CONVENTIONAL UNIT CELLS OF THE 14 BRAVAIS LATTICES

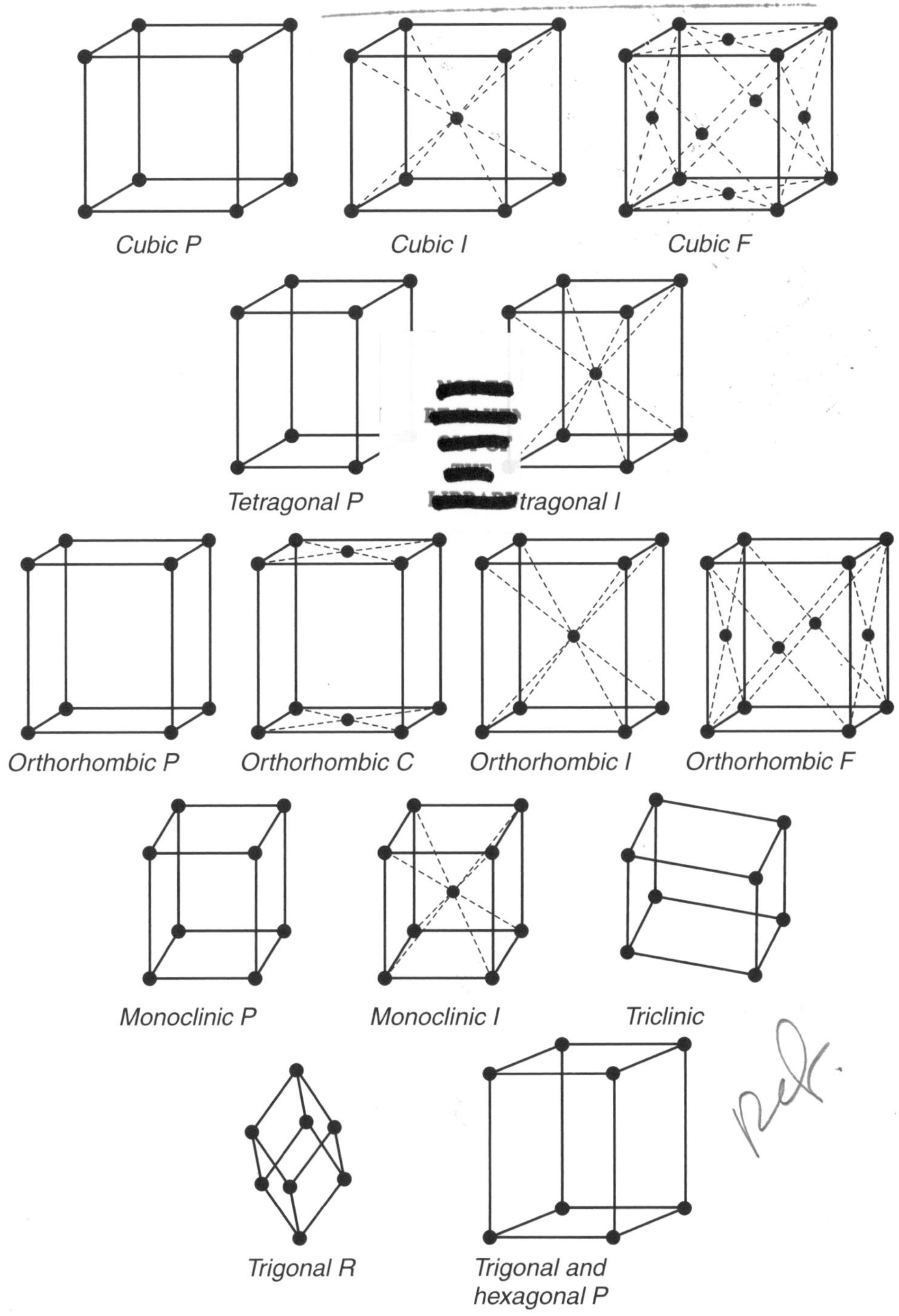

ENCYCLOPEDIC DICTIONARY OF CONDENSED MATTER PHYSICS

ФИЗИКА ТВЕРДОГО ТЕЛА

Энциклопедический Словарь

Главный редактор

В. Г. БАРЬЯХТАР

Киев Наукова думка

FIZIKA TVERDOGO TELA
ENCIKLOPEDICHESKII SLOVAR

Original Edition (in Russian):

Naukova Dumka, Kiev, Ukraine
Volume 1, 1996
Volume 2, 1998
Institute of Physics, Institute of Magnetism

Editor-in-Chief of the Original Edition:

V. G. BAR'YAKHTAR

Department of Physics and Astronomy
National Academy of Sciences of Ukraine

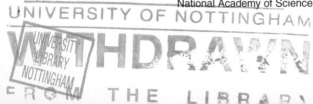

ENCYCLOPEDIC DICTIONARY OF CONDENSED MATTER PHYSICS

VOLUME 1

A – M

EDITED BY

CHARLES P. POOLE JR.

Department of Physics and Astronomy
University of South Carolina
Columbia, SC, USA

ELSEVIER
ACADEMIC
PRESS

2004

Amsterdam – Boston – Heidelberg – London – New York – Oxford
Paris – San Diego – San Francisco – Singapore – Sydney – Tokyo

ELSEVIER B.V.
Sara Burgerhartstraat 25
P.O. Box 211,
1000 AE Amsterdam
The Netherlands

ELSEVIER Inc.
525 B Street, Suite 1900
San Diego, CA 92101-4495
USA

ELSEVIER Ltd
The Boulevard, Langford Lane
Kidlington, Oxford OX5
UK

ELSEVIER Ltd
84 Theobalds Road
1GB London WC1X 8RR
UK

This is a translation from Russian.

First edition 2004

Library of Congress Cataloging in Publication Data: A catalog record is available from the Library of Congress.

British Library Cataloguing in Publication Data: A catalogue record is available from the British Library.

ISBN 0-12-088398-8 volume 1

ISBN 0-12-088399-6 volume 2

ISBN 0-12-561465-9 set

⊗ The paper used in this publication meets the requirements of ANSI/NISO Z39.48-1992 (Permanence of Paper).

Printed in The United Kingdom.

EDITOR'S PREFACE TO THE ENGLISH EDITION

At the present time Condensed Matter Physics is the largest of the main subdivisions of physics. For several years the Encyclopedic Dictionary of Solid State Physics has been one of the principal reference works in this area, but unfortunately until now it was only available in the Russian language. Its appearance in English is indeed long overdue. It has been a great pleasure for me to undertake the task of editing the English language edition. This two-volume work contains a great deal of material ordinarily not found in the solid state texts that are most often consulted for reference purposes, such as accounts of acoustical holography, bistability, Kosterlitz–Thouless transition, and self-similarity. It provides information on materials rarely discussed in these standard texts, such as austenite, ferrofluids, intercalated compounds, laser materials, Laves phases, nematic liquid crystals, etc. Properties of many quasi-particles are presented; for example, those of crowdions, dopplerons, magnons, solitons, vacancions, and weavons. The volumes are fully cross-referenced for tracking down all aspects of a topic, and in this respect we have emulated the excellent cross-referencing systematics of the Russian original. Long lists of alternate titles for looking up subjects can considerably shorten the time needed to find specialized information.

I wish to thank the members of the Review Panel of the English Language Edition who read over the initial draft of the translation, offered many suggestions and comments for improving the translation, and also provided valuable advice for refining and updating the contents of the articles: Victor Bar'yakhtar, Alexei A. Maradudin, Michael McHenry, Laszlo Mihaly, Sergey K. Tolpygo and Sean Washburn. Thanks are also due to Horacio A. Farach, Vladimir Gudkov, Paul G. Huray, Edwin R. Jones, Grigory Simin, Tangali S. Sudarshin, and especially to Ruslan Prozorov, for their assistance with the editing. The translators did an excellent job of rendering the original Russian text into English, and the desk editor Rimantas Šadžius did an outstanding job of checking all the many cross references and the consistency of the notation. The Academic Press Dictionary of Science and Technology has been an invaluable source of information for clarifying various issues during the editing.

I wish to thank my wife for her patience with me during the long hours expended in the editing process. The page proofs of this work arrived at my home on the octave day of our 50th wedding anniversary, which we celebrated on October 17, 2003, with our five children and fifteen grandchildren.

Charles P. Poole Jr.

FOREWORD

An Encyclopedic Dictionary of Condensed Matter Physics can serve a diverse set of audiences. Professional researchers and university and college teachers interested in learning quickly about a topic in condensed matter physics outside their immediate expertise can obtain the information they seek from such a source. Doctoral students embarking on research in condensed matter physics or materials science can find it helpful in providing explanations of concepts and descriptions of methods encountered in reading the literature dealing with their research subject and with related subjects. Undergraduate students with no previous background in condensed matter physics will find it a particularly valuable complement to their textbooks as they take their first courses in the field. Educated laypersons can turn to such an encyclopedia for help in understanding articles published in popular scientific journals.

Until the publication of the Russian edition of the Encyclopedic Dictionary of Solid State Physics, with Professor Viktor G. Bar'yakhtar as its General Editor, no such resource was available to the scientific community. Its more than 4300 entries were written by more than 700 contributors, specialists in their branches of condensed matter physics, many of whom were well known to an international audience.

The task of making this encyclopedia accessible to non-Russian readers was undertaken by Academic Press, with Professor Charles P. Poole Jr. in the role of the General Editor for the English translation. Both the publisher and editor deserve the gratitude of the condensed matter community for bringing this project to fruition. All involved in its production hope that that community will find it as useful as its creators intended it to be.

Irvine, California
October 29, 2003

Alexei A. Maradudin

FOREWORD OF THE EDITORIAL BOARD OF THE ORIGINAL RUSSIAN EDITION

This publication is an attempt at a systematized presentation of modern knowledge within all areas of solid state physics from the most fundamental ideas to applied questions. There has been no similar encyclopedic compilation of solid state physics available in the world until now. The Dictionary contains more than 4000 articles. Among them there are relatively long articles devoted to the basic concepts and phenomena of solid state physics, moderate-length articles on particular problems and physical concepts, and, finally, brief definitions of terms used in the literature. Considerable attention is paid to technical applications of solid state physics. Brief historical information is given in many articles such as the names of authors and the dates of discoveries. In a number of cases the history of the question is covered in the article. Together with the description of classical topics, special attention is paid to those new ideas that arose only recently in solid state physics and, therefore, often still do not have commonly accepted definitions. Many articles are provided with illustrations which supplement the text material to help the reader to acquire more complete and more visual information.

More than 500 scientists participated in the preparation and the writing of the articles of the Dictionary; these articles were written by specialists working in the particular fields that they cover.

The goal of the publication is to provide needed reference material to a wide audience of readers, which will help them to understand the basic principles of solid state physics. Therefore, the authors endeavored to probe deeply into some salient details of the subject, while preserving a broad perspective throughout the text of the article. The content of articles is interconnected by a system of cross-references. The appended Subject Index will help the reader to orient himself easily in the scope of the various articles.

To our deep sorrow, during the period of preparation of the manuscript of this Dictionary, several outstanding scientific members of the Editorial Board Valentin L'vovich Vinetskii, Yakov Evseevich Geguzin, Emanuil Aizikovich Kaner, Boris Ivanovich Nikolin, Kirill Borisovich Tolpygo, Leonid Nikandrovich Larikov and Alexander Bronislavovich Roitsyn have passed from among us. The recently expired well-known scientist and prominent publishing specialist Aleksandr Aleksandrovich Gusev, who shared with us his vast knowledge and encyclopedic expertise, provided invaluable aid in the publication of the dictionary.

Victor G. Bar'yakhtar, Editor-in-Chief
Ernest A. Pashitskii, Vice-Editor-in-Chief
Elena G. Galkina, Secretary

We started as the team (EDITORIAL BOARD OF THE ORIGINAL RUSSIAN EDITION):

V.G. Bar'yakhtar, Editor-in-Cief
V.L. Vinetskii, Vice-Editor-in-Chief
E.A. Pashitskii, Vice-Editor-in-Chief
E.G. Galkina, Secretary

A.S. Bakai
Ya.E. Geguzin
M.D. Glinchuk
Yu.V. Gulyaev

B.A. Ivanov
M.I. Kaganov
E.A. Kaner
A.P. Korolyuk

LIST OF CONTRIBUTORS

F.Kh. Abdullaev
A.A. Adamovskii
S.S. Afonskii
I.A. Akhiezer
A.F. Akkerman
A.B. Aleinikov
V.P. Alekhin
A.S. Aleksandrov
L.N. Aleksandrov
B.L. Al'tshuler
E.D. Aluker
S.P. Anokhov
L.I. Antonov
K.P. Aref'ev
I.E. Aronov
M. Ashe
Yu.E. Avotin'sh
V.G. Babaev
Yu.Z. Babaskin
S.G. Babich
Yu.I. Babii
A.S. Bakai
Yu.V. Baldokhin
O.M. Barabash
S.D. Baranovskii
Z.V. Bartoshinskii
V.G. Bar'yakhtar
V.G. Baryshevskii
E.M. Baskin
F.G. Bass
A.N. Bekrenev
I.P. Beletskii
Ya.A. Beletskii
S.I. Beloborod'ko
E.D. Belokolos
V.F. Belostotskii
A.V. Belotskii
M.V. Belous
N.P. Belousov
K.P. Belov
D.P. Belozerov
A.I. Belyaeva
S.S. Berdonosov
B.I. Beresnev
L.I. Berezhinskii
V.I. Bernadskii
I.B. Bersuker
V.N. Berzhanskii
P.A. Bezirganyan
V.F. Bibik
I.V. Blonskii
E.N. Bogachek
A.N. Bogdanov
Yu.A. Bogod
S.A. Boiko
V.S. Boiko
S.I. Bondarenko
V.N. Bondarev
A.S. Borovik-Romanov
A.A. Borshch
Yu.S. Boyarskaya
A.M. Bratkovskii
O.M. Braun
S.L. Bravina
M.S. Bresler
A.B. Brik
M.S. Brodin
V.A. Brodovoi
L.T. Bugaenko
V.N. Bugaev
M. Buikov
L.L. Buishvili
E.I. Bukhshtab
L.N. Bulaevskii
B.M. Bulakh
A.A. Bulgakov
S.I. Bulychev
A.I. Buzdin
A.I. Bykhovskii
A.M. Bykov
Yu.A. Bykovskii
G.E. Chaika
A.A. Chel'nyi
S.P. Chenakin
V.T. Cherepin
N.F. Chernenko
Yu.P. Chernenkov
A.A. Chernov
S.I. Chugunova
B.A. Chuikov
K.V. Chuistov
A.A. Chumak
V.E. Danil'chenko
V.D. Danilov
L.I. Datsenko
A.B. Davydov
A.S. Davydov
S.A. Dembovskii
V.L. Demikhovskii
S.A. Demin
E.M. Dianov
V.V. Didyk
I.M. Dmitrenko
N.L. Dmitruk
V.N. Dneprenko
V.S. Dneprovskii
V.N. Dobrovol'skii
R.D. Dokhner
I.F. Dolmanova
F.E. Dolzhenkov
V.F. Dorfman
S.N. Dorogovtsev
N.V. Dubovitskaya
E.F. Dudnik
V.A. Durov
V.V. Dyakin
M.I. Dykman
A.I. Efimov
A.L. Efros
Yu.A. Ekmanis
M.A. Elango
V.M. Elinson
P.G. Eliseev
V.A. Elyukhin
Yu.P. Emets
V.V. Emtsev
V.Z. Enol'skii
I.R. Entin
I.R. Entinzon
E.M. Epshtein
A.I. Erenburg
A.S. Ermolenko
E.D. Ershov
A.M. Evstigneev
M.I. Faingol'd
V.M. Fal'chenko
V.L. Fal'ko
B.Ya. Farber
M.P. Fateev
I.M. Fedorchenko
O.P. Fedorov
A.G. Fedorus
Ya.A. Fedotov
I.V. Fekeshgazi
A.V. Filatov
B.N. Filippov
V.A. Finkel'
V.M. Finkel'
A.Ya. Fishman
N.Ya. Fogel'
E.L. Frankevich
L.I. Freiman

V.M. Fridkin
I.Ya. Fugol'
B.I. Fuks
A.M. Gabovich
S.P. Gabuda
A.Yu. Gaevskii
Yu.A. Gaidukov
V.P. Galaiko
R.V. Galiulin
Yu.M. Gal'perin
M.Ya. Gamarnik
G.D. Gamulya
E.M. Ganapol'skii
V.G. Gavrilyuk
Ya.E. Geguzin
B.L. Gel'mont
D.S. Gertsriken
I.A. Gilinskii
I.A. Gindin
S.L. Ginzburg
E.I. Givargizov
E.I. Gladyshevskii
A.A. Glazer
L.I. Glazman
A.I. Glazov
M.D. Glinchuk
E.Ya. Glushko
V.M. Gokhfel'd
V.L. Gokhman
V.A. Golenishchev-Kutuzov
A.V. Golik
M.F. Golovko
T.V. Golub
D.A. Gorbunov
L.Yu. Gorelik
Yu.I. Gorobets
V.I. Gorshkov
V.V. Gorskii
V.G. Grachev
S.A. Gredeskul
V.P. Gribkovskii
P.P. Grigaitis
V.N. Grigor'ev
O.N. Grigor'ev
Yu.M. Grin'
V.V. Gromov
V.V. Gudkov
A.V. Gulbis
A.P. Gulyaev
Yu.V. Gulyaev
M.E. Gurevich
Yu.G. Gurevich
K.P. Gurov
R.N. Gurzhi
M.B. Guseva
A.N. Guz'
V.M. Gvozdikov
A.I. Ignatenko
V.V. Il'chenko
E.M. Iolin
I.P. Ipatova
S.S. Ishchenko
Z.A. Iskanderova
V. Iskra
V.G. Ivanchenko
P.G. Ivanitskii
A.L. Ivanov
B.A. Ivanov
M.A. Ivanov
N.R. Ivanov
G.F. Ivanovskii
P.F. Ivanovskii
V.G. Ivantsov
R.K. Ivashchenko
O.M. Ivasishin
Yu.A. Izyumov
G.A. Kachurin
A.N. Kadashchuk
A.M. Kadigrobov
M.I. Kaganov
E.B. Kaganovich
Yu.M. Kaganovskii
N.P. Kalashnikov
P.A. Kalugin
G.G. Kaminskii
E.A. Kaner
L.N. Kantorovich
A.S. Kapcherin
A.A. Kaplyanskii
A.L. Kartuzhanskii
A.L. Kasatkin
V.N. Kashcheev
N.I. Kashirina
G.A. Katrich
B.V. Khaenko
I.B. Khaibullin
V.A. Kharchenko
E.I. Khar'kov
S.S. Khil'kevich
A.I. Khizhnyak
G.E. Khodenkov
A.R. Khokhlov
G.A. Kholodar'
V.A. Khvostov
K.I. Kikoin
S.S. Kil'chitskaya
O.V. Kirichenko
A.P. Kirilyuk
P.S. Kislyi
A.E. Kiv
Ya.G. Klyava
V.M. Knyazheva
E.S. Koba
L.S. Kogan
V.V. Kokorin
Yu.A. Kolbanovskii
G.Ya. Kolbasov
A.G. Kol'chinskii
V.L. Kolesnichenko
V.V. Kolomiets
V.I. Kolomytsev
E.V. Kolontsova
V.G. Kon
A.A. Konchits
S.N. Kondrat'ev
M.Ya. Kondrat'ko
V.A. Kononenko
V.I. Konovalov
A.I. Kopeliovich
V.A. Koptsyk
T.N. Kornilova
Yu.V. Kornyushin
E.N. Korol'
N.S. Kosenko
A.M. Kosevich
O.G. Koshelev
A.S. Kosmodamianskii
E.A. Kotomin
N.Ya. Kotsarenko
Yu.N. Koval'
V.V. Koval'chuk
A.S. Kovalev
O.V. Kovalev
M.Yu. Kovalevskii
Yu.Z. Kovdrya
I.I. Kovenskii
V.I. Kovpak
V.I. Kovtun
A.V. Kozlov
N.N. Krainik
V.V. Krasil'nikov
A.Ya. Krasovskii
L.S. Kremenchugskii
V.P. Krivitskii
M.A. Krivoglaz
V.N. Krivoruchko
A.A. Krokhin
S.P. Kruglov
I.V. Krylova
A.Yu. Kudzin
L.S. Kukushkin
I.O. Kulik
V.A. Kuprievich
A.I. Kurbakov
A.V. Kurdyumov
V.G. Kurdyumov
G.G. Kurdyumova
M.V. Kurik
M.I. Kurkin
N.P. Kushnareva
V.A. Kuz'menko
R.N. Kuz'min
V.N. Kuzovkov
A.I. Landau
V.F. Lapchinskii
L.N. Larikov
F.F. Lavrent'ev
O.D. Lavrentovich
B.G. Lazarev
L.S. Lazareva
Ya.S. Lebedev
B.I. Lev
A.P. Levanyuk
A.A. Levin
I.B. Levinson
V.A. Likhachev
M.P. Lisitsa
V.M. Lisitsyn
V.A. Lisovenko
F.V. Lisovskii
P.G. Litovchenko
V.G. Litovchenko
V.S. Litvinov
V.A. Lobodyuk
V.M. Loktev
V.F. Los'

D.V. Lotsko
V.P. Lukomskii
G.A. Luk'yanov
V.A. L'vov
S.N. Lyakhimets
V.G. Lyapin
B.Ya. Lyubov
I.F. Lyuksyutov
R.G. Maev
V.M. Maevskii
L.L. Makarov
N.M. Makarov
V.I. Makarov
D.N. Makovetskii
L.A. Maksimov
V.V. Malashenko
B.Z. Malkin
G.D. Mansfel'd
V.G. Manzhelii
V.G. Marinin
S.V. Marisova
V.Ya. Markiv
Yu.V. Martynenko
T.Ya. Marusii
S.V. Mashkevich
V.S. Mashkevich
T.V. Mashovets
V.V. Maslov
O.I. Matkovskii
Z.A. Matysina
V.F. Mazanko
I.I. Mazin
E.A. Mazur
B.V. Mchedlishvili
M.V. Medvedev
V.K. Medvedev
A.I. Mel'ker
G.A. Melkov
V.S. Mel'nikov
F.F. Mende
Yu.Ya. Meshkov
G.A. Mesyats
V.P. Mikhal'chenko
V.S. Mikhalenkov
V.V. Milenin
V.K. Milinchuk
Yu.V. Mil'man
M.G. Mil'vidskii
V.N. Minakov
E.V. Minenko
D.N. Mirlin
I.A. Misurkin
V.Ya. Mitrofanov
N.V. Morozovskii
V.V. Moshchalkov
N.P. Moskalenko
S.A. Moskalenko
V.N. Murzin
V.P. Naberezhnykh
T.A. Nachal'naya
V.M. Nadutov
E.L. Nagaev
V.E. Naish
N.G. Nakhodkin
V.D. Natsyk
A.G. Naumovets
O.K. Nazarenko
E.I. Neimark
I.M. Neklyudov
I.Yu. Nemish
V.V. Nemoshkalenko
S.A. Nepiiko
V.A. Nevostruev
S.A. Nikitov
M.Yu. Nikolaev
B.I. Nikolin
A.B. Nikol'skii
G.K. Ninidze
A.I. Nosar'
N.N. Novikov
N.V. Novikov
V. Novozhilov
S.G. Odulov
V.I. Okulov
E.F. Oleinik
N.V. Olechnovich
S.I. Olikhovskii
B.Z. Ol'shanetskii
A.N. Omel'yanchuk
A.N. Orlov
S.P. Oshkaderov
M.E. Osinovskii
S.S. Ostapenko
E.L. Ostrovskaya
D.E. Ovsienko
R.P. Ozerov
G.A. Pacharenko
P.P. Pal'-Val'
S.V. Panyukov
I.I. Papirov
E.S. Parilis
A.M. Parshin
E.A. Pashitskii
L.A. Pastur
R.E. Pasynkov
E.P. Pechkovskii
G.P. Peka
V.I. Peresada
S.P. Permogorov
V.G. Peschanskii
I.S. Petrenko
E.G. Petrov
V.A. Petrov
Yu.N. Petrov
G.A. Petrunin
V.Yu. Petukhov
A.S. Pikovskii
G.E. Pikus
A.N. Pilyankevich
V.I. Pipa
F.V. Pirogov
G.S. Pisarenko
E.A. Pisarev
L.P. Pitaevskii
A.B. Plachenov
B.T. Plachenov
Yu.N. Podrezov
A.E. Pogorelov
V.Z. Polinger
Yu.N. Polivanov
V.V. Polotnyuk
S.S. Pop
Yu.M. Poplavko
L.E. Popov
V.E. Pozhar
L.I. Pranyavichus
A.I. Prikhna
V.E. Primachenko
V.G. Prokhorov
G.I. Prokopenko
P.P. Pugachevich
V.V. Pustovalov
V.I. Pustovoit
A.P. Rachek
V.A. Rafalovskii
V.V. Rakitin
E.I. Rau
B.S. Razbirin
A.R. Regel'
V.R. Regel'
V.Yu. Reshetnyak
Yu.Ya. Reutov
I.S. Rez
I.M. Reznik
Yu.A. Reznikov
D.R. Rizdvyanetskii
A.B. Roitsyn
O.V. Romankevich
A.E. Romanov
A.S. Rozhavskii
S.S. Rozhkov
V.V. Rozhkov
E.Ya. Rudavskii
A.V. Rudnev
I.N. Rundkvist
S.M. Ryabchenko
P.V. Ryabkov
K.P. Ryaboshapka
V. Rybalka
A.I. Ryskin
V.I. Ryzhkov
A.M. Sabadash
V.A. Sablikov
A.V. Sachenko
M.V. Sadovskii
O.G. Sarbei
V.M. Schastlivtsev
Yu.V. Sedletskii
S.I. Selitser
A.V. Semenov
Yu.G. Semenov
A.I. Senkevich
T.N. Serditova
T.N. Sergeeva
S.E. Shafranyuk
S.I. Shakhovtsova
A.M. Shalaev
B.N. Shalaev
B.D. Shanina
B.Ya. Shapiro
I.G. Shaposhnikov
O.D. Shashkov
V.G. Shavrov
D.I. Sheka
V.I. Sheka
R.I. Shekhter
V.A. Shenderovskii

S.I. Shevchenko
A.Ya. Shik
V.B. Shikin
Yu.M. Shirshov
S.V. Shiyanovskii
G.N. Shkerdin
O.A. Shmatko
A.S. Shpigel'
T.D. Shtepa
E.I. Shtyrkov
Yu.N. Shunin
A.K. Shurin
K.K. Shvarts
E.A. Silin'sh
A.F. Sirenko
A.I. Sirko
A.A. Sitnikova
F.F. Sizov
E.V. Skokan
V.V. Skorokhod
V.V. Sleptsov
A.A. Slutskin
A.A. Smirnov
L.S. Smirnov
G.A. Smolenskii
O.V. Snitko
V.I. Sobolev
A.I. Sokolov
I.M. Sokolov
A.S. Sonin
E.B. Sonin
A. Sorokin
M.S. Soskin
A.L. Sozinov
V.B. Spivakovskii
I.E. Startseva
A.N. Starukhin
E.P. Stefanovskii
M.F. Stel'makh
I.A. Stepanov
I.A. Stoyanov
M.I. Strashnikova
V.I. Strikha
V.I. Sugakov
A.P. Sukhorukov
S.V. Svechnikov
V.M. Svistunov
I.I. Sych
A.K. Tagantsev
G.A. Takzei
A.P. Tankeev
V.V. Tarakanov
V.V. Tarasenko
G.G. Tarasov
Yu.V. Tarasov
I.A. Tarkovskaya
V.A. Tatarchenko
V.A. Tatarenko
I.I. Taubkin
V.A. Telezhkin
Yu.S. Tikhodeev
L.V. Tikhonov
E.A. Tikhonova
B.L. Timan
S.F. Timashev
B.B. Timofeev
V.B. Timofeev
Yu.A. Tkhorik
K.B. Tolpygo
S.K. Tolpygo
P.M. Tomchuk
V.I. Tovstenko
V.I. Trefilov
O.A. Troitskii
V.A. Trunov
V.A. Tsekhomskii
B.E. Tsekvava
R. Tsenker
B.S. Tsukerblat
V.M. Tsukernik
L.T. Tsymbal
N.V. Tsypin
L.I. Tuchinskii
A. Tukharinov
E.A. Turov
Yu.D. Tyapkin
V.M. Tyshkevich
V.A. Tyul'nin
K.M. Tyutin
A.A. Urusovskaya
A.I. Ustinov
B.K. Vainshtein
S.B. Vakhrushev
M.Ya. Valakh
V.I. Val'd-Perlov
A.A. Varlamov
I.A. Vasil'ev
L.I. Vasil'ev
M.A. Vasil'ev
F.T. Vas'ko
A.V. Vedyaev
B.G. Vekhter
O.G. Vikhlii
V.L. Vinetskii
E.A. Vinogradov
B.B. Vinokur
V.M. Vinokurov
Yu.P. Virchenko
A.S. Vishnevskii
I.M. Vitebskii
N.A. Vitovskii
V.I. Vladimirov
V.V. Vladimirov
N.A. Vlasenko
A.D. Vlasov
K.B. Vlasov
O.G. Vlokh
E.D. Vol
A.F. Volkov
P.Yu. Volosevich
V.A. Voloshin
V.D. Volosov
S.V. Vonsovskii
D.L. Vorob'ev
S.A. Vorob'ev
Yu.V. Vorob'ev
A.M. Voskoboinikov
B.E. Vugmeister
D.A. Yablonskii
Yu.N. Yagodzinskii
L.N. Yagupol'skaya
Yu.I. Yakimenko
B.I. Yakobson
M.V. Yakunin
V.A. Yampol'skii
I.K. Yanson
A.M. Yaremko
A.A. Yatsenko
A.F. Yatsenko
I.A. Yurchenko
Yu.F. Yurchenko
P.A. Zabolotnyi
V.A. Zagrebnov
Yu.R. Zakis
E.Ya. Zandberg
I.M. Zaritskii
E.V. Zarochentsev
I.K. Zasimchuk
E.E. Zasimchuk
S.S. Zatulovskii
E.A. Zavadskii
P.Yu. Zavalii
Yu.S. Zharkikh
I.S. Zheludev
G.N. Zhizhin
A.F. Zhuravlev
S.N. Zhurkov
O.S. Zinets
D.N. Zubarev
M.D. Zviadadze
L.O. Zvorykin
B.B. Zvyagin
I.P. Zvyagin

INTRODUCTION

We introduce to the reader the Encyclopedic Dictionary of Solid State Physics[1]. This Dictionary was planned several years ago, and the aim of its compilers was not only to provide a wide audience of readers with comprehensive and detailed information about different processes, phenomena, and effects in the solid state, but also to enlarge our understanding of the modern terminology in which this information is presented. The latter was very important for the compilers, since they were united by the idea that the achievements of solid state physics during recent decades were due not only to the fantastic development of experimental techniques and the level of experimental skill, but were also due to the creation of a unified system of concepts and terms adequately reflecting the observed regularities and the depth of understanding of the physical processes taking place in solids. The balanced rhythm of the modern language of solid state physics is determined by the theory of symmetry. It governs all theoretical constructions and the coefficients of the equations describing equilibrium and near-equilibrium states of a solid. The dynamic theory of a crystal lattice in the harmonic approximation, which is based on the discrete symmetry of ordered atomic arrangement in a crystal, is a classical example how symmetry considerations are utilized. Another "hidden" symmetry is used, e.g., to describe a number of properties of magnetic, ferroelectric, and superconducting materials. A particular kind of symmetry based on the regularities in diverse phase transitions is apparent in modern scaling theory. The reader will find applications of the theory of symmetry in many of the articles.

The majority of the phenomena of recent interest in solids were associated with low-level excited states of a substance. It appeared that while atoms or molecules are structural elements of a crystal lattice (and, generally, of any condensed state of a substance), certain collective excitations are the elements of every type of motion (elements of dynamics) in a solid. In view of the macroscopic homogeneity of the unbounded model of a solid, or the periodicity of an unbounded crystal lattice, such collective motions must take the form of plane waves, with a dispersion law that satisfies strict symmetry requirements. The quantization of elementary excitations leads to the image of a "corpuscle", or a quasi-particle. The concept of a quasi-particle and the idea of almost ideal quantum gases composed of various quasi-particles, enrich the language of solid state physics and allow one, within a unified system of terms and concepts, to treat physical phenomena that are quite diverse and have, at first glance, little in common. Therefore, it is not surprising that in the articles devoted to phonons, magnons, excitons, and to electrons or polarons as well, we find very similar equations and conclusions. The universality of the physical language reflects the unity of Nature itself.

The passage to the study of highly excited states of a material led to the creation of the nonlinear dynamics of solids. New contributions to the lexical reserve of the language of solid state physics originated from nonlinear dynamics. For example, the concepts of topological and dynamic solitons were put forward, dynamic solitons appearing in solid state physics as a new type of collective excitation of a condensed medium.

Not all parts of solid state physics are presented here with the same completeness and the same amount of detail. The branches reflecting the scientific interests of solid state physicists from the Ukraine and other former Soviet Union republics are the most extensively described. Although general problems of solid state

[1] This is the title of the book in Russian (Editor's note).

physics are less systematically presented, the following areas are covered rather deeply and broadly: physical properties of metals, including superconductivity, electrical and optical properties of semiconductors and insulators, magnetism and magnetic materials, mechanical properties of crystals, and physics of disordered systems. One may also notice an emphasis on the applied physics of plasticity and strength. This is because of the fact that, although physics is an objective international science, some of its presentations or overviews found in textbooks, monographs, or encyclopedic publications unintentionally introduce subjective ideas and personal preferences of authors.

The Dictionary includes articles reflecting the latest achievements in solid state physics, sometimes with results that have not yet appeared in monographs. This relates both to experimental discoveries and theoretical ideas. There are articles on high-temperature superconductivity, the quantum Hall effect, Kosterlitz–Thouless phase transitions in two-dimensional systems, autowave processes, fractals, etc. Novel nontraditional materials and techniques for investigation are described: amorphous substances, liquid crystals, quasi-crystals, polymers, low-dimensional systems, etc.

In addition let us point out to the reader the "quantum-mechanical emphasis", which may be noted in many articles devoted to diverse macroscopic phenomena and properties of solid bodies. Traditionally, use is made of the point of view that the quantum statistics of quasi-particles undoubtedly defines such macroscopic low-temperature properties of crystals as specific heat and thermal conductivity, electrical conductivity and magnetism. However, some authors only assign to quantum mechanics the role of describing microscopic laws explaining the dynamics of atomic particles. The discovery of high-temperature superconductivity transferred the quantum manifestations of particle dynamics into the region of moderately high temperatures. State-of-the-art experimental techniques make it possible to observe a number of phenomena involving macroscopic quantum tunneling. Thus, macroscopic solids fairly often exhibit some particular quantum features.

Finally let us stress the role of models in solid state physics. As is well known, a model is a much simplified speculative scheme which allows a researcher, on the one hand, to propose a simple qualitative explanation of a property or phenomenon under study, and, on the other hand, to give a quantitative description of properties and to perform a complete theoretical calculation of characteristics of the phenomenon being studied. Models are utilized very often in solid state physics. Apparently, the most ancient model is that of an "absolutely solid body". The model of perfect gas of quasi-particles is frequently employed; Ising's model is well known in the study of magnetism. The model of a two-dimensional lattice of classical spins in the theory of phase transitions proposed by Onsager allowed him to give the first statistical description of second-order phase transitions. The Frenkel–Kontorova and Peierls models are widely used in dislocation theory. References to them can be found in many of the articles included in this work.

The need for this publication arises from the recent tremendous progress in this field of physics, especially by the extension of its areas of practical applications. There have appeared new fields of electronics (optical electronics, acoustoelectronics); semiconductor electronics has changed and now employs an increasingly wider range of complex structures such as metal–insulator–semiconductor devices and computers with superconducting elements (cryotrons); and memory elements based on new physical principles have been developed.

There are many promising new solid state materials based on garnets, rare-earth metals, magnetic semiconductors, amorphous metals, semiconductors, and, finally, the discovery of high-temperature superconductors in 1986.

The compilers of this Dictionary hope that this publication will be useful for researchers working in solid state and other fields of physics (nuclear physics, plasma physics, etc.) and also for a wide audience of other readers, including, but not limited to, engineers investigating and utilizing solids, as well as doctoral and advanced students at universities.

V.G. Bar'yakhtar

A.M. Kosevich

HOW TO USE THE ENCYCLOPEDIC DICTIONARY

In this Dictionary, we follow standard fundamental rules generally adhered to in encyclopedic publications.

1. Articles are arranged alphabetically; if a term (typeset in boldface) has a synonym, then the latter is given in the entry after the main meaning of the term (after comma, in usual print).
2. Some terms in the titles of articles adopted from other languages are followed by a brief etymological note (with clear abbreviations: abbr., abbreviated, fr., from; Gr., Greek; Lat., Latin; Fr., French; Germ., German, etc.).
3. The names of scientists appearing in the title or text of an article are commonly followed by the date of obtaining the result (effect, phenomenon) under consideration.
4. We use a system of cross-references (in *Helvetica italics*) to other articles. Additional terms explained within the same article are typeset in *Times italic* and, together with the main entries, are listed in the Subject Index at the end of Volume 2. Many terms synonymous to those in the article titles may be also found in this Subject Index.
5. Units of physical quantities and their symbols are given in the systems of units most commonly used in solid state physics. For the relations which convert these units into SI, see the tables included in the article *Units of physical quantities.*
6. The figures are explained either in their captions or in the text of the article.
7. Usually, the notation used in formulae is explained in the text of that article. However, some symbols have the same meaning throughout the Dictionary (unless otherwise specified): c, speed of light; h and $\hbar = h/2\pi$, Planck constant; T, absolute temperature; k_B, Boltzmann constant; λ, wavelength; ν and $\omega = 2\pi\nu$, frequency. The notation of some elementary particles is also standardized: ν, photon and gamma quantum; e, e^-, electron; p, proton; n, neutron; d, deuteron; π and π°, pions; $\mu^\pm$, muons; $K^\pm$ and K°, K-mesons.

Aa

ABLATION (fr. Lat. *ablatio*, removal, taking away)

The removal of mass from the surface of a solid brought about by a heated gas flow sweeping past it.

This is observed during the motion of material bodies and space crafts through the atmosphere at high velocities. Ablation occurs as a result of erosion, melting, and sublimation. It is used to cool key parts of super- and hypersonic aircrafts, spacecrafts, and rocket engines by coating them with layers of an ablatable material (e.g., *ceramics*, or pyrolytic *graphite*). The ablation material must have high *melting* and *sublimation* temperatures and a large *specific heat*.

ABRASIVE MATERIALS, abrasives (fr. Lat. *abrasio*, scraping)

Fine-grained solid materials intended for working (polishing, finishing) a material surface.

The working consists in cutting and scraping the surface by a great number of randomly oriented sharp edges of abrasive grains. Typical abrasive materials are diamond (natural or synthetic), *carborundum*, the cubic modification of boron nitride, and *boron carbide*. Abrasive materials are characterized by hardness, strength, granularity (describing the magnitude and dispersion of the grain size); knowing these parameters permits the selection of abrasive materials for particular purposes, and provides the expected precision of the working. Depending on their relative abrasive abilities, abrasive materials can be classified in terms of the following scale: diamond (1.00), cubic boron nitride (0.78), *boron carbide* (0.60), *silicon carbide* (0.50), and titanium carbide (0.42).

ABRIKOSOV–SUHL RESONANCE

(A.A. Abrikosov, H. Suhl, 1965)

The abnormal increase of the *electron spin-flip scattering* amplitude of conduction electrons on a paramagnetic impurity when the energy ε of the scattered electron approaches the Fermi energy ε_F of the *metal*.

The cause of the Abrikosov–Suhl resonance is the creation of a long-lived bound state of a conduction electron with a paramagnetic impurity due to the indirect interaction of *conduction electrons* with each other via the impurity spin when the spin–paramagnetic impurity *exchange interaction* constant is negative, i.e. $J < 0$. The Abrikosov–Suhl resonance manifests itself below the *Kondo temperature* T_K by an anomalous temperature dependence of the behaviour of the resistivity and the magnetic susceptibility of a metal (see *Kondo effect*). The expression for the resistivity of a metal is derived by the selective summation of the "dangerous" terms of the perturbation expansion in the parameter $|J|/\varepsilon_F \ll 1$ (diagrams calculated in the so-called parquet approximation), and has the form

$$\rho = \rho_s^{(0)}\left[1 + \frac{J\nu(\varepsilon_F)}{n}\ln\left(\frac{\varepsilon_F}{k_B T}\right)\right]^{-2},$$

where $\rho_s^{(0)}$ is the resistivity calculated in the first *Born approximation*, $\nu(\varepsilon_F)$ is the conduction *density of electron states* at the *Fermi energy*, and n is the metal density. The denominator vanishes at the Kondo temperature

$$T_K = \frac{\varepsilon_F}{k_B}\exp\left[-\frac{n}{|J|\nu(\varepsilon_F)}\right].$$

In the temperature range $T \leqslant T_K$, perturbation theory is no longer valid, and a more detailed analysis shows that a spin complex of a conduction electron with an impurity is created, where at

$T = 0$ the impurity spin is shielded by the conduction electrons. In the vicinity of T_K the spin parts of the thermodynamic and kinetic characteristics of a metal sharply change their temperature dependences, remaining finite over the entire temperature range. The exact solution of the Kondo problem was obtained by P. Wiegmann and N. Andrei (1980).

ABRIKOSOV VORTICES, vortex filaments (A.A. Abrikosov, 1956)

Vortex current structures of *type II superconductors* in magnetic fields with strength higher than the *lower critical field* $B_{c1}(T)$, but lower than the *upper critical field* $B_{c2}(T)$, i.e. in the domain of the mixed state (see also *Shubnikov phase*).

First predicted theoretically and then observed experimentally Abrikosov vortices are topologically stable singlet states, involving the penetration into a superconductor of minimal magnetic flux quantities, equal to the flux quantum Φ_0 (see *Quantization of flux*). One can think of these vortices as quantum "rods" of magnetic flux encircled by superconducting current flow and pierced by magnetic field lines (Fig. 1). For an isolated

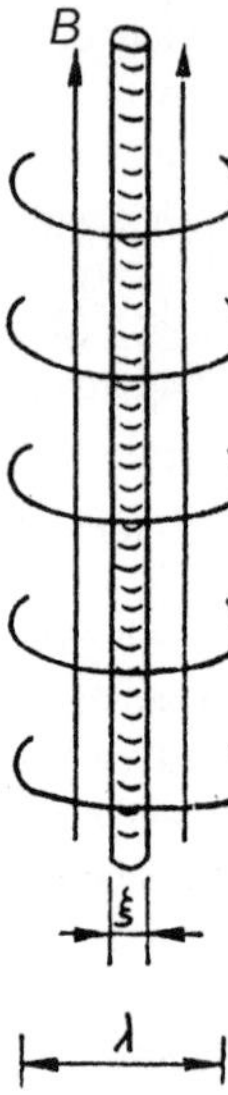

Fig. 1. Abrikosov vortex. The arrows show the direction of the supercurrent flow. The normal phase core region is cross hatched. The coherence length ξ and penetration depth λ are indicated.

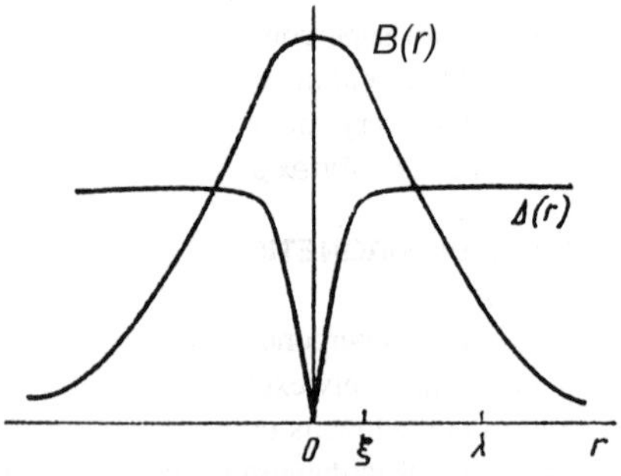

Fig. 2. Behaviour of the order parameter Δ and the magnetic field strength B in the region of an isolated vortex. Within the core ($r < \xi$) the order parameter $|\Delta|$ decreases rapidly, reaching zero at the center.

Abrikosov vortex, the outer diameter has a radius equal to the London penetration depth λ_L (see *Penetration depth of magnetic field*), and inside there is a core of radius equal to the *coherence length* ξ, in which superconductivity is suppressed by the high density of encircling shielding current that supports the magnetic field (Fig. 2). One circuit around the vortex axis changes the phase of the *order parameter* Δ by 2π. The magnetic field is able to penetrate the superconductor in the form of vortices because of the negative *surface energy* of the N–S boundary in a type II superconductor. When an externally applied magnetic field $\boldsymbol{B}$ is present there is a gain in energy via the formation of a microscopic inhomogeneous region with a maximal surface. The extent of the splitting is determined by the flux quantization condition, whereby the minimal *domain* size pierced by the magnetic field has the value $(\Phi_0/B)^{1/2}$. Since this minimal size cannot be smaller than ξ, it leads to the limitation that superconductivity is only possible for magnetic fields smaller than the value $B_{c2} = \Phi_0/2\pi\xi^2$ (upper critical field). In an ideal type II superconductor, Abrikosov vortices arrange themselves in a triangular lattice due to their mutual repulsion (see *Vortex lattices in superconductors*). When transport current flows perpendicular to the magnetic field direction Abrikosov vortices can drift in the direction perpendicular to $\boldsymbol{J}$ and $\boldsymbol{B}$ due to the *Lorentz force* $\boldsymbol{J} \times \boldsymbol{B}$, which causes the appearance of electrical resistance in the superconductor (see *Resistive state*). For a resistanceless supercurrent to flow in a superconductor (called a hard superconductor) the Abrikosov vortices should be held in position by pinning sites or defects of the crystal lattice (see *Kosterlitz–Thouless transition*, *Vortex pinning*).

ABSOLUTE DIAMAGNETISM, ideal diamagnetism

The state of a superconductor in which a magnetic field is completely excluded from its interior at a temperature below its *critical temperature* T_c.

This occurs when the applied magnetic field $\boldsymbol{B}$ is lower than the *thermodynamic critical magnetic field* $B_c(T)$ for a *type I superconductor*, and lower than the *lower critical field* $B_{c1}(T)$ of a *type II superconductor* (see *Meissner effect*). This implies that in the superconducting state the *magnetic susceptibility* χ attains its highest negative value $\chi = -1$ in SI units ($\chi = -1/4\pi$ in Gaussian units). Absolute diamagnetism is caused by the persistent supercurrents which circulate in the outer surface layer of the material and shield the interior from the penetration of magnetic flux (see *Penetration depth of magnetic field*). This diamagnetism results in repulsion between a superconductor and a magnet, and the levitation of a magnet above a superconductor.

Absolute diamagnetism is utilized in the manufacture of frictionless bearings, and the operation of trains levitated above their tracks.

ABSORPTION (fr. Lat. *absorptio*, suck)

Absorption is the uptake or extraction of ingredients dissolved in or mixed with the bulk of a gas or liquid by another solid or liquid material called an *absorbent*.

Absorption is one of the cases of *sorption*. In contrast to the surface process of *adsorption*, absorption occurs in the bulk of the absorbent. Absorption can be complicated by the chemical interaction of the absorbing substance with the absorbent (*chemisorption*).

Absorption is used in industry to separate the components of a gas mixture by selective absorbents, possibly obtaining them in pure form by *desorption* with absorbents that can be reused afterwards. Examples are (1) the absorption of butadiene during the manufacturing of synthetic *rubber*; (2) removing gases from harmful admixtures (e.g., H_2S, SO_2, CO_2, CO); (3) obtaining a gas product such as sulphuric acid by absorption of SO_3, (4) the extraction of hydrocarbon gases (natural and synthetic) from their mixtures (e.g., propanes, butanes and pentanes).

ABSORPTION COEFFICIENT, absorption factor, absorptance, absorptivity

The ratio of energy (particle) flux, absorbed by a given sample, to the incident flux. The term "absorption factor" is sometimes used in reference to an *absorption index* that is defined as the natural logarithm of the reciprocal of the absorption factor.

ABSORPTION EDGE

See *Intrinsic light absorption edge*.

ABSORPTION INDEX

A quantity α, which is the reciprocal of a distance, by which radiation flux in a parallel beam is attenuated by the factor of $1/e$ as a result of its absorption by the material (cf. *Absorption coefficient*). It is introduced to characterize radiation intensity $I = I_0 e^{-\alpha x}$ which decays exponentially with the distance (see *Beer–Lambert law*). The value of the absorption index in *crystals* may depend on the direction of propagation, and the type of polarization of the radiation (see *Dichroism of crystals*, *Pleochroism*).

ABSORPTION, LOW-FREQUENCY

See *Low-frequency enhanced magnetic resonance absorption*.

ABSORPTION, NONRESONANT

See *Nonresonance absorption*.

ABSORPTION OF ELECTROMAGNETIC WAVES in superconductors

The decrease in intensity of radiation, which is due to the partial transfer of radiation energy to the *normal vibrations* in the superconductor and, finally, to the crystal lattice. The rate of absorption of electromagnetic energy by a unit surface area of a superconductor is proportional to the active component of its *surface impedance*. At $T = 0$ the absence of this electromagnetic wave absorption exhibits a frequency threshold, i.e. it exists for $\hbar\omega \geqslant \Delta$ (where Δ is the *energy gap*), when an electromagnetic field quantum possesses enough energy to break a *Cooper pair*. Measuring this threshold is one of the methods for determining Δ.

In the *mixed state* of *type II superconductors* the absorption of electromagnetic waves is related to energy dissipation in the lattice of *Abrikosov vortices* and their viscous motion. In this case the measurement of electromagnetic absorption allows one to determine the *magnetic flux resistance* and the so called *pinning frequency*. In electromagnetic fields of high amplitude nonlinear effects arise which are related to the probability of generating vortices by the wave field, *kinetic decoupling of Cooper pairs*, and many-quantum absorption processes.

ABSORPTION OF LIGHT

See *Light absorption.*

ABSORPTION OF SOUND

See *Sound absorption.*

ACCEPTOR

An impurity or structural *defect* with a *local electronic level* in the *band gap* of a semiconductor which can receive electrons from the valence band.

An acceptor center is charged negatively if its local states are filled with electrons, and it is neutral if the states are empty. A predominance of acceptors over donors produces p-type conductivity (see *Holes*) with a temperature dependence determined by the height (typically <0.1 eV) of the energy level in the gap above the top of the valence band. Acceptors can also affect the carrier mobility and other nonequilibrium processes. Acceptors are divided into shallow and deep types. The former, hydrogen-like centers, are well described in the effective mass approximation. The most typical example is a Group III atom like In or Ga in a Group IV atomic semiconductor like Si or Ge. Deep acceptor states in Si and Ge arise from impurities of Groups I (except for Li which is interstitial) and II, as well as to *amphoteric centers*: vacancies, Te, W, and several *transition metals.* In an n-type semiconductor the presence of an acceptor compensates the charge of a donor.

ACCOMMODATION on a surface

The process of transferring part of the kinetic energy of a gas atom or molecule to induce the excitation of quasi-particles (*phonons*, *plasmons*, *electron–hole pairs*) at the impact of a gas with a *solid surface.*

It is characterized by the *accomodation coefficient* $\alpha = \Delta E/E$, where E is the incident particle energy, and ΔE is the energy transferred. The accommodation coefficient depends on the energy E, the temperature of the solid body, and the mechanism of the interaction between the particle and surface. Typically, neutral particles excite phonons in solids, while charged ones can also excite surface plasmons. When the particle is chemically active, the excitation of *electron–hole pairs* occurs. If the particle energy after the collision is less than its potential energy at infinity, the particle "sticks" to the surface and adheres to it in the adsorbed state (see *Adsorption*). The latter process is characterized by a *sticking coefficient*, defined as the ratio of the number of adhering particles to the total number of incident particles per unit time. Accommodation plays an important part in the processes of adsorption, *heat transport*, *friction*, and heterogeneous catalysis.

ACCUMULATION of current carriers

Increase of carrier concentration in part of a bulk *semiconductor* as a result of an externally applied field.

This process most frequently occurs in a potential well formed near a potential barrier, when the external field is oriented so that it drives the carriers toward the barrier. When the field increases, the accumulation increases and becomes more noticeable until it is limited by the lowering of the barrier height by the external field. When current flows in a semiconductor with bipolar electrical conductivity (see *Electrical conductivity*), the accumulation of both electrons and holes can simultaneously occur at the same location due to their mutual attraction. In this case, the size of the accumulation domain can reach several *diffusion lengths.* Near a semiconductor surface, either electrons or holes can accumulate depending on the sign of the charges outside of it (for example, depending on the sign of the voltage applied to the metal electrode of a *metal–insulator–semiconductor structure*). In more precise terminology the aggregation of majority carriers near a surface is called *accumulation*, while that of minority carriers is called *inversion*.

ACOUSTIC ANALOG OF COTTON–MOUTON EFFECT

Consists in the *acoustic wave ellipticity* and the *rotation of acoustic wave polarization plane* (originally linearly polarized) as it propagates normal to a static applied magnetic field (magnetic induction) $\boldsymbol{B}$. The polarization plane of the initial wave should not be parallel to or normal to $\boldsymbol{B}$. This effect is caused by linear magnetic double refraction (due to different *phase velocities* of transverse waves polarized parallel and normal to $\boldsymbol{B}$) and linear magnetic dichroism (differential absorption of transverse waves polarized parallel and normal to $\boldsymbol{B}$, denoted as $\alpha_{\parallel}$ and $\alpha_{\perp}$, respectively). For $\alpha_{\parallel} = \alpha_{\perp}$ the dependence of the ellipticity ε on the distance l covered by an acoustic wave in an elastically-anisotropic medium (see *Anisotropic medium*) is represented by a saw-toothed curve, and the dependence of the rotation angle of the polarization plane Φ on l jumps by $\pi/2$ at points where $\varepsilon = \pm 1$. A difference between $\alpha_{\parallel}$ and $\alpha_{\perp}$ smooths out the jogs on the curves $\varepsilon(l)$ and $\Phi(l)$. This effect provides information about magnetoelastic constants. It was first observed by B. Luthi (1963) in iron-yttrium *ferrite*, and later it was seen in rare-earth garnets, *magnetite*, nickel, and $RbNiF_3$ crystals.

ACOUSTIC ANALOG OF FARADAY EFFECT

Rotation of acoustic wave polarization plane and *acoustic wave ellipticity* during its propagation parallel to an applied static *magnetic field* $\boldsymbol{B}$ along a *symmetry axis* of rotation not lower than third order (see *Crystal symmetry*). This acoustic analog of the *Faraday effect* arises from the interaction of the elastic subsystem with either the spin subsystem, or with the of *conduction electron* subsystem. This interaction may have either a nonresonant, or a resonant character. In the case of a nonresonance interaction, and also of a weakly coupled resonant interaction (characterized by small variations of dispersion curves of normal modes near resonance) the rotation of the polarization plane is caused by *magnetic circular birefringence* (different *phase velocities* of oppositely circularly-polarized waves). By this, the rotation angle of the plane of polarization φ, as in the Faraday effect for linearly polarized light, depends on the distance l traversed by the acoustic wave in the elastically anisotropic medium: $\varphi = (1/2)(k'_+ - k'_-)l$, and the ellipticity ε is linearly connected with the *magnetic circular dichroism* (or differential adsorption of the circularly-polarized waves), having the asymptotic behaviour with increasing l

$$\varepsilon = \frac{\exp(-k''_+ l) - \exp(-k''_- l)}{\exp(-k''_+ l) + \exp(-k''_- l)},$$

where $k'_\pm$ and $k''_\pm$ are the real and imaginary components, respectively, of the complex wave vectors $k = k' + \mathrm{i}k''$ of the circularly polarized acoustic waves. In case of strong coupling (e.g., under conditions of *helicon–phonon resonance*) the ellipticity and rotation of the polarization plane results from interference not only of two, but of three circularly polarized waves, with φ and ε having oscillation dependences on l and $\boldsymbol{B}$. The acoustic analog of the Faraday effect was observed in metals under conditions of *Doppler-shifted cyclotron resonance* (Al, Cu, W), of *doppleron–phonon resonance* (In, W), of helicon–phonon resonance (K, In), and in *magnetic substances*: yttrium iron garnet and some other *iron garnets*.

ACOUSTIC BREAKDOWN

Change of electron path topology in *metals* in a magnetic field when acted on by an intense ultrasonic wave.

During an intraband acoustic breakdown the periodic strain acting on the sonic wave splits the energy band of a metal into several subbands, each of which is related to a specific electron path in the external magnetic field. An interband acoustic breakdown appears when the *quasi-momentum* of the sonic wave is close to the minimal distance between the electron paths in *momentum space* with no sound present. The interband acoustic breakdown always manifests itself in combination with *magnetic breakdown*: in the presence of sound, the transitions linked with the magnetic breakdown take place in much lower magnetic fields, and can lead to changes in the electron path topology. The acoustic breakdown causes the appearance of new periods of Shubnikov–de Haas oscillations (see *Shubnikov–de Haas effect*), as well as to changes of the more gradually varying part of the *electrical conductivity* tensor in strong magnetic fields.

ACOUSTIC CONTACT

A transition layer of a thickness significantly smaller than the sound wavelength, used for transmission of acoustic energy from a radiator to a solid-state sound wave guide, or between two sound guides. Depending on operating conditions, and frequency and polarization of the sound, the following typical acoustic contacts are used: the welding layer of metallized surfaces, silicone liquids, compounds with a rubber base, stilbene, crystal salol, epoxy resin, etc.

ACOUSTIC CYCLOTRON RESONANCE

Enhancement of the collision-free absorption of ultrasound in *metals* located in a magnetic field arising under the condition $\omega = n\Omega$, where $n = 1, 2, \ldots, \omega$ is the sound frequency and Ω is the *cyclotron frequency*. The enhancement arises because of the synchronization between the sound oscillations and the cyclotron motion of the electrons in the magnetic field. See also *Magnetoacoustic effects*.

ACOUSTIC DEFECTOSCOPY

See *Defectoscopy*.

ACOUSTIC DIFFRACTION

The phenomenon of acoustic wave bending around material bodies, obstacles and medium inhomogeneities with a size less than or comparable with the wavelength.

Acoustic diffraction exhibits all the features characteristic of wave motion. Diffraction effects play an important part in acoustics, manifesting themselves in emission, propagation and detection of the oscillations. As the wavelength decreases, acoustical diffraction becomes less noticeable, and wave propagation occurs in the accordance with the laws of *geometrical acoustics* (see *Acoustic wave scattering in solids*). This permits the use of acoustic oscillations for nondestructive testing (see *Defectoscopy*, *Acoustic microscopy*). Acoustic diffraction calculations are usually based on the *Huygens–Fresnel principle* according to which any acoustic field can be treated as resulting from interference of secondary wavelets emitted by virtual sources on the wavefront. See also *Raman scattering of sound by sound*, *Acousto-optic diffraction*.

ACOUSTIC DOUBLE MAGNETIC RESONANCE

Phenomena associated with the simultaneous interactions of electronic, paramagnetic, nuclear magnetic and other systems in solids with two alternating resonance fields, at least one of them being an acoustic field. The electron–electron, electron–nuclear and nuclear–nuclear acoustic double magnetic resonances (ADMR) can be distinguished. The most studied among electron–electron ADMRs are those with hypersonic and ultra-high frequency electromagnetic fields. If the frequencies of hypersound Ω_H and of the electromagnetic field Ω_E conform to the relation $\Omega_H \approx \Omega_E \approx (E_2 - E_1)/\hbar$ where E_1 and E_2 are the energies of quantum levels of the solid-state magnetic system, then the ADMR is displayed in the form of the interaction of hypersonic and electromagnetic fields with the amplitude, width, and shape of the lines corresponding to *electron paramagnetic resonance* and *acoustic paramagnetic resonance*. This provides information about *spin–lattice interaction* constants, *magnetic relaxation* times, etc. If the paramagnetic system is multilevel (e.g., a three-level system satisfying $E_3 - E_1 \gg E_2 - E_1$) then using an electromagnetic field tuned in frequency to the transition $E_3 \leftrightarrow E_1$ it is possible to invert the *level populations* of E_2 and E_1. As a result the amplification of hypersound at the frequency $\Omega_H \approx (E_2 - E_1)/\hbar$, in the presence of hypersonic resonance, generates phonons without an external acoustic field (see *Phaser*). Also ADMR is possible when both frequencies are acoustic and $\Omega_H \gg \Omega_E$. More complex versions of ADMR are resonances of the multiquanta phonon–photon and phonon–phonon types. With the help of *phonon–photon ADMR* the phase-sensitive detection of hypersound is carried out. In *electron–nuclear ADMR* the acoustic and magnetic fields effecting the magnetic system of an impurity with a nonzero nuclear spin bring about the polarization of nuclei, the value and sign of which is controlled by the frequency and intensity of the applied fields. Using electron–nuclear and also nuclear–nuclear ADMR allow one to investigate the dynamical processes in solids under less severe *selection rules* than is the case for purely electromagnetic *double resonances*.

ACOUSTIC EMISSION

Appearance of acoustic waves (see *Acoustic vibrations*) in solids as a result of the local dynamic reorganization of their internal structure.

Pulses of acoustic emission emerge in regions of rapid *slip*, *twinning of crystals*, *failure* (formation of *cracks*), *martensitic transformations*, or other lattice *phase transitions*, as well as in the domains of magnetic or electric structure variation, accompanied by lattice deformation (*striction*) (see *Magnetostriction*, *Electrostriction*). Such zones are referred to as sources (or sites) of acoustic emission. Common examples of acoustic emission are the crackling of bending tin and zinc rods, in both cases, the acoustic noise is the result of twinning. The detection of acoustic emission signals is carried out, as a rule, by piezoelectric transducers (see *Piezoelectricity*); also used are laser interferometers and other sensors, enabling the detection of rapid displacements of the local features of a solid surface. In some cases special sound-guides from the body under study to the sensor are applied. The processing of acoustic emission signals and the clarification of the characteristics of the acoustic noise is carried out by special electronic equipment, capable of measuring the number of signals per unit of time, the amplitude and duration of the signals, their form and their spectral content. For large enough bodies it is also possible to localize the spatial position of the emission sources. Acoustic emission parameters characterize the solid body's response to the mechanical, thermal, magnetic, electric, and other factors. The acoustic emission in macroscopic tests is typically a stochastic pulse process arising from the operation of individual sources distributed throughout the body bulk, and to some extent varying randomly in time. If the damping time of the signals and the duration of transient processes during their recording are longer than the time between the appearance of pulses, then the phenomenon is called *continuous acoustic emission*. A sequence of clearly separated signals with large amplitudes is called *discrete acoustic emission* (sometimes the term *explosive acoustic emission* is used). The energy of individual signals of the discrete emission exceeds the continuous emission signal energy by some orders of magnitude, and can be as high as roughly 0.1 J under material failure. The frequency spectrum of the emission ranges from sonic frequencies to frequencies of about 100 MHz. In common use as information parameters of acoustic emission are acoustic emission characteristics: namely the number of pulses of the discrete acoustic emission in the selected time interval; the total acoustic emission count, i.e. the number of continuous acoustic emission pulses which exceed a certain discrimination (limiting) level during the observation time; the acoustic emission count rate, or the number of continuous acoustic emission pulses per unit of time in the selected small interval; the acoustic emission signal energy, or the energy measured at the point of observation; the acoustic emission signal amplitude, or the maximum value of the emission pulse envelope energy; the acoustic emission voltage, or the mean square value of the electric voltage in the selected small time interval. In macroscopic tests of solids one seeks to establish the dependences of the above parameters on time, or to obtain the time dependence of parameters specifying the testing regime: *mechanical stress* or *strain*, temperature, magnetic and electric fields, etc. The results of acoustic emission detection under *plastic deformation* and material failure are represented by *acoustic emission diagrams* which provide a graphic image of the acoustic emission count rate N dependence on the deformation rate ε $(N(\varepsilon))$. The comparison of this diagram to the *deformation diagram* gives a qualitative insight into the intensity of the structure variation processes at different stages of deformation. As a rule, the acoustic emission diagrams exhibit one or several sharp spikes near the *yield limit* of a material, and a spike near the *ultimate strength* just before failure. *Strain hardening* is typically followed by a reduction in count rate. An important feature of acoustic emission from the bodies under strain is the *Kaiser effect*, whereby after stopping the deformation process, unloading the sample and then loading it again, the acoustic emission appears only when the load exceeds the level reached before. This emission is absent at the stage of elastic strain (see *Elasticity*), but appears and rapidly grows at the stage of microplasticity. The *acoustic yield point* of the material is the stress beyond which the sharp growth of acoustic emission count rate begins.

Experimentally, acoustic emission signals that accompany the individual acts of inelastic deformation are also observed: during the formation of twins, martensite inclusions (see *Martensitic polytypes*), reemission of the sound by an individual slip band, appearance of microcracks. The theoretical description of acoustic emission in the framework of *continuum mechanics* is based on the assumption that the structural changes taking place in the sources (*spontaneous transformation strain*) are matched to the ambient material without upsetting the macroscopic continuity, but cause the emergence of time-varying forces localized at the source (*internal stresses*), which excite the acoustic signals by their action. In some cases, use is made of formal modeling of the source by a system of point sources (dipoles, quadrupoles, etc.) varying in time according to a given law. Then, the problem is reduced to finding wave solutions of the dynamic equations of *elasticity theory*, which are satisfied by these forces. The dynamics of reconstructing the structure at the semimicroscopic level is described in terms of elementary “carriers”. In the cases of slipping, twinning, and martensitic transformations, the part played by such carriers is related to corresponding *perfect dislocations*, twin dislocations, and transformation dislocations (see *Dislocations*). The *phase interfaces* and the sides of the cracks can also be treated as the elementary “carriers”. The physical description of an acoustic emission source is reduced to establishing the relationships between spontaneous structure transformation, the strain and dynamic characteristics of the carriers, and the formulation of the dynamic equation of elasticity theory involving the distribution of the forces determined by the “carrier” fluxes. The end purpose of the physical theory of acoustic emission is to ascertain the connection between the wave solutions of the dynamic equation and the spatial position, speed and accelerations of “carriers”. In the framework of the approach, a great number of elementary mechanisms of acoustic emission connected with the dislocation motion is described: *annihilation of dislocations* of opposite signs, escape of dislocations to the crystal free surface (*transition radiation by dislocations*), the work of the dislocation *Frank–Read source*, the oscillatory motion of straight-line dislocations, and the dislocation segments fixed at ends. Also described are the acoustic signals generated by the motion of simple dislocation accumulations, sides of cracks, and *nucleation* in a *first-order phase transition*. An important trend in the theory of acoustic emission is the analysis of the transient procedures in recording and processing of acoustic emission signals, as well as the statistical and correlation analysis of pulse fluxes. The detection and analysis of acoustic emission yield important information on the structure and its change in time for solids exposed to various factors. That is why the phenomenon of acoustic emission is widely used as the basis for various *nondestructive testing techniques*, in particular for control of *corrosion cracking*, and technological processes such as diffusion saturation, *welding*, *hardening*, cutting. The use of acoustic emission control requires comparison of acoustic emission parameters with those of emerging and developing *defects*, especially at the stage when their further development becomes critical for the object under investigation. The acoustic emission method is also used in physical research for studying plastic deformation processes, failure, and phase transformation kinetics.

ACOUSTIC GENERATOR

See *Acoustic wave transducer.*

ACOUSTIC GYROMAGNETIC PHENOMENA

in magnetic materials

Rotation of acoustic wave polarization plane and the appearance of ellipticity of the initially plane-polarized transverse elastic wave propagating along the *magnetization* direction.

The phenomena take place in a medium with non-zero magnetization $\boldsymbol{M}$ (spontaneous or induced by an external magnetic field). They are conditioned by the emergence of an antisymmetric component of the dynamic *elastic modulus* (if $\boldsymbol{M}$ is parallel to the z axis, then the components $c_{yzxz} = -c_{xzyz}$). The first effect is related to the real part of the antisymmetric components, resulting in a difference between the *phase velocities* of right and left circularly polarized waves. The second effect arises from the imaginary part of these components, which causes the damping of the right- and left-polarized waves to be different (*magnetic circular dichroism*). Both effects increase sharply (in a resonant manner) with the frequency approaching *magnetoacoustic resonance*, as has been proven experimentally.

ACOUSTIC HOLOGRAPHY

A direct analog of optical *holography*, i.e. an interference mode of recording three-dimensional images using reference and object waves. The recording conditions are selected so that the image can be reproduced using the diffraction of light in the visible range. A phase interface between a liquid and a gas can serve as a recording medium. Under the action of the acoustic wave, the interface surface undergoes a deformation, and the relief which appears represents a *phase hologram* for the optical radiation. The use of *cholesteric liquid crystals* for recording the acoustic holograms is also available. At the locations of the antinodes of the acoustic standing wave the arrangement of liquid crystal molecules becomes disordered, conserving its initial ordered condition at the nodes. As a result, the extinction coefficient becomes space-modulated, and an *amplitude hologram* is detected.

The possibility of directly measuring the acoustic wave phase with the help of linear detectors (in contrast to optical detectors capable of recording intensity waves only) provides some unusual recording methods in acoustic holography. For example, during the recording of an acoustic hologram, the reference wave can be introduced in the form of an electrical signal that is summed with the acoustic signal transformed into an electric one. The instantaneous distribution of the acoustic waves in space is registered with the help of a rapidly moving range of detectors which record the complete information about the structure of the *wave front* scattered by an object. The reference wave is not needed for this recording method.

ACOUSTIC IMPEDANCE

A complex-valued quantity Z_a characterizing the resistance of a medium to the excitation or propagation of acoustic waves. Acoustic impedance is equal to the ratio of the complex amplitudes of sound pressure and the volume speed. The latter equals the product of the area S where acoustic impedance is determined times the normal component of the velocity of the oscillating medium particles, averaged over S. The real part of the acoustic impedance is associated with the losses in the system itself plus the radiation loss. The imaginary part is due to the reaction of the forces of elasticity and inertia. One can also define the *specific acoustic impedance* z_a and the *mechanical impedance* Z_m, and these are related to the acoustic impedance by $Z_m = Sz_a = S^2 Z_a$. At great distances from the source, Z_a coincides with the wave resistance of the medium, equal to ρv, where ρ is the medium density and v is the *sound velocity*. The ratio of the amplitudes of transmitted and reflected plane waves from the interface of two media, $z_a^{(1)}/z_a^{(2)} = \rho^{(1)}v^{(1)}/(\rho^{(2)}v^{(2)})$, also determines the matching conditions for acoustic waveguides. The efficiency of sound transducers, the power of emitters, etc., depend on the acoustic impedance match.

ACOUSTIC METHODS of studying solids

Determining properties of solids by measuring their acoustic characteristics. The acoustic methods are especially important for the study of opaque objects.

These methods cover a wide range of frequencies and are used, in particular, as *nondestructive testing techniques* for characteristics of solids. *Defectoscopy* can be complemented by investigating objects using an acoustical microscope, and this is especially effective for non-transparent media, or searching for inhomogeneities in transparent ones (see *Acoustic microscopy*). In addition to determining the macroscopic parameters of a medium, acoustic methods are successfully applied in investigations of its microscopic properties. *Acoustic paramagnetic resonance* and *acoustic nuclear magnetic resonance* provide information on the spin–phonon interaction (see *Spin–phonon Hamiltonian*), and the character of *crystal fields* (their symmetry, the presence of defects). Metals can also be studied. The investigation of the *acousto-electronic interaction* provides characteristics of the *electron–phonon interaction*, and the electrical conductivity of a medium. The observation of nonlinear acoustic effects enables the study the *phonon–phonon interaction* and related phenomena: *thermal expansion* and *thermal conductivity* of materials.

Ultrasonic spectroscopy is an important method of acoustic investigation. It can provide the spectral dependences of acoustic parameters, such as velocity, absorption, scattering of ultrasound. This method yields information on such physical characteristics of solids as *relaxation time*, electron *effective mass*, *Fermi surface* shape, as well as on the presence of structural inhomogeneities, in particular, *magnetic domain structure*.

ACOUSTIC MICROSCOPY

A technique which provides magnified acoustic images of small objects and measures the distribution of local physico-mechanical properties.

In acoustical microscopy, ultrasound and hypersound are used as probing radiation. As distinct from *optical microscopy* where the image is formed by *refractive index* variations, the contrast of an acoustic image results from changes in the acoustic impedance and in the sound absorption coefficient. Acoustic images have high contrast for both optically transparent and opaque objects. There are several schemes for acoustical microscopy implementation, including scanning-raster, scanning-laser, photo- and electronic acoustical microscopy. The latter two are based on the detection of an acoustic signal excited in the object by a focused optical or electron beam scanning its surface. In *scanning-laser* acoustic microscopy the object under study is placed in a liquid and exposed to an acoustic plane wave. After the interaction of the wave front with the object at the liquid boundary, the surface relief that is formed is read by a laser beam and forms an image on the display after signal processing. The *scanning-raster* approach is based on exposing the object under study to a focused electron beam created by an acoustical lens; with the object located near the focus of the acoustic microscope. It is possible to obtain the acoustic image by detecting the transmitted, reflected or scattered radiation, and scanning the focal region near the sample. In addition, this scanning-raster method permits the measurement of the local speed of the Rayleigh waves on the object's surface, and, hence, the distribution of the *elastic moduli* and their anisotropy (see *Elastic modulus tensor*, *Surface acoustic waves*). The resolution depends on the ultrasound frequency, and is typically comparable to optical microscopy resolution. The highest resolution achieved in cryogenic-scanning-raster microscopy is 25 nm. Acoustic microscopy is useful for studying the structure and *defects* of crystals and polycrystals, mono- and polycrystal *films*, many-layer systems, *coatings* and their *adhesion*, the structure of *composite materials*, *alloys* and *minerals*, biological tissues and cells; nondestructive testing of microelectronic elements, etc.

ACOUSTIC NUCLEAR MAGNETIC RESONANCE (ANMR)

Selective absorption of ultrasonic oscillation energy by nuclear *spins* in *solids*, occurring when the ultrasound frequency coincides with the energy level spacing of the nuclear spins in the external magnetic or intracrystalline electric field.

The *crystal lattice vibrations* at the ultrasonic frequency lead to the appearance of intracrystalline oscillating electric or magnetic fields which induce electric quadrupole or magnetic dipole transitions between nuclear spin levels. Therefore, ANMR can be thought of as an acoustic analogue of *nuclear magnetic resonance* (NMR). The ANMR is detected by the absorption of ultrasound at *resonance*, or by changes in the NMR signal via acoustical saturation of the spin levels. Further developments include acoustic nuclear–nuclear and electron–nuclear double resonances (see *Double resonances*), nuclear polarization (see *Nuclear orientation*), and *spin echo*.

ANMR has been studied in *insulators*, *semiconductors*, *magnetic materials*, and *metals*. These measurements provide values of the spin–phonon interaction tensors (see *Spin–phonon Hamiltonian*) and make it possible to clarify the *crystal field* theory and spin system dynamics descriptions. A method of crystal *defectoscopy* based on ANMR has been developed. For metals and high-conductivity semiconductors ANMR is the only method for investigating a spin system in regions not limited by the skin depth (see *Skin-effect*). These magnetoacoustic methods are useful for the study of ferro- and antiferromagnets. See also *Acoustic paramagnetic resonance*.

ACOUSTIC PARAMAGNETIC RESONANCE (APR)

Resonant absorption of hypersound by *paramagnetic centers* in a crystal.

This consists in the transfer of sound wave energy to these centers by the modulation of the *crystal field* acting upon them. APR is realized under conditions similar to EPR (see *Electron paramagnetic resonance*), when a quantum of the hypersound elastic vibrational energy is equal to the Zeeman splitting of the electron spin energy levels of the paramagnetic centers. The mechanism of APR or the transfer of acoustic vibrational energy to unpaired spins is the same as the mechanism involved in EPR relaxation via single-phonon processes. They both involve the *electron–phonon interaction*, and the intensity of an APR line, or the hypersound resonant absorption coefficient, is proportional to the strength of this interaction. The *selection rules* of the allowed transitions in the hypersound wave field, which are determined by the electron–phonon interaction, have a quadrupole character. Therefore, studying APR can provide direct information on the form of the electron energy spectrum of a paramagnetic center, and its electron–phonon interaction. In this way, one removes the known limitations of the EPR method, which involve the strict selection rules of allowed magnetic dipole transitions. The APR of hypersound waves can supplement information gained by EPR, and elucidate particular features of the paramagnetic center spectrum.

The APR measurements are conducted with pulse hypersound APR spectrometers (see *Acoustic spectrometer*) which record the resonant absorption of hypersound in dielectric crystals containing paramagnetic centers at low temperatures. It is possible to retrieve the electron spectral structure of the ground and excited states of the center by using the position of the APR line in the magnetic field, the anisotropy of the line, and the temperature dependence of its width and intensity. The APR method is also employed in studies of the electron–phonon interaction (the coupling between the paramagnetic center and the lattice vibrations). To this end, the APR line intensity is measured for different orientations of the wave vector of the longitudinal and transverse hypersound wave, and of the magnetic field relative to crystallographic directions. The complete set of electron–phonon interaction tensor components can be obtained. Investigations of the electron energy spectrum by APR make use of non-Kramers paramagnetic centers via forbidden magnetic dipole transitions, and in particular, those centers with an orbital ground state degeneracy in a cubic crystal field, plus strong electron–phonon coupling. The hypersound APR spectroscopy method (see *Acoustic spectroscopy*) has been employed to investigate non-Kramers spectra of the iron group, as well as spectra from centers in insulators and semiconductors with strong electron–phonon coupling. In the case of these centers, the APR line intensity dependence on the temperature, and the orientation of the magnetic field relative to crystallographic directions, provides the structure of the electron or electron-vibrational (vibron) levels. It is possible to evaluate the magnitude of the electron–phonon coupling, and assess the character of the relaxation processes arising from this coupling.

An intense hypersound propagating in a crystal with paramagnetic centers induces *non-linear APR*, which is accompanied by a number of specific phenomena such as *resonant self-channeling* and *self-focusing of the hypersound*. These occur because the change in the paramagnetic center exerts an influence back on the hypersound wave which has caused these changes.

The investigations of the APR of paramagnetic centers in crystals, and especially centers with strong electron–phonon coupling, are important because transition series ions often form centers of this type, and APR data can correlate with some optical and electrical properties of these materials.

ACOUSTIC PHONONS

See *Phonons*.

ACOUSTIC PLASMON (D. Pines, 1956)

The quantum of low-frequency *collective excitations* in some plasmas which have the linear acoustic *dispersion law* $\omega_q = qu$ $(q \to 0)$, where q is the plasmon wave vector and u is its phase velocity.

The branches of acoustic plasmons exist in multicomponent or low-dimensional Fermi systems with Coulomb interactions between particles (e.g., in a degenerate electron–hole plasma, or in a space-quantized electron gas in an *inversion layer* of a semiconductor) having a large difference between the *effective masses* of the "light" and "heavy" *current carriers*, or a strong anisotropy of the electron spectrum. For a two-component isotropic system subject to the conditions $m_1^* \ll m_2^*$ and $v_{F1} \gg v_{F2}$, where m_i^* and v_{Fi} are the effective mass and *Fermi velocity* of the ith component, respectively, regardless of the charge carrier sign, the acoustic plasmon spectrum in the momentum range $q \ll k_{F1}$ has the following form at sufficiently high densities:

$$\omega_q \cong q\left(\frac{\omega_{p2}^2 + \frac{3}{5}q^2 v_{F2}^2}{\kappa_1^2 + q^2[1 - \kappa_1^2/(8k_F^2)]}\right)^{1/2}, \qquad (1)$$

where k_{F1} is Fermi wave number of the light fermions, ω_{p2} is the plasma frequency of the *heavy fermions*, and κ_1^{-1} is the Thomas–Fermi *screening radius* of the "light" component (setting $\hbar = 1$). At $q \geqslant \omega_{p2}/v_{F2}$, the acoustic plasmon branch falls into the region of strong quantum *Landau damping* due to the decay into "heavy" electron–hole pairs, while at $q < \omega_{p2}/v_{F2}$ it lies in the region of relatively weak damping due to the decay into "light" Fermi excitations. The *phase velocity* and *group velocity* of acoustic plasmons at $q \to 0$ are

$$u = \frac{\omega_{p2}}{\kappa_1} = v_{F1}\left(\frac{m_1^* n_2}{3m_2^* n_1}\right)^{1/2} = v_{F2}\left(\frac{m_2^*}{3m_1}\right)^{1/2}\left(\frac{n_2}{n_1}\right)^{1/6}, \qquad (2)$$

where n_i is the concentration of the ith component, so the weak damping condition of the long-wave acoustic plasmons has the form $v_{F2} \ll u \ll v_{F1}$, i.e.

$$\frac{m_2^*}{m_1^*} \gg \max\left\{\frac{n_2}{3n_1}, 3\left(\frac{n_1}{n_2}\right)^{1/3}\right\}. \qquad (3)$$

After the transition of one or both components into the superconducting state (see *Superconductivity*), the acoustic plasmons dispersion relation remains practically the same, and only the acoustic plasmon damping in the energy range $\omega < 2\Delta_i$ vanishes, where Δ_i is the gap in the quasi-particle spectrum of the ith superconducting component.

The acoustic plasmons can be viewed as a Goldstone mode (see *Goldstone theorem*) in systems with a long-range Coulomb interaction, which arises due to the *spontaneous symmetry breaking* of the initial Hamiltonian as a consequence of the difference in the effective masses of the particles. That is, when the "light" fermions have the time to shield the "adiabatically slow" (low-frequency) density oscillations of the "heavy" fermions, and they effectively block the long-term interaction which prevents the acoustic branch (*hydrodynamic sound*) from appearing. The "heavy" component can also be acted upon by the charged bosons, e.g., *bipolarons* of large enough mass.

In a *two-dimensional electron gas*, the *plasmon* spectrum is quasi-acoustic with a square-root behaviour ($\omega_q \sim \sqrt{q}$) and weak damping in the case of one component, and it is linear ($\omega_q \sim q$ at $q \to 0$) in the case of two components. In layered *quasi-two-dimensional crystals* and chain *quasi-one-dimensional crystals* with the Coulomb interaction between conducting layers and chains, the plasmon spectrum is three-dimensional even when disregarding the electron tunneling (see *Tunneling phenomena in solids*) between neighboring layers and chains, but it is anisotropic, $\omega_q \sim (q_\parallel/q)\omega_p$, where $q = (q_\parallel^2 + q_\perp^2)^{1/2}$, $q_\parallel$ and $q_\perp$ being the longitudinal and transverse components of the momentum, respectively, so that $\omega_q \to 0$, but $q_\perp \neq 0$. The virtual low-frequency acoustic plasmons can play an important role in weakening the Coulomb repulsion between electrons and thereby favor superconductivity in anisotropic (multiband) and low-dimensional metals (see *Plasmon mechanism of superconductivity*).

ACOUSTIC RADIATION BY DISLOCATIONS

Generation of acoustic waves in a crystal, which accompanies *dislocation* motion and various dislocation reactions. This is one of the most important mechanisms of *acoustic emission* by a under *strain* crystal.

Acoustic radiation by dislocations in a homogeneous medium arises in nonstationary processes, as well as at the initiation and *annihilation of dislocations*. In inhomogeneous media, elastic waves are also excited under stationary motion of dislocations at points where changes occur in *elastic moduli* (*transition radiation* by dislocations). A particular case of transition radiation is observed at the escape of a dislocation to the free surface of a body. A dislocation segment (see *Dislocation string*) fixed at its ends can vibrate under the action of external oscillating stresses in the manner of a tight string. While moving, a dislocation also experiences the action of the periodic Peierls potential relief (see *Peierls–Nabarro model*) and alternating fields of *internal stresses*; it can be pinned by *impurity atoms*, and break away from them. All of these nonstationary processes make contributions to the radiation of elastic waves.

ACOUSTIC RESONANCE, X-RAY

See *X-ray acoustic resonance*.

ACOUSTICS

See *Geometrical acoustics*, *Nonlinear acoustic effects*, *Quantum acoustics*.

ACOUSTIC SELF-INDUCED TRANSPARENCY

Hypersound pulse propagation without resonant absorption of energy by impurity *paramagnets* and other resonant species, brought about by the coherent self-action of the hypersound via the resonant electron system of impurity centers.

The impact of a hypersound pulse on these resonant centers causes the inversion of the populations of their energy levels E_1 and E_2, with the subsequent stimulated emission of coherent *phonons* into the region behind the pulse. Because of this, the pulse energy remains constant, but its velocity and envelope shape change noticeably. Acoustic self-induced transparency may be realized only for pulses which are shorter than the relaxation time of the impurity centers. The region of the input pulse

$$\Theta = \hbar^{-1} \int_{-\infty}^{+\infty} \left| F_{12}(t) \right| \mathrm{d}t$$

(F_{12} is an *electron–phonon interaction* Hamiltonian matrix element) must exceed π. When $\Theta = 2\pi$ the pulse is stationary. If $\Theta > 4\pi$, the input hypersound pulse splits into 2π-pulses. Acoustic self-induced transparency has been observed experimentally in dielectric and ferroelectric crystals, alloyed by paramagnetic ions of the iron group, satisfying the conditions of *acoustic paramagnetic resonance* at liquid helium temperatures.

ACOUSTIC SOLID-STATE RESONATOR

A device involving *acoustic vibrations* in a mode close to that of a standing wave, viz. when the distance between the end plates of an acoustic resonator is an integral multiple of half-wave lengths.

Typically, an acoustic resonator is made from a piezoelectric material (see *Piezoelectricity*), and the oscillations are excited by an electric field. Piezoelectric acoustic resonators are electrical two-pole and multipole elements with the *impedance* exhibiting resonant behavior near resonant frequencies. These resonators exhibit a high stability of the natural oscillation frequency, a long-term one of $\sim 10^{-6}$–10^{-8} Hz, and a short-term one of 10^{-8}–10^{-10} Hz and higher. These are used to stabilize oscillator frequencies (of resonators), and form amplitude responses of filters (filter resonators).

ACOUSTIC SPECTROMETER

An experimental installation for measuring *sound absorption* and *sound velocity dispersion*.

An acoustic spectrometer comprises the following basic components: a source of electromagnetic ultra-high frequency (UHF) oscillations, a component to pass on these oscillations to an *acoustic wave transducer* with a connection to the sample under study, a UHF section from the sample to the receiving end of the spectrometer, and a recording device. Acoustic spectrometers typically employ pulse and pulse-phase measurement techniques, those with continuous oscillations being less common. As a rule, the spectroscopic measurements determine the absorption and sound velocity dispersion as functions of external parameters such as magnetic field, temperature, pressure, etc., at a fixed UHF frequency. In pulse acoustic spectrometers, these quantities are the pulse amplitude and its traveling time through the sample under study, while in pulse-phase acoustic spectrometers the pulse that passes through the sample is compared to a signal coherent with it. The changes in amplitude and traveling time of the pulse through the sample are determined by beating the measured pulse with a reference signal. Pulse and pulse-phase acoustic spectrometers operate in both "transmission" and "reflection" modes. In the first case, two acoustic wave transducers are placed on opposite sides of the sample, whereas in the second one, the same transducer is used for exciting and recording the acoustic oscillations. The pulse duration must be much less than its transit time through the sample. Continuous wave acoustic spectrometers operate in transmission only, and are used in measurements requiring low sensitivity to absorption factor variations. To increase the acoustic spectrometer response, use is made of low-frequency modulation of external parameters, with subsequent synchronous demodulation or digital accumulation of the signal.

ACOUSTIC SPECTROSCOPY of solids

A collection of methods for investigating the resonant and oscillatory effects induced by the interaction of acoustic oscillations with different subsystems (electron, nuclear, spin, etc.) in a solid.

Acoustic spectroscopy of normal metals. The acoustic spectroscopy of pure metals in external magnetic fields at low temperatures is the most diverse case since a large number of non-damping low-frequency collective waves can exist in metals. Owing to the cyclotron character of conduction electron motion and its interaction with acoustic waves in pure metals at low temperatures, the oscillating and resonant features of the *sound absorption* and the *sound velocity dispersion* depend on the external magnetic field. If the conditions $kl, \Omega\tau \gg 1$ are met (l, τ are the length and time of the mean free path of conduction electrons, $\boldsymbol{k}$ is the wave vector of the acoustic wave, and Ω is the electron *cyclotron frequency*), one can observe a *magnetoacoustic effect* whereby particular electrons most effectively interact with the sound, as indicated by the magnitude and direction of the constant magnetic field relative to $\boldsymbol{k}$, and by the value of parameter $\omega\tau$ (ω is the frequency of the sound wave). Particular phenomena appear, such as *giant quantum oscillations* of sound absorption and velocity dispersion, *geometrical resonance*, *acoustic cyclotron resonance*, *Doppler-shifted cyclotron resonance*, coupled acoustic and electromagnetic waves. These magnetoacoustic effects can provide extremum cross-sections, *Fermi surface* geometry, cyclotron masses, *Fermi velocities* and other energetic and kinetic parameters of conduction electrons (see also *Electron–phonon interaction*, *Electromagnetic generation of sound*, *Acoustic wave ellipticity*, *Rotation of acoustic wave polarization plane*). Acoustic spectroscopy is also an effective technique for studying an *energy gap* and its dependence on temperature and crystallographic direction.

Acoustic spectroscopy of paramagnetic centers in insulator and semiconductor crystals is based on the application of hypersound waves for investigating the electron energy spectrum structure and the electron–phonon interaction of paramagnetic centers based on hypersound *acoustic paramagnetic resonance* (APR) spectroscopy. The resonant absorption of hypersound corresponding to APR in dielectric or semiconductor crystals containing paramagnetic centers is associated with induced transitions between the levels of the electron paramagnetic centers' energy spectrum, and the APR line intensity (resonant absorption value) is proportional to the electron–phonon coupling of the center. That is why the APR characteristics such as the *resonant magnetic field* and *line width*, as well as their dependence on the temperature, relative orientation of the magnetic field and crystallographic directions, yield direct and accurate data on the positions of the ground and excited energy levels of the paramagnetic center, and its interaction with selected crystal lattice vibrations (electron–phonon interaction). Hypersound APR spectroscopy is an effective technique for studying the energy spectrum of non-Kramers centers with forbidden magnetic-dipole transitions and, especially, paramagnetic centers with orbital degeneracy in cubic *crystal fields*, involving a strong electron–phonon–ion interaction.

Acoustic spectroscopy of magnetically ordered crystals. These crystals exhibit elementary excitations of magnetization, such as *magnons* or *spin waves*. The magnetic subsystem energy is determined by the electric (exchange) and magnetic (dipole–dipole and spin–orbit) interactions which depend on the distances between atoms. Magnons interact with *phonons* and undergo mutual scattering. This interaction is especially pronounced at crossings of magnon and phonon spectral branches in the region of *magnetoacoustic resonance*. In this case, coupled magnetoelastic waves having the amplitude equally dependent on the magnetization and elastic strain propagate in the crystal. Hypersound investigations of magnetoacoustic resonance, which underlie acoustic spectroscopy of magnetically ordered crystals, provide direct data on the structure of the magnon spectrum. One measures the resonant absorption value and the hypersound velocity change in a magnetic field as a function of the temperature for several relative orientations of the external magnetic field and the crystal. Acoustic spectroscopy is highly effective for the study of *phase transitions* in magnetically ordered crystals.

ACOUSTIC SPIN ECHO

Coherent acoustic response of an electric or nuclear spin system in a paramagnetic crystal to the application of an acoustic or acoustico-electromagnetic pulse sequence during its resonant interaction with a spin system.

Through the spin–phonon interaction (see *Spin Hamiltonian*), the acoustic pulse creates an in-phase precession of *dipole moments* or *quadrupole* moments which, after the end of the pulse, become dephased in the time T_2 under the influence of relaxation processes. If the excitation pulse length is less than T_2, the spin system dynamics is independent of the excitation field amplitude at that given time, but it is determined by its value in the preceding period. Therefore, the application of the second pulse after the time $\tau < T_2$ causes the reversal of the dephasing process so that the initial phased state is recovered after the time interval 2τ (so-called *phase memory effect*). In materials with a strong spin–phonon interaction acoustic oscillations are generated (*acoustic echo-response*) at the moment of full phasing, in addition to the generation of electromagnetic field pulses. Acoustic spin echo is the acoustic analogue of the magnetic resonance *spin echo* and *photon echo*. Acoustic spin echo enables one to measure the *relaxation time* and the spin–phonon interaction. Spin memory accounts for such nonlinear effects as *acoustic self-induced transparency* and *induction*.

ACOUSTIC VIBRATIONS of crystal lattice

A type (branch, mode) of small oscillations of *crystal* atoms, when a displacement of the center of mass of a *unit cell* occurs; sometimes they are called *translational vibrations*.

The *dispersion law* (the dependence of the frequency ω on the wave vector $\boldsymbol{k}$) of acoustic vibrations for small values of k is linear: $\omega = v_{\text{ph}}k$, where v_{ph} is the *phase velocity* of the atomic displacement waves. It is an ordinary sound wave (hence the term "acoustic branch"), i.e. a matter density wave with frequency 0–10^{13} s^{-1}. At high frequencies, the dispersion law becomes nonlinear. In a three-dimensional crystal, there are always three branches of vibrations: with $\boldsymbol{k}$ directed along the symmetric direction in the crystal, one longitudinal and two transverse modes exist (the same as for an elastic-isotropic medium), the displacement vectors in transverse waves being mutually perpendicular. The quanta of vibrations are referred to as *acoustic phonons*. See also *Crystal lattice dynamics* and *Phonons*.

ACOUSTIC WAVE AMPLIFICATION by current carrier drift

An effect which is the reverse of the electronic attenuation of acoustic waves. Amplification of acoustic waves by charge carrier drift occurs in *semiconductors* and layered structures of the piezodielectric-semiconductor type when the drift velocity of free current carriers exceeds the *phase velocity* of the acoustic wave. For this amplification to occur the acoustic waves must interact with free charge carriers. The strongest and most extensively studied interaction mechanisms are those involving the piezoelectric effect (see *Piezoelectricity*) and the *strain potential*. The *piezoelectric interaction* is dominant in *piezosemiconductors* at low frequencies (<10 GHz). When travelling through a piezoelectric crystal, an acoustic wave is accompanied by a wave of electric potential φ. The free charge carriers (electrons) screen the wave field and become clustered in such a way that a *space charge* travelling wave arises. As this occurs energy exchange between the wave and the electrons takes place. The physical features of this exchange depend essentially on the ratio between the acoustic wave length λ, and the length l of the charge carrier mean free path. If $l/\lambda \ll 1$, then electrons

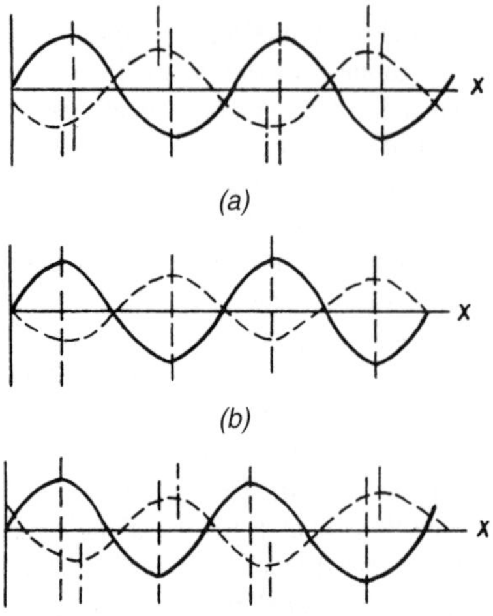

Fig. 1. Phase relationships between potential wave oscillations (solid lines) and electron concentration oscillations (dashed lines) for three successive times (a), (b) and (c).

at wavelength λ undergo a large number of collisions. Due to momentum relaxation, the motion of electrons with the wave is not inertia-free: as a result, the peak values of the electron concentration lag behind the potential minima, and the majority of grouped electrons become located at those positions of the charge pattern where the wave field accelerates electrons in the wave propagation direction (see Fig. 1(a)). As this takes place, the energy of the wave is transferred to electrons, and the wave decays.

The generation of a *current carrier drift* in the wave propagation direction in a semiconducting crystal brings about a change of the value of the absorption. Once the drift velocity v_d equals the acoustic wave velocity v_s, the concentration maxima coincide in space with the negative potential peaks, and the electron absorption discontinues because the wave no longer expends energy on inducing electron transport (Fig. 1(b)). When $v_d > v_s$, the peak values of electron concentration are ahead of the negative peaks of the potential (Fig. 1(c)). The electrons, which had been accelerated by the drift field are now decelerated by the alternating wave field. The energy, which is acquired by electrons from the electron drift generator, is transferred to the wave during the deceleration. The wave amplitude increases, and the wave builds up. For low wave amplitudes (in the linear amplification mode when $e\varphi/(k_B T) \ll 1$), the wave builds up exponentially as it propagates. For amplification, when $v_d > v_s$, the *electron absorption coefficient* becomes negative, and for $l/\lambda \ll 1$ it is given (in cm^{-1}) by

$$\alpha_e = \frac{K^2 q}{2} \frac{\omega\tau_M(1 - v_d/v_s)}{\omega^2\tau_M^2(1 - v_d/v_s)^2 + (1 + q^2 r_D^2)}, \tag{1}$$

where K is the electromechanical coupling constant for the given direction of wave propagation, τ_M is the *Maxwell relaxation* time, r_D is the Debye *screening radius*, ω is the wave frequency, and $q = 2\pi/\lambda$. Eq. (1) provides, for the most part, an adequate description of experiments on acoustic wave amplification in piezosemiconductors which exhibit moderate electron mobility: CdS, CdSe, ZnO, GaAs, etc. ($\mu = 50$–$5500\ cm^2 \cdot V^{-1} \cdot s^{-1}$). The general nature of the relation between the electron absorption coefficient and the drift velocity is shown in Fig. 2. Experimentally obtained electron amplification coefficients are of the order of tens and even hundreds of dB/cm.

The nature of the interaction changes with increased frequency and with increased *mean free path* of the carriers. As the wavelength approaches the carrier mean free path the energy exchange between free carriers and the waves ceases to be dependent on collisions. In the case of a purely

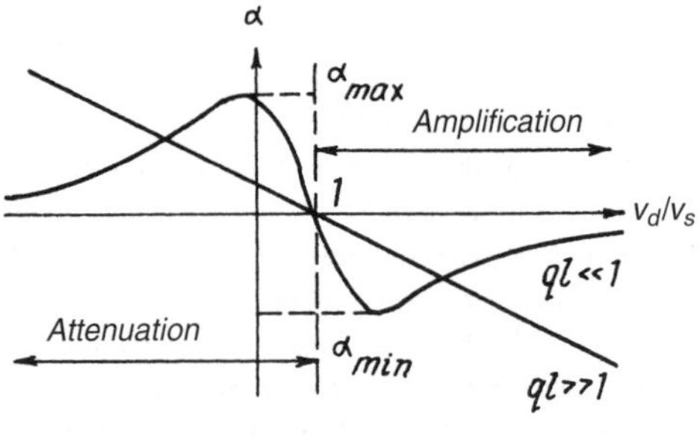

Fig. 2. Dependence of the absorption coefficient α on the ratio v_d/v_s of the electron drift velocity to the wave velocity for the two limits $ql \ll 1$ (curve) and $ql \gg 1$ (straight line). Note that for both limits at $v_d = v_s$ the coefficient α changes sign.

collisionless interaction ($l/\lambda \gg 1$), the energy exchange with the wave involves only those carriers (e.g., electrons), which exhibit the projections v_{Tx} of the thermal velocity on the wave direction x close to the acoustic wave velocity (see *Landau damping*). The role played by the remaining electrons is limited to shielding the wave piezoelectric potential. In the process of energy exchange, the carriers characterized by $v_{Tx} < v_s$ gain energy from the wave, whereas those of $v_{Tx} > v_s$ give up energy to the wave. In the absence of drift, as well as at $0 < v_d < v_s$ (or under the conditions of back drift), most of the carriers belong to the former group (i.e. exhibit $v_{Tx} < v_s$), and hence the wave is attenuated. Under supersonic conditions favoring the drift of carriers ($v_d > v_s$), the electron velocity distribution exhibits a distinguishing inversion, namely, the electrons that give up their energy to the wave exceed the number of those that gain energy, and hence the sound wave is amplified. The energy of the source of the drift field is transferred to the waves, and the electron absorption coefficient (in cm^{-1}) changes sign at $v_d = v_s$:

$$\alpha_e = \frac{K^2 q}{2} \frac{\sqrt{\pi}(v_s/v_T) q^2 r_D^2}{(1 + q^2 r_D^2)^2} \left(1 - \frac{v_d}{v_s}\right). \qquad (2)$$

The amplification coefficient is a linear function of the drift velocity. The collision-free mode of interaction is easily observed in semiconductors featuring a high electron mobility, e.g., in n-InSb, at liquid-nitrogen and liquid-helium temperatures at sound frequencies $\geqslant$200–500 MHz.

As the frequency increases ($\geqslant$10 GHz), the piezoelectric coupling mechanism gives way to the *strain-potential mechanism*, which occurs in all semiconductors, not only in piezoelectric ones. At high frequencies and low temperatures the electron de Broglie wavelength in certain semiconductors becomes comparable to the sound wavelength. In this case quantum effects are of considerable importance in the *acousto-electronic interaction*. These effects manifest themselves in certain features of the screening of the electric field, and the attenuation and "shutdown" of acousto-electronic amplification at $\hbar q \geqslant 2p$, where p is the Fermi momentum or thermal momentum, depending on the electron statistics used. The effect of amplification of acoustic waves by charge carrier drift may be utilized in *acousto-electronic amplifiers*, which are the solid-state analogues of travelling-wave tubes. These devices convert a signal into an acoustic form, amplify it, and then convert it back into electromagnetic oscillations. The advantages of acousto-electronic amplifiers are low values of mass and overall dimensions. These devices, however, are not free from serious drawbacks, such as: high heat dissipation, high supply voltages, low efficiency, losses by energy conversion, high noise level, etc. Some of these disadvantages may be eliminated either fully or partially by application of amplifiers that operate by *surface acoustic waves*. The latter devices are based on laminated structures of a piezodielectric-semiconductor, the surface acoustic waves being initiated and propagated on the piezoelectric surface, which adjoins the semiconductor. The *acousto-electronic interaction in laminated structures* is induced by virtue of ac fields that penetrate from the piezoelectric into the semiconductor. Amplification of acoustic waves by current carrier drift was first observed by A.R. Hutson, J.H. McFee, and D.L. White (1961) in CdS crystals.

ACOUSTIC WAVE CONICAL REFRACTION

See *Conical refraction of acoustic waves.*

ACOUSTIC WAVE ELLIPTICITY

Quantitative characteristic of an elliptically polarized acoustic wave. The magnitude (absolute value) of the acoustic wave ellipticity (ε) is the ratio of the semiminor to the semimajor axes of the ellipse along which an element of volume of the medium of a propagating sound wave moves. The sign of the ellipticity indicates the direction of the motion. The limiting cases of ellipticity, namely $\varepsilon = +1$ and $\varepsilon = -1$, correspond to clockwise and counterclockwise circular polarizations while $\varepsilon = 0$ corresponds to a linearly polarized wave. The acoustic wave ellipticity is always accompanied by a rotation of the acoustic wave polarization plane. The acoustic wave ellipticity may be either related to the internal structure of the medium, or be due to the interaction of the matter with a magnetic field (see *Acoustic analog of Faraday effect, Acoustic analog of Cotton–Mouton effect, Rotation of acoustic wave polarization plane*).

ACOUSTIC WAVE FOCUSING

Concentration of the *sound* energy in a small region by the formation of a convergent *wave front.* In an isotropic medium the convergent wave fronts are portions of spherical and cylindrical surfaces. The minimum sizes of focused regions are of the order of the sonic wave length. Focusing is used for its local effect in small regions of space in the medium under investigation. Acoustic lenses and reflectors of corresponding shape may be used as focusing elements.

ACOUSTIC WAVEGUIDE

A region of a medium relatively small in cross-section compared to its length, which localizes acoustic wave energy for the transmission and processing acoustic signals.

The energy localization inside the wave-guiding structure is influenced by the acoustic nonuniformity of the medium. In the case of a smooth nonuniformity, the velocity profile of an acoustic plane wave must have a minimum inside the structure in order to focus acoustic waves into the waveguide, taking into account wave front curvature. In the case of a sharp nonuniformity of the acoustical properties at the boundary, almost all the energy is localized inside the waveguide, and is retained there owing to the repeated reflections from the boundaries. The interference of the waves produces a complicated field pattern in the waveguide arising from the superposition of normal modes of acoustic oscillations.

Most widespread are *planar acoustic waveguides* which are inhomogeneous structures on the surface of *substrates*: *strip waveguides* formed from applied strips of another material, *topographic waveguides* involving grooves and ridges etched on the substrate, and *diffusion waveguides* wherein the inhomogeneities arise from the implantation of impurities in the substrate. There are also *fiber acoustic waveguides.*

Distinctive features of the waveguide propagation are the presence of many *acoustic vibration* modes with strong dispersion. The latter is widely used for the development of dispersion devices for acoustic signal processing: *ultrasonic delay lines*, filters, waveguide lenses. The pronounced energy localization inside a waveguide enables the directing and transmitting of acoustic signals: without loss from the wave divergence, multiplexing acoustic channels in space, and increasing the efficiency of the non-linear wave interaction.

ACOUSTIC WAVE REFRACTION

Variation of the direction of wave propagation in a nonuniform medium caused by the dependence of the *phase velocity* on the coordinates. The term "refraction of acoustic waves" refers to cases where *geometrical acoustics* is applicable, and it means the refraction of sonic beams at a planar or smoothly bent (on the scale of wavelengths) interface between two media. In the medium with a continuous dependence of the phase velocity on the coordinates the beam direction varies continuously, which involves *refraction of waves.* At the interface between two anisotropic media (see *Bulk acoustic waves*, *Elastic modulus tensor*) one incident plane harmonic elastic wave in the general case excites three waves – one quasi-longitudinal and two quasi-transverse refracted, and three reflected waves (see *Reflection of acoustic waves*). The wave vectors of all these waves lie in the plane of incidence, determined by the normal to the interface and by the incident wave vector. In anisotropic cases refraction of the wave vector in the reflection of energy (sonic beam) from the interface can also occur (see *Group velocity*).

ACOUSTIC WAVE ROTATION

See *Rotation of acoustic wave polarization plane.*

ACOUSTIC WAVE SCATTERING in solids

The process of formation of weak additional sound waves, differing in their characteristics from the incident wave, as a result of *acoustic diffraction* in a medium with inhomogeneities or obstacles. The process of scattering may be either coherent or incoherent, depending on whether the frequency of the vibrations is conserved or not during the scattering. In either case the scattering results in the attenuation of the sound wave. An example of *incoherent acoustic wave scattering* is the observation of hypersound in a dielectric crystal at low temperatures when the condition $\omega\tau > 1$ is satisfied (ω is the hypersound angular frequency, and τ is a thermal phonon lifetime). Absorption in that case is due to a three phonon scattering process whereby the incident phonon flux forms a hypersonic wave of thermal phonons with long lifetimes, a result which arises from *anharmonic vibrations* of the *crystal lattice.* Irreversible absorption of the hypersound takes place, with the incident energy transferred to the thermal phonons. For *coherent acoustic wave scattering* off static inhomogeneities in a solid, scattered waves of the same frequency appear, with characteristics that depend on the type of incident wave, the inhomogeneities, and the boundary conditions at the surface. An example of such a coherent process may be found in the scattering of sound from crystalline inhomogeneities, such as the temperature-independent residual attenuation of hypersound in dielectric crystals at low temperatures.

ACOUSTIC WAVE, SUBSURFACE

See *Subsurface volume acoustic waves.*

ACOUSTIC WAVE, SURFACE

See *Surface acoustic waves.*

ACOUSTIC WAVE TRANSDUCER

A device which transforms an electromagnetic vibration into acoustic ones, and vice versa. The majority of acoustic wave transducers are reciprocal, i.e. they may serve both as emitters and as receivers of acoustic vibrations. In ultrasonic and hypersonic investigations of *bulk acoustic waves* the most common transducers make use of the piezoelectric effect (see *Piezoelectricity*) or the phenomenon of *magnetostriction.* In resonance piezoelectric transducers, which have a thickness equal to an odd number of half wavelengths, monocrystalline quartz or lithium niobate (see *Niobates*) plates are used as operating elements at frequencies up to 1 GHz. Alternative materials are *barium titanate*, lithium tantalate, bismuth germanate, and *piezoelectric ceramics.* At frequencies in the UHF range the resonance acoustic wave transducers are applied at the samples under investigation, or at the sound wave guides in the form of *thin films* of ZnO, CdS and other *piezoelectric materials.* The alternating electric field is applied and recorded at the piezoelectric element either by the direct insertion of the latter into the electromagnetic circuit, or with the help of UHF resonators or transmission lines. A *thin magnetostriction transducer* of acoustic waves is a *magnetic film* applied at the end of the sample. The film magnetized by an applied magnetic field is deformed by the oscillating magnetic component of the UHF field. Improved matching of the *impedance* of the acoustic wave transducer with the UHF guides and resonators may be achieved with the aid of piezoelectric transducers based on the depletion layers of *piezosemiconductors*, formed with many-layered and many-element constructions. For the excitation and recording of *surface acoustic waves* one may employ *counter-pin transducers*, in which the distribution of the electromagnetic field and the corresponding deformation at the sound guide surface is pregiven or fixed. These acoustic wave transducers are also used for excitation and recording of volume modes, spread at an angle to the surface at which the transducer is placed. This mode of exciting volume oscillations is used in dispersion delay lines when the effect of the *sound velocity dispersion* is reflected by the passage through different distances of vibrations with different wave lengths. For measurements of the absolute values of *crystal lattice vibrations capacitance transducers* have found their wide application, with their design in the shape of flat plate capacitors, one plate of which is rigidly connected with the sample. See also *Phonon generation, Electromagnetic generation of sound.*

ACOUSTO-ELECTRIC DOMAIN

A stationary or moving region of strong electric field and an elevated intensity of acoustic noise which forms in *semiconductors* as a result of the amplification of thermal *crystal lattice vibrations* by the electric current at a charge carrier drift velocity faster than that of the sound (see *Acousto-electric effect*).

ACOUSTO-ELECTRIC EFFECT

Establishment of a constant current or electromotive force caused by the transfer of momentum and energy from a traveling ultrasonic wave to conduction electrons (see *Acousto-electronic interaction*).

The capture of electrons by the local electric fields arising in the conductor under the action of an ultrasonic wave leads to their being carried along in the direction of the sound propagation, and an *acousto-electromotive force* arises along the sample. When the sample is introduced into a closed electric circuit an electric current begins to flow through it. The current density is $j_{\text{A.E.}} = -2\mu\Gamma I/v$ (G. Weinreich, 1957), where Γ is the coefficient of sound absorption by the *conduction electrons*, I is the ultrasonic wave intensity, v is the *sound velocity*, $\mu = e\tau/m$ is the mobility, τ is the mean free path time. As seen from the expression for $j_{\text{A.E.}}$, the sign of the acousto-electric current or acousto-electromotive force $E_{\text{A.E.}} = j_{\text{A.E.}}\sigma$ (σ is the conductivity) depends on the sign of Γ, and can change depending on the ultrasonic wave propagation direction, its damping, or its amplification. During the propagation of a *surface acoustic wave* a transverse acousto-electric effect is observable. When a transverse acousto-electric current directed into the sample bulk arises along with the longitudinal one; this is linked with the wave localization on the surface. See also *Acoustomagnetoelectric phenomena, Surface acoustic wave.*

ACOUSTO-ELECTROMAGNETIC EFFECT

The appearance of a magnetic moment in a semiconductor crystal under application of a sufficiently strong electric field causing the amplification of acoustic noise (*phonons*). The generated flux of acoustic energy results in the movement of charge carriers (see *Acousto-electric effect*), and in some cases the charges may not move along the electric field direction (e.g., in an anisotropic *crystal* where the direction of maximal noise amplification may not coincide with that of the applied electric field). As a result, a circular electric current begins to flow around the sample, and hence the appearance of a magnetic moment. Acoustic energy flux injected into a sample from outside can induce a magnetic moment in the absence of the external electric field (*acoustomagnetic effect*). The nonlinearity of charge carrier motion is related both to the *anisotropy of crystals*, and the acoustic energy flux nonuniformity. This phenomenon is also observable in metals. The acoustomagnetic field arises, in particular, during the propagation of *surface acoustic waves*. In this case the acting force field is always nonuniform because the oscillatory displacement of the particles decays into the bulk of the sample.

ACOUSTO-ELECTRONIC AMPLIFIER

See *Acoustic wave amplification.*

ACOUSTO-ELECTRONIC INTERACTION (AEI)

Interaction of acoustic waves with charge carriers (*conduction electrons* and *holes* in *metals*, *semimetals* and *semiconductors*).

In metals, due to the Pauli's principle, the interaction with acoustic waves (*sound*) involves only electrons with energies close to *Fermi energy* E_{F}. There are two main mechanisms of AEI: (1) the *deformation mechanism* involves a change in *periodic potential* of the crystal lattice and a local change of the dispersion law of *quasi-particles* $E(\boldsymbol{p})$ in the electron *Fermi liquid*, (2) the *electromagnetic mechanism* is due to forces exerted on the electrons by electric and magnetic fields arising from the displacement of ions during the passage of the ultrasonic wave. For longitudinal sound, the fields are electrostatic and are screened by free electrons (see *Electric charge screening*) to a great extent. This leads to renormalization of the *strain potential* $\Lambda_{ik}^{(p)} = \lambda_{ik}(\boldsymbol{p}) - \langle\lambda_{ik}(\boldsymbol{p})\rangle$, where the angle brackets indicate averaging the bare deformation potential λ_{ik} over the *Fermi surface.* The quantity Λ_{ik} can be interpreted as a distinctive phonon electric-charge in the metal. During propagation of the transverse sound wave the electrons are affected by the vortex electric fields (see *Acousto-electric effect*). In an external magnetic field $\boldsymbol{B}$, an additional, so-called *induction interaction* of the electrons with the acoustic waves arises, its strength being $e(\boldsymbol{v} \cdot [\boldsymbol{U} \times \boldsymbol{B}])$, where $\boldsymbol{v}$ is the electron velocity, and $\boldsymbol{U}$ is the ion displacement velocity. This interaction involves the additional parameter $(qR_{\text{L}})^{-1}$, where q is the wave vector of the sound wave, and $R_{\text{L}} = p_{\text{F}}/eB$ is the electron Larmor radius. The induction interaction prevails over the deformation in sufficiently strong magnetic fields when $qR_{\text{L}} < 1$. Besides, in the presence of a magnetic field AEI leads to various *magnetoacoustic effects*, in which electromagnetic resonances, ballistic and collective effects add their characteristic features.

The great difference between the electron and ion masses leads to the fact that the *sound velocity* is many times less than the *Fermi velocity* of the electrons. This circumstance (*adiabaticity*) plays an important role in the interaction of electrons with sound. Adiabaticity supports introducing the concept of electrons and phonons as weakly interacting *quasi-particles*, and makes it possible to neglect inertial terms (*Stewart–Tolman effect*) in the AEI energy. In pure metals with a long *mean free path* l ($ql \gg 1$) the adiabaticity underlies the fact that only electrons with a velocity $\boldsymbol{v}$ almost perpendicular to the sound wave vector can effectively interact with the sound (*autophasing effect*).

In semiconductors, the AEI is mainly determined by the strain interaction related to the local change in the *band gap* under the action of the *strain* arising when a sound wave travels through the crystal. In *ionic crystals* and semiconductors with a large number of charged *impurity atoms* in the lattice, the electromagnetic mechanism of AEI is also possible.

In superconductors, the AEI at temperatures lower than *critical temperature of superconductors* differs in some features from the AEI in normal metals. Due to the Cooper pairing of electrons (see *Cooper pairs*), the *phonon* absorption probability decreases and, as a consequence, a sharp (exponential) decrease of the ultrasound absorption factor occurs. Besides, the *Meissner effect* precludes the electromagnetic mechanism of AEI.

ACOUSTO-ELECTRONICS

A branch of electronics dealing with the effects of excitation and propagation of acoustic waves in condensed media, the interaction of the waves with electromagnetic fields and electrons, and the development of *acousto-electronic devices* based on these effects.

The diversity of effects utilized in the design of acousto-electronic devices suggests the division of acousto-electronics into the high frequency (microwave) acoustics of solids (effects of excitation, propagation and reception of high frequency acoustic waves and hypersonic waves), acousto-electronics proper (interaction of acoustic waves with *conduction electrons* in the bulk and on the surface of solids, see *Acousto-electronic interaction*), and *acousto-optics* (interactions between light and acoustic waves). Acousto-electronics was formed as an independent part of electronics in the decade of the 1960s with the start of intensive studies stimulated by the discovery of the *acoustic wave amplification* by conduction electrons in cadmium sulfide crystals. The rapid development of acousto-electronics was caused by the necessity to create simple, reliable and small-scale radio signal processing devices for radio-electronic equipment. At the present time acousto-electronic devices perform the following signal transformations: (1) in time (delaying and changing the duration of signals), (2) in phase and frequency (frequency and spectrum conversion, phase shift), (3) in amplitude (amplification, modulation), and (4) those involving more complicated functional transformations (coding and decoding, signal convolution and correlation). In some cases, the acousto-electronic methods of signal transformation are simpler (as compared, e.g., to electronic methods) and, sometimes, they constitute the only possible choice. The use of these devices is predicated on the low propagation velocity of acoustic waves (compared to that of electromagnetic waves), weak absorption of acoustic waves in crystals with a high acoustic Q-factor (quality factor), as well as favorable interactions of these waves with electromagnetic fields and electrons. In acousto-electronic devices, both *bulk acoustic waves* and *surface acoustic waves* (SAW) are used. The manufacture of these devices employs *piezoelectric materials* and laminated structures composed of piezoelectric and semiconductor layers.

Many acousto-electronic devices convert high frequency electric signals into acoustic waves (*acoustic wave transducers*) to propagate in a sound guide and then be converted back again into a high frequency signal. Components used for excitation and reception of bulk acoustic waves are mainly *piezoelectric transducers*: piezoelectric plates (at frequencies up to 100 MHz) and film transducers (at frequencies above 300 MHz), while those employed for the excitation and reception of a SAW are opposite-pin transducers.

The *ultrasonic delay lines* operating in the frequency range from several MHz to several tens of GHz, as well as quartz resonators for the radio frequency generator stabilization (see *Acoustic solid-*

Table 1. Physical characteristics of some acousto-electronic materials

Material (name, formula)	Orientation	Coefficient of electromechanical coupling	Speed of sound, m/s	Existence of cut with zero temperature coefficient of delay	Acoustic loss in air (dB) at 100 MHz	Comments
α-quartz, SiO_2	ST, x	0.0017	3138	yes	0.03	Industrial product
Lithium niobate, $LiNbO_3$	y, z	0.045	3488	no	0.04	Industrial product
Lithium tantalate, $LiTaO_3$	x, 112°	0.05	3295	yes	0.03	Industrial product
Lithium tetraborate, $Li_2B_4O_7$	x, 6.7°	0.014	3808	no	0.04	Laboratory product
$Bi_{12}GeO_{20}$	[110], [111]	0.015	1681	no	0.0145	Industrial product
$Bi_{1.2}Sr_{0.8}TiSi_2O_8$	[110]	0.015	2640	yes	0.03	Laboratory product
$Pb_2KNb_5O_{15}$	z, 74.4°	0.073	2505	yes	0.05	Specialized laboratory product
$K_{2.98}Li_{1.55}Nb_{5.11}O_{15}$	[100]	0.09	3100	yes	0.05	Specialized laboratory product
Tl_3VS_4	[110], 24°	0.062	1010	yes	0.95	Specialized laboratory product
$Pb_{0.91}In_{0.03}[Zr_{0.44}Ti_{0.53}(Li_{3/5}W_{2/5})_{0.08}]O_3$	–	0.01	2270	no	0.10	Industrial product (high temperature and pressure)
$Pb[(Mn_{1/3}Sb_{2/3})_{0.08}Zr_{0.42}Ti_{0.50}]O_3$	–	0.01	2300	no	0.10	Industrial product (high tempetature and pressure)

state resonator), are among the most widely used acousto-electronics devices for bulk waves.

The most common are *acousto-electronic devices based on the SAW* due to small loss for wave excitation, the possibility of wave propagation control at any point of a sound guide (along the wave propagation path), and the feasibility for designing devices with controllable frequency, phase and other responses. Such devices include resonators on the SAW, which are used as narrow-band acousto-electronic filters and are also introduced into a generator circuit for its frequency stabilization; delay lines (for long signal delay in memory elements); various filters based on the SAW (e.g., band filters, matched filters); coding and decoding devices and so on. The interaction of acoustic waves with conduction electrons in semiconductors and laminated structures of the piezoelectric–semiconductor type leads to such phenomena as electron absorption or amplification of acoustic waves, the *acousto-electric effect*, and others. These effects are useful in acousto-electronic generators. The nonlinear effects arising under the interaction of acoustic waves in piezoelectrics, semiconductors, and laminated structures of a piezoelectric-semiconductor type find wide use in signal convolution and correlation devices. The interaction of light and acoustic waves in condensed media forms the basis for the operation of *acousto-optic devices* (deflectors, modulators, filters etc.). They can be used to control the amplitude, polarization, and spectral content of optical radiation, as well as its propagation direction (see *Acousto-optics*).

ACOUSTO-ELECTRONICS MATERIALS

Materials used to create working parts for the branch of *functional electronics*, which involves devices and systems for processing large and extra-large (order of gigabits) information flows. The physical basis for the operation of such devices is provided by the interaction of space and *surface acoustic waves* of various types (Rayleigh, Gulyaev–Bleustein, Love, etc.) with electromagnetic radiation fluxes at various frequencies – from 10 to 10^{11} Hz. The main physical characteristics of some important acousto-electronics materials are given in Table 1. Some layered systems for devices are based on surface acoustic waves,

Table 1. Physical characteristics of some layered acousto-electronic materials

Material	Coefficient of electro-mechanical coupling	Dielectric constant	Attenuation (dB) at 1 GHz	Carrier mobility, $cm^3 \cdot V^{-1} \cdot s^{-1}$
Ideal material	>0.010	<10	<1.0	>5000
Lithium niobate, $LiNbO_3$	0.045	36	1.0	–
$Bi_{12}GeO_{20}$	0.013	38	1.6	–
Lithium tetraborate, $Li_2B_4O_7$	0.012	10	(1.0)	–
Quartz, SiO_2	0.012	4.5	2.6	–
Aluminum nitride on colorless sapphire	(0.010)	(10)	(<1.0)	–
Zinc oxide on silicon	0.010	9	(3.0)	200
Cadmium sulphide on silicon	0.005	9.5	3.5	340
Gallium arsenide on silicon	0.002	11	4.2	8500
Silicon (substrate)	–	12	1.6	1500

and possibilities offered by applying a *piezoelectric material* layer to a *semiconductor* substrate (or vice versa) are realized in these systems. Parameters for the materials of such systems are given in Table 2.

ACOUSTOLUMINESCENCE

Luminescence in a *solid* arising under its excitation by an acoustic wave with an energy flux density in excess of a certain threshold value.

The threshold energy flux density (typically 1–10 W/cm^2) decreases as the density of structural *defects* (e.g., *dislocations*) rises. The luminescence is nonuniform throughout the bulk, and is concentrated in regions of defect accumulation. The spectrum of acoustoluminescence at the energy flux threshold is close to that of *photoluminescence* of the corresponding material, it contains bands associated with intrinsic structural defects and impurity centers, and it is stable in time (some hours). The increase of energy flux density by a factor of 1.5 to 2 relative to the threshold leads to a sharp change in the acoustoluminescence spectrum. A large number of narrow lines having half-widths $\leqslant 1$ meV appear instead of wide bands with half-widths of some tens of meV. The restructuring of the spectrum is accompanied by an instability in the amplitude and sign of the acousto-electric current. Under these conditions the luminescence is unstable in time. Alongside of the bulk luminescence there is a surface luminescence with a spectrum which depends on the composition and pressure of the ambient gaseous medium. See also *Acoustoluminescence*.

ACOUSTOMAGNETOELECTRIC PHENOMENA

Kinetic effects in *semiconductors* and *metals* arising from the entrainment of *conduction electrons* by an ultrasonic wave in the presence of an external magnetic field. As is the case with *galvanomagnetic effects*, acoustomagnetoelectric phenomena are due to charge carrier trajectories bending in the magnetic field (see *Lorentz force*). Identified as acoustomagnetoelectric phenomena are the acoustic analogs of the Hall effect, Ettingshausen effect, and so on. Their distinctive feature is that the ultrasonic flux rather than the external electric field creates the initial current. There also exists the *acoustomagnetoelectric effect* consisting in the appearance of an electromotive force in the direction perpendicular to the ultrasonic flux under the action of the magnetic field, when there is no acousto-electric current present (see *Acousto-electric effect*). In bipolar semiconductors and *semimetals*, the acoustomagnetoelectric effect is caused by the separation of the electron–hole flux drawn along by the ultrasonic wave in the magnetic field. In monopolar semiconductors and metals, the acoustomagnetoelectric effect is associated with the fact that carriers with different energies are entrained by the ultrasonic wave and deflected differently by the magnetic field. Therefore, the magnitude and even the sign of the monopolar effect are sensitive to the carrier scattering mechanism. Acoustomagnetoelectric phenomena are also possible in a planar configuration when the directions of the ultrasonic flux, magnetic field, and resultant electromotive force lie in the same plane. Acoustomagnetoelectric phenomena have been observed in Bi, Te, InSb, and *graphite*.

ACOUSTO-OPTIC DEFLECTOR

See *Light deflectors*.

ACOUSTO-OPTIC DEVICES

A class of active functional devices based on light diffraction by acoustic waves (see *Acousto-optics*, *Acousto-optic diffraction*).

These devices are used for measuring optical radiation characteristics, controlling its parameters, and processing optical, radio-frequency and acoustical signals. The basic element of an acousto-optic device is an acousto-optic cell, a *crystal* where the ultrasound is excited by the *acoustic wave transducer*, and where the acousto-optic interaction takes place. Since the cell has two inputs (optical and acoustical), two classes of acousto-optic devices can be distinguished.

One type for controlling light beam parameters includes adjustable *optical filters*, *light deflectors* and monochromatic radiation modulators (see *Modulation spectroscopy*). They have the advantages: fast response (adjustment time 1–30 μs), large light-gathering capability (angular aperture 1–15°, spatial aperture 1–10 cm^2), considerable resolution (number of resolvable points is 10^3–10^4), and operational flexibility. The adjustments of the transparency window filters and the beam scanning deflectors are implemented by changing the ultrasound frequency, whereas the light intensity modulation is carried out by controlling the sound power. These elementary acousto-optic devices can be components of more complicated instruments: spectrometers, group switches, scanners, etc.

Another class of acousto-optic devices is intended for processing signals carried by the ultrasound, e.g., radio-frequency signals. In spectrum analyzers the angular dependence of the diffracted monochromatic beam repeats the ultrasound spectrum, while in correlators the time dependence of the intensity of the radiation diffracted in two acousto-optic cells replicates the correlation function of the acoustic wave amplitude in the cells. These acousto-optic devices are components of various systems for radio signal processing.

Of particular importance are acousto-optic matrix processors intended for algebraic matrix operations over numeric data arrays because the latter can be entered into the processor through the acoustic and optical input.

ACOUSTO-OPTIC DIFFRACTION, diffraction of light by sound

A class of phenomena involving the deviation of light from a straight path in a medium under the action of an ultrasonic wave.

By changing the *refractive index* of the medium in a periodic manner the ultrasonic wave acts on the light as a phase *diffraction grating* moving at the *sound velocity*. When a monochromatic light beam passes through the region of a medium where an ultrasonic wave propagates, then beams of diffracted light appear. The frequencies and wave vectors of the incident (subscript I) and diffracted (subscript D) light waves are interrelated through the acoustic wave parameters ω_{ac} and k_{ac}:

$$\omega_D = \omega_I \pm m\omega_{ac}, \qquad k_D = k_I \pm mk_{ac},$$

where the order m of the diffraction is an integer. The following basic types of acousto-optic diffraction are known:

1. *Raman–Nath diffraction* (sometimes called *Debye–Sears diffraction* or *Lucas–Biquard diffraction*), which is equivalent to the diffraction by thin holograms, arises for normal light incident on a low-frequency ultrasonic beam when the relation $k_{ac}^2 L/k_I \ll 1$ holds, where L is the length of the domain of interaction of the light and ultrasound in the direction of the first diffraction maximum. After passing through the ultrasonic wave, the light beam exhibits a sinusoidal phase variation in the plane perpendicular to the direction of its propagation. As a result, a series of diffracted beams appears with frequencies ω_D characterized by $m = 1, 2, \ldots,$ propagating at angles $\theta = \arcsin(m\lambda/\Lambda)$, where λ and Λ are

the wavelengths of the light and sound, respectively. The Raman–Nath diffraction is negligible in the hypersound frequency domain where the above inequality is not satisfied.

2. At high frequencies when the inequality $k_{\mathrm{ac}}^2 L/k_{\mathrm{I}} \gg 1$ holds, then only the zero and first order maxima are observed for the diffracted light. This phenomenon is usually known as *Bragg diffraction*, because it arises only for the light incident on the ultrasonic beam at the *Bragg angle* $\theta_{\mathrm{B}} = \arcsin[\lambda/(2\Lambda)]$. This is equivalent to diffraction by a thick hologram. The diffraction efficiency is specified here by the formula

$$\frac{I_{\mathrm{D}}}{I_{\mathrm{I}}} = \sin^2\left[\frac{k_{\mathrm{I}}L}{4(M_2 I_{\mathrm{ac}})^{1/2}}\right]$$

and can be as high as 100%, where $M_2 = p^2 n^6/\rho v^3$ is the acousto-optic quality factor of the crystal determined by its density ρ, refraction index n, photoelastic constant p, and sound speed v.

Under acousto-optic diffraction in anisotropic media, the polarization of the incident and diffracted light are generally not the same. Therefore, under this diffraction with a rotated plane of polarization in an anisotropic medium (anisotropic diffraction), as distinct from diffraction in an anisotropic medium without any polarization plane rotation or in an isotropic medium, the angles of incidence and diffraction are generally not the same. For oblique light incidence on a solid surface, along which a *surface acoustic wave* propagates, the diffraction of the light is observed in both the transmitted and reflected beams. This is explained by the fact that, instead of photoelastic changes of dielectric constant, there is a periodic bending of the surface, which corresponds to the normal component of the displacements in the surface acoustic wave, and creates a varying difference in path-length for the light waves. In a conducting medium the acoustic wave creates two diffraction gratings with the same spatial period, typically phase-shifted, for the electromagnetic wave. One of the gratings corresponds to a change in dielectric constant owing to the deformation of the lattice by the acoustic wave. The other one constitutes an electron diffraction grating due to the local change in electron concentration under the action of the ultrasonic wave. See also *Brillouin scattering, Raman scattering of light.*

ACOUSTO-OPTICS

Study of the diffraction of optical radiation by acoustic waves in solids, and the development on this basis of devices for controlling light flux parameters and processing optical and radio signals (see *Acousto-optic diffraction, Acousto-optic devices*).

The basis of *acousto-optic diffraction* in crystals is the photoelastic effect (see *Photoelastic tensor*). The elastic wave creates a periodic disturbance of the crystal *dielectric constant* which acts as a phase *diffraction grating* for the light. The frequencies and wave vectors of the incident and diffracted light waves are coupled through the parameters of the acoustic wave. The spectral and angular selectivity of acousto-optic diffraction can be controlled by changing the sound frequency, i.e. the period of the acoustic diffraction grating. The sound intensity determines the energy distribution between the initial and diffracted rays. These effects underlie acousto-optic methods and *visualization of acoustic fields*, measurement of different acoustic characteristics of materials, and a wide range of acousto-optic devices.

ACOUSTOTHERMAL EFFECT

Appearance of a temperature difference in a thermally isolated body as a consequence of a heat transfer by a traveling acoustic wave.

The acoustothermal effect is a result of the departure of the sound wave regime from adiabatic behaviour. It is more pronounced in materials with a high thermal conductivity (metals and metal alloys). If the *thermal conductivity* κ rises with increasing pressure, P, the sample-averaged gradient of temperature ∇T is oriented opposite to the wave propagation direction

$$\langle\nabla T\rangle = -\frac{1}{2}\frac{\partial\kappa}{\partial P}\alpha\langle T\rangle\frac{K^2 b^2\omega^4}{v\omega^2\kappa^2 + v^5 c^2},$$

where α is the volume *thermal expansion* coefficient, K is the bulk modulus, b, ω and v are the amplitude, angular frequency and *phase velocity* of the wave, respectively; and c is the *specific heat* per unit volume. With the *strain* amplitude in the wave of the order of 10^{-4} in metals, $\langle \nabla T \rangle$ is several hundred K/m.

ACTINIDES (fr. Gr. $\alpha\kappa\tau\iota\varsigma$, beam, and $\varepsilon\iota\delta o\varsigma$, type)

A family of 14 radioactive elements with a partially filled $5f^n$ electron shell and an atomic number in the range 90 to 103, located in the seventh period of the periodic system of elements after actinium Ac.

Like Ac itself, these elements fit into Group III of the system. The actinides include, in atomic number order, the following: *thorium* Th, *protactinium* Pa, *uranium* U, *neptunium* Np, *plutonium* Pu, *americium* Am, *curium* Cm, *berkelium* Bk, *californium* Cf, *einsteinium* Es, fermium Fm, mendelevium Md, nobelium No and lawrencium Lr. The first three actinides (Th, Pa, U) are found in nature (U and Th – in sizeable quantities). The remaining actinides were artificially synthesized during the years from 1940 until 1963 with the help of nuclear reactions, and Np and Pu were later detected in radioactive ores. Like the *lanthanides*, the actinides are often listed in a row at the bottom of the periodic table. The symbol An is sometimes used as a common designation for an actinide. All these elements have similar physical and chemical properties, and their electronegativity values range from 1.1 to 1.2. Free actinide atoms have the inner electron configuration $[\mathrm{Rn}]5f^n6s^26p^6$ plus one $6d$ and two $7s$ valence electrons. The latter is responsible for the usual formation of the species An^{3+} in ionic compounds, although other valence states do form. In the neutral atoms of actinides and ions with the same charge (e.g., An^{3+} and An^{4+}), the radius gradually decreases in proportion to the increase of the atomic number; this is the so-called *actinide compression* (e.g., the U^{3+} radius is 0.104 nm, Np^{3+} is 0.102 nm and Pu^{3+} is 0.101 nm).

Of all the actinides, U and Pu find the widest practical application. The isotopes ^{233}U, ^{235}U and ^{239}Pu serve as a nuclear fuel in nuclear reactors, and act as explosives in nuclear weapons, while ^{232}Th and ^{238}U can serve as source material for the production of fissionable ^{233}U and ^{239}Pu. Several actinide nuclides emit α-particles in their radioactive decay (^{238}Pu, ^{242}Cm, and others).

ACTINIUM, Ac

Radioactive chemical element of Group III of the periodic system, with atomic number 89.

It has no stable isotopes. The mass number of the longest living isotope is 227. The electron configuration of the valence outer shells is $6d^17s^2$. Successive ionization energies are 6.9, 12.06, and 20 eV. The neutral atom atomic radius is 0.203 nm; and the Ac^{3+} ionic radius is 0.118 nm. The oxidation state of ionic compounds is +3, and the electronegativity is ~1.05. Actinium in a free form is a silvery *metal*. It has a face centered cubic lattice with parameter $a = 0.5311$ nm at 293 K, and crystallographic space group $Fm\bar{3}m$, O_h^5. The density is 10.06 g/cm^3 at 298 K, $T_{\text{melting}} = 1323$ K, $T_{\text{boiling}} = 3473$ K. The heat of melting is 10.5 kJ/mole and the isothermal bulk modulus is 25 GPa at room temperature. Chemically, it is the highest analogue of lanthanum; and a dangerous radioactive poison. The mixture of ^{227}Ac with beryllium is used for the preparation of radiation sources.

ACTIVATED ADSORPTION

A type of *adsorption* for which molecules (atoms) colliding with a surface must overcome a potential barrier.

The height of the barrier determines the adsorption energy. Due to the long-range character of *van der Waals forces*, physical adsorption is the first and necessary stage of the adsorption interaction during the approach of the adsorbing particle to the surface. It requires no *activation energy* and precedes the subsequent *chemisorption*. Depending on the energy where the potential energy curves of physical and chemical adsorption cross each other, the transition to the chemisorption state can either be activated or non-activated. The *activated transition* may occur in the case of dissociative chemisorption of molecules, when the "stretching" of the molecule occurs during its motion in the repulsive range of the physical *adsorption* process, which precedes the dissociation. Like a chemical reaction, activated adsorption takes place in accordance with the *Arrhenius law*, in which the exponent contains the activation energy corresponding to the transition from the state of physical adsorption to that of chemisorption. According to reaction rate theory, the activated adsorption time scale is determined by the concentration of adsorbed molecules in the transient activated complex. This complex forms at the potential energy maximum along the most energetically favorable path of chemisorption.

ACTIVATION ANALYSIS

A method for determining the chemical composition of a substance, together with the impurities that are present, with the aid of parameters induced by radioactivity.

The sample is exposed to a particular type of *nuclear radiation* that induces nuclear reactions in atoms of the material. Then, the type of induced radioactivity is determined experimentally, along with the secondary particle energy and the half-life. Using these data with the aid of nuclear reaction tables one can deduce the isotopes of the chemical elements that underwent the radioactive transformations. The intensity of the secondary radiation and the cross-sections of the corresponding nuclear reactions provide the initial concentrations of these isotopes. The sensitivity of the method varies over a wide range for isotopes of different elements. A sufficiently large nuclear reaction cross-section is required, and the half-life should be not be too long and not too short. Activation analysis is efficient for determining the gold, platinum, oxygen, boron content in *silicon*; the niobium in tantalum, and others. Activation analysis establishes the total impurity content in the sample, but is insufficient for deducing physical states in a crystal (*phase*, *doping atoms* versus *substitutional atoms*, etc.). Depending on the type of activating radiation, one can distinguish *neutron activation analysis*, *proton activation analysis* and *gamma activation analysis*.

ACTIVATION ENERGY

Basic characteristic of a thermally activated process which determines the rate of the process at a given temperature T, and its sensitivity to a variation of the temperature. The activation energy U is related to the probability W of the occurrence of an elementary event whose successive occurrences implement the corresponding process through the expression

$$W = W_0 \exp\left(-\frac{U}{k_B T}\right).$$

Examples of *thermally activated processes* are the following: *diffusion* of impurity atoms or vacancies; *electrical conductivity* of semiconductors; decay of *metastable states* at *first-order phase transitions* by the formation of nuclei of the new phase; *plastic deformation* of a crystal due to the motion of *dislocations* overcoming the *Peierls barriers* (see *Peierls relief*) or local defects. Such processes, referred to as elementary events (elementary structural rearrangements), are the passage of a diffusing atom between neighboring equivalent positions in a *crystal lattice*, the birth of a current carrier in the *conduction band* of a semiconductor, the breaking loose of a dislocation from its pinning center, the formation of a *critical embryo* of a new phase, and so on. Therefore, one can characterize the rate of a thermally activated process by the number of elementary events per unit time $\dot{n}$:

$$\dot{n}(T) = \dot{n}_0 \exp\left(-\frac{U}{k_B T}\right), \qquad (1)$$

where $\dot{n}_0$ is a normalization constant. This expression, called the *Arrhenius law*, is used in chemical kinetics to describe thermally activated reactions.

This specific equation for the activation energy, and its relationship with the parameters describing the structure and the state of a material, depend on the type of process under consideration. The detailed expression for Eq. (1), and the calculation of the activation energy U for the majority of cases require solving a complicated statistical mechanical problem. A simple qualitative interpretation of Eq. (1) based on the thermodynamic theory of *fluctuations* consists of the following: an individual event of the thermally activated process includes overcoming a certain energy barrier of height U; the exponential probability law for its occurrence as a function of the temperature is a consequence of exponential *Gibbs distribution* for the energy states of the subsystem where the given elementary process takes place.

The value of the activation energy can depend on external parameters a which determine the material condition: the *mechanical stress*, the electric field strength, and so on ($U = U(a)$). This circumstance allows one to control the rate of a thermally activated process, not only through a temperature change, but also by varying the external parameters a: $\dot{n} = \dot{n}(T, a)$, where

$$\dot{n}(T, a) = \dot{n}_0 \exp\left[-\frac{U(a)}{k_B T}\right]. \qquad (2)$$

One of the experimental methods for determining the activation energy is the measurement of the rate of a process at various temperatures while keeping the parameter a constant, then drawing a plot of $\ln \dot{n}$ versus $1/T$, and finding the slope of the resulting straight line. Another method is based on the measurement of the sensitivity of the rate of a thermally activated process to small variations of the temperature δT (again keeping a constant), with the subsequent application of the following formula deduced from Eq. (2):

$$U = k_B T^2 \left[\frac{\ln \dot{n}(T + \delta T, a) - \ln \dot{n}(T, a)}{\delta T}\right] \approx k_B T^2 \left(\frac{\partial \ln \dot{n}}{\partial T}\right)_a. \qquad (3)$$

The measurement of the activation energy and the analysis of its dependence on external parameters is called *thermal activation analysis*. This provides information on the physical nature of the elementary reactions involved in the thermally activated process (see also *Activation volume*).

ACTIVATION VOLUME

A quantity with the dimension of volume, which characterizes the sensitivity of the rate of activation analysis in a solid relative to the change of the applied *mechanical stress* (i.e. a change of *stress tensor* σ or pressure P).

Initially, the term activation volume arose in the description of the pressure effect on point defect *diffusion* in a solid, and it had the sense of the volume change involved in the transition of a *point defect* from an initial equilibrium state to a so-called *activated state*. At present it is used in a more general sense, and reflects the dependence of the *activation energy* of any thermodynamically activated process on σ or P. For example, in the case of the diffusion of *vacancies* the parameter is uniform pressure, while in the case of the motion of *dislocations* it is the shearing stress acting in its *slip plane*. In such cases the rate of the process $\dot{n}$ (or the number n of elementary acts per unit time) is described by the formula

$$\dot{n}(T, \sigma) = \dot{n}_0 \exp\left(-\frac{V(\sigma)}{k_B T}\right). \qquad (1)$$

The activation volume v is introduced formally as the negative derivative of the activation energy volume with respect to the stress (pressure), $v = -\mathrm{d}V(\sigma)/\mathrm{d}\sigma$. The activation energy change under a small stress change $\delta\sigma$ can be expressed via the activation volume using the relation $V(\sigma + \delta\sigma) = V(\sigma) - v(\sigma)\delta\sigma$. It should be noted that the activation volume does not always have the sense of a physical volume. The experimental determination of an activation volume is based on the measurement of the activation energy for different values of the stress, following numerical differentiation of the dependence so obtained, or on the change in the process rate sensitivity to small stress variations at a fixed temperature using the following formula derived from Eq. (1):

$$v = k_B T \frac{\ln \dot{n}(T, \sigma + \delta\sigma) - \ln \dot{n}(T, \sigma)}{\delta\sigma} \simeq k_B T \left(\frac{\partial \ln \dot{n}}{\partial \sigma}\right)_T. \qquad (2)$$

Measurement of the activation volume and analysis of its dependence on the value of σ (or P) is one of the important components of the procedure of *thermo-activation analysis*, and it provides information on the energy and geometric parameters of activation processes.

ACTIVE CARBONS

Highly porous carbon materials containing 85–97% *carbon*.

These materials are obtained by carbonization and activation of carbon-containing substances such as naturally occurring wood, mineral *coals*, peat, oil and coal cokes, as well as synthetic *polymers* such as phenol-formaldehyde, styrene-divinylbenzene, and other resins. Carbonization products, the so called *carbonizates*, comprise an ordered part (graphite-like carbon crystallites) and an amorphous part (high carbon content radicals), associated mainly with the prismatic faces of *crystallites*. Upon activation (850 to 950 °C), the oxidizing gases (CO_2, H_2O et al.) interact first of all with the amorphous carbon, and then with the carbon of the crystallites. The burning-out of impurities and partially of carbon from the contacting and coalescent carbonizate crystallites results in the formation of microporous regions containing *pores* of two types: intracrystallite ones with dimensions up to 1.5 nm, which are flat slots with walls formed by the basic faces of the crystallites; and intercrystallite ones with the walls in the form of prismatic faces. A portion of the external surface of the microporous areas forms the surface of larger mesopores (1.5 to 100 nm) and macropores (200 nm). The dimensions of single zones in active carbons are between 10 and 60 nm, while their number per unit of mass for the majority of active carbons is the order of 10^{16}. The number of micropores in a single microporous area is as high as several thousand.

Typical of active carbons is the *turbostrate structure* characterized by the absence of strict periodicity of the elementary layers. The distances between the latter are unequal, and fluctuate within the range 0.334 to 0.365 nm, in contrast to the equidistant graphite structure which has an inter-layer separation of 0.33554 nm. The carbon atoms in the crystallites of the turbostrate structure are displaced relative to the lattice plane within the range 0.014 to 0.017 nm. Active carbons possess high *electrical conductivity* due mainly to π-electrons in the planes of hexagonal rings along the system of conjugated bonds. The electric resistance of the active carbons depends strongly on the pretreatment temperature and its duration, and can be 10^8 to 10^{10} Ω·cm for the carbonizates, and 0.1 to 10 Ω·cm for well-activated carbons. This is associated with structural changes: of the conductive systems in cokes (carbonizates) separated from one another by compounds with an aliphatic character which burn out during an oxidizing thermal treatment. The ordering of the structure is associated with the reduction of the electric resistance.

Active carbons, in comparison with other transitional forms of carbon, are characterized by a broad spectrum of atomic carbon hybrid electronic states – sp^2, sp, $sp^2\pi$, etc. *Dangling bonds* at the surface of active carbons may bind oxygen, nitrogen, hydrogen, sulfur, etc., with the formation of strong surface compounds. Of great importance are surface compounds with oxygen. Especially prepared *oxidized carbons* form a separate class of active carbons with specific properties such as an extremely high cation-exchange selectivity. Surface compounds with nitrogen and sulfur exert a strong effect on the properties of active carbons.

Active carbons, because of their well developed porosity, have important applications as adsorbents and catalysts: for air and gas purification, for extraction of solvents, for preparation of drinking water, for purification of sewage, in the production of food, and for the extraction of valuable components (e.g., rare and noble metals) from their complex mixtures. Oxidized carbons are used for the acquisition and analysis of particularly pure materials for electronic engineering, and they are used in medicine for detoxification of blood and other biological liquids (hemo-, lympho-, plasmo-, enterosorbtion), during treatment of many diseases, and also as carriers of medicinal preparations.

ACTIVE CENTERS

These centers are *impurity atoms* injected into *solids* to produce changes in their electrical, optical, magnetic and other properties. An injected impurity which brings about no change is called a *passive impurity*. The impurity activity in a given material is determined not only by its chemical nature, but also by its position in the crystal lattice, and this depends on the manner and conditions of its injection, and the subsequent processing of the sample. Activation or passivation of impurities can be caused by *heat treatment*, the effect of nuclear radiation, pressure, and other factors. For instance, electrically inactive *interstitial atoms* of oxygen in *silicon* transform under high-temperature *annealing* into donor centers (called *thermo-donors*). The injection of Au, Cu and other atoms into photoelectrically inactive silicon crystals results, after subsequent exposure to radiation, in the appearance of *photoconductivity*. Impurity activation can arise from the change in position of the impurity atom as a result of irradiation. Active centers are not necessarily impurities, but can be intrinsic crystal lattice *defects*, as well as complexes of *intrinsic defects* and impurity atoms.

Table 1. Important properties of active solid state laser materials doped with neodymium

Material	Ruby (aluminum oxide doped with chromium)	Aluminum-yttrium garnet	Gallium-scandium-gadolinium garnet	Yttrium aluminate	Silicate glass	Potassium-neodymium phosphate glass
Property	trigonal, optically uniaxial	cubic, isotropic	cubic, isotropic	orthorhombic, optically biaxial	amorphous, isotropic	amorphous, isotropic
Wavelength of emitted radiation λ, μm	0.694	1.06	1.06	1.06	1.05	1.05
Density, g/cm^3	3.92	4.55	6.55	5.35	2.55	2.85
Melting or softening temperature, °C	2040	1970	1850	1875	435	400
Index of refraction (for emission wavelength λ)	1.755–1.763	1.82	1.943	1.94–1.97	1.56	1.555
Thermal conductivity, $W{\cdot}cm^{-1}{\cdot}K^{-1}$	0.42	0.13	0.07	0.11	0.013	0.01
Concentration of activators, $ions/cm^3$	$1.6{\cdot}10^{19}$	$3{\cdot}10^{19}$	$2{\cdot}10^{20}$	$2{\cdot}10^{20}$	$5{\cdot}10^{20}$	$3{\cdot}10^{21}$
Cross-section of basic generator transition σ, cm^2	$2.5{\cdot}10^{-20}$	$8{\cdot}10^{-19}$	$1.5{\cdot}10^{-19}$	$2.5{\cdot}10^{-20}$	$3{\cdot}10^{-20}$	$3.8{\cdot}10^{-20}$
Excited state lifetime, μs	3000	230	200	200	300	330
Emission linewidth, nm	0.53	0.45	1.4	1.0	26	15
Activator distribution coefficient	0.7–0.8	0.18	0.28	0.8	–	–

ACTIVE LASER ELEMENT

An optical element in *quantum amplifiers* or generators which attains population inversion (see *Level population*). Active elements in use with *solid-state lasers* are ruby, neodymium glass, *yttrium–aluminum garnet* with a neodymium admixture, *alkali-halide crystals* with color centers, semiconducting *heterostructures*, etc.

ACTIVE MEDIUM

A material disturbed from its *thermodynamic equilibrium* state in such a way that, when waves of a certain type (electromagnetic, sonic) propagate in it, their intensity increases in some frequency range. In other words, energy is pumped from the active medium to the waves. The active medium, in its turn, receives its energy from an external source, e.g., an intense light beam (optical pumping), an incident electron beam, etc. See also *Laser active medium*.

ACTIVE MEDIUM OF LASER

See *Laser active medium*.

ACTIVE SOLID-STATE LASER MATERIALS

Monocrystals and glasses (see *Vitreous state of matter*) used as laser *active media*. The basic properties of active solid-state laser materials most widely used in industrial applications are given in Table 1. Characteristics of materials intended for use in minilasers are given in Table 2. Such materials are expected to permit pumping by minilamps and/or semiconductor light diodes, so that such minilasers may be built into systems of *fiber optics*.

ACTIVE SPECTRUM in radiation physics

Stationary space-energy distribution of fast particles within solids arising from nuclear radiation. The active spectrum is essentially different from the spectrum of the primary radiation source because it is formed as a result of interaction of the

Table 2. Important properties of minilaser materials

Material, chemical formula	Concentration of neodymium, ions/cm^3	Pump level absorption coefficient, cm^{-1}	Excited state lifetime, τ, μs	Effective cross-section of basic generator transition, arbitrary units	Relative efficiency, arbitrary units
$Y_{0.7}Nd_{0.3}Al_5O_{12}$	$0.14 \cdot 10^{21}$	1.1	230	1	1
Silicate glass doped with Nd	$0.28 \cdot 10^{21}$	2.0	300	0.05	0.14
$NdAl_3B_4O_{12}$	$5.5 \cdot 10^{21}$	90	19	2	6.7
$Nd_{0.16}Y_{0.84}Al_3B_4O_{12}$	$0.88 \cdot 10^{21}$	15	48	2	57
NdP_5O_{14}	$4 \cdot 10^{21}$	40	120	0.4	38
$Nd_{0.15}La_{0.85}P_5O_{14}$	$3 \cdot 10^{21}$	30	240	0.4	56
$LiNdP_4O_{12}$	$3 \cdot 10^{21}$	30	120	0.6	48

primary radiation with the irradiated system; it is the active spectrum that causes radiation effects in solids. This spectrum is characterized by the set of functions $N_i(\varepsilon, \boldsymbol{r})$, where ε is the particle energy, $\boldsymbol{r}$ is the space coordinate, and the dimensionality of N is $MeV^{-1} \cdot cm^{-2} \cdot s^{-1}$. The expression

$$N_i(\varepsilon, \boldsymbol{r}) \Delta\varepsilon \Delta A \Delta t \qquad (1)$$

gives the number of fast particles with energies in the range from ε to $\varepsilon + \Delta\varepsilon$, which intersect at right angles any region of the large cross-section of the sphere ΔA centered at the point $\boldsymbol{r}$ during the time interval from t to $t + \Delta t$ (see Fig.). Vectors 1–3

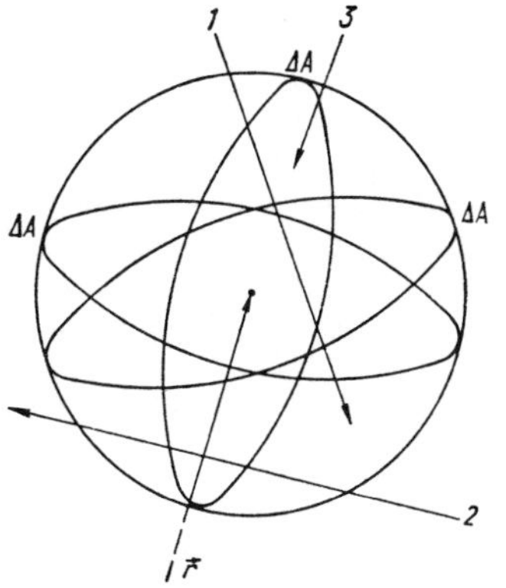

Construction for the determination of the active spectrum at the point $\boldsymbol{r}$. Incoming particles 1 and 2 contribute to the value $N(\varepsilon, \boldsymbol{r})$ of Eq. (1), whereas particle 3 does not do so.

designate the trajectories of the motion of three particles, whose energies during the motion in the time interval Δt lie within the range $(\varepsilon, \varepsilon + \Delta\varepsilon)$. During the interval Δt particles 1 and 2 intersect the area ΔA at a right angle and contribute to Eq. (1). Particle 3, though it succeeded in penetrating the sphere, did not proceed far enough to pass through the large cross-section area at right angles, and hence it does not make any contribution to the value (1).

The active spectrum, as well as the *absorbed dose*, has a macroscopic value, unambiguously defined only for such ΔA and Δt at which it is possible to neglect fluctuations of the value $N(\varepsilon, \boldsymbol{r})$. For a sufficiently large volume the *Bragg–Grey approximation* is valid, namely that the energy given by the particle to the ionization and excitation of atoms remains entirely within the volume. In this case the Bragg–Grey correlation holds between the absorbed dose of radiation $D(\boldsymbol{r})$ and the active spectrum:

$$D(\boldsymbol{r}) = T \int_{\varepsilon} N(\varepsilon, \boldsymbol{r}) S(\varepsilon)\, d\varepsilon, \qquad (2)$$

where $S(\varepsilon)$ is the average mass braking capability, which describes the *charged particle energy loss* during passage through the material (dimensions are $MeV \cdot g^{-1} \cdot cm^{-2}$); T is the time of the radiation, with $T \gg \Delta t$. Knowledge of the active spectrum is necessary for calculating the basic parameter of the radiative processes, namely the yield G. Fast particles in particular energy ranges contribute to different radiation induced processes (ionization of different atomic shells, removal of an atom from its site, etc.). Thus, besides the spectrum, it is necessary to know the cross-sections of the processes $\sigma(\varepsilon)$ as a function of the incoming particle energy. It is possible to calculate the yield G of the radiation induced formation of *Frenkel defects* within a crystal irradiated by accelerated electrons, with the aid of the formula

$$G(\boldsymbol{r}) = \int_{\varepsilon} N(\varepsilon, \boldsymbol{r})\sigma_{\mathrm{F}}(\varepsilon)\,\mathrm{d}\varepsilon, \qquad (3)$$

where $\sigma_{\mathrm{F}}(\varepsilon)$ is the cross-section of Frenkel defect formation upon collision of fast incoming electrons of energy ε with crystal atoms.

ACTIVITY of particles (quasi-particles)

A quantity that represents the effective concentration of a solute in an ideal solution. It has the dimension of particle concentration, and is used in the equation of the law of mass action in cases when the *Boltzmann distribution* does not apply. The classical expression for the relationship between the particle concentration n and the chemical potential μ is

$$\mu = \mu_0 + k_{\mathrm{B}}T \ln n, \qquad (1)$$

where $\mu_0 = \text{const}$. This equation does not hold in the presence of degeneracy (e.g., when *Fermi–Dirac statistics* or *Bose–Einstein statistics* apply), or when the interaction (electric, elastic, etc.) between particles and *quasi-particles* leads to a redistribution and spatial nonuniformity of their concentration, with a component of shielding. Under these conditions a quantity a called the activity is introduced instead of n to retain the form of Eq. (1). The ratio $A = a/n$, called the *activity coefficient*, is a measure of the effectiveness of particular particles (atoms, ions, etc.) in undergoing reactions or other actions. The concept of activity is widely used in physical and chemical research. In descriptions of non-equilibrium processes the form of Eq. (1) is sometimes retained, but with the chemical potential μ associated with the *Fermi level* replaced by a *Fermi quasi-level* counterpart that corresponds to a non-equilibrium concentration n.

ADDITIONAL LIGHT WAVES, Pekar waves, supplementary light waves

Waves arising from effects of *spatial dielectric dispersion* in the range of exciton resonances in crystals. In contrast to classical *crystal optics*, which implies that in a given direction $\boldsymbol{s}$ and at a frequency ω two mutually orthogonally polarized waves can propagate (*birefringence*), the appearance of additional light waves means that there can be more than two such waves with the same $\boldsymbol{s}$ and ω. In the case of isolated exciton transitions, there are two waves with the same polarization but with different complex *refractive indices*, i.e. with different velocities and attenuation coefficients. The number of additional light waves is equal to the number of *excitons* that are at *resonance*. The dispersion of additional light waves is shown in Fig. for the cases of positive (a) and negative (b) exciton *effective masses*, where a dashed line indicates curves that do not take into account the spatial dispersion.

In the presence of additional light waves the Maxwell boundary conditions imposed on the fields for reflection and transmission at the interface between two media are insufficient for the determination of the amplitudes of all the waves. Therefore, it is necessary to introduce additional boundary conditions that are obtained from the law of exciton wave reflection from a crystal surface $[\boldsymbol{P}_{\mathrm{ex}}(\boldsymbol{r},t)]_{z=0} = 0$, where $z = 0$ is the surface, and $\boldsymbol{P}_{\mathrm{ex}}(\boldsymbol{r},t)$ is the partial exciton contribution to the polarization $\boldsymbol{P}(\boldsymbol{r},t)$. For excitons of large radius, due to the image forces in the vicinity of the surface, there appears a "dead" layer with a thickness of $l \approx 2$–3 exciton radii into which the excitons cannot penetrate; and one should impose the additional boundary conditions on the surface $z = l$. There also exist other boundary conditions.

The additional light waves exist in a narrow spectral range in the vicinity of an exciton resonance at low temperatures. When moving away from the resonance toward higher or lower frequencies a transition takes place to the classical theory of birefringence because the amplitudes of the additional light waves go to zero. When the temperature is raised and the *exciton–phonon interaction* constant Γ is increased, then a rearrangement of the dispersion curves takes place; and when Γ reaches a certain critical value Γ_{cr} they acquire the classical form.

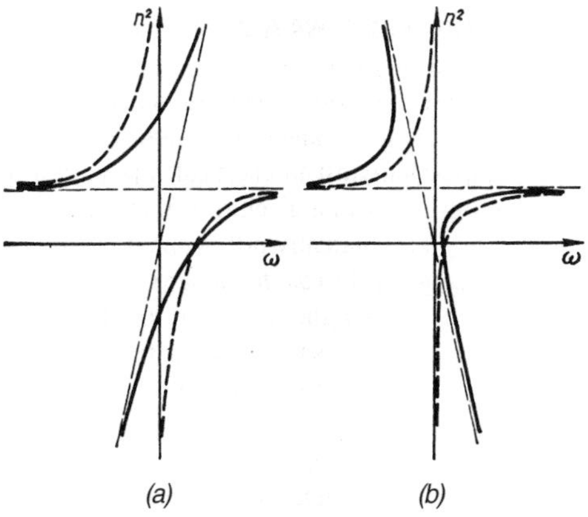

Additional light waves.

The existence of additional light waves is supported by many experiments. For example, two equivalently polarized waves with different refractive indices were observed after the transmission of light through a wedge-shaped crystal of CdS.

ADHESION (fr. Lat. *adhaesio*, sticking)

Coupling or attachment of two dissimilar bodies brought into contact.

In the case of bodies of the same physicochemical nature, the phenomenon is referred to as *cohesion*. Adhesion is caused by interatomic (intermolecular) surface attraction, the nature of the latter being related to the deformation of the atomic electron shells at the interface of the bodies. Various mechanisms can be operative, such as polarization attraction via the van der Waals interaction (see *Van der Waals forces*) with characteristic energies $\varepsilon \approx 10^{-3}$–$10^{-1}$ eV/atom, and the interchange or sharing of valence electrons (chemical interaction, $\varepsilon \approx 1$–10 eV/atom). High *adhesion durability* (several units of eV/atom) is exhibited by contacts of metal with most (but not all) other metals and semiconductors, as well as by those of polymeric films with different materials. The durability of contacts of *metals* with *insulators* and *graphite* is much lower (10^{-1} eV/atom). Adhesion plays an important role in some interface phenomena such as *friction*, *wetting*, and the heterogeneous growth of *crystals*. Adhesion finds practical use in soldering, welding, and gluing.

ADIABATIC APPROXIMATION

A specific method of approximate solution of quantum mechanical problems, based on the distinction between "fast" and "slow" variables.

In solid state physics, this approximation is used to separate the electronic ($\boldsymbol{r}$) and nuclear (ionic, $\boldsymbol{R}$) variables in the Schrödinger equation. A solid body is an electron–ion system bound together by strong interactions, in which the characteristic frequencies of the electron subsystem are much higher than those of the vibrating ions. This makes it a good candidate for applying *adiabatic perturbation theory*. The basic parameter that permits this application is the mass ratio $\kappa = (m/M)^{1/4}$, where m is the electron mass and M is the mass of an ion. The kinetic energy of the ions is κ^2 times less than the characteristic electron energy so it is neglected in the zeroth-order approximation, i.e. the Schrödinger equation is solved for the electrons with the ions fixed in position at their equilibrium configuration. The resulting wave function $\Psi_n(\boldsymbol{r}, \boldsymbol{R})$ and energy $E_n(\boldsymbol{R})$ of electrons in the state n depend on the coordinates $\boldsymbol{R}$ of all the ions. At the next stage of the problem the ion movement is solved for; the motion, as a rule, being oscillations of the ions around their equilibrium positions. In the adiabatic approximation, the motions of the ions do not induce transitions in the electron subsystem, but there can be a non-dissipative transfer of energy between the electronic and ionic subsystems. The adiabatic approximation makes it possibile to treat the electronic states of the system independently, to a considerable extent, of the oscillations of the lattice, and hence, every subsystem makes an additive contribution in the total energy. In this approximation, the oscillations of ions occur in the field of an effective potential (*adiabatic potential*) which equals to the sum of the electron energy $E_n(\boldsymbol{R})$ and ion–ion interaction potential. The adiabatic description assumes that terms higher than κ^5 are neglected. Higher-order terms describe interactions between the electron and ion systems (*electron–phonon interaction*) and higher-order anharmonicity (*phonon–phonon interaction*).

The main *adiabaticity criterion* is that the electron spectra have no excitations with energies of the order of nuclear oscillation frequencies: i.e. that $\hbar\omega_{\mathrm{D}} \ll |E_n - E_m|$, where E_n and E_m are the energies of the free and occupied electron states, and ω_D is the characteristic nuclear vibrational frequency. This criterion is well satisfied in *insulators* and *semiconductors* due to the presence of a *band gap* E_g in the electron spectrum. For this case the adiabaticity parameter is $\hbar\omega_D/E_g$. This criterion can be violated in *narrow-gap semiconductors*. In *metals*, the excitation energy is of the order of the Fermi energy E_F (see *Fermi energy*), so for electrons far below the Fermi surface the criterion is satisfied due to the smallness of the ratio $\hbar\omega_D/E_F$. The delocalized electrons in the energy range $\sim\hbar\omega_D$ near the *Fermi surface* are non-adiabatic. The relative smallness of the volume occupied by such states permits the use of the adiabatic approximation for bulk characteristics of a metal such as the binding energy and the phonon spectrum.

ADIABATIC DEMAGNETIZATION COOLING,
magnetic cooling

Decrease of crystal lattice temperature, T, arising from *demagnetization* without heat exchange with the surrounding medium of paramagnetic particles (spins) contained in the crystal lattice.

This was predicted independently by P. Debye and W. Giauque (1926). The first experiments were carried out in 1933–1934. It is used to obtain temperatures considerably lower than 1 K (e.g., $\sim 10^{-3}$ K). Transition metal ions of the *iron* group or *rare-earth elements* can be used as paramagnetic particles, as well as other nuclei with non-zero magnetic moments. The cooling is performed in several stages: (1) the sample is precooled to an initial moderately low temperature (~ 1 K) by other methods; (2) at these temperatures, the sample is magnetized isothermally in a field $B \sim 1$–10 Tesla; (3) then the sample is *adiabatically demagnetized* and its temperature falls rapidly far below 1 K. The physical reason for this decrease is associated with the transfer of *entropy* from the magnetic subsystem to the crystal lattice (and vice versa). At stage 2 the spin alignment of the magnetic subsystem occurs, diminishing its entropy and increasing the crystal lattice subsystem entropy. Since step 2 is isothermal there is no change in the sample temperature. Since stage 3 is isentropic the demagnetization is accompanied by withdrawal of the heat from the lattice, and the sample temperature decreases.

ADIABATIC FAST PASSAGE

Manner of scanning the applied magnetic field B_{app} through the Larmor magnetic resonance condition $B_{app} = B_0 = \omega_0/\gamma$ in a time subject to two criteria: (1) the effective magnetic field vector direction, $\boldsymbol{B}_e = \boldsymbol{B}_0 - \boldsymbol{B}_{app} + \boldsymbol{B}_1$, changes so slowly that the *magnetization* vector, $\boldsymbol{M}$, remains approximately parallel to $\boldsymbol{B}_e$; and (2) the passage through the *resonance* is so rapid that the *relaxation* processes have no time to deviate $\boldsymbol{M}$ from the direction of $\boldsymbol{B}_e$. The mathematical statement of these two criteria in solids is as follows:

$$\frac{1}{T_1} \ll \frac{1}{B_1}\left|\frac{\mathrm{d}\boldsymbol{B}_{app}}{\mathrm{d}t}\right| \ll |\gamma \boldsymbol{B}_1|,$$

where $\boldsymbol{B}_1$ is the alternating magnetic field vector which oscillates at the fixed frequency ω_0; $\boldsymbol{B}_0$ is the resonant field, γ is the gyromagnetic ratio, and T_1 is the *spin–lattice relaxation* time.

ADIABATIC PARAELECTRIC COOLING

Decrease of the temperature of a dielectric (*insulator*) via a decreasing electric field due to adiabatic depolarization; the electric analog of adiabatic demagnetization of a spin system (see *Adiabatic demagnetization cooling*).

Cooling by adiabatic depolarization occurs in systems with impurity electric dipoles, such as $KCl{:}OH^-$, $KCl{:}Li^+$, $RbCl{:}CN^-$. In these systems, the temperature was observed to decrease to 4 mK. Systems of reorientable electric dipoles having large *dipole moments* and a small *inversion splitting* are favorable for obtaining low temperatures. The measurement of the temperature change (ΔT) dependence on the electric field strength and on the initial temperature elucidates some basic characteristics of these dipoles, such as the impurity concentration and the magnitude of the *dipole moment*, as well as the inversion splitting of the impurity energy levels.

ADIABATIC POTENTIAL

The energy of a many-atom system as a function of nuclear coordinates, obtained in the solution of the electronic part of the Schrödinger equation with the nuclei fixed in position. If the criterion for the validity of the *adiabatic approximation* is met, the adiabatic potential has the sense of the potential energy of the nuclei in the field of the electrons (this interpretation is inapplicable in the case of electronic degeneracies or pseudodegeneracies, when the adiabatic potential is especially complicated and leads to a whole range of new properties of the system). See also *Vibronic instability*, *Jahn–Teller effect*.

ADIABATIC PROCESS

A thermodynamic process taking place without the *heat transport* to and from the surrounding medium.

It is represented by an adiabat on a *phase diagram*. It occurs in a thermally isolated system or one enclosed in an-insulated environment. It also can occur in a high-speed process, such as the rearrangement of electron subsystem of a crystal, or a molecule undergoing rapid rearrangement of its configuration (see *Adiabatic approximation*).

ADIABATIC SHIFT in plastic deformation

Formation of localized zones of enhanced plastic deformation in a material under stress (see *Localization of plastic deformation*).

The adiabatic shift phenomenon is due to the fact that almost 90% of the plastic deformation energy transforms to heat, and the *yield limit* of most metals diminishes as the temperature increases. If the heat generation under plastic deformation occurs faster than its removal by thermal conduction, pronounced localized heating occurs near *slip bands*, which leads to the appearence of the adiabatic shift. The transition from uniform deformation via *shear strain* to adiabatic shift is determined by the magnitude and the *plastic deformation rate*. The temperature in the adiabatic shift domain depends on the amount and the rate of local strain, as well as on the thermal, elastic and strength characteristics of a material. The plastic deformation in the adiabatic shift zone is superplastic (see *Superplasticity*), and the deformation rate is high. The adiabatic shift is observed in *explosion deformation*, during collisions between bodies, pressure processing, mechanical processing, static *pressing*, grinding, etc. When *steels* are heated above a certain temperature there is a transformation of the ferrite into *austenite*. The formation of adiabatic shift zones in materials, whose strength diminishes as the temperature increases, can lead to catastrophic *failure* along these bands.

ADIABATIC SURFACE SCATTERING

Scattering of quantum particles or waves, taking place in situations when the scattering potential on the surface varies slowly from point to point. Such scattering occurs at the multiple near-specular reflection of particles from the boundary, when they undergo finite periodic motion in the direction transverse to the surface (e.g., electrons in a conductor with a constant magnetic field parallel to the boundary, or current carriers in a thin plate with mirror faces). Under these conditions, the state of such motion is "quantized", i.e. the corresponding energy levels are discrete. The adiabaticity of the scattering potential manifests itself in the correlation of successive collisions of the particles with the surface. The strong correlation of neighboring collisions causes the scattering to preserve the adiabatic invariant of the periodic motion. It reduces of the phase volume of states, into which the particle can transit after collision, which in turn diminishes the scattering intensity.

ADSORPTION (fr. Lat. *ad*, on, and *sorbeo*, am taking in)

Accumulation of material (*adsorbate*) from a gaseous or liquid phase on a *solid surface* or liquid surface (*adsorbent*) due to its unusually strong attraction and retention near the *phase interface*.

Adsorption is due to forces between the molecules (atoms) that form on the adsorbent surface and those arriving from the gaseous medium or solution. Depending on the nature of these forces, two types of adsorption are distinguished, namely physical and chemical (*chemisorption*). Physisorption is caused by *van der Waals forces* of intermolecular interaction, and chemisorption occurs under the action of *chemical bonding* forces. Depending on the forces acting, there are also two types of heats of adsorption. In case of *physical adsorption*, the heat of adsorption slightly exceeds that of *condensation* and is approximately 10 kJ/mole, while the heat of chemisorption is typically the same order of magnitude as that of molecular bonding, namely ~300 kJ/mole. Chemisorption involves saturation of the free valence bonds of surface atoms, and this occurs via the formation of chemical compounds with the adsorbent on the surface. Valence bonds become saturated after laying down a *monolayer* of adsorbent.

The compounds that are formed are not necessarily stoichiometric. Their components are atoms or individual radicals of molecules dissociated during the process of adsorption. *Sorption* is a more general concept which includes *adsorption* and *absorption* (see *Activated adsorption*).

ADSORPTION, ACTIVATED

See *Activated adsorption.*

ADSORPTION, CHEMICAL

See *Chemisorption.*

ADSORPTION FATIGUE

A nonreversible change in the condition of a solid body subject to cyclic strains, revealed in an adsorption-active medium when *cracks* appear (see *Strength reduction through adsorption*). The essential feature of the process of the development of adsorption fatigue is that the *adsorption* of the substance on the crack surface promotes its development under extension and prevents its closing at compression. Adsorption fatigue is clearly evident in brittle materials; the degree of its appearance diminishes as the cyclic deforming frequency increases, and it rises as the amplitude of the deformation rises. The adsorption fatigue effect can be weakened by the creation of *residual stress* of compression in the solid.

ADSORPTION-INDUCED STRENGTH REDUCTION

See *Strength reduction through adsorption.*

ADSORPTION POTENTIAL

The potential energy $V(r)$ of interaction between a particle being adsorbed (atom, molecule) and a *solid surface.*

This energy depends on their separation distance, and is numerically equal to the work required to transfer the particle to infinity (see *Adsorption*). The curve defining the adsorption potential consists of two regions, one corresponding to the short-range forces of repulsion, the other arising from long-range forces of attraction. The distance dependence is often approximately described by the *Lennard-Jones potential* $V(r) = -A/r^6 + B/r^{12}$, where A and B are positive, and r is the distance of the particle from the surface. A more general description of the absorption potential is obtained by summing over over the contributions of each *pairwise interaction* of the particle being adsorbed with each atom of adsorbent. This potential depends not only on the distance between the particle and the surface, but also on its particular position relative to the *substrate* atoms.

ADSORPTION, RADIATION-INDUCED

See *Radiation-induced adsorption.*

AEROSOL PARTICLES in the atmosphere

An *aerosol* is a suspension of very fine particles of a solid, or very fine droplets of a liquid in a gas.

The radii of aerosol particles in the Earth's atmosphere are 10^{-3} to 10^{-4} m, their concentration in polluted air may reach 10^{11} m^{-3}. The aerosol particles enter the atmosphere or are formed in it as the result of chemical reactions between gases (the anthropogenic sources contribute 20%). Natural sources of these particles are air erosion of soil and rocks, forest fires, volcanic eruptions, sea foam spraying, reactions between atmospheric gases resulting in the formation of sulfates, nitrates and ammonium salts. The contribution from extraterrestrial sources is about 0.05%. The typical particle lifetime in air depends on its size and its altitude in the atmosphere: particles live up to 2 days in intermediate layers (to 1.5 km), and as long as 2 weeks in the middle and upper troposphere. The concentration drops exponentially with the altitude in the lowest atmospheric layer, remaining constant in higher layers. The density of the distribution in size is of the form r^{-n} (where r is the particle radius, and $n = 4$ for Junge's law). The *atmospheric aerosol* plays a significant role in the optics and energetics of the atmosphere, with some particles (~1%) acting as centers for heterogeneous vapor *condensation* (condensation nuclei). A few aerosol particles in a cubic meter could be the centers of ice nucleation (nuclei of *ice* formation or freezing). Experimental studies of the aerosols make use of impactors, millipore filters, condensation counters, diffusion chambers and thermoprecipitators.

AFFINITY, ELECTRON

See *Electron affinity*.

AGING

See *Alloy aging*, *Magnetic aging*, *Strain aging*, *Recovery of aged alloys*.

AGING OF POLYMERS

Change of properties of a *polymer*, e.g., mechanical or electrical, with time; this change results from concurrently proceeding chemical and physical processes. Oxidation processes due to interaction with atmospheric oxygen and ozone are accelerated under exposure to light or radiation, when internal stresses are present in the polymeric material. In combination with various mechanisms of *polymer degradation*, internal stresses may result in the propagation of a network of cracks. The course of polymer aging is influenced by microscopic structural inhomogeneity, and by the change of this characteristic with time, by migration of the *polymeric material* components (plasticizers, antioxidants, light stabilizers (UV absorbers), etc.) either to the surface of the polymer, or into regions that separate individual structural components of polymeric material from one another. Acids, alkalis, oxidizing agents, *surface-active agents*, and moisture are capable of changing the polymer structure, and affecting the strength of the adhesion contact between the components of the polymeric *composite material*. The rate of polymer aging changes with time in accordance with the mechanism of the process; in particular, if the major mechanism of aging is oxidative breakdown, then the rate of aging increases with decreasing concentration of antioxidant, and with growing concentrations of products which define the auto-catalytic nature of the process.

AHARONOV–BOHM EFFECT (Y. Aharonov, D. Bohm, 1959)

A macroscopic manifestation of the nonlocal dependence of the phase Θ of the wave function $\Psi = |\Psi|e^{i\Theta}$ of a charged particle with a charge e on the electromagnetic field vector potential $\boldsymbol{A}(\boldsymbol{r}, t)$

$$\Theta(\boldsymbol{A}) = \Theta_0 + \frac{e}{h}\int \boldsymbol{A}\,\mathrm{d}l, \qquad (1)$$

where the integration is carried out along the quasi-classical trajectory of the particle. The nonlocal dependence (1) is a consequence of the local gauge symmetry of the electromagnetic interaction in quantum mechanics (see *Gauge invariance*). In view of this, the electromagnetic field must affect charged particles even when the magnetic ($\boldsymbol{B} = \nabla \times \boldsymbol{A}$) and electric ($\boldsymbol{E} = -(\partial \boldsymbol{A}/\partial t)$) field strengths are zero, i.e. when the *Lorentz force* is absent, but the time-independent vector potential $\boldsymbol{A}(\boldsymbol{r})$ is nonzero. In order to observe the dependence of the phase Θ on $\boldsymbol{A}$, Bohm and Aharonov proposed an experiment involving the interference of coherent electron beams passing on both sides of an infinite solenoid. The constant magnetic flux Φ is concentrated inside the solenoid, while outside the magnetic field $\boldsymbol{B}$ is zero, but $\boldsymbol{A}(\boldsymbol{r}) \neq 0$. Since, according to Stokes' theorem,

$$\Phi = \int_S \mathrm{d}S\,\nabla \times \boldsymbol{A} = \oint_L \mathrm{d}l\,\boldsymbol{A}, \qquad (2)$$

where the integration is carried out over the cross-sectional area, S, of the solenoid, and along a closed contour, L, encircling the solenoid, the phase difference of the electron wave functions in the two beams is

$$\Theta_1 - \Theta_2 = \frac{e}{h}\oint_L \boldsymbol{A}\,\mathrm{d}l = \frac{\Phi}{\Phi_0'}. \qquad (3)$$

Here $\Phi_0' = h/e$. Thus, the electron beam interference pattern observed beyond the solenoid (e.g., on a fluorescent screen) exhibits a shift that results from a variation of Φ. The first experiment of this kind using a tiny ($R \approx 14$ μm) solenoid (Chambers, 1960) confirmed qualitatively the theoretical predictions. Subsequent more precise experiments were carried out using long and very thin ($R \sim 1$ μm) filaments of ferromagnetic materials (see *Ferromagnet*), toroidal solenoids having almost no stray fields, and *superconducting quantum interference devices* in which the interference of wave functions of *Cooper pairs* with charge $2e$ takes place. The double charge of a Cooper pair is responsible for the factor of 2 in the denominator of the fluxoid (flux quantum) definition $\Phi_0 = \Phi_0'/2 = h/2e$ (see *Flux quantum*).

AIRY STRESS FUNCTION (G.B. Airy, 1863)

A function used to express the solution of the problem of planar *strain* of a homogeneous, linear elastic solid. *Planar strain* is the strain at which one of the components of the *displacement vector* $\boldsymbol{u}$, called u_z, is equal to zero, and the other two components u_x and u_y do not depend on z. In this case the equations for the equilibrium of the elastic body can be written in the form

$$\frac{\partial \sigma_{xx}}{\partial x} + \frac{\partial \sigma_{xy}}{\partial y} + \rho X = 0,$$

$$\frac{\partial \sigma_{yy}}{\partial y} + \frac{\partial \sigma_{xy}}{\partial x} + \rho Y = 0,$$

where σ_{ij} are the components of the *stress tensor*, ρ is the density, X and Y are components of the *bulk force* density vector. If this vector can be derived from a potential V, i.e. $X = -\partial V/\partial x$, $Y = -\partial V/\partial y$, then there exists a function χ called the Airy stress function, in terms of which one can express the components of the stress tensor that are the solutions of the problem at hand:

$$\sigma_{xx} = \frac{\partial^2 \chi}{\partial x^2} + \rho V,$$

$$\sigma_{yy} = \frac{\partial^2 \chi}{\partial y^2} + \rho V,$$

$$\sigma_{xy} = -\frac{\partial^2 \chi}{\partial x \partial y}.$$

Under certain conditions the Airy stress function is biharmonic.

ALBEDO

The reflecting ability of a matte surface. The *plane (Lambert) albedo* is the ratio of the light flux scattered by a flat element of surface to the incident flux. This is measured by an albedometer. The *apparent albedo* is the ratio of the brightness of a flat surface element illuminated by a beam of parallel rays to that of an absolutely white surface oriented normal to the incident beam. The albedo for monochromatic radiation is called *spectral albedo*. A similar physical quantity in neutron optics, *neutron albedo*, is the probability of the neutron reflection that results from multiple scattering in the material. The concept of albedo is used, e.g., in optical engineering calculations, in temperature measuring by the radiometer method, and in neutron diffusion theory.

ALFVEN WAVES

See *Solid-state plasma.*

ALGEBRA, COMPUTER

See *Computer algebra.*

ALKALI-HALIDE CRYSTALS

Ionic crystals of compounds of *alkali metals* and halogens. They crystallize mainly in the NaCl-type FCC *crystal lattice* (except CsCl, CsBr, CsI which adopt the CsCl BCC structure). NaCl is a typical representative of alkali halides: the *ionicity* is 0.94; interionic spacing is 0.282 nm; binding energy per pair of ions is 7.93 eV; *melting temperature* 1074 K; *Debye temperature* 281 K; *specific heat* 48.86 $\mathrm{J{\cdot}mol^{-1}{\cdot}K^{-1}}$ (250 K); thermal conductivity 0.07 $\mathrm{W{\cdot}cm^{-1}{\cdot}K^{-1}}$ (273 K); static *dielectric constant* is 5.9, optical dielectric constant 2.25. Alkali halides are one of the most important model systems in solid state physics. During investigations of these crystals many fundamental phenomena were discovered, such as *self-localization* of electron excitations, the existence of free and self-localized states of *quasi-particles*, the decay of electron excitations per pair of defects, *light absorption* and *luminescence*, relations of intraband transitions in *insulators*.

The *valence band* (width 2–5 eV) develops from p-states of the halogen, the *conduction band* (somewhat wider) arises from the empty s-state of the alkali metal. The broad *band gap* (6–14 eV) causes the transparency in the visible and UV spectral regions. In the IR range the transparency region is limited by the lattice absorption (excitation of optical *phonons* $\nu \sim 10^{13}$ Hz).

In the UV region the *transparency* limit is determined by the exciton absorption bands situated on the long-wave edge of the fundamental absorption. These bands involve the excitation of the halogen s-state to the p-state (*Frenkel exciton*). In the transparent region absorption bands caused by structural or impurity defects (*color centers*) may be observed.

A strong *electron–phonon interaction* results in a sharp difference between free and relaxed quasi-particle state properties. Effective masses are: for the electron band $m_e^* \approx 0.5m_e$, and for the polaron $m_p^* \approx m_e$. The drift mobility is about 10^4 $cm^2 \cdot V^{-1} \cdot s^{-1}$ at 5 K, and about 10 $cm^2 \cdot V^{-1} \cdot s^{-1}$ at 300 K. For holes and excitons *relaxation* leads to the self-localization of quasiparticles in the regular lattice (*holes* as a molecular halogen ion X_2^-, *exciton* as an excited quasimolecule $(X_2^{2-})^*$).

Holes travel a distance of about 10^2–10^3 nm in a time of about 10^{-11} s before their self-localization. For the self-localization of an exciton the *activation energy* of about 10 meV is required, therefore, the lifetime of a free exciton and its shift depend on the temperature. With the increase in temperature the thermally activated motion of self-localized holes and excitons is observed, with activation energies 0.18–0.54 eV for holes and approximately half that value for excitons. It is likely that for excitons this motion represents a series of reorientation jumps of $(X_2^{2-})^*$, terminating in a radiative or nonradiative annihilation, or capture by a defect. For holes there is a combination of reorientation jumps of X_2^- and "great leaps" for a distance about 100 nm. This hopping, which provides the major part of the contribution to the hole motion, is due to a thermally activated transition of the V_k-center in a hole band, or a hole polaron of large radius.

The *recombination* of an electron with a self-localized hole, or a nonradiative exciton annihilation, takes place with a high probability of the formation of a *Frenkel defect* pair in the anion sublattice. This is responsible for the low *radiation resistance* of alkali halides. Luminescence of alkali halides is related to both radiative exciton annihilation (free and self-localized exciton) and radiative transitions in local centers (*luminescent center*). An impurity *ion* or *impurity atom* (e.g., Ag^+, In^+, Tl^+, Pb^{2+}), or a structural defect, is typically a nucleus of such centers.

The luminescence of local centers in alkali halides is observed both at the photoexcitation of absorption bands of these centers (the *intra-center luminescence*), at the excitation in a region of fundamental absorption, and by interactions involving incident high-energy particles and quanta. In the latter case the energy transfer from the lattice to the luminescence centers is provided by electron and hole migration (*recombination luminescence*), or by exciton migration. At $T \geqslant 200$ K the *ionic conductivity* arises mainly from the motion of cation *vacancies*, while for $T \geqslant 300$ K it is due to the motion of anion vacancies.

The main technical applications of alkali halides are for optical materials, for the detectors of *nuclear radiations* (scintillators NaI doped with Tl, CsI with Tl, CsI with Na, and thermoluminescence dosimeters based on LiF), and *lasers* based on color centers.

ALKALI METALS, alkali elements

Chemical elements of Group I of the periodic table (Li, Na, K, Rb, Cs, radioactive Fr). Owing to the high radioactivity of Fr its properties have been rarely examined. The distribution of Na and K in the Earth's crust is rather high (2.64 and 2.5 mass %); the remaining alkali metals are rare and widely dispersed. All alkali metals have low electronegativities, with *electronegativity* values ranging from 0.96 (Li) to 0.69 (Fr). The outer electronic shell of an alkali atom contains a single s-state electron (configuration s^1), with a stable inert gas shell (s^2 or s^2p^6) lying below it. The atomic radius expands from Li (0.155 nm) to Fr (0.280 nm). At the same time, the number of electron shells increases, resulting in a decrease of the electrostatic attraction force of the outer valence electron to the nucleus, and a decrease of the ionization energy from 5.392 eV for Li to 3.894 eV for Cs. This is why alkali metals easily give up their valence electron and transform into ions M^+.

Alkali metals are light *metals*: density increases from 0.539 g/cm^3 (Li) to 1.9039 g/cm^3 (Cs). They melt easily: $T_{melting}$ changes from 453.69 K (Li) to 301.75 K (Cs). These metals have the ability to color a colorless flame: Li into red, Na into yellow, K and Cs into violet, Rb into a purple-red color. Alkali metals are similar in their physical and chemical properties, all being very active chemically. They interact with oxygen and water at room temperature. Hydroxides of alkali metals with the general formula MOH are all strong bases, and are highly water soluble. They are often referred to as *caustic alkalis*. Most alkali metals salts are very water soluble, exceptions being the Li compounds LiF, Li_2CO_3, Li_3CO_4, and also perchlorate salts $MClO_4$, where M = Li, Na, K, Rb, Cs, Fr, NH_4.

Uses of alkali metals in the free state are severely limited: Na and K are used as thermo-carriers; Cs and other alkali metals as getters in vacuum equipment, etc.

ALLOTROPY (fr. Gr. $\alpha\lambda\lambda o\varsigma$, other, and $\tau\rho o\pi\eta$, turn)

The ability of a chemical element to exist as two or several distinct forms (*allotropic forms*) having different physical and chemical properties. The allotropy can be caused by various processes such as the formation of molecules having different numbers of atoms (e.g., oxygen O_2 and ozone O_3), *crystals* of a different structure (e.g., *graphite*, *diamond*, *carbyne*, *fullerene* forms of *carbon*), or high pressure (diamond). In the case of carbon the allotropy is a kind of *polymorphism*.

ALLOWED SPECTRAL LINES

Spectrum of lines of *allowed transitions*, i.e. those which satisfy quantum-mechanical selection rules, and have intensities that far exceed the intensities of *forbidden quantum transitions*.

ALLOWED TRANSITIONS

Transitions which are permitted by the selection rules of quantum mechanics. The dimensionless squared magnitude of the matrix element of the periodic perturbation that induces transitions between particular initial and final states prescribes the probability of such a transition to occur, and for an allowed transition this matrix element is close to unity. The relative intensity of transitions in *magnetic resonance* is proportional to $|\langle in_i|g\mu_B J|fn_f\rangle|^2 \boldsymbol{B}_1^2$, where $\boldsymbol{B}_1$ is an alternating microwave or radio-frequency field, g is the dimensionless g-factor, μ_B is the Bohr *magneton* (the nuclear magneton is used for NMR), J is the *spin*, $n_{i,f}$ are the numbers of *photons*, i and f are the quantum numbers of the spin system initial and final states. If one denotes the states of an isolated spin by the value of its projection M along the applied magnetic field $\boldsymbol{B}_0$, then the transitions with $\Delta M = \pm 1$, $n = 1$ are allowed in strong magnetic fields for $\boldsymbol{B}_1 \perp \boldsymbol{B}_0$. For very weak applied magnetic fields (e.g., weaker than a crystal field term that causes magnetic quantum number splittings) the intensities of other transitions (forbidden in high fields) sometimes approach the value unity as well, that is they also become allowed, e.g., $\Delta M = \pm 2, \pm 3, \ldots$ or $\Delta M = 0$ (integer spin and $\boldsymbol{B}_1 \parallel \boldsymbol{B}_0$). When spins interact with each other, transitions involving a change in the projection of just one of the two are considered allowed transitions as well: $\Delta M_1 = \pm 1$, $\Delta M_2 = 0$, $\Delta n = 1$ or $\Delta M_1 = 0$, $\Delta M_2 = \pm 1$, $\Delta n = 1$.

ALLOY AGING

Change of *crystal structure* and physical properties of *alloys* as a result of the decomposition of a supersaturated *solid solution* (see *Alloy decomposition*), which is generated when an alloy undergoes rapid cooling (*quenching*) from the single-phase region of the *phase diagram* to the two-phase (low-temperature) one. The process of alloy aging takes place when the mutual *solubility* of alloy components in each another decreases because of a lowered temperature. The resulting supersaturated solid solution is in a *metastable state* and attains equilibrium, i.e. reaches a state characterized by a minimum of free energy (see *Thermodynamic potentials*), by precipitating the excess amount of alloying element in the form of second-*phase* particles (see *Alloying*). This process is referred to as *natural aging* if it occurs at room temperature, as distinguished from the *artificial aging* process, which requires heating of the hardened alloy. Natural alloy aging is common to *aluminum alloys* (e.g., Al–Cu, Al–Zn, Al–Ag, etc.), whereas the artificial variety is characteristic of alloys featuring higher heat resistance (e.g., Cu–Be, Cu–Ti, Ni–Be, Co–Be, Co–Ti, Co–Cu, Ni–Al, Ni–Ti, Fe–Be, Fe–Cu, etc.).

Alloy aging is a *first-order phase transition* involving *nucleation* of a new phase followed by growth of the nucleus. The late stages of alloy aging are characterized by the *coalescence* process, with larger particles growing at the expense of smaller ones. The formation of the new phase nucleus calls for overcoming a potential barrier due to the generation of a new volume which exhibits a new crystal structure (different from original matrix), composition and phase boundary. When the matrix and precipitated phase have different specific volumes, an additional contribution to the new volume is made by the elastic energy, which is due either to elastic coherent stresses arising in the case of a coherent interface, or to fields of stresses due to interphase dislocations in the case of an incoherent interface (see *Coherent precipitates*). Two decomposition types are distinguished. If the structures of the matrix and of the precipitated phase have much in common, then the process is called *isomorphic decomposition* (see *Isomorphism*), otherwise it is referred to as *nonisomorphic decomposition*. The potential barrier may be substantially lowered if the nucleus of the new phase arises at *grain boundaries*, *dislocations*, *stacking faults*, i.e. a discontinuous-type heterogeneous nucleation takes place. In the event that the nuclei are formed at point defects, the process is called "continuous homogeneous nucleation". Another mechanism of alloy aging is *spinodal decomposition*.

The form and nature of the spatial arrangement of the precipitated particles in an aged alloy are determined by the following factors: type of decomposition (continuous or discontinuous), magnitude of coherent interphase strains, value of the difference between the *elastic moduli* of the phases, type of *anisotropy of elasticity* exhibited by the matrix, volume fraction of precipitates in the alloy. Provided a certain balance is reached between the above parameters (elastically isotropic matrix, high coherent strains, and volume fraction of precipitates), the particles of the coherent phases in an alloy can achieve a regular distribution in the form of a *macrolattice*, forming a so-called *modulated structure* during the process of coalescence. The shape of precipitate particles depends on the coherent strains and on the ratio between the elastic moduli of the matrix and of the precipitated phase. There exist spherical, laminar-shaped, and needle-shaped precipitate particles.

The process of aging does not ordinarily involve the immediate formation of the precipitated phase: the alloy passes through a sequence of intermediate metastable states which correspond to certain free energy minima. At the initial stages (pre-precipitation), either zones enriched with the alloying element (*Guinier–Preston zones*), or dispersed particles of the coherent intermediate phase are formed. The dimensions of the latter range from 0.5 to 1.0 nm. At subsequent stages their structure and composition approximate the equilibrium values of the equilibrium-precipitated phase. The realization of one or another intermediate state is determined by the kinetics of alloy aging (i.e. depends on the *activation energy* of the process). Aging is accompanied by considerable changes in alloy properties: increase of *strength* and *high-temperature strength*, *coercive force*, magnetic energy, *electrical conductivity*, etc. It is for this reason that alloy aging is used in industry for the purposes of upgrading alloy characteristics.

ALLOY BRITTLENESS

High susceptibility to *brittle failure*, appearing at the *alloying* of metals, and at the formation of *alloys*. The brittleness of alloys appears due to:

(1) *Hardening*, introduced by atoms of the doping element. The increase of flow stress σ_S within the *solid solution* limits results in a decrease of plastic characteristics (elongation δ, transverse narrowing ψ), and in an increase of the *cold brittleness* temperature T_b. Alloys where a *rhenium effect* occurs are an exception.
(2) *Dispersion hardening* (see *Precipitation-hardened materials*). When phases of refractory compounds are introduced into the alloy as dispersed particles, then the mean free path of *dislocations* is restricted. Also this favors their conglomeration at barriers, leads to an increase of σ_S, and to a decrease of δ and ψ, thereby inducing an increase of T_b.
(3) Formation of intermetallic brittle phases (see *Intermetallic compounds*) (σ-, χ-phases, etc.). The reasons are the same as in case of disperse hardening, but they are compounded by the primary release of these phases along *grain boundaries*.
(4) Decay of solid solutions oversaturated by interstitial impurities (*dispersion hardening*). The reasons are the same as in two previous cases (see *Alloy decomposition*).
(5) Solid solution decay of the stratification type (see *Stratification-type failure*). At the decay extremely small (of the order of nanometers) solid solution regions based on each alloy component are formed (e.g., in the Cr–Fe system: α-phase based on Fe, and α'-phase based on Cr). The presence of fine-grained phases presents difficulties for the formation and motion of dislocations, and thereby decreases δ and ψ and increases σ_S and T_b.

ALLOY DECOMPOSITION

A *first-order phase transition* involving the separation of an excess doping element in the form of a new phase, or the separation of a pure component from a supersaturated *solid solution* of the *alloy* in which one of the components exhibits a drop in its *solubility* in the other at lower temperatures. The separated phase either may exhibit a *crystal structure* similar to that of the alloy or the matrix (*isostructural decay*), or it can have a different structure (*heterostructural alloy decomposition*). As for the acting mechanism, the alloy can separate via a *spinodal decomposition*, or through the formation and growth of an embryo of a new phase. One may also identify a *heterogeneous alloy decomposition* when *defects* of a crystal lattice take part in the *nucleation* (except for *point defects*), and a *homogeneous alloy decomposition* when the embryos form without any input from defects. As for their morphology, one may distinguish a *continuous alloy decomposition* that proceeds uniformly throughout the bulk of the crystal grain, and a *fragmentary alloy decomposition* (*cellular alloy decomposition*) which starts at *grain boundaries* and propagates into the bulk grain in the form of pseudopearlite cellular colonies.

ALLOY DEHARDENING

See *Dehardening by alloying.*

ALLOY DISORDERING

A process inverse to *alloy ordering*, when atoms of each type redistribute so that the parameters of *long-range and short-range order* decrease in their absolute values. Disordering of alloys may occur with increasing temperature and corresponding change of pressure under the effect of various radiations, and also during *plastic deformation.*

ALLOY HETEROGENIZATION

A process of obtaining metal *alloys* with a structure consisting of two or more phases (*heterogeneous structures*). According to the equilibrium *phase diagram* of an alloy the phase composition may change under a variation of temperature or pressure. For example, if there is a single-phase (homogeneous) *solid solution* in thermodynamic equilibrium at a high temperature, then lowering the temperature could convert it to a nonequilibrium state. In this case, using a rapid-rate of cooling (*quenching*) a homogenous (in the macroscopic sense) solid solution with the high temperature composition becomes fixed at room temperature, despite its lack of thermodynamic equilibrium. During subsequent heating and exposure to air this homogeneous solid solution could undergo a decay with the precipitation of inclusions of new phases, i.e. it proceeds toward heterogenization. This heterogenization is *alloy aging.* At aging of, for example, an alloy on a base of aluminum (duralumin, 4% Cu + others) has its *strength* significantly increased. *Nickel alloys* doped with aluminum, chromium, and titanium noticeably improve their *high-temperature strength* as a result of the formation of a system of isolated regions of a new phase (*gamma phase*). Heterogenization is extensively used for obtaining *hard magnetic materials.* Heating and subsequent exposure are often carried out in a strong magnetic field (*thermomagnetic treatment*). Inclusions of a strongly magnetic phase grow in the magnetic field so the direction of their stretching coincides with the field direction. Rod-like precipitates result in *magnetic anisotropy* and an increase in the *coercive force.*

Heterogeneous structures are produced not only by the *thermal treatment* of alloys, but also with the help of other diverse techniques applied during the manufacture of neutron glasses, *composite materials* (*impregnation* of high-strength fibers, networks, etc., arranged in a particular way).

ALLOYING

Introducing *atoms* of various chemical elements into materials (*insulators*, *metals*, and *alloys*) so as to impart certain properties. The formation of metal alloys with various properties that differ from those found in pure metals (increased *strength*, *high-temperature strength*, *corrosion resistance*, highly increased *coercive force*, fine dispersion structure, etc.). Concentrations of alloying elements in metals and alloys vary typically from tenths of a percent to tens of percents.

ALLOYING ELEMENT

An auxiliary *alloy* added to *liquid metals* or alloys to alter their chemical composition and improve their qualities. The doping element (see *Alloying*) is better assimilated if one uses an alloyed element, rather than the element in its pure form. Alloying elements are obtained by melting together the necessary components, or by reducing ores, concentrates and oxides. Alloying elements find their widest application in ferrous *metallurgy*, mainly for the *modification* and doping of *steel* and *cast iron*.

ALLOY ORDERING

The process of redistribution of atoms in *alloys*, whereby atoms of each type begin to occupy their individual *crystal lattice* sites so long-range order becomes established (see *Long-range and short-range order*). As this takes place, the local increase (or reduction) of the concentration of atoms in the *coordination spheres* around a central (arbitrarily chosen) also atom takes place; this local change of concentration corresponds to the presence of short-range order along with the long-range order (see *Correlation in alloys*). The phenomenon of alloy ordering is of common occurrence; it is observed in both *substitutional alloys* and *interstitial alloys* (*vacancies* at the sites and interstices of the lattice may also be involved), in oxides, in two-dimensional structures consisting of atoms adsorbed on the crystal surface, in ferromagnetic (see *Magnetic materials*) and *antiferromagnetic metals* (where differently oriented atomic spin moments become ordered), etc. Consider the simplest case of a *binary alloy* consisting of N_A atoms of type A and N_B atoms of type B that are distributed over respectively N_1 positions of type 1 and N_2 positions of type 2 (positions of types 1 and 2 being, respectively, the sites and interstices of the lattice). In this case, the relative concentrations of atoms A and B may be introduced: $c_A = N_A/N$, $c_B = N_B/N$, where $N = N_A + N_B = N_1 + N_2$. The long-range order is characterized by the long-range order parameter η (see *Order parameter*), which is given by the formula $\eta = (p_A^{(1)} - c_A)/(1 - \nu)$, where $p_A^{(1)}$ is the *a priori* probability of site 1 being occupied by atom A, and $\nu = N_1/N$. The name "*disordered alloy*" is commonly given to alloys that exhibit no long-range order (short-range order may exist, however), i.e. $p_A^{(1)} = c_A$ and $\eta = 0$. For a completely *ordered alloy* the stoichiometric composition is $c_A = \nu$, so the long-range order parameter $\eta = 1$. Also other types of long-range order parameters (as well as short-range order ones) may be introduced.

The most direct and widely used methods for experimentally investigating ordered structures (*superlattices*) in alloys are radiographic (see *X-radiography*) and neutron diffraction (see *Neutron diffractometry*) methods, which allow the determination of the arrangement of atoms in various superlattice structures (Fig. 1). It has been found that η decreases with increased temperature. For many alloys, η becomes equal to zero at a certain absolute temperature $T = T_0$ (*ordering temperature*) below the *melting temperature*. Here T_0 is called the *order–disorder phase transition* temperature, which may occur at either a *second-order phase transition* (e.g., in alloys like β-brass) or a *first-order phase transition* (in alloys of $AuCu_3$ and AuCu type). The temperature dependence of the long-range order parameter for these cases is given in Fig. 2. The value of the ordering temperature T_0 depends upon the alloy

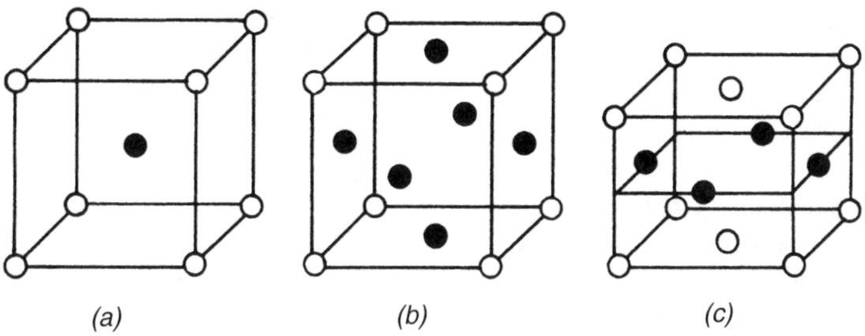

Fig. 1. Distribution of atoms A (open circles) and B (closed circles) in perfectly ordered alloys with three common structures of the types (a) β-brass, (b) $AuCu_3$ and (c) AuCu.

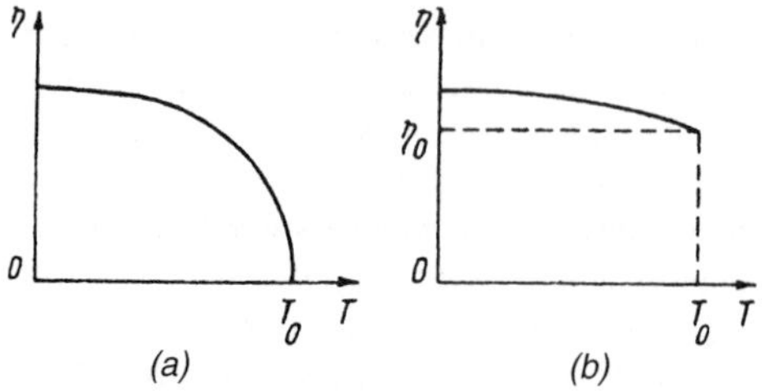

Fig. 2. Temperature dependence of the order parameter $\eta(T)$ for a phase transition of (a) second order and (b) first order.

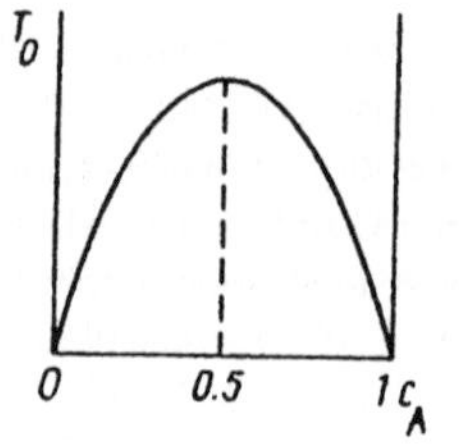

Fig. 3. Dependence of the ordering temperature T_0 on the relative concentration $c_A = N_A/N$ of A type atoms in the alloy β-brass.

composition. The short-range order does not die out at $T = T_0$, but reduces gradually with increasing T. Certain alloys (e.g., Fe–Al) exhibit successive formations of various superlattice structures at various temperatures; i.e. these alloys feature not only order-disorder transitions between superlattice structures, but also transitions of the order-order type. Various *phase transitions* involving a change of atomic order have been found in the subsystem of interstitial atoms in interstitial alloys (e.g., in metal–hydrogen systems).

The theory of atomic ordering is ordinarily based on one of two approaches: the thermodynamic approach or the statistical one. The thermodynamic theory treats alloy ordering in the neighborhood of the T_0 temperature on the basis of the general *Landau theory of second-order phase transitions.* Another thermodynamic approach (E.M. Lifshits, 1941, 1944) consists in analyzing the symmetry of the atomic probability density function using *group theory* methods, which allows the determination of structures prone to second-order transitions.

The statistical theory of alloy ordering is based on employing simplified models of alloy atomic structure. It provides a means of determining the equilibrium values of order parameters in alloys over the whole range of variation of these parameters. The commonly used model is based on the *pairwise interactions of atoms* in the context of the configuration approximation. Even when the only interatomic interactions considered are those between nearest neighbor atoms (*Ising model*), the determination of thermodynamic functions remains a complicated task. The exact solution of this statistical problem has been obtained for one- and two-dimensional crystal lattices (*Onsager method*). The difficulties associated with the three-dimensional problem necessitate the application of approximate methods, of which the simplest is the *Gorsky–Bragg–Williams approximation* (V.S. Gorsky, 1928; W.L. Bragg, E.J. Williams, 1934–1935), which completely ignores correlation in alloys. The application of this method gave rise to the theory of ordering in binary A–B alloys with two equivalent atom positions. In the context of this theory the equilibrium equations for determining the T- and c_A-dependences of the equilibrium value of η have been derived. The general forms of these equations, which are applicable to all types of structures, are given by

$$k_B T Z = \frac{1}{\nu(1-\nu)N}\frac{\partial E}{\partial \eta},$$

$$Z = \ln\frac{[c_A + (1-\nu)\eta](1 - c_A + \nu\eta)}{[1 - c_A - (1-\nu)\eta](c_A - \nu\eta)},$$

where E is the pairwise interaction energy of an atomic system. The $\eta(T)$ dependences are presented in Fig. 2, and the corresponding phase transitions follow from these expressions. One can also obtain the $T_0(c_A)$ dependence, which takes the form given in Fig. 3 in the case of alloys of the β-brass type. Generalization of the alloy ordering theory to alloys with inequivalent sites, and to interstitial alloys with inequivalent interstices, brings the equilibrium equation to the following form: $k_B T Z = R_1\eta + R_2$, where R_1 and R_2 are constants, and $R_2 = 0$ for the case of equivalent positions of atoms. The latter equation suggests the occurrence of *order–order phase transitions*, which do not involve the generation of new superlattice structures. The alloy ordering theory is extended also to cases when the type of the resulting superlattice structures is not known in advance (method of branch points, and method of static concentration of waves); it allows the prediction of the types of superlattice structures arising from the disordered state of an alloy of a given structure. The theory of alloy ordering under *pressure* is developed by taking into account the relationship existing between the energies associated with the interatomic interaction and the interatomic distance.

Allowance for correlation in alloy ordering theory has been made using various approaches, of which the *Kirkwood approximation* (J.G. Kirkwood, 1938) and the *quasi-chemical approximation* should be mentioned. The former is based on expanding the free energy in a power series of $w/(k_B T)$, where w is the ordering energy. The latter approximation treats groups of neighboring atoms as quasi-molecules. Allowance made for correlation results in the $T_0(c_A)$ curves being contracted to the medium region of the *phase diagram*, in the increase of the η_0 jump at first-order transitions, in a more correct determination of the phase transition type in alloys of the AuCu type, and in other refinements of the theory.

In a theory of alloy ordering, consideration must be given to the electronic subsystem of the crystal. Alloy ordering has a pronounced effect on the mechanical properties of alloys, strongly affecting the *elastic moduli*, critical *shear* stress, *creep*, *high-temperature strength* (e.g., the hardness of the alloy CuPt more than doubles on ordering). Alloy ordering is responsible for abrupt changes in atomic *diffusion coefficients* and the specific electrical resistance (see *Resistivity of ordered alloys*), as well as for profound changes in thermal properties (characteristic anomalies of *specific heat* in the neighborhood of T_0, abrupt changes of alloy energy during the course of first-order transitions). The assumption of the existence of ordered structures is based on the observation of minima on the plots of $\rho(c_A)$ for the Au–Cu alloys. Quantum perturbation theory (with no regard for correlation) provides the dependence of residual resistivity on c_A and η under conditions of alloy ordering:

$$\rho_0 \sim \left[c_A(1-c_A) - \nu(1-\nu)\eta^2\right].$$

Alloy ordering strongly affects the energy spectrum of electrons (theoretically, there is a certain probability of generating a gap in the electron energy spectrum (see *Band gap*) which may in principle induce *metal–insulator transitions*), therefore, alloy ordering exerts a strong influence on *galvanomagnetic effects*. The Hall constant of the alloy $AuCu_3$ even exhibits a change of sign. Alloy ordering has a pronounced effect on magnetic properties (e.g., the alloy Ni_3Mn is weakly magnetic in the disordered state, and becomes a *ferromagnet* in the ordered state).

ALLOYS

Macroscopically homogeneous many-component systems. The main method of alloy production is *crystallization* from the melt. There are also other techniques of alloying: *sintering*, *diffusion* of the alloying elements into a solid; accumulative crystallization of vaporized components; electroplating; *ion implantation*, etc. In certain cases the concentrations of the components are in a strictly defined ratio, i.e. the alloy is a chemical compound (with only slight deviations from *stoichiometric composition*); e.g., *brass* structures of the β (CuZn), γ (Cu_5Zn_8) and ε ($CuZn_3$) types, or compounds $AuCu_3$, AuCu, etc. The alloy name is often determined by the name of its major component (*copper alloys*, *titanium alloys*, *aluminum alloys*, etc.). Certain groups of alloys go under conventional names: cast irons and steels (Fe–C), brasses (Cu–Zn), bronzes (remaining Cu-based alloys). Alloys are divided into single-phase (homogeneous) and multiphase (heterogeneous) types. The following individual alloy phases are distinguished: *solid solutions* formed when atoms or ions of components share a common lattice (characteristic of one of the components) when mixed in arbitrary proportions; *intermetallic compounds* characterized by certain relations among the constituent elements; crystal lattices of intermetallic compounds that differ from those of the constituent elements. Alloys are grouped according to common inherent features: nonferrous metal alloys, refractory metal alloys, ferrous metal alloys, semiconducting alloys, etc. The most widely used types are metal alloys which are divided into casting alloys and wrought alloys depending on the technique of production of semifinished items or complete articles. In terms of applications, alloys are classified into constructional alloys, tool alloys, alloys with special electrical properties, solders, etc. In terms of the principal operating characteristic, *high-temperature alloys*, *thermal-environment resistant alloys*, wear-resistant (see *Wear*) ones, corrosion-resistant (see *Corrosion resistance*) alloys, extra-hard alloys, etc., are distinguished. The large group of alloys with special characteristics due to their definite chemical composition, lack of harmful (exerting strong adverse effect on properties) impurities, are called *precision alloys*. This group includes soft magnetic alloys and hard magnetic alloys (see *Soft magnetic materials*, *Hard magnetic materials*), alloys with specified electric, thermal, elastic properties (including alloys exhibiting particular temperature dependences of these properties), as well as superconducting and bimetallic alloys. High purity of alloys is achieved by using components of extra-purity, and performing the smelting in vacuo, inert, or refining (hydrogen-containing) media, by application of *zone refinement*, electroslag remelting, vacuum refining, methods of floating zone melting. In order to increase the anisotropy of properties, directional crystallization and *thermomagnetic treatment* are used. The finishing stage of the development of alloy properties is heat treatment, which facilitates eliminating the nonuniform distribution of alloying components within the boundaries of individual grains, establishing the needed dislocation structure (see *Dislocations*), phase composition, distribution of atoms within separate phases (see also *Alloy ordering*, *Alloy aging*, *Order parameters in alloys*). The thermodynamic description of alloys usually involves the temperature T and composition (molar ratios of components) as independent state variables. The characteristic functions are *entropy* S, enthalpy H, and *Gibbs free energy* $G = U - TS + pV$ (U is internal energy, V is volume) (see *Thermodynamic potentials*). The components of alloys are described using partial molar derivatives, e.g., *chemical potential* $\mu = \partial G/\partial n_i$.

ALLOYS OF METALS

For different kinds see *Aluminum alloys*, *Beryllium alloys*, *Chromium alloys*, *Copper alloys*, *Germanium–silicon alloys*, *Iron alloys*, *Lead alloys*, *Magnesium alloys*, *Molybdenum alloys*, *Nickel alloys*, *Noble metal alloys*, *Rhenium-bearing alloys*, *Tantalum alloys*, *Titanium alloys*, *Tungsten alloys*, *Heusler alloys*, *Mishima alloys*.

ALLOY STRATIFICATION

A *phase transition* consisting in *isostructural decay* of a supersaturated *solid solution* (see *Alloy decomposition*) into two or perhaps three solutions with differing concentrations of doping elements. Experiments have demonstrated alloy stratification in systems exhibiting immiscibility regions in their solid states (such as Au–Pt, Au–Ni, Cu–Ni–Fe). A spinodal mechanism of the phenomenon is suggested (see *Spinodal decomposition*).

ALLOY STRUCTURE AND PROPERTIES

For structure and properties of different kinds of alloys, see *Amorphous metals and metallic alloys*, *Binary alloys*, *Eutectic alloys*, *Eutectoid alloys*, *High-coercivity alloys*, *Interstitial alloys*, *Monotectic alloys*, *Peritectic alloys*, *Superconducting alloys*, *High-temperature alloys*, *Substitutional alloys*, *Thermal-environment resistant alloys*, *Ferromagnetism of metals and alloys*, *Radiation physics of metals and alloys*, *Failure of amorphous metallic alloys*, *Order parameters in alloys*, *Resistivity of ordered alloys*.

ALNI (fr. aluminum and nickel)

Hard magnetic materials of the system Fe–Ni–Al with Cu commonly added. The optimal composition is close to Fe_2NiAl. The products from these alloys are manufactured by *casting* and grinding. A high-coercivity state results from the decomposition of an oversaturated *solid solution*. Properties of some alni alloys are given under the entry *Hard magnetic materials*.

ALNICO (fr. aluminum, nickel and cobalt)

Hard magnetic materials based on the system Fe–Co–Ni–Al. They are commonly used with the addition of up to 4% Cu. They are similar in many aspects to *alni* alloys. Alloys with contents of Co 18 to 25% are sensitive to a *thermomagnetic treatment* resulting in the formation of *anisotropic magnetic materials*. Their properties are given in the table under the entry *Hard magnetic materials*.

ALPHA PHASE (α-phase) (introduced by F. Osmond, 1885)

A *phase* on the basis of the low-temperature modification of a polymorphous chemical element (see *Polymorphism*). The term alpha phase was introduced to designate the low-temperature modification of *iron* and related *solid solutions*; later it was extended to other polymorphous elements. In binary systems, the alpha phase indicates the solid solution of the second component in the substrate metal. The temperature and concentration regions of the alpha phase are determined by the thermodynamic potentials of its initial and concurrent phases. If the alpha phase exists over wide concentration limits, then *alloy ordering*, *alloy stratification*, and magnetic transformations may occur at its cooling.

ALUMINOSILICATES

A group of *minerals*: various salts of silicic acid where *aluminum* occupies the same position in the *crystal structure* as *silicon*. Also present are anion complexes, namely SiO_4 tetrahedra, and also AlO_4. The number of Si^{4+} ions replaced by Al^{3+} ions does not exceed 50%. The ratios Si : Al equal 3 : 1 and 1 : 1 are found. The ratio (Si + Al) : O in anion radicals is also 1 : 2. Aluminosilicates can have chain, laminar or framework structures. In minerals with the chain structure, Al can replace up to 1/4 of the Si, and in the laminar and three dimensional framework ones – up to 1/2 of Si. In this case, the charge balance is restored by the introduction of ions of monovalent (Na, K) or divalent (Ca, Ba) elements into the structure. The most common aluminosilicates are feldspar, *zeolites*, scapolite, micas, chlorites, and others.

ALUMINUM, Al

A chemical element of Group III of the periodic table with atomic number 13 and atomic mass 26.98154.

The electron configuration of the outer shell is $3s^2 3p^1$. Successive ionization energies are 5.99; 18.83; and 28.44 eV. Atomic radius is 0.143 nm; radius of Al^{3+} ion is 0.051 nm. The main oxidation state is +3, with +2 and +1 possible at high temperatures. Electronegativity is 1.5. Free aluminum is a silvery-gray *metal*, under normal conditions covered with a thin oxide film. It has one stable isotope ^{27}Al. It is the most abundant metal in the earth's crust. It cannot occur free in nature due to its high chemical activity. The crystal lattice of aluminum is face centered cubic, with lattice parameter $a =$ 0.4049 nm. Its density is 2.70 g/cm^3 (at 293 K), $T_{melting} = 933$ K, $T_{boiling} = 2793$ K. Heat of melting is 10.55 kJ/mole; heat of vaporization is 291 kJ/mole, specific heat is 25.1 $J{\cdot}kg^{-1}{\cdot}K^{-1}$ (at 273 K). Debye temperature is 663 K; thermal linear expansion coefficient is $24.56{\cdot}10^{-6}$ K^{-1} from 293 to 473 K. Specific thermal conductivity is 225 $W{\cdot}m^{-1}{\cdot}K^{-1}$, electric resistivity is 2.66 $\mu\Omega{\cdot}cm$ (at 273 K). The superconducting transition temperature is 1.2 K. Aluminum is weakly paramagnetic. Its elastic modulus is 68.6 GPa, Brinell hardness for calcined aluminum is 167 MPa. Aluminum is easy to process by pressing, rolling, forging, stamping and drawing. Its ultimate tensile strength is 78.5 to 98.1 MPa, yield limit upon stretching is 29.4 MPa, its rigidity is 25.5 GPa. After cold rolling, its ultimate strength grows to 177 to 245 MPa. Adiabatic elastic moduli of the aluminum monocrystal: $c_{11} =$ 106.43; $c_{12} = 60.35$; $c_{44} = 28.21$ GPa (at 300 K). Pure aluminum has a high reflection factor which makes it suitable for the manufacture of reflectors. Aluminum and *aluminum alloys* are used as *construction materials* in machinery, buildings, aircraft, civil engineering, etc.

ALUMINUM ALLOYS

Alloys based on aluminum.

They are characterized by low density, as well as high *corrosion resistance*, *thermal conductivity*, *electrical conductivity* and specific *strength*. There are deformable and cast aluminum alloys.The *deformable alloys* possess high *plasticity*, they can be welded (see *Welding*), and machined, do not become brittle (see *Embrittlement*) at low temperatures. They are separated into hardenable and nonhardenable types by a thermal treatment. The mechanical properties of nonhardenable alloys can be improved by *alloying* and *work hardening*, while those of the hardenable alloys are improved by *quenching* and aging (see *Alloy aging*). To increase their plasticity the alloys are annealed (see *Annealing*). Basic alloying elements of deformable alloys are Cu, Mg, Mn, Si, Zn, and Cr. *Cast alloys* possess fluidity, high density, and low linear shrinkage. They do not tend to form voids or cracks. The main alloying elements of cast alloys are: Cr, Cu, Mg, Mn, Ni, Ti, and Zr. Aluminum alloys are used as *antifriction materials.* In volume of production and range of application aluminum alloys are second after ferrous metals.

ALUMINUM NITRIDE, AlN

The only chemical compound of *aluminum* with nitrogen. It has a hexagonal wurtzite-like (ZnS) lattice of space group $P6_3mc$ (C^4_{6v}); crystal lattice parameters are $a = 0.3112$ nm, $c = 0.4981$ nm. Aluminum nitride decomposes at 2400 °C; density is 3.26 g/cm^3; heat conductivity of polycrystalline samples varies from 50 to 150 W·m^{-1}·K^{-1} depending on purity; heat conductivity of monocrystals is 270 to 360 W·m^{-1}·K^{-1}; microhardness is 1.2 GPa. Aluminum nitride is an insulating material with electrical resistivity $\rho > 10^{15}$ Ω·m and band gap width exceeding 6.2 eV. It reacts with water, diluted by mineral acids; it does not interact with melt Al, Ga, Cu, cryolite, *gallium arsenide.* Aluminum nitride is obtained by nitration of Al, by reduction of Al with simultaneous nitration, by thermal decomposition of aluminum-containing compounds ($AlCl_3 \cdot 6NH_3$). It is used as a fireproof and electrical *insulating material.*

ALUMINUM–YTTRIUM GARNET

See *Yttrium–aluminum garnet.*

AMBIPOLAR DIFFUSION

The *diffusion* of quasi-neutral disturbances of an electron–hole plasma density n driven by the gradient $\partial n/\partial x$ (see *Solid-state plasma*). If the electron and hole diffusion rates differ then some spatial charge separation occurs. An electric field arises inside the disturbed density region to retard the fast carriers and accelerate the slow ones. Therefore, the carriers of either type, while diffusing in the direction of their concentration gradient, simultaneously drift in the *self-consistent* (so-called *imbedded*) *electric field* of the separated charges. With this field accounted for, it is possible to describe the density cluster dynamics by just one *ambipolar diffusion coefficient*

$$D = \frac{n_e\mu_e D_h + n_h\mu_h D_e}{n_e\mu_e + n_h\mu_h},$$

where $n_{e,h}$, $\mu_{e,h}$, $D_{e,h}$ are the concentrations, mobilities and *diffusion coefficients* of electrons and holes. Non-degenerate carriers obey the Einstein relation $D_{e,h} = (T/e)\mu_{e,h}$, where e is the elementary charge, T is the temperature expressed in energy units, and as a result, the previous equation simplifies to

$$D = \frac{n_e + n_h}{n_e/D_h + n_h/D_e}.$$

AMBIPOLAR DRIFT

The drift of the quasi-neutral perturbation of *charge carrier* density under the action of an external electric field: $E = -q\partial n/\partial x$. Since the electric force acts in opposite directions on electrons and holes, an additional electric field arises due to charge separation in the density perturbation. In addition to the drift fluxes from each-type carrier density irregularity, fluxes in the disturbed field arise which to some extent compensate the former. When the undisturbed electron and hole concentrations are equal ($n_e = n_h$), complete compensation takes place, and the bunches become immobile. The ambipolar drift velocity is given by $v = \mu E$, where $\mu = \mu_e\mu_h(n_e - n_h)/(n_e\mu_e + n_h\mu_h)$ is the *ambipolar mobility*, where $\mu_{e,h}$ are the electron and hole mobilities. It follows from the formula that in the limiting cases $n_e \gg n_h$ and $n_e \ll n_h$ the ambipolar mobility μ is equal to μ_h and $-\mu_e$, respectively, i.e. it coincides with the mobility of the minority carriers. The expression for the ambipolar drift mobility is valid if the carrier mobilities are isotropic and independent of n_e and n_h and there is no heating of the electrons and holes by the field. Otherwise v can differ from zero even when $n_e = n_h$.

AMERICIUM, Am

A chemical element of Group III of the periodic table with atomic number 95 and atomic mass 243.0614.

Fifteen radioactive isotopes are known with mass numbers 232, 234 to 247. There are no stable isotopes. Electronic configuration of filling and outer shells is $5f^76d^07s^2$. The ionization energy is 5.99 eV. Atomic radius 0.174 nm; radius of Am^{3+} ion is about 0.10 nm, of Am^{4+} ion 0.09 nm, of Am^{5+} ion 0.086 nm, of Am^{6+} ion

0.080 nm. Oxidation state is +2 to +7, that of +3 being the most stable in solution. Americium in a free form is a silvery *metal*, luminous in the dark due to its intrinsic α-radiation. Depending on temperature and pressure, americium exists in three crystallographic modifications, the room temperature form being the α-modification with a double hexagonal close-packed structure. Americium is used for manufacturing sources of neutrons, α-particles and γ-radiation (used in the defectoscopes and densimeters) as well as for the production of heavier *transuranium elements*. Americium and its compounds are extremely toxic.

AMORPHIZATION

The transformation, without melting or sublimation, of a *solid* from the crystalline state characterized by long-range order to an amorphous one which lacks long-range order.

Since the *amorphous state* of a solid is a nonequilibrium one, the process of amorphization necessarily includes the transition to a nonequilibrium state. Amorphization may be brought about by intense penetrating radiation which generates many structural *defects* in the solid, under strong *plastic deformation*, or under the special thermal *annealing* of a mixture of two crystalline materials. For example, if multilayer metal ribbons Ni–Zr–Ni–Zr$\cdots$ are annealed at the temperature 300 °C, the amorphous material Ni–Zr appears after 260 hours. The amorphization in this case is due to the diffusional mobility of one type of atom (Ni). *Amorphous solid solutions* are also obtainable by mixing the atoms of multilayer crystalline films under ion or electron exposure. The strong plastic deformation of some alloys (e.g., Ni–Ti) results in their amorphization. Amorphization also occurs at some chemical reactions in crystalline bodies; e.g., the oxidation of silicon films in an atmosphere of oxygen leads to the formation of *amorphous films* of silicon dioxide. Different methods of amorphization of solids are under development to provide bulk samples of amorphous materials (mainly metallic ones), which are not available via the methods of *quenching* or *sedimentation*.

AMORPHON

A type of *noncrystalline cluster*, the element of a structure of amorphous material having tetrahedral bonding (e.g., *amorphous silicon*, germanium, etc.), which is a regular dodecahedron with pentagonal faces, and atoms located at its apices.

AMORPHOUS EPITAXY, artificial epitaxy

Formation of single crystal (usually semiconductor) layers on amorphous substrates. As distinct from the *epitaxy* carried out on single crystal substrates, a strip-like microscopic relief is produced on the surface of an amorphous insulator or semiconductor by photolithography. The relief depth is usually to 0.3 μm with a step of a few microns. During the heating of this structure, e.g., by laser or light pulses, the relief-induced nonuniformity of the heat absorption in the layer leads to a periodic temperature distribution. Therefore, the heterogeneous nucleation of the crystal phase (see *Crystallization*) in the forming melt turns out to be ordered, resulting in a directional growth of *crystallites*. Profiled cuts are made in the substrate to serve as heat sinks. The oriented block formation is facilitated by the anisotropy of *surface energy* of the melt–substrate boundary, which determines the rate of crystal nucleation during the crystallization, and by that the crystal growth rates. Use is also made of the effect of oversaturation at the artificial epitaxy, and the formation of a liquid layer under the crystallites (*rheotaxy*). These methods were applied to the systems Si/Si_3N_4, $GaAs/SiO_2$, Ge/SiO_2, and are used for producing A_2B_6, A_3B_5 films and other substances (see *Semiconductor materials*). Artificial epitaxy, in addition to the "penciled" type (*graphoepitaxy*), includes methods leading to the outgrowth of monocrystalline portions from the orienting matrix (Si, GaAs) through the windows in an *insulator* (*selective epitaxy*), and between dielectric layers (SiO_2, glass) from the end-face orienting crystal. The orienting action of the substrate can be used through an amorphous layer or a gap (*reproductive epitaxy*). The layers obtained using the artificial epitaxy method have the following defects: *dislocations*, *grain boundaries*, extended *stacking faults*. Yet, one can attain the required degree of structural perfection, and the electrophysical properties that suffice for making the elements of *integrated circuits*. The artificial epitaxy method has been developed as a basis for the technology of multilayer three-dimensional *microelectronics*.

AMORPHOUS FILMS

Layers of solid materials in the *amorphous state* or *vitreous state of matter* with thicknesses up to several tens of millimeters. Physical properties (structural, mechanical, electrical, magnetic, optical) of amorphous films differ a great deal from those of crystalline *films* and from bulk amorphous materials.

Amorphous films are obtained mainly by rapid *condensation* (10^3 to 10^7 K/s) of the source material onto a substrate via *spraying* in vacuo or in an inert atmosphere, by rapid cooling of a thin layer of melt of the material, or by electrical deposition. Methods are available for producing amorphous films by *amorphization* of crystalline films. The best studied are amorphous films of *semiconductor materials* (silicon, chalcogenides of arsenic, germanium, phosphorus, antimony, etc.); of *insulators* (SiO_x, Al_2O_3, Si_3N_4, etc.); of *metals* (Be, Cr, Mn, Fe, Co, etc.); of *alloys* (metals of Group VII and *transition metals* with B, C, P, Ge, Si, alloys of Fe, Co, Ni, Cu with transition metals, alloys of Te with Cu, Ag, In, Tl, etc.). The structure and stoichiometric composition of amorphous films are thermodynamically unstable. Amorphous films subject to heat or radiation can undergo *crystallization*, changes in chemical composition (evaporation through the surface, reactions with the ambient medium), and deviations from uniformity (*granular films*). There are various applications in electronics (magnetic shields, electronic switches with memory, *field-effect transistors*, etc.), in *optoelectronics* (optical memory including *optical storage disks*), audio and video information storage (screens of color kinescopes, photoresistors in xerography, etc.) (see also *Metallic glasses*, *Amorphous semiconductors*). Many amorphous materials deposited from vapors onto a substrate at cryogenic temperatures exhibit *superconductivity* (e.g., granular superconductors, superconducting amorphous films of Ga, $Tl_{90}Te_{10}$, Be, $Sn_{90}Cu_{10}$).

AMORPHOUS MAGNETIC MATERIALS

Amorphous magnetic materials are those which exhibit *amorphous magnetism*.

These materials come in the form of *metallic glasses* and coordination compounds. Pure *transition metals* obtained as amorphous *thin films* have low *crystallization* temperatures (about several tens of Kelvin degrees). In order to hinder the generation and growth of the crystalline phase, and the increase of the crystallization temperature, one can introduce into metallic *alloys* either elements with strong *covalent bonds* such as B, P, Si, C, Ge (up to 15 to 20 atomic %), or elements, differing appreciably in their atomic size (*rare-earth elements*, Nb, Zr, Hf), up to 10 atomic %. Amorphous magnetic materials are obtained in the form of ribbons, wires, films, or foils by the following methods: cooling of the melt jet at the rotating disc or between cooled rollers, drawing of the melt in the form of microwire, ion-plasma and laser spraying, chemical and electrochemical deposition, and *ion implantation*.

Magnetic ordering of different types (see *Amorphous magnetism*, *Spin glass*) is observed in amorphous magnetic materials. Such parameters as *magnetization*, *exchange interaction* parameter, and local *magnetic anisotropy* are determined in amorphous magnetic materials as average values, and also by the values of their root-mean-square *fluctuations* and by their correlation radii. The type of magnetic order in an amorphous magnetic material may differ significantly from that of its crystal analogue. Amorphous magnetic materials have a unique combination of physical properties: high *strength* (350 to 400 kg/mm^2) and *hardness* (1000 kg/mm^2); good *plasticity*; resistance against all types of *corrosion* (it is enhanced with proceeding along the sequence P, C, Si, B); high saturation induction (up to 17.5 T) and high *magnetic permeability* (up to $6\cdot 10^5$); low *coercive force*; rectangular shaped *hysteresis loop*, low sensitivity of magnetic properties to mechanical factors; high electric *resistance* (3 to 4 times higher than that of crystal analogs), low specific loss for *magnetic reversal*. The last is associated with the absence of long-range crystalline order (see *Long-range and short-range order*), and hence, of crystallographic magnetic anisotropy, and of those common defects which inhibit remagnetization such as *crystallite* boundaries, *dislocations*, and *stacking faults.* Immediately after their synthesis, amorphous magnetic materials have high internal stresses bringing about the appearance of magnetoelastic anisotropy. To eliminate these stresses, thermal and thermomagnetic *annealing* is performed at temperatures below the crystallization temperature, and alloys with a zero *magnetostriction* coefficient are used. One disadvantage of amorphous magnetic materials is the presence in them of structural *relaxation.*

Amorphous magnetic materials have prospective applications in electrical and electronic engineering. The replacement of conventional electrical sheet *steels*, precision alloys of the permalloy type (see *Soft magnetic materials*) and ferrites in different electromagnetic devices and transformers with amorphous magnetic materials manufactured by a simpler and cheaper process allows reducing the dimensions of the devices, the energy loss, and the cost. In electronic engineering, the low-coercive field amorphous magnetic materials find wide application in the production of magnetic heads, delay lines, and magnetic shields. Ferromagnetic films (see *Magnetic films*) of amorphous magnetic materials of a transition metal or rare-earth metal type, with high single-axis magnetic anisotropy (up to $5{\cdot}10^5$ erg/cm^3) induced during the process of synthesis, can be used as inexpensive media for the recording devices in *magnetic bubble domains.* With the films Co–Gd, the diameter of magnetic bubble domains achieves submicron dimensions with mobility 20 m$\cdot$s$^{-1}\cdot$Oe^{-1}. The Fe–Tb films are prospective materials for contact printing and optical *information recording* with magneto-optic reading. Their advantages are low Curie temperature (310 to 370 K) and, hence, low optical energy required for the *thermomagnetic information recording*, strong *magnetooptical effects*, absence of the intercrystallite boundaries which bring about additional scattering and absorption of light, and, as a result, high signal-to-noise ratio in the reading of information.

AMORPHOUS MAGNETIC SUBSTANCES

Involves solid substances having long-range magnetic order, but lacking long-range crystalline order (see *Long-range and short-range order*).

There are three classes of amorphous magnetic substances. The first such class includes highly disordered *ferromagnets* with ferromagnetic *exchange interactions* which fluctuate noticeably in magnitude. The class is exemplified by *metallic glasses* $Fe_{80}B_{20}$, $Fe_{40}Ni_{40}B_{20}$. Most physical properties of these systems differ only slightly from those of ordered ferromagnets. The main distinction is that amorphous ferromagnets are characterized by local frozen magnetization m_i and its first moments $M = \langle m_i \rangle_c$ and $q = \langle m_i^2 \rangle_c$ (averaging over the configurations of exchange integrals). As a result, some interesting new phenomena arise which are also present in other classes of amorphous magnetic substances, e.g., the critical magnetic scattering of neutrons by the configuration fluctuations of magnetization, and fluctuations decaying with distance following the exponential dependence $\exp(-\kappa r)$ rather than the *Ornstein–Zernike law*.

The second class of amorphous magnetic substances comprises systems with concurrent ferro- and anti-ferromagnetic exchange interactions, exemplified by the *alloys* $Eu_xSr_{1-x}S$ and metallic glasses $Fe_xMn_{1-x}P_{16}B_6Al_3$. The exchange interaction can have different signs at different bonds, and depending on these signs these materials may be *paramagnets*, ergodic ferro- or antiferromagnets, nonergodic ferro- or antiferromagnets (*asperomagnets*) as well as *spin glasses.* These states are characterized by different sets of first and second moments, possible local frozen-in magnetization, and the presence or absence of *nonergodicity.* The most interesting property of this class of substances is the emergence of nonergodic phases; experimentally the phases are associated with various irreversible phenomena.

The third class of amorphous magnetic substances comprises *alloys* and the metallic glasses which include *rare-earth elements* and noble metals of the DyCu and TbAg type. These exhibit ferromagnetic interactions and high local *magnetic anisotropy* with random directions of anisotropy axes. The last property is the salient one characterizing the physics of this class of magnetics. Amorphous magnetic substances are asperomagnets (typically, nonergodic) or spin glasses, and they are used as *hard magnetic materials*; *magnetic bubble domains* can be observed inside them. Despite the dissimilarities of their interactions, this class of materials is quite close to the previous one in some of its properties, including various characteristics of the local frozen magnetization and its different manifestations of nonergodicity.

The magnetic branch of the boson spectrum of amorphous ferro- and antiferromagnets in the long-wavelength range corresponds to *magnons* which obey the *dispersion law* $\omega = Ak^2$. Typically, the magnon stiffness coefficient A in the amorphous magnetic phase is less than that in the corresponding crystalline phase. In the wave vector range $k \sim a^{-1}$ (a is the atomic distance), the frequency function $\omega(k)$ has a minimum which corresponds to a *magnetic roton*, the excitation specific to amorphous magnetic materials.

AMORPHOUS MAGNETISM

The field of the physics of magnetic phenomena dealing with the magnetic properties of atomically disordered solids (see *Amorphous magnetic substances*, *Amorphous magnetic materials*, *Spin glass*).

AMORPHOUS METALLIC ALLOYS

See *Amorphous metals and metallic alloys*, *Failure of amorphous metallic alloys*, *Plastic deformation of amorphous metallic alloys*.

AMORPHOUS METALS AND METALLIC ALLOYS

Solid *metals* and metallic *alloys* of different chemical composition without periodic structure, but with topological and compositional (see *Disordered solids*) short-range order. Long-range order (see *Long-range and short-range order*) is absent, as in liquids. Amorphous metallic alloys form a new type of metallic material which is of interest for both fundamental solid state physics and applications. These alloys have high mechanical *viscosity*, high *yield limit*, high *magnetic permeability* and low *coercive force*, high *corrosion* resistance, and an electric resistance which depends weakly on the temperature. Amorphous metallic alloys are obtainable by *amorphization* of crystalline phases, by condensation of vapors onto a cold substrate, with the help of chemical and electrochemical deposition, by plasma spraying, and with the aid of rapid cooling (*quenching*) of the liquid melt. Amorphous metallic alloys obtained by quenching are called *metallic glasses*.

AMORPHOUS PHOTOCONDUCTIVITY

See *Photoconductivity of amorphous materials*.

AMORPHOUS SEMICONDUCTORS

Materials in the *amorphous state*, possessing the properties of *semiconductors*.

They are obtained by deposition of atomic or molecular fluxes and ionized gases onto substrates, by *amorphization* of crystalline solids, and by rapid cooling of melts. Amorphous semiconductors obtained by cooling of melts are called *vitreous semiconductors* (see *Vitreous state of matter*). According to their composition and structural properties, amorphous semiconductors (ASCs) are divided into tetrahedral, chalcogenide, pnictide based, oxide, and organic types. *Tetrahedral* ASCs include simple materials such as *amorphous silicon* and germanium; binary materials (GaAs type) and more complex ones (of the type of solid solutions Ge_xSi_{1-x}, $Si_x(GaSb)_{1-x}$, etc.). The atoms in these materials are fourfold coordinated, each having four *covalent bonds* which are saturated. *Chalcogenide* ASCs include glasses based on Se and Te, and also binary, ternary, and quaternary systems (sulftellurides), e.g., AsS, AsSe, AsTe, AsSSe, AsGeSeTe, AsSbSSe, etc. There are materials of the $CdAs_2$ and $CdGeAs_2$ types, which can be considered as mixed tetrahedral ASCs and *pnictides*. Among *oxide* amorphous semiconductors are Y_2O_5 and other compounds based on *transition metal* oxides.

Atomic structure and structural defects. Amorphous semiconductors have no long-range order (see *Long-range and short-range order*) in the location of atoms, but they do have short-range order, and often the *intermediate order* due to the saturation of interatomic bonds. These semiconductors have structures of the *random continuous network* type. Different cluster models as well as the *Polk model* were suggested as the model structures of tetrahedral ASCs. The structural defects of ASCs include empty sites, networks, *interstitial atoms*, *impurity atoms*, *dangling bonds*, and two-dimensional defects of the boundary type, typical of amorphous bodies (see *Polycluster amorphous solids*). Multicentered, "stretched" and distorted bonds, atomic configurations of low rigidity, and "soft" anharmonic interatomic potentials with low *elastic moduli* may exist in amorphous semiconductors. *Three-centered bonds* probably play an

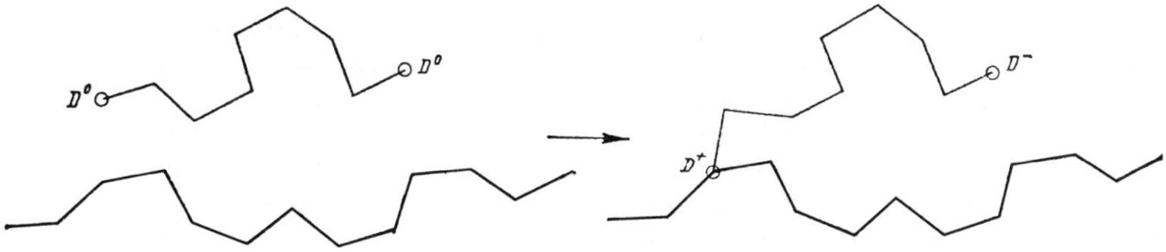

Fig. 1. Sketch of the rearrangement of atomic configurations under the $2D^0 \to D^+ + D^-$ transformation.

important role in chalcogenide amorphous semiconductors. The concentration of broken bonds of tetrahedral amorphous semiconductors, and of three centered bonds in chalcogenide ones, according to different measurements, may be as high as 10^{20} to 10^{21} cm^{-3}. The *dangling bonds* in tetrahedral ASCs are mostly electrically neutral (each of them has spin $1/2$) and thus they can be detected by *electron paramagnetic resonance.* Pairs of electrically neutral broken bonds in chalcogenide ASCs transform, as a rule, into charged ones according to the scheme $2D^0 \to D^+ + D^-$ (Fig. 1). In such a transformation, two electrons are coupled at one of the defects, thus forming D^-, while the Coulomb repulsion between the electrons is compensated by the rearrangement of the ambient atomic configuration, and D^+ is built into the network with an increase of coordination number (see Fig. 2). The charged defects D^+ and D^- are usually called *variable valence pairs.* In binary ASCs defects of the "irregular bonding" type are observed, e.g., Se–Se and

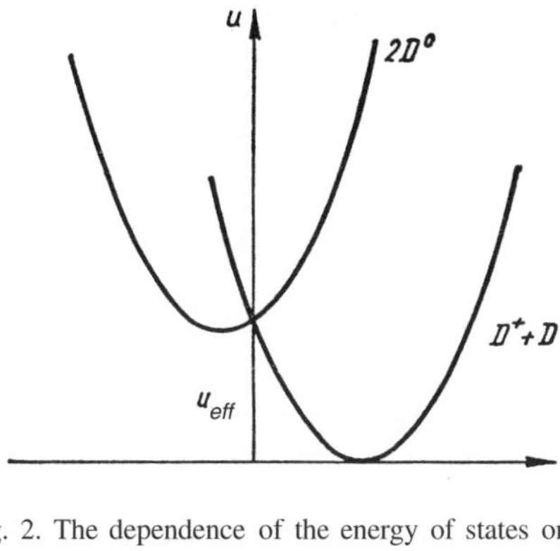

Fig. 2. The dependence of the energy of states on the configuration variable.

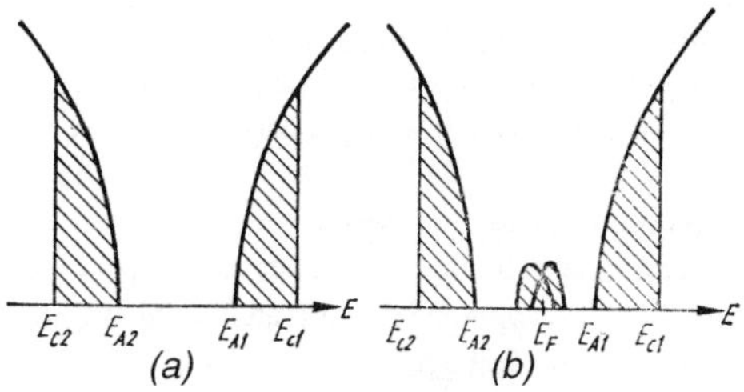

Fig. 3. Diagram of the density of electronic states of amorphous semiconductors while neglecting the levels associated with defects (a), and while taking them into account (b). The regions of the spectrum corresponding to localized states are shaded. E_{c1}, E_{c2} are the mobility edges of the conduction and valence bands, and E_{A1} and E_{A2} are the edges of "tails" in these regions.

As–As in As_2Se_3, the peroxide bridge O–O and "oxygen vacancy" Si–Si in SiO_2, etc. Large-scale fluctuations of the concentrations of atoms participating in the material composition and density, crystalline inclusions, and pores fall into the category of three-dimensional defects.

The *electronic structure* of amorphous semiconductors resembles, due to the presence of short-range order (in accordance with the *Ioffe–Regel hypothesis*), that of crystalline semiconductors, i.e. it is close to the band type. Due to the topological disorder of a random continuous network, the "tails" of *valence bands* and of *conduction bands* are found to extend to the *pseudogap* and contain energy levels of localized states (see Fig. 3(a)). The hypothesis on the existence of "tails" of the electron distributions in the pseudogap of ASCs was first formulated by A.I. Gubanov and later elaborated by N. Mott and E. Davis (see *Mott–Davis model*). Also located in the gap are electron levels associated with defects, viz., the levels of broken bonds D^0 and of charged D^+ and D^- states (Fig. 3(b)). The presence of two unfilled electronic levels lying below the single-electron states in chalcogenide ASCs due to the effective attraction U_{eff} between electrons and of free electronic levels at the broken bonds in tetrahedral ASCs brings about the so-called *Fermi level pinning* in the middle of the gap. Doping with donor (acceptor) impurities fails to shift the Fermi level E_F from the middle of the pseudo-gap to the *mobility edge* of electrons (holes) at doping with donor (acceptor) impurities because the electrons (holes) fill the empty levels of defects in the middle of the pseudo-gap. Experimental measurements reveal rapidly decaying "tails" of the valence and conduction bands, and the presence of several peaks in the of *density of states* in the pseudo-gap, apparently associated with different types of structural defects.

A possible way to change electronic properties of tetrahedral ASCs with *doping* is accomplished by preliminary hydrogenization, or the introduction of hydrogen, which saturates the broken bonds and thus counteracts the *Fermi level* pinning in the middle of the gap. Changes in the electronic properties of ASCs without the preliminary saturation of broken bonds can be brought about by the introduction of relatively higher concentrations of doping elements (see also *Doped amorphous semiconductors*).

Electrical conductivity. In amorphous semiconductors the Fermi level lies in the energy gap. The transfer of electrons to the external electrical field may occur either by crossing over to the conduction band above the mobility edge E_{c1}, or by hopping (jumping) from the level of one defect to an empty level of another, adjacent defect (see *Hopping conductivity*). The prevalence of either mechanism of conductivity depends on the electronic structure of ASCs and on the temperature. Prevailing at high temperatures and small gaps between E_F and E_c is the transfer connected with activation transitions of electrons to the conduction band; here

$$\sigma = \sigma_0 \exp\left(-\frac{E_c - E_F}{k_B T}\right),$$

where $\sigma_0 \sim 10^3$ to 10^4 $\Omega^{-1}\text{cm}^{-1}$. At lower temperatures, the electrons from the pseudo-gap may transfer to delocalized states of the conduction band, arriving first at the localized states (the "tail") of this band. In this case

$$\sigma = \sigma_1 \exp\left(-\frac{E_A - E_F + W}{k_B T}\right),$$

where typically $\sigma_0 \leqslant 10$ $\Omega^{-1}\text{cm}^{-1}$, W is the activation energy of the transitions between localized states within the tail of the conduction band. At low temperatures and large energy gap widths, jump conductivity prevails. This latter conduction mechanism is characteristic of tetrahedral ASCs, while the former one applies to chalcogenide ASCs. The mobility of electrons and holes is usually low, depending on the electric field strength and on the sample thickness. The conductivity of such amorphous semiconductors can be changed by 7 to 8 orders of magnitude with the help of doping.

Many chalcogenide amorphous semiconductors have the property of *conductivity switching*, i.e. the ability to transform reversibly from a state with high electrical resistance to one with low resistance under the influence of a strong electric field with intensity 10^5 V/cm (and higher). The switching commonly occurs during short (10^{-10} s) time intervals, and the underlying cause of the phenomenon is twofold. First, it can occur as a result of *injection* of additional electrons (or holes) into ASCs with an associated increase of the concentration of carriers in the delocalized state. It can also be due to ohmic heating of the sample in the area of a current channel. The formation of *current channels* with small cross-sections occurs due to the local heating of the sample upon the reduction of resistance, so that the switching in this case involves the rearrangement of the material, uniform in cross section, by the flow of electrons into narrow current channels. In some chalcogenide ASCs, the state of low electrical resistance persists for a long time, which is explained by the partial *crystallization* of the sample in the region of a current channel. The return to the state with high electrical resistance occurrs by the transmission of a short pulse of current. Amorphous semiconductors exhibit the phenomenon of photoconductivity (see *Photoconductivity of amorphous materials*).

Optical properties. The absorption spectra of light in amorphous semiconductors reflect the structure of the electron density of states. The coefficient of absorption of the photon $\alpha(\omega)$ with energy $\hbar\omega$ is proportional to the density of electrons which are capable, having absorbed the energy, of transferring it to unfilled levels. Accordingly, it is high when $\hbar\omega$ is approximately equal to or greater than the width E_0 of the pseudo-gap. In this case $\alpha(\omega) \propto c_0(\hbar\omega - E_0)/(\hbar\omega)$, where $c_0 \sim 10^5$ to 10^6 cm^{-1}eV^{-1}. For a photon energy slightly below E_0, the absorption coefficient exponentially decreases along with $\hbar\omega$ (*Urbach rule*): $\alpha(\omega) \propto \exp[-\beta(E_0' - \hbar\omega)]$, where E_0' differs little from E_0, and the constant β depends weakly on the temperature. An explanation of the Urbach rule given by J. Dow and D. Redfield (see *Dow–Redfield model*) is based on local fluctuations of the potential, which occur due to the topological distortions of the network of atoms. The absorption of the photons with energies considerably lower than E_0 is associated with transitions of electrons to localized states within the pseudo-gap.

The phenomenon of *photoluminescence*, which was discovered in chalcogenide and tetrahedral ASCs, is inherent in ASCs, the number of radiated light quanta being comparable to that of absorbed phonons (or an order of magnitude less). The luminescence spectra, usually wide, lie in the interval of energies below E_0 and lack *fine structure*. Chalcogenide ASCs exhibit a reduction of *luminescence* in time during the excitation process, accompanied by the detection of electron paramagnetic resonance. This phenomenon of photo-induced electron paramagnetic resonance is explained by the redistribution of D^0, D^+ and D^- defects under the action of light.

Applications. Amorphous semiconductors find a variety of applications along with crystalline semiconductors. In electrical photography, they are used as photosemiconductor layers with high electrical resistance and low charge mobility. They serve as materials for *nonsilver photography* which includes a stage of chemical development; in *holography*, for manufacturing of masks, microfiches, etc., where use is made of the unique property of chalcogenide ASCs to change optical parameters under the irradiation with strongly absorbed light, known as the *phenomenon of photostructural transformations*. These amorphous semiconductors form the basis of elements of electronics and fiber optics in the IR range, including the wave length 10.6 μm. Fast switches and memory arrays based on switching phenomena are available. Hydrogenated silicon and other tetrahedral ASCs are used in the manufacture of solar cells, effective electroluminophors, videocons, etc.

AMORPHOUS SILICON (α-Si)

Silicon is one of the most comprehensively investigated *amorphous semiconductors*. It has tetrahedral coordination, and is characterized by the same short-range order (see *Long-range and short-range order*) as crystalline *silicon*: the coordination number is close to four, interatomic distances in α-Si differ from corresponding crystalline values by several percent, and *fluctuations* of tetrahedral angles are about 10°. Usually α-Si is obtained in the form of *films* by spraying or decompositing silane SiH_4 in a high-frequency glow discharge. The latter method is used to obtain *amorphous films* of hydrated amorphous silicon (α-Si:H) with hydrogen content of several atomic percent.

In the *pseudogap* of the non-hydrated material, the density of the localized states is high (10^{16} to 10^{19} eV^{-1}cm^{-3}). As a result, the *Fermi level* undergoes almost no shift upon *doping*. *Dangling bonds* are considered to be the main type of *defect* that creates *deep levels* in the mobility gap. For these deep defects, the intracenter energy of electron correlation is positive. The concentration of defects changes under *annealing*, and also under the *Stebler–Vronsky effect*, which consists in the reversible (after annealing) reduction of the conductivity of films after long-term preliminary irradiation with white light. The direct current conductivity σ at low temperatures ($T \leqslant 300$ K) is described by the *Mott law* $\ln\sigma \sim T^{-1/4}$, and the alternating current *electrical conductivity* depends on the frequency according to the power law $\mathrm{Re}\,\sigma(\omega) \sim \omega^s$, where typically $s = 0.8$ to 0.95; the thermal electromotive force at low temperatures is independent of temperature. All this indicates that the *hopping conductivity* mechanism with a variable jump length is realized in amorphous silicon.

The hydrated material differs by a low density of localized states in the pseudo-gap. By bringing the hydrogen *concentration* in films to 5–15 atomic %, the density of localized states can be reduced to 10^{15} to 10^{16} $eV^{-1}cm^{-3}$. The material can be doped by adding phosphine PH_3 or diborane B_2H_6 to the source gas mixture during the fabrication of films. In both doped and undoped films the resistance usually depends exponentially on the inverse temperature.

The optical width of the band gap of α-Si:H depends on the hydrogen content, and is usually 1.6 to 1.8 eV. Films of α-Si:H exhibit a high coefficient of light absorption (in the maximum region of the solar spectrum, it exceeds the absorption coefficient of crystalline silicon by approximately an order of magnitude) and have high photosensitivity (ratio of *photoconductivity* to dark conductivity is as high as 10^6–10^7). The above-mentioned properties of α-Si:H, and the possibility of obtaining large-area multilayer structures on its basis within the scope of a continuous automated process, make α-Si:H favorable for the production of inexpensive *solar cells.* Maximum efficiencies achieved for such structures of small area exceed 13%. Amorphous silicon is used in electric photography and *microelectronics*, in particular for creating thin-film transistors acting as drivers in flat *liquid-crystal displays.* See also *Amorphous semiconductors, Doped amorphous semiconductors.*

AMORPHOUS SOLID

See *Polycluster amorphous solids.*

AMORPHOUS STATE

A state of a solid without any long-range order (see *Long-range and short-range order*) of the atomic arrangement. As distinct from liquids and gases, the mean atomic positions in amorphous solids do not change in time, i.e. the changes in atomic configurations due to the thermal *fluctuations* occur randomly about fixed lattice sites and can be ignored in the structure description. Solids can be transformed to the amorphous state by several different methods: deposition of molecular, atomic, ion and plasma fluxes on substrates, rapid cooling (quenching) of a liquid, *amorphization* of crystalline solids. Amorphous solids obtained by cooling of liquids are called *glasses*. The solid material in the amorphous state is a thermodynamically unstable or metastable system that usually becomes crystalline upon *annealing.* However, polymeric glasses consisting of *high molecular weight compounds* typically do not yield to *crystallization.* The physical properties of amorphous solids differ substantially from those of crystalline solids with the same composition. See also *Vitreous state of matter, Amorphous metals and metallic alloys, Amorphous semiconductors, Metallic glasses.*

AMORPHOUS SUPERCONDUCTORS

Superconducting materials with an amorphous structure.

These materials are, typically, amorphous metallic *films* obtained by deposition on cold substrates and *metallic glasses* produced by rapid cooling (*quenching*) from the melt. Many compounds which are nonsuperconducting in the crystalline phase become superconducting in the amorphous state, e.g., Be, Bi, transition metal alloys such as $Zr_{1-x}Ni_x$, $Zr_{1-x}Co_x$, and others. Conversely, many crystalline compounds lose superconductivity upon amophization, e.g., compounds with the A-15 structure (e.g. Nb_3Sn), and *high-temperature superconductors.*

Amorphous superconductors based on non-transition metals have a strong *electron–phonon interaction*, while those based on *transition metals* and their alloys have a weak or intermediate strength electron–phonon interaction, even though in the crystalline phase these metals have a trend toward strong coupling. At temperatures below the *critical temperature of superconductors* T_c, amorphous superconductors are the most favorable materials for studying *two-level systems* in metals. In these materials, due to their short *coherence length* ξ, the domain of strong thermodynamic *fluctuations* covers a considerably wider temperature range than in ordered systems. In the case of extensive disorder (at the threshold of *Anderson localization*), thermodynamic fluctuations bring about a reduction of T_c. Here, statistical spatial fluctuations of the superconductor *order parameter* become essential; they may involve the percolation nature of the superconducting *phase transition.* A reduction of T_c in amorphous superconductors can also occur as a result of: (a) the enhancement of the Coulomb repulsion at the slow diffusion motion of electrons; (b) renormalization of the electron *density of states* due to interference of the non-elastic electron–electron interaction, and the elastic scattering of electrons by impurities and defects (*Altshuler–Aronov effect*), and (c) weakening of the electron–phonon interaction due to the "inefficiency" of the low frequency phonons with a wave length in excess of the electron *mean free path*. The *upper critical fields* B_{c2} of amorphous superconductors are comparable to the maximum critical fields of crystalline *type II superconductors*. The temperature dependence $B_{c2}(T)$ of amorphous superconductors reveals, in contrast to that of crystalline superconductors, that over a wide temperature range there are regions of zero and even positive curvature, which are attributed to the effects of Anderson localization and the electron–electron interaction. The current carrying capacity of amorphous superconductors in a magnetic field is considerably lower than that of crystalline materials due to high homogeneity of amorphous systems, and to the paucity of effective centers where *vortex pinning* might have occurred. The main mechanism of pinning in amorphous superconductors is *collective pinning*, and this is incapable of providing high values of *critical current*. Amorphous superconductors have a high mechanical *strength* and *plasticity*, and a high stability of their critical parameters with respect to radiation damage.

AMPHOTERIC CENTER

A *defect* having both donor and acceptor levels (see *Acceptor*, *Donor*) in the *band gap* of a semiconductor. Depending on the electron ocupancy, an amphoteric center can be (alternately) in three or more charge states including: negative, positive and neutral. An amphoteric center is a common type of local electric center in semiconductors. For example, atoms of Au in Ge crystals produce four energy levels; the lowest is a donor and the remainder are acceptors. See also *Local electronic levels*.

AMPLIFIER, ACOUSTIC WAVE

See *Acoustic wave amplification*.

AMPLIFIER, ACOUSTOELECTRONIC

See *Acoustic wave amplification*.

AMPLITUDE-DEPENDENT INTERNAL FRICTION

Change of *internal friction* with the magnitude of the applied stress.

This results from a nonlinear relation between the *stress* and the *strain* (*static hysteresis*). Hysteresis energy loss in nonmagnetic materials is mainly determined by the movement of *dislocations*. The displacement of domain walls and the rotation of domains cause static hysteresis in *ferromagnets* and *antiferromagnets*, as well as in *ferroelectrics*. Relative strain amplitudes during studies of amplitude dependent internal friction range within 10^{-7}–10^{-2}, while the frequencies lie in the range 0.1–10^6 Hz. The most widely studied amplitude-dependent internal friction is that caused by dislocations. In its initial stage this value of friction arises most often by the dislodging of dislocations from their pinning points (see *Granato–Lücke model*), and at its subsequent second stage by the action of *Frank–Read sources*, as well as by the *cross slip* of the dislocations. See also *Internal friction*.

AMPLITUDON

Energy quantum of a normal vibration of a *crystal* in an incommensurate phase (see *Ferroelectricity*) that corresponds to a change in the modulus of the *order parameter* for a symmetric normal–*incommensurate phase* transition. In the simplest case, the order parameter for this transition is one of the normal coordinates (or more precisely, a linear combination of these coordinates) of the crystal, corresponding to the wave vector within the *Brillouin zone* of the symmetric phase. Such a normal coordinate is a complex quantity, i.e. one characterized by its modulus and phase. In other words, the order parameter is transformed according to a one-dimensional complex representation. The structure of the incommensurate phase is often expressed in terms of a periodic spatial nonuniformity of the order parameter of a symmetric, commensurate phase transition. The oscillations of the amplitude of such a "wave" correspond to amplitudons. Since oscillations of only the modulus of the order parameter are involved, the amplitudon is quite similar to a *soft mode* phonon in an ordinary *structural phase transition*. In the general case, the order parameter of the symmetric incommensurate phase type transition may correspond to a non-unidimensional representation of the symmetry group of the symmetric phase, so that several amplitudon branches may exist in this phase. During vibrations corresponding to amplitudons, quantum effects are usually not too noticeable.

ANALYSIS, CHEMICAL

See *Chemical methods of analysis.*

ANALYTICAL MECHANICS (introduced by J.L. Lagrange)

Theoretical mechanics based on mathematical equations without the aid of geometric constructions.

Modern analytical mechanics applies a set of effective and quite general methods, based in part on minimization principles, for the solution of mechanical problems; in particular utilizing symmetry for finding the various *integrals of motion*. One expresses the equations of motion in the form of *Lagrange's equations* in the presence of holonomic constraints, and adopting *generalized coordinates* can make it possible to eliminate some unnecessary coordinates.

The Lagrange function L (the difference between the kinetic and potential energies of the system), is expressed via the *generalized coordinates* q_k and *generalized velocities* $\dot{q}_k$. The n Lagrange's equations have the form

$$\frac{\mathrm{d}}{\mathrm{d}t}\frac{\partial L}{\partial \dot{q}_k} - \frac{\partial L}{\partial q_k} = 0, \quad \text{where } k = 1, 2, \ldots, n.$$

If one defines the *generalized momenta* as $p_k = \partial L/\partial \dot{q}_k$, and composes the *Hamiltonian* $H(\ldots, p_k, \ldots, q_k, \ldots, t) = \sum p_k \dot{q}_k - L$, the system of $2n$ first-order equations $\dot{q}_k = \partial H/\partial p_k$, $\dot{p}_k = -\partial H/\partial q_k$ (*Hamilton's canonical equations*) is obtained. Another approach is to define *Hamilton's principal function* S in terms of the *action integral* $S = \int_{t_0}^{t} L(\ldots, q_k, \ldots, \dot{q}_k, \ldots, t)\,\mathrm{d}t + \text{const}$. While evaluating the integral, the time dependences of q_k and $\dot{q}_k$ are assumed to correspond to the actual motion between the initial and final $q_k(t)$ values of the system coordinates. Then the function S depends on these coordinate values, and satisfies the *Hamilton–Jacobi partial differential equation*

$$\frac{\partial S}{\partial t} + H\left(\ldots, \frac{\partial S}{\partial q_k}, \ldots, q_k, \ldots, t\right) = 0,$$

with the momenta in the Hamiltonian being replaced by the partial derivatives $p_k \to \partial S/\partial q_k$. After integration, the Hamilton–Jacobi equation provides $n+1$ arbitrary constants, $S = (t, q_1, \ldots, q_0, \alpha_1, \ldots, \alpha_0) + \alpha_{n+1}$, and it is possible to obtain the solutions of the equations of motion by means of the set of partial differentiations $\partial S/\partial \alpha_k = \beta_k$, where β_k are additional constants. All q_k are expressed in terms of them as functions of time and the constants α_k. Instead of solving the differential equations, one can use the *least action principle* and seek the extremum trajectory from the condition $\delta \int L\,\mathrm{d}t = 0$ (*Hamilton's principle*), or from other principles related to it. The application of variational calculus methods often allows solving the dynamic problem approximately, or investigating the character of the solution. If some variables from the set $q_1, \ldots, q_k$ are explicitly absent from α or H (they are called *cyclic coordinates*), then the integrals of the motion can be found directly. For example, if $\partial H/\partial q_k = 0$, then the corresponding canonical momentum p_k is conserved. When the Hamiltonian has no explicit dependence on the time t, and the forces are derivable from a conservative potential (i.e. work is independent of the path), then the Hamiltonian is the total energy: $H(p, q) = E$.

ANDERSON–BOGOLYUBOV MODE in superconductors

Oscillations of the *order parameter* phase when a *superconductor* is treated as a neutral superflowing *Fermi fluid* (see *Superfluidity*). The Anderson–Bogolyubov mode corresponds to a *Goldstone excitation* arising from a *spontaneous symmetry breaking* at the transition to the superconducting state. At zero temperature and small wave vectors $\boldsymbol{k}$, the Anderson–Bogolyubov mode has a linear *dispersion law* $\omega = vk$, where in a pure case $v = v_F/\sqrt{3}$ (v_F is the electron *Fermi velocity*). At the excitation of this mode, an electron density disturbance takes place, and the spectrum of such oscillations in real superconductors arises due to

Coulomb interactions involving the plasma frequency (see *Collective excitations in superconductors*).

ANDERSON HAMILTONIAN

See *Anderson model.*

ANDERSON LOCALIZATION

The disappearance of *electrical conductivity* in a system of noninteracting electrons when the impurity concentration becomes sufficiently high.

This can explain *metal–insulator transitions* at a low temperature in *amorphous semiconductors*, *doped amorphous semiconductors*, and other *disordered solids.* Anderson localization is also studied in disordered spin systems (disappearance of *spin diffusion*), and for normal modes of disordered lattices (*fractons*). Under consideration when investigating Anderson localization is the *correlation function* of the density–density type $\langle n(\boldsymbol{r}, t)n(\boldsymbol{r}, 0)\rangle$, or the *Green's function* $G_{\boldsymbol{rr}}(t)$. If these functions do not vanish in the limit $t \to 0$, the electron "remembers" its initial state and, hence, is in a localized state. Serving as a *localization criterion* in the energy representation is the presence of a pole of the Fourier-transformed Green's function $G(\omega, t)$ at zero frequency. In the initial formulation, use was made of a model in which an electron jumps from site to site, the energy of a site having random values in the range $(-W, W)$, with the tunneling amplitude depending on the distance according to a power law $r^{-\alpha}$ (see *Anderson model*). If the tunneling amplitude decays rapidly enough ($\alpha > 3$), all the states are localized provided the following inequality holds:

$$t < \gamma W, \qquad \gamma \approx 1, \tag{1}$$

where t is the mean tunneling amplitude, or a quantity proportional to it, the width of the *conduction band* that would be created for $W \ll t$. The value of γ is determined numerically. In the *Lifshits model*, the disorder is not specified by the spread in the site energies, but by the randomly positioned short-range impurities. In this model, Anderson localization takes place when the *mean free path* decreases to the order of the *de Broglie wave* length. If condition (1) for complete localization is not satisfied then the states in the central region of the band are not localized, but the ones close to the top and bottom of the band still are. In this case, the electron system is in the insulating state if the Fermi energy E_F is lower than the threshold energy (see *Mobility edge*) $E_\mathrm{c} \approx W$, and it is in the conducting state, if $E_\mathrm{F} > E_\mathrm{c}$. The electrical conductivity behaves in the close vicinity of the metal–insulator transition point as follows:

$$\begin{aligned} &\sigma \sim (t - \gamma W)^\nu \quad \text{or} \\ &\sigma \sim (E_\mathrm{F} - E_\mathrm{c})^\nu, \qquad \nu = 0.5. \end{aligned} \tag{2}$$

If the random potential in which the electrons move changes only slightly on a scale of the order of the electron wavelength, then the quantum value of the *critical index* $\nu = 0.5$ "works" in a very narrow region, whereas in a wider region near the *critical point* the index ν takes on the quasi-classical value 1.7 predicted by *percolation theory.*

All the preceeding remarks concern ordinary three-dimensional systems. It has been proven that in the one-dimensional case an electron has only localized states for any value of the disorder parameter W. Apparently this is also valid in the two-dimensional case, except for the only infinitely extended state in the middle of the band.

Closely related to Anderson localization are such phenomena as *weak localization* and *mesoscopics*, in which the interference of *electron scattering* processes by several impurities, although weak, still plays a significant role. It should be noted that Anderson localization is described in the single particle approximation. At finite temperatures, the inelastic particle interaction becomes significant and cancels Anderson localization. However, it can be shown that *self-localization* takes place even in an *ideal crystal* if inequality (1) is satisfied, where the quantity W has the sense of a characteristic level difference at neighboring sites. In an ideal crystal, this difference of levels is not due to impurities, but to the intrinsic interaction of particles with one another. Autolocalization has been observed in studies of the *diffusion* of ^{3}He atoms in solid ^{4}He. See also *Elementary excitation spectra of disordered solids.*

ANDERSON MECHANISM OF HIGH-TEMPERATURE SUPERCONDUCTIVITY

(P.W. Anderson, 1987)

A mechanism of *superconductivity* in systems with "*resonating*" *valence bonds* (RVB).

This magnetic mechanism of Cooper pairing involves a fundamentally new type of ordering for two-dimensional systems with an antiferromagnetic Heisenberg interaction (see *Antiferromagnetism*). Proceeding from the experimental data on the insulating (La_2CuO_4) and conducting ($La_{2-x}(Ba, Sr)_xCuO_4$) phases of lanthanum compounds, Anderson proposed several hypotheses regarding their properties, and a possible superconductivity mechanism of *high-temperature superconductors*. The presence of bivalent copper Cu^{2+} with one hole in the unfilled $3d$-shell classifies La_2CuO_4 as a magnetic *insulator* adequately described by Hubbard's Hamiltonian (see *Hubbard model*) with a strong intra-atomic Coulomb repulsion hindering the free motion of the carriers (holes), and causing their localization (*Mott metal–insulator transition*). The Mott–Hubbard insulator ground state with one electron (hole) per atom (half-filled band) is equivalent to Heisenberg's antiferromagnet (see *Heisenberg magnet*) with one spin $S = 1/2$ per site. A second Anderson suggestion based on H. Bethe's works on one-dimensional *magnetic materials*, as well as on L. Pauling's *chemical bond* theory, matches with the essentially quasi-two dimensional character of the structure of superconducting cuprates. The suggestion means that under certain conditions quantum fluctuations, even at $T = 0$, are able to completely destroy antiferromagnetic ordering in the CuO_2 planes, so that the complete wave function of the ground state of an individual plane can be represented by a linear combination of the wave functions of "valence-coupled" pairs of the localized carriers:

$$\Psi_{RVB} = \sum_{\{n,m\}} C\{n,m\} \prod_{n,m} (\alpha_n \beta_m - \beta_n \alpha_m),$$

$$|\Psi_{RVB}|^2 = 1, \qquad (1)$$

where $\boldsymbol{n}, \boldsymbol{m}$ are the bidimensional site vectors, $\{\boldsymbol{n}, \boldsymbol{m}\}$ represent all possible configurations with contributions determined by the coefficients $C_{\{n,m\}}$, and α_n, β_n are functions of "up" and "down" spin projections. In this state, the mean value of a spin at each site is zero, but the phase coherence is retained. The lowest excitations above the ground state (1), known as *spinons*, are *dangling bonds* or individual unpaired spins. The latter are the topological defects (*kinks*), their thermodynamically equilibrium number being determined by the temperature. The kink displacement corresponds to the "valence bond transfer" from one site to another, with no change in energy and no transfer of charge taking place. It is in this sense that the state (1) corresponds to "resonating valence bonds", while spinons ($S = 1/2$) turn out to be neutral, zero mass fermions.

The transition from the insulating RVB-state to the metallic one is caused by introducing new holes into the system (by doping or a change in stoichiometry). The "extra" (free) holes are coupled to a spinon forming a new charged Bose-particle called a *holon*. The motion of holons both inside and between planes, their scattering by spinons, and the phenomenon of *holon confinement* determine the finite resistance in the normal metallic phase, as well as the attraction with the formation of holon *Cooper pairs*, thereby facilitating a transition to the superconducting state at temperatures on the order of Heisenberg *exchange interaction* constant. The holon gas does not undergo a *Bose–Einstein condensation*. There is a relationship between the wave function (1) and the condensate wave function of the *Bardeen–Cooper–Schrieffer theory*

$$\Psi_{BCS} = \prod_{k} (u_k + v_k a^+_{k\uparrow} a^+_{-k\downarrow})|0\rangle,$$

$$|\Psi_{BCS}|^2 = 1, \qquad (2)$$

where u_k, v_k are coefficients, and $a^+_{k\sigma}$ ($\sigma = \uparrow, \downarrow$) are Fermi creation operators of carriers from the vacuum state $|0\rangle$. The wave function (2), projected on the pairs and then transformed into real coordinate space, acquires the functional form (1). Another feature of the Anderson mechanism lies in the possibility of a qualitative description of some unusual properties of high-temperature superconductors (such as the anomalous temperature dependence of the heat capacity, the resistance in the normal phase, anisotropy of conductance, etc.). Experiment shows that cuprate metal oxides in the normal state are ordinary Néel *antiferromagnets*. However, some arguments favor inducing the RVB-state in metal phases under conditions of strong spin interactions of free and localized carriers.

ANDERSON MODEL (P.W. Anderson)

(1) In *disordered systems theory* – a strong coupling model with diagonal disorder described by *Anderson's Hamiltonian*

$$H = J \sum_{\langle \boldsymbol{n},\boldsymbol{m}\rangle} |\boldsymbol{n}\rangle\langle\boldsymbol{m}| + \sum_{\boldsymbol{n}} \varepsilon_{\boldsymbol{n}} |\boldsymbol{n}\rangle\langle\boldsymbol{m}|,$$

where $\boldsymbol{n}$, $\boldsymbol{m}$ are the sites of a d-dimensional lattice, $\langle \boldsymbol{n},\boldsymbol{m}\rangle$ are pairs of the nearest neighbors, J is the transfer integral, and the energies $\varepsilon_{\boldsymbol{n}}$ at different sites $\boldsymbol{n}$ are independent, uniformly distributed random quantities. The pioneer work of Anderson gives arguments in favor of the fact that for $d = 3$, if $\varepsilon_{\boldsymbol{n}}$ are uniformly distributed in the interval $[-W/2, W/2]$ and the quantity W/J exceeds some critical value $(W/J)_{\mathrm{c}}$, then all the states of this Hamiltonian are localized (*Anderson localization*). The Anderson model provides an effective analysis of the transition from quasi-Bloch (propagating) states to localized ones, and clarifies the presence and position of the *mobility edge*, characteristics of the fluctuation *elementary excitation spectra of disordered solids*, and other properties of *disordered solids*.

(2) In the *theory of magnetism* this approach provides a model Hamiltonian to describe the formation of the local *magnetic moment* of an impurity atom in a nonmagnetic metal as the impurity level changes relative to the *Fermi level*. The model admits an exact solution by means of the Bethe substitution (*Bethe ansatz*).

(3) In the *theory of metal-oxide superconductors* – the model provides an explanation of resonating valence bonds (see *High-temperature superconductivity*, *Anderson mechanism of high-temperature superconductivity*).

ANDERSON THEOREM (P.W. Anderson, 1959)

A theorem stating that the *electron scattering* by nonmagnetic impurities in superconductors has a slight effect on the *critical temperature* only. This theorem is based upon the invariance of the Hamiltonian with respect to time inversion. The Anderson theorem is not necessarily valid outside the framework of the *Bardeen–Cooper–Schrieffer theory*, namely, if one takes into account the weak localization of carriers, anisotropy of the *Fermi surface*, fine structure of the *density of electron states*, the presence of a dielectric gap on a part of the Fermi surface, etc.

ANDREEV REFLECTION of quasi-particles (A.F. Andreev, 1964)

A specific reflection of *quasi-particles* (electrons and *holes*) from the interface between the superconducting (S) and normal (N) phases of a *metal* (NS-boundary).

During this process the *quasi-momentum* $\boldsymbol{p}$ of a reflected particle is conserved, but the signs of the charge and all three velocity components change. When an electron of energy ε (measured from the *Fermi level*) and velocity $\boldsymbol{v} = \partial\varepsilon/\partial\boldsymbol{p}$ is incident from the normal metal side on the NS interface, the reflected quasi-particle is a hole with the same energy ε and velocity $\boldsymbol{v}$. The reflection of holes proceeds in a similar way. The value of the Andreev reflection coefficient R depends on the quasi-particle's energy: $R(\varepsilon) = 1$ for $0 < \varepsilon < \Delta$ (Δ is the value of the *energy gap* in the superconducting phase); $R(\varepsilon) \to 0$ for $\varepsilon \gg \Delta$. The essence of Andreev reflection is that at the NS-boundary the incident electron enters the superconducting condensate, forming a *Cooper pair* with an electron having the opposite momentum. A hole arising in the process moves back into the normal region. When the normal metal layer is situated between superconducting slabs (e.g., in the *intermediate state* of a superconductor or in SNS-junctions), Andreev reflection gives rise to distinctive quantization of the low-energy ($\varepsilon < \Delta$) excitations of the N-layer (*Andreev quantization*). This quantization corresponds to a periodic motion of quasi-particles inside the N-layer, with the quasi-particle being an electron for half of the period and a hole for the other half.

Andreev reflection manifests itself in a number of features of the thermodynamic and kinetic properties of the *intermediate state* of a superconductor. In particular, it leads to geometrical oscillation effects in ultrasonic attenuation, and to an additional thermal resistance of the intermediate state during thermal flux propagation normal to N- and to S-layer, etc. As distinct from thermal flux propagation, the passage of an electric current through the NS-boundary at low temperatures ($k_{\mathrm{B}}T \ll \Delta$) is not associated with overcoming the excess resistance of the S-boundary due to the quasi-particle reflection, since the conversion of the normal excitation current into that of Cooper pairs takes place under Andreev reflection. Andreev reflection has been directly observed experimentally in studies of the *radio-frequency size effect*, and of transverse magnetic focusing of electrons in NS-structures.

ANGULAR DEPENDENCE OF SPECTRA

of magnetic resonance

The dependence of the *resonant magnetic field* B_r (in case of *electron paramagnetic resonance*, EPR) or resonance frequencies (in case of *electron–nuclear double resonance*, ENDOR; *nuclear magnetic resonance*, NMR) on the relative orientation of the constant external magnetic field $\boldsymbol{B}$ and the crystal axes, i.e. on the values of the polar angle θ and azimuthal angle φ that specify the $\boldsymbol{B}$ direction in a coordinate system relative to the crystal. The parameters of a generalized *spin Hamiltonian* are obtainable by comparing experimentally observed angular spectral dependences to those calculated from the Hamiltonian. Examining angular spectra in three mutually perpendicular planes permits determination of all parameters of the effective spin Hamiltonian (see *Spin Hamiltonian parameters, Invariants*), and the *local symmetry groups* of its interactions. An example of the angular dependences of the magnetic resonance spectra of the Gd^{3+} ion at a site of symmetry C_2 is given in Fig. which plots the values of the resonant *magnetic field* B_r versus the azimuthal and polar angles θ and φ, respectively, for the indicated transitions $M \to M-1$, where M is the eigenvalue of the projection of the electronic spin $S = 7/2$ on $\boldsymbol{B}$. When the symmetry and arrangement of crystallographic axes (see *Crystal symmetry*) are unknown, they can be determined from EPR (NMR) spectra recorded for rotations around three mutually perpendicular axes fixed in the crystal. In certain cases, the values of θ and φ also affect the *line width* and *line intensity*.

An EPR spectrum is customarily called an *anisotropic EPR spectrum*, if the positions and intensities of its lines depend on the orientation of the sample relative to the external magnetic field. For spin $S \leqslant 1$ and in the absence of anisotropic *hyperfine interactions*, the anisotropic spectrum is observed only for centers belonging to non-cubic spin Hamiltonian symmetry groups, whereas in the presence of anisotropic hyperfine interactions

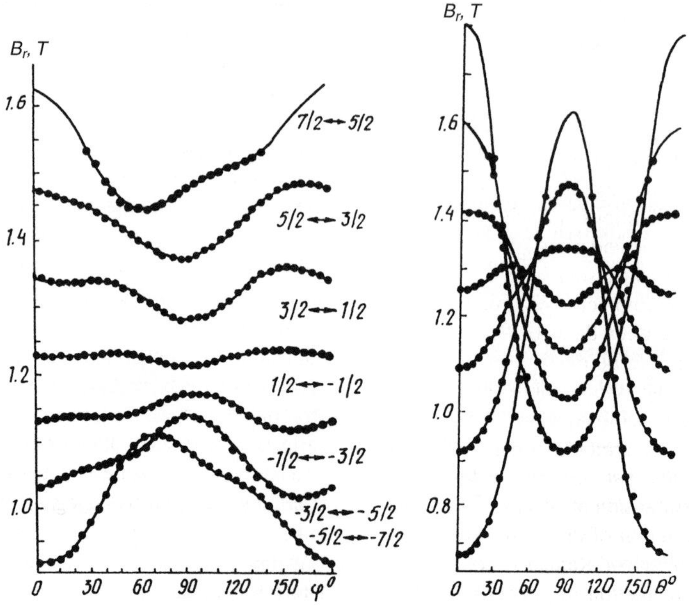

Angular dependence of EPR spectra of Gd^{3+}.

with surrounding nuclei an anisotropic spectrum can be observed even for cubic symmetry groups. If the positions and intensities of EPR spectral lines are independent of the orientation of the sample under investigation relative to the direction of the external constant magnetic field, then the spectrum is called an *isotropic EPR spectrum*. The concepts "anisotropic spectrum" and "isotropic spectrum" are valid also for other resonance phenomena such as NMR.

ANGULAR MOMENTUM, moment of momentum

Vector product $\boldsymbol{r} \times m\boldsymbol{v}$, where $\boldsymbol{r}$ is the position vector, $\boldsymbol{v}$ is the velocity of a material point of mass m. The angular momentum of an arbitrary system of points with masses m_i, which move with velocities $\boldsymbol{v}_i$, is the sum of vector products $\sum_i \boldsymbol{r}_i \times m\boldsymbol{v}_i$, where $\boldsymbol{r}_i$ is the position vector of the ith mass point relative to an origin of coordinates 0. The angular momentum depends on the choice of origin.

ANHARMONICITY, VIBRONIC

See *Vibronic anharmonicity*.

ANHARMONIC VIBRATIONS of crystal lattices

Deviation from the simple harmonic nature of atomic vibrations in a *crystal*, i.e. a nonlinear relationship between the displacements of atoms (ions) from their equilibrium positions and the restoring forces.

While the potential energy of a crystal in the *harmonic approximation* includes only terms of second order in the relative atom displacements, taking anharmonic vibrations into account leads to the appearance of cubic and higher order terms in the potential energy expansion in powers of the displacements. Anharmonicity of vibrations involves interaction between the normal modes (see *Normal vibrations*) of a harmonic crystal, which corresponds to taking into account both elastic and inelastic *phonon* collisions from the quantum point of view. Anharmonic vibrations are responsible for the *thermal conductivity* of insulators, the *thermal expansion* of bodies, the change in the dielectric function of the crystal with temperature, IR absorption and Raman scattering, etc. The anharmonic interaction of phonons leads to their scattering and decay which, in turn, determines the width (lifetime) and shift of the energy levels. Accounting for the anharmonicity is a typical many-body problem, with its solution providing the answers to some fundamental questions of solid state physics, such as the possibility of forming two-phonon coupled states (*biphonons*), the role of *soft modes* in *structural phase transitions*, etc. Slightly and highly anharmonic crystals can be distinguished. *Quantum crystals*, systems with structural phase transitions, in particular *ferroelectrics*, fall into the latter type. The presence of *solitons*, which can contribute to the *energy transfer* and to the process of the restoring the equilibrium state after an external perturbation, is possible in an anharmonic crystal.

ANISOTROMETER

See *Magnetic anisotrometer*.

ANISOTROPIC MAGNETIC MATERIALS

Magnetic materials having different magnetic properties along different directions.

The most common anisotropic magnetic materials have one *easy magnetization axis*. Along this axis, high *magnetization* and low values of the *coercive force* are observed in *soft magnetic materials* and the maximal *remanence* is realized in *hard magnetic materials*. The magnetic anisotropy is natural for monocrystalline magnetics. The widely used polycrystalline anisotropic magnetic materials are obtained by a variety of methods. One of them is the production of crystallographic *texture*, i.e. preferable orientation of easy magnetization axes of individual *crystallites* along one common direction. Thus, in sheets of anisotropic electrical steel, the crystalline structure is produced by cold rolling and annealing, and in magnetically hard materials it is obtained by *powder metallurgy* which involves pressing in a magnetic field. Another method of obtaining anisotropic magnetic materials involves producing *magnetic texture* by means of *thermomagnetic treatment* or *thermomechanical treatment*. In some anisotropic magnetic materials, both the crystallographic and magnetic texture are used simultaneously. Thus, e.g., the best magnetic properties are realized after thermomagnetic treatment of alloys of the *alnico* type, in which the columnar crystallographic texture is produced by directed *crystallization*.

ANISOTROPIC MEDIUM

A medium with some properties that depend on the spatial directions in it.

Anisotropic media necessarily possess long-range order, involving translational and/or orientational order (see *Long-range and short-range order*). The natural *anisotropy of crystals* is related to the symmetry of the ordered arrangement of atoms (molecules) in space, and is the more pronounced, the lower the *crystal symmetry*. Many mechanical (*anisotropy of elasticity*, *plasticity*, *hardness*, etc.), thermal (*thermal expansion* and others), electric (*electrical conductivity*, *dielectric constant*, etc.), magnetic (*magnetic anisotropy*), and optical (*birefringence*) properties of monocrystals are anisotropic. Only a few crystal properties (e.g., *density* and *specific heat*) do not depend on direction (isotropy). Ordinary liquids, in the absence of external factors, are isotropic like gases, due to the absence of long-range order in the relative positions and orientations of the constituent molecules (or atoms). The *anisotropy of liquid crystals* is due to the asymmetry and the long-range correlation of relative molecular orientations. The *quantum liquid* ^{3}He exists in two antiferromagnetic modifications: a highly anisotropic A-phase with weak ferromagnetic properties (in which superfluid motion is not always possible in some directions) and a weakly anisotropic B-phase. The inhomogeneity of a body in its surface layer results in local anisotropy of physical properties. Some chemical characteristics of a medium can also be anisotropic, e.g., the oxygenation and etching rates. The anisotropic properties of a medium can be characterized by corresponding tensor coefficients (susceptibilities).

ANISOTROPY FACTOR of a cubic crystal

This factor $H = 2c_{44} + c_{12} - c_{11}$ depends on the elastic constants c_{ij} (see *Elastic moduli*). For an isotropic medium, $H = 0$ (see *Isotropy*). The anisotropy elasticity ratio defined by $A = 2c_{44}/(c_{11} - c_{12})$ has the value $A = 1$ for an isotropic medium. The quantity A can be either less than or greater than unity. For example, A has the values 1.00 for W, 1.21 for Al, 2.36 for Fe, 3.21 for Cu, and 0.69 for Cr. H is measured in Pa, and A is dimensionless.

ANISOTROPY FIELD

See *Magnetic anisotropy*.

ANISOTROPY, GROWTH

See *Growth anisotropy*.

ANISOTROPY, MAGNETIC

See *Magnetic anisotropy*.

ANISOTROPY OF CRYSTALS

Dependence of the physical properties of *crystals* on the direction.

The anisotropy is caused by the periodic arrangement of atoms in a crystal, and is related to the crystal lattice symmetry (see *Crystal symmetry*). Physical properties are the same only in equivalent crystallographic directions. According to the *Neumann principle* (see *Crystal physics*), the symmetry of any physical property cannot be lower than the symmetry of the *point group* of the crystal, but can be higher than that. Physical properties of crystals are described by tensors of rank from zero to four. Examples of properties described by a zero-rank tensor (scalar) are *density* and *specific heat* (they are isotropic for every crystal). A first-rank tensor (vector) describes, for instance, the *pyroelectric effect*, while *thermal conductivity*, *thermal expansion* and *magnetic susceptibility* are described by tensors of rank two. Tensors of third rank describe the piezoelectric effect (see *Piezoelectricity*) and *electrooptical effect*; most elastic properties of a crystal are described by a tensor of rank four. The symmetry of these properties depends on the crystal lattice symmetry. For crystals of low symmetry (triclinic, monoclinic), all these properties are anisotropic. Some of these properties for high symmetry crystals are isotropic, e.g., in a cubic crystal, all the properties described by a tensor of rank two are isotropic.

ANISOTROPY OF ELASTICITY

Dependence of the elastic properties of a material on the spatial direction in it.

The anisotropy of elasticity is characterized by the symmetry of its *elastic modulus* tensor, and by the values of its components. The *elastic modulus tensor* for an isotropic medium has the form

$$C_{iklm} = \lambda\delta_{ik}\delta_{lm} + \mu(\delta_{il}\delta_{km} + \delta_{im}\delta_{kl}),$$
$$i, k, l, m = x, y, z, \qquad (1)$$

where λ and μ are *Lamé coefficients*. Any deviation of a tensor C_{iklm} from the form (1) corresponds to the presence of anisotropy of elasticity. In *Voigt notation*, the tensor C_{iklm} is replaced by the matrix c_{IJ} $(I, J = 1, 2, \ldots, 6)$ which, for the lowest symmetry, has 21 independent elements. The symmetry of interatomic interactions in a crystal imposes restrictions on the number of independent elements of c_{IJ}. If *central forces* act between atoms and every site of a crystal lattice is a *center of symmetry*, six *Cauchy relations* are satisfied: $c_{12} = c_{66}$; $c_{13} = c_{55}$; $c_{23} = c_{44}$; $c_{14} = c_{56}$; $c_{25} = c_{46}$; $c_{36} = c_{45}$, which, for instance, in crystals with a *heteropolar bond* reduce to one relation: $c_{12} = c_{44}$. In hexagonal close-packed crystals no site is a center of symmetry, and the Cauchy relations do not hold even if the interatomic forces are central. The deviations from the Cauchy relations give an insight into the contribution of anharmonicity to c_{IJ}, and about nonpairwise (many-particle) forces of interaction. Noncentrality of the interaction has a strong effect on the temperature dependence $c_{IJ} = c_{IJ}(T)$, as well as on the value c_{II}. The elements of c_{IJ} must satisfy a number of relations which are corollaries of the *mechanical stability conditions for a crystal*. For example, in a hexagonal close-packed crystal

$$c_{II} > 0 \quad (I = 1, \ldots, 6);$$

$$c_{11}^2 - c_{12}^2 > 0;$$

$$c_{33}(c_{11} + c_{12}) - 2c_{13}^2 > 0;$$

$$c_{11}c_{33} - c_{13}^2 > 0.$$

Cubic crystals exhibit *anisotropy of shift*, which can be characterized by the factor $A_{\text{shift}} = 2c_{44}/(c_{11} - c_{12})$. In addition to the shift anisotropy with the factor $A_{\text{shift}} = c_{44}/c_{66}$, crystals of hexagonal, trigonal and tetragonal symmetry have also an anisotropy of linear compressibility (with the factor $A_{\text{compr}} = (c_{11} + c_{12} - 2c_{13})/(c_{33} - c_{13})$ for hexagonal symmetry). Near the polymorphic transformation temperature, some combinations of elasticity constants can diminish abnormally (*softening of normal modes*). The defects in a crystal cause the components of the tensor of effective elastic moduli to decrease by several percent. When studying crystal distortions localized near defects, one needs to know elastic moduli tensors of higher orders. The effective elastic moduli in a polycrystalline material are found by averaging over orientations of grains.

ANISOTROPY OF g-FACTOR

A property which reflects the dependence of the splitting of electronic (nuclear) energy levels in a magnetic field (see *Landé g-factor*) on the mutual orientation of the *crystal* and the field. It manifests itself in the *angular dependences of spectra* of magnetic resonance, and is described by the second-rank g-factor tensor g_{pq} $(p, q = x, y, z)$ with unequal principal values. The magnitude of the tensor in the principal axis system for the polar angles θ, φ is given by $g = [(g_{xx}^2 \sin^2\varphi + g_{yy}^2 \cos^2\varphi)\sin^2\theta + g_{zz}^2 \cos^2\theta]^{1/2}$. For axial symmetry $g = (g_\perp^2 \sin^2\theta + g_\parallel^2 \cos^2\theta)^{1/2}$, where $g_\perp = g_{xx} = g_{yy}$ and $g_\parallel = g_{zz}$. Anisotropy is present in *paramagnetic centers* with a *local symmetry* lower than cubic. Ordinarily the tensor is symmetric, with $g_{pq} = g_{qp}$.

ANISOTROPY OF LIQUID CRYSTALS

Dependence of physical properties of *liquid crystals* on the direction of a measurement.

In oriented samples the direction of the orientational order is specified by the *director*. If $n_\parallel$, $\chi_\parallel$, and $\varepsilon_\parallel$ are respectively the *refractive index*, diamagnetic susceptibility and *dielectric constant* along the director, and $n_\perp$, $\chi_\perp$ and $\varepsilon_\perp$ are the same quantities in a transverse direction, then $\Delta n = n_\parallel - n_\perp$, $\chi_{\text{a}} = \chi_\parallel - \chi_\perp$ and $\varepsilon_{\text{a}} = \varepsilon_\parallel - \varepsilon_\perp$ are referred to as *optical anisotropy*, *diamagnetic anisotropy* and *dielectric anisotropy*, respectively. For example, for thermotropic liquid crystals $\Delta n \sim 0.1$–0.4; $\chi_{\text{a}} \sim 10^{-7}$ emu, $\varepsilon_{\text{a}} \sim 5$–$30$.

The sign of the anisotropy depends on the chemical structure of liquid crystals. For nematics (see *Nematic liquid crystal*), $\chi_{\text{a}} > 0$, $\Delta n > 0$, $\varepsilon_{\text{a}} > 0$ and $\varepsilon_{\text{a}} < 0$. For cholesterics (see *Cholesteric liquid crystal*), $\chi_{\text{a}} > 0$ and $\chi_{\text{a}} < 0$, $\Delta n < 0$, $\varepsilon_{\text{a}} > 0$ and $\varepsilon_{\text{a}} < 0$. For smectics (see *Smectic liquid crystal*), $\chi_{\text{a}} > 0$, $\Delta n > 0$, $\varepsilon_{\text{a}} > 0$ and $\varepsilon_{\text{a}} < 0$. Typical values for *lyotropic liquid crystals* are: $\chi_{\text{a}} \sim 10^{-8}$ emu, $\Delta n \sim 10^{-3}$. For lyotropic *discotics* $\chi_{\text{a}} > 0$ and $\chi_{\text{a}} < 0$, $\Delta n > 0$, for lyotropic *calamitics* $\chi_{\text{a}} > 0$ and $\chi_{\text{a}} < 0$, $\Delta n < 0$. Because of the anisotropy, the description of elasticity and viscosity of liquid crystals requires using several coefficients.

ANISOTROPY OF MECHANICAL PROPERTIES

Differences in mechanical characteristics: *elasticity*, *plasticity*, and *durability*, along various directions of a solid.

There are two types of anisotropy of mechanical properties: the natural or fundamental (primary) anisotropy, and the induced (secondary) type. The former is determined by the crystal symmetry (see *Anisotropy of crystals*) and can be easily observed. The latter is acquired during the process of mechanical and thermal treatment of a metal. There are two origins of secondary anisotropy: textural (see *Texture*) and structural. The textural type arises as a result of reorientation of *crystallites* along preferred crystallographic directions. The structural secondary anisotropy is caused by the oriented arrangement of grains and subgrains, inclusions of the second phase, and *pores* (see also *Structural texture of metals*). The elastic anisotropy for a number of textures can be derived analytically (see *Anisotropy of elasticity*). *Anisotropy of plasticity* and *anisotropy of durability* have more involved dependences than the anisotropy of elasticity, and are determined experimentally in mechanical tests.

ANISOTROPY, OPTICAL

See *Optical anisotropy.*

ANNEALING

Heat treatment of materials at elevated temperatures aimed at investigating or improving their properties. Material annealing can lead to *phase transitions*, *recrystallization*, *polygonization*, *homogenization*, relaxation of *internal stresses*, removal of aftereffects of cold *plastic deformation* (*strain hardening*), annihilation and rearrangement of *defects*, and so on. The results of annealing depend significantly on its kinetics: the rates of heating and cooling and the time of exposure at a given temperature. To investigate annealing of radiation-induced and other defects, use is made of *isothermal annealing* (long time exposure at constant temperature), and *isochronal annealing* (exposure during a given time interval at successively increasing temperatures). Various stages of isochronal annealing correspond to the presence of defects of different kinds. The annealing of the defects of a particular type can lead to either their annihilation or their transformation into defects of more complicated types. The kinetics of isothermal annealing yield information on the nature of the reactions that result in the annealing.

ANNEALING, IRRADIATION

See *Irradiation annealing.*

ANNEALING OF SEMICONDUCTOR

See *Pulsed annealing of semiconductors.*

ANNEALING, RECRYSTALLIZATION

See *Recrystallization annealing.*

ANNIHILATION OF DISLOCATIONS

Merger and mutual destruction of two *dislocations* or elements of one *dislocation loop* with opposite signs (see *Dislocation sign*).

The annihilation of dislocations leads to the disappearance (mutual compensation) of *elastic fields of dislocations.* If parts of a dislocation with opposite signs come closer together as a result of *dislocation slip* in one common plane, then annihilation restores the perfect crystalline structure at their meeting point (see Fig. 1). The annihilation of *screw dislocations* occurs with the restoration of the structure even when they initially were not in the same *slip plane.* Nonscrew dislocations of opposite signs lying in different, but not too widely separated, slip planes form a dipole due to their mutual attraction. Its subsequent annihilation is possible by means of a head-on *encounter* with the creation of *interstitial atoms* or *vacancies*

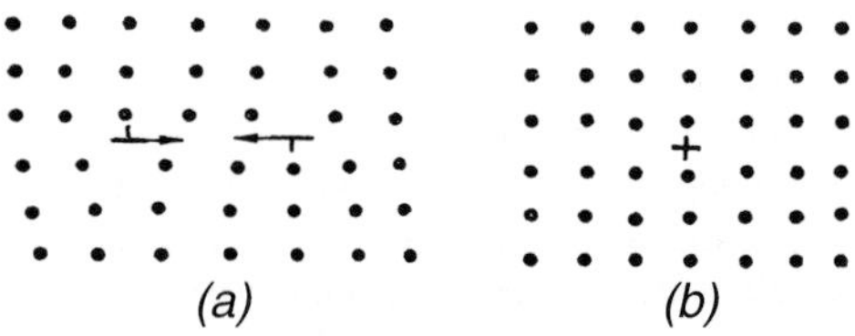

Fig. 1. Annihilation of edge dislocations of opposite signs, which move in the same slip plane: (a) approach of the dislocations; (b) restoration of the perfect crystal structure upon annihilation.

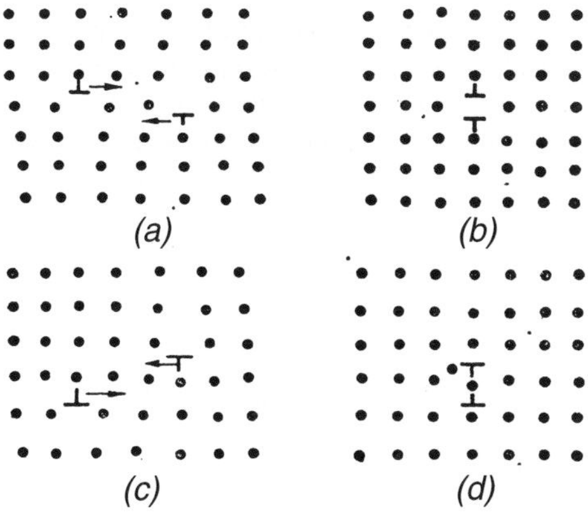

Fig. 2. Annihilation of edge dislocations of opposite signs, which move in neighboring slip planes: (a, c) approach of the dislocations; (b) formation of vacancy at the location of annihilation; (d) formation of interstitial atom at the location of annihilation.

(Fig. 2). When dislocations of opposite signs in different slip planes approach each other and reach a characteristic distance r_a referred to as the *annihilation radius*, then the region of the crystal between them loses its stability, and annihilation occurs dynamically, rather then by diffusion. Studies of the dislocation annihilation process on an atomic scale by mathematical modeling provide the value $r_a \approx (5\text{–}10)b$, where b is the *Burgers vector*. The annihilation of dislocations plays an important role as an effective channel for reducing the dislocation density (see *Dislocation density tensor*) during the annealing process. This annihilation is accompanied by the emission of sound, and it is one of the most important mechanisms of *acoustic emission*.

ANODIZATION, anode oxidation

The creation of a layer of oxide on the surface of a *metal* at a positive potential during electrolysis. Anodization is used to change the color, durability, hardness, electrical conductivity, catalytic activity, and other properties of surfaces of various metals and alloys. There is a great variety of electrolytes, and either direct or alternating current can be used. Depending on this, one obtains either a thin (less than 1 μm), or a thick (tens to hundreds of μm) enamel-like *coating*.

ANOMALOUS AVERAGES

The values of products of two (or more) creation $a^+_{k\sigma}$ (or annihilation $a_{k\sigma}$) operators of half-integer *spin* particles (fermions), averaged over the lowest level (ground) quantum-mechanical states (at absolute zero temperature $T = 0$), or the thermodynamic states (at $T \neq 0$) of a system. The anomalous averages were introduced by N.N. Bogolyubov (1958) to describe superconducting (superfluid) Fermi systems in the *self-consistent field* approximation. They provide the complex-valued *order parameter* (wave function) of a macroscopic coherent quantum state (see *Macroscopic quantum coherence*).

In a superconductor at a temperature below T_c, the anomalous averages of the $\langle a^+_{k\sigma} a^+_{-k-\sigma}\rangle$ and $\langle a_{-k-\sigma} a_{k\sigma}\rangle$ types are nonzero. They describe the condensate of *Cooper pairs* of electrons with antiparallel momenta ($\boldsymbol{k}' = -\boldsymbol{k}$) and spins ($\sigma' = -\sigma$), which is analogous to a *Bose condensate*. In the *superfluid phases of* ^{3}*He* at temperatures below 1 K ($T < 0.003$ K) and under a pressure ($p > 30$ atm), the *triplet pairing* of ^{3}He atoms with parallel spins ($\sigma' = \sigma$) is possible along with the singlet Cooper pairing ($\sigma' = -\sigma$), when the anomalous averages $\langle a^+_{k\sigma} a^+_{-k\sigma}\rangle$ and $\langle a_{k\sigma} a_{-k\sigma}\rangle$ do not vanish. In anisotropic crystals with a degenerate electron spectrum $\varepsilon(\boldsymbol{k}) = -\varepsilon(\boldsymbol{k} + \boldsymbol{Q})$ and flattened, nested regions of the *Fermi surface* separated by the wave vector $\boldsymbol{Q}$, a spatially nonuniform state can arise. This state results from electron–hole pairing or a structural *Peierls instability*, and has nonzero anomalous averages of the type $\langle a^+_{k+Q,\sigma} a_{k,\pm\sigma}\rangle$ and $\langle a^+_{k,\pm\sigma} a_{k-Q,\sigma}\rangle$ which describe *charge density waves* and *spin density waves*.

ANOMALOUS DIFFUSION

A *diffusion* process in which the time evolution of the concentrations of diffusing particles fails to follow the Fick law, occurring asymptotically faster or slower than in the normal case.

As for any diffusion process, anomalous diffusion is associated with a *random walk* of particles. However, in contrast to an ordinary random walk, the mean square displacement of a particle in the time t exhibits the behavior $\langle r^2(t)\rangle \sim t^{\nu}$, with $\nu \neq 1$. The causes of anomalous behavior are associated with the disorder of the system where the random walk occurs. For example, anomalous diffusion behavior is observed in systems with a distribution of activation energies, when the distribution of the wait time of the next jump becomes very broad and the mean value of this time diverges; in a random walk in *fractal* systems the diffusion is slower ($\nu < 1$); in a random walk with flights, when a particle can travel a great distance during a short time ($\nu > 1$), etc. To characterize anomalous diffusion, one often does not use the power index ν, but rather a related quantity called an *anomalous diffusion index* that characterizes properties of diffusion and random walk in a fractal system. In the normal case, for instance, on any Euclidean lattice, the square of the recession of a particle in random walk for sufficiently long time t is proportional to this time: $\langle L^2(t)\rangle \sim t$. In contrast to such universal behavior, the diffusion in a fractal system occurs more slowly, $\langle L^2(t)\rangle \propto t^{2/(2+\theta)}$, where $\theta > 0$ is the anomalous diffusion index. The quantity θ for fractal lines is related to the fractal dimension of the line D: $\theta = 2D - 2$, although in other cases the quantity θ is some independent characteristic of a fractal system. The index θ for an infinite cluster at the percolation threshold is related to well-known critical indices of *percolation theory* $\theta = (\tau - \beta)/\nu$, where τ is the critical index of the percolation system conductivity, β is the critical index of the extent of the infinite cluster, and ν is the critical index of the correlation length (see *Correlation length*).

ANOMALOUS DISPERSION OF LIGHT

Dispersion of light characterized by the anomalous dependence of the *refractive index* n of a material on the frequency ν (wavelength λ) of the light: n decreases as ν grows (λ decreases). The anomalous dispersion of light occurs in the absorbtion bands of materials.

ANOMALOUS ELASTICITY

See *Superelasticity*.

ANOMALOUS MAGNETORESISTANCE

The dependence of a conductor resistivity on the external magnetic field, which is not describable within the classical picture of electron orbit bending (see *Galvanomagnetic effects*, *Magnetoresistance*). In *magnetic materials*: ferromagnetic and antiferromagnetic *metals*, *semimagnetic semiconductors*, such departures from classic dependences (both in sign and in characteristic field values) are related to the influence of the magnetic field on the spin subsystem. In the case of nonmagnetic conductors, the anomalous magnetoresistance is caused by quantum-interference corrections to the conductivity, their contribution being determined by the coherence length (phase memory time) of the carriers. Such corrections arise both at *weak localization* (causing *negative magnetoresistance*), and due to interference of Coulomb and impurity scattering when the magnetoresistance is positive.

ANOMALOUS MASS TRANSFER

Transfer of inherent and *impurity atoms* in crystals at a rate several orders of magnitude greater than that of *diffusion*. It is observed during *pulse exposure* causing *plastic deformation* with a rate greater than 1 s^{-1}. It results from *dynamic recovery* of nonequilibrium *point defects* formed during the process of plastic deformation under *forging*, *explosion*, *pulse heating* (thermoplastic deformation), in a pulsed magnetic field (due to the action of ponderomotive forces). The mechanism of anomalous mass transfer consists in the creation of *interstitial atoms* under the nonconservative motion of dislocation thresholds at *screw dislocations* with a velocity higher than the critical one. The interaction of the moving dislocations with interstitial atoms causes the drift of the latter. Hence, the anomalous mass transfer can be treated as diffusion of nonequilibrium interstitial atoms under the conditions of drift (in the presence of driving forces of transfer processes). The distribution of the concentration C of diffusing atoms is described by the equation

$$C(x,t) = C_0(4\pi Dt)^{-1/2}\exp\left[-\frac{(x-Vt)^2}{4Dt}\right] - C_0\frac{V}{4D}\exp\left(\frac{Vx}{D}\right)\operatorname{erfc}\left(\frac{x+Vt}{2\sqrt{Dt}}\right),$$

where C_0 is the concentration at the instant of time $t = 0$; D is the *diffusion coefficient*, and V is the drift velocity in the x direction. The temperature dependence of the anomalous mass transfer rate is weaker than that of the diffusion coefficient, and cannot in general be described by a simple exponential over a wide temperature range. The critical rate of deformation is defined by the nonlinear dependence of the *yield limit* on the deformation rate. Below the critical rate, the energy of external stresses causing the dislocation slip is insufficient for the formation of interstitial atoms. Therefore, only dislocation thresholds that create *vacancies* move conservatively. At fast rates, both vacancies and interstitial atoms are created. The latter prevail at small dislocation densities, so anomalous mass transfer diminishes with the extent of the plastic deformation. It diminishes also when an elevated concentration of excess vacancies is created in a *crystal* before the pulse arrival. The rise of the speed of *dislocation* displacement due to the swift rise of the external stresses leads to the diminishing of the vacancy contribution to the anomalous mass transfer. The dynamic *multiplication of dislocations* is superimposed, and the dragging along of interstitial atoms by *shock waves* begins to play a significant role in the anomalous mass transfer. This transfer is faster in metals with face-centered cubic and hexagonal close-packed lattices than it is in those with a body-centered cubic lattice. The effect of anomalous mass transfer acceleration is more pronounced for substitutional impurities than for interstitial ones. As a result of the anomalous mass transfer, nonequilibrium *solid solutions* can also be created from the component with low mutual solubility in the solid phase. Under the *low-temperature deformation* of crystals with *ionic bonds*, other mechanisms of anomalous mass transfer appear. Thus, rather small deformation of LiF crystals at liquid helium temperatures leads to *helium* penetration into the surface layer of a crystal due to the dragging along of helium atoms by the dislocation thresholds, and their motion along the dislocations. Indentation (see *Microhardness*) of MgO crystals leads to directed plastic deformation (without the creation and motion of dislocations due to those of *crowdions*) (see also *Anomalous diffusion*). Engineering applications of anomalous mass transfer are diverse. Thus, hot forging or rolling of ingots of *alloys* at a deformation rate greater than the critical one allows considerable acceleration of the *homogenization* of dendrite *liquation*. Percussion *welding* in the solid phase forms welded joints of heterogeneous metals during a time too short for the creation of *intermetallic compounds* responsible for the *embrittlement* of the weld. The effective hardening of the surface layer of machine parts is achieved by means of electric spark *alloying*. Taking anomalous mass transfer regularities into account leads to improving the alloyed surface layer obtained by laser processing (see *Laser technology*). Anomalous mass transfer occurs also at shot-blasting or *ultrasonic treatment* of a metallic *coating* on the surface of products. The mechanical mixing of surface layers of contacting metals under shock loading known as *mechanical diffusion* does not fall into the category of anomalous mass transfer.

ANOMALOUS PASSAGE OF X-RAYS, Borrmann effect (G. Borrmann, 1941)

Propagation of X-rays in *monocrystals* along directions satisfying the *Bragg law*, when the effective (so-called interference) absorption coefficient is considerably smaller than the photoelectric absorption coefficient μ.

This is observed in the case of *Laue diffraction* for sufficiently thick crystals at absorption levels $\mu t \geqslant 3$–5 (t is the sample thickness). The anomalous passage of X-rays is the result of multiple scattering by atomic planes of a crystal, and the interference of the scattered waves, which leads to the creation of a total wave field with a spatially modulated amplitude. If the antinodes of this field lie to the side of atomic planes, it is absorbed very weakly (much more weakly than in ordinary interference). As a result, the X-rays can travel in crystals for relatively great distances (up to 1–10 mm). The anomalous passage was first observed in perfect crystals; and has subsequently been seen in mildly distorted crystals. It occurs for the diffuse background as well, which leads to considerable, often asymmetric deformation of the *diffuse scattering of X-rays* profile compared to the kinematic case (amplification of central part of the profile relative to its wings). See also *Dynamic radiation scattering*, *Two-wave approximation*.

ANOMALOUS PENETRATION
of the electromagnetic field into a metal

This occurs in an external magnetic field, and is caused by the specific dynamics of the *conduction electrons*. The effects of anomalous penetration are subdivided into two groups. The first group, that of plasma or collective phenomena, comprises the aggregate of weakly decaying waves in the degenerate electron plasma of a metal (see *Helicon*,

Magnetic plasma waves, *Cyclotron waves*, *Quantum waves*, *Doppleron*, etc.). The second group includes ballistic effects of anomalous penetration which are caused by a trajectory transfer of an electromagnetic wave from the skin depth layer δ into the bulk metal by numerous groups of electrons, involving individual motions of the charged particles and quasi-particles. Because of the *skin-effect*, the most effective interaction of the electrons with the electromagnetic field occurs for those electrons which move along the wavefront, i.e. parallel to the surface of the *metal* (so-called *effective electrons*). Electrons moving at large angles with the surface spend less time in the skin layer, and thus are less effective in gaining energy from the wave. In a magnetic field the direction of the electon velocity changes periodically due to the cyclotron motion. Therefore, the most intense interaction with the electromagnetic wave occurs in the vicinity of points of electron trajectories where the velocity is parallel to the surface (so-called *effective points*). Due to the cyclotron motion the number of these points is infinite, and under the *anomalous skin-effect* $\delta \ll D$ (D is the characteristic length of a trajectory), some of these points are located outside the skin layer. Their existence is related to resonant excitation of the electron collective motion. When moving inside this layer the electron acquires a velocity increase $\Delta \boldsymbol{v}$ and creates the current $\Delta \boldsymbol{j} = -e\Delta \boldsymbol{v}$, where e is the magnitude of the electronic charge. At the next effective point lying in the depth of a metal, the electron again moves parallel to the sample surface and reproduces the current increment $\Delta \boldsymbol{j}$ there. This is what constitutes anomalous penetration of the ballistic type: bursts of current and magnetic field spaced by the distance D arise inside the metal. The various types of electron trajectories lead to different mechanisms of anomalous penetration.

In a magnetic field $\boldsymbol{B}$ parallel to a metal surface, due to the complicated structure of the Fermi surface, there exists *anomalous penetration along the chain of (closed) electron trajectories* with extremal diameters D_{ext} (see *de Haas–van Alphen effect*). The chain of trajectories from the central section of the *Fermi surface* is created in a slightly tilted $\boldsymbol{B}$. At a greater inclination of $\boldsymbol{B}$ there occurs a kind of *focusing of effective electrons* drifting into the interior of the metal. These electrons provide bursts of anomalous penetration at depths that are multiples of the displacement along the normal to the sample surface, extremal over the cyclotron period. This mechanism of anomalous penetration is related to a sharp rise of the Fourier component of the conductivity due to *Doppler-shifted cyclotron resonance*. The following effects arise from the drift type of anomalous penetration: focusing of electrons from the vicinity of the reference point of the Fermi surface, bursts on the helical paths of the Fermi surface boundary cross sections, trajectory transfer by effective electrons on an open Fermi surface, etc. Finally, in a magnetic field perpendicular to the metal surface, the ballistic effects of anomalous penetration are associated with the magnetic focusing of drifting or ineffective, electrons, the so called *Gantmakher–Kaner effect*.

The various phenomena of anomalous penetration of trajectories of weakly damped waves are closely interrelated, and can transform into each other when the frequency ω or the magnetic field $\boldsymbol{B}$ is changed. The physical pattern of anomalous penetration depends on the nature of the reflection of electrons from the metal surface. The anomalous penetration of radio waves manifests itself experimentally as a *radio-frequency size effect*. It is used to investigate Fermi surface parameters, the temperature dependence and anisotropy of the *mean free path*, and mechanisms of surface *electron scattering*.

ANOMALOUS PENETRATION OF ULTRASOUND into a metal

A significant nonlocal phenomenon appearing at high frequencies ω, when the damping depth of a sound wave (calculated within the framework of a local theory), $L \cong v_{\text{F}}/\omega$, is smaller than the mean free path of the carriers, $l = v_{\text{F}}\tau$ (the latter is the penetration depth of electron density oscillations caused by sound, and interacting with it in a conductor; see *Anomalous skin-effect*). Thus, the anomaly criterion is $\omega\tau \gg 1$ for the asymptote of the acoustic field $u(x,t)$ in the bulk of a *metal* ($x \gg L$), when it is a relatively weak but slowly damped quasi-wave (non-exponentially damped in the collision-free limit) propagating with the electron velocity v_{F}.

The appearance of *electron-acoustic quasi-waves* is associated with the degenerate statistics of charge carriers in metals, i.e. with the presence in the electron spectrum of a *Fermi velocity* in every direction. Anomalous penetration of ultrasound occurs for an arbitrary electron *dispersion law*, and for any polarization of the sound. In the typical case of a convex *Fermi surface* and monochromatic longitudinal vibrations, we have

$$u(x,t) \cong Cu(0,t)\frac{s^2}{v_{F0}\omega x}\ln^{-2}\left(\frac{\omega x}{v_{F0}}\right) \times \exp\left(\frac{\mathrm{i}\omega x}{v_{F0}} - \frac{x}{l_0}\right), \quad x \geqslant \frac{v_{F0}}{\omega}.$$

Here C is a numerical factor of the order of unity determined by the form of the electron spectrum, and the nature of their reflection from the metal boundary ($x = 0$); s is the *sound velocity*, and $v_{F0} = (v_x)_{max}$ is the electron velocity at the reference point on the Fermi surface. The quasi-waves are generated also by other particular values of v_x on the Fermi surface; the power law of asymptotic decay $u(x) \sim x^{-3/2}$ as $l \to \infty$ is characteristic for an internal extremum of v_x on a nonconvex Fermi surface.

With respect to short (compared with τ) acoustic pulses, the anomalous penetration of ultrasound is revealed in the formation of an acoustic signal "forerunner" by electrons. In the absence of a *magnetic field* the quasi-wave *precursor* travels with the velocity v_{F0}. Hence, the observation of the anomalous penetration of ultrasound in the pulse mode could make it possible, by varying the geometry and width of samples, to measure directly both electron velocities and local values of their mean free paths over the entire Fermi surface of the metal under study. Observed in the magnetic field $\boldsymbol{B}$ in the pulse mode is the ultrasound field transfer by *conduction electrons* along the electron orbit chain (the analogue of bursts of the electromagnetic field). Having effectively interacted with the ultrasonic field in the region where $v_x = s$, the fast electron carries away the information on the sound pulse to create its "image" at its path locations where $v_x = s$ again. As a result, the forerunners and afterrunners of the sound signal arise, the distance between them (in the absence of electron drift along x) being a multiple of the extremal diameter of an electron orbit. To identify the effect unambiguously, the generating ultrasound pulse width must be much less than the Larmor diameter. Otherwise, the effect appears as oscillations of the sound speed as the magnetic field changes. The shape of the forerunners repeats that of the generating pulse.

The analogue of the anomalous penetration of ultrasound is the *anomalous penetration of an alternating longitudinal* (i.e. normal to the interface) *electric field*. Though its main part decays exponentially into the metal for a distance that approximates a Debye radius, the nonuniformity of the electron density remains down to macroscopic depths $\cong l$. Owing to the deformation interaction of a nonequilibrium electron gas with the ionic lattice vibrations, this leads to a new (linear in the applied field amplitude) mechanism of *electron-acoustic transformation*, where efficiency grows as $\omega\tau$ increases. Such an effect related to the anomalous penetration of ultrasound was first observed in a *tin* single crystal at liquid helium temperatures.

ANOMALOUS PHOTOCONDUCTIVITY

Photoconductivity which is characterized by unusual behavior, such as: independence of the photocurrent from the photoexcitation intensity over a wide range, marked dependence on the spectral content of the light, and long-term relaxation of the photoconductivity after extinction of the light. Anomalous photoconductivity was discovered and investigated in detail on layers of amorphous selenium treated in mercury vapors. It is explained by the existence of deep *traps for mobile particles* with small probabilities of both thermal excitation of electrons to the *conduction band*, and *recombination* with a free *hole*. The traps are assumed to be *clusters* containing about 10^3 atoms, and localized in a region near the surface of the sample. Anomalous photoconductivity is sometimes referred to as the phenomenon of current reduction under illumination, but in this case the term *negative photoconductivity* is more often used.

ANOMALOUS PHOTOVOLTAIC EFFECT

Appearance of a voltage V_p during illumination of a sample by light of energy in excess of the *band gap* E_g.

Under ordinary mechanisms of the photoelectromotive force: $eV_p \leqslant E_g$ (see *Photoelectric phenomena*). *Anomalous photovoltages* are observed in semiconductor films (e.g., CdTe), in lithium *niobate* crystals, selenites and other materials exhibiting *unipolarity* along some direction. The values of V_p in the anomalous photovoltaic effect can be as high as many kilovolts. This effect can occur either with the absence of a *center of symmetry* of the crystal lattice, or with a certain nonuniformity of the distribution of *defects* in the sample. In the first case, the appearance of a photovoltage provides for the transfer of free *current carriers* in one of the directions along the unipolar axis during the process of photoexcitation (such a mechanism was discovered experimentally in lithium niobate crystals), or is responsible for the unipolar character of *current carrier scattering* by oriented dipoles created by lattice *defects.* In the second case, the cause of the anomalous photovoltaic effect can be the formation of interconnected microscopic sources of electromotive forces, e.g., *semiconductor junctions.* This mechanism has been used to explain the anomalous photovoltaic effect in semiconductor films.

ANOMALOUS SKIN-EFFECT

A variety of phenomena of electromagnetic field damping in *metals* (and plasma) under the condition that the *mean free path* l of conduction electrons considerably exceeds the skin layer depth, δ. In contrast to the normal (classical) skin-effect when $l \ll \delta$ and electromagnetic absorption is determined by collisions of *conduction electrons* with scatterers, the mechanism of absorption in the anomalous skin-effect has the collision-free character of *Landau damping.* The screening of the external wave by the conductor in this effect is caused by the sliding (so-called effective) electrons that move almost in parallel to the surface at angles on the order of δ/l. The salient features of the anomalous skin-effect are the non-exponential decay of the amplitude of an electromagnetic wave deep inside the metal, independence of the skin layer depth δ on l (in the normal skin effect $\delta \propto l^{-1/2}$), a different frequency dependence of the *surface impedance* $Z_a \propto \omega^{2/3}$ (whereas $Z_n \propto \omega^{1/2}$) and the change of the relation between its real and imaginary parts. The value of the impedance in the anomalous skin-effect depends on the character of *electron scattering* at the surface: under diffuse reflection, and the impedance is 12.5% greater than under specular reflection. The anomalous skin-effect has been observed in pure *monocrystals* of metals at low (helium) temperatures in the frequency range from several to hundreds of GHz. It was discovered by H. London (1940) in *tin.*

ANOMALOUS THERMAL EXPANSION

Excessive *thermal expansion* of the *lattice* surface (in the first 1–2 monolayers), which takes place in *semiconductors, metals* and simple chemical compounds. For example, the ratio of the linear thermal expansion coefficient (normal and tangential) of the surface to its bulk value for a Si crystal is approximately 8, for Ge: 4, for GaAs: 3, for NaF and LiF: 2, for W: 2–3. Anomalous thermal expansion was established by various methods: *low-energy electron diffraction,* scattering of H and D atoms, spin polarization of reflected low-energy electrons. This expansion does not harm the crystal surface (for moderately high temperatures) due to the *relaxation* (compression) of the surface layers of the lattice, and the presence of monatomic steps and other defects on the crystal surface. For instance, under heating to 500 K the lattice parameter of bulk LiF will rise by 1.5%, while the increase on the surface is 3%. The extra expansion (1.5%) of the surface is compensated by surface relaxation or structural *defects.*

ANOMALY, KOHN

See *Kohn anomaly.*

ANTHRACENE, $C_{14}H_{10}$ (fr. Gr. $\alpha\nu\vartheta\rho\alpha\varsigma$, coal)

An aromatic conjugated, condensed three ring hydrocarbon with molecular weight 178.22.

A crystal of anthracene (specific weight $1.25\ \text{g/cm}^3$; $T_{\text{melting}} = 217.8\,°\text{C}$; $T_{\text{boiling}} = 355\,°\text{C}$) consists of anthracene molecules bonded by weak *van der Waals forces*; it is a typical representative of *molecular crystals.* Crystals of anthracene belong to the prismatic *monoclinic system* with space group $P2_1/c$ (C_{2h}^5) and two molecules per *unit cell.* The monoclinic axis b, the *crystal symmetry* axis, coincides with the middle axis of the refractive index ellipsoid. Crystals of anthracene can be produced by *sublimation*, as well as by *crystallization* from the melt (Bridgeman's method), or from solution. Using *monocrystal* growth by sublimation one obtains thin (from a fraction to hundreds of microns) colorless foils, their blue *fluorescence* being readily observable under ordinary illumination (*quantum yield* of fluorescence close to unity). The plane of these foils contains the a and b axes of the crystal. Anthracene is easily soluble in hot benzene, poorly soluble in alcohol, ether, and is insoluble in water. It is crystallized from solution to form colorless prisms. Using the Bridgeman method, one can obtain large crystals with any developed plane. Crystals of anthracene begin to absorb light in the near ultraviolet region. The *Davydov splitting* of the 0–0 band of the first electron transition is equal to 230 cm^{-1}. Under irradiation with ultraviolet light anthracene can be dimerized or oxidized to *anthraquinone*. A crystal of anthracene is an *organic semiconductor*, and it is the basis of several large groups of *dyes*. Many valuable natural substances are anthracene derivatives (alizarin, cochineal, chrysophanic acid, etc.). Anthracene *films* produced by spraying can be used as nonsilver carriers for *information recording*.

ANTICORROSIVE COATINGS

Layers of material applied to metal parts for *corrosion protection*.

The protection is assured either by insulation of the construction metal from the corrosive environment, or due to the protective or inhibitory action of the coating upon the metal being protected. In terms of composition, anticorrosive coatings are classified into nonmetallic, metallic and composite; and in terms of their mode of action they can be insulating, protecting and inhibiting. Concerning their durability, coatings are protective (for a given service life) and preserving (temporary protection). Some important properties of the coatings are their chemical resistance, porosity (permeability), *adhesion* to the basic metal, and deformability. The tin, lead, copper, nickel, chrome, and titanium coatings are classified as corrosion resistive (insulating) metallic coatings, while zinc, cadmium, aluminum coatings on steel are protective types.

Concerning their mode of application metallic coatings are subdivided into electroplated, hot-dip (of the melted metal) and metallized ones. The nomenclature of metallized coatings is quite diverse, and is differentiated by the deposition method (electric arc, gas-flame, plasma, and gas-thermal metallization). To create coatings or corrosion layers on the metal surface, one can also use electrolysis from melted salts, deposition from gas-vapor phase, and implantation of metal ions or noble gases. To improve the operating protective properties of the surface layer, use is often made of diffusive heating that provides for *diffusion* of coating components into the underlying metal.

The greatest scope of application (about 70%) involves protective polymeric (and film) lacquer paint coatings, protective fettle coatings, protective enamel and other inorganic nonmetallic coatings differing in their chemical composition, technology of application and range of use. Composite metal-polymer coatings, inhibited metallic and nonmetallic coatings that provide better reliability and protection duration have gained wide use as well.

ANTIEMISSION MATERIALS

Metals, *alloys*, chemical compounds exhibiting low intensity of electron emission or its absence. The antiemission materials include *gold*, *platinum*, *titanium*, *zirconium*, *carbon*, carbides of *tungsten* and *molybdenum*, titanium silicides, and others. They are used for *coating* the working surfaces of instruments, as well as their nonoperational surfaces to suppress electron emission during their useful lifetime.

ANTIFERROELECTRICITY (fr. Gr. $\alpha\nu\tau\iota$, against, and ferroelectricity)

A set of phenomena related to the appearance of ordered antiparallel electric dipole moments in the crystal lattice of antiferroelectric crystals.

The spontaneous polarizations (see *Ferroelectricity*) of neighboring ionic chains or dipole groups of the same type are oriented in opposite directions, so the total polarization is zero. Phases having the free energy (see *Thermodynamic potentials*) close to that of pyroelectric and ferroelectric states are usually considered favorable for forming antiferroelectric phases, because the *phase transitions* to paraelectric and ferroelectric states are possible from these phases under a change of temperature, application of electric field, pressure, and substitutions for some lattice ions. The electric field decreases the temperature of an antiferroelectric phase transition and increases the *dielectric constant* ε. There are antiferroelectric phases with their structure change explainable by condensation of several *soft modes* having wave vectors on the *Brillouin zone* boundary. The phase transition from the para- to the antiferromagnetic phase is accompanied by a change in the lattice superstructure, reduction of ε, spontaneous *strain*, changing the birefringence and *quadrupole* bond constant, elastic and electrooptical properties, etc. More than 60 antiferroelectric compounds are known with many different structures. Some of them have a *displacive type phase transition* (e.g., $PbZrO_3$, $NaNbO_3$, $PbMg_{1/2}W_{1/2}O_3$), and others with the *perovskite* type structure undergo an *order–disorder phase transition* (also $NH_4H_2PO_4$, $Cu(HCOO)_2{\cdot}4H_2O$). There are incommensurate nonpolar phases which are antiferroelectric in the crystals $NaNO_2$, Rb_2ZnCl_4 and others. Antiferroelectrics are used as electrooptic materials and as components of various solid solutions, which are the bases of capacitor, *piezoelectric materials* and electrostriction materials for optics, electronics, radio engineering, and so on.

ANTIFERROELECTRICS

Insulators with internal oppositely directed electric dipole moments due, for example, to rows of ions displaced in opposite senses. A transition from the high-temperature paraelectric phase to the low-temperature ordered, antiferroelectric state occurs via a *structural phase transition*, which is accompanied by a considerable anomaly in the *dielectric constant*. The transition temperature T_C typically depends strongly on the applied electric field, so antiferroelectricity can be induced by the application of a field, as well as by lowering the temperature. Since the transition is typically a *first-order phase transition*, a jump-like change in the polarization P is observed as the field E changes, and the dependence $P(E)$ has the shape of a (possibly double) *hysteresis loop* (see *Hysteresis*). Typical antiferroelectrics are $NH_4H_2PO_4$ ($T_C = 148$ K), $PbZrO_3$ ($T_C = 506$ K), and $NaNbO_3$ ($T_C = 627$ K).

ANTIFERROMAGNET

A material with antiferromagnetic ordering of *magnetic moments* of atoms and ions (see *Antiferromagnetism*).

A substance becomes antiferromagnetic when cooled below its *Néel point* T_N, and generally remains in this state down to $T = 0$ K. Elemental antiferromagnets are: solid oxygen (α-modification) at $T < 24$ K; Mn (α-modification) with $T_N = 100$ K; Cr ($T_N = 310$ K); a number of rare-earth metals (with T_N from 12.5 K for Ce to 230 K for Tb). Chromium has a sinusoidally *modulated magnetic structure*, and the heavy rare-earth metals also have complicated magnetic structures provided by neutron diffraction (see *Magnetic neutron diffractometry*). In the temperature range between T_1 and T_N ($0 < T_1 < T_N$), they are antiferromagnetic, and below T_1 they become *ferromagnetic*. Chlorides of Fe, Co and Ni were the first known antiferromagnets, and by now thousands of them are known. The chemical formula includes at least one transition ion. Antiferromagnets include numerous simple and complex oxides of transition elements such as *ferrite spinels*, *iron garnets*, *orthoferrites* and orthochromites, as well as fluorides, sulfates, carbonates, etc. There are a few antiferromagnetic organic compounds, as well as metallic alloys, in particular alloys of the iron and platinum groups. The temperature dependence of the *specific heat* has a maximum at T_N which is typical of a *second-order phase transition*. Noted among cubic antiferromagnets is the family of rare-earth ferrite garnets in which Fe is replaced by Al or Ga. Dysprosium-aluminum garnet $Dy_3Al_5O_{12}$ is of special interest because it is an antiferromagnetic *Ising magnet* in which the anomalous behavior close to the *triple critical point* has been studied in detail.

The detailed investigation of hydrous copper chloride ($CuCl_2{\cdot}2H_2O$) led to the discovery of *antiferromagnetic resonance*, *magnetic phase transition* – reversal of sublattices (*spin-flop transition*) in a magnetic field, and weak antiferromagnetism.

Noted among *uniaxial crystals* with anisotropy of the easy-axis type is the group of fluorides (MnF_2 and others), which served to clarify optical absorption spectra and provided the discovery of exciton-magnetic excitations, two-magnon absorption, and *Raman scattering of light* by magnons. The optical absorption spectra were also investigated with binary fluorides like $KMnF_3$, $CsMnF_3$. *Brillouin scattering* of light by magnons was observed in antiferromagnetic $FeBO_3$ and $CoCO_3$. In uniaxial CoF_2 the piezomagnetic effect was discovered (see *Piezomagnetism*), and in Cr_2O_3 the *magnetoelectric effect*. In uniaxial crystals with anisotropy of the easy-plane type (α-Fe_2O_3, $MnCO_3$, $CoCO_3$, NiF_2) the *weak ferromagnetism* due to the *Dzialoshinskii interaction* was discovered. Of special interest among substances with weak ferromagnetism are the orthoferrites ($YFeO_3$ and others) where *orientational phase transitions* (the change of the antiferromagnetic ordering axis) are observed, and transparent $FeBO_3$ with T_N above room temperature which displays a magnetoelastic interaction. The strongest magnetoelastic interaction among antiferromagnets is observed in α-Fe_2O_3. This is the first compound to exhibit a wide gap in the *spin wave* spectrum that equals $\Delta E = \hbar\gamma(2B_{ME}B_E)^{1/2}$ and is caused by the effective field of magnetoelastic anisotropy B_{ME} (B_E is the exchange field) (see *Magnetoacoustic resonance*). Some antiferromagnets contain more than two *magnetic sublattices*. For example, uranium oxide (UO_2) is a four-sublattice antiferromagnet characterized by three *antiferromagnetism vectors* that form an orthogonal vector triplet. There is a group of antiferromagnetic semiconductors (chalcogenides of Mn, Eu, Gd and Cr) with very strong *magnetooptical effects*.

The low-dimensional antiferromagnets are of particular theoretical interest. The already mentioned Fe and Co fluorides as well as some binary fluorides $BaCoF_4$, Rb_2CoF_4 are two-dimensional antiferromagnets. The well-known metal-oxide compounds displaying properties of *high-temperature superconductivity* (La_2CuO_4, $YBa_2Cu_3O_{7-\delta}$) in their insulating phase are many-sublatticed quasi-two-dimensional antiferromagnets. One-dimensional antiferromagnets include $KCuF_3$, $CuCl_2$, $RbNiCl_3$ and others. Especially strong interactions of electron and nuclear spin system oscillations were observed in $KMnF_3$, $MnCO_3$, and $CsMnF_3$. Anhydrous sulfates of copper and cobalt as well as CoF_2 exhibit antiferromagnetic ordering induced by a magnetic field at temperatures above T_N due to the Dzialoshinskii interaction. The values of T_N of most antiferromagnets lie below room temperature.

ANTIFERROMAGNETIC DOMAIN

A region of an *antiferromagnet* with a constant value of the *antiferromagnetism* vector $\boldsymbol{L} = \boldsymbol{M}_1 - \boldsymbol{M}_2$. The directions of $\boldsymbol{L}$ in different antiferromagnetic domains are different. Antiferromagnetic domains are separated by antiferromagnetic *domain walls*. It is not always true that the division of antiferromagnets into *domains* results in a decrease of the energy of the *demagnetization fields*. The appearance of these domains either in the *intermediate state* of magnets, or for antiferromagnets with a weak *ferromagnetism*, is thermodynamically favorable. In other cases, these are thermodynamically unfavorable. In addition in some cases, e.g., for domains with zero *magnetization* and differing in the sign of $\boldsymbol{L}$ (antiferromagnetic domains with 180° neighbors), states in different domains, within the macroscopic framework, are considered identical because they differ from one another only in the redesignation of the *magnetic sublattices*. This is a substantial distinction of the properties of antiferromagnetic domains in comparison to, e.g., those of *ferromagnets* with different directions of magnetization.

The presence of antiferromagnetic domains in real antiferromagnets can be linked with the fact that a *magnetic phase transition* to an antiferromagnetic state due to nonuniformity can occur independently in various finite regions of a crystal. If these regions differ in the sign of $\boldsymbol{L}$, then the spreading of the transition over the whole sample causes antiferromagnetic domains to appear. In addition, the division of antiferromagnets into domains can reduce the energy of elastic *strain*, or (at moderate temperatures, near the *Néel point*) become thermodynamically favorable due to the contribution of entropy to the free energy of the antiferromagnet. These domains significantly affect such properties as the electrical conductivity of a metal antiferromagnet, and *sound absorption* in some antiferromagnets. The presence of antiferromagnetism hinders the observation of a series of effects sensitive to the sign of $\boldsymbol{L}$, e.g., *piezomagnetism*. Antiferromagnetic domains of the non-180° kind (so-called *T-domains* or *twin domains*) are accompanied by a distortion of the crystal lattice, thus providing a way to detect, for example, a conversion of an antiferromagnet to a single domain state, and to study it by various methods. The observation of the 180° antiferromagnetic domains (*S-domains*, i.e. spin-rotation domains) was carried out by using the *linear magneto-optic effect*. Those domains, in which this effect is allowed by the symmetry in the external magnetic field, possess optical properties dependent on the direction of $\boldsymbol{L}$, which allows one to employ optical methods for *magnetic domain characterization*.

ANTIFERROMAGNETIC METALS

Solids with metallic conductivity exhibiting *antiferromagnetism*. They include a number of $3d$-, $4f$-, and $5f$-*metals*, their *alloys* and some of their compounds. When analyzing the physical mechanism responsible for antiferromagnetism in the metals, it is important to know to what extent the electrons responsible for the magnetic ordering (magnetic electrons) are localized. Accordingly, one can speak about *band antiferromagnetism* (see *Band magnetism*) and the *antiferromagnetism of magnetic ions*.

In *band antiferromagnets* the magnetic electrons form a relatively wide, partially filled band with the *exchange splitting* $\Delta\varepsilon \leqslant \zeta$, where ζ is the chemical potential referenced to the bottom of a band. These materials are described in terms of a static *spin density wave* with a wave vector $\boldsymbol{k}$ that can be either commensurate or incommensurate with a *reciprocal lattice* period (see *Incommensurate structures*, *Modulated magnetic structures*). A strict theory exists solely for the case of weak band antiferromagnetism ($\Delta\varepsilon \ll \zeta$). An example is pure Cr with the *Néel point* $T_N = 312$ K. Below this temperature, the magnetic moment per atom is $\mu = 0.59\mu_B$ (μ_B is Bohr *magneton*), much less than that of an isolated Cr atom ($\mu = 3\mu_B$). The spin density wave period is incommensurate with the reciprocal lattice period $\boldsymbol{k} = (2\pi/a)(1-\delta, 0, 0)$, where a is the lattice constant. At $T = 118$ K, $\delta = 0.049$ and depends slightly on the temperature; at $T = T_N$, $\delta = 0.037$. *Chromium alloys* can exhibit both commensurate and incommensurate spin density waves, the origin of latter being determined by the specific nature of the *Fermi surface*, which includes congruent hole and electron octahedra whose component parts can be made coincident with themselves via translation by a reciprocal lattice vector $\boldsymbol{k}$. It is the *triplet pairing* of electrons (holes) of congruent parts that manifests itself as a spin density wave with wave vector $\boldsymbol{k}$. The presence of congruent parts of Fermi surfaces leads to instability of the paramagnetic state with respect to the emergence of a spin density wave (*spin density wave instability*). Weak band antiferromagnets include compounds with metallic conduction CrB_2 ($T_N = 85$ K), VX, VX_2, V_3X_4, V_5X_8, where X ≡ S or Se (for VS $T_N = 1040$ K, for V_3S_4 $T_N = 9$ K, for V_3Se_4 $T_N = 16$ K, for V_5S_8 $T_N = 29$ K, for V_5Se_8 $T_N = 35$ K). The weak band antiferromagnet MnSi possesses a helical structure (see *Helicomagnet*) with a long wavelength (180 Å) and hence it may be considered as a weak band antiferromagnet with an inhomogeneous ground state in the form of a helical spin density wave. The examples of *strong band antiferromagnets* are: α-Mn ($T_N = 95$ K); γ-Mn ($T_N = 450$ K), γ-Mn being metastable in the region of antiferromagnetism (it is stable only at $T > 1352$ K); γ-Fe ($T_N = 8$ K; γ-Fe is stable in the range 910–1400 °C). They all support commensurate spin density waves.

Antiferromagnetism of magnetic ions takes place in rare-earth metals due to the smallness of the radius of $4f$-shells whose electrons are responsible for the magnetic ordering (see *Magnetism of rare-earth metals*). Rare-earth metals are commonly classified into light (from Ce to Eu) and heavy (from Gd to Lu) ones. The light rare-earth metals (except Nd) at T_N are antiferromagnets with a collinear structure. In the heavy rare-earth metals (except Gd, Yb, Lu) and in Nd, an antiferromagnet–ferromagnet transition is observed at a lower temperature $T = T_c < T_N$ in addition to the paramagnet–antiferromagnet transition at $T = T_N$. In the antiferromagnetic range T_N–T_c, magnetic moments of rare-earth metals create complicated modulated magnetic structures (helical, cycloidal, sinusoidal). The complicated magnetic structures arise also in alloys of rare-earth metals with each other and with Y (a paramagnet in the pure state). The Néel temperatures of the light rare-earth metals are as follows: for Ce $T_N = 12.5$ K; for Pr $T_N = 20$ K; for Sm $T_N = 14.8$ K; for Eu $T_N = 90$ K; for Nd $T_N = 19$ K with ferromagnetism observed below $T_c = 7.5$ K.

Intermediate type of antiferromagnetic metals. The magnetic order of a number of antiferromagnetic metals is determined by the presence of two factors: spin density wave instability (near congruent parts of Fermi surface) and localized magnetic moments (FeRh, Pt_3Fe, alloys: Cr–(Fe, rare-earth metal), Mn–rare-earth metal, Y doped with rare-earth metal). Small additions of rare-earth metals to paramagnetic Y cause a helical antiferromagnetic structure to appear. In compounds FeRh, Pt_3Fe, one part of the Fermi surface exhibits spin density wave instability, while another forms local moments.

Antiferromagnetic actinides: pure metallic curium ($T_N = 52$ K) and several compounds with metallic conductivity: UP (130 K), UAs (182 K), USb (213 K), UBi (290 K) (T_N given in parentheses).

ANTIFERROMAGNETIC RESONANCE

The selective absorption by an *antiferromagnet* of electromagnetic wave energy with a frequency close to the natural frequencies of precession of magnetization vectors of magnetic sublattices; a variety of electron *magnetic resonance.* Antiferromagnetic resonance involves the excitation of resonant coupled oscillations of magnetization vectors of magnetic sublattices of the antiferromagnet relative both to each other and to the direction of the applied magnetic field $\boldsymbol{B}_0$ (see *Antiferromagnetism*). The frequency of these oscillations is determined by the value of effective magnetic fields acting on magnetic moments of sublattices: $\boldsymbol{B}_E$, the effective field of the *exchange interaction* of sublattices, $\boldsymbol{B}_A$ the field of *magnetic anisotropy*, and $\boldsymbol{B}_0$. The dependence of the resonance frequency ω on the effective magnetic field in antiferromagnets is complicated, and depends on the crystal structure. When the external magnetic field is applied parallel to the *easy magnetization axis* then two antiferromagnetic resonance frequencies lying in the range 10–100 GHz correspond to one value of $\boldsymbol{B}_0$. The study of antiferromagnetic resonance makes it possible to determine the value of effective magnetic fields in antiferromagnets.

ANTIFERROMAGNETISM

Magnetically ordered state of a crystalline material in which neighboring atomic *magnetic moments* are aligned antiparallel, and the total magnetic moment of a magnetic unit cell of the crystal equals zero (or a very small fraction of an atomic moment).

Antiferromagnetism occurs at temperatures below the *Néel point* T_N. More generally, antiferromagnetism refers to the set of physical properties associated with the antiferromagnetic state. The *exchange interaction* seeking to align the *spins* and hence the magnetic moments antiparallel to each other is responsible for the antiferromagnetism. The idea that the exchange interaction can cause antiferromagnetism was put forward by L. Néel (1932) and independently by L. Landau (1933), who, in addition, described theoretically the *phase transition* from the paramagnetic to the antiferromagnetic states. At $T > T_N$, when the thermal energy ($E = k_B T$) exceeds the exchange energy (μB_E, where μ is the atomic magnetic moment and B_E is the exchange field), the material exhibits paramagnetic behavior (see *Paramagnetism*). The temperature dependence of the *magnetic susceptibility* χ of such a material at $T > T_N$ follows the *Curie–Weiss law* $\chi = C/(T - \theta)$ with a negative Weiss constant θ, and C is a constant. At $T = T_N$, antiferromagnetism appears in a material. In most cases the phase transition at T_N is a second-order transition accompanied by the characteristic anomalies of magnetic susceptibility, *specific heat, thermal expansion* coefficient, *elastic moduli,* etc.

The magnetic ordering is characteristic of an *atomic magnetic structure* with a symmetry described by point and space *magnetic symmetry groups.* The magnetic structure of an *antiferromagnet* is conveniently described by two, four, or more interpenetrating *magnetic sublattices* with magnetizations $\boldsymbol{M}_j$ that mutually compensate each other. As an example, Fig. 1 shows two antiferromagnetic structures: (a) MO, where M = Mn^{2+}, Ni^{2+}, Fe^{2+}; (b) MF_2, where M = Mn^{2+}, Co^{2+}, Fe^{2+}. The difference is that the first one has a cubic magnetic unit cell with lattice parameter a_M twice that of the crystallographic unit cell (a_C), while the magnetic unit cell in the second (tetragonal) structure coincides with the crystallographic

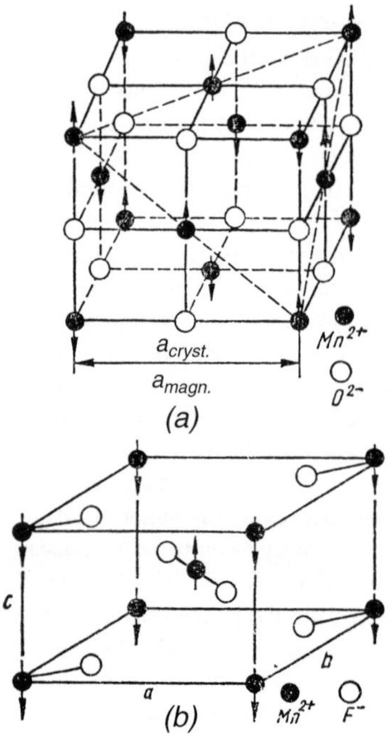

Fig. 1. Magnetic structure the antiferromagnets (a) MnO and (b) MnF_2.

one. In most antiferromagnetic materials with no external magnetic field present, $\sum \boldsymbol{M}_i = 0$. However, there exists a large group of antiferromagnetic materials for which this condition is not strictly met. They are materials with *weak ferromagnetism* (*weak ferromagnets*), in which $|\sum \boldsymbol{M}_i| \ll \sum |\boldsymbol{M}_i|$. There are also more complicated antiferromagnetic structures, which can not be described in terms of sublattices. They are *helical and sinusoidal antiferromagnets* (see *Modulated magnetic structures*, *Spin density wave*). In both types a preferred direction exists. In the helical structure, the spins of all magnetic atoms in each crystallographic layer perpendicular to the preferred direction are parallel to each other. However, the direction of magnetization of each layer is turned relative to the neighboring ones by some angle $\Delta\varphi = \pi/n$, where n is typically a noninteger number. In sinusoidal structures, the magnetizations of layers are not parallel to each other, but their value varies according to the law $M_k = M_0 \sin(k\Delta\varphi)$. Both structures are incommensurate in the sense that the magnetic and crystallochemical translational periods are not multiples of each other. In most antiferromagnetic materials, the directions of magnetization vectors $\boldsymbol{M}_i$ relative to crystallographic axes are determined by much weaker interactions than the exchange (E_{ex}) interactions, causing the anisotropy of antiferromagnetic materials (E_a) (see *Magnetic anisotropy*). The axis along which the magnetizations of sublattices are aligned is known as the *easy axis*. Antiferromagnetic materials with $E_a \geqslant E_{ex}$, referred to as *metamagnetics*, do not exhibit a *spin-flop transition* (see *Magnetic phase transitions*). Also referred to as metamagnetics are *layered antiferromagnets* in which the exchange interaction E_{ex} inside a layer is ferromagnetic and much stronger than the antiferromagnetic interaction between the layers. A direct method to determine the antiferromagnetic structure (including the direction and even the temperature dependences of the sublattice magnetizations) is *magnetic neutron diffractometry*. The intensity of magnetic diffraction peaks is proportional to the square of the sublattices magnetization. In real antiferromagnetic substances *antiferromagnetic domains* exist.

In weak magnetic fields, the magnetization $\boldsymbol{M}$ in antiferromagnetic materials, as well as that in a *paramagnet*, depends linearly on the magnetic field ($\boldsymbol{M} = \chi \boldsymbol{B}$); however, the temperature dependences of χ in these substances differ significantly. In *uniaxial crystals* the salient features of antiferromagnetism (e.g., anisotropy of magnetic susceptibility) are exhibited most clearly. In particular, the longitudinal (along the antiferromagnetism axis) magnetic susceptibility $\chi_{\|}$ decreases sharply as temperature T decreases, while the transverse one $\chi_{\perp}$ is almost independent of the temperature (Fig. 2). Thus, there is a strong anisotropy of the magnetic susceptibility. If the magnetic field directed along the easy axis attains the critical value $B_{c1} = (2B_A B_E)^{1/2}$ (B_A is the effective anisotropy field), magnetic sublattices flip over in the direction perpendicular to the magnetic field. This *spin-flop transition* occurs because at $B > B_{c1}$ the difference of magnetic energies in the easy axis and flipped over states becomes greater than the anisotropy energy E_A: $(1/2)(\chi_{\perp} - \chi_{\|})B^2/\mu > E_A$. In samples of finite

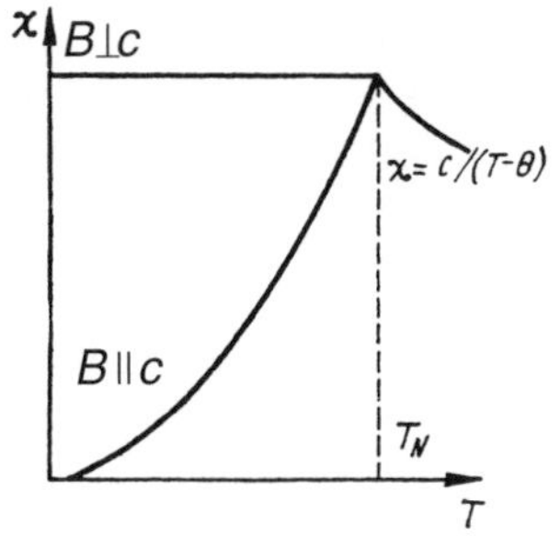

Fig. 2. Anisotropy of the magnetic susceptibility χ in an axial antiferromagnet.

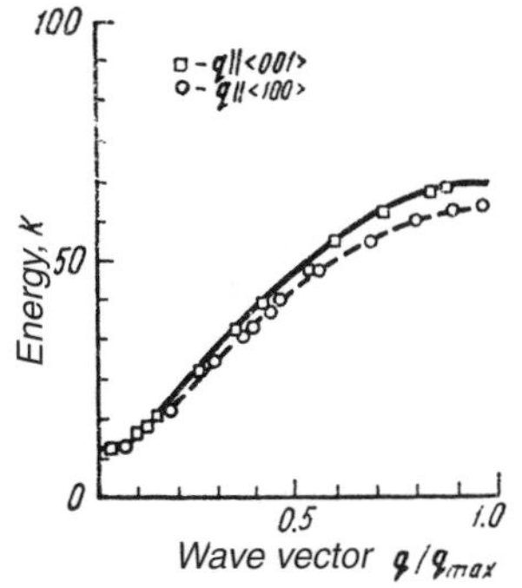

Fig. 3. Spectrum of spin waves in MnF_2.

size, the *intermediate state* with alternating layers of flipped over and normal phases is formed in the narrow region of fields near B_{c1} due to the energy of demagnetization. There exists a second critical field $B_{c2} = 2B_E$ at which the antiferromagnetism is destroyed completely, and all magnetic moments align strictly parallel to the applied field (i.e. a *spin-flip transition* occurs). Other magnetic phase transitions are observed in antiferromagnets as well.

A theoretical description of antiferromagnetism can be given within both phenomenological and microscopic models. A phenomenological description based on the *Landau theory of second-order phase transitions* uses the expansion of the free energy in even powers of components of magnetization vectors of magnetic sublattices, taking into account the *crystal symmetry*. This theory predicts symmetries of magnetic phases, as well as the shape of *phase diagrams* in the vicinity of T_N. The microscopic theory uses a *Heisenberg Hamiltonian* written in terms of *spin operators*, and leads to the energy of the "ground" state and a spectrum of *spin waves*. In contrast to *ferromagnets*, the microscopic theory of *Heisenberg magnets* faces a fundamental difficulty because the state with the strict antiparallel spin alignment is not a true ground state of the system in the presence of an antiferromagnetic exchange interaction.

At a finite temperature T, the *quasi-particles* that destroy antiferromagnetic order are *magnons* (*spin waves*). The existence of several sublattices leads to the existence of several branches, one for each sublattice, of spin waves. Two of them are quasi-acoustic, and the rest are of the exchange type. The *quasi-acoustic spin waves* satisfy a linear *dispersion law*. In most antiferromagnetic materials, however, their spectrum has an energy gap, with the width determined by the expression $\hbar\omega = \mu_B(2B_A B_E)^{1/2}$, where μ_B is the Bohr *magneton*. The existence of the linear dispersion law is confirmed by the linear temperature dependence of the specific heat, and more reliably by *inelastic neutron scattering*. Fig. 3 shows the spin wave dispersion relationships of MnF_2. In the scale used, the gap is barely observable. The gap in the spin wave spectrum is studied by means of *antiferromagnetic resonance* without any particular difficulties in *easy-plane antiferromagnets* where an almost gapless branch is observed, in addition to the magnon branch with a large gap ($\approx\mu_B(2B_A B_E)^{1/2}$). The study of this nearly gapless branch allows the subtle effects of the dynamic interaction of the spin system with other degrees of freedom of the crystal (*phonons*, oscillations of nuclear spins, etc.) to be revealed. This becomes possible because of the "exchange amplification" of the frequency of coupled vibrations in antiferromagnets. Branches of the *exchange spin waves* are relatively less studied; spin-wave analogues of optical phonons have a wide gap $\approx\mu_B B_E$, and rather weak dispersion.

Many antiferromagnetic materials are transparent in the optical and infrared regions. The distinctive features of the optical properties of antiferromagnetic materials are the abnormally high spontaneous magnetic birefringence arising in the easy-plane, and the change of birefringence in the easy-axis antiferromagnetic materials. Single-magnon scattering also turns out to be stronger than what could be expected from the magnetic dipole–dipole transitions alone. Exciton–magnon excitations are observed in the absorption spectra.

ANTIFERROMAGNETISM VECTOR

A macroscopic parameter of *antiferromagnets* and *ferrimagnets*; the order parameter at the *magnetic phase transition* to the antiferromagnetic state (see *Antiferromagnetism*). The vector of antiferromagnetism is defined as such a linear combination of the magnetization vectors $\boldsymbol{M}_\alpha$ of individual *magnetic sublattices*, that it differs from a sum $\boldsymbol{M} = \sum_{\alpha=1}^{k} \boldsymbol{M}_\alpha$ which defines the total *magnetization*, where $\alpha = 1, 2, 3, \ldots, n$, and n is the number of sublattices. To describe the *atomic magnetic structure*, $n - 1$ antiferromagnetism vectors $\boldsymbol{L}_i$ are introduced. The expression for $\boldsymbol{L}_i$ depends on the magnetic structure, e.g., in the simplest two-sublattice antiferromagnet (or ferrimagnet), only one such antiferromagnetism vector exists, $\boldsymbol{L} = \boldsymbol{M}_1 - \boldsymbol{M}_2$. Sometimes, normalized antiferromagnetism vectors are used, $\boldsymbol{l} = \boldsymbol{L}/|\boldsymbol{L}|$, $\boldsymbol{l}^2 = 1$. It is possible to write dynamic equations for the antiferromagnetism vector to define linear and nonlinear *spin waves* in antiferromagnets and ferrimagnets (see *Landau–Lifshits equation*).

ANTIFRICTION MATERIALS

Materials having low values of the *friction* coefficient.

Besides low values of the friction coefficient (0.004–0.10 with lubricant, and 0.12–0.20 without lubricant), antifriction materials are characterized by considerable wear resistance, adequate mechanical *strength* and *plasticity*, *corrosion resistance*, and absence of sticking. These materials slightly wear the adjoining surfaces. Depending on the operating conditions antifriction materials with special properties are created to be used with lubrication by oil or water, without a lubricant, in the atmosphere or in vacuo, at a high temperature, in chemical environments, etc. Classified among antifriction materials are the following: cast materials (*babbitts*, *bronzes*, and *brasses*); baked ones manufactured by *powder metallurgy*, including composite materials on the base of iron or bronzes; combined (mostly metal-fluoroplastic) materials; *antifriction cast irons* and *antifriction steels* (copper steel, graphitized steel). The cast antifriction materials are manufactured by *melting* source components with subsequent casting. As to the structure type, they are subdivided into those with harder inclusions of *intermetallic compounds* placed in a relatively light matrix (e.g., aluminum bronze, tin bronze, tin babbitt), and those with soft inclusions with metallic structural components in a stronger *solid solution* matrix (e.g., lead bronze, *aluminum alloys* containing *tin*, *lead* and cadmium). A typical feature of baked (metalloceramic) antifriction materials is the presence of pores (8–27% of the volume) which are impregnated with a fluid lubricant. The metallic base of such materials is *iron*, *steel*, *copper*, bronze, and *nickel*. In addition to these, baked materials contain components (*graphite*, sulfides, selenides and others) to serve as a *dry lubricant*. Baked antifriction materials in most cases have higher tribotechnical properties than the ones obtained by *casting*, and can provide unique properties unattainable in cast alloys.

ANTIMONY (Lat. *stibium*), Sb

A chemical element of Group V of the periodic system with atomic number 51 and atomic mass 121.75. It is related to As and Bi in its properties. Natural antimony has 2 stable isotopes Sb^{121} (57.25%) and Sb^{123} (42.74%). Important radioactive isotopes are Sb^{122} ($T_{1/2} = 2.8$ days), Sb^{124} ($T_{1/2} = 60.2$ days), Sb^{125} ($T_{1/2} = 2.0$ years). Electronic configuration of the outer shell is $5s^2 5p^3$. Successive ionization energies are 8.64, 16.5, 25.3, 44.0, 55.4 eV. Atomic radius is 0.161 nm; radius of Sb^{3+} ion is 0.090 nm, of Sb^{5+} 0.062 nm, of Sb^{3-} 0.208 nm. Work function is 4.08 eV. Oxidation states are −3, +3, +5. Electronegativity is ≈2.0.

Antimony is a silvery-white *metal* with strong brilliance. In a free form antimony exists in several crystalline and amorphous modifications. Under normal conditions, the rhombohedral crystalline modification is stable with parameters $a = 0.45064$ nm and $\alpha = 57.1°$. Under increased pressure the cubic and hexagonal modifications appear. Besides, three amorphous modifications are known: yellow, black and explosive ones. The rhombohedral modification is characterized by gray star-like crystals. For this modification the density is 6.690 g/cm^3; $T_{\text{melting}} = 903.5$ K,

$T_{\text{boiling}} = 1907$ K. Heat of melting is 20 kJ/mole, heat of evaporation is 124 kJ/mole; specific heat $c_p = 25.2$ kJ·mole^{-1}·K^{-1}. Thermal conductivity is 18.84 W·m^{-1}·K^{-1}, temperature coefficient of linear expansion of polycrystalline antimony is $9.2 \cdot 10^{-6}$ K^{-1} (at 273 K), $10.3 \cdot 10^{-6}$ K^{-1} (at 873 K). Debye temperature is 204 K. Brinell hardness of technical grade antimony is 320 to 580 MPa; of zone refined antimony it is 260 MPa. Shear modulus is 20 GPa, normal elasticity modulus is 56 GPa.

Antimony is a diamagnet; magnetic susceptibility of polycrystalline antimony is $0.81 \cdot 10^{-9}$ (at 293 K); electrical resistivity 0.39 μΩ·m (273 K), temperature coefficient of electrical resistivity is $5.1 \cdot 10^{-3}$ K^{-1}. Compounds of antimony with indium, gallium, aluminum, cadmium and other metals (so-called *antimonides*) possess semiconductor properties, with indium antimonide InSb possessing the highest carrier mobility among all known *semiconductor materials*. Antimony is also used as a component of various *alloys*, it is contained in lead storage battery plates (4 to 12% by mass). Antimony oxide Sb_2O_3 is introduced into the composition of glasses characterized by a low refractive index.

ANTIREFLECTION COATING OF OPTICAL ELEMENTS

A means of increasing light transmission of optical elements by coating their surfaces with nonabsorbtive *films*. The optical paths through such films should be comparable to the radiation wavelength. Without an antireflection coating the loss of light during reflection from their optical surfaces may reach 10% or more. The antireflection effect involves interference of light beams reflected from the front and back sides of the dielectric layers that form the coating. Provided the angles of incidence remain close to normal the antireflection coating has a maximum effect. The thickness of the thin coating film should be an odd number of quarter wavelengths, and the *refractive index* of the film, n_2 should satisfy the relation $n_2^2 = n_1 n_3$, where n_1 and n_3 are the indices of refraction of the two media adjacent to the film (the first is often air, $n_1 = 1$). For optical elements made of glass with a low value of n_2 a single layer antireflection coating is often insufficient, whereas two-layer coating films almost totally eliminate *reflection of light* from the surfaces of the optical elements. Note that this is true in a rather narrow spectral range only. To attain antireflection protection over a broad spectral range triple and multiple layer *coating* systems are commonly used. Antireflection coating plays a very important role in lowering optical losses in multilens optical systems, in laser technology where coating makes it possible to lower feedback losses from the optical cavities, in opto-acoustic instruments, etc.

ANTIRESONANCE, MAGNETIC

See *Magnetic antiresonance.*

ANTIRESONANCE, OPTICAL

See *Optical antiresonance.*

ANTI-SCHOTTKY DEFECT

A two-component *defect in crystals* of binary (MX) or more complex composition. Its components are I_M and I_X, the amounts of *interstitial atoms* of chemical elements M and X which themselves constitute the regular lattice of the crystal. Anti-Schottky defects can be equilibrium *stoichiometric defects* arising at elevated temperatures, e.g., because of "vaporization" of M and X atoms from the surface to the inside of the MX crystal. The creation of anti-Schottky defects is hampered by the necessity of inserting the larger M and X atoms into the smaller-sized interatomic "cavities" (i.e. interstitial sites). These defects have the name anti-Schottky because a Schottky defect is a vacancy formed by transferring an atom from the bulk of the crystal to its surface.

ANTISCREENING FACTOR

A parameter characterizing *antiscreening* (the phenomenon which is inverse to *electric charge screening*).

If P_0 is a response arising as a result of some influence in the absence of a medium, and P is the same response in the presence of the medium, then one can write $P = P_0(1 - \gamma)$. If the dimensionless parameter γ is less than zero it is commonly referred to as an antiscreening factor. The *Sternheimer antiscreening factor* (R.M. Sternheimer, 1951) γ_S describes the enhancement of an intracrystalline electric field gradient at a nucleus due to the polarization of closed atomic shells by external electric charges. The value of $|\gamma_S|$ can be rather large (50–100).

ANTISITE DEFECT

An atom of one sublattice of a two- or multicomponent *crystal*, occupying a site of another sublattice; actually, a *substitutional atom*, denoted as X_M (atom X at a site of sublattice M). Adjacent defects X_M and M_X represent an *antistructural defect* pair. For unequal concentrations of M_X and X_M the antisite defects determine an excess of component M or X, i.e. a deviation from *stoichiometric composition*.

ANTI-STOKES LUMINESCENCE

Photoluminescence, or luminescence at wavelengths less than that of the exciting light, which is in contradiction with *Stokes' rule*.

Anti-Stokes luminescence is common in the region of overlap between the excitation and the luminescence wavelengths. The emission of a quantum with an energy exceeding that acquired in the absorption occurs due to the additional involvement of the thermal motion of the luminescent system. This results in the *Vavilov law*, which reflects the fact that the probability of such a process decreases rapidly as the involved energy rises, and it depends on the thermal motion intensity (temperature). Of greater importance are the fundamental violations of Stokes' law when the emitted luminescence lies in the region of wavelengths less than that of the exciting light. Such effects are possible by making use of several exciting light quanta (most often, of two) in the emission process, by the simultaneous absorption of two photons by one electron at a high excitation intensity, by sequential absorption of two quanta by the electron, and by pooling the energy of two excited electrons to produce the emission (*cooperative luminescence*), etc.

ANTI-STOKES RAMAN LINES

Lines in the spectrum of *Raman scattering of light* that are located on the high-frequency side of the exciting line.

Anti-Stokes light scattering lines are caused by annihilation of the excitation quantum in the system by the inelastic scattering of an incident *photon* on it (a *phonon*, *magnon*, *plasmon*, etc., is typically annihilated in a crystal). The intensity of anti-Stokes lines depends on the initial *level population* (excited) and decreases with a lowering of the temperature. The presence in a spectrum of Stokes and anti-Stokes pairs of lines, symmetric relative to exciting one, is often used to discriminate between the processes of Raman scattering and *luminescence*, while their intensity ratio characterizes the scattering medium temperature.

ANTISTRUCTURAL DEFECT

A *defect* in a binary or more complex *crystal* containing two chemical elements M and X.

The defect consists of two components – *substitutional atoms* M_X (atom M at site X) and X_M (atom X at site M). The formation of antistructural defects in a nonmetallic crystal can be accompanied by changing the charge state of its components. For example, atoms M and X in the normal positions at their "own" sites M_M and X_X can be neutral, while as components of an antistructural defect, if M is a *metal*, M_X tends to become an *acceptor* and X (metalloid) a *donor* X_M. Antistructural defects either exist in the equilibrium state at elevated temperatures (e.g., in crystals Bi_2Te_3, Mg_2Sn, and others) or are created as nonequilibrium metastable defects, e.g., as a result of *nuclear radiation* (see *Antisite defect*).

ANTISYMMETRIC VIBRATIONS of crystal lattice

Normal vibrations which reverse their sign under the application of an appropriate symmetry operation (see *Crystal symmetry*). In *ionic crystals* antisymmetric vibrations involve a change of the electric *dipole moment*.

ANTISYMMETRY of crystals, Shubnikov symmetry

One of the particular cases of *colored symmetry* (*hypersymmetry*). In ordinary *crystal symmetry*, only spatial (position or orientation) symmetry is considered, while in hypersymmetry, another attribute (conventionally, color or spin direction) is assigned a value at every point of the *crystal*. If the additional attribute can take on only two alternative values (black and white), such a particular case of hypersymmetry is referred to as antisymmetry, in other words as *black and white symmetry*. If there are many values then one speaks about *color symmetry*. Another term for antisymmetry is *Shubnikov symmetry* after A.V. Shubnikov who introduced the concept. In analogy to the ordinary crystal symmetry groups, there exist 122 point, 36 translational and 1651 space *Shubnikov groups*. All of them have been derived and classified, a system of notation has been developed, and they are used in concrete physical studies. Examples of the physical realization of crystal antisymmetry can vary. Treated as the color change operation may be, for example, a change in the chemical type of atoms in an ordered *binary alloy*, or in the sign of the displacement of atoms and of the electric charge, and others. Solid state physics makes the widest use of this approach for the Shubnikov classification of the magnetic symmetry of crystals, where the color change operation has the sense of the *time inversion* operation R, and, as a consequence, the inversion of the *magnetic moments* and *magnetic fields* (see *Magnetic symmetry group*). Operations of antisymmetry, which occur in magnetic crystal symmetry and are combinations of rotation or translation with R, are referred to as *antirotations* and *antitranslations*.

ANYONS (fr. any)

Particles that obey a statistics which is intermediate between the Bose (see *Bose–Einstein statistics*) and Fermi (see *Fermi–Dirac statistics*) types. Such particles can exist in two-dimensional space since the topology of this space admits the presence of indeterminate wave functions for (identical) particles a with strong short range repulsion (hard core). These functions obey the permutation rules $P_{jk}\psi = \exp(\mathrm{i}\pi\alpha)\psi$, where P_{jk} means the permutation of jth and kth particles, and α is any real number which is sometimes called a *statistical parameter*. The value $\alpha = 0$ corresponds to bosons and $\alpha = 1$ denotes fermions (see *Second quantization*), while intermediate values are indicative of an *intermediate statistics*. The possibility of the existence of an intermediate type statistics was first pointed out by J.M. Leinaas and J. Myrheim (1977), and the term "anyon" was introduced by F. Wilczek (1982). Intermediate statistics can also be obeyed by particles in one-dimensional space (to which a term "anyon" is inapplicable, as a rule), and by distinguishable particles (anyons of many types). The permutation conditions for the latter have the form $P_{jk}^2\psi = \exp(\mathrm{i}\pi\alpha)\psi$, and at $\alpha \neq 0, 1$ one can speak about an *in-between statistics*.

The ambiguous nature of anyon wave functions results in the fact that the quantum-mechanical problem of many anyons cannot be solved even for the simplest case when there is no interaction. The usual procedure of seeking solutions in terms of single-particle wave functions is not applicable here.

The statistics of all elementary particles is determined by their spin in accordance with the Pauli principle. Therefore (since real space is three-dimensional), elementary particles with an intermediate statistics cannot exist in free space; this statistics can be valid only for *quasi-particles* in quasi-two-dimensional and quasi-one-dimensional systems. In particular, the elementary excitations in the *fractional quantum Hall effect* (see *Quantum Hall effect*) obey intermediate statistics, i.e. they are anyons. At a filling factor $\nu = 1/q$ where q is an odd integer, the statistical parameter of these excitations is $\alpha = \nu$. The appearance of the effect at filling factors $\nu = p/q$ kept constant is explained quite well in the framework of so-called *hierarchic theory*. This implies that there appear excitations over a "primary" electron gas that represent the anyon gas. Other excitations can arise which also represent the anyon gas but with another statistical parameter, and so on infinitely. One can demonstrate that under particular conditions an anyon gas is a *superconductor*. As far as it is known *high-temperature superconductivity* is of a quasi-two-dimensional nature, and an hypothesis about an anyon *mechanism of superconductivity* has been suggested, but there is no experimental confirmation of it. See also *Anderson mechanism of high-temperature superconductivity*.

APW METHOD

See *Augmented plane waves method*.

ARMCO IRON (ARMCO is an acronym of American Rolling Mill Corporation)

Highly purified *iron* with high *plasticity*, enhanced *electrical conductivity*, high magnetic saturation and resistance against *corrosion*. ARMCO iron falls into *soft magnetic materials*, its magnetic properties depending upon the quantity of impurities, sizes of crystal grains, elastic and plastic stresses. ARMCO iron is used for the manufacture of electrical engineering products to be operated in constant and slowly varying magnetic fields. In *metallurgy*, it is employed as the basic element in the manufacture of many magnetic alloys and as the charge in the production of alloyed *steel*.

AROMATIC COMPOUNDS

A group of organic compounds possessing a system of conjugated π-bonds.

The term originated from the specific odor of natural substances containing aromatic compounds. These compounds are exemplified by the *polycyclics* – benzene (C_6H_6), naphthalene ($C_{10}H_8$), anthracene ($C_{14}H_{10}$), tetracene ($C_{18}H_{12}$), pentacene ($C_{22}H_{14}$), and others, with a common formula $C_{4n+2}H_{2n+4}$, where n is the number of benzene-type rings. The molecules of polycyclics are planar and form *molecular crystals* in the solid phase. Most of them belong to the *monoclinic system* with two molecules in a unit cell. Electronic properties of aromatic compounds are determined mainly by p_z electrons of C atoms forming a delocalized π-orbital. The π-electrons excited by light undergo a π–π^* transition and account for the absorption spectral shift from the ultraviolet to the visible part of the spectrum in the above-indicated sequence of polyacenes. The mobility of π-electrons determines important physical and chemical properties of aromatic compounds: *photoconductivity* of their crystals, migration of electron excitation energy, catalytic activity. Aromatic compounds are model systems for *organic semiconductors*. The dark *electrical conductivity* of pure *crystals* is low, $\sigma \sim 10^{-2}$–$10^{-12}\ \Omega^{-1}\text{cm}^{-1}$. The typical mobility of excess electrons and holes moving in narrow bands takes on values of the order of $1\ \text{cm}^2{\cdot}\text{V}^{-1}{\cdot}\text{s}^{-1}$.

ARRHENIUS LAW (S. Arrhenius, 1889)

Relation stating the temperature dependence of the *reaction rate constant* K

$$K \sim K_0 \exp\left(-\frac{U}{RT}\right),$$

where U is the *activation energy* of the reaction, and R is the universal gas constant. The parameter K_0 has a temperature dependence weaker than exponential. The value of K is determined from the expression for the reaction rate $n\text{A}+m\text{B}+l\text{C}+\cdots$ which is proportional to the product of the appropriate powers of concentrations (denoted by brackets): $v = K[\text{A}]^n[\text{B}]^m[\text{C}]^l \cdots$. The Arrhenius law was deduced empirically and can be justified from the *Boltzmann distribution*. The figure schematically presents the dependence of the energy of the reacting particles on their generalized coordinate. The activation energies U_1 of the reaction $1 \to 2$ and U_2 of reaction $2 \to 1$ are determined by the energies of the initial states 1, 2 and the height of the potential barrier separating them. The barrier energy U varies from several electron-volts to almost zero; it can equal zero, e.g., in ionic-molecular reactions or in the reactions with mechanisms based on the *tunnel effect*. Such reactions can occur at temperatures close to absolute zero. Reactions in *solids* with participation of the inherent and *impurity atoms*, *conduction electrons*, and various *quasi-particles* also follow the Arrhenius law. In them, a part of the activation energy U_D can arise from the *diffusion* of reagents, determined by the temperature dependence of the *diffusion coefficient* $D = D_0 \exp(-U_D/RT)$. See also *Interdiffusion, Activation energy, Thermal activation*.

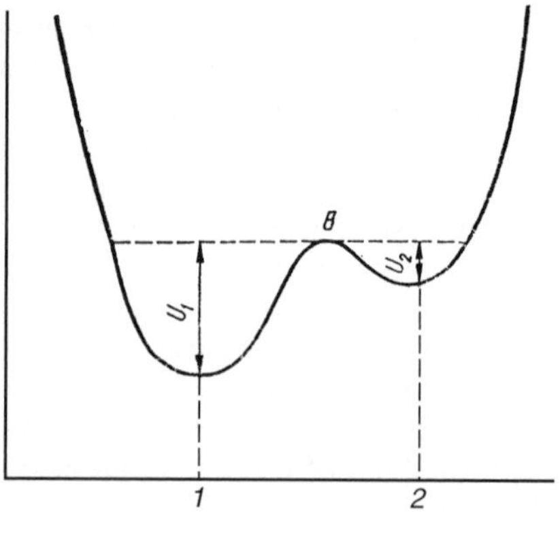

Activation energies U_1 and U_2 for a double minimum potential.

ARTIFICIAL EPITAXY

See *Amorphous epitaxy*.

ASBESTOS (fr. Gr. $\alpha\sigma\beta\varepsilon\sigma\tau o\varsigma$, unquenchable)

Fine-fiber *minerals* capable of splitting into flexible and thin fibers (up to 0.5 μm thick).

The most significant is serpentine asbestos. There are such varieties as chrysotile-asbestos and picrolite. Chrysotile-asbestos $Mg_6Si_4O_{10}(OH)_8$ has a lamellar-tubular structure with a monoclinic lattice. Felted and parallel-fiber aggregates are typical of it. The cleavage is imperfect along (110) at the angle 130°. Density is 2.5 to 2.65 g/cm^3. Mohs hardness is 2.0 to 3.0. Impurities: Fe (up to 3%), Ca, Mn, Ni, Al, Na. Water content is 13.0 to 14.8%. Refractive indices: $n_g = 1.522$ to 1.555; $n_p = 1.508$ to 1.543. The fibers of asbestos are rupture-strong (285 to 365 kg/mm^2). Asbestos is fireproof (T_{melting} is 1500 °C); it has high sorption ability (see *Sorption*), swelling ability; it conducts heat poorly; it is stable against alkalis; upon heating, it loses water and transforms to forsterite. Asbestos is used in textile, cement, paper, radio industry, road surfacings, etc.

ASCENDING DIFFUSION

A process of transport, whereby the flow of particles is directed along the concentration gradient. In contrast to conventional diffusion, in this case the difference in concentration of the diffusing component does not diminish but rather grows. The possibility of ascending diffusion arises from the fact that the driving force of the diffusion is the gradient of the chemical potential which, in general, can differ from the concentration gradient. Different fields which contribute to the chemical potential in addition to the *entropy* of mixing, can cause the ascending diffusion; examples are thermal fields (*thermodiffusion*), electrical fields, vacancy fields, phase non-uniformity of the material, and the field of elastic stresses. These fields make the potential barrier that the diffusing particle encounters an unsymmetrical one, and the result is a preferred drift in the direction of the field. In a general phenomenological theory ascending diffusion can be described in a systematic uniform manner, but the detailed mechanisms operating in the above mentioned cases are different.

The *Gorsky effect* (V.S. Gorsky, 1935), which involves ascending diffusion caused by local gradients of elastic stresses in *crystals*, is specific for solids. For example, during the bending of a *plate* made of metal alloys the concentration of atoms with the greater atomic radius grows at the exterior side of the flexure (a partial *stress relaxation* accompanies this process). By providing for the *segregation* of atoms of different types, ascending diffusion, evidently, plays an important role in the phase separation in *alloys*, and at the formation of atmospheres of the *impurity atoms* in the strained regions of a crystal, e.g., interstitial atoms preferentially concentrate in the stretched region of the field of stresses at dislocations (see *Cottrell atmosphere*).

ASPEROMAGNET

An *amorphous magnetic substance* in which the random frozen *magnetization* $\boldsymbol{m}_i$ has a first moment $\boldsymbol{M} = \langle \boldsymbol{m}_i \rangle_c$ and two second moments, $q_\parallel = \langle (\boldsymbol{m}_i^\parallel)^2 \rangle_c$ and $q_\perp = \langle (\boldsymbol{m}_i^\perp)^2 \rangle_c$.

The averaging is performed over configurations, the indices $\parallel$ and $\perp$ denoting projections of $\boldsymbol{m}_i$ onto the vector $\boldsymbol{M}$ or onto the plane perpendicular to $\boldsymbol{M}$, respectively. The *asperomagnetic state* is observed in two types of amorphous magnetic materials: in *alloys* and *metallic glasses* containing *rare-earth elements* and noble metals, such as DyCu, TbAg; in systems of the *spin glass* type. In the first case, the origin is the large value of the *magnetic anisotropy* constant (typical of rare-earth elements) and the random orientations of the anisotropy axis (typical of alloys and glasses). In the second case, the cause is the presence of concurrent random ferro- and antiferromagnetic interactions. Asperomagnets often exhibit nonergodic behavior (see *Nonergodicity*).

ASTERISM

The characteristic oblong shape of *X-ray diffraction* spots from deformed *monocrystals* exposed to X-radiation of a continuous type.

In the case of transmission images (*Laue patterns*), the diffraction spots are elongated toward the center of the picture. Asterism arises due to the reflection of the X-ray beam from the bent part of a crystal. The horizontal diameter of the spot in angular units is $\varphi_1 = 2\alpha$ and the vertical one is $\varphi_2 = -2\alpha \sin\theta$, where θ is the *Bragg angle* (see *Bragg law*) and α is the flexure angle of the reflecting plane. At small θ, $\varphi_2 \ll \varphi_1$ which explains elongation of spots at asterism, and proves that the length of asterism bands can serve as a measure of *flexure.* In reflection imaging (*epigrams*), the directional stretching of diffraction spots is observed according to the flexure axis orientation relative to the initial beam.

ATHERMIC CREEP

A type of low-temperature logarithmic *creep*, independent of temperature. Its rate after the instant jump of the initial *strain* ε_0 at the moment of load application quickly decreases in time, $\dot{\varepsilon} = \alpha/t$. Typical for this creep is the time dependence of strain caused by the motion of dislocations, $\varepsilon = \varepsilon_0 + \alpha \ln(\nu t + 1)$, α and ν being constants. A feature of athermic creep is the smallness (0.1 to 1%), in comparison with ε_0, of the strain under creep. The deceleration of athermic creep is associated either with the increasing density of support of intersecting dislocations or with the "depletion" of mobile dislocations. At temperatures close to absolute zero ($\sim$1 K), the barriers of the support type are overcome with the help of a quantum-mechanical tunnel effect (see *Low-temperature yield limit anomalies*). According to Mott, in this case α is independent of temperature, $\alpha \propto 2\pi h/(2Mu)^{1/2}$, where h is the *Planck constant, M* is the effective mass of dislocations, u is a quantity on the order of the energy of the step transformation at the dislocation ($3\cdot 10^{-19}$ J). Within the range of temperatures between 1 K and $T \sim 0.1\Theta$, where Θ is the *Debye temperature*, the crossing of the support may be activated by zero-point vibrations of dislocation segments. The factor

$$\alpha = \frac{1}{2}\frac{k_B\Theta}{h\nu}\left(1 + \left(\frac{T}{\Theta}\right)^2\right)$$

insignificantly changes with temperature at $T < \Theta$. The value of ε at the athermic creep is not large, but it needs to be accounted for in the operation of precision parts and products.

ATHERMIC PLASTICITY

Lack of sensitivity of the characteristics of a *plastic deformation* process of a solid to changes in temperature.

Athermic plasticity characterizes the regularities of plastic deformation of crystals at low and very low temperatures. It emphasizes that the thermoactivating character of elementary dislocation processes in determining the deformation kinetics, such as overcoming Peierls barriers by *dislocations* (see *Peierls–Nabarro model*) or local barriers determined by the *defects* of crystalline structure, is of limited applicability for the description of low-temperature plasticity. Athermic plasticity is not fully understood; it has been proposed as (i) playing a crucial role in quantum fluctuations that arise during the motion of dislocations through barriers, and (ii) playing a role in the replacement of a thermoactivated barrier penetration by an activation-free process of the over-barrier motion, which occurs owing to the high level of effective forces acting upon dislocations, and the low level of dynamic friction restraining the dislocations (see *Internal friction*), typical for low-temperature deformation.

ATMOSPHERE

See *Cottrell atmosphere, Snoek atmosphere, Suzuki atmosphere.*

ATOM

An atom consists of a positively charged nucleus with a size on the order of 10^{-15} m and a surrounding electronic cloud with a size of about 10^{-10} m (see *Atomic radius*).

An atom as a whole is electrically neutral; the positive charge of the nucleus Ze (Z is the atomic number and e is the elementary electron charge) is compensated by the negative charge of its Z electrons. An atom can accept or donate one or more electrons to become, respectively, a negatively or positively charged *ion.* The complicated structure of the electron cloud is determined mainly by Coulomb interactions, the attraction of electrons to the nucleus and their mutual repulsion, as well as by the Pauli exclusion principle. Relatively small but fundamentally important contributions to the energy arise from the *spin–orbit interaction* and the *spin–spin interaction.*

A *single-electron approximation* is widely used to describe an atom: the wave function of the electrons is a product of one-electron wave functions (*Hartree method*) or a linear combination of such products which satisfies the requirement of interchange invariance (*Hartree–Fock method*).

In this approximation one can speak about one-electron energy levels and corresponding electron shells of an atom. To describe them approximately, the model of a *hydrogen-like atom* is employed, with the shells characterized by the principal quantum number $n = 1, 2, 3, \ldots$, and their substructure by the orbital quantum numbers $l = 0, 1, 2, \ldots, n - 1$ and $m = 0, \pm 1, \ldots, \pm l$. The electrons of the K-shell with $n = 1$ are closest to the nucleus with the deepest (most negative) energy levels. Then follow the $L, M, N, \ldots$ shells with $n = 2, 3, 4, \ldots$. In the ground states of the atoms H and He only the K-shell is occupied. As Z grows, the orbits with higher quantum numbers become filled. The electron configuration of an atom is specified by symbols that indicate the values of quantum numbers n and l of filled shells in ascending order, and the number of electrons in them; the values $l = 0, 1, 2, 3, 4, \ldots$ being denoted with corresponding letters $s, p, d, f, g, \ldots$. For example, the configuration $1s^2 2s^2 2p^6$ of the neon atom has two electrons with $n = 1$ and $l = 0$, plus two electrons with $l = 0$, and six electrons with $l = 1$ at $n = 2$. Each l shell contains $2(2l + 1)$ electrons when it is full.

Another approximate method for describing an atom (*Thomas–Fermi method*) is based upon a structureless cloud of atomic electrons with their density distribution determined by the *Poisson equation*.

The outer valence electrons determine the chemical properties of atoms, their ability to form this or that compound, the structure, binding energy, and other characteristics of compounds. Although atoms are the structural elements of solids, the distribution of the atomic electrons in a solid differs from that of a free atom. The configurations of valence electrons undergo considerable changes, depending on the type of bonding force between atoms, the number and positions of nearest neighbors, etc. In a crystalline solid, the states of valence electrons depend on whether the atoms occupy sites of the regular lattice, or they are *doping atoms* or *substitutional atoms*. The presence of the two latter types introduces additional levels in the energy spectrum of *electrons in crystals* and influences the optical, electrical and other properties of a *crystal*. The inner atomic shells also undergo some change, but to a much lesser extent. This is substantiated by data of *X-ray emission spectroscopy*, by the observed differences of characteristic X-ray frequencies for the inner shells of atoms of a given chemical element incorporated into different chemical compounds. It is not feasible to solve exactly the problem of the states of atoms forming (comprising) a solid. The approximation often used in condensed matter theory is based on reducing the number of independent variables per atom down to 3. Here, the atom is represented as a structureless entity, while interactions involving atoms are described by various, mostly empirically chosen, potentials with limited applicability (see *Interatomic interaction potentials*). In a more precise approach the valence electrons are singled out, and the nucleus with the deep electron shells is treated as a point particle with an effective charge.

The potential of an atom is determined by the distribution of its electron density. An approaching test particle perturbs this distribution in a manner that depends on its nature and its speed relative to the atom. Atom–atom potentials differ for different pairs of atoms. In the quantum description one cannot speak about each particular electron being affiliated with one particular atom of a solid. The delocalization of electrons among many atoms causes the splitting of atomic energy levels and the creation of *bands of allowed energies* in crystals. Such broadening is rather small for electrons in deep levels. For those in outer atomic shells, broad (several electron-volts) allowed *energy bands* arise, and are particularly prominent in metals, with their structure determining the electric, optical, and other properties of the crystal.

ATOM DISPLACEMENTS

See *Static displacements of atoms*, *Thermal displacements of atoms*.

ATOMICALLY CLEAN SURFACE

A surface free of impurities and having a pristine *crystal structure*.

To obtain and preserve an atomically clean surface, various methods of cleaning in a super-high vacuum (10^{-8} Pa) are used. The main methods are high temperature heating, bombardment with ions of noble gases with annealing and chemical pretreatment to obtain products that are removable at moderate temperatures. An absolutely clean surface is impossible to obtain: the sources of contamination are the impurities in a *crystal* bulk and the *adsorption* of particles from residual gases. Therefore, the surface is considered to be clean, if the concentration of impurities is so small that it fails to affect phenomena under study. *Auger electron spectroscopy* is the most suitable method of monitoring surface cleanliness (the sensitivity is 10^{-2}–10^{-3} monolayers). The investigation of an atomically clean surface structure is performed by *low-energy electron diffraction* and high-energy electron diffraction methods (see *Electron diffraction*). From an analysis of the data from structural investigations, the atomically clean surfaces of crystals, especially semiconductor ones, can be reconstructed. *Surface structures* can form with periods other than those in atomic planes in a crystal bulk parallel to the surface. A special notation is adopted for these surface structures. They include the chemical symbol of the crystal, the orientation of a face, and the ratio of the *unit cell* vectors of surface structure to the corresponding vectors in the bulk, e.g., Si (111) – 7 × 7, Au (100) – 5 × 1 and so on. According to modern concepts, the cause of the reconstruction of the upper atomic planes of a crystal is the tendency to reduce the excessive free energy of a *surface lattice* relative to that in the bulk. The supposition that the reconstruction is determined by the uncontrollable contamination of a surface failed to be confirmed experimentally. The difficulties connected with clarification of the nature of surface reconstruction are complicated by the fact that the problem of the determination of the coordinates of the unit cell atoms for the surface structures with large unit cells is yet to be solved because of the complexity of the calculations in the structural analysis. Therefore, various hypothetical models are used in discussions of this reconstruction, their ability to approximate reality being tested using data obtained by various methods. All the varieties of models can be divided into two classes: those based on small static displacements of atoms from their normal positions in the surface lattice, and those involving the concept of orderly-spaced defects at the surface (vacancies, adatoms, etc.). There has been considerable progress in understanding surface structures because of data obtained from *scanning tunneling microscopy*. According to estimates based on different data, the reconstruction encompasses several atomic planes of a crystal. There are only a limited number of orientations of planes which coincide with atomic planes having the same *crystallographic indices* and characteristic surface structures. Such surfaces are referred to as *atomically smooth*. In other cases, energy can be gained by the decay of a flat surface to micro *facets*, or by the formation of a stepped surface including terraces of low-index facets connected by equispaced *atomic steps* as high as one or more interplane distances. Besides the structure, there are also other important properties of a surface such as the electronic energy band structure and dynamical characteristics (frequencies of atomic vibrations, their vibration amplitudes) that are different from the analogous properties in the bulk. Therefore, the near-surface region is considered as a special *surface phase* with its own conditions of equilibrium. The atomically clean surfaces of some crystals of semiconductors and metals with particular orientations are the place where *order–disorder phase transitions* as well as order–order ones occur, while there are no phase transitions in the bulk. The order–order reconstruction can occur as transitions between surface structures, both with unit cells lying in one plane, and with changes of facet structure – *morphological transitions*.

ATOMIC CHARGE

See *Effective atomic charge*.

ATOMIC FORM FACTOR

A parameter of the elastic scattering of a plane wave (X-ray, electron, neutron) by an isolated atom, which determines the amplitude of the scattered wave as a function of the scattering angle.

In the scattering of waves in *crystals*, the amplitudes of scattering by individual atoms are amplified in some directions, while they are decreased in others; the total result being determined by the values of the atomic form factors of the atoms forming the crystal. A general expression for an atomic form factor is

$$f(s) = K \int \rho(\boldsymbol{r}) \exp(\mathrm{i}\boldsymbol{s} \cdot \boldsymbol{r})\, \mathrm{d}r',$$

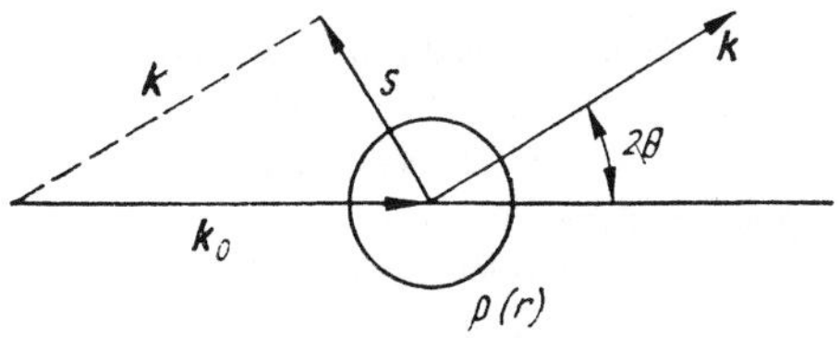

Atomic form factor.

where $s = k - k_0$, and k_0, k are the wave vectors of the incident and scattered waves, respectively (see Fig.), and K is a factor dependent on the type of radiation. The X-ray scattering is determined by the electron density $\rho_X = |\psi(r)|^2$, and values of $f_X(s)$ are on the order of 10^{-11} cm. The electron scattering is determined by the electrostatic potential of the atom $\varphi(r)$; $\rho_{el}(r) = \varphi(r)$, with values of $f_{el}(s)$ of the order of 10^{-8} cm. Neutron scattering may be thought of as taking place at the point potential of the atomic nucleus, with values of $f_n(s)$ of the order of 10^{-12} cm, and no angular dependence.

ATOMIC MAGNETIC STRUCTURE

Position and spatial orientation of atomic *magnetic moments*, or of the magnetic moment density of delocalized electrons (the latter pertains to *band magnetism*). The presence of a magnetic structure in a material implies that the time averages of the above quantities differ from zero. These averages are zero in materials having no magnetic structure (an external magnetic field is presumed absent). The emergence of a magnetic structure in a material is always accompanied by symmetry lowering, if for no other reason than that the time reversal operation is not a symmetry transformation. In this connection, the state of a material exhibiting a magnetic structure is called a *magnetically ordered state*, and the process of transforming to a magnetically ordered structure is a *magnetic phase transition* of the disorder-to-order type.

A magnetic structure is commonly found in those crystalline substances where the magnetic atoms tend to occupy symmetrical crystallographic sites, while the mean magnitudes of atomic magnetic moments are oriented in a definite manner with respect to one another and to the crystallographic axes. The magnetic structure of most magnetic crystals can be classified as one of three basic types that correspond to *ferromagnetism*, *antiferromagnetism*, and *ferrimagnetism*.

A *ferromagnetic structure* refers to one with all the magnetic moments aligned in the same direction. This structure is characterized by the magnitude and direction of the *magnetization* M.

The simplest *antiferromagnetic structure* is a set of two interpenetrating *magnetic sublattices*, with their respective magnetization vectors, M_1 and M_2, equal in magnitude and opposite in direction, viz., the net vector $M = M_1 + M_2$ is zero. The two-sublattice antiferromagnetic structure is described by the magnitude and direction of the *antiferromagnetism vector*, $L = M_1 - M_2$. The condition $M_1 = -M_2$ necessarily assumes that atoms belonging to different magnetic sublattices are the same, and their crystallographic positions are equivalent, since a difference in the magnetic moments inevitably results from the dissimilarity of the atoms, or their positions. More complicated antiferromagnetic structures have been discovered in many crystals. For example, in Mn_3NiN, the magnetic moments are directed along the sides of an equilateral triangle (a *non-collinear antiferromagnetic structure* with three sublattices), while in $YMnO_3$, they follow the directions of the sides of two equilateral triangles rotated by 60° relative to one another (a six-sublattice structure). *Noncoplanar antiferromagnetic structures* are also encountered, e.g., the magnetic moments in the four-sublattice antiferromagnet UO_2 lie along the body diagonals of a cube.

Any particular magnetic structure of the ferro- or antiferromagnetic type forms in a substance as a result of electrostatic exchange or direct interactions, with the *exchange interactions* far exceeding, as a rule, those of the direct type (i.e. *magnetic dipole interactions* and *spin–orbit interactions*). In quite a number of materials exchange interactions establish an antiferromagnetic structure (a bi- or polysublattice one), whereas the other interactions (see *Dzialoshinskii interaction*) bring about a small distortion of this structure, which causes a nonzero ferromagnetism vector, $M \ll L$, to appear. Such materials are called *weak ferromagnets*, and the corresponding magnetic structures are called *weak ferromagnetic structures*.

The competition of the strong electrostatic and weak direct forces may also have other consequences i.e. it may produce a long-period *modulated magnetic structure* in a crystal. Here, the exchange forces order the magnetic moments according to either a ferro- or antiferromagnetic pattern, while the relativistic forces lead to the vectors $\boldsymbol{M} = \boldsymbol{M}(\xi)$ or $\boldsymbol{L} = \boldsymbol{L}(\xi)$ gradually changing along a certain direction ξ that is called the *axis of the modulated structure*. In the simplest modulated structures, $\boldsymbol{M}$ (or $\boldsymbol{L}$) just rotates around the axis (*long period helices*); there are also more complicated modulated magnetic structures. The formation of a magnetic helix in some crystals is due to the exchange interaction. The helix period is of the same order of magnitude in this case as the distance between neighboring magnetic atoms. A magnetic helix differs fundamentally from a noncollinear antiferromagnetic structure (e.g., of $YMnO_3$ type) in that the angles between the magnetic moments may not correspond to $360°/n$ values ($n = 1, 2, 3, \ldots$). The helix period is not commensurate with the interatomic distances ξ_{lm} in this case.

Another common type of magnetic ordering (*ferrimagnetic structure*) occurs in *ferrimagnets*, where the exchange forces position the atomic magnetic moments in a manner resulting in nonzero values of both $\boldsymbol{M}$ and $\boldsymbol{L}$. In particular, a ferrimagnetic structure appears when the magnetic atoms occupy crystallographic positions of two different types, with each type constituting a magnetic sublattice. Magnetic structures of a special type, impossible in magnetic crystals, have been discovered in chemically disordered *solid solutions* (alloys) (see *Amorphous magnetic substances*, *Spin glass*, *Asperomagnet*). A particular form of modulated magnetic structure, the *spin density wave* whose period may be incommensurable with the *crystal lattice* period, may appear in a system of delocalized electrons.

All the above magnetic structures have been observed in a number of materials, such as those involving $4f$-elements and their alloys (see *Magnetism of rare-earth metals*), as well as *actinide*-containing compounds. There is still another form of magnetic structure called the antiphase domain structure which possesses an alternating sequence in space of the groups of n_1 atoms with the magnetic moments oriented along a certain axis, and of the groups of $n_2 \neq n_1$ atoms with opposite magnetic moment orientations (e.g., in thulium, $n_1 = 4, n_2 = 3$).

ATOMIC ORBITALS

See *Orbitals*.

ATOMIC ORDER on a crystal surface

The ordered arrangement of surface *atoms* (*ions*, molecules) that corresponds to a minimal free energy (see *Thermodynamic potentials*) of a finite *crystal*. The order is determined by the interaction of surface atoms with one another and with bulk atoms. It can differ from the arrangement in atomic planes inside a crystal, but parallel to the surface (see *Surface superlattice*, *Surface reconstruction*). The possible symmetry types of periodic surface structures are described with five parallelogram systems of equivalent points (two-dimensional analog of *Bravais lattices*, see *Two-dimensional crystallography*) with primitive *unit cells* in the shape of a general parallelogram, rectangle, square, rhombus with angle 120°, rhombus with angle unequal to 120° (in this case, one usually chooses a centered rectangle as a nonprimitive cell). Sometimes, the equilibrium state of a crystal surface is in correspondence with the ordered arrangement of *vacancies*, adsorbed atoms, *atomic steps* and terraces on it, e.g., a system of equidistant atomic steps of the same height on the facets with high Miller indices (see *Crystallographic indices*) in *silicon* and *germanium*. The results of surface structure studies by *scanning tunneling microscopy*, and by *low-energy electron diffraction*, indicate the presence of atomic order on a crystal surface.

ATOMIC RADII

The approximate sizes or radii r_a of particular *atoms* in *crystals*, assuming them to be spheres touching each other. A typical interatomic spacing between two atoms in a crystal is the sum of their atomic radii. In neutral atom crystals such as solid argon the electron density distribution of outer shells of atoms displays a relatively weak dependence on their nearest neighbors. The value of the outer shell radius averaged over the sphere, which characterizes the atomic size, varies with changes in the chemical composition of the nearest neighbors by a few percent when the number of neighbors (referred to as the *coordination number k*) is the same, and it varies by up to 10–12% in compounds with different coordination numbers. Common coordination numbers are $k = 12$, 8, 6, and 4 for face centered cubic (fcc), body centered cubic (bcc), simple cubic (sc) and diamond (ZnS) structures, respectively. Thus, an approximate *additivity rule* exists: $d(\mathrm{AB}) = r_a(\mathrm{A}) + r_a(\mathrm{B})$, where d is the closest distance between atoms A and B in the compound AB, and the r_a are their atomic radii. In view of the approximate character of the additivity rule, there is some indeterminacy in the choice of numerical values of atomic radii, and several systems of atomic radii exist. All of them, however, predict values of interatomic distances in crystals with a good accuracy, which has been confirmed with many compounds. The *metallic atomic radii* r_M for $k = 12$, 8, 6, and 4 are in the ratio 1 : 0.98 : 0.96 : 0.88. The atomic radii for Li, K, Al, Fe, Au are 1.55, 2.36, 1.43, 1.26, 1.44 Å. To describe ionic compounds *ionic radii* r_i are used; with the closest distance between two ions of opposite sign assumed to equal the sum of their ionic radii. Underlying the determination of r_i are the radii of fluoride (F^-) and oxygen (O^{2-}), which are adopted to be $r_i(F^-) = 1.33$ Å, $r_i(O^{2-}) = 1.32$ Å in the Goldschmidt system and $r_i(F^-) = 1.36$ Å, $r_i(O^{2-}) = 1.40$ Å in the Pauling system.

The concept of atomic radii can be extended to other cases. *Covalent atomic radii* r_c are characterized by a noticeable dependence on the bond multiplicity. The value r_c for the atoms C, N, O, S reduces by 12–14%, if two pairs of electrons take part in the bond, and by 20–22%, if three, relative to a single electron pair covalent bond. The covalent atomic radii of C, Cl, Br are 0.77, 0.99, 1.14 Å. In *molecular crystals*, the additivity condition is approximately met for distances d between the nearest atoms of neighboring molecules. One can introduce a system of intermolecular atomic radii r_V (also referred to as *van der Waals atomic radii*). Due to the relative weakness of their bonding forces, the values of r_V are greater than those of atomic radii for strong bonding. For He, C, Cl, Kr, $r_V = 1.40$, 1.70, 1.78, 2.02 Å.

ATOMIC STEPS on a surface

Raised boundary regions in a *monocrystal* that are transitions between terraces – flat regions of atomically smooth surface (see *Atomically clean surface*).

The steps are as high as one or several interplanar distances. As distinct from analogous steps of macroscopic size with a height of several micrometers and more, depending on the extent of a surface treatment, the atomic steps cannot be observed with an optical microscope, but are detectable using *low-energy electron diffraction* and are observable with an electron microscope, a scanning tunneling microscope, or an atomic force microscope. In experiments involving the reflection of heavy *ions* with energy 1–50 keV under grazing incidence on a monocrystal surface with a variation of the angle of incidence, the scattering by the atomic steps "upward" and "downward" (relative to the incident beam direction) produces characteristic features in the energy spectrum of the scattered ions, and marked differences in the extent of their ionization. This allows the use of the scattering of medium energy electrons to diagnose the step structure of a solid surface at the atomic level.

ATOMIC STRUCTURE OF A SURFACE

See *Surface atomic structure.*

ATOM–SURFACE INTERACTION

The interaction of *atoms* that is defined by characteristic features of their position on the surface.

The *crystal* surface has an excess of energy compared to the bulk. The coordination number of surface atoms is less than that of the atoms in the bulk (see *Coordination sphere*), which causes the appearance of uncompensated elastic forces acting on the surface atoms. The formation of a surface can be accompanied also by the appearance of atomic dipole repulsive moments caused by the shift of electron clouds into the depth of a crystal due to the electric potential jump on the surface. The effect of these factors is that the dynamical characteristics of the surface atoms are changed compared to those in the bulk: growth of atomic vibration amplitude, decrease of effective *Debye temperature*, and so on. Reconstruction of the upper atomic planes caused by the tendency to reduce the free *surface energy*, and the phenomenon of a surface *phase transition* in the absence of one in the bulk (see *Atomically clean surface*) are typical for a number of crystal surfaces. The screening of the Coulomb interaction of adsorbate atoms by charge carriers in a nearby-surface *space charge region* can appreciably modify the atom–surface interaction.

ATTENUATED TOTAL INTERNAL REFLECTION, ATR method

Technique for studying optical properties of solids (see *Optics of solids*), based on the phenomenon of *total internal reflection* of light, which occurs when a light wave penetrates from an optically denser medium 1 into medium 2 of lower optical density ($n_2 < n_1$). Provided that the angle of incidence φ exceeds the critical value $\varphi_{crit} = \arcsin(n_2/n_1)$ there forms a nonuniform (decaying) field $E(z) = E_0 \exp(-z/d_{penetr})$ (z is normal to the interface between the media), with an effective penetration depth d_{penetr} that is close to the radiation wavelength. Now, assuming that medium 2 is absorbing, the *reflectance* is $R \neq 1$, which indicates the presence of *attenuated total internal reflection*. The characteristics of spectra recorded by this reflection $(1 - R)$ method, and by the standard absorption method, are similar to each other. That is why attenuated total internal reflection is used to study strongly absorbing media, and in particular optical properties of their surfaces (e.g., perturbed unstructured layers, adsorption films, transition layers of various origins). To enhance the contrast accumulated during repeated internal reflections, specially designed prisms are used that provide multiple reflections which effectively extend the optical path of the light wave through medium 2. The propagation of light through medium 2 may result in its polarization, so one may speak about the propagation of a volume *polariton* wave. From the law of the conservation of the tangential component of the wave vector k_x it is possible to determine the dispersion of volume polaritons in medium 2: $k_x = (\omega/c)[\varepsilon_2(\omega)]^{1/2}$ where $\varepsilon_2(\omega)$ is the relevant *dielectric constant*. If medium 2 appears surface active in the respective frequency range, i.e. $\varepsilon_2(\omega) < 0$, one may apply a modified type of attenuated total internal reflection to excite *surface polaritons*. To that end, one introduces an air (or vacuum) gap between the prism and the material under study [*Otto modification* (A. Otto, 1968)], or places the material on the base of the prism as a thin film [*Kretschmann modification* (E. Kretschmann, 1968)].

ATTO...

A decimal prefix denoting a 10^{-18} part of the initial unit of a given physical quantity. For example, 1 as (attoseconds) $= 10^{-18}$ s (seconds).

AUGER EFFECT in semiconductors (P.V. Auger, 1925)

A semiconductor analogue of the Auger effect in an atom; a double electron transition with one electron crossing to a lower level, and another one to a higher level which is often in the continuum, outside the atom. The presence of *energy bands* of two kinds of *current carriers*, and the variety of structural *defects* in semiconductors, leads to the wide diversity of the Auger effect. The role of this effect in nonequilibrium processes in semiconductors may be quite different: *recombination*, generation, *impact ionization* of local centers, participation in *scattering* processes, *relaxation* of the energy and momentum of charge carriers. This effect can also be a cause of the strength and of the temperature extinction of *luminescence*. The set of Auger effect variants is shown in Figs. (a–h). There are processes with participation of a multiply charged local center (a), singly charged local center (f, g), two singly charged local centers (b, c, d, h), and free charge carriers (e). Included among Auger processes are also all processes that are inverse to those represented in this figure. A salient feature of an Auger processes is the pronounced dependence of its probability on the distance R between the interacting carriers ($\propto R^{-6}$, corresponding to the energy of one dipole in the field of another). In the case of an Auger interaction between free carriers, this means that the rate varies as the square of their concentration. Thus, in process (e), the hole *lifetime* with respect to the Auger process is proportional to the inverse square of the electron concentration, which means the dependence of the rate on R^{-6}. Both the theory of the Auger effect and experimental data demonstrate that the probability of this process in semiconductors is weakly

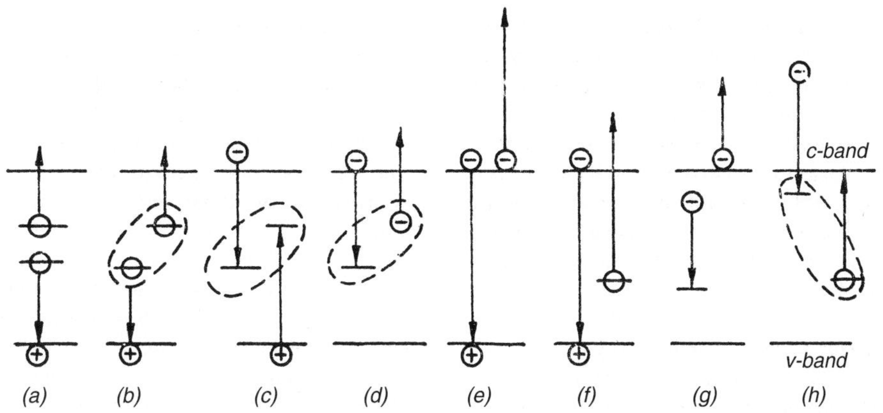

Auger effect processes in semiconductors.

dependent on the temperature. The dependence of the probability on the energy exchanged between the carriers is a power function with an exponent between 2 and 5. When the distance between the electrons involved in the Auger interaction mechanism is about 10 nm in order of magnitude (corresponding to a free charge carrier concentration about 10^{18} cm^{-3}), the cross-sections of the Auger processes can be quite large (close in a crystal *unit cell* size, i.e. $\sim 10^{-15}$ cm^2). In this case, the Auger processes can dominate the recombination and scattering, thus determining the lifetime of the carriers, the probability of radiative and *nonradiative quantum transitions*, the *relaxation times* of energy and momentum, and other parameters of the material.

AUGER ELECTRON SPECTROSCOPY

A method of analysis of the elemental and phase composition of surface layers of *solids*, based on the *field ionization* of atoms excited by a beam of primary electrons with energies 1 to 10 keV. The sensitivity of the method is 10^{10}–10^{12} atom/cm^2, and the depth of the analyzed layer is 0.5 to 2.5 nm. The mechanism for the emergence of *Auger electrons* is a complicated multistage process that includes the ionization of an inner atomic level x, the filling of the produced *vacancy* by an electron from a higher atomic level y, the nonradiative transfer of the released excess energy to an electron at level z, and the emission of the Auger electrons. The identification of chemical elements is based on comparison of calculated and experimentally found values of Auger electron energy $E_{xy} = E_x - E_y - E_z + U_{\text{eff}}$. Here, E_x, E_y, and E_z are, respectively, the binding energies of electrons at x, y, and z levels (generally designated by X-ray terms) of the neutral atom participating in the Auger process; U_{eff} is a factor to allow for the presence of two vacancies in atomic shells. Auger lines of different elements are well resolved. The change in the electron density at an atom results in a shift of the Auger line, the magnitude and sign of the shift being characteristic of the atom's chemical state. Analysis of the width and shape of the Auger lines, whose formation is participated in by *valence band* electrons, yields information on the *band structure* of a solid. The total Auger electron current is proportional to the number of vacancies at atomic levels formed per unit time within the thickness of the region of the exit from the surface. This circumstance underlies the application of Auger electron spectroscopy to the quantitative analysis of solids. In the scanning mode of the electron beam deflection, Auger electron spectroscopy is used for studying the planar distribution of chemical elements with a spatial resolution of a few tens of nanometers. In combination with *ion etching*, it is applicable for determining the depth profiles of element distributions.

AUGER ION SPECTROSCOPY

Determination of the elemental and chemical composition, the structure, and the state of a *solid* and its surface from Auger electron spectra (see *Auger electron spectroscopy*) excited by ion bombardment. An important feature which distinguishes Auger ion spectroscopy from Auger electron spectroscopy is its selectivity, i.e. Auger peaks of a particular kind specific for a given solid are typically observed in Auger spectra excited by *ion bombardment.* In addition, these peaks appear only under bombardment by certain ions. In contrast to the electron bombardment case, the character of ion-stimulated spectra is determined by characteristics of the ionization process in the inner shells, and the emission of Auger electrons at atomic collisions. In particular, the ion Auger spectra have a great number of satellites, and represent quasi-molecular spectra with a typical continuous background. Yet many of Auger peaks are narrower, since they are emitted by free atoms dislodged from the solid. Furthermore, the *recoil atoms* that emerge under ion bombardment and emit the Auger electrons are very fast moving, and induce an extra "Doppler" broadening (see *Doppler effect*). Under ion bombardment of *monocrystals*, one can observe both azimuthal and polar *orientation effects* in the Auger spectra, which are related to the crystal structure of the surface. Also classified as Auger ion spectroscopy is *ion-neutralization spectroscopy* that deals with the analysis of the spectrum of secondary electrons emitted during the *Auger neutralization* of slow positive ions (including multiply charged ones) at the *solid surface.* Auger ion spectroscopy is a convenient method for diagnostics of the solid surface, and is effective when one has to detect atoms of a certain element or a group of elements on the surface, or during a *layer-by-layer analysis* of a thin layer. This is particularly convenient in devices where ion beams are used for implantation, *etching*, *spraying*, *epitaxy*, etc.

AUGER NEUTRALIZATION

The process whereby an *ion* near a *solid surface* traps an electron (or several electrons in the case of multiply charged ions), with subsequent transfer of the resulting energy of neutralization to another (or other) electron(s) in the solid by means of the Auger process (see *Auger effect*). A condition for Auger neutralization is that the ionization potential of the ion V_i must exceed the *work function* of the solid φ ($V_i > \varphi$), and for electron potential emission we require $V_i > 2\varphi$. The Auger neutralization rate W depends on the scattering of ions to the surface S in accordance with $W = A\exp(-S/a)$ where $A \sim 10^{16}$ s^{-1}, $a \approx 0.2$ nm. A consequence of intensive Auger neutralization is a low ($\approx 1\%$) degree of ionization of atoms with high ionization potentials, e.g., inert gases reflected from a *metal* surface.

Auger neutralization underlies the *ion-neutralization spectroscopy* of a solid surface that allows one to determine the electron density in the *conduction band*, the position of the *Fermi level*, and other characteristics of the electronic structure of a solid surface. The Auger neutralization of multiply charged ions proceeds stepwise, as a series of successive acts of Auger neutralization and *Auger relaxation* in the system of levels of ions with decreasing charge, with a mean step of about 25–30 eV. As a result of the neutralization, the number of emitted Auger electrons is proportional to the resultant energy of the ion neutralization, and can be as high as a few tens of electron volts. On the surface of nonmetals this effect leads to the accumulation of a positive charge (considerably larger than the initial charge) in a microscopic region, and could possibly lead to a *Coulomb explosion.* This causes *sputtering* of the solid, and represents a mechanism for metal surface erosion under action of slow multiply charged ions. Auger neutralization of multiply charged ions on the metal surface can serve for their identification on the background of singly charged ions due to stimulated electron emission.

AUGER RELAXATION

Nonradiative *relaxation* of an excited *atom* near a *solid surface*, an analog of the Penning effect in gases. During Auger relaxation, the excitation energy is transferred by the Auger process to an electron of the solid, which can be emitted outward. Along with *Auger neutralization*, Auger relaxation is a mechanism of potential emission of electrons. This relaxation is possible for both metastable excited atoms coming to the surface, and *ions* that trap electrons from the solid at excited levels. The

Auger relaxation time is 10^{-14} to 10^{-15} s, which is six orders of magnitude less than the time of radiative *relaxation*. This circumstance suppresses strongly the radiation of the excited atoms near the metal surface. On the surface of nonmetals or metals covered by adsorbed layers, the rate of the Auger relaxation can be slightly smaller.

AUGMENTED PLANE WAVES METHOD, APW method (J.C. Slater, 1937)

One of the main *band structure computation methods*. It considers crystal potentials in the form of a *muffin-tin potential* (MT potential) and is based on the variational principle. Its basis functions are the APWs. Outside the MT-sphere an APW is a plane wave, $e^{i(\boldsymbol{k}+\boldsymbol{G})\boldsymbol{r}}$ ($\boldsymbol{G}$ is a *reciprocal lattice* vector). Inside the MT-sphere the APW method employs a linear combination of solutions of the Schrödinger equation in a spherically symmetric potential. The coefficients for that combination are selected to satisfy the continuity condition for APWs at the MT-sphere boundary. The amount of computations needed follows from the necessity of solving a nonlinear secular equation of high order. The order of the equation, i.e. the number of APWs, is usually quite large, from several dozen for the simplest lattices to several hundred for polyatomic structures. For the linearized APW method, see *Linear methods of band structure computation*.

AUSTENITE (named after W. Roberts of Austen)

A high-temperature *gamma phase* that is a *solid solution* of carbon (or nickel, manganese, nitrogen) in γ-iron with a face-centered cubic lattice (see *Face-centered lattice*).

Alloying with these elements stabilizes γ-iron. *Alloys* in which the γ-state is present at room temperature are called *austenite alloys*. They are easily deformable and possess high *corrosion resistance*. Carbonaceous austenite is stable under cooling to 725 °C; at lower temperatures, a phase transition occurs which can proceed either in a diffusion or a nondiffusion manner, depending on the cooling rate or the isothermal exposure. At cooling rates of a few tens of a deg/s or at isothermal exposure in the temperature range 450–723 °C, the austenite decomposes into a ferrite–cementite mixture (see *Cementite*) and forms structures of a pearlite class (*pearlite*, *sorbite*, *troostite*) that differ in the degree of dispersion of cementite plates. At temperatures below 450 °C, when the diffusion rate of iron and alloying elements is insufficient, austenite transforms into *bainite*; the transformation needs only the diffusion of carbon. At lower temperatures and at a cooling rate greater than 100 deg/s, austenite transforms into *martensite* in a diffusionless manner (see *Martensitic transformation*). When overheated above the *critical point*, perlite transforms into austenite at a varying rate; however, the newly formed austenite is heterogeneous even within a single grain volume. To obtain a homogeneous austenite, it is necessary to operate at a temperature above the completion of the perlite–austenite transformation, which produces *homogenization*. The rate of the homogenization is largely determined by the initial structure of the *steel*. The trends of the processes of decomposition (see *Alloy decomposition*) and formation of austenite underlie the processes taking place during the *heat treatment* of steel, with the treatment providing the required combination of strength and durability properties of steel.

AUTOEPITAXY, homoepitaxy

The oriented and ordered build-up of a material on a monocrystal surface of the same material.

Autoepitaxy is observed in the deposition of material from liquid solutions or melts, by chemical reactions at *crystallization from the gas (vapor) phase*, and at *condensation* from vapors in vacuo. In autoepitaxy the *crystal lattice* of the building-up *phase* is oriented relative to the lattice of the *substrate* monocrystal. Important influences on autoepitaxy are the chemical purity and the distribution of defects on the monocrystal surface, the chemical composition of the environment, the temperature, the goodness of the vacuum chamber, and the chemical composition of the residual atmosphere.

The phenomenon of autoepitaxy is used to prepare *thin films* of *germanium*, *silicon*, *gallium arsenide* and other semiconductors on monocrystal substrates of the same materials (see also *Epitaxy*).

AUTOIONIC MICROSCOPY

The same as *Field ion microscopy*.

AUTOIONIZATION

The same as *Field ionization*.

AUTOPHOTOELECTRONIC EMISSION

Electronic emission by a conducting body being irradiated by light in the presence of a strong external electric field (of the order of 10^9 V/m).

The mechanism of autophotoelectronic emission is the *field electron emission* excited by light. This emission appears most prominently in p-type semiconductors at low temperatures. The light excites the electrons from the *valence band* to the *conduction band*, with the long wavelength (red) limit of photosensitivity determined by the width of the *band gap*. The external electric field penetrates deeply inside the cathode and facilitates the transport of electrons to the surface and their escape into the vacuum. The *quantum yield* can reach 0.5–0.8.

Autophotoelectronic emission is used in converters of images in infrared rays to visible images. The cathodes in such converters are multitip (up to 10^6 tips/cm^2) matrix systems of p-type semiconductors (*silicon*, *germanium*, and others).

AUTORADIOGRAPHY, radioautography

Analysis of the radioactivity distribution in a material by means of *autoradiograms* which are photographic images produced by radioactive emission.

Autoradiography is based on the ability of some substances (typically, silver bromide) to undergo chemical transformations under the action of light, X-rays, gamma rays, and charged particles. Using contact contrast analysis, the emission intensity and the radioactive isotope distribution in the material are determined by the extent, density, and distribution of the blackening of the various regions of the autoradiograms. To obtain and analyze the autoradiogram, a sample containing the radioactive isotope is brought into contact with the photoemulsion and kept in position there for some time (up to several days), depending on the emission type and intensity. Then the autoradiogram is developed and analyzed by means of an optical or *electron microscope*. In the latter case, the object under study is a carbon *replica* with emulsion deposited on its surface. Autoradiography is used in the *tracer* analysis method for studying the distribution of doping elements and impurities in various materials, for measuring their mobilities, etc.

AUTOWAVES

Nonlinear waves in an active medium, in which the energy dissipation is compensated by an applied external source; the analog of the *self-oscillations* for distributed oscillatory systems.

As distinct from auto-oscillations, when the system oscillates as a whole, in autowave processes the excitation is transferred between neighboring elements of the medium (the term *autowaves* was introduced by R.V. Khokhlov). There is a great variety of autowave process types in non-equilibrium systems. They are described by systems of non-linear differential equations; a typical form of the equation for the variables U_i, e.g., temperature, particle concentration (*quasi-particles* in solids), electric field, etc., is

$$\frac{\partial U_i}{\partial t} = D_i \nabla^2 U_i + F_i(U),$$

where t is the time, and ∇^2 is the Laplacian. The first term describes the evolution of U_i relative to the spatial spreading of this nonuniformly distributed variable (if U_i is the particle concentration, then D_i is the *diffusion coefficient*, if U_i is the temperature, then D_i is the *thermal conductivity coefficient*, etc.). The functions $F_i(U)$, where U is a totality of the all the individual U_i, describe the non-linearity of the medium, and the interaction between different components of U. These functions include the permanent energy source, which covers the energy dissipation expenditure. There are also more complicated sets of equations describing autowave processes.

Self-organization is possible (see *Synergism*) in autowave media, i.e. the spontaneous regularization of both the temporal evolution of the system (rise of periodic or more complex oscillations) and the spatial distribution of its characteristics (emergence of *dissipative structures*). However, complete randomization of autowaves often occurs. Autowave processes are widespread in nature, occurring in many biological systems, and in physical and chemical transformations (e.g., combustion and other reactions, gas discharges). In solids, autowave modes are often realized in the excitation of the electron subsystem: in *semiconductors* by an electric field, by heating, by the injection of charge carriers (*thermoelectric domains*, *Gunn effect*, *recombination waves*, electron–hole drops, etc.); in *phase transitions*, e.g., a metal–semiconductor transition, in *ferroelectric phase transitions*, in active non-linear optical solid media (*lasers*, *real-time holography*), etc. The autowave processes in solids are characterized by a variety of spatial–temporal scales (*diffusion length* 10^5–10^{-7} cm, frequency 10^{-2}–10^{10} Hz), and mechanisms of the excitation transfer (fluxes of *electrons* and *holes*, *excitons*, *magnons*, *phonons*, *photons*).

Applications of autowaves are extensive, including, for example, optical, radio-, and UHF amplifiers and generators; generation of ultra- and hypersound; recording and processing of information; image recognition; processes (propagation of nerve impulses and regimes of cardiac muscle operation), etc.

AVALANCHE OF PHONONS

See *Phonon avalanche.*

AVOGADRO NUMBER (A. Avogadro, 1811)

The number of molecules (*atoms*, *ions*) in one mole, identical for any substance in any *state of matter*, $N_A = 6.02214 \cdot 10^{23}$ mole^{-1}. One *mole* (gram-molecule) is the mass of a substance in grams, numerically equal to its atomic or molecular weight. The unit of molecular weight is $1/12$ of the mass of the carbon isotope ^{12}C. In the SI system, the mole is accepted as the unit of the quantity of a substance.

AXIAL SYMMETRY

The symmetry of a body, which comes into coincidence with itself when rotated through an arbitrary angle about some axis. Such a *rotation symmetry axis* (∞) pertains to cones and cylinders with the *limiting symmetry groups* ∞, $\infty 2$, ∞m, $\frac{\infty}{m}$, $\frac{\infty}{m}m$ corresponding to them.

AZBEL–KANER RESONANCE

The same as *Cyclotron resonance in metals.*

Bb

BACKSCATTERING of high-energy particles

Occurs under bombardment of a *solid surface*, and results from collisions of incoming particles with atoms of the target. In this case the direction of the normal component of the particle momentum may change, which causes some of the particles that have penetrated into the target to escape back through the surface under bombardment. Contributions to backscattering come from both single scattering through large angles, and from multiple scattering. The energy and angular distribution of the scattered particles carry information about the masses of the target atoms; it is the collisions with these atoms in the near-surface region that determine the energy loss by the incident particles. This effect is used for the analysis of the elemental composition of materials, including their impurities. *Ions* with energy $E \leqslant 1$ keV are almost completely backscattered from most solids. At $E \geqslant 10$ keV, a significant number of ions penetrates into the target, thus reducing the efficiency of the backscattering, especially for heavy ions. The energy of the escaped atoms decreases with increasing depth z where the reversal in direction occurrs, thus allowing an estimate of z. The efficiency of the backscattering method for determining the content of *defects in crystals* (especially impurity and intrinsic *interstitial atoms*) grows significantly for irradiation in the *channeling* regime. Here, one can determine the positions of interstitial atoms in the channels of a monocrystal target with a good accuracy, and detect the disordering of *crystal lattices* at a depth of several micrometers from a surface that is irradiated with ions.

BAINITE, acicular troostite

A structural component of *steel*, a highly disperse mixture of ferrite and *iron* carbide. It was named after E. Bain. In carbon steels, bainite forms during the isothermal hardening process as a result of decomposition of the *austenite* (see *Alloy decomposition*) in the temperature range 250–450 °C. In some alloyed steels (see *Alloying*), bainite can appear during cooling in air. The conditions of decomposition (mainly, thermal conditions) of the overcooled austenite,determine the formation of *higher bainite* or *lower bainite* with specific morphologic and crystalline features. The transition between them occurs at 350 °C. The bainite structure allows one to process steels that have high static, dynamic and cyclic *strength* as well as wear resistance and *elasticity*.

BALLISTIC AFTEREFFECT

An inertia effect occurring when a *magnetic bubble domain* is displaced under the action of a pulse of a nonuniform magnetizing field, and continues its motion at the initial speed after the pulse termination.

The path covered can be much greater than the domain displacement during the pulse action. The ballistic aftereffect is observable only when the magnetizing field gradient along the domain diameter exceeds a certain critical value. The cause of the ballistic aftereffect is as follows: in the overcritical regime, there appears a great number of horizontal and vertical *Bloch lines* in the *magnetic domain wall* of the magnetic bubble domain; these lines are dynamically stabilized and accumulate considerable mechanical momentum. After turning off the nonuniform magnetic field, the *magnetization* distribution in the domain wall becomes unstable, and the process of "untwisting" of the Bloch lines begins. The decrease of accumulated mechanical momentum is compensated for by the motion of the domain as a whole. The magnetic bubble domain eventually comes to a stop due to viscous friction forces (see *Domain wall drag*).

BALLISTIC PHONONS

Nonequilibrium acoustic *phonons* propagating unhindered in crystalline solids, usually in straight line paths without scattering by lattice *defects*, free current carriers and *impurity atoms*, and also free of phonon–phonon scattering at low temperatures. A typical distance that phonons of frequency $<10^{12}$ Hz can travel ballistically in high quality single crystals at low temperatures ($T \leqslant 4.2$ K) is 0.1–1 cm. A source of ballistic phonons can be a thin film heater placed at one side of a crystal. A thin film thermoresistor (bolometer) placed on the opposite side usually can serve as a detector of ballistic phonons.

BALLISTIC REGIME

The conditions under which the free motion of *quasi-particles* in a solid takes place under the action of external fields only.

The ballistic regime can be achieved in a region with dimensions less than a *mean free path* of quasi-particles. In the ballistic regime the sample size can act as the effective mean free path. Under conditions of the ballistic regime, *size effects* become highly pronounced, e.g., space quantization of the electron motion in finite size samples. Appearing in the ballistic regime are physical phenomena determined by the dynamics of free particle motion and therefore sensitive to the type of their *dispersion law*, for example, the longitudinal and transverse focusing of electrons in the magnetic field in a metal at low temperatures, or the focusing of phonons in insulating *monocrystals*.

BALLISTIC TRANSPORT

Charge carrier transfer in crystals with characteristic dimensions smaller than or comparable to the mean free path of the *current carriers*.

Characteristic dimensions l can be exemplified by the thickness of the active region in *semiconductor diodes*, the thickness of the base in *bipolar transistors*, the gate length or the source–drain distance in *field-effect transistors*, etc. For $l \leqslant l_0$ (where l_0 is the *mean free path* of current carriers), the carrier motion in the active region of semiconductor elements occurs free of scattering, or with very few scattering events. The former case is referred to as ballistic transport, and the latter *quasi-ballistic transport*. The realization of the condition $l \leqslant l_0$ depends on the character and dimension of the semiconductor structure, the carrier energy, the temperature, the concentration of *impurity atoms* and *defects*, the *electron–phonon interaction* constants, and the applied electric field. Depending on these factors, the critical value l, i.e. the length at which the carrier motion takes on the character of ballistic transport, varies from several tenths of a micron to tens of microns. In pure *metals* at low temperatures, this length can reach several millimeters.

A salient feature of ballistic transport is the nonequilibrium of current carriers due to a high operating voltage U as compared to the "diffusion" voltage k_BT/e. Under conditions of ballistic transport, the potential distribution and the *current–voltage characteristic* are determined by the *injection* of electrons (holes) that results in the violation of Ohm's law. In the ordinary case, the dependence of the current density j on the voltage U is described by the "three halfs law" ($j \propto U^{3/2}$) as in the case of a vacuum thermoemission *diode*. There are also nonmonotonic and N-shaped current–voltage characteristics observed.

Ballistic transport can be implemented also by the optical excitation of carriers. Under such conditions, the optically excited charge carriers follow a non-Maxwellian velocity distribution: there exists a prevailing direction of their motion determined by the momentum of the exciting photons. This provides a flux of excited carriers, its parameters depending on features of the *band structure* of the semiconductor, as well as the spectral composition and the polarization of the exciting light. This flux manifests itself in some special *photoelectric phenomena* (ballistic photoelectromotive force, ballistic photomagnetic effect, and so on). The ballistic transport possesses some features suggesting the possibility of producing generators, amplifiers and other devices of submillimeter range as well as the high-speed elements of digital circuits with operation times of 10^{-11}–10^{-12} s. The high-frequency properties of semiconductor structures under the conditions of ballistic transport offer a possibility for the control of the real and imaginary parts of the impedance at certain frequencies; they are also highly nonlinear.

BAND BENDING

A change in the position of an *energy band* edge (see *Band structure*) near an energy surface region (or in the vicinity of a *heterojunction*) due to the electrostatic potential arising from charge redistribution at the boundary. Band bending occurs in *semiconductors* and *semimetals* with small concentrations of free carriers, so the *screening radius* is large compared to the lattice constant. The bending is utilized in various devices, modulating the properties of the near-surface region by the charges from *dangling bonds* at the interface, by impurities and defects that are present in this region, and also by an external field. If the potential difference at the boundary exceeds the *band gap* width the band bending results in the appearance of an *inversion layer* which is widely used in *microelectronics*. The bending in the near-surface electrical field also affects the optical properties of materials, which allows one to utilize the sensitive electrooptical modulation method for investigating the band structure and surface properties of semiconductors.

BAND GAP, forbidden energy zone

The range of energies where there are no allowed electron states in an *ideal crystal* with energy bands. For *semiconductors* and *insulators*, the energy region between the highest level (top) of the *valence band* and the lowest level (bottom) of the *conduction band* is normally meant when referring to the band gap. The energy difference ΔE between these extreme energy values is called the band gap width. In *many-valley semiconductors* the location in *k-space* of the conduction band minimum may not coincide with that of the valence band maximum, and in these cases the minimal ΔE value is called an indirect band gap width. The widths ΔE in different materials vary from zero (see *Gapless semiconductors*) to several electronvolts, e.g., up to 10–20 eV in insulator crystals. They are typically much smaller in semiconductors, being 1.43 eV for the direct band gap of GaAs, and 1.11 eV for the indirect band gap of Si. See also *Band theory* of solids.

BAND MAGNETISM

Term used for describing *ferromagnetism* or *antiferromagnetism* of delocalized electrons, or more precisely, of the electrons which form a broad partially filled band with $\Gamma, \zeta \geqslant \Delta\varepsilon$, where Γ is the band width, $\Delta\varepsilon$ is the exchange splitting at the *Fermi level*, and ζ is the chemical potential referenced to the band edge. The term *itinerant magnetism* refers to a phenomenon opposite to the case of *magnetic ions* ($\Gamma < \Delta\varepsilon$). The $3d$-metals, their alloys, and a group of intermetallic compounds (see *Ferromagnetism of metals and alloys*) are classified as *band ferromagnets*. Band magnetism was originally described by E.C. Stoner (1938) who generalized the *Weiss theory* for the case of delocalized electrons with the *dispersion law* for the energy $\varepsilon = \varepsilon(\boldsymbol{p})$, where the momentum is $\boldsymbol{p} = \hbar\boldsymbol{k}$, and $\boldsymbol{k}$ is the wave vector. The *exchange interaction* of electrons is replaced by the action of a *molecular field* $U = -\theta\sigma\kappa$, where θ is the magnetic field parameter; σ is the *Pauli matrix*, $\boldsymbol{\kappa} = \boldsymbol{m}/(\mu_B n_m)$; μ_B is the Bohr *magneton*; $\boldsymbol{m}$ is the *magnetization*, and n_m is the density of electrons responsible for the magnetism (magnetic electrons). The self-consistency equations in an external magnetic field $\boldsymbol{B}$ at temperature T are described by *Stoner relations*

$$\boldsymbol{m} = \mu_B \int \frac{d^3\boldsymbol{p}}{(2\pi\hbar)^3}\{n_+(\boldsymbol{p}) - n_-(\boldsymbol{p})\},$$

$$n_m = \int \frac{d^3\boldsymbol{p}}{(2\pi\hbar)^3}\{n_+(\boldsymbol{p}) + n_-(\boldsymbol{p})\}, \qquad (1)$$

$$n_\pm(\boldsymbol{p}) = \left\{\exp\left[\frac{\varepsilon(\boldsymbol{p}) \mp \theta\kappa \mp \mu_B B - \zeta}{k_B T}\right] + 1\right\}^{-1}.$$

The condition for ferromagnetism in the *Stoner model* is the *Stoner criterion* $g\theta > n_m$ (or equivalently $\kappa > 0$); where g is the *density of electron states* at the *Fermi surface*. The drawbacks of the Stoner model have not yet been completely overcome. One of them is the disregard of the Bloch factor in the wave function of the electrons, the factor having the period of the *crystal lattice*, which precludes the quantitative treatment of the situation $\zeta > \Delta\varepsilon$, and thus restricts quantitative results to the case of so-called *weak band ferromagnets* ($\kappa \ll 1$). Another drawback involves the molecular field approach that is inapplicable in the neighborhood of the *Curie point* T_C (near point $T = T_C$, $B = 0$ in the (T, B) plane).

Following from Eqs. (1) is the so-called *Wohlfarth isotherm* (E.P. Wohlfarth, 1968), which is the dependence of the magnetization of a band ferromagnet on B and T:

$$\left(\frac{m(B,T)}{m_0}\right)^3 - \frac{m(B,T)}{m_0}\left(1 - \frac{T^2}{T_C^2}\right) = \frac{2\chi_0 B}{m_0}, \quad (2)$$

where $m_0 \equiv m(0,0)$, and χ_0 is the longitudinal (in $\boldsymbol{B}$) static *magnetic susceptibility* at $T = 0$. Eq. (2) is quantitatively valid for weak band ferromagnets, e.g., $ZrZn_2$, ScIn, Ni_3Ga, and a number of Fe–Ni–Mn alloys. The quadratic dependence of the spontaneous magnetization on the temperature, which follows from Eq. (2), viz. $m_0^2 - m^2(T) \propto T^2$, was also experimentally observed in *strong band ferromagnets* ($3d$-metals and alloys).

The theory of *magnons* in band ferromagnets is based on the work of A.A. Abrikosov and I.E. Dzyaloshinskii. The temperature dependence of the magnon spectrum in band ferromagnets was studied by V.P. Silin and others.

If the Stoner criterion (1) is not met then the metal is paramagnetic. If, however, $|\kappa| \ll 1$, then *strong band paramagnetism* occurs (for band antiferromagnetism see *Antiferromagnetic metals, Spin density wave*).

BAND STRUCTURE

This is the energy spectrum (plots of average energy E versus wave number k) of single-electron states represented in the form of a *dispersion law* of electrons $E_{\boldsymbol{k}\lambda}$, where $\boldsymbol{k}$ is the *quasi-wave vector*, λ is a quantum number called the *band index* or *band number*. Occasionally, electronic wave functions are also included in this definition. The band structure is a key notion of the *band theory of solids*. See also *Band structrure computation methods*.

BAND STRUCTURE COMPUTATION

See *Linear methods of band structure computation*.

BAND STRUCTURE COMPUTATION METHODS

Procedures and algorithms for solving the single-particle Schrödinger equation in a *periodic potential* by using modern computers. The construction of a crystalline potential, self-consistent in the sense of *density functional theory*, is an important part of band structure computation. Pseudopotential methods generally incorporate a weak pseudopotential; and full-electron methods place no restrictions on the potential strength.

Early band structure computation methods had been developed even before high-speed computers were available. These were the *augmented plane waves method* (APW), *Korringa–Kohn–Rostoker method* (KKR), *orthogonalized plane waves* method (*pseudopotential method*, version I), and *tight binding method*. The APW and KKR methods are designed for a potential of a special form. The further development of band structure computation methods was chiefly determined by the emergence of *linear methods of band structure computation*, e.g., linear combination of atomic orbitals (LCAO), linear APW, etc., and generalizations of the methods of the APW and KKR families to potentials of arbitrary form. Standing apart are methods related to the tight binding approach, viz. the *linear combination of atomic orbitals* method (Gaussian orbitals and Slater orbitals). One can notice their connection to the methods of the KKR theory. The time required for computing the *band structure* is determined, mainly, by the number of atoms in a *unit cell*. In most of the methods the computational time grows approximately proportional to the cube of this number.

BAND STRUCTURE, SURFACE

See *Surface band structure*.

BAND THEORY of solids

A theory describing a crystalline metal or *solid* as a system of noninteracting electrons, moving in a *periodic potential*. The potential is assumed to arise from nuclei or positive ions screened by electrons. The effects of mutual interactions are considered to be included in the effective potential. The Bragg scattering of electrons in the crystal potential results in a complicated dependence of electron energies on *quasi-momenta* (so-called *band structure*), with the appearance of *band gaps*, i.e. energy gaps where the electrons have no allowed states.

In the ground state, at zero temperature, all electron states with energies lower than the chemical potential (referred to in band theory as the *Fermi energy* E_F, or *Fermi level*) are occupied,

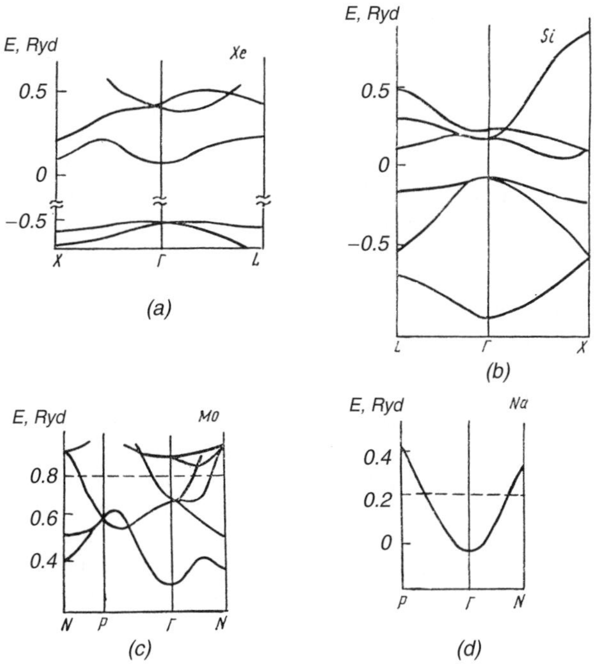

Energy bands of (a) the insulator crystalline xenon, (b) the semiconductor silicon, (c) the transition metal molybdenum, and (d) the nearly free electron metal sodium.

and those with higher energies than the chemical potential are empty. In *semiconductors* and *insulators* the chemical potential lies within the gap, so there are no allowed states with the energy E_F itself. There are such states in a *metal*, and they form a surface in ***k**-space*, called the *Fermi surface*. An important parameter of the electronic spectrum is the *density of electron states*, which indicates the number of allowed states per unit energy at a given energy. The *density of states* at the Fermi level is of special importance.

The main task in band theory is to solve the one-electron Schrödinger equation in a crystal potential with translational symmetry. The latter imposes certain constraints on a solution ψ_k with quasi-momentum $\boldsymbol{k}$: $\psi_{\boldsymbol{k}}(\boldsymbol{r} + \boldsymbol{R}) = \psi_{\boldsymbol{k}}(\boldsymbol{r})e^{i\boldsymbol{k}\boldsymbol{R}}$, where $\boldsymbol{R}$ is a lattice translation period (*Bloch theorem*). Such solutions are called *Bloch wave functions*. The requirements of *crystal symmetry* also impose other constraints on the wave functions and the electron energy spectrum, and these are often used to simplify the solution of the Schrödinger equation. However, in the vicinity of nuclei the influence of crystal symmetry is not pronounced, and the potential is close to an atomic one. Hence, it is possible to speak about electronic states (or bands) of predominantly s-, p- or d-type character, similar to the classification of states according to the atomic angular momentum.

Band theory permits the classification of solids according to the qualitative features of their band structure. In insulators the allowed bands do not overlap, and some are completely full (*valence bands*), while the remainder are empty (*conduction bands*). Semiconductors differ from insulators only by their small (fraction of an electronvolt) width of the forbidden gap. In metals the Fermi level lies within the allowed band, or more commonly, the allowed bands overlap one another. In *semimetals*, the overlap of these bands also occurs, but to a lesser extent. There is also a separate class of *gapless semiconductors*, distinguished by the width of the forbidden gap becoming zero at one or several points of the *Brillouin zone*.

Quantitative band theory is based on the numerical solution of the Schrödinger equation in a crystal. Initially used for this purpose were the *nearly-free electron approximation* and the *tight binding method*, which are valid within the limits, respectively, of a weak crystal potential, and of a potential consisting of isolated wells separated by high barriers. Later the *pseudopotential method* was developed, which expanded the applicability of the method of nearly-free electrons. More widely used are non-pseudopotential methods, capable of dealing with more generalized types of crystal potentials. With the appearance of *density functional theory*, a formal basis emerged for the transformation in the effective potential from a system of strongly interacting electrons to that of noninteracting particles, i.e. to band theory proper. *Linear methods of band structure computation* have been elaborated, which permit solving the Schrödinger equation in a crystal 2 to 3 orders of magnitude faster than by using classical methods of computation. Typical examples of band structures are shown in Fig. (see previous page): (a) an insulator (crystalline Xe) with a narrow valence band, a wide gap, and bands well described by the tight binding method; (b) a semiconductor (Si) with its indirect gap width small compared to that of the valence band; (c) a transition metal (Mo) with valence electrons comprising the group of bands formed predominantly by s- and d-electrons; where the Fermi level lies inside the d-band with a high density of states; and (d) a simple metal (Na) with s-electrons forming an almost parabolic energy band with a low density of states, in accordance with the model of nearly free electrons, where X, Γ, L, N, and P are symmetry points of the Brillouin zone.

BAND THEORY RELATIVISTIC EFFECTS

See *Relativistic effects in band theory of metals.*

BAND-TO-BAND TUNNELING

See *Interband tunneling.*

BARDEEN BARRIER (J. Bardeen)

A potential barrier which emerges at the boundary between two bodies in contact (e.g., two *solids*, a solid and a liquid or gas) in the presence of a significant concentration of *surface electron states* at the interface. In the limit of high concentrations, the Bardeen barrier height does not depend on the *contact potential difference* due to screening of the Coulomb interaction by the charges in the surface electron state. In practice, the *Fermi level* of a Bardeen barrier is fixed at the surface (at the interface). In most cases, this term is used with respect to contacts between a semiconductor and a metal (called a *Schottky barrier*), an insulator, or another semiconductor.

BARDEEN–COOPER–SCHRIEFFER THEORY, BCS theory (J. Bardeen, L. Cooper, J. Schrieffer, 1957)

The first and the currently accepted microscopic theory of *superconductivity* based on the presence of so-called *Cooper pairs* in a superconductor at low temperatures $T < T_c$ (where T_c is the critical temperature). The Cooper pairs consist of electrons with antiparallel momenta and spins (see *Cooper effect*) that form as a result of their attraction via the *electron–phonon interaction*. The starting point of the theory is the model Hamiltonian

$$H_{BCS} = \sum_k 2\varepsilon_k c_{k\sigma}^+ c_{k\sigma} + \sum_{k,k'} V_{kk'} b_{k'}^+ b_{k'}, \quad (1)$$

where ε_k is the initial energy of the conduction electrons with momentum $\hbar\boldsymbol{k}$, $c_{k\sigma}^+$ and $c_{k\sigma}$ are the creation and annihilation operators of electrons, respectively, b_k^+ and b_k are the same for the Cooper pairs (see *Second quantization*), and $V_{kk'}$ are the matrix elements of the effective electron–electron interaction. The minimization of the energy of the ground state of the electron subsystem of the superconductor yields the following expression for the *quasi-particle* spectrum through the use of the variational method within the BCS framework:

$$E_k = \left[(\varepsilon_k - \mu)^2 + \Delta_k^2\right]^{1/2}, \quad (2)$$

where μ is the chemical potential (*Fermi level*) of the degenerate electrons in the metal, and Δ_k is the *energy gap* determined by the following nonlinear equation:

$$\Delta_k = -\sum_{k'} V_{kk'} \frac{\Delta_{k'}}{2E_{k'}} \tanh \frac{\Delta_{k'}}{2k_B T}. \quad (3)$$

A nontrivial solution of Eq. (3) $\Delta_k \neq 0$ exists only for the condition $V_{kk'} \equiv V < 0$, i.e. so in a certain range of energy (momentum) of the *Fermi surface* of a metal, the effective attraction due to the electron–photon interaction exceeds the screened Coulomb repulsion (see *Electric charge screening*). In the weak coupling limit Eq. (3) yields the exponential BCS formula for the gap Δ at $T = 0$:

$$\Delta = 2\hbar\omega_{\rm D} \exp\left\{-\frac{1}{N(0)V}\right\}, \qquad (4)$$

(where $\omega_{\rm D}$ is the Debye phonon frequency, and $N(0)$ is the electron density of states on the Fermi surface in a normal metal), and for the critical temperature $T_{\rm c}$ of the transition to the superconducting state we have:

$$T_{\rm c} = \frac{2{\rm e}^{\gamma}}{\pi}\Theta_{\rm D} \exp\left\{-\frac{1}{N(0)V}\right\}, \qquad (5)$$

where $\Theta_{\rm D} = \hbar\omega_{\rm D}/k_{\rm B}$ is the *Debye temperature*. Eqs. (4) and (5) give the *BCS relationship* $k_{\rm B}T_{\rm c} = {\rm e}^{\gamma}\Delta_0/\pi = 0.567\Delta_0$, where $k_{\rm B}$ is the *Boltzmann constant* and $\gamma = 0.577\ldots$ is the Euler–Mascheroni constant. The BCS relationship is valid solely for superconductors with weak coupling, whereas for strongly-coupled superconductors $k_{\rm B}T_{\rm c} < \gamma\Delta(0)/\pi$. The latter is due to the effects of the retardation of the electron–phonon interaction, which cause the gap Δ to depend on the energy of the quasi-particles and, hence, the value of Δ at $T = 0$ increases compared to the value $\pi k_{\rm B}T_{\rm c}/{\rm e}^{\gamma}$.

The BCS theory has explained the principal physical properties of superconductors in terms of the presence of the energy gap in the quasi-particle spectrum. This includes magnetic vortices, flux quantization, the exponential dependence of the electron *specific heat* on the temperature and its jump in value at the *phase transition* point $T = T_{\rm c}$; the temperature dependence of the *thermodynamic critical magnetic field*; the features of *ultrasound attenuation*; relaxation of nuclear spins, and the absorption of electromagnetic waves in superconductors; the current–voltage characteristics of the superconducting *tunnel junctions*, and so on. The originators of the BCS theory received the Nobel prize in 1972.

BARIUM, Ba

A chemical element of Group II of the periodic system with atomic number 56 and atomic mass 137.33. It consists of a mixture of seven stable isotopes, with ^{138}Ba predominating (71.66%). The electronic configuration of the outer shell is $6s^2$. Successive ionization energies are 5.21 and 10.00 eV. Atomic radius is 0.221; ionic radius of Ba^{2+} is 0.134 nm. Oxidation state is +2. Electronegativity is ≈0.94. In a free form barium is a silvery-white soft *metal*. It has a body-centered cubic lattice with parameter $a = 0.5019$ nm at room temperature; Space group is $Im\bar{3}m$ (O_h^9). Density is 3.76 g/cm^3; $T_{\rm melting} = 998$ K; heat of melting is 7.66 kJ/mole; heat of sublimation is 205 kJ/mole; $T_{\rm boiling} = 1912$ K; heat of evaporation is 175 kJ/mole; specific heat is 284.7 kJ·kg^{-1}·K^{-1}; Debye temperature is 110 K; coefficient of linear thermal expansion is ≈19.5·10^{-6} K^{-1} (at 273 to 373 K); adiabatic bulk modulus is 9.575 GPa; Young's modulus is 15.57 GPa; rigidity is 6.33 GPa; Poisson ratio is 0.229 (at 293 K); Brinell hardness 0.042 GPa; Mohs hardness is 2; ion-plasma frequency is 4.52 THz, Sommerfeld linear electronic specific heat coefficient γ is 2.72 mJ·mole^{-1}·K^{-2}; electric resistivity is 357 nΩ·m (at 293 K); temperature coefficient of the electrical resistance is 0.00649 K^{-1}; work function of a polycrystal is 2.49 eV; molar magnetic susceptibility is 20.4·10^{-6} cgs units; nuclear magnetic moment of ^{137}Ba is 0.931 nuclear magnetons. Barium is used in alloys with lead, aluminum, and magnesium. Barium and its compounds are added to the materials intended for protection against radioactive and X-ray radiation.

BARIUM TITANATE, $BaTiO_3$

A colorless solid substance; $T_{\rm melting} \sim 1618$ K; it has the structure of *perovskite* (tetragonal *crystal lattice* with lattice constants $a = 0.3992$ nm, $c = 0.4036$ nm); density is 6.08 g/cm^3. Each titanium is surrounded by a somewhat distorted octahedron of oxygens. Chemically it is very resistant: it does not change in air or react with water, diluted acids or alkali; it dissolves in hot strong sulfuric acid. It is an *insulator* with a very high *dielectric constant* (1000 at 20 °C), and one of the most important *ferroelectrics* with a high piezoelectric modulus (see *Piezoelectricity*) and specific optical properties. It is obtained by caking and melting together of titanium dioxide with *barium* oxide, hydroxide or carbonate, or by heating up to 800 to 1200 °C of mixture obtained by common deposition of carbonates or oxalates from solutions. It is used in the form of *monocrystals* and *solid solutions* for manufacturing of capacitors, piezoelectric cells, pyroelectric collectors of radioactive energy, etc.

BARKHAUSEN EFFECT (H.G. Barkhausen, 1919)

A jump-like change of *magnetization* M in magnetically ordered materials under continuous variation of external conditions (e.g., applied magnetic field B).

Barkhausen discovered that clicks could be heard in a telephone supplied by an electron amplifier and connected with a coil containing a ferromagnetic sample during very slow (quasi-static) magnetizing of the latter. The main cause of a Barkhausen effect is the various inhomogeneities of the crystalline structure of a *ferromagnet* (inclusions, *residual stresses*, *dislocations*, *grain boundaries*, the heterogeneity of the composition, and other *defects* in the crystals). When moving in the magnetic field, the magnetic domain walls attach to these inhomogeneities until the increasing field reaches a certain (critical) value. Then the domain wall is released (*Barkhausen jump*), so that its motion turns out to be jump-like, in contrast to a smooth movement in a *perfect crystal*. These jump-like changes are irreversible. Another mechanism of the Barkhausen effect is possible: the nucleation of new phase centers with their subsequent growth. The Barkhausen effect is best observable in *soft magnetical materials* at the steepest part of the *magnetization curve* (of stepped shape in that region, see Fig.). This effect

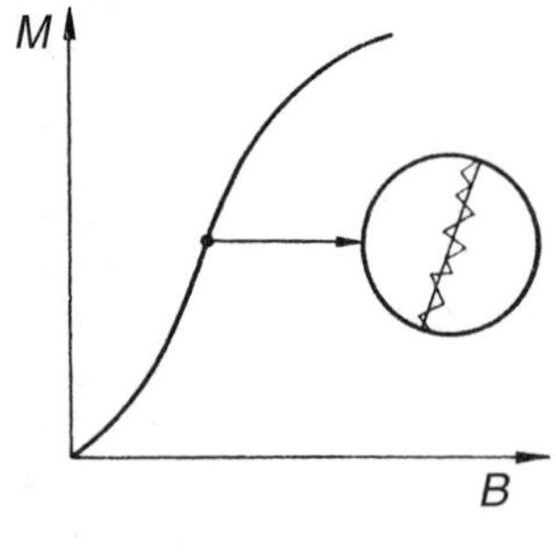

Barkhausen effect.

is highly sensitive to various external factors and structural changes in the sample. The connection of the Barkhausen effect with the *magnetic domain structure*, the main elements and the shape of the hysteresis loop, allows it to be used as a method of investigation of the domain structure of ferromagnets. By analogy with ferromagnets, the appearance of *repolarization jumps* in ferroelectric crystals under variation of an applied electric field is also referred to as a Barkhausen effect.

BARKHAUSEN NOISE

Irregular sequences of jumps in the the *magnetic moment* of a ferromagnetic sample placed in a periodically varying magnetic field (see *Barkhausen effect*). The component is due to fluctuations of specific parameters of *Barkhausen jumps* (the shape and duration, the critical field, the magnetic moment varying at the jump, the number of jumps in a given range). Studies of Barkhausen noise yield substantial information on complicated phenomena involving the rearrangement of the *magnetic domain structure* in *ferromagnets*.

BARNETT EFFECT (S. Barnett, 1909)

Magnetizing of *ferromagnets* under a change in their mechanical moment.

Barnett confirmed *Ampère's hypothesis* that magnetism can arise from rotational motion of charges in a material (*molecular current hypothesis*). In a body with no *atomic magnetic structure*, the intrinsic *magnetic moment* is proportional to the angular velocity and oriented along the rotation axis. In some cases, the Barnett effect allows one to determine the *gyromagnetic ratio* or *Landé g-factor* of atoms of some materials. In particular, this ratio for Fe, Co, Ni and their alloys corresponds to the electron magnetic moment and indicates that *ferromagnetism* is mainly due to the electron *spin*. The Barnett effect, along with the converse *Einstein–de Haas effect*, are called *magnetomechanical phenomena*. *Magnetic resonance* methods are much more precise for determining g-factors.

BARODIFFUSIONAL EFFECT

An effect of hydrostatic *pressure* on the diffusion *healing of defects*, or the retardation of pore growth (at low pressures), and on the decrease of the *diffusion coefficient* in crystals (at *high pressures*).

At low pressures, barodiffusional effects are caused by the influence of pressure on the equilibrium between the pore and the vacancy subsystem (see *Vacancy*) taking into account oversaturation $\Delta = c_\mathrm{v} - c_{\mathrm{v}0}$ in a *crystal* (where c_v and $c_{\mathrm{v}0}$ are the actual and equilibrium concentrations of vacancies, respectively). Under the characteristic pressure $P_k = 2\alpha/R_{0k}$ (where α is the surface tension, R_{0k} is the critical radius of a viable pore embryo in the absence of pressure), the nucleation and the growth of pores are forbidden. At $R_{0k} = 10^{-5}$ cm, $P_k \approx 10^2$ atm. Such pressure has a significant effect on the kinetics of the diffusional *coalescence* of the pore ensemble due to the transformation of some of the growing pores into dissolving ones. Under high pressures, the barodiffusional effect is caused by the decrease of c_v and by the increase of the energy outlay for an elementary migration event with growing pressure. There is an exponential diffusion coefficient dependence

$$D_P = D_{P0}\exp\left(-\frac{\omega_\mathrm{A}P}{k_\mathrm{B}T}\right),$$

where $\omega_\mathrm{A} = \omega_\mathrm{v} + \omega_\mathrm{m}$ is the *activation volume*, $\omega_\mathrm{v} < \omega$ is the vacancy volume dependent on the extent of relaxation of the atoms surrounding the vacancy, ω is the atomic volume, ω_m is the volume of the activated complex at a crossing point between two neighboring positions. It follows from the dependence $D_P(P)$ that at a characteristic pressure $P^* = k_\mathrm{B}T/\omega_\mathrm{A}$ the value of D_P reduces e times. For metals, $P^* \approx 10^4$ atm.

BARRETT FORMULA (J.H. Barrett)

An expression describing the temperature dependence of the *dielectric constant* of ferroelectrics in the *paraelectric phase* in a temperature (T) range that includes the quantum region:

$$\varepsilon(T) - \varepsilon_\infty = C\left\{\left(\frac{\hbar\Omega_0}{2k_\mathrm{B}}\right)\coth\left(\frac{\hbar\Omega_0}{2k_\mathrm{B}T}\right) - T_\mathrm{c}\right\}^{-1},$$

where Ω_0 is the frequency of zero-point vibrations, T_c is the "high-temperature" value of the Curie–Weiss temperature. At $T > \hbar\Omega_0/k_\mathrm{B}$, the Barrett formula transforms into the classical expression since $\varepsilon(T)$ satisfies the *Curie–Weiss law*. The system behavior at $T \leqslant \hbar\Omega_0/k_\mathrm{B}$ depends on the ratio of T_c and $\hbar\Omega_0/(2k_\mathrm{B})$. If $T_\mathrm{c} > \hbar\Omega_0/(2k_\mathrm{B})$, the system has a low-temperature *phase transition* at $T = T_\mathrm{cL} < T_\mathrm{c}$. There is no ferroelectric phase transition, if $T_\mathrm{c} < \hbar\Omega_0/(2k_\mathrm{B})$, in this case the crystal is a *virtual ferroelectric*. The Barrett formula is used to explain the observed low-temperature values $\varepsilon(T)$ in $SrTiO_3$ and $KTaO_3$, but is only valid when the *soft mode* frequency is high relative to acoustical frequencies. In the general case, taking into account the interaction of *optical vibrations* and *acoustic vibrations* results in a characteristic modification of the Barrett formula in a low-temperature range.

BARRIER

See *Bardeen barrier, Lomer–Cottrell barrier, Potential barrier, Schottky barrier, Local barriers, Surface barrier, Surface barrier structures.*

BARRIER CAPACITANCE, junction capacitance

Electric capacitance caused by the existence of a potential barrier in a *semiconductor junction*, in a *heterojunction*, at a *metal–semiconductor junction*, and so on. Such a system possesses a high-ohmic layer that is depleted of current carriers and situated between high-conduction layers, i.e. is similar to a capacitor. The electric charge in the depletion layer is concentrated at *local electronic levels*. The barrier capacitance value is determined by the thickness of the depletion layer and the *dielectric constant* of the crystal. A layer with an increased current carrier concentration also has a barrier capacitance, e.g., an *accumulation* layer in the *metal–insulator–semiconductor structure.*

BARRIER INJECTION TRANSIT TIME DIODE (BARRITT diode)

A superhigh-frequency (SHF) *semiconductor diode* consisting of a semiconductor plate with contacts made of *semiconductor junctions* (p^+–n–p^+, p^+–n_1–n_2–p^+, where $n_1 > n_2$, and other structures), *Schottky barriers* (m–n–m structures, m is *metal*), or their combinations. With the bias voltage applied to the diode, one contact operates in the forward direction and the other in the reverse direction. A BARRITT diode operates in a punch-through mode where the depletion regions of the forward- and reverse-biased contacts overlap, so that a small change of the voltage across the diode results in the current growing by several orders of magnitude. Dynamic *negative resistance* in the BARRITT diode arises in a frequency range ω where the phase delay of carrier transit through the depletion layer of the reverse-bias junction is $\theta = \omega\tau \approx 3\pi/2$ (τ is carrier transit time). The barrier drift time through the depletion layer of the reverse biased junction depends on the width of the depletion layer and the saturation velocity. The effect is due to the phase shift between the SHF electric field and the conduction current. The required phase shift is caused: (i) by the phase delay of the *injection* of the minority charge carriers through the forward biased *p–n* junction, and (ii) by the finite time of the *current carrier drift* through the depletion layer. BARRITT diodes are used for signal generation and amplification in the frequency range from 3 to 15 GHz. The main advantage of these diodes is a low noise level.

BARRITT DIODE

The same as *Barrier injection transit time diode.*

BASAL PLANE of crystals

A plane containing equivalent crystallographic (coordinate) axes x and y. Ordinarily it is the plane that forms the bottom of the conventional unit cell, with the index (001) in tetragonal and (0001) in trigonal and hexagonal crystals (see *Crystallographic system*). The spatial arrangement of the faces and the sides of crystalline polyhedra is described with the help of symbols relative to these axes. The faces that intersect only one axis and possess numbers (100), (010) and (001) (see *Crystallographic indices*) are called *basal (coordinate) faces*.

BASALT

An extrusive and congealed dark gray or black volcanic rock consisting mainly of plagioclase, pyroxenes and often olivine, typically with an admixture of magnetite or ilmenite.

Color is dark sulfur-like to black. Basalt forms frequently have a rod-like or a "pencil"-like separation. Holocrystalline basalt is called dolerite; basalt of vitreous texture is called hyalobasalt. The almond-shaped basalts, or mandelstones, are porous varieties. The *pores* (voids) are of a rounded, ellipsoidal, more rarely elongated, tubular shape and are filled with *minerals*, mostly, agate, chalcedony, opal, quartz, or sometimes, amethyst, zeolites, calcite, and so on. The density is 2.8–3.2 $\mathrm{g/cm^3}$; the resistance to compression in a dry state is 300–500 $\mathrm{kg/cm^2}$ (3–5 MPa); electric resistivity $\rho = 10^3$–10^5 Ω·m; porosity is 7.9%, water absorption is 1.07%, and magnetic susceptibility $\chi = (125\text{–}15500)\cdot 10^6$ CGS units. The basalts are used as building, electric insulation, and acid-resistant materials.

BAUSCHINGER EFFECT (J. Bauschinger, 1886)

A partial *dehardening* of single and polycrystalline solids observable at the change in the sign of the *strain*. The effect manifests itself as a reduction of some initial mechanical properties (*elastic limit*, *yield limit*, and so on) when, prior to a test, a sample was loaded in the opposite direction until an insignificant residual strain appears. The Bauschinger effect is observable both in static tests and in the course of fatigue loading (see *Fatigue*). At the same time, its value depends essentially on the kind of material, of the material pretreatment, and the strain conditions. Numerous attempts have been made to explain the Bauschinger effect on the basis of dislocation theory (see *Dislocations*); however, a satisfying agreement with the experimental data has been achieved only in individual cases.

BEER–LAMBERT LAW (also called Bouguer–Lambert–Beer law, P. Bouguer, J.H. Lambert, A. Beer)

A law determining the dependence of the intensity I of a parallel monochromatic light beam on the thickness d of the absorbing matter:

$$I = I_0\,\mathrm{e}^{-kd}, \tag{1}$$

where I_0 is the incident light intensity, k is the *absorption coefficient* ($\mathrm{cm^{-1}}$ or $\mathrm{m^{-1}}$). The latter is related to the dimensionless absorption index κ as $k = 4\pi\kappa/\lambda$, where λ is the wavelength. Eq. (1) is used to evaluate the absorptivity of a solid or liquid layer. For a dilute solid or liquid solution of concentration n, Eq. (1) takes the form:

$$I = I_0\,\mathrm{e}^{-\chi n d}, \tag{2}$$

where χ is the molecular absorption index, cm^2. In the particle beam radiation of solids the term *absorption cross-section* is often employed for χ, and k is referred to as the absorption factor. Sometimes the Beer–Lambert law is expressed in terms of common logarithms (i.e. a power of 10) as

$$I = I_0 \cdot 10^{-\varepsilon c d}, \qquad (3)$$

where c is the molar concentration, and ε is the decimal molar absorption index, also called the *absorption or extinction coefficient.*

BEILBY LAYER (G.T. Beilby, 1901)

A glass-like *surface layer* that appears during mechanical processing (grinding, *polishing, forging,* punching, cutting, and so on) at the surface and over the *grain boundaries* of crystalline solids (*metals*, quartz, various *minerals* and chemical compounds). A Beilby layer has its normal crystal structure obliterated. Long term grinding or polishing is unable to remove the layer, but rather causes its thickness to grow. A Beilby layer is removable by chemical methods. Its thickness ranges from a several to 100 or more microns. A Beilby layer along grain boundaries imparts great *strength* to a material, it is characterized by pronounced *hardness*, and specific optical, electric, and magnetic properties. A Beilby layer crystallizes under heating at a temperature between 0.4 and 0.6 that of the *melting temperature.*

BENDING VIBRATIONS of a crystal lattice

Vibrations of chains (layers) of *crystal* atoms in a direction normal to the chains (layers). Bending vibrations exist in those crystals where the bonding between the chains (layers) is weak compared to that of the atoms within the chain (layer). The linear *dispersion law* for bending vibrations is valid only for the lowest values of the wave vector along the chain (layer). At large values of the wave vector, a quadratic dispersion law similar to that for the vibrations of strings and *membranes* applies.

BERKELIUM, Bk

An *actinide* element of Group III of the periodic system with atomic number 97 and atomic mass 247.07.

There are 11 known radioactive isotopes with mass numbers 240, 242 to 251 (no stable isotopes have been discovered). Electronic configuration of outer shells is $5f^7 6d^2 7s^2$. Ionization energy is 6.30 eV; atomic radius is 0.174 nm; radius of Bk^{3+} ion is 0.0955 nm, of Bk^{4+} ion is 0.0870 nm. Oxidation state is +3 (most stable in solution) and +4 (strong oxidizer). Electronegativity is 1.2. In a free form, berkelium is a silvery-white *metal.* Stable at a temperature below ~1250 K is α-Bk with double hexagonal close-packed structure (hexagonal layers alternate ABAC...; space group $P6_3/mmc$ (D_{6h}^4), basic parameters $a = 0.3416$ nm, $c = 1.1069$ nm at room temperature). Stable above ~1250 K up to $T_{\text{melting}} \approx 1290$ K is β-Bk with hexagonal close-packed structure (space group $Fm\bar{3}m$, O_h^5; parameter $a = 0.4997$ nm at room temperature after hardening). Density is 14.80 g/cm^3; T_{boiling} is about 2880 K. At low temperatures, α-Bk becomes antiferromagnetic with *Néel point* $T_N \approx 22$ to 34 K). In β-Bk, no magnetic ordering has been detected. Berkelium isotopes are used for research in nuclear physics and radiochemistry. Berkelium is highly toxic.

BERNAL'S FLUID MODEL

See *Random close-packing model.*

BERYLLIUM, Be

A chemical element of Group II of the periodic system with atomic number 4 and atomic mass 9.01218.

Natural beryllium consists of the stable isotope ^{9}Be with electronic configuration $1s^2 2s^2$. The ionization energies are 9.323 or 18.21 eV. Atomic radius is 0.113 nm, Be^{2+} ion radius is 0.035 nm. It is chemically active; oxidation state is +2. Electronegativity is 1.47. In a free form, beryllium is a light bright gray *metal,* which is brittle under normal conditions. The properties of beryllium depend on its purity, grain size, texture and degree of anisotropy that depend on its pretreatment. Some properties change with an increase of pressure and temperature, and upon strong neutron irradiation ($>10^{23}$ neutrons/m^2). Beryllium crystallizes in two allotropic modifications. Below 1527 K, α-Be is stable (hexagonal close-packed lattice with parameters $a = 0.2283$ nm, $c = 0.3584$ nm at 300 K and impurity content no more than 1%; space group $P6_3/mmc$, D^4_{6h}). Above 1527 K up to $T_{\text{melting}} = 1560$ K, β-Be is stable (body-centered cubic with parameter $a = 0.2546$ nm at 1527 K; space group $Im\bar{3}m$, O^9_h). Bond energy of pure beryllium is 3.33 eV/atom at 0 K; density is 1.848 g/cm^3 (at 293 K); melting heat is 11.7 kJ/mole; sublimation heat is 314 kJ/mole; $T_{\text{boiling}} = 2743$ K; evaporation heat is 309 kJ/mole; specific heat is 1.80 kJ·kg^{-1}·K^{-1}; Debye temperature is 1160 K. Coefficient of linear thermal expansion is $9.2 \cdot 10^{-6}$ K^{-1} along the principal axis $\underline{6}$ and $12.4 \cdot 10^{-6}$ K^{-1} perpendicular to it (at 300 K). Heat conductivity reduces from 230 W·m^{-1}·K^{-1} to 178 W·m^{-1}·K^{-1} in the temperature range 273 to 323 K; adiabatic coefficients of elastic rigidity of α-Be monocrystal: $c_{11} = 288.8$, $c_{12} = 20.1$, $c_{13} = 4.7$, $c_{33} = 354.2$, $c_{44} = 154.9$ GPa at 298 K (within the range 423 to 1530 K $c_{13} \leqslant 0$ GPa). Bulk modulus is ≈109 GPa, Young's modulus is ≈304 GPa, rigidity is ≈143 GPa; Poisson ratio is ≈0.056 (at 298 K); ultimate tensile strength is 0.196 to 0.539 GPa; relative elongation is 0.2 to 2%. Pressure treatment considerably improves mechanical properties; ultimate strength in stretching direction reaches 0.392 to 0.784 GPa; yield limit is 0.245 to 0.588 GPa; and relative elongation up to 4 to 12%; mechanical properties perpendicular to stretching undergo minimal change. Brinell microhardness is 1.18 GPa; impact strength is 0.001 to 0.005 GPa; transition temperature of commercially pure beryllium from brittle to plastic state is 473 to 673 K. Highly pure beryllium (above 99.9%) with grain size <10 μm looses its red-shortness, becomes easily forgeable and rollable and, at temperatures within the range 873–973 K, becomes superplastic. Coefficient of self-diffusion of α-Be is $32.9 \cdot 10^{-12}$ m^2/s along the $\underline{6}$ principal axis and $53.4 \cdot 10^{-12}$ m^2/s perpendicular to it at 1370 K. Effective cross-section of the thermal neutron trapping is 0.009 barn/atom, their scattering cross-section is 6.1 barn/atom. The properties of irradiated beryllium are restored upon annealing. Ion-plasma frequency is 49.22 THz, Sommerfeld coefficient γ of the linear low temperature electronic specific heat is ≈0.19 mJ·mole^{-1}·K^{-2}. Electric resistivity of α-Be is 35.4 nΩ·m along the principal $\underline{6}$ axis and 31.3 nΩ·m perpendicular to it at 298 K; temperature coefficient of electrical resistance is 0.00628 K^{-1} (at 293 K). The Hall constant is $+24.3 \cdot 10^{-11}$ m^3/C at room temperature ($+77 \cdot 10^{-11}$ m^3/C for the magnetic field applied along the $\underline{6}$ principal axis in α-Be at room temperature). Work function from polycrystal is 3.92 eV; coefficient of secondary electron emission is 0.4 to 0.5 for the primary electron energy 0.07 to 0.2 keV; superconducting transition temperature is 0.064 K. Beryllium is diamagnetic with the molar magnetic susceptibility $9.02 \cdot 10^{-6}$ cgs units at 300 K (the value along the $\underline{6}$ axis is higher than that in the perpendicular direction); nuclear magnetic moment of the ^{9}Be isotope is 1.18 nuclear magnetons. Thus, Be advantageously combines small atomic mass and low density, high specific heat, vaporizing heat, thermal conductivity, heat resistance, elastic modulus, hardness, strength and capability of retaining the stability of dimensions, high electrical conductivity, corrosion resistance, small thermal neutron trapping cross-section and satisfactory stability under irradiation, high permeability for X-rays and intense emission of neutrons under the bombardment by α-particles. Due to all this, beryllium is used in aviation, rocket and space engineering, in underwater instrument making, in reactor material science, radioelectronics, X-ray engineering, metallurgy. Its low plasticity (due in part to impurities) is the main obstacle to its use as a structural material. The processing of beryllium is complicated because of the high toxicity of its volatile compounds in the form of powder and dust.

BERYLLIUM ALLOYS

Alloys based on *beryllium*. The main advantages of beryllium alloys are high specific *strength* and specific rigidity up to temperatures of 600 to 800 °C, high *specific heat* and low neutron trapping cross-section. The main disadvantages are low *plasticity* at room and cryogenic temperatures, and toxicity. Hardness varies with *low-alloyed* (see *Alloying*), and *disperse-hardened alloys* (including those with intermetallide hardening). There are structural low-alloyed ones with the *solid solution* structure, two-phase composite ones with pliable base, and those with a *disperse structure*. The most common are beryllium alloys with *aluminum* (38%), which crystallize in the range 1280 to 650 °C. At temperatures below 645 °C there is a brittle beryllium phase, and pliable aluminum-beryllium *eutectics* (98–99.5% of Al) are formed in such alloys. The strength and plasticity of beryllium alloys are improved by alloying them with *magnesium* (5–8%). Beryllium alloys with a pliable base are easily amenable to *welding*, stamping, *pressing*, rolling (Be–Al alloys at temperatures of 600 to 650 °C, and Be–Al–Mg alloys at temperatures less than 450 °C).

BETA PHASE (β-phase) (introduced by F. Osmond, 1885)

The second phase in polymorphous chemical elements (see *Polymorphism*).

This phase forms from the alpha phase (α-phase) at increasing temperatures. For iron, the term β-phase means the paramagnetic phase (see *Paramagnetism*) with a body-centered cubic lattice that is really the β-phase existing above the *Curie point*. In *alloys*, the β-phase is a *solid solution* based on the beta-modification of a solvent element. The region of existence depends on the nature of doping elements (see *Doping*). For example, the additives of *molybdenum*, *vanadium*, *tantalum* or *chromium* to the element *titanium* expands this region, while the additives of *aluminum*, *boron* or *oxygen* narrow it by increasing the polymorphic transformation temperature. The type and the transformation mechanism of a β-phase are determined by the nature of the components, the chemical composition of alloys, the pretreatment and the cooling rate. In systems without a polymorphic transformation, the term beta phase on the *phase diagram* denotes the second phase on the concentration coordinate. In a number of univalent systems and in some transition metal alloys (e.g., with beryllium, cadmium, aluminum, antimony), the beta phase is one of the types of *electron compounds* with a body-centered cubic lattice (Cu_3Al, CuZn, etc.), a hexagonal close-packed lattice (AgCd, Au_5Sn, etc.), or the complicated cubic lattice of the β-Mn type (Ag_3Al, Au_3Al, etc.). The beta-phase stability is determined by the electron concentration (number of valence electrons in an atom of the crystal) $e/a = 3/2$ (see *Hume–Rothery phases*).

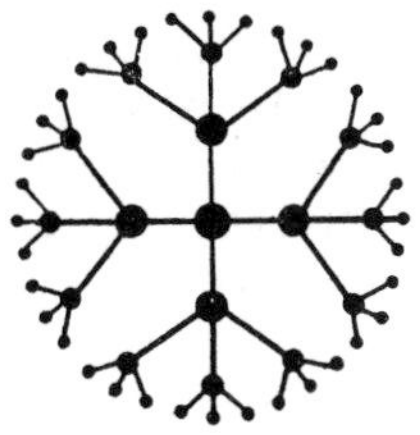

Bethe lattice.

BETHE LATTICE (H.A. Bethe, 1935)

A model structure of an atomic lattice where every site has the same coordination number. All interatomic bonds are equivalent; they exist only between atoms of the first *coordination sphere*; in addition, there are no cycles formed by the bonded atoms. The figure shows a portion of a Bethe lattice with the coordination number 4. The Bethe lattice was introduced to describe thermodynamic properties of Heisenberg ferromagnets (see *Heisenberg model*). This lattice is used when solving problems involving *elementary excitation spectra of disordered solids*, and it allows one to obtain exact solutions. It is also applied to simulate solids by the molecular dynamics method.

BETHE–PEIERLS APPROXIMATION (H. Bethe, 1935; R. Peierls, 1936)

One of the simplest *cluster approximations* used for calculating statistical sums and thermodynamic parameters of *magnetic substances*.

A *cluster* in a Bethe–Peierls approximation is chosen as a central site with the *coordination sphere* of its nearest neighbors surrounding it. Then, the contribution to the statistical sum of the interaction between the central *spin* and its nearest environment is taken into account exactly, while that of the interaction of spins of the first coordination sphere with spins of the remaining coordination spheres is approximated by an effective field. The ordered phase exists as long as the effective field does. For ferromagnetic ordering the field is determined from the condition of the equality of the thermodynamic mean values of the central and peripheral spins of the cluster (a condition of self-consistency). In the *Ising model* for the *Bethe lattice*, the Bethe–Peierls approximation provides exact results. This approximation partially accounts for the *fluctuations* in a spin system, and can describe a number of physical effects not accounted for by the *molecular field* approximation. These are the absence of the long-range order (see *Long-range and short-range order*) at $T \neq 0$ in one-dimensional systems, the existence of short-range order above the transition temperature to an ordered phase, and the appearance of the percolation concentration threshold (see *Percolation theory*) in disordered dilute magnetic materials at $T = 0$. However, like the molecular field approximation, the Bethe–Peierls approximation fails to describe collective excitations (spin waves, magnons etc.) in magnetic materials with vector spins.

BETHE SPLITTING

The splitting of degenerate energy levels of atoms, ions, or molecules under the action of crystalline electric fields of various symmetries (see *Crystal field*).

The latter is lower than the spherical symmetry typical of a free atom so that there is a total or partial removal of the *degeneracy* in a *crystal*, and several closely spaced bands (lines) appear in the absorption or emission spectra instead of a single line. Their maximum number corresponds to the degree of degeneracy. In a weak crystalline field, the *spin–orbit interaction* is retained and the Bethe splitting is less than the multiplet one. In an intermediate field, the Bethe splitting exceeds the multiplet one and remains less than the distance between neighboring multiplets. In this case, the spin–orbit coupling is broken. The Bethe splitting in a strong field exceeds the distance between neighboring multiplets. This uncouples (quenches) the total *orbital angular momentum* from the spin angular momentum, and it can also uncouple the individual spin orbital angular momenta from each other. Bethe splitting is observed in *phonon optical spectra* as well.

BETTI RECIPROCITY THEOREM (E. Betti)

A statement of *elasticity theory* according to which the following equality holds for an elastic body in the equilibrium state when no *bulk forces* are present:

$$\int_S (\boldsymbol{F}\boldsymbol{u}')\,\mathrm{d}S = \int_S (\boldsymbol{F}'\boldsymbol{u})\,\mathrm{d}S.$$

Here $\boldsymbol{F}$ and $\boldsymbol{F}'$ are *surface force* densities, $\boldsymbol{u}$ and $\boldsymbol{u}'$ are the corresponding body displacement vectors, and S is the elastic body surface area. In other words, the work that would be done by $\boldsymbol{F}$ acting through $\boldsymbol{u}'$ equals the work that would be done by $\boldsymbol{F}'$ acting through $\boldsymbol{u}$. The theorem is a generalization of the Maxwell reciprocity theorem for a linear elastic response.

BIAXIAL CRYSTALS

Crystals which have two *optical axes* with the property that light travelling along either of these directions is not subjected to *birefringence* or *polarization*. The angle between the two optical axes (2θ) is called the *optical axis angle*. Biaxial crystals are optically positive if the bisector angle $2\theta < 90°$ and negative if $2\theta > 90°$ (see also *Uniaxial crystals*).

BICRITICAL POINT

A point on the *phase diagram* of a crystal where two curves of *second-order phase transitions* and one curve of a *first-order phase transition* meet. The latter curve usually separates the regions of existence of two different ordered phases (see *Multicritical point*).

BIEXCITON, exciton molecule

A bound state of two *excitons* in *semiconductors* and *insulators*.

A biexciton in a semiconductor is a stable four-particle complex containing two electrons and two *holes*. Since the effective masses of an electron m_e and a hole m_h in semiconductors are close to each other, the biexciton stability problem has no small parameter for carrying out a power series expansion. Therefore, the *adiabatic approximation* widely used in molecular spectroscopy is inapplicable in this case; and the calculations are usually carried out with the help of the variational method. In the limit $m_e/m_h = 1$, the biexciton stability problem is analogous to the *positronium* molecule problem. The binding energy of the positronium molecule is extremely small: $\Delta_M \approx 0.02E_x$ where E_x is the binding energy of the positronium atom (electron–positron bound state analogous to hydrogen atom). For example, the biexciton binding energy in Ge and Si is $\Delta_M \leqslant 1$ meV. Variational calculations of this binding energy in semiconductors with different energy band structures demonstrate that biexcitons are stable for arbitrary values of the ratio m_e/m_h. In the limit $m_e/m_h \to 0$, corresponding to a hydrogen molecule, $\Delta_M = 0.35E_x$.

Since the energy gain due to biexciton formation from two excitons is insignificant, from the thermodynamic viewpoint biexciton formation is likely only at low enough temperatures. The probability of formation is in proportion to the square of the exciton concentration; and the estimates show that this concentration should be very large. The most direct method for the detection and investigation of biexcitons is the analysis of their *recombination* emission spectra. The most probable process of radiative biexciton decay is via a single *electron–hole pair* recombination accompanied by the creation of a photon and a recoil exciton. The radiative decay of biexcitons was detected in the indirect-gap semiconductors Ge and Si.

In direct-gap semiconductors, the recombination times are extremely short (10^{-9} s or less). Hence, the processes of binding excitons into biexcitons and the nonradiative decay of the latter takes place under highly nonequilibrium conditions with respect to both the phonon system and between the electron–hole components of the system itself. Therefore, to study the biexciton properties in direct-gap crystals experimentally, nonlinear optical methods were developed that are most efficient under highly nonequilibrium conditions. The methods include the two-phonon resonance excitation of biexcitons, the induced one-phonon transformation of an exciton into a biexciton, as well as the two-phonon resonance *Raman scattering of light* (or *hyper-Raman scattering*) with participation of biexciton states. The efficiency of these processes in direct-gap semiconductors is associated with a very large value of the nonlinear susceptibility at two-phonon resonance in the biexciton state. This is the basis for nonlinear effects of the multiwave mixing (see *Frequency mixing*) and *bistability* with participation of biexcitons. With the help of these methods, biexcitons were detected and studied in a number of direct-gap semiconductors (CuCl, CuBr, CdS, etc.).

BIMETALS, bimetallic materials

Materials consisting of two different, tightly bonded *metals* or *alloys*.

A distinctive feature of bimetals is a combination of various properties such as *strength* and *thermal-environment resistance*, wear resistance, or low electric resistance. Joining of bi- or multilayer metals occurs via the action of interatomic forces which bring the pure metal surfaces to a separation close to an interatomic one. The bimetals are obtained either using *plastic deformation* (rolling, *pressing*, pressure *welding*, *drawing*, *diffusion welding*) or without it (*casting*, electroslag welding, metallization by *spraying*, and so on). In industry, bimetals are used for preserving expensive metals or alloys, as well as materials with special properties. Bimetals containing metals (alloys) with different *thermal expansion* coefficients are called *thermobimetals*.

BINARY ALLOYS, double alloys

Alloys formed from two component systems.

Binary alloys are used when studying *phase diagrams* of the phase equilibrium and the physical, chemical, thermodynamic and other properties of alloys composed of two components. For example, binary alloys of the iron–carbon system, binary alloys based on nickel, and so on. These alloys include *single-phase alloys* and *two-phase alloys*, *intermetallic compounds*, *interstitial alloys*, and *substitutional alloys*, as well as various metallic materials regardless of the mode of their manufacture. A common feature is their two-component composition. For two-component alloys (differing in concentration) the term *binary alloy system* is used (e.g., the melt diagram of the binary system Al–Si). Alloys containing three or four components are called *tri-component alloys* and *tetra-component alloys*, respectively. Alloys composed of four and more components are called *multicomponent alloys*.

BIOLOGICAL CRYSTALS

Crystals built of biological molecules (proteins, nucleic acids or virus particles).

Due to the enormous dimensions of biological macromolecules containing 10^3–10^4 atoms, biological crystals have very large *crystal lattice* periods (5–20 nm) compared to ordinary crystals,

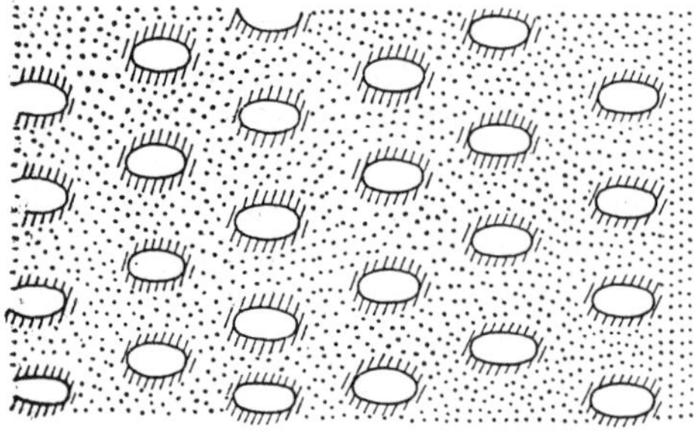

Packing of molecules in a protein crystal.

the lattice periods in a virus reaching 100 nm or more. An important feature of biological crystals is that they contain, in addition to the macromolecules themselves, some of the host solution (35–80%), from which the molecules were crystallized, typically water with various ions (see Fig.). Many biological crystals exist solely in equilibrium with such solutions. Desiccation brings about the denaturation (structural decay) of molecules and the crystal as a whole. Regularity of the arrangement of molecules in biological crystals is determined by the electrostatic interaction of the charged atomic groups at the surface of the molecules, with *van der Waals forces* and *hydrogen bonding* playing a major role. The solvent molecules adjacent to the surface are arranged in order; those in the intermolecular space are disordered. The methods of *crystallization* of biological crystals are based on the variation of the temperature and oversaturation. To change the solubility, some particular salts, or organic solutes, etc., are introduced into the solution.

The enormous dimensions of biological crystals allow one to directly observe their packing in the crystal lattice with the help of *electron microscopy. X-ray diffraction* also provides data on the most complicated spatial configuration of the molecules composing the crystals, but the enormous number of reflections obtained ($\geqslant$100,000) render the process of decoding them extremely complex.

In addition to the true three-dimensional periodic biological crystals, there exist biological crystals with different types of ordering. Thus, deoxyribonucleic acid (DNA) forms textured gels, i.e. *liquid crystals.* Their X-ray structural analysis allowed a spatial model of DNA to be constructed, and the nature of the genetic information transfer to be established. Some globular proteins (catalase and others) crystallize with the formation of tubes with monomeric walls built of molecules arranged according to spiral symmetry. Research of crystals of spherical viruses with the help of electron microscopy and X-ray analysis allowed one to establish the character of the mutual packing, and the structure of the protein molecules composing the virus shell. The latter are arranged in a virus particle according to icosahedral symmetry, with an axis of 5th order (see *Crystal symmetry*).

BIPHONON

A *quasi-particle* representing a bound state of two *phonons*, arising due to the *anharmonic vibrations.*

Both phonons move in the crystal as an integral whole, and are characterized by the same value of the wave vector. A biphonon influences *phonon optical spectra of absorption* and spectra of *Raman scattering of light* as a sharp maximum situated at an energy below the lower edge of the corresponding two-phonon band. The latter characterizes the crystal excitation region for the simultaneous creation of two phonons (the so-called sum process), or at the creation of a phonon of higher energy with simultaneous destruction of another one of lower energy (the so-called difference process). The separation of a biphonon peak from the two-phonon band and its intensity are determined by constants characterizing the anharmonicity, and by the density distribution of two-phonon states. Biphonons are frequently observable in phonon spectra at low temperatures.

BIPOLAR DIFFUSION

The same as *Ambipolar diffusion.*

BIPOLARON

A system consisting of two *conduction electrons* bound by a strong interaction with the medium; in other words, two bound *polarons.*

The appearance of bipolarons is possible in crystals, amorphous materials, and liquids. If the macroscopic polarization dominates the interaction with the medium then the large *dielectric constant* of the medium favors the creation of bipolarons. The electron motion in a bipolaron is correlated; an essential feature is their space separation inside the polarization potential well. The possibility of forming bipolarons was first established analytically in *ionic crystals* and then extended to amorphous insulators, metals and other materials. In bipolarons the electrons with opposite spins are bound, which leads to the absence of *paramagnetism* of the free charge carriers. The experimental evidence of the existence of bipolarons was obtained in a series of oxide crystals with variable *valence* (e.g., Ti_4O_7), and in some compounds of linear organic molecules. The space–time properties and energy scale of a bipolaron are essentially different from those in a *Cooper pair*. See also *Bipolaron mechanism of superconductivity*.

BIPOLARON MECHANISM OF SUPERCONDUCTIVITY

A possible mechanism to explain supercurrent flow through a crystal without resistance at low temperatures due to the *superfluidity* of a charged *Bose gas* or a *Bose fluid* of bipolarons.

The elementary excitation spectrum in such a system of charged bosons possesses, due to the long-range Coulomb repulsion, a finite energy gap $E_0 = \hbar\omega_p$ at zero momentum (where ω_p is the frequency of the collective *plasma oscillations* in the *bipolaron* gas). This also satisfies the *Landau superfluidity criterion*, which in this case corresponds to superconductivity. The bipolaron mechanism of superconductivity differs from the Cooper pairing of the *Bardeen–Cooper–Schrieffer theory* (BCS) in that bipolarons of small and intermediate radii are separated in space, and are not expected to decay above the transition from the superconducting to the normal state. To estimate the critical temperature of such a transition, one can use the expression for the *Bose–Einstein condensation* temperature $T_0 = 3.31\hbar^2 n_0^{2/3}/m_B^*$ (where n_0 is the concentration and m_B^* is the effective mass of the bipolarons) of an almost ideal charged Bose gas of high density under the condition $n_0^{1/3} a_B^* > 1$ (where $a_B^* = \varepsilon_0\hbar^2/(4e^2 m_B^*)$ is the effective Bohr radius of a bipolaron with charge of $2e$, and ε_0 is the static *dielectric constant* of the crystal). An estimate $T_0 \geqslant$ 2–50 K was obtained for bipolarons of intermediate radius in *ionic crystals* with $\varepsilon_0 \gg 1$ of $SrTiO_3$ type at $m_B^* \leqslant 10m_0$ (m_0 is free electron mass) and $n_0 \approx 10^{17}$–10^{20} cm^{-3}, while $T_0 \geqslant$ 8–80 K for bipolarons of small radius with $m_B^* \approx (100$–$1000)m_0$ at $n_0 \geqslant 10^{22}$ cm^{-3}. The bipolaron mechanism of superconductivity has been proposed theoretically for *heavy fermions* ($CeCu_2Si_2$, UBe_{13}, UPt_3), semiconducting *perovskites* $BaPb_{1-x}Bi_xO_3$ and possibly the new *high-temperature superconductor* cuprates.

BIPOLAR TRANSISTOR

A *semiconductor device* with its active zone containing two *semiconductor junctions* separated by a thin region called a base. The outer zones having the same conduction type are called the emitter and collector (see Fig.). In an n–p–n type *transistor* the minority current carriers of the base (electrons) diffuse through the p-type base region and penetrate the n-type collector region; in a p–n–p type bipolar transistor the minority current carriers (*holes*) diffuse through the n-type base region. A bipolar transistor is designed in such a way that small variations of the electric current or voltage could cause significant current variations between the emitter and the collector. As to their design, bipolar transistors can be either homostructural, heterostructural, or have graded bands. Bipolar transistors serve for the generation and amplification of electromagnetic oscillations, and for the control of voltage and current.

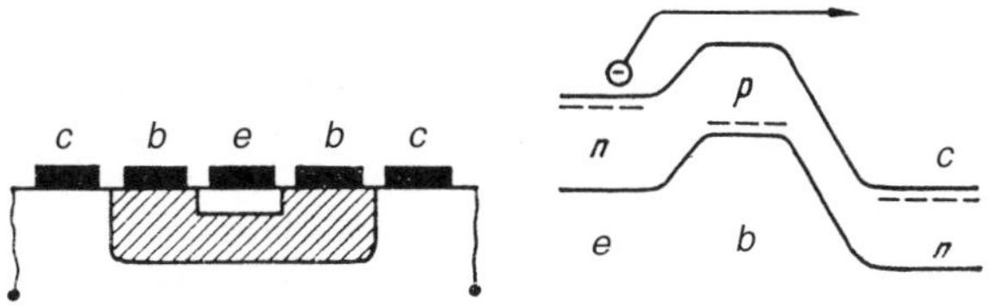

Structure (left) and energy diagram (right) of an n–p–n bipolar transistor showing the emitter (e), base (b) and collector (c) regions. The donor and acceptor energy levels are indicated by dashed horizontal lines.

BIRADICAL

A molecule containing two spatially separated radical centers (unpaired electrons), with the singlet–triplet energy splitting of the orbital state lower than or close to $k_B T$. The appearance of biradicals in a solid causes changes in an *electron paramagnetic resonance* spectrum due to *exchange interactions* and *dipole–dipole interactions.* In chemical reactions, a biradical behaves as two independent, but joined together, *free radicals.*

BIREFRINGENCE

Refraction of light in *anisotropic media* (e.g., *crystals*), which results in its splitting into two beams with different polarizations. It was discovered by E. Bartholin (1669) in calcite crystals. Birefringence is caused by the tensorial dependence of the *polarization vector* of the anisotropic medium on the light wave electric field intensity. The result is that the *polarization of light* in the medium and the phase velocities of the waves (*refractive indices*) both depend on the direction of propagation. In optically inactive crystals, upon birefringence the two waves are locally polarized in mutually perpendicular planes (in the general case of an optically active crystal the waves are elliptically polarized) and they have different *phase velocities.*

In single-optical-axis crystals one of the waves (ordinary ray) is polarized perpendicular to the *main cross-section*, i.e. the planes which pass through the direction of the light wave and the *optical axis of a crystal*, and the other wave (extraordinary ray) is polarized parallel to it. Refraction of the *ordinary wave* is similar to refraction (see *Refraction of light*) in an isotropic medium since the ray vector coincides with the wave vector and the refraction index does not depend on the direction of propagation. For the *extraordinary wave* the common law of refraction is not satisfied: the wave and ray vectors do not coincide, and the refraction index n_0 depends on the direction of propagation. For light propagation along the optical axis there is no birefringence. Biaxial crystals are characterized by three refraction indices, and they have two directions (two optical axes), along which there is no birefringence. The phase velocities of both waves depend on the direction of propagation, and both waves, upon refraction, may emerge from the plane of incidence.

Birefringence is possible not only in naturally anisotropic media, but also in media with an induced anisotropy caused by external fields: electrical (*Pockels effect, Kerr effect*), magnetic (*Voigt effect*), and acoustical. Phenomena similar to birefringence have also been observed outside the optical range of the electromagnetic spectrum. See also *Self-induced light polarization change.*

BISMUTH (Lat. *Bismuthum*), Bi

Chemical element of Group V of the periodic table; atomic number 83, atomic mass 208.98.

Natural bismuth consists of one stable isotope ^{209}Bi. Electronic configuration of outer shells is $6s^2 6p^3$. Ionization energies are 7.289, 16.74, 25.57, 45.3, 56.0 eV. Atomic radius is 0.182 nm, radius of Bi^{3+} ion is 0.120 nm, of Bi^{3-} ion is 0.213 nm. Electronegativity is ≈1.95. In the free form Bi is a silvery *metal* with a pink tint. Crystalline lattice is rhombohedral with the following parameters: $a = 0.47457$ nm and $\alpha = 57°14'13''$. Its density is 9.80 g/cm^3 (at 293 K); density of the liquid form is 10.27 g/cm^3 (at 544.4 K); $T_{melting} = 544.4$ K, $T_{boiling} = 1833$ K. Specific heat is 0.129 (at 293 K), in liquid state 0.142 (at 544.4 K), 0.148 (at 673 K), 0.166 kJ·kg^{-1}·K^{-1} (at 1073 K); thermal coefficient of linear expansion is 13.3·10^{-6} K^{-1}, volume expansion of metal upon solidifying +3.32% (at 544.4 K); thermal conductivity coefficient is 8.374 W·m^{-1}·K^{-1} (at 293 K), 15.491 (at 673 K); heat of melting is 11.38 kJ/mole; heat of evaporation 179 kJ/mole. Specific electrical resistance is 1068 μΩ·m (at 293 K), 1.602 (at 373 K), 1.289 (at 573 K), 1.535 (at 1053 K). The transition temperature T_c of a superconducting thin film state is ≈7 K. Under the influence of a magnetic field the electrical resistance of Bi increases more than that of other metals so it is used for the measurement of strong magnetic fields. The tensile strength is 4.9 to 19.6 MPa, elastic modulus is 31.37 GPa, rigidity 12.35 GPa, Brinell hardness is 94.2 MPa, Mohs hardness is 2.5. Adiabatic elastic moduli of Bi in the hexagonal system, $\bar{3}m$ (D_{3d}) class (room temperature): $c_{11} = 63.7$, $c_{12} = 24.88$, $c_{13} = 24.7$, $c_{14} = 7.17$, $c_{33} = 38.2$, $c_{44} = 11.23$ GPa. In the temperature range 423 to 523 K, bismuth compresses in a comparably easy manner. Bismuth is a diamagnetic metal with magnetic susceptibility $-1.35{\cdot}10^{-9}$ (upon melting the susceptibility decreases 12.5 times). Cross-section for thermal neutron capture is not high in Bi (0.034 barn). Bismuth is used for the production of low-melting *alloys* (e.g., Wood's metal alloy with melting point of 343 K). Bismuth alloys are used in the helices of devices for measuring strong magnetic fields, and it also serves as a heat-transfer agent in nuclear reactors.

BISTABILITY

A property whereby a system has two stable states in a certain region of variation of its parameters. The existence of bistability is connected with nonlinear properties of the system, and produces *hysteresis* during a fast enough variation of the parameters. It is exemplified by *optical bistability*, *polarization optical bistability*, an N- or S-shaped *current–voltage characteristic*, the magnetic polarization of a *ferromagnet*, in a magnetic field, etc. Nonlinear systems may exhibit more than two stable states (*multistability*).

BITTER PATTERN

See *Magnetic powder pattern*.

BLACK-AND-WHITE GROUPS

The same as *Shubnikov groups*.

BLAHA–LANGENECKER EFFECT (F. Blaha, B.-Z. Langenecker, 1935)

Reduction of the force needed for plastic deformation of a specimen under simultaneous action of static and alternating loads.

The effect is observed in technological processes involving the *shape-changing of metals* (such as drawing, punching, rolling, and so on). The effect is especially noticeable when using ultrasound (see *Ultrasonic treatment*); in this case, the reduction of static stress depends on the ultrasound intensity, the nature and the condition of the material, the temperature, and other conditions. The effect of the ultrasound is explained by a local absorption of the energy of the high-frequency oscillation by various defects of the crystalline structure (see *Defects in crystals*), which facilitates the multiplication of *dislocations* and *vacancies*, as well as their motion in the crystal.

BLISTERING

An appearance of blisters on a *crystal* surface under bombardment with atoms (ions) of inert gases, protons, and other particles.

The blisters "ripen" with an increase in dosage (see *Dosimetry*) up to a stage when the blister dome breaks due to the increased pressure inside it. A new generation of blisters can develop on the uncovered area. The subsequent stages of the blistering are shown in the figure. The phenomenon of blistering is caused by migration of irradiation-induced *vacancies* and vacancy *pores* towards the surface. The mechanical stresses arising at embedding of the bombarding ions into the subsurface layer can also induce blistering. If the *defect* density is sufficiently uniform along the surface, the *scabbing* of thin plates off the surface of the irradiated material takes place instead of blister formation. The blistering and flaking phenomena lead

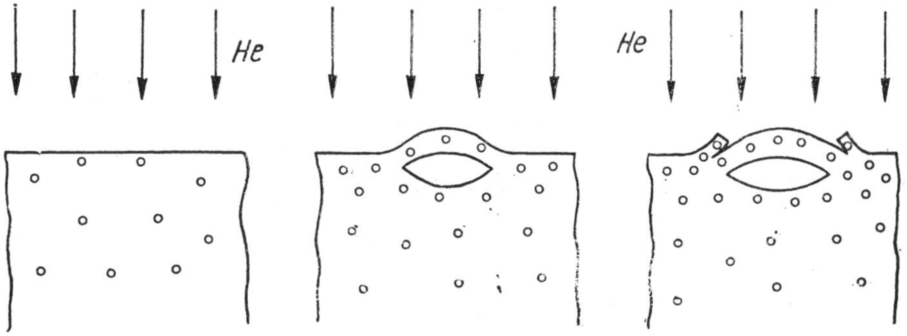

Blistering process.

to strong *surface erosion* of irradiated material at high irradiation doses.

BLOCH EQUATIONS (F. Bloch, 1946)

Phenomenological equations describing the magnetic properties of ensembles of electrons ($\gamma < 0$) and nonzero-spin nuclei ($\gamma > 0$) in external magnetic fields:

$$\frac{\mathrm{d}\boldsymbol{M}}{\mathrm{d}t} = \gamma(\boldsymbol{M} \times \boldsymbol{B}) - \frac{M_x \boldsymbol{i} + M_y \boldsymbol{j}}{T_2} - \frac{M_z - M_0}{T_1}\boldsymbol{k}, \qquad (1)$$

where M_x, M_y, M_z are the components of *magnetization* vector $\boldsymbol{M}$; M_0 is the equilibrium value of M_z; $\boldsymbol{i}, \boldsymbol{j}$ and $\boldsymbol{k}$ are the unit vectors of the laboratory coordinate system; $\boldsymbol{B} = \boldsymbol{k}B_0 + \boldsymbol{i}2B_1 \cos\omega t$, B_0 is the constant magnetic field component directed along z axis; ω is the frequency; and B_1 is the amplitude of alternating magnetic field ($B_0 \gg B_1$); γ is the *gyromagnetic ratio*. The constant T_1 is called the *longitudinal or spin–lattice relaxation time* and T_2 is the *dephasing time*, also referred to as the *transverse or spin–spin relaxation time*. In the general case, T_1 is unequal to T_2 ($T_1 \geqslant T_2$) since T_1 characterizes the rate of establishing thermal equilibrium between the *spins* and the *lattice*, while T_2 is related to rate of establishing thermal equilibrium within the spin system, and this latter is equivalent to the decay rate of the phase coherences between precessing spins (as a result of the interaction between each other and with their environment, the spins precess in different local fields).

The Bloch equations are based on the following reasoning. In an arbitrary uniform magnetic field $\boldsymbol{B}$ the equation of motion of the magnetization vector $\boldsymbol{M}$ of the ensemble of free spins has the form $\mathrm{d}\boldsymbol{M}/\mathrm{d}t = \gamma(\boldsymbol{M} \times \boldsymbol{B})$. It is postulated that: (i) in a constant magnetic field B_z, the longitudinal magnetization M_z relaxes to M_0 in an exponential manner, i.e. satisfies the equation $\mathrm{d}M_z/\mathrm{d}t = -(M_z - M_0)/T_1$; (ii) transverse components M_x and M_y decay exponentially to zero as

$$\frac{\mathrm{d}M_x}{\mathrm{d}t} = -\frac{M_x}{T_2}, \qquad \frac{\mathrm{d}M_y}{\mathrm{d}t} = -\frac{M_y}{T_2}; \qquad (2)$$

(iii) in an external magnetic field (comprised of a strong constant and weak alternating components) the relaxation motion of the magnetization vector is superimposed on the motion of free spins.

Under the conditions of *resonance*, the stationary solutions of Eqs. (1) lead to the following expressions for the real (χ') and imaginary (χ'') parts of the *magnetic susceptibility*:

$$\chi' = (\omega_0 - \omega)T_2\chi'',$$

$$\chi'' = \frac{1}{2}\gamma M_0 T_2\left[1 + (\omega - \omega_0)^2 T_2^2 + S\right]^{-1},$$

where $\omega_0 = \gamma B_0$, χ'' determines the *Lorentzian shape* of the line with a half-width at a half-height of $\Delta = (1 + S)^{1/2}/T_2$; and the saturation factor (see *Saturation effects*) $S = \gamma^2 B_1^2 T_1 T_2$ indicates the dependence of the line shape on the alternating field intensity. In the absence of saturation ($S \ll 1$), the half-width is determined by the spin–spin relaxation time T_2. The Bloch equations are also used for the analysis of pulsed experiments. One can obtain Eqs. (1) from the general theory of relaxation (see *Bloch–Redfield theory*). For *two-level systems* they are valid if alternating fields are small and there are no spin–spin interactions. In solids the value of T_2 is often estimated from the width. In this case Eqs. (1) lead to a discrepancy with the experimental data in the limit of strong saturation. Despite the limits to their applicability, the Bloch equations play an important role in elucidating many aspects of *nuclear magnetic resonance* and *electron paramagnetic resonance* in solids.

BLOCH FUNCTION

See *Bloch theorem*.

BLOCH–GRÜNEISEN FORMULA (F. Bloch, E. Grüneisen, 1930)

An interpolating expression describing the temperature dependence of that part of electric resistance of *metals* ρ_T that is caused by *electron scattering* by lattice vibrations (*phonons*):

$$\rho_T = 4\left(\frac{T}{\Theta}\right)^5 \rho_0 \int\limits_0^{\Theta/T} \frac{z^5\,\mathrm{d}z}{(\mathrm{e}^z - 1)(1 - \mathrm{e}^{-z})}, \qquad (1)$$

where Θ is the *Debye temperature*, and ρ_0 is a constant independent of temperature. In the simplest case of a spherical *Fermi surface*, the value of ρ_0 is

$$\rho_0 = \frac{3\pi}{\hbar}\frac{m^2}{e^2 n^2}\frac{\varepsilon^2}{M_0 v_0 k\Theta}, \qquad (2)$$

where m is the electron *effective mass*, M_0 is the atomic mass of a metal atom, v_0 is the volume per atom, n is the density of *conduction electrons*, ε is a particular energy parameter (in the simplest case, the *strain potential*) characterizing the electron–phonon interaction; $\varepsilon \sim 1$–10 eV. At high temperatures ($T \gg \Theta$), Eq. (1) has the form $\rho_T = (T/\Theta)\rho_0$, i.e. it describes an ordinary linear growth of the metallic resistance with increasing temperature. At low temperatures ($T \ll \Theta$), Eq. (1) yields the well-known *Bloch law* $\rho_T \sim T^5$, where $\rho_T = 497.6(T/\Theta)^5\rho_0$. For a majority of simple metals the parameter Θ evaluated from Eq. (1) is in good agreement with the Debye temperature obtained from measurements of the specific heat.

BLOCH LAW, three halves power law (F. Bloch, 1930)

The temperature T dependence of spontaneous *magnetization* M_s of *ferromagnets* in the temperature range significantly lower than *Curie point* Θ, but higher than the *magnon* activation energy:

$$M_s = M_{s0}\left[1 - \alpha\left(\frac{T}{\Theta}\right)^{3/2}\right],$$

where M_{s0} is the maximum value of M_s at $T = 0$ K, and α is a constant which is specific for a given substance. This formula represents the first terms of the expansion of $M_s(T)$ in powers of T/Θ. The subsequent terms of the series are negligible (at $T < \Theta$) and, as experiments show, the *Bloch law* is satisfied well up to $T \approx \Theta/2$. The decrease of M_s with increasing T is a result of the disturbance of the ordered orientation of atomic (spin) *magnetic moments* in ferromagnets due to the thermal excitation of magnons. Their number grows as $T^{3/2}$ with increasing temperature, in accordance with the equation.

BLOCH LINE

A linear inhomogeneity of a 180° *magnetic domain wall* (*Bloch wall*) that separates wall regions with different directions of magnetization rotation (see Fig.). The z axis is the Bloch line. A Bloch line cannot end anywhere in the domain wall, it either forms a closed ring, or terminates at the surface of the *magnetic substance*, i.e. the medium. The domain wall thickness decreases near the Bloch line, as shown. Bloch lines are distinguished according to their orientation. *Vertical Bloch lines* are directed normally to the surface of the magnetized material, and exist in domain walls of materials with *magnetic bubble domains*. They can be detected with the help of direct methods, such as electron microscopy or, in some cases by magneto-optic methods (see *Magnetic domain characterization*), and by the deformation of a moving domain wall at the location of an accumulation of vertical Bloch lines. Magnetic bubble domains with a large number of vertical Bloch lines (see *Rigid magnetic bubble domain*, *Dumbbell domain*) possess anomalous dynamic properties. S. Konishi (1983) proposed to use these lines for solid-state systems capable of superdense magnetic *information recording*. *Horizontal Bloch lines* that are directed parallel to the sample surface, as well as Bloch lines of a more involved shape, can arise during the motion of a twisted domain wall (see *Domain wall twisting*). Horizontal Bloch lines

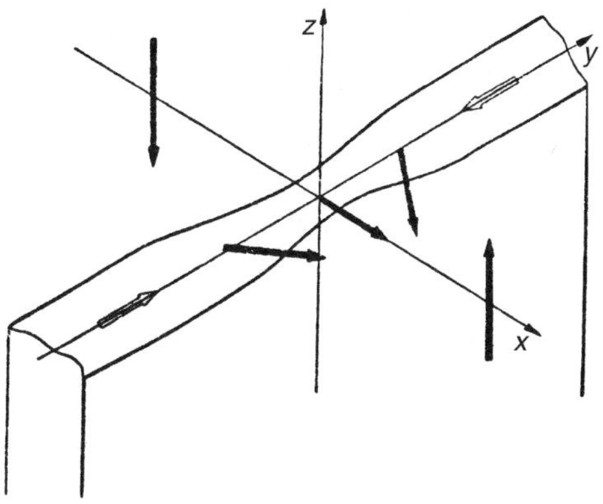

Distribution of magnetization indicated by solid arrows outside and within a domain wall with vertical (along z) Bloch lines; light arrows indicate directions of magnetization rotation.

play an important role in the formation of the phenomenon of *ballistic aftereffect* in the motion of magnetic bubble domains.

BLOCH POINT

A point inhomogeneity, a singular point of the *magnetization* $\boldsymbol{M}(\boldsymbol{r})$ at the merging of two *Bloch lines* of different signs in a magnetic *domain wall*. The magnetization on a closed surface surrounding a Bloch point takes on all possible values. Thus, $\boldsymbol{m} = \boldsymbol{M}/\boldsymbol{M}_S$ is equal to $+\boldsymbol{e}_z$ and $-\boldsymbol{e}_z$ far from the domain wall, to $+\boldsymbol{e}_y$ and $-\boldsymbol{e}_y$ in the middle of the domain wall far from the Bloch line, and to $+\boldsymbol{e}_x$ and $-\boldsymbol{e}_x$ in the middle of Bloch line segments of opposite sign (see Fig.). Hence, when approaching a Bloch point, the direction of magnetization $\boldsymbol{m}$ depends on the direction of approach, thus producing a "hedgehog"-type singularity of the magnetization field (see *Topological inhomogeneity*). The following distribution corresponds to this singularity at $a < r < \Delta$ (a is the interatomic distance, Δ is the domain wall thickness):

$$\boldsymbol{m} = \pm\frac{\boldsymbol{r}}{r}, \quad (1)$$

where the different signs determine the sign (i.e. *topological charge*) of a Bloch point. There are other possible distributions that can be obtained

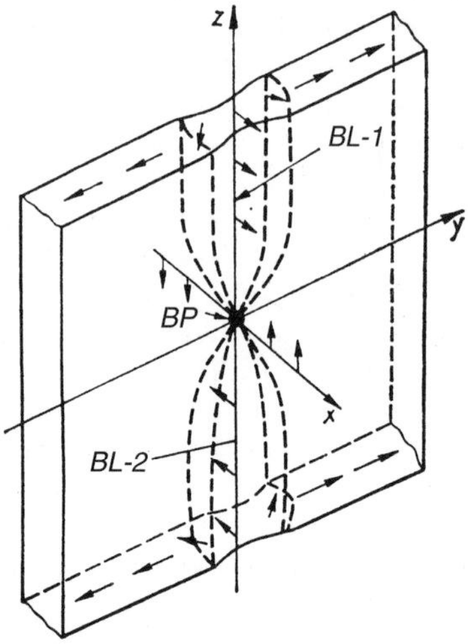

Distribution of magnetization indicated by arrows in a domain wall with a Bloch point BP crossed by Bloch lines BL-1 and BL-2.

from Eq. (1) by a uniform rotation. (The Bloch point in Fig. takes the form $\boldsymbol{m} = \boldsymbol{r}/r$ after the rotation of $\boldsymbol{m}$ by $\pi/2$ about the y axis.) In materials with strong uniaxial *magnetic anisotropy*, the loss of domain wall energy due to the presence of a Bloch point is not significant. Bloch points must always arise under certain distributions of the magnetization. Bloch points and their displacements under the action of an appropriately directed magnetic field (along x axis) are detectable in domain walls of materials with *magnetic bubble domains* by the deviation of the direction of domain motion from the direction of the magnetic field gradient (see *Rigid magnetic bubble domain*).

BLOCH–REDFIELD THEORY

A magnetic resonance theory describing the time evolution of a spin system, located in an external magnetic field, and interacting with its molecular environment (with the *lattice*).

The main result of the theory is the Markovian *kinetic equation* for the *spin density* matrix. A consistent quantum-statistical derivation of the kinetic equation was first presented by F. Bloch and R.K. Wangness, then Bloch generalized the theory for alternating magnetic fields of arbitrary magnitude. Later A.G. Redfield put forth a semiclassical theory of *relaxation* where an approach to the calculation of the kinetic equation coefficients was treated for the first time. The basic physical concepts of the Bloch and Redfield theories are the same. Later, both theories were described within the framework of a general formalism by other authors.

Let a closed region comprise a spin system and a lattice with the Hamiltonian $H = H_s + H_s(t) + H_{sf} + F$ where H_s is the time-independent Hamiltonian of the spin system, $H_s(t)$ is the interaction operator of the spin system with alternating magnetic fields, $H_{sf} = \sum_\alpha u_\alpha v_\alpha$ is the spin–lattice interaction operator, u_α are the lattice operators, v_α are the *spin operators*, F is the Hamiltonian of the lattice. The exact equation of motion for the *density matrix* ρ of the closed system, $\mathrm{i}\,\mathrm{d}\rho/\mathrm{d}t = [H, \rho]$, is iterated in the interaction representation, and the infinite series obtained is truncated at the second term (*spin–lattice interaction* considered as small perturbation). Then the spin density matrix evolves with the expression: $\sigma = \mathrm{Tr}_q\,\rho$ where the trace (Tr) of the matrix is taken over the lattice states. The kinetic equation for elements σ_{mn} has the following form:

$$\frac{d\sigma_{mn}}{dt} = i[\sigma, H_s + H_s(t)]_{mn} + \sum_{kl}(2\Gamma_{mkln}\sigma_{kl} - \Gamma_{klmk}\sigma_{ln} - \Gamma_{lnkl}\sigma_{mk}). \quad (1)$$

Here the indices m, n, k, l enumerate the energy levels of Hamiltonian H_s;

$$\Gamma_{mkln} = \pi \sum_{\alpha\beta} j_{\alpha\beta}(\omega_{mk}) v_{\beta mk} v_{\alpha ln} \Delta(\omega_{mk} + \omega_{ln}),$$

where $j_{\alpha\beta}(\omega_{mk})$ is the Fourier transform of the lattice correlator:

$$j_{\alpha\beta}(\omega_{mk}) = \frac{\sum_{r,r'} u_{\alpha r'r} u_{\beta rr'} e^{-\beta F_r} \delta(\omega_{mk} + \omega_{rr'})}{2\pi \mathrm{Tr}_q e^{-\beta F}},$$

where $\omega_{r'r}$ are the lattice eigenfrequencies; ω_{mk} are the eigenfrequencies of the spin system; $v_{\beta mk}$ and $v_{\alpha ln}$ are the matrix elements with respect to eigenfunctions of Hamiltonian H_s; $u_{\alpha r'r}$ and $u_{\beta r'r}$ are the matrix elements of eigenfunctions of operator F; $\beta = \hbar/(k_B T)$, where T is the lattice temperature, F_r is the energy of rth lattice level. One must retain on the right-hand side of Eq. (1) those Γ_{mkln} for which $|\omega_{mk} + \omega_{ln}| \ll \tau_c^{-1}$. This is formally achieved by the introduction of the Heaviside step function $\Delta(x) = 1$ at $|x| < \tau_c^{-1}$ and $\Delta(x) = 0$ otherwise, where τ_c is the lattice *correlation time* (perturbations in the lattice disappear within time τ_c).

The derivation of the equation requires a number of assumptions:

(i) the lattice stays close to thermal equilibrium;

(ii) the influence of the spin system on the lattice is negligible, since the lattice has many more degrees of freedom than the spin system ($\rho = \sigma\rho_f$, where ρ_f is the equilibrium density matrix of the lattice);

(iii) there exists a time interval Δt such that $\Delta t \gg \tau_c$ and simultaneously $\Delta t \ll \tau_r$, where τ_r is the shortest *relaxation time* of the spin system (if the latter condition is met then the density matrix σ changes insignificantly within the time Δt_l, thus allowing one to apply perturbation theory);

(iv) alternating magnetic fields are small; $\gamma B_1 \ll \tau_c^{-1}, \gamma B_0$, where B_1 is the amplitude of the external alternating magnetic field, B_0 is the amplitude of the externally applied magnetic field, and γ is the *gyromagnetic ratio*;

(v) the interaction between the lattice and the spin system results in a small shift of the levels of the latter that is ignored in Eq. (1); and

(vi) the relaxation coefficients Γ are time-independent and comparable to the reciprocals of the relaxation times.

It follows from the relation

$$\Gamma_{mkln} = \Gamma_{lnmk} \exp(\beta\omega_{ln}) \approx \Gamma_{lnmk} \exp(\beta\omega_{km})$$

that in the absence of alternating magnetic fields, the solutions of Eq. (1) are elements of the equilibrium spin density matrix, i.e. due to the relaxation the spin system equilibrates with the lattice. Eq. (1) describes the *magnetic resonance* and the relaxation of magnetically dilute spin systems.

BLOCH THEOREM

One of the most important theorems in solid state physics, which states that the intrinsic states of a particle in a periodic potential $V(\boldsymbol{r})$ satisfying condition $V(\boldsymbol{r} + \boldsymbol{t}) = V(\boldsymbol{r})$ (where $\boldsymbol{t}$ is the constant distance vector called the *direct-lattice vector*) can be characterized by two quantum numbers, viz., the band number λ and a real-valued vector $\boldsymbol{k}$ called the *quasi-wave vector* (*lattice wave vector* or *crystal momentum*). The wave function $\Psi_{\boldsymbol{k}\lambda}(\boldsymbol{r})$ called the *Bloch function* satisfies the condition $\Psi_{\boldsymbol{k}\lambda}(\boldsymbol{r} + \boldsymbol{t}) = \Psi_{\boldsymbol{k}\lambda}(\boldsymbol{r}) \exp(i\boldsymbol{k}\boldsymbol{t})$; and can be written as $\Psi_{\boldsymbol{k}\lambda}(\boldsymbol{r}) = u_{\boldsymbol{k}\lambda}(\boldsymbol{r}) \exp(i\boldsymbol{k}\boldsymbol{t})$ where $u_{\boldsymbol{k}\lambda}$ has the symmetry of the potential, so that $u_{\boldsymbol{k}\lambda}(\boldsymbol{r} + \boldsymbol{t}) = u_{\boldsymbol{k}\lambda}(\boldsymbol{r})$. The function $\Psi_{\boldsymbol{k}\lambda}(\boldsymbol{r})$ can also be written as a linear combination of *Wannier functions* $\Phi_\lambda(\boldsymbol{r})$, i.e. $\Psi_{\boldsymbol{k}\lambda}(\boldsymbol{r}) = \sum_{\boldsymbol{R}} e^{i\boldsymbol{k}\boldsymbol{r}} \Phi_\lambda(\boldsymbol{r} - \boldsymbol{R})$, where the set of points $\boldsymbol{R}$ transforms to itself upon displacement by the vector $\boldsymbol{t}$, i.e. it possesses the symmetry of the potential.

BLOCH WALL

See *Magnetic domain wall.*

BLOCH WAVE FUNCTION

See *Bloch theorem.*

BLOCK STRUCTURE

Partitioning a *crystal* into regions (*blocks*) slightly disoriented relative to each another (disorientation angles are on the order of a few minutes of arc). Other terminologies are mosaic structure and *blocks of mosaics* (see *Mosaic crystals*).

A block structure comprises disordered regions separated by closed *small-angle boundaries* in the form of *dislocation interlacements*, skeins, tangles, nets and *dislocation walls* of various structural and dislocation compositions. The regions contain both a *cellular structure* and a variety of dislocation structures such as individual *dislocations*, dislocation dipoles, multipoles, clusters, interlacements, wisps, nets, walls, and so on. The boundaries may be thin or thick, dense or rarefied, etc. The nets, boundaries, and walls may be regular or irregular; the networks may be hexagonal, tetragonal, etc. Incomplete (i.e. either forming or decaying) boundaries are a common occurrence. Depending on the dislocation composition and structure, the boundaries may be twisting or inclination ones, or a mixture of different varieties. The blocks may be equiaxial or nonequiaxial textured (see *Texture*). Block structures with small-angle boundaries composed of fairly regular nets and walls are often called *polygonized structures*. If the block size is commensurable with the grain size in a *polycrystal*, or with a single crystal size, the blocks are often called *subgrains*. The block size can vary from 0.1 to 10 μm.

BLUE BRITTLENESS

Lowering of the *plasticity* of unhardened construction *steel* at temperatures of blue color iridescence (300–500 °C). It is responsible for the hardening of ferrite at temperatures about 300 °C under conditions of static loading, and the formation of aggregations of impurity atoms which enhance its brittleness at temperatures about 400–500 °C under conditions of dynamic loading. In this case the plasticity is reduced owing to the formation of *Cottrell atmospheres* around dislocations because of the sharp growth of the *diffusion* rate as the temperature rises. In addition, under conditions of shock loading, the velocity of dislocation motion exceeds that of impurity atom atmospheres. The latter start to lag behind the dislocations, forming elongated aggregations of impurity atoms, which extend in the direction of the steel plastic flow. The presence of such aggregations in steel ferrite increases the *alloy brittleness*. Blue brittleness is reduced in steel by introducing elements linking interstitial atoms in *carbides* and *nitrides*.

BLUE PHASES of liquid crystals

Phases existing in a narrow temperature range (1–0.1 °C in order of magnitude) between a liquid phase and a cholesteric phase (see *Cholesteric liquid crystal*), possessing a spatial three-dimensional periodicity, and formed by bilaterally asymmetric elongated molecules. Three blue phases are known, designated as BPI, BPII, and BPIII, with no translational long-range order present in the highest-temperature phase BPIII of the group (see *Long-range and short-range order*).

In view of the bilateral symmetry of phases BPI and BPII, a coil twisting is locally advantageous in two directions (in spite of the single direction where it takes place in a cholesteric). Therefore, a three-dimensional lattice of linear defects (*disclinations*) with a period 0.5 μm appears in the system. Thus, blue phases are the analogues of ordinary solid crystals where the structural elements are not the individual atoms or molecules, but the cells of the disclination network. There also exist some other models of blue phases. The positions of the molecules themselves are not fixed; only their orientations are conserved at every point. Despite the lack of any direct proof of the existence of the disclination lattice in a blue phase, there are some indirect effects arguing in favor of the model, such as the increase of the lattice period with a decrease in the temperature. Due to a periodic structure, the BPI and BPII are able to form well-faceted *monocrystals*, and also to selectively scatter light in a visual spectral range, thus acquiring the blue color that gives them their name. Two optical characteristics of the BPI and BPII phases are the strong *rotation of the light polarization plane* (arising due to a coil structure) and the lack of linear double refraction (due to the cubic form of the disclination network cells). Although blue phases have been well known since the discovery of *liquid crystals* in 1888, nowadays these phases are among the least studied liquid crystals.

BODY-CENTERED LATTICE, I-lattice

A *lattice* whose unit cell parallelepiped contains an extra site in the cell center (see *Bravais lattices*) as well as the sites at the vertices. In addition to the three basic *translations* $\boldsymbol{a}$, $\boldsymbol{b}$, $\boldsymbol{c}$, the lattice possesses a further translation $\boldsymbol{t} = (\boldsymbol{a} + \boldsymbol{b} + \boldsymbol{c})/2$. The latter can appear only in rectangular parallelepipeds; therefore, body-centered lattices exist only in orthorhombic, tetragonal, and cubic *crystallographic systems*. In triclinic and trigonal crystal systems the I-lattice reduces to a smaller primitive lattice (see *Primitive parallelepiped*). In a monoclinic crystal system, this is a base-centered lattice with the *unit cell* volume and other basis translations half as large. In the hexagonal lattice case, the appearance of the translation $\boldsymbol{t}$ leads to a lowering of the symmetry to that of a monoclinic base-centered lattice. Considering the point symmetry of the sites, the body-centered lattice belongs to the body-centered Bravais groups.

BOGOLYUBOV CANONICAL TRANSFORMATIONS

See *Canonical transformation method*.

BOGOLYUBOV–DE GENNES EQUATIONS
(N.N. Bogolyubov, 1958; P.G. de Gennes, 1963)

Equations of the *mean field approximation* in the theory of *superconductivity* that determine the energy eigenvalues ε_λ and the corresponding two-component wave functions $\binom{u_\lambda(\boldsymbol{r})}{v_\lambda(\boldsymbol{r})}$ of Bogolyubov quasi-particles (see *Bogolyubov method*) in a spatially inhomogeneous state of a superconductor described by the *order parameter* $\Delta(\boldsymbol{r}) = g\langle\Psi_s\Psi_{-s}\rangle$. Here $\Psi_s(\boldsymbol{r})$ is the annihilation operator of an electron with spin s, and g is the matrix element of the effective electron–electron interaction in *Bardeen–Cooper–Schrieffer theory*. These equations have the form:

$$\widehat{H}_0 u_\lambda(\boldsymbol{r}) + \Delta(\boldsymbol{r}) v_\lambda(\boldsymbol{r}) = \varepsilon_\lambda u_\lambda(\boldsymbol{r}),$$

$$-\widehat{H}_0^* v_\lambda(\boldsymbol{r}) + \Delta^*(\boldsymbol{r}) u_\lambda(\boldsymbol{r}) = \varepsilon_\lambda v_\lambda(\boldsymbol{r}),$$

where

$$\widehat{H}_0 = \frac{1}{2m}\left(-\mathrm{i}\hbar\nabla - e\boldsymbol{A}(\boldsymbol{r})\right)^2 + U(\boldsymbol{r}) - E_\mathrm{F}$$

is the one-electron Hamiltonian of the normal state of a *metal*, m is the electron mass, $\boldsymbol{A}(\boldsymbol{r})$ is the magnetic field vector-potential, $U(\boldsymbol{r})$ is the effective one-particle potential for the electrons in the metal, and E_F is the *Fermi energy*. In the thermodynamic equilibrium state, the value of $\Delta(\boldsymbol{r})$ satisfies the self-consistency condition

$$\Delta(\boldsymbol{r}) = g\sum_\lambda u_\lambda(\boldsymbol{r}) v_\lambda^*(\boldsymbol{r}) \tanh\left(\frac{\varepsilon_\lambda}{2k_\mathrm{B}T}\right).$$

The Bogolyubov–de Gennes equations were used to solve spatially nonuniform problems of superconductivity theory such as, e.g., calculation of the *upper critical field* $B_{c2}(T)$, finding the excitation spectrum in the *vortex lattices*, determining the boundary conditions in *proximity effect* theory, elucidating particular features of the reflection of quasi-particles from a "normal metal–superconductor" interface (see *Andreev reflection* of quasi-particles), finding the excitation spectrum in S–N–S contacts, etc.

BOGOLYUBOV METHOD in the theory of superconductivity (N.N. Bogolyubov, 1958)

A field theory method for calculating the ground state energy and the spectrum of single particle and *collective excitations in superconductors*.

The initial Hamiltonian of the electron and phonon subsystems of a *metal* was diagonalized using the unitary transformations of creation and annihilation operators of bosons and fermions (see *Canonical transformation method*) and the principle of renormalization, i.e. the compensation of divergent *Feynman diagrams* by a perturbation theory series. The Bogolyubov method is an alternative to the variational method in *Bardeen–Cooper–Schrieffer theory* and, unlike the latter, allows a consistent treatment of collective effects and the Coulomb interaction in superconductors. In particular, within the framework of the *Fröhlich Hamiltonian* for the *electron–phonon interaction*, this method was used to obtain in the weak coupling limit, for the first time, an equation for the *energy gap* parameter Δ in the spectrum of a superconductor taking into account *effects of retardation.* In addition, with the help of the Bogolyubov method, a further weakening of the screened electron–electron repulsion was shown to exist (see *Bogolyubov–Tolmachyov logarithm*), which facilitates Cooper pairing due to the electron–phonon interaction. When the Bogolyubov method was applied to collective excitations involving undamped branches in the energy range $\omega < 2\Delta$ with a linear dispersion law $\omega_q = qv_F/\sqrt{3}$ (where v_F is the *Fermi velocity*), they were found to exist in neutral superfluid Fermi systems. That branch represents a *Goldstone mode.* In charged Fermi systems with long-range Coulomb repulsion (electrons in a metal), the branch transforms into *plasma oscillations* with the frequency $\omega_q \neq 0$ at $q \to 0$.

BOGOLYUBOV–TOLMACHYOV LOGARITHM

(N.N. Bogolyubov, V.V. Tolmachyov, 1958)

The logarithm of the ratio of the *Fermi energy* in *metals* $E_F \sim$ 1–10 eV to the mean energy of *phonons* $\hbar\omega_D \sim$ 0.1–0.01 eV (ω_D is the *Debye frequency*) that characterizes the extent of the effective weakening of the Coulomb repulsion between the electrons in a superconductor:

$$\widetilde{V}_c = \frac{V_c}{1 + N(0)V_c \ln[E_F/(\hbar\omega_D)]},$$

where V_c is the matrix element of the screened Coulomb interaction that is averaged over the Fermi surface, and $N(0)$ is the *density of states* at the *Fermi level.* The extra (relative to screening) weakening of the repulsion is caused by the decreased probability of the appearance of two electrons within the limits of the screening region due to Coulomb scattering. It facilitates electron *Cooper pair* formation via an effective attraction, which results from *virtual phonon* exchange.

BOHR MAGNETON

See *Magneton.*

BOHR RADIUS

The radius a_B of the orbit of an electron in the ground state of a hydrogen *atom.*

More generally, the distance between the center of the hydrogen atom and the maximum of the quantum-mechanical probability for detecting an electron; or the distance through which the electron wave function reduces by the numerical factor of e = 2.71828. The Bohr radius has the value $a_B = 4\pi\varepsilon_0\hbar^2/(me^2) = 5.2918\cdot10^{-11}$ m, where m and e are the electron mass and charge, respectively, $\hbar = h/(2\pi)$ is the reduced *Planck constant.* In the physics of insulators and semiconductors the effective Bohr radius a_B^* is used which differs from the Bohr radius a_B with m replaced with the *effective mass* of an electron in a crystal and e replaced by the effective charge $e^* = e/\sqrt{\varepsilon}$, where ε is the relative *dielectric constant.* The effective Bohr radius determines the approximate dimensions and energy, $E_n = -e^{*2}/(8\pi\varepsilon_0 a_B^* n^2)$ (where n is the principal quantum number), of shallow hydrogen-like centers and *Wannier–Mott excitons* in semiconductors. It is a convenient scale for expressing the dimensions of localized electron states in nonmetallic crystals.

BOHR–VAN LEEUWEN THEOREM (N. Bohr, 1911; generalized by J. van Leeuwen, 1919)

A theorem of classical statistical physics, which states that the *magnetization* of a system of charged particles in a constant external magnetic field under the conditions of statistical equilibrium is equal to zero. The theorem demonstrates that it is impossible, within the framework of the classical statistical mechanics of charged particles, to explain *ferromagnetism, paramagnetism* and *diamagnetism.* As was shown later, the magnetism of a substance is caused by quantum properties of the particles comprising it.

BOLOMETER

A device for measuring the radiated energy (power) in the infrared and optical wavelength ranges. It is in common use for various detection schemes with narrow ranges of extreme sensitivity to particular excitations.

It was first used by S. Langley (1880) for measurements in the IR range of the solar spectrum. The principle of bolometer operation is based on the change of the resistance of the thermosensitive element under the action of radiation. Bolometers are nonselective *optical radiation detectors.* To obtain equivalent spectral sensitivity in a given range, absorbing *coatings* are applied to the sensitive elements of the bolometer. There are metallic, semiconductor (thermistor) and insulator bolometers. The use of cryogenic effects allowed producing cooled bolometers based on the conduction of *shallow levels* in semiconductors, as well as superconducting (isothermal and nonisothermal) bolometers.

BOLTZMANN CONSTANT (L. Boltzmann)

One of the fundamental physical constants, a parameter of the statistical *Boltzmann distribution* and *Gibbs distribution*. Through them, the Boltzmann constant k_B enters various thermodynamic relations, for example, it relates the thermodynamic probability W of a system state to the *entropy* S through the expression $S = k_B \ln W$, and determines the mean thermal energy $k_B T/2$ of a particle per degree of freedom. The gas constant R involved in the equation of state for one mole of the *ideal gas*, $pV = RT$ (p is pressure, V is gas volume), is related to the Boltzmann constant by $k_B = R/N_A$, where N_A is the *Avogadro number*. The standard value based on the 1986 CODATA recommendations is $k_B = 1.38066 \cdot 10^{-23}$ J/K.

BOLTZMANN DISTRIBUTION

A *distribution function* of coordinates and momenta for particles or *quasi-particles* of an ideal nondegenerate gas in the state of thermodynamic equilibrium. The probability $\mathrm{d}W$ of a particle to have a momentum in a range $\boldsymbol{p} \Rightarrow \boldsymbol{p} + \mathrm{d}\boldsymbol{p}$ and a coordinate in a range $\boldsymbol{r} \Rightarrow \boldsymbol{r} + \mathrm{d}\boldsymbol{r}$ is related to the Boltzmann distribution $f_n(\boldsymbol{r}, \boldsymbol{p})$ by

$$\mathrm{d}W(\boldsymbol{r}, \boldsymbol{p}) = f_n(\boldsymbol{r}, \boldsymbol{p})\,\mathrm{d}\boldsymbol{r}\,\mathrm{d}\boldsymbol{p},$$

where in the classical case we have the approximation

$$f_n(\boldsymbol{r}, \boldsymbol{p}) = A \exp\left[-\frac{\varepsilon(\boldsymbol{p}) + U(\boldsymbol{r})}{k_B T}\right]. \tag{1}$$

Here $\varepsilon(\boldsymbol{p})$ is the kinetic energy, $U(\boldsymbol{r})$ is the potential energy, k_B is the *Boltzmann constant*, and A is the normalization factor found from the condition that the probability of a particle to be present in the phase space equals unity, i.e. $W = \int \mathrm{d}W = 1$. Sometimes the following expression obtained by integrating Eq. (1) over all momenta $\boldsymbol{p}$ is referred to as the Boltzmann distribution function:

$$f(\boldsymbol{r}) = A' \exp\left[-\frac{U(\boldsymbol{r})}{k_B T}\right],$$

while Eq. (1) is called the *Maxwell–Boltzmann distribution*. Integration of Eq. (1) over coordinates yields the *Maxwell distribution* of velocities. In a quantum description of particles characterized by a set of single-particle energy levels E_i, the Boltzmann distribution takes the form

$$\overline{n}_i = \exp\left[\frac{\mu - E_i}{k_B T}\right], \tag{2}$$

where $\overline{n}_i$ is the average number of particles in the state with energy E_i, μ is the chemical potential of the system that acts as the normalization factor determined by setting the sum $\sum_i n_i$ equal to the total number of particles, N. The Boltzmann distribution is the high temperature limiting case of the more accurate Bose–Einstein and Fermi–Dirac distribution functions (see *Bose–Einstein statistics*, *Fermi–Dirac statistics*) that take into account the quantum character of particle motion, and this limiting case is valid for an ideal gas with a small density of particles at high temperatures. In the physics of *semiconductors*, the Boltzmann distribution is widely used in the statistics of a gas of quasi-particles (*electrons* and *holes*) to determine their equilibrium concentrations and other specific parameters.

BOLTZMANN EQUATION (L. Boltzmann, 1872)

An integro-differential *kinetic equation* that provides the time (t) evolution of the one-particle distribution function (f) of a low-density monatomic gas. The Boltzmann equation has the form

$$\frac{\partial f}{\partial t} + \frac{1}{m}(\boldsymbol{p}, \nabla_{\boldsymbol{x}} f) + (F, \nabla_{\boldsymbol{p}} f) = \rho L_{\boldsymbol{p}}(f, f),$$

where $f(\boldsymbol{x}, \boldsymbol{p}, t)$ is the probability density function in the phase space of coordinates and momenta of a particle $(\boldsymbol{x}, \boldsymbol{p})$, m is the particle mass, F is the external force field applied at the point $(\boldsymbol{x}, \boldsymbol{p})$ of the phase space, and ρ is a dimensionless parameter proportional to the mean density of gas particles. The collision operator (or *collision integral*) L_p has the following form:

$$L_p(f,f) = \frac{1}{m}\int \mathrm{d}\boldsymbol{p}' \int \mathrm{d}\sigma\, |\boldsymbol{p}-\boldsymbol{p}'| \times \left\{ f(\boldsymbol{p}^*) f(\boldsymbol{p}'^*) - f(\boldsymbol{p}) f(\boldsymbol{p}') \right\},$$

where the explicit dependence of the function f on the coordinates and time is omitted, and $\mathrm{d}\sigma$ is the differential cross-section which includes conservation laws of both energy and momentum. Momenta $\boldsymbol{p}^*$, $\boldsymbol{p}'^*$ can be found from solving the scattering problem for two particles with incident momenta $\boldsymbol{p}$ and $\boldsymbol{p}'$ and a certain impact parameter, colliding in accordance with the laws of classical mechanics. When deriving the Boltzmann equation, the evolution of the function $f(\boldsymbol{x},\boldsymbol{p},t)$ within a small time interval is assumed to be determined by its value at a given instant of time t by means of the free motion of gas molecules and their pairwise collisions, provided the interaction time during the collision of two gas particles is much smaller than their mean free path time. The operator L_p is constructed taking into account the number of particles in a phase volume element. The Boltzmann equation is derived from the Liouville equation in the lowest-order approximation using an expansion in powers of the density. If a system is in statistical equilibrium the collision integral vanishes, and the *Maxwell distribution* will be a solution to the Boltzmann equation.

BOND

See *Chemical bonds, Covalent bond, Hydrogen bond, Ionic bond, Ionic-covalent bond, Metallic bond.*

BOOJUM

A *point defect* in the field of a *director* at the boundary of a nematic (see *Nematic liquid crystal*).

Boojum-induced elastic distortion spreads over the entire bulk medium. Displacement of a boojum into the bulk is energetically unfavorable because a boojum in this case would be a point defect connected to the surface, i.e. a *disclination.* The director distribution in the vicinity of a boojum depends on the orientation of nematic molecules near the boundary. When this orientation changes, e.g., as shown in the figure, from a tangential orientation to a normal one, a boojum can either disappear or transform into a *hedgehog* and move from the surface to the interior. Experimentally, boojums are observable in plane *textures of liquid crystals* in the form of defect points at the sample surface as well as in spherical nematic drops. The boojum was first studied by N.D. Mermin (1976) in superfluid ^{3}He-A. The nature of the ordering in the latter is similar to that in nematic liquids (see *Superfluid phases of ^{3}He*).

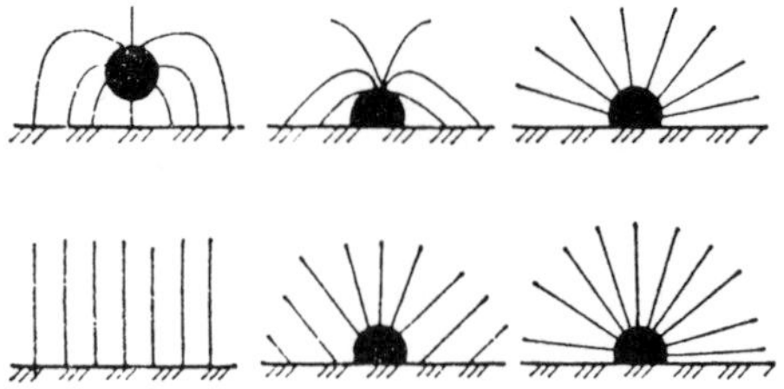

Change in the director distribution during the transformation of a boojum into a hedgehog (above), and into a uniform state (below), during the variation of the director angle relative to the surface. Each distribution is axially symmetric, and the figure presents sections in a vertical plane.

BORN APPROXIMATION (M. Born, 1926)

A quantum mechanical perturbation theory method of calculating the probability of scattering of a particle (*quasi-particle*) off a distance-decaying potential $V(\boldsymbol{r})$. It is based on using plane waves for both the initial wave function prior to scattering, and the final wavefunction far beyond the range of $V(\boldsymbol{r})$ after scattering. The *scattering amplitude* in first order perturbation theory (*first Born approximation*) is

$$A_{kk'} = -\frac{m}{2\pi\hbar^2}\int \mathrm{e}^{-\mathrm{i}(\boldsymbol{k}-\boldsymbol{k}')\boldsymbol{r}} V(\boldsymbol{r})\,\mathrm{d}^3x,$$

where m is the scattered particle mass, and $\boldsymbol{k}$ and $\boldsymbol{k}'$ are the wave vectors of the incoming and scattered states of the particle, respectively. Subsequent orders of perturbation theory take into account the deviation of the wave function of the particle from a plane wave, which is caused by the interaction potential. The criterion of the Born approximation is $|V(R)| \ll \hbar^2/(mR^2)$, where R is the range of the interaction potential. For an electron scattering off a Coulomb potential, the criterion of the Born approximation is $v \gg \overline{v}$ (v is the relative velocity of colliding particles, $\overline{v}$ is the velocity of electron motion in the Bohr orbit (see *Bohr radius*), $\overline{v} = Ze^2/(4\pi\varepsilon_0\hbar)$ (Z is the atomic number, e is the electron charge). The Born approximation finds wide use in the theory of current-carrier scattering and scattering of other microscopic particles in solids, in the treatment of ionization processes, e.g., light absorption by local electron centers (see *Local electronic levels*) in nonmetallic *crystals*, etc.

BORN–HABER CYCLE (M. Born, F. Haber, 1919)

A method of approximate determination of the atomic bond energy in crystals (mainly *alkali-halide crystals*) from experimental data on reaction energies whose sequence (cycle) results in the formation of a *crystal* from its gaseous atoms.

The bonding energy at the temperature of absolute zero is determined as a sum of energies needed for (i) heating the crystal to room temperature (1.5 to 3 kcal/mole); (ii) separating the crystal into the pure metal and the diatomic halide gas (60 to 150 kcal/mole); (iii) sublimation of the metal (20 to 40 kcal/mole); (iv) dissociation of the halogen into a monatomic gas (25 to 30 kcal/mole); (v) cooling the gases to absolute zero (to 2.9 kcal/mole); (vi) removing the valence electrons from alkali atoms (90 to 120 kcal/mole); (vii) formation of negative halide ions from atoms and electrons (70 to 80 kcal/mole). As a result of summing, the bonding energy turns out to equal ~180 kcal/mole for LiI, ~185 kcal/mole for NaCl, ~170 kcal/mole for KCl, ~160 kcal/mole for KBr, ~145 kcal/mole for CsI, and so on. Application of methods similar to the Born–Haber cycle allows one to estimate the energies of various reactions in *solids*, which can be included into one or another cycle, using the known energies of other components of the cycle, e.g., from the energy of formation of Frenkel pairs (see *Unstable Frenkel pair*) or Schottky pairs in different charge states (see *Schottky defect*).

BORN–VON KÁRMÁN BOUNDARY CONDITIONS (M. Born, Th. von Kármán, 1912)

Periodic boundary conditions imposed on the Bloch function of a particle in a periodic potential, or on a plane wave

$$\varphi(x + L_x, y, z) = \varphi(x, y + L_y, z) = \varphi(x, y, z + L_z) = \varphi(x, y, z),$$

where $L_i = N_i a_i$, $i = x, y, z$, are the Cartesian coordinates; a_i and N_i are the *crystal lattice* constant and the number of lattice periods in the direction i, respectively. Born–von Kármán boundary conditions allow one to determine the number of allowed states of electrons with different *quasi-wave vectors* $\boldsymbol{k}$ in a unit volume of the reciprocal space. Then, an extra (in addition to the *Bloch theorem*) condition $\psi_{k\lambda}(\boldsymbol{r} + \boldsymbol{L}) = \psi_{k\lambda}(\boldsymbol{r})$ (where $\boldsymbol{L}$ is a linear combination of vectors $\boldsymbol{L}_1, \boldsymbol{L}_2$, and $\boldsymbol{L}_3$ with integer coefficients) is imposed on the wave function. Hence, the number of allowed $\boldsymbol{k}$ values equals the number of *unit cells* in the volume $V = L_1 \times L_2 \times L_3$ and the density of permitted states in ***k**-space* equals $V/(2\pi)^3$. None of the observed values, including the *density of electron states* referred to one unit cell, or per unit volume of direct space, are dependent on the selection of L_1, L_2, L_3 or V.

BORON, B

A chemical element of Group III of the periodic system with atomic number 5 and atomic mass 10.81.

The isotope composition of natural boron: ^{10}B (19.57%) and ^{11}B (80.43%); the radioactive isotopes ^8B, ^{12}B and ^{13}B are known (half-lives are fractions of a second). Outer electronic shell configuration is $2s^2 2p^1$. Successive ionization energies are 8.298; 25.16 and 37.93 eV. Atomic radius is 0.091 nm; ionic radius of B^{2+} is 0.023 nm. The most typical oxidation state is +3, and in compounds with metals −3 (borides). In the free state boron exists in the form of brown fine-crystal powder (so-called *amorphous boron*) and of dark-gray crystals (*crystalline boron*). The most important boron modifications are simple α-rhombohedral B (I), tetragonal B (II), and complicated β-rhombohedral B (III). The α-rhombohedral boron lattice parameters are: $a = 0.506$ nm, $\alpha = 58.1°$, density is 2.365 g/cm^3; the tetragonal boron lattice parameters are $a = 1.012$, $c = 1.414$ nm, and the density is 2.365 g/cm^3; for β-rhombohedral boron we have: $a = 1.014$ nm, $\alpha = 65.3°$, density is 2330 kg/m^3. The modifications B (I) and B (II) under heating above 1773 K irreversibly transform to B (III). T_{melting} of B (III) is 2346 K; T_{boiling} is around 3973 K; thermal coefficient of the linear expansion is

$(1.1–8.3)\cdot 10^{-6}$ K^{-1} (from 293 to 1023 K). Thermal conductivity coefficient of crystalline boron is 25.96 $W\cdot m^{-1}\cdot K^{-1}$ (at room temperature); specific heat of amorphous boron is 12.97 $J\cdot mole^{-1}\cdot K^{-1}$, specific heat of rhombohedral boron is 11.1 $J\cdot mole^{-1}\cdot K^{-1}$. Standard enthalpy is 5.895 $J\cdot mole^{-1}\cdot K^{-1}$ for crystalline boron and 6.548 $J\cdot mole^{-1}\cdot K^{-1}$ for amorphous boron; Debye temperature is 1220 K.

Crystalline boron is a *semiconductor*, the forbidden gap width determined from electrical and optical measurements is 1.42 and 1.53 eV, respectively. Boron is a *diamagnet*, with specific magnetic susceptibility $-0.62\cdot 10^{-9}$ (at room temperature). It is characterized by high hardness. At low temperatures, boron is a brittle material; above 1800 °C, it begins to deform plastically.

Chemically, boron is a nonmetal. Elementary boron has low chemically activity: in the air it burns at 973 K, forming B_2O_3; with nitrogen above 1473 K, it forms *boron nitrides*; with carbon above 1573 K, it forms *boron carbides*. It does not interact directly with hydrogen, but a whole number of hydride compounds B_nH_m is known (see *Hydrides*), obtained indirectly using silicon. Above 1273 K, there are silicon borides; formed with phosphorus and arsenic above 1173 K are phosphides and arsenides. It finds wide application in the production of materials (*boroplastics*), high-temperature strong and superhard materials (e.g., *borazon*), neutron-adsorbing materials for nuclear power engineering, *semiconductor materials*, corrosion resistant and antifriction *coatings*.

BORON CARBIDE, B_4C

The only chemical compound of *boron* with *carbon* with a broad homogeneity range (molar content of C is 8.6 to 21%). It forms a rhombohedral lattice of *space group* $R3m$ (C^5_{3v}), and the parameters of the crystal lattice in the homogeneity range vary from $a = 0.5602$ to 0.5616 nm, $c = 1.2066$ to 1.2120 nm. $T_{melting} = 2400\,^{\circ}C$, density is 2.51 g/cm^3, *thermal conductivity* is 121.4 $W\cdot m^{-1}\cdot K^{-1}$ at 300 K, *microhardness* is 45.0 GPa, *elastic modulus* is 450 GPa. Boron carbide is a *semiconductor* of p-type, *band gap* width is 0.84 eV, electrical resistivity is 0.01 $\Omega\cdot m$. Boron carbide does not decompose in mineral acids and alkalis. It is obtained through interaction of B_2O_3 with coal in arc furnaces, by synthesis from the elements, by *crystallization* from melts, from their gaseous compounds. Boron carbide is used as an *abrasive material*, for the borating of metals and alloys, for adsorption of neutrons.

BORON NITRIDE, BN

Chemical compound of *boron* with nitrogen, with three modifications: α-BN, β-BN, γ-BN. The α-BN compound is hexagonal with space group $P6_3/mmc$ (D^4_{6h}), crystal lattice parameters: $a = 0.2504$ nm, $c = 0.6661$ nm; β-BN is cubic with sphalerite-type space group $F\bar{4}3m$ (T^2_d), $a = 0.3615$ nm; γ-BN has the hexagonal wurtzite-like structure, space group $P6_3mc$ (C^4_{3v}), $a = 0.255$ nm, $c = 0.423$ nm. The α-BN compound transforms into β-BN under *high pressure* and temperature ($P[\text{kbar}] = 11.5 + 0.025T$). In the course of heating under normal conditions, β-BN transforms to α-BN at temperatures above 1400 °C; α-BN transforms to γ-BN at shock compression 120 to 130 kbar. The density is 2.20 g/cm^3 for α-BN, 3.45 g/cm^3 for β-BN and 1.8 g/cm^3 for γ-BN; the thermal conductivity is 21 $W\cdot m^{-1}\cdot K^{-1}$ for α-BN and 1800 $W\cdot m^{-1}\cdot K^{-1}$ for β-BN. Microhardness of β-BN is 60 GPa; elastic modulus is 953 GPa. Boron nitride is an insulating material with electrical resistivity $\rho > 10^{15}$ $\Omega\cdot m$ and *band gap* width of β-BN ~10 eV. Boron nitride is obtained by nitration of boron. The α-BN form is used as fireproof and electrical insulating materials; β-BN is a tool and *abrasive material*.

Cubic boron nitride is a *superhard material* having no natural analogs. It was obtained by Robert H. Wentorf, Gr. (1956) under high pressure (above 4.0 GPa) and temperature (above 1470 K) from hexagonal boron nitride in the presence of alkali and alkali-earth metals, lead, antimony, tin and *nitrides* of the said metals. The rules of formation of binary compounds with a tetrahedral configuration are as follows. The components should belong to groups equidistant from Group IV of the periodic system; the average number of valence electrons in this compound should equal four. The majority of *binary alloys* with sphalerite and wurtzite structures satisfy these rules. Cubic boron nitride with sphalerite-type (β-BN) and wurtzite-type (γ-BN) lattices are characterized by the tetrahedral arrangement of atoms with each atom of one type surrounded by four atoms of the other type. In β-BN and γ-BN lattices the four B–N bonds have the same length 0.157 nm, with 109.5° angles between them. The crystal lattice of β-BN is similar to that of *diamond* with spacing 0.3615 nm. Cubic boron nitride is an insulator with a density $\approx$3.465 g/cm^3. Nanotubes can be made from boron nitride.

BORRMAN EFFECT

The same as *Anomalous passage of X-rays*.

BORRMAN TRIANGLE (G. Borrman)

A triangle in the scattering plane with one apex lying in the entry surface of a *crystal*, the two side lines directed along the wave vectors of the incident wave $\boldsymbol{K}_0$ and the Bragg-diffracted wave $\boldsymbol{K}_d$ (see *Bragg diffraction*), respectively, and the base formed by the segment of the exit surface of the crystal intercepted by these lines. If an X-ray beam is incident along $\boldsymbol{K}_0$ at a certain point on the entry surface, and the crystal is oriented for Bragg reflection, then the wave field inside a thin crystal propagates only within the Borrman triangle. Inside a Borrman triangle the response functions also have finite values (see *Dynamic radiation scattering*).

BOSE CONDENSATE

A macroscopically large number of particles with integer *spin* (*bosons*) in an ideal or slightly nonideal *Bose gas*, which are in their lowest state with zero point energy and momentum at a temperature below the *Bose–Einstein condensation* temperature. Since the Pauli exclusion principle does not apply, bosons can accumulate in the lowest energy quantum state, all with the same energy forming a coherent condensate. The latter is described by a common complex wave function $\Psi = |\Psi|\mathrm{e}^{\mathrm{i}\Theta}$ with a magnitude $|\Psi|$ and phase Θ that are constant or slowly varying in time. This fundamental property of bosons underlies the phenomena of *superfluidity* in liquid ^{4}He (see *Bose fluid*) and *superconductivity* in metals (see *Cooper pairs*). Related to this is the possibility of laser generation (see *Lasers*) of a macroscopically large number of photons with the same wavelength under the conditions of a nonequilibrium inverted *level population*, as well as the possibility of such an accumulation of photons in a resonator. An analog of a Bose condensate is the ground state in systems with *spontaneous symmetry breaking*, which is described by a complex *order parameter*, in particular, the so-called *Higgs field* (see *Higgs mechanism*).

BOSE CONDENSATION

The same as *Bose–Einstein condensation*.

BOSE–EINSTEIN CONDENSATION, Bose condensation

The property of Bose particles (bosons) with integer spins to accumulate in macroscopically large amounts in a single quantum state described by a unique (coherent) wave function (see *Bose condensate*). Bose–Einstein condensation is a property of three-dimensional Bose systems only. It is missing in two- and one-dimensional systems.

Bose–Einstein condensation of excitons and biexcitons is a variety this condensation for the case of *quasi-particles* (*excitons* and *biexcitons*, between which repulsion forces dominate). Bose–Einstein condensation of excitons interacting with *phonons* leads to renormalization of the energy spectrum of both subsystems and alters the shape of *light absorption* and *luminescence* bands. The bands contain additional sharp peaks generated by a condensed phase and sloping wings arising from condensate-free excitons. The Fermi nature of electrons and holes forming excitons leads to the non-Boseity of exciton operators. Excitons metallize when their concentration n_{ex} becomes comparable to the inverse of the volume a_{ex}^{-3} of the exciton, where a_{ex} is its Bohr radius. The Bose–Einstein condensation of excitons is realizable under the condition $na_{\mathrm{ex}}^3 < 1$ only.

Bose–Einstein condensation of dipole-active excitons and photons (or *polaritons* with wave vector $k_0 \neq 0$) is of special interest. In this case, the Bose-condensed state can be considered as an intense coherent wave described by a coherent Glauber state with a macroscopic amplitude. Equations describing inhomogeneous in space and time weakly of excitons and of photons have been derived; the diagram techniques for nonequilibrium states has been developed to describe coherent quasi-particles existing simultaneously with a condensate. On this basis, the features of coherent nonlinear propagation of light in densely condensed media are assigned to the exciton spectral range. The values of wave vectors and frequencies have been found, where, due to the absolute and convective instability of the spectrum, generation and amplification of waves is possible. Rearrangement of the energy spectrum of the phonoriton type (see *Phonoriton*) is due to peculiarities of the distribution functions of incoherent quasi-particles and conditions of stimulated *Brillouin scattering* of light. The role of quantum fluctuations and the influence of incoherent polaritons upon the statistical properties of an intense polariton wave have been clarified. A stimulated Bose–Einstein condensation of excitons and polaritons under the action of coherent resonant laser radiation, and a stimulated Bose–Einstein condensation of biexcitons obtained by two-photon excitation with counter-directed or parallel laser beams, were experimentally carried out.

BOSE–EINSTEIN STATISTICS (Sh. Bose, A. Einstein, 1924)

The statistics describing the distribution in quantum states of a system of very many identical particles with zero or integer (in $\hbar$ units) *spin*.

Particles as well as *quasi-particles* that obey Bose–Einstein statistics are called *bosons*; including *photons, phonons, magnons, excitons, bipolarons*, π-mesons, and so on. The wave function of bosons is symmetric with respect to the exchange of any pair of particles. A particular feature of bosons is the possibility for any number of them to be in the same quantum state. In view of this property, it follows from the *Gibbs distribution* that the average number of particles $\overline{n}_i$ in quantum state i with energy ε_i for a boson *ideal gas* (*Bose gas*) in *thermodynamic equilibrium* is

$$\overline{n}_i = \left\{\exp\left(\frac{\varepsilon_i - \mu}{k_B T}\right) - 1\right\}^{-1}, \qquad (1)$$

where k_B is the *Boltzmann constant*, μ is the chemical potential determined by the total number N of particles in the system, and $N = \sum_i \overline{n}_i(\mu)$. In a system with a variable number of particles, $\mu = 0$; under constant N, $\mu < 0$. Eq. (1) is called the *Bose–Einstein distribution*. At low temperatures, a finite portion of the particles pass to the lowest states with $\varepsilon = 0$; their number $N_0 = N[1 - (T/T_0)^{3/2}]$ where

$$T_0 \cong 3.31\left(\frac{N}{gV}\right)^{2/3}\frac{\hbar^2}{mk}$$

is called the *condensation temperature*, g is the *degeneracy* factor, m is the particle mass, V is the system volume. Other particles are distributed over states according to Eq. (1) at $\mu = 0$. Such a transition is called a *Bose–Einstein condensation* (Bose condensation). The Bose–Einstein distribution determines the efficiency of scattering of particles moving in a solid (*electrons* and *holes, neutrons, photons* etc.) by *photons, nonradiative quantum transitions* in solids, *phonon–phonon interactions* and also values of *electrical conductivity* and *thermal conductivity, specific heat*, acoustical properties, parameters of optical bands, spectral composition of black body radiation, and so on.

BOSE FLUID

A *quantum liquid* that condenses at low temperatures and consists of interacting *bosons*, i.e. particles with integer (in $\hbar$ units) intrinsic *angular momentum* (spin). An example of a Bose fluid is liquid *helium* (^{4}He) at $T < 4.2$ K. Below some critical temperature (see *Lambda point*) Bose fluids can transform into the superfluid state (see *Superfluidity*).

BOSE GAS (Sh. Bose)

A quantum system of weakly interacting particles (*quasi-particles*) with integer spins (bosons) subject to *Bose–Einstein statistics* with the *distribution function* (at spin $\sigma = 0$)

$$n_k = \left[\exp\left(\frac{\varepsilon_k - \mu}{k_B T}\right) - 1\right]^{-1},$$

where ε_k is the energy of the kth quantum state or of a free particle with momentum $\boldsymbol{k}$, μ is the chemical potential ($\mu \leqslant 0$), T is the absolute temperature. A Bose gas of neutral particles is ideal for a small density n, when $n^{1/3}r_0 \ll 1$, where r_0 denotes the action radius of exchange or *van der Waals forces*, and a Bose gas of electrically charged bosons (e.g., *bipolarons*) becomes weakly nonideal at a high density, when $n^{1/3}a_0 \gg 1$, where a_0 denotes the *Bohr radius* of particles. In a Bose gas with a given (constant) number of particles, under a pronounced decrease of temperature, *Bose–Einstein condensation* takes place, while in a Bose gas with a variable number of particles (quasi-particles) determined by the thermodynamical (thermal) equilibrium condition (e.g., the *phonon* gas in a *crystal* or in liquid *helium*), the Bose–Einstein condensation does not occur.

BOSE STATISTICS

The same as *Bose–Einstein statistics.*

BOTTLENECK OF PHONONS

See *Phonon bottleneck.*

BOUNDARY DIFFUSION

Diffusion over two-dimensional structural defects like *grain boundaries* and subgrains, i.e. over elements of imperfection of the crystal structure which separate parts of a bulk metal from *crystal lattices* of different space orientation, different *crystal structure*, or different material composition. Among the various types of boundary diffusion, the greatest interest involves material transfer over grain boundaries in polycrystalline materials. Boundary diffusion deviates from the usual Arrhenius linear dependence (see *Interdiffusion*) of $\log D$ versus $1/T$ where D is the *diffusion coefficient*; lower temperatures and grain pulverization result in greater departures from the *Arrhenius law.* The boundary diffusion coefficient in *metals* is usually several orders of magnitude higher than that for bulk diffusion. A model for quantitatively estimating diffusion parameters put forward by I.C. Fisher (1951) provides the expression

$$D_{\text{GB}}\delta = \frac{0.21(D/t)^{1/2}}{\tan^2\alpha},$$

where D_{GB} is the diffusion coefficient over the grain boundaries, t is the diffusion time, δ is the width of the boundary layer, and α is the slope of the straight line characterizing the dependence of the logarithm of the concentration on the diffusion depth. According to various authors the width of the boundary layer lies in the range from 0.3 nm to a few μm. The boundary layer has a so-called *diffusion width*, i.e. a region in which the mobility of atoms is higher than in the bulk. This width is approximately double the distance from the grain boundary to the point where the diffusing atom concentration is reduced by the factor of $e = 2.718$ from its maximum value. For some metals (e.g., for self-diffusion in *nickel*) this distance is 4 to 5 μm. Diffusion over a phase boundary is of great importance for *phase transition* kinetics. On a phase boundary, like on a grain boundary, a change of the crystal lattice state, *defect* clustering, and chemical *segregation* are possible factors that enhance the atomic mobility compared to that in the bulk. The principal methods for investigating grain diffusion are macro-, micro- and electron microscopic *autoradiography*, and the method of secondary ion–ion emission.

BOUNDARY EFFECTS in liquid crystals

Effects caused by the action of anisotropic surface forces. In technological applications and in laboratory studies *liquid crystals* are in contact with substrates and hosts (glass, *polymers*, isotropic liquids, cleavages of *monocrystals*, and so on). Anisotropic interphase interactions caused by the mutual action of various molecular forces (dispersion, dipole–dipole, steric) orient the molecules of a liquid crystal either normal to the boundary (*homeotropic orientation*), at an angle (*conical or tilted orientation*), or in the plane of the boundary (*planar orientation*). A quantitative measure of the orienting action of the substrate (i.e. of the external medium) on the liquid crystal is the cohesion energy (*anchoring energy*) W. The magnitude of W determines the work needed to turn the *director* (molecules) of the liquid crystal from its equilibrium position at the boundary through a certain angle. If W is comparable to or greater than the energy of intermolecular interactions in the liquid crystal then the cohesion is called strong; otherwise it is called weak. Owing to long-range orientational ordering (see *Orientational order*), the conditions of cohesion at the surface affect the bulk properties of the liquid crystal. For example, in the case of weak cohesion the threshold tension of a *Frederiks transition* is reduced; at the same time, the response time decreases with a simultaneous increase in the steepness of the voltage–contrast characteristic in a liquid crystal cell, and this allows one to increase the information capacity of *liquid-crystal displays.* Boundary effects influence not only in the type of orientation of the molecules on the surface, but also the *degree of order* in the interior. Near the surface the degree of order differs from that in the bulk (the former being higher, as a rule). It is experimentally possible to form, at the boundary of a *nematic liquid crystal*, a thin layer of *smectic liquid crystal* that is characterized by extra translational ordering of the molecules, and by a higher degree of orientational ordering of their long axes. The spatial inhomogeneity of the degree of ordering near the liquid crystal surface results in a macroscopic electric *polarization* of the subsurface layer which is referred to as *order electricity* (compare with the *flexoelectric effect*).

BOUND WATER, combined water

Water interacting with the surface of mineral particles, and thereby acquiring some special properties. It freezes at temperatures below 0 °C. The "bonding" mechanism of water in rocks involves the processes of physical and chemical *adsorption*, capillary *condensation* (see *Capillary phenomena*), and osmosis. The action of surface forces in *minerals* results in the perturbation of *hydrogen bonds* between the molecules of water, which is then transferred in succession from molecule to molecule of water. Bound water is usually separated into adsorbed, capillary, and osmotic types. Adsorbed water is more strongly held at the surface of minerals, and is therefore considered strongly bound, while capillary and osmotic water are both weakly bound. Strongly bound water is nonuniform. One may identify "island" water adsorption (bonding energy within 40–130 kJ/mol) and multilayer adsorption (bonding energy within 0.4–40 kJ/mol). Capillary and osmotic water are contained in rocks at humidities exceeding the maximum rock *hygroscopicity*. They both differ in their properties from either free or strongly bound water. Osmotic water may only be contained in rocks if some ions are present. It is the type most weakly bound to the surface, with its bonding energy remaining below 0.4 kJ/mol. The freezing temperature for capillary and osmotic water varies from −0.2 to −(2–3) °C.

BOUSSINESQ EQUATION (J. Boussinesq, 1871)

An equation of the form $U_{tt} - U_{xx} - 6(U^2)_{xx} - U_{xxxx} = 0$, where $U(x,t)$ is the velocity field; and x and t are, respectively, the space and time coordinates written in a dimensionless form (subscripts denote partial differentiation).

This equation was introduced by Boussinesq to describe the propagation of waves of small but finite amplitude when the dissipation is negligible, and characteristic wavelengths significantly exceed the reservoir depth. In solid-state physics, the Boussinesq equation is also referred to as a *nonlinear string equation* in connection with its application to the theory of nonlinear waves of a one-dimensional lattice and a nonlinear string. The Boussinesq equation is more general than the *Korteweg–de Vries equation* (see *Soliton*) because it takes into account motion in both directions. The Boussinesq equation is completely integrable by the inverse spectral transformation associated with a scattering problem for a third order operator. Therefore, it has all attributes pertinent for integrable systems: an infinite series of polynomial conservation laws, Bäcklund transformation, and many-soliton solutions. The simplest single-soliton solution of the Boussinesq equation is

$$U(x,t) = \frac{1}{4}(c^2 - 1)\mathrm{Sech}^2\left(\frac{x \pm ct}{\Delta}\right),$$

where c is the soliton velocity, and $\Delta = 2(c^2 - 1)^{-1/2}$ is its width. It is a single parameter family of stationary solitary waves. There are several modifications of the Boussinesq equations which cannot be integrated. In particular, $U_{tt} - U_{xx} - 6(U^2)_{xx} - U_{xxtt} = 0$ is the adjusted Boussinesq equation used to describe longitudinal acoustic waves in *rods*; and $U_{tt} - U_{xx} - 6(U^3)_{xx} - U_{xxtt} = 0$ is the modified adjusted Boussinesq equation used in the theory of nonlinear seismic waves.

BOUSSINESQ–GALYORKIN SOLUTION (J. Boussinesq, B.G. Galyorkin)

A general solution of the equilibrium equations of an elastic medium in the absence of *bulk forces* expressed via an arbitrary biharmonic vector. If the equilibrium equation is written in the form $(1 - 2\nu)\nabla^2 \boldsymbol{u} + \mathrm{grad\,div}\,\boldsymbol{u} = 0$, where $\boldsymbol{u}$ is the displacement vector, and ν is the *Poisson ratio*, then the Boussinesq–Galyorkin solution has the form:

$$\boldsymbol{u} = \nabla^2 \boldsymbol{f} - \left(\frac{1}{2}(1 - \nu)\right) \times \mathrm{grad\,div}\,\boldsymbol{f},$$

where f is an arbitrary biharmonic vector, i.e. a vector satisfying the equation $\nabla^2\nabla^2 f = 0$.

BRAGG DIFFRACTION (W.L. Bragg)

A type of diffraction of X-rays, neutrons, electrons, etc., by crystals when the reflection from each plane of atoms is specular, and the rays reflected from successive planes constructively interfere with each other. Symmetric and asymmetric Bragg diffraction are distinguished in which the reflecting planes (hkl) are parallel or not parallel to the entrance plane of the crystal, respectively. The most important effect in Bragg diffraction is the total interference reflection of the transmitted wave in a small angular interval (for X-rays, a few seconds of arc) due to the *extinction* effect when the wave field penetrates a crystal to a depth of the order of the *extinction length*. The reflectivity of a crystal of finite thickness in Bragg diffraction has some subsidiary maxima in its angular dependence, which are the analogues of the pendulum solution in *Laue diffraction*. Bragg diffraction is mostly employed to check the structural perfection of the near-surface layers of crystals. Unlike Laue diffraction, Bragg diffraction is not limited by sample thickness t; and for thick samples $t \gg t_n$, where t_n is the absorption length, the diffraction pattern is simpler, and it is more readily interpreted.

BRAGG DOUBLE REFLECTIONS

This diffraction occurs when an X-ray, electron, neutron, etc., beam reflected in accordance with the *Bragg law* from a family of parallel planes in a *crystal* reflects again from the planes of the same family.

The direction of propagation and the intensity of Bragg double reflections are determined by the *crystal structure* of a crystal, and by the experimental conditions. While the first and the second reflections occur in crystal regions disoriented by the angle δ, the direction of Bragg double reflections differs from that of the primary radiation by the angle $\varepsilon = \delta \sin\theta$ where θ is the Bragg angle. There are three kinds of Bragg double reflections in *polycrystals*: intrablock, intragrain and intergrain ones. At large scattering angles ($\varepsilon \geqslant 10°$), these reflections are veiled by the diffuse background due to Compton scattering (see *Compton effect*) and thermal scattering. At small scattering angles ($\varepsilon \leqslant 1°$), the contribution of Bragg double reflections is significant and exceeds the thermal and the Compton scattering by several orders of magnitude. The *small-angle scattering of X-rays* due to Bragg double reflection allows one to determine the size of grains and blocks in polycrystals, their orientation distribution, and the angular block disorientation (see *Bragg multiple scattering*).

BRAGG LAW, Bragg relation (W.H. Bragg, W.L. Bragg, G.V. Vulf, 1913)

The main equation of *X-ray diffraction*. X-ray diffraction from the atomic lattice of a monocrystal can be formally described as reflection from atomic planes. According to the Bragg law, the angle of reflection (*Bragg angle*) θ between the atomic plane and diffracted beam satisfies the condition

$$2d_{hkl} \sin\theta = n\lambda,$$

where d_{hkl} is the interplane distance between the series of parallel atomic planes with *crystallographic indices* (hkl), λ is the X-ray wavelength, and $n = 1, 2, 3, \ldots$ is the order of the reflection.

BRAGG MULTIPLE SCATTERING

Scattering of radiation in a *crystal* under the conditions when the Bragg relation (see *Bragg law*) is valid for two or n reflections simultaneously. One can represent the conditions of the Bragg multiple scattering with the help of an *Ewald sphere* in ***k****-space*: the Ewald sphere with radius $|\boldsymbol{k}_0| = 1/\lambda$ corresponds to the incident radiation with wavelength λ. Vectors $\boldsymbol{k}_0$, $\boldsymbol{k}_1$, $\boldsymbol{k}_2$ corresponding to the incident wave and waves reflected from *reciprocal lattice* sites (1) and (2) for the case of triple Bragg scattering are shown in the figure ($n = 3$, including the primary beam). Here, $|\boldsymbol{k}_0| = |\boldsymbol{k}_1| = |\boldsymbol{k}_2|$. By rotating the crystal about the scattering vector (straight line H_{01} connecting sites (0) and (1)), one can remove site (2) from the Ewald sphere, i.e. violate the Bragg condition for the site and thus exclude the triple dynamic scattering. At this, site (1) remains at the sphere and the Bragg condition applies to it (see *Bragg double reflections*). The greater the dimensions of a crystal *unit cell* and the shorter the wavelength λ (i.e. the larger the Ewald sphere radius),

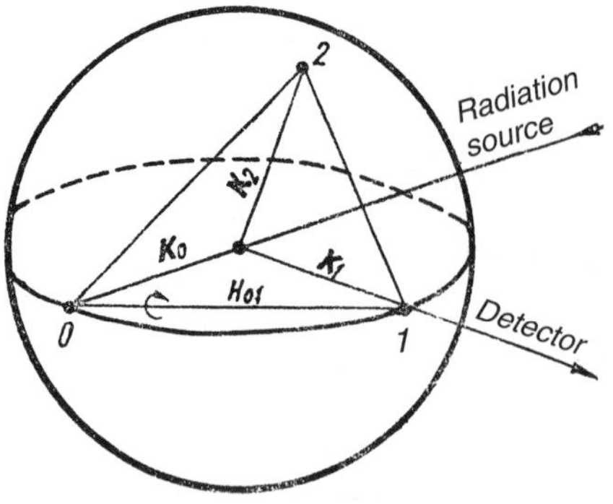

Bragg multiple scattering.

the greater the number of reciprocal lattice sites that can be on the sphere. That is, the more difficult it will be to find a position of the crystal where there is no Bragg multiple scattering. The evaluation of the contribution of the Bragg multiple scattering to the intensity of the measured reflection is quite a difficult problem to solve, considering that it is complicated by the crystal mosaicity, the divergence, and the nonmonochromaticity of the incident radiation. Bragg multiple scattering allows one to observe reflections in those directions where the Bragg scattering is prohibited by the *crystal symmetry* conditions; this is the so-called *Renninger effect* (M. Renninger, 1937). Bragg multiple scattering in crystals can also play a positive role, e.g., it underlies one of the methods of high-precision determination of crystal lattice parameters.

BRAGG RULE

This rule involves *charged particle energy loss* during the passage through a material with the chemical formula $A_lB_mC_n\ldots$ and it is expressed by the approximate relation

$$-\frac{dE}{dZ}=l\left(-\frac{dE}{dZ}\right)_A+m\left(-\frac{dE}{dZ}\right)_B+n\left(-\frac{dE}{dZ}\right)_C+\cdots,$$

where $(-dE/dZ)_i$, $i = A, B, C$, are the energy losses per unit length due to the ith component of the material.

BRAKING RADIATION

See *Bremsstrahlung.*

BRANCHING OF FRACTURES

A process of the separation (bifurcation or polyfurcation) of a main-line *fracture* into two or more secondary ones. One can distinguish quasistatic branching of fractures caused, in particular, by processes of *plastic deformation* at its vertex, and dynamic branching which is observed when the fracture attains about 60% of the speed of transverse elastic waves. Due to this the stresses in front of the fast moving fracture stop being a maximum, and the *failure* becomes more probable in other directions. Dynamic branching of fractures has been observed in hardened glasses (see *Vitreous state of matter*), *steels*, Plexiglas and epoxy resins. Branching arises in the core layers of a sample at the front of the main-line fracture, and is directed along the tangent to the curved segment of the fracture trajectory. Thus, a secondary fracture is oriented at an angle of approximately 30° with respect to the primary one. After each act of branching the speed of the main-line fracture decreases. Therefore, multiple branching produces destructions that move in jumps with high accelerations, up to 10^8 m/s^2, thus managing to branch at small distances (e.g., in the millimeter range for hardened steel), break loose, pick up critical speed, and then branch again. The branching is stimulated by internal residual stresses that arise in the sample being fractured.

BRASS

Alloy of *copper* and *zinc* belonging to the class of compounds with *metallic bonding.* It exists over a wide range of concentration. At low Zn content the *alpha phase* (face-centered cubic lattice) is formed. At higher Zn content the stability of the new phases (the so-called "electronic phases") is determined by the average concentration n of valence electrons per atom via the proper matching of the Fermi surface and the Brillouin zone. The three electronic phases are the *beta phase* (compound CuZn, body-centered lattice, $n = 3/2$, below 460 °C ordered according to the CsCl type), *gamma phase* (compound Cu_3Zn_8, complex cubic lattice, ordered, contains 52 atoms, $n = 21/13$) and *epsilon phase* (compound $CuZn_3$, hexagonal

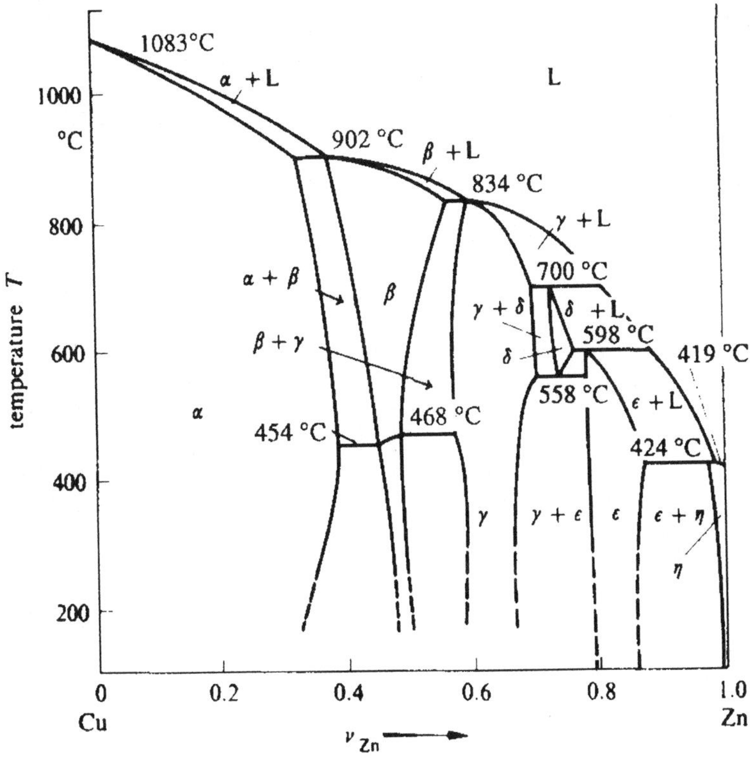

Brass phase diagram.

close-packed lattice, $n = 7/4$). Unusual mechanical properties are found in *β-brass*, caused by a *martensitic transformation* (*shape memory effect*, *superelasticity*, high damping). The figure shows the phase diagram.

BRAVAIS LATTICES (A. Bravais, 1848)

The classification of three-dimensional *lattices* into 14 types that takes into account the full point symmetry groups (*holohedries*) and the parallel translation groups.

If a *Bravais coordinate system* (chosen along the most symmetrical directions of the lattice) coincides with the principal lattice system then the *Bravais parallelepiped* constructed on it contains lattice sites only at its apices, i.e. is a *primitive parallelepiped*. Such lattices are called *primitive or P-lattices*; all other lattices are referred to as *centered lattices*. There are three ways to center the Bravais parallelepipeds, I – body-centered, F – face-centered, and C – base (or side)-centered. Two lattices are included in the same *Bravais type* if their coordinates are the same and the parallelepipeds are centered identically. The Bravais classification has also a group theory basis: two lattices are included in the same Bravais type if their full symmetry groups are isomorphous (see *Isomorphism*). These groups (14 out of 73 symmorphic *space groups*) are called *Bravais groups* and are designated by the symbols of the space groups. Each of 14 Bravais types is presented in Table 1 (see next page) by a corresponding Bravais parallelepiped with points at the apices, and where appropriate, with centering points indicated (see inside front cover).

BREAKDOWN of solids

A sharp increase of electric current density in *insulators* and semiconductors that occurs when a certain critical level of the applied electric field strength, E_{crit}, is reached. The value E_{crit} is called the *electric strength*. The value of E_{crit} falls within

Table 1. Bravais lattices

System	Lattice type			
	Primitive	Base-centered	Body-centered	Face-centered
Triclinic	 P			
Monoclinic	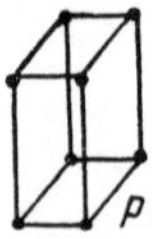 P	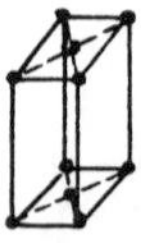 C		
Orthorhombic	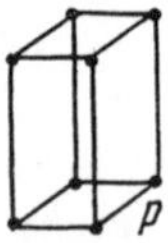 P	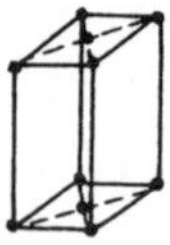 C	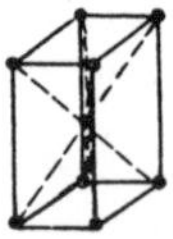 I	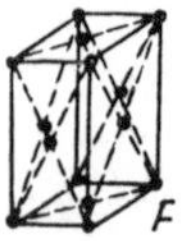 F
Trigonal (rhombohedral)	 P			
Tetragonal	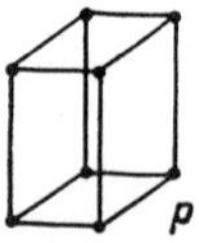 P		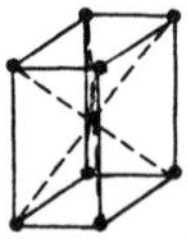 I	
Hexagonal	 P			
Cubic	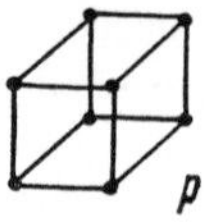 P		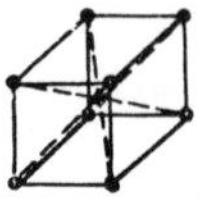 I	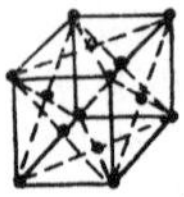 F

10^5–10^7 V/cm for various dielectrics and starts from 10^5 V/cm for semiconductors. During *impact ionization* it may be as low as 1 to 10 V/cm. One may distinguish between the electric and thermal mechanisms of breakdown. The mechanism of *electric breakdown* is typically characterized by an avalanche increase of the number of free carriers (electrons or *electron–hole pairs*), for example due to ionization by collision. *Current filamentation* may also occur, i.e. a conducting channel in which the current density is higher than its average across the sample may form. This effect results from an instability of the spatially homogeneous distribution of carriers when the sample *current–voltage characteristic* has an S-shape. In the case of an electric breakdown the value of E_{crit} only weakly depends on the sample temperature, thickness, and the duration of electric field pulse, τ, provided $\tau = 10^{-7}$–10^{-6} s. *Thermal breakdown* results from melting of the material via Joule heating when its thermal equilibrium is disturbed (thermal breakdown without a disturbance of thermal equilibrium is also known, the so-called *type II thermal breakdown*). A combined mechanism of breakdown is possible, when the Joule heating of a particular region of the sample works to lower the value of E_{crit} necessary for an electrical breakdown in that region. The application of a strong field may result in a gradual change of the composition and structure of an insulator, so that both eventually work to lower the value of E_{crit}. These effects could, for example, lead to the formation of *color centers* in *alkali-halide crystals* or to erosion of *polymer* surfaces. Such processes are called *electric aging of insulators*. In contrast, dielectric breakdown in *semiconductors* may not necessarily be accompanied by changes in the crystal lattice. In certain cases, when the conduction electron concentration is limited and the field strength, E is moderate, the initial process of breakdown does not result in the failure of the material, and may be repeatedly reproduced (see *Impurity breakdown, Thermoelectric instability of photocurrent, Streamer breakdown*).

For different varieties, see also *Acoustic breakdown, Electric breakdown, Magnetic breakdown.*

BREAKDOWN NOISE

See *Telegraph noise.*

BREAKING STRAIN

Strain that results in the break-up of a body into several macroscopic parts (*failure*). The breaking strain produces such a degree of deformation that it results in failure. This strain is one of the principal characteristics of a material, delimiting its *margin of plasticity.* It is determined during the *mechanical testing of materials.*

Breaking strain depends on the type of *state of stress* (distension, compression, *flexure* etc.), i.e. on the "rigidity" of various types of mechanical tests, and on the *softness factor* $a = t_{max}/S^n_{max}$, which is the ratio of the maximum tangential tension, t_{max} to the maximum normal reduced stress, S^n_{max}. With an increase in the material capability to withstand stress the *plastic deformation* increases. Breaking strain also depends on the type of *crystal lattice*, the structure, testing temperature, environment, etc. When determining the level of breaking strain the scheme reproducing both the tensed and the deformed states of the specimen during tests should be as close as possible to the actual operating conditions of the material.

BREAKING STRESS, failure stress, failing stress

A stress reached during distension tests, which result in the breakup of the specimen into two or more parts. Breaking stress is defined by the ratio of the load to the specimen cross-section at the moment of *failure.* In brittle metals (see *Alloy brittleness*) this stress is achieved via the opening of brittle cracks when the external stress remains below the *yield limit*, and macroscopic *plastic deformation* is absent. When plastic metals subjected to stresses exceed the yield limit, there develops a *tough failure* or a *quasi-brittle failure* in them, while the failure stress exceeds the yield limit because of *strain hardening.* Failure stress in brittle metals depends only weakly on the temperature, approximately as the *elastic modulus*, while its dependence in plastic materials is controlled mainly by the temperature dependence of the yield limit, and may happen to be stronger.

BREIT–RABI FORMULA (G. Breit, I. Rabi)

An exact solution of the secular equation for the *spin Hamiltonian* of a paramagnetic ion with electron effective *spin* $S = 1/2$, nuclear spin I, and isotropic $\boldsymbol{g}$, $\boldsymbol{g}_n$ and hyperfine ($\boldsymbol{A}$) tensors. The energy levels are as follows:

$$E(F, m_F) = -\frac{\Delta}{2(2I+1)} - g_n\beta_n B m_F \pm \frac{1}{2}\Delta\left[1 + x^2 + \frac{4xm_F}{2I+1}\right]^{1/2},$$

where $F = I + 1/2$, $m_F = m \pm 1/2$, $m = -I, \ldots, +I$; $\Delta = a(I + 1/2)$ is the splitting in zero magnetic field $\boldsymbol{B}$; a is the isotropic *hyperfine interaction* constant, $x = (g\beta + g_n\beta_n)B/\Delta$, and β and β_n are the Bohr and nuclear *magnetons*, respectively.

BREMSSTRAHLUNG, braking radiation, collision radiation

An emission of *photons* by electrically charged particles decelerating in an electric field (or in a magnetic field, called *magnetic Bremsstrahlung* or *synchrotron radiation*). The intensity of Bremsstrahlung is proportional to the square of the particle acceleration, i.e. proportional to Z^2/M^2 where Z is the atomic number, and M is the mass of the nucleus being decelerated by the electric field. Owing to the small mass of electrons and *positrons*, the intensity of their Bremsstrahlung is $3 \cdot 10^6$ times greater than that for protons. This leads to a substantial contribution to the Bremsstrahlung by the energy loss of fast electrons and positrons in a medium (see *Charged particle energy loss*). The spectrum of Bremsstrahlung is continuous. The maximum value of the energy of the emitted photon is $\hbar\omega_M = E - mc^2$ (here E is the total energy, m is the electron mass, and c is the speed of light).

In a *crystal* the constructive interference of photons scattered by closely situated atoms of the crystal influences some specific features of the Bremsstrahlung, in particular the intensity of the Bremsstrahlung increases (see *Coherent Bremsstrahlung*). These changes are especially significant in the case of the *channeling* of electrons and positrons (see *Quasi-characteristic radiation*). See also *Radiation by charged particles.*

BREMSSTRAHLUNG, COHERENT

See *Coherent Bremsstrahlung.*

BRILLOUIN FUNCTION (L. Brillouin, 1926)

Function $B_j(x)$ introduced to describe *paramagnetism* of noninteracting atoms. For atoms with the angular momentum j (in units of *Planck constant* $\hbar$), the function is

$$B_j(x) = \frac{2j+1}{2j}\coth\left(\frac{2j+1}{2j}x\right) - \frac{1}{2j}\coth\left(\frac{x}{2j}\right).$$

The Brillouin function determines the mean value of the projection of j on the direction of the magnetic field H (directed along the z axis) $\langle j_z\rangle = jB_j(x)$, $x = g_j|\mu_B|H/(k_BT)$, where μ_B is the Bohr *magneton* and g_j is the *Landé g-factor.* The Brillouin function is the basis for the quantum theory of paramagnetism of substances with uncompensated atomic magnetic moments, and of magnetically ordered materials (when described in the framework of the *molecular field* approximation; in this case, H is the effective magnetic field).

BRILLOUIN SCATTERING, also called Mandelshtam–Brillouin scattering

Inelastic *light scattering* by thermal acoustic *phonons* in condensed media with the light frequency changed by a value equal to a phonon frequency. (Analogous processes by optical phonons are called *Raman scattering of light*). The scattering is caused by fluctuating inhomogeneities of the *refractive index*, which are related to propagation of thermal (Debye) excitations of a medium. From the classical point of view, Brillouin scattering can be interpreted as result of the *diffraction of light* by thermal elastic waves. By virtue of the diffracted light interference, only waves with wavelength Λ, which satisfies the Bragg condition $\lambda_l = 2n\Lambda\sin(\theta/2)$ (where n is the refractive index of the medium, λ_l is wavelength in vacuum, θ is angle of scattering), are responsible for scattering in a given direction. Furthermore, due to the *Doppler effect*, the light with frequency ω_l, which is scattered by elastic waves, undergoes a frequency change $\omega_s = \omega_l \pm \omega_0$, defined by the relationship $\omega_0/\omega_l = \pm v/(\Lambda\omega_l) = \pm 2n(v/c)\sin(\theta/2)$, where c is the speed of light in vacuo, and v is the *sound velocity* in the medium. Elastic waves with frequency $\sim 10^{10}$ Hz (hypersonic waves) are of importance in the light scattering. The *line width* is determined by hypersound attenuation.

The scattering efficiency is of the order of 10^{-6} scattered photons per incident photon. However, if the intensity of the incident light exceeds some threshold value ($\sim 10^{10}$ W/cm^2), an avalanche-like growth of the amplitudes of sound and of scattered waves occurs, a process called *stimulated Brillouin scattering*. Brillouin scattering is used for studying material properties: sound wave propagation velocity and decay rate, molecular structure, intermolecular interactions, etc. Stimulated Brillouin scattering is used for the generation of intense hypersonic waves, and in experiments on *wave front reversal* of light waves. Resonance Brillouin scattering is an effective method for investigating excitonic *polaritons.*

BRILLOUIN ZONE

A limited region in the reciprocal space (***k***-*space*) of wave vectors inside which the *quasiparticle* energies in the crystal take on the entire spectrum of permitted values. The overall $\boldsymbol{k}$-space can be partitioned into an infinite number of Brillouin zones of the same dimensions, and equal in volume to a *reciprocal lattice* cell. The use of a single Brillouin zone is due to the crystal periodicity since the quasi-particle energy $E(\boldsymbol{k})$ is periodic in the quasi-momentum space with the periodicity of a reciprocal lattice vector $\boldsymbol{G}$: $E(\boldsymbol{k}) = E(\boldsymbol{k} + \boldsymbol{G})$. A region with its center at the point $\boldsymbol{k} = 0$ is chosen as the first Brillouin zone, or simply referred to as the Brillouin zone. It is delimited by planes passing through the mid-points of the lines (normally to them) that connect point $\boldsymbol{k} = 0$ with neighboring sites of the reciprocal lattice. A Brillouin zone is analogous to a *Wigner–Seitz cell* in real (coordinate) space. Its boundaries are determined by the reciprocal space form of the Bragg equation $2\boldsymbol{k}\cdot\boldsymbol{G} = \boldsymbol{G}^2$, involving a reciprocal lattice vector. At the boundaries the quasi-particle energy $E(\boldsymbol{k})$ reaches an extremum. To describe some physical phenomena it is convenient to use the extended Brillouin zone scheme which makes use of the periodicity of the quasi-particle spectrum in $\boldsymbol{k}$-space.

BRITTLE CRACK

A *crack*, whose propagation is not followed by a *plastic deformation*. During brittle crack propagation the *theoretical strength* is reached at its vertex. The breaking of interatomic bonds occurs without activation. Elastic energy, stored in the material during a deformation process, converts into *surface energy* at its propagation. The critical stress for the brittle crack propagation is determined by the Griffith equation (see *Griffith theory*).

BRITTLE FAILURE

A type of solid *failure* involving purely elastic *strains*. Ideally only materials with no capability for *plastic deformation* (e.g., glass, ceramics, diamonds, some minerals) undergo brittle failure under conditions of ordinary tension. Metals, due to the presence of *dislocations*, behave as ideally brittle only in special circumstances when the failure is initiated in extremely small regions. These regions have an ideal (defect free) structure in *perfect crystals* (whiskers), or in microregions near the tip of atomically-sharp *submicrocracks*, where a fracture of atomic bonds occurs through *microspalling* (without attendant plastic deformation). However, in so far as the origin of microcracks in metals is possible only as a result of the displacement and *pile-up of dislocations*, so a failure in metallic materials can only be of a quasi-brittle type (see *Quasi-brittle failure*) since a condition of purely elastic deformation does not appear at their failure.

In practice when a failure of products, constructions and installations (e.g., hulls of ships) occurs, it is considered to be the brittle type, if an overall deformation of the product was elastic before its failure, in spite of the fact that a large local plastic deformation took place in the focus of failure. Hence in practice the term "brittle failure" does not have a precise meaning. However, it can correspond to a complicated type of failure combining features of overall elastic and local plastic deformation. The latter forms only in small regions near cracks, or in action zones of large residual stresses of a welding or quenching origin. An important feature of such a failure is that the average (nominal) failure stress should be less than the *yield limit* of the material. Excessively high strength of the material, low temperature (*cold brittleness*), high velocity of deforming (a shock load), the presence of *cracks* or other concentrators of stresses, the presence of large residual macrostresses of the first type; all of these factors contribute to the significance of brittleness in engineering practice.

BRITTLE FRACTURE

A *fracture* arising from *brittle failure*. This type of *failure* is associated with failure by *spalling*, when a progression of the crack is not accompanied by *plastic deformation*, and failure occurs over the plane of *crystal cleavage* through the breaking of atomic bonds. Actual brittle failure has been found in *covalent crystals* such as *diamond* and *silicon carbide*. A brittle *transcrystalline failure* is possible also in other crystals with a high degree of covalent bonding, where *dislocations* are practically "frozen in". The typical feature appearing on the surface of a brittle fracture is a *step of spalling* arising at the intersection of a *screw dislocation* and a crack front. A junction of the splitting steps makes up a stream-like pattern. The same features of fracture are typical for *quasi-brittle failure* through splitting accompanied by plastic deformation of the region adjacent to the crack. Such a failure is typical for crystals with body centered cubic lattices at ambient temperatures.

Intercrystalline brittle failure occurs at sharply weakened intergranular linkages. It is observed along grain boundaries through the catastrophic crack progress due to the *segregation* of impurities, and also to the presence of *internal stresses*. Intercrystalline failures along boundaries of grains and subgrains are distinguished. When a failure runs along grain boundaries a fracture is completely free from a stream-like pattern, linkage points of three (or more) boundaries of grains appear, and often a "bubble" due to the segregation of impurities is observed on *facets* of split grains.

BRITTLENESS AND TOUGHNESS CRITERION

This is a condition determining the relation of parameters responsible for the type of *failure* of a solid under load. There is no unique brittleness and toughness criterion, since these criteria differ in various areas of science involving failure problems. In solid state physics, the brittleness and toughness criterion is connected with the possibility of plastic *stress relaxation* via the generation of *dislocations* by a vertex of a perfectly sharp *crack* in a *crystal*. This possibility depends on the relation between *theoretical strength* against shear τ_{theor} and rupture σ_{theor} and maximum tangential $\tau_{\max}$ and tensile stress $\sigma_{\max}$ at the vertex of the crack. If $\tau_{\text{theor}}/\sigma_{\text{theor}} < \tau_{\max}/\sigma_{\max}$, the crystal is tough (FCC metals: *copper*, *silver*, *gold*); otherwise, it is brittle (diamond and BCC metals: *tungsten*, *iron*, *sodium*). In physical *materials science*, as a brittleness and toughness criterion, the one based on the *microspalling criterion* my be chosen: if $j > K_{\text{t}}$, then the failure is brittle, if $j < K_{\text{t}}$, then it is tough (more precisely, a *quasi-brittle failure*). According to the fractographic classification of destruction types, one does not succeed in giving a clear distinction between brittleness and *toughness* in metals, and brittleness and toughness criteria are qualitative: if in a *fracture* of brilliant *facets* of a spalling no appreciable traces of *plastic deformation* are seen, then the failure is supposed to be of a brittle (or quasi-brittle) type; if typical *pores* or pits are observed (a fibrous break), the failure is supposed to be tough. In mechanics, the engineering criterion of brittleness is used, whether the product is destroyed by the average (nominal) stress σ_{n} below the *yield limit* σ_{y}, i.e. when $\sigma_{\text{n}} < \sigma_{\text{y}}$, which can be satisfied only in the presence of stress concentrators (see *Stress concentration*).

BRITTLENESS, BLUE

See *Blue brittleness*.

BRITTLENESS OF ALLOYS

See *Alloy brittleness*.

BROADENING OF A LINE

See *Line (level) width*.

BROKEN BOND

The same as *Dangling bond*.

BROOKS–HERRING FORMULA (H. Brooks, C. Herring)

This expression determines the energy dependence of the mean free path time $\tau(E)$ of charge carriers of energy E in *semiconductors* when the scattering takes place predominantly from ionized impurities (low temperatures, high impurity concentrations). The Brooks–Herring formula is

$$\tau(E) = \frac{\varepsilon^2(2m^*E^3)^{1/2}}{\pi e^4 N\Phi(x)},$$

where ε is the dimensionless *dielectric constant*; m^* is the *effective mass* of the carriers, N is the impurity concentration, $\Phi(x) = \ln(1+x) - x/(1+x)$ where $x = 8m^*E/(\hbar^2 q^2)$, q is the inverse Debye *screening radius*. It follows from the Brooks–Herring formula that scattering by ionized impurities becomes more efficient at small carrier energies and, consequently, at low temperatures. The Brooks–Herring formula was derived in the *Born approximation* of collision theory taking into account the screening of impurities by free carriers. In its derivation the impurities are assumed to be arranged in a *crystal lattice* in a disordered way (see *Current carrier scattering*).

BUBBLE DOMAIN

See *Magnetic bubble domain*.

BUBBLE DOMAIN LATTICE

See *Magnetic bubble domain lattice*.

BUCKYBALL

See *Fullerene.*

BULK ACOUSTIC WAVES

Propagating periodic *strains* of an elastic medium obeying the wave equation $\rho\partial^2u_i/\partial t^2 - c_{ijkl}\partial^2u_k/\partial x_j\partial x_l = f_i$, where ρ is the medium density, $\boldsymbol{u}$ is the displacement of the medium element, c_{ijkl} are the components of the *elastic modulus tensor*, and $\boldsymbol{f}$ is the bulk density of the external (nonelastic) forces.

In contrast to liquids and gases, *solids* can support the propagation of traverse waves in addition to longitudinal waves; the displacements in the former being normal to the propagation direction. Three natural modes of acoustic waves exist in an isotropic body (see *Isotropy*): one *longitudinal acoustic wave* with the velocity $v_\| = (c_{1111}/\rho)^{1/2}$, and two degenerate *transverse acoustic waves* with $v_\perp = (c_{1212}/\rho)^{1/2}$. In an anisotropic medium (such as a *crystal*) pure longitudinal and traverse waves can propagate only in preferred directions which have sufficiently high symmetry. For an arbitrary propagation direction, there are three modes of bulk waves with different *phase velocities* (a typical *sound velocity* in solids is $v \approx 10^5$ cm/s).

Acoustic waves can be classified by their frequency. Waves in the audible and infrasonic range ($f < 10^4$ Hz) lack important applications in solid state physics due to their long wavelength. Waves with frequencies from 20 kHz to 1 GHz are called ultrasonic (*ultrasound*), and are widely used in physics and engineering. The frequency range 10^9 to 10^{12} Hz, called hypersonic (*hypersound*), exhibits quite sizable attenuation (attenuation coefficient $\kappa > 1$ cm^{-1}). Acoustic waves with frequencies above 10^{12} Hz are called *terasound*, and exhibit a significant dispersion as their wavelength becomes comparable to the *crystal lattice* period. *Sound velocity dispersion* and anisotropy of the medium cause the *group velocity* characterizing the propagation of perturbation waves, which has a narrow spectral-angular distribution ($\boldsymbol{v}_{\rm gr} = \partial\omega/\partial\boldsymbol{k}$), to differ from the phase velocity describing the phase propagation ($\boldsymbol{v}_{\rm ph} = \boldsymbol{k}\omega/k^2$) in both magnitude and direction.

BULK CHARGE

The same as *Space charge.*

BULK FORCE, volume force

A force distributed over the volume of a material. It is characterized by a *volume force density* f, i.e. a force acting upon a unit body volume. An example is a *mass force* f that acts on mass and is associated with the acceleration a of a body: $f = -\rho a$, where the *density* ρ can depend on the position, and vary throughout the body. The force of gravity $f_{\rm G}$ (at the Earth's surface $f_{\rm G} = \rho g$, where g is the acceleration of gravity) is a special case of a mass force. Other volume forces are: a thermoelastic force (see *Elasticity theory, Thermoelasticity*); the forces of electric and magnetic fields acting upon charges and *magnetic moments*, respectively; and the force acting upon the body from the *conduction electrons* that drift in an electric field E, $f = enE$ (here n is the number of conduction electrons per unit volume, e is the electron charge).

BULK MODULUS, compressibility modulus

One of the *elastic moduli* of an isotropic medium, K; it is expressed via the *Lamé coefficients* as $K = \lambda + (2/3)\mu$. Its reciprocal value is called the *compressibility*. In crystals of cubic symmetry, $K = (c_{11} + 2c_{12})/3$ (here c_{ik} are the second-order elastic moduli in *Voigt notation*). The bulk modulus is measured in pascals. K is 0.77 GPa for Al; 1.37 for Cu; 1.78 for Fe; 0.75 for Ge; 0.24 for NaCl; 4.00 for *diamond.*

BULK PHOTOVOLTAIC EFFECT

The appearance of a stationary photocurrent under uniform illumination of a macroscopically homogeneous piezoelectric (see *Piezoelectricity*) or *ferroelectric* crystal. A bulk photovoltaic effect was discovered in the late 1960s which exhibited a linear dependence of the photocurrent on the light intensity (see *Glass' constant*). The effect leads to the generation of photoelectric voltages which can exceed the band gap width by several orders of magnitude, and are limited only by the *photoconductivity* of the crystal. The direction and magnitude of the photocurrent depend on the orientation of the light polarization. Linear and circular effects are known. Phenomenologically, it is convenient to establish the characteristics of the bulk photovoltaic effect by expanding the photocurrent I_i in a power series of the electric field E_j as

$$I_i = \sigma_{ij}E_j + \alpha_{ijk}E_jE_k^* + \gamma_{ij}i\left[EE^*\right]_j + \cdots$$

The first term with the second-rank tensor σ_{ij} describes photoconductivity in an external field E_j. The second term describes the *linear bulk photovoltaic effect* (i.e. in linearly polarized light); E_j and E_k being the projections of the light polarization vector. The components of the third-rank tensor α_{ijk} of this effect differ from zero for only 20 piezoelectric symmetry groups. The third term describes the *circular photovoltaic effect.*

The microscopic theory of the bulk photovoltaic effect takes into account two basic mechanisms. The first, so-called ballistic, mechanism is related to the asymmetry of the *distribution function* of the nonequilibrium carriers in momentum space. This asymmetry, in its turn, is caused by the asymmetry of the generation, recombination, and scattering of the nonequilibrium carriers in a crystal without a center of symmetry. The second, so-called shift mechanism, is due to displacement of the center of gravity of the wave packet of an electron during its optical transition between bands, or between an impurity center and a band in a crystal without a center of symmetry. The two mechanisms describe both intrinsic and impurity bulk photovoltaic effects, and predict similar magnitudes of the bulk photovoltaic effect. Therefore, the experimental differentiation of these mechanisms is a complicated task. At present, the linear and circular effects are observable in a large number of ferro- and piezoelectric crystals, including ferroelectric *niobates* and tantalates, Te, GaAs, GaP, ZnS, and some other *piezosemiconductors.*

BULK SOUND HARDENING, sonic strengthening

An increase of the *yield limit, microhardness* and other mechanical properties, observable during propagation of high-intensity elastic waves (sonic waves) in solids, mainly in *metals.* For every material in equilibrium there is a critical vibration amplitude, which depends on the ambient temperature and frequency. It is starting from this amplitude that the density of *dislocations, vacancies* and other *defects* of the crystal structure increases. The shape and the size of a mechanical part remains practically unchanged in this case, which is an advantage of this method of hardening (see *Strain hardening*). Bulk sound hardening can arise due to the intensification of thermally activated processes. Using the latter, one can control processes of homogenization, *recovery, recrystallization,* and *phase transitions.* A specific feature of the action of intensive vibrations on a material is the nucleation and growth of microcracks (see *Crack*) at early stages of loading, a *fatigue* type phenomenon. Depending on the conditions of treatment, and the initial state of materials, *acoustic dehardening* may occur, e.g., with sound waves incident on initially deformed metals with a face-centered cubic lattice.

BURGERS CONTOUR

A contour constructed in a regular *crystal lattice* by sequentially tracing a closed line from one atom to another.

It is used to classify *defects* in crystals, and is of particular importance for the description of *linear defects* or *dislocations* (Fig. (a)). If a dislocation appears in a crystal enclosed with such a contour, the contour remains open. In order to close a Burgers contour, it must be complemented with a vector called a *Burgers vector* that closes the break (Fig. (b)). A Burgers vector $\boldsymbol{b}$ characterizes a dislocation unambiguously. Linear *needle-shaped defects* of different types unconnected with translational *slip* of a part of a crystal

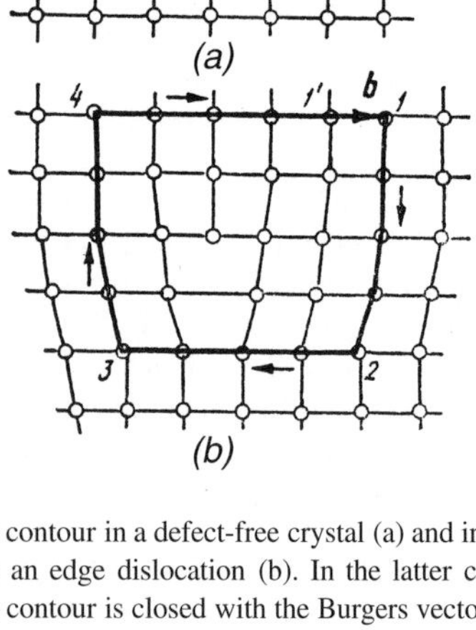

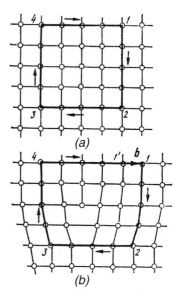

Burgers contour in a defect-free crystal (a) and in a crystal with an edge dislocation (b). In the latter case, the Burgers contour is closed with the Burgers vector $\boldsymbol{b}$.

relative to another part (chain of *vacancies*, chain of *impurity atoms*, etc.) never cause a break of a Burgers contour. Sometimes a closed Burgers contour is constructed within a crystal with a dislocation (Burgers contour in the Frank interpretation). Then, upon removal of the dislocation from the crystal, the Burgers contour turns out to be open, and it is closed by a Burgers vector. When dealing with small-angle and large-angle *grain boundaries*, use is also made of a special Burgers vector constructed in the reference lattice (contour in Frank's interpretation). This contour is constructed with the help of similar (according to the crystallographic direction) vectors drawn in neighboring grains so that the vectors originate from the same point located at the boundary. The Burgers contour is closed in this case with the closure vector $\boldsymbol{B}$.

BURGERS EQUATION (J.M. Burgers, 1940)

An equation of the form $U_t + UU_x - \mu U_{xx} = 0$, where $U(x,t)$ is the velocity field; x, t are space and time coordinates, subscripts indicate partial differentiation, and μ is the *viscosity* coefficient. The equation is a one-dimensional reduction of the Navier–Stokes equations (see *Continuum mechanics*), and is the simplest member from among the most important equations of the physics of nonlinear waves. Burgers equation gives a good description of the dynamics of weakly nonlinear waves in continua, including solids, when only first order dissipation effects are taken into account. The complete solution of Burgers equation is connected with the *Cole–Hopf transformation* (J.D. Cole, 1951; E. Hopf, 1950): $U = -2\mu(\ln\varphi)_x$, which transforms the former into the linear *diffusion* equation: $\varphi_t - \mu\varphi_{xx} = 0$. If there is no dissipation ($\mu = 0$), then the nonlinearity causes a gradual increase of the wave front slope, and finally leads to the formation of a break (*simple Riemann wave*). The latter, increasing in the presence of viscosity ($\mu > 0$), leads to the intensification of dissipation effects that seek to decrease the profile slope. It is the mutual compensation of these contradicting trends, represented in Burgers equation by the last two terms, that brings about the possibility of the propagation of waves with a stationary profile, including a *shock wave* with a finite shock width, directly proportional to μ and inversely proportional to the shock amplitude.

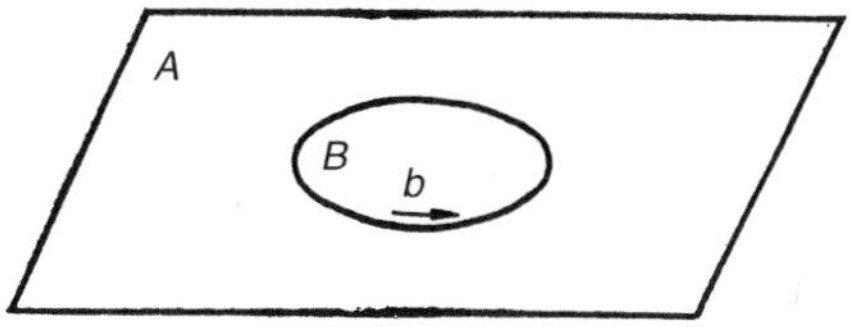

Closed dislocation line in a crystal: A is the region unaffected by the shift; B is the region where a shear deformation has occurred; Burgers vector $\boldsymbol{b}$ is indicated.

BURGERS VECTOR (J.M. Burgers, 1939)

The most important translational characteristic of linear *defects* in *crystals* is the Burgers vector ($\boldsymbol{b}$) which closes an open *Burgers contour* around a *dislocation* line. When there is no dislocation, the contour is closed. The magnitude and direction of a Burgers vector determine how a portion of material on one side of the dislocation line is shifted relative to the material on the other side (see Fig.). A Burgers vector and a tangent vector of the dislocation line determine the *slip plane* of a dislocation element. In magnitude, a Burgers vector can equal either a whole repetition period of the crystal lattice in the vector direction (for *perfect dislocations*) or equal a fractional part of it (for *partial dislocations*). A Burgers vector is constant along the entire length of a dislocation line, whatever its configuration, and remains constant during the dislocation motion. The sum of Burgers vectors of dislocations meeting at a dislocation site equals zero if all the dislocation lines are directed either to or from the site (*Frank networks*). Elastic *strains* and stresses produced by a dislocation in a crystal are proportional to the magnitude of the Burgers vector, while the strain energy is proportional to its magnitude squared. When dealing with small-angle and wide-angle *grain boundaries* in crystals, a *closure vector* $\boldsymbol{B}$ is introduced, which is similar to a Burgers vector and is caused by the presence of dislocations at the boundary. This vector carries the information on the angle of disorientation of neighboring grains.

BURSTEIN–MOSS EFFECT (E. Burstein, T.S. Moss, 1954)

An upward shift in energy of the *intrinsic light absorption edge* in *semiconductors* accompanying an increase of the free current carrier concentration.

The Burstein–Moss effect is due to filling the bottom of the *conduction band* by electrons in n-type semiconductors or the top of the *valence band* by holes in p-type semiconductors. Then, only states above the *Fermi level* E_F are available for transitions involving photoexcited carriers. As a result, direct optical band-to-band transitions at $T = 0$ K are possible with the energy of quanta satisfying the inequality

$$\hbar\omega \geqslant \hbar\omega_{\min} = E_g + \frac{\hbar^2 k_F^2}{2m_n}\left(1 + \frac{m_n}{m_p}\right),$$

where E_g is the *band gap*, m_n and m_p are the effective masses of the electron and the hole, respectively; and k_F is the wave vector corresponding to the Fermi level. At $T > 0$ K and $m_n/m_p \ll 1$, the magnitude of the Burstein–Moss effect is approximately equal to $\Delta\hbar\omega = E_F - 4k_B T$ in direct-gap semiconductors, and in semiconductors with small m_n (e.g., InSb), it already appears at carrier concentrations $\geqslant 10^{17}$ cm^{-3}. The Burstein–Moss effect is used in studies of the energy *band structure* of semiconductors.

BURSTING

See *Penetration deformation.*

C

CADMIUM, Cd

Chemical element of Group II of the periodic system with atomic number 48 and atomic mass 112.41. Natural cadmium consists of stable isotopes with mass numbers: 106 (1.215%), 108 (0.875%), 110 (12.39%), 111 (12.75%), 112 (24.07%), 113 (12.26%), 114 (28.86%), 116 (7.58%). The electronic configuration of the outer shells is $4d^{10}5s^2$. Successive ionization energies are 8.994, 16.908 eV. Atomic radius is 0.148 nm; radius of Cd^{2+} ion is 0.097 nm. Oxidation state is +2. Electronegativity is ≅1.5.

In free form, cadmium is a silvery-white forgeable and ductile *metal*. It has a hexagonal close-packed crystal lattice; $a = 0.297912$ nm, $c = 0.561827$ nm (at 298 K). Density is 8.65 g/cm^3 at 293 K; $T_{\text{melting}} = 594$ K; $T_{\text{boiling}} = 1030$ K. Bond energy is 1.16 eV/atom at 0 K. Heat of melting is 6.38 kJ/mole; heat of sublimation is 112.2 kJ/mole; heat of evaporation is 99.9 kJ/mole; specific heat is 0.231 kJ·kg^{-1}·K^{-1}; Debye temperature is ≅200 K; coefficient of the linear thermal expansion of cadmium monocrystal is $54.0\cdot10^{-6}$ K^{-1} along the principal axis $\underline{6}$ and $+19.6\cdot10^{-6}$ K^{-1} perpendicular to it (at 300 K); coefficient of thermal conductivity is 83.7 W·m^{-1}·K^{-1} along the principal axis $\underline{6}$ and 105 W·m^{-1}·K^{-1} perpendicular to it (at room temperature). Adiabatic coefficients of elastic rigidity of cadmium monocrystal: $c_{11} = 114.5$, $c_{12} = 39.5$, $c_{13} = 39.9$, $c_{33} = 50.85$, $c_{44} = 19.85$ (in GPa) at 300 K; bulk modulus is ≅58 GPa; Young's modulus is ≅65 GPa; shear modulus is ≅25 GPa; Poisson ratio is ≅0.31 (at 300 K), ultimate tensile strength is ≅0.08 GPa and relative elongation is ≅32%. Brinell hardness is ≅0.18 GPa. Cadmium is easily forged and rolled. Surface tension of liquid cadmium is 0.628 J/m^2 (at 623 K). Self-diffusion coefficient of atoms in cadmium polycrystal at T_{melting} is $1.7\cdot10^{-12}$ m^2/s. Effective high-energy (3 to 10 MeV) neutron cross-section for absorption and scattering is 4.3 barn, and 2610 barn for thermal (0.025 eV) neutrons. Ion-plasma frequency is 8.54 THz; low-temperature linear electronic specific heat (Sommerfeld coefficient) is 0.70 mJ·mole^{-1}·K^2; electrical resistivity is 83.6 nΩ·m (at 293 K) along the principal axis $\underline{6}$ of a monocrystal and 68.7 nΩ·m in a perpendicular direction at room temperature, thermal coefficient of electrical resistance in these directions is 0.0041 and 0.00405 K^{-1}, respectively; Hall constant is $\cong +5.7\cdot10^{-11}$ m^3/C at room temperature. Work function of cadmium polycrystal is 4.1 eV. Critical temperature of cadmium transition to the superconducting state is 0.517 K; critical magnetic field is 2.8 mT (at 0 K). Cadmium is a diamagnetic material with molar magnetic susceptibility $-19.7\cdot10^{-6}$ CGS units (at 300 K); nuclear magnetic moment of ^{111}Cd isotope is −0.592 nuclear magnetons. Cadmium is used in nuclear power engineering and as a component of alloys for solders, printing blocks, jewelry, electrodes.

CADMIUM SULFIDE

A semiconductor compound of the $A^{II}B^{VI}$ group (see *Semiconductor materials*), which crystallizes in the cubic *sphalerite structure* (*zinc blend structure*), space group $F\bar{4}3m$ (T_d^2), and in the hexagonal wurtzite structure, space group $P6_3mc$ (C_{6v}^4). The *unit cell* of sphalerite contains four molecules with lattice parameter $a = 0.5835$ nm, while that of wurtzite contains two molecules, with lattice parameters $a = 0.41368$ nm, $c = 0.6713$ nm, and the ratio of anion and cation radii is 0.52. For both structures the tetrahedral positions of the atoms is characteristic, the crystals have no *center of symmetry*, they are polar; they form polytypes (see *Polytypism*) as derivatives of basic structures. Below the pressure 3 GPa a transition to the structure of sodium chloride is observed. The cubic modification is stable at considerably lower temperatures than those of growing *monocrystals*. Data given for the hexagonal wurtzite modification are as follows: temperature coefficients of linear expansion with temperature reduction from 600 to 150 K are reduced from $5.9{\cdot}10^{-6}$ to $0.5{\cdot}10^{-6}$ K^{-1}, density is $4.825\ g/cm^3$, melting temperature is 1748 K, heat of formation $\Delta H^{\circ}_{f,298.15} = -157$ kJ·mole^{-1}, standard entropy $S^{\circ}_{298.15} = 71$ J·mole^{-1}·K^{-1}, specific heat $C^{\circ}_{p,298.15} = 47.4$ J·mole^{-1}·K^{-1}, heat conductivity 0.20 W·cm^{-1}·K^{-1}, *Debye temperature* is 219.3 K. *Covalent bonding* prevails with some admixture of *ionic bonding*. The *phonon spectrum* has nine optical and three acoustic branches. The band gaps are direct, the *valence band* presents three closely located subbands: Γ_9 (the upper one), Γ_8, Γ_7; the final *conduction band* has symmetry Γ_7. The *band gap* width: $E_{g,1.8\,K} = 2.58$ eV, $E_{g,300\,K} = 2.53$ eV. Average values of temperature coefficient of E_g variation at constant pressure $(dE_g/dT)_P$ in the range 90 to 300 K and coefficient (dE_g/dP): $-5.2{\cdot}10^{-4}$ eV/K and $3.3{\cdot}10^{-6}$ eV/bar. The *spin–orbit interaction* constant is 0.065 eV, the valence band splitting due to the *crystal field* action is 0.027 eV, *electron affinity* is 4.73 eV, work function is 7.26 eV. *Effective masses* of electrons, holes: $m_n^* = 0.205$, $m_p^* = 0.7$ $(\perp c)$, $m_p^* = 5$ $(\parallel c)$. Refraction index dispersion and birefringence are characterized by the values: $n_e = 2.726$, $n_o = 2.743$ at $\lambda = 515$ nm; $n_e = 2.312$, $n_o = 2.996$ at $\lambda = 1500$ nm. Static and high-frequency *dielectric constants* are: $\varepsilon_s = 8.64$ $(\parallel c)$, $\varepsilon_s = 8.28$ $(\perp c)$, $\varepsilon_\infty = 5.24$.

The Cd–S system in the phase equilibrium diagram has one compound CdS with a narrow range of homogeneity. Inherent atomic defects to a large extent determine the *diffusion* and *solubility* of impurities, the structure of local centers, optical and electric properties. Elements of Group III Al, Ga, In, can replace the cations, elements of Group VII Cl, Br, I, replace the anions, *vacancies* of anions, interstitial cations are shallow hydrogen-like donors; elements of Group I Cu, Ag, replace the cations, cation vacancies behave as deep *acceptors*. The association of the point defects is characteristic. Conductivity may change from a semimetal to insulator type, no p-type conductivity is observed due to the self-compensation phenomenon. Mobility of electrons and holes: 350 cm^2·W^{-1}·s^{-1} and 30 cm^2·W^{-1}·s^{-1}. Minority carrier *lifetimes* (holes) are 10^{-6} to 10^{-8} s, diffusion lengths of holes are short (7 to 0.5 μm).

The presence of direct band gaps facilitates effective luminescent and laser radiation. Blue radiation is caused by the annihilation of free and associated *excitons*, green, orange, red and infrared radiations are associated with impurity *recombination*. At $T \leqslant 15$ K energy of excitons $E^A_{ex} = 2.550$ eV, $E^B_{ex} = 2.564$ eV, $E^C_{ex} = 2.632$ eV, bonding energies of excitons: $E_A = E_C = 0.0294$ eV, $E_B = 0.0295$ eV. Centers of radiative recombination are centers of photosensitivity. Having the high photosensitivity in a wide spectral range: from UV to near IR, in the X-ray spectral range CdS is characterized by high sluggishness, low temperature and time stability of photocurrent. The absence of a center of inversion and a strong *electron–phonon interaction* determine the piezoelectric properties. The capability of fabrication by simple methods in the form of powders, tablets, and caked layers of *thin films* with controllable properties is a pronounced advantage. The uniqueness and variety of its physical properties allows its use as a model material of the physics of semiconductors and quantum electronics: in photo-, opto-, acoustoelectronics, optics, laser engineering. Cadmium sulphide is widely used as a crystallophor for the creation of the screens of electron tubes, including color television, as a photoconductor for the creation of *photoresistors*, of nuclear particles radiation sensors, as an element of *heterojunctions* for solar phototransformers.

CALAMITICS

Uniaxial *liquid crystals* with structure units (molecules or micelles) of oblong cylindrical shape (cf. *Discotics*).

CALCIUM, Ca

Chemical element of Group II of the periodic system with atomic number 20 and atomic mass 40.078, it belongs to alkaline-earth metals. Natural calcium is a mixture of 6 stable isotopes with mass numbers 40, 42–44, 46, 48; among them, ^{40}Ca (96.94%) is the commonest and ^{46}Ca (0.003%) is the rarest. Electronic configuration of outer shell is $4s^2$. Atomic radius is 0.197 nm. Successive ionization energies are 6.133, 11.872, 50.914 eV. It is chemically active; oxidation state +2. Electronegativity is 1.04.

In free form, calcium is a silvery-white *metal*; it darkens in the air due to formation of products of calcium interaction with oxygen, nitrogen, water vapors, etc., at the surface. From room temperature to 716 K, α-Ca is stable with face-centered cubic lattice (with parameter $a =$ 0.556 nm); at temperatures above 716 K and up to $T_{\text{melting}} = 1120$ K β-Ca is stable with a body-centered cubic lattice (with parameter $a =$ 0.448 nm), $T_{\text{boiling}} = 1770$ K. Density of α-Ca is 1.55 g/cm^3. Heat of melting is 8.4 kJ/mole; heat of evaporation is 152 kJ/mole; specific heat $c_p = 25.9$ J·mole^{-1}·K^{-1}. Work function 2.75 eV. Debye temperature is 220 K; temperature coefficient of linear expansion is $2.2\cdot10^{-5}$ K^{-1} (at 273 to 573 K); heat conductivity is 125 W·m^{-1}·K^{-1}; electrical resistivity is $3.8\cdot10^{-2}$ μΩ·m (at 273 to 373 K); temperature coefficient of electrical resistance is $4.57\cdot10^{-3}$ K^{-1}. Hall constant is $-1.78\cdot10^{-10}$ m^3/C; magnetic susceptibility is $\chi = +1\cdot10^{-9}$. Modulus of normal rigidity is 21 to 28 GPa, tensile strength is 59 GPa, elastic limit is 4 MPa, limiting yield stress is 37 MPa. Brinell hardness is ~250 MPa.

Metallic calcium is used in *metallurgy* for reduction of U, Th, Zr and other rare metals from their compounds; as deoxidizing and degassing agent in melting of special steels, bronze, etc. Calcium is used as a getter in electronic devices. *Monocrystals* of the calcium compound, fluorite CaF_2, find wide use in optical and laser devices.

CALIFORNIUM, Cf

Chemical element of Group III of the periodic system with atomic number 98 and atomic mass 251.08, belongs to *actinides*. There are no stable isotopes, and 17 radioactive isotopes are known with mass numbers 240 to 256. Electronic configuration of outer shells of atom is $5f^9 6d^1 7s^2$. Atomic radius is 0.169 nm; ionic radius of Cf^{3+} is 0.906 nm. Oxidation state is +3 (less often, +2, +4, +5, +6). Electronegativity is 1.23.

In free form, californium is a silvery *metal*. It exists in two allotropic modifications: low-temperature (α-Cf) and high-temperature (β-Cf). The first has a hexagonal close-packed crystal (space group $P6_3/mmc$ (D^4_{6h}), $a = 0.3988$ nm, $c = 0.6887$ nm at room temperature); according to other data, it has a double hexagonal close-packed crystal lattice ($a = 0.339$ nm, $c = 1.101$ nm). The other modification has a face-centered cubic lattice (space group $Fm\bar{3}m$ (O^5_h), $a = 0.5743$ nm and $a = 0.494$ nm upon hardening by quenching at ~1000 K and 870 K, respectively) at temperatures above 860 K to $T_{\text{melting}} = 1170$ K. $T_{\text{boiling}} \approx$ 1500 K. Density of α-Cf is 15.1 g/cm^3. Below the Curie point $T_C = 51$ K, α-Cf transforms from a paramagnetic state to a ferromagnetic one; in β-Cf, no magnetic ordering has been found. Californium isotopes (in particular, ^{252}Cf) are used as neutron sources in *activation analysis*, in medicine, etc. Californium is highly toxic.

CALORIMETRIC COEFFICIENTS

The thermodynamic coefficients of the form $(\partial Q/\partial\mu)_\nu$ where μ and ν take on the values of any of the thermodynamic parameters P, V, T (*pressure*, volume, *temperature*) of a system. The quantity ∂Q is the amount of heat transferring the system into a new thermodynamic state under a variation of μ by $\partial\mu$ at $\nu =$ const. There are 6 principal calorimetric coefficients:

$(\partial Q/\partial T)_V = C_V$,
specific heat at constant volume,
$(\partial Q/\partial T)_P = C_P$,
specific heat at constant pressure,
$(\partial Q/\partial P)_T$, heat of isothermal compression,
$(\partial Q/\partial P)_V$, heat of isochoric compression,
$(\partial Q/\partial V)_P$, heat of isobaric expansion,
$(\partial Q/\partial V)_T$, heat of isothermal expansion.

The calorimetric coefficients can be expressed via the derivatives of the thermodynamic state functions: enthalpy H and internal energy U (see

Thermodynamic potentials):

$$\left(\frac{\partial Q}{\partial T}\right)_V = \left(\frac{\partial U}{\partial T}\right)_V,$$
$$\left(\frac{\partial Q}{\partial T}\right)_P = \left(\frac{\partial H}{\partial T}\right)_P,$$
$$\left(\frac{\partial Q}{\partial P}\right)_T = \left(\frac{\partial H}{\partial P}\right)_T - V,$$
$$\left(\frac{\partial Q}{\partial P}\right)_V = \left(\frac{\partial U}{\partial P}\right)_V,$$
$$\left(\frac{\partial Q}{\partial V}\right)_P = \left(\frac{\partial U}{\partial V}\right)_P + P,$$
$$\left(\frac{\partial Q}{\partial V}\right)_T = \left(\frac{\partial U}{\partial V}\right)_T + P.$$

In addition to calorimetric coefficients, thermodynamics also makes use of six *thermal coefficients*: $(\partial V/\partial P)_T$, $(\partial P/\partial T)_V$, $(\partial T/\partial V)_P$, $(\partial V/\partial P)_S$, $(\partial P/\partial T)_S$ and $(\partial T/\partial V)_S$, where S is the entropy.

There are 9 independent thermodynamic relationships between these 12 thermodynamic coefficients so only 3 thermodynamic coefficients can be selected as independent ones to be evaluated theoretically or determined experimentally. In principle all other coefficients can be calculated via these three, chosen as independent. This circumstance allows one to calculate coefficients that are difficult to measure experimentally through easily measurable ones. For instance,

$$C_V = C_P + T\left(\frac{\partial P}{\partial V}\right)_T \left(\frac{\partial V}{\partial T}\right)_P^2,$$
$$\left(\frac{\partial V}{\partial P}\right)_T \left(\frac{\partial P}{\partial T}\right)_V \left(\frac{\partial T}{\partial V}\right)_P = -1,$$
$$\left(\frac{\partial H}{\partial V}\right)_T = T\left(\frac{\partial P}{\partial T}\right)_V + V\left(\frac{\partial P}{\partial V}\right)_T.$$

CALORIMETRY (fr. Lat. *calor*, heat, and Gr. $\mu\varepsilon\tau\rho\varepsilon\omega$, am measuring)

Methods for measuring thermal effects accompanying different processes. Calorimetric methods are used in the interval 0.05–3500 K and attain an accuracy 0.01%. Calorimetry at temperatures above 400 K is conventionally called a high-temperature type.

Instruments for calorimetric measurements are called *calorimeters*. Their design is determined by the temperature interval and desired accuracy. The two most widely used types in studies of solids are *adiabatic calorimeters* where the heat exchange between calorimeter and medium is negligibly small, and *dynamic (heat-conducting) calorimeters* based upon detection and control of the heat flux between the sample and the environment.

Adiabatic calorimeters with a pulse heat input are very accurate for measurements made close to equilibrium. They are used at temperatures from 0.05 to 1000 K to determine the *specific heat*, thermodynamic functions, latent heats of *phase transitions*, the *critical index* α of phase transitions; density of states of electrons at the Fermi level $N(E_F)$ by the electronic specific heat coefficient γ; obtaining direct information on energy levels of valence electrons by the Schottky effect; investigation of energy properties of *vacancies* and coordination polyhedra. A drawback of this method is the long time required for making the measurement.

Dynamical calorimeters, in particular the Calve type, are used at temperatures from 77 to 1880 K. The temperature variation rate is from 1 K/h to 64 K/min. The important advantage of calorimeters of this type is the possibility of contact between the sample and the environment, control of the pressure or gas phase during the experiment, and access to the center of the calorimeter when necessary. The field of application is the determination of heat capacities, thermodynamic functions, latent heats of phase transitions, heats of oxidation–reduction reactions, heats of gas–solid reactions, the energy related to the *strain* of a solid, and heats of *adsorption* and *desorption*. At temperatures above 1800 K the specific heat of solids is determined by the *offset method*. Benzoic acid and corundum have been adopted as standard substances in calorimetry.

CANONICAL TRANSFORMATION METHOD (N.N. Bogolyubov, 1947)

A method utilizing a unitary transformation of particle creation a_k^+ and annihilation a_k operators in order to reduce the initial complex Hamiltonian of a many-particle quantum system, involving interacting particles expressed in a *second quantization* representation, to a simpler Hamiltonian in diagonal form for noninteracting *quasi-particles* with a restructured energy spectrum.

If the initial Hamiltonian has the form

$$H = \sum\left[A_k a_k^+ a_k + \frac{1}{2}B_k a_k a_{-k} + \frac{1}{2}B_k^* a_k^* a_{-k}^*\right], \tag{1}$$

then by introducing new operators according to the relations

$$a_k = \lambda_k c_k + \mu_k^* c_{-k}^+, \qquad a_{-k}^+ = \lambda_k^* c_{-k}^+ + \mu_k c_k \tag{2}$$

the Hamiltonian assumes the form

$$H = \sum \varepsilon(k) c_k^+ c_k + E_0, \qquad \varepsilon(k) = \left[A_k^2 \mp |B_k|^2\right]^{1/2}. \tag{3}$$

The parameters of the canonical transformation obey the normalization condition

$$|\lambda_k|^2 \mp |\mu_k|^2 = 1.$$

In formulae for the energy of the quasi-particles and the transformation parameters the upper sign corresponds to bosons and the lower sign corresponds to fermions.

The canonical transformation for spin waves was applied by Holstein and Primakoff (1940) to a ferromagnetic material, and it was applied by Bogolyubov (1947) to bosons for the case when many of them are in the state of a Bose–Einstein condensate (see *Bose–Einstein condensation*). The method was exploited by Bogolyubov to construct his microscopic theory of superconductivity (1947). It results in the following elementary excitations:

$$E(k) = \sqrt{k^2 u^2(k) + \left(\frac{k^2\hbar^2}{2m}\right)^2}, \tag{4a}$$

$$u(k) = \sqrt{\frac{NV(k)}{m}}, \tag{4b}$$

where m is the boson mass, N is the number of particles in a unit volume, and $V(k)$ is the Fourier transform of the *pairwise interaction* potential. For particles with repulsion, when $V(k) > 0$, Eq. (2) yields at $k \to 0$ the acoustic *dispersion law* of quasi-particles $E(k) \approx ku(0)$, which is a necessary but insufficient condition for the *superfluidity* of neutral Bose systems (see *Landau superfluidity criterion*). When attraction dominates at large distances with $V(0) < 0$, the spectrum (2) at $k \to 0$ becomes unstable (imaginary), which corresponds to an exponential growth of long-wavelength perturbations in time, and a spontaneous contraction (collapse) of the system. For charged bosons (e.g., *bipolarons*) with a long-range Coulomb repulsion $V(k) = 4\pi e^2/k^2$ (e is the boson charge), the spectrum (2) of quasi-particles at $k = 0$ has a finite energy gap $E(0) = \hbar\omega_p$ (ω_p is the plasma frequency) and satisfies the superfluidity criterion, which corresponds to superconductivity (see *Bipolaron mechanism of superconductivity*).

In the case of electrons (fermions) interacting under the conditions of Cooper pairing (see *Bardeen–Cooper–Schrieffer theory*) one can use Hamiltonian (1), where k denotes both the momentum and the spin of the electron, and A_k and B_k are given by

$$A_k = \xi(k), \qquad B_k = \Delta(k).$$

Here $\xi(k)$ is the energy measured from the Fermi level, and $\Delta(k)$ is the non-zero energy gap below the transition temperature from the normal to the superconducting (superfluid) state of a Fermi system. In the *self-consistent field* approximation, which is asymptotically accurate in the thermodynmic limit (see *Bogolyubov method in the theory of superconductivity*), transformations (2) lead to a reconstructed spectrum of quasi-particles in a superconductor with the energy $\varepsilon(k) = [\xi^2(k) + \Delta^2(k)]^{1/2}$.

CAPACITANCE

A quantitative measure of the ability of an object or material to retain electric charge q. Charged particles of the same sign attempt to go out of a charged body due to electrostatic repulsion. Mobile charge carriers in a conducting medium spread over its surface and provide an equal value of electric potential φ everywhere on the surface. If the conductor is electrically charged then its potential differs from that at infinity where by convention $\varphi_\infty = 0$. The value φ is proportional to q; and it specifies the conductor capacitance by the ratio $C = q/\varphi$. The greater the value of C, the greater the charge on the conductor at a given φ. The magnitude C of an isolated conductor is determined by its geometric dimensions and shape; e.g., the capacitance of a sphere is equal to its radius (in Gaussian system of units). In a system of conductors, a correspondence between their charges and potentials is determined by the linear relations $q_i = \sum_j C_{ij}\varphi_j$ (C_{ij} is called the *capacity coefficient* of the ith and jth conductors).

In a system metal–insulator–metal (*condenser*), the capacitance is determined by the relation $C = q/\Delta\varphi$ where q is the value of charges of opposite sign on every capacitor plate; $\Delta\varphi$ is the potential difference between the plates. For a planar capacitor $C = \varepsilon A/d$ where A is the area of each plate, d is the distance between them, and ε is the relative (dimensionless) *dielectric constant* of the medium between the plates. For a very small d, the capacity of the system can be very large. The systems metal–semiconductor, *semiconductor junction*, and so on are similar to a capacitor, and can also have a significant capacitance (see *Barrier capacitance*). Variations of the thickness of the dielectric layer under actions of any kind provide information on the distribution of *electrically active extended defects* in a system (see *Capacitance–voltage characteristic*, *Capacitance spectroscopy*).

Josephson junctions, with typical capacitances $C \sim 10^{-15}$ F, exhibit voltage changes $\Delta V = e/C \sim 0.016$ mV for individual electron tunneling. This is called a Coulomb blocade, and it appears on a current versus voltage characteristic curve (I versus V, or $\mathrm{d}I$ versus $\mathrm{d}V$ plot) as a sequence of steps called a *Coulomb staircase*.

CAPACITANCE, BARRIER

See *Barrier capacitance*.

CAPACITANCE SPECTROSCOPY

A collection of methods for determining the energy spectrum of *deep levels* in semiconductors from changes in their *barrier capacitance*, C, during variation of the applied voltage. A simple version of capacitance spectroscopy is based on measuring steady-state *capacitance–voltage characteristics*. When a reverse bias (i.e. a bias in the cut-off direction) is applied to a system containing a *semiconductor junction* or a *Schottky barrier*, the thickness L of the semiconductor region with depleted *current carriers* increases, and the capacitance of the depletion layer decreases. More information can be obtained by methods of *nonstatic capacitance spectroscopy* of deep levels. It is based on the measurement of the capacitance relaxation kinetics for different combinations of factors applied to fill or ionize the deep levels (such as forward and reverse bias pulses of various durations, photoionizations, and temperatures). Using these methods provides separate measurements of concentrations of deep levels with different ionization energies, and also determines the cross-sections of charge carrier trapping at the levels. The efficient *Lang method* measures the differences between C at two times, t_1 and t_2, during the course of the relaxation of C with the emission of electrons or holes from individual levels into a corresponding allowed band. There are modifications of this method when the relaxation of current rather than capacitance is measured; and taken together they constitute the *relaxation spectroscopy of deep levels*.

CAPACITANCE, SURFACE

See *Surface capacitance*.

CAPACITANCE–VOLTAGE CHARACTERISTIC

Dependence of the sample capacitance C on the applied voltage V.

Measurement of the capacitance–voltage characteristic is one of the most widespread methods of investigating semiconductor *electron–hole transit*, *heterojunctions*, *Schottky diodes* and *metal–insulator–semiconductor structures*. From the shape of a C–V characteristic it is possible to determine the distribution of doping impurity in the structure. In particular, in sharp p–n junctions and Schottky diodes the capacitance–voltage characteristic plotted as C^{-2} versus V is a straight line with a slope characterizing the *doping* of the semiconductor, and the cutoff voltage is equal to the equilibrium height of the potential barrier at the junction (*contact potential difference*). The characteristic shape of the C–V characteristic for M–I–S structures is shown in the figure.

The presence of *deep levels* within the bulk or at the surface of the sample, which recharge during voltage variations, causes additional features of the capacitance–voltage characteristic. Besides, the shape of the C–V characteristic has a strong dependence on the frequency f of the probing signal, because at high frequencies there is not enough time to recharge the deep levels, and to

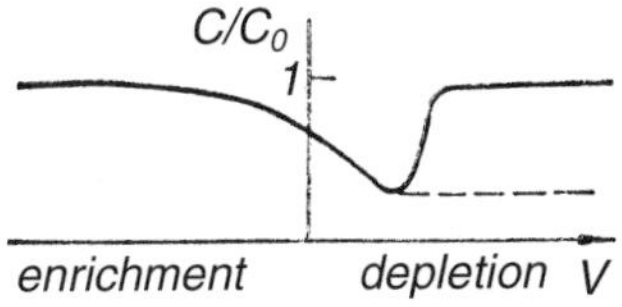

Dependence of the capacitance on the voltage in an M–I–S structure at a low (solid line) and a high (dashed line) frequency, where C_0 is the insulator capacitance.

form the *inversion layer* in the M–I–S structure. This allows one to use the C–V characteristic and its frequency dependence to obtain information about the concentration, energy spectrum and trapping profiles of the bulk and surface levels (see also *Varicap, Capacitance spectroscopy*).

CAPILLARY PHENOMENA

Physical phenomena arising from surface tension. Capillary phenomena are most easily observed in thin tubes ($R \leqslant 10^{-3}$ m). The fundamentals of capillarity theory laid by P.S. Laplace (1806–1807) include two basic laws:

(1) the excess pressure under the curved surface of liquid is

$$\Delta p = \pm\sigma\left(\frac{1}{R_1} + \frac{1}{R_2}\right), \qquad (1)$$

where σ is the surface tension coefficient, R_1 and R_2 are the radii of curvature of two mutually perpendicular sections of the surface at the point in question; and

(2) the *edge angle* θ_0 of a *solid surface* by a liquid is independent of the shape of the surface.

Of importance in the theory is the *capillary constant* a with the dimensions of length:

$$a = \sqrt{\frac{\sigma}{\Delta\rho g}}, \qquad (2)$$

where $\Delta\rho$ is the difference between the densities of the phases in contact, and $g = 9.8\ \mathrm{m/s^2}$.

For cylindrical tubes the *Borelli–Jurin equation* holds

$$h = \frac{2\sigma\cos\theta_0}{gR\rho}, \qquad (3)$$

where h is the height of the rise of the wetting liquid and ρ is the liquid density. Eq. (3) describes the impregnation of pore structures, and substantiates the role of capillary forces in the root feeding of plants utilized for soil tillage. Eq. (1) explains the presence of a considerable capillary counter-pressure in a system of alternating drops of liquid and gas bubbles in capillaries (gas embolism, decompression sickness in divers). See also *Thermocapillary effect, Electrocapillary phenomena.*

CARBIDES (fr. Lat. *carbo*, coal, and Gr. *ειδος*, type)

Chemical compounds of *carbon* with *metals* and some nonmetals. We can distinguish *covalent carbides* (beryllium carbide, *boron carbide, silicon carbide*), metal-like, and salt-like carbides. *Metal-like carbides* include those of scandium, yttrium, lanthanum, lanthanides, *transition metals* of Groups IV to VII of the periodic system, and metals of the iron family. Carbides of *transition metals* (TiC, Cr_3C_6, Cr_3C_2, Mo_2C, WC and others) are interstitial phases and structures close to interstitial, with the carbon atoms occupying octahedral or tetrahedral sites (voids) of close-packed metal lattices. Such carbides exhibit metallic characteristics of conductity, high melting (decomposition) temperature, hardness and chemical stability. *Salt-like carbides* include those of s-state metals (alkali, alkaline-earth metals and aluminum). The best known among them is *calcium carbide* CaC_2, a colorless crystal, with density 2.21 g/cm^3, $T_{\mathrm{melting}} \sim 2300\,^\circ$C. Carbides are obtained from elementary matertials in vacuo or in a reducing gas atmosphere, as well as by interaction of metallic oxides with carbon in vacuo or in a reducing atmosphere, by interaction of metals or their oxides with carbon-containing gases, and by interaction of gaseous compounds of some metals (halides) and carbon in the hydrogen atmosphere. The favorable properties of carbides are used in nuclear and space engineering, microelectronics, metal processing and many other areas.

CARBIDE TRANSFORMATIONS in steel

Formation and rearrangement of iron *carbides* during the course of different treatments of *steel.* Carbides are one of the basic phases in steel; they may form upon cooling of steel heated above some critical point, or upon *tempering* of hardened steel (see *Quenching*). Above 300 °C a stable *iron carbide, cementite,* is present in carbon steel. At the tempering of alloyed steels, the temperature of cementite (Fe_3C) formation increases. Redistribution of *carbon* and alloying elements between phases leads to *alloying* of iron carbide. With a certain temperature of tempering, degree of steel alloying, and ratio of carbide-forming element content to that of carbon, the formation of special carbide is possible. Destabilization of cementite (or special carbide) occurs at complete replacement of one atom of iron in cementite (or the basic element in special carbide) with the alloying element. In multicomponent alloying such replacement occurs with several alloying elements simultaneously. Carbide transformations that run according to the mechanism of *alloy aging* and direct precipitation of carbides from ferrite, bypassing cementite, are used to harden structural, stamp, tool, high-speed and certain austenite steels, as well as to maintain hardening in long-term operation of boiler and turbine, motor equipment, of industrial holders and fastening parts, etc., at increased temperatures.

The specifics of redistribution of carbon and alloying elements between phases in multicomponent alloying causes the carbide transformations to run at a lower number of alloying elements than in single or simple alloying (and the same total degree of alloying). Here, the ferrite contains more alloying elements, which permits a more complete use of their effect upon physical mechanical, technological and service properties of steel. The composition and type of the carbide crystal lattice, distribution of carbides within the structure, and the mechanism of their formation are investigated using chemical *phase analysis*, *X-ray structure analysis*, microscopic diffraction and magnetic analyses, *Mössbauer spectroscopy*, *internal friction*, and electron microscopy with the help of *replicas* and foils.

CARBON, C

The chemical element of Group IV of the periodic table with atomic number 6 and atomic weight 12.011. The mass numbers of the two stable carbon isotopes are 12 (^{12}C, 98.892%) and 13 (^{13}C, 1.108%). The ^{14}C isotope is unstable (half-life 5730 years). The wide diversity of properties of carbon compounds is due to the electronic configuration of the carbon atom: $1s^2 2s^2 2p^2$. Interaction with other atoms results in rearrangement of the density of the electron clouds and the concentration of electron density in certain directions, such as along chemical bonds (see *Hybridization of orbitals*). Depending on the way atomic orbitals are combined, sp-, sp^2- and sp^3-hybridizations are distinguished. Each hybridization corresponds to a particular spatial arrangement of a bond. Oxidation states of carbon in its compounds are -4, -2, $+2$, and $+4$; electronegativity value is ≈ 2.58; atomic radius 0.077 nm, ionic radii: 0.016 nm for C^{4+} and 0.26 nm for C^{4-}.

There exist five allotropic (see *Allotropy*) modifications of solid carbon: *diamond* (sp^3-hybridization of orbitals), *graphite* (sp^2-hybridization), *fullerene* (C_{60} and analogues, sp^2-hybridization), *carbyne* (sp-hybridization), and *amorphous carbon* (mixed type of bonding). Carbyne is a quasi-one-dimensional semiconductor, the width of the *band gap* of which is $\sim$1 eV. Amorphous carbon is a polyphase structure, which consists of the finest inclusions of polymorphic modifications of carbon. Depending on the mode of preparation, carbon may also contain chemically bound hydrogen atoms, which saturate the *dangling bonds* of the carbon structure. The principal carbon-containing *minerals* are salts (*carbonates*) of carbonic acid H_2CO_3. Carbon atoms are the chief constituents of *coal*, oil, peat and other materials. Carbon is a nonmetal and behaves as an active reducing agent on heating. It forms *carbides* with various elements, e.g., *boron carbide*, *silicon carbide* SiC, *cementite* Fe_3C. Carbon is used as a reducing agent, as a component of various *alloys*, as fuel, for production of electrodes, etc.

CARBON, ACTIVE

See *Active carbons*.

CARBON FILM, DIAMOND-LIKE

See *Diamond-like carbon films*.

CARBYNE (fr. Lat. *carbon*, coal)

The third allotropic modification of *carbon*, which consists of rectilinear carbon chains bound into crystals through *van der Waals forces*. The rectilinear shape of the chains is due to sp-hybrid bonding of the carbon atoms. In actual samples the individual chains are cross-linked, with perhaps some periodicity, and this involves a transition of some carbon atoms into sp^2- or sp^3-hybrid type bonding. The presence of carbyne can be detected by the methods of *electron diffraction* and *X-ray diffraction*.

Crystalline particles of carbyne are colorless, and no large *monocrystals* have been prepared. Carbyne is hard to obtain in a pure form (free from admixture of laminated modifications of carbon). Density and other physical properties depend on the method of manufacture and purification. Carbyne has a higher *specific heat* than *graphite* and *diamond*, and it surpasses graphite in *hardness*. Heat of combustion is 84 to 372 kJ/mole, which is significantly lower than that of graphite and diamond, which attests to the thermodynamic stability of carbyne.

There are 3 bands in the Raman spectrum of carbyne at 1360, 1580 and 2140 cm^{-1}, the last of them being the characteristic one related to the vibrations of conjugate C–C bonds. Carbyne is an n-type *semiconductor* with a *band gap* of 0.64 eV; electrical resistivity is 10^6 Ω·cm (at 293 K). Carbyne was originally obtained and characterized by the methods of oxidative polycondensation of acetylene. It forms upon *condensation* from vapor-gas phase, plasma, an ion-molecular beam under the conditions that are softer compared to those for forming diamond, and of laminated *polymers* of carbon. Natural carbyne was discovered in 1968 in a meteorite crater and received the name *chaoite* (white veins and disseminations in graphite). In its chemical inertness it surpasses graphite and approaches diamond. Defect-free *thread-like crystals* of carbyne are believed to have very high rupture strength.

CASCADE, COLLISION

See *Collision cascade*.

CASTING

The process of forming pieces (casts) by filling a hollow casting form or mold with melted material (*metals*, their *alloys*, plastics, certain rocks, metallurgy slags). The principal advantages of casting include the possibility of obtaining casts of a complicated configuration that is as close in size and shape to the final product as possible. Mold casting of *steel*, *cast iron*, nonferrous alloys (from 5–10 to 100 g and more in mass) is used to produce various parts, from time-piece components to heavy-duty constructions for engineering, metallurgy, power and other branches of machine building. Steels and alloys are cast into molds (plate casts, forging casts) or poured continuously into water-cooled crystallizers (for a continuously operating casting installation). More than 50 types of casting techniques are available. Traditional technologies include pouring molten metal from dippers or crucibles driven by gravity into sand casting forms. Special casting techniques make it possible to improve the quality of casts, perform precision casting in both size and shape, and achieve special physical and mechanical characteristics. Such technologies involve various external factors which act upon the crystallizing metal and provide for using special casting forms, models or casting techniques. Among the latter are casting under high or low pressure, centrifugal casting, electric slag casting, continuous or suspension casting, casting into melted-out models, casting into chill molds, ceramic and shell forms, etc. Art casting is done using melted-out models to produce monumental sculptures, architectural decorations, and jewelry.

CASTING MATERIALS

Macroscopically inhomogeneous substances obtained by fusion of two and more mutually soluble *metals*, nonmetals, or organic substances, and by subsequent *solidification* of the liquid phase in a mold of a particular shape. Depending on the temperature range of metastability of the liquid phase, the structure of casting materials may be crystalline or amorphous. In a general case, the conditions of solidification in the amorphous or crystalline states may be controlled by the cooling rate, by the purity of the materials, by their solubility, and by their crystallization activity.

The properties of casting materials crystallizing during the course of solidification are not only determined by their chemical composition, by the method of fusion, thermal and other types of deformationless treatment, but are also dependent on the thermal conditions of solidification, on the nucleation rates of centers, and the linear growth rates of *crystallization* nuclei. These factors determine the degree of development of physical, chemical and structural inhomogeneities in casting materials. The processes of their solidification and crystallization, and hence their properties, may be significantly changed by *modification*, electromagnetic mixing and other methods (see *Casting*).

CAST IRON

Alloy of *iron* with *carbon* (usually 2 to 4% C), also containing (Si, P and S, Mn), and sometimes additional alloying elements (Al, V, Cr, Ni, etc.); it solidifies with the formation of *eutectics*. Unlike *steel* due to its low *plasticity*, cast iron is machinable under pressure. It is characterized by a high liquid *yield*. Cast irons are subdivided into the following classes: *pearlite*, *ferrite*, perlite-ferrite, *sorbite*, *troostite*, *bainite*, *martensite*, troostite-ferrite, and *austenite*. According to its microstructure the following types are distinguished: *gray cast iron* (in the form of cementite or iron carbide Fe_3C), and *forgeable cast iron* obtained by annealing *white cast iron* (flocculent graphite). The properties of cast iron are improved by *alloying*, *modification*, *thermal treatment*. Cast iron is the primary product obtained from the processing of iron ore materials. It is used in chemical, petroleum, and agricultural machine building, in tool, car and tractor fabrication, in civil engineering, etc.

CATALYSIS, RADIATION

See *Radiation catalysis*.

CATASTROPHE, POLARIZATION

See *Polarization catastrophe*.

CATEGORY (fr. Gr. $\kappa\alpha\tau\eta\gamma o\rho\iota\alpha$, accusation)

A level of crystallographic classification by the number and order of special (or individual) directions, i.e. directions in the *crystal lattice* which do not repeat under a symmetry operation (see *Crystal symmetry*). There are three categories: the highest one (no special direction and some symmetry axes of order higher than 2); the medium type (one special direction along a unique symmetry axis of order 3, 4, or 6); the lowest one (no special direction and no symmetry axes of order higher than 2). The highest category involves crystals of the *cubic system*, the medium one includes those of the *hexagonal system*, *trigonal system* and *tetragonal system*, and the lowest category encompasses structures of the *orthorhombic system*, *monoclinic system* and *triclinic system*. Depending on the category the nature of the anisotropy, and many physical properties of a solid change characteristically (see *Anisotropy of crystals*).

This classification term "category" is not commonly used in the Western literature.

CATHODE

See *Cold cathode*, *Field-emission cathode*, *Thermionic cathode*, *Photocathode*.

CATHODE SPUTTERING

See *Sputtering*.

CATHODOLUMINESCENCE

A type of *radioluminescence* excited by an electron beam ("cathode rays"). Depending on the electron energy E, one distinguishes low-voltage ($E = 10$–100 eV) and high-voltage ($E > 0.5$ keV) cathodoluminescence. At high-voltage the electrons overcome the surface barrier, penetrate to a depth proportional to $E^{1.4}$, knock out secondary electrons that also ionize atoms, create *electron–hole pairs*, or directly excite the ions of the activator (Mn^{2+}, Tb^{3+} and others). The minority carriers migrate across the lattice and are trapped at *luminescent centers*, accompanied by radiative *recombination*. At low voltage the excitation by primary electrons only covers several near-surface atomic layers, and the *diffusion* of nonequilibrium carriers produces a zone of fluorescence spreading into the depth of the sample. The forming negative charge is neutralized at high voltage luminescence mainly through the emission of electrons from the surface of a *luminophor*, while at low voltage the neutralization is due to the conductivity of the latter. In its spectrum, cathodoluminescence is similar to other types of *luminescence*. Its brightness in a certain spectral region is proportional to E and the current density in the beam. The energy yield is 10–30% for high voltage and 0.5–2% for low voltage. Cathodoluminescence is employed in picture tube screens and in low-voltage indicators that are used in devices for the visualization of information, TV sets and other electronic devices. Cathodoluminescence underlies the operation of *lasers* excited by an electron beam.

CATLOW–DILLER–NORGETT POTENTIAL

(C.R. Catlow, K.M. Diller, M.J. Norgett, 1977)

A central pair-potential that describes short-range ion-to-ion interactions in *ionic crystals* using the *shell model*. An analysis of the relation between the equilibrium interionic distance in a *crystal* and its elastic constants on the one hand, and the interionic interaction potentials yielded by this model for *alkali-halide crystals* (such as LiCl, NaCl, KCl, RbCl) on the other hand, demonstrated that the ion-to-ion interaction is attractive at long distances r and repulsive at short distances. Starting with that observation, Catlow, Diller and Norgett offered the following two proposals for the short-range potential of (identical) second nearest neighbors in alkali halides.

I. The *Buckingham potential* $V(r) = A\exp(-r/\rho) - Cr^{-6}$ contains an exponential repulsive term with parameters A, ρ determined using an electron gas model, and an attractive van der Waals potential with a parameter C (see *Van der Waals forces*). For example, the F^-–F^- ionic interaction in a LiF crystal is described by $A = 1128$ eV, $\rho = 0.02753$ nm, and $C = 12.71$ eV·nm^5. The parameters for the anion–anion (C_{--}) and cation–cation (C_{++}) interactions are arbitrarily assumed to be equal to each other, $C_{--} = C_{++} = C/2$.

II. A combined ("spline") potential consists of expressions used in proposal I for attraction at long distances and repulsion at short ones, plus the third and fifth power polynomials for intermediate distances (see Fig.):

$$V(r) = -Cr^{-6}, \quad r > r_a, \tag{1}$$

$$V(r) = \sum_{i=0}^{3} a_i r^i, \quad r_a > r > r_m, \tag{2}$$

$$V(r) = \sum_{i=0}^{5} b_i r^i, \quad r_m > r > r_b, \tag{3}$$

$$V(r) = A\exp\left(-\frac{r}{\rho}\right), \quad r_b > r. \tag{4}$$

Matching the functions at points r_a, r_b, and at the minimum point r_m is carried out by invoking continuity of the functions and their first derivatives. The parameters A and ρ coincide, while C considerably exceeds the value expected from proposal I. For example, $C = 26.8$ eV, while $r_a = 0.3457$ nm, $r_b = 0.2$ nm, $r_m = 0.2833$ nm for the F^-–F^- interaction. These parameters are self-consistently adjusted over the entire alkali halide series.

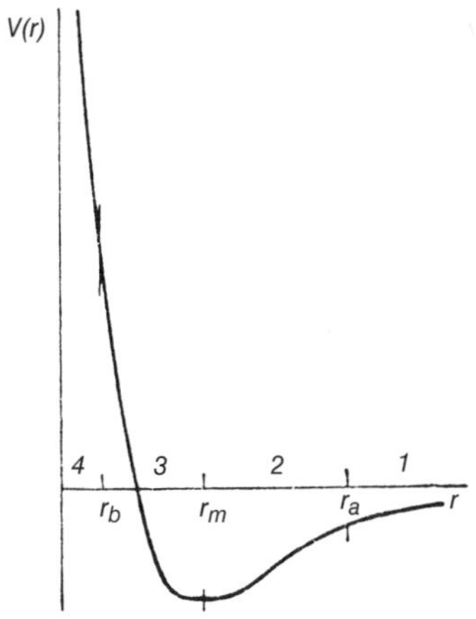

Schematic represention of the potential described in proposal II for the four regions: (1) $r > r_a$, (2) $r_a > r > r_m$, (3) $r_m > r > r_b$, and (4) $r_b > r$.

Having fixed the potential parameters for second nearest neighbors, one may update the respective parameters for first neighbors, that is for adjacent ions, that potential being the common *Born–Mayer potential*, $V(r) = A\exp(-r/\rho)$. Next the parameters of the ions are determined using the shell model. The shell charge and the nucleus–shell bonding constant are found by comparing the static and the high frequency dielectric constants and the transverse optical frequencies with experimental data for all the crystals, so that the polarization of a crystal may be adequately described by the result.

The Catlow–Diller–Norgett potential, as presented in II above, has a number of advantages over other potentials used for ionic crystals. It covers every individual ion and is easily applied to modeling alkali halides. It offers a consistent picture of the ion-to-ion interaction for all the alkali halides. In combination with the Coulomb potential it adequately describes the properties not only of an *ideal crystal*, but of a real one as well. In the latter the ion-to-ion distances may differ considerably from the equilibrium ones in the vicinity of *defects*. Thus it becomes possible to study both intrinsic crystalline and impurity defects in various ionic crystals (see *Interatomic interaction potentials*).

CAUCHY RELATIONS (A. Cauchy)

Additional relations between the 36 crystal *elastic moduli* c_{ik} following from the symmetries of interatomic interactions, where $i, k = 1, 2, \ldots, 6$. If the interatomic forces are pairwise and central then each term is at a *center of symmetry*, the undeformed crystal is stress free, and the *harmonic approximation* is valid. Then in the general case there are 21 independent modulus constants c_{ik} (in *Voigt notation*) by virtue of the symmetry condition $c_{ik} = c_{ki}$. The six Cauchy relations $c_{12} = c_{66}$, $c_{13} = c_{55}$, $c_{23} = c_{44}$, $c_{14} = c_{56}$, $c_{25} = c_{46}$, and $c_{36} = c_{45}$ reduce the number of independent elastic moduli to 15. For cubic crystals there are three independent elastic constants (c_{11}, c_{12}, c_{44}), and one Cauchy relation ($c_{12} = c_{44}$). These results hold well for some *alkali-halide crystals*, but not for most other cubic crystals. In metallic crystals the Cauchy relations usually do not hold (e.g., in noble metals $c_{12}/c_{44} \sim 2$–4). In elastically isotropic crystals obeying the Cauchy relations $c_{11} = 3c_{44}$, and the *Poisson ratio* is $\nu = 1/4$. For orthorhombic crystals there are 9 independent constants and 3 Cauchy relations, corresponding to 6 independent moduli when the Cauchy relations hold.

CAVITATION RESISTANCE

The ability of a material to resist the effect of *cavitation* (formation of vapor bubbles) that appears in a liquid at the passage of a powerful sonic or ultrasonic wave (*acoustic cavitation*), and also during high-speed flow past bodies (*hydrodynamic cavitation*). The collapse of a cavitation bubble at the compression phase produces a local hydraulic shock which affects the surface of a tool, an article, or a part. Cavitation erosion is especially strong for brittle materials such as rocks, concrete, and *ceramics*. *Metals* possess considerably higher cavitation resistance that reduces noticeably in hostile media and suspensions with abrasive particles, and upon increasing the pressure within the volume of a liquid. Ultrasound is widely applied in some technological processes where cavitation plays an important role, and this has prompted the search for materials with high cavitation resistance. Its measure is the loss of mass per unit area during a preset period of time. The cavitation intensity is evaluated by the destruction of a thin aluminum foil placed in the cavitation area. There are also other methods to determine cavitation resistance in investigations of hydrodynamic cavitation, such as rotational, jet-impact, and flow-through ones.

CAYLEY TREE (A. Cayley, 1889); also called a Bethe lattice

A branched structure in which each site has f nearest neighbors, and no two branches are permitted to cross each other. It is cycle-free, and also self similar since each branch is itself an entire tree. Such graphs are used for percolation studies, in cluster techniques to provide a convenient boundary condition for a *cluster*, to calculate the number of isomers of saturated hydrocarbons, and to describe *disordered solids*.

CELLULAR STRUCTURE

Regions 1 μm in size which are free of *dislocations*, or which contain individual dislocations, dislocation dipoles, multipoles, interlacements, networks, etc., and are separated by *dislocation interlacements*, balls, or rods. The cell boundaries can be either dense or spread out, either narrow or wide, and so on. The cells can be equiaxial or nonequiaxial. Sometimes the cellular structure is called *block-cellular structure* (see *Block structure*).

CEMENTATION

Coating a metal, usually iron or steel, by heating it with another substance so that the added material, usually a powder, diffuses into the surface; it is a type of *chemical heat treatment*. The coating material might be a metal like *aluminum* or *zinc*, or perhaps *carbon*. Diffusion saturation of a steel product surface by carbon typically produces a saturated layer that is 0.5 to 2 mm thick, with a carbon content of 0.8–1.1%. Cementation with subsequent *quenching* and low *annealing* provides articles with a high surface hardness, wear resistance and increased fatigue strength, leaving their core tough. Cementation may be carried out in gaseous, liquid, and solid media containing carbon. It is used in the production of tools.

CEMENTITE (fr. Lat. *caementum*, broken stone)

A structural component of *steel* and *cast iron* containing the elements *iron* and *carbon*. Cementite is the most stable *iron carbide* (Fe_3C), containing 6.7 wt.% C (see *Carbide transformations*). It is formed upon the cooling of iron–carbon alloys (see *Iron alloys*) if the conditions are favorable for *crystallization* to occur in the metastable *phase diagram* of these alloys (iron–cementite). Cementite is often a component of *pearlite*, *sorbite*, *troostite* and *ledeburite*. The *crystal lattice* of cementite is orthorhombic; the unit cell contains 12 atoms of Fe and 4 atoms of C. Lattice constants are: $a = 0.4514$ nm, $b = 0.5079$ nm, $c = 0.6730$ nm. According to some indications cementite is close to an interstitial phase (see *Interstitial alloys*). It forms substitutional *solid solutions*: carbon may be replaced by nitrogen or oxygen, iron by *manganese*, *chromium* or *tungsten* (alloyed cementite). Their dispersity, the form of cementite crystals, and their crystal structure associated with a ferrite environment (see *Ferrites*), to a large extent determine the *strength* and *plasticity* of iron–carbon alloys after a *heat treatment*. Cementite is hard and brittle, and there is practically no elongation if it fails during the course of tensile tests. The melting temperature is ~ 1873 K; at this high temperature it decomposes into *austenite* and *graphite*. It is ferromagnetic with the *Curie point* of 486 K.

CENTER

See *Active centers*, *Amphoteric center*, *Color center*, *Dilatation center*, *Luminescent centers*, *Paramagnetic center*, *Positron center*, *Positronium center*.

CENTER OF INVERSION

See *Center of symmetry*.

CENTER OF MASS, center of inertia

A point in a solid body relative to which the total first moment of the all the components of the mass vanish, i.e. $\sum_i m_i \boldsymbol{r}_i = 0$, where $\boldsymbol{r}_i$ is the radius-vector of the ith mass relative to the center of mass. In the absence of torques a solid body moves as if the resultant sum of all applied forces $\boldsymbol{R} = \sum_i \boldsymbol{F}_i$ were acting at the center of mass, and the total mass of the body $M = \sum_i m_i$ were concentrated there. This definition implies that the center of mass is, in a uniform gravitational field, the *center of gravity*, as well as the point where the resultant force of gravity $M\boldsymbol{g}$ is acting.

CENTER OF PERCUSSION

The point of a body about which it starts to rotate when an impulse (impulsive force) is applied at another point in the body.

CENTER OF SYMMETRY, center of inversion

A special point of a figure possessing the property that during the inversion operation ($\boldsymbol{r} \to -\boldsymbol{r}$) through that point as origin of coordinates, all the elements (points) of the figure transform into themselves ($\boldsymbol{r}$ is the coordinate of an arbitrary point of the figure). The presence of a center of symmetry is equivalent to the presence of an *inversion axis* of the first order. Many properties of the solid state depend on the presence or absence of a center of symmetry; see, e.g., *Electric field effects in radio-frequency spectroscopy*.

CENTI... (fr. Lat. *centum*, a hundred)

Prefix to the name of a physical value to form a fractional unit equal to $1/100$ of the source unit. Abbreviation symbol is c. For instance, 1 cm (centimeter) $= 0.01$ m.

CENTRAL FORCES

In solids these are forces of interaction between pairs of atoms (ions) acting along the straight line connecting them. In such a case these atoms are considered to be point masses. Formally the concept "central forces" is based on "rigid" point atoms (ions). When interatomic interactions in solids are described exclusively by means of central forces, this constitutes a coarse and phenomenological approximation. The central force is a special case of a *pairwise interaction* between atoms, when its *adiabatic potential* U has the form

$$U = \frac{1}{2}\sum_{ij} \varphi(|\boldsymbol{R}_i - \boldsymbol{R}_j|),$$

where $\boldsymbol{R}_i$ is the radius-vector of the lth atom, and φ is a function which depends on the type of interaction. For a gravitational field involving two masses $\varphi = -Gm_1m_2/|\boldsymbol{R}_1 - \boldsymbol{R}_2|$, corresponding to an inverse square central force law, where G is the gravitational constant.

CENTRAL PEAK

A sharp increase of the magnitude of the order parameter *fluctuation* at a frequency approaching zero in the neighborhood of the *critical point* T_c of a *structural phase transition*. Initially a central peak was observed in 1971 in the energy spectrum of a beam of monochromatic *neutrons* scattered by a $SrTiO_2$ crystal in the vicinity of $T_c = 105$ K. Central peaks have also been observed experimentally in spectra of *Raman scattering of light*. Depending on the nature and mechanism of its formation one should distinguish three types of central peaks: static intrinsic, static dynamic, and impurity types. A *static intrinsic central peak* is associated with the relaxation of lattice distortions of crystals without impurities. The mechanisms of *dynamic intrinsic central peak* formation are much more diverse. These involve fluctuations of the entropy (*ferroelectrics*: $SrTiO_2$, $KTaO_3$, KDP type), the presence of an overdamped *soft mode* at $T \to T_c$ (K_2SeO_4), two-phonon anharmonic interactions (germanate of Pb, $SrTiO_2$), *diffusion* of mobile ions in *solid electrolytes* (Ag, F, β-alumina), formation of a paraphase, and relaxation at $T \to T_c$ of *clusters* of the prior ordered *phase*.

The *impurity central peak* is associated with the relaxation of *defects* which differ from the point of view of a given central peak formation by the symmetry of their local surroundings, and by the rigidity of their coupling with a *crystal lattice*. Direct experiments on KPD-type crystals have shown that the central peak disappears at the annealing of defects. In the phenomenological theory the integrated intensity of a dynamic intrinsic central peak at $T \to T_c$ increases as the *susceptibility* (at a given external field), and its width is 1 to 2 orders less than an impurity central peak width. For example, in $NaNO_3$ the central peak width is $\sim 10^{13}$ s^{-1}, which is an order of magnitude less than the central peak width in KH_2PO_4. A central peak was found when quasi-one-dimensional magnetic materials (ferromagnetic $CrNiF_3$, antiferromagnetic $KMnF_3$, etc.) were examined at temperatures higher than the three-dimensional ordering point. This central peak is due to the contribution of nonlinear spin excitations, i.e. *magnetic solitons*.

CERAMIC FERROELECTRIC

See *Ferroelectric ceramics*.

CERAMIC GLASS

See *Glass ceramics*.

CERAMIC PIEZOELECTRIC

See *Piezoelectric ceramics*.

CERAMICS (fr. Gr. κεραμικος, made of clay)

A baked (sintered) material; pottery. A class of compact materials, based on the compounds that the nonmetallic elements of Groups III and IV of the periodic system form with one another and/or with *metals*, produced by technological processes that allow for *mass transport*, as well as for the binding of the components. Classified as *ceramic materials* are: glass (see *Vitreous state of matter*), monocrystals of *refractory materials*, and *composites* with a ceramic matrix (see *Composite materials*). Traditional (Table 1) and novel ceramics may be distinguished.

The vast majority of ceramic materials are mono- or heterophase polycrystalline materials based on high-melting compounds. A wide diversity and intricacy of the electronic structure of high-melting compounds comprising different types of interatomic bonds (covalent, ionic, metallic) results in a unique combination of properties (high energy of fragmentization, *hardness*, refractory properties, large *elastic moduli*, a collection of distinctive electrophysical properties, etc.). Their spatial distribution of interatomic bonds is related

Table 1. Types of traditional ceramics

- Pieces of Art
- Household pottery and china (tableware and tiling, crockery, bathroom particulars, etc.)
- Glass (a variety of items and materials for all areas of application)
- Binding materials (cement, gypsum, lime, mortar, etc.)
- Construction materials (pipes, bricks, roof tiles, etc.)
- Fire proofing materials (bricks, crucibles, etc.)
- Abrasive materials

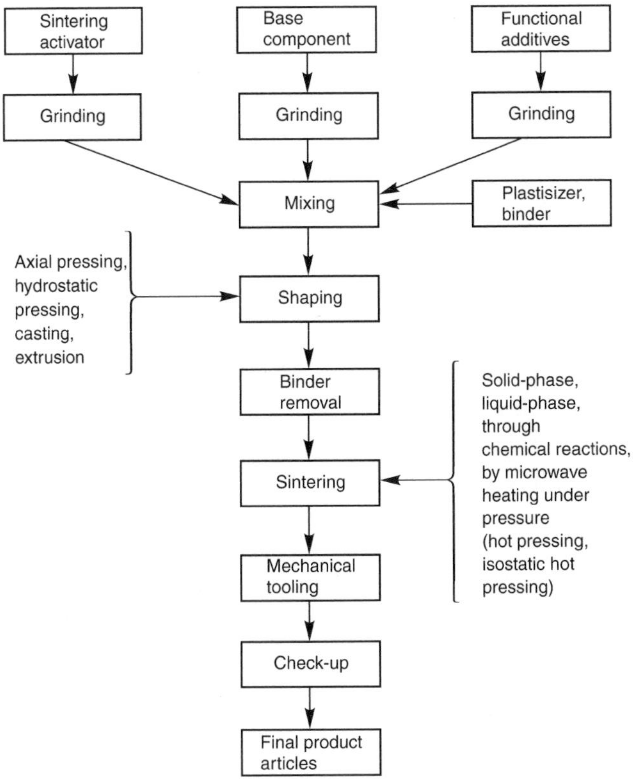

Fig. 1. Typical scheme of the production technology of ceramics.

to diamond-like crystals with predominantly *covalent bonds* (*carbides*, *silicon nitride* and *boron nitride*), and these substances all have similar structures constructed on the principle of diversified sequential packing of identical layers. These compounds are characterized by the existence of a substantial number of polytypes (see *Polytypism*); the latter have also been found in *diamond.*

Besides covalent and *ionic crystals*, there are many ceramic material components from a large group of metal-like high-melting compounds, such as interstitial solutions that form easily in *transition metals* by light elements with relatively small *atomic radii.* Most of these *phases* are compounds of variable proportion having wide ranges of homogeneity, and a great variety of structure types (e.g., CaFe, ZnS, CaB_6, etc.).

Conventionally the properties of a ceramic are defined not only by the properties of its components, but to a substantial degree, by the technology of production (Fig. 1). Also widely employed are the methods of *self-propagating high-temperature synthesis*, *chemical sedimentation*, *infiltration from the gas phase*, etc. The stability of the *crystal lattice* and the low mobility of defects in high-melting compounds results in low rates of the diffusive and viscous processes responsible for mass transport, and for settling at the *solid-phase sintering*. For that reason, additives to activate the *sintering* are introduced into the batch charge, and this is then subjected to *explosion* processing to augment the number of defects, or there could be a physical and chemical *modification* of the surface, etc. The activating additives, usually oxides, secure the *liquid-phase sintering* regime, which not only facilitates the rearrangement of component particles, but also results in their *recrystallization* via the liquid phase. This ac-

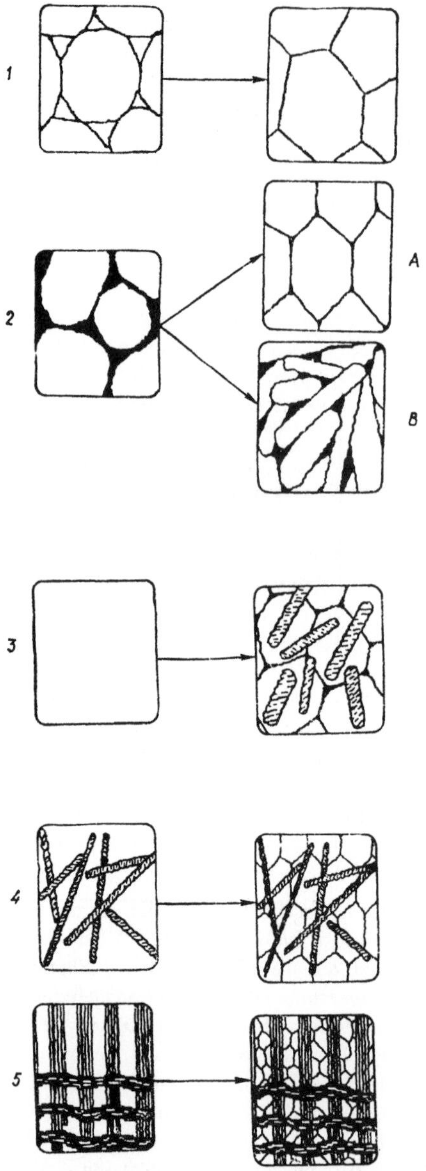

Fig. 2. Microstructure of ceramics fabricated by way of the processes: (1) solid-phase sintering, (2) liquid-phase sintering with the formation of (A) solid solutions, and (B) systems comprising a crystalline matrix with glassy phases at grain boundaries; (3) crystallization of eutectic liquids; obtaining composites of crystalline matrix with thread-like crystals (4), or with fibers (5), by solid- or liquid-phase sintering, or by the chemical sedimentation and infiltration from the gas phase.

Table 2. Properties of several types of alumina

Type of ceramic	Strength, GPa	Type of defects
An ideal crystal, theoretical strength	50	–
Whisker thread-like crystals	15	Screw dislocations along the axis of growth, and surface defects
Single crystals	7	Surface and bulk defects of crystal lattice
Hot-pressed ceramic	0.8–1.0	Same as above, plus grain boundaries and additional phases
Sintered technical ceramic	0.2–0.3	Same as above, plus pores

celerates the sintering process, while permitting a substantial decrease in the temperature, but the ceramic thus fabricated will be interlayered with an amorphous phase at grain boundaries. Typical microstructures obtained with such processes are presented in Fig. 2. The type of microstructure that is realized, as well as the character of the defects and inhomogeneities in the ceramic, determine its mechanical behavior and its complex of physicochemical properties resulting, for instance, in a variation in *strength* by over two orders of magnitude (see Table 2).

The wide variety of types of structure, and sometimes the attainment of record values of the physical properties of the phase components of the ceramics, provides us with the opportunity for fabricating unique materials for a large number of purposes. Along with the attainment of high strength (more than 1500 MPa), a high *fracture toughness* can be achieved to compensate for the traditional brittleness.

The "viscous" state is not only reached by means of a *"skeletal" reinforcement*, but also because the processes of elastic energy dissipation are being purposely initiated at the vertex of a spreading *crack*; the effects of such processes are similar to the action of plastic relaxation at crack vertices in metals. In ceramics, the energy dissipation is conditioned by the development of microcracks, by first- and second-order *phase transitions*, etc. The stability to cracking of such

Table 3. Fields of application of novel (refined, advanced) ceramics

- In electrical engineering and circuit electronics (substrates, varistors, magnetic, ferroelectric and high-temperature superconducting ceramics, transducers, sensors, etc.)
- Structural ceramics (engine parts in the aerospace and automobile industries, facing materials, fairings, thermoresistive bearings, heat exchangers, special fireproof materials, etc.)
- In medicine (implants, bio-ceramics, etc.)
- In nuclear technology (nuclear fuel, materials for nuclear reactors, etc.)
- Wear resistant ceramics
- Low weight armor
- Fiber and laser optics, etc.

a ceramic may be enhanced from a typical level of K_{1c} = 1–3 MPa·m$^{1/2}$ up to values in the range of 10–30 MPa·m$^{1/2}$, viz., up to a magnitude characteristic of hard *alloys* and some metallic *construction materials*.

The discovery in 1986 of *high-temperature superconductivity* in copper oxides, and the raising of the *critical temperature of superconductors* up to 90–135 K for ceramics of the type of $YBa_2Cu_3O_{7-\delta}$ (123 phase) and $HgBa_2Ca_2Cu_3O_{8+\delta}$, respectively, opened up essentially new opportunities in *superconductor electronics*. All the superconducting cuprate ceramics are *type II superconductors*, possess a perovskite-like structure (see *Perovskites*), and have a carrier current density smaller than that of metals and alloys, the carriers generally being holes (in some cases electrons). The 123 phase, the first one discovered with T_c above 77 K, has granular particles of 1–10 μm dimensions that are in weak electrical contact with each other in the ceramic body, with the electromagnetic properties of the ceramic strongly influenced by intercrystallite grain-boundary layers.

On the whole, there is a broad range of applications of these novel ceramics (Table 3), and these materials are used in many fields of technology.

CERIUM, Ce

Chemical element of Group III of the periodic system with atomic number 58 and atomic mass 140.12; it is the lightest *rare-earth element* aside from lanthanum. Natural cerium has 4 stable isotopes ^{136}Ce (0.193%), ^{138}Ce (0.25%), ^{140}Ce (88.48%) and ^{142}Ce (11.07%). The outer shell electronic configuration is $4f^25d^06s^2$. Ionization energies are 6.9, 12.3, 19.5, 36.7 eV. Atomic radius is 0.181 nm; radius of Ce^{3+} ion is 0.107 nm, of Ce^{4+} ion is 0.094 nm. Oxidation states are +3, +4. Electronegativity is 1.08.

Cerium is a silvery-blue *metal*; in air it is covered by a grayish-white oxide film. Four crystalline modifications of cerium are known (α-Ce, β-Ce, γ-Ce, δ-Ce). The crystal lattice of α-Ce is face-centered cubic with $a = 0.485$ nm (at 77 K); β-Ce has a double hexagonal close-packed lattice (α-*lanthanum* type), $a = 0.3673$, $c = 1.1802$ nm (at 298 K); γ-Ce is face centered cubic with $a = 0.51612$ nm (at 293 K); δ-Ce has a body centered cubic lattice, $a = 0.411$ nm (at 1041 K).

At room temperature γ-Ce is stable. During the course of cooling, starting from the temperature 263 K, γ-Ce partially transforms to β-Ce; the transformation is of a martensitic nature. At the temperature ~95 K the non-transformed part of γ-Ce transforms to α-Ce. Below 77 K β-Ce also transforms to α-Ce, but this transition is not completed even at the temperature of liquid helium 4.35 K. The complete transition may be achieved only by plastic deformation of the sample at 77 K. The transitions $\gamma \leftrightarrow \alpha$ and $\gamma \leftrightarrow \beta$ are characterized by large hysteresis depending on the temperature and pressure. At atmospheric pressure the reverse transformations $\alpha \to \gamma$ and $\beta \to \gamma$ start from the respective temperatures ~160 K and ~373 K. If the pressure is over 2500 atm β-Ce cannot exist at any temperature. The behavior of Ce strongly depends on its history and purity. The transformation $\gamma \leftrightarrow \delta$ takes place at 998 K. Density is 6.678 g/cm^3 at 293 K, $T_{\text{melting}} = 1071$ K, $T_{\text{boiling}} = 3740$ K. Heat of melting is 9.211 kJ/mole, heat of evaporation is 390.18 kJ/mole, heat of sublimation is 398 kJ/mole; specific heat is 27.9 J·mole^{-1}·K^{-1} at 298 K; coefficient of thermal conductivity is 10.89 W·m^{-1}·K^{-1} at 300 K, linear expansion coefficient is 8.5·10^{-6} K^{-1} at 298 K. Electrical resistivity is 0.753 μΩ·m at 298 K, temperature coefficient of electrical resistance is 8.7·10^{-4} at 273 K. Above 12.5 K cerium is paramagnetic with molar magnetic susceptibility +2300·10^{-6}.

Thermal neutron cross-section is 0.63 barn. Debye temperature is 139 K. At room temperature the elastic modulus is 29.99 GPa, shear modulus is 11.99 GPa, Poisson ratio is 0.248; ultimate tensile strength of cast sample is 0.1206 GPa, relative elongation is 32.6%. Pure cerium is a forgeable and tough metal. At room temperature without preheating it is possible to manufacture cerium sheets by forging, and wire by pressing. It is used as an alloying addition to the light metals, and also as a getter.

CERMET (fr. ceramic metal)

A composite material consisting of *ceramic* grains dispersed in a metallic matrix. It is formed by techniques of *powder metallurgy.*

CESIUM, Cs

Chemical element of Group I of the periodic table with atomic number 55 and atomic mass 132.9054. There is one stable isotope ^{133}Cs, and 22 radioactive isotopes, among which ^{137}Cs with a half-life 27 years has found the broadest applications. Outer shell electronic configuration is $6s^1$. Ionization energies are 3.893, 25.1, 34.6 eV. Atomic radius is 0.262; radius of Cs^+ ion is 0.167 nm. Oxidation state is +1. Electronegativity is 0.68.

Cesium is a silvery-white *metal.* It has a body-centered cubic lattice with $a = 0.605$ nm (at 448 K). Density is 1.9039 g/cm^3 (at 273 K) and 1.880 g/cm^3 (at 300 K); $T_{\text{melting}} = 301.75$ K; $T_{\text{boiling}} \approx 960$ K. Heat of melting is 15.8172 kJ/kg, heat of evaporation is 613.2 kJ/kg; specific heat at constant pressure is 32.2 J·mole^{-1}·K^{-1} at 298 K; thermal conductivity ratio is 16.7 to 27.2 W·m^{-1}·K^{-1} at 301.6 K; linear expansion coefficient is $9.7\cdot10^{-5}$ K^{-1} (273 to 300 K). Electrical resistivity is 0.183 μΩ·m (at 273 K) and 0.2125 μΩ·m (at 300 K). Mohs hardness is 0.2; Brinell hardness is 0.015. Normal elastic modulus is 1.716 GPa at room temperature. Debye temperature is 42 K. Ion-plasma frequency is 1.724 THz; linear low-temperature electronic specific heat coefficient is ≈3.20 mJ·mole^{-1}·K^{-2}. Electronic work function of cesium polycrystal is ≈1.85 eV. Cross-section of thermal neutron trapping is 29 barn. Cesium is paramagnetic with magnetic susceptibility $0.1\cdot10^{-6}$ (at 291 K).

Cesium is used in the production of photocells (it exceeds all the other metals in light sensitivity), of gas absorbers, and of low work function coatings for emitters. The application of "cesium plasma" is prospective for ionic reactor motors.

CHAIN REACTIONS

Reactions of active particles (*free radicals* in chemical reactions, neutrons n in nuclear reactions), when the elementary interaction produces one or a number of additional active particles (radicals, A·). In the first case the reaction is called an ordinary, non-branched chain reaction, in the second case it is a branched one. The chlorination process of H_2: $Cl\cdot + H_2 \to HCl + H\cdot$, $Cl_2 + H\cdot \to HCl + Cl\cdot, \ldots$, etc., is an example of a non-branched chain reactions. The process of H_2 oxidation represents an example of a branched chain reaction: $H\cdot + O_2 \to \cdot OH + O{:}$, $O{:} + H_2 \to \cdot OH + H\cdot$, $\cdot OH + H_2 \to H_2O + H\cdot, \ldots$, etc. In such a manner, the primary species A· may give rise to a large number (from ten to 10^7) of transformations of inactive particles (e.g., molecules). The number of such transformations per primary active species A· is called the *chain length.* The breaking of a chain (chain reaction termination) may be due to the presence of an impurity which interacts with A· so an additional A· is not produced. Also it may be due to the departure of A· from the reaction region.

In the simplest situation (homogeneous case) the chain reaction rate is determined by the equation:

$$v = K_1 C_1 P_0 \times \exp\left\{\int_0^t \left[(\alpha - 1)K_1 C_1 - K_2 C_2\right]\mathrm{d}t - \beta t\right\},$$

where P_0 is the concentration of the primary species A· created by the initiating chain reaction, K_1 and K_2 are the velocity constants of chain reactions and the chain breaking reaction; C_1 and C_2 are the concentrations of initial substances of the chain reactions, and of the chain breaking centers; α is the number of A· species arising in the elementary act of chain initiation; β is a constant related to the probability of A· leaving the reaction region. The exponential term in this equation reflects the time change of the radical concentration; and the relations between the terms in the exponent index determine the kinetic character of the chain reactions. So, if the first term in the exponent index turns out to be larger than the sum of the others, then the chain reactions velocity increases exponentially with respect to time, and a chain explosion is possible.

In solids chain reactions have been found in organic matrices. The matrix itself and also the components diluted in it can participate in these reactions. For example, in organic polymeric matrices which contain halide complexes of Au, Pt, Cu, Bi chain reactions can be initiated by the action of light or ionizing radiation on the system. As a result, atomic Cl· appears (e.g., $[AuCl_4]^- + Cl\cdot \to [AuCl_2]^- + Cl_2 + Cl\cdot$), and the chain length in such systems may reach 10^4. The use of chain reactions for the photographic recording of information provides new opportunities for the creation of sensitive media with high resolution.

Free radicals giving rise to chain reactions may be generated at the heterogeneous *burning* of solids and its mixtures in the reaction zone. In such a case the very existence these radicals determines the burning kinetics. There is some evidence that at burning, chain reactions also occur in the condensed phase.

Nuclear chain reactions take place in uniform solid specimens of U-235 or Pu-239 (nuclear bomb), and also in the case of a nonuniform distribution of U-235 in some other material (graphite, heavy or light water, or of course U-239) which form the nuclear reactor. The first nuclear chain reaction was realized by E. Fermi in 1942 in a uranium-graphite reactor. Up to the present time only uranium nuclear reactors are in use, and they satisfy the following conditions. The elementary act of fission involves a neutron n striking a U-235 nucleus in accordance with the reaction

$$\text{U-235} + n \to \text{A} + \text{B} + kn + Q,$$

where A and B are nuclear fragments (daughter nuclei), k is the neutron multiplication factor, and the chain reaction proceeds because the average value of $k \sim 2.5$ for the elementary act of fission. The kn secondary neutrons in turn induce further fissions The energy Q of the reaction is harnessed in nuclear power plants for heating cities, producing electricity, etc. and the number of neutrons generated is greater than one (i.e. $k > 1$). The cross-section of the fission reaction of U-235 increases with the decrease of the neutron velocity v as $1/v$, and there are resonances in the range of neutron energy from a fraction of an electronvolt to perhaps a hundred electronvolts. As a result, the so-called "thermal neutrons" play the main role in fission reactions. At these low energies the cross-section for the U-235 fission reaction is much larger than the cross-sections for other reactions with neutrons. Since there is a large probability that many neutrons will escape from a very small sample without reacting, there is a critical mass for the chain reaction to take place, and this is approximately 50 kg for U-235 and 10 kg for Pu-239. If the average neutron multiplication factor $\langle k \rangle$ is too large, the number of neutrons in the specimen can grow exponentially, leading to an explosion. To prevent this, neutron absorbing rods (e.g., graphite) are inserted into the reactor to remove excess neutrons and keep $\langle k \rangle \sim 1$ so the chain reaction will proceed slowly, and generate heat gradually.

CHALCOGENIDE MATERIALS (fr. Gr. $\chi\alpha\lambda\kappa o\varsigma$, copper and $\gamma\varepsilon\nu\nu\alpha\omega$, engender)

Compounds of elements of Group VI of the periodic table (O, S, Se, Te, Po, collectively called chalcogens) with other elements. The class of *chalcogenides* (oxides, sulfides, selenides, tellurides) contains the X^{-2} anion of Group VI elements. In view of their differences in properties and fields of application, oxides are often unfairly excluded from the list of chalcogenide materials, and polonium compounds are rarely included because they are strongly radioactive. In modern areas of engineering chalcogenide materials are often used in the form of *thin films* or *monocrystals*, which are obtained through precipitation of aqueous solutions, decomposition of chelate compounds on a *substrate*, crystallization from gaseous (vapor) phase, etc. Typical ionic lattices of chalcogenides are characteristic of these elements, which exhibit very low values of ionization potentials. Many chalcogenide materials form lattices of the chalcopyrite type, the spinel type, and others. They are often nonstoichiometric compounds. Chalcogenides comprise a broad class of compounds, including *insulators*, *semiconductors* and *conductors* of electric current. Chalcogenide materials are used in opto- and microelectronics, power engineering, in metallurgy, as photosensitive elements, transparent contacts, contactless switches, image converters, light indicators and electric phosphors, resistive salts, memory storage cells, etc.

Historically the term chalcogenide was used for ores and *minerals* containing copper and sulfur: chalcopyrite (copper pyrite) $CuFeS_2$, the most important of all copper ores; chalcocite Cu_2S; chalcanthite $CuSO_4 \cdot 5H_2O$ (blue vitriol), chalcostibite $CuSbS_2$, and others.

CHANNELED PARTICLE FRICTION

See *Friction of channeled particles.*

CHANNELED PARTICLE SELF-ACCELERATION

A phenomenon observed during the passage of a beam of relativistic charged particles through a *monocrystal* in parallel to crystallographic planes or axes. Charged particles travelling in the *channeling* mode may radiate short-wave *photons.* By this action the channeling charged particles may absorb quanta emitted by other charged particles of the beam, and thereby increase their initial energy.

Quasi-bound motion of charged particles in the planar or axial channneling mode may be regarded as one-dimensional or two-dimensional atoms. Thus in analogy with an atomic system, the process of self-acceleration of channeled particles may be called a quasi-Auger effect. For ultrarelativistic electrons entering a monocrystal parallel to crystallographic planes, the main radiation mechanism is electromagnetic *quasi-characteristic radiation* associated with the transitions of electrons from the free state to the channeling one. An electron in the quasi-bound channeling state may transfer to another quasi-bound state or escape into the continuum spectrum as a result of electron trapping (this exchange takes into account the possible contributions of not only real, but also virtual photons). The kinematics of this process shows that in the final state it is possible to detect charged particles with double their initial energy.

CHANNELING in monocrystals

The transversely restricted (relative to the initial direction) motion of charged particles along open channels between neighboring close-packed atomic rows or planes in crystals. The condition for channeling is a small value of the angle subtended by the particle direction while entering the channel to the corresponding axis (for *axial channeling*) or plane (for *planar channeling*). This means that the particle being channeled experiences repeated glancing impacts with atoms of restraining chains (planes). These bend the trajectory in such a way that encounters with a small impact parameter become rare. The classical description of channeled *ions* proposed by J. Lindhard is based on using a potential averaged over the atoms of chains (planes), which is referred to as a *string potential*, or a *continuous potential*. In this case the angle ψ of the particle velocity vector relative to the axis of a chain or plane must be smaller than a critical value $\psi_k = (U_0/E)^{1/2}$, where E is the particle energy, and U_0 is the height of the potential barrier created by the atoms of a chain or a plane. If ψ is greater than ψ_k, called the *Lindhard angle*, the particles are no longer confined to the regime of channeling (see also *Dechanneling*). The manner of reflection of a particle from the walls of a channel suggests that a positively charged channeling particle spends most of its time moving past regions of a crystal with relatively low electron density. Therefore, the *trapping* of a particle into the channeling mode causes a reduction of its ionization energy loss and increases its effective mean free path in a *monocrystal.* In the channeling regime there is a considerable reduction of the probability of processes with small impact parameters relative to the sites of a crystal lattice (see *Orientation effects*).

The occurrence of channeling was observed first for positively charged ions, and later for electrons and *positrons.* It is of interest that the channeling phenomenon was "discovered" twice. In 1912, the German physicist J. Stark took notice of the fact that charged particles can penetrate to considerably greater depths along some preferential axes. However, this effect was forgotten for a long time until a "second discovery" in the early 1960s. It was found by computer simulation that incident bombarding ions which have abnormally long paths through a monocrystal when they enter the crystal almost parallel to a main crystallographic direction. The channeling of relativistic charged particles gives rise to an intense electromagnetic *quasi-characteristic radiation* in the gamma ray and X-ray bands. See also *Hyperchanneling, Molecule channeling, Neutral particle channeling, Quantum states of channeled particles, Okorokov effect, Channeled particle self-acceleration, Surface channeling, Surface conductive channel, Current filamentation.*

CHANNELING OF ION BEAMS

This takes place along the directions of atomic chains and planes in a *monocrystal* because the ions scattered at small angles to these directions cannot penetrate deep into the *crystal* (see Fig. 1).

Shadow traces of the atomic chains and planes appear (the so-called *shadow effect*) at the target detecting the scattered *ions*. The emission of products of a nuclear reaction between the incident ions and the crystal atoms also produces the shadow effect. The shadow disappears, if the time between the trapping of the ion nucleus by the crystal atom nucleus and the emission of the reaction product is long enough for the activated atom to displace noticeably from the blocking chain (plane). This allows one to estimate experimentally the nuclear reaction time. One can observe the blocking by the shadow effect resulting from the embedding of a radioactive isotope into a *monocrystal*. Fig. 2 shows the angular distribution of the intensity of emitted α-particles in a tungsten crystal upon the introduction of the ^{222}Rn isotope; the minimum corresponds to the $\langle 111 \rangle$ direction. Heating the crystal results in washing out the shadow due to thermal displacements of the atoms.

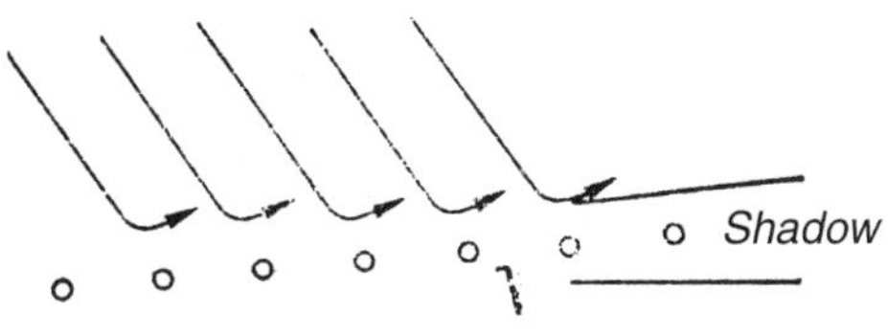

Fig. 1.

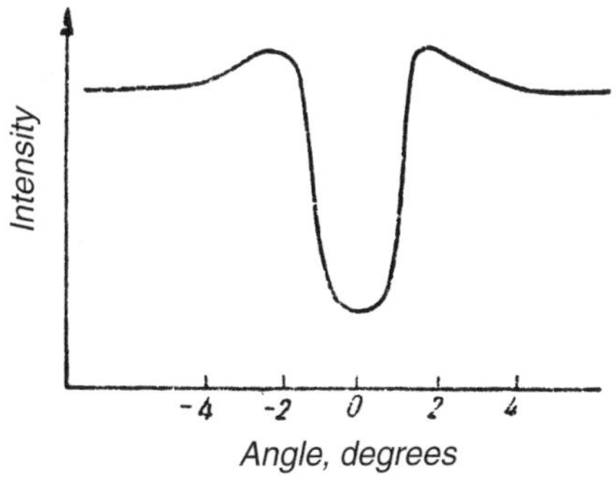

Fig. 2.

CHARACTERISTIC ENERGY LOSS

Discrete energy losses, which are independent of the energy of incident primary electrons, and are characteristic of a given *solid*; they are observed in reflection and transmission spectra of secondary electrons. Characteristic energy loss is energy expended on the excitation of (i) electron transitions from inner core levels or from *valence band* levels to unoccupied levels, vibrational energy levels, and (ii) collective longitudinal oscillations of a *solid-state plasma* (*plasmons*). The values of characteristic energy losses of the first kind (30–10^5 eV) are easily determined from the binding energies (E_b) of electrons in atoms, and are used in *ionization spectroscopy* for the analysis of solids. Different solids exhibit different binding energies because of many-particle effects (*relaxation*, multiplet splitting, *configuration interaction*, *shake-off* and *shake-up*, chemical shifts). Characteristic energy losses of the second kind (1–30 eV) are more difficult to interpret, since they are due to processes involving levels of the valence band and upper unoccupied levels. Vibrational characteristic energy losses are related to transitions between the vibrational states of surface atoms (molecules) of a solid, or of adsorbed layers; these characteristic energy losses are a source of information on phonon involvements. Volume and surface plasmon characteristic energy losses ($\hbar\omega_P$ and $\hbar\omega_S$, respectively) are determined by the concentration of valence electrons n: $\hbar\omega_P = (4\pi e^2 n/m)^{1/2}$, $\hbar\omega_S = \hbar\omega_P/\sqrt{2}$, and they manifest themselves as distinct lines in the spectra of materials with weakened valence electron bonding. Other materials (e.g., *transition metals*) exhibit diffuse plasma characteristic energy loss lines. See also *Characteristic energy loss spectroscopy.*

CHARACTERISTIC ENERGY LOSS SPECTROSCOPY

Branch of physics concerned with the discrete energy losses that are experienced by a monochromatic primary electron beam interacting with a *solid.* Characteristic energy loss spectroscopy examines samples using either the transmitted beam or the reflected beam of light. The techniques are applied to the investigation of inner core levels in atoms of solids, shifts of these levels caused by changes of the chemical environment, electron transitions from valence band levels to unoccupied levels, transitions among vibrational states of atoms of adsorbed layers at *solid surfaces*, generation of collective longitudinal *plasma oscillations* (*plasmons*). These studies provide information on the composition and electronic properties of the material as a whole, and in particular of its surface. The major fields of application are the spectroscopy of *surface plasmons* and *vibrational characteristic energy loss spectroscopy*, which allows one to perform a complete analysis of the surface. As distinct from *optical spectroscopy*, vibrational characteristic energy loss spectroscopy affords: (1) operation over a broad energy band (0.001–1 eV), which includes phenomena inaccessible by optical methods; (2) excitation of optically forbidden transitions, and (3) an increase of the scattering cross-section. There exists the following classification of spectroscopic methods according to the type of physical process involved in the energy losses: *ionization spectroscopy* of valence and inner core levels, *plasmon characteristic energy loss spectroscopy* and high-resolution vibrational spectroscopy. The intensity of spectral lines due to *characteristic energy losses* is not very high, and decreases with an increase of the probing beam energy; therefore, electrons with energies below 2 keV are used, so the probing depth in this case does not exceed 2 nm. Thus, characteristic energy loss spectroscopy is the preferred method for examining near-surface layers of solids.

CHARACTERISTIC LENGTH OF MAGNETIC MATERIAL

Parameter l_W of uniaxial magnetic materials, with the dimensionality of length, which is equal to the ratio of the surface energy density of a Bloch *domain wall* to the maximal volume energy density of the *demagnetization field* $\mu_0 M^2/2$. In materials with a quality or Q-factor (see *Magnetic films*) exceeding 1 the characteristic length determines the minimal diameter of a *magnetic bubble domain* (or minimal spacing of the band *domain structure*), that may occur in thin layers (films) along the *easy magnetization axis*, which is perpendicular to the developed surface. The largest possible diameter of an isolated magnetic bubble domain is $\approx 3.9 l_W$, and the layer thickness in this case must be $\approx 3.3 l_W$.

CHARACTERISTIC LENGTHS IN SUPERCONDUCTORS

Coherence length ξ, *penetration depth of magnetic field* λ, *length of imbalance relaxation* λ_E (see *Penetration depth of electric field*), *Josephson penetration depth.*

The *superconducting phase transition* is characterized by the generation of bound states, i.e. *Cooper pairs* of electrons which are bosons, and are subject to *Bose–Einstein condensation.* The coherence length ξ is the effective radius of a Cooper pair, and it determines the characteristic distances over which the complex *order parameter* of the superconductor $\psi(r)$ (effective wave function of Cooper pair) changes to an appreciable extent when the superconductor is in a certain nonequilibrium state; e.g., at the penetration of *Abrikosov vortices* into a superconductor (ξ is the radius of the normal state vortex core), at the interface of a superconductor with a normal metal (*proximity effect*), etc. Typical values of ξ range from 10^{-6} m in pure *metals* to 10^{-9} m in *alloys* and compounds.

The coherence length of a pure (type I) superconductor is $\xi_0 = \hbar v_F/\pi\Delta_0$, and in the case of a (type II) superconducting alloy with electron *mean free path* l_l, which is small compared with ξ_0, the coherence length at the temperature 0 K is of the order of $\xi(0) \approx (\xi_0 l_l)^{1/2}$. In the neighborhood of the transition point, ξ tends to infinity as $\xi(T) \approx \xi(0)/(1-t)^{1/2}$, where

$t = T/T_c$. In a *type II superconductor*, the magnitude of ξ is related to the value of the *upper critical field* B_{c2} through the expression $B_{c2} = \Phi_0/(2\pi\xi^2)$ ($\Phi_0 = h/(2e)$ is the *flux quantum*).

The *Meissner effect* in superconductors arises from the fact that external magnetic field creates a current (*Meissner current*) in the surface layer which screens the bulk of the superconductor from the penetration of the field. The characteristic thickness of this layer is called the penetration depth λ. In the case of a pure *type I superconductor*, the *London penetration depth* is determined from the *London formula* $\lambda_0 = [mc^2/(4\pi ne^2)]^{1/2}$ and is of the order of 10^{-7} m. In the case of a superconducting alloy, $\lambda > \lambda_0$, with the value $\lambda \approx \lambda_0(\xi_0/l_l)^{1/2}$ at $\xi_0 \gg l_l$. In the neighborhood of T_c, λ has the temperature dependence $\lambda(T) \approx \lambda(0)/(1-t)^{1/2}$. The penetration depth λ may be determined experimentally from the change of the superconductor inductance, which is due to partial penetration of the field into the bulk of the solid.

The imbalance relaxation length λ_E determines the characteristic distances of nonequilibrium processes in superconductors. The simplest case is the current flow through a galvanic contact between a normal metal and a superconductor. In the normal metal, the current is carried by electrons, whereas in the superconductor it is carried by Cooper pairs. The conversion of normal excitations into pairs in the normal metal takes place at the distance λ_E from the interface. The electric field $\boldsymbol{E}$ differs from zero in this region, therefore λ_E may be considered as the depth of penetration of a longitudinal (in contrast to the Meissner effect) "dissipative" electric field. The nonequilibrium state of the superconductor exists within the distance $\sim\lambda_E$ of the interface. This state is characterized by an imbalance in the occupancy of electron-like ($P > P_F$, where P_F is the Fermi momentum) and hole-like ($P < P_F$, where P is excitation momentum) spectrum band branches. The imbalance relaxation is related to inelastic collisions (i.e. electron–phonon types), characterized by a mean free path length $l_\varepsilon = v_F\tau_\varepsilon$, which are inefficient at low temperatures, and therefore lead to rather high values of λ_E (e.g., $\lambda_E \sim 10$ μm). According to theory, $\lambda_E = (D\tau_Q)^{1/2}$, where $D = v_F l_l/3$ is the electron *diffusion coefficient*, and $\tau_Q = [4T/(\pi\Delta)]^{1/2}\tau_\varepsilon$ is the *period of the imbalance relaxation*. In the neighborhood of T_c, λ_E varies as $\lambda_E(T) \approx \lambda_E(0)/(1-t)^{1/4}$. The value of λ_E determines the characteristic dimensions of the region of inhomogeneity (so-called *phase slip centers*) in the *resistive state* of superconducting films.

The Josephson penetration depth has the value $\lambda_J = [\Phi_0/(2\pi\mu_0 J_c d_{eff})]^{1/2}$ where J_c is the critical current and d_{eff} is the effective thickness of the junction. A Josephson junction is considered short when its length is less than λ_J, and it is long when it has a greater length.

CHARACTERISTIC SURFACE, ellipsoid

A second-order surface, which represents the orientation dependence of a particular physical property of a *crystal*, which is described by a second-rank tensor.

The equation of the characteristic surface, when reduced to the principal axes of the *crystallophysical coordinate system* (e.g., an orthogonal crystal coordinate system compatible with nonorthogonal lattice constant directions) is of the form

$$s_1x_1^2 + s_2x_2^2 + s_3x_3^3 = 1,$$

where s_1, s_2 and s_3 are the principal values of the tensor s_{ij}. If all s_{ii} are positive, then the characteristic surface is (in the general case) a three-axis ellipsoid; with its semiaxes equal, respectively, to $(s_1)^{-1/2}$, $(s_2)^{-1/2}$ and $(s_3)^{-1/2}$, and proportional to the values of the particular physical property represented by the s_{ij} in question along the principal axes. For a random direction, specified by radius vector $\boldsymbol{r}$, we have $r_{ij} = (s_{ij})^{-1/2}$.

In accordance with the *Neumann principle* (which states that the *symmetry group* of a certain physical property of a crystal must include the symmetry elements of the *point group* of the crystal), the symmetry of the characteristic surface corresponds to the symmetry of the physical property. For all positive values of s_{ii} the characteristic surface is a sphere for cubic crystals which belong to the highest symmetry *category*. The characteristic surface of the crystals which belong to the medium category, those with tetragonal, hexagonal or trigonal structures, has the shape of spheroid (prolate or oblate ellipsoid), whereas crystals of the lowest symmetry category which are orthorhombic, monoclinic, or triclinic, are described by a characteristic surface in the form of a three-axis ellipsoid. An example of a characteristic surface associated with optical anisotropy represents the dependence of the *refractive index* n on the direction in the crystal; depending on the shape of this surface crystals are classified as optically isotropic, *uniaxial crystals* or *biaxial crystals.*

The main property of the characteristic surface may be stated as follows: if vectors $\boldsymbol{A}$ and $\boldsymbol{B}$ are interconnected by the relation $B_i = s_{ij} A_j$ and the radius vector $\boldsymbol{r}$ of the characteristic surface tensor s_{ij} is parallel to the vector $\boldsymbol{A}$, then the direction of the vector $\boldsymbol{B}$ coincides with the normal to the plane, which is tangent to the characteristic surface at the contact point of the vector $\boldsymbol{r}$. Characteristic surfaces apply in the linear or harmonic response range of physical properties.

Besides the characteristic surface, the concept of an *indicating surface* is sometimes introduced:

$$\frac{x_1^2}{s_1^2} + \frac{x_2^2}{s_2^2} + \frac{x_3^2}{s_3^2} = 1.$$

Radius vectors, which terminate at this surface, designate the absolute values of the physical properties in a given direction.

CHARACTER OF GROUP REPRESENTATION MATRIX

The trace or sum of the diagonal elements of a group representation matrix. The character is independent of the choice of coordinate system, and all irreducible representations of the same class have the same character. The character table of the irreducible representations has orthogonality properties (see *Group theory*).

CHARGE CARRIER DRIFT

See *Current carrier drift.*

CHARGE COUPLED DEVICES

A broad class of devices (*charge transfer devices*, *charge coupling devices*, and related *charge-injection devices*) that allows for the scanning of an optical image in a manner similar that of a cathode ray tube (*vidicon*). The main purpose of such devices is the transmission of TV signals of various spectral bands: they are also used in random-access memories. The mechanism of charge coupling action can, to some extent, be represented by a combination of a *bipolar transistor* (motion of minority *current carriers*) and a *field-effect transistor* (controls carrier motion by changing displacement voltage at the gate). The electrodes (gates) have a mosaic structure, and the distance between the edges of adjacent elements does not exceed the *diffusion length* of nonequilibrium carriers. The application of a bias voltage to a single gate element creates a "potential well" under this element for minority carriers. The latter accumulate under the gate in quantities proportional to their rate of generation. In a two-dimensional array of such elements, each "operating element" with the bias that corresponds to the accumulation of minority carriers has unbiased elements as neighbors on every side. Upon supplying the bias to an element adjacent to the operating one, a "potential well" appears under it. If the well is deep enough the packet of minority carriers from the "operating element" well moves into it. The sequence of voltage pulses at the gate elements is set so that the packet shifts by the length of one element with each pulse, thus the charge of every "operating" element can be led sequentially through to the output element, producing an output signal proportional to this charge.

CHARGE DENSITY OSCILLATIONS

See *Plasma oscillations.*

CHARGE DENSITY WAVE in metals

Periodic redistribution of electronic, ionic and total charge in space, caused by small periodic displacements of ions near their equilibrium positions in the *crystal lattice.* The charge density wave (CDW) state is detected via the scattering of X-rays, fast electrons and neutrons, which exhibits diffraction peaks from the source lattice and weaker peaks from their "satellites" (see *X-ray structure analysis*, *Electron diffraction analysis*). A charge density wave can appear when the metal is cooled below some critical temperature. The *phase transition* to the CDW state is characterized by changes in the temperature dependence of the resistance, Hall constant, magnetic susceptibility, and a modification of the electronic spectrum of the metal. The charge density wave period can be commensurate with the period of the source lattice, and in these cases one can talk about *commensurate* CDWs which differ from *incommensurate* ones. As a rule in the cases of incommensurability the lattice period depends on the temperature, and structural transitions to the commensurable CDW are possible.

The transition to a charge density wave state is found in metals with strong anisotropy of the electronic spectrum. This anisotropy can have a two-dimensional character, when the electrons move freely in a plane (their Wannier wave functions at different points overlap), but between the planes their motion is restricted (due to the weak overlapping of the electronic *Wannier functions*). Related to such compounds are, e.g., the layered compounds of dichalcogenides of transition metals like TaS_2, $NbSe_2$ (see *Quasi-two-dimensional crystals*). Anisotropy of a one-dimensional type is realized in compounds with a chain structure, e.g., in organic conductors (see *Quasi-one-dimensional crystals*).

The origin of transitions to the CDW state for all these systems is associated with the features of the geometry of the *Fermi surface*. The theory shows that if sufficiently large sections of this surface are nested, meaning that they are can be overlapped with each other by translation along a reciprocal lattice vector $\boldsymbol{Q}$, then the polarizability of the system of electrons in the periodic electric field of the lattice $E = \mathrm{Re}\,(E_0\,\mathrm{e}^{\mathrm{i}\boldsymbol{Q}\boldsymbol{r}})$ ($\boldsymbol{r}$ is the radius-vector of the lattice point) is high and the lattice becomes unstable with respect to the appearance of periodic distortions with wave vector $\boldsymbol{Q}$. The distortions are formed below the critical temperature and cause an energy gap to appear in the electronic spectrum at the nested sections of the Fermi surface, i.e. with the complete or partial loss of metallic properties. The degree of nesting, as well as the extent of the metallic property loss grows with increasing anisotropy of the electronic spectrum. In quasi-two-dimensional (layered) compounds the appearance of an energy gap at the whole Fermi surface is impossible, and they preserve their metallic properties at transitions to CDW state, or in case of large displacements of atoms they become *semimetals*. In quasi-one-dimensional compounds the region of Fermi surface nesting is more extensive, and the energy gap may occur on the whole Fermi surface. Then quasi-one-dimensional compounds in a state with a charge density wave become insulators (see *Peierls transition*).

CHARGE DENSITY WAVE IN SUPERCONDUCTOR

See *Superconductor with charge (spin) density waves*.

CHARGED PARTICLE ENERGY LOSS in a solid

This loss occurs during the interaction of incoming charged particles with *atoms* and electrons of a target. It is characterized by the average *braking ability* equal to $-\mathrm{d}E/\mathrm{d}x$, where E is the particle energy and coordinate x is its initial direction, and by *straggling* of the charged particle energy loss (by the root mean square scattering of the energy loss value). This energy loss determines the values of *charged particle path* in the material. Calculation of the energy loss is a complex many-particle problem, and to solve it requires the adoption of a series of approximations which differ for high- and low-energy particles, for long-range and short-range collisions, and also for the different correlations of masses and charges of interacting particles, taking into account the so-called *density effect* associated with the polarization of the medium by the ionizing particle, of the resonance recharging of the moving *ions*, etc. The main mechanisms of charged particle energy loss are the nuclear decelleration or braking caused by elastic collisions (taking place without variation of internal energy), collisions of atoms (ions); electron braking caused by the inelastic collisions leading to the excitation and ionization of the atomic electron shells, and also of the free electrons of the irradiated material (see *Characteristic energy loss*). Besides, there is the charged particle energy loss from radiation – *Bremsstrahlung* and *Cherenkov radiation*; the loss is substantial for light particles, especially for electrons and *positrons*. The value of this loss depends on the energy, charge, and mass of the moving particles, as well as on the charge, mass, and density of the target atoms. For motion close to the direction of low-index crystallographic axes and planes, the energy loss strongly depends on the direction of the incident beam (see *Orientation effects*; *Channeling*); in other cases the target may be approximately considered as an amorphous one with isotropic loss. In the approximation of the *Lindhard–Scharf–Schiott theory* charged particle energy loss of ions in an amorphous medium is expressed in terms of dimensionless variables for the energy $\varepsilon = aEM_2/[e^2 Z_1 Z_2(M_1 + M_2)]$, and coordinate $\rho = 4\pi x a^2 N M_1 M_2/(M_1 + M_2)^2$, where M and Z are the masses and atomic numbers of the incident ions (subscript 1) and of the atoms of the medium (subscript 2), N is the density of the atoms of the medium, $a = 0.8853 a_B (Z_1^{2/3} + Z_2^{2/3})^{-1/2}$, a_B is the *Bohr radius.* Taking into account the structure of the atomic electron shells supplements the results of the Lindhard–Scharf–Schiott theory concerning an oscillating (within limits of tens of percents) dependence of dE/dx on Z_2 caused by the contribution of the valence atomic shells to the charged particle energy loss of the electrons.

The charged particle energy losses of positrons and electrons of not too high energies $E < E_{cr}$ when the role of braking radiation is not yet appreciable, are determined by nuclear and electron braking, with values having qualitative regularities close to the charged particle energy loss of ions, but differing from the ion values mainly due to the mass difference between the electrons (taking into account relativistic corrections) and ions. The value of E_{cr} may be evaluated from the correlation $Z_2 E_{cr}[\text{MeV}] \approx 800$. For different energy ranges, masses, particle charges, and target parameters, there are different approximations for the calculation of the charged particle energy loss, and there are also many available experimental data.

CHARGED PARTICLE PATHS in condensed media

Characteristics of charged particle trajectories that are limited by the *charged particle energy loss* during collisions with atoms and electrons of the medium. The particle path in a medium is a broken line since each collision results in a change of particle momentum and direction. The total vector path, $\boldsymbol{R}$, is the distance between the particle entry point and its terminal point where it has already expended most of its energy and stopped. The projective path R_p presents a projection of $\boldsymbol{R}$ onto the direction $\boldsymbol{x}$ of the initial particle momentum (e.g., the latter may be normal to the bombarded plane, yz). The transverse path $R_\perp$ characterizes the change of the initial coordinate of the particle in the yz plane at the time the particle stops. Finally, one may consider the total or effective path R_L which is equal to the full length of the distance travelled by the particle taking into account all deviations from a linear track. The number of particle collisions per unit path length, the impact parameters for each collision, and, hence, the trajectory of each individual particle are probability variables, even for identical initial conditions of particle motion, so there exists a certain distribution of charged particle paths over their lengths. When bombarding a sample with a beam of ions of the same chemical element with equal initial energies the profile of stopped ions appears to exhibit a certain broadening. Such a distribution is often characterized by the moments $M_1 = R_p = \langle x \rangle$; $M_2 = \Delta R_p^2 = \langle (x - \langle x \rangle)^2 \rangle$, ..., $M_k = \langle (x - \langle x \rangle)^k \rangle$ $(k = 3, 4, \ldots)$, $\langle x^k \rangle = \int (r\cos\theta)^k p(r, \theta; E)\,dr$, where $p(r, \theta; E)$ is the probability for a particle entering the crystal with energy E at the zero coordinate point along the direction $\boldsymbol{n}$ to stop at point $\boldsymbol{r}$, so that $\theta = \arccos(\boldsymbol{nr}/r)$. The value ΔR_p presents the statistical average scatter (*straggling*) of particle paths, M_3 that is a measure of distribution asymmetry. The accepted measure of asymmetry is the so-called distribution *skewness*, $S_k = M_3/M_2^{3/2}$. In the case of $S_k < 0$, the path distribution falls off more sharply from its maximum in the bulk sample, and when $S_k > 0$ a sharper fall-off is at the surface of the sample. The pathways become longer in the case of *channeling.* Electron pathways in the medium typically feature much sharper trajectory breaks compared to the pathways of heavy particles. Because of multiple collisions, the electron quite soon "forgets" its initial direction of propagation. Moreover, electrons penetrate the medium in a diffusion manner. To retrieve the pathways experimentally electrons of varying initial energies are shot through thin plates. The maximum plate thickness at which a throughput electron current is still recorded corresponds to the maximum path length for an electron with a given energy penetrating the material. A normal pathway is one with a plate thickness such that the initial electron beam intensity decays by a factor of 6 when passing through it. An empirical dependence for electrons with initial energies $E \geqslant 0.6$ MeV is $R_\mathrm{p}[\mathrm{mg/cm^2}] = 526E[\mathrm{MeV}] - 94$.

CHARGE, MAGNETIC

See *Magnetic charge*.

CHARGE-TRANSFER COMPLEXES

Organic compounds consisting of molecules of two types: one acts as an electron *donor* and the other as an electron *acceptor*, the charge being redistributed between the neighboring donors and acceptors. Charge-transfer complexes have an electric *dipole moment* even if they are created from nonpolar molecules, and they exhibit paramagnetic properties, whereas the initial donor and acceptor molecules are diamagnetic. If an electron in the ground state of the organic solid is completely transferred from donor to acceptor, such a compound is called an *ion-radical salt* (limiting bond case in charge-transfer complexes). Besides weak charge-transfer complexes of ion-radical salts, there are many such compounds with intermediate strength bonds. The properties of charge-transfer complexes are strongly dependent on the extent of charge transfer, and on the magnitude of the donor–acceptor interaction. The electrical conductivity of these complexes at room temperature varies over a wide range: from 10^{-15} up to 10^2–10^3 $\Omega^{-1}\cdot\mathrm{cm}^{-1}$. The temperature dependence of the *electrical conductivity* is of the semiconductor type; hence, most charge-transfer complexes are *organic semiconductors*. These are exemplified by compounds of different molecules with the molecular acceptor *tetracyanoquinodimethane* (TCNQ). Because of the strong acceptor properties, the planar structure, and the high symmetry of the TCNQ molecule, one can obtain charge-transfer complexes with a rather high conductivity (as high as 10^4 $\Omega^{-1}\cdot\mathrm{cm}^{-1}$). A large variety of organic charge-transfer complexes have been synthesized that exhibit properties ranging from those of *semiconductors* to *metals* and superconductors (see *Superconductivity*). The latter two categories include mainly ion-radical salts. Molecular charge-transfer complexes are of interest as photosemiconductors (see *Photoconductivity*).

CHEMICAL BOND PREDISSOCIATION

The state of a *chemical bond*, usually an excited one, having an energy that exceeds the energy of atoms in the dissociated state. The latter is separated from the initial state by a barrier which may be overcome by tunneling (see *Tunneling phenomena in solids*). The simplest example of chemical bond predissociation is the excited hydrogen molecule H_2 with a dissociation energy $\approx$4.4 eV, whereas the excitation energy upon transition of one of the atoms to the level $2p$ is $\sim$10 eV. Another example is the state of an α-particle in a radioactive nucleus, e.g., of Ra or U, whose average lifetime is, respectively, 1600 and $4.4\cdot 10^9$ years.

CHEMICAL BONDS in solids

Bonds between atoms of a *solid*; isolated atoms decrease their free energy during the redistribution of their electron densities when they join together through chemical bonding. The nature of a chemical bond is determined by the nature of the constituent atoms, and by the type of structure adapted by the solid. There exist *molecular crystals*, whose structural elements are molecules; under evaporation this *crystal* type decomposes into molecules. Crystals can be formed from monatomic molecules of noble gases (He, Ne, Ar, Kr, Xe, Rn), diatomic molecules (H_2, O_2, NO, CO), triatomic molecules (H_2O, H_2S, NO_2), and molecules of a more complex composition (NH_3, CH_4, etc.). In these cases every molecule has saturated bonds, and the attraction of molecules toward each other is through *van der Waals forces* or *hydrogen bonds*, as in H_2O, H_2S, NH_3. A certain (additional) role is played by Coulomb forces, if the atoms have a partial charge, as, e.g., in CO_2 and H_2O molecules. In cases when the molecules contain atoms of Groups I, II, III, together with those of Groups V, VI, VII of the periodic table, i.e. they are *polar molecules*, then Coulomb forces become dominant to the extent that individual molecules are not observed in the crystal, and its structural elements are *ions*. Typical examples are provided by *alkali-halide crystals*, and oxides of alkaline earth metals. In these compounds every ion is usually symmetrically surrounded by ions of opposite sign (six for FCC NaCl, eight for BCC CsCl); and the stronger *ionic bond* is formed. For these cases the action of van der Waals forces between ions becomes of secondary importance.

The opposite case is provided by crystals of elements of Group IV: *diamond, silicon, germanium,* gray *tin.* There are no charges on the atoms of these crystals, and the formation of *covalent bonds,* or *homopolar bonds,* is due to the sp^3-*hybridization of orbitals* (atomic orbitals), and to the attraction of electrons to the nuclei (skeletons) of neighboring atoms. As only four hybrid orbitals may be formed by a combination of s- and p-orbitals, the number of closest neighbors (*coordination number*) for these crystals is limited to four. Each atom of one sublattice is tetrahedrally surrounded by four atoms of the other sublattice. The binary $A^{(n)}B^{(8-n)}$ type *crystallization* of elements (sphalerite lattice) is formed analogously, e.g., BN, GaP, ZnS, HgTe, CuCl. However, the atoms of the metal and the metalloid in these compounds possess fractional charges of opposite sign, with the metal atom positively charged. This is a combined (hybrid) *polar-covalent bond* (*ionic-covalent bond*). The smaller the number n of the periodic table group, the more ionic the bond.

In *metals* and *alloys* the so-called *metallic bond* also occurs: it is formed through delocalization of valence electrons. At the formation of the crystal the outer (valence) s- and p-electrons (elements of Groups I to III, and transition elements) leave their atomic nuclei and migrate between ions over the entire lattice. As this takes place, their kinetic energy decreases (in accordance with the uncertainty relation). It is this energy decrease that is responsible for the formation of a metallic bond; *alkali metals* and alkaline earth metals possess relatively weak metallic bonds (1 to 2 free electrons per atom), whereas elements such as Fe, Ti, Pt exhibit rather strong metallic bonds.

For a fairly broad class of hydrogen-containing organic compounds there is a noticeable contribution to the crystal formation energy from *hydrogen bonds.* Typical examples are provided by *ice,* solid ammonia, and the ferroelectric crystal KH_2PO_4. Because of its comparatively small mass, a positively charged hydrogen atom exhibits quantum behaviour and spends some time near an electronegative atom (see *Electronegativity*) of one of the neighboring molecules, which results in a decrease of the overall system energy. Many solids exhibit several types of chemical bonds, usually with one or two bonding types dominant.

CHEMICAL DIFFUSION

The same as *Heterodiffusion.*

CHEMICAL HEAT TREATMENT

Heat treatment in a chemically active medium involves the diffusive saturation of a metal surface with chemical elements of the medium. Both nonmetals (B, C, N, etc.) and metals (Al, Ti, Cr, Zn, etc.) are used as diffusible elements. In a number of cases simultaneous saturation with metals and nonmetals is used, including multicomponent saturation. During chemical heat treatment a diffusion layer is formed as a result of prior processes of dissociation in the active external medium, *adsorption* of dissociated atoms on the surface of the base metal, and *diffusion* of the saturating element. Chemical reactions bring about dissociation with the generation of active atomic states of chemical elements (e.g., ammonia decomposition under nitriding). Adsorption of dissociated atoms may take place either through the action of *van der Waals forces* of attraction, with the generation of an adsorption layer (*physical adsorption*), or because of chemical bonding with atoms of a metal surface (*chemisorption*). Simultaneously with these processes, the main chemical heat treatment process goes on, namely the diffusion of adsorbed atoms into the depth of the metal. The result is the formation of a diffusion layer with a certain distribution of diffusible elements relative to the depth. The concentration changes monotonically with depth, so a continuous sequence of *solid solutions* forms with the base metal. The chemical composition of the layer is more complicated if the solid solutions have limited solubility, or if chemical compounds form. In this case the layer includes different phases, in accordance with the *phase diagram* of the saturating element with the base metal at the diffusive saturation temperature, and the concentration dependence on the depth can undergo abrupt changes. The duration of diffusive saturation is determined by the depth of the diffusion layer, which increases quadratically with time. The main methods of chemical heat treatment with respect to the type of active medium are: saturation from powders and pastes, gaseous media, melts of metals and salts, as well as *evaporation* of the diffusible element in vacuo. Chemical heat treatment involving a single element includes *nitriding*, *aluminizing*, *boronizing*, *siliconizing*, *titanizing*, *zinc plating*, *cementation*. Saturation with two elements takes place at *aluminoboronizing*, *aluminosiliconizing*, *chromosiliconizing*, *nitrocementation*, *cyaniding*. This treatment involving several diffusible elements includes saturation with boron and silicon together with other metals (Ti, Cr, Mo, W, etc.), as well as saturation with aluminum and chromium together with other metals (Ti, V, Mn, Zn, etc.).

Chemical heat treatment is used in industry to increase the reliability and service life of metallic working parts of devices and engines. Depending on the chemical composition of the metallic work piece and the nature of the saturating elements, treated articles may acquire increased *hardness*, *strength*, *corrosion resistance* and *cavitation resistance*, wear resistance (see *Wear*), etc.

CHEMICAL METHODS OF ANALYSIS

Methods of analysis based on the use of chemical reactions. Methods of *quantitative analysis* involve the determination of the masses of the reaction products (*gravimetry*); determination of the volume of the solution of one substance required to react quantitatively with a given solution volume of another substance (*titration analysis*), determination of thermal effects of the reaction (*calorimetry*), of its rate (kinetic methods), etc. Methods of *qualitative analysis* are based on carrying out chemical reactions, which indicate either the presence of ions of certain type in the solution, or the presence of certain atoms in organic compounds. Chemical methods of analysis have been largely superseded by more accurate physical and *physicochemical methods of analysis*.

CHEMICAL POTENTIAL (introduced by J.W. Gibbs, 1873–1878)

Thermodynamic state function, which defines the change of *thermodynamic potential* for the change in the number of particles of the ith type N_i in open multicomponent systems with varying numbers of particles, which are described by the grand canonical distribution (see *Gibbs distributions*). The chemical potential μ of the ith component equals the partial derivative of every thermodynamic potential (e.g., *Helmholtz free energy* F) with respect to N_i while holding other thermodynamic quantities constant (e.g., $\mu_i = \partial F/\partial N_i$ at $T = \text{const}$, $V = \text{const}$ and $N_j = \text{const}$ for $j \neq i$). The chemical potential for constant pressure conditions is closely related to *Gibbs free energy* $G = \sum_i \mu_i N_i$, and in the case of a single-component system $\mu = G/N$, i.e. the chemical potential is equal to Gibbs energy per particle.

For quantum systems of bosons, which obey *Bose–Einstein statistics*, the chemical potential is negative at temperatures higher than the *Bose–Einstein condensation* temperature T_0, and is equal to zero below T_0, and in systems with varying numbers of particles (*phonons* in solids, *photons* in equilibrium with thermal radiation). For a system of fermions, which obey *Fermi–Dirac statistics*, the chemical potential is positive at temperatures below the degeneracy temperature (Fermi temperature) T_F, it coincides with the *Fermi energy* E_F at absolute zero of temperature; and for $T > T_F$ the chemical potential satisfies the inequality $\mu < E_F$.

In external potential fields (gravitational, electric) the role of the chemical potential is played by the quantity $\mu_i = \mu_i^0 + U_i(\boldsymbol{r}) = \text{const}$, where μ_i^0 is the chemical potential in the absence of the field, and $U_i(\boldsymbol{r})$ is the potential energy of particles of the ith type in the external field. For charged particles in an electrostatic field $\boldsymbol{E} = -\nabla\varphi(\boldsymbol{r})$ the expression for the *electrochemical potential* $\mu_i = \mu_i^0 + e_i\varphi(\boldsymbol{r}) = \text{const}$ holds true. Equal values of chemical potential in different *phases* define equilibrium conditions for phase equilibria (the so-called *Gibbs' phase rule*) and chemical reactions. The chemical potential in a solid determines the electronic *work function* of a crystal for *field electron emission*, *thermionic emission* and *photoelectron emission*, *electron affinity*, *contact potential difference*, and other properties.

CHEMICAL SHIFT

Term used to describe changes (depending on the specific nature of the material under investigation) of the positions of energy levels or spectral lines which can arise from the local electronic or chemical environment. The values of chemical shifts are usually given as parameters corresponding to the spectrum (energy state) of a given material. Generally the chemical shift is measured either in the same units as the basic quantities which characterize the spectrum, or in dimensionless units.

The term is most widely used in high resolution *nuclear magnetic resonance* (NMR) spectroscopy, in *Mössbauer spectroscopy*, in the physics of *semiconductors* to describe the location of impurity energy levels, etc.

In high-resolution NMR the chemical shift is the shift of the *resonance* signal. The mechanism producing this shift is as follows: an atom (molecule) is placed in an applied magnetic field $\boldsymbol{B}_0$; and as a result of *Larmor precession* of electrons oriented parallel to and antiparallel to $\boldsymbol{B}_0$ the field $\boldsymbol{B}^*$ arises (see *Diamagnetism*), so the nucleus experiences the action of an effective magnetic field $B = B_0 - B^* = B_0(1 - \sigma)$, where σ is called the *screening constant*; usually $\sigma \sim 10^{-6}$. For nuclei of a given type, which are in different electronic surroundings, i.e. either in different molecules or at nonequivalent sites of one molecule, the resonance will be observed for different values of B_0 (at a given frequency). In practice the NMR chemical shift is measured in relative units $\delta = (B - B_{\rm st})\cdot 10^6/B_0$, where B and $B_{\rm st}$ are the respective resonance values of the external field for the sample under investigation and a standard substance. The chemical shift is one of the main characteristics of high resolution NMR spectra, which are observed for liquids (and occasionally gases). The value of this shift provides information on the electronic structure of molecules, and the nature of *chemical bonds*. The development of a recent trend in spectroscopy, namely *high-resolution NMR spectroscopy* of solids, permits the measurement of chemical shifts in solids.

In the *Mössbauer effect* the chemical shift (also called *isomer shift*) makes its appearance if the source and absorber differ chemically.

In the physics of *semiconductors* a chemical shift manifests itself in the difference between the energy states of impurity levels, expressed in terms of the *effective mass*, and the energy states calculated from theory, taking into account corrections to the potential due to the specific atom (ion). For example, some donor levels exhibit fairly large chemical shifts in their ground state energies due to correlation and exchange effects involving the donor and core electrons.

CHEMILUMINESCENCE

Luminescent excitation that results from chemical reactions. The process of chemiluminescence involves the emission of radiation either by products of a chemical reaction, or by other components which become excited by absorbing energy transferred to them from reaction products. Chemiluminescence in solids arises at a boundary with a gaseous or liquid phase as a result of heterogeneous reactions (see *Heterogeneous system*), which proceed either spontaneously (*intrinsic chemiluminescence*) or under action of various external factors: e.g., electrical, electrolysis (*electroluminescence*). A specific case of chemiluminescence is *bioluminescence* (e.g., luminescence of rotten wood). Chemiluminescence constitutes the basis of action of chemical *lasers*, and is used for the analysis of gases and *solid surfaces*.

CHEMISORPTION, chemical adsorption

Chemical binding of particles of a substance (atoms, molecules, radicals) by a *solid surface*. It differs from *physical adsorption* by its stronger bonding, that is by a higher value of the binding energy (0.5–10 eV). The bond between the adsorbed particle and the solid can have both covalent and ionic components; the former due to overlap of the electronic wave functions of the adsorbed particle and the *substrate* atoms, and the latter due to the transfer of charge from the adsorbed particle to the solid material. An adsorbed ion in conjunction with charges which screen it can form a dipole, which results in the adsorbed film changing the electronic *work function* of the solid. Particles chemically adsorbed on the surface of a solid interact with one another, and sometimes generate an ordered two-dimensional lattice of adatoms (see *Monolayer*). Chemisorption plays an important role in the oxidation of metals, heterogeneous catalysis, crystal growth, electronics, etc.

CHEMISTRY

See *Crystal chemistry*, *High-energy chemistry*, *Mechanochemistry*, *Mesonic chemistry*, *Quantum chemistry*, *Radiation chemistry*, *Solid state chemistry*.

CHERENKOV RADIATION (S.P. Vavilov, P.A. Cherenkov, 1934); referred to as Vavilov–Cherenkov radiation in the Russian literature

Photons emitted by charged particles moving in a medium with a speed v which exceeds the *speed of light* $u = c/n$ in this medium, where n is the *refractive index*. In condensed matter with high values of n Cherenkov radiation is observed at electron energies of about 10^5 eV, and at protons energies of 10^8 eV. This radiation is explained via Huygens principle by noting that waves which are emitted at points along the charged particle trajectory for the condition $v > u$ will overtake waves emitted earlier by the particle. As a result of their interference, a light cone is formed having a vertex that coincides with the instantaneous position of the emitting particle, with θ the angle between the direction of the particle's movement and the direction of propagation of the outgoing radiated wave satisfying the relation

$$\cos\theta = \frac{1}{n}\frac{c}{v}.$$

The spectral distribution of energy $E(\omega)$ for Cherenkov radiation is determined by the relation

$$\frac{\mathrm{d}E}{\mathrm{d}\omega} = \frac{e^2}{c^2}\omega\left(1 - \frac{c^2}{v^2 n^2}\right)$$

for the condition $v > c/n(\omega)$. There is no radiation for frequencies ω which do not satisfy the above condition. Cherenkov radiation has found wide application in *Cherenkov counters* used for determining the speeds, charges, and masses of fast charged particles (electrons, positrons, mesons, protons, and antiprotons); and for studying optically *anisotropic media*, including waveguides.

CHI PHASE (χ-phase)

An *intermetallic compound* with a cubic lattice isomorphic to that of cubic α-Mn. It is formed mainly by *transition metals*. The best known chi phases are those involving *rhenium* with *refractory materials*. In alloys of rhenium with molybdenum, tungsten, niobium and tantalum the chi phase contains about 75 at.% Re. In the systems Mo–Re and W–Re the chi phase is obtained by the *peritectic reaction* (see *Peritectic alloys*) at 1850 °C in the system Mo–Re and at 2150 °C in the system W–Re; and this reaction only takes place for a narrow region of homogeneity. In the systems Nb–Re and Ta–Re the χ-phase melts without decomposition and exhibits a broad range of homogeneity. In systems formed by Re with Ti, Zr and Hf the composition of the χ-phase is approximately Me_5Re_{24}; it is formed by the peritectic reaction, and has a moderate range of homogeneity (Me–Ti, Zr or Hf). The chi phase has also been observed in binary systems Al–Re, Nb–Os, Tc–Zr. Chi phases are assumed to be "electron compounds" (see *Hume-Rothery phases*). A number of chi phases transform to the superconducting state, with the highest transition temperature $T_c = 12.9$ K for $Nb_{0.24}Tc_{0.76}$ (see *Superconducting intermetallic compounds*).

CHIRALITY (fr. Gr. $\chi\varepsilon\iota\rho$, hand)

Handedness, or belonging to one of two ("left-hand" or "right-hand") non-superimposable mirror image configurations. See also *Enantiomorphism*.

CHIRALITY IN LIQUID CRYSTALS

Absence of planes of symmetry in the molecules of *liquid crystals*. Chirality in liquid crystals leads to instability of a uniform (homogeneous) state, and the formation of spatially-periodic (spiral) structures (right-handed or left-handed) in some liquid crystal phases (*cholesteric liquid crystals*, *blue phases*, *smectic liquid crystals* of the C type); see *Chirality*, *Enantiomorphism*.

CHIRALITY OF MAGNETIC BUBBLE DOMAINS

The sign of the relative rotation of the *magnetization* in the center of a domain wall for a *magnetic bubble domain* with *topological index* $S = 1$, which does not contain *Bloch lines*. The magnetization in a *magnetic domain wall* is either aligned along the vector tangent to the lateral surface of a domain wall, or is opposed to it. Hence two possible states of a magnetic bubble domain, labeled $\chi^{\pm}$, are more favorable (in the context of magnetostatics) than all other states. By convention the label χ^{+} is assigned to a clockwise rotation of the magnetization when observing along the rotation axis direction. The twisting deflections of χ^{+} and χ^{-} are equal in magnitude.

It should be noted that, by virtue of *domain wall twisting*, the above definition is valid only for the cross-section of the magnetic bubble domain by a plane that is at equal distances from the *magnetic film* surfaces. Transitions between the χ^+ and χ^- states of magnetization rotation usually occur when horizontal Bloch lines are present.

CHOLESTERIC LIQUID CRYSTAL, cholesteric, chiral nematic, cholesteric phase (fr. Gr. $\chi o\lambda\eta$, bile, $\sigma\tau\varepsilon\rho\varepsilon o\varsigma$, solid, and $\chi\varepsilon\iota\rho$, hand)

An analog of a nematic phase (see *Nematic liquid crystal*) for a material containing chiral molecules, that is molecules which differ from their mirror image structures (see *Chirality in liquid crystals*). When chiral molecules are dissolved in a nematic liquid the structure undergoes a helical distortion to form a helical phase called a cholesteric. This phase was originally observed in the ethers of cholesterol, hence its name. A cholesteric, like a nematic, lacks long-range order, and the molecular orientation has a preferred axis denoted by a *director* $\boldsymbol{n}$ which is perpendicular to the helical z axis. The local symmetry is close to $\infty 22$ (or $C_{\infty h}$), with the spatial distribution of the apex positions of the director vector $\boldsymbol{n}$ tracing out a spiral or helix. The spatial period or step along the helical axis is one half the pitch of the spiral because director orientations $\boldsymbol{n}$ and $-\boldsymbol{n}$ are equivalent, as is the case in nematics. The step of the spiral L depends on the chemical composition and on the temperature, and typical values lie in the range from 350 nm to 650 nm, corresponding to visible and near UV wavelengths. Cholesteric liquid crystals possess unique optical properties such as *Bragg diffraction* of light beams at the spiral structure, enhanced optical activity, etc. These properties are widely used in various fields of engineering, in particular in the medical area.

CHRISTOFFEL SYMBOLS (E.B. Christoffel, 1869)

The Christoffel symbol is $\Gamma^i_{kl} = g^{in}\Gamma_{n,kl}$, where the *metric tensor* g_{ik} defines $\Gamma_{i,k,l} = (\partial_k g_{il} + \partial_l g_{ki} - \partial_i g_{kl})/2$, and $\partial_i = \partial/\partial x^i$. The Christoffel symbols do not form a tensor; in particular by choosing a particular system of coordinates one may reduce all the Christoffel symbols to zero at any given point. In curvilinear coordinates in a flat space the Christoffel symbols are used to form covariant derivatives of tensors. These symbols are used in *elasticity theory* to characterize changes of the basis vectors e_i of the local coordinate system deformed together with the body: $\mathrm{d}e_i = \Gamma^l_{ki} e_l\,\mathrm{d}x^k$. Christoffel symbols and g_{ik} are both expressed through the *strain tensor* u_{ik}: $g_{ik} = \delta_{ik} + 2u_{ik}$ (δ_{ik} is the unit tensor), $\Gamma_{i,kl} = \partial_k u_{il} + \partial_l u_{ki} - \partial_i u_{kl}$. To take into account *dislocations* and *internal stresses*, the Christoffel symbols are replaced, within the framework of the mechanics of solids, by the coefficients of linear (affinity) coherence, Γ_{ikl}. The number of independent values of Γ_{ikl} is equal to 15 (in contrast to 9 for the previous case). Γ_{ikl} may be separated into symmetric ($\Gamma_{i(kl)}$) and antisymmetric ($\Gamma_{i[kl]}$) parts with respect to the indices k, l: $\Gamma_{ikl} = \Gamma_{i(kl)} + \Gamma_{i[kl]}$. The quantity $\Gamma_{i(kl)}$ characterizes the pure deformation of the trihedron defined by the basis vectors e_i, and $\Gamma_{i[kl]}$ characterizes the rotation of the trihedron during its transition from point to point. The quantity $\Gamma_{i[kl]}$ is directly linked to the density of dislocations in the material; the sum of the *Burgers vectors* of dislocations passing through a small area S is proportional to the convolution of the tensor of the area S and $\Gamma_{i[kl]}$.

CHROME PLATING

Application of chromium *coatings* to the surface of materials to increase their *hardness*, wear resistance (see *Wear*), *corrosion resistance* and *friction* resistance, and also for decorative purposes. The thickness of the coatings varies from 1 μm to 3 mm, their hardness ranges from 500 to 1200 kg/mm^2. *Electrolytic chrome plating* is carried out from aqueous chromium-oxide solutions either with a sublayer of copper and nickel, or without it. The appearance, structure, physicochemical and mechanical properties of the coatings may be changed by variation of the electrolyte composition, and the method of electrolysis. Also in some areas of engineering *diffusion chrome plating* is used, preferably from the lowest halide compounds of chromium, by means of thermal decomposition or reduction in a gaseous, solid or liquid medium, with successive *diffusion* of chromium into the depth of the parts being coated.

CHROMIUM, Cr

A chemical element of Group VI of the periodic table with atomic number 24 and atomic mass 51.996. There are four stable isotopes: ^{50}Cr, ^{52}Cr, ^{53}Cr, ^{54}Cr. The outer shell electronic configuration is $3d^5 4s^1$. Successive ionization energies are 6.764, 16.49, 31, 51, 73, 90.6 eV. Atomic radius is 0.125 nm; radius of Cr^{2+} ion is 0.084 nm, of Cr^{3+} is 0.063 nm. Covalent radius is 0.118 nm. Oxidation states are +2, +3, +6. Electronegativity is 1.55.

Under normal conditions chromium is a silvery-white *metal*. The crystal lattice is body-centered cubic, space group $Im\bar{3}m$ (O_h^9), with parameter $a = 0.28848$ nm (at 298 K). Density is 7.194 g/cm^3 (at 298 K); $T_{melting} = 2176$ K, $T_{boiling} = 2742$ K. Heat of melting is $\approx$14.0 kJ/mole, heat of evaporation is 393.34 kJ/mole; specific heat is 460 $J \cdot kg^{-1} \cdot K^{-1}$ (at 298 K). Debye temperature is 630 K. Linear thermal expansion coefficient is $\approx 6.5 \cdot 10^{-6}$ K^{-1} (at 298 K); coefficient of thermal conductivity is 67 $W \cdot m^{-1} \cdot K^{-1}$ (at 298 K). Adiabatic elastic moduli $c_{11} = 350.0$, $c_{12} = 67.8$, $c_{44} = 100.8$ (in GPa). Young's modulus is 279.65 GPa (at 298 K); shear modulus is 102.02 GPa (at 298 K), Poisson ratio is 0.31. Ultimate strength is 84 MPa (at 298 K, after recrystallization). Relative elongation is 0% (at 298 K, after recrystallization). Vickers hardness is 100 to 200 HV. Linear low-temperature electronic specific heat coefficient is 1.4 $mJ \cdot mole^{-1} \cdot K^{-2}$. Electrical resistivity is 129 $n\Omega \cdot m$. Temperature coefficient of electrical resistance is $3 \cdot 10^{-3}$ K^{-1} (293 to 373 K). Hall constant is $3.63 \cdot 10^{-10}$ m^3/C (at 293 K). Coefficient of absolute thermal electromotive force is 16.2 μV/C. Chromium is an antiferromagnet with Néel temperature $T_N = 311$ K. Molar magnetic susceptibility is $3.6 \cdot 10^{-6}$ CGS units. Nuclear magnetic moment is 0.474 nuclear magnetons. Reflection coefficient of 500 nm wavelength light is 67%. Electronic work function of polycrystal is 4.68 eV. Chromium is employed to give glass an emerald green color, and it finds wide use as a catalyst.

CHROMIUM ALLOYS

Chromium-based *alloys*. Doping Cr with small additions (up to 1%) of *yttrium*, *rare-earth elements* and *titanium* causes the binding of these atoms into refractory compounds (see *Refractory materials*), and refining the metal matrix decreases the *cold brittleness* temperature slightly, and increases its *plasticity*. At the same time the *corrosion resistance* increases. Strong doping of Cr with iron, cobalt, nickel, ruthenium, and rhenium changes the electronic structure of chromium, and in some cases its *crystal structure* and phase composition. It provides for an appreciable decrease of the cold brittleness temperature. However, in such a case *alloy brittleness* appears, due to the formation of a *sigma phase*. Chromium alloys are smelted in archer or induction furnaces in an inert gas atmosphere, or fabricated by the method of *powder metallurgy*. A *thermomechanical treatment* of low-doped chromium alloys, when a fine-grained and cellular dislocation structure arises (see *Cellular structure*), permits an increase of the *strength* and, at the same time, the cold brittleness temperature. Low-doped chromium alloys are used as *thermal-environment resistant materials*, and as *high-temperature materials* at 1000–1100 °C, with a short-term heating capability up to 1500–1600 °C. Highly-doped alloys are used in the chemical engineering industry, in the production of components for magnetohydrodynamic generators, foundry castings, etc.

CINNABAR (fr. Gr. $\kappa\iota\nu\nu\alpha\beta\alpha\rho\iota$), HgS

A *mineral*, the red-colored stable modification of *mercury* sulfide. The Mohs hardness is 2–2.5, and the density is 8.0–8.2 g/cm^3; the crystal family is trigonal, and the *space group* is $P3_12_1$ (D_3^4), $a = 0.4146$ nm, $c = 0.9497$ nm. Another modification of mercury sulfide, black metacinnabarite with the ZnS type structure, is unstable and gradually transforms into cinnabar. This most important mercury mineral is used to produce the *metal*. Artificially, cinnabar is prepared by rubbing mercury with sulfur with subsequent treatment with alkalis. The heat of formation equals −59 kJ/mol (at $T = 298$ K). Cinnabar exhibits a high volatility ($t_{sublimation} = 580$ °C). The *crystal lattice* energy is 3,578 kJ/mol. The crystals are composed of polymer chains –Hg–S–Hg–S– with the parameters: d(HgS) = 0.236 nm, $\angle$HgSHg = 105°, $\angle$SHgS = 172°. Artificial cinnabar is a *semiconductor material* used for photoresistors, as a catalyst, and as a pigment.

CIRCULAR PHOTOVOLTAIC EFFECT

Photovoltaic effect observed in doubly refracting crystals (see *Gyrotropic media*) illuminated by circularly or elliptically polarized light (see *Polarization of light*). There is a difference between the circular photovoltaic effect and other photovoltaic effects: the resulting current is proportional to the degree of circular polarization of the radiation P_{circ} and changes its direction in accordance with the sign of the polarization. The "circular" photovoltaic effect for elliptically polarized light is described by the following relation:

$$j_\alpha = \gamma_{\alpha\beta} i[\boldsymbol{E} \times \boldsymbol{E}^*]_\beta,$$

where $\boldsymbol{j}$ is the current density, and $\boldsymbol{E}$ is the amplitude of the light electric field (see also *Bulk photovoltaic effect*). For circular polarization this becomes

$$i[\boldsymbol{E} \times \boldsymbol{E}^*] = |E|^2 P_{\text{circ}} \boldsymbol{q}/q,$$

where $\boldsymbol{q}$ is the wave vector of the light. Initially the circular photovoltaic effect was observed in *tellurium*, then it was studied in a variety of *crystals* including a number of *ferroelectrics*. The origin of this effect arises from the asymmetry of the distribution of the electrons excited by the light. This asymmetry is due to the correlation between the electron angular momentum (spin) and its linear momentum in gyrotropic crystals. The inverse effect, the so-called *optical activity* (see *Rotation of light polarization plane*) induced by the current and resulting from the spin orientation during the current flow, has also been observed in tellurium. In nongyrotropic crystals, which possess no center of symmetry, the circular photovoltaic effect does not occur, while the phenomenon of optical activity may appear from uniaxial deformation or an applied magnetic field. Such a magneto-induced circular galvanomagnetic effect was observed in p-type GaAs. In crystals of low symmetry, including those possessing a center of symmetry, the *circular drag effect* described by the relation

$$j_\alpha = T_{\alpha\beta\gamma} i[\boldsymbol{E} \times \boldsymbol{E}^*]_\beta q_\gamma$$

may be observed.

In contrast to the circular photovoltaic effect, the circular drag effect is due to the transfer of phonon momentum to the electrons. The current in this effect changes its sign, not only when the sign of the circular polarization changes, but also when the light propagation direction changes. The circular photovoltaic effect may used in inertialess receivers of polarized laser radiation, in elements of optical memory on photorefractive crystals (see *Photorefraction*), and also as a mechanism of four-wave mixing of isofrequency waves in various nonlinear *optical frequency converters*.

CLATHRATE COMPOUNDS (fr. Gr. $\kappa\lambda\eta\theta\rho\alpha$, adler-tree)

Phases comprising at least two compounds, with the molecules of one enclosed in the molecular voids or *lattice* cavities of the other. They bear a close similarity to interstitial *solid solutions*.

As a rule, only *van der Waals forces* act between the molecules of a clathrate compound. The binding energy of each enclosed molecule is about $\approx$50 kJ/mol. This relatively high value for van der Waals interaction is explained by the molecule and void geometry fitting well into one another, and by the many points of bonding by the molecule. Depending on the void type, clathrate compounds are categorized as lattice or molecular types. *Lattice clathrate compounds* can only be solids, and are further classified by the shape of their voids as channel, cage (originally only this type was termed clathrate), and laminated varieties. Examples of "host" compounds that form cage clathrates are uria, thiouria, hydroquinone, phtalocyanines, etc. There is a wide range of corresponding "guest" compounds: hydrocarbons, alcohols, many gases. Cyclic molecules usually form the *molecular clathrate compounds*, such as cyclodextrin and other *macrocyclic compounds*. Compounds of relatively small dimensions can be enclosed in the voids of such "hosts". The clathrate compounds exhibit, in the general case, a variable composition. They are employed mainly to isolate and purify "guest" molecules.

CLAUSIUS–MOSSOTTI EQUATION (R. Clausius, O. Mossotti, 1879)

Relation between the relative (dimensionless) *dielectric constant* of isotropic nonpolar *insulators*, $\varepsilon(\omega)$, at a frequency ω (see *Temporal dielectric dispersion*) and the *polarizability*, $\alpha(\omega)$, of the atoms that make up the *crystal lattice*:

$$\frac{\alpha(\omega)N}{3\varepsilon_0} = \frac{\varepsilon(\omega) - 1}{\varepsilon(\omega) + 2}, \qquad (1)$$

where N is the number of atoms per unit volume, and SI units are used. This expression is sometimes written in terms of the electric susceptibility $\chi(\omega) = \varepsilon(\omega) - 1$. If the square of the refractive index, $n(\omega)$, in the transparency region is substituted for $\varepsilon(\omega)$ then Eq. (1) is called the *Lorentz–Lorenz relation*. Eq. (1) differs from the frequently used expression $\varepsilon = 1 + \alpha N/\varepsilon_0$ which is applicable in the case of a *linear response* of a medium (crystal) to a macroscopic electric field averaged over the medium, since the Clausius–Mossotti equation is derived taking into account the local electric field that acts on an atom in the crystal, and this differs from its mean value. The Clausius–Mossotti equation for cubic crystals coincides with the expression for ε derived from the rigorous *Ewald method*. See also *Langevin–Debye equation*.

CLEMENTI FUNCTION (E. Clementi)

A wave function of a certain nl-shell of an atom (or ion) whose radial part is represented as a superposition of Slater functions (see *Slater rules*), the parameters of the latter being derived from minimizing the total energy of a given multi-electron state. Clementi functions can be used as the first approximation for a non-empirical computation of *crystals*.

CLOSE-PACKED STRUCTURES, closest-packed structures

Geometrical arrangement of rigid spheres of the same radius, which provides the densest filling of space. There is only one way of attaining such a configuration in a plane, which consists in arranging a *lattice* (the sites formed by the centers of balls) based on an equilateral triangle configuration, as shown in Fig. 1. The three-dimensional closest-packed sphere structure is constructed from the above planar arrangement by superimposing layers, one onto another, in such a way that the upper layer of a pair lies with its atoms set in the recesses of the layer below; the density of packing (packing fraction) in this case being $\pi/\sqrt{18} \approx 0.74$. There are two ways to do this, and either one is effectively equivalent. The third layer also has two possibilities, but they are not equivalent. One way for positioning the third layer is to place its spheres in recesses between second layer atoms that are not directly above the

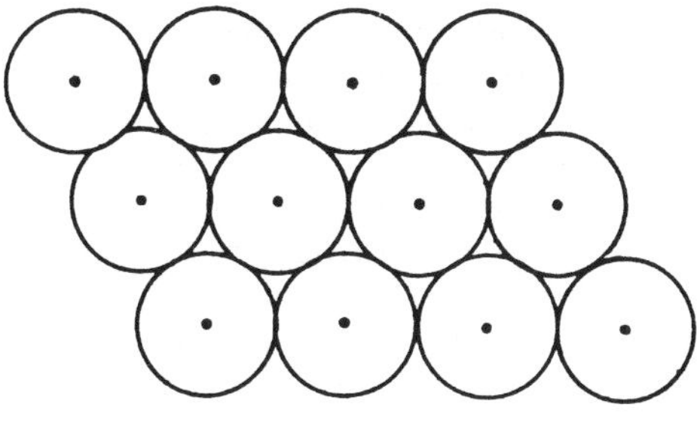

Fig. 1. Planar triangular close packing arrangement of spheres.

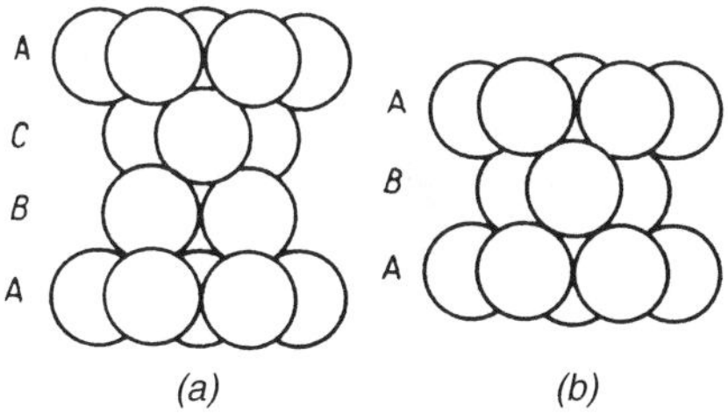

Fig. 2. Three-dimensional close packing arrangements of spheres: (a) face-centered cubic, and (b) hexagonal close-packed.

atoms of the first layer, and this gives what is called an ABC arrangement of layers, as shown in Fig. 2(a). None of the three layers is directly above another. Another way to position the third layer is with its atoms in recesses directly above the first layer, thereby forming what is called the ABA sequence of layers, as shown in Fig. 2(b). Each subsequent layer can be positioned in one of two ways, leading to an infinity of possible closest packed structures if the number of layers becomes very large, and the sequencing of the A, B, and C type layers is random subject to the condition that two identical layers cannot be adjacent to each other. If the sequence is a periodic one, then the corresponding closest-packed sphere structure is spoken of as *periodic highest-density spherical packing*. According to the *Belov theorem*, periodic closest-packed sphere structures may exhibit only eight different *space groups* of symmetry: $P3m1$, $R3m$, $P\bar{3}m1$, $R\bar{3}m$, $P\bar{6}m2$, $P6_3mc$, $P6_3/mmc$, $Fm3m$.

The packing consisting of only two alternating layers $ABAB\ldots$ is referred to as highest-density hexagonal packing or *hexagonal close-packed* (HCP). The three-layered $ABCABC\ldots$ arrangement is given the name highest-density cubic packing or *cubic close packing* (CCP or FCC). The centers of the spheres of the latter structure form a *face-centered cubic lattice*. The tendency of atoms to attain closest packing results in the fact that most metallic atoms crystallize in one of the close-packed arrays which is almost always the hexagonal (e.g., Mg, Nd, Zn) or the face-centered cubic (e.g., Ag, Al, Cu) one. In addition bulky components (e.g., anions oxygen or sulfur) of various widely occurring chemical compounds also form close-packed arrays, and the diversity of structures stems from the fact that there are many ways of arranging the smaller cations throughout the smaller tetrahedral and larger octahedral cavities of the closest-packed structure.

CLOSE PACKING, RANDOM

See *Random close-packing model.*

CLUSTER

A stable aggregation or group of atoms or molecules distinguished by certain physical or chemical features from other similar stable micro-objects. Several particles constitute the lower limit on the size of a cluster, while the upper limit can often be deduced from one of the following two traits: (1) the addition of one more particle to the aggregation appreciably affects the system properties; or, (2) the number of particles at the surface, i.e. those which differ in properties from the bulk due to their proximity to the surface, is comparable to the total number of internal particles in the cluster. These conditions usually limit the number of particles in a cluster to the range 10^2–10^3. Larger agglomerates of particles in a material are called *phase precipitates*. Examples of clusters are: agglomerations of *impurity atoms* or intrinsic defects (see *pile-up of dislocations* and *point defect clustering*) arising in the bulk of a solid due to their mutual attraction and to the *diffusion* of particles (during the growth of crystals or during *heat treatment*), under the action of external energetic particles (see *Collision cascade* and *Subcascades*), under the action of light (in photographic emulsions), etc. The atomic or molecular aggregates that comprise a *small particle* can be formed during the coagulation of particles, e.g., those dispersed in air (*aerosol particles*), during the sputtering of materials under *ion bombardment* or other high-energy factors, during *evaporation*, etc. Surface clusters can be formed by deposition of particles on the surface of a solid. The stability of a cluster can depend on the number n of constituent particles, and sometimes it depends on whether n is odd or even. Thus it was established by means of *mass spectrometry* that when Li atoms are evaporated into an inert gas an appreciable amount of Li_2, Li_6 and Li_8 appears, whereas Li_3, Li_5, Li_7, and Li_9 are either entirely absent, or are present in only trace amounts. During the ion sputtering of Ag crystals three times more Ag_n clusters form with odd n compared to the number with even n.

A method termed the *cluster approximation* is often employed in numerical computations of the energy spectra of ideal crystals, or of crystals with defects, with a cluster defined here as a lattice of a finite number of atoms. Atom–atom potentials and cluster boundary conditions are chosen to provide for a high cluster symmetry, for correct values of the *crystal lattice constants*, and for stability. Systems of particles can also have fractal properties (see *Fractal*) arising from special rules followed in building the system. For instance, a *percolation cluster* is a set of certain lattice sites obtained by successive translations of a given length from an arbitrarily chosen starting site to other randomly chosen sites; the dimensions of such a cluster are not limited to integers (see *Percolation theory*). In quantum scattering theory coupled states of several particles are termed clusters; with every cluster considered as a separate scattering channel.

CLUSTER APPROXIMATION

The breaking down of a system into subsystems (see *Cluster*) that interact weakly with each other. This approximation is employed in thermodynamics, statistical physics, and quantum scattering theory. It involves expressing the wave function of a multiparticle system far away from the scattering region as a superposition of states corresponding to individual isolated particles and/or groups of them. The cluster approximation in the theory of lattice *defects* in crystals is based on singling out a region in the crystal that incorporates the defect and several *coordination spheres* surrounding it. The local defect properties (binding energy, energy level position, wave function of the localized state, lattice configuration) are then computed over this region using one of the standard molecular theory methods. The interaction of the cluster with the rest of the crystal is accounted for by using a *self-consistent field* approach.

Sometimes the term cluster approximation is employed to designate an approximate method for computing the parameters of an *ideal crystal* by representing the crystal as a cluster consisting of a relatively small number of atoms. Certain boundary conditions are imposed upon the outermost atoms to ensure the stability of such a configuration. See also *Watson sphere*.

CLUSTER, CYBOTAXIC

See *Cybotaxic clusters*.

CLUSTER DEPOSITION

Formation of heterogeneous *films* composed of *clusters* of particles. The films form via the deposition from a flow of clusters, e.g., produced by *ion* sputtering of different targets, or by homogeneous gas phase nucleation, or due to a heterogeneous nucleation at the *substrate* surface. A cluster may contain from several to a few thousand particles. An assortment of cluster deposition technology methods provides a range of structures and cluster properties which determine the film formation kinetics and film properties. Using the method of cluster deposition at low temperature one can obtain the *epitaxial* deposition of multicomponent compounds such as semiconductors, or one can grow *diamond-like carbon films*, oxides, *nitrides*, *carbides*, etc.

CLUSTER, NONCRYSTALLINE

See *Noncrystalline clusters*.

CLUSTER SPUTTERING

Sputtering of a solid under the action of *ion bombardment* during conditions when *clusters* of different size composed of the solid's own atoms plus impurity atoms are predominant in the flux or flow of sputtered material. Not only neutral clusters, but also positive and negative ions, polyatomic and molecular ions and their fragments, are found in the sputtered flux during the cluster sputtering of *metals*, *semiconductors*, *insulators*, organic compounds, *polymers* and biomolecules. The mass-spectra of sputtered clusters range from dimers through aggregates of tens and hundreds of atoms, up to large bioorganic molecules and their clusters. The energy distribution of sputtered clusters reaches 100–200 eV, the average being 2–10 eV per cluster of ions and 10–20 eV for neutral clusters. The sputtering probability sharply drops with an increase in the number of atoms in the cluster, and it can undergo oscillations associated with the binding energy. In particular, there may be less charged metallic clusters having an even number of atoms than those having an odd one. The mechanisms of the cluster sputtering of solids are not yet understood. Theories of direct knock-out (for small clusters), of *recombination*, and of *shock waves*, have been proposed to explain the phenomenon. The latter theory allows for the emergence of large clusters and of complete bioorganic molecules (such as insulin) under the impact of a single heavy accelerated ion, or of a fragment from nuclear fission. Cluster sputtering plays an important role in *secondary ion mass-spectrometry*, and it also finds applications in designing sources for accelerated cluster and macromolecular ion formation, including ionized bioorganic molecules.

COADSORPTION

A cooperative adsorption of adsorbates of differing chemical composition. In the case of physisorption the mutual influence of these adsorbates is associated with the difference in the heats of adsorption, which may result in the particles with a greater heat of adsorption driving out those with lesser heats of adsorption. In the case of *chemisorption*, the mutual effects of the adsorption of different species is associated with the formation of chemical bonds of the adsorbates with the substrate; this can result in a promotion or inhibition of the sorptional activities of neighboring centers. Such an influence may stretch over the distance of several lattice constants, and may result in the emergence of not only a short-range, but also a long-range order. The phenomenon of *heterogeneous catalysis* is based on the process of coadsorption of molecules of differing chemical composition.

COAGULATION

See *Coalescence.*

COAL

Carbonaceous porous combustible stratified rock; with a surface of fracture which is either lustrous or lusterless. Coal contains a maximum of 30% of incombustible components (ash). Coal is formed by carbonization (*metamorphic transformation*) of the remains of decomposed plants under prolonged exposure to temperature, ionizing radiation of natural radionuclides, etc. According to the *degree of carbonization*, brown coal (lignite) and black coal are distinguished. The *carbon* content increases from ~65% (by mass of organic components of coal) for soft lignites to >91% for anthracites; accordingly, the oxygen content decreases from 30 to 2%, and the hydrogen content decreases from 8 to <4%. The percentages of sulfur and nitrogen are little affected by the extent of carbonization, and are respectively 0.5–3% and 0.5–2%. Coal can also include P, Ge, Ga, B, V, W, U, and other elements. The X-ray diffraction pattern is indicative of the presence of both the amorphous (see *Amorphous state*) and the *crystal structures* of coal. The carbonization process involves a structural-molecular rearrangement of the organic material; this rearrangement being responsible for the extreme values of the following characteristics of coal at intermediate stages of carbonization (carbon content ~85%): *elastic modulus*, *microhardness*, *yield*, *density*, heat of combustion, caking ability, volume and radius of *pores*, extracting characteristics, etc. At these intermediate stages coals gain the capacity to form coke on heating. The electrical conductivity of coal ranges from 10^{-2} to 10^{-10} $(\Omega\cdot\text{cm})^{-1}$. Coals which contain >91% C exhibit an abrupt increase of aromaticity, and an abrupt decrease of electric resistance. Coals contain *paramagnetic centers*, with concentration reaching 10^{20} spin/g.

COALESCENCE (fr. Lat. *coalesco*, grow together, be coupled)

The spontaneous merger of *small particles* of a solid that results in the formation of a single larger particle whose shape minimizes its free energy. The phenomenon of the formation of *pores* in a solid is also known. The coalescence of particles may occur in sols, aerosols, and on the surface of a solid. Coalescence is distinct from *coagulation*, the latter being the mere sticking together of colliding particles. Coalescence is associated with diffusion processes that take place under the influence of *surface tension* forces. It will be slowed down by charged particles. *Surface coalescence* is conditioned by strains in the *substrate*, by its surface roughness, and by its polycrystalline structure.

COATING

Application of a layer, which has protecting, decorating, conducting, etc., properties, to the surface of a material. *Metals* and *alloys*, as well as nonmetallic substances, are used as coating materials. Application of a metallic coat is performed through hard facing with a thin layer of salts in oil (baking coating), chemical deposition, galvanization (electrolytic deposition, see *Electroplating*), dip coating (dipping into melted metal or into aqueous salt solutions in the presence of oxygen), *vacuum evaporation*, cathode *sputtering* (sprayed metal coating), etc. Nonmetallic coats (varnishes, enamel and oil paints) are applied to surfaces through brushing, spraying under pressure, fusion. Colloid mixtures of *graphite* with varnish, or of a metal with varnish, are used as nonmetallic conducting coatings. Anodic *oxidation* is used in order to coat the surface with an oxide *film*; carbon films are obtained through decomposition of a dried layer of carbonic acid through heating to 473 K, or high vacuum evaporation.

COATING, ANTICORROSIVE

See *Anticorrosive coatings.*

COBALT (Lat. *cobaltum*), Co

Chemical element of Group VIII of the periodic system with atomic number 27 and atomic mass 58.9332. The naturally occurring stable isotope ^{59}Co is very close to 100% abundant. The most important artificial isotope is radioactive ^{60}Co with a half-life of 5.24 years. The outer shell electronic configuration is $3d^74s^2$. Successive ionization energies are 7.865, 17.06, and 33.50 eV; electronegativity is 1.7. Atomic radius equals 0.125 nm; the Co^{2+} ionic radius equals 0.072 nm, that of Co^{3+} is 0.063 nm. The oxidation state is +2, +3, rarely, +1 and +4.

Cobalt in the free state is a *metal* of silver-steel color. It exists in two polymorphic configurations: α-Co with a hexagonal close-packed lattice (space group $P6_3/mmc$ (D_{6h}^4), $a = 0.2507$ nm and $c = 0.40686$ nm at 298 K), and β-Co with a face-centered cubic lattice (space group $Fm\bar{3}m$ (O_h^5), $a = 0.3548$ nm). The temperature of the start of the transformation of α-Co into β-Co by heating is 750 K; the temperature of the start of the reverse transformation by cooling is 676 K. The density of α-Co is 8,183.6 g/cm^3 (at 293 K), that of β-Co is 8,739 g/cm^3 (at 298 K), the binding energy is −4.387 eV/atom at 0 K and −4.4 eV/atom at room temperature; $T_{\text{melting}} = 1768$ K, $T_{\text{boiling}} = 3203$ K. The heat of fusion is 16.2 kJ/mol. The specific heat is 414 J·kg^{-1}·K^{-1} at standard pressure and temperature. The Debye temperature is 385 K. The coefficient of linear thermal expansion parallel to the c axis is $14.62\cdot10^{-6}$ K^{-1}, and perpendicular to the c axis, it is $10.96\cdot10^{-6}$ K^{-1}. The thermal conductivity is 142 W·m^{-1}·K^{-1} at 163 K and 70.9 W·m^{-1}·K^{-1} at 290 K; the adiabatic elastic rigidity coefficients of α-Co crystal are: $c_{11} = 306.28$, $c_{12} = 165.12$, $c_{13} = 101.9$, $c_{44} = 75.33$, and $c_{33} = 357.4$ GPa at 298 K. Young's modulus of α-Co is 220.9 GPa, that of β-Co is 199.7 GPa; the modulus of rigidity of α-Co is 84.5 GPa, that of β-Co is 74.7 GPa; Poisson ratio of α-Co is 0.307, and that of β-Co is 0.337 (at 298 K). The ultimate tensile strength is 260 MPa (at 298 K), and the relative extension is 5% (at 293 K). The Brinell hardness is 124 MPa for cobalt subjected to 1.2–3 GPa treatment. The self-diffusion coefficient equals $5.91\cdot10^{-26}$ m^2/s. The Sommerfeld low-temperature electronic specific heat coefficient (linear term) equals 4.73 mJ·mol^{-1}·K^{-2}. Electrical conductivity is 55.7 nΩ·m (at 293 K); the temperature coefficient of electric resistance equals $604\cdot10^{-5}$ K^{-1}. The Hall coefficient is $0.36\cdot10^9$ m^3/C. The absolute thermal electromotive force coefficient equals $18.5\cdot10^{-6}$ V/K. The optical reflectivity for 5.0 μm wavelength is 92.9%. The work function for a polycrystal equals 4.41 eV. Cobalt is a *ferromagnet* with *Curie point* $T_C = 1393$ K. The nuclear magnetic moment of ^{59}Co equals 4.639 nuclear *magnetons*, and the atomic magnetic moment equals 1.715 Bohr magnetons. The thermal neutron absorption cross-section is 34.8 barn.

Cobalt finds its use as a component of hard magnetic, corrosion resistant and *high-temperature alloys* and *coatings*, as well for glass dyeing; the *intermetallic compound* $SmCo_5$ is used for the fabrication of *permanent magnets* (see *Samarium–cobalt magnet*), and radioactive ^{60}Co is employed in medicine as a source of γ-radiation.

COERCITIMETER

Device for measuring the *coercive force* or coercivity B_c of magnetic materials. The most common types are coercitimeters for coercive force measurement by magnetization $_MB_c$ or by induction $_BB_c$. For an $_MB_c$ measurement the sample is magnetized to saturation and then demagnetized up to the point of zero magnetization, and the magnetic field strength that brings this about corresponds to $_MB_c$. For measuring B_c by induction the sample is made part of a closed *magnetic circuit*, and the strength of the demagnetization field at which the induction B in the sample goes to zero is accepted as the value B_c (i.e., $_BB_c$). For small values of the coercive force $_MB_c \approx {}_BB_c$. See *Magnetometry*, *Magnetic structure analysis*.

COERCIVE FORCE, coercive field, coercivity (fr. Lat. *coercitio*, confinement)

Value of an applied magnetic field B_c which reduces the magnetization of a magnetized *magnetic substance* to zero. In many magnetic materials the magnetization process is not reversible, but takes place around a hysteresis loop. An applied magnetic field B_m produces a state of magnetization M_m in the material. The removal of the applied field does not reduce the magnetization to zero, but rather to the remanent magnetization value M_{rem}. A reverse direction field $-B_c$ must be applied to reduce the magnetization to zero. Increasing the reversed applied field to $-B_m$ induces the reversed magnetization $-M_m$, and most of a *hysteresis* loop is thereby traced out. When B_m is small the *coercive field* B_c is also small, where of course $B_c < B_m$, and increases in B_m cause B_c to increase. Very large applied fields B_m induce a saturation magnetization M_s, so further increases in B_m have no effect. The coercive field B_c measured at saturation is called the *coercivity*, and the associated hysteresis loop is called an extremal hysteresis loop. The coercivity of various materials varies over a very wide range from 10^{-4} mT to 10 T. (Some authors define $H_c = B_c/\mu_0$, where 1 mT $\cong$ 795.8 A/m.) The coercivity arises from the back magnetization process, and involves the *magnetic anisotropy* energy, the *magnetostriction*, and the saturation magnetization. In a particular material the coercivity can differ considerably depending on its structural state; it depends also on the direction in anisotropic materials, on the temperature, and on the presence of external stresses. The extremal coercivity value for a given material equals its anisotropy field and can be achieved, e.g., in fine monocrystal ferromagnetic *single-domain particles*, in which back magnetization is brought about by irreversible rotation of the spontaneous magnetization vector. High coercivity values can be obtained in close to perfect crystals where, near saturation, the formation and growth of back magnetization nuclei are impeded. In most *ferromagnets* the coercivity is determined by the critical field for irreversible displacement of *magnetic domain walls*, and this can be impeded by various inhomogeneities such as internal stress gradients, foreign inclusions, structural defects, and so on.

COHERENCE

See *Macroscopic quantum coherence*, *Structural coherence*.

COHERENCE FACTORS

Renormalization coefficients $F_{1,2}(\boldsymbol{k},\boldsymbol{k}')$ appearing in the *Bardeen–Cooper–Schrieffer theory* (BCS theory) which connect the matrix elements $B_1(\boldsymbol{k}',s' \mid \boldsymbol{k},s)$ for single-particle transitions $\boldsymbol{k},s \to \boldsymbol{k}',s'$ in the system of non-interacting *Bogolyubov quasi-particles* (see *Bogolyubov method*), and the matrix elements $B_2(\boldsymbol{k}',s' \mid -\boldsymbol{k},-s)$ for the processes of the annihilation of quasi-particle pairs in the states $\boldsymbol{k}',s'$ and $-\boldsymbol{k},-s$. The latter appear in a superconductor under the effect of an external perturbation with corresponding matrix elements $B_0(\boldsymbol{k}',s' \mid \boldsymbol{k},s)$ for the single-electron transitions $\boldsymbol{k},s \to \boldsymbol{k}',s'$ in the normal state of a *metal*: $B_{1,2} = F_{1,2}B_0$.

According to the BCS theory,

$$F_1(\boldsymbol{k},\boldsymbol{k}') = U_{\boldsymbol{k}'}U_{\boldsymbol{k}} \pm V_{\boldsymbol{k}'}V_{\boldsymbol{k}},$$

$$F_2(\boldsymbol{k},\boldsymbol{k}') = U_{\boldsymbol{k}'}V_{\boldsymbol{k}} \pm U_{\boldsymbol{k}}V_{\boldsymbol{k}'},$$

where the $U_{\boldsymbol{k}}, V_{\boldsymbol{k}}$ are the Bogolyubov transformation coefficients (see *Canonical transformation method*), with the sign determined by the nature of the external perturbation.

Coherence factors are needed for theoretical descriptions of the different kinetics in superconductors associated with transitions in the system of Bogolyubov quasi-particles. This provides, e.g., in the BCS theory, an explanation of the difference between the type of temperature dependence of the *nuclear spin relaxation time* and that of the coefficient of ultrasound adsorption in a superconductor.

COHERENCE LENGTH in superconductors

Length scale for the variation of the *order parameter* (see *Ginzburg–Landau theory of superconductivity*)

$$\xi(T) = \frac{\xi_0}{\sqrt{2}(1 - T/T_c)^{1/2}}, \qquad \xi_0 = \frac{\hbar v_F}{\pi\Delta_0},$$

where v_F is the *Fermi velocity*, $\Delta_0 = E_g/2$ is the superconducting *energy gap* at $T = 0$, and T_c is the critical temperature of the *second-order phase transition* from the normal to the superconducting state. According to the scale invariance hypothesis, at the point $T = T_c$ the coherence length ξ goes to infinity as $\xi(T) \sim (1 - T/T_c)^{-\nu}$. The *critical index* is $\nu = 1/2$ in the *self-consistent field* approximation. The coherence length also determines the effective size of *Cooper pairs* with the binding energy 2Δ. In pure superconductors with $T_c \approx$ 1–10 K, for $T \ll T_c$, ξ is equal to $\xi_0 \approx 10^{-5}$–10^{-4} cm in order of magnitude. In "dirty" superconductors with short electron mean free paths $l \ll \xi(T)$ the effective coherence length is $\xi_{eff} \cong (\xi l)^{1/2}$. In highly disordered superconducting alloys (compounds) and in *amorphous superconductors* $\xi \leqslant 10^{-6}$ cm, which provides high values of the *Ginzburg–Landau parameter* $\kappa = \lambda/\xi$ and the *upper critical field* $B_{c2}(T)$, where λ is the magnetic field penetration depth.

COHERENT BREMSSTRAHLUNG

A type of *Bremsstrahlung* (braking radiation) specific for *monocrystals*. The Bremsstrahlung of charged particles in single crystals at high particle energies effectively differs from that of an isolated atom. For relativistic particles (total energy E far exceeds rest energy, Lorentz factor $\gamma = [1 - (v/c)^2]^{-1/2} E/(m_0c^2) \gg 1$), there is a *formation zone* one *radiation length* l long in which the Bremsstrahlung process takes place which can substantially exceed an atomic size, i.e. the creation of a radiated *photon* takes a long enough a time so that a particle can traverse a macroscopic distance. The time of photon generation is long compared with the duration of a particle's impact upon a nucleus, therefore the process characteristics are determined mainly by interference during the radiation process, rather than by the nature of the collision with an individual nucleus. The formation zone in monocrystals may be substantially longer than the separations between the atoms of the crystal lattice. In this case all the atoms in the zone take part in the process of coherent radiation, with the Bremsstrahlung intensity in monocrystal increasing, relative to that of an amorphous substance, proportional to the number of atoms located in this zone:

$$N_{\text{eff}} \propto \frac{l}{a} \propto \frac{2E(E - \hbar\omega)}{m^2c^3\omega a}.$$

The presence of interference in the monocrystal Bremsstrahlung arises from simple kinematic considerations. Suppose a particle entering a crystal at a small angle, θ, relative to a crystallographic axis encounters atoms along its path spaced a distance a/θ apart (a is the lattice constant), and subsequent acts of Bremsstrahlung emission take place. Suppose the energy of the radiation of quanta is small, and the radiation only weakly affects the particle motion (its momentum is virtually unchanged at radiation). The quanta are emitted mainly in the forward direction of the particle motion (within a cone whose angle is inversely proportional to the Lorentz factor γ). If t_0 is the time of emission at the first atom, the

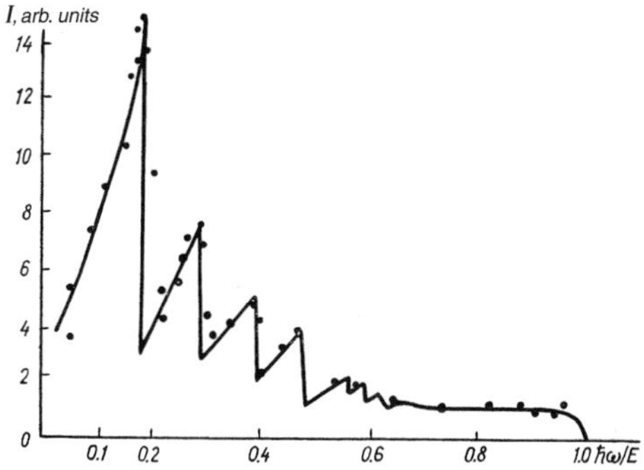

Intensity I of Bremsstrahlung of electrons in a diamond single crystal as a function of the energy $\hbar\omega$ of the emitted quanta at the incidence angle $\theta = 4.6$ mrad relative to the (110) axis (an electron beam of initial energy 1 GeV is directed along the (001) plane); shown are theoretical calculations (solid lines) and experimental data (dots).

Bremsstrahlung emission at the second atom will occur at the time $t_0 + a/(v\theta)$. As the first emitted quantum reaches the second atom at the moment $t_0 + a/(c\theta)$, the phase difference between the two must be equal to $2\pi n$ (where n is an integer) for the interference to amplify the radiation:

$$\omega\frac{a}{\theta}\left(\frac{1}{v} - \frac{1}{c}\right) = 2\pi n.$$

With a fixed particle energy, E, and a fixed magnitude of quantum energy, $\hbar\omega$, there is a particular angle $\theta_{\max}$ at which constructive interference is observed:

$$\theta_{\max} = \frac{a}{2\pi nl},$$

where l is the radiation length. There may be different numbers of lattice sites inside the Bremsstrahlung region, depending on the angle of incidence of the charged particle relative to the crystallographic planes or axes, which results in an appearance of maxima and minima in the Bremsstrahlung spectrum presented as a function of the angle at which fast particles enter the monocrystal. The intensity of coherent Bremsstrahlung in the *Born approximation* is proportional to the square of the number of atoms within the radiation length. If the Born approximation is inapplicable, the dependence of coherent emission intensity I on the number of atoms within the formation zone is weaker, having a logarithmic dependence (see Fig.). The coherent Bremsstrahlung exhibits a rather high degree of polarization, and this results in intense beams of polarized γ-quanta being produced by electron accelerators.

COHERENT CHANNELING

See *Okorokov effect.*

COHERENT CONDENSATE

See *Bose condensate.*

COHERENT POTENTIAL METHOD (P. Soven, D. Taylor, 1967)

An analogue of the mean field method used for computing the electronic structure and electronic properties of disordered alloys (see *Disordered solids*) based on the formalism of multiple scattering theory. The applicability of this method is not related to limitations on the magnitude of the impurity scattering potential, or of the *alloy* component ratios. A *scattering matrix* (t*-matrix*) is assigned to each individual atom in this approach. Systems of atoms are described on the basis of the scattering characteristics of individual atoms using a *Green's function* that depends upon the positions of the scattering atoms. It is assumed that the energy spectrum of the ensemble as a whole can be "macroscopically" described with the mean Green's function

$$\overline{G}(z) = \left[z - \varepsilon(\boldsymbol{k}) - \sum_{\boldsymbol{k}}(z)\right]^{-1},$$

where $\sum_{\boldsymbol{k}}(z)$ is a self-energy term (see *Mass operator of quantum mechanics*), and $\varepsilon(\boldsymbol{k})$ is the *dispersion law* for an electron in an unperturbed translationally-invariant crystal. To complete the Green's function definition, the system is presumed to behave as if one and the same *coherent potential*, $\sigma(z)$, is assigned to each lattice site, and an effective potential, $\widetilde{w}_{\boldsymbol{n}} = w_{\boldsymbol{n}} - \sigma(E)$, is introduced for every site to replace the random potential energy of an electron $w_{\boldsymbol{n}}$ (see *Random potential*). The self-consistency condition is taken as follows:

$$\langle t_{\boldsymbol{n}} \rangle = \left\langle \frac{\widetilde{w}_{\boldsymbol{n}}}{1 - \widetilde{w}_{\boldsymbol{n}} \langle \boldsymbol{n} | \overline{G}(z) | \boldsymbol{n} \rangle} \right\rangle = 0.$$

This implicit equation defines $\sigma(E) \equiv \sum_{\boldsymbol{k}}(E)$, and permits the determination of the coherent potential as a function of a spectral variable z. This method is believed to provide one of the best single-site approximations for obtaining the spectral characteristics of disordered systems. The small parameter of the method for the calculation of the energy spectrum is the quantity $(a/R_0)^3$, where a is a lattice constant, and R_0 is a characteristic damping or jump length of the hopping integral. A similar small parameter appears when computing corrections to the expression for the electrical conductivity of a disordered *binary alloy* within the formalism of the coherent potential method. The physical sense of this parameter can be explained as follows: if a sphere of radius R_0 encloses a large number of sites then one can neglect the fluctuations of the number of atoms of both types A and B inside this sphere.

COHERENT PRECIPITATES

New *phases* formed at phase boundaries which display a *structural coherence* there. Coherent precipitates arise as a result of the decomposition of supersaturated *solid solutions* (see *Alloy decomposition*), or from other *phase transitions*: a *martensitic transformation*, a eutectic (see *Eutectic alloys*), etc. Coherent precipitates may be wholly or partially coherent, depending on the type of structural coherence. They form during the early stages of phase transformations; later on, the structural coherence will be destroyed due to structural *dislocations* which form at the *phase interface* to make the precipitates non-coherent. The appearance of coherent precipitates with an appreciable mismatch of *crystal lattice constants* $\delta = (a_{\rm p} - a_{\rm m})/a_{\rm m}$, where $a_{\rm p}$ and $a_{\rm m}$ are lattice constants of the precipitate and matrix, respectively, gives rise to coherent elastic stresses, and to an increase of the elastic energy of the system by a magnitude $\varepsilon_{\rm e} \propto \delta^2$. That results in an increase of the *strength* of an alloy, of its *coercive force*, and of the magnetostatic energy (see *Magnetostatics*). In aging alloys (see *Alloy aging*), the elastic interaction of coherent precipitates may produce *modulated structures*. Coherence disruption, and an appearance of *non-coherent precipitates*, will usually result in deterioration of the alloy strength characteristics.

COHERENT SCATTERING REGIONS

Largest regions of a *crystal* from which *elastic scattering of radiation* (X-rays, neutrons, etc.) by individual atoms or *unit cells* is coherent. The dimension D of a coherent scattering region can be defined as the distance at which the phase shift of the waves scattered by atoms (cells) reaches π as compared to an *ideal crystal.* The value of D depends on the reflection indices hkl of the planes.

In the case of an ideal *mosaic crystal,* D coincides with the size l of mosaic blocks. Provided the disorientation of adjacent blocks is $\psi \gg \lambda/l$ (λ is the radiation wavelength), the value of D may be estimated from the width $\delta\theta$ of the X-ray line at its half-height: $D \approx \lambda/(\Delta\theta \cdot \sin\theta_B)$, where θ_B is the Bragg angle. In the opposite case (when $\psi \leqslant \lambda/l$), the regions of coherent scattering may cover several blocks ($D \propto \lambda/\varphi > l$). The blocks of the mosaics do not exclusively determine the range of coherent scattering in a real crystal. *Stacking faults* (leading to an effective reduction of the coherent scattering regions), small-angle boundaries, degree of uniform bending of *crystallites,* and internal *microstresses* in crystallites also play roles. To separate the effect of mosaics from that of microstresses, one may apply harmonic analysis to calculate the normalized Fourier coefficients A_n in the reflection curve (the profile of the X-ray line), $P(2\theta)$. The values of A_n can be written as the product $A_n = A'_n A''_n$, where A'_n, A''_n are the contributions from regions of coherent scattering and of microstresses, respectively. In the xyz coordinate system, with xy planes coinciding with reflecting planes, one may define the dependence of A_n on the reflection index $00l$ and then, extrapolating $A_n(l)$ to $l = 0$, find $A'_n = A_n(0)$. The derivatives of A'_n characterize the average size D_0 of the coherent scattering region, and the distribution P_n of the regions of coherent scattering transverse and along the z axis direction: $-(dA'_n/dn)_{n=0} = c/D_0$ (c is the unit period along the z axis), $d^2A'_n/dn^2 = P_n$.

COHESION (fr. Lat. *cohaesus*, stuck)

Linkage (binding) between two *solid* bodies of similar chemical composition in contact with each other. It results from a balance between long-range repulsive and short-range attractive forces. The *cohesion energy* is an important characteristic associated with *chemical bonds.* It is defined as the work required to move far apart the units of a bound structure. It can be experimentally found by *evaporation* or *sublimation.* The largest cohesion energy values are found in condensed phases (liquids and solids) where the chemical bonds are mainly ionic, covalent or metallic in character. The cohesion energy in real gases is several orders of magnitude smaller.

The *cohesion strength* depends on the mutual orientations of the crystal lattices of the bodies in contact, on their geometrical perfection, and on the degree of their surface cleanliness. The cohesion strength is calculated from the expression $P = (\alpha_{10} + \alpha_{20} - \alpha_{12})/a$ where α_{10} and α_{20} are the specific energies of the surfaces in contact, α_{12} is the specific energy of the grain boundary interface, and a is an interatomic distance. If the contacting surfaces are aligned relative to the symmetry elements then $\alpha_{10} = \alpha_{20} = \alpha$, $\alpha_{12} = 0$, and $P_{max} = 2\alpha/a$. The actual strength is smaller than P_{max} due to the lack of smoothness of an actual contact. Showing up in the cohesion linkage zone are stresses arising from mismatches in the crystal geometry of the surfaces in contact.

COINCIDENT-SITE LATTICE

An auxiliary *lattice* considered in the description of particular *grain boundaries.* A coincident-site lattice is that sublattice of the *complete superposition lattice* whose sites are occupied by atoms of both lattices of a bicrystal, i.e. the sites coincide in each of the crystals that are separated by the boundary. The reciprocal relative fraction of coincident sites is designated by Σ. The symmetry of the coincident-site lattice is determined by the symmetry of the pair of conjugated crystals, by the axis of their relative orientation, and by the value of the corresponding rotation angle. See *Grain boundary, Grain boundary lattice.*

COLD BRITTLENESS

The change of the *failure* mechanism from *tough failure* to *brittle failure* (or *quasi-brittle failure*), which occurs with a decrease of temperature within a comparatively narrow temperature range. This change of failure mechanism involves abrupt reduction of *plasticity*-related characteristics and *strength.* According to the Ioffe theory, cold brittleness is the consequence of an abrupt increase of the *yield limit* σ_s with decreasing temperature, the temperature dependence of the *breaking stress* σ_f being comparatively weak. Cold brittleness occurs below the temperature at which the curves of $\sigma_s(T)$ and $\sigma_f(T)$ versus T intersect. Modern microscopic dislocation theories attribute the sharp temperature dependence of the yield point to the presence of a covalent component in the interatomic bond (see *Covalent bond*), and to the high value of the Peierls–Nabarro stress (see *Peierls–Nabarro model*) at the *plastic deformation of covalent crystals*. Hence the cold brittleness arises due to the abrupt decrease of the mobility of *dislocations* with decreased temperature, and to the decrease of the rate of *stress relaxation* at the apexes of *cracks* and at other stress concentrators (see *Stress concentration*).

The reciprocal of cold brittleness is given the name *cold reluctance*. This term is descriptive of the ability of materials to retain plasticity and failure toughness with decreasing temperature.

The susceptibility of a material to cold brittleness is determined by the *cold brittleness temperature* (or *cold brittleness threshold*) T_c. Upper and the lower cold brittleness points (T_{cb}^u and T_{cb}^l) are commonly distinguished. The lower point T_{cb}^l is the minimal deformation temperature at which *plastic deformation* precedes the failure, and the failure is totally tough in its nature. The value of T_{cb}^u depends heavily on the deformation rate $\dot{\varepsilon}$, increasing with its increase. Various methods of *hardening of materials* (raising the *dislocation* density): *solid-solution hardening*, *dispersion hardening* (see *Precipitation-hardened materials*), etc. result in an increase of the value of T_{cb}^l. Only a reduction of the crystal grain size or the formation of a disoriented dislocation *cellular structure* can lead to the reduction of T_{cb} and the improvement of strength properties. As the failure mechanism changes from *transcrystalline failure* to *intercrystalline failure*, the cold brittleness temperature may abruptly increase.

COLD CATHODE

Unheated cathode emitting electrons by means of *field emission of electrons*. These are *field-emission cathodes*, and there are also various types of cathodes emitting nonequilibrium electrons excited by an internal field: dispersed metal films, semiconductor films, *p–n* junctions biased in the gate direction (see *Semiconductor junction*), transistor structures, M–I–S structures (see *Metal–insulator–semiconductor structures*), M–S structures (M is metal, I is insulator, S is semiconductor), semiconductor structures with a negative *electron affinity*, etc.

COLD ELECTRONS

An electron gas in the *conduction band* of semiconductors (*holes* in the *valence band*) with an average energy $\bar{\varepsilon}$ appreciably less than the thermodynamic equilibrium value $\bar{\varepsilon}_0$. This phenomenon of electron cooling is due to various mechanisms related to the motion of the current carriers in an electric field, or to optical transitions of the charge carriers. For instance, electron cooling will occur if the overall current in a semiconductor is directed opposite to the internal electric field in a *space charge region*, or because of another inhomogeneity of the semiconductor. In electric fields, where the current is directed along the field, a cooling of the carriers is possible when the dominant scattering of the electron energy occurs by optical *phonons*, and the momentum scattering is by ionized impurities under conditions when the electron–electron interaction is negligible. While acted on by an electric field an electron is able to emit an optical phonon, and thereby diminish its energy (i.e. cool), and the scattering by ionized impurities promotes the accumulation of such electrons. Hence it is possible to cool carriers by light through the agency of direct band-to-band transitions. For this purpose it is necessary (when the energy of scattering via inter-electron collisions is negligible compared to such scattering of photocarriers by optical phonons) to excite the photoelectrons with a fixed initial energy ε_i greater than $\bar{\varepsilon}_0$ by utilizing light of such a wavelength that, after the emission of n optical phonons, the photocarriers should have a final energy $\varepsilon_f < \varepsilon_i$.

COLD WORKING

Industrial term no longer used much. It means *hardening* of a metal under the action of cold *plastic deformation*, but not *strain hardening* (see also *Work hardening*). Cold working in optimal mode increases *hardness* and *yield limit*, resistance to fatigue deformations, and a decrease of *plasticity* and *impact strength*. It is used in technology for *surface hardening* of work pieces with the help of steel- or cast iron-shot blasting impact treatment, rolling, etc.

COLLAPSE OF ELECTRON ORBITS

A sharp decrease of the average radius of an electron orbit in an excited *atom* occurring for a small change of the nuclear charge (atomic number), Z. The collapse of electron orbits takes place when the screened electrostatic potential $-U(r)$, acting on the electrons and the orbital or "centrifugal" potential $\hbar^2 l(l+1)/(2mr^2)$ (where r is the distance from the electron to the nucleus, l is the orbital quantum number, and m is the electronic mass) approximately compensate for each other. Then the effective potential

$$U_{\text{eff}} = -U(r) + \frac{\hbar^2 l(l+1)}{2mr^2}$$

can exhibit two valleys or minima, one near the atom nucleus (at $r \approx a_B$, where a_B is the *Bohr radius*), and the other at the atom periphery (at $r \approx 10a_B$), separated by a positive potential barrier. The deepening of the internal well with increasing Z at a given l results, at a certain critical value of $Z = Z_c$ (which depends on l, and also on the electron configuration and on the level), in the redistribution of electric charge density from the outer valley into the inner one, which is interpreted as an electron collapse. This collapse of electron orbits is possible for electrons with $l \geqslant 2$. In the *isoelectronic sequence* of argon-like *ions*, the collapse of $3d$-electrons in the $2p^5 3d$ configuration takes place at $Z_c = 19$, viz., at the transformation from Ar to K^+, while in the isoelectronic sequence of xenon-like atoms the collapse of the $4f$-electron in the $3d^9 4f$ configuration takes place at $Z_c = 55$, i.e. at the transformation from Xe to Cs^+. If the atom or ion is present in a solid then the orbits that did not collapse strongly overlap with neighboring atomic *coordination spheres*, and the corresponding electrons often become delocalized. Photoabsorption spectra reflect the participation of such states, and provide information on the electronic structure of such a solid (see *Fine structure of X-ray absorption spectra*).

COLLECTIVE EXCITATIONS

Motions in a system with a large number of interacting particles associated with the excitation of some collective degrees of freedom, and manifesting themselves in a cooperative dynamics of all the particles (see *Phonons*). The collective excitations are not necessarily associated with mechanical displacements of the particles: they may be, e.g., electronic excitations in atoms (see *Excitons*) or excitations of magnetic (spin) degrees of freedom (see *Magnons*). The collective excitations in condensed media propagate as an ensemble of waves, each of them representing a certain kind of motion characterized by a velocity and an energy. In quantum theory collective excitations are regarded as *quasi-particles*, their dynamics being described by the corresponding equations of motion. The collective excitations for weak interactions in a system, when the *harmonic approximation* is valid, obey the superposition principle and play the part of elementary excitations. Introducing the concept of collective excitations facilitates the treatment of a number of physical properties of weakly excited condensed media. Explaining some of these properties requires taking into account the interactions of the collective excitations, which is equivalent to introducing an anharmonicity into their description. In macroscopic systems with a high degree of excitation the collective excitations become transformed, and they can appear as solitary waves (solitons) or *shock waves*.

COLLECTIVE EXCITATIONS IN SUPERCONDUCTORS

Mutually coupled oscillations of the *order parameter* and of an electric field. At a temperature near the critical (transition) temperature, the existence of weakly damped oscillations of this type with an acoustic spectrum is possible. Such oscillations (experimentally discovered by P.L. Carlson and A.M. Goldman in 1975) show up at frequencies which exceed both the reciprocal energy relaxation time, and the reciprocal lag time for the establishment of an equilibrium distribution of quasi-particles over the branches of the energy spectrum. This collective mode represents oscillations of the population differences between the branches of the energy spectrum. Simultaneously, a longitudinal electric field appears in the superconductor, along with currents of normal excitations and of the superconducting condensate directed opposite to one another, so that there is no net current. The phase velocity of such excitations equals: $c = [2\Delta D\chi(2\pi T\tau)]^{1/2}$, where $D = v_F l/3$ is the electron diffusivity (diffusion coefficient), Δ is the *energy gap* in the superconductor, τ is a characteristic time of scattering from impurities, T is the temperature, χ is the *Gorkov function* introduced in the theory of *superconducting alloys*, v_F is the *Fermi velocity*, and l is the electron *mean free path*.

Another type of oscillation existing in superconductors is an oscillation of the order parameter amplitude which has a resonance character (*Schmid mode*, A. Schmid, 1975). This is associated with the possibility of a virtual decay of *Cooper pairs*, which requires an energy input of 2Δ. This process involves violation of the energy *conservation law* and, in accordance with the quantum-mechanical uncertainty principle relating energy and time, must proceed for times of the order of $\hbar/(2\Delta)$. The amplitude of the order parameter perturbation oscillates with a frequency $\omega = 2\Delta/\hbar$, and tends toward its steady-state value with a power law dependence on the time. Inelastic scattering mechanisms can bring about an exponential decay of such oscillations.

In a hypothetical superconductor treated as a superfluid *Fermi liquid* there are weakly damped oscillations of the phase of the order parameter which have an acoustic spectrum (*Anderson–Bogolyubov mode*). However, an excitation of the electron density accompanies these oscillations, hence, their frequency is increased up toward the plasma frequency in real superconductors because of Coulomb interactions.

COLLECTIVE VARIABLES METHOD

A method of statistical physics that introduces long-range interactions into the thermodynamic and structural properties of multiparticle systems. The description of interacting particles (e.g., atoms, *ions* or molecules) using collective variables is based on an extended phase space consisting of individual particle coordinates and collective variables associated with the waves of density fluctuations of a certain system parameter (see *Fluctuations of atomic positions*). Depending on the specific problem, the latter may be the density of the number of particles, density of electric charge, *spin density*, etc. In the simplest case of a single-component system, the collective variables will be defined by means of the Fourier components of the operator of the number density of particles:

$$\widehat{\rho}_k = \frac{1}{\sqrt{N}} \sum_{j=1}^{N} \exp(\mathrm{i}\boldsymbol{k}\cdot\boldsymbol{r}_j), \quad k \neq 0, \qquad (1)$$

where $\boldsymbol{r}_j$ are the coordinates of the jth particle, and N is their total number. The condition $k \neq 0$ excludes the constant term $\widehat{\rho}_{k=0} = \sqrt{N}\delta_{k,0}$. The transition to an expanded phase space that includes individual coordinates, $\{\boldsymbol{r}\} = (\boldsymbol{r}_1, \ldots, \boldsymbol{r}_N)$, and collective variables, $\{\rho_k\}$, is effected via a transfer function which is a product of corresponding δ-functions. Hence, e.g., the system configuration integral can be represented as follows:

$$Q_N = Q_N^0 \int (\mathrm{d}\rho) J(\rho) \exp\bigl(-\beta\theta(\rho)\bigr), \qquad (2)$$

where Q_N^0 is the configuration integral of the subsystem with short-range interactions, the function

$$J(\rho) = \frac{1}{V^N Q_N^0} \int (\mathrm{d}r) \exp\bigl(-\beta U_0(\rho)\bigr) \times \prod_{k\neq 0} \delta(\rho_k - \widehat{\rho}_k), \qquad (3)$$

is the Jacobian of the transition to collective coordinates, the differential factors have the form

$$(\mathrm{d}r) = \prod_{j=1}^{N} \mathrm{d}\boldsymbol{r}_j, \qquad (\mathrm{d}\rho) = \prod_{k\neq 0} \mathrm{d}\rho_k,$$

where $\beta = 1/(k_B T)$ is the reciprocal temperature of the system, $U_0(r)$ is the potential energy of the short-range interactions, and the function $\theta(\rho)$, given by

$$\theta(\rho) = \sum_{k\neq 0} \frac{N}{V} \nu(k)(\rho_k \rho_{-k} - 1) + \frac{N^2}{2V} \nu(k=0), \qquad (4)$$

is the representation in collective variables of the potential energy of long-range pair inter-particle interactions, $\nu(k) = \int \mathrm{d}r\, \mathrm{e}^{-\mathrm{i}rk} \Phi(r)$ is a Fourier component of the long-range potential $\Phi(r)$, and V is the system volume.

A similar transformation to collective variables also can be brought about by means of an integral transform of the Gaussian factor $\exp[-\beta\theta(\rho)]$.

The method of collective variables is used to describe the properties of a variety of real systems. Thus, an ion-molecular approach in the theory of electrolytic solutions, based on an equivalent account of the interactions of all the particles in solution, has been developed within the framework of collective variables. This method, supplemented by the *density matrix* shift transformation, has been successfully employed in the theory of quantum Bose- and Fermi-systems, in particular for constructing a theory of liquid ^{4}He and describing the electron gas in metals (see *Bose–Einstein statistics* and *Fermi–Dirac statistics*).

Variables associated with the *order parameter* of the system have been introduced into the set of collective variables in the theory of *second-order phase transitions*, and the explicit form of the transfer Jacobian, $J(\rho)$, is of a non-Gaussian character for three-dimensional systems. In another approach, the collective motions of systems of interacting particles are treated by the *canonical transformation method*, whereby a transition is made to collective coordinates and momenta that correspond to collective oscillations, or to the elemental excitations associated with them. Since the collective coordinates and momenta are not conventional variables, a transition to them involves additional conditions that conserve the total number of degrees of freedom in the system.

COLLISION CASCADE

The avalanche-like process of dislodging (displacement of) atoms (ions) from the sites of a *crystal lattice* (or of an amorphous *solid*). The collision cascade arises after energy transfer to the target atom by an external particle as a result of sequential collisions of atoms in the vicinity of the initially displaced one.

If the energy of the initially displaced atom is ε, the collision cascade occupies the spatial domain of size $l(\varepsilon)$, where $l(\varepsilon)$ is the mean free path as a function of its energy. The cascade time is 10^{-14}–10^{-13} s; where the cascade over a domain of enhanced concentration of radiative defects arises with its central part enriched with *vacancies*, and the peripheral one enriched with *interstitial atoms*. The primary relaxation of a cascade takes $\sim 10^{-13}$–10^{-11} s, with some interstitial atoms returning to their initial sites. Then the secondary stage of relaxation occurs, controlled by thermally activated processes (primarily by *diffusion*). On irradiation with dense beams or with radioactive decay, spatially uniform collision cascades can arise, their only common quantitative characteristic being the *cascade function* or the *radioactive damage function* $\nu(\varepsilon)$. It equals the total number of *atoms* (*ions*) displaced from the sites of the solid body lattice by the initial atom with energy ε.

Low-energy cascade: $\varepsilon < \varepsilon_c$, where ε_c is the critical energy, with negligible retardation of the moving atom by electrons below it. All the energy of the atom goes to displace atoms from lattice sites, so that

$$\nu(\varepsilon) = \frac{\alpha\varepsilon}{\varepsilon_d} \quad (\varepsilon < \varepsilon_c),$$

where ε_d is the energy for creating a stable vacancy–interstitial atom pair, and α is a factor on the order of unity. As commonly assumed, $\varepsilon_d \sim 25$–80 eV, $\alpha \sim 0.4$–1. Theoretical values of ε_c are 0.1 MeV for Fe and 3 MeV for U.

For a *high-energy cascade*: $\varepsilon > \varepsilon_c$. The retardation of atoms down to energy ε_c occurs mainly due to electrons of the medium, i.e. without dislodgment of new atoms. Therefore,

$$\nu(\varepsilon) = \frac{\alpha'\varepsilon_c}{\varepsilon_d} \quad (\varepsilon > \varepsilon_c),$$

where α' is a factor on the order of unity (it can differ somewhat from α). See also *Seeger zone*.

COLLISION IONIZATION

See *Impact ionization*.

COLLISION RADIATION

See *Bremsstrahlung*.

COLOR CENTERS, F-centers

Point defects with optical absorption in the transparent region of a defect-free *crystal*. Color centers are related to either impurities (*impurity color centers*), or intrinsic structural defects in crystals (*intrinsic color centers*). The main methods to stimulate the formation of color centers are

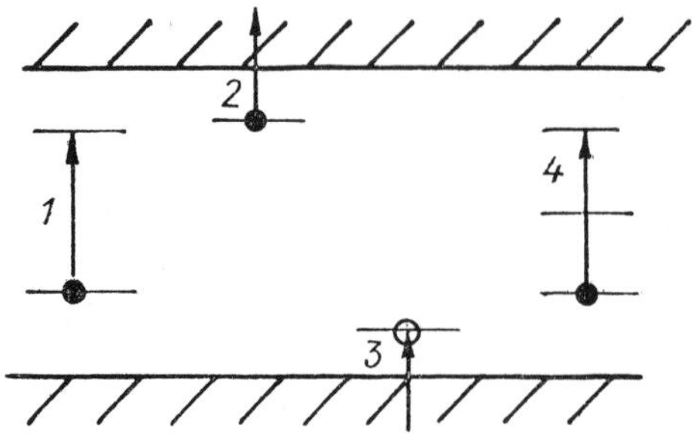

Color centers.

doping, deviation from *stoichiometric composition*, and irradiation by ionizing radiation. Electronic transitions that determine the color center absorption are shown schematically in the figure. Transitions 2 and 3 produce, as a rule, wide absorption bands; transitions 1 give either narrow or broad bands, depending on the magnitude of the *electron–phonon interaction*. Transitions 4 are observed only under high-power pulse excitation with the aid of time-resolved spectroscopy. Reverse transitions may proceed with photon emission; in this case a color center is simultaneously a *luminescent center*. Many color centers involve a point defect that has trapped an electron (*electron color centers*) or a hole (*hole color centers*). The association of elementary defects (*vacancies*, *interstitial atoms*, *impurity atoms*, or impurity ions) results in the formation of aggregate color centers. An increase in temperature and illumination in the absorption bands of color centers may lead to the release of trapped electrons or holes, i.e. to the destruction (discoloration) of color centers. The most widely studied color centers are in *ionic crystals*, in particular, in *alkali-halide crystals*. For example, F-centers (anionic vacancy with a trapped electron, named from the German word "Farbe" for "color") produce an intense absorption band in the visible region, which corresponds to a transition similar to the $1s$–$2p$ transition in a hydrogen atom (F-band), and weaker K-, L-bands refer to transitions to higher F-center excited states. Also H-centers (interstitial atoms of halide X^0), and V_k-centers (self-localized holes X_2^-), produce broad bands in the ultraviolet region. In *covalent crystals*, the absorption of color centers typically occurs as narrow lines, e.g., $GR1$ and $ND1$ centers in *diamond* (neutral and negatively charged vacancies). Color centers are also found in glasses (see *Vitreous state of matter*), e.g., a non-bridging oxygen atom in vitreous quartz (band with a maximum at 4.75 eV). The main applications of color centers in practice are *lasers*, *photopigments*, *dosimetry*, and diagnostics of *minerals*.

COLORED SYMMETRY

A theory of generalized symmetry of material objects (media and structures), which combines the global symmetry of the system as a whole with the *local symmetry* of its structural elements. Depending on the method of combination of the groups of global (external) G- and local (internal) P-symmetries, the following *colored symmetry groups* are distinguished: (1) P-symmetry groups, which are subgroups $G^{(p)} \subseteq P \times G$ of the direct products of groups P and G; (2) Q-symmetry groups, which are subgroups $G^{(q)} \subseteq P\lambda G$ of the semidirect products of P and G; (3) W_p-symmetry groups $G^{(W_p)} \subseteq P_{wr}G = P^G \times G$, which are subgroups of the wreath product (for groups $G^{(W_q)}$ – of twisted wreath product) of P and G, where P^G is the direct product of groups P, taken n_G times, $n = |G|$ is the order (or power) of the group G: $P^G = P^{g_1} \times P^{g_2} \times \cdots \times P^{g_n}$, and the elements $g_i \in G$ indicate the position of each individual group P in the n-fold direct product. These groups differ in the manner of their effect on a mass point, and in the multiplication law of their elements.

In terms of physical interpretation, the lowest groups $G^{(p)}$ and $G^{(q)}$, isomorphic to the groups G, describe the magnetic exchange symmetry of the structures of magnetically ordered crystals. Groups $G^{(W_p)}$ and $G^{(W_q)} \leftrightarrow G$, involving the effect of translation on a mass point (depending on its coordinate), describe the symmetry of *real crystals*, including *quasi-crystals* (icosahedral phases), and polysystem *molecular crystals* with so-called "super-Fedorov" symmetry.

COLORED SYMMETRY GROUP

In the general case the subgroups of the direct products ($P \times G$) and the semidirect products ($P\lambda G$) of groups of external (global) G and internal (local) P symmetries, as well as subgroups of the wreath product $P_{wr}G = P^G\lambda G$, which operate over subspaces of geometrical coordinates

$r \in R(3)$, and of their nongeometrical ("colored") counterparts $\overset{\circ}{S} \in S(d)$. According to the laws of multiplication and to the action of combined elements (p, g) or $\langle \ldots p(\boldsymbol{r}_k) \ldots | g\rangle$, $p \in P$, $p(\boldsymbol{r}_k) \in P^{q_k r_1} = P^{r_k} \subset P^{r_1} \times \cdots \times P^{r_k} \times \cdots = P^G$ of the material point $(\overset{\circ}{S}, \boldsymbol{r})$, groups of four types are singled out; of *global action*:

$$P\colon\ (p, g)\big(\overset{\circ}{S}, \boldsymbol{r}\big) = \big(p\overset{\circ}{S}, g\boldsymbol{r}\big) = \big(\overset{\circ}{S}{}', \boldsymbol{r}'\big),$$

$$(p_i, g_j)(p_k, g_l) = (p_i p_k, g_i g_l) \in G^{(p)} \subset P \times G,$$

$$Q\colon\ (p, g)\big(\overset{\circ}{S}, \boldsymbol{r}\big) = \big(pg\overset{\circ}{S}, g\boldsymbol{r}\big) = \big(\overset{\circ}{S}{}', \boldsymbol{r}'\big),$$

$$(p_i, g_j)(p_k, g_l) = \big(p_l\big(g_j p_k g_j^{-1}\big), g_j g_l\big) \in G^{(q)} \subset P\lambda G$$

and of *local action* – depending on the coordinate $\boldsymbol{r}_k = g_k \boldsymbol{r}_1$ of the point $(\overset{\circ}{S}, \boldsymbol{r}_k)$:

$$W_p\colon\ \langle \ldots p(\boldsymbol{r}_k) \ldots | g\rangle\big(\overset{\circ}{S}, \boldsymbol{r}_k\big) = \big(p(\boldsymbol{r}_k) g\overset{\circ}{S}, g\boldsymbol{r}_k\big),$$

$$\langle \ldots p_j(\boldsymbol{r}_k) \ldots | g_j\rangle \langle \ldots p_l(\boldsymbol{r}_k) \ldots | g_l\rangle = \langle \ldots p_j(\boldsymbol{r}_k) p_l(\boldsymbol{r}_k) \ldots | g_l g_l\rangle,$$

$$W_q\colon\ \langle \ldots p(\boldsymbol{r}_k) \ldots | g\rangle\big(\overset{\circ}{S}, \boldsymbol{r}_k\big) = \big(p(\boldsymbol{r}_k) g\overset{\circ}{S}, g\boldsymbol{r}_k\big),$$

$$\langle \ldots p_j(\boldsymbol{r}_k) \ldots | g_j\rangle \langle \ldots p_l(\boldsymbol{r}_k) \ldots | g_l\rangle = \big\langle \ldots p_j(\boldsymbol{r}_k)\big(g_j p_l(\boldsymbol{r}_k) g_j^{-1}\big) \ldots \big| g_j g_l\big\rangle.$$

The colored symmetry groups are closely connected with the theory of so-called fiber SR-spaces ($SR \subset S(d) \times R(3)$, or $SR \subset S^R(d) \times R(3)$), and $n = (d+3)$-dimensional groups, which are their generalized projections onto the three-dimensional space. The colored symmetry groups are used to describe magnetically ordered crystals (see *Magnetic symmetry group*). See also *Colored symmetry*.

COLOR GROUP

See *Shubnikov groups*.

COMBINED HEAT TREATMENT

The heat treatment of materials including combinations of heat with chemical and mechanical modes. The choice of combined techniques takes into account the loading conditions, detailed geometry, sequence of operations being carried out, and values of the technological parameters being used.

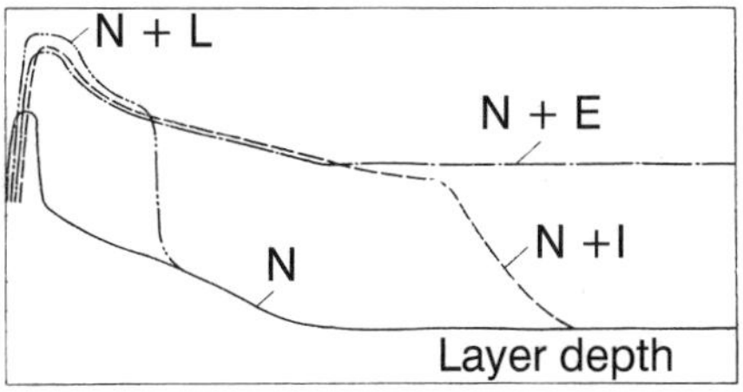

Curves of hardness versus layer depth for different combined heat treatment techniques: nitration or nitrocementation chemical heat treatment (N), and electrocontact (E), induction (I), and laser (L) high-speed hardening.

Multiple combined *heat treatment* techniques allow the variation of properties corresponding to different requirements for surface layers and internal bulk (core) material. Schematic curves of hardness as a function of layer depth are presented in Fig. as an example, where the lower solid curve is a thermo-chemical treatment employing nitration or nitrocementation (N), and the remaining curves represent high-speed hardening via electrocontact (E), induction (I), and laser (L) processes.

It is possible to improve hardness and other mechanical properties in near-surface regions, and depending on the methods of heating, at different distances from the surface, taking into account structure and microstructure variations. It is also possible to obtain gradients of particular properties needed for specific loading conditions. As examples of how metals can undergo combined heat treatments we refer to prior *quenching* with *tempering* before thermo-chemical treatment at low temperatures; *cementation*; *nitrocementation* with hardening, etc.

COMMENSURABLE STRUCTURE

See *Commensurate structure*.

COMMENSURATE STRUCTURE,

commensurable structure

A long-period structure, which arises from periodic distortions of an initial lattice due to displacements of atoms from equilibrium positions. The commensurate structure is generally formed during lowering of the temperature at a *phase transition*. In a number of cases the system undergoes a sequence of phase transitions to various commensurate states (with sequential variations of the *superlattice* constant), until reaching the final commensurate *phase* which has a period a' that is a multiple of the initial lattice constant a, i.e $a' = na$, where n is an integer (so-called *"devil's staircase"*).

The transition from the initial structure to a commensurate one proceeds in certain materials through an intermediate temperature *incommensurate structure*, with a superstructure constant incommensurate with the lattice constant of the initial phase (see *incommensurate–commensurate phase transition*). The commensurate structure may be represented as modulated by shear waves, e.g., *charge density waves* (*spin density waves*), in one, two or three directions; with the modulation period determining the displacements of atoms from their initial equilibrium positions. The periodicity of the new structure is fully commensurate with that of the initial structure: $\boldsymbol{Q} = n\boldsymbol{q}$, where $\boldsymbol{Q}$ and $\boldsymbol{q}$ are wave vectors of the commensurate and the original structures, respectively, and n is an integer. The translational symmetry is not violated in the generation of the commensurate structure. The commensurate structure can be detected by diffraction methods, e.g., through an electron diffraction observation of extra reflections corresponding to basic structure reflections for multiple distances, and by changes of other physical characteristics: electrical *resistance*, *magnetic susceptibility*, *Young's modulus*, etc. The commensurate structure occurs in dichalcogenides of transition elements MX_2 (M = Ta, Nb; X = Se, Te, S), in crystals of the A15 and cubic Cu_2Mg types, in the premartensitic region of temperatures in a number of compounds based on equiatomic TiNi, in Cr, etc.

COMPACTION

See *Pressing*.

COMPENSATION

See *Conductivity self-compensation*, *Ionic conductivity compensation*, *Magnetic compensation*.

COMPENSATION RULE, Meyer–Neldel rule (W. Meyer, H. Neldel)

A relationship between the *activation energy* ΔE and the preexponential factor σ_0 in the temperature dependence of the electrical conductivity σ of semiconductor:

$$\sigma(T) = \sigma_0 \exp\left(-\frac{\Delta E}{2k_B T}\right). \qquad (1)$$

The analysis of many experimental data obtained for semiconductors possessing *current carriers* with small mobilities has shown that

$$\lg \sigma_0 = \alpha \Delta E + \beta. \qquad (2)$$

Here α and β are parameters characteristic of the material. Eq. (2) represents the compensation rule which has been widely investigated for polymeric semiconductors. This is explained with the help of the charge transport tunnel model (see *Tunnel effect*).

COMPLEX ELASTIC MODULI

Tensor factors in *Hooke's law* written for harmonically time-varying stresses and *strains* in dissipative media (see *Dissipative structures*): $\sigma_{ik}^{\omega} = c_{iklm}(\omega)\varepsilon_{lm}^{\omega}$, where σ^{ω} and ε^{ω} are Fourier components of stresses and strains, respectively. The imaginary part of the complex elastic modulus, called the *loss modulus*, is responsible for the damping of elastic waves in a medium.

COMPLEXY

A thermodynamic equilibrium *defect* in a solid solution consisting of an *impurity atom* complex located at a group of lattice sites whose number differs from the number of complexy atoms. The concept of complexy was introduced by M.A. Krivoglaz (1970). It can be energetically advantageous to form a complexy from atoms with strongly differentiated atomic radii if this results in a partial removal of stresses and a decrease of the overall elastic energy. A complexy is less advantageous from the entropy standpoint than isolated impurity atoms, and the complexy concentration in a *solid solution* decreases with temperature in contrast to the behaviour of *point defects*. Complexy formation radically affects the temperature and concentration dependences of *crystal lattice constants*, *diffusion coefficients*, solubility of impurity atoms, *diffuse scattering of X-rays*, and other solution characteristics. A noticeable change in lattice parameters should be observed in *alloys* containing complexies, and these parameters are detected by *dilatometric analysis* and *X-ray diffraction* techniques. Complexies have been detected experimentally in a lead–gold alloy.

COMPOSITE MATERIALS (fr. Lat. *compositio*, combination)

Materials consisting of two or more components with a distinct boundary between them, and possessing properties not found in the individual components taken separately. Anisotropic and isotropic (quasi-isotropic) composite materials are distinguished. Among the first are *reinforced composite materials* including matrix and reinforcing components in the forms of fibrils, wires, *thread-like crystals*, strips, and grids oriented in it in a predefined fashion. There are metallic, polymeric and ceramic composite examples differentiated by the matrix type. Among the second are *precipitation-hardened materials*, *pseudoalloys*, and randomly reinforced composites. The dispersion-hardened varieties have a matrix structure consisting of a basic material with fine dispersed hardening inclusions; the pseudo-alloys can have the structure of interpenetrating skeletons, a statistical powder mixture, or a matrix structure depending on its composition and formation process; randomly reinforced materials have a matrix structure with no preferred orientation direction of fibrils or inclusions. Reinforcing fibrils and inclusions endow composite materials with *strength*, hardness, and the matrix with bonds between components, load transmission on fibrils, and *corrosion resistance*. Properties of composite materials depend significantly on phase boundary strength, hardening components, their size and quantity. One of the main requirements imposed on composite materials used for construction is the lack of significant diffusion (see *Diffusion*) and chemical interactions of the components during the preparation and high-temperature use. By varying the structure and composition it is possible to modify the *strength* characteristics over a wide range, *fracture toughness*, *electrical conductivity*, *thermal expansion* factor, and other properties.

COMPOSITE MATERIALS, STRENGTH

See *Strength of composite materials*.

COMPRESSIBILITY, volume compressibility factor

Relative variation of the volume of a body under the effect of total *pressure*. One can distinguish the *isothermal volume compressibility*, $\kappa_T = -(\partial v/\partial p)_T/v = -((\partial \ln(\partial \Phi/\partial p)_T)/\partial p)_T$ (Φ is *thermodynamic potential*) and the *adiabatic volume compressibility*, $\kappa_S = -(\partial v/\partial p)_S/v = -(\partial v/\partial p)_T/v - Tv\alpha^2/c_p$ (where c_p is the *specific heat* of the body under constant pressure, α is the volume coefficient of *thermal expansion*). The isothermal compressibility exceeds the adiabatic one for the majority of solids by not more than several percent. The compressibility has the units $\mathrm{m^2/N} = 1/\mathrm{Pa}$ (typical value is $\sim 1/\mathrm{GPa}$).

COMPRESSIBILITY MODULUS

The same as *Bulk modulus*.

COMPRESSION, UNIFORM

See *Uniform compression*.

COMPTON EFFECT (A. Compton, 1922)

The scattering of an X-ray or *gamma quantum* off a free (or weakly bound) electron to which the *photon* transfers a part of its energy and momentum. The change of the photon wavelength λ at scattering is determined by the energy and momentum *conservation law* and is equal to $\Delta\lambda = \lambda' - \lambda_0(1 - \cos\theta)$, where λ' is the wavelength of the scattered photon, $\lambda_0 = h/(mc)$ is the *Compton wavelength* equal to $2.426 \cdot 10^{-12}$ m, θ is the angle between the directions of motion of the incident and scattered photons. The scattering intensity evaluated by quantum theory provides the angular distribution of scattered photons (mainly forward scattering) and the scattering cross-section σ which depends on the photon energy ε, with $\sigma(\varepsilon)$ decreasing with ε. In the limit $\varepsilon \to 0$ we have $\sigma \to \sigma_\mathrm{T} = 8\pi r_0^2/3$ (r_0 is *classical electron radius*, $r_0 = 2.8 \cdot 10^{-15}$ m). The *Thomson's scattering cross-section* σ_T was derived by J. Thomson in the classical theory of electromagnetic wave scattering. The Compton effect can occur for scattering off other charged particles as well, but the effect is seen only for high-energy photons scattering off particles with large masses. If the electron involved in the Compton scattering has a high velocity then an increase in photon energy is possible at scattering (*inverse Compton effect*). The Compton effect gives a strong contribution to the generation of fast electrons in gamma ray irradiated solids, and these generated electrons create radiation defects through impacts with target atoms.

COMPUTATIONAL MICROTOMOGRAPHY

A method of scanning an object with a probe of penetrating radiation and, with the aid of a computer, subsequently reconstructing an image which contains details of the internal microstructure of the object in a series of thin layer cross-sections. Ordinary transmission images, being two-dimensional shadow projections of the actual three-dimensional object, do not give significant information about its internal structure, because they display information averaged according to the thickness. In contrast to this the digitized images called *tomograms* (see *Tomography*) provide considerably sharper differentiation of areas with differing *absorption coefficients*, and better separation of the images of overlaying structures.

To apply computational microtomography the object is transilluminated with a narrow beam of the penetrating radiation (e.g., X-rays, electrons, ultrasound, or IR) and the radiation that passes through it is recorded by corresponding detectors. By sequentially scanning the beam and/or turning the sample, a set of projection data necessary for reconstructing the tomogram is obtained. The result is a sequence of two-dimensional halftone cross-section images, and the data from these cross-sections can be used to reconstruct a three-dimensional image which reflects the true microstructure of the object.

Microtomography has been successfully used in transmission *electron microscopy* for the investigation of molecules, bacteriophages and membranes (with spacial resolutions of angstroms), in *scanning electron microscopy*, and in optical and *X-ray microscopy* for investigating the properties and structure of *solids* with dimensions of about 1 mm and resolutions of micrometers. The technique is also used in the non-destructive *defectoscopy* of components of *solid-state quantum electronics*, providing visualization and identification of the subsurface micro-structure and defects. See also *Tomography of microfields*.

COMPUTER ALGEBRA

A process in applied mathematics and computer engineering, that uses computers for the automation of symbolic derivations. Computer algebra is designed for the execution of cumbersome (routine) analytic manipulations that cannot be readily carried out manually: multiplication (division) of polynomials, cancellation of similar terms, formation of an arbitrary algebraic expression with respect to a dependent variable, substitution of variables, integration, differentiation, etc. This approach can facilitate and improve the accuracy of subsequent numerical calculations, provide analytical relationships between various parameters, and overcome difficulties involved in the subtraction of nearly equal large magnitude numbers.

COMPUTERS in solid state physics

Various methods for applying computers to solve problems in the field of solid state physics. In experimental solid state physics a computer performs the following three functions: control of apparatus (e.g., the control of devices designed for molecular beam *epitaxy* when synthesizing *superlattices*); data analysis (such as deciphering various spectra); planning experiments. In an optimal situation, all these functions are coordinated, so that the computer performs quite a complicated range of research tasks. In analytical solid state physics most of the important applied problems, as well as an ever-growing share of the fundamental problems, are solved using computers. Two approaches are dominant in this field: the numerical solution of equations, and the direct modeling of observed solid state phenomena with the help of computers (see also *Computer simulation*).

The first approach of solving systems of equations is not specific for solid state physics, and its efficiency is determined, to a great extent, by the capabilities of the computer. For single-processor computers the absolute or theoretical limit of their productivity is about 10^9 operations/s, but in practice the limit is lower by one or two orders of magnitude. For a relatively simple problem, such as a study of the initial stages of *monocrystal growth* from an atomic beam at a perfect monocrystal face, it is necessary to analyze a Markovian chain consisting of hundreds or (at low saturation in the vicinity of the critical point) a few thousand nonlinear differential equations. The determination of the distribution function of the atomic groupings for any given set of initial conditions requires the performance of about 10^8 operations/s. For a reasonably complete analysis involving the entire range of available growth conditions one needs to perform 10^{10} or more operations/s. The problem becomes too difficult for multicomponent crystals, and indeed insoluble for single-processor computers when the final growth stages are included in the calculation. Multiprocessor computers are also inefficient in the framework of the first approach, because extensive parallelism of the data processing is hardly achievable.

In contrast, the second approach of modeling observed phenomena is quite compatible with the architecture of multiprocessor computers. The real space of the crystal can be divided into a network of conventional cells (hence the term *network models*) and a step-by-step analysis of changes in the states of cells and their interactions can be carried out. In the extreme case, a cell is represented by an individual atom (e.g., *Ising model*). The organization of this modeling system can be either linear, two-dimensional, or three-dimensional. A conformity between the computational process and the processes occurring in solids is provided by means of both the programs and the apparatus. The former are universal (because the program approach allows one to model any physical process using any computer) but they do increase the duration of each step of the modeling. The latter (i.e. the apparatus part) specifies the computer for each individual problem, and provides for its rapid solution. In the extreme case, the relations between the processors of the modeling system are isomorphous to the relations in the modeled solid; and the productivity increases as the number of processors increases. There exist modeling systems (so-called distributed matrix processors, "hypercube") consisting of 10^5 (in order of magnitude) processors. Theoretically it is possible to produce such systems containing up to 10^{12} modeling processor units with a clock period to 10^{-9} s, which permit the solution of a large part of the problems of solid state physics in real time. Further progress depends, in principle, on creating new solid and molecular modeling systems that implement the probability functions at the quantum level.

In terms of the complexity of the modeling, the problems of solid state physics could be classified as follows: (i) *crystallization* and recrystallization in the absence of continuous *defects*; (ii) the same processes with the participation of dislocations and other continuous imperfect defects; (iii) electronic, thermal, and diffusion (see *Diffusion*) processes with changes in the location of internal boundaries (e.g., *semiconductor junctions*); (iv) non-structural *phase transitions*; (v) first-order *structural phase transitions* with alteration of the lattice symmetry; (vi) second-order structural phase transitions; (vii) modeling processes on a "rigid" lattice with a quantum-mechanics analysis of elementary events; (viii) subsequent quantum-mechanical analysis of processes in solids with changes of lattice symmetry. At present, models of the former four types are really quite well-developed, and the efficacy of the latter four types depends on future progress.

COMPUTER SIMULATION, computer modeling

Investigation of structures and phenomena by their information simulation on computers. Direct simulation is achieved by an organization of elements of the computer and the topology of communications between them such that the properties of data-regulated processes are similar to the corresponding characteristics of the object. In this case, the maximum performance of the simulation process is achieved, but each structure and each type of phenomena need a special machine (or its complicated adjustment). Alternatively simulation is carried out by numerical treatment of mathematical equations describing the object at various boundary and initial conditions. The similarity to the object is achieved only by means of software. The hardware of simulation is universal but its efficiency (even at the maximum speed of the elements) is rather limited. Historically, the first simulation was performed by analog machines for physical processes formally described by the same equations as the phenomena under study. Now there are analog machines based on various physical phenomena (transients in electric circuits, optical and acoustic waves, processes in solids and on their surface, etc.). They allow one to carry out a parallel multichannel simulation in the real-time mode of a physical process achieving an effective capacity up to and even more than 10^{10}–10^{11} operations per second. The physical structure and manufacturing techniques of such devices are relatively simple, and the energy consumption is minimal. However, besides the specialization, their shortcoming is low accuracy of calculations and instability of the simulation process. Computer simulation by numerical solution of the equations describing a physical object possesses quite opposite features. There are two approaches to computer simulation combining the advantages of digital and analog principles. The first is based on multiprocessor computers (see *Computers in solid state physics*). In this case, a similarity is reached between the organization of flows of the internal information exchange in a computer and mass–energy flows in a modeled system. The performance of computer simulation increases by several orders; under optimal conditions, proportional to the number of processors in the system. Another approach is based on joining digital and analog processors in one device (so-called *hybrid devices*) in which, at each stage of simulation, the optimal distribution of data fluxes between digital and analog subsystems is realized. The analog subsystem provides the maximum speed of analysis of special problems arising during simulation, and the digital one provides stability and universality of simulation, and the accuracy of the calculated data. Both approaches (digital multiprocessing and digital–analog hybrid) are based on the development of *microelectronics* and an increased integration of computer elements. In prospect, supermultiprocessor (up to 10^9–10^{11} modeling cells) hybrid devices will unite the advantages of all the above-mentioned approaches to computer simulation.

CONCENTRATION

The relative content of a component in a system (solution, mixture, melt). There are many different ways of expressing this: mass, moles, volume fraction, relation of individual mass or individual moles to the total. These concentrations can be expressed as fractions, as mass or volume percentages (%), or for high dilutions as, e.g., parts per million (ppm) or parts per billion (ppb).

Other ways for expressing the amount of a component and the amount of a total system are: *mass concentration* – the relation of the component mass to the system volume (kg/m^3, more often g/l); *molar concentration* (*molarity*) – the relation of the component quantity in moles to the system volume (mol/m^3, more often mol/l); *molal concentration* (*molality*) which is the relation of the component quantity in moles to the solvent mass (mol/kg); *normal concentration* (*normality*, n) – the number of gram-equivalents of the substance in one liter of solution; *titer* – the number grams of the component dissolved in 1 ml of the solution. The last two quantities are common in analytical chemistry. Furthermore, in the literature concentration is sometimes expressed as mass of the component being dissolved in 100 g of solvent, or as number of moles of the component in 1000 moles of solvent. The SI system recommends using only mass and molar concentrations.

CONCENTRATION-DEPENDENT PHASE TRANSITION

A *phase transition* characterized by the spontaneous onset of long-range order at $T = 0$ (see *Long-range and short-range order*) at a change in *concentration* of the repetative units forming the crystal. Depending on the physical nature of the associated *order parameter* that arises, such a phase transition can be ferromagnetic, ferroelectric, structural, etc. Let us consider for definiteness the ferromagnetic concentration-dependent phase transition in a system in which magnetic atoms replace nonmagnetic ones in a random fashion. Because the *exchange interaction* rapidly decreases with distance it is possible to consider as a first approximation that only nearest neighbor magnetic atoms are interacting. At absolute zero temperature all magnetic moments coupled by exchange are oriented parallel. If the concentration x of magnetic atoms is small they form *clusters* isolated from each other with dimensions which do not vary with an increase of the crystal volume (*finite clusters*). The average magnetic moment of the whole crystal is equal to zero in this case since various clusters have oppositely directed

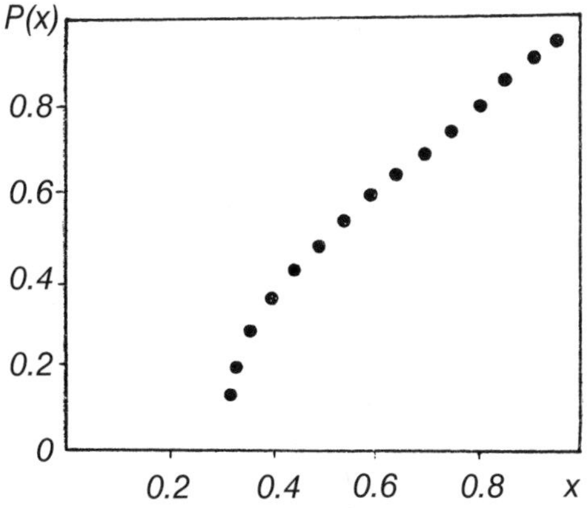

The dependence of the probability $P(x)$ for the formation of an infinite cluster on the concentration of magnetic atoms x for a simple cubic lattice.

moments. With an increase in x the characteristic size of terminal clusters increases, and at a certain critical concentration x_C a cluster forms whose number of particles is proportional to the system volume, the so-called *infinite cluster*. The magnetic moment of a crystal $M(x)$ for $x > x_c$ differs from zero, being equal to $M(x) = M(1)P(x)$, where $M(1)$ is the magnetic moment of the ordered crystal, and $P(x)$ is the probability of the formation of an infinite cluster (one occupation per lattice site). Thus, at $x = x_c$ and $T = 0$ a phase transition from the paramagnetic state to the ferromagnetic state takes place, an effect closely related to percolation (see *Percolation theory*).

Examples of these phase transitions in materials with interacting nearest neighbors are: *semimagnetic semiconductors* $Cd_{1-x}Mn_xTe$, $Cd_{1-x}Mn_xSe$, *insulators* $SnTe_xSe_{1-x}$, etc. The theoretical dependence of $P(x)$ on the concentration x of magnetic ions for a simple cubic lattice ($x_c = 0.31$) is given in the figure. At a finite temperature T the phase transition persists, and its transition temperature $T_c(x)$ approximately satisfies the condition $T_c(x) = P(x)T_c(1)$, where $T_c(1)$ is the temperature of the usual phase transition in an ordered crystal.

CONDENSATION

A transition of a material from the gaseous state to a condensed (liquid or solid) one due to cooling or compression. Condensation is a *first-order phase transition*. Vapor condensation is only possible at subcritical conditions. Condensation to the liquid phase occurs if the temperature and pressure are above their *triple point* values, and it occurs directly into the solid phase if they are lower than the triple point values. The quantity of heat released at condensation is the same as that which was supplied at *evaporation* or *sublimation*, processes which are the inverse of condensation. At constant temperature condensation occurs until a vapor–condensed phase equilibrium is established which is described by the Clausius–Clapeyron equation. If the saturated vapor is a mixture of gases then this equation holds for each individual component.

For the determination of condensed phases that are in equilibrium with a vapor of complex composition it is possible to make use of the complete *phase diagrams*. During condensation metastable phases are often formed in which the system can persist for a long time before reverting back to a more stable (at the given conditions) state (steel hardening, amorphous metal formation on cooling). Condensation throughout a volume containing vapor and on a surface are distinguished. During condensation on a cooled surface the velocity of condensing molecules has an important significance which depends on several things: the difference between the vapor and the surface temperatures, heat exchange processes, condensation surface (solid or liquid, both continuous and dispersed), the presence of noncondensible gases, etc. The mere presence of a supersaturated vapor is not sufficient for inducing condensation within the vapor volume in the absence of condensed phase particles; it is necessary to have condensation centers available (impurity particles, ions). The formation of an equilibrium nucleus of a new (condensed) phase can be unstable, and is describable by the *Kelvin equation*. The condensation of a vapor into a crystalline phase occurs during *crystallization*, *epitaxy*, or physical and chemical precipitation from the vapor phase.

CONDENSED MATTER INTERFACE

A layer which separates two materials of different chemical composition that are in contact. One can distinguish geometrical and physical condensed matter interfaces. The concept of a *geometrical interface* is used in describing the physical properties of two bodies in contact when the layer thickness is very small compared to the bulk dimensions of these bodies so it has a negligible influence on the bulk properties. A *physical condensed matter interface* is a transition layer of finite thickness between bulk materials whose chemical composition differs from the chemical composition of the media in contact, and hence it can influence their properties. One can distinguish *sharp and smooth interfaces*. A transition layer between contacting media with a thickness the order of magnitude of an interatomic distance is called sharp, and it can be obtained by the molecular *epitaxy* method. A transition layer much thicker than the atomic scale is called smooth.

CONDENSED STATE OF MATTER

Solid and liquid states of matter. The condensed state is characterized by high matter *density* in comparison with a gas, and by a rather low speed of motion of its component *atoms*. The regular arrangement of atoms or molecules on a three-dimensional lattice with long-range order (see *Long-range and short-range order*) determines the *crystal structure* that corresponds to an equilibrium state. There is only short-range ordering in the arrangement of atoms in a liquid. The structure of amorphous solid bodies is equivalent to that of a liquid; and from the thermodynamic point of view the *amorphous state* represents a supercooled liquid with a very long *relaxation time* for transition to the equilibrium crystal state. Macroscopic liquids and amorphous materials are isotropic. Materials intermediate between crystal and amorphous solids are *polymers*, *liquid crystals*, *quasi-crystals*, and the *gaseous crystalline state*. Polymers consist of periodic linear chains at sites where particular atom groupings are allocated. The *chain molecules* can be bound together in different ways. The parallel arrangement of polymeric chains defines the polymer anisotropy (see *Anisotropic medium*). The properties of liquid crystals are determined by the ordered orientational arrangement of their molecules which are often elongated or disk like in shape, and this results in the *anisotropy of liquid crystals*. There is only short-range order in the spatial arrangement of the molecule centers.

CONDENSON (fr. Lat. *condensatio*, densification, thickening)

A self-localized state of an electron in an elastically deformed medium, or a hypothetical *quasiparticle* in a *covalent crystal* in which all atomic bonds are covalent. An electron induces dipoles in neutral surrounding atoms of a medium, and then by its electric field attracts them. As a result of small atom displacements, an increased density region is formed which represents a potential well for the electron (see *Electrostriction*) where it is localized. A macroscopic consideration of a condenson which takes into account a *strain potential* shows that if self-localized condenson states are at all possible in homopolar crystals, they will have a small radius. A macroscopic condenson can be realized in three-dimensional homopolar crystals with a "loose" crystal lattice due to weak medium anharmonicities. In one-dimensional crystals a condenson can also exist in an harmonic elastic medium. A similar situation is realized, in particular, with the application of a *quantizing magnetic field* to a crystal. In crystals with negative *electron affinity* energy (He, Ne) a reciprocal type of electron *self-localization* is possible with the formation of a lattice disperse region or even the appearance of a small cavity inside which the electron cloud ($|\Psi|^2$) is mainly localized.

CONDON APPROXIMATION

See *Franck–Condon principle*.

CONDUCTANCE, ELECTRICAL

See *Electrical conductance*.

CONDUCTANCE, HIGH-FREQUENCY

See *High-frequency conductance*.

CONDUCTION BAND

An *energy band* of a crystal comprising electronic states with energies higher than those of filled *valence bands*. Weak external effects on the conduction electrons can cause a change in their average linear momentum, i.e. induce the appearance of current.

To create current carriers in the conduction band of intrinsic semiconductors and *insulators* it is necessary to overcome an energy barrier, with a minimum value called the *fundamental gap* (see *Semiconductors*). The gap ranges from about 0.1 to several electronvolts in semiconductors, and can reach about 10 eV in insulators. Typical physical mechanisms of current carrier generation are optical and thermal excitation in semiconductors, and ion collisions as well as electrical breakdown in insulators.

The parameters that describe the conduction band are the same as those for the valence band: width, *effective mass*, presence of *Van Hove singularities* and their symmetry, and spin–orbital splitting (see *Spin–orbit interaction*). Methods for studying the conduction band include *light absorption, cyclotron resonance, photoeffect, Hall effect* (see *Galvanomagnetic effects*), etc.

CONDUCTION ELECTRONS

Elementary excitations (*quasi-particles*) of the electron *Fermi liquid* of metals which carry electric charge in transport. Every quasi-particle has the same charge and spin as a free electron (exceptions are double charges ($2e$) in superconductors, and fractional charges in the *quantum Hall effect*). In the absence of external fields, the conduction electrons are characterized by the *quasi-momentum* $\boldsymbol{p}$ and the energy band number s. Their energy $\varepsilon_s(\boldsymbol{p})$ is periodic (*dispersion law*) in respect to $\boldsymbol{p}$ with a period $2\pi\hbar\boldsymbol{b}$ where $\boldsymbol{b}$ is a reciprocal lattice vector. A deviation of $\varepsilon_s(\boldsymbol{p})$ from the quadratic isotropic dispersion law of free electrons serves as a source of numerous diverse effects which are particularly pronounced in strong magnetic fields. At a temperature T much less than the Fermi temperature $T_F = \varepsilon_F/k_B$ (here ε_F is the *Fermi energy*, and for a typical *metal*, $T_F \sim 10^5$ K), the conduction electrons form a weakly-nonideal gas. A specific contribution of the electron–electron interaction to the inverse lifetime of the conduction electrons with $\xi \equiv |\varepsilon_s(\boldsymbol{p}) - \varepsilon_F| \ll \varepsilon_F$ is of the order of $\nu = (\varepsilon_F/\hbar)(\xi/\varepsilon_F)^2$. At $\xi \to 0$, the ratio $\hbar\nu(\xi)/\xi \to 0$ which means that conduction electrons at low levels of excitation $\xi \ll \varepsilon_F$ behave as long-lived quasi-particles. This is the basis for the *Landau theory of Fermi liquid* describing the electronic properties of normal metals.

CONDUCTIVE CHANNEL, SURFACE

See *Surface conductive channel*.

CONDUCTIVITY

See *Electrical conductivity, Effective electrical conductivity, Electronic conductivity. Fröhlich conductivity, Hopping conductivity, Intrinsic conductivity, Ionic conductivity, Surface electrical conductivity*.

CONDUCTIVITY SELF-COMPENSATION

A process of *electrical conductivity* compensation in doped *semiconductors* when additional compensating defects are formed as a result of introducing free current carriers (*conduction electrons, holes*) into the semiconductor. The formation of the inherent compensating defect during the self-compensation may be stimulated by the introduction of impurity donor or acceptor atoms, as well as by *injection* of carriers through the contact or by their formation under the effect of light on an intrinsic semiconductor. Self-compensation of conductivity is effective in the case when some of the inherent defect types in semiconductors have sufficiently deep acceptor levels (for n-type semiconductors) or analogous donor levels (for p-type semiconductors) that the energy outlay W for thermal formation of the defects of the given type is almost completely compensated by the energy gain E in the trapping of a free carrier to the level of the forming defect. As the defect formation leads to the growth of crystal entropy S, at a sufficiently high temperature T and small energy difference $(W - E)$ the minimum free energy of the crystal $F = E - TS$ (E is the Helmholtz energy of the system) may be attained at defect concentrations N_D comparable to the concentration of the carriers being introduced. Here, in the state of thermodynamic equilibrium, the actual concentration of current carriers may be significantly lower than the concentration of impurities of donor and acceptor types with *shallow levels* (low ionization energies). It is not possible to obtain high conductivity over the entire series of semiconductor materials (e.g., n-Cu_2O, p-CdS, alkali-halide crystals, etc.). The kinetics of self-compensation of conductivity is determined by the probability of the elementary act of defect formation, stimulated by the trapping of the charge carrier at a *deep level* during the course of defect generation. If conductivity self-compensation occurs under conditions involving the exchange of atoms between the crystal and its environment (high temperatures, pressures), then *vacancies* or impurity *interstitial atoms* may be the compensating defects.

CONDUCTIVITY, THERMAL

See *Thermal conductivity* and *Thermal conductivity coefficient.*

CONDUCTIVITY, THERMALLY STIMULATED

See *Thermally stimulated conductivity.*

CONDUCTORS, electrical conductors

Materials with low specific electrical resistance (resistivity, usually below $10^{-6}\ \Omega\cdot$m). Conductors include, for example, *metals*, metal *alloys*, *semimetals* (including *graphite*). The *current carriers* (charge carriers) in solid conductors are ordinarily *conduction electrons.* Typical for a conductor in its normal (not superconducting) state is the linear dependence of the current density on the strength of the applied electric field (*Ohm's law*). Identifying conductors as a separate class of materials can be somewhat subjective since conductivity depends on many factors, including temperature. For example, many (poor) conductors transform to a superconducting state at extremely low temperatures.

CONFIGURATION INTERACTION

Interaction of multiparticle states (configurations) described by wave functions constructed from single-particle states (see *Slater determinant*). The matrix elements of this interaction operator were found by J.C. Slater (1929) for orthogonal, and by P.O. Löwdin (1955) for nonorthogonal single-electron functions. The configuration interaction decreases the total energy of a fermion system by the so-called *correlation energy.* It is necessary to take the configuration interaction into account for systems with uncompensated electronic spin (atoms, ions, radicals, impurities with partially filled d- and f-shells in crystals, etc.), that is systems with unfilled shells. From considering a set of multiparticle configurations as the basis for solving the multielectronic Schrödinger equation, and utilizing its linearity, one arrives at the method of *second quantization* which is a more convenient one for treating a large number of particles. Applying the configuration interaction in solids results in the narrowing of the electronic bands obtained from solutions of the Hartree–Fock equations (see *Hartree–Fock method*).

CONFOCAL DOMAIN, focal conic domain

A domain *defect* in a structural liquid crystal (*smectic liquid crystals* A, C, *cholesteric liquid crystals*) which has the form of a cone of revolution. Because of the different orientations of molecules on the liquid crystal boundary the layers inside the confocal domain are curved with their shapes determined by the condition of maintaining an invariable width of each layer. The layers with *Dupin cyclides* fulfill this condition, as shown in Figs. 1 and 2. This results in the appearance in

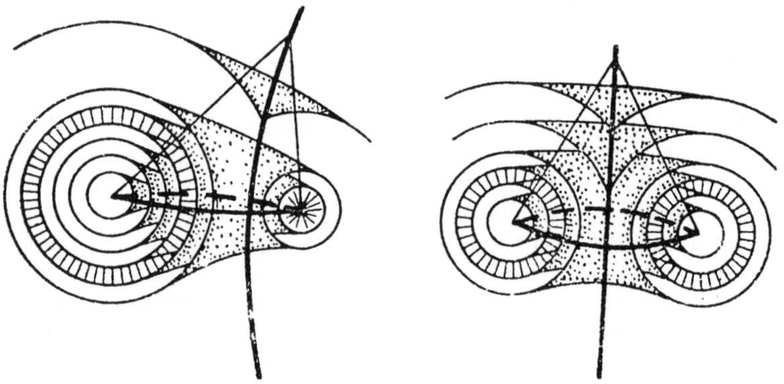

Fig. 1. Configuration of smectic layers in a confocal domain of a general form (left) and of a toroidal form (right).

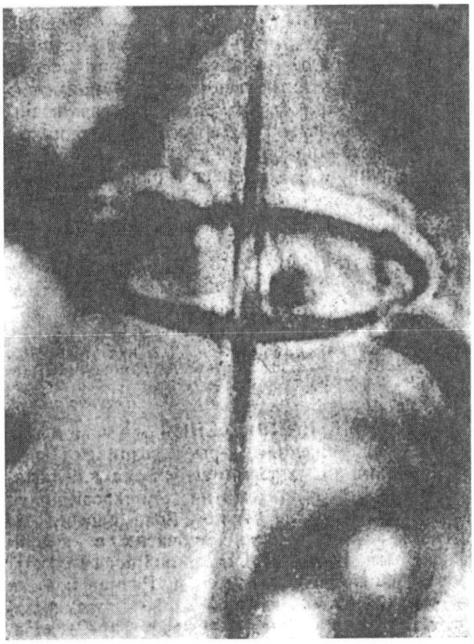

Fig. 2. Micrograph of an isolated domain in a type A smectic liquid crystal.

the confocal domain of a pair of defect lines on which the layer wrapping undergoes a discontinuity. More often one observes a confocal domain with a pair of confocal ellipses and hyperbolae situated in mutually perpendicular planes. The ellipse is located on the perimeter of the confocal domain base, and the hyperbola passes through the confocal domain vertex. In a domain with a torus configuration of layers the ellipse degenerates into a circumference and the hyperbola into a line. Confocal domains with defects of a parabolic type are known as well.

The confocal domain is the main element for extensively filling the layers in a liquid crystal; in particular they form *confocal textures* which are most commonly observed in smectics and cholesterics. At the filling, a system of adjacent confocal domains in contact with each other and of sequentially decreasing sizes is formed; the smaller *domains* being inserted into the gaps between the larger ones. The minimal confocal domain size is determined by the conditions of molecular orientation near the liquid crystal surface (see *Boundary effects in liquid crystals*). Interstitial spaces free of confocal domains are filled with layers of a flat or spherical shape. The only discontinuities in the layer wrapping for this filling are confocal pairs.

Confocal domain formation plays an important role in *phase transitions* and *relaxation* of mechanical stresses. Polarizing microscopic analysis of confocal domain structure in smectic textures has formed the basis for clarifying the one-dimensional stratified structure of these media.

CONFORMATIONS (fr. Lat. *conformatio*, shape, arrangement)

Different spatial configurations of a polyatomic molecule which can transform between each other without breaking valence bonds, by executing rotational displacements around single bonds, or by changing bond angles. For example, the ethane molecule C_2H_6 which has a C_3 symmetry axis joining the carbon atoms, changes its energy by twisting one CH_3 group relative to the other in an approximately sinusoidal manner with energy maxima corresponding to D_{3h} point group symmetry when the C–H bonds of the two CH_3 groups are aligned, and energy minima corresponding to D_{3d} symmetry when they are staggered. An example of conformations involving *crystals* is *twinning of crystals* in calcite $CaCO_3$ wherein it is the possible to transform the unit cell from the form of a skewed prism into a direct prism, and then into a plane-reflected one, via a *shear strain*. The conformation concept is used also in a quasi-molecular definition of impurity-defect complexes and *clusters*, which are models of localized regions of a *solid*. Conformation analysis has been widely applied to cluster models for the determination of macrostructures of random semiconductor (see *Disordered solids*) surfaces and boundary layers. The local minima on a molecular potential energy surface correspond to stable conformations. The transition between local minima is carried out along a curve designated by K. Fukui (1970) as the *path of minimum energy*. Regions of multidimensional potential energy surfaces, so-called *conformation maps*, are useful in practice. In conformation analysis the *conformer* concept is introduced. It represents a set of conformations which form in the space of independent geometrical variables a continuous region including coordinates of the minimum *adiabatic potential*, with the region under consideration laying below the lowest adjacent first-order saddle point.

CONICAL REFRACTION OF ACOUSTIC WAVES

A special type of *acoustic wave refraction* on a flat *crystal* face oriented normally to an *acoustic axis*, i.e. in a direction along which a great number of quasi-transverse waves with identical velocity can propagate. An exception is the acoustic axis being coincident with a 4th or 6th order symmetry axis. Conical refraction was observed for the first time during the propagation of acoustic waves in *nickel* crystals (J. de Klerk, M.J.P. Musgrave, 1955) and in dihydroammonium phosphate crystals (K.S. Aleksandrov et al., 1963).

Interior and exterior conical refraction are distinguished. For the *interior conical refraction* the whole cone of acoustic beams corresponds to one wave normal, while for the *external conical refraction* the whole cone of wave normals corresponds to one beam direction. The angular opening is determined by the crystal *elastic moduli* and can reach tens of degrees as distinct from *conical refraction of light waves.* For example, it is 61°30′ for a calcite crystal. The special case of conical refraction in the direction of an acoustic axis located in the plane of elastic symmetry of a hexagonal crystal, when the refracted beams are distributed in a fan with a flat section, was predicted by A.G. Khatkevich (1962). Under the conditions of conical refraction several effects have been observed: the reflection of a linearly polarized acoustic wave from a free crystal face parallel to the wave vector of an incident wave accompanied by a 90° rotation of its polarization plane, acoustic wave refraction on the internal surface of a twinned crystal (see *Twinning of crystals*), in quartz, and in a number of other ways. At ultrahigh frequencies the observation of conical refraction is complicated by acoustical activity phenomena (see *Rotation of acoustic wave polarization plane*).

CONICAL REFRACTION OF LIGHT WAVES

A special type *refraction of light* in optically *biaxial crystals* observed during propagation in the direction of any binormal *optical axis* (see *Crystal optics*). In conical refraction the thin polarized light beam which is incident on a face of the plane-parallel lamina makes up the beam in the form of a full cone whose vertex is located at the point of the beam incidence on the face. Thus linearly polarized waves with continuously varying (in the range of 180°) polarization azimuth propagate in directions of cone generators. Conical refraction was predicted by W. Hamilton (1832) and observed by H. Lloyd (1833). Internal and external conical refraction can be distinguished. *Internal conical refraction* occurs in light beam transmission in the binormal direction, in this case the vertex is located on the forward lamina face; on emerging from the lamina the beam makes up a hollow cylinder. *External conical refraction* occurs for light transmission in the biradial direction, in this case the cone vertex is located on the output lamina face. If the light beam direction does not coincide with the optical axis direction then ordinary *birefringence* occurs.

CONOSCOPY (fr. Gr. κωνος, a cone, and σκοπεω, am looking, am observing)

A method of investigating optical properties of oriented anisotropic media (birefringent *crystals*) with the help of interference figures (so-called *conoscopy figures*) observed in convergent polarized light beams. In conoscopy polarized microscopes, conoscopes, and gas lasers with polarized devices are used.

A conoscopy figure is obtained for light beams convergent on a birefringent crystal (*uniaxial crystal* or *biaxial crystal*) using a crossed polarizer and analyzer. The type of the figure depends on the number of *optical axes of a crystal*, the angles between them, the presence or absence of optical activity (see *Rotation of light polarization plane*), crystal lamina orientation or their combination,

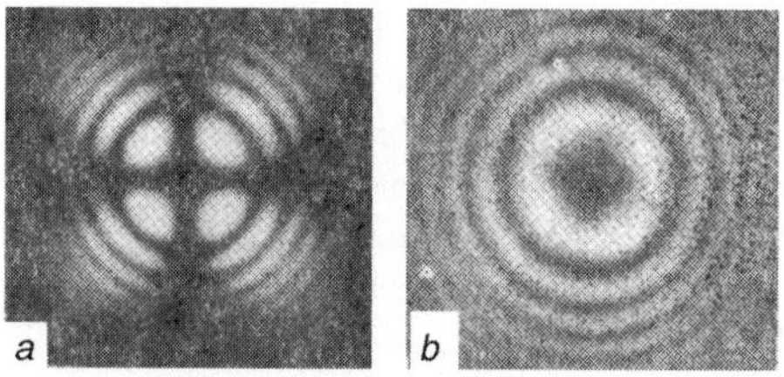

Conoscopy figures obtained with light beams converging on a single crystal with crossed polarizer and analyzer: (a) for linearly polarized light, and (b) for circularly polarized light.

and *polarization of light*. For linearly polarized light, conoscopy figures consist of isochromates and isogyres as shown in Fig. (a). The *isochromates* form a family of concentric interference rings or bands, each of which corresponds to a particlar phase difference of the interfering waves; in white light the bands have a rainbow coloring. The *isogyres* are mutually perpendicular brush-like dark bands corresponding to directions of light transmission through the crystal in which the linear polarization is not affected. With circularly polarized light the isogyres are absent, as shown in Fig. (b).

CONSERVATION LAWS of quasiparticles

Regularities whereby the values of some physical parameters of materials do not change in time. Most conservation laws existing in physics (energy, momentum, angular momentum, etc.) are applicable to solid state physics, provided the necessary conditions hold.

The states of elementary excitations (*quasi-particles*) in crystals are characterized by a vector, similar to the linear momentum, the *quasi-momentum* $\hbar\boldsymbol{k}$. It is defined to within an arbitrary vector of the *reciprocal lattice* ($\hbar\boldsymbol{k}$ and $\hbar\boldsymbol{k}'$ describe physically equivalent states, if $\boldsymbol{k} - \boldsymbol{k}' = 2\pi\boldsymbol{b}$, where $\boldsymbol{b}$ is a vector of the reciprocal lattice). In the interaction of elementary excitations (the collision of quasi-particles), the *law of conservation of quasi-momentum*, similar to that of linear momentum, holds:

$$\sum_i \hbar\boldsymbol{k}_i - \sum_i \hbar\boldsymbol{k}'_i = 2\pi\hbar\boldsymbol{b}.$$

Here $\hbar\boldsymbol{k}_i$ and $\hbar\boldsymbol{k}'_i$ are the quasi-momenta of quasi-particles before and after a collision. To avoid ambiguity, the quasi-momenta of all colliding quasi-particles (before and after a collision) should be "located" in the same *Brillouin zone*. Therefore, if the number of colliding quasi-particles is not large, the vector of the reciprocal lattice, involved in the conservation law of quasi-momentum, does not exceed one or two elementary vectors of the reciprocal lattice. Collisions with $\boldsymbol{b} = 0$ are referred to as *normal collisions* (emphasizing the satisfaction of the law of linear momentum conservation), while collisions with $\boldsymbol{b} \neq 0$ are called *Umklapp processes* (see Fig.). A periodic *crystal lattice* is a space where quasi-particles exist. The law of conservation of quasi-momentum is a corollary of the symmetry (periodicity) of this space, just as the law of conservation of linear momentum reflects the symmetry (uniformity) of empty space. Related to the uniformity of time is the *law of conservation of quasi-particle energy* in a crystal

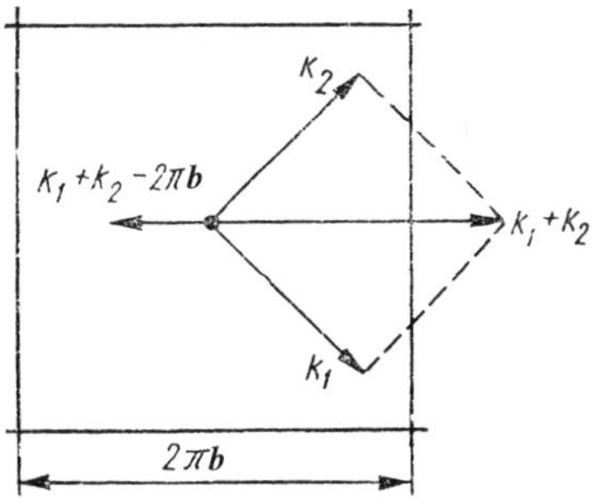

Schematic diagram of an Umklapp process.

$$\sum_i \varepsilon(k_i) = \sum_i \varepsilon(k'_i),$$

where $\varepsilon(k_i)$ and $\varepsilon(k'_i)$ are energies of quasi-particles before and after a collision, respectively. Related to the law of conservation of angular momentum is the symmetry of the *stress tensor* in nonpolarized media.

The essential information on the interaction of quasi-particles (as well as ordinary particles) can be obtained on the basis of a conservation law analysis of their kinematics. In particular, this is how *selection rules* are established that forbid processes where the conservation laws do not hold. The complexity of the *dispersion laws* of quasi-particles and the presence of Umklapp processes make the kinematics of quasi-particle collisions very complicated: mutual particle transformations, creation and annihilation, as well as breakup and fusion of particles, are possible. Even in the case of an elastic collision of quasi-particles (e.g., electron scattering by an *impurity atom* in a crystal), some unusual (for particles) situations are possible, e.g., the transition from one energy band (minimum) to another band (minimum), or

Table 1. Criteria for materials of the first wall of a tokamak

Criteria	Materials
I. Radiation damages and lifetime	
(a) swelling	Ti, V, Mo, stainless steel
(b) blistering	C, Nb, V, Ti, stainless steel
(c) surface atomizing	V, Ti, Al, C
II. Compatibility with cooler and tritium	
(a) Li	Ti, V, Nb, Mo, stainless steel
(b) He	Stainless steel, Ti, Mo, Al, C
(c) water	Stainless steel, Al, Ti
(d) tritium	Mo, Al, stainless steel
III. Mechanical and thermal properties	
(a) yield point	Mo, Nb, V, Ti, stainless steel
(b) ultimate stress limit σ_T	Stainless steel, Ti, Al
(c) creep limit	Mo, V, Ti, stainless steel
(d) thermal stress parameter M	Mo, Al, Nb, V
IV. Processibility and welding	Stainless steel, Al, Ti
V. Developed industry and data base	Stainless steel, Al, Ti, C
VI. Cost	C, Al, stainless steel, Ti
VII. Induced radioactivity	V, C, Ti, Al

scattering when the particles scattered in a given direction have different quasi-momenta.

CONSTRUCTION MATERIALS

Materials for production of engine parts, building constructions, etc., for which shape maintenance under mechanical forces is necessary. *Metals* and their *alloys* such as steel, natural and synthetic *polymers*, and *ceramics* are used as construction materials. Composite construction materials play a special role, and for usage in aviation and astronautics a reduction in weight without sacrificing strength is important. The main quality factors of these materials are *strength*, *elastic modulus*, *hardness*, *impact strength*, and temperature dependence of physico-mechanical properties.

CONSTRUCTION MATERIALS FOR THERMONUCLEAR REACTORS

Materials of thermonuclear reactor (TR) elements in contact with the plasma: first wall, aperture, and *divertor*. The other parts of the reactor construction are exposed mainly to neutron irradiation so the choice of materials for them is less complicated. In the advanced TR design called the *tokamak* the main factors which affect the first wall are: (1) neutron flux with energy up to 14 MeV and power up to 2.5 MW/m^2, (2) heat flux with power up to 0.25 MW/m^2, (3) flux of neutral *atoms* of hydrogen isotopes D and T generated at the ion recharge, $j \sim 10^{20}$–$10^{21}\,\text{m}^{-2}\cdot\text{s}^{-1}$. The main factors which affect the aperture and divertor receiving plates are (1) flow of D^+ and T^+ ions and heavy impurities with energies up to 20–50 eV, $j \sim 10^{23}\,\text{m}^{-2}\cdot\text{s}^{-1}$, (2) heat flux equal to 10–20 MW/m^2.

In addition in a contingency the first wall, aperture, and divertor should be able to sustain a discharge disruption and release of all the energy accumulated in the plasma (1–2 MJ/m^2) during a time of 20 ms. For uncontrollable modes of TR operation the aperture and divertor can erode because of the development of unipolar arcs, i.e. electrical discharges between the plasma and plasma trap chamber.

For design and economy reasons the first wall cannot be changed, and should operate in an industrial reactor for ~30 years. Aperture and divertor receiving plates are changeable. Table 1 cites main criteria for choice of materials for the first wall of a tokamak listed according to their priority. The thermal stress parameter M

$$M = \frac{2\sigma_T k(1-\nu)}{\alpha E},$$

where k is the thermal conductivity, α is the thermal expansion, E is *Young's modulus*, and ν is the *Poisson ratio*, characterizes the thermal stress from the temperature gradient: the greater the M value the greater is the allowable wall thickness. The tokamak operates in the pulsed mode, and the international reactor INTOR pulse lasts 200 s with a 20 s pause between. Thermal load cycling promotes the aging of materials. The most important features for the material choice for the aperture and divertor receiving plates are thermal and chemical properties, and surface erosion. For a hybrid reactor combining a thermonuclear and fission reactor which uses thermonuclear neutrons, construction materials will have less plasma applied loads because the high power of thermonuclear energy is not required, but the materials are subjected to radiation exposures typical of those from atomic fission reactors.

CONSTRUCTION MATERIAL STRENGTH

The strength of construction materials or their models having design shapes and geometrical sizes typical of actual construction elements. In contrast to the material *strength* which is evaluated experimentally by testing relatively small material samples on devices of simplified shape, the constructional strength is evaluated by testing material samples with geometrical shapes and sizes that are typical of the majority of actual parts of engines or other devices. Values of constructional strength are much lower than those of material strength which is generally obtained for simplified laboratory samples. This difference arises mainly from the design shapes which often involve sharp transitions from one section size to another, the presence of holes which induce some *stress concentration*, and other factors. In addition, the surface quality of actual samples can differ considerably from the surface of laboratory ones because the handling technology of actual device production, as a rule, differs from the technology used for preparing simplified laboratory samples. The strength indices of actual samples are strongly influenced by their production technology, especially by *welding* which causes inhomogeneities in the material, and also the tangent stress due to so-called welding stresses.

Constructional strength is appreciably degraded by the decline of performance factors arising from exposure to repeatedly changing loads. There are always a number of factors under actual working conditions which cause the concentration of stress to be the main origin of the appearance of fatigue *cracks* on exposure to repeatedly changing stresses.

CONSTRUCTION PLASTICS

Plastics (see *Polymeric materials*) designed for the production of parts and components capable of keeping their shape during exposure to mechanical loads. Polyolefines, polyformaldehydes, polyamides, fluoroplasts, polyarylates, etc. are used in various industries as constructional plastics. The main processing methods for converting these plastics into components are pressure die *casting*, *hydroextrusion*, and *pressing*. In most cases constructional plastics are multicomponent systems: antioxidants, lighting stabilizing agents, plasticizing agents, oilers, bulking agents, etc., which are loaded as additional binders that impart specific properties (e.g., *fiber glass*, *carbon*, *boron nitride*, fire-retardant additives, etc.).

CONTACT ANGLE

Angle between surfaces of a fluid and a solid body (or another fluid) in contact. The contact angle θ is counted from the fluid. The equilibrium contact angle is referred to as the *edge angle* and is denoted by θ_0. During contact formation the angle varies from $\theta = \pi$ up to $\theta \approx \theta_0$ during 0.02–0.03 s. On rough solid surfaces the contact angle usually decreases, so that for $\theta < \pi/2$, $\cos\theta = \gamma\cos\theta_0$, where γ is the surface roughness coefficient. The contact angles for fluid inflow on a substrate $\theta_{\rm in}$, and for outflow $\theta_{\rm ex}$, differ from θ_0. This is caused by the roughness and heterogeneity of the actual surface. For *substrate* backings with a high *surface energy* the property of "*autophobia*" is possible in contact with polar fluids: the fluids on the substrate generate adsorption films on which they form a contact angle $\theta > 0$. It is known that $\theta_{\rm in} > \theta_0 > \theta_{\rm ex}$, and this effect is referred to as *wetting hysteresis*. With increasing temperature $\theta_{in} \to \theta_0$ and $\theta_{ex} \to \theta_0$, so the hysteresis decreases (see *Wetting*).

CONTACT CORROSION, galvanic corrosion

A type of electrochemical *local corrosion* resulting from the contact of two or more metallic materials with different initial *corrosion* potentials, situated in a conductive corrosion environment. The contact corrosion intensity is determined by a polarization cathode and anode resistance, the ratio of their dimensions, and decreases with increased distance from the contact point. The *metal* quantity being dissolved because of contact corrosion is equivalent to the current. Methods for protecting against contact corrosion consist of sensible equipment designing, eliminating or isolating contacts of metals having much different corrosion potentials, using protectors, etc.

CONTACT POTENTIAL DIFFERENCE

Difference of electric potentials between surfaces of conductors brought into contact at thermodynamic equilibrium. It is defined by the difference in the *work functions* of the media in contact divided by the electron charge. The contact potential difference arises from the requirement of electrochemical potential equality for media in contact; it is established because of electron exchange between the conductors. Thus the surfaces of both media become charged: negative for the conductor with the greater work function, and positive for the other. The width of the boundary (contact) layer in which the electric field differs from zero is defined by the shielding distance in the conductor, with values 10^{-10}–10^{-9} m for metals and up to 10^{-6} m for semiconductors. The potential difference can reach several volts, and depends on the physico-chemical condition of the boundary region. In multiple layer heterophase structures representing a chain of conductor contacts the overall potential difference does not depend on the intermediate links, but equals the potential difference of the extreme outer conductors. This potential difference is important for the formation of electrovacuum devices and solid contact structures: *semiconductor junctions, heterogeneous structures, metal–semiconductor junctions*, etc.

CONTACT PROBLEM OF ELASTICITY THEORY

Determination of the *strain* and *stress* distribution in a system of solid bodies having a general contact surface. For the case of two elastic bodies in three-dimensional space this problem was set up and solved first by H. Hertz (1882) by considering that the contact area consists of small, strain-free surfaces around the contact point referred to as second-order surfaces. With the help of an electrostatic analogy he determined the pressure function at the contact point, and derived explicit formulae relating the *magnitude of the impression* h, the *impressing force* F, and the *main radii of curvature* of strain-free surfaces at the contact point, from which a relationship of the type $h = \text{const} \cdot F^{2/3}$ followed. Among contact problems of the theory of elasticity are *stamp problems* in which one of contacting bodies is an elastic halfplane, and the other is an *perfectly rigid body* impressed into the halfplane by fixed forces. In the simplest case of a flat die uniformly pressed onto the boundary of a half-plane on the segment $(-a, a)$ of the Ox axis the contact pressure distribution $P(x)$ is determined by the formula $P(x) = F/[\pi(a^2 - x^2)^{1/2}]$ where F is the pressing force. One approach to solving such a problem involves reducing it to a boundary value problem of the theory of analytical functions.

CONTACT STRESSES, mechanical

Stresses which appear at mechanically interacting solid deformable bodies, or around them (e.g., at the compression of bodies in contact). Knowledge of contact stresses is important for the *strength* calculation of bearings, gear and worm transmissions, ball and cylinder compactors, colliding bodies, etc. The contact surface size is often small in comparison with the body dimensions, with the contact stresses rapidly decreasing with distance from the initial contact point. The stress distribution at the contact area and in its neighborhood is nonuniform, and the peak shearing stresses which characterize the strength of compressed bodies can appear at a particular distance from the contact point. In Fig., the upper part shows two spheres in contact, subject to the compression force P, with A indicating the point of peak stress. The lower part shows how the stress p in the interface decreases with the radial distance r

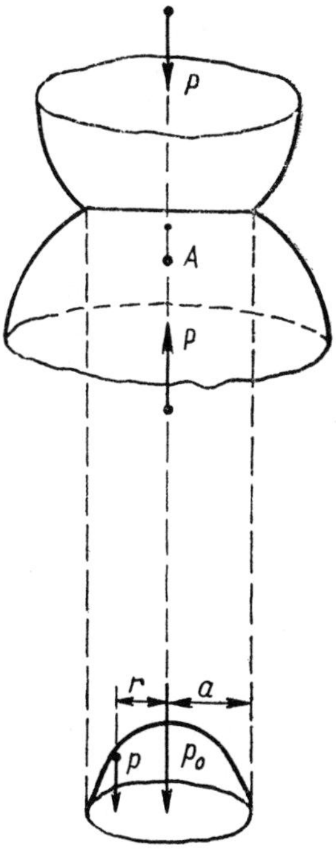

Upper part: two spheres in contact subject to the compression force P, with the point of peak stress designated by A. The region of contact has the radius a. Lower part: the stress p variation with the radial distance r from its peak value p_0 in the center.

from its peak value p_0 in the center of the contact area. See also *Contact problem of elasticity theory.*

CONTINUOUS SYMMETRY TRANSFORMATION GROUPS, Lie groups (S. Lie)

Transformation groups with continuous symmetry. Infinite groups in which a finite number of parameters, n, corresponds to each element of the group. Here, the group operations (i.e. mappings $(g, h) \to gh$ and $g \to g^{-1}$ for elements g and h of a group G) must be continuous and infinitely differentiable functions in the parameter space. A number n is called the *group dimensionality*. In general, the n-dimensional Lie group is a topological space (differentiable manifold), the neighborhood of each of its points being *homomorphic* to a certain region in Euclidean space $\mathbb{R}^n$. For example, an n-dimensional sphere $\mathbb{S}^n$ is not a Euclidean space, however, any part of such a sphere is homomorphic to a certain region of $\mathbb{R}^n$.

Lie groups are among the most important transformation groups used in physics. For example, various *conservation laws* are related to such groups. The Lie groups most often found in physics are the *orthogonal groups* $O(n)$ and the *unitary groups* $U(n)$ of real- and complex-valued $n \times n$ matrices, which preserve their scalar product in the real, $\mathbb{R}^n$, and complex, $\mathbb{C}^n$, Euclidean spaces, respectively. Their subgroups consisting of matrices with a unit determinant (det $= +1$) are called *special orthogonal groups* and *special unitary groups*, denoted by $SO(n)$ and $SU(n)$, respectively. Spherical harmonics transform by $SO(n)$, and spinors ($S = 1/2$) by $SU(2)$.

CONTINUUM INTEGRAL

See *Path integral.*

CONTINUUM MECHANICS, mechanics of continua

Branch of mechanics devoted to describing the properties and movements of matter (solids, liquids, gases, plasma, etc.) in the *continuum approximation.* This approximation involves long wave length vibrations ($\lambda \gg$ lattice constant) with small propagation constants $k = 2\pi/\lambda$. The mechanics of continua may be divided into the *mechanics of deformable solids* (including *elasticity theory*, *plasticity* theory, theory of *failure*, etc.), hydrodynamics, gas dynamics, etc. The continuum approximation applies to any dynamic system that consists of a large number of particles that may be considered as a homogeneous medium. This description is predicated on introducing physically infinitesimal volumes, whose size (l_{mic}) far exceeds characteristic distances between particles (l_{part}). Averaging microscopic mechanical characteristics of the system (forces, coordinates, momenta of particles, energy, etc.) yields macroscopic quantities (density, velocity field) that have a continuous distribution in space. The equations of motion of particles undergo a similar averaging; to obtain closed form equations for the above macroscopic quantities. To accomplish this it is necessary, as a rule, to perform a certain "uncoupling" of cross-terms of dynamic equations. The latter procedure imposes a limitation on l_{mic}, i.e. $l_{mic} \ll l_{mac}$, where l_{mac} is the characteristic length of variations of averaged macroscopic quantities. One can obtain in this manner, e.g., *hydrodynamic equations* from quantum equations of motion of the *density matrix*, or equations of elasticity theory from classical equations of motion of *crystal* ions. Thus the validity criterion for the present description of a dynamical system of a large number of particles is of a self-consistent nature, and is stated in the form of an inequality $l_{part} \ll l_{mic} \ll l_{mac}$. This criterion corresponds to sufficiently homogeneous systems (gases, liquids). In the case of inhomogeneous systems that contain many (macroscopic) structural elements this description may prove to be too detailed, and it may be necessary to perform a second averaging of macroscopic equations of motion over volumes of a larger characteristic size l' that satisfies the inequality $l_{str} \ll l' \ll l'_{m}$. Here, l_{str} is the characteristic size of structural elements or the mean distance between them, and l'_{m} is the characteristic length of variations of doubly averaged quantities. Typical examples are the averaged description of the motion of multiphase media with a heavy mixing of phases (i.e., attenuation of *sound in solids* with many *pores*, or the elastic properties of a *composite material*), *plastic flow* of solids under conditions of a high density of *dislocations* and dispersed particles, and the possible propagation of a main *crack* in a material with a number of microcracks and cavities in the zone of failure.

Two equivalent methods are used for describing the motion of macroscopic particles of continuous media. The *Lagrange method* involves tracking the location of a particle $\boldsymbol{r} = \boldsymbol{r}(\boldsymbol{a}, t)$ at every instant of time t; vector $\boldsymbol{a}$ specifies the particle, and is usually taken to be the particle location at the initial instant of time t_0. The *Euler method* involves tracking the velocities, accelerations, etc. of various particles that pass through a certain point $\boldsymbol{r}$ of the space. Ordinarily the Eulerian description is used when the mechanical state of the medium is characterized by fields of various types: displacement field $\boldsymbol{u}(\boldsymbol{r}, t)$, velocity field $\boldsymbol{v}(\boldsymbol{r}, t)$, density field $\rho(\boldsymbol{r}, t)$, etc. The fields usually satisfy second-order partial differential equations (*Navier–Stokes equations* of hydrodynamics, equations of elasticity theory, etc.). Along with $\boldsymbol{u}$ and $\boldsymbol{v}$, these equations involve the force characteristics of the medium, described most often by the field of the *stress tensor* $\sigma_{ik}(\boldsymbol{r}, t)$. Evaluating the system of equations of the mechanics of continua requires specifying the constitutive relations (*rheological equations*) that establish a link between σ_{ik} and $\boldsymbol{u}$, $\boldsymbol{v}$. In the case of a solid under smooth and rather small *strain*, this link may be considered linear and local (i.e. $\sigma_{ik}(\boldsymbol{r}, t)$ depending on the values of $\boldsymbol{u}$, $\boldsymbol{v}$, taken at the same point $\boldsymbol{r}$ and at the same instant of time t). More complicated relationships between σ_{ik} and $\boldsymbol{u}$, $\boldsymbol{v}$ are also possible (linear, nonlocal, with allowance for the history of deformation of the medium), represented by complicated functions $\sigma_{ik} = f_{ik}(\boldsymbol{u}, \boldsymbol{v})$ or by functionals $\sigma_{ik} = \Phi\{\boldsymbol{u}(\boldsymbol{r}, t), \boldsymbol{v}(\boldsymbol{r}, t)\}$. For instance, we have for a linear nonlocal medium

$$\sigma_{ik}(\boldsymbol{r}, t) = \iint A_{ik}^{lm}(\boldsymbol{r}, t; \boldsymbol{r}', t') u_{lm}(\boldsymbol{r}', t')\, \mathrm{d}\boldsymbol{r}'\, \mathrm{d}t'.$$

Equations of continuum mechanics in the linear local approximation often proceed from general considerations (law of conservation of particle number, symmetry conditions, etc.). They contain a number of unknown coefficients (*elastic moduli*, coefficient of *viscosity*, etc.) that should be determined experimentally or calculated from a microscopic theory.

The above procedure for obtaining macroscopic equations from microscopic ones involves not only purely spatial averaging, but also statistical averaging (for quantum dynamic systems both kinds of averaging are performed simultaneously through the density matrix). Statistical averaging may fail to produce closed mechanical equations, since the continuum equations may contain thermodynamic and chemical quantities (temperature, concentration, chemical potentials, etc.) along with parameters characterizing internal degrees of freedom of microparticles or other subsystems of the body (electric, magnetic, and others). In this case, the mechanical problem should be solved together with the thermal, chemical, etc. ones by adding equations of *thermal conductivity*, *diffusion*, etc. to continuum mechanics equations. Typical examples are diffusive-viscous creep of polycrystals (see *Diffusion mechanism of crystal creep*), dynamics of *ferroelectrics* and *ferromagnets*, and propagation of *shock waves* in crystals. Besides that, empirical or calculated relationships may be available, which describe the dependence of mechanical parameters (e.g., elastic constants) on temperature, concentration, *magnetization*, etc. The most consistent description of related mechanical, thermal, electrical, and magnetic processes can be made within the framework of the thermodynamics of *irreversible processes*, which includes much of continuum mechanics as a special case. In certain situations, however, one can confine oneself to the solution of a purely mechanical problem. For instance, in a static state during a very slow deformation of a solid, the temperature has time to equilibrate and the deformation process is described by equations of elasticity theory with isothermal elasticity moduli. Also, under very rapid deformation (e.g., propagation of high frequency sound), *heat transport* between neighboring parts of a body does not take place, and the process is described by the same equations, but with adiabatic elastic moduli.

CONTINUUM THEORY OF DEFECTS

Theory of *defects* in a broad sense based on the continuum approximation in analysis. From a more specialized viewpoint the continuum approach to defects is founded on the assumption that they fill the physical space of the crystal. The subjects of this analysis are usually *dislocations*, *disclinations*, *dispyrations*, and *planar defects*, although the methodology can be extended to other problems which are not closely related to the presence of defects: e.g., this theory is used for the calculating thermal stresses, electrostrictive or magnetostrictive mechanical effects (see *Electrostriction*, *Magnetostriction*), elasto-plastic fields generated by *phase transitions* or *twinning of crystals*, etc. This continuum theory is constructed on rather general physico-mechanical principles.

It includes the determination of the tensor of densities of defects, the balance conditions for these densities, and the continuity condition (existence of solutions). In this theory the connection between defect motion and the plastic fields generated by this motion (see *Plastic deformation*) is analyzed. The forces acting in this process and other factors are evaluated. The system of equations in the continuum theory involves a fundamental relation that connects the mobility of defects and the forces acting on them. This fundamental equation is considered to be known because it cannot be derived, apparently, within the framework of the continuum approximation.

CONTRACTION, TRANSVERSE

See *Transverse contraction.*

CONVOLUTION OF LINE SHAPE

See *Line shape convolution.*

COOLING AGENT, refrigerant

A material (working medium) used to cool a substance (cooling agent) or maintain it at a low temperature (cryo-reservoir). The most widely used refrigerants are liquids and gases with low *condensation* and *crystallization* temperatures (^{3}He, ^{4}He, H_2, Ne, N_2, CO_2/acetone, etc.), paramagnetic salts (magnesium cerium nitrate $Ce_3Mg_3(NO_3)_{12}\cdot 24H_2O$, ferric chromium alum and ferric aluminum alum), van Vleck *paramagnets* ($PrNi_5$, $PrPt_5$, $PrTl_5$, $PtCu_6$, etc.), metals with paramagnetic nuclei (Cu, Ag, In, etc.).

COOLING DUCT

A heat exchanging device to provide the thermal connection of an object being cooled with a *cooling agent* (refrigerant), or with a *cryostat* (low-temperature reservoir). In the simplest case a cooling duct is a rod (tube, stripe, etc.) of a material with good *thermal conductivity*, which provides a reliable thermal connection with the parts of an installation containing the object being cooled, as well as the cooling agent, or the low temperature reservoir. Often (especially at low temperatures) the cooling duct is in the form of a large number of thin wires (usually made of *copper*) which improves the *heat transfer* efficiency by increasing the contact area. To arrange for the adjustment of the temperature the cooling duct can be provided with a heater, and when it is necessary to rapidly change the thermal connection then *heat switches* are used. The associated properties of a thermal switch and a cooling duct are combined in a thermosiphon, which is an hermetically sealed tube filled with a gas under pressure, providing for its liquefaction in the desired temperature range.

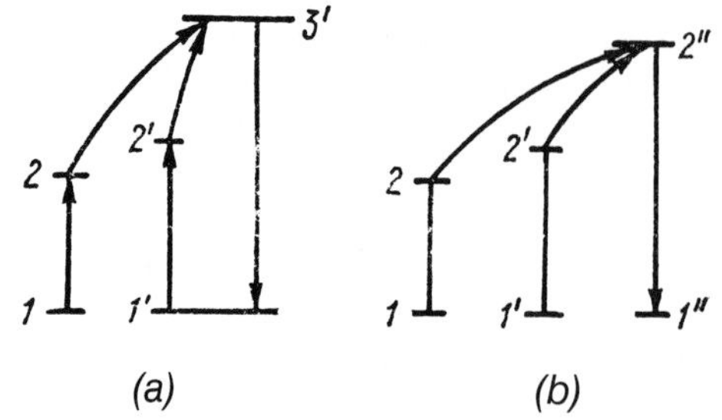

Cooperative luminescence.

COOPERATIVE LUMINESCENCE

Luminescence caused by radiative transitions from excited states in crystals that result from the delocalization of energy from individual excited states (e.g., photoexcited states of impurity centers); it is a special case of *anti-Stokes luminescence.* Cooperative luminescence violates the *Stokes' rule* and *Vavilov law.* The phenomenon was discovered by C.A. Parker (1962) and F. Muel (1962) in *dye* solutions. In the solid state it was observed by P.P. Feofilov, V.V. Ovsyankin (1966) and separately by F. Auzele (1966) in crystals activated with rare-earth ions, and in broad-band semiconductors with an adsorbed dye. An example of the mechanism of cooperative luminescence is the absorption of light by local crystal centers (transitions $1 \to 2$ and $1' \to 2'$), the addition of photoexcitation energy to induce the transition from one of these excited states to the $3'$ level, followed by the subsequent radiative transition $3' \to 1'$ (Fig. (a)). This luminescence process can also occur through excitation energy transport from two local centers (2 and $2'$ levels) to a third (level $2''$), with the subsequent radiative deexcitation of the latter ($2'' \to 1''$) (Fig. (b)).

COOPER EFFECT (L. Cooper, 1956)

Instability of the ground state of a fermion system with respect to the spontaneous formation of bound pairs of fermions with antiparallel momenta and *spins* (see *Cooper pairs*), provided that an arbitrarily weak attraction $V_0 < 0$ exists between these particles in a narrow energy band near the *Fermi surface.* The binding energy of a selected pair of fermions (lowering of its energy with respect to the *Fermi level*, E_F) is equal to $\Delta E = \hbar\omega_0 e^{-2/g}$, where $g = N(E_F)V_0$ is the dimensionless coupling constant, $N(E_F)$ is the *density of states* at the Fermi surface, and $\hbar\omega_0 \ll E_F$. In *momentum space* the Cooper effect is similar to the quantum-mechanical formation of a bound state that develops when a particle of mass m moves in a two-dimensional rectangular potential well of depth U_0 and width a. Within the limit of weak coupling (shallow and narrow well) the energy of the lowest level depends exponentially on the interaction constant: $E_0 \propto U_0 e^{-1/g_0}$, where $g_0 \propto U_0 a^2 m/\hbar^2$.

While treating the problem initially and selecting an isolated pair of fermions (i.e. electrons) from the Fermi sea, L. Cooper ignored the principle of particle identity, but took into account the Pauli principle. A more systematic self-consistent treatment of the problem was provided by J. Bardeen, L. Cooper and J. Schrieffer (1957) (see *Bardeen–Cooper–Schrieffer theory*).

COOPER PAIR, KINETIC DECOUPLING

See *Kinetic decoupling of Cooper pairs.*

COOPER PAIRS

Correlated bound states of fermion pairs (electrons, sometimes holes) with opposite *quasi-momenta* and antiparallel (singlet) or parallel (triplet) *spins*, that form due to an attractive interaction (see *Cooper effect*). Such an attraction in *metals* (*semimetals*, doped semiconductors etc.) between conduction electrons (*holes*) near the *Fermi surface* at low temperatures can be produced by the exchange of *virtual phonons.* For a strong enough *electron–phonon interaction* the attraction may override the screened Coulomb repulsion. The Cooper pairs formed by this attraction have the binding energy $2\Delta \ll E_F$ (Δ is the *energy gap*, E_F is the *Fermi level*), and are characterized by a correlation radius larger than the average electron-to-electron distance $r = 10^{-9}$ m. This is the so-called *coherence length* $\xi = 10^{-8}$ m, so that the wave functions of Cooper pairs strongly overlap. As a result a coherent condensate of Cooper pairs forms within the volume of the crystal (see *Bose condensate*) that is described by a complex wave function $\psi = |\psi|e^{i\theta}$ (the *order parameter*) with a phase θ that is, generally, coordinate dependent. Since the spin of a Cooper pair is an integer (0 or 1) it is a Boson particle, and a *Bose–Einstein condensation* called a Cooper pair condensate can form in which all of the pairs are in the same quantum-mechanical ground state. However, Cooper pairs have been called "bad" bosons since their creation $B_p^+ = a_{p\sigma}^+ a_{-p,-\sigma}^+$ and annihilation $B_p = a_{-p,-\sigma} a_{p\sigma}$ operators are more complicated than the usual creation $a_{p\sigma}^+$ (annihilation $a_{p\sigma}$) operators of a fermion with momentum $\boldsymbol{p}$ and spin $s = \pm 1/2$ (see *Second quantization*). Cooper pairs have the more complex commmutator relations

$$[B_p B_{p'}^+ - B_{p'}^+ B_p] = (1 - n_{p\sigma} - n_{-p,-\sigma})\delta_{pp'},$$

instead of the usual Bose operator commutation rules. Here $\delta_{pp'}$ is the Kronecker symbol and $n_{p\sigma} = a_{p\sigma}^+ a_{p\sigma}$ is the number operator, which is 1 for an occupied state and 0 for an unoccupied one. The Cooper pair condensate state is separated by a gap of 2Δ from excited states (excitations above and below the Fermi surface), the latter having a continuous spectrum $\varepsilon_p = \pm(\Delta^2 + v_F^2(p_F - p)^2)^{1/2}$ that satisfies the *Landau superfluidity criterion* (v_F and p_F are the Fermi velocity and Fermi momentum, respectively). Thus, the formation of Cooper pairs in a charged electron *Fermi liquid* (or Fermi gas) in metals results in *superconductivity*, while in a neutral Fermi system (^{3}He, neutron matter) a process involving rotons results in the related phenomenon of *superfluidity* (see *Superfluid phases of ^{3}He*).

COORDINATION COMPOUNDS

These are compounds of complex composition in which a central atom or ion (complexing agent) has molecules and ions called *ligands* bound to it by donor–acceptor bonds (see *Donor–acceptor pairs*). The central metal atom or ion is generally a transition metal with electron-*acceptor* properties. Ligands representing electron *donors* can be neutral molecules (H_2O, NH_3, CO, etc.) or anions ($C_2O_4^{2-}$, CN^-, etc.). The central atom together with its ligands form the "complex" or inner coordination sphere of complex compounds, and examples are $[Fe(CN)_6]^{3-}$, $[Co(NH_3)_6]^{3+}$ and $[Pt(NH_3)_2Cl_2]$. The inner sphere is usually enclosed in square brackets. The complex can be neutral or charged, in the latter case it is a cation (+) or anion (−), and counterions (monatomic or complex) comprise the outer *coordination sphere.* To separate the *complex ion* from its component ions such as SO_4^{2-} or NH_4^+ one can assume that the reaction of complex formation from the central atom (ion) and ligands can occur under normal chemical conditions, and that the central ion is capable of existing independently in solution in a solvated form. This limitation is of a conditional character since there is no essential difference between such ions and, for example, PF_6^- and VF_6^- in their *chemical bond* nature, and the majority of their physico-chemical characteristics.

Each ligand is attached to the central atom through one, two, or several atoms occupying, respectively, one (monodentate ligand), two (bidentate ligand) or several (polydentate ligand) coordination sites. The total number of coordination sites in a complex is its *coordination number*, and it is determined by the central atom oxidation state and by the nature of the ligand. Complexes with coordination number six and octahedral symmetry are the most widespread ($[PtCl_6]^{2-}$, $[Ru(NH_3)_6]^{2+}$); others are commonly four-fold coordinated with square planar ($[PtCl_4]^{2-}$) or tetrahedral ($[CoCl_4]^{2-}$) symmetry. Coordination polyhedra can be distorted through ligand inequivalence, electronic effects (see *Jahn–Teller effect*) or *crystal lattice* influences.

COORDINATION SPHERE

A sphere centered at a particular atomic site with surrounding atoms of the *crystal* located on its surface. Coordination spheres are numbered in the ascending order of their radii: the first (closest), then the second, and so on. The first sphere contains the z nearest neighbors of the given atom, where the *coordination number* z is is a measure of the packing density of the atoms; the most loosely packed semiconductor structure is the *diamond* type (Si, Ge) with $z = 4$, in NaCl type *ionic crystals* $z = 6$, in BCC metals and those with the fluorite structure (*ceramics*) $z = 8$, in FCC and HCP metals $z = 12$. In *ideal crystals* the coordination sphere radii coincide with peak positions of the radial distribution function (RDF), with the number of atoms in the nth sphere proportional to $4\pi r_n^2 S_n$, where S_n is the space under the nth RDF surface. Using this correspondence, the coordination sphere concept can be extended to liquids and amorphous solids (see *Amorphous state*), understanding as the nth coordination sphere the spherical layer with atom positions within the range Δr_n of radius values r_n corresponding to the magnitude of nth RDF peak, and considering the number of atoms in the nth sphere as $4\pi r_n^2 S_n \rho$, where ρ is the average atom density. For an amorphous solid, with increasing r the RDF peaks rapidly broaden and become flattened, so the coordination sphere concept is used only for small n (usually $n = 1$–3). Thus, in virtue of the statistical nature of the RDF the number of atoms in the sphere (the coordination number) can be nonintegral. The coordination sphere concept is useful in the analysis of structures with short-range order (see *Long-range and short-range order*), for self-ordered alloys, for the approximate construction of a crystal potential for *band structure* calculations, and so on.

COPPER (Lat. *cuprum*), Cu

Chemical element of Group IB of the periodic table with atomic number 29 and atomic weight 63.546. Stable isotopes ^{63}Cu (69.1%) and ^{65}Cu (30.9%). Nine radioactive isotopes are known. Outer shell electronic configuration $3d^{10}4s^1$. Successive ionization energies (eV): 7.72, 20.29, 36.83. Atomic radius is 0.127 nm. Ionic radii are (nm): Cu^{+} 0.096, Cu^{2+} 0.072. Covalent radius 0.117 nm. Oxidation states +1, +2, more rarely +3. Electronegativity is 1.75.

Copper is a red-tinted *metal* of face-centered cubic structure, space group $Fm\bar{3}m$ (O_h^5); $a =$ 0.361479 nm. Density is 8.932 g/cm^3; $T_{\text{melting}} =$ 1356 K, $T_{\text{boiling}} = 2843$ K, heat of melting 13.0 kJ/mole, heat of sublimation 340 kJ/mole, heat of evaporation 307 kJ/mole (under standard conditions). Specific heat is 390 J·kg^{-1}·K^{-1} at 298 K. Debye temperature is 339 K; linear thermal expansion coefficient 16.8·10^{-8} K^{-1}. Coefficient of thermal conductivity is 385–402 W·m^{-1}·K^{-1} (at 273 K) and 5 W·m^{-1}·K^{-1} (at 20 K). Adiabatic elastic moduli of crystal: $c_{11} = 168.8$, $c_{12} =$ 121.8, $c_{44} = 75.5$ GPa at 298 K. Young's modulus is 110–145 GPa, shear modulus is 42–55 GPa, Poisson ratio is ≈0.33. Rupture strength is ≈230 MPa, relative elongation 60%. Brinell hardness is 35 HB (at 293 K). Coefficient of self-diffusion is 9.1·10^{-14} m^2/s (at 1233 K). Low-temperature linear electronic heat capacity coefficient 0.687–0.695 mJ·mole^{-1}·K^{-2}. Electric resistivity is 15.5 nΩ·m. Electric resistance temperature coefficient is 440·10^{-5} K^{-1} (at 273–373 K). Hall constant is −5.2·10^{-11} m^3/C. Absolute thermal electromotive force coefficient +1.7·10^{-6} V/K. Reflection factor for 5 μm optical wave is 98.9%. Work function of a copper polycrystal is 4.4 eV. Copper is diamagnetic with magnetic susceptibility (5.4–10)·10^{-6} CGS units. Ion plasma frequency is 7.655 THz. About 50% of copper produced is used in the electrical power industry, and over 30% is used for *copper alloys*. It is widely used for chemical equipment.

COPPER ALLOYS

Copper forms the following systems with other elements: (a) infinite *solubility* in solid and liquid states: with Ni, Rh, Pt, Pd, Mn, Au; (b) infinite solubility in liquid and limited solubility in solid state: with Li, B, Ag and Bi; (c) peritectic type of interaction (see *Peritectic alloys*): with Ir, Nb, Co, Fe; (d) limited solubility in liquid and solid phases: with Pb, Tl, Cr, V); (e) intermediate compounds with many elements (for dissimilar electrochemical properties Cu forms chemical compounds; for the same properties, an electron compound is formed). Cu dissolves up to 31.9 Zn, 20 Li, 16.6 Be, 16.4 Ga, 15.8 Al, 12 Ge, 11.15 Si, 10 In, 8 Ir, 8 Ti, 8 Sb, 7.7 Sn, 7 Mg, 6.85 As, 4.5 Fe, 3.4 P, 2.1 Cd (molar content, %). Some copper alloys have special names: *bronzes* (with Al, Sn and some others), *brasses* (with Zn), *German silver* (with Ni). On addition of Zn, Sn or Al to Cu, *α-solid solutions* appear first; then, as the concentration increases, β-, γ-, δ-phases appear, and so on. All industrial brasses and bronzes are of two types differing in structure and properties: single-phase α-alloys (plastic, easily deformable, *strength* and *hardness* slightly exceeding those of copper) and two-phase alloys (α-solid solution + another phase). The latter exhibit higher strength and hardness but lower *plasticity*. Of practical use are brasses that contain up to 43% Zn, tin bronzes with up to 22% Sn, aluminum bronzes that contain up to 11% Al. *Tin bronzes* exhibit enhanced strength, hardness, *corrosion* resistance (α-bronzes that contain up to 7–8% Sn). *Aluminum bronzes* are stronger and more plastic, corrosion- and *wear*-resistant; two-phase alloys are commonly used (≈ 10% Al). *Quenching* and *tempering* (because of *phase transitions* in the solid state) may change their properties essentially. *Silicon bronzes* exhibit less specific weight and high mechanical properties. If silicon content is 2–5%, they are mostly single phase and often used with Mn, Fe, Zn additions. *Beryllium bronze* (2–3% Be) exhibits high *elasticity*, corrosion resistance. In *cadmium bronze* Cd improves the strength of Cu. *Lead bronze* separates into layers already in the liquid phase; at Pb content of 20–30%, it is used as a bearing alloy. Cu forms the following compounds with Ni: *copper–nickel* (32% Ni), *constantan* (40–45% Ni) that exhibits high electrical resistance, *bullet German silver* (20% Ni). Single phase alloys with Mn possess enhanced electrical resistance: *manganin* (12% Mn + (2–4)% Ni) and *double manganin* (20% Mn + 5% Al) with electrical resistance twice that of manganin.

CORBINO DISK (O.M. Corbino, 1911)

A metal or semiconductor disk with a hole in the center, with a contact located on the inner wall of the hole, and another at the periphery. It is used for investigating *galvanomagnetic effects*. The plane of the disk is normal to an external magnetic field $\boldsymbol{B}$. When electric current flows between the contacts, due to the sample axial symmetry the electric field has only a radial component and the Hall field is absent. As a result, there is no (partial) compensation for the curvature of charge carrier paths in the magnetic field by the Hall field, and the relative resistance $\Delta R/R$ in the magnetic field appears to be much greater than the *magnetoresistance* value $\Delta\rho/\rho_0$ (ρ_0 is the resistance with the field absent, and $\Delta\rho$ is the change brought about by the field) measured in a long (thread-like) sample. In n-InSb at 300 K in a field $B = 1$ T, $\Delta R/R = 17.7$, and $\Delta\rho/\rho_0 = 0.48$. These values are connected by the relationship

$$\frac{\Delta R}{R_0} = \frac{\Delta\rho}{\rho_0} + \frac{(\mu_{\mathrm{H}} B/c)^2}{1 + (\Delta\rho/\rho_0)}$$

(μ_{H} is the Hall mobility of charge carriers) from which it follows that when B increases the ratio $\Delta R/R$ does not tend to saturation. Therefore, the Corbino disk is utilized in devices for strong magnetic field measurement (see *Magnetometry*).

CORRELATION in alloys

The interconnections in the equilibrium distribution of *alloy* component atoms on *crystal lattice* sites. They depend on the differences in the interaction energies of different atoms since each atom tends to surround itself by atoms of another type in ordered alloys, or by the same atoms in dissociating alloys (see *Alloy ordering*, *Alloy decomposition*). Quantitative measures of this correlation are correlation parameters, or short-range order parameters. The latter characterize mainly the extent of the formation of pairs of dissimilar atoms, or pairs of identical ones.

CORRELATION ENERGY

The difference between the ground state energy of an interacting particle system and the minimum energy of the system determined by a *self-consistent field* method. The correlation energy concept is most often used for the description of an electron gas. For the self-consistent *Hartree–Fock method* the correlation energy is the true energy minus the kinetic, electrostatic and exchange energies. In this case the correlation energy is related to the decrease in energy due to the vanishing of the probability density for the presence of two electrons at the same point because of the singularity of the Coulomb interaction. A rigorous calculation of the correlation energy always represents a complicated problem, especially when dealing with a crystal or a polyatomic molecule. In such cases the wave function is represented by a superposition of antisymmetrized products of one-particle states (in Hartree–Fock theory only one such term is retained), see *Configuration interaction*, or the wave function is formed on the basis of two-, three-, and so on molecular *orbitals* (e.g., *geminals*). The correlation energy of a homogeneous electron gas is the best studied. Here the energy, according to Hartree–Fock theory, corresponds to the first order of the perturbation theory of the interelectronic interaction, and the correlation energy is the sum of the remaining terms of a series, each term of which diverges. Several ways are known for summing these series that result in the expansion of the one electron correlation energy

$$\frac{2\hbar^2}{me^4}\varepsilon_{\mathrm{corr}} = 2(1 - \ln 2)\ln r_{\mathrm{s}} - 0.096 + Ar_{\mathrm{s}}\ln r_{\mathrm{s}} + Br_{\mathrm{s}} + \cdots,$$

where $r_s = [3n/(4\pi)]^{1/3}$ is the average distance between electrons in *Bohr radii* $[\hbar^2/(me^2)]$, and n is the number of electrons per unit volume. This expansion in powers of r_s cannot be rigorously justified for real valence electron densities for which $1.5 < r_s < 6$. Nevertheless, the known results for the correlation energy are in good agreement with those obtained with the aid of the variational principle, that are not dependent on the smallness of r_s. The correlation energy of inhomogeneous systems is more often calculated within the framework of density functional theory (see *Density functional theory*) using the approximation of local homogeneity. This gives

$$E_{\text{corr}} = \int \varepsilon_{\text{corr}}\big(n(r)\big)n(r)\,\mathrm{d}^3r.$$

The *exchange interaction* energy is not always separated from the correlation energy, and then one speaks about an *exchange-correlation energy*.

CORRELATION ENERGY OF A CENTER

The difference U between the energies E_1 and E_2 in the sequential *trapping* of the first (1) and the second (2) electron on *alternative levels* of a local electronic center (see *Local electronic levels*) in a non-metallic *crystal*. Because of the mutual repulsion of electrons the relation $U = E_1 - E_2 > 0$ (*positive correlation energy*) is the usual one. The energy level positions and the charge states of the center for $U > 0$ are shown in Fig. (a, b) for a doubly-charged and an *amphoteric center*, respectively. The deformation of the atomic environment of the center caused by the strong interaction of the electrons with the atomic subsystem can give rise to a significant increase of the potential well and the binding energy of the second electron on the center. In this case the relation $U < 0$ (*negative correlation energy*) becomes possible. Centers with a negative correlation energy are filled pairwise by electrons under the condition of *thermodynamic equilibrium*. In particular, if an amphoteric center concentration with $U < 0$ dominates, then one half of the centers are filled with pairs of electrons and have a negative charge, and the other half have a positive charge. The *Fermi level* E_F thus is located in the middle between the *donor* and *acceptor* levels (Fig. (c)), and does not vary with a change in the concentration of the centers (so-called pinning). The amphoteric interstitial boron

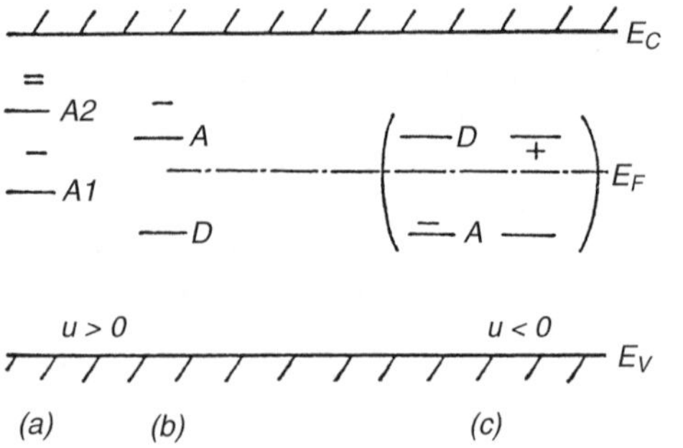

Donor (D) and acceptor (A) energy levels between the top of the valence band (E_V) and the bottom of the conduction band (E_C) for positive ($U > 0$) and negative ($U < 0$) correlation energies, where E_F is the Fermi energy.

atom in *silicon* is an example of a center with a negative correlation energy, another example is the Si V *vacancy* in silicon, where $U < 0$ for the trapping of the second electron on V^-, and for the trapping of the second hole on V^+.

CORRELATION FUNCTION

The average of the product of several variables or functions corresponding to different points in space–time. Correlation functions describe the correlation of dynamic variable values. A correlation function or *correlator* of nth order is the quantity $\langle\xi_1(X_1)\xi_2(X_2)\ldots\xi_n(X_n)\rangle$, where $X_i = (x_i, y_i, z_i, t_i)$, x, y, z are the space coordinates, t is the time; ξ_i is a dynamic variable or function, and $\langle\rangle$ stands for a certain averaging. In classical physics $\xi(X)$ represents a random (often complex) field, and the averaging is carried out over all possible fields under consideration. In the particular case when ξ_i depends only on time $\xi_i = \xi_i(t)$ represents a random quantity, and the averaging is carried out over time. In quantum physics the $\xi_i(X_i)$ represent Heisenberg operators describing quantum fields, and the averaging is carried out over some state of the fields specified by the statistical operator (*density matrix*) ρ: $\langle\xi_1(X_1)\ldots\xi_n(X_n)\rangle = \mathrm{Tr}\{\rho\xi_1(X_1)\ldots\xi_n(X_n)\}$. The Heisenberg operators can depend only on time, thus $\xi_1(t_1)\ldots\xi_n(t_n) = \mathrm{Tr}\{\rho\xi_1(t_1)\ldots\xi_n(t_n)\}$. An important case arises when the ξ_i represent operators for the creation or annihilation of field quanta (see *Second quantization*), or of particles (*quasi-particles*), or the quantum system as a whole in a certain state: $\xi_i = a_i^+$ or a_i. In solid state physics these are creation and annihilation operators of electrons (*holes*), *excitons*, *phonons*, *magnons*, *photons*, or states of impurity centers (see *Impurity atoms*). In solid state quantum theory *pair correlations* $\langle\xi_1(t_1)\xi_2(t_2)\rangle$, or *second-order correlation functions* $\langle\xi_1(X_1)\xi_2(X_2)\rangle$, are the most widely used. When considering the pair correlator it is convenient to use average and difference times: $t = (t_1 + t_2)/2$, $\tau = t_1 - t_2$, hence $\langle\xi_1(t_1)\xi_2(t_2)\rangle = f(\tau, t)$. Of fundamental importance is the spectral representation $f(\omega, t) = \int e^{i\omega\tau} f(\tau, t)\,d\tau$ which defines the system spectral characteristics. In the stationary case the dependence on t is absent. The pairwise temporal correlator $f = \langle a^+(t_1)a(t_2)\rangle$ in its spectral representation describes, through the frequency or the energy $E = \hbar\omega$, the occupation of the state to which the operators a^+, a are related. To evaluate the correlation function it is usually more effective to use the related Green's functions.

CORRELATION FUNCTION OF RANDOM FIELD

A statistical characteristic of a random field (see *Random potential*) in a disordered system defined by the equality

$$\psi_n(\boldsymbol{r}_1, \ldots, \boldsymbol{r}_n) = \langle V(\boldsymbol{r}_1) \ldots V(\boldsymbol{r}_n)\rangle,$$

where $n \geqslant 2$ is the order of these functions. Here $\langle 1 \ldots n\rangle$ means averaging over the random field $V_{\boldsymbol{r}}$, and the origin of the energy scale is selected so that $\langle V\rangle = 0$. Most widely used is the pair correlation function of a random field ψ_2 in terms of which many experimentally measured quantities can be expressed, such as the square of a random field fluctuation $\langle V^2(\boldsymbol{r})\rangle$. In the *Gaussian random field* case all functions ψ_n ($n \geqslant 3$) are expressed in terms of ψ_2. In a macroscopically homogeneous disordered system the functions $\psi_n(\boldsymbol{r}_1, \ldots, \boldsymbol{r}_n)$ depend only on $(n - 1)$ differences $\boldsymbol{r}_1 - \boldsymbol{r}_n, \boldsymbol{r}_2 - \boldsymbol{r}_n, \ldots$; and in particular, $\psi_2(\boldsymbol{r}_1, \boldsymbol{r}_2) = \psi_2(\boldsymbol{r}_1 - \boldsymbol{r}_2)$.

An important property of random fields in real systems is the attenuation of correlations in systems at infinity. With the distancing of the points $\boldsymbol{r}_1$ and $\boldsymbol{r}_2$ from each other the correlation of the quantities $V(\boldsymbol{r}_1)$ and $V(\boldsymbol{r}_2)$ decreases. This provides an opportunity for the factorization of the average of the types $\langle V(\boldsymbol{R}_1) \ldots V(\boldsymbol{R}_n)V(\boldsymbol{r}'_1) \ldots V(\boldsymbol{r}'_m)\rangle = \psi_m(\boldsymbol{r}'_1, \ldots, \boldsymbol{r}'_m)\psi_n(\boldsymbol{r}_1, \ldots, \boldsymbol{r}_n)$, if $\boldsymbol{R}_i = \boldsymbol{r}_i + \alpha$ and $|\alpha| \to \infty$. In particular, $\psi_2(\boldsymbol{r}) \to 0$ as $|\boldsymbol{r}| \to \infty$. The distance $|\boldsymbol{r}_1 - \boldsymbol{r}_2|$ at which the *pair correlation function* $\psi_2(\boldsymbol{r}_1, \boldsymbol{r}_2)$ becomes and remains a negligible quantity is called the *correlation radius* or correlation length. This radius is an important characteristic length scale of a random field, which determines the behavior of macroscopic quantities that depend on the field. See *Correlation length*.

CORRELATION LENGTH, correlation radius

The characteristic dimension of the region within which correlation between physical variables is appreciable. At distances exceeding the correlation radius these variables become statistically independent of each other, and their *correlation function* becomes equal to the product of their averages. Consider, e.g., a ferromagnet that is far from its *critical point* (transition temperature). The correlation radius is of the order of an interatomic distance. Thermal motions and quantum *fluctuations* destroy the correlation of spin orientations at distances noticeably larger than this radius. As the critical point T_c is approached from above, the correlation radius grows, accompanied by the stronger and stronger cooperative effects that eventually result in the formation of a new ordered *phase* at $T = T_c$, and the material is ferromagnetic below T_c.

CORRELATION TIME

Time interval τ during the course of which the value of the second-order *correlation function* decreases by the factor of $e = 2.718$. The stationary random quantity $u(t_1)$ (e.g., thermal displacement of the crystal atom) correlates with its value $u(t_2)$ at a later time t_2 if the ensemble average value of the product $c(t_1, t_2) = \langle u(t_1)u(t_2)\rangle \neq 0$. The quantity $c(t_1, t_2)$ is called the *second-order correlation function*. For stationary processes, e.g., lattice vibrations, $c(t_1, t_2)$ depends only on the time difference $\tau = t_1 - t_2$, and the dependence $c(\tau)$ is such that $|c(\tau)| \leqslant |c(0)|$. For some classes of random quantities $c(\tau)$ has an exponential dependence on τ: $c(\tau) = c(0)\exp(-\tau/\tau_c)$, where

τ_c is the correlation time, or the time constant associated with the relaxation decay of correlations. The quantity $c(\tau)$ does not always have an exponential dependence on τ. For example, the single-phonon correlation function $\langle u_i(t_1)u_j(t_2)\rangle$ is not exponential in the low-temperature region where the anharmonicity of vibrations is negligible (see *Anharmonic vibrations*), and its correlation time corresponds to the reciprocal width at half-maximum of the Fourier transform $c(\omega) = \int_{-\infty}^{\infty} c(\tau)\exp(\mathrm{i}\omega\tau)\,\mathrm{d}\tau$.

CORROSION (fr. Lat. *corrodo*, gnaw away)

Physicochemical interaction of a *solid* and a medium (liquid, gas) resulting in the degradation of the material operating properties. During corrosion the oxidation of source material or its individual components (*anode process*) occurs with a simultaneous *cathode process*, i.e. the reduction of medium components by the uptake of electrons released at the anode reaction. Corrosion takes place irregularly on the surface due to the chemical and structural inhomogeneity of an actual solid body, and it is accompanied by product formation (including compact non-conducting depositions) from the elements present in the solid composition and surrounding medium. The local corrosion rate is limited by the rate of the slowest process, in accordance with the mixed electron potential on the surface (*corrosion potential*) which is being generated for conditions during which the rates of anode and cathode processes are equal. On the basis of the type of external exposure on the solid, which accelerates the breakdown process, we can distinguish *electrochemical corrosion* (during electric current flow), *stress corrosion*, and *photocorrosion* (during illumination). In the latter case, for oxidation–reduction processes charge carriers are excited in the solid, and corrosion products participate in generating a new phase on its surface. There are external and internal *photoeffects*, such as the photoactivation of anode dissolution of n-type *silicon*, at a rate which is limited by the generation of holes at the Si surface layer.

The nature of the surrounding medium determines different corrosion types: electrochemical, gas, atmospheric, soil, biological, sea, and so on. During high-temperature *gas corrosion* solid oxidation products (scale) are usually formed. In such conditions the oxidizing agent diffuses through the scale to the solid surface, and solid components diffuse in the reverse direction. If the transport of components through the scale is impeded its primary layers show a protective, anticorrosion effect at the subsequent contact with the medium. The corrosion rate can be varied by introducing into the corrosion active medium some chemical compounds, e.g., *inhibitors* which decrease the rate of corrosion, or *promoters* which increase it. The protective mechanism of inhibitor molecules usually consists in their ability to diminish the limiting step rate of the corrosion process at *adsorption* on a solid surface, or to promote the formation of a protective coating which prevents corrosion (see *corrosion protection*). Volatile inhibitors which protect engineering industry products from atmospheric corrosion have gained wide-spread acceptance: corrosion inhibitors are added to lubricants, oils, fuel, brake fluids, and emulsions that considerably decrease material losses from corrosion processes. The formation of *anticorrosive coatings* can also be promoted by *alloying* materials (steels) with special additives (Cr, Al, Si).

CORROSION, CONTACT

See *Contact corrosion.*

CORROSION CRACKING

Brittle failure of a metal in a corrosive medium with a static field of tensile stresses present in the lattice. The onset of corrosion cracking requires the presence in the medium of a component which creates the conditions for sharp localization of the corrosion process on the metal surface, and at structural defects. Corrosion cracking takes place in solutions of electrolytes (sulfide, chloride cracking, caustic embrittlement of steel, hydrogen cracking of *steel* in acid, etc.); in gases (hydrogen, ammonia), and in molten metals (mercury). A measure of a *metal* stability against corrosion cracking is its limiting tension loading with a particular base test, or more precisely, the critical coefficient of stress intensity (cracking stability) K_{SI} in the corrosion *crack* vertex below which the crack does not grow. The rate of crack growth is determined by the anode (active) metal dissolving rate in the crack vertex, the possible decrease of adsorption *strength*, hydrogen saturation of the metal volume before the crack, and the value of the stress intensity. The relative role of the corrosion factors differs for different metal–solution systems. In a hydrogen atmosphere at high temperature with cathode hydrogen saturation the main role is played by the hydrogen saturation of the metal, i.e. *hydrogen brittleness*.

CORROSION FATIGUE

The degradation of a metallic *crystal lattice* in a corrosive medium at exposure to a variable field of tension stresses. Corrosion fatigue is a measure of the limit of permissible cyclic *metal* loads in a given corrosive medium based on the number of selected load cycles, or on the critical intensity coefficient of cyclic tension stresses below which the growth of cracks is not observed.

The corrosion fatigue limit is much lower than that of metal *fatigue* in an inert medium. An increase of metal *strength* through a long-term base test, as a rule, does not cause an increase in corrosion fatigue in a given metal–medium system. Depending on the frequency and loading level, we can distinguish low-cycle and high-cycle corrosion fatigue. The types of loading are the stretching from zero, expansion–contraction, arching (pure and cantilever), torsion with different forms of loading cycles (sinusoidal, rectangular, trapezoidal, symmetric, asymmetric). Taking into account the presence of residual technology loading in real systems, the most general is asymmetric loading, which is also the most dangerous for a metal–medium system that is susceptible to *corrosion cracking*. Methods for limiting the increase in corrosion fatigue are the creation of compression stresses in the surface layers of a construction, surface *alloying* or use of protecting *coatings*, change of corrosion medium composition, lowering of the temperature, application of inhibitors, and electrochemical protection.

CORROSION, INTERCRYSTALLINE

See *Intercrystalline corrosion*.

CORROSION OF METALS

Metal destruction due to chemical or chemicophysical interaction with a corrosive medium. A common cause of the corrosion of a metal (M) is a chemical oxidizing reaction, e.g., by dry oxidizing gases in accordance with the following reaction:

$$nM + mO_2 = M_nO_{2m}. \qquad (1)$$

During the corrosion of metals in electrolytic media (electrolytic solutions, including those impregnated by porous bodies (e.g., soils), electrolytic melts) M^{z+} cations transfer into the medium to form compounds with different solubilities. There are corresponding characteristic reactions in acidic and neutral aqueous solutions:

$$M + zH_3O^+ = M^{z+} + 0.5zH_2 + zH_2O, \qquad (2)$$

$$M + 0.25zO_2 + 0.5zH_2O = M^{z+} + zOH^-. \qquad (3)$$

A metal–medium system, as a rule, is thermodynamically unstable, and hence it is an environment for the spread of corrosion. Despite successes in the development of materials with increased *corrosion resistance*, *anticorrosive coatings*, and other *corrosion protection* methods, the problem of metal corrosion persists because of the deficiencies of materials, the greater severity of the conditions imposed on metal usages, and other reasons. Economic losses from the corrosion of metals result from the decreased machinery and communications productivity, direct metal losses, and so on. In the case of the formation of thin (<5 nm) surface *films* M_nO_{2m} reaction (1) is associated with the tunneling of metal electrons through the boundary film–gas, and is limited by the tunneling itself or by cation M^{z+} migration arising from the appearance of an electric field in the film. In the case of a thick film (scale), because of its density the reaction rate is more often determined by cation M^{z+} *diffusion* to the boundary film–gas, the diffusion of O^{2-} ions to the metal–film boundary, or their cross-diffusion. Reactions (2) and (3) usually go like an electrochemical process, i.e. without any direct contact between the metal atom being ionized and the ion or oxidizer molecule being adsorbed at another surface location, with the electron transfer from metal to oxidizer taking place at the metal *conduction band*. Such a process includes two reactions associated with the condition of system electroneutrality, but which are kinetically independent, namely anode metal ionization (e.g., $M = M^{2+} + 2e$) and cathode reduction of oxidizer particles on the surface (e.g., $2H_3O^+ + 2e^- = H_2 + 2H_2O$). To inhibit the corrosion of metals it is sufficient to inhibit either of these two reactions. The sharp inhibition of the anode corrosion reaction of metals when forming protective (usually oxide) films (*surface passivation*) made possible the use of many thermodynamically unstable metals (*stainless steel*, *titanium*, *aluminum*, and so on). Other specific types of corrosion can result from the simultaneous influence of effects on the metal from the medium, and from mechanical factors (see *Local corrosion*, *Corrosion cracking*, *Corrosion fatigue*, *Corrosion under stress*).

CORROSION PROTECTION

Procedures for protecting *metals* against corrosive breakdown (see *Corrosion*). The following principles underlie it: provision for chemical stability of a *construction material*; reduction of corrosive medium antagonism; preclusion of contact with a hostile medium by application of coatings (see *Anticorrosive coatings*); regulation of the potential of a metallic structure surface. The chemical stability of a metal under particular conditions is assured by the selection of a metal possessing the necessary properties. *Iron* and *iron alloys* are the most available and economically effective. In highly hostile media *steels* alloyed with nickel and chromium are used. *Titanium* and *titanium alloys*, *aluminum* and *aluminum alloys*, *copper* and *copper alloys*, as well as *tantalum* and *tantalum alloys* are also used. Chemical stability can be enhanced by *alloying*, by increasing the purity of the main additions, or by the creation of the necessary structural state of the material. Reduction of the corrosive medium hostility is achieved by removing dissolved oxygen or other gaseous and corrosion-active substances from it, or by introducing some inhibitors, or additives for passivation of the metal surface (see *Surface passivation*). Protection methods based on inorganic or organic *coatings* are also widely used. In view of the dependence of the rate of corrosion on the potential, the potential of the surface is chosen to ensure that the material either will not be destroyed by cathode polarization (cathode protection), or that it will be passivated by anode polarization (anode protection).

CORROSION RESISTANCE

The measure of material stability under particular corrosion conditions expressed using a ten point scale in terms of coefficients of corrosion rate dependence on depth (mm/year), and on rate losses of mass ($g{\cdot}m^{-2}{\cdot}hour^{-1}$). Corrosion resistance is not an absolute property, but rather depends on the composition and parameters of the corrosive medium (temperature, pressure, rate, external loads, and so on), thermodynamic properties of the material, possible kinetic decelerations in the system material–corrosion medium (*surface passivation*), external polarization, and also on the design features of the product. An improvement in corrosion resistance can be brought about by *alloying*, increasing the metal purity by eliminating non-metallic inclusions, application of a special surface treatment, use of protective or conservation *coatings*, change of parameters, composition, or adjusting the oxidation–reduction characteristics of the corrosion medium, electrochemical protection, application of corrosion inhibitors.

CORROSION UNDER STRESS

A physico-chemical interaction in a *metal* having a static or cyclicly varying stress field with corrosive medium components. During simultaneous exposure of the metal to mechanical stresses and a corrosive medium, the rate of a general surface *corrosion* changes slightly in comparison with the possibly more rapid corrosion localization at the most highly stressed points: at rising *defects* of surface layers near *grain boundaries*, at nonmetal inclusions, at the end points of *dislocations*, at glide plane steps. These factors result in the loss of *strength* or density of compactness of structure because of corrosion *cracks*. A measure of a materials stability against corrosion under stress is the *corrosion-mechanical strength*, which is the limit of a prolonged static or cyclic strength in a given corrosive medium expressed in terms of exposure by time or by number of loading cycles.

COSTA-RIBEIRO EFFECT

See *Thermodielectric effect*.

COTTON–MOUTON EFFECT

See *Voigt effect*.

COTTON–MOUTON EFFECT, ACOUSTIC ANALOG

See *Acoustic analog of Cotton–Mouton effect*.

COTTRELL ATMOSPHERE (A.H. Cottrell)

Nonuniform distribution of *point defects* around a dislocation line caused by a hydrostatic component of the elastic *dislocation* stress field. At thermal equilibrium the *defect* (*dilatation center*) concentration in a Cottrell atmosphere or Cottrell medium, taking into account that one lattice site should be occupied by no more than one defect, is described by an expression similar to the Fermi–Dirac distribution (see *Fermi–Dirac statistics*):

$$c(r,\theta)=\left\{1+\exp\left[\frac{W(r,\theta)-G_0}{k_BT}\right]\right\}^{-1},$$

where k_BT is the temperature expressed in energy units, $G_0=k_BT\ln[c_0/(1-c_0)]$, and c_0 is the concentration far from the dislocation. The elastic interaction energy $W(r,\theta)$ in the simplest case equals the work required to replace an atom of the matrix at the point (r,θ) (see Fig.) by a dilatation center (impurity atom) with an excess

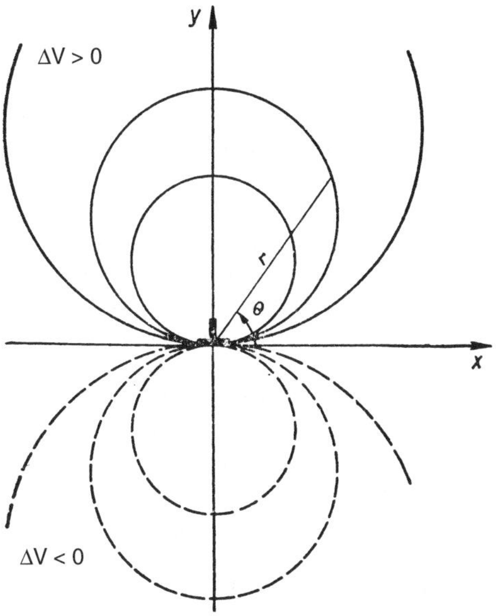

Equipotential lines $W(r,\theta)=\text{const}$ for an edge dislocation. Each circle is for a particular W_0 value. The array of lines for $c(r,\theta)=\text{const}$ is similar.

volume ΔV. For an *edge dislocation* in an elastically isotropic medium (see *Isotropy of elasticity*) the elastic interaction energy is

$$W(r,\theta)=\frac{Gb}{3\pi}\frac{1+\nu}{1-\nu}\frac{\sin\theta}{r}\Delta V,$$

where b is the *Burgers vector*, G is the *shear modulus*, and ν is the *Poisson ratio*, respectively. Above the Cottrell medium condensation temperature, T_c, defined by the relation

$$\exp\left(\frac{W_0-G_0}{k_BT_c}\right)\sim 1,$$

where $W(r,\theta)$ is a maximum at $r=b$, $\theta=\pi/2$

$$W_0\approx\frac{1+\nu}{1-\nu}\frac{G\Delta V}{3\pi},$$

the Cottrell medium becomes diluted:

$$c(r,\theta)\approx c_0\exp\left[\frac{W(r,\theta)-G_0}{k_BT}\right]\ll 1,\qquad T>T_c.$$

The formation of a Cottrell medium causes dislocations to become pinned. To separate dislocations from the impurity medium the stress $\sigma_c\approx W_0/b^3$ is needed, and $\sigma_0\approx 1$ GPa for typical values $W_0\approx 0.1$ eV.

COULOMB EXPLOSION in a solid

Expulsion of positive *ions* driven by Coulomb repulsion when several nearby atoms are simultaneously ionized. The Coulomb explosion may be caused by the simultaneous ionization of two *atoms*. The formation of an electron vacancy in an internal electron shell of an atom affected by ionizing radiation, or the appearance of a slow multicharge ion inside the *solid* or at its surface, can result in cascades of Auger transitions (see *Auger effect*). These may be both "vertical" in a single atom, and "horizontal" (when the growing positive charge is distributed among neighboring atoms, and ultimately results in a Coulomb explosion as well). A Coulomb explosion in a solid produces *point defects*, defect *clusters* can also form, and the surface of the solid may suffer *sputtering*. Coulomb explosions contribute to the degradation along the trajectories of heavy accelerated ions or fragments of nuclear fission. The Coulomb explosion in *metals* is not effective because of the rapid "dispersal" of microscopic positive charge in a metallic conductor. The rates at which charges are neutralized in *insulators* and *semiconductors* with their low concentration of current carriers are much slower, and the influence of the Coulomb explosion on the formation of defects may be much more effective.

COULOMB GAP

A gap in the *density of states* $g(E)$ of electron excitations around the *Fermi level* μ in a disordered system with localized electrons, e.g., the gap in weakly doped compensated or *amorphous semiconductors* that results from the long range Coulomb interaction. An important characteristic of the Coulomb gap is its energy Δ. Provided that the Coulomb interaction energy between two localized *current carriers* separated from each other by an average distance (averaged over all localized carriers) remains small in comparison with the characteristic spread of the energy levels, the value of Δ is of the order of $e^3 g_0^{1/2}/\varepsilon^{3/2}$ in a three-dimensional system. Here ε is the *dielectric constant*, and g_0 is the density of states at the Fermi level in the absence of the electron–electron Coulomb interaction (see Fig.). Such a situation is found in a crystalline semiconductor for both strong and weak compensation, and also in amorphous semiconductors. In the case of medium compensation in a weakly doped semiconductor Δ is comparable to the overall width of the *impurity band*, and the Coulomb gap is more strongly expressed. The density of states at the Fermi level vanishes only for zero temperature. With increasing temperature the Coulomb gap "closes" and totally vanishes as soon as the thermal energy $k_{\mathrm{B}}T$ reaches several tenths of the value of Δ. This gap manifests itself in experiments that measure the *hopping conductivity* versus hopping distance in regimes where that the principal contribution to the electrical conductivity arises from states in the energy band that are within the energy range Δ from the Fermi level. A technique to observe the Coulomb gap directly is *tunneling spectroscopy*.

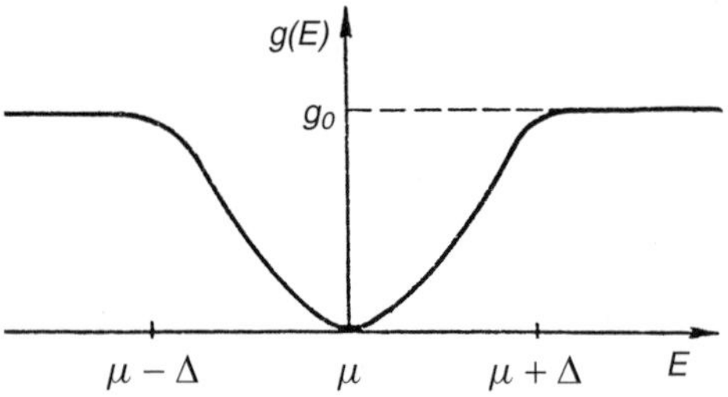

Density of states $g(E)$ of an amorphous semiconductor near the Fermi level μ.

COULOMB PSEUDOPOTENTIAL

See *Morel–Anderson Coulomb pseudopotential.*

COVALENT BOND, homopolar bond

A particular type of *chemical bond* formed between similar atoms or atoms whose properties differ very little from each other, when the pairing (overlap) of the electron shells of neighboring atoms takes place without any appreciable exchange of the electric charges of the two atoms. A typical example of a covalent bond is the bond in a hydrogen molecule. The two-electron wave function of such a molecule can be written down in the simplest *Heitler–London approximation* as follows:

$$\psi(r_1\tau_1, r_2\tau_2) = \left[\psi_a(r_1)\psi_b(r_2) + \psi_a(r_2)\psi_b(r_1)\right] \times \chi_a(\tau_1\tau_2), \quad (1)$$

where ψ_a and ψ_b are the $1s$-type one-electron wave functions centered on the a and b atoms, respectively, and χ_a is the antisymmetric wave function of the spins. The difference between the eigenenergy obtained from solving the Hamiltonian of the molecule constructed using the wave functions (1) and twice the electron energy in an isolated atom corresponds to the covalent bond energy. A similar construction is used for the wave function of two electrons of a σ-bond between nearest neighbor atoms in *homopolar crystals* (diamond, gray tin, germanium, silicon) as well as in the $A^{III}B^{V}$ and $A^{II}B^{VI}$ compounds (ψ_a and ψ_b are distinguishable in this case, appearing as sp^3 hybrid functions of the valence electrons of the atoms involved). If the crystal is formed from atoms belonging to different groups of the periodic system then there can be a greater or lesser admixture of *ionic bond* character in the bond, and this will be termed an *ionic-covalent bond.*

COVALENT CRYSTALS

Crystals held together by covalent interatomic chemical bonds. Covalent crystals are most often formed by elements of Group IV and adjacent groups of the periodic system, and are characterized by a tetrahedral (sp^3) *hybridization of orbitals*, with the *chemical bonds* formed by singlet pairs of electrons localized in the space between nearest neighbor atoms (see *Covalent bond*). Because these bonds display distinct direction and strength, covalent crystals posses a high hardness and elasticity, some of them being brittle. Covalent crystals usually exhibit a high *thermal conductivity*. The most typical covalent crystal is *diamond* (C); others are silicon (Si), germanium (Ge), gray tin (α-Sn), and a range of compounds formed from elements alongside group IV of the periodic system. These are: the $A^{III}B^{V}$ compounds, e.g., *borazon* (BN), GaAs, GaSb, InAs and AlP; $A^{II}B^{VI}$ compounds, such as beryllium oxide (BeO), zinkite (ZnO), sphalerite (ZnS), CdTe, etc. All these compounds are *semiconductors*, with their energy *band gap* varying between 0.2 to 2–4 eV. As the elements move apart horizontally in the periodic system, like in the $A^{I}B^{VII}$ compounds: CuCl, CuBr and AgI, the covalent nature of the bond weakens and it becomes partially ionic in character. Moving vertically down the periodic table some metallic features appear: for instance, crystals of white tin (β-Sn) are virtually metallic.

Covalent crystals of triple and more complex compounds which likewise possess tetrahedral atomic coordination also exhibit some degree of metallic behavior; examples are chalcopyrite ($CuFeS_2$), stannite (Cu_2FeSnS_4), $CdSnAs_2$, etc. Covalent crystals with octahedral coordination are exemplified by PbS, PbSe, SnTe, Bi_2Fe_3, Bi_2TeS_2, etc. Many crystals are heterodesmic (see *Heterodesmic structures*), i.e. the atoms in their crystalline structure have bonds of different types. Thus, *graphite* crystals are covalently bonded within the hexagonal layers, while the bonds between the layers are of the van der Waals type. The structure of elements close to Group IV, such as P, S, Se or Te, may be described in a similar way: the atoms form covalently bonded groups, but the groups themselves are bonded together by *van der Waals forces*. Many covalent crystals are widely used in technology: native and synthetic diamonds as abrasives and *silicon* crystals of the highest purity serve as the basis of semiconductor electronics together with the covalent crystals Ge, GaAs, InSb, etc.

COVALENT CRYSTALS, PLASTIC DEFORMATION

See *Plastic deformation of covalent crystals*.

COVALENT RADIUS

A parameter of the size of an *atom* in covalent compounds (see *Covalent crystals*). The covalent radius is defined as half the distance between two identical atoms held together by a covalent *chemical bond*. The length of a single chemical bond between differing atoms can be evaluated as a sum of the covalent radii of the corresponding atoms. A commonly employed system is the set of *Pauling covalent radii* which are defined for atoms in tetrahedral crystals such as diamond (C) and sphalerite (ZnS) (see *Atomic radius*).

COVARIANCE

The quality of physical quantities found in reference frames of a certain theory being transformed to other frames according to representations of the *invariance* group of this theory. For instance, in special relativity *Lorentz covariance* presumes using physical quantities that transform as relativistic vectors (e.g., 4-vectors), tensors, spinors, etc. via Lorentz transformations. *Gauge covariance* is referred to in the case of a gauge transformation (see *Gauge invariance*). Covariance also implies the conservation of the form of an equation with respect to the transformations of the invariance group. For instance, the covariance of the form of the equations of general relativity implies that the equation appears the same in any four-dimensional reference frame, even though the specific equation coefficients can (covariantly) change with a space–time transformation.

CRACK

A disturbance of continuity of a solid which is localized along a certain surface; a plane defect with broken molecular bonds along its edges. There is a singularity at the crack tip which under mechanical loading leads to *stress concentration* proportional to $1/\sqrt{r}$ where r is the distance from the crack tip. Depending on the shape of the displacement between the edges, cracks are classified as rupture cracks, and longitudinal and transverse *shear* cracks. In respect to stability, cracks are divided into equilibrium (i.e. subcritical which do not grow spontaneously) and nonequilibrium (i.e. overcritical which can propagate under action of the elastic potential accumulated by the system). In respect to the involvement of *plastic deformation*, cracks are divided into brittle (i.e. not connected with material yield) and tough (see *Brittle failure*, *Tough failure*) with a certain portion of *plasticity*. There are microscopic and macroscopic cracks. Classed among the former are sizes of cracks from the atomic scale to the subgrain scale. Referred to the latter are cracks whose dimensions are greater, by at least an order of magnitude, compared to the structural elements of the solid such as subgrains, *crystallites*, etc.

The first stage of *failure* is *microcrack* nucleation. There are two kinds of microcracks differing in the shape: elastic (force) cracks and dislocation cracks. The former are like an incision along an atomic plane with broken molecular bonds from one tip to another, and the surfaces are opened by the applied stresses. A particular feature of such a crack is its instability. It can spontaneously collapse after the stress release, so that it cannot exist in an unloaded body. A *dislocation microcrack* is an empty nucleus of a superdislocation with a large *Burgers vector*. It is mechanically stable, and can occur in an unloaded body. The nucleation of this type of crack is determined by collective dislocation processes. There is a series of crack nucleation mechanisms: *dislocation* junction, braked shear, opening of bent *slip band*, interaction of dislocations at crossing *slip planes* and at broken walls of subgrains, generation according to the scheme of opposite clusters and interacting twin layers, and so on.

The second stage of microcrack development (subcritical growth) is also fully determined and controlled by dislocation phenomena. Its essence is the "falling" of dislocations in the crack that induces the growth of its Burgers vector. There are several mechanisms for this process: the crack is initiated by the slip band approaching a tip; two slip bundles adjoin the crack; an accumulation in the crack of dislocations from the slip bands stopped by boundaries. At this stage of failure a substantial role is played by rotation processes when individual grains and their groups are able to rotate freely about the crack tip (see *Plastic twisting deformation*). Subcritical growth of the macroscopic crack proceeds via the same mechanisms with the added feature that the plastic deformation involving a large volume leads to the appearance of a system of secondary microcracks and *pores* prior to the main crack formation. The failure proper is then effected through a merging of the crack system into a common one by the tough rupture of bridges.

The third stage of crack development, involving overcritical growth, proceeds with completely or almost completely suppressed plasticity at the vertex. Here, the spreading of the crack exhibits a distinctly elastic and overstructural character, with the limiting velocities of growth determined by the elastic constants of the medium, and by the sample geometry. The maximum possible velocities of failure, according to different evaluations, vary from 0.53 of the transverse elastic wave velocity to 0.38 of the longitudinal wave velocity. The limiting velocity of the crack, equal to the velocity of surface Rayleigh waves (see *Surface acoustic waves*) is considered to be the preferable index. In the case when additional feeding of the crack vertex by elastic energy is not mediated by the elastic waves that spread through the destructible material, but is performed directly, e.g., by plasma expansion at the vertex of crack under the effect of a laser beam, the velocities of crack propagation may be considerably higher, and may even exceed the *sound velocity*.

See also *Brittle crack*, *Nucleating crack*, *Submicrocracks*, *Tough crack*, *Wedge-shaped crack*.

CRACKING

The process of extraction of products with lower molecular mass from hydrocarbons, mainly of petroleum origin. The first and foremost high tonnage industrial application of the process is extraction of motor fuels. Many processes of cracking are run in the presence of solid catalysts (*catalytic cracking*). The catalysts most often used are highly porous synthetic aluminosilicates containing Al_2O_3 and SiO_2 in various proportions.

Complex sets of reactions involved in catalytic cracking always include the formation of new solid phases which poison the catalysts. These are cokes, and their mass can reach several per cent of the catalyst mass. To remove cokes the silica-aluminas are oxidized with air, or an air–vapor mixture (process of catalyst regeneration).

CRACKING CORROSION

See *Corrosion cracking.*

CRACK RESISTANCE

The same as *Fracture toughness.*

CRAIG EFFECT (D. Craig, 1955)

Changes in the spectra of electronic and vibrational excitations of a molecule located in a crystal matrix. The influence of the local *crystal field* and intermolecular interactions results in a partial splitting of degenerate states, and changes of *selection rules*, thereby producing a larger number of observable bands in the optical spectra, and also changes their relative intensities and polarization properties. These phenomena were initially encountered during applied studies of *molecular crystals.*

CREEP

Property of a solid to increase *strain* or to deform during the course of time under the action of a constant external *stress* or load, usually at an elevated temperature. The relationship between strain and time is given by the *creep curve*, which is arbitrarily divided into characteristic regions. At the moment of applying stress an instantaneous strain is induced; then comes the stage of initial creep or *transient creep*. The rate of deformation decays at this stage. At the second stage (*secondary creep*) the rate of deformation is constant. The third stage is characterized by an increase of the creep rate, then the process ends with *failure*. The shape of a creep curve depends on the temperature, stress, initial structure, and stability of the initial structure. Depending on these factors, there are four kinds of creep with different deformation mechanisms.

(1) The first kind of creep is observed if the applied stress is lower than the critical *shear* stress. This type of creep is called *elastic creep* or *reversible creep*. It is adequately described by the equation $\varepsilon = \varepsilon_l + \varepsilon_0[1 - \exp(-t/\tau)]$, where ε_l is a purely elastic strain (see *Elasticity*), t is the time after loading, ε_0 and τ are constants. The second term of this equation stands for inelastic strain, which tends to ε_0. After load release the strain completely regains its value due to inelastic creep. One of the mechanisms, which provides an explanation for inelastic creep, proposes the diffusive transfer of impurity *interstitial atoms* into crystal lattice interstices, which are stretched under the action of the applied stress (see *Diffusion*). The strain ε_0 is proportional to the applied stress, and to the concentration of impurity interstitial atoms. After release of stress interstitial atoms tend to become uniformly arranged at interstices, and the strain is restored. Inelastic creep may also be caused by short-distance transfer of *dislocations*, which does not involve abstraction from pinnings.

(2) At stresses that are higher than the critical shear stress, the second kind of creep is observed in the low-temperature region, called *logarithmic creep* or *low-temperature creep*, and described by the equation: $\varepsilon = \alpha \ln(1 + \nu t)$, where $\alpha \geqslant 0$ and $\nu \geqslant 1$ are constants. This relation is valid up to very slight strains ($\varepsilon \leqslant 0.2\%$). So far it is unknown whether logarithmic creep is followed by secondary creep, because tests to confirm this are very time consuming to carry out.

(3) As the temperature exceeds T_G (for *metals* $T_G \approx 0.5T_{\text{melting}}$ and decreases with increased load) the third kind of creep is observed: *high-temperature creep*. The mechanism of high-temperature creep is related to the creeping of dislocations, and is controlled by processes that are determined by the rate of self-diffusion (motion of thresholds at *screw dislocations*, rate of creeping of dislocations around obstacles, etc.). High-temperature creep is described by the relation: $\varepsilon = \varepsilon_0 + \beta t^m + Kt$, where ε_0 is the instantaneous strain, β and m are time-independent constants, $1/4 < m < 2/3$ (according to Andrade (1910) $m = 1/3$), K and Kt are, respectively, the rate and strain at the first stage. At this stage *hardening* dominates over *dehardening*, the density of dislocations increases, and the size of subgrains decreases; during the steady state the size of subgrains remains constant, whereas disorientation of subgrains increases in proportion to the increase of strain.

At the second stage of creep the duration and rate of deformation depend on the stress and temperature. The equation $K = A\exp[H/(k_B T)] \times \sinh(B\sigma)^n$, where T is temperature, σ is stress, A, B, n are constants, describes the temperature dependence and stress dependence of K. The *activation energy* H monotonically increases with the increase in temperature; for $T > T_G$ it is a constant equal to the self-diffusion activation energy. This means that at low temperatures diffusion-related mechanisms of *plastic deformation* are activated. It has been found that slippage along the *grain boundaries* contributes significantly to the creep at the second stage. The dependence of $\dot{\varepsilon}$ is $\propto d^2$ at low temperatures and $\propto 1/d$ at high temperatures, while at intermediate temperatures it takes the form $k/d + Kd^2$, where k and K are constants, and d is the grain size. The influence of grain boundaries at elevated temperatures may be explained by *localization of plastic deformation* in border zones.

The third stage of creep is prepared by structural changes, which take place at preceding stages. It is thought that acceleration of creep occurs because of an increase of stress that is related to the generation of a *neck*, the decrease of the actual cross-sectional area caused by accumulation of microcracks (see *Cracks*) and *pores*, or processes of *recovery* and *recrystallization*.

(4) The fourth kind of creep, which is called *diffusive creep*, is observed at low temperatures and low stresses, and does not involve the transfer of dislocations, but is a result of directional diffusive *mass transport*. The rate of such creep, due to the self-diffusion or flux of *vacancies* between external surfaces of the sample, has been calculated in works of F.R.N. Nabarro, C. Herring (1950) and R.L. Coble (1963). Taking into account the increase of creep rate caused by grain boundary diffusion, we obtain the equation:

$$\dot{\varepsilon} \approx \frac{5D}{d^2}\frac{\sigma\Omega}{k_B T}\left(1 + \frac{\pi\delta D_g}{Dd}\right),$$

where D and D_g are the coefficients of bulk and grain boundary self-diffusion, respectively, δ is the effective thickness of a grain boundary, and Ω is the atomic volume.

Creep resistance of materials is characterized by the *creep limit*, i.e. the stress at which a certain strain is achieved in a given time. Creep resistance depends on the type of interatomic bond, type of *crystal lattice*, stacking fault energy, grain structure, and other characteristics of the material. High creep resistance is the main factor determining the *high-temperature strength* of materials. See also *Athermic creep*, *Diffusion mechanism of crystal creep*, *Radiation-induced creep*.

CREEP RESISTANCE

See *Creep*.

CRITERION OF LOCALIZATION

See *Anderson localization*.

CRITICAL CURRENTS in superconductors

The minimum value of the current I_c or current density j_c at which the superconducting state is destroyed. Several types of critical currents are identified. In *type I superconductors* the value of I_c depends on the shape (geometry) of the sample. In particular, for a long enough cylindrical superconductor it is obtained from the condition that the azimuthal component of the magnetic field of the longitudinal current, $B_f = I/(2\pi a)$, reaches the value of the *thermodynamic critical magnetic field* $B_c(T)$ at the surface of a sample of radius a. When $I > I_c = 2\pi a B_c(T)$ the superconductor transforms to the *intermediate state* with a finite resistance. In *type II superconductors* the critical current density is determined by the pinning strength of the *Abrikosov vortices*, or by the fluidity limit of the *vortex lattice* (see *Vortex pinning*). There is also a *decoupling critical current* at which the *Cooper pairs* are broken and the supercurrent flow is terminated. The current velocity $v = j/(eN_0)$ (here j is the current density, N_0 is the concentration of electrons in the *metal*) exceeds the critical velocity $v_c = \Delta/p_F$, where Δ is the superconductor *energy gap*, and p_F is the Fermi momentum (see *Kinetic decoupling of Cooper pairs*). Such currents may be reached in very thin superconducting films and filaments, the magnetic field of the current at their surface being lower than $B_c(T)$ in type I superconductors, or such that their thickness is smaller than the London *penetration depth of the magnetic field* of a type II superconductor. However, recalling the presence of inhomogeneities, it becomes clear that the destruction of the superconducting state, and the recovery of the *resistive state* in thin films and filaments are usually controlled by vortex pinning at these inhomogeneities. Finding ways to increase critical currents is one of the most important tasks of applied *superconductivity*.

CRITICAL DISORIENTATION OF DISLOCATION SUBBOUNDARY

The threshold value of the disorientation angle in a *cellular structure* at which, similar to *grain boundaries*, cell boundaries start to resist the movement of *dislocations*. The concept of critical disorientation of dislocation subboundaries resulted from comparing the mechanical properties of *metals* during the formation of the sublattice dislocation structure (either cellular or fragmented) that develops in the process of *plastic deformation* (see *Fragmentation*). While the overall density of dislocations remains unchanged with growing plastic deformation, they undergo redistribution in the process. The number of dislocations in the body of the subgrain diminishes, their density at *subboundaries* increases and the angle of disorientation of the adjacent cells grows, so that *dislocation slip* to large distances becomes more difficult. The critical value of the angle of disorientation, θ_{cr}, usually remains within several degrees, and may be estimated from the expression

$$\theta_{cr} = \left(\frac{K_y}{G}\right)^2 \frac{(1-\nu)^2 2\pi n}{0.45b},$$

where G is the *shear modulus*, ν is the *Poisson ratio*, K_y is a constant value from the Hall–Petch equation (see *Petch relation*), b is the *Burgers vector*, n is the ratio of the effective slip length L to cell size l. Following the growing plastic deformation the number of subboundaries in the already formed dislocation cells increases, such that their disorientation angle is either equal to or already exceeds the value of θ_{cr}. This results in a significant strengthening of the material, while the temperature of *cold brittleness* simultaneously decreases, which means that the formation of a disordered cellular dislocaton structure significantly improves the overall mechanical properties of the material. For such a structure to form certain conditions have to be met, and the *strain* should take place within the temperature range of warm deformations.

CRITICAL EMBRYO, critical bubble

Seed or embryo of a stable thermodynamic *phase* in unstable equilibrium with its surrounding metastable phase (see *Metastable state*). The radius of a spherical critical embryo is given by

$$r_E = \frac{2\gamma v}{\mu_M(P) - \mu(P)} = \frac{2\gamma T_c(v_M - v)}{q(T - T_c)}.$$

Here γ is the coefficient of *surface tension*; v, v_M are the atomic volumes; μ and μ_M are the chemical potentials (see *Thermodynamic potentials*) for the embryo and the metastable phase, respectively; P is the external pressure, T_c is the *phase equilibrium* temperature; q is the latent *heat of the phase transition* from the metastable phase to the embryo phase. Only embryos above the critical size continue to grow. The process of growth, stable with respect to fluctuations, sets in at $r > r_E + [k_B T/(4\pi\gamma)]^{1/2}$. For a *supercritical embryo* ($r > r_E$) to form it has to penetrate the thermodynamic potential barrier $4\pi r_E^2 \gamma/3$. See also *Nucleation* at phase transitions, *Crystallization*.

CRITICAL EXPONENTS, critical indices

Power law indices or exponents defining the dependency of dynamic and thermodynamic parameters such as the external field h, etc., on the reduced temperature, $\tau = T/T_c - 1$, etc. around a *second-order phase transition* point (T_c). Critical exponents only characterize the principal, leading asymptotes of these quantities. When τ, h change the system may pass through several regimes or phases, each of them described by its specific set of critical exponents. For example, when approaching the *triple critical point* the system passes through a transition region in which its behavior changes from that typical of a second-order phase transitions to a tricritical (around that point) behaviour.

Critical exponents used to describe thermodynamic (statistical) properties of the system define the temperature dependences of the *specific heat* $c_p \propto |\tau|^{-\alpha}$, the *order parameter* $M \propto (-\tau)^{\beta}$, the *susceptibility* $\chi \propto |\tau|^{-\gamma}$, and the *correlation length* $r_c \propto |r|^{-\nu}$, where the exponents $\beta, \gamma, \nu > 0$. The critical exponents of anomalous dimensions for fluctuations of the order parameter characterize the law by which the correlation function decreases with distance r for $r \ll r_c$, $G(r) \propto r^{-(d-2+\eta)}$, where d is the dimensionality. For $T = T_c$ the critical exponents describe the behavior of the system in the external field h: $c_p \propto h^{-\varepsilon}$, $M \propto h^{1/\delta}$, $r_c \propto h^{-\mu}$ $(\delta, \mu > 0)$. Critical exponents are related to each other via certain algebraic expressions called *scaling relations*: $\alpha + 2\beta + \gamma = 2$, $\beta\delta = \beta + \gamma$, $\varepsilon(\beta + \gamma) = \alpha$, $\mu(\beta + \gamma) = \nu$, $\nu(2 - \eta) = \gamma$, $\gamma d = 2 - \alpha$ (see *Scaling invariance hypothesis*). Critical exponents also describe the kinetic (dynamic) properties of the system around T_c. For example, the *relaxation time* $t(k, T)$ of fluctuations of the order parameter with wave vector k has the form $t(0, T) \propto |\tau|^{-y}$ and $t(k, T_c) \propto k^{-z}$ in the *critical region*. Scaling relates these exponents to each other: $y = z\nu$. Critical exponents are universal values, i.e. they do not depend on the nature of the particular *phase transition*, or on the specifics of the physical system. They depend, instead, only on the number of components of the order parameter, and on the dimensionality d of the space. *Critical fluctuations* become insignificant for $d > d_c$, where d_c is the so-called *upper critical dimension*, so that the critical exponents coincide with those provided by the *Landau theory of second-order phase transitions*. At $d = d_c$ fluctuations generate multiplicative logarithmic corrections to the results from the Landau theory. If $d < d_c$ the corresponding critical exponents differ from those prescribed by the *phenomenological theory of phase transitions*. Their values may be (approximately) calculated using the techniques of high temperature expansions within the *renormalization group method*, in particular through expansions in powers of $\varepsilon = d_c - d$ (*epsilon expansion*), in terms of the spherical model, employing the $1/n$ expansion approach, and by other techniques.

CRITICAL FLUCTUATIONS

Particularly strong thermodynamic fluctuations of the *order parameter* and other physical variables observed near *second-order phase transitions*. The *correlation length* of these fluctuations and their respective *relaxation times* grow without limit when approaching the transition point, thus reflecting destabilization of the corresponding *phase*, and the stronger role of the cooperative effects in forming the dynamics and thermodynamics of the *crystal*. Much effort has been expended toward explaining the strong nonlinear interactions of critical fluctuations (see *Renormalization group method, Epsilon expansion*). Experimental studies of critical fluctuations by neutron diffraction, optical, radiospectroscopic, and other techniques yield rich data on the behavior of materials in the *critical region*.

CRITICAL MAGNETIC FIELDS in superconductors

Values of applied magnetic field strength (that depend on *superconductor* shape) above which the character of the penetration of the magnetic field into the superconducting material changes, or the sample reverts to the normal (nonsuperconducting) state. One distinguishes between the *thermodynamic critical magnetic field* B_c in *type I superconductors* (see *Meissner effect*), and the *lower critical field*, B_{c1}, the *upper critical field*, B_{c2}, and the *third critical field* B_{c3} in *type II superconductors*. See also *Mixed states*.

CRITICAL PHENOMENA

Specific effects observed in the neighborhood of *second-order phase transition* points. One distinguishes between static and dynamic critical phenomena. Among the first are the anomalies of the thermodynamics of *crystals*, and features in the behavior of static *kinetic coefficients*. Examples of static critical phenomena are: sharp increase in the *magnetic permeability* of ferromagnets as the phase transition point T_c is approached; anomalous temperature dependence of the *specific heat*, *elastic moduli*, etc.; increase in the electrical conductivity of normal metals near *superconducting phase transitions*; and temperature anomalies in the specific heat of *ferroelectrics*. Dynamic critical phenomena involve the dynamic susceptibility, a sharp increase in *relaxation times* as T_c is approached (*critical slowing down*), anomalies in *Raman scattering of light*, in *inelastic neutron scattering*, in X-ray scattering, and in Mössbauer spectra (see *Mössbauer effect*).

Sometimes critical phenomena are understood in a narrower sense, as effects produced specifically by *critical fluctuations* of the order parameter. The susceptibility χ, the *order parameter* M, and the specific heat C, all behave as power law functions of the reduced temperature $\tau = (T - T_c)/T_c$

around the critical point T_c: $\chi \propto |\tau|^{-\gamma}$, $M \propto |t|^{\beta}$, $c \propto |t|^{-\alpha}$. Second-order *phenomenological theory of phase transitions* predicts the values $\alpha = 0$, $\beta = 1/2$, $\gamma = 1$ for these power law indices called *critical exponents*, but it does not take fluctuations into account. Experiments usually yield the ranges $-0.1 < \alpha < 0.1$, $0.3 < \beta < 0.4$, and $1.2 < \gamma < 1.4$ for three-dimensional (non-layered, non-quasi-one-dimensional) systems. Specific values of the critical indices depend on the number of components of the order parameter.

CRITICAL POINT

A point associated with a *second-order phase transition*. When the transition occurs during a temperature change the critical point corresponds to a *critical temperature*. Examples are the *Curie point* of a ferromagnet, and the *Néel point* of an antiferromagnet. The critical point can also be the value of an electric or magnetic field which induces a change in phase, such as the upper and lower critical magnetic fields of type II superconductors.

CRITICAL POINT, TRIPLE

See *Triple critical point*.

CRITICAL PORE SIZE

Size of a *pore* in unstable equilibrium with its surrounding environment. Smaller pores shrink and heal (see *Healing of defects*), while larger sizes grow. The critical radius of a spherical pore r_{cr} depends on the coefficient of *surface tension* γ, the density of *point defects*, their mobility, the nature of their elastic interaction with the pore (see *Elastic interaction of defects*), the gas pressure inside the pore and outside the sample, and other factors. For example, we have $r_{cr} = 2\gamma/[P + k_B T N_0 \ln(1 + c/c_0)]$ in the presence of extra *vacancies* (vacancy concentration c, equilibrium concentration c_0). Here P is the difference between the pressures inside the pore and outside the sample, k_B is the *Boltzmann constant*, N_0 is the number of atoms per unit volume of material. In polarized substances (with electric or magnetic moments) the pore tends to stretch along the moment direction. The equilibrium ratio of pore axes is determined by competition between the growing *surface energy* and the diminishing field energy as the lines of force flatten, so that ratio grows for larger pore volumes. As the pore enlarges it sucks in more and more energy from the field, and it eventually reaches the critical size $V = V_{cr}$, where $V_{cr}^{1/3} = 0.476\gamma(\ln(0.309/N^{7/6}))^{1/6}/(M^2 N^{7/6})$. The principal axes of the demagnetization factor tensors of the sample and pore (see *Demagnetization*) are assumed to coincide. Here M is the moment per unit volume, and N is the principal value of the sample demagnetization tensor along the pore axis. As a rule, the *segregation* of impurities in the pore surface layer lowers the surface tension and results in a lower r_{cr}.

CRITICAL REGION, critical temperature range

A fluctuational range in the neighborhood of a phase transition, located in the *phase diagram* adjacent to the *second-order phase transition* line, or to the *critical point*, where the thermodynamics and dynamics of the system are strongly affected by *critical fluctuations* of the order parameter. The size of the critical region is characterized by the *Ginzburg number*, and depends on the type of interatomic (electron, spin) interaction potential associated with the phase. The behavior of the system in the critical region may differ both quantitatively and qualitatively from that predicted by the *Landau theory of second-order phase transitions*.

CRITICAL STATE

The limiting equilibrium state of a two-phase system at which the physical differences between the two *phases* vanish. For example, the α- and γ-phases of cerium with the same face-centered cubic lattice become identical in the critical state. The "liquid–vapor" case is quite well known. The critical state on a *phase diagram* corresponds to a *critical point* or critical line. When approaching that point *critical fluctuations* of the density of the substance become very much larger, while the latent heat of a *first-order phase transition* approaches zero. At the critical point itself the phase transformation occurs simultaneously throughout the entire system, similar to a *second-order phase transition*. The values of the state parameters for the critical state are called *critical parameters*, e.g., *critical temperature*, critical pressure, critical volume.

CRITICAL STRAIN, critical degree of deformation

The extent of *strain* that results in a qualitative change in the behavior of a material under a given external action (e.g., deformation which produces sample *failure*), or an irreversible change of the sample shape (i.e. *plastic deformation*). Sometimes one may speak about a critical deformation with respect to certain physico-chemical characteristics of a solid, in particular about a critical plastic deformation resulting in *recrystallization*, etc.

CRITICAL SUPERCOOLING FIELD

The value of the *magnetic field* B_s at which a sample in a normal (i.e. nonsuperconducting) *metastable state* transforms into a superconducting state. A type I superconductor can be supercooled, meaning that it can remain in the normal state when the temperature is below the superconducting transition temperature T_c, and the applied magnetic field is also below the *thermodynamic critical magnetic field* B_c. A further lowering of the applied field to the value B_s brings about a *phase transition* to the superconducting state. The effect is only observed in pure (type I) superconductors, and its magnitude is characterized by the ratio $\alpha = B_s/B_c < 1$. In some superconductors (Al, In, Sn, etc.) the extent of the overcooling may be quite significant, and indium has been supercooled down to $\alpha < 0.1$.

CRITICAL TEMPERATURE

A temperature corresponding to a *critical point*, i.e. a point on a *phase diagram* at which the equilibrium curve for two *phases* in contact comes to an end. The concepts of critical temperature and critical point were introduced by D.I. Mendeleev (1860). The critical point is an isolated point of a *second-order phase transition*, and more generally T_c is often understood as the temperature of second-order phase transitions, including cases when the phase diagram contains a line or even a range of continuous *phase transitions*. The *thermodynamic potentials* of the system, which are functions of thermodynamic variables (such as temperature, volume, electric or magnetic polarization), feature singularities at $T = T_c$. While approaching T_c their derivatives, such as *susceptibility*, *specific heat*, etc., may grow without limit. In the neighborhood of T_c the temperature dependences of thermodynamic variables are described by power law functions (see *Critical phenomena*, *Scaling invariance hypothesis*). Several examples of critical temperatures are: transformation of a *metal* to a superconducting phase, transitions to magnetically ordered or ferroelectric states (*Curie point*), and transformation of a liquid to a superfluid state.

CRITICAL TEMPERATURE OF SUPERCONDUCTORS

Transition temperature T_c from the normal state of finite resistance R to the superconducting state of zero resistance, which is specific for each superconductor (see Fig.). In the absence of a magnetic field the transformation corresponds to a *second-order phase transition*. For the *phonon mechanism of superconductivity* the value of T_c depends on the magnitude of the *electron–phonon interaction* constant λ. This interaction involves the attraction between pairs of electrons in the vicinity of the *Fermi surface* with the formation of *Cooper pairs*, despite repulsion from the Coulomb interaction. The calculation of T_c from first principles is so complex that T_c is usually estimated from its dependence on various parameters in terms of simplified models. The most useful approach involves the *Bardeen–Cooper–Schrieffer theory* (BCS) which yields the following expression for T_c:

$$T_c \approx \Theta_D e^{-1/g}.$$

Here Θ_D is the *Debye temperature*, and g is a dimensionless positive coupling constant, with the BCS model value given by $g = N(E_F)(V_{ph} - V_c)$, where $N(E_F)$ is the *density of states* of electrons at the *Fermi level* E_F, V_{ph} and V_c are the

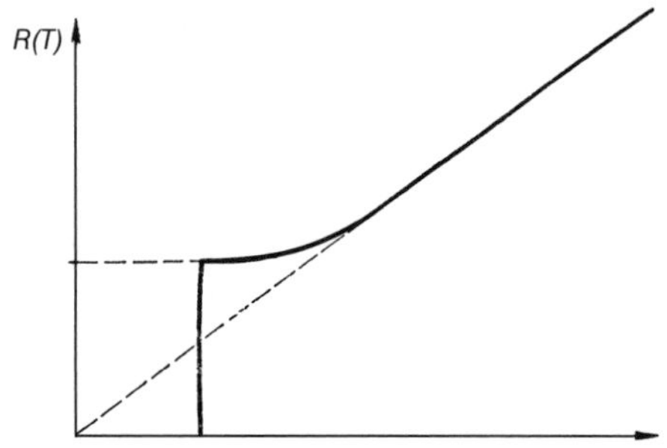

Critical temperature of superconductors.

matrix elements of the electron–phonon interaction (second order in perturbation theory) and the Coulomb repulsion, respectively. Note that $g > 0$ for $V_{ph} > V_c$, corresponding to the *BCS criterion for superconductivity*. A more accurate accounting for the Coulomb repulsion in the energy range near E_F results in its significant weakening (the logarithm $\ln[E_F/(\hbar\omega_D)]$ becomes large (see *Bogolyubov–Tolmachyov logarithm*) and a softening of the superconductivity criterion

$$V_{ph} > V_c^* \equiv \frac{V_c}{1 + N(E_F)V_c \ln[E_F/(\hbar\omega_D)]}.$$

McMillan (1968) was the first to make approximate calculations of T_c for the cases of intermediate and strong coupling in the limit $\lambda \geqslant 1$. He used the integral *Eliashberg equations* and obtained:

$$T_c = \frac{\Theta_D}{1.45} \exp\left\{-\frac{1.04(1+\lambda)}{1-\mu^*(1+0.62\lambda)}\right\},$$

where $\mu^* = N(E_F)V_c^*$ is the *Morel–Anderson Coulomb pseudopotential*.

CROSS-DEFORMATION INTERACTION,

electron–phonon interaction in one-dimensional metals

Interaction between *conduction electrons* and long-wave *phonons* in conductors exhibiting one-dimensional *electrical conductivity* (one-dimensional conductors). It is produced by modulation of a random static scattering field of impurities or other *defects* by a sound wave. It is non-diagonal in the electron momenta at the *Fermi surface* of a one-dimensional metal, i.e. the sign of the electron velocity can change due to this interaction. It is similar to the common *deformation interaction* (A.I. Akhiezer, 1938) that is described by introducing an additional term $H_d = \lambda_{ik}u_{ik}$ into the Hamiltonian of an electron, where the *strain potential* tensor λ_{ik} is of the order of the *Fermi energy*, e_F, and u_{ik} is the *strain tensor*. The cross-deformation interaction results in the appearance of a new term in the Hamiltonian, its structure schematically represented by $H_{cd} = \Delta_{ik}u_{ik}$, where Δ_{ik} is the complex *cross-deformation potential* tensor. In the case of the weak scattering of electrons at impurities (*defects*), the cross-deformation interaction appears much weaker than deformation scattering, meaning $|\Delta_{ik}|/\lambda_{ik} \propto (e_F\tau)^{-1/2}$ for $e_F\tau \gg 1$, where τ^{-1} is the frequency of electron–impurity scattering. However, since the deformation potential in one-dimensional metals does not depend on the *quasi-momentum* of the electron at the Fermi surface, and the electron *energy bands* do not overlap, the Coulomb interaction between the electrons almost completely screens the potential λ_{ik} (to the accuracy of terms in $(qa)^2$ of the order of magnitude $(qa)^2 \ll 1$, where q is the wave number of sound, and a is the crystal lattice constant). Therefore, alongside the *inertial interaction* resulting from the Stewart–Tolman effect, the cross-deformation interaction is the principal type of interaction of electrons with long-wavelength sound waves in one-dimensional metals (see *Electron–phonon interaction*).

CROSS-LINKING OF POLYMERS

See *Polymer cross-linking*.

CROSSOVER TRANSITIONS

See *Direct transitions*.

CROSS-RELAXATION

Relaxation processes in spin and other quantum systems with non-equidistant energy levels that control the reestablishment of equilibrium between these levels after excitation by a radiation source. For example, consider a spin system with two resonant frequencies ω_a and ω_b, so that the difference between them does not exceed their line widths. The process of cross-relaxation develops via mutual spin flips, the so-called *spin-flip process*, resulting from the *magnetic dipole interaction* between the spins. It can also involve spin exchange. The extra energy $(\omega_b - \omega_a)$ released during cross-relaxation is compensated by the increase of the dipole–dipole interaction energy of the spin system.

Cross-relaxation processes may involve two qualitatively different types of *spin* systems: resonant frequencies ω_a and ω_b for different types of spins (Fig. (a)); and frequencies ω_a and ω_b for one and the same spin type (Fig. (b)). The latter case can occur in a system with $S > 1/2$ under the influence of a crystalline electric field. The rate

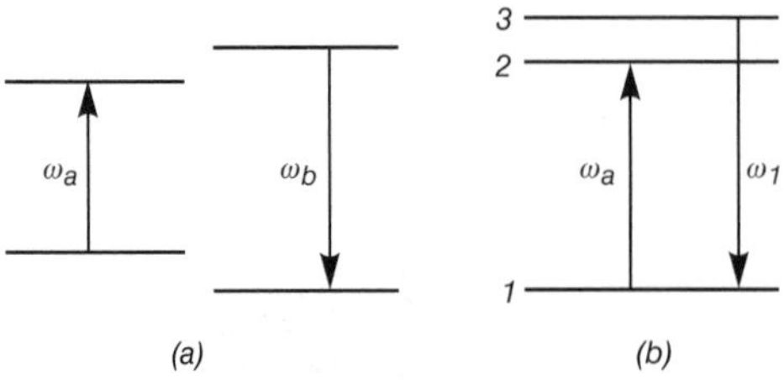

Examples of cross-relaxation processes.

of cross-relaxation depends strongly on the difference between the energies of the respective levels, and becomes negligible when this difference exceeds the average dipole–dipole interaction energy. This phenomenon was initially observed during *nuclear magnetic resonance* in LiF where the rate of transfer of *resonance* saturation from the Li^7 nuclei to the F^{19} nuclei depends on the value of the applied magnetic field. The process of cross-relaxation can decrease the *spin–lattice relaxation* rate of a slowly relaxing system of spins by coupling them with another spin system that relaxes more rapidly via a much shorter spin–lattice *relaxation time*. Cross-relaxation is thus a mechanism for thermal contact between the two spin systems. For example, slowly relaxing nuclear spins can transfer their energy to the lattice much more rapidly if they undergo cross-relaxation with rapidly relaxing electronic spins. Beside these two-spin cases, cross-relaxation processes may involve larger numbers of spins, such as during *dynamic nuclear polarization*.

CROSS-SECTION of a process, effective cross-section

A factor characterizing the averaged rate of change in the state of particles as a result of their interaction (scattering, dispersion, chemical, nuclear and other reactions). Let the velocity of one of two interacting particles relative to each other be v. Consider a system of N (cm^{-3}) fixed and n (cm^{-3}) particles moving with relative velocity v in some initial state, with their flux characterized by the intensity $\Phi = nv$ particles·$\text{cm}^{-2}\cdot\text{s}^{-1}$. The cross section σ of the process of transition from the initial to the final state caused by particle interactions is determined from the relation $dn/dt = \sigma \Phi N = \gamma n N$, where t is the time, and $\gamma = \sigma v$ is the probability of an elementary act of the process. The cross-section σ at a given Nnv defines the fraction of particles undergoing the change: $dn/n = \sigma N v\, dt$.

The change caused by *scattering* is usually characterized by a *scattering angle* θ, by the change of scattering energy of the particles, or by the value of their orbital angular momentum l after scattering. In the case of nuclear or chemical reactions, the system after the reaction is defined by the final reaction products and their states. If it is necessary to determine the rate of reaction product formation by characteristics from an infinitesimal interval of values, then *differential cross-sections* are used. For example, scattering in the interval of angles $[\theta, \theta + d\theta]$, and scattering with power transmission from the incident particle to the scatterer in the energy (temperature) interval $[T, T + dT]$, are characterized by the differential cross-sections $d\sigma/d\theta$, $d\sigma/dT$, etc. The corresponding *total cross-sections* are $\sigma = \int (d\sigma/d\theta)\, d\theta$ and $\sigma = \int (d\sigma/dT)\, dT$, where the latter integration is from $T = 0$ to $T = T_{max}$, representing the maximal energy transferred in the interaction.

The cross-section is defined by quantum-mechanical probabilities of transitions between initial and final states of the system under the influence of particle interactions, or by reasoning from the classical mechanics equations of motion. Both the scattering cross-section and cross-sections of various reactions often differ appreciably from the geometrical cross-sections of the interacting particles, or from dimensions defined by the radius of action of a short-range force, and they can be much larger or much smaller than the geometrical dimensions. Thus, the elastic scattering of particles on a ball of radius r is characterized in classical mechanics by the cross-section $\sigma = \pi r^2$ corresponding to the maximal geometrical ball cross-section. Quantum theory takes into account the wave properties of scattering particles, and provides the expression $\sigma = 4\pi r^2$ for $\lambda \gg r$ (λ is the *de Broglie wave* length, $\lambda = h/p$, p is the scattered particle linear momentum). For $\lambda \ll r$ the cross-section is $\sigma = 2\pi r^2$, and the contribution to the total cross-section of the order πr^2 gives the small angle scattering. For resonant scattering the elastic scattering cross-section reaches the value $\sigma_l = 4\pi\lambda^2(2l + 1)$, and for slow particles with λ considerably exceeding the radius of action of the force it is much greater than the geometrical dimensions of the system. This effect has a quantum origin. The opposite situation with the cross-section much smaller than geometrical dimensions is realized in reactions where there is an energy potential barrier between the interacting particles. Then, on the average, repeated approaches of reacting particles are necessary to overcome the barrier. A similar situation often occurs in *quasi-chemical reactions* involving *point defects* in a *solid.*

CROSS SLIP

A transition of helical sections of *dislocations* from one *slip plane* with the normal $\boldsymbol{n}$, to another one with normal $\boldsymbol{m}$, if the line of the intersection of these planes is parallel to the *Burgers vector* of the dislocation $\boldsymbol{b}$, i.e. $[\boldsymbol{n} \times \boldsymbol{m}] \| \boldsymbol{b}$. A slip plane is the closest packed plane of a crystal on which dislocations glide. The cross slip appears, first of all, near the obstacles for the usual *dislocation slip.* If such an obstacle is local, then after the motion within the cross slip plane the dislocation may return to the initial plane, i.e. the process of *double cross slip* takes place. It is experimentally established that in face-centered cubic single crystals the beginning of the third stage of *hardening* is associated to the fact that as a result of the *stress concentration* by single or multiple steps there appear areas of dislocation lines which may perform cross slip. The process of cross slip determines the *creep* of metals in the interval of moderate temperatures (e.g., for *aluminum* single crystals within the temperature range 0 to 100 °C).

CROWDION (fr. crowd)

A specific quasi-one-dimensional *point defect* of the type of an *interstitial atom* such that adjacent atoms are preferentially displaced along one of the directions of dense packing. The extra atom in a crowdion in a fully packed row displaces 5 to 10 atoms along the line by individual minimal distances (much shorter than the interatomic distance a, see Fig.). H. Paneth (1950) introduced the concept of a crowdion to describe self-diffusion in *alkali metals* with BCC lattices (with crowdion direction $\langle 111 \rangle$). Experimental data are available proving the existence of crowdions in FCC metals (Cu, Ni) at low temperatures (along the $\langle 110 \rangle$ direction). A crowdion may form during radiation.

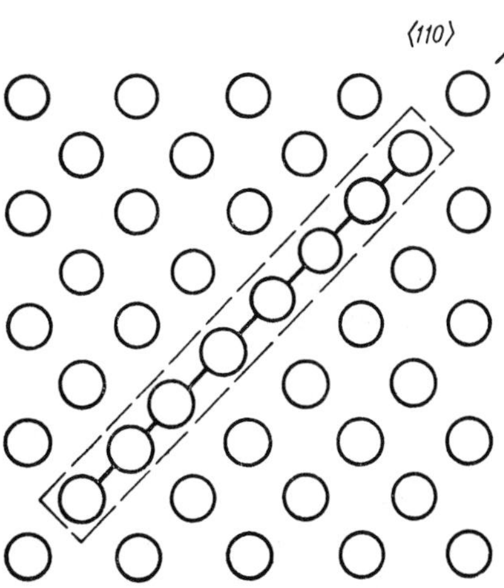

Crowdion.

A *dynamic crowdion* appears when a pulse sufficient to displace a neighboring atom is transmitted along a fully packed atomic row to one of the atoms of the crystal. A relay displacement then takes place, and the crowdion propagates along the direction of the transferred pulse. A site *vacancy* remains at the site of the initially dislodged atom. A dynamic crowdion is able to propagate a distance of the order of $10^3 a$ from the position of the initially dislodged atom. A static crowdion may remain stable in some crystals within a certain range of temperature. It propagates as a whole, and the process involves one-dimensional *diffusion* along the atomic line. Although a crowdion can travel a long distance, the displacement of an individual atom at each step in the displacement process is the same (equal to a). In its quantum-mechanical description the crowdion is a *quasi-particle* (a crowdion wave).

CRYOCRYSTALS

Solidified gases, i.e. *crystals* formed by atoms or small molecules with weak intermolecular interaction forces, which therefore feature low *crystallization* temperatures. Their relative simplicity is convenient for studies of the general properties of the crystalline state of matter.

Atomic cryocrystals – solid Ne, Ar, Kr, Xe are closest to theoretical models of crystals with central force interactions. They form face-centered cubic lattices containing identical atoms interacting via valence forces of attraction, and dispersion forces of repulsion. Their low-frequency collective excitations are *phonons*, and their high-frequency ones are *excitons*.

Solid N_2, O_2, CO, CO_2, CD_4, etc., are *molecular cryocrystals*. Their molecules may be considered nondeformable since their intramolecular forces are much stronger than the intermolecular ones. Due to their rigidity these molecules rotate as whole units. Non-central interaction forces (dispersion, exchange, and multipole) result in certain orientations of these molecules in the *crystal lattice*. *Phase transitions* mainly arise from changes in that orientational order. Their collective excitations are phonons (both acoustic and optical); *librons* (which appear only when some long range orientational order exists); bound *intramolecular vibrations*; excitons, and their combinations. If the central interaction dominates over the non-central variety (e.g., in solid CD_4), the breakdown of long-range orientation order (*orientation melting*) occurs within the solid phase. Regions of short-range orientational order may still persist in orientationally disordered crystals. With increasing temperature these regions shrink, and librons transform into weakly perturbed rotations. Of particular interest is *solid oxygen*, a molecular *magnetic substance* with particular structural features and collective excitations (*magnons*, *biexcitons*) that result from the uncompensated electron spin of its paramagnetic O_2 molecule.

Quantum crystals form a separate group of cryocrystals (^{3}He, ^{4}He, H_2, HD, D_2 and CH_4). The dynamics of the translational and rotational motions of these molecules defies a classical description, being controlled by the laws of quantum mechanics instead.

The weak intermolecular interactions, small size, and light mass of these molecules all combine to produce properties typical of cryocrystals: low crystallization temperature, *strength* and *thermal conductivity*, and high *plasticity*, *compressibility*, vapor pressure, and *thermal expansion* coefficient.

CRYOGENIC TECHNOLOGY

Branch of technology involving the development, manufacture and use of instruments and machines that produce and maintain low temperatures, and that can liquefy and separate various gases. Industrial applications of cryogenic technology are numerous, since pure oxygen obtained by liquefying air and separating its gases at low temperature is widely used in *metallurgy* and other industries. Liquid nitrogen, also obtained from liquid air, is more and more widely used for the long-term preservation of food products and biological materials, to operate transport refrigerators, for the needs of cryofracturing of various substances, etc. Various options of hydrogen liquefiers (up to 10^4 l/h), helium liquefiers (100 l/h), and refrigerators which reach millikelvin temperatures using ^{3}He dissolved in superfluid ^{4}He are manufactured.

Cryogenic technology applies low-temperature methods to other technical areas, such as superconductivity. Using the phenomenon of *superconductivity* one may develop economically efficient systems (e.g., large-scale magnetic devices with *superconducting magnets* for nuclear magnetic resonance imaging, accelerators, and thermonuclear installations). One may attain high specific power output for electric motors and generators, achieve very high sensitivities in measuring devices using the *Josephson effect*, etc. Projects are in progress to build superconducting electric power lines and transportation systems.

CRYOGENIC TEMPERATURES

Low temperatures, in the range from 120 K to below 1 K (see *Low-temperature physics*).

There exist various methods of deep freezing, or reaching these temperatures: cooling a gas when throttling (*Joule–Thompson effect*), isoentropic gas expansion in a *gas expander*, *desorption cooling*, expansion liquefaction, vapor pumping over a *cooling agent*, and so on. Combinations of two or more methods are often used.

The liquefaction of nitrogen, hydrogen and helium provides the principal refrigerants in technical and scientific research. Vapor pumping over these liquids produces low temperatures in the range of 120–0.25 K ($N_2 \sim$ 120–63 K; $H_2 \sim$ 20–10 K; ^{4}He $\sim$ 4–0.7 K; ^{3}He $\sim$ 3–0.25 K). These refrigerants are employed in practice for cooling various electrical engineering devices (superconducting magnets, power transmission lines, electric generators, electric motors, transformers, etc.), electronic devices (radiation detectors, first stages of amplifiers, computer elements, etc.) and other units. Deep freezing is used for gas blend separation, gas purification, cooling of food, and for other purposes in medicine, construction, engineering, and so on.

CRYOSORPTION (fr. Gr. $\kappa\rho\upsilon o\varsigma$, cold, and Lat. *sorbeo*, absorb)

The process of absorption of gas from the environment by a strongly cooled solid or a cryogenic liquid called a *cryosorbent*. The removal from the gas phase by the liquid cryosorbent is called *absorption*. This process is explained by the finite solubility of certain gases in cryogenic liquids. A solubility coefficient α (in m^3 per ton of cryosorbent at $0\,^\circ C$ and 760 mm Hg) is introduced. The value of α is a function of the partial pressure, P, and the amount of the absorbed gas is given by the expression $\sigma = \alpha P$. The coefficient α has a different dependence on the temperature for different materials. It decreases with the temperature if gaseous H_2 or He is absorbed by liquid nitrogen; while for most gases absorbed by liquid methanol it grows with the drop in temperature. The absorption process is used in several separation cycles. During cryosorption of vapors by porous solids capillary *condensation* may occur. Cryosorption of vapors from certain multiphase gas systems may proceed by way of freezing the atoms of highly volatile fractions into a solid layer of a weakly volatile fraction precipitated at a cooled surface.

A special case of cryosorption is *adsorption* of a material from the gas phase by a surface layer of a cooled adsorbent. The amount of gas, V, adsorbed by a unit mass of adsorbent depends on the nature of the adsorbent, and that of the adsorbed substance, and it grows for lower temperatures and higher gas pressures. The curve describing the dependence of V on P at $T = \text{const}$ is called the *adsorption isotherm*. The amount of adsorbed gas is linear with pressure at low pressures, following *Henry's law*: $V = kP$.

As a rule, several processes proceed in parallel during cryosorption. Cryosorption is widely used in systems of vacuum cryogenic evacuation (see *Low-temperature pumping methods*).

CRYOSTAT (fr. Gr. $\kappa\rho\upsilon o\varsigma$, cold, and $\sigma\tau\alpha\tau o\varsigma$, standing, unmoving)

A device for cryostatting, i.e. for maintaining a constant cryogenic temperature between 0 and 120 K. Cryostats are used to study physical properties of materials, the phenomenon of *superconductivity*, to cool photodetectors, and to pursue other tasks at low temperatures. The simplest cryostat consists of two vacuum concentric bottles. The inner bottle, where the specimen under study is placed, is filled with a cryogenic liquid boiling at low temperature, e.g., with liquid *helium* (4.2 K). The outer bottle playing the role of thermal isolation is filled with a liquid boiling at a higher temperature, e.g., liquid nitrogen (77 K). Cryostats are often used to cool and store the specimen at a specific temperature within a wide range (from room to a very low value, e.g., 4.2 K). Such a device is not a cryostat in a strict sense, or is not one at every temperature. However, if its minimum attainable temperature falls within the cryogenic range, it is commonly called a cryostat.

In most cryostats the sources of cold that absorb the heat removed from the specimen are liquefied gases. The most widely used are liquid nitrogen (N_2), helium (^{3}He, ^{4}He), hydrogen (H_2) and, less commonly, oxygen (O_2). Special cryostats used to cool photosensors sometimes contain solid noble gases. More recently compact refrigerators have been developed that can continuously pump a liquid or gaseous *cooling agent* into a cryostat. Together the two units form a joint cryostatting system. To reach superlow temperatures ($T < 1$ K) complex multistage devices are employed which achieve *cryogenic temperatures* of a specimen using liquid gases (^{3}He–^{4}He solution refrigerators), or through adiabatic demagnetization (see *Adiabatic demagnetization cooling*).

Depending on the type of study, the cryostat may include optical windows (*optical cryostat*), a superconducting solenoid, a very high frequency lead, or other devices to probe the specimen in various ways.

Many types of cryostats are available, varying in purpose, cooling technique, temperature range covered, permissible thermal load, accuracy, reliability, stability, character of probing, etc.

Two main approaches have been followed in building cryostats:

1. Complex design for highly specific tasks, e.g., achieving extra low temperatures, cooling superconducting devices, cooling radiation sensors, solving problems in material development, etc.
2. Simple and cheap universal design for scientific and technological applications, to cover a wide temperature range, have short priming time, quickly reach functional range, have low maintenance cost, match well with peripheral measurement devices, permit manipulation of the cooled specimen, etc.

Choosing the technique of heat evacuation from the specimen under study is of particular importance, especially for high heat influx. Several common techniques for heat removal are: thermal radiation exchange, direct mechanical contact with the cooled surface, heat exchange through a gas carrier, the Swenson effect, and forced freezing of a target with a cold gas jet. These are all applicable over a wide temperature range, and have already reached advanced stages of development.

The most widely applied cryostats make use of a single cryogenic fluid, and are of two types, namely liquid immersion and stream flow. To absorb thermal energy from the specimen *liquid cryostats* ordinarily use the heat of evaporation of a liquid coolant, but this only supports a temperature close to the boiling point of the coolant. Over a wide temperature range heat can be absorbed and removed via the *specific heat* of a gaseous coolant. The temperature of a flowing gas is easily varied, with the coolant brought directly to the specimen in the amount needed to compensate for the heat input from probing at a given temperature. *Stream cryostats* are of a more general nature, and permit using all the five known methods of heat evacuation immediately from the cooled specimen. Stream cryostats have several important advantages over the liquid type: they can achieve any temperature above the boiling point of the coolant; the coolant is very efficiently utilized; the temperature can be changed rapidly; the cryostat may be arbitrarily oriented in space; these cryostats are easily made small, etc. This small size capability is particularly important when matching cryostats to standard experimental devices, such as spectral instruments, optical or electronic microscopes, etc.

CRYOTRON

A superconducting device used as a switch, or a memory cell, in computers. Cryotrons employ the phenomenon of *superconductivity* turn-off by a magnetic field. The current in an input circuit magnetically controls the superconducting-to-normal transition in one or more output circuits. In practice a cryotron is a superconducting rod, or more commonly, a *thin film* which may be in either the normal (switched-off) or superconducting (switched-on) state. A different *superconductor* with a higher critical field controls it. The cryotron is switched from one state to another by varying the current through a control circuit. Cryotrons have the advantage that in their superconducting state the dissipation of energy in them is zero. An important characteristic is the switching time, which in more advanced designs may be as short as 10^{-10} s.

CRYSTAL BASAL PLANE

See *Basal plane of crystals.*

CRYSTAL CHEMISTRY

Branch of *crystallography* studying the locations of atoms and particles in *crystals*, and the nature of their *chemical bonds*. Crystal chemistry brings together the results from experimental X-ray and other diffraction techniques used to study the atomic structure of crystals (see *X-ray structure analysis*, *Electron diffraction analysis*, *Neutron diffractometry*), and combines them with both classical and quantum theories of chemical bonding to compute the energy of crystal structures, taking into account the *crystal symmetry*. Crystal chemistry approaches make it possible to explain, and in certain cases to predict, the positions of atoms and molecules in a *crystal lattice*, and their nearest neighbor distances, starting from the chemical composition of the material.

Chemical bonding between atoms in crystals results from the interaction between their outer valence electrons. Depending on the nature of the bonding, the equilibrium distance between atoms is usually within 0.15–0.4 nm. When atoms are brought closer together than this a strong repulsion develops. Therefore, to a first approximation, certain "sizes" or characteristic radii may be attributed to atoms for each type of bonding. Then one may proceed from a physical model of a crystal envisaged as an atomic electron-dominated system, to a geometrical model that looks upon a crystal as an ordered arrangement of incompressible balls.

A full crystal chemistry description includes specifying the size of the *unit cell*, the *space group* of the crystal, its individual atom coordinates, distances between them, and the types of chemical bonds that are present. In addition, the near environment of separate atoms, the characteristic atomic groupings, thermal vibrations of the atoms, etc., have to be taken into account.

In terms of the type of their chemical bonding, crystals may be divided into four major groups: *ionic crystals* (e.g., NaCl), *covalent crystals* (e.g., *diamond*, *silicon*), metallic crystals (*metals* and *intermetallic compounds*), and *molecular crystals* (e.g., naphthalene).

A certain structure is typical for each of these types of crystals. However, when thermodynamic conditions change these may sometimes alter (*polymorphism*). As a rule, the simpler the chemical formula of the compound, the more symmetric its structure. Crystals with identical chemical formulae (in the ratios of constituent atoms) may have identical *crystal structures* (identical structural types), despite possible differences in their bonding type (*isostructure*). Alkali halides of the NaCl type and certain oxides (e.g., MgO), as well as several alloys (e.g., Ti–Ni), are isostructural. Isostructural crystals with the same kind of bonding are called isomorphic (see *Isomorphism*). In many cases a continuous sequence of *solid solutions* exists between isomorphic crystals.

The geometric model of crystals involves the concept of effective radii of atoms, molecules and ions (*crystal chemical radii*), so that their bonding distance is equal to the sum of such radii (additivity of crystal chemical radii). Using experimental data one can tabulate crystal chemical radii for all the types of bonding (see *Atomic radii*, *Ionic radius*). The principal geometric concept in crystal chemistry is the *theory of dense packing* that explains the positioning of atoms in certain metallic and ionic structures graphically. For ionic structures one may talk about "holes" in the packing of larger anions that are occupied by smaller cations of similar ionic radii.

The structural unit in molecular crystals is a molecule. Crystal chemistry of organic compounds considers the rules of dense packing of molecules, the relation between the symmetry of molecules and the symmetry of the crystal, and the types of organic structures involved. Specific crystal patterns are identified in the structures of *polymers*, *liquid crystals*, and *biological crystals*.

The coordination number K (see *Coordination sphere*) and the type of coordination polyhedron characterize the chemical bonding of a given atom, as well as the structure of a given crystal as a whole. For example, Be (with a few exceptions) and Ge form a tetrahedral environment ($K = 4$), Al and Cr have an octahedron ($K = 6$) for their coordination polyhedron, and Pd and Pt are often square planar coordinated ($K = 4$). Low coordination numbers point to a significant role of directional *covalent bonds*, and large K values (e.g., 6, 8, 12) indicate an appreciable role of ionic and metallic bonding. Bonds of various types coexist in many crystal structures (graphite, MoS_2, etc.). Such structures are called *heterodesmic structures* in contrast to *homodesmic ones*, the latter featuring uniform bonding (diamond, metals, NaCl, crystals of inert elements), and the former mixed bonding.

The laws of thermodynamics determine whether or not a particular crystal structure will form. The structure most stable at a given temperature T is the one that has the lowest *Helmholtz free energy* $F = U - TS$, where U is the bonding energy of the crystal (energy needed to break up the crystal into separate atoms or molecules) at $T = 0$ K, and S is its *entropy*. The higher the free energy the stronger the bonding in the crystal. It is usually within 40–80 kJ/mol for crystals with covalent bonding, is somewhat lower for the molecular crystals, and is lowest for van der Waals bonding (4–15 kJ/mol). In practice, theoretically calculating the free energy and predicting the structure of crystals is only possible for comparatively simple cases. Such calculations can be carried out within the framework of the *band theory* of solids. In some cases semiempirical expressions for the potential energy of interaction of atoms in crystals can yield fairly accurate results for certain types of bonding (see *Interatomic interaction potentials*).

Table 1. Meaning of the first (I), second (II) and third (III) terms in the International Symbols of the 32 crystallographic point groups of the seven crystal systems (see figure).

Crystal system	Position of each term in the symbol		
	I	II	III
Triclinic	Unit symbol corresponding to any direction	–	–
Monoclinic	Twofold axis and/or reflection plane	–	–
Orthorhombic	Twofold axis (x) or reflection plane	Twofold axis (y) or reflection plane	Twofold axis (z) or reflection plane
Trigonal	Threefold symmetry axis	Possible twofold axis or reflection plane	–
Hexagonal	Sixfold symmetry axis	Possible twofold axis or reflection plane	Possible twofold axis or reflection plane
Tetragonal	Fourfold symmetry axis	Possible twofold axis or reflection plane	Possible twofold axis or reflection plane
Cubic (isometric)	A symmetry element	Diagonal threefold axis	Possible twofold axis or reflection plane

CRYSTAL CLASSES

Crystallographic classes or symmetry classes are categories of crystallographic *point groups* of symmetry (introduced in 1830 by J. Hessel and, independently, by A.V. Gadolin in 1867). To denote crystal classes symbols are used (see *Crystal symmetry*) that are based on group theory theorems for the combination of symmetry operations. According to these, the product of two symmetry elements produces another element of the symmetry group. Table 1 presents notational rules for crystal classes that are found in various *crystallographic systems*. For example, the point group $\bar{4}3m$ corresponding to the mirror plane class of the cubic system, denoted by T_d (dihedral-tetrahedral) in the Schönflies system, has a fourfold inversion axis ($\bar{4}$), a threefold rotation axis (3) and a mirror plane (m). The axial monoclinic point group (Schönflies C_2) has only a twofold axis. The coordinate and diagonal elements of symmetry are understood as planes and axes passing along the coordinate planes (axes) and the bisectors of the angles between them, respectively. The crystal classes are separated into primitive (single symmetry axis), centric (symmetry axis plus *center of symmetry*), planar (n symmetry planes along the axis), axial (having n twofold axes normal to the symmetry axis), etc. Elements of the crystal classes of the 32 crystallographic point groups are presented in the figure.

CRYSTAL CLEAVAGE

The capacity of *monocrystals* to be split (cleaved) along certain crystallographic planes (*crystal cleavage planes*) on exposure to a mechanical force or impact. The cleavage is related to the specific features of the *crystal structure.* The *crystal* is split along directions or planes of smallest *crystallographic indices*, which involves the breakage of the weakest bonds. Thus, the *graphite* structure is susceptible to actions counteracting *van der Waals forces* perpendicular to (0001), which involves the separation of adjacent sheets of carbon atoms from one another. Most cubic crystals cleave along cubic planes. The crystal cleavage plane of BCC *metals* is often (100), whereas hexagonal crystals cleave along the *basal plane.* The orientation of the cleavage planes of certain crystalline materials (e.g., sphalerite) is determined by the charge of adjacent separable atomic sheets. Cleavage along planes of opposite charge requires more energy than splitting along planes containing ions of the same sign. The following degrees of cleavage of crystals are distinguished: highly perfect cleavage (cleaved into thinnest layers), perfect cleavage (crystal split

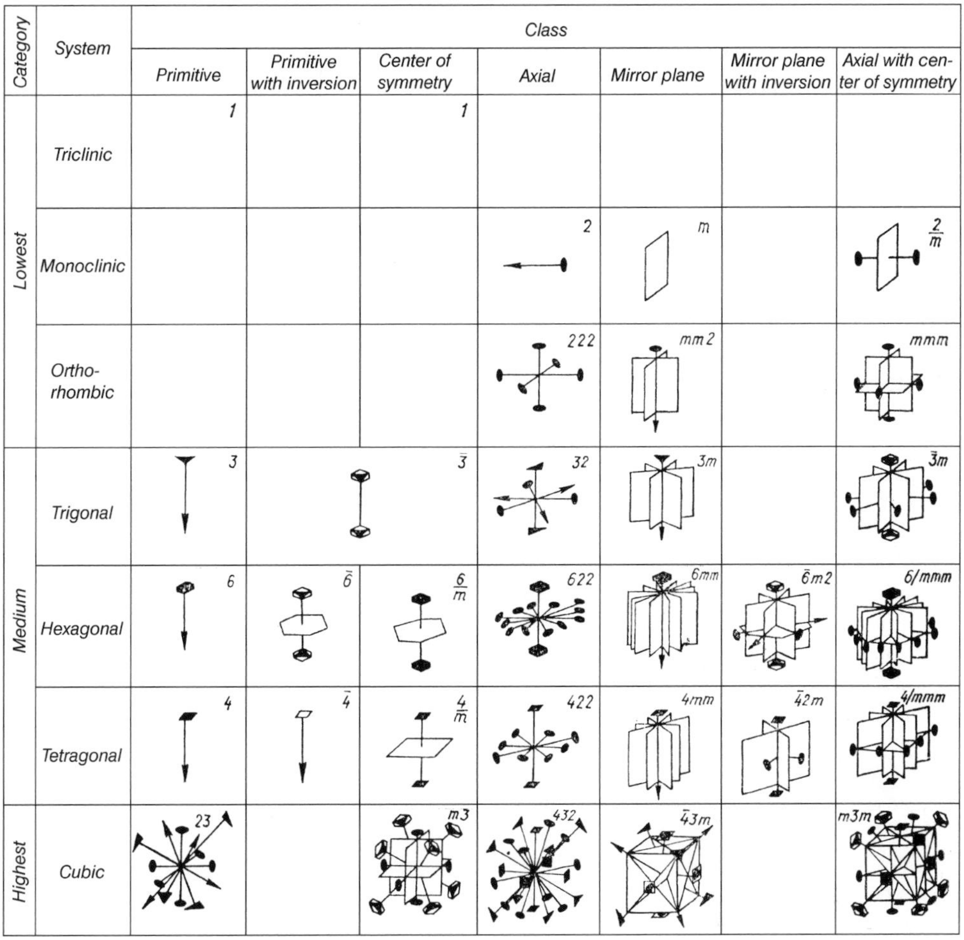

Category	System	Class: Primitive	Primitive with inversion	Center of symmetry	Axial	Mirror plane	Mirror plane with inversion	Axial with center of symmetry
Lowest	Triclinic	1		1				
	Monoclinic				2	m		$\frac{2}{m}$
	Ortho-rhombic				222	mm2		mmm
Medium	Trigonal	3		$\bar{3}$	32	3m		$\bar{3}m$
	Hexagonal	6	$\bar{6}$	$\frac{6}{m}$	622	6mm	$\bar{6}m2$	6/mmm
	Tetragonal	4	$\bar{4}$	$\frac{4}{m}$	422	4mm	$\bar{4}2m$	4/mmm
Highest	Cubic	23		m3	432	$\bar{4}3m$		m3m

into bars along cleavage plane), average degree of cleavage (cleavage facets accompanied by randomly shaped facets), imperfect cleavage (fractography patterns exhibit mostly macroscopically irrational *fracture* regions), completely imperfect cleavage (practically no cleavage: fracture is conchoidal).

Cleavage is one of the characteristic properties of a crystalline material. It is helpful for the determination of a crystal disorientation. Surfaces of high perfection, exposed on splitting along the crystal cleavage plane, are convenient for study. The phenomenon of crystal cleavage is used in jewelry trades to prepare a gem for its *crystal faceting*.

CRYSTAL DETECTOR, crystal counter

An instrument to record and analyze spectroscopic properties of nuclear particles and *photons*. The crystal counter is an electric capacitor (usually parallel plate type), filled with a single crystal *insulator* (e.g., diamond or CdS). A charged particle entering the dielectric produces ionization along its path, and free *current carriers* appear which are driven to the capacitor electrodes by the external electric field. A short electric pulse results, its amplitude proportional to the energy released by the particle in the *crystal*. After amplification it is recorded by a counting device or amplitude analyzer.

Simplicity of design and maintenance, small size (below 1 mm^2) and the fact that certain crystals, e.g., diamond, can function at high temperatures, make crystal counters convenient *nuclear radiation detectors* that are widely used in physics, dosimetry, various industrial applications, etc.

CRYSTAL DIFFRACTION SPECTROMETERS

A class of X-ray spectral instruments for studying absorption and emission spectra of matter and the structural features of *monocrystals*. Crystal diffraction spectrometers may be single-axis (sample crystal), double-axis (collimator crystal plus sample), and triple-axis (collimator, sample, and analyzer crystals) types. As the number of elements in the instrument grows, its spectral dispersion and angular divergence of the radiation diminish, so that it becomes possible to better expand the diffracted intensity over the angle. *Single-axis diffraction spectrometers* are only capable of measuring the integrated intensity of the diffraction. *Double-axis instruments* can record *X-ray spectra* or *rocking curves* for monochromatic radiation (i.e. dependence of diffracted intensity on X-ray angle of incidence). They also provide for antiparallel and parallel positioning of the crystals so that the resulting dispersion is either doubled after the beam passes through the sample crystal, or it reverts to zero. The first position is used for spectral studies, and the second (zero dispersion) for structural studies during which the width and intensity of the diffraction maximum are measured. *Triple-axis spectrometers* permit the separation of the coherent component of the integrated intensity from the diffuse one. See also *X-ray spectrometer*.

CRYSTAL FACETING, crystal cut

Outward shape of crystals specific for a given *crystal class*. Faceting conforms strictly to the point group symmetry relationships of a crystal (see *Crystal symmetry*), and only includes faces of certain categories, the so-called *simple forms* (see *Crystal polyhedron*). As a rule, the simple forms that are possible within a given *point group symmetry* tend to display various combinations one with another. The shapes of individual crystals are closely related to their structure. Materials without clearly preferred directions form, as a rule, uniformly developed *isometric polyhedra*. The *acicular crystals* are typical of materials with needlelike or chain structures; *plate-like crystals* are typical of materials with laminated structures, and so on. In most cases, the only crystal faces that participate in the crystal faceting are those parallel to *crystal lattice* planes of the structure. The latter possesses high *reticular density*, i.e. their unit area maximizes the number of particles (*Bravais law*). In an ideal case the faces of each simple form must have the same configuration and dimensions. The shape of a real individual crystal is determined by the specific conditions of its growth: temperature, pressure, concentration, *impurity atoms* content, etc. When characterizing the crystal faceting, the terms *form* and *habit* are commonly used. The former term concerns a general view of a polyhedron (isometric, acicular, fibriform, etc.), while the latter serves to describe the outward appearance of the individual crystal with the help of dominant simple forms (octahedral, prismatic, dipyramidal, and so on).

CRYSTAL FIELD, internal crystal field, crystalline electric field

The electric field at a point within a *crystal*, produced by its atoms or ions. This field determines energy level splittings of atomic electronic configurations, and the periodic potential of the crystal field determines the *band structure* of the crystal for an ideal lattice corresponding to the crystallographic *space group*. An impurity ion (or local ionic center) has an energy spectrum determined by the local crystal field symmetry. The internal electric fields of defects bring about an *inhomogeneous broadening* of the spectral lines of resonant transitions. The *line shape* (profile) depends on the nature of the defects and their distribution within the crystal. To compute the shape of the resonance lines one can use *statistical theory of line shape*, or possibly the method of moments.

Crystal field theory is applied in various branches of physical science (optics, radiospectroscopy, etc.) to calculate the energy transitions responsible for spectra. It considers various models of the field, and basically accounts for the spatial motions of electrons and nuclei. Energy level splittings produced by the crystal field generally lie in the optical frequency range (10^{13}–10^{15} Hz). Accounting for the spin variables along with the spatial degrees of freedom results in an additional splitting or shifting of the levels. The crystal field affects the spin through the *spin–orbit interaction*. Spin level splittings may be treated using a framework similar to crystal field theory called the *spin Hamiltonian* technique. The energy level spacings associated with electron spin interactions lie in the frequency bands of electron paramagnetic resonance (10^8–10^{11} Hz), while those resulting from the nuclear spin are 2 to 3 orders of magnitude smaller, the same as the frequency range of nuclear magnetic resonance.

CRYSTAL FIELD GRADIENT

A symmetric second-rank zero trace tensor with components equal to the second derivatives of the potential $V(\boldsymbol{r})$ of the electric field at the site of an atom or ion in a *crystal*: $V_{\alpha\beta} = (\partial^2 V/\partial x_\alpha \partial x_\beta)|_{\boldsymbol{r}=0}$. The crystal field gradient defines the energy H of the quadrupole moment $Q_{\alpha\beta}$ of a nucleus (see *Quadrupole*), located at the site:

$$H = \sum_{\alpha\beta} Q_{\alpha\beta} V_{\alpha\beta} = \sum_m Q_m^{(2)} V_{-m},$$

where $Q_m^{(2)}$ are the components of the tensor operator of the nuclear quadrupole moment:

$$V_0 = \frac{1}{2} V_{zz},$$

$$V_{\pm 1} = \pm\left(\frac{1}{6}\right)^{1/2} (V_{xz} \pm \mathrm{i} V_{yz}),$$

$$V_{\pm 2} = \left(\frac{1}{24}\right)^{1/2} (V_{xx} - V_{yy} \pm 2\mathrm{i} V_{xy}).$$

The crystal field gradient of the tensor $V_{\alpha\beta}$ in a system with two principal axes is characterized by two independent parameters $eq = V_{zz}$ and $\eta = (V_{xx} - V_{yy})/V_{zz}$, where $|V_{zz}| \geqslant |V_{yy}| \geqslant |V_{xx}|$, the dimensionless asymmetry parameter η is limited to the range $0 \leqslant \eta \leqslant 1$, and the Hamiltonian of the quadrupole interaction assumes the form

$$H = e^2 q Q\left[4I(2I-1)\right]^{-1} \times \left\{3I_z^2 - I(I+1) + \eta\left(I_x^2 - I_y^2\right)\right\},$$

where q is the quadrupole moment of the nucleus with a spin $I \geqslant 1$. The crystal field gradient for cubic symmetry is zero. When calculating a crystal field gradient one has to take into consideration the fields of point charges (lattice summation carried out by the *Ewald method*), and the *multipole* moments of ions and conduction electrons. To take into account the polarization of the full electron shells of an ion in the *crystal field*, its field gradient at the nucleus is written in the form $(1-\gamma)V_{\alpha\beta}$, where γ is the *antiscreening factor*. For most ions the value of γ is within the range from -10 to -100.

CRYSTAL FIELD SPLITTING

See *Bethe splitting*.

CRYSTAL FIELD THEORY (H. Bethe, 1929)

Theory describing electronic states of *ions* with unfilled d- and f-shells in *crystals*. The scope involves such areas as optical spectroscopy, magnetic properties, and mechanisms of the *electron–phonon interaction*, in crystals containing ions of the three transition metal groups, the *lanthanides*, and the *actinides*. Crystal field theory treats the splitting of electronic energy levels of localized d- and f-electrons by an effective internal crystalline field. The Hamiltonian is solved using perturbation techniques for a single configuration, with the states of a single electron in the central field of a free ion taken as the zero-order approximation. The perturbation operator defined in the space of the wave functions of the $(nl)^N$-shell equals the sum of the energies of the electrostatic interaction, H_{ee}, the *spin–orbit interaction*, H_{so}, and the energy of the electron in the crystal field, H_{c}. The Hamiltonian, H_{c} is averaged over the radial distribution of electron density, and has the form:

$$H_{\mathrm{c}} = \sum_i^N \sum_{kq} B_q^k C_q^{(k)}(i),$$

where $C_q^{(k)}(i)$ are spherical operators in the space of angular variables of the ith nl-electron, and B_q^k are the *crystal field parameters*, quantitative characteristics of the potential of the electric field at the paramagnetic ion site in an equilibrium configuration. They have dimensions of energy, are generally complex, and satisfy the relations

$B_q^{k*} = (-1)^q B_{-q}^k$. The number of independent parameters depends on the symmetry of the crystal field, while the structure of the operator H_c is determined by the choice of local coordinate system. For example, a *cubic crystal field* has the octahedral *point group of symmetry* O_h. In the system of Cartesian coordinate axes coinciding with the C_4 symmetry axes the Hamiltonian of the electron interaction with the cubic crystal field is given by

$$H = B_4\left[C_0^{(4)} + \sqrt{\frac{5}{14}}\left(C_4^{(4)} + C_{-4}^{(4)}\right)\right]$$
$$+ B_6\left[C_0^{(6)} + \sqrt{\frac{7}{2}}\left(C_4^{(6)} + C_{-4}^{(6)}\right)\right].$$

The even component of the crystal field potential is the same in the case of tetrahedral (T_h, T_d) coordination. Often the cubic symmetry component dominates at a low-symmetry site. The parameter $10Dq = (10/21)B_4$ equals the difference between the two d-electron energies in a crystal field of cubic symmetry (doublet e_g and triplet t_{2g} states), and provides a measure of the field "strength" for ions with partially filled d^n-shells. The f^n-shell multiplet splittings in cubic symmetry are defined by the parameters B_4 and B_6, and the structure of the wave functions of the respective states depends on only a single parameter (x), proportional to the ratio B_4/B_6. The effects produced by the spatial correlation of electrons, and the change of the electrostatic interaction parameters in the crystal field, are described by a two-term operator:

$$H_c^{(2)} \sum_{i>j} \{C^{(k_1)}(i) \times C^{(k_2)}(j)\} B_{q'}^{(k)} B_q^{(k_1 k_2)k},$$

that has a complex tensor structure. The parameters B_q^k, $B_q^{(k_1k_2)k}$ can be evaluated from optical or magnetic measurements, or computed within various crystal field models. Quantum-mechanical calculations taking into account the electron structure of the nearest neighbors of a paramagnetic ion are sometimes called *ligand field theory*. When calculating the energy spectrum of the $(nl)^N$ electron shell one should distinguish between strong crystal fields (ions of 4d and 5d transition groups, $H_c > H_{ee} > H_{so}$), intermediate fields (ions of the 3d iron group, $H_c \sim H_{ee} > H_{so}$), average fields ($H_{ee} > H_c > H_{so}$), and weak fields (rare-earth elements, $H_{ee} > H_{so} > H_c$). Various options for constructing the matrix of the H_c operator depend on the choice of the basis for the zeroth approximation (e.g., single electron wave functions, functions of terms $|\gamma LSM_L M_S\rangle$, or functions of multiplets $|\gamma LSJM_J\rangle$). Piezospectroscopic or electric field effects in the spectra of paramagnetic crystals can result from changes in the crystal field during corresponding external actions on the lattice. Modulation of the crystal field by lattice vibrations is the principal mechanism of the electron–phonon interaction.

Crystal field models help with computational techniques. The simplest and historically the first *point charge model* envisaged the crystal lattice as a collection of point charges $Q_\lambda e$ located at lattice sites (λ labels the site), with the charges themselves prescribed by the valences of the respective ions. To determine the field of point charges one has to calculate the lattice sums:

$$B_q^k = -(-1)^q\left(\frac{4\pi}{2k+1}\right)^{1/2} e^2\langle r^k\rangle$$
$$\times \sum_\lambda \frac{Q_\lambda Y_{-q}^k(\theta_\lambda \varphi_\lambda)}{R_\lambda^{k+1}},$$

where Y_q^k are spherical harmonics, r_λ, θ_λ, φ_λ are the spherical coordinates of site λ (the paramagnetic ion is at the coordinate origin), and $\langle r^k\rangle$ are the moments of the electron radial density in the unfilled shell. Discrepancies of an order of magnitude often exist between calculated and measured crystal field parameters. To account for this one may consider changes in the radial density of the d- and f-electrons in the crystal field, fields of point *multipole* moments (dipoles, *quadrupoles*), the distribution of charges of lattice ions, screening of f-electrons by the outer filled electron shells, etc. The electronic structure of selected complexes that include the paramagnetic ion and its nearest neighbors (*ligands*) immersed in the crystal is treated by cluster models (see *Cluster*). Electron energies in the complex are calculated using a one-electron Hamiltonian that operates in the space of molecular orbitals (LCAO method, see *Molecular orbital*), or in the space of mixed configurations. Charge transfer from the ligands to the paramagnetic ion is also taken into account (the *Heitler–London approximation*). To simplify estimates of matrix elements of the one-electron Hamiltonian (see *Single-electron approximation*) one uses either the *Mulliken–Wolfsberg–Helmholtz method* (R.S. Mulliken, M. Wolfsberg, G. Helmholtz, 1952) or the local exchange potential (X_α-technique). The dominant role in forming the energy spectrum of the cluster and the corresponding potential of the effective crystal field involves effects produced by the overlapping of the spatial distributions of charges of the ligands and the paramagnetic ion. Among such effects are the *Kleiner correction* to the Coulomb field of the ligands, the *exchange interaction*, the mutual nonorthogonality of wave functions of the ligands and the metal ion, covalency, virtual excitations of the cluster accompanied by charge transfer from the ligands to the paramagnetic ion, followed by the formation of holes in its filled electronic shells.

One may explicitly determine the interrelations between the composition and structure of the crystal lattice, on the one hand, and the energy spectrum of the paramagnetic ions on the other, using semiphenomenological models that involve a limited number of parameters determined from experimental data. The energy of an nl-electron in an axially symmetric field of the ligand λ is equal to $H_{c(\lambda)} = \sum_k \overline{B}_0^k(R_\lambda)C_0^{(k)}$ (quantization axis parallel to $\overline{r}_\lambda$). Assuming that individual ligand fields may be summed, and using the power series approximation $\overline{B}_0^k \propto R_\lambda^{-t_k}$, one may express the terms of the crystal field through four ($k = 2$, 4 for $l = 2$) or six ($k = 2$, 4, 6 for $l = 3$) parameters of the "superposition model" $\overline{B}_0^k$, t_k. To parameterize the spectra of paramagnetic ions using the *model of angular overlap* and the *model of exchange charges* it is sufficient to represent the functions $\overline{B}_0^k(R_\lambda)$ by linear combinations of squared overlap integrals of the wave functions of the d- (or f-) electron with the ligand wave functions. Satisfactory models are still lacking for rare-earth metals and actinide compounds. The so-called *odd components of the crystal field* are treated separately since they mix together the states of different parity that correspond to the ground state ($(nl)^N$, $1 = N < 2(2l + 1)$), and to the excited ($(n'l')^{4l'+1}(nl)^{N+1}$, $(nl)^{N-1}(n'l')$) *electronic configurations* of the ions of rare-earth and transition metals. The energy of a localized electron in an odd crystalline field corresponds to the Hamiltonian $H = \sum B_k^{2p+1}C_k^{2p+1}$, defined in the subspaces of one-electron functions $\langle n'l'm_l'|$, $|nlm_l\rangle$ with their parameters B_k^{2p+1} (where $0 \leqslant p \leqslant (l + l' - 1)/2$) different from zero when the *point group* of symmetry of the ion does not include inversion. The odd components control the intensities of the dipole transitions between the states of an unfilled $(nl)^N$-shell (forbidden for free atoms), as well as the linear *electric field effects* in optical spectra and the spectra of paramagnetic crystals.

When perturbation theory is applicable one can introduce an effective operator $(HA + AH)/\Delta$ instead of the odd single particle operator A in the space of states of the basic electron configuration. Here Δ is the average energy difference between the excited and ground state configurations. Additional terms in the effective operator of the dipole moment proportional to the squared electron radius (r^2) are called *pseudo-quadrupole moments*. Calculating the Stark structure of the spectrum of the unfilled shell, the effective even field H^2/Δ is allowed along with the even component of the crystal field. The odd components, induced by odd vibrations, define the intensities of the electron–vibrational transitions between the states of the same parity for ions in the crystal lattice sites, the latter also serving as the *centers of symmetry*.

CRYSTAL GRAINS

See *Polycrystal*.

CRYSTAL, IDEAL

See *Ideal crystal*, *Perfect crystal*.

CRYSTAL LATTICE

A geometrically ordered arrangement of atoms (ions, molecules) typical of the crystalline state of matter (see also *Lattice*). The geometry of an ideal crystal involves a spatial periodicity (*translational symmetry*, see *Crystal symmetry*) characterized by three noncoplanar basis vectors $\boldsymbol{a}_j$ ($j = 1$, 2, 3) such that the displacement of the lattice by a vector $n_1\boldsymbol{a}_1 + n_2\boldsymbol{a}_2 + n_3\boldsymbol{a}_3$ (n_1, n_2, n_3 are integers) brings it into itself. If every $\boldsymbol{a}_j$ is selected to be the shortest possible displacement in its respective direction, the parallelepiped formed by the three basis vectors is called the *unit cell* or the *primitive parallelepiped.* By simply replicating the unit cell in space one may generate the entire *crystal structure.* The choice of the set of $\boldsymbol{a}_j$ vectors is not unique; yet the unit cell volume is independent of such a choice. If a unit cell can be chosen so that it contains only a single atom then it is called a *primitive cell.* All other unit cells are called *nonprimitive cells.* A complex crystal lattice may be envisaged to consist of one or several *Bravais lattices* fitted into each other, each such Bravais lattice being built of atoms of a single type. The presence of a crystal lattice explains the phenomenon of *anisotropy of crystals* and *crystal faceting.* Experimentally, crystal lattices can be identified and classified by X-ray, electron, and neutron diffraction techniques.

CRYSTAL LATTICE BASIS, unit cell basis

Set of atoms and their coordinates with the same location at each point of a *Bravais lattice.* By adding the basis to the Bravais lattice, one obtains a *crystal lattice* structure.

CRYSTAL LATTICE CONSTANTS

The magnitudes of three noncoplanar *translations* (a, b, c), by which the *Bravais lattice* may be constructed for a given crystal. The lattice is to satisfy three main conditions: (1) the symmetry of the lattice *unit cell* must correspond to the microscopic *crystal symmetry*; (2) the unit cell is to possess the highest possible number of right angles (or the highest possible number of equal angles and equal sides); (3) the unit cell is to have a minimal possible volume (with due regard for the conditions (1) and (2)). The set of crystal lattice constants (a, b, c) and the angles between them: α (between b and c), β (between a and c) and γ (between a and b), is called the *crystallographic basis.* There are seven types of crystallographic basis sets that correspond to the seven *crystallographic systems*, as follows:

- $a = b = c$, $\alpha = \beta = \gamma = 90°$, *cubic system*;
- $a = b \neq c$, $\alpha = \beta = \gamma = 90°$, *tetragonal system*;
- $a = b \neq c$, $\alpha = \beta = 90°$, $\gamma = 120°$, *hexagonal system*;
- $a = b = c$, $\alpha = \beta = \gamma \neq 90°$, *trigonal system*;
- $a \neq b \neq c$, $\alpha = \beta = \gamma = 90°$, *orthorhombic system*;
- $a \neq b \neq c$, $\alpha = \beta = 90°$, $\gamma \neq 90°$, *monoclinic system*; and
- $a \neq b \neq c$, $\alpha \neq \beta \neq \gamma$, *triclinic system.*

CRYSTAL LATTICE DYNAMICS

The branch of solid-state mechanics, which studies the internal movements within a *crystal*, taking into account the discreteness of its atomic structure. It includes the classical and quantum mechanics of collective movements of atoms in an *ideal crystal*, the theory of interactions of the crystal with penetrating radiation, and also the *dynamics of imperfect crystals.*

Due to the strong chemical bonding the internal motion cannot be described as the superposition of independent displacements of separate atoms. If the displacements are small and the *harmonic approximation* applies then *normal vibrations* occur, each of which involves the participation of all the atoms. The normal vibration has a form of a plane wave characterized by the *wave vector* $\boldsymbol{k}$, expressed within the first *Brillouin zone*, which determines the direction of propagation of the wave front, with wave length $\lambda = 2\pi/|\boldsymbol{k}|$, and a *polarization vector* which indicates the direction of the displacement of atoms within the wave. During the course of a normal vibration all the atoms of the crystal vibrate harmonically near their equilibrium positions at the same frequency ω which is determined by the *dispersion law* that connects the values of ω and $\boldsymbol{k}$:

$$\omega^2 = f_\alpha(\boldsymbol{k}), \quad \alpha = 1, 2, \ldots, 3q,$$

where α denotes the branch of the dispersion law and q is the number of atoms in the crystallographic *unit cell.*

In three dimensions there are at least three branches of vibrations. In the long-wave approximation ($\lambda \gg a$, a is the interatomic distance) all the atoms in the unit cell vibrate with the same phase, corresponding to *acoustic vibrations* (Fig. 1(a)). In this limit $\lambda \gg a$ they are transformed into the usual sound waves in a solid with the linear dispersion law $\omega = c_s k$, $s = 1, 2, 3$. Acoustic vibrations cover the frequency band from 0 to 10^{13} Hz. At high frequencies the dispersion law is no longer linear. However, in the

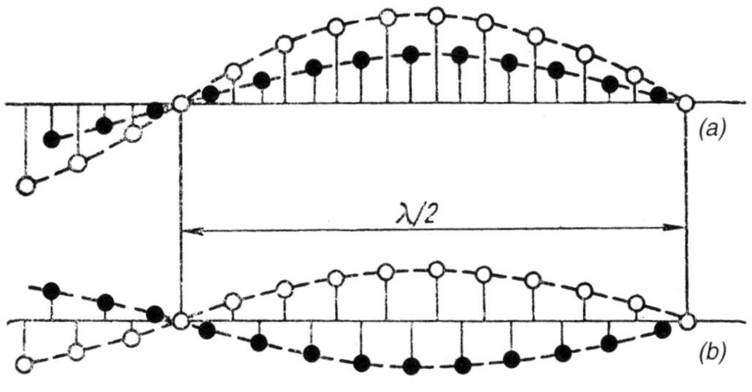

Fig. 1. Diagram of a long-wave vibration of a one-dimensional crystal: (a) acoustic mode; (b) optical mode.

Debye model it is assumed that acoustic vibrations satisfy the linear dispersion law at all frequencies within the range $0 < \omega < \omega_D$, where the *Debye frequency* ω_D is the maximum possible vibrational frequency (10^{13} Hz), and serves as the most important parameter of the spectrum of crystal vibrations. In a complex *crystal lattice* ($q > 1$), $3q - 3$ branches of *optical vibrations* are possible, and at $\boldsymbol{k} = 0$ the center of mass of the unit cell remains stationary with the motion reduced to relative displacements of atoms within the unit cell (Fig. 1(b)). In the limit $\boldsymbol{k} \to 0$ the optical vibration frequencies $\omega(0)$ are different from zero. Ordinarily the frequency bands of optical vibrations are located above those of acoustic vibrations, with a forbidden region, or *band gap*, separating them. It is also possible for acoustic and optical frequency bands to overlap.

There are crystals in which some optical frequencies strongly depend on external parameters (temperature, pressure, magnetic field, etc.) and at specific values of these parameters they may tend to zero (*soft modes*). As a result, static relative displacements of atoms can occur, i.e. the unit cell is restructured by undergoing a *structural phase transition*. Optical vibrations of *ionic crystals* can strongly interact with an electromagnetic field producing coupled polarized vibrations of the lattice and the field. This phenomenon permits the excitation of optical vibrations of ionic crystals by incident oscillating electromagnetic fields (e.g., light waves), generally in the IR frequency range, which is the reason for calling these frequencies optical.

Many lattice modes can be excited simultaneously with different amplitudes. The total number of possible independent vibrations equals the number of mechanical degrees of freedom of all the atoms in the crystal, and their distribution in frequency is determined by the frequency *distribution function* $\nu(\omega)$ (phonon density of states). The form of $\nu(\omega)$ strongly depends on the dimensionality of the crystal. In a three-dimensional lattice at low frequencies ($\omega \ll \omega_D$) for each acoustic branch $\nu_3(\omega) \propto \omega^2/{\omega_D}^3$. For increasing frequency the behavior of $\nu(\omega)$ changes: it tends to zero at the band edges, and within the bands it exhibits *Van Hove singularities* (Fig. 2). The overall vibrational density of states is obtained by summing the distributions of the individual branches.

The quantum dynamics of a crystal lattice is well described by the harmonic approximation: to each wave of normal vibration with frequency ω and wave vector $\boldsymbol{k}$ there is a corresponding *quasi-particle* with the energy $\varepsilon = \hbar\omega$ and quasi-momentum $\boldsymbol{p} = \hbar\boldsymbol{k}$, with the number of quasi-particles proportional to the wave intensity. These quasi-particles are the elementary excitations of crystals and are called *phonons*; they obey *Bose–Einstein statistics* and form an *ideal gas* in the harmonic approximation. At sufficiently low temperatures when the crystal is weakly excited mechanically, its thermodynamic properties arise from these elementary excitations, and in particular the lattice portion of the crystal energy coincides with energy of the phonon gas.

The quantum nature of the crystal motions is manifested through the presence of so-called

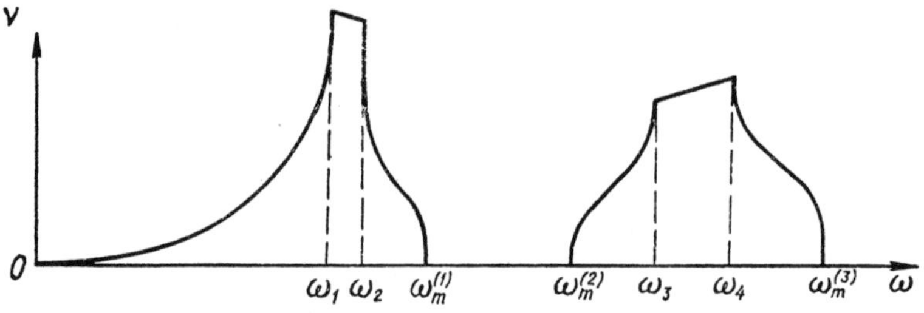

Fig. 2. Plot of the frequency distribution function in a three-dimensional crystal for the acoustic ($0 < \omega < \omega_m^{(1)}$) and optical ($\omega_m^{(2)} < \omega < \omega_m^{(3)}$) branches, where $\omega_1, \omega_2, \omega_3, \omega_4$ are the frequencies of Van Hove singularities.

zero-point vibrations of atoms at $T = 0$ K. The amplitude of these vibrations is usually considerably smaller than the interatomic distance, but in crystals consisting of light atoms it may not be so very much smaller (*quantum crystals* He, H_2, Ar). In solid He the zero-point vibrations are so pronounced that the solid state exists at $T = 0$ K only under a pressure of more than 25 atm. At lower pressures it "melts" and transforms into a *quantum liquid.* Other crystals simply melt with an increase in temperature. *Melting* involves the disruption of the crystal structure and the long-range order, and it occurs when the average amplitude of vibrations of atoms exceeds some critical value.

With increasing amplitudes of vibration, anharmonic behavior begins to appear. *Anharmonic vibrations* cause interactions of phonons and play an important role in kinetic processes such as *thermal conductivity* and *sound absorption* in crystals. The interaction of phonons with other *elementary excitations* (electrons, *excitons, spin waves*) influences the dynamics of the crystal lattice through the electrical conductivity, and the optical and relaxational properties of the crystals. The dynamic interaction of a crystal with photons (including X-rays and γ-quanta), neutrons and accelerated charged particles not only reflects the structural features of the crystalline state, but also provides experimental methods for investigating the dynamics of crystal lattices. For example, *inelastic neutron scattering* can furnish the dispersion law and polarization of the vibrations; and the *Mössbauer effect* can provide estimates of the root-mean-square displacements of vibrating atoms.

CRYSTAL LATTICE VIBRATIONS

Coherent oscillatory motion associated with a *crystal lattice.* The theory of crystal lattice vibrations uses the *Born–Oppenheimer adiabatic approximation* (M. Born and R. Oppenheimer, 1927), whereby the relatively light and fast moving electrons adiabatically follow the relatively slow motion of heavy atomic nuclei or ions. This justifies presenting the potential energy of a crystal as a function of atomic coordinates only. Assuming small vibrational amplitudes, the potential energy is expanded in a power series in the displacements of atoms from their equilibrium positions. Taking into account only quadratic terms of the expansion corresponds to the *harmonic approximation,* and leads to linear equations of motion. If $\boldsymbol{u}_s(\boldsymbol{n})$ is the displacement of the sth atom in the unit cell of vector-number $\boldsymbol{n}$, its variation with time is given by the equation:

$$\frac{m_s \mathrm{d}^2 u_s^l(\boldsymbol{n})}{\mathrm{d}t^2} + \sum_{n's'k} \beta_{ss'}^{ik}(\boldsymbol{n}, \boldsymbol{n}') u_{s'}^k(\boldsymbol{n}') = 0,$$

where m_s is the mass of the sth type of atom, i and k are Cartesian indices, and $\beta_{ss'}^{ik}(\boldsymbol{n}, \boldsymbol{n}')$ are *force constants. Normal vibrations* emerge as eigensolutions of the lattice vibration equations. There are 3 *acoustic vibration* modes and $3q - 9$ *optical vibration* modes for $q > 3$ (q is the number of atoms in the crystallographic *unit cell*). Thus the superconductor $Tl_2Ba_2CaCu_2O_8$ has 3 acoustic modes and 36 optical modes of vibrations (16 Raman active and 20 infrared active).

CRYSTALLINE STATE, GASEOUS

See *Gaseous crystalline state.*

CRYSTALLITE, MICRO-

See *Microcrystallites.*

CRYSTALLITES

Small crystal grains (see *Polycrystal*) in various polycrystalline formations: metal ingots, processed *metals* and *alloys*, rocks, *minerals*, *ceramics*. Fine *monocrystals* lacking a clear crystallographic polyhedral shape, and large microcrystals with poorly developed structure (*dendrites*), plastic deformed microcrystals, etc., are also called crystallites. Depending on the nature and conditions of formation, the average size of crystallites varies within the limits 5–10 μm, with their orientation either arbitrary or following certain patterns (*textures*). As far as their shape is concerned, crystallites may be equiaxial, columnar, platelike, etc. The crystallite structure strongly affects the *strength*, *plasticity*, and other properties of a substance as a whole. *Quenching* the liquid phase yields particularly fine crystallites, down to the so-called *X-ray amorphous structure* range (crystallites below 3 nm in size).

CRYSTALLIZATION

The process of transforming a substance from its liquid, amorphous or gaseous phase into a *crystal*. For a crystal to form it is necessary for the chemical potential μ of the crystal *phase* (which depends on pressure p, temperature T, and concentration c) to be lower than the respective chemical potential of the initial phase. Since a crystal phase usually corresponds to lower temperatures, higher pressures and concentrations in the *phase diagram* than those found at the point of phase equilibrium, one needs to produce overcooling ΔT, excess pressure Δp, or supersaturation Δc in the initial phase. The relation between ΔT, Δp, Δc and the change in chemical potential, $\Delta\mu$, may be expressed as

$$\Delta\mu = \Delta H\frac{\Delta T}{T},\qquad \Delta\mu = k_B T\frac{\Delta p}{p},\qquad \Delta\mu = k_B T\frac{\Delta c}{c}, \tag{1}$$

where ΔH is the change in enthalpy per particle during the phase transition, and k_B is the *Boltzmann constant.*

Crystallization may occur via random formation and growth of crystal precipates (embryos) in various parts of the total volume of the initial phase, a process called *mass crystallization.* During the formation of a precipitate the binding energy of particles entering it (either atoms or molecules) is released. Energy is expended to form the precipitate surface. For a spherical precipitate of radius r the total change of *thermodynamic potential* is

$$w(r) = -\frac{4\pi}{3}r^3\frac{\Delta\mu}{v} + 4\pi r^2\alpha, \tag{2}$$

where v is the volume per individual atom or molecule, and α is the interphase energy per unit surface area of the precipitate (coefficient of *surface tension*). The function $w(r)$ is nonmonotonic, and reaches a maximum at $r = 2\alpha v/\Delta\mu$. The corresponding radius of the precipitate is called the critical radius r_c, and the energy of its formation is $w(r_c) = (16\pi/3)(\alpha^3 v^2/\Delta\mu^2)$. *Precipitates of a critical radius* (see *Critical embryos*) are in a state of unstable equilibrium with their environment. An increase or decrease of the precipitate radius results in its growth or decay, respectively. When a precipitate forms in an initial phase containing no extraneous microimpurities, the process of its formation is called *homogeneous precipitation.*

When the initial phase contains microimpurities, critical precipitates may form at their surface (or on the walls of the container). In that case, the formation process is called *heterogeneous precipitation.* The ratio between the work required to form precipitates of a critical radius heterogeneously and homogeneously is given by

$$\psi(\theta) = \frac{1}{4}(1-\cos\theta)^2(2+\cos\theta), \tag{3}$$

where θ is the *edge angle* depending on the coefficient of surface tension at the system interfaces. These are the initial phase–microimpurity; microimpurity–crystal; and crystal–initial phase interfaces. *Wetting* lowers the work of precipitate formation. The number of precipitates of a critical radius per unit volume of the initial phase is proportional to the probability of their formation: $\exp[-w(r_c)/(k_B T)]$. The respective proportionality coefficient is close to the number of molecules in a unit volume, n. The *rate of formation of precipitates* I, i.e. the number of precipitates formed in a unit volume per unit time, may be determined within the formalism of the thermodynamic theory of their formation. One has to multiply the density of precipitates of a critical radius by the number of atoms (molecules) striking their surface per unit time:

$$I = \begin{cases} 4\pi r_c^2 pn(2\pi mk_B T)^{-1/2}, \\ \quad \text{precipitates form from gas phase}, \\ 4\pi r_c^2 nva \exp\left(-\dfrac{E}{k_B T}\right), \\ \quad \text{precipitates form from condensed phase}, \end{cases}$$

where m is the mass of the molecule, a is the atom-to-atom distance, v is the frequency of atomic oscillations, and E is the *activation energy* for particles adding to the precipitate. To account for the evolution of the system of precipitates of radii below r_c in the initial phase one adds an extra term in the expression for I, namely:

$$z = \left[\frac{4r_c\alpha}{9k_B T n_c}\right]^{1/2},$$

where n_c is the number of particles in a precipitate of a critical radius. With decreasing temperature (at relatively small overcooling) I grows as precipitates grow, since their formation energy diminishes; on the other hand, the energy of thermal motion and the mobility of particles in the initial phase simultaneously decrease, so that the rate at which they add to the surface of the precipitate decreases as well. Competition between these two factors produces a maximum of I at a certain temperature, after which it starts to decrease at lower temperatures. The value of I is remains negligibly small at comparatively small supercooling (or supersaturating) of the initial phase, so that very few precipitates form in volumes of appreciable size. In other words, the initial phase remains in a *metastable state*. The range of respective supercooling (supersaturating) is called the *metastability range*.

As a result of the mass crystallization a *polycrystal* forms, with fine crystals growing from separate precipitates (*crystallites*) that fill the volume of the initial phase with their crystallographic axes randomly oriented. The structure of the polycrystal depends on the values of I and v. Its granularity roughens for larger I and smaller v. See also *Monocrystal growth*.

CRYSTALLIZATION CENTER

See *Crystal nuclei*.

CRYSTALLIZATION FROM THE GAS (VAPOR) PHASE

The formation of *crystals* by the transformation of a material from its gas (vapor) phase to a solid. When crystals grow from the vapor an adsorption layer differing from the crystal in composition and properties appears at the crystal–medium interface (see *Adsorption*). Moreover, *surface diffusion* of the adsorbed particles at the growing faces plays a decisive role there.

Crystallization from the gas phase has been used to produce massive *monocrystals*, *plates*, bands, acicular crystals, *thread-like crystals*, and epitaxial *films*. *Condensation* techniques and chemical reactions are also employed to grow crystals. When using the first it is assumed that material reaches the growing crystal as an atomic or molecular vapor, or as associations of such particles. Depending on the technique used to bring the material to the crystallization zone one may speak about molecular beams in vacuo (see *Molecular beam epitaxy*), *cathode sputtering* (see *Sputtering*), *volume vapor phase* in an isolated system, and crystallization in an *inertial gas flow*. To obtain high quality crystals and films chemical reaction techniques (e.g., as chemical transport, decomposition of compounds, chemical synthesis in the vapor phase) appear to be quite efficient. In certain cases when filaments or massive crystals are grown (as well as epitaxial films), the vapor–liquid–crystal transport mechanism is employed. Vapor material reaches the growing crystal through a liquid layer of different composition formed either unintentionally or intentionally on the growing surface by introducing impurities into the crystallization zone. It may result from deviations from a *stoichiometric composition*, or present a liquid phase intentionally formed at the substrate as a smelt droplet of a prescribed composition. The phase-to-phase energy of the crystal–vapor interface is almost an order of magnitude less than the respective energy of the crystal–vapor interface. Two-dimensional precipitates (embryos) form at the liquid–crystal interface at a rate that is much higher than such a rate at the vapor–crystal interface. This feature combines with other factors (such as a higher condensation coefficient at the liquid surface, low resistance to diffusion in the smelted droplets, etc.) to sharply increase the growth rates.

CRYSTALLIZATION, HYDROTHERMAL

See *Hydrothermal crystallization.*

CRYSTALLIZATION OF POLYMERS

First-order phase transition in polymeric systems. *Crystallization* of high molecular weight polymers is more complex than crystallization of low molecular weight substances because of the influences of various parts of the *polymer* chains (main chain links and side branches). Long chain polymers have a *primary structure* from the linking together of their repeat units or *monomers.* A *secondary structure* can arise from coiling or linking nearby parts through weaker hydrogen bonds, and a *tertiary structure* can form if larger sections fold into themselves to form a globular or condensed conformation. A crystal structure can form from stretched out and aligned macromolecular chains, from regularly arranged molecular chains folded "upon themselves", from adjacent polymers interlinked in a *quaternary structure,* etc. These various *conformations* of the macromolecule as a whole are found in a crystal with unchanged sublattice parameters (repeatedly packed chain links or monomers), and an unchanged translation period along the primary chain axis.

There are globular crystals, supercrystals, with long-range ordering in the packing of the macromolecules (see *Long-range and short-range order*), e.g., globular proteins or their elements (blocks, co-polar blocks). Such longer range ordering may be present along with short range ordering in the packing of repeated molecular chain links. Deformation processes affecting the conformations of macromolecules can affect the ability of some polymers to crystallize (polyisobutylene is an example).

CRYSTALLIZATION, SELF-SUSTAINING

See *Self-sustaining crystallization.*

CRYSTALLIZATION WAVES

A particular form of slowly damping oscillations of the interface between solid and liquid *helium* at low temperatures (below the *lambda point* 2.17 K) where liquid ^{4}He is a superfluid *quantum liquid* (see *Superfluidity*) and solid ^{4}He is a *quantum crystal.* When a crystallization wave propagates through this system it is accompanied by periodically alternating *melting* and *crystallization.* When crystallization waves are mechanically excited, the surface of a ^{4}He crystal goes into a dynamic state of quantum roughness, similar to a thermodynamic equilibrium state of atomic roughness found at the surface of a classical crystal (see *Phase transitions at a surface*).

The spectrum of crystallization waves at low frequencies has the form:

$$\omega^2 = \widetilde{\alpha}\frac{\rho_L k^3}{(\rho_S-\rho_L)^2} + g\frac{\rho_L k}{\rho_S-\rho_L} - \mathrm{i}\omega k\frac{\rho_L\rho_S}{K(\rho_S-\rho_L)},$$

where ω and k are the frequency and the wave number of the crystallization waves; ρ_L and ρ_S are the densities of the liquid and the solid, respectively; g is the free fall acceleration (9.8 m/s^2); K is the kinetic coefficient of growth (proportionality coefficient between rate of growth of a rough crystal face, and difference between crystal and liquid chemical potentials); $\widetilde{\alpha} = \alpha + \partial^2\alpha/\partial\varphi^2$ is the surface rigidity of a given face of the crystal (α is the specific surface energy, and φ is the crystallographic angle corresponding to the direction $\boldsymbol{k}$). The value of K for helium is particularly large below 0.7 K so propagating crystallization waves are only weakly damped.

CRYSTALLOGRAPHIC CLASSES

The same as *Crystal classes.*

CRYSTALLOGRAPHIC INDICES

A triplet (quadruplet) of numbers defining the position of points (sites), directions and planes in a *lattice.* Let the radius-vector from the origin to a given site in the *crystal lattice* be projected on to the a, b, c axes in a (in general, oblique) coordinate system so that these projections are equal to ax, by, cz. Then x, y, z are the crystallographic indices of that site, and the direction of

Table 1. Crystallographic classification scheme

System	Number of space groups	Example of point group	
		International	Schönflies
Triclinic (anorthic)	2	$\bar{1}$	C_i
Monoclinic	13	$2/m$	C_{2h}
Orthorhombic	59	mmm	D_{2h}
Trigonal	25	$\bar{3}m$	D_{3d}
Hexagonal	27	$6/mmm$	D_{6h}
Tetragonal	68	$4/mmm$	D_{4h}
Cubic (isometric)	36	$m\bar{3}m$	O_h

a vector to that site is denoted by $[xyz]$. A vector parallel to the x coordinate axis is in the [100] direction, etc. Crystallographic planes are denoted by what are called *Miller indices* (hkl). Consider a plane which intersects the a, b, c axes at the points $h'a$, $k'b$, $l'c$, and write down the reciprocals of these intercepts $1/h'$, $1/k'$, $1/l'$. The Miller indices are the smallest three integers h, k, l having these same ratios. For example, if $h' = 4$, $k' = 1$ and $l' = 2$ then $(hkl) = (142)$. The three integers which define orientation directions in a crystal, and the positions of its verges, are called the *Weiss indices*, and they are written in the form $[p_1 p_2 p_3]$. As for the *hexagonal system* and the *trigonal system*, their directions and planes are sometimes presented in a four axes system (three equivalent directions are then positioned at 120° to each other in the basal plane of the crystal). The Bravais indices $[mnrp]$ and $(hkil)$ satisfy the conditions $r = -(m+n)$, $i = -(h+k)$. When using the Bravais indices one may clearly see the equivalence of the respective directions and planes in the crystal.

CRYSTALLOGRAPHIC SYSTEM

Crystallographic classification scheme for the 230 space groups that takes into account their characteristic symmetries (see *Crystal symmetry*), and uses the ratios of their *symmetry axes*. There exist seven independent combinations of the *rotation symmetry axes* of the nth order ($n = 1$, 2, 3, 4, 6), and seven crystallographic systems corresponding to them: *cubic system*, *hexagonal system*, *trigonal system*, *tetragonal system*, *orthorhombic system*, *monoclinic system* and *triclinic system*. The concept of crystallographic system coincides with the concept of *syngony* for every crystallographic system, with the exception of the hexagonal and the trigonal systems which together comprise the hexagonal syngony or family. Table 1 outlines the crystallographic classification scheme, and lists a high-symmetry point group for each system.

CRYSTALLOGRAPHIC TEXTURE

See *Texture*.

CRYSTALLOGRAPHY (fr. crystal, and Gr. $\gamma\rho\alpha\varphi\omega$, am writing)

The science describing *crystals* and the crystalline state of matter. Six basic branches are recognized within crystallography: *geometric crystallography* (*crystal symmetry*, *crystal structures*, and *crystal morphology*); *crystallogenesis* (crystal formation and growth, see *Crystallization*); *crystal physics*; *crystal chemistry*; *applied crystallography* (methods and techniques for crystal growth and use in technical applications); *space crystallography* or growing crystals in zero gravity (in orbit) and studying their properties. Geometric crystallography gave birth to the hypothesis of ordered three dimensional positioning of particles (atoms and molecules) composing the crystal which form the *crystal lattice*. The mathematical apparatus of crystallography is based on Euclidean geometry, group theory, and tensor calculus.

Structural crystallography studies the atomic and molecular composition of crystals using *X-ray structure analysis*, *electron diffraction analysis*, and *neutron diffractometry*, all based on the diffraction of waves and particles within the crystals. *Optical spectroscopy*, *electron microscopy*, and resonance methods, etc., are sometimes used. As a result, the crystal structures of over a hundred thousand chemical substances have been determined. When studying the processes of formation and growth of crystals one uses the basic principles of thermodynamics, the laws of *first-order phase transitions*, and of surface phenomena. One also accounts for the interaction of crystals with their ambient environment, and for anisotropies. Industrial processes used to grow diamonds, rubies, and other crystals have been perfected (see *Synthetic monocrystals*), and provide a basis for quantum and semiconductor electronics, optics, acoustics, etc. Crystallography also studies various perturbations in an ideal crystal lattice, such as *defects* and *dislocations*, that originate during the course of crystal growth due to various external influences, and influence many properties of crystals. The structure and properties of various aggregates formed from microcrystals (*polycrystals*, *textures*, *ceramics*), as well as substances with atomic ordering similar to that found in crystals (*liquid crystals*, *polymers*), are also studied. Consistent symmetrical and structural patterns studied within the framework of crystallography find their application when treating general structural features and properties of various non-crystalline states of condensed matter, such as amorphous solids and liquids, polymers, macromolecules, supermolecular structures, etc. (*generalized crystallography*). Studies of biological crystals have been successful in determining the structures of large protein and nucleic acid molecules.

Modern crystallography typically concentrates on the atomic and defect structure of crystals, processes of their formation and growth, and searches for new properties of crystals. These problems involve comprehensive tasks aiming at discovering new materials with important physical properties. Results obtained from crystallographic studies are routinely applied to physics, mineralogy, chemistry, molecular biology, etc.

CRYSTALLOGRAPHY, INTERNATIONAL TABLES

See *International Tables for Crystallography*.

CRYSTALLOGRAPHY, TWO-DIMENSIONAL

See *Two-dimensional crystallography*.

CRYSTAL MORPHOLOGY (fr. Gr. $\mu o \rho \varphi \eta$, appearance, shape, and $\lambda o \gamma o \varsigma$, word)

Outward shape of *crystal* that grows from a solid, liquid or gaseous phase. It depends on the nature of the substance, on growth conditions, and other factors. During *crystallization* from the melt with a small deviation from equilibrium, crystals of some materials grow in the shape of regular *crystal polyhedra* (Fig. 1), while crystals of other materials grow in a shape approximating a sphere (Fig. 2(a)). A faceted shape (see *Crystal faceting*) is typical of materials that exhibit a high *entropy* of fusion $\Delta S = L/T > 4R$ (L is the latent heat of melting, R is the gas constant), and a high anisotropy of *surface energy*. A rounded shape is typical of *metals* and other materials that exhibit a low entropy of fusion ($\Delta S < 4R$) and a low *surface energy* anisotropy. Metals grown from a gaseous phase or from solution may exhibit a faceted shape. A smooth shape loses stability and develops bulges (Fig. 2(b)) at increased overcooling of the melt (in metals, this process occurs at smaller values of overcooling, than in crystals with a high entropy of fusion). The presence of impurities intensifies this process owing to emerging structural overcooling. Thermal and structural overcooling first results in the formation of bulges (*cellular morphology*), which then evolve into *dendrites* (Fig. 2(c, d)). Under sufficiently high overcooling, the dendritic shape degenerates into a needle one without lateral branches. When crystallization takes place from a viscous medium,

Fig. 1. Crystal polyhedron.

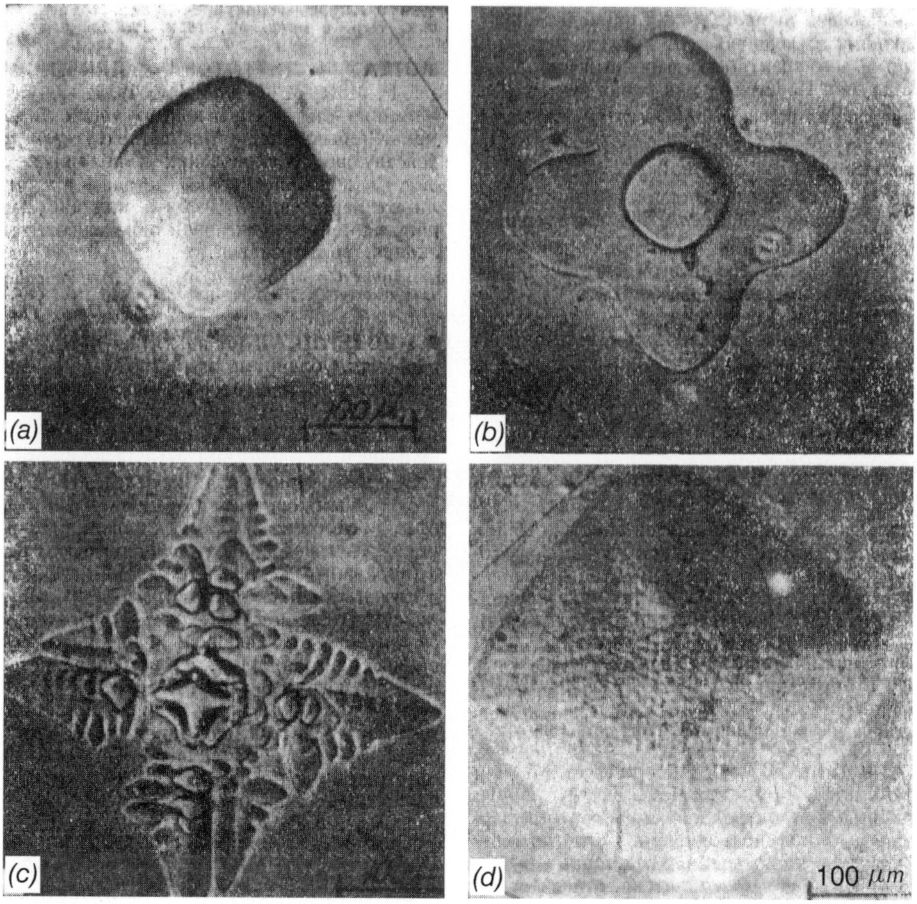

Fig. 2. Crystal morphologies.

spherulites form, which are monocrystalline filaments or *crystallite* aggregates of radial-beam shape.

CRYSTAL NUCLEI, crystallization centers

Crystal phase clusters of critical (and larger) sizes with the capability of growing (see *Crystallization*) in a melt. The critical size of a crystal nucleus is associated with the equality of the bulk and the surface parts of its free energy (which decreases as the melt is overcooled). Crystallization centers may appear spontaneously on solid surfaces, where this process is facilitated by the formation of crystal nuclei.

CRYSTAL OPTICS

The branch of optics investigating light propagation in *crystals*. Among phenomena of crystal optics, there are *birefringence*, interference and *polarization of light*, optical activity (see *Rotation of light polarization plane*), *conical refraction of light*, *pleochroism*, etc. These phenomena can appear or vary under the action of external electric and magnetic fields, mechanical stress, acoustic waves, and this is the subject of electron, magnetic, piezo-, and acoustic optics, respectively.

The classical approach to the phenomena of crystal optics is based on the electrodynamics of continuous media, *crystal symmetry*, tensor calculus; it is based on the fact that the period of a *crystal lattice* ($\sim$1 nm) is much shorter than the wavelength of light ($\sim$500 nm), and therefore, a crystal can be considered as a homogeneous anisotropic medium. This is expressed by a tensor form of material equations connecting the *electric flux density* D_i and *magnetic flux density* B_i with the corresponding fields E_j and H_j: $D_i = \varepsilon_{ij} E_j$ and $B_i = \mu_{ij} H_j$ (here ε_{ij} and μ_{ij} denote the polar tensors of rank two of *dielectric constant* and *magnetic permeability*). In most problems of crystal optics, one considers nonmagnetic crystals ($\mu_{ij} = 1$) and concentrates upon the properties of the tensor ε_{ij}. In the absence of *spatial dielectric dispersion*, the dissipation of energy, and an external magnetic field, by the symmetry principles of kinetic coefficients, the tensor ε_{ij} is symmetric and real. Therefore, for a fixed frequency of light ω, in anisotropic media, the directions of vectors $\boldsymbol{D}$ and $\boldsymbol{E}$ do not coincide. This determines the anisotropy of the optical properties of crystals, i.e. the dependence of the *phase velocity* $v = c/n$ and of *refractive index* $n = \sqrt{\varepsilon}$ on the direction. For the totality of all directions at an arbitrary point of a crystal, the magnitudes of n form an ellipsoidal surface, which is called the *optical indicatrix*. The symmetry axes of the optical indicatrix form three mutually orthogonal principal directions in the crystal (*crystallophysical axes*), along which the directions of vectors $\boldsymbol{D}$ and $\boldsymbol{E}$ coincide. In the rectangular Cartesian coordinate system with axes coinciding with the crystallophysical axes, the equation of the optical indicatrix is of the form

$$\frac{x^2}{n_x^2} + \frac{y^2}{n_y^2} + \frac{z^2}{n_z^2} = 1,$$

where n_x, n_y, and n_z are the principal values of n. Crystals of a *cubic system* are optically isotropic, i.e. their ellipsoid degenerates into a sphere; the crystals of medium systems are optically uniaxial (the indicatrix is an ellipsoid of revolution); the crystals of lower systems are optically biaxial (the indicatrix is a triaxial ellipsoid). The *optical axis of a crystal* corresponds to the normal of a circular section of the optical indicatrix.

The laws of propagation of plane harmonic waves in dielectric crystals follow from the wave equation together with the tensorial relation between the vectors $\boldsymbol{D}$ and $\boldsymbol{E}$. Thus, we get the *dispersion equation*, which corresponds to a homogeneous system of three linear equations for the components of the vector $\boldsymbol{E}$:

$$\left[\frac{\varepsilon_{ij}(\omega)\omega^2}{c^2} - k^2\delta_{ij} + k_i k_j\right] E_j(\omega) = 0,$$

where k, k_i, and k_j are the magnitudes of the wave vector and its corresponding components, $\delta_{ij} = 0$ for $i \neq j$, and $\delta_{ij} = 1$ for $i = j$. This system has a nontrivial solution $\boldsymbol{E} \neq 0$ if and only if the determinant in the parentheses is not zero. The latter condition leads to the *Fresnel equation*

$$\frac{k_x^2}{v^2 - v_x^2} + \frac{k_y^2}{v^2 - v_y^2} + \frac{k_z^2}{v^2 - v_z^2} = 0,$$

where v is the magnitude of the phase velocity, the direction of which coincides with the direction of the wave vector $\boldsymbol{k}$; v_x, v_y, and v_z are the principal phase velocities of the wave for the corresponding directions of the vector $\boldsymbol{D}$ when the light propagates along one of the principal directions.

The dispersion and Fresnel equations are quadratic with respect to k and v, i.e. at light propagation in anisotropic crystals, there arise two waves (see Fig.) with different values of the magnitude of the wave vector and different phase velocities (refractive indices) v_1 and v_2 coincide only along the directions of the optical axes. They correspond to two mutually orthogonal values of the vectors $\boldsymbol{D}$ and $\boldsymbol{E}$.

If from some point of the crystal one locates all the vectors corresponding to v_1 and v_2 in all the directions $\boldsymbol{m}$, then the endpoints of these vectors form two surfaces called the *surfaces of normals*. In *uniaxial crystals*, one of these surfaces is a sphere (characterizing the ordinary ray v_0), and the other one is an ellipsoid of revolution (characterizing the extraordinary ray v_e) tangent to the sphere at its intersection points with the optical axis. In *biaxial crystals*, both surfaces are ellipsoids intersecting at 4 points lying pairwise on two optical axes (*binormals*). The ray vectors form similar surfaces called *ray surfaces* (or *wave surfaces*). Thus, in anisotropic crystals, in an arbitrary direction $\boldsymbol{m}$, there may propagate two plane waves polarized in two mutually orthogonal planes. The

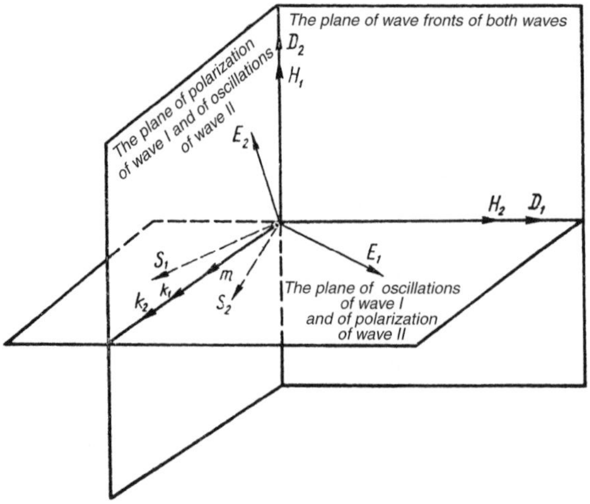

Crystal optics.

directions of vectors $\boldsymbol{D}_1$ and $\boldsymbol{D}_2$ of these waves coincide with axes of the ellipse of the central section of the optical indicatrix by the plane orthogonal to $\boldsymbol{m}$, and the lengths of the ellipse axes determine their refractive indices. Consequently, the optical indicatrix and surfaces of normals are mutually dependent.

The birefringence in the crystals because of the different refraction indices of two polarized waves propagating in the same direction implies a path difference. If, in a convergent light, one reduces the polarization of these waves into one plane (using some polarization devices), this results in the interference of the light. Such *interfering conoscopic images* allow one to investigate crystal-optical properties of crystals.

In the presence of a spatial dispersion of first order with respect to k or an external magnetic field, and also in magnetic crystals, the tensor ε_{ij} is complex and, in general, nonsymmetric; moreover, $\varepsilon_{ij}(\omega, k, H) = \varepsilon_{ij}(\omega, -k, -H)$. In particular, in such crystals two elliptically polarized waves propagate with orthogonal axes of the ellipse coinciding with the axes of the section ellipse of the optical indicatrix. The directions of by-pass of vectors $\boldsymbol{E}$ are opposite in both waves. Along the direction of the optical axis, and in cubic crystals, the polarization ellipses degenerate into circles, i.e. the waves are circularly polarized and have different velocities. Moreover, in uniaxial crystals, the surfaces of normals do not osculate at the points of intersection with the optical axis. The ellipticity of waves brings about optical activity of crystals, which is expressed, e.g., as *gyration*, *electrogyration*, *Faraday effect*, etc. The complexity of the tensor ε_{ij} and the elliptical two-ray refraction are also caused by light absorption in crystals. The directions of by-pass of ellipses in this case coincide. In such crystals, the absorption of differently polarized waves is different, and occasionally *circular dichroism* takes place in optically active crystals.

Crystal optics adjoins *crystallography*, solid state physics, and mineralogy. With the appearance of laser technology, new branches of physical optics and optical tool making arose that are fundamentally based on crystal optics: coherent and *nonlinear optics*, *holography*, *optoelectronics*, *integrated optics*, etc. The control of optical (in particular, laser) radiation is based on the phenomena of crystal optics.

CRYSTAL, PERFECT

See *Perfect crystal*, *Ideal crystal*.

CRYSTAL PHYSICS, physical crystallography

Branch of *crystallography* involving the physical properties of *crystals* and crystalline aggregates of *polycrystals*, as well as changes in their properties arising from external influences. Crystal physics explains *anisotropy of crystals*, and basic variations in the properties of crystals due to the regular distribution of particles (molecules, atoms, and ions) in the *crystal lattice*, and the types of their interrelations. The mathematical apparatus of matrix algebra, tensor calculus, and group theory provides a quantitative description of the physical properties of crystals. Directionally independent properties such as density are characterized by scalar variables. To describe physical properties involving relationships between two vectors (e.g., between either *polarization* or electric current density and electric fields, etc.) or between pseudo-vectors (e.g., magnetic induction and magnetic field strength) one uses second-rank tensors (e.g., *dielectric constant* and *magnetic permeability* tensors). Many physical fields in crystals (e.g., electric and magnetic fields, mechanical strains) are themselves tensor fields. The relation between the physical fields and the properties of crystals may be described by tensors of higher rank, as is the case with the piezoelectric effect (see *Piezoelectricity*), *electrostriction*, *magnetostriction*, *elasticity*, *photoelastic tensor*, etc.

It is convenient to represent dielectric, magnetic, elastic, and other physical properties of crystals in the form of *characteristic surfaces* such as ellipsoids. The radius-vector that traces out such a surface indicates the value of a particular property along its given direction (see *Crystallographic indices*). The symmetry of any property of a crystal may not be lower than the symmetry of its external shape (*Neumann principle*, F. Neumann). In other words, a symmetry group G_1 that describes any physical property of a crystal includes necessarily the symmetry elements of its associated point group G, that serves as its supergroup: $G_1 \supset G$. The symmetry elements determine the orientations of the principal axes of characteristic surfaces, as well as the number of tensor components that determine a given physical property. In many cases these are surfaces of ellipsoids with their principal axes related to the symmetry axes of the crystal (see *Crystal classes*). Crystals of low symmetry have their physical properties described by second-rank tensors with three principal values (plus the orientation of the tensor principal axes). Physical properties of cubic symmetry crystals that are described by second-rank tensors are directionally independent or isotropic. Crystals of intermediate symmetry (tetragonal, trigonal and hexagonal) have these properties characterized by the symmetry of their rotation ellipsoid, i.e. their second-rank tensors have two independent components, one of which describes the property along the principal axis, and the other along any direction orthogonal to this axis. To describe the property along an arbitrary direction one needs to know these two principal components plus the direction cosines of the direction.

Physical properties described by higher-rank tensors are characterized by a larger number of parameters: e.g., elasticity properties, described by a fourth-rank tensor, need three such variables in a cubic crystal, and two independent variables in an isotropic medium. The elastic properties of a triclinic crystal require 21 independent tensor components for their description. To determine the number of independent components of tensors of higher rank (5th, 6th, etc.) one can make use of the corresponding *point group of symmetry*. To obtain a complete description of the physical properties of crystals and *textures* one may make use of radiofrequency, acoustical, optical, and other techniques.

Crystal physics studies both phenomena specific for anisotropic media (*birefringence*, *rotation of light polarization plane*, direct and inverse piezoeffects, electro-optic, magneto-optic, and piezooptic effects, generation of optical harmonics, etc.), as well as properties more often associated with isotropic media (electric conductivity, elasticity, etc.). In crystals these latter properties may also exhibit features resulting from the crystal anisotropy.

Many phenomena within the scope of crystal physics are related to changes in the crystal symmetry under various thermodynamic conditions. The *Curie principle* may be used to predict changes in both the point and the space symmetry groups of crystals undergoing *phase transitions*, e.g., entering ferromagnetic or ferroelectric states (see *Ferromagnetism*, *Ferroelectricity*). Crystal physics studies various defects of the crystal lattice (*color centers*, *vacancies*, *dislocations*, *grain boundaries*, *stacking faults*, *domains*, etc.) and their effects upon the physical properties of crystals (*plasticity*, *strength*, *luminescence*, electrical conductivity, mechanical Q-factor, etc.). Among the tasks of crystal physics is the search for new crystalline materials with specific prospective properties needed for practical technological applications.

CRYSTAL POLYHEDRON

A *monocrystal* grain under equilibrium conditions (see *Monocrystal growth*) consisting of a certain combination of faces, edges and vertices. The number of faces, edges and vertices should follow the *Euler theorem* (L. Euler): the sum of the number of faces and vertices equals the number of edges plus two ($F + V = E + 2$). The so-called *simple forms* participate in *crystal faceting*. These are polyhedra whose faces can be obtained from one another via symmetry operations that are specific for the *crystal class* of the given crystal; they usually form various combinations with each other. There are 47 geometrically distinct simple forms known; this number increases to 193 when taking into account enantiomorphous (mirror reflection) pairs (see *Enantiomorphism*), or to 146 in case one accounts for differences in point symmetry. Only certain simple forms are permissible in each *point group*. Specific forms of crystal polyhedra depend on their structure and conditions of *crystallization*.

CRYSTAL, QUANTUM

See *Quantum crystals*.

CRYSTALS (fr. Gr. $\kappa\rho\upsilon\sigma\tau\alpha\lambda\lambda o\varsigma$, ice)

Solids with ordered three-dimensional periodic spatial atomic structure. The formation medium may be an overcooled vapor or liquid of a pure substance, a supersaturated solution or flux of one material in another (the solvent), or a crystal medium of a polymorphous modification or composition different from that of the newly formed crystal. Some features of crystals depend on their formation process. If the material is supplied evenly and uniformly from the ambient environment then a polyhedral shape results, limited by flat faces intersecting each other along straight edges. These faces, often mirror smooth in perfect crystals, are tilted toward each other at certain angles characteristic for the particular crystal type (*law of constancy of angles*). The ratios between the intercepts on the axes for different tilted faces of a crystal can always be expressed by rational numbers such as $1:2$ or $2:3$, and never by irrational ones such as $1:\sqrt{2}$ (*law of rational indices*).

Other general macroscopic properties of crystalline matter include *crystal homogeneity* (independence of the choice of measurement point), *anisotropy of crystals* (dependence on direction), and *crystal symmetry* (spatial transformations that bring the crystal into itself).

There are 32 classes of morphological symmetry of crystals, each corresponding to its specific set of symmetry elements: axes, planes, center of symmetry, etc. Three-dimensional atomic structure periodicities are described in terms of a three-dimensional space lattice (the *crystal lattice*), a three-dimensional periodic system of points that may be formed starting from an arbitrary point by simply repeating the lattice by a series of parallel *translations*. There exist a total of 14 types of such lattices called *Bravais lattices*.

The crystal appearance (*crystal morphology*) is characterized by the crystal shape and faceting, i.e. the set of faces bounding the crystal, and by the ratio of the sizes of these faces (see *crystal faceting*). Such facets and their relative sizes characterize the crystal morphology or *crystal habit*.

The shapes of crystals reflect their *crystal structure*, and the nature of the interactions between their atoms. In particular, these factors affect the shape of the forming crystal through *kinetic phenomena* occurring at the surface of a growing crystal. Such phenomena depend on the level of supersaturation, temperature, composition of the ambient medium, *thermal conductivity*, *diffusion*, and the presence of impurities. The shape acquired by the crystal during its growth (the *growth shape*) is very sensitive to the conditions of crystallization, and reflects the mechanism of growth. Therefore, the growth shape implies certain conclusions beyond the conditions of growth. During the artificial growth of crystals the crystallization parameters may be chosen accordingly. Although a typical crystal shape is a polyhedron, rounded growth shapes or strongly perturbed shapes may form under certain ambient conditions (*thread-like crystals*, *plates*, *dendrites*, etc.). Some crystals can form in two different modifications, a right-hand and left-hand one (*enantiomorphism*). Substances are found that may crystallize in the shape of twins (*twinning of crystals*). In cases when crystal faceting is suppressed by the limited availability of feeding material, *forced growth shapes* develop. The crystal can then acquire the shape of the surface limiting its growth (e.g., crucible walls, smelting isotherm in a smelt, an isoconcentration saturation surface in a gas phase). Most *monocrystals* are the result of forced growth. Technologies for growing crystals with forced growth shapes find wider and wider applications. Almost all semiconductor crystals are produced that way, as well as most piezoelectric, ferroelectric, and optical crystals (see *Monocrystal growth*).

If crystal generation and growth occur simultaneously throughout the whole volume a *polycrystal* forms, that is an aggregate of randomly oriented fine crystals of various sizes and shapes called *crystallites* or *crystal grains*. Sometimes crystallites follow a certain predominant orientation indicating the presence of a *texture*. During mass crystallization when embryos appear only in certain regions of the generating medium, the collective growth of crystals ensues, characterized by the interaction of the growing crystals with each other, to produce a geometric selection. Crystals having an optimal orientation for the given ambient conditions then survive. Such a competition results in the formation of a pole structure, typical for metal ingots, in which all the crystals are elongated and almost parallel to each other. Examples of such collective growth are *druses* (crystal agglomerations in which the vertices of separate crystals face in more or less the same direction) and *parallel inshoots* (crystals growing together to merge in a parallel orientation).

The ideal structure of a crystal always undergoes certain perturbations (accumulates *defects*) because growth conditions change, impurities are trapped, and various external influences are always at work. Real crystals often display *macrodefects* (gas–liquid, solid inclusions, *cracks*, etc.) which worsen their properties. A typical *structural defect* is mosaicity (see *Mosaic crystals*) which results from various defects (*internal stress*, block boundaries, twins (see *Twinning of crystals*), *stacking faults*, *dislocations*, etc.) that appear during crystal growth. The presence of foreign atoms and ions (impurities) in a crystal during its growth may result in the development of zonal, sectional, and other structures. Substituting crystal atoms (*substitutional impurities*) or imploding between the atoms of the crystal (*implosion impurities*) can produce *point defects* (see *Impurity atoms*). Other point defects perturbing the lattice periodicity are *vacancies* and *interstitial atoms*.

The level of deficiency of the crystal structure affects both crystal growth (e.g., the dislocation mechanism of growth) and various crystal properties that are structure sensitive (*plasticity*, *strength*, *electrical conductivity*, *photoconductivity*, *light absorption* capacity, etc.). Therefore, to grow crystals with prescribed properties, one must intentionally introduce desired defect structures. Examples are donor and acceptor impurities which determine the electrical properties of semiconductors. Added impurities (*doping*) are widely used to produce semiconducting crystals with various types and levels of conductivity, to produce laser crystals with a prescribed spectral transmission band, etc. The development of electronics, quantum electronics, computer technology, and various other industries is directly linked to using crystals and complex crystal structures based on them (e.g., epitaxial layers, *heterostructures*, microchips, *superlattices*).

CRYSTAL STABILITY CONDITIONS

Conditions imposed on the coefficients of the expansion of the crystal potential energy U in terms of atomic displacements u_{si}:

$$U = U_0 + \sum_{s,i} A_{si}u_{si} + \frac{1}{2}\sum_{s,t,i,j} A_{stij}u_{si}u_{tj} + \cdots,$$

where U_0 is the crystal energy at equilibrium (the indices s and t are used for labeling atoms; $i, j = x, y, z$). From the condition that the resultant of all forces acting on the atoms at equilibrium vanishes, it follows that $A_{sl} = 0$. From the invariance of U under *translation* and rotation of the crystal as a whole it follows that $\sum_t A_{stij} = 0$, $\sum_t (A_{stij}X_{tk} - A_{stik}X_{tj}) = 0$ (X_{si} designates the sth atom coordinates). Crystal symmetry properties impose additional restrictions on the harmonic *force constants* A_{stij}, and the application of these relationships considerably reduces the number of independent force constants. Besides that, some restrictions on the force constants follow from the condition of crystal equilibrium ($U = \min$ at equilibrium), and these are in the form of inequalities arising from the requirement of positive definiteness for the quadratic form (third term) of the expansion of U. Similar conditions may be derived for the anharmonic force constants, which serve as the coefficients of additional terms of the expansion of U.

CRYSTAL STRUCTURE

The structure of an actual crystalline *solid.* The concept of crystal structure starts with the concept of *crystal lattice*, i.e. of the parameters of a *unit cell*, such that any ideal crystal (*monocrystal*) may be reproduced by repeating its unit cell in space. A full description of a crystal structure also includes specifying the positions of atoms, ions and molecules inside the cell (including the orientations of molecules). Beside the locations of the atoms, other important factors are the distribution of the electron density in the unit cell, the atomic and *ionic radii*, *chemical bonds* (see *Crystal chemistry*), coordination numbers (see *Coordination sphere*), etc.

Ferroelectric and magnetic *crystals* are also characterized by the direction of their electric and magnetic moments, respectively. The corresponding lattices may not necessarily coincide with the crystallographic ones, and they may possibly be mutually incommensurate.

The crystal structure of an actual crystal is quite variable, thus opening ways for numerous practical applications. Specific branches of the physics of solids are dedicated to their description (see *Crystal physics*). The concept of crystal structure also includes various lattice *defects*: point defects (such as vacancies, interstitional or substitutional impurity atoms, atomic clusters); linear defects (dislocations); two-dimensional defects (stacking faults, twins, etc.). When studying the crystal structure of *solid solutions* one has to know the distribution of atoms (ions, molecules) of various types in their sublattices. Crystal structure also includes describing the physical composition and parameters of structural elements (grains, cells, and blocks) and various other factors.

The main techniques for studying crystal structures are X-ray diffraction structural analysis, neutron diffraction, and electron diffraction.

CRYSTAL SYMMETRY, microscopic crystal symmetry

Geometrical regularity in the spatial arrangement of particles (atoms, *ions*, molecules, complexes), composing a *crystal.* One can characterize the symmetry of a crystal by the totality of all spatial transformations $(x', y', z') = f(x, y, z)$, which bring the crystal into coincidence with itself. From all conceivable transformations it is sufficient to restrict attention to linear ones, and not necessarily to all of these, but only to deformation-free types, i.e. to linear orthogonal transformations that preserve the lengths of segments and angles between them: $(x', y', z') = \|\alpha\|(x, y, z) + (x_0, y_0, z_0)$, where $\|\alpha\|$ is the transformation matrix. The determinant $D = |\alpha|$ in this case can have only two values $D = \pm 1$. The transformations with $D = 1$ are called proper transformations, and all of them involve rotational spatial displacements (i.e. rotations) around some axis through an angle φ, while transformations with $D = -1$ are called improper; the unit improper transformation being *inversion* $(x', y', z') = (-x, -y, -z)$. All improper transformations are combinations of proper transformations with inversion. In general a linear transformation can involve a shift or *translation*, designated by the term (x_0, y_0, z_0) through a distance appropriate to the crystal structure lattice spacings. If the term $(x_0, y_0, z_0) = 0$ there is no translation, and the transformation is either a proper rotation for $D = +1$, or an improper rotation for $D = -1$.

The three-dimensional periodicity of a crystal automatically sharply restricts the possible angles φ of allowed proper (and improper) rotations to five values: 0°, 60°, 90°, 120° and 180°. It is customary to express the allowed angles φ by the expression $\varphi = 2\pi/n$, and in the international notation to designate proper rotations by the five integers: $n = 1, 2, 3, 4, 6$. Improper rotations are designated by $\bar{1}$, $\bar{2} = m$, $\bar{3}$, $\bar{4}$, $\bar{6}$, where m is the symbol for a *mirror reflection* in a plane. In the older *Schönflies notation* the proper and improper rotations are designated, respectively, by C_1, C_2, C_3, C_4, C_6, I, $S_2 = \sigma$, S_3, S_4, S_6. There are two special combined rotation-translation *symmetry operations*: (a) a *screw rotation* which is a rotation by $2\pi/n$ combined with a translation by half a lattice spacing along the rotation axis, and (b) a *glide reflection* which is a reflection in a plane combined with a translation by half a lattice spacing parallel to the plane.

The complete set of all available symmetry transformations of a crystal form the *space group* of the crystal (called *Fedorov group* in the Russian literature). It is discrete, but with an infinite set of elements, because translation through all possible periods is endless. There exist 230 possible space groups, and they are described explicitly in the *International Tables for Crystallography.*

Also important in the physics of crystals are finite groups containing only proper (and possibly improper) rotational elements (without translations) which are called *crystal classes* or *point groups*, and their number is only 32. In three dimensional space there are only 14 *translational groups* which contain no rotational elements. A set of points invariant under translation form a crystal *lattice* (*Bravais lattice*). Each space group has associated with it a point group and a crystal lattice.

When describing the symmetry of magnetic (magnetically ordered) crystals (see *Magnetism*), the situation becomes complicated, since transforming a crystal into itself requires matching not only coordinates of individual identical atoms, but also the directions of their atomic *magnetic moments* (*spins*). The Shubnikov extension of symmetry theory to the symmetry of magnetic crystals involves the addition of the non-spatial operation R of spin reversal (see *Shubnikov groups*). The addition of this new symmetry element permits one to construct 1651 space (Shubnikov) groups, 122 point groups and 36 translational *magnetic symmetry groups* (see also *Antisymmetry*). A further generalization brought about the appearance of the theory of colored symmetry, and the creation of many additional *colored symmetry* groups. All these types of groups find use in solid state physics.

Concerning macroscopic physical properties (optical, electrical, mechanical, magnetic, etc.), crystals behave as media that are homogeneous, but anisotropic (see *Anisotropy of crystals*); and the discreteness of their atomic structure is not significant. Because of this only the rotational symmetry of a crystal, i.e. its point group, is of importance for the macroscopic directionally dependent properties described by tensors of different ranks. Then the distinction between crystals by their differences in symmetry finds its expression in the intrinsic symmetry of tensors for different physical properties. The literature provides information on the specific forms of various tensors of many ranks for all 32 of the crystallographic point groups.

CRYSTAL SYSTEM

The classification of *crystals* in accordance with their crystallographic coordinate system, i.e. by the metric represented by the set of independent linear (a, b, c) and angular (α, β, γ) *crystal lattice constants.* It is possible to choose seven different unit cells with the same number of crystal systems corresponding to them (see Table 1). The concept of *syngony* (in common use in the Russian literature) coincides with the concept of crystal system for all crystal systems except the hexagonal.

Table 1. Unit cell shapes and metric parameters for the seven crystal systems

Crystal system	Shape of unit cell	Metric parameters
Triclinic	Oblique-angled parallelepiped	$a, b, c, \alpha, \beta, \gamma$
Monoclinic	Straight prism with parallelogram in base	a, b, c, β
Orthorhombic	Right-angle parallelepiped	a, b, c
Tetragonal	Right-angle parallelepiped with square in base	a, c
Hexagonal	Prism with base in rhombic shape, $\gamma = 120°$	a, c
Trigonal	Rhombohedron	a, α
Cubic	Cube	a

CRYSTAL, THREAD-LIKE

See *Thread-like crystals.*

CUBIC SYSTEM

A *crystallographic system* defined by the presence of four third-order *symmetry axes* (C_3) in *crystals* with cubic symmetry. The *unit cell* of the system is a cube with lattice parameter a. The coordinate axes of a cubic system contain 4, $\overline{4}$, and 2 (i.e. C_4, S_4, and C_2) symmetry elements. This system includes three *Bravais lattices* (primitive, body-centered, face-centered), five *point groups* (T, T_h, T_d, O, O_h), and 36 *space groups* (see *Crystal symmetry*).

CUMULATIVE EFFECT

Accumulation is a process of building up the energy flux density in a medium by focusing the flows of energy. It is achieved either through specific shaping of an elongated energy source, or through the interaction of *shock waves.* This effect is usually produced using explosives. In solids this may be realized in several ways, as follows. *Cumulative jets* are formed by compressing metal linings placed into the cumulative cavities in explosive charges. The shape of the lining may be hemispherical, conical, parabolic, hyperbolic, etc. Compression concentrates the principal part of the explosive energy into a thin internal layer of the lining so that this layer forms a cumulative jet. The mass of metal transformed into the jet may reach 6–20% of the total mass of the lining, and the energy density in the jet may reach 10^6 MJ/m^3.

Shock wave collision. During a frontal collision of shock waves in a solid (see *Explosion*) each element of the solid is sequentially compressed first by the incident and then by the reflected shock wave. As a result, pressures in the area of the reflected shock wave exceed those in the incident wave by a factor of several times, while temperatures remain comparatively low, that is lower than those resulting from compression by a single shock wave. During a glancing collision of shock waves that meet at a certain angle a triple shock wave configuration forms, and pressures in its forward wave significantly exceed those produced during a frontal collision. By means of this technique a pressure of 1.8 TPa was achieved in copper samples.

Cumulation in a medium with varying density. When a shock wave exceeding a critical value passes through a system of gradually tapering alternating layers of light and heavy material its propagation develops into a periodic self-reinforcing process (see *Self-similarity*) that results in the unlimited growth of pressure at its front.

CUNICO

Hard magnetic material based on the Cu–Ni–Co system. A high-coercive state is achieved upon *quenching* from 1100 °C and *tempering* at 625 °C. Cunico *alloys* are readily plastically deformable, so they are employed for the manufacture of *permanent magnets* of a complicated shape. For properties, see Table 1 in *Hard magnetic materials.*

CUNIFE

Hard magnetic material based on the Cu–Ni–Fe system. A high-coercive state is achieved upon *quenching* from 1100 °C, *tempering* at 700 °C, applying cold *strain* with high compression, and tempering at 650 °C. The strain induces uniaxial *magnetic anisotropy* along the rolling or *drawing* direction. Cunife *alloys* are used as thin wires or stampings. For properties, see Table 1 in *Hard magnetic materials.*

CURIE GROUP

See *Limiting symmetry groups.*

CURIE LAW (1895)

Temperature dependence of the *magnetic susceptibility*, $\chi = C/T$, in many *paramagnets*, where C is the Curie constant ($C > 0$). All gases in which the atoms, molecules, or ions have permanent *magnetic moments* follow the Curie law, as well as dilute liquid solutions of *transition metal* ions, as well as *crystals* of compounds of these elements with non-metallic bonding. It holds for temperatures at which the thermal energy significantly exceeds the energy of interaction of neighboring ionic magnetic moments with each other, and with the external magnetic field.

CURIE POINT, Curie temperature

The point (temperature T_C) of a *second-order phase transition*. Examples of a Curie point are the temperatures T_C of *magnetic phase transitions*, *ferroelectric phase transitions*, *structural phase transitions*, and other *phase transitions* in solids. Special names are used for certain transitions, such as the *Néel point* for the antiferromagnetic Curie point.

CURIE PRINCIPLE (P. Curie, 1984)

This principle deals with the symmetric aspect of the causality principle. Namely, the symmetry of a system of causes is (abstractly) preserved in the symmetry of the outcomes. This principle is a philosophical generalization of the *Neumann–Minnegerode–Curie principle* which states that the symmetry of physical properties of a system made of a material object (e.g., a *crystal*) and of an external action is not lower than the intersection (a general subgroup) of the groups of symmetry of the object and action. The Curie principle holds for all deterministic systems, to which the notion of symmetry can be applied.

A modern formulation of the Curie principle establishes a *homomorphism* relation in the particular case of an *isomorphism* between the symmetry groups of a system of causes $G_{\rm sys}^{\rm c}$ and a system of outcomes $G_{\rm sys}^{\rm out}$:

$$G_{\rm sys}^{\rm out} = \bigcap G_j^{\rm out} G_{\rm out}^{\rm sym} \quad \leftrightarrow \quad \bigcap G_j^{\rm c} G_{\rm c}^{\rm sym} = G_{\rm sys}^{\rm c}.$$

For a homogeneous system, parts of which are connected by some symmetrizer $G^{\rm sym}$, its symmetry is defined by the intersection of the symmetry groups of its parts, $\bigcap G_j = G_1 \cap G_2 \cap \cdots$, multiplied by $G^{\rm sym}$ (the *Shubnikov theorem*). For an inhomogeneous system, $G^{\rm sym}$ reduces to identity and the system symmetry is defined by the intersection $\bigcap G_j$ (the *Curie theorem*). Taking, e.g., the interaction of an object with an external action as the cause of a physical phenomenon, we can find the symmetry group $G_j^{\rm ph}$ of a systematic phenomenon (for $G_{\rm c}^{\rm sym} = 1$), or in a particular case,

$$G_{\rm object} \cap G_{\rm action} = G_{\rm sys}^{\rm c}.$$

For subsystem properties, the case

$$G_j^{\rm out} \leftrightarrow G_j^{\rm ph} \leftarrow G_{\rm sys}^{\rm c}$$

is possible if $G_{\rm c}^{\rm sym} \neq 1$ (the *Wigner theorem*).

This systematic formulation explains violations of the Curie principle associated with the incompleteness of a given system of causes, an inexact definition of symmetry of the system components, or with using the Curie theorem instead of the Shubnikov theorem.

CURIE–WEISS LAW (P. Curie, P. Weiss, 1907)

The temperature dependence of the *susceptibility* χ of ordered systems above the point of their transition to an ordered phase at the *Curie point* T_C has the form:

$$\chi = \chi_0 + \frac{C}{T - \theta},$$

where C is the *Curie constant*, χ_0 is the temperature independent part of the susceptibility, and θ is the characteristic temperature. For the *magnetic susceptibility* χ of *paramagnets*, or *ferromagnets* in their paramagnetic phase above T_C, as well as for the dielectric susceptibility of *ferroelectrics* and other materials, θ is called the *Weiss temperature*, or sometimes the *Curie–Weiss temperature*. The value of θ depends on particle interactions, and may differ appreciably from T_C (see also *Dielectric constant*). In the case of *antiferromagnets* (or antiferroelectrics) and *virtual ferroelectrics* the value of θ may become negative. The Curie–Weiss law holds for the range of temperatures in which the fluctuations are not *critical fluctuations*, that is for $(T - T_C)/T_C \gg Gi$, where Gi is the *Ginzburg number*.

CURIUM, Cm

A chemical element of Group III of the periodic system with atomic number 96 and atomic mass 247.07; it is a member of the *actinide* family. 15 isotopes are known with mass numbers 238 to 252, none of them being stable. The electronic configuration of the outer shells is $5f^7 6d^1 7s^2$. The ionization energy is 6.09 eV. Atomic radius is 0.1749 nm; radii of Cm^{3+} and Cm^{4+} ions are 0.0946 and 0.0886 nm, respectively. Oxidation state is +3 (more rarely, +4, +6).

In the free form, curium is a soft silvery-white *metal*. It exists in two allotropic modifications: a low-temperature α-Cm below 873 K, and a high-temperature form β-Cm. The α-Cm variety has a double hexagonal close-packed crystal lattice with alternation of layers according to the ABAC... scheme; space group $P6_3/mmc$ (D_{6h}^4), $a = 0.344$ nm, $c \approx 1.16$ nm. The β-Cm form has a face-centered cubic lattice, space group $Fm\bar{3}m$ (O_h^5), $a = 0.5039$ nm upon hardening. Density of α-Cm is 13.51 g/cm^3; $T_{\text{melting}} =$ 1620 K, T_{boiling} is about 3380 K; specific heat $c_p = 27.6$ J·mole^{-1}·K^{-1}; heat of melting is 14.64 kJ/mole, heat of sublimation is 372.6 kJ/mole. Below the Néel point $T_{\text{N}} = 52$ K, α-Cm transforms from the paramagnetic state to the antiferromagnetic one (see *Antiferromagnetism*); below the critical temperature $T_{\text{c}} = 205$ K, β-Cm is a *ferrimagnet*, or has a noncollinear magnetic structure. Targets from isotopes ^{246}Cm and ^{248}Cm are used in the synthesis of the heavier *transuranium elements* by bombardment with the multicharged ions. High heat release in the preparations from the isotopes ^{244}Cm and ^{242}Cm caused by their α-decay suggests the possibility of using these nuclides to produce small-size nuclear electric current sources that persist for several years (e.g., in onboard space systems).

CURRENT CARRIER DRIFT, charge carrier drift

A directed movement of charge carriers in *semiconductors* under the action of external fields imposed on their disordered (thermal) motion. The current density resulting from the charge carrier drift in the electric field $\boldsymbol{E}$ (*drift current*) is $\boldsymbol{j} = \sigma\boldsymbol{E}$, where the electrical conductivity $\sigma = e(\mu_{\text{e}}n + \mu_{\text{h}}p)$, n and p are the concentrations of the *conduction electrons* and *holes*, respectively, and μ_{e} and μ_{h} are their mobilities (see *Mobility of current carriers*). The total conduction current density is the sum of the drift current, the diffusion current, and the thermoelectric current arising from the presence of a temperature gradient. The charge carrier drift can also result from the carrying along of carriers by either an ultrasonic wave (see *Acousto-electric effect*) or an electromagnetic wave (radio electric effect, light electric effect). In this event the drift involves nonequilibrium carriers, and therefore it becomes complicated by the appearance of bulk charges whose field should be taken into account in addition to the external field and the *recombination of current carriers*. The motion of injected nonequilibrium carriers (see *Injection of current carriers*) in the external electric field is described by so-called *ambipolar mobility*

$$\mu_{\text{a}} = \frac{\mu_{\text{e}}\mu_{\text{h}}|n-p|}{\mu_{\text{e}}n + \mu_{\text{h}}p},$$

which, in general, differs from μ_{e} and μ_{h}. For an intrinsic semiconductor $n = p$ and $\mu_{\text{a}} = 0$; for an n-type semiconductor $n \gg p$, $\mu_{\text{a}} = \mu_{\text{h}}$; and for a p-type semiconductor $p \gg n$, $\mu_{\text{a}} = \mu_{\text{e}}$, which means that in an impurity semiconductor μ_{a} coincides closely with the mobility of the minority carriers. The velocity of the grouping of nonequilibrium carriers in an external field $\boldsymbol{E}$ is equal to $\mu_{\text{a}}\boldsymbol{E}$.

An important characteristic of charge carrier drift is its *drift length* l, which is the distance that the carriers traverse in going from their point of generation (see *Generation of current carriers*) to their recombination point. This length has the value $l = \mu E\tau$, where τ is the lifetime of the nonequilibrium carriers. The method for measuring the drift length is the same as that for measuring the *diffusion length*.

In anisotropic crystals the direction of the drift can differ from that of the electric field (since mobilities are tensors). In strong electric fields the drift can be anisotropic even in isotropic (cubic) *many-valley semiconductors*. In the presence of a transverse magnetic field the drift direction does not coincide with the electric field direction.

In a strong magnetic field $\boldsymbol{B}$ normal to the external electric field $\boldsymbol{E}$ satisfying the condition $cB \gg E$, the charge carriers drift in the direction normal to both $\boldsymbol{E}$ and $\boldsymbol{B}$ with the velocity $v = E/B$ that does not depend on the carrier mobility. A circular motion of the carriers at the *cyclotron frequency* is superimposed on this drift.

CURRENT CARRIER GENERATION

See *Generation of current carriers*.

CURRENT CARRIER MOBILITY

See *Mobility of current carriers*.

CURRENT CARRIER RECOMBINATION

See *Recombination*.

CURRENT CARRIERS, charge carriers

Mobile charged particles and *quasi-particles.* In an external electric field, the random motion of the current carriers acquires a directional component (see *Current carrier drift*), so that an *electric current* flows. The current carriers in *metals* are usually electrons of the *conduction band,* but when this band is more than half full, the current carrying role is better described in terms of *holes* in this band. In metals the concentration of current carriers is about 10^{22}–10^{23} cm^{-3}, i.e. is of the order of the number of atoms in the crystal itself, while the mobility μ of *conduction electrons* usually lies within 10^5–10^7 $cm^2{\cdot}V^{-1}{\cdot}s^{-1}$). However, in super-pure metals, it may increase by 3–4 orders of magnitude (see *Mobility of current carriers*). The carriers of current in *semiconductors* are usually electrons in the conduction band and holes in the *valence band.* Their concentrations, n (electrons) and p (holes), may vary by many orders of magnitude from sample to sample of the same material, depending on the concentration of *impurity atoms* (*donors, acceptors*), intrinsic *defects,* the temperature, and external actions. *Dark current carriers* are those which exist in semiconductors in the absence of photoexcitation; *equilibrium current carriers* are those present when thermodynamic equilibrium exists between allowed *energy bands* and *local electronic levels.* For the case $n > p$ at equilibrium, the *majority current carriers* are electrons, otherwise they are holes. *Nonequilibrium current carriers* appear when the electron subsystem is variously excited. The nonequilibrium concentration of *minority current carriers* may be easily raised far above its equilibrium value. It is controlled by the efficiency of the injection of minority current carriers into the band, and by their *lifetime* in that band. In semiconductors, the mobility of carriers usually lies within the range 10^2–10^4 $cm^2{\cdot}V^{-1}{\cdot}s^{-1}$. The electron (hole) component of conductivity may include impurity conductivity only, without any input from the allowed band (see *Impurity band, Hopping conductivity*), and the carrier mobility μ may then remain much lower than 1 $cm^2{\cdot}V^{-1}{\cdot}s^{-1}$. The interaction of free electrons or holes with the surrounding atoms of the crystal may result in the formation of polarons and fluctuation states of the current carriers (see *Polaron, Fluctuon*). In *insulators* the role of charge carriers may be played by interstitial ions of either intrinsic or impurity atoms, or by charged vacancies (see *Ionic conductivity*).

CURRENT CARRIER SCATTERING in crystals

A transition of an *electron* (or *hole*) between states described by Bloch wave functions under the action of a perturbation of the periodic *crystal field.* There are two types of such perturbations that influence the mechanism of the scattering. The first type includes static defects of the crystal lattice: *impurity atoms, vacancies, dislocations, crystallite* boundaries, and the external surface of the specimen (see *Surface scattering*). The second type includes dynamic perturbations: *phonons,* other electrons, *magnons,* etc. The interaction of *current carriers* with the perturbation of the periodicity is described by a Hamiltonian $H' = H - H_0$, where H is the Hamiltonian of the actual crystal and H_0 is the Hamiltonian of an *ideal crystal.* This results in electronic transitions between the states with *quasi-wave vectors* $\boldsymbol{k}$ and $\boldsymbol{k}'$ and energies $\varepsilon(\boldsymbol{k})$ and $\varepsilon(\boldsymbol{k}')$. Provided $|\varepsilon(\boldsymbol{k}') - \varepsilon(\boldsymbol{k})| \ll \varepsilon(\boldsymbol{k})$ then we have *quasi-elastic (elastic) scattering*. To each current-carrier scattering mechanism there corresponds a separate Hamiltonian H', a probability $W(\boldsymbol{k}, \boldsymbol{k}')$ of the transition from state $\boldsymbol{k}$ to state $\boldsymbol{k}'$, and a *relaxation time* for quasi-momentum τ_p, provided such a variable may be introduced. Calculations of $W(\boldsymbol{k}, \boldsymbol{k}')$ are usually carried out within first-order perturbation theory ($|H'| \ll H$, the *Born approximation*), and $\Delta t \ll \tau_p$, where Δt is the action time for the perturbation in the elementary scattering event. Scattering hampers directional motions of electrons either in the external electric field or under the effect of a temperature gradient, these motions generating the electric current and the energy flux, respectively. The quantity $W(\boldsymbol{k}, \boldsymbol{k}')$ enters the collision integral in the Boltzmann *kinetic equation* (see *Boltzmann equation*), its solution permitting the determination of kinetic coefficients. Among the latter are, in particular, the *mobility of current carriers* μ and the *Hall constant* (see *Galvanomagnetic effects*). The solution of that equation simplifies if the variable τ_p may be introduced. For the simultaneous action of several scattering mechanisms the values $W_i(\boldsymbol{k}, \boldsymbol{k}')$ and τ_{pi}^{-1} are summed: $1/t_p = \sum_i (1/\tau_{pi})$. In some cases this summation process extends to $1/\mu = \sum_i (1/\mu_i)$ and $\rho = \sum_i \rho_i$ (*Matthiessen rule* for resistivities ρ_I), which hold for τ_{pi} independent of $\boldsymbol{k}$. Provided $W(\boldsymbol{k}, \boldsymbol{k}')$ depends on the angle between $\boldsymbol{k}$ and $\boldsymbol{k}'$ only, the microparticle scattering is called *isotropic scattering*, and when it also depends on the directions of $\boldsymbol{k}$ and $\boldsymbol{k}'$ it is *anisotropic scattering*. The relaxation time is a tensor in the latter case.

Current carrier scattering in semiconductors. Besides the mechanism of scattering proper, this effect is controlled by the *dispersion law* for the current carriers. When $\boldsymbol{k}$ and $\boldsymbol{k}'$ belong to the same valley the process is called *intravalley scattering*, otherwise it is designated *intervalley scattering*. If one may express $\tau_p \propto e^s$ and $\mu \propto T^\alpha$ then the quantities s and α control the scattering process.

During the *scattering of electrons by lattice vibrations (phonons)* a phonon with an energy $\hbar\omega(\boldsymbol{q})$ and a quasi-wave number $\boldsymbol{q}$ (ω is the frequency) is either emitted or absorbed in an elementary scattering event. The *conservation laws* of energy $\varepsilon(\boldsymbol{k}) = \varepsilon(\boldsymbol{k}') \pm \hbar\omega(\boldsymbol{q})$ and quasi-momentum $\hbar\boldsymbol{k} = \hbar\boldsymbol{k}' \pm \hbar\boldsymbol{q} + \hbar\boldsymbol{b}$ are then obeyed. The "minus" sign here corresponds to the absorption of a phonon, while "plus" corresponds to its emission, and $\boldsymbol{b}$ is a reciprocal lattice vector. Processes with $\boldsymbol{b} \neq 0$ are the so-called *flip-over* or *Umklapp processes*. It is important to account for them when $\hbar(\boldsymbol{k} - \boldsymbol{k}')$ lies outside the first *Brillouin zone*, where $\hbar\boldsymbol{b}$ brings $\hbar\boldsymbol{q}$ back into the first zone. Umklapp processes are insignificant for valleys in the center of the first Brillouin zone (see *Many-valley semiconductors*). *Acoustic vibrations* of atoms deform the crystal, changing the energy of the bottom of the *conduction band*, and produce a scattering *acoustic strain potential* $H'_{ac} = \varepsilon_1 \Delta v/v_0$, where ε_1 is the *strain potential* constant (sometimes the strain potential itself), and $\Delta v/v_0$ is the relative change of a given semiconductor volume, e.g., a *unit cell*.

The above expression for H'_{ac} holds for an isotropic electron dispersion law only, and for an extremum at the center of curvature of the first Brillouin zone. In the general case the strain potential constant is a tensor (*strain potential tensor*) that relates H'_{ac} to the *strain*. When $T > T_0 = m_n u^2/k_B \approx 1$ K, where m_n is the effective mass of the electron, u is the *sound velocity*, k_B is the *Boltzmann constant*, the scattering is quasi-elastic, so that only the phonons with $q < 2k$ take part in it. The Planck distribution function for them may be approximated by the classical expression $k_B T/\hbar$ that corresponds to an equilibrium energy distribution over the degrees of freedom, so that these are *equidistributed phonons*. In the simplest case of an isotropic parabolic dispersion law for $\varepsilon(\boldsymbol{k})$ we have $s = -1/2$, $\alpha = -3/2$, $r = 3\pi/8 = 1.18$.

Optical vibrations of atoms produce a scattering *optical strain potential* (scattering by *nonpolar optical phonons*). When $k_B T \gg \hbar\omega_0$, where ω_0 is the frequency of an optical phonon, the scattering is elastic, $s = -1/2$, $\alpha = -3/2$, $r = 3\pi/8 = 1.18$. At all other temperatures the scattering becomes inelastic. Provided, however, that the dispersion law is isotropic (or close to isotropic and averaged over directions), the equality $W(\boldsymbol{k}, \boldsymbol{k}') = W(\boldsymbol{k}, -\boldsymbol{k}')$ makes it possible to introduce τ_p. If $k_B T \ll \hbar\omega_0$, $\mu \propto \tau_p \propto \exp[\hbar\omega_0/(k_B T)]$, $r = 1$. Only these two cases of scattering by phonons are possible in nonpolar crystals such as Ge and Si. In polar crystals (e.g., $A^{III}B^{V}$) scattering by the *polar optical potential* resulting from the lattice polarization (see *Polarization of insulator*) contributes to the optical vibrations of the atoms (*scattering by polar optical phonons*). Such scattering may be stronger than scattering from acoustic and optical deformation potentials. It is elastic for $k_B T \gg \hbar\omega_0$ and $s = 1/2$, $\alpha = -1/2$, $r = 45\pi/128 = 1.105$, and becomes inelastic at other temperatures. However, when $k_B T \ll \hbar\omega_0$ one may introduce τ_p since electrons can only absorb an optical phonons in a weak electric field. Moreover, having absorbed the optical phonon, the electron emits it almost instantaneously (the probability ratio then becomes $\exp[\hbar\omega_0/(k_B T)] \gg 1$. The electron energy undergoes almost no change during such a composite scattering event, while $\boldsymbol{k}$ does change. We have $\mu \propto \tau_p \propto \exp[\hbar\omega_0/(k_B T)]$, $r = 1$ in the absence of degeneracy.

In crystals with partial *ionic bonding* acoustic vibrations may also produce, beside the deformation potential, a *piezoelectric scattering potential*, provided the crystal possesses no *center of symmetry* (e.g., in $A^{III}B^{V}$, $A^{II}B^{VI}$ semiconductors, see *Piezoelectricity*). A *piezoelectric constant* enters the formulae describing scattering by that potential, and $s = 1/2$, $\alpha = -1/2$, $r = 45\pi/128 = 1.105$ for the case.

At low temperatures the phonon gas is rarefied and the main effect may consist in *scattering by impurity atoms*; the scattering in that case is elastic because of the great difference between the masses of the electron and the atom. Scattering from ionized impurities is similar to the Rutherford scattering of α-particles. The *Conwell–Weisskopf formula* (E.M. Conwell, V.F. Weisskopf, 1950) obtained for μ for that kind of scattering does not take into account the screening by free electrons of the field produced by the impurity ions (see *Electric charge screening*), while the *Brooks–Herring formula* does. In both cases, and to within the accuracy of certain factors slowly changing with temperature T and the impurity concentration N, we have $\tau_p \propto e^{3/2}/N$, $\mu \propto T^{3/2}/N$, $r = 315\pi/512 = 1.93$. *Scattering by neutral impurities* is similar to electron scattering by atoms of hydrogen: $s = 0$, $\alpha = 0$, $r = 1$. The corresponding formula for τ_p is called the *Erginsoy formula* (C. Erginsoy, 1950). In the case of scattering from *point defects* with a short-range δ-function potential $\alpha = -1.2$, $s = -1.2$. In all the above theories the scattering of current carriers is assumed to be *incoherent scattering*, with the assumption of no interference between electron waves scattered by different impurity atoms. This is generally true for a random distribution of impurity atoms. The above dependences of the mobility μ on the temperature T are usually not observed experimentally, since they were obtained from idealized models. The reason for such a pronounced disagreement between theory and experiment may lie in a dependence of m_n and of the deformation potential on the temperature T. Also, the band may deviate from a parabolic shape, the electrons may exhibit partial *degeneracy*, etc. For example, in the case of scattering by phonons in *germanium*, $\alpha = -1.66$ for electrons and $\alpha = -2.33$ for holes.

Electron–electron scattering does not change the total quasi-momentum of the electron gas, and hence has not affect on either the current or μ. However, provided some other scattering mechanism is simultaneously at work, characterized, as it were, by a $\tau_p(\varepsilon)$, both the current and μ will change because of energy redistribution among the electrons. This effect is characterized by the *electron–electron scattering* time, τ_{ee}. When the electrons collide with each other, a noticeable exchange of both quasi-momentum and energy takes place, because of the equality of their masses, and τ_{ee} controls the rate of both processes. In the absence of degeneracy we have $\tau_{ee} \propto n^{-1}$, where n is the electron concentration. The *electron–hole scattering* is significant for high concentrations of current carriers (high temperatures, high levels of *injection of electrons*). It is similar to scattering by ionized impurities when the ionic charge is opposite to that of the scattered carriers.

Scattering by dislocations occurs due to the interaction of current carriers with charged dislocation filaments, and the deformation potential produced by the field of strained dislocations. When the scattering center has a *magnetic moment* then *electron spin-flip scattering* becomes possible, which leads to the *Kondo effect*. The *energy relaxation time*, τ_e, may often be represented in the form $t_e \propto \mathrm{e}^l$, where the value of l depends on the scattering mechanism. We have $l = -1/2$ for the acoustic deformation potential, $l = 1/2$ for the optical, $l = 3/2$ for the polar optical, and $l = 1/2$ for the acoustic piezoelectric potential. The inequality $\langle\tau_p\rangle \ll \langle\tau_e\rangle$ holds for the averages of τ_p and τ_e in the case of quasi-elastic scattering.

Scattering of current carriers in metals. This manifests itself in the principal kinetic properties of the metal, its *electrical conductivity* and *thermal conductivity*. The electrical resistance is related to the relaxation of the electron quasi-momentum, while the thermal resistance arises from both momentum and energy relaxation. When the momentum relaxation prevails, the *Wiedemann–Franz law* holds: $\kappa/(\sigma T) = \pi^2 k_{\mathrm{B}}^2/(3e^2)$, where κ is the *electronic thermal conductivity* and σ is the electrical conductivity. The scattering of current carriers in metals is also influenced by the shape of the *Fermi surface*. In the case of *electron–phonon scattering*, determining the explicit form of the Hamiltonian H' is a difficult problem, since the displacement of ions leads to a significant restructuring of the electronic wave functions. At the same time semiempirical techniques are available that make it possible to relate matrix elements of the Hamiltonian to the structure of the Fermi surface. The most well-known one is based on combining the *model of rigid (undeformable) ions* with the *pseudopotential method.* It is not difficult to demonstrate, without falling back to model considerations, that the squared matrix element of the Hamiltonian H' is proportional to the phonon momentum to first order for the case of scattering from long-wave vibrations.

The mechanism of *electron–phonon relaxation* depends qualitatively on the relation between the momentum of thermal phonons $\hbar q$ and the typical size P_{F} of the Fermi surface (note that an intermediate range $P_{\mathrm{F1}} < \hbar q < P_{\mathrm{F2}}$ is also realized in Bi, P_{F1} and P_{F2} being the major and the minor semiaxes of the ellipsoidal Fermi surface). Let us introduce the maximum energy of the phonons that can interact with electrons: $\theta = uP_{\mathrm{F}}$. The ratio θ/k_{B} in typical metals is of the order of the *Debye temperature.* At high temperatures we have $k_{\mathrm{B}}T \gg \theta$ and the phonons with energy of the order of θ and momentum $\hbar q \approx P_{\mathrm{F}}$ are important. The number of such phonons and, accordingly, the probability of electron–phonon collisions, are proportional to the temperature T. The electrical resistance is, correspondingly, $\rho \propto T$. Since the characteristic energy scale of the change in the Fermi distribution function for electrons (see *Fermi–Dirac statistics*) $k_{\mathrm{B}}T$ is much larger than the phonon energy, such collisions are quasi-elastic, and the Wiedemann–Franz law holds. At low temperatures $k_{\mathrm{B}}T \ll \theta$ and phonons with energy of the order of $k_{\mathrm{B}}T$ and momentum $\hbar q = k_{\mathrm{B}}T/u = P_{\mathrm{F}}k_{\mathrm{B}}T/\theta$ start to play the decisive role. In this particular case energy relaxation occurs during a single collision event. A significant change in the electron momentum $\hbar\omega$ requires $[P_{\mathrm{F}}/(\hbar q)]^2 = [\theta/(k_{\mathrm{B}}T)]^2$ collisions. Such a relaxation is essentially a *diffusion* of electrons over the Fermi surface that proceeds with a typical step of $\hbar q \ll P_{\mathrm{F}}$. Due to the laws of conservation of energy and momentum such an electron may only interact with those phonons that have their momenta normal to the electron velocity vector, the number of such phonons being proportional to T^2. In addition, the squared matrix element of the *electron–phonon interaction* is proportional to the first power, $q \propto T$. As a result, the energy relaxation time at low temperatures is $t_e' \propto T^{-3}$, while the transport (pulse) time is proportional to $\tau_p \approx \tau_e[\theta/(k_{\mathrm{B}}T)]^2 \approx T^{-5}$. The corresponding electrical resistivity follows the *Bloch law*, $\rho \propto T^5$. Naturally, the Wiedemann–Franz law now no longer holds. Strictly speaking, the above refers to compensated metals (with equal numbers of electrons and holes), or to metals with open Fermi surfaces. For an uncompensated metal the relaxation of total momentum at low temperatures is controlled by electron–phonon collisions, accompanied by *Umklapp processes.* Moreover, such processes are only possible in the regions where closed Fermi surfaces come closest to each other, the so-called *hot spots* or *hollows*. The transport relaxation time, roughly speaking, is equal to the sum of the time span of transitions across the Fermi surface to one of the hollows by diffusion, plus the time an electron takes to flip over or undergo an Umklapp process in a hollow. Usually the first time span significantly exceeds the second (Umklapp processes proceed relatively readily), so that the Bloch law remains valid. Umklapp processes may control momentum relaxation only in the range of extremely low temperatures (about 1 K), resulting in an exponential temperature dependence of the electrical resistivity against the background of the predominant influence of electron–impurity scattering.

Electron–electron collisions. Only electrons from regions where the Fermi distribution is blurred or diffuse may take part in such collisions, due to the Pauli principle and the conservation of energy law. In other words, a given electron may only collide with a small fraction of the conduction electrons, of the order of $k_{\mathrm{B}}T/\varepsilon_{\mathrm{F}}$ (ε_{F} is the *Fermi energy*), since the number of available states is proportional to $k_{\mathrm{B}}T/\varepsilon_{\mathrm{F}}$. (After one of the electrons enters the diffuse range, another appears automatically, due to energy conservation during their collisions.) Thus, taking into account the Pauli principle lowers the probability of an electron–electron collision by a factor of $[\varepsilon_{\mathrm{F}}/(k_{\mathrm{B}}T)]^{-2}$ (see *Fermi liquid*). One should also take account of the fact that the *screening radius* in a typical metal is of the order of the lattice constant, and as a result, the *mean free path* for electron–electron collisions is $l_{ee} \approx a[\varepsilon_{\mathrm{F}}/(k_{\mathrm{B}}T)]^2$. Umklapp processes during electron–electron collisions are possible when the Fermi surface crosses a plane normal to a certain vector $\boldsymbol{b}$ of the reciprocal lattice located a distance $b/4$ from the center of the Brillouin zone. The law $\rho \propto T^2$ might manifest itself at low enough temperatures when one could neglect the input from electron–phonon scattering, but there is no reliable experimental evidence for this. Meanwhile, electron–electron collisions play a role during the absorption of electromagnetic waves in metals in the near-IR spectral range. Having absorbed an electromagnetic quantum with energy $\hbar\omega \gg k_B T$ the electron may interact with other electrons in a layer of thickness $\hbar\omega$. Calculations result in a substitution in the expression for the probability of electron–electron collisions: the $k_B^2 T^2$ factor is replaced by $k_B^2 T^2 + [\hbar\omega/(2\pi)]^2$. This does not apply at $T = 0$ where everything is in the ground state.

Scattering from static defects is usually elastic, hence the collision probability does not depend on the temperature, and the Wiedemann–Franz law holds. Local defects such as impurity atoms significantly perturb the crystal field at distances of the order of a lattice constant a. The field of a charged impurity is screened by electrons, the screening radius (Debye radius) in a typical metal also being of the order of a. Therefore, the scattering cross-section is of the order of a^2, and depends on the relatively weak influence of the local defects. The mean free path is $l_{ed} \propto aC^{-1}$, where C is the number of local defects divided by the number of cells in the crystal lattice.

CURRENT, CRITICAL

See *Critical currents.*

CURRENT, ELECTRIC

See *Electric current.*

CURRENT FILAMENTATION, current channeling

Electric current flow in a sample with a nonuniform distribution over the cross-section, i.e. high densities along some channels (filaments), and low or zero densities along neighboring parallel paths. The phenomenon of current filamentation may result from an S-shaped *current–voltage characteristic.* Such a characteristic can arise from certain mechanisms of current carrier scattering in solids. In addition an S-shaped characteristic in a uniform semiconductor may arise from the release of Joule heat (*overheating instability*); when a temperature increase caused by the current flow leads to additional electron transitions from *local electron levels* in the *conduction band* then the current can increase still further, with a further rise in the lattice temperature. The channeling often brings about irreversible processes in the current channel, which could put the device out of operation. If the channeling associated with an S-shaped characteristic happens at high frequencies it may be used for the generation of super-high frequency oscillations (harmonics), as occurs in various types of diodes (e.g., *trapped plasma avalanche triggered transit diode, tunnel diode, impact ionization avalanche transit time diode*). Besides, the *pinch effect* may be responsible for current channeling in *semiconductors* (see *Magnetic pressure*). Also this channeling may be observed in pure metals at low temperatures through the magnetodynamic mechanism of nonlinearity (see *Magnetodynamic nonlinearity*).

CURRENT, PERSISTENT

See *Persistent current.*

CURRENT STATES IN METALS

A hysteresis effect (in an external magnetic field B_0) of electric current rectification, and the excitation of a static *magnetic moment* in the sample when irradiating the latter by an rf (radio wave) of a high amplitude B_{rf}. This takes place at the quasi-static limit $\omega\tau \ll 1$ (here ω is the rf frequency; τ is the electron characteristic lifetime) of the *anomalous skin-effect*, with B_0 parallel to the rf magnetic field vector B_{rf}. The current states are excited when B_{rf} exceeds a threshold value $B_{cr} \sim 0.1$–1 mT that is determined by ω, the mean free path l, and the character of the electron *surface scattering.* For diffusion scattering $B_{cr} \propto \omega^{-1/3} l^{-2}$. In the ideal case of specular reflection the current states do not emerge. At the regime of strong nonlinearity $B_{rf} > B_{cr}$, the magnetic moment does not vanish even at $B_0 = 0$. For growing B_{rf}, the *hysteresis* loop approaches a universal curve which is determined only by amplitude B_{rf}, and does not depend on the electrodynamic properties of the metal. The excitation of current states is stimulated by sharp conductivity jumps during the rf period which are due to the appearance and disappearance of trapped electrons in the skin-layer (see *Magnetodynamic nonlinearity*).

CURRENT STATES IN SUPERCONDUCTORS

States with a finite current I at zero voltage which correspond to an infinite electrical conductivity. The current states are metastable under the condition $I > I_c$ where I_c is the temperature-dependent critical current which goes to zero at the transition point T_c. In this case, in the vicinity of T_c, we have $I_c = \text{const}(T_c - T)^{3/2}$ (*Bardeen law*). The energy E is related with the current as $E = L_k I^2/2$, where L_k is the kinetic inductance of the superconductor. Strictly speaking, a current state as a *metastable state* has a finite lifetime, but for macroscopic superconducting rings the persistent current decay times are so long that it is reasonable to speak about an "infinite conductivity". The existence of current states is the consequence of the *macroscopic quantum coherence* of electronic states of the superconductor which are described by a unified wave function $\psi = F e^{i\varphi}$. In this case the current state is given by the magnitude of phase gradient $\nabla\varphi$ according to

$$j = n_s e v_s,$$

$$v_s = \frac{1}{2m}(\hbar\nabla\varphi - 2eA),$$

where v_s is the *superfluid velocity* (velocity of motion of condensate); n_s is the temperature-dependent concentration of superconducting electrons ($n_s/2$ is the concentration of *Cooper pairs*). For a pure superconductor $n_s = n$ at $T = 0$, where n is the total concentration of electrons; for a *superconducting alloy* with a mean free path l less than the *coherence length* ξ_0, $n_s = nl/\xi_0$ at $T = 0$. As v_s grows, the value n_s decreases so that the dependence $j(v_s)$ takes the shape of a curve with a maximum ("*depairing curve*"). The *critical depairing current* j_c is determined as the maximum of the dependence $j(v_s)$, and is estimated as $j_c \sim n_s(v_s = 0)ev_c$ (here $v_c = \Delta/p_F$ is the critical velocity). This estimate gives a critical current density value of $j_c \sim 10^9$ A/cm^2 which is achieved in thin superconducting films or filaments (with thickness or diameter less than ξ_0). In bulk samples the depairing current is not reached, as a rule, and the critical current is determined by the decay of the superconductive state via the intrinsic magnetic field of the current in *type I superconductors* (*Silsbee rule*, F.B. Silsbee, 1916), or by the Abrikosov mechanism of *vortex pinning* in *type II superconductors*. The current states also occur at superconducting *tunnel junctions* (dc *Josephson effect*). In this case the factor which determines the current through this transition is the phase difference $\varphi_1 - \varphi_2$ of the condensate wave functions of the two superconductors: $I = I_c \sin(\varphi_1 - \varphi_2)$.

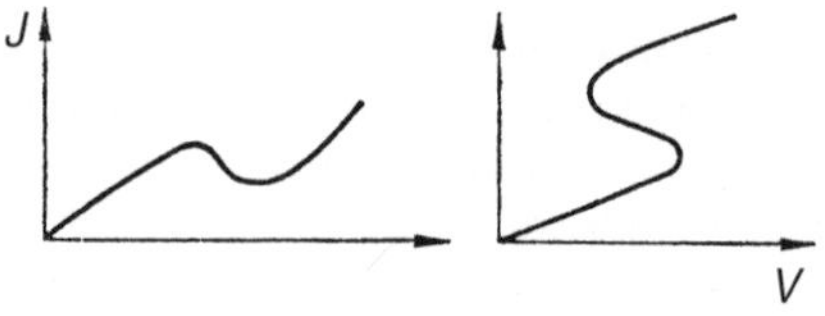

N-shaped (left) and S-shaped (right) current versus voltage characteristics.

CURRENT–VOLTAGE CHARACTERISTIC

The dependence of the electrical current I on the voltage V applied to an electric circuit element. If the element resistance does not depend on the current then the current–voltage characteristic (CVC) is a straight line (*Ohm's law*). The I versus V characteristics of some circuit elements can have sections which are nonlinear with various shapes. In semiconductors the appearance of non-linear sections can result from the limitations of current through the contact arising from the dependence of the emission ability of the contact on the presence of regions of space charge, and from the dependence of the current carrier mobility on the electric field strength.

In some cases a current–voltage characteristic has sections of negative slopes, corresponding to negative resistance dV/dI (see Fig.). Negative resistance means that the conduction occurs under unstable conditions. Fig. left: an I versus V characteristic with an N-shape which is seen in tunnel *diodes* and in semiconductors containing *traps* with a specific dependence of the carrier *trapping* on their energy. Right: an S-shape characteristic which can result from an "overheating" mechanism, associated with the *electron heating*, and also from an avalanche breakdown due to *impact ionization* of impurity atoms.

CURVATURE TENSOR

Riemann tensor, a 4th rank tensor that characterizes the deviation of the geometry of space from the Euclidean type. The mixed curvature tensor, R^i_{klm}, is defined through the *Christoffel symbols* Γ^i_{kl} as follows: $R^i_{klm} = \partial_l \Gamma^i_{km} - \partial_m \Gamma^i_{kl} + \Gamma^i_{nl}\Gamma^n_{km} - \Gamma^i_{nm}\Gamma^n_{kl}$ (here $\partial_i = \partial/\partial x^i$, repeated indices imply summation). The covariant curvature tensor $R_{iklm} = g_{in}R^n_{klm}$ (where g_{in} is the metric tensor) satisfies the relations $R_{iklm} = -R_{kilm} = -R_{ikml} = R_{lmik}$; $R_{iklm} + R_{ilmk} + R_{imkl} = 0$. A property of this tensor is that there exists a coordinate system such that the tensor is identically zero when the space is locally Euclidean (i.e. g_{ik} may be reduced to the metric tensor of flat space with the corresponding signature via a coordinate transformation in a finite neighborhood of an arbitrary point). The inverse statement is also valid. Thus, the vanishing of the curvature tensor R^i_{klm} provides an invariant criterion for the Euclidean nature of space that is independent of the choice of coordinate system. This property of the curvature tensor is used in *elasticity theory* to derive the conditions for the *strain compatibility* that are equivalent to the vanishing of the curvature tensor formed from the Christoffel symbols which are calculated from the displacement field. When the solid contains *dislocation* and *disclination* types of *defects*, which give rise to incompatible deformations, the condition of compatibility $R^i_{klm} = 0$ is violated. The solid is then internally stressed, i.e. there are tensions in it even in the absence of external loads (see *Internal stresses*). Such a solid is described by a *non-Riemann geometry*, with a curvature tensor that differs from zero, and another tensor called the *Einstein tensor* $R_{km} = R^n_{knm}$ is formed from it. Note that $R_{km} = B_{km}$, where the so-called "*material tensor*" B_{km} describes the distribution of additional matter that produces incompatible deformations within the body. For example, this relation describes internal stresses in a solid that result from a prescribed density of dislocations (see *Dislocation density tensor*).

CYANIDING

Diffusion saturation of the surface of steel and cast-iron products by nitrogen and carbon simultaneously, generally in melts containing cyanide salts. The thickness of the cyanided layer is 0.05 to 2 mm. Cyaniding increases the surface hardness of the products, their wear resistance, and their endurance limit. Tool *steels* are cyanided at 450 to 600 °C, structural steels below 950 °C. Cyaniding is distinguished by *cementation* and *nitriding* (see *Chemical heat treatment*) depending on the higher velocity of the process. After the high-temperature cyaniding the products are subjected to hardening and to low-temperature annealing.

CYBOTAXIC CLUSTERS

Domains with a fluctuation origin having a smectic structure found in a *nematic liquid crystal*, existing in the vicinity of a nematic–smectic *phase transition*. Cybotaxic clusters present in the nematic phase influence the longitudinal bending and twisting strains. Smectic layers in the cluster should be equidistant. As a result, the *Frank elasticity constants* K_2 and K_1 in the neighborhood of the transition in the *smectic liquid crystal* phase increase directly as the cluster size (see *Elasticity of liquid crystals*).

CYBOTAXIS

Name of a hypothetical three-dimensional arrangement of particles (atoms, molecules) in fluids which are in a state intermediate between one that is completely disordered and one that is crystalline. The term "cybotaxis" was introduced by D. Stewart and R. Morrow (1927). The atoms and molecules in cybotaxes have mobility, but their motion and relative arrangement are not disordered as in the gaseous state. A cybotaxis has no sharply delineated boundary. The lifetime of cybotaxes depends on the composition of the fluid, and the temperature. The basis for introducing the cybotaxis concept was X-ray structure analysis data that reveal that at temperatures slightly above the *melting* point the arrangement of atoms and molecules in liquids is similar to the arrangement in crystalline *solids*. It is possible to connect the formation of locally regular *clusters* in glasses which inherit the structural elements of smelts on *quenching*, with the presence of cybotaxes in smelts. Cybotaxes are often associated with localized arrangements of molecules in liquid crystals. See also *Polycluster amorphous solids*, *Lebedev crystallite hypothesis*, *Liquid state models*.

CYCLIC HEAT TREATMENT

A *heat treatment* of metals and alloys involving periodic heating and cooling of the material. This is employed to induce structural changes that lead to the improvement of mechanical properties. One heating and one cooling event constitute a unit cycle of a cyclic heat treatment. The cycle configuration (i.e. temperature versus time dependence), its parameters (maximum and minimum temperatures of the cycle, heating and cooling rates, exposure time at the maximum temperature of the cycle) and the number of cycles can vary, depending on the type of material being treated, its original condition, and the needed structural changes.

A cyclic heat treatment can be carried out so that no *phase transitions* take place in the temperature range of the cycle. In this case structural changes in the cyclic process account for the *plastic deformation* due to the thermal stresses which arise as a result of temperature gradients induced by heating and cooling the material. Substantial changes take place in the intragrain dislocation structure, grain interfaces (see *Grain boundaries*), regions of grains adjacent to interfaces, and the quantity and morphology of particles of phases which gain *strength*. In particular, the *durability* of metals and alloys under the conditions of high-temperature *creep* increases significantly.

If a cyclic heat treatment involves a phase transition within its temperature range of thermal cycling (e.g., the $\alpha \leftrightarrow \gamma$ transformation in iron and its alloys), the structural changes of a cyclic heat treatment arise as a result of both the plastic deformation caused by thermal stresses and the phase transitions. At low rates of heating and cooling (or for a small difference between T_{max} and T_{min}) phase transitions play the main role, and the presence of these transitions can result in the production of tiny uniform sized grains which increase the *plasticity* and strength of the material.

CYCLIC STRENGTH

Resistance to fatigue *failure* of materials under the application of cyclic loads. Quantitatively cyclic strength may be estimated by the *endurance limit*. As a rule, cyclic strength is somewhat lower than *static strength*. First of all, cyclic strength is determined by the maximum cyclic stress and its amplitude. The greater the average cyclic stress, the lesser the stress amplitude needed for material destruction at the same test base. The character of the stress change between σ_{max} and σ_{min} has a negligible effect on the cyclic strength. That is why cycles of complicated shapes may always be reduced to ordinary ones, and standard fatigue tests are carried out with the use of the simplest cycle configuration.

An increase of the cycle frequency, other things being equal, usually gives rise to an increase in the cyclic strength, especially at high temperatures. So long as fatigue *cracks* are forming from regions of *stress concentration* on the surface, their state strongly influences the resulting cyclic strength. As the temperature increases, a shift of *fatigue* curves towards lower stresses is observed. When temperature changes take place, fatigue disturbances may also arise because of the onset of thermal stresses. Cyclic strength rises with the increase of metal purity with respect to interstitial elements, with the decrease of grain size, and with the formation of a cellular dislocation structure (see *Fragmentation*). It also depends on the presence and size of the second phase of dispersed particles.

The environment in contact with the surface of the crack affects the crack growth due to physico-chemical processes taking place at its peak (see *Fatigue crack growth*). The reliable determination of the cyclic strength is only possible when one takes into account the nature of the product, and the details of the fatigue.

CYCLOTRON FREQUENCY, gyromagnetic frequency

Gyromagnetic frequency Ω of charged particle rotation in a constant magnetic field $\boldsymbol{B}$ in a plane perpendicular to the field. For a free charged particle the cyclotron frequency Ω is determined from the balance of the *Lorentz force* $e\boldsymbol{v} \times \boldsymbol{B}$ by the centrifugal force mv^2/r, and it is given by $\Omega = eB/m$, where e and m are, respectively, the charge and mass of the free particle, and $v = \Omega r$.

The cyclotron frequency determines the energy difference ΔE between the energy levels of a particle in a magnetic field: $\Delta E = \hbar\Omega$. In *crystals* the motion of particles is more complicated since they interact with the ions of the lattice. In a constant magnetic field the energy of the electron (or hole) E and the projection p_B of its *quasi-momentum* $\boldsymbol{p}$ along the direction $\boldsymbol{B}$ are conserved. Hence in *momentum space* the motion takes place along the intersection of the *isoenergetic surface* $E(p) = E$ with the plane $p_B = \text{const}$. If this curve is closed then the motion is periodic with the frequency $\Omega = eB/m^*$, where m^* is the charge carrier cyclotron *effective mass*.

CYCLOTRON PARAMETRIC RESONANCE

Selective absorption of electromagnetic radiation by a *semiconductor* located at the maximum magnetic field of a stationary wave (frequency ω) inside a resonator, when a longitudinal constant magnetic field B_0 is applied, and the *cyclotron frequency* Ω is a multiple of half the pump wave frequency ω. Cyclotron parametric resonance is due to instability, arising from time modulation of the electron cyclotron rotation frequency (see *Cyclotron resonance*). The stabilization of this instability results from nonlinear processes, arising at large electron energies, that lead to a highly nonequilibrium and anisotropic distribution of carriers formed at the cyclotron parametric resonance. The presence of this distribution results in the generation and amplification of electromagnetic and acoustic vibrations in semiconductors over a broad frequency range. In a thin sample at cyclotron parametric resonance the electron distribution function is independent of the coordinates. In a bulk semiconductor, when the sample size exceeds the wavelength of the pumping wave in the resonator; then the heterogeneity of the wave magnetic field plays an important role. The emerging nonequilibrium distribution becomes rather heterogeneous. It results in gradients of the average energy and concentration, and finally it gives rise to a static electric field E_{ind} (hence a static electromotive force) and a heterogeneous Hall effect with a resonance character. The main cyclotron parametric resonance appears when the cyclotron frequency $\Omega_0 = eB_0/m$ (e and m are particle charge and mass) is half the modulation frequency $\omega/2$ ($\Omega_0 \sim \omega/2$ for linear polarization of the pump field in a rectangular resonator, and $\Omega_0 \sim \omega$ for circular polarization). Cyclotron parametric resonance, like any other parametric resonance of mechanical or electrical vibrations, occurs in resonance zones of instability, and diminishes with the increase of the resonance number n ($\Omega = n\omega/2$, $n = 1, 2, 3, \ldots$). This resonance can be observed when the electrons make many revolutions in B_0 before they collide with other particles and scatter, i.e. $\Omega_0 \gg \nu_1$, where ν_1 is the a relaxation rate of the electron pulse. In *solids* collisions of the *conduction electrons* with crystal lattice defects ($\nu_1 = 10^9$–10^{11} s^{-1}), and scattering by thermal vibrations (*electron–phonon interaction*) are the most important. The last process limits the observation region of cyclotron parametric resonance: it is at low temperatures (1 to 10 K), where collisions with thermal phonons become rather infrequent. The actually achieved *relaxation times* provide the lower limit of the frequency region ($\omega > 10^9$ Hz) where cyclotron parametric resonance can be detected at the field value $B_0 = 0.1$ to 1 T. This resonance is a threshold-type effect which can be observed when the pump levels are rather high, i.e. $\Omega_1 > (\nu_e \nu_1)^{1/2}$, $\Omega_1 = eB_1/m$, where B_1 is the amplitude of the alternating magnetic field, and ν_e is the relaxation rate of the electron energy, $\nu_e \approx (10^{-4}$–$10^{-3})\nu_1$. This condition is valid in fields $B_1 \sim 0.01$–0.1 mT. The energy of the resonance particles increases exponentially with time, and can attain large values (electron gas overheating can reach 100 K). It results from the fact that the electron distribution becomes rather nonequilibrium and anisotropic. The energy increase is not infinite, but is limited by nonlinear processes. These are processes such as the nonquadratic spectrum of conduction electrons, which results in the dependence of the mass (and the cyclotron frequency) on the energy, and also affects the energy dependence of the electron relaxation rates.

Cyclotron parametric resonance can be used as a method of high-frequency resonance pumping of electrons in conductors to produce an inverted population distribution of charge carrier energies.

CYCLOTRON–PHONON RESONANCE

A sharp increase of the electromagnetic wave absorption in *semiconductors*, when the condition $\omega = \omega_0 + n\omega_{\text{L}}$ is satisfied, where ω is the electromagnetic wave frequency, ω_0 is the optical *phonon* frequency, ω_{L} is the Larmor precession frequency, and n is an integer. The resonance is observed at high frequencies ($\omega\tau \gg 1$) and in a *quantizing magnetic field* ($\omega_{\text{L}}\tau \gg 1$, $\hbar\omega_{\text{L}} \gg k_{\text{B}}T$); τ is the time of the electron mean free path. The resonance is associated with transitions of electrons between Landau levels with the simultaneous absorption of a photon and an optical phonon. The singularity of the absorption coefficient at its approach to resonance has a logarithmic nature. Cyclotron–phonon resonance has often been studied experimentally. It is used to determine semiconductor parameters, and in particular the frequency of the optical phonon.

CYCLOTRON RESONANCE

Resonant absorption of radiofrequency or microwave energy by an electron or other charged particle in a static magnetic field when its gyrofrequency equals the frequency of a perpendicular electric field. It corresponds to the diamagnetic resonance absorption of an electromagnetic wave, accompanied by the transition of an electron (or hole) from one stationary orbit to another (mainly, to a neighboring one) in a crystal situated in an external constant magnetic field $\boldsymbol{B}$. By its nature this is an electric dipole resonance. In a solid a charge carrier can move freely longitudinally along $\boldsymbol{B}$. The transverse motion takes place in a fixed orbit with energy equal to $\hbar\omega_c n$ (in an elementary Brillouin zone with energy $E(k) = \hbar^2 k^2/(2m^*)$), where $\omega_c = eB/m^*$ is the *cyclotron frequency* and n is a positive integer (*Landau levels*). In the case of absorption, the change of carrier energy is $\hbar\omega_c$ for the ground state absorption, and a multiple of $\hbar\omega_c$ for harmonics arising when the dependence $E(k)$ is more complex. The nature of $E(k)$ can be deduced by methods of group theory in solid state physics. A main feature of cyclotron resonance is the possibility of determining the *effective mass* m^* directly through a measurement of the absorption frequency. When $E(k)$ is anisotropic then ω_c depends on the angle between $\boldsymbol{B}$ and the *symmetry axes* of the crystal. The main condition for observing cyclotron resonance is the requirement that the conduction electron be able to complete an orbit before colliding ($\omega\tau > 1$), where impurities, heterogeneities, and *crystal lattice vibrations* determine the scattering time τ. For complicated forms of the energy $E(k)$, in particular for the presence of both light and heavy holes in cubic *semiconductors*, the equations of motion have an obvious classical interpretation. The theory uses a quantum-mechanical calculation based on the *Luttinger Hamiltonian* matrix. In addition to cyclotron resonance there are spin resonances of free carriers (*combined resonance* and *paramagnetic resonance*). These types of absorption, together with the investigation of magnetooptical interband transitions, provide a reliable starting point for understanding energy band structures (see *Band theory*).

CYCLOTRON RESONANCE, ACOUSTIC

See *Acoustic cyclotron resonance*.

CYCLOTRON RESONANCE IN METALS,

Azbel–Kaner resonance

In metals *cyclotron resonance* amounts to an increase in the *high-frequency conductance* at frequencies ω which are multiples of the electron cyclotron resonance frequency in a constant magnetic field parallel to the sample surface. This phenomenon is due to the repeated synchronous acceleration of electrons during the part of the orbit located within the skin depth δ (see *Skin-effect*). In 1956 M.Ya. Azbel and E.A. Kaner theoretically predicted cyclotron resonance in metals, and E. Fawcett discovered it in Sn and Cu. It has been observed in over 32 *metals*.

The electron acceleration mechanism for cyclotron resonance in metals is as follows (see Fig.). For a magnetic field $\boldsymbol{B}$ parallel to the metal surface all electrons move in circular orbits, and periodically with the *cyclotron frequency* Ω they return to and pass through the skin δ where they are accelerated by the electromagnetic wave. If the resonance condition $\omega = n\Omega$, $n = 1, 2, 3, \ldots$, is satisfied then the acceleration of the electrons takes place on each cycle of the trajectory, just as it occurs in a cyclotron with a single accelerating region. The skin depth δ in a metal plays the role of the latter. The metal region outside the skin acts as the dees

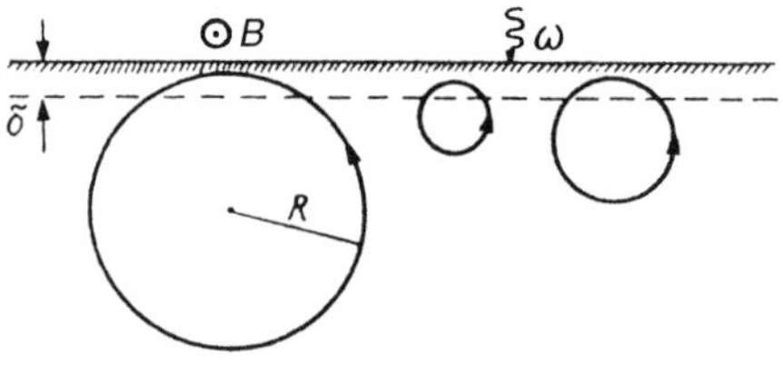

Cyclotron resonance in metals.

of the cyclotron. Cyclotron resonance is observed in pure single crystal metals at helium temperatures when the rate of electron collisions is less than Ω. It is observed only for the conditions of the *anomalous skin-effect*, when the radius of the orbit R and the electron mean free path l are much larger than the skin depth δ.

Cyclotron resonance in metals differs significantly from *diamagnetic cyclotron resonance* in semiconductors. The latter is also called cyclotron resonance when it takes place in a gas-discharge plasma. Diamagnetic resonance occurs in a homogeneous radio-wave field. Therefore: (a) it is observed at the fundamental frequency (first harmonic) $\omega = \Omega$, but not at multiple frequencies, as is the case of cyclotron resonance in metals; (b) it is a maximum when the polarization of the rf electric field is perpendicular to the vector $\boldsymbol{B}$, whereas cyclotron resonance in metals exists at any polarization; (c) it is characterized by a sharp increase of the electromagnetic absorption at resonance, while absorption decreases at cyclotron resonance in metals. The last circumstance is associated with the fact that a sharp increase of the current at cyclotron resonance in metals results in its resonance screening from the the external wave, in the increase of the reflection coefficient, and, therefore, in a decrease of the energy absorption. The line shape of cyclotron resonance in metals is rather asymmetric due to the presence of intrinsic *cyclotron waves*, which have their spectrum localized near harmonics $n\Omega$ of cyclotron frequencies. In the case of a nonquadratic electron *dispersion law*, which is typical for most metals, the cyclotron frequency Ω depends on the value of the electron momentum projection along the direction $\boldsymbol{B}$, hence it differs for different electrons. In this case cyclotron resonance in metals occurs at the extreme frequencies $n\Omega_{\text{ext}}$, since for electrons with Ω_{ext} there is a singularity in the *density of states*. Owing to the decrease of the resonance electron number the amplitude of the cyclotron resonance in metals decreases, and the lines broaden and become more asymmetric in comparison with the case $\Omega = \text{const}$. In so doing the resonance of electrons from the central cross-section, and from the vicinity of special points of the *Fermi surface*, becomes sensitive to the wave polarization. A small slope of the field $\boldsymbol{B}$ relative to the surface leads firstly to the splitting of the resonance peaks, then to the doubling of their number ($\Omega = n\Omega/2$), and to the inversion of the lines. In numerous experiments on cyclotron resonance in metals an important fact has been established: *electron scattering* on the surface heterogeneities of the sample is almost of the mirror-type. It results in a considerable decrease of the amplitude of the cyclotron resonance in metals, since there arises a new group of electrons sliding along the surface near the mirror boundary of the skin depth. Their contribution to the current is rather more than the contribution of resonance particles, and cyclotron resonance in metals arises from corrections to the smooth nonresonance part of the *surface impedance* of metals. A complete self-consistent theory of cyclotron resonance in metals at arbitrary reflections of electrons from the metal surface over a wide range of frequencies and magnetic fields has been developed.

CYCLOTRON RESONANCE, QUANTUM

See *Quantum cyclotron resonance.*

CYCLOTRON WAVES

Intrinsic electromagnetic excitations of degenerate electron–hole *solid-state plasmas* in a magnetic field $\boldsymbol{B}$. The spectrum of cyclotron waves is localized near the *cyclotron resonance* frequencies $\omega = n\Omega$ $(n = 1, 2, 3, \ldots)$, for a wave vector $\boldsymbol{k} \perp \boldsymbol{B}$. Its origin is the collective precession of electrons with the *cyclotron frequency* Ω, in the absence of *magnetic collision-free damping*. The distinguishing features of this spectrum are due to the nonlocal properties of the conductivity operator, and depend on the electron *dispersion law*. For the square law of dispersion there are three independent waves in the vicinity of each cyclotron resonance, namely the ordinary, extraordinary, and longitudinal ones, differing in their spectrum and polarization. The spectrum of the transverse cyclotron waves begins at the point $k = 0$ from the frequency $n\Omega$, reaches a minimum (about $n\Omega - \Omega/2$) at $k \sim R^{-1}$ (R is the cyclotron radius), and then asymptotically approaches the initial value $n\Omega$. The attenuation of cyclotron waves is determined by the rate of electron relaxation. Parameters of the *Fermi liquid* interaction in alkali halides have been measured by cyclotron waves.

Dd

DAMPING, COLLISION-FREE

See *Magnetic collision-free damping.*

DAMPING, LANDAU

See *Landau damping.*

DANGLING BOND, broken bond

An unsaturated *covalent bond*; a bond from one atom without a terminus, a kind of structural *defect* in a solid involving a covalent or *ionic-covalent bond.* Broken bonds are inherent in both crystalline and amorphous solids (see *Amorphous state*), and can be arranged in a type of *random continuous network*; this is common in organic and polymeric glasses (see *Vitreous state of matter*). Associated with broken bonds in *amorphous semiconductors* and *insulators* are features of the electronic spectrum that significantly affect the processes of electron transfer and photostimulated phenomena. In many cases broken bonds sufficiently high in concentration can be observed with the help of *electron paramagnetic resonance* since they involve unpaired electrons. The density of broken bonds in amorphous solids can reach 10^{20} to 10^{21} cm^{-3}. Examples of broken bonds are unsaturated bonds of surface atoms. Since the energy can be lowered by saturation (closure) of broken bonds, efforts are made toward *surface reconstruction* of covalent crystals, and this is an important factor in *chemisorption* on crystals.

DAVYDOV SPLITTING (A.S. Davydov, 1948)

Splitting of electronic, electronic-vibrational and vibrational states in *molecular crystals* caused by the presence of inductive coupling between molecules and by the migration of excitation energy in a crystal. Davydov splitting is possible when there are two or more identical molecules with different orientation in a *unit cell.* It is observable by the splitting of bands of absorption and *luminescence*, the components of which appear to be polarized in orthogonal directions.

Davydov splitting happens in the spectra of inorganic crystals with ionic or mixed-type bonds in which one can distinguish groups of strongly bound atoms with different orientation in the lattice (K_2CrO_4, $CaWO_4$, etc.). Davydov splitting is especially apparent in *semiconductors* with a layered and chain-like structure (As_2S_3, CdP_2, ZnP_2, etc.) where the unit cell is composed of atoms belonging to two or more translationally inequivalent chains or layers, between which, like in molecular crystals, there is a weak van der Waals bond (see also *Fermi–Davydov combined resonance, Van der Waals forces*).

Davydov splitting in *liquid crystals* is close in nature to Davydov splitting in crystals with several molecules in the unit cell and is of the order of $\Delta\nu_e \sim 10^{-13}$ s^{-1} for electronic transitions or $\Delta\nu_v \sim 10^{-11}$ s^{-1} for vibrational transitions. In liquid crystals, there appears an additional contribution to the splitting, which is associated with the orientation dependence of the dispersion interaction. It is comparable in its magnitude with the Davydov splitting, and observable in impurity absorption bands.

DE BOER PARAMETER (J. de Boer, 1948)

Ratio of the amplitude of the *zero-point vibrations* of the atoms in a crystal to the crystal lattice spacing, denoted by Λ_B. This provides the stability criterion of the classical crystal state, namely, $\Lambda_B \ll 1$. However, there are crystals for which $\Lambda_B \sim 1$. Since the existence of the crystalline state in such a situation is itself problematic, the definition of the de Boer parameter should not contain the lattice period. It is determined through the parameters of the pairwise interaction of atoms, parameters of the *Lennard-Jones potential*. The de Boer parameter for a series of elements has the values:

^{3}He	^{4}He	H_2	D_2	Ne	Ar	K	Xe
0.48	0.43	0.28	0.20	0.09	0.03	0.02	0.01

Crystals for which $\Lambda_B \sim 1$ are called *quantum crystals*.

DE BROGLIE WAVE (L. de Broglie, 1924)

The quantum mechanical attribution of wave properties to microparticles (electrons, atoms, molecules, etc.). According to the assumption of L. de Broglie a moving particle with the energy E and momentum p has associated with it a wave with a frequency $\nu = E/h$ and a wavelength $\lambda = h/p$. The de Broglie wave length corresponding to the thermal motion of the particles increases with the decrease in temperature T. At a sufficiently low temperature when λ becomes comparable to the distance between the particles of a system, then quantum effects begin to play a role (see *Quantum liquid, Quantum crystal*).

DEBYE FREQUENCY

See *Specific heat.*

DEBYE–HÜCKEL THEORY (P. Debye, E. Hückel, 1923)

Thermodynamic theory of electrolytes taking into account the screening of the electrostatic potential (see *Electric charge screening*) by the mobile ions of the solution. It follows from Debye–Hückel theory that an ion located within the solution is surrounded by a coordination sphere of ions of opposite sign, its radius being $l_D = \sqrt{\varepsilon k_B T/(8\pi e^2 I)}$, where ε is the dielectric constant, I is "ionic force" of the solution, $I = (1/2)\sum_i n_i Z_i^2$ where n_i is the concentration of ions of the ith type, and Z_i is the *valence* of the ith ion. The coefficient of activity of the ith ion is equal to $\exp[-Z_+Z_-e^2/(2k_B T\varepsilon l_D)]$, where Z_+, Z_- are valences of ions of the corresponding sign of the charge. In the Debye–Hückel theory the possibility of forming a bound pair of oppositely charged ions is also taken into account (in this case the components of the pair provide the mutual screening). The relation between the concentration of pairs and the concentration of individual screened ions is determined by the total concentration of ions in the solution. This theory is used for the description of defects in nonmetallic crystals, e.g., for the calculation of *diffusion coefficients*, of *ionic conductivity*, etc.

DEBYE–SCHERRER METHOD, powder X-ray method, polycrystal method (P. Debye, P. Scherrer)

Method of obtaining the diffraction diagram of a polycrystalline material (see *Polycrystal*) by X-ray scattering. In this method a monochromatic narrow beam of X-rays is incident on a sample in the form of a cylinder or a polished section from a polycrystal. All the crystallites in the sample are assumed to have random spatial orientations, and the Bragg condition is satisfied for each set of suitably oriented crystallographic planes. For each allowed crystallite orientation diffracted X-rays will emerge along a cone coaxial with the incident X-ray beam with an aperture angle 4θ, where θ is a Bragg angle (see *Bragg law*). The result is a family of cones corresponding to diffracted X-rays with various allowed aperture angles 4θ for the various allowed reflections. The diffraction picture in the Debye–Scherrer method may be recorded on a cylindrical or flat photographic film (*debyegram*). In the *X-ray diffractometry* of polycrystalline samples the photographic film is replaced by a rotating detector of X-rays.

DEBYE TEMPERATURE (P. Debye, 1912)

The characteristic *crystal* temperature Θ_D corresponding to the *Debye frequency* ω_D ($k_B\Theta_D = \hbar\omega_D$), where ω_D is the maximum frequency of the *crystal lattice vibrations* of the crystal ($\omega_D \propto v_s/a$, where v_s is the *sound velocity* and a is the crystal lattice constant). In the *Debye model* the crystal vibration frequencies ω are limited by the Debye frequency ($\omega \leqslant \omega_D$), and the linear *dispersion law* $\omega = v_s|\boldsymbol{k}|$ is assumed to be valid, where $\boldsymbol{k}$ is the phonon wave vector which takes on $3N$ discrete values (N is the number of atoms in the crystal), the largest value of $\boldsymbol{k}$ being π/a. The numerical values of the Debye temperature depend on the atomic mass, and on the strength of the elastic bond between atoms (e.g., for Pb $\Theta_D \sim 90$ K, for diamond $\Theta_D \sim 2340$ K). Knowledge of the value of Θ_D allows one to obtain important information about the vibrational spectra of a crystal, and about the characteristics of its chemical bonds. The notion of a Debye temperature is essential for treating the thermodynamics of crystal; it is the approximate temperature boundary which separates the low-temperature ($T \ll \Theta_D$) and high-temperature ($T \gg \Theta_D$) lattice properties of the crystal, where T is the crystal temperature. Using only the single parameter Θ_D together with some approximation expressions it is possible to describe the behavior of a whole series of observed physical characteristics of crystals over a broad temperature range below as well as above the Debye temperature. Examples of these are: *specific heat*, electrical *resistance* (*Bloch–Grüneisen formula*), *elastic moduli*, *Debye–Waller factor* for the scattering of X-rays, neutrons, electrons, etc.; properties involving the *Mössbauer effect*, IR absorption, etc.

DEBYE THEORY OF SPECIFIC HEAT

See *Specific heat*.

DEBYE–WALLER FACTOR (P. Debye, I. Waller)

The temperature factor in the expression for the probability of the radiation of *photons* from crystals, or of elastic processes for the scattering of particles or photons, which takes into account thermal and static displacements of atoms from their sites on a *crystal lattice*. The Debye–Waller factor is written in the form e^{-2W}, where $e^{-W} = \langle\langle \exp[i\boldsymbol{Q}(\boldsymbol{u}_{th} + \boldsymbol{u}_{st})]\rangle\rangle$, $\langle\langle \ldots \rangle\rangle$ means double (thermodynamic and configurational) averaging, and $\boldsymbol{Q}$ is the scattering vector or the wave vector of the radiated photons. If the displacements are counted with respect to the sites of the "average" lattice, W assumes the approximate form $W \approx (1/2)\langle\langle [\boldsymbol{Q}(u_{th} + u_{st})]\rangle\rangle^2$. In the isotropic approximation $2W_\alpha = (Q^2/3)\langle u_\alpha^2\rangle$. For a monatomic crystal, assuming the *Debye model*, it has the temperature dependence

$$2W_{th} = \frac{3Q^2}{8Mk_B\Theta_D}\left\{1 + \frac{2\pi^2}{3}\left(\frac{T}{\Theta_D}\right)^2 + \cdots\right\},$$

where Θ_D is *Debye temperature*, and T is absolute temperature of the crystal. In the general case W_{th}. depends on the *force constants* of the crystal, on the atomic masses and on temperature. The *static Debye–Waller factor* depends on the elastic constants of the crystal, and on the characteristics of defects (concentration, dimensions, correlation in their location, etc.). Usually $W_{th} \leqslant 1$, and for room temperature $W_{th} \sim 0.01$ to 0.1. For *impurity atoms*, small *clusters* and *dislocation loops* $W_{st} \leqslant 1$; for *dislocations* $W_{st} \gg 1$ and the effects of coherent dissipation disappear in the kinematic approximation; but taking into account the dynamic effects of scattering W_{st} in the case of dislocations requires renormalizing, while, e.g., for the scattering of X-rays the renormalized Debye–Waller factor $W_{st} < 1$ for a dislocation density less than 10^5 to 10^7 cm^{-3}.

The Debye–Waller factor determines the reduction of the coherent scattering intensity $I(\boldsymbol{Q})$ of X-rays, neutrons and other radiations in crystals: $I(\boldsymbol{Q}) = I_0(\boldsymbol{Q})e^{-2W}$, where I_0 is the intensity of scattering by the fixed ideal (average) lattice. A Debye–Waller factor also describes the efficiency of *gamma ray* radiation in a crystal without recoil (see *Mössbauer effect*); in this case it is sometimes called the *Lamb–Mössbauer factor*.

DECA... (fr. Gr. $\delta\varepsilon\kappa\alpha$, ten), da

Decimal prefix which denotes a 10-fold increase of the value of the basic unit. For instance, 1 dal (dekaliter) = 10 l.

DECAY, ISOSTRUCTURAL

See *Isostructural decay*.

DECHANNELING

Exit from the *channelling* mode; influenced by the excitation of *phonons*, electrons, *plasmons*, etc., by the channeling particles. Motion of the channeled particles in the transverse direction can be considered as a drift in the field of a random force, caused by deviations from the average continuous potential. In this case the energy of the transverse motion of the channelized *ions* $E_\perp$ increases in proportion with their movement along the channels (upon reduction of their total kinetic energy), and upon reaching the critical value $E_{\perp c} = E_0\Psi_k$ the dechanneling occurs; E_0 is the initial channelized particle energy, Ψ_k is the *Lindhard angle* (see *Channeling*). The distance at which the density of channeling particles is reduced by one half is called the *dechanneling length*. This length for protons with energies from 1 to 10 MeV in silicon lies within the range from 2 to 18 μm. Dechanneling is enhanced by the presence of *defects* in *crystals*, especially by impurity or inherent *interstitial atoms*, located near the axes of the channels, and also by *dislocations*; in alloys it depends on the degree of ordering, etc.

DECI... (fr. Lat. *decem*, ten), d

The decimal fraction prefix meaning 1/10 of a particular unit. For instance, 1 dg (decigram) = 0.1 g.

DECOMPOSITION OF ALLOYS

See *Alloy decomposition.*

DECOMPOSITION, SPINODAL

See *Spinodal decomposition.*

DECORATION

See *Surface decoration.*

DEEP ENERGY LEVELS AND GREEN'S FUNCTIONS

See *Green's functions and deep energy levels.*

DEEP LEVELS in semiconductors

Local electronic levels in semiconductors whose ionization energies are comparable with the *band gap* width. Deep levels exhibit a diversity of properties due to their depth, and their physicochemical characteristics. The atoms Zn, Cd, Mn, Fe, Co, and Ni in Si and Ge are double (M^{2+}) acceptors; Cu and Ag are triple acceptors, and the impurity Au in Ge produces three acceptor and one donor level. Deep levels are responsible for a series of special electric, optical, photoelectric and other properties in *semiconductor materials.* The direct introduction of impurities is the main method for obtaining material with the desired properties. Deep level centers possess abnormally large cross-sections for *trapping of current carriers*, so they can serve as centers of rapid *recombination.* An attempt to formulate a quantitative theory of deep levels encounters some significant obstacles associated with the need for a microscopic description of the *defect* portion of the crystal and with the multi-electron nature of the problem (i.e. simultaneously taking into account both the electrons of the defects and the valence electrons of neighboring atoms). For calculating local level parameters given in the literature, various quantum chemical methods are used (see *Quantum chemistry of solids*), in particular, a Lifshits method of regular perturbations (*Green's functions*), a modified method of *effective mass*, and some others. These methods are arbitrary in their choice of the periodic field potential perturbed by a defect in an *ideal crystal*, and as a rule are not very accurate.

DEFECT

For different varieties see *Defects in crystals, Defects in liquid crystals, Anti-Schottky defect, Antisite defect, Antistructural defect, Electrically active extended defects, Frenkel defect, Grain-boundary defects, Intrinsic defects, Macroscopic defects, Monopole-type defect, Needle-shaped defect, Planar defect, Rod-shaped defects, Schottky defect, Stacking faults, Stoichiometric defects, Surface defects, Swirl defects, Thermal defects.*

DEFECT CLUSTERING

See *Point defect clustering.*

DEFECT ELASTIC POTENTIAL

See *Elastic potential of a defect.*

DEFECT INTERACTION

See *Elastic interaction of defects, Strain interaction of defects.*

DEFECTON

A *quasi-particle* which arises from a light *defect* of a crystal lattice as a result of quantum tunneling (see *Tunneling phenomena*) through the periodic relief of a *crystal.* A defecton is characterized by a *quasi-momentum* which determines its energy, usually lying within a narrow allowed energy band (*defecton band*). A *vacancion* (vacancy wave) in solid *helium* and an *impurity atom* of ^{3}He in crystalline ^{4}He (*impuriton*) exemplify defectons. Defectons take part in *quantum diffusion* which has an unusual temperature dependence, and is very sensitive to the concentration of defects. Due to the small width of the defecton band a non-uniform *strain* of a crystal brings about the localization of defectons, and their transformation into ordinary *point defects.*

DEFECTOSCOPY, flaw detection

The totality of methods which permit the detection of *defects* (*cracks*, *pores*, gas blisters, extraneous inclusions, *liquation*, *segregation* of impurities, etc.) in the structure of a solid. There are several types of defectoscopy.

Magnetic defectoscopy records the spreading out of magnetic fields in the regions of defects, or measures the magnetic properties which change in the presence of defects in the material (ferromagnetic materials or ferromagnetic inclusions in nonmagnetic materials). There are the following variants of magnetic defectoscopy: *magnetic-powder method* where the scattering fields are detected at the surface of a magnetized sample of small (several μm) particles of ferromagnetic powders or suspensions (recording is visual); the *magnetic-luminescent method* where luminofores are used as powder (particle location recorded using UV radiation); the *magnetographic* method uses superposition at a ferromagnetic film, which is then studied with the aid of a tape-recorder head and an electron-beam indicator; the *ferroprobe method* using different ferroprobes (in particular, tape-recorder heads, magnetic diodes, induction coils, etc.) for recording the scattering fields.

Electrical defectoscopy records the spreading out of the electric fields near defects, thermal electromotive force, electrical *resistance*.

X-ray defectoscopy detects defects by the variation of the contrast from X-rays passing through the sample. This is applicable to thick samples, and uses photographic film or special sensors which record the X-ray intensity at every point of the space, subject it to analysis, and transmit the results to a video device screen.

Gamma defectoscopy uses the variation of the intensity of *gamma rays* passing through the material in the region of defects, and is applicable to thick samples.

Beams of high-energy electrons generated by accelerators can detect hidden defects in critical materials.

Neutrons can detect defects in massive structures and engineering devices, as well as the concentration of inhomogeneities in ordinary bodies.

Defectoscopy based on dynamic light scattering in *nematic liquid crystals* is used for the control of insulating layers on *metal–insulator–semiconductor structures* (integrated circuits, conductive plates with dielectric coating, etc.). This method can detect 1 μm pores in *insulators*, and it provides a nondestructive control of a structure with dielectric insulation, giving information about the type of defects present. The most widespread use of *acoustic defectoscopy* is the *shadow method* where the reduction in amplitude or the variation of phase of the incident wave, scattered by the acoustic non-uniformity, is recorded, and the *echo-method* which measures the time of return of a short acoustic pulse to the piezo-transformer after reflection from a defect (see *Piezoelectricity*). The minimum dimensions of recorded defects and the accuracy of their localization by these methods is of the order of the ultrasound wave length. Also there is a series of special methods for measuring specific parameters of samples. For instance, the thickness of plates is controlled by the value of their resonance acoustic frequencies, the quality of lamellar structures by their *acoustic impedance*, and the presence of defects in strongly sound-adsorbing bodies through the spectrum of their free (low-frequency) oscillations.

Physico-chemical methods of investigation include the *capillary method*, *luminescent analysis*, the *chromatic method* which employs the interaction of surface-active liquids (see *Surface-active agents*) with defects located at the surface or in regions close to the surface. To investigate microdefects of the structure, the method of *etching* with subsequent microscopic or electron-microscope analysis is used (see *Electron microscope*). Upon the interaction of the crystal surface with a chemically-active medium the points of exposure of the defects, e.g., of *dislocations*, are "etched" more intensely, and this is then detected by optical or *scanning electron microscopy*. Dislocations and *dislocation loops* in foils are detected by transmission *electron microscopy*. *Electron–positron annihilation* (see *Positron spectroscopy*) detects and identifies microdefects due to the ability of positrons to localize and annihilate in the defective areas of the crystal. The processing of positron annihilation spectra or the analysis of their lifetime gives information about the type and distribution of microdefects in crystals. The use of different types of penetrating radiations and their transformation into optically visible images with the help of special electronic devices called *image transformers* is a type of defectoscopy. X-rays, γ-quanta, β-radiation of isotopes, beams of high-energy electrons and ions, neutrons, electromagnetic radiation in the range from the near infrared to millimeter and submillimeter waves, are used as penetrating radiations (see *Optical defectoscopy*).

DEFECT RESISTANCE

A property of a material whereby it resists damage and possible *failure* through the effect of small concentrators of stresses (see *Stress concentration*). Examples are scratches, small *cracks*, rough nonmetallic inclusions and other *defects*, which are fractions of a millimeter in size, i.e. no more than 1 or 2 orders of magnitude larger than the size of a polycrystalline metal grain (see *Polycrystal*). The *defect resistance parameter* is a quantitative measure of the defect resistance:

$$D_2 = \log K_{\text{v}} - n_\tau$$

where $K_{\text{v}} = R_{\text{ms}}/\sigma_\tau$ is the *viscosity* coefficient of the metal, R_{ms} is the *microspalling resistance*, σ_τ is the *yield limit*, and n_τ is the exponent in the parabolic equation governing the initial stage of *strain hardening* of the metal, $\sigma_e = Ae^n$ (A is a constant, e is the actual *strain* or deformation):

$$n_\tau = \frac{\log(S_{\text{v}}/\sigma_\tau)}{\log(e_{\text{e}}/e_\tau)},$$

where S_{v} is the true *ultimate strength*, e_{e} is the uniform true strain, and e_τ is the strain at the yield limit. The criterion of defect resistance $D_2 = \log(\sigma_1/\sigma_i)$ (where the ratio σ_1/σ_i is the rigidity of the *state of stress* at the vertex of the defect, σ_1 is the main normal stress, σ_i is the intensity of stresses) expresses the condition for the destruction of a metal containing a small defect within an area of low *plastic deformation* of the magnitude $e \approx 0.02$. The parameter of defect resistance as a characteristic of steel toughness supplements the index of *fracture toughness* (crack resistance), because the latter is not applicable to cracks and defects of submillimeter size.

DEFECTS, CONTINUUM THEORY

See *Continuum theory of defects*.

DEFECTS, HEALING

See *Healing of defects*.

DEFECTS IN CRYSTALS

Disturbances of the periodicity of atom (ion) positions within a *crystal* (*structural defects*). They are subdivided into point, linear, surface (or planar) and bulk types. *Point defects* are one dimensional imperfections (of atomic scale size) distributed in three dimensions. Examples are *vacancies*, *interstitial atoms* and some of their combinations (*bivacancies*, complexes, etc.), as well as *impurity atoms* which may be located at regular sites of the crystal lattice (*substitutional atom*) or between regular sites (*doping atom*). Unlike other types, point defects can be in thermodynamic equilibrium at the lattice temperature.

Linear defects are small in two dimensions, and can have a considerable length in the third. The character of linear defects, and of the displacement fields created by them, is described by the *Burgers contour*. This contour may either close on itself or be open, and if it is open then the Bravais lattice vector connecting the starting and ending points is called the *Burgers vector*. *Needle-shaped defects* characterized by a closed contour are chains of vacancies or interstitial atoms. They are not very stable, and easily decompose by thermally activated or tunnel *diffusion* of atoms. *Dislocations* correspond to open Burgers contours.

Lattice disturbances which are elongated in two dimensions and small in the third are called surface defects. They form at crystal locations where adjoining lattices differ from one another by space orientation or by atom positionings. Some characteristic types are *stacking faults*, *grain boundaries*, and subgrain boundaries (see *Subboundaries*).

Bulk defects are disturbances which are not small in any of their three dimensions. Aggregates of point defects, *pores* and inclusions of a second phase are included in this type.

The influence of defects on the physical properties of crystals is extremely diverse. It is determined by the character of the binding forces within the crystal, by their energy spectrum (*metals*, *semiconductors*, *insulators*), by the nature of the defects, etc. In semiconductors and insulators, where good electrical, optical or photoelectric properties are crucial (see *Photoelectric phenomena*) point defects play an important role, and dislocations may also make a considerable contribution to the electrical *resistance*. In *construction materials* where high strength is essential (see *Strength*), the role of surface and bulk defects is important. An atmosphere of point defects can strongly influence the motion of dislocations (see *Cottrell atmosphere*).

DEFECTS IN LIQUID CRYSTALS

Inhomogeneities of the orientational (see *Orientational order*) and/or translational ordering of molecules of *liquid crystals*. Molecular size and macroscopic defects are distinguished. Those of molecular size are formed by additions into the liquid crystal, e.g., molecules of *dyes*, of *ions*, of *surface-active agents*, or as a result of light- or radiation-induced transformations of molecules already present. Macroscopic defects include the following: *point defects* (*hedgehogs*, *boojums*), linear ones (*dislocations*, linear *solitons*), *planar defects* (*domain walls*) and bulk defects (particle-like solitons). Macroscopic defects are called topologically stable if no continuous transformation is able to transform the field of the order parameter to the uniform state.

The classification of macroscopic defects is based on the calculation of the *homotopic groups* $\pi_i(R)$ of the space of rank R (region of degenerate states) of the medium. Each class of defects corresponds to an element of the group, identical within the accuracy of continuous transformations of the order parameter; a state without a defect corresponds to the unit defect. This formalism permits one to associate with each class of defects some topological invariant, called the *topological charge*, which is a generalization of the *Burgers vector* in the theory of dislocations. The continuous variation of the topological charge of an isolated defect is impossible. In the general case the homotopic group $\pi_i(R)$ classifies a defect of dimensionality $d' = d - i - 1$, where d is the dimensionality of the medium. In a three-dimensional ($d = 3$) medium group $\pi_0(R)$ describes walls ($i = 0$, $d' = 2$), $\pi_1(R)$ describes lines ($d' = 1$), and $\pi_2(R)$ describes points ($d' = 0$). The processes of decomposition and merging of the defects are limited by the multiplication law of the group elements. If the group is Abelian (all elements commute) the result of such processes is always unambiguous, while for a non-Abelian group the result depends on the path along which the decomposition (merging) of the defects occurs.

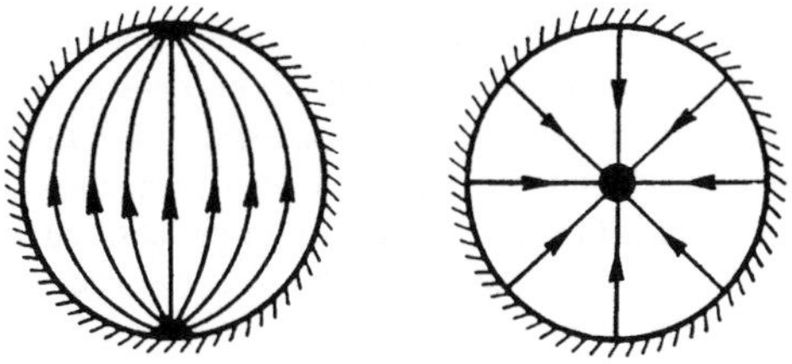

Defects in liquid crystals.

Macroscopic defects cause distortions of the orientation of liquid crystal molecules within the whole volume of the sample. Since it is the orientation of the molecules which determines the distribution of the *optical axes* (see *Director*) of the liquid crystal, the defects are easily subjected to investigation by the methods of polarization microscopy (see *Optical polarization method*).

The formation and motion of defects play important roles in practically all processes of the liquid crystal response to external disturbances, such as the *stress relaxation*, the reconstruction of structure under the effect of an electromagnetic field, at *phase transitions*, etc. As a rule, some *topological defects* are present even in the equilibrium state of a liquid crystal, and their stability is provided by the boundary conditions (see Fig.). Thus, for example, in a spherical drop of a nematic material (see *Nematic liquid crystal*) with normal boundary conditions in equilibrium there always exists a point defect within the volume, the hedgehog (Fig., right), and at the tangential lines there are two boojums or surface point defects (Fig., left).

Due to the wide variety of defects, and the possibility of performing a broad range of experiments, defects constitute convenient models for the development of topological representations in condensed matter physics (see *Topological inhomogeneity*).

DEFECTS IN PHASE TRANSITIONS

Phase transition defects are of two types, those involving a *random local field*, and those involving a *random local temperature*, depending on their influence upon the coefficients of the density $\eta(r)$ or the thermodynamic potential $\Phi(r)$ describing the *phase transition*. Within the phenomenology of the *Landau theory*, the density in the absence of defects has the form:

$$\Phi(r) = \frac{A}{2}\eta^2(r) + \frac{B}{4}\eta^4(r) + \cdots + \frac{D}{2}\big(\nabla\eta(r)\big)^2 + \cdots - h_0\eta, \qquad (1)$$

where for simplicity the long-range forces are not taken into account, and the one-component *order parameter* $A = A_0(T - T_0)$ is included; B, D, A_0 are medium constants, and h_0 is the external field which is conjugated to the order parameter. Point defects of the random local field type add the following contribution to $\Phi(r)$:

$$\Delta\Phi_h = \sum_i h_i(r - r_i)\eta(r), \qquad (2)$$

and defects of the random local temperature type lead to the additional contribution

$$\Delta\Phi_T = \frac{1}{2}A(r - r_i)\eta^2(r). \qquad (3)$$

The origin of the terms, defects of random local field and random local temperature, is associated with the fact that the "averaged over configuration" values $\Delta\overline{\Phi}_h(r) = \overline{h}\eta(r)$, $\Delta\overline{\Phi}_T(r) = (1/2)\overline{A}\eta^2(r)$ describe renormalization by the defects of the value of the external field h_0 and of the phase transition temperature T_0.

The local field defects bring about strong local distortions of the crystal matrix, because it follows from Eq. (1) that (not too close to T_0) $\eta(r) \propto (1/r)\exp(-r/r_c)$ with $r^2{}_c = D/A \to \infty$ at $T \to T_0$. These defects eliminate the long-range order (see *Long-range and short-range order*), leading to a complex *domain* structure in systems with infinitely degenerate η, or to long-term *relaxation* in systems with finite degenerate η. Such polarized defects create an average field within the crystal, and induce $\eta \neq 0$ at any temperature T, i.e. the phase transition, speaking strictly, is absent (blurred).

The random local field defects bring about a shift of T_0 and lead to anomalies of different physical values in the vicinity of T_0. In the area of similarity theory these defects cause a variation of the *critical indices*. The reorienting random local field defects are a special class of defects which, unlike frozen defects, bring about an increase in the temperature of pure crystal phase transitions.

DEFECT STRESS SCALE

See *Stress scales of defect structures.*

DEFECT STRUCTURE ANALYSIS

See *Microprobe X-ray defect structure analysis.*

DEFECT SYMMETRY GROUP

This involves both the symmetry of a *defect* itself as well as the *symmetry group* of the crystal where the defect appears. Translational symmetry (see *Translation*) disappears when a single defect is introduced into a *crystal*. The *point group of symmetry* of a crystal with a defect reflects the symmetry of the *point defect*, and is a subgroup of the point group of the *ideal crystal*. If we disregard the crystal distortion induced by a defect then the symmetry of this defect is the same as the symmetry of the site that it occupies. The symmetry groups of the various points of the *unit cells* for all 230 *space groups* are given in the *International Tables for Crystallography*. For an *impurity atom* (or a *vacancy*) at a site of a *Bravais lattice* the defect point symmetry is the same as the crystal point symmetry. The presence of a pair of point defects, *dislocation loops* in cubic crystals, or particles of a new phase of different symmetry all result in a lowering of the symmetry compared with that of the host crystal.

DEFECT THRESHOLD ENERGY

See *Threshold energy of defect formation.*

DEFORMATION AT LOW TEMPERATURES

See *Low-temperature deformation.*

DEFORMATION BY EXPLOSION

See *Explosion deformation.*

DEFORMATION DIAGRAM, stress–strain curve

Graphic representation of the dependence of the deforming force (stress) on the material response (*strain*). The deformation diagram reflects the results of testing solids by different methods such as stretching, compressing, *flexure* and torsion. As deformation parameters one can select absolute values of elongation, of *shearing* angle, of torsion angle (see *Rod twisting*), of deflection, or preferably one can work with *relative deformations* which are ratios of sample dimensions during loading to those before the test. Relative deformations are usually subdivided into two types: a *conditional deformation* which is the the sum of successive variations of size all relative to the same initial dimension before the test; and a *true deformation* which is the sum of small variations of size, with each of them relative to the current value of the dimension. The deforming effort is characterized also by either the absolute value of the projection of the force or torsion along some direction (direction of deformation), or by the specific parameter of *stress*, i.e. by the force effecting a unit area of some cross-section of the body. The *normal stress*, namely the component of force perpendicular to the cross-section area, and the *tangential stress*, due to the projection of the force along (parallel to) the area of the cross-section, are singled out. The stresses are usually divided into *conditional and true stresses*: the former determined by the relation of the actual force to the initial area of the cross-section, and the latter by the same relation to the current area. To plot a deformation diagram it is possible to use different combinations of the above mentioned deformation and deforming force parameters. For the analysis of physical dependencies of the deformation process the diagram using the coordinates "true deformation–true stress" is the most significant. This diagram provides data for the calculation of many basic mechanical parameters of the material (e.g., *elastic moduli, yield limit, ultimate strength, strain hardening* factor, *margin of plasticity*). The analysis of a deformation diagram also gives information about the physical mechanisms of *plastic deformation*, of *hardening* and of *degeneracy*. Alternate terminology for a deformation diagram is *deformation curve, hardening curve* and *diagram of loading*.

DEFORMATION KINETICS

The development and interplay of elemental acts of deformation that produce *strain* in a material. Generally, the strain-inducing stress is a function of temperature, strain, and rate of deformation. The simplest *kinetic equation* presenting a microscopic description of *plastic deformation* in terms of the *dislocation model of plastic deformation* takes the form: $\dot{\varepsilon} = b\rho v$, where $\dot{\varepsilon}$ is the rate of deformation, b is the *Burgers vector*, ρ is the density of mobile *dislocations*, and v is the mean velocity of their motion. The deformation kinetics is accompanied by an irreversible displacement of dislocations. In addition to the displacement of dislocations, one has to consider the *multiplication of dislocations* and interactions between them, as well as the effects of other lattice *defects*, in particular, of *disclinations*. Two competing processes proceed simultaneously when a crystal is undergoing a deformation: *strain hardening* and *recovery*. During hardening the stresses required to maintain a constant rate of deformation increase, while the rate of deformation decreases, if a constant external stress is applied. Recovery decreases the rate of hardening during active deformation ($\dot{\varepsilon} =$ const), and produces a steady-state *creep* under a constant stress. Existing theories describe the deformation kinetics at the first stages of deformation, in particular, the appearance of a *yield cusp* (upper and lower *yield limits* on the *deformation diagram*, or stress–strain curve). Still, the deformation kinetics at large plastic deformations cannot yet be unambiguously described in terms of existing theories.

DEFORMATION, PLASTIC

See *Plastic deformation, Discontinuous plastic deformation, Localization of plastic deformation, Plastic twisting deformation, Temperature conditions of plastic deformation.*

DEFORMATION RATE

See *Plastic deformation rate.*

DEFORMATION TENSOR

See *Strain tensor.*

DEFORMATION WORK

See *Work of deformation.*

DEGENERACY

There are two main uses for this term.

1. The first use is the presence in a quantum-mechanical system of several different states which have the same value of some physical quan-

tity, such as energy, *angular momentum*, etc. The number of such states is called the *multiplicity* or *degeneracy factor*. Usually degeneracy is associated with the symmetry of a system; lowering by external factors (fields) brings about the partial or complete removal of degeneracy, e.g., it leads to the splitting of energy levels (see, e.g., *Zeeman effect*, *Stark effect*).

2. The second use of this term concerns the state of a gas of identical particles, whereby their quantum properties become revealed below what is called the *degeneracy temperature* T_0. For Bose particles (*bosons*) degeneracy corresponds to *Bose–Einstein condensation*, and in the case of an ideal *Bose gas*

$$T_0 = \frac{3.3\hbar^2 N^{2/3}}{mk_B(gV)^{2/3}},$$

where N is the number of gas particles, V is the volume, m is the particle mass, and g is the multiplicity. For Fermi particles (*fermions*) the state of degeneracy corresponds to the almost complete filling by the particles of states with energies less than the *Fermi energy*, and to the very low occupancy of higher energy states. For an ideal Fermi gas

$$T_0 = \frac{(6\pi^2 N)^{2/3}\hbar^2}{2mk_B(gV)^{2/3}}.$$

Conduction electrons in *metals* ($N/V \propto 10^{22}$ to 10^{23} cm^{-3}) are degenerate up to temperatures of about 10^4 K. In *semiconductors* with high *doping* almost complete degeneracy of electrons in the *conduction band* (or *holes* in the *valence band*) can take place at room temperature or below.

DEGENERACY, SPIN

See *Spin degeneracy*.

DEGENERATE SEMICONDUCTOR

A *semiconductor* with the charge carrier energy distribution described by *Fermi–Dirac statistics*. The Fermi level is located either within the *conduction band* or *valence band*, or is situated in the *band gap* within $\sim k_B T$ from the edge of one of these bands. Inherent semiconductors become degenerate at high temperatures when $k_B T$ becomes comparable to the gap width ε_g, and when these materials have narrow gaps (HgSe, HgTe) then one or both types of carrier is already degenerate at room temperature. In doped semiconductors the *conduction electrons* (holes) become degenerate at high donor (acceptor) concentrations. For high optical excitation or strong *injection* of charge carriers the degeneracy of nonequilibrium carriers is possible.

For an arbitrary degree of *degeneracy* or multiplicity the thermodynamic and kinetic characteristics of an equilibrium electron–hole semiconductor system are expressed through the Fermi–Dirac integral:

$$F_n(z) = \int_0^\infty \frac{x^n\,dx}{\exp(x-z)+1},$$

where $z = \xi/(k_B T)$, and ξ is the chemical potential. For strong degeneracy, $\exp[\xi/(k_B T)] \gg 1$, these formulae are considerably simplified, and for doped degenerate semiconductors they have the same form as for *metals*.

The degeneracy of charge carriers is especially noticeable in those kinetic effects, which arise from the thermal spread in the energy distribution of the carriers. These include the *magnetoresistance* effect, electronic thermal conductivity, *Peltier effect*, *Ettingshausen effect*, and other effects in semiconductors with isotropic energy spectra (see *Thermomagnetic phenomena*, *Thermoelectric phenomena*). In completely degenerate semiconductors at $T = 0$ K these effects are absent, because, according to the Pauli principle, in *transport phenomena* only the charge carriers located at or near the *Fermi surface* and possessing the same energy take part. At $T \neq 0$ K these effects are present but they are not appreciable, their value being approximately $\xi/(k_B T)$ or $[\xi/(k_B T)]^2$ times (depending on the effect under consideration) less than in non-degenerate semiconductors.

The features of degenerate semiconductors are most clearly evident in the presence of quantized magnetic fields (see *De Haas–van Alphen effect*, *Shubnikov–de Haas effect*, *Quantum Hall effect*, etc.).

DEGREE OF ORDER of liquid crystals (V.N. Tsvetkov, 1942)

The quantity which designates the fraction of *liquid crystal* molecules with their long axes aligned along the *director* $\boldsymbol{n}$. The degree of order S, or the *order parameter*, is given by the following equation:

$$S = \frac{1}{2}\langle 3\cos^2\theta - 1\rangle,$$

where θ is the angle between the long axis of an individual molecule and $\boldsymbol{n}$, and the angle brackets designate averaging over the orientations of the molecules. In the isotropic liquid crystal phase $S = 0$, for ideal ordering $S = 1$, and in *nematic liquid crystals* $S \approx 0.6$. The degree of order may be determined, e.g., from NMR spectra.

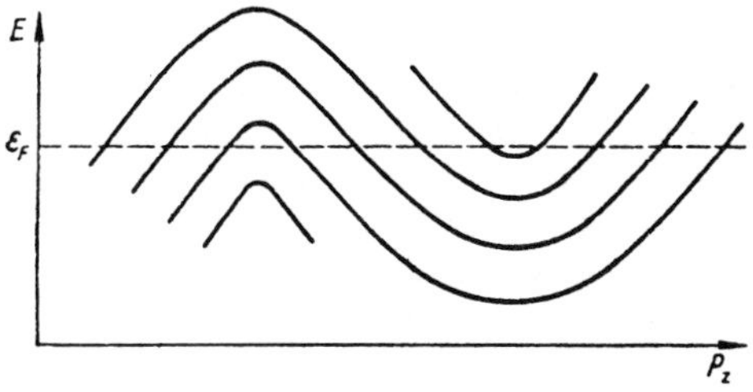

The curves show the quantum terms in the magnetic field $E_n(p_z)$, and the dashed line indicates the Fermi energy. At $T = 0$ all the states with $E_n < \varepsilon_F$ are occupied by electrons, and those with $E_n > \varepsilon_F$ are all vacant.

DE HAAS–VAN ALPHEN EFFECT (W. de Haas, P. van Alphen, 1931; predicted by L.D. Landau, 1930)

The oscillating dependence of thermodynamic characteristics of *metals*, *semimetals* and *degenerate semiconductors* on the magnetic field $\boldsymbol{B}$, caused by the quantization of energy levels of the conduction electrons, with a fixed projection of quasi-momentum (p_z) along the direction $\boldsymbol{B} = (0, 0, B)$ (*Landau quantization*).

The de Haas–van Alphen effect is explained in terms of the presence of a selected *Fermi energy* ε_F in the spectrum of excitations of an electron *Fermi liquid*. When the magnetic field $\boldsymbol{B}$ is scanned the various n terms of magnetic quantization with the energy $E_n(p_z)$ and momentum p_z in turn intersect ε_F, and the density of one-electronic states having the energy ε_F oscillates and passes through a sharp peak each time one of the n curves $E = E_n(p_z)$ is tangential to the dashed horizontal line $E = \varepsilon_F$ (see Fig.). It is these oscillations which constitute the de Haas–van Alphen effect. Usually the number of terms (N_0) with energy $E_n < \varepsilon_F$ is very high: $N_0 = p_F^2/e\hbar B$ (p_F is the electron momentum at the energy ε_F) and the structure of the *de Haas–van Alphen oscillations* is determined by the quasi-classical *Lifshits–Onsager quantization rules* for electrons with an arbitrary *dispersion law* $\varepsilon(\boldsymbol{p})$ ($\boldsymbol{p}$ is the *quasi-momentum*) in the magnetic field: $S(E, p_z)/e\hbar B = 2\pi(n + 1/2)$, $n \gg 1$. Here S is the cross-section area of the *isoenergetic surface* $\varepsilon(p) = E$ at the plane $p_z = \text{const}$. The oscillations are periodic with respect to the reciprocal magnetic field with the period $\Delta(1/B) = 2\pi e\hbar/S_{\text{ext}}$, where S_{ext} is the extremal cross-section of the *Fermi surface* $\varepsilon(\boldsymbol{p}) = \varepsilon_F$ (the points of extrema $E_n(p_z)$ and $S(\varepsilon_F, p_z)$ coincide at $E_n = \varepsilon_F$). If the Fermi surface has several extremal cross-sections, then the oscillating parts of the thermodynamic characteristics contain harmonics with periods $\Delta(1/B)$, responding to all the extremal areas. The experimental study of the dependence of $\Delta(1/B)$ on the direction $\boldsymbol{B}$ gives detailed information about the Fermi surface geometry, and in some cases provides data sufficient for its complete reconstruction.

Since the quasi-classic parameter N_0^{-1} is negligible in magnitude the quantum oscillations of the thermodynamic potential of the electronic system have the relative amplitude $A \propto N_0^{-5/2}$. As a rule it is not the potentials themselves that are determined, but rather their derivatives (*magnetic susceptibility*, *compressibility*, etc.). In this case A may grow by several orders. At finite temperatures, $T \neq 0$, the boundary of the energy distribution of the electrons is blurred, which reduces the value of A. In the temperature region $k_BT \gg \hbar\Omega$ (Ω is the *cyclotron frequency* corresponding to the given extremal cross-section of the Fermi surface) the relative amplitude of the lth harmonic is proportional to the exponential factor $A_T = \exp[-2\pi^2 lk_BT/(\hbar\Omega)]$. In addition, the broadening of the electronic terms caused by the *electron scattering* further reduces A. This effect is taken into account by the so-called *Dingle factor* $A_D = \exp(-2\pi l\nu/\Omega)$, where ν is the reciprocal lifetime of the electronic state, averaged over the extremal cross-section of the Fermi surface. In order to make the factors A_T, A_D of the order of unity, low temperatures (He range) and pure samples are needed. At $T \to 0$ not only do *point defects* (*impurity atoms*, *vacancies*) influence the value of A_D, but also the weak deformational fields of *dislocations*, usually unimportant for the kinetics of the conduction electrons. In some cases the contribution of dislocations to A_D may dominate.

A systematic theory of the de Haas–van Alphen effect was formulated by I.M. Lifshits and A.M. Kosevich (1955). In the presence of *magnetic breakdown* the de Haas–van Alphen effect is qualitatively more complicated in comparison with the quasi-classical one, but the Lifshits–Kosevich theory can be generalized to this case.

DEHARDENING, softening

Decrease in *strength* of a material that had been brought into a hardened, thermodynamically nonequilibrium state by the effect of some external field (temperature, gravitational, mechanical, electric, magnetic, etc.). Such a softening or dehardening proceeds by the transformation of the material into a more stable state with lower energy. Dehardening may occur either during the process of the dehardening action itself (*dynamic dehardening*) or subsequent to it (*static dehardening*). To activate the latter in practice the material is usually heated.

In the case of crystalline solids dehardening proceeds through the formation, displacement (redistribution) and annihilation of *point defects*, of linear, surface and volume crystal defects, and their complexes. Depending on the type of defects responsible for the dehardening the latter may exhibit different kinetics, which variously affect the physical and mechanical properties of the material. In most cases dynamic dehardening improves the material *plasticity*, and only weakly affects the specimen shape as the latter changes under the effect of the external field. Dehardening that is related to the formation and spread of cracks lowers the material plasticity.

DEHARDENING BY ALLOYING, softening by alloying

A decrease of the critical *shear* strength in *metals* upon the addition of a small amount of an alloying element (e.g., when *alloying* tantalum with rhenium). Dehardening by alloying is observed at room and lower temperatures when the thermally activated *dislocations* penetrate the Peierls barriers (see *Peierls relief*), thereby, to a large extent, determining the value of the critical shear stress. One may relate dehardening by alloying with the lowering of the Peierls–Nabarro strength (see *Peierls–Nabarro model*). The latter may be explained by the easier formation of double *kinks on dislocations* at those points where the atoms of the alloying element are located. Dehardening by alloying is also observed when highly active elements are added to body-centered cubic metals. Such elements bond the *doping atoms* into a second *phase*, clean the metal matrix, and thus lower the impurity *hardening*. In that case, the effect of lowering the critical shear stress is observed over a wide temperature range.

DELAY LINE, ULTRASONIC

See *Ultrasonic delay lines*.

DELTA PHASE (δ-phase)

The fourth allotropic modification (see *Allotropy*) of polymorphous metals (see *Polymorphism*) (e.g., δ-Mn, δ-Ce), and also *solid solutions* based on them. *Iron* is an exception, because the solid solution on its third high-temperature modification (see *Gamma phase*) is called the delta phase or δ-Fe. Like α-Fe it has a body-centered cubic lattice and exists from 1400 °C up to the melting temperature 1540 °C. *Alloying* iron with elements of Groups III, IV, V and VI broadens the temperature range of the delta phase existence. In some systems (e.g., Fe–Si, Fe–Cr, Fe–P) the stabilization of the δ- and α-phases (see *Alpha phase*) brings about a merging of their phase regions in their *phase diagram*. As a result, such phases over their entire temperature range of existence only have a body-centered cubic lattice. This phenomenon is used to create *soft magnetic materials* (e.g., silicon *steels*, high-chromium steels, stable against corrosion under stress, etc.). In phase diagrams of other systems, the fourth intermediate phase (counting from the first toward the second component) is called the delta phase. In systems with interstitial elements (C, N, etc.) the monocarbides with the NaCl type *crystal structure* are designated as delta phase.

DEMAGNETIZATION

A process whereby the *magnetization* of a magnetic material is reduced. A sample with a magnetic susceptibility located in an externally applied *magnetic field* B_{app} has a magnetization given by $M = (B_{app}/\mu_0)[\chi/(1+\chi N)]$ where N is the demagnetization factor which depends on the sample shape and orientation in the field, and can assume values $0 \leqslant N \leqslant 1$. For an ellipsoidal shape with principal directions x, y, z the corresponding demagnetization factors obey the normalization relation $N_x + N_y + N_z = 1$, so $N = 1/3$ in all directions for a sphere. For an ellipsoid of revolution with the external field applied along its symmetry axis we have $N \approx 1 - \delta$ for a strongly oblate (disk-shaped) sample, and $N \approx \delta$ for a strongly prolate (needle-shaped) sample, where $\delta \ll 1$.

This discussion assumes that there is no hysteresis, so that the magnetization M vanishes when the applied field is reduced to zero. When hysteresis is present there is *partial demagnetization* even when there is no shape demagnetization ($N = 0$) and the external magnetic field is lowered to zero. When B_{app} reaches zero the *remanence* M_{rem} remains (see also *Hysteresis*). Complete demagnetization in the presence of hysteresis can be achieved in three different ways:

1. Demagnetization by a constant magnetic field. After the applied field reaches zero it is increased in the reverse direction until it reaches the value $-B_c$, called the *coercive force* (coercive field). The specimen is now demagnetized to zero ($M = 0$) in an applied reverse field. To completely demagnetize the sample in zero field it can be cycled through its hysteresis loop many times with gradually decreasing peak field values.
2. Demagnetization by an alternating magnetic field. The specimen is subjected to an alternating applied field that is gradually diminished to zero from an amplitude that can (or may not) initially exceed B_c.
3. Complete demagnetization by heating. The specimen is heated to a temperature above the *Curie point* and subsequently cooled in an environment free of any magnetic fields.

DEMAGNETIZATION COOLING

See *Adiabatic paraelectric cooling.*

DEMAGNETIZATION FACTOR TENSOR

A second-rank symmetric tensor $\widehat{N}$ defining the relationship $\boldsymbol{H}_D = -\widehat{N}\boldsymbol{M}$ between the demagnetization field $\boldsymbol{H}_D$ and the magnetization vector $\boldsymbol{M}$ of a specimen of permeability $\hat{\mu} = \mu_0(1+\chi)$ placed in a uniform external magnetic field $\boldsymbol{B}_0 = \mu_0\boldsymbol{H}_0$, where $\chi = M/H_i$ is the dimensionless magnetic susceptibility of the material, H_i is the *internal field*, and SI units are being used. The demagnetization factor tensor $\widehat{N}$ coincides with the *depolarization factor tensor* for a specimen of identical shape and orientation of *dielectric constant* $\hat{\varepsilon}$ placed in a uniform external electric field $\boldsymbol{E}_0 = \boldsymbol{D}_0/\varepsilon_0$. Moreover, all the results valid for the demagnetization factor tensor hold also for the depolarization factor tensor, provided the notations are changed, respectively, from the magnetic to electric case: $\boldsymbol{H} \to \boldsymbol{E}$, $\boldsymbol{B} \to$ induction $\boldsymbol{D}$; permeability tensor $\hat{\mu} \to \hat{\varepsilon}$; $\mu_0\boldsymbol{M} \to$ polarization $\boldsymbol{P}$; $\boldsymbol{H}_D \to$ depolarization field $\boldsymbol{E}_D$. The source of the latter field lies with the electric charges induced by the applied field $\boldsymbol{E}_0$ at the specimen surface.

For a specimen of arbitrary shape the demagnetization factor tensor is coordinate dependent, due to the nonuniform nature of the demagnetization fields, so its calculation becomes problematic. The most important shape permitting an exact analytic solution is an ellipsoid arbitrarily oriented with respect to $\boldsymbol{B}_0$. Within a specimen of that shape the internal field values $\boldsymbol{B}_i$, $\boldsymbol{M}$, and $\boldsymbol{H}_i = \boldsymbol{H}_0 + \boldsymbol{H}_D$ appear to be uniform, but they may have different directions in space. In the important special case when the principal directions of the tensor $\hat{\mu}$ coincide with the ellipsoid principal axes and the applied field $\boldsymbol{B}_0$ is oriented along one of those axes, then the internal fields $\boldsymbol{B}_i$, $\boldsymbol{M}$, and $\boldsymbol{H}_i$ are all parallel to the applied field $\boldsymbol{B}_0$. The demagnetization factor tensor is in diagonal form with its principal N_i all positive, obeying the normalization relation $\sum_{i=1}^{3} N_i = N_x + N_y + N_z = 1$. These *demagnetization factors* are given by the integral expression:

$$N_i = \frac{a_1 a_2 a_3}{2} \int_0^{\infty} \frac{dq}{(a_i^2 + q) f(q)}, \qquad (1)$$

where $f(q) = [(q + a_1^2)(q + a_2^2)(q + a_3^2)]^{1/2}$. For ellipsoids of revolution ($a_1 = a_2 \neq a_3$) the elliptic integrals (1) reduce to elementary functions.

The internal fields are related to the applied field through the expression

$$NB_{\mathrm{i}} + (1 - N)\mu_0 H_{\mathrm{i}} = B_0 \tag{2}$$

with the following individual values:

$$B_{\mathrm{i}} = B_0 \frac{1+\chi}{1+N\chi}, \quad H_{\mathrm{i}} = \frac{H_0}{1+N\chi}, \quad M = \frac{\chi H_0}{1+N\chi}, \tag{3}$$

where $\boldsymbol{B}_0 = \mu_0 \boldsymbol{H}_0$. From symmetry considerations it is clear that $N = 1/3$ for a sphere. In the limiting cases of an infinitely oblong and oblate ellipsoids of revolution (a long cylinder and a flat disc (plate)), the same considerations yield $N = 1/2$ for the case when $\boldsymbol{B}_0$ is normal to the cylinder axis, $N = 0$ for $\boldsymbol{B}_0$ parallel to that axis, while $N = 1$ for $\boldsymbol{B}_0$ normal to the plate plane.

The demagnetization factor tensor is used to calculate internal magnetostatic fields and also external fields produced by *permanent magnets* of various shapes, and likewise for electrostatic calculations. The demagnetization factor tensor plays a significant role when determining frequencies of *ferromagnetic resonance* and *ferrimagnetic resonance*, and it is important for *electron paramagnetic resonance* for $g\mu_{\mathrm{B}}B/(k_{\mathrm{B}}T) \geqslant 1$, where $g\mu_{\mathrm{B}}B$ is the Zeeman splitting (see *Zeeman effect*). The demagnetization factor tensor also plays an important role in the theory of the *intermediate state* of superconductors and magnetic materials in an external magnetic field, while the dielectric analogue theory is used to explain the interaction of electromagnetic waves with *small particles* of various shapes in the Rayleigh approximation (see *Rayleigh scattering of light*). In the latter case they often introduce a generalization of the concept of *depolarization factors*, considering particles placed in a medium with the dielectric constant $\varepsilon_{\mathrm{m}} > \varepsilon_0$ so that besides the purely geometric factor (1) there appears a dependence on the dielectric properties of both the particle and the medium. If the principal axes of tensors $\hat{\varepsilon}$ and $\widehat{N}$ coincide, the generalized factor has the form:

$$\widetilde{N}_{\mathrm{i}} = \frac{\varepsilon_{\mathrm{i}} - \varepsilon_{\mathrm{m}}}{\varepsilon_{\mathrm{i}} - \varepsilon_0} \frac{N_{\mathrm{i}}}{\varepsilon_{\mathrm{m}}/\varepsilon_0}. \tag{4}$$

A generalization similar to (4) may also be introduced for the demagnetization factor tensor with allowance made for the above change in notation.

DEMAGNETIZATION FIELDS

Magnetic fields within a *magnetic substance* (that is either magnetized or placed in an external magnetic field), which seek to lower its *magnetic moment*. The sources of demagnetization fields are either surface or bulk (ficticious) *magnetic charges*. To describe demagnetization fields one traditionally uses the coefficients of the *demagnetization factor tensor* $\widehat{N}$ that relates the demagnetization field $\boldsymbol{H}_{\mathrm{D}}$ with a specimen magnetization $\boldsymbol{M}$, so that $\boldsymbol{H}_{\mathrm{D}} = -\widehat{N}\boldsymbol{M}$. The value of the tensor $\widehat{N}$ significantly depends on the specimen shape and may be calculated exactly for an ellipsoidal shape only. In the latter case, both $\boldsymbol{H}_{\mathrm{D}}$ and $\widehat{N}$ remain uniform within the specimen bulk and the tensor $\widehat{N}$ is diagonal with the normalization condition $N_x + N_y + N_z = 1$. For a specimen of an arbitrary shape, its demagnetization field is nonuniform. Either empirical or approximate relationships are then used to describe the tensor $\widehat{N}$. When exciting a nonuniform precession of the magnetization M in a variable magnetic field charges form that result in the appearance of variable demagnetization fields. Such fields appreciably affect the spectrum of magnetostatic vibrations and waves.

DENDRITES (fr. Gr. $\delta\varepsilon\nu\delta\rho o\nu$, tree)

Crystals (or aggregations of crystals) of a branched tree-like configuration. They are formed during *crystallization* from the melt, solution, vapor, and some *metallic glasses*. The formation of dendrites is explained by the instability of the source crystal convex shape under the conditions of oversaturation or overcooling of the medium. At the first stage projections appear which develop into axes (trunks, first-order branches) at which branches of second and higher orders form (see *Crystal morphology*). The directions of *dendrite* growth usually coincide with the axis of a pyramid with its sides formed by planes of greatest reticular density: $\{100\}$ in a face-centered or body-centered cubic lattice, $\{10\bar{1}0\}$ in a hexagonal close-packed lattice, and $\{110\}$ in a body-centered tetragonal lattice. In the structures formed by these growths (*dendrite structure*) rounded and faced dendrites are singled out. The former are characteristic of most crystals of pure metals, and of some alloys and organic materials which grow from the alloy and have low melting *entropies*. *Edged dendrites* have been found during crystallization from the melt of many nonmetallic materials (semiconductors, oxides, semimetals), and also upon crystallization from vapor and solution.

DENSITY

One of the fundamental physical characteristics of a body, which is numerically equal to the mass of a unit volume: $\rho = \mathrm{d}m/\mathrm{d}V$, where $\mathrm{d}m$ is the mass of a differential element of this substance of volume $\mathrm{d}V$. The *mean density* of a body is defined as the ratio of the mass of body m to its volume V, i.e. $\rho_{\mathrm{m}} = m/V$. The ratio of the densities of two materials under certain standard physical conditions is referred to as *relative density*. In the case of porous and loose materials, two types of density are distinguished: *true density* (not including cavities present in the body) and *apparent density* (ratio of mass of body to total volume occupied by it). The SI unit for density is $\mathrm{kg/m^3}$, the CGS unit is $\mathrm{g/cm^3}$. As a rule the density increases with increasing *pressure* and decreases with increasing temperature (due to *thermal expansion*), except near phase transitions. The value of the density undergoes an abrupt change as a material undergoes a change of state (see *States of matter*): it sharply decreases upon the transition into the gaseous state, and increases on *solidification* (there exists an anomaly in the behavior of the density of water and *cast iron*: it decreases at the liquid–solid transition). The value of the density of a body is determined by precise determination of the mass and volume of this body by application of various types of *densimeters*. Also used in density determinations is the dependence of density on the velocity of sound propagation, the intensity of γ- and β-radiation passing through a substance, etc.

DENSITY FUNCTIONAL, NONLOCAL

See *Nonlocal density functional.*

DENSITY FUNCTIONAL THEORY

Method involving a formal, accurate reduction of the problem of the ground state of a many-particle system with strong interactions to a single-particle problem. Along with *Fermi liquid* theory and many-particle perturbation theory (method of *Green's functions*), density functional theory provides a formalism for explaining many-particle systems. The basis for density functional theory is the *Hohenberg–Kohn theorem* (P. Hohenberg, W. Kohn, 1964), which states that the energy of a system of interacting fermions (usually electrons) in the ground state is a function of the total particle density only; an external field appears in this functional only as the term $\int V_{\mathrm{ext}}(\boldsymbol{r})\rho(\boldsymbol{r})\,\mathrm{d}\boldsymbol{r}$, and the functional minimum in the given field corresponds to the true ground state in this field. Usually the following energy contributions are explicitly separated out in the *density functional*: the kinetic energy of noninteracting fermions of the same density $\rho(\boldsymbol{r})$, the energy of interaction with the external field, and the Coulomb (Hartree) energy. The remaining contributions to the energy are called exchange-correlation (E_{XC}) ones (in the context of many-particle theory this term has a somewhat different meaning; the difference has to do with introduction of part of the kinetic energy into E_{XC}). To determine the density distribution, which minimizes the above functional, in the form of the density of a certain system of noninteracting particles under the influence of the effective potential, the effective potential itself is expressed through an externally applied potential, the Hartree potential and the exchange-correlation potential $V_{\mathrm{XC}}(\boldsymbol{r}) = \delta E_{\mathrm{XC}}\{\rho\}/\delta\rho(\boldsymbol{r})$. The resulting set of self-consistent equations is referred to as the *Kohn–Sham equations* (W. Kohn, L. Sham, 1965). In the limit of nearly constant density, the *local density approximation* is valid, according to which $E_{\mathrm{XC}} = \int \rho(\boldsymbol{r})\varepsilon_{\mathrm{XC}}(\rho(\boldsymbol{r}))\,\mathrm{d}\boldsymbol{r}$, where $\varepsilon_{\mathrm{XC}}(\rho)$ is the density of the *exchange-correlation energy* (see *Correlation energy*) of the uniform electron gas of density ρ. Most density functional theory calculations are carried out in the context of this approximation. The accuracy of this approximation is quite high for the calculation of observable properties of real solids (equilibrium volume, compressibility, etc.), but there exists a number of problems (e.g., description of the interaction of an electron with a metal surface), which require greater accuracy. Available for application are also *nonlocal density functionals*, in which

E_{XC} is an integral function of the electron density in a certain region (mean density approximation, weighted density approximation), and gradient density functionals. The simple expansion, with the gradient-dependent part of E_{XC} proportional to the square of the density gradient, is now rarely used. Widely accepted are *Langreth–Mehl processes* (D.C. Langreth, M. Mehl, 1983) and the generalized gradient expansion, with its complicated dependence of E_{XC} on the density gradient. Density functional theory may be generalized to treat magnetic systems (*spin density* functional) at finite temperatures, and in time-independent external fields. There exists a relativistic formulation of density functional theory.

Density functional theory provides a formal basis for the *band theory* of solids, since it permits the treatment of the *band structure* of noninteracting particles under the influence of an effective periodic potential; the properties of the ground states of this calculated band structure are in full agreement with the properties of the ground states of the crystal under consideration. In combination with *band structure computation methods*, the density functional theory allows the calculation of the following parameters: equilibrium *crystal lattice constants* (to an accuracy of several percent), elastic moduli and phonon frequencies (accurate within 10–20%), magnetic moment in *magnetic substances* (correct to several percent), distribution of electron and spin densities, relative stability of *crystal structures*, etc.

The formulation of density functional theory using the Kohn–Sham equation results in the generation of an additional spectrum which, in general, does not coincide with the actual spectrum of single-particle excitations of the system (since density functional theory is a ground state theory). At the same time, the assumption of a close relationship between these spectra has become the usual practice; this assumption allows one to calculate observed X-ray, optical and other spectra, kinetic characteristics, *electron–phonon interaction*, etc. Thus, density functional theory provides an opportunity to derive system response functions; in the context of this theory these functions are adequately represented by the above-mentioned additional single-particle spectrum, and the second-order variational derivative of E_{XC}. The errors involved in calculations of various quantities, based on the local density approximation theory, are (compared to experiment) from several percent (equilibrium lattice constant, magnetic moment of *ferromagnets*, etc.) to tens of percents for elastic moduli, etc.

DENSITY MATRIX, statistical operator (J. von Neumann, L.D. Landau, 1927)

Operator $\widehat{\rho}$ useful for the calculation of the mean value of any physical quantity in quantum statistical mechanics. The density matrix plays the same role in quantum statistical physics as the *distribution function* does in classic statistical mechanics. Ordinary quantum mechanics deals with systems characterized by a definite state vector $|\psi\rangle$. The mean value $\overline{A}$ of any physical quantity, represented by an operator $\widehat{A}$, in this state is $\overline{A} = \langle\psi|\widehat{A}|\psi\rangle$. Quantum-statistical mechanics describes systems that cannot be represented by a state vector, but may be described by giving probabilities $W_1, W_2, \ldots$ for existing in states $|\varphi_1\rangle, |\varphi_2\rangle, \ldots$, respectively. These systems are called *statistical mixtures*. The mean value in such systems is defined by the formula

$$\overline{A} = \sum_k W_k \langle\psi_k|\widehat{A}|\psi_k\rangle, \qquad \sum_k W_k = 1,$$

which may be written in the following contracted form:

$$\overline{A} = \mathrm{Tr}(\widehat{\rho}\widehat{A}), \qquad \widehat{\rho} = \sum_k W_k |\psi_k\rangle\langle\psi_k|,$$

where Tr is the trace of the operator (i.e. sum of its diagonal elements). In order to write the operator $\widehat{\rho}$ in the form of a matrix ρ_{mn} in some representation, one has to select an appropriate complete set of wave functions $\{|\varphi_i\rangle\}$. Then $|\psi_k\rangle = \sum_i a_{ki}|\varphi_i\rangle$ and $\rho_{mn} = \langle\varphi_m|\widehat{\rho}|\varphi_n\rangle = \sum_k W_k a^*_{km} a_{kn}$. The probability of finding a system in a ("pure") state $|\varphi_m\rangle$ from the overall set $\{|\varphi_i\rangle\}$ is given by the diagonal matrix element ρ_{mm}. In the general case, the system is in the "pure" state, if its matrix operator satisfies the condition $\widehat{\rho}^2 = \widehat{\rho}$. To find the operator $\widehat{\rho}$, nonequilibrium quantum statistical mechanics uses the quantum Liouville equation (*Neumann equation*) which plays the same role as the Schrödinger equation for the time evolution of "pure" states (see *Quantum kinetic equation*). When studying systems that interact with the environment the reduced density matrix is used. It is obtained through averaging the total density matrix of the whole system (quantum system plus environment) over variables of the latter. The reduced density matrix plays an important role in investigating the system response to external disturbances, as well as in studies of such *irreversible processes*, as *transport phenomena*, *relaxation* processes, etc.

DENSITY OF ELECTRON STATES

One of the major characteristics of the electron spectrum of *solids* and other many-electron systems. The density of electronic states $g(E)$ equals the number of electronic states within a unit interval of energy. If the state characterized by *quasi-wave vector* $\boldsymbol{k}$ in the band λ has the energy $E_{\boldsymbol{k}\lambda}$, then the density of electronic states per *unit cell* is given by the expression $g(E) = (1/N)\sum_{\boldsymbol{k}\lambda}\delta(E - E_{\boldsymbol{k}\lambda})$ (see *Density of states*). If the electron wave functions are expanded in terms of a certain set of basis functions, then the partial densities of electronic states may be introduced, which define the relative contribution of a given basis function (e.g., orbital state) over a certain energy interval. Of particular importance is the electronic density of states at the *Fermi level*. It determines kinetic, superconducting, and other properties of a material; and it bears a direct relation to the stability of a *crystal lattice* and to the thermodynamic properties of electrons.

DENSITY OF STATES, spectral density of states

Number of the energy levels of a system per unit energy interval (in more exact terms, the ratio of the number of levels within a certain energy interval to the value of the said interval, when the latter tends to zero). Phonon, electron, magnon, exciton and other densities of states are distinguished in solids. These different types of densities of states correspond to different types of *elementary excitation spectra of disordered solids*. The density of states determines the spectral (*cross-section* of absorption and scattering of radio frequency and X-ray radiation, light, neutrons, gamma quanta, ultrasound), thermodynamic (*specific heat*, *thermal expansion*, etc.), and a number of kinetic properties of a material.

Of particular importance for *metals* is the *density of electron states* at the *Fermi surface*. Methods of determining the density of states are highly diversified, and depend on the type of spectrum. Thus, the best results for the phonon and magnon cases of different materials over a wide range of energies are obtained by application of *inelastic neutron scattering*, whereas the electronic density of states may be determined by employing X-ray photoemission (see *X-ray emission spectroscopy*).

In periodic structures (*crystals*) the states are described by the *quasi-wave vector* $\boldsymbol{k}$, and the energy spectrum is determined by the *dispersion law* $\varepsilon_\lambda(\boldsymbol{k})$, where λ is the number of the spectral band or branch. For these structures, the density of states per unit cell may be represented as

$$g(E) = \frac{1}{N}\sum_{\lambda\boldsymbol{k}}\delta\bigl(E - \varepsilon_\lambda(\boldsymbol{k})\bigr) = \frac{v_0}{(2\pi)^3}\sum_\lambda\int\frac{\mathrm{d}S}{\hbar v_\lambda},$$

where N is the total number of cells, $\delta(x)$ is the Dirac delta function, v_0 is the unit cell volume, and integration with respect to $\mathrm{d}S$ is taken over the surface in *$\boldsymbol{k}$-space* (within the *Brillouin zone*), on which $\varepsilon_\lambda(\boldsymbol{k}) = E$, and $v_\lambda = (1/\hbar)|\nabla_{\boldsymbol{k}}\varepsilon_\lambda(\boldsymbol{k})|$ is the group velocity at this surface.

For the case of the *phonon spectrum* in three-dimensional crystals, the spectral density of states behaves as $g(E) \propto E^2$ at low values of the energy due to the linear form of the *dispersion law* (for two-dimensional systems $g(E) \propto E$, whereas for one-dimensional ones the density of states is constant in this energy region). *Van Hove singularities* occur in the density of states in the neighborhood of edges of phonon, electron, and other types of spectra, as well as within the allowed spectral regions, and they feature zero group velocity. In the case of three-dimensional systems, $g(E) \propto (E - E_{\mathrm{c}})^{1/2}$ in the neighborhood of these points, whereas one- and two-dimensional systems show sharper singularities.

In imperfect crystals, on the one hand, the density of states peaks in the neighborhood of energies of impurity localized states (see *Local electronic levels*), which become more and more broadened with increasing concentration of defects, and on the other hand, the smearing of Van

Hove singularities in the neighborhood of spectral edges takes place, and "tails" of the density of states arise in the previously forbidden spectral region. As this takes place, the type of states changes: the continuous spectrum, in which the states are spread throughout the whole crystal (i.e. delocalized) goes over into a spectrum of discrete states, the wave function of which is localized within a limited spatial region. These two types of states are separated by the *mobility threshold* (see *Mobility edge*, *Anderson localization*). The kinetic properties of excitations on different sides of this threshold are quite different, but the density of states in this region remains fairly smooth. At a certain concentration of defects, e.g., *impurity atoms* of large radius, delocalized states may appear also in the neighborhood of impurity levels. The excitations of low energy in glasses and amorphous compounds (see *Amorphous state*) are due to quantum transitions between specific two-level states (see *Two-level systems*). The density of states in this case does not tend to zero as it does in the case of ordinary phonon states, but turns out to be constant at low values of energy. This is responsible for the specific features of the thermodynamic and kinetic properties of these materials at low temperatures (see *Low-temperature anomalies in amorphous solids*).

DEPLETION of charge carriers

Depletion in a bulk semiconductor (or part of it) of charge carriers due to *charge carrier drift* in an external electric field. Depletion is a phenomenon opposite to *accumulation*. It occurs in a region adjacent to a potential barrier (e.g., near a semiconductor contact). The barrier limits the amount of carrier flux entering this region. If the carriers depart from it at a high rate due to their drift in an external field, the higher the drift rate, the more depleted the region becomes. With a growing field the size of the region increases. When the current flows in the semiconductor with bipolar conduction the region can be simultaneously depleted of charge carriers of both signs as a result of the *Maxwell relaxation* of uncompensated charge of free carriers.

DEPOLARIZATION FIELD

See *Photodomain effect*.

DESORPTION

Removal of adsorbed substances from the surface of an adsorbent, a process which is opposite to *adsorption*. In respect to the effect which causes the desorption the following types are singled out: *thermal desorption*, *electron-stimulated desorption*, *photodesorption*, *field desorption*, and desorption resulting from surface bombardment by fast ions and atoms (cathode *scattering*). The composition of desorption products may differ from the composition of the adsorbed material. In addition to adsorbed molecules and their parts, molecules chemically bound to the adsorbent may be subjected to desorption. Some material might also accumulate at the adsorbent surface (e.g., carbon during electron-stimulated desorption of carbon-containing adsorbates), or adsorbent material itself might be carried away.

DESORPTION COOLING

Process whereby cooling occurs via adiabatic gas *desorption*. It is carried out by pumping-out the thermoisolated adsorbent (see *Adsorption*), which has previously adsorbed some quantity of gas. As a result of desorption the temperature of the adsorbent and of the remaining gas is reduced.

DESORPTION, ELECTRON-INDUCED

See *Electron-induced desorption*.

DESORPTION, THERMAL

See *Thermal desorption*.

DETECTORS OF OPTICAL RADIATION

See *Optical radiation detectors*.

DEUTERATED CRYSTALS

Replacement of protons by deuterons within crystals such as those of *ferroelectrics* with *hydrogen bonds*. In a series of cases it produces a considerable isotopic shift of the *Curie point* T_C. The shift reaches its greatest value for potassium dideuterophosphate D-KDP with $T_C = 213$ K, compared with $T_C = 123$ K for undeuterated KDP (potassium dihydrogen phosphate, KH_2PO_4). Due to the anomalous growth of the electrooptical coefficient and the respective reduction of the controlling stresses near T_C, crystals of D-KDP are used as targets of space-time modulators of light, and they are also widely used in *electrooptical modulators*, shutters and deflectors of different laser devices. The complete deuteration of crystals of *triglycinesulfate* increases the Curie temperature from 322 K to 333 K. Simultaneously with the increase of T_C the *light absorption* of the crystals D-KDP and cesium dideuteroarsenate (D-CDA) decreases to $\lambda = 1.06$ μm, which promotes their application in the wide-aperture elements of neodymium *lasers* at installations for controllable laser thermonuclear synthesis. Some ferroelectrics with hydrogen bonds, e.g., Rochelle salt, exhibit only a small shift of T_C upon deuteration.

DEVICES, SEMICONDUCTOR

See *Semiconductor devices, Microwave semiconductor devices, Photocontrolled microwave semiconductor devices.*

"DEVIL'S STAIRCASE"

A temperature- or pressure-stimulated stepwise change of the wavevector $\boldsymbol{k}_s$ of a superlattice in a crystal whose compositional *phase diagram* exhibits a sequence of thermodynamically stable *phases* with various integral (in regard to *reciprocal lattice* vectors) values $\boldsymbol{k}_s$. The possibility of a stepwise change of the vector $\boldsymbol{k}_s$ that determines the period of a superlattice modulated phase with a gradual monotonic change of the external thermodynamic parameters was proposed by I.E. Dzyaloshinskii (1964). For such a stepwise change the vector $\boldsymbol{k}_s$ runs through a series of sequential magnitudes corresponding to nonsymmetric ("random") points of the *Brillouin zone*. In principle there are no limits to restrict the number of steps of a "devil's staircase", however, for stability with regard to thermal and quantum fluctuations only those phases appear which have small translational periods. As the period grows, the region of existence of the corresponding phase shrinks. Stepwise changes of the vector $\boldsymbol{k}_s$ were observed in the crystal CeSb at temperatures below $T_N = 16$ K.

DIAMAGNET

A material which becomes magnetized in the direction opposite to that of an external *magnetic field*; a material without magnetic ordering, where *paramagnetism* is either absent or is weaker than *diamagnetism*. The *magnetic susceptibility* of diamagnets is negative and low, and the molar magnetic susceptibility χ is of the order of -10^{-5} to -10^{-6}. Many liquids and gases, the molecules (or atoms) of which do not have inherent *magnetic moments* (inert gases, H_2, N_2, F_2, H_2O, some organic compounds) are diamagnets, and the strongest diamagnetism appears in *aromatic compounds*. Many *molecular crystals* formed by such molecules are diamagnets, as well as the series of covalent and *ionic crystals*, containing no atoms of the transition and rare-earth elements (NaCl, CsCl, Ge, Si). *Metals* are diamagnets when *Landau diamagnetism* and diamagnetism of the ionic cores prevail over the Pauli paramagnetism (e.g., *copper, gold, zinc, graphite*). *Type I superconductors* with $\chi = -1$ in mks units ($\chi = -1/(4\pi)$ cgs) are considered as ideal diamagnets (see *Absolute diamagnetism*). Anisotropy (tensor character) of the magnetic susceptibility, which is especially strong for molecular crystals of aromatic hydrocarbons and metals with an anisotropic electronic spectrum (graphite), is characteristic of many crystalline diamagnets.

DIAMAGNETIC DOMAINS

Uniformly magnetized macroscopic regions of a diamagnetic *metal* into which it spontaneously decomposes in an external magnetic field. The appearance of the diamagnetic domains is caused by the fact, that at low temperatures the relationship between the applied magnetic field intensity H and the internal magnetic field B involves the de Haas–van Alphen oscillations (see *de Haas–van Alphen effect*) (see Fig.). The sections of the curve $H(B)$ with $\partial H/\partial B < 0$ reflect a thermodynamically unstable state (see *Schoenberg effect*) which brings about *second-order phase transitions* with a jumpwise variation of the *magnetization* M. The magnetization jump $\Delta M = M_2 - M_1$ occurs at $H = H_c$ and has the magnitude $\Delta M = (B_2 - B_1)/\mu_0$ as a result of the equality of the areas from B_1 to B_2 that are shaded in Fig. This equality gives for the value of the integral $\int_1^2 H(B)\,dB = H_c(B_2 - B_1)$. For a sample of ellipsoidal shape with the dimensionless demagnetization factor N ($0 \leqslant N \leqslant 1$), placed in the uniform magnetic field $B_0 = \mu_0 H_0$, the jumps of magnetization are absent, and upon the variation of the

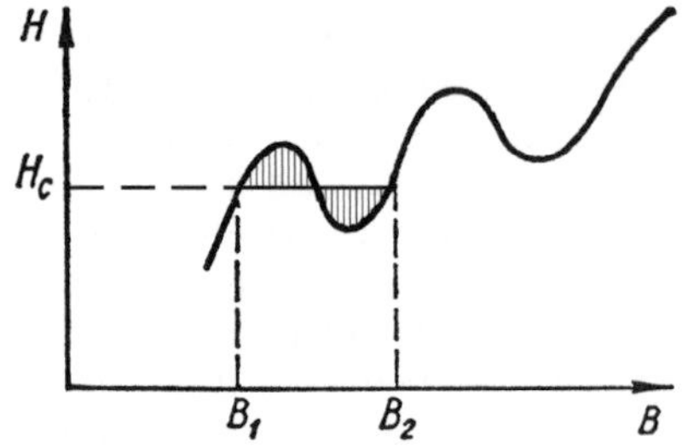

applied field H_0 within the range $NB_1 + (1 - N)\mu_0 H_c < \mu_0 H_0 < NB_2 + (1 - N)\mu_0 H_c$ the internal field $\overline{B}$ averaged over the sample grows linearly with the increase of H_0, changing from B_1 to B_2. In this case, as only the phases with $B = B_1, B_2$ are thermodynamically stable, the sample is decomposed into interlacing *domains* with the internal fields B_1 and B_2. The share of phases with each B value is determined by H_0. In the simplest case of a flat-parallel plate with thickness d, oriented perpendicular to H_0, we have $N = 1$ and there appears a periodic picture of interlacing domains with the shape of flat layers with thickness of approximately $d^{1/2}$, parallel to H_0.

DIAMAGNETISM (fr. Gr. $\delta\iota\alpha$, prefix in this case meaning opposed, and magnetism)

A property whereby a material magnetizes antiparallel to an applied *magnetic field* $\boldsymbol{B}$. Diamagnetism is inherent to all substances. Materials without magnetic order in which diamagnetism prevails over *paramagnetism* are called *diamagnets.* In terms of classical physics diamagnetism may be explained by electromagnetic induction whereby an applied magnetic field acts on closed electric currents circulating in atoms and molecules of a solid, causes electrons to precess, and thereby induces *magnetic moments.* The latter, in accordance with the Lenz law, are antiparallel to the external field. In a quantum-mechanical description diamagnetism is caused by the shifting and splitting of atomic energy levels $\Delta\varepsilon$ under the effect of an applied magnetic field. The induced *magnetization* is $M = \chi H$, where the *magnetic susceptibility* χ in the case of diamagnetism is small and negative (typical values are -10^{-6} to -10^{-5}). For non-interacting atoms χ arises from the susceptibilities χ_a of individual atoms, $\chi_a = -e^2\langle r^2\rangle/(6mc^2)$, where $\langle r^2\rangle$ is the mean square distance of the electron from the nucleus. This formula provides a description of the diamagnetism of monatomic gases and liquids; when atoms are combined into molecules (or *crystals*) the value of χ usually decreases, but it may also increase. The latter occurs for the delocalization of electrons in molecules such as aromatic hydrocarbons, and in some crystals.

In *metals* and *semiconductors* diamagnetism is determined by quantization of the movement of electrons in the magnetic field (see *Landau diamagnetism*). At low temperatures in pure metals the relationship between M and H becomes complex (non-linear, alternating) (see *de Haas–van Alphen effect*). Diamagnetism caused by surface currents appears in *superconductors,* and in bulk samples of *type I superconductors* an internal magnetic $\boldsymbol{B}$ field is excluded (see *Meissner effect*) corresponding to $\chi = -1$ in mks units ($\chi = -1/(4\pi)$ cgs). This value of χ is the lowest one that is possible.

DIAMAGNETISM, ABSOLUTE

See *Absolute diamagnetism.*

DIAMAGNETISM, FLUCTUATION

See *Fluctuation diamagnetism.*

DIAMAGNETISM, LANDAU

See *Landau diamagnetism.*

DIAMAGNETISM, PERFECT

See *Absolute diamagnetism.*

DIAMOND

An allotropic modification of *carbon.*

Occurs in nature (a native nonmetallic *mineral*) which can also be prepared synthetically. The strength of the diamond lattice determines its extreme physical properties and chemical inertness. Colorless (in pure form), it crystallizes in a face-centered cubic lattice ($a_0 = 0.357$ nm), with coordination number $z = 4$, density 3.515 g/cm^3, space group $Fd3m$, and 8 atoms per *unit cell* at the positions: (0 0 0), (1/2 1/2 0), (0 1/2 1/2), (1/2 0 1/2), (1/4 1/4 1/4), (3/4 3/4 1/4), (1/4 3/4 3/4), (3/4 1/4 3/4). The structure is a three-dimensional lattice of rigid C–C sp^3-hybrid *covalent bonds* 0.15445 nm long; with each carbon atom centered in a tetrahedron of four others

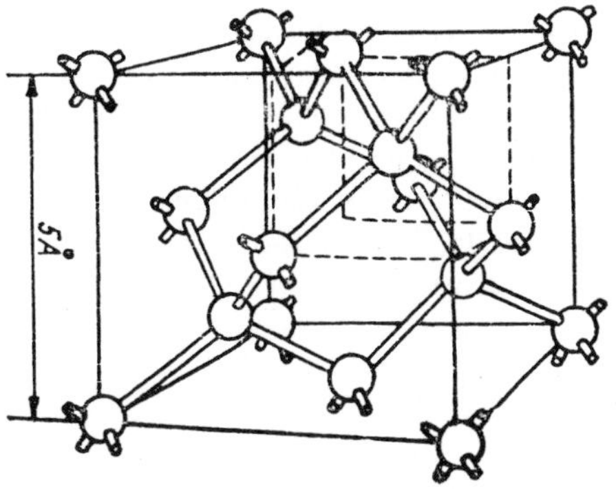

Crystallographic structure of diamond.

(see Fig.). Crystals have common shapes of octahedron, rhombododecahedron, and cube as well as their combinations. The *crystal cleavage* along (111) is perfect.

Diamond is metastable, and at atmospheric pressure its enthalpy exceeds that of *graphite* over the entire temperature range. The high activation barrier $Q_{act} \approx 470$ kJ/mole favors the onset of diamond *graphitization* above 1300 K. The transformation is rapid at 2000 K; and the transition heat is 1.898 kJ/mole. The triple point of diamond–graphite–liquid is at 11 GPa and 3700 K. Standard entropy $S(T = 298\text{ K}) = 2.37\text{ J}\cdot\text{mole}^{-1}\cdot\text{K}^{-1}$; isobaric specific heat at 298 K $c_p = 6{,}600\text{ J}\cdot\text{mole}^{-1}\cdot\text{K}^{-1}$; and the Debye temperature $\theta_D = 1890$ K is very high.

Diamond is the hardest of all naturally occurring materials with a Mohs scale hardness of 10 which varies with the particular crystal face. Diamond is brittle and cracks under a sudden blow. Young's modulus is $9.36\cdot10^7$ Pa; bulk modulus is $6.0\cdot10^7$ Pa; ultimate strength is $1.76\cdot10^5$ Pa; thermal linear expansion coefficient is $1.06\cdot10^{-6}$. Refraction index is 2.417 for $\lambda = 0.589$ μm (yellow Na-D line), angle of total internal reflection is 24°50′ (Na). Diamond does not interact with acids or alkali solutions; it oxidizes in air at 1120 K, in oxygen at 990 K. At high temperatures etched by alkali melts, oxidizing salts, metals; reacts with water vapors, with carbon dioxide.

Classification of synthetic and natural diamonds is based on their optical properties; the most common being the classification according to the IR-absorption spectra. Type I crystals absorb in the two ranges 700 to 1660 cm^{-1} and 1600 to 3300 cm^{-1}, and type II crystals absorb only in the latter range 1600 to 3300 cm^{-1}. The 1600 to 3300 cm^{-1} absorption characteristic of all diamonds is due to intrinsic lattice vibrations, while that within the range 700 to 1600 cm^{-1} is determined by the impurity content. The most typical impurity in diamonds is nitrogen (up to 0.25% in type I and up to 0.001% in type II). Type Ia comprises natural diamonds containing nitrogen in the clustered (N_2) nonparamagnetic form; type Ib comprises natural and synthetic diamonds containing nitrogen in the disperse paramagnetic form (EPR g-factor = 2.0). Type IIa comprises natural and synthetic diamonds, characterized by high thermal conductivity, 10 $\text{kW}\cdot\text{m}^{-1}\cdot\text{K}^{-1}$ (6 $\text{kW}\cdot\text{m}^{-1}\cdot\text{K}^{-1}$ for type I) and by photoconductivity. Diamonds of types I and IIa are *insulators* with the energy gap width $\Delta E = 5.5$ eV; for type I: $\rho = 10^{16}$ Ω·cm, and for type II: 10^{14} Ω·cm. Type IIb comprises natural and synthetic diamonds which are p-type semiconductors due to the impurity boron. Typical is hole conductivity with the hole binding energy 0.37 eV. Implantation of lithium produces semiconducting diamond layers with donor ionization energy 0.1 eV. Implanted nitrogen acts as a donor with deep local electronic levels.

There are four types of natural diamonds:

(1) *Balas* has a globular shape with crystal faces at the surface. It is colorless or blackish-gray, semitransparent and opalescent, nontransparent black, with color due to graphite inclusions. The density 3.512 g/cm^3 decreases with increasing quantity of inclusions. Also made artificially.

(2) *Bort* is an aggregate of intergrown granular crystals, typically of irregular shape. The color is gray to black; due to graphite or mineral impurities. The weight of ashes after burning can reach several percent.

(3) *Carbonado* consists of microgranular aggregates of irregular shape with micrometer sized grains; the color is dark gray, black; less often grayish-green, brown, violet. Massive samples with matte or glossy surface occur as well as porous, slag-like ones; with pores often filled with minerals. Density 3.0 to 3.4 g/cm^3

decreases with increasing porosity. Also obtained artificially.

(4) *Impact diamonds*, found in meteorites and in meteorite craters, are cryptocrystalline textured aggregates containing considerable (sometimes up to 50%) hexagonal phase or *lonsdalite*. Density is 3.3 to 3.5 g/cm^3. They have a high specific surface area (20 to 40 m^2/g), high abrasive capacity, and enhanced birefringence (0.0005 to 0.0100). Artificial analogues include *explosive diamond* obtained by exposing graphite to a *shock wave*.

Synthesis of diamonds. (1) At *high pressure* by the direct transformation of graphite to diamond above 12 GPa and 1300 K. (2) Transformation of graphite to diamond via the melt of a metal (Mn, Fe, Co, Ni, etc., and their alloys). *Crystallization* of the dissolved carbon takes place above 4 GPa and 1500 K. (3) Explosive method exposes graphite to a shock wave with $P > 49$ GPa. (4) Detonation method using explosive gas containing extra carbon, at pressure 13 to 25 GPa and temperature 3000 to 4000 K. Rapid adiabatic expansion of explosion followed by rapid cooling yields 20 to 90% diamond. (5) Synthesis from atomic carbon (via epitaxial growth, deposition from ionic beam, condensation of vapors under laser heating, and ionized carbon deposition from subnormal discharge plasma).

Both natural and synthetic diamonds are used as extremely hard abrasive material for manufacturing of grinding, polishing, cutting, drilling and other tools; in electronic engineering as heat eliminating elements, luminescent radiation sources, temperature sensors, photoreceivers, nuclear radiation detectors. The semiconducting properties of diamond are used in solid-state physics; its hardness and transparency are utilized in high-pressure apparatus.

DIAMOND-LIKE CARBON FILMS

Metastable carbon or hydrogen-containing carbon *films* with *hardness*, *density*, optical and electrical properties, and chemical stability close to *diamond*, whereas the structure of films differs from that of diamond. The structure of diamond-like films is amorphous, though in some cases the presence of the diamond phase crystals is observed. Diamond-like films as a rule are a heterophase, containing regions of short-range order (*clusters*) with sp^3-, sp^2- and sp-hybridization C–C and C–H bonds, i.e. the films may contain the phases of diamond, *graphite*, and *carbyne*, respectively. The phase C_8 has been also discovered. The formation of other metastable allotropic modifications of carbon (see *Allotropy*) is possible as well. The dimensions of clusters, their composition and correlation with different kinds of bonds *hybridization* are determined by the contents, average energy and density of the particle flux, from which the formation of film occurs, and also by the nature of the substrate and its temperature. Hydrogen content may reach 25%. The density of films lies within the range 1.0 to 3.5 g/cm^3; refraction index is 1.6 to 2.7, optical width of the *band gap* is 0.8 to 2.6 eV. Diamond-like films may be used as durable, masking, passivating, anti-reflection coatings of optical elements for different ranges of wavelengths, as coatings for *optical storage disks*, UV range lasers, and also for the design of fundamentally new active elements for *microelectronics* and *optoelectronics*. Density of states at the boundary of a semiconductor upon application of a diamond-like film can be 10^{10} to $5\cdot10^{11}$ cm^{-2}.

DIAZOTYPE MATERIALS

See *Nonsilver photography*.

DICHROISM OF CRYSTALS (fr. Gr. $\delta\iota\kappa\rho o\nu o\varsigma$, two-colored)

Dependence of *light absorption* on its polarization (see also *Pleochroism*); in the narrower sense the term is sometimes applied to *uniaxial crystals*. The latter, due to the anisotropy of absorption, have two "main" colorations: the "basic" one and the "axial" one for white light propagating, respectively, along and perpendicular to the *optical axis*. *Linear dichroism* (difference of absorption of radiation linearly polarized in mutually perpendicular directions) and *circular dichroism* (difference of absorption of right-and left-hand circularly polarized radiation) are singled out. Circular dichroism of crystals is associated with *spatial dielectric dispersion*.

DIELECTRIC CONSTANT, dielectric function, permittivity

A physical value characterizing the response of a medium (e.g., a *solid*) to the electric component $\boldsymbol{E}$ of an external electromagnetic field. In free space the electric displacement $\boldsymbol{D}$ is given by $\boldsymbol{D} = \varepsilon_0 \boldsymbol{E}$ where ε_0 is the dielectric constant of the free space. In anisotropic media, e.g., in *crystals*, this vector relation becomes a tensor expression $D_\alpha = \sum \varepsilon_{\alpha\beta}\varepsilon_0 E_{\alpha\beta}$ involving the dimensionless relative dielectric constant or *permittivity* tensor $\varepsilon_{\alpha\beta}$, where $\alpha, \beta = x, y, z$. When there is no time dependence (static) and the electric field is uniform in space, the permittivity is constant, and causes the weakening of the internal effective field due to the separation of charges of opposite signs (*polarization of medium*). For electric fields $\boldsymbol{E}(t, \boldsymbol{r})$, varying in time and non-uniform in space, it is necessary to take into account the *temporal dielectric dispersion* and *spatial dielectric dispersion* of the dielectric constant, i.e. the dependence of $\varepsilon_{\alpha\beta}$ on the frequency ω and wave vector $\boldsymbol{k}$. In the general case this tensor $\varepsilon_{\alpha\beta}(\boldsymbol{k}, \omega)$ (as well as its inverse $\varepsilon_{\alpha\beta}^{-1}(\boldsymbol{k}, \omega)$) is complex, and its real and imaginary parts satisfy *Kramers–Kronig relations.* The imaginary part brings about the attenuation of electromagnetic waves. The *relative dielectric function* may be expressed in terms of the complex electrical conductivity tensor $\sigma_{\alpha\beta}$ connecting the current density $\boldsymbol{j}$ and electric field $\boldsymbol{E}$ in the medium:

$$\begin{aligned} \varepsilon_{\alpha\beta}(\boldsymbol{k}, \omega) &= \delta_{\alpha\beta} + \frac{4\pi \mathrm{i}}{\omega}\sigma_{\alpha\beta}(\boldsymbol{k}, \omega), \\ j_\alpha(\boldsymbol{k}, \omega) &= \sigma_{\alpha\beta}(\boldsymbol{k}, \omega)\boldsymbol{E}_\beta(\boldsymbol{k}, \omega), \end{aligned} \tag{1}$$

or through the *polarizability* tensor (*dielectric susceptibility*) of the medium $\kappa_{\alpha\beta}$, connecting the polarization vector $\boldsymbol{P}$ with the field $\boldsymbol{E}$:

$$\begin{aligned} \varepsilon_{\alpha\beta}(\boldsymbol{k}, \omega) &= \delta_{\alpha\beta} + 4\pi\kappa_{\alpha\beta}(\boldsymbol{k}, \omega), \\ \boldsymbol{P}_\alpha(\boldsymbol{k}, \omega) &= \kappa_{\alpha\beta}(\boldsymbol{k}, \omega)\boldsymbol{E}_\beta(\boldsymbol{k}, \omega), \end{aligned} \tag{2}$$

since $\boldsymbol{P}$ and $\boldsymbol{j}$ are related through the expression $\partial \boldsymbol{P}/\partial t = \boldsymbol{j}$. In an isotropic medium, where $\boldsymbol{E}$ and $\boldsymbol{j}$ may be separated into longitudinal ($\nabla \times \boldsymbol{E}_\mathrm{l} = \nabla \times \boldsymbol{j}_\mathrm{l} = 0$) and transverse ($\nabla \cdot \boldsymbol{E}_\mathrm{t} = \nabla \cdot \boldsymbol{j}_\mathrm{t} = 0$) components, the tensor $\varepsilon_{\alpha\beta}(\boldsymbol{k}, \omega)$ assumes the form

$$\varepsilon_{\alpha\beta}(\boldsymbol{k}, \omega) = \varepsilon_\mathrm{l}(\boldsymbol{k}, \omega)\frac{k_\alpha k_\beta}{k^2} + \varepsilon_\mathrm{t}(\boldsymbol{k}, \omega)\left[\delta_{\alpha\beta} - \frac{k_\alpha k_\beta}{k^2}\right], \tag{3}$$

where ε_l and ε_t are the longitudinal and transverse dielectric constants. As a consequence, the Fourier components of the charge and current densities induced within the medium by external charges $\rho_0(\boldsymbol{r}, t)$ and currents $\boldsymbol{j}_0(\boldsymbol{r}, t)$ are given by

$$\begin{aligned} \rho(\boldsymbol{k}, \omega) &= \frac{1}{\varepsilon_\mathrm{l}(\boldsymbol{k}, \omega)}\rho_0(\boldsymbol{k}, \omega), \\ \boldsymbol{j}(\boldsymbol{k}, \omega) &= \frac{\omega^2 - k^2c^2}{\omega^2\varepsilon_\mathrm{t}(\boldsymbol{k}, \omega) - k^2c^2}\boldsymbol{j}_0(\boldsymbol{k}, \omega), \end{aligned} \tag{4}$$

and the zeroes (poles) of the denominators of the expressions (4) $\varepsilon_\mathrm{l}(\boldsymbol{k}, \omega) = 0$ and $\varepsilon_\mathrm{t}(\boldsymbol{k}, \omega) - k^2c^2/\omega^2 = 0$ determine the spectrum of the longitudinal and transverse electromagnetic collective excitations (*normal vibrations*) of the medium (e.g., *plasmons*, *excitons*, optical *phonons*, etc.).

In the high-frequency limit as $\omega \to \infty$ and $\boldsymbol{k} \to 0$ in any medium

$$\varepsilon_\mathrm{l}(0, \omega) = \varepsilon_\mathrm{t}(0, \omega) = 1 - \frac{\omega_\mathrm{p}^2}{\omega^2}, \tag{5}$$

where $\omega_\mathrm{p} = \sqrt{4\pi e^2 n/m}$ is the plasma frequency, m is the electron mass, and n is the concentration of all the electrons of the atoms present. In a *metal* located in a uniform high-frequency electromagnetic field ($\boldsymbol{k} \to 0$) the *conduction electrons* undergo scattering by vibrations (phonons) and point defects with a characteristic time τ, to give for the transverse dielectric constant

$$\varepsilon_\mathrm{t}(0, \omega) = 1 - \frac{\omega_\mathrm{p}^2}{\omega(\omega + \mathrm{i}/\tau)}, \tag{6}$$

so that for $\omega \to 0$ we have $\varepsilon_\mathrm{t}(0, \omega) = 1 + 4\pi \mathrm{i}\sigma_0/\omega$, where $\sigma_0 = e^2 n\tau/m$ is the static electrical conductivity of the metal. At $\omega = 0$ (but $\boldsymbol{k} \neq 0$) the static longitudinal dielectric constant of an isotropic metal, semiconductor or *semimetal* has the form

$$\varepsilon_\mathrm{l}(\boldsymbol{k}, 0) = \varepsilon_\mathrm{i} + \frac{4\pi e^2}{k^2}\Pi(\boldsymbol{k}), \tag{7}$$

where $\Pi(\boldsymbol{k})$ is the static polarizability (*polarization operator*) of the free charge carriers (electrons, holes) and ε_i is the relative dielectric constant of the lattice caused by the polarization of the bound electrons in the shells of atoms, and by the displacement of ions from their equilibrium positions in *ionic crystals* (or by the orientation of molecular *dipole moments* in polar crystals). Eq. (7) describes the effects of screening the electric fields and the Coulomb interaction of the charges (see *Electric charge screening*).

In dielectrics (*insulators*) the difference between the local electric field at the point where an atom (ion) is located and the average macroscopic field leads to the *Clausius–Mossotti equation*, and in crystals with molecules possessing electric dipole moments it leads to the Onsager–Kirkwood relations and to the *Langevin–Debye equation.*

In the CGS system the relative and absolute dielectric constants or permittivities are equal to each other and are all dimensionless. In the SI system the absolute ε_a and relative (dimensionless) ε are related by the expression $\varepsilon_a = \varepsilon\varepsilon_0$ where ε_0 is is the *dielectric constant of vacuum* with the value $8.8542{\cdot}10^{-12}\,\mathrm{F/m}$.

DIELECTRIC DISPERSION

See *Spatial dielectric dispersion.*

DIELECTRIC DISPERSION, TEMPORAL

See *Temporal dielectric dispersion.*

DIELECTRIC FUNCTION

See *Dielectric constant.*

DIELECTRIC LOSS

A part of the electric field power transferred to an *insulator* (dielectric) located in an electric field, and dissipated in the form of heat. The value of the dielectric loss is characterized by the *dielectric loss tangent angle* δ in the expression $\tan\delta = \varepsilon''/\varepsilon'$, where ε' and ε'' are the real and imaginary parts of the *dielectric constant* (or permittivity), respectively. There are various mechanisms of dielectric loss in solid dielectrics which lead to specific frequency and temperature dependences of $\tan\delta$. Losses via the resonance excitation of bound electrons of dielectrics or of oscillating ions occur at a series of resonance frequencies with a weak temperature dependence. In the vicinity of a resonance the dielectric loss exhibits a maximum. In the case of inertial (delayed response) mechanisms of polarization, the value of the dielectric loss is determined by the *relaxation time* τ of the medium. The *Debye theory of dielectric relaxation* applied to relaxation processes acting when an applied electric field is switched on provides the expression

$$\tan\delta = \frac{(\varepsilon_s - \varepsilon_\infty)\omega\tau}{\varepsilon_s + \varepsilon_\infty\omega^2\tau^2},$$

where ε_s and ε_∞ are respectively the static ($\omega \to 0$) and high-frequency dielectric constants, and $\tan\delta$ has a maximum at the electromagnetic field frequency $\omega_M = (1/\tau)(\varepsilon_s/\varepsilon_\infty)^{1/2}$. The small dielectric loss at low frequencies $\omega \ll \omega_M$ is explained by the slow response of the medium, and at high frequencies it is explained by the fact that the inertial response does not have time to follow the changes in the field when $\omega \gg \omega_M$. For relaxation processes involving the reorientation or displacement of ions, the temperature dependences of $\tau(T)$ and $\omega_M(T)$ are significant. Some relaxation mechanisms result in deviations from the Debye theory. The dielectric loss caused by the electrical conductivity, σ, of a dielectric in the absence of other types of loss is determined by the relation $\tan\delta = \sigma/(\varepsilon_0\omega)$ (mks units) and grows in proportion to σ as the temperature increases. For *ferroelectrics*, specific loss associated with the reorientation of *domains* in the electric field, as well as the presence of anomalies in the vicinity of the *phase transition* temperature, can play a significant role.

DIELECTRIC MEASUREMENTS, dielectrometry

Measurements of electric polarization and of technical parameters of *insulators* at frequencies below the optical range. Between 0 to 10^8 Hz capacitor methods of measurement are used: the relative (dimensionless) *dielectric constant* $\varepsilon = Cd/A$ is determined from the geometric dimensions of a parallel plate capacitor with capacitance C, area A and thickness d, filled with the dielectric under study. The *dielectric loss* (loss tangent, or $\tan\delta$) is found from the variations of the phase φ of alternating current passing through the capacitor. Typical equipment includes bridges, Q-meters, impedance measuring devices, phase meters and dielectrometers. Within the range 10^8 to 10^{11} Hz there are waveguide (microwave) methods of determining ε and $\tan\delta$, as well as resonator ones. The waveguide methods are based on measuring the standing wave ratio, the phase of the reflected signal, or that of the signal passing through the investigated dielectric sample. In resonator methods the change in the frequency and the Q (loss or quality) factor of a resonator are measured when the sample of dielectric is introduced. Measuring devices include generators and lines, frequency meters, reflectometers, phase meters, etc. At the frequencies 10^{11} to 10^{15} Hz quasi-optical (beam) methods are employed, which permit the determination of the wavelength of electromagnetic radiation in the dielectric and its attenuation there: *Fabry–Perot interferometers*, goniometers (see *Goniometry*), refractometers (see *Refractometry*). Fourier spectrometers (see *Fourier spectroscopy*) can analyze the response of dielectrics to a pulsed electric field.

DIELECTRIC RELAXATION

See *Maxwell relaxation.*

DIELECTRICS

See *Insulators.*

DIELECTRIC VISCOSITY

Phenomenon of a time lag in the variation of electric parameters (*polarization of insulator, dielectric constant,* etc.) in ferroelectric materials depending on the intensity of the external electric field. Viscous phenomena in *ferroelectrics* are similar to the those of *magnetic viscosity* and superviscosity in *ferromagnets.* As a parameter of the switching properties of ferroelectric crystals during the rearrangement of the *magnetic domain structure* the *dielectric viscosity coefficient* β is introduced, and it determines the relationships between the value of the external electric field and the velocity of processes involving the variation of dielectric parameters under the effect of this field. The equation describing the transition of ferroelectricity in the equilibrium state with polarization P_0 under the effect of the constant electric field with intensity E reflects the role of the dielectric viscosity during the processes of polarization:

$$P(t) = P_0\left[1 - \exp\left(-\frac{Et}{\beta}\right)\right],$$

where t is time and $P(0) = 0$. The value of β depends on the temperature of the ferroelectric, and in the region of high fields it begins to depend on E.

DIELECTRONICS, dielectric electronics

A branch of physics and engineering which deals with transformation of electric, acoustical, mechanical and electromagnetic waves by the utilization of particular attributes of *insulators* (dielectrics) such as piezoelectric, pyroelectric, electrooptic, and other properties. The response of a dielectric to an external action or perturbation can entail either a transformation of the perturbing energy into another kind of energy, such as electrical or optical energy (or in the reverse direction), and so on, or it can involve the nonlinear optical, electric, magnetic, or acoustical properties of dielectrics. Quite a number of discrete and integrated devices of *microelectronics, acoustoelectronics,* and *optoelectronics* are widely used in data processing systems, and in information transfer systems. Dielectronic-based devices are employed in the energetics for, e.g., the transformation of nuclear radiation energy or solar energy into electricity, etc.

In the narrow sense, the word "dielectronics" is the name of a branch of solid state physics that deals with "*semi-insulators*". These materials have properties intermediate between insulators and highly conductive semiconductors, with their specific electrical resistivity in the range 10^3–10^{10} Ω·cm. Classified among them are strongly compensated *semiconductors,* materials which have a relatively wide *band gap.* The nonlinear electrical properties of semi-insulators are controlled through the *injection of current carriers* by means of contacts; their *current–voltage characteristics* are sensitive to the presence of intrinsic and extrinsic defects in the sample.

DIFFRACTION

See also *Acoustic diffraction, Acousto-optic diffraction, Electron diffraction, Laue diffraction, X-ray diffraction.*

DIFFRACTION GRATING

Periodic structure of many regularly arranged elements at which the *diffraction of light* takes place. Diffraction gratings may be flat or concave, reflecting or transmitting. The grating decomposes the incident light beam into its component wavelengths according to the equation $d(\sin\alpha + \sin\beta) = m\lambda$, where d is the grating period, α and β are the angles of incidence and diffraction; λ is the wavelength, and $m = 0, \pm1, \pm2\ldots$ is the order of diffraction. The basic parameters of a diffraction grating are: *angular dispersion* $\Delta\beta/\Delta\lambda = m/(d\cos\beta)$; *resolution* $R = \lambda/\Delta\lambda = mN$ (N is number of lines of the diffraction grating); *free spectral region* $F_\lambda = \lambda/m$. The *diffraction efficiency* is the fraction of the intensity of a monochromatic light beam diffracting in the required order. To increase the number of lines of a diffraction grating a step like structure of grooves can be used (*echelette*). *Holographic gratings*, which record in photosensitive media upon the interference of laser beams, may be thin (thickness $l \ll d$) or bulk ($l > d$), and are of high quality. The diffraction efficiency may be close to 100%. Gratings are used in spectral devices, *dispersion laser resonators*, as sensors of linear and angular motions, as beam dividers, etc.

DIFFRACTION METHODS OF INVESTIGATION

Methods of investigating materials based on the scattering (diffraction) of X-rays, electrons, neutrons and other microparticles by the atoms of the object under investigation. They are subdivided into indirect and direct ones. With the help of indirect methods based on the measurement of the parameters of the scattered radiation intensity, the atomic structure of *crystals*, the character and density of *defects* in crystals are investigated, a *phase analysis* is performed; the structure of amorphous and liquid substances is studied. The indirect diffraction investigations are carried out using special equipment – *diffractometers* (X-ray and neutron types), *electronographs*, etc. For recording the intensity of the scattered radiation the following equipment and materials are used: photographic films and plates, and also ionization and scintillation counters (detectors) with the intensity recorded on ruled paper, or with the output of the counter signals to printing devices. *X-ray structure analysis*, *electron diffraction analysis*, *neutron diffractometry* and methods of analysis using the *orientation effects* on the scattering of charged particles are related to indirect methods. The first three methods have a sufficiently common basis. During elastic scattering of incident radiation by periodically spaced atoms of the *crystal lattice* there appear sharp maxima of the scattered radiation and the diffuse background, caused by inelastic (Compton) scattering. By the analysis of the positions and intensities of the diffraction maxima the type of crystal lattice, *unit cell* size and atomic coordinates may be determined. Crystal lattice defects change the positions, intensities and shapes of the profiles of the diffraction maxima; and can add extra *diffuse scattering* which helps to provide information about the defects in the crystal. Indirect methods are also used for investigating the structure of amorphous and liquid substances. With the help of electron diffractometry it is convenient to study the thin surface layers of a material, and with the help of neutron diffraction the compositions of elements with close atomic numbers, and the ordering of alloys. A method using the *orientation effects* of charged particle scattering has found wide application. For the analysis by the *method of orientation effects* particles with energy from several electronvolts to kilo- and megaelectronvolts are used. *Proton microscopes* are employed for investigation of various structural defects. The method is especially effective for the study of the localization of impurities within a unit cell of the lattice, as well at a small distance below the crystal surface. Different modifications of this method are based on the *channeling* of particles, on the effect of shadows, and on the double orientation.

The direct methods are used for the investigations of defects of crystalline material (e.g., for dislocation structure studies). Some examples of such methods: *transmission electron microscopy* and *X-ray topography*. Upon using these methods the local variations in the intensity of the diffraction maxima, caused by the defects of the crystalline structure, are analyzed. The direct diffraction methods are also used for investigations of *magnetic domain structures* in *ferromagnets* and *ferroelectrics*.

DIFFRACTION OF LIGHT

Deviation of the light waves in media with optical non-uniformities of the order of a wavelength and more (e.g., at the boundaries of bodies, slits, etc.); a particular case of the diffraction of electromagnetic waves (see *Diffraction of waves*). The approximate theory of diffraction of light was formulated by A. J. Fresnel (1816). Diffraction manifests itself in the formation of a diffraction picture (alternation of intensity maxima and minima). The *Fresnel band* $(z\lambda)^{1/2}$ (λ is the wave length, z the distance from the object of size a from which diffraction takes place to the observation point) is an important notion in the theory of diffraction. For $a \sim (z\lambda)^{1/2}$ diffraction occurs for converging beams (*Fresnel diffraction*, A.J. Fresnel), and for $a \ll (z\lambda)^{1/2}$ it occurs for approximately parallel beams (*Fraunhofer diffraction*, J. von Fraunhofer). Diffraction determines the resolution of optical devices. The properties of diffraction from volume *diffraction gratings* form the basis of *holography*. In solid-state physics there is an important role of the diffraction of light from dynamic optical non-uniformities, including those caused by ultrasound (*acousto-optical diffraction*) by the self-dispersal of radiation within the medium (*self-effect of light waves* in non-linear optics, *self-diffraction of light* in *real-time holography*), and by the spontaneous and stimulated inherent elastic vibrations of solids and liquids (see *Brillouin scattering*). In lasers diffraction determines the minimum (diffraction) divergence of beams of the stimulated radiation $\varphi \propto \lambda/a$, and many properties of optical *resonators*. Operation of a diffraction grating, of a zone plate, of acoustooptical modulators and deflectors (see *Light deflectors*) is based on the diffraction of light.

DIFFRACTION OF RADIATIONS in ultrasonic field

This is usually investigated in perfect *monocrystals* with application to X-rays and neutrons. The effect of ultrasound depends on the correlation between the ultrasound wavelength λ and the *extinction length* Λ of radiation in crystal. For $\lambda \gg \Lambda$ ultrasound increases the intensity of the diffracted beams due to the widening of the angular interval of the rays, satisfying the *Bragg law*, and it also suppresses the *anomalous passage of X-rays*. The suppression is sharply enhanced under conditions of *X-ray acoustic resonance* at $\lambda \approx \Lambda$. If $\lambda < \Lambda$ ultrasound causes transitions between the branches of the dispersion surfaces which determine the dynamics of radiation in the crystal, the *pendular oscillations of intensity* of the diffracted beams due to the scattering at the ultrasonic superlattice. In the case of neutron diffraction it is necessary to take into account not only the exchange of momentum, but also of energy (about 10^{-7} eV) of the ultrasonic wave. The high sensitivity of scattering of ultrasound is used for the detection of the structure of vibrational modes in piezoacoustic converters (see *Acousto-electronics*).

DIFFRACTION OF WAVES

The deviation from rectilinear propagation of waves in a medium containing non-uniformities or obstacles. The mathematical description of the wave fields (sound, X-rays, neutrons, electrons) follows from the Huygens principle (see *Acoustic diffraction*), and in this sense it is similar to the description of the *diffraction of light*. Wave diffraction is usually discussed for the case of an appreciable deviation from the laws of geometrical optics. For example, upon the scattering of a plane wave of monochromatic light from the object (or from a non-uniformity of a medium) with the characteristic dimension a the angular deviation of the beams is $\theta \sim \lambda/a$ (λ is the wavelength). Therefore, at a screen located at the distance l from the object there appear linear displacements of beams in the transverse direction of the order $x \sim l\theta \sim \lambda l/a$. For the formation of a sharp geometrical image of the object the condition $x \ll a$ or $l \ll a^2/\lambda$ should be satisfied, which is the criterion for the applicability of geometrical optics. For $l \sim a^2/\lambda$ (*Fresnel diffraction*) the linear displacements of the diffracted beams at the screen are of the order of a. At the screen different beams, deviated from rectilinear propagation, may converge and interfere, so that a complicated distribution of diffraction radiation intensity is observed at the screen, that is not unambiguously associated with the optical properties of the scattering object (e.g., at the image center of a nontransparent object a bright spot (*Arago spot*) may be observed; the *Arago–Poisson–Fresnel effect*, 1811). For $l \gg a^2/\lambda$ (*Fraunhofer diffraction*) at every point of the screen the beams, deviated approximately in the same direction, converge and interfere; and hence (at least for small deviation angles) the intensity distribution at the screen reproduces a two-dimensional Fourier-image (of obstacles or non-uniformities), because the coordinates

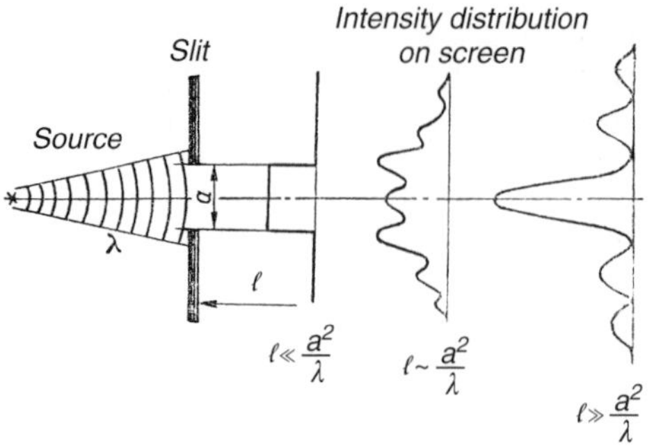

Source, slit, and distribution of intensity at the screen. The intensity distribution is shown for geometric (left), Fresnel (center) and Fraunhofer (right) diffraction.

of the points are linearly connected with the transverse components of the diffracted radiation wave vector. The figure shows, from left to right, these three cases of wave propagation or diffraction for the example of light passing through a slit of width a, namely: geometric ($l \ll a^2/\lambda$), Fresnel ($l \sim a^2/\lambda$) and Fraunhofer ($l \gg a^2/\lambda$) (see also *Diffraction of light*).

In solid state physics wave diffraction plays a double role. In microscopy it is desirable to obtain a sufficiently sharp image of the object, so wave diffraction is harmful, and restricts the microscope resolution to a value of the order of λ. Fresnel diffraction is seldom used due to the difficulties of its interpretation, its main use is for the observation of *crystal lattice* defects by the method of X-ray contrast, when it is impossible to obtain the geometric image. Fraunhofer diffraction is used to determine the locations of atoms in solids by *X-ray structure analysis*, and also to investigate lattice *defects*. In the latter case the diffracted X-radiation consists of a coherent (Bragg) component, which reflects the properties of the "average" lattice of a defective crystal (e.g., relative variation of the lattice parameter $\Delta a/a$, thermal and static *Debye–Waller factors*), and an incoherent (diffuse) component, whose distribution provides information about concentration, type, size, and internal structure of defects. In the case of amorphous solids the scattering intensity comprises only the incoherent part, from which it is possible to determine the radial distribution function, total and partial structure factors (see *Structure amplitude*), coordination numbers, etc. In diffraction from crystals, kinematic and dynamic modes may be realized. In the kinematic case the individual scattering of a wave by the atoms and the subsequent interference of the secondary waves are taken into account. The kinematic approximation is valid for ideally-mosaic (see *Mosaic crystals*) or sufficiently strongly distorted crystals, when the condition $l_0 \ll \Lambda$ is met (l_0 is the characteristic size of mosaic blocks, or the *coherent scattering regions*, Λ is the *extinction length*). In the opposite case ($l_0 \geqslant \Lambda$) *dynamic diffraction* takes place, i.e. through multiple coherent scattering the plane waves entering the crystal are transformed (at the distance $\sim\Lambda$) into Bloch waves (see *Bloch theorem*), whose diffraction in the crystalline medium is accompanied by a series of new qualitative effects (see *Elastic scattering of radiation*, *Dynamic radiation scattering*, *Extinction*).

DIFFRACTION POLARIZATION of X-rays

Variation of polarization in the course of *X-ray diffraction* within the crystal, associated with the difference of the diffraction properties of the crystal for different polarizations of the incident beam. The character of diffraction polarization depends on the conditions of diffraction and on the structural perfection of the *crystal*. For perfect crystals this polarization is described by the theory of *dynamic radiation scattering*. Upon two- or many-wave diffraction (see *Two-wave approximation*) in such crystals the effects of *birefringence* and dichroism (see *Dichroism of crystals*) are weakened in proportion to the crystal deviation from perfection. For three- or many-wave diffraction the effect of *rotation of light polarization plane* is observed. In distorted crystals the diffraction leads to depolarization of X-ray beams. Natural (non-polarized) radiation is polarized upon diffraction. The degree of its polarization and the character of its variation within the limits of the Bragg scattering curve depend on the defectiveness of the crystal (see *Defects in crystals*). On the basis of diffraction polarization effects the methods of X-ray *polarimetry* have been developed.

DIFFRACTION SPECTROMETERS, CRYSTAL

See *Crystal diffraction spectrometers*.

DIFFRACTOMETRY, GAMMA RAY

See *Gamma diffractometry*.

DIFFRACTOMETRY, NEUTRON

See *Magnetic neutron diffractometry, Neutron diffractometry.*

DIFFRACTOMETRY, X-RAY

See *X-ray diffractometry, Energy-dispersion X-ray diffractometry.*

DIFFUSE PHASE TRANSITION

A *ferroelectric phase transition* taking place within a relatively wide temperature range, called the *Curie range*, instead of at a certain specific temperature. In contrast to common *ferroelectrics*, temperatures at which the anomalies in various properties of the material occur are distributed within that range. Piezoelectric properties are observed at temperatures higher than the temperature T_m of the maximum *dielectric constant* ε. Meanwhile, the specific heat, the optical absorption edge, the index of refraction, and the electrooptic properties all change slowly instead of sharply when passing through the Curie range.

In one of the most actively studied substances with a diffuse phase transition, $PbMg_{1/3}Nb_{2/3}O$ (PMN), the Curie range spreads out to cover more than 100 °C. The dielectric constant ε of PMN passes through a maximum at $T = 0\,°C$, and shifts to larger temperatures for higher measurement frequencies (see Fig.). Such a relaxation behavior of ε is typical for all materials with the diffused transition, therefore they are sometimes called *relaxor ferroelectrics* or *ferroelectrics with a diffuse phase transition*. This transition is commonly observed in oxide compounds with a *perovskite* structure in which ions of different types randomly occupy identical points in the lattice, as, e.g., the ions of Mg and Nb in PMN, or the ions of Sc and Ta in PST ($PbSc_{1/2}Ta_{1/2}O_3$).

In PMN type materials (1 : 2 group) the extent of the diffusion of the phase transition does not depend on their preparation technique, while materials of the PST type may only be obtained under certain technical conditions, and attain an arbitrary degree of diffusion even down to the normal ferroelectric transition. The diffuse phase transition in *solid solutions* (e.g., such as $BaTi_{1-x}Sn_xO_3$, $Pb_{1-y}La_yZr_{1-x}Ti_xO_3$) is only observed for certain values of x and y.

The sources of random fields that result in a diffused phase transition are fluctuations of the

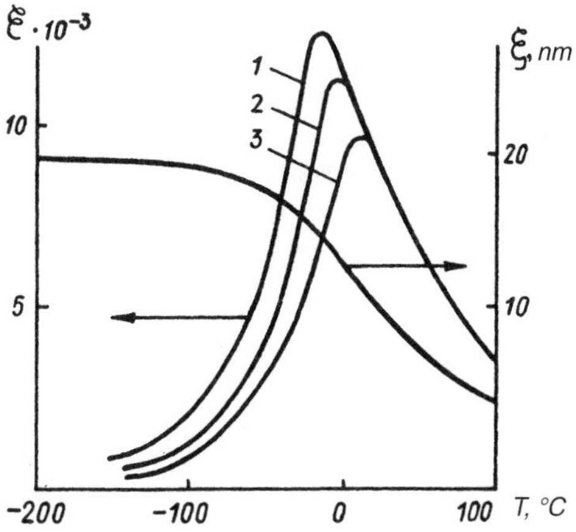

Temperature dependence of the correlation radius ξ and of the dielectric constant ε of $PbMg_{1/3}Nb_{2/3}O_3$ at various frequencies: (1) 0.4 kHz; (2) 45 kHz; and (3) 4500 kHz.

material composition. It was suggested initially that such fluctuations lead to a distribution of local temperatures of *phase transitions* in crystal microvolumes, with long-range ferroelectric order forming individually in each such microvolume. It was experimentally demonstrated later that these clusters only have short-range ferroelectric ordering. This picture corresponds to a *dipole glass* that is an electric analogue of a spin glass. At low temperatures, $T < T_g$ (the *freezing temperature* T_g usually satisfies the condition $T_g < T_m$), the system appears to be in a nonequilibrium frozen state in which the electric dipole moments of its polar clusters are randomly oriented. In all these materials a distribution of *relaxation times* typical of glasses is observed, the temperature dependence of the maximum time being described by the *Vogel–Fulcher law* (see *Spin glasses*).

The external electric field orients polar clusters and induces the ferroelectric state. As a result, cooling the material to $T < T_g$ in an external electric field (field cooling, FC) or in the absence of any field (zero field cooling, ZFC) produces either normal or relaxor *ferroelectricity*, respectively. The polarized specimen features a residual piezoelectric effect (see *Piezoelectricity*) and birefringence, both of which vanish if the specimen is heated in the absence of an electric field (zero field heating, ZFH) at temperatures $T < T_m$. The *correlation length* ξ grows at lower temperatures up to $T \approx T_g$. This radius remains constant below that temperature ($\xi_{max} = 20$ nm in PMN, see Fig.), in contrast to the situation in common ferroelectrics.

High values of certain parameters found in materials featuring a diffuse phase transition make it possible to use them in capacitors for electrostriction (see *Electrostriction*), and for other *electrooptic materials.*

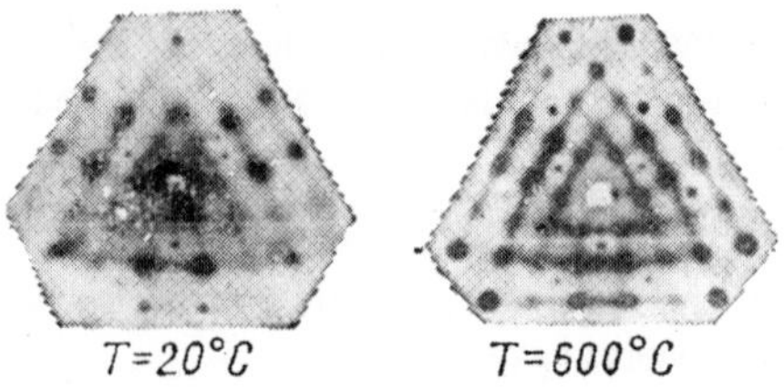

Diffuse scattering of X-rays (K_α, Mo) arising from thermal atomic displacements in the monocrystal $LiNbO_3$ at 20 °C (left) and at 600 °C (right).

DIFFUSE SCATTERING of X-rays

Scattering of X-rays in solids from atoms displaced from regular *crystal lattice* sites. In an *ideal crystal* the displacements are caused by thermal vibrations. The intensity and distribution of the diffuse scattering depend on the type of structure, on the forces of the interatomic interactions, and on the temperature (see Fig.). In real crystals the displacements of atoms are caused by the their vibrations and by *defects*, and the diffuse X-ray scattering is determined by the nature of these defects, by their concentration and location, and their shape and dimensions. Diffuse scattering is used for investigating interatomic interactions, the phonon spectrum, the elastic characteristics of crystals, its defects, regions with an electron density that differs from the average value, and features formed through various effects and processes (irradiation, thermal and mechanical treatment, *alloy decomposition*, *alloy aging*, etc.). The utilization of diffuse scattering is one of the most effective methods for a detailed analysis of variations in the configurations of atoms, and of the forces of interatomic interaction, because this scattering is sensitive not only to the entire spectrum of frequencies of vibrating atoms, but also to other factors influencing the spectrum. Investigation of diffuse scattering facilitates the detection of many types of defects and determines their parameters, including plots of how local electron density differs from that of an ideal perfect crystal, and of the type of *Guinier–Preston zones* formed by precipitation during the decomposition of solid solutions and the aging of alloys. By studying the temperature dependence of the intensity of the diffuse scattering it is possible to determine the nature of the displacements of atoms in crystals with defects (of the dynamic or static type).

The diffuse scattering method provides more complete information when monochromatic radiation is used to investigate *monocrystals*, although *polycrystals* are also widely used, especially for deducing the degree and type of short-range order (see *Long-range and short-range order*) associated with the locations of atoms. From the analysis of how the intensity of the diffuse scattering by monocrystals depends on the distances to the sites of $\boldsymbol{k}$*-space* it is possible to determine the characteristics of defects (*point defects*, aggregations, *dislocation loops*) and their dimensions. The investigation of diffuse scattering over broad regions of $\boldsymbol{k}$-space allows the determination of the form of defects, dimensions, how they disturb the surrounding crystal lattice, and also the location of the defects in the volume of the crystal. To interpret the effects of diffuse scattering it is expedient to investigate also the vicinity of the zero node by the method of the *small-angle scattering of X-rays.*

The uniqueness of the interpretation of the effects of diffuse scattering depends on the extent of the region of $\boldsymbol{k}$-space that is studied and analyzed, as well on the degree of monochromaticity of the radiation being used. For the investigation of defects it is necessary to take into account, and if possible, to exclude thermal effects by working at low temperatures, something that is especially important for small concentrations of defects.

DIFFUSION

Phenomenon of disordered (often thermal) motion of the particles of a medium (atoms, molecules, ions, electrons, and in a condensed medium *quasi-particles*), which leads to the mixing of particles, to the spatial homogenization of concentrations, or to the appearance of directed particle fluxes. The following types of diffusion are singled out: *self-diffusion* (in a medium of similar diffusing particles, e.g., *tracer* atoms), *heterodiffusion*, and *interdiffusion*. The rate of diffusion is provided by the *diffusion coefficient* D, which linearly relates the flux density j of diffusing particles to the gradient of their concentration through the *first Fick law*

$$j = -D \operatorname{grad} n, \qquad (1)$$

where n is the number of particles per unit volume.

In gases diffusion occurs through random collisions of diffusing particles with the gas molecules, and D is determined by the expression

$$D = \frac{1}{3}\lambda v = \frac{1}{3}\frac{\lambda^2}{\tau_c},$$

where λ is the *mean free path*, v is the average speed of the thermal motion, τ_c is the average time between collisions; and D is inversely proportional to the gas pressure with a weak temperature dependence ($\propto T^{1/2}$).

In condensed systems (liquids, solids) diffusion occurs via random jumps of atoms between states corresponding to absolute or relative minima of their potential energy (e.g., sites and interstices of the *crystal lattice*). The corresponding coefficient of diffusion is

$$D \approx \frac{a^2}{\tau},$$

where a is the characteristic length of an elementary jump of a particle (of the order of the average distance between the molecules), τ is the average time between jumps which at high temperatures depends exponentially on the temperature: $\tau = \tau_0 \exp[Q/(k_B T)]$, $\tau_0 \approx 10^{-12}$ to 10^{-13} s is the characteristic period of the particle vibrations, and Q is the *activation energy*. Thus, D depends exponentially on the temperature: $D = D_0 \exp[-Q/(k_B T)]$; $D_0 \approx a^2/\tau_0$, which is consistent with the fluctuation nature of the particle jumps. At low temperatures the jump between adjacent energy minima may be performed by quantum tunneling; and the temperature dependence of D is different (see *Quantum diffusion*). In crystals diffusion is a second-rank tensor D_{ik}, and in cubic crystals this tensor D_{ik} reduces to a scalar.

The kinetics of diffusion is determined by the *second Fick law*

$$\frac{\partial n}{\partial t} = -\operatorname{div} j.$$

The diffusion flux also appears under the influence of any external or internal fields which determine the gradient of the chemical potential μ of the particles – the "motive force" of diffusion:

$$j = nv = -n\frac{D}{k_B T}\nabla\mu. \qquad (2)$$

It can be seen that the drift velocity v of the particles is proportional to the motive force ($-\nabla\mu$) and to the mobility b which is related to D through the *Einstein relation* $b = D/(k_B T)$. In gases or dilute solutions $\nabla\mu = k_B T \nabla n/n$, and Eq. (1) follows from (2) as a particular case. In the general case the motive force of diffusion is determined by the concentration gradients as well as by the types of interactions of the diffusing particles with the fields. For example, for diffusing particles with charge q in an electric field E the diffusion flux is

$$j = -D\nabla n + \frac{D}{k_B T} nqE. \qquad (3)$$

The second term in Eq. (3) determines the flux of particles under the influence of the electric field. In many cases (e.g., gases, liquids, solid electrolytes) the directed diffusion drift of charged particles in the electric field determines the *electrical conductivity* of the medium, and this conductivity σ is proportional to D through the *Nernst–Einstein relation*:

$$\sigma = \frac{Dnq^2}{k_B T}.$$

This expression permits the determination of D from the known value of σ (and vice versa). If the temperature varies throughout the medium then $\nabla(\mu/T)$ serves as the motive force of particle diffusion. The current density j is the sum of the usual flux and the *thermodiffusion* flux (*Soret effect*)

$$j = -D\nabla n - \frac{DQ^* n}{k_B T^2}\nabla T,$$

where the dimensionless quantity $k_T = Q^*/(k_B T)$ is called the *thermodiffusion relation*, and Q^* is the heat transferred by the particles per unit act of diffusion. In the course of the diffusion of atoms along grain boundaries (*boundary diffusion*), along the surface (*surface diffusion*), along dislocations (*tubular diffusion*), a speeding up of the diffusion rate compared with that of bulk diffusion is observed. The diffusion process may take place through the formation or decomposition of phases (*reactive diffusion*).

Diffusion in alloys. The specifics of diffusion in *alloys* is associated with the presence of several diffusing substances. The density of the diffusion flux of αth substance in the one-dimensional case is written in a form that generalizes Eq. (1),

$$j_\alpha = -\sum_\beta L_{\alpha\beta} \frac{\partial \mu_\beta}{\partial x}, \tag{4}$$

where $L_{\alpha\beta}$ is the corresponding kinetic coefficient, and μ_α is the chemical potential of the αth substance. In the majority of cases it is assumed that $L_{\alpha\beta} = 0$, $\alpha \neq \beta$; then $L_{\alpha\alpha}$ determines the partial diffusion coefficient D_α. The differences between the partial diffusion coefficients in binary *substitutional alloys* leads to a series of new physical effects. If, for example, $D_A > D_B$ in a sample enriched with atoms A then *vacancies* will be formed, which may either combine in pairs (*Frenkel effect*) or disappear at *dislocations*, which leads to the motion of atomic planes and markers associated with them (*Kirkendall effect*).

The Frenkel and Kirkendall effects exclude the exchange, ring, and interstitial mechanisms of diffusion and point to the vacancy mechanism. The concentration c_A under the conditions of interdiffusion in A–B alloys for the one-dimensional case is determined from the expression

$$\frac{\partial c_A}{\partial t} = \frac{\partial}{\partial x}\left[(D_A c_B + D_B c_A)\frac{\partial c_A}{\partial x}\right],$$

where D_A and D_B are the partial diffusion coefficients of the components. Thus, the coefficient of interdiffusion determined in the experiments is given by $D_{AB} = D_A c_B + D_B c_A$.

See also *Ambipolar diffusion, Anomalous diffusion, Ascending diffusion, Boundary diffusion, Tubular diffusion, Radiation-induced diffusion, Reactive diffusion, Spin diffusion* and *Nuclear spin diffusion, Surface diffusion.*

DIFFUSION ANNEALING

See *Homogenization.*

DIFFUSION COEFFICIENTS

Coefficients of proportionality between the value of the flux and the concentration gradient of particles; the main parameter of the kinetic process of the diffusional dispersal of particles within a medium. Depending on the *diffusion* conditions this process is characterized by various diffusion coefficients. For self-diffusion and for *heterodiffusion* in an isotropic medium the diffusion spreading is described by the *isotopic diffusion coefficient* (*self-diffusion coefficient*) D^*, which coincides with the diffusion coefficient describing a *random walk* of the particles (*Brownian motion*). For the *interdiffusion* of two components A and B the diffusion of each individual particle is characterized by the *partial diffusion coefficients* D_i ($i = A, B$), which differ from the isotopic one by a factor that takes into account the non-ideality of the solutions: $D_i = D_i^*(1 + \partial \ln \gamma_i / \partial \ln c_i)$, where the γ_i are *activity coefficients*, and c_i are the concentrations of the components. The process of equalizing the concentration by interdiffusion is determined by the *chemical diffusion coefficient*, or by the *interdiffusion coefficient* $\widetilde{D}$ which is related to the partial diffusion coefficients by the relation $\widetilde{D} = D_A c_B + D_B c_A$. If a system with two (or more) components has internal fields proportional to the gradients of the component concentrations then the *intrinsic diffusion coefficients* D' are expected to have the form $D' = D_A^* D_B^* / (c_A D_A^* + c_B D_B^*)$. They are sometimes called *internal* or *ambipolar diffusion coefficients.*

In the gas-kinetic approximation, in order of magnitude we have $D \sim l^2/\tau$, where l is the average displacement of the diffusing particle between two successive collisions, and τ is the average time between collisions. For particles diffusing in a condensed medium $D = \eta a^2 \nu$, where a is the average length of a jump (of the order of an interatomic distance), ν is the frequency of jumps, and η is a numerical coefficient of the order of unity which depends on the structure of the material.

DIFFUSION-FREE TRANSFORMATION

A phase transformation that takes place in solids without diffusion at low temperatures ($< 0.4T_{\text{melting}}$) through *shears* of atom positions or atomic planes (see *Martensitic transformation*).

DIFFUSION LENGTH

A mean distance L_D reached by a diffusing particle or *quasi-particle* away from its initial position within its lifetime τ; here $L_D = (D\tau)^{1/2}$, and D is the *diffusion coefficient.* Owing to the random nature of the *diffusion* motion and its wandering (zig-zagging) path, the diffusion length is much shorter than the overall trajectory length of a diffusing particle. Charged particles, e.g., current carriers in *semiconductors* in an electric field E, drift and diffuse at the same time. The distance traversed by these drifting particles during a time τ is specified by a *delay length* which can be longer or shorter than L_D, and in a strong electric field the drift can dominate. For this case the diffusion is mainly transverse, and within a time τ the current carriers drift over the distance $L_E = \mu E\tau$ called the *drift length* (here μ is the *ambipolar mobility*) (see *Ambipolar diffusion*).

DIFFUSION MECHANISM OF CRYSTAL CREEP, diffusion creep

Microscopic (atomic) mechanism of irreversible variation of shape of a crystal subject to constant external loading, based on the diffusion transfer of material in the solid. It is carried out at sufficiently high temperatures and low loadings. The creep of monocrystalline samples is associated with the presence of *point defects*, i.e. *vacancies* and *interstitial atoms* capable of diffusional motion. The diffusion flux of point defects usually starts at separate surfaces or lines (sources) and finishes at others (drains). The external surfaces of the sample, *grain boundaries* in *polycrystals*, *edge dislocations* and other extended effects may serve as sources and drains of point defects. The deposition of a microscopic number of vacancies (interstitial atoms) in the plane of a drain decreases (increases) the number of regular sites of the *crystal lattice*, which leads to shortening (elongation) of the sample in the direction perpendicular to this plane. In the plane which is the source of point defects the inverse process takes place. The action of such processes occurring under the effect of the external forces acting in all the sources and drains induces changes of the sample shape, while conserving the continuity (uniformity) of the material. A non-uniform distribution of external forces at the surface of a sample usually causes the appearance of diffusion flux. A sketch of single-axis loading which generates a directed diffusion flux of vacancies and interstitial atoms, leading to elongation of the sample, is shown in Fig. The velocity of relative elongation appears to be proportional to pDL^{-2}, where D is the *diffusion coefficient* of the point defects, and L is the size of the sample. The diffusion mechanism of crystal creep under such conditions had been described by F.P.N. Nabarro (1948) and C. Herring (1950). The *crystallite* grain size plays the role of L in a *polycrystal.* This diffusion mechanism in polycrystals was investigated by I.M. Lifshits (1963). The presence of prismatic *dislocation loops* within the crystal provides sources and drains, and diffusion may transfer material from one group of *dislocations* to another. Shear loading brings about the diffusion transfer of "favorably" oriented dislocation loops together with the reduction in size of "unfavorably" oriented loops. In this *dislocation-diffusion creep mechanism* the product of the distance between the centers of prismatic loop dislocation generation and the average radius of the emerging loop play the role of the parameter L^2. For this mechanism the characteristic temperature dependence of the plastic creep velocity results from the exponential dependence of the diffusion coefficient: $D \propto \exp[-Q/(k_BT)]$, where Q is the *activation energy* of point defect diffusion.

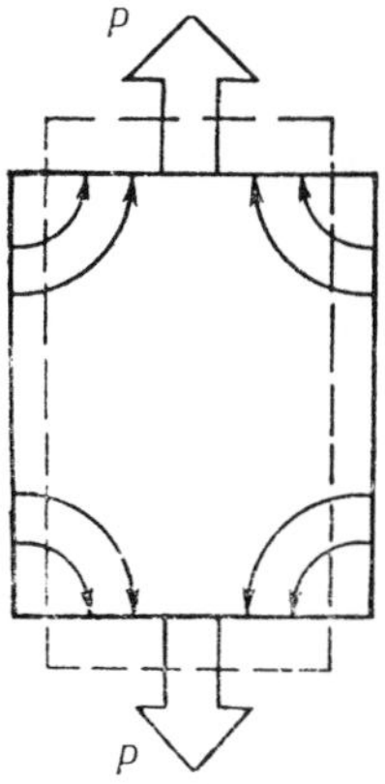

Sketch of the diffusion flux of material (shown by the arrows) arising from single-axis loading.

DIFFUSION POROSITY

The aggregate of *pores* within a crystal, formed due to the precipitation of excessive *vacancies* and their combining by the process of diffusion (see *Diffusion*). In crystals with pronounced *surface tension* anisotropy the pores are faceted ("crystals" of cavities or pores). When the reserve of excess vacancies is restricted (e.g., crystal hardened by heating) and is determined by the value of the initial over-saturation $\Delta = c_{V_1} - c_{V_0}$ (c_{V_1}, c_{V_0} are initial and equilibrium vacancy concentrations) then pore formation will reach completion during the time $\tau = \overline{R}^2/(\Delta D_V)$, where the average pore size $\overline{R}$ is determined by the condition $4\pi\overline{R}^3/3 = \Delta/n_0$, n_0 is the number of pores per unit volume, and D_V is the self-diffusion coefficient of the vacancies. If the oversaturation by vacancies Δ within the lattice persists as a long-lasting source (evaporation of volatile component, *interdiffusion*, radiation effect), the kinetics of growth will depend on the link which limits the process). Microcavities and microcracks of noncontinuity at the boundaries may form nuclei for the formation of diffusion pores. Diffusion porosity makes its appearance during the thermal treatment of alloys, the preparation of coatings, the irradiation of crystals, etc.

DIFFUSION WELDING

Method of pressure *welding* carried out by heating below the *melting* temperature of the welded metals (without using solders) with the application of *pressure* sufficient for creating the necessary *plastic deformation* of the welded parts. The connected parts subject to compressing pressure approach each other at distances determined by the interatomic forces. A welded joint is formed via *interdiffusion* through the surface of the interface in the solid or liquid state (welding-soldering by melted intermediate layer). The formation of a satisfactory welded joint requires the proper choice of three main parameters – temperature, pressure, and time lag at the given temperature and pressure. To connect similar metals the temperature as a rule is $(0.5$ to $0.7)T_{\text{melting}}$, for welding non-uniform metals this temperature is equal to the lowest T_{melting}. The time lag usually does not exceed 1 to 50 min. The ability to form compounds of metallic materials with *ceramics*, *graphite*, powder materials, and optical fibers shows the advantages of diffusion welding. The formation of intermetallic and carbide phases in the region of the joint is a deficiency of diffusion welding. To avoid the embrittling phases (see *Embrittlement*) other types of *pressure welding* can be used such as *magnetic-discharge welding*, *cold welding* or *percussion welding in vacuo*.

DIFFUSION ZONE

The region of a solid where there is a non-uniform concentration distribution arising from the *interdiffusion* of two or more components of the system. This non-uniformity sustains the diffusional movement of atoms, leading to the establishment of a phase composition determined by the compositional phase diagram. The time dependence of the diffusing component concentration $c(t)$ in the diffusion zone for a the two-component system is determined by the two *Fick equations* (A. Fick, 1855):

$$j = -\frac{\widetilde{D}}{\omega}\frac{\partial c}{\partial x},$$

$$\frac{\partial c}{\partial x} = \frac{\partial \widetilde{D}}{\partial x}\frac{\partial c}{\partial x} + \widetilde{D}\frac{\partial^2 c}{\partial x^2},$$

where j is flux, $\widetilde{D}$ is the coefficient of interdiffusion (see *Diffusion coefficients*), c is the concentration, x is the coordinate, and ω is the atomic volume. In time, under isothermal conditions the characteristic linear dimension l_D of the diffusion zone increases according to the law $l_D \approx 2(\widetilde{D}t)^{1/2}$, where t is the time of diffusion annealing (see *Homogenization*). The spontaneous broadening of the diffusion zone is a consequence of the tendency to decrease the free energy of the system caused by the diffusional mixing of atoms. At the same time accompanying processes may take place which lead to a temporary, local increase of the free energy. Among them are pore formation (see *Frenkel effect*), creep of the substance (see *Kirkendall effect*), the appearance of elastic stresses, and the formation of dislocation structures (see *Dislocation*). The appearance of these factors in the diffusion zone indicates a lack of equilibrium corresponding to the presence of intermediate states of the system on its way toward reaching equilibrium.

DIFFUSIVITY, temperature conductivity factor

The parameter that determines the dependence of the *temperature* of a material on the coordinates and the time during nonstationary thermal processes; it serves as a measure of the rate of heat diffusion. In the model of an isotropic homogeneous body with thermal characteristics independent of temperature, the diffusivity is a constant a which enters the *thermal conductivity* equation: $a\nabla^2 T = \partial T/\partial t$. A formal expression for the diffusivity is $a = \lambda/(C\rho)$ where λ is the *thermal conductivity coefficient*, C is the *specific heat* (at constant pressure), ρ is the *density* of the material.

In the presence of appreciable temperature gradients in a body it is necessary to take into account the dependence of the thermal characteristics of the material on the temperature. The thermal conductivity equation conserves its form only in the case of the replacement of the temperature by new variable $\Theta = (1/\lambda_0)\int \lambda\, dT$, and we have $a\nabla^2\Theta = \partial\Theta/\partial t$, where a is expressed through the variable Θ, and is a function of the temperature. The dimensionality of a is $[a] = [L^2][t^{-1}]$; the unit of measurement is m^2/s.

DILATATION (fr. Lat. *dilato*, making wider)

The relative local elastic change of volume upon deformation. The element of volume dv after deformation becomes dV:

$$\frac{dV}{dv} = \left|\frac{\delta X_i}{\delta x_k}\right| = \left|\delta_{ik} + \frac{\partial U_i}{\partial x_k}\right|$$

(X_k and x_k are the Cartesian coordinates of the body points before and after its deformation, and U_k is the displacement vector). When account is taken to second order of the smallness of the *strain tensor*

$$u_{ik} = \frac{1}{2}\left(\frac{\partial u_i}{\partial X_k} + \frac{\partial u_k}{\partial X_i}\right),$$

we obtain the following expression for the dilatation:

$$\delta = \frac{dV - dv}{dv} = \sum_i u_{ii} - \frac{1}{2}\sum_{ik} u_{ik}u_{ki} + \frac{1}{2}\left(\sum_i u_{ii}\right)^2.$$

For elastically isotropic bodies, taking into account only the terms that are linear in u_{ik}, we have:

$$\delta = \frac{1}{3K}\sum_i \sigma_{ii},$$

where σ_{ik} is the deformation tensor, and K is the modulus of uniform compression. In isotropic crystals with point defects and dislocations the dilatation is everywhere zero in the linear deformation tensor approximation except at places where defects are located. Dilatations different from zero can occur for such defects when account is taken of the crystal anisotropy, or when deformation tensor terms which are second order in smallness make a contribution.

DILATATION CENTER

A *point defect* which inserts an "extra" infinitesimal volume ΔV into a *crystal*. A macroscopic analog of a dilatation center is a spherical cavity in an elastic medium into which an elastic globule of larger volume is inserted. Its elastic field is equivalent to the field of three pairs of forces acting along the coordinate axes and concentrated in the center. Elastic *strains* and *stresses* decrease inversely with the cube of the distance from the center. A dilatation center simulates *interstitial impurities* ($\Delta V > 0$) (see *Impurity atoms*), *substitutional impurities* (any sign of ΔV is possible) and *vacancies* within the crystal ($\Delta V < 0$).

DILATOMETER (fr. Lat. *dilato*, extend, and Gr. $\mu\epsilon\tau\rho\epsilon\omega$, am measuring)

A device for measuring variations of the dimensions of bodies (e.g., length) under the influence of external actions, internal physico-chemical processes, or both. The most wide spread types are optico-mechanical (sensitivity 10^{-6} to 10^{-7} cm), interference (10^{-7} to 10^{-8} cm), electronic with inductive, capacitive, electro-mechanical, galvanomagnetic (see *Galvanomagnetic effects*), or *magnetostriction* sensors (10^{-7} to 10^{-9} cm), radioresonance (10^{-12} cm), and X-ray (10^{-5} to 10^{-6} cm). Dilatometers are used for determining the coefficient of *thermal expansion* of materials, and for *dilatometric analysis*.

DILATOMETRIC ANALYSIS

The analysis of physical and physico-chemical properties of solids according to the variation of their volume or sizes. Dilatometric analysis is based on the tendency of a solid to change its volume (dimensions) under the influence of heat, pressure, electric and magnetic fields, ionizing radiation and other external effects, and also due to internal physico-chemical processes. The variations of volume are characterized by the *bulk expansion coefficient*. Dilatometric analysis is most widely used for investigations of *thermal expansion*. Processes taking place in solids are characterized by variations of volume (bulk effects) and of the *temperature coefficient of bulk expansion*. Absolute measurements (with respect to a fixed standard) and differential ones (the standard undergoes changes similar to the test body) are performed during the course of dilatometric analysis. This analysis is used to determine the true and average temperature coefficients of expansion, their phase composition at critical points, the evaluation of their structural state, and for the identification of ongoing physico-chemical processes and their rates.

DILATON

A non-linear excitation that occurs under the action of a stretching load in anharmonic atomic chains, or in structures fabricated from such chains (e.g., in fibrillar crystals of *polymers*). A dilaton is a dynamic structural *defect* which can alleviate *failure*. Two types of dilatons are distinguished: a *subcritical dilaton* and *supercritical dilaton*. The first is a deformation fluctuation created by a group of interatomic bonds, each of which is stretched by an amount ε_{I} which lies within the range $\varepsilon < \varepsilon_{\mathrm{I}} < \varepsilon_{\mathrm{cr}}$, where ε is the stretching strain, and $\varepsilon_{\mathrm{cr}}$ is the *strain* corresponding to the *ultimate strength* of the bond. The supercritical dilation has in the center of the group an anomalously strained interatomic bond with the strain $\varepsilon_{\mathrm{II}} > \varepsilon_{\mathrm{cr}}$. Both types of dilatons are particular solutions of the same equation of motion in which the cubic and quartic anharmonic terms are retained in the expansion of the potential energy.

DINGLE FACTOR (R.B. Dingle, 1952)

An exponential multiplier of the form $[-2\pi m^*/(e\tau B)]$, (where τ is the average time of collision, B is the magnetic field; m^* is the *effective mass*), which describes the amplitude reduction of quantum oscillations of thermodynamic and kinetic parameters of *metals* due to conduction *electron scattering* (see *de Haas–van Alphen effect*, *Shubnikov–de Haas effect*).

DIODE

A two-electrode electronic gas-discharge, vacuum tube or *semiconductor device* with the property of *unipolarity* (unidirectionality) of its *electrical conductivity*. Nowadays *semiconductor diodes* play a dominant role in science and engineering; and other (non-linear) properties not associated with their unipolarity have found various practical applications. A *semiconductor junction* or *metal–semiconductor junction* forming a *Schottky barrier* transmits current well in one direction and has a high electrical resistance in the opposite or blocking direction. Because of this, the rectifying action a diode can be used as a *semiconductor gate* to rectify low-frequency current up to 1 kA (*power gate*) and for voltage stabilization (*supporting diode*). *Pulse diodes* perform transformations of ultrahigh frequency radiation, e.g., for switching, or for detection associated with the variation of the radiation frequency. *Demodulation* or the removal of a modulated low-frequency signal from a high frequency carrier wave is based on this process. Diodes with *negative resistance*, or *generation type semiconductor diodes*, have been used for generating, amplifying, switching, multiplying (increasing the frequency) and other functions in the ultra-high frequency range, up to hundreds of gigahertz. *Gunn diodes* which have no *p–n* junction, or Schottky barrier have found wide application; their action is based on the presence of a negative resistance arising from the characteristics of the energy band structure of the *many-valley semiconductor*. The action of a *varactor* (also called *varicond*) is based on the dependence of the capacitance, present in the semiconductor diode due to a *p–n* junction or Schottky barrier, on the value of the applied external voltage. The non-linear dependence of the capacitance on the voltage forms the basis of the action of *parametric diodes* which are used for the amplification and conversion of ultra-high frequencies. In *photodiodes* the separation of non-equilibrium electrons and holes by the field of a *p–n* junction causes the appearance of a photo-electromotive force; and this is utilized in highly sensitive detectors of radiation in the infrared, visible and near ultraviolet regions of the electromagnetic spectrum (see, e.g., *Optical radiation detectors*). Solar batteries operate on such a principle. The action of *light emitting diodes* is based on the emission of photons during the *recombination* of electrons and holes in regions of p–n hetero- and homojunctions, and it can be very intense in direct-band semiconductors upon the application of a reversed bias (reversed electric field). In *magnetodiodes* use is made of the dependence of the current carrier mobility on the applied magnetic field. The various types of *semiconductor diodes* are differentiated according to the physical phenomena which determine their operation principles, as well as according their design; see *p–i–n diode*, *Barrier injection transit time diode*, *Impact ionization avalanche transit time diode*, *Trapped plasma avalanche triggered transit diode*, *Tunnel diode*, *Schottky diode*, *Varactor*.

DIPOLAR INSTABILITY

See *Vibronic instability*.

DIPOLE–DIPOLE INTERACTION

The interaction energy of two (electric) *dipole moments* $\boldsymbol{p}_1$ and $\boldsymbol{p}_2$, described by the expression

$$\Phi(\boldsymbol{R}) = \frac{\boldsymbol{p}_1 \cdot \boldsymbol{p}_2}{R^3} - 3\frac{(\boldsymbol{p}_1 \cdot \boldsymbol{R})(\boldsymbol{p}_2 \cdot \boldsymbol{R})}{R^5},$$

where $\boldsymbol{R}$ is the coordinate vector connecting the two dipoles. The dipole–dipole interaction involves one of the terms that arises from the multipolar expansion of the potential produced by a local distribution of electric charges. It is the energy of one dipole interacting with the electric field produced by another dipole. The formula is equally valid for electric as well for *magnetic dipoles* μ_1 and μ_2, and in particular it applies to spin magnetic moments in magnetic materials. The dipole–dipole interaction plays an important role in determining the internal long-range forces in *ionic crystals*, in *crystal lattice dynamics*, in the theory of *magnetism*, and also in the theory of the electric polarization of crystals, especially ferroelectric crystals.

DIPOLE–DIPOLE RESERVOIR

An energy subsystem of interacting *magnetic dipoles* $H_D = -2\sum_{i>1}\mu_i\mu_j\sum_{\alpha,\beta} S_{\alpha i} D^{ij}_{\alpha\beta} S_{\beta j}$, where $S_{\alpha i}$ and $S_{\beta j}$ are the projections of ith and jth spins on the axes $\alpha,\ \beta = x,\ y,\ z$, respectively, and μ_i, μ_j are their magnetic moments (see *Dipole–dipole interaction*). This energy is part of the total energy $H = H_0 + H_D$ of the ensemble of interacting spins, where H_0 is the main energy term and H_D determines the splitting of this main energy into sublevels with the magnitude $\hbar\omega_D$. The dipolar Hamiltonian term H_D also determines the rate of establishing a quasi-equilibrium distribution of particles in these sublevels. If this rate exceeds that of the *spin–lattice relaxation* of individual spins or exceeds the rate of heat exchange ω_x with the lattice, then one can say that H_D acquires its own temperature T_D which is not necessarily the same as that of the lattice, and hence we have a dipolar *heat sink*. The concept of a dipole–dipole reservoir can also apply to interacting electric dipoles.

DIPOLE-FORBIDDEN TRANSITION

Quantum transition between states for which the Hamiltonian matrix element of the electric *dipole moment* vanishes. Such a transition could occur via other mechanisms such as magnetic dipole or electric quadrupole. The dipolar prohibition of the transition is associated with symmetry, and in *solids* it is strictly forbidden only for specific lattice configurations (usually equilibrium configurations are assumed).

DIPOLE GLASS

Electrical analogue of magnetic *spin glass*. A dipole glass is the low-temperature state of randomly localized electric *dipole moments* mutually acted upon by forces arising from the *dipole–dipole interaction*. The randomness of the dipole moment orientations in the glass state is caused by the alternating character of the dipole–dipole interaction potential which depends on the relative locations and orientations of nearby dipoles.

Alkali-halide crystals with non-central ions or with dipole moments, e.g., KCl:Li^+, KCl:OH^- are typical dipole glasses. Also there are *ferroelectric–antiferroelectric* mixed crystals which at particular compositions exhibit properties similar to those of spin glasses, and they are related to dipole glasses. For example, in the mixed crystals $Rb_{1-x}(NH_4)_xH_2PO_4$ the spin-glass like behavior is caused by the random freezing of protons at positions along hydrogen bonds. The term *structural glass* is often used, because the forces of interaction of atoms in such systems are different from dipolar interaction forces. Also the *orientational* or *quadrupole glasses* of the type $(KCN)_x(KBr)_{1-x}$, where the random freezing of the quadrupole moment of CN^- occurs, are related to the structural glasses.

DIPOLE, INDUCED

See *Induced dipoles.*

DIPOLE LAYER on a surface

A system of electric dipoles on a surface due to the *adsorption* of atoms or molecules. A dipole moment per unit area np appears (n is concentration of adsorbed particles per unit area, p is their *dipole moment*) as a result of the orientation of the intrinsic dipoles of the adsorbed particles, of their polarization, or of charge transfer between the adsorbed particles and the surface. A dipole layer alters the *work function* φ of the surface: $\Delta\varphi = 4\pi np$.

DIPOLE, MAGNETIC

See *Magnetic dipole.*

DIPOLE MOMENT

A characteristic of a system of electrically charged particles is the vector $\boldsymbol{p} = \sum e_i \boldsymbol{r}_i$, called the dipole moment, where e_i is charge of the ith particle, $\boldsymbol{r}_i$ is its coordinate, and the sum is taken over all the particles of the system. For a charge distribution of density $\rho(\boldsymbol{r}')$ the dipole moment is $\boldsymbol{p} = \int \boldsymbol{r}'\rho(\boldsymbol{r}')\,\mathrm{d}^3\boldsymbol{r}'$. If the total charge of the system is zero then its electric field at a distance which considerably exceeds the dimensions of the system is approximately determined by the dipole moment. The value of the dipole moment $\boldsymbol{p}$ does not depend on the choice of origin of the coordinates. The electric dipole moment per unit volume is called the electric polarization, with the symbol $\boldsymbol{P}$ (see *Polarization of insulator*). It may be formed by the spatial separation of opposite polarity mobile charges, as well by the presence of elementary dipoles. The dipole moment of two charges q equal in magnitude and opposite in sign separated by the distance l is given by $\boldsymbol{p} = q\boldsymbol{l}$; where the vector $\boldsymbol{p}$ is directed from the negative to the positive charge.

Variations of the electrical dipole moment of a system system of charges, e.g., brought about by their motion, results in the emission of electromagnetic waves. The value of a dipole moment is measured in debyes, where one debye (1 D) is the dipole moment of two unit electric charges $+e$ and $-e$ separated by the distance of one angstrom ($e = 1.6022\cdot 10^{-19}$ C).

DIPOLE RELAXATION of an insulator

Return of a polarized *insulator* to its equilibrium non-polar state after switching off an applied electric field. In the electric field the randomness of the heat induced motion of weakly-bound electrons, ions, and dipoles is weakened, and some of them appear to be "fixed" by the electric field in positions corresponding to the polarized state. Unlike liquids and gases where a dipole is a rotator with unlimited rotation angles, in a crystal the dipole reorientation capability is restrictive: to only a limited number of stable orientations separated by potential barriers with heights that depend on the applied field. The following events may bring about dipole relaxation: jumps of weakly-bound ions in the vicinity of a *defect* with opposite charge; redistribution of the electron density in the vicinity of an anion *vacancy* (assuming degenerate electron states); jumps of protons in polar crystals between local energy minima along hydrogen bonds, and other possible mechanisms. The *relaxation time* τ usually (for weak dipolar interactions) depends exponentially on the temperature. At 300 K for various dielectrics $\tau = 10^{-4}$ to 10^{-10} s.

DIRAC MONOPOLE (P. Dirac, 1931)

See *Magnetic monopole.*

DIRECTOR

A unit vector $\boldsymbol{n}$, whose directions $+\boldsymbol{n}$ and $-\boldsymbol{n}$ are equivalent. The term is used for the description of the molecular ordering in *liquid crystals*, where $\boldsymbol{n}$ points along the preferred direction of the axes of the molecules. For the description of the most typical liquid crystal, which is the single-axis *nematic liquid crystal* type, it is sufficient to introduce only one director, because in this medium only one of the three molecular axes is oriented along a common direction, the other two being randomly oriented. In the two-axis nematics it is necessary to specify the triplet of directors $\boldsymbol{n}$, $\boldsymbol{l}$, and $\boldsymbol{n} \times \boldsymbol{l}$ because the three molecular axes are orientationally ordered. To describe C-type *smectic liquid crystals* the notion of a ***c**-director* is introduced. Smectic C has a lamellar structure and the director $\boldsymbol{n}$ does not point along the normal to the layers. The $\boldsymbol{c}$-director is introduced to designate the direction of the common slope of the molecules in the plane of the layers. It is a regular vector because the states with $+\boldsymbol{c}$ and $-\boldsymbol{c}$ are non-equivalent.

Through the components of the director and its coordinate derivatives it is possible to express all the characteristic quantities of a liquid crystal: its *dielectric constant* tensor, its elastic energy, its moment of forces, etc.

DIRECT RELAXATION PROCESS

A relaxation transition between two levels with a splitting of $\hbar\omega$ accompanied by either emission or absorption of a single resonance frequency phonon (a *single-phonon process*). For a direct relaxation process occurring by way of transitions between the spin sublevels the rate of *spin–lattice relaxation* is $\tau_1^{-1} = a\coth[\hbar\omega/(2k_{\mathrm{B}}T)]$, where a is a coefficient depending on the value of the matrix element of the *spin–phonon Hamiltonian*, the resonance frequency (or the external magnetic field) and certain constants of the material (material density and the speed of sound). When $\hbar\omega/(2k_{\mathrm{B}}T) \ll 1$ we have $\tau_1^{-1} \propto T$, an important feature of the direct relaxation process that may be tested experimentally. We have $\hbar\omega/(2k_{\mathrm{B}}T) \gg 1$ in the low-temperature limit, so that τ_1^{-1} does not depend on T, but rather it is controlled by the rate of spontaneous emission of *phonons* from the upper spin state. Another important aspect of the direct relaxation process that makes it possible to clarify the mechanism of spin–lattice relaxation is the magnetic field B dependence of τ_1^{-1}. Consider the case of the *Kronig–Van Vleck mechanism* (see *Van Vleck mechanism*) where the electric *crystal field* is modulated by the lattice vibrations. In the high-temperature limit we have $a \propto B^2$ for ions with an even number of electrons, and for transitions between *Kramers doublets* where the number of electrons is odd we have $a \propto B^4$. Direct relaxation usually dominates at low temperatures ($T < 10$ K).

DIRECT TRANSITIONS, crossover transitions, vertical transitions

Band-to-band *electron* (*hole*) transitions in crystals resulting from the absorption of light quanta, such that they occur without any appreciable change in the electron *quasi-momentum* (to the accuracy of the photon momentum, the latter being negligibly small within the scale of the *Brillouin zone*). See *Semiconductors*.

DISCLINATION MODELS OF PLASTIC DEFORMATION

They regard *plastic deformation* as a consequence of the motion of disclinations. Plastic deformation may be realized by three main mechanisms: diffusion, shear (translational), and rotational. *Point defects*, mainly *vacancies* (see *Diffusion mechanism of crystal creep*) are the carriers of diffusion plasticity, *dislocations* are the carriers of shear plasticity, and *disclinations* are the carriers of rotational plasticity. Disclination models of plastic deformation describe the generation and development of plastic rotations in solids. A.F. Ioffe found at the beginning of the 1920s that during the course of plastic deformation mutual rotations of crystal microvolumes take place. We now know that these rotations not only accompany deformation, but constitute an essential and often determining factor. The rotation of a microvolume cannot be smooth because of its interaction with surrounding volumes. Like *shear strain* it leads to the generation and motion of linear defects or disclinations; see Fig. (b). Disclinations with ratings (*Frank vectors*) $|\omega| = 0.1$ to $5°$ are observed in solids. After the rotation of the adjacent microvolumes to such angles their ideal conjugation is impossible. A flat *defect* formed along the plane of a turn together with a linear defect – the line of the rotation jump – is called a *partial disclination*. The stresses from a single disclination do not attenuate, and so screened disclination structures – *disclination loops* and *disclination multipoles* – are usually encountered. A dipole is an example of such a structure (see Fig. (c)), creating elastic fields which at great distances are equivalent to the fields of an *edge dislocation* with *Burgers vector* $b \approx 2\Delta\omega$. *Dislocation walls* are flat defects of partial disclinations which in turn are defects of ordered dislocational structures and of flat boundaries (at Fig. (c) the ends of torn dislocational walls). Disclinations can differ according their rating, geometry, and scale. The smallest disclination loops have diameters of several angstroms units, and describe, e.g., the rotations of molecules in *polymers*. Disclinations with dimensions 0.1 to 1 μm perform the rotations at the level of blocks and small fragments, 1 to 10 μm – of rough fragments and small grains, 10 to 1000 μm – of grains and their groups. Sometimes rotation is also observed at the macrolevel, associated with the passage of disclinations of the order of the body size. The most frequently encountered *twisting structures* (see *Plastic twisting deformation*) are the *fragmentation* and *stripe structures*. The fragmented structure consists of equal-axes cells or *fragments* disoriented with respect to each other. After starting a mutual rotation the angles increase during deformation until they reach a critical value with the generation of *cracks*. A hierarchy of fragments according to their size and the spread of their angles is often observed, with some splitting into smaller ones. The *fragmentation* kinetics is described by models for the generation and motion of dislocation loops. The paths of loops do not exceed the dimensions of the corresponding fragment. Stripe structures are formed at the onset of rotation-current instability (see *Slip bands*). Their kinetics follows models for the generation and motion of disclination quadrupoles, dipoles and their ensembles (see Fig. (c)). For example, dipoles of size $2\varDelta = 0.1$ to 1 μm can move 100 μm to several millimeters, possibly intersecting *grain boundaries*. The collision of the disclination dipole with a strong obstacle may bring about the formation of micro- or macrocracks. The generation of disclination multipoles and loops occurs under conditions of high dislocation density or flat boundaries, of high local stresses, and especially their gradients, creating torsion moments in microvolumes. Therefore, rotational structures form in the presence of strongly non-uniform external forces (active process), retarded shears (relaxation process), and non-uniform strain (passive process). Rotational structures occur in almost every important type of deformation. These are drops (macrostriped structures); plastic treatment of metals (striped and fragmented); *creep* (many-scale fragmented structures at low stresses and striped structures at high stresses); *fatigue* (fragmented and striped structures of stable strips of slipping); *internal friction* (fragmented structures of surface layers); *strain hardening* at stage III, etc. In some cases strain processes may be described within the framework of a *dislocation model of plastic deformation*, as well as within the framework of a disclination model of plastic deformation.

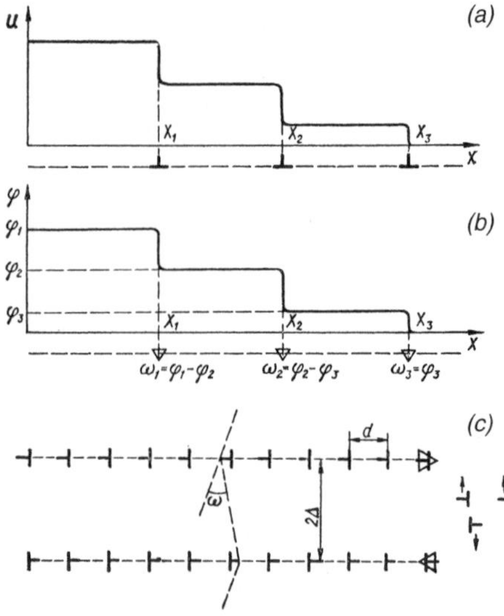

Linear defects in solids: (a) dislocations at the points X_1, X_2 and X_3 of the displacement jumps $\boldsymbol{u}$; (b) disclinations at the points X_1, X_2 and X_3 of the jumps in turning angle; (c) dipole of partial dislocations, which causes the layer turn between dislocation walls to the angle $\omega \approx b/d$.

DISCLINATIONS (fr. Lat. *dis*, apart, violation, and Gr. $\kappa\lambda\iota\nu\omega$, slope)

Defects in solid and *liquid crystals* which are lines along and near which the orientational ordering of the atomic planes or molecules is violated. In the *Volterra model of defects* (singularities) of an elastic continuum disclinations are formed as a result of a mutual rotation of the

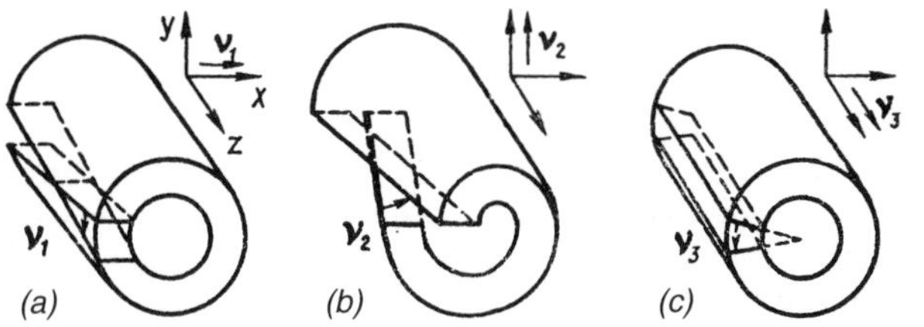

Fig. 1. Torsional disclination (a, b) and wedge disclination (c) along the axis of a continuous cylinder.

edges of a cut of an initially continuous cylinder to some pregiven vector angle (*Frank vector*; see *Somigliana–Volterra dislocations*) as shown in Fig. 1. Negative and positive disclinations are distinguished, respectively, by the rotation of the planes toward each other and in the opposite direction. If a Frank vector is parallel to the disclination line, it is called a *wedge disclination* (*disclination of slope*), while if it is perpendicular to it then it is a torsional disclination. The space containing a disclination acquires the Riemann–Cristoffel curvature (see *Curvature tensor*), and space with translational *dislocations* has a Cartan torsion. In crystalline media complete and partial disclinations are singled out depending on whether the rotation angle satisfies the angular symmetry of the medium or not. *Partial disclinations* are not isolated defects, but serve as the edges of defect planes. Partial disclinations violate the stacking regularity of the sites of the crystal lattice in one or several planes, adjacent to the disclination line. A torn slope boundary is an example of a partial disclination. In crystals with helical axes of symmetry *complete disclinations* are combined with translational dislocations having a *Burgers vector* equal to the pitch of a screw. Such a defect is called a *dispyration*. Any complete disclination violates the natural *crystal symmetry*. Thus positive wedge 60-degree disclinations transform sixfold rotation axes into pseudo-fivefold ones, and the negative ones bring about a transformation to pseudo-sevenfold axes. Disclinations move either conservatively, or non-conservatively. In the latter case they emit (absorb) lattice dislocations. Disclinations appearing as linear singularities are sources of stress. The stress field near rectilinear disclinations are at the line of the disclination as well as at infinity. The field of stress of disclinational dipoles at large distances from them is an elastic one, and its dependence on the distance is similar to that of the field of stress of a dipole of dislocations (see *Elastic field of dislocations*). In crystals of macroscopic size R isolated rectilinear disclinations are seldom found because their elastic energy is proportional to R^3. However, disclinations may be present in crystals in the form of small loops, dipoles, or networks, forming *grain boundaries*, whose energy is proportional to r, where r is the radius of the loop or the distance between adjacent disclinations. The other possibility for reducing the high disclination elastic energy is by limiting the crystal size, e.g., using two-dimensional crystals or liquid crystals.

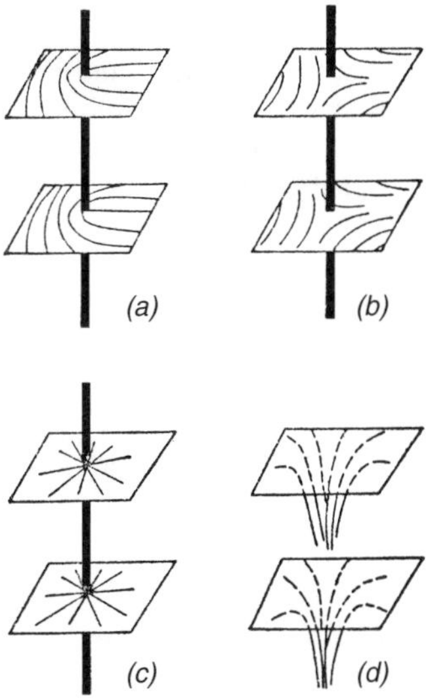

Fig. 2. Disclinations in nematics: (a, b) distribution of director near disclinations with $m = 1$ and $m = -1$, respectively, (c, d) disclination with $m = 2$ which is unstable with respect to a transition to the uniform state.

In liquid crystals disclinations have the form of ruptures of vector fields, which describe the orientational ordering of molecules, as shown for nematics (see *Nematic liquid crystals*) in Fig. 2. These disclinations appear in the form of flexible mobile filaments within the volume and at the surface of practically any sample, a circumstance which led to the introduction of term "nematic" itself. Due to the fact that nematics have a second-order *symmetry axis*, the rotation angle of a disclination is restricted to the set $m\pi$, where m is an integer. All disclinations with odd m are topologically stable, so to bring about their transition to the uniform state requires an energy that is considerably in excess of their self-energy. Those with even m are unstable and readily transform to the uniform state, as shown in Fig. 2. Disclinations are the dominating *defects in liquid crystals*, where they are observable by an optical microscope. They are present in biological objects, small crystals, and thin crystalline films. Disclinations are used to explain the structure of glasses (see *Vitreous state of matter*), and sometimes also for interpretating processes of *plastic deformation* and *fracture* of materials.

DISCONTINUOUS PLASTIC DEFORMATION

Process of irregular, unstable plastic flow (see *Plastic deformation*) of crystalline materials. Under conditions of active deformation (extension or contraction of samples with constant velocity) spiked or jumpwise *strain* appears on the *deformation diagram* (stress versus strain plot, $\tau(\varepsilon)$) as a more or less irregular alternation of spontaneous increases and decreases of deforming stress taking place at a constant average deformation rate, so the diagram acquires a spiked nature (Fig. 1).

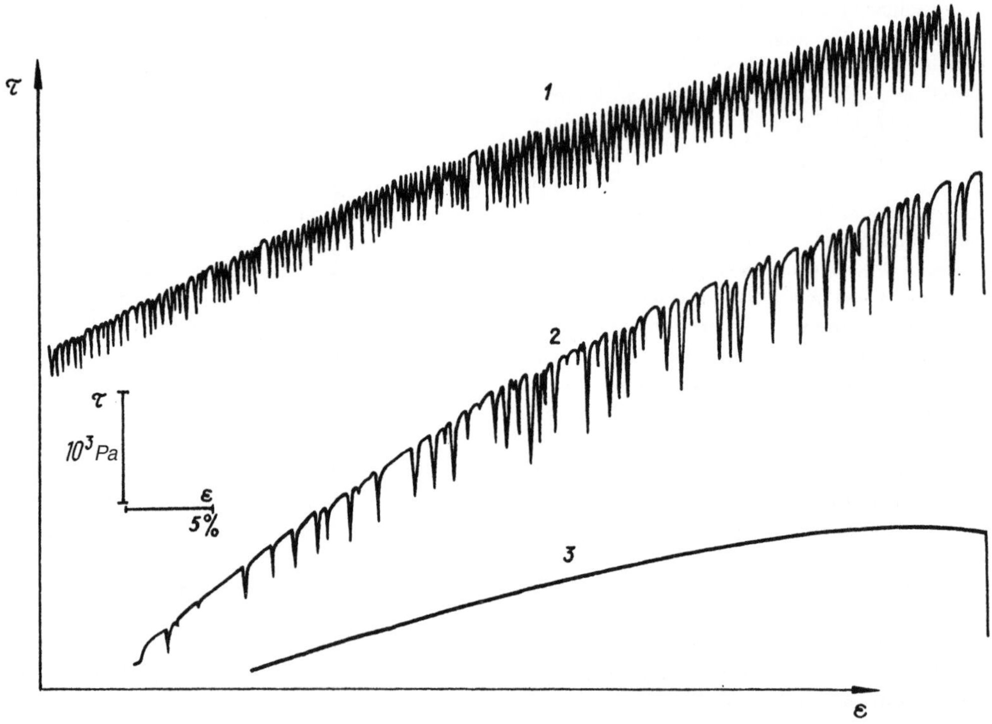

Fig. 1. Stress τ versus strain (deformation) ε plot showing regions of deformation of aluminum monocrystals with various purity values 99.5% (1), 99.99% (2) and 99.9997% (3), illustrating spiked (jumpwise) stress (1, 2) and its disappearance at high purity (3). The temperature of the experiment was 1.5 K.

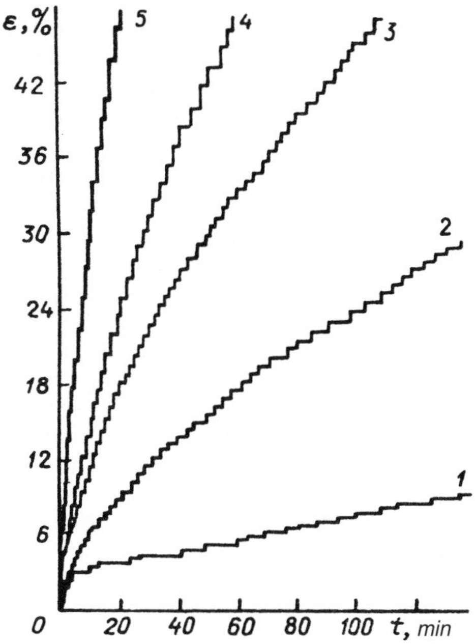

Fig. 2. Stepwise creep of zinc monocrystal during slipping under exposure to constant stress that exceeds the yield limit at the temperature 300 K. Various creep curves, or plots of shear deformation ε versus time t (curves 1 to 5), correspond to different values of deforming stress.

Under *creep* conditions this irregular strain appears as an alternation of cycles of higher-speed and lower-speed creep, so the creep curve acquires a step-like nature (Fig. 2). Stepwise strain at room temperature and high temperature is mainly observed in *alloys* [*Portevin–Le Chatelier effect* (A. Portevin, F. Le Chatelier, 1923)] and is explained by the specificity of the dislocation diffusion processes. In the low temperature range not only alloys, but also many pure mono- and polycrystalline metals are susceptible to stepwise strain, therefore it is considered one of the most characteristic properties of low-temperature *plasticity*. At low temperatures the susceptibility of a material to step strain can be regulated by a change of shape, size, or surface state, experimental conditions (deformation velocity, hardness of deforming device, temperature, and requirements of heat-sink cooling), internal state (change of defect structure, or sample transition from normal to superconducting state, see *Superconductivity*), and a number of other factors. The variety of the forms and conditions of its observation provides the basis for explaining stepwise strain by more than a single physical mechanism. Several mechanisms are known and actively discussed: *mechanical twinning*; periodic formation of dislocation aggregations and breaking of barriers by them; interactions of volumetric or surface dislocation sources with a spectrum of initial stresses; onset of nonlinear waves of dislocation density in sufficiently dense dislocation ensembles; warming up of a sample or individual parts of it and thereby perturbing the thermal stability of strain processes; nonmonotonic dependence of dislocation drag on the velocity; geometric *dehardening* due to the increasing narrowing of the sample cross-section all along its length or in separate parts.

The theoretical formulation of a step strain process assumes the solution of two main problems. The first problem involves the derivation of an instability criterion involving the particular inequality that connects the sample features with the experimental conditions that exist when the strain process deviates from a stable mode. The second, generally more complicated, problem is associated with deducing the mechanisms that stabilize the strain after the loss of stability, and delineating the time evolution of the process within each separate cycle (step). One approach for explaining low-temperature stepwise strain adapts the hypothesis of thermal instability, whereby the two problems mentioned above are formulated and solved for specific approximation conditions. Qualitative suggestions for instability criteria have been proposed for various mechanisms.

DISCOTICS (S. Chandrasekhar, 1977)

Liquid crystals formed by molecules in the form of discs. The existence of several liquid-crystalline phases of *discotics* has been experimentally demonstrated: of nematic phase, of orthogonal columnwise and of tilted columnwise ones.

In the *discotic nematics* (see *Nematic liquid crystals*) the molecules freely slip with respect to one another, due to which translational ordering is absent, but there is orientational ordering of the normals to the planes of molecules along the common direction of the *director* $\boldsymbol{n}$, as shown in Fig. 1.

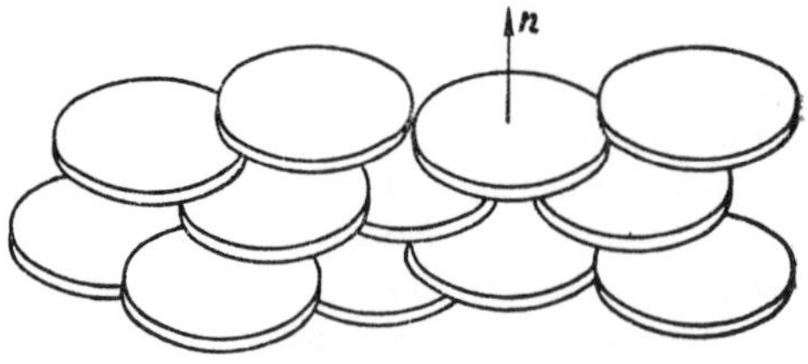

Fig. 1. Ordering of molecules in a discotic nematic.

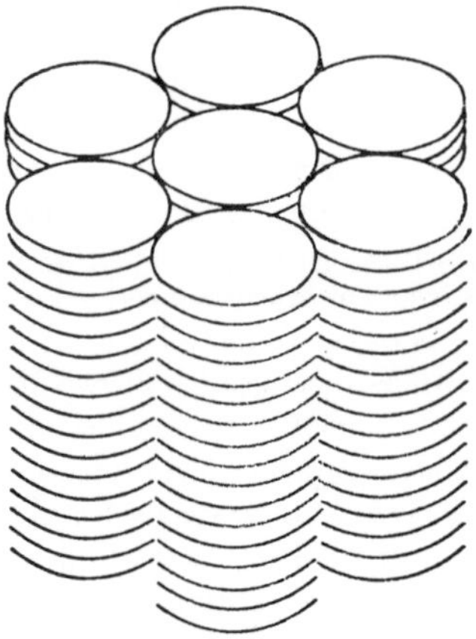

Fig. 2. Ordering of molecules in the orthogonal columnwise phase of a discotic.

The *columnwise phases of discotics*, or the discotics themselves, are two-dimensional translationally ordered lattices of long columns formed by the stacks of disc shaped molecules, as shown in Fig. 2. Along the axes of the columns there is no long-range order in the location of the mass centers of the molecules, and the system is liquid in one dimension and *solid* in two others (compare with *smectic liquid crystals*). This is because the interaction of the central parts of the molecules (usually rigid) is considerably weaker than the interactions of the lateral parts which contain flexible hydrocarbon chains. If the molecular planes are perpendicular to the plane of the column the discotic is called *orthogonal discotic*, and if they subtend some angle to this plane it is a *slope discotic*. Depending on the two-dimensional lattice type, *hexagonal discotics* and *rectangular discotics* are distinguished.

The investigation of the properties of discotics has wide practical applications because it is known that during the formation of coke and artificial fibers from oil and coal pitch at high temperatures (400 to 550 °C) the so-called *carbonic phases* are formed, which are analogues of discotic nematics, and consist of large aromatic polynuclear molecules that are plate-like in shape. The quality of the final products is to a large extent determined by the character of the ordering in this liquid-crystalline carbonic phase.

DISLOCATION ACOUSTIC RADIATION

See *Acoustic radiation by dislocations.*

DISLOCATION ANNIHILATION

See *Annihilation of dislocations.*

DISLOCATION BARRIER, LOCAL

See *Local barriers.*

DISLOCATION CONTRAST

Local variation of the intensity on an electron-microscope or X-ray topogram (see *X-ray topography*), caused by the field of *dislocation* distortions. The value of the local *reciprocal lattice* vector $\boldsymbol{b}_h(\boldsymbol{r}) = \boldsymbol{b}_h - \nabla(\boldsymbol{b}_h \boldsymbol{u})$ ($\boldsymbol{b}_h$ is a reciprocal lattice vector of the *ideal crystal*, $\boldsymbol{u}$ is the atomic displacement) changes due to the distortions, which leads to the variation of local conditions for Bragg reflection (see *Bragg law*), and therefore to the variation of the intensity of transmitted and diffracted radiation. As a result, the intensity distribution has an interference nature typical of dislocation contrast; from the number and configuration of the light and dark interference fringes it is possible to determine the dislocation orientation within the crystal and the direction of *Burgers vector.* Images of rectilinear dislocations have the form of thin fringes, whose width is inversely proportional to the half-width of the Bragg reflection, of the order of an *extinction length.*

DISLOCATION DENSITY TENSOR

Second-rank tensor which characterizes the distribution of dislocation lines and their *Burgers vectors* in a crystal. In the case of a continuous medium, it describes the continuum distribution of dislocations of infinitely small Burgers vectors of the form

$$\mathrm{d}\boldsymbol{b} = \int_{\mathrm{d}c} \mathrm{d}\boldsymbol{U} = \int_{\mathrm{d}c} \widehat{\beta}\,\mathrm{d}\boldsymbol{r} = \widehat{\alpha}\,\mathrm{d}\boldsymbol{s},$$

where $\mathrm{d}c$ is the infinitely small *Burgers contour*, $\mathrm{d}\boldsymbol{s}$ is the infinitely small surface which is bounded by $\mathrm{d}c$, β is the *distortion* tensor having components $\beta_{ik} = \partial U_i/\partial x_k$, and the tensor $\widehat{\alpha} = \mathrm{rot}\,\widehat{\beta} = \nabla \times \widehat{\beta}$ is referred to as the dislocation density tensor. By definition the divergence of a curl is zero, hence $\mathrm{div}\,\widehat{\alpha} = \nabla \cdot \widehat{\alpha} = \mathbf{0}$, i.e. the dislocation lines cannot end inside the crystal. An isolated dislocation line may be characterized in terms of δ-functions in the x and y coordinate directions, the dislocation density tensor being in this case $\nabla \cdot \widehat{\alpha} = \boldsymbol{\tau} \cdot \boldsymbol{b}\,\delta(x)\delta(y)$, where $\boldsymbol{\tau}$ is the unit vector tangent to the dislocation line, and $\boldsymbol{b}$ is the Burgers vector. The state of *internal stresses* is defined by the incompatibility tensor $\widehat{\eta}$, which is expressed through $\widehat{\alpha}$ as $\widehat{\eta} = (1/2)(\widehat{\overline{\alpha}} \times \nabla \times \nabla \times \widehat{\alpha})$, where $\widehat{\overline{\alpha}}$ is the transposed dislocation density tensor. Thus, the specification of the dislocation density tensor completely defines the field of internal stresses. The dislocation density ρ is experimentally determined as the overall length of dislocation lines in a unit volume: $\rho = L/V$. The value of ρ is commonly expressed in cm^{-2}.

DISLOCATION–DISLOCATION INTERACTION

The forces which affect a *dislocation* arising from other dislocations.

The dislocation–dislocation interaction can be long-range (through the fields of elastic, electric or magnetic stresses) or short-range (accompanied by reconstructions in the cores of dislocations at their intersection). The strength of a dislocation–dislocation interaction depends on the form of dislocation lines and their spatial distribution. The elastic dislocation–dislocation interaction involving distances between them which are larger than the core radius r_0 are determined by the field of elastic stresses created by these dislocations. The stresses are described by the elastic *stress tensor* $\widehat{\sigma}$, with components σ_{ij} which determine the force per unit length F acting on the dislocation. The components of the vector F are written in the form $F_i = \varepsilon_{ilm}\tau_l\sigma'_{mk}b_k$, where $\sigma'_{mk} = \sigma_{mk} - (1/3)\delta_{mk}\sigma_{ll}$ are the components of the deviator tensor (see *Stress deviator*); ε_{ilm} are the components of the so-called *Levi-Civita tensor* among which only six differ from zero, namely $\varepsilon_{123} = \varepsilon_{231} = \varepsilon_{312} = +1$ and $\varepsilon_{132} = \varepsilon_{213} = \varepsilon_{321} = -1$; τ_l is the lth component of the unit vector of the normal to the dislocation line, b_k is the kth component of the *Burgers vector* of the dislocation.

The force of interaction between two parallel rectilinear *edge dislocations* with Burgers vectors $\boldsymbol{b}_1, \boldsymbol{b}_2$ is not central; and its projection on the direction of the Burgers vector $\boldsymbol{b}_1$ is equal to

$$F_b = \frac{\mu(\boldsymbol{b}_1, \boldsymbol{b}_2)\cos\varphi\cos 2\varphi}{2\pi(1-\nu)r},$$

where r, φ are the polar coordinates of the second dislocation with respect to the first, in which the axis from which the angle φ is measured lies along the direction $\boldsymbol{b}_1$. F_b is equal to 0 at $\varphi = \pi/2$ (stable equilibrium for the case $\boldsymbol{b}_1\boldsymbol{b}_2 > 0$ corresponds to this) and $\varphi = \pi/4$ (this corresponds to stable equilibrium for dislocations of different signs, $\boldsymbol{b}_1\boldsymbol{b}_2 < 0$). The field of characteristic elastic stresses arising upon the bending of a dislocation line induces the appearance of the *dislocation self-action force*. The presence of broken or uncompensated interatomic bonds means that close to dislocation lines, within a region several angstroms in size, electrical space charge can accumulate. In *metals* this charge is strongly screened, and in *insulators* and semiconductors a stable redistribution of space charge can occur. The electrostatic field associated with this charge takes part in the *electrostatic dislocation–dislocation interaction*. In *ionic crystals* containing dislocations, ions with charges of different signs contribute to this interaction. It is convenient to determine the effective charges at dislocations by counting the *dangling bonds*. For example, elementary steps at a pure edge dislocation have a charge of $\pm e/2$. For a medium with dielectric constant $\varepsilon < 6.12$ the electrostatic interaction of steps is stronger than their elastic interaction. In crystals with magnetostriction properties (see *Magnetostriction*) a *magnetic dislocation–dislocation interaction* is possible. At high temperatures the *diffusion dislocation–dislocation interaction* can arise through the flux of *vacancies* or *interstitial atoms* between the dislocations.

DISLOCATION ELASTIC FIELD

See *Elastic field of dislocation.*

DISLOCATION EXCITON

An *exciton* localized near a dislocation line by the *elastic field of dislocations*, which creates the potential well for the exciton. The electronic excitations (excitons) in such states may travel along the *dislocation* line, undergoing small transverse motions in the plane perpendicular to it. The quantum state of a rectilinear dislocation is characterized by a wave vector directed along the dislocation line, and by a discrete set of quantum numbers, related to the transverse motion. In semiconductors there are excitons of several varieties, which differ by the type of bonding of the electron and hole forming the exciton with the dislocation core: excitons localized within the dislocation core, or far from it in the field of the *strain potential*; excitons of a mixed type with one charge within the dislocation core and the other in the deformation potential field. *Kinks on dislocations* (*dislocation steps*) on the dislocation line create traps for the dislocation exciton and disrupt the extension of the longitudinal motion. The exciton introduces specific features into optical and kinetic properties of semiconductors.

DISLOCATION-FREE CRYSTALS

Highly perfect monocrystalline samples that contain no *dislocations.* Dislocation-free crystals can be obtained by eliminating the causes that induce mechanisms of *nucleation* and *multiplication of dislocations* in the growing and subsequent cooling of a *monocrystal.* When producing dislocation-free crystals, the important factors are the material purity, the crystal thickness, and the degree of oxidation of its surface; the most significant feature is the thermal stress induced due to a nonuniform temperature distribution in the crystal. When priming is used the *subboundaries* and dislocations could grow from the priming. The most advanced results are achieved in obtaining the dislocation-free crystals of Si and Ge. With the help of the Czochralski method (see *Monocrystal growth*), the dislocation-free crystals of Si and Ge 80 mm and 35 mm in diameter, respectively, were obtained. The growth of perfect metallic crystals exerts severe requirements on the heating conditions, since in metals the *plasticity* decreases at a lower rate, and the temperature range where the thermoelastic stresses induce the nucleation and the multiplication of the dislocations is much wider. Nevertheless, one can obtain dislocation-free crystals of Cu up to 6 mm in diameter and 60 mm in length by the Czochralski method, provided the temperature regime is maintained with high precision, a thin ($\leqslant$1 mm) neck is created, and a heater of special design and a system of heat screens are used. These are to be combined with other technological procedures to provide for the necessary conditions for break-off of the crystal from the melt and for its subsequent cooling.

So-called *thread-like crystals* (cat's whiskers, fibers) up to 20 μm in diameter, with a ratio of length to diameter not less than 10^3, have also been obtained free of dislocations (see *Quasi-one-dimensional crystals*). Such crystals possess a number of interesting physical properties, in particular their *strength* being close to the theoretical value.

DISLOCATION GRIDS

A system of mutually intersecting *dislocation* lines. At the point of intersection, a reorganization can be observed resulting in the appearance of sites whose equilibrium position can be approximately determined by the relation $\sum_{i=1}^{n} \tau_i b_i^2 = 0$. Here τ_i is the unit vector tangent to the ith dislocation line, entering the site; b_i is the magnitude of its *Burgers vector* (summation carried out over all dislocations passing through the given site). If the attractive (or repulsive) forces act between the different dislocations near the site, the dislocation reactions may result in the appearance of grids of dislocations with hexagonal cells. Within the boundaries of purely planar torsion (110) in body-centered cubic crystals, hexagonal grids with $\tau = (1/\sqrt{6})\langle 112\rangle$ and $b = (1/2)\langle 111\rangle$ appear. In face-centered cubic crystals two types of lattice points are distinguished. The lattice points are called *K-sites*, if they are formed by three dislocations so that one character A, B or C in notation based on the *Thompson tetrahedron* is located in the sector. If a pair of characters A, B or C separated by an extension of opposite dislocations is located in each sector, then the lattice points are called *P-sites*. The symmetric K-sites, denoted by $K_{\rm s}$, are characterized by a clockwise sequence of characters A, B, C, and the asymmetric $K_{\rm a}$ sites have a counterclockwise sequence. The lattice points with a counterclockwise sequence A, B, C, A, B, C are called P_s, and the ones with a clockwise sequence are called P_a. If only subtraction *stacking faults* are possible, then K-sites are collapsed and P-sites are split. The K-sites can split if it is possible to form defects of implantation stacking faults. In annealed crystals (see *Annealing*) stable three-dimensional *Frank networks* of dislocations are possible.

DISLOCATION INTERACTIONS

See *Microcrack–dislocation interaction*, *Point defect–dislocation interaction*, *Vacancy–dislocation interaction*.

DISLOCATION INTERLACEMENTS

A relatively isolated group of *dislocations*, with lines forming a more or less complex interlacing. The following types of dislocation interlacements can be singled out: those containing an excess of dislocations of the same sign, which involve fields of elastic stresses decreasing with the distance r as r^{-1} (see *Elastic field of dislocations*), and those with their total *Burgers vector* equal to zero. The latter elastic stresses decrease with distance at least as fast as r^{-2}. The boundaries of disoriented *block structures* may consist of closed dislocation entanglements of the same sign, located in a relatively broad band. Local non-uniformities of the dislocation density (dislocation interlacements, tangles and bundles) form during early stages of *plastic deformation* at the second stage of the evolution of dislocation structures, with a dislocation density of 10^8 to 10^{11} cm^{-2}.

A special type of dislocation interlacement consisting of a relatively isolated series of parallel dislocations belonging to one system of *dislocation slip* and located in adjacent *slip planes*, constitutes a *pile-up of dislocations*, which under the effect of an applied stress become grouped in front of some obstacle. Such aggregations are *concentrators of stresses* (see *Stress concentration*), creating elastic stresses which are proportional to the applied stress, and to the number of dislocations near the obstacle. As a result of plastic creep (transverse slip and *climb of dislocations*), at the head of an aggregation near an obstacle there may form *dislocation tangles*, mixed up dislocation lines. Burgers vectors of dislocations in the tangle are usually of the same sign, and thus the long-range elastic stress far from the aggregation may be described as the field of a *superdislocation* with the total Burgers vector of the tangle. However, at the head of the relaxed aggregation the elastic stress is lower due to its "bluntness". The interlacements may consist of a system of *dislocation multipoles* (dipoles, quadrupoles, octupoles, etc.), formed from equal numbers of dislocations with mutually opposite Burgers vectors. There are the following special types of dislocation interlacements:

(1) bulk *dislocation grids*;
(2) walls of *dislocation loops* in a layer much thinner than the other two dimensions (all the loops need not belong to one system);
(3) *dipole dislocation walls*.

DISLOCATION KINK

See *Kinks on dislocations*.

DISLOCATION LOOP

A *dislocation* whose line forms a closed curve. If all the points of this line are confined to a plane then the dislocation loop is called planar. We can distinguish a planar slipping dislocation loop with its *Burgers vector* $\boldsymbol{b}$ lying in the same plane, and a planar prismatic dislocation loop with its Burgers vector perpendicular to this plane.

At large distances $r \gg r_L$ from a dislocation loop of size r_L the displacement vector u_i and the stress tensor σ_{ik} decrease with distance in accordance with the laws $|u_i| \sim b/r$ and $|\sigma_{ik}| \sim Gb/r^2$, where G is the shear modulus, and b is the magnitude of the Burgers vector. The energy W of the dislocation loop is of the order $W \sim Gb^2 r_L \ln(r_L/r_C)$, where r_C is the radius of the dislocation core.

DISLOCATION MODEL OF PLASTIC DEFORMATION

Explains *plastic deformation* caused by the motion of *dislocations* within a crystal, including conservative motion (slipping) in the *slip plane*, and non-conservative motion of *edge dislocations*

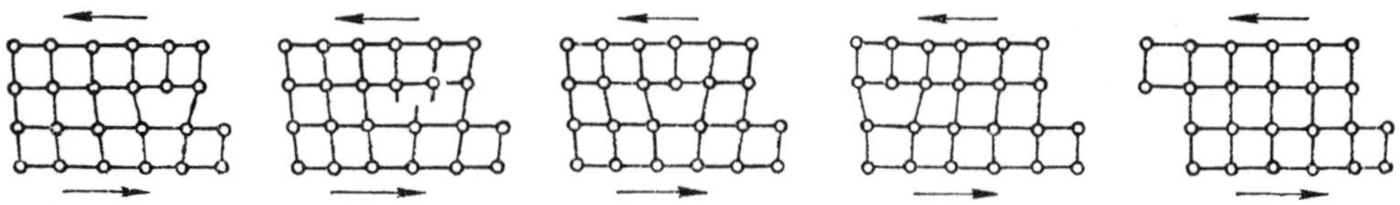

Diagram sketching the motion of an edge dislocation.

along the normal to the slip plane (climb). A dislocation can be considered as the boundary between regions shifted during the course of deformation, and the unshifted (undisplaced) regions of the crystal, so the motion of dislocations in the slip plane brings about *plastic deformation* by shearing. Upon the application of an external stress τ due to the presence of an *elastic field of dislocation*, there arises a force which affects each segment of dislocation length: $\boldsymbol{F} = \tau \boldsymbol{b}$, where $\boldsymbol{b}$ is the *Burgers vector*. During the dislocation motion in the slip plane interatomic bonds are broken and rebound only near the dislocation core, but not along the entire slip plane (Fig.). The *shear* in the slip plane occurs sequentially in proportion to the dislocation motion. During the passage of the dislocation through the crystal the magnitude of the displacement is equal to the Burgers vector $\boldsymbol{b}$. The atoms near the dislocation core are almost symmetrically displaced to opposite sides from the extra semiplane, which involves equal and oppositely directed forces for the shear. For some crystals this produces the minor stress (several orders below the *theoretical strength*) necessary for moving dislocations in a perfect crystal lattice at 0 K, the so-called *Peierls–Nabarro stress* τ_P. This stress is $\tau_P \approx G\exp[-2\pi w/b]$, where G is the *shear modulus* and w is the dislocation width (see *Peierls–Nabarro model*). Upon the slipping of dislocations, rotation (twisting) and bending of interatomic bonds take place, so the Peierls–Nabarro stress is increased in *covalent crystals* with directed interatomic bonds (small w), where dislocations may be almost immobile at room temperature (see *Plastic deformation of covalent crystals*). At temperatures above 0 K the motion of dislocations may be accelerated through the thermally activated overcoming of Peierls barriers (see *Peierls relief*), with the accumulation of dislocations lying at the energy minima (*Peierls valleys*) of double bends. Quantum-mechanical effects appear necessary for overcoming the Peierls barriers by dislocations at temperatures close to 0 K (see *Low-temperature yield limit anomalies*). Screw sections of dislocations can easily change their slip plane, which causes the transfer of dislocations to other planes via *double transverse slipping*. The *multiplication of dislocations* during the course of plastic deformation is performed via the action of *Frank–Read sources*, multiple slipping, and other mechanisms. The *climb of dislocations* occurs as a result of the diffusion of *vacancies* or *interstitial atoms* to the extra semiplane at the dislocation core (or due to their outflow), and this becomes significant at high temperatures and long loading hold times (see *Diffusion*). The processes of crystal hardening by *doping* of the solid solution, by the introduction of dispersed particles of a second *phase* (see *Precipitation-hardened materials*), and also by *strain hardening*, are explained by the dislocation model of plastic deformation. This model has been experimentally confirmed by transmission *electron microscopy*, by *X-ray topography*, by the method of *etching* of pits, and others. See also *Disclination models of plastic deformation.*

DISLOCATION PILE-UP

See *Pile-up of dislocations.*

DISLOCATIONS (fr. Lat. *dislocatio*, displacement, movement)

Defects of a crystal lattice which distort the correct arrangement of atomic planes. Dislocations differ from other defects in crystals because they involve a considerable disruption of the regular interlacing of atoms concentrated within a close vicinity of some line passing through the crystal, and this line is the edge of an incomplete displacement of one part of the crystal with respect to another part through one of the lattice *translation* vectors.

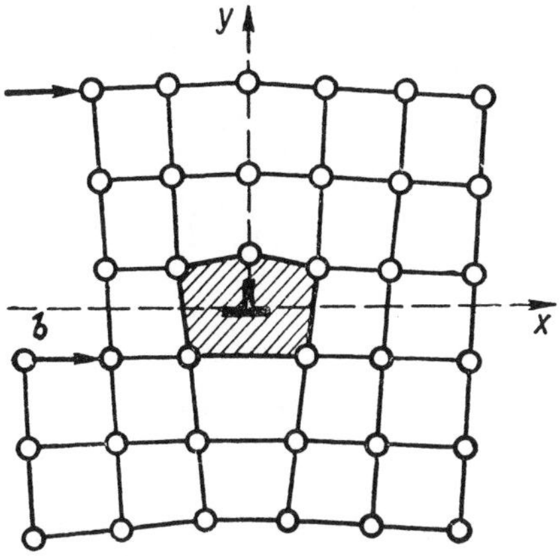

Fig. 1. Sketch of the positions of atoms in the core of an edge dislocation.

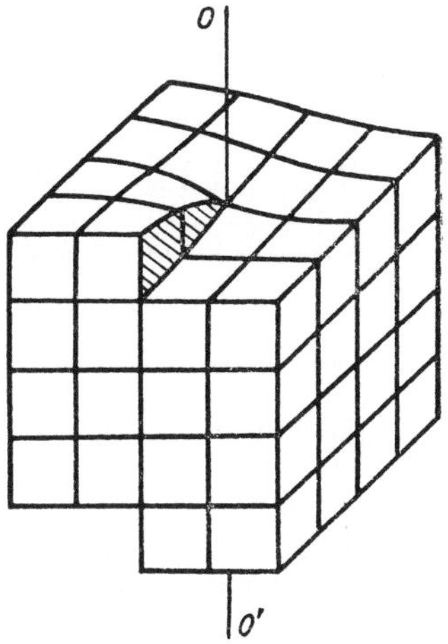

Fig. 2. View of the positions of atoms (cubes) in the neighborhood of a screw dislocation.

Edge dislocations and *screw dislocations* are the simplest types. In an *ideal crystal* adjacent atomic planes are rigorously parallel, and if one of these planes is split within the crystal, as shown in Fig. 1, an edge dislocation occurs, with the edge of the "extra" vertical semiplane as its axis (z). The five irregularly arranged atoms forming the contour of the shaded pentagon called the *dislocation core* are pictured in Fig. 1. Regions of irregularly spaced atoms extend several interatomic distances along the dislocation line x from the core, so the atom displacements are perpendicular to the dislocation axis z. The figure also indicates how this edge dislocation may be obtained as a result of an incomplete displacement of the upper part of the crystal by one lattice spacing along the x direction at the intersection of the dislocation axis.

The screw dislocation shown in Fig. 2 may be represented as the result of a displacement through a lattice period of one part of the crystal with respect to another part along some semiplane, with the edge of the semiplane playing the role of the dislocation axis. Therefore, the displacement that generates the screw dislocation is parallel to its dislocation axis. In the case of a screw dislocation the atomic planes, being only approximately parallel, are joined into one helical surface. If the axis of a helical dislocation reaches the external surface of the crystal, then at this surface a characteristic step is formed with a height of one atomic layer. This step becomes evident during the process of *crystallization* because atoms deposited from the vapor or melt are connected to the step at the surface of the growing crystal, bringing about an intensive helical growth of the crystal.

Between the limiting cases of edge and screw dislocations various intermediate cases are possible, including curved dislocation lines. If the displacement generating a linear dislocation is directed at an arbitrary angle to its axis, it is called a *mixed dislocation*. The displacement generating a dislocation is designated by a vector that is constant along the dislocation line (*Burgers vector* $\boldsymbol{b}$), and coincides with one of translational periods of the crystalline lattice. The plane passing through the vector $\boldsymbol{b}$ and tangent to the dislocation line at the point under consideration, is called the *slip plane* of the given dislocation element. The possible systems of slip planes are determined by the structure of the crystal. Dislocation lines cannot end within the crystal, but must be either closed (*dislocation loops*), or emerge from the surface of the crystal (*crystallite*), or branch, thereby forming a *dislocation grid*. The number of dislocations is characterized by their density (see *Dislocation density tensor*), meaning the average number of dislocation lines intersecting a unit area of space. This density varies from 10^2–10^3 cm^{-2} in the most perfect *monocrystals* to 10^{11}–10^{12} cm^{-2}

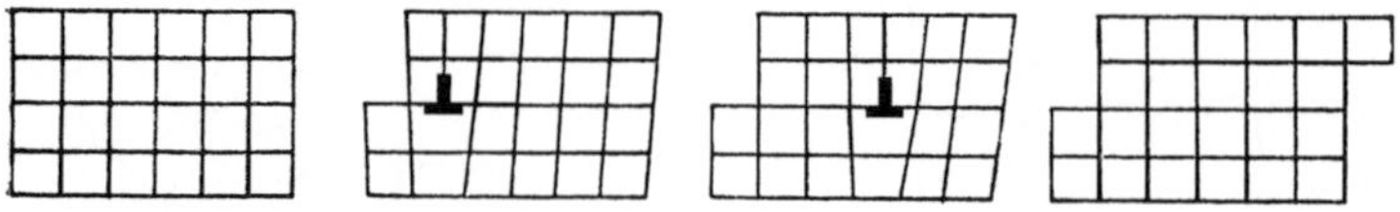

Fig. 3. Illustration of a plastic deformation that results from sequential dislocation slip.

in appreciably distorted (cold-rolled) metals. If the dislocation core is enclosed by some tube, then outside this tube the crystal may be considered ideal, subjected only to elastic deformation. Therefore, distortions far away may be analyzed by the methods of *elasticity theory*. Dislocations generate elastic deformations and stresses in their neighborhood so they are sources of *elastic fields of dislocations* within the crystal. These fields are reminiscent of the magnetic field lines encircling an electric current. The stress field determines the inherent elastic dislocation energy proportional to b^2 with a value of about 10^{-4} erg per cm of its length, and it also involves interactions between dislocations. A dislocation as a source of an elastic field is subjected to forces arising from the shear stresses within the crystal, reminding one of the Lorentz force of a magnetic field on a conductor carrying electric current. The value of the force $f = b\sigma$ applied to a unit length of a dislocation involves a component of the *strain tensor* σ_{ik}. For example, for an edge dislocation with its axis lying parallel to the z axis and the vector $\boldsymbol{b}$ directed along the x axis the force component is $f_x = b\sigma_{xy}$. The external force fairly easily sets dislocations into motion. A moving dislocation is capable of emitting elastic (sonic) waves, which explains the emission of sound during plastic deformation (see *Acoustic radiation by dislocations*).

Shear forces can move dislocations and thereby cause *plastic deformation*. Plastic deformation brought about by a sequence of individual dislocation *slips* is sketched in Fig. 3. Plasticity is associated with different types of this motion, as well as with the motion of *pile-up of dislocations*. Slipping occurs at comparably low external loadings, and the stresses at the onset of slipping determine the microscopic *elastic limit* σ_s of a monocrystal. The value of σ_s is 10^2 to 10^4 times smaller than the *shear modulus* of a monocrystal. This small value explains why the shear *strength* of real crystals with dislocations is several orders of magnitude smaller than that of monocrystals without dislocations. Some dislocation motions (*climb*) displace an edge dislocation to a direction perpendicular to the slip plane, and the growth or dissolution of atomic rows at the edge of an "extra" semiplane is accompanied by the formation or disappearance of vacancies in the vicinity of the dislocation core. This climb provides the mechanism for *dislocational diffusion* (see *Diffusion mechanism of crystal creep*, *Dislocation model of plastic deformation*).

The development of dislocational plasticity is determined by the mobility of dislocations, and the intensity of their generation. The mobility of isolated dislocations in a very pure monocrystal depends on the interatomic bonding forces, and on the interaction with *phonons* and electrons (in *metals*). The mobility of dislocations in non-ideal crystals is reduced by their interaction with each other, and with other defects which can retard or stop them (*dislocation barriers*). Impurities can block dislocation motion and pin a dislocation line at some points. At high temperatures the barriers can be overcome by thermal activation, while at low temperatures quantum tunneling is possible (see *Tunnel effect*).

Dislocations mainly affect the mechanical properties of solids (*elasticity*, *plasticity*, strength, *internal friction*), for which their presence is often the determining factor. The elastic fields induce an increased concentration of impurities near the axis and the formation of a *Cottrell atmosphere*, which may bring about the coagulation of impurities at the dislocations (in transparent crystals it "decorates" dislocations). Dislocations change the optical properties of the crystals, which is the basis of the method of observing isolated dislocations in transparent materials. At points where dislocation lines emerge at the external surface of the body the chemical strength of the crystal is weakened, and special reagents can destroy the neighborhood of a dislocation axis, forming visible pits. This is the basis of the *selective etching method* which is the the main method for observing individual dislocations in bulk samples of non-transparent materials. A system of *dangling bonds* at the dislocation core distinguishes the dislocation line from electrical, magnetic, and optical points of view. Dislocations may carry or capture electrical charge, and possess a *magnetization* that differs from that of the bulk crystal. They can also increase the electrical *resistance* of conductors and change the concentration of free electrons in semiconductors. They play an important role in magnetic crystals by determining different relaxation phenomena. While moving, a dislocation can emit or adsorb vacancies, thereby changing their total number within the crystal. The dynamic formation of charged vacancies in *ionic crystals* and semiconductors may be accompanied by a glow (*luminescence*) of the crystal. The rate of diffusion of point defects along the dislocation axis is, as a rule, greater than the rate of their *diffusion* through the volume of the regular crystal. The coefficient of linear diffusion along a dislocation can exceed the coefficient of bulk diffusion by several orders of magnitude. Therefore, dislocations play the role of "drain tubes" along which point defects can easily travel relatively long distances within the crystal.

See also *Edge dislocation, Epitaxial dislocations, Frank dislocation, Helical dislocation, Partial dislocation, Perfect dislocation, Prismatic dislocations, Screw dislocation, Somigliana–Volterra dislocations, Surface dislocations.*

DISLOCATION SIGN

Characteristic (sign) attributed to elementary (linear) *dislocations*, determined by the orientation of the *Burgers vector* $\boldsymbol{b}$, and of the vector tangent to the dislocation line. In the case of linear dislocations lying in parallel *slip planes* and having parallel lines, the dislocation sign is associated with the direction of the Burgers vector. To determine the direction of $\boldsymbol{b}$ a rule is used which, not withstanding a certain arbitrariness, gives an unambiguous result in successive applications. Initially a positive direction is chosen (arbitrarily) along the dislocation line; i.e. the direction of a unit vector τ tangent to this line is assigned at a point, and then a closed *Burgers contour* in a crystal with dislocations is drawn so that, when sighting along the positive τ direction, the tracing of the boundary is in the clockwise sense. The resulting Burgers vector $\boldsymbol{b}$ connects the end with the beginning of the gap in the Burgers contour traced in the initial ideal lattice. This method of defining $\boldsymbol{b}$, suggested by Frank and Bilby, is called the *rule of "end to beginning along a right-hand screw"*. As a result, we have $\boldsymbol{b} = \oint_c (\partial u/\partial l)\,dl$, where the integral along the closed contour c that encloses the dislocation line is taken in the direction determined by a right-hand screw relative to τ. If the positive direction along this line is changed, then the Burgers vector reverses its sign. Sometimes a definition of $\boldsymbol{b}$ is used whose sign is changed to the opposite, so it is necessary to check the rule for assigning the sign to $\boldsymbol{b}$. It is important that the Burgers vector remain constant along the dislocation. In the simplest case of a linear *edge dislocation*, a positive (+) sign is assigned to the dislocation, for which the extra half-plane is situated in the upper part of the crystal, and the negative (−) sign corresponds to the extra half-plane located below. "Upper" and "lower" are relative concepts that are determined by the crystal orientation relative to an observer. The sign of elements of a closed *dislocation loop*, having a unique Burgers vector, is determined by the direction of the vector tangent to a given element.

DISLOCATION SLIP, dislocation glide

Dislocation motion along the plane, determined by a line of dislocations and its *Burgers vector* $\boldsymbol{b}$ (*slip surface*). The slip of dislocations is accompanied by the shear of one part of the crystal relative to another through $\boldsymbol{b}$, i.e. the slip of dislocations causes the *shear strain* of the crystal. All possible slip planes and their directions form the *slip system*. The presence of *pencil slip* causes twisting slip lines to appear on surfaces of deformed body-centered cubic *metals*, in contrast to the long linear slip lines in face-centered cubic and hexagonal close-packed crystals. Screw dislocation components (see *Screw dislocations*) in inhomogeneous internal stress fields can exhibit double *cross slip*.

DISLOCATION STEP

A dislocation segment connecting two sections of an *edge dislocation* lying in different parallel *slip planes*. If the distance between the slip planes equals one lattice parameter the dislocation step is called a unit step, and for larger separations it is called a *superstep*. The step at a screw *dislocation* sector coincides with the *kink on dislocation* because in this case the plane where the dislocation is located, is its slip plane. It is possible to ascribe a sign (positive or negative) to a dislocation step to distinguish steps which transfer the dislocation line from the given slip plane to the plane located above, and to that below. Between two steps at the dislocation line there is a Coulomb type interaction whereby steps of the same sign repel and those of opposite sign attract. In the elastic-isotropic continuum approximation (see *Isotropy of elasticity*) the modulus of the interaction energy of two steps, spaced a distance L apart, is given by the expression

$$W = A\frac{Gb^2a^2}{8\pi L},$$

where b is the magnitude of *Burgers vector*, a is the size of the dislocation step in the direction perpendicular to the slip planes, G is the shear modulus, and A is a constant of the order of unity, the exact value of which depends on the type of dislocation. There are several mechanisms for forming a dislocation step. They may occur at the intersection of two dislocations moving in intersecting slip planes, so the steps on one dislocation are formed parallel to the Burgers vector of the other one. The magnitude and direction of the Burgers vector of the dislocation step are the same as those of dislocation sections located above (or below) the slip plane. A pair of steps of opposite signs appears in the so-called *cross slip* of edge dislocations. Since in the general case the slip plane of the dislocation step does not coincide with the slip plane of the main part of the dislocation line, the dislocation slip as a whole can occur only during non-conservative motion (climb) of the step, accompanied by the appearance within the crystal of chains of *vacancies* or *interstitial atoms*. Such a step motion may only occur when thermal activation strongly contributes to the motion of the dislocations with steps. Due to the diffusion flow (see *Diffusion*) of vacancies (or of interstitial atoms) the step may move along the dislocation line, i.e. the dislocation step is a source (or sink) for *point defects* of the crystal lattice. When the Burgers vector of the dislocation being intersected is coplanar with the slip plane of the initial dislocation then the step located at it will have the same slip plane, and such a dislocation with its step may move conservatively. At the intersection of two dislocations, each of which is split into *partial dislocations*, there appears a constriction at the *stacking fault* connecting the partial dislocations, because the formation of the step brings about the merging of the partial dislocations accompanied by the annihilation of the stacking fault at the step. The intersection of split dislocations involves a stress that is considerably higher than that at the intersection of unsplit dislocations. In an *ionic crystal*, at a dislocation with a Burgers vector of the $\langle 011\rangle$ type and a line along $\langle 100\rangle$, at a dislocation step with a bend height of one interatomic distance there appears an effective (either positive, or negative) charge, equal to the half of the charge of the boundary ions. A neutral step appears at such a dislocation, if the step length is equal to two interplanar distances. Dislocation steps play a very important role in the kinetics of crystal *strains* since their appearance and motion to a large extent determine the aggregation of point defects (vacancies and interstitial atoms) during the process of *plastic deformation*.

DISLOCATION STRING

Mechanical model which treats a *dislocation* in terms of a flexible elastic string with a linear mass density M and intrinsic linear energy density (linear stretching) C. This model is used for the description of the conservative movement (*slip*) of weakly bent dislocations. If the linear deviation of the dislocation line $u\ (x,t)$ from some static configuration $u_0(x)$ projected along the x axis is cho-

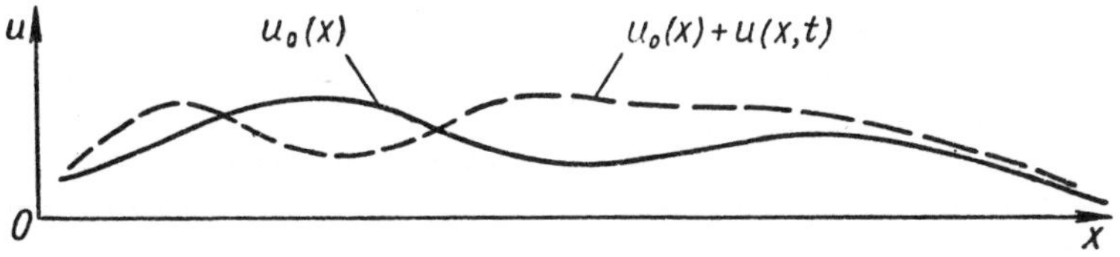

Static $u_0(x)$ and instantaneous $u_0(x) + u(x,t)$ configuration of a weakly bent dislocation line.

sen as the dynamic variable (see figure), then the equation for the dislocation motion in this string approximation has the form

$$M\frac{\partial^2 u}{\partial t^2} - C\frac{\partial^2 u}{\partial x^2} + B\frac{\partial u}{\partial t} = F(u,x,t).$$

The first term in the left represents the force of inertia, the second represents the stretching force, and the third is the force arising from viscous friction. The total force per unit length $F(u,x,t)$ on the right-hand side includes external *mechanical stresses*, *internal stresses*, and the Peierls potential (see *Peierls relief*). The parameters M and C provide the approximate *effective mass* and self-energy of the dislocation, with values determined by the *Burgers vector* of the dislocation $\boldsymbol{b}$, and by the material constants of the crystal; they can be estimated from the expressions $M \propto \rho b^2$ and $C \propto Gb^2$, where ρ is the density of the crystal, and G is the *shear modulus* along the direction of slipping. The coefficient B of the equation characterizes the viscous retardation of the dislocation by elementary excitations of the crystal (by *phonons*, *conduction electrons*, etc.) or by mobile *point defects*. The string equation replaces the more complex rigorous equation of dislocation motion where the inertial force and the self-action force have a nonlocal nature. The condition for the applicability of the *string model* is limited by the inequalities $\partial u/\partial x \ll 1$ and $\partial u/\partial t \ll c$, where c is the speed of sound. The string model is widely used for the theoretical analysis of the effect of dislocations on the *specific heat*, *thermal conductivity*, and *electrical conductivity*, the acoustical properties of crystals, and also for investigating the processes of dislocation motion through Peierls barriers, and through *point defects* of the crystal structure.

DISLOCATION SUPERCONDUCTIVITY

Non-uniform superconducting state which occurs near a dislocation network at a temperature and magnetic field exceeding the corresponding critical values for the same material without *dislocations*. The supercurrent that flows within such a structure is anisotropic with respect to the axes of the dislocations. This superconductivity, which occurs in the vicinity of individual dislocations, and also of their aggregates, has a localized nature. The *critical magnetic field* in this case is strongly anisotropic with respect to the dislocation axes.

Dislocations in superconductors very strongly affect the value of the *critical current*, which due to the *vortex pinning* at the dislocations considerably exceeds its value in "pure" superconductors.

The critical magnetic field of superconductors with dislocations is significantly increased by the variation of the properties of the material, as well as due to the variation of the *mean free path* of the electrons, which becomes anisotropic. The presence of a gap in the electron spectrum changes the force resisting dislocation motion, and helps to increase the *plasticity* of the material in the superconducting state.

DISLOCATION TENSION, LINEAR

See *Linear dislocation tension.*

DISLOCATION TRANSITION RADIATION

See *Transition radiation by dislocations.*

DISLOCATION WALLS, dislocation boundaries

One-dimensional ensembles of straight, parallel, regularly arranged *dislocations* of the same type with parallel *slip planes*. Regular equilibrium dislocation walls of *edge dislocations* do not create long-range stress fields since the latter exponentially attenuate at distances of the order of the average spacing between dislocations within the wall. Non-equilibrium dislocation walls, however, may serve as a source of considerable stress. A wall formed by parallel *screw dislocations* creates stresses that extend over distances comparable with the characteristic sizes of dislocation walls. If the wall is torn it becomes the source of a stress field that is similar to the stress field from a *disclination*, corresponding to the disorientation of a dislocation wall. These walls play an important role in the *strain hardening* of monocrystals because they are sources of both dislocations and the barriers to them. Dislocation walls have photoresistive (see *Photoconductivity*) and semiconducting properties in semiconductor crystals.

DISORDERED SOLIDS

A generic name for *solids* lacking any type of either short-range or long-range order. Routinely, one has to specify which particular type of *disorder* is implied. The destruction of long-range order in the positions of atoms in a solid, regardless of the types of atoms in particular positions, is called a *topological disorder*. When the term "disordered solid" is used without specifying the type of disorder, a topological disorder is usually implied. Disordered solids featuring this disorder are classified as amorphous solids (see *Amorphous state*). The breakdown of order in the relative positioning of atoms of differing types is called *compositional disorder*. Topologically ordered *binary alloys* may exhibit this type of disorder, with each type of atom randomly occupying the sites of a regular crystal lattice. The breakdown of the relative ordering of *magnetic moments* of different atoms, or atomic complexes that display certain short-range order, represents a *magnetic disorder*, while that of the relative orientation of molecules or atomic complexes is an *orientational disorder*. An example of solids featuring orientational order while having no topological order or translational invariance of their structure is represented by *quasicrystals*. The breaking of topological, compositional, magnetic, orientational, and other types of symmetry or order may arise from thermal *fluctuations* or by external actions, such as electromagnetic fields, mechanical strains, etc. The emergence of order, its breakdown and change in type, all correspond to *order–disorder phase transitions* in solids. See also *Long-range and short-range order*, *Phase transitions*.

DISORDERED SOLIDS, SPECTRA

See *Elementary excitation spectra of disordered solids*.

DISORDER, ELECTRON-INDUCED

See *Electron-induced disordering*.

DISORDERING OF ALLOYS

See *Alloy disordering*.

DISPERSE STRUCTURE

The structure of liquid or solid systems consisting of *dispersed phase* particles distributed in various ways within the volume of a matrix *phase*. The inclusions may be solid, liquid, gaseous. They are introduced into the matrix as a result of *phase transitions* (*alloy aging*, *martensitic transformation*, eutectoid *alloy decomposition* (see *Eutectoid alloys*), eutectic *crystallization*, devitrification of *amorphous metals and metallic alloys*, etc.), by internal oxidation, by introduction of the high-melt particles into the melt (*composite materials*), and as a result of the action of nuclear radiation or *ion implantation*. The properties of the inclusions depend on the degree of dispersity of particles, on the energy of their interaction with the medium, on shape, *state of matter* and structure (e.g., presence of atomic ordering of the inclusion lattice). Depending on their size, the inclusions are classified as rough (low) dispersed (>0.1 μm) and fine (highly) dispersed (0.001 to 0.1 μm). Their shape may be threadlike (needle-like), plate-like, or spherical. Disperse particles have excess *surface energy*, which implies thermodynamic instability and a disposition toward *coalescence* (*coagulation*). Inclusions formed as a result of phase transformations have a shape and orientation with respect to the crystal lattice or matrix that is conditioned by the ratio of the surface energy to the energy of the internal stresses, which arise during the formation of a disperse structure. The matrix containing the disperse structure is characterized by increased *strength* and *high-temperature strength* (see *Alloy aging*, *Precipitation-hardened materials*). The distribution of particles over the volume of the matrix may be classified as regular or irregular. A regular, quasi-periodic distribution of particles is called a *modulated structure*. The disperse particles create a *macrolattice*, with a *unit cell* which may be cubic, tetragonal etc., depending on the symmetry of the elastic fields around the particles (see *Crystal symmetry*) within the matrix volume. For a low bond energy between the particles (less than their thermal energy, in order of magnitude), the so-called *thixotropic disperse structures* form, which have low thermodynamic stability, but recover in time (e.g., disperse structures formed in aqueous solutions of *aluminum* or *iron* hydroxide). (Thixotropic gels liquify when shaken and solidify when left standing.) Non-thixotropic disperse structures are more stable, but can be destroyed irreversibly. These are related to disperse structures of ceramics, binding materials (cements) and of the dispersion-hardening alloys.

DISPERSION (fr. Lat. *dispersus*, scattering, dispersion)

In different areas of the natural sciences dispersion determines the characteristics of the scatter of values of a physical parameter. Its meaning varies somewhat with the parameter under consideration, and some examples will be given. In mathematical statistics the *dispersion σ of a random variable parameter x* determines the intensity of *fluctuations*: $\sigma^2 = \langle x - \langle x \rangle \rangle^2$, where the angular brackets $\langle \ldots \rangle$ denote an average value. The wave *dispersion relation* is the dependence of the frequency $\omega(\boldsymbol{k})$ on the wavenumber k which provides the magnitude $k = 2\pi/\lambda$ and the direction of motion of the wave, where λ is the wavelength. The *phase velocity* $v_{\text{ph}} = \omega/k$ and the group velocity $\boldsymbol{v}_{\text{g}} = \text{d}\omega/\text{d}\boldsymbol{k}$ depend on the dispersion law (see *Sound velocity dispersion*, *Dispersion of light*). In the presence of dispersion the propagation of an anharmonic wave is accompanied by a variation of its shape. A *dispersive medium* is one whose *dielectric constant* $\varepsilon = \varepsilon' + \text{i}\varepsilon''$, *magnetic permeability* $\mu = \mu' + \text{i}\mu''$ (or susceptibility $\chi = \chi' + \text{i}\chi''$) depend on the frequency (wavelength). When these quantities are complex, the real (primed) part is called *dispersion* and the imaginary (double primed) part is called *absorption*. The real part determines the frequency and the imaginary part determines the losses in a resonator. The singularities of propagation and absorption of elastic (including sonic) and light waves in dispersive media depend on ω and $\boldsymbol{k}$ (e.g., susceptibility, elastic constants, etc). *Spatial dispersion* is caused by the fact that the response of the dispersive medium to an external disturbance at an arbitrary point $\boldsymbol{r}$ depends on the value of the disturbing field not only at this point, but also at surrounding points (nonlocal response). *Temporal dispersion* is caused by the time lag of the response, i.e. by its dependence on the value of the disturbing field not only at the moment of observation, but also at earlier times.

In addition, the dependence of the frequency ω or the energy E of a *quasi-particle* in a solid on its *quasi-momentum* $\hbar\boldsymbol{k}$ is called dispersion, or more accurately, a *dispersion law*. The dispersion law $E(\boldsymbol{k})$ determines the structure of the *energy bands* of electrons, *holes*, *excitons*, and the spectrum of *phonons*, *plasmons*, *magnons*, and other quasi-particles.

DISPERSION, ANOMALOUS

See *Anomalous dispersion of light*.

DISPERSION DIELECTRIC CONSTANT ANALYSIS

One of the methods for determining the frequency dependence of the *dielectric constant* of solids from experimental data about the response of the body to an applied alternating electric field. The simplest model to explain the frequency dependence of the reflectivity coefficient is that of a *solid* consisting of a set of the non-interacting dispersion oscillators. By variation of oscillator parameters (frequencies, attenuation coefficients, and also their number), it is possible to attain good agreement between model calculations and experiment.

DISPERSION GRINDING

Fine grinding of a solid to form *disperse structures*. The specific amount of work spent for the dispersion depends on the cohesion parameters (see *Cohesion*), on the structural characteristics of the pulverized body, on the *surface energy*, and on the extent of the grinding. Introduction of *surface-active agents* into the system reduces the energy expenditure during the course of dispersion, and increases the dispersity of the ground phase. Depending on the composition and properties of the dispersed phase and of the dispersing medium, and also on the method of dispersion, the lower boundary of particle sizes may range from a fraction of a micrometer to several dozen μm. In industrial and laboratory practice the dispersion is carried out using mills of various types: ball mills, vibratory mills, jet mills, etc.

DISPERSION HARDENING

Improving the *strength* (*hardness*) of metallic oversaturated *solid solutions* by their decomposition (see *Alloy decomposition*), and by the separation of the dispersed phase enriched by the alloying element (see *Alloying*). The degree of dispersion hardening depends on the bulk share (quantity) of the separation phase, on the size and shape of its particles, on the character of the interface between the phases (coherent or non-coherent), on the value of coherent stresses, on the internal structure (atomically ordered or disordered), on the value of the *elastic moduli* of the matrix, and on the extent of the separation. Several theories describe the hardening of alloys and the *yield limit* at different stages of decomposition (see *Alloy aging*). Dispersion hardening at the early stages is well described by the *Mott–Nabarro theory*, which explains it in terms of the interaction of internal coherent stresses around the particles and moving *dislocations*. The hardening at later stages of decomposition is satisfactorily described by the *Orowan theory* of precipitation strengthening (E. Orowan, 1948), where the yield limit of the aged alloy depends only on the distance between its non-coherent particles.

DISPERSION LASER RESONATOR

An optical *resonator* with losses or Q-values which depend on the frequency ν. This dependence is achieved by the introduction of dispersion elements into the resonator (Fig. 1): refracting prism (a), holographic *diffraction grating* (b), *Fabry–Perot interferometer* (c), interference-polarization filters (d). The selection of modes is determined by the propagation of the radiation beams at different angles relative to the axis of the resonator (see Fig. 1(a, b)), or by the spectral dependence of the transparency of the dispersion elements (see Fig. 1(c, d)). The adjustment of the radiation spectrum ν_L of the *laser* with the help of a dispersion

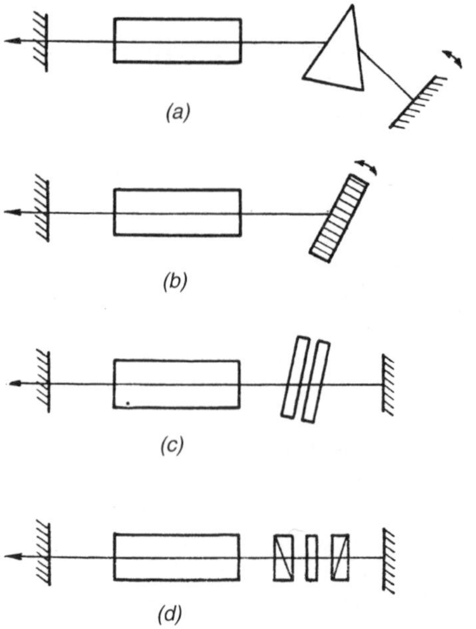

Fig. 1. Dispersion laser resonators.

laser resonator is carried out by moving the curve of losses $\gamma(\nu)$ along the amplification curve $\alpha(\nu)$ of the *active laser element* (Fig. 2), for $\delta\nu_\gamma < \delta\nu_\alpha$, where $\delta\nu_\gamma$ and $\delta\nu_\alpha$ are the half-widths of the $\gamma(\nu)$ and $\alpha(\nu)$ curves, respectively. For $\delta\nu_\gamma \ll \delta\nu_\alpha$ the generation may be of a single frequency. The adjustment of the resonator during the power generation provides for the control the spectral-kinetic parameters of the radiation (see *Tunable lasers*). Using the principle of a dispersion laser resonator, *refractive lasers* of the solid and liquid *active media* types (crystals with color centers, organic com-

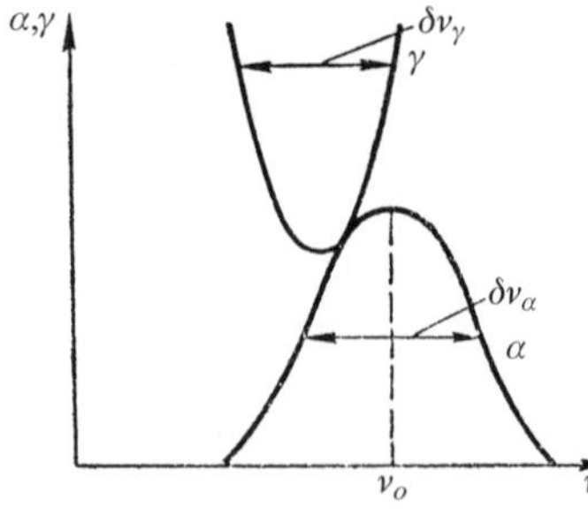

Fig. 2. Amplification curve of the laser active element.

pound solutions, etc.) have been designed with a broad band adjustable output frequency.

DISPERSION LAW

Relationship either between energy ε and momentum $\boldsymbol{p}$ (*quasi-momentum* in the periodic *crystal lattice* potential) of elementary excitations (*quasi-particles*), or between the frequency ω and the wave vector $\boldsymbol{k}$ of waves (sonic, electromagnetic, spin, etc.) propagating in many-particle systems (e.g., *solids*). The dependence of $\varepsilon(\boldsymbol{p})$ on $\boldsymbol{p}$, or $\omega(\boldsymbol{k})$ on $\boldsymbol{k}$, can be rather involved, and sometimes it is multivalued (see *Semiconductors*).

DISPERSION LAW, KANE

See *Kane's dispersion law*.

DISPERSION OF LIGHT

Decomposition of a light beam into its component frequencies during the *refraction of light*, *interference of light* or *diffraction of light*. It results from the dependence of the *refractive index* of a material on the wavelength (frequency) of the light. The dispersion of light was discovered by T. Hariot (approx. 1605), then rediscovered by J.M. Marci (1658), and experimentally studied by I. Newton (1672). In the classic electronic theory of light dispersion introduced by H. Lorentz an electron bound to an atom, ion or molecule in the *crystal lattice* is modeled by a harmonic oscillator. In the field of the light wave $E\mathrm{e}^{\mathrm{i}\omega t}$ with frequency ω the oscillator performs forced vibrations near its equilibrium position, and this induces an ac polarization of the crystal at this frequency. When the frequency ω approaches the natural frequency ω_0 of the electron's vibration, there is a *resonance* that enhances the *light absorption*. The frequency dependences of the refractive index n and the *absorption index* κ of nonconducting crystals are determined by the relations

$$n^2 - \kappa^2 = 1 + \frac{Ne^2}{m\varepsilon_0}\left[\frac{\omega_0^2 - \omega^2}{(\omega_0^2 - \omega^2)^2 + \gamma^2\omega^2}\right], \quad (1)$$

$$2n\kappa = \frac{Ne^2}{m\varepsilon_0}\frac{\gamma\omega}{(\omega_0^2 - \omega^2)^2 + \gamma^2\omega^2}, \quad (2)$$

where γ is the *friction* coefficient, N is the concentration of electron oscillators, m is the mass, and

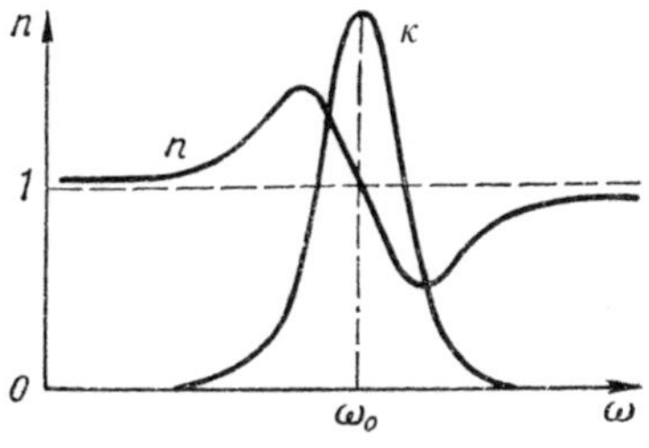

Dispersion of light.

ε_0 is the dielectric constant of vacuum. If several electrons, bound by different elastic forces, take part in the process, the right-hand sides of Eqs. (1) and (2) are replaced by summations.

Typical dispersion n and adsorption κ curves for an individual oscillator are shown in the figure. In the region between the peaks n decreases with increasing ω (region of *anomalous dispersion of light*), and outside this region n grows with the frequency. The absorption is strong in the anomalous dispersion region near ω_0. The frequency range $(\omega_0^2 - \omega^2)^2 \gg \gamma^2\omega^2$ corresponds to the region of transparency ($\kappa \approx 0$). In this region

$$n^2 = 1 + \frac{Ne^2}{m\varepsilon_0}\frac{1}{\omega_0^2 - \omega^2}. \quad (3)$$

For *metals* the classic *Drude–Zener–Kronig theory* (P. Drude, C. Zener, R. Kronig) of light dispersion is based on the free electron model. For this model Eqs. (1) and (2) reduce to the form

$$n^2 - \kappa^2 = 1 - \frac{Ne^2}{m\varepsilon_0}\frac{1}{\omega^2 + \gamma^2}, \quad (4)$$

$$2n\kappa = \frac{Ne^2}{m\varepsilon_0\omega}\frac{\gamma}{\omega^2 + \gamma^2}. \quad (5)$$

Ideal metals with infinitely long average electron *mean free paths* should not absorb light, so $\gamma = 0$ and

$$n^2 = 1 - \frac{Ne^2}{m\varepsilon_0\omega^2}. \quad (6)$$

If $0 < Ne^2/(m\varepsilon_0\omega^2) < 1$, then $0 < n < 1$. For normally incident light the metal is transparent, but beginning with small angles *total internal reflection* occurs, characteristic of short wavelengths. For long wavelengths when $Ne^2/$

$(m\varepsilon_0\omega^2) > 1$, n has an imaginary value, and total internal reflection occurs for all angles of incidence. As the free motion of *conduction electrons*, limited by their mean free path, only lasts for very short time intervals, the above extreme cases are not easily observed. Nevertheless, the experiments of R. Wood with thin layers of *alkali metals* demonstrated their transparency in the UV spectral range and their strong absorption in the visible and infrared. The quantum theory of light dispersion leads to similar results.

DISPERSION OF MAGNETIC ANISOTROPY in magnetic materials

Dependence of the constants of the *magnetic anisotropy*, and the direction of the *easy magnetization axis* on the crystal structure spatial coordinates. The dispersion of anisotropy constants is called *amplitude dispersion*, and that of the easy magnetization axes is called *angular dispersion*. In *magnetic films* the angular dispersion leads to the appearance of the so-called *magnetization ripples* (or ripple structures).

DISPERSITY

A parameter of a *disperse structure* associated with the dimensions of inclusions (e.g., precipitated particles) in a medium. The term dispersity often denotes the average size of scattered material. More precisely, dispersity is characterized by the distribution function of the dimensions of dispersed inclusions, or by the distribution function of the dispersive phase volume according the sizes of particles $F(V)$. The probability $\mathrm{d}W$ to find an inclusion with a volume between V and $V + \mathrm{d}V$ is given by the expression $\mathrm{d}W = F(V)\,\mathrm{d}V$. A monodispersed system is characterized by the presence of one narrow maximum $F(V)$, corresponding to the most probable dimension of the inclusions. For polydispersive systems, where the function $F(V)$ has the form of a diffuse curve, perhaps with a succession of local maxima, it is more correct to characterize dispersity by the specific surface area of the dispersed inclusions.

DISPLACED LAYER STRUCTURES

Crystals with high concentrations of parallel *stacking faults* of various kinds, which break the regular alternation pattern of atomic layers in a *lattice*. These include *polytype defects* formed by the displacement of layers to alternate structure positions (e.g., from layers A or B of the wurtzite hexagonal type structure to layers A, B or C of the sphalerite FCC structure), and *turbostratum stacking faults* where the layers are displaced to random locations. The first type of disordering is typical of SiC, ZnS, and other tetrahedral structures tending toward *polytypism*. In this case no turbostratum defects are formed. The largest amount of layer disordering in tetrahedral structures was detected in *lonsdalite*, a hexagonal modification of *diamond*. The second kind of disordering takes place exclusively in strongly anisotropic laminated structures, such as *graphite*, graphite-like *boron nitride*, and some clay *minerals*. The concentration of polytype and turbostratum stacking faults is determined from characteristic features in diffraction patterns of layer-disordered structures.

DISPLACEMENT, ELECTRIC

See *Electric flux density*.

DISPLACEMENT SPIKE

The region of a *solid*, which contains atoms subjected to inelastic displacements and vacant lattice sites, which are generated as a result of propagation of the *collision cascade* of atoms produced by irradiation with energetic particles. Two zones are distinguished within the displacement spike: the depletion zone, which is located at the center of the cascade and contains an excess quantity of vacant sites, and the enriched zone, which occupies the peripheral region of cascade and contains an excess quantity of *interstitial atoms*. See also *Seeger zone*.

DISPLACEMENT VECTOR, vector of displacement, vector of a deformation

The difference $\boldsymbol{u}$ of radius vectors of a point in the solid without ($\boldsymbol{x}$) and with ($\boldsymbol{x}'$) a *strain*; $\boldsymbol{u} = \boldsymbol{x}' - \boldsymbol{x}$.

DISPYRATIONS

The arrangement of Volterra dislocations (see *Somigliana–Volterra dislocations*) which simultaneously combine *disclinations* and translational *dislocations* subject to the condition that the *Frank dislocation* vector of disclination is parallel to the *Burgers vector* of the dislocation. Perfect and partial dispyrations are distinguished. *Perfect dispyrations* can only appear in *crystals* with screw *symmetry axes*, and for the formation of a linear *defect* the operation of translation together with rotation is required. If the onset of a dispyration is associated with an arbitrary screw motion along the edge of the cut then a *partial dispyration* is created. Within the crystal it is necessary for a *stacking fault* to be adjacent to it. In a continuum the separation of a dispyration does not make practical sense, because all the elastic field parameters may be obtained by simply summing those from the disclinations and dislocations that form the dispyration. Sometimes what is observed is a special continuum of dispyrations limited to the screw parameter. A medium with dispyrations possesses Riemann curvature (see *Curvature tensor*). The existence of dispyrations in ordinary crystals has not been experimentally proven, although they probably play some role in organic and biological materials, and they seem especially important for membrane biology (see *Membrane*). There is a possibility of forming non-crystalline structures within glasses (see *Vitreous state of matter*), where they add a stable curvature.

DISRUPTED SURFACE LAYER

A layer at a *solid surface* saturated with structural *defects*. It forms during mechanical processing (cutting, polishing, finishing, shot-blasting, broaching, rolling, etc.). Depending on the form of the structural defects: *macrocraks* and *microcracks*, large- and small-angle *grain boundaries* and *subgrain* boundaries, *vacancies*, *micropores*, and *dislocations*, the disturbed (or perturbed) surface layer is subdivided into various zones. As a rule, there are three such zones: that of surface roughness, that of macro- and microcracks, and a transition layer saturated with microdefects. The disturbed surface layer is usually several microns to several hundred microns thick. This layer lends fatigue strength and wear resistance to many industrial materials. Its presence sometimes increases and sometimes decreases their *corrosion resistance*. The disturbed layer plays a negative role in processes of epitaxial deposition of materials onto substrates with such a layer.

DISSIPATION FUNCTION

Rate of *energy dissipation* (usually per unit volume). This is associated with the transformation of part of the energy of ordered motion (e.g., motion of a pendulum) into energy of disordered motion (thermal energy). *Dissipation forces* bring about this dissipation (e.g., *friction* forces).

DISSIPATIVE STRUCTURES

Ordered spatial structures (stationary or changing in time), which appear in open non-equilibrium systems as a result of *self-organization* processes. The name dissipative structure emphasizes the fact that the system is maintained in the ordered state by *energy dissipation*. As a rule, these structures are stable non-uniform distributions of composition, electron density, electric field, temperature and other physical values, which appear as a result of the development of instabilities in the uniform state with respect to the spatial *fluctuations* of these values. Examples of dissipative structures are: in chemical kinetics – complex spatial distributions of chemical components in the *Belousov–Zhabotinsky reaction*; in hydrodynamics – convection cells (*Bénard cells*), turbulence, etc; in solids – structures of *domains* in semiconductors with N-shaped *current–voltage characteristics* (*Gunn diodes*), *current filamentation* in semiconductors with S-shaped current–voltage characteristics. A particular case is the non-uniform structures which occur during the *relaxation* of a strongly unbalanced system to thermodynamic equilibrium, which during the course of relaxation appear to be "frozen" due to the slowing-down of the kinetics. Another case is long-period structures which appear at the *spinodal decomposition* of a solid solution. Non-uniform structures which are thermodynamically balanced (such as an *intermediate state*, domain magnetic structure, et al.) are not related to dissipative structures. Dissipative structures are studied by *non-equilibrium thermodynamics*. In many cases the structure type can be determined by the analysis of the stability of the uniform state with respect to low level fluctuations, performed by linearizing the equations of the system kinetics. Beyond the limits of applicability of the linearized equations these structures may be studied using non-linear differential equations, or by numerical methods.

DISTORTION

Disturbance of the normal arrangement of atoms of a *crystal* in the vicinity of a *defect* which lowers the *crystal symmetry*; a parameter of the non-uniformity of displacements upon deformation; in optics – degradation of the geometrical symmetries of an object and its image, caused by optical system aberrations.

DISTORTION, JAHN–TELLER

See *Jahn–Teller distortion*.

DISTORTION TENSOR

Gradient of the *displacement vector* U: $U_{ik} = \partial u_k/\partial x_i$ (x_i are Cartesian coordinates of points of the body before deformation). The distortion tensor in the general case is non-symmetrical. Its symmetric part gives the *strain tensor* of linear *elasticity theory* $u_{ik} = (U_{ik} + U_{ki})/2$, and the antisymmetric part is the *rotation tensor* in the linear approximation $\Omega_{ik} = (U_{ki} - U_{ik})/2$. This tensor accompanies the *rotation vector* $\omega_i = (\nabla \times u)_i/2 = -e_{ijk}\widetilde{\Omega}_{jk}/2$ (here e_{ijk} is the third-rank unit antisymmetric tensor). The distortion tensor and rotation vector are both dimensionless.

DISTRIBUTION FUNCTION

Basis for the statistical description of systems consisting of a large number of identical objects; it defines the probability density used for determining various parameters of a system. In particular, the distribution function defines the probability density of finding a particle (*quasi-particle*) of a many-particle system in a certain state; and using this function the statistical averages of concentrations, momenta and other characteristics of a macroscopic system can be determined. For a classical system in *thermodynamic equilibrium* the *Gibbs distribution function* is used, while in particular quantum mechanical cases a Fermi–Dirac (half-integer spin) or Bose–Einstein (integer spin) distribution applies (see *Fermi–Dirac statistics*, *Bose–Einstein statistics*), *Maxwell distribution* and *Boltzmann distribution*. *Nonequilibrium distribution functions* may be calculated from the *kinetic equations*.

DIVACANCY, sometimes called bivacancy

A complex consisting of two vacancies situated at neighboring *crystal lattice* sites. The appearance of divacancies is especially probable in crystals containing excessive nonstoichemical or nonequilibrium vacancies at not too high temperatures determined by the dissociation energy of the divacancy (as a rule, this is on the order of decimal parts or units of electron-volts). A divacancy can diffuse as a whole over a crystal with a *diffusion coefficient* which is much smaller than that for an isolated vacancy. Well studied are divacancies in silicon crystals where they are one of the commonest radiation *defects*. A divacancy in silicon is an *amphoteric center*, mobile at temperatures $T \geqslant 200$–$300\,^\circ$C (the diffusion *activation energy* is 1.3 eV, the dissociation energy is 1.6 eV). In crystals of a complicated composition, several kinds of divacancies consisting of atoms of various elements in the crystal can exist.

DOMAIN

Regions in an otherwise crystallographically and chemically homogeneous *solid*, or other *condensed state of matter*, which differ one from another in magnetic, electric, and other properties, or in an *order parameter*. There are *equilibrium domains* which appear when an overall heterogeneous state is thermodynamically more favorable than a homogeneous one (*magnetic domains*, domains in superconductors, ferroelectrics, normal metals involving the *de Haas–van Alphen effect*, etc.; see also *Intermediate state*). There are also *static nonequilibrium domains* such as, for example, elastic twins (see *Elastic twinning*), *antiferromagnetic domains*, domains in liquid crystals, and domains which appear in nonequilibrium systems subject to external pumping (see *Gunn effect*). Domains are separated from each other by *domain walls*.

See also *Acousto-electric domain*, *Antiferromagnetic domain*, *Diamagnetic domains*, *Dumbbell domain*, *Ferroelectric domains*, *Honeycomb domain structure*, *Kittel's domain structure*, *Landau–Lifshits domain structure*, *Opposed domains*, *Structural domain*, *Thermoelectric domain*.

DOMAIN BOUNDARY

The same as *Domain wall*.

DOMAIN, MAGNETIC

See *Magnetic domain.*

DOMAIN SPLITTING IN FERROELECTRICS, trilling in ferroelectrics

Splitting of a *ferroelectric* into *domains* with directions of spontaneous polarization at an angle of 120° with respect to one another. This corresponds to the presence of three crystallographically equivalent directions in the initial high-temperature phase. The *domain structure* of the threefold split type may occur in crystals which undergo a change of symmetry from C_{3v} to C_s or from C_{3v} to C_1; the vector of the resulting spontaneous polarization lies in the σ_v plane, and is directed transversely to the threefold axis of the phase of C_{3v} symmetry. See also *Twinning of crystals, Twinning structure.*

DOMAIN WALL in ferroelectrics

An interface between *ferroelectric domains* which is a few lattice constants thick. The center of a 180° domain wall has zero polarization. The appearance of a domain wall is associated with a localized energy loss arising from elastic *strain* and electrostatic interactions between dipoles; the establishment of a system of domain walls can provide a state of minimum free energy. The orientation of a wall does not involve any spontaneous deformation accompanying the polarization of *domains* in the wall. The thicknesses and energies of domain walls differ significantly in different crystals. A 180° domain wall in *barium titanate* has a thickness of about $(5–20)\cdot 10^{-8}$ cm and an energy of about 10 erg/cm^2; while in Rochelle salt these values are approximately $200\cdot 10^{-8}$ cm and 0.012 erg/cm^2, respectively. Domain walls can move under the influence of an external electric field, and their lateral motion provides one of the main mechanisms for the process of repolarization.

DOMAIN WALL in liquid crystals

A layer that separates two regions (*domains*) of a *liquid crystal* differing in the direction of the uniform orientation of their molecules; at the wall the *director* $\boldsymbol{n}$ changes its direction from one orientation in the first domain to another in the second domain. A rotation of the director is performed

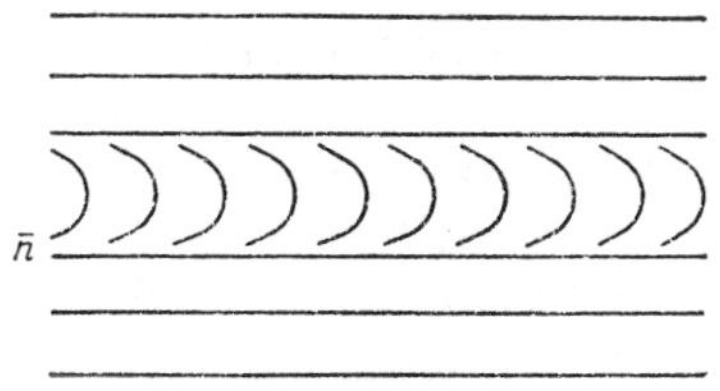

(a)

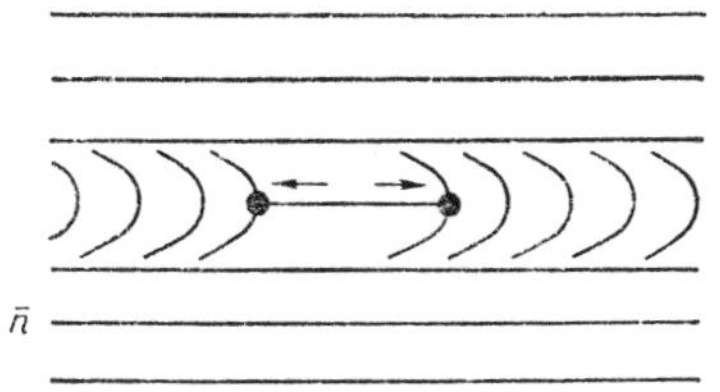

(b)

Domain walls in liquid crystals.

about an axis that is either parallel to the wall (Fig. (a)), or normal to it. In this way the walls in liquid crystals are similar to classical *Bloch domain walls* and *Néel walls* in *ferromagnets* (see *Magnetic domain walls*). However, there is a difference: because liquid crystal walls can be cut at linear defects, i.e. at *disclinations*. This topological feature has an important physical consequence: it is not necessary to reorient the director in the entire region of a domain in order to remove a wall from the liquid crystal. It is sufficient to make a disclination ring which, by expanding, annihilates the wall (Fig. (b)). This process is energetically favorable because it brings about a uniform distribution of $\boldsymbol{n}$. Domain walls often appear when applying external electric or magnetic fields (see *Frederiks transition*). For their description the term *planar soliton* is sometimes used.

DOMAIN WALL BRAKING

See *Domain wall drag.*

DOMAIN WALL DISPLACEMENT PROCESSES in magnetization

Initial stage of *magnetization process* in multidomain *ferromagnetic* materials, consisting in the displacement of magnetic *domain walls* between domains: the favorably oriented *magnetic domain* volume increases because neighboring domains have energetically less favorable magnetizing orientations. In this case the domains have magnetized directions close to that of the overall magnetizing field. One can distinguish reversible and irreversible displacement processes. At low applied fields the initial gently sloping part of the *magnetization curve* (region of initial *magnetic susceptibility*) corresponds to *quasi-elastic reversible displacement processes*. On this portion of the curve *magnetic hysteresis* is almost absent. At higher fields *irreversible displacement processes* become prominent in the next part of the magnetization curve associated with a maximum magnetic susceptibility. In this region intermitent irreversible displacements of domain walls take place (see *Barkhausen effect*), which constitute the main mechanism responsible for magnetic hysteresis in *ferromagnets*. The displacement process contribution to the magnetization is determined by the size and form of the *magnetic texture*, by characteristics of *defects in crystals*, by domain wall structure.

DOMAIN WALL DRAG in a magnetic material, domain wall braking

This braking or frictional drag is determined by the energy dissipation of the moving *domain wall*, and is described by the frictional force which acts on and retards the motion of the wall: $\boldsymbol{v}\cdot\boldsymbol{F}_{\text{b}} = -\text{d}\sigma/\text{d}t$, where σ is the wall energy. The domain wall drag is caused by: *defects in crystals* (*dislocations*, impurities); the presence of *dynamic braking* (viscous friction) due to the transfer of the domain wall energy to various subsystems of the magnetic system which are close to the state of *thermodynamic equilibrium* (to *magnons*, *phonons*, *quasiparticles*, magnetic *impurity atoms*, and so on).

The domain wall drag determines the character of the *magnetic domain wall dynamics*: the velocity of its stationary uniform motion under action of the external force (*magnetic pressure* P_{m}), or the dissipative characteristics at its nonuniform motion. At small velocities $\boldsymbol{v}$, the force is $\boldsymbol{F}_{\text{b}} = F_0\boldsymbol{v}/|\boldsymbol{v}| + \eta\boldsymbol{v}$, where $\boldsymbol{F}_0$ is the force of friction at rest relative to the *coercive force* of the magnetic material, η is the *viscosity* coefficient. Therefore, the velocity is $\boldsymbol{v} \propto (\boldsymbol{P}_{\text{m}} - \boldsymbol{P}_{\text{m}}^0)$, where $\boldsymbol{P}_{\text{m}}$ is related to $\boldsymbol{F}_0$ (see *Domain wall mobility*). When describing domain wall dynamics in the framework of the *Walker solution*, we have $\eta = \lambda v\sigma(0)/\Delta$, where Δ is the wall thickness, an extended phenomenological description gives $\lambda \to \lambda_{\text{eff}}$, where λ_{eff} involves exchange coupling, and relaxation arising from the spin orbit and dipole–dipole interactions (see *Landau–Lifshits equation*). When taking into account the *domain wall twisting* or the presence of *Bloch lines*, the value of η increases, in some cases, quite considerably.

In materials which contain *magnetic ions* with strong spin–orbit coupling (e.g., in rare-earth *iron garnets* and *orthoferrites*) the domain wall braking results from the wall energy transfer to rare-earth ions. In this case, an ordinary phenomenological description based on the relaxation term of the Landau–Lifshits equation is in good agreement with experiment. These materials exhibit, as a rule, large *magnetic losses* and the value $\lambda = 1$ to 10^{-2}. In magnetic materials with small losses, and lacking the ions indicated above, the main contribution to the domain wall drag arises from the interaction of the wall with thermal quasi-particles (especially with magnons). In this case an ordinary phenomenological description does not agree with, experiment: e.g., the values of the relaxation constant λ determined from measurements of the wall mobility, and from the width of the *magnetic resonance* line, can differ considerably from each other. Using a microscopic description, the drag force of the wall is expressed in terms of the probability of processes involving the interaction of thermal magnons or photons with the moving wall.

For magnon relaxation processes there is a reduction of the domain wall drag or, which is equivalent, a rise of the domain wall mobility with decreasing temperature; as a rule $\eta \propto T^n$ or $\mu \sim T^{-n}$, where $n = 1$ to 4. Such behavior is readily observed in magnetic materials with small losses. *Relaxation* from impurity ions grows with the decrease in temperature, and becomes substantial at low temperatures (see Fig.).

Upon increasing the domain wall velocity, the dependence $F(v)$ becomes nonlinear, resulting in a nonlinear dependence $v(P_{\text{m}})$. If the wall velocity is close to the sound velocity v_{s} then *phonon Cherenkov radiation* emerges, and induces substantial growth of $F_{\text{b}}(v)$ at $v \cong v_{\text{s}}$. In this case, there appears an anomaly in the dependence $v(B)$

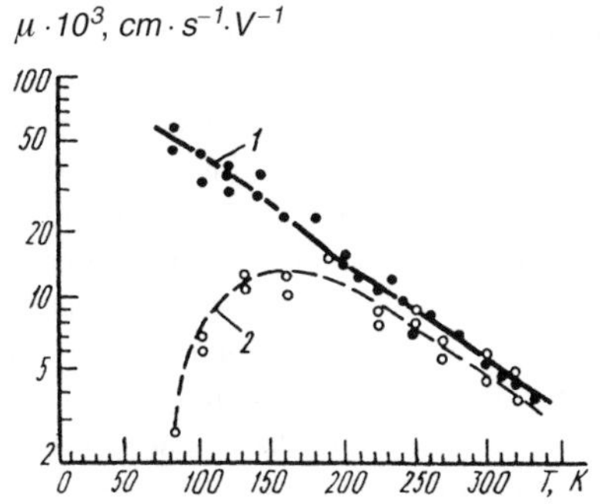

Temperature dependence of the mobility $\mu(T)$ for yttrium orthoferrite $YFeO_3$. Curve 1 refers to a very pure sample for which magnon processes of domain wall drag dominate, and curve 2 corresponds to a sample with impurities whose contribution leads to a reduction of the mobility at low temperatures. From F.C. Rossol, Phys. Rev. Lett., Vol. 24, p. 1021 (1970).

which has the form of a region with small *differential mobility*. This effect can appear only in magnetic materials in which the *domain wall velocity limit* v_c is greater than the *sound velocity*, e.g., in orthoferrites.

DOMAIN WALL DYNAMICS

See *Magnetic domain wall dynamics.*

DOMAIN WALL, MAGNETIC

See *Magnetic domain wall.*

DOMAIN WALL MOBILITY

Ratio between the rate v of *magnetic domain wall* motion and the driving *magnetic field* B, which is defined within the linear interval of the $v(B)$ dependence: $\mu = v/B$. The motion is limited by the force of viscous friction, which acts upon the domain wall. In *magnetic substances*, which exhibit an appreciable *coercive force*, domain wall mobility is defined as follows: $\mu = v/(B - B_s)$, where B_s is the starting field of the domain wall. Sometimes the *differential domain wall mobility* is introduced, which is defined by the equation $\mu_d = dv/dB$ (see *Magnetic domain wall dynamics*, *Domain wall drag*).

DOMAIN WALL PEAK VELOCITY

The maximal velocity of the uniform steady (without change of structure) motion of a twisted domain wall (see *Domain wall twisting*) under the action of an applied magnetic field. As the velocity reaches its peak value, the breakdown of the steady-state mode occurs (see *Slonczewski limiting velocity*). The value of this peak domain wall velocity is below the *Walker critical velocity* (see *Walker solution*), which is the "peak velocity" for a spatially uniform boundary. According to J. Slonczewski (1973), the breakdown of the steady-state mode is caused by structural instabilities in the twisted boundary, which give rise to dynamic horizontal *Bloch lines*. By this mechanism, the peak velocity is inversely related to the film thickness. The question of experimental confirmation of the above inverse relation, as well as the question of whether the whole mechanism is suitable or not, is still not clearly understood. Calculations involving the *Slonczewski equations* are suggestive of the important role of the deflections of the boundary plane across the whole width of the film.

The instabilities in the twisted boundary are eliminated and the value of the peak velocity of the domain wall increases in the presence of a strong enough magnetic field applied to the film plane, or when high rhombic *magnetic anisotropy* is present.

DOMAIN WALL TWISTING

The effect of *demagnetization field* on the structure of a *magnetic domain wall* in uniaxial *magnetic films* with quality factor $Q \gg 1$. It results from the component of the demagnetizing field perpendicular to the plane of the wall arising from the *magnetic charges* of the domains on both surfaces of the film. The demagnetizing field competes with the local magnetostatic field at the boundary, hindering the transition from a *Bloch wall* to a *Néel wall*. A twisted domain wall consists of two nearly Néel-type sections of opposite polarities adjacent to each surface of the film, smoothly connected by a quasi-Bloch region in the middle of which the boundary is of the purely Bloch type. The size of each isolated Néel domain wall section makes up about 1/8 of the film width, and decreases with a decrease of the film width or the *magnetic domain structure* period. The twisted domain wall corresponds to a minimum of the energy in films with perpendicular *magnetic anisotropy*. However, the contribution to the wall energy from twisting is small ($\sim 1/Q$) when $Q \gg 1$. The twisted wall can contain a number of statically stable horizontal *Bloch lines* localized in the quasi-Bloch region. The superposition of magnetic fields in the basal plane is usually accompanied by the appearance of Bloch lines defining processes of domain wall remagnetization.

In terms of dynamics, the two transition regions from a Néel to a quasi-Bloch section differ by an increased instability relative to the formation of dynamic horizontal Bloch lines. Some dynamic properties of twisted walls are related to this factor (see *Domain wall peak velocity, Slonczewski limiting velocity*).

DOMAIN WALL VELOCITY LIMIT

Maximum possible velocity of stationary *domain wall* motion in a *magnetic substance*. For different types of domain walls this velocity varies over wide limits, and values from several meters per second to 20 km/s have been experimentally observed. For thin *magnetic films* of ferromagnets or of *ferrites* with an anisotropy axis perpendicular to the film surface, and with low *magnetic anisotropy* in the basal plane, the domain wall is usually not one-dimensional (see *Domain wall twisting*). In this case its velocity limit is determined by the *domain wall peak velocity* (see *Slonczewski limiting velocity*). In these magnetic materials the domain wall velocity limit does not exceed several tens of meters per second. If the wall twisting is less significant (in the presence of strong magnetic anisotropy within the plane of the field, upon the switching of the external magnetic field within the film plane), the velocity limit sharply increases (up to several hundred meters per second), and may approach the limiting Walker velocity (see *Walker solution*). The domain wall velocity limit also increases at the approach to the point of *magnetic compensation* of a ferrimagnet. For *ferromagnets* and *ferrimagnets* a quantitative theoretical explanation of the observed values has not yet been formulated. In weak ferromagnets (see *Antiferromagnet*), experiments in quantitative agreement with theory show that the domain wall velocity limit coincides with the *spin wave* velocity in the linear region of the *dispersion law*, reaching values of 10^4 m/s (20 km/s for orthoferrites, 14.5 km/s for iron borate $FeBO_3$). See also *Magnetic domain wall dynamics*.

DONOR

An impurity or structural *defect* that introduces into the *band gap* of a semiconductor one or more energy states which can supply the *conduction band* with electrons. A donor is electrically neutral when its local states are filled with electrons; and it is charged positively if the local states have given up their electrons. If a *donor impurity* is dominant then the semiconductor has n-type conductivity in the temperature range determined by the level depth of the donor states. Donors participate in the scattering of charge carriers since they can be either neutral, singly or multiply charged centers. Conventionally, donors are divided into small hydrogen-like ones (well described by an *effective mass*), and deeply lying ones. The most characteristic examples of small donors are atoms of Group V (e.g., P, As, Sb) in atomic semiconductors of Group IV (e.g., Si, Ge). Deep donor states in Si and Ge are introduced by Mn, impurities of Group VI, and some *amphoteric centers*, e.g., Au, some transition *metals*, and *vacancies* (in Si).

DONOR–ACCEPTOR PAIRS

Systems composed of two oppositely charged centers. In the simplest case, such a dipole system contains a positively charged *donor* D^+ and a negatively charged *acceptor* A^-. The nature of the pair changes with changes in the distance between the components. A bound state (D^+, A^+) can form unless the separation between these centers is too large. A great deal of experimental information on donor–acceptor pairs in semiconductors has been obtained by optical methods (*luminescence, infrared spectroscopy*) and *electron paramagnetic resonance*.

DOPED AMORPHOUS SEMICONDUCTORS

Amorphous semiconductors obtained by introducing impurities (*doping*) that strongly affect their electronic structure. Thin films of amorphous Si doped with elements of Groups III and V are produced by decomposing mixtures of SiH_4 with PH_3, AsH_3 or B_2H_6 in a gas discharge. Introducing 10^{-2}–10^{-1}% of an impurity increases the *electrical conductivity* by 7 to 8 orders of magnitude (from 10^{-9}–10^{-10} to 10^{-2} $(\Omega\cdot\text{cm})^{-1}$, see Fig.). In the case of impurities from Group V, 20 to 40% of the atoms introduced are *donors*, i.e. they

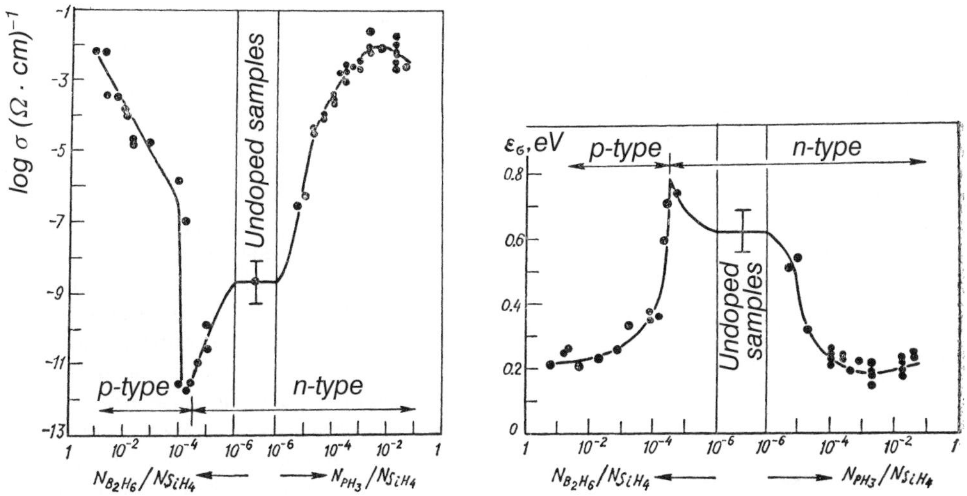

Electrical conductivity at room temperature (left), and conduction activation energy (right), of amorphous Si as functions of composition of the gas mixture used to form the films.

have fourfold coordination in Si. Boron impurities result in hole conduction. *Ion implantation* is used to dope *amorphous silicon* with H as a donor impurity. The conductivity of amorphous silicon increases from 10^{-17}–10^{-18} to 10^{-11} $(\Omega\cdot\text{cm})^{-1}$ when doped (in a melt) with $2\cdot 10^{-3}\%$ of O as an acceptor impurity, and up to 10^{-9} $(\Omega\cdot\text{cm})^{-1}$ when $5\cdot 10^{-2}\%$ of donor Cl is added. Chalcogenide glasses significantly increase their conductivity when As_2S_3, As_2Se_3, and As_2Te_3 are doped with Cu, Ag, In, Ti impurities. GeS reacts similarly to Ca and Ag impurities. The mechanism for the increase in the conductivity of doped amorphous semiconductors consists in reconstruction of their atomic structure and accompanying changes in the ratio of their negatively and positively charged free bonds. Transition series and rare earth ions added to oxide glasses produce new *local electronic levels* in the forbidden gap of the latter. The levels influence optical absorption, *luminescence*, and *electron paramagnetic resonance* spectra.

Due to low production costs, and the possibility of varying their physical properties over wide limits, doped amorphous semiconductors present a promising class of materials for various engineering applications. These doped semiconductors have been successfully used for *solar cells* (*p–n* junction in amorphous Si), *solid-state lasers*, etc.

DOPING

Introducing impurity atoms into *semiconductors* to increase the *electrical conductivity* of the material, i.e. to produce *donor* centers in electron semiconductors and *acceptor* ones in hole semiconductors. The corresponding elements are called *doping impurities*. Pairs of elements forming centers of opposite types are called *compensating impurities*. Both doping and compensating impurities also affect the photoelectric, optical, magnetic and other properties of semiconductors. A particular feature of semiconductors is the high sensitivity of their properties (especially electric) to the introduction of impurities. As for pure materials, their electric properties start to alter noticeably at impurity concentrations of 10^{10}–10^{11} cm^{-3}. *Monocrystals* may be doped during the growth process in the following ways; add elements to the melt from which the *solid* crystallizes; stimulate thermally activated *diffusion* from a *solid surface* coated with a doping element beforehand; perform *ion bombardment* (see also *Ion implantation, Ionic synthesis*); subject the doped material to *nuclear radiations* that precipitate nuclear reactions in it (see *Nuclear doping*). Diffusion from the surface and ion implantation are used to dope surface layers so as to achieve surface *hardening* of metal parts, and to change the distribution of electrical resistance in electronic semiconductor devices.

DOPING ATOM

An *impurity atom* occupying a lattice site (substitutional), or at an off-site position (interstitial) in a *crystal lattice*, e.g., in the empty space between surrounding atoms located at the neighboring sites of a crystal lattice (see *Interstitial atom*, *Substitutional atom*, *Dumb-bell interstitial defect*, *Crowdion*).

DOPING, ION IMPLANTATION

See *Ion implantation doping.*

DOPING, NUCLEAR

See *Nuclear doping.*

DOPING, SURFACE

See *Surface doping.*

DOPPLER EFFECT in solid state physics (Ch. Doppler, 1842)

A change of frequency ω perceived by an observer when moving toward or away from an oscillating source. The frequency of the waves emitted by a radiation source which moves in a medium depends significantly on the velocity v of the radiator and the phase velocity v_{ph} of the waves. Any system can serve as a radiation source if it possesses its own inherent oscillation frequency, e.g., atoms, molecules, electrons oscillating under the action of alternating external fields, and channeled (see *Channeling*) relativistic particles in a crystal. Not only can light waves be emitted in a medium, but also plasma (longitudinal) waves (see *Plasma oscillations*), sound waves, etc., that is all possible types of waves that can exist in a given crystal. From the *conservation laws* of energy and momentum for the source before and after emission the following equations are obtained for the frequency ω of a quantum emitted at an angle θ relative to the direction of motion of the source (in what follows, for definiteness we will speak of the emission of photons):

$$\omega = \begin{cases} \dfrac{\omega_0\sqrt{1-\beta^2}}{1-\beta n(\omega)\cos\theta}, & \beta n(\omega)\cos\theta < 1, \\[2ex] \dfrac{\omega_0\sqrt{1-\beta^2}}{\beta n(\omega)\cos\theta - 1}, & \beta n(\omega)\cos\theta > 1, \end{cases}$$

where ω_0 is the intrinsic frequency of the radiator at rest in the fixed coordinate system, $\beta = v/c$ (v is the radiator speed), and $n(\omega)$ is the *refractive index* (the medium is assumed to be transparent and isotropic). For particles channeling in a crystal, ω_0 is determined by the difference in the energies of the transverse motion. In the motion of a radiator with a velocity whose projection on the direction of propagation of the photon is smaller that the phase velocity of light in the medium (e.g., $v\cos\theta < v_{\text{ph}}$), the photon emission event takes place at the transition of the system from a more excited state to a less excited state (*normal Doppler effect*). In the opposite case ($v\cos\theta > v_{\text{ph}}$), the photon emission is accompanied by excitation of the system (*anomalous Doppler effect*). The equations for ω in a medium with dispersion are not linear with respect to ω. As a result, for a given ω_0, v, and θ the emission of quanta with different frequencies is possible (*complex Doppler effect*). If the radiator moves in a vacuum there is no dispersion, and a quantum of a single frequency is emitted (*simple Doppler effect*). A source with an arbitrary intrinsic frequency cannot overtake the irradiated energy at any speed and in any refracting medium. It is for this reason that there is always a wave in the emission spectrum for which $v\cos\theta/W(\omega_\theta) < 1$, where $W(\omega_\theta)$ is the group velocity of the waves radiated at the angle θ. The inverse condition $v\cos\theta/W(\omega_\theta) \geqslant 1$ determines the range of existence of a complex emission spectrum. Therefore, the anomalous Doppler effect is always complicated since, in addition to a frequency for which $v\cos\theta/W(\omega_\theta) > 1$, there is always a normal component for which $v\cos\theta/W(\omega_\theta) < 1$. In this case the frequency of the normal component is higher than the anomalous one. As a result, the normal component that overtakes the radiator carries away more energy per unit frequency than the anomalous component. The expressions obtained are valid for an optically transparent body. Taking into account absorption, photon emission accompanied by system excitation also occurs in a medium with $n < 1$. Both complex and anomalous Doppler effects are observable in the X-ray range. In view of a small departure of the X-ray refractive index from unity, to observe the effect in this case one should use channeled relativistic particles For a non-stationary medium (whose parameters change with time) a frequency change can occur even when both the radiator and the acceptor are motionless, and this is called the *parametric Doppler effect*.

DOPPLERON

A low-frequency, weakly decaying, slow electromagnetic wave that can appear and propagate in a *metal* in the presence of a permanent magnetic field $\boldsymbol{B}$. A doppleron is one of five known intrinsic electromagnetic excitations in a dense electron metal plasma (see *Solid-state plasma*). Its existence is caused by the *Doppler-shifted cyclotron resonance* which is a consequence of the Fermi degeneracy of *conduction electrons* in metals. Therefore, a doppleron as a particular feature of metals, in contrast to a *helicon*, has no analogues in an equilibrium gas plasma. Being a *collective excitation* of a metal plasma, a doppleron nevertheless is closely connected with the local characteristics of the *Fermi surface*, and this determines the diversity of information contained in it. On the one hand, there is a possibility of restoring the local conduction in a magnetic field, and on the other hand, one can extract accurate information about the Fermi surface curvature, the mean free path of resonance groups of carriers, the state of the metal surface, and so on. In the case of complicated Fermi surfaces, the existence of several doppleron modes is possible, including multiple dopplerons for a non-axial Fermi surface.

Doppleron propagation in pure metals is predominantly along the axis of high *crystal symmetry* $\boldsymbol{c}$ ($\boldsymbol{k} \| \boldsymbol{B} \| \boldsymbol{n} \| \boldsymbol{c}$) in strong magnetic fields $\boldsymbol{B}$ at low temperatures when $\omega \ll \nu \ll \Omega$, where ν is the collision frequency of carriers, Ω is the *cyclotron frequency*, $\boldsymbol{n}$ is a unit vector normal to the metal surface, ω and $\boldsymbol{k}$ are the frequency and the wavevector of the doppleron wave, respectively. Attenuation is characteristic of both collision (with photons and lattice defects) and collision-free types (see *Magnetic collision-free damping*). The latter is determined by the resonance cyclotron absorption of the wave by other groups of carriers at the Fermi surface, and by the magnetic *Landau damping* that suppresses the dopplerons when $\boldsymbol{B}$ is bent away from $\boldsymbol{c}$. The propagation of dopplerons is also possible along open Fermi surfaces in a magnetic field parallel to the metal boundary.

Specific features of dopplerons are the circular polarization of the wave in the plane of $\boldsymbol{B}$, the dependence of the wavelength on the magnetic field, the threshold (in regard to frequency and magnetic field) nature of the spectrum, and, as a rule, anomalous dispersion ($\mathrm{d}\omega/\mathrm{d}k < 0$). These features can appear, for example, in compensated metals, in the long-wave limit, or at certain singularities at the Fermi surface such as saddle points and local minima of curvature.

DOPPLERON–PHONON RESONANCE

A resonance in an applied magnetic field involving the interaction of *dopplerons* with *sound*. It is similar to a *helicon–phonon resonance*. The doppleron–phonon resonance exists in metals under the conditions needed for the existence of a doppleron; involving the same circular polarization in the range of intersection of the doppleron wave and sound wave spectra (assuming no coupling between elastic and electromagnetic subsystems), i.e. for coinciding *phase velocities* of the ultrasonic and doppleron waves. The wave coupling results in the removal of degeneracies in the resonance region. However, in the case of the weak bond that is specific for the doppleron–phonon resonance, a mode repulsion does not take place: the sound wave brings elastic energy to the resonance range, while the doppleron wave brings electromagnetic energy. The damping of both the waves in the resonance gives rise to a resonance maximum. Since the doppleron is circularly polarized the resonance does not exist for a longitudinal ultrasonic wave, appearing only for circular polarization of the transverse wave. As a result, for linearly polarized transverse elastic waves in the region of resonance a *rotation of light polarization plane* and ellipticity may arise. The transformation of electromagnetic waves into ultrasonic waves and vice versa can occur in the resonance range. For non-axial *Fermi surfaces* a multiple doppleron–phonon resonance is possible. For surface (Rayleigh) sound waves, a resonance similar to the doppleron–phonon one for bulk sound has been found (*subsurface doppleron–phonon resonance*) when the external field is parallel to the metal surface.

DOPPLER-SHIFTED CYCLOTRON RESONANCE

A resonance interaction of the *conduction electrons* that are drifting along a magnetic field $\boldsymbol{B}$ with an electromagnetic wave propagating in a *metal*. This Doppler-shifted cyclotron resonance appears when the electron velocity $\tilde{\boldsymbol{v}}$ along the wave vector $\boldsymbol{k}$ averaged over the cyclotron period differs from zero, i.e. if $\boldsymbol{k}$ is not perpendicular to $\boldsymbol{B}$, or in the presence of open cyclotron orbits on the Fermi surface, and $kl \gg 1$, where l is the mean free path of the free carriers. The propagation of an ultrasonic wave in a metal is known as a *Doppler-shifted acoustic cyclotron resonance*, and it satisfies the following condition in the free-electron model:

$$|\omega \pm k_B \tilde{v}_B| = \omega_c, \tag{1}$$

called the *Kjeldaas edge* (1959). This means that electrons drifting toward the wave shift their frequency to the cyclotron value ω_c owing to the *Doppler effect*. In connection with the high electron velocities ($\approx 10^8$ cm/s) on the Fermi surface, the Doppler shift of the frequency is very large, $\omega \ll k_B \tilde{v}_B$, and the resonance is observed at frequencies $\omega \ll \omega_c$. The condition (1) takes the form

$$\omega_c = |k_B \tilde{v}_B| \tag{2}$$

or

$$u(p_B) = \lambda_B, \qquad u(p_B) = eB\frac{\partial s}{\partial p_B}, \tag{3}$$

i.e. those electrons are involved in the resonance whose displacement along $\boldsymbol{B}$ within a cyclotron period, $u(p_B)$, is equal to the wavelength λ, where p is the electron momentum, and s is the area of the cross-section of the Fermi surface defined by the plane $p_B = \text{const}$. Thus, the Doppler-shifted cyclotron resonance, similar in many ways to the usual temporal *cyclotron resonance*, is, in fact, a space resonance.

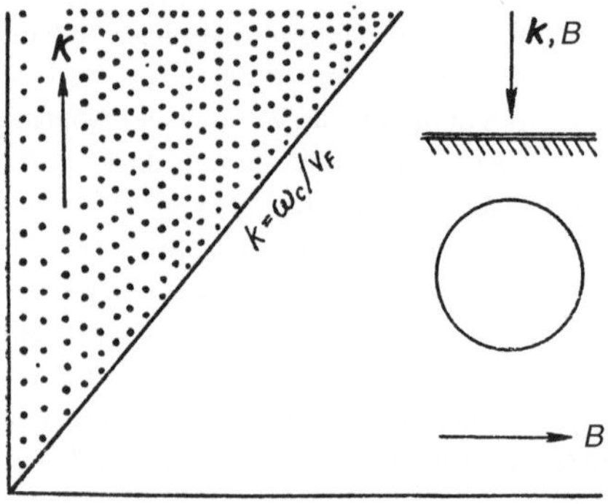

Doppler-shifted cyclotron resonance.

The spectrum of this resonance (see Fig.) in the coordinates $(\boldsymbol{k}, \boldsymbol{B})$ represents a straight line that passes through the origin, and separates the regions (shaded) where the collision-free Doppler-shifted cyclotron absorption is present, and where it is absent. The latter is caused by electrons with values of $\tilde{v}_B$ varying from the *Fermi velocity* v_F at the base point to zero at the central cross-section of the Fermi surface. The Doppler-shifted cyclotron resonance results from the electron *Fermi–Dirac statistics* and the presence of a sharp velocity distribution of electrons.

In the case of an anisotropic Fermi surface, the Doppler-shifted cyclotron resonance can exist for a few groups of carriers. Distinguished (or resonant) are the groups with extremal values of $u(p_B)$, because in this case a large number of carriers is formed which move in phase, and at the same time do not contribute to the wave absorption. The contribution of other carriers is insignificant due to their out-of-phase motion, and can be regarded as providing a background signal.

For a non-axial Fermi surfaces the condition for the Doppler-shifted cyclotron resonance takes the form

$$|k_B \tilde{v}_B| = n\omega_c, \tag{4}$$

$$u(p_B) = n\lambda, \tag{5}$$

where the integer n is determined by *selection rules*. For example, for the resonance with longitudinal sound $n = m\alpha$; and in the case of transverse sound $n = m\alpha \pm 1$, where $m = 0, \pm 1$, and α is the order of the crystal *symmetry axis*.

The nature, shape, and magnitude characterizing the resonance as a function of the sound absorption coefficient Γ versus B are dependent on the type of ultrasonic wave (longitudinal or transverse), its frequency and polarization, the multiplicity of the resonance n, the type of resonance singularity $\partial s/\partial p_B$, and so on. Doppler-shifted cyclotron resonance can manifest itself (converted to a function of ω, B, etc.) in the form of either an edge, an absorption step, or a *resonance* or *antiresonance*. At sufficiently high ω and l, the line of the Doppler-shifted cyclotron resonance can be split (see Eq. (1)). This resonance allows one to determine the curvature of the Fermi surface $(\partial s/\partial p_B)_{\text{extr}}$, the *effective mass* of the carriers of resonance groups, the mean free path, and so on. See also *Doppleron*.

DOSE, SMALL

See *Small-dose effect*.

DOSIMETRY

A determination of the effects of *nuclear radiation* on irradiated objects. A quantitative measure of these effects depends on the case (*solids* experience a change of resistance, optical absorption, strength, etc.). However, independently of secondary effects, the primary cause of radiation-induced effects and the most general physical value that characterizes the interaction of radiation with the matter is the *absorbed energy*. The *absorbed energy dose D* is the macroscopic value that best characterizes the absorbed energy:

$$D = \frac{\Delta E}{\Delta m}, \tag{1}$$

where ΔE is the energy absorbed in the bulk, and Δm is the mass of the substance. In SI units the absorbed radiation dose in joules per kilogram is called a grey (Gy): 1 Gy = 1 J/kg. Let E_{in} be the sum of the energies of all the particles that during time Δt enter volume ΔV which contains the mass Δm, and E_{out} be the energy sum of all particles that leave this volume during the same time. If there are no sources of ionizing radiation, and no reactions, then the absorbed energy $\Delta E = E_{\text{in}} - E_{\text{out}}$ will be expended entirely for ionization, excitation, and elastic interactions. If nuclear reactions and ionizing radiation sources are present in volume ΔV then ΔE should take into account the total energies from the nuclear transformations E_{n} and the sources of radiation energy E_{s}:

$$\Delta E = E_{\text{in}} - E_{\text{out}} + E_{\text{n}} + E_{\text{s}}. \tag{2}$$

The macroscopic values determined in this fashion, i.e. the absorbed energy and the *dose*, do not reflect the stochastic nature of the interaction of ionizing radiations with matter. Therefore, when determining the dose, Δm and Δt should be, on the one hand, sufficiently large (to exclude the effect of fluctuations), and, on the other hand, sufficiently small so as to be able to disregard macroscopic quantitative and qualitative changes in the incoming radiation (i.e. *active spectrum*) and in the radiation-damaged material under consideration.

A determination of the absorbed radiation dose is related to the measurement of any dose-dependent radiation effect in matter. However, a radiation-stimulated process convenient for measurement may be lacking in the irradiated material for which the absorbed dose has to be determined. In such cases a standard material is introduced at the detection point, and the absorbed dose is deduced from the radiative reactions in the extra material The *Bragg–Grey probe method* which is widely used to investigate irradiation damage by both non-charged and charged high-energy particles is based on this principle. A probe, that is a material introduced at the detection point, must satisfy the *Bragg–Grey conditions*: (i) the active spectrum at the probe should be the same at that detection point in the absence of the probe; (ii) charged particles generated in the probe by indirectly ionizing non-charged particles should not contribute significantly to the irradiation dose absorbed in the probe. These conditions are sufficiently satisfied if the linear dimensions of the probe are small compared to the mean free path $\overline{R}$ of the charged particles (e.g., $0.1\overline{R}$). If the Bragg–Grey conditions are satisfied, and the mean energy $\overline{\varepsilon}$ of the active spectrum of charged particles (electrons) at the detection point is known, then with the help of the Bragg–Grey equation one can find the dose D_m in the material with a good accuracy:

$$D_m = D_p \frac{S_m(\bar{\varepsilon})}{S_p(\bar{\varepsilon})}, \quad (3)$$

where D_p is the experimentally found dose in the probe; $S_m(\bar{\varepsilon})$ and $S_p(\bar{\varepsilon})$ are respectively the mean braking abilities of the material and of the probe, which represent the mean *charged particle energy loss* per unit of path divided by the material density. A more exact characteristic of the radiation in the medium (in comparison with the absorbed dose) can be obtained from the determination of the active spectrum. Knowledge of this spectrum allows one to calculate the absorbed dose and the yield of various radiation-stimulated reactions.

DOUBLE LAYER, ELECTRIC

See *Electric double layer.*

DOUBLE RESONANCES

Resonances in systems oscillating at two frequencies arising from external effects. In solids where the constituent microparticles have a set of discrete energy levels $\varepsilon_i, \varepsilon_j, \varepsilon_k, \ldots$ the two resonances are due to transitions (indicated in Fig. by arrows) between pairs of levels with different energy separations corresponding to the frequencies ν_{ij} and ν_{jk}. Usually in double resonance experiments there is a coupling between the transitions through interactions which permit one resonance to be recorded through the other one. The two pairs of levels may have similar or different natures (e.g., magnetic sublevels with one pair involving an electron flip and the other involving the flip of a nuclear spin), close frequencies or frequencies in different ranges of the electromagnetic spectrum, those that belong to spins of different types (electrons and atomic nuclei), or to identical particles which differ by some parameter. Double resonances may arise from the same type of interaction (e.g., electromagnetic wave) or from different types (e.g., electromagnetic and acoustic waves). The most common are magnetic in nature: *electron–nuclear double resonance* (ENDOR), *acoustic double magnetic resonance*, and *optical–magnetic resonance*, and these are all various methods of *magnetic resonance spectroscopy*.

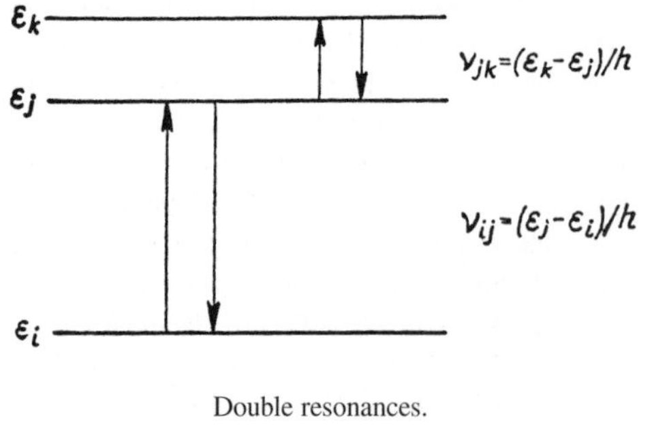

Double resonances.

In *double magnetic resonance* there is selective absorption (radiation) at two resonance frequencies of electromagnetic energy by a spin located in a constant magnetic field. Observation of double magnetic resonance simplifies the spectrum of one resonance by suppressing the interaction between the spins through high power irradiation at the second resonance frequency, and it also increases the sensitivity or resolution of the low-frequency resonance by recording it as part of the high frequency spectrum.

In *nuclear–nuclear double resonance* (NNDR) the *nuclear magnetic resonance* (NMR) experiment is carried out simultaneously at two frequencies which belong to atomic nuclei with different energy spacings between their spin sublevels. It is used to investigate substances with nuclei (1 and 2) coupled together though the *spin–spin interaction*, which causes each individual resonance to influence the other. *Homonuclear NNDR* involves two nuclei of the same type and *heteronuclear NNDR* involves nuclei of different types. In the former case the resonance at nucleus 1 is detected at low power, and then the irradiation of resonance 2 at high power suppresses their interaction. This simplifies the spectrum of nucleus 1 and enhances the amplitude of the detected NMR signal, a phenomenon called the *Overhauser effect*. A second experiment, called *internuclear double resonance* (INDOR), is a method of recording the NMR of nucleus 2 in a manner similar to double electron–nuclear resonance. The technique of NNDR permits one to study the NMR of rare isotopes or of nuclei which provide weak signals disturbed by interaction with other nuclei, to determine the signs of the spin–spin interaction constants, and to obtain information about processes of nuclear spin relaxation.

In *electron–electron double resonance* (ELDOR) the *electron paramagnetic resonance* (EPR) is performed simultaneously at two microwave frequencies (of observation and pumping) corresponding to the resonance lines of two pairs of spin levels. These lines may involve transitions between the spin sublevels of one *paramagnetic center*, or between two different centers coupled by the *spin–spin interaction*. In ELDOR the investigated substance is placed into a special bimodal resonator where it is subjected to the effect of two microwave fields of different frequencies. Variations are observed in the EPR signal at the observation frequency when the power level is changed at the pumping signal frequency. This experiment provides information about spin–spin interactions, *cross-relaxation*, and mechanisms of EPR line broadening.

DOUBLET

The term is employed in two senses: a double degeneracy of levels (e.g., a *Kramers doublet*), and two closely situated lines in the spectrum of a resonance absorption which are usually of the same origin (e.g., *hyperfine structure* in an *electron paramagnetic resonance* spectrum when the nuclear spin $I = 1/2$). In a similar way, one can speak about a *triplet* (three), a *quadruplet* or *quartet* (four), and a general expression for such terms is *multiplet* (many). For a single line or an absence of degeneracy it is common to use the term *singlet*.

DOW–REDFIELD MODEL (J. Dow, D. Redfield, 1972)

A description of the *light absorption* in the case of direct (i.e. without emission and absorption of *phonons*) exciton transitions in a strong uniform electric field E. It was shown that the light absorption coefficient $\alpha(\hbar\omega)$ near the absorption edge energy ε_0 depends on the photon energy $\hbar\omega$ according to an exponential law, i.e. as

$$\alpha(\hbar\omega) \sim \exp\left[-\frac{\beta(\varepsilon_0 - \hbar\omega)}{E}\right],$$

where β is a constant, and E is the electric field strength. The light absorption at $\hbar\omega < \varepsilon_0$ becomes significant when E reaches a value of 10^6 to 10^7 V/cm. The Dow–Redfield model was applied to explain the *Urbach rule* that involves an exponential law for the $\alpha(\hbar\omega)$ dependence near the light absorption edge in *amorphous semiconductors*. It is assumed that there exist strong local (on the scale of atomic dimensions) electric fields in amorphous semiconductors. These fields are due to a local distortion of the atomic structure of the short-range order inherent in amorphous materials.

DRAG OF DOMAIN WALL

See *Domain wall drag.*

DRAWING in metallurgy

A method based on metal *pressure* treatment, whereby stock of fixed cross-section is introduced into the channel of the drawing instrument (die) and is pulled (drawn) through it. The shape of the channel cross-section is the same or close to that of the metal being pulled, and it smoothly decreases from the point of metal entry to its exit. During the process of drawing the stock is deformed and its cross-section varies, at the end attaining the shape and dimensions of the narrowest section of the channel.

DRIFT, AMBIPOLAR

See *Ambipolar drift.*

DRIFT, MAGNETIC

See *Magnetic drift.*

DRIFT OF CURRENT CARRIERS

See *Current carrier drift.*

DROP-IMPACT EROSION

Failure of a solid surface under the action of nonstationary fields of stresses arising at impact from high-speed drops of a liquid. The threshold speed v_k, below which no failure occurs, is related to the drop diameter d by $v_k d = \text{const}$. The damage results from the stress wave propagation, jet-type spreading of a drop, and penetration of a liquid inside the *defects* of a solid. The mean value of the pressure in the contact zone is given by $P = \rho c v$, where ρ is the liquid density and c is the speed of sound in the liquid. The maximal pressure

$3P$ is observed after a time $t = 2dv/c^2$ from the instant of contact, and is concentrated on the contact zone boundary. Corresponding to this instant is the beginning of the drop spreading over the surface at the speed $(3–4)v$. Under a single impact at transonic or supersonic speed depressions appear in *metals* and *alloys*, sometimes with a crater at the center. In nonmetals, circular *cracks* arise with a subsurface zone of failure at the central axis. Under multiple impacts we define drop-impact erosion at the stage of steady-state failure by the ratio of mass loss to the area and the amount of liquid striking the solid surface. The dependence of erosion E on the drop speed v and the angle θ of the drop speed direction to the normal to the surface is given by the relation $E \propto (v\cos\theta - v_k)^n$. The value of n depends on the properties of the material and the method of testing, and varies within the range 2.5 to 6. See also *Surface erosion.*

DRUDE–LORENTZ–SOMMERFELD MODEL

See *Macroscopic description.*

DUMB-BELL DOMAIN

A kind of *rigid magnetic bubble domain*, oval-shaped with a neck. Its specific feature is the great number (>100) of vertical *Bloch lines* in a magnetic *domain wall* giving the predominant contribution to the wall energy. In the case of a high density of vertical Bloch lines, a corresponding diameter of a cylindrical magnetic domain can become greater than the diameter of the elliptic instability that is the cause of the formation of an "oval" domain. The appearance of a neck on the oval is due to the dependence of the equilibrium distance between the vertical Bloch lines on the curvature of the domain wall. A dumb-bell domain is stable over a certain range of magnetizing field variation, at the lower boundary of which the dumb-bell domain transforms into a *strip domain*; and at the upper boundary it either collapses (in the case of very high density of vertical Bloch lines), or transforms into an *elliptic domain.* The field strength of the dumb-bell domain collapse (unlike the field strength of the collapse of circular rigid cylindrical magnetic domains with a small vertical Bloch line density) decreases with an increase in the number of vertical Bloch lines on its surface. Under a periodic pulsed modulation of the magnetizing field, a dumb-bell domain undergoes a rotation whose direction is determined by the sign of the vertical Bloch lines. See also *Magnetic bubble domains.*

DUMB-BELL INTERSTITIAL DEFECT

A configuration of an *interstitial atom* in a monatomic crystal where the interstitial atom displaces one of the adjacent lattice atoms from its lattice position position O and forms with it a pair with the center of gravity at O. The energy E_i^+ of a dumb-bell depends on the orientation $\langle ijk\rangle$ of its axis. In many face-centered cubic crystals this is a minimum for orientations $\langle 100\rangle$ while in body centered cubic crystals the minimal value is attained for orientations $\langle 110\rangle$. Computer calculations gave the following results: Cu: $E_i^+ \geqslant 3$ eV, α-Fe: $E_i^+ = 3$–4 eV. A migration of a dumb-bell involves the displacement of a neighboring atom from its lattice position followed by the formation with it a new dumb-bell with another orientation. The other atom of the initial dumb-bell returns simultaneously to its original lattice position. With the help of computer calculations, using various interatomic potentials (see *Interatomic interaction potentials*) one can estimate the dumb-bell migration energy, which corresponds to the migration energy of an interstitial atom; for Cu $E_i^m = 0.05$–0.13 eV, in agreement with the experimental value 0.12 eV. The turning energy of a dumb-bell without a displacement of its center of gravity is 3 to 4 times higher. In *substitutional alloys* interstitial atoms can form *mixed dumb-bells* consisting of atoms of two kinds.

DURABILITY

Time during which a solid preserves its properties. In the case of *strength*, durability is an important characteristic for long-term loading when a *creep* develops in a solid. Here durability is the time before *fracture.* The higher the temperature and the higher stress, the shorter the time t_f before fracture:

$$t_f = A\exp\left[\frac{U_0 - \gamma\sigma}{k_BT}\right].$$

On the other hand, the rate of stationary creep $\dot{\varepsilon}$ increases with increasing stress and temperature:

$$\dot{\varepsilon} = B\exp\left[-\frac{Q_0 - \gamma_1\sigma}{k_BT}\right].$$

In these dependences A and B are constants; γ and γ_1 determine the measure of the stress action on the height of the barrier (*activation volumes*), and U_0 and Q_0 are the *activation energies* of fracture and quasi-viscous flow, respectively. It has been shown that $U_0 = Q_0$ and $\gamma = \gamma_1$. Therefore, processes that develop at the second stage of creep and at fracture have the same nature, i.e. fracture occurs at the second stage of creep, which confirms the earlier finding

$$t_f \dot{\varepsilon} = AB = \text{const}.$$

The validity of this relation was established using *alloys* with various compositions and structures. However, it should be realized that this expression starts to be violated at elevated temperatures and small σ values due to the change of the fracture mechanism. This relation suggests the possibility of predicting the durability of a loaded solid long before it begins to fracture from short-time tests of creep. There are also other methods of durability prediction. Examples are parametric methods based on semi-logarithmic diagrams of durability versus T or $1/T$ dependences. However, in order to ensure the validity of the extrapolation it is important not to try and predict durability beyond an order of magnitude of the time for a short term test.

DYES

Organic compounds of low molecular weight with semiconductor-type *electrical conductivity* and *photoconductivity*, capable of dyeing other materials (see *Organic semiconductors*). The classic representatives of dyes with semiconductor properties are phthalocyanines, molecules which are planar, rather large, and contain a system of conjugated bonds. They are close in structure to the porphyrin molecule. In terms of conductivity, dyes fall into electron (crystal fuchsin – violet, aqueous fuchsin – blue) and hole (uranine, erythrosin, phosphine, etc.) types. Dyes are used to build solar energy transformers. They are also used as photosensitizers of organic and inorganic materials, and as active *laser* media (*dye lasers*).

DYNAMIC CHAOS

Irregular, quasi-random vibrations in otherwise deterministic dynamical systems. The stochastic character of a process is not specified by its *fluctuations* (noise), but rather by its inherent complex dynamics. Dynamic chaos, or chaos for short, possesses the properties of a random process: the autocorrelation function is reduced (see *Correlation function*) and the power spectrum is continuous. The instability of trajectories in the phase space of the dynamical system serves as the criterion of chaos. The *Lyapunov exponent*, often taken for the index of disorder, is the quantitative index of this instability. In Hamilton's dynamical systems chaos occurs as a set of non-zero measure within the *phase space* where the motion has the property of *ergodicity*. There is, in particular, a strict proof of chaos in the system of elastically colliding spheres, which permits substantiation of the ergodic hypothesis for this system. In typical situations, within the phase space there are regions of chaos which coexist with regions of regular (quasi-periodic) behavior. *Hamilton's dynamic chaos* appears in the dynamics of non-linear chains (lattices) of atoms, during the movement of particles in external periodic fields, and in problems of celestial mechanics.

In dissipative dynamical systems chaos corresponds to a set of non-zero measure in phase space referred to as the *strange attractor*. As a rule, the transition from regular behaviour to chaos involves changes in one or more control parameters. Some scenarios for this transition (e.g., through an infinite succession of *bifurcations* or period doublings) involve universal numerical constants, and they are described with the help of *renormalization group methods*. Chaos has been experimentally observed in such systems as *lasers*, spin-wave (see *Spin waves*) turbulence, non-equilibrium electron-hole plasmas in solids (see *Electron–hole liquid*), non-linear electronic chains, etc. In quantum-mechanical systems the local instability of motion is absent, thus, strictly speaking, there is no dynamic chaos. However, quantum systems that are stochastic in the classical limit do possess some features of dynamic chaos.

DYNAMIC COOLING

See *Dynamic nuclear polarization*.

DYNAMIC HOLOGRAPHY

See *Real-time holography*.

DYNAMIC MIXED STATE

A state that occurs in *type II superconductors* and all superconducting thin films upon the application of a magnetic field exceeding the *lower critical field*, $B > B_{c1}$, and a current exceeding the critical *vortex pinning* current. The field penetrates into the superconductor in the form of vortices, which effects the *Lorentz force* arising from the superfluid component of the transport current. When this force exceeds the vortex pinning force then the vortices begin to move and dissipate energy. They are generated at one edge, move across, and disappear at the other edge of the superconductor. Since encircling a vortex changes the *order parameter* phase by 2π, their motion involves time variations of phase and the appearance of a potential difference, i.e. a finite resistance (superconductor *resistive state*). In superconducting thin films at some critical speed of vortex motion the uniform dynamic mixed state becomes unstable to the formation of lines of *phase slip*.

DYNAMIC NUCLEAR POLARIZATION

The totality of methods for orienting the nuclear spins of a material along a specific direction (ordinarily along a constant applied magnetic field $\boldsymbol{B}_0$) under the action of ultra-high frequency electromagnetic fields. Unlike static methods of nuclear polarization (see *Nuclear orientation*), the dynamic methods do not require extremely low temperatures or very strong fields $\boldsymbol{B}_0$; and with their use almost 100% nuclear polarization can be achieved. The basic methods of dynamic nuclear polarization in solids are the *Overhauser effect*, *solid effect* and dynamic cooling.

The *dynamic cooling method* involves the following: the width of the electron paramagnetic resonance line arises from the dipole–dipole (d–d) interaction of the electronic spins, and saturating, non-strictly resonant ($\Delta \equiv \omega - \omega_s \neq 0$) microwave power is applied to the sample. For this case, in the elementary act of absorption of the quantum of energy $\hbar\omega$ by the electronic subsystem the d-d energy E_D of spins with the inverted temperature $\beta_d = 1/(k_B T_d)$ play the role of compensating for the unbalanced energy $\hbar\Delta$. This causes the system to cool ($\omega < \omega_s$) or heat up to a negative temperature ($\omega > \omega_s$). In both cases there occurs a reduction of $|\beta_d|$ by the factor ω_s/ω_d, where $\hbar\omega_d$ is the "quantum" of energy E_D (for a sufficiently high spin concentration ω_d is comparable to the electron paramagnetic resonance line width). The nuclear spins I of a matrix with $\omega_I \leqslant \omega_d$ are in effective heat contact with the cooled subsystem D, so the absolute value of their temperature is reduced by the factor ω_s/ω_d, which brings about the nuclear polarization.

DYNAMIC RADIATION SCATTERING

Diffraction of radiation (X-rays, electrons, neutrons, gamma quanta etc., see *Diffraction of waves*) taking into account the multiplicity of scattering and of the interference interaction of the scattered waves. Dynamic radiation scattering is observed only in sufficiently perfect *crystals*; the condition for it to occur is $l \geqslant \Lambda$ (l is the size of the *coherent scattering region*, Λ is the *extinction length*). In other words for the condition $l \geqslant \Lambda$ it is necessary to study multi-wave scattering, which provides more information about defects in a material than single wave scattering. The multi-wave scattering is divided into two-wave and many-wave (three-wave, etc.) types. The total wave field within the crystal is a Bloch wave (see *Bloch theorem*), for the wave vector $\boldsymbol{K}_\delta$ with $\delta = 1, 2, \ldots, n$ for the nth order dispersion equation of wave diffraction. If the radiation beam propagates at the Bragg angle (see *Bragg law*) the various $\boldsymbol{K}_\delta$ differ by a value of the order of $1/\Lambda$, so the total field is space-modulated with the period Λ. This provides a series of interference scattering effects in ideal and weakly-distorted crystals, such as the *anomalous passage of X-rays*, *pendular oscillations of intensity*, dynamic *extinction*, etc. The *two-wave approximation* is usually realized in the case of X-rays, neutrons, and channeling charged particles.

For distorted crystals the amplitude of the wave field $A(\boldsymbol{r})$ is ordinarily represented as the superposition of smoothly modulated plane waves: $A(\boldsymbol{r}) = \sum_{\boldsymbol{G}} A_{\boldsymbol{G}}(\boldsymbol{r}) \exp[2\pi i(\boldsymbol{K} + \boldsymbol{G})\boldsymbol{r}]$ ($\boldsymbol{K}$ is the wave vector, $\boldsymbol{G}$ are the *reciprocal lattice* vectors for the strong reflections). Substituting $A(\boldsymbol{r})$ into the corresponding wave equation and considering that amplitudes $A_{\boldsymbol{G}}(\boldsymbol{r})$ are slowly changing, the following system of equations is obtained:

$$\frac{\partial A_{\boldsymbol{G}'}(\boldsymbol{r})}{\partial S_{\boldsymbol{G}}} = \sum_{\boldsymbol{G}'} B_{\boldsymbol{G}\boldsymbol{G}'}(\boldsymbol{r}) A_{\boldsymbol{G}'}(\boldsymbol{r}), \qquad (1)$$

where S_G is the coordinate along $\boldsymbol{K}+\boldsymbol{G}$, and the coefficient $B_{GG'}$ contains information about the distortions. These equations are solved by computer or analytically using successive approximations, depending on the distortions. In particular cases Eq. (1) transforms to the system of equations for the many-wave diffraction of electrons, and to the system of *Takagi equations* (S. Takagi, 1962) for the two-wave diffraction of X-rays ($\boldsymbol{G}$ assumes two values: 0, $\boldsymbol{H}$). These equations provide practically complete solutions for electron-microscope or X-ray contrast of defects (*clusters*, *dislocation loops*, etc.) or their complexes. For very smooth distortions the approximation of geometrical optics applies, and Eqs. (1) may be transformed into the usual equations for beam trajectories.

The exact solution of system (1) in the two-wave approximation is obtained for the important case of X-ray diffraction in a uniformly bent crystal. This allows one to follow the variation of the mode of X-ray scattering with the growth of the degree of distortions (the transition from purely dynamic scattering to the mode characteristic for a *mosaic crystal*, and from this to the kinematic mode), to clarify the limits of applicability of the geometric optics approximation, to investigate the focusing of plane wave X-rays, or of point sources by bent crystals, etc. The other case of regular distortions involves deformation fields created by ultrasonic waves (see *Diffraction of radiations* in ultrasonic field). For the case of statistical distortions the *statistical dynamic theory of two-wave diffraction* of X-rays in the limit $l_c \ll \Lambda$ (l_c is the correlation length of the distortion field) is applicable, and closed form equations of energy transfer of the type (1) for the coherent and incoherent components of the scattered intensity are obtained. These equations describe the following effects: weakening of the reflecting ability of the distorted atomic surfaces; increasing of the absorption coefficient of coherent and non-coherent radiation due to the scattering from the random distortions, and the redistribution of the coherent component intensity to an incoherent one. Besides this, the Takagi equations have been successfully applied for the determination of deformations in thin ($<\Lambda$) *disrupted surface layers*, and in perfect crystals from X-ray diffraction data.

The equations of dynamic radiation scattering may be written in the Fourier representation for the amplitudes, which is especially convenient for consideration of diffraction of radiation distortions in crystals with randomly distributed defects. The dependences of integrated reflections of X-rays on the thickness of the plane-parallel sample, calculated by this method, correlate well with experiment. According to the experimental data it is possible to determine the statistical *Debye–Waller factor* L (with accuracy $\sim 10^{-3}$) and the effective addition $\Delta\mu$ to the adsorption coefficient due to the *diffuse scattering*, and these provide such parameters of defects as their concentrations, and the average size of clusters or dislocation loops.

DYNAMIC RECOVERY

The *recovery* of the properties of crystals in the course of *plastic deformation*. This includes the dynamic *restoration of crystals*, dynamic *polygonization* and dynamic recovery with reorientation. The initial stage of dynamic recovery involves *healing of defects* of a point type (*interstitial atom*, *vacancy*) formed during the course of plastic deformation, and it is accompanied by *anomalous mass transfer*. The partial *annihilation of dislocations* and their redistribution during the course of deformation bring about dynamic polygonization. These processes play an important role in *creep* with recovery, and also at the third stage of the *strain hardening* of the crystals. The dynamic recovery with reorientation occurs in crystals with a high degree of plastic deformation under conditions which bring about dynamic polygonization. Reorientation of separate sections of a deformed crystal accomplishes their *dehardening*, and prevents the formation of *cracks*. Dynamic recovery is observed also at temperatures of 0.1 times the absolute *melting temperature*, and is enhanced by increasing the strain temperature.

DYNAMIC SCALING, dynamic similarity hypothesis, dynamic scale invariance

In the theory of *critical phenomena* it is a generalization of the *scaling invariance hypothesis* to the function, which determines the critical behavior of kinetic (dynamic) characteristics of the system: *thermal conductivity*, coefficient of *sound absorption*, etc., which have anomalies at the point of a *second-order phase transition*. It was initially introduced in the works of R.A. Ferrell, N. Menyhard et al. (1968), and also of B.I. Halperin,

P.C. Hohenberg (1969). By analogy with the static *similarity* hypothesis, wherein the critical behavior of the system is determined by a single parameter with the dimensions of length, namely the *correlation length*, in the case of dynamic scaling it is the characteristic parameter with the dimension of frequency which determines the dynamics of a system in the vicinity of the phase transition point. In particular, this involves the so-called *critical retardation of the kinetic processes*.

DYNAMICS OF IMPERFECT CRYSTALS

This field of *crystal lattice dynamics* is devoted to the study of internal motions of real crystals (i.e. those containing *defects*). It includes the theory of lattice vibrations with defects, and the dynamics of defects.

Vibrations of a crystal with defects. The *normal vibrations* of a real crystal are not plane waves as they are in an *ideal crystal*. New vibrations may appear that are completely localized in the vicinity of defects (*local vibrations*). Frequencies which lie above the limiting frequency of an ideal crystal or are found in a *band gap* between bands correspond to these vibrations. If there are many *point defects* of the same type then resonance transitions of the local vibrations from one defect to another are possible. In this case the defective crystal has an *impurity band* of vibrational frequencies. The local vibrations of extended defects (e.g., *dislocations* or *stacking faults*) spread along them without penetrating into the crystal volume, corresponding to an inherent spectrum of dislocational or surface vibrations, different from those of the bulk. In addition to these local vibrations so-called *quasi-local vibrations* are possible which embrace the crystal as a whole, but with amplitudes which considerably exceed those of the regular lattice vibrations. The frequencies of such vibrations are found in the continuous spectrum of the ideal crystal, but most often in the vicinity of the edges of these bands. Local and quasi-local vibrations show themselves also in the appearance of additional lines in infrared adsorption spectra, at singularities of neutron scattering, and of Mössbauer spectra.

Dynamics of defects. Point defects such as *impurity atoms*, *vacancies* and *interstitial atoms* are capable of moving through the crystal by means of *diffusion*, and also via quantum tunneling (see *Tunnel effect*) from some position to an adjacent one. As a result of this latter possibility the crystal can transform to having the *quasi-particle*, called a *defecton*, moving slowly through it. An interstitial atom migrates with mechanical motion even in a classic crystal if it is in the so-called *crowdion configuration* (see *Crowdion*). Often the mechanical movement (*slip*) is characteristic of the specific linear defect called a *dislocation*. The displacement of its line along the slip plane does not violate the continuity of the crystal and thus it occurs comparatively easily. It appears that the dislocation motion is always associated with an uncontrollable variation of the shape of the crystalline sample, therefore the dislocation is the unit carrier of crystal *plasticity*. The atomic rearrangements that accompany dislocation motion do not require very high loads, and this is why *plastic deformation* of a crystal starts at stresses that are far below the theoretical *strength* of the crystal. The motion of dislocations is accompanied by the emission of elastic (sound) waves, which is similar to the manner in which moving (accelerated) electric charges bring about the emission of electromagnetic waves. Besides this, while interacting with the lattice vibrations the dislocation takes part in the oscillator dissipative motion, and thereby contributes to the *internal friction*.

Two-dimensional defects of the twin type (see *Twinning of crystals*), *cracks* or martensite inclusions may manifest themselves as dynamic formations. Together with dislocations, they play some role in the plasticity and strength of crystals.

DYNAMICS OF LIQUID CRYSTALS

Hydrodynamics of *anisotropic liquids*. In the case of nematic liquid crystals, besides the common hydrodynamic variables like speed $v(\boldsymbol{r},t)$ and pressure $p(\boldsymbol{r},t)$ there is also the *director* $\boldsymbol{n}(\boldsymbol{r},t)$. Therefore, the usual hydrodynamic equations should be supplemented with the equation of motion for the director. In addition it is necessary to take into account that the shear stresses within the nematic (stress tensor components) may be of three types:

(1) purely fluid stresses associated with the gradients of the velocity of the liquid ∇v,
(2) stresses not depending on the orientation or speed of the rotation of $\boldsymbol{n}$, and

(3) stresses involving the interaction of the orientation of molecules $\boldsymbol{n}$ with the non-uniform flow (∇v).

To treat these shear stresses coefficients of *viscosity* are introduced, called *Leslie coefficients*. The viscosity (or friction) associated with the motion of the director may be of two types: purely orientational and hydrodynamic, corresponding to the second and third types of shear stresses. The dynamics of *cholesteric liquid crystals* and *smectic liquid crystals* is similar to that of nematics (nematodynamics), but its formulation appears to be more complex (see also *Nematic liquid crystals*).

DYNAMICS OF QUANTUM VORTICES

The equation of *quantum vortex* motion within a superconductor has the form $M\partial \boldsymbol{v}/\partial t + \eta \boldsymbol{v} = \boldsymbol{f}_{\mathrm{L}}$, where M is the vortex mass per unit length, η is the *viscosity* coefficient, and $\boldsymbol{f}_{\mathrm{L}}$ is the force (*Lorentz force*) exerted on the vortex magnetic field $\boldsymbol{B}$ in the core by a transport current $\boldsymbol{j}$: $\boldsymbol{f}_{\mathrm{L}} = \boldsymbol{j} \times \boldsymbol{B}$. Ordinarily it is possible to neglect the inertial term because the motion is so slow and the viscosity associated with energy dissipation by the current flow in the normal state core is very high. Therefore, the velocity of *Abrikosov vortices* $v = |\boldsymbol{j} \times \boldsymbol{B}|/\eta$ is proportional to the average current which leads to Ohm's law for the resistivity generated by moving vortices $\rho_{\mathrm{f}} = \rho B/B_{c2}$.

The motion of the *vortex lattice* in a somewhat irregular (non-uniform in thickness) film due to the interaction of the vortices with the nonuniformities produces radiation that appears at the frequency $\omega = 2\pi v/a$, where a is the lattice period. Irradiating the system by an external high-frequency field causes singularities to appear in the current–voltage characteristic, similar to Shapiro steps (see *Josephson radiation*) in superconducting tunnel junctions (see *Tunnel effects in superconductors*), thereby producing high-frequency interference in superconducting films.

The dynamics of vortices in superconducting tunnel junctions (*Josephson vortices*) is characterized by a low viscosity, so inertial effects play an important role and induce a variety of dynamic phenomena. The motion of vortices along the surface of the tunnel junction induces radiation with the frequency $\omega = 2e\overline{V}/\hbar$, where $\overline{V}$ is the average voltage at the barrier, which can be looked upon as corresponding to the *dc Josephson effect*. The interaction of vortices with the fields within a superconductor resonator (in the high density limit they are considered as a "*density wave of Josephson current*") brings about the appearance of *Fiske steps* (or *Shapiro steps*) on the current–voltage characteristic of the junction. The interaction of two fluxons of opposite polarity ("vortex–antivortex" system) can involve the formation of a dynamically bound state, etc.

DYNAMIC SPIN HAMILTONIAN

See *Spin–phonon Hamiltonian*.

DYNAMIC TRANSFORMATION OF MAGNETIC BUBBLE DOMAINS

Irreversible variation of the *domain wall* structure of *magnetic bubble domains* (cylindrical regions of magnetization) during the process of motion. It is observed when the drop of the magnetic biasing field at the domain diameter exceeds some critical value, i.e. for a sufficiently high speed of motion. To characterize the state of a magnetic bubble domain the notions of *chirality of magnetic bubble domains* and average *topological index* $\overline{S}$ are used. The variation of chirality and of the average index can occur only in the mode of non-stationary motion (spins within the domain wall precess) due to the breaking through the dynamic horizontal *Bloch lines* to the surface of the plate (film), to annihilate or generate pairs of *Bloch points* in the process of interaction with *defects in crystals*, or under the effect of thermal excitation. It is possible to control the processes of magnetic domain transformations using a pulsed magnetic field, oriented parallel to the developed surface of the film, which finds its application in the storage devices at the hexagonal lattices of magnetic bubble domains with information coding according the average index $\overline{S}$ (see *Magnetic bubble domain devices*).

DYSON EQUATION (F. Dyson, 1948)

The equation connecting the full *Green's function*, the zero-order Green's function and the *mass operator of quantum mechanics*. The equation has the form $G = G_0 + G_0 M G$, where G is the full (exact) Green's function, G_0 is the zero-order (without taking account of the interaction) Green's function, M is the *self energy operator* (sometimes called mass operator). Using the diagram technique (see *Feynman diagrams*) the Dyson equation is obtained by summing an infinite series of diagrams. The solution of this equation with respect to the full Green's function is

$$G = \frac{1}{G_0^{-1} - M}.$$

This solution is formal as the self energy operator in its turn contains Green's functions, but it is useful for investigating general properties of Green functions, as well as for their approximate calculation. In the spectral representation $G(\omega)$ is given by

$$G(\omega) = \frac{1}{\omega - \Omega - M(\omega)},$$

where Ω is determined by the zero-order Hamiltonian.

DYSON–MALEYEV REPRESENTATION
See *Spin operator representations.*

DYSON SHAPE (F. Dyson, 1955)
The *line shape* of electron paramagnetic resonance arising from *conduction electrons* in metals. It is determined by the following parameters: electron mean free path l, thickness of skin-depth δ (see *Skin-effect*), time of electron passage through it, and times of *spin–lattice relaxation* and *spin–spin relaxation.* The shape differs for regions of normal ($l < \delta$) anomalous ($l > \delta$) skin effect, and in the general case the shape is asymmetric.

DYSPROSIUM, Dy
A chemical element of Group III of the periodic table; atomic number 66, atomic mass 162.50; it is a *lanthanide.* 21 isotopes are known with mass numbers from 147 to 167. Natural dysprosium consists of the mixture of 7 stable isotopes with mass numbers 156, 158, 160 to 164. Outer electronic shell configuration is $4f^{10}5d^06s^2$. Successive ionization energies are 5.93, 11.67, 22.8 eV. Atomic radius is 0.177 nm; radius of the ion Dy^{3+} is 0.091 nm. Oxidation states are +3 (the most characteristic), +4. Electronegativity is ≈1.19.

In the free form dysprosium is a silvery-gray *metal.* There are α and β modifications. α-Dy is hexagonal close packed with parameters $a = 0.3591$ nm, $c = 0.5649$ nm at room temperature (space group $P6_3/mmc$, D_{6h}^4). Above 1661 K up to $T_{\text{melting}} = 1681$ K β-Dy is stable with body centered lattice and parameter of cubic conventional unit cell $a = 0.398$ nm (extrapolated value) after hardening (space group $Im\bar{3}m$, O_h^9). The density is 8.54 g/cm^3, $T_{\text{boiling}} \approx 2700$ K. Heat of melting is 16.6 kJ/mole, heat of sublimation 280.5 kJ/mole and heat of evaporation is 283 kJ/mole, specific heat is 0.170 kJ·kg^{-1}·K^{-1} at room temperature; Debye temperature ≈185 K; linear thermal expansion coefficient of a monocrystal is $16.6\cdot10^{-6}$ K^{-1} at 400 K along the main axis $\underline{6}$ and $6.0\cdot10^{-6}$ K^{-1} in the perpendicular direction, and of the dysprosium polycrystal $8.6\cdot10^{-6}$ K^{-1} (at room temperature); coefficient of thermal conductivity is 10.1 W·m^{-1}·K^{-1}. Adiabatic coefficients of elastic rigidity of an α-Dy monocrystal: $c_{11} = 74.66$, $c_{12} = 26.16$, $c_{13} = 22.33$, $c_{33} = 78.71$, $c_{44} = 24.27$ GPa at 298 K; adiabatic modulus of uniform compression is 41.1 GPa, isothermal modulus of uniform compression is 39 GPa; Young's modulus is 63.08 GPa; tensile strength is 0.263 GPa, elongation is 6%; Vickers hardness is 42 HV. Dysprosium is easily subjected to mechanical treatment. Effective thermal neutron cross-section is 1050 barn. Resistivity of a monocrystal at 298 K is of 774 nΩ·m along the main axis $\underline{6}$ and 1003 nΩ·m in the perpendicular direction; temperature coefficient of electrical resistance of polycrystalline dysprosium is 0.00119 K^{-1}. Work function of a dysprosium polycrystal is 3.09 eV. In dysprosium two temperature regions of magnetic ordering are observed: in the temperature range T_c to T_N – helicoidal *antiferromagnetism* with screw axis along the main axis $\underline{6}$, and in the range 0 K to T_c – collinear *ferromagnetism*, and in both magnetic structures the magnetic momenta are located in the basal planes of α-Dy. Transition from the paramagnetic to the anti-ferromagnetic state at the *Néel point* $T_N = 178.5$ K is a *second-order phase transition* (according to the other data, a "weak" *first-order phase transition*). At the *Curie point* $T_C = 86$ K there is a first-order phase transformation to the low-temperature ferromagnetic state, and when the α-Dy crystal is subjected to jumpwise small distortions its symmetry is lowered to orthorhombic. According to other data, the dysprosium transition from the antiferromagnetic state to the ferromagnetic state upon cooling is observed at 88.0 K, and the reciprocal transformation upon heating at 90.6 K. Magnetic susceptibility is 0.102 CGS units, nuclear magnetic moment of the 24.97% abundant isotope ^{163}Dy is −0.53 nuclear magnetons. Dysprosium is a component of some magnetic alloys (with iron, nickel, copper and aluminum), of phosphors, fireproof ceramic materials, et al.

DZIALOSHINSKII INTERACTION
(I.E. Dzialoshinskii, 1957)

A part of the *spin–spin interactions* and *spin–orbit interactions* in two-sublattice *antiferromagnets*, leading through a phenomenological description to the appearance of terms of the form $w_D = D_{ik}M_iL_k$, where $\boldsymbol{M}$ is the *magnetization*, and $\boldsymbol{L}$ is the *antiferromagnetism vector*. The form of the tensor D_{ik} is determined by the symmetry of the crystal which enters the expression for the magnetic energy. The added terms are introduced for the explanation of the weak ferromagnetism of antiferromagnets (see *Antiferromagnetism*). As a rule, the Dzialoshinskii interaction is small due to the smallness of the spin–orbit interaction in comparison with the *exchange interaction*. The case when $w_D = D[\boldsymbol{M} \times \boldsymbol{L}]_z$ is in practice the most important, where z is the symmetry axis direction, because in this case the Dzialoshinskii interaction is enhanced due to the exchange interaction (T. Moriya, 1960), an enhancement which is also referred to as *antisymmetric exchange*. The Dzialoshinskii interaction is found only in crystals whose *magnetic symmetry group* allows the presence of the tensor D_{ik}.

Ee

EASY MAGNETIZATION AXIS

The direction of the *magnetization* vector of a ferromagnet, $\boldsymbol{M}$, which corresponds to the minimum of the *magnetic anisotropy* energy, or the direction of possible alignment of the vector $\boldsymbol{M}$ in an infinite *magnetic material* with neither an external magnetic field nor an elastic *strain* present. If the magnetic anisotropy is not unidirectional, the easy magnetization axes are axial directions, so their number is always even, depending on the symmetry and nature of the magnetic anisotropy. In the case of unidirectional anisotropy, the easy magnetization axes represent polar directions. The direction corresponding to the absolute minimum of the magnetic anisotropy energy is sometimes referred to as the *easiest magnetization axis*. See also *Hard magnetization axis*.

ECHELETTE

See *Diffraction grating*.

ECHO, ELECTROACOUSTIC

See *Electroacoustic echo*.

ECHO, PHONON

See *Electroacoustic echo*.

ECHO, SPIN

See *Spin echo*.

EDGE ANGLE, wetting angle

The angle θ_0 between the contacting surfaces of a liquid and a *solid* (or another liquid), measured from the side of the liquid that is in a state of thermodynamic equilibrium (see Fig.). Depending on the value of θ_0, there are wetting and non-wetting liquids (see *Wetting*). When the surface of the solid is deformed, the equilibrium of a droplet is characterized by two angles, θ_0 and α, as shown in the figure. These angles are related to surface tensions σ_L, σ_S, and σ_SL of the liquid, solid, and solid–liquid interface, respectively (see *Surface tension*), by the following expressions:

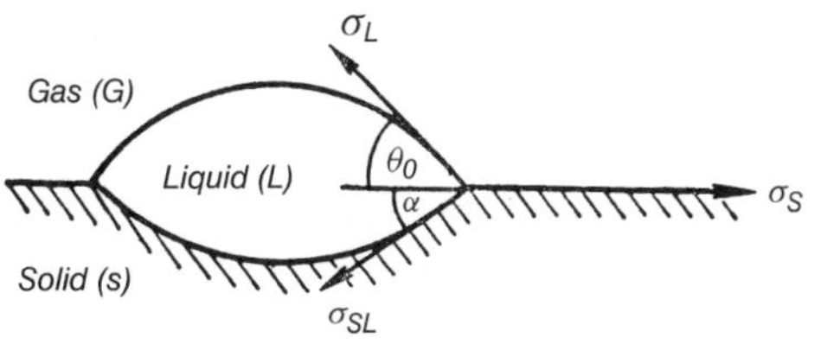

Edge angle.

$$\sigma_\mathrm{S} = \sigma_\mathrm{L}\cos\theta_0 + \sigma_\mathrm{SL}\cos\alpha,$$

$$\sigma_\mathrm{L}\sin\theta_0 = \sigma_\mathrm{SL}\sin\alpha.$$

By measuring θ_0 and α and knowing σ_L, one may calculate σ_S and σ_SL.

EDGE DISLOCATION

A *dislocation* line which is normal to a *Burgers vector*. In an infinite isotropic elastic medium (see *Isotropy of elasticity*) a linear edge dislocation creates a plane *strain* inversely proportional to the distance from it, and independent of the position along the dislocation. The Burgers vector in a crystal coincides with one of the lattice vectors. An edge dislocation can be represented as an edge of an incomplete (removed) lattice plane normal to the Burgers vector. The plane surface parallel to the Burgers vector and passing through an edge dislocation line is referred to as a *glide plane*. An edge dislocation can move along this plane, remaining parallel to itself without any *diffusion* of atoms (con-

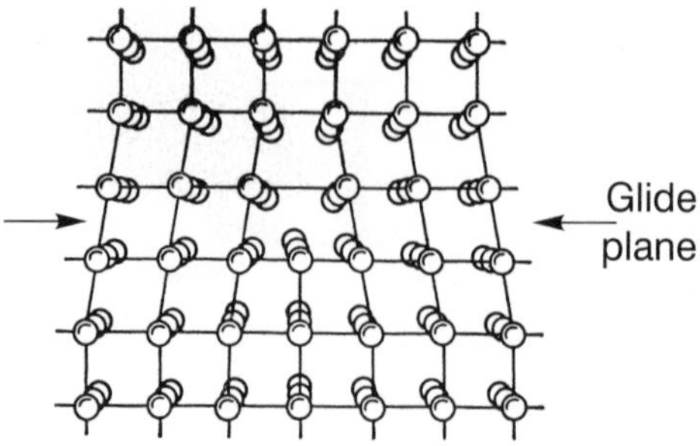

Edge dislocation.

servative motion). The figure shows a model of an edge dislocation in a simple cubic lattice.

EFFECTIVE ATOMIC CHARGE

Difference in electron density between a crystal and a free atom (ion) integrated over the volume per atom. Due to the ambiguity of the shape of the indicated volume, an effective atomic charge is an estimated value. Sometimes the concept of a *dynamic effective charge* is used; this is equal to the ratio between the *dipole moment* of displaced *atom* and the magnitude of this displacement. In this case, the polarization of the atomic electron shell is also taken into account. The dynamic charge is dependent on the wave vector of the vibrational motion, and is different for the various phonon branches.

EFFECTIVE ELECTRICAL CONDUCTIVITY

Coefficient of proportionality between applied external electric field $\boldsymbol{E}$ and an average value of current density $\boldsymbol{J}$ (see *Electrical conductivity*) in an isotropic heterogeneous conductor: $\langle \boldsymbol{J}\rangle = \sigma\langle \boldsymbol{E}\rangle$. In the case of a weakly heterogeneous medium, $\langle(\sigma - \langle\sigma\rangle^2\rangle \ll \langle\sigma\rangle^2$ and the effective conductivity is given by $\sigma_e = \langle\sigma\rangle - \langle(\sigma - \langle\sigma\rangle)^2\rangle/(3\langle\sigma\rangle)$. In the case of a strongly heterogeneous medium, σ_e is calculated from *percolation theory*. In the general case, $\langle\sigma^{-1}\rangle^{-1} \leqslant \sigma_e \leqslant \langle\sigma\rangle$ and $\langle\langle\sigma^{-1}\rangle_{\parallel}^{-1}\rangle_{\perp} \leqslant \sigma_s \leqslant \langle\langle\sigma\rangle_{\perp}^{-1}\rangle_{\parallel}^{-1}$ where the brackets $\langle\ldots\rangle_{\parallel}$ and $\langle\ldots\rangle_{\perp}$ mean averaging over Cartesian coordinates parallel and normal to the current direction $\boldsymbol{J}$, respectively. In semiconductors, an increase of $\langle\boldsymbol{E}\rangle$ results in changes in the nonuniformity of the spatial distribution of *current carriers*, the presence of a σ_e versus $\langle\boldsymbol{E}\rangle$ dependence, and a nonlinear *current–voltage characteristic*. In terms of macroscopic considerations, *anisotropic conductors* are described by an *effective electrical conductivity tensor* that is not symmetric in the presence of a magnetic field.

EFFECTIVE ENHANCEMENT OF SPACE DIMENSIONALITY

A phenomenon in which physical characteristics of low-symmetry crystals associated with dipole-active excitations are the same as those of *crystals* in a d-dimensional space with $d > 3$. Long-wave dipole-active excitations in crystals (polar optical *phonons*, *excitons*, *plasmons*) are accompanied by a longitudinal macroscopic electric field. In low-symmetry crystals (*uniaxial crystals* and *biaxial crystals*), along with transverse-longitudinal splitting of the energy spectrum, this field brings about *directional dispersion* which means the dependence of the energy of such excitations on the direction of a quasi-momentum $\boldsymbol{k}$ with respect to the *optical axes of a crystal*: $E = E(\boldsymbol{k}/k)$ as $\boldsymbol{k} \to 0$. For example, the energy spectrum of excitations with the *polarization vector* parallel to the optical axis of a uniaxial crystal, has the form

$$E(\boldsymbol{k}) = E_0 - \frac{\hbar^2\boldsymbol{k}^2}{2m} - \alpha\sin^2\psi, \qquad (1)$$

where ψ is the angle between $\boldsymbol{k}$ and the optical axis. For such excitations, the main contribution to the density of states

$$\rho(E) = \frac{1}{(2\pi)^3}\int \delta\big(E - E(\boldsymbol{k})\big)\,\mathrm{d}^3\boldsymbol{k}$$

is that of excitations with $\theta \approx 0, \pi$. The volume element is $\mathrm{d}^3\boldsymbol{k} = 2\pi k^2\,\mathrm{d}k\sin\psi\,\mathrm{d}\psi$. We must take $\sin\psi \approx \psi$ and thus obtain the quasi-five-dimensional volume element $\mathrm{d}^3\boldsymbol{k} = 2\pi k^2\,\mathrm{d}k\,\psi\,\mathrm{d}\psi$. Then the density of states has also the form corresponding to the dimension $d = 5$ of the space:

$$\rho(E) = \mathrm{const}(E_0 - E)^{3/2}\theta(E_0 - E), \qquad (2)$$

where $\theta(E)$ is a step function. In another geometry, the effect of dimension $d = 4$ may arise. Such an enhancement of dimension changes the character of singularities of the density of states that are different from the *Van Hove singularities*. When *ferroelectric phase transitions* are studied, due to the effective enhancement of the space dimensionality, *fluctuations* near the point of the phase transition appear to be damped. This alters the localization laws of elementary excitations at *point defects*, at plane *stacking faults*, and at composition fluctuations in *solid solutions*.

EFFECTIVE IMPURITY-ATOM CHARGE

Difference between the electric charge of an *impurity atom* and an intrinsic atom in a solid. The value of the effective impurity-atom charge arises from the redistribution of the electron density of a chemically foreign atom when it is embedded in a solid. The same difference with respect to an elementary charge is called *impurity valence*. This differs from the *valence* of an isolated atom, and from a difference in valences of impurity and intrinsic isolated atoms. In the general case, this is specified by a non-integral or fractional number (so-called *fractional valence*).

EFFECTIVE MASS

A value that characterizes *quasi-particles* with a quadratic energy *dispersion law* $E(\boldsymbol{p})$ where $\boldsymbol{p}$ is the *quasi-momentum*. Formally, it is the quantity m^* present in the formula $E(\boldsymbol{p}) = \boldsymbol{p}^2/(2m^*)$ that is called the effective mass. A differential definition is $m^* = \hbar^2(\partial^2 E/\partial k^2)^{-1}$, where $\boldsymbol{p} = \hbar\boldsymbol{k}$. The concept of effective mass results from solving dynamical problems (Newton equation, Schrödinger equation, or Euler–Lagrange equation) in which the process of taking into account the interaction of a given particle with the environment is reduced to the substitution of a certain effective mass for its ordinary mass. For example, the effective mass of a *polaron* is related to the *electron–phonon interaction*. In the *band theory* of charge carriers in a crystal, the concept of effective mass is introduced near the extrema of the band energy $E_n(\boldsymbol{p})$ where n is the band number. If there is a minimum (electrons) or a maximum (holes) at a particular point $\boldsymbol{p}_0$, $E_n(\boldsymbol{p})$, then in the vicinity of $\boldsymbol{p}_0 = \hbar\boldsymbol{k}_0$ we have the expression

$$E_n(k) = E_n(k_0) + \frac{\hbar^2 m_{ij}^{-1}(k_i - k_{i0})(k_j - k_{j0})}{2}, \tag{1}$$

where the second-rank tensor m_{ij}^{-1} is called the *inverse effective mass tensor*. In its principal axes system the tensor m_{ij}^{-1} has three components m_i^{-1} ($m_i > 0$ is associated with electrons, and $m_i < 0$ corresponds to holes). The anisotropy of the dispersion law (1) as well as the presence of a few maxima result in a great diversity of *galvanomagnetic effects* in *semiconductors* and *metals*. There also exist other kinds of effective masses, e.g., a *cyclotron effective mass* which is measured with the help of a *cyclotron resonance* experiment, and the effective masses of *rotons*, *excitons*, etc.

EFFECTIVE RYDBERG (after J.R. Rydberg)

Ionization energy of a small Coulomb impurity center with a hydrogen-like energy spectrum in a *semiconductor*. For a nondegenerate band with an isotropic parabolic *dispersion law* in the vicinity of a minimum, the effective Rydberg $\mathrm{Ry}^* = [m^*/(m\varepsilon^2)]\mathrm{Ry}$, where m^* is the *effective mass*, m is the mass of a free electron, ε is the dimensionless static *dielectric constant* of the crystal, Ry is a *Rydberg* with the value $\mathrm{Ry} = me^4/(8\varepsilon_0^2 h^2) \approx$ 13.60 eV corresponding to the ionization energy of a hydrogen atom. Values of the effective Rydberg in *germanium* and *silicon* are of the order of 0.01 eV. The expression "effective Rydberg" is also used in the broader sense to designate the binding energy of a system with an electrostatic interaction (sometimes with a more general type of interaction), in particular to describe *Wannier–Mott excitons*. For such excitons, $\mathrm{Ry}^* = [\mu/(m\varepsilon_\infty^2)]\mathrm{Ry}$, where μ is the electron and *hole* reduced effective mass, and ε_∞ is the dimensionless optical dielectric constant.

EFFECTS OF RETARDATION

Effects stemming from the finite velocity of propagation (light velocity) of an interaction implemented through the electromagnetic field. Unlike the instantaneous Coulomb interaction which depends only on the positions of charges, the retarded interaction depends also on the distribution of their velocities. It is caused by the solenoidal portion of the electromagnetic field, and corresponds to the exchange of virtual transverse *photons*. The effects of retardation can be described by the dimensionless parameter $\alpha = \tau/T$, where T is the characteristic time of the motion of the interacting particles, and τ is the time of the field propagating between them. For instance, if applied to atomic electrons, $1/T \sim e^2/(4\pi\varepsilon_0 a\hbar)$ is the Bohr frequency of the electron; e is its charge; a is the radius of its orbit; $\tau \approx a/c$, and $\alpha = e^2/(4\pi\varepsilon_0\hbar c)$ is the fine structure constant. In the case of a crystal with excitonic excitation there is an interaction between electrons that are much farther from one another than in an atom, at a distance on the order of an *exciton* wavelength λ, i.e. $\tau \approx \lambda/c$. Also, the time for an appreciable displacement of charge is $T \approx 1/\omega$, where ω is the frequency of the exciton wave. The result is $\alpha \approx 1/n$, where n is the *refractive index* of the electromagnetic wave accompanying the exciton. Thus, the effects of retardation are significant as long as n (or the wave vector $k = n\omega/c$) is not too large. The latter holds in the region of the intersection of the spectral curves of photons and excitons, where effects of retardation are appreciable, and a mixed electromagnetic wave or *polariton* (*photoexciton*) appears.

EINSTEIN–DE HAAS EFFECT (A. Einstein, W.J. de Haas, 1915)

Variation of the mechanical angular momentum of a *magnetic material* as a result of its *magnetization* in an external magnetic field. For a magnet, the induced mechanical momentum is proportional (neglecting *magnetic anisotropy*) to the variation of its *magnetic moment*, and it is directed along the magnetization axis. Experimentally a twisting motion is observed for a sample freely suspended in an alternating magnetic field with a frequency close to the frequency of the intrinsic oscillations of the sample. The Einstein–de Haas effect is a complement of the the *Barnett effect* (slight magnetization induced by rapid rotation), and it is an example of a *magnetomechanical phenomenon*.

EINSTEINIUM, Es

Chemical element of the *actinide* group of the periodic system with atomic number 99 and atomic mass 252.0828. There are 14 isotopes known with mass numbers from 243 to 256, none of which are stable. Outer shell electronic configuration is $5f^{11}6d^07s^2$. Radius of Es^{3+} ion is 0.0953 nm. Oxidation state is +3, more rarely +2. Electronegativity is 1.27.

Einsteinium is a *metal* characterized by high volatility, it has a face-centered cubic lattice, space group $Fm\bar{3}m$ (O_h^5); $a \approx 0.558$ nm at room temperature. The existence of a metastable modification of einsteinium is possible with a hexagonal close-packed lattice, space group $P6_3/mmc$ (D_{6h}^4), having a two atom basis and a primitive unit cell volume 0.08916 nm^3 (at room temperature); with $T_{\text{melting}} = 1130$ K. Einsteinium is paramagnetic, with no known magnetic ordering. Targets of isotope ^{253}Es with 20.47 day half-life are used for synthesis of heavier transplutonium elements.

EINSTEIN RELATION (A. Einstein, 1905)

An expression which relates the *diffusion coefficient* D_i of particles i (electrons, *holes*, *ions*, etc.) possessing an electric charge e_i and the mobility μ_i in an electric field E: $e_i D_i = k_B T \mu_i$. The Einstein relation is valid when the *Boltzmann distribution* is satisfied by the particles. In the case of electric field heating of *current carriers* then the Boltzmann distribution applies using the electron (hole) temperature T^*, so the Einstein relation remains valid when T^* is substituted for T.

ELASTIC ENERGY, strain energy

The energy due to the elastic *strain* of a solid body (see *Elasticity theory*). The elastic *free energy* density F can be expressed in terms of the *distortion tensor* u_{ik}:

$$F = \frac{1}{2}\int c_{ik,lm} u_{ik} u_{lm}\,\mathrm{d}V,$$

where $c_{ik,lm}$ is the fourth-rank *elastic modulus tensor* of the crystal,

$$u_{ik} = \frac{1}{2}\left(\frac{\partial u_i}{\partial x_k} + \frac{\partial u_k}{\partial x_i}\right),$$

and $\boldsymbol{u}_i = \boldsymbol{u}_i(x,t)$ is a displacement vector at the point x. This free energy can also be presented in terms of the *strain tensor* σ_{ik} as these quantities are connected with each other by *Hooke's law*: $\sigma_{ik} = c_{ik,lm} u_{lm}$.

ELASTIC ENERGY, STORED

See *Stored elastic energy*.

ELASTIC FIELD OF DISLOCATION

Field of *internal stresses* (and *strains*) created by a *dislocation* in a solid. The nature of the stress field depends strongly on the degree of curvature of the dislocation line in the case of a bent dislocation, and on the magnitude of the angle α between the *Burgers vector* and the dislocation line direction for a linear dislocation.

In an elastically-isotropic infinite medium (see *Isotropy of elasticity*), only *shear* stresses are produced by a *screw dislocation* ($\alpha = 0$). In the case of a screw dislocation that is directed along the z-axis of an orthogonal xyz coordinate system we have the stress tensor components

$$\sigma_{xz} = -\frac{Gb}{2\pi}\frac{y}{x^2+y^2},$$

$$\sigma_{yz} = \frac{Gb}{2\pi}\frac{x}{x^2+y^2},$$

where G is the *shear modulus*, and b is the magnitude of the Burgers vector.

An *edge dislocation* ($\alpha = 90°$) generates both shear and normal stresses. Consider an edge dislocation aligned along z, with the Burgers vector directed along the Ox-axis:

$$\sigma_{xx} = -\frac{Gb}{2\pi(1-\nu)}\frac{y(3x^2+y^2)}{(x^2+y^2)^2},$$

$$\sigma_{yy} = \frac{Gb}{2\pi(1-\nu)}\frac{y(x^2-y^2)}{(x^2+y^2)^2},$$

$$\sigma_{xy} = \frac{Gb}{2\pi(1-\nu)}\frac{x(x^2-y^2)}{(x^2+y^2)^2},$$

where ν is *Poisson ratio.* When the dislocation line makes an arbitrary angle α with Burgers vector $\boldsymbol{b}$, the total stress field is defined as the sum of stress fields produced by an edge dislocation with the Burgers vector $b\sin\alpha$ and a screw dislocation with the Burgers vector $b\cos\alpha$.

The stresses generated in a solid by a line dislocation decay away from dislocation as $1/r$, where r is the distance from dislocation. The field of stresses produced by a bent dislocation is of a more complicated nature. The free surface of a solid of finite sizes exerts a certain influence on the stresses around the dislocation by virtue of the action of *image forces.* This influence is stronger, the closer the dislocation is to the solid surface. The calculations of internal stresses, which are produced by dislocations in an elastically anisotropic medium, are rather complicated. However, the results based on the assumption that the medium under consideration is isotropic are often appropriate for most practical applications.

ELASTIC INTERACTION OF DEFECTS in a crystal

The coordinate-dependent contribution to the interaction of *defects* in a crystal lattice. This interaction is conditioned by nonuniform fields of static elastic *strains*, which are produced by defects. Generally, the elastic interaction energy between two defects falls off as the distance between these defects becomes sufficiently long (in comparison to their characteristic sizes and the radius of interatomic interactions in the lattice). For a broad variety of *point defects*, dislocation lines, dislocation and disclination dipoles and loops (see *Dislocation*, *Disclination*, *Dislocation loop*), and particles of a precipitated phase in the lattice, the energy decays with distance as a power function, in particular as $\propto 1/r^3$ for the case of point defects in a slightly elastical isotropic medium. When the spacing between the defects is of the order of their sizes, the energy of the elastic interaction of defects may be determined by calculations carried out in the context of a microscopic (discrete) model. The value of the defect elastic interaction energy depends on the elastic properties of the crystal lattice, on the orientation of defects with respect to one another, and on the orientation of the radius vector $\boldsymbol{r}$ connecting the defects with respect to the crystallographic axes of the lattice. By virtue of its long-range action, the elastic interaction of defects plays a noticeable role in processes involving the approach to equilibrium or to metastable configurations in a crystal with defects.

ELASTICITY

The property of a solid to change its shape and size under action of external loads, and to resume its original configuration upon removal of the load. An elastic *strain* is completely reversible; in other words, the positions of all points of the solid during the loading process coincide with the positions of these same points during the course of the load removal. Elastic deformations in a crystalline solid may be related not only to the action of external forces, but also to the presence of structural defects (*dislocations*, *disclinations*, second phase particles, etc.). Elasticity of solids results from interatomic (intermolecular) interaction forces which tend to restore the body to its initial equilibrium state that had been previously disturbed by external forces which altered the original mutual arrangement of atoms.

ELASTICITY ANISOTROPY

See *Anisotropy of elasticity.*

ELASTICITY ISOTROPY

See *Isotropy of elasticity.*

ELASTICITY OF LIQUID CRYSTALS

The capability of *liquid crystals* to resist a change of translational order and *orientational order* on being subjected to external actions (mechanical loads, electric and magnetic fields), and the ability to return to their initial configuration spontaneously when the said actions have ceased. The specific nature of liquid crystals consists in the fact that no forces counteracting the separation of neighboring points arise in these materials at deformation. Owing to this, *Young's modulus* and the *shear modulus* of *nematic liquid crystals* are both equal to zero. Molecules of nematics are aligned with a particular common direction specified by the *director* $\boldsymbol{n}$, and deformation of these materials brings about only rotational moments that obstruct the change of the $\boldsymbol{n}$ direction. Similar to *Hooke's law*, the free-energy density for slow spatial changes of the director, called the *Frank energy*, is a quadratic function of the deformations $\nabla\boldsymbol{n}$:

$$f = \frac{1}{2}K_1(\nabla\boldsymbol{n})^2 + \frac{1}{2}K_2\big(\boldsymbol{n}\times(\nabla\times\boldsymbol{n})\big)^2 + \frac{1}{2}K_3\big(\boldsymbol{n}\times(\nabla\times\boldsymbol{n})\big)^2, \qquad (1)$$

where the factors K_1, K_2, K_3, called *Frank moduli*, or *Oseen–Frank moduli* (C.W. Oseen, F.C. Frank), associated respectively with transverse bending, twisting, and longitudinal bending strains (see Fig.), have the values $K_1 \sim K_2 \sim K_3 \sim 10^{-11}$ N for the majority of nematics. For

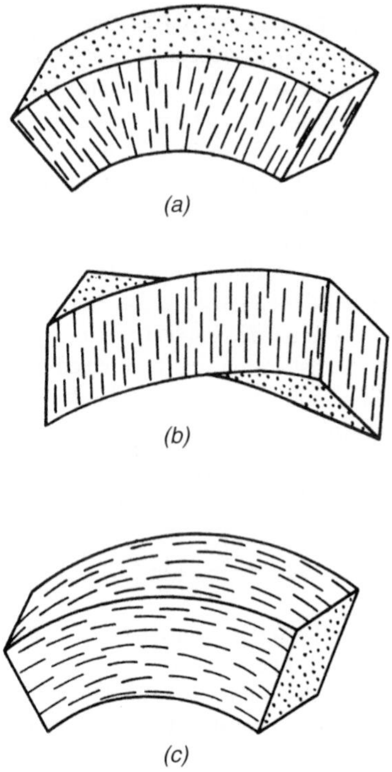

Deformations involving (a) transverse bending, (b) twisting, and (c) longitudinal bending.

p-azoxyanisole, e.g., $K_1 = 0.7\cdot10^{-11}$ N, $K_2 = 0.43\cdot10^{-11}$ N, and $K_3 = 1.7\cdot10^{-11}$ N at 120 °C. Frank moduli are temperature-dependent, e.g., K_2 and K_3 increase indefinitely in the neighborhood of the nematic–smectic transition (see *Cybotaxic clusters*).

The molecules of *cholesteric liquid crystals* do not exhibit bilateral symmetry, and hence the following term $K_2\boldsymbol{q}(\boldsymbol{n}\times\operatorname{rot}\boldsymbol{n})$ linear in the deformation ($\boldsymbol{q}$ is the cholesteric spiral wave vector) appears in Eq. (1). This cross-product term is present because the equilibrium state of a cholesteric is a spontaneously curled structure.

Other properties of *smectic liquid crystals* of A-type also exhibit specificity. On the one hand, the requirement of equidistant layers causes the moduli K_2 and K_3 to become infinite (the modulus K_1 remains approximately equal to that of a nematic). On the other hand, there arises the necessity of introducing an elastic modulus B, which is similar to Young's modulus, and characterizes the change of interlayer distance under the action of an external force. In this case, the free-energy density is given by

$$f = \frac{1}{2}K_1(\nabla \boldsymbol{n})^2 + \frac{1}{2}B\left(\frac{\partial u}{\partial z}\right)^2, \qquad (2)$$

where $B \sim 10^6\ \mathrm{N/m^2}$, and u is the displacement of the smectic layers. In the case of smectic C the free energy (2) must include a term for the energy of the C-director elastic distortions, as well as terms to take into account the interactions of the deformations of layers and the C-director (see *Wave-like modulation*). In the presence of an electric field $\boldsymbol{E}$ and a magnetic field $\boldsymbol{B}$ the energy terms Π_{e} and Π_{m} are added to the free energy f:

$$\Pi_{\mathrm{e}} = -\frac{\varepsilon_{\mathrm{a}}}{2}(\boldsymbol{n}\cdot\boldsymbol{E})^2,$$

$$\Pi_{\mathrm{m}} = -\frac{\chi_{\mathrm{a}}}{2}(\boldsymbol{n}\cdot\boldsymbol{B})^2.$$

These terms depend on the *dielectric constant* ε_{a} and the *magnetic susceptibility* χ_{a} (see *Anisotropy of liquid crystals*).

ELASTICITY THEORY

Branch of *continuum mechanics* concerned with reversible *strains* of solids. Elasticity theory is used in solid state physics for describing structural distortions which vary only slightly over distances on the order of interatomic spacings. This theory provides the basis for a continuum theory of *dislocations*, *grain boundaries*, *cracks*, *pores*, *point defects*, for physical theories of *plasticity*, *strength*, *failure* and acoustical phenomena.

The basic concepts of elasticity theory are related to *internal stresses* and deformations, described respectively by the *stress tensor* σ_{ik} and the *strain tensor* u_{ik}. In linear elasticity theory the deformation tensor is given by $u_{ik} = (1/2)(\partial u_i/\partial x_k + \partial u_k/\partial x_i)$, where u_i is a *displacement vector*, and the stress tensor σ_{ik} is related to u_{ik} by *Hooke's law*: $\sigma_{ik} = c_{iklm}u_{lm}$, where c_{iklm} is the *elastic modulus tensor*. The condition for applicability of linear elasticity theory is that the stresses be small compared with *elastic moduli*. Elasticity theory assumes that the equilibrium state of an elastic solid at given temperature T is unambiguously determined by the deformation tensor u_{ik} (deformations are coordinate-dependent), and hence the free energy of a deformed solid is a functional of the deformations. The free energy of a unit volume of the deformed solid may be represented as a series in terms of powers of u_{ik}, where only quadratic terms (*harmonic approximation*) of this series are employed in linear elasticity theory:

$$F = F_0(T) + \frac{1}{2}c_{iklm}^{T}u_{ik}u_{lm}, \qquad (1)$$

where $F_0(T)$ is the free energy of the undeformed solid, and the terms c_{iklm}^{T} are temperature-dependent isothermal elastic moduli. The remaining (anharmonic) terms of the series are used for describing effects that are left out in the linear approximation (e.g., volume effects in a crystal with dislocations, nonlinear effects occurring at the propagation of elastic waves, etc.).

Thermodynamic definitions of stresses and elastic moduli in linear elasticity theory are as follows:

$$\sigma_{ik} = \left(\frac{\partial F}{\partial u_{ik}}\right)_T,$$

$$c_{iklm} = \left(\frac{\partial^2 F}{\partial u_{ik}\partial u_{lm}}\right)_T.$$

The dynamic equation of elasticity theory is of the form:

$$\frac{\rho\partial^2 u_i}{\partial t^2} = \frac{\partial \sigma_{ik}}{\partial x_k} + f_i, \qquad (2)$$

where ρ is the *density* of the material, and f_i is the force acting on a unit volume of the solid. This equation presumes the applicability of Hooke's law with certain elastic moduli, and it involves adiabatic elastic moduli in the case of high-frequency (sound) vibrations. The boundary conditions for Eq. (2) are established either by specified *surface forces* $\sigma_{ik}n_i = f_i^{\mathrm{s}}$ (n_i is the unit vector of the outward-directed normal to the surface, f_i^{s} is the force acting on a unit surface area of the deformed solid), or by specified shifts of the outer surface of a solid. The applied force f_i may be that of gravity ρg_i (where g_i is the gravitational acceleration), or the force arising from the difference between the local temperature and the volume-averaged temperature of the body under nonuniform heating. The distribution of forces f_i may be generated by *defects* in the solid when the net applied force (summation taken over the volume), and the total moment of the forces both equal zero.

The static equilibrium equation (*equation of a solid in equilibrium*), which follows from the dynamic equation of elasticity theory, admits unique solutions under specified boundary conditions. The majority of technical applications of elasticity theory are based on applying these solutions. The wave solutions, which describe sound vibrations in solids, are of greatest interest among the various dynamic solutions of the elasticity theory equation. The main characteristic of wave solutions is the *dispersion law* (the relationship between wave frequency and wave vector). There exist three branches of sound vibrations in solids (three different sound waves associated with each value of a wave vector). In the case of an infinite isotropic (see *Isotropy*) body, these three branches correspond to one longitudinal wave, and two transverse ones.

All elasticity theory expressions are long-wave limits of the corresponding results of *crystal lattice dynamics*.

ELASTICITY THEORY CONTACT PROBLEM

See *Contact problem of elasticity theory*.

ELASTICITY THEORY INVARIANTS

The deforming state of a solid is characterized by the *strain tensor*

$$u_{ik} = \frac{1}{2}\left[\frac{\partial u_i}{\partial x_k} + \frac{\partial u_k}{\partial x_i} + \frac{\partial u_j}{\partial x_i}\frac{\partial u_j}{\partial x_k}\right],$$

where u_i is a *displacement vector*, or by the *stress tensor* σ_{ik}. For the tensor u_{ik} (σ_{ik}) of an isotropic body there are some invariants independent of the choice of a local coordinate system (i.e. quantities that do not change under rotations of the coordinate system through uniform angles, or under uniform displacements). The invariants I are: to first order $I_{1,u} = u_{ii}$, $I_{1,\sigma} = \sigma_{ii}$, to second order $I_{2,u} = u_{ik}u_{ki}$, $I_{2,\sigma} = \sigma_{ik}\sigma_{ki}$, and to third order $I_{3,u} = \det|u_{ik}|$, $I_{3,\sigma} = \det|\sigma_{ik}|$. These invariants are sufficient for the construction of the *Helmholtz free energy* F of an isotropic solid body

$$F = \frac{1}{2}Ku_{ii} + \mu\left(u_{ik} - \frac{1}{3}\delta_{ik}u_{jj}\right)^2.$$

Table 1. Number of second-order elasticity invariants for various crystallographic systems

Crystallographic system	Number of invariants
Triclinic	21
Monoclinic	13
Orthorhombic	9
Tetragonal	7, 6
Rhombohedral	7, 6
Hexagonal	5
Cubic	3
Isotropic	2

Here K is the *bulk modulus*, and μ is the *shear modulus*. The tensor $u'_{ik} = (u_{ik} - \delta_{ik}u_{jj}/3)^2$, called the *strain deviator*, characterizes a *shear strain*, and the tensor $\sigma'_{ik} = (\sigma_{ik} - \delta_{ik}\sigma_{jj}/3)^2$ is called the *stress deviator*. These tensors have nonzero invariants to second and third order (the first-order invariants for both u'_{ik} and σ'_{ik} are zero):

$$I'_{2,u} = u'_{ik}u'_{ki} = u_{ik}u_{ki} - \frac{1}{3}u_{jj}^2,$$

$$I'_{3,u} = \det|u'_{ik}|,$$

$$I'_{2,\sigma} = \sigma'_{ik}\sigma'_{ki} = \sigma_{ik}\sigma_{ki} - \frac{1}{3}\sigma_{jj}^2,$$

$$I'_{3,\sigma} = \det|\sigma'_{ik}|.$$

The deformed state in a crystal is characterized by the strain tensor, where the scale of inhomogeneities involved in the deformations is much larger than the lattice parameters. The number of invariants in a crystal exceeds the number in an isotropic body. This is because, due to its symmetry, a crystal only has certain discrete axes of rotation and certain discrete displacements (see *Crystal symmetry*). For example, to construct the Helmholtz free energy F one needs the invariant to second order in u_{ik}. For a cubic crystal these invariants are:

$$I_{2,1} = u_{xx}^2 + u_{yy}^2 + u_{zz}^2,$$

$$I_{2,2} = u_{xx}u_{yy} + u_{yy}u_{zz} + u_{zz}u_{xx},$$

$$I_{2,3} = u_{xy}^2 + u_{yz}^2 + u_{zx}^2.$$

The general method of any invariant construction makes use of the second-order invariants of irreducible representations of the symmetry group.

The number of second-order independent invariants for the various *crystallographic systems* are given in Table 1.

ELASTICITY THEORY, NONLINEAR

See *Nonlinear elasticity theory.*

ELASTICITY THEORY PLANE PROBLEM

See *Plane problem of elasticity theory.*

ELASTICITY THEORY POTENTIALS

The scalar potential φ of the irrotational part, and vector potential $\boldsymbol{\psi}$ of the solenoidal (circulation) part of a *strain* field $\boldsymbol{u}$, permit one to represent $\boldsymbol{u}$ in the form $\boldsymbol{u} = \nabla\varphi + \nabla \times \boldsymbol{\psi}$. If the mass density of the *bulk force* $\boldsymbol{F}$ can also be represented as the sum of irrotational and solenoidal parts, $\boldsymbol{F} = \nabla\Phi + \nabla \times \boldsymbol{\Psi}$, then the equations of motion of an isotropic elastic body

$$\mu\nabla^2\boldsymbol{u} + (\lambda + \mu)\,\mathrm{grad\,div}\,\boldsymbol{u} + \rho\boldsymbol{F} = \rho\frac{\partial^2\boldsymbol{u}}{\partial t^2}$$

are satisfied if the potentials φ, $\boldsymbol{\psi}$ are the solutions of wave equations

$$c_{\mathrm{L}}^2\nabla^2\phi - \frac{\partial^2\phi}{\partial t^2} = -\Phi, \quad c_{\mathrm{L}}^2 = \frac{\lambda + 2\mu}{\rho},$$

$$c_{\mathrm{T}}^2\nabla^2\boldsymbol{\psi} - \frac{\partial^2\boldsymbol{\psi}}{\partial t^2} = -\boldsymbol{\Psi}, \quad c_{\mathrm{T}}^2 = \frac{\mu}{\rho}.$$

Here λ, μ are the elastic *Lamé coefficients*, ρ is the *density*, and c_{L}, c_{T} are the speeds of propagation of the longitudinal and transverse waves.

ELASTIC LIMIT

The maximum stress causing *strain* of a body, such that after its removal the body can still recover its initial size, i.e. it is the largest or limiting stress for which a strain is still elastic (see *Elasticity*). The elastic limit may coincide with the *proportionality limit*, or it may be higher than it. In the first case within the region of elastic strain *Hooke's law* is valid, in the second case at stresses close to the elastic limit a deviation is observed from the linear dependence between applied stress and relative elastic strain.

ELASTIC MODULI

Characteristics of elastic properties of materials under small *strains*. Elastic moduli are coefficients of expansion which form the *stress tensor* σ_{ij} in terms of components of the *strain tensor* u_{ki}: $\sigma_{ij} = c_{ijkl}u_{kl} + b_{ijklmn}u_{kl}u_{mn} + \cdots$ (summation is over repeated indices that denote Cartesian coordinates). Considering only the first term of this expansion, we obtain *Hooke's law*: $\sigma_{ij} = c_{ijkl}u_{kl}$. Elastic moduli c_{ijkl}, $i, j, k = 1, 2, 3$, form a tensor of fourth rank, which is called the second-order *elastic modulus tensor*. Elastic moduli with $2n$ indices form a tensor of rank $2n$, which is called an elastic modulus tensor of nth order.

Elastic modulus tensors exhibit a certain symmetry with respect to permutation of indices: if sequential indices are grouped into pairs, then the tensor is symmetrical for index permutations within these pairs, e.g.: $c_{ijkl} = c_{klij} = c_{jikl} = c_{ijlk}$. In certain cases, e.g., in *magnetic materials*, this permutation symmetry may be partially broken. For an isotropic medium (see *Isotropy*), the tensors are constructed of unit tensors δ_{ij}. For example, the second-order elastic modulus tensor for an isotropic medium is $c_{ijkl} = \lambda\delta_{ij}\delta_{kl} + \mu(\delta_{ik}\delta_{jl} + \delta_{il}\delta_{jk})$, and thus has two linearly independent moduli λ, μ, called *Lamé coefficients*. The third-order elastic modulus tensor b_{ijklmn} for an isotropic medium has three linearly independent moduli.

For a *crystal*, elastic modulus tensors have a more complicated structure that depends on the *point group* of the crystal symmetry. The number of linearly independent components, and the canonical form, of the tensor for a crystal are determined either by the so-called method of direct inspection, or by the application of *group theory*. In particular, the number of linearly independent elastic moduli of nth order is described by the equation

$$k = \frac{1}{N_F N_P}\sum_{g\in F}\sum_{p\in P}\chi(Q^{l_1})\chi(Q^{l_2})\cdots\chi(Q^{l_r}),$$

where F is the point group of crystal symmetry, N_F is the order of the group, χ is the character of the vector representation of group F, P is the tensor symmetry group with respect to index permutations, N_P is the order of the group P, $l_1, \ldots, l_r$ are the lengths of permutation cycles

$p \in P$. For instance, the number of linearly independent second order elastic moduli equals 21 for the *triclinic system*, 13 for the *monoclinic system*, and 9 for the *orthorhombic system*; for the *tetragonal system*, the number of moduli is 7 for symmetry classes 4, $\bar{4}$ and $4/m$, and 6 for classes 422, $\bar{4}2m$, $4mm$ and $4/mmm$; for the *trigonal system*: 7 for classes 3 and $\bar{3}$, and 6 for classes 32, $3m$, $\bar{3}m$; the number of linearly independent second-order elastic moduli is 5 for the *hexagonal system* and 3 for the *cubic system*. The number of linearly independent elastic moduli of third order equals 56 for triclinic, 32 for monoclinic, and 20 for orthorhombic; for the tetragonal system: 16 for classes 4, $\bar{4}$, $4/m$ and 12 for classes 422, $\bar{4}2m$, $4mm$, $4/mmm$; for the trigonal system: 20 for classes 3, $\bar{3}$ and 14 for classes 32, $3m$, $\bar{3}m$; for the hexagonal system: 12 for classes 6, $\bar{6}$, $6/m$ and 10 for classes 622, $\bar{6}m2$, $6mm$, $6/mmm$; for the cubic system: 8 for classes 23, $m\bar{3}$ and 6 for classes 432, $\bar{4}3m$, $m3m$. The components of higher-order elastic modulus tensors of crystals are determined in a similar manner.

If the deformation occurs so slowly that the material temperature remains unchanged, the moduli are called *isothermal elastic moduli* (see *Elastic moduli, relaxed*), and if it is so rapid that no heat exchanges between nearby regions of the body then the moduli are called *adiabatic elastic moduli*. Isothermal $\lambda^{(i)}$, $\mu^{(i)}$ and adiabatic $\lambda^{(a)}$, $\mu^{(a)}$ Lamé coefficients for an isotropic body are related as follows: $\lambda^{(i)} < \lambda^{(a)}$, $\mu^{(i)} = \mu^{(a)}$. If the rate of deformation is comparable to that of relaxation processes (see *Stress relaxation*) that take place in the body, the relation connecting stress and strains at a given instant of time is replaced by the *rheological equation* whereby stresses and strains are related not only at a given time, but also at prior times. Elastic moduli of a *polycrystal* depend on the location in the medium. However, if the characteristic deformation length significantly exceeds the sizes of *crystallites*, then it is convenient to treat the polycrystal as a homogeneous medium with constant effective elastic moduli, derived from elastic moduli of individual crystallites by averaging over various orientations (e.g., *Voigt averaging*, *Reuss–Voigt–Hill approximation*). The above effective medium method is used for calculating effective characteristics of mixtures, *composite materials*, etc. The units of elastic moduli are gigapascal, GPa.

ELASTIC MODULI, COMPLEX

See *Complex elastic moduli*.

ELASTIC MODULI, RELAXED AND UNRELAXED

Factors of proportionality between stress and deformation under an infinitely slow (rapid) deformation process when relaxation processes in the medium have sufficient (inadequate) time to take place. A *relaxed elastic modulus* is an isothermal elastic modulus, used if the *relaxation* of the medium has time to take place, and, therefore, equilibrium with respect to all degrees of freedom is established. An *unrelaxed elastic modulus* is an adiabatic elastic modulus, used if the system undergoes no *irreversible processes* under deformation. To describe the relation between stress and deformation in partial relaxation, the *rheological equation* of the medium is used. For a standard linear material (*Zener model*), the rheological equation is of the form $\sigma + \tau_\varepsilon(\mathrm{d}\sigma/\mathrm{d}t) = M_\mathrm{R}(\varepsilon + \tau_\sigma(\mathrm{d}\varepsilon/\mathrm{d}t))$, where σ and ε are the rates of variation of stress and strain, and τ_σ and τ_ε are the *relaxation times* at constant stress and constant strain, respectively. When we expand stress and deformation into Fourier integrals, we obtain the following equation for the ratio of their complex Fourier amplitudes (this ratio is the *complex elastic modulus* M^*): $M^* = \sigma(\omega)/\varepsilon(\omega) = M_\mathrm{R}(1 + \mathrm{i}\omega\tau_\sigma)/(1 + \mathrm{i}\omega\tau_\varepsilon)$. For $\omega = 0$ we have $M_\mathrm{R} = \sigma(0)/\varepsilon(0)$, where M_R is the relaxed elastic modulus, and for $\omega = \infty$ we have $M_\mathrm{UR} = \sigma(\infty)/\varepsilon(\infty) = M_\mathrm{R}\tau_\sigma/\tau_\varepsilon$, where M_UR is the unrelaxed elastic modulus. Typically, $M_\mathrm{R} < M_\mathrm{UR}$.

ELASTIC MODULUS TENSOR

Tensor of fourth rank c_{ijkl}, which relates the *stress tensor* σ_{ij} and the *strain tensor* u_{ki} in the linear *Hooke's law* $\sigma_{ij} = c_{ijkl}u_{kl}$. The elastic modulus tensor possesses the following symmetry properties with respect to the permutation of indices: $c_{ijkl} = c_{jikl} = c_{ijlk} = c_{klij}$. Values of tensor components for *crystals* often lie in the range 10^{11}–10^{13} dyn/cm^2.

The number of linearly independent components of this tensor is determined by the *crystal system* and cannot exceed 21. An isotropic medium (see *Isotropy*) is described by two *elastic moduli*, a cubic crystal by three, and so on. Sometimes, the *Voigt notation* is used, where the first pair of tensor indices is replaced with one index according to the scheme: $11 \to 1$; $22 \to 2$; $33 \to 3$; $23, 32 \to 4$; $31,\ 13 \to 5$; $12,\ 21 \to 6$. A similar replacement is also done for the second pair; as a result, the matrix c_{ik}, $i, k = 1, 2, 3, \ldots, 6$, is introduced instead of the fourth-rank tensor; this matrix has the same number of independent elements, as does the tensor c_{iklm}.

The tensor s_{ijkl}, that is reciprocal to c_{ijkl}, is called the *compliance tensor*, $u_{ij} = s_{ijkl}\sigma_{kl}$. The concept of elastic modulus tensor is important not only in static elasticity theory, but also in dynamic *elasticity theory*, since the elastic modulus tensor enters the *wave equation* to describe the propagation of *elastic waves* in crystals:

$$\rho = \frac{\partial^2 u_i}{\partial t^2} = c_{ijkl}\left(\frac{\partial^2 u_i}{\partial x_j \partial x_k}\right),$$

where ρ is the crystal density. When the nonlinear Hooke's law applies ($\sigma_{ij} = c_{ijkl}u_{kl} + c_{ijklqr}u_{kl}u_{qr} + \cdots$), the nonlinear acoustic equations (see *Nonlinear acoustic effects*) involve a tensor of sixth rank c_{ijklqr}, the so-called third-order elastic modulus tensor.

ELASTIC POLARIZABILITY OF DEFECTS

The ability of *defects* to produce additional mechanical stresses $\Delta\hat{\sigma}^{\mathrm{d}}(\hat{\varepsilon})$ in a *crystal* which has been exposed to external mechanical stresses and related *strains* $\hat{\varepsilon}$. The initiation of $\Delta\hat{\sigma}^{\mathrm{d}}(\hat{\varepsilon})$ in the presence of defects is related to the fact that the magnitudes of interatomic interaction forces in a crystal undergo additional changes when the $\hat{\varepsilon}$ field is established in this crystal. The induced stresses $\Delta\hat{\sigma}^{\mathrm{d}}(\hat{\varepsilon})$ die out as the field of external deformations is turned off. In the field of small crystal deformations (smooth field of atomic displacements), the $\hat{\varepsilon}$-dependence of $\Delta\hat{\sigma}^{\mathrm{d}}(\hat{\varepsilon})$ for a single defect is linear:

$$\Delta\hat{\sigma}^{\mathrm{d}}_{ij}(\hat{\varepsilon}) \approx V^{-1}\alpha_{ijkl}\varepsilon_{kl}, \qquad |\varepsilon_{kl}| \ll 1,$$

where V is the crystal volume, and indices i, j, k, l represent the coordinate axes x, y, z. The fourth-rank elastic polarizability tensor of the defects $\hat{\alpha}$ exhibits permutation symmetry identical to that of the coefficients of the elastic rigidity tensor, and of the tensor of *elastic moduli* c_{ijkl} with respect to permutations of the indices i, j, k, l, and hence it has the same number of independent components. It is related to the change Δc_{ijkl} of the c_{ijkl} moduli. For the case of N_{d} uniformly distributed *point defects*, with a small relative concentration $c = N_{\mathrm{d}}/N$ ($c \ll 1$), we have $\Delta c_{ijkl} \approx V_{\mathrm{c}}^{-1}c\alpha_{ijkl}$, where V_{c} is the primitive *unit cell* volume, and N is the number of such cells in the crystal. The components of $\hat{\alpha}$ are of the order of 1–10 eV, and higher.

ELASTIC POTENTIAL OF A DEFECT

A phenomenological parameter that characterizes the displacement field in a medium containing a *point defect*. In an unbounded isotropic medium the elastic potential determines the displacement $\boldsymbol{u}(\boldsymbol{r})$ produced by a single point defect at the point characterized by radius vector $\boldsymbol{r}$, where $\boldsymbol{u}(\boldsymbol{r}) = -A\nabla(1/r)$, and A is called the elastic potential of the defect. This elastic potential is analogous to the value of an electrostatic point charge that produces around itself an electric field $E = -\nabla V$, where $V(r) = k/r$ is the Coulomb potential. In contrast to the electrostatic charge case, A is determined by the elastic properties of the medium, and the force constants of the defect.

ELASTIC RECOVERY

Spontaneous decrease of elastic macrostresses that occurs in a solid on the removal of an external load. After being exposed to elastic deformation (see *Elasticity*), the material completely resumes its original shape during the course of elastic recovery (e.g., a spring returns to its original shape). If the applied load exceeds the *yield limit*, then only partial recovery of the original shape takes place, the degree of recovery being lower, the higher the degree of *plastic deformation*. In such cases, elastic recovery can be observed by cutting the deformed solid into parts, or by partial dissolution of the solid (see *Recovery*).

ELASTIC SCATTERING

See *Elastic scattering of radiation*.

ELASTIC SCATTERING OF RADIATION

The elastic or kinematic scattering of radiation (X-rays, gamma rays, electrons, neutrons, etc.) by condensed media in which a single interaction of each quantum with the matter is dominant, and the scattered "particle" undergoes no change in energy or frequency. The condition for elastic scattering by crystalline *solids* is the inequality: $\lambda \leqslant l \leqslant \Lambda$, where λ is the radiation wavelength, l is the dimension of the *coherent scattering region*, and Λ is an *extinction length*. The opposite case ($l \geqslant \Lambda$) corresponds to the regime of inelastic or *dynamic radiation scattering*. A useful aspect of elastic scattering is that the scattering intensity, I, is related in a straightforward way to the atomic structure of the material under investigation. Thus for Fraunhofer *diffraction* when the scattered radiation is recorded at a large effective distance from the object, I reproduces the Fourier transform of the distribution of nuclei, of electron density, or of electric potential in the material, viz., it actually yields the Fourier transform of the spatial distribution of the atoms. Based on this principle is the crystallographic structure analysis of solids which can determine the lattice type, and the number and positioning of atoms in a *unit cell*. For Fresnel diffraction the detector or photographic plate is situated closer to the object under study, and an image of the distribution of strains in the object can be recorded for this type of radiation (the *kinematic contrast*) which provides information on crystal structure defects, such as their type, positioning, density, etc.

For X-rays, Λ amounts to 1–10 μm, hence the kinematic regime is often realized in practice. In the case of *electron diffraction*, Λ is smaller, therefore, the dynamic scattering regime shows up more often; nevertheless, in many cases the scattering pattern can be reduced to an elastic type by introducing an effective potential instead of the actual one, by taking into account the electron beam absorption, etc. The elastic scattering regime can be easily obtained for thermal neutron scattering and gamma ray scattering; so these radiations can provide data that are easy to interpret.

ELASTIC STRAIN, FINITE

See *Finite elastic strain*.

ELASTIC TWINNING

The initial stage of mechanical *twinning of crystals*, at which a wedge-shaped twin is generated in a crystal. This twin undergoes a reversible change of size when the external load is varied: it increases with increased load and decreases when the load is reduced, "exiting" the crystal when the load is removed. Elastic twinning is commonly observed near the vicinity of a crystal surface in nonuniform elastic fields (e.g., under a localized load, see the figure).

The phenomenon of elastic twinning was discovered by R.I. Garber (1938). This process has been observed in calcite, sodium nitrate, antimony, bismuth, iron silicate, etc. It has been concluded that elastic twinning is an essential stage of mechanical twinning, common to all crystals. The elastic twin is extremely sharp-tipped and, as a rule, very thin; the ratio between its thickness and height is of the order of 10^{-4}–10^{-3}, which enables one to treat elastic twinning as a planar agglomeration of twinning *dislocations*. The physical causes of elastic twinning, i.e. the factors responsible for the initiation of the restoring force, are elastic energy due to inhomogeneous deformation about the twin wedge, and the *surface energy* of the twin boundary (see *Twinning structure*). The direction of twin propagation is defined by the relation between the external load, the *surface tension* forces which act on the tip of twin and tend

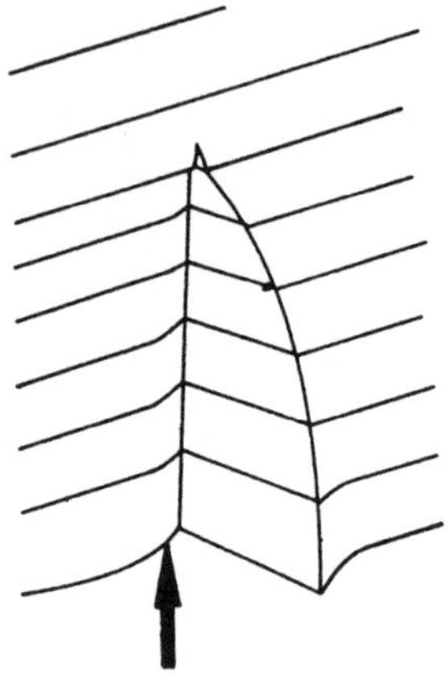

Schematic representation of a wedge-shaped twin near the crystal surface, which is brought about by application of a localized load (arrow).

to reduce its length L, and the decelerating forces (similar to dry *friction* forces) which retard the motion of the twin boundary.

The equilibrium value of L monotonically increases with increased external load. In the case of a crystal of finite dimensions under considerable load, the twin becomes unstable and "slips through" the crystal, extending all the way through it, forming a plane-parallel residual twin interlayer. If the concentrated applied load provides the twin with some stability, then the twin may remain in the crystal after the load has been completely removed (*wedged twin*). This is related to the fact that the total surface tension force, which acts as a buoyant one, is proportional to $L^{-1/2}$, while the retarding force is L-independent. The presence of the dry friction forces is the reason for a hysteretic dependence of L on the external load (see *Hysteresis*). The temperature dependences of the friction and the surface tension γ determine the conditions under which elastic twinning and allied phenomena are realized. The dynamic motion of elastic twins is accompanied by *acoustic emission*.

The quantitative description of elastic twinning, which permits detailed correlation with experimental results, is given in the context of dislocation theory, the latter being extended to related phenomena: thermoelastic *martensitic transformation*, *superelasticity* effects, behavior of *domains* in *ferroelastics*.

ELASTOMERS

Polymers and polymer-based materials which possess highly-elastic properties over a wide temperature range of their application. Typical elastomers are *rubbers*, or materials with rubber-like properties. See also *Polymeric materials*.

ELASTOOPTIC TENSOR

See *Photoelastic tensor*.

ELECTRETS

Amorphous and crystalline *insulators* with permanent electric polarization P; electrical analogue of a permanent magnet. In other words, the electric polarization (so-called residual polarization; see *Polarization of insulator*) remains a long time after removing the external action (electric field, mechanical stress, and so on) that produced it. Electrets in the polarized state possess all the properties of *pyroelectric materials*. When cutting an electret along its center (neutral) line, two electrets are formed, as is the case of magnets. The name "electret" originally appeared in formal analogy with the term "magnet". One should differentiate electrets from *ferroelectrics* which are electric analogues of *ferromagnets*. In ferroelectrics the macroscopic electric moment is a result of a *structural phase transition* accompanied by the formation of regions of spontaneous polarization (see *Ferroelectricity*), namely *domains*, whose orientation induced by the electric field results in the *residual polarization*.

The *electret state of an insulator* is determined by the long-term persistence of an electric *dipole moment* per unit volume in it. Electrets bring about an electric field in their surroundings. A dipole moment can be induced by positioning an insulator which contains polar molecules (i.e. molecules with intrinsic dipole moments) in an electric field at an elevated temperature, thereby orienting the initially randomly directed molecules, with subsequent cooling. The partially oriented polar molecules become quenched (frozen in position); then the external field is switched off. The electron–hole mechanism for the appearance of the electret state is the spatial redistribution of charge carriers that leads to their separation. Electrons and holes captured at a deep local electron level aligned in accordance with the direction of the polarizing electric field, i.e. nonuniformly over the sample, with sufficient cooling conserve their induced dipole moment for a long time. In crystals with a lattice possessing a spontaneous dipole moment, the presence of free charges results in *electric charge screening*. The lack of free electrons (either external or internal ones) or other mechanisms of shielding (e.g., the formation of oppositely directed *domains*) leads to the electret state in such crystals. In the case of orientational polarization as well as at the displacement of charge carriers from inside the sample to the surface, the direction of the induced field is opposite to the inducing field direction (*electrets with heterocharge*). In the case of the external injection of charge carriers, both directions coincide (*electrets with homocharge*). The electret can be either electrically neutral overall, or it can have a certain charge. Depending on their method of formation, such as heating and cooling, strong electric field, illumination, *nuclear ra-*

diation, etc., one can distinguish *thermoelectrets*, *electroelectrets*, *photoelectrets* and *radioelectrets*.

A number of methods are known for the preparation of electrets. Thermoelectrets are obtained by cooling a melt of dielectric (such as wax, paraffin wax, resins) in a strong electric field. This was the method used by the Japanese physicist M. Eguchi who had made the first electret (1929). Highly mobile polar molecules of the melt quite easily acquire a predominant alignment in an electric field. This orientation sometimes remains stable for several years after the solidification, i.e. the electret state is metastable but quite persistent. Residual polarization of the lattice can appear as a result of the orientation in the electric field of so-called *quasi-dipoles*: two vacancies of different sign, or an impurity ion with a vacancy of opposite sign (e.g., in *alkali-halide crystals*, in rutile). When evolving toward a *thermodynamic equilibrium* state, the electret gradually loses its induced dipole moment. This process is accompanied by a *depolarization current* (or, at heating, by a *thermodepolarization current*).

Electret properties are well pronounced in wax, paraffin wax, glass ceramics, corundum Al_2O_3, rutile TiO_2, and so on. In good electrets, such as carnauba wax, the polarization persists for several decades. An attainable potential difference between opposite surfaces of polymer films a few microns thick is a few hundred volts. Electrets are applied as *membranes* in microphones, telephones, vibration and piezoelectric transducers, and in the area of electric photography (photoelectrets).

ELECTRIC-ACOUSTIC INTERACTION

See *Acousto-electronic interaction.*

ELECTRICAL CONDUCTANCE

Reciprocal $G = 1/R$ of the electrical resistance of a material. Sometimes the dimensionless variable G/G_0 is called the conductance, where $R_0 = h/(4e^2)$ is the quantum of resistance, and $G_0 = 1/R_0$.

ELECTRICAL CONDUCTIVITY, electric conduction, conductivity

Measure of the ability a material to allow the passage of (to transmit) an *electric current* in response to the application of an electric field; a coefficient that specifies this property. The electric conduction current is a result of the *current carrier drift* in an electric field. In an alternating electric field, a *displacement current* which is not connected with the carrier drift arises. In weak electric fields of strength $\boldsymbol{E}$, current density $\boldsymbol{J}$ and $\boldsymbol{E}$ are related linearly in accordance with *Ohm's law*: $J_i = \sigma_{ik} E_k$, where σ_{ik} is the *(electrical) conductivity tensor*; $i, k = x, y, z$ are Cartesian coordinates. Ohm's law can be rewritten in the inverted form $E_i = \rho_{ik} J_k$ where ρ_{ik} is the *resistivity tensor*. In a magnetic field $\boldsymbol{B}$ we have $\sigma_{ik}(\boldsymbol{B}) = \sigma_{ki}(-\boldsymbol{B})$ by the *Onsager principle* of the symmetry of kinetic coefficients (1931). More generally, in a magnetic field the tensor σ_{ik} is divisible into symmetric and asymmetric parts S_{ik} and A_{ik}, respectively: $\sigma_{ik} = S_{ik} + A_{ik}$, where $S_{ik}(\boldsymbol{B}) = S_{ik}(-\boldsymbol{B})$, $A_{ik}(-\boldsymbol{B}) = -A_{ik}(\boldsymbol{B})$. The tensor S_{ik} describes the effect of magnetoresistance, A_{ik} is related to the Hall effect (see *Galvanomagnetic effects*). *Crystal symmetry* reduces the number of independent tensor components. In particular, in a cubic crystal with $\boldsymbol{B} = 0$ the conductivity is a scalar, i.e. isotropic: $\boldsymbol{J} = \sigma \boldsymbol{E}$, $\boldsymbol{E} = \rho \boldsymbol{J}$ where σ is the *electrical conductivity*, $\rho = 1/\sigma$ is the *specific resistance* (*resistivity*).

Electronic conductivity by the transport of electrons (or holes) has been explained in terms of *band theory*. In an electric field the electrons are accelerated, thereby passing to higher energy levels in the band. According to the Pauli principle, this is possible only if the band is not completely filled with electrons, as is the case in *metals*. In metals free current carriers exist at any temperature, and their concentration does not depend on the temperature. If the upper filled (valence) band (at temperature $T \to 0$ K) has no unoccupied levels then the crystal is a *semiconductor* or an *insulator*. An equilibrium conduction in such a crystal arises through the thermal excitation of electrons, e.g., from the *valence band* to the *conduction band*, in this case $\sigma = e\mu_n n + e\mu_p p$ where e is the electron charge, μ_n and μ_p are the mobilities, and n and p are the concentrations of electrons and holes in the conduction and valence bands, respectively. In an intrinsic semiconductor ($n = p$), the conduction results from the *intrinsic conductivity*. In doped semiconductors the excitation of electrons and holes to bands from *local electronic levels* of defects takes place and brings about an *impurity conductivity*. If the excitation is induced by an external action, the resulting conduction is called *nonequilibrium conduction*. Depending on the kind of action, one distinguishes photo-, X-ray, gamma conduction, and so on. Electron (hole) *injection* also induces a nonequilibrium conduction. For $\mu_n n \gg \mu_p p$ the conduction is called (n-type) *electronic conductivity* in contrast to (p-type) *hole conductivity* at $\mu_n n \ll \mu_p p$. If the carriers of one sign contribute to the conduction much more than the carriers of the opposite sign, the conduction is predominately *monopolar conduction*; in the case when both contributions are comparable then one speaks about *bipolar conduction*.

In *highly-doped semiconductors* (see *Doping*) and amorphous semiconductors, a type of *hopping conductivity* might predominate. In cases when the percolation properties of the conductive medium play the dominant role then the electrical conductivity is described in terms of the percolation mechanism.

If the period of an alternating electric field is of the order or less than the quasi-momentum *relaxation time*, then the current transport is characterized by a *high-frequency conductance* that has a complex value and depends on the frequency (see *High-frequency conductance*).

ELECTRICAL CONDUCTIVITY, SURFACE

See *Surface electrical conductivity*.

ELECTRICAL CONDUCTORS

See *Conductors*.

ELECTRICALLY ACTIVE EXTENDED DEFECTS

Dislocations and *grain boundaries* which possess *local electronic levels*. Such *defects* can have a dominant effect on the concentration of *current carriers*, and the electrical properties of a material. In a monopolar semiconductor with shallow impurity levels of *point defects* and *deep levels* at dislocations and grain boundaries, the number of localized current carriers increases as the density n_{d} (per unit area) of extended defects increases, and when a particular (critical) density n_{dc} is reached all the existing current carriers become localized on them. At the same time, the degree of filling the levels of extended defects remains low due to the Coulomb repulsion of charges localized at them. When the energy bonding the carrier with the defect, $E \gg k_{\mathrm{B}}T$, does not depend on the temperature T or the density of levels localized on the defect, then the critical density is given by $n_{\mathrm{dc}} = [3ne^2/(\varepsilon E)]\ln[0.163\varepsilon E/(e^2 n^{1/3})]$ where n is the current carrier concentration, e is the electron charge, and ε is the *dielectric constant*. Typical values of n_{dc} are 10^{11}–10^{12} m^{-2}. The charge per unit length of a dislocation is $q_{\mathrm{L}} = en/n_{\mathrm{d}}$ for $n_{\mathrm{d}} \geqslant n_{\mathrm{dc}}$ (n_{d} is the dislocation density per unit area), and $q_{\mathrm{L}} = en/n_{\mathrm{dc}}$ for $n_{\mathrm{d}} \leqslant n_{\mathrm{dc}}$. Complete localization of carriers at grain boundaries corresponds to a specific defect size $l_{\mathrm{c}} = (6\varepsilon E/\pi e^2 n)^{1/2}$. Typical values for l_{c} are 1 to 10 μm. The charge per unit boundary area $q_{\mathrm{A}} = 0.577enl_{\mathrm{c}}$ at $l \gg l_{\mathrm{c}}$ (here l is the grain size), and $q_{\mathrm{A}} = enl/3$ at $l \leqslant l_{\mathrm{c}}$. In the case of $n_{\mathrm{d}} = n_{\mathrm{dc}}$ or $l = l_{\mathrm{c}}$ in n-type semiconductors, then all the electrons in the *conduction band* are localized at defects, and are capable of transferring to defects in the *valence band* as their density increases; this leads to an inversion of the conductivity type. In the case of localization of all the carriers at dislocations and grain boundaries, the *Mott law* of conductivity is valid. The electrostatic field around the charged dislocation line causes the *band bending* that makes possible the inversion of the conduction type in the range adjacent to the dislocation axis. Thus, a *semiconductor junction* can arise. In the case of amphoteric extended defects (see *Amphoteric center*), a pinning of the *Fermi level* between the donor and acceptor levels is possible (e.g., for dislocations in p-type *silicon* and *germanium*).

ELECTRICAL PROBES

Contacts attached to a sample for determining the potentials of various points at its surface. In many cases the electrical probes are prepared in the form of thin pointed wires made of a hard conducting material. Thin electrical probes branching from the surface, manufactured by a common technological process and made of the same *semiconductor material* as the sample, are also used. The probes terminate at large contact areas to which the wires of the control circuit are attached.

Samples of these probes are made in the form of dumb-bells, and in the form of a clover leaf. The input resistances of the devices connected to probes are chosen so that the current uptake by them does not distort the electric field in the sample, and there is a negligibly small potential drop at the transition resistance between the sample and the probe. Probe methods used for specific resistance measurements of semiconductors include: the *two-probe method*, the *four-probe method*, and the *van der Pauw method* (1958). In these methods the specific resistance value is determined from the potential drop between two probes at a known current. The four-probe method is applicable for determining the specific resistance of semiconductor bars; the van der Pauw method is useful for plates of arbitrary shape. If a contact between the probe and the sample is undesirable, one can use a *vibrating capacitive probe* which utilizes an electrode vibrating above some point of the sample surface. If the potentials of this point and the probe are different, an alternating current runs in the circuit between the point and the probe. The lack of current flow means that the potential of the point is equal to the potential of the probe. A vibrating capacitive current is used when measuring the *contact potential difference*. In the case of an alternating electric field in a sample (e.g., the *Gunn effect*), immobile capacitive probes are employed.

ELECTRIC BREAKDOWN in a solid

Sharp increase of electric current through a non-conducting (insulator) or weakly conducting (semiconducting) crystal in a strong electric field exceeding a certain critical field E_{cr} (see *Breakdown of solids*, *Impurity breakdown*, *Streamer breakdown*). This disruptive process involves complete failure under electrostatic stress, and can do irreversible damage.

ELECTRIC CHARGE SCREENING, electric charge shielding

Effective reduction of the Coulomb interaction of charges in a medium as a result of its polarization. In *insulators* the electric charge screening is reduced by the factor $\varepsilon_0/\varepsilon$(here ε is the static *dielectric constant* of the material, and ε_0 is its free space value) in the electrostatic interaction for distances exceeding the lattice constant value owing to the polarization (displacement) of bound charges, i.e. *electrons* and *ions*.

In any material (*metal*, *semimetal*, *semiconductor*, electrolyte, and gas-discharge plasma) with a finite concentration of free *current carriers*, the electric charge screening is determined by the redistribution of the current carrier density in the field of a given charge. In the simplest approximation of a constant (independent of energy and momentum) *linear response* of the medium this is described by the (Yukawa) exponential law as

$$V(r) = \frac{e^2}{r}\exp\left(-\frac{r}{r_e}\right), \tag{1}$$

where $V(r)$, called the *Yukawa potential*, is the potential energy of two charges of value e separated by a distance r; here r_e is the *screening radius* that is equal to either the *Debye–Hückel shielding radius* (see *Debye–Hückel theory*) in a classical (nondegenerate) plasma of a semiconductor or gas discharge and in weak solutions of electrolytes, or it is the *Thomas–Fermi shielding radius* in a degenerate electron plasma (Fermi gas or *Fermi liquid*) in metals, doped semiconductors, and semimetals.

A more detailed quantum-mechanical calculation demonstrates that the static linear response (*polarization operator*) of the ideal Fermi gas depends on the transfer of momentum q, and has a singularity at the point $q = 2k_F$ (k_F is the Fermi momentum, setting $\hbar = 1$) which corresponds to a sharp discontinuity of the fermion *distribution function* at the *Fermi surface* (see *Fermi–Dirac statistics*), and depends on the type and dimensionality of the electronic spectrum. So for an isotropic three-dimensional (3D) system with a quadratic *dispersion law* and an isotropic (spherical) Fermi surface, the static polarization operator Π in *momentum space* takes the form:

$$\Pi_{3D}(q) = N_{3D}(0)\left[1 + \frac{4k_F^2 - q^2}{4k_F q}\ln\left|\frac{q + 2k_F}{q - 2k_F}\right|\right], \tag{2}$$

where $N_{3D}(0)$ is the *density of electron states* at the *Fermi level*. As follows from Eq. (1), $d\Pi_{3D}(q)/dq$ at $q = 2k_F$ has a logarithmic singularity, which leads to the fact that the Fourier component of the shielded potential of the Coulomb interaction

$$V(q) = \frac{4\pi e^2}{q^2 + 4\pi e^2 \Pi_{3D}(q)}, \tag{3}$$

when transformed to coordinate space, is no longer described by an exponential law, but rather by an alternating in sign oscillating function of distance r which decays at $r \gg (2k_F)^{-1}$ according to the power law $\cos(2k_F r)/r^3$ (see *Friedel oscillations*). For the case of a two-dimensional (2D) spectrum (cylindrical Fermi surface) in layered metals (see *Quasi-two-dimensional crystals*), or in *inversion layers* of semiconductors, we have instead of Eq. (2) the following expression:

$$\Pi_{2D}(q_\parallel) = N_{2D}(0)\left[1 - \mathrm{Re}\sqrt{1 - \left(\frac{2k_F}{q_\parallel}\right)^2}\right], \tag{4}$$

where $q_\parallel$ is the transferred 2D-momentum. A square root singularity $\Pi_{2D}(q_\parallel)$ at the point $q_\parallel = 2k_F$ leads to the oscillating dependence of the shielded interaction on r with the asymptotic of Friedel oscillations in the form $\sin(2k_F r)/r^2$ in the layer plane.

In chain crystals (see *Quasi-one-dimensional crystals*) with planar areas on the Fermi surface, and a one-dimensional (1D) spectrum of electrons, we have

$$\Pi_{1D}(q_x) = N_{1D}(0)\frac{2k_F}{q_x}\ln\left|\frac{q_x + 2k_F}{q_x - 2k_F}\right|, \tag{5}$$

where q_x is the component of the momentum q in the direction of the chains (i.e. along the x axis). The logarithmic singularity $\Pi_{1D}(q_x)$ at the point $q_x = 2k_F$ leads to alternating-sign Friedel oscillations in the form $\cos(2k_F x)/x$. Similar Friedel oscillations appear in the case of the shielded Coulomb interaction of charged particles adsorbed on the metal surface (see *Adsorption*). As a result of the space separation of the external charge and the electron cloud that screens it inside the metal, simultaneously with the oscillating component, the dipole–dipole repulsion (see *Image forces*) is present in the electrostatic potential of the *lateral interaction* of adsorbed particles:

$$V(R) = \frac{A}{R^3} + \frac{B}{R^n}\mathrm{e}^{-2k_F x}\cos(2k_F R + \theta), \tag{6}$$

where A and B are constants, x is the distance from the adsorbed atom to the surface, θ is the phase of the Friedel oscillations, and n varies from 3 to 1, depending on the shape and orientation of the Fermi surface with respect to the metal surface.

ELECTRIC CURRENT

Directed motion of electrically charged particles, i.e. mobile current carriers (*electrons, holes, ions*).

Free carriers undergo random thermal motion. In the absence of an external action, such as an applied electric field, the particle fluxes moving through an arbitrarily chosen area in opposite directions are, on the average, balanced, i.e. their algebraic sum equals zero (no current). When an electric field is applied the equality of contercurrent fluxes is disturbed. An excess charge carrier flux (directed along the field in an isotropic medium) determines the magnitude of the electric current. The moving current carriers are scattered by collisions in the medium (see *Current carrier scattering*), so when a constant field is switched on, the current rapidly builds up to a final steady-state value limited by the scattering. Relaxation processes cause the energy obtained by charge carriers from the field in a solid to be transferred to the atomic system (vibrational motions) of the solid, and it appears as heat. The scattering determines the resistance of the electric current (*electric resistance*). A particular case is the effect of superconductivity for which the resistance vanishes.

Another cause of electric current flow may be the presence of a gradient of a mobile-current-carrier concentration. For example, one can induce a concentration gradient of electrons and holes in a semiconductor by exciting them from their ground states by a spatially nonuniform (or nonuniformly absorbed) light beam. Under these conditions, a *photoelectromotive force* and corresponding current can arise in a closed circuit (*photovoltaic effect*). The energy expended producing the flow of current comes from the energy of the absorbed photons. In non-metallic crystals without a center of inversion, uniform illumination can induce an electric current in the absence of an electric field due to the *unipolarity* of the photoexcitation and scattering processes. The appearance of a current is also possible as a result of the difference in the mobilities of the individual charge carriers (e.g., electrons and holes), or due to nonuniform heating of the conductor (*thermal electromotive force*).

The mechanism of transport of charged particles may be related not only with their quasi-free motion (delocalized states of electrons and holes), but can also involve jumps of current carriers between localized states (e.g., electronic *hopping conductivity* and *ionic conductivity*). In this case the current carrier spends most of its time as a stationary particle localized at a defect, but from time to time this carrier makes a jump to a similar neighboring site. In an electric field the jump probability (frequency) along the field direction is higher than in the opposite direction; and this determines the magnitude of the current.

Quantitatively the current is characterized by a *current density* j, i.e. by the charge transported per unit time unit through a unit area normal to the current flow. The direction of positive charge flow is accepted to be the direction of the current. The value of the current $I = JS$ (S is the traverse area) is measured in amperes, 1 A = 1 C/s, and the current density J is measured in $\mathrm{A/m^2}$. The current density includes an applied electric field term $\boldsymbol{J}_{\text{drift}} = \sigma \boldsymbol{E}$, where $\sigma = en\mu$ is the specific conductivity (reciprocal to the resistivity or specific resistance); e is the electron charge, the concentration of current carriers n is averaged over the length $\mathrm{d}x$, μ is the mobility; and $E(x)$ is the electric field strength. The action of concentration gradients, mobility, and temperature leads to a second term J_{diff}. These terms are referred to as *drift current* and *diffusion current*, respectively.

ELECTRIC DIPOLE MOMENT OF A PARAMAGNETIC CENTER

Matrix element of the electric *dipole moment* operator $\boldsymbol{D}$ evaluated using wave functions ψ_i of the main electronic term, as follows:

$$(D_\beta)_{ij} = \int \Psi_i^* \widehat{D}_\beta \Psi_j \, \mathrm{d}\tau. \qquad (1)$$

Electric dipole moments of *paramagnetic centers* are used for describing *electric field effects* in *electron paramagnetic resonance* since the initial operator for the interaction of paramagnetic centers with the electric field $\boldsymbol{E}$ has the form $(\boldsymbol{E} \cdot \boldsymbol{D})$. Taking into account the symmetry of the localized paramagnetic center and its behavior under *time inversion* makes it possible to identify matrix elements which vanish, and to establish interrelationships between non-zero matrix elements without actually calculating them. In particular, for paramagnetic centers with a *center of symmetry* the wave functions of the main electronic term have the same parity, and due to the polarity of the $\boldsymbol{D}$ vector operator we have $(\boldsymbol{D}_\beta)_{ij} = 0$ for any i and j. As a result, linear electric effects do not occur. If the states ψ_i of the main term are pure spin states then only diagonal matrix elements of operator $\boldsymbol{D}$ should exist, and they are all equal to each other so electric effects are still lacking. Therefore, for the appearance of an electric effect it is necessary to have an admixture of orbital states in the main term. In other words, two circumstances play a significant role in the appearance of the electric effects: the *spin–orbit interaction*, and an interaction with the crystal resulting in the contribution of an admixture of orbital states and states with different parity.

The matrix elements $(\boldsymbol{D}_\beta)_{ij}$ are related to the parameters of the of the *electric–spin interaction* Hamiltonian, and they differ (in contrast with the *spin Hamiltonian* method) in their description of electric effects in terms of the *perturbation matrix method*.

ELECTRIC DOUBLE LAYER

Electric charges of opposite signs (ions, polar molecules), distributed along the boundary at the interface of two phases. An electric double layer appears at the surface of a *solid* (electrode), submerged in a solution of electrolyte. We can distinguish a dense part of this double layer [*Helmholtz layer* (G. Helmholtz, 1879)] located directly at the surface of the electrode and a diffusion part [*Gouy–Chapman layer* (C. Gouy, 1910; D. Chapman, 1913)] located within the solution behind the Helmholtz layer (see also *Semiconductor–electrolyte boundary*). The effective distance between the layers is the order of a molecular dimension so this double layer has a high electrical capacitance (about 10 to 100 $\mu\mathrm{F/cm^2}$), it is electrically neutral, and within the layers the electric field intensity may reach very high values. The formation of an electric double layer influences electrokinetic phenomena, as well as the course of electrochemical reactions (e.g., in the chemical sources of current, in electrolysis, etc.). For investigations of the double layer structure different methods are used: measuring the capacitance with an alternating current, adsorption methods including radioisotope investigations, optical methods (e.g., *electroreflection*, *Raman scattering of light*, etc.).

ELECTRIC FIELD EFFECTS in radio-frequency spectroscopy

Various actions of an electric field on a *paramagnetic center*, or at a nucleus, in a spectrum of *electron paramagnetic resonance* (EPR), *nuclear magnetic resonance* (NMR), and *electron–nuclear double resonance* (ENDOR). The electric field can be either constant or alternating, and either external or internal (*crystal field*). The most notable indication of an electric field effect is a splitting of an EPR line by an externally applied electric field (see Fig.). Other cases correspond to: transitions between magnetic sublevels under the action of the electric or magnetic component of an electromagnetic wave, these are called *electric dipole transitions*; transitions induced by electric or magnetic components of the electric wave in the system of energy level splittings produced by the external electric field; *magnetic flux density* and *spin echoes*, that is the excitation of the electric *dipole moment* precession of paramagnetic centers by a single pulse or a series of pulses of the electric or magnetic component of the electromagnetic wave with a subsequent response of the system to this excitation; the modulation of the *spin echo* signal by an external electric field; the paramagnetic *magnetoelectric effect*, and the EPR line broadening (see *Line (level) width*) induced by the electric field of crystal defects. As a rule, classed among electric field effects in EPR is a linear electric field effect which appears as a linear dependence of the EPR line on the electric field strength. A necessary condition for observing this is the presence of an electric dipole moment on the paramagnetic center in its ground state, which is only possible when the location of the paramagnetic center does not coincide with a center of inversion. The origin of the electric field effects is the influence of the electric field on the energy levels of paramagnetic centers (displaces and splits them, and induces transitions between them). A *spin Hamiltonian* which describes these effects is proportional to the components of the electric field E_i. Its most important terms can be represented as

$$W_E = \sum_i E_i \left\{ \sum_{j\leqslant k} R_{ijk}(S_j S_k + S_k S_j) + \sum_{jk} S_k (T_{ijk} B_j + F_{ijk} I_j) \right\}, \quad (1)$$

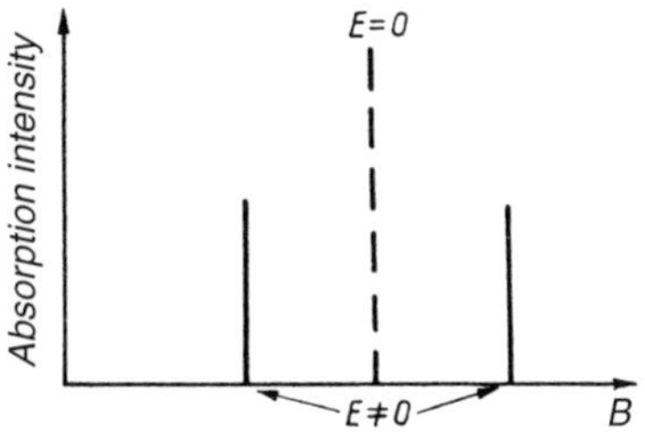

Splitting of an EPR line into two components under the action of an electric field (B is the magnetic field strength; E is the electric field strength).

where S_k and I_j are the operators of the electron and nuclear spins of the paramagnetic center, and the tensors R, T and F characterize the influence of the electric field on the energy of interaction of the paramagnetic center with the intracrystalline field, the Zeeman energy (see *Zeeman effect*), and the *hyperfine interaction*, respectively. The spin Hamiltonian can also include higher power terms of the operators S_i for these interactions, as well as additional terms to describe them. The number of tensor components to be included is determined by the *local symmetry* of the paramagnetic center (the higher the symmetry, the smaller the number). The operator W_E containing the components E_i and S_k describes a complicated mechanism of interactions of the electric field with the spin system, one called the *electric–spin interaction*. In EPR (excluding *zero-field magnetic resonance*) the degeneracy of the energy levels has already been raised by the external magnetic field, so the extra electric field only shifts the levels of the lines corresponding to the EPR spectrum. However, if there exist *inversion-inequivalent positions* in the crystal, the shift can appear as a splitting (due to the shift of lines in opposite directions for various nonequivalent positions). Since this arises from the shifting rather than the splitting of terms it is often called the *Stark pseudoeffect*.

As a rule, electric field effects are studied in the linear range of the line shift dependence on the electric field strength. However, at a sufficiently high field strength (~1000 kV/cm) the interaction energy of the paramagnetic centers with the electric field can be comparable with the Zeeman energy or the intracrystalline electric field energy. In this case, in addition to first-order perturbation theory terms (linear in E_i), nonlinear effects are also possible: a *quadratic effect*, and even a *cubic electric field effect*, despite the linearity in the field of the original operator W_E. In contrast to this situation, there are possible cases when the original operator already contains terms $\sim E_i E_j$ (or $E_i E_j E_k$), which appear in EPR spectra, and are specific for strongly polarized crystals such as *ferroelectrics*. As a result, the observation of electric field effects for them is available even when the site of the paramagnetic center is a center of inversion.

The most important applications of electric field effects are the following: unambiguous determination of the symmetry group of paramagnetic centers, and uncovering the nature of defects and the character of their distribution over a crystal.

Electric field effects in nuclear magnetic resonance and electron–nuclear double resonance are similar to those of electron paramagnetic resonance, consistent with the use of the same phenomenological description for EPR and NMR spectra: the nuclear Zeeman energy corresponds to the electron Zeeman energy, and the quadrupole interaction energy corresponds to the intracrystalline electric field. The energy operator for the interaction of the nucleus with the electric field differs from the electronic operator only by the interchange $\boldsymbol{S} \leftrightarrow \boldsymbol{I}$ where $\boldsymbol{S}$ and $\boldsymbol{I}$ are the electron and nuclear spins, respectively. The most specific features of the effects are the shift (splitting) of lines in the NMR and ENDOR spectra under the action of a constant electric field, and the quantum transitions between nuclear magnetic levels under the action of an alternating electric field. To observe the linear electric field effect, the site of the nucleus should not be a center of inversion. In the presence of inversion-nonequivalent positions, the shift of lines of the NMR (ENDOR) spectrum appears as a splitting. The observation of the electric field effect in ENDOR provides new approaches for studying paramagnetic centers. Thus, ENDOR expands significantly the class of objects that might be investigated, since nuclei of paramagnetic centers are generally situated at sites that are not centers of inversion; this is true whether or not the paramagnetic center possesses inversion symmetry. (An exception is when the impurity ion nucleus belongs to a paramagnetic ion at an inversion center.)

ELECTRIC FIELD GRADIENT

A tensor of the second rank with components $q_{jk} \equiv \partial E_j/\partial x_k$ characterizing the spatial nonuniformity of an electric field, $\boldsymbol{E}$. The electric field vector can be written as the gradient of a scalar potential $\boldsymbol{E} = -\nabla V$, or in component form $E_j = -\partial V/\partial x_j$. In the absence of charges Laplace equation holds $\nabla^2 V = 0$, and there is a system of *principal axes of a tensor* where the q_{jk} tensor is diagonal and traceless, i.e. $q_{xx} + q_{yy} + q_{zz} = 0$. If the choice of axes is made so that $|q_{xx}| \leqslant |q_{yy}| \leqslant |q_{zz}|$ then we can define $q_{zz} \equiv q$ and $(q_{xx} - q_{yy})/q_{zz} \equiv \eta$, where the quantity η is called the *asymmetry parameter* of the electric field gradient. For axial symmetry $q_{xx} = q_{yy}$ and $\eta = 0$.

ELECTRIC FIELD, LOCAL

See *Local electric field.*

ELECTRIC FIELD MODULATION METHOD

One of the methods employed to observe *electric field effects* in *electron paramagnetic resonance* (EPR) spectroscopy. It is often used in cases when electric effects are small, or other special conditions arise to hinder the direct observation of the shift of EPR spectral lines in an electric field. The method consists in comparing EPR lines that are recorded with the help of two different techniques: an ordinary technique (using magnetic field modulation) and a special technique (using electric field modulation). The comparison provides the shift of lines due to the applied electric field.

ELECTRIC FIELD PENETRATION DEPTH

See *Penetration depth of electric field.*

ELECTRIC FIELD, QUANTIZED

See *Quantizing electric field.*

ELECTRIC FLUX DENSITY, electric displacement

A vector $\boldsymbol{D}$ which characterizes the electric field, and equals a sum of electric field strength $\boldsymbol{E}$ (the main field characteristic) and *dielectric polarization* $\boldsymbol{P}$ (see *Polarization of insulator*). In the CGS system of units $\boldsymbol{D} = \boldsymbol{E} + 4\pi\boldsymbol{P} = \varepsilon\boldsymbol{E}$; in SI units $\boldsymbol{D} = \varepsilon_0\boldsymbol{E} + \boldsymbol{P} = \varepsilon_0(1+\chi_e)\boldsymbol{E} = \varepsilon\boldsymbol{E}$, where $\chi_e = P/(\varepsilon_0 E)$ is the electric susceptibility, ε_0 is the *dielectric constant* of free space with the value $\varepsilon_0 = 1/\mu_0 c^2 = 8.8542\cdot10^{-12}$ F/m, and μ_0 is the *magnetic permeability of vacuum*.

The relation between $\boldsymbol{D}$ and $\boldsymbol{E}$ is called a material equation. In a static electric field ε is a tensor for *anisotropic media*, in particular, *crystals*. For an alternating field in a medium with space and time *dispersion*, the relation between $\boldsymbol{D}$ and $\boldsymbol{E}$ is not local, and ε is a function of the frequency ω and wave vector $\boldsymbol{k}$.

ELECTRIC PULSE MACHINING

One of the technological methods of material processing based on using an electric pulse discharge between the processed item (as a rule, anode) and the other electrode (i.e. instrument, cathode) separated from it by a high electric potential. The electric discharge is accompanied by heat generation and a directed explosion which produces *surface erosion* of the material. Electric pulse machining is employed mainly for electrically conducting materials (for cutting, hardening the instrument, *alloying* the surface, and so on).

ELECTROACOUSTIC ECHO, phonon echo

Phenomenon involving the restoration of the coherence of acoustical vibrations at certain instants of time after their impression onto a *piezoelectric material* via pulses of alternating electric field. After application of radio-frequency pulses at times $t = 0$ and $t = \tau$, the acoustical vibrations excite the piezoelectric, but they lose their phasing during the time interval $0 < t < \tau$ and soon decay. As a result of the nonlinear interaction between the vibrations induced by these two pulses, there appear reversed vibrations from the second pulse with phases opposite in direction to the phases of vibrations produced by the first pulse. These phases of reversed vibrations coincide after an additional time interval τ which equals the interval between the first and the second pulses. Hence at $t = 2\tau$ the averaged electric field produces the signal of a *two-pulse echo*. The vibrational phases of a third pulse applied at moment T become coincided at $T + \tau$, and the signal of a *three-pulse echo* appears. Electroacoustic and other nonlinearities of different orders take part in the parametric processes of the birth of reflected vibrations occurring in the dielectric material; in piezoelectric powders where the echo appears at the acoustic resonance of powder particles, the main contribution is often provided by elastic nonlinearities. The *relaxation time* of the two-pulse echo T_2 is inversely proportional to the *absorption coefficient* of the acoustical vibrations. It has been found that in piezoelectric crystals and powders the three-pulse echo might be observed when applying the third pulse after attenuation of the sound induced by the first and the second pulses, i.e. there exists a long-term persistent memory with a storage time of a few months. The long-term storage is due to nonuniform distribution of electric charge and residual strain in the bulk crystal; the latter takes place under the action of a resulting stress induced by the nonlinear effects from the pair of recording pulses. In powders, a contribution to a record can also be provided by particle orientation processes stimulated by ponderomotive forces.

The electroacoustic echo is used for studying *phase transitions* and sound attenuation in solids. The method may be promising for storage and correlation processing of radio-frequency signals.

ELECTROCAPILLARY PHENOMENA

Phenomena caused by a change of the surface tension σ of a liquid at the boundary between two phases, where a discontinuity occurs in the electric potential φ at the interface. An example is a change of the shape of this interface. The electrocapillary phenomenon was discovered by G. Lippmann (1875). The *Gibbs equation* for the electrocapillarity takes the form:

$$\mathrm{d}\sigma = -q\,\mathrm{d}\varphi - \sum_i \Gamma_i\,\mathrm{d}\mu_i, \qquad (1)$$

where q is the surface electric charge density, Γ_i is the adsorption coefficient of ith component of the solution (excess of ith component in bulk of surface layer expressed in moles per unit interface area), μ is the chemical potential. At constant composition we obtain the *Lippmann–Helmholtz equation* for the surface charge density σ:

$$\frac{d^2\sigma}{d\varphi^2} = -C, \quad (2)$$

where C is the differential capacitance of the *electric double layer* at the conductor boundary expressed per unit area. At a smooth surface of metal in aqueous solutions of electrolytes, $C = 0.1$ to $0.4\ \mathrm{F/m^2}$, in melts of alkali-halide salts $C \approx 0.4\ \mathrm{F/m^2}$. From Eq. (2) we have

$$\sigma = \sigma_0 - \frac{C(\varphi - \varphi_0)^2}{2}. \quad (3)$$

This is the *electrocapillary curve equation*, where the factor φ_0 takes into account measuring the potential relative to an auxiliary unpolarizable standard electrode. In accordance with Eq. (3), in the case of electrochemical polarization of the metal–electrolyte interface the measurement of σ over the range of $e\varphi$ from 100 to 200 $\mathrm{mJ/m^2}$ is feasible. Electrocapillary phenomena are employed for degreasing metal surfaces in electrolytes, for electrical amalgamation of gold, *coating* deposition on conducting and insulating surfaces in electrolytes, for accelerating the drop refining of metals in slag, etc.

ELECTROCHROMIC EFFECT

A change of the frequency (spectral) dependence of the light transmitted through a material (ordinarily a solid, sometimes a liquid) brought about by the application of an electric field, which is perceived visually as a color change. An electrochromic effect can be either due to the polarization of molecules, due to the orientation of dipoles or *domains* (e.g., in *ferroelectrics*), or caused by the *injection* of electrons and holes from electrodes. It can also arise from the injection of hydrogen ions or *alkali metal* ions (e.g., lithium) through special electrodes in the form of solid or liquid electrolytes. An electrochromic effect obtained by polarizing large domains or by injection usually remains in place after the field is switched off. Sometimes it can be "erased" by applying a field reversed in direction.

ELECTROCRYSTALLIZATION

Transition of a material from an ionized state in solution (or melt) to a crystalline state as a result of an electrochemical reaction. Electrocrystallization is the basis for all types of electrical deposition of metals, as well as the formation of oxide layers and barely-soluble compounds at the anode. An example is the formation of electrolytic protective-decorative *coatings* in chemical fabrication processes. Electrocrystallization differs from ordinary crystallization from vapor or solution by the fact that the charge transfer from the electrode to the ion either occurs prior to the development of a crystalline structure, or both processes take place simultaneously. The emergence of nuclei of a new phase (see *Crystal nuclei*) during the electrocrystallization requires a certain oversaturation that is determined by the overpotential at the electrode. The higher the overpotential, the greater the number of nuclei emerging per unit time at a given area. The nuclei increase in size as a result of the layer-by-layer growth of faces. The process can include the formation of bidimensional nuclei, or it can develop through layering-spiral crystal growth at *screw dislocations*. As a result of linear growth in crystals, their coagulation takes place with the formation of a continuous layer of electrolytic coating.

The ratio between the rate of nucleation of new crystals and their linear growth rate determines the morphology of the surface of crystalline deposits, and their properties. The size of crystals in the coating, their habit and their orientation, depend on the overpotential at the cathode. A particularly strong influence on the nucleation and growth of crystals is exerted by *surface-active agents* and other impurities in the reaction zone. The *adsorption* of surface-active agents reduces the *surface energy*, decreases the work of formation of new crystals, and lowers the linear growth rate. As a result, monocrystal coatings (sometimes brilliant) are formed. Crystalline deposits obtained from the baths with additives of surface-active agents are less porous and hence possess superior protective properties. Such coatings themselves, however, are more active electrochemically since they contain some impurities, i.e. their lattice is characterized by a high *dislocation* density.

ELECTROGYRATION

Appearance or change of particular optical properties (e.g., optical activity) under the action of an electric field on a crystal (see *Gyration*), an effect that was first observed in quartz. A quantitative description of the electrogyration can be obtained by expanding the electrooptical tensor g_{ij} in a power series of the electric field components E_i:

$$g_{ij} = g_{ij}^0 + \gamma_{ijk} E_k + \beta_{ijkl} E_k E_l + \cdots,$$

where g_{ij}^0 is this tensor in the absence of an electric field (in centrosymmetric crystals $g_{ij}^0 = 0$), γ_{ijk} and β_{ijkl} are axial tensors of the third and fourth ranks which describe a *linear electrogyration* (*Pockels effect*) and a *quadratic electrogyration* (*Kerr effect*), respectively. The quadratic effect occurs in crystals of all symmetry classes with the exception of the cubic classes $m3m$, $\bar{4}3m$, and 432; while the linear effect occurs only in crystals without a center of symmetry (no inversion operation). Twenty of the 32 crystal classes can exhibit the Pockels effect (crystals in these same classes can also exhibit piezoelectricity). As a rule, optical activity is measured experimentally by the linear Pockels effect in a parallel applied electric field, or by the quadratic Kerr effect in a perpendicular field. Crystals with inversion symmetry, as well as liquids, do not exhibit the linear Pockels effect, but they can display the quadratic Kerr effect. When there is inversion symmetry the larger Pockels effect masks the weaker Kerr effect for low values of the applied electric field. Ferroelectric crystals exhibit temperature-dependent electrooptical anomalies in the neighborhood of a *phase transition*. At a transition from a centrosymmetric phase to a non-centrosymmetric phase optical activity can arise spontaneously; and repolarization of the crystal is accompanied by *hysteresis*. Electrooptical effects are employed to investigate *ferroelectric phase transitions*, for controling light polarization, in optical modulators, controlled filters, and *optoelectronics* devices. See also *Electrooptical effects*.

ELECTROHYDRODYNAMIC INSTABILITIES

in liquid crystals

Instabilities in *liquid crystals* arising as a result of charge (ion) transport in an applied electric field E. The ion current causes the liquid to flow, and the intrinsic anisotropy of the electrical conductivity $\sigma_a = \sigma_\parallel - \sigma_\perp$ and the *dielectric constant* $\varepsilon_a = \varepsilon_\parallel - \varepsilon_\perp$ in the liquid crystal (see *Anisotropy of liquid crystals*) brings to the convection some properties specific for liquids. In *nematic liquid crystals*, the electrohydrodynamic instabilities are especially pronounced for $\varepsilon_a < 0$ and $\sigma_a > 0$ in the original planar texture. A method for observing these instabilities is described in *Electrooptical effect*. The instabilities appear when the electric potential U reaches a certain threshold

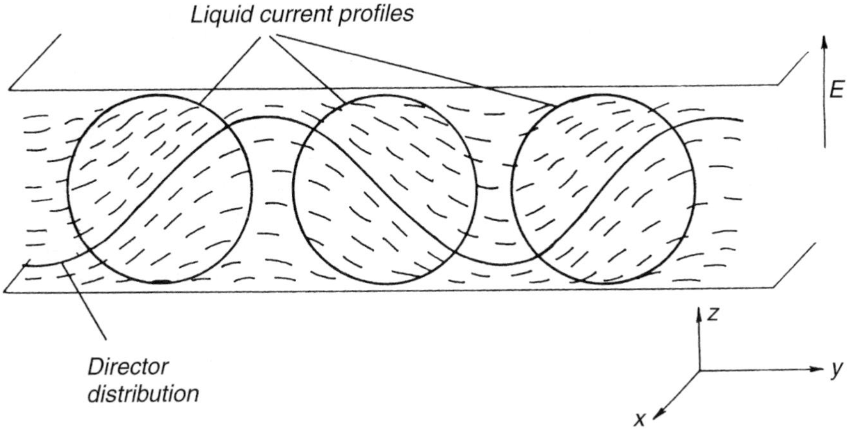

Electrohydrodynamic instabilities in liquid crystals.

value $U_c = 5$–10 V. They are seen under a microscope as alternating dark and light bands caused by the periodic variation of the orientations of liquid crystal molecules, and hence variations in the distribution of liquid crystal optic axes (director) (see Fig.). For a further rise in the applied potential the motion in the liquid crystal takes on a turbulent character, with a sharp increase of the light scattering at spatial distortions of the optical axis. This effect, called *dynamic light scattering*, is utilized in the production of *liquid-crystal displays.*

ELECTROLUMINESCENCE

Luminescence excited by an electric field capable of changing the potential and/or kinetic energy of electrons and *holes*. The mechanism of excitation is divided into injection and prebreakdown types of electroluminescence.

Injection electroluminescence arises when applying the electric potential to an *semiconductor junction*, *Schottky barrier*, or *metal–insulator–semiconductor structure* which lowers the potential barrier and stimulates the *injection* of minority carriers, whose *recombination* with majority carriers leads to the emission of *photons*. Electroluminescence was first observed by O.V. Lossev (1923) in crystals of SiC (see *Lossev effect*). More recently the effect has been observed in many *semiconductors* (Si, Ge, $A^{III}B^{V}$, $A^{II}B^{VI}$, and so on). Its specific features are the following: low exciting electric potential 1.5 to 5 V; proportionality of the radiation intensity to between the first and second power of the current over a broad region; internal *quantum yield* close to unity; fast response ($<10^{-7}$ s). Injection electroluminescence is the basis for the action of *light emitting diodes* and *semiconductor lasers.*

Prebreakdown electroluminescence arises in strong electric fields ($>10^5$ V/cm) as a result of: (a) *impact ionization* (excitations) of lattice or impurity atoms by the free carriers, in most cases by electrons which have obtained sufficient kinetic energy from the field (impact electroluminescence); (b) *field ionization* or thermal field ionization (*tunneling electroluminescence*). It is observed in inversely shifted *p–n* junctions and Schottky barriers, capacitors filled by crystal-phosphor powder dispersed in insulating powder and subjected to an alternating electric field, e.g., ZnS:Cu, Cl (*Destriau effect*); thin-film structures as metal–semiconductor–metal, metal–semiconductor–insulator–metal, and metal–insulator–semiconductor–insulator–metal, where the *n*-layer of high-ohmic (10^5–10^{12} Ω·cm) semiconductor or semi-insulator contains luminescent centers, e.g., ZnS:Mn, CdF_2:Eu, SrS:Ce, Cu, Br, etc.; moving high-field domains, acoustic-electric domains, Gunn domains (see *Gunn effect*) etc. (*domain electroluminescence*). Depending on the sample design, prebreakdown electroluminescence can be excited by either a constant or an alternating field.

Impact electroluminescence is the most widespread and important type of prebreakdown electroluminescence. It includes the following main stages (see the figure):

(a) either the injection of free carriers into strong fields from electrodes $(1, 1')$ or surface states (2), or their generation by the impact or tunnel ionization of deep donors, traps, and so on $(3, 3')$;
(b) acceleration of carriers. Effective electroluminescence is possible if every electron is capable of being accelerated to the required energy, i.e. its energy increment during a mean *mean free path* exceeds losses due to the scattering on phonons;
(c) excitation (4) or ionization $(4')$ of luminescence centers directly at an inelastic collision or through some intermediate stages, e.g., the *impact ionization* of the lattice atom (5) with subsequent trapping of the hole onto the center (6);
(d) radiation $(7, 7', 8)$. Particularly effective are intra-center radiative transitions $(7')$, e.g., in transition-metal and rare-earth ions (Mn^{2+}, Tb^{3+}, and some others) since in a strong electric field the probability of radiative recombination is sharply reduced due to forcing the electrons and holes to separate at opposite sample boundaries.

The strong field can cause: an increase of the probability of radiative dipole-forbidden transitions, a change of the band shape and structure due to the *Franz–Keldysh effect* and *Stark effect,*

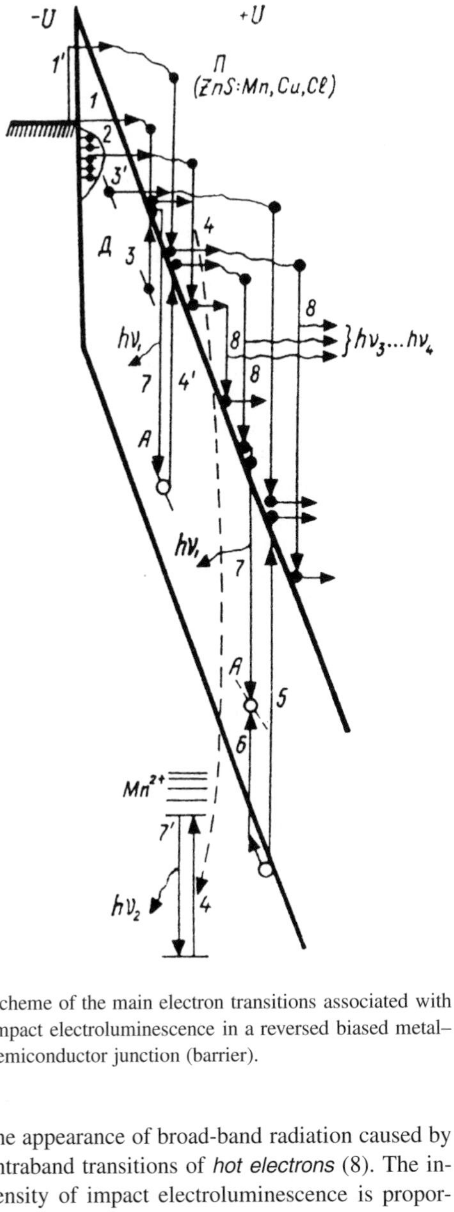

Scheme of the main electron transitions associated with impact electroluminescence in a reversed biased metal–semiconductor junction (barrier).

the appearance of broad-band radiation caused by intraband transitions of *hot electrons* (8). The intensity of impact electroluminescence is proportional to the carrier current density, the luminescence center concentration, and the *quantum yield* of the radiative transition. Its internal yield can be greater than unity. Impact electroluminescence is the basis for the operation of various indicators and screens used in devices to display information. Especially promising are thin-film electroluminescence devices based on metal–insulator–semiconductor–insulator–metal structures which exhibit high brightness (to 10^4 cd/m^2), longevity ($>10^4$ hours), and adequate efficiency (4–8 lm/W).

ELECTROLYTE, SOLID

See *Solid electrolytes.*

ELECTROLYTIC POLISHING

Phenomenon of smoothing an anode surface during the electrolysis of electrically conducting media which causes them to shine. The electrolytic polishing process provides a luster to the surface without deforming the adjacent layer below; in addition it reduces the *friction* coefficient and the electronic emission, increases the *corrosion resistance* and, consequently, improves several important working characteristics of hardware. Electrolytic polishing is also widely utilized in *structure studies* (see also *Polishing*).

ELECTROMAGNETIC-ACOUSTIC EFFECT

See *Acousto-electromagnetic effect.*

ELECTROMAGNETIC GENERATION OF SOUND

Exciting acoustic vibrations by electromagnetic (EM) waves incident on a *metal* surface. The EM generation of sound in metals involves forces exerted on a crystal lattice by *conduction electrons.* Along with the direct action of the electromagnetic field on lattice ions, an essential contribution comes from the action on the lattice of forces exerted by electrons whose equilibrium has been disturbed by the electric field $\boldsymbol{E}(\boldsymbol{r},t)$ of the EM wave. A complete set of equations that describes the EM generation of sound comprises Maxwell's equations, the *kinetic equation* for the electron *distribution function*, and the elasticity equation. The latter has the form of the equation of forced vibrations:

$$\rho\frac{\partial^2 u_i}{\partial t^2} - K_{iklm}\frac{\partial^2 u_m}{\partial x_k \partial x_l} = F_i,$$

where $\boldsymbol{u}$ is the ion displacement, ρ is the metal mass density, K_{iklm} is the adiabatic *elastic modulus tensor* (without taking into account electron renormalization), $\boldsymbol{F}$ is the force exerted by the electrons on the lattice (a functional of the field

$\boldsymbol{E}(\boldsymbol{r},t)$). Various mechanisms of the electron–lattice interaction make additive contributions to $\boldsymbol{F}$. For non-magnetic metals these are the *inertia mechanism*, the induction mechanism, and the strain mechanism. The contribution of the inertia force term $\boldsymbol{F}_1 = -(m_0/e)\partial \boldsymbol{J}/\partial t$ is usually small in comparison with other mechanisms, where $-e$ and m_0 are the charge and the mass of a free electron, respectively; and $\boldsymbol{J}$ is the total current density taking into account the electrical neutrality.

The *induction mechanism* is determined by the average *Lorentz force* $\boldsymbol{F}_2 = [\boldsymbol{j} \times \boldsymbol{B}_0]$ where $\boldsymbol{B}_0$ is the external uniform magnetic field.

The *strain mechanism* is due to the electron–lattice interaction through a *strain potential* $\hat{\lambda}(\boldsymbol{p})$:

$$F_{3i} = \frac{\partial}{\partial x_k}\langle \Lambda_{ik}\chi\rangle,$$

$$\Lambda(\boldsymbol{p}) \equiv \hat{\lambda}(\boldsymbol{p}) - \frac{\langle\hat{\lambda}\rangle}{\langle g\rangle},$$

where χ is the nonequilibrium addition to the electron distribution function; the angular brackets denote averaging over the *Fermi surface*; and g is the *density of states*. The role of various components of the force $\boldsymbol{F}$ depends on external parameters: the wave frequency, the temperature, the magnetic field, and the metal characteristics. For *semimetals* the EM generation of sound is due to a strain force $\boldsymbol{F}_3$. For typical non-magnetic metals with a normal (classical) *skin-effect*, the EM generation of sound is caused by the induction mechanism. At low temperatures under the conditions of the *anomalous skin-effect*, this generation of sound takes place due. mainly, to the strain force $\boldsymbol{F}_3$. The efficiency of this generation depends essentially on the character of the electron reflection from the metal surface: for boundaries with diffusive scattering the EM generation is more significant than it is for mirror reflecting boundaries. The transformation of a metal to the superconducting state strongly reduces the intensity of the sound generation since this transformation causes a pronounced decrease of the *penetration depth of electromagnetic field* into the metal. A particular feature of the EM generation is the strain mechanism which involves various resonance effects intrinsic in electromagnetic and *magnetoacoustic effects* in metals: *cyclotron resonance*, geometrical oscillations, and so on. The intensity of the sound increases in the manner of a resonance phenomenon when the sound interacts with collective vibrations of the metal electron–hole plasma: *helicons*, *dopplerons*, and *cyclotron waves*.

In magnetic metals, in addition to the above mentioned mechanisms, the magnetostriction electromagnetic generation of sound occurs (see *Magnetostriction*). This is caused by forces that act on the lattice due to the *magnetoelastic interaction* of the *magnetization* with the lattice. The induction mechanism of sound generation is modified by magnetization currents. The magnetoelastic interaction can be enhanced by *ferromagnetic resonance*.

The highest intensity of EM sound generation is reached in *films* by exciting standing sound waves. This generation method is employed in practical applications in laboratory practice and engineering. It does not require direct contact with the material, in contrast to routine transformers, and it has no major restrictions in frequency and temperature.

ELECTROMAGNETIC WAVES WITH DISCRETE SPECTRUM IN A METAL

Electromagnetic waves which are weakly attenuated only in the neighborhood of discrete values of the wave vectors k_n. Such waves are observed in *metals* in magnetic fields satisfying the condition $kR \gg 1$, where R is the cyclotron radius of the electron orbits, and k_n is given by the expression $k_n = \pi(n + 1/4)/R$, where $n = 0, 1, 2 \ldots$. The physical cause of the weak attenuation at these discrete k_n values is associated with oscillations of the dissipative conductivity σ_{diss} as a function of the magnetic field $\boldsymbol{B}$. There are minima at $k = k_n$ where σ_{diss} becomes small in comparison with the Hall conductivity. The effective *dielectric constant* of the electron gas in this case becomes mainly real-valued, which makes possible the propagation of weakly decaying electromagnetic waves.

ELECTROMECHANICAL EFFECTS

Reciprocal causality and correspondence between electrical and mechanical states (properties) in solids. In *ferroelectrics* the electromechanical effects are much more pronounced compared to ordinary *insulators*, piezoelectrics, and *semiconductors*; this is characterized by abnormally large values of piezoelectric constants (e.g., piezomoduli d_{ik} and coefficients of electromechanical relations k_{ik}, see *Piezoelectricity*) and strong dependence of material parameters (*dielectric constant* ε_{ij}, elastic compliances $S_{kk'}$, and piezomoduli d_{ik}) on temperature T, electric field E, and mechanical stress σ (or uniform pressure). The main cause of strong electromechanical effects is the existence of a *ferroelectric phase transition* that, in a certain range of variation of T and σ_k, stimulates the appearance (disappearance) or change of magnitude or direction of the spontaneous polarization vector $\boldsymbol{P}_s$ which is accompanied by a significant deformation of the crystal lattice (see *Ferroelectricity*). In the vicinity of a phase transition, i.e. in the neighborhood of the loss of ferroelectric thermodynamic stability, one can observe an abnormally high pliability (strong response) to an external electric and mechanical action, a gain of reciprocity of thermal, elastic (mechanical) and electric properties, and their nonlinearity. Outside the phase transition region these phenomena are less pronounced, however, both the pliability and the electromechanical effects in ferroelectrics remain greater than they are in ordinary crystals. Similar phenomena can be observed in *ferroelastics* and, in general, in the majority of cases of *structural phase transitions* which are related to a symmetry change of a crystal lattice. A quantitative description of the indicated effects can be obtained by differentiating the thermodynamic potential $\Phi(T, D_i, \sigma_k)$ with respect to D_i and σ_k (here D_i is the *electric flux density* (polarization P_i), and σ_k are the mechanical stresses) to obtain the fundamental equations:

$$\frac{\partial \Phi}{\partial D_i} = \alpha_{ij} D_j + \eta_{ijl} D_j D_l + \beta_{ijlm} D_j D_l D_m - g_{ik}\sigma_k - \kappa_{ijk} D_j \sigma_k = \frac{E_i}{4\pi},$$

$$\frac{\partial \Phi}{\partial \sigma_k} = S_{kk'}\sigma_{k'} + g_{ik} D_i + \kappa_{ijk} D_i D_j = u_k,$$

where u_k is the *strain tensor*. The summation is carried out over repeated indices $i, j, l, m = 1, 2, 3$; $k, k' = 1, 2, \ldots, 6$; the quantities α_{ij}, $S_{kk'}$, g_{ik}, κ_{ijk} are, respectively, the tensors of reciprocal electric susceptibility, elastic compliance, reciprocal piezoeffect, and *electrostriction*. To simplify these expressions, the series is limited to terms of order D^4. In the vicinity of a ferroelectric phase transition, i.e. when approaching the *Curie point* $T \to \theta(\sigma_k)$, some diagonal components of the inverse susceptibility α_{ij}, or principal minors of the determinant of the set of equations vanish. This is equivalent to the loss of thermodynamic stability of the crystal lattice, and is related to the sharp increase under action of external electric and mechanical forces of those components of electric flux density $D_i(T, E, \sigma)$ and strain $u_k(T, E, \sigma)$ which are the solutions of the set, i.e. the dielectric constant ε_{ij}, piezomoduli d_{ik}, and so on. At the same time, the fundamental equation and corresponding properties of the ferroelectric become essentially nonlinear. Outside the neighborhood of the phase transition, due to low values of some constants α_{ij} (low stability) and a high electrostriction (i.e. constants κ_{ijk}), the above-indicated effects in ferroelectrics are also abnormally large. In ferroelectrics which lack the piezoeffect ($g_{ik} = 0$) in the para-range, i.e. at $T > \theta$ and $\boldsymbol{P}_s = 0$, in the ferro-range ($T < \theta$ and $\boldsymbol{P}_s \neq 0$) the effect is caused entirely by the so-called linearized electrostriction, i.e. it arises from a term $\kappa_{ijk} D_i D_j \sigma_k$ in the thermodynamic potential; in other words, the effect is due to the presence of a relation $u_k(E) = \kappa_{ijk} D_{is} D_j(E) = d_{ik} E_i$ where $D_{is} = 4\pi P_{is}$ is the spontaneous electric flux density. Strong electromechanical effects in ferroelectrics determine a number of specific complex nonlinear dependences $\boldsymbol{P}_s(T, \sigma_k)$, $\varepsilon_{ij}(T, \sigma_k, E_i)$, $S^E_{kk'}(T, \sigma_k, E_i)$, $d_{ik}(T, \sigma_k, E_i)$; the phenomenon of electromechanical *hysteresis*, dependences of *dielectric loss* and mechanical loss on values of static and dynamic electric and mechanical fields, as well as anomalies of the piezoresistive effect and *electrical conductivity* in ferroelectrics-semiconductors (see *Photoferroelectrics*).

ELECTRON AFFINITY

A measure of the tendency of an atom (or molecule) in its ground state to bind to a free electron, thereby becoming a *negative ion*. For an *energy affinity* $\chi > 0$, i.e. the generation of a negative ion involves the release of energy and is therefore energetically favorable, then the corresponding atom is said to have a *positive electron affinity*. Positive values of χ are exhibited by atoms, e.g., hydrogen, carbon, oxygen, halogens, and by various molecules. The order of χ ranges from 0.1 to 1 eV.

In the context of the physics of *semiconductors* and other condensed media, the term "electron affinity" is sometimes used in reference to the height of the *potential barrier* to electron transfer into the vacuum (an alternative term is "*outer work function*"). The quantity χ may assume various numerical values, depending on whether the electron is removed from the semiconductor to infinity, or transferred directly into the region near the *semiconductor surface* where *image forces* play a role. The value of χ also depends on whether the electron is removed from a uniform neutral semiconductor, or from one with a *electric double layer* or charged surface. Special treatment of the semiconductor surface allows establishing an electric field in the near-surface layer, which lowers the χ barrier, thus promoting electron escape from the crystal. The potential barrier height may be lowered to such an extent that the χ level will be located below the bottom of the *conduction band*, in the quasi-neutral region, deep within the bulk semiconductor (*negative electron affinity*). This negative value of the electron affinity is characteristic of certain condensed media, e.g., liquid helium. A negative electron affinity is favorable for high emissive capacity of materials under conditions of the X-ray photoeffect, electronic excitation, etc.

ELECTRON BEAM PUMPED LASER

Solid-state lasers (mainly, *semiconductor lasers*) in which bombarding the semiconductor with fast electrons produces high concentrations of excess carriers and, hence, conditions suitable for generating coherent radiation. Electron beam pumped lasers contain an active element – a semiconductor target with an optical *resonator* cavity, a source of electrons, and a system for their acceleration and focusing. During the interaction of electrons with the solid medium, up to 30% of their initial energy is spent on internal ionization, i.e. on the generation of *electron–hole pairs*. A high electron energy is selected below the radiation defect formation threshold to obviate degradation from *cathodoluminescence*. Electron beam pumping is an important all-purpose way to produce laser emission in a semiconductor, and it is applicable in a wide range of *active media* in the wavelength range from 0.3 to 30 μm. There are also heterostructure lasers. Advantages of electron beam pumped lasers include rapid scan of the emitting spot, and high peak output power densities. Laser cathode-ray tubes are used for projection television, data presentation, and optical memory systems.

ELECTRON BEAM TECHNOLOGY

Technological processes for the treatment of solids by an electron beam in vacuo. Included among methods of electron beam technology are *welding*, evaporation, fusion, and *thermal treatment* (including cutting, milling and drilling of materials). An electron beam is formed in a vacuum of 10^0–10^{-3} Pa in a device called an *electron gun*; the beam is accelerated in an electrostatic field, focused, bent (deflected) with the help of magnetic fields when needed, and directed into a vacuum working chamber. It is in the chamber that objects involved in the electron-beam process are arranged: in the case of welding, heat treatment and trimming these are the parts to be welded or treated; in the case of fusion these constitute a crucible and the material to be melted in it; in the case of evaporation these are the material to be evaporated and those parts (*substrates*) at whose surface the *sputtering* is performed. The vacuum pumping system is an important part of any device of electron beam technology. In terms of the density of gas flow during the heating and melting of a solid, the output of the vacuum pumping systems in the range of pressure 10^{-2}–10^{-3} Pa is 10^2–10^5 l/s. Typical electron beam parameters are the following:

(i) for welding: power $W \sim 10^3$–10^5 W, energy $U \sim 20$–150 keV;
(ii) for evaporating $W \sim 10^3$–10^4 W, energy $U \sim 10$–20 keV;
(iii) for fusion: $W \sim 10^3$–10^6 W, energy $U \sim 20$–30 keV;
(iv) for trimming $W \sim 10^2$–10^3 W, energy $U \sim 50$–150 keV.

Electron-beam welding is employed in the instrument-making industry, for electrovacuum production, energy engineering, and the aircraft and automotive industries. *Electron-beam evaporation* is used for producing the metallic and nonmetallic protective *coatings* at working blades of gas-turbine apparatus, for obtaining some special multilayer and dispersed materials. *Electron-beam melting* is applied for refining metals and alloys, obtaining multilayer plate preforms made of different metals, and in the production of original semiconductor materials. An *electron-beam heat treatment* is used for modifying physical and mechanical properties of materials.

ELECTRON COMPOUNDS

See *Hume-Rothery phases.*

ELECTRON DIFFRACTION

Elastic (without energy loss) scattering of electrons by individual atoms (molecules) or in media (gases, liquids, amorphous and crystalline solids). For the structural analysis of solids collimated beams of electrons are used after passage through a specific drop in potential V, and the directions and intensities of the scattered beams are recorded far from the object under investigation (Fraunhofer *diffraction of waves*). Depending on the value of V the following types are distinguished: *low-energy electron diffraction* (LEED, $V \sim 10$ to 10^2 V), *diffraction of electrons of medium energies* ($V \sim 10^2$ to 10^3 V), and *high-energy electron diffraction* ($V \sim 10^4$ to 10^6 V). In the LEED method the depth of penetration of electrons into the solid is several angstroms so this method is used for investigations of the structure of the surface layers of solids. Using high-energy electron diffraction it is possible to study the structure of samples with thickness 10^{-7} to 10^{-4} cm, and (by reflection) surface layers of about 10 nm. Since for high-energy electron diffraction the wave length is fractions of an angstrom this is a highly sensitive method for investigating solids – the type of *unit cell*, *mosaic crystals*, *texture*, phase composition, *defects*, etc. Electron diffraction is the basis of *electron diffraction analysis.*

ELECTRON DIFFRACTION ANALYSIS

A method for investigating the structure of solids based on *electron diffraction.* In this method the accelerated and focused electron beam from an electron gun is directed onto a sample, and as a result of its scattering, a diffraction pattern (Fig.) is produced. One can differentiate the diffraction from macroscopic regions of the sample which are hundreds of micrometers in size, and *microdiffraction* with a resolution down to 5 nm. From the nature of the pattern, which is a section of the *reciprocal lattice*, one can determine the *crystal structure*, the phase composition, the shape and dimensions of particles, and other structural features of a sample under investigation. This technique is often used in conjunction with the electron-microscopic observation of surface features (*electron micrograph*). The main applications of electron diffraction analysis are the investigation of the structure of materials containing light elements, thin films, ultradispersed bodies, and processes in solids that develop at the submicroscopic level. A particular branch of electron diffraction analysis is *low-energy electron diffraction* (LEED) which utilizes an accelerating potential ~10–100 V, in sharp contrast to ordinary electron diffraction analysis where the accelerating potential is ~100 keV. This low-energy version provides information about the structure of monomolecular layers, adsorbed layers, etc.

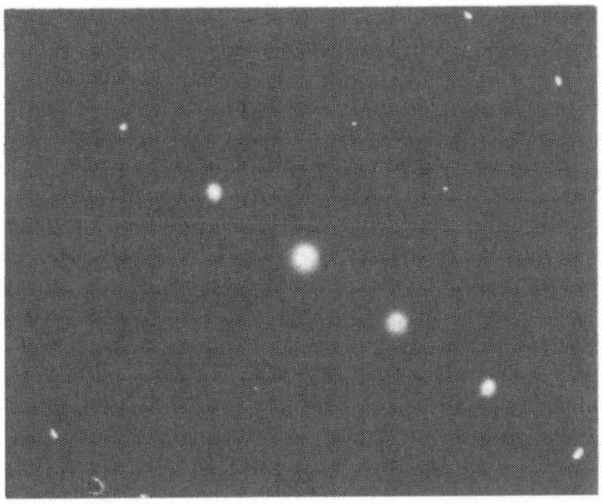

Electron diffraction pattern obtained from a thin foil of stainless steel.

ELECTRONEGATIVITY

Measure of ability of an atom in a molecule or condensed matter to attract electrons involved in *chemical bond* formation. Robert Mulliken defined the electronegativity of an *atom* in terms of its *ionization energy* and its *electron affinity*, and Linus Pauling used bond dissociation energies to devise an electronegativity scale. Tabulated values are arrived at using averages of data from many compounds, and they range from 0.79 for Cs to 3.98 for F. Atoms with a high electronegativity easily add electrons to form anions, and atoms with low electronegativity easily release a valence electron to form a cation. The electronegativity value is of significance when determining the type of chemical bond that exists between atoms which form molecules, amorphous and crystalline solids.

ELECTRON EMISSION, SECONDARY

See *Secondary electron emission.*

ELECTRON GAS, TWO-DIMENSIONAL

See *Two-dimensional electron gas.*

ELECTRON HEATING

Build-up of electron kinetic energy above its *thermodynamic equilibrium* value as a result of pumping additional energy into the electron subsystem of a crystal by means of an electric current, laser irradiation, etc. *Electron heating by a field* consists in increasing the kinetic energy during the electron drift in an electric field E (see *Current carrier drift*). When $\bar{\varepsilon} - \bar{\varepsilon}_0 \ll \bar{\varepsilon}_0$, where $\bar{\varepsilon}$ is the average electron energy, and $\bar{\varepsilon}_0$ is its value in thermodynamic equilibrium, the electrons are called *warm electrons*, while for $\bar{\varepsilon} - \bar{\varepsilon}_0 \gg \bar{\varepsilon}_0$, they are called *hot electrons*. Electrons acquiring additional energy from the heating electric field within their *mean free path* pass it on it to the crystal lattice during their collisions with *phonons*, *impurity atoms* and other *defects* of the lattice. Conditions favorable for electron heating by a field include a long mean free path, low temperature, elastic scattering of electrons by crystal defects, and high strength of the electric field E. To avoid raising both the lattice temperature T and $\bar{\varepsilon}_0$ the electric field it is often applied to the material in short pulses.

The degree of electron heating may be characterized by the *electron temperature*. When the field is weak, the average electron momentum $p \sim E$ and $\bar{\varepsilon} - -\bar{\varepsilon}_0 \propto E^2$, so that the change in electron energy during the course of the transport phenomena may be neglected as of second order, and the electron characteristics do not depend on the field itself. In contrast to this, when the field is strong it results in certain specific effects among which are the dependence of the electron mobility on the field strength, saturation of electron drift velocity, *hot electron generation*, the appearance of a thermal electromotive force for nonuniform field electron heating, ionization of the impurity states by electrons, avalanche ionization, N- and S-shaped current–voltage characteristics, regions of negative differential conductivity, domains of strong field, *current filamentation*, the *Gunn effect*, the *Sasaki effect*, a many-valued electron distribution, and sign change of the bipolar drift velocity.

Electron heating by a field occurs in semiconductors, but it plays no role in *metals* because of the high value of the Fermi energy of the electrons that provide the *transport phenomena*. Under certain conditions optical transitions of electrons, or the electric field itself, may produce *cold electrons*, with an energy $\bar{\varepsilon} < \bar{\varepsilon}_0$.

ELECTRON–HOLE LIQUID

A dense electron–hole phase in which a gas of *high-density excitons* in *semiconductors* is condensed at fairly low temperatures. In an electron–hole liquid the excitons loose their individuality, and these almost free *current carriers* (electrons and holes) persist by means of exchange and correlation forces. The mean distance between the carriers is close in order of magnitude to the Bohr exciton radius a_B, and their equilibrium concentration is $n_0 \sim a_B^{-3}$. For example, in Ge and Si the corresponding free carrier equilibrium concentration is $n_0 \sim 10^{17}$ cm^{-3} and $n_0 \sim 10^{18}$, respectively. The electron–hole liquid differs from an ordinary *electron–hole plasma* in semiconductors analogous to the way a liquid metal differs from an electron–ion plasma: the liquid is sustained by the internal Coulomb forces and possesses a certain equilibrium density $n_0(T)$ dependent on the temperature. Therefore, the electron–hole liquid, in contrast to its plasma counterpart, does not spread over the whole sample, but rather it occupies only that portion of the bulk which can be filled uniformly with the temperature-dependent density $n_0(T)$ consistent with the number of nonequilibrium electrons and holes that had been introduced into the sample.

The condensation of the free exciton gas into the electron–hole liquid is a *first-order phase transition*. Upon reaching the mean concentration of excitons equal to the temperature-dependent value $n_c(T)$ (at low temperatures $n_c(T) \ll n_0(T)$), a layering into two phases takes place in the bulk, i.e. a separation into portions filled by the exciton gas, and others filled by drops of electron–hole liquid (*electron–hole drops*) with the density n_0. At further increasing the average bulk concentration of nonequilibrium electrons and holes, the volume occupied by the electron–hole liquid grows, but its density at a given temperature $n_0(T)$ does not change until the electron–hole liquid fills the whole sample. The condensation of an exciton gas into an electron–hole liquid was first observed in Ge crystals. The presence of this liquid in crystals can be established from radiative *recombination* spectra, from plasma and magneto-plasma resonance absorption, and from *light scattering*. The electron–hole liquid is stable in semiconductors with a strong *effective mass* anisotropy and a high degree of orbital degeneracy of the energy spectrum (many-valley type) (see *Many-valley semiconductors*).

ELECTRON–HOLE PAIR

An electron in the *conduction band* associated with a *hole* in the *valence band*. It can exist either as separated particles (the Coulomb attraction is negligibly small) or as a bound pair in the form of a *Wannier–Mott exciton*. Electron–hole pairs are formed as a result of either thermal excitation of the electron subsystem, or under the action of light, *nuclear radiation*, electron (hole) *injection*, *shock waves*, etc.

ELECTRON–HOLE PLASMA

A system of free electrons and holes at sufficiently high densities and temperatures where there is a dissociation of the *high-density excitons* produced in *semiconductors* by light. Unlike an *electron–hole liquid* which appears at high densities and low temperatures due to the *Mott metal–insulator transition* in the exciton system, a gas-like electron–hole plasma fills the entire crystal bulk. The behaviour of the charged gas particles in a plasma is dominated by the electromagnetic interactions between them. A nonequilibrium electron–hole plasma can be formed by the action of strong very-high-frequency fields, and by electron injection through a contact in a strong electric field (see *Solid-state plasma*).

ELECTRONIC CONDUCTIVITY

This nonstandard term is somewhat arbitrarily defined as referring to the *electrical conductivity* of a material caused by (i) electron and (or) hole transport, as in metals, as distinct from *ionic conductivity* where the current carriers are ions, as takes place in liquids, (ii) jump or *hopping conductivity* when charges hop from ionic site to ionic site, and (iii) in *semiconductors* there is *n-type conductivity* when the majority *current carriers* are *conduction electrons*, as distinct from *p-type conductivity* when the majority carriers are holes. The n-type occurs in semiconductors when the *donor* concentration exceeds the *acceptor* concentration, and p-type occurs when the acceptors dominate.

ELECTRONIC CONFIGURATION

Complex of occupied one-electron states of a system with designated quantum numbers which determine the electron energies. The electron configuration of an atom $(n_1l_1)^{N_1}\ldots(n_il_i)^{N_i}$ is given by the *occupation numbers* of N_i of electron shells with the principal quantum number n_i and *orbital angular momentum* l_i. As a rule, only partially occupied, non-inner core, shells with $1 \leqslant N_i < 2(2l_i+1)$ are designated. The energy levels (terms) of the system with a given electron configuration are determined by inter-electron (Coulomb and *exchange interactions*) and other interactions (in particular, *spin–orbit interactions*). Single-electron states (*orbitals*) in molecules and isolated complexes in condensed media (solutions, glasses, crystals) are classified by irreducible representations of their symmetry *point group*. An electron configuration of a molecule is given in the form $(n_1\Gamma_1)^{N_1}\ldots(n_i\Gamma_i)^{N_i}$ where $n_i \geqslant 1$ is the occupation number of the state of a given symmetry listed in the order of increasing energy; the symbol Γ_i denotes the irreducible representation with the degeneracy (multiplicity) g_Γ, where $1 \leqslant N_i \leqslant 2g_\Gamma$. In *crystal field theory* the electron configurations of complexes of d-state transition metals with cubic symmetry are constructed from orbitals $e(d_{z^2}, d_{x^2-y^2})$ and $t_2(d_{xz}, d_{yz}, d_{xy})$

which, respectively, constitute basis functions of the two-dimensional E_g (Γ_{3g}) and three-dimensional T_{2g} (Γ_{5g}) irreducible representations of the octahedral group O_h. The electron configurations $e^m t_2^h$ are mixed as a result of the interelectronic interaction (configuration overlapping). In the valence bond method (*Heitler–London approximation*), the electron configurations of complexes and molecules are constructed from hybrid atomic orbitals; the LCAO method (see *Linear combination of atomic orbitals*) uses linear combinations of orbitals of atoms which comprise a molecule. The molecular orbitals are categorized by both their change in energy at bond formation (bonding, nonbonding, and antibonding orbitals) and the symmetries of their charge density (σ- and π-bonds). See also *Group theory*.

ELECTRONIC LEVEL, LOCAL

See *Local electronic levels*.

ELECTRONICS

See *Acousto-electronics*, *Acousto-electronics materials*, *Dielectronics*, *Functional electronics*, *High-energy electronics*, *Microelectronics*, *Nanoelectronics*, *Optoelectronics*, *Solid-state quantum electronics*, *Superconductor electronics*, *Vacuum electronics*.

ELECTRON-INDUCED DESORPTION

Removal of adsorbed material, comprising ions or neutral particles, from the adsorber surface under electron bombardment of the surface. An initial stage of the process of electron-induced *desorption* is the excitation of an electron from a bonding orbital or an inner shell of the adsorber or adsorbed atom. The energy of the incident electron needed for this excitation determines the energy threshold of the process. Then the adsorbed atom passes to a new equilibrium state, and if it acquires sufficient kinetic energy it can leave the surface. If its energy is insufficient for desorption (e.g., a metallic bond with short specific *relaxation times*) but sufficient for a transition to a neighboring center of adsorption, one can observe an *electron-induced disordering* of the adsorbed layer.

ELECTRON-INDUCED DISORDERING

Disturbance of order in the atomic structure of an adsorbed submonolayer *film* under low-energy electron irradiation (as a rule, with energies $E < 100$ eV). The term is applied only to adsorbed films. Many similar phenomena in a bulk solid correspond to the *formation of radiation defects*. Electron-induced disordering is so far known only for films made of Li, ^{2}H, and ^{1}H where it was observed by *low-energy electron diffraction* (LEED) at low temperature. The cross-section of electron-induced disordering in these films is $\sigma = 10^{-18}$–10^{-17} cm^2. The value of σ is sensitive to the isotope mass M (σ decreases with the increase of M). For some films (e.g., hydrogen on molybdenum), a competitive process of *electron-induced ordering* develops in parallel with the electron-induced disordering. The nature of this disordering as well as ordering seems to be related to the nonequilibrium vibrational excitation of adsorbed atoms whereby they become capable of migrating over the surface. The generation of structural defects results in electron-induced disordering, whereas defect annealing leads to electron-induced ordering. This disordering can exist with a threshold (in Li films $E \geqslant 54$ eV) as well as without a threshold (e.g., in hydrogen films the effect utilizes the energy yield from the electron passage). Accordingly, the excitation of atomic vibrational energy that exceeds the migration barrier can be either transferred directly to these levels (H), or transferred through a stage of electron excitation with subsequent Auger relaxation (Li).

ELECTRON–ION EMISSION

Emission of surface atoms (ions) as a result of electron bombardment of a *solid surface*. Of particular interest is electron–ion emission with the electron energy below a threshold value. In this case the emission of atomic particles takes place due to excitation of the electron subsystem of the *solid*, and this emission is of especial significance since it provides information on the mechanisms of subthreshold defect formation (see *Threshold energy of defect formation*). In the case of electron-ion emission, these mechanisms differ in their dependence on the type of *chemical bond* in a solid.

In ionic crystals the ionization mechanisms of electron-ion emission have been observed experimentally. These experiments confirmed the known fact that the inefficiency of ionization mechanisms in the bulk of *alkali-halide crystals* is due to the high symmetry of their *crystal lattice*, and the relatively compact arrangement of atoms in such lattices. It had been shown by C.C. Parks et al. (1983) that the electron-ion emission at the (100) surface of NaF crystals is caused by ionization of the K-shell of Na. In the case of alkali-halide crystals, this emission is due to the interatomic interaction forces, and involves the spatial arrangement of an overcharged ion. In the case of a surface with a prevalence of *covalent bonds*, the electron–ion emission occurs either as a result of local excitation of chemical bonds with the potential barrier reduced to the point where thermal fluctuations can eject an electron to the vacuum state outside, or the emission results from the disappearance of the barrier when an electron changes to an antibonding orbital state. Mechanisms of electron–ion emission can also induce stimulated *desorption* from a solid surface of any type.

ELECTRON MICROSCOPE

An electronic optical device in which, as a result of the interaction of a primary beam of accelerated electrons with a sample, the image of its surface or its internal structure is formed with a resolution δ down to 0.1 nm. Electron microscopes are divided into (i) magnetic (electromagnetic and magnetostatic), (ii) electrostatic, and (iii) combined magnetic/electric types; into devices with resolution $\delta = 0.1$–1.0 nm, $\delta = 1$–3 nm, and $\delta > 3$ nm; and into (i) transmission, (ii) reflection, (iii) emission, (iv) scanning, (v) specular, and (vi) shadow apparatus depending on the kind of radiation that forms the image (see also *Electron microscopy*).

A *transmission electron microscope* allows one to study the internal structure of objects with a resolution down to 1 nm. This class of electron microscopes is divided into three subgroups according to the energy value U of the applied accelerating potential: (i) $U < 100$ keV, (ii) $U = 100$–500 keV, and (iii) $U > 500$ keV. An electron microscope equipped with attachments can be used for crystallographic and phase analysis carried out at various temperatures, for studying samples under strain, and for the determination of their elemental composition. To carry out research on these systems the sample should be thin enough to permit the passage of electrons.

Several varieties of electron microscopes will be described.

The working principle of a *reflection electron microscope* involves the reflectance of the incident electrons, $\delta = 30$–35 nm.

In an *emission electron microscope* the image is formed by electron emerging from the sample surface, $\delta = 20$ nm.

In a *scanning electron microscope* an electron probe scans over the sample surface, thereby forming its image, $\delta = 3$ nm. This device can work by means of secondary reflections and transmitted electrons at $U < 50$ keV. The analyzers provide the determination of the elemental composition of a sample.

A *specular electron microscope* furnishes an image of a potential relief (i.e. the distribution of potential near the sample surface), $\delta = 8$–15 nm (in practice this value may be closer to 100 nm).

A *shadow electron microscope* is designed for studying surface relief.

ELECTRON MICROSCOPE, SCANNING

See *Scanning electron microscope.*

ELECTRON MICROSCOPY

Set of methods for investigating the structure of solids (from the mesoscopic to the atomic-molecular level) and their local chemical composition with the help of electron-optical devices called *electron microscopes*.

Transmission electron microscopy can attain a resolution up to 0.1 nm when passing the electron beam through a sample 10 nm thick. With the help of high-resolution electron microscopy, the atomic structure of solids becomes available for observation. Using thicker sample-foils (0.1 to 1.0 μm thick) one can study imperfections of the crystalline structure of solids which become detectable through their phase contrast. The use of a narrow electron beam (about 1 nm in diameter) scanning over the foil surface makes it possible to combine the electron-microscopic study of the solid bulk structure with the analysis of its local chemical composition (see *Local analysis*). Spectrometers of various kinds are used for investigating this chemical composition: *X-ray spectrometer* with energy dispersion, X-ray absorption

A micrograph of the defect structure of a thin film of stainless steel obtained in the transmission mode of an electron microscope. Magnification 25,000.

near edge structure spectrometer, and characteristic electron energy loss spectrometer (see *Characteristic energy loss spectroscopy*). Application of the above-mentioned spectrometers allows one to study the local elemental or chemical composition of samples with regard to elements from ^{3}Li to ^{92}U, although the very lightest elements are more difficult to measure quantitatively. The modern electron microscope provides microstructural resolution and the determination of the local chemical composition in combination with the identification of the crystal structure. To study solids via transmission electron microscopy thin electron-transparent samples 10 nm to a few micrometers thick can be prepared by electrochemical methods, *ion etching*, thermal evaporation, and so on. These methods are used for investigating the defect structure (see Fig.) and *phase transitions* in solids. To study the structure of complex compounds such as *alloys* and *minerals* which contain dispersed particles, a direct method is often supplemented with a semidirect one by obtaining a replica from the sample surface with embedded particles of various phases.

To study the surface of bulk samples, a *scanning electron microscope* (raster electron microscope) is employed. To analyse the surface topography and composition *Auger electron microscopy* can be used. This allows one to study thin surface layers in localized regions about 50 nm in size. This method is particularly convenient for determinating the distribution of elements over the solid surface. Auger electron microscopy requires high purity of the surface under investigation, which can be achieved only under the conditions of an extremely high vacuum.

An important application is the dynamic study of solid structures during variations of the temperature, mechanical stress, and magnetic field, as well as during the accumulation of radiative damage in material subjected to bombardment by high-energy electrons (1 MeV in order of magnitude).

ELECTRON–NUCLEAR DOUBLE RESONANCE (ENDOR) (G. Feher, 1956)

Selective absorption of electromagnetic wave energy which induces quantum transitions between spin sublevels in a system of electrons coupled with nuclei, carried out simultaneously at two frequencies (electronic ν_e and nuclear ν_n); it is a version of magnetic *double resonance*. It measures *nuclear magnetic resonance* (NMR) transitions through their influence on the *electron paramagnetic resonance* (EPR) signal. For the observation of ENDOR the substance containing *paramagnetic centers* is irradiated by the oscillating microwave magnetic field at the frequency ν_e, while at the same time the radio-frequency NMR field is applied and its frequency is swept. At the coincidence of the energy of the radio-frequency quantum with the difference of energies between two nuclear sublevels nuclear spin transitions are induced which change some parameter of the EPR signal (intensity, width and shape of line, resonance frequency). Usually the variation of the microwave adsorption at some point of the EPR line is recorded. This variation, which has a form of resonance line, is called the *ENDOR signal*, and the frequencies ν_n at which it occurs are called the *ENDOR frequencies*. The main advantage of ENDOR as a method of measuring nuclear spin transitions lies in its combination of the high resolution of NMR with the high sensitivity of EPR. Figs. (a)–(d) present two simple situations which provide simple visual explanations of the mechanisms of ENDOR. Figs. (b), (c) correspond to so-called *non-stationary ENDOR* when the spin *relaxation times* are very long. Here the equalizing of the populations of the electronic sublevels is achieved with the help of a very high

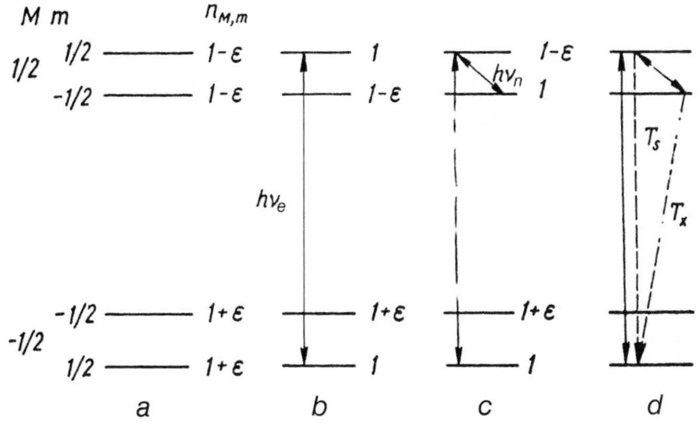

Energy levels of the system of an electron and a nucleus with spins $S = I = 1/2$ in a constant applied magnetic field $\boldsymbol{B}$ taking into account the Zeeman and hyperfine interactions: (a) levels in thermodynamic equilibrium; (b,c) variation of populations under effect of the microwave frequency and the radio-frequency acting in the absence of spin relaxation; (d) ENDOR in the presence of relaxation, the fastest processes being associated with levels, affected by the input power, as shown by the dashed line; T_s and T_x are characteristic relaxation times.

power microwave input signal which saturates and thereby reduces the EPR signal to zero amplitude. The subsequent application of the radiofrequency and an *adiabatic fast passage* through resonance cause the inversion of the nuclear sublevel populations which induces the nonstationary ($\mathrm{d}n_{M,m}/\mathrm{d}t \neq 0$) emission of the EPR signal; M, m are the projections of the spins on the direction of $\boldsymbol{B}$, $n_{M,m}$ are the relative populations of levels ($h\nu_n \ll h\nu_e$ and $\varepsilon \ll 1$, $\varepsilon = h\nu_e/(2k_{\mathrm{B}}T)$), $h\nu_e$ and $h\nu_n$ are the energies of the microwave and radio-frequency quanta which cause the corresponding electronic and nuclear transitions. In *stationary ENDOR* (Fig. (d)) the intensity of the partially saturated EPR signal is determined by the competition between the comparatively fast processes of relaxation with induced transitions. The passage through nuclear resonance is performed sufficiently slowly, so that the system remains close to equilibrium: $\mathrm{d}n_{M,m}/\mathrm{d}t = 0$. The enhancement of the EPR signal results from the shortening of the spin relaxation time due to switching from the T_x process to the T_s process by the applied radio-frequency, and as a result, the extent of the EPR saturation is reduced.

Electron–nuclear double resonance may arise from the nucleus of an *impurity atom* which constitutes the paramagnetic center (*ENDOR at doping nucleus*), as well as from the nuclei of atoms (*ligands*) adjacent to the paramagnetic center (*ligand ENDOR*) and from nuclei distant from the paramagnetic center (*distant ENDOR*). There is no direct interaction of the electrons of the paramagnetic center with such nuclei. Its mechanism is associated with the space diffusion of nuclear polarization from adjacent nuclei, polarized through the *solid effect* or through the *dipole–dipole reservoir*, to the distant nuclei. The ENDOR of this type has a signal at the nuclear Larmor frequency ν_n^* which, after turning off the incident radio-frequency, falls off with a time scale characteristic of nuclear *spin–lattice relaxation*. The ENDOR of non-ordered systems (polycrystals, amorphous bodies, glasses) is also observed at frequencies close to ν_n^*, but its mechanism is the same as for the adjacent nuclei. If the radio-frequency field increases the intensity of microwave adsorption, the ENDOR is called *positive*, and in the case of reduction the ENDOR is called *negative*. *Radio-frequency selective saturation* is a specific kind of ENDOR using pulsed methods of detection.

In ENDOR the replacement of the microwave or radio-frequency electromagnetic waves by sonic vibrations is possible, to provide *acoustic-magnetic* (or *magnetoacoustic*) ENDOR, and when the signal is recorded through optical transitions it is called optically detected ENDOR. In ENDOR information is extracted from the frequency, width, shape and amplitude of the resonance lines, from their angular dependences on the $\boldsymbol{B}$ field, from ENDOR dynamics (dependences on intensities of microwave or radio-frequency fields), and also from the variations of signal as a result of external effects.

ENDOR is one of the most effective and accurate methods of measuring *hyperfine interactions* and of clarifying the structure of defects (impurity, radiative, etc.) in solids. It allows one to determine the spatial distribution of *spin densities* $e|\psi|^2$ of electrons at paramagnetic centers (ψ is the unpaired electron wave function). It is used also for determining electric quadrupole and *magnetic dipole* moments of atomic nuclei, and spin relaxation parameters in electronic–nuclear systems. ENDOR provides information about fundamental parameters of crystals: their *band structure*, intracrystalline fields, *strain potentials*, and local properties in the vicinity of paramagnetic centers.

In the presence of two radio-frequency sources a *triple electron–nuclear–nuclear resonance* can be carried out. It is similar to ENDOR in its mechanism and properties, the difference being the presence of two radiofrequency sources. The extra radiofrequency permits one to determine the relative signs of hyperfine coupling constants, to separate different processes of spin relaxation, and to determine their characteristic times.

ELECTRON ORBIT COLLAPSE

See *Collapse of electron orbits.*

ELECTRON PAIR, UNSHARED

See *Unshared electron pairs.*

ELECTRON PARAMAGNETIC RESONANCE

(EPR), also called electron spin resonance (ESR)

Resonance absorption (emission) of electromagnetic waves of the radio-frequency or microwave (superhigh-frequency) ranges (10^9–10^{12} Hz) by *paramagnets* whose paramagnetism is caused by unpaired electrons; this is a particular case of *magnetic resonance.* The effect was discovered by E.K. Zavoisky (1944). A paramagnetic spin in a constant magnetic field forms a system of energy levels E_i which differ in the magnetic quantum number M of its *magnetic moment.* Under the action of electromagnetic waves, magnetic dipole transitions can be induced between these levels, called magnetic or *spin sublevels*, with the absorption (emission) of an energy quantum of frequency $\nu_{ij} = |E_i - E_j|/h$. In the particular (but quite common) case of a single electron with spin $s = 1/2$, located in a magnetic field B, the following relation is satisfied:

$$h\nu = g\beta B, \qquad (1)$$

where g is the electron g-factor, β is the Bohr magneton. The frequency ν defined by Eq. (1) is called the *Zeeman frequency*, energy levels $E_{\pm 1/2} = \pm g\beta \boldsymbol{B}/2$ responding Eq. (1) are called *Zeeman energies*, and the spacing between these levels $E_{\pm 1/2}$ is called the *Zeeman splitting*. As a rule, $B = 0.1$ to 5 T and $g \approx 2$, which yields $\nu = 3\cdot 10^9$–$1.5\cdot 10^{11}$ Hz.

The EPR experiments are carried out with the help of a special apparatus called a *magnetic resonance spectrometer* which can operate in the centimeter or millimeter ranges of wavelengths. Microwave (radar) instrumentation is used, systems of waveguides and *resonators* with generators (klystron) and detectors (crystal diode). A sample several cubic millimeters in volume is located in a resonator where a component of the electromagnetic wave (usually a magnetic one), which induces the transitions, has a maximum value. The resonator is situated between the pole pieces of an electromagnet (ordinary or superconducting) which provides the magnetic field that produces the Zeeman splitting of the levels. The resonance condition (1) is satisfied by scanning the magnetic field strength $\boldsymbol{B}$ while keeping the frequency at a constant value ν. The magnitude of the magnetic field at *resonance* depends, in general, on the orientation of the vector $\boldsymbol{B}$ with respect to the sample. A signal with a bell shape or its derivative is observed with the help of an oscilloscope or recorder. In a crystal positioned in the external magnetic field, a paramagnetic particle with spin $s = 1/2$ has two levels ($E_{1/2}$ and $E_{-1/2}$). These levels vary in proportion to the value of the magnetic field strength $\boldsymbol{B}$, and in many single crystals they are dependent on the orientation of vector $\boldsymbol{B}$

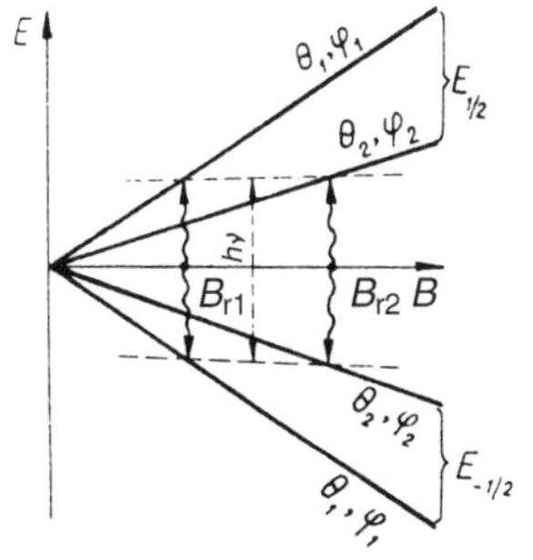

Energy levels E in a magnetic field B for electronic spin $s = 1/2$ and two orientations of θ and φ.

relative to the crystallographic axis given by the polar coordinate angles θ and φ through the following expression for the g-factor in Eq. (1):

$$g = \left[\left(g_1^2 \cos^2 \phi + g_2^2 \sin^2 \phi\right) \sin^2 \theta + g_3^2 \cos^2 \theta\right]^{1/2}, \qquad (2)$$

where g_1, g_2 and g_3 are the three principal values of the g-factor tensor. The figure shows the energy levels for two orientations of θ and φ. At resonance, e.g., the magnetic field magnitudes B_{r1} and B_{r2} indicated on the figure correspond to the orientation angles θ_1, φ_1 and θ_2, φ_2, respectively, for the same Zeeman splitting energy $h\nu$ in Eq. (1). The spectrum is a single line which appears at the position B_{r1} for the angles θ_1, φ_1, and at the magnetic field position B_{r2} for the orientation θ_2, φ_2. For the special case of axial symmetry with $g_1 = g_2 = g_\perp$ and $g_3 = g_\parallel$, the angular dependence given by

$$g = \left[g_\perp^2 \sin^2 \theta + g_\parallel^2 \cos^2 \theta\right]^{1/2} \qquad (3)$$

only depends on the azimuthal angle.

A Hamiltonian that describes an EPR spectrum is usually called a *spin Hamiltonian*. The spin Hamiltonian is the complete Hamiltonian averaged with respect to the coordinate part of the wave function. This depends on only spin operators and belongs to a certain electron configuration. The spin Hamiltonian that takes into account the principal types of interaction has the form:

$$\widehat{W} = \widehat{W}_Z + \widehat{W}_{cf} + \widehat{W}_{hf} + \widehat{W}_{ss} + \widehat{W}_p + \widehat{W}_E. \quad (4)$$

These terms, from left to right, mean respectively: the interaction of a paramagnetic spin with the external magnetic field (*Zeeman interaction*), the interaction with crystalline electric fields, the interaction with the magnetic moments of surrounding nuclei (*hyperfine interaction*), the mutual interaction of paramagnetic particles (*exchange interaction, dipole–dipole interaction, spin–spin interaction*), the effective interaction with applied external pressure (deformations), and the interaction with an external electric field. The result of averaging over the space coordinates is included in Eq. (4) in the form of parameters called *spin operators*. The energy structure and wave functions are found by solving a set of equations which correspond to Eq. (4). The number of equations (dimensionality of secular determinant) is equal to $\left[\prod_{k=1}^{n}(2s_k + 1)\right]\left[\prod_{r=1}^{p}(2I_r + 1)\right]$ where n and p are the number of electron spins (s) and nuclei (I). As a rule, s_k and I_r assume values from 1/2 to 7/2; $n = 1, 2$; $p = 1$–10. A purpose of the phenomenological description of the problem is finding (for a certain transition) the expression for the resonant field B_r as functions of the parameters of the spin Hamiltonian and the angles of orientation of the external fields and the pressure. By comparing calculated field positions $(B_r)_{theor}$ with experimentally measured ones $(B_r)_{exp}$ one establishes the validity of of Eq. (4), and can evaluate the parameters of the spin Hamiltonian. The purpose of the microscopic theory is to calculate the spin Hamiltonian parameters from "first principles" and establish the interrelationships between the EPR characteristics and the properties of the solid.

The magnetic fields from the magnetic moments of surrounding electronic and nuclear spins change the local field at each observed electronic spin, and lead to homogeneous line broadening. Imperfections of the crystal can lead to *inhomogeneous broadening* of the EPR lines. These factors can be described by the terms of Eq. (4), but other approaches have been used, such as the *statistical theory of line shape*. An important role in the phenomenon of EPR is played by relaxation processes which seek to restore the equilibrium that is disturbed by the action of the microwave radiation. Due to the interactions between the electronic and nuclear spins, and between the electronic spins and the *lattice*, the relaxation proceeds continuously (*spin–lattice relaxation* and *spin–spin relaxation*). These processes arise from transitions between the levels which compete with the microwave-induced transitions. If these are dominant then saturation (equalizing of *level populations*) occurs, which results in a broadening and decrease in amplitude of the EPR signal. The relaxation processes are specified by *relaxation times* and described by kinetic equations and equations for energy level populations.

The phenomenon of the EPR has found a wide application in the physics of solids as a method for studying crystal lattice defects, and site symmetries. Owing to the existence of *low-symmetry effects* and *electric field effects*, the determination of the local symmetry of a defect is possible from EPR data. A combination of the EPR method with the *nuclear magnetic resonance* method, i.e. *electron–nuclear double resonance*, allows one to probe nearest neighbor paramagnetic particles and their localization in the lattice. Thus the nature of defects becomes available for study. EPR is also applied in studies of *solid surfaces*, *phase transitions*, and disordered systems. An important technical application is the design and production of *masers* (see *Quantum amplifier*).

ELECTRON PARAMAGNETIC RESONANCE FINE STRUCTURE

See *Fine structure of electron paramagnetic resonance (EPR) spectra.*

ELECTRON–PHONON INTERACTION

Interaction of free charge carriers (*conduction electrons* and *holes*) with quanta of *crystal lattice vibrations.* In *insulators* and *semiconductors* with predominately *covalent bonding* in the long-wave limit the electron–phonon interaction with acoustical *phonons* can be described in terms of a *strain potential*, whereas in *ionic crystals* and in *metals* the basis of the electron–phonon interaction is the Coulomb (electric) interaction of electrons with ions. The electron–phonon interaction determines many kinetic and thermodynamic properties of solids, and this is described by the *Fröhlich Hamiltonian.* This interaction leads to *current carrier scattering* in crystals. In the case of *semiconductors* one should distinguish the mechanisms of scattering off acoustical phonons and optical (polar and nonpolar) phonons. Correspondingly, the *relaxation times* in regard to energy and momentum are specified by different temperature dependences. The strong electron–phonon interaction in ionic and polar crystals can lead to electron (hole) *self-localization*, i.e. to the formation of *polarons* (electron–phonon interaction with optical phonons) or *condensons* (interaction with acoustical phonons). In metals an important role involves the particular case of the electron–phonon interaction, that is the *acousto-electronic interaction.* In second-order perturbation theory the electron–phonon interaction leads to an effective attraction between electrons. Such an attraction in metals in the vicinity of the *Fermi surface*, arising due to the exchange of virtual phonons, can lead to the Cooper pair coupling of electrons (see *Cooper pairs*), and the transition from a normal state to a superconducting state (see *Phonon mechanism of superconductivity*). The electron–phonon interaction in metals is specified by the *electron–phonon interaction constant* λ as

$$\lambda = \frac{N(E_F)|\langle \boldsymbol{p}|\nabla V_{ei}|\boldsymbol{p}'\rangle|^2}{M\langle\omega^2\rangle},$$

where $N(E_F)$ is the *density of states* at the *Fermi level* E_F per unit spin, $\langle \boldsymbol{p}|\nabla \boldsymbol{V}_{ei}|\boldsymbol{p}'\rangle$ is the matrix element of the gradient of the shielded pseudopotential of the electron–ion interaction calculated using Bloch functions (see *Bloch theorem*) of electrons in the *conduction band*, M is the ion mass, and $\langle\omega^2\rangle$ is the square of the photon frequency averaged over the spectrum. The expression in the numerator is sometimes referred to as the *Hopfield parameter* (J.J. Hopfield, 1969). A specific (average) transferred momentum of the electron–phonon interaction ($\boldsymbol{q} = \boldsymbol{p} - \boldsymbol{p}'$) is of the order of the Fermi momentum p_F, i.e. the effective wave length of the virtual phonons is of the order of $h/p_F \approx 10^{-8}$ cm. For such short-wave phonons the approximation of uncorrelated local vibrations of ions (atoms) at crystal lattice sites is valid with a high accuracy. In *superconducting alloys* and compounds consisting of atoms of different kinds, an important role is played by the interstitial correlations of ion displacements which are responsible, in particular, for the concentration dependence of λ (and, consequently, the *critical temperature* T_c of the transition to the superconducting state) for variations of the alloy composition, or for deviations of the compound from stoichiometry. In this case, an overall electron–phonon interaction constant cannot be represented by a sum of partial constants for individual components, but rather it has a more complicated nature that is, e.g., clear from the nonlinear dependence of T_c and electronic *specific heat* γ on the composition of *solid solutions* of transition metals. The difficulties involved in calculating this interaction constant for complex compounds from first principles became evident with the discovery of *high-temperature superconductivity* in the cuprates.

ELECTRON–PHONON INTERACTION IN ONE DIMENSION

See *Cross-deformation interaction.*

ELECTRON–PHOTON SPECTROSCOPY

Method of investigation of *solid surface* properties based on measurements of the spectral, polarization, angular dependence, etc., characteristics of *photons* emitted during electron bombardment (see *Inverse photoelectron emission*). Electron-photon spectroscopy is utilized for studies of the energy band structure of the surface layers of solids (bulk samples and *films*, including *insular films*); this provides, in particular, data on the *density of electron states*, the frequencies of surface *plasma oscillations*, and exciton transitions. This spectroscopy is employed to study *luminescent centers*, adsorption–desorption processes, mechanisms involving the interaction of electrons with the surface, as well as to determine the degree of *surface roughness*, and so on.

In electron–photon spectroscopy the electrons with the energies exceeding 10^3 eV can probe the surface under study in a superhigh vacuum (needed to ensure purity). The emission characteristics are analyzed with the help of optical methods using spectral devices.

ELECTRON-PLASTIC EFFECT (discovered in 1963)

Characterizes the effect of a directed electron flux (electron drift) on the plastic flow of a material (see *Plastic deformation*). In the case of metals, the electron-plastic effect causes repeated acceleration of the flow rate and reduction of the material resistance to deformation. This effect is opposite to the electron-strain effect. This effect takes place in the presence of two factors: (i) mechanical stress that exceeds the *yield limit*; (ii) the drift of free electrons inside the deforming metal. In its turn, the electron drift can be produced by (1) the application of a potential difference across a sample to induce an *electric current*; (2) the injection of accelerated electrons into a sample; (3) irradiation by UV, X-ray or γ-radiation to bring about the *photoeffect* and *Compton effect*, as well as the formation of electron–positron pairs; (4) exposure of the sample to high power light radiation (subsurface drift); (5) the *Stewart–Tolman effect.* In these cases one can distinguish the electron-plastic effects of current, radiative, electromagnetic, light and inertial origin. The basis of the electron-plastic effect is the transfer of momentum and energy from electrons to the deforming *crystal lattice*, mainly at defect points such as *dislocations*, dislocation stops, and *point defects.* The electron-plastic effect depends linearly on the current density; the effect is polar, and in this respect it differs in principle from the *Joule–Lenz effect.* The force of drifting electrons acting on a unit dislocation length is determined by the expression $F = (j/e)(1 - (v_h/v_e))mv_F b$ where e and m are the electron charge and mass, respectively; v_F is the electron velocity on the *Fermi surface*; j is the current density; b is the *Burgers vector.* The condition for the appearance of the electron-plastic effect is the excess of the electron drift velocity v_e over the *phase velocity* of elastic dislocation waves v_d, or simply over the dislocation velocity (in the case of uniform strain), i.e. $v_e > v_d$. Since $v_e = j/(en)$ (n is electron concentration per unit volume), to produce the effect it is necessary to provide powerful electron fluxes (i.e. a huge j). In practice, during rolling, *drawing*, punching and flattening one can use short-time powerful pulses with magnitude $j = 1000$ A/mm^2 and duration 10^{-5}–10^{-2} s. This facilitates the metal treatment with pressure (the treatment effort is reduced by 35–50%); after the treatment the metal plasticity increases considerably; the physical and mechanical properties of the material, such as residual *plasticity* and degree of *texture* perfection, become improved, and some undesirable phase transitions (e.g., $\gamma \to \alpha$ transformation of *steel*) become suppressed. An alternative view of the mechanism of the electron-plastic effect regards current pulses as inducing temperature flashes in the vicinity of structural defects, thereby promoting a release of dislocations and facilitating the plastic strain. It is also supposed that local heat yield stimulates the dynamic *recrystallization* in the material at a sufficiently accumulated deformation and an optimal overall input of electromagnetic field energy.

ELECTRON–POSITRON ANNIHILATION METHOD

See *Positron spectroscopy.*

ELECTRON PROBE

A focused electron beam with an energy from units to hundreds of keV which is employed in modern *microelectronics* technology, and electron-probe methods for investigating *solids* (see *Microprobe*).

During the bombardment of a solid by an electron beam the incoming (primary) electrons can induce chemical or physical changes in the target material. A portion of the electron beam is reflected, and the part that penetrates produces secondary electrons, X-rays, nonequilibrium charge carriers, phonons, etc. This allows one to use an electron probe for various purposes. Electron-probe-induced heating is applied for special treatments of the *solid surface*, and (in industry) for the *evaporation*, fusion, and *welding* of metals. A typical application of non-thermal *electron-probe processing* is electron *lithography* in microelectronics, and the formation of clear and latent submicron structures. Local charging of the *insulator* surface, the effects of photo- and cathode-luminescent storage, electron-beam-induced polymerization of molecules at free supporting films are utilized for the purpose of information recording and storage. An electron probe is widely used in various electron-beam tubes, in *electron microscopy*, X-ray microanalysis (see *X-ray microscopy*), Auger spectroscopy, and in a series of secondary-emission methods for studying the properties and structure of solids. The investigation of the surface topography and the crystal structure of solids, as well as chemical element analysis, and the measurement of electric and magnetic microscopic fields, are carried out with the help of the electron probe method.

ELECTRON RELAXATION TIMES in a crystal

The characteristic time interval during which an electron property (e.g., momentum, energy, phase of wave function) changes by an amount comparable with its initial value due to interactions with other electrons or with the *crystal lattice.* During current flow the electrons obtain momentum and energy from the electric field, and these are redistributed within the electron system and passed on to the lattice. The rate of the first process is determined by the *time of electron–electron scattering* (*electron–electron relaxation*) τ_{ee}, which is the average time between collisions of electrons. These collisions are inelastic and involve the exchange of momentum and energy. In the case of a non-degenerate (see *Degeneracy*) electron gas the order of magnitude of τ_{ee} can be estimated from the expression

$$\tau_{ee} = \frac{\kappa^2 m^{1/2}}{ne^4}(k_B T_e)^{3/2} \ln^{-1}\left(\frac{mk_B T_e}{\hbar^2} L_D^2\right),$$

where κ is the *dielectric constant*, m is the *effective mass*, T_e is the *electron temperature*, L_D is the *screening radius* and n is the concentration of electrons; and approximately $\tau_{ee} \sim 1/n$. The momentum transfer from electrons to the lattice is described by the *momentum balance equation* which is the conservation law for the electron momentum associated with the drift (see *Current carrier drift*) in the electric field $\boldsymbol{E}$. In the simplest and the most common case of a uniform electron gas under stationary conditions $-e\boldsymbol{E} = mv/\langle\tau_p\rangle$, where v is the average drift velocity of electrons in the electric field, $\tau_p(\varepsilon)$ is the *relaxation time of the momentum of the electrons*, ε is the electron energy; $e\boldsymbol{E}$ is the force acting on the electron, and $mv/\langle\tau_p\rangle$ is the average momentum transferred by one electron to the lattice per unit of time. The symbol $\langle\,\rangle$ denotes averaging over electron states with non-equilibrium distribution functions. The value of τ_p coincides with the *relaxation time* in the *Boltzmann equation*, and it determines the *mobility* of electrons $\mu = e\langle\tau_p\rangle/m$. Other names given to τ_p are *transport relaxation time* and *mean free path time*. The *mean free path* l is defined as $l = v_T\tau_p$, where $v = v_T$ is the thermal speed of the non-degenerate electron gas, and $v = v_F$ is the Fermi velocity in a degenerate Fermi gas. Energy exchange between the electron gas and the lattice is described by the *energy balance equation* which is the energy conservation law for a gas of electrons drifting in the electrical field. For a uniform gas under stationary conditions $e\mu E^2 = (\langle\varepsilon\rangle - \langle\varepsilon_0\rangle)/\langle\tau_\varepsilon\rangle$, where $\tau(\varepsilon)$ is the *relaxation time of the electron energy*, also called the energy relaxation time, $\langle\varepsilon_0\rangle$ is the equilibrium value of $\langle\varepsilon\rangle$; in the absence of degeneracy $\langle\varepsilon_0\rangle = (3/2)k_\mathrm{B}T$, where T is the lattice temperature. The left-hand side of this equation is the energy obtained from the field by one electron per unit time, and the right-hand side shows the energy transferred to the lattice per unit time. For the quasi-elastic collision of electrons $\langle\tau_p\rangle \ll \langle\tau_\varepsilon\rangle$, and often $\tau_p \sim \varepsilon^s$, $\tau_\varepsilon \sim \varepsilon^l$, where the exponents s and l are numerical coefficients. Upon scattering on the acoustic *strain potential* we have $s = -1/2$ and $l = -1/2$; on the optical deformation potential (see *Acoustic vibrations*) $-1/2,\ 1/2$; on the polar optical potential $1/2,\ 3/2$; on the acoustic piezoelectric potential (see *Piezoelectricity*) $1/2,\ 1/2$; on ionized impurities $s = 3/2$; and on neutral impurities $s = 0$ (see *Impurity atoms*).

The relaxation times τ_{ee}, τ_p and τ_ε play an important role in the theory of *hot electrons*. The relaxation of the non-stationary space charge in an electron gas is characterized by the *Maxwell relaxation* time, and by the frequency of the *plasma oscillations*.

ELECTRONS in a crystal

The electrons of *atoms* which compose a *crystal*. The states of the inner atomic shell electrons in crystals deviate only slightly from their states in gaseous atoms. In contrast to this, the valence shell electrons of atoms in a crystal determine the character and the magnitude of the *chemical bonding* forces as well as the principal physical properties, i.e. the electric, magnetic, optical, thermal and mechanical properties. The basis for describing the structure of their energy spectrum and the nature of the quantum states of electrons in crystals is *band theory* that explains the division of solids into *metals*, *insulators* and *semiconductors*. The band electrons interact with each other, with *phonons*, and with various lattice *defects*. In metals the *conduction band* is partially full, and the electron system behaves as a degenerate *Fermi liquid* even at room temperature. Elementary excitations of this liquid with the same charge and *spin* as electrons are called *conduction electrons*. The Fermi liquid in crystals can also possess elementary excitations with a positive charge, i.e. *holes*. The main characteristics of metals can be described by the *dispersion law* of conduction electrons (or simply, electrons) $\varepsilon_\mathrm{c}(p)$ in close proximity with the *Fermi surface*. In this spectral range (only in it) the attenuation of electron states due to the electron–electron interaction is insignificant, and disappears completely at $\varepsilon_\mathrm{c}(p) = \varepsilon_\mathrm{F}$, which provides some justification for the widely accepted representation of the system of electrons in a metal as an ideal gas. At present, with the help of resonance, galvanomagnetic, and other experiments, the Fermi surface topology with its main features, the velocity, and the *effective mass* tensor of electrons at the Fermi surface, has been determined for almost all metals.

The system of mobile conduction electrons plus the background of compensating stationary positive charge forms a so-called *solid-state plasma* that exhibits a set of related collective properties: weakly damped high-frequency vibrations (*plasmons*, *helicons*, etc.) can be excited in it. A static charge is screened the more effectively the greater the density of the electron spectrum at the Fermi surface, and in good metals the *screening radius* is of the order of the lattice constant. In some cases correlation effects induced by mutual electron interactions prevent an interpretation in terms of band theory (e.g., *Wigner crystal*). Fermi-liquid correlations between the electron spins sometimes lead to the appearance of Bose-type excitations in a metal which have a *spin wave* nature.

The interaction of electrons with *crystal lattice vibrations* (see *Electron–phonon interaction*) in metals is insignificant because the specific *phonon* energy is small compared to ε_F. Phonon exchange causes the attraction between electrons that induces the formation of Cooper pairs and the transition of a metal to the superconducting state (see *Superconductivity*) if this attraction exceeds the Coulomb repulsion.

In insulators and semiconductors at $T = 0$, the *valence band* is fully occupied, and the conduction band is empty. The *band gap* width E_g in insulators has specific atomic values ($\sim$10 eV), with much smaller values in semiconductors ($E_\mathrm{g} = 0$

to 3.5 eV) (see *Gapless semiconductors*). Of great significance for the electronic properties of these crystals are the states corresponding to the bottom (electrons) of the conduction band and the top (holes) of the valence band, as well as *local electronic levels* in the forbidden gap (donor or acceptor ones) which are, as a rule, associated with either *impurity atoms* (*donors* or *acceptors*) or intrinsic defects in the crystal. The density of charge carriers in semiconductors is several orders of magnitude less than it is in metals, the frequencies of *plasma oscillations* in them are much lower than in metals, and the shielding radius is greater.

An electron and a hole attract each other and can form a bound state of the hydrogen-like type, the so-called *Wannier–Mott exciton*. A related excitation in a *molecular crystal* is the *Frenkel exciton*. The Wannier–Mott exciton radius is usually one to two orders of magnitude larger than the lattice constant, so at a certain exciton density attainable using laser irradiation (see *Laser*) their wave functions overlap, thereby making it possible to form either an *electron–hole plasma* or a new phase, that is an *electron–hole liquid* with metallic properties.

In semiconductor *heterostructures* the motion of electrons is quasi-two-dimensional. Such systems located in a strong *quantizing magnetic field* exhibit a series of unusual properties such as, e.g., the *quantum Hall effect*.

There is a small group of crystals, called *semimetals*, in which the conduction band weakly overlaps the valence band. The electrical conductivity of such systems is of a metallic character, but much less so than is the case for a true metal, because the number of charge carriers is 3–4 orders of magnitude less than that in typical metals.

In the case of strong scattering of the conduction electrons the band theory approximation no longer holds. Transits of electrons occur via occasional jumps between localized states of the polaron type, or states localized at either impurity atoms or intrinsic defects. The interaction of conduction electrons with phonons (which is quite strong) results in the formation of autolocalized states (see *Self-localization*) in which the electron deforms the lattice and becomes confined in a newly-formed potential well. In *ionic crystals* this species (*polaron*) is produced by the interaction of electrons with the polarized optical vibrations, whereas in homopolar (covalently bonded) crystals this is due to acoustical bending vibrations (*condenson*).

High-energy electrons which penetrate into a crystal as a result of external irradiation are braked (decellerated) by their interaction with the target atoms. The energy transferred to the crystal in this manner causes ionization, and the excitation and displacements of atoms from their equilibrium positions. A subsequent *relaxation* leads to a partial return of the excited electrons and atoms to their equilibrium states; however, some of the relaxation can be to *metastable states*, e.g., atoms can occupy interstitial positions while their original lattice sites remain unoccupied (vacancies), and the electrons can populate local electronic levels of defects. Permanently displaced atoms constitute what might be called *electron-induced disordering*. The paths of high-energy electrons in crystals are characterized by *charged particle energy losses* and *mean free paths* which depend on both the incoming electron energy, and the crystal characteristics. In conducting materials the excess electrons are easy emitted from the irradiated sample, which thereby remains electrically neutral. In insulators there can be an accumulation of a noticeably negative charge due to electrons trapped in the sample if the sample thickness exceeds the electron mean free path.

ELECTRON SCATTERING

See *Current carrier scattering*.

ELECTRONS, HOT

See *Hot electrons*.

ELECTRON SPECTROSCOPY

Study of the energy distribution (spectra) of electrons emitted by atoms, molecules, and *solids* under action of incident electromagnetic radiation, primary electrons or *ions* that strike the target, or electric fields. The methods of electron spectroscopy are used for investigating the energy levels of atoms, molecules and macroscopic systems composed of them, as well as the quantum transitions between these levels, to obtain information about the structure and the properties of materials. Core electron levels and valence electron levels, *collective excitations* (*plasmons*) and *vibrational spectra* are also studied by electron spectroscopy. The limiting positions in the observed spectra provide the energies of individual groups of emitted electrons, whereas the spectral line intensities furnish the relative numbers of these electrons.

Electron spectroscopy originated from the classical experiments of J. Franck and G. Hertz (1912) where the discrete atomic energy levels were determined from the energy loss of electrons passed through a low-density gas. These data in combination with *optical spectroscopy* data provided experimental justification for the Bohr theory (1913), then the experimental underpinning for quantum mechanics which, in its turn, became a theoretical basis of modern electron spectroscopy.

Depending on the type of excitation, one can distinguish electron-electron, ion-electron, tunneling (autoelectron), photoelectron, and *beta spectroscopy*. The β-spectroscopy deals with the spectra of electrons and positrons ejected during a nuclear beta-decay, as well as *conversion electrons* (emitted during the energy transfer from an excited atomic nucleus to a surrounding atomic electron shell), and electrons generated during the interaction of *gamma rays*, X-rays, and other high-energy radiations with matter. Depending on the kind of emission, electron spectroscopy is divided into *tunneling spectroscopy* (or electron spectroscopy of inelastic tunneling), *characteristic energy loss spectroscopy* (EELS), *Auger electron spectroscopy*, *Auger ion-electron spectroscopy*, electron spectroscopy of the appearance and disappearance of threshold potentials, *photoelectron spectroscopy*, X-ray-electron spectroscopy (*electron spectroscopy for chemical analysis*), and the "*e–2e experiment*". Characteristic energy loss spectroscopy is divided into *ionization spectroscopy* which deals with energy loss spectra for core level excitation, and high-resolution characteristic energy loss spectroscopy (vibrational electron spectroscopy) where the vibrational states of the surface and the molecules adsorbed at it are studied. There are integral and angle-resolved types of electron spectroscopy which differ in the collection angle of the emission. In some cases, the scattered emitter electrons are divided according to their spin polarization. In the *e–2e* experiment, the coincidence technique is used for identification of two emitted electrons with particular energies and momenta excited by the primary electron. A comparison of experimental distributions of emitter electrons with calculated distributions provides the dispersion dependence of the electrons in a solid, information about the orientation of adsorbed molecules with respect to the surface, and about their molecular orbitals involved in the bonding to the surface. The most important areas of application of electron spectroscopy are the determination of the cross-sections of excitation and ionization of atoms and molecules, the energy structure of core and valence levels in solids, the energy structure of surface and adsorbed layers, the *qualitative analysis* and the *quantitative analysis* of the material and its surface. Especially widespread is Auger spectroscopy which has high sensitivity (10^{10} to 10^{12} atoms/cm^2).

The wide application of photoelectron and *X-ray-electron spectroscopy* is stimulated by the availability of *synchrotron radiation*, i.e. a continuous photon source with quantum energies from 10 eV to 10^4 eV. Tunneling spectroscopy and *vibrational electron spectroscopy* allow one to study the vibrational terms of adsorbed molecules in the region of wavelengths $\lambda \leqslant 0.3$ mm which is inaccessible for optical spectroscopy, and to gather information about the microscopic structure of adsorbates. Tunneling spectroscopy is also used in studies of the distinct features of the electronic structure of *superconductors* (see *Tunneling spectroscopy of superconductors*). When using the photon or electron excitation in electron spectroscopy, the depth of the layer under study depends on the conditions of the experiment and the emitter electron energies. For energies below $\sim$100 eV there is a minimum of a few atomic layers, i.e. the method is applicable for surface analysis. Combining electron spectroscopy with *ion etching* allows one to carry out a *layer-by-layer analysis*. The fine structure of the electron spectra contains information on the density of free and occupied states (see *Density of states*), the structure of atomic and molecular orbitals, the spin–orbit and multiplet splittings, the *Jahn–Teller effect*, as well as multielectron processes like *shake-off*, *shake-up*, and so on. Elastically and inelastically scattered electrons also provide information on the atomic structure of condensed systems and surface defects. The electron spectroscopy of appearance potentials and of surfaces, and *extended X-ray absorption fine structure* (EXAFS) (see *Fine structure*) allow one to determine radial distribution functions (see also *Emission spectral analysis*).

ELECTRON SPIN-FLIP SCATTERING

Quantum transitions of *electrons in crystals* between two states differing in the direction of their *spin* that occur during the interaction of electrons with *quasi-particles* of other types (*phonons*, *holes*), impurities or defects. This spin-flip scattering with an exchange of spin orientations occurs either due to the *exchange interaction* of electrons with magnetic impurities (see *Exchange scattering*), or current carriers of other groups, or under the effect of impurities or *crystal lattice vibrations* when the *spin–orbit interaction* is sufficiently strong. This type of scattering of electrons controls the intensity and shape of *electron paramagnetic resonance* and *combined resonance* lines, the degree of polarization of particles during optical orientation of free current carriers in semiconductors, and the scale of spin-dependent kinetic effects. The resonance growth of the scattering cross-section in the vicinity of the *Fermi energy* level, produced by the exchange interaction of conduction electrons with magnetic impurities in *metals* (the s–d interaction), is the source of the *Kondo effect*.

ELECTRON SPIN RESONANCE (ESR)

The term "electron spin resonance" is widely used in the Western, but not in the Russian literature. See *Electron paramagnetic resonance*.

ELECTRON STATE DENSITY

See *Density of electron states*.

ELECTRON STATE, SURFACE

See *Surface electron states*.

ELECTRON-STRAIN EFFECT

Appearance of electron drift (see *Current carrier drift*) under the action of high-rate *plastic deformation* of a material. It was discovered in 1977. In the case of *metals* this phenomenon is opposite to an *electron-plastic effect*. The mechanism of the electron-strain effect, in contrast to that of the electron-plastic effect, consists of the carrying along of free electrons by dislocations and other movable *defects* in regions of plastic flow. The electron-strain effect takes place under the conditions of a high-rate directed *strain* of a metal (with the rate 10^2–10^3% per second). It can be observed when the *twinning of crystals* or the motion of *slip bands* take place, as well as when moving a *neck* or a strain region over the extended billet. The electron-strain effect is detected either by the presence of microscopic currents in the strain region, or from the potential difference that compensates these currents. This effect can appear at modern drawing and rolling machines. In particular it can cause breakage of steel and copper wire under high-rate, very fine *drawing*.

ELECTRON TEMPERATURE

A value of temperature which specifies the mean energy of free electrons (*holes*) when they are overcooled or are heated up (see also *Electron heating*). Its significance and properties depend on the ratio between the mean *relaxation time* and those of the quasi-momentum τ_p, energy τ_E, electron–phonon scattering $\tau_{\text{ph-e}}$, and electron–electron scattering τ_{ee} (see *Current carrier scattering*). Under field-stimulated warming-up, while drifting in the electric field with the mean velocity v_{d}, the electrons obtain from the field *quasi-momentum* $\boldsymbol{p}$ and energy E which they redistribute between themselves and then pass on to the lattice. For the usual case $\tau_{\text{ee}} \ll \tau_{\text{ph-e}}$ the internal interaction within the electron subsystem dominates over its interaction with the lattice. Therefore, in view of the lack of the equilibrium between the electrons and the lattice, a state of equilibrium is established within the electron subsystem over the momentum $\boldsymbol{p}$ and the energy E characterized by an average electron velocity u, and the electron temperature T_{e} which is higher than the lattice temperature. This approximation is often called the *hydrodynamic approximation*. For this case $\tau_p < \tau_E < \tau_{\text{ee}}$. This is often introduced with the following relation: mean value $\langle E\rangle = (3/2)k_{\text{B}}T_{\text{e}} + E_{\text{d}}$ where $E_{\text{d}} = m_n v_{\text{d}}^2/2$ is the drift energy, and m_n is the electron *effective mass*. In the case when $m_n v_d^2/2 \ll k_B T_e$, the energy acquired by the electrons is distributed uniformly over the degrees of freedom, and the system as whole slowly drifts. In the case of inelastic scattering $\tau_p \approx \tau_E$, the energy is comparable with the random motion energy, and T_e is determined from $k_B(T_e - T) \approx E_d$.

ELECTROOPTICAL EFFECTS

Variation Δn of the *refractive index* n of light in crystals located in an electric field of strength E. It arises from the orientation in the field of the atoms or polar molecules that compose these crystals. In crystals without a *center of symmetry* a *linear electrooptical effect* (in respect to the field strength) takes place: $\Delta n \sim E$ (*Pockels effect*). The applied electric field transforms an optically isotropic piezocrystal into an uniaxial one, and an uniaxial crystal transforms into a biaxial crystal. Therefore, a consequence of electrooptical effects is the appearance of induced birefringence.

Electrooptical effects are the physical basis for the action of an electrooptic modulator of coherent radiation (*Pockels cell*) with a very short relaxation time (about 10^{-13} s). Electrooptical effects are used also for the modulation of laser *resonators* designed to provide single pulses of high-power coherent radiation (giant pulses). A disadvantage of modulators based on electrooptical effects is the high working voltage (tens to hundreds of kilovolts). In centrally symmetric crystals a *quadratic electrooptical effect* can appear, with $\Delta n \sim E^2$ (*Kerr effect*).

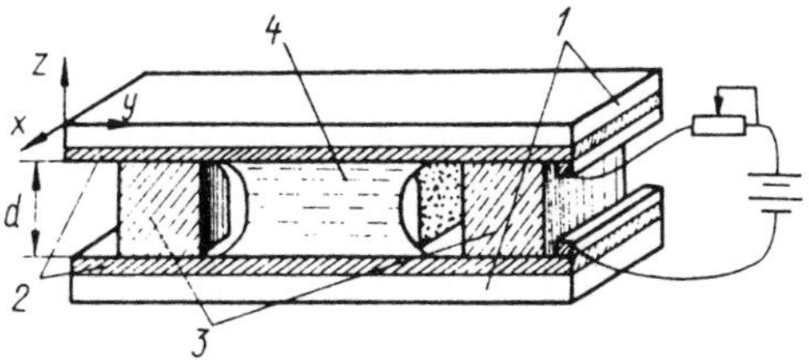

Electrooptical effects in liquid crystals.

ELECTROOPTICAL EFFECTS IN LIQUID CRYSTALS

Variation of the passage of light through a sample of a *liquid crystal* to which a constant or alternating electric field is applied, due to the orienting action of the external field on the *director* (optic axis) (see *Twist effect, Guest–host effect, Frederiks transition*).

A device for observing these effects is sketched in the figure. The unit consists of glass plates 1 with transparent electrodes 2 (e.g., made of SnO_2) and spacers 3 (made of an insulator such as mica or teflon) which match the thickness to the liquid crystal layer 4. As a rule, this thickness $d = 10$ to 100 μm. The observations can be performed using either transmitted or reflected light (lower electrode can be reflecting). These effects are widely used in practice (see *Liquid-crystal displays, Defectoscopy*).

ELECTROOPTICAL MODULATORS

Bulk or integral-optic devices employed in quantum electronics and power optics up to thermonuclear synthesis for the purposes of amplitude and phase modulating laser beams; their action is

Table 1. Characteristics of electrooptical modulators

Properties of modulators	Type of modulator			
	No. ML3	No. DP663	No. ML5	No. ML7
			Lithium niobate	Galium arsenide
Half-wave voltage, V (at $\lambda = 0.63$ μm)	730	370	185	3000 (at $\lambda = 10.6$ μm)
Transmission band, μm	0.35–1.1	0.35–1.1	0.5–4.5	1.0–20
Band width, MHz	0–100	0–100	0–200	0–25
Aperture, mm	3	2.5	2	3
Contrast	100	135	120	135:1
Normalized control power, W/MHz	3.68	1.26	0.19	51.8

a result of the influence of the electric field on the optical properties of the material. The main characteristics of some bulk electrooptical modulators are presented in Table 1. Nonlinear optical materials have also been used.

Integral-optic modulation is mainly carried out on plates made of lithium niobate with waveguides formed by titanium diffusion or cold welding; the controlling power decreases to 1 μW/MHz as the upper limit of the frequency range is extended to tens of GHz. This completely satisfies the requirements of multichannel fiber-optic communication and control systems, including the integration of user terminals with national and international networks of databases. See also *Electrooptic materials*.

ELECTROOPTIC MATERIALS

Crystals with comparatively strong *electrooptical effects*. There are linear and quadratic (with respect to the electric field) electrooptical effects (see *Pockels effect*, *Kerr effect*). Of particular practical significance for the transformation and modulation of radiation is the Pockels effect which is inherent in *piezoelectric materials*. The most important electrooptic materials are the so-called KDP compounds of the group potassium dihydrogenphosphate KH_2PO_4. There are several isomorphic room temperature crystals containing Rb, Cs and NH_4 instead of K, arsenates containing AsO_4 instead of PO_4, and deuterium-substituted compounds such as, e.g., $KDHPO_4$. At low temperatures several of them are *ferroelectrics*, and others are *antiferroelectrics*. Concerning their structure, all of them belong to a tetragonal-scalenohedral type of symmetry with space group $I\bar{4}2d$, D_{2d}^{12} (V_d^{12}).

Cubic sphalerite-type crystals can also possess electrooptical properties. Classed among them are *semiconductors* of the types $A^{II}B^{VI}$ (ZnS, ZnSe, ZnTe, and CdTe) and $A^{III}B^{V}$ (GaAs and GaP), copper and silver halides. Technological problems make it difficult to obtain large, defect-free crystals of the majority of these compounds, and this significantly restricts their application as basic elements of *electrooptical modulators*. To the class of electrooptic materials there also belong *cubic system* crystals like eulytite $[Bi_4(SiO_4)_3]$, sodalite $[Na_4Al_3(SiO_4)_3Cl]$, and pleomorphic sillenite ($n Bi_2O_3 \cdot m R$) where R is an oxide of germanium, titanium, silicon, aluminum, zinc, lead, etc., and n and m are integers. The *molecular crystals* $(CH_2)_6N_4$, ammonium chloride, and sodium chlorate are also electrooptic.

Non-oxide type ferroelectrics and antiferroelectrics include the well-known electrooptic materials Rochelle salt and *triglycinesulfate*. A numerous group of electrooptic materials is composed of octahedral-oxygen cooordinated ferroelectrics and antiferroelectrics: crystals of the perovskite type ($CaTiO_3$, $BaTiO_2$), double oxides like $PbO \cdot Nb_2O_5$, lithium niobate and tantalate ($LiNbO_3$ and $LiTaO_3$). Lithium niobate (see *Niobates*) is used in *nonlinear optics* as a frequency transformer of coherent radiation, and for the production of modulators.

ELECTROPLATING

A technique for depositing *metals* (more rarely non-metals) on the surfaces of solids with the help of electrolysis. There are two approaches, adding a thin metal protective or decorative coating to a surface, and obtaining precise, easily detachable metallic replicas from a host surface. A simple electrodeposition system consists of an electric circuit with two electrodes (a cathode and an anode) and an electrolyte containing the ions of the metal (or metals) to be discharged at the cathode. The structure and the properties of electrolytically deposited metals depend on the *crystallization* conditions, and these are mainly determined by the electrolyte composition and the electrolysis regime. Solutions of complex, in particular cyanic, salts that produce a fine-structure deposit are widely used. The loss of ions discharged at the cathode is compensated by ions dissolving and thereby entering the electrolytic solution at the anode. For applications using anodes that do not readily dissolve, the ion loss at the cathode is compensated for by a periodic introduction into the electrolyte of salts containing these lost ions.

The electrodeposition method is important in the chemical and electronics industries, and is used for putting *coatings* on machinery, as well as for adding conduction contacts to insulating *substrates*.

ELECTROREFLECTION

A method of *modulation spectroscopy* in which the light reflection coefficient R is modulated by an electric field. The electroreflection signal is $\Delta R/R = [(\alpha - \mathrm{i}\beta)\Delta\varepsilon]$, where the so-called *Seraphin coefficients* (B.O. Seraphin) α and β depend on the optical constants of the reflecting system, and $\Delta\varepsilon$ is the *dielectric constant* variation in the electric field. Physical mechanisms of electroreflection are the *Franz–Keldysh effect* involving the action of the electric field on the exciton states or the plasma of the free charge carriers, and the *Stark effect* which involves the modulation of the energy band populations, and other effects. The Franz–Keldysh effect is the lengthening of the wavelength of the optical absorption edge of a semiconductor in an applied electric field. In strong electric fields, an electroreflection spectrum contains Franz–Keldysh oscillations, and this is equivalent to a spectrum of $\partial^3\varepsilon/\partial\omega^3$ (here ω is the modulation frequency). Therefore, the electroreflection method provides greater resolution than other modulation methods (such as modulating the wavelength, temperature or pressure) where the spectra are of the type $\partial\varepsilon/\partial\omega$. The measurement of electroreflection yields the energies of critical points of the *band structure*, *effective masses*, and broadening parameters that correspond to them, the plane-band potential, estimates of the concentration of free charge carriers, the third-order nonlinear optical susceptibility, and some other parameters. The method of electroreflection is used in studies of the band structure of *solids*, electrical-physical processes near the surface in semiconductors, *metal–insulator–semiconductor structures*, *homojunctions* and *heterojunctions*, and in electrochemical studies of the system metal–electrolyte. See also *Modulation spectroscopy*.

ELECTROSTRICTION

Strain of an *insulator* that varies as the square of electric field strength, E^2. Electrostriction is caused by the *polarization of insulator* in an applied electric field; this is inherent in all insulators, i.e. solid, liquid and gaseous ones. One should differentiate electrostriction from an inverse piezoeffect which is linear in the field E (see *Piezoelectric materials*).

In isotropic media, including gases and liquids, electrostriction is observed as a density variation under the action of the electric field, and described by the formula:

$$\frac{\Delta V}{V} = AE^2, \qquad (1)$$

where $\Delta V/V$ is the relative bulk *strain*, $A = (\beta/2\pi)d(\partial\varepsilon/\partial d)$ where β is the *compressibility*, d is the *density*, and ε is the *dielectric constant*. For organic liquids (xylene, toluene, nitrobenzene) $A \sim 10^{-12}$ CGS units. In anisotropic crystals electrostriction can be described by the relationship between two second-rank tensors, the tensor of the square of the electric field strength and the *strain tensor*:

$$\varepsilon_{ij} = \sum_m \sum_n R_{ijmn} E_m E_n, \qquad (2)$$

where ε_{ij} is the strain tensor component; and E_m and E_n are the components of the electric field vector. The factor R_{ijmn} is the *electrostriction coefficient*. The number of independent electrostriction coefficients depends on the *crystal symmetry*. For the lowest crystallographic symmetry, namely triclinic, the electrostriction tensor has 36 independent coefficients. Ordinarily $R_{ijmn} \sim 10^{-14}$–10^{-10} CGS units, and in a field $E \sim 300$ V/cm the value $\varepsilon_{ij} \sim 10^{-6}$.

Sometimes a large electrostriction is ascribed to *ferroelectrics*. Actually this is a *reciprocal piezoeffect*. However, in the ferroelectric case where the volumes of differently polarized *domains* are the same, the deformation does not depend on the field direction. Under the action of an alternating electric field with the frequency ω, an insulator actuated by the electrostriction effect oscillates with the frequency 2ω (this is specific for all the quadratic effects). Electrostriction can be used to transform electric vibrations into sound oscillations.

ELECTROTHERMAL GRADIENT EFFECT

The appearance of electric charge and an electric polarization P_i at planes normal to the direction of a temperature gradient that is present in a *crystal* of arbitrary symmetry. The polarization is proportional to the temperature gradient as follows:

$$P_i = b_{ij}\frac{\partial T}{\partial x_j},$$

where b_{ij} is the polarization coefficient tensor.

The electrothermal gradient effect is caused by a nonuniformity of the thermal expansion, together with a contribution from phonon states. As a result, this effect is observable not only in *ferroelectrics* where it reaches a relatively large magnitude in the vicinity of the *Curie point* T_C, but also in centrally symmetric crystals.

ELEMENTARY EXCITATION SPECTRA OF DISORDERED SOLIDS

Energy spectra of low-lying states of *disordered solids*, which do not possess translational symmetry. Elementary excitation spectra of disordered solids cannot be described in terms of *quasi-momentum*, and hence are more complicated in nature than corresponding spectra of ordered solids. First of all, there are branches of the spectrum which correspond to a single-particle pattern, i.e. the analogues of *quasi-particles* of an ordered solid. However, apart from those states with an amplitude, which is of the same order of magnitude over the entire body of the solid (analogues of Bloch functions of an ordered system), there also exists a macroscopic number of localized states in a disordered solid (see *Anderson localization*). If the degree of disorder is considerable, then the contribution of localized states may be rather high (it should be mentioned that all states of one- and two-dimensional systems are localized, see *One-dimensional models of disordered structures*, *Scaling theory of localization*), and in this case localized states have a pronounced effect on the mechanisms of *kinetic phenomena*. Thus, the static conductivity of a three-dimensional macroscopic disordered sample is zero at zero temperature if the *Fermi level* falls within the region of localized states (it is always zero if the sample is one-dimensional or two-dimensional). Such a sample is the so-called *Anderson insulator*. Its static conductivity can arise only through a thermal (or other, e.g., tunneling) activation mechanism at $T = 0$. Beside quasi-particles, other types of excitations may exist in disordered solids. For example, correlation effects were found to play an important role in the system of interacting electrons in a random field of *impurity atoms*. In the latter case, two neighboring energy states of the system correspond to different arrangements of electrons over the impurity sites, and the *density of states* is different from that of a single-particle pattern: the so-called *Coulomb gap* appears in the spectrum. Of crucial importance may also be excitations that are due to the rearrangement of the particles which constitute the framework. These excitations are related to quantum transitions of the whole system among the configurations, which differ from one another only by the positions of a small number of inner core particles. If the energies of these configurations are close together and the energy barrier separating them is not too high, then the time of a transition is short compared with the characteristic observation period, and hence the excitations of this type are in equilibrium. The existence of the simplest excitations of this type, which are called *two-level excitations*, has been established for several classes of disordered solids (in insulating and *metallic glasses*, semiconductors, etc.), and results in a variety of thermodynamic and kinetic features of such systems (see *Two-level systems*).

The simplest quantity which is characteristic of the spectrum of single-particle excitations in disordered solids is the density of states $\rho(E)$. This is a *self-averaging quantity* in disordered systems, which are spatially uniform only on the average, and it exhibits a decrease of correlations at distant points. Therefore, spectra of all forms of a macroscopic system coincide, since they consist of points at which the nonrandom function $\rho(E)$ differs from zero.

The *spectral limits* consist of nonrandom points E_g; and the density of states goes to zero on crossing these limits. The spectral limits are differentiated into fluctuation limits and stable ones. The spectral region in the neighborhood of a *fluctuation limit* (fluctuation region of the spectrum) owes its existence to improbable fluctuations, i.e. to non-typical arrangements of impurities; the spectrum in these regions arising from states that are localized at those fluctuations. The location of the fluctuation limit depends on the specific form of disorder (type of impurities, correlations in their arrangement, etc.). The spectrum of energy states in the neighborhood of *stable limits* arises from all kinds of factors, and the density of states converges asymptotically to that of a certain effective ordered system. The location of a stable limit, and the energy dependence of the asymptotic density of states in its neighborhood, are insensitive to the particular type of disorder. Examples of stable limits are the high-energy region of the electron spectrum, and the bottom of the acoustic band of the *phonon spectrum*. The spectral limits are in many cases singular points with respect to the density of states. Other singular points appear only in the limit of minor disorder (the local level E_0 (see *Local electronic levels*), the spectrum limit of the initial disordered system). The rearrangement of quantum states and the systematics of these states take place in the neighborhood of singular points. The disturbance due to disorder in the neighborhood of these points is not small in terms of the effect that it exerts on the states and on the spectrum. The presence of small parameters, in addition to those relative to $|E - E_g|$ and $|E - E_0|$, often allows the study of these new systematics, and permits one to obtain the asymptotic density of states (see *Lifshits model, Fluctuation levels*).

The localized states of a macroscopic disordered system correspond to a *high-density discrete spectrum*, whereas the delocalized states correspond to a continuous energy spectrum. Moderately disordered three-dimensional systems exhibit a region of delocalized states, which is either adjacent to the stable limit, or located at the center of the band. The probability of such states existing with localized particles of equal energy is rather low, because the generation of discrete levels in the continuous spectrum takes place as a result of fine interference effects, and is therefore highly unstable. Consequently, there exist nonrandom states E_c (*mobility limits*, see *Mobility edge*) along the energy axis, which separate the region of localized states from that of delocalized ones. The latter have, in order of magnitude, one and the same amplitude on a set that is a macroscopic part of the whole volume. Such a set, commonly called an *infinite cluster*, may possess a multiply-connected, highly branched (fractal) structure (see *Fractal*). Therefore, the delocalized states of disordered systems bear little resemblance to Bloch states of ordered systems, at least in the neighborhood of the mobility limit.

In the three-dimensional case of a system with a fluctuation limit E_g at sufficiently high energy (in the neighborhood of the stable boundary, $E \to +\infty$), the motion of a particle is quasi-classical; it may be scattered by individual impurities although, as a rule, only through small angles, so a large number of collisions is needed to substantially change the particle quasi-momentum. As a result, the initial phase is completely erased from the system memory, and hence the particle motion may be described in terms of classical theory. Since the energy is assumed to be much higher than the potential variation relief heights, such a "diffusion" motion takes place in the classically allowed region and, consequently, the particle is in an *infinite or delocalized state*, i.e. it belongs to the continuum spectrum. Furthermore, all states in the proximity of the fluctuation limit are highly localized. Thus, the spectrum in this case necessarily contains both continuous and discrete components, which are separated by the mobility limit E_c. The relative contributions of these components to the spectrum are determined by the degree of disorder of the system.

The qualitative pattern of states in the neighborhood of the mobility limit E_c becomes clear when described for the sake of simplicity in quasi-classical terms. If a random field due to disorder is at the same time sufficiently smooth, then the *quasi-classical approximation* is also applicable at energies below the peak of the *random potential* amplitude. The regions of classically allowed motion are in this case sufficiently high, and hence the particle motion at this energy may be described as largely classical in nature. Then the question of whether the quantum state of a certain energy is localized or delocalized, and whether the corresponding particle motion is finite or infinite, reduces to a question of the geometry of the classically allowed region: whether it is divided into individual parts of finite dimensions, which are separated from one another, or take up the whole space of the system. The value of the energy, at which the transition between the above-described geometric patterns takes place, is called the *percolation level* (in the quasi-classical case it corresponds to the mobility limit) and the broad range of problems, which arise in this connection, is the subject matter of *percolation theory*. When the band is finite in width and both limits are fluctuation limits, then there exist two mobility limits $E_{c1} < E_{c2}$, the positions of which will move into the band with an increased degree of disorder, and finally merge: $E_{c1} = E_{c2}$. Such a confluence results in the disappearance of delocalized states from the spectrum: the metal turns into an Anderson insulator (*Anderson transition*). A completely different type of situation occurs in a one-dimensional system. In this case, the momentum of the particle, which is elastically scattered by an impurity, is either conserved or reversed. Therefore, the scattering cannot be considered to be slight, the particle motion is not quasi-classical even at high energies, and an important role is played by the interference of incident and scattered waves, which requires a sophisticated consideration of multiple scattering effects. As a result all states of a one-dimensional system become localized at the introduction of even a small amount of disorder.

One of the best known *localization criteria* of states in the neighborhood of the energy E is the positive nature of the *Anderson function*,

$$p(\boldsymbol{r}, E) = \left\langle \sum_{\text{discr}}^{\infty} \delta(E - E_n)\left|\psi_n(0)\psi_n(\boldsymbol{r})\right|^2 \right\rangle$$

the physical meaning of which is simple: the integral of this function over the volume

$$p(E) = \left\langle \sum_{\text{discr}} (E - E_n)\left|\psi_n(0)\right|^2 \right\rangle$$

is the density of localized states. In the case of a one-dimensional system we have $p(E) = \rho(E)$ over the entire energy range, and the asymptote of the Anderson function of weak scattering by a single impurity is given by $p(x, E) \propto \exp[-|x|/(4l)]$ with $|x| \gg l$, where $l(E)$ is the *localization length*. Consider the case of a three-dimensional system and a semi-infinite band $[E_g, +\infty]$, with $p(E) = \rho(E)$ below the mobility limit ($E < E_c$), while for $E > E_c$ the Anderson function goes to zero. Another criterion for localization is related to the behavior of an important physical quantity: the static electrical conductivity $\sigma_{dc}|_{T=0}$, which is considered as a function of the *Fermi energy*. If $E_F < E_c$, then the Fermi energy falls within the region of localized states and $\sigma_{dc}|_{T=0} = 0$. As E_F increases, the insulator (Anderson type)–metal transition takes place for $E_F = E_c$ and $\sigma_{dc}|_{T=0}$ becomes different from zero at $E_F > E_c$. To clarify the behavior of the conductivity in the neighborhood of the mobility limit see the *Scaling theory of localization*.

ELEMENTARY PARALLELEPIPED

See *Primitive parallelepiped.*

ELIASHBERG EQUATIONS (G.M. Eliashberg, 1960)

A set of interconnected nonlinear integral equations obtained on the basis of *Green's functions in superconductivity*. The Eliashberg formalism contains an equation for the renormalized, anomalous *intrinsic-energy part* (function of *energy gap*) of the spectrum of *quasi-particles* $C(\omega)$, and an equation for the function of the renormalization spectrum $Z(\omega)$ of a superconductor with a strong *electron–phonon interaction*. Since the mean *phonon* energy in *metals* is much lower than the electron *Fermi energy*, the Eliashberg equations involve a single-variable (energy) function; at $T = 0$ these take the form

$$C(\omega) = \frac{1}{Z(\omega)} \int_{\Delta}^{\infty} d\omega' \operatorname{Re}\left\{ \frac{C(\omega')}{\sqrt{\omega'^2 - C^2(\omega')}} \right\} \times K_+(\omega, \omega'),$$

$$\left[1 - Z(\omega)\right]\omega = \int_{\Delta}^{\infty} d\omega' \operatorname{Re}\left\{ \frac{\omega'}{\sqrt{\omega'^2 - C^2(\omega')}} \right\} K_-(\omega, \omega').$$

Here

$$K_\pm(\omega, \omega') = \int_0^{\infty} d\Omega\, g(\Omega) \left[\frac{1}{\Omega + \omega' + \omega + i\delta} \pm \frac{1}{\Omega + \omega' - \omega + i\delta} \right],$$

where $g(\Omega)$, often called the *Eliashberg function*, is the characteristic function of the electron–phonon interaction given by the expression

$$g(\Omega) = \int_{S_F} \frac{dS_p}{v_p} \int_{S_F} \frac{dS_{p'}}{v_{p'}} \sum_j \left|M^j_{pp'}\right|^2 \times \delta\left(\Omega - \omega^j_{p-p'}\right) \left\{ \int_{S_F} \frac{dS_p}{v_p} \right\}^{-1},$$

where $M^j_{pp'}$ is the matrix element of the electron–phonon interaction, ω^j_q is the *dispersion law* of the jth branch in the phonon spectrum, v_p is the electron velocity at the Fermi surface (integrated or averaged over the entire Fermi surface S_F). The Eliashberg function $g(\Omega)$ is often represented as $g(\Omega) = \alpha^2(\Omega)F(\Omega)$, where $\alpha(\Omega)$ is the electron–phonon coupling strength, and $F(\Omega)$ is the phonon density of states. In the vicinity of the *critical temperature* ($T \to T_c$), when the superconductor energy gap parameter diminishes to zero ($\Delta \to 0$), the linearized Eliashberg equations in the framework of Green's function temperature technique, with the Coulomb repulsion taken into account, assume the following form:

$$Z(\mathrm{i}\omega_n)C(\mathrm{i}\omega_n) = \pi T_c \sum_{\omega_m} \left[Q(\mathrm{i}\omega_m - \mathrm{i}\omega_n) - \mu^*\right] \frac{C(\mathrm{i}\omega_m)}{|\omega_m|},$$

$$\omega_n\left[1 - Z(\mathrm{i}\omega_n)\right] = -\pi T_c \sum_{\omega_m} Q(\mathrm{i}\omega_m - \mathrm{i}\omega_n)\,\mathrm{sgn}\,\omega_m,$$

where

$$Q(\mathrm{i}\omega_m - \mathrm{i}\omega_n) = \int_0^\infty \mathrm{d}\Omega\, \alpha^2(\Omega)F(\Omega) \frac{2\Omega}{\Omega^2 + (\omega_m - \omega_n)^2},$$

where μ^* is the *Morel–Anderson Coulomb pseudopotential*, $\omega_n = (2n - 1)\pi T_c$ are the Matsubara "frequencies", and $n = 0, \pm 1, \pm 2, \ldots$.

A precise numerical solution of the Eliashberg equations was first obtained by D.J. Scalapino, J.W. Wilkins, and J.R. Schrieffer (1963) who uncovered an essential dispersion (energy dependence) of both real and imaginary parts of the complex gap parameter $C(\omega)$, and called attention to related anomalies of the tunneling *density of states*. Further, with the help of a numerical solution of the inverse problem, the method of reconstructing the Eliashberg function $g(\omega)$ from the tunneling current–voltage characteristics of a superconductor was successfully developed (see *Tunneling spectroscopy of superconductors*).

ELLIPSOMETRY, optical polarization method

Branch of optics which studies *solid surfaces* and various multilayer *coatings* on solids by analyzing the polarization of a reflected or transmitted light beam (*reflection ellipsometry* and *transmission ellipsometry*).

The classical configuration of an *ellipsometer*, an instrument for the analysis of elliptically polarized light, includes a light source (mercury lamp, *laser*, or monochromator), polarizer, table with sample under study, compensator (quarter-wave plate), and analyzer. Incident linearly polarized light (see *Polarization of light*) reflected by a sample becomes, in general, elliptically polarized. This is characterized by the polarization angles Ψ and Δ which describe the amplitude ratios and the phase difference of the parallel (p) and perpendicular (s) components. The main equation of ellipsometry establishes the relation between angles Ψ and Δ, and the optical properties of the reflecting system. This equation is based on the *Fresnel equations*; specific terms depend on the choice of the model for the surface. It is generally solved by numerical methods with the aid of a computer. Since each ellipsometric measurement yields two magnitudes Ψ and Δ, one can find two unknown parameters. In the case of a multilayer reflection system, when it is necessary to determine a large number of parameters (*refractive indices* and *absorption coefficients* of the substrate and film situated on it, film thickness, and so on), one should perform many measurements under various conditions: at several angles of incidence (*multiangle ellipsometry*); for several thicknesses of a homogeneous film (*multithickness ellipsometry*); in various ambient media (*immersion ellipsometry*). To obtain the optical parameters of the system from measured angles Ψ and Δ it is necessary to solve a reciprocal problem. The use of modern high-speed computers removed the ambiguity problem involved in the interpretation of ellipsometric data, thereby putting on a firm basis the investigation of the surface state (roughness, adsorption coatings and coatings of other kinds, subsurface disturbed layers, etc.), the determination of the thickness and the optical parameters of thin ($\geqslant$1 nm) films, substrate parameters, and direct studies of various physicochemical processes at the solid surface. The sensitivity of the method increases sharply under resonance conditions when surface polaritons are excited with the enhancement of the electromagnetic wave field near the surface. This is achieved by using the *attenuated total internal reflection* method, or nonplanar (randomly rough or regularly profiled) surfaces. Advantages of ellipsometry are its lack of direct contact with the surfaces, and its unique sensitivity to superthin layers situated at the surface (to $\sim$0.1 of a monolayer).

ELLIPTICITY OF ACOUSTIC WAVE

See *Acoustic wave ellipticity.*

ELONGATION MODULUS

See *Young's modulus.*

EMBRITTLEMENT

A sharp reduction of *plasticity* of metals and *alloys.* Embrittlement can be a consequence of a high level of impurities (see *Impurity atoms*) and nonmetallic inclusions, as well as a result of *plastic deformation* or *heat treatment* of metals and alloys. The most common cause of embrittlement is the *segregation* of impurities at *grain boundaries,* and the weakening of the grain-to-grain bonding *strength.* In this case, the *failure* is of an intercrystallite character (see *Crystallites*). A major hazard is the segregation of impurities with low *surface energy* at grain boundaries, since this leads to the lowering of the failing stress. To lessen the segregation of harmful impurities at grain boundaries, an appropriate heat treatment can be carried out. Such a treatment either stimulates the transfer of the impurity to the grain body, binds the impurity to chemical compounds, or else leads to the competitive segregation of elements, unable to reduce the surface energy, at the grain boundaries. Embrittlement in alloys may be a consequence of solid solution decomposition (see *Alloy decomposition*) with the formation of *Guinier–Preston zones* (clusters of precipitate without definite crystalline structure) followed by the precipitation of intermetallic compounds. In *metals* with a body-centered cubic lattice embrittlement can result from the blocking of *dislocations* by impurities with the establishment of atmospheres of various kinds (see, e.g., *Cottrell atmosphere*). Embrittlement can be caused by the formation of unfavorable *texture* and macrodefects such as slivers, laps, pits, etc. during the course of plastic deformation. In some cases, the embrittlement can be explained as a rise of the *cold brittleness* temperature above room temperature as a result of structural factors.

EMBRITTLEMENT, IRRADIATION

See *Irradiation embrittlement.*

EMERSLEBEN TRANSFORM (O. Emersleben, 1952)

Mode of transformation of lattice sums such as $\sum_{l'} 1/|r_{sl} - r_{s'l'}|^n$ (here l, l' are cell numbers, s, s' are the numbers of atoms in a cell) into two rapidly convergent series in terms of the direct and reciprocal lattices; a method similar to the *Ewald method* for integers $n \geqslant 1$. This is applied, in particular, for the calculation of the *crystal lattice* energy related to *van der Waals forces* for n equal to 6, 8, 10, or the lattice energy associated with the potential $A/r^m - B/r^n$, $m > n$ (m, n are positive integers; $m = 12$, $n = 6$ for the *Lennard-Jones potential*).

EMISSION ELECTRONICS

See *Vacuum electronics.*

EMISSION OF HOT ELECTRONS

See *Hot electron emission.*

EMISSION OF PARTICLES

See *Particle emission.*

EMISSION SPECTRAL ANALYSIS

Totality of methods for determining the chemical composition of a substance by studying its optical radiation (emission) spectrum. Qualitatively, emission spectral analysis involves the detection and identification of the spectral lines of the element of interest in the emission spectrum of a material. For this purpose, there exist special tabulations and atlases of the spectral lines of elements. This method permits the determination of almost all the elements of the Mendeleev periodic system. In practice for the emission spectral range (160 to 1000 nm) in and near the visible region the elements of Groups I–II with low atomic number Z are found with the best sensitivity, while halogens have the lowest sensitivity. Quantitatively emission spectral analysis is based on the relationship between the intensity of the analytical line I_A^i of the defined element A and its content C_A in a sample, established according to the semiempirical *Lomakin–Scheibe relation* (B.A. Lomakin, G. Scheibe, 1932) on reference samples of a composition similar to that of the one under analysis: $I_A^i = aC_A^b$ (a and b are constants depending on the properties of the analytical line and the light source). A comparison line is usually chosen from an element already in the sample, or the line of an element artificially introduced into the sample to serve as an *internal standard*. The values I_A and I_{stand} are selected such that the ratio of their intensities should be insensitive to a variation of the discharge conditions. The relative error of emission spectral analysis for contents $\geqslant 10^{-2}\%$ is 1 to 10%.

Devices for emission spectral analysis are differentiated by their spectral wavelength band, the type of dispersing element, and the light detection method. Spectrographs with interchangeable diffraction gratings and photoelectric quantometers, provided with microcomputers, have particular advantages.

In emission spectrum analysis, about ten types of light sources and their modifications are used. The best sources for the analysis of solid nonconducting materials are arc discharges; for the analysis of monolithic metals and alloys – glow discharges in a hollow cathode after Grimm, and low-voltage discharges with the frequency $\leqslant$500 Hz in an argon jet; for the analysis of particularly pure substances (after dissolving them and concentrating the analyzed impurities) – a high-frequency induction-bound plasma; for the analysis of the barely-excitable elements – glow discharges in hollow cathodes and high-frequency induction-bound plasmas. In use for *layer-by-layer analysis* of solid materials, and for the analysis of their inclusions with a volume 200 to 10^6 μm^3, are the corresponding pulsed discharges in a cooled hollow cathode and lasers. The method of *laser fluorescent analysis*, where a *laser* with a tunable frequency serves as the light source, is the most prospective and efficient method of emission spectral analysis. This method possesses extremely high selectivity and sensitivity, unachievable by pre-laser analytical methods, which might possibly permit the detection of single atoms (in a pregiven quantum state) in bulk material illuminated with the laser radiation.

ENAMELING

Application of a thin glassy, opaque ceramic *coating* to a metal surface. Enamel coatings are notable for *hardness*, *wear* resistance, *abrasive resistance* (see *Abrasive materials*), *corrosion resistance*, *thermal-environment resistance*, color resistance, and weathering resistance. These coatings are single or multiple layers 0.07–0.2 mm thick applied by pouring, spraying, and other methods, followed by further sintering and glazing in special furnaces. Sometimes a paint or varnish is used that produces an enamel-like coating.

ENANTIOMORPHISM, also called enantiomerism (fr. Gr. εναντιος, opposite and μορφη, form)

Property of molecules and crystals to form mirror image structural units called enantiomers. One of the enantiomers is called left, with the other being designated right. Enantiomorphism influences morphological, optical, and other structural characteristics. Thus the left and right modifications of the structure of quartz rotate the polarization plane (see *Rotation of light polarization plane*) of light passing along the optical axis in opposite directions. Enantiomorphism is also inherent in many organic materials. As a rule, living systems are constructed of only one of the two enantiomorphisms of biological molecules, such as the exclusive use of left-handed aminoacids by proteins. See also *Crystal symmetry*, *Chirality*.

ENDOSCOPY

Methods for studying the internal structure and properties of opaque objects. X-ray, gamma ray and ultrasonic methods, as well as *tomography* and magnetic resonance imaging (MRI), provide information on internal structure. However, the term endoscopy generally refers to the use of endoscopes which are inserted into, and permit the visual examination of, parts of the interior of a body.

ENDOSCOPY, EPR

See *EPR imaging*.

ENDOSCOPY, NMR

See *Magnetic resonance imaging*.

ENDURANCE LIMIT

Maximum value of a stress cycle which a sample endures without *failure*, or without intolerable *strain* loading during the course of a preset number of cycles (usually 10^6 cycles; see *Cyclic strength*). To define the endurance limit a series of tests is performed with different loads, and the number of cycles of loading until failure is determined. The stress values are plotted on a diagram that shows their dependence on the number of cycles N which at a given stress level the sample had endured until failure (*fatigue curve* or *Wöhler curve*). For pure *metals*, or *polymers*, or in case of *corrosion fatigue*, the Wöhler curve monotonically decreases with the number of cycles; for less pure samples and for *alloys* the Wöhler curve approaches a straight line, parallel to the abscissa. The endurance limit is determined by the nature of the interatomic bonds, by the *crystal lattice* type, by the structural state, and by other parameters of the material; it also depends on the conditions of testing (amplitude and frequency of loading, sample size, testing medium). *Fatigue* tests exhibit considerable scatter of the results, so it is important to use statistical methods in evaluating them. The endurance limit of the sample in the presence of a *crack* may be characterized by the smallest value of the cyclic stress intensity ratio that brings about a further spreading of the crack. The endurance limit in the presence of a crack may be determined with acceptable accuracy by retesting the same sample. The ability of a metal to function after the appearance of a fatigue crack is called its *viability*.

ENERGY BANDS in crystals

Energy spectrum which results from the quantum equations of motion which are invariant under crystal lattice *translations*. Band spectra are inherent in electrons, *phonons*, *magnons*, *excitons*, etc. However, as a rule, the terms energy bands and band theory are used when describing electrons in a crystal. The formation of electron energy bands is connected with the Bragg reflection of electrons in crystals (see *Bragg law*).

Electronic energy bands represent the overall spectrum of solutions to the Schrödinger equation for an electron moving in the periodic potential of a *crystal lattice*. The electron bands "genetically" originate from the energy levels of the free atoms that come together to form a crystal. In the crystal, as a result of the interaction between the atomic electronic shells, each atomic level broadens and transforms into a *band*, consisting of a great number of levels. This number is equal to the degree of degeneracy of the atomic level multiplied by the number of equivalent atoms in the crystal; it means that in practice the band is occupied continuously (quasi-continuously). A band spectrum of an electron in a crystal is represented by alternating bands of allowed and forbidden energies (*allowed band* and *forbidden band*, or *band gap*). Even in the case of strong localization of the atomic wave functions one can speak about narrow bands. For example, in *rare-earth elements* atomic f-levels form a narrow f-band. The quantum theory of metals is based on the *single-electron approximation* which postulates that the electron moves in the periodic potential arising from nuclei and other electrons in the crystal. Stationary states of such electrons are described by *Bloch functions* $\psi_{nk}(r)$ (see *Bloch theorem*), and the energy spectrum (or *dispersion law*) $\varepsilon_n(\boldsymbol{k})$ where $\boldsymbol{k}$ is the wave vector, and n is the band number. The periodicity of the crystal potential causes the periodicity of the energy $\varepsilon_n(\boldsymbol{k})$ in $\boldsymbol{k}$*-space* corresponding to $\varepsilon_n(\boldsymbol{k}+\boldsymbol{b}) = \varepsilon_n(\boldsymbol{k})$, where $\boldsymbol{b}$ is a *reciprocal lattice* vector. Many physically different values of $\boldsymbol{k}$ are situated in the first *Brillouin zone*. To every value of $\boldsymbol{k}$ there corresponds an infinite number of discrete energy levels denoted by the index n. For every n the electron energy assumes values within a certain finite range. The form of the dispersion law depends upon the *crystal symmetry*, and the symmetry of the wave functions of the individual atoms that compose the crystal. The bands are often separated by energy gaps, but they can also partially overlap. In the case of band overlapping a superposition (*degeneracy*) of values $\varepsilon_n(\boldsymbol{k})$ and $\varepsilon_{n'}(\boldsymbol{k})$ at the same points of space is possible. The locations of such points are related to the crystal symmetry. The occupation of energy levels by electrons obeys *Fermi–Dirac statistics*. In pure semiconductors and insulators the *Fermi level* falls within the forbidden gap that separates the *valence band* from the *conduction band*. In metals and doped semiconductors the Fermi level is in an allowed band. In the vicinity of a band minimum or maximum the *isoenergetic surfaces* of electrons ($\varepsilon_n(\boldsymbol{k}) = \text{const}$) assume the shape of spheres or ellipsoids, which means that the dispersion law of electrons is quadratic with respect to the *quasi-momentum* $\hbar\boldsymbol{k}$. Hence, near the band extrema a band electron is similar to a free electron, but with an *effective mass* m^* defined by the expression

$\varepsilon_n(\boldsymbol{k}) = \hbar^2\boldsymbol{k}^2/2m^*$. In diamond (Ge) and zinc-blende (GaAs, ZnS, etc.) type tetrahedral semiconductors the effective mass ratio $m^*/m_0 < 1$ for electrons and holes (m_0 is the free electron mass). In the general case the effective mass is anisotropic, or more precisely, the dispersion law is characterized by the inverse effective mass tensor.

ENERGY-DISPERSION X-RAY DIFFRACTOMETRY

Method for investigating the structural state of materials. Energy-dispersion X-ray diffractometry is similar to the *Laue method*, and consists of irradiating the sample by an X-ray beam with a continuous spectrum followed by detection of the diffracted radiation with wavelengths satisfying the *Bragg law* by a semiconductor detector (Si(Li) or Ge(Li)). The exposure is carried out with a stationary radiation source, detector, and sample.

The semiconductor detector produces electrical signals with amplitudes proportional to the quantum energy of the diffracted radiation. After an analog–digital transformation, the signals are accumulated by a multichannel pulse analyzer, and reproduce the picture of the diffraction spectrum. The interplanar distance d (nm), γ-quantum energy E (keV) and diffraction angle θ are related by the expression $d = 0.6199/(E\sin\theta)$. Energy-dispersion X-ray diffractometry allows one to perform a simultaneous synthesis of the data from the required spectral range within a few minutes at the resolution $\Delta d/d \approx 10^{-4}$. It can be used for the purpose of a rapid analysis, for studying the intensity distribution of diffracted irradiation with high values of the scattering vector, and the influence of various external actions.

ENERGY GAP in superconductors

Basic characteristic of the energy spectrum of a superconductor associated with the minimal energy needed for the appearance of quasi-particles. The energy of quasi-static excitations ε_k (in the isotropic model) at $T < T_c$ has the form $\varepsilon_k = [\xi_k^2 + \Delta^2(T)]^{1/2}$ where $\Delta(T)$ is the *energy gap*, and ξ_k is the electron energy relative to the Fermi surface. In the *Bardeen–Cooper–Schrieffer theory* (BSC theory) $2\Delta(T)$ is the energy needed for breaking up a *Cooper pair*. The energy gap grows with decreasing temperature from the value $\Delta(T_c) = 0$ at the transition temperature to the superconducting state T_c to its final value $\Delta(0)$ at $T = 0$. In the BCS model this has the value $\Delta(0) = 1.76k_BT_c$. In a superconductor with strong electron–phonon coupling (Pb, Nb, etc.) the ratio $2\Delta(0)/(k_BT_c)$ can exceed the theoretical value. A consequence of the existence of the energy gap is the exponential dependence of the specific heat and the electronic absorption coefficient of ultrasound, as well as some other characteristics of superconductors, on the temperature at $T < T_c$. The most direct method of measuring the energy gap utilizes the *tunnel effect*. At $T = 0$, the contact tunnel current between a superconductor and a normal metal appears only at the voltage $eV = 2\Delta$. In the case of a contact between different superconductors with gaps Δ_1, Δ_2, the current produces a jump (at zero temperature and $T > 0$) at $eV = -\Delta_1 + \Delta_2$.

ENERGY LOSS, CHARACTERISTIC

See *Characteristic energy loss, Characteristic energy loss spectroscopy.*

ENERGY LOSS OF CHARGED PARTICLES

See *Charged particle energy loss.*

ENERGY OF SURFACE

See *Surface energy.*

ENERGY TRANSFER in crystals

The vector of an energy flux density (EFD) $\boldsymbol{S}$ is numerically equal to the quantity of energy transferred per unit time through a unit surface area which is perpendicular to the direction of the energy flow at a given point. In a crystal (solid state body) there are as many different EFD vectors as there are types of quasi-particles existing in it. A general expression for EFD by a j-type quasi-particle is $S_j = \sum_j \varepsilon_j(\boldsymbol{k})v_j(\boldsymbol{k})n_j(\boldsymbol{k})$. Here $\varepsilon_j(\boldsymbol{k})$, $v_j(\boldsymbol{k})$, and $n_j(\boldsymbol{k})$ are, respectively, the energy, group velocity, and distribution function of j-type quasi-particles for the same state $\boldsymbol{k}$. The EFD associated with lattice vibrations, electromagnetic waves, magnetic vibrations, thermal energy, etc., can be presented in terms of macroscopic quantities: *stress tensor* σ_{ji}, *displacement vector* $\boldsymbol{u}_i$, electric and magnetic field strengths $\boldsymbol{E}$ and $\boldsymbol{B}$, respectively, exchange constant A, magnetization $\boldsymbol{M}$, thermal conductivity tensor $\chi_{ik} = \chi_{ki}$. Formulae for these EFD relations are:

(a) *Umov vector* of EFD in lattice (1874):

$$S_i = -\sigma_{ij}\frac{\partial u_j}{\partial t};$$

(b) *Poynting vector* for EFD of electromagnetic field (J. Poynting, 1884):

$$S_i = [\boldsymbol{E} \times \boldsymbol{H}]_i;$$

(c) *Akhiezer vector* for EFD of magnetic waves (spin waves, 1958):

$$S_i = -A\frac{\partial^2 M}{\partial t \partial x};$$

(d) thermal energy flux:

$$S_i = -\chi_{ik}\frac{\partial T}{\partial x_k};$$

here T is the temperature; and for a crystal of cubic symmetry and an isotropic medium $\chi_{ik} = \chi\delta_{ik}$.

At *diffusion*, including *thermodiffusion*, every migrating particle carries the transferred heat q, so the energy flux density S_i is given by $S_i = qJ_i$, where J_i is the particle flux density. The quantity q may assume both positive and negative values, and in the context of a microscopic theory q is the height of a potential barrier which is to be overcome by the particle during its migration.

ENERGY TRANSFER BY EXCITONS

The transfer of excitation energy from *excitons* to local impurity and defect centers, as well as to the *surface states* of a crystal. The efficiency of such energy transfer depends on the mobility of the excitation in the crystal. The nature of energy transfer by excitons is determined by the relationship among the chief kinetic temporal parameters: the lifetime of an exciton in a perfect lattice τ_0, the time of exciton scattering (by *phonons*, local centers, etc.) τ_s, and the lifetime with respect to trapping the excitation on an *acceptor* τ_l. If $\tau_l < \tau_s$, τ_0, then the transfer is performed coherently at a rate equal to the *group velocity* of an exciton in a band. If the scattering time τ_s turns out to be the smallest of these parameters, then the process is diffusive in nature. The mean free path of an exciton in the case of strong scattering is of the order of a lattice constant ($l \propto a$), and energy transfer by excitons takes place in the form of random jumps or hopping from one lattice site to another. Such a pattern is characteristic of the exciton motion in crystals with narrow bands, or at the presence of impurities or defects in large quantities. The value of the *diffusion coefficient* D in this case is $D \sim 10^{-3}$ cm^2/s. If excitons are scattered weakly, $l \gg a$, then their motion is related to energy bands, and the diffusion coefficient is several orders of magnitude higher: from 1 cm^2/s (*alkali-halide crystals* and *cryocrystals*) to 10^3 cm^2/s (semiconductors with "light" excitons). For the *diffusion* of localized excitons $\ln D \propto T^{-1}$. In the case of free excitons, $D \propto T^{-1/2}$ if the transfer is performed by thermalized excitons, and $D(T) = \text{const}$, if the transfer occurs via "hot" excitons. If there is a strong exciton–exciton interaction in the crystal, then the probability of *energy transfer* by many-particle bound states: *biexcitons*, exciton drops, etc. should be taken into account.

ENHANCED RAMAN SCATTERING OF LIGHT

A comparatively strong *Raman scattering of light* by various molecules adsorbed on a *metal* surface. The scattering cross-section per molecule is 5 to 6 orders of magnitude higher than the scattering cross-section of the same molecule in solution. This is observed on the rough surface of a metallic electrode in an electrochemical cell or in a vacuum, on a metal surface with especially prepared relief, in tunnel structures, in metallic *granular films*, or on colloidal particles. Despite the great diversity of the objects involved, the observed phenomena have some common features. First, enhanced Raman scattering exists for molecules in sharply heterogeneous conditions. The most favorable dimension of a heterogeneity (size of roughness, metallic islands, colloid particles) for enhanced scattering lies in the range 10 to 100 nm. Second, the dependences of the scattering cross-section on the exciting-light frequency have the form of smooth curves which is evidence of the non-resonant nature of the giant enhancement.

A possible cause of Raman scattering enhancement is the growth of the *polarizability* of an adsorbed molecule due to its interaction with a metal substrate (molecular-chemical amplification). Estimates show that in this case the amplification does not exceed 10–10^2. A second possible cause of giant enhancement is related to the electric field amplification in the vicinity of a metal surface (electrodynamic amplification). Field amplification occurs due to its increase near surface protrusions, in microscopic cracks, etc., which is connected, e.g., with the excitation of localized *surface plasmons.* An enhanced Raman scattering occurs in media with a *dielectric constant* varying from negative to positive values (*anomalous dispersion region* in crystals), as well as in the transition regions in *semiconductors* between the those with low and high free carrier concentrations. This scattering process has important practical significances for surface quality control, heterogeneous catalysis, and investigations of biological objects.

ENTHALPY, heat content

A thermodynamic potential H defined by the expression $H = U + PV$, where U is the internal energy, P is the pressure, and V is the volume of the system. Enthalpy is called heat content because the increase in enthalpy in a constant pressure process is equal to the quantity of heat absorbed. See *Thermodynamic potentials.*

ENTRAINMENT OF ELECTRONS BY PHONONS

Directional drift of free *current carriers* in *semiconductors* or *metals* that occurs provided there is a nonequilibrium *phonon* flow in these materials to entrain or carry along the electrons. This directional drift of charge carriers is brought about by the momentum transfer from phonon to electron that takes place via the *electron–phonon interaction.* The entrainment process is most complete when the prevailing electron–phonon collisions are elastic ones, i.e. the collisions between electrons and phonons proceed with conservation of *quasi-momentum.* During the process of entrainment, the drift distribution (see *Current carrier drift*) of *quasi-particles* is established, which specifies the drift velocity common to both types of quasi-particles. These conditions are difficult to attain in typical metals which are readily susceptible to *Umklapp processes* at electron–phonon collisions down to low temperatures. More characteristic of pure metals at temperatures $T \leqslant T^*$ (the T^* temperature amounts to approximately one tenth of the *Debye temperature*) is the situation of incomplete entrainment. At $T \gg T^*$, the frequency of phonon–phonon collisions involving transport with anharmonic coupling exceeds the frequency of phonon–electron collisions, and hence the entrainment of electrons becomes attenuated. As shown by Peierls, the electric resistance of a metal (see *Electrical conductivity*) under conditions of complete entrainment of electrons by phonons decays exponentially with decreasing temperature for uncompensated metals with closed *Fermi surfaces* (this electric resistance behaviour is only known for Na and K). The entrainment of electrons by phonons is of considerable importance in *thermoelectric phenomena* and *thermomagnetic phenomena* (the theory was formulated by L.E. Gurevich in 1945). The sign of the phonon contribution to the thermoelectromotive force is determined by the sign of the current carriers (electrons and holes) in semiconductors, and by the orientation of the Fermi surface curvature in metals. The entrainment of electrons by phonons increases steeply in a *quantizing magnetic field.* The phenomenon of entrainment of electrons by phonons is also referred to as *phonon wind.*

ENTRAINMENT OF ELECTRONS BY PHOTONS

Generation of an electron flow in a solid conductor, which occurs due to momentum transfer from the directional flux of *photons* to electrons. The phenomenon of entrainment or carrying along of electrons by photons is observed in the optical and very high frequency energy bands of *semiconductors, semimetals* and certain *metals.* This phenomenon is most extensively studied in semiconductors (Ge, Si, compounds of $A^{III}B^{V}$ type), where the entrainment of bound electrons (*photoionization*), or of *conduction electrons* and *holes* occurs. A considerable part of the momentum of photons, which is passed on to the solid as a whole, is originally taken up by the mobile charge carriers, resulting in their induced motion. Since the photon momentum is equal to the sum of momenta acquired by the lattice and the electrons, it is possible for the momentum acquired by an electron to be opposite in sign to that of the photon. The entrainment of electrons by photons is detected by the presence of a current (*entrainment current*) or electromotive force. This phenomenon is used for determining time characteristics of pulse *laser* radiation, and for detecting IR radiation.

ENTROPY

One of the thermodynamic variables that characterizes the state of a macroscopic body or system of bodies, and the direction of flow of nonequilibrium processes; a measure of the disorder in a system, i.e. a measure of its distribution over available states. The ordinary symbol for entropy is S. The variation of the entropy $\mathrm{d}S$ is related to the incremental amount of heat received by a body $\mathrm{d}Q$ by the expression $\mathrm{d}S = \mathrm{d}Q/T$. According to the *first law of thermodynamics* (law of conservation of energy) the variation of the internal energy of a body (see *Thermodynamic potentials*) is given by $\mathrm{d}E = T\,\mathrm{d}S + \mathrm{d}W$, where $\mathrm{d}W$ is the work done on the body. The entropy is a function of the state variables, which means that after a nondissipative cyclic process the body entropy returns to its initial value. The entropy can also be chosen as an independent variable specifying the system state through the expression $\mathrm{d}H = T\,\mathrm{d}S + V\,\mathrm{d}P$ for the enthalpy H. The entropy is related to the Helmholtz free energy F and to the Gibbs free energy G through partial derivatives:

$$S = -\left(\frac{\partial F}{\partial T}\right)_V = -\left(\frac{\partial G}{\partial T}\right)_P.$$

Entropy is an extensive variable, so the entropy of a system composed of different bodies is the sum of the entropies of individual bodies. More specifically, if the observation time exceeds the *relaxation time* in separated portions of the body but is less than the time for establishing equilibrium in the system as a whole, then one can speak about the entropy of separated portions of the body, and about its additivity as a whole. According to the *second law of thermodynamics*, the entropy of a closed system can either grow (irreversible process) or remain constant (reversible process). As a result, without doing work over the system, the heat can pass only in the direction from bodies at higher temperatures to bodies at lower temperatures.

If an irreversible process takes place in the body itself (e.g., a chemical reaction), its entropy also increases; in an equilibrium state the entropy is a maximum with respect to the parameters which specify the state of the body. If the body is thermally isolated, and the environmental conditions vary quite slowly, then during such a process, called an *adiabatic process*, the entropy of the body remains unchanged, i.e. the process is reversible.

According to the *third law of thermodynamics* (*Nernst heat theorem*), the entropy of a body at absolute zero temperature diminishes to zero. At fixed values of pressure and volume, the entropy of the body takes the form

$$S = \int_0^T \frac{1}{T} C_{P,V}\,\mathrm{d}T, \tag{1}$$

where $C_{P,V}$ is the *specific heat* at constant pressure, or constant volume, respectively.

In statistical physics the entropy is a measure of the system disorder, and this is determined by the statistical weight or the degeneracy (multiplicity) of a given equilibrium state of the system: $S = \ln \Delta\Omega$, where $\Delta\Omega$ is the number of quantum-mechanical states (in the vicinity of the average energy $\overline{E}$) within a range with a dimension of the order of the energy *fluctuation* of the system under consideration. Using a knowledge of the system distribution of states function W_n (here n is the number of the state) the expression for the entropy can be written as

$$S = -\sum_n W_n \ln W_n, \qquad \sum_n W_n = 1. \tag{2}$$

Within the framework of the canonical *Gibbs distribution*, the expression for W_n takes the form

$$W_n = \frac{1}{Z}\exp\left(-\frac{E_n}{k_\mathrm{B}T}\right),$$

where

$$Z = \sum_n \exp\left(-\frac{E_n}{k_\mathrm{B}T}\right)$$

is the *partition function*, E_n is the energy of the nth state of the system, and k_B is the *Boltzmann constant*.

The law for the increase of entropy in statistical physics means that the evolution of the system is directed toward the most probable and uniform distribution of energy over individual levels. An approach for justifying the law of increasing entropy involves the *Boltzmann H-theorem*.

EPITAXIAL DISLOCATIONS, mismatch dislocations

Dislocations that emerge during the growth of a crystal on a substrate made of another crystalline material. Epitaxial dislocations appear at the interface between the substrate and the layer which is growing in an epitaxial fashion (see *Epitaxy*) as a result of incomplete matching of their structures.

EPITAXIAL IRON GARNET FILMS

Magnetic films of 0.1 to 100 μm thick, monocrystalline layers of pure or mixed *iron garnets* obtained by *epitaxial* growth (from gaseous or liquid phase) on substrates of nonmagnetic garnets (gadolinium–gallium garnet $Gd_3Ga_5O_{12}$, calcium–gallium–germanium garnet $Ca_3Ga_2Ge_3O_{12}$, and some others). Depending on its purpose, various epitaxial iron garnet films are used. The most widespread are films of *yttrium iron garnet* $Y_3Fe_5O_{12}$ which are employed in superhigh-frequency equipment; films with *magnetic bubble domains* of various compositions (e.g., $(Sm,Lu)_3Fe_5O_{12}$, $(Eu,Tm)_3(Fe,Ga)_5O_{12}$ and others) for storage devices (see *Magnetic bubble domain devices*), as well as films of bismuth-containing or praseodymium-containing iron garnets with a large specific Faraday rotation (10^4 deg/cm and more) for magnetooptic applications (see *Faraday effect*, *Magnetooptics*). To obtain high-quality films it is necessary to provide good alignment of the lattice parameters of the film with those of the substrate within an accuracy of ~0.1 nm. The *magnetic anisotropy* of epitaxial iron garnet films can differ essentially from the cubic (crystallographic) magnetic anisotropy of bulk *monocrystals* because of the presence of *growth anisotropy* and magnetostriction contributions (due to mismatch of film and substrate lattice parameters). Thus, in films with magnetic bubble domains the dominant uniaxial magnetic anisotropy is the normal anisotropy that is introduced during the growth process.

EPITAXY

Oriented growth of crystals on monocrystalline substrates. The term epitaxy was proposed by L. Royer (1928). Initially *epitaxial films* made of various materials served only as objects of scientific investigation. In 1951 the first epitaxial germanium *p–n* junctions (see *Semiconductor junction*) were fabricated, and within a decade epitaxial semiconductor film growth techniques became highly developed for engineering applications. Since 1960 when epitaxial transistors began being produced epitaxy has become one of the basic technological methods of semiconductor electronics. Depending on the kind of device, epitaxial films can be obtained by the methods of gas, liquid, or solid phase epitaxy, *molecular beam epitaxy*, vacuum deposition, as well as laser, cathode, and magnetron sputtering (see also *Autoepitaxy*).

The methods of *gas phase epitaxy* are based on the use of chemical reactions (synthesis, dissociation, and disproportionation) which bring about the formation of the epitaxial layer at the *substrate* surface. For *liquid-phase epitaxy* (*rheotaxy*), the material of the layer separates out on the substrate in the form of a solution or melt which provides direct contact with the substrate. In the case of *solid-phase epitaxy*, the material of the epitaxial layer diffuses towards the substrate through a layer of crystalline material. Finally, for molecular beam epitaxy and other vacuum methods, the atoms of deposited substances are transported to the substrate after evaporation or sputtering from vacuum sources.

The initial period of epitaxial growth involves a nucleation stage, and this is described by a thermodynamic theory based on the minimum of the free energy of the critical nucleating center. Under the conditions of strong oversaturation when the nucleating center consists of only a few atoms the theory loses its validity. It is for this case that D. Walton and T. Rhodin (1962) proposed a statistical mechanical model where, instead of utilizing free energies, statistical sums and potential energies are used. Of importance for the theory of epitaxy is the problem of mismatch between the crystal lattices of the film and the substrate. Royer put forward a scale criterion which suggests the existence of a threshold mismatch. According to the modern viewpoint, in the case of moderate mismatching (<0.1) the film grows initially on the substrate in a pseudomorphous manner (interphase boundary is coherent) which corresponds to the free energy minimum. Thicker films have a larger elastic strain energy related to the stress of mismatching, so after reaching a critical thickness it becomes possible to relax the inner tension through the formation of *dislocations* or in other ways (domainization, *twinning of crystals*), i.e. a breakdown of coherence takes place. Another type of epitaxial growth is the *recrystallization* of amorphous or polycrystalline films during their heating by optical radiation, a laser beam, or an electron beam.

See also *Amorphous epitaxy, Gas phase epitaxy, Liquid-phase epitaxy, Molecular beam epitaxy.*

EPR IMAGING, EPR endoscopy

Set of methods for studying internal spatial characteristics of a subject using *electron paramagnetic resonance* (EPR). A typical task of EPR imaging is the investigation of the *paramagnetic center* distribution in a bulk sample. In addition, EPR imaging studies the macroscopic and *point defect* distribution, the variation of spatial characteristics under external action (irradiation, annealing), and so on. Various methods for spatial focusing are employed, e.g., by utilizing plane resonators with narrow gaps between walls, or loop gap resonators which concentrate the microwave field, etc. However, the most common method used in EPR imaging (taken from *magnetic resonance imaging*, or MRI) is the detection of EPR signals in systematically nonuniform magnetic fields, i.e. $\boldsymbol{B}_0(x, y, z)$ has a known configuration. The essence of this method is the following. First we consider an EPR signal in a uniform magnetic field $\boldsymbol{B}_0$ (no x, y, z dependence) with the frequency dependence $I(\omega) = I_0 g(\omega - \omega_0)$, where $g(\omega - \omega_0)$ is the line shape function (with peak frequency ω_0 and width δ), I_0 is the signal amplitude: $I_0 = \int_V k\rho(x, y, z)\,dx\,dy\,dz$, k is a dimensional factor, and the function $\rho(x, y, z)$ provides the distribution of paramagnetic centers over the volume V. In a nonuniform magnetic field the EPR signal is determined by the convolution integral

$$I(\omega) = \int_V k\rho(x, y, z) g\big[\omega - \omega_0(x, y, z)\big]\,dx\,dy\,dz, \tag{1}$$

where $\omega_0(x, y, z) = \gamma B_0(x, y, z)$ is the resonance frequency which depends on the position, and γ is the *gyromagnetic ratio*.

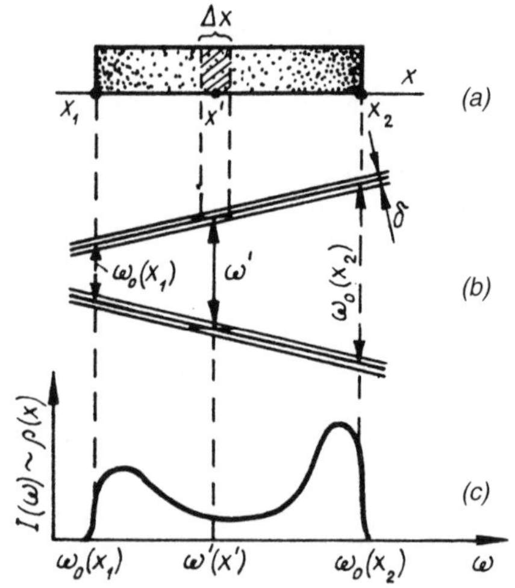

A sample with a nonuniform distribution of paramagnetic centers represented by dots (a), energy levels of paramagnetic centers in a nonuniform magnetic field (due to ∇B_0) with energy level spacings that depend on the coordinate x (b), and the resulting EPR signal (c).

Let $\nabla B_0 = \partial B_0/\partial x$, corresponding to the case $\omega_0(x) = \gamma B_0(x)$ in which the resonant frequency has no y or z dependence. If $\omega_0(x_2) - \omega_0(x_1) = \gamma(\partial B_0/\partial x)(x_2 - x_1) \gg \delta$ (see Fig.) then at a certain arbitrary frequency $\omega = \omega'$, a contribution to the EPR signal will arise from those centers with coordinates $x = x' \pm \Delta x/2$ and arbitrary coordinates y and z. In other words, the signal has no y or z dependence. The value Δx which determines the resolving ability of EPR imaging satisfies the relation $\Delta x \approx (\delta/\gamma)(\partial B_0/\partial x)^{-1}$. In this case the profile of the EPR signal determines the projection of *spin density* $\rho(x, y, z)$ on the axis Ox along which ∇B_0 is directed, i.e. $I(\omega) \sim \rho(x)$. Using various orientations of ∇B_0 one can sequentially obtain projections of $\rho(x, y, z)$ for different directions, and then find the overall $\rho(x, y, z)$ with the help of reconstructive *tomography*.

The spin density $\rho(x, y, z)$ can be found with the application of diverse nonuniform external actions such as pressure, temperature, electric field, etc. If ω_0 is dependent on the electric field, then for a nonuniform field $E(x, y, z)$ one can find $\rho(x, y, z)$ using Eq. (1) where $\omega_0(x, y, z)$ is determined by the *electric field gradient* (but not by the magnetic field gradient). With the help of subsequent localization of the field E in limited portions of the sample (e.g., using pointed electrodes) one can "extract" from an overall EPR signal of the sample the signal from the region of the field localization. This problem of obtaining the EPR signal from a region of field localization can be also solved by using an alternating electric field as a modulating field (i.e. in the absence of magnetic field modulation). In addition to studying centers with an ω_0 dependence on E, this method also applies for studying reorienting centers for which the field E changes the intensity of various EPR lines. For a modulating field E, the EPR signal is produced only by portions of the sample where the field E is localized.

EPSILON EXPANSION (ε-expansion)

Method of analysis of the system behavior in a *critical region*, based on calculations of its characteristics (*critical exponents*, ratios of critical amplitudes, and so on) in the form of a power series in $\varepsilon = d_c - d$, where d is the system dimensionality, and d_c is the so-called *upper critical dimensionality* corresponding to the minimal value of d at which the critical fluctuations become insignificant, i.e. the *Landau theory of second-order phase transitions* is valid. This was proposed by K.G. Wilson and M.E. Fisher (1972), and represents the most widespread (but not the only) version of analytical mean field perturbation theory used for studies of *critical phenomena*. The epsilon series expansions are asymptotic for $d_c = 4$ for the majority of models describing magnetic, structural, superconducting, and other *phase transitions*. Nevertheless the cutoff of these series at the second or third term with the substitution $\varepsilon = 1$ yields results that are in quite good agreement with experiment. For example, in the *Ising model* (uniaxial ferromagnet) the epsilon expansion of the critical index for the susceptibility γ has the form:

$$\gamma = 1 + \frac{1}{6}\varepsilon + \frac{25}{324}\varepsilon^2 + \cdots;$$

the substitution $\varepsilon = 1$ yields $\gamma = 1.24$, and experimental values are in the range 1.20–1.25. Taking into account additional terms of the expansion worsens the result; however, the use of methods for summing asymptotic series provides exact values of critical indices not only at $\varepsilon = 1$, but also at $\varepsilon = 2$. In the theory of systems with continuous symmetry groups another version of the epsilon expansion is often adopted, corresponding to an expansion in powers of $\varepsilon = d' - d'_c$, where d'_c is the *lower critical dimensionality*, that is the maximum dimensionality at which the thermal fluctuations of the *order parameter* destroy the ordered phase, i.e. $T_c = 0$ K. This variant had been proposed in 1975 by A.M. Polyakov, and it allows one to find critical characteristics of three-dimensional systems as a "rise" in dimensionality from $d'_c = 2$ (*Heisenberg model*) or $d'_c = 1$ (Ising model). In addition to its application in the theory of phase transitions, the epsilon expansion is used in *percolation theory*, the theory of localization, and in some other areas of condensed matter theory.

EPSILON PHASE (ε-phase)

Metallic phase with a two-layered hexagonal close-packed lattice. Depending on the conditions of formation, equilibrium and metastable (martensite) ε-phases can be distinguished. The equilibrium ε-phase appears during the course of *crystallization* in the form of an electronic compound (see *Hume-Rothery phases*) with an electron concentration (ratio of number of electrons to number of atoms) $e/a = 7/4$ (e.g., in alloys Cu–Zn and other alloys based on copper), or as the polymorphous modification of some metals at *high pressures* (see *Polymorphism*). For example, in pure iron an ε-phase appears at pressure exceeding $135 \cdot 10^8$ Pa as the third polymorphous modification: from the *alpha phase* it appears under pressure, and from the *gamma phase* it appears upon cooling. A martensite ε-phase is formed without diffusion from *austenite* at low temperatures in alloys with low *stacking fault* energy (e.g., Fe–Mn, Fe–Mn–C, Fe–Ru). Upon heating in the range of temperatures $\sim$430 to 670 K the ε-phase transforms into austenite, i.e. an inverse *martensitic transformation* $\varepsilon \to \gamma$ takes place. The *crystal lattice* of ε-phase *martensite* is naturally oriented with respect to the austenite lattice: $(0001)_\varepsilon \parallel (111)_\gamma$ and $[1\bar{2}10]_\varepsilon \parallel [0\bar{1}1]_\gamma$. As a result, the ε-phase crystals are located in the form of plates parallel to the $\{111\}$ octahedral planes of austenite. The microstructure of alloys in the martensite ε-phase is similar to the *Widmanstätten structure* where sheets of ε-phase are interlaced with sheets of residual austenite. The *hardness* of the martensite ε-phase is higher (to 40%) than the austenite hardness, the *yield limit* is 2 to 3 times higher, and the *plasticity* is lower than the corresponding values for austenite. The maximum amount of martensite ε-phase in iron–manganese alloys is 70 to 80%, and at *strain* it reaches 100%. Upon *plastic deformation* or *hydroextrusion* the ε-phase transforms into *alpha martensite*. The repeated transition of austenite into the ε-phase and the inverse transformation lead to *phase wear hardening*, i.e. a type of *hardening* of the alloy. Multiple $\gamma \leftrightarrow \varepsilon$ transitions in iron–manganese alloys do not change the hexagonal symmetry of the ε-phase, and the addition of 0.25 to 0.60% of *carbon* to them leads to the formation of an 18-layered martensite phase. The ε-phase can be detected with the help of *X-ray structure analysis* and *metallographic analysis*.

EQUATIONS OF STATE

Equations which describe the relationship between the variables that characterize the state of a *thermodynamic system*, such as a gas, a liquid, or a *crystal*. *Thermal equations of state* relate intensive quantities (independent of system volume) to extensive ones (proportional to the system volume), and to the *temperature* T, e.g., $f(P, V, T) = 0$, where P is *pressure*, V is volume. *Caloric equations of state* represent *thermodynamic potentials* (free energy, enthalpy, etc.) or *entropy* as functions of pressure and temperature (or V and T, etc.). The relation between thermal and caloric equations of state is established by the application of fundamental laws of thermodynamics. Equations of state, which serve as a necessary connecting link that allows applying thermodynamic laws to particular systems, cannot be derived from fundamental thermodynamic laws alone. There are no universal equations of state for solids, rather the role of these equations is played by certain approximate expressions obtained on the basis of simple theoretical models or experimental data. These theoretical models are, e.g., the Debye theory of *specific heat*, and the lattice model of crystal vibrations developed either with or without regard for anharmonicity (see *Crystal lattice vibrations*). The role of empirical laws may be played by, e.g., the temperature and pressure dependences of *elastic moduli*, or by other relationships, depending on the process under investigation.

EQUILIBRIUM OF PHASES

See *Phase equilibrium*.

EQUILIBRIUM SEGREGATION

A *metastable state* of a solid containing impurities and their traps (*dislocations*, *stacking faults*, *grain boundaries*, *phase interfaces*, etc.). In such a state, the content of *impurity atoms* differs in the material matrix and at its structural defects. It is determined by the conditions of minimum *thermodynamic potential* yielded by the impurity concentrations at traps, and by the free parameters of structural defects (e.g., by the width of the stacking faults between *partial dislocations*). Depending on the value of the bonding *enthalpy* between the impurities and the traps one may consider competitive *segregation* of the various impurities to a given trap, and competition between various traps in the process of pinning a given impurity. In case the number of atoms of an impurity is smaller than or comparable to the number of the positions available for them at traps, such competition may lead to nonmonotonic temperature dependences of the content of a given impurity at a given trap. Such dependences exhibit a minimum at a temperature $T = T_{\min}$ and a maximum at a temperature $T = T_{\max} > T_{\min}$. Equilibrium segregation lowers the impurity content in the matrix and, hence, lowers the temperature of dissolution of carbides, nitrides, and other phases. Equilibrium impurity segregation affects the strength and plastic properties of materials (see *Strength*, *Plasticity*), and the local electrophysical properties of semiconductors.

EQUILIBRIUM, THERMODYNAMIC

See *Thermodynamic equilibrium*.

ERBIUM, Er

Chemical element of Group III of the periodic system with atomic number 68 and atomic mass 167.26; belongs to *lanthanides*. There are 23 known isotopes with mass numbers from 151 to 173; naturally occurring isotopes are: ^{162}Er (0.14%), ^{164}Er (1.61%), ^{166}Er (33.6%), ^{167}Er (22.95%), ^{168}Er (26.8%), ^{170}Er (14.9%). Outer shell electronic configuration is $4f^{12}5d^06s^2$. Atomic radius is ≈0.1754 nm; radius of Er^{3+} ion ≈0.0886 nm. Oxidation state +3, electronegativity ≈1.2.

Erbium is a silvery-white *metal*. At low temperatures erbium has a hexagonal close-packed crystal lattice, space group $P6_3/mmc$ (D_{6h}^4), with parameters $a = 0.35592$ nm, $c = 0.5585$ nm (at room temperature). At ≈1690 K the polymorphous transformation of erbium into a body-centered cubic modification is possible, space group $Pn\bar{3}n$ (O_h^2), with parameter $a = 0.3940$ nm (extrapolated value) after hardening. Density is 9.051 g/cm^3 at 293 K, $T_{\text{melting}} \approx 1790$ K, $T_{\text{boiling}} \approx 2865$ K. Heat of melting is 17.2 kJ/mole, heat of sublimation is 280.5 kJ/mole, heat of evaporation is 271 kJ/mole; specific heat is 0.168 kJ·kg^{-1}·K^{-1} at room temperature. Debye temperature is ≈180 K; linear heat expansion coefficient of erbium polycrystal is ≈$10.8\cdot10^{-6}$ K^{-1}; thermal conductivity coefficient is 9.6 W·m^{-1}·K^{-1}. Adiabatic elastic moduli rigidity of erbium polycrystal: $c_{11} = 83.88$, $c_{12} = 29.26$, $c_{13} = 24.21$, $c_{33} = 84.32$, $c_{44} = 26.45$ (GPa) at 300 K; adiabatic bulk modulus is 44.913 GPa, Poisson ratio is 0.238. Vickers hardness is ≈55 HV; tensile strength is 0.280 GPa. Activation energy of self-diffusion of atoms in erbium polycrystal at the melting temperature is 300 kJ/mole. Effective thermal neutron trapping cross-section is 162 barn. Resistivity of erbium monocrystal (at 298 K) is 750 nΩ·m along the principal axis $\underline{6}$ and 440 nΩ·m in the perpendicular direction, and that of erbium polycrystal is ≈940 nΩ·m, thermal coefficient of electric resistance is 0.00201 K^{-1}. Work function of erbium polycrystal is 3.12 eV. At the Néel point $T_N = 84.4$ K erbium transforms from paramagnetic state to antiferromagnetic state with sine wave-modulated component of magnetic moment along the principal $\underline{6}$ axis and with disordered components of magnetic moment in the basal plane. Below the critical temperature $T_c = 52.4$ K the modulation of the magnetic moment along the principal axis $\underline{6}$ is conserved, but there appears helicoidal ordering of the perpendicular component of the magnetic moment in the basal plane (of the complex helical type). At the Curie point T_C (18.0 K < T_C < 19.6 K) a first-order phase transition yields ("freezes out") the sinusoidal modulation of the magnetic moments with respect to the $\underline{6}$ axis: the magnetic moments are oriented at the angle ~29.6° to the $\underline{6}$ axis, and the perpendicular components conserve the helicoidal ordering in the basal planes (i.e. cone-like structure of the ferromagnetic type of spiral is formed). Magnetic susceptibility of erbium is $+44500\cdot10^{-6}$ CGS units; nuclear magnetic moment of isotope ^{167}Er (22.95% abundant) is 0.48 nuclear magnetons. Pure erbium is used for the purposes of investigations, and its alloys (with cobalt, iron and nickel) are used for manufacturing *permanent magnets*, etc.

EROSION, DROP-IMPACT

See *Drop-impact erosion.*

EROSION, SURFACE

See *Surface erosion.*

ESAKI DIODE

See *Tunnel diode.*

ETCHING

A method of chemical or electrochemical treatment of the surface of solid materials aimed to remove dirt and oxides, to reveal the material structure, to produce a desired microstructure of the surface, to remove the *surface layer* disturbed by the mechanical processing, and so on. *Chemical etching* is brought about by the direct interaction of the material surface with the etching agent solution, or with an appropriate reagent in the gaseous phase. At *electrochemical etching*, the treated surface is submerged into an electrolyte solution and used as an electrode when passing direct or alternating current through the solution. See also *Selective etching, Ion etching.*

ETCHING GROOVE, THERMAL

See *Thermal etching groove.*

ETCHING, ION

See *Ion etching.*

ETCHING, SELECTIVE
See *Selective etching*.

EULER ANGLES, Eulerian angles

Three angles ψ, θ, and φ which allow one to unambiguously define the orientation of a solid (rigid body) having internal coordinates x, y, z with respect to a fixed (laboratory) system of coordinates ξ, η, and ζ (see Fig.). In one common convention the Eulerian angles are the angle ψ between the ξ axis and a line ON of intersection of planes $(\xi\eta)$ and (xy). This line ON is called the *line of nodes*, and ψ is the *precession angle*. The angle between planes $(\xi\eta)$ and (xy) (or between axes ζ and z) is the *nutation angle* θ. The angle between the x axis and the line of nodes ON is the *spinning angle* or *rotation angle* φ. When the solid undergoes rotational motion all the angles can be changing; and the individual *angular velocities* corresponding with each of these angles are as follows: $\dot{\psi}$ is directed along axis ζ, $\dot{\theta}$ is directed along line of nodes ON, $\dot{\varphi}$ is directed along axis z. The angular velocity ω of the solid is the vector sum $\dot{\psi}+\dot{\theta}+\dot{\varphi}$ with the components

$$\omega_x = \dot{\psi}\sin\theta\sin\varphi + \dot{\theta}\cos\varphi,$$

$$\omega_y = \dot{\psi}\sin\theta\cos\varphi - \dot{\theta}\sin\varphi,$$

$$\omega_z = \dot{\psi}\cos\theta + \dot{\varphi}.$$

These are the kinematic *Euler equations*. The angle θ varies in the range from 0 to π, and angles ψ and φ vary from 0 to 2π. (The standard convention in the Western literature is to interchange the symbols φ and ψ so that φ is the precession angle, and ψ is the spinning angle.)

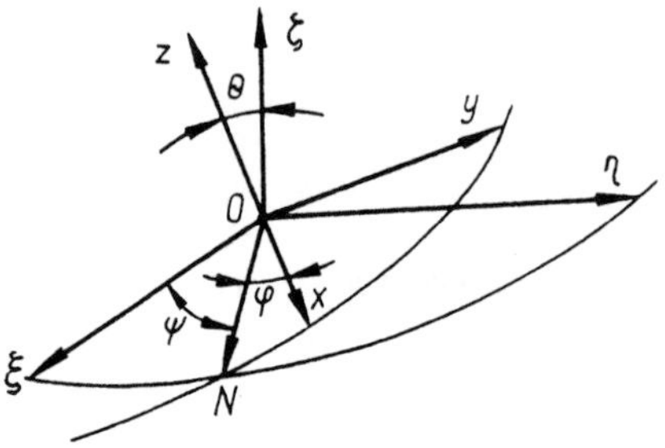

Euler angles.

EULERIAN COORDINATES (L. Euler, 1765)

Three components of the velocity of a system aligned parallel to fixed coordinate axes. These coordinates are used in *continuum mechanics*.

EUROPIUM, Eu

Chemical element of Group III of the periodic table, atomic number 63 and atomic weight 151.96. Belongs to *lanthanides*. Natural europium consists of two stable isotopes ^{151}Eu (47.8%) and ^{153}Eu (52.2%). Outer shell electron configuration is $4s^2p^6d^{10}f^75s^2p^66s^2$. First three ionization energies (eV): 5.664, 11.25, 24.7. Atomic radius is 0.202 nm. Ionic radii are (in nm): Eu^{2+} 0.112, Eu^{3+} 0.098. Europium possesses the lowest work function (2.54 eV) among all lanthanides. Electronegativity is ≈1.08.

Free europium is a soft light-gray *metal* possessing a body-centered cubic structure with parameter $a = 0.4572$ nm. Density is 5.245 g/cm^3 at 298 K, $T_{\text{melting}} = 1100$ K, $T_{\text{boiling}} \approx 1760$ K. Latent heat of melting is 9.2 kJ/mol, latent heat of evaporation 146 kJ/mol, specific heat 27.6 J·mol^{-1}·K^{-1}. Debye temperature is 121 K. Electrical resistivity is 0.813 $\mu\Omega$·m at 25 °C. Temperature coefficient of resistance is $4.80\cdot10^{-3}$ K^{-1} (273–373 K). Europium is a strong *paramagnet* with magnetic susceptibility $\chi = +22\cdot10^{-8}$ CGS units. Temperature coefficient of linear expansion is $2.6\cdot10^{-5}$ K^{-1}. Europium has the lowest hardness among all the lanthanides, with the Brinnel value of 98–147 MPa.

Naturally occurring isotopes possess high thermal neutron capture cross-sections; therefore, europium (in the form of an oxide) is used as an effective neutron absorber. Microscopic amounts of europium serve as activators when added to *luminophors* based on compounds of yttrium (used in color TV sets) and other elements. A *laser*, which is produced using europium-activated ruby, provides radiation in the visible region of the spectrum. Europium is introduced into specialized types of glasses.

EUTECTIC

A liquid system (melt or solution) under a particular pressure in equilibrium with solid phases, the number of which equals the number of components of the system. The *crystallization* of such a system, in accordance with the *Gibbs' phase rule*, takes place at a constant temperature, like the crystallization of pure substances. In this case a mechanical mixture of solid phases of the same composition (*solid eutectic*) is formed. For a given system, the *melting temperature* of a solid eutectic is lower than the melting temperature of a mixture of any other composition. See also *Eutectic alloys*.

EUTECTIC ALLOYS

Alloys crystallizing from a liquid (melt) with the simultaneous formation of two solid *phases* α and β, if the alloys have two components. In three-component eutectic alloys (or more complicated types), three or more solid phases could appear simultaneously from a liquid. Eutectic alloys possess a lower (compared to their components) *melting temperature*. As a rule, the *crystallization* of eutectic alloys occurs by cooperative growth of crystals of both solid phases, which results in the formation of bicrystal colonies (grains). Eutectic alloys are used as *casting materials*, *solders*, wear-proof alloys (see *Wear*) for *coatings* and surfacings, and for obtaining eutectic composite materials and amorphous alloys (see *Amorphous metals and metallic alloys*) by quenching them from the liquid state. See also *Phase diagram*.

EUTECTOID ALLOYS

In two-component eutectoid *alloys* a transformation of one solid *phase* β into other solid phases α and γ can take place during cooling. In three-component eutectoid alloys a solid phase can transform into three other phases. The product of the transformation, called a *eutectoid*, has as a rule a layered structure. An example of a eutectoid alloy is the blend of *ferrite* and *cementite* (Fe_3C) which, while cooling, forms *austenite* (interstitial *solid solution* of carbon in γ-Fe).

EVAPORATION

The transition from liquid (or solid) *state of matter* to a gaseous one (vapor) by adding latent heat to the liquid. The direct evaporation of solids is called *sublimation*.

EVAPORATION IN VACUUM

See *Vacuum evaporation*.

EWALD METHOD (P. Ewald, 1921)

Method for calculating lattice sums of an electrostatic potential at position j for charges of magnitude $Z_i e$ at positions i the distances r_{ij} away using mathematical techniques which bring about rapid convergence of the series. The Coulomb potential $\sum_i Z_i e/r_{ij}$ arising from *crystal lattice* ions is reduced to two absolutely, uniformly, and rapidly converging series, i.e. those over the direct and the *reciprocal lattices*. In this way, one achieves a sharp reduction of the computational work needed for determining both the potentials, and the field strength with its gradient. The Ewald method allows one to determine the Hertz vector of a system of oscillating dipoles, as well as the highest multipole moments (see *Multipole*) associated with crystal lattice sites. In particular, the Ewald method is efficient for calculating both the *Madelung constant* and the electrostatic part of the energy of an *ionic crystal*. Kornfeld extended the Ewald method for the calculation of dipolar and quadrupolar arrays.

EWALD SPHERE

See *X-ray structure analysis*.

EXA... (fr. Gr. $\varepsilon\xi$, six; means the sixth power of a thousand)

A prefix for the name of a unit of physical value which equals 10^{18} of the original value. Denoted by E. Example: 1 Em (exameter) $= 10^{18}$ m $= 10^{15}$ km $= 0.009465$ light year.

EXCESS NOISE

See *Flicker noise*.

EXCHANGE-COUPLED PAIRS

The simplest complex involving two paramagnetic ions. The presence of an exchange-coupled (exchange bound) pair affects the optical properties, the *spin–lattice relaxation*, the *specific heat*, and the *magnetic susceptibility* of crystals. Exchange-bound pairs may couple via various kinds of exchange interactions: isotropic or anisotropic, antisymmetric, biquadratic, etc.

EXCHANGE FIELD of a magnetic material

The characteristic *magnetic field* associated with the *exchange interaction*. The order of magnitude of the magnetic exchange field B_e is $J/(2\mu_B)$, where J is the exchange integral, and μ_B is the Bohr magneton. For magnets with several *magnetic sublattices*, B_e is a basic material constant defined as the magnetic field magnitude at which the *spin-flip magnetic phase transition* occurs, e.g., where the *magnetizations* of sublattices become aligned. The exchange field for various *antiferromagnets* varies from a few dozen tesla (for low *Néel points* T_N of several kelvins) to a few kilotesla (for T_N up to kelvins in the hundreds). The value of B_e is relatively small for metamagnetic media (several tesla).

EXCHANGE INTEGRAL

See *Exchange interaction*.

EXCHANGE INTERACTION (W. Heisenberg, 1926; W. Heitler, F. London, 1927; P.A.M. Dirac, 1929)

A specific mutual interaction between identical particles whereby the system energy depends on its *resultant spin*. The exchange interaction is a purely quantum effect, involving *Fermi–Dirac statistics* or *Bose–Einstein statistics* (for fermions and bosons, respectively). It can involve particles interacting either with each other (e.g., via the Coulomb interaction) or with external fields. The Hamiltonian of the exchange interaction H_e for two electrons with spins $\boldsymbol{S}_1$ and $\boldsymbol{S}_2$ has the form:

$$H_e = -J\boldsymbol{S}_1 \cdot \boldsymbol{S}_2, \qquad (2)$$

where J is called the *exchange integral*. It is the difference between the interaction energies of electrons, U_S, for the two states with different total spin values $\boldsymbol{S}$ ($S = 1$ or $S = 0$), $J = U_0 - U_1$. Within the framework of nonrelativistic quantum mechanics (expressing the wave function as a product of spin and space coordinate terms), the origin of the exchange interaction, i.e. why $U_1 \neq U_0$, is found in the *Pauli principle*. According to it, the wave function for fermions (electrons in our case) is antisymmetric with respect to the interchange of particle variables (both coordinate and spin). Since the spin part of the wave function is symmetric for $\boldsymbol{S} = 1$, its coordinate part should be antisymmetric and, accordingly, for $S = 0$, its spin part is antisymmetric and its coordinate part is symmetric. The difference in the symmetries of the coordinate parts of the wave functions results in a difference between the corresponding particle interaction energies (with each other and with other charges, such as atomic nuclei in a molecule). The exchange integral may be either positive (parallel spin orientation, $\boldsymbol{S} = 1$, is more favorable in energy) or negative (state $\boldsymbol{S} = 0$ is favored in energy). Here are some of the simplest examples. The Coulomb repulsion of several electrons in a partly filled atomic shell (e.g., a transition atom d- or f-shell) is minimized by the "most antisymmetric" wave function, so that $J > 0$ and the shell has its maximum possible spin value (*Hund rules*). For electrons of *covalent bonds* (e.g., in the molecule H_2), a configuration with a symmetric wave function is energetically favored, so the electron density clusters in the region between two charged nuclei. Because of that, $J < 0$ and the net spin of a covalent bond is zero. The exchange interaction of electrons of different atoms is appreciable when the atomic wave functions of these electrons overlap. The exchange integral is of the order of $\xi(e^2/a_B)$ where a_B is the Bohr radius, and the factor ξ which describes the extent of overlap of the wave functions (usually $\xi \leqslant 10^{-2}$) falls off exponentially with increasing interatomic distance. The exchange interaction of electrons localized on different atoms is often reduced to that of atomic spins. Equation (1) also describes the exchange interaction between the total spins of unfilled atomic shells, $\boldsymbol{S}_1$ and $\boldsymbol{S}_2$, but such a treatment is not rigorous in the case $S > 1/2$. To pursue a rigorous approach, additional terms must be introduced, e.g., biquadratic in the atomic spins (so-called *biquadratic exchange interaction*). The exchange interaction is often related only to the electrostatic electric interaction of electrons with each other and with atomic nuclei. In that case it includes the scalar products of spins only, and remains invariant with respect to arbitrary spin rotations. When one considers other types of interactions (e.g., *spin–orbit interaction*) anisotropic terms of the form $J_{\alpha\beta}S_\alpha S_\beta$ may appear in the exchange interaction Hamiltonian, where $\alpha, \beta = x, y, z$ (see *Spin Hamiltonian*). In the case when $J_{\alpha\beta}S_\alpha S_\beta \neq J\boldsymbol{S} \cdot \boldsymbol{S}$, one speaks of the *anisotropic Dzialoshinskii–Moriya exchange interaction* (see *Dzialoshinskii interaction*), the so-called *antisymmetric exchange interaction*. This is described by antisymmetric tensor components $J_{\alpha\beta}$, $J^{(a)}_{\alpha\beta} = -J^{(a)}_{\beta\alpha}$; in order of magnitude $J^{(a)} \sim (\Delta g/g)J$, where Δg is the change of the g-factor due to the spin–orbit interaction. The *symmetric exchange interaction* is described by symmetric components $J^{(s)}_{\alpha\beta} = J^{(s)}_{\beta\alpha}$, and determines both the ordinary Heisenberg exchange interaction, and the *exchange anisotropy*. The *Ising model* is a limiting case of the anisotropic exchange interaction. Non-Heisenberg exchange interaction terms are particularly large for *rare-earth element* and *actinide* ions. There is also an *indirect exchange interaction* involving either an intermediate (nonmagnetic) ion, or *conduction electrons* (see *Ruderman–Kittel–Kasuya–Yosida interaction*). The exchange interaction plays an important role in the binding energy of atoms, molecules, and solids (especially *metals* and *covalent crystals*); it explains trends in the spectroscopy of solids (see *Exchange splitting*, *Exchange-coupled pairs*), and underlies the physics of *magnetism*.

EXCHANGE INTERACTION, INDIRECT

See *Indirect exchange interaction*.

EXCHANGE MAGNETOSTRICTION

See *Magnetostriction*.

EXCHANGE SCATTERING

Scattering of electrons (see *Current carrier scattering*) with a change in the projection of their spin angular momentum. The presence of exchange scattering involving identical particles makes it necessary to antisymmetrize their electronic wave functions. To describe the scattering of delocalized band electrons by local *paramagnetic centers* when the wave vector of the scattering electron is smaller than the reciprocal radius of the local center state, the exchange scattering makes use of a zero-radius potential $\widehat{V} = [2\pi\hbar^2/m^*]a\boldsymbol{S}\cdot\boldsymbol{S}_e\delta(\boldsymbol{r})$ (where $\boldsymbol{S}_e$ is the spin of the scattering electron, m^* is its *effective mass*, $\boldsymbol{S}$ is the spin of the paramagnetic center, and a is the exchange scattering length that is either considered as a phenomenological parameter, or is calculated by taking into account the specific electronic structure of the paramagnetic center.

Exchange scattering is responsible for such phenomena as the spin *relaxation* of paramagnetic centers, *conduction electrons*, and *excitons* in semiconductors (see *Paramagnetic relaxation*), electronic polarization, the formation of localized *magnetic moments* in metals, and the *indirect exchange interaction* of atoms through conduction electrons (see *Ruderman–Kittel–Kasuya–Yosida interaction*), as well as some other phenomena which involve participation of electron spins.

EXCHANGE SPLITTING

An energy difference between levels differing in the value of the *resultant spin* S for a doublet, triplet, or more complicated complex involving exchange-interacting paramagnetic ions. For an isotropic *exchange interaction* characterized by the parameter J, the position of the energy level is found to be $E_S = JS(S+1)/2$, and this difference is easy to calculate. The quantity J was first measured and calculated for a helium atom (ortho- and parahelium), and for a hydrogen molecule.

EXCIMER

An excited dimer formed by two atoms or two molecules in an excited state. Such an elementary excitation in an organic compound is characterized by *chemical bond* formation between neighboring molecules as a result of the excitation of one of them, when no bond exists before the excitation. The crystals where the excimers are formed exhibit monomer absorption (*monomer*) with broad structureless bands in *fluorescence* spectra which are specific for pairs of molecules called *dimers* (see *Exciplex*). A typical example of an organic compound where excimers form is *pyrene* ($C_{16}H_{10}$). There are four molecules in the unit cell of pyrene with the molecules arranged in pairs, thereby satisfying the conditions for the formation of an excimer state. Excimers are observable in gaseous, liquid and solid states of organic molecules. Critical parameters for excimer formation are the distance between molecules, and the possibility of their interaction at the optical excitation of one them. Excimers play a significant role in the optical properties of many organic compounds, in particular, in processes of photogeneration of *current carriers* in them. Excimers are used in excimer *lasers*.

EXCIPLEX

An excimer (excited dimer) compound formed as a result of an interaction between two different molecules, one of which is in an excited state (dimer in excited state, see *Excimer*). Such an excited molecular charge complex is unstable in the ground state. It may form from an excited state acceptor molecule interacting with a ground state donor, and vice versa. Exciplex *fluorescence* arises in lattices constructed from physical *dimers*, *donor* and *acceptor* molecules, which do not interact in their ground (unexcited) states.

EXCITATIONS, COLLECTIVE

See *Collective excitations*.

EXCITON

An elementary collective electronic excitation in nonmetallic crystals, free of current, and existing in the crystal as a wave of excitation. The exciton concept was introduced by Ya.I. Frenkel (1931). Exciton energy states involve a band with a particular *dispersion law* that depends on the crystal properties. A specific feature of the band structure of dipole-active excitons is its splitting into branches of longitudinal and transverse excitons (with regard to the angle between the vectors of the wave and the dipole moment). The value of the transverse–longitudinal splitting is proportional to the square of the *dipole moment* of the exciton transition, and it determines the range of existence of long-wave excitons (those with small wave vector magnitude). In optical processes the main excitons that participate are those with small wave vectors which correspond to the magnitude of the light wave E vector. In most cases excitons are induced by an electromagnetic wave. If the exciton transition has a large *oscillator strength*, the quanta of the electromagnetic field in the medium (*photons*) are mixed with the quanta of "mechanical" (electronic) excitations, thereby forming exciton-induced *polaritons* (*photoexcitons*). In this case there is no longer a sharp difference between the excitons and the photons in the crystal. If the oscillator strength is small then such mixing is insignificant.

There are two kinds of excitons: the *Frenkel exciton* (*small-radius exciton*) and the *Wannier–Mott exciton* (*large-radius exciton*). A Frenkel exciton is basically an excited electronic state of an ion. It can move around the crystal by transferring the excitation from ion to ion, so no ions need to change places. A example of a Wannier–Mott exciton is an electron in the conduction band bound to a hole in the valence band by the Coulomb interaction $V(r) = -e^2/(4\pi\varepsilon r)$, where the dielectric constant of the crystal ε is greater than that ε_0 of free space, typically by a factor of 10. The energy levels are the same as the Rydberg series levels of a hydrogen atom, but with energy values reduced by the ratio $(\varepsilon_0/\varepsilon)^2$:

$$E = E_g - \frac{me^4}{2\hbar^2(4\pi\varepsilon)^2 n^2},$$

where E_g is the energy gap, and n is the principal quantum number. The exciton Bohr radius $a_B = 4\pi\varepsilon\hbar^2/(me^2)$ exceeds that of a hydrogen atom by the ratio $\varepsilon/\varepsilon_0$. There also exist excitons of an intermediate type with a binding energy $E_b \sim 1$ eV and a small radius ($R = 10^{-7}$–10^{-8} cm) in *alkali-halide crystals*, and in crystals of noble gases. The exciton radius is the mean distance between the excited electron and its non-excited state that remained vacant (hole) and bound in the exciton wave.

An exciton existing in the form of a wave (so-called *free exciton* or *coherent exciton*) can be trapped by a lattice structure defect and bond with it, thus forming a *bound exciton*. The energy levels of bound excitons are discrete and situated lower than the band of free excitons for the same value of binding energy with defects.

The exciton absorption spectra in perfect crystals are represented by a series of discrete bands whose width and shape are determined by the structure of the exciton band, and the character of the interaction of the excitons with *photons* and *phonons*. When Wannier–Mott excitons are excited with the help of a low-power light source, their concentration is small (to 10^{-14} cm^{-3}); this state can be regarded as a gas of non-interacting *quasi-particles*. Under intensive laser excitation *high-density excitons* are produced. In this case the density is so high that the exciton size is comparable with the inter-exciton distance; their interaction becomes significant and leads to the appearance of exciton molecules (*biexcitons*), which either condense or decay with the formation of an electron–hole plasma, or drops of an *electron–hole liquid*. Excitons play an essential role in processes of excitation energy transport in crystals (see *Energy transfer by excitons*), in radiation processes, for laser generation in homogeneous semiconductors, and so on.

See also *Biexciton, Dislocation exciton, Frenkel exciton, High-density excitons, Photoexciton, Polyexcitons, Self-localized exciton, Surface exciton, Trion, Triplet excitons, Wannier–Mott exciton, X-ray excitons.*

EXCITON, ENERGY TRANSFER

See *Energy transfer by excitons*.

EXCITON–IMPURITY COMPLEX

Bound state of two or more *excitons* with a small neutral impurity center (donor or acceptor) in indirect-gap *semiconductors* (e.g., Si and Ge). The stability of exciton–impurity complexes results from the degeneracy of the bands. In a direct-gap semiconductor with simple bands, at the center point Γ of the Brillouin zone, there is a $1s$ orbital state which can only hold two electrons with opposite spins. In indirect gap Si which has six electronic valleys situated at the points Δ, twelve electrons with antiparallel spin orientations or Bloch functions can coexist in this state. The *valence band* of Si at an extremum point is fourfold degenerate; hence four holes with angular momenta projections j_z equal to $\pm 3/2$, $\pm 1/2$ can be in the same orbital state. In radiative *recombination* spectra at high excitation densities the exciton–impurity complexes are observed as a series of narrow lines. The long-wave lines of the series appear at high levels of excitation. An individual series is associated with each impurity center.

In the *single-electron approximation* the electronic structure of an exciton–impurity complex is described in terms of a *shell model*: the electrons and the holes of the complex fill the electron and hole shells in accordance with the Pauli principle, similar to the way electron shells are filled in many-electron atoms. The shell degeneracy is determined by the multiplicity (by principal and orbital quantum numbers). The electrons and holes are in a self-consistent field whose symmetry coincides with the symmetry of the impurity center. Therefore, the structure and the degeneracy of the shells are the same as in simple donors and acceptors.

The description of the energy spectrum and shell filling based on the single-electron approach is a crude approximation that fails to explain the fine structure of the spectra of the exciton–impurity complex caused by the electron–electron (hole) correlation.

EXCITON LUMINESCENCE

Luminescence in which the radiation arises from transitions involving excitonic energy levels. The *exciton* luminescence is observed in most crystal phosphors and organic *luminophors* in the form of spectrally narrow (in comparison with other kinds of luminescence) bands (lines) with line positions and shapes which depend on the temperature, the level, and the mode of excitation, the presence of defects in the material, the action of external electric and magnetic fields, deformations, and so on.

A *radiative quantum transition* that produces exciton luminescence is exciton *recombination* occurring with the emission of light. The energy of the emitted quantum is close to the energy of the corresponding exciton. Many possible exciton states (exciton bands; bound excitons; excitons due to *self-localization*) result in the emission of a few bands (lines) specific for the exciton luminescence of each individual material. The excitons of the exciton band, under action of the excitation, can spread the energy over a broad continuous range. However, the corresponding exciton luminescence bands are usually narrow. This is due, first, to *selection rules* which provide a sufficient emission probability only for those electrons with a certain energy, and second, to the nonuniform distribution of excitons over the energies of the band. If the time delay between the excitation and the emission is sufficient for a great number of collisions of excitons with *phonons* to take place, then the excitons become thermalized, i.e. the energy no longer depends on that acquired at excitation, they will occupy a range of states with the thermal energy k_BT while passing their excess energy to the lattice; consequently, the exciton luminescence will consist only of quanta with the same energy. The dipole approximation selection rules only permit radiative transitions of excitons with the *quasi-momenta* close to zero. The exciton states related to these quasi-momenta are all within a narrow range of the possible energies that will appear in the emission.

In the general case, both the kinetics of the change of the exciton energy after its excitation (i.e. its *relaxation*), and the selection rules (taking into account interactions of excitons with phonons, *plasmons* and other units in the crystal) are more complicated. For example, for a strong interaction of excitons with the crystal electromagnetic field, the relaxation of exciton *polaritons* is characterized by a sharp dependence of the *mean free path* on the polariton energy, and the probability of emission is determined, in addition, by the conditions of transformation of polaritons into phonons at the surface. The exciton energy distribution is determined in this case by a combination of probabilities from different modes of relaxation. The dependence of the probability of emission on the exciton energy also becomes complex and nonmonotonic. All these factors affect the shape of the exciton luminescence bands, thus making it possible to determine from experimental data the relaxation mechanisms, their probabilities, and their selection rules.

The application of external fields can change the internal energy of excitons, thereby causing them to shift, split, and broaden the exciton luminescence bands. For high excitation levels when a high-density of excitons and other contributors (*current carriers*, nonequilibrium phonons) appear, their interaction also changes the internal energy of the excitons, and correspondingly, the exciton luminescence spectrum. As a result, the changes involved in combining excitons in "exciton molecules" (*biexcitons*, drops of *electron–hole liquid* and exciton liquid) can be detected. The change of the exciton energy, or the decay of the exciton states as a result of the interaction with the current carriers (*electric charge screening*), is also accompanied by corresponding changes in the exciton luminescence spectra (appearance or disappearance of bands, their transformation into other types of bands, e.g., into an emission band of an *electron–hole plasma*, etc.).

In most cases the strong temperature dependence of *luminescence quenching* is specific for the exciton luminescence. In this case processes competing with exciton luminescence may be either dissociation of excitons into pairs of nonbound current carriers, or recombination with the emission of many phonons.

EXCITON MECHANISM OF SUPERCONDUCTIVITY

Hypothetical non-phonon mechanism of *high-temperature superconductivity* which assumes that *Cooper pairs* of degenerate free carriers (conduction electrons) form through the exchange of virtual quanta of *collective excitations* of bound electrons in atoms and molecules of solids (i.e. by *excitons*). The mechanism had been put forward by W. Little and independently by V.L. Ginzburg (1964). Little considered analytically a quasi-one-dimensional model of conductive polymer chains in organic (metal-organic) compounds, in which, as a result of the Cooper interaction, the conduction electrons polarize the electron shells of molecules (radicals), thereby reducing their mutual repulsion. The result of this polarization is an effective interaction between the electrons in the chain which takes the form:

$$V(q,\omega) = \frac{4\pi e^2}{q^2\varepsilon(\omega)},\qquad \varepsilon(\omega) = 1 + \frac{\Omega_p^2 f_0}{\omega_0^2 - \omega^2}, \tag{1}$$

where $\hbar\omega$ and $\hbar q$ are the transferred energy and momentum, respectively; $\varepsilon(\omega)$ is the *dielectric constant* of organic molecules surrounding the conductive chain; ω_0 and f_0 are respectively the frequency and oscillator strength of the optically allowed dipole transition; Ω_p is the plasma frequency of bound electrons. In the frequency range

$$\omega_0 < \omega < \left(\omega_0^2 + \Omega_p^2 f_0\right)^{1/2} \tag{2}$$

the interaction $V(q,\omega)$ of Eq. (1) is negative corresponding to an attractive interaction involving the exchange of virtual excitations (excitons) between molecules. It was supposed that this attraction, owing to the high energy of the excitons

($\hbar\omega_0 \geqslant 1$ eV), could lead to superconductivity with a high critical temperature ($T_c \geqslant 100$ K) if one accepts the validity of the BCS formula for the transition temperature $k_B T_c \approx \hbar\omega_0 \exp(-1/\lambda_{ex})$ (see *Bardeen–Cooper–Schrieffer theory*) with the assumption that the dimensionless electron–exciton interaction constant is not small, i.e. $\lambda_{ex} \geqslant 1/2$. As the range of attraction (2) is a narrow peak situated far from the *Fermi level*, under realistic conditions the optimal values are $\lambda_{ex} \leqslant 0.1$, and consequently T_c cannot exceed 1 K. At present a series of quasi-one-dimensional (chain) organic crystals with metallic conductivity (see *Quasi-one-dimensional crystals*) that transform into the superconduction state at $T \approx 10$ K have been synthesized (see *Organic conductors and superconductors*), but there are no convincing experimental proofs of the operation of the exciton mechanism of superconductivity in such systems. V.L. Ginzburg considered a possibility for this mechanism in quasi-two-dimensional layered structures of the "sandwich" type (semiconductor–metal–semiconductor) when the Cooper pairing of electrons in ultrathin metal film of $d \approx 1$ μm thick takes place due to the attraction caused by their interaction with virtual excitons in the semiconductor plates. There have been reports about the observation of superconductivity in monatomic films made of silver and gold at the surface of germanium monocrystals, or in multilayer structures Au–Ge, but it was not been demonstrated that the exciton mechanism acts in this case. In recent years the interest in this mechanism has grown in connection with the cuprate *high-temperature superconductors* ($T_c \approx 30$ to 165 K) which have a layered crystal structure (see *Quasi-two-dimensional crystals*) and a complicated strongly anisotropic band spectrum whose particular feature is intensive (corresponding with high oscillator strength) interband transitions which probably enhance the Cooper coupling and thereby raise T_c. There is one more branch of excitations of the boson (exciton) type in copper-containing layered metal-oxides which is associated with a particular electronic structure of bivalent copper (configuration $3d^9$). The carriers are capable of exchanging these excitations. Since in a cubic *crystal field* the d-state of the Cu^{2+} ion forms an orbital e_g-doublet, this ion is a *two-level system* with the splitting of the levels due to a low-symmetry component of the *ligand* field. When traveling through the lattice, the quasi-free hole described by hybridized p–d-states can excite a transition between the two e_g-levels of the hole localized at the Cu^{2+} ion. This transition corresponds to the formation of a quadrupole-type *Frenkel exciton*. At the same time d–d transitions can also be excited that are linked with the change of the *spin* projection, and which in turn, could be one of the causes of breaking the long-range antiferromagnetic order in the presence of charge carriers. By treating the d-excitons as carriers of the inter-hole interaction, one might approach a theory similar to the Bardeen–Cooper–Schrieffer theory with a relatively high superconducting transition temperature due to the high frequency (to 0.5 eV) of the d–d excitations.

EXCITON MOLECULE

See *Biexciton*.

EXCITON–PHONON INTERACTION

A dynamic process perturbing exciton states. This is described by a matrix element $V_{\text{e-ph}}$ which depends on the mechanism involving:

(i) the interaction of *excitons* with acoustical *phonons* by means of the *strain potential*:

$$V'_{\text{e-ph}} = \sqrt{\frac{\hbar|q|D^2}{2V\rho u}}; \qquad (1)$$

(ii) the interaction of excitons with acoustical phonons resulting in a piezoelectric effect (non-centrally symmetric crystals):

$$V''_{\text{e-ph}} = \sqrt{\frac{\hbar}{2\rho u}\frac{4\pi e\varepsilon_{ij}}{\varepsilon_0 q^{1/2}}}(q_e - q_h); \qquad (2)$$

(iii) the interaction of excitons with phonons of the Fröhlich kind (see *Fröhlich Hamiltonian*):

$$V'''_{\text{e-ph}} = \frac{\hbar q}{2\mu}\sqrt{\pi\left(\frac{\varepsilon_\infty}{\varepsilon_0} - 1\right)} \times \sqrt{\left(\frac{m_e - m_h}{m_e + m_h}\right)^2 \frac{a_{ex}^3}{V}\hbar\omega}, \qquad (3)$$

where ρ is the crystal *density*, u is the sound velocity, D is the deformation potential, V is the crystal volume, q is the phonon momentum, $q_{e,h}$ are the momenta of the electron and *hole* which compose the exciton, ε_{ij} are the components of the *piezoelectric tensor*, ε_∞ and ε_0 are the high-frequency and static *dielectric constants*, respectively, $m_{e,h}$ are the effective masses of the electron and hole, μ is the reduced exciton mass, a_{ex} is the *Bohr radius* of the exciton, $\hbar\omega$ is the phonon energy.

There are three kinds of electron–phonon interactions: those with strong, intermediate, and weak strength. This interaction is the basis of exciton *light absorption*, the transport of the excitation energy throughout the crystal (see *Energy transfer by excitons*), and exciton *photoluminescence*. These factors can lead to exciton band broadening, changes in their spectral line positions and shape, the appearance of split states of *self-localized excitons*, and can stimulate nonradiative exciton relaxation and *luminescence quenching*.

EXCITON SPECTROSCOPY

Branch of *optical spectroscopy* devoted to the study of the energy spectrum of excitons as well as their interactions with each other, with electrons (*holes*), *phonons*, *magnons*, with light, and with defects of the *crystal structure*. This includes the analysis of spectra of *light absorption* and *reflection of light*, photoluminescence, photoconduction, resonance *Raman scattering of light* and *Brillouin scattering* spectra in the spectral range of exciton resonance at *cryogenic temperatures* using polarized light (see *Polarization of light*), and the effects of external electric, magnetic, and deformation fields.

From the spectral positions of exciton bands one finds the energy parameters of excitons: energy of formation E_{ex} and binding energy ε_{ex}. Asymmetrically shaped absorption bands (deviations from *Lorentzian shape*, *Gaussian shape*, Voigt contour, etc.) and their changes with temperature allow one to determine the strength of the exciton–phonon bond (see *Exciton–phonon interaction*), and the dominant mechanisms of energy dissipation of the excitonic excitation; the *absorption coefficient* yields the *oscillator strength* of the transition.

EXCITON STARK EFFECT

See *Stark effect of excitons*.

EXOELECTRONIC EMISSION

Low-temperature emission of electrons from a disrupted *solid surface* which is induced by a mechanical disturbance, radiation, or some other action, or by physicochemical processes such as *adsorption*, *desorption*, *oxidation*, *corrosion*, catalysis, and so on; this emission accompanies the restoration of the equilibrium state. *Exoelectronic emission* results from the above-mentioned interactions due to the breakage of chemical bonds and charge separation ("excitation"), followed by processes of restoring the broken bonds and *recombination* of the charge (*relaxation of excitation*). The relaxation events are associated with the release of energy that had been expended for the electron yield from the surface. Clean surfaces of metals and semiconductors do not exhibit exoelectronic emission. Processes of *failure* of dielectric or oxide surface layers on metals and semiconductors under the external action are accompanied by the emission, not only of electrons, but also of ions (*exoionic emission*). In this connection the term "exoemission" is sometimes used. Exoelectronic emission has been applied in industrial engineering and technology as an efficient method for controlling the initial stages of the fracture of solids under the action of applied factors, such as mechanical and thermomechanical action, ionizing radiation, corrosion, etc., as well as for monitoring the technological processing of materials.

EXOIONIC EMISSION

Low-temperature emission of charged particles (*ions*) from a disrupted *solid surface*, induced by mechanical or radiation actions, or by physicochemical processes at the surface, such as *adsorption*, *desorption*, dehydration, *corrosion*, catalysis, etc. Exoionic emission can result from the emission of ions of the host material (emitter) or of impurity ions. A possible mechanism of this process is the emission of charges resulting from the energy released during the restoration of chemical bonds, and when a disturbed surface reverts to its equilibrium state. In many cases exoionic emission accompanies the phenomenon of *exoelectronic emission*, and obeys the same laws. The effect is used for the detection and investigation of initial stages of the breakdown of metals and nonmetals, including *polymeric materials* and *composite materials*, and for the investigation of adsorption processes and chemical reactions at the surface.

EXOPHOTON EMISSION

Emission of *photons* from a *solid surface* disturbed by either mechanical processes, radiation, or another action such as bombardment by atomic, molecular, or ionic beams, as well as by some physicochemical process (*adsorption*, *recombination* of active particles, etc.). Exophoton emission takes place as a result of energy released during the recombination of active particles which are formed under the above-indicated processes and interactions. Exophoton emission is a surface effect which is strongly influenced by the pressure and the chemical nature of gases in the ambient atmosphere. The photon emission centers are situated either at the surface or in a thin near-surface layer of the solid. This emission can be employed for studying *surface defect* formation, material *failure*, and physicochemical processes at the surface (including *strain* and fracture of metals, disintegration of materials, *sputtering*, *chemisorption*, and surface chemical reactions). Exophoton emission accompanies *exoelectronic emission* in many cases, and both processes are described by the same laws.

EXPANSION, HIGH-TEMPERATURE

See *High-temperature expansion*.

EXPANSION, THERMAL

See *Thermal expansion*.

EXPERIMENTAL METHODS OF PHOTOELECTRON SPECTROSCOPY

Experimental techniques for determination of the spectral characteristics of *photoelectron emission* in *photoelectron spectroscopy*, including *quantum yield*, energy, degree of spin polarization, and angular distribution of photoelectrons N, etc. Varying the test conditions and recording different quantities, one may carry out various methods of photoelectron spectroscopy, which are sensitive to diverse physical properties of solids. The experimental parameters may be:

(1) energy $h\nu$, polarization P, and angle of incidence α of the photon;
(2) kinetic energy E, quasi-momentum k, spin σ, emission angle, initial state energy E_i and final (excited) state energy E_f of the electrons;
(3) crystallographic structure, orientation of the surface, *electron affinity* and electron *work function*, surface concentration of adsorbed atoms, atomic structure of adsorbed *films*, temperature and other characteristics of the material under investigation.

The diversity of experimental parameters determines the diversity of photoelectron spectroscopy methods: Ultraviolet Photoemission Spectroscopy (UPS), X-ray Photoemission Spectroscopy (XPS) and synchrotron photoemission spectroscopy are named in accordance with the energies of the exciting radiation. The experimental parameter of Angle-Integrated Photoemission Spectroscopy (AIPES) is the angle of emission; whereas Angle-Resolved Photoemission Spectroscopy (ARPES) and normal Photoemission Spectroscopy (PES) take into account the angular resolution. The spectroscopic methods which deal with energies of initial and final electron states are Constant-Final-State Spectroscopy (CFS), Constant-Initial-State Spectroscopy (CIS), Normal Constant-Final-State Spectroscopy (NCFS), and Normal Constant-Initial-State Spectroscopy (NCIS). The values of quantum yield are measured using Y-spectroscopy techniques: Yield Spectroscopy of total quantum yield (YS) and Partial Yield Spectroscopy (PYS), in which the photoelectrons are distinguished with respect to energy and angle of emission. There is also *Electron Spectroscopy for Chemical Analysis* (*ESCA*). Some of these acronyms are, and some are not commonly used in the Western literature.

The analysis of the dependences of the yield $Y(h\nu, P, \sigma, \alpha)$ and number $N(E, k, \sigma, h\nu, P)$ of photoelectrons may provide information on the density of empty and occupied (band and inner core) electron states, dispersion of energy bands, symmetry of wave functions and spin state of electrons, geometric structure of the surface, orientation and symmetry of orbitals of the surface atoms and molecules, composition in atomic percent and chemical composition. It is also used for examining the positions of localized adsorbed atoms and molecules and the orientation of their orbitals in the electron spectrum. The methods of photoelectron spectroscopy are used for determination of the potential barrier characteristics: work function, electron affinity, location of *Fermi level*, band edges and bendings at the surface. Information on scattering mechanisms can also be obtained.

EXPLOSION

A process of energy release within a restricted volume during a short time interval.

As a result of an explosion, the material filling the volume where energy is released is transformed to a strongly heated gas under very high pressure. During the course of the explosion a *shock wave* is formed and spreads into the environment. An explosion in a solid medium is accompanied by its collapse and destruction. Different types of explosions can be distinguished by the physical nature of the energy source, and by the manner of its release. An explosion can occur due to the emission of chemical or nuclear energy, releasing the compressed gas energy, or transforming the magnetic, electric, radiation or kinetic energy into heat energy. A *heat explosion* can occur when thermal equilibrium cannot be established between the reacting material and its environment. At sufficiently large values of *activation energy* the speed of the chemical reaction ω increases with increasing temperature T in accordance with *Arrhenius law*. A *chain explosion* occurs when a chemical reaction develops as a branched *chain reaction* during the course of which large concentrations of active particles or radicals appear (comparable with source material concentrations). The thermal and chain modes of propagating the explosion can occur during nuclear transformations, i.e. the reactions of nuclear fission and fusion. An explosion can be caused by sharp external effects, by shock, by friction, etc. The cause through shock is probably the result of local heating of the material. A shock wave causes the specific type of explosive transformation which does not appear simultaneously along the whole charge, but rather propagates in the material at a constant speed, i.e. *detonation* occurs. Processes during which no internal energy of the material is released, but rather the energy of an external source (e.g., the collision of bodies moving with large speeds; or the matter where high-power laser radiation is focused, etc.) are also related to explosions. In scientific investigations involving explosions extremely high values of pressure, temperature and density are achieved. It is used for the attainment of *superstrong magnetic fields*, for the instigation of *phase transitions* (see *Phase transitions under pressure*), and for the formation or synthesis of new substances (see *High pressures*).

EXPLOSION, COULOMB

See *Coulomb explosion*.

EXPLOSION DEFORMATION, explosion strain

Deformation of a material due to the passage of *shock waves* initiated by an *explosion*. The magnitude of *strain* ε is determined by the expression $\varepsilon = (4/3)\ln(V/V_0)$, where V and V_0 respectively are the specific volumes of the material before and after the passage of the shock wave. Explosion deformation is distinguished from quasi-static deformation by several special features: pulse wave character, shortness of the effect, high effective tangential stresses, lack of significant residual macroscopic deformation. These features condition multiple *slip* even at small degrees of deformation, initiation of deformational twinning (see *Twinning of crystals*) at room temperature for face-centered cubic metals, twisting deformation mechanisms (see *Plastic twisting deformation*), and the formation of a *substructure* with a high density of *defects*. The possibilities of an explosion deformation are greater than that of a quasi-static one. At present shock waves are investigated in materials with a *pressure* within the wave front of about a TPa.

EXPLOSION NOISE

See *Telegraph noise*.

EXPLOSIVE ELECTRON EMISSION

Emission of an intense electron flow, brought about by transforming cathode metal material from the condensed phase (see *Condensed state of matter*) to a dense plasma (see *Solid-state plasma*) as a result of local heating. The metal-to-plasma transition can be initiated by the *explosion* of the metal due to its heating by a *field electron emission* current of high density ($j = 10^8$–10^9 A/cm^2), due to the impact of particles at the cathode, due to radiation flux incident on the cathode, etc. The initial explosion and subsequent explosive electron emission are accompanied by the near-cathode generation of plasma, which expands with the speed $v \approx 10^6$ cm/s, and constitutes a single peak of current

$$I = 3.7\cdot 10^{-5}\cdot U^{3/2}\frac{vt}{(d - vt)},$$

where U is the voltage between the cathode and anode during the course of the emission, d is the distance between them, and t is the time. This electron emission is accompanied by the carrying off of the material from the cathode. This latter effect can be reduced by decreasing the electron current, but if this current becomes less than some critical value the electron emission stops. Explosive electronic emission is used in strong-current amplifiers of electrons, and in pulse sources of high-intensity X-rays. It also plays a fundamental role in vacuum discharges, in some types of gas discharges, and in the generation of shot noise.

EXPONENT, CRITICAL

See *Critical exponents.*

EXTERNAL FRICTION

A dissipative process at the area of contact between two bodies pressed together as they slide past each other, accompanied by heat release, and the possibility of electrification (*triboelectricity*), erosion (surface *wear*), etc.

The work of *friction* (W_{fr}) is determined by thermal ($\sim$99.9% of W_{fr}) and non-thermal losses connected with the increase of irregularities of the layers near the surface, and the possible destruction of surface material. External friction can be characterized by the *friction force* $F = dW_{fr}/dl$ which appears in the plane of contact of the sliding bodies, and is directed opposite to direction of the displacement dl. For small relative displacements Δl which are less than a "zero" value Δl_0 the friction force grows from zero to a maximum, and beyond this value ($\Delta l > \Delta l_0$) it no longer depends on Δl. Typical zero displacements Δl_0 lie in the interval 10–0.01 μm. In the general case there are two types of relevant factors: those depending on the specific load normal to the friction surface (P) and those independent on the load $F = A + BP^{\chi}$, where A and B are independent of the pressure. Sometimes the two limiting cases are considered, corresponding to $\chi = 1$, $A \neq 0$ (Coulomb's law) and $\chi = 1$, $A = 0$. In practice it is convenient to use the *friction coefficient* $f = F/P$. The *Michin–Kragelsky equation* is the best known expression for this coefficient: $f = \tau_0/P_r + \beta + k\alpha_H\sqrt{h/R}$, where P_r is the actual load at the contact. The two first terms constitute the molecular component of the friction coefficient, and the last represents the mechanical (deformation) component; β and k are numerical parameters. The mechanical component is the resistance arising from pressing together the material, defined by the coefficient of hysteresis loss α_H and by the intrusion depth h of the unevenness, simulated by a spherical *indenter* of radius R. This component is very sensitive to the variation of the *elastic modulus* of the softer material of the friction pair, and can be 50 to 0.01% or less than the coefficient of friction. The first part of the molecular component τ_0/P_r depends on the load P_r and is determined by the value of specific *adhesion* τ_0. For laminated compounds (*graphite*, NB, MoS_2) which are frequently used as *antifriction materials*, τ_0 is determined by the value of the lamination energy along the basis plane (E_{001}) and by the characteristic dimension of the *crystallite* (L_a): $\tau_0 \approx E_{001}\rho_{001}/(L_a N_a)$, where ρ_{001} is the atomic density of the basis plane. For metallic pairs the specific adhesion τ_0 and the coefficient of molecular bond hardening are determined experimentally. Boundary and *hydrodynamic regime* of friction are distinguished. In the first case the structure and properties of the surface of the friction pair material have an essential influence which determines the value of actual area of contact, and the value of the *adsorption* into the contact gap of lubrication materials which can be in either the gaseous or liquid states (*liquid friction*) in the contact gap. Upon adsorption the lubrication materials at the contact form wedging layers which create normal pressure up to $152\cdot 10^6$ Pa (1500 atm) that provides the sliding. Here the contact gap can be regarded as a slot micropore with a characteristic dimension of about 10^{-9} m. In the boundary lubrication mode the coefficient of friction does not depend on the lubricant *viscosity* (η) and does not depend (or depends weakly) on the sliding speed (V). In

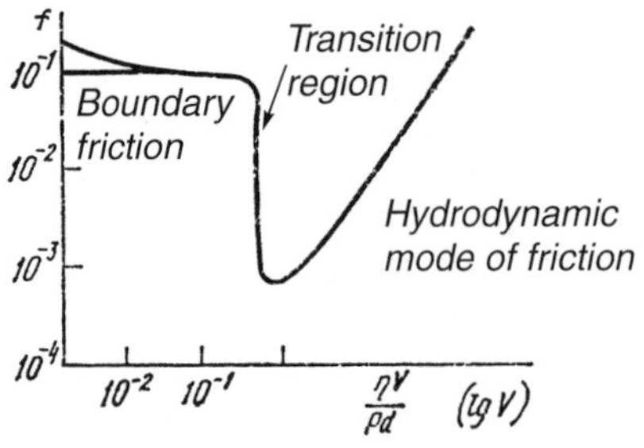

Dependence of the coefficient of friction f on the sliding speed V, and on the Sommerfeld criterion $\eta V/(Pd)$.

the second case the coefficient of friction is determined by the hydrodynamic properties of the lubricating liquid and depends on the dimensionless *Sommerfeld criterion* $S_m = \eta V/(Pd)$, where d is the bearing length. At the transition from boundary friction to the hydrodynamic mode (Fig.) the coefficient of friction decreases and then increases in value proportional to the liquid viscosity. The gas phase composition strongly influences the friction coefficient between the two surfaces.

At the working-in of the friction pairs the roughness corresponding to the minimal coefficient of friction is set spontaneously. For metallic systems *fragmentation* of the surface layer is observed in the friction process, and for laminated compounds a film forms with its basis planes oriented with respect to the friction surface. When the load increases, the depth of the surface layer changes, it breaks into fragments, mutual disorientation and rotation of these fragments begins, and with further increases in the load the external friction transforms to *internal friction*, i.e. there is no jump of speed at the transition from one body to another.

Sliding friction and *rolling friction* can be distinguished, each characterized by its own coefficient. For rolling friction $f_r = M/N$, where N is the normal load, and M is the moment of resistance to rolling. Rolling friction is less than sliding friction. It is conditioned by losses due to elastic *hysteresis*, by the material *work of deformation* at the formation of a bead in front of the rolling body, and overcoming of the forces of molecular interaction between the contacting surfaces. At moderately high speeds resistance to rolling increases and the force of sliding friction becomes less than that of rolling friction.

EXTINCTION

Reduction of the intensity of a radiation beam that propagates in a medium. For weakly diffracting radiation (e.g., light propagating in an almost-transparent body), a single mechanism of the extinction is associated with the inelastic processes of the radiation interaction with the matter, which is accompanied by energy loss. In this case the radiation intensity I decays with the depth of penetration l in an exponential manner (Beer's law): $I = I_0 \exp(-\mu l)$, and the extinction is characterized by the absorption coefficient μ which describes the inelastic processes. If the medium is inhomogeneous ("turbid") then elastic scattering on the medium inhomogeneities takes place. In this case the intensity is expressed as $I = I_0 e^{-(\mu+\nu)l} + \Delta I$, where the first term describes the intensity in the direction of the propagation of the initial beam, $\Delta I(l)$ is the intensity of the radiation scattered in other directions ("diffuse background"), and ν is the additional attenuation of the transmitted beam caused by diffusion scattering. The measurement of ν provides information on the characteristics of the medium inhomogeneities. In the case of interference during the propagation of the radiation (X-ray, neutron, electron, or any other) in crystals, the extinction acquires a more complicated form. For directions approaching the Bragg direction (see *Bragg law*), at a distance close in order of magnitude to the *extinction length* Λ, the beam intensity transforms partially into the intensity of the diffracted beam. As a result, an interference wave field appears, and in the case of *Bragg diffraction*, the intensity $I(z)$ decays exponentially with the depth (i.e. $I \propto \exp(-z/\Lambda)$). This phenomenon is called *dynamic extinction*. In the case of *Laue diffraction*, $I(z)$ is described by a more complicated expression that includes the specific effects of dynamic radiation scattering (e.g., *anomalous passage of X-rays*, and so on) which cannot be reduced to the ordinary extinction of the exponential type. In addition, in real crystals with a mosaic structure one can observe a phenomenon related to the extinction when the light propagates in a medium with random inhomogeneities. In view of the difference in orientation between mosaic blocks, a part of the transmitted beam, when diffracting, irreversibly goes out of the range of the diffraction path and never returns to the transmitted beam. This results in angular broadening and attenuation of the transmitted beam, which in some cases can be described by introducing an effective absorption coefficient μ_e. From the measurement of μ_e one can make judgments about the statistical properties of the mosaic structure (see *Mosaic crystals*).

EXTINCTION LENGTH

Crystal thickness at which the intensity of a diffracted wave (X-ray, neutron, electron) goes from a maximum to its first zero (see *Laue diffraction*). The extinction length Λ is the period of the interference beat of the intensity of the diffracted (and transmitted) wave over the crystal thickness. The beat is due to the difference between the wave vectors $\boldsymbol{k}_{h1}$ and $\boldsymbol{k}_{h2}$ (or $\boldsymbol{k}_{01}$ and $\boldsymbol{k}_{02}$ in the case of a transmitted wave) of the components (plane waves) forming it along the normal to the entry surface of the crystal (see *Two-wave approximation*). The extinction length depends on the wavelength λ and the geometric conditions of the diffraction. A typical value of the extinction length in the case of X-ray diffraction in perfect semiconductor crystals is $\Lambda_{perf} \approx 10$ μm; in the case of weakly absorbing crystals (e.g., Si) Λ is much smaller than the absorption length. In single crystals with randomly distributed defects $\Lambda_{real} = \Lambda_{perf}\,e^L$ (L is the static *Debye–Waller factor*).

EXTRACTION of charge carriers

Diminishing of charge from a *semiconductor* (insulator) by the withdrawal of current carriers (electrons) from a metallic contact. Extraction is a phenomenon opposite to *injection*. It takes place when the external electric field is parallel to the field of the near-contact potential barrier, thereby increasing its height, and causing the drift flux to prevail over the diffusion flux. The main trends of extraction as well as injection are determined by the field of the charge that emerges from the bulk semiconductor. Since the sign of this charge is opposite to the sign of the carriers drawn out of the contact, the charge-generated field hinders the withdrawal of carriers. Differences in mechanisms of bulk charge formation (as in the case of injection) make it necessary to distinguish its monopolar, bipolar, stationary and non-stationary varieties. One should differentiate extraction from other mechanism for draining the semiconductor of charge carriers – *depletion*.

EXTRAORDINARY RAY, extraordinary wave

One of the two light waves propagating through a *uniaxial crystal*, the extraordinary (e-) ray has its polarization direction in the plane of the principal section of the crystal (see *Crystal optics*). This ray may lie outside the plane of incidence, and Snell's law does not apply to it. The velocity v_e and *refractive index* n_e of the e-ray depend on the direction of propagation. Two extraordinary rays arise in a *biaxial crystal*. The extraordinary and *ordinary rays* may become separated in space.

Ff

FABRY–PEROT INTERFEROMETER (Ch. Fabry, A. Perot, 1899)

Optical system in which a beam of light is split into multiple parallel beams which are repeatedly reflected between two parallel mirror surfaces, and are then transmitted and recombined to form an interference pattern. The incident collimated beam of light undergoes multiple reflections of constant path difference $\Delta = 2nh\cos\theta$, where h is the distance between the mirrors, n is the *refractive index* of the medium between the mirrors, θ is the angle of reflection. Many Fabry–Perot interferometers are optical *laser* resonators with the radiative (i.e. lasing) medium located between the mirrors. Since the *transmittance* depends on the refractive index of the medium between the mirrors, a Fabry–Perot interferometer is used for constructing bistable optical devices, with the *nonlinear medium* confined between the mirrors (see *Optical bistability*). The function of this interferometer is commonly carried out by a single crystal, with the *reflection of light* taking place on its parallel planar faces.

FACE-CENTERED LATTICE (F-lattice)

A *Bravais lattice* for which the translations are written in terms of the vectors $\boldsymbol{a}$, $\boldsymbol{b}$, $\boldsymbol{c}$, $(\boldsymbol{a}+\boldsymbol{b})/2$, $(\boldsymbol{b}+\boldsymbol{c})/2$, and $(\boldsymbol{c}+\boldsymbol{a})/2$; the *crystal lattice basis* is [000], [1/2 1/2 0], [1/2 0 1/2], [0 1/2 1/2]. There are only two face-centered lattices among the 14 Bravais lattices, one in the *cubic system* and the other in the *orthorhombic system*.

FACETS

Microfaces at *crystal* surfaces with orientations different from that of the surface. If the instability of a surface with a given orientation towards decomposition into microfaces leads to a reduction of the surface *free energy* (see *Surface energy*), this may be the cause of facet formation. If this condition is fulfilled only within a restricted temperature range, then varying the temperature through this range produces reversible reconstructions of flat surfaces to surfaces edged by facets (e.g., some atomically-pure surfaces of *germanium* and *silicon*). Facets may appear at the surface as a result of different types of surface etching (chemical, ion, thermal) due to different speeds of dissolving and evaporating along different crystal directions (facets at edge of etching figures), and also at surfaces of growth (see *Monocrystal growth*). In addition to mechanisms associated with the surface structure and features characterizing its growth, impurities can play an important role in facet formation.

FAILURE

A multistage process of weakening and breaking interatomic bonds that results in the separation or splitting of a macroscopic specimen into two or more parts. The following principal stages are identified during failure.

Stage I. Accumulation of lattice energy by *plastic deformation* accompanied by an increase in *dislocation* and *disclination* density. Substructures may form during its course that lead to failure: *pile-up of dislocations* and *dislocation walls*, *subboundaries*, twins (see *Twinning of crystals*), etc. Plastic deformation also produces a change in shape that decreases the specimen strength. During *uniform strain*, the overall material cross-section decreases, while during a local deformation it sharply decreases only in the region of the neck.

Stage II. Generation of microcracks (see *Cracks*) and their growth to critical size in accordance with the *Griffith theory*. There are several dislocation and disclination mechanisms of microcrack generation, with examples presented in

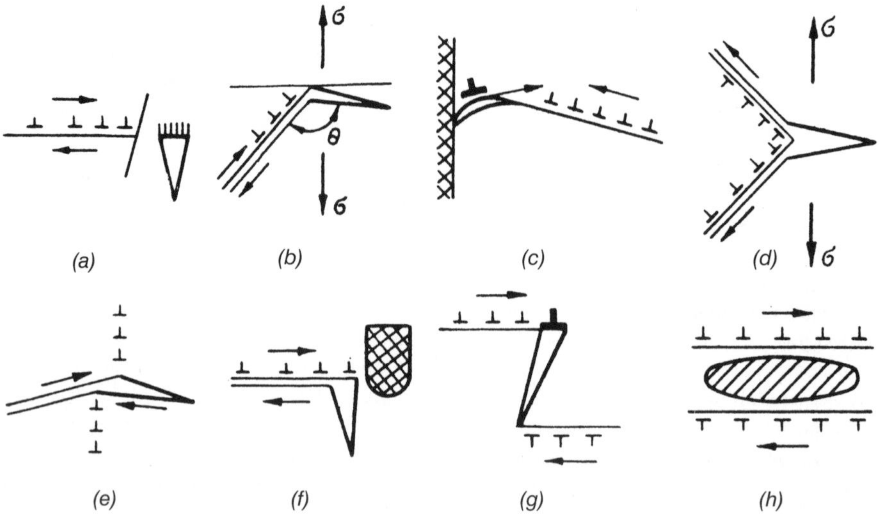

Mechanisms for the generation of microcracks, as described in the text.

the figures. The first three mechanisms are based on the concept of *slip bands* blocked by obstacles, which are regarded as flat disclination heaps that produce high *stress concentrations* in their head areas. Such a concentration may result in either a restart of a *slip* or an opening of a crack. Fig. (a) shows a mechanism for the merging of dislocations in the head of a dislocation heap resulting in the formation of an embryo microcrack. The mechanism of hampered *shear* (*Stroh theory*) is presented in Fig. (b). The crack here develops at an angle of approximately $110°$ relative to the *slip plane*, as determined by the distribution of stress at the head of the dislocation heap. The *Rozhanski–Gilman mechanism* (Fig. (c)) is based on the concept of the bending of the slip plane in which the edge dislocations accumulate. The crack develops from the point of braking in the slip plane so as to meet that pile-up. The *Cottrell mechanism* (Fig. (d)) assumes a crossing of two slip lines $\{110\}$ in body-centered cubic metals accompanied by the formation of dislocations in the $\{100\}$ plane with its *Burgers vector* $a[001]$, as prescribed by the reaction $(a/2)[\bar{1}\bar{1}1] + (a/2)[111] \to a[001]$. The merging of those dislocations into bands initiates the formation of microcracks. There are more advanced comprehensive models available for describing the generation of cracks during the interaction of two different dislocation pile-ups (Fig. (e, f)), or during the interaction of dislocation pile-ups with inclusions (Fig. (g)) or *pores* (Fig. (h)) in a crystal. Even partial disclinations (or their dipoles) are a significant source of *internal stress*, and may also result in the generation of microcracks. The probability of opening microcracks at various interfaces (*grain boundaries*, *subboundaries*) is often greater due to the presence of segregated impurities (see *Segregation*), and to the availability of energy stored during plastic deformation.

Stage III. Crack growth, the final break-up and disintegration of the specimen into two parts. The description of processes happening during this stage generally focuses on the concepts of linear mechanisms of failure that entail the concept of a critical coefficient of stress intensity (*crack resistance*). The micromechanism of failure is controlled by the character of the interatomic bonds, temperature, type of *state of stress*, and specimen structure. Micromechanisms of failure may be characterized by the *fracture toughness* and the structural indicator of failure. By taking into account the failure viscosity one may separate the types of failure into *brittle failure*, *quasi-brittle failure*, and *tough failure*. Meanwhile, according to structural indicators, *transcrystalline failure* or *intercrystalline failure* may take place. These factors of *fracture toughness* and structural indicators may be found in any combination, and they produce the following six basic types of failure:

1. Brittle transcrystalline failure, a catastrophic failure of crystals along their cleavage planes (see *Crystal cleavage*) with the formation of typical spalling streams without any noticeable plastic deformation in the crack apex.
2. Quasi-brittle transcrystalline failure, a failure by *spalling* accompanied by plastic deformation.
3. Tough transcrystalline failure, a typical micromechanism for it being the generation and merging of microcavities accompanied by the formation of pitted relief. The major crack then propagates by way of successive lacerations of the neck between the cavities formed in front of the apex of the growing crack. Variations of that type of failure include *rupture* (transverse narrowing is then $\psi = 100\%$) and *shear failure* of microcrystals.
4. Intercrystalline brittle failure propagates as a spalling along the grain and subgrain boundaries.
5. Quasi-brittle intercrystalline failure involves spalling along the grain boundaries while a preceding plastic deformation forms characteristic steps at fracture surfaces (see *Fracture*).
6. Tough (pit) intercrystalline failure during which the major crack propagates via successive lacerations of necks between the cavities formed as a result of plastic deformation situated in subboundary volumes. In that case the *localization of plastic deformation* in the subboundary volumes results from the presence of an intergrain boundary layer that is softer than the internal volumes of the grains.

In certain cases, particularly in the presence of *structural texture of metals*, anisotropies are observed in various types of failure. A combination of several varieties of failure then merge as the crack develops in various directions. With any increase in temperature one may observe a succession of micromechanisms of failure, extending from brittle fragmentation to viscous failure. However, a change in the structural type of failure may disrupt that tendency. Studies of failure mechanisms provide data for developing materials with improved mechanical and service characteristics.

See also *Brittle failure*, *Intercrystalline failure*, *Quasi-brittle failure*, *Stratification-type failure*, *Tough failure*, *Transcrystalline failure*.

FAILURE KINETICS

The development and interrelationships between elemental actions that specify mechanical *failure* and the dependence of *strength* on temperature, and on the time during which the stress continues. It has been experimentally established that the time τ which elapses until failure (see *Durability*), for solids with various types of interatomic bonding and lattice structure, obeys the equation: $\tau = \tau_0 \exp[(U_0 - \gamma\sigma)/(k_B T)]$, where σ is the tensile stress, and τ_0, U_0 and γ are parameters of a material. The exponential dependence of τ on the temperature indicates that thermal activation underlies the failure. The durability of a body under load depends on thermal fluctuations resulting in the generation of stable material discontinuities called *cracks*, and their accumulation up to a critical concentration in the failure zone. Large-scale density fluctuations serve as a source for the formation of micro-discontinuities in a continuous body. The initiation of a crack may result from a fluctuation of the number of *phonon* collisions over their mean free path, Λ, resulting in a kinetic instability with ensuing phonon pumping, tensile *strain*, and break-up of interatomic bonds inside the fluctuation region. The size of a *nucleating crack*, r, will be greater than or equal to Λ according to this model, and its *activation energy* U is given by

$$U = \frac{a^3}{G^2\kappa} - \frac{\Lambda a^2\sigma}{G},$$

where a is an interatomic distance, G is the Grüneisen constant (see *Grüneisen tensor*), and κ is the *bulk modulus*; these relations agree with experiment. Cracks are generated randomly over the material because of the random nature of thermal fluctuations, so their accumulation is accompanied by a spontaneous clustering that results in the formation of enlarged cracks which then grow to generate the failure itself. The transformation from delocalized to localized crack formation occurs when the mean distance between the cracks reaches a critical relative value K^* (expressed in the units of crack size, K), with $K^* \geqslant$ e (e is the base of natural logarithms) and K^* slightly increases with the body volume. Hence, the bulk concentration of cracks cannot exceed a value $C^* = (K^* r)^{-3}$. In practice this concentration limit is reached with r varying from 10^{-8} m in *polymers* to 10^4 m at earthquakes. Crack enlargement is accompanied by changes in the energy release characteristics (e.g., growth in amplitude of acoustic signals produced at crack formation). One can utilize the latter effect to predict failure kinetics behavior, e.g., to identify a material approaching the pre-break-up state, and to evaluate its durability.

FAILURE OF AMORPHOUS METALLIC ALLOYS

Failures characteristic of an isotropic continuous metal when the direction of the major destroying *crack* is controlled only by the nature of the load and by the temperature. At low temperatures, within the range of *cold brittleness*, one observes brittle failure of amorphous metallic alloys with the plane of the crack normal to the axis of the distending stress σ (*plane of maximum normal stress*). Amorphous metallic alloys may become plastic at high temperatures, so that the major crack may then open along the surfaces oriented at $\sim 45°$ to the distension axis, as shear stresses produce preliminary localized plastic deformations (see *Localization of plastic deformation*) that reach their maxima there. The stress responsible for the final rupture of the specimen into two parts during the failure is the normal component of the stresses at those surfaces σ_n ($\sigma_n < \sigma$).

FANO RESONANCE (U. Fano, 1961)

Appearance in absorption, reflection, scattering and *luminescence* optical spectra of the effect of a resonance interaction of two states, overlapping in energy, one of which is characterized by a discrete energy transition, and the other by a broad, continuous spectrum. A quantum-mechanical treatment of such an interaction was initially performed by G. Breit and E. Wigner (1936) in connection with the scattering of neutrons. The resonance was theoretically considered by U. Fano (1961) in the context of atomic spectra.

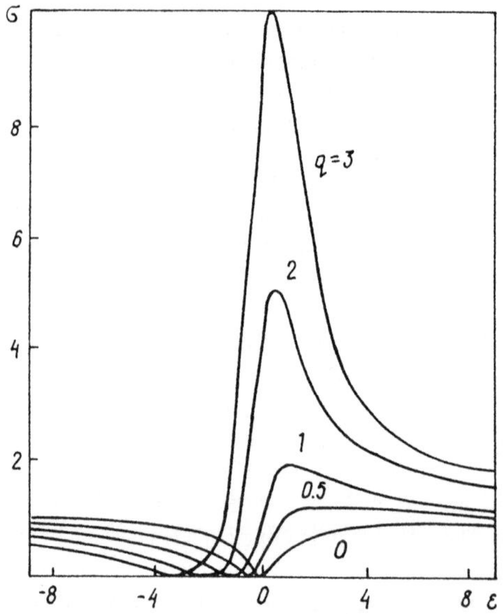

Dependence of the spectral contour or cross-section σ on the dimensionless energy parameter ε (Fano resonance) for several values of the parameter q.

The interference between states of discrete and continuous excitation produces the characteristic, and in general asymmetrical spectral *Fano contour*, as described by the formula

$$\sigma = \frac{(q+\varepsilon)^2}{1+\varepsilon^2} = 1 + \frac{q^2-1}{1+\varepsilon^2} + \frac{2q\varepsilon}{1+\varepsilon^2},$$

where σ is the cross-section or spectral line contour of the process under investigation (e.g., of *light absorption*), ε is the dimensionless energy in binding energy units, q is a parameter depending on the correlation of the probabilities of transitions from the ground state to the discrete and continuous excited states. The terms on the right side of the equation represent, from left to right, the background absorption, the symmetrical Lorentz peak, and the interaction between discrete and continuous transitions. For $q \gg 1$ the Fano contour $\sigma(\varepsilon)$ transforms to a Lorentzian curve (see *Lorentzian shape*), for $q \approx 1$ the dependence has a dispersion form, and for $q \ll 1$ an inverted spectrum with a minimum at $\varepsilon = 0$ (see Fig.) is observed. The minimum at $\varepsilon = -q$ is called the *antiresonance valley*.

Numerous appearances of a Fano resonance in the spectra of solids are known. They were observed in the exciton spectra of absorption and luminescence of *semiconductors*, in electron absorption spectra of the impurity centers in crystalline matrices, in the spectra of *Raman scattering of light* of photon and electron excitations in crystals. Investigations of the Fano resonance in solids gives information about the energy spectrum of excitations, about their life times, and their interaction mechanisms.

FARADAY, Faraday number, F

A fundamental physical constant numerically equal to the amount of electric charge which liberates 1 mole of a monovalent substance at each electrode after passage through an electrolytic solution. It has the value $F = N_{\mathrm{A}}e = 9.6485\cdot 10^4$ C/mole, where N_{A} is the *Avogadro number*, and e is the elementary electric charge.

FARADAY EFFECT, Faraday rotation (M. Faraday, 1845)

Rotation of light polarization plane of electromagnetic radiation (light) propagating within a material parallel to the direction of a constant applied magnetic field (see *Magnetooptical effects*). After passage through a layer of nonmagnetic material with the thickness d in a magnetic field with intensity B, the plane of polarization rotates through an angle $\theta = VBd$, where V is the *Verdet constant* which depends on the properties of the material, on the wavelength of the light λ, etc. The Faraday effect is caused by the difference in the speeds of the right- and left-circularly polarized components of the electromagnetic radiation which propagates longitudinally along an applied magnetic field; the superposition of these components forms linear-polarized radiation. In magnetically ordered materials the Faraday effect is determined not only by B, but also by the spontaneous magnetization.

The sign of the Faraday effect does not depend on the direction of propagation with respect to the B direction, which follows from the pseudovector character of the magnetic field intensity. This distinguishes the Faraday effect from the rotation of light polarization plane in optically active media (helical symmetry). In transparent materials the Faraday effect is expressed in terms of off-diagonal components of the *magnetic permeability* tensor μ_{ij}, which are proportional to the odd powers of B, and thus they are unsymmetrical. At high B intensities, especially near resonant frequencies, there is a contribution from terms with higher powers of B than the first one.

It is used for the determination of the *band structure* parameters of semiconductors, the properties of impurity states, the structure of molecules, the *magnetization* of solids. The *effective masses* of the carriers in many semiconductors have been determined by measuring the Faraday effect of free carriers in the IR range of the spectrum. The Faraday effect is used in light modulators.

FATIGUE

Progressive accumulation of damage in a material under action of alternating stresses. This results in a change of properties, generation and propagation of *cracks*, and *failure* of the material. The tendency to counteract the development of fatigue is referred to as the *fatigue resistance* of a material. By analogy with the term *long-term strength*, this property is sometimes called *cyclic strength* (see also *Endurance limit*).

The problem of fatigue is one of the most complicated and important ones for those branches of physics and engineering dealing with the *strength* of solids. *Materials science* and the mechanics of deformable bodies offer various models for the origin and propagation of fatigue damage, but the quantitative characteristics of fatigue are best obtained experimentally by comparing materials and structural members under various conditions which depend on the task to be achieved. The data provided by relatively simple fatigue tests can be used to calculate the characteristics of fatigue for complicated cases of loading, to determine the *durability* of structural members at a given load, or conversely, to determine the permissible load for a given durability. Many of these problems have been solved by analyzing the mechanics of cracks (see *Fatigue crack growth*).

Numerous methods have been developed for increasing fatigue resistance by applying special treatments to those parts of machines and constructions which are exposed to the greatest stresses. From the materials science standpoint, the major goal is the preparation of materials (both homogeneous and composite) which offer enhanced resistance to the initiation and propagation of fatigue cracks. The problem is that the attainment of high material static *strength* is not necessarily accompanied by a corresponding increase of its cyclic strength.

See also *Adsorption fatigue*, *Corrosion fatigue*, *Thermal fatigue*.

FATIGUE CRACK GROWTH

Cyclic nonstationary process of increasing *crack* length, which causes *failure* of a material under cyclic loading (see *Cyclic strength*). At low-cycle *fatigue* the growth of cracks is of a quasi-static nature in every loading cycle, leading to the advancement of the crack, and to the formation of a *fracture* with a *neck*. After many fatigue cycles the local dislocation structure changes (see *Dislocation*) in the surface layer at the stage of crack generation, and near the vertex of the advancing crack at the growth stage prior to the fatigue failure. The most characteristic defects which prepare for the formation of the fatigue crack are *intrusions* and *extrusions*, or microcavities and microbulges at the surface, formation in the surface layer of stable *slip bands*, where irreversible *dehardening* of the material occurs. The *Cottrell–Hull model* (A.H. Cottrell, D. Hull, 1957), where two surface dislocation sources in crossing slip systems are effective, is the most widely used model for explaining the mechanism of formation of extrusions and intrusions. At the first stage of fatigue crack growth, its propagation occurs at the angle of 45° to the direction of the load application axis, or in the direction of action of the maximum tangential stresses. The duration of this stage of growth is no more than 5% of the total time until failure. At the second stage characterized by a stable growth speed the generated crack, having achieved its macroscopic size, changes its trajectory and spreads perpendicular to the axis of application of the stresses. The rate of the crack advancement is determined by the structure and phase composition of the material, by the magnitude of the amplitude of cyclic loading, and also by the testing medium. The sections of flexure being formed at the stage of stationary growth are characterized by specific grooves, reflecting the cyclic interlacing of the sharpening and blunting of the advancing crack. Evolution of the material structure in the vicinity of the growing *fatigue crack* leads to the appearance of secondary microcracks, located parallel to the fatigue grooves. The third, and concluding, stage of fatigue crack growth proceeds with acceleration after reaching the critical length of the fatigue crack. In this section, the crack growth proceeds at high speed. Failure in this region may have a tough, as well as a brittle nature (see *Tough fracture*, *Brittle fracture*).

FAULTING

A *fault* is a fracture in which movement of material takes place along its plane. Faulting involves the *plastic deformation* of crystals associated with the nonuniform distribution of *slip* (gliding) within the sample. Nonuniformity of sliding upon faulting during the course of *strain* (usually upon compression) is caused by the loss by the *crystal*, at first of elastic stability, and then of plastic stability. The sliding develops at regions with orientations that are more favorable for it. At the moment of faulting, temporary unloading of the sample takes place, which is manifested through the sharp falling of the level of stresses seen on a *deformation diagram*. Faulting is observed in those crystals where it is possible to create a state of deformation which hinders the development of gliding: in the case of hexagonal metals, it is necessary to perform compression or stretching approximately along the basal plane, and in the case of CsCl type (BCC) crystals the compression or stretching is carried out close to one of the mutually orthogonal gliding directions of the type ⟨100⟩. At the earliest stages of sliding, there appear sections of *stress concentration*, most often in the *fractures* at the surface, where upon crank flexure the lattice is subjected to the greatest deviation from the source orientation. In these areas short *slip bands* are generated, spreading rapidly perpendicular to themselves toward the center of the sample. Several such bands may be generated in different parts of the sample, and then multi-layer *faults* or several separate bands of faults are formed. As faulting is associated with the behavior of large collective bodies of *dislocations*, it may be described in terms of the *disclination model of plastic deformation*.

FAULTS

Regions being formed during the course of *plastic deformation*, generally in the form of interlayers, where the lattice is turned through some angle with respect to the remainder of the crystal. Unlike the *twinning of crystals*, the angle of lattice rotation is not constant, but varies from 0 to 90° and more, depending on the degree of *strain*. Movement of material can take place along the plane of a fracture that constitutes a fault. The boundaries of faults are not distinct, and they exclude regions of sharp *fracture* of the crystal. In the remaining parts of the faults, there is observed a smooth rotation of the lattice from its source orientation to its maximally rotated state. These were originally described by O. Lehman (1885) and O. Mügge (1898) who observed faults in a broad group of *mineral* crystals (gypsum, molybdenum, etc.). Later, they were obtained upon plastic deformation of *zinc* and *cadmium*, of CsCl-type *alkali-halide crystals*, etc. (*bismuth*, galena, the polycyclic hydrocarbon naphthalene). In appearance, faults in crystals are displayed in different ways. Most often (especially at comparatively low deformations under conditions of compression) they cause the elbowed *flexure* of the samples. At higher degrees of compression, the multiple elbowed flexure occurs in two or more directions, as a result of which thickening appears near the edges or in the center of the sample. In these cases faults constitute the totality of many narrow bands with discretely changing orientation of the lattice. Studies of the formation conditions of the fault bands show that most often they appear during compression under conditions that make the *slip* difficult. Bands of faults are regions of *localization of plastic deformation* by slip (see *Faulting*).

FAULT, STACKING

See *Stacking faults*.

FEDOROV SYMMETRY GROUP

Name for a crystallographic *space group* used in the Russian literature.

FEMTO... (fr. Danish, Norwegian *femten*, fifteen)

Prefix to the label of a physical quantity to denote a value equal to 10^{-15} of the basic value. The abbreviated designations begin with "f", e.g., 1 fC (femtocoulomb) $= 10^{-15}$ C.

FERMI CONSTANT

See *Hyperfine interaction*.

FERMI–DAVYDOV COMBINED RESONANCE

A phenomenon caused by the effect of intermolecular interactions that leads to an intermolecular resonance (*Davydov splitting*) with a number of components, a distribution of intensities, and a polarization of the quantum mechanical transitions resulting from a *Fermi resonance*. It may appear preferentially in vibrational and electronic-vibrational spectra of crystals in the presence of van der Waals bonding (see *Van der Waals forces*). A typical example of this phenomenon is a methyl iodide crystal whose vibrational spectrum, instead of displaying two liquid phase Fermi components, exhibits four components, close together in intensity, and polarized respectively in mutually perpendicular directions (A.S. Davydov, 1951).

FERMI–DIRAC STATISTICS, Fermi statistics (E. Fermi, P. Dirac, 1926)

The statistics describing the distribution, according to their quantum states, of a large number of identical particles with half-integer spin. Particles and *quasi-particles* obeying Fermi–Dirac statistics are called *fermions*; examples of these particles are: electrons (also *holes* in *solids*), *muons*, protons, *neutrons*, neutrinos, etc. The wave function of a fermion is antisymmetric with respect to the interchange of any pair of particles (see *Second quantization*). The impossibility of more than one particle occupying the same quantum state (*Pauli exclusion principle*) is characteristic for fermions. From the *Gibbs distribution*, taking this property into account, the expression for the average number of particles $\bar{n}_i$ in quantum state i with energy ε_i for an ideal gas of fermions in thermodynamic equilibrium is given by

$$\bar{n}_i = \left[\exp\left(\frac{\varepsilon_i - \mu}{k_B T}\right) + 1\right]^{-1}, \qquad (1)$$

where k_B is the *Boltzmann constant*, and μ is the *chemical potential* of the system, which depends on the total number of particles $N = \sum_i \bar{n}_i(\mu)$. Eq. (1) is called the *Fermi–Dirac distribution*. For the limit $\exp[(\varepsilon_i - \mu)/(k_B T)] \gg 1$ it closely approximates a *Boltzmann distribution*. At $T = 0$ all the fermion states below some energy ε_F are occupied, all the states lying above ε_F are empty; and the value of ε_F coincides to the value of μ:

$$\varepsilon_F = \frac{p_F^2}{2m}, \qquad p_F = \left(\frac{6\pi^2 N}{gV}\right)^{1/3}\hbar, \tag{2}$$

where m is the fermion mass, N is number of particles in the volume V, and g is the *degeneracy*.

The quantity ε_F is called the *Fermi energy*, and p_F is the *Fermi momentum*. In *metals* and *degenerate semiconductors* with a nonquadratic dispersion law the dependence $\varepsilon_F(p_F)$ has a more complex form than Eq. (2) (see *Fermi surface*), and is often anisotropic. For $T > 0$ a certain number of fermions has an energy greater than ε_F; empty states vacated by these elevated fermions at $\varepsilon < \varepsilon_F$ are called *holes*.

FERMI ENERGY

The value of energy, below which all the states of a system of particles obeying *Fermi–Dirac statistics* (*fermions*), are occupied at absolute zero temperature, and all higher states are vacant. The existence of a Fermi energy is a consequence of the *Pauli exclusion principle* according to which there can be no more than one fermion particle in a state (or two electrons with opposite spins).

For an ideal degenerate gas of fermions the Fermi energy coincides with the value of the chemical potential at $T = 0$ K, and may be expressed as follows in terms of the number N of gas particles in a unit volume:

$$\varepsilon_F = \frac{(2\pi\hbar)^2}{2m}\left[\frac{3N}{(2S+1)4\pi}\right]^{2/3},$$

where m and S are the mass and *spin* of the particle, respectively. The quantity $p_F = (2m\varepsilon_F)^{1/2}$ is the *Fermi momentum*. At $T = 0$ K all the states with momenta $p < p_F$ are occupied, and those with $p > p_F$ they are empty. In other words, at $T = 0$ K fermions in an isotropic momentum space occupy all the states within the sphere $p^2 = 2m\varepsilon_F$ (*Fermi sphere*), with the radius p_F in *momentum space* ($\boldsymbol{k}$-space). The value $v_F = p_F/m$, called the *Fermi velocity*, determines the upper limit of fermion speeds at $T = 0$ K.

The gas of *conduction electrons* in *metals* and in *degenerate semiconductors* may be characterized by other than a quadratic dispersion law. Then at $T = 0$ K the conduction electrons fill momentum space in regions of a more complex shape (see *Fermi surface*). The location of the Fermi energy in a partly filled electron band is called the *Fermi level*. In the case of *semiconductors* the position on the energy scale of the *chemical potential* at a finite temperature is also called the Fermi level. It is necessary to remember that this designated Fermi level may be located within the gap or band of forbidden energies, and hence not coincide with any actually existing energy level.

FERMI LEVEL

See *Fermi energy*.

FERMI LEVEL PINNING, Fermi level limiting position

Establishment of a fixed value of the *Fermi level* μ_F of a semiconductor in the presence of electrically active centers or defects, a value which does not change with further additions of these centers or defects. Fermi level pinning occurs when donors and acceptors are introduced into the sample in equal concentrations $N_d = N_a$ (e.g., levels of *vacancies* and *interstitial atoms* with formation of *Frenkel defects*). Upon reaching a sufficiently high concentration, when the numbers of charged donors and acceptors $N_d^+ = N_a^-$ dominate the *electroneutrality condition*, μ_F does not depend on N_d, N_a, but rather is equal to the limit value $\mu_F = (\varepsilon_a + \varepsilon_d)/2$, where ε_a and ε_d are the positions of *donor* and *acceptor* levels (measured from a common origin). A Fermi level limiting position is also realized when there is a dominant concentration of defects of one type, and these defects are *amphoteric centers*, especially if they are centers with a negative *correlation energy*.

FERMI LIQUID

Quantum liquid, whose elementary excitations (*quasi-particles*) have half-integer spin (*fermions*), and due to the Pauli exclusion principle (see *Fermi–Dirac statistics*) they have a characteristic *Fermi momentum* p_F determined by the density of particles of the liquid $p_F = (3\pi^2 N/V)^{1/3}\hbar$ (V is the volume of liquid, N is the number of fermions). The Fermi momentum determines the *Fermi energy* ε_F, from which the energies of quasi-particles $\xi = |\varepsilon(p) - \varepsilon_F|$ are derived. The excitations of a Fermi liquid are generated in pairs: a hole quasi-particle with $p < p_F$ and a corresponding electron quasi-particle with $p > p_F$. A strong interaction between the quasi-particles is characteristic of a Fermi liquid, so the single-particle state is not the natural state of a many-particle system, and it breaks down after a finite time. Due to the features of the Fermi spectrum (presence of filled states), regardless of the interaction strength, the decay time τ_{dec} is long for the single-particle states near the *Fermi surface* $\varepsilon(p) = \varepsilon_F$ (at sufficiently low temperatures): $\tau_{dec} \sim \varepsilon_F \hbar/\xi^2$ (at $T \neq 0$, $\xi \sim k_B T$). Thus the conditions for applying Fermi liquid theory are: $\xi \ll k_B T \ll k_B T_f = \varepsilon_F$. Conduction electrons in a *metal*, the liquid helium isotope ^{3}He, and nuclear matter serve as examples of Fermi liquids. For conduction electrons in a metal $T_f \sim 10^4$ to 10^6 K.

The theory of a Fermi liquid was formulated by L.D. Landau on the basis of two assumptions: (1) the interactions of quasi-particles do not disturb the systematics of levels near the Fermi surface, and in the presence of interactions the original fermions are replaced by quasi-particles; (2) interactions of particles with one another may be described with the help of a *self-consistent field* by the introduction of a *distribution function* for the quasi-particles $\hat{n} = n_{\alpha\beta}(p, r, t)$. The function n is a 2×2 operator-matrix in spin space. The quasi-particle energy depending on the spin $\varepsilon_{\alpha\beta}(p, r, t)$ is determined as the first variational derivative of the total system energy $E[\hat{n}]$:

$$\delta E = \mathrm{Tr}_\sigma \int \hat{\varepsilon}(p, r, t)\, \delta\hat{n}\, \mathrm{d}\tau\, \mathrm{d}^3 r, \qquad \mathrm{d}\tau = \frac{\mathrm{d}^3 p}{(2\pi\hbar)^3}, \tag{1}$$

where p is the momentum, and $\hat{\varepsilon} = \hat{\varepsilon}_0 + \delta\hat{\varepsilon}$, so

$$\delta\hat{\varepsilon}(p, r, t) = \mathrm{Tr}_{\sigma'} \int \hat{f}(p, r; p', r') \times \delta\hat{n}(p', r', t)\, \mathrm{d}\tau'\, \mathrm{d}^3 r'. \tag{2}$$

The quantity $\hat{f}$ called the *Landau correlation function* is a 4×4 matrix in spin space which predetermines a set of phenomenological parameters that are taken from experiment. In metals, where the *correlation length* and wavelength of electrons are of the order of interatomic distances, it is considered to be independent of the coordinates.

In the presence of only the *exchange interaction* the matrix $\hat{f}$ is expressed with the help of two scalar functions $\eta(p, p')$ and $\xi(p, p')$:

$$\hat{f}(p, p') = \eta(p, p') + \sigma\sigma'\xi(p, p') \tag{3}$$

(σ_i are *Pauli matrices*).

In an isotropic system the functions $\eta(\boldsymbol{p}, \boldsymbol{p}')$ and $\xi(\boldsymbol{p}, \boldsymbol{p}')$ depend only on the angle θ between the vectors $\boldsymbol{p}$ and $\boldsymbol{p}'$, and may be expanded in terms of Legendre polynomials. Dimensionless *Landau parameters* A_l and B_l are coefficients of the expansion the Legendre polynomials of the functions $\eta(\theta)$ and $\xi(\theta)$ multiplied by $g(\varepsilon_F)(2l+1)^{-1}$, where $g(\varepsilon_F)$ is the *density of states* of the quasi-particles at the Fermi surface. The function $\hat{f}$ satisfies the stability requirements for the ground state of a Fermi liquid with respect to both deformations of the Fermi sphere, and small spin vibrations. This imposes specific conditions at the parameters A_l and B_l (*Pomeranchuk criterion*, I.Ya. Pomeranchuk, 1958):

$$\begin{aligned} 1 + A_l &> 0, \\ 1 + B_l &> 0. \end{aligned} \tag{4}$$

Inequality (4) for B_0 expresses a condition for a Fermi liquid paramagnetic state with respect to a transition to the ferromagnetic one (*Stoner criterion*, see *Ferromagnetism of metals and alloys*). The Fermi distribution of quasi-particles follows from the requirement that the overall Fermi entropy be maximal (for a fixed number of particles, $\delta N = 0$, and a fixed energy, $\delta E = 0$):

$$\hat{n}(\hat{\varepsilon}) = \left(\exp\frac{\hat{\varepsilon} - \varepsilon_F}{k_B T} + 1\right)^{-1}, \tag{5}$$

where the single-particle energy $\hat{\varepsilon}$ is a functional of $\hat{n}$. The *specific heat* C depends linearly on the temperature, as also is the case for a Fermi gas, $C = (\pi^2/3)g(\varepsilon_F)k_B T$. Within the framework of the isotropic model the *magnetic susceptibility* of a Fermi liquid $\chi = g(\varepsilon_F)\mu_B^2(1 + B_0)^{-1}$ contains the parameter B_0 (at $B_0 < 0$ the condition

$|B_0| \to 1$, $\chi \to \infty$ corresponds to the transition to the ferromagnetic state). The *sound velocity* is expressed with the help of the parameter A_0: $s^2 = [p_F^2/(3m_0m^*)](1 + A_0)$. When Galilean invariance is satisfied the quasi-particle *effective mass* m^* is related to the source particle mass m_0 by the expression $m^* = m_0(1 + A_1)$. In the presence of an external electromagnetic field with vector and scalar potentials $\boldsymbol{A}(\boldsymbol{r}, t)$ and $\varphi(\boldsymbol{r}, t)$, respectively, the quasi-particle energy $\hat{\varepsilon}(\boldsymbol{p})$ is replaced by the expression

$$\begin{aligned} \hat{\varepsilon} &= \hat{\varepsilon}\left(p - \frac{e}{c}\boldsymbol{A}(\boldsymbol{r}, t)\boldsymbol{r}, t\right) \\ &\quad - \mu_B \sigma \boldsymbol{B} + e\varphi(\boldsymbol{r}, t), \\ \boldsymbol{B} &= \nabla \times \boldsymbol{A}. \end{aligned} \tag{6}$$

In a neutral Fermi liquid two modes of sonic vibrations are possible. The low-frequency ones ($\omega\tau \ll 1$, τ is the mean free path time of the quasi-particles, $\tau \sim 1/T^2$) correspond to ordinary sound, which is sometimes called first sound. The velocity s of this sound, which involves density variations, is determined by the *compressibility*, and in the isotropic case it is expressed through parameters A_0 and A_1 (see above). At reduced temperatures ($\omega\tau \sim 1$) ordinary sound strongly attenuates. An undistorted Fermi surface shape is characteristic of ordinary sound: both the Fermi momentum and the density of the liquid vary, and the Fermi surface vibrates as a whole with an amplitude whose value is associated with the speed of the liquid motion within the wave. High-frequency vibrations ($\omega\tau \gg 1$, collision-free mode) are called second sound or *zero sound*. This sound is sometimes referred to as a temperature wave which does not involve density variations. The velocity $u = \omega/k$ of zero sound is determined from the equation

$$(\boldsymbol{k}\boldsymbol{v} - \omega)\delta\hat{n} - \boldsymbol{k}\boldsymbol{v}\frac{\partial n_0}{\partial \varepsilon}\int \eta(\boldsymbol{p}, \boldsymbol{p}')\,\delta n'\delta\tau' = 0, \tag{7}$$

where $\boldsymbol{v} = \partial\varepsilon/\partial\boldsymbol{p}$ is the quasi-particle velocity. In the general case the presence of several branches of *zero sound vibrations* is possible, each propagating with different speeds. Unlike ordinary sound, zero sound is associated with periodic distortions of the Fermi surface shape. If in this case the Fermi surface deformations are not axially symmetrical, then the vibrations take place without any variation of the liquid density. In the simplest case $\eta(\theta) = A_0 g^{-1}(\varepsilon_F)$ the inequalities $u > v$, $A_0 > 0$ are the conditions for the existence of zero sound, and the Fermi surface is axially symmetric and drawn toward the direction of the wave propagation. *Spin waves* (in a neutral Fermi liquid) are vibrations of the spin characteristic of a Fermi liquid associated with the second term in Eq. (3). In the absence of an external magnetic field the expression for these vibrations is obtained from Eq. (7) by replacing η by ξ. In the particular case $\xi = B_0 g^{-1}(\varepsilon_F)$, for the propagation of a spin wave $B_0 > 0$ is necessary. For liquid ^{3}He $B_0 < 0$, which supports the impossibility of propagating waves of this type.

In the case of a charged Fermi liquid (e.g., electrons in a metal) vibrations of the electron density lead to the origin of the electric fields that alter the conditions for the existence of zero-sound vibrations. This means that vibrations of the electron density at frequencies $\omega < \omega_p$ in a metal are absent (ω_p is the plasma frequency). Although zero-sound vibrations are possible in a Fermi liquid in which there is no variation of the electron density, zero sound has not yet been detected experimentally in an electron liquid (see *Spin waves in nonmagnetic metals*). See also *Microscopic theory of Fermi liquid.*

FERMI MOMENTUM

See *Fermi energy* and *Fermi–Dirac statistics.*

FERMION, HEAVY

See *Heavy fermions.*

FERMIONS IN A SOLITON FIELD

A specific state of a fermion, in particular, of a *conduction electron*, described by some one-dimensional models of interacting fermion and boson fields with a *spontaneous symmetry breaking* due to the double degeneracy of the ground state of the free boson field. These two states are associated with the spatially nonuniform solution of the classic equation for a boson field, called a *soliton* or *kink*. In a fermion–soliton system there appear two associated states, degenerate in energy, and carrying the charges of $\pm e/2$ (e is unit fermion charge), depending on whether the given state is occupied or empty. These states are interpreted as soliton states with a non-integer charge.

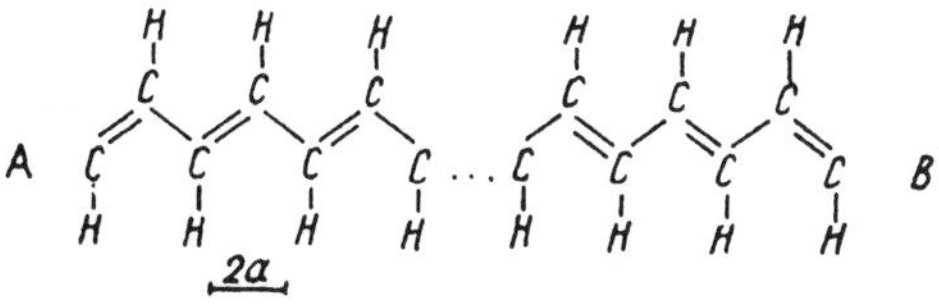

Fermions in a soliton field.

The phenomenon of *fractional charge* was predicted by R. Jackiw and C. Rebbi (1976) in the model system described above. As an example of a real system exhibiting a similar phenomenon, consider a *polymer* – the *polyacetylene* $(CH)_x$ which may exist in two different forms: trans-$(CH)_x$ and cis-$(CH)_x$. The solitons in polyacetylene are the *domain walls* (kinks) separating these two possible structures of the dimerized chains $(CH)_x$ (i.e. chains with interlacing single and double bonds). The figure shows the chain dimerized in the trans-$(CH)_x$ form, with a domain wall $(\cdots)$ separating the two possible configurations A and B.

In the non-dimerized structure the equilibrium separation is $a = 0.122$ nm between adjacent CH-groups. The nondimerized structure has mirror symmetry, and a spontaneous breaking of this symmetry due to *Peierls instability* in this quasi-one-dimensional electron–phonon system leads to the illustrated dimerized picture with the lattice period $2a$ (there occur small ~ 0.004 nm bond length variations). On the assumption of a half-filled electron band the dimerized chain in the ground state is an insulator, each state having two electrons with up and down *spins*. In the uniform configuration (i.e. in phase A or in phase B) the occupied states are separated from the unoccupied ones by the energy gap 2Δ. Upon the introduction of a domain wall in the system a radical rearrangement of the spectrum takes place: in the center of the gap there appears a bound state ψ_0, the so-called *zero mode*, and in both the *valence band* and the *conduction band* there appears a vacant or "semi-state" (for one direction of spin). As the total number of electrons is conserved, the missing electron in the valence band appears in the state ψ_0. In the valence band all the electrons are paired, and the individual electron with unpaired spin 1/2 is in the state ψ_0. As a result of this, a soliton with charge $Q = 0$ and spin 1/2 is formed. The extra spin in the system is compensated by the spin of an antisoliton, to which the remaining semi-state in the valence band corresponds. The existence of soliton–antisoliton pairs is expected by *Kramers theorem* and by topological considerations.

If the state ψ_0 is empty or doubly occupied, then the soliton spin is zero, and the charge $Q = \pm e$. Charged and neutral solitons are formed, respectively, in doped and undoped polymers, and are recognized by their electrical and magnetic properties.

FERMI QUASI-LEVELS

Energy levels associated with the filling of allowed *energy bands* by charge carriers in *semiconductors* under nonequilibrium conditions. In the state of thermodynamic equilibrium the distribution of carriers according to their energies is described by *Fermi–Dirac statistics*, and it is determined by the temperature T and the *Fermi energy* ε_F. Irradiation of semiconductor or *injection of charge carriers* disturbs this equilibrium, and a quasi-equilibrium state may appear. If the *relaxation times* of the momentum and energy for electrons and holes are much shorter than the time of their *recombination*, then within each allowed energy band an equilibrium distribution of energies is established corresponding to a lattice temperature. Due to the correlation of electrons and holes, however, this is not an overall equilibrium situation. It means that there is no definite *Fermi level* for the system as a whole, but rather in each band the Fermi distribution for electrons and holes corresponds to its own individual "intrinsic" Fermi level:

$$\Psi_e(\varepsilon) = \left[1 + \exp\left(\frac{\varepsilon - \varepsilon_F^e}{k_B T}\right)\right]^{-1},$$
$$\Psi_h(\varepsilon) = \left[1 + \exp\left(\frac{\varepsilon_F^h - \varepsilon}{k_B T}\right)\right]^{-1}, \quad (8)$$

where ε_F^e, ε_F^h are the distances between the Fermi quasi-levels and the band edges.

Under conditions when it is possible to introduce Fermi quasi-levels, the correlations which associate the concentration of electrons n_e and holes n_h with the positions of corresponding Fermi quasi-levels have a form which is similar to that under equilibrium conditions:

$$n_e n_h = n_i^2 \exp\left(\frac{\varepsilon_F^e - \varepsilon_F^h}{k_B T}\right), \qquad (9)$$

where n_i is the equilibrium concentration of carriers of each sign in the intrinsic semiconductor. Eq. (2) is a generalization of the *law of mass action* for the quasi-equilibrium system. If there is a fast enough exchange of charge carriers between the *conduction band* (or *valence band*) and the *local electronic levels* in the *band gap*, then it is possible to introduce a generalized Fermi quasi-level for them.

FERMI RESONANCE (E. Fermi, 1931)

Anharmonic interaction of vibrational states of a crystal having randomly coinciding or closely spaced frequencies, which leads to the mixing of their respective wave functions, and to the formation of two new states with other frequencies and intensities within the spectrum. Usually in the simplest case vibrations of different orders resonate, e.g., one fundamental frequency and the first harmonic of another fundamental frequency. As a result, the spacing between bands in the adsorption spectrum or in the spectrum of *Raman scattering of light* increases (new states are repelled from each other), and the intensity of the harmonic approaches that of the fundamental. There are more complex cases, e.g., the interaction of two ground state vibrations with the harmonic of at third. In the lattice spectra of *ferroelectric* crystals (e.g., quartz, $AlPO_4$) so-called *soft modes* of vibration are observed, with frequencies that tend to zero at the approach to the *phase transition* temperature. If the harmonic of an acoustic vibration is below the frequency of the "soft" optical vibration, then heating causes the distance between them to decrease to such extent that, due to enhancement of the anharmonic interaction, the frequency of the soft mode within some neighborhood stops to decrease, whereupon the harmonic frequency becomes lower, and its intensity grows.

A Fermi resonance often appears in the spectra of impurity crystals and *solid solutions*, which permits one to vary the composition (or concentration) and properties of impurities, and thereby change the frequency of lattice or *local vibrations* within wide limits. Thus, there is a probability of crossing the single-phonon band with the two-phonon excitations (see *Biphonon*), with the appearance of a Fermi resonance. This happens in such particular systems as $Zn_xCd_{1-x}S$, $In_xGa_{1-x}P$, $ZnTe_xSe_{1-x}$, etc. Attempts to explain the novel features of these *vibrational spectra* without invoking the notion of Fermi resonance have been unsuccessful.

A simpler case of a Fermi resonance interaction type appears when the frequency of a local vibration enters the region of a two-phonon spectrum of the main lattice. In ZnSe:Mg crystals the local vibrations created by the *impurity atoms* of magnesium serve as an example of this. Due to the high probability of the excitation of vibrational excitons of small radius in *molecular crystals*, a Fermi resonance in its pure form does not occur in them, but the presence of a *Fermi–Davydov combined resonance* is a characteristic singularity in most crystals of this type.

FERMI STATISTICS

The same as *Fermi–Dirac statistics*.

FERMI SURFACE

An *isoenergetic surface* in the space of *quasi-momenta* (***k**-space*) which separates the filled electron states from the unoccupied states; it is one of the main constructs of the electronic theory of *metals*. Due to degeneracy of the electron system, the location of the Fermi surface plays a major role in determining many properties of metals. Their velocities are in directions normal to Fermi surface: $\boldsymbol{v} = (\partial\varepsilon/\partial\boldsymbol{k})_F$. As is the case for all isoenergetic surfaces in $\boldsymbol{k}$-space, the Fermi surface has a a periodicity

$$\varepsilon_F = \varepsilon(\boldsymbol{k}) \equiv \varepsilon(\boldsymbol{k} + \boldsymbol{b}), \qquad (1)$$

where ε_F is the Fermi energy, and $\boldsymbol{b}$ is an arbitrary vector of the *reciprocal lattice*. If a periodic Fermi surface splits into separate regions (sheets, pockets), each of which is located within a *unit cell* of $\boldsymbol{k}$-space, such a Fermi surface is called a *closed Fermi surface*, and if the Fermi surface ranges continuously through $\boldsymbol{k}$-space, it is called an *open Fermi surface*. Also *electron Fermi surfaces* and *hole Fermi surfaces* are distinguished. In the former case the normal (velocity) is directed outside the surface, in the latter case it is

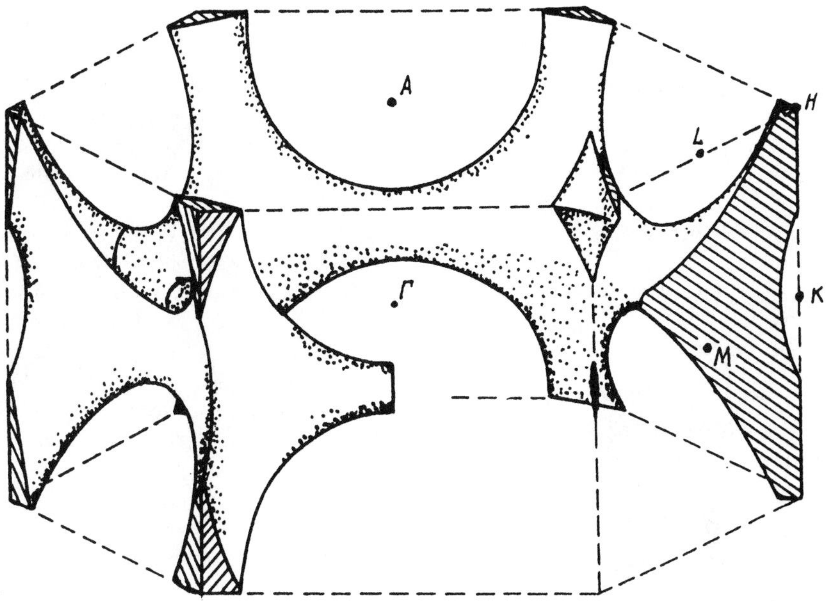

Fig. 1. Hexagonal close packed structure Fermi surface of zinc showing the symmetry points Γ, A, K, L, and M: the first Brillouin zone exhibits hole "pyramids" at H points; the second zone comprises a hole "monster"; the third zone involves electron needles at K points (the electron lens centered at the center Γ point is not shown).

inside the surface. A Fermi surface can have several convolutions (see Figs. 1–4). Among them closed and open, electron and hole varieties are encountered. The volume of each enclosure of a Fermi surface Ω_i (calculated for one unit cell of $\boldsymbol{k}$-space) is associated with the density of electrons: $n_i = 2\Omega_i/(2\pi\hbar)^3$, whereas

$$\sum_i n_i = n, \qquad (2)$$

where n is the density of *conduction electrons* of the given metal ($n = z/V_0$, z is the number of valence electrons in the unit cell of volume V_0). The volume of a hole-type Fermi surfaces is usually associated with the density of holes $n_i^{(h)}$, so Eq. (2) is naturally modified by this. Metals (Be, Bi, P, Sb, etc.) where the volume of the hole Fermi surfaces equals the volume of the electron Fermi surfaces are called *compensated metals*. For them the equation $n_e = n_h$ ($n_{e,h}$ are the total densities of electrons and holes, respectively) replaces the condition (2).

In an applied magnetic field $\boldsymbol{B}$, under the action of the Lorentz force $e\boldsymbol{v} \times \boldsymbol{B}$ a conduction electron between collisions moves along a trajectory in $\boldsymbol{k}$-space which is at the intersection of the Fermi surface with the plane $\boldsymbol{k}\cdot\boldsymbol{B} = \text{const}$. The projection of the electron trajectory in $\boldsymbol{r}$-space to the plane $\boldsymbol{r}\cdot\boldsymbol{B} = \text{const}$ is similar to its trajectory in $\boldsymbol{k}$-space. This circumstance served as a basis for devising various methods for investigating the conduction electron motion (on Fermi surfaces of metals) according to their properties in a magnetic field (see *Galvanomagnetic effect, De Haas–van Alphen effect, Shubnikov–de Haas effect, Size effects*, etc.). Fermi surfaces of all the monatomic metals and of many *intermetallic compounds* are known with good accuracy. The branch of the electronic theory of metals which focuses attention on geometrical properties of Fermi surfaces (topology, including presence of singularities), and on the relationships of these geometrical properties with physical properties, is referred to as *fermiology*.

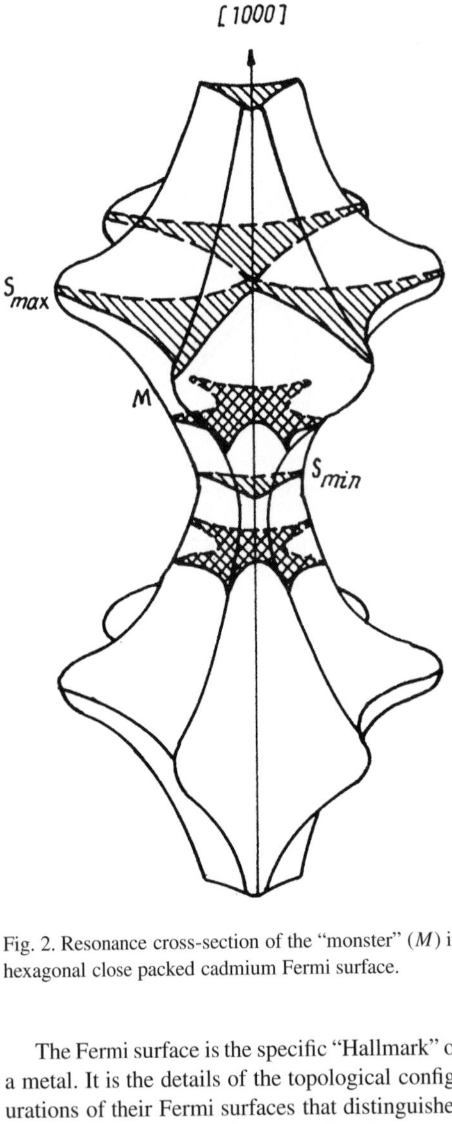

Fig. 2. Resonance cross-section of the "monster" (M) in hexagonal close packed cadmium Fermi surface.

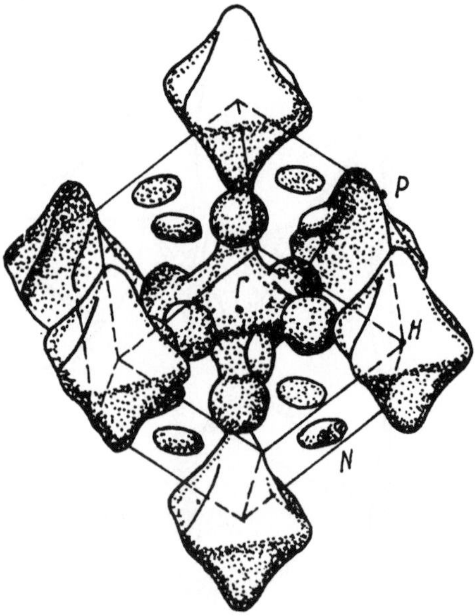

Fig. 3. Lomer model of the body-centered cubic structure Fermi surface of molybdenum and tungsten.

The Fermi surface is the specific "Hallmark" of a metal. It is the details of the topological configurations of their Fermi surfaces that distinguishes one metal from another. The variations in properties which occur in metals are naturally reflected in their Fermi surfaces. Thus, at the transition from the paramagnetic state to the ferromagnetic one Fermi surface is split to two: electrons with different projections of spins belong to separate surfaces; at the *superconducting phase transition* a gap forms above the Fermi surface, a region of forbidden energy states which arises due to the appearance of *Cooper pairs*; upon the onset of a metal–insulator transition the conduction electrons vanish (i.e. they bond to the metal atoms), and the Fermi surface disappears with them. There are transitions (called electron-topological transitions or *Lifshits transitions*) which arise directly from a variation of the Fermi surface topology. Under the influence of external pressure, or by a variation of the density of conduction electrons (due to donor or acceptor impurities) the *Fermi energy* may "pass" through a value of the band energy ε_c at which an isoenergetic surface changes its connectedness (see *Van Hove singularities*). At $\varepsilon_F = \varepsilon_c$ the following may happen: either a new convolution of the Fermi surface appears (disappears), or a bridge near the Fermi surface is severed (created). A variation of the connectedness is accompanied by a singularity of the *density of states* $\nu(\varepsilon)$ at the energy equal to Fermi energy ε_F: $\delta\nu(\varepsilon_F) \sim |\varepsilon_c - \varepsilon_F|^{1/2}$, and also at the temperature $T = 0$ by a free energy singularity $\delta F \sim |\varepsilon_c - \varepsilon_F|^{5/2}$. At $T \neq 0$ the singularity is diffused, but for $k_B T \ll \varepsilon_F$ the diffusiveness is negligible, and the presence of the singularity results in anomalous behavior of some thermodynamic and kinetic characteristics of the metal.

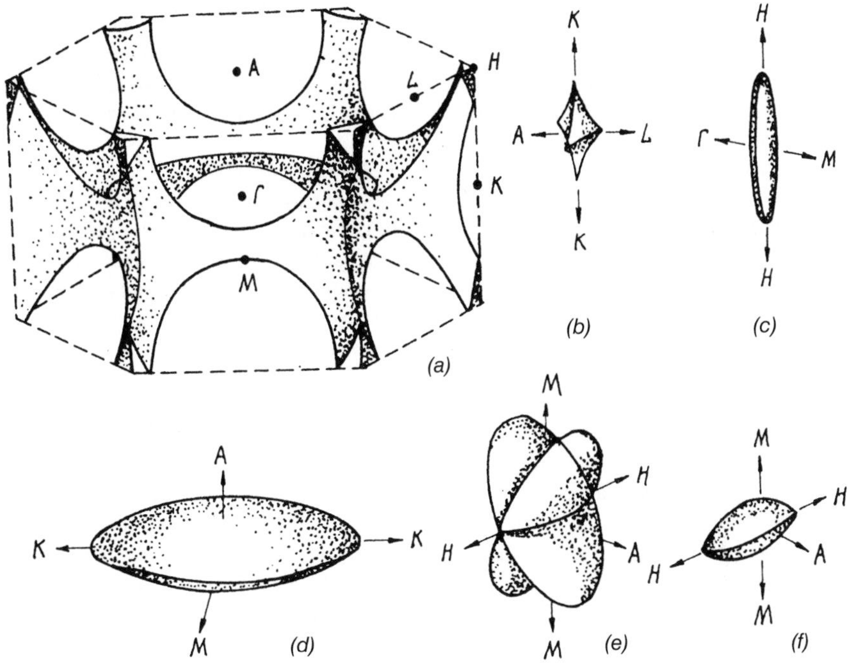

Fig. 4. Fermi surface of divalent metals with a hexagonal close-packed lattice according to the model of almost free electrons: (a) "monster", (b) "pyramid", (c) "needle", (d) "lens", (e) "star", (f) "cigar".

The Fermi surface generalizes the notion of a Fermi sphere to the case of a degenerate electron gas. The deviation of the Fermi surface from a spherical shape arises from the influence of the electric fields of the background of ions where the conduction electrons travel. Taking into account interactions between electrons (transition from Fermi gas to Fermi liquid) does not alter the representation of the Fermi surface as the boundary between filled and empty states, but when taking these interactions into account it is not the particles but rather the quasi-particles which fill the states for $p < p_F$ ($\varepsilon < \varepsilon_F$) that are being considered (see *Fermi liquid*).

FERMI SYSTEM

A quantum system involving many particles with half-integer spin (*fermions*) which obey the Pauli exclusion principle. Elementary excitations in Fermi systems are described by *Fermi–Dirac statistics*. See also *Fermi liquid*.

FERMIUM, Fm

See *Transuranium elements*.

FERMI VELOCITY

See *Fermi energy*.

FERRIMAGNET

Double sublattice or multisublattice magnetically ordered crystal with antiparallel orientation of *magnetization* vectors of inequivalent *magnetic sublattices*. Many oxide crystals (*ferrites*) with the spinel structure (*ferrite spinels*, cobaltite, chromite, manganite), garnets (*iron garnets*) or *perovskites* (*orthoferrites* with spins of ions of *rare-earth elements* ordered at low temperatures), and compounds with hexagonal structures (e.g., *magnetoplumbites*) are ferrimagnets. Ferrimagnets also include some *amorphous magnetic materials* which are related and contain several different types of magnetic atoms (e.g., alloys of d- and f-metals). Also a negative *indirect exchange interaction* between sublattices, which in ferrites acts through oxygen ions and in alloys acts through *conduction electrons*, can lead to ferrimagnetic ordering. Ferrimagnets form the most numerous group among magnetically ordered media, with applications in various areas of engineering. See also *Ferrimagnetism*.

FERRIMAGNETIC RESONANCE

Selective absorption by a *ferrimagnet* of electromagnetic wave energy within the radio frequency and the infrared ranges; it is a type of electron *magnetic resonance*. The presence of two or more *magnetic sublattices* in a ferrimagnet leads to the presence of multiple branches of ferrimagnetic resonance. Each branch is characterized by the dependence of the resonance frequency ω on the value of the external magnetic field $\boldsymbol{B}$ corresponding to the excitation of a specific type of resonance vibration of *magnetization* vectors of the sublattices with respect to one another. The low-frequency branch of ferrimagnetic resonance corresponds to the excitation of the precession of the spontaneous magnetization vector of the ferrimagnet $\boldsymbol{M}_s$ in the effective field $\boldsymbol{B}_{\text{eff}}$ which is determined by the external field, by the fields of *magnetic anisotropy*, and by the *demagnetization fields*, as in the case of ferromagnetic resonance. The uniform precession takes place in such a manner that the magnetization vectors of the sublattices remain antiparallel, so $\omega = \gamma_{\text{eff}} B_{\text{eff}}$. This type of ferrimagnetic resonance is analogous to *ferromagnetic resonance*, with its one particular value of the *gyromagnetic ratio* γ_{eff}. In the simplest case of a ferrimagnet with two sublattices of magnetizations M_1 and M_2 we have

$$\gamma_{\text{eff}} = \frac{M_1 - M_2}{M_1/\gamma_1 - M_2/\gamma_2},$$

where γ_1 and γ_2 are the gyromagnetic ratios of the sublattices.

The high-frequency branches of ferrimagnetic resonance correspond to precessions of sublattice magnetization vectors that violate their antiparallelism condition. These branches are sometimes called *exchange resonances*. Their frequencies are proportional to the exchange fields which act between the sublattices: $\omega = \gamma\alpha M$, where α is an *exchange interaction* constant. These frequencies are located in the IR range of the electromagnetic spectrum. Ferrimagnetic resonance can occur in ferrimagnets with noncollinear sublattice magnetization vectors, and it can also occur near the compensation point of the *ferrite*.

FERRIMAGNETISM

A phenomenon involving antiparallel ordering of the *magnetization* vectors of inequivalent *magnetic sublattices*, which leads to the presence of an uncompensated net *magnetic moment*. A necessary condition for the appearance of *ferrimagnetic ordering* is the existence of a negative *exchange interaction* between the *magnetic ions* (atoms) associated with the different sublattices. Ferrimagnetism is observed not only in crystalline magnetic materials, but also in amorphous ones, e.g., in alloys of d- and f-metals, in the temperature range from 0 to the *Néel point* T_N. Ferrimagnetic ordering allows the presence of *magnetic compensation* at the temperature $T_k < T_N$, where the resultant magnetic moment becomes zero. *Intrinsic ferrimagnetism* and *induced ferrimagnetism* are distinguished. The existence of intrinsic ferrimagnetism is caused by the crystallographic structure itself, i.e. by the presence of inequivalent positions for the magnetic ions of the same or different types; induced ferrimagnetism appears only in media with magnetic ions of different types due to the formation of a *superlattice*. It is necessary to distinguish ferrimagnetism from *weak ferromagnetism* (see *Antiferromagnetism*), when the resulting magnetic moment is caused by small deviations from antiparallelism of the magnetization vectors of two equivalent magnetic sublattices.

FERRITES

1. Ordinarily the term "ferrite" is used to designate double or more complex crystalline oxide compounds which contain iron sesquioxide Fe_2O_3. In the general case the chemical formula of ferrites may be written in the form $n_1Fe_2O_3{\cdot}n_2M_aO_b{\cdot}n_3R_cO_d$, where the n_i are integers, and M and R are metal ions. *Ferrite spinels* correspond to the case $n_1 = n_2 = 1$, $n_3 = 0$, $a = b = 1$ (e.g., Fe_2MgO_4); *orthoferrites* constitute the case $n_1 = n_2 = 1$, $n_3 = 0$, $a = 2$, $b = 3$; *iron garnets* comprise the case $n_1 = 5$, $n_2 = 3$, $n_3 = 0$, $a = 2$, $b = 3$ (e.g., gadolynium iron garnet $Gd_3Fe_5O_{12}$; the names take their origin from

the crystalline structures of single crystals formed by the corresponding compounds); for $n_1 = 2k_1 \geqslant 6$, $n_2 \geqslant 0$, $n_3 \geqslant 1$ and $a = b = c = d = 1$ the corresponding oxide compounds are called *hexaferrites*; for the orthoferrites and iron garnets M is an ion of a *rare-earth element* or *yttrium*. The great majority of ferrites are *ferrimagnets* with two or more *magnetic sublattices*. *Magnetic ions* which enter the tetrahedral or (d)-sublattice are surrounded by four oxygen ions, the octahedral or $[a]$-sublattice is formed by the magnetic ions surrounded by six O^{2+} ions, in iron garnets rare-earth element ions are located in dodecahedral coordination (12 nearest O^{2+} ions), thus forming a $\{c\}$-sublattice. The *exchange interaction* between different magnetically active (metallic) ions acts through oxygen ions, i.e. it constitutes an *indirect exchange interaction*. Both intrasublattice, as well as intersublattice, exchange interactions in ferrites are negative; the $[a]$–(d) interaction is the dominant one, which leads to ferrimagnetic ordering. The total *magnetization* of ferrites is the vector sum of the sublattice magnetization vectors.

Ferrites may form *solid solutions*, i.e. they allow the replacement of the initial metallic ions by other dia- or paramagnetic ions with close *ionic radii*. Undiluted from the magnetic point of view, ferrites possess a high *Néel point* (up to 800 K); the replacement of iron ions by diamagnetic ones leads to a reduction of the Néel temperature. Many ferrites have a *magnetic compensation* point at which the net magnetization goes to zero.

Ferrites belong to the class of magnetic insulators, and they have narrow *ferromagnetic resonance* line widths ΔB. For example, yttrium iron garnet $Y_3Fe_5O_{12}$ exhibits record low values of $\Delta B \approx 0.01$ mT. Due to this fact ferrites find their application in UHF engineering for the design of filters, gates, circulators, etc. Ferrites and the more complex compounds based on them are also used in radio engineering as cores, magnetic circuit elements, etc.; for creating *permanent magnets* with a relatively low *coercive force*. Because of their extensive technical applications the term "ferrite" is often applied to any ferromagnetic insulator. *Epitaxial iron garnet films* are characterized by a pronounced *magnetic anisotropy*, and they are used in *magnetic bubble domain devices*.

2. The term "ferrite" is also a name used for an allotropic modification of *iron* with a body-centered cubic lattice (space group $Fm\bar{3}m$, O_h^5; $a = 0.286653$ nm at 298 K) which exists at temperatures below 1184 K. This ferrite is ferromagnetic up to the temperature $T_C = 1040$ K (*alpha phase*), and above the *Curie point* it transforms to the paramagnetic state (β-phase). The high-temperature phase (1665 to 1810 K) of iron (*delta phase* or *delta ferrite*) with a body-centered cubic lattice (space group $Im\bar{3}m$, O_h^9, $a = 0.29322$ nm at 1667 K, paramagnet) is also called ferrite.

3. Another use of the term "ferrite" is to designate one of the phases in iron alloys, steels, and cast irons, which is a many-component solid solution of interstitial (usually C, N) and substitutional (Si, Al, Mn, Cr, Ni, etc.) elements in α-iron. The proportion of this ferrite may reach 95% depending on the alloy composition. The properties of these ferrites depend on the type and concentration of the doping elements, on the dimensions of the crystalline grains, the density of dislocations, the presence of dispersed particles.

FERRITE SPINELS

Oxide monocrystals (*ferrites*) with the structural formula $MO{\cdot}Fe_2O_3$ (M = Mg, Ca, Zn, Cd, Mn, Fe, Co, Ni, Cu, or their combination), isomorphous to the spinel mineral $MgO{\cdot}Al_2O_3$. They belong to the *cubic system*, space group $Fd\bar{3}m$ (O_h^7). The lattice constant is 0.833 to 0.851 nm; density is 4.75 to 5.42 g/cm^3; *magnetic moment* per formula unit at $T = 0$ is $(1.1\text{–}5.0)\mu_B$, where μ_B is the Bohr *magneton*; *Néel point* is from 9.5 to 670 K. The O^{2-} ions in ferrite form a face-centered cubic lattice, where metallic *magnetic ions* occupy the tetrahedral ((d)-sublattice) and octahedral ($[a]$-sublattice) coordinations. *Indirect exchange interactions* in ferrite spinels are negative; the $[a]$–(d) interaction is the strongest one, which in the great majority of cases leads to collinear *ferrimagnetism*.

FERROELASTICS (term first introduced by K. Aizu, 1969)

Insulating *monocrystals* in which particular regions (*ferroelastic domains*) differ in the spontaneous *strain* of their crystal lattice relative to some initial structure. When the sample temperature decreases, spontaneous deformation can develop as a result of a *structural phase transition* from the initial *paraelectric phase* to a *ferroelastic phase* of lower symmetry. The decomposition of the crystal into separate domains corresponds to minimizing the elastic energy of the crystal. In contrast to linearly elastic materials (see *Hooke's law, Elasticity*), the dependence of the deformation of a ferroelastic on the applied mechanical stress has the shape of a *hysteresis* loop. Moreover, at a certain value of stress, called the coercive stress, the crystal undergoes a transition to a single-domain state, accompanied by a change of sign of its spontaneous deformation. Among known ferroelastics are crystals of $KH_3(SeO_3)_2$ (transition temperature to the ferroelastic state $T_c = -62\,^\circ C$), $KD_3(SeO_3)_2$ ($T_c = 12\,^\circ C$); Nb_3Sn, V_3Si, $DyVO_4$, $TbVO_4$, and $RbMnCl_3$. Some ferroelastics are *ferroelectrics* at the same time. Ferroelastics are promising materials for building acousto-electric and acousto-optical devices (see *Acousto-electronics, Acousto-optics*).

FERROELECTRIC CERAMICS

A polycrystalline material produced using ceramics technology, with a basic crystalline phase exhibiting ferroelectric properties. *Ferroelectrics of the oxygen-octahedron type* are the ones most easily mastered in state-of-the-art ceramics technology. Their chemical composition and crystalline structure vary widely. Both simple and complex ferroelectric compounds may serve as the basis for ferroelectric ceramics, as well as *solid solutions* based on them. Such are, first of all, materials with an ABO_3 chemical formula and a *perovskite*-type structure. Numerous *ferroelectrics* crystallize in that structure, the most important of them being *barium titanate*, $BaTiO_3$. Some ferroelectric ceramics have the tetragonal potassium–tungsten bronze type of structure with the common chemical formula: $A_6B_{10}O_{30}$ or AB_2O_6. Others have a layered structure and common chemical formulae: $A_2B_2O_7$, $A_2Bi_2B_3O_{12}$, $ABi_2B_2O_9$, $A_3Bi_2B_4O_{15}$, etc. Another group alternates perovskite-type layers of oxygen octahedra with layers of $(Bi_2O_2)^{2+}$, where the number of perovskite layers usually varies from two to five. There are also materials featuring a structure similar to pyrochlore, with the common chemical formulae: $A_2B_2O_7$, $A_2(B^IB^{II})O_6$, those with a pseudoilmenite structure, ABO_3, etc. Since *ceramics* are agglomerates of randomly arranged *crystallites* they appear isotropic in their properties. These properties depend quite strongly on the ratios between the different crystalline phases, the ratio between the relative distribution of crystalline phases and the intercrystallite layers, on the stoichiometric composition of the basic phase, the nature of impurities and doped additions that are introduced, the choice of production and processing technologies and regimes. For example, a $BaTiO_3$ fine crystallite ceramic with grain size below 1 μm exhibits a relative *dielectric constant* $\varepsilon \sim 3000$–5000 which is 2–3 times higher than that found in common ceramic samples with grain sizes of 20–100 μm. The dielectric constant of these materials varies only slightly with temperature, which is promising for building compact ferroelectric ceramic capacitors. Adding oxides of certain elements (e.g., Sn, Zr, Bi, Ca and other compounds based on them) results in a significant broadening diffusion of the *phase transition*. Compounds, solid solutions, and heterogeneous mixtures with a *diffuse phase transition* serve as a basis for numerous ferroelectric ceramic capacitor materials with significantly nonlinear properties within broad temperature (-60 to $85\,^\circ C$) and frequency (10^3–10^7 Hz) ranges.

The parameters of certain ferroelectric capacitor materials are summarized in Table 1. Materials with high nonlinearity produced by *orientational polarization* (see *Polarization of insulator*) are widely used to produce various nonlinear ferroelectric elements. Solid solutions based on $BaTiO_3$, $SrTiO_3$, $PbTiO_3$ yield a basis for producing capacitor ferroelectric elements controlled by either an alternating or a direct electric field, the so-called *variconds*. Using ferroelectric ceramics with intergranular layers, such that the composition and structure of phases at the grain surface and inside it are different, makes it possible to reach ultrahigh values of ε (up to 50,000). Ferroelectric materials reduced in an atmosphere of dry hydrogen or doped with oxides of *rare-earth elements* feature a low specific resistance (10^{-3}–10^2 Ω·m) and are *semiconductors*. Using semiconducting ferroelectric ceramics of $BaTiO_3$ with

Table 1. Ferroelectric ceramics

Ceramic type	Dielectric constant ε at a temperature of $20 \pm 5\,^{\circ}$C	Relative variation in dielectric constant $\Delta\varepsilon$, %	Relative variation in reversible dielectric constant $\Delta\varepsilon_r$, %	Dielectic loss tangent at two temperatures 25 °C	85 °C	Resistivity ρ, GΩ·m at $100 \pm 3\,^{\circ}$C	Electrical strength E, MV/m
T-900	900	±30	±10	0.002	–	0.1	5
T-2000	2000	±20	±10	0.025	–	0.1	4
SM-1	4000	±70	±30	0.03	0.02	0.1	3
T-8000	8000	–	±30	0.003	0.02	0.1	3

Notes:
1. The relative changes in the dielectric constant $\Delta\varepsilon$ are its changes in the temperature interval $-40\,^{\circ}$C to $+85\,^{\circ}$C compared to ε at a temperature of $20 \pm 5\,^{\circ}$C.
2. The relative change in the reversible dielectric constant $\Delta\varepsilon_r$ is the change in ε with the applied electric field growing from 0 to 0.5 MV/m.

surface barriers or insulating layers several micrometers thick one may manufacture low-voltage capacitors with a maximum specific capacity of about 3 μF/cm^3. Semiconducting ferroelectric ceramics display a positive thermal resistance factor, and are used to produce a particular type of thermal resistors – *posistors*. The nonlinear *current–voltage characteristics* of semiconducting ferroelectric ceramics make it possible to build *varistors* with relative resistance variation to voltage variation ratios that exceed the value of 3. Ferroelectric materials with a high pyrofactor value serve as a basis for various pyroelectric transformers. The operative spectral range of such devices is quite wide, stretching from the far IR to short-wave X-rays and γ-rays. Some of the materials found among pyroelectric ferroelectric ceramics (see *Pyroelectric materials*) are solid solutions of $PbTiO_3$–$PbZrO_3$ (PZT); their composition usually corresponding to the *morphotropic region*, in which the tetragonal and rhombohedral phases coexist. Their pyrofactor, as well as many other factors reach their maximum values there. The extreme compositions of either $PbTiO_3$ or $PbZrO_3$ differ in their properties: the former is ferroelectric (tetragonal structure), while the latter is antiferroelectric (rhombohedral structure). The morphotropic region in the mixture ($PbTi_{1-x}Zr_xO_3$) is of the order of Δx and reaches 0.15 at around the 50% composition. One may separate the ferroelectric ceramics used to fabricate electrooptical devices into specific types. The electrooptical ferroelectric ceramic most thoroughly studied and most widely used is zirconium–titanium–lead (PZT), modified with *lanthanum* (PLZT). To obtain highly uniform and defect-free electrooptical ferroelectric ceramics, one uses the techniques of joint chemical precipitation from solutions, followed by hot *pressing*. Such electrooptical ferroelectric ceramics feature high transparency, and exhibit pronounced *electrooptical effects*.

Using ferroelectric ceramics one may produce polar-anisotropic *textures* (with a predominant orientation of the elementary electric moments of crystal grains, or of separate grain regions). Such textures exhibit piezoelectric properties. Ceramics with high piezoelectric parameters are usually called *piezoelectric ceramics*. To produce piezoelectric ceramics, solid PZT solutions as well as solid solutions based on $BaTiO_3$ and $NaNbO_3$ are used, with compositions close to the ferroelectric morphotropic region. Certain piezoelectric ceramic materials with properties meeting various piezoelectric instrumentation requirements have been perfected. High workability, availability of raw materials, possibility of task-specific alteration of their electrophysical properties, and reproducibility in samples of varying size and shape, provide ferroelectric materials with a wide range of applications in various branches of technology.

FERROELECTRIC DOMAINS

Separate regions differing in their spontaneous polarization direction, into which an *ideal crystal* breaks up at a transition to the ferroelectric state in the absence of external influences. Since the polarization direction (see *Polarization of insulator*) is determined by the crystal symmetry, the *magnetic domain structure* is characteristic for each particular *crystal lattice*. For example, in $BaTiO_3$ the angle between ferroelectric domain polarization vectors can equal either 90° or 180° (called 90°- and 180°-domains, respectively) whereas in *triglycine-sulfate* and *ferroelectrics* such as KH_2PO_4 only 180°-domains are found. A typical width of a ferroelectric domain is of the order of 10^{-4} cm. From the crystallographic point of view, ferroelectric domains are similar to twins, where the elements of *twinning of crystals* are symmetry elements lost at the transition to the ferroelectric phase. Therefore, the macroscopic crystal symmetry of the crystal broken into ferroelectric domains resembles the symmetry of the non-ferroelectric phase.

The formation of ferroelectric domains is due to the decrease of crystal free energy in the many-domain state. To determine the ferroelectric domain geometry one should minimize the total crystal free energy, including the depolarization energy that depends on the sample shape, and the *domain wall* energy. If the depolarization field is compensated (e.g., in good conducting ferroelectrics), the free energy minimum can correspond in the absence of defects to that of a single-domain configuration.

In imperfect crystals, defects of various types influence the formation of ferroelectric domains. For example, in the case of a *first-order phase transition* the defects determine both the locations of new phase seed nuclei, and the spontaneous polarization direction in them. A mismatch of the polarization directions in individual seed nuclei can also result in the formation of ferroelectric domains.

The geometry of ferroelectric domains is rather complicated, and depends on many factors, including the crystal symmetry, electrical conductivity, configuration of defects, values of spontaneous polarization, elastic and dielectric susceptibilities, and also the prior history of the sample.

FERROELECTRIC DOMAIN SPLITTING

See *Domain splitting in ferroelectrics.*

FERROELECTRIC DOMAIN STRUCTURE

A system of uniformly polarized regions in a *ferroelectric* material (*domains*). The cause of the formation a domain structure during the process of transition from a *paraelectric phase* to a *ferroelectric phase* (a polar one) is the appearance of the polarization field due to a jump of spontaneous electric polarization. The depolarization field increases the crystal free energy (see *Thermodynamic potentials*). In addition to the *spontaneous polarization shielding* by free charges, the decisive role belongs to the division of the sample into domains. The shape and the equilibrium dimensions of domains are determined by a competition between the depolarization field energy and the energy of the *domain walls* (boundaries) that divide the domains. The relative orientations of polar axis in neighboring domains as well as the direction of the domain wall are determined by the *crystal symmetry* group in the paraelectric and ferroelectric phases. There are single-axis and multi-axis ferroelectrics that have either two antiparallel directions or several directions of spontaneous polarization, respectively (see *Ferroelectricity*). In these crystals domain structures occur with antiparallel orientation of the spontaneous polarization, as well as domains with spontaneous polarization directions at certain angles (90°, 60° etc.) with respect to each other. These angles are often used to designate the corresponding domains and domain walls. For layer-shaped samples the concepts of *s-domains* (direction of spontaneous polarization normal to the plane) and *a-domains* (parallel arrangement of polar axis in the plane) are introduced. The domain structure can be changed by applying an external electric field. Various methods are employed for domain structure detection. The most efficient are optical methods based on the difference between optical parameters of the crystal along different crystallographic directions. For multi-axis crystals, a widely used method is the observation of transmitted polarized light. In crystals such as lead germanate, the domain structure is observed through the dependence of the *rotation of light polarization plane* on the polar axis direction.

FERROELECTRICITY

The complex of effects associated with the presence of a *spontaneous polarization* P_S in a material, or the final value of the *polarization of an insulator* that arises irrespective of the presence of an external electrical field E, but can change its direction under action of this field. The value of P_S for some ferroelectrics can range up to hundreds of $\mu C/cm^2$. Upon heating P_S decreases, and at the Curie point T_C it disappears when a *phase transition* to the *paraelectric phase* takes place, and the crystal lattice symmetry changes. In the *polar phase* below T_C the ferroelectric is divided into separate regions called *domains* with differing directions of P_S so that the overall polarization of the crystal remains close to zero. The P_S originates from the displacements of ions from their initial lattice sites in *displacive type ferroelectrics*, and it arises from the ordering of atomic groups with electrical dipole moments in *order–disorder type ferroelectrics* (see *Ferroelectrics*). At a *second-order phase transition* the second derivative of the thermodynamic potential exhibits anomalies: also the dielectric *susceptibility*, *dielectric constant*, *specific heat*, coefficient of *thermal expansion*, *elastic modulus*, piezomodulus, pyroelectric coefficient, electrooptic coefficients, etc. At a *first-order phase transition*, besides the distinct changes of the above mentioned quantities, the first derivative of the thermodynamic potential has a discontinuity: P_S, the volume, the entropy, and the birefringence are discontinuous, latent heat is released, etc. In ferroelectrics with a domain structure (multidomain ferroelectric) an applied field E brings about its polarization not only due to elastic displacements of atoms and their electronic shells, but also due to the reorganization of the *magnetic domain structure*, which proceeds by way of the formation and penetration of domain germ nuclei with P_S close to the direction of E, and transverse motions of *domain walls*. The dependence of the ferroelectric polarization on E exhibits *hysteresis* at increasing and subsequently decreasing E, therefore, after removal of the field the crystal retains a *remanent polarization*. The *ferroelectric phase transition* is the result of instability of the crystal lattice with respect to a certain *soft mode* of lattice vibrations. Namely, in an intrinsic ferroelectric phase transition (where the *order parameter* is proportional to P_S) the frequency of the polar transverse optical mode tends to zero as it approaches the transition to the ferroelectric phase, and the mode condenses at the center of the *Brillouin zone* at T_C causing P_S to appear. At an extrinsic phase transition the transition parameter is not proportional to P_S, but rather to a quantity with other transformation properties, and P_S appears as a secondary effect because of the interaction of the soft mode responsible for the transition with the polar mode. The transition parameter can be nonuniform across the crystal and vary with a certain period, generally incommensurat with the lattice parameter. Such phases are referred to as *incommensurat phases*. Several hundred compounds are known to possess ferroelectric properties. Among them ferroelectrics with the *perovskite* type structure are the most important for engineering applications. Ferroelectrics are used as capacitor, piezoelectric, pyroelectric, electrooptic, and electrostrictive compounds.

FERROELECTRICITY INDUCED BY IMPURITIES

Ferroelectric ordering of dielectric crystals caused by the introduction of impurities. It was observed in $KTaO_3$ crystals with Na, Nb, Li impurities, in PbTe with Ge impurities, and in other strongly polarized *insulators* possessing phonon *soft modes*. A characteristic of *ferroelectricity* is the concentration dependence of the *phase transition* temperature T_C, and at an impurity concentration of several percent $T_C \sim 100$ K. The simplest physical mechanism for inducing ferroelectricity is the changing of the *unit cell* dimensions by inserting impurity ions. Since in weak but strongly polarized lattices there is almost complete neutralization of the long-range attractive forces between ferroactive ions in different cells, and a short-range repulsion of adjacent ions, even a small change of lattice constant can disturb this balance and lead to complete neutralization. This makes the frequency of the soft mode vanish, that is, it induces a *ferroelectric phase transition.* In the above compounds, adding impurities can result in a decrease of the unit cell dimensions. However, decreasing the unit cell size under hydrostatic *pressure* causes a decrease in the *dielectric constant*, and the crystal is no longer ferroelectric.

The *ionic radius* of the impurities causing the decrease of the lattice constant of the crystal is smaller than that of matrix atoms. In this situation, the impurity ions, as a rule, become *noncentral ions* and generate impurity electric *dipole moments* in an *ionic crystal*. The electric *dipole–dipole interaction* forces result in the *ferroelectric ordering of dipoles*. The *spatial dielectric dispersion* that is usually significant in media with large polarizability enhances the possibility of ferroelectric phase transitions in similar compounds. The presence of spatial dispersion characterized by the correlation radius of the polarization r_c changes the effective potential of the dipolar coupling, so that interaction between impurities that are located at distances $r < r_c$ from each other appears to be chiefly ferroelectric, and more long-range than the usual dipole–dipole interaction. Owing to this long-range action, configuration fluctuations of local fields (destroying the ferroelectric phase transition in weakly polarized, e.g., in *alkali-halide crystals*) decrease, and long-range order can appear in the system under certain conditions (see *Long-range and short-range order*). Such a phase transition is accompanied by the emergence of a spontaneous polarization (see *Ferroelectricity*) connected with displacements of lattice atoms, since oriented dipole moments polarize the lattice.

The theoretical description of phase transitions in ferroelectrics with impurities (disordered ferroelectrics) differs from that of phase transitions in ordered crystals similar to $BaTiO_3$ or KH_2PO_4. The *molecular field* theory is applicable to the latter but not to the former, since the local field fluctuations are large in impurity systems, and this violates the conditions required for applying molecular field theory.

The thermodynamic mean $\langle\bar{l}\rangle$ of the electric dipole moment directions of impurities, averaged over the configurations of impurities (indicated by the overhead bar) $L = \langle\bar{l}\rangle$ defines the relative number of coherently oriented electrical dipole moments, and can be chosen as the *order parameter* of the ferroelectric phase transition. The quantity L obeys the self-consistent equation

$$L = \int \mathrm{d}E\, f(E, L)\langle l\rangle_E, \qquad (1)$$

where $f(E)$ is the random field distribution function (see *Random potential*) of E exerted on the ith dipole by the nearby impurities: $f(E) = \overline{\langle\delta(E - E_i)\rangle}$, $\delta(x)$ is the Dirac delta function, and $\langle l\rangle_E$ defines the mean value of the electric dipole moment of the random field. The molecular field approximation corresponds to the substitution $f(E) = -\delta(E - \langle\overline{E_i}\rangle)$. Under this approach, ferroelectric order always exists at low temperatures. However, thermal fluctuations can result in the smearing of the δ-function, so the solution with $L \neq 0$ does not always exist. It is precisely the thermal fluctuations that lead to the suppressing of the ferroelectric phase transition in alkali-halide crystals. Eq. (1) allows one to calculate the critical impurity concentration n_{cr} above which the ferroelectric phase transition induced by off-center ions ($n_{cr}r_c^3 \sim 10^{-2}$) is activated. At concentrations lower than this critical value, the *dipole glass* state exists.

FERROELECTRICITY, VIBRONIC THEORY

See *Vibronic theory of ferroelectricity*.

FERROELECTRIC LIQUID CRYSTALS

Liquid crystals with spontaneous polarization (see *Ferroelectricity*) that varies under the influence of an external electric field. These crystals were predicted by R. Meyer (1975) and detected by him and his collaborators in 1977. Unlike in *solids*, spontaneous polarization in liquid crystals does not arise from dipole–dipole interactions, but rather is usually due to a short-range steric interaction. Dipolar ordering can arise in a smectic C*-phase (see *Smectic liquid crystal*), consisting of chiral (see *Chirality*) molecules with long axes inclined at an angle relative to the smectic layers. In this case, the rotation of molecules around long axes is restricted, and the spontaneous polarization P is directed perpendicular to the plane of the molecule tilt. Hence, dipolar ordering arises as a side effect of a steric interaction. The value $P \approx 10^{-6}$–10^{-4} C/m^2 is proportional to the *order parameter* of the molecular short axis. Possible uses of ferroelectric liquid crystals are in information handling devices, imaging, and recording, because they ensure liquid crystal device performance times in the range 10^{-6}–10^{-3} s (see *Optical techniques of information recording*).

FERROELECTRIC PHASE

Crystallographic modification of a *ferroelectric* existing in a particular temperature range, and exhibiting spontaneous polarization which is temperature-dependent, and can be reoriented by an externally applied electric field (see *Ferroelectricity*). The ferroelectric phase passes into the *paraelectric phase* above a temperature T_C called the *Curie point* by analogy with *ferromagnets*. Most ferroelectrics have only one Curie point. The transition to the ferroelectric phase takes place at a temperature below the *melting* point. For some ferroelectrics (e.g., Rochelle salt, $NaKC_4H_4O_6{\cdot}4H_2O$) the ferroelectric phase is bounded by upper (24 °C) and lower (−12 °C) Curie points. Materials in the ferroelectric state generally exhibit a ferroelectric hysteresis loop (see *Polarization switching*), and in the case of unipolar ferroelectrics there is a *pyroelectric effect*.

FERROELECTRIC PHASE TRANSITION

A *structural phase transition* in crystals involving the appearance of *spontaneous polarization* P (electric dipole moment per unit volume), and the *pyroelectric effect* which appears if the sample is heated in the temperature range below the transition ($T < T_C$) when an electric field is present in it. At the transition some ions are displaced from their equilibrium positions in the structure of the more-symmetric phase (paraphase, or *paraelectric phase*) so that an electric dipole moment, and hence bound surface charges, appear in the crystal. This charge is usually compensated by incoming free charges, and since its value depends on the lattice constant, the compensation is disturbed by heating. The origin of the ferroelectric phase transition involves the softening and loss of stability of certain oscillatory or relaxation lattice modes. The transition is accompanied by features of the static *dielectric constant* ε and a number of other variables. If at the transition a polar mode loses stability then it is ordinarily convenient to select the polarization P as the *order parameter*, and the transition is called an *intrinsic ferroelectric phase transition*. The polar mode is softened at the *Brillouin zone* center, and the temperature dependence $\varepsilon(T)$ has a singularity of the type $C/(T - T_C)$ (*Curie–Weiss law*). An *extrinsic ferroelectric phase transition* is related to the loss of stability of a nonpolar mode with which the polar one interacts; as a result of this effect, spontaneous polarization appears below T_C. Thus ε is not a critical susceptibility, and one ordinarily observes only a "weak" singularity on the diagram of the $\varepsilon(T)$ dependence, e.g., a discontinuity. A ferroelectric phase transition can be either first order or second order, and can also be differentiated into a *displacive type transition* or an order–disorder type transition. Let $V(Q)$ be a single-particle potential for a change in the position Q of the ion which is related to the transition, so that Q is a relevant collective coordinate. Usually $V(Q)$ is of the form of two troughs or wells separated by a potential barrier of height ΔV. If ΔV exceeds the interaction energy U_{int} of displaced ions in different elementary unit cells then the polarization P is induced at a temperature below T_C because of a preferential redistribution of ions in various potential wells, and an *order–disorder phase transition* takes place. If $\Delta V < U_{int}$ the presence of the barrier is not relevant, at the transition point the displacement of the participating ions begins, and a displacive type transition occurs. An example of a proper displacive type ferroelectric phase transition occurs in *barium titanate* $BaTiO_3$ from a cubic phase to a tetragonal one at 393 K. In triglycinesulfate (TGS), $(CH_2NH_2COOH)_3H_2SO_4$ and potassium dihydrogen phosphate (KDP), KH_2PO_4, intrinsic order–disorder ferroelectric phase transitions are realized. The Curie–Weiss constant C for order–disorder transitions is significantly smaller than that for transitions of the displacive type: $C = 170{,}000$ K for $BaTiO_3$, 3200 K for TGS and 3500 K for KDP. A whole sequence of ferroelectric and other structural phase transitions sometimes take place in a single compound (e.g., $BaTiO_3$). Usually, the *critical region* at a ferroelectric transition is sufficiently narrow; so it is meaningful to take into account only the first-order fluctuation corrections of the *Landau theory of second-order phase transitions*. However, the introduction of defects into the crystal can expand the width of the fluctuation region, and modify the transition or bring about its disappearance.

FERROELECTRICS

Crystalline *insulators* (or semiconductors) which exhibit *ferroelectricity.* Ferroelectrics may be divided into two types: displacive ferroelectrics and order–disorder ferroelectrics, depending on the shape of the single-particle potential of their ions.

Displacive type ferroelectrics are characterized by low anharmonicity of their single-particle potential, so that their ferroelectrically active ions are found in a single-well potential in the paraelectric phase. Within the *harmonic approximation*, the dynamics of their lattice may be described in terms of phonon modes. Their crystal lattice is stable with respect to minor deformations provided all the normal phonon modes have real frequencies. In certain ionic or partially *ionic crystals*, the frequency $\omega_0(q)$ of one of the polar lattice modes may turn purely imaginary in the harmonic approximation, thus resulting in a *ferroelectric phase transition* of the displacive type (if $q = 0$). The imaginary nature of $\omega_0(q)$ is a corollary of the fact that the harmonic force constant in ionic crystals, and therefore $\omega_0^2(q)$, contains two contributions with opposite signs – short-range repulsive forces between ions and long-range Coulomb forces. If the square of the harmonic frequency $\omega_0^2(q)$ is negative for a given normal mode, as a result of compensation between short-range repulsive and long-range attractive forces, anharmonic interactions will stabilize the system above the *Curie point* T_C. Taking into account the oscillation anharmonicity effects in the quasi-harmonic approach yields a value of the observed quasi-harmonic frequency that is real and positive, however, it depends on the temperature, and the anharmonic contribution drops as the temperature decreases. The frequency of the so-called *soft mode* decreases down to zero at the approach to the Curie point, with the *Cochran relationship* (W. Cochran) observed, $\omega_0^2(q) = k(T - T_C)$. The low-temperature phase structure is determined by frozen displacements corresponding to the soft mode.

Order–disorder type ferroelectrics are characterized by strong anharmonicity of the single-particle potential (see Fig.). Owing to this effect, ions (or radicals) determining the crystal polarization can be located in two or more equilibrium positions, displaced relative to centrally symmetric

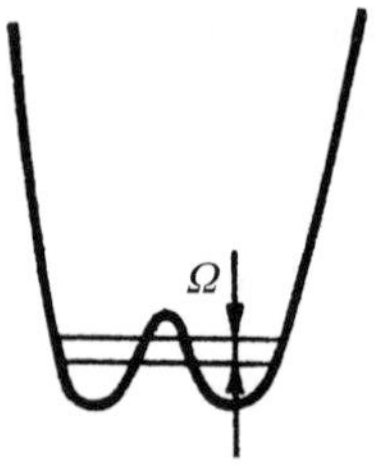

Sketch of single-particle potential in ferroelectrics of the order–disorder type (Ω is the tunnel splitting).

positions in the unit cell. Above the ferroelectric phase transition temperature T_C these positions are statistically uniformly populated, however, below T_C a spontaneous population asymmetry appears, resulting in polarization of the crystal. Such cases are encountered frequently, e.g., in crystals with *hydrogen bonds* such as potassium dihydrogen phosphate (KDP), KH_2PO_4, where the proton can move between two equilibrium positions (see Fig.) in a hydrogen-bond potential. The proton position on the bond is usually described in terms of a pseudospin operator that is the usual $S = 1/2$ *spin operator* used for quantum *two-level systems.* As in displacive type ferroelectrics, the phase transition in systems of the order–disorder type can be expressed in terms of a soft mode. It is not *phonons*, however, but rather unstable *pseudospin waves* that are represented in this case of weak oscillatory excitations. The pseudospin mode is present in order–disorder type ferroelectrics, in addition to usual phonon modes arising from harmonic lattice oscillations. There is another important distinction between displacive and order–disorder systems in respect to soft mode dynamics. In displacive type systems, the soft mode always has a resonant nature, although it can be highly damped. In order–disorder type systems the imaginary part of the soft mode frequency is always nonzero, and the real part is not zero except for a strongly split ground states determined by tunneling (as shown in Fig.); otherwise the system response has a relaxation but not a resonant nature. This takes place in ferroelectrics similar to $NaNO_2$ or *triglycinesulfate* that are characterized by negligible tunneling. The resonant nature of the response has been detected in crystals similar to KH_2PO_4 characterized by strong tunneling.

FERROELECTRICS, QUANTUM

See *Quantum ferroelectrics*.

FERROELECTRICS, VIRTUAL

See *Virtual ferroelectrics*.

FERROFLUID, magnetic liquid

A liquid with pronounced magnetic properties, primarily with a relatively large *magnetic susceptibility*, χ (viz., with χ larger than is typical for *paramagnets* or other weak magnetic materials, which typically have $\chi \sim 10^{-9}$–10^{-4}). Homogeneous liquid *ferromagnets* have yet to be found, and ferrofluids are only exemplified by colloidal solutions of small (10^{-4}–10^{-6} cm) ferromagnetic particles (as a rule, *single-domain particles*; sometimes magnetic liquid particles exhibit *superparamagnetism*). Ferrofluids with χ up to 10^{-1} are stable colloidal systems, which reach a saturation magnetization of up to 3.5 mT with a magnetizing field up to 1 T. Ferrofluids are used to make magnetically controlled mechanical devices (magnetic clutches and couplings, etc.), in magnetic *defectoscopy*, and for *magnetic domain characterization* by the *magnetic powder pattern* method.

FERROICS

General name for materials which, below the *Curie point* or in some temperature range, exhibit spontaneous reorientable physical quantities described by tensors of first rank (spontaneous polarization (see *Ferroelectricity*), spontaneous *magnetization*), second rank (spontaneous *strain*) or higher rank (e.g., *piezoelectricity*, elastic constants (see *Elastic moduli*), etc.). In the ferrophase ferroics may become subdivided into orientation states (domains) with different directions of the vector *order parameter* (director). In specific physical situations the entire crystal of a ferroic may be in an uniform orientational state (with a uniform direction of the vector order parameter). If switching a crystal from one orientational state to another is carried out under the influence of one of the physical fields, the material corresponds to *first-order ferroics*, which include *ferroelectrics*, *piezoelectric materials* and *ferroelastics*. If the switching of the crystal may be performed only under an action involving two fields (e.g., electric field and mechanical stress, magnetic field and mechanical stress, magnetic field and electric field, etc.), such materials correspond to *second-order ferroics* (e.g., NH_4Cl, CoF_2, Cr_2O_3, SiO_2, etc.). There also exist ferroics of higher orders.

FERROMAGNET

A material exhibiting *ferromagnetism*: below the transition temperature (*Curie point* T_C) the average values of the *magnetic moments* of ferromagnetic materials are oriented parallel to each other, thereby producing strong *spontaneous magnetization* M_s. It is necessary to differentiate ferromagnets from the other magnetically-ordered materials with $M_s \neq 0$, e.g., from *ferrimagnets*, and from magnetic materials with *modulated magnetic structures*, although some of their macroscopic parameters may be similar (see *Atomic magnetic structure*). A large number of ferromagnets is known (see Table 1), many of which are *metals*, their *alloys*, or *intermetallic compounds*.

Among the chemical elements the $3d$- and $4f$-metals included in the table are ferromagnets (see *Ferromagnetism of metals and alloys*). Also some insulating ferromagnets are known, such as the *ionic crystals* EuO, EuS, $CrBr_3$, etc., which usually exhibit an *indirect exchange interaction* through the ions S, O, or Se, as is the case for ferrimagnets. Insulator ferromagnets are often transparent, with *magneto optical effects* observed in them. Ferromagnets are widely used in modern engineering (see *Magnetism*, *Magnetic materials*).

Table 1. Curie temperature T_C and spontaneous magnetization M_s of some ferromagnets

Material	T_C, K	$M_s(T=0)$, G
Fe	1043	1740
Co	1338	1446
Ni	627	510
Gd	293	2100
Tb	219*	2713
Dy	85*	2910
MnAs	313†	870
$CrBr_3$	33	270
EuO	69	1910

* Point of transition to a simple helical structure.

† First-order phase transition with pronounced hysteresis.

FERROMAGNETIC RESONANCE

Selective absorption by a *ferromagnet* of the electromagnetic wave energy at frequencies (usually microwaves), which coincide with the intrinsic precession frequency of the ferromagnet *magnetic moment* (see *Larmor precession*); a variety of electron *magnetic resonance*. Ferromagnetic resonance involves the resonance precession of the magnetization M_s of a ferromagnet in the effective magnetic field B_{eff}, which is determined by the magnetic anisotropy, by the magnetoelastic interaction constants, and by the demagnetization fields. The latter dependence means that B_{eff} depends on the shape of the sample (see *Demagnetization factor tensor*). The presence of the domain structure complicates ferromagnetic resonance.

Unlike *electron paramagnetic resonance*, in the case of ferromagnetic resonance the electron spin moments (see *Spin*) of the sample atoms, due to their strong exchange interaction with each other, perform vibrations in unison, and the presence of spontaneous magnetization in the sample causes the local magnetic field to differ from the external applied field. Some generators and amplifiers in the UHF range, frequency transformers, and other UHF devices are based on the phenomenon of ferromagnetic resonance.

FERROMAGNETISM (fr. Lat. *ferrum*, iron, and *magnetism*)

Magnetically ordered state of a material (*ferromagnet*) where the average values of the *magnetic moments* of all the atoms (ions) are oriented in parallel due to the *exchange interaction*. The ordering occurs at temperatures T below the Curie point T_C (see *Magnetic phase transitions*). For ferromagnets the existence of *spontaneous magnetization* $M_s(T)$ in each individual *domain* in the absence of an external *magnetic field* $\boldsymbol{B}$ is a characteristic property, the value of M_s reaching a maximum at $T = 0$, and going to zero at $T = T_C$. The dependence of the overall or net magnetization M on the external field $\boldsymbol{B}$ (*magnetization curves*) has a fairly complex nature due to its existence in the equilibrium (demagnetized) state of the *magnetic domain structure*. In this state at $\boldsymbol{B} = 0$, prior to having placed it in an applied magnetic field, the total magnetic moment of the sample arises from the cancellation of the magnetic moments of different *magnetic domains*, and the average magnetization $\langle M\rangle$ is zero. Small ferromagnetic particles may be too small to split into domains, so they become *single-domain particles*. The single-domain state, which is a *metastable state*, may also exist in bulk ferromagnetic samples.

In single-domain samples, and in individual component domains of bulk samples, the *magnetization* is oriented along an *easy magnetization axis*. Upon magnetization along this axis, due to the growth and merging of energetically favorable domains, the magnetization of the sample $\langle M\rangle$ may grow very fast and even reach the nominal saturation value of M_s in very weak fields, in this case the *magnetic susceptibility* $\chi = \langle M\rangle/H$ reaches values up to 10^6. The same situation may also occur in *polycrystals*, high values of χ make ferromagnetism useful in electrical engineering applications (see *Soft magnetic materials*). During the magnetization of a ferromagnetic *monocrystal* along a *hard magnetization axis* domain *magnetization rotation processes* play the main role, and the susceptibility is lower, of the order of M_s/H_A, where H_A is the anisotropy field (see *Magnetic anisotropy*), i.e. the values of χ are quite low for some anisotropic magnetic materials, but they may also be higher for weakly anisotropic materials. In ferromagnetic materials magnetized up to saturation the variation of the magnetization takes place due to the *paraprocess* (the corresponding susceptibility is low). The above-mentioned magnetizations are related to the magnetic susceptibility relative to the internal magnetic field $\boldsymbol{H}_i$, which differs from the external magnetic field $\boldsymbol{H}_e$ due to *demagnetization fields*. The connection between $\boldsymbol{H}_i$ and $\boldsymbol{H}_e$ is determined by the sample shape, their values being close only for particular types of geometry (see *Demagnetization factor tensor*).

Upon increasing the magnetic field the value of $\langle M\rangle$ tends to M_s, i.e. the magnetization curve is sharply nonlinear. Ferromagnetism is characterized by *magnetic hysteresis* so $\langle M\rangle$ depends on the history of the sample, i.e. the connection of $\langle M\rangle$ with H_e and other external parameters (temperature, pressure, etc.) is multivalued. Due to this fact a ferromagnetic material can possess a magnetic moment (remanent magnetization) even for the applied field $H_e = 0$, hence it may be used as a *permanent magnet* (see also *Hard magnetic materials*). During a variation of the magnetization of a ferromagnetic material its dimensions and size vary slightly (see *Magnetostriction*), and in the adiabatic case the temperature may also change (see *Magnetocaloric effect*). In addition anomalies are observed in some nonmagnetic properties of the material: elastic moduli, specific heat, thermal expansion coefficient, etc.; the maxima of these singularities occur in the vicinity of T_C or at the positions of other phase transitions.

FERROMAGNETISM OF METALS AND ALLOYS

Phenomenon of *ferromagnetism* observed in three of the $3d$-*transition metals* (Fe, Co, Ni), in six of the $4f$-transition metals (*lanthanides, rare-earth elements*: Gd, Tb, Dy, Ho, Er, Tm), in a series of alloys (ordered as well as amorphous), and in *intermetallic compounds*. They may be classified according to the electronic structure of the component atoms:

(1) Alloys of the d- and f-transition metals with each other: (a) pure ferromagnetic transition metal alloys (Fe–Ni, Co–Ni, Fe–Co–Ni, Fe–Gd, Gd–Dy, etc.); (b) transition metal ferromagnetic materials with antiferromagnetic and paramagnetic metals (Fe–Cr, Ni–Ti, Co–Pt, Gd–V, Y–Ni, etc.); (c) transition antiferromagnetic metals with transition paramagnetic metals (Cr–Pt, Mn–Pt, etc.).
(2) Alloys of transition metals with normal (nonmagnetic) elements: (a) transition ferromagnetic metals with normal elements (Ni–Cu, Co–Ag, Fe–Al, etc.); (b) transition $3d$-antiferromagnetic metals (Mn, Cr) with normal metals (*Heusler alloy* types): Cu_2MnM, where M = Sm, Al, Ge, Zn, As, In, etc., and also Mn_4N, Mn–Au, Mn–Bi, Mn–Te, etc.; (c) transition paramagnetic metals with normal elements (Zr–Zn, Sc–In, Au–V, etc.).

In the ferromagnetic alloys and intermetallides a transition metal having an unfilled internal d- or f-shell with an uncompensated spin and orbital magnetic moment (*Hund rules*) can be one of the components. The d-electron wave functions in the crystal overlap considerably, which can lead to the bonding or delocalization of these electrons. For deep lying f-electrons there is no such overlap, so they are localized within the atom of the crystal in almost the same way as in an isolated atom.

In the theory of ferromagnetic metals there are two main approaches: for d-electron ferromagnetic metals there is the *model of delocalized electrons* or *band magnetism* (Frenkel, Bloch, Stoner, Moriya, et al.), for f-electron ferromagnetic metals there is the *model of localized electrons* (see *Heisenberg magnetism*). The fact that in the ferromagnetic metals Fe, Co, Ni the average magnetic moment (in Bohr *magneton* units μ_B: $P_S = \langle P_S \rangle = M_S/(n\mu_B)$, where n is the number of atomic magnetic moments per unit volume) is less than that in the isolated atoms, and in addition they have fractional values such as 2.2, 1.7, 0.6 instead of the corresponding 4, 3, 2 values in the isolated atoms, gives support to the model of delocalized electrons. Further support is the high value of the low-temperature electronic *specific heat* of transition metals compared to normal metals. In the case of d-electron metals and alloys the magnetic properties are determined only by the spin magnetic moment of the d-level, because the orbital moments are "quenched" and the coupling between spin and orbital components of the d-electron angular momentum in the level is broken by the strong crystalline electric field (*ligand field*), therefore the exchange interaction acts between the individual electrons with spin $J = S = 1/2$. Fig. 1 shows that for the d-electron ferromagnetic metals the theoretical dependence of the magnetization on the temperature $M_S(T)$ calculated for $J = 1/2$, coincides most closely with experiment.

The magnetism of the delocalized electrons may be described by the *single-electron approximation*. A free electron in the periodic *crystal field*, taking into account only its interaction with the electric field of the ion cores (neglecting dynamic electron correlation) has a standard band theory spectrum which consists of bands of finite width, separated by the forbidden gap ΔE, with a complex *dispersion law* $E(\boldsymbol{p})$, only coinciding with the quadratic form near "the bottom" and "the top" of the bands. The external magnetic field $\boldsymbol{H}$ in a paramagnetic metal changes the distribution of electrons with the spin projection values, equal to $+1/2$ or $-1/2$ (i.e. up "+", or down "−" spins): the levels which were degenerate at $H = 0$ are

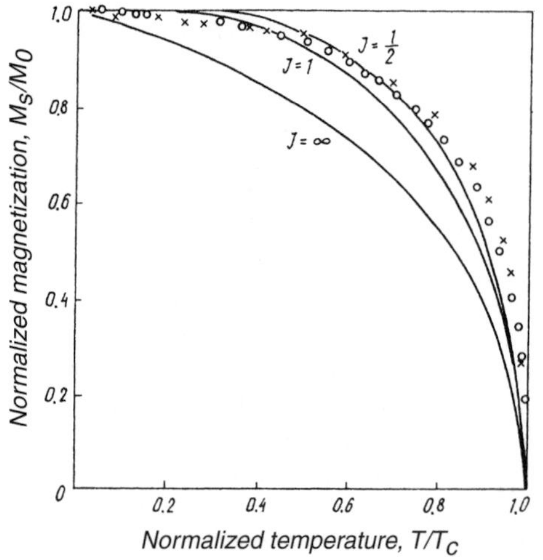

Fig. 1. Dependence of the normalized magnetization on the normalized temperature. Theoretical curves are shown for spin $J = 1/2$, 1 and ∞.

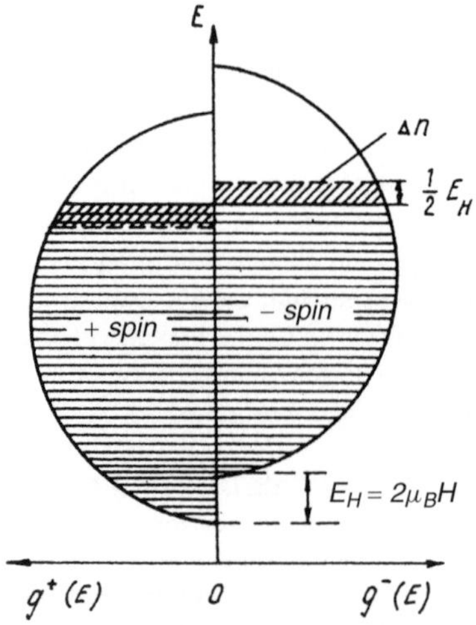

Fig. 2. Shift of the density of states curves for spin up (+ spin) and spin down (– spin) electrons in an applied magnetic field.

now split to two with the separation $E_H = 2\mu_B H$, and the density of state curves $g^+(E)$ and $g^-(E)$ for "+" and "−" electrons will be shifted by E_H, as shown in Fig. 2.

Electrons from the upper part of the subband "−" with thickness $E_H/2$ will transfer to the upper part of subband "+". Their number is $\Delta n = (1/2)g(E_F)E_H = g(E_F)\mu_B H$, where E_F is the *Fermi energy*, and $g(E_F)$ is the *density of electron states* at the *Fermi surface*. Thus the magnetization of an electron gas has the value $\Delta M = 2\mu_B \Delta n = 2g(E_F)\mu_B^2 H$. Within the framework of this approximation it is possible to describe Pauli *paramagnetism*. The *Pauli–Dorfman susceptibility* (W. Pauli, 1927; Ya.G. Dorfman, 1924) of a Fermi gas is $\chi_{pm} = \Delta M/H = 2\mu_B^2 g(E_F)$. In a ferromagnetic metal the same shift of "+" and "−" subbands occurs spontaneously, and at $H = 0$ under the effect of the *molecular field* $H_{mol} = N\Delta M$, where N is the molecular field constant, which is the main assumption of the *Frenkel–Stoner theory* (Ya.I. Frenkel, 1928; E.C. Stoner, 1936). By this the Fermi kinetic energy increases: $\Delta E_{kin} = (1/2)\Delta n E_H = (\Delta M)^2/(2\chi_{pm})$, the exchange energy reduction is $\Delta E_{exc} = -\Delta M H_{mol}/2 = -N(\Delta M)^2/2$ and the total energy balance is $(\Delta M)^2(1 - N\chi_{pm})/(2\chi_{pm})$, with its sign depending on the correlation between χ_{pm} and N^{-1}, the ferromagnetic state appears at $\chi_{pm} > N^{-1}$ or at $g(E_F)A > 1$, and $A = N/(2\mu_B^2)$ (the *Stoner criterion*). Therefore, an appreciable exchange energy A and a high value of $g(E_F)$ are necessary for ferromagnetism. It is also necessary to determine the origin of A. It is either due to direct d–d exchange, to indirect exchange between the different groups of d-electrons, or through conduction s-electrons. The development of the *Stoner model*, which is correct in principle, but only qualitative, occurs by specifying the calculations of the dispersion law and of the density of states $g(E)$, and by taking into account the electronic correlation and the fluctuations of the spin and charge density of the magnetic ions. One of the difficulties of the Stoner model lies in the impossibility of explaining the temperature dependence of χ_{pm} in many ferromagnetic metals, which at $T > T_C$ is close to the *Curie–Weiss law*, rather than following the weak quadratic dependence of all paramagnetic metals. Also some properties of weakly ferromagnetic metals, such as those of *strong band paramagnets* of the $ZrZn_2$ type with $M_s/\mu_B \gg n$, which have a very small exchange displacement of the "+" and "−" electron subbands (see Fig. 3) compared to E_F, are not understood.

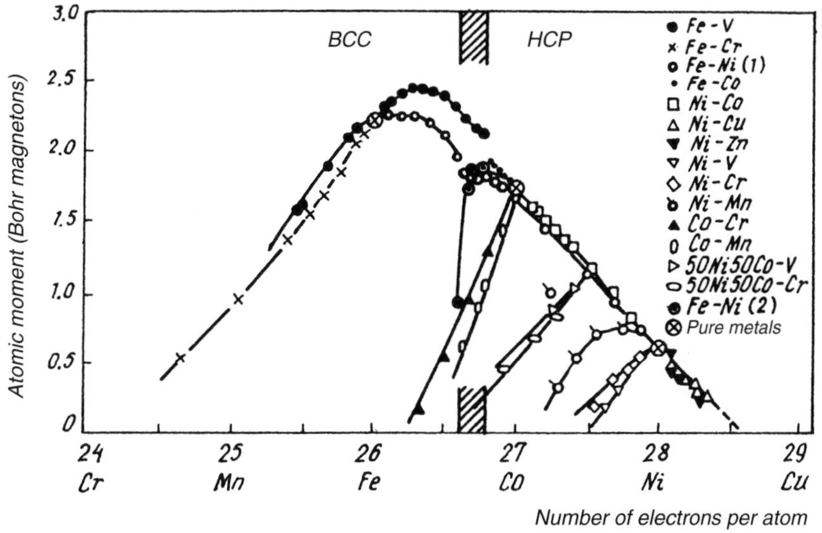

Fig. 3. Dependence of the atomic saturation magnetic moment on the number of electrons per atom.

At $T < T_C$ the magnetic properties are well explained by a correlated electron model, although at $T > T_C$ the Curie–Weiss law is closely obeyed. The same fact is observed for paramagnetic metals, close to ferromagnets (of Pt type). T. Moriya developed a *general theory of spin fluctuations* which provides a unified description of ferromagnetic metals, ferromagnetic semiconductors and insulators, including in the limit the models of delocalized and localized electrons (*Moriya theory*). For a general approach to spin fluctuations the *Rhodes–Wohlfarth curve* (P.R. Rhodes, E.P. Wohlfarth, 1963) (see Fig. 4) provides the dependence of $P_c/P_s = M_c/M_s$ on T_C for ferromagnetic metals, alloys and intermetallides. Here P_s and P_c are the average atomic magnetic moments in the units of μ_B, calculated respectively from M_s at $T = 0$, and from the constant C in the Curie–Weiss law under the assumption of localized magnetic moments, that is $C = n\mu_B^2 P_c(P_c + 2)/(3k_B)$.

In the model of localized electrons $P_c/P_s = 1$, and in the model of delocalized electrons $P_s \to 0$, which follows from the general theory of spin fluctuations, and agrees with experiment (see Fig. 4). The values of P_c/P_s are widely distributed along the length of the curve. This points to the fact that general theory of spin fluctuations explains the paradox concerning the Curie–Weiss law in the model of delocalized electrons, and provides a unified picture of ferromagnetic metals. In metals close to ferromagnets (Pd and Pt), due to the exchange amplification in the paramagnetic state, local spin fluctuations (*paramagnons*) are formed which tend to be conserved, i.e. to pass over to the state of "nuclear paramagnetism" or to that of a paramagnet with exchange amplification. Yu.P. Irkhin et al. showed that the strong dependence $\chi_{pm}(T)$ at $T > T_C$ in the ferromagnetic metals may be explained by the sharp dependence of the density of states $g(E)$ near E_F. This is important in the general theory of spin fluctuations because the spin fluctuations themselves depend strongly on the type of $g(E)$. It is also necessary to take into account the hybridization of s- and d-bands (*s–d exchange*); there are some reasons to think that the d-band in ferromagnetic metals is split to two bands.

The reduction of ΔM_s with the increase in temperature at low temperatures follows from the appearance of magnetic excitations, i.e. *spin waves* (*magnons*). In the case of ferromagnetic metals, in addition to the exchange of magnons, Stoner type excitations are also possible. Magnons lead to the *Bloch law* dependence $\Delta M_s \propto T^{3/2}$,

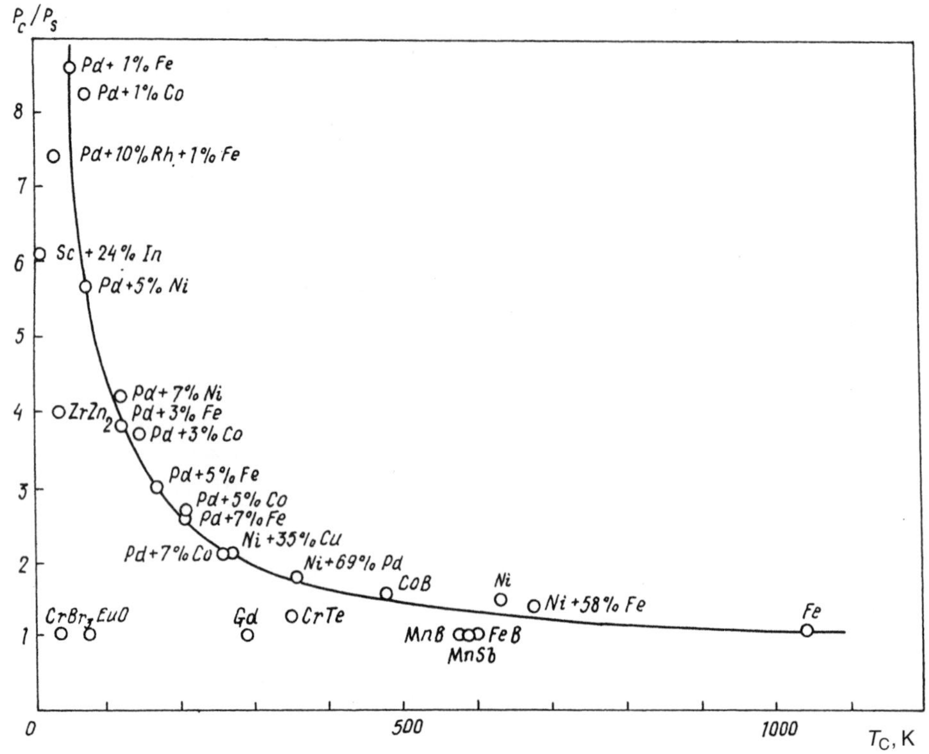

Fig. 4. Dependence of the ratio P_c/P_s of various materials on the Curie temperature T_C.

and Stoner excitations lead to the parabolic dependence $\Delta M_s \propto T^2$. For weak ferromagnetic metals the model of delocalized electrons leads to the dependence $\Delta M_s \propto T^2$ at not very low temperatures, $T < 0.5T_C$.

The magnetic properties of ferromagnetic metals, alloys and intermetallides are caused by rare-earth atoms, whose partly filled inner $4f$-shells which are shielded by surrounding outer closed $5s^2 5p^6$ shells determine the magnetic moment. Unlike the d-metal case, these shells are not subjected to the effect of the surrounding ligand fields, which quenches the orbital angular momentum in d-metals. In $4f$-metals the magnetic coupling is small compared to the electrostatic fields, and even relative to the spin–orbit interaction within the shell, therefore the magnetic moment of a rare-earth ion is determined by the sum of its spin and orbital momenta. Thus for $4f$-shells it is not possible to use the single-electron approximation, but it is necessary to consider the many-electron system of a $4f$-shell. Due to weak overlapping of the wave functions of $4f$-electrons of adjacent layers of the crystal lattice direct f–f exchange is vanishingly small. Moreover, the values of T_C in ferromagnetic metals of the rare-earth elements reach 300 K (in Gd), which is brought about by the indirect *exchange interaction* with adjacent $4f$-shells through the conduction electrons, the so-called *s–f exchange* (Shubin, Vonsovski, Zener, *Ruderman–Kittel–Kasuya–Yosida interaction*). The exchange integral A_{ind} has an alternating sign, is a variable function of the distance r between electrons with the dependence $\sim r^{-3}$, and its maximum value is of the order of A_{sf}^2/E_F. The oscillating characteristic of the Ruderman–Kittel–Kasuya–Yosida (RKKY) exchange interaction has been experimentally verified by *nuclear magnetic resonance*. The long-range character of this interaction leads to the appearance of a helical or spiral magnetic atomic structure which is observed in all the rare-earth ferromagnetic metal elements except Gd. I.Ye. Dzialoshinskii developed a general theory of this phenomenon based on the theory of *second-order phase transitions*, taking into account the effect of the band structure of the rare-earth elements (shape of *Fermi surface*).

Ferromagnetic alloys possess a great variety of properties. Their values of T_C and M_s depend essentially on the concentration of magnetic components, and on the structure. The *Slater–Polling curve* shown in Fig. 3 provides the dependence of average atomic magnetic moment on the number of electrons per atom, and we see that for Ni–Co alloys the experimental points lie at the straight line between the points with coordinates $(1.7, 27)$ and $(0.6, 28)$, and the simple mixing predicted by the model of localized electrons is observed. For other alloys this model may not be observed, e.g., for Ni–Cu. For this case it is not possible to use the model of localized electrons, but rather the model of delocallized electrons to account of the shape of $g(E)$. Partial atomic magnetic moments of the components of alloys are determined by the method of *magnetic neutron diffractometry*. The magnetic properties of ferromagnetic intermetallides, i.e. of chemical compounds with integer proportions of the components, are more complex. In the series of ferromagnetic alloys (Cu–Mn, Fe–Al, Ni–Mn, et al.), particular compositions exhibit the *spin glass* state. With the increase in concentration of the magnetic component ferromagnetic *clusters* may appear instead of individual disordered spins, and these may form an antiferromagnetic system with each other; this state is called *mictomagnetism*. There are also other magnetic states which can appear in *amorphous magnetic materials*. In addition there are intermetallides based on the *actinides* ($UNiS_4$, URu_2Si_2, etc.) in which ferromagnetism sometimes coexists with superconductivity (URu_2Si_2, etc.). Also, there are ferromagnetic intermetallides which exhibit the weak ferromagnetism of delocalized electrons in a specific temperature range, known for the lowest critical point (Y_2Ni_7).

FERRONS

Quasi-particles which may be formed in *magnetic semiconductors*. Ferrons are *conduction electrons* self-localized within regions with increased *magnetization* values. *Self-localization* can arise when the *exchange interaction* of conduction electrons with magnetic atoms causes the electronic energy to attain a minimum value at complete ferromagnetic ordering. The conditions for the existence of ferrons are especially favorable in antiferromagnetic semiconductors. In a nondegenerate antiferromagnetic semiconductor each electron may individually create a ferromagnetic microregion, and cause it to be stable by its localization there. In these semiconductors collective ferron states are possible, when several electrons with spins, parallel to each other, are self-localized simultaneously in the same ferromagnetic microregion. As a result, the crystal appears to be in a heterophase antiferromagnetic–ferromagnetic state.

In isotropic three-dimensional antiferromagnetic crystals the conditions for the existence of ferrons are fulfilled at Néel temperatures (see *Néel point*) $T_N < 10$ K, and the number of magnetic atoms in a ferromagnetic region per electron may reach 10^3 to 10^4. The conditions for the existence of ferrons in two-dimensional and *quasi-two-dimensional crystals* are far more favorable (they are possible even in antiferromagnetic semiconductors with $T_N \sim 100$ K). The stability of ferrons is increased by polarization of the crystal lattice by the self-localized electron, and as a result, the self-localized state becomes a ferron–polaron one. By increasing the temperature up to $T \sim T_N$ ferrons are destroyed; they also may be destroyed by applying a magnetic field which transforms the entire crystal into the ferromagnetic state.

The effective mass of ferrons is very high, and grows with an increase of the spin S of the magnetic atoms. Thus the conditions for *Anderson localization* of ferrons, caused by impurities and defects within the crystal, are very favorable. For high spin it is possible to consider ferrons immobile, and their destruction should lead to the transition from the insulating to the conducting state, observed experimentally in EuSe and EuTe. See *Magnetic semiconductors*.

FEYNMAN DIAGRAMS (R. Feynman, 1949)

A graphical technique used for the visual representation of perturbation theory expansions. This diagrammatic technique is widely employed in modern quantum solid state physics. Its idea is as follows: cumbersome analytical expressions of separate terms entering the perturbation series are represented in a pictorial and concise graphical manner, utilizing a construction called a Feynman diagram. After establishing the correspondence between the elements of this diagram (edges, vertices, and so on) and the elements of the analytical expressions (matrix elements, multipliers, energy differences, and so on), certain rules of graphical operations (summing, multiplication etc.) are introduced, so that any Feynman diagram composed of these elements can be uniquely converted back into its analytical counterpart. The characteristics of these rules are different for boson and fermion operators (see *Second quantization*), or for *spin operators*. The rules are specific for the particular type of particle interactions in the system under study.

A remarkable feature of Feynman diagrams, which underlies the efficiency of this diagrammatic technique, is the fact that a particular rule for "summing" these diagrams can be correlated with the actual summation of some series of terms (either finite or infinite) entering the perturbation theory expansion. The Feynman diagram depicting this sum is composed of elements that, in their turn, also denote certain summations. This makes it possible to simultaneously handle blocks composed of many elements, instead of being restricted to treating the individual elements one-by-one. Although we are unable, as a rule, to exactly sum an entire perturbation series, the partial sums obtained by operating with Feynman diagrams provide a considerably more accurate final answer. By using Feynman diagrams, a more rigorous reasoning and a more thorough understanding became possible for a number of fundamental concepts in modern solid state physics, that hitherto had been only intuitive. In particular, Feynman diagrams were largely instrumental in substantiating the theory of *Fermi liquids* (L.D. Landau), of Bose systems with a *Bose condensate* (N.N. Bogolyubov), the *molecular field* theory of magnetism, and many others.

FEYNMAN INTEGRAL

See *Path integral.*

FIANITES (named after FIAN, Physical Institute of the USSR Academy of Sciences, where they were originally synthesized in 1970–1972)

Transparent synthetic crystals $(Zr,Hf)O_2$. The cubic modification is stabilized by Y_2O_3. Small amounts of the impurities (Ce, Eu, Tb, Ho, Ti V, Mn, Fe, Co, Cu) give fianites various colors and hues: yellow, yellow-brown, red, pink, violet, green, white, etc. The melting temperature of the fianites is from 2873 to 3023 K; Mohs hardness is 7.5 to 8.5; density is 6 to 10 g/cm^3. Refractive index approaches that of diamond: 2.10 to 2.25, dispersion is 0.059 to 0.065. Manufacturing is carried out by high-frequency fusion in a cold container. Fianite is used as a material for optical *laser* devices, and for high-quality optical lenses and "windows". It also serves for jewelry in imitation of precious stones.

FIBER GLASS

The finest filaments which are produced from molten glass. The diameter of these threads ranges from 1 to 30 μm. Two types of fiber glasses are distinguished: continuous fiber (up to several kilometers in length) and staple fiber, which consists of short segments (up to 50 cm long). Continuous glass is similar to silk in appearance, whereas a staple fiber resembles cotton. Glass fiber materials exhibit an uncommon combination of properties: high flexural (see *Flexure*), tensile, and compression strength; *thermal resistance*; low *hygroscopicity*; insensitivity to chemical and biological attacks; high light translucence. The above properties favor using fiber glasses as electric-, thermal- and sound-insulating materials; glasses of high purity serve as materials for optical *light pipes* (see *Fiber optics*) for transmission of information in the form of light pulses. A light pipe consists of a glass light-guiding filament (core) embedded in a cladding (usually another type of glass) of lower *refractive index* so that *total internal reflection* takes place at the interface between the core and the cladding, and the light propagates along the core. The following techniques of fiber glass production are known in modern engineering: drawing fibers from billets (billet method), extruding the molten glass through spinnerets by application of a reel block or a bobbin (spinneret method), the centrifugal method (separation of glass mass jet by centrifugal force), the technique of glass blowing, which consists in moulding the glass mass using a jet of vapor or air with a high nozzle speed.

FIBER OPTICS

A branch of optics involving the propagation of light along fiber light guides, and related phenomena. Fiber optics appeared in the 1950s, and during the first 20 years of its development single fiber light guides or bunches of them with regular and irregular packing and lengths up to several meters were used. They were made from multicomponent *optical glasses*, and had a transmission factor in the visible region of only 30–70% at the length of one meter. The numeric aperture of light guides can reach 0.5–1. Light guides are mostly used in instrument making, particularly in technical and medical endoscopy for illuminating not so easily accessible regions (nasopharynx, stomach, cavities in the machines, etc.) and for transmission of images. In the latter case a regular packing of many fibers is used. The resolution is determined by the number of fibers, the diameter of the light guide core and the quality of their fabrication, and it is usually about 10 to 50 lines per mm.

Fiber light guides of quartz based glass with small optical losses (~1 dB/km) have been developed for the near-IR region. These guides have minimal optic losses in the range of wavelengths 1.3 to 1.6 μm, as shown in Fig. 1. Their transmission factor is about 50% at the length of several kilometers. As a result, new applications of fiber optics had appeared, such as fiber telecommunication systems, transmission of telemetric information, and sensors for monitoring different physical values (magnetic field, temperature, acoustic waves, rotation speeds, etc.). An infrared guide in its simplest version is a long filament with a core from highly-transparent glass, surrounded by a shell from material with a lower *refractive index* (Fig. 2(a)). Usually glass is used for the shell, but different *polymers* can also be employed. Fiber light guides are divided into multi- and single-mode ones depending on the number of modes which can be propagated along them. The core diameter is usually 5 to 8 μm for a single mode in the near-IR range, and from several tens to several hundreds micrometers for multimode ones. The difference between the refractive indices of the shell and core materials, providing light channeling by means of *total internal reflection* at their boundary, is, as a rule, 1 to 2% for multimode light guides, and several tenths of a percent for the single-mode ones.

In the mid-1980s, *multimode light guides* with a refraction index profile distributed over the cross-section (*gradient light guides*) and *single-mode light guides* (Fig. 2(b)) became widely used.

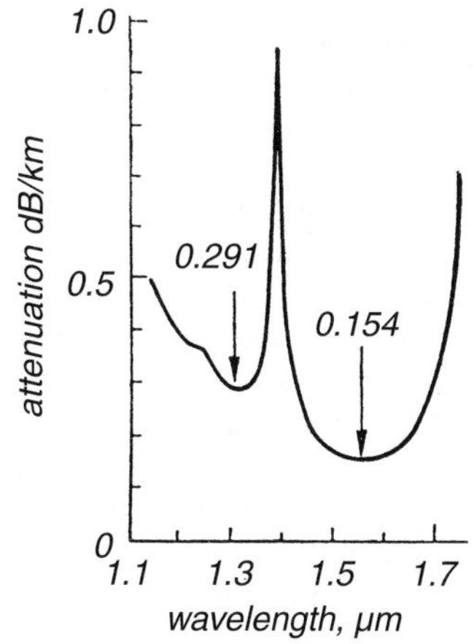

Fig. 1. Optical loss spectrum of a single-mode fiber light guide with SiO_2 core and SiO_2:F coating, showing the attenuation, dB/km (vertical scale), plotted versus the wavelength λ, μm (horizontal scale).

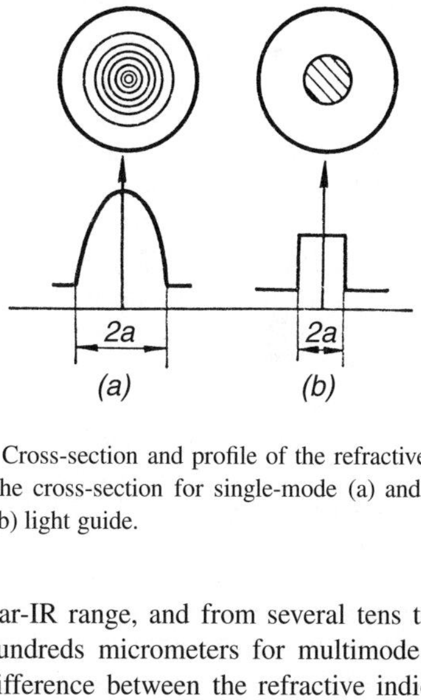

Fig. 2. Cross-section and profile of the refractive index along the cross-section for single-mode (a) and multimode (b) light guide.

For a refraction index profile close to parabolic mode-to-mode dispersion is sharply reduced and an information transmission band about 1 GHz wide was achieved in the near-IR spectral range. Fiber light guides based on quartz glass have a breaking *strength* up to 5 GPa.

The technology for manufacturing light guides from quartz glass is based on chemical deposition of the light guide material from the vapor phase, with volatile halides and oxygen serving as source materials. From among the volatile halides usually the chlorides of silicon, germanium, etc. are used. The quartz glasses of the core and shell are distinguished by different addition compounds (GeO_2, P_2O_5, B_2O_3, F^-).

FIELD DESORPTION

See *Field emission of atoms.*

FIELD EFFECT

Bringing about a change of the near-surface *electrical conductivity* by the application of an electric field normal to a *solid surface* (see Fig.). As a rule, the field effect is observed in semiconductors, but its appearance in a metallic phase is also possible (in particular, in *thin films*). An induced electric charge is distributed between the bulk semiconductor (*space charge region*) and the *surface states.* The concentration of free electrons (holes) in the space charge region is determined by its surface potential φ_S (surface band bending $e\varphi_S$) (see *Surface phenomena in semiconductors*) and the position of the *Fermi level* E_F in the bulk semiconductor. An external field establishes three states of the space charge region (and, correspondingly, three regimes of the field effect):

(i) the enhancement with majority carriers (in comparison with the bulk of the n-type semiconductor this means a strong bending of the bands downward, i.e. $e\varphi_S > 0$);
(ii) the formation near the surface of the depletion layer where the bands are bent up; in this case the conductivity relative to the field effect decreases with the subsequent upward bending;
(iii) as the concentration of minority carriers (with respect to the bulk) increases near the surface, a point is reached where the minority carriers near the surface exceed in concentration the majority carriers in the bulk, and an *inversion layer* emerges. When the surface bending becomes so pronounced that the valence band crosses the level E_F (not shown), then a degenerate electron gas forms (in certain situations this includes two-dimensional quantization of the carrier band spectrum). The formation of the inversion layer again produces an increase of the conductivity, but this time due to minority carriers. Thus a dependence of the conductivity on the applied voltage V has the shape of a curve with a minimum. From the analysis of this curve

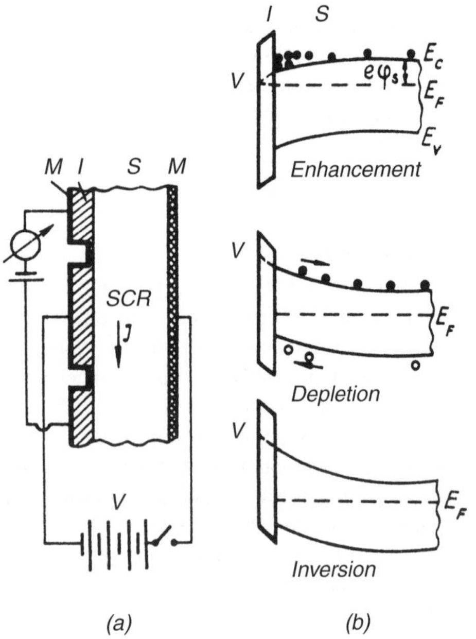

(a) Electrical connections of a metal (M), insulator (I), semiconductor (S), metal structure showing the applied voltage V and the current density J in the space charge region (SCR). (b) Plot of the voltage V as a function of the distance x from the insulator (I) into the semiconductor (S) in the subsurface regions of enhancement, depletion, and inversion. The energies of the bottom of the conduction band E_C, the top of valence band E_V, the Fermi energy E_F, and the surface band bending energy $e\varphi_S$ are indicated. In Fig. (b) electrons (solid circles) are shown just above the bottom of the conduction band, and holes (open circles) are shown just below the top of the valence band.

one can determine φ_S as a function of V, as well as the distribution of the induced charge between the space charge region and the surface states.

The study of the kinetics of the formation of electrical conduction under the conditions of the field effect provides the nonequilibrium parameters of the *surface states*. For the regime of enrichment these are the kinetic parameters involved in the trapping of carriers, in the case of the inversion layer regime these are the generation–recombination parameters of the surface minority carriers. By measuring the field effect relaxation one can divide the surface centers into fast (with relaxation time $<10^{-3}$ s) and slow ones ($\geqslant 10^{-1}$ s). A long-time constant relaxation of a non-exponential nature is due to surface heterogeneity. Slow relaxation can be caused by ionic processes (ion diffusion), adsorption–desorption processes, and field-stimulated chemical reactions.

FIELD-EFFECT TRANSISTOR (FET)

Semiconductor *transistor* of planar configuration, in which the flow of majority carriers goes from the emitting electrode (source S) to the collecting electrode (drain D) through a layer of doped semiconductor material. A change of voltage at the middle electrode (gate G) modulates the width of the region under the gate, thus changing the conduction characteristics of the material (see *Field effect*). Small changes of voltage at the gate cause pronounced changes of current in the source–drain circuit (see Fig.). The width and level of doping of the semiconducting material determine the limiting value of the current in the transistor. The upper frequency limit of a field-effect transistor increases with a decrease of the gate length and an increase of mobility of the carriers in the material. According to constructional features field-effect transistors are divided into three groups:

(1) Transistors with a gate in the form of a p–n junction or *Schottky barrier* at the boundary between the metal and semiconductor; there is one conductivity type in the space between source and drain.
(2) Transistors with a gate in a *metal–insulator–semiconductor structure* (MIS structure); in particular, with an oxide as insulator (*metal–oxide–semiconductor structure*, MOS structure), the regions of emitter and collector, and the gate region are oppositely doped (see *Doping*) as shown in the figure.
(3) Transistors with a heterostructure gate.

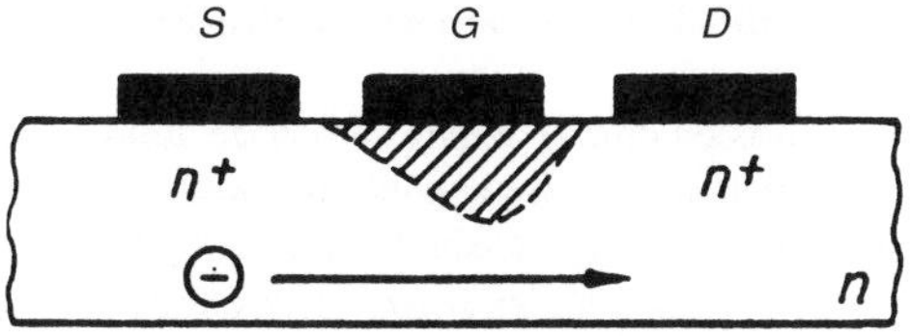

Sketch of a field-effect transistor (FET).

FIELD ELECTRON EMISSION

The emission of electrons by conducting bodies under the action of a strong ($E \sim 10^9$–10^{10} V/m) external electric field.

The mechanism of field electron emission is electron tunneling (see *Tunnel effect*) through the potential barrier near the conductor surface. Under field emission from a *metal* to the vacuum, the emission current density j over a wide range (10^{-1}–10^{11} A/m^2) is well described by the approximate Dushman equation

$$j = \left(\frac{4\pi me}{h^3}\right)(k_B T)^2 e^{-\varphi/k_B T},$$

where φ is the work function of the emitter, in eV. The energy spectrum of electrons in field emission from a metal is quite narrow (half-width $\sim$0.1 eV), with a shape which depends on the energy distribution of electrons inside the emitter, and on the characteristics of the quasi-levels of adsorbed atoms. Field electron emission from semiconductors has several characteristic features (nonlinearity of the $\log(j/E^2)$ versus $1/E$ dependence, photo- and thermosensitivity, the "contraction" of the emission cone from the needle-shaped emitter at high voltages) arising from the smaller electron concentration and the penetration of the field inside the semiconductor. If the emitter temperature increases and, accordingly, the electric field strength decreases, the field electron emission transforms into *thermal electron field electron emission*, and then to the *thermionic emission* amplification by the field due to the Schottky effect (see *Schottky diode*). At $j \sim 10^{12}$ A/m^2, a rapid destruction (explosion) of the emitter tip occurs, a cathode plasma torch arises, the emission current rises by about a factor of 100, and field electron emission transforms to *explosive electron emission* which is the initial stage of vacuum breakdown.

For applications see *Field-emission cathode.*

FIELD-EMISSION CATHODE

A cold cathode that emits electrons by means of *field electron emission.*

Field-emission cathode materials are various metals (clear or coated with activating layers), and metal-like compounds of semiconductors. A typical field-emission cathode is a needle produced by chemically or electrochemically *etching* a wire or a rod. The limiting field electron emission current density of a needle is determined by the heating of the needle tip by the current, and it has the value $\sim 10^{11}$ A/m^2. With typical needle tip radii $r \sim 0.1$–1 µm, the limiting field electron emission current is 0.01–1 A. To obtain higher currents the emitting area can be increased by increasing r and thereby increasing the operating voltage (e.g., $V \approx 5Er$ and $E \sim 10^9$–10^{10} V/m), or by using many-needle systems, blades, etc. The stability of a field-emission cathode is degraded by the adsorption of impurities which change its work function, and by ionic bombardment which pits its surface. Ways to increase the stability are to improve the vacuum; to use materials that adsorb residual gases; to separate the electron and ion trajectories; and to use field electron emission of microcells where, due to small (1–5 µm) interelectrode distances and low (20–50 eV) electron energy, ionic bombardment is strongly suppressed.

Field-emission cathodes are used in *field emission microscopy*, in vacuum *microelectronics*, and in infrared receivers (see *Autophotoelectronic emission*). Electron guns with a field-emission cathode are 10^3–10^4 times brighter than those with heated cathodes, and are used in *electron microscopy* and *scanning electron microscopy*, microanalysis, *X-ray microscopy*, etc.

FIELD EMISSION MICROSCOPY, field electron microscopy

A method for obtaining a factor of 10^5–10^6 times magnification of the image of the surface of a *solid* by using *field electron emission* from the surface.

The sample which serves as the cathode in the field of an *electron microscope* is a needle of a conducting material having a tip radius of curvature $r \sim 10^{-7}$–10^{-8} m. The anode is a luminescent screen placed at a distance of $R \sim 10$ cm from the tip. A high vacuum (residual pressure 10^{-7}–10^{-8} Pa) is maintained inside the microscope. Several thousand volts are applied between the screen and tip, which induces an electric field of strength 10^9–10^{10} V/m near the tip surface, as needed for the field electron emission. The emitted electrons are accelerated radially and produce on the luminiscent screen an image of the needle tip magnified approximately R/r times. The resolution of 2–3 nm is limited by *electron diffraction* of the tangential components of the electron velocity. The device thus cannot quite resolve the atomic structure of the surface. The contrast of the field electron image is determined by the local curvature (determining the local electric field) and the *work function* of different parts of the tip surface. Due to its small size, the tip is typically a *monocrystal.* The field electron image of a tip, smoothed and cleansed from impurities by high-temperature heating, exhibits the lattice symmetry of the tip material. Such a surface is used to study the changes in work function of different faces under *adsorption* of different materials, as well as the *surface diffusion* of adsorbed atoms, *phase transitions* in the adsorbed layers, field desorption of atoms (see *Field emission of atoms*), etc. Processes involving *crystal* shape changes under thermo-field treatments are also investigated by means of field electron emission.

FIELD EMISSION OF ATOMS, field evaporation of atoms

Emission or evaporation of surface atoms of a solid or liquid electrical conductor under the action of a strong electric field (of intensity $\sim 10^{10}$ V/m). This emission of adsorbed layers is also called *field desorption.* Field evaporation is the limiting case of positive *surface ionization* in strong electric fields; its mechanism consists in thermal evaporation of ions through the surface energy barrier, which is lowered by the field. Emission of multiply charged ions (charge of emitted ion may reach 6) is a distinctive property of field evaporation. These ions are believed to be generated by *post-ionization*, i.e. additional field ionization of outgoing ions at short distances from the surface. Field evaporation of *refractory materials* is enhanced at adsorption of certain molecular gases (H_2, N_2); as this takes place, emission of ions of complexes (hydrides, nitrides) is observed. Surface evaporation of semiconductors and *liquid metals* produces a large quantity of polyatomic ions.

Field evaporation is used in *field ion microscopy* for primary cleaning of the surface of the sample, for step-by-step removal of layers to investigate the internal composition of the sample, at the operation of field ion microscope in the mode of an atomic probe, etc. Field evaporation is the basis of operation of a *field desorption microscope*, and *liquid metal sources of ions.*

FIELD EMISSION OF ELECTRONS, cold emission of electrons

Electron emission by conducting bodies at low temperatures under the influence of external and internal electric fields. Cold emission involves *field electron emission* and emission of nonequilibrium electrons (including hot ones) excited by internal electric fields. Generally the term cold emission does not apply to various types of electron emission arising from irradiation of a cathode (*photoelectron emission*, *secondary electron emission*, *ion–electron emission*, etc.) or from other types of action upon it (*exoelectronic emission*, etc.).

FIELD EMISSION, THERMAL

See *Thermal electron field emission*.

FIELD IONIZATION, autoionization

The ionization of atoms and molecules in a strong electric field.

When a bound atomic electron resides in a potential well and an electric field is applied, a potential barrier is formed at one side of the well. The basic mechanism of field ionization is the tunneling of an electron (see *Tunnel effect*) through the barrier away from the atom. The probability of tunneling depends exponentially on the barrier height and thickness, and these depend on the electric field strength, E, and on the electron energy. A pronounced field ionization of simple gas molecules occurs at $E \approx (2\text{–}6)\cdot 10^{10}$ V/m, and that of alkali atoms and excited molecules occurs at 10^8–10^9 V/m. The area of the potential barrier near the metal surface becomes reduced due to image forces, and the probability of field ionization with the same E increases noticeably. The field ionization starts at a critical distance x_{cr} between the atom and surface, at which the energy level of an atomic electron rises above the *Fermi level* in the metal owing to the applied electric field, and the possibility for an electron tunneling into the empty energy levels in the *metal* is enhanced. If $x > x_{cr}$, the tunneling probability decreases due to the widening of the potential barrier, and thus the ionization zone half-width is a few tenths of an angstrom. The field ionization phenomenon is used in *field emission microscopy*, in the construction of ion sources for mass-spectrometers (see *Mass spectrometry*), in scanning tunneling microscopes, etc. A considerable increase of the ion current is achieved in cryogenic ion sources where the field ionization of gas molecules is concentrated at the surface of the tip.

FIELD ION MICROSCOPY

Method of investigation of atomic structure of pure *metals* and their *alloys*. The method is based on *field ionization* of noble gas atoms near the surface of the sample under investigation. The autoionic image of the surface is produced by an *ion projector* for a sample in the form of a needle with the radius of the curvature of the point in the range 10–100 nm. Because of its high magnification ($>10^6$) an ion projector provides information on the arrangement of individual atoms, and structural details of atomic sizes on the surface of the point. Field ion microscopy is used to study *point defects* of a crystal structure (*vacancies*, intrinsic and *impurity atoms*), *grain boundaries*, *stacking faults*, twin boundaries, *segregations* of impurities at defects and various surface phenomena, e.g., *surface diffusion*, *adsorption*, *corrosion*, *oxidation*, deposition of *thin films*. Limitations of field ion microscopy are related to the necessity to use needle-shaped samples, which must be cooled to helium temperatures, and the fact that the surface of many metals is severely damaged or destroyed by strong electric fields. Field ion microscopy does not provide a direct identification of the chemical nature of observed atoms. This problem is solved by using a combination of an ion projector with a mass spectrometer (see *Mass spectrometry*).

FILAMENTARY CRYSTALS
See *Thread-like crystals.*

FILM GROWTH, RADIATION-INDUCED
See *Radiation-induced film growth.*

FILM IONIC MODIFICATION
See *Ionic modification of films.*

FILMS
Layers of a particular material, commonly grown on a *substrate* for different purposes. As a rule, the properties of a film differ from those of the bulk material: there exists a *transition region between film and substrate*, the film shows size effects (see *Thin films*) and exhibits internal *mechanical stresses.* The first applications of films in engineering were for protective and decorative *coatings.* Films form the basis of solid-state electronics, hence numerous preparation techniques have been developed, and electrical, optical, mechanical, and other properties of films have been widely studied. The development of production methods like *epitaxy* and *photolithography*, followed by the attainment of microminiaturization, reproducibility of parameters, and high reliability of radio electronic circuits and devices based on them, promoted the wide application of films in electronic engineering. Modern methods of film technology, including *molecular beam epitaxy* and electron, ion, and X-ray *lithography*, continually increase the number of active elements that can be deposited per unit surface area of film.

See also *Thin film growing*, *Thin films*, *Amorphous films*, *Diamond-like carbon films*, *Epitaxial iron garnet films*, *Granular films*, *Insular films*, *Langmuir–Blodgett films*, *Magnetic films.*

FILMS, GRANULAR
See *Granular films*, *Insular films.*

FILM VIBRATIONS
See *Vibrations of rods and plates.*

FILTER, NUCLEAR
See *Nuclear filters.*

FINE STRUCTURE of electron paramagnetic resonance (EPR) spectra
An observable line splitting for $S \geqslant 1$ *paramagnetic centers* (here S is the electron or effective spin) due to the indirect influence of the crystalline electric field (molecule, *ligand*) upon the position of spin levels, via the interaction of the electric field with the *orbital angular momentum*, and of the latter with the spin. Fine structure energy splittings are often comparable to Zeeman level splittings for paramagnetic materials in magnetic fields close to one tesla in magnitude. This structure is described by including in the spin Hamiltonian the irreducible tensor $T_{lm}(S)$ or Stevens' operators $O_l^m(S)$ (see *Irreducible tensor operators*) with even l (values lm allowed for particular symmetry groups of the spin Hamiltonian are listed in *Spin Hamiltonian*). *Spin Hamiltonian parameters* for operators with $l = 2$ are sometimes represented in the form of a Cartesian $\boldsymbol{D}$-tensor. Typical forms of the spectra demonstrating the fine structure and the dependences of splitting on the orientation of the applied magnetic field are shown in the figures in *Hyperfine structure* and *Angular dependence of spectra.*

FINE STRUCTURE of X-ray absorption spectra
A system of extrema in X-ray absorption spectra. The greatest interest is associated with the structure located in the range of threshold ionization energy maxima of internal atomic shells in solids (so-called *absorption edges*) which are related, mainly, to *X-ray excitons*, and the structure whose energy exceeds the energy of edges by 10–1000 eV. The latter structure is called *extended X-ray absorption fine structure* (EXAFS); this results from the interference of the photoelectron wave emerging from the region of the ionized atom, with the wave of the same electron reflected from the neighbors of this atom. Such structure contains information about the spatial distribution of atoms in the solid, and this is employed for the structural analysis of crystalline

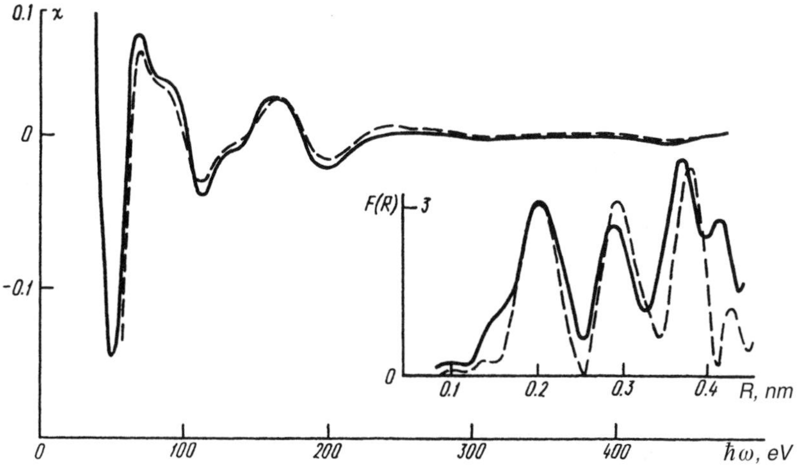

Oscillating part of K-absorption spectrum of Fe in deoxyhaemoglobin, and (insert) its Fourier transform which reflects the radial distribution of the Fe neighbors (solid lines are experimental data; dashed lines correspond to analytical calculation).

and noncrystalline solids, as well as *polymers* and macromolecules (see the figure).

FINITE ELASTIC STRAIN, finite elastic deformations

Elastic *strains* with values comparable to unity. *Rubber* is an example of a medium in which large deformations are possible (see *Elasticity*). The theory of finite elastic deformations is expressed in the language of differential geometry. If a medium before deformation is characterized by the metric tensor g_{ij} and after deformation by G_{ij} then the *strain tensor* is $u_{ij} = (G_{ij} - g_{ij})^2$. Let the deformation vector $\boldsymbol{v} = v_i \boldsymbol{G}^i$ (see *Displacement vector*), where the $\boldsymbol{G}^i$ are basis vectors after deformation, then $u_{ij} = (D_j v_i + D_i v_j - D_i v^r D_j v_r)^2$, where the symbol D_j means covariant differentiation: $D_j v_i = \partial v_i / \partial q^i - \Gamma^s_{ji} v_s$, the q^i are *Lagrangian coordinates* of a point in the medium, and the Γ^s_{ji} are *Christoffel symbols.* Deformation tensor invariants (see *Invariants of elasticity theory*) I_1, I_2, I_3 are defined as the coefficients in the expansion of the determinant $|g^{rm} G_{ms} + \delta^r_s| = \lambda^3 + I_1\lambda^2 + I_2\lambda + I_3$ in powers of λ. Thus, $I_1 = g^{rm} G_{rm}$, $I_2 = G^{rm} g_{rm}$, $I_3 = G/g$. Here $G = |G_{ij}|$, $g = |g_{ij}|$. For an incompressible medium $I_3 = 1$. If a deformable medium before and after deformation can be located in a Euclidean space, the components of *curvature tensor* calculated with the help of the metric tensor g_{ij} or G_{ij} are equal to zero. This imposes on the deformation tensor components some constraints that are called *compatibility criteria.* The application of these criteria means that *disclinations* and *dislocations* do not appear during deformation of the medium. The equations of motion of an elastic medium for finite elastic deformations have the form $D_i \sigma^{ij} + \rho F^j = \rho a^j$, where σ^{ij} is the *stress tensor*, ρ is the medium density after deformation, F^j is the *bulk force* density per unit mass, and a^j is the acceleration of a unit mass of the medium. The stress tensor σ^{ij} is symmetric if the moments of stresses acting on the medium are equal to zero, otherwise it can be asymmetric. If the *surface forces* $\boldsymbol{P}$ are defined on the surface of an elastic body characterized by a normal vector $\boldsymbol{n}$ then the boundary conditions look like $\sigma_{ij} n_j = P^i$. The deformed medium is called an *elastic medium* if a unique dependence exists between stresses and deformations, and it is called a *hyperelastic medium* (see *Hyperelastic material*) if it can be described by an *elastic energy* W which is a unique continuous scalar function of the deformation tensor u_{ij}.

It is possible to consider W as a polynomial of the components u_{ij} without loss of generality, as it is possible to approximate a continuous function to any precision by a polynomial, and the definition of W is reduced to the creation of a polynomial basis of deformation tensor invariants. For a known deformation energy the deformation tensor is determined by the formula $\sigma^{ij} = (\delta W/\delta u_{ij} + \delta W/\delta u_{ji})^2$. For a uniform isotropic body W is a function of I_1, I_2, I_3 only. In practice the following expressions for the deformation energy of a uniform isotropic body are commonly used: *harmonic (semilinear) energy*, Murnaghan deformation energy (F.D. Murnaghan, 1967), Rivlin–Saunders deformation energy (R.S. Rivlin, D.W. Saunders, 1951), etc. Solving *elasticity theory* problems for finite elastic deformations can be carried out by the method of successive approximations when the deviations from linear behavior are small, or by the so-called *semi-inverse method* when the deformed state is defined beforehand with an accuracy to within constants and some unknown functions, and it is necessary to determine these constants and functions to satisfy the boundary conditions. Finite elastic deformation theory can be converted into linear elasticity theory when it is possible to neglect higher powers of the deformation vector components and their derivatives in comparison with their first powers.

FINITE ZONE POTENTIAL

A potential $u(x)$ of a one-dimensional Schrödinger operator $H = -\partial_x^2 + u(x)$ (where ∂_x denotes differentiation with respect to coordinate x) with the spectrum as a sum of a finite number of intervals of a double absolute continuum. An arbitrary N-zone potential, its respective eigenfunctions, boundaries of spectra, *density of states*, etc. are expressed via the N-dimensional Riemann Θ-function:

$$\Theta(z) = \sum_{n\in\mathbb{C}^N} \exp\left[\mathrm{i}\pi(Bn, n) + \mathrm{i}2\pi(z, n)\right],$$

where $z \in \mathbb{C}^N$, B is a symmetrical complex $N \times N$ matrix possessing a positive definite imaginary part. An example of the N-zone potential is

$$u(x) = C - 2\mathrm{d}_x^2 \ln \Theta(\alpha x + \beta),$$

where $C \in \mathbb{R}$, $\alpha, \beta \in \mathbb{C}^N$ are constants. With the eigenfunctions in hand, making use of the theory of residues, one is able to determine matrix elements of any single-particle operator, and hence, to compute any single-particle property of an electron within the finite zone potential, e.g., the *electron–phonon interaction* constant. Generally, the finite zone potential is a quasi-periodic function and becomes periodic only when its periods satisfy the mutual commensurability condition. In the case of a smooth *periodic potential*, the *band gaps* in the spectrum of the respective Schrödinger operator rapidly shrink with energy, so any periodic potential could be approximated by the finite zone potential within any desired accuracy by neglecting small band gaps. This potential is an exact solution to the Peierls problem of the self-consistent state of conduction electrons within the lattice (see *Peierls transition*), yielding *polarons* and *charge density waves* as particular cases. The finite zone potentials are a periodic general case of *non-reflective potentials*.

FIRST-ORDER PHASE TRANSITION

The *phase transition* of a material from one thermodynamic *phase* to another, accompanied by a certain quantity of heat (*heat of phase transition*) being either absorbed or evolved. Examples of first order phase transitions are *melting* and *crystallization*, certain ferroelectric and magnetic transitions, etc. If the symmetry of the system does not change at the transition, then the curve of the first-order transition on the *phase diagram* may end at the *critical point*. The free energy and the *thermodynamic potential* of the system are continuous at the first-order transition point, but their first derivatives have discontinuities at this point. The *order parameter*, *susceptibility*, *specific heat* and other quantities undergo an abrupt change at a first-order transition. An abrupt decrease or increase of the temperature may result in thermal *hysteresis*, and the substance may reach a supercooled or superheated *metastable state* (pure water, e.g., may be supercooled down to $-40\,^\circ$C without crystallizing to *ice*). The presence of thermal hysteresis is often used for identifying first-order phase transitions.

FISKE STEPS in tunnel junctions (M.D. Fiske, 1964)

Stepwise dependence of the current on the voltage at a Josephson *tunnel junction* or *Josephson junction*, caused by the excitation of standing electromagnetic waves matched to the Josephson current. At a voltage step V of the junction in a constant magnetic field with intensity B the current density has the form $j = j_c \sin(\omega t - kx)$, where $\omega = 2eV/\hbar$, $k = 2e\Lambda B/\hbar$, and Λ is the *penetration depth of magnetic field* into the superconducting junction. A step appears at the current–voltage characteristic of the tunnel contact when ω coincides with one of the resonant frequencies of the coupled stripline superconducting resonator. The variation of the applied field B causes an oscillating dependence of the Fiske step height on B.

FISSURES, tears

Macrodefects appearing during *plastic deformation* of solids, and *metals* in particular. Fissures are tears at edges and surfaces of billets (ingots) that appear during the course of their rolling, *forging*, stamping and piercing, and result from deformations that are pronounced in both degree and rate, that is, they go beyond what is technically acceptable. Further *strain* after the appearance of fissures leads to *failure* of the billet. Most often fissures form in the course of processing materials of low *plasticity* by pressure. Such materials are best processed under *uniform compression* (see *Hydroextrusion*). The material plasticity increases with increasing processing time. However, there may be a danger of *metal overheating* (that may lead to the formation of extensive crystallization) and even of burning (*oxidation* and sometimes even melting of *grain boundaries*). It all results in a significant decrease of plasticity and *viscosity*. In certain cases billets with minor fissures may be used after the perturbed surface layer is cut away.

FIXED POINT

A point in a space that remains invariant with respect to a prescribed set of transformations of that space. In *critical phenomena* theory such transformations are elements of a renormalization group, operating either in the *Gibbs distribution* space for the configurations of fluctuating fields of dynamic variables describing the *order parameter*, or in the space of corresponding effective Hamiltonians. The fixed point is associated with a self-similarity distribution which describes the system at a *critical point*, determines the nature of a *phase transition*, and allows evaluating the *critical indices* (see *Renormalization group method*).

FLAKING

See *Scabbing*.

FLASH HARDENING

A technical operation consisting essentially in the fast cooling of a melt. Physically flash hardening may be provided in several ways. One approach is to form a thin layer or a small droplet of melt fast enough while maintaining its good thermal contact with a thermally highly conductive substrate, or some tempering medium to carry away the heat, and ideally the rate of cooling may reach 10^6–10^{10} K/s. High initial supercooling of the melt and quick removal of heat from the *crystallization* front and from the cooling solidified material then combine to produce a heavily supersaturated *solid solution*. New metastable *phases* may also appear in this manner (see *Metastable state*). One may achieve highly disperse microstructures with a specific *crystal morphology*, or freeze an *amorphous state* into metals and alloys. Flash hardening techniques may be divided into three basic groups. *Sputtering* techniques envisage that droplets of melt cool and solidify during their flight through a gaseous medium. The second group of techniques provides hardening at certain cooling surfaces or in flowing media. In the latter case, the melt is entering a heat receptor continuously. By injecting the melt into a cooler crucible one may produce solid filaments. Extracting the melt or punching its droplets between two colliding pistons may form microwires and narrow bands encased in glass coating; the so-called *Taylor–Ulitowski techniques* (J.F. Taylor, 1924; A.V. Ulitowski, 1957). Next come the techniques of *melt spinning* whereby one pressure-feeds the melt onto a rotating cooler. A widely used version of that technique consists in rolling the melt between the two rotating coolers. The droplet may be shot onto a substrate (the so-called *splatting*). Splatting provides cooling rates up to 10^{10} K/s. The techniques belonging to the third group of methods (*welding*) envisage that a localized surface layer of limited thickness cools and hardens due to heat evacuation into the bulk volume of metal or metal substrate. Such techniques are employed during electric arc welding. They are also at work during *electric pulse machining*, electron beam or laser beam processing of surfaces.

The output products of flash hardening have the form of powder particles of various shapes, flakes, beads, filaments, or surface coatings. To obtain bulk samples and pieces with flash-hardened surfaces one has to gradually build up thicker layers of modified metal by repeatedly processing their surface with electron or laser beams. Next, one has to apply high hydrostatic pressure to compact the products of flash hardening. To achieve the needed physical and chemical properties of the piece, and attain its required operational qualities, one has to select and apply appropriate technological schemes.

FLEXIBILITY OF POLYMER CHAIN

Ability to change the *conformation* of a polymer chain through thermal motion (*equilibrium* or *thermodynamic flexibility*), or under the action of an external, in particular mechanical, action (*kinetic flexibility*). A conformational change results from a rotation about a single bond of the main chain. The mechanical and relaxational properties of *polymers* are related to the flexibility of macromolecules. The following parameters are used as characteristics of the equilibrium flexibility: the ratios of mean-square distances between the ends of a chain in a solvent, and the free rotation of its segments; the magnitude of a Kuhn segment (a statistical segment); the persistence length of a macromolecule. The kinetic flexibility determined by the magnitude of a kinetic segment is not a constant for an isolated macromolecule, but depends on the rate of application of an external load (or the oscillation frequency). In the absence of a solvent the flexibility of amorphous polymers manifests itself in a *vitrification* region and at higher temperatures.

FLEXOELECTRIC EFFECT

Appearance of macroscopic electrical polarization in a crystal during *flexure* deformation. The inverse flexoelectric effect involves the flexure of a crystal under the effect of an electric field. The direct and inverse flexoelectric effects appear particularly strong in *liquid crystals*. The presence of a constant electric *dipole moment* in the molecules, and the asymmetry of their molecular shape, is a physical cause of the flexoelectric effect in liquid crystals. In the absence of *strains* the orientation of the dipoles "forward" and "backward" is equally probable so there is no sample polarization (see Fig. (a)). Upon the flexure of a liquid crystal (Fig. (b)), due to the asymmetry of the shape of molecules the dipole orientation appears in the preferred direction, and therefore a macroscopic polarization $\boldsymbol{P}$ develops. A quadrupole mechanism for the appearance of the flexoelectric effect is also possible.

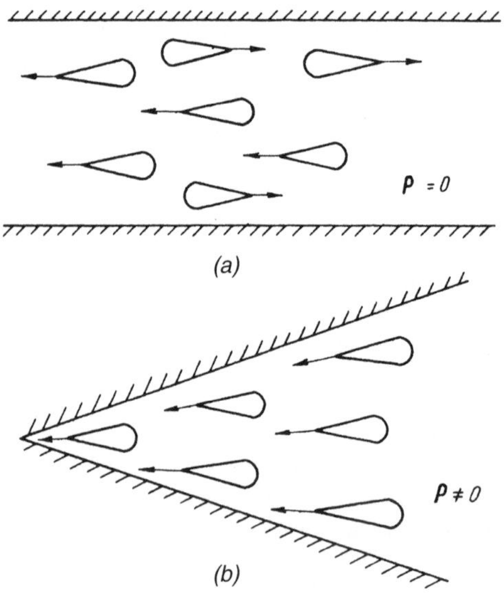

Flexoelectric effect.

FLEXURAL WAVES

Elastic waves in long *rods* or *plates* when the vibrations are accompanied by *flexure* (bending) strain. The displacements in flexural waves are normal to the rod axis or the plate plane. The flexural wave length, λ, must be much greater than the body thickness. Two kinds of flexural waves with *dispersion laws* of the form

$$\omega_i = \left(\frac{EI_i}{\rho S}\right)^{1/2} k^2$$

are possible in rods (ω is the frequency, $k = 2\pi/\lambda$ is the wave vector, E is *Young's modulus*, ρ is the matter density, S is the cross-section area, $i = 1, 2$, I_i are the *principal moments of inertia*). Only one type of flexural wave with a similar dispersion law

$$\omega = \left(\frac{h^2 E}{12\rho}\right)^{1/2} \left(1 - \nu^2\right)^{-1/2} k^2$$

(h is the plate thickness; ν is the *Poisson ratio*) can propagate in plates. These waves exhibit *dispersion*, and in both cases their phase velocities are $c_{ph} \approx \sqrt{\omega}$, the group velocities being $c_{gr} = 2c_{ph}$.

FLEXURE

A strained state (see *Strain*) which emerges in a specimen under action of forces and moments directed normally to its axis, and it is accompanied by distortion of the axis. Normal stress σ developing in a cross-section of the sample results in a moment M of force, normal to the axis, referred to as a *bending moment*. Static bending tests are commonly employed for assessment of mechanical properties of a material. Two loading schemes can be used in bending tests of a specimen at rest on stationary supports:

(i) The load is applied by a concentrated force at the midpoint between the supports.
(ii) The load is applied by two equal forces at two points equally distant from the supports.

The first scheme leads to a distribution of bending moments along the sample length in the shape of an isosceles triangle, the moment attaining its maximum at the midpoint of the span between the supports. Under the second scheme, the bending moment distribution follows the trapezium law, whereas the magnitude of the moment in the interval of "*pure flexure*" is constant. Bending tests are useful to determine mechanical properties of brittle materials, to evaluate the temperature of their transition from the brittle state to the viscous state (see *Cold brittleness*), and to study *dislocation* mobility at a given level of stress and temperature.

FLICKER NOISE, excess noise

Fluctuations of the dependence of the spectral density S_f on the frequency of the form $1/f^{\alpha}$, where the exponent α is close to unity (usually $0.8 < \alpha < 1.4$). This is observed in diverse types of conductors: in carbon resistors, in semiconductors, in devices based on *semiconductor junctions*, in continuous metallic *insular films*, etc. The quantity S_f decreases with an increase in frequency, and at high frequencies flicker noise is hardly noticeable in the background of other noise sources. The highest frequencies at which it has been observed are $\approx 10^6$ Hz. The lower boundary of flicker noise, where the dependence $S_f \sim 1/f^{\alpha}$ no longer holds, has not been reached experimentally, although measurements have been carried out down to extremely low frequencies ($\sim 10^{-6}$ Hz). Existing theories of flicker noise do not explain the absence of the lower limit, although in some systems they predict the possibility of noise with a dependence close to $1/f$ over a broad spectral range. The spectral density of electromotive force noise of the $1/f^{\alpha}$ type is generally proportional to the square of the current flow, indicating that it arises from a broad spectrum of two-level systems wth extremely low level cutoffs so the $1/f$ noise can persist to arbitrarily low temperatures.

FLUCTUATION DIAMAGNETISM

Anomalous diamagnetism of *superconductors* above the *superconducting phase transition* temperature T_c, arising from the fluctuation formation of *Cooper pairs*. The fluctuation diamagnetic *susceptibility* χ_{fl} is much lower in magnitude than the susceptibility of a perfect superconductor $\chi = -1$ in SI units ($-1/4\pi$ CGS, see *Meissner effect*). Fluctuation diamagnetism becomes more pronounced with the reduction of the system dimensionality (in *thin films*, *thread-like crystals* and *small particles*), and also upon contamination of the material by impurities (see *Fluctuations in superconductors*). Measurements of χ_{fl} in small superconducting particles with the size $d \ll \lambda$ (λ is *penetration depth of magnetic field*) allow one to determine the width of the *critical region*, and its dependence on d. The value of χ_{fl} may be differentiated from the background diamagnetic susceptibility of a metal by its strong temperature dependence, which follows a power law.

FLUCTUATION-DISSIPATION THEOREM

Theorem of statistical physics, establishing the connection between the *fluctuations* of a system in its equilibrium state and its nonequilibrium properties. It was proposed by H.B. Callen and T.A. Welton (1951). The reaction of a system to some perturbation under the action of a force f (depending on time t as $\cos\omega t$) which enters as the additional term $-f\hat{x}$ in the system Hamiltonian ($\hat{x}$ is an operator corresponding to the physical value of x), leads to a variation of the average value $\overline{x}$ of the quantity x to $\delta\overline{x} = \alpha(\omega) f$, where $\alpha(\omega)$ is called the generalized *susceptibility* of the system which determines its nonequilibrium properties. According to the fluctuation-dissipation theorem, the Fourier transform of the correlation function

$$\varphi(\omega) = \int_{-\infty}^{+\infty} \frac{1}{2}\left[\overline{\Delta x(t)\Delta x(0)} + \overline{\Delta x(0)\Delta x(t)}\right] e^{i\omega t}\, dt$$

is associated with $\alpha(\omega)$ by the relationship $\varphi(\omega) = \hbar \coth[\hbar\omega/(2k_B T)] \operatorname{Im}\alpha(\omega)$, where Im denotes imaginary part. The fluctuation-dissipation theorem is important for the evaluation of noise in quantum systems. It is applicable to many systems in which correlations are studied, and frictional or fluctuating forces are operating.

FLUCTUATION LEVELS

Energy levels of *quasi-particles* in the fluctuation region of a spectrum (see *Elementary excitation spectra of disordered solids*). The quantum states corresponding to fluctuation levels are localized; they appear at low-probability *fluctuations* associated with the potential wells of quasi-particles, and they are concentrated at these fluctuations. In a macroscopic (infinite) system the spectrum of fluctuation levels becomes dense, but the distance separating the centers of the localized of states corresponding to two fluctuation levels (i.e. between corresponding potential wells) grows exponentially in proportion to the convergence of their energies. As a result, the movement in each such well, which "seems" to the quasi-particle to be infinitely deep, is almost independently quantized. The density of fluctuation levels at any particular energy is, as a rule, exponentially small, and it is in fact formed by a single *optimal fluctuation*, for which the energy E corresponds to the ground state in the fluctuation potential well. Nevertheless, within the scope of this exponential smallness in different parts of the fluctuation region there may appear energy dependences of the fluctuation level density, which reflect a different structure of the optimal fluctuations forming the spectrum at these regions. The presence of fluctuation levels is a characteristic singularity of the elementary excitation spectra in *disordered solids*. The identification of this region of the spectrum, and its detailed investigation, was initially carried out by I.M. Lifshits. A series of important and exacting results substantiating the picture of such a fluctuation spectrum was obtained by J. Fröhlich and his colleagues.

FLUCTUATION REGION

The same as *Critical region*.

FLUCTUATIONS in solids

Random deviations of physical values x_i, characterizing a solid, from their average values $\bar{x}_i$. The random, undetermined character of the fluctuations means that in the same physical system subject to constant external conditions the value of x_i depends on time ($x_i = x_i(t)$), but it is impossible to predict the explicit form of this dependence from the experimental conditions. A description of the state of a solid based on a restricted set of values x_i corresponding to the temperature, *magnetization*, *polarization*, concentration of current carriers, etc., can only be approximate. The transition from a fundamental description based on mechanical equations of motion for the particles belonging to the body, to a simpler description based on actual measurements, requires that approximations be made, which obscure detailed information about the initial state of the particles when their motion is averaged over short intervals of space or time. Thus the fluctuation $\delta x_i = x_i - \bar{x}_i$ provides a measure of the incompleteness of the statistical description of the system. Fluctuation process may be characterized with the help of a *correlation function* $S(t,t') = \overline{\delta x_i(t)\,\delta x_i(t')}$, where the top bar means averaging over all possible values of $\delta x_i(t)$ at $t < t'$, or of $\delta x_i(t')$ at $t' > t$. In stationary systems $S(t,t')$ depends only on the time difference $t - t'$: $S(t,t') = S(t - t')$. The function $S(t,t')$ indicates how the presence of a fluctuation δx_i at a moment t influences its value at the later instant t', where $t < t'$. The function $S(t,t')$ declines with the growth of $|t - t'|$, eventually vanishing with the characteristic memory time of small fluctuations. Instead of a time correlation function it is possible to use the spectral presentation determined by the expression

$$S_\omega = \int_{-\infty}^{+\infty} \mathrm{d}t\, \mathrm{e}^{\mathrm{i}\omega t}\, \overline{\delta x(t)\, \delta x(0)},$$

where S_ω is called the *spectral density of fluctuations* of the value x. It is often possible to correlate fluctuations with values characterizing the *linear response* of the system to an external field. For example, the spectral densities of fluctuations of heat flow, and of electric currents in a state of *thermodynamic equilibrium*, determine the values of the corresponding *thermal conductivity* and *electrical conductivity*, respectively. It is possible, by measurements of S_ω, to find the values mentioned above. In addition, fluctuations are the cause of noise appearing in weak signal amplifiers, and also in various generators.

FLUCTUATIONS, CRITICAL

See *Critical fluctuations.*

FLUCTUATIONS IN SUPERCONDUCTORS

Random deviations of the *order parameter* in superconductors from its average value which minimizes the free energy F. The probability W of the appearance of fluctuations is determined by *Gibbs factor* $W \approx \exp(-\Delta E/T)$; it depends on the volume of the fluctuations ΔV, and on the free energy of the condensate in the superconductor $\Delta F \approx \Delta V B_c^2(T)/(2\mu_0)$, where B_c is the thermodynamic critical field. The extent of the fluctuations is of the order of the *coherence length* $\xi(t) = \xi(0)(1 - T/T_c)^{-1/2}$. Here

$$\xi(0) = \min\left|\left(\frac{\hbar v_F}{k_B T_c}\right), \left(\frac{\hbar v_F}{k_B T_c} l\right)^{1/2}\right|,$$

where v_F is the Fermi velocity, l is the electron mean free path, and T_c is the *critical temperature.* Fluctuations are strongest in superconductors of reduced dimensionality $n < 3$ (n is a number of dimensions along which the size of superconductor exceeds $\xi(T)$), and they show their presence within a narrow temperature range δT near the *phase transition* point at $T = T_c$. The width of the *critical region* where fluctuations are strong is determined by the *Ginzburg number*

$$\varepsilon_n \equiv \frac{\delta T}{T_c} = \left[\frac{\varepsilon_F}{T_c} \frac{k_F^{-3}}{\xi^n(0) d^{3-n}}\right]^{1/(2-n/2)},$$

where $d < \xi(T)$ is the smallest dimension of the material, ε_F and k_F are the Fermi energy and momentum, respectively. In pure (type I) three-dimensional superconductors the width of the critical region $\varepsilon_3 \approx 10^{-15}$ is not experimentally measurable. However, dirty superconductors of reduced dimensionality (films, filaments, granules, lamellar and chain crystals) exhibit values of $\varepsilon_n \sim 10^{-1}$ to 10^{-3}, and fluctuation effects are experimentally observed. Their presence is evident in the experimentally measurable effects *fluctuation diamagnetism*, *paraconductivity*, *specific heat* anomalies. Above the critical temperature the fluctuations lead to the possibility of combining electrons into nonequilibrium (fluctuation) *Cooper pairs*, with life times near T_c of $\tau = \pi\hbar/[8k_B(T - T_c)]$. Due to the relatively small size ξ of a Cooper pair and the large number of particles with a negligible probability of forming one, the fluctuations of a superconducting region should be involved in the pair generation process.

Fluctuations influence various physical parameters of superconductors: specific heat, conductivity, magnetic *susceptibility*, resistance of *tunnel junctions*, *sound absorption* factor, etc. Several diverse factors contribute to this. The first concerns the influence of fluctuation-generated Cooper pairs themselves on these physical characteristics of a normal metal. Thus in superconductors at temperatures somewhat above the critical temperature the mechanism of *paraconductivity*, predicted by A.G. Aslamazov and A.I. Larkin (1968), results in the additional transport of electrical charge by fluctuation-generated Cooper pairs. The corresponding correction factor to the conductivity appears to be singular near T_c: $\Delta\sigma/\sigma_n \propto [T_c/(T - T_c)]^{2-n/2}$, where n is the dimensionality of the superconducting system. The second contribution to the physical parameters of superconductors above T_c is caused by variations of the single-electron *density of states.* It is not so significant very near T_c because it has a much weaker dependence on $T - T_c$ than the first one, but it is comparable to it in respect to the temperature range associated with its contribution ($\Delta T \geqslant T_c$). Finally, in sufficiently dirty superconductors ($l < \xi$) above T_c there is a third fluctuation contribution, the so-called *Maki–Thompson contribution* (K. Maki, 1968; R.S. Thompson, 1970).

This is caused by quantum interference associated with the scattering of the electrons that form the fluctuation Cooper pairs. At the same impurities, and close to T_c, it has the same (and in films to some extent an even stronger) temperature dependence as the *Aslamazov–Larkin contribution*. The Maki–Thompson contribution in low-temperature superconductors depends essentially on the presence in them of paradestructive mechanisms (of the magnetic field, paramagnetic impurities, etc.). Below T_c fluctuation phenomena appear to be more diverse, involving the modulus and phase of the order parameter, or fluctuations of scalar and vector potentials. These fluctuations also influence other properties of superconductors, and in particular they lower the value of T_c, as expected from the *Bardeen–Cooper–Schrieffer theory*.

FLUCTUATIONS, MESOSCOPIC

See *Mesoscopic fluctuations*.

FLUCTUATIONS OF ATOMIC POSITIONS, fluctuation wave

Variations in *static displacements of atoms*, and of displacements in the concentration of *solid solutions*, involving *long-range and short-range order* parameters, and also other internal parameters, which can have equilibrium and spatially nonuniform distributions in the volume of a crystalline material. Fluctuation inhomogeneities of the internal parameters play a determining role in the *diffuse scattering of X-rays*, *neutrons* and electrons. They are especially high in systems located near the point of a *second-order phase transition*.

FLUCTUON

Self-localized *conduction electron* in a disordered or partially ordered nonmetallic medium. It is a region of variation (*fluctuation*) of the medium parameter which creates the potential well where an electron is localized, and it supports by its field the persistence of the fluctuations. A fluctuon may be regarded as the bound state of an electron and a fluctuation (unstable in the absence of an electron). In solutions fluctuons may be formed in a region of varying concentration, in gases – in regions of varying density. In para- and ferromagnetic semiconductors (see *Magnetic semiconductors*) a fluctuon is associated with the fluctuation of magnetization. It forms a ferromagnetic region in a *paramagnet*, and a zone of increased magnetization in a *ferromagnet*. In the vicinity of a *first-order phase transition* there may appear fluctuating regions of the second phase near which the electrons may be localized, thus forming the particular case of fluctuons – *phasons*.

When particular conditions are satisfied fluctuons can trap a large number of atoms. Fluctuons are thermodynamically favorable only in a limited range of temperatures surrounding the *Curie point* in a ferromagnet and the critical point in solutions or gases, but usually not including low temperatures. The narrowness of the temperature range where free electrons can transform into fluctuons, a process which is considered as a *diffuse phase transition* in the electronic subsystem, is a characteristic feature of fluctuons.

Fluctuons move in an external electric field since they are *current carriers*; and they have an unusual mechanism of mobility, associated with the *diffusion* of atoms or spins, or with viscous flow in the medium. Bound fluctuons may appear near impurity centers. The transformation of electrons into fluctuons can strongly influence the electrical, magnetic, optical, and other properties of semiconductors.

FLUORESCENCE

Luminescence during the excitation with a short-duration afterglow (unlike *phosphorescence*). It is characteristic of *molecular crystals*, excitons, plasmas, and some other types of luminescence in crystallophors (see *Luminophors*). Fluorescence may arise from electronic, as well as vibrational *radiative quantum transitions*. Most elementary processes which lead to luminescence have been investigated in detail, and the term fluorescence refers to those cases in which the system returns to its initial state almost instantaneously after the radiative quantum transition takes place, so that the maximum duration of afterglow is mainly determined by the probability of this radiative transition taking place.

FLUORIDES

Compounds of fluorine with other elements. This halogen forms compounds with all elements except the light noble gases helium, neon and argon. Fluorides are divided into covalent and ionic ones, although there is no rigid line of demarcation between these two groups. The most characteristic examples of covalent fluorides are compounds with nonmetallic elements (B, C, Si, N, P, As, O, S, Se, Te, Cl, Br, I). These compounds are characterized by low melting and boiling points (most are gases, some are volatile liquids). In the solid state these fluorides possess molecular lattices of low strength. Another group of covalent fluorides involves polyvalent *metals* (platinum metals, V, Nb, Sb, Ta, Mo, W, Tc, Re, U, Np, Pu) and noble gases (krypton and xenon). Most of these fluorides are volatile solid substances, which form a molecular crystal lattice of low strength. Molecules of certain fluorides of this group are held together by bridges of fluorine atoms. Fluorides of various transition and nontransition group metals, which exist in low oxidation states (usually from +2 to +4), are intermediate in properties between the two above groups. All of them are low-volatility high-melting compounds of polymeric structure (see *Polymers*). The driving force for the polymerization is the tendency of atoms to increase their coordination numbers. The process of polymerization involves the formation of bridges of fluorine atoms. Both chain and three-dimensional polymeric structures are known.

Typical representatives of ionic fluorides are those of alkali and alkaline-earth metals. They exhibit the highest melting and boiling points among all fluorides, and possess ionic lattices of high strength. Ionic fluorides are soluble in water, because hydration energies of ions of alkali metals (except Li^+) and the F^- ion exceed the crystal lattice energies of the corresponding fluorides.

Interhalogen compounds of fluorine with chlorine are used as rocket fuel oxidizers. Ionic fluorides are used for growing *monocrystals* of high-melting materials, in optics (CaF_2 and LiF are transparent to UV and visible radiation), as heat-transport media and working media in atomic-power engineering (fluorides of lithium, beryllium, thorium, uranium), in electrochemical production of aluminum and other active metals, as fluxes for welding *refractory materials*, as thermoluminescent and lyoluminescent ionizing radiation dosimeters (fluorides of calcium, magnesium, lithium).

FLUXOID

See *Flux quantum*.

FLUXON

See *Josephson vortex*.

FLUX QUANTIZATION

See *Quantization of flux*.

FLUX QUANTUM, fluxoid

The minimal value of magnetic flux Φ_0 that can be frozen in a superconducting ring carrying an electric current. The value of the flux quantum is $\Phi_0 = h/2e = 2.06785 \cdot 10^{-15}$ T·m^2. See *Quantization of flux*.

FOCUSON, focusing collision

One of the types of short-lived excited states of an atomic chain in a *crystal*. If an atom of the chain receives an impulse (e.g., from a foreign particle flying by) at a sufficiently low angle to the chain direction, then it transfers the momentum to a neighboring atom in the chain at a smaller angle, thereby returning after the collision to its original lattice site. A succession of such elastic collisions leads to a considerable amount of excitation transfer along the chain (of the focuson passage). In an undistorted crystal the focuson transit does not induce structural changes. If there is a *defect* along the focuson path (*impurity atom*, *dislocation*, *solid surface*, etc.), the spreading of the focuson may lead to some atomic reconstruction in the vicinity of a defect at a considerable distance from the place where the focuson originated. The condition for focuson formation in the standard model of a solid, where the atoms are represented by hard spheres of radius R, has the simple form $d < 4R$, where d is the interatomic distance. As R decreases with the growth of the energy of the atom, focusons are formed with very low amounts of energy transferred to the primary atom, of the order of hundreds of electronvolts. At the small angles and high energies of the primary collision, when the primary knocked-out atom replaces an adjacent one in the chain, the effect of collision focusing leads to the formation of a *dynamic crowdion* (see *Crowdion*), which proceeds along the chain, ending with the appearance of a static configuration of widely separated Frenkel pairs (see *Frenkel defect*). At sufficiently high angles of momentum transfer to the primary atom, a *defocusing of atomic collisions* takes place.

FOKKER–PLANCK EQUATION (A.D. Fokker, 1914; M. Planck, 1917)

Equation for the propagation density of probabilities $f(Q,t)$, where $Q=(q_1,\ldots,q_n)$ is a set of dynamic variables. A general representation of the Fokker–Planck equation has the form

$$\frac{\partial f}{\partial t}=-\sum_{i=1}^{n}\frac{\partial}{\partial q_i}\left[a_i(t,Q)f\right]+\frac{1}{2}\sum_{i,k=1}^{n}\frac{\partial^2}{\partial q_i\,\partial q_k}\left[b_{ik}(t,Q)f\right].$$

The coefficients $a_i(t,Q)$ are called *drift coefficients*, and $b_{ik}(t,Q)$ are *matrix diffusion ratios*.

Many equations of physical kinetics are reducible to the Fokker–Planck equation, such as

(a) *self-diffusion* and interdiffusion processes (see *Interdiffusion*) in systems with large numbers of particles (gases, liquids) are described by the *Fick equation*

$$\frac{\partial f}{\partial t}=D\sum_{i=1}^{3}\frac{\partial^2 f}{\partial x_i^2},$$

where x_1, x_2, x_3 are Cartesian coordinates of a designated particle, $f=f(x_1,x_2,x_3,t)$ are probability density distributions of the particle location within the space, and D is the coefficient of diffusion;

(b) the process of *Brownian motion* of a foreign particle in a liquid is described by the *Smoluchowski equation*

$$\frac{\partial f}{\partial t}=-\sum_{i=1}^{3}\left(\frac{F_i}{\gamma}\frac{\partial f}{\partial x_i}\right)+D\sum_{i=1}^{3}\frac{\partial^2 f}{\partial x_i^2},$$

where F_1 is the force of gravity which acts on the particle, γ is a coefficient of *friction*;

(c) a *relaxation* distribution process involving the momentum of a heavy gas impurity in an atmosphere of light gas molecules in *thermodynamic equilibrium* with each other, is described by the equation

$$\frac{\partial f}{\partial t}=-\gamma\sum_{i=1}^{3}\frac{\partial}{\partial P_i}(P_i;f)+\frac{\sigma}{2}\sum_{i=1}^{3}\frac{\partial^2 f}{\partial P_i^2},$$

where P_1, P_2, P_3 are the Cartesian components of the heavy-particle momentum vector, $f=f(P_1,P_2,P_3,t)$ are probability distribution densities for the heavy particles, and σ is the strength of the effect that the light gas molecules have on the heavy particle.

For the applicability of the Fokker–Planck equation to the description of the time variation of the distribution density $f(Q,t)$ it is necessary to fulfill the following conditions (see *Kinetic equation*):

(1) the variation of $f(Q,t)$ in a short time interval Δt is totally determined by the function $f(Q,t)$ itself, and by the forces applied at the moment t, i.e. there is no "memory" of the character of the system evolution in the past,

(2) the complete evolution process of the set of physical values $(q_1,\ldots,q_n)$ consists in a large number of individual elementary acts, which occur at random times, and have random durations. The process of variation of the distribution of values q_i is far slower than the elementary processes which generate it, i.e. the average duration of events comprising the microscopic mechanism of evolution is much less than the characteristic times of variation of the values q_i;

(3) the average variations of the values $q_1,\ldots,q_n$, on which $f(Q,t)$ depends in every elementary act, are low compared to their individual characteristic values.

For all of the above-mentioned physical phenomena these conditions are fulfilled because the described processes proceed much slower than the elementary acts (collisions of particles) generating them, and the average relative change of the dynamic variables (coordinates, momenta) at each collision is very small. It is necessary to take into consideration that the actual processes involved in the variation of the physical values have a statistical nature, hence conditions 1 and 2 are only approximately satisfied, and therefore describing the physical values with the help of the Fokker–Planck equation is only an approximation, whose accuracy depends on the accuracy of fulfilling these conditions. This is clearly demonstrated by the example of *diffusion* processes of atoms in solids, and especially by the passage of gases through pressed powders when a random arrangement of the medium reinforces the random variation of the particle motion. Since this state varies so slowly, there remains some memory of the past evolution of the particle, which is expressed in the statistical connection between the separate elementary acts associated with the particle motion, and this connection may become pronounced to such an extent that condition 1 would not be satisfied. In particular, there is a so-called *percolation threshold* (see *Percolation theory*) depending on the parameters characterizing the medium, above which diffusion ordinarily cannot occur, and hence the Fokker–Planck equation only applies in the above mentioned physical systems when the parameters of the medium are much less than their threshold values.

FORBIDDEN BAND

See *Band gap.*

FORBIDDEN ENERGY ZONE

See *Band gap.*

FORBIDDEN QUANTUM TRANSITIONS

Conventional term for transitions with an intensity much less than that of *allowed transitions.* They include the transitions with $\Delta M \neq \pm 1$, $\Delta n = 1$, and, in strong magnetic fields, those transitions with the simultaneous changing of the projections of two spins $\Delta M_1 = \Delta M_2 = \pm 1$, $\Delta n = 1$, or $\Delta M_1 = -\Delta M_2 = \pm 1$, $\Delta n = 1$, where Δn is the change in the number of photons (see Fig. in *Hyperfine structure*). Included also in the forbidden category are many-quantum transitions with $\Delta n = 2, 3, \ldots$. With intersecting energy levels forbidden quantum transitions become comparable to the allowed ones in their intensity. The position, width and intensity of lines of forbidden quantum transitions can provide additional information not available from allowed transitions, e.g., details about low-symmetry *spin Hamiltonian parameters.* Quantum transitions between some states, forbidden via one type of interaction, might be easily observable in other circumstances. For instance, electric dipole-forbidden transitions under *light absorption* can be allowed for *Raman scattering of light,* and vice versa (see *Mutual exclusion*).

FORBIDDEN TRANSITION, DIPOLE

See *Dipole-forbidden transition.*

FORCE CONSTANTS

Coefficients in the quadratic term for the series expansion of the crystal potential energy U for displacements of atoms from equilibrium positions $u_s(\boldsymbol{n})$ ($\boldsymbol{n}$ labels the unit cells in the lattice, s labels the atoms in the *unit cell*):

$$U = U_0 + \frac{1}{2}\sum \beta^{ik}_{ss'}(\boldsymbol{n},\boldsymbol{n}')u^i_s(\boldsymbol{n})u^k_{s'}(\boldsymbol{n}'),$$

where the summation is carried out over all indices (here i, k are the indices of Cartesian coordinates). The symmetrical and real force constants $\beta^{ik}_{ss}(\boldsymbol{n},\boldsymbol{n}')$ define the type of *crystal lattice vibrations* and the spectrum of their frequencies:

$$\beta^{ik}_{ss'}(\boldsymbol{n},\boldsymbol{n}') = \beta^{ki}_{s's}(\boldsymbol{n}',\boldsymbol{n}).$$

They obey the following relations:

$$\sum_{ns} \beta^{ik}_{ss'}(\boldsymbol{n},\boldsymbol{n}') = 0,$$

$$\sum_{n's'} \beta^{ik}_{ss'}(\boldsymbol{n},\boldsymbol{n}')R^l_{s'}(\boldsymbol{n}') = \sum_{n's'} \beta^{il}_{ss'}(\boldsymbol{n},\boldsymbol{n}')R^k_{s'}(\boldsymbol{n}').$$

Here $R^k_s(\boldsymbol{n})$ is the kth component of the coordinate of the sth atom in the $\boldsymbol{n}$th cell. The correlations between force constants may be obtained from the symmetry of each specific lattice (note that some terms may vanish). In an infinite (uniform) crystal we have

$$\beta^{ik}_{ss'}(\boldsymbol{n},\boldsymbol{n}') = \alpha^{ik}_{ss'}(\boldsymbol{n}-\boldsymbol{n}').$$

In the limiting case of long-wavelength *acoustic vibrations* the equations of crystal lattice vibrations are equivalent to the dynamic equations of motion of *elasticity theory,* where the *elastic moduli* are expressed with the help of the force constants. For instance, in a monatomic crystal

$$c_{iklm} = -\frac{1}{2V_0}\sum_n \alpha^{ik}(\boldsymbol{n})R^l(\boldsymbol{n})R^m(\boldsymbol{n}),$$

where V_0 is the unit cell volume.

FORCE, IMAGE

See *Image forces.*

FORCE, SURFACE

See *Surface force.*

FORERUNNER

See *Precursor.*

FORGEABILITY, malleability

The ability of *metals* (*alloys*) to change their shape or size via *forging* and volume pressing. When comparing the feasibility to process metals with pressing one should take into account both their *plasticity* (transverse contraction, ψ, and extension before *failure*, δ) and *strength* (σ_t). Forgeability is characterized by a *malleability index K* with its value given by the expressions $K_\psi = \psi/\sigma_t$ or $K_\delta = \delta/\sigma_t$. The value of K depends upon the metal identity, chemical composition and structure, as well as on temperature, *plastic deformation rate*, and the *state of stress*. To enhance its forgeability a metal should be cleansed of chemical elements which form low-melting compounds, as well as of oxygen, hydrogen and nitrogen; the processing by pressure should be carried out at elevated temperatures; and surface *defects* should be eliminated. To determine the material forgeability, one gauges the magnitudes of strength and plasticity at varying temperatures and rates of deformation.

FORGING

A technique of metal processing by *pressure* whereby a mechanical part is shaped into a required configuration and dimensions by means of repeated impacts of a hammering tool. The metal is heated at forging to enhance its *plasticity* and to reduce the pressures required. Forging is performed using hammers and presses. The most important smithy operations include: setting, broaching, punching, chopping, bending and twisting. The parts obtained with forging stand out by their enhanced *toughness* and *impact strength* compared to parts that are cast (see *Casting*). The forging regime is chosen so it not only ensures the material shaping, but also provides for a favorable grain structure and *dislocation substructure* in order to enhance the material strength and plasticity.

FORM FACTOR

In *X-radiography* and *neutron diffractometry*, and in atomic and nuclear physics the electromagnetic form factor is a function characterizing the space charge distribution (*electric form factor*) or *magnetic moment* distribution (*magnetic form factor*) within the atom, elementary particle or nucleus. An *X-ray form factor* corresponds to an electric form factor (it describes the electromagnetic wave interference within the region occupied by the overall electronic cloud of the atom), and a *neutron form factor* corresponds to a magnetic form factor (it describes the neutron wave interference within the limits of the atom's outer electronic shells with *spin density*). These form factors are measured with the help of the elastic scattering of X-rays, neutrons, electrons (or muons) in crystals or other materials.

In X-ray crystallography an atomic form factor is defined by the expression $f(\boldsymbol{K}) = -(1/e)\int e^{i\boldsymbol{K}\cdot\boldsymbol{r}}\rho(\boldsymbol{r})\,d^3\boldsymbol{r}$, where $\rho(\boldsymbol{r})$ is the charge distribution in the atom, and $\boldsymbol{K}$ is a reciprocal lattice vector. Form factors for strongly inelastic processes can be defined. The weak form factors which characterize weak interactions of elementary particles are introduced analogously to electromagnetic form factors, and it is not only the vector currents, but also the axial currents that take part in this kind of interaction. Weak interaction processes are associated with vector and axial form factors. The spatial variation of the matrix elements of an atomic or nuclear interaction can be described by a form factor.

FORM FACTOR, ATOMIC

See *Atomic form factor.*

FOURIER SPECTROSCOPY of solids

Method for recording absorption, reflection and emission spectra in the long-wavelength, medium and short-wavelength IR spectral regions. Optical–infrared Fourier spectroscopy has progressed very rapidly since *Fourier spectrometers* have demonstrated their advantages over classical (frequency scanning) ones; the method features high intensity, monochromatic sources, precise

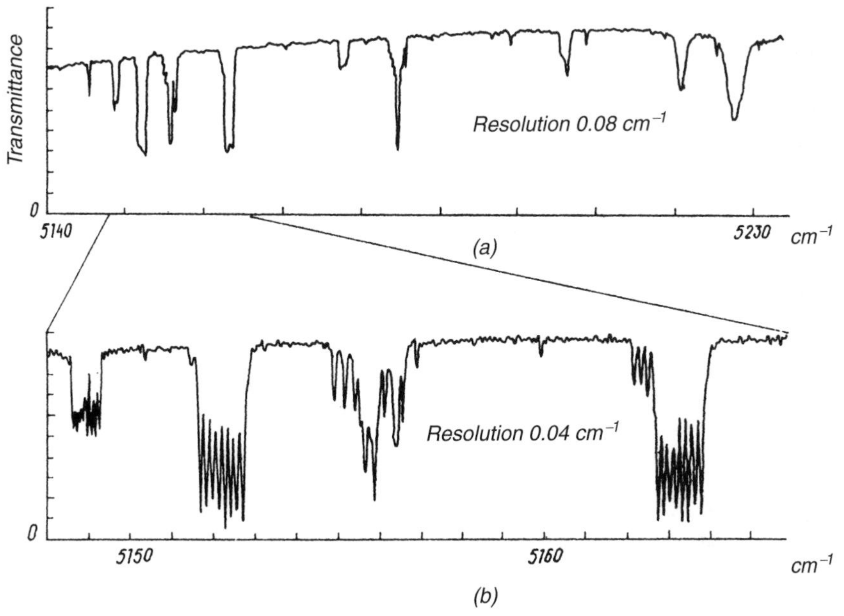

Infrared spectrum of Ho^{3+} doped $LiYF_4$ (a), and enlargement of the low-frequency region (b).

wavelength determination (Fourier spectrometers incorporate a laser interferometer), use of computers for data acquisition and processing.

The required spectrum is obtained by calculating the Fourier transform of the measured interferogram (autocorrelation function) at the output of a double beam *interferometer* (often Michelson interferometer). The resolvable spectral interval is inversely proportional to the maximum range in the optical path achievable by the interferometer, and the overall spectral range is inversely proportional to the smallest resolvable interval of the interferogram sampling. Fourier spectrometers exhibit resolving powers up to 10^{-3} cm^{-1}, and the observable spectral region extends from 5 to $5 \cdot 10^4$ cm^{-1}, i.e. from submillimeter wavelengths to the UV. Of prime importance for the phonon spectroscopy of crystals is the broad interval of observable wavelengths, and the high photometric accuracy with which reflection spectra are obtained. The optical constants of materials may be obtained from these spectra through the use of the *Kramers–Kronig relations*. The investigation of the temperature dependence of the IR reflectivity permits one to evaluate the *energy gap* of *high-temperature superconductors*. With the use of Fourier spectrometers, total internal reflection spectroscopy has become a leading method for investigating surface molecular adsorbate structure; this method provides information on the composition, type of *chemical bond*, structure of layers and their rearrangement at *phase transitions*. This spectroscopy is nondestructive in nature, involves no surface reactions (catalysis, corrosion), and does not disrupt the structure of adsorbates and their two-dimensional patterns. The method of probing the surface by electromagnetic waves is two orders of magnitude more sensitive for monomolecular layers. This technique is carried out using a Fourier spectrometer over a broad range of wavelengths. The implementation of the method of photothermal ionization of fine *donors* in semiconductors (*Lifshitz method*) using Fourier spectrometers increases the sensitivity for the determination of impurity content to the level of 10^6 atoms per cm^3, which is still unattainable by atomic emission spectroscopy. The measurements of optical absorption in the long-wavelength IR spectral region provide important information on the nature of two-dimensional layers of space charge in metal–oxide–semiconductor structures: *effective mass*, characteristics of the scattering processes, band structure of *surface states*, and elementary excitations. A Fourier thickness gauge is used for exercizing control over the thickness of epitaxial layers. High-resolution Fourier spectroscopy (0.001 cm^{-1}) may be successfully used to study *hyperfine interactions* of atomic impurity centers in crystals, since nonuniform widths of bands are 0.03–0.06 cm^{-1}, whereas hyperfine splitting reaches 2 cm^{-1}, and the overall Stark multiplet spans 100–150 cm^{-1} (see *Stark effect*). A portion of an absorption spectrum of the crystal $LiYF_4$–Ho^{3+} at 6 K in the region of the transition ${}^5I_8$–${}^5I_7$ of the impurity ion Ho^{3+} is given in Fig.: (a) represents a Stark multiplet, and the enlargement of (b) shows part of this multiplet (resolution 0.04 cm^{-4}); the eight components of the hyperfine splitting ($I = 7/2$) are distinctly visible in all the spectral bands.

FRACTAL

Term (combined from words: fraction and fracture) coined by Benoit B. Mandelbrot (1977) to describe objects that have a non-integer or fractional dimension; in addition to this property they also exhibit self-similarity. In terms of mathematics, for a one-dimensional Euclidean space ($D_E = 1$) fractals are sets of points along a line with a fractional dimensionality D_H in the range

$$D_T < D_H < D_E, \tag{1}$$

where $D_T = 0$ is the intuitive topological dimensionality of a point. The quantity D_H is called the *Hausdorff–Bezikovich dimensionality* (F. Hausdorff, 1918; A. Bezikovich) or the *fractal dimensionality*. For a two-dimensional Eucklidean space $D_E = 2$ fractals are sets of lines ($D_T = 1$) with a nonintegral dimension D_H in the range $1 < D_H < 2$, and so forth for fractals in higher-order Euclidean spaces. It is characteristic of a fractal that the dimension D_H is not related to topology, but rather to the way in which the set under consideration is constructed.

To obtain a formula for dimensionality consider a line segment of length a_0 divided into N line segments each of length a_0/N, a square a_0 on the side divided into N squares each with a side

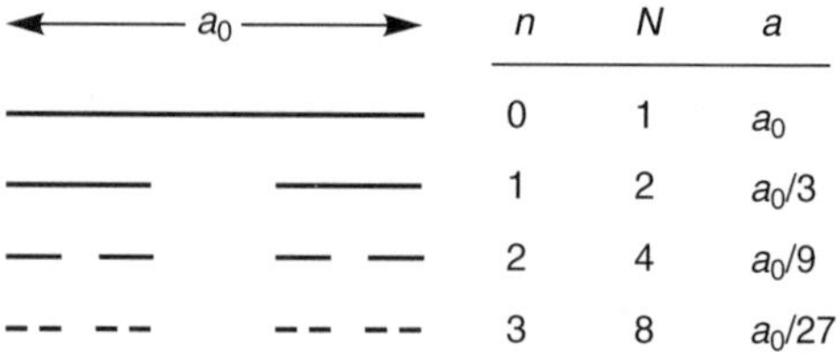

Cantor set of line segment fractals.

of length $a \ll a_0$, or a cube of side a_0 divided up into N small cubes each with dimension $a \ll a_0$. In each case the number of objects N is given by the equation

$$N(a) = \left(\frac{a_0}{a}\right)^D, \tag{2}$$

where the dimension $D = 1, 2, 3$ for the line segment, the square and the cube, respectively. Taking the logarithm of both sides of this equation gives for the dimension

$$D = D_H = \frac{\log N(a)}{\log(a_0/a)}. \tag{3}$$

This expression is not only valid for the example, but it is also the Hausdorff or fractal definition of dimensionality D_H. As an example of its significance, consider the Cantor set of line segments depicted in Fig. An initial line segment of length a_0 in one-dimensional Euclidean space ($D_E = 1$) is divided into two segments of length $a = a_0/3$ with a space of length $a_0/3$ between them. Fig. shows one further subdivision of this type. If this subdivision process is repeated n times we obtain $N(a) = 2^n$ micro line segments or points, each of length $a = 3^{-n}a_0$. The points have the topological dimension $D_T = 0$. Inserting these expressions into Eq. (3) gives $D_H = \log 2/\log 3 = 0.6309$, a non-integer value which shows that the Cantor set is a fractal object. It is clear that this Hausdorff dimensionality obeys inequality (1). The Cantor set also exhibits self-similarity since, if a portion of the set after many subdivisions is magnified, it will resemble the set after only a few subdivisions.

Fractals occur in various physical objects such as the coastline of a beach, geothermal rocks, cloud formations, woody plants, and speech forms. Fractals are often associated with the presence of random processes or chaos, such as the kinetics of the generation of *clusters*, the growth of *dendrites*, turbulent flow, problems of percolation through disordered lattices, etc. See *Percolation theory*.

FRACTAL DIMENSIONALITY

See *Fracton dimensionality*.

FRACTOGRAPHY

Physical method of investigating the structure of *fractures* for the purpose of determining the causes and nature of the *failure* in connection with the structure of the material under investigation. Fractography permits the qualitative and quantitative determination of the origin of failure, i.e. the location and dimensions of the defect, the direction of *crack* propagation, and an evaluation of the energy consumption during the process. Fractography provides the basis for classifying failure mechanisms according to the criteria of *fracture toughness* and fracture structure (*brittle fracture*, quasi-brittle fracture, *tough fracture*, etc.). The main instrument used in these investigations is the scanning electron microscope (see *Scanning electron microscopy*), which provides images of fractures over a broad range of magnification from 20 to 2000. Fractography methods are used to inspect the quality of materials with respect to *strength* and *reliability of materials*, and to detect inhomogeneities in the structure of *alloys*.

FRACTON DIMENSIONALITY, spectral dimensionality

A characteristic D_s of a *fractal system*, which is determined by the *fractal dimensionality* D and *anomalous diffusion* index Θ of the system: $D_s = 2D/(2+\Theta)$. The name of this quantity arose from the behavior of the *density of states* of a fractal system (see *Fractal*). For example, the density of vibrational states in a system with isotropic elastic forces (see *Isotropy of elasticity*) is given by the equation $N(\omega) \propto \omega_s^{D_s-1}$ (the density of vibrational states for ordinary lattices under the same conditions is given by $N(\omega) \propto \omega^{d-1}$, where d is the space dimension). The corresponding vibrational excitations are often localized and are called *fractons*. Systems which appear homogeneous at large scales exhibit a transition from fracton behavior to ordinary phonon behavior (see *Phonons*) at sufficiently low frequencies. This behavior of the density of states accounts for certain features of the *specific heat* of polymeric resins. The same quantity determines the spectrum of electronic states of the fractal system within the tight binding approximation (see *Tight binding method*): $N(\varepsilon) \propto \varepsilon^{(D_s/2)-1}$ for $\varepsilon \to 0$.

The value of D_s is responsible for certain features of *diffusion* and *random walk* in fractal systems. The random walks are unrestricted (non self-avoiding) for $D_s \leqslant 2$, and are self-avoiding for $D_s > 2$ (see *Migration*). The value of D_s also determines the kinetics of diffusion-controlled reactions in fractal systems (chemical reactions in a porous matrix, *luminescence* of mixed *molecular*

Table 1. Asymptotic behavior of certain types of diffusion-controlled reactions in fractal systems

Type of reaction	Asymptotic limit	Comments
$A + B \to B$ Extinguishing the agitation in flux saturation and mobility	$n_A \sim \exp(-a n_B t)$, $D_s > 2$, $n_A \sim \exp(-a n_B t^{D_s/2})$, $D_s < 2$	n_B is the concentration flux, a is a numerical coefficient. Intermediate asymptote
The same	$n_A \sim \exp(-a n_B^{2/(2+D_s)} t^{D_s/(D_s+2)})$,	Actual asymptote after long time
$A + A \to 0$ Annihilation of triplet excitons	$n_A \sim t^{-1}$, $D_s > 2$, $n_A \sim t^{-D_s/2}$, $D_s < 2$	–
$A + B \to 0$, $n_A(0) = n_B(0)$ Electron–hole recombination, annihilation of radiation defects	$n_A = n_B \sim t^{-1}$, $D_s > 4$, $n_A = n_B \sim t^{-D_s/4}$, $D_s < 4$	Actual asymptote after long time

crystals, with the concentration of one component close to the percolation threshold, etc.). The time dependence of concentrations (n_A, n_B) of excitations and reagents for different types of reactions are given in Table 1.

In the case of an infinite percolation cluster the *Alexander–Orbach relation*, $D_s \approx 4/3$, is satisfied for a space of any dimensionality (S. Alexander, R. Orbach, 1982). See also *Percolation theory*.

FRACTURE

The surface damage resulting from the *failure* of a mechanical part, a semifinished item, or a specimen. The study of fracture is one of the ancient methods of metal quality control performed either visually or with the help of an optical microscope; at the present time one makes use of a transmission electron microscope (*replica* method) or a scanning microscope. The type of the fracture depends on the variety of crystal lattice, the metal structure, the testing (or operating) temperature, the loading rate, and the strain state. Using structural indicators, a distinction can be made between *transcrystalline failure* and *intercrystalline failure*; and using a *fracture toughness* criterion one distinguishes between *brittle failure*, *quasi-brittle failure* and *tough fractures*. Some fractures may exhibit these signs in various combinations. In visual observation, the brittle fracture surface obtained by *spalling* is shiny and bright; that of a ductile fracture is dull gray.

FRACTURE BRANCHING

See *Branching of fractures*.

FRACTURE, BRITTLE

See *Brittle fracture*.

FRACTURE TOUGHNESS, fracture ductility, crack resistance

The ability of a material to resist the spreading of large macroscopic *cracks* within itself. Fracture toughness depends on the level of intermolecular bonding forces, on *plastic deformation* in the crack, and on barriers in the way of its spreading. The mechanism of plastic flow in front of the crack vertex involves dislocations at low levels, and rotations at high levels of plastic deformation (see *Plastic twisting deformation*, *Dislocation model of plastic deformation*). Macroscopic obstacles (boundaries of composite layers, pliable insertions, welded joints, riveted stringers, slots, local thermal and elastic fields) as well as microscopic ones (interblock boundaries and *grain boundaries*, triple grain junctions, *slip bands* and *faults*, second *phase* inclusions, etc.) may serve as barriers. At the beginning of the spreading of a crack the external stress should break the bonds at its opening, with accounts for the work expended in plastic deformation and overcoming barriers. As a quantitative measure of fracture toughness three main parameters, fixed at the moment of the crack movement under the load, are commonly used: (1) G_{1c}, the *critical velocity of energy release*, J/m^2, (2) K_{1c}, critical *stress intensity factor*, MPa·m$^{1/2}$, (3) δ_c, *critical crack opening*, mm. For quasi-elastic (low plasticity) materials (see *Quasi-brittle failure*), the values G_{1c}, K_{1c} and δ_c are interrelated as

$$G_{1c} = \sigma_s \delta_c = \frac{K_{1c}^2(1-\nu^2)}{E},$$

where σ_s is the *yield limit* of the material, ν is the *Poisson ratio*, and E is *Young's modulus*. The parameter K_{1c} is determined experimentally by tensile tests of special samples with an applied fatigue crack (see *Fatigue*) and a fixed minimum load, whereby the crack increases via flat deformation, i.e. by flat break-off under the effect of normal stresses. The factor K_{1c} is almost independent of the sample dimensions (for sufficiently large dimensions) and at a given temperature and deformation speed it is a constant determined by the structure of the material. For construction steels of average strength ($\sigma_b = 400$ to 600 MPa) typical values are $K_1 \approx 25$ to 30 MPa·m$^{1/2}$ (~80 to 100 kg·mm$^{-3/2}$). Fracture toughness is not applicable for describing stress resistance involving small cracks and defects of submillimeter dimensions (see *Defect resistance*).

FRAGMENTATION

Process of formation of a fragmented structure in crystalline materials, i.e. the generation of small structural units which are called *fragments*. There is no unambiguous definition of the concept "fragment". Originally the term referred to a structural unit immediately below *polycrystal* grain in the size scale. In this context, fragments of polycrystal grains and *monocrystals* result from various processes: *crystallization*, *recrystallization*, *phase transition* in the solid state, *plastic deformation* under conditions of both *creep* and resistive load, including pressure treatment of metals. In all cases, except for resistive plastic deformation, the disorientation angles between the fragments are $\theta \approx 0.2$–$2°$, and a *block structure* is often observed inside the fragments. The sizes of fragments range from decimal fractions of a micrometer to several millimeters (in monocrystals).

Substructures with dislocation boundaries, which arise under active plastic deformation, can be called *cellular structures.* Slightly ($\theta \leqslant 2$–$4°$) and highly (larger θ, up to tens of degrees) disoriented cellular structures are distinguished. The dimension of cells ranges from 0.2 μm to several micrometers. Sometimes the cells (particularly highly disoriented ones) are assumed to be types of fragments. Two size ranges of fragments are distinguished: (1) dimensions ranging from 0.5 to 1 μm, $\theta = 1$–$15°$; (2) dimensions ranging from several micrometers to several tens of micrometers, with disorientation angles of tens of degrees, and narrow "knife-edge" boundaries. In certain cases (particularly at moderate temperatures) the fragmented structure (of the second size range) has a streaky appearance. Fragments of this size often serve as boundaries for the texture component region of a deformation *texture.* In the context of this approach, it is appropriate to use the term "cell" when they are slightly disoriented with wide dislocation boundaries. These various structural scale levels may occur both simultaneously and separately. The tendency toward fragmentation increases with the elevation of the *strain* temperature, increase of the number of *slip systems*, increase of the *stacking fault* energy. Fragmentation exerts a profound influence on mechanical properties, and in a number of cases it results in an increase of the *strength* of a material with a concurrent decrease of the temperature of the brittle–ductile transition.

FRANCIUM, Fr

Radioactive element of the Group I of the periodic table with atomic number 87 and atomic weight 223.0197; belongs to *alkali metals.* Francium is the least stable among all the radioactive elements occurring in nature. The only natural radionuclide ^{223}Fr (produced at one of the stages of ^{227}Ac decay) has a half-life of approximately 21.8 min. The surface layer of the Earth's crust (1.6 km thick) contains only 24.5 g of Fr. The electronic configuration of the outer shell is $7s^1$. Atomic radius 0.280 nm; radius of the ion Fr^+ 0.180 nm. The work function is 1.8 eV, and the electronegativity is 0.7.

The following characteristics of francium are deduced through extrapolation calculations. It is expected that francium exhibits a body-centered cubic lattice; $T_{melting} \approx 293$ K, $T_{boiling} \approx 913$ K, heat of melting 2.1 kJ/mole, heat of sublimation 69 kJ/mole; specific heat 32 J·mole^{-1}·K^{-1}; resistivity 0.45 μΩ·m (at 290 K). ^{223}Fr is used in radiochemical determinations of the radioactive element *actinium.*

FRANCK–CONDON PRINCIPLE (J. Franck, E. Condon, 1925)

A principle which governs the distribution of intensity in the optical transition band of an electronic-vibrational system such as a molecule or impurity center of a crystal. The Franck–Condon principle is an analogue of *selection rules* for coupled electronic-vibrational (i.e. vibronic) transitions. However, this principle does not impose strict exclusions on the possibility of particular transitions, like selection rules do; rather it indicates which transitions are more likely. The classical Franck–Condon principle states that during the period of an electronic transition in a molecule (or a crystal) the atomic nuclei do not appreciably change their coordinates or momenta. It is also assumed that the transition occurs at the turning points (points of slowest speed) of the classical motion of the nuclei. The semiclassical Franck–Condon principle, though accepting the claim that the coordinates and momenta of nuclei remain fixed during the course of the vibronic transition (which is inconsistent with the uncertainty relation), allows for transitions at all values of the vibrational coordinates R of the atomic nuclei, with probabilities that are determined by the quantum-mechanical distribution of R. A rigorous calculation of the probability of a vibronic transition (quantum-mechanical Franck–Condon principle) involves the assumption that the matrix element of the electronic operator of the *dipole moment* is only slightly dependent on the coordinates of the nuclei; thereby justifying the expansion of this operator in a power series in terms of R in the neighborhood of the equilibrium positions of the nuclei. If we restrict our consideration to the zeroth-order term of this series (*Condon approximation*), then the probability of the electronic-vibrational transition is determined by the overlap integral of the vibrational wave functions of the initial and final states of the system under consideration. In particular, if the adiabatic potentials of both states are identical, then in the context of this approximation, those transitions which involve a change of the vibrational state of the lattice are forbidden, which follows from the orthonormality of the vibrational wave functions.

FRANK DISLOCATION

A pinned dislocation which generates additional dislocations when subjected to an applied stress. The site of a Frank dislocation is called a *Frank–Read source.*

FRANK NETWORKS

Networks generated by dislocation lines, which belong to three (or more) different *slip systems* in the case when the *dislocations* form stable dislocation sites (see *Dislocation grid*), at which their *Burgers vectors* sum to zero. If the dislocation densities in these systems are equal, then there is a possibility of generating stable Frank networks with nearly equal cell sizes. These Frank networks, which exhibit cells several micrometers in size, are observed in polygonized *alkali-halide crystals.* Since individual units of Frank networks belong to different slip systems, the network sites may turn out to be fixed points of dislocations; under the action of applied external stresses the segments of these networks act as *Frank–Read sources.*

FRANK–READ SOURCE (F.C. Frank, W.T. Read, 1950)

A source of a *Frank dislocation*, which is a pinned dislocation that generates additional dislocations when subjected to an applied stress. It is a source of *multiplication of dislocations* at the *plastic deformation* of crystalline materials. There exist two types of Frank–Read sources: two-pole sources and one-pole sources.

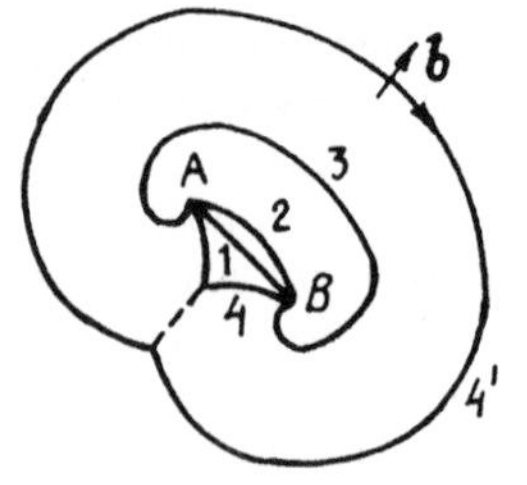

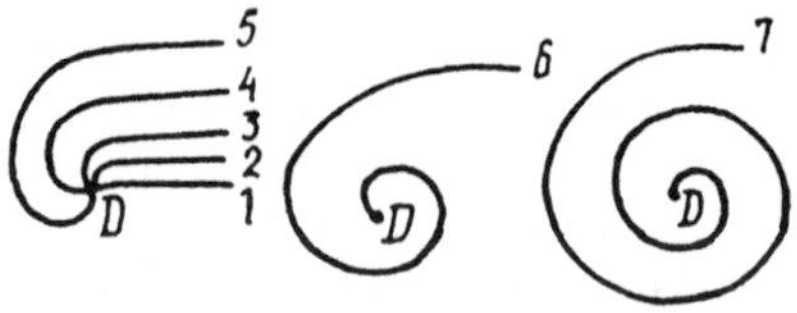

Frank–Read source.

A *two-pole Frank–Read source* is the dislocation segment AB (see Fig. (a), where $\boldsymbol{b}$ is *Burgers vector*, and the arrow indicates the direction of the dislocation line), which has ends that are fixed by the nodes of the *dislocation grid*, by impurities, or by segments of the dislocation line which leave the *slip plane.* This segment buckles at slipping under the action of an applied *shear* stress. If the stress is so strong that it cannot be compensated by the stretching force of the dislocation line, then the dislocation segment goes through the successive stages 1–2–3–4 depicted in Fig. (b). Dislocation line segments of opposite signs meet each other and annihilate (annihilation proceeds at the dashed line in Fig. (a)) at the fourth stage (see *Annihilation of dislocations*). This results in the formation of the *dislocation loop* ($4'$), and the recovery of the initial segment, which is capable of producing new dislocation loops.

In the process of development of a *single-pole Frank–Read source* the original line *dislocation* turns into a spiral, which increases the length of the dislocation line in the bulk of the crystal, i.e. leads to the growth of the dislocation density. Seven arbitrarily chosen successive stages of the spiral development are shown in Fig. (b), where

D is the point of the dislocation pinning. Frank–Read sources have been repeatedly observed in crystals.

FRANZ–KELDYSH EFFECT (W. Franz, L.V. Keldysh, 1958)

The shift of the *intrinsic light absorption edge* of a *semiconductor* toward lower frequencies, which takes place in the presence of an external applied electrical field. This effect was experimentally discovered by V.S. Vavilov and I.K. Britsyn in *silicon*. In the absence of an electric field, the frequency corresponding to the absorption edge is given by the formula $\omega = E_g/\hbar$, where E_g is the width of the *band gap*. In an electric field E the absorption edge becomes blurred because of electron tunneling (see *Tunnel effect*) from the valence band to the conduction band, and the absorption of light of frequencies $\omega < E_g/\hbar$ becomes possible. The effective shift of the band edge is given by the formula $\Delta E \sim (e^2 E^2 \hbar^2/m^*)^{1/2}$, where m^* is the *effective mass* of the electron. Accompanying a change of the absorption coefficient edge there is a change of the refractive index. The Franz–Keldysh effect is used for the modulation of optical radiation (see *Modulation spectroscopy*).

FREDERIKS TRANSITION (V.K. Frederiks, 1927)

The change of orientation of the *director* ***n*** in *liquid crystals*, as a result of the action of magnetic or electric fields. The physical cause of the Frederiks transition is that the liquid crystals tend to reach the field configuration ***n*** of the lowest free energy. In the case of a *nematic liquid crystal* with finite and infinite values of energies of adhesion of molecules to the supporting surfaces the Frederiks transition is a threshold process. The physical cause of the existence of a threshold lies in the fact that the change of the director orientation (the director orients along the external field direction) is hindered by elastic deformations in the neighborhood of the supporting surfaces. As the field increases, the undistorted texture ***n*** turns into a distorted one through a *second-order phase transition*. The critical field values for the Frederiks transition for magnetic and electric fields are, respectively:

$$B_c = \frac{\pi}{d}\left(\frac{K_{ii}}{\chi_a}\right)^{1/2},$$

$$E_c = \frac{\pi}{d}\left(\frac{4\pi K_{ii}}{\varepsilon_a}\right)^{1/2},$$

where d is the sample thickness, K_{ii} are elastic constants which assume different values depending on the geometry of the experiment, χ_a and ε_a are the respective values of the diamagnetic and dielectric anisotropy of the liquid crystals (see *Anisotropy of liquid crystals*).

Typical values are $B_c \sim 0.1$–0.5 T, $E_c d = V_c \sim 1$–5 V. In the case of an inclined orientation of the director with respect to the field direction the Frederiks transition is a non-threshold type. An analogous effect may be observed in *smectic liquid crystal* C. In the case of *cholesteric liquid crystals*, in which $\chi_a > 0$ and $\varepsilon_a > 0$ and the direction of the field is perpendicular to the spiral axis, the Frederiks transition involves uncoiling the spiral structure and a cholesteric–nematic transition.

FREE ENERGY

The quantity of energy available to perform work, or the capability of a system to do useful work. The *Helmholtz free energy* F for constant volume processes, $F = U - TS$, and the *Gibbs free energy* G for constant pressure processes, $G = H - TS$, are defined thermodynamically in terms of the energy U, the enthalpy H, the temperature T, and the entropy S. See *Thermodynamic potentials, Gibbs free energy, Helmholtz free energy*.

FREE RADICALS

Paramagnetic compounds with unpaired electrons at outer atomic or molecular *orbitals*, and which usually display high reactivity. Some free radicals are stable (have long lifetimes), but most are unstable. Stable free radicals may be isolated as chemically pure substances, and unstable free radicals can be obtained and studied in solid matrices or in solutions. One of the principal methods used to detect and study free radicals is *electron paramagnetic resonance* (EPR). Free radicals having two or more uninteracting unpaired electrons are called *biradicals* or *polyradicals*, respectively. If they are charged they are called *radical ions*. Free radicals form from corresponding molecules under the effect of electromagnetic radiation, high-energy particle fluxes, at catalyst surfaces, during heating, etc. Free radicals are widely found in nature: at the Sun, in stars and comets, in interstellar space, etc. Respiration and photosynthesis as well as many individual processes, such as combustion, polymerization, etc., involve free radicals.

FREE SPIN COMPLEXES

The same as *High-spin complexes*.

FREE VOLUME MODEL

A theoretical model used to describe the physical properties of liquids and solids in the *amorphous state*. The model is based on the assumption that the amorphous state is completely determined by the distribution of volume in it. The *free volume* v_f per given atom is the difference between the volume of the cell v_n occupied by the given atom and the volume of that atom itself, $v_a \propto a^3$ (here a is the atomic diameter). The basic assumptions underlying the model are as follows:

(a) free volume may redistribute itself between cells without any changes in either the total volume or the total free energy of the solid;
(b) *diffusion* of atoms takes place along fluctuationally developing cavities, their volume exceeding a certain critical value v^* that is close to v_a (these form during the course of free volume redistribution);
(c) the total free volume changes with changes in temperature, and under the effect of tensions that produce *plastic deformations*.

For a liquid in thermodynamic equilibrium the distribution function $p(v_f)$ of free volume was found to have the form:

$$p(v_f) = \left(\frac{\gamma}{\bar{v}_f}\right)\exp\left(-\frac{v_f}{\bar{v}_f}\right),$$

where γ is a geometric factor of the order of unity, the average free volume $\bar{v}_f$ may be represented in the form of a polynomial expansion over temperature T as $\bar{v}_f = \xi(T - T_0)$; here ξ is the *thermal expansion* coefficient, T_0 is a phenomenological parameter. An expansion of the diffusion coefficient D follows from the formula for $p(v_f)$ and assumption (b) above:

$$D \approx \frac{1}{6} au \exp\left(-\frac{\gamma v^*}{\bar{v}_f}\right),$$

where u is the thermal motion velocity of atoms. From the above formula, there follows an expression for the *viscosity* coefficient, $\eta = Ak_BT/(aD)$, where A is a constant of the order of unity. In this form the above formula coincides with the empirical expression for the viscosity coefficient known as the *Vogel–Fulcher–Tamman formula*: $\eta = B\exp[C/(T - T_0)]$, where B and C are parameters. Within the model of free volume, this formula is satisfied by *amorphous metals and metallic alloys* around the *vitrification* temperature, so that the model may be applied to *metallic glasses*.

Using the model of free volume, a phenomenological description has been suggested for plastic deformations, diffusion, and glass–liquid transitions in metallic glasses. The applicability of the model to such glasses, however, has not yet been proven.

FRENKEL DEFECT

Crystal structure *defect* which arises when an atom migrates from its regular lattice site to an interstitial one. Thus, Frenkel defects consist of two components: the *vacancy* and the *interstitial atom*, therefore this defect is also called a *Frenkel pair* (*interstice–vacancy pair*). In nonmetallic crystals the components of a Frenkel defect may exist in various charge states, e.g., in *alkali-halide crystals* the neutral Frenkel pair consists of *color centers* of type F (halogen vacancy with an electron) and H (interstitial halogen atom); a charged Frenkel pair consists of color center α (halogen vacancy without an electron) and I (negatively charged halogen ion at interstice). In crystals of complex composition there can exist several types of Frenkel defects which are associated with different sublattices; e.g., the binary crystal MX contains pairs V_M–I_M and V_X–I_X (V are vacancies in sublattices M and X, I are interstitial atoms of M and X). Frenkel defects can be generated at sufficiently high temperatures as a result of thermal *fluctuations*. The probability that an atom escapes the potential barrier of height U, which separates the site and the interstice, is of the order of $\nu\exp[-U/(k_BT)]$, where ν is the oscillation frequency of the atom at the lattice site. Frenkel defects are also produced at elastic collisions between crystal atoms and high-energy particles at exposure of the crystal (e.g., alkali-halide) to *nuclear radiations*, and they can arise from excitation and subsequent annihilation of *electron–hole pairs* or *excitons*. The energy U for the thermal generation of a Frenkel pair (of the order of eV) is often several times less than the energy released at the impact generation of defects. This difference is due to different patterns of *strain* of the atomic environment associated with the slow thermal and rapid impact mechanisms of defect generation.

Separated and bound Frenkel defects are distinguished. A *separated Frenkel pair* consists of isolated vacancies and interstitial atoms, which do not interact because they are too far removed from each other. The evolution of the *bound Frenkel pair*, the members of which are close to each other, may lead either to its annihilation or to its dissociation into isolated defects. The character of this process has a pronounced effect on the efficiency of the exposure of the crystal to irradiation with various energies of defect-producing particles. Frenkel defects and more complicated objects, which include Frenkel defects and other types of crystal defects, may have a profound effect on the crystal properties. See also *Unstable Frenkel pair*.

FRENKEL EFFECT

Generation of *pores* (cavities) in the *diffusion zone* during the course of diffusion homogenization; the pores are generated because of the coagulation of surplus *vacancies* in systems where *diffusion* occurs by a vacancy mechanism. The presence and oversaturation of surplus vacancies are maintained by the inequality of counter diffusion flows of the system components. In the simplest case of diffusion homogenization, the Frenkel effect may be observed also in a single-component system, in which the directional flux of vacancies is maintained by the structural nonuniformity of the sample. The strength or effectiveness of the source of surplus vacancies, which are responsible for the generation of pores, is given by the equation $g_v = \operatorname{div} I_v$, where I_v is the flux of vacancies, which is equal to the difference between the fluxes of the system components. The role of centers of pore nucleation is played by intrinsic *defects* of the structure (boundaries, *cracks*) and the defects, which arise at relaxation of the diffusion-induced stresses (see *Stress relaxation*). The Frenkel effect in multicomponent systems is accompanied by swelling of the diffusion sample by an amount which equals the volume of the pores produced. The swelling manifests itself clearly at sintering of the mixture of powdered mutually insoluble substances. This effect may be prevented by the application of low ($\sim 10^2$ atm) *uniform compression* (see *Barodiffusional effect*). In alkali-halide crystals in an external electric field, the Frenkel effect shows itself in the form of *vacancion breakdown*.

FRENKEL EXCITON (Ya.I. Frenkel, 1931)

An *exciton* consists of a single excited atom with a finite lifetime in a lattice (or a single excited molecule in a *molecular crystal*). It was introduced to explain photoelectrically inactive *light absorption*. By virtue of its interaction with the environment, the excitation may be transferred (without energy loss) to neighboring (identical) atoms (or molecules). If the excitation is not accompanied by a lattice deformation, then it can travel throughout the crystal, being characterized by a certain wave vector $\boldsymbol{k}$ (or by momentum $\boldsymbol{p} = \hbar\boldsymbol{k}$) and by the *dispersion law* $\varepsilon(\boldsymbol{k}) = p^2/2m^*$ with the *effective mass* $m^* = \hbar^2(\partial^2\varepsilon/\partial k^2)^{-1}$. In this case a Frenkel exciton is a free or *traveling exciton*. When there is an interaction with the environment then a deformation of a certain region of the lattice (which provides obstacles for migration of the excitation) may take place; in this case a *self-localized exciton* is formed, which may travel throughout the crystal carrying along this region of deformation. Therefore, a self-localized exciton is characterized by a substantially larger effective mass than the free exciton. There exists a possibility of deexcitation of the crystal, which involves the transfer of the Frenkel exciton energy to *photons* (luminescence) and *phonons*.

FRENKEL–KONTOROVA MODEL (Ya.I. Frenkel, T.A. Kontorova, 1937)

A one-dimensional microscopic model of a *crowdion* or *dislocation* in a crystal. An infinite chain of atoms linked together by elastic (spring) forces is considered to be located within the range of an externally applied periodic potential. In the context of a linear (with respect to relative shifts) approximation the equation of atomic motion takes the form

$$\frac{d^2u_n}{dt^2} = \alpha(u_{n+1} + u_{n-1} - 2u_n) - a\omega_0^2 \sin\frac{2\pi u_n}{a},$$

where u_n is the shift in position of nth atom, a is the equilibrium interatomic distance, ω_0 is the frequency of uniform small-amplitude vibrations of the chain, and α is the elastic interatomic interaction constant. The dislocation satisfies the following boundary conditions:

$$u_n = a, \quad n \to -\infty,$$

$$u_n = 0, \quad n \to +\infty.$$

Assuming that the length of the region of localization of the inhomogeneous *strain* is much greater than a, we take the limit to the equation for the continuous function $u(x)$:

$$\frac{\partial^2 u}{\partial t^2} = c^2\frac{\partial^2 u}{\partial x^2} - a\omega_0^2 \sin\frac{2\pi u(x)}{a}.$$

This *sine-Gordon equation* is widely used in one-dimensional nonlinear dynamics. The solution of this equation with prescribed boundary conditions corresponds to the solitary wave of the nonlinear deformation field $\partial u/\partial x$, and it is the simplest type of *soliton*. The dislocation described in this continuum approximation may be displaced by an infinitesimal force. There is an initial or threshold stress which must be exceeded to start the dislocation motion (*Peierls stress* for dislocation in a one dimensional crystal; see *Peierls–Nabarro model*).

FRENKEL PAIR

See *Unstable Frenkel pair.*

FREQUENCY MIXING in a solid

A nonlinear process during which a pump wave (electromagnetic) field containing two or more frequency components (e.g., ω_1 and ω_2) is transformed in a *nonlinear medium* into waves with frequencies that are linear combinations of the pump wave frequencies (e.g., $\omega_3 = \omega_1 \pm \omega_2$, $\omega_4 = 2\omega_1 \pm \omega_2, \ldots$). The wave transformation takes place with conservation of the total electromagnetic field energy, so the nonlinear medium does not absorb or radiate energy. From a quantum-mechanical point of view the frequency mixing is a *multiphoton process* in the solid state, with the same starting and ending energy conditions. When looked upon as a macroscopic phenomenon, the mixing of frequencies is characterized by a nonlinear susceptibility of the second, third, and higher order (see *Nonlinear optics of solids*).

Because of the frequency dispersion of the *refractive index*, the spatial distribution of the initial nonlinear wave polarization in the medium does not coincide with that of the electromagnetic wave excited at the new "mixed" frequency. Therefore, phenomena like interference are observed: the characteristic direction of the mixed frequency radiation has a sharp minimum and maximum, and the dependence of its power on the path length in the medium (in the direction of maximum intensity) has an oscillating nature. The frequency mixing processes of greatest practical importance are the optical generation of sum and difference frequencies $\omega_3 = \omega_1 \pm \omega_2$ used for frequency conversion (e.g., for generating the second harmonic), and four-wave frequency mixing of the types of $\omega_4 = 2\omega_1 \pm \omega_2$ that is the basis for a nonlinear spectroscopy method corresponding to *active spectroscopy* of *Raman scattering of light.* See also *Optical parametric oscillator.*

FRESNEL EQUATIONS (A.T. Fresnel, 1823)

Relations which connect the amplitudes, phases and polarizations of reflected and the refracted light waves. The reflected and refracted waves arise at the passage of light through the plane boundary between two transparent media, and they are described in terms of corresponding parameters of the incident wave. Fresnel obtained these relations using the now discredited concept of a vibrating elastic ether. The Fresnel equations follow from the electromagnetic theory of light, and may be obtained from the solution of Maxwell equations with boundary conditions at the interface between two homogeneous isotropic media with *refractive indices* n_1 and n_2. For plane waves the amplitudes of the electric vectors of the incident (i), reflected (r) and transmitted (t) waves are respectively resolved into components, which are parallel (subscript p) or perpendicular (subscript s) to the plane of incidence, and the Fresnel equations take the form

$$T_p = \frac{2\sin\theta_t \cos\theta_i}{\sin(\theta_i + \theta_t)\cos(\theta_i - \theta_t)} I_p,$$

$$T_s = \frac{2\sin\theta_t \cos\theta_i}{\sin(\theta_i + \theta_t)} I_s,$$

$$R_p = \frac{\tan(\theta_i - \theta_t)}{\tan(\theta_i + \theta_t)} I_p,$$

$$R_s = -\frac{\sin(\theta_i - \theta_t)}{\sin(\theta_i + \theta_t)} I_s,$$

where θ_i, θ_t are respectively the angles of reflection and *refraction of light*, which are related to the refractive indices n_1 and n_2 by *Snell's law*: $n_1 \sin\theta_i = n_2 \sin\theta_t$. It follows from the above equations, that the phase of every component of the reflected and transmitted wave is either equal to the phase of the corresponding component of the incident wave or differs from it by π. Therefore, e.g., if the incident wave is linearly polarized, then the reflected and refracted waves are also linearly polarized.

The applicability of the Fresnel equations is limited by the requirement that the irregularities of the surface must be small compared to the wavelength, and refraction indices of the media must not depend on the amplitude of the E-vector of the light wave. In the case of very intense light sources (laser light) the reflection and refraction are accompanied by a number of new effects, and the amplitudes of the waves at the boundary between the two media no longer satisfy the Fresnel equations (see *Nonlinear optics*).

FRICTION

A dissipative process arising as a result of the resistance to motion of a body or a system of bodies, or a relative displacement of parts of the same system which leads to scattering and the loss of energy. The sliding motion of bodies in contact leads to *external friction*; and this situation in respect to relative displacements of parts of the same system is referred to as *internal friction*. In a mechanical system friction causes the total mechanical energy E to decrease with its partial transformation into other types of energy, including heat energy, energy accumulated in material defects, and so on. As a rule, one introduces *dissipative forces* (*friction forces*) which counteract the relative body motion. In the case of the dependence of these forces on generalized velocities $\dot{q}$, it is convenient to consider a *dissipation function* $\Phi = (1/2)\sum_{ij} a_{ij}\dot{q}_i\dot{q}_j$, where q_j are the generalized coordinates of the system, a_{ij} are the coefficients (which, in general, depend on q_i) representing a measure of the scattering of the total mechanical energy per unit time, and $E = -2\Phi$. In this case, when writing the equations of motion of the mechanical system it is necessary to take into account the terms proportional to $\dot{q}$ which represent the frictional forces. Friction in momentum space is the process of a particle decellerating (braking) under the action of small forces uncorrelated in regard to either direction nor magnitude. *Diffusion* in momentum ($\boldsymbol{P}$) space (see *Fokker–Planck equation*) can be described by a kinetic-type expression

$$\frac{\partial}{\partial t} f = b \cdot \mathrm{grad}_p\left[\boldsymbol{P} f + \mathrm{grad}_p(k_B T f)\right],$$

where f is the particle distribution function over momenta, and the first term of the right-hand side determines the force acting upon a particle which is directed opposite to its motion, and proportional to its speed. It can be shown that the time averaged momentum increment is $\langle \Delta\boldsymbol{P}\rangle = -b\boldsymbol{P}\tau$, where τ is the correlation time of the forces acting upon the particle, and b is the *friction constant*. Such a consideration is used, e.g., when calculating the coefficient for the transfer and dissipation of the energy of channeled particles (see *Friction of channeled particles*). By introducing the dissipation term the *kinetic equation* can be regarded as including *dynamic friction* forces. The energy dissipation produced by *dipole moment* oscillations can be described as the action of a *radiation friction* force.

FRICTION CONSTANT

A coefficient in the *Fokker–Planck equation* that determines the force acting on a moving particle, proportional to the particle velocity and directed oppositely to the direction of motion. By considering the particle motion as a random process, and introducing the total force $F(t)$ acting on the particle at every instant of time t, it is possible to determine the friction constant b:

$$b = \frac{1}{mk_B T} \int_{-\tau}^{t} \langle F(t+\tilde{t}) F(t) \rangle \mathrm{d}\tilde{t},$$

where m is the particle mass, τ is the *correlation time* of the forces acting on the particle, and $\langle F(t+\tau)F(t)\rangle$ is the *correlation function* of the effective forces. Here the averaging $\langle \ldots \rangle$ is performed over all the particles of the system excluding the probe particle under consideration, and in accordance with the assumption that the random nature of the process does not depend on the time, and determines the subsequent increment of momentum $\langle \Delta P \rangle$. During the time τ, $\langle \Delta P \rangle = -bP\tau$, where P is the momentum of the particle. The friction constant is similar to the coefficient of mobility in the *diffusion* equation, and is associated with the *diffusion coefficient* in momentum space $D_p = 2bmk_B T$. The friction constant b is used to obtain expressions for the transport coefficients in liquids and gases, in investigations of the energy dissipation during the channeling of particles (see *Friction of channeled particles*), etc.

FRICTION, EXTERNAL

See *External friction.*

FRICTION, INTERNAL

See *Amplitude-dependent internal friction*, *Internal friction.*

FRICTION OF CHANNELED PARTICLES

A characteristic of the process whereby channeled particles dissipate energy by scattering off electrons and thermally displaced atoms (see *Thermal displacements of atoms*) of the crystal. When describing the energy change of the particle (in a direction normal to the forward motion, $\Delta E_\perp$) per unit channel length z, one can introduce an index of dynamic braking $\langle \Delta E_\perp / z \rangle$ where the averaging $\langle \ldots \rangle$ is carried out over the entire region available for a particle with a given transverse energy $E_\perp$. Considering this value as the dynamic friction coefficient, one can write down a *kinetic equation* (similar to *Fokker–Planck equation*) where the above-defined term is related to the change of transverse energy due to braking (see *Channeling*, *Neutral particle channeling*).

FRIEDEL OSCILLATIONS (J. Friedel, 1952)

Periodic space oscillations of the electron density n_e, which occur with the period $L = \pi / k_F$ (where k_F is *Fermi wave number*), and are due to excessive electric charge in the bulk of a metal or at its surface. The magnitude of Friedel oscillations has a power-law-type dependence on the distance r from the charge. The particular form of the decay law depends on the nature (dimensionality) of the electron configuration, and on the shape of the *Fermi surface.* In the case of a point charge in an isotropic metal with a spherical Fermi surface, the electron density perturbation decreases at large distances $r \gg k_F^{-1}$ in inverse proportion to the cube of the distance $\propto \cos(2k_F r)/r^3$. In anisotropic layered (see *Quasi-two-dimensional crystals*) or chain (see *Quasi-one-dimensional crystals*) metals with nearly cylindrical or slightly corrugated flat Fermi surfaces the asymptotes of Friedel oscillations are of the form $\propto \cos(2k_F r)/r^2$ in the plane of the layers or $\propto \cos(2k_F z)/z$ in the direction of the chains. The generation of Friedel oscillations is due to the diffraction of *de Broglie waves* of electrons with the wavelength $\lambda_D = 2\pi / k_F$ (corresponding to *conduction electrons* in the neighborhood of the Fermi surface) by *point defects* (impurity atoms, vacancies), and by extended *defects in crystals* (dislocations, phase boundaries, twinning surfaces). Friedel oscillations play an important role in the screened Coulomb interaction of atoms, which are adsorbed by the metal surface, and in the indirect interaction of spins (see *Ruderman–Kittel–Kasuya–Yosida interaction*).

FRIEDEL SUM RULE

A relation which permits the determination of the change in the *density of electron states* ΔN in a solid due to impurities introduced into it. It is derived in terms of a wave function which is a sum over products $\Psi_l(r)P_l(\cos\theta)$ of radial functions and Legendre polynomials. The Friedel sum rule given by

$$\Delta N = \frac{2}{\pi} \sum_l (2l+1)\delta_l$$

expresses ΔN in terms of the phase shifts δ_l of the wave functions of electrons at the *Fermi surface* which are scattered by impurities. Ordinarily the phase shifts are negligible for l greater than 3 or 4.

FRÖHLICH CONDUCTIVITY

Contribution to the conductivity of a quasi-one-dimensional semiconductor, which takes place at temperatures below that of the *Peierls transition*, and is due to the motion of an electronic *charge density wave* (CDW) in the electric field. The Fröhlich conductivity may be observed when the magnitude of the electric field exceeds a certain threshold value E_T which is determined by the pinning of the charge density wave at lattice defects, and by commensurability effects. Typical values of E_T for a number of quasi-one-dimensional inorganic compounds ($NbSe_3$, TaS_3, $K_{0.3}MoO_3$), in which the motion of a CDW is observed, lie in the range between 0.01 and 1 V/cm. It is assumed that the phenomenon of Fröhlich conductivity is related to certain unusual properties of the semiconductors with charge density waves: the sharp frequency dependence of the conductivity in a weak alternating field at frequencies $\sim$ 10–100 MHz; the nonlinear nature of static conductivity in fields with magnitudes that exceed E_T at $T < T_P$ (T_P is the temperature of the Peierls transition); the generation of narrow-band noise at a frequency which is proportional to that of the transport current carried by the charge density wave; the generation of Shapiro steps (see *Josephson effects*) on the current–voltage characteristic in the regime of Fröhlich conductivity, etc.

FRÖHLICH HAMILTONIAN (H. Fröhlich, 1950)

Hamiltonian of the *electron–phonon interaction* in the *quasi-momentum* space in the solid, which is written in the formalism of *second quantization*:

$$H_F = \sum_{k,k',q,j,\sigma} g^j_{kk'}(q) a^+_{k',\sigma} a_{k,\sigma} (b^+_{qj} + b_{-qj}) \times \sqrt{\frac{\hbar\omega^j_q}{2V}} \Delta(q + k' - k), \qquad (1)$$

where $g^j_{kk'}(q)$ is the vertex part (triple pole) of the electron-phonon interaction, $a^+_{k\sigma}$ ($a_{k\sigma}$) and b^+_{qj} (b_{qj}) are respectively creation and annihilation operators of the electron of quasi-momentum k and spin σ, the phonon of momentum q belongs to the jth branch of frequency ω^j_q, V is the normalizing volume, and $\Delta(q)$ takes into account the quasi-momentum conservation law. The constant g in covalent crystals (*insulators*, *semiconductors*) is proportional to the *strain potential*. The *Fröhlich constant* in ionic crystals is equal to

$$g_F(q) = \frac{2e}{q}\sqrt{\pi\left(\frac{1}{\varepsilon_\infty} - \frac{1}{\varepsilon_0}\right)}, \qquad (2)$$

where ε_0 and ε_∞ are the static and high-frequency *dielectric constants*, respectively. In the case of *metals*, one should include in Eq. (1) the matrix element of the electron–phonon interaction for $g^j_{kk}(q)[\hbar\omega^j_q/(2V)]^{1/2}$:

$$M^j_{kk'}(q) = -\mathrm{i}\left(\frac{n}{M\omega^2_q}\right)^{1/2} \langle k'|\nabla V_{ei}|k\rangle e^j_q, \qquad (3)$$

where V_{ei} is the pseudopotential of the electron–ion interaction, e^j_q is the polarization vector of the jth phonon branch, $q = k - k'$, M is the ion mass, n is the concentration of free electrons. The Fröhlich Hamiltonian describes processes of single-phonon electron scattering, and the related effective attraction between electrons in second-order perturbation theory.

FROZEN CONDUCTIVITY, residual conductivity, stimulated conductivity, metastable conductivity

A physical phenomenon in semiconductors, consisting in the persistence of an elevated value of electrical conductivity for a long time after the external action (e.g., illumination) is terminated. Frozen conduction is not a common property of semiconductors; it appears only after special techniques of sample preparation. A fast *relaxation* from the frozen conduction state to the equilibrium state can be caused by increasing the temperature, by applying a strong electrical field, or by illuminating with light having quanta of energy less than the width of the *band gap* (thermal, field, and IR extinction of frozen conduction).

There are several mechanisms of frozen conductivity. In inhomogeneous semiconductors (including, in particular, strongly compensated semiconductors), the separation of nonequilibrium carriers by the fields of inhomogeneities gives rise to a sharp increase of life-time, which is displayed via the long-term relaxation of conductivity, and hence frozen conduction. Frozen conduction induced by light can result from *photochemical reactions* in the bulk of the sample, or on its surface. It can also be caused by optical ionization of deep centers which strongly interact with the crystal lattice. The *trapping* of carriers at such centers requires lattice reconstruction, and proceeds extremely slowly.

Frozen conduction can be used in the systems of *information recording.* The phenomena involved in frozen conduction can also be responsible for degradation effects in photoelectronic devices.

FRUSTRATION

The situation in a *crystal*, when conflicting spatial requirements prevent complete atomic or magnetic ordering from being established. This can occur, e.g., when a regular lattice of antiferromagnetically interacting spins cannot establish an ordered arrangement in which each spin has only oppositely directed nearest neighbors. (see *Long-range and short-range order*).

FULDE–FERRELL–LARKIN–OVCHINNIKOV STATE (P. Fulde, R. Ferrell, A.I. Larkin, Yu.A. Ovchinnikov, 1964)

Nonuniform superconducting state which arises in a uniform magnetic field B. The superconducting *order parameter* Φ in the Fulde–Ferrell–Larkin–Ovchinnikov state varies periodically in space with a period which is close to the *coherence length* ξ_0 in the superconductor. The *phase transition* in the presence of B from the normal state to the Fulde–Ferrell–Larkin–Ovchinnikov state at zero temperature $T = 0$ is a *second-order phase transition*, taking place in the field $B = 0.76\Delta_0/\mu_B$ (μ_B is Bohr *magneton*, Δ_0 is the value of the energy gap Δ at $T = 0$ and $B = 0$). When B reaches the *Clogston–Chandrasekhar paramagnetic limit* (A.M. Clogston, B.S. Chandrasekhar) $B = \Delta_0/(\sqrt{2}\mu_B)$ (see *Paramagnetic limit*) the Fulde–Ferrell–Larkin–Ovchinnikov state gives way to the uniform superconducting state through a *first-order phase transition.* This state can only be observed in pure (type I) superconductors, because in the case of an impure superconductor the periodic changes of Φ are suppressed by the diffusional motion of electrons of mean free path $l \ll \xi_0$.

FULLERENE, buckyball (named for architect Buckminster Fuller, inventor of geodesic domes built from straight bars)

Molecule of general formula C_n (n is an even number) which has the shape of a closed hollow (in most cases convex) polyhedron; constitutes the fourth allotropic modification of *carbon.*

The best known fullerene molecules are C_{70}, and especially C_{60}. The molecule of the formula C_{60} exhibits the highest symmetry among all molecules known to the present day. This molecule exists in the shape of a truncated icosahedron consisting of 20 hexagons and 12 regular pentagons, resembling a soccer ball (see Fig.). Carbon atoms in the fullerene molecule C_{60} are close to sp^2-hybridized, and each of them forms one double bond C=C and two single bonds C–C. The length of the single bonds (located along pentagonal edges) is 0.146 nm, and the length of the double bonds (at boundary between two hexagons) is 0.140 nm. The average bond length, 0.144 nm, is close to that of graphite (0.142 nm). These bonds are located at the distance of approximately 0.351 nm from the polyhedron center. Fullerene molecules are diamagnetic, i.e. their ground electron state is a singlet; the fullerene of the formula C_{84} is gyrotropic (see *Gyrotropic medium*).

The fullerene of formula C_{60} exhibits in its IR spectrum four bands at 527, 576, 1182 and

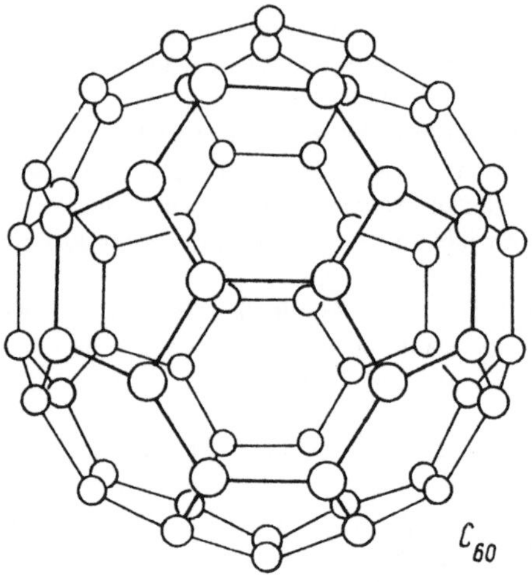

Structure of C_{60} fullerene molecule. The alternating single–double bonds are not indicated.

1428 cm^{-1} due to allowed electric dipole *intramolecular vibrations*. The ^{13}C NMR spectrum of this fullerene is a singlet (chemical shift $\delta = 142.68$ ppm), demonstrating that all 60 carbon atoms are equivalent.

Fullerenes of all compositions (including *hyperfullerenes* of formula C_n, where $n > 70$ and up to ≈ 200) are formed in the gaseous phase at the evaporation of *graphite* through laser or electric arc heating, and exhibit unambiguous signals of highly stable ions C_n^+ or C_n^- in mass spectra (see *Mass spectrometry*). Fullerenes are soluble in aromatic, saturated and chlorinated hydrocarbons; they may be separated from solution in the form of a solid phase, which is sometimes called *fullerite*.

The study of physical and physicochemical properties of fullerenes is important for the understanding of various processes, in particular catalysis, oxidation, combustion. Investigation of molecules of the formula C_n is of much interest also from the point of view of cosmology and astronomy, because certain spectral bands, which had not been previously identified and are now ascribed to fullerene, have been spectroscopically observed in the atmospheres of carbon stars, and the tails of comets.

Many unique opportunities for the application of fullerenes are in sight; fullerenes may provide the basis of new lubricants, molecular *traps* for atoms (including radioactive atoms), new medicines, etc.

The possibility for the existence of a chemically stable cluster with the formula C_{60} had been predicted by E. Osawa (1970) and, independently, by D.A. Bochwar and E.N. Galperin (1973). This fullerene was first observed in mass spectra in 1985 by H.W. Kroto and R.E. Smalley, who subsequently proposed the name "fullerene" for this new class of molecular carbon compounds. Fullerene was, for the first time, obtained in a pure state through a synthesis carried out by W. Krätschmer and D. Huffman (1990).

Fullerides are chemical compounds which include the molecule fullerene as one of the constituents. In particular, fullerides are found in the form of stable complexes of the formula $C_{60}M_n$ (M = Na, K, Cs, Ca, Sr, Ba, La, U, etc.) in the mass spectra of vapors obtained through evaporation of a mixture of graphite and salts of the metals. The metal atom in this complex is generally outside the fullerene molecule, but it can be inside it, or may be incorporated in the fullerene lattice. Chemical reactions, which result in the generation of fullerides, ordinarilly begin with the elimination of one or several pairs of carbon atoms. On the other hand, the generation of, e.g., the fulleride of the formula $C_{60}H_{36}$ takes place through rearrangement of intrinsic π-bonds of the C_{60} molecule. The broad diversity of fullerides, especially metal-containing ones with the formula $[(C_6H_5)_3P]_2MC_{60}$ or $\{[(C_2H_5)_3P]_2M\}_6C_{60}$, where M = Pt, Ir, Pd, etc., has opened up new trends in the physics and chemistry of organic compounds.

Fullerite is the solid phase of fullerene. The most attention has been given to the study of fullerites based on C_{60} molecules. Fullerites can be precipitated from solution, or generated through *condensation* of graphite vapor with the subsequent extraction with a hydrocarbon, chromatographic treatment, and *crystallization*. Fullerite is a semitransparent crystalline material of dark-yellow, brown or dark-brown color (color depends on C_{60}/C_{70} ratio in the composition). A C_{60} crystal has a cubic close-packed (FCC) structure with lattice constant 1.417 nm and density 1.72 g/cm^3. This substance is stable in air and sublimes at 434 °C.

Solid fullerenes are molecular *insulators* with the value of the *band gap* $\leqslant 2$ eV; the interaction forces between the molecules are mostly van der Waals in nature (see *Van der Waals forces*) with a slight intermixture of covalent bonding (see *Covalent bond*), which leads to weak anisotropy of the intermolecular forces. Within the dipole approximation, the lowest optical transitions are forbidden in both fullerites and fullerenes.

The fullerites of formula C_{60} or C_{70} may be easily turned into semiconductors (of both n- and p-types) and metals through *doping*. Both semiconducting and metallic properties of fullerites are due to inclusion of atoms of alkali (Na, K, Rb, Cs) or alkaline-earth (Sr, Ba) metals into the cavities (octahedral, tetrahedral sites) of the face-centered lattice (or lattice of another type); outer shell (valence) electrons of these atoms are transferred into lowest empty bands of fullerites; and these bands become the conduction bands of the material.

At compositions close to stoichiometric, ion-radical salts of the general formula M_3C_{60} ($M = K$, Rb, Cs) exhibit not only metallic, but also superconducting properties, which are unique for three-dimensional organic compounds. In particular, the *critical temperature* of the doped triple fullerite Rb_2CsC_{60} is $T_c \approx 31$ K. There is a well-defined correlation between T_c and the atomic radius of the doping metal: the critical temperature increases with the increase of the doping alkali metal atomic radius; this enhancement of T_c is caused by the increase of spacing of the fullerite lattice, and the corresponding increase of the *density of states* at the Fermi level (which is in full conformity with the *Bardeen–Cooper–Schrieffer theory*). However, the attempts to observe the phenomenon of *superconductivity* in doped fullerites based on C_{70} and other fullerenes have not yet been successful. Solid fullerenes resemble *high-temperature superconductors* in a number of superconducting properties (values of correlation length, critical fields and currents, etc.), but are sharply different from them in crystal and electronic structure. Solid fullerenes and fullerides are considered as promising materials for various applications: semiconductor and *superconductor electronics*, light-weight rechargeable batteries of high capacity, transistors, heterojunction diodes, catalysis, hydrogen storage, etc.

FUNCTIONAL ELECTRONICS

Trend in *microelectronics*, in which the role of the main data carrier is not played by the electric state of a certain circuit cell (e.g., logic gate or memory storage register), as in the cases of *integrated circuits* (IC), and very large integrated circuits (VLIC), but rather by a dynamic inhomogeneity, i.e. a local nonequilibrium state in a continuous homogeneous medium. Examples are *magnetic bubble domains*, *surface acoustic waves*, magnetostatic waves (see *Spin waves*), charge "pockets" and packets in *charge coupled devices*, double *Bloch lines*, repolarization domains in *ferroelectrics*, etc. The application of dynamic inhomogeneities is possible not only in solids, but also in liquid media (e.g., *liquid crystals*), in *gels*, and in vacuum-gaseous media.

The devices of functional electronics may operate due to a single particular physical effect and a single type of dynamical inhomogeneity, or they may involve several continuous media with the integration of different physical effects in a single device, i.e. a system of media with various dynamical inhomogeneities which interact with each other. Functional electronics, as a circuitry-less branch of electronics; it is not in opposition to traditional circuit engineering (e.g., IC) since they can exist in parallel and complement each other. An advantage of functional electronics is that it utilizes no intercellular connections, which impose strict limitations on the speed of response and the reliability of integrated and large integrated circuits with minimal dimensions 1 μm or smaller (*nanotechnology*). The possibility of carrying out not only sequential input (output) and transport of data, but also "instantaneous" transport of large data files, is another advantage of functional electronics. This may be considered as data processing at the level of elementary functions of higher order than those of IC and VLIC. This principle is a promising one for processing large data files in real time, e.g., when solving problems of pattern recognition and identification. Examples of functional electronics devices are *magnetic bubble domain devices*, solid-state photoelectric converters with charge couplers (see *Charge coupled devices*), a variety of acousto-electronic devices (see *Acousto-electronics*) using surface acoustic waves, electron–optical image converters (*electrooptical modulators*), liquid-crystal diodes (LCD) (see, e.g., *Liquid-crystal displays*).

Gg

GADOLINIUM, Gd

An element of Group III of the periodic table belongs to *lanthamides*, atomic number 64, atomic mass 157.25; possesses 20 stable isotopes with mass numbers from 143 to 162. Natural gadolinium is mainly a mixture of isotopes with mass numbers 154 to 158 and 160, plus weakly radioactive ^{152}Gd which has a half-life of $1.1 \cdot 10^{14}$ years. Electronic configuration of outer electronic shells $4f^7 5d^1 6s^2$, successive ionization energies 5.98, 12.1, 20.6 eV. Atomic radius 0.179 nm, Gd^{3+} ionic radius ≈ 0.0954 nm. Oxidation state is +3, more rarely +2 and +1. Electronegativity is ≈ 1.20.

In free form a silvery-gray *metal*. It exists in α and β modifications. α-Gd has hexagonal close packed lattice with parameters $a = 0.3634$ nm, $c = 0.5781$ nm (space group $P6_3/mmc$, D_{6h}^4) under normal conditions. Above 1535 K up to $T_{\text{melting}} = 1585$ K, α-Gd is stable. β-Gd with the body centered cubic lattice ($a = 0.405$ after hardening; space group $Im\overline{3}m$, O_h^9). Density is 7.886 g/cm^3 at 293 K. $T_{\text{boiling}} = 3553$ K. Binding energy is 4.14 eV at 0 K. The latent heat of $\alpha \to \beta$ transformation is 3.9 kJ/mole, latent heat of melting is ≈ 12 kJ/mole, latent heat of sublimation is 301.5 kJ/mole, latent heat of evaporation is ≈ 330 kJ/mole. Specific heat is ≈ 0.20 kJ$\cdot$kg$^{-1}\cdot$K^{-1} at room temperature. Debye temperature is ≈ 180 K; linear thermal expansion coefficient of monocrystal is $12.5 \cdot 10^{-6}$ K^{-1} along the principal axis $\underline{6}$ and $6.3 \cdot 10^{-6}$ K^{-1} in perpendicular directions (at 600 K); thermal conductivity is ≈ 10 W$\cdot$m$^{-1}\cdot$K^{-1} at room temperature. Adiabatic elastic moduli of α-Gd monocrystal: $c_{11} = 66.7$, $c_{12} = 25.0$, $c_{13} = 21.3$, $c_{33} = 71.9$, $c_{44} = 20.7$ GPa (at 298 K); adiabatic compressibility is 37.8 GPa (at 298 K), isothermal compressibility is 35.48 GPa (at zero external pressure); Young's modulus is 56.19 GPa; endurance limit is 0.1824 GPa; yield limit is 0.182 GPa (at 293 K). Brinell hardness is 0.588 GPa. Gadolinium is easily mechanically processible. Activation energy of self-diffusion of atoms (at T_{boiling}) is 136 kJ/mole (i.e. 1.41 eV/atom). Natural gadolinium has very high effective thermal neutron capture cross-section 45,000 barn (around 160,000 and 70,000 barn for ^{157}Gd and ^{155}Gd, respectively). Sommerfeld coefficient (of low temperature electronic specific heat) 19.5 mJ$\cdot$mole$^{-1}\cdot$K^{-2} in the temperature range 2.5 to 10 K. At 298 K electric resistivity of α-Gd monocrystal along the principal axis $\underline{6}$ is 1220 n$\Omega\cdot$m, and in perpendicular directions it is 1390 n$\Omega\cdot$m, temperature coefficient of electrical resistance of gadolinium polycrystal is 0.00176 K^{-1}. Work function for polycrystal is 3.07 eV.

Gadolinium is the only *rare-earth element* which transforms from the paramagnetic to the ferromagnetic state directly at the Curie point $T_c = 293.2$ K (*second-order phase transition*). Gadolinium is a collinear ferromagnet below T_c and above $T_c' = 232$ K. At low temperatures the magnetic moments precess around a six-fold axis, forming a cone-like ferromagnetic structure (phase transition is of "order–order" type). Magnetic susceptibility $75{,}500 \cdot 10^{-6}$ cgs; nuclear magnetic moment of isotope ^{157}Gd (15.64% abundant) is -0.34 nuclear magnetons. Gadolinium is a prospective material for control rods (neutron absorbers) of nuclear reactors. Gadolinium is used for the *alloying* of steel, titanium, and magnesium; as a component of the *alloys* (with iron, nickel, and cobalt), possessing high magnetic induction and magnetostriction, for the luminophors and other materials. Some strongly paramagnetic gadolinium salts are used for cooling to extremely low temperatures, $\sim$0.001 K (see *Adiabatic demagnetization cooling*).

GALLIUM, Ga

A chemical element of Group III of the periodic table with atomic number 31 and atomic mass 69.723. The electronic configuration of the outer shells is $3d^{10}4s^24p^1$. Successive ionization energies are 5.998, 20.514, 30.71 eV. Atomic radius of neutral Gd is 0.139 nm, radius of Ga^{3+} ion is 0.062 nm. Work function is 4.19 eV. Oxidation state is +3, more rarely +2 and +1. Electronegativity is 1.65.

In free form, gallium is a silvery-white *metal*. Under normal conditions the α-form is stable (α-Ga or GaI) with an orthorhombic crystal lattice with parameters $a = 0.4529$ nm, $b = 0.4519$ nm, $c = 0.7657$ nm. The peculiarity of this lattice lies in the fact that the atoms are arranged as pairs, as diatomic molecules Ga_2. Under the pressure 1.2 GPa and temperature 275.5 K GaI transforms to GaII (tetragonal lattice with parameters $a = 0.396$ nm, $c = 0.437$ nm); at the pressure 3 GPa and temperature 323 K the form GaIII has been detected. For α-Ga $T_{\text{melting}} = 302.9$ K, $T_{\text{boiling}} \approx 2600$ K; density of solid gallium is 5.094 g/cm^3 (at 303 K), density of liquid gallium is 6.095 g/cm^3 (at 303 K); latent heat of melting is 5.59 kJ/mole, latent heat of evaporation is 256 kJ/mole; specific heat $c_p = 25.9$ J·mole^{-1}·K^{-1}. Debye temperature is 333 K; electrical resistivity is 0.272 μΩ·m (at 273 K); thermal coefficient of electrical resistivity is $8\cdot10^{-6}$ K^{-1}. In the solid state gallium is a *diamagnet* with magnetic susceptibility $-0.31\cdot10^{-9}$ CGS units (at 290 K). Hall constant is $-0.63\cdot10^{-10}$ m^3/C. Transition temperature to the superconducting state is 1.085 K. Thermal coefficient of linear expansion is $2\cdot10^{-5}$ K^{-1} (at 223 to 293 K), thermal conductivity is 41 W·m^{-1}·K^{-1} (at 273 K). Brinell hardness is 25 MPa.

Metallic gallium is used for manufacturing high-temperature (873 to 1573 K) thermometers, manometers, thermoregulators, it substitutes for toxic mercury in diffusion pumps. Compounds of gallium with elements of Group V of the periodic system (GaP, GaAs, GaSb, and some others) are *semiconductors* used in the high-temperature rectifiers, transistors, solar batteries, and IR radiation detectors. Mirrors made from gallium possess high reflectivity. Gallium is added to special types of glasses.

GALLIUM ARSENIDE

A binary semiconducting compound (see *Semiconductor materials*) with the *sphalerite* (*zinc blende*, ZnS) structure with each As atom centered in a regular tetrahedron formed by four Ga atoms, and vice versa.

Thus each Ga atom is in the same position relative to its closest neighbors as As. The GaAs *crystal lattice* can be thought of as two interpenetrating face-centered cubic sublattices built of Ga and As atoms. The two sublattices are shifted along the cube diagonal relative to each other by 1/4 of its length. The *space group* is cubic $F\bar{4}3m$ (T_d^2), which has three 2-fold (π) *symmetry axes*, four 3-fold ($2\pi/3$) symmetry axes and six symmetry planes, and in addition it lacks a *center of symmetry*. The distance between the positions of Ga and As nearest neighbors is 2.44 Å, the sum of their *atomic radii* Ga (1.26 Å) and As (1.18 Å). The lattice constant is 5.653 Å, the density 5.32 g/cm^3, Mohs hardness is 4.5, specific heat under constant pressure is 2.76 cal·mole^{-1}·K^{-1} at 80 K, heat conductivity is 0.37 cal·cm^{-1}·s^{-1}·K^{-1}. *Debye temperature* is 362 K, static dielectric constant is 12.9, the optical (high-frequency) permittivity 11.6, the melting temperature 1510 K, electron affinity 4.07 V, work function 4.7 eV, temperature coefficient of linear expansion $\Delta L/L\Delta T = 5.9\cdot10^{-6}$ deg^{-1}. The energy minima $E(\boldsymbol{k})$ ($\boldsymbol{k}$ is *quasi-wave vector*) in the conduction band are located along the crystallographic directions (100) (X-valleys), (111) (L-valleys) and near the center point (000) (Γ). The energy distance between the Γ-minimum and the maximum of the valence band equals the width of the gap $E_g = 1.43$ eV at $T = 300$ K. The temperature dependence $E_g(T) = E_g(0) - \alpha T^2/(T + \beta)$, $E_g(0) = 1.519$ eV; $\alpha = 5.405\cdot10^{-4}$ eV/K ($100 < T < 500$ K), $\beta = 204$ K. At $T = 300$ K, E_g grows with pressure P, $\mathrm{d}E/\mathrm{d}P = 12.6\cdot10^{-6}$ eV·kg^{-1}·cm^2. Energy gaps between the Γ-points and the two other valleys are $E_{\Delta k} = 0.28$–0.36 eV, $E_{\Delta L} \approx 0.45$ eV, respectively. The *effective mass* of electrons near the bottom of the *conduction band* is

$0.068m_0$ (m_0 is the free electron mass). The dependence $E(\boldsymbol{k})$ in the *valence band* has one doubly degenerate maximum. Effective masses of the light and heavy holes are equal to $0.12m_0$ and $0.5m_0$, respectively. The effective *density of states* at $T = 300$ K is $4.7{\cdot}10^{17}$ cm^{-3} in the conduction band and $7{\cdot}10^{18}$ cm^{-3} in the valence band; the intrinsic concentration of electrons is $1.1{\cdot}10^{7}$ cm^{-3}. The mobility of electrons is 9000 cm^2·V^{-1}·s^{-1}, that of holes is 400 cm^2·V^{-1}·s^{-1}. The main *acceptors* are atoms of Group II of the periodic system that substitute for Ga and As atoms. The main *donors* are atoms of Group VI at As sites. Atoms of Group IV can be donors, if they occupy the places of Ga, or acceptors if they belong to the As sublattice. The most important acceptors are Zn and Cd. At the melting temperature their solubilities are Zn $\geqslant 10^{20}$ cm^{-3}, Cd $\geqslant 10^{19}$ cm^{-3}. At low concentrations, Mg is a donor, and at high concentrations ($\geqslant 10^{18}$ cm^{-3}) it is an acceptor. Selenium, tellurium and sulfur are shallow donors (see *Shallow levels*). Oxygen creates deep donor levels (see *Deep levels in semiconductors*). Elements of the transition group Cr, Mn, Fe, Co, Ni are deep acceptors. GaAs is the most widely used of all $A^{III}B^{V}$ compounds. Manufactured on its basis are *transistors*, *diodes* of ultra-high frequency range, tunnel, switching, emission diodes, devices of the ultra-high frequency range with application of bulk effects, the devices with negative differential resistance, parametric amplifiers, heterostructures, functional *integrated circuits*, microwave varactors, injection *lasers*, infrared lasers, solar converters.

GALVANOMAGNETIC EFFECTS

Manifestations of the action of a magnetic field $\boldsymbol{B}$ on the conduction properties of a solid. These effects are due to the *Lorentz force* $\boldsymbol{j} \times \boldsymbol{B}$ which deflects the moving charge (*current carriers*) from their direction of motion. This leads to a dependence of the resistivity ρ on $\boldsymbol{B}$ (*magnetoresistance*) and to the *Hall effect*, that is the appearance with an electric Hall field $\boldsymbol{E}_{\mathrm{H}}$ which is orthogonal to both $\boldsymbol{B}$ and the electric current density $\boldsymbol{J}$ that compensates the action of the Lorentz force. The experimental study of galvanomagnetic effects stimulated the development of the concept of *quasi-particles* in crystals, i.e. elementary excitations with energy ε that are periodic functions of their *quasi-momenta* $\boldsymbol{P}$. The observation of *Hall fields* of opposite signs justified the concept of *holes*, excitations with a specific charge $e/m^* > 0$ (m^* is the cyclotron *effective mass*). Current carriers with specific charges of opposite sign moving in directions opposite to each other, deflect in the same direction. Hence, in an intrinsic semiconductor with an equal mobility of electrons and holes, and in a compensated *metal* with the same number of electrons N_1 and holes N_2, the Hall field is very small or does not appear at all. The Hall field can also be removed by the proper selection of the sample geometry (so-called *Corbino disk geometry*).

In experimental studies of galvanomagnetic effects, a current with a given magnitude and direction flows along a conductor, and the electric field components in three orthogonal directions are measured. There are a *transverse galvanometric effect* ($\boldsymbol{j} \perp \boldsymbol{B}$) and a *longitudinal effect* ($\boldsymbol{j} \parallel \boldsymbol{B}$), the latter being quite small. A dimensionless parameter determining the role of the magnetic field is the ratio r_c/l where r_c is the radius of the current carrier orbit (electron radius), and l is the mean free path. The dependence of ρ on r_c/l at various temperatures approximately scales to a single curve, and this is called the *Kohler rule*. The bending of current carrier trajectories in a weak magnetic field ($r_c \gg l$) leads, as a rule, to the appearance of a small increase in resistance which is quadratic in the magnetic field. For $r_c \ll l$ the behavior of the resistance provides information on the carrier dispersion law. The main contribution to the *electrical conductivity* comes from electrons with energy very close to the *Fermi energy*. When moving in a magnetic field these electrons travel along curves laying on the *Fermi surface* $\varepsilon(\boldsymbol{p}) = \varepsilon_{\mathrm{F}}$ in the plane with $\boldsymbol{p} \cdot \boldsymbol{B} = \mathrm{const}$; the trajectories of the electrons in the coordinate space in the plane perpendicular to $\boldsymbol{B}$ are similar to the trajectories on the Fermi surface.

In metals with closed Fermi surfaces the magnetic field confines the *current carrier drift* to the plane perpendicular to $\boldsymbol{B}_0$. The drift in this plane is due to collisions. However, in non-compensated metals ($N_1 \neq N_2$), the resistivity $\rho(B)$ is the same order of magnitude as $\rho(0)$, growing and reaching saturation at $B \to \infty$. Even in very clean single crystals there is no pronounced anisotropy of the galvanomagnetic trajectories. This is due to the fact that the carrier drift velocity in the electric field $\boldsymbol{E}$ perpendicular to $\boldsymbol{B}$, which has the form $\boldsymbol{u} = (\boldsymbol{E} \times \boldsymbol{B})/B^2$, is orthogonal to both $\boldsymbol{E}$ and $\boldsymbol{B}$. The associated Hall current $\boldsymbol{j}_{\mathrm{H}} = (N_1 - N_2)e\boldsymbol{u}$ in the specimen is balanced by the Hall field $E_{\mathrm{H}} = -RB_j$; where the Hall constant R is given by $R = 1/[(N_2 - N_1)e]$. As a result, the bending of the trajectories is small, and the conductivity remains high. In the compensated metals ($N_1 = N_2$), the Hall field and the electric field along the current are of the same order, and ρ grows without limit as B^2.

A dependence of the mobility of current carriers in semiconductors on their energy results in the appearance of a temperature gradient in the direction of the field, corresponding to a *Hall–Ettingshausen effect*. It is caused by a sorting of current carriers by energy in the magnetic field. The force corresponding to the *Hall field* counteracts the Lorentz force acting on the carriers moving with speeds close to an average speed. For charges having a mobility higher than average, the Hall electric force is less than the Lorentz force and the charges deflect toward one side of the specimen, while the current carriers with a smaller mobility deflect toward the opposite side. The appearance of a thermoelectric field in addition to the Hall field equalizes these two fluxes. In *thin films*, the magnitude of the thermoelectric field caused by the Ettingshausen effect can appreciably exceed the Hall field. Its direction depends on the trend, either the mobility grows or declines with the energy. The appearance of the two counteracting fluxes of "hot" and "cold" current carriers in the direction of the Hall field, i.e. the lack of compensation of the Hall and Lorentz forces for carriers with different mobilities, leads to an increase of the sample resistance in the magnetic field even for the case of an isotropic dispersion law for the current carriers.

In metals with open Fermi surfaces, the trajectories of the current carriers can be either open or closed, depending on the direction of $\boldsymbol{B}$ with respect to the crystallographic axes. A sharp anisotropy of the transverse resistance at $r_{\mathrm{c}} \ll l$ has been observed in single crystals of gold, zinc, cadmium, and other metals; this is connected with the appearance or disappearance of open cross-sections of the Fermi surface at changes in the orientation of the vector $\boldsymbol{B}$ relative to the crystal symmetry axes. A presence of an insignificant layer of open Fermi surface by the plane $\boldsymbol{pB} = \mathrm{const}$ with a relative thickness $\delta \gg (r_{\mathrm{c}}/l)^2$ causes an essential change of the magnetoresistance at $B \to \infty$ ($\rho \sim B^2$ if the field direction is not orthogonal to the opening direction).

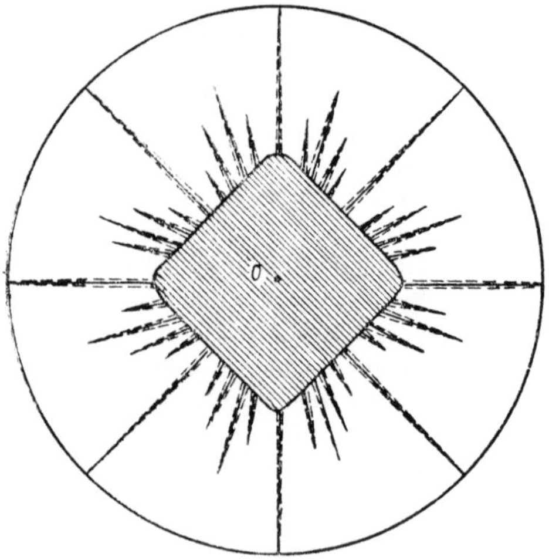

Stereographic projection in the magnetic field direction which presents an open cross-section of the Fermi surface of tin (continuous lines and cross-hatched region, O indicates fourth-order axis).

Data from galvanomagnetic effect studies acquired during the sixties provided the Fermi surface topology of almost all the metals. Most metals possesses open Fermi surfaces (see Fig.); only tungsten, molybdenum, indium, aluminum, alkali metals and *semimetals* of the bismuth group have closed Fermi surfaces. In polycrystalline samples the averaging carried out over randomly oriented *crystallites* of metals with open Fermi surfaces provides a dependence of the resistance on B that is close to being linear (*Kapitza law*, 1928); in metals with a closed Fermi surface the transverse resistance at $r_{\mathrm{c}} \ll l$ either reaches saturation or is proportional to B^2 if the numbers of electrons and holes are compensated.

In imperfect crystals containing expanded *defects*, the bending of the current lines can lead to a linear growth of not only the transverse but also the longitudinal magnetoresistance observable in some metals (aluminum, copper, alkali metals). In small monocrystalline samples the resistance anisotropy associated with the dimensions and the shape of conductor is easily differentiated from the sharp anisotropy at $r_c \ll l$ caused by the appearance or disappearance of a layer of an open section of a Fermi surface; hence the so-called *scale effects* are not obstacles for the reconstruction of Fermi surface topology with the help of galvanomagnetic measurements. *Quantization of electron energy* is usually displayed in strong magnetic fields in the form of small amplitude oscillations superimposed on a smooth "classical" dependence of the galvanomagnetic characteristic on B (*Shubnikov–de Haas effect*).

GAMMA, γ

Magnetic unit designating a hundred-thousandth fraction of an Oersted ($1\ \gamma = 10^{-5}$ Oe $= 7.958 \cdot 10^{-4}$ A/m); it is also a rarely used fractional unit of mass ($1\ \gamma = 10^{-9}$ kg).

GAMMA DEFECTOSCOPY

See *Defectoscopy*.

GAMMA DIFFRACTOMETRY

A method of investigating the structure and the properties of *monocrystals* based on *Bragg diffraction* of γ-radiation. Gamma diffractometry uses intense sources of highly monochromatic ($\Delta\lambda/\lambda \leqslant 10^{-6}$ at 300 K for ^{118}Au) γ-quanta that emerge from the decay of an element (Au, Sm, Tb, Ir, Re) activated by thermal neutrons. *Gamma diffractometers* (see the figure) differ significantly from X-ray apparatus in dimensions and design. A short-wave-length γ-ray ($\lambda = 0.003$ nm, $E = 412$ keV for ^{198}Au) from source 1 passes through collimating device 2 that forms a primary beam with an angular divergence of a few seconds of arc. A sample is positioned on a four-circle goniometer 3, and detector 4 records an angular distribution of the radiation scattered by the crystal.

Gamma diffractometry has some important advantages in comparison with routine *X-ray diffrac-*

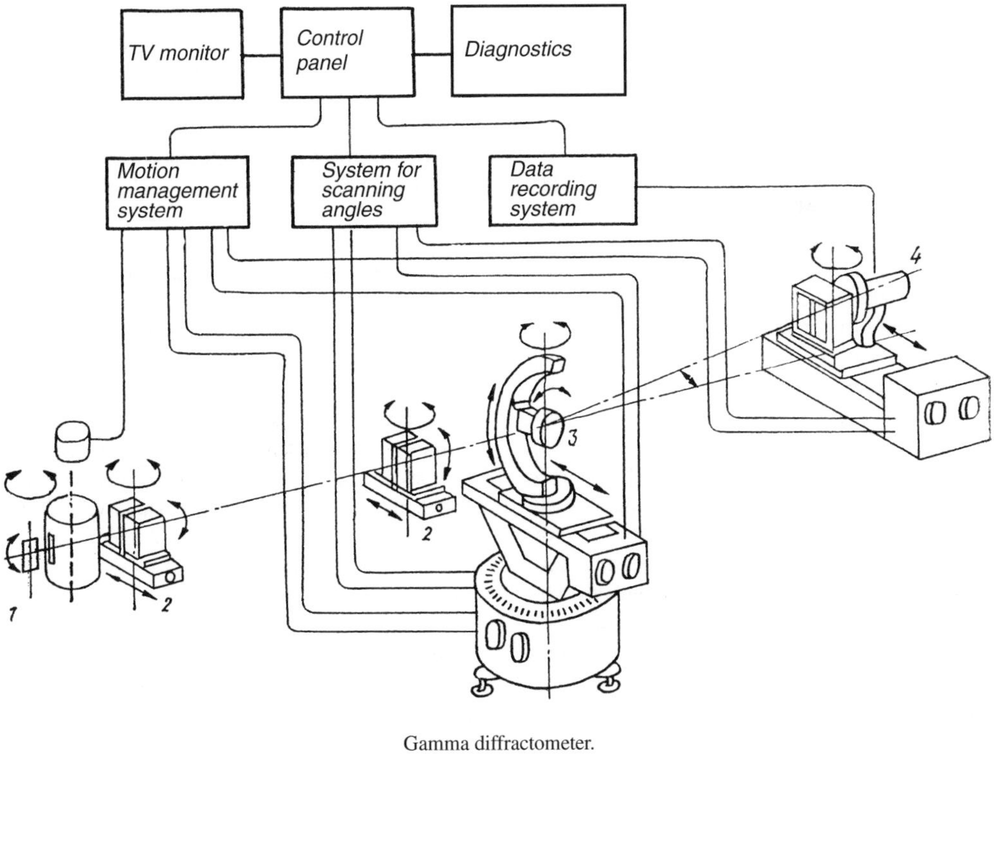

Gamma diffractometer.

tometry: a high energy and angular resolution; a significant penetration ability of the radiation allowing one to study large crystals without damaging them; a possibility to measure the integrated reflection coefficient on an absolute scale; and a high sensitivity to the presence of strain gradients and other distortions in a monocrystal lattice. A gamma diffractometer can be used to study *mosaic crystals* serving as neutron monochromators. This method has been applied to optimize the growth of crystals (see *Crystallization*); to study the distortions produced by microscopic defects in defect-free Si; to study the influence of isovalent doping and neutron irradiation on the properties of the lattice of Si; for investigations of *phase transitions*, and for determining the atomic structure of materials. A particular case of gamma-diffractometry is *Mössbauerography*.

GAMMA PHASE (γ-phase) (introduced by F. Osmond, 1885)

The third phase in polymorphous metals (see *Polymorphism*) formed under increasing temperature; for example, γ-Mn, γ-U, γ-Ce, and also *solid solutions* formed on this basis. The concept of a gamma phase was introduced to designate a polymorphous face-centered cubic modification of *iron*, however, as it has turned out more recently, a gamma phase in iron and its alloys represents a solid solution on the basis of a second allotropic modification (see *Allotropy*) referred to as *austenite*. The region of occurrence of a gamma phase may change depending on the doping elements (see *Doping*) used in practice for obtaining austenite or ferrite *steels*. A transformation $\alpha \Leftrightarrow \gamma$ in iron alloys is the basis of a *thermal treatment* stimulating a variation of their *strength* and *plasticity* over wide ranges. In alloys without polymorphous transformations, a gamma phase in a *phase diagram* designates a third equilibrium phase on a concentration axis. In a series of systems of *copper* and *silver*, a gamma phase represents a kind of electron compound with a complicated cubic lattice whose stability is determined by the *electron concentration* (number of valence electrons per atom) $e/a = 21{:}13$ (see *Hume–Rothery phases*).

GAMMA QUANTUM

See *Gamma ray*.

GAMMA RAY, γ-quantum

A high-energy *photon*, a quantum of extremely short wavelength electromagnetic radiation. There is no sharp boundary separating *gamma rays* (γ-rays) from lower energy X-rays; but it is generally accepted to occur in the energy range $E \sim 10$–100 keV, or at wavelengths $\lambda \sim 10^{-11}$–10^{-10} m. Gamma quanta emerge during nuclear reactions, decay of elementary particles, braking (Bremsstrahlung) of high-speed electrons and *positrons*.

A quantitative measure of the γ-ray energy transfer in matter is the absorption or scattering cross-section per atom μ_a expressed in the units cm^2/g. The figure shows how μ_a depends on the γ-ray beam energy for various target materials. The main interactions involving gamma quanta passing through matter are photoabsorption (*photoeffect*), Compton scattering (see *Compton effect*), and the formation of *electron–positron pairs* (pair production). A measure of the efficiency of these individual processes are the partial cross-sections per atom μ_a, where $\mu_a = \mu'_a(\text{photo}) + \mu'_a(\text{Compton}) + \mu'_a(\text{pairs})$. The attenuation of a narrow beam of gamma radiation in a material is described by the exponential decay expression $J(x) = J(0)\exp(-\mu x)$, where J is the beam intensity at point x; the coordinate $x = 0$ denotes the surface under irradiation; and the absorption coefficient μ is given by $\mu = \mu_a N$, where N is the concentration of atoms (g/cm^3) of the absorbing substance.

At a *photoeffect*, the absorption of gamma quanta by electrons takes place. In this case the energy of an excited electron is given by $E_e = E_\gamma - E_i$, where E_i is the energy of the electron in its initial ith state in an atomic shell, and E_γ is the gamma-quantum energy. The photoeffect provides the principle contribution to the magnitude of μ at comparatively low incident γ-ray energies; for example, in Al this is up to 100 keV, in Pb to 1 MeV. The dependence of $\mu'_a(\text{photo})$ on E_γ is not monotonic: with the increase of E_γ there are jumps in energy coinciding with the energies of the $\ldots M$, L, K electron shells of *atoms*. With a further increase in the energy E_γ the energy losses of γ-rays via the Compton effect begin to play a principle role, and at even higher energies (~30 MeV in Al, 10 MeV in Pb) the main losses

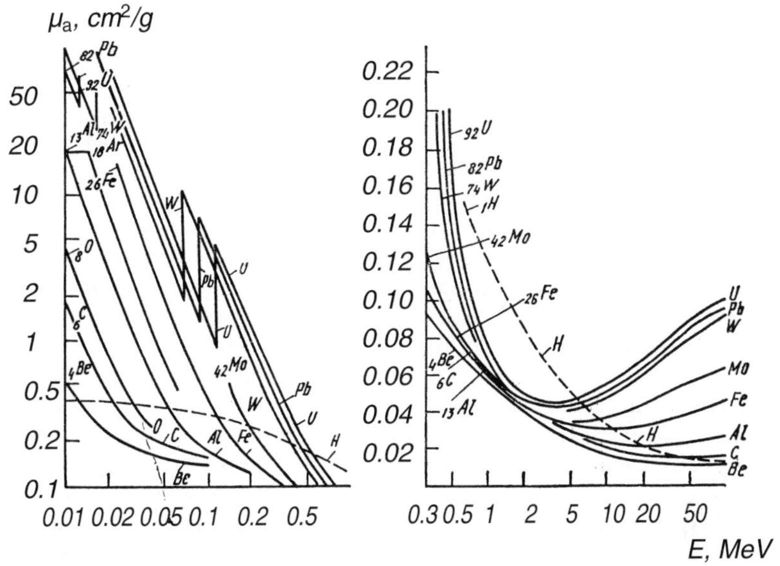

Dependence of the absorption cross-section μ_a on the incident gamma ray energy for a number of target materials. The right-hand figure is a semi-logarithmic plot covering a wide energy range, and the left-hand figure is a log–log plot for low energies.

are due to the formation of electron–positron pairs near atomic nuclei, and this is possible beginning from $E_\gamma \geqslant 2mc^2$ (where m is the electron mass, c is the speed of light). The scattering of gamma quanta on atomic nuclei is much weaker than on electrons. An interesting feature is *resonance scattering of gamma quanta* whose energy coincides with a difference between the ground state and any excited state of a nucleus. A process of resonance scattering consists of absorption and subsequent emission of resonance gamma quanta. Due to the recoil of the nucleus absorbing and emitting a photon, the latter is shifted in frequency toward lower values. In solids, the recoil energy is transferred to the photon. However, in a *Mössbauer effect*, the recoil is absorbed by the solid as a whole, and the frequency shift when absorbing or emitting a gamma quantum is insignificant.

The incoming gamma ray photons possess energies sufficient for knocking out atoms from their positions in the lattice, and collisions with them can produce radiation *defects*. In addition, the defect nucleation events can be stimulated by Auger processes (see *Auger effect*) occurring as a result of knocking out electrons from deep lying (inner) atomic electron shells. Nuclear reactions between high-energy gamma quanta and atoms in solids can lead to a *nuclear doping* of materials.

GANTMAKHER–KANER EFFECT

A *radio-frequency size effect* in a magnetic field $\boldsymbol{B}$ normal to a metal surface, caused by a penetrating component of the electromagnetic field, the so-called *Gantmakher–Kaner* mode. Its existence is connected with *ballistic transport* along the constant magnetic field of the perturbation in the electron system induced by the applied radiofrequency wave. Associated with it is an oscillating dependence of the electric field $\boldsymbol{E}$ in the metal on the coordinate $z \| \boldsymbol{B}$ due to the electrons with the extremal displacement along $\boldsymbol{B}$ during a cyclotron period: $u_0 = (1/eB)(\mathrm{d}S/\mathrm{d}p_z)_{\text{ext}}$ (where S is the area of the cross-section of the *Fermi surface* normal to $\boldsymbol{B}$, and p_z is the longitudinal (parallel to $\boldsymbol{B}$) component of the electron *quasi-momentum*). For $z \gg u_0$ this dependence has the form $E(z) \sim (\delta/z)^\alpha \exp(\mathrm{i}k_{\text{GK}}z)$ (where δ is the skin depth, $k_{\text{GK}} = 2\pi/u_0$, α is an exponent depending on the shape of the Fermi surface). The presence of the preexponential factor introduces a difference between the coordinate dependence of a Gantmakher–Kaner mode and the coordinate dependence of the field of weakly-attenuating electromagnetic waves (*helicon*, *doppleron*). This preexponential factor is associated with a branch point singularity in the non-local conductivity as a function of a complex wave number. Experimentally the Gantmakher–Kaner effect is observed as an oscillatory dependence of the surface *impedance* of a metal plate on its thickness, and on the magnitude of the external magnetic field. The oscillations induced by this effect exist over a wide range of sufficiently strong magnetic fields, with $\Omega \gg \nu$, where Ω is the *cyclotron frequency* and ν is the electron *relaxation* frequency). Their principal features are the following: (a) constant period in respect to the magnetic field; (b) small amplitude and its independence of the magnetic field in the range of the *anomalous skin-effect*; (c) the existence of oscillations for both senses of circular polarizations of an incident electromagnetic wave on the metal. The temperature dependence of the amplitude of the Gantmakher–Kaner oscillations allows one to determine the temperature dependence of the frequency ν.

GAPLESS SEMICONDUCTORS

An intermediate class of substances between *metals* and *semiconductors*.

They can be defined as metals with a pointwise *Fermi surface* or as semiconductors with a zero *band gap*. There are two kinds of zero-gap semiconductors differing by the mechanism causing the gap to disappear. The first kind includes those in which the gap vanishes due to the accidental coincidence of the energies of the edges of the *conduction band* and the *valence band*. In such zero-gap semiconductors, the zero-gap condition could be destroyed under the action of any disturbance including those that do not change the *crystal symmetry*. This kind of zero-gap semiconductors is exemplified by Bi_xSb_{1-x}, $Pb_{1-x}Sn_xTe$ and $Pb_{1-x}Sn_xSe$ alloys with certain ratios between components x. The gap in these alloys vanishes at four equivalent points L of the *Brillouin zone*, whereas the energy of electrons and holes depends linearly on the wave vector.

In zero-gap semiconductors of the second kind, the gap vanishes because of the crystal lattice symmetry. The so-called inverse zone scheme in these zero-gap semiconductors is possible if both the conduction band and the valence band belong to the same irreducible representation of the *crystal lattice* symmetry group. In all the zero-gap semiconductors of the second kind, the wave functions of the conduction band in the center of the Brillouin zone are of p-type symmetry. Due to a strong *spin–orbit interaction*, the sixfold degeneracy inherent in a p-symmetry state (taking into account the spin) is lifted, and the latter splits into fourfold and doubly degenerate states in accordance with the total angular momentum composition law. The fourfold degenerate state in an ordinary semiconductor gives rise to bands of light and heavy holes, while that in a zero-gap semiconductor of the second kind splits into the conduction band and the valence band. The zero-gap semiconductors of the second kind crystallize in a sphalerite-type lattice; they include gray tin (α-Sn) and the chalcogenides HgTe, HgSe and β-HgS. This zero-gap state disappears only under disturbances that lower the crystal lattice symmetry. Hydrostatic pressure does not change the symmetry; hence, the energy gap width for excitation of free current carriers remains equal to zero in the zero-gap semiconductors of the second kind. Both uniaxial compression and an applied magnetic field induce the appearance of an energy gap.

The electron mass is less than the hole mass in all known zero-gap semiconductors of the second kind. At finite temperatures, the electrons and the holes are created; the *Fermi level* is situated in the conduction band, the electrons are degenerate and the holes are nondegenerate. In this case, the Hall coefficient (see *Galvanomagnetic effects*) is a power function of the temperature. The electric properties of the zero-gap semiconductors are extremely sensitive to the presence of impurities. The donors are ionized at any temperature, so there are free electrons in the zero-gap semiconductors even at very low temperatures. There can also be resonance acceptor levels on the background of continuous spectrum of the conduction band. The small *level width* is related to the small ratio of the electron and the hole masses. In contrast to ordinary semiconductors, there is no low-temperature impurity compensation while the Fermi level is lower than the *acceptor* level. The compensation occurs with the temperature increase when the electrons pass to the acceptor levels. Due to the absence of an energy gap, there should be no ohmic region on the *current–voltage characteristic* at $T \to 0$ since in the electric field electrons and holes are created in equal numbers. An ohmic region is possible only at finite temperatures or in the presence of impurities that lead to the appearance of free current carriers in the absence of an electric field. In a magnetic field, an energy gap appears in zero-gap semiconductors, which leads to the activated creation of free carriers. Hence, the character of the Hall coefficient temperature dependence changes in a magnetic field. Zero-gap semiconductors, as well as the semiconductors with small gaps ($E_g \leqslant 0.1$ eV), possess quite high sensitivity to external factors (electromagnetic field, pressure, temperature, etc.). This provides wide opportunities for the application of the zero-gap semiconductors as resistors, photodiodes, radiation generators, and so on.

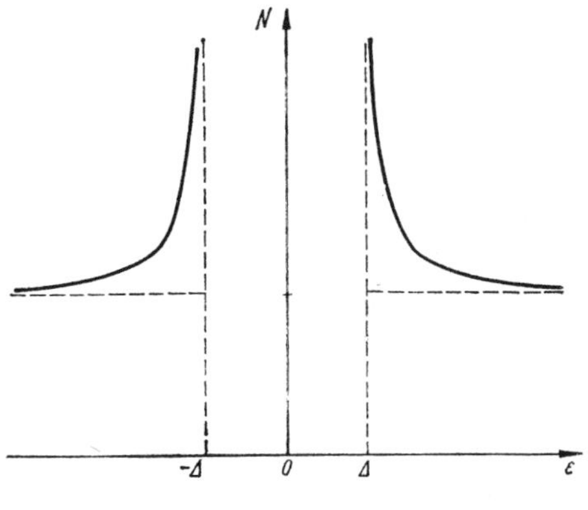

Fig. 1. The density of states of a BCS superconductor.

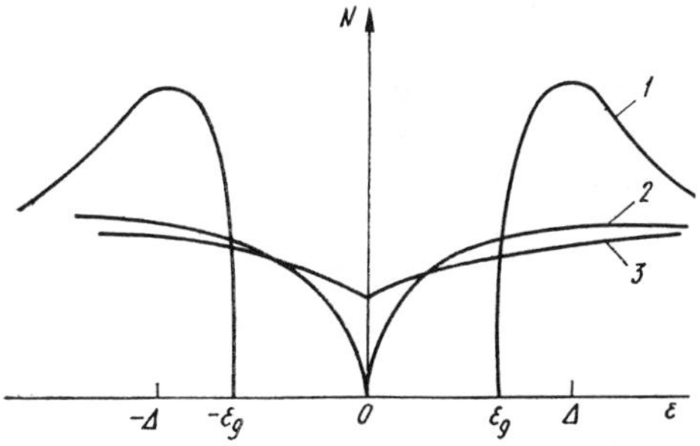

Fig. 2. The density of states of a superconductor with paramagnetic impurities. (1) $n < 0.9n_c$ (energy gap $\varepsilon_g < \Delta$); (2) $n = 0.9n_c$; (3) $0.9n_c < n < n_c$ (zero-gap state, $\varepsilon_g = 0$).

GAPLESS SUPERCONDUCTORS

Superconducors with a nonzero density of states $N(\varepsilon)$ for single particle excitations at and in the vicinity of the Fermi level, as distinct from ordinary superconductors where the density of states is zero within the the *energy gap* Δ in the spectrum of quasi-particles (see *Bardeen–Cooper–Schrieffer theory*), and has singularities at $\varepsilon = \pm\Delta$ (see Fig. 1). Gapless superconductors are exemplified by superconducting alloys with paramagnetic impurities having uncompensated magnetic moments. At high enough impurity concentrations, $N(\varepsilon)$ does not vanish at any energy (zero-gap state) although it does dip in the vicinity $|\varepsilon - \varepsilon_F| \sim k_B T_c$ of the Fermi energy (Fig. 2).

At the first glance, the absence of the gap contradicts the *Landau superfluidity criterion* for the critical condensate velocity $v_c = \Delta/p_F$ (where p_F is the Fermi momentum). However, this criterion is applicable only to homogeneous systems with translational *invariance*. For example, in a bimetallic plate consisting of a normal metal and a superconductor the current flowing along the superconducting layer parallel to the surface would be superconducting despite the fact that density of states of the system as a whole has no gap (see *Proximity effect*). Superconducting alloys with paramagnetic impurities are characterized by two *relaxation times*, one without a change of the electron spin direction (τ_1) and the other with a change of the spin projection (τ_2). *Electron spin-flip scattering* occurs with probability of $1/\tau_s = 1/(2\tau_1) - 1/(2\tau_2)$, and corresponds to the decay of a *Cooper pair* and a lowering of the critical temperature T_c. The value of T_c is related to τ_s by the Abrikosov–Gorkov relationship

$$\ln\frac{T_c}{T_{c0}} = \psi(1/2) - \psi(1/2 + \rho),$$

where T_{c0} is the critical temperature of the pure metal, $\rho = \hbar/(2\pi\tau_s k_B T_c)$, and $\psi(x)$ is the Euler digamma function. The value of ρ as well as the scattering frequency $\tau_s{}^{-1}$ are both proportional to the paramagnetic impurity concentration. According to this equation, T_c vanishes at a critical value of ρ corresponding to a critical paramagnetic impurity concentration n_c. In the concentration range $0.9n_c < n < n_c$, the energy gap vanishes (curve 3 in Fig. 2). At $n < 0.9n_c$, the "normal" gap is restored although $N(\varepsilon)$ has no singularity (i.e. divergence of $N(\varepsilon)$) at the gap edge. In the range $0.9n_c < n < n_c$, the dynamic characteristics of gapless semiconductors are similar to those of normal *metals*; in particular, their specific heat has a linear temperature dependence. However, the electrodynamics of zero-gap superconductors retain the same salient features that are typical of superconductors with a gap (BCS theory). Thus, the *Meissner effect* and the finite supercurrent are retained. The Josephson current in alloys with paramagnetic impurities also remains finite in the zero-gap range. Superconductivity with a zero gap is a feature which can occur in high-T_c superconductors when the wave function symmetry is d-wave.

GARNET FILMS

See *Epitaxial iron garnet films.*

GARNET, IRON

See *Iron garnets.*

GARNET, YTTRIUM–ALUMINUM

See *Yttrium–aluminum garnet.*

GASEOUS CRYSTALLINE STATE

A *state of matter* of a solid in which the centers of molecules form a regular *crystal lattice*, while the molecular orientational axes are randomly distributed in space with spherical or axial symmetry. In parallel with *liquid crystals*, a gaseous crystalline state is an example of a mesomorphous phase (see *Mesophases*), in which reorientations of molecules can take place without a noticeable change in the packing density. A special characteristic of a gaseous crystalline state is the presence of scales of order, long-range order (see *Long-range and short-range order*) in the arrangement of the centers of the molecules, and the absence of order in their relative orientations. A transition between a gaseous and an ordinary crystalline state occurs via a *first-order phase transition* referred to as *Frenkel orientational melting*. A gaseous crystalline state manifests itself by singularities of physical quantities (e.g., *dielectric constant*) or a *heat of phase transition* in the vicinity of the orientational melting point. This state is observed in some *molecular crystals*.

GAS EXPANDER

A piston or turbine device designed to allow an adiabatically expanding gas to cool itself and perform external work. The action of a gas expander is based on the *adiabatic process*. Gas expanders are used in *cryogenic technology*, and are employed for the liquefaction and separation of gases (nitrogen, *oxygen*, *helium*, etc.).

GAS, IDEAL

See *Ideal gas.*

GAS PHASE EPITAXY

An epitaxial deposition (see *Epitaxy*) on a monocrystalline substrate of layers of the same (*autoepitaxy*) or different (*heteroepitaxy*) material from a gaseous phase. Gas phase epitaxy represents one of the most widely used methods of forming semiconductor epitaxial *films* on conducting and insulating *substrates* during the manufacture of *semiconductor devices* and *integrated circuits*. It has the following advantages: simplicity and high productivity of the equipment; high deposition rates in comparison with vacuum vaporization; relatively low deposition temperatures; convenience for *doping* and control of stoichiometry (see *Stoichiometric composition*); a possibility of depositing layers with a complicated geometry; a possibility to clean the surface by chemical *etching* directly before the layer deposition. The main disadvantages of the method are the involvement of some toxic, explosive or *corrosive* gaseous products, the difficulty of producing multilayer structures with layers of controlled composition, and the uncertainty about precisely preserving the *superlattice* period.

In the design of apparatus for gas phase epitaxy, one can distinguish a closed process and an open process. In the closed process a substrate, a source of a stream of the transport material (e.g., iodine), and a source of the depositing material are positioned in different temperature zones of a break-sealed or dismountable ampoule filled, as a rule, with hydrogen or inert gas. The temperatures of the source and the substrate are chosen to facilitate the appropriate movement of products of disproportionation or decomposition reactions. For example, the reaction $GeI_4 \Leftrightarrow Ge + 2I_2$ near the source should take place in such a manner that there is a continuous transfer of *germanium* from the source to the substrate. In an open process, the substrate is washed by the transport gas flux (as a rule, this is hydrogen) with additives of halogens, arsenes, or phosphenes of the depositing substance which are decomposed or reduced by hydrogen on the substrate surface. Disproportionation reactions are also used in this process in the presence of a source. Doping of the deposited layers is performed by a doping source which adds gaseous compounds of the doping impurity in the transport gas flux (in the open process). An accompanying (usually undesirable) process is *autodoping*, i.e. a diffusion transfer of impurities from the substrate to the growing epitaxial film.

GAUGE INVARIANCE

The *invariance* of field equations (of observable quantities) with respect to symmetry transformations unrelated to the properties of space–time (so-called *internal symmetries*). An example of gauge invariance is electrodynamics where the electromagnetic field $F_{\mu\nu} = \partial_\mu A_\nu - \partial_\nu A_\mu$ remains unchanged under the transformation

$$A_\mu \to A'_\mu = A'_\mu - \partial_\mu \theta, \tag{1}$$

where θ is an arbitrary function of coordinates and time $x^\mu = \{x^0 = t, \boldsymbol{r}\}$, $\partial_\mu \equiv \partial/\partial x^\mu$,

$$A^\mu = \{A^0 = \Psi, A\},$$

Ψ and $\boldsymbol{A}$ are the scalar and vector potentials (hereinafter, the system of units with $c = \hbar = 1$ is used). The above transformation, referred to as a *gauge transformation*, acquires fundamental importance for a theory in the presence of charged particles. In the electrodynamics of zero-spin charged particles (see *Spin*) described by the complex-valued scalar field $\varphi(x)$ (the field φ can be considered also as the *order parameter* in a *superconductor*), the Lagrangian $L = L(\varphi, \partial_\mu \varphi)$ for the free field φ displays certain properties of symmetry with respect to the charge transformations of the unitary $U(1)$ group:

$$\varphi \to \varphi' = e^{ie\theta(x)}\varphi, \tag{2}$$

where e is the charge, and θ is a parameter. This transformation is referred to as a global one if $\theta = \text{const}$, and a local one if θ depends on x. If one replaces ordinary derivatives ∂_μ with covariant ones D_μ (where $D_\mu = \partial_\mu + ieA_\mu$, and A_μ is referred to as the *gauge field*) in a Lagrangian L that is invariant relative to the global transformation, i.e. $L(\varphi, \partial_\mu\varphi) = L(\varphi', \partial_\mu\varphi')$ for $\theta = \text{const}$ (in this case the charge conservation law is fulfilled), one obtains a new Lagrangian $L_G = L_G(\varphi, D_\mu\varphi)$ which is invariant with respect to the local transformation $\varphi \to \varphi'$ provided the gauge field A_μ transforms according to Eq. (1), i.e. $L_G(\varphi, (\partial_\mu + ieA_\mu)\varphi) = L_G(\varphi', (\partial_\mu + ieA'_\mu)\varphi')$. Transformations (1) and (2) are referred to as *local gauge transformations*, and the invariance of L_G relative to this transformation is called *local gauge invariance*. The requirement of local gauge invariance leads to the appearance of new fields A_μ and the construction of a new Lagrangian (to be supplemented with a Lagrangian of the gauge fields). From the physical point of view, the gauge field is the carrier of the interaction of charged fields. In the case of electrodynamics the gauge field is a field of *photons*. In the case at hand, the symmetry group $U(1)$ (see *Group theory*) of the Lagrangian is Abelian. The scheme of constructing a theory with a non-Abelian symmetry group remains unchanged. The gauge theories of fields with a given symmetry group (see *Continuous symmetry transformation groups*) are built from a Lagrangian that is invariant relative to the global transformations of the given group (this yields *conservation laws* as required from the physical point of view).

Abelian gauge theory is in full correspondence with the *Ginzburg–Landau theory of superconductivity* (GL theory) where the *order parameter* can be treated as a complex-valued scalar *Higgs field*, and the vector-potential as an Abelian gauge field in the Abelian Higgs model (see *Higgs mechanism*). The GL theory was the first gauge theory of a field with a *spontaneous symmetry breaking*. Abelian and non-Abelian gauge fields appear also in the theory of condensed matter, such as *liquid crystals* and *spin glasses*.

GAUSSIAN DISTRIBUTION, normal distribution (K.F. Gauss)

A *Gaussian* function

$$y(x) = \frac{1}{\sigma(2\pi)^{1/2}} \exp\left(-\frac{x^2}{2\sigma^2}\right)$$

normalized to unity that gives the frequency of occurrence (probability of appearance) of a parameter x (σ characterizes the deviation of x from its most probable value at $x = 0$). The Gaussian distribution is used in solid state physics when analyzing errors of measurements (as a normal law of error distribution), to describe the shape of a resonance line in spectroscopy (see *Gaussian shape*), to account for the distribution of fields of *defects in crystals*, and for characterizing *fluctuations*.

GAUSSIAN SHAPE, Gaussian

A spectral line contour conforming to the expression $I(x) = I_m \exp[-x^2 \ln 2/\delta^2]$ where I_m and δ are the amplitude maximum (for $x = 0$) and the line half-width at a half-height ($I(x) = I_m/2$ for $x = \delta$), respectively. When the line is recorded in the form of the first or higher derivatives of initial curves, the Gaussian is described by the corresponding derivative of this expression. In solid state spectroscopy, the Gaussian shape accompanies line broadening mechanisms of the statistical type (e.g., the hyperfine structure envelope for lines of *electron paramagnetic resonance*). As a rule, the Gaussian shape characterizes *nuclear magnetic resonance* lines in solids.

GEL

A colloidal system consisting of a solid dispersed phase (see *Disperse structure*) suspended in a liquid (*lyogel*) or in a gas (*aerogel*). A gel emerges as a result of the processes of structure formation and *sol* coagulation. A sol-to-gel transition is not referred to as a *phase transition*; under the action of external (mechanical or physicochemical) factors a reverse gel-to-sol transition (*peptization*) is possible, where a sol is more liquid-like and a gel is more solid-like. Gels possess mechanical properties specific for solids such as *plasticity* and elasticity; and in addition gels exhibit a property of thixotropy, i.e. the ability to restore its spacial structure after a mechanical *fracture*. Being a nonequilibrium system, a gel changes its properties with time; aging and the loss of the thixotropic property associated with phase layering take place. The removal of the disperse medium leads to the formation of *xerogels*, i.e. disperse solids possessing an extended surface area reaching 500 m^2/g and more. A particular feature of xerogels is their porosity, with the *pore* size varying over a wide range. The value of the specific surface and the xerogel porosity are to a great extent established at the stage of the sol-to-gel transition (drying temperature, nature of washing liquid, and so on). Xerogels are employed as adsorbents and heterogeneous catalysts.

GEMINALS (fr. Lat. *gemini*, twins)

Wave functions of two interacting electrons $\psi(r_1, r_2)$ that cannot be reduced to a simple or antisymmetrized product of one-electron functions $\varphi_1(r_1)$, $\varphi_2(r_2)$. Geminals are used in the theory of molecules and molecular crystals when greater precision is required for the description of a many-electron system compared to the *Hartree–Fock method*. In this case it is necessary to take into account the correlation motion of at least all the electron pairs that provide a significant contribution to a bonding energy of molecules and crystals. In the simplest cases of a two-electron system (H_2 molecule and He atom), a successful selection of geminals allows one to calculate the energy with an accuracy sufficient for comparison with spectroscopically determined values. For systems with the number of electrons 2N larger than 2 it is convenient to represent a wave function using an antisymmetrized product

$$\psi(r_1, \ldots r_{2N}) = \prod_{i=1}^{N} \psi(r_{2i-1}, \ldots, r_{2i}).$$

In this way one can achieve a greater accuracy for ψ than that obtained with a *Slater determinant*. For even greater accuracy, one can use a linear combination of such $\psi(r_1, \ldots, r_{2N})$ with different geminals. In this case, the method of geminals competes successfully with a *many-configurational approximation*. From a technical viewpoint, a calculation with geminals is more complicated than one with *orbitals*, but it can be simplified by imposing the requirement of strong orthogonality on different geminals. This, however, results in systematically overestimating the calculated energy.

GEMS, jewels, tinted stones

General term to designate precious and semiprecious minerals, and also artificial stones which, when cut and polished, are used for jewelry and ornamental purposes. *Diamond*, beryl and corundum (ruby and sapphire) are *precious stones*. For almost all these gems the color is due to the presence of impurities.

Beryl is a *mineral* $Be_3Al_2[Si_6O_{18}]$. Its composition is as follows: BeO 14.1%, Al_2O_3 19%, SiO_2 66.9%. Its *crystal system* (syngony) is hexagonal; symmetry type is dihexagonaldipyramidal. Its bases are hexadic rings $[Si_6O_{18}]$, located one over another and fixed by the atoms of Be and Al, which are located respectively in tetrahedral and octahedral coordinations of oxygen atoms. Unit cell parameters: $a = 0.921$ to 0.924 nm, $c = 0.919$ to 0.922 nm. Density is 2.6 to 2.9 g/cm^3. Transparent beryls of different coloring have specific names: *emerald* – brightly green (addition of Cr), *aquamarine* – bluish, *vorobievite* – pink (addition of Cs), *heliodor* – yellow (addition of FeO). Dielectric constant is 4.48 to 7.5. Magnetic susceptibility $(0.4\text{–}0.8)\cdot 10^{-6}$ cm^3/g. Refractive indices: $n_o = 1.568$ to 1.602, $n_e = 1.564$ to 1.595; birefringence $n_o - n_e = 0.004$ to 0.008. Indices of refraction and birefringence grow with increasing alkali concentration. Melting temperature is 1723 K. Mohs hardness is 7.5 to 8, cleavage is imperfect. In UV radiation it glows yellow. It does not dissolve in acids. Emerald and aquamarine are precious stones. Due to its beautiful coloring, transparency and brilliance, beryl is a popular gem-cutting material. Nonjewelry beryl is the basic ore of *beryllium* used in engineering (beryllium bronzes, superlight alloys, reflectors of neutrons in atomic reactors, etc.).

Corundum is the mineral α-Al_2O_3. Crystallographic system (syngony) is trigonal, type of symmetry is ditrigonal-scalenohedral. Crystals are columnar or barrel-like, they are represented by a combination of ditrigonal prisms, dipyramids, rhombohedron and pinacoid. *Rubies* (Cr doped) and *sapphires* (Co, Cr, Ti doped) are the gem varieties of corundum. Corundum occupies second place by its hardness (9 according to the Mohs scale) and abrasive ability after diamond. Density is 3.99 to 4.05 g/cm^3. No cleavage is observed. Refraction indices: $n_e = 1.765$ to 1.776, $n_o = 1.757$ to 1.768, $n_e - n_o = 0.008$ to 0.009. There is strong pleochroism for ruby: yellow-red, red; for sapphire: yellow, light-yellow. Ruby luminesces in UV rays with bright ruby-red color, and sapphire by blue, orange or purple color. Since 1958 ruby has been is used in *quantum amplifiers* (masers) and in *solid-state lasers*. Transparent varieties of corundum, besides their use as gems, are used in precision devices and watches as bearings and step bearings for the rotating parts, the nontransparent ones are used as *abrasive materials*.

[Following the modern classification scheme we proceed by listing less valuable stones after more valuable ones.]

Chrysoberyl is a mineral $BeAl_2O_4$ with additions of iron and chromium. It crystallizes in the rhombic system. Mohs hardness is 8.5. Microhardness is 16,592 to 18,896 MPa. Density is 3.631 to 3.835 g/cm^3. Refractive indices $n_g = 1.753$ to 1.758, $n_m = 1.747$ to 1.749, $n_p = 1.744$ to 1.747; birefringence is 0.009 to 0.011. Dispersion is 0.015. Color: yellow-green, yellow, more rarely brown, red, violet. The *alexandrite* variety of chrysoberyl, which contains up to 1% of chromium, has emerald-green coloring in the daylight and a violet coloring under artificial light. The selective transmission of bluish-green (460 to 530 nm) and red (620 nm up to the boundary of the visible spectrum) beams is the cause of the color change. In UV light rays sometimes a red glow may be detected. Density is 3.644 to 3.663 g/cm^3.

Opal is a mineral, solid hydrogel of *silica*, containing variable amounts of water (6 to 10%). It is formed by colloidal silica of globular structure with different size and disordered localization of globules; gem grade opal has regularly packed globules of the same size with diameters from 150 to 400 nm. Mohs hardness is 5.5 to 6. Density is 1.9 to 2.3 g/cm^3, it depends on the additions and water contents. Color: white, due to foreign additions it may be yellow, grayish brown, gray, green, blue, black. Sometimes it opalesces with iridescent play of colors (noble opal). It is isotropic, refractive index is 1.455 to 1.460. Its dielectric constant is 6.2 to 9.7. Diamagnetic susceptibility is $-0.29\cdot 10^{-6}$ cm^3/g. It is soluble in HF.

Garnets are a group of minerals with a common formula $A_3B_2Si_3O_{12}$, where A = Ca, Fe^{2+}, Mg, Mn^{2+}; B = A, Fe^{3+}, Cr, Ti, V, Zr. The terminal members of this group are *pyrope* $Mg_3Al_2Si_3O_{12}$, *almandine* $Fe_3Al_2Si_3O_{12}$, *spessartine* $Mn_3Al_2Si_3O_{12}$, *grossular* $Ca_3Al_2Si_3O_{12}$, *andradite* $Ca_3Fe_2Si_3O_{12}$ and *uvarovite* $Ca_3Cr_2Si_3O_{12}$. Crystallographic system (syngony) is cubic, hexoctahedral type of symmetry. Crystal lattice parameters: $a = 1.1459$ nm (pyrope), 1.1526 nm (almandine), 1.1621 nm (spessartine), 1.1851 nm (grossular), 1.2048 nm (andradite), 1.205 nm (uvarovite). The structure is formed by isolated groups $[SiO_4]$ located along the helical axis of the fourth order, which explains the rhombic dodecahedral and tetragonal-trioctahedral habit of crystals. Density and refractive indices are the following: pyrope, 3.71 g/cm^3, $n = 1.705$; almandine, 4.32, $n = 1.83$; spessartine, 4.18, $n = 1.80$; grossular, 3.53, $n = 1.735$; andradite, 3.83, $n = 1.895$; uvarovite, 3.78, $n = 1.86$. Upon melting together the density drops and the refraction index also becomes lower. Color: dark-red, red, pink, orange-yellow, black, green; there are no blue garnets. Magnetic susceptibility of garnets is proportional to the content of Fe, specific magnetic susceptibility is $(60–160)\cdot 10^{-6}$ cm^3/g. Dielectric constant 4.25 to 4.51 (almandine), 4.70 to 4.87 (pyrope), 5.22 (grossular), 6.40 (andradite). Mohs hardness is 6.5 to 7.5, it is the highest for almandine, pyrope and spessartine (7 to 7.5). Cleavage is imperfect. Only andradite is soluble in HCl, the others only after melting. The transparent varieties of garnets are jewels; garnets may be used as abrasive materials, artificial garnets are *ferrimagnets*.

Topaz is a mineral $Al_2[SiO_4](F,OH)_2$. Fe, Cr, Mg, Ti, V, Ge and other elements occur as additions. Crystallographic system (syngony) is rhombic, rhombic dipyramidal type of symmetry; crystal lattice structure is insular, it consists of the columns of Al-octahedra connected by Si-tetrahedra; crystal lattice parameters: $a = 0.465$ nm, $b = 0.880$ nm, $c = 0.840$ nm. Density is 3.4 to 3.6 g/cm^3. Color: colorless, yellow, wine-yellow, green, bluish-green, blue, violet-blue, pink, red-grayish brown. Chromophores (Cr, V, Fe, Ti) and crystal lattice defects are responsible for coloring. Refraction indices: $n_g = 1.638$ to 1.616; birefringence is 0.009 to 0.011. Colored topaz exhibits pleochroism: along n_g it is violet-pink, along n_m yellowish-pink, along n_p brownish-yellow. It luminesces in cathode radiation by red, blue light, in UV rays by red, yellow, pale green light. It is easily electrified by friction, compression, heating. Dielectric constant is 6 to 7; diamagnetic susceptibility is $0.42\cdot 10^{-6}$ cm^3/g. Mohs hardness is 8, cleavage is perfect in one direction.

Zircon is a mineral $ZrSiO_4$. Composition: Zr 49.5% (ZrO_2 67%, SiO_2 33%). The additions are possible: Hf, Th, U, Ca, Al, Fe. Crystallographic system (syngony) is tetragonal, structure is insular with radical-ionic lattice of anion groups SiO_4 and cations Zr^{4+} surrounded by 8 oxygen ions. Parameters of the unit cell $a = 0.6586$ to 0.6622 nm, $c = 0.594$ to 0.6025 nm. Density is 4.7 g/cm^3, density of metamict zircons is 3.9 to 4.2 g/cm^3, compressibility coefficient is $(47.6–48.3)\cdot 10^{-13}$ cm^2/dyn. Zircon may be transparent, translucent and nontransparent. Color: grayish-brown, yellow, barely red, blue, black, pink, lilac, green, powder may be white and yellow. The coloring varies depending on the effect of light, heating, X-rays. Dielectric constant is 3.6 to 5.2, that of the transparent ones 8 to 12. Magnetic susceptibility is $(-0.19$ to $0.90)\cdot 10^{-6}$ cm^3/g. Refraction index $n_g = 1.968$ to 2.015, $n_m = 1.923$ to 1.960; birefringence $n_g - n_m = 0.045$ to 0.058. Mohs hardness is 7 to 7.5, cleavage is imperfect. Transparent zircons have a brilliant glow, the translucent and nontransparent ones have glass glow. Zircon luminesces under the UV radiation with yellow, orange, green light, in cathodoluminescence with green light; it may exhibit thermoluminescence.

Olivine is a mineral. For the continuous isomorphous series the end pont formulae are $(Mg,Fe)_2[SiO_4]$, the end terms of the series are forsterite Mg_2SiO_4 and fayalite Fe_2SiO_4. The composition: Mg up to 34.3%, Fe up to 54.8% (MgO up to 57.3%, FeO up to 70.5%, SiO_2 up to 42.7%), additions of Mn, Ni, Co, Ca, Ti. Crystallographic system (syngony) is rhombic, rhombic pyramidal type of symmetry, parameters of the unit cell: $a = 0.4700$ to 0.4815 nm, $b = 1.0197$ to 1.0431 nm, $c = 0.5937$ to 0.6068 nm, cell parameters lengthen with increasing Fe content. The structure is insular, with $(Mg,Fe)O_6$ octahedra forming zigzag-like chains along the c axis, which are held together by SiO_4 tetrahedra. Density is 3.2 to 4.4 g/cm^3. Color: olive- or bottle-green, yellow, grayish brown, more rarely violet;

the powder is white. It is transparent to translucent. Refractive indices: $n_g = 1.670$ (forsterite), $n_m = 1.651$ to 1.869, $n_p = 1.635$ to 1.827. Dielectric constant is 3.3 to 9.1 for different terms of the isomorphous series and 6.8 to 7.1 for forsterite; it is a paramagnetic material, with the magnetic susceptibility up to $18 \cdot 10^{-6}$ cm^3/g (for the sample with contents of Mn 0.07% and of Fe 8.0%) depending linearly on the total contents of these elements. Mohs hardness is 6.5 to 7, cleavage is average. Decomposes in HNO_3, forsterite gleams in UV beams with crimson-colored light. Transparent, beautifully colored olivine (chrysolite) is a jewel.

Quartz (*silica*) and its varieties may be subdivided into the crystalline forms (rock crystal, smoky quartz (morion), pink quartz, amethyst etc.) and also the hidden-crystalline chalcedonies (sard (or carnelian), chrysoprase, agate, onyx etc.).

Amethyst is the violet variety of quartz. Color from red-violet to blue-lilac. The type of coloring is caused by the structural addition of Fe: Si^{4+} may be substituted by Fe^{3+} with charge compensation by Na ions. Radiation also influences the amethyst coloring. Coloring disappears upon heating up to 543 to 773 K, and is restored under the effect of X-rays and gamma radiation. Upon heating above 793 K the coloring irreversibly disappears. Intensely colored stones possess weak paramagnetism unlike diamagnetic colorless quartz.

Chrysoprase is the most valuable variety of chalcedonies. The color is apple green. The coloring depends on the addition of nickel, which contents varies from 0.3 to 3.3%. The color is, as a rule, nonuniform – from white to deep green. Green coloring may be explained by the color of gel.

[There follow stones known as jewelery, artificial or natural.]

Lazurite is a mineral, *aluminosilicate* of Na and Ca, $(Na, Ca)_8[AlSiO_4]_6(SO_4, S)_2$, additions of Se, Cl, CO_3. The crystals belong to the cubic system (syngony), they are usually met in the form of dense aggregates. Lattice parameter $a = 0.905$ nm. Density is 2.8 to 2.42 g/cm^3. Color: blue, violet-blue, light blue, greenish-blue, the powder is blue; the coloring depends on the presence of sulfur, which in the brightly blue lazurite is 0.5 to 0.7%. Refraction index $n = 1.50$. In UV rays it sometimes luminesces with weak orange color. Upon heating the intensity of blue coloring increases. Mohs hardness is 5.5 to 6, cleavage of the crystals is perfect. Under the effect of acids lazurite decomposes, releasing sulfurous gas.

Malachite is a mineral $Cu_2[CO_3](OH)_2$. The composition Cu 57.4% (CuO 71.9%, CO_2 19.9%, H_2O 8.2%), additions of Fe, Ca. Crystallographic system (syngony) is monoclinic, prismatic symmetry type, unit cell parameters: $a = 0.948$ nm, $b = 1.203$ nm, $c = 0.321$ nm. Density is 3.9 to 4.1 g/cm^3. Color: bright-green, blackish-green; streak is pale green; luster is from glass to diamond. Refraction indices: $n_g = 1.909$, $n_m = 1.875$, $n_p = 1.655$; birefringence is 0.254. Magnetic susceptibility $(10–25) \cdot 10^{-6}$ cm^3/g. Upon heating there appears endothermic effect of dehydration and dissociation within the range 533 to 773 K. Mohs hardness is 3.5 to 4, cleavage is average. It decomposes in HCl with release of CO_2. It is a copper ore, and is used as a manufacturing and decorative stone.

Nephrite is hidden-crystalline matted-fibrous aggregate of amphibole of the tremolite-actinolite series, related to the prismatic type of symmetry, with general formula $Ca_2(Mg, Fe)_5Si_8O_{22}(OH)_2$. Color: white (*tremolite-nephrite*), green of different tints – apple-green, grayish, light bluish, grass-green, dark green; intensive green coloring is caused by the presence of chromium (*actinolite-nephrite*). It is translucent in plates of thickness up to 2 cm. Density is 2.9 to 3 (tremolite-nephrite) and 3.1 to 3.3 g/cm^3 (actinolite-nephrite), Mohs hardness is 5.5 to 6.5. It is distinguished by high strength and toughness.

Rhodonite, or *manganese spar* is a mineral, silicate of Mn, $(Mn, Ca) \cdot SiO_3$, additions of Fe, Mg, Al, Zn. Crystallographic system (syngony) is triclinic, pinacoidal symmetry type. Unit cell parameters: $a = 0.779$ nm, $b = 1.247$ nm, $c = 0.675$ nm. The crystal lattice structure is represented by simple chains of silicon–oxygen tetrahedra with 5-fold period of identity. Density is 3.4 to 3.7 g/cm^3. Color: pink, crimson, scarlet; in the thin layer sometimes translucent. Refraction indices: $n_g = 1.730$ to 1.744; $n_m = 1.726$ to 1.735; $n_p = 1.721$ to 1.728; birefringence is 0.011 to 0.013. Distinct yellow- and pink-red pleochroism. Glows in cathode radiation with red color; in UV radiation with red, pink, orange light. Dielectric constant is 4.68 to 7.56, it increases with increasing Ca contents up to 11.2; magnetic susceptibility is $53.4 \cdot 10^{-6}$ (at 24% of Mn) and $87.7 \cdot 10^{-6}$ cm^3/g (at 36% of Mn and 1% of Fe). Mohs hardness is 5.5 to 6.5, cleavage is perfect.

Charoite is a mineral $(Ca, Na, K, Sr, Ba)_3$-$[Si_4O_{10}](OH, F) \cdot H_2O$ (it is around 56.5% of SiO_2, 1.01 to 1.85% of Al_2O_3, 20.5% of CaO, 8.2 to 10.5% of K_2O, and also there are oxides of Ba, Sr, Na (0.9 to 3.5%), fluorine and water). The isomorphous additions: Pb, Mn, La, Zr. Unit cell parameters: $a = 3.182$ nm, $b = 0.713$ nm, $c = 2.210$ nm, $\beta = 94.15°$. Crystallographic system (syngony) is monoclinic, cleavage is perfect in three directions. Angle between the cleavage planes is 124°. The color is lilac, of other tints up to violet; luster is glass-like. The spectrum of optical adsorption shows that the coloring of charoite is determined by the presence of Mn ions. Mohs hardness is 5 to 5.5. Density is 2.54 to 2.68 g/cm^3. Microscopically in the passing light charoite is colorless along n_g and pink along n_p; refraction indices: $n_g = 1.559$, $n_m = 1.553$, $n_p = 1.550$; birefringence is 0.009. It is not soluble in acids.

Cordierite or *dichroite*, *iolite*, or "*lynx sapphire*" is a mineral, Mg *alumosilicate*; $(Mg,Fe)_2$-$Al_3[AlSi_5O_{18}] \cdot nH_2O$, additions of Mn, Ca, Ti, Na. Crystals of rhombic crystal system (syngony), rhombic dipyramidal type of symmetry; crystal lattice consists of the rings of silicon–oxygen tetrahedra located in parallel to {0001}, thus forming channels along the c axis where alkali cations or H_2O molecules may be located. Crystal lattice parameters: $a = 1.708$ nm, $b = 0.710$ nm, $c = 0.933$ nm. Density 2.54 to 2.75 g/cm^3. Color: grayish-blue, blue up to bluish-violet, greenish-blue, colorless, yellow, grayish-brown. Refraction index: $n_g = 1.538$ to 1.568, $n_m = 1.532$ to 1.560, $n_p = 1.527$ to 1.558; birefringence is 0.009 to 0.010, sometimes 0.003, rarely 0.012. Magnetic susceptibility of cordierite is proportional to its content of Fe^{2+}: $6 \cdot 10^{-6}$ cm^3/g at 2% of Fe^{2+}, $\sim 10 \cdot 10^{-6}$ at 4% of Fe^{2+}, and $2.4 \cdot 10^{-6}$ at 8.5% Fe^{2+}. Mohs hardness is 7 to 7.5, cleavage is perfect.

[We follow by listing stones of organic origin: amber, pearl, coral, agate, which are related to jewels.]

Pearl is formed by the internal layer of shells of bivalve and gastropod mollusks (*mother of pearls*). It consists of very thin crystals of *aragonite* (orthorhombic form of calcium carbonate) 82 to 96% and of an organic substance (protein) conchiolin 10 to 14%. It has an iridescent luster. Density is 2.6 to 2.8 g/cm^3, Mohs hardness is 3.5 to 4.5. Small crystals of aragonite are oriented perpendicular to the surface of the internal layer of the shell. The luster of pearl is explained by refraction and reflection of beams in the prismatic layers of aragonite. Color of the sea pearls is white, creamy, pink, blue, red, black.

Corals are the external skeletons of marine invertebrates (polyps) forming tree-like branching constructions. In the composition of corals calcite with the addition of Mg carbonate prevails, sometimes there are oxides of Fe, Mn, etc. There are small quantities of organic material. The color of coral depends on the additions captured during the course of their growth, it may be white, pink, flesh-pink, red, dark red, sometimes light blue and black (containing almost 100% organic material). Density from 2.7 to 1.32 g/cm^3 (decreases with content of organic material); Mohs hardness is 3 to 3.5. Refractive index is 1.65 to 1.49 for pink coral, 1.56 for the black type. In UV rays coral luminesces with violet or red light.

Amber is a *mineral* of the organic substance class, fossil resin of coniferous trees, mainly of the Paleogene period. Chemical composition: C 76 to 81%, H 10.0 to 10.5%; O 7.5 to 13.0%, N and S tenths of a percent. Amber is amorphous, it is a volume polymer. Color: watery-transparent (barely), milky-white, reddish-brown (oxidized amber), usually it is yellow. Depending on its turbidity the following types of amber are differentiated: cloudy (semitransparent), hybrid (translucent in thin chips), bony and foamy (nontransparent). Amber exhibits a specific IR spectrum in the range of 700 to 1900 cm^{-1}, which facilitates reliable identification. Mohs hardness is 2.0 to 2.5. Density is 1.00 to 1.10 g/cm^3. It is easily machinable (with exception of the foamy type). It melts with decomposition at 573 to 613 K; without contact with air it softenes at 413 K, small pieces may then be pressed into large blocs of so-called pressed amber; turbid varieties transform to the transparent state. Amber is a good *insulator*. It is formed upon petrifaction of resin as a result of oxidative polycondensation of the saline acids and terpenes. It is used as a manufacturing stone, and also in the production of paint and varnish material. Pressed amber is used for the fabrication of electrical insulators.

GENERATION OF CURRENT CARRIERS

Excitation of electrons to the *conduction band* and holes to the *valence band* where they function as current carriers. This generation of current carriers is stimulated by the thermal motion of the crystal lattice atoms (*thermal generation*) and also by some external factors: illumination (*optical generation*), particle beam irradiation, action of strong fields, etc. A measure of this process is the rate of generation, i.e. the number of carriers produced per unit volume per unit time. Thermal carrier generation in an equilibrium semiconductor is balanced by their recombination (see *Recombination of current carriers*), hence, the heat generation rate G is equal to the recombination rate: $G = n_0/\tau$, where n_0 is the equilibrium concentration of carriers, and τ is the lifetime of the nonequilibrium current carriers.

In the case of optical generation the nonequilibrium carrier concentration can exceed the equilibrium value by many orders of magnitude. The interband *light absorption* that occurs when the quantum of light energy $\hbar\omega$ exceeds the *band gap* width ε_g results in the generation of electron–hole pairs ($G_e = G_h$), while impurity absorption leads to the generation of electrons ($G_e \neq 0$, $G_h = 0$) or holes ($G_e = 0$, $G_h \neq 0$). The rate of the optical generation of current carriers at $\hbar\omega > \varepsilon_g$ depends on the light intensity. At low intensities, this dependence is usually linear, described by the formula

$$G = \eta\alpha I_0 \exp(-\alpha x), \tag{1}$$

where I_0 is the beam density of light quanta (number of quanta arriving per unit area unit per unit time), α is the light absorption coefficient, x is the depth of the penetration, η is the *quantum yield* (frequency dependent efficiency coefficient for producing charge carriers). At $\hbar\omega \leqslant \varepsilon_g$, $\eta \leqslant 1$ because intraband light absorption does not produce new carriers. At $\hbar\omega > 2\varepsilon_g$ it is possible to have $\eta > 1$ since a single photon can excite more than one electron.

For $\hbar\omega \gg \varepsilon_g$ (X-ray or gamma radiation), the generation of current carriers consists of primary ionization events at which high energy ($\sim\hbar\omega$) carriers appear, and multiple *impact ionization* processes which produce electron–hole pairs. In this case $1 \ll \eta < \hbar\omega/\varepsilon_g$. The latter inequality is connected with the requirement of momentum conservation in the elementary events of electron–hole pair creation and interaction with lattice vibrations. For $\hbar\omega \gg \varepsilon_g$, an approximate formula $\eta \approx \hbar\omega/(3\varepsilon_g)$ is often used. The carrier generation proceeds in a similar way when, instead of photons, high-energy charged particles are involved (electrons, protons, α-particles, etc.). At high light intensity (laser radiation) when multi-quantum light absorption processes are operative, the dependence of the rate of carrier generation on the intensity becomes nonlinear (see *Multiphoton processes*, *Semiconductor lasers*).

Current carrier generation also occurs in the presence of a strong electric field as a result of impact ionization and tunneling transitions of electrons from impurities, and from the valence band to the conduction band (so-called *Zener breakdown*).

GENERATION–RECOMBINATION NOISE

A type of *noise in semiconductors* caused by a stochastic aspect of the *generation* and the *recombination* of current carriers. If a variation of the number of carriers δN decays in time following an exponential law, the spectral density of fluctuations (see *Fluctuations in solids*) of the number of carriers is

$$S_\omega = 2\overline{\delta N}^2 \frac{\tau}{1 + \omega^2\tau^2},$$

where $\overline{\delta N}^2$ is the self-*correlation function*, and τ is the carrier recombination time. In the low-frequency limit ($\omega\tau \leqslant 1$), S_ω grows with increasing τ. This fact plays an important role in the design of photosensitive detectors based on the *photoconductivity* effect. Since the low-frequency sensitivity of such devices increases with the growth of τ, it is necessary in this case to take into account the increase of the noise.

Generation–recombination noise appears in a variety of *semiconductor devices*, but the spectral characteristics of this noise do not always conform the expression given above because the number of current carriers in a limited volume can change as a result of *current carrier drift* and *diffusion*.

GENERATION–RECOMBINATION PARTICLE COAGULATION

The formation of diverse clusters in a system consisting of two types of complementary particles such as *vacancies* and *interstitial atoms* in a crystal, and the generation–recombination processes that take place for a random spatial distribution of these particles. The probability of *recombination* of a pair of particles is negligible if their separation r is sufficiently large. One can introduce the concept of a *recombination sphere* of an isolated particle, i.e. a sphere of such a radius a that the probability of recombination equals one for $r < a$, and equals zero for $r > a$. A homogeneous arrangement of two types of particles with equal concentrations n of each leads to $n \leqslant n_{\mathrm{m}} \leqslant a^{-3}$. In the generation–recombination process, as a result of the formation of clusters, the value n sometimes exceeds n_{m}. A generation–recombination particle coagulation can take place both for fixed particles of both kinds, and for the *diffusion* of isolated particles. If a single particle happens to be inside the recombination sphere during a particle birth, the probability of its recombination $W \approx 1$; if there are V particles, $W = 1/V$ for each particle. Thus, near-neighbor particles of the same kind are shielded during the recombination; the particle lifetime with respect to recombination increases with the growth of the number of its neighbors of the same kind. Under the condition of spatially-homogeneous generation and increased "longevity" of particles situated in dense *clusters*, there appears a trend toward coagulation. This process of coagulation has been studied in detail using computer-simulation experiments. These processes develop when crystals are exposed to defect-producing radiations (e.g., nuclear radiation), sometimes concurrently with the action of processes of *coalescence* of particles of the same kind.

GENERATION TYPE SEMICONDUCTOR DIODES

Diodes with *negative resistance* used for the generation of microwave electromagnetic oscillations. They are divided into three groups: drift (*impact ionization avalanche transit time diodes*, *tunnel diodes*, and *barrier injection transit time diodes*), *trapped plasma avalanche triggered transit diodes*, and intervalley-transferred-electron or *Gunn diodes*. In drift and trapped plasma diodes, under the action of constant or alternating fields, *injection of current carriers* into a drift space takes place; in the drift space the current carriers move most of the time in a braking microwave field; as a result the *current carrier* energy obtained from the constant electric field transforms partially to the electromagnetic energy of the microwave oscillations. Generation type semiconductor diodes produce oscillations with a frequency up to 400 GHz. See also *Tunnel diode*.

GENERATOR, THERMOMAGNETIC

See *Thermomagnetic generator*.

GEOMETRICAL ACOUSTICS

An approach for describing *sound* propagation by disregarding diffraction phenomena; an analogue of geometrical optics. In geometrical acoustics the acoustic energy propagation is described in terms of non-interacting sound beams which move along straight paths in a homogeneous medium (see *Conical refraction of acoustic waves*, *Reflection of acoustic waves*, *Acoustic wave refraction*, *Bulk acoustic waves*).

GEOMETRICAL RESONANCE, Pippard oscillations

The oscillating dependence of sound attenuation and *sound velocity* on the inverse magnetic field $\boldsymbol{B}$ for the conditions $kR \gg 1$ and $\boldsymbol{k} \perp \boldsymbol{B}$ (where $\boldsymbol{k}$ is the sound wave propagation vector, and R is the cyclotron radius), and in the presence of closed cyclotron orbits during the absorption of ultrasound in *metals*. A geometrical resonance predicted by H. Bömmel (1955) and A.B. Pippard (1957) is caused by a periodically arising (with variation of $1/B$) multiplicity of the value of the extremal displacement of the electron along the propagation vector $\boldsymbol{k}$ between the positions of a strong interaction of the electrons with the sound ($\boldsymbol{k} \cdot \boldsymbol{V} = 0$, $\boldsymbol{V}$ is the electron speed) over the distance of a sound wavelength. An oscillation period of a geometrical resonance is $\Delta(1/B) = 2\pi e/(kP_{extr})$; which depends on the extremal magnitude P_{extr} between the points of noncollision absorption at a *Fermi surface* measured in a direction normal to the vectors $\boldsymbol{k}$ and $\boldsymbol{B}$. The measurement of $\Delta(1/B)$ at various orientations of the vector $\boldsymbol{B}$ relative to crystallographic directions allows one to determine the Fermi surface topology.

GERMANIUM, Ge

A chemical element of Group IV of the periodic system with atomic number 32 and atomic mass 72.59. Its contents in the Earth's crust is $7{\cdot}10^{-4}\%$ in the mixture of 5 stable isotopes: ^{70}Ge (20.55%), ^{72}Ge (27.35%), ^{73}Ge (7.78%), ^{74}Ge (36.5%), ^{76}Ge (7.86%). Outer electron shell configuration is $4s^2 4p^2$. Successive ionization energies are 7.899, 15.934, 34.2, 45.1 eV. Atomic radius is 0.139 nm; radius of ions Ge^{2+} is 0.065 nm, of Ge^{4+} is 0.044 nm. Oxidation state is generally +4, occasionally +2.

In a free form, germanium is a solid with a metallic shine. It exists in one amorphous and several crystalline modifications. The crystal modification that is stable under normal conditions has a cubic structure of the *diamond* type with the lattice parameter $a = 0.56575$ nm, space group $Fm\bar{3}m$ (O_h^5). The density of solid germanium is 5.327 g/cm^3 (at 298 K), of liquid germanium is 5.557 g/cm^3 (at 1273 K); $T_{melting} = 1211$ K, $T_{boiling} = 2973$ K. Melting heat is 443 kJ/kg; heat of evaporation 4700 kJ/kg; coefficient of thermal conductivity is 60.7 W·m^{-1}·K^{-1}; coefficient of linear expansion is $4.5{\cdot}10^{-6}$ K^{-1} (in the range 73 to 273 K). Mohs hardness is 6.25.

Germanium is a typical *semiconductor* with the band gap 0.78 eV (at 273 K); temperature coefficient of the band $3.9{\cdot}10^{-4}$ eV/K; electrical resistivity of extremely pure germanium 0.60 Ω·m; mobility of electrons 3800 cm^2· W^{-1}·s^{-1}, of holes 1900 cm^2·W^{-1}·s^{-1} (at 298 K). Conduction band has four equivalent energy minima (valleys) located near the boundary of the *Brillouin zone*, the surfaces of constant energy as a function of quasi-wave vector are ellipsoids of rotation with their symmetry axes along the $\langle 111 \rangle$ directions. *Effective mass* of electron: longitudinal one is $1.64m$, transverse $0.082m$ (m is mass of free electron). Valence band energy maximum is in the center of the Brillouin zone at $\boldsymbol{k} = 0$, the band is threefold degenerate, effective mass of the heavy holes is $0.28m$, of the light holes is $0.044m$, due to *spin–orbit interaction* the third band is split off to 0.28 eV, effective mass of holes in the split-off band is $0.077m$.

The impurities of Groups III and V in germanium are hydrogen-like *acceptors* and *donors*, respectively. Shallow donor states are introduced also by lithium. The deep acceptors are zinc, beryllium, cadmium, mercury, copper, silver, gold. The deep donors are oxygen, sulfur, selenium, tellurium. Germanium is an *amphoteric impurity* in semiconductors of the A^3B^5 type. Heat treatment and the effect of nuclear radiations bring about the so-called *electron–hole* (n–p) *conversion* of germanium, or changing the type of conductivity, caused by the binding of atoms of the Group V into the electrically inactive complex with nonequilibrium vacancies. By irradiation with electrons with an energy of about 1 MeV at $T < 65$ K metastable pairs may be created: vacancy + interstitial atom (*unstable Frenkel pair*). Upon irradiation by the slow neutrons the most important reaction of transmutation occurs: ^{74}Ge(n, γ)–^{75}Ge which disintegrates with a half-life 82 min to form ^{74}As. This process is the basis of *neutron doping* of germanium. Monocrystals are obtained by the Czochralski method or by zone melting (see *Monocrystal growth*) to produce germanium with the highest purity of all the elements with only 10^9 cm^{-3} of electrically active impurity atoms, so germanium can be used for manufacturing *nuclear radiation detectors* and also of *photodiodes*, IR radiation receivers (transparency range 1.5 to 20 μm), and avalanche *diodes*.

GERMANIUM–SILICON ALLOYS

A continuous series of substitutional *diamond* structure *solid solutions* with unlimited mutual solubility of the components. Due to *liquation* and weak *interdiffusion*, the production of uniform *monocrystals* in the range of concentration ratio $0.08 < \mathrm{Si/Ge} < 0.95$ is technologically complicated. Monocrystals of germanium–silicon alloys are characterized by a high density of structural defects (*dislocations*) and non-uniformity of composition. The lattice parameter and the density of the *alloys* depend linearly on the composition over its entire range. The *band structure* of these alloys has been calculated theoretically, and also reconstructed from experimental measurements of optical adsorption and reflection spectra, as well as from thermal conductivity, electrical, and other properties. The dependence of the *band gap* width on the alloy composition is described by two linear regions with a point of inflection, corresponding to $Si/Ge = 0.15$. Variations in the alloy composition bring about the restructuring of the energy spectrum of the *conduction band* and the *valence band*. The functional dependences of the *kinetic coefficients* point to the non-uniformity of the alloys, which is determined by the characteristics of their structure: by the difference of *ionic radii* and by fluctuations of component distributions.

Germanium–silicon alloys are used for manufacturing thermoresistors, thermooscillators and memory chips; they are favorable for the design of receivers of IR radiation with the long-wave limit of sensitivity 14 μm, of tensometers, and of thermal neutron monochrometers.

GERMAN SILVER

Copper alloy with *nickel* (13.5 to 16.5%) and *zinc* (18 to 30%), sometimes with additions of *lead* (1.6 to 2.0%), and other metals. The density of German silver is 8.7 g/cm^3, electrical resistivity is $26\cdot10^{-8}$ Ω·m; thermal linear expansion coefficient at temperatures 293 to 373 K is equal to $16\cdot10^{-6}$ K^{-1}; coefficient of *thermal conductivity* at 293 K is 89.5 $W\cdot m^{-1}\cdot K^{-1}$; normal *elastic modulus* is 115 GPa; *shear modulus* is ≈45 GPa; limit of linearity is 120 to 140 MPa in the annealed state and 480 to 590 MPa in the strained state; elasticity limit upon *flexure* $\sigma_{0.005} = 560$ MPa. German silver is used in instrument making for components of timepieces and fine mechanical devices, the contact plates of automobile mechanical relays, and the spring parts of electromechanical television devices and computers.

GETTERING in solids

A process of extraction of impurities or structural *defects* from a *solid* through contact with an active material (getter), a passive part (drain), or a vacuum. The term gettering is taken from a vacuum technology process which involves cleansing a gaseous medium from undesirable components during either *adsorption*, *absorption*, or bonding chemical reactions at the surface or in the bulk of disperse layers of chemically active substances. The gettering process in solids includes three main stages: (a) activation of particles involved in gettering; (b) *mass transport*; (c) reaction or interaction with a getter accompanied by fixation or removing the particles from the sample. Depending on the type of activation, gettering is divided into the thermal-, chemical-, and radiation-activated types. The highest temperature process is the thermally activated one.

A radiation-activating process is possible even at cryogenic temperatures. With respect to mass transport gettering is divided into diffusion (including *radiation-induced diffusion*), drift (via electric field or mechanical force) and dynamic (with mass transport, relay-race method, acoustic wave including impact wave, and dilation collective transport, see *Dilatation*). With respect to position we can differentiate: (a) bulk gettering (inclusions of gettering phase embedded in the bulk, e.g., in *silicon* these are SiO_2, SiC, Si_3N_4, *rare-earth elements*, *metals*, phosphorosilicates, and so on); (b) surface gettering (gettering layers Si_3C_4, SiO_2, $SiO_{2-x}P_x$, $SiO_{2-x}Pb_x$ and others deposited on the surface, mass transport is in the bulk); (c) planar gettering (mass transport at the surface or over the interface separating the body and the contacting phase) (see the figure). In view of facilitated migration at the surface, this is

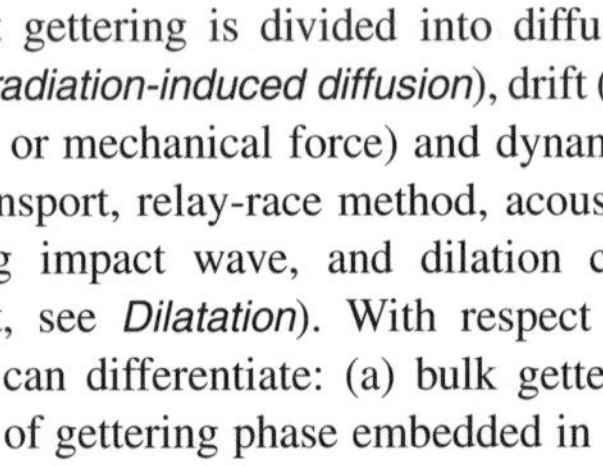

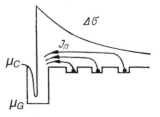

Model of a getter, where $\Delta\sigma$ is the gradient of the mechanical pressure, J_n is the current of the getter particles in the region of flow, μ_G is the chemical potential of the particles in the getter region, and μ_C is the chemical potential of the particles in the cleansing region.

characterized by a strong acceleration of the gettering process (by several orders of magnitude) and, therefore, will function at low temperature. In terms of the bonding mechanism we can distinguish phase gettering (absorption due to increased solubility of an impurity in a getter) and structural gettering (absorption due to defect structure and the capability of absorbing *point defects* which bond impurities). Gettering is widely used in *microelectronics*; making it possible to increase the lifetimes of minority *current carriers* and prevent the formation of defects under various kinds of treatment.

g-FACTOR

See *Landé g-factor, Anisotropy of g-factor, Nuclear g-factor shift.*

GIANT MAGNETOSTRICTION

Magnetostriction characterized by an abnormally high value of λ, viz., $\lambda = \Delta l/l > 10^{-3}$, found in many magnetically ordered *rare-earth elements*, their alloys and compounds. Table 1 shows the magnitudes of giant magnetostriction in the field of *magnetic saturation* for such *magnetic substances*; $\Delta l/l$ for *nickel* and *iron* are listed for the sake of comparison. Other giant magnitudes are the magnetostriction-related effects: *mechanostriction*, ΔE*-effect* (effect of magnetic field on modulus of elasticity), *spontaneous magnetostriction*, etc. Giant magnetostriction has been found in a number of ferromagnetic compounds of uranium (U_3As_4 and U_3P_4) and other *actinides*. Some paramagnetic rare-earth compounds ($Tb_3Ga_5O_{12}$, $TbLiF_4$ and $Dy_2Ti_2O_7$) exhibit a giant magnetostriction much greater than that of nickel in the ferromagnetic state, which exemplifies the phenomenon in paramagnets ($\Delta l/l \sim 10^{-3}$–10^{-4}). Microscopic mechanisms underlying giant magnetostriction are the interaction of an anisotropic f-orbital electron cloud with the *crystal field*, and the sharp dependence of the exchange energy (in some crystallographic directions) on interatomic distances. Another occasion for giant magnetostriction is the presence of a phase transition in the lattice of a magnetic material, such as a martensitic transition.

Table 1. Values of giant magnetostriction in various substances

Magnetic substance	λ $(\Delta l/l)_s \cdot 10^{-6}$	Temperature of measurement, K	Crystallographic axis
Fe	−10	300	Polycrystal
Ni	−37	300	Polycrystal
Ni	−60	78	$B \parallel [111]$
Tb	1,230	78	Polycrystal
Tb	5,460	4.2	$B \parallel a$ axis
Tb	22,000	4.2	$B \parallel c$ axis
$TbFe_2$	4,700	300	$B \parallel [111]$
$Tb_3Fe_5O_{12}$	2,460	4.2	$B \parallel [111]$
Dy	1,400	78	Polycrystal

GIANT QUANTUM OSCILLATIONS

A quantum effect involving sharp changes of absorption coefficient Γ of slow waves in pure *metals* at low temperatures in a *magnetic field*. Giant quantum oscillations have been observed for *sound absorption* and electromagnetic waves (*helicons*). The effect is due to the quantization of the energy of *conduction electrons* in a magnetic field and a periodic (on the scale of inverse magnetic field B^{-1}) switching on of the (noncollision) interaction with the wave of those electrons at the *Fermi surface* $\varepsilon(\boldsymbol{p}) = \varepsilon_F$, whose velocity v_z along the magnetic field direction $\boldsymbol{B}$ coincides with the wave *phase velocity* v_{ph}. As a result of the quantization (for a quadratic isotropic *dispersion law*), the electron energy assumes the form (see Fig.):

$$\varepsilon_n(p_z) = \hbar\Omega\left(n + \frac{1}{2}\right) + \frac{p_z^2}{2m}.$$

Here $n = 0, 1, 2, \ldots$ is the Landau level number, $\Omega = eB/m$ is the *cyclotron frequency*, e and m

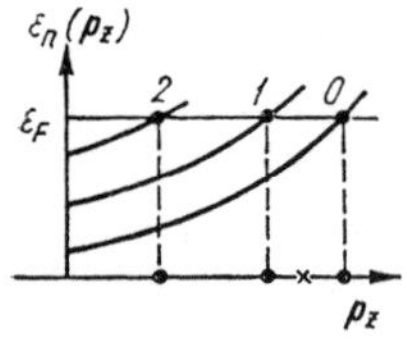

Curves show the dependence of the electron energy ε_n on the projection p_z of the quasi-momentum $\boldsymbol{p}$ along the direction of the magnetic field $\boldsymbol{B}$ for different n. Points on the $p_z = mv_z$ abscissa axis corresponding to quantized velocities v_{zn} are shown for v_{z0}, v_{z1} and v_{z2}.

are the electron charge and mass, respectively, and p_z is the projection of the *quasi-momentum* of electrons on the direction of the magnetic field. At $T = 0$ the absorption as a function of B^{-1} takes the form of a periodic system of high and low rectangular "pulses" separated by wide gaps with zero absorption. At a finite temperature, the maxima decrease and become rounded. A particular feature of these oscillations is the large magnitude of the ratio $\Gamma_{\max}/\Gamma_{\min} \gg 1$ under the conditions of quasi-classical quantization of electron states when the electron motion has a one-dimensional character. Giant quantum oscillations is one of the few *macroscopic quantum effects* in solids (quantized flux in superconductors is another).

GIBBS DISTRIBUTION FUNCTION, also Gibbs' ensemble (J.W. Gibbs, 1901)

Functions describing the probability that a system consisting of many particles be in a certain state. Gibbs ensembles are characterized either by coordinates or momenta of particles in the case of a classical description, or by quantum numbers in the case of a quantum description. In solid state physics, a *Gibbs' canonical ensemble* and a *Gibbs' grand canonical ensemble* characterize, respectively, the behavior of a system in equilibrium which exchanges energy or both energy and particles with a heat bath. The probability that a system with a constant number of particles (in the general case, of various kinds) is in a certain lth state with energy E_l is

$$P_l = \frac{1}{Z}\exp\left(-\frac{E_l}{k_B T}\right)$$

(Gibbs' canonical distribution), where Z is the *partition function* determined from the normalization condition $\sum_l P_l = 1$:

$$Z = \sum_l W_l \exp\left(-\frac{E_l}{k_B T}\right) = \exp\left(-\frac{F}{k_B T}\right),$$

where W_l is the number of different states with the same energy E_l, and F is the Helmholtz free energy. The value Z can be represented in operator form by the trace

$$Z = \mathrm{Tr}\left[\exp\left(-\frac{\widehat{H}}{k_B T}\right)\right],$$

where $\widehat{H}$ is the Hamiltonian operator, and the operation $\mathrm{Tr}[\widehat{A}]$ means a trace or summing over all the diagonal elements of the operator $\widehat{A}$ calculated using a complete set of wave functions of the system.

The equilibrium properties of a system with a variable number of particles (i.e. the system can exchange particles with a *heat sink*) are described by the Gibbs' grand canonical ensemble. The probability that the system under consideration contains N_m particles of kind m and is in state l with energy $E_{l\ldots N_m\ldots}$ is

$$P_{l\ldots N_m\ldots} = \exp\left\{\frac{1}{k_B T}\left[\Omega - \sum_m \mu_m N_m - E_{l\ldots N_m\ldots}\right]\right\},$$

where Ω, called the grand potential, or grand *thermodynamic potential*, is determined from the normalization condition that the total probability $P_{l\ldots N_m\ldots}$ be unity:

$$\Omega = -k_B T \ln \sum_{N_m}\left\{\exp\left(\frac{\sum_m \mu_m N_m}{k_B T}\right) \times \sum_l \exp\left(-\frac{E_{l\ldots N_m\ldots}}{k_B T}\right)\right\},$$

where μ_m are the values of the chemical potential of particles of the mth kind. The average statistical numbers N_m of particles of type m for a given system are determined with the help of the derivative $N_m = -\partial\Omega/\partial\mu_m$. If each of the different subsystems are in equilibrium (individual parts of the body, various phases, etc.), then each of the chemical potentials should be the same for the whole system.

GIBBS FREE ENERGY

See *Free energy*.

GIBBS' PHASE RULE (J.W. Gibbs, 1876)

A law of thermodynamics stating that for a system with a number f ($f \geqslant 0$) of thermodynamic degrees of freedom (e.g., number of particles, temperature and pressure, or temperature and volume, etc.) that can be changed while keeping constant the number ϕ of phases existing in *phase equilibrium*, the number f is determined by the expression $f = c - \phi + k$, where c is the number of system components and k is the number of parameters determining the state of a single phase. In almost all cases of interest there are only two parameters so $k = 2$ (e.g., temperature and pressure, or temperature and volume, etc.), and Gibbs' phase rule is written $f = c - \phi + 2$. At equilibrium, the temperature, the pressure, and the chemical potential (see *Thermodynamic potentials*) are equal for different phases. Gibbs' phase rule is valid for large phase dimensions when one can neglect surface phenomena, and in the absence of semipermeable walls. It follows from the phase rule that for single-component substances ($c = 1$, for example, water) the maximum number of phases in equilibrium for the condition $f = 0$ is $\phi_{\max} = 3$ (triple point: vapor, water, *ice*); and the number for the condition $f = 1$ is $\phi = 2$ (e.g., melting line with water and ice in equilibrium). For a two-component system with no degrees of freedom we have $c = 2$, and $f = 0$ which gives $\phi_{\max} = 4$ phases in equilibrium. For example, the two-component system of water and a soluble salt like NaCl with $f = 0$ has four phases in equilibrium (e.g., precipated salt, floating ice, liquid salt solution, and water vapor).

GIGA...

A decimal multiple prefix signifying a 10^9 times increase of the initial unit of a physical quantity. For example, 1 GHz (Gigahertz) $= 10^9$ Hz.

GINZBURG–LANDAU–ABRIKOSOV–GORKOV THEORY (GLAG theory)

See *Ginzburg–Landau theory of superconductivity*.

GINZBURG–LANDAU PARAMETER

A dimensionless parameter κ equal to the ratio of the *penetration depth of magnetic field* $\lambda(T)$ to the *coherence length* $\xi(T)$ in a superconductor that characterizes the relation between the *thermodynamic critical magnetic field* $B_c(T)$ and the *upper critical field* $B_{c2}(T)$:

$$\kappa = \frac{\lambda(T)}{\xi(T)} = \frac{B_{c2}(T)}{B_c(T)\sqrt{2}}.$$

In a *type II superconductor* $\kappa > 1/\sqrt{2}$ and $B_{c2}(T) > B_c(T)$, so that there exists a region of *mixed state* due to the negative surface energy α between normal and superconducting phases. In a *type I superconductor* $\kappa < 1/\sqrt{2}$ and α is a positive quantity. For such a type of superconductor there exists an intermediate state in an external magnetic field.

In "dirty" superconductors (alloys) with a high concentration of impurities and defects in the crystal lattice and, correspondingly, with a small electron *mean free path* $l \ll \xi$, the effective coherence length $\widetilde{\xi}_{\text{eff}} \approx l$ is small in comparison with λ so that $\kappa \gg 1$, the material is type II, and $B_{c2}(T) \gg B_c(T)$. High values of κ and $B_{c2}(T)$ are achieved also in a series of superconducting compounds based on *transition metals* due to unusually large values of λ caused, in particular, by the small width of the d-band and the large effective mass of the charge carriers. Maximal values $\kappa \sim 10^3$ are achieved by the cuprate high-temperature superconductors (see *High-temperature superconductivity*).

GINZBURG–LANDAU THEORY OF SUPERCONDUCTIVITY (V.L. Ginzburg, L.D. Landau, 1950)

A phenomenological theory of a superconducting state obtained by a *second-order phase transition* from a normal (non-superconducting) state in the presence of a magnetic field. The Ginzburg–Landau (GL) theory of superconductivity is based on the *Landau theory of second-order phase transitions*, the hypothesis of *macroscopic quantum coherence* of electrons in superconductors, and Maxwell's equations. In the framework of this theory the free energy density of a superconductor in a magnetic field $\boldsymbol{B}$ in the vicinity of critical temperature T_c takes the form:

$$F_s = F_n + \alpha(T)|\psi|^2 + \frac{\beta}{2}|\psi|^4 + \frac{1}{2m^*}\left|(-i\hbar\nabla - e^*\boldsymbol{A})\psi\right|^2 + \frac{B^2}{2\mu_0}, \quad (1)$$

where F_n is the Helmholtz free energy in the normal state, ψ is the complex coherent wave function (*order parameter*) of the superconducting charge carriers called *Cooper pairs* with effective mass $m^* = 2m$ and charge $e^* = 2e$ (m and e are free electron values), and $\boldsymbol{A}$ is the vector-potential of the magnetic field ($\boldsymbol{B} = \nabla \times \boldsymbol{A}$). The coefficients α and β obtained by Gorkov (1959) from the *Bardeen–Cooper–Schrieffer theory* are given by

$$\alpha(T) = \frac{\hbar^2}{m^*\xi_0^2}\left(\frac{T}{T_c} - 1\right); \qquad \beta = \frac{0.098}{N(0)}\left(\frac{\hbar^2}{m^*\xi_0^2 k_B T_c}\right)^2, \quad (2)$$

where $\xi_0 = \hbar v_F/(\pi\Delta_0)$ is the *coherence length* or the size of a Cooper pair, Δ_0 is the *energy gap* in the spectrum at temperature $T = 0$, v_F is the *Fermi velocity*, $N(0)$ is the *density of states* at the *Fermi surface*, and the superelectron density $n_S = 2|\psi|^2$. The condition of the free energy minimum (1) provides the first Ginzburg–Landau equation:

$$\frac{1}{2m^*}(-i\hbar\nabla - e^*\boldsymbol{A})^2\psi + \alpha(T)\psi + \beta|\psi|^2\psi = 0, \quad (3)$$

which is supplemented by the quantum-mechanical expression for the supercurrent density, called the second Ginzburg–Landau equation:

$$J_S = -\frac{ie^*\hbar}{2m^*}(\psi^*\nabla\psi - \psi\nabla\psi^*) - \frac{(e^*)^2}{m^*c}|\psi|^2\boldsymbol{A} \quad (4)$$

and by Maxwell's equations.

The *Ginzburg–Landau–Abrikosov–Gorkov (GLAG) theory* provides solutions of the nonlinear equations (3) and (4) in an applied magnetic field which describe the macroscopic electromagnetic and thermodynamic properties of *superconductors*. In particular, this theory provides values of the *upper critical field* $B_{c2}(T)$ and the *thermodynamic critical magnetic field* $B_c(T)$, the *penetration depth of magnetic field* $\lambda(T)$, the coherence length $\xi(T) = \hbar[2m^*\alpha(T_c - T)]^{-1/2}$, and the *Ginzburg–Landau parameter* $\kappa = \lambda(T)/\xi(T)$. This theory also accounts for the structure of Abrikosov vortices in the mixed (vortex) state, *critical currents* and *critical magnetic fields* in thin superconductor films and wires, the *Little–Parks effect* and the *quantization of flux* in multiply-connected superconductors, the nucleation of incipient superconductivity at the sample surface, and so on.

The behavior of a superconductor in an external magnetic field $\boldsymbol{B}$ depends essentially on the value of *Ginzburg–Landau parameter.* In pure elemental superconductors $\kappa < 1$ (with the exception of niobium), but in dirty superconductors (*alloys*) with an electron free path length $l \ll \xi(0)$ κ can be much more than unity. In *type I superconductors* with $\kappa < 1/\sqrt{2}$, the *surface energy* α_s between the boundary of the superconducting and normal phase is positive. They undergo a *first-order phase transition* to the normal state when the applied magnetic field reaches the value of the thermodynamic critical magnetic field B_c. In *type II superconductors* with $\kappa > 1/\sqrt{2}$, the value $\alpha_s < 0$, and in the range of applied fields $B_{c1} < B_{app} < B_{c2}$ the material is in the *mixed state.* Upon reaching the lower critical field B_{c1} ($B_{c1} \approx B_c \ln\kappa/(\sqrt{2}\kappa)$ at $\kappa \gg 1$), it becomes energetically favorable to form filamentary inclusions of the normal phase called *vortices* (see *Abrikosov vortices*) each of which carries a *flux quantum* $\Phi_0 = h/(2e)$. The state of minimal energy is a regular triangular lattice of vortices. Their number per unit area of cross-section $\nu = B/\Phi_0$ increases as the field increases, and when its value approaches $B_{c2} = \sqrt{2}\kappa B_c$, the distance between adjacent vortices becomes close to $\xi(T)$. In fields exceeding the upper critical field B_{c2} bulk superconductivity disappears by way of a second-order phase transition, but it persists in a surface layer of thickness ξ up to the *third critical field* $B_{c3}(T) = 1.69B_{c2}(T)$.

In the presence of an electric current density $\boldsymbol{J}$, the Lorentz force $\boldsymbol{J} \times \boldsymbol{B}$ acts on the vortex filaments. In an ideal homogeneous material, the vortex motion under action of this force is hindered by viscous friction only, and a finite resistance appears in the superconductor. In practice many vortices are held or pinned at defects and structural heterogeneities (see *Vortex pinning*), and this pinning has a dominant influence on the critical current of a type II superconductor. See also *Meissner effect* and *Gauge invariance*.

GINZBURG NUMBER, Ginzburg–Levanyuk parameter (V.L. Ginzburg, A.P. Levanyuk, 1960)

A dimensionless parameter designated by Gi (or τ_G) which characterizes the width $\tau = |T/T_c - 1|$ of a *critical region* near the *critical temperature* T_c of a *second-order phase transition* outside which the *Landau theory of second-order phase transitions* is applicable. The Ginzburg number Gi is connected with the coefficients of the Ginzburg–Landau theory expansion, in the absence of a magnetic field (i.e. the vector potential $\boldsymbol{A} = 0$), of the Helmholtz free energy $F(h, T)$ in terms of the order parameter ψ

$$F = F_n + \alpha(T)|\psi|^2 + \frac{1}{2}\beta|\psi|^4 + g\left(\frac{\partial\psi}{\partial h}\right)^2,$$

where $h = x/\xi_0$, and ξ_0 is the *coherence length.* For the usual approximation, $\alpha = \alpha_0(T - T_c)$, the Ginzburg number has the value

$$Gi = \tau_G = \frac{T_c\alpha_0\beta^2}{4g^3},$$

and Landau theory is applicable at $\tau > \tau_G$.

In the region $\tau < \tau_G$ the thermodynamics (and dynamics) of the system is significantly affected by *critical fluctuations* of the order parameter. The number Gi is determined by the nature of the interaction forces between particles (atoms, molecules, electrons, their spins, and so on) causing the phase transition. For a particular material Gi is the order of magnitude of $(r_0/r_c)^6$, where r_0 is the particle interaction radius, and r_c is the *correlation length* of *fluctuations* at a temperature far from the phase transition point, when $|T - T_c| \sim T_c$, i.e. when $\tau \sim 1$. A criterion for the applicability of the thermodynamic theory of phase transitions (*self-consistent field* theory) up to $T \approx T_c$ is the inequality $Gi \ll 1$. For example, for *superconductors* r_c coincides with the coherence length of the *Bardeen–Cooper–Schrieffer theory* $\xi_0 = \hbar v_F/(\pi\Delta_0) \geqslant 10^{-6}$ cm, where Δ_0 is the value of the energy gap, and r_0 is close in order of magnitude to the mean interatomic distance ($d \sim 10^{-8}$ cm), so that $Gi \ll 10^{-12}$, thus providing a measure of the precision of the *Ginzburg–Landau theory of superconductivity.* In contrast, in liquid helium (see *Superfluidity*) in the vicinity of the phase transition point (*lambda point*) to the superfluid state, $T_\lambda = 2.17$ K, $Gi \sim 1$, and the self-consistent field theory is not applicable.

GLAG THEORY

See *Ginzburg–Landau theory of superconductivity.*

GLASS

See *Vitreous state of matter, Dipole glass, Fiber glass, High molecular weight glasses, Ideal glass, Metallic glasses, Optical glass, Photochromic glass, Spin glass, Structural glasses.*

GLASS CERAMICS, pyroceramics

Polycrystalline bodies of fine-grained structure, which are obtained by controlled directional *crystallization* of glasses of special composition. The composition of the glass (see *Vitreous state of matter*) for the production of glass ceramics is chosen in accordance with the *phase diagram* in such a way that the end product of the crystallization exhibits certain predetermined properties. In order to carry out uniform bulk crystallization and obtain a microcrystallite structure (dimensions range from fractions of a μm to 1 μm) the glass is often doped with a crystallization catalyst (metals Au, Ag, Pt, Cu, oxides TiO_2, Cr_2O_3, P_2O_5, SnO_2, fluorides CaF_2, Na_2SiF_6, etc.). The transformation of glasses into glass ceramics is performed through a two-stage *heat treatment.* The temperature of the first stage fits the optimal conditions for the generation of crystal nuclei and their growth up to critical sizes, whereas the temperature of the second stage is adequate for a maximal crystal growth rate (curing time determined by requirements imposed on the phase composition of glass ceramics). Owing to their fine-grained structure, glass ceramics feature high strength characteristics, high thermal shock resistance, and low thermal expansion, whereas their other specific properties depend mainly on the quantity (60–90%) and type of the resulting crystalline phases, as well as on the composition of the residual glass phase. *Glass-ceramic photomaterials*, obtained by S.D. Stookey (1957), form a special class of glass-ceramics produced from light-sensitive glasses containing Au, Ag, Cu by exposing the latter to the action of UV or ionizing radiation followed by heat treatment. This causes the segregation of colloid particles of the metal, which catalyze the further crystallization. The selective irradiation of such glasses can provide photographic images. Of high practical value are *ash glass-ceramic materials*, *petrol glass-ceramic materials*, and *slag glass-ceramic materials*; these types being obtained from industry wastes and rocks.

GLASS' CONSTANT (A.M. Glass, 1974)

A parameter characterizing the magnitude of a *bulk photovoltaic effect*. The photocurrent density J is proportional to the incident light intensity I, and the proportionality coefficient for heterogeneous absorption has the form $k\alpha$:

$$J = k\alpha I, \qquad (1)$$

where α is the *light absorption* coefficient independent of I; and the parameter k is called the Glass' constant. Eq. (1) is for the isotropic case. Customary units are $[J] = \text{A/cm}^2$, $[I] = \text{W/cm}^2$, $[\alpha] = \text{cm}^{-1}$, and $[k] = \text{A·cm·W}^{-1}$. The value of the Glass' constant depends upon the mechanism of the bulk photovoltaic effect, and often involves both intrinsic and extrinsic defects. Therefore, this value can depend strongly on the individual properties of the sample, and change under various kinds of influences, e.g., under the action of high-energy irradiation, annealing, and so on. In addition, Glass' constant depends on the frequency and polarization of the exciting light. In the crystal $LiNbO_3$, the value k in the vicinity of the maximum of the absorption spectrum is close to 10^{-9} A·cm·W^{-1} in order of magnitude, while in crystals of sillenite $Bi_{12}SiO_{20}$ it reaches $\sim 10^{-8}$ A·cm·W^{-1}.

GLASS-REINFORCED PLASTICS

Composite materials consisting of fiberglass fillers (fibers, multiple fibers, tissues, etc.) and polymeric binders (polyester, epoxy, organic-silicon, or other structure). Under operating conditions the fillers carry the main load, while the binders cement together separate filaments, and contribute to a uniform load distribution. Glass-reinforced plastics are divided into the three following groups: *glass textolites* (laminated sheet materials) obtained by hot pressing previously impregnated glass fabric; *oriented glass-reinforced plastics* obtained by winding, drawing, or packing *fiber glasses* and multiple glass fibers, combined with simultaneous impregnation of these materials with resin; *fiberglasses* obtained, as a rule, by continuous impregnation followed by refinement of filaments and multiple fibers; *glass-reinforced plastics based on linens and mats* available from pressing under low pressures; and *soft rolled glass-reinforced plastics* produced by continuous impregnation and heat treatment of various roll glass-fiber materials under standard pressure. Glass-reinforced plastics find wide application in the building trade (roofing and facing materials, damp-proofing, sound insulators), in the chemical industry (materials for pipes, containers, etc.), in electrical engineering (insulators, collectors, etc).

GLIDE-REFLECTION PLANE, mirror glide plane

Crystallographic symmetry operation combining *translation* along a plane for a distance equal to the half of the lattice translation period t, and reflection across this plane (see *Symmetry plane*). The *glide vector* $t/2$ is sometimes called the *glide component*. The designation of the glide symmetry plane depends on the glide direction: if the glide vector is parallel to the a, b, or c axis, then the plane is designated respectively a, b, or c; if the glide vector is directed along the body diagonal or the base diagonal and equals half of the corresponding diagonal, then the glide-reflection plane is designated n and referred to as the *diagonal glide plane* or *clino-plane*; if the glide vector is aligned with the base diagonal of the *face-centered lattice*, or with the body diagonal of the *body-centered lattice* and equals $1/4$ of the corresponding diagonal, then the plane is designated d and named *diamond plane*.

GLOBAR

A radiator consisting of a rod made of compressed refractory material (*silicon carbide*) which glows during the passage of an electric current, and can be used as a source of infrared radiation.

A typical working temperature of a globar is close to 1500 K; with the maximum of the radiation intensity close to the wavelength of about 2 μm.

GOLD (Lat. *Aurum*), Au

A chemical element of Group I of the periodic table with atomic number 79 and atomic mass 196.967. Natural gold consists of one stable isotope ^{197}Au. Outer shell electronic configuration is $4f^{14}5d^{10}6s^1$. Successive ionization energies are 9.266, 20.5, 30.5 eV. Electron affinity energy is 2.31 eV. Atomic radius of Au is 0.144 nm, radius of Au^+ ion 0.137 nm, of Au^{3+} ion 0.085 nm. Oxidation state is +3, more rarely, +1 and +2. Electronegativity is ≈1.5.

Gold is a soft yellow *metal*. It has a face-centered cubic crystal lattice with parameter $a = 0.40783$ nm and density 19.32 g/cm^3 (at 293 K). $T_{\text{melting}} = 1340$ K, $T_{\text{boiling}} = 2950$ K. Latent heat of melting 12.5 kJ/mole, heat of vaporization 349 J/mole, specific heat 25.4 J·mole^{-1}·K^{-1}. Coefficient of linear expansion is 14.2·10^{-6} K^{-1} (at 273 to 373 K), thermal conductivity is 311 W·m^{-1}·K^{-1}. Electrical resistivity is 2.25 μΩ·cm, thermal coefficient of resistance is 3.96·10^{-3} K^{-1} (at 273 to 373 K). Gold is diamagnetic. Elastic modulus is 77 GN/m^2; the tensile strength for annealed Au is 100–140 MPa. Brinell hardness is 177 MPa (for gold annealed at 673 K), Mohs hardness is 2.5. Adiabatic elastic moduli of gold monocrystals are $c_{11} = 192.9$, $c_{12} = 163.8$, $c_{44} = 41.5$ (GPa) at room temperature.

Gold is used in engineering in the form of *alloys* with other metals, which improves its hardness and strength. It is used in production of chemically stable equipment, in electrical engineering, etc. Coating of surfaces with thin layers of gold provides high *corrosion* resistance and high reflectivity.

GOLDSTONE EXCITATIONS

See *Goldstone theorem* and *Spontaneous symmetry breaking*.

GOLDSTONE MODES

See *Goldstone theorem* and *Spontaneous symmetry breaking*.

GOLDSTONE THEOREM (J. Goldstone, 1961)

A statement concerning the excitations of a system with *spontaneous symmetry breaking*. According to the Goldstone theorem, a system with such a broken symmetry (see *Continuous symmetry transformation groups*) possesses excitations whose frequency ω goes to zero as the wave vector $\boldsymbol{k}$ is reduced to zero. These zero-gap (massless) excitations are called *Goldstone modes* (*Goldstone bosons*). In the case of a *Heisenberg magnet*, the symmetry group of the Hamiltonian is the three-dimensional rotation group; the symmetry group of the ground state (there is a continuum of states with the same energy) is the group of two-dimensional rotations about a certain arbitrary direction of the spontaneous magnetization $\boldsymbol{M}$. The symmetries relative to rotations about the other two axes (orthogonal to each other and the axis $\boldsymbol{M}/|\boldsymbol{M}|$) are broken. The appearance of *Goldstone excitations*, i.e. small oscillations about the indicated axes, can be associated with the tendency of the system to restore its symmetry by undergoing transitions from one degenerate state to another, which requires an infinitesimal energy (hence, $\omega \to 0$ when $\boldsymbol{k} \to 0$). The *dispersion law* of the Goldstone modes and their number are determined by the equations of motion in each particular case. The Goldstone excitations in a *ferromagnet* are *spin waves* or *magnons* with a square dispersion law ($\omega \propto k^2$). In many non-relativistic systems the Goldstone theorem fails to hold. In particular, the presence of long-range forces (e.g., the Coulomb forces in superconductors) results in the appearance of a gap in the spectrum of Goldstone modes (e.g., a finite plasma frequency ω_p when $\boldsymbol{k} \to 0$; see *Plasma oscillations*). A gap also arises in the presence of interactions affecting the degeneracy of the ground state. For example, both a magnetic field and magnetic crystalline anisotropy (see *Magnetic anisotropy*) lead to the appearance of a gap in the magnon spectrum. A violation of the Goldstone theorem is also possible due to the *Higgs mechanism*.

GONIOMETER, X-RAY

See *X-ray goniometer*.

GONIOMETRY

A method of measuring the angular coordinates of crystal faces and processing the results to determine the *crystal symmetry*, the *crystallographic indices* of faces, and other geometric factors. As a rule, spherical coordinates φ and θ of normal directions to faces (φ is the longitude, θ is the latitude angle) are used. The most prevalent two-circle or theodolite *goniometers* have two mutually perpendicular axes connected with limbs for reading φ and θ, an optical collimator for illuminating the crystal, and an optical tube. Prior to a measurement, the crystal is arranged at the point of intersection of the goniometer axes, and adjusted in position where one of its specific directions coincides with the axis of angle φ. The measuring procedure includes a subsequent rotation of the crystal about the goniometer axes, a detection of the beams reflected from its faces (reflections) with the help of the optical tube, and recording the corresponding angles φ and θ. The apparatus has an accuracy of about $\pm 1'$.

GORKOV–NAMBU METHOD (L.P. Gorkov, 1958; Y. Nambu, 1960)

A matrix formalism used for constructing a microscopic theory of superconductivity. In this formalism, the two-component operators of the electron field are introduced:

$$\Psi(x) = \begin{pmatrix} \Psi_\uparrow(x) \\ \Psi_\downarrow(x) \end{pmatrix}, \quad \Psi^+(x) = \big(\Psi_\uparrow(x), \Psi_\downarrow(x)\big), \quad x = (r, \tau), \tag{1}$$

where the arrows indicate the spin direction. Their product averaged over the *Gibbs distribution* determines a matrix *Green's function*:

$$\widehat{g}(x, x') = -\langle T_\tau \Psi(x)\Psi^+(x')\rangle, \tag{2}$$

where T_τ is the symbol for ordering over the thermodynamic time parameter τ, where $0 \leqslant \tau \leqslant 1/T$. The diagonal matrix elements of Green's function (2) are the ordinary electron Green's functions, and the non-diagonal ones are the *Gorkov anomalous functions* describing the *Bose condensate* of Cooper pairs. The Gorkov–Nambu method is quite convenient because the perturbation theory series for the matrix Green's function of a superconductor has the same form as for the ordinary Green's function of a normal metal. This method is used for superconductors with strong coupling. The intrinsic-energy part is equal to a sum of contributions of the *electron–phonon interaction* and the Coulomb interaction, and the *Eliashberg equations* are used for its determination. The method takes into account not only singlet coupling (antiparallel spins) of the *Cooper pair* electrons, but also *triplet pairing* involving parallel spins.

GORSKY–BRAGG–WILLIAMS APPROXIMATION

See *Alloy ordering*.

GRABNER EFFECT (L. Grabner, 1960)

Appearance of an electric field in the direction of a magnetic field, whose vector $\boldsymbol{B}$ lies in the sample plane and is perpendicular to the electric current vector. This effect, which was first studied in n-Ge, is entirely due to the anisotropy of the crystal properties. The dependence of the magnitude of the effect on the magnetic field has a complicated form: in a weak magnetic field it is proportional to B^n ($n \geqslant 1$); in the range of intermediate magnetic fields it passes through a maximum, and finally in classically strong magnetic fields it decays as B^{-1}.

GRAIN BOUNDARY

Region of a *polycrystal* with a specific atomic structure that connects two *crystallites* situated next to each other. A plane grain boundary between crystallites (1 and 2 in Fig. 1) in a bi- or polycrystal is, in general, characterized by nine parameters. Three of them determine the orientation in space of the coordinate system corresponding to one of the grains, two specify the normal to the boundary plane (grain boundary orientation); three prescribe the mutual rigid displacement of crystals 1 and 2 at the distance $r < a_0$ (where a_0 is the interatomic distance), and the final parameter designates the *chirality* of crystals without a *center of symmetry*. A turn through an angle of rotation $|\theta|$ transforms crystal 1 into crystal 2 (θ is the disorientation of the grains). If the rotation axis is in the grain boundary plane ($\boldsymbol{\theta} \perp \boldsymbol{n}$) than this is a *boundary of inclination*; if $\boldsymbol{\theta} \parallel \boldsymbol{n}$ this is a *twist boundary*; and for intermediate values of $\boldsymbol{\theta}$ and $\boldsymbol{n}$ it is a *mixed boundary*. If the bisector of angle θ is in the grain boundary plane than the boundary is symmetric, otherwise it is asymmetric. The position of a grain boundary in a polycrystal can be revealed by the method of *etching* at a plane edge; *grain boundary defects* are determined from transmission *electron microscopy* studies; the arrangement of atomic planes adjoining along a grain boundary is determined with the help of a high-resolution transmission electron microscope or a field ion microscope (see *Ion projector*). The effective thickness of a boundary (relaxed layer) does not exceed (2–3)a_0. As a rule, the boundary energy per unit area has a magnitude $\gamma < \gamma_0$, where γ_0 is the surface free energy of the crystal, but in the case of low-index angles θ there are dips on curves of $\gamma(\theta)$ at some special angles $\theta = \theta_i$ ($i = 1, 2, \ldots$), in particular for the case of twinning misalignments (see *Twinning of crystals*). Such grain boundaries are referred to as *special boundaries*, and others are *ordinary boundaries*. A model of the atomic structure of special grain boundaries is based on the concept of a *coincident-site lattice*.

Fig. 1 shows an arrangement of atoms in a plane in the vicinity of a symmetrical *grain boundary* between identical simple cubic lattices with inclination $\Sigma = 5$, rotation axis [001], tan $(\theta/2) = 0.5$, rotation angle $\theta = 53.13°$. Darkened circles denote the coinciding nodes (occupied sites for both lattices). In the plane of the figure the vertical distance between darkened circles is $\sqrt{5}a_0$, and this is the repeat unit distance vertically along the boundary between crystals 1 and 2. Computer calculations and observations demonstrate that the structure shown in Fig. 1 does not correspond to a minimum of boundary energy density $\gamma(r)$. The boundary energy would be reduced by a slight shift (rigid displacement) of crystal 2 relative to crystal 1. A further reduction of γ is possible if one allows all the atoms of the boundary layer to move ("to relax") independently. Such displacements take place at distances not more than (1–2)a_0 from the grain boundary. Calculations carried out for inclination and twist boundaries in face-centered cubic crystals demonstrate that grain boundaries with $\Sigma > 100$ consist of small structural elements of two different types (special elements) which alternate in a manner that minimizes the number of similar neighboring elements.

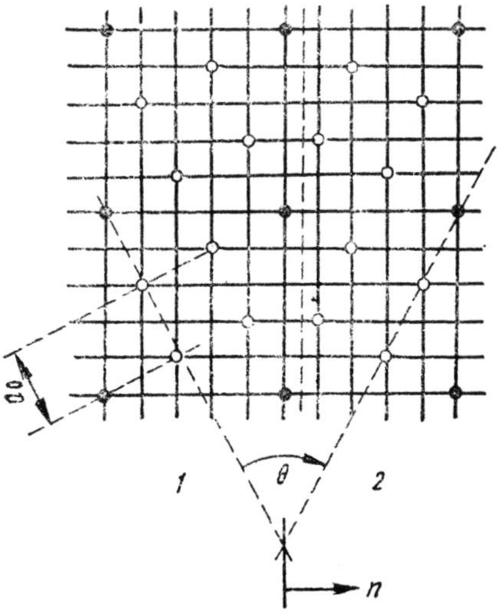

Fig. 1. Grain boundary.

Some special boundaries consisting of particular elements of the same kind are called *particular grain boundaries*.

Thus, the model of structural elements is applicable to ordinary grain boundaries of inclination and twisting, and, probably also, to mixed grain boundaries. There are models of the atomic structure of grain boundaries in *lattices* with a more complicated basis; in particular, in crystals with a *diamond*-like (zinc sulphide) lattice where some structural units contain *dangling bonds*. For further details see *Grain boundary lattice*.

GRAIN-BOUNDARY DEFECTS

Disturbances of the structure of a *complete superposition lattice*. The mutual shift of aligned crystal grains along their boundary by a lattice translation vector $\boldsymbol{b}$ leaves the alignment pattern unchanged, but only displaces it along the sites of the former complete superposition lattice. If the shift does not occur over the entire *grain boundary* area, but is truncated along some line, this line is a linear grain-boundary defect, also called a *grain-boundary dislocation* with *Burgers vector* $\boldsymbol{b}$. If $\boldsymbol{b}$ lies in the plane of the boundary, the grain-boundary dislocation is of the slipping type; if $\boldsymbol{b}$ has a component normal to the boundary, then it is a sessile dislocation (one that cannot migrate

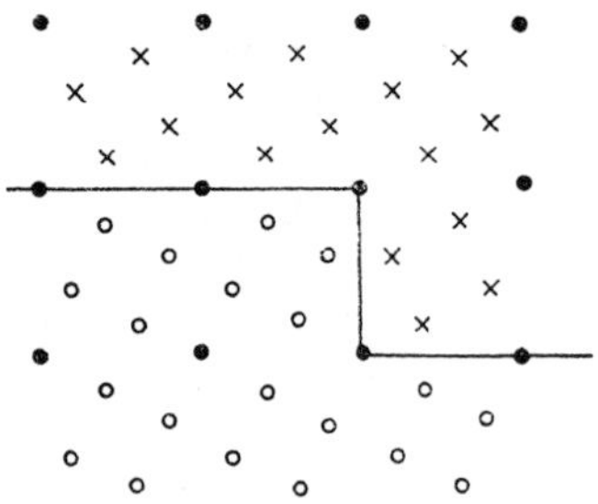

Fig. 2. Grain-boundary defects. Shelf.

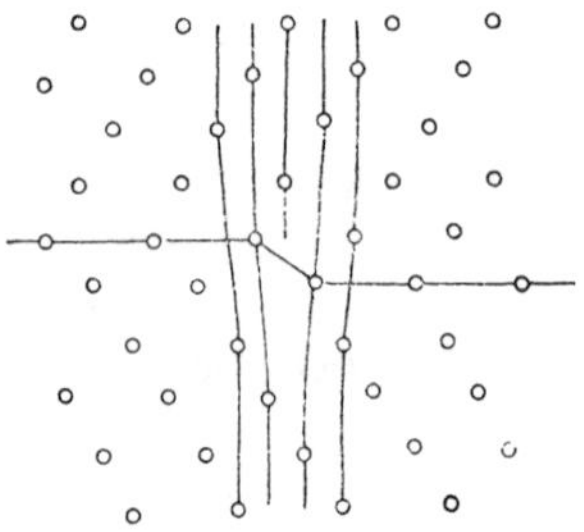

Fig. 3. Grain-boundary defects. Twinning dislocation.

easily). The shift by the vector $\boldsymbol{b}'$ of the shifted *grain boundary lattice* is not a vector of the complete superposition lattice; it disturbs the superposition pattern and the atomic neighborhoods at the boundary. As a result, a *grain-boundary stacking fault* appears at the boundary. If the shift by $\boldsymbol{b}'$ covers only a part of the boundary area rather than all of it, the boundary of the shift is a partial grain-boundary dislocation. It separates the grain-boundary stacking fault and the defect-free boundary, or it separates two such defects of different packing types. Another linear grain-boundary defect is the *shelf* type sketched in Fig. 2. The boundary passes along the shelf from one plane of the grain-boundary shift lattice to a parallel one. The shelf may be related to a grain-boundary dislocation, and the *twinning dislocation* sketched in Fig. 3 is an example (see *Twinning of crystals*). Since the defect-free lattice is a sublattice of the complete superposition lattice, the Burgers vector of the lattice *dislocation* is a vector of the fully aligned lattice, and the lattice dislocation may be built into the boundary. Typically, such a boundary-trapped lattice dislocation restructures by splitting into several dislocations with smaller Burgers vectors. *Grain junctions* are also linear grain-boundary defects. A junction can contain a *junction dislocation* and a *junction disclination*. According to computer simulation studies, a *vacancy* situated in a boundary retains its individuality unless it adjusts itself to a sessile grain-boundary dislocation. In contrast to lattice vacancies, grain-boundary vacancies at different points of boundary structural elements have different energies.

GRAIN-BOUNDARY EFFECTS

Properties of monocrystals and phenomena involving them, which are due to *grain boundaries*. To some extent all properties specific to polycrystals are associated with the presence of grain boundaries. The *recrystallization* rate is controlled by the mobility of boundaries $\mu = v_m/F = \mu_0 \exp[-U/(k_B T)]$, where v_m is the migration rate, and F is the motive force. Plots of the dependences of μ and μ_0 on the disorientation angle θ exhibit maxima, and the plot of the *activation energy* U dependence on θ exhibits minima. An elementary act of migration involves many atoms, and one possible mechanism is the stepwise activated motion of a grain-boundary dislocation (see *Grain-boundary defects*). Self-diffusion and impurity diffusion of *substitutional atoms* (see *Diffusion*) along grain boundaries occurs, as a rule, according to a vacancy (see *Vacancy*) mechanism. The *boundary diffusion* coefficient $D_b = D_{b0} \exp[-U_b/(k_B T)]$ is greater than D_v, its volume diffusion counterpart. D_b for special boundaries is less than it is for ordinary ones. Since the potential relief of crystals in the boundary plane and along the normal to it is nonuniform, the measured activation energy U_b is an effective value and relates to the effective width δ of the grain boundary, which equals several interatomic spacings. Typically, what is measured experimentally is the product $D_b\delta(U_b \approx (1/2)U_v)$, where U_v is the activation energy of volume diffusion. Some types of *impurity atoms* are attracted to grain boundaries (*horophilic impurities*), while others are repelled by them (*horophobic impurities*). The incorporation of impurities in grain boundaries (intercrystallite internal *adsorption*) may cause *embrittlement* of a polycrystal, and accelerate intergrain *corrosion*. Facilitated sliding along grain boundaries with participation of grain-boundary dislocations increases the *creep* rate and *plasticity* (see *Superplasticity*).

The restructuring of atomic configurations of grain boundaries by an alternating voltage results in *internal friction* Q^{-1}, characterized by a grain-boundary peak on the temperature function $Q^{-1}(T)$. In semiconductor *polycrystals* (e.g., polysilicon), many grain boundaries contain *dangling bonds* that are centers for *recombination of current carriers*.

GRAIN BOUNDARY LATTICE

An auxiliary *lattice* for describing the location of atoms in a bicrystal with a specific boundary oriented in such a way that the surface density σ^{-1} of coincident sites (see *Coincident-site lattice*) in its plane (or parallel to it) is smaller than its maximum possible value $(\sigma^{-1})_{\max}$. Coincident sites are sites of the *complete superposition lattice* that are occupied by atoms of both individual bicrystal lattices. At $\sigma^{-1} = (\sigma^{-1})_{\max}$ the grain boundary lattice coincides with the complete superposition lattice so that all of the sites are coincident.

Fig. 4 shows a boundary between two simple cubic lattices $\Sigma = 5$ with the rotation axis [001], $\tan(\theta/2) = 0.5$, rotation angle $\theta = 53.13°$. The sites of the atoms of the lattice of crystal No. 1 at the bottom are designated by (∘), those of the other crystal No. 2 at the top are denoted by (×). The boundary region is delimited by coincident sites, indicated by darkened circles, ABC at the bottom and $GFED$ at the top. Sites of both lattices, which

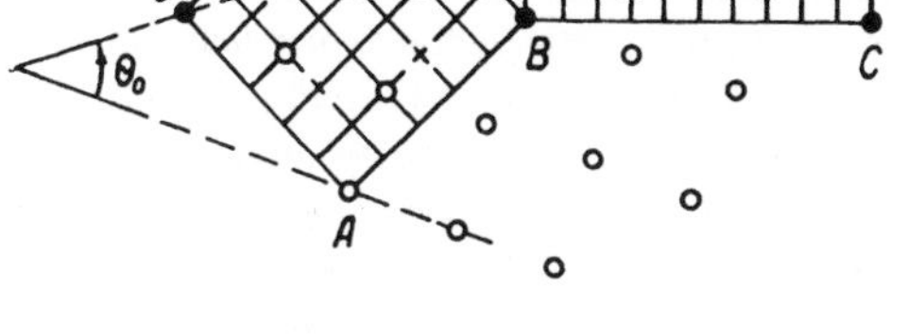

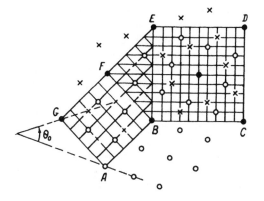

Fig. 4. Grain boundary lattice.

constitute the complete superposition lattice are shown within the boundary region. In actuality the boundary will tend to have the same density of atoms as the two individual crystal lattices, so it will have a grain boundary lattice with many less atoms than those corresponding to the sites of the complete superposition lattice. The atoms in the boundary region will arrange themselves in a configuration of lowest energy which involves a tendency to align themselves with the atoms outside the boundary.

Another view of a grain boundary arrangement with the same angle θ is shown in Fig. 1, see *Grain boundary*.

GRANATO–LÜCKE MODEL (A. Granato, H. Lücke, 1956)

A model of the dissipation of energy of elastic vibrations (see *Internal friction*) which arises from *dislocation* motion. The Granato–Lücke (GL) model is based on an analogy between dislocation motion and a string vibration, including the tension, viscous and inertial forces. It assumes that a *monocrystal* contains a *dislocation grid* consisting of *screw dislocations* and *edge dislocations*. A *dislocation line* possesses two kinds of *pinning centers*, sturdy ones (nodes of the dislocation network) and weak ones (*impurity atoms* and other *point defects*). Correspondingly, two kinds of dislocation segments, L_S and L_W, are considered; $L_S \gg L_W$. The GL model takes into account different length distributions of line segments L_W for zero temperature. When the magnitudes of the applied stress σ are small (magnitude independent losses), line segments L_W vibrate between pinning points (like in an elastic string). As long as the motion is hindered by friction there is a phase lag between the *strain* and the stress; hence energy loss Δ appears as well as variations in the *elastic modulus*. It follows from the GL model that for amplitude-independent internal friction at low frequencies ($\omega \sim 1$ kHz) we have $\Delta \sim \Lambda L_W^4$, where Λ is the dislocation density (see *Dislocation density tensor*); while at high ω there is a peak in the losses $\Delta(\omega)$ of the resonance or relaxation type which depends on the magnitude of the damping constant. For further increases in σ the dislocations tear off from the weak pinning centers, so their motion is limited by the strong pinning at network nodes, and there appears a frequency dependence $\Delta(\omega)$ (*amplitude-dependent losses*, or *hysteresis losses*). The GL model has some limitations; and it can be further developed by taking into account thermal activation and the point defect distribution. The description of *dislocation internal friction* based on this model has been applied successfully to metals, semiconductors, and ionic crystals.

GRANITE (fr. Lat. *granum*, grain)

A construction stone; a crystalline rock which consists of quartz (25 to 30%), feldspars (plagioclase and microcline, 60 to 65%), colored *minerals* (up to 10 to 15%), micas (biotite, muscovite), amphibole, and less often pyroxene. Granite is classified by the contents and character of colored minerals: *alaskite* (without dark-colored minerals); *leucocratic granite* with low amounts of dark-colored minerals; *biotite granite* – the most common type, contains 6 to 8% of biotite; and *double-micaceous granite* with biotite and muscovite, which is widely distributed in the Earth's crust. The rock colors light-gray, yellowish, pinkish-gray, and less often greenish, are to large extent determined by the coloring of the feldspars which are dominant in the rock composition. Density is 2.6 to 2.7 g/cm^3, the compression strength ranges from 100 to 300 MPa, with most values in the range 120 to 180 MPa. Porosity varies from 0.3 to 3.84%, with an average value of 1.3%; adsorption of water 0.1 to 0.6%; electrical resistivity $\rho = 10^3$ to 10^5 Ω·m; magnetic susceptibility χ varies from 0 to $4500 \cdot 10^6$ cgs units.

GRANULAR FILMS

Films consisting of individual small particles (granules) with sizes from 10^{-3} to 1 mm. Their formation proceeds with the distruction of the macroscopic structure of a *thin film*, and usually occurs at the *condensation* of a thin layer material on a substrate with defects (local *crystallization* centers on a vitreous substrate, an adsorbed impurity on a crystalline substrate, etc.), or at heating or aging of such thin (usually amorphous) films accompanied by partial heterogeneous vaporization or *diffusion* of the material. See also *Insular films*.

GRANULAR SUPERCONDUCTORS

Heterogeneous superconductors consisting of separate superconducting granules (grains) connected by weak Josephson coupling through an intervening thin non-superconducting medium (see *Josephson junction*). The latter is a specially prepared insulator host, or oxide on the surface of metal granules, or pores filled by gas, or insulating surface layers as in the case of the *ceramic* $BaPb_{1-x}Bi_xO_3$. Granule dimensions range from a few nanometers to tens of microns. From the viewpoint of their superconducting properties, granular superconductors can be classified as three-dimensional and two-dimensional. Superconductivity of individual granules (if these are sufficiently small) can be suppressed by fluctuations of the *order parameter* (efficient "zero-dimensionality"). There are often two superconducting *phase transitions* in granular superconductors: the transition within individual granules at the temperature $T = T_G$, and the transition of the superconductor as whole at $T = T_W < T_G$ as a result of the establishment of phase coherence throughout the system of Josephson junctions. The latter phase transition at $T = T_W$ can be absent in the case of small granules due to a large value of the Coulomb energy $U \approx e^2/C$ which should be overcome during the tunneling of a *Cooper pair* (here e is the electron charge, C is the capacitance of a granule), and the space quantization of the electron spectrum. Two-dimensional granular superconductors exhibit a topological phase transition (see *Kosterlitz–Thouless transition*) and suppression of superconductivity due to a *weak localization* of charge carriers. A granular superconductor is an example of a disordered *Josephson medium* or superconducting glass.

GRAPHITE (fr. Gr. $\gamma\rho\alpha\varphi\omega$, am writing)

One of the crystalline allotropes of *carbon*. Mohs scale hardness is one unit, density is 2.33 g/cm^3. Color is from black to steel gray. It is fireproof with high *thermal resistance*. It crystallizes with horizontal layering. Equilibrium "graphite–vapor" at atmospheric pressure (0.1 MPa) occurs at 3270 °C. With the growth of pressure up to 10 MPa the equilibrium temperature increases up to 3700 °C, the of *triple point* (graphite–liquid–vapor) is 3750 °C under the pressure 12.5 MPa.

Table 1. Some properties of different classes of materials obtained on the basis of graphite

Properties	From roasted coke	From unroasted coke	Pyrographite	Glass–carbon	Carbon–carbon
Density, g/cm^3	1.6 to 1.7	1.72 to 1.98	2.15 to 2.23	1.45 to 1.52	1.4 to 1.45
Compression strength, MPa	33.0/33.0	98	295 to 345	225	410
Elastic modulus E, GPa	6.4/4.9	10.3	26.5	26.5 to 34.0	132
Coefficient of thermal expansion $\alpha \cdot 10^{-6}$, K^{-1} (at 20 °C)	3.7/4.1	6.6/6.0	10/0.55	–	–
Thermal conductivity, $W \cdot m^{-1} \cdot K^{-1}$ (at 20 °C)	120/103	94	70/227	–	–
Typical crystallite dimension in plane of basis L_a, nm	40	60	100	5 to 10	–

Note: The numerators provide values of properties in the direction parallel to the axis of compression, and the denominators denote those in the perpendicular (transverse) direction.

Natural graphite is in the form of sheets without distinct crystalline shape, and their aggregates. It is transparent in the IR range. It is stable towards heating in the absence of air, and it burns in air. It has 10 to 20% impurities (hydrogen, nitrogen, oxides and dioxides of carbon, methane, ammonia, hydrogen sulfide, water). It is used as a lubricant after chlorine purification at about 2000 °C. Graphite polycrystalline materials are related to *semimetals* with zero (or close to zero, depending on impurities and submolecular structure) overlapping of the *valence band* and *conduction band*. The electric resistivity depends on porosity, grain dimensions, *crystallite* dimensions and the *thermal treatment* temperature, thus varying for different brands of graphites by factors of 10 or even 100. In monocrystals it is highly anisotropic; in the plane of the layers it has a metallic character (0.385 μΩ·m), and in the perpendicular direction (52.0 μΩ·m) graphite is a *semiconductor*. The resistivity of polycrystalline materials parallel to the basal plane (perpendicular to the axis of *pressing*) is several dozen times lower than along the axis.

The strength parameters of graphite are also anisotropic. Thus, e.g., for pyrographite obtained by deposition from the gaseous phase, the average limiting compression strength σ_{comp} parallel to the deposition plane has a value 312 MPa, and perpendicular to this plane the value is 61 MPa. The strength parameters of anisotropic graphites depend on the submolecular structure associated with the treatment temperature, the method of production, etc. For polycrystalline graphite the anisotropy is insignificant.

Besides conventional graphites, natural and artificial ones obtained from roasted and unroasted coke, there is also a whole series of carbon materials based on graphite which differ from each other in many of their properties. Quasi-monocrystals of graphite used as monochromators, possess sharply expressed anisotropy (disorientation of crystallite packing does not exceed $1'$ to $5'$) and the characteristic crystallite dimension in the basal plane is about 1 μm. At the same time on the basis of the so-called nongraphitizing resins *glass carbon* is obtained, which is a completely anisotropic gas-impermeable material, stable in corrosive media. Glass carbon has a globular structure with closed porosity ~30%; and a characteristic crystallite dimension of about 5 to 10 nm.

Materials based on graphite are differentiated by their supramolecular structure, porosity, crystallite size and manner of packing. The wide spectrum of properties of these materials is shown in Table 1 for the cases of: material obtained on the basis of roasted coke; material on the basis of unroasted coke; pyrographite – obtained by deposition from the gaseous phase; glass carbon – on the basis of phenol-formaldehyde resin; composition materials of the carbon–carbon type, obtained on the basis of carbon fibers.

The various properties of graphite determine their areas of application. High electrical conductivity, thermal stability in an oxygen-free atmosphere and inertness with respect to the whole series of corrosive media allow the use of graphite for manufacturing heaters, electrodes, crucibles and cuvettes. The main consumers are the *metallurgy*, chemistry and semiconductor industries. Due to its low cross-section for thermal neutron capture (~0.0045 barn) and its ease of machining, graphite is used as moderator of neutrons and a *construction material* for thermonuclear reactors. Together with conventional graphite materials, carbon becomes more and more widely used: carbon composites, carbon plastics, structural materials in rocket and aircraft industry, sporting goods, in medicine for implantable non-detachable prostheses, etc. Graphite high-modulus fibers and graphite "whiskers" have values of the elasticity modulus ~500 GPa.

The structure of graphite is a typical lamellar type (see *Quasi-two-dimensional crystals*). The layers form a series of parallel flat planes, which consist of regular hexagons with carbon atoms at their vertices (see the figure). In the plane of the hexagonal lattice (basal plane) the atoms are bound together by strong *covalent bonds* formed by overlapping of sp^2-hybrid orbitals (see *Hybridization of orbitals*). The shortest distance between the atoms in the plane is $a = 0.1415$ nm, and the energy of the covalent bonds is 710 kJ/mole. The order of the location of planes along the c axis and the symmetry in this direction determine the *space group*. There are two

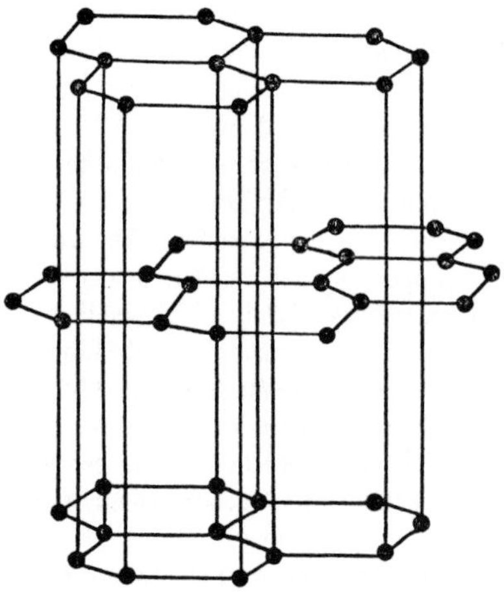

Graphite.

modifications. The hexagonal structure of graphite is related to the space group $C6/mmc$ (D_{6h}^4), with four atoms per *unit cell*, which is a prism with the height 0.671 nm and a rhombus in the base. The sides of the rhombus are 0.246 nm with the angle 60°. The displacement of layers in relation to each other is 0.1418 nm, and every third layer repeats the first one. The rhombohedral modification corresponds to the space group $R\bar{3}m$ (D_{3d}^5). The packing of layers is such that every fourth layer repeats the first one. The rhombohedral structure is usually found in natural graphite, where its content may reach 30%; it almost never appears in artificial graphite.

The forces of interaction of the basal planes are weak *van der Waals forces*. The binding energy between the planes is 4.2 to 18.2 kJ/mole and the distance between them is 0.3354 nm. Due to the low values of the bonding forces between the basal planes the latter may be the main planes of displacement and crystallite stratification, and cleavages are possible along them. The ease of displacement along the basal planes has for a long time been incorrectly associated with the good antifriction parameters of the material (L. Bragg hypothesis) similar to those of a deck of playing cards. The lubricating ability of graphite is explained by the formation in the contact gap of cleaving layers, adsorbed at the surface of molecules (see *External friction*).

During the production of artificial graphite, at the stage of annealing, a material is obtained with the turbostrate structure, characterized by the absence of ordering in the direction of the c axis. The characteristic interplane distance for turbostrate carbon is more than that for the ordered modification, and has a value 0.34 nm. The amount of turbostrate carbon that is produced decreases in proportion to the increase of the treatment temperature.

GRAPHITIZATION

The formation of *graphite* like material in carbide-containing metal *alloys* (mainly based on iron) whose *carbides* are unstable at atmospheric pressure. Graphitization occurs at increasing temperature, and is represented by nucleation and growth of graphite nuclei in a metallic alloy host (due to the diffusion influx of *carbon* atoms of the dissolving carbide, and self-diffusion of metal atoms from the graphite surface). The higher the temperature, the higher the rate of graphite formation. Graphitization is stimulated by preliminary *quenching* of alloys, deformation, irradiation by high-energy particles, and introduction of silicon, or more rarely, aluminum. The process is hindered by the elements which increase the sturdiness of carbides (e.g., chrome and manganese). Sometimes the term graphitization is used for the formation of graphite in alloys which contain no carbides. In such alloys graphite separates out from the carbon-supersaturated solution during solidification and further cooling. In addition, variations in the structure of carbon-based materials are also called graphitization.

GRAPHS in solid state physics

Geometrical figures consisting of a number of points (vertices) and segments (edges) connecting these points. Properties of graphs and various problems related to graphs are studied in the *theory of graphs*, which is a branch of discrete mathematics. Their first applications in physics occurred in the 19th century when G.R. Kirchhoff employed them for the calculation of the electric circuits and A. Cayley used them for calculating the number of isomers of the various saturated hydrocarbons, so-called "*trees*" or connected, non-oriented graphs without cycles. At present graphs are commonly used in various areas of the solid state physics for designing diverse structures, processes, interactions, and so on.

Concrete applications of graph theory involve the topology of disordered structures, the calculation of electron and *phonon* properties of *disordered solids* using model lattices (see *Bethe lattice*, *Cayley tree*), the development of the *Ising model*, and the problem of dimers in a theory of *adsorption* (number of nearest neighbor compounds is determined for a distribution of diatomic molecules on a doubly-periodic lattice-surface). A theory of graphs is used for the graphic representation of a perturbation theory series, for example, for *Green's functions* (see *Feynman diagrams*). In a Feynman diagram the segments represent "quasiparticles" in motion, and the vertices represent events; every segment has its own "momentum". More generally, diagrams are planar graphs with particular physical properties assigned to the vertices and edges. The *lattice gas* model widely used in solid state physics assumes that the atoms and molecules of the gas occupy positions that are not random, but rather are situated at *crystal lattice* sites, each site having at most a single particle. This model is convenient for studying the randomness of substitutions in disordered systems. Configuration integrals over particle coordinates transform into lattice sums which, in some simple cases, can be calculated exactly. Diagram representations of expansions of interaction integrals and Green's functions as sums lead to the necessity of summing the corresponded series only over connected diagrams, based on a fundamental theorem of graph theory (diagrams). The resistance of a random network of resistors can be treated from the viewpoint of the theory of graphs (*percolation theory*, *seepage theory*), and by analogy with the Kirchhoff problem one can study the passage of electric current over a tree with random links.

GRAVIMETRIC ANALYSIS, weight analysis

A *quantitative analysis* of the chemical composition of a material through measurements of its mass by weighting. Gravimetric analysis is based on the mass conservation law during chemical reactions, and on the law of the constancy of composition of chemical compounds. As a rule, the analysis includes weighting the sample (or measuring its volume), dissolving it, quantitatively isolating it from the solution in its elemental condition or as a barely soluble compound, transformation (by drying or heating) of the product into a stable compound with an exactly known composition, and weighting the final product. The amounts of the desired components in the analyzed product are calculated.

GREEN'S FUNCTIONS in solid state physics (H. Green)

Green's functions are constructed from two-time *correlation functions*. It is in terms of two-time Green's functions (or correlation functions) that one can express observed macroscopic characteristics of solids (see *Macroscopic description* in solid state physics), microscopic properties (e.g., positions and widths of energy levels, see *Microscopic description* in solid state physics), and kinetic coefficients (e.g., *electrical conductivity, susceptibility*). For example, an electrical conductivity tensor has the form

$$\sigma_{xx'}(\omega) = -\frac{ie^2 n}{m\omega}\delta_{xx'} + \frac{1}{\omega}\tanh\frac{\omega}{2k_B T} \times \int_{-\infty}^{\infty} \langle j_x(0) j_{x'}(t) + j_{x'}(t) j_x(0)\rangle e^{-i\omega t}\, dt,$$

where n is the electron concentration, $j(t)$ is the Heisenberg operator of the current density; and averaging is performed over an ensemble in the equilibrium state at temperature T.

In solid state physics there are Green's functions built on creation and annihilation operators (see *Second quantization*) of various particles: *electrons, phonons, magnons, photons* (*partial Green's functions*), states of a quantum (sub)system (e.g., an impurity center), and the system as a whole (*global Green's functions*). The technique of unlinking and the diagram technique are employed. The *technique of unlinking* is based on obtaining (with the help of time differentiation) a chain of equations containing more and more complicated Green's functions, and the subsequent closing of the chain by unlinking, i.e. by the appropriate substitution of a mean value of a product of a certain number of operators by a mean value of a product of a lesser number of operators. The *diagram technique* is based on the expansion of a Green's function in a power series of interactions of the particles or parts of the system, bringing the individual members of the expansion into correspondence with certain *Feynman diagrams*, and the summation of an infinite (this is the heart of the method) number of the most essential diagrams (so-called *partial or selective summation*). In those cases when one succeeds in applying a diagram technique the method becomes the most effective.

Quite general results can be obtained with the help of non-equilibrium Green's functions built on creation and annihilation operators. Let a_α^+ and a_α be the operators of creation and annihilation of either particle (fermion or boson, see *Fermi–Dirac statistics* and *Bose–Einstein statistics*) or a subsystem as a whole in state α. The diagram technique is built for the system of functions

$$G_{\alpha\beta}^{>}(t_1,t_2) = \langle a_\alpha(t_1) a_\beta^+(t_2)\rangle,$$

$$G_{\alpha\beta}^{<}(t_1,t_2) = \langle a_\beta^+(t_2) a_\alpha(t_1)\rangle,$$

$$G_{\alpha\beta}^{+}(t_1,t_2) = -i\Theta(t_1-t_2) A_{\alpha\beta}(t_1,t_2),$$

$$G_{\alpha\beta}^{-}(t_1,t_2) = i\Theta(t_2-t_1) A_{\alpha\beta}(t_1,t_2).$$

Here $A = G^{>} + \eta G^{<}$, $\eta = 1$ for fermions, -1 for bosons, and 0 for the subsystem; $\langle\ldots\rangle$ means averaging over the non-equilibrium state. In a spectral representation, i.e. a Fourier representation relative to the time difference $t_1 - t_2$, the following relations (in a matrix form) are valid:

$$G^{\gtrless}(\omega) = G^{+}(\omega)\Sigma^{\gtrless}(\omega)G^{-}(\omega),$$

$$G^{\pm}(\omega) = \frac{1}{\omega - \Omega - \Gamma^{\pm}(\omega)},$$

$$\Sigma^{>} + \eta\Sigma^{<} = i(\Gamma^{+} - \Gamma^{-}) \equiv \Gamma,$$

where Ω is determined by a zero-order Hamiltonian, Γ and Σ are eigenvalue energy functions, $\Gamma^{\pm}$ is the mass operator, $\Sigma^{>}$ and $\Sigma^{<}$ correspond to leaving and entering the state, respectively. The expression $\Gamma = \Sigma^{>} + \eta\Sigma^{<}$ has a simple physical sense: the width of a level for the system as a whole is equal to the rate of particle interchange; in the case of fermions (bosons) this is a sum (difference) of the rates of leaving and entering. The result for the case of bosons means a narrowing of the line with the increase of input; and this effect is implemented in a *laser*.

The possibility of introducing exact eigenvalue energy functions is important for calculating Green's functions. The diagram technique was developed for these functions, and the following two expressions are valid:

$$\int \frac{d\omega}{2\pi} A_{\alpha\alpha}(\omega) = 1,$$

$$\int \frac{d\omega}{2\pi} G_{\alpha\alpha}^{<}(\omega) = \langle n_\alpha\rangle,$$

where n_α is the population number operator of state α, $A_{\alpha\alpha}(\omega)$ describes the shape of the level responding to state α, and $G_{\alpha\alpha}^{<}(\omega)$ is the spectral distribution of the level population. For equilibrium states, the usual single-particle Green's functions are commonly used:

$$G^0_{\alpha\beta}(t_1,t_2) = \Theta(t_1 - t_2)G^>_{\alpha\beta}(t_1,t_2) + \eta\Theta(t_2 - t_1)G^<_{\alpha\beta}(t_1,t_2),$$

where Θ is the step function, and the temperature Green's functions are obtained from the usual ones by the exchange $t \to -\mathrm{i}\hbar\tau$, $0 \leqslant \tau \leqslant (k_{\mathrm{B}}T)^{-1}$, where $-\mathrm{i}\tau$ is the imaginary "time" parameter.

GREEN'S FUNCTIONS AND DEEP ENERGY LEVELS

A method for calculating one-electron energy levels and wave-functions of non-metallic crystals with *defects* that introduce energy levels into the *band gap* of the main crystal. In this method a one-electron *Green's function* is used, i.e. the operator $\widehat{G}_0(E) = (E - \widehat{H}_0)^{-1}$, where E is the energy, and $\widehat{H}_0$ is the effective one-electron Hamiltonian of the crystal. In practical applications the operator $\widehat{G}_0$ is replaced by the matrix G_0. The latter is calculated in a basis of localized orbitals, as a rule atomic orbitals based on the *band structure* and Bloch functions of an ideal lattice (see *Bloch theorem*). The enormous dimensionality (order of the *Avogadro number*) of matrix G_0 does not cause any complications since, in fact, it is only necessary to know a comparatively small submatrix $\widetilde{G}_0$ of matrix G_0 for the portion of the crystal lattice around the defect. The defect is considered as a perturbation of an ideal lattice, and is described by the perturbation operator $\widehat{V}$ with the matrix V expressed in a basis of localized orbitals. In the latter only a submatrix $\widetilde{V}$ is non-zero for the perturbed region near the defect. The unknown levels in the gap are then determined from the equation $\mathrm{Det}\,\|1 - \widetilde{G}_0(E)\widetilde{V}\| = 0$.

In addition to these equations corresponding to localized electronic states (see *Local electronic levels*), the defect-induced *resonance states* can be calculated with the help of Green's functions. These states appear as peaks and valleys on the curve $\Delta n(E) = n(E) - n_0(E)$, where $n_0(E)$ and $n(E)$ are, respectively, the densities of states in an ideal crystal and in a crystal with a defect. To find these, the energy E in the Green's function is considered as a complex number with an infinitesimal imaginary part. Then use is made of a known relation between the density of states and the imaginary part of the Green's function, plus a relation between the Green's function of an ideal lattice and a perturbed Green's function of a lattice with a defect (*Dyson equation*).

The method of Green's functions was initially developed by I.M. Lifshitz in 1940s for vibrations of a lattice with defects, and then reformulated by Slater and Koster (1954) for electronic structure (see *Koster–Slater model*). At present, various versions of the method as well as the molecular *cluster* method are widely used in calculations of *point defects* (*vacancies*, substitutional impurities, *interstitial atoms*) and complicated defects in *insulators* and especially in *semiconductors*, and also in calculations of electronic states in crystals with a surface which is considered as a two-dimensional defect. The method of Green's functions differs from the cluster method in correctly taking into account the "infinite" crystal lattice, but this becomes labor-consuming when calculating the complete energy and studying multielectron effects.

GREEN'S FUNCTIONS AND MAGNETISM

Methods of quantum field theory applied to *magnetism*. A principal element is a *spin Green's function* in terms of which one can describe in a self-consistent manner the main characteristics of a quantum magnetic system: the *susceptibility*, the *magnetization*, the *thermodynamic potential*, the *kinetic coefficients*, the spectrum of intrinsic excitations, etc. (see *Green's functions*). There are retarded, advanced, causal, and spin temperature Green's functions. Their analytical properties are determined both by general requirements of statistical physics, and by symmetry properties of a system Hamiltonian. Like the case of Bose and Fermi operator Green's functions (*Fermi–Dirac statistics* and *Bose–Einstein statistics*), the poles of retarded spin Green's functions determine the energy and attenuation of the intrinsic vibrations of the spin system. One of the calculation modes consists in breaking the infinite chain of the equations for Green's functions so that in a particular approximation the chain is reduced to a closed one. A calculation of the spin Green's function can also be carried out by using a representation of *spin operators* in terms of Bose or Fermi operators (see *Spin operator representations*). The *Wick theorem* was formulated and the diagram technique was developed for use with *spin* operators. However, since a spin operator commutator is also an operator, the algebraic grounds for such a mode of Green's function calculation consists in a subsequence of more complicated relations (in comparison with Bose and Fermi operators). The perturbation approach to spin type Green's functions is sufficiently straight forward. There exist several variants of the perturbation technique (expansion over reciprocal spin, *magnon* density, reciprocal radius of interaction) with each suitable for a particular application, and they provide descriptions of high-frequency, thermodynamic and kinetic properties of magnetic systems.

GREEN'S FUNCTIONS AND SUPERCONDUCTIVITY

A field theory treatment of the superconducting properties of *conduction electrons* in *metals* with the help of the *Dyson–Gorkov equations* for normal and anomalous Green's functions (see *Green's functions, Gorkov–Nambu method*). In the case of singlet Cooper pairing (see *Cooper pairs*), these equations in momentum–energy space take the form:

$$\begin{aligned}&\big[\omega-\xi(\boldsymbol{k})-\Sigma(\boldsymbol{k},\omega)\big]G(\boldsymbol{k},\omega)\\&\quad-\Delta(\boldsymbol{k},\omega)F^{+}(\boldsymbol{k},\omega)=1,\\&\big[\omega+\xi(\boldsymbol{k})+\Sigma(-\boldsymbol{k},-\omega)\big]F^{+}(\boldsymbol{k},\omega)\\&\quad-\Delta^{+}(\boldsymbol{k},\omega)G(\boldsymbol{k},\omega)=0,\end{aligned}\tag{1}$$

where $\xi(\boldsymbol{k})$ is the energy of electrons with *quasi-momentum* $\boldsymbol{k}$ in the *conduction band* measured from the *Fermi level*, Σ and Δ are, respectively, the normal and anomalous *eigenenergy parts* (*mass operators of quantum mechanics*) of electrons which are determined by the following integral equations:

$$\begin{aligned}\Sigma(p)&=\mathrm{i}\int\frac{\mathrm{d}^4p'}{(2\pi)^4}\Gamma(p,p';p,p')G(p'),\\\Delta(p)&=\mathrm{i}\int\frac{\mathrm{d}^4p'}{(2\pi)^4}\Gamma(p,p';p,p')F(p'),\end{aligned}\tag{2}$$

where $\Gamma(p_1,p_2,p_3,p_4)$ is the *vertex part* (*four pole*) of the effective non-local (retarded) electron–electron interaction, $p\{\boldsymbol{k},\omega\}$ is the energy–momentum four-vector (here and below $\hbar=1$). The four-pole Γ contains the *electron–phonon interaction* due to the exchange of virtual phonons (see *Virtual transitions*) and the Coulomb repulsion shielding due to the exchange of virtual plasmons (see *Electric charge screening*). The electron–phonon interaction is described by the phonon Green's function which is determined by the *Dyson equation*:

$$D(p)=D_0(p)+D_0(p)\widetilde{\Pi}(p)D(p),\tag{3}$$

where D_0 is the bare (un-renormalized Green's function), $\widetilde{\Pi}$ is the self-energy part of the phonons:

$$\begin{aligned}\widetilde{\Pi}(p)=-2\mathrm{i}g\int\frac{\mathrm{d}^4p'}{(2\pi)^4}G(p')G(p'-p)\\\times\widetilde{g}(p',p'-p;p),\end{aligned}\tag{4}$$

where g and $\widetilde{g}$ are the bare and renormalized vertex parts (*three pole*) of the electron–phonon interaction. The shielded Coulomb interaction in the *random phase approximation* takes the form:

$$\widetilde{V}(\boldsymbol{k},\omega)=\frac{V(\boldsymbol{k})}{1-V(\boldsymbol{k})\Pi(\boldsymbol{k},\omega)}\equiv\frac{4\pi e^2}{k^2\varepsilon(\boldsymbol{k},\omega)},\tag{5}$$

where

$$\varepsilon(\boldsymbol{k},\omega)\equiv1-\frac{4\pi e^2}{k^2}\Pi(\boldsymbol{k},\omega)$$

is the *dielectric constant*, and

$$\Pi(p)=-2\mathrm{i}\int\frac{\mathrm{d}^4p'}{(2\pi)^4}G(p')G(p'-p)$$

is the electron *polarization operator*. Eqs. (1)–(5) together with the expression

$$\begin{aligned}\widetilde{g}(\boldsymbol{k},\omega)&=g(\boldsymbol{k})+V(\boldsymbol{k})\Pi(\boldsymbol{k},\omega)\widetilde{g}(\boldsymbol{k},\omega)\\&\equiv\frac{g(\boldsymbol{k})}{\varepsilon(\boldsymbol{k},\omega)}\end{aligned}\tag{6}$$

form a closed set of nonlinear equations that is used for the description of physical properties of *superconductors* with strong coupling (see *Eliashberg equations*).

GREEN'S TENSOR

A tensor of rank two, G_{ik}, whose elements are the proportionality coefficients between the components of the *displacement vector* u_i and the components of the force F_k applied to a certain point of an infinite elastic medium: $u_i = G_{ik}F_k$. The components of Green's tensor are homogeneous functions of the coordinates to first order. For an isotropic medium the tensor is given by

$$G_{ik} = \frac{1+\nu}{8\pi E(1-\nu)}\left[(3-4\nu)\delta_{ik} + n_i n_k\right]\frac{1}{r},$$

where ν is *Poisson ratio*, E is *Young's modulus*, δ_{ik} is the unit tensor (Kronecker delta), n_i $(i = 1, 2, 3)$ are components of a unit vector, and r is the distance from the force application point.

GRIFFITH THEORY (A.A. Griffith, 1920)

A theory that describes possible reasons for the lack of conformity between the actual *strength* and theoretical strength of crystalline and vitreous bodies (see *Vitreous state of matter*). Griffith suggested that when a body is *fracturing*, the stress need not be equal to the theoretical strength over the whole body bulk but only at the tip of a sharp and narrow *crack*. Such a crack may become unstable under relatively low nominal stress. His reasoning was based on a thermodynamic approach that correlates an increment of crack length with a change of the potential energy of the stressed system. If there is an external elongation stress it produces a penetrating crack of length L which is small compared to the dimensions of the body, and a zone free of elastic stress arises in the region adjacent to the crack. Increasing the crack size increases the volume of the *relaxation* zone and, consequently, it increases the relaxed elastic energy. The concomitant formation of crack voids requires an expenditure of energy for breaking chemical bonds. As the crack size increases, the energy balance of the system changes, and the energy needed for forming new interfaces increases linearly with the crack length L, while the energy gain increases in proportion to L^2. As a result at a certain critical value L_{cr} the energy expenditure reaches a maximum, beyond which the total potential energy of the system begins to decrease continuously. It means that the gain of elastic energy in the bulk in the vicinity of the crack exceeds the expenditure for atomic bond breakage, so that the crack can propagate under the constant external stress without external energy input. Since the energy gain increases as the square of the crack length, the crack spontaneously accelerates either up to the total fracture of the body, until it enters a region without tensile stress, or until it enters a region where there is compression stress in a plane normal to the crack. The *Griffith critical stress* needed for crack development under load in the case of a planar stressed state is the following:

$$\sigma_{\text{G}} = \sqrt{\frac{2\gamma E}{\pi L}},$$

where E is *Young's modulus*, and γ is the *surface energy*. The thermodynamic criteria for the crack growth are necessary conditions only. To carry out the process it is necessary to provide at the crack tip a local stress that is within an order of magnitude of the theoretical tensile strength of the metal. The *Griffith criterion* could be both a necessary and a sufficient condition for fracture of a crystalline solid if the crack tip radius is comparable in size to an interatomic distance.

GRINDING

Treatment of sample surfaces by abrasive devices (see *Abrasive materials*). The grinding of metallic wares is accomplished on grinding machines by rotating abrasive wheels or whetstones. In the case of surfaces that are difficult to treat, electrochemical (electrolytic) grinding wheels are provided with a constant electric current, and an electrolyte is used for the cutting section. Grinding may be carried out by a soft abrasive (see *Polishing*), by abrasive powder suspended in a liquid, and also by means of a vibrational treatment. For the grinding of stones one may employ carborundum plates, whetstones, etc., of various grain sizes.

GRINDING, DISPERSION

See *Dispersion grinding*.

GROUP CHARACTER

See *Character of group representation matrix*.

GROUP

For different kinds of mathematical groups used in solid state physics, refer to *Colored symmetry group, Continuous symmetry transformation groups, Defect symmetry group, Limiting symmetry group, Local symmetry group, Magnetic symmetry group, Point groups, Shubnikov groups, Space group, Symmetry group, Wave vector group, Wittke–Garrido groups.*

GROUP THEORY in solid state physics

A mathematical formalism which allows one to exploit fundamental features of the *crystallography*, of *crystal lattice dynamics*, of *current carriers*, the environment in the vicinity of *defects in crystals* (impurities, *dislocations*), etc., on the basis of the *crystal symmetry.* Taking into account this symmetry simplifies equations and clarifies the physical picture. Areas of application of group theory are the following: discerning properties of tensors in crystals; theory of *crystal lattice vibrations*; *phase transitions* in multicomponent systems, characteristics of electron motions: it is possible to determine general properties of equations without knowing their solutions, to describe some characteristics of electron dynamics (and *holes* when treating semiconductors), and also the response to an external perturbation. Group theory formulates precisely the *conservation laws*, the *selection rules*, and the multiplicity of energy level degeneracies (see *Band theory*).

The complex of symmetry operations g_α under which the system Hamiltonian is invariant, form a group of transformations of the wave function solutions ψ_i of the stationary state Schrödinger equation, corresponding to the expression $g_\alpha\psi_i = \sum\psi_j\tau_{ij}(g_\alpha)$. A collection of the matrices τ_{ij} corresponding to all the elements of the group g_α is called a *representation of the group*. The matrices of the representations $\tau(g_\alpha)$ can be determined by group theory methods without having a concrete expression for ψ_i. These methods allow one to predict the behavior of symmetric systems in external fields, and to deduce a possible form of the equations of motion (e.g., in the *effective mass* approximation *spin Hamiltonian* approach). These equations are determined by the particular Hamiltonian H. Both the construction of the Hamiltonian and its form (as a rule, a matrix form) are essentially based on the characteristics of the representation of the group (in particular, the dimensionality of H coincides with the dimensionality of the matrices of this representation). There can be several such representations; therefore, group theory can offer more than one possible model of H. Further insights are obtained by comparing the predictions of theory with experiment. Particularly efficient investigations under this approach involve magnetooptic studies which include both band-to-band and intraband transitions (e.g., *light absorption, paramagnetic resonance, cyclotron resonance, combined resonance*, and *Raman scattering of light*). In combination with an external electric field and pressure these experiments permit one to select a model of the effective Hamiltonian, and to determine some of its parameters (e.g., the constants of each invariant). The same Hamiltonian is used to calculate the energy spectrum of localized states in the theory of *kinetic phenomena.* For the most widely studied semiconductors the number of independent techniques for determining the parameters of H exceeds the number of parameters, and the agreement between the results supports the reliability of the conclusions of the group-theoretical approach.

GROUP VELOCITY, ray velocity

The velocity of propagation of the envelope of a collection of monochromatic waves close to each other in frequency, i.e. a *wave packet*. The concept of group velocity is applicable when the propagation velocity does not depend on the intensity, and the frequency spectrum can be characterized by its envelope. For media with *dispersion* the term group velocity is applicable for short enough times within which the wave-packet remains cohesive within a given accuracy. In this case, the group velocity is the velocity of translation of the envelope $u = \partial\omega/\partial k$.

Inside the envelope, the wave propagates with a certain *phase velocity* $v = \omega/k$, where $k = 2\pi/\lambda$ is the mean value of the wave number, ω is the mean angular frequency, and λ is the mean wavelength of the waves in the packet. In accordance with a *Rayleigh criterion*, the group velocity is $u = v + k\partial v/\partial k = v - \lambda\partial v/\partial\lambda$. The case $u > v$

corresponds to normal dispersion, $u < v$ to anomalous dispersion, and $u = v$ indicates the absence of dispersion. The direction of the group velocity for an electromagnetic wave coincides with the direction of the Poynting vector $\boldsymbol{S} = \boldsymbol{E} \times \boldsymbol{H}$ whose magnitude equals the the energy S passing in a unit time through a unit area normal to the energy flux direction. In *anisotropic media* the group velocity direction, in general, does not coincide with the phase velocity direction, and we have the relation $u \cos \alpha = v$, where α is the angle between the directions of the group velocity and the phase velocity.

GROWTH ANISOTROPY

Differences in properties of solids along different directions, induced during the growth process. It is most prominently represented by the presence of *magnetic anisotropy* in magnetically ordered media, i.e. in the surface regions of large *monocrystals*, and in thin monocrystalline or *amorphous films*. In amorphous films growth anisotropy often plays a dominant role (see *Epitaxial iron garnet films, Metallic glasses*). The existence of a special direction, such as the normal to the surface of a growing crystal or film, is a physical characteristic of magnetic growth anisotropy. The contribution of growth anisotropy to the free energy f of a medium depends on the orientation of the *magnetization* vector with respect to the growth direction $\boldsymbol{n}$, i.e.

$$f_{\text{g.a.}} = k_{ij}\beta_i\beta_j + k_{ijkl}\beta_i\beta_j\alpha_k\alpha_l + \cdots,$$

where $k_{ij}, k_{ijkl}, \ldots$ are the components of the tensors of the growth magnetic anisotropy constants, β_i and α_k are the respective direction cosines of the magnetization vector $\boldsymbol{M}$ with respect to $\boldsymbol{n}$ and to the edges of the crystal *unit cell* (in amorphous films there is no contribution of crystallographic anisotropy). Growth anisotropy is metastable; and it disappears during long-term, high-temperature *annealing*.

GROWTH OF MONOCRYSTALS

See *Monocrystal growth*.

GROWTH PYRAMID, growth sector

The region in a *monocrystal*, which is generated during the process of *crystallization* (crystal growth) off an individual face. If the *crystal* has the shape of a polyhedron (see *Crystal polyhedron*), its body consists of growth pyramids centered at the nucleus or seed, from which the crystal has grown. If a certain face is generated at a later stage of crystal growth, then the vertex of the corresponding growth pyramid is at the point, at which the face had originated. A particular case of a growth pyramid is a tubular domain, which is left in a crystal pulled from the melt by an atomically smooth (singular) face, and differs from other tubular (cylinder-shaped) domains that frame it, and have formed on atomically rough surrounding regions of the crystallization front. The growth pyramid is often bounded by nonplanar surfaces, which are constituted by traces of the motion of edges that frame the face, which forms this pyramid. The crystal consisting of growth pyramids exhibits the so-called *sector structure* (the existence of these sectors is the consequence of the lack of thermodynamic equilibrium in the processes of crystal growth; it breaks the ideal microscopic *crystal symmetry* to a variable degree with respect to different properties). The growth pyramids differ from one another in properties because of the difference in concentrations of impurities (sometimes by several fold), inclusions, dislocations, etc. (see *Defects in crystals*). Hence they deviate in values of *crystal lattice constants* (up to $10^{-3}\%$). Therefore, there occur *internal stresses* at the interfaces between the growth pyramids in a crystal. Even the growth pyramids of crystallographically equivalent faces can differ in properties, for the defects "memorize" not only the index of a face, but also the direction of its growth. For instance, the so-called *anomalous pleochroism* in the growth pyramids of the minor rhombohedron of quartz is due to the fact that the substitution of Si^{4+} for $Al^{3+} + Na^{+}$ proceeds with different probabilities in three silicon–oxygen tetrahedra which are inequivalent with respect to a particular face.

GRÜNEISEN LAW

See *Thermal expansion*.

GRÜNEISEN TENSOR (E. Grüneisen, 1908)

A tensor γ_{ij} which is a generalization of the *Mie–Grüneisen parameter* $\gamma_S = -\mathrm{d}\ln\omega_S/\mathrm{d}\ln V$

which is a measure of the frequency shift induced by the change of volume V for an arbitrary *strain*. The value of γ_S can be estimated from experimental data as a mean value

$$\gamma = \frac{1}{3N}\sum_{S=1}^{3N}\gamma_S = \frac{V}{C_V}\beta K,$$

where β is the thermal expansion coefficient (see *Thermal expansion*), K is the *bulk modulus*, C_V is the *specific heat* at constant volume, and N is the total number of particles. The absolute values of the dimensionless parameter γ, which are sometimes called *Grüneisen constants*, vary from 0.5 for *covalent crystals* to 3.0 for crystals with weak *van der Waals forces*, thus reflecting to some extent the anharmonicity of vibrations (see *Anharmonic vibrations*). Generalizing γ_S for any kind of deformation leads to the relation

$$\gamma_{ij}^{(S)} = -\frac{d\ln\omega_S}{d\ln u_{ij}} \quad \text{and} \quad \gamma_{ij} = \frac{V}{C_V}c_{ijkl}\alpha_{kl},$$

where u_{ij}, c_{ijkl}, and α_{kl} are the tensors of deformation, elasticity and thermal expansion, respectively. Thus, γ_{ij} is a symmetric tensor of rank two whose components within the accuracy of the known scalar V/C_V are equal to the components of the thermal stress tensor $c_{ijkl}\alpha_{kl}$ in a "squeezed" crystal. The Grüneisen tensor is applied in studies of thermal expansion, the dependencies of c_{ijkl} on temperature and pressure, the behavior of solids in the state of hydrostatic and impact compression, the lattice thermal conductivity of *semiconductors* and *insulators*, the scattering of various ionizing radiations and particles on *crystal lattice vibrations*, the absorption and attenuation of ultrasonic waves as a result of their interaction with *phonons*, quantum zero point motion, and so on.

GUEST–HOST EFFECT

An *electrooptical effect* involving the change of the intensity of coloration, color, or luminosity of a system consisting of a "dichroic *dye* in a *liquid crystal*" (guest–host) at a *Frederiks transition*. Due to the geometric anisotropy, the dye molecules are oriented in the liquid crystal host in accordance with its *director* orientation. Therefore, when an electric field is applied to the system with a strength higher than the threshold value of the Frederiks transition, the dye molecules reorient in common with the host molecules. If the dye has a dichroic absorption band in the visible range, then this reorientation leads to a change of the color intensity of the system. In the case of two mutually perpendicular polarized absorption bands in the dye spectrum, a switching of color is observed. When using luminescent dichroic dyes, the luminosity intensity of the system changes. The guest–host effect is employed in colored *liquid-crystal displays*.

GUINIER–PRESTON ZONES, G-P zones (A. Guinier, G.D. Preston, 1938)

Regions or clusters without a definite crystal structure that emerge from a supersaturated *solid solution* during the initial stages of precipitation. They involve a process of solid solution decomposition (see *Alloy decomposition*) that is accompanied by concentration variations in small regions of the crystalline host. In a general case, the formation of these zones induces atom displacements relative to the sites of the initial lattice (see *Alloy aging*). At early stages of aging the X-rays scattered from Guinier–Preston zones of Al–Cu *monocrystals* indicate the presence of layer-like units 0.5 nm thick and 5–10 nm wide oriented parallel to {100} planes of the host cubic lattice. The formation of Guinier–Preston zones can result in the strengthening of alloys.

GUNN DIODE

A *semiconductor diode* with a negative differential resistance arising in an active layer of a semiconductor as a result of the *Gunn effect*. The transformation of energy from a source to the microwave oscillations in a *diode* is carried out in various ways depending on the level and the profile of the impurity distribution in a sample, the active layer dimensions, the cathode contact properties, the offset value of the diode, and the type of external circuit. A Gunn diode is used for the generation, amplification, and transformation of microwaves in the frequency range 4–200 GHz; it has a low noise level (see *Gunn effect*).

GUNN EFFECT (J.B. Gunn, 1963)

The appearance of a portion of a semiconductor *current–voltage characteristic* with a *negative*

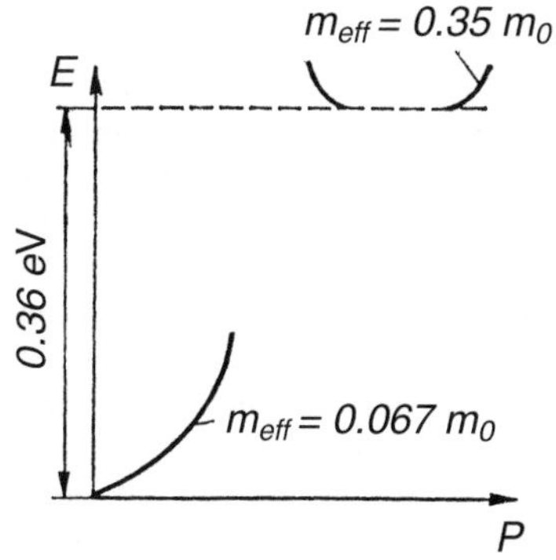

Fig. 1. Dependence of the energy E on the quasi-momentum P for electrons in gallium arsenide; m_{eff} is the electron effective mass and m_0 is the free electron mass.

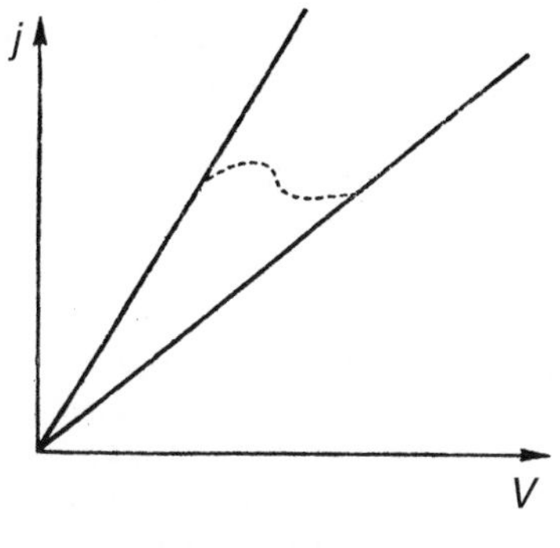

Fig. 2. Dependence of the electric current j on the voltage V (current–voltage characteristic) for two bands: the upper straight line is for the lower band in which the effective mass is smaller, and the lower line is for the upper band with larger effective mass.

resistance caused by a specific electron spectrum. A typical *semiconductor* in which one can observe the Gunn effect is *gallium arsenide.* As shown in Fig. 1, the spectrum consists of two electron bands or valleys with different curvatures and hence different electron effective masses, the mass being larger in the upper band on the right than in the lower one on the left. It follows from obvious physical considerations supported by calculations that for a given electric field strength the larger the efficient mass, the lower the electric current.

Fig. 2 shows the current–voltage characteristics of a semiconductor with a two-band electron

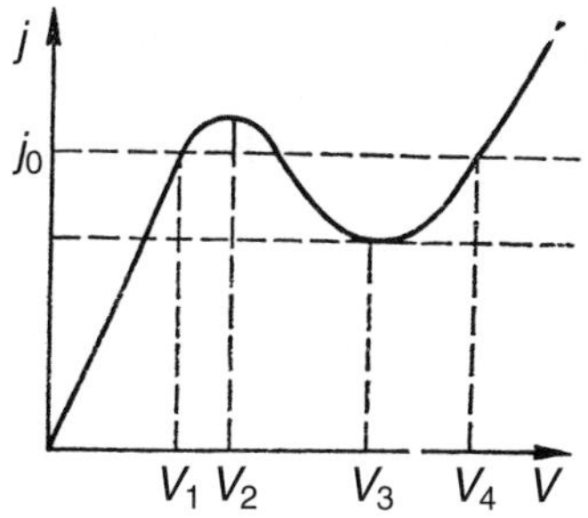

Fig. 3. N-shaped current–voltage characteristic, as described in the text.

spectrum. The upper linear plot corresponds to all the electrons being in the lower band, while the lower plot is the case for all the electrons in an upper band. With the increase of the electric field the electron energy grows, and some of the electrons are excited from the lower to the upper band (Fig. 2, dotted line). The typical current–voltage characteristic sketched in Fig. 3 is an N-shaped curve. The falling portion of the characteristic is unstable, and only the parts to the left of V_2 and to the right of V_3 can be realized. As seen in Fig. 3, two values, V_1 and V_4, of the field strength correspond to the same value of current j_0. Experimental studies show that the Gunn effect arises with the periodic appearance in a sample of a region with an increase in field strength, a so-called *Gunn domain.* The domain is formed on the cathode under action of the electric field, moves along the sample, and disappears upon reaching the anode. Thus, bursts of field strength arise periodically in the sample. This system works as a generator of electromagnetic waves and is called a Gunn generator. The electromagnetic field frequency ω can be estimated from the formula $\omega = v/L$, where L is the sample length, v is the electron drift velocity. The Gunn generators work in the range from short centimeter to long millimeter wavelengths. The Gunn effect is the basis for semiconductor devices called *Gunn diodes.*

GYRATION

In general, a rotation described by some characteristic scale. In the field of optics a first order *spatial dielectric dispersion* in a *crystalline* material exhibiting elliptic (circular) *birefringence.* It is characterized by the variation of the phase and the nature of the *polarization of light* passing through a crystalline plate; a particular case of gyration is *optical activity*. Gyration is associated with an axial tensor of gyration g_{ik}; it can exist only in acentric crystals, i.e. crystals which lack a center of symmetry (with the exception of classes $\overline{6}$, $\overline{6}m2$, $\overline{4}3m$). Gyration is observed as a *rotation of the light polarization plane* in cubic crystals, as well as along the *optical axis* of anisotropic crystals. In crystalline classes $\overline{4}$ and $\overline{4}2m$, a rotation about an axis is forbidden; in other directions, gyration usually manifests itself by alternations of the polarization ellipse. A special case is gyration in crystals of planar classes $3m$, $4mm$, $6mm$ for which g_{ij} is fully antisymmetric; in these crystals the ordinary wave is polarized linearly and the extraordinary one is polarized elliptically. See also *Electrogyration, Self-induced light polarization change.*

GYROMAGNETIC FREQUENCY

See *Cyclotron frequency.*

GYROMAGNETIC MEDIUM

See *Gyrotropic medium.*

GYROMAGNETIC PHENOMENA

See *Acoustic gyromagnetic phenomena, Magnetomechanical phenomena.*

GYROMAGNETIC RATIO, sometimes called magnetogyric ratio

The ratio of the *magnetic moment* of a fundamental particle (or of a system of fundamental particles: atoms, ions, atomic nuclei, etc.) to its angular momentum. The ratio is defined by the equation $\gamma = g\mu/\hbar$, where g is the dimensionless *Landé g-factor* (or other), and μ is the *magneton* (Bohr magneton for an electron, and nuclear magneton for a nucleon or nucleus). The gyromagnetic ratio defines the action of an applied *magnetic field* B on a substance that has both a magnetic moment and an angular momentum. It provides for the splitting of atomic and molecular energy levels in an external magnetic field (see *Zeeman effect*), and *nuclear magnetic resonance* and *electron paramagnetic resonance* at frequencies ω_{r} determined by the equation $\omega_{\mathrm{r}} = \gamma B$.

GYROSCOPE

An axially symmetric solid rapidly rotating about its symmetry axis that can change its direction relative to its container while preserving its direction in the external inertial frame of reference. In most cases the bearings on the axis ends are positioned on opposite sides of a ring that can rotate inside another ring about an axis normal to the gyroscope axis. If this ring can also can rotate then the gyroscope axis can be oriented in any direction relative to its container. In the absence of external torques the gyroscope axis conserves its direction, or undergoes a *precession* if the conserved *angular momentum* does not coincide with the gyroscope rotation axis. Under the application of external forces, the gyroscope axis turns to be parallel to the external torque direction (*gyroscopic effect*). Gyroscopes are widely used in technology, e.g., as a compass, as stabilizers of various kinds, and for other automatic controlling devices.

GYROSCOPE, QUANTUM

See *Quantum gyroscope.*

GYROSCOPE WHEEL

The internal movable part of a *gyroscope*, usually a symmetrical body rotating around its *symmetry axis*, with its point of rest lying below the center of gravity, is called a gyroscope wheel. When the axis is oblique, it performs precessional and nutational motion (see *Euler angles*). At high rotational speeds when the latter motion is so slow that it is practically imperceptible, then the motion is called *pseudoregular precession*. Due to *friction* at the point of rest there appear additional torques tending to vertically orient the gyroscope wheel axis, resulting in the so-called *sleeping gyroscope wheel* motion.

GYROTROPIC MEDIUM, optically active medium

A continuous (crystalline, amorphous, liquid and other) medium characterized by nonsymmetric tensors of response to an electromagnetic field: *dielectric constant* ε_{ik}, magnetic susceptibility χ_{ik}, etc. A gyrotropic medium is called *gyroelectric* if $\varepsilon_{ik} \neq \varepsilon_{ki}$, and *gyromagnetic* $\chi_{ik} \neq \chi_{ki}$. Sometimes, the term "*gyrotropy*" is used to designate a property inherent in a gyrotropic medium. Physical examples of materials that are gyrotropic can be distinguished on the basis of *Onsager relations* (see *Onsager theory*) according to which, e.g.,

$$\chi_{ik}(\boldsymbol{k}, \omega, \boldsymbol{B}_0) = \chi_{ki}(-\boldsymbol{k}, \omega, \boldsymbol{B}_0), \qquad (1)$$

where $\boldsymbol{B}_0$ is the *magnetic flux density*. On the strength of this relation (or an identical one for ε_{ik}), the gyrotropy of the medium can be caused by either a *spatial dielectric dispersion* (dependence on $\boldsymbol{k}$), or by the presence of a magnetic field $\boldsymbol{B}_0$.

If the gyrotropy is due to spatial dispersion, the corresponding gyrotropic medium is called a medium with natural *optical activity* (see *Gyration*). Based on Eq. (1), a tensor form typical for this is the following:

$$\varepsilon_{ik} = \varepsilon_{ik}^{(0)} + i\varepsilon_{ikj}\left(\widehat{g}\boldsymbol{k}\right)_j, \qquad (2)$$

where $\varepsilon_{ik}(0) = \varepsilon_{ki}(0)$, ε_{ikj} is the totally antisymmetric third-rank tensor, $\widehat{g}$ is the *gyration tensor*, and an identical relation is valid also for χ_{ik}. Apparently, on the basis of Eq. (2), a gyrotropic medium must be invariant with respect to some mirror reflections. This property can be caused by the symmetry of a gyrotropic medium: by the presence of a *screw axes* in the crystal (see *Crystal symmetry*) as well as by a screw-like character of a molecular ordering in *cholesteric liquid crystals*, and spins in magnets with *modulated magnetic structures*. It can manifest itself in both amorphous and liquid phases of a matter if the lack of bilateral symmetry is inherent in particles composing the material. Weak but observable gyrotropic effects appear due to the violation of space parity in weak interactions.

In the presence of a magnetic induction $\boldsymbol{B}_0 = \mu_0(\boldsymbol{H}_0 + \boldsymbol{M}_0)$, the medium can be gyrotropic even without taking into account space dispersion. In these cases, *magnetic gyrotropy* is said to be present, and the gyrotropic medium is called *magnetogyrotropic*. In magnets, the magnetogyrotropy is present even without the external magnetic field $\boldsymbol{B}_0$ due to the *magnetization* $\boldsymbol{M}_0$ (in magnets with a complicated type of magnetic ordering, e.g., in *antiferromagnets*, the contribution of the *antiferromagnetism vector* $\boldsymbol{L}$ can be significant). A magnetogyrotropic medium can be both gyroelectric (e.g., a plasma in a magnetic field, including a *solid-state plasma*) and gyromagnetic (magnets: *paramagnets* and materials with magnetic ordering) at the same time. A gyrotropic medium is characterized by a variety of particular features in regard to electromagnetic wave propagation (see *Rotation of light polarization plane*).

Hh

HADFIELD STEEL (after English metallurgist R.A. Hadfield)

High-manganese austenite *steel* which contains 11 to 15% Mn and 1 to 1.5% C. It can be strongly hardened under dynamic loads. After slow cooling the structure consists of *austenite* and double *carbides* of the cementite type $(Fe, Mn)_3C$. To obtain a single-phase austenite structure it is heated up to 1050–1100 °C for several hours with subsequent cooling in water. Hadfield steel is used as a resistant to wear material under conditions of moderate wear, but it has low resistance to abrasive *wear*. It has good *yield* (in the liquid state) which allows its use as a casting material (see *Casting*). A strong tendency to *strain hardening* complicates rolling and forging. Poor *thermal conductivity* is a cause of forming hot *cracks* in casting and cracks upon the accelerated heating for *quenching*.

HAFNIUM, Hf

A chemical element of Group IV of the periodic table with atomic number 72 and atomic mass 178.49. Natural hafnium consists of 6 stable isotopes with the mass numbers 174, 176 to 180. Electronic configuration of outer shells is $4f^{14}5d^26s^2$. Successive ionization energies are 7.5, 15.0, 23.3, 33.3 eV. Atomic radius is 0.159 nm; radius of Hf^{4+} ion is 0.078 nm. Oxidation is state is +4 (less common +3, +1). Electronegativity is 1.23.

In the free form hafnium is a silvery-white *metal*; it is pliable, easily subjected to cold and hot processing (rolling, forging, stamping, milling, drilling). It exists in α- and β-modifications. α-Hf has hexagonal close-packed lattice with parameters $a = 0.3188$ nm, $c = 0.5042$ nm at 273 K; space group $P6_3/mmc$ (D_{6h}^4). Above 2013 K up to $T_{\text{melting}} \approx 2500$ K β-Hf is stable, body-centered cubic with lattice parameter $a = 0.351$ nm, space group $Im\bar{3}m$ (O_h^9). Binding energy of pure hafnium is 6.35 eV/atom (at 0 K). Density is 13.3 g/cm^3, $T_{\text{boiling}} = 4870$ K; latent heat of melting is 21.8 kJ/mole; latent heat of sublimation is 712 kJ/mole; latent heat of evaporation is 650 kJ/mole; specific heat 143 J·mole^{-1}·K^{-1} (at 298 K); Debye temperature is 252 K for hafnium with purity 99.96%; linear thermal expansion coefficient is $5.9\cdot10^{-6}$ K^{-1} (in the temperature range 273 to 1273 K); thermal conductivity coefficient with 2 at% Zr decreases from 22.32 to 20.52 W·m^{-1}·K^{-1} (in the range of temperatures 323 to 723 K); adiabatic coefficients of elastic rigidity of α-Hf monocrystal: $c_{11} = 181.1$, $c_{12} = 77.2$, $c_{13} = 66.1$, $c_{33} = 196.9$, $c_{44} = 55.7$ GPa (at 298 K); bulk modulus is 108.9 GPa, Young's modulus is 83.4 GPa; rigidity is 30.4 GPa, Poisson ratio is 0.37 (at 298 K); Vickers hardness is 152 in HV units for hafnium iodide (0.72% of Zr); thermal neutron capture cross-section is 105 barn; ion-plasma frequency is 13.3 THz. Sommerfeld linear temperature term in low-temperature specific heat is 2.15 mJ·mole^{-1}·K^{-2}; electrical resistivity is 306 nΩ·m (at 295 K), temperature coefficient of electrical resistivity is 0.00351 K^{-1} (273 to 1073 K); work function of polycrystal is 3.53 eV; superconducting transition temperature is 0.13 K; hafnium is paramagnetic with molar magnetic susceptibility $+70\cdot10^{-6}$ cgs; nuclear magnetic moment of ^{177}Hf isotope (18.39% abundant) is 0.61 nuclear magnetons.

Hafnium is used in nuclear power engineering (control rods of reactors, screens protecting from neutron radiation) and in electronic engineering (cathodes, getters, electrical contacts), manufacture of *high-temperature alloys*. A solid solution of hafnium and tantalum carbides is the most refractory *ceramic* material ($T_{\text{melting}} = 4270$ K).

HAGEN–RUBENS APPROXIMATION (E. Hagen, H. Rubens, 1903)

Theory of optical absorption in *metals* and narrow-band *semiconductors* at low frequencies $\omega \ll \omega_\tau$ ($\omega_\tau = 1/\tau$ is the collision frequency, τ is the electron mean free path time); according to this theory, optical properties of materials depend only on their static conductivity σ_0. For this case the *refractive index* $\tilde{n} = n + \mathrm{i}\chi$, where $n \approx \chi = [\sigma_0/(2\varepsilon_0\omega)]^{1/2}$, $\varepsilon_0 = 8.85 \cdot 10^{-12}$ F/m is the vacuum *dielectric constant*. In this case, the *reflectance* $R \approx 1 - 2/n$. The deviation from total reflection is given by the Hagen–Rubens formula $1 - R \approx (8\varepsilon_0\omega/\sigma_0)^{1/2}$, i.e. the lower the conductivity of metals and semiconductors, the lower is their reflecting power. In the Hagen–Rubens approximation, the *absorption coefficient* $\alpha \approx (2\mu_0\sigma_0\omega)^{1/2} \propto \omega^{1/2}$ ($\omega \ll \omega_\tau$), where $\mu_0 = 4\pi \cdot 10^{-7}$ H/m is the *magnetic permeability of vacuum*. At high frequencies, $\omega \gg \omega_\tau$, the *Drude formula* is valid, according to which $\alpha \propto \omega^{-2}$.

HALL EFFECT

See *Galvanomagnetic effects*.

HARDENING of materials

A process for increasing the *strength* of a material. *Strain hardening* is widely used for increasing the strength of metals. The *strain* may give rise to a "*forest of dislocations*" (uniformly distributed dislocation structure), or to a *cellular structure* (see *Fragmentation*). In the former case, the hardening achieved is higher ($\Delta\tau \approx \alpha G b \sqrt{\rho}$, where $\alpha = 0.1$–1.5, G is the *shear modulus*, b is *Burgers vector*, ρ is the *dislocation* density), but it is accompanied by a decrease of *plasticity*. The hardening achieved in the latter case is somewhat lower, but the structure obtained is more stable, the hardening process being accompanied by a decrease of the *cold brittleness* temperature, and an increase of plasticity at low temperatures. At high enough degrees of reduction, when the disorientation of individual dislocation cells relative to one another exceeds a certain critical value (~2–3° of arc), the cell boundaries start to behave as *grain boundaries*, acting as barriers to dislocation motion. In this case, the hardening process is described by the *Petch relation*, where the value of the cell size is substituted for the grain size. Strain hardening is carried out during the process of plastic working (*pressing*, *forging*, rolling, *drawing*, etc.). The nature of the dislocation structure is determined by the degree and the temperature of the strain.

The hardening due to the formation of dispersed particles of the second phase (see *Precipitation-hardened materials*, *Dispersion hardening*) may be either direct or indirect. Direct hardening is related to the direct interaction between dislocations and dispersed particles, the latter acting as barriers for dislocations, which slip during the course of *plastic deformation*. Indirect hardening is related to the fact that the presence of dispersed particles of the second phase brings about an increase of stability of the nonequilibrium structural state, and an increase of the *recrystallization* temperature. Precipitation-hardened systems are obtained by means of *alloy aging* (age hardening), by application of *powder metallurgy* methods, etc.

Hardening through *alloying* is widely used in engineering; the method being based on the interaction between dislocations and the atoms of dissolved elements (see *Solid-solution hardening*).

In the case of *alloys* subject to *phase transitions*, the hardening is achieved through phase recrystallization during the course of a *heat treatment* (*quenching*, *tempering*, etc.). The hardening in this case is due to a directional change of the phase composition, grain refining, phase peening (see *Strain hardening*, *Work hardening*).

See also *Dispersion hardening*, *Flash hardening*, *Irradiation hardening*, *Linear hardening of crystals*, *Solid-solution hardening*, *Bulk sound hardening*, *Shock hardening*, *Surface hardening*, *Surface sound hardening*.

HARD MAGNETIC MATERIALS, high-coercive materials, hard magnetics

Materials characterized by a large *coercive force* ($B_c = 10^{-2}$–1 T). The origin of high B_c in hard magnetic materials lies in the lag of *magnetic reversal* in ferromagnets, arising from three main mechanisms: (i) *irreversible magnetization rotation processes*; (ii) lag in the formation and growth of remagnetization nuclei; and,

Table 1. Principal characteristics of the most important hard magnetic materials (1 kA/m corresponds to 1.26 mT and 12.6 Oe). Note that $B_c = \mu_0 H_c$

Material	Composition	Magnetic properties						Remarks
		H_c		B_r		$(BH)_{max}$		
		Oe	kA/m	G	T	10^6 G/Oe	kJ/m^2	
Alni	15.5% Al, 25% Ni, 4% Cu, the rest Fe	500	40.0	5000	0.5	0.9	7.2	
Alnico	9% Al, 14% Ni, 24% Co, 4% Cu, 0.3% Ti, the rest Fe	550	44.0	12,300	1.23	4.0	32	Thermo-magnetic treatment
Alnico	9% Al, 15% Ni, 25% Co, 4% Cu, 0.8% Nb, the rest Fe	775	62	12,800	1.28	6.6	52.8	Thermo-magnetic treatment, Columnar structure
Tikonal	7.6% Al, 14% Ni, 35% Co, 3.5% Cu, 5% Ti, 0.8% Nb, the rest Fe	1560	125	11,200	1.12	12.0	96	Thermo-magnetic treatment, Columnar structure
Vicalloy 2	52% Co, 13% V, the rest Fe	370–470	29.6–37.6	9000–9500	0.9–0.95	1.0–1.75	8–14	
Cunife 2	50% Cu, 20% Ni, 2.5% Co, the rest Fe	260	20.8	7300	0.73	0.35–0.40	2.8–3.2	Anisotropic
Cunico 2	35% Cu, 24% Ni, 41% Co	450	36.0	5300	0.53	0.5	4.0	
Pt–Co	76% Pt, 24% Co	4800	384	64,000	6.4	9.2	73.6	
Fe–Co–Cr	63% Fe, 25% Cr, 12% Co	630	50	14,500	1.45	7.7	61	Anisotropic
Mn–Al–C	70% Mn, 29.5% Al, 0.5% C	2700	216	6100	0.61	7.0	2.8	Anisotropic
Barium ferrite	$BaO{\cdot}6Fe_2O_3$	1450	116	4080	4.08	3	24	Anisotropic
Sm–Co	$SmCo_5$	9500	760	9800	9.8	24	190	Anisotropic
Sm–Co–Fe–Cu–Zr	$Sm(Co,Fe,Cu,Zr)_8$	10,000	800	12,000	1.2	33	260	Anisotropic
Nd–Fe–B–Co	$Nd_{15}Fe_{62.5}B_{5.5}Co_{16}Al_1$	11,000	880	13,200	13.2	41	324	Anisotropic

(iii) pinning of *domain walls* at various crystal irregularities and defects. Hard magnetic materials find their use in *permanent magnets*, hysteresis motors, and magnetic recording. Employed in the last two cases are *magnetic materials* with moderate $B_c = 10^{-3}$–10^{-2} T, while hard magnetic materials with $B_c = 10^{-2}$–1 T are used for permanent magnets. The appraisal of a hard magnetic material is mainly based on its *demagnetization curve*, i.e. on the portion of the major *hysteresis* loop lying in the 2nd quadrant.

The most important parameters of a hard magnetic material, along with B_c, are the remanence, B_r, and the maximum value of the product of fields on the demagnetization curve, $(BH)_{max}$, the so-called *energy product*.

The high-coercive state in hard magnetic materials is formed following a certain production cycle (sometimes, a rather complex one). In the context of present day technology hard magnetic materials can be classified into the following groups (see Table 1):

(1) *steels* hardened into *martensite*, displaying relatively low B_c, and rarely used;
(2) rigid cast alloys like *alni*, *alnico* and *ticonal*, displaying a broad spectrum of magnetic properties; these alloys are the most common materials for permanent magnets;
(3) ductile alloys like *vicalloy*, *cunife*, *cunico*, the alloys Fe–Co–Cr, Mn–Al–C, and also *noble metal alloys*: Pt–Co, Pd–Fe and Pt–Fe;
(4) powdered hard magnetic materials produced by powder *pressing* followed by *heat treatment*. Metalloceramic, and metalloplastic hard magnetics are distinguished here, along with *oxide magnets* and *powder magnets*.

Metal-ceramic hard magnetic materials are fabricated from a metallic powder by compaction without any binder, with a subsequent high-temperature sintering. Magnets of the largest energy capacity, based on the *rare-earth elements* (e.g., the *samarium–cobalt magnets*), are metalloceramic magnets. Metalloplastic hard magnetics are fabricated by compaction of powder with an insulating binder that polymerizes at a moderate temperature (see *Magnetodielectrics*).

HARD MAGNETIZATION AXIS

The direction of the *magnetization* vector of a ferromagnet, $\boldsymbol{M}$, which corresponds to the maximum of the *magnetic anisotropy* energy. To magnetize an infinite *magnetic substance* (with no elastic *strains* present) along the difficult magnetization axis, it is necessary to subject the crystal to the action of an external magnetic field with a strength greater than that needed to magnetize it along any other direction. If the magnetic anisotropy is not unidirectional, the difficult magnetization axes are axial directions; so their number is always even. The direction corresponding to the absolute maximum of the magnetic anisotropy energy is sometimes referred to as the *most difficult magnetization axis*. See also *Easy magnetization axis*.

HARDNESS

The ability of materials to resist *plastic deformation* or *brittle failure* under local loading. Its magnitude and dimensionality depend on the method and conditions of measurement. The most widespread method is the static indention of an especially loaded solid *indenter* into the material to be studied. In this case, the hardness is determined by a ratio between the load and the imprint area. Applied also are tests with dynamic indenting, pendular methods, methods of elastic recoil, sclerometric methods (scratching tests), and so on. There are established correlations of hardness with various mechanical characteristics of materials, such as *elastic moduli*, *yield limit*, *ultimate strength*. See also *Microhardness* and *Macrohardness*.

HARDNESS DIAGRAM

See *Indenter–hardness diagram*.

HARDNESS, MAGNETIC

See *Magnetic hardness*.

HARDNESS, SECONDARY

See *Secondary hardness*.

HARD-SPHERE MODEL

Model for calculating physical properties of solids and liquids, which consists in replacing the actual atomic volume with solid spheres of diameter b. Then, the two-body repulsive potential becomes a potential of absolutely hard spheres

$$\Phi(R) = \begin{cases} \infty, & R < b, \\ 0, & R > b, \end{cases}$$

where R is the interatomic distance (see *Interatomic interaction potentials*). This replacement helps to solve a number of problems in a purely geometric manner. For instance, if spheres of equal diameter are arranged in close-packed layers one atop another, then, depending on the periodicity in the positions of the layers, different structures (*crystal lattices*) arise. In particular, two-layer periodicity results in a hexagonal close-packed structure, while three-layer periodicity provides a face-centered cubic structure. Violations of periodicity cause *stacking faults*. One can obtain more complicated structures by taking spheres of different diameters in certain proportions. The volume of one and the same atom depends on the type of solid crystal; and metallic, ionic, van der Waals, and covalent diameters (radii) of atoms have been tabulated (see *Atomic radii*).

The hard-sphere model simplifies the description of processes for the generation of radiation damage (see *Radiation physics*, *Subcascades*). Indeed, the scattering laws for hard spheres are the simplest; the probability of scattering through a certain angle (and, accordingly, of energy to be transferred to a crystal atom) is independent of the incident atom energy. In order to make the potential more realistic and allow for the anisotropy of scattering, the diameter of colliding spheres (collision diameter) is considered to be a variable that depends on the energy E of the incident particle. This diameter is found from the condition of the equality of the real repulsion potential and the relative kinetic energy of the colliding atoms (two-body problem). For instance, in a solid consisting of atoms of one type, the repulsion is describable by the *Born–Mayer potential*, with $b = a\ln(2A/E)$, where a and A are constants of the Born–Mayer potential (see *Interatomic interaction potentials*).

HARMONIC APPROXIMATION

Expansion of the vibrational energy of a solid in terms of displacements of atoms from their equilibrium positions, and retaining only the first-order or linear terms (quadratic in the displacements) provides the following expression:

$$E = \frac{1}{2}\sum_{l\alpha} M_l \left(\frac{\partial \overline{U}_\alpha^l}{\partial t}\right)^2 + \frac{1}{2}\sum_{l\alpha,l'\alpha'} V_{\alpha\alpha'}(R_{ll'})U_\alpha^l U_{\alpha'}^{l'},$$

which leads to the oscillator energy sum

$$E = \sum_{k,\lambda} \hbar\omega_\lambda(k)\left[n_\lambda(k) + \frac{1}{2}\right].$$

Here $R_{ll'}$ is the radius vector connecting atoms l and l'; M_l is the lth atom mass; U_α^l is the displacement of lth atom from its equilibrium position along the αth coordinate axis; $V_{\alpha\alpha'}(R_{ll'})$ is the coefficient in the first non-vanishing term of the energy expansion, t is the time; $\omega_\lambda(k)$ is the frequency of λ-polarized *phonons* with *quasi-momentum* $\hbar\boldsymbol{k}$; and $n_\lambda(k)$ are the occupation numbers of phonons. In the harmonic approximation, the Helmholtz free energy of a solid is

$$F = k_B T \sum_{k\lambda} \ln\sinh\left[\frac{\hbar\omega_\lambda(k)}{k_B T}\right] = k_B T \int g(\omega)\ln\sinh\left(\frac{\hbar\omega}{k_B T}\right)\mathrm{d}\omega,$$

where k_B is the *Boltzmann constant*, and $g(\omega)$ is the phonon *density of states*. If the dependence of the phonon frequency on the unit cell volume is taken into account in this expression, then one obtains a quasi-harmonic approximation taking into account anharmonic terms in E which are even with respect to displacements (see *Anharmonic vibrations*). The harmonic approximation is the most widely used approach for treating acoustic and optical vibrations, as well as the thermodynamic properties of solids.

HARTREE–FOCK METHOD (D. Hartree, 1928; V.A. Fock, 1930)

Quantum-mechanical method of description and calculation of systems of interacting fermions, which is widely applied to many-electron systems: atoms, molecules, crystals. Sometimes the term "Hartree–Fock method" is used in reference to a purely multiplicative approximation for the wave function, a set of coupled equations to solve, one for each electron, containing Coulomb ($\int \psi_k^* V \psi_k \, d\tau$) and exchange ($\int \psi_j^* V \psi_k \, d\tau$) integrals, and occasionally exchange terms are neglected. In the usual Hartree–Fock method the wave function of a system of N electrons has the form of a *Slater determinant*, the product of one-electron wave functions (spin-orbitals) $\psi_1, \psi_2, \ldots, \psi_N$. As a rule, the spin-orbitals are assumed to be orthonormal, and are determined through the variational principle. The conditions for a stationary energy functional lead to the problem of determining pseudo-eigenvalues of the *Fock operator* (*Fockian*): $F(P)\psi_k = \varepsilon_k \psi_k$, where $F(P) = h + J(P) - K(P)$, h includes the operator of the kinetic energy of the electrons and the potential energy of attraction of the electrons to the nuclei. The Coulomb term $J(P)$ and exchange operator $K(P)$ are defined on spin-orbitals ψ_k through the *Fock–Dirac density matrix*:

$$P = \sum_{1 \leqslant k \leqslant N} |\psi_k\rangle\langle\psi_k|.$$

The term $J(P)$ includes the diagonal elements of P, which represent the density of particles. In practice, the solutions are obtained through iterative procedures and self-consistency. In deriving the variational equations, the self-action terms are formally included in operators F and K; these terms compensate each other, and this justifies the formation of the Hermitian Fockian, which is common for all spin-orbitals. This is significant for fulfilling the conditions of spin-orbital orthonormality. The initial *Hartree method* used a wave function product of spin-orbitals without antisymmetrization. In this case exchange terms do not appear, and each spin-orbital ψ_k is related to its operator F_k, which does not contain the self-action term of the kth electron in $J(P)$.

In the Hartree–Fock approach the spin-orbitals are specified in the form $\varphi_m\alpha$ and $\bar{\varphi}_m\beta$, where α and β are spin wave functions corresponding to the two spin directions. In the restricted Hartree–Fock method, the functions of the space coordinates (*orbitals*) φ_m and $\bar{\varphi}_m$ are assumed to have an identical form (spin limitation) and to transform with respect to the irreducible representation of the symmetry group of the system under consideration (symmetry limitation). For crystals φ_m can be expressed in Bloch form. The nonlinearity of the Hartree–Fock equations is responsible for their plurality of solutions. The solutions of the restricted Hartree–Fock method may be unstable, i.e. may correspond to a saddle point of the energy surface. Then the decrease in energy at the removal of spin or symmetry restrictions (unrestricted Hartree–Fock method) can be accompanied by the appearance of components of different wave function multiplicities of the system, or by a violation of its symmetry (solutions of the charge density wave or spin density wave type in crystals).

The Hartree–Fock equations are integrodifferential because of the nonlocality of $K(P)$. Even in the simplest cases they cannot be solved analytically. The spin-orbitals in crystals are constructed through the *linear combination of atomic orbitals* (LCAO) approximation, or through expansion in a plane wave basis set, with coefficients determined by the variational method. Atomic orbitals are constructed from Slater (exponential) or Gaussian functions, or through the use of solutions of the atomic problem with the so-called *muffin-tin potential*. The necessary matrix elements are found, as a rule, by numerical integration, but sometimes (e.g., with the aid of Slater's basis set) they may be found analytically.

Besides this nonempirical (*ab initio*) approach, based on an algebraic modification of the Hartree–Fock method, various semiempirical and simulation methods have been developed for calculating the electronic structure of molecules and crystals. In these methods certain parts of molecular integrals are neglected, so the remainder are calculated using simplified equations, or are considered as parameters (see *Hoffmann method, Mulliken–Wolfsberg–Helmholtz method*). A drawback of the Hartree–Fock method is that it neglects electron correlations, which are due to Coulomb interactions between electrons; these correlations are especially important for electrons of opposite spin. The error of the Hartree–Fock method with respect to the determination of the binding energy (*correlation energy*) of atoms and molecules reaches 1–2 eV per electron pair. To take into account electron correlations, which are not always included in perturbation theory, it is sufficient to employ self-consistent approaches, which make use of the basic set of several Slater determinants (many-configuration approach; see *Many-configuration approximation*).

HASIGUTI PEAK (R.R. Hasiguti, 1960)

Maximum point of the temperature dependence of the *internal friction* of crystals with a face-centered cubic structure, which contain *point defects*: *impurity atoms*, *vacancies*, *interstitial atoms*. It is observed in samples which are deformed at some temperature and quickly cooled (quenched). The Hasiguti peak is caused by thermally activated disconnections of *dislocations* from point defects under the action of oscillating stresses which accompany low-frequency vibrations of the sample. There are several variations of Hasiguti peak theory (Hasiguti himself proposed five models), which specifically take into account the motion of dislocations, and interactions between dislocations and point defects (see *Point defect–dislocation interaction*).

HEALING OF DEFECTS

Processes involving redistribution and annihilation of *defects* (see *Annihilation of dislocations*) in crystals. Healing of defects includes *restoration of crystals*, *elastic recovery*, *recovery* of properties, *polygonization*, and *recrystallization*. Practically complete healing of atomic scale defects (nonequilibrium *point defects*, *dislocations*) occurs upon heating (*recrystallization annealing*), while partial healing of defects is also possible during the course of *plastic deformation* (*dynamic recovery*, *dynamic recrystallization*). To heal defects of a macroscopic scale (macrodefects, *pores* in solids, *cracks*), the heating is combined with *high pressure* processing. The process of *hydroextrusion* can bring about the dynamic healing of defects of a microcrack type. Through this healing, the totality of structurally sensitive properties of the solid is changed. The *mechanical properties of solids* most affected are the decrease of the *yield limit* and *strength* (see *Dehardening*), and the increase of the *plasticity*. The healing of defects is accompanied by the release of *stored elastic energy*, and by an increase in *density*. The change in *electrical conductivity* during healing of defects depends on the electronic structure. The annihilation of defects in *metals* increases this conductivity, whereas the inverse phenomenon is observed in semiconductors and *insulators*. *Type II superconductors* change their parameters upon the healing of defects due to the decrease in the *vortex pinning*. Healing of defects affects *ferromagnets* also: the *coercive force* decreases, but the *magnetic permeability* increases. The healing of defects can influence magnetic properties indirectly through the change in *texture*.

HEALING OF SOLIDS with cracks, restoration of solids with cracks

Recovery of continuity and *strength* of a damaged solid, i.e. healing of an already existing crack (see *Healing of defects*). The latter may go either via diffusion (see *Diffusion*) or via direct recovery of molecular bonds between the lips of the opened crack (in the absence of *mass transport*). An example of this restoration is the collapse of the *Rosette channels* when twinning interlayers leave the volume of a *monocrystal*.

The strength of a restored crystal typically reaches only 70–80% of its initial strength. This is due to the bonding being only partially recovered through the body cross-section. The healed crack contains various discontinuities, e.g., hollow channels along the *spalling* steps and fronts formed at crack stops.

HEAT CAPACITY

See *Specific heat*, *Surface specific heat*.

HEAT CONDUCTION

See *Thermal conductivity*.

HEAT CONDUCTIVITY FACTOR

See *Thermal conductivity coefficient*.

HEAT CONTENT

See *Enthalpy*.

HEAT EMISSION

See *Heat loss.*

HEAT EXCHANGE

See *Heat transport.*

HEATING PULSE

See *Pulse heating.*

HEAT LOSS, thermal loss, heat release, heat emission, heat removal

A *heat transport* between a solid surface to a medium (liquid, gas), called a carrier of heat, which is in contact with the surface. The heat loss occurs through convection, *thermal conductivity*, and radiation. There exist heat losses at the free and induced movement of the heat carrier, as well as at a change of the *state of matter*. The heat loss intensity is characterized by a thermal loss factor $\alpha = \delta\Phi/(\Delta T \mathrm{d}A)$ where $\delta\Phi$ is the heat flux through a surface element of area $\mathrm{d}A$, ΔT is the temperature difference (sometimes called temperature pressure) between the medium and the surface. In SI units the thermal loss factor α is expressed in the units $W{\cdot}\mathrm{m}^{-2}{\cdot}\mathrm{K}^{-1}$. Heat loss can be regarded as part of a more general process of *heat transfer*. The heat transfer ΔQ at the temperature T is accompanied by a change of entropy ΔS, where $\Delta S = \Delta Q/T$. The loss of heat by a material lowers its temperature, and also its entropy. The mechanisms and kinetics of heat loss at a change of state of matter of heat carriers (including cases of boiling and condensation) depend on the conditions of *wetting* the confining walls by the liquid.

HEAT OF MELTING

The amount of heat L_{melting}, called latent heat, transferred to a crystalline solid during the equilibrium isobaric–isothermal process of its transformation into the liquid state; a particular case of the latent *heat of phase transition*. There is a *specific heat of melting* (measured in J/kg, or kcal/kg) and a *molar heat of melting* (J/mole). The heat of melting is expended for weakening the interatomic (intermolecular) bonds in the material to break them apart, resulting in *melting*. The heat of melting of metals as well as their *melting temperature* T_{melting} increases as the electron concentration grows (with increasing interatomic energy).

Many *covalent crystals* melt with the complete severing of their directed two-electron bonds and convert to the metallic state (Si, Ge, Sb, Bi, etc.); the heat of melting of these crystals is a maximum (see Table 1). Amorphous and vitreous materials, which do not possess a permanent T_{melting} and soften as the temperature rises, also lack a definite heat of melting (see *Vitreous state of matter*). Those alloys which solidify over a range of temperature (between the liquidus and solidus

Table 1. Latent heat of melting for some substances

Material	T_{melting}, °C	L_{melting}, kcal/kg	L_{melting}, kJ/kg
Hg	−38.86	2.82	11.9
H_2O (ice)	0.0	79.72	334
Na	97.8	24.4	102
Sn	231.9	14.4	60.2
Pb	327.4	5.9	24.7
Zn	419.5	24.4	102
Al	660.4	94.5	385
Ag	961.9	25.0	105
Au	1064.49	15.3	64
Cu	1084.5	49.0	205
Si	1415.0	337.0	1409
Co	1494.0	62.1	264
Fe	1539.0	63.7	266
Cr	1890.0	62.1	264

temperatures, see *Phase diagram*) can be specified by the ratio of their heat of melting to their melting temperature range. The heat of melting is related with T_{melting}: as T_{melting} increases (with the same kind of interatomic bonds), the heat of melting grows. The ratio $L_{\text{melting}}/T_{\text{melting}} = \Delta S_{\text{melting}}$ is the *entropy of melting*. The heat of melting can be found from the *Clausius–Clapeyron equation* $\mathrm{d}P/\mathrm{d}T = L_{\text{melting}}/(T\,\Delta V)$, or directly by means of the methods of *calorimetry*.

HEAT OF PHASE TRANSITION

An amount of heat Q_{L} called *latent heat* transmitted to (or removed from) material undergoing a *first-order phase transition*. At a *second-order phase transition* the latent heat is zero. An equilibrium isobaric-isothermal transition at a given pressure takes place at a constant temperature T_{c} that is the *phase transition* temperature. The latent heat equals the product of the phase transition temperature and the entropy difference ΔS between the two phases involved into the transition, i.e. $Q_{\text{L}} = T_{\text{c}}\Delta S$. The latent heat can be designated as specific or molar when referred to 1 g (CGS), 1 kg (SI), or 1 mole of matter, respectively (see *Heat of melting*, *Heat of polymorphic transformation*).

HEAT OF POLYMORPHIC TRANSFORMATION

An amount of heat generated (absorbed) at an equilibrium isobaric-isothermal transformation of matter from one polymorphic modification to another (see *Polymorphism*). This is a particular case of the latent *heat of phase transition*.

HEAT OF SOLUTION

An amount of latent heat generated (absorbed) by material dissolving in solution. The amount of heat absorbed per molecule of dissolved matter when dissolving under constant pressure is $q = k_{\text{B}}T^2\partial\ln c_0/\partial T$ where c_0 is the solubility, that is the concentration of saturated solution in *thermodynamic equilibrium* with the dissolving matter. This expression is valid for $c_0 < 1$. If $q < 0$ then heat is evolved during the dissolving process.

HEAT RADIATION

See *Thermal radiation*.

HEAT-REFLECTING COATINGS

Coatings put at the surface of *construction materials* to reflect the incident *infrared radiation* from a heat source. In the IR spectral range of electromagnetic waves, the reflection ability of solids is determined mainly by the presence of free *conduction electrons*. Therefore, *metals* possess the highest reflection ability. Al, Au, Ag and Cu have the highest IR radiation reflection coefficients, and its magnitude increases with the increase of wavelength λ; at $\lambda = 10$ μm the reflection coefficient reaches 98%. *Films* 100–200 nm thick made of the metals listed above, with their oxides deposited on the film surface using chemical or electrical deposition, or *sputtering* under high vacuum, are the most widespread heat-reflecting coatings. These coatings are applied in engineering, and in the power, aircraft, and rocket-space industries in order to regulate the heating cycle of construction surfaces operating under predominantly radiative *heat transport* conditions. For protection against *thermal radiation* one often uses films and glasses with oxide or metal coatings 1–5 nm thick that reflect up to 70% of the incoming heat.

HEAT RELEASE

See *Heat loss*.

HEAT REMOVAL

See *Heat loss*.

HEAT SINK

A macroscopic system designed for the transfer to and storage of thermal energy. The heat energy is a portion of the internal energy of a body (see *Thermodynamic potentials*). As a rule, any large system in a thermal equilibrium state can serve as a heat sink. For equilibrium systems the concept of *temperature* is introduced, and the ability to transfer and to accumulate thermal energy is characterized by the thermal conductivity and the *specific heat*. Ordinarily we assume that the processes of energy exchange with the heat sink do not appreciably affect its equilibrium condition. If its temperature remains constant, the heat sink is called a *thermostat*. The amount of thermal energy stored in a heat sink increases with the growth of the number of degrees of freedom, and an increase in temperature. Thermal equilibrium is often, but not always, present in a heat sink. A *phonon sink* is a particular case of a heat sink when the *phonon* energy (of lattice vibrations) is the only energy being utilized.

HEAT SWITCH

A device which provides a sharp change of *heat transfer* between objects (e.g., between the consecutive circuits of a cooling system). The most widespread types of heat switches are the following: mechanical heat switches where the heat linkage is through mechanical contact (e.g., by the compression of elastic plates which detach at the end of the compression); gaseous heat switches where the heat transfer takes place through the gas *thermal conductivity* (the gas is evacuated to break the contact); superconducting heat switches which utilize the large difference in the thermal conductivity of metals in normal and superconducting states (see *Superconductivity*). The latter is widely used in obtaining temperatures below 1 K owing to the simplicity of controlling such a heat switch: its transition from the superconducting to the normal state is easily brought about by applying a magnetic field that exceeds a critical value.

HEAT TRANSFER

Heat transport between two carriers of heat through a solid wall or interface separating them. A heat transfer process includes a *heat loss* from a hot liquid or gas to a wall, *thermal conductivity* in the wall, and the heat loss from the wall to a colder movable medium outside. The magnitude of the heat transfer is characterized by a heat transfer coefficient $k = \delta Q/(\Delta T\, \mathrm{d}S)$ where δQ is the heat flux through a surface element of area $\mathrm{d}S$, and ΔT is the temperature difference (sometimes called temperature pressure) between the medium and the surface. The value $R = 1/k$ is called the *overall thermal resistance*. In most cases k is determined by testing.

HEAT TRANSPORT, thermal transport, heat exchange

A process of transport of energy in the form of heat through a medium with a nonuniform *temperature* distribution, or other nonuniformities of physical fields. Heat transport in solids is carried out by the heat propagation inside a body (see *Heat transfer*), and by the interaction of its surface with the environment (see *Heat loss*). The main kinds of heat transport are *thermal conductivity*, convection, *thermal radiation*, and heat transfer at *phase transitions* (*heat of phase transition*). The heat transport depends on the geometric dimensions and the shape of the body, its temperature and thermal conductivity, the degree of brightness of its surface, and also on the temperature and the physical properties of the external medium. Its intensity serves as a quantitative measure of the transport of heat, and this is expressed through a thermal transport coefficient with the dimensionality $\mathrm{W{\cdot}m^{-2}{\cdot}K^{-1}}$. The heat transport at phase transformations in solids is due to the change in enthalpy (see *Thermodynamic potentials*); this is determined by both the thermal conductivity and the ability of a body to accumulate heat, i.e. by its *diffusivity*.

HEAT TREATMENT

Modification of the structure, phase composition, and related physical and mechanical properties of *metals* and *alloys* by subjecting them to heating and cooling. The temperature and its rate of change can vary over wide ranges, determined by the combination of a number of material characteristics.

There is a hardening and dehardening treatment, and also treatments for imparting some technological characteristics to a metal. The hardening effect is achieved by raising the degree of precipitation of hardening *phases* (see *Precipitation-hardened materials*) which are associated with the host, or with the formation of other concentration and structural inhomogeneities enabled to play a role as obstacles for *dislocation* motion, e.g.: *grain boundaries*, *segregations*, twins (see *Twinning structure*), and so on. In the case of a dehardening heat treatment, the role of the above-mentioned factors is less. The principal kind of hardening heat treatment of *steels* is their *quenching* with subsequent *tempering*, and quenching for the next aging. Dehardening can involve various kinds of *annealing* applied to remove the consequences of cold *plastic deformation* (*strain hardening*).

To improve the technological properties of metals, special kinds of heat treatment are used.

Classified among them are spheroidizing, graphitizing, and homogenizing annealings, *patenting*, normalizing, and full *graphitization* (ferritization) of *cast iron*. Particular kinds of heat treatment include *decarburizing annealing* carried out either in air or in a hydrogen atmosphere, and *oxidation* in order to form an oxide film at the surface. The oxidizing of *iron*-based alloys combined with tempering and *burnishing*, improves the esthetical appearance of the metal. Combinations of heat treatment with other methods of material modification are widely used: *combined heat treatment*, *thermomechanical treatment*, *chemical heat treatment*, and so on. The term heat treatment is applied to practically all metal alloys when they undergo phase and structural transformations during heating and cooling. Related research approaches and investigations in the field of practical applications are the subject of such sciences as *metal science*, the *physics of metals*, the heat treatment and technology of metals.

See also *Mechanical heat treatment of metals*, *Chemical heat treatment*, *Combined heat treatment*, *Cyclic heat treatment*, *Rapid heat treatment*, *Rapid electric heat treatment*.

HEAVY FERMIONS in metals

Charge carriers of half-integer spin in solids of abnormally large *effective mass* $m^* = (10^2\text{–}10^3)m_0$. Heavy fermions exhibit a narrow, high amplitude peak in the *density of electron states* $g(E)$ in the neighborhood of the *Fermi level* E_F in a number of metallic compounds based on *rare-earth elements* (Ce, Eu, Sm, Tm, Yb) and the *actinides* (U). Compounds of this type are referred to as *heavy fermion systems*. The nature of heavy fermions is still not completely understood. It is generally assumed that for the majority of heavy-fermion systems ($CeAl_3$, $CeCu_2Si_2$, $CeCu_6$, etc.) the peak in $g(E)$ is the many-particle *Abrikosov–Suhl resonance* generated at E_F due to the antiferromagnetic *exchange interaction* between the conduction electrons and the localized magnetic moments of f-ions that make up the regular lattice. By analogy with a separate magnetic impurity in a metal (*Kondo impurity*), the latter lattice consisting of localized spins that interact with conduction electrons is given the name *Kondo lattice*. The properties exhibited by heavy-fermion systems are unusual for metals (see Table 1): the enormous value of the electronic *specific heat* coefficient, the enhanced Pauli *paramagnetism*, high value of the thermoelectric coefficient, the appreciable contribution of electron–electron scattering to the resistivity, extremely low value of the *Fermi velocity* ($\sim 10^5\text{–}10^6$ cm/s).

Of chief interest are *superconducting heavy fermion systems* represented by $CeCu_2Si_2$, UBe_{13}, UPt_3, URu_2Si_2. Although exhibiting low values of the critical temperature ($T_c = 0.5\text{–}0.9$ K), these superconductors feature extremely high *upper critical fields* B_{c2}, extremely high values of the dB_{c2}/dT derivative (at $T \to T_c$) (see Table 1), and are also highly sensitive to the presence of both magnetic and nonmagnetic impurities. The electronic heat capacity, *sound absorption*, *spin–lattice relaxation* time, and other characteristics of superconducting heavy-fermion systems behave at $T = T_c$ as power law functions of temperature rather than as exponential functions. This suggests that the superconducting gap $\Delta(k)$ goes to zero for certain directions of the wave vector $\boldsymbol{k}$. These anomalous properties of superconducting heavy-fermion systems are indicative of the peculiar nature of their superconductivity. In particular, ordinary BCS-pairing of two electrons with opposite spins (see *Bardeen–Cooper–Schrieffer theory*)

Table 1. Characteristics of heavy fermion superconducting systems, and their comparison with superconducting tin

Compound	$\gamma(T \to 0)$, mJ·mole^{-1}·K^{-1}	T_c, K	$B_{c2}(0)$, kG	$-\dfrac{dB_{c2}}{dT}$ $(T = T_c)$, kG/K
$CeCu_2Si_2$	1050	0.4–0.6	10–18	100–250
UBe_{13}	1100	0.86	150	250–400
UPt_3	450	0.54	16	63
Sn	1.78	3.733	0.3	0.14

in heavy-fermion systems may turn out to be less efficient than p-state or d-state pairing, which gives rise to a Cooper pair with a nonzero orbital moment $l \neq 0$ exhibiting singlet ($s = 0$) or triplet ($s = 1$) types of superconductivity (see *Triplet pairing*). Table 1 compares the superconducting characteristics of three heavy-fermion systems with those of the "normal" superconductor Sn.

HECTO...

A decimal multiple prefix, now rather antiquated, signifying a 10^2 times increase of the initial unit of a physical value. For example, 1 hW (hectowatt) $= 10^2$ W.

HEDGEHOG

A *point defect* in a vector-type field describing the long-range *orientational order* of a condensed medium. Hedgehogs are generated in isotropic *ferromagnets* in the magnetization vector field, as well as in *nematic liquid crystals*, and in type A *smectic liquid crystals* in the *director* field. Elastic distortions due to a hedgehog can spread over the entire bulk of the medium. The simplest hedgehogs are a *radial hedgehog* and a *hyperbolic hedgehog* where every possible spatial orientation of the vector occurs only once (see the figure); this allows a *topological invariant (charge)* N equal to unity to be correlated with each of them. When the direction of the arrows is reversed, N takes on the value of -1. In ferromagnets the coalescence and decay of hedgehogs are describable as simple addition of the charges N. For nematics the result of hedgehog merging and decay is ambiguous, since the sign of N is indeterminate due to the nonpolar nature of the nematic; e.g., two hedgehogs with equal charges $|N| = 1$ may merge either to annihilate ($|N| = 0$) or to form a hedgehog with $|N| = 2$. This depends on the merging path and on the presence of linear *disclinations* in the sample: in tracing around them, the director orientation changes by 180°. In nematics and smectics A hedgehogs similar to those sketched in the figure have been observed experimentally.

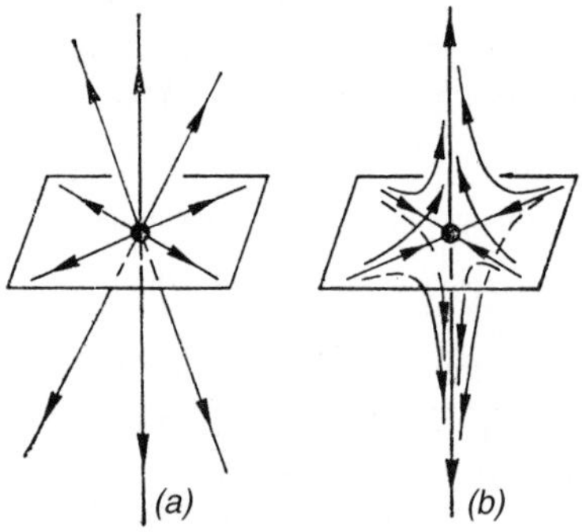

Radial (a) and hyperbolic (b) hedgehogs.

HEISENBERG HAMILTONIAN

A model *spin Hamiltonian* describing the *exchange interaction* in *insulators* containing atoms or ions with magnetic moments

$$\mathcal{H} = -J \sum_{(\boldsymbol{m},\boldsymbol{n})} \widehat{S}_{\boldsymbol{m}} \widehat{S}_{\boldsymbol{n}},$$

where $\widehat{S}_{\boldsymbol{m}}^{\alpha}$ ($\alpha = x, y, z$) is the operator of αth projection of the spin at the $\boldsymbol{m}$th site of a *crystal lattice*; J is the exchange integral for the atoms at sites $\boldsymbol{m}$ and $\boldsymbol{n}$. The Heisenberg Hamiltonian is a many-particle generalization of the *Heitler–London approximation* for a hydrogen molecule. The standard approximation is to take into account only the interactions of nearest neighbors, and to assume that they have a constant exchange integral $J_{mn} = J$, so the summation is taken over pairs of nearest neighbors. For $J > 0$ this Hamiltonian describes a ferromagnetic material (see *Ferromagnetism*) whose total spin in the ground quantum state has the maximum value NS (N is the number of atoms in lattice, S is the spin of an individual atom), i.e. classically all the spins are parallel. For $J < 0$ the Heisenberg Hamiltonian describes an antiferromagnetic material (see *Antiferromagnetism*) with minimal total spin in the ground state (0 or S, depending whether N is even or odd), for which the wave function and the ground state energy are known only approximately (exactly in the one-dimensional case for spin $S = 1/2$). Since the exchange interaction involved in the Heisenberg Hamiltonian is isotropic, the total spin of the system is conserved, and the stationary states can be classified in accordance with the projection of the total spin along a coordinate axis. A state with a single spin deviation from its ground state constitutes an elementary excitation of a ferromagnet, i.e. a *magnon*, which can propagate along the spin system, and states with several spin deviations are *spin complexes* which describe excited states of magnons.

HEISENBERG MAGNET

A *crystal* with magnetic ordering described by the *Heisenberg Hamiltonian*. In the case of a positive exchange integral $J > 1$ (see *Exchange interaction*), a Heisenberg magnet is a *ferromagnet*, otherwise (negative exchange integral, $J < 1$) it is an *antiferromagnet*. If the signs of the exchange integrals for the nearest and next to nearest neighbors differ, the magnetic structure is more complicated, for example, helical (see *Modulated magnetic structures*). The isotropic character of the exchange interaction affects the energy spectrum and macroscopic properties. The elementary excitations, i.e. *magnons*, in a Heisenberg magnet are *Goldstone modes* (zero-gap). Hence, the low-temperature additives to thermodynamic values are temperature-dependent in the form of a power (but not exponential) law. There is no long range ordering at finite temperatures in low-dimensional (one- and two-dimensional) Heisenberg magnets due to the formation of topological defects. In three dimensions there is a finite critical temperature (*Curie point* or *Néel point*) with a magnitude determined by the exchange integral.

HEITLER–LONDON APPROXIMATION

(W. Heitler, F. London, 1927)

A method of constructing the wave functions of the electrons of a molecule from products of wave functions of electrons of individual atoms, taking into account the identity of the electrons. It was first applied for the calculation of the energy spectrum of the hydrogen molecule H_2. According to the principle of identity, the coordinate part of the wave functions of H_2 in the zeroth-order approximation of perturbation theory can be either a symmetric (singlet) state, or an antisymmetric (triplet) state, with respect to the permutation of electrons:

$$\Psi_s = \alpha_s\big(\psi_A(\boldsymbol{r}_1)\psi_B(\boldsymbol{r}_2) + \psi_A(\boldsymbol{r}_2)\psi_B(\boldsymbol{r}_1)\big),$$

$$\Psi_t = \alpha_t\big(\psi_A(\boldsymbol{r}_1)\psi_B(\boldsymbol{r}_2) - \psi_A(\boldsymbol{r}_2)\psi_B(\boldsymbol{r}_1)\big),$$

where $\Psi_{A(B)}(\boldsymbol{r}_j)$ is the wave function of electron $\boldsymbol{r}_j$; and $\alpha_{s,t}$ is the normalization coefficient. The Heitler–London approximation is employed in solid state physics to study impurities and *excitons* in molecular crystals. In a wider sense this approximation provides wave functions of the crystal in the form of a product of the wave functions of all the molecules in the system, and in a limited sense it is associated with neglecting the contributions to the energy of crystal states involving two or three excited molecules. This means that in the following Hamiltonian for low-density *Frenkel excitons* ($<10^{17}$ cm^{-3}):

$$\widehat{H} = \sum_{n,f} E_f B_{nf}^+ B_{nf} + \sum_{nf,mf} M_{nm}^f B_{mf}^+ B_{nf} + \frac{1}{2}\sum_{nf,mf} M_{nm}^f \big(B_{nf}^+ B_{mf}^+ + B_{nf} B_{mf}\big),$$

the third term is neglected. Here B_{nf}^+ and B_{nf} are the creation and annihilation operators of the excitations of an f-state molecule at lattice site n; E_f is the energy of the excited molecule without accounting for the transfer of the excitation to other sites; and M_{nm}^f is the matrix element of the resonance excitation transfer. The Heitler–London approximation is applicable when the width Δ of an exciton band is small in comparison with energy E_f. For some molecular crystals this condition is satisfied (e.g., for the lowest singlet excitations of anthracene and naphthalene $\Delta \approx 500$ and 150 cm^{-1}, respectively, and $E_f \approx 25{,}100$ and $31{,}500$ cm^{-1}).

HELICAL DISLOCATION

Dislocation lines of a specific configuration in the form of a coil arising from the creeping of a dislocation segment (see *Dislocations*) of the Burgers kind, or of a mixed kind due to the absorption or emission of *vacancies*. Helical dislocations were often observed in experiments involving the high oversaturation of intrinsic *point defects* emerging at high temperatures during quenching, or when subject to reactor irradiation (see *Radiation physics*). The figure shows an example of the formation of a helical dislocation. There exist right- and left-screw types differing in the direction of the winding of the coil. In a crystal oversaturated by intrinsic point defects, the process of formation of the helicoids can closely correlate with the nucleation and the decay of prismatic

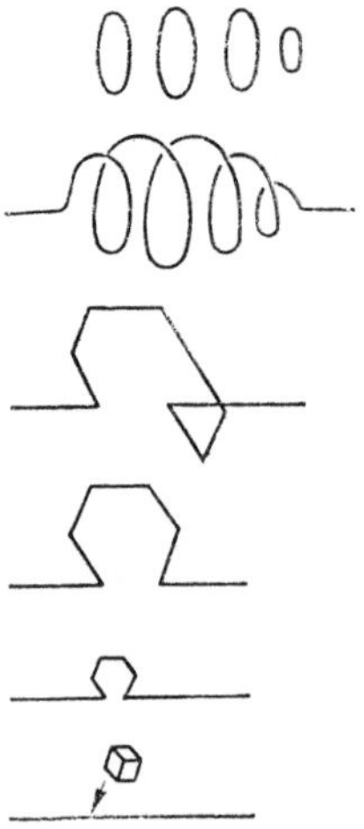

Successive stages, from bottom to top, in the formation of a helical dislocation due to the attachment of vacancies to a screw dislocation, and the eventual break-away from the loops of the helix.

loop types (see *Prismatic dislocations*), and as seen in the figure, inter-transformations are possible between them. In addition, helical dislocations can play an essential role in processes of dislocation structure rearrangement at various stages of the development of *plastic deformation*, thus stimulating the formation of *dislocation clusters*.

HELICAL MAGNETIC STRUCTURE

See *Modulated magnetic structures.*

HELICOMAGNET

A magnetically ordered material (see *Magnetism*) with a helical magnetic structure (a simple or ferromagnetic coil). See also *Modulated magnetic structures.*

HELICON, helical wave

A weakly attenuating electromagnetic wave in a *solid-state plasma*, propagating either along or at a small angle θ with respect to an external magnetic field $\boldsymbol{B}$. The name corresponds to a circular polarization of the wave in a plane normal to the wave vector $\boldsymbol{k}$. A helicon exists in the region of strong magnetic fields when the cyclotron frequency $\Omega = eB/m$ of the conduction carriers (electrons and holes) greatly exceeds the relaxation frequency ν (e and m are, respectively, the charge and the mass of the carriers). In this case the carriers are "magnetized": their diffusion in the plane normal to the vector $\boldsymbol{B}$ proceeds very slowly since $\nu \ll \Omega$, and the dissipative (diagonal) components of the static tensor of the transverse conductivity of a metal are much smaller than non-dissipative (non-diagonal) components equal to $\pm(n_e - n_h)e/B$ (where n_e and n_h are, respectively, the electron and hole concentrations). Therefore, the electric field oriented normally to $\boldsymbol{B}$ does almost no work over the electron system: in directions normal to $\boldsymbol{B}$, the metal behaves as an *insulator.* As a result, a helicon can propagate in metals.

A particular feature of a helicon is the quadratic dependence of the frequency ω on wave vector $\boldsymbol{k}$:

$$\omega = \frac{k^2 B\cos\theta}{4\pi e(n_1 - n_2)}.$$

A helicon is a slow wave, since its *phase velocity* $v_{\rm H} = \omega/k$ in typical metals at $B \cong 1$ T and $\omega \cong 10^7\ {\rm s}^{-1}$ is approximately 10^4 cm/s, which is significantly less than the Fermi velocity $v_{\rm F} \cong 10^8$ cm/s. The attenuation of a helicon is caused by the *current carrier scattering* by lattice defects and phonons, as well as by collisionless absorption of the wave energy (*magnetic collision-free damping*). Correspondingly, a damping decrement Γ consists of two terms:

$$\Gamma = \frac{\nu}{\Omega}\sec\theta + \frac{3\pi}{16}k\frac{v_{\rm F}}{\Omega}\sin^2\theta.$$

In *semiconductors* there is usually no *spatial dielectric dispersion* ($\nu \gg kv_{\rm F}$) since the first term plays the main role. In pure metals at low temperatures, a strong space dispersion occurs ($\nu \ll kv_{\rm F}$), and the second term dominates. In this case, the condition $kv_{\rm F}/\Omega \ll 1$ provides a lower limit to the magnetic field $\boldsymbol{B}$ where the attenuation of helicons is small.

HELICON–PHONON RESONANCE

A resonance interaction with respect to an external magnetic field $\boldsymbol{B}$ of a *helicon* with *sound*, resulting in the excitation of coupled waves. During the propagation of a helicon along the vector $\boldsymbol{B}$, a helicon–phonon resonance with the transverse sound occurs for coinciding *phase velocities* $v_{\rm H}$ and $s_{\rm t}$ of the helicon and the sound, respectively.

In typical metals this condition can be satisfied at a frequency of tens of megahertz, and in fields up to 10 T. The wave arises due to the *electron–phonon interaction*. Owing to a helicon–phonon resonance, the spectrum of collective vibrations becomes reconstructed: with increasing magnitude of the wave vector a phonon branch of the spectrum transforms into a helicon branch and a helicon branch transforms into a phonon one.

HELIUM, He

A chemical element of Group VIII of the periodic table with atomic number 2 and atomic mass 4.002602. Electron configuration is $1s^2$. Ionization energy is 24.59 eV (the highest of all elements). Radius of helium atom 0.122 nm. The element is chemically inert. Natural helium has stable isotopes ^{4}He (99.999862%) and ^{3}He.

Helium is a light colorless monatomic gas; its density is 0.1785 kg/m^3 (at 273 K and $1.013\cdot10^5$ Pa); it is barely soluble in water. Heat conductivity is 0.1438 W·m^{-1}·K^{-1} (at 273 K), viscosity is 18.60 μPa·s (at 273 K). Dielectric constant is 1.000074 (at 273 K and 1 atm pressure). ^{4}He is weakly diamagnetic, $\chi = -0.78\cdot10^{-13}$ m^3·kg^{-1}. Index of refraction for the yellow Na line $n_D = 1.000034$; $T_{\text{boiling}} = 4.22$ K, the lowest for all liquids. *Liquid helium* is a transparent colorless liquid. Helium isotopes remain liquid at the pressure of saturated vapors up to arbitrarily low temperatures (see *Low-temperature physics*), which is associated with the weakness of the interaction between helium atoms and the high amplitude of their *zero-point vibrations*.

Since at very low temperature the de Broglie wave length of the helium atoms is comparable with their interatomic distance, helium is a *quantum liquid*: more specifically, ^{3}He forms a *Fermi liquid*, and ^{4}He forms a *Bose fluid*. Liquid helium undergoes a *second-order phase transition* to the superfluid state (see *Superfluidity*). The superfluid phase of ^{4}He (called He II) exists at temperatures below the lambda (λ) point 2.172 K, and the *superfluid phases* of ^{3}He (A–, B–, A_3–^{3}He) exist below $2.7\cdot10^{-3}$ K. Solutions of ^{3}He–^{4}He form quantum liquids.

Solid helium is a colorless transparent *crystal*, which exists only at *high pressures* ($P >$ 2.5 MPa). It experiences polymorphic transitions (see *Polymorphism*). Solid helium has a low density (up to 0.19 g/cm^3), high plasticity (up to $3.5\cdot10^{-8}$ Pa^{-1}); yield limit at shear deformations of about 10^3 Pa; the Debye temperature is very low (up to $\theta_{\text{D}} = 25$ K). It is the best known representative of a *quantum crystal*.

Helium is used for reaching low and superlow temperatures (4.2 to 10^{-3} K), for *thermometry* in the range of temperatures 1 to 80 K, for providing a protective inert medium at melting, cutting and welding of metals and alloys, growing of semiconductor *monocrystals*, etc.

HELIUM-3, SUPERFLUID

See *Superfluid phases of ^{3}He.*

HELMHOLTZ FREE ENERGY

See *Free energy*.

HERZBERG–TELLER INTERACTION

(G. Herzberg, E. Teller, 1933)

The part of the interaction between the electron motion and the nuclear motion in polyatomic systems, which is treated by the separation of the electron and the nuclear variables in an adiabatic wave function (see *Adiabatic approximation*). The presence of a Herzberg–Teller interaction explains (i) a possibility of electronic-vibrational transitions involving odd vibration sublevels of an excited electronic state in the case of non-totally-symmetric vibrations; (ii) the lack of reflection symmetry in the intensity distribution of conjugated optical bands of the *light absorption* and light *fluorescence* of impurity molecules; (iii) temperature and other dependences of the optical spectra of crystals (see *Optical spectroscopy*).

HETERODESMIC STRUCTURES

(fr. Gr. $\varepsilon\tau\varepsilon\rho o\varsigma$, other and $\delta\varepsilon\sigma\mu o\varsigma$, bond)

Crystalline structures which contain two or more types of *chemical bonds*. Included among heterodesmic structures are both monatomic *crystals* (e.g., *graphite*) and crystalline compounds, for example, various *molecular crystals*, *coordination compounds*, *polymers*. It is specific for heterodesmic structures to include portions (layers, chains, molecules) in which the atoms are linked by strong bonding such as a *covalent bond*, *ionic bond*, *metallic bond*; as well as having parts connected with each other by weak bonds (*van der Waals bond*, *hydrogen bond*). Crystals with a single kind of bonding are referred to as homodesmic structures (*diamond*, *silicon*, *germanium*, *alkali-halide crystals*, *metals*, etc.).

HETERODIFFUSION, chemical diffusion

A flux of foreign atoms diffusing into a homogeneous medium. In this case the diffusion mobility (see *Diffusion*) of the foreign atoms is determined not only by the parameters of the medium, but also by the interactions of the diffusing atoms which depend on their concentration. The *heterodiffusion coefficient* is a function of concentration. It is defined in the same way as the partial *diffusion coefficients* which arise in the theory of *interdiffusion* formulated within the framework of the thermodynamics of *irreversible processes*. The driving force of heterodiffusion is the gradient of the chemical potential. Introducing the chemical potential μ as $\mu = k_B T \ln(c\gamma)$ (c is the atomic concentration of a component, and γ is its thermodynamic activity coefficient) the expression for the heterodiffusion coefficient takes the form:

$$D_\mathrm{h} = D^*\left(1 + \frac{\partial \ln \gamma}{\partial \ln c}\right),$$

where D^* is the self-diffusion coefficient.

HETEROEPITAXY

Oriented *crystallization* of a material on the crystal surface of another material. This phenomenon is widely found in both natural and artificial crystallization processes. The degree of ordering of a *solid* formed as a result of heteroepitaxy can vary from a predominant orientation of *polycrystal* grains to a structurally perfect *monocrystal*. The latter case is possible only for crystallogeometric compatability of a *substrate* and the growing material, at least in the plane of growth. As no two materials have precisely the same crystal lattices the inevitable differences between their parameters leads to the appearance of dislocations (see *Epitaxial dislocations*) which can hinder a *film* from attaining a particular thickness. A critical thickness at which dislocation failure occurs depends on the relative misalignment of the *crystal lattice constants* and their *elastic moduli*. Heteroepitaxy plays the following roles in solid-state-device technology: design of semiconductor *integrated circuits* on dielectric substrates (heteroepitaxy occurs for *semiconductors* on metals and *metals* on semiconductors); design of devices on *heterojunctions* (e.g., *lasers*) with a double *injection of current carriers*; design of new solids with alternating ultrathin layers made of semiconductor or other materials with various *band structures* (artificial *superlattices*). The synthesis of a superlattice can be a chain of sequentially alternating processes of a heteroepitaxy.

HETEROGENEOUS STRUCTURE

A structure of a heterogeneous material; elements of a heterogeneous structure are distributed irregularly (distinguished from *homogeneous structure*). A heterogeneous structure can be exemplified by the structure of a system consisting of parts differing in physical properties or chemical compositions (*phases*), a so-called *heterophase structure*. The parts are separated by interfaces involving a sharp change in material concentration, *crystal lattice* parameters, or other characteristics. We can distinguish macro- and microheterogeneity of the electron distribution in the system. Such a structure can be formed through thermal, radiation and other kinds of material treatments. For example, a single-phase *alloy* which is thermodynamically unstable at a certain temperature can decay during *annealing* to phases with different chemical compositions. In the process of heterogeneous structure formation, the *grain boundaries* of a host (main) phase as well as the *shear lines* and twinning boundaries (see *Twinning of crystals*) play a significant role which is connected with the decrease of the energy of the interfaces due to the accumulation of *impurity atoms*. A macroscopic heterogeneity manifests itself, for example, through various shrinkage cavities, *cracks*, martensitic phases, and so on.

Producing a heterogeneous distribution can provide materials with desired combinations of properties (see *Alloy heterogenization*), such as *composite materials* and semiconductor *heterostructures*.

HETEROGENEOUS SYSTEM

A system consisting of physically or chemically heterogeneous (differing) regions, separated spatially but in contact. In many cases heterogeneous systems are of natural origin (for example, *minerals*), and synthetic counterparts are widely used in science and technology. Such a system may consist of phases differing in their *states of matter*, for example, gas–liquid, gas–solid, colloidal particles in solution, and so on. Included among heterogeneous systems are solids having a *heterophase structure*, for example, *ferroelectrics* and *magnetic substances* with a *magnetic domain structure*; metal–semiconductor systems (see *Metal–semiconductor junction*), metal–insulator–semiconductor, etc. A system consisting of two or more contacting *semiconductors* of different chemical compositions corresponds to a solid heterogeneous structure. Due to a difference between polarity and between energy band widths, *heterojunctions* can form in such structures (e.g., n–p and multicomponent heterojunctions); *quantum wells* and more complicated multiwell structures based on them; *superlattices* consisting of alternating layers of different chemical composition. Heterostructures and heterojunctions based on them are particularly important components of diverse devices of *microelectronics*.

HETEROJUNCTION

A contact of two *semiconductors* differing in chemical composition and physical properties. Depending on the type of conduction of the semiconductors in contact, there are distinguished *anisotropic heterojunctions* (different types of conduction) and *isotropic heterojunctions* (similar type of conduction) (see *Inversion layer*, *Quantum well*). Combinations of various heterojunctions form a *heterostructure*. At the formation of a heterojunction, due to a difference between *work functions* of the semiconductors in contact (see Fig. (a)) or due to the presence of a charge in the *surface states* (see Fig. (b)), *band bending* φ_1 and φ_2 as well as potential barriers for electrons and holes arise near the interfaces.

Due to differing electron affinity, jumps in energy of *conduction band* and *valence band* of semiconductors, $\Delta\varepsilon_c$ and $\Delta\varepsilon_v$, respectively, occur. Under constant applied voltage, a direct

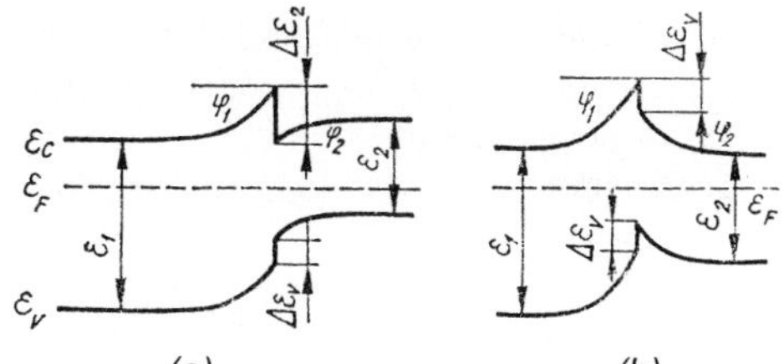

Sketches of potential energy changes (solid lines) and Fermi energy level position (ε_F, dashed lines) near the interface for two types of heterojunctions, namely, a junction involving a difference in work functions (a) and a junction involving surface states (b). The energies of the top of the valence band ε_v, the bottom of the conduction band ε_c, and the band bendings φ_1 and φ_2 at the interfaces are indicated.

current passes through a heterojunction as a result of the difference of the potential barriers for electrons and holes, producing a *current–voltage characteristic* which is sharply asymmetric. Different heights of barriers for holes and electrons allow one to obtain either hole or electron currents. Under application of an alternating electric field the heterojunction is characterized by a parallel combination of resistive and reactive (as a rule, capacitive) impedances whose magnitude depends on the applied constant electric field. The complicated energy structure of a heterojunction provides a wide spectrum of applications in semiconductor electronics for *lasers*, *light emitting diodes*, photodetectors, etc.

HETEROJUNCTION, ISOPERIODIC

See *Isoperiodic heterogeneous systems.*

HETEROJUNCTION LASER

A *semiconductor laser* (the most common *injection laser*) based on a *heterostructure* that incorporates one or more *heterojunctions* adjacent to an active region. Potential barriers limit the region of recombination of excess *current carriers* (active region), preventing their migration to neighboring regions (*electronic confinement*). On the other hand, optimization of optical properties can be achieved by preparing a dielectric waveguide (*optical confinement*). A comparatively low generation threshold (below 100–200 A/cm^2)

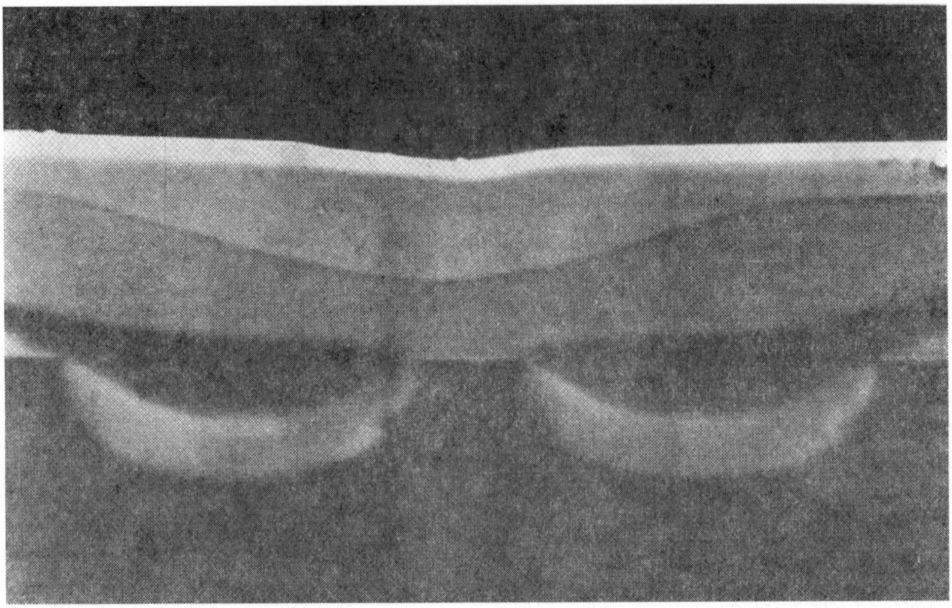

A heterostructure for a heterojunction laser.

Table 1. Some characteristics of heterojunction lasers

Active medium	Emitting material	Substrate	Wavelength range, μm	Comments
InGaAsP	InGaP	GaAsP	0.64–0.8	–
GaAs	CaAlAs	GaAs	0.85–0.88	Ordinary heterostructure
			0.77–0.80	Quantum-well heterostructure
InGaAsP	InP	InP	1.06–1.67	–
GaSb	GaAlSbAs	GaSb	1.75–1.8	–
InGaSbAs	GaAlSbAs	GaSb	1.8–2.4	–
PbSnTe	PbTe	PbSnTe	8–16	Working temperature 10–77 K

and high efficiency (up to 50%) at room temperature are achievable. A basic model of a heterolaser is a *planar heterostructure* with an active (narrow-band) layer sandwiched between two passive emitter (wider-band) layers. Optimized multilayer heterostructures with gradient or stepwise waveguide which includes an active layer are also employed.

Strip-geometry (V-groove) heterolasers with a contact in the form of a strip at the planar heterostructure serves as a waveguide configuration which provides the proper distribution of optical amplification (under the contact) and absorption (out of the contact area). Heterostructures have been designed with a refractive index gradient both in side directions and in vertical cross-section, corresponding to a two-dimensional dielectric waveguide (*refractive confinement*). The figure provides a microphotograph of an overgrown heterostructure based on InGaAsP/InP for a 1.3 μm wavelength heterolaser. The threshold current in such heterolasers can be reduced to 5–10 mA at room temperature.

An important class is represented by the *quantum-well heterolasers* in which the active layer (or active few layers) thickness is comparable with or less than the de Broglie wavelength of the current carriers. In such a *quantum well*, the quantization of levels takes place with a substantial modification of the energy spectrum of the active region, and with the appearance of new *selection rules* for the optical transitions. In addition, in such heterolasers the active region volume is quite small, close to 10^{-12} cm^3, which allows one to significantly reduce the threshold current. On the basis of these structures not only single-strip lasers, but also multiple-element lasers with more than 1 W continuous radiation have been designed.

The initial developments involving physics and engineering of heterolasers were concerned exclusively with the heterosystem GaAlAs/GaAs. To expand the wavelength range improved heterosystems became necessary, and these are mainly quaternary ones such as InGaAsP/GaAs, InGaAsP/InP, and GaAlSbAs/GaSb. Some characteristics of heterolasers based on various materials are given in Table 1. In their manufacturing various methods of *epitaxy* are used.

HETEROSTRUCTURE

A semiconductor structure formed by a few *heterojunctions*. The geometry of these junctions is established during the growth of a heterostructure in contrast to *heterogeneous systems* for which the shape of heteroboundaries is determined by the conditions of phase equilibrium, or by the random conditions existing during production. Several parameters of the component materials (e.g., positions of the band extrema, *effective masses*, *dielectric constants*, etc.) provide possibilities for controlling the properties of the charge carriers and the propagation of the electromagnetic waves. The reflection and refraction of electromagnetic waves propagating in a heterostructure is accompanied by the appearance of waveguide modes localized in layers with a large *refractive index*. Due to a difference between the *crystal lattices* forming the *heterojunctions* the phonon spectrum of a heterostructure changes, and local phonon modes can arise which can influence the nonequilibrium charge carriers. The scattering of electrons and holes (as well as their distribution in the structure) can also be controlled by selective *doping* when donors and acceptors are introduced only in selected regions of the structure.

Characteristics of the motion of charge carriers in heterostructures are determined by relative energy positions of band extrema (i.e. by the type of heterojunction) and by the geometry of these structures. In structures based on GaAs–AlAs, GaAs–GaP and some others the gap of a *narrow-gap semiconductor* is situated on the background of a wide-band material. This band diagram (i.e. a dependence of band extremum positions on a coordinate) corresponds to a *heterostructure of the first kind. Heterostructures of the second kind* (e.g., InP–$Al_{0.48}In_{0.52}As$) correspond to a band diagram where the breaks of c- and v-bands at the heterojunction are of the same sign. The case of non-overlapping band gaps (e.g., in the system InAs–GaAs) is referred to as a *heterostructure of the third kind*. More complicated variants of band diagrams occur for heterostructures formed by materials of different symmetry (e.g., heterojunctions with band inversion, or those between a *gapless semiconductor* and an ordinary semiconductor), or in the case of the structures formed by *semimagnetic semiconductors* (spin heterostructures) or materials in which the band extrema are situated at different points of a *Brillouin zone* (different-valley heterostructures); various combinations of these above variations also occur.

To account for the change in the geometry of structures various cases of low-dimensional electron states as well as tunnel processes in heterostructures (see *Tunneling phenomena*) became available. In Fig. 1 the band diagrams, positions of the energy levels, and the wave functions are shown for various heterostructure types: (a) single *quantum well*, (b) multiple quantum wells, (c) *superlattice*, and (d) selective-doped single heterojunction. Two-dimensional electron states are formed in narrow-gap materials; and in the case of a superlattice a mini-band energy spectrum appears due to a tunneling between these states. Structures with tunneling-transparent barriers (Fig. 2) are used for transverse charge carrier transfer. At these two-barrier structures separated by a narrow quantum well, an increase of the transparency for the energies close to the resonance level is observed (*resonance tunneling effect*).

In addition to the above-described plane heterostructures, *quantum wires* and *quantum dots* have been intensively investigated (see Fig. 3). In

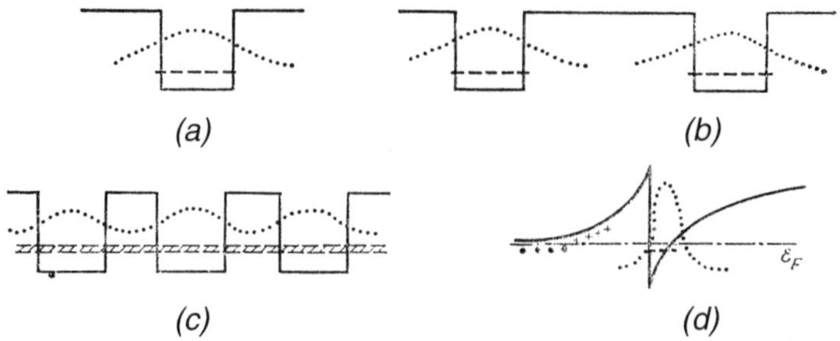

Fig. 1. Diagrams of (a) single quantum well, (b) multiple quantum wells, (c) superlattice, and (d) doped single heterojunction. The first three figures show the potentials (solid lines), energy levels (a, b) and miniband (c), and wave functions (dotted lines) plotted versus position. The fourth sketch is a plot versus the Fermi energy ε_F.

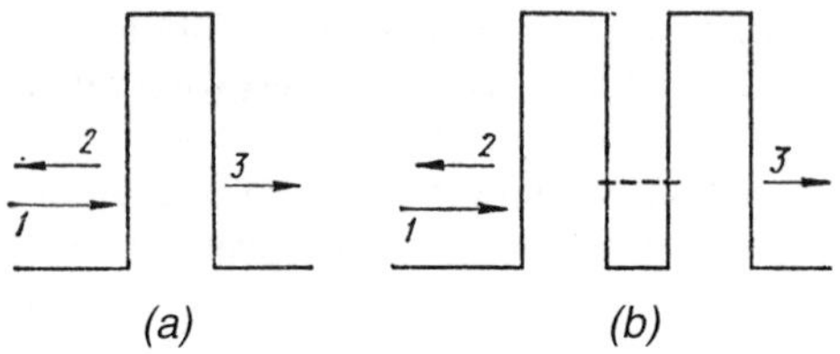

Fig. 2. Tunnelling processes through a single (a) and a double (b) barrier with the incident (1), reflected (2) and transmitted (3) currents indicated.

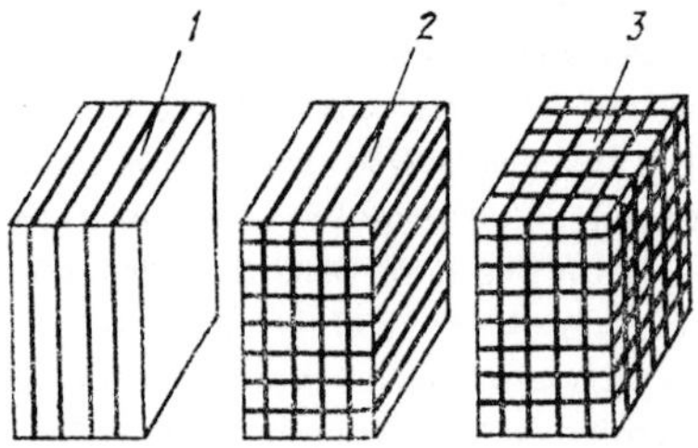

Fig. 3. Heterojunctions of lower dimensions: quasi-two-dimensional (1: quantum well), quasi-one-dimensional (2: quantum wire), and quasi-zero-dimensional (3: quantum dot). The potential barriers are darkened.

quantum wires the charge carrier motion is limited in two directions by the potential barriers of heterojunctions, or the electrostatic potential emerged on the surface, so that a one-dimensional energy spectrum appears. Therefore, the galvanomagnetic and optical properties of the quantum wires differ from the case of two-dimensional systems. Wholly localized states of charge carriers (i.e. zero-dimensional energy spectrum) in quantum dots are studied using optical methods (see *Nanoelectronics*).

To obtain semiconductor heterostructures the method of growing ultrathin semiconductor layers (*liquid-phase epitaxy* or *molecular beam epitaxy*, chemical deposition from a gas, see *Thin film growing*) as well as various methods of surface treatment (*ion implantation*, *etching* and so on) and control are used. The appearance of heterostructures not only opened broad opportunities for investigating atomic and electronic properties of interfaces, and the physics of a *two-dimensional electron gas*, but it also prompted the design of new semiconductor devices and the improvement of already existing ones in *microelectronics* and *optoelectronics*. Heterostructures are applied in manufacturing of semiconductor *lasers* and *light emitting diodes* as well as for various *optical radiation detectors* (avalanche *photodiodes*, phototransistors, infrared detectors, etc.) and *solar cells*. New kinds of transistors are designed on the basis of heterostructures.

HEUSLER ALLOYS (F. Heusler, 1898)

Three-component *alloys* possessing an ordered body-centered cubic lattice with ions of *copper* (*silver*, *gold*, *palladium*, etc.) at the vertices of a cube, and ions of *manganese* and *aluminum* (*tin*, *antimony*, *indium*, *gallium*, etc.) alternating in the centers of cubes. Heusler alloys can be *ferromagnets* or *antiferromagnets*.

HEXAFERRITES

Oxide ferrimagnetic monocrystals of the hexagonal or trigonal types (see *Crystal symmetry*) with a composition $n_1\text{BO}\cdot 2n_2\text{MO}\cdot 2n_3\text{Fe}_2\text{O}_3$ where $n_1 \geqslant 1$, $n_2 \geqslant 0$ and $n_3 \geqslant 3$ are integers, B = Ba, Sr, Pb, and M = Mn, Co, Ni, Cu, Zn, Mg, and so on. The *unit cell* consists of a sequence of alternate blocks arranged normally to a principal *symmetry axis*: two-layer block S with a cubic spinel-like structure (see *Ferrite spinel*), three-layer block R with a hexagonal structure, and four-layer block T with a hexagonal structure. The best known hexaferrites are the following: with structure M or *magnetoplumbites* (block arrangement is SRS^*R^*, *space group* $P6_3/mmc$, D^4_{6h}); with structure W (block arrangement is $S_2RS_2^*R^*$, space group is $P6_3/mmc$, D^4_{6h}); with structure Y (block arrangement is $STS^*T^*S^*T^*$, space group is $R\bar{3}m$, D^5_{3d}); with structure X (block arrangement is $(SRS_2^*R^*)_3$, space group $R\bar{3}m$, D^5_{3d}); with structure Z (block arrangement is $SRSTS^*R^*S^*T^*$, space group $P6_3/mmc$, D^4_{6h}); with structure U (block arrangement $(SRS^*R^*S^*T^*)_3$, space group $R\bar{3}m$, D^5_{3d}); and an asterisk means a turn of a block through 180° with respect to a principal symmetry axis. In some hexaferrites (e.g., those with structure Y) helical magnetic ordering can occur (see *Modulated magnetic structures*). A particular feature of hexaferrites is a strong, single-axis *magnetic anisotropy*.

HEXAGONAL SYSTEM

A *crystallographic system* characterized by the presence of a single axis of symmetry 6 (or $\bar{6}$) and referred to as hexagonal *crystal class*. In a hexagonal system, a sixfold coordinate axis (6 or $\bar{6}$) is parallel to the z direction, $a = b \neq c$, $\alpha = \beta = 90°$, $\gamma = 120°$. The hexagonal system includes one *Bravais lattice*, seven *point groups* and 27 *space groups*.

HIGGS EFFECT

See *Higgs mechanism*.

HIGGS MECHANISM, Higgs effect (P. Higgs, 1964; first proposed by P.W. Anderson)

An analogue of the *Meissner effect* in gauge field theories with *spontaneous symmetry breaking*. The simplest theoretical field model is *Higgs' Abelian model*, in which a complex scalar field, called the *Higgs field*, interacts with an Abelian *gauge field* (like an electromagnetic field). In the static case the Lagrangian of this model, e.g., correlates with the free energy in the *Ginzburg–Landau theory of superconductivity* for *type II superconductors*, where the *order parameter* plays the role of the Higgs field. The Higgs mechanism generates a mass by the gauge field: in terms of electromagnetism this means a finite depth of penetration of an electromagnetic field into the condensate (Meissner effect). Higgs mechanism is based on a spontaneous breaking of continuous symmetry and local *gauge invariance* (in the Abelian case – the ordinary gauge invariance of electromagnetism). Spontaneous symmetry breaking is related to the appearance of *Goldstone excitations* (*bosons with zero mass and zero spin*, see *Goldstone theorem*). In the case of the electromagnetism of a *superconductor* a measure of the reciprocal effective mass of a "photon" (accurate to a dimensional factor) is the penetration depth of an electromagnetic field into a superconductor. In particle physics the Higgs mechanism is that aspect of the gauge theory of the weak force that provides mass for the W and Z bosons. The discovery of the Higgs mechanism turned out to be of decisive importance for Weinberg and Salam's 1967 derivation (Nobel Prize, 1979) of their theory involving the unification of the weak and electromagnetic interactions.

HIGH-COERCIVITY ALLOYS

Metallic ferromagnetic *alloys*, which possess, after specific pretreatments, a high *coercive force* and can be used as hard magnetic materials. Among these are the widely used dispersion-hardened cast alloys of the *Alni* (Al–Ni) and *Alnico* (Al–Ni–Co) types, the deformable alloys *Vicalloy*, *Cunife* (Cu–Ni–Fe), *Cunico* (Cu–Ni–Co), and related alloys of the Fe–Co–Cr system. The high coercitivity state in these alloys is due to the presence in them of small single-domain formations (see *Single-domain particles*), with their remagnetization brought about by uniform rotation of the spontaneous magnetization (see *Magnetization rotation process*). The ordered alloys Pt–Co, Pt–Fe, Mn–Al–C are the other class of high-coercivity alloys, in which the *magnetic reversal* processes are difficult, mainly due to pinning of the *magnetic domain walls* at the anti-phase boundaries of the ordering domains. The properties of many alloys are given in Table 1 in *Hard magnetic materials*.

HIGH-DENSITY EXCITONS

Situation arising at high levels of optical radiation in *crystals* when interactions between *excitons* become appreciable. This high exciton density favors the formation of exciton molecules (*biexcitons*), and the appearance of coherent nonlinear phenomena and new luminescence bands. At densities $n_{ex}a_{ex}^3 \geqslant 1$ (here a_{ex} is the exciton radius) *Wannier–Mott excitons* metallize and transform into an *electron–hole plasma*. The presence of this plasma at low temperatures in crystals like Ge and Si with a complex electron band structure leads to the formation of an *electron–hole liquid* and electron–hole drops. It was found experimentally that crystalline Si exhibits the presence and inter-transformation of excitons, biexcitons and electron–hole drops. The quantum statistics of excitons and properties close to *Bose–Einstein condensation* are found in the crystal Cu_2O.

HIGH ELECTRON MOBILITY TRANSISTOR (HEMT)

A *field-effect transistor* whose channel contains a *heterojunction* interface between wide- and narrow-gap *semiconductors*. In a narrow potential well ($\sim$10 nm) formed at the heterojunction interface, both the *quasi-momentum* and the electron energy are quantized, and there appears a finite *density of states*, independent of the energy, which in combination with selective (modulated) *doping* leads to the suppression of scattering mechanisms, and a sharp rise of the *mobility of current carriers*. The mobility growth is promoted by a thin undoped layer formed at the heterojunction interface. The current carriers in the channel form a *two-dimensional electron gas* with the carrier mobility reaching $2\cdot10^6$ cm$^2\cdot$V$^{-1}\cdot$s^{-1} at 4 K, and $5\cdot10^5$ cm$^2\cdot$V$^{-1}\cdot$s^{-1} at 77 K. In the case of a small drift space in the device, this leads to a very large drift velocity, up to $4\cdot10^7$ cm/s, a shortening of the delay time to 1 ps, and a raising of the frequency limit to hundreds of GHz. This two-dimensional electron gas exhibits the *quantum Hall effect* and the *Shubnikov–de Haas effect*. In metrology this makes it possible to measure the fine structure constant, and to define the international ohm standard.

HIGH-ENERGY CHEMISTRY

The branch of physical chemistry dealing with high temperature activation processes caused by electronic, atomic, molecular, ionic, neutron, photon impacts, absorption of *photons* and *phonons*, application of electric and magnetic fields, mechanical stress, shear, shock. A common feature of these processes is that the energy which is absorbed from the external source creates in the medium nonequilibrium concentrations of intermediate highly reactive particles and *quasiparticles*: *holes* and *conduction electrons*, localized holes and electrons, *ions*, ionic *clusters*, *excitons* and other excited states, *free radicals*, etc. At the instant of formation these particles are very far from thermal equilibrium with the medium, and in a number of cases they take part in chemical reactions until they slow down, cool down, and approach thermal equilibrium. Thus high-energy chemistry involves the study of thermal transformations in chemical systems of all types in the absence of thermodynamic equilibrium. Within this scope the following branches are included: photochemistry, laser chemistry, *radiation chemistry*, plasma chemistry, chemistry of shock waves, as well as chemistry of sound (phonons), chemistry of infrared and extra-high frequency radiations (at high densities of corresponding actions), chemistry resulting from nuclear transformations, chemistry of hot atoms, chemistry of "exotic" atoms (*positronium*, *muonium*, etc.), chemistry of processes of dissolving defect-containing substances, chemistry of molecular beams, photoelectrochemistry, emission of ions and electrons, electric discharge in condensed medium, etc. The main tasks of high-energy chemistry are: development of methods of production of new substances and materials including those resistant to superthermal activation of various kinds, determination of the resistance of substances and materials towards various types of corrosive actions, the study of the mechanisms of chemical processes, of the composition and physicochemical properties of intermediate particles and quasiparticles, etc.

HIGH-ENERGY ELECTRONICS of the solid state

A discipline at the border of solid-state physics and electronics, studying the properties of *solids* upon the passage of high density electron beams through them. Such investigations underwent intensive development during the 1970s after the introduction of high-current pulsed electron beam accelerators. These were based on *explosive electron emission*, with a pulse period up to 1 ns, and with electron energies from 0.3 to 0.5 MeV, regulated by a current I up to 5,000 A with a density up to 2,000 A/cm^2. These accelerators had been developed in Russia and the USA. Upon irradiation by strong-current beams and high-energy hole plasmas with densities of about 10^{13} to 10^{14} cm^{-3} some specific new effects appeared. One such effect is a new type of *luminescence*, caused by band-to-band transitions of electrons and holes. This emission has a wide spectral band and a short *relaxation time*, much less than a picosecond. The electrical conductivity of the irradiated body even in a weak external electric field is determined by the strongly excited conduction electrons and holes with energies of about an electron volt beyond the edges of the corresponding bands. The specific resistance of an *insulator* penetrated by the beam drops by 10^{11}–10^{15} orders of magnitude, and can reach values of about 10 Ω·cm, and a current density of thousands of A/cm^2. An electron–hole plasma, which is formed by the pulsed strong-current irradiation, greatly influences the *brittle failure* of crystals. Pulsed accelerators can be used to investigate the fine details of the process of *Frenkel defect* formation via the excitation of the electronic subsystem in non-metallic crystals (see *Pulse spectroscopy*).

HIGH-FREQUENCY CONDUCTANCE

A complex *kinetic coefficient* $\sigma(\omega)$ in the generalized Ohm's law that is frequency (ω) dependent and defines the linear relation between the AC current density $j(\omega)$ and the strength of the applied AC electric field $E(\omega)$:

$$j(\omega) = \sigma(\omega)E(\omega). \tag{1}$$

Isotropic conductors (*metals*, *semiconductors*) usually have their high frequency conductivity quite well described by the classical *Drude formula* (P. Drude, 1900):

$$\sigma(\omega) = \sigma_0 \frac{1 + \mathrm{i}\omega\tau}{1 + \omega^2\tau^2}, \tag{2}$$

where $\sigma_0 = \mathrm{e}^2 n\tau/m^*$ is the static *electrical conductivity*, n and m^* are the concentration and the effective mass of free current carriers (*conduction electrons*, *holes*), and τ is the characteristic time of scattering of carriers by each other, acoustic *phonons*, crystal lattice *defects*, *impurity atoms*, etc.

Eq. (1) holds in a local approximation, provided the characteristic distance over which the amplitude of the external electric field E changes (e.g., the depth of the field penetration into the metal through the *skin-effect*) is much larger than the effective *mean free path*, $l_{\mathrm{eff}} = |l/(1 - \mathrm{i}\omega t)|$, where $l = v\tau$ and v is the average carrier velocity.

HIGH-FREQUENCY SIZE EFFECTS

Dependence of the *impedance* and/or the transparency of an electrically conducting medium (metal or semiconductor) on its dimensions (see *Size effects*). During bilateral antisymmetric excitation of electromagnetic waves under the conditions of the normal *skin-effect* (the skin depth δ exceeds the electron mean free path, l), the impedance is at its maximum when δ equals a half of the plate thickness, d, i.e. the entire sample volume placed in the resonant cavity takes part in lowering the cavity quality factor Q. For $\delta > d/2$, the intensities of the waves reflected from and passing through the plate fall off with decreasing frequency ω due to the superposition of waves inside the plate. The *Fischer–Kao size effect* (H. Fisher, Y.-H. Kao, 1969), which provides a direct determination of the skin depth by a radio-frequency measurement, has been observed in *semiconductors*. In metals this effect takes place even at room temperatures up to frequencies of $\omega \leqslant 10^{14}$ Hz. It may be employed to control the state of the specimen surface, and when the surface is improved through mechanical (polishing) or chemical processing the maximum of the impedance shifts to lower frequencies. At liquid helium temperatures l is quite large (reaching 0.1–1 cm in pure metals) and the *anomalous skin-effect* occurs ($\delta \ll l$) over a wide frequency range, including radio-frequencies. For $\delta \ll l$ high frequency size effects develop when d is comparable to l. They are related to the excitation of weakly damped waves that are produced by electrons, and that carry information on the state of the field within the narrow skin depth layer to distances of the order of l into the specimen.

In a magnetic field $\boldsymbol{B}$ tilted by an angle $\theta \leqslant d/l$ from the surface of a thin metal layer the intensity of waves reflected from that layer and passing through it will oscillate with changing $\boldsymbol{B}$, provided that the radius r of the electron orbit satisfies $r \ll d$ (analogue of *Sondheimer oscillations*). Let $\theta = 0$ and $\delta \ll 2r < l \ll d$. Then, provided that there are no open electron trajectories, the electromagnetic field penetrates into the bulk in the form of narrow bursts, having a width of the order of δ only, while the distance between such bursts remains equal to the extremal diameter of the electron orbit. The transparency of a thin plate sharply increases when a burst of high frequency field approaches the face opposite the skin-layer. This effect, in its turn, results in an oscillatory dependence of the plate impedance and transparency on $\boldsymbol{B}$ (see Fig. 1). Both the widths of such bursts $(\delta_1, \delta_2, \ldots)$ and their shape significantly depend on l and on the state of the specimen surface. The larger l, the narrower the burst, so that in plates with rough surfaces one may deduce the mean free path of those charge carriers that exhibit high-frequency size effects from the width of the corresponding high-frequency lines. The burst approaching a smooth enough surface area to within $\Delta \ll r$ is transformed quite appreciably. The lower Δ is, the more often the electrons almost specularly reflected by the surface enter such bursts, and thus the more appreciably they screen the electromagnetic field. Within the range $\delta < \Delta \ll r$ the burst narrows in proportion to $(\Delta/r)^{1/6}$. At the same time its amplitude is proportional to $(\Delta/r)^{1/3}$ and is sensitive to the state of both faces of the plate. In the case of *multiple mirror reflection* the trajectory of an electron travelling through

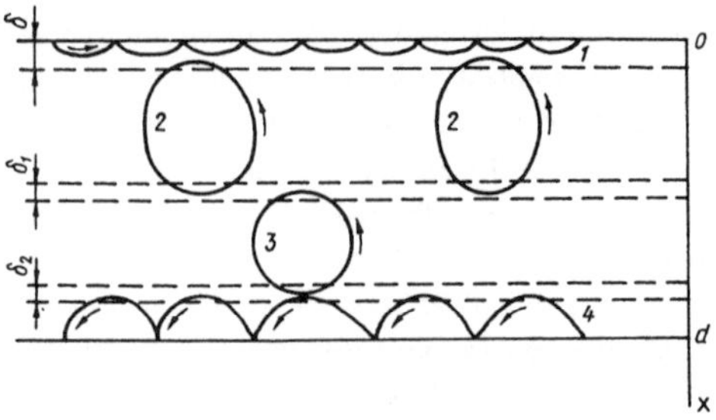

Fig. 1. Specular reflection of charge carriers by the conductor surface: electrons 1 (remaining in the skin layer) produce a strong high-frequency current screening the electromagnetic field; electrons 2 and 3 "pull" the high-frequency field into the bulk of the metal; electrons 4 screen the field in the burst nearest to the $x = d$ face.

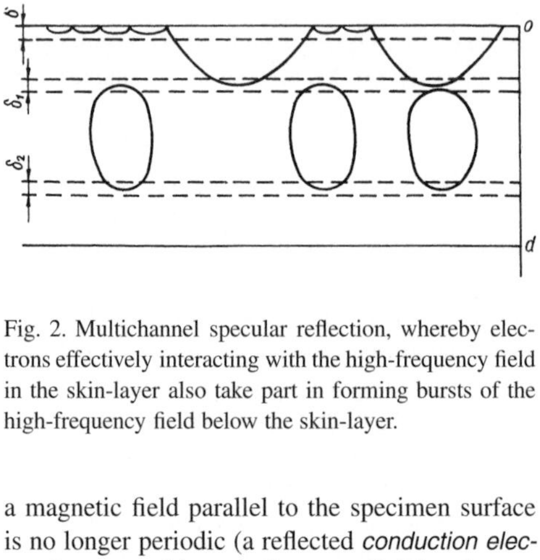

Fig. 2. Multichannel specular reflection, whereby electrons effectively interacting with the high-frequency field in the skin-layer also take part in forming bursts of the high-frequency field below the skin-layer.

a magnetic field parallel to the specimen surface is no longer periodic (a reflected *conduction electron* has several inequivalent states). The electron alternatively "glides" along the conductor surface, remaining in the skin-layer, or it enters the bulk of the material, producing a burst of high-frequency field at a distance of the order of r (see Fig. 2). The further transfer of high-frequency fields produced by that burst into the bulk material is carried out by those electrons exhibiting an extremal diameter of their orbits. As a result, there appears an additional series of high-frequency size effect lines absent in plates with rough surfaces. By studying these additional lines, one may determine the mutual position of separate indentations of the *Fermi surface* in *momentum space*. Each high-frequency field burst is formed by a small group of conduction electrons, the diameters of their orbits scattering within the order of δ around each other. Meanwhile, the field of a burst in the radio-frequency region is at least $(r/\delta)^{1/2}$ times lower than the field in the skin-layer.

Proceeding to the UHF range one may find bursts of significantly higher intensities. Under the conditions of *cyclotron resonance in metals*, when the energy spectrum of charge carriers is anisotropic, the surface is rough, and $\omega\tau > r/\delta$, the field in the burst coincides with that in the skin-layer in its order of magnitude ($\tau = l/v$ is the charge carrier mean free time). This is because of the scatter of the orbit diameters of the resonance electrons, such orbits forming in the skin-layer when $r(\omega\tau)^{-1}$, is smaller than or of the order of δ. At even higher frequencies, $\omega > v(r\delta)^{-1/2}$, the wave phase changes several times while the electron flies through the skin-layer, and instead of bursting into the metal the high frequency field penetrates it via a different mechanism, that of electron transit. Note that the latter is also sensitive to the state of the material surface. Hence, if the electrons are almost specularly reflected then the transparency of a thin ($d \ll l$) plate acquires a resonance character as long as the electrons repeatedly return into the skin layer. In the case of a weak magnetic field ($r \gg d$), the electrons carrying information on the field into the skin layer perform essentially a nonlinear unwinding of the high-frequency temporal oscillations of the field into the respective spatial oscillations. At $\omega \gg v(rd)^{-1/2}$ this results in an oscillatory dependence of the transparency of a thin plate. It is then a function of $B^{-1/2}$, the period of this function being dependent on ω, d, and the local properties of the Fermi surface.

A *size cyclotron resonance* takes place within the $d < 2r < l$ range. Instead of the "cut-off" of the cyclotron resonance frequencies at the plate depth there appear new resonance frequencies at a multiple of the rotation frequency of those electrons in the magnetic field which have their orbit diameters equal to d. Studies of cyclotron resonance for various values of d make it possible to determine not only particular *effective masses*, but all the effective masses of charge carriers at the Fermi surface. At these frequencies the impedance has a weaker, logarithmic frequency dependence, instead of the fractional power law dependence.

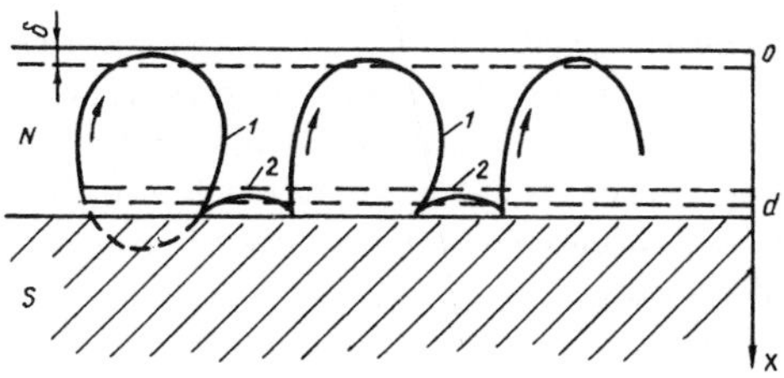

Fig. 3. The Andreev reflection of charge carriers from the N–S (normal metal–superconductor) interface. Trajectory 2 complements trajectory 1 to form an open electron orbit in a bulk specimen, and there is no "cut-off" of the resonance frequency at $r < d < 2r$.

In a plate with smooth enough faces charge carriers repeatedly return to the skin-layer due to multiple specular reflections off the face opposite to the skin-layer, and these also take part in the resonance. Dimensional cyclotron resonance is also possible in weak enough magnetic fields (however, strong enough to satisfy the condition $r < l^2/d$). By measuring the width of the resonance lines one may determine the dependence of the specular reflection probability on the angle of incidence of the electrons on the conductor surface. In a thin layer of a normal metal superposed on a superconducting substrate, the "cut-off" of part of the cyclotron resonance takes place at appreciably lower values of B, when the half-diameter of the orbits of the resonance electrons equals the depth d of the normal metal layer. When $r < d < 2r$ then *Andreev reflection* of charge carriers by the normal metal–superconductor (N–S) interface takes place, with the sign of the velocity of those carriers reversing itself ($\boldsymbol{v} \to -\boldsymbol{v}$). In that case, the frequency of a charge carrier entering the skin-layer appears to be the same as that in the bulk material (see Fig. 3). There forms a narrow burst of high-frequency field in the presence of the magnetic field. As B decreases the burst approaches the surface of the layer in which the electromagnetic wave is excited, so that at $r = d$ a signal associated with high-frequency size effects is detected. Now, for the condition $r > d$ a significant fraction of electrons, having effectively interacted with the high-frequency field in the skin-layer, and having subsequently experienced Andreev reflection from the N–S interface, must eventually collide with the layer surface. Meanwhile, resonance effects are only possible when there is almost specular reflection from that surface. Experimental studies of these effects make it possible to directly observe the Andreev reflection of electrons.

HIGHLY-DOPED SEMICONDUCTORS

Semiconductors whose concentration of impurities N is so high that weak compensation causes electronic states at the *Fermi level* to become delocalized, and the low-temperature electrical conductivity assumes a metallic, non-activated nature. Experiments show that the transition from activated conductivity to the metallic type with the growth of N (*Mott metal–insulator transition*) takes place at the concentrations that fulfill the condition $Na_B^3 \approx 0.02$, where a_B is the *Bohr radius* of the impurity atom. The theory of highly-doped semiconductors usually rests on the assumption that the inequality $Na_B^3 \gg 1$ holds. In this case, in the *weakly-compensated highly-doped semiconductors* the electrons are degenerate, and the structure of the *density of states* is described by the linear theory of screening (see *Electric charge screening*). For this situation the frequency dependence of the band-to-band light adsorption coefficient at frequencies below the threshold provides the density of states of minority carriers. With the growth of compensation (or temperature), when the Fermi level lies deep in the *band gap*, the direct association of the frequency dependence of the absorption coefficient with the density of states vanishes.

A special class arises with *strongly-compensated highly-doped semiconductors*, where the electron density at sufficiently low temperatures appears to be highly nonuniform in space. The screening of the potential associated with fluctuations of impurity concentrations in such a system is nonlinear. The greatest spatial R and energy γ scales of fluctuations, which are not screened by the electrons, are equal in order of magnitude to the values $R = N_t^{1/3}/n^{2/3}$, $\gamma = e^2 N_t^{2/3}/(\varepsilon n^{1/3})$, respectively. Here N_t is the overall concentration of impurities, n is the difference in concentration of basic and compensating impurities, ε is the *dielectric constant* of the system. In such semiconductors electrons form drops isolated from one another. Within each drop the conductivity has a metallic nature, but the drops occupy only a small fraction of the volume. They are separated from one another by potential barriers with height γ and thickness R, and at not very low temperatures conduction takes place by the relegation of electrons to the processes of percolation (see *Percolation theory*), i.e. it has an activated nature. The transition from metallic conductivity to the activated variety with the increasing extent of compensation (*Anderson transition*) in highly-doped semiconductors has been experimentally observed in n-Ge, GaAs, CdTe and n-InSb. Such a transition may occur under the effect of a magnetic field, because the magnetic field, by restricting the motion of electrons in directions transverse to the field direction, helps to bring about their localization (see *Anderson localization*). The limiting case of strongly-compensated highly-doped semiconductors is sometimes investigated theoretically for the purpose of modeling properties of *amorphous semiconductors*.

HIGH MELTING TEMPERATURE MATERIALS

See *Refractory materials*.

HIGH MOLECULAR WEIGHT COMPOUNDS

Compounds containing in a single molecule a large number (from several thousands to many millions) of chemically bound atoms, e.g., proteins, starch, polyethylene, and other *polymers*. In the case of flexible-chain polymers the location of the boundary between the low molecular weight and the high molecular weight categories depends on the *flexibility of the polymer chain*. In the case of rigid-chain molecules forming three-dimensional networks the transition from the low to the high molecular weight categories is blurred.

HIGH MOLECULAR WEIGHT GLASSES

Amorphous *polymers* in the solid state. The properties of high molecular weight glasses (e.g., *specific heat*) depend on the temperature of the transition to the *vitreous state of matter* (the temperature of *vitrification*), and on the chacteristics of the vitrification process. In particular they depend on the rate of cooling during vitrification, the type of solvent, and the velocity of its evaporation upon the formation of the polymer glass from the solution. Concerning mechanical effects, the transition to the glass-like state depends on the loading speed and on the frequency of vibrations (mechanical vitrification). High molecular weight glasses are capable of extensive *strain* when loaded in the thermal range below vitrification (induced high elasticity).

HIGH-PERMEABILITY MAGNETIC MATERIALS

Materials which have high values of *magnetic saturation*, and which can be easily remagnetized (see *Magnetic reversal*) and *demagnetized*. This is achieved by forming alloy compositions with a highly uniform *atomic magnetic structure*, and close to zero constants of *magnetic anisotropy* (K_1) and of *magnetostriction* (λ_s).

There are several systems of high permeability magnetic materials:

1. In the Fe–Ni alloy system there are three types of compositions, used for production of high-permeability magnetic materials, called *permalloys*. At 70–85% of Ni with the help of *annealing* in the range of temperatures 1100–1300 °C in vacuo (or hydrogen) and cooling at a regulated rate, a state with K_1 and λ_s, close to zero, is obtained, in this case with values of *magnetic permeability* (initial and maximum) $\mu_a = 2{\cdot}10^4$ to $2{\cdot}10^5$, $\mu_{max} = 10^5$ to 10^6, but these alloys have low saturation induction $B_s =$ 0.5 to 0.8 T. To facilitate the process of *thermal treatment* of such alloys and obtain certain combinations of magnetic, electrical and mechanical properties, they are alloyed with the following elements which influence the ordering processes: Mo, Cr, Cu, V and Si. The best properties in this group are exhibited by *supermalloy* containing 80% Ni, 5% Mo, 15% Fe and with $\mu_{max} > 10^5$. The alloys with 55 to 68% Ni after *thermomagnetic treatment* and *thermomechanical treatment* acquire a *magnetic texture* with an *easy magnetization axis*, and for magnetizing along this direction they have a rectangular *magnetic hysteresis* loop and $\mu_{max} > 800{,}000$. The compositions with 45–50% Ni have an increased $B_s = 1.3$ to 1.5 T. During the course of secondary *recrystallization* at 1150 to 1200 °C a crystallographic *texture* appears in them, which provides high-permeability magnetic materials with μ_a and μ_{max} within the limits 2,000 to 5,000, and 20,000 to 300,000, respectively. All the Fe–Ni alloys are pliable, they can be subjected to cold rolling up to micron thicknesses, to cutting and welding; after thermal treatment in the state with maximum permeability they are very sensitive to mechanical stresses and deformations, under the influence of which their magnetic properties sharply deteriorate.

2. Alloys containing 33 to 40% Ni, 25 to 30% Co, additions of: Cr, Si, Mo, Cu, and residual Fe. With the help of a thermomagnetic treatment they exhibit a rectangular hysteresis loop with $B_r/B_s =$ 0.85 to 0.98 and $\mu_{max} = 4{\cdot}10^4$ to $8{\cdot}10^5$.

3. Alloys based on the Fe–Al system containing about 12% Al have a high *hardness*; they are used for manufacturing magnetic circuits and products which are routinely subjected to mechanical loads.

4. Some *metallic glasses*, obtained by *quenching* from the liquid state, have high values of μ_a and μ_{max}. For example, Fe–Ni glasses with 45% Ni after *annealing* in a longitudinal magnetic field have $\mu_{max} = 310{,}000$. Some Co-based glasses after *tempering* in vacuo at 300 °C have $\mu_a = 30{,}000$ and $\mu_{max} = 370{,}000$.

HIGH-PRESSURE CELLS

Devices for the creation of static *pressure* in excess of 100 MPa.

High-pressure cells with a gaseous or liquid pressure-transmitting medium are capable of producing a hydrostatic pressure of up to 3 GPa. These cells are made in the shape of hollow thick-walled (often multilayer) cylinders, closed (if the pressure is created by means of a pump) or with a movable piston (if they are driven by a press). A higher (but quasi-hydrostatic) pressure (up to 170 GPa) is produced in solid-state high-pressure cells, i.e. with a solid pressure-transmitting medium. These cells are press-driven and consist of a container (material: pyrophyllite, talc, lithographic stone) with the material to be compressed inserted into it, and force bearing hardened instrumental *steel*, hard *alloy* or *diamond* placed around the container. The shape of the force applicators are chosen so that a cavity capable of reducing its volume is created around the container. This cavity may be either initially closed (high-pressure cells with gliding punch), or with clearances left between power details (high-pressure cells with compressible burr). When the force applicators come closer together, a part of the container's material is forced out into these clearances to form the burr locking the cavity, and clamp together the surfaces of power details adjacent to the cavity. The lateral surfaces of the force applicators are clamped together by pressing them into steel cartridges. The steel solid-state high-pressure cells create a pressure up to 5 GPa, and the diamond ones up to 170 GPa. For synthesis of diamond, cubic *boron nitride* and sintering of *composite materials* on their base, the industry employs solid-state high-pressure units for pressures 5–8 GPa in the shape of two anvils with depressions, together with a cylinder with flared holes at both ends and two punches (see also *High pressures*, *High pressure measurements*).

HIGH PRESSURE MEASUREMENTS

Measurements of pressure carried out with the help of various *manometers* whose operation is based on the pressure dependence of many diverse characteristics of matter, viz., thermodynamic, electric, magnetic, spectral, and so on. Manometers must be calibrated by comparison either with results of absolute measurements using piston gauges, or with secondary data recommended by the special group of the International Association on Development of Research in the field of *high pressures* (1986). Below 1.4 GPa, the pressure measurements can be based on *mercury* melting temperature data. Manometer calibration can also be based on characteristics of *phase transitions* at room temperature, calculated data relating the *pressure* (in kbar) to the *compressibility* of NaCl, or the spectral shift of the R_1 line in the spectrum of ruby at room temperature: $2.746\Delta R_1(\text{Å}) = P(\text{kbar})$ (this equation is valid up to 200 kbar). The Raman spectrum of ruby is currently the most popular calibration method.

Measurements of high pressures are often reported in non-standard units, such as: kilobar and megabar, as well as derived SI units: megapascal (10^6 Pa), gigapascal (10^9 Pa), and terapascal (10^{12} Pa) (see *Units of physical quantities*). Their interrelationships are given below:

	kbar	Mbar	MPa	GPa	TPa
kbar	1	10^{-3}	10^2	10^{-1}	10^{-4}
Mbar	10^3	1	10^5	10^2	10^{-1}
MPa	10^{-2}	10^{-5}	1	10^{-3}	10^{-6}
GPa	10	10^{-2}	10^3	1	10^{-3}
TPa	10^4	10	10^6	10^3	1

HIGH PRESSURES in solid state physics

Conventional name for a range of pressures at which the distance between the atoms in solids decreases more through compression than it does during cooling of the same body at a normal *pressure*. Pressures above 100 MPa are classed as high pressures.

In solid state physics the high pressures that are generally employed are of the static type. In this case the value of the attainable pressure is determined by the design characteristics of the *high-pressure cell* and chamber. Depending on the application, the specifications required for particular investigations can put limits on the range of working pressure. For example, a high pressure chamber can be weakened by windows incorporated for the admission of light, by conductoring wires inserted for resistance measurements, or by the incidence of ionization radiation. For magnetic investigations the chambers should be manufactured from non-magnetic materials; for chambers working at high or low temperatures appropriate materials must be selected. The pressure, together with the temperature, both enter the main thermodynamic relations, and can appear as differentials of energy, *entropy* and volume. The dependence of the pressure on the volume and temperature is described by an *equation of state*. Several approximate and empirical equations of state have been proposed which give satisfactory agreement with experimental data, at least for pressures up to 20 GPa.

Under compression changes can occur in various interactions, and in particular, interactions between electrons of the same configuration, interactions with nuclei, as well as inter-configurational interactions. The values of these interactions sharply differ for "inner" and "outer" electron shells. For instance, the variation of electrostatic interactions of f-electrons can bring about a shift of energy levels up to approximately 10 cm^{-1}/GPa, and the shift of outer configuration levels can be an order of magnitude higher.

Using such relations and the fact that high pressures can provide a large variation of inter-atomic distances, it is possible to single out these interactions and to acquire some information about them which cannot be obtained by the other methods. Quantitative investigations of this type have been carried out using *magnetoacoustic resonance*, *tunneling spectroscopy*, *electron paramagnetic resonance*, *nuclear magnetic resonance*, etc. It often happens that at the attainment of a particular degree of compression qualitative changes in properties occur, the characters of interactions change, and *phase transitions* are observed. For example, metallic cesium undergoes an electronic phase transition (see *Phase transitions under pressure*) in the vicinity of 4.5 GPa; its volume significantly decreases without a change of the close packing of the atoms or the lattice symmetry. Such transitions are studied by measuring the electric resistance, because they often involve a metal-to-insulator transformation (and also *superconductor*). Modern experimental installations permit one to make simultaneous measurements of electric resistance under high pressures (up to 50 GPa) and low temperatures (below 1 K). Different spectral investigations (*absorption*, *luminescence*, *Raman scattering of light*, etc.) performed at low temperature (4.2 K using "piston–cylinder", and 25 K using "diamond anvil" type chambers) and high pressure (up to 6 GPa in two-step "piston–cylinder", and up to 16 GPa in "diamond anvil" chambers) are devoted to the study of *structural phase transitions* and *electronic phase transitions*.

During magnetic investigations at high pressures (up to 1 GPa) and low temperatures (below 0.05 K) some rare-earth substances characterized by magnetic ordering loose this ordering down to the lowest temperatures. Investigations of the *Mössbauer effect* have shown, in particular, that these materials do not change their valence up to very high degrees of compression. At present the Mössbauer effect can be studied (at room temperature) at pressures up to 70 GPa.

HIGH-SPIN COMPLEXES

Complexes of *transition metals*, in which the electrostatic interaction of electrons within a partially full d-shell of the central transition atom exceeds that of the crystal field of the *ligands*. High-spin complexes are formed preferentially with strongly electronegative ligand atoms and groups utilizing strongly ionic type bonds, without essential participation of d-orbitals. In high-spin complexes the structure of the d-shell is not disturbed by the influence of weak ligand fields, and the number of unpaired d-electrons, which has a maximal value according to the *Hund rule* for the main term of the isolated central atom, remains the same in complexes with ligands. *Low-spin (spin-paired) complexes* are also distinguished; they are complexes of transition metals, where the maximum possible spin of unpaired electrons in the unfilled d-shell of the central atom is not realized due to the strong interaction with the ligands. This interaction brings about the splitting of the d-level of the central atom into sublevels, the multiplicity of which is determined by the crystal field symmetry (see *Crystal field*). Differences in the total *spin* of the ground state, depending on the ligand type, are possible for the electronic configuration d^n of the central atom with $n = 4, 5, 6, 7$ in the case of octahedral complexes, and with $n = 3, 4, 5, 6$ in the case of tetrahedral complexes. In low-spin complexes the energy gain from placing electrons at the lower energy sublevels exceeds the loss in exchange energy. These complexes are characteristic of the second, and especially of the third series of transition elements; they are formed preferably with weakly-negative ligands, bound to metals by *covalent bonds*.

HIGH-TEMPERATURE ALLOYS

Alloys based on nickel, iron–chromium–nickel, cobalt, or a mixed base, known for their *high-temperature strength*. The main features characterizing high-temperature alloys are high resistance to *creep* and *fatigue*, and *long-term strength*.

Deformable and cast high-temperature alloys are distinguished. The heat resistance of the *deformable high-temperature alloys* is due to the structure, where the particles of *intermetallic compounds*, borides and *carbides* are evenly distributed. Such a structure appears as a result of *heat treatment* (homogenizing and aging, see *Homogenization*) bringing about the *alloy heterogenization* of the microstructure. Besides, this appears during the course of *alloying* with refractory chemical elements (*tungsten, molybdenum, vanadium*) and strengthening elements (e.g., *titanium, aluminum, niobium, boron*). Finally, it can be created by reduction of the content of *lead, tin, antimony, bismuth* and sulfur, and by addition of refining elements (*calcium, cerium, barium* and boron). The long-term heating during the course of hardening causes the depletion of alloying elements (titanium, aluminum, etc.) in the upper layer of the products. The defective layer is removed by cutting, *polishing* or electrochemically. The *thermal-environment resistance* of high-temperature alloys in the air, in the products of fuel combustion, and other gas media depends on the content of alloying elements, and it is determined by the operating temperature, operational life, and alloy composition. Heat resistance reduces sharply in a medium of combustion products of raw oil, black oil, of combustion products of fuels with addition of sea water, whereas *chromium* and aluminum improve it. If the high-temperature alloy products are intended for foundry use at 800 °C, their surface is additionally subjected to a diffusion thermochemical treatment (calorizing, chromium calorizing, enameling, deposition of refractory oxides, etc.). Deformable high-temperature alloys are fused in vacuo by the methods of HF induction, including plasma overheating, complying with strict requirements on the purity of alloys.

Cast heat-resistant alloys differ from deformable alloys by their dendriticity (see *Dendrites*). The inter-dendrite spaces contain, as a rule, fusible element admixtures, oxides, sulfides, etc., so that it is best to perform melting and pouring under vacuum or in argon, using extremely pure charge materials. Homogenization can partly correct for structural inhomogeneity. The products made of cast high-temperature alloys are calorized (aluminized) in dry mixtures or by dipping into liquid aluminum alloy with subsequent diffusion heating at high temperatures. Calorizing (chromium calorizing) improves the oxidation resistance of surface layers as well as the thermal stability of the products. High-temperature alloys are used to manufacture parts of steam and gas turbines, and internal combustion engines operating at high temperatures, air jet engines, and in rocket engineering.

HIGH-TEMPERATURE EXPANSION

A method for the approximate calculation of thermodynamic functions (see *Thermodynamic potentials*) and average values of physical parameters, based on their expansion in an infinite series with respect to the small parameter $\delta E/(k_B T)$, where δE is the energy scale characteristic for the given problem. Often only the term linear in $1/T$ is taken into account. This is acceptable if the energy spectrum of the system is restricted to the narrow interval $\Delta E \ll k_B T$. In the general case this approximation can be applied to a group of levels which lie within the relatively narrow energy gap $\Delta E_1 \ll k_B T$ and are separated from other levels by an appreciable energy interval $\Delta E_2 \gg k_B T$. The high-temperature expansion is used in the theory of *magnetism* because spin systems have a restricted spectrum. In particular, the *Curie law* for the magnetization of an ideal *paramagnet* is well described by this approximation. In non-equilibrium thermodynamics this approximation allows one to linearize the *kinetic equations* for reciprocal temperatures of quasi-equilibrium subsystems.

HIGH-TEMPERATURE MATERIALS

Materials known for their enhanced *high-temperature strength*. High-temperature materials also exhibit high resistance to *creep* and *failure* at elevated temperatures. They include *high-temperature alloys* based on iron, nickel and cobalt, *refractory materials*, and also complex *composite materials*: metal–oxide, metal–carbide and metal–intermetallic compounds with various high-temperature properties. The properties of high-temperature materials depend on the *melting temperature* of their base, on homologous temperatures, diffusion mobility (see *Diffusion*), atomic bond energy in *alloys*, and on their structure. Within each group of materials and each base (aluminum and magnesium *steels*, *nickel alloys*, cobalt and refractory alloys), mechanical properties (*long-term strength, Young's modulus, impact strength*, tensile *ultimate limit*, etc.) vary over a wide temperature range depending on the *alloying* of the base, on melting methods, *heat treatment*, and gas medium composition.

HIGH-TEMPERATURE STRENGTH

The ability of a material to resist *plastic deformation* and *failure* under the conditions of high temperature, and constant and alternating stresses. The main physical parameters which determine the high-temperature strength of materials are the magnitude and direction of interatomic bonds in crystal lattices of their component *phases*, and also the *crystal structure*. The magnitude and direction of interatomic bond forces determine the upper limit of temperature, where the material still has a high resistance to plastic deformation. Depending on their purpose, heat-resistant materials may be unalloyed *metals*, *solid solutions*, dispersion-hardened alloys (see *Dispersion hardening*) and *composite materials*.

Heat treatment and mechanical heat treatment are capable of producing a certain structure and dislocation substructure in the material, which increase the resistance to plastic deformation at high temperatures. The higher the density of crystalline structure *defects*, the more disperse and uniformly distributed the secondary phases, and also the higher the resistance of the material to plastic deformation. Since thermodynamic stability differs for different structure types, the longer the material must operate at high temperature the more stable the strengthing structural elements should be. In the case of a very long service life (over 100,000 hours), the material is subjected beforehand to special stabilizing types of treatment. Depending on this expected life, on the nature of the loaded state: static, dynamic or cyclic, and also on the value of permissible deformation, different parameters are used to define high-temperature strength of the material. *Hot hardness* and *short-term strength* at high temperatures serve for estimating the high-temperature strength of the material in the case of short-term operation, and also for a tentative appraisal in the case of long-term service. *Creep* is often observed in the material after the long-term action of static stress at a high temperature. To evaluate destruction resistance, the *long-term strength limit* is used, which is defined as the minimal stress, at which the material fails during a certain time period at a specified temperature, e.g., σ_{1000}^{500}, σ_{3000}^{700}, σ_{10000}^{100}, where the subscript shows the time to failure (hrs), and the superscript gives the temperature (°C). If the maximum permissible lengthening serves to limit the material service life, then the *conditional creep limit* is used, i.e. the stress, at which a specified deformation is attained during the predetermined time period, e.g., $\sigma_{0.2/500}^{100}$, $\sigma_{1.0/1000}^{1200}$, etc. (superscript gives temperature (°C), subscript % *strain*/time to develop deformation). When the deformation in the material is preset as constant, then the *stress relaxation* occurs at a high temperature during the course of time. Elastic strain is gradually transformed, by virtue of creep, into plastic deformation. In this case, high-temperature strength is determined by the *relaxation strength* which is evaluated by the stress, σ_τ, remaining in the material for the time period τ, or by a decrease in stress $\Delta\sigma_\tau$, for the same time period: $\Delta\sigma_\tau = \sigma_0 - \sigma_\tau$, where σ_0 is the initial stress. If the stress changes in a cyclic manner then one can observe the creep via the cyclic stress, *dynamic creep* and high-temperature *fatigue* in the material. Creep at cyclic stress can occur if the value of stress changes according to a rectangular cycle with a long period (tens of minutes to tens of hours) as a superposition of cyclic stress σ_a over average stress σ_m. When compared to creep at constant stress equal to the maximum level in the cycle, the creep resistance at cyclic stress often proves to be lower. In the case of dynamic creep, the tension also changes in a cyclic way, but the frequency is considerably higher (from units to tens of hertz) and the value of the stress change is considerably below the average stress in the cycle. Reversing the sign of the load at a high temperature leads to fatigue in the material. Fatigue resistance is expressed in terms of the number of cycles until the appearance of a macrofracture (see *Fractures*), or until the failure of the deformation amplitude at a preset temperature.

HIGH-TEMPERATURE SUPERCONDUCTIVITY

A phenomenon of *superconductivity* observed at temperatures much higher than those of ordinary superconductors ($T > 30$ K) and discovered in 1986 by J.G. Bednorz and A.K. Müller and C. Chu in metal oxide ceramics (see *High-temperature superconductors*). The highest transition temperature into the superconducting state now stands at $T_c \approx 145$ K for the mercury cuprate compound $HgBa_2Ca_2Cu_3O_{8+x}$, which also exhibits $T_c \approx 166$ K at a pressure of 30 GPa.

In 1946 R.A. Ogg Jr. observed a sharp drop in the resistance of a metal–ammonium solution $Na:NH_3$ frozen in liquid nitrogen ($T = 77$ K). Based on his experiments, Ogg suggested that bound electron pairs with zero spin form as a result of the polarization of dipole molecules of NH_3 (prototype of singlet *Cooper pairs* or *bipolarons*), and he attributed zero resistance (i.e. superconductivity) to the *superfluidity* of such charged bosons. Ogg's experimental results, however, were later found to be due to the precipitation of sodium fibers from the solution.

Almost twenty years later the idea of high-temperature superconductivity appeared in studies by W. Little (1964) who suggested its possibility in *organic polymers* at $T_c \sim 10^2$–10^3 K. Similarly, V.L. Ginzburg pointed to a possibility of high-temperature superconductivity at a metal–insulator interface or in "sandwich" systems of the insulator–metal–insulator or semiconductor–metal–semiconductor type. These studies considered a non-phonon mechanism of Cooper pairing of *conduction electrons* (see *Cooper effect*), which might occur due to exchange of high-energy excitations of the Bose type, i.e. *excitons* (see *Exciton mechanism of superconductivity*). V.L. Ginzburg and D.A. Kirzhnits wrote a book entitled *High Temperature Superconductivity* in 1977. The above ideas stimulated experimental research in systems of low dimensionality, such as the *quasi-one-dimensional crystals* (both organic and inorganic) of the TTF-TCNQ, KPC type, the $(SN)_x$ polymer, $Nb(Se, S)_3$, or the quasi-two-dimensional crystals of the type of $Nb(Se, S)_2$ and $Ta(Se, S)_2$. However, the maximum values of $T_c \approx 10$ K could only be reached in the $(ET)_2Cu(NCS)_2$ compound.

In 1967–1968 J. Garland, H. Fröhlich, and E.A. Pashitsky independently suggested a mechanism of high-temperature superconductivity in the *transition metals* and in *degenerate semiconductors, semimetals* and layered semiconducting structures where there seemed to be a possibility of another type of collective boson excitation exchange of an electronic nature, namely *acoustic plasmons*. Such excitations had been predicted earlier by D. Pines (1956) for a two-component *Fermi liquid* with "light" and "heavy" charge carriers (see *Plasmon mechanism of superconductivity*). B.T. Geilikman (1966) treated a similar mechanism of electron superconductivity involving the screened Coulomb interaction in transition metals and alloys. No account, however, was taken of the effects of retardation that develop during the exchange of acoustic plasmons.

After the discovery of high temperature superconductivity in 1986, numerous new theoretical concepts appeared to explain this unique phenomenon. Such concepts may be separated conditionally into the traditional ones that treat this or that mechanism of Cooper pairing (see *Bardeen–Cooper–Schrieffer theory*), and the nontraditional varieties that offer entirely new approaches to the solution of the problem of high-temperature superconductivity. Among the first are the models with strong *electron–phonon interaction* that are based on the *Eliashberg equations* for the *energy gap in superconductors* and the critical temperature. Various non-phonon mechanisms of superconductivity are considered, in particular those stemming from the exchange of dipole and quadrupole permitted excitons (see *Exciton mechanism of superconductivity*) or of collective excitations of charge density (e.g., of plasmons, see *Plasmon mechanism of superconductivity*). Such excitations may take place in layered (quasi-two-dimensional) metals and in narrow-band metals with charge carriers almost localized at lattice points. In 1962 I.Ya. Privorotsky suggested another mechanism of Cooper pairing for the case of triplet p-wave pairing of electrons that could go via the exchange of virtual antiferromagnetic *magnons* (see *Magnon mechanism of superconductivity*). D. Pines (1990) applied this mechanism to a modified case of singlet d-wave coupling of holes in CuO_2 cuprate layers to explain the nature of high temperature superconductivity. According to a concept suggested by A.A. Abrikosov (1993), an important role in Cooper pairing that takes place in high-temperature superconductors based on layered copper metal-oxide compounds is played by the quasi-one-dimensional saddle anomalies (lines of saddle points) found in the quasi-two-dimensional *dispersion law* (band spectrum). These are experimentally observable using angle resolved *photoelectron spectroscopy*. Such anomalies are characterized by *Van Hove singularities* in their *density of electron states*.

Other nontraditional theories of high temperature superconductivity which have been proposed include the *model of resonating valence bonds* (see *Anderson mechanism of high-temperature superconductivity*), various modifications of the Hubbard and the Heisenberg models (see *Hubbard model* and *Heisenberg magnet*), the model of a charged bipolaron gas superconductivity (see *Bipolaron mechanism of superconductivity*), the *anyon model* – a model of quasi-particles that follows a statistics intermediate between the Boson and Fermion types in strongly correlated two-dimensional systems, etc. Currently, there is no comprehensive unified commonly accepted theory of high-temperature superconductivity, although the BCS theory does quite well in explaining most known properties.

HIGH-TEMPERATURE SUPERCONDUCTORS

Solid-state materials (chemical compounds such as *alloys*, *solid solutions*, especially *ceramics*) with rather high *critical temperatures* (T_c) of their *superconducting phase transition* (assumed $T_c > 30$ K). The first high-temperature superconductor was discovered by J.G. Bednorz and K.-A. Müller (1986) in the form of ceramic samples of a LaBaCuO metal-oxide compound, $T_c \approx 30$ K. At present, several types of copper-oxygen containing (cuprate) high-temperature superconductors are known: $La_{2-x}M_xCuO_4$ (M = Ba, Sr, Ca), the maximum value of the critical temperature, $T_{c\,max} \approx 40$ K; $YBa_2Cu_3O_{6+x}$ (123) and $YBa_2Cu_4O_{7+x}$ (124), $T_{c\,max} \approx 90$ K; $Bi_2Sr_2Ca_{n-1}Cu_nO_x$, $T_{c\,max} \approx 110$ K (for $n = 3$); $Tl_mBa_2Ca_{n-1}Cu_nO_x$, $T_{c\,max} \approx 115$ K (for $m = 1$ and $n = 5$), or $T_{c\,max} \approx 125$ K (for $m = 2$, $n = 3$); $HgBa_2Ca_{n-1}Cu_nO_x$, $T_{c\,max} \approx 135$ K (for $n = 3$) and $T_{c\,max} \approx 160$ K under pressure (30 GPa). All such compounds feature a perovskite-type structure (see *Perovskites*) and contain from 1 to 5 of the so-called cuprate layers of CuO_2 in their *unit cell*, the latter serving as the basic conducting and superconducting channel in a high-temperature superconductor. Layers of CuO_2 alternate with layers of metal ions (Ca^{2+}, Sr^{2+}, Y^{3+}) and oxide layers (LaO, BaO, SrO, BiO, TlO, HgO) in their *crystal lattice*. Figs. 1–3 demonstrate the crystalline structures of certain high-temperature superconductors. Fig. 1 shows the structure of $LaCuO_4$ (a) and $YBa_2Cu_3O_7$ (b); Fig. 2(a) shows the structure

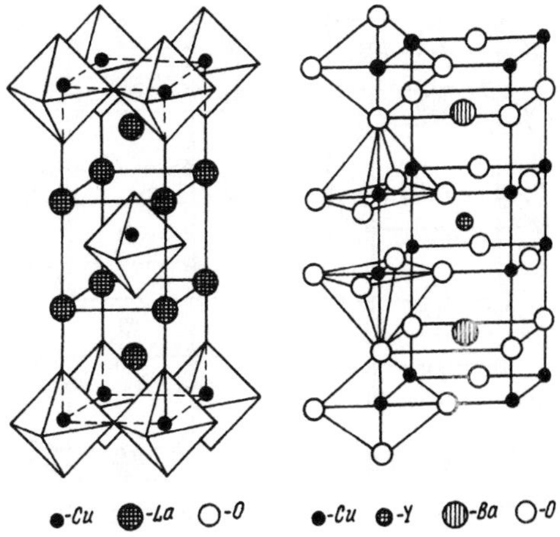

(a) (b)

Fig. 1.

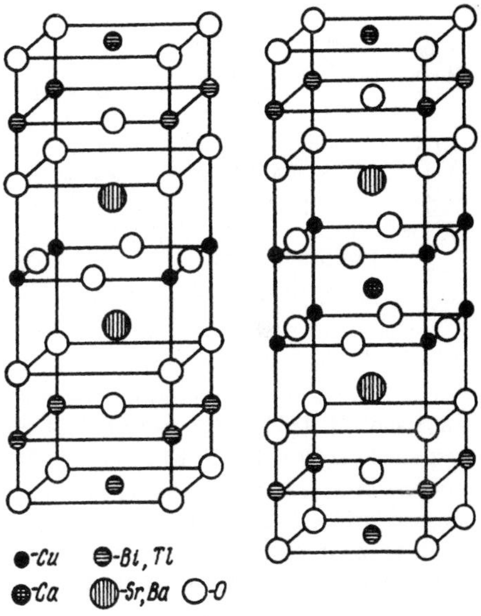

(a) (b)

Fig. 2.

of $Bi_2Sr_2CuO_6$ and $Tl_2Ba_2CuO_6$ with a single cuprate layer, $n = 1$. Fig. 2(b) shows the structure of $Bi_2Sr_2CaCu_2O_8$ and $Tl_2Ba_2CaCu_2O_8$ with two cuprate layers, $n = 2$. Fig. 3(a) shows

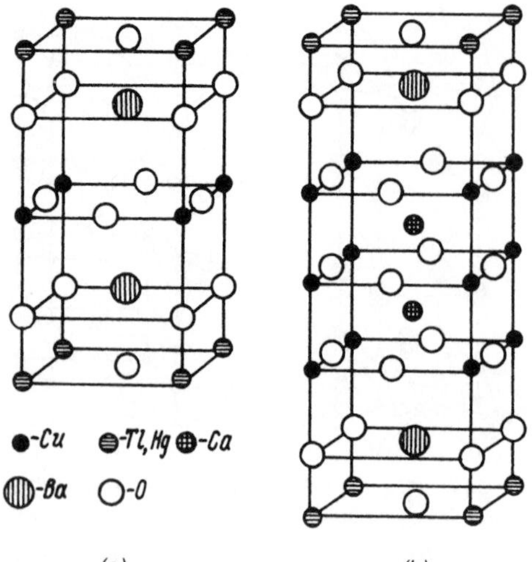

Fig. 3.

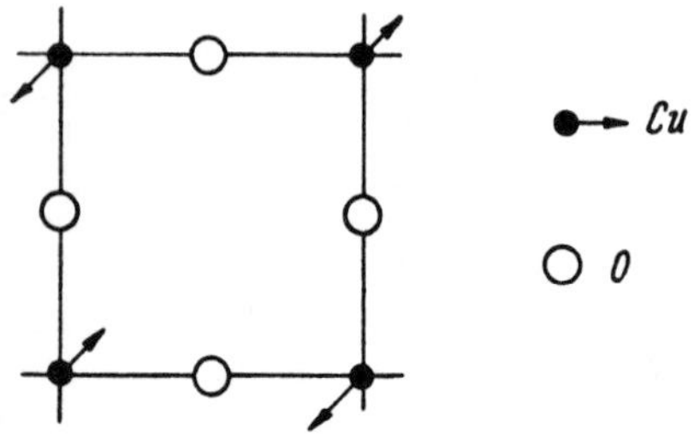

Fig. 4.

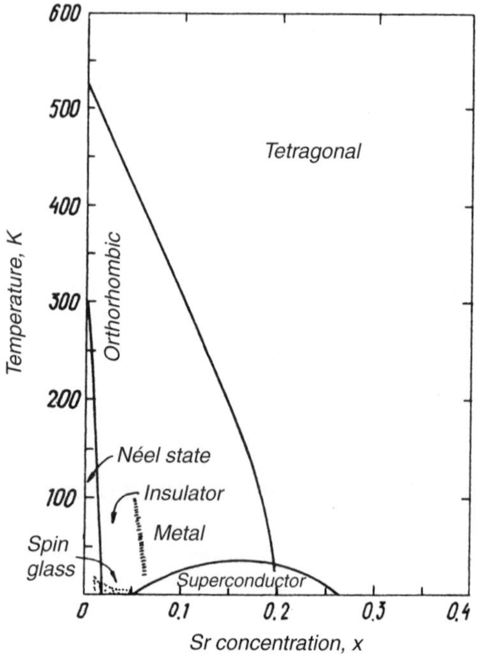

Fig. 5.

the structure of $TlBa_2CuO_6$ and $HgBa_2CuO_6$ with a single layer of CuO_2 ($n = 1$). Fig. 3(b) shows the structure of $TlBa_2Ca_2Cu_3O_{10}$ and $HgBa_2Ca_2Cu_3O_{10}$ with three layers of CuO_2 ($n = 3$). Using the method of molecular epitaxy (see *Molecular beam epitaxy*) a metastable layered system was constructed $(Ca_{1-x}Sr_x)_{1-y}CuO_2$ that contains an "infinite" number of cuprate layers of CuO_2, separated by intermediate layers of Ca^{2+} and Sr^{2+} ions, with the critical temperature $T_c \approx 110$ K.

Depending on the contents of the intermediate layers, the valence (charge) state of the Cu and possibly O ions also changes, as well as the magnetic spin state of the Cu ions. For example, $La_{2-x}M_xCuO_4$ (M = Sr, Ba), $x < 0.05$ (and $YBa_2Cu_3O_{6+x}$, $x < 0.5$) are both antiferromagnetic *insulators* (see *Antiferromagnetism*), with *Néel points* found at $T_N = 300$ K and 400 K, respectively, for $x = 0$. They also feature a two-dimensional ordering of half-integer spins ($S = 1/2$) of their Cu^{2+} ions that form a double sublattice collinear structure (Fig. 4). As x grows, large-scale macroscopic ordering in the CuO_2 planes deteriorates, although a high degree of localization of spin states and a near-neighbor antiferromagnetic ordering both persist. Charge states appear of O^- instead of O^{2-}, as well as extra (doped) *holes*, first localized at atoms of oxygen. Next, as their threshold concentration is reached (at $x \approx 0.05$ per single layer of CuO within the volume of the unit cell) these holes delocalize and the system as a whole transforms to its metallic state (see *Mott metal–insulator transition*, *Anderson localization*). At $T < T_c$ the same system transforms to its superconducting state (see *High-temperature superconductors*). Fig. 5 presents a typical compositional *phase diagram*, drawn for the $La_{2-x}Sr_xCuO_4$

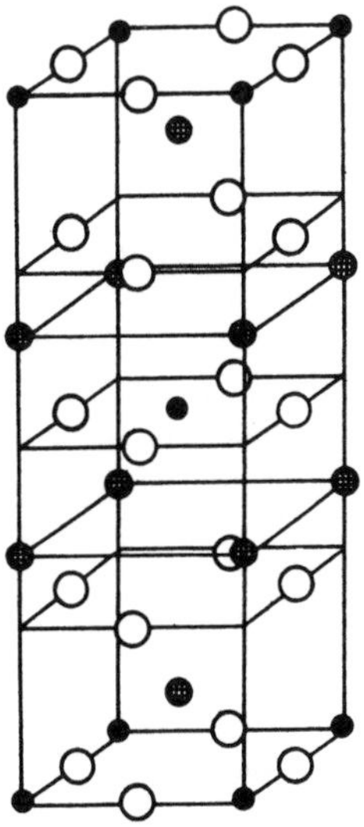

Fig. 6.

system, in the temperature (T) versus doped carrier concentration (x) space. It provides an example of how various phases or states of cuprate high-temperature superconductors can be related to each other. Besides the cuprate compounds described above that feature hole conductivity in their metallic state, there also exist compounds of the $Nd_{2-x}Ce_xCuO_4$ type and $Pr_{1-x}Ce_xCuO_4$ type which exhibit electronic conductivity. However, their critical temperatures are significantly lower, $T_c < 25$ K. Their crystalline structure is shown in Fig. 6.

One of the most prominent features of high-temperature superconductors are the nonmonotonic dependences of their critical temperature T_c on x, and on the number n of CuO_2 cuprate layers in their crystal unit cells. There is also the absence of an appreciable isotope effect in all components of the cuprate metal-oxide compounds, with the exception of oxygen and (possibly) copper. According to the classical *isotope effect*, $M^{\alpha} T_c = \text{const}$, where M is the isotope mass and $\alpha = 1/2$ for the simplified BCS model involving phonon-coupled Cooper pairs. The oxygen isotopic shift of the critical temperature T_c is also anomalous when O^{16} is replaced by O^{18}, in both its dependence on the concentration of carriers, and the value of T_c. As T_c increases for higher or lower x, the index α of the oxygen isotopic effect drops all the way to zero, and even becomes negative, in contrast to regular *superconductors* with an *electron–phonon interaction*. In the latter, the index of their isotopic shift in T_c decreases for lower T_c. This feature may point to a non-phonon mechanism of high-temperature superconductivity.

At $T < T_c$ high-temperature superconductors typically display anomalous thermodynamic and kinetic properties. For example, they feature a non-exponential temperature dependence of their electronic *specific heat*, the rate of *spin–lattice relaxation*, and the damping coefficient of *ultrasound*. The high-temperature superconductors are type II with abnormally high values of their transverse *upper critical field*, $B_{c2}^{\perp}(T = 0) > 100$ T. This is typical for layered superconductors with weak Josephson binding between the layers (see *Josephson effect*). In its turn, the layered structure results in strong anisotropy of the critical fields and currents (see *Critical currents in superconductors*).

The normal metallic state of high-temperature superconductors above T_c also exhibits certain unusual properties, among which are a linear temperature dependence of their electrical *resistance* and an inverse value of their Hall coefficient (see *Galvanomagnetic effects*) over a broad temperature range. Moreover, they feature nonlinear temperature dependences of their electron specific heat, and the rate of their spin–lattice relaxation, pronounced deviations from the Drude formula in the frequency dependence of their *high-frequency electrical conductivity*, etc. All these anomalies of normal state physical properties of the cuprate superconductors (as well as the phenomenon of high-temperature superconductivity itself) have as yet found no comprehensive theoretical explanation.

Some authors include among the family of high-temperature superconductors the $Ba_{1-x}M_x$-BiO_3 metal-oxide compound ($M = K, Rb$) with a cubic perovskite structure and the maximum value of $T_c \approx 30$ K, and perhaps *fulleride*, a $C_{60}M_x$ solid *fullerene* doped with alkali metals ($M = K$, Na, Rb, Cs), its $T_{c\,max} \approx 47$ K for $C_{60}Cs_3$.

Studies of physical properties of high-temperature superconductors, both already known and newly synthesized, are ongoing, and have flourished during their active development phase. One should not completely rule out the possibility of discovering "room temperature" high-temperature superconductors ($T_c > 300$ K). It is hard even to grasp the perspectives of practical applications of such an eventuality.

HIGH-TEMPERATURE X-RADIOGRAPHY

Use of *X-ray diffraction* to investigate solid-state physics and material science problems at high temperatures (500 to 3000 K). One main application is the investigation of *first-order phase transitions* and *second-order phase transitions* at high temperatures, including *polymorphism*, the *crystal structure* of the high-temperature modifications, determination of the *phase transition* temperature, finding volume and *orientation relations* at the transformations, and elucidating the kinetics of different transformations in the solids. The X-ray diffraction data can provide high-temperature phase diagrams of multicomponent system equilibria. This technique is also used for studying heat expansion of crystals at high temperatures (*X-ray dilatometry*). Although the accuracy of X-ray dilatometry is not very high, it has the advantage of allowing one to measure the *thermal expansion* tensor of anisotropic *crystals*. The X-ray diffraction data can provide estimates of the equilibrium concentration of *point defects*. Measurements of the temperature dependence of the intensity of X-ray diffraction lines, together with heat diffusion dissipation at high temperatures, can provide the characteristic *Debye temperature*. X-ray diffraction intensity measurements are also used to study *alloy ordering processes*.

HILBERT RELAXATION TERM

See *Landau–Lifshits equation*.

HOFFMANN METHOD, extended Hückel method (R. Hoffmann, 1963; E. Hückel, 1931)

A semiempirical method, no longer much in use, for the quantum mechanical calculation of the electron structure of multiatomic molecules. In the Hoffmann method the single particle states of a molecule (*molecular orbitals*) are determined in the form of linear combinations of valence orbitals of component atoms, whose coefficients are components of particular vectors C_i of the matrix of the single-electron Hamiltonian H from the eigenvalue equations

$$HC_i = \varepsilon_i S_i C_i, \qquad C_i^+ SC_j = \delta_{ij},$$

where S is the matrix of orbital overlap, C_i^+ is the transposed column-vector C_i; ε_i are the energies of the one-electron levels; above the lowest levels of the range the electrons are distributed in accordance with the Pauli principle. The matrix elements of H are determined by the formulae

$$H_{\mu\mu} = -I_\mu,$$
$$H_{\mu\nu} = kS_{\mu\nu}\frac{H_{\mu\mu} + H_{\nu\nu}}{2}, \qquad \mu \neq \nu,$$

where I_μ is the ionization potential which responds to removal of an electron from a valence atomic orbital μ, and k is a numerical coefficient (as a rule, $k = 1.75$). A principal drawback of the Hoffmann method is its neglect of the Coulomb interaction between electrons; however, owing to its simplicity, it has been used to assess the electron density distributions and the *spin density* in the ground state of molecules and radicals, and to estimate the energy of *conformations*, internal rotation barriers, and characteristics of deformation vibrations. The method *is not* satisfactory for determining bond lengths.

HOLE

A *quasi-particle* introduced to account for the cooperative motion of electrons in *energy bands* of solids. Holes are found in the *valence band* of semiconductors, and in the *conduction band* of hole-type *metals* and *semimetals*. The introduction of holes replaces taking into account the actual motion of all the band electrons in external electric ($\boldsymbol{E}$) and magnetic ($\boldsymbol{B}$) fields by consideration of the motion of fictitious particles, i.e. holes, each of which has a positive charge e equal in magnitude to electron charge; each hole occupies one of the free (unoccupied) states of the band and has the *quasi-wave vector* $\boldsymbol{k}$ of this state. As this takes place, the following circumstances should be noted: (1) the concentration of holes is equal to the concentration of free-electron states in the band, and can be calculated with the help of the statistical equations which describe these states;

(2) the motion of holes produces current, which equals the current carried by actually existing electrons; (3) the change of state of a hole is described by quasi-classical equations for positively charged particles with positive effective masses:

$$\frac{1}{m_{ij}} = \frac{1}{\hbar^2}\frac{\partial^2 \varepsilon(\boldsymbol{k})}{\partial k_i \partial k_j},$$

where the m_{ij} are components of the effective mass tensor, and $\varepsilon(\boldsymbol{k})$ is the electron energy.

If, as it is the case for Ge or Si, the valence band width matches or overlaps the maximum of two or more electronic spectrum branches $\varepsilon_1(k)$ and $\varepsilon_2(k)$, which have different values of effective mass, then holes are classified into *heavy (slow) holes* and *light (mobile) holes*, respectively.

In a certain region of ***k-space***, i.e. in the cone of negative effective masses in Ge, Si, $A^{III}B^{V}$ semiconductors, the effective masses of holes and, respectively, the differential conductivity may have negative values. This provides an opportunity for the design of extra-high frequency amplifiers and generators.

Holes are responsible for *hole (p-type) conduction* of semiconductors, metals and semimetals. In metals and semimetals the *Fermi level* may intersect several energy bands with overlapping energy values. As this takes place, the role of *current carriers* can be played by holes in one energy band and by electrons in another. If the concentrations of electrons and holes are equal, then the metal (semimetal) is called a *compensated metal* (e.g., Bi).

The term hole is also used in a sense which is unrelated to an energy maximum. Free regions, which arise under the *Fermi surface* at $T \neq 0$, are also called holes (*Fermi holes*). A field $\boldsymbol{E}$, which acts upon a metal, shifts the distribution of electrons (or holes in a hole-type metal). This may be interpreted as the generation of *kinetic holes* in a portion of the surface area in the neighborhood of the Fermi surface, and of *kinetic electrons* at the opposite $\boldsymbol{E}$ region. The absence of an electron at the *local electronic level* of a defect is sometimes treated as the localization of a hole on it. The term *X-ray hole* is also used, meaning a hole which arises through ionization of deep lying electron shells of atoms.

HOLE BURNING

Formation of a dip in the contour of an *electron paramagnetic resonance* (EPR) (or *nuclear magnetic resonance*) spectral line due to the action of a short saturating microwave pulse applied to the spin system. The hole burning is observed immediately following the termination of the saturating pulse by recording the EPR signal using low-power (non-saturating) continuous wave (cw) microwaves incident on the system. The hole burning occurs because the energy absorbed in the rapid saturation of particular resonating spins is only slowly passed on to nearby spins, so the decrease of the signal intensity due to the *saturation effects* occurs only at one particular location on the line. In the presence of coherent effects or of *selective saturation* the process of hole burning becomes more complicated. This procedure is used to clarify the mechanism of EPR line broadening and the efficiency of *spin diffusion*.

HOLE CONDUCTIVITY

Electrical conductivity of a semiconductor in which the majority *current carriers* are *holes*. Hole conductivity takes place in semiconductors when the concentration of *acceptors* exceeds that of *donors*. In the *band theory* of solids, hole conductivity is called p-type conductivity.

HOLMIUM, Ho

A chemical element of Group III of the periodic table with atomic number 67 and atomic mass 164.93; it belongs to the *lanthanide* group; 21 isotopes are known with mass numbers from 150 to 170, among which only ^{165}Ho is stable. Electron configuration of the outer shells is $4f^{11}5s^25p^65d^06s^2$. Successive ionization energies are 6.02, 11.8, 22.8 eV. The atomic radius is 0.176 nm, the radius of the Ho^{3+} ion is 0.091 nm, the oxidation state is +3 (more rarely +2), electronegativity is 1.2.

In the free form holmium is a silvery-white *metal* with α- and β-modifications known. α-Ho has a hexagonal close-packed lattice with parameters $a = 0.3578$ nm, $c = 0.5618$ nm at room temperature; the space group is $P6_3/mmc$ (D_{6h}^4). Above 1700 K up to $T_{\text{melting}} = 1740$ β-Ho has a body centered cubic lattice with lattice parameter $a = 0.396$ nm (extrapolated value) after annealing;

space group is $Im\overline{3}m$, O_h^9. The binding energy of holmium is 3.0 eV/atom (at 0 K). The density is 8.78 g/cm^3 (at 293 K), $T_{\text{boiling}} \sim 2900$ K. Latent heat of melting is ~15 kJ/mole; latent heat of sublimation is 280.5 kJ/mole; latent heat of evaporation is ~260 kJ/mole, vapor pressure is 42.1 Pa (at the melting temperature); entropy in the standard state is 75.45 J·mole^{-1}·K^{-1} (at 298.15 K) at room temperature; Debye temperature is ~175 K; linear thermal expansion coefficient of monocrystal at 300 K is $13\cdot10^{-6}$ K^{-1} along the principal axis $\underline{6}$ and $5.0\cdot10^{-6}$ K^{-1} in the perpendicular direction; coefficient of thermal conductivity of holmium polycrystal is 7 W·m^{-1}·K^{-1} (at 100 K) and ≈16 W·m^{-1}·K^{-1} at room temperature. Adiabatic coefficients of elastic rigidity of α-Ho monocrystal: $c_{11} = 76.12$, $c_{12} = 26.0$, $c_{13} = 20.72$, $c_{33} = 80.15$, $c_{44} = 25.92$ GPa (at 300 K); isothermal bulk modulus is 39.7 GPa; Young's modulus is 67.17 GPa; ultimate tensile strength is 0.280 GPa; yield limit is 222 MPa (at 293 K). Brinell hardness is 0.49 GPa. Holmium is easily machined. Effective thermal neutron capture cross-section is 65 barn. Electrical resistivity of holmium monocrystal is 610 nΩ·m along the principal axis $\overline{6}$ and 1020 nΩ·m in the perpendicular direction (at 298 K), temperature coefficient of electrical resistance of holmium polycrystal is 0.00171 K^{-1}. Work function of an electron from holmium polycrystal is 3.09 eV. Below the Néel point $T_N = 132$ K, but above the Curie point $T_C = 19.6$ K holmium is a helicoidal antiferromagnet with a distorted simple spiral magnetic structure (the principal axis $\underline{6}$ of α-Ho serves as the helicoid axis, and the magnetic moments are located in the basal planes of α-Ho). Below T_C holmium has a cone-like helicoidal (spiral type) ferromagnetic structure with projections of the magnetic moments helicoidally ordered in the basal planes of α-Ho, and the moments themselves deviated by the angle ≈80° from the principal axis $\underline{6}$ (see *Atomic magnetic structure*). Magnetic susceptibility of holmium is $68{,}200\cdot10^{-6}$ CGS units; nuclear magnetic moment of the isotope ^{165}Ho is 3.31 nuclear magnetons. Holmium is a component of magnetic *alloys* (with iron, cobalt, nickel) with high magnetic induction and *magnetostriction*, and it is also a part of composition of some *luminophors*, materials for microelectronics and special glasses.

HOLOGRAPHIC GRATING
See *Diffraction grating.*

HOLOGRAPHIC LASER, dynamic grating laser, degenerate four-wave mixing laser

A class of *lasers* based on the generation and amplification of light waves through energy exchange with incident pumping waves on light-induced dynamic gratings, while satisfying synchronization conditions (see *Nonlinear optics*, *Real-time holography*). A working principle of a holographic laser, sketched in the figure, is the following: the pumping field of frequency ω_P and the radiation of frequency ω_S scattered in nonlinear medium 1 interfere and write in it the volume gratings 2 (*real-time holograms*) which are either immobile (for $\omega_P = \omega_S$) or move (for $\omega_P \neq \omega_S$); the recording beams diffract on gratings, which leads to amplification of the scattered radiation; a major role in the generation is played by the radiation at frequency ω_R for which the threshold pumping is minimal in a given resonator. The *resonator* of a holographic laser can be of either a linear or a ring type, and it can be formed from both ordinary mirrors 3, 4 or phase-conjugate mirrors (see *Wave front reversal*). The amplification band width of a holographic laser is $\delta\omega = 2(\omega_P - \omega_R) = \tau^{-1}$, where τ is the grating relaxation time. A combination of ordinary and nonlinear media is possible in a holographic laser.

A holographic laser can function on stimulated scattering in media with a nonlocal response (the extrema of the field and grating are shifted) where a single pumping beam is sufficient (Fig. (a)); in parametric four-wave mixing in media with a local response (the extrema of the field and grating coincide) when a pair of beams, those of pumping ω_{P1} and generation (radiation) ω_{R1}, write the grating while the second pumping beam ω_{P2} diffracts on it to produce the second generated (radiated) beam ω_{R2} (Fig. (b)).

In solids, which comprise the main *active media* of holographic lasers (photorefractive crystals, see *Holography in solid media*, *Semiconductors*) a gain of up to 10^2 cm^{-1} in order of magnitude can be achieved. A holographic laser provides the possibility of generating beams with a diffraction divergence on inhomogeneous media and of achieving wave front reversal of laser beams. Designs for optical communication systems based on

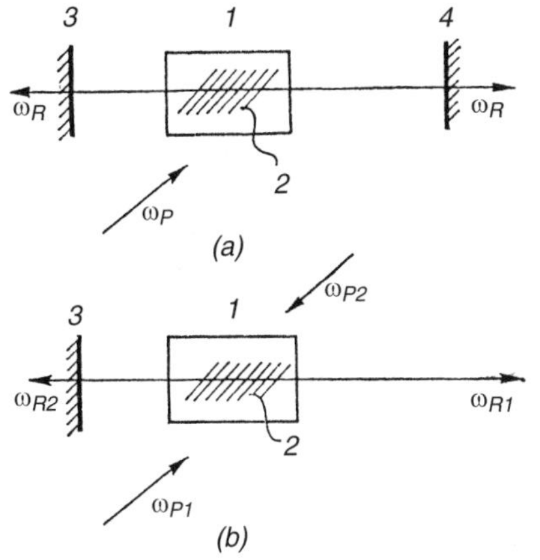

Sketch of operating principle of holographic laser based on (a) a single incident pumping beam ω_P, and (b) two incident pumping beams ω_{P1} and ω_{P2}.

inhomogeneous media, new methods of interferometry and optical activity, and optical processors have been based on holographic lasers.

HOLOGRAPHY in solid media

A technique of recording and restoring an interferometric wave field (in most cases, electromagnetic) based on alteration of the characteristics of solids under the action of an appropriate field. In particular, under the action of light on many solids, their optical characteristics change both at the exciting light frequency and in other spectral ranges (*refractive index*, *absorption coefficient*). In most cases, for a radiation of moderate power that does not produce any bulk or surface damage, these changes are reversible, i.e. disappear spontaneously or can be deleted with the help of another irradiation or heat, or by application of a field, and so on. This property of solid materials is used for either real-time or reversible cyclic holographic recording (see *Optical techniques of information recording*).

A solid detecting specimen for holography can have a thickness d much greater than the spatial period of a recording interference pattern Λ. As a result, a *hologram* becomes spatial (three-dimensional), and during its restoration all extra orders of diffraction (with the exception of standard and object beams) are attenuated by the interference (*Bragg condition of restoration*). When a high-quality restoration is not required, the diffraction efficiency of a hologram can reach 100%. In this case the angular selectivity of the hologram $\Delta\Theta$, i.e. a permissible divergence from a Bragg angle, is related directly to the layer thickness where $\Delta\Theta = \Lambda/d$, and for a medium with $d = 1$ mm and $\Lambda = 1$ μm, $\Delta\Theta = 10^{-3}$. High selectivity permits the detection in a single specimen of hundreds and thousands of holograms of complicated images, and this is of considerable interest for optical storage systems (see *Optical storage disks*).

Physical processes determining the optical nonlinearity of solids are extraordinarily diverse. Many investigations are carried out for media where light causes the appearance of either local or *collective excitations*, photoinduced charge transfer, and *photochemical reactions.* In the medium with impurity centers whose energy levels are situated in the gap of the host crystal, using light with its frequency close to a *resonance* enables one to saturate a transition between a pair of levels, thus stimulating strong changes of absorption and dispersion (*resonance nonlinearity*). In a semiconductor crystal the excitation near the *intrinsic light absorption edge* enables the production of a noticeable concentration of nonequilibrium free carriers. Their appearance shifts the plasma frequency (see *Plasma oscillations*) thus modifying the crystal dispersion over quite a broad range (*Drude–Lorentz nonlinearity*). Optical properties of crystals are also changed as a result of the photogeneration of *excitons* and other excitations.

A particular class of recording solid materials with a pronounced optical nonlinearity are represented by *photorefractive crystals* (see *Optical recording media*). Under the action of light (in some cases with the application of an external field), in these materials a space separation of charges takes place; and as a result, a static electric field appears that changes the refractive index due to the *electrooptical effect.* In crystals with strong *photoconductivity*, the ultimate variation of the refractive index is determined, not by the light wave intensity, but rather by the value of the electric field producing the charge separation (i.e. the external field, the "internal" photovoltaic field, or the diffusion field). The lack of a *center of symmetry* in photorefractive crystals determines the possibility of a special nonlocal nonlinear response providing a stationary amplification of the weak light wave on account of the diffraction of the strong beam at the emerging hologram (see *Real-time holography*).

HOLOGRAPHY, ACOUSTICAL

See *Acoustical holography.*

HOLOGRAPHY, POLARIZATION

See *Polarization holography.*

HOLOHEDRIES

Centrosymmetric lattice *point groups of symmetry* of crystal systems; highest point symmetry group of a crystal family. There is a single holohedry in one dimension. There are four holohedries in two dimensions, three of which are symmetry groups of parallelograms (oblique, rectangular, square), and the fourth of a hexagon (hexagonal). In three dimensions there are seven holohedries, six of which are the allowed symmetry groups of parallelepipeds (triclinic, monoclinic, orthorhombic, tetragonal, trigonal, cubic), and the seventh is the symmetry group of a hexagonal prism (hexagonal). A holohedry can also be defined as a symmetry group of the *Wigner–Seitz cell.* The holohedries of direct and *reciprocal lattices* are the same.

HOLSTEIN–PRIMAKOV REPRESENTATIONS

See *Spin operator representations.*

HOLTSMARK SHAPE, Holtsmarkian (T. Holtsmark, 1919)

A *line shape* described by the function

$$I_\Omega(\omega-\omega_0)=(\pi\Omega)^{-1}\int_0^\infty \cos(\omega-\omega_0)\frac{\rho}{\Omega}\times\exp\left(-\rho^{3/2}\right)\mathrm{d}\rho,$$

where ω_0 is the frequency of the resonance line in the absence of broadening mechanisms, characterized by the parameter Ω. The Holtsmark line shape is an intermediate one between the *Lorentzian shape* and *Gaussian shape.* The function I appears in the first-order *statistical theory of line shape of magnetic resonance* when the field components f_k, which cause a frequency shift, fall off with distance from the resonance center as r^{-2}. An example is the electric fields of charged *point defects* in crystals, which lead to *inhomogeneous broadening* of an electron paramagnetic resonance line. This line shape is rarely used.

HOMEOTROPY

A homogeneous state of a *liquid crystal* layer with a *director* normal to principal surfaces. In this case the direction of the director is called the *optical axis* of the crystal.

HOMOEPITAXY

See *Autoepitaxy.*

HOMOGENEOUS BROADENING

This is the broadening of spectral lines resulting from irreversible phase relaxation, i.e. related to the finite lifetimes of particles in various energy levels. Its effect on general broadening may be determined in the most direct manner by the *spin echo* technique (in the case of *magnetic resonance*), or by using a light echo, since only irreversible phase relaxation contributes to decaying echo signals. While the resonance is saturated by a monochromatic variable field, all the portions of the uniformly broadened line behave identically: in contrast to the case of inhomogeneous broadening, their amplitudes monotonically decrease. The corresponding mechanisms include processes of spin–lattice relaxation and *ion–lattice relaxation*; relaxation processes due to internal molecular motion, by atomic diffusion, etc. The inhomogeneous contribution is always present during dipole–dipole broadening due to the presence of static local fields that significantly influence the situation, particularly in magnetically diluted systems. See also *Inhomogeneous broadening.*

HOMOGENEOUS STRUCTURE

See *Heterogeneous structure.*

HOMOGENEOUS SYSTEM

A *thermodynamic system* whose properties (composition, density, pressure, etc.) vary continuously (remaining close to mean values) in space. Owing to this continuous variation such a system (in contrast to a *heterogeneous system*) has no parts confined by interfaces on which even a single property could vary by a jump, i.e. the system is truly a single-phase (see *Phase*). *Solid solutions* are examples of homogeneous systems.

HOMOGENIZATION, homogenizing annealing

A *thermal treatment* aimed to increase the homogeneity (more exactly, microhomogeneity) of an *alloy* by means of diffusion. To accelerate the *diffusion*, the *homogenization* is performed at an elevated temperature: for *iron alloys* (*steel*) it is usually a prolonged exposure during a few hours at 1150–1250 °C. Simultaneously with producing a chemical homogenization of the microscopic bulk, the homogenizing treatment leads to the growth of crystalline grain (see *Polycrystal*) which deteriorates the alloy properties. After homogenization a secondary heat treatment at lower temperatures (850–950 °C for steels) disperses the grain and regenerates the alloy. The most important application of homogenization is the production of doped alloys (see *Alloying*). Due to dendrite *liquation* the ingots of these alloys possess a lowered *plasticity* at high temperature, so in most cases the homogenization of such alloys is necessary prior to *forging*.

HOMOJUNCTION

A contact between regions of the same material (e.g., GaAs) with different types of conduction (n–p junction) or different impurity concentrations of one type (n–p' and p–p' junctions). Since a p–n junction is characterized by a potential barrier for electrons (holes) whose height depends on the applied voltage it exhibits unipolar conduction (conduction in one direction only). It can be used for rectification of alternating current; in some devices it serves for the amplification and transmittal of electrical signals (see *Transistor*), and for the transformation of optical signals into electrical ones (*photodiodes*). Homojunctions of the types n–n' and p–p' with increased concentrations of current carriers in n' or p' regions relative to those n or p regions, respectively, are usually used for the fabrication of ohmic contacts for semiconductors. They are prepared by methods involving *diffusion*, fusion, *ion implantation doping*, laser irradiation and epitaxial growth. Depending on the length of the intermediate region between the layers which differ in conductivity one can distinguish sharp and smooth homojunctions. Electrophysical properties of homojunctions are: electric field distribution $E(x)$, potential barrier height Φ, length of the *space charge region* ω, breakdown voltage V_b, the capacitance $C(V)$, and the rectification coefficient $K(V)$ dependence on the applied voltage V. All these characteristics can be calculated if the laws of the doping impurity distribution, *band gap* width E_g, *dielectric constant* of the semiconductor ε, and *impact ionization* coefficients of electrons and holes α, $\beta(E)$ are known. The quantities $E(x)$ and ω are found from the *Poisson equation*. The potential barrier height for a p–n junction is given by $\Phi = (E_g - \mu_n - \mu_p)/e$, where e is the electron charge, $\mu_{n,p}$ are the distances from the *Fermi level* to the *conduction band* in an n-type semiconductor, and to the *valence band* in a p-type semiconductor, given by

$$\mu_{n,p} = k_B T \ln\left\{\frac{h^3 N_{d,a}}{2(2\pi m_{n,p} k_B T)^{3/2}}\right\},$$

and $N_{d,a}$ are the concentrations of the doping impurities (*donors* in n-type and *acceptors* in p-type semiconductor), $m_{n,p}$ are the *effective masses* of the electrons and holes, respectively.

HOMOLOGY (fr. Gr. $o\mu o\lambda o\gamma\iota\alpha$, similarity)

Correspondence between elements that play similar roles in distinct geometrical figures (or mathematical functions). Homology may be looked upon as a generalization of symmetry which allows one to classify low-symmetry figures and crystals in more detail than by their group symmetry classes (see *Crystal symmetry*). For example, a rectangle and a figure composed of two identical circles tangent to each other possess the same symmetry. However, the rectangle can be transformed into itself by an oblique reflection in one of its diagonals in the direction of the second, which differentiates the first figure from the second (according to E.S. Fedorov, this is a *visual symmetry*).

V.I. Mikheyev proposed to use a homology for indexing the debyegrams (Debye–Scherrer X-ray powder patterns) of low-symmetry crystals (see *Debye–Scherrer method*). The divergence between their structure and the structure of *ideal crystals* is small, as a rule. While the faces of a cube in crystals belonging to a *cubic system* produce a single line in a debyegram, *crystals* close to cubic produce two or three such lines. A small difference in the interplanar distances of these three families of atomic lattices leads to the appearance of separate groups of split lines in a debyegram. By proceeding from the kind of splitting and the number of lines in a group one can determine the type of *strain* of the crystal.

If a figure Φ with a symmetry group G is subjected to a homogeneous deformation σ (compression or shear), then it will go over into $\Phi' = \sigma(\Phi)$, and G will isomorphically reflect itself to the group H of affine transformations Φ' in accordance with the law: $H = \sigma G \sigma^{-1}$. In this case, H is called the *homology group* of Φ', and the transformations from H are called the *homology transformations* of Φ'. If G is a point, or a space, or a discrete (e.g., crystallographic) group, then H is the same. There exist four kinds of *rotation axes of homology* of kth order: λ_k is the oblique elliptic axis, λ'_k is the direct elliptic axis, λ''_k is the oblique circular axis, and G_k is the symmetry axis; there are four kinds of *inverse-rotation axes of homology* with the same names and designations, and two kinds of *planes of homology*: p is the plane of oblique reflection, P is the *symmetry plane*. There are two kinds of axes of the second order: the direct and oblique ones. The *center of homology* is the *center of symmetry*. With the help of various affine deformations, 215 point crystallographic groups of homology emerge from 32 *point groups of symmetry*, and 1848 space groups of homology emerge from 230 space groups. These point groups describe the external shape of crystals while the space groups describe their structure.

HOMOPOLAR BOND

See *Covalent bond*.

HONEYCOMB DOMAIN STRUCTURE

Magnetic domain structure of a thin film with a two-dimensional hexagonal lattice. The *domains*, which are located at the sites of this lattice, are similar in shape to honeycomb cells. A honeycomb domain structure is the limiting case of a very close packed *magnetic bubble domain lattice*.

HOOKE'S LAW (R. Hooke, 1660)

Expression of a linear dependence between the *stress tensor* σ_{ik} and the *strain tensor* u_{ik}: $\sigma_{ik} = c_{iklm} u_{lm}$, where c_{iklm} is the *elastic modulus tensor* (elastic stiffness tensor) of second rank. The inverse relation has the form $u_{ik} = s_{iklm} \sigma_{lm}$, where s_{iklm} is the second-rank *elastic compliance tensor*. Tensors c_{iklm} and s_{iklm} are symmetric relative to a transposition of indices in every pair, and of pairs of indices between themselves. They obey the relation $c_{iklm} s_{lmpq} = (\delta_{ip}\delta_{kq} + \delta_{kp}\delta_{iq})/2$. In the case of an isotropic medium (see *Isotropy*), Hooke's law has the form

$$\sigma_{ik} = K u_{ll} \delta_{ik} + 2\mu \left(u_{ik} - \frac{u_{ll}\delta_{ik}}{3} \right),$$

where K is the *bulk modulus* and μ is the *shear modulus*. A reciprocal relation has the form:

$$u_{ik} = \frac{\sigma_{ll}\delta_{ik}}{9K} + \frac{\sigma_{ik} - \sigma_{ll}\delta_{ik}/3}{2\mu}.$$

If the applied stress is a tensile stress with only one component σ_{33} then this expression becomes

$$u_{ik} = \frac{\sigma_{33}\delta_{ik}}{9K} + \frac{\sigma_{ik} - \sigma_{33}\delta_{ik}/3}{2\mu}$$

with the particular values

$$u_{33} = \frac{\sigma_{33}}{9K\mu}(\mu + 3K),$$

$$u_{11} = u_{22} = \frac{\sigma_{33}}{9K\mu}\frac{(\mu - 3K)}{2},$$

and $u_{ik} = 0$ for $i \neq k$. *Young's modulus* E, which is defined as σ_{33}/u_{33}, has the value

$$E = \frac{9K\mu}{\mu + 3K}.$$

The *Poisson ratio* $\rho = -u_{22}/u_{33}$ is the ratio of lateral contraction strain u_{22} to the linear stretching strain u_{33}. It is given by the expression

$$\rho = \frac{1}{2}\frac{3K - 2\mu}{3K + \mu},$$

and hence Young's modulus can be written $E = 3K(1 - 2\rho)$.

HOPPING CONDUCTIVITY

A mechanism of conductivity working through localized states. Such a mechanism operates through activated tunneling of charge carriers (*electrons, holes, polarons*) from an occupied state with energy ε_i to a vacant state with energy ε_j. This mechanism bypasses activation of current carriers into the band of delocalized states. The energy deficit $\varepsilon_j - \varepsilon_i$ is compensated by the absorption of *phonons*. It entails a temperature dependence for the activation of jump conduction. At moderately low temperatures when "hops" between adjacent levels dominate, the activation energy is constant, in other words, $\ln \rho \propto T^{-1}$ (ρ is the resistivity). With decreasing temperature hops only take place between levels that are close to each other in energy (although spatially they may be quite far apart). Such a regime is called the variable range. Then $\ln \rho \propto T^{-n}$, $0 < n < 1$ (see *Mott law*). In strong electric fields, when the energy accumulated during average hops becomes equal to the activation energy in zero field, an activation-free conductivity develops with a characteristic dependence on the field strength, with $\ln \rho \propto E^{-n}$. Hopping conductivity is observed in *disordered solids*, in which the quantum states localized at various points in the system have different energies. In doped or compensated *semiconductors* the value of the hopping conductivity displays a strong exponential dependence on the impurity concentration. Calculating the concentration, temperature, and field strength dependences of hopping conductivity may in certain cases be reduced to calculating the *Miller–Abrahams resistance network* (A. Miller, E. Abrahams) plotted over random points (see *Percolation theory*).

HOT ATOMS

Recoil atoms in nuclear transmutations that acquire an extra kinetic energy ~10–100 eV due to the emission of neutrons, protons, α or β particles, or *gamma rays* by a radioactive nucleus. These atoms can effectively participate in chemical reactions in solids (see *Radiation chemistry*). The rate of such reactions is independent of temperature due to nonequilibrium energy distribution of hot atoms. At any temperature hot atoms are used in the synthesis of labeled compounds, the separation and enrichment of isotopes, etc.

HOT ELECTRON EMISSION

Emission of very fast, *hot electrons* from a *solid* to its surroundings. The majority of the free electrons in a solid have an energy close to the mean energy. Hot electrons have energies much higher than the mean; for a conductor this is an energy far above the *Fermi energy*, and for a semiconductor it is an energy far in excess of the *conduction band* energy. The number of electrons whose energy ε is significantly higher than the mean energy at thermodynamic equilibrium decays exponentially with their increase in energy, i.e. in proportion to the Boltzmann factor $\exp[-\varepsilon/(k_B T)]$, where T is the temperature of the solid (lattice). Among these fast electrons there exist some whose energy exceeds the *work function* φ. A portion of these electrons (whose velocities are perpendicular to the surface) contribute to the *thermionic emission* current, with a magnitude determined by the *Richardson equation* $I = B(T)\exp[-\varphi/(k_B T)]$.

The dependence of the preexponential factor $B(T)$ on the temperature is described by a power law. The presence of hot electrons causes the thermal emission current to sharply increase. In the simplest situation when (owing to the dominant role of interelectron collisions) the electron gas can be characterized by a certain *electron temperature* for the emission current, we again have the Richardson formula, but with the electron temperature replacing the lattice temperature. In this case the electron temperature, with its dependence on the heating electric field strength, can be several times or even an order of magnitude greater than the lattice temperature, and the current of hot electron emission will exceed by many orders of magnitude the thermal emission value. However, in most cases the distribution of fast electrons in a semiconductor differs from the exponential (Maxwell) law, and depends on the magnitude of heating field, the scattering mechanisms, and the *band structure* of the given material. Therefore, it is not possible to write for the hot electron emission current a general formula (like the Richardson formula) with a single parameter that would be the work function of a given material.

Hot electrons can be obtained from semiconductor–metal *insular films* on an insulator *substrate*, as well as from "sandwiches" of, e.g., the metal–insulator–metal type.

The phenomenon of hot electron emission has prospects for the production of *cold cathodes*. The advantage of these in comparison with heated cathodes is obvious. The evaporation of cathode material does not take place at a cold lattice with hot electrons. In addition, such a cathode is almost noninertial (no time is needed to preheat it). Insular metal films are, in addition, promising for electronics from the viewpoint of the continuous trend toward the microminiaturization of electronic devices.

HOT ELECTRON GENERATION

Almost unbounded increase of the energy of *conduction electrons* in a constant electric field. The hot electron generation occurs when the rate at which the energy increases exceeds the rate of energy dissipation through interaction with the lattice. A similar result is predicted by a theory which assumes a fixed scattering mechanism whereby, e.g., energy and momentum are dispersed through the agency of optical *phonons* or ionized impurities (see *Current carrier scattering*). In actual situations the increase of energy is limited by the increasing role played by other scattering mechanisms, e.g., *intervalley scattering* of electrons involving transitions to valleys of higher energy, or *impact ionization*. In a single-valley semiconductor (see *Many-valley semiconductors*), the hot electrons may cause *electron gas overheating*, with the result that the differential conductivity may take on a negative value (and the *current–voltage characteristic* may become S- or N-shaped); the latter corresponding to *current instability due to overheating*.

HOT ELECTRON PHOTOLUMINESCENCE, hot photoluminescence

Radiation, which is produced at *recombination* of photoexcited electrons, the kinetic energy of which far exceeds $k_B T_L$ (hence the name *hot electrons*), where T_L is crystal lattice temperature. Thorough investigations of hot electron photoluminescence are performed using direct band gap semiconductors of the $A^{III}B^{V}$ group (GaAs, InP) as samples. The spectrum of hot electron photoluminescence provides a type of time evolution of energy *relaxation* processes. Excitation of *electron–hole pairs* by linearly polarized light gives rise to the *optical alignment* of electrons and holes with respect to velocities, which manifests itself in the appearance of polarization in hot electron photoluminescence. The study of depolarization of hot electron photoluminescence in a magnetic field allows one to determine characteristic electron *relaxation times* (e.g., for the processes of emission of longitudinal optical phonons, intervalley transitions). Spectra of hot electron photoluminescence provide a number of important parameters of *band structure* (the width of the valence band). The final stage of energy relaxation involves thermalization, which takes place due to electron–electron collisions: a Fermi–Dirac distribution of electrons of temperature T_e becomes established. At low temperatures, T_e often noticeably exceeds T_L even at moderate levels of excitation. The magnitude $T_e - T_L$ is determined by the ratio between the pumping power and the rate of transmission of energy from the electron subsystem to the lattice.

HOT ELECTRONS

Conduction electrons in a solid with an increased (owing to excitation from an external source) mean energy. Any input of energy into the electron subsystem (due to electric current, laser irradiation) disturbs the thermodynamic equilibrium between the electron and lattice subsystems. The electrons become "hot", i.e. their mean energy exceeds the equilibrium value corresponding to the lattice temperature. The extent of *electron heating* is greater, the higher the input power and the lower the magnitude of the *electron–phonon interaction*. The heating of electrons is easily observed in semiconductors with weak electron–phonon coupling (Ge, Si, InSb, GaAs, and so on). The materials with p-conductivity allow the formation of *hot holes*.

The energy distribution of hot electrons can differ considerably from the equilibrium one: when heated, the electrons "populate" new parts of the energy band (see *Intervalley redistribution*). Therefore, there appears a number of new phenomena dependent on the band structure, and the prevailing scattering mechanisms (see *Current carrier scattering*) in semiconductors with hot electrons. Examples of such new phenomena are: the non-linear current dependence on the electric field; the deviation of the electric current direction from the electric field direction (*Sasaki effect* in *many-valley semiconductors*); the current oscillation under an applied constant voltage (*Gunn effect*), and others. Some of these effects have found their way into applications. Solid-state microwave generators designed on the basis of the Gunn effect (see *Gunn diode*) are especially widely used. In *metals*, the heating of electrons does not play a significant role with the exception of *insular films* or ultralow temperatures. In *insulators* the *injection* of hot electrons can be of importance.

HOT ELECTRON TRANSISTOR

A super high frequency (SHF) transistor with an active region layered structure into which electrons are injected which possess an energy equal to or exceeding the equilibrium energy; these so-called *hot electrons* are accelerated and propelled through this region in a ballistic fashion with a speed higher than the average electron speed. There are hot electron transistors based on metals and on semiconductors. The metal-based type has a construction which is transparent for the electron flux (see Fig.). A semiconductor-based transistor has a planar construction in which *heterojunctions* or various combinations of doped layers are used to produce the needed energy profile of the *conduction band*.

Hot electron transistors are used for the generation or amplification of oscillations in the range from hundreds of megahertz to hundreds of gigahertz, as well as being utilized in some logic devices.

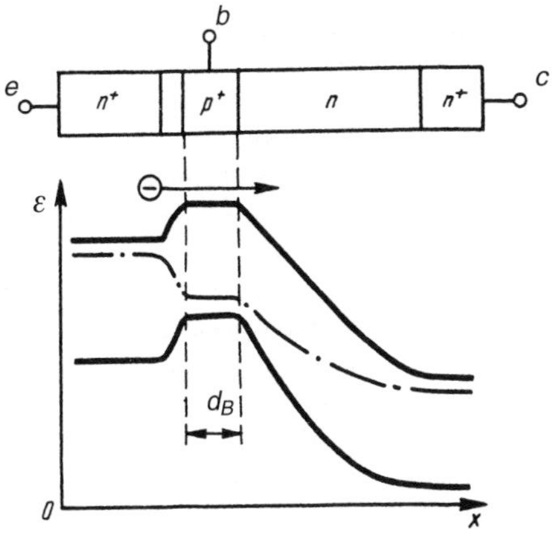

Sketch of the structure of a transistor with a barrier volume of thickness d_B (top), and its energy diagram (bottom), showing the Fermi level (dash-dotted curve) between the bottom of the conduction band (upper curve) and the top of the valence band (lower curve), where e is the emitter, b is the base, c is the collector, ε is the electron energy, and x is the coordinate.

HOT LUMINESCENCE of crystals

One of the components of resonance secondary luminescence (see *Secondary luminescence*) of intrinsic or impurity electron states having a complex energy spectrum (energy bands or vibrational sublevels). A hot luminescence appears because of light emission prior to establishing thermal quasi-equilibrium between the sublevels of an excited electron state and the *crystal lattice*. A particular case is a situation when, prior to light emission from an excited electronic level, quasi-equilibrium is established which is characterized by a temperature which exceeds the lattice temperature (*electron temperature*). Hot luminescence has been investigated for the cases of vibrational sublevels of the excited electron states of multiatomic impurity ions, for free electrons in semiconductors, and for photoexcited *hot electrons*. Measurements have been made of the *relaxation times* of picosecond states during stationary excitation. Studies of the intensity and the polarization provide detailed information on the processes of energy relaxation in the excited state (see *Relaxation of excitation*). When free excitons are excited by monochromatic light, a part of the hot luminescence can be regarded as resonance light scattering through real intermediate states with conservation of the wave-vector at all intermediate stages. In this case the spectrum has some features of a *Raman scattering of light*, although its intensity and time-decay are determined by the lifetime of the intermediate state.

HOT MAGNONS

Magnons whose energy distribution corresponds to a temperature which exceeds that of the *crystal lattice.* In ferromagnetic semiconductors (e.g., EuO) a situation could occur in which the *conduction electrons* scatter strongly off magnons (see *Magnetic semiconductors*) which are only weakly coupled to *phonons.* In this case, the *electron heating* by the external electric field leads to a heating of the magnons due to the transfer of energy from electrons to magnons. The phonons serve as a heat bath because the lattice is maintained at a constant temperature. Due to the weak coupling, the magnon temperature can substantially exceed the phonon (lattice) temperature. Heating the magnons causes a dependence of magnetic characteristics (e.g., *magnetization*) on the electric field. Heating both the electrons and the magnons can be accomplished with the help of laser irradiation (see *Laser*). When several coherent light beams are incident on a ferromagnetic semiconductor, *superlattices* can arise both from *hot electrons* and from the hot magnons (superlattices in concentration, effective temperature, and other parameters) due to interference effects.

HOT PHOTOLUMINESCENCE

See *Hot electron photoluminescence.*

HUBBARD MODEL (J. Hubbard, 1963–1965)

One of the fundamental models which treats a system of strongly interacting electrons in a crystal. The one-state Hubbard model is the simplest model of correlated electrons. This model describes the motion of electrons on a lattice, taking into account only the local Coulomb interactions between the electrons at a particular lattice site.

The *Hubbard Hamiltonian* contains only two parameters: the hopping amplitude matrix element t, which corresponds to jumping from one lattice site to a neighboring one, and the Coulomb repulsion term U at a site:

$$H = -t\sum_{ij\sigma} C_{i\sigma}^{+} C_{j\sigma} + U\sum_{i} n_{i\uparrow} n_{i\downarrow}. \quad (1)$$

Here $C_{i\sigma}$ ($C_{i\sigma}^{+}$) are Fermi annihilation (creation) operators of an electron at the site i with projection σ, and $n_{i\sigma} = C_{i\sigma}^{+} C_{i\sigma}$ is the electron number operator at the site with a given spin projection (up $\uparrow$ or down $\downarrow$). Instead of the parameter t, it is more convenient to use the range of the interaction region $W = 2zt$, where z is the number of nearest neighbors.

For $U \ll W$ the Hamiltonian (1) corresponds to a *Fermi liquid*, and in the opposite "tight binding" limit ($U \gg W$) the Hamiltonian (1) is consistent with a strongly correlated system. In the latter case it is convenient to operate on the premise of the atomic limit, which allows us to discard the kinetic (hopping) term t of the Hamiltonian. In this limit every site has two electron levels, separated by U. When the kinetic term is taken into account these levels become smeared and turn into bands, so at high values of U the correlation splitting of the interaction zone into two Hubbard subbands of one- and two-particle states is to be expected. In the initial approach, the *Green's function* splitting (*"Hubbard 1 state" approximation*) leads to band formation, which, however, in the limit of low values of U lacks physical sense. The more sophisticated (and more realistic) *"Hubbard 3 state" approximation* gave correct results: the splitting is observed when U exceeds a certain critical value of the order of W. Thus, in the case of half-filling (number of electrons per site $n = 1$), the system is a *metal* for low values of U and an *insulator* for high values of U. The metal–insulator phase transition takes place at intermediate values of the parameter U/W, so there is no small parameter in the theory. The region of intermediate values of this parameter may be handled in two ways: the nearly free electron limit ($U \ll W$) and the tight binding limit ($U \gg W$). The crossover of these modes can be studied using various methods: analytical (variational method, mean-field theory, etc.) and numerical (exact diagonalization of small *clusters*, quantum *Monte Carlo method*).

Different methods of investigating the intermediate region sometimes yield erratic results, therefore, of prime importance are exact solutions, which are known for a number of particular cases of the the Hubbard model. Thus, the exact solution of one-dimensional Hubbard model ($d = 1$) for all values of U was found by E. Lieb and F.Y. Wu (1968), and investigations in the limit of infinite spatial dimensionality were begun by W. Metzner and D. Vollhardt (1989). In the context of many-particle systems theory, the limit $d = \infty$ corresponds to the *mean field approximation.* It has been shown that in the limit of $d = \infty$ the statistical mechanics of the Hubbard model is equivalent to that of a certain auxiliary problem involving an impurity center in the *Anderson model*, which has been more widely studied than the Hubbard model. The evolution of the quasi-particle spectrum in the Hubbard model during a variation of the parameter U is more complicated than that predicted by an approximate approach like the "Hubbard 3 state" approximation. Depending on the U/W ratio, there may occur a one-band and a two-band spectrum, as well as a spectrum with a gap, in which there is a sharp Kondo resonance at the *Fermi surface* (see *Kondo effect*). Within the context of this approach it is possible to describe a whole set of interrelated phenomena: the metal–insulator transition, the generation of localized magnetic moments, and the disturbance of the Fermi liquid pattern by an increase of the parameter U.

The exact solution of the one-dimensional Hubbard model, which is obtained through the application of the Bethe ansatz, shows that the ground state is insulating for all values of U. The third exact result in the Hubbard model, known as the *Nagaoka theorem* (Y. Nagaoka, 1966), states that as U tends to infinity, a single hole in the half-filled system makes the ground state ferromagnetic. This allows one to expect that for a large and finite U value the ground state will be ferromagnetic over a wide range of electronic concentrations.

In the limit of $U \gg W$, one may change from Hamiltonian (1) to the effective Hamiltonian of the *t–j model* which replaces the Coulomb repulsion term U by the Heisenberg term $J \sum S_i S_j$:

$$H = -t \sum_{ij\sigma} (1 - n_{i\sigma}) C_{i\sigma}^{+} C_{j\sigma} (1 - n_{j\sigma}) + J \sum_{ij} S_i S_j, \qquad (2)$$

where $C_{i\sigma}$ ($C_{j\sigma}^{+}$) are operators of annihilation (creation) of an electron on the lattice site provided that there is no other electron at that site, S is the spin operator at the site, $n_i = \sum_\sigma n_{i\sigma}$ is the number of electrons at the site i. The factors $1 - n_{j\sigma}$ arise from the Pauli exclusion principle, which forbids electrons with spin projection σ from occupying a site which is already occupied by another electron of spin σ. The Hamiltonian (2) describes the electron motion over the lattice for the above condition, and the *exchange interaction* at the neighboring sites of the antiferromagnetic integral $J = 2t^2/U$. When the band becomes half-filled, the first term in Eq. (2) vanishes, and we arrive at the Heisenberg model (see *Heisenberg Hamiltonian*) of spin $S = 1/2$ and an antiferromagnetic interaction. If the number of electrons per lattice site $n = 1$, then the ground state of the system is that of an antiferromagnetic insulator. Deviations of n from unity may result in the generation of charge carriers in the system; and these charge carriers undergo a strong interaction with the magnetic order. The carrier (electron or hole) becomes surrounded by a cloud of spin deviations, and is a *quasi-particle* called a *magnetic polaron.*

The Hubbard model is used as a test model for describing *band magnetism* in metals, a metal–insulator phase transition, and various interrelationships between *magnetism* and the electrical properties of materials.

The Hubbard model is often used to describe narrow-band metals with d- and f-electrons, and compounds based on these metals. In recent years, the Hubbard model and its limiting case, the t–j model ($d = 2$, electron concentration near $n = 1$), have been used to describe cuprate *high-temperature superconductors.*

HUME-ROTHERY PHASES

Phases formed in *alloys* of Cu, Ag, Au and some *transition metals* with *metals* of Groups IIB, IIIB and IVB of the periodic system of elements. The *crystal structure* of a Hume-Rothery phase is determined by the electron concentration, that is the number of valence electrons per atom of the crystal lattice. An *alpha phase* (α-phase) with a face-centered cubic lattice covers *solid solutions* with the electron concentrations up to 1.4. *Beta phases* (β-phases) are disordered solid solutions with a body-centered lattice (e.g., CuZn, Cu_3Al, Cu_5Sn, etc.) that are characterized by an electron concentration equal to 1.5. Compounds CoAl, NiAl, and FeAl with a body-centered cubic lattice can be referred to as β-phase under the supposition of a zero *valence* for the transition metals. As the temperature decreases, *alloy ordering* (CuZn) or eutectoid *alloy decomposition* (see *Eutectoid alloys*) (Cu_3Al, Cu_5Zn) take place. *Gamma phases* (γ-phases) with a complicated ordered cubic lattice with 52 atoms per unit cell (Cu_5Zn_8, Cu_9Al_4, $Cu_{31}Si_8$, and some others) are characterized by an electron concentration equal to 1.62. *Epsilon phases* (ε-phases) with a hexagonal close-packed structure ($CuZn_3$, Au_5Al_3, Cu_3Sn, and so on) are characterized by the electron concentration equal to 1.75. The β-, γ-, and ε-phases are called *electronic compounds*. The phase boundaries between α- and β-phases in *brass* were calculated from the analysis of a difference of vibrational components of the *entropy* in the face-centered cubic and body-centered cubic lattices. As the temperature decreases the contribution of the entropy to the total free energy decreases, and the β-phase lattice becomes unstable.

HUME-ROTHERY RULES (W. Hume-Rothery)

Semiempirical rules which determine the conditions for two elements to form *solid solutions* over a wide range of concentrations (on phase diagrams). These conditions can be reduced to the following: if the difference in atomic size (scale factor) of the components which form an *alloy* exceeds 10–15%, the *solubility* in the solid state becomes limited. The formation of stable intermediate compounds would narrow the ranges of restricted solid solutions, so solubility is greater when the two elements have similar electronegativities (*electronegative valence effect*). Solubility is highest if the elements have the same valence; and if the valences differ then solubility in the element with smaller *valence* always exceeds that in the element with larger valence (*relative valence effect*). For complete solid solution solubility (over the entire range) all these rules should be satisfied.

HUND RULES (F. Hund, 1927)

Empirical rules determining the relative orientations of spin and *orbital angular momenta* of electrons in an atom. According to these rules a partially full electron shell with a given principal quantum number n and orbital quantum number l is characterized by:

(1) a total spin quantum number S with the highest value permitted by the Pauli exclusion principle;
(2) a total orbital quantum number L with the highest value permitted after the choice of (1);
(3) a total quantum number J of the resultant angular momentum $\boldsymbol{J} = \boldsymbol{L} + \boldsymbol{S}$ with the value (magnitude) $J = |L - S|$ for a less than half-full shell, and $J = L + S$, if the shell is more than half-filled.

In accordance with the Hund rules, the atomic *magnetic moment* of an atom with a partly occupied internal electronic shell ($3d$, $4d$ and $5d$; $4f$ and $5f$) differs from zero. This fact has a dominant role in the phenomenon of *magnetism* in solids.

HUSIMI CACTUS (K. Husimi, 1950)

Graph of a special type. Each block of this graph represents a unique ordinary cycle, and none of its edges may belong to more than one ordinary cycle. Initially the Husimi cactus was studied in connection with the computation of cluster interaction integrals in the condensation theory of J.E. Mayer (1938). The Husimi cactus is used in solid state theory as a pseudo-lattice in various problems involving elementary excitations. All sites of such a lattice (pseudo-lattice) are equivalent, and a transition from one site to another one is determined by simple recurrence correlations (see *Bethe lattice*). A Husimi cactus is constructed using regular triangles, quadrangles, pentagons, etc. with their apices in contact. These appear as graphical representations of ordinary cycles.

HYBRIDIZATION OF ORBITALS

Formation of mixed (hybrid) *orbitals* at *chemical bonding*. Hybridization utilizes the *Pauling's maximum overlap principle*. The most common occurrence involves the mixing of s and p (p_x, p_y, p_z) orbitals; and when transition ions are involved in the bonding, d orbitals also participate. For example, a better approximation to the energy of the hydrogen molecule than the *Heitler–London approximation* is obtained if the two-electron wave function is built up from $1s$ orbitals with a small admixture of $2p$ orbitals oriented towards each other to achieve a larger overlap. Hybrids of $2s$ and $2p$ orbitals of carbon atoms provide the strong in-plane chemical bonds in crystals of *graphite* and the tetrahedral bonding in *diamond*.

In the first case, three combinations (unnormalized) $s + \sqrt{2}p_x$, $s - (1/\sqrt{2})p_x \pm \sqrt{3/2}p_y$ produce three orthogonal trigonal orbitals in the xy-plane with interaxial angles 120°. In the second case, four orbitals of a carbon atom on the first sublattice $s + p_x + p_y + p_z$, $s + p_x - p_y - p_z$, $s - p_x + p_y - p_z$, and $s - p_x - p_y + p_z$ are oriented along the cube diagonals and maximally overlap with the orbitals of the nearest neighbors belonging to the second sublattice whose orbitals differ from the above set by a sign change of all the p-type functions. Similar hybridizations occur for the $3s$ and $3p$ orbitals in a *silicon* crystal, the $4s$ and $4p$ orbitals in a *germanium* crystal, and the $5s$ and $5p$ orbitals in a crystal of gray *tin* (α-Sn). With increasing nuclear charge Z and quantum number n of the outer shells, the difference between the energies of ns and np orbitals increases, and the formation of such sp^3 hybrid states requires more energy than is gained by the *crystal* formation. Therefore, there is no hybridization of orbitals of this kind in the next (after $n = 5$) element lead; Pb is a *metal* similar to white tin (β-Sn). Similar hybridization of orbitals takes place in various compounds III–V, II–VI and I–VII. However, due to the difference in the number of valence electrons, in parallel with *covalent bonding*, *ionic bonding* acquires a significant role in this case (e.g., in CuCl).

HYBRID VALENCE

See *Intermediate valence*.

HYDRATION OF SURFACES

Attachment of water molecules to the surface of a solid. Hydration is a particular case of *solvation* (attachment of any solvent). Unlike the *hydrolysis* of a surface, hydration is not accompanied by the formation of hydrogen and hydroxyl *ions*. The adsorbed water molecules are oriented in a particular way and interact with ions of the *solid*, thus influencing the value of the potential jump in a double ion layer, and the electromotive force of the electrochemical system. Concerning the capacity to hydrate, i.e. to form a water bond, there are distinguished *hydrophilic* (water attracting) and *hydrophobic* (water repelling) surfaces, each characterized by an *edge angle* (*wetting angle*). Small additions of *surface-active agents* can strongly affect the wetting angle. Water losses of solids at heating, vacuum vaporization, etc., are called *dehydration*.

HYDRIDES

Compounds of chemical elements with hydrogen. There are simple and complex hydrides. Simple hydrides are divided into ionic (salt-like), metal-like (metallic) and volatile (covalent). Classed among ionic hydrides are those of alkali and alkaline earth metals (e.g., NaH, CaH_2) that are formed under compression of the *crystal lattice*. Metal-like hydrides are predominantly of varying composition such as solid solutions of hydrogen in metals (*lanthanides*, Ti, Fe, Pt, etc.). Their particular features are varying composition, as well as significant *electrical conductivity* and *thermal conductivity*. Covalent hydrides are formed by metalloids of Groups IV, V, VI and VII of the periodic system, as well as by *boron*. *Boron hydrides* have the most complicated structures; the presence of double three-centered bonds, as shown in the figure, is assumed to explain their electron deficiency.

Complex hydrides are those with complex anions, for example, lithium and aluminum boron hydrides $LiBH_4$ and $Al(BH_4)_3$, respectively, and lithium aluminohydride $LiAlH_4$. The properties of simple and complex hydrides depend on the position of the hydride-forming element in the periodic table, and the extent of the *electronegativity* of hydrogen. Hydrides can be components of complicated equilibrium systems, and many hydride ductility and solubility diagrams (see *Phase*

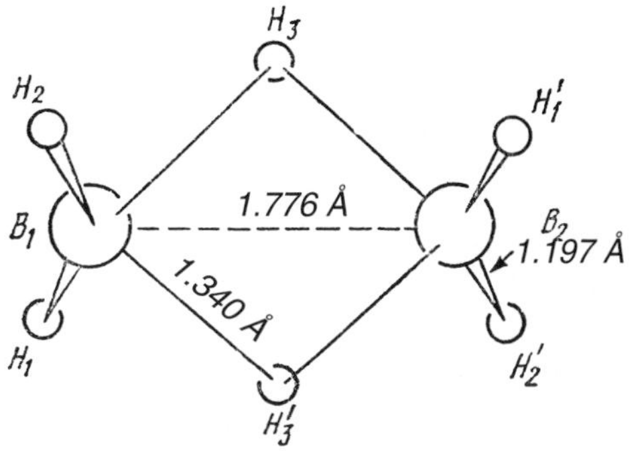

Sketch of a boron hydride molecule.

diagram) are known. With the help of thermal degradation of volatile hydrides of boron, silicon, arsenic and other elements, with prior refinement by distillation and *condensation*, the highest purity *semiconductor materials* are obtained in bulk or in the form of *films*. Some metallic hydrides have been sources of highly purified elements (e.g., silicon and germanium); lithium and sodium hydrides are used in organic synthesis as reducing agents and catalysts. Some hydrides are favorable for use as jet fuels, and sources for the rapid generation of hydrogen.

HYDRODYNAMIC REGIME

A situation in a gas of *quasi-particles* arising when a size or other specific distance in a solid is significantly greater than the *mean free path* between the normal collisions of quasi-particles but significantly less than the mean free path between collisions involving momentum (or *quasi-momentum*) loss (collisions with impurities or Umklapp processes, see *Conservation laws*). Each small element of a body in the hydrodynamic regime has an equilibrium state characterized by a local density (temperature) and a velocity of ordered quasi-particle motion. *Kinetic phenomena* in this regime can be described in terms of a gas flowing along a crystalline sample similar to the flow of a viscous liquid. An example is the flow of a *phonon* gas in crystalline ^{4}He. Propagation of sound waves in the gas of quasi-particles (see *Second sound*) is possible in the hydrodynamic regime.

HYDROEXTRUSION

Shape modification of *metals* and *alloys* performed by extruding the samples through a scaling jet out of a *high-pressure cell*. This technological process includes two main physical effects of the high pressure on the deforming material: (1) plasticizing of crystalline solids under the high hydrostatic pressure; (2) preservation of improved mechanical properties in the material treated under a high pressure (see *Hydrostatic treatment of materials*).

The following problems are solved by the use of hydroextrusion: (a) improved shaping of brittle metals and alloys; (b) obtaining objects with complicated, shaped profiles; (c) obtaining objects with improved complex working properties.

HYDROGEN BOND

A chemical bond between two *atoms* A and B involving an intervening hydrogen atom. The hydrogen is more strongly bound with one of the atoms, e.g., A, and the bond is designated as R_1A–H...BR_2, where the radicals R_1, R_2 involve other atoms of the system. The interaction within the hydrogen bond arises from electrostatic attraction and is a result of the partial redistribution of electron density to the side of atom B during its approach to the A–H group. The hydrogen bond energy is defined as the difference of the energy of the electron subsystem of the complex as a whole and the sum of the energies of its parts R_1A–H and BR_2. It varies (from fractions of to tens of kcal/mole), and the hydrogen bond length (i.e. the distance A–B) varies correspondingly from more than 0.3 to 0.26 nm. Hydrogen bonds determine the structure of *ice*, solid methane and other *molecular crystals* constructed from heteropolar molecules and of many *polymers* (in particular, deoxyribonucleic acid and protein). See also *Chemical bond in solids*.

HYDROGEN IN METALS

Significant quantities of hydrogen can dissolve in some metals.

This was originally discovered for metallic *palladium* by T. Gram (1866). At a low external gas pressure p_{H_2} the equilibrium concentration of hydrogen in metals is proportional to $\sqrt{p_{H_2}}$ (*Sievert's law*) which is evidence for the dissociation of H_2 molecules in the solution. In disordered solutions, which are called *alpha phases*, the hydrogen atoms statistically occupy particular interstitial sites of the *metal* lattice. By this the delocalization of electrons takes place and different types of bonding are possible: in compounds with alkali metals hydrogen plays the role of a halogen, and in transitional metals *metallic bonds* prevail, with the hydrogen electron occupying states of the metallic matrix near the *Fermi energy*. The H atoms cause deformation of the metallic lattice which results in hydrogen–hydrogen attraction. Experiments have revealed the specific *condensation of a lattice gas* of hydrogen in metals and of *phase transitions* "gas–liquid–solid" associated with this phenomenon. Due to the long-range character of the attraction the properties of such transitions may depend on the shape of the sample. At relatively high hydrogen atom contents a reconstruction of the metallic sublattice takes place and stoichiometric compounds called *hydrides* are formed.

The *diffusion* of atomic hydrogen in metals has a very high rate due to its small size (sometimes it is possible to speak about screened protons), the lifetime of which, e.g., in *vanadium* is 10^{-12} s at room temperature, so that the *diffusion coefficients* generally correspond to liquid-state ones. Thus the isotope effect, high diffusion coefficients and deviations from *Arrhenius law* at low temperatures are evidence of a quantum character of diffusion of hydrogen in metals (see *Quantum diffusion*).

Applications involving phenomena associated with hydrogen in metals are high: hydrogen *embrittlement* of metals, materials for retarders in nuclear reactors, hydrogen filters, hydrogen accumulation for power systems, etc. *Superconductivity* occurs in the hydride compounds Th_4H_{15} and PdH ($T_s \approx 10$ K), but it has not been observed in the hydrides of superconducting metals (e.g., Ta).

HYDROGEN, METALLIC

See *Metallic hydrogen.*

HYDROPHILIC AND HYDROPHOBIC BEHAVIOR

See *Lyophilic and lyophobic behavior.*

HYDROSTATIC TREATMENT OF MATERIALS

A treatment based on *plastic deformation* of a solid under high hydrostatic pressure where the *plasticity* of all crystalline bodies sharply increases. In some cases, the increased plasticity (as well as the *strength*) is restored after removal of the pressure. It is also important for improved properties to be preserved in crystalline bodies after a *thermal treatment* so that *construction materials* will acquire enhanced working characteristics (e.g., plastic *zinc*, *molybdenum*, *beryllium*, etc.). There are technological processes which produce plastic shape modification under high hydrostatic pressure resulting in an improved working quality that needs no further complicated mechanical processing. Examples are *extrusion* (see *Hydroextrusion*), *drawing*, and bulk and sheet punching performed in a medium under high pressure. These processes solve practical problems in the engineering industry and *metallurgy*.

HYDROTHERMAL CRYSTALLIZATION

A method of growing the *crystals* of inorganic substances from water solution at increased temperature and pressure (up to supercritical). To perform a hydrothermal crystallization, one makes use of the solubility increase of a series of non-organic compounds by facilitating the flow transport reactions which provide a *mass transport* of the crystallizing material and its deposition on a priming device. This crystallization is carried out in autoclaves with a volume from 100 cm^3 (for laboratory studies) to $\approx$12 m^3 for industrial production. Working regimes vary from 0.1 to 300 MPa, and from 300 to 1100 K. The duration of technological cycles can vary from a few days in research laboratories to one and a half years (for example, when growing large scale *monocrystals* of optical quartz). During a hydrothermal crystallization, a particular temperature gradient is maintained between a bottom zone where there is a charge of optimal composition and chemical purity, and a growing zone where precisely oriented, especially prepared primings are positioned on crystal-holders. The *crystal growing rate* is controlled both by the magnitude of the temperature gradient and by the convection process with the help of diaphragms established inside the autoclave and, in many cases, perforated, and jet-directing devices which govern the hydrodynamic characteristics of the mass transport process in the liquid and fluid phases. The crystal growing rate can vary up to 1 mm per day and more. Hydrothermal crystallization has been applied for growing quartz single crystals (industry yield exceeds hundreds of tons a year) as well as for obtaining synthetic monocrystals of calcite and zincite.

HYGROSCOPICITY

A property of materials whereby they absorb moisture from the air. Hygroscopicity is inherent in wetting materials (hydrophilic materials, see *Lyophilic and lyophobic behavior*) with a capillary-porous structure where the moisture is taken up in thin capillaries, as well as in water soluble materials, especially chemical compounds which form hydrates with water. The amount of moisture absorbed by the substance (*hygroscopic humidity*) increases with the increase of the moisture content

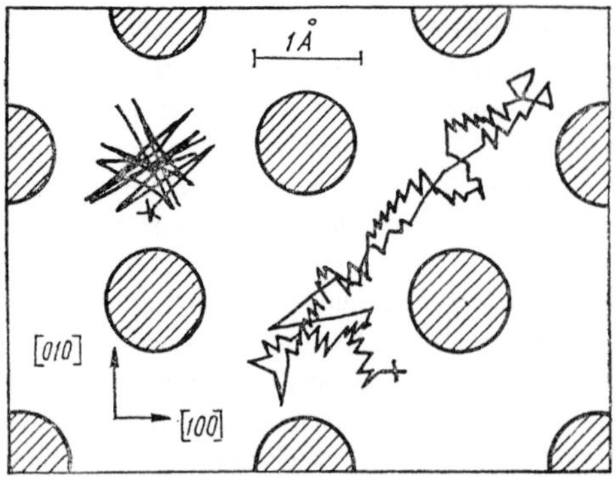

Fig. 1. Trajectories (generated by computer modeling) of 5 keV copper ions travelling along a channeling axis in a copper single crystal parallel to the [001] direction (i.e. perpendicular to the plane of the figure). The trajectories are projected on the (001) plane of a body centered cubic monocrystal of copper. The point of entry is marked by a cross.

of the air, and reaches a maximum for a relative humidity of 100%.

HYPERCHANNELING

One of the types of channeling of charged particles in *monocrystals*; a kind of *axial channeling* arising when the entrance angle is very small relative to the axis. The following varieties of trajectories are possible for axial channeling: limited to a single channel (shown in the left upper corner in Fig. 1), and unlimited (freely roaming between channels during ion passage through the crystal and shown to the right in Fig. 1). When the channeling is limited to a single channel the ions undergo a large number of glancing collisions, and they move in regions of low potential which are surrounded by relatively tightly packed rows of atoms. In this case, the critical angle for channeling is smaller than the Lindhard angle (see *Channeling*) because the magnitude of the potential barrier is determined not at the center of the atomic chain, but by the potential at the boundary between the channels, indicated in Fig. 2 by the point P.

HYPERELASTIC MATERIAL

A material which exhibits zero dissipation of *strain energy*. Thermodynamic properties of a hyperelastic material can be described by components of a *strain tensor* involving, in combination with the temperature, a closed system of thermodynamic variables. All truly elastic bodies possess a hyperelasticity (see *Finite elastic strain*).

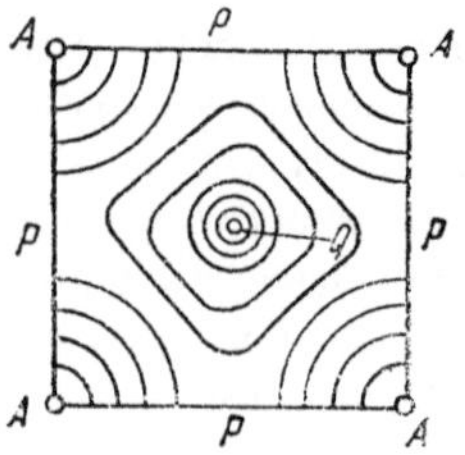

Fig. 2. Arrangement of continuous equipotential lines around atomic chains A for copper, where Q indicates the location of a potential minimum and P denotes saddle points.

HYPERFINE INTERACTION

The interaction of the *magnetic moment* of the nucleus μ_n with the magnetic moments of electrons (*orbital angular momentum* l and spin moment S). The hyperfine interaction results in the *hyperfine structure* of spectral lines (hence the name). The effect of a hyperfine interaction upon nuclei is described by *hyperfine fields* at the nuclei, $\boldsymbol{B}_n$. Due to the quantum nature of electron motions in solids, $\boldsymbol{B}_n$ may be written in the form

$$\boldsymbol{B}_n = -2\mu_\mathrm{B}\left[\frac{\boldsymbol{l}}{r^3} - \frac{\boldsymbol{S}}{r^3} + \frac{3\boldsymbol{r}(\boldsymbol{r}\boldsymbol{S})}{r^3} + \frac{8}{3}\pi\delta(\boldsymbol{r})\right],$$

where $\boldsymbol{r}$ is the distance from the electron to the nucleus. The last term in the expression for $\boldsymbol{B}_n$ is called the Fermi contact field. It exists only for those electrons that have a charge density at the nucleus different from zero. The average values of the operator $\langle \boldsymbol{B}_n \rangle$ are related to the magnetic moment of the atom $\boldsymbol{M}$ by the expression

$$\langle \boldsymbol{B}_n^\alpha \rangle = \sum_\beta A_{\alpha\beta} \boldsymbol{M}_\beta, \quad \alpha, \beta = x, y, z,$$

where the symmetry of the tensor $\widehat{A}$ is determined by the local symmetry of the atomic environment. Meanwhile, the values of *hyperfine coupling constants* $A_{\alpha\beta}$ depend on the form of the electron wave functions, and correlate with experimental data provided by, e.g., the techniques of the *nuclear magnetic resonance* (NMR) or nuclear gamma ray resonance (*Mössbauer effect*). The fields $\boldsymbol{B}_n$ only weakly affect the NMR frequencies (*Knight shift*, *chemical shift*) of *paramagnetic centers* in which the values of $\boldsymbol{M}$ are usually low due to thermal motions. In magnetically ordered materials with small thermal oscillation amplitudes $\boldsymbol{M}$, the average values of $\langle \boldsymbol{B}_n \rangle$ may reach 10–100 T, so the hyperfine interaction is the main factor controlling the behavior of nuclear spins. There result certain features in the NMR, such as enhancement effects, *nuclear spin waves*, etc. Measurements of $\langle \boldsymbol{B}_n \rangle$ provide information on the electric properties of solids, and the distribution of magnetic atoms in them.

The isotropic part of the hyperfine interaction (*contact interaction*) is described by the *spin Hamiltonian*, $a\boldsymbol{J} \cdot \boldsymbol{I}$, where $\boldsymbol{I}$ is the nuclear *spin*, and $\boldsymbol{J}$ is its electronic counterpart. The isotropic hyperfine interaction parameter (*Fermi constant*) is $a = 16\pi g\beta g_n\beta_n(|\psi_\uparrow|^2 - |\psi_\downarrow|^2)/3$ where $|\psi_\uparrow|^2$ is the sum of the electron densities at the atomic (ion) shells, with spins aligned upward ($\uparrow$) at the nucleus, and analogously for $|\psi_\downarrow|^2$. Note that the nucleus is considered to be a point charge. Next come $g(g_n)$ and $\beta(\beta_n)$. These are the electron (nuclear) g-factor and the magneton, respectively. Depending on the filling of the electron shell and the value of the nuclear moment, the value of an atomic moment a may vary from several kHz to tens of GHz. It is measured using the techniques of *electron paramagnetic resonance* (EPR), NMR, nuclear γ-resonance, *electron–nuclear double resonance* (ENDOR), *selective saturation* of EPR, and *spin echo*. These measurements yield information on the spins and magnetic moments of the nuclei, on the isotropic hyperfine interaction of the *paramagnetic center* with surrounding nuclei, and also on the spatial distribution of its electron cloud. The *isotropic hyperfine interaction* serves as a measure of the covalence of complexes. Different isotopes and different charge states of ions also have differing values of a. The isotropic hyperfine interaction is a factor enhancing the effect of radio frequency fields upon the nucleus (sometimes that enhancement reaches several thousand). Disregarding the approximation of a point nucleus results in a weak dependence of the isotropic hyperfine interaction on the distribution of *magnetization* and the density of the electron wave function within the volume of the nucleus. As a result, the isotropic hyperfine interaction $a\boldsymbol{I} \cdot \boldsymbol{S}$ and $a'\boldsymbol{I}' \cdot \boldsymbol{S}$ for two isotopes with nuclear magnetic moments μ and μ' yields an anomaly in the hyperfine structure characterized by the parameter Δ in the ratio $a/a' = (1 + \Delta)\mu/\mu'$. The value of Δ may exceed 10^{-2} for heavy nuclei, with $\Delta < 10^{-5}$ for the light ones.

Depending on the mutual orientation of the crystal axes, and on the total electron $\boldsymbol{J}$ and nuclear $\boldsymbol{I}$ spins, the anisotropic part of the hyperfine interaction has the form $\boldsymbol{J}\widehat{A}\boldsymbol{I}$. It is characterized by the parameters $B_{pq} = A_{pq} - \mathrm{Tr}\,\widehat{A}$, and is determined by averaging the *dipole–dipole interaction* between the electron moment, distributed with the probability $|\psi|^2$ over space, and the point nuclear moment. For distances between $\boldsymbol{J}$ and $\boldsymbol{I}$ that exceed the radius of the electron cloud, the *anisotropic hyperfine interaction* is close to the dipole–dipole interaction of point magnetic dipole moments $\beta\boldsymbol{q}\boldsymbol{J}$ and $\beta_n\boldsymbol{q}_n\boldsymbol{I}$. The parameters B_{pq} are found from the *angular dependences of spectra* of EPR and ENDOR.

HYPERFINE STRUCTURE

The splitting of spectral lines due to intrinsic nuclear spin interactions with a paramagnetic ion. The *hyperfine interaction* of the magnetic moment (spin) of the nucleus $\boldsymbol{I}$ with the magnetic moment of the net electronic angular momentum $\boldsymbol{J}$, as well as the Zeeman interaction of the nuclear spin with an external magnetic field $\boldsymbol{B}$, and the interaction of the quadrupole moment of the nucleus with the *electric field gradient*, can all be described by combinations of terms in the generalized spin Hamiltonian of the form $\boldsymbol{J}^k\boldsymbol{I}^l\boldsymbol{B}^n$ (here $k, l, n = 0, 1, 2, \ldots$, $k + l + n$ is an even number), with their symmetry corresponding to the point symmetry of the *paramagnetic center*. It is a "fingerprint" of an ion, since it contains the information on its nuclear spin and magnetic moment, its hyperfine and quadrupole interactions, and the relative abundance of the given isotope (see Fig. 1).

The *superhyperfine structure* (SHFS) of an *electron paramagnetic resonance* (sometimes called "*ligand* hyperfine structure") is the splitting that results from the interaction of the net magnetic moment of the electronic angular momentum

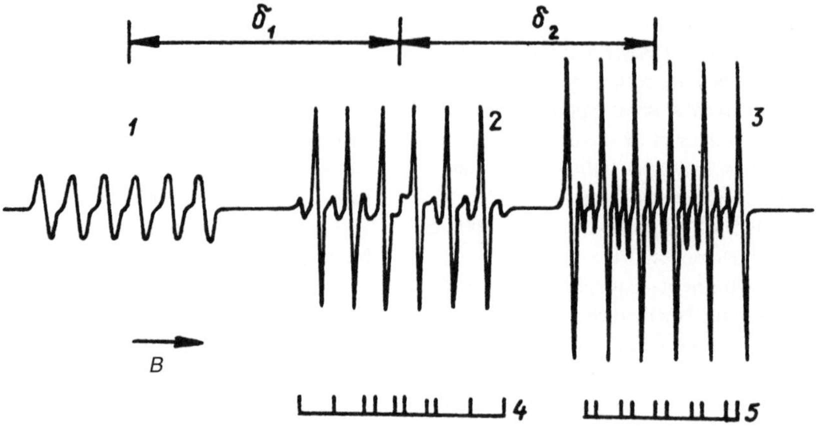

Fig. 1. A typical spectrum of Mn^{2+}: $(1, 2, 3)$ are the sextets of hyperfine components ($M = 5/2 \to 3/2$, $3/2 \to 1/2$, $1/2 \to -1/2$, respectively) of allowed transitions that comply with the following selection rules for changes in the electronic, M, and nuclear, m, spins: $\Delta M = \pm 1$, $\Delta m = 0$. The two higher field sextets ($M = -3/2 \to -5/2$ and $-1/2 \to -3/2$) symmetric with sextets (1) and (2) are not shown. Panels (4, 5) identify the forbidden transitions with $\Delta M = \pm 1$, $\Delta m = \pm 1$ which appear between the stronger allowed lines of sextets (2) and (3), respectively. The terms δ_1 and δ_2 identify the splittings that characterize the fine structure.

at a paramagnetic defect $\boldsymbol{J}$ with the spins of the nearby nuclei $\boldsymbol{I}_i$. To study this one may use the techniques of *electron–nuclear double resonance*, electron *spin echo*, and *selective saturation*. Such a structure is described by a sum of operators of the generalized spin Hamiltonian of the form $\boldsymbol{J}^k \boldsymbol{I}_i^l \boldsymbol{B}^n$ mentioned above. Note that the symmetry of the interactions with the ith nucleus arise from its *local symmetry group*. As for the simplest groups of equivalent nuclei with isotopic hyperfine interactions, their SHFS consists of a set of components with a binomial ratio of intensities (provided one neglects the difference between the transition probabilities for $I_i \geqslant 1$), see Fig. 2.

The spectrum becomes more complex when there is a strong quadrupole interaction, and also in the case of anisotropic hyperfine interactions. SHFS yields information on the type and location of the nearby nuclei, on the distribution of electron density, and on the isotopic composition of the crystal.

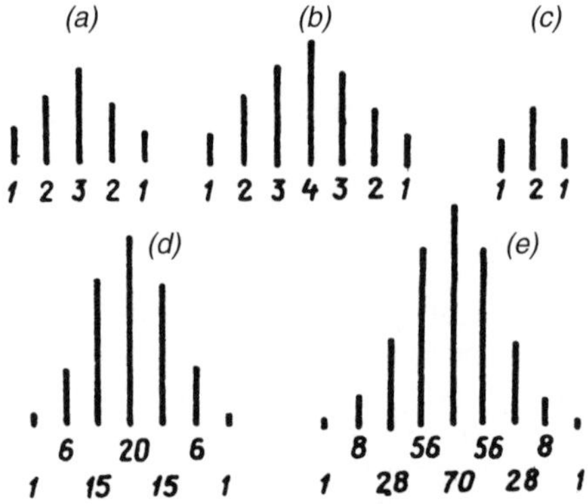

Fig. 2. Superhyperfine structure for equivalent nuclei: patterns (a, b, c) correspond to pairs of nuclei with $I_i = 1, 3/2, 1/2$, respectively, and (d, e) are hyperfine patterns for six and eight equally coupled $I_i = 1/2$ nuclei, respectively.

HYPER-RAMAN SCATTERING, two-photon Raman scattering

A process of non-linear interaction of light with matter leading to inelastic scattering. As a result, the irradiation of matter by monochromatic rays of frequency ω_l produces scattered radiation at the frequency $\omega_s = 2\omega_l \pm \omega_0$, where ω_0 is the frequency of a quantum of elementary excitation (vibration) of the matter. According to quantum theory, the hyper-Raman scattering process can be interpreted as an annihilation of two incident photons accompanied by the generation of a scattered photon, and the creation (Stokes process) or annihilation (anti-Stokes process) of a quantum of an elementary excitation (phonon) of matter.

The magnitude of this scattering is proportional to the square of the intensity of the incident light and, as a rule, is extremely small. Therefore, it is necessary to use pulse-periodic *lasers* for excitation, and photon counting systems with gating for its detection. The efficiency of the process becomes significantly enhanced when the frequencies ω_l and/or $2\omega_l$ approach the frequency of an intrinsic transition in the medium where a *resonance hyper-Raman scattering* takes place.

Important advantages of hyper-Raman scattering result from *selection rules* which permit this process to occur for excitations that are inactive in both Raman and infrared spectra (*silent lattice vibrations*). In this scattering IR-active excitations are permitted; consequently, *polaritons* (quanta of interaction between phonon and photon) in media of any symmetry are also permitted, and there are favorable conditions for studying both phonon and exciton polaritons. When the polariton frequency is far removed from an intrinsic resonance frequency of the medium (polariton energy corresponds to "photon" portion of dispersion curve) the hyper-Raman scattering on polaritons is also called *hyperparametric light scattering*, *four-photon light scattering*, or *light on light scattering in matter*. See *Raman scattering of light*.

HYPER-RAYLEIGH SCATTERING

A process of nonlinear interaction of a sufficiently intense light wave of frequency ω_0 with matter, which results in the appearance of scattering at frequencies $2\omega_0$, $3\omega_0$, and so on, with the magnitude proportional to, respectively, I_0^2, I_0^3 and so on (here I_0 is the intensity of the incident light wave). See also *Rayleigh scattering of light*, *Light scattering*.

HYPOELASTIC MATERIAL (introduced by C. Trusdell, 1955)

A material in which the components of *stress* variation rates under a *strain* are homogeneous functions of the deformation rate components. A hypoelastic material is non-viscous and its infinitesimal deformations are reversible relative to the initial stress. This idea is a generalization of an intuitive concept of an elastic material for a case of finite deformations.

HYSTERESIS

Dependence of a physical state on the history of controlling parameters, rather than just on their values. Such a multivalued correspondence of the state with the parameter values is observed in any process which exhibits *relaxation*, whose time constant (relaxation time) determines the time for reaching the equilibrium state under the variation of external conditions. The ambiguity in the value of the state of the system parameter depends on how the external condition parameter relates to the system *relaxation time*. In the case of a *first-order phase transition* one can observe a difference between the temperatures of the *phase transition* occurring during cooling and heating. This phenomenon is called *temperature hysteresis*, and it is a consequence of the possible existence of *metastable states* within a certain temperature range where the *thermodynamic potentials* of the system are double-valued. For a given material this temperature range depends on the rate of variation of the temperature. For complete *thermodynamic equilibrium* in the system (which is never reached due to the lag of the *nucleation* processes and other causes), the temperature range of the hysteresis must be zero, and the phase transition should occur at the temperature corresponding equilibrium of thermodynamic potentials. Another kind of hysteresis phenomenon is observed in many physical systems with magnetic or electric polarizations, when a transition from one phase to another includes a state with domains (an *intermediate state*), and the ambivalence of the dependence of values does not disappear under infinitesimally slow variations of the external conditions. Examples of this are *ferroelectric hysteresis* and *magnetic hysteresis* observable in *ferroelectric* and magnetically ordered materials, respectively, and also *elastic hysteresis*, i.e. a lag in the phase of a *strain* relative to that of a *stress* (see, e.g., *Shape memory effect*, *Martensitic transformation*). The phenomenon of elastic hysteresis is inherent in all *solids*. In experiments, these hysteresis phenomena appear as loop-like dependences (hysteresis loops) of values characterizing the system state (polarization, *magnetization*, *internal stress*, etc.) under a cyclic variation of corresponding values specifying the external conditions of a *hysteresis loop* (electric field, magnetic field, applied stress). Hysteresis is also typical of certain states of condensed media with a nonergodic behavior (*spin glasses* and other *amorphous magnetic substances*, *dipole glasses*, and so on).

HYSTERESIS IN PHASE TRANSITIONS

A double-valued dependence of an *order parameter* (e.g., the *magnetization* at *magnetic phase transitions*) on physical values specifying the external conditions (temperature, pressure, external field, etc.) associated with a phase transition. The hysteresis appears due to the presence of *metastable states* at *first-order phase transitions*. It is observable under sufficiently fast changes of the external conditions owing to the slow response of the system, and its persistence in a state with an order parameter value corresponding to a relative minimum rather than the absolute minimum of the *thermodynamic potential* (see *Hysteresis*, *Phase transitions*).

HYSTERESIS LOOP

See *Hysteresis*, *Magnetic hysteresis*.

HYSTERESIS, MAGNETIC

See *Magnetic hysteresis*.

Ii

ICE

Solid *phase* of water, most important rock-forming *mineral*, and monomineral rock of cryolithosphere. Ten crystalline modifications of ice are known in addition to amorphous ice. Under natural conditions there is only one modification (ice I), whose *crystals* have the hexagonal space lattice of the *trigonal system* with six H_2O molecules forming a regular hexagonal cell with dimension $b = 0.9$ nm along the axis. The coordination number is four with 0.276 nm between the closest centers of the molecules. Six cavities or voids located around each molecule at the distance 0.347 nm, with dimensions exceeding those of the molecules, form channels associated with alternating hexagonal cells. The spatial lattices of other ice modifications belong to the rhombic (ice II), tetragonal (ice III, VI, VIII and IX), monoclinic (ice V) or cubic (ice IV) *crystal class*. Each modification has its own range of stability. Under ordinary atmospheric pressure in the temperature range -130 to $120\,^\circ$C, ice IV with cubic symmetry can exist. It forms by *condensation* of vapors on a substrate cooled to $-125\,^\circ$C, and at lower vapor condensation temperatures amorphous ice forms (see *Amorphous state*). With increasing of temperature, these forms transform to common ice I. The density of ice at $0\,^\circ$C and a pressure 0.1 MPa equals 0.9167 g/cm^3, which is 9% lower than that of water. Crystals of ice exhibit *anisotropy of mechanical properties*, *optical anisotropy*, anisotropic electrical and other properties. A low coefficient of visible light adsorption (*transparency*, see *Light absorption*) and very weak *birefringence* characterize ice which is optically positive and uniaxial. The *melting temperature* of ice under pressure of 0.1 MPa is adopted as $0\,^\circ$C. At reduced pressure the melting temperature lowers by $0.075\,^\circ$C per 1 MPa. Heat of melting at $0\,^\circ$C is 334 J/g, and heat of *sublimation* is 2834 J/g. *Thermal conductivity* of ice near $0\,^\circ$C is 2.2 W·m^{-1}·K^{-1} which is approximately 4 times more than the value for water, and it increases as the temperature is lowered. Ice has a high electrical resistivity ($\rho \sim 10^8$ to 10^9 Ω·m). It exhibits plastic properties (see *Plasticity*), and *plastic deformations* which depend on the temperature, on the character of the loading, and on the deformation rate. The *viscosity* of ice varies from 10^3 to 10^8 MPa·s. Elastic properties of ice are displayed only under very short-term loads. With the reduction of temperature, *elasticity* and *brittleness* of ice grow; and with its increase, the plasticity grows. *Ultimate strength* of ice under compression is about $25\cdot10^5$ N/m^2, shear strength $6\cdot10^5$ N/m^2, rupture strength 1 to $2\cdot10^5$ N/m^2, *flexure* strength is 20 to 30% less due to plastic deformations.

Subterranean ice. Ice in the Earth's lithosphere is present in rock of the cryolithosphere as a mineral, piercing the rock mass in the shape of thin layers, lenses, small scattered crystals (ice-cement) and also as a monomineral rock, thus forming separate large aggregations with thicknesses between 0.3 m and 60 m. Overall subterranean ice occupies about 2% of the total volume of ice of the cryosphere ($\approx 0.5\cdot10^6$ km^3). Subterranean ices are subdivided as to their mechanism of their formation into vein, injection, segregation, cave and buried types.

Ice in the atmosphere is found in the form of cloud particles and precipitation, usually of the ice I crystal modification. At temperatures above that of homogeneous liquid water freezing (above $-41\,^\circ$C) heterogeneous ice nucleation takes place, a process dependent on the properties of atmospheric aerosols (see *Aerosol particles*). Water vapor diffusion controls the growth of ice crystals in the atmosphere up to the radius of 10^{-3} m, and is responsible for the formation of regular shaped crystals. About 80 different shapes of ice crystals are found in the atmosphere, the basic types being dendrites, plates, prisms, needles, etc. Further growth and enlargement proceeds via coagulation of crystals with crystals (snow flakes), crystals with droplets (fine hail, up to $0.5 \cdot 10^{-2}$ m radius and hail, up to 10^{-1} m radius). The formation and fallout of precipitation in the middle and polar latitudes is closely related to the presence of ice in the clouds.

ICE BOMB (B.G. Lazarev, L.S. Lazareva, 1939)

A simple device used to produce high uniform pressures (up to 0.2–1 GPa) at low temperatures without employing any external compressor. Several materials feature higher densities in their liquid state than in the solid one (e.g., water by 8.3%, gallium by 3%). Therefore, under chilling and freezing in a confined space, they develop a high pressure that is sustained at low temperatures in the solid phase (up to 0.173 GPa for water and up to 1 GPa for gallium). That pressure is transferred to the sample placed inside the "bomb" container. A uniform pressure p is developed during slow chilling of a beryllium bronze bomb (features high strength, high thermal conductivity, weak magnetism). By attaining *high pressures* this way without compressors the ice bomb opened up a broad new field in the studies of solids, low-temperature physics at high pressures. Ice bombs provide a quantitative experimental base for studying the thermodynamics of *type I superconductors*, such as the pressure dependences $\mathrm{d}T_\mathrm{c}/\mathrm{d}p$ and $\mathrm{d}B_\mathrm{c}/\mathrm{d}p$ of the *critical temperature of superconductor* T_c, and the *thermodynamic critical magnetic field* B_c. Studies have been made of electron-topological transitions of metals under elastic strain (see *Lifshits transition*). The effect of pressure (up to 1.1 GPa) on the *magnetization* of ferromagnets was studied using gallium. Due to the small size of the bomb and its stand-alone nature, it is convenient for high pressure studies at temperatures below 1 K (down to about 0.05 K). The ice bomb initiated the process of designing new devices in which the agent to transfer high pressure is a solid (see also *High-pressure cells*).

IDEAL CRYSTAL

A theoretical model of a *crystal* with a perfect periodic lattice consisting of immobile identical atoms. The ideal crystal model is widely used in the theory of solids. Real crystals always contain a certain number of *defects* and thus are not ideal. The closest to ideal crystals are *monocrystals*. See also *Perfect crystal*.

IDEAL GAS

Model of a gas which neglects the potential energy of interaction between its particles. This gas obeys the ideal gas law $PV = nRT$, where P is the pressure, V is the volume, n is the number of moles, R is the gas constant, and T is the absolute temperature (in kelvins). The ideal gas model is widely used in solid state physics to describe a system of *quasi-particles* at low density. The interaction of ideal gas particles takes place only during binary collisions. The particles are assumed to be nondeformable with no internal degrees of freedom. An ideal gas of quantum particles (quasi-particles) is described by *Fermi–Dirac statistics* for particles with half-integral spin (e.g., conduction electrons in a metal), by *Bose–Einstein statistics* for particles with integral or zero spin, and by the *Boltzmann distribution* for classical cases.

IDEAL GLASS

An idealized (simplified) model used in the theory of solids to describe the *vitreous state of matter* (by analogy with an *ideal crystal*). In the case of amorphous solids with *covalent bonds*, a body whose atomic structure represents a random network without *dangling bonds* is considered to be an ideal glass. Unlike an ideal crystal which is defined unambiguously by its lattice symmetry and the free energy minimum condition (as temperature tends to zero), an ideal glass structure lacks an unambiguous definition. For *metallic glasses*, an ideal glass is defined in a model where all the interatomic cavities are tetrahedral, and the local coordination numbers Z of the atoms vary over a rather wide range $9 \leqslant Z \leqslant 16$.

As is well known, three-dimensional Euclidean space cannot be filled with regular tetrahedra without gaps. For example, when packing five

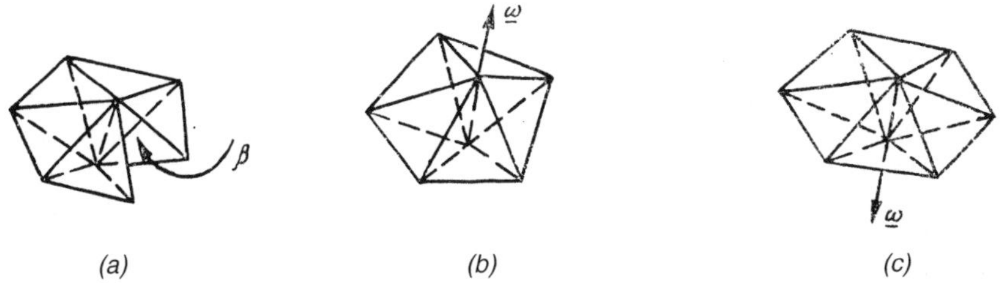

Disclination and tetrahedral packing.

tetrahedra with a common edge there arises a deficiency of space angle $\beta = 4\pi - 20\arcsin(1/\sqrt{3})$ (Fig. (a)). Elimination of the deficiency is possible either with the help of elastic closure of the slit, which is equivalent to the introduction of a positive wedge *disclination* along the common edge of tetrahedra with the magnitude $\omega = \beta/2 \cong 7^\circ\,20'$ (Fig. (b)), or by adding an extra regular tetrahedron into the existing slit (Fig. (c)). The latter operation leads to the appearance of a negative disclination with $\omega \cong 2\pi/5 = 72^\circ$. Therefore, tetrahedra with local tetrahedral packing turn out to be distorted, and the packing geometry can be described by the number q of tetrahedra adjoined along the linking edge. In the examples given, $q = 5$ and $q = 6$.

An *ideal glass with tetrahedral packing* is an atomic configuration in three-dimensional space, which is transformed with the help of a relaxation process to a condition involving more or less differing bond lengths. For an ideal glass, mean values of $\overline{q}$ and $\overline{Z}$, related by the expression $\overline{q} = 6 - 12/\overline{Z}$, are equal to $\overline{q}_i = 2\pi/\arccos(1/3) \cong 5.104$ and $\overline{Z}_i \cong 13.397$. Values close to $\overline{Z}_i$ are specific for coordination numbers of the so-called *Frank–Kasper phases* (F.C. Frank, J.S. Kasper) in complex *alloys* with a large number of atoms in a unit cell (e.g., $\overline{Z} = 13.358$ for $Mg_{32}Zn_{49}$). In the framework of the present approach, Frank–Kasper phases can be treated as near-ideal glasses that degenerate to crystals due to their chemical composition. One can define various *defects* with respect to ideal glasses. Disclinations that are the lines of distorted atomic coordination in local tetrahedral packing play a principal role in glasses with local tetrahedral packing.

IDEAL SOLID SOLUTION

A homogeneous, multicomponent ($i = 1, 2, \ldots, n$) solid *phase* of varying composition, its formation being accompanied by the same changes of *thermodynamic potentials* as in mixing *ideal gases*. In contrast to ideal gases, interactions between particles in an ideal solid solution do exist but undergo no changes with variations of composition, while the partial enthalpies H_i and *entropies* S_i (both vibrational and rotational) are conserved. Only the configuration component of the entropy S changes. Hence, the changes of thermodynamic potentials are $\Delta H = 0$, $\Delta S = -R\sum_{i=1}^{n} x_i \ln x_i$ and $\Delta G = -T\Delta S$ (R is the gas constant, x_i is the relative concentration of the ith component, and G is the Gibbs free energy). These solutions are idealized states; close to them are isotopic *solid solutions*. More common are *regular solid solutions* with $\Delta H \neq 0$, $\Delta S = -R\sum_{i=1}^{n} x_i \ln x_i$. Real solid solutions deviate from both ideal and regular solutions in that the entropy increases, $S > S_{id}$, with the formation of defects (e.g., *Schottky defect*), and the entropy decreases $S < S_{id}$ with the reduction of the particle oscillation frequencies under ordering.

IMAGE FORCES in electrostatics

Electrostatic attractive or repulsive forces affecting an electric charge near a *solid surface* or near the interface between different media. Forces of the image type arise because of the polarization (redistribution of natural charges) in a medium acting on a test charge e at rather long (macroscopic) distances x from the interface. These image forces are determined from the classical potential:

$$W(x) = \frac{e^2}{4x}\frac{\varepsilon_1 - \varepsilon_2}{\varepsilon_1 + \varepsilon_2},$$

where ε_1 is the relative *dielectric constant* of the medium with the charge e, and ε_2 is the dielectric constant of the medium bordering it. In particular, if the charge is in a vacuum ($\varepsilon_1 = 1$) close to a metal surface ($\varepsilon_2 \to \infty$), then image forces are a maximum and correspond to *reflecting image forces* with the potential $W(x) = -e^2/(4x)$, which arises because of the redistribution of free conduction electron density in the metal. The divergence of $W(x)$ at the point $x = 0$ is eliminated by systematically taking into account the effects of *spatial dielectic dispersion* at distances comparable to the *screening radius*, and also by accounting for quantum effects. For rapidly moving charges the image forces are attenuated due to retardation effects arising from the inertia of natural charges in a medium. For *metals* this happens at times of flight comparable to the *plasma oscillation* period of about 10^{-15} s. Under the action of image forces *surface electron states* can arise, which are localized near the interface, e.g., on the surface of liquid helium (see *Levitating electrons*). Image forces play an important role in the *transverse interaction* of adsorbed atoms or molecules at a solid state surface (see *Adsorption*). In particular the interaction between adsorbed atoms with charge Q, located a distance x from the metal surface and separated from each other by the distance $R > x$, has the following form for $R \gg x$ as a result of reflecting image forces:

$$W(R) = -\frac{Q^2}{\sqrt{R^2 + 4x^2}} + \frac{Q^2}{R} \approx \frac{P^2}{2R^3},$$

where $P = 2xQ$ is the effective *dipole moment*, i.e. $W(R)$ exhibits a dipole–dipole type repulsion (see *Dipole–dipole interaction*).

Image forces in imperfect crystals are those taking into account boundary conditions in a crystal with *defects*. In an infinite crystal, *displacement vectors* $\boldsymbol{u}$ due to defects decrease, according to the continuum theory of elasticity, as the distance from the defect grows. However, in a crystal of finite dimensions such a solution does not obey the surface boundary conditions (e.g., condition of absence of stress on a free surface). To meet such conditions image forces are introduced, which are virtual forces acting from the sites of imaginary external defects inducing so-called displacements of "images". In an infinite medium, displacements of the image converge to zero, while in a spatially-restricted body of arbitrary shape they depend on this shape, and on the arrangement of real defects relative to its boundary. Despite the comparatively small value of corresponding deformation "images", bound by surface image forces (and inversely proportional to the cube of the characteristic crystal size L), they are evenly (if defects do not concentrate near the surface layer) spread throughout the crystal volume $\sim L^3$. Therefore, image forces do make a finite contribution to the change of some characteristics (e.g., volume) upon the introduction of defects, which is comparable to changes arising from the emergence of static deformation fields that decrease at long distances from defects.

The energy contribution to the total *strain interaction of defects* related to image forces is generally an "attractive" one and varies very smoothly even at distances r between defects comparable with L, i.e. it is almost independent of r for a macroscopically uniform average distribution of defects. It is particularly insensitive to fluctuational redistributions of their concentration at distances $r \ll L$.

IMAGING, EPR

See *EPR imaging*.

IMAGING, NMR

See *Magnetic resonance imaging*.

IMPACT IONIZATION, collision ionization

Ionization, which occurs when an ionizable ion acquires energy as a result of a collision with another particle (electron, *ion*, *atom*). In the context of solid state physics, the term impact ionization is used in reference to ionization brought about by applying a strong electric field to the sample. *Conduction electrons* are accelerated by the field and give up part of their kinetic energy at collisions with atoms of the solid. In a strong enough field, the energy acquired by electrons may exceed the *band gap* width E_g, in which case an electron can jump from the *valence band* to the *conduction band* as a result of collisions between conduction electrons and vibrating atoms. The approximate criterion for impact ionization is the equality of the energy E_g gained during the passage along a *mean free path* length l, i.e. during the average interval between two successive collisions of an electron with *phonons* or lattice defects (see *Defects in crystals*), i.e. $eE = E_g/(\alpha l)$, where E is electric field intensity, and α is a numerical factor specifying the amount of energy lost by a high-speed electron in a collision. The process of impact ionization results in an avalanche-like increase of the concentration of free *current carriers*. For field intensities from 10^5 to 10^6 V/cm, this increase is limited by the *recombination* of electrons and holes produced by impact ionization. At higher fields (of the order of 10^7 V/cm), however, impact ionization can bring about the electrical breakdown of a semiconductor (see *Breakdown of solids*).

IMPACT IONIZATION AVALANCHE TRANSIT TIME DIODE, IMPATT diode

A *semiconductor diode* with a dynamic *negative resistance* in the UHF range that forms due to an avalanche breakdown (see *Semiconductors*) precipitated by a phase shift between the voltage and the current at the device output. This effect arises from the finite inertia of the avalanche multiplication of carriers and the finite time, τ, that it takes for these carriers to traverse the carrier-depleted layer (transit time). The efficient transfer of IMPATT energy input into UHF oscillations of frequency f occurs when $f \cdot \tau \approx 0.5$. IMPATT devices are used to generate, amplify, and transform oscillations in the range from units to hundreds of GHz. They remain the most powerful solid-state sources of UHF oscillations in the 20–400 GHz band.

IMPACT MALLEABILITY

See *Toughness*.

IMPACT STRENGTH

The *strength* of materials under short-time impact loading, one of the forms of *dynamic strength*. Impact strength determines the resistance to *failure* under short-time (of the order of 10^{-3} s and less) loads of high severity. The action of these loads is characterized by wavelike propagation, and hence by transient stress distribution over the bulk of material under load, as well as by the high-rate deformation of the material which, in a number of cases, enhances its strength properties (*yield limit*, *ultimate strength*, see *Shock hardening*). As a result of a nonuniform distribution of stresses and stress concentrations in certain regions, the *plasticity* of a structural member under impact load is lower than that under static load (see *Construction material strength*). The impact strength of a structural member is improved by a design plan that allows for increasing the uniformity of stress distribution throughout the bulk of material.

IMPATT DIODE

The same as *Impact ionization avalanche transit time diode*.

IMPEDANCE (fr. Lat. *impedio*, am hindering)

A complex-valued quantity that is an analogue of the electrical resistance for harmonic (AC) processes. The impedance has the complex form $Z = R + iX$, where the real part R is the resistive or lossy term, and the imaginary part X is the reactive term. A distinction is drawn between the impedance of an AC circuit element (two-terminal impedance) and the impedance of a (planar) surface in a monochromatic electromagnetic field (*surface impedance*). The concept of impedance was introduced into electrodynamics by O. Heaviside and O. Lodge; that of surface impedance belongs to S. Schelkunoff (1938). Impedance characteristics are used not only in electrodynamics, but also in descriptions of transmission lines for wave disturbances of any nature (e.g., *Acoustic impedance*). In AC electrical circuit theory the impedance Z of a resistance R, inductance L and capacitor C in series has the magnitude $Z = [R^2 + (X_L - X_C)^2]^{1/2}$, where $X_L = \omega L$ and $X_C = 1/(\omega C)$ are the inductive and capacitive reactances, respectively, $\omega = 2\pi\nu$ is the frequency, and $\omega_0 = 1/(LC)^{1/2}$ is the resonant frequency. The voltage V and current I are complex quantities related by Ohm's law $V = IZ$, and the phase angle ϕ between the current and voltage is given by the expression $\tan\phi = (X_L - X_C)/R$. The losses in the circuit arise from the resistance R.

IMPEDANCE, SURFACE

See *Surface impedance*.

IMPREGNATION (fr. Lat. *im*, in, and *praegnans*, pregnant)

A technological process of filling the microcavities and/or *pores* of solids with various liquid compounds (except for metal melts) in order to impart some specific properties to these solids. The process comprises introducing liquid into material, drying the latter, and adding, if necessary, a heat treatment. Impregnation allows one to produce some kinds of self-lubricating materials, to render wood and fabric moisture-resistant, and so on. See *Soaking, Infiltration.*

IMPURITON

A *quasi-particle* formed by a light impurity in a *crystal lattice* during the process of quantum tunneling (see *Tunneling phenomena in solids*). An impuriton is a special case of a *defecton.* Examples of impuritons are atoms of H in *metals* at low temperatures, or atoms of ^{3}He in a ^{4}He crystal.

IMPURITY ATOM CHARGE

See *Effective impurity-atom charge.*

IMPURITY ATOMS

The atoms of chemical elements contained within a solid without being part of its chemical composition. The impurity atom concentration is considerably smaller than the concentration of the basic material atoms. Despite this, they strongly influence the physical properties of many solids. Control of the impurity content of materials can be very important. It includes removal of impurity atoms that enter from the raw materials or from the elements of technological devices, and also adding impurities by *doping. Semiconductors* are especially sensitive to the presence and variety of the impurity atoms: thorough purifying of some of them (e.g., of *silicon*) permits one to detect their effect on the electrical properties of semiconductors in concentrations 10^{11} to 10^{12} cm^{-3}. Doping *metals* and *alloys* to control their strength parameters requires an impurity atom concentration that amounts to several percent. The spatial distribution of the impurity atoms, which may be highly nonuniform, e.g., in case of *ion implantation,* appreciably influences the material properties.

IMPURITY BAND

A band of permitted energies, which appears within the *band gap* of a semiconductor as a result of the overlapping of the closely-spaced *local electronic levels* of impurities. As the *impurity atom* concentration, $N_{\rm i}$, increases, the following limiting cases of semiconductor *doping* are realized. At a low concentration $N_{\rm i}$, the following inequalities hold:

$$N_{\rm i}^{-1/3} \gg r_0, \qquad (1)$$

$$N_{\rm i}^{-1/3} \gg a_{\rm W}, \qquad (2)$$

where r_0 is the *screening radius,* and $a_{\rm W}$ is the radius of the wave function of the impurity atom electron. These are *weakly doped semiconductors.* Local electric fields around the impurity atoms and the wave functions of the electrons localized on them do not overlap, the local electronic levels are discrete, and the activation energies of the impurity atoms are the same. Upon increasing $N_{\rm i}$, inequality (1) becomes violated, and due to the interaction of the electric fields the discrete levels are shifted with respect to each other: the so-called *classical level broadening* takes place. Upon further increases of $N_{\rm i}$, condition (2) is also violated and the levels broaden into an impurity band (*quantum level broadening*). If the impurity band is separated from the *valence band* and *conduction band* by regions (gaps) of forbidden energies, then the materials are called *intermediately doped semiconductors*; at the merging of the impurity band with one of the other bands they are called *highly-doped semiconductors.*

IMPURITY BREAKDOWN in semiconductors

Sharp increase in semiconductor *electrical conductivity* in an external electric field. The increase results from *impact ionization* by collisions of either conduction electrons or holes with impurity atoms. At liquid helium temperatures the impurity breakdown is observed in crystals of *germanium* placed in electric fields of several volts per cm. Due to the limited concentration of *impurity atoms* their full ionization does not result in a true voltage *breakdown* that would entail sharp changes in the material properties. Therefore, the impurity "breakdown" is reversible. Provided the external field increases sufficiently, a *thermal breakdown* due to Joule heating may ensue.

When a sample possesses trapping levels, conditions appropriate for impurity breakdown may be produced by filling the levels with current carriers. For example, one may irradiate a sample with photons of energy exceeding the *band gap* width (so-called induced impurity breakdown).

IMPURITY CONDUCTIVITY

Low-temperature *hopping conductivity* involving impurities in amorphous or doped (impurity) semiconductors. The hops (jumps) may be accompanied by the absorption or emission of *phonons.* At moderately low temperatures the hop occurs by thermal activation through a potential barrier that separates the impurities. At low temperatures, quantum tunneling appears to be essential. See also *Electrical conductivity, Semiconductors.*

Sometimes the term impurity conductivity is used in another sense. When the doping of a semiconductor is sufficiently high and the current carriers that determine conductivity are formed by ionization of the impurities, these carriers can move in an *impurity band.*

IMPURITY DISTRIBUTION

See *Interphase impurity distribution coefficient.*

IMPURITY-INDUCED FERROELECTRICITY

See *Ferroelectricity induced by impurities.*

IMPURITY INTERACTION

See *Magnetic impurity interaction.*

IMPURITY STATE, SURFACE

See *Surface impurity states.*

INCLINED VORTICES in superconductors, Kulik's vortices

A specific kind of *vortex lattice* formed in planar superconducting layers with a thickness of the order of the *coherence length* $\xi \sim 10^{-5}$–10^{-6} cm. The layer is located in the neighborhood of the surface and retains superconducting properties under the action of an applied field exceeding the field needed to destroy bulk *superconductivity.* The vortex state is realized at all angles of magnetic field inclination, except for $\theta = 0$ (i.e. strictly parallel to surface) in the field interval from $B_{c2}(\theta)$ to $B_{c3}(\theta)$ (see *Surface superconductivity*). The vortex state is a lattice that is oriented in the direction of the projection of the magnetic field on the metal surface; the lattice has the structure of triangles elongated in this direction. Inclined vortices determine the current-carrying capacity of the surface layer, which is due to the pinning of inclined vortices by defects of the surface structure (see *Vortex pinning*).

INCLUSION, MACROSCOPIC

See *Macroscopic inclusions.*

INCOMMENSURATE–COMMENSURATE PHASE TRANSITION

A *phase transition* in a crystal, which displays a *superlattice*; the transition is accompanied by a reduction of the incommensurability parameter to zero (see *Incommensurate structures*) and the resulting restoration of translational symmetry. An incommensurate–commensurate phase transition is, as a rule, a *first-order phase transition.* The transition may be detected through diffraction methods; changes of certain physical properties (electric resistance, *magnetic susceptibility,* etc.) are also indicative of an incommensurate–commensurate phase transition. The transition is observed in dichalcogenides of *transition metals* with the formula MX_2 (M=Ta, Nb; X=Se, Te, S), over the premartensitic temperature range of compounds based on TiNi; in *ferroelectrics* K_2SeO_4, $(ND_4)_2BeF_4$, etc.

INCOMMENSURATE MAGNETIC STRUCTURE

See *Modulated magnetic structures.*

INCOMMENSURATE STRUCTURE

A superstructure in a *crystal,* with a period incommensurate with the lattice constant (may be found in more than one crystallographic direction). It originates when atoms are displaced off their positions in the initial structure in such a way that it becomes impossible to find a translation of the lattice that would superpose the crystal with itself. An incommensurate structure fails to fall into any of 230 crystallographic *space groups.* The transition to this structure occurs as a *phase transition* when the temperature is lowered. The modulation responsible for this structure may be produced not only by displacement waves, but also by *charge density waves* or *spin density waves.*

Wave vectors of the incommensurate and the initial structures, $\boldsymbol{Q}$ and $\boldsymbol{q}$, are related to each other as $\boldsymbol{Q} = n(1+\delta)\boldsymbol{q}$ where δ is the nonintegral *incommensurability parameter*, and n is an integer. Incommensurate structures are identified either by using diffraction techniques (e.g., by appearance of anomalous satellites during electron (neutron) diffraction analysis), or by tracking the changes in certain macroscopic properties (e.g., electric resistance, etc.). Such structures have been identified in certain laminated dichalcogenides of the MX_2 type (M = Ta, Nb; X = Se, Te, S), in compounds based on TiNi, Cr, in *quasi-one-dimensional crystals*, etc.

INCOMPATIBILITY TENSOR

A second-rank tensor $T = \mathrm{Ink}\,A$ is the curl of the transposed curl of a given second rank tensor A. In coordinate notation,

$$T_{ik} = e_{imn} e_{kpq} \frac{\partial^2 A_{qn}}{\partial x_m \partial x_p},$$

where e_{ijkl} is the alternating sign tensor. If A is symmetric, then $\mathrm{Ink}\,A$ is also symmetric. Hence, under the condition $\mathrm{Ink}\,A = 0$, the tensor A can be written in the form of the deformation of some vector u_i:

$$A_{ik} = \frac{1}{2}\left(\frac{\partial u_i}{\partial x_k} + \frac{\partial u_k}{\partial x_i}\right).$$

This property of an incompatibility tensor is used for an invariant formulation of the conditions for the *strain compatibility*.

INCOMPLETE MAGNETIC DOMAIN,
non-see-through domain

One of several possible types of *magnetic domains* in highly anisotropic plates with a single magnetic axis (*magnetic films* with quality factor exceeding unity), such that their *easy magnetization axis* remains normal to their developed surfaces. Characteristic of such domains is the lack of contact of a *magnetic domain wall* with either one or both surfaces of the plate. In the first case one speaks of subsurface incomplete domains, and in the second case one speaks of internal volume domains. Incomplete magnetic domains are stable only in plates that are inhomogeneous along their thickness, and are localized in that region of the plate where the plate *magnetization M* is at its maximum, or where the surface energy density of the domain walls σ is at its minimum. When the distribution profiles of either M or σ exhibit several respective minima or maxima through the plate, there can be a variety of subsurface and internal volume *magnetic domain structures* of various types.

INCONGRUENT MELTING

Melting when the liquid phase (the melt) differs in chemical composition from the solid phase. Incongruent melting is observed in solid solutions and in phases of some compounds of varying composition. This melting can complicate the processes of growing monocrystals with a given *stoichiometric composition* (see *Monocrystal growth*). The composition stability of a growing crystal is ensured by the change of melt composition calculated from the compositional *phase diagram*.

INDENTER

A hard tip of a device serving for local loading of materials in testing their *hardness*. The most common indenter shapes are as follows. A hard ball is used as the indenter in testing the *Brinell hardness*, with the relation $d/D \approx 0.375$ maintained (where d is the indentation diameter, D the ball diameter). A cone with an apex angle of 120° and a rounded tip is used in the *Rockwell hardness* test. A regular tetrahedral pyramid with an angle of 136° between opposite faces is used in testing *Vickers hardness*. A regular trihedral pyramid with an angle of $\sim$77° between a face and the axis is used for *Berkovich hardness*. A tetrahedral pyramid with a rhombic base is used in testing *Knopp hardness* with the ratio between the long and short diagonals of the imprint equal to seven. Bicylindrical Yegorov's indenter and a flat die type are also known. Equivalence of the indenters of various shapes (i.e. a possibility to obtain the same hardness values) is due to the same value of the elastic *strain* and *plastic deformation* under the indenter. Indenter material must possess high values of hardness and an *elastic modulus* exceeding those of the material tested. Local loading with the rigid indenter allows one to determine the hardness as well as the *fracture toughness* (crack resistance),

Young's modulus, *breaking stress*, *cold brittleness temperature*, and other mechanical properties of a material.

INDENTER–HARDNESS DIAGRAM

A graphic presentation of the dependence of the loading of the *indenter* P and the depth of its penetration (indentation) h during hardness checking. The indenter–hardness diagram exhibits the following stages: (1) active loading at a controlled rate (load increased from zero to maximum), (2) imprint hold time under pressure, and (3) active unloading and possible repeated loading with passage through a *hysteresis* loop. The appearance of the indenter–hardness diagram depends on the indenter shape, e.g., the pressing of the prism with a flat end produces a diagram which resembles a single-axis compression. The *hardness* H_h, calculated from the depth dependence $P(h)$ of the applied load, is a measure of the resistance to elastico-plastic deformations that come into play, unlike the Vickers or Brinell hardness H (see *Macrohardness*) which to a greater extent reflects the resistance to *plastic deformation*. The correlation between H_h and H is an important structure-sensitive parameter of the measurement, depending on the interatomic bond type, on the plasticity, and on the degree of material loosening. The information content of the indenter–hardness diagram is especially high when testing surface layers of coatings and powder compacts (see *Powder metallurgy*). The resolution reaches 10 nm, with the loading resistance limited by elementary mechanisms of *mass transport*. When P is reduced the quantities H, H_h and their nonuniformity grow due to the statistical nature of *strength*. Modern theories permit a separation by this method of the elastic and plastic deformations of the material under the indenter, and also permit a determination of *Young's modulus*. The *indenter hardness test*, its nonuniformity, the ratio H/H_h, and the roles of elastic (reversible) deformation all come under the scope of "*kinetic hardness*" or "*microhardness*".

INDEX OF REFRACTION

See *Refractive index*.

INDIRECT ELECTRON SPIN–SPIN COUPLING

The coupling of electron shells of *paramagnetic centers* caused by various fields of elementary excitations in crystals: *phonons*, *plasma oscillations*, *conduction electrons*, *spin waves*, and so on. This coupling of paramagnetic ions through a phonon field was first considered by K. Sugihara (1959). He made use of an *electron–phonon interaction* mechanism which changes the orbital motion of valence electrons, taking into account spin–orbit coupling which brings about the correlation of the spin variables of paramagnetic centers with the phonon fields. In the second order of perturbation theory the spin variables of the paramagnetic centers are "mixed", the system energy involves two paramagnetic centers, and phonon field terms appear which depend on the spin variables of the paramagnetic centers, and their reciprocal distances. These terms correspond to the energy of the indirect electron spin–spin coupling, and written in operator form they represent the effective Hamiltonian of the indirect electron spin–spin coupling of the paramagnetic centers. Couplings of this type are anisotropic in the spin variables, have a long-range nature, and are comparable in magnitude with the magnetic *dipole–dipole interaction*. For S-state ions the analogous mechanism is less important. This indirect coupling can be initiated by any elementary excitation field which can influence the orbital electron motion in any manner. In addition to the spin–orbit indirect coupling mechanism described above, an alternative mechanism is possible that involves exchange coupling caused by fields of elementary excitations.

INDIRECT EXCHANGE INTERACTION

Exchange interaction of spin degrees of freedom of localized electrons (or atomic nuclei) through the perturbation of another subsystem such as interstitial (diamagnetic) ions located among magnetic ions in magnetic *insulators*, or conduction electrons in *metals* and *semiconductors*.

In magnetic insulators containing ions of nontransition series metals, where the direct overlap of transition metal ion d-orbitals is small, the main mechanism of exchange coupling is *Kramers–Anderson indirect exchange* (*superexchange*). The specificity of the superexchange arises from the nature of the localized wave functions of the magnetic electrons, represented by the superposition of atomic d-orbitals (f-orbitals for rare-earth compounds) with the s- and p-wave functions of the interstitial anions (F^-, O^{2-}, S^{2-}, Se^{2-} ligands), and the influence of the overlap between them.

For d-orbital transition ions with octahedral or tetrahedral coordination the parameter I_{ab} of the isotropic exchange coupling $I_{ab}(S_a S_b)$ is determined by the superposition of the contributions of unpaired magnetic electrons in d-states (e_g or t_{2g}) split by the crystal field:

$$I_{ab} = (n_a n_b)^{-1} \sum_{i_a, j_b} I_{i_a, j_b} n_{i_a} n_{j_b},$$

where n_a and n_{i_a} are the numbers of unpaired electrons on the center, and on the ith orbital of this center. In the Anderson theory (P.W. Anderson) two basic superexchange mechanisms are considered: (1) potential ferromagnetic exchange (direct exchange coupling of orthogonal molecular orbitals of magnetic ions), and (2) second-order antiferromagnetic kinetic exchange generated by the virtual transport of unpaired electrons from one center to another. The dependence of the magnitude and sign of the exchange parameters $I_{i,j}$ (and hence the dependence of I_{ab}) on the filling of d-states, and the angle of the superexchange metal–ligand–metal bond are explained empirically by the *Goodenough–Kanamori–Anderson rules* (J.B. Goodenough, J. Kanamori, P.W. Anderson), which we summarize as follows:

1. If half filled orbitals on adjacent ions overlap then the exchange coupling parameter $I_{i,j}$ is antiferromagnetic and relatively strong. In the case of a 180° bond the e_g-state coupling is the strongest. For a 90° bond relatively strong exchange coupling takes place in octahedral coordination for the d_z-state electrons of one center, and for d_{xy} on the other.
2. If there is no overlap between half filled orbitals then the exchange is ferromagnetic and, as a rule, weaker.
3. For each magnetic ion the magnitude of the exchange parameters I_{ab} and angle of the bond increase with the replacement of one ligand by another in the series F^-, O^{2-}, S^{2-}, Se^{2-}, this increase being caused by an increase of *covalent bonding*.

The application of these rules provides the following relations involving the exchange parameters I_{ab} between ions with various numbers of d-electrons on the centers:

180° exchange:

- $I(d^8-d^8) > I(d^5-d^5) > I(d^3-d^3) > 0$ (antiferromagnetic bonds),
- $I(d^3-d^8) < I(d^3-d^5) < 0$ (ferromagnetic bonds);

90° exchange:

- $I(d^3-d^5)$, $I(d^5-d^5) > 0$ (antiferromagnetic bonds),
- $I(d^8-d^8) < 0$; $I(d^3-d^3)$ can have different signs.

In substances with metallic conductivity the indirect exchange coupling of electronic or nuclear spins occurs through *conduction electrons* (see *Ruderman–Kittel–Kasuya–Yosida interaction* (RKKY interaction), *Indirect nuclear spin–spin coupling*).

INDIRECT EXCITATION

A many-wave effect in dynamic scattering of X-rays by *monocrystals*. It occurs in cases when three or more points of the *reciprocal lattice* all lie on the Ewald sphere (see *X-ray structure analysis*), e.g., points (000), (hkl), and $(h'k'l')$, with (hkl) corresponding to a forbidden reflection. If the reflection $(h-h', k-k', l-l')$ is not forbidden, a wave scattered in the direction $[hkl]$ appears, since in this case the reflection $(h'k'l')$ plays the role of a new zero point, and the wave field at first scatters in the direction $[h'k'l']$, and then in the direction $[hkl]$. See also *Many-wave approximation*, *Dynamic radiation scattering*.

INDIRECT NUCLEAR SPIN–SPIN COUPLING

Coupling of nuclear spins through orbital and spin magnetic moments of nearby electrons. Indirect nuclear spin–spin coupling takes place because of the polarization of electronic shells by magnetic fields arising from nuclear moments. Nuclear spins, because of their coupling with orbital and spin electron moments, affect the state of surrounding electron shells, which changes their effect on other nuclei, thus giving rise to an indirect nuclear coupling. In addition to the *chemical shift*, indirect nuclear spin–spin coupling effects appear in high-resolution *nuclear magnetic resonance* (NMR) spectra, determining line positions. Often indirect coupling constants are comparable to chemical shifts, which complicates the interpretation of NMR spectra, and suggests the use of stronger magnetic fields. In *metals* and *superconductors* indirect nuclear spin–spin coupling is caused by the coupling of nuclei with *conduction electrons*, and is known as *Ruderman–Kittel isotropic coupling* (see *Ruderman–Kittel–Kasuya–Yosida interaction, RKKY interaction*). In *ferromagnets* anisotropic indirect nuclear spin–spin coupling is referred to as *Suhl–Nakamura coupling* (H. Suhl, T. Nakamura), and it takes place because of the coupling of nuclei with *magnons*. Indirect nuclear spin–spin coupling in diluted *paramagnets* is caused by dipole coupling of nuclei with impurity electron spins, and has the same order of magnitude as chemical shifts.

INDIRECT TRANSITIONS

Optical transitions of *electrons in a crystal* when their quasi-momentum changes appreciably. Indirect transitions determine the low-energy part of the *intrinsic light absorption edge* in *indirect band gap semiconductors* (e.g., Ge, Si), where the absolute extrema of the *valence band* and the *conduction band* occur at different values of the quasi-momenta $\boldsymbol{k}$. Due to the *conservation law* of quasi-momentum and the low value of the *photon* momentum, indirect transitions are only possible with the participation of a *phonon* (or some other particle), or in the field of a defect. Indirect transitions can be described by second-order perturbation theory, and the corresponding values of the *absorption coefficient* are 2–3 orders of magnitude lower than those for *direct transitions*. The spectral dependence and temperature sensitivity of an indirect transition absorption edge are determined by the symmetry of the bands involved in the transition, and by whether or not the transition is accompanied by the emission or absorption of a phonon.

INDIUM, In

Chemical element of Group III of the periodic system with atomic number 49 and atomic mass 114.82. It is a mixture of isotopes ^{113}In (4.33%) and ^{115}In (95.67%). Outer shell electronic configuration is $4d^{10}5s^25p^1$. Successive ionization energies are 5.786, 18.869, 28.03 eV. Atomic radius is 0.158 nm; radius of In^{3+} ion is 0.081 nm, radius of In^+ is 0.130 nm. Oxidation state is +3, less often +1, +2. Electronegativity is 1.49.

In a free form, indium is a silvery-white soft *metal*. It has body-centered tetragonal crystal lattice with parameters $a = 0.32512$ nm, $c = 0.49467$ nm at room temperature, space group $I4/mmm$ (D_{4h}^{17}). Density is 7.31 g/cm^3 at 293 K, $T_{\text{melting}} = 429.9$ K, $T_{\text{boiling}} = 2348$ K. Binding energy is -2.6 eV/atom at 0 K. Heat of melting 3.27 kJ/mole, heat of sublimation 217.7 kJ/mole, heat of evaporation 232 kJ/mole; specific heat is 234.5 J·kg^{-1}·K^{-1} (at 273 to 423 K); Debye temperature is 121.8 K; linear thermal expansion coefficient along principal crystallographic axis is $-7.5\cdot10^{-6}$ K^{-1} and perpendicular to it $+50\cdot10^{-6}$ K^{-1}; coefficient of thermal conductivity is 71 W·m^{-1}·K^{-1}; adiabatic coefficient of elastic rigidity of indium crystal: $c_{11} = 45.4$, $c_{12} = 40.1$, $c_{13} = 41.5$, $c_{33} = 45.2$, $c_{44} = 6.51$, $c_{66} = 12.1$ GPa at 298 K; adiabatic bulk modulus is 42.58 GPa, Young's modulus is 15.8 GPa, shear modulus 5.23 GPa, Poisson ratio is 0.441 (293 K); breaking strength is 0.0023 GPa. Brinell hardness is 0.009 GPa; low-temperature electronic specific heat γ (Sommerfeld coefficient) 1.69 mJ·mole^{-1}·K^{-2}; resistivity is 81.9 nΩ·m (at 273 K), thermal coefficient of resistivity is 0.0049 K^{-1}; Hall constant is $+0.1597\cdot10^{-9}$ m^3/C; optical reflection factor for 5.0 μm wavelength is 96.6%; polycrystal workfunction is 3.8 eV; superconducting transition temperature $T_c = 3.40$ K, critical magnetic field is 29.3 mT (at 0 K). Indium is diamagnetic with molar magnetic susceptibility $-12.6\cdot10^{-6}$ CGS units (at 298 K); nuclear magnetic moment of ^{115}In is 5.507 nuclear magnetons.

The basic area of application of indium and its compounds (InSb, InAs and InP) is that of *semiconductor materials*. Thus, InSb is used in the detectors of infrared radiation; InAs is used in measuring devices for magnetic field intensities.

Doping of Si and Ge with microscopic quantities of indium provides hole conduction and *p–n* junctions. Besides, indium is used as sealing, soldering and corrosion-resistant materials in the electronics industry. Indium *coatings* possess high reflectivity and may be used for manufacturing mirrors and reflectors.

INDUCED DIPOLES

Dipoles produced at locations of atoms and molecules by an external electric field. The physical cause for the *dipole moments* to appear is the displacement of ions from their equilibrium positions, and the shifted polarizations of atomic electron shells. Induced dipoles are responsible for the polarization of a dielectric material. In electric fields of complicated configurations, e.g., in *crystal fields*, the induction of higher order multipoles such as quadrupole moments (see *Quadrupole*) is possible.

INDUCED LIGHT SCATTERING

Light scattering of high intensity off a solid, with the simultaneous creation in the solid, under the influence of the incident radiation, of coherent oscillations or wave-like excitations. The scattered radiation contains Stokes' components with lower frequencies, and anti-Stokes' components with higher frequencies than the incident beam. From the classical point of view the Stokes' components of the induced light scattering can be interpreted as arising from elementary excitations in the medium which subtract energy from the incoming Raman wave and produce a lower frequency scattered wave. Induced light scattering is exemplified by *Raman scattering of light*, as well as by *Brillouin scattering* of light. This scattering is used for transformations of laser radiation, for generating ultra-sound and other types of excitations in solids, and for investigation of the internal structure of matter (nonlinear spectroscopy).

INDUCED OPTICAL ANISOTROPY

Optical anisotropy arises in optically isotropic media in the presence of external fields which induce a preferred direction in these media. The decrease of symmetry causes a difference of optical properties of the media along the preferred direction (*optical axis*) and in the plane perpendicular to it. On removal of the field, the optic axis, as a rule, vanishes (because of relaxation processes). The following fields may be used for inducing an optic axis: electric field $\boldsymbol{E}$ (see *Kerr effect*), magnetic field $\boldsymbol{B}$ (see *Voigt effect*, *Faraday effect*), field of elastic forces, electromagnetic field (see *Self-induced light polarization change*). In the context of phenomenological *crystal optics*, induced optical anisotropy may be treated as a relationship between the *dielectric constant* tensor $\varepsilon_{ij}(\omega, \boldsymbol{k})$ and external fields $\boldsymbol{E}$, $\boldsymbol{B}$, and also *mechanical stresses* σ_{ij}. If the external fields are rather weak, then the tensor ε_{ij} may be expanded in a power series in $\boldsymbol{E}$, $\boldsymbol{B}$, σ_{ij}, and taking into account *crystal symmetry* considerably decreases the number of independent tensor components that appear in the expansion. Starting from this expansion the polarization of normal electromagnetic waves is determined. Molecules in solids are anisotropic and are characterized by a *polarizability* tensor. When the molecules are arbitrarily spatially oriented, then the medium as a whole is macroscopically uniform. On the application of an external electric field, the molecules can orient themselves with the axis of highest polarizability along the field, a process responsible for the initiation of anisotropy. Three polarizability contributions are usually singled out: an electronic one due to displacement of electronic shells from nuclei in the external field; an ionic one associated with shifts of ions relative to each other; and an orientational (dipole) one due to change of orientation of elementary dipoles in the external field. Each of these polarizability components has its characteristic time of dielectric relaxation; and these times differ sharply. Thus, e.g., orientation relaxation in solids is characterized by *relaxation times* in the range from 10^{-10} s to several hours, depending on the physical nature of the induced optical anisotropy and the frequency of the initiating radiation. If molecules possess permanent *magnetic moments* they may be oriented by a constant applied magnetic field.

INDUCED RADIATION

Emission of electromagnetic waves by an excited quantum system under the influence of external radiation, such as the stimulated emission of a laser. The probability of induced radiation is proportional to the external radiation intensity.

The frequency, propagation direction, phase and polarization of the emitted and inducing radiations coincide with each other. If external *photons* of frequency $\nu = (E_2 - E_1)/h$ enter a medium where the upper energy level E_2 is more populated than the lower level E_1 then the process of induced radiation can develop into an avalanche. The phenomenon of induced radiation is used in lasers for the generation and amplification of electromagnetic waves (see *Quantum amplifier, Quantum radio-frequency generator*).

INDUCTANCE

A physical quantity, L, that characterizes magnetic properties of an electrical circuit. The inductance is numerically equal to the ratio of *magnetic flux* Φ threading a contour (e.g., loop or coil of wire) to the strength of the current I in the contour producing this flux, $L = \Phi/I$. The inductance depends on the size and shape of the contour, on the *magnetic permeability* μ of the conductors composing it, and on the magnetic susceptibility χ of the environment. The magnetic energy storage in an inductive circuit is $LI^2/2$. A changing current through an inductive circuit produces the voltage $V = -L\,\mathrm{d}I/\mathrm{d}t$.

Inductance coils with cores of soft magnetic materials are used to increase the inductance. Owing to the dependence of the magnetic permeability of *ferromagnets* on the magnetic field strength, the inductance of such coils depends on I. The inductance of a solenoid situated in a medium with relative permeability μ is $L = \mu N^2 S/l$, where μ_0 is the permeability of free space, N is the number of turns of the coil, S and l are the coil cross-section and length, respectively. The SI unit of inductance is a henry, H, where $1\ \mathrm{H} = 1\ \mathrm{T{\cdot}m^2/A}$.

INDUCTION

This term involves *electric flux density* D, *magnetic flux density* B (electromagnetic field characteristics); and also electromagnetic induction, electrostatic induction.

INDUCTION, NUCLEAR

See *Nuclear induction.*

INELASTIC NEUTRON SCATTERING in solids

Scattering accompanied by energy and momentum exchange between the *neutron* and the scattering system during the course of their interaction. *Thermal neutrons* are commonly used because their energy and momentum is comparable to those of elementary (vibrational) excitations in the *condensed state of matter.* The presence of the neutron magnetic moment makes it possible to study the magnetic properties of solids. The types of inelastic neutron scattering include coherent and incoherent, nuclear and magnetic (see *Neutrons*). Coherent inelastic neutron scattering is governed by the collective dynamics of all the particles, corresponding to the interaction between a neutron and the *collective excitations* of a lattice (*phonons, magnons*) when both the energy and the momentum of the interacting particles are conserved. Experiments on this scattering by *monocrystals* yield complete information on phonons (*nuclear neutron scattering*) and magnons (*magnetic neutron scattering*) in a crystal, including phonon and magnon dispersion curves, largely unattainable by other techniques. During incoherent inelastic neutron scattering, the particles interact with individual nuclei. However, due to the strong bonds in a lattice, other nuclei also affect the process of such scattering, and a large group of particles takes part in it. Therefore, incoherent scattering may also be treated as an interaction of the neutron with quanta of elementary excitations when only the energy of the colliding particles is conserved. Incoherent inelastic neutron scattering enables retrieval of data on *phonon spectra* of crystals. Compared to other (e.g., optical) techniques, using neutron scattering allows one to carry out experiments over a wide range of wave vectors at very low vibration frequencies. Besides, this technique is not limited by *selection rules* since all the vibrations are active. Because of the large element-to-element difference in the neutron scattering cross-sections, this inelastic scattering permits one to study the dynamics of atoms of a specific type in complex compounds. With the appearance of powerful neutron sources (such as high flux reactors, accelerators) the capabilities of this technique are greatly expanding. For example, its sensitivity (energy resolution of 1 μeV ~ $0.8 \cdot 10^{-2}$ cm^{-1}) is comparable to that provided by optical techniques, and its high intensity makes it possible to conduct studies with small samples (down to 10^{-2} cm^3), including surfaces. Inelastic neutron scattering data demonstrate its unique capabilities for studying phonons, *structural phase transitions*, magnons, collective excitations in liquids, diffusion of *hydrogen in metals*, rotational excitations and tunneling in *molecular crystals*, *crystal field* effects in compounds of rare-earth elements, dynamics of *disordered solids*, etc. See also *Neutron spectroscopy*.

INFILTRATION

Process of penetration or permeation. Penetration of liquids or solids through microscopic cavities and/or *pores* under action of capillary forces. The phenomenon of infiltration underlies the technological processes of *impregnation* and *soaking*.

INFORMATION RECORDING in solids

Processes based on using external effects (deformation, electromagnetic fields, chemical reactions, etc.) for storing information via residual changes of physicochemical properties of solids. The information recording technique and the information carrier determine, to a large extent, the design of a computer, or of an information system in general (data input and output, storage system, processor, etc.). At present, the following techniques of information recording in solids are extensively employed: electrical, magnetic, optical, and molecular.

Electrical information recording is widely used in processors and in storage systems in the form of semiconductor *transistors* and large-scale integrated circuits (LSIC). The LSICs are the main processors in microcomputers. A representative LSIC contains about 10^6 elements per cm^2 and has a response time of 1–10 ns. The operating principle of semiconductor transistors involves switching between two states $(1; 0)$. In recent years *metal–insulator–semiconductor structures* (or metal–oxide–semiconductor structures, MOS structures) have found wide use for processing information, and especially images. MOS structures in the form of multicomponent arrays are used in devices with charge transfer. There, the action of light on a semiconductor releases charge carriers which, after amplification, are collected at the metallic layer of the MOS structure. The visualization of images in MOS arrays is possible using an electroluminophor. Devices with charge transfer are used in TV engineering, for visualization and amplification of IR images, etc. To process information, the phenomenon of low-temperature superconductivity can also be employed.

Magnetic information recording via *ferrites* is most widely used in random access and read only memories (RAM, ROM) of large computers. The principle of recording in binary is based by reversing the magnetization direction; and to accomplish this each magnetic memory element is supplied with two current circuits. Early computers used ferrite storage systems in the form of plates with holes (0.5 mm) with magnetizing wires passing through them. In order to design more compact storage systems, ferrite cores (dimensions 0.5×0.1 mm^2) or *thin films* (6 μm; core diameter 2 mm) can be used. Periods of information retrieval 100 ns long and total memory volumes of 10^{11} bits are achieved with ferrite memories. Information recording in magnetic materials requires rather high values of magnetizing currents (power ~10 W). Thin *magnetic films* on an aluminum substrate are used in ROM. Good prospects for magnetic memory systems are *thermomagnetic information recording* systems, and systems using *magnetic bubble domains*.

Optical information recording is widely used in devices for the reproduction of sound and video images (see *Optical storage disks*, *Optical recording media*).

Molecular information recording is now at the research stage. The huge information capacity of protein molecules (*organic polymers*), genetic molecules (DNA, RNA), etc., can be used for it. When they are utilized in the form of thin films on a solid substrate, one can class them as solid systems. Attempts have been made to use various tautomeric reactions, polymerization, reactions of enzymes with a substrate, etc., for information recording. To indicate a record, one can use changes in electrical, optical and other properties.

INFORMATION RECORDING, THERMOMAGNETIC

See *Thermomagnetic information recording.*

INFRARED RADIATION (IR radiation)

Electromagnetic radiation in the wavelength range from $\lambda = 0.74$ μm ($4.05 \cdot 10^{14}$ Hz) to 1,000 μm ($3 \cdot 10^{11}$ Hz) conventionally divided into near (overtones, 0.74–2.5 μm), medium range (fundamental modes, 2.5–50 μm) and long-wave (far) (50–1000 μm) spectral bands. This range covers most vibrational states of atoms, ions, and molecules of *solids*, as well as a significant part of the crystal electronic excitation states. Infrared radiation is absorbed and emitted by such elementary excitations in solids as *polaritons*, *phonons*, *excitons*, *plasmons*, *magnons*, and others.

The study of solids (without sample damage) using infrared transmission and reflection spectra is the most widespread application. These spectra provide information on the energy spectrum of elementary excitations, their *oscillator strengths* and *lifetimes*, as well as on the interaction characteristics of similar or different *quasi-particles*, e.g., plasmon–phonon and *phonon–phonon interactions.*

All solids with the exception of super-pure *diamond* crystals absorb IR radiation in some spectral range. All *metals* strongly reflect IR radiation, with the reflectivity growing monotonically with increasing wavelength. Infrared radiation is strongly absorbed by atmospheric water vapor, carbon dioxide, ozone, and other air molecules. Infrared radiation was first discovered by M. Pictet (1790) and rediscovered by W. Herschel (1860).

INFRARED SPECTROSCOPY of solids

A branch of the *optical spectroscopy* of crystals that employs emission, absorption and reflection spectra in the infrared range. For nonmetallic *solids*, the spectral dependence of infrared absorption is determined by the energy spectrum of mainly vibrational excitations, with contributions from electronic and rotational excitations. For highly conducting metals, an almost total reflection is observed over the entire infrared range. Infrared spectroscopy is especially informative when studying the energy spectrum of vibrational excitations since most frequencies of the *intramolecular vibrations* and optical *phonons* correspond to the infrared range. The only vibrations that appear in the spectra of *perfect crystals* are those involving a change in the *dipole moment.* The number of vibrational bands in the spectrum of a pure substance as well as their frequency, intensity, width, shape and polarization properties are determined by the number of atoms forming a *unit cell*, by the cell symmetry, and the nature of the binding forces; they are also dependent on external parameters (temperature and pressure). Therefore, infrared spectroscopy is used as a structural analysis method for the identification of material compositions. It is especially useful for identifying the presence and the local environment of various functional groups like amino $-NH_2$, hydroxyl –OH, nitro $-NO_2$, etc., in organic molecular crystals. To interpret complicated spectra unambiguously, one obtains data at several temperatures using polarized radiation.

The presence of *impurity atoms* and other defects in solids significantly affects their vibrational infrared spectra. Typical effects are the broadening of bands, changes in their intensity and polarization properties. The appearance of some extra defect-induced bands of *local vibrations* and *quasi-local vibrations* is also possible. Under noticeable structural disordering, which is typical for amorphous materials, the spectrum appears as a set of broad overlapping bands, and thus reflects the features of the phonon *density of states.*

Electronic excitations in *insulators* and *semiconductors* reveal continuous, wide-band, or discrete spectral dependences of optical constants in the infrared range. Narrow-band spectra are due to transitions between discrete energy levels, e.g., for impurity centers and *excitons.* The *current carrier* excitations, where the initial or final state corresponds to an energy band, appear as frequency bands whose shape characterizes the band density of states. Under excitations within a single band, there is a monotonic variation of the spectral characteristics with the wavelength of the infrared radiation. The analysis of infrared spectra provides data on the energy spectrum of the excitations, the type and the concentration of impurity atoms, the concentration and the *effective mass* of free current carriers, their scattering by phonons, impurities, and other defects. The experimental research is generally performed with the help of spectrophotometers, or tunable sources of monochromatic infrared radiation such as *lasers*.

INHOMOGENEOUS BROADENING

Spectral line broadening resulting from the spread of resonance frequencies from various atoms. When resonance frequencies lie close to each other there appears an unresolved spectrum, with an envelope having the shape of an inhomogeneously broadened line. This shape is a convolution of a *uniformly broadened line*, i.e. a shape obtained by integrating the spectral line produced by an isolated atom or a group of atoms with the same resonance frequency over a spectral distribution of resonance frequencies. In the case of *magnetic resonance*, individual components of the line contour are called *spin packets*, and a set of *spins* in identical local fields contributes to each such packet. The model of spin packets is particularly useful when describing interactions of a spin system with a microwave field, its frequency ω_p fitting into the contour of an inhomogeneous line of *electron paramagnetic resonance*. According to such a model, only spins belonging to the packet of frequency $\omega_i = \omega_p$ are directly affected by the field. If one scans a strongly saturated electron paramagnetic resonance line using an additional weak microwave frequency there appears a characteristic dip or trough in that line, a "burned-out hole" centered at the frequency ω_p (see *Hole burning*). The width of the hole depends on the extent of the overlap of the neighboring packets, and on processes spreading the saturation along the overall inhomogeneous line contour. Two mechanisms for spreading saturation are *cross-relaxation* or spectral diffusion, and packet-to-packet spin "jumps" due to fluctuations of local fields. In the first case, the saturation is transferred from the "burned-out hole" first to packets closest in frequency, and then it spreads further by the process of *diffusion*. The role of the characteristic length in frequency space is played by the quantity $\Delta_D = (T_1 D_\omega)^{1/2}$, where T_1 is the *spin–lattice relaxation* time, and D_ω is the spectral *diffusion coefficient*. The *diffusion spin length* Δ_D defines the frequency range over which the spin excitation spreads within its lifetime (until absorbed by the lattice). This value also determines the half-width of the "burned-out hole" in the case of spectral diffusion. The second mechanism makes it equally probable for an excited spin to enter any packet. The "burned-out hole" does not broaden with time; but as the microwave power grows, the whole line saturates evenly.

The spread of resonance spin frequencies that produces inhomogeneous broadening may be related to microscopic inhomogeneities of the crystal, to the presence of defects and impurities (all result in random changes of the internal field), or to a *hyperfine interaction* of electrons with spins of the surrounding nuclei. Nonuniformity of the external magnetic field, the polycrystalline nature of the sample, the presence of slightly disoriented macroscopic blocks in an imperfect *monocrystal* (so-called mosaic structure), and the dipolar interaction between spins with differing Larmor frequencies (see *Larmor precession*) all result in inhomogeneous line broadening. To calculate the shape of a spectral distribution of resonance frequencies, the *statistical theory of line shape* is sometimes used. See also *Homogeneous broadening*.

INJECTION of current carriers

Introduction of nonequilibrium (excessive) current carriers into a semiconductor (insulator) from an external source that could be either a contact (semiconductor or metal contact) or an incident electron beam. The remaining semiconductor bulk may be activated by a limited region (usually a subsurface one) of enhanced generation of carriers by light, fields, etc.

Injection from a contact arises when an external electric field reduces the contact potential barrier compared to its equilibrium value that is the *contact potential difference*. When two solids with different work functions Φ_1 and Φ_2 ($\Phi_1 > \Phi_2$) are in contact, there is an exchange of electrons. The subsurface layers of both bodies become charged, the solid with Φ_1 being charged negatively and the solid with Φ_2 positively. These charges produce a potential barrier whose height at thermodynamic equilibrium is $(\Phi_1 - \Phi_2)/|e|$. The equilibrium barrier field produces drift fluxes of electrons and holes that compensate exactly for their diffusion fluxes. If the external field is directed against the barrier field, it reduces the barrier height so the drift fluxes become less than the diffusion ones, thus inducing the flux of excessive electrons directed into the solid with Φ_1, while those of holes are directed into the solid with Φ_2. Injection efficiency of the contact with a semiconductor is described by the *injection factor*. This is the ratio between the current of carriers entering the semiconductor from the contacting conductor through the near-contact potential barrier and the total current through the contact. The injection factor of a perfect injecting contact is close to unity.

Under *monopolar injection* from the contact an excess of charge carriers in the semiconductor bulk grows with time. It produces a field that precludes further penetration of carriers into the semiconductor, and it also makes the potential barrier higher and reduces the injection. The charge stops growing when the injection ceases completely. If there is a second contact through which the carriers can drain from the semiconductor then stationary monopolar injection is possible. Its level is controlled by the competition between the injection-inducing external field due to the voltage applied to the semiconductor, and the anti-field produced by the excessive carrier charge. Hence, in a semiconductor with a high free-carrier concentration one can obtain a high-level injection (significant concentration growth in the whole semiconductor) only for strong external fields in samples with very small distances between the contacts. If the equilibrium concentration is small due to low *doping*, the conditions needed to produce stationary monopolar injection are made easier.

This conclusion is, however, not valid for semiconductors with a high charged impurity concentration (i.e. *donors* and *acceptors* in equal numbers compensating each other); here low free-carrier concentration is due to weak *generation of current carriers* from impurity levels to allowed bands (at low temperature and/or due to high impurity depth). For semiconductors with low generating rates the appearance of even small (comparable to equilibrium) concentrations of excessive carriers leads to the accumulation of a large charge of excess carriers at impurities due to the increase of the *trapping* rate; and this charge inhibits injection. The accumulation of a large impurity charge is very prolonged, running slower for a lower free carrier concentration. Therefore, in these semiconductors efficient nonstationary injection is possible even for quite slow processes; such injection noticeably increases the material conductivity compared to its stationary value. The excessive carriers are able to penetrate into the semiconductor depth up to the length determined by *current carrier drift* during trapping.

If one of the contacts supplies the semiconductor with excessive electrons and the other one with holes, then *bipolar (double) injection* occurs. In this case, the excessive minority carriers introduced into the semiconductor stimulate, by their field, the influx of excessive majority carriers up to almost complete neutralization of their charge. Due to the lack of charge hindering the injection, efficient stationary double injection is easily achieved even at high semiconductor conductivity. The excessive carriers under stationary conditions can also penetrate into the semiconductor depth a long distance from the contacts, their path being determined by their lifetime. The penetration depth in weak fields is equal to the *ambipolar diffusion* length; this depth in strong fields is determined by the *ambipolar drift*.

One should differentiate injection from a related phenomenon called *accumulation*. The operation of most *semiconductor devices* is based on injection.

INJECTION LASER

The commonest kind of *semiconductor laser* (*diode laser, laser diode*). Its particular feature is current carrier injection through a nonlinear electrical contact (e.g., the *semiconductor junction*) used as the pumping mechanism. The advantages of this type of laser are the small size of its active element, simplicity of its design, economy of operation (efficiency up to 50%), possibility of direct modulation of radiation over a wide frequency range ($\sim 10^{10}$ Hz), simplicity of frequency tuning and selection of the *active medium* to cover a wide wavelength range (0.6–45 μm), durability (to 10^6 h), low-voltage power supply (1–2 V), compatibility with other *semiconductor devices*, and possibility of its incorporation in integrated circuits. A simple version of an injection laser is a crystal *diode* (Fig. 1) with two plane parallel faces forming a plane laser resonator. So-called

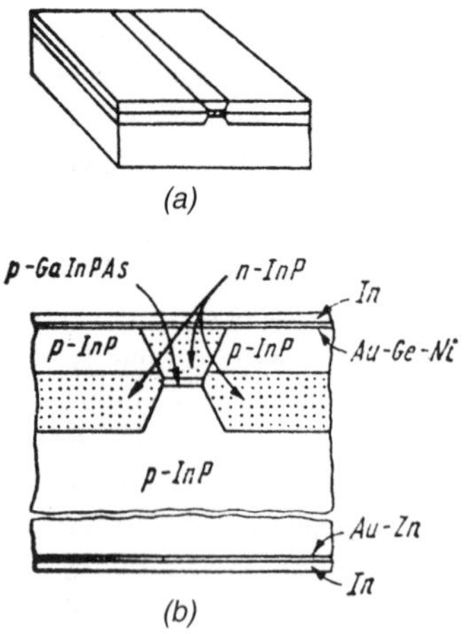

Fig. 1. Diagram of a crystal diode (a) and cross-section (b) of a stripe active region (p-InGaAsP) in a nucleating heterostructure based on InGaAsP/InP.

stripe-geometry injection lasers are the commonest. The active medium in these lasers forms a stripe reaching from one mirror to another, with an active volume 10^{-11}–10^{-10} cm^3. Power generation occurs when a direct current is passed through the diode. The threshold current in the lowest conductive injection lasers is 5–10 mA at room temperature. A typical radiation power of the stripe-geometry injection laser operating in a single-mode regime is 1 to 10 W. In some phased many-stripe arrays (gratings), the continuous radiation power reaches 1 W and more. In the pulse mode, an injection laser with a wide multimode active medium can emit radiation up to 100 W in power. Active media for injection lasers, so-called *direct-gap semiconductors*, are binary compounds and their *solid solutions* of $A^{III}B^{V}$ and $A^{IV}B^{VI}$ types (Fig. 2). The commonest injection laser called a *heterojunction laser* employs a design based on a *heterogeneous structure* formed by a combination of *heterojunctions* and *p–n* junctions. Available commercial injection lasers are based on heterosystems GaAlAs/GaAs (wavelength 0.78–0.85 μm) and InGaAsP/InP (wavelength 1.06–1.67 μm). Lasers operating at shorter wavelengths are used in disk systems to read (in some cases, to record) audio- and video-information. The injection laser radiation is focused to a sharp optical "needle" (beam cross-section is less than 1 μm^2)

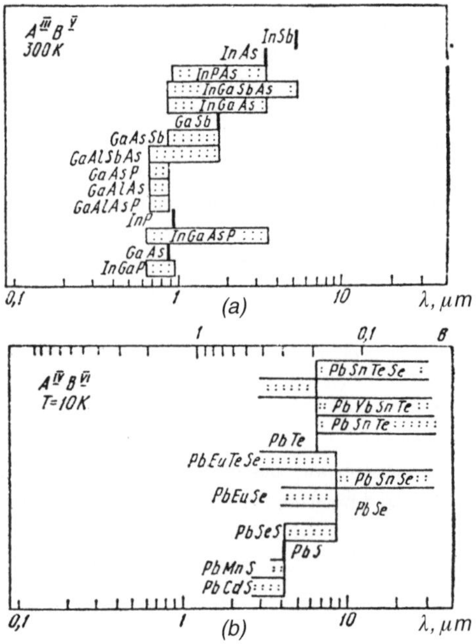

Fig. 2. Spectral ranges covered by injection lasers based on binary compounds and solid solutions: (a) $A^{III}B^{V}$ and (b) $A^{IV}B^{VI}$.

thus enabling the retrieval of information recorded with high density. Injection lasers operating at longer wavelengths, in particular at wavelengths 1.3 μm and 1.55 μm, find wide use in fiber-optic communication systems for data transmission, including telecommunication. The latter application is possible due to the high transparency of the fiber light guides at the wavelengths indicated above. The transmission range without retransmission is 30 to 200 km depending on the laser power, the frequency range, and the reception mode (either direct photodetection or optical heterodyning). In demonstration experiments at a long range greater than 100 km, the frequency band could as wide as 1 GHz. Other injection lasers include diodes based on $A^{IV}B^{VI}$ compounds (see *Semiconductor materials*) such as PbSnTe and other mixed crystals. These adjustable injection lasers serve as sources for high resolution spectroscopy in the IR range. They are also used to detect harmful impurities in air, and to find their individual concentrations by measuring the optical absorption at characteristic wavelengths.

INORGANIC POLYMERS

Simple and complex inorganic substances consisting of atoms or groups of atoms held together by a continuous system of *covalent bonds* (nonpolar, or with polar component) or coordination bonds which form a three-dimensional spatial, two-dimensional laminar, or one-dimensional chain structure. The principal structure-forming factor is the skeleton, i.e. the system of atoms directly linked by *chemical bonds* to form one-, two-, or three-dimensional frameworks. Materials with ionic structures are not considered *polymers*, since the ionic bond is not directional as it arises from electrostatic attraction alone. However, there is no sharp dividing line between ionic and polymeric structures, since there is none between ionic and covalent-polar bonds (e.g., consider the progression from the ionic structure of $BaCl_2$ and $SrCl_2$ through intermediate $CaCl_2$ and $MgCl_2$ to polymeric $BeCl_2$ with extended chains of atoms). Substances with a metallic structure are also not polymers, but there is no distinct border between polymeric and metallic structures (e.g., the sequence from the fourfold coordinated covalent *diamond* structure of C, Si, Ge to the approximately six-coordinated white tin to the FCC structure of metallic *lead*). While there is a monomer existing in its individual state that corresponds to almost every carbon-containing *organic polymer*, there are no monomer analogues for most inorganic polymers that would exist under ordinary conditions. The polymer is more stable than the monomer because the compound is coordination-unsaturated in its monomer state. Polymer formation is explained by the increase of its coordination number (see *Coordination sphere*) to its optimum value. For example,

$$2n[\text{I–Hg–I}] \rightarrow \left(\begin{array}{cc} \diagup \text{I} \diagdown & \diagup \text{I} \diagdown \diagup \\ \quad \text{Hg} & \text{Hg} \\ \diagdown \text{I} \diagup & \diagdown \text{I} \diagup \diagdown \end{array} \right)_n .$$

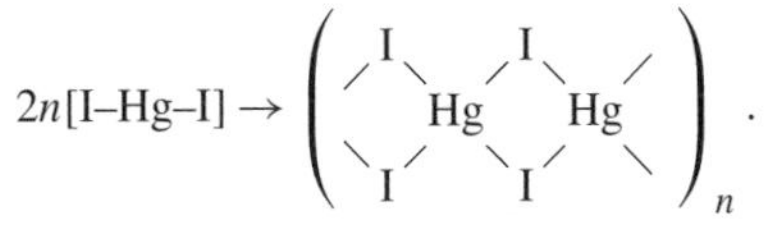

Three- and two-dimensional structures are more common among inorganic polymers, which distinguishes them from organic polymers in which one-dimensional chain structures play the dominant role. Inorganic polymers are classified as homochains, with skeletons consisting of atoms of the same element, and heterochains, with atoms of different elements (usually there

are two of them) alternating in the skeleton. In contrast to organic polymers, heterochain structures are more often found in inorganic polymers. Many simple substances, such as plastic sulfur, selenium, tellurium, black and red phosphorus, arsenic, graphite, diamond, carbyne, silicon, boron and others feature homochain polymeric structures. Among them only sulfur, selenium, and the carbyne allotropic variety of carbon form one-dimensional chain structures. Most binary compounds, halides, chalcogenides, nitrides, phosphides, arsenides, carbides, silicides, borides, many minerals, and other substances feature a heterochain polymeric structure.

The spatial arrangement of the skeletons of inorganic polymers determines their physical properties. Polymers with a chain structure (elastic sulfur, phosphonitrylhalides, etc.) are close to organic polymers in certain properties (e.g., elasticity). Polymers with laminar or three-dimensional structures typically exhibit pronounced hardness and brittleness, as well as high softening, melting and decomposition temperatures (e.g., diamond, quartz, etc.).

Information concerning the polymeric structure of compounds can be obtained directly from X-ray structural analysis and various spectroscopic techniques, and also from indirect data on their physical and chemical properties, as well as from theoretical deductions based on the nature of their chemical bonds.

INSTABILITY OF SUPERCONDUCTORS

See *Structural instability of superconductors*.

INSTANTANEOUS AXES

Axes used to describe an arbitrary displacement of a rigid body. According to *Chasles' theorem* the most general displacement of a rigid body can always be carried out by a translation plus a rotation through a certain angle about a certain axis. Continuous motion may be expressed as a sequence of infinitely small translations plus rotations about corresponding axes called instantaneous axes. If a body rotates about a fixed point 0, the totality of instantaneous axes forms a cone with its vertex at 0. It is convenient to imagine two reference frames: one at rest, usually aligned along the angular momentum, and the other fixed in the rigid body, and usually aligned along a principal axis of the body. The totality of instantaneous axes forms cones in both spaces; the motion of the solid being a result of the rolling of the body cone about the space one, the two cones being in contact along instantaneous axes at every instant of time.

INSULAR FILMS, granular films

Systems of non-contacting *small particles* on a *substrate*, obtained by spraying or by *dispersion* within thin continuous films as a result of heating. Of interest are electrophysical, emission, optical, and magnetic properties of granular or insular films. The electrical conductivity of these films, which is of a tunneling nature, is sensitive to minor changes in the distances, and the potential barrier height between the islands or grains. This property is employed to produce strain gauges, vapor pressure pickups for various substances, and temperature-sensitive elements. By changing the substrate and island materials, and the morphology of the islands, one can obtain positive, negative or even zero temperature coefficients of resistance. The system of islands covered by the adsorbate can provide an N-shaped *current–voltage characteristic*, and thus switching and storage elements. The films where the island size varies monotonically along a coordinate axis exhibit conductivity asymmetry. Small metal particles in close thermal contact with an insulator substrate are capable of withstanding high power flux without destruction. This property, in combination with the reduction of the magnitude of the *electron–phonon interaction* as the particle size decreases, allows heating the electron gas of small metal particles to incite *hot electrons*. The appearance of hot electrons (during passage of a current or under laser irradiation) causes electron emission (see *Hot electron emission*) and glow of such films. During the transition from continuous films to granular or insular ones, optical constants change, the dynamic *polarizability* increases, and the *magnetooptical effect* becomes stronger.

INSULATING MATERIALS, electrical insulating materials

These are *insulators*, or dielectric materials with very high resistance, which are used to isolate conductors electrically, as well as capacitor elements in electrical circuits. They are employed in high-voltage generators, transformers, cable manufacturing, etc. Examples of good electrical insulating materials are resin, wax, varnish, various *polymeric materials*, *ceramics*, *mica*, and so on. The presence of defects in a real dielectric together with thermal excitations determine the final value of resistance, and hence there can exist a small current leakage through an insulating material. The resistivity of electrical insulating materials reaches 10^{18}–10^{19} $\Omega\cdot$cm. These materials are also specified by the value of their *dielectric strength*, that is the critical electric field E_{th} at which *breakdown* occurs. The value of E_{th} in solid electrical insulators is 10^4 to 10^5 V/cm. Special requirements are imposed on the insulating materials employed in *microelectronics* (sapphire, quartz, etc.) in connection with very thin insulating layers ($\approx 10^{-6}$ cm).

INSULATOR POLARIZATION

See *Polarization of insulator.*

INSULATORS, dielectrics (fr. Gr. $\delta\iota\alpha$, through, and $\eta\lambda\epsilon\kappa\tau\rho o\nu$, amber)

Term introduced by M. Faraday for substances into which an external electric field penetrates. If the material contains free current carriers the applied electric field $\boldsymbol{E}$ causes them to move, and they establish a spatial charge distribution that compensates for the field within the conductor. In the absence of the free carriers only partial compensation of the field is possible, caused by small displacements of the bound charges (see *Polarization of insulator*). Thus the other definition of insulators as materials with low *electrical conductivity* is closely connected with the first one. Materials with resistivity in the range 10^8 to 10^{17} $\Omega\cdot$cm (i.e. insulators) are referred to as dielectrics in the Russian literature. Many *solids* as well as liquids and gases possess the properties of insulators. In real insulators there are always some free charge carriers, but at their low concentration their screening of the electric field appears to be weak. Low concentrations of free electrons in crystals with large *band gaps* $E_{\mathrm{g}} \geqslant 3$ eV (i.e. $E_{\mathrm{g}} \gg k_{\mathrm{B}}T$) produce a low value of the conductivity. The efficiency of compensation of the external field $\boldsymbol{E}$ in insulators is determined by the *dielectric constant* (also called permittivity) ε of the substance; thus in Coulomb's law

$F = e_1e_2/(4\pi\varepsilon r^2)$ the charges e, placed into the dielectric medium, could be replaced by effective charges $e^* = e/\sqrt{\varepsilon/\varepsilon_0}$.

Based upon the polarization mechanisms insulators are subdivided into three classes:

(1) *nonpolar insulators*, where only polarization of the electronic shells is present (crystals of O_2, inert gases, diamond, etc.). At frequencies far from those of resonance absorption ε does not depend on the frequency and $\varepsilon/\varepsilon_0 = n^2$ (n is the *refractive index*);

(2) *ionic insulators*, which consist of oppositely charged ions capable under the effect of the electric field to undergo only small displacements from their equilibrium positions (*alkali-halide crystals*, oxides, sulfates, metal phosphates etc.). For them there is a significant difference between the static and high-frequency dielectric constants;

(3) *polar insulators* or *dipole dielectrics* consisting of polar molecules (having intrinsic *dipole moments*). Another class of crystals has two or more different equilibrium positions for ions in the *unit cell*. This group contains water (in the solid and liquid state), solutions of polar molecules, electrolytes; materials of the second group with lattice defects, etc. For these dielectrics the value of ε is associated with the dipole moment by the Onsager relations (see *Onsager theory*). At low temperatures reorientation of the dipole moments is "frozen" and the behavior of the dielectric constant becomes similar to the behavior of ε for ionic insulators.

Dielectrics or insulators of the types of *ferroelectrics*, *piezoelectrics*, pyroelectrics (see *Pyroelectricity*) possess their own specific properties. Insulators are used in quantum electronics, in particular, in *optoelectronics* for the creation of the *active laser elements*, in holographic devices, in infrared techniques, in capacitors; as insulating materials etc. (see also *Dielectronics*).

INTEGRATED CIRCUIT

A tiny solid-state electronic device produced by incorporating a given arrangement of circuit elements in a single microcrystal. An integrated circuit (IC) is manufactured on a semiconductor substrate using layer-by-layer planar technology (see *surface smoothing*) to form all the elements into a single technological unit. Layers with a given topology of circuit components (*transistors*, condensers, *diodes*, *resistors*, etc.) are sequentially formed on the surface of crystals made of materials with various electrical properties, and are arranged under one another. Combinations of such layers produce various circuit elements. After forming the multilayer structure the circuit elements are interconnected with a surface conductive layer of the required configuration. The basic material for chip production is *silicon* single crystals as well as *gallium arsenide* single crystals featuring high carrier mobility. Materials used as insulator layers include silicon dioxide SiO_2, sapphire Al_2O_3, spinels AB_2O_4, *iron garnets*, and others. According to the application, the IC can be divided into digital and analogue types. Depending on the packing density (number of devices per unit area on a crystal), the integrated circuits can be classified into *small-scale IC* containing 10–12 devices on a crystal with the surface area within a mm^2 or cm^2; *mid-scale IC* containing 12 to 100 devices; *large-scale IC* containing more than 100 devices; *very-large-scale IC* that contains 10 000 and more digital devices. The ICs are used for microminiaturization of radioelectronic many-purpose systems like microcomputers for automatic control (in space complexes, in robotics) as well as for data transmission and data processing. See also *Microelectronics*.

INTEGRATED OPTICS

A branch of optoelectronics oriented to designing solid-state devices for optical information processing that are manufactured with the help of planar technology (see *Surface smoothing*) developed for *integrated circuits*. The application areas of these integrated optics are optical selectors, retransmitters, commutators, spectral multiplexes of optical channels in fiber-optic communication lines, spectral analyzers in radio-beam systems, optoelectronic and fiber-optic sensor devices, integrated optoelectronic circuits, planar optoelectronic instruments and systems.

The constructive base of integrated optics (see Fig. 1) is the *dielectric waveguide* (DW) that can be planar (a), three-dimensional (b) or stripe (c).

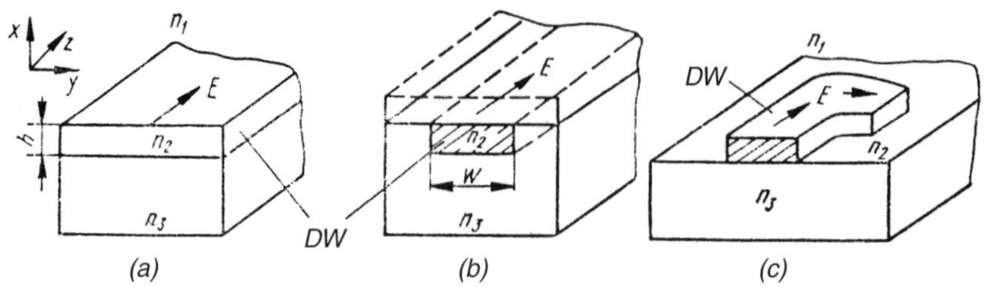

Fig. 1. Dielectric waveguides (DW).

Three-dimensional dielectric waveguides provide additional limitation of the optical radiation in the traverse direction. This increases the integration level and efficiency of integrated optics circuits and simplifies the interfacing of the dielectric waveguides to other integrated optics components and the fiber-optical cable (see *Fiber optics*). The dielectric waveguide is an optical line with low loss ($\leqslant$1 dB/cm) at the operating radiation wavelengths (from 0.62 μm to 1.6–2 μm and more). The waveguide modes, e.g., the fundamental modes of integrated optics devices, are solutions to the wave equation

$$\nabla^2 E(\boldsymbol{r}) + k_0^2 n_2^2(\boldsymbol{r}) E(\boldsymbol{r}) = 0, \quad k_0 = 2\pi/\lambda_0,$$

in the form

$$E(\boldsymbol{r}, t) = E(x, y) \exp\bigl[\mathrm{i}(\omega t - \beta z)\bigr],$$

where $\beta = \omega/v_{\rm ph}$ is the propagation constant; ω and $v_{\rm ph}$ are the angular frequency and the *phase velocity* of the electromagnetic wave; n_2 is the *refractive index* of the dielectric waveguide carrier layer; and $\lambda_0 = 2\pi c/\omega$ is the wavelength in free space. A waveguide mode with the field varying harmonically inside the dielectric waveguide and decaying exponentially in a transverse direction outside of the waveguide satisfies the condition $n_2 < n_1, n_3$ (see Fig. 1). The absence of the radiation in the substrate (as a rule, $n_3 \geqslant n_1$) for a given mode configuration is controlled by the following inequality:

$$\Delta n = n_2 - n_3 > \frac{(2m-1)^2 \lambda_0}{32 n_2 n_1^2},$$

whereby we have $\Delta n \approx 10^{-2}$–10^{-3} for the basic modes and $\lambda_0 = 1$ μm. The bending of a dielectric waveguide introduces some extra loss (0.1–0.08 dB/rad for glasses), which restricts the permissible relative bending radii R/λ that are the greater, the smaller $\Delta n/n_2$. Level frequency and phase characteristics of a dielectric waveguide are easiest to achieve when using a single waveguide type of transmission. Single- and few-mode regimes limit the transverse dimensions of a dielectric waveguide (its format) to the order of the of transmitted radiation wavelength (0.3–3 μm in height, 1–5 μm in width), and they should be maintained within the error of a few hundredths of μm. A particular waveguide mode is ensured by preset refractive index profiles. There are dielectric waveguides with stepwise and continuous (graded) refractive index variations. In a gradual microwaveguide it is easier to provide single-mode operation that results in much less loss. At a given frequency the number of waveguide modes is limited by the relation

$$m \leqslant 1/a + ah/\lambda_0\bigl[2n_3(n_2 - n_1)\bigr]^{1/2},$$

where $a = 2$ and 4 for the stepwise and gradient dielectric waveguides, respectively.

For passive applications there are dielectric waveguides on glass substrates where a given profile $n_2(x, y)$ is produced with the help of the ion exchange method, *ion implantation*, hydrolysis, or the deposition of *thin films* (including chalcogenide and polymeric ones). To control the light beam, the waveguide material must have certain electrooptic, acousto-optic, and other active properties determining the dependence of its refractive index on the controlling action. Active substrates in dielectric waveguides are exemplified by a crystal of lithium *niobate* with the required refractive index profile in the crystal produced by the diffusion of Ti (transmission range is $\lambda \leqslant 3$ μm, with SiO_2 antireflection

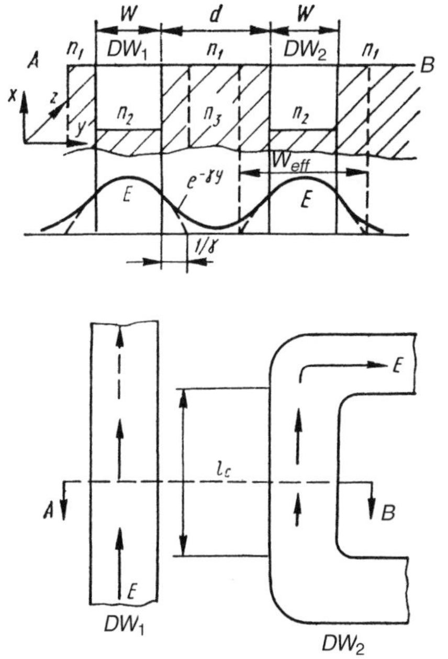

Fig. 2. Two dielectric waveguides DW_1 and DW_2 a distance d apart.

coating at end faces). There are semiconductors (mainly $A^{III}B^V$) where the waveguide layer is formed by the *space charge region* (heterostructures AlGaAs/GaAs, InP/InGaAsP and others; transmission range $\lambda \leqslant 10$ μm, Si_3N_4 antireflection coating). Radiation escape from the waveguide sets the lower bound for its cross-section: the radiation propagates along the waveguide surface so that its self-contained nature and isolation are disturbed. The same effect allows one to produce a distributed communication in the waveguide that results from the interference of symmetric (or asymmetric) waves propagating along dielectric waveguides situated at a distance d one from another (where d is of the same order of magnitude as the field screening length $1/\gamma$ outside the waveguide, see Fig. 2). Owing to the distributed communication, the pumping of the radiation power from one waveguide to another (with identical waveguide characteristics) takes place over a finite length l_c. Values of l_c (from 0.5 to 10 mm) depend on the device type, the ratio between the supplied and consumed power, topological features, and the technological spread of parameters of the distributed communication elements. Integrated optics is used in modulators, selectors, and couplers. Components for radiation input to the waveguide are prisms, tapering films (wedges), and *diffraction gratings*. The two latter are used also to establish a coupling between the integrated optics circuits and the fiber-optic cable. The formation of a given phase front of the optical beam, Fourier transformation and other operations in integrated optics are performed by planar focusing elements that form an integral whole with the dielectric waveguide. If the wave elements of integrated optics are applied to connect the optoelectronic elements (such as light sources and photodetectors), they form *integrated-optical circuits* that are the analogues of functional SHF circuits. In the latter the signal generation, transmission, detection, modulation, filtration, and transformation are carried out on the basis of various types of interaction of radiation with matter.

Of special importance for integrated optical circuits (both monolithic and hybrid) is the physical integration in addition to the technological one. Integrated optical circuits are developed mainly on a $A^{III}B^V$ semiconductor base.

INTERATOMIC INTERACTION POTENTIALS

The approximate representation of the *adiabatic potential* of an atomic system (equal to the electron energy for random but fixed values of the coordinates of their nuclei) in the form of a sum of pairwise terms depending on the distances R between the atomic nuclei.

Historically these terms initially appeared during the description of two interacting atoms considered within the formalism of the classical mechanics *two-body problem* involving the motion of two oppositely charged masses m_1 and m_2 attracted to each other by the Coulomb interaction. In this formalism the two particle problem dependence $R(x)$ is reduced to the problem of the one-dimensional motion of a single particle with the reduced mass $\mu = m_1 m_2/(m_1 + m_2)$ in the field with the effective potential energy

$$\Phi(R) = \Phi_0(R) + \frac{L^2}{2\mu R^2}.$$

Here $\Phi_0(R)$ is the pairwise potential of the interacting particles separated by the distance R, L

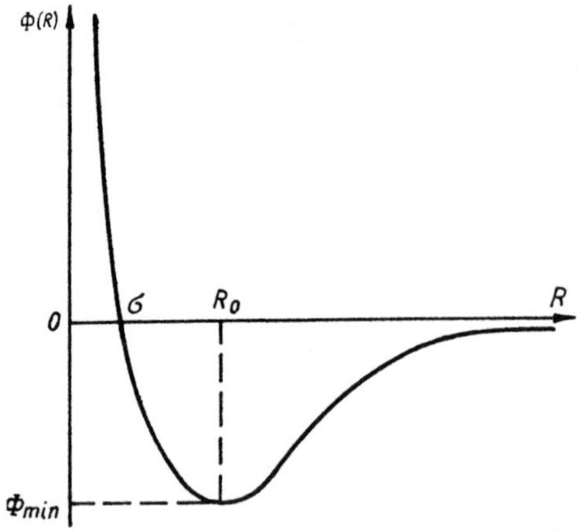

Coulomb interaction potential.

is the *angular momentum*, and $L^2/(2\mu R^2)$ is the centrifugal energy.

For the Coulomb interaction of two point charges q_1, q_2 of opposite signs we have

$$\Phi(R) = \frac{L^2}{2\mu R^2} - \frac{|q_1 q_2|}{R}.$$

At $R_0 = L^2/(\mu|q_1 q_2|)$, $\Phi(R)$ has a minimum equal to $\Phi_{\text{min}} = -[\mu/(2L^2)](q_1 q_2)^2$ (see Fig.). By generalizing the expression of the second equation, G. Mie (1903) suggested that the pairwise interatomic potential of a *solid* be written in the form

$$\Phi(R) = \frac{A}{R^n} - \frac{B}{R^m},$$

where $A, B > 0$, and $n > m > 0$.

With the development of quantum mechanics the *Mie potential* obtained a partial explanation. For instance, for the van der Waals interaction (involving polarization) it appeared possible to calculate B and to show that $m = 6$. *Crystals* constructed from atoms of inert gases, whose closed electron shells have almost no overlap, are the simplest solids. Here the interaction remains the van der Waals type, so $m = 6$; and from subsequent work it was empirically found that n can have values from 9 to 14. For convenience of calculations it is accepted that $n = 12$, and the 6–12 potential, which is called the *Lennard-Jones potential*, has the canonical form

$$\Phi(R) = 4\varepsilon\left[\left(\frac{\sigma}{R}\right)^{12} - \left(\frac{\sigma}{R}\right)^{6}\right],$$

where $\varepsilon = B^2/(4A)$, and $\sigma = (A/B)^{1/6}$. The parameter ε gives the depth of the potential well, and the parameter σ, which is found from the condition $\Phi(\sigma) = 0$, is called the repulsion diameter. For *ionic crystals*, where the repulsion is mainly caused by the exchange interaction, M. Born and J. Mayer (1932) proposed the following formula:

$$\Phi_{\text{rep}}(R) = A\exp\left(-\frac{R}{a_{\text{BM}}}\right),$$

where A and a_{BM} are constants, the latter being called the screening constant. If instead of the constant A we assume a Coulomb potential and slightly alter the screening constant, then the *Yukawa potential* or *Bohr potential* (screened Coulomb potential) is obtained:

$$\Phi_{\text{rep}} = \frac{Z_1 Z_2}{R}\exp\left(-\frac{R}{a_{\text{B}}}\right),$$

where Z_1, Z_2 are the charge numbers, and a_{B} is the Bohr screening constant. This potential is widely used in *radiation physics of solids*. The detailed description of the interaction of ions in ionic crystals gives the *Catlow–Diller–Norgett potential*. Combining the repulsive term in the Born–Mayer form with the van der Waals attraction gives the *Buckingham potential*, in terms of which the pairwise interaction in *molecular crystals* is described.

In the case of *covalent crystals* the classical *Morse potential* is used (P.M. Morse, 1929):

$$\Phi(R) = \Phi_0\big[\exp\big(-2\alpha(R - R_0)\big) - 2\exp\big(-\alpha(R - R_0)\big)\big].$$

At present for many metals the oscillating pairwise potentials that are used are found by the *pseudopotential method*. For body-centered cubic transition metals, in the past a number of different empirical potentials have been proposed, and among the best known is the potential suggested by R.A. Johnson (1964) for α-iron:

$$\Phi(R) = \begin{cases} a_1(a_2 - R)^3 - a_3 R + a_4, & R < R_{\text{m}}, \\ 0, & R > R_{\text{m}}. \end{cases}$$

Here a_i are constants found with the aid of elastic moduli, and R_{m} is a cut-off radius.

For covalent and molecular crystals interatomic potentials are also obtained on the basis of pseudopotential theory. However now it is necessary to introduce additional assumptions (e.g., the *model of charges on bonds*) in order to take into account the directionality of the interatomic bonds in these crystals. From an empirical approach, in addition to the pairwise potentials three-particle terms are introduced, depending on the angles between the bonds. By combining different angular functions with the previously considered pairwise potentials it is possible to obtain (usually in the *harmonic approximation*) the potential energy of a *crystal lattice*: *Keating potential*, *Vukcevich potential* (M.A. Vukcevich). This facilitates the calculation of properties of crystals that depend on small displacements of atoms from their equilibrium positions (elastic constants, phonon spectra). The many-parameter potential functions for describing the transition to the liquid state can be found by the same empirical approach.

INTERBAND TUNNELING, band-to-band tunneling, Zener tunneling (C. Zener, 1934)

Tunneling of electrons from the *valence band* to the *conduction band* of a semiconductor, which takes place in a strong electric field. The energy diagram of a semiconductor, shown in Fig. (a), changes in an electric field: the electron energy is augmented with the electrostatic energy of the applied field $\mathcal{E}$. As a result, the plot of the electron energy E versus the coordinate x becomes inclined as shown in Fig. (b), and electron tunneling (see *Tunneling phenomena in solids*) between states of equal energy in nearby occupied and empty bands becomes possible. The probability of tunneling is proportional to $\exp[-\pi m^{1/2}E_g^{3/2}/(2\hbar e\mathcal{E})]$, where m is the reduced *effective mass* of the electron and hole, E_g is the width of the *band gap*, and e is the electron charge. In parallel with tunneling, electron oscillations also occur inside the band, caused by reflection of electrons from band boundaries (see *Quantizing electric field*). In strong fields, interband tunneling may cause breakdown (*Zener breakdown*). Band-to-band tunneling provides the basis for *tunnel diodes*, where an electric field is induced in the region of the *semiconductor junction*.

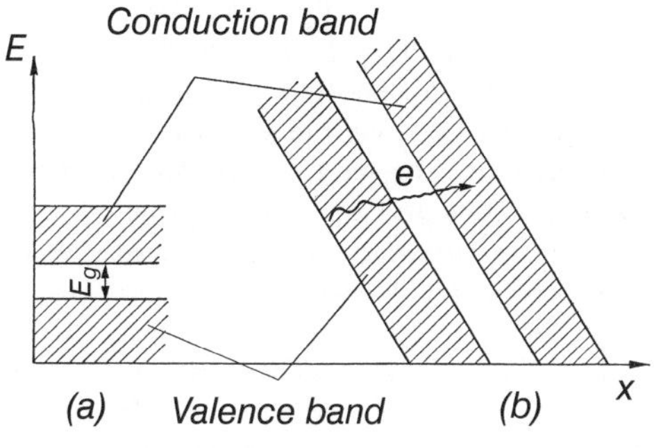

Interband tunneling.

INTERCALATED STRUCTURES

Crystals of a laminated type with *intercalants*, i.e. extra atoms, atomic groups, or molecules, introduced between the layers. The most commonly studied intercalated structures are compounds based on laminated dichalcogenides, *transition metals* compounds, and *graphite*-based compounds. Typical examples of the first type are the *metals* $K_{0.4}MoS_2$, $Fe_{0.05}TaS_2$, $Ca_x(NH_3)_yMoS_2$, and $TaS_2(Py)_{1/2}$ where Py is the pyridine (C_5H_5N) molecule. The source crystals are semiconducting MoS_2 and metallic TaS_2 or $NbSe_2$. Graphite is also easily intercalated by metals and molecules. *Intercalated compounds* possess physical properties that often differ greatly from those of source crystals and intercalants due to electron transfer from the intercalated atoms to the layers. For example, MoS_2 is a semiconductor, but upon intercalation with *alkali metals* it becomes a metal, and superconductivity has been observed below 6 K (see *Superconductivity*). In a similar way, neither graphite nor alkali metals exhibit superconductivity, but the intercalated compounds superconduct below 5.5 K, due to the appearance of new electron states occupied by electrons that had left the alkalis. At the same time, transition metal dichalcogenides intercalated by molecules exhibit very little change in their properties if these properties arise from electron motion inside the layers. However, as a rule, intercalated compounds differ from source crystals by a stronger anisotropy of the electronic and mechanical properties. For example, in $TaS_2(Py)_{1/2}$ the electrical conductivity across the layers at 20 K drops by 5 orders of magnitude compared to the source crystal TaS_2, while this conductivity along the layers is only two times lower due to reduction of the carrier concentration at the layer spacing. Therefore, intercalated compounds provide so-called *quasi-two-dimensional systems* (see *Quasi-two-dimensional crystals*) where the electron motion is close to being two-dimensional. The intercalated structures are also interesting from the physical viewpoint since they provide the potentiality to obtain new superconductors with a possible hypothetical *exciton mechanism of superconductivity.* These structures admit the existence of combinations of layers with different electronic properties in the same crystal, e.g., metallic layers associated with semiconducting ones, or superconducting layers with magnetic ones ($Fe_{0.05}TaS_2$ is such a system). In this sense these structures are similar to metallic *superlattices.*

INTERCALATION

Introduction of extra atoms, or atomic groups, or molecules (*intercalants*) into the space between crystal layers. Usually intercalation is found in crystals with laminated structures and weak chemical or van der Waals binding of layers (see *Van der Waals forces*). Easily intercalated structures are exemplified by dichalcogenides of transition metals of the TaS_2 type with mainly van der Waals interactions between the layers, as well as by *graphite.* The intercalation process could be carried out by immersion of the source crystal into the gas or liquid phase of intercalants. During the intercalation procedure, the layers of the original crystal move apart to accommodate the intercalants, and the expansion can be appreciable. For example, in the compound TaS_2(octadecylamine)$_{1/3}$ the TaS_2 layers are 0.3 nm thick, are spaced 0.3 nm apart, and separate to a distance of 5 nm to accommodate two layers of decylamine linear molecules 2.5 nm long. The intercalation process is energetically favorable because the bond between the intercalants and the layers proves to be stronger than that of the layers between each other. Ordinarily this is due to partial or complete transfer of electrons from the intercalants to the layer. The intercalants form regular structures between the layers, with the structures depending on the extent of the intercalation. For example, many structures are known for graphite, corresponding to different stages of intercalation.

INTERCRYSTALLINE CORROSION

Selective *corrosion* of boundary layers of grains (*crystallites*) of metallic materials, which spreads inside the metal along intercrystalline boundaries; a kind of *local corrosion.* The tendency toward intercrystalline corrosion often results from heating which causes microstructural transformations along *grain boundaries.* Extra phases (often *chromium*-based *carbides*) are generated, enriched with certain alloy components, and next to them bands of solid solution are formed, which are depleted of the above-mentioned components (often Cr). The most typical intercrystalline corrosion occurs near welding seams (so-called *knife-line corrosion*). Depending on the properties of the hostile medium and the chemical composition of structural components along the grain boundaries, intercrystalline corrosion may develop in the form of selective corrosion of either depleted bands, excessive phases, or both. The mechanical properties of a metal that shows a tendency toward intercrystalline corrosion, or has been exposed to it, become degraded.

INTERCRYSTALLINE FAILURE

Nucleation and propagation of the main crack (*Griffith crack*, see *Griffith theory*) along boundaries of adjacent *crystallites. Grain boundaries* or cell boundaries (see *Cellular structure*), and those of twins (see *Twinning of crystals*) and martensitic crystals, could serve as interfaces between crystallites. The best studied is the failure along grain boundaries in polycrystalline alloys. In limiting cases, intercrystalline breakdown may occur in either a brittle manner (see *Brittle failure*), i.e. without formation of a strain relief on the surface, or in a tough or viscous manner (see *Tough failure*) according to the mechanism of nucleation and merging of pits over a sufficiently wide range of temperatures and *strain* rates. The physical cause of intercrystalline breakdown is the decrease of metal–vacuum *surface energy* γ_0 during crack formation at an interface in a *polycrystal* by the value γ_b of the energy of existing boundaries. In the presence of surface-active impurities (see *Surface-active agents*) or doping agents (see *Doping*) that cause weakening of the binding forces at boundaries, the quantity γ_0 changes to γ_c. Here, γ_c is the surface energy of the crack that has crossed the interface which existed in the crystal, and where *segregations* of dissolved atoms have significantly changed γ_0 and, accordingly, the surface energy of the intercrystalline failure $\gamma_{\text{eff}}^{\text{b}} = \gamma_c - \gamma_b/2$. The impurity *doping atoms* nitrogen and *oxygen* in body-centered cubic transition metals noticeably reduce the value γ_0. In a similar way, phosphorus and sulfur affect *iron*, sulfur affects *nickel* and *copper*, etc. In deformed metals the intercrystalline breakdown along cell boundaries (see *Cellular structure*) can result from the increase of γ_b due to the increased density of unrelaxed *dislocations* at the *subboundary*. The value $\gamma_{\text{eff}}^{\text{b}}$ decreases sharply in this case. The *strength* σ_{cr} under intercrystalline breakdown is described by the dependence $\sigma_{\text{cr}} \geqslant \beta\varepsilon\gamma_{\text{eff}}^{\text{b}}/d$, where β is a constant; ε is the strain related to the amount of mobile dislocations; and d is the structural element size.

The relationship between the *cold brittleness* temperatures under *transcrystalline failure* and intercrystalline failure can be illustrated by a diagram (see Fig.). *Alloying* the *solid solution* causes a monotonic increase of the cold brittleness temperature $T_{\text{c.g.}}$ of the host grain. At small contents of surface-active agents (especially interstitial impurities of body-centered transition metals), the sharp increase of the cold brittleness temperature $T_{\text{c.b.}}$ occurs near boundary regions due to the appearance of segregation (Fig. from 1 to A). The

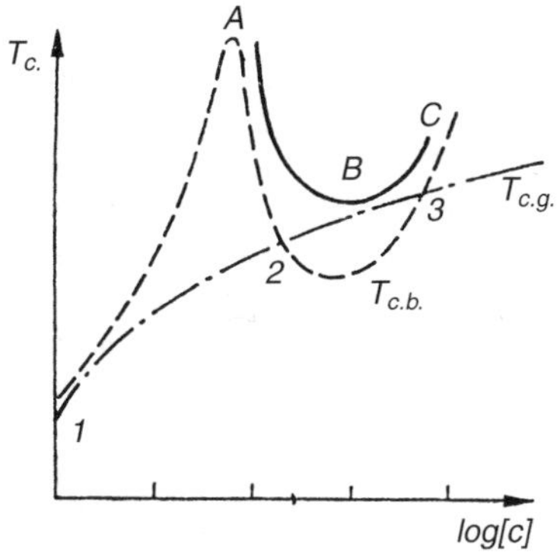

Sketch of variation of temperatures of intercrystalline ($T_{\text{c.b.}}$) and transcrystalline ($T_{\text{c.g.}}$) embrittlement under doping with carbon.

increase of the interstitial impurity content (or change of limit of possible oversaturation at *heat treatment* or *thermomechanical treatment*) causes the formation of *disperse structures*. The latter hinder the nucleation of *intercrystalline cracks* which results in a decrease of $T_{\text{c.g.}}$ (from A to 2). Along line segment 2 to 3 brittle failure either reverts to transcrystalline (dashed curve with crossover in Fig.) or remains intercrystalline (upper solid line curve with no intersection), but the cold brittleness temperature unambiguously decreases. A pronounced rise in temperature takes place at concentrations above point 3 when structure-free layers of excessive brittle *phase* (interstitial phase, *intermetallic compounds*, etc.) appear at the grain boundaries, these layers having very high cold brittleness temperatures. The formation of continuous interlaying of the second phase along grain boundaries embrittles the alloy in a deleterious manner. At the decline of γ_c ($\gamma_{\text{eff}}^{\text{b}} \to 0$), failure occurs below the *yield limit* of the alloy. This phenomenon is called *intercrystalline brittleness* (such as oxygen-induced brittleness of polycrystalline *molybdenum alloys*, *steel* temper brittleness, antimony-induced brittleness of copper, etc.). Doping and heat treatment can stimulate impurity redistribution between the grain bulk and the grain boundary, thus changing both the effective surface energy of formation of an intercrystallite crack, and the conditions for crack *relaxation*. This, in turn, causes a change in the cold brittleness temperature at the boundaries and in the bulk of the grains.

INTERDIFFUSION

Mutual penetration of atoms and molecules across an interface of contact between two materials (solid, liquid, gas) of different compositions, caused by thermal motion of the atoms or molecules.

Interdiffusion is established with respect to the displacement of the front of constant concentration. Because of counter-diffusion flow the mechanical equilibrium in the system is disturbed, but then it is immediately restored by the mechanical shift of layers, and this reestablishes the front of constant concentration (see *Kirkendall effect*). The theory of interdiffusion for two-component solid metallic systems takes into account the Kirkendall effect and the vacancy mechanism of *diffusion*. The two resulting atomic fluxes are equal in magnitude, opposite in direction, and proportional to the atomic concentration gradient. The proportionality ratio is called the *interdiffusion coefficient* D. This coefficient depends on the temperature, pressure and composition. The dependence of D on the temperature follows approximately, but not precisely, the *Arrhenius law*: $D = D_0 \exp[-Q/(k_B T)]$, where D_0 is a constant, and Q is the *activation energy* of interdiffusion. The dependence of D on the composition is different for different systems, and has been especially investigated under isothermal conditions. It is often possible to determine the distribution of one of the components (so-called *concentration curve*) in the *diffusion zone*, and to find the functional dependence on the concentration with the aid of a special calculational method (*Matano–Boltzmann method*). The theory gives the relationship between D and the partial diffusion coefficients D_1 and D_2 (see *Heterodiffusion*) of the two components: $D = C_1 D_2 + C_2 D_1$, where C_1, C_2 are atomic concentrations of the components. This same relation is also valid for liquids and gases. The shape of the concentration curve depends of the *phase diagram* of the two-component system. If at the given temperature the whole diffusion region corresponds to a continuous series of *solid solutions* then the curve is continuous, while if there are different *phases* then the curve is piecewise continuous, with each region corresponding to a separate phase. In the latter case during the course of interdiffusion the layers of different phases appear and grow, and two-phase areas are absent. For polycomponent systems the investigation of diffusion becomes considerably more complicated, and there exist only approximate ways for calculating D in three-component systems.

INTERFACE

See *Condensed matter interface*, *Metal–insulator interface*, *Phase interface*.

INTERFERENCE COATINGS

Single- or multilayer thin-film systems applied to the surface of solids to decrease or increase the *reflectance* through interference effects. They are

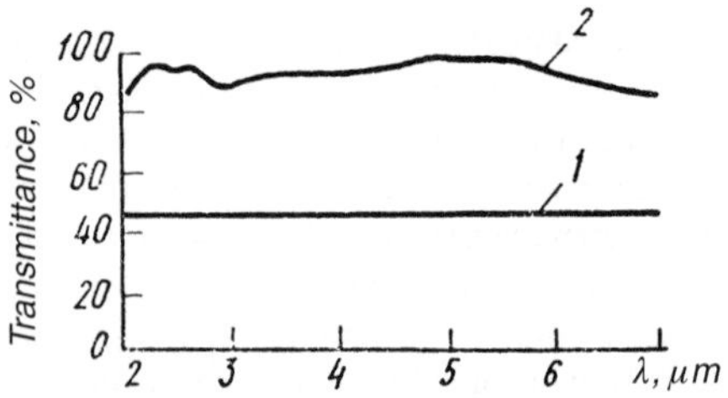

Transmittance of Ge plate without antireflection coating (1), and with three-layer antireflection coating of $Si + CeO_2 + MgF_2$ films (2). All layers are $\lambda/4$ thick for $\lambda = 3.5$ μm.

used as antireflection coatings, for producing dielectric mirrors with high reflection coefficients, interference filters, radiation polarizers, and beam splitting systems. Of widest use are multilayer systems with alternating layers of transparent materials with low and high *refractive indices*; the optical thickness of such layers equals or is a multiple of $\lambda/4$. Computer aided design allows developing interference coatings having layers of unequal thickness with more efficient parameters. Application of antireflection coatings is especially needed in multicomponent objective lenses where the reflection loss can reach 70%, and also in semiconductor receivers and sources of radiation made of high-refraction materials (see the figure).

Multilayer dielectric mirrors (unlike metallic ones) reflect almost all the incident radiation over a broad spectral range without absorption loss. Such mirrors are indispensable for manufacturing Fabry–Perot type resonators in laser engineering, because metal mirrors degrade easily at high radiation intensity. Two *dielectric mirrors* with a layer of a high-refractive index material (with optical thickness a multiple of $\lambda/2$) between them compose an *interference light filter*. Light beam splitters are formed from the same dielectric mirrors with a reflection coefficient close to 50%, and variable over a broad range in special cases. The splitters are used to divide a light beam into two beams with intensities equal to each other, or in a certain ratio.

INTERFERENCE, QUANTUM

See *Quantum interference phenomena*.

INTERFEROMETER

An instrument operating on the basis of the interference of light. The incident light beam in an optical interferometer is divided in two (or more) beams that propagate along different optical paths, and then are reunited in the plane where the interference occurs. A crystal or other substance is placed in the path of one beam, while a comparison standard is in the other beam. By measuring the path difference Δ between the interfering beams, one can obtain data on surface defects of the *crystals* under study, their optical homogeneity, their *dielectric constant* ε or *refractive index* $n = \sqrt{\varepsilon}$. Interferometers of various designs are used to measure geometrical dimensions, such as the diameter of balls, the surface roughness of optical components, semiconductor washers, etc.

INTERFEROMETER, FABRY–PEROT

See *Fabry–Perot interferometer.*

INTERFEROMETRY, X-RAY

See *X-ray interferometry.*

INTERMEDIATE ORDER

Ordering or presence of symmetry elements at sites of atoms separated by distances much longer than the average interatomic distance in amorphous bodies (see *Amorphous state*). One may assume the presence of intermediate order from the behavior of an atomic radial distribution function, $g(r)$. If we define the *correlation length* of an atomic structure r_c as such a value of r that the difference $(g(r) - 1)$ becomes very small for $r > r_c$ while the variations of that difference remain within the measurement accuracy for $r < r_c$, then one may state that a single-phase intermediate order exists in the body for $r_c \gg a$, where a is the average interatomic distance. The characteristic dimension of ordered regions is r_c. The intermediate order may be nonuniform, with the correlation radius in the atomic arrangement varying from one part of an amorphous body to another.

Direct observations with the help of high-resolution instruments (such as scanning tunneling microscopes, autoionic microscopes, etc.) reveal the presence of regions of ordering, with a size $\sim$2.5 nm in certain amorphous alloys. Indirect experimental data are also available that suggest the presence of intermediate order in certain glasses (see *Vitreous state of matter*) and amorphous semiconductors (see *Long-range and short-range order*).

INTERMEDIATE STATE of a solid

A spatially inhomogeneous state of a *solid* formed in thermodynamic equilibrium during a *first-order phase transition*, and induced by an external magnetic field (electric field) $\boldsymbol{B}$ $(\boldsymbol{E})$ (see *Phase transitions in an external field*). The formation of an intermediate state is a common property of such *phase transitions*. Usually the intermediate state consists of alternating macroscopic homogeneous regions of competing phases, i.e. *domains* that are separated from each other by inhomogeneous transition regions (*domain walls*). The domain size significantly exceeds the thickness of the domain walls, and remains much smaller than the size of the specimen. The principal factor stabilizing the inhomogeneous state is the demagnetizing (depolarizing) field (see *Demagnetization fields*) produced by the surface magnetostatic (electrostatic) charges. These charges form due to a jump in the normal component of the *magnetization* vector (electric polarization), $\boldsymbol{M}$ $(\boldsymbol{P})$ at the specimen surface. When the external field changes the domains change their size to equalize the internal field with the equilibrium field of the competing phases $(\boldsymbol{B}_n)$. The principal equation describing the evolution of the initial state during the change of external field has the form

$$\boldsymbol{B} = \boldsymbol{B}_n + \widehat{N} \sum_{\nu=1}^{n} \xi_\nu \boldsymbol{M}_\nu(\boldsymbol{B}_n),$$

where $\widehat{N}$ is the *demagnetization factor tensor*, ξ_ν is the volume fraction of the νth phase, and n is the number of phases remaining in equilibrium in the field $\boldsymbol{B}$.

The term "intermediate phase" was first introduced by R. Peierls (1936) in studying the thermodynamics of *superconductors* in the region where they undergo a first order phase transition back into their normal state, induced by an external field. Peierls demonstrated that a specimen of finite dimensions developed a certain transition (buffer) region within a certain range of magnetic field strengths, and he called that range the intermediate state. The intermediate state structure in a superconductor was clarified by L.D. Landau (1937), who demonstrated that it represents a thermodynamically stable domain structure consisting of alternating patches of metal in the normal and the superconducting states.

The intermediate state has been found and studied in the region of first order phase transitions in *antiferromagnets*, as well as in *ferrite spinels* and *orthoferrites*, where it develops during various *magnetic phase transitions*.

INTERMEDIATE STATE OF SUPERCONDUCTORS

Thermodynamically stable state of a *type I superconductor* of finite dimensions in a certain range of external magnetic fields. Such fields have values close to those that bring about field induced phase transitions back to the normal state. It is a structure consisting of alternating macroscopic *domains* of the normal and superconducting phases in the shape of plates parallel to the external field, with a characteristic transverse size much smaller than the specimen size. Such a structure forms due to the positive free surface energy.

When affected by the magnetic field produced by current flowing through the specimen, a *dynamic intermediate state* develops in which the domains are normal to the direction of the current flow, and move continuously through the specimen. They are generated at one of its surfaces and end at the opposite one.

When a superconducting sample is placed in an external homogeneous field $\boldsymbol{B}_0$, that field is pushed out of the sample due to the *Meissner effect*. The field force lines then concentrate at the surface meaning that the field in that range exceeds $\boldsymbol{B}_0$ and may reach the value of the *thermodynamic critical magnetic field*, $\boldsymbol{B}_c$, where $B_0 < B_c$. In such a case the superconducting state should start to disintegrate in the neighborhood of the maximum concentration of flux lines, with the concomitant penetration of the magnetic field into the specimen. However, the normal phase appears to remain in the range of magnetic field strengths that are lower than the critical field, so there is an apparent contradiction. To overcome it R. Peierls (1936) and F. London (1936) both independently suggested that the intermediate state is one in which the Meissner effect takes place only partially throughout the material. Then, as the external field $\boldsymbol{B}_0$ gradually increases, the value of magnetic induction, $\boldsymbol{B}$, inside the sample linearly grows from zero at $B_0 = B_c^*$ (at the transition from the superconducting to the intermediate state) to $\boldsymbol{B}_c$ when $B_0 = B_c$ (at the transition from the intermediate to the normal state). Meanwhile, accounting for *demagnetization field*, we have $B_c^* < B_0 < B_c$ for the whole intermediate state range, with the Maxwell field inside the specimen $H = B_c/\mu_0$. The value of B_c^* depends on the value of B_c and the shape of the material.

For an arbitrary sample shape the normal and the superconducting phases may coexist because of the inhomogeneity of the demagnetizing fields. However, they are always separated from each other by an intermediate state region. In the case of an ellipsoid placed in a uniform external field, the induction and the field inside it are always uniform. Therefore, the entire ellipsoidal specimen transforms into the intermediate state as a whole, with a single value of $H = B_c^*/\mu_0$ throughout it. When the direction of $\boldsymbol{B}_0$ coincides with the ith principal axis of the ellipsoid, the directions of $\boldsymbol{B}$ and $\boldsymbol{H}$ also coincide with it, so that $B_c^* = B_c(1 - N_i)$, where N_i are the dimensionless *demagnetizing factors* ($i = 1, 2, 3$) (see *Demagnetization factor tensor*). Considering the extreme cases of the field parallel to the axis of a long cylinder ($N = 0$) we find the intermediate state virtually absent, while in the case of the field normal to a flat plate ($N = 1$) the intermediate state develops starting from a close to zero value of B_0.

The above description explains the basic experimental results, but says nothing about the intermediate state structure, and it fails to clarify the physical content of $\boldsymbol{H}$ and $\boldsymbol{B}$ inside the specimen. The Landau (1937) theory of the intermediate state represented the macroscopic field $\boldsymbol{H}$ inside the specimen as the result of averaging microscopic fields over the normal phase domains, while the induction, $\boldsymbol{B}$, was arrived at by similar averaging over the entire specimen. The fraction of the normal phase in the intermediate state equals $(B_0 - B_c^*)/(B_c - B_c^*)$. The spatial period d of the domain structure is determined by minimizing the free energy of the magnetic field forced outside the specimen by the Meissner effect, plus the *surface energy* of the domain walls, so there is a competition between the bulk term $[B_c^2/(2\mu_0)]\Delta V_n$

and the surface free energy at the interface between the normal and the superconducting phases. Here ΔV_n is the additional volume generated by the expansion of normal domains lying close to the surface. For a type I superconductor its surface free energy is positive and equals $[B_c^2/(2\mu_0)]\xi S$, where ξ is the *coherence length*, S is the overall domain boundary surface, and $d \propto (\xi L)^{1/2}$ ($\xi \ll d \ll L$), where L is the size of the specimen along the field. For *type II superconductors* the negative surface free energy dominates, equaling $-[B_c^2/(2\mu_0)]\lambda S$ (where λ is the *penetration depth of magnetic field*). The domain structure is absent, and instead of the intermediate state there is a partial Meissner effect in the *mixed state*.

INTERMEDIATE VALENCE, hybrid valence

A special state of matter characterized by a *resonance* between the energy states of the localized (bound) and the collective or delocalized (band) electrons. States of intermediate valence are typical for certain *rare-earth elements* (cerium, samarium) and several of their chalcogenides (see *Chalcogenide materials*) as well as doped compounds of rare-earth elements and *alloys* (see *Alloying*). In these materials the states in which their f-electrons are localized at the rare-earth ion ($4f^n$ configuration) appear to be close in energy to those of a type $4f^{n-1} + d$ electron in the *conduction band*. Therefore, the f-electrons acquire a partially band nature, and the average number of electrons on the center ("*valence*") appear to be fractional (hence the term *interconfiguration fluctuation states*).

A qualitative scheme of energy levels illustrating the development of intermediate valence (see Fig. 1) demonstrates how an external action (e.g., *pressure*) changes the position of the f-level, E_0, with respect to the d-band parabola (E_F is the *Fermi level*). As the external conditions change (e.g., temperature, pressure, composition), *phase transitions* develop, related to filling of electron levels in compounds with intermediate valence. Such transitions are accompanied by certain anomalies in lattice *elasticity*, by jumps in the *magnetic moment*, *specific heat* and crystal volume, and by sharp changes in conductivity. As a result of the mutual interaction of rare-earth ions via the *phonon* field, a spatial ordering of ions of differing valence becomes possible, so that the so-called

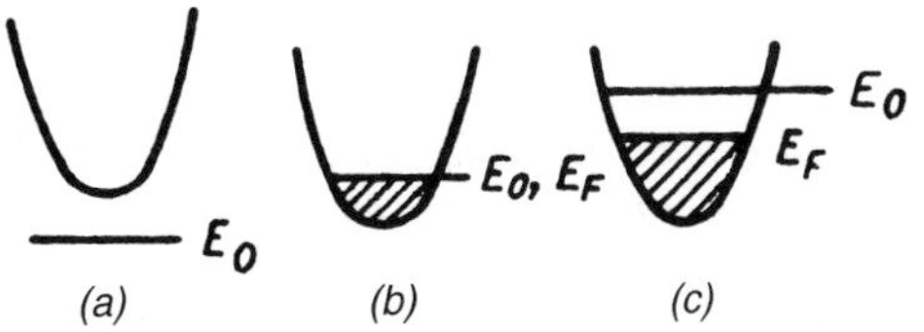

Fig. 1. Energy level scheme for (a) $4f^n$ configuration with whole valence, (b) $4f^{n-1} + d$ state with whole valence, and (c) intermediate valence phase. Different valence states of an R ion correspond to (a) and (b), e.g., R^{2+} and R^{3+}.

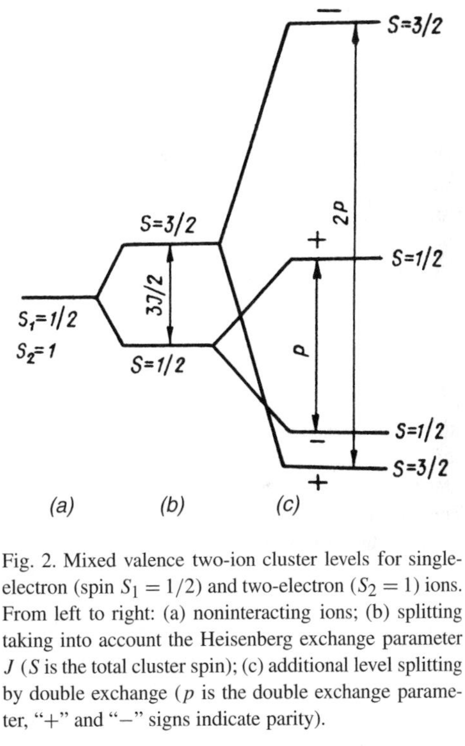

Fig. 2. Mixed valence two-ion cluster levels for single-electron (spin $S_1 = 1/2$) and two-electron ($S_2 = 1$) ions. From left to right: (a) noninteracting ions; (b) splitting taking into account the Heisenberg exchange parameter J (S is the total cluster spin); (c) additional level splitting by double exchange (p is the double exchange parameter, "+" and "−" signs indicate parity).

valence density waves or *charge density waves* form. It is suggested that one might employ phase transitions in intermediate valence compounds to record and store data. Intermediate valence states of certain substances are closely connected with their catalytic properties. *Molecular crystals* with intermediate valence contain *clusters* consisting of a few ions (2, 3, or 4) in various degrees of oxidation as their structural elements. The migration of an "extra" electron (double exchange) equalizes the valences. The magnetic properties are defined by the relation between the Heisenberg *exchange interaction* (that results in either *ferromagnetism* or *antiferromagnetism*) and double exchange. For strong double exchange the cluster has a maximum spin, as shown in Fig. 2 for a two-ion cluster with total spin $\boldsymbol{S} = \boldsymbol{S}_1 + \boldsymbol{S}_2$ obtained by vector addition. The interaction among clusters may stabilize the charge-ordered phase, and two phase transitions become possible under certain conditions, the ordered state forming within a finite temperature range $T > 0$.

INTERMETALLIC COMPOUNDS, intermetallides

Chemical compounds of *metals* involving other metals or semimetal compounds. The intermetallic compounds exist in the solid state and have mostly metallic type *chemical bonds*. About 14,000 intermetallic compounds are known; and some types comprise more than 20 compounds (there are 27 compounds in the Nd–Ni–Ga system). The most completely investigated intermetallic compounds are the borides (about 750 compounds), aluminides (1100), and silicides (1300). In view of their specific properties, the intermetallic compounds of the *rare-earth elements* (4500 compounds) are intensively studied. The synthesis of these compounds is carried out by fusion, sintering, and electrolysis. Their composition, structure, and properties depend on the nature of the components, electronic structure of the atoms, their dimensions and *electronegativity*. The compositions are quite diverse; and in most cases these compounds fail to conform to the *valence* rules. Many of them have wide regions of homogeneity (daltonides and berthollides (nonstoichiometric)). The heterodesmic (with more than one type of chemical bonding) intermetallic compounds (see *Heterodesmic structures*) containing semimetal elements include framework, laminated, chain, and insular compounds. Their crystal structures fall into almost 1,000 structural types grouped into 17 classes according to their coordination characteristics; the classes differ in coordination polyhedra down to the size of atoms. Intermetallic compounds possess a close packed structure, and are characterized by complete or partial ordering. These compounds find their use as catalysts, semiconductors, superconductors, as well as magnetic hydrogen-absorbing, thermal emission, and other special materials. They are involved in compositions of the most important *alloys* such as structural, refractory, hard, high-temperature, and corrosion-resistant alloys. See also *Superconducting intermetallic compounds.*

INTERMETALLIC SUPERCONDUCTOR

See *Superconducting intermetallic compounds.*

INTERNAL ENERGY

See *Thermodynamic potentials.*

INTERNAL FRICTION

The property of solids to irreversibly transform applied elastic oscillation energy into heat energy, and the various mechanisms involved in this transformation. W. Thomson, Lord Kelvin, observed this in 1865.

The increased interest in internal friction is caused, mainly, by two reasons. The first is associated with the need to account for the dissipated energy in construction material during the calculation of oscillations of mechanical systems operating in the resonance mode. The second reason is connected with the creation of a precision analytical method of investigating physical and chemical phenomena in *crystals*, which is based on internal friction. The high structural sensitivity and its non-destructive character are the advantages of this internal friction method. Internal friction can be demonstrated with the help of a diagram of *stress* σ (see *Stress tensor*) versus *strain* ε. In the ideally elastic (see *Elasticity*) region the plot of $\sigma(\varepsilon)$ is a straight line. In the presence of internal friction the plot becomes a *hysteresis loop* (see Fig.) with an area equal to the loss of energy per unit volume during one cycle of oscillations ΔW. Internal friction can be measured in different ways: through the increase in temperature (for very high values of the internal friction), using the tangent of the shear angle between σ and ε ($\tan\theta$), from the amplitudes of free and forced oscillations (Δ), and through measuring the absorption of elastic waves in the material (α) (see *Sound absorption*). For small values of the *internal friction* ($\ll 0.1$) measurements of it are connected with each other in the following way: $Q^{-1} = \tan\theta = \Delta W/(2\pi W) = \Delta/\pi = \alpha\lambda/\pi$, where W is the

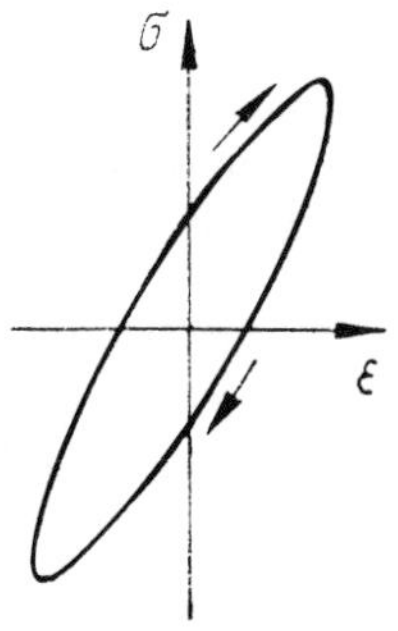

Internal friction.

total oscillation energy, Δ is the logarithmic decrement of the oscillations (natural logarithm of the ratio of two sequential amplitudes), α is the attenuation constant, and λ is the wavelength of the elastic wave. Methods of measuring internal friction are divided into three basic groups: (1) the *torsional pendulum method* with the range of measurement frequency ($\omega = 0.1$ to $5 \cdot 10^2$ Hz); (2) the *method of the resonance rod* (transverse oscillations of $\omega = 5 \cdot 10^2$–10^4 Hz, longitudinal ones of $\omega = 10^4$–10^6 Hz); and (3) the *method of ultrasonic pulses* ($\omega = 10^6$–10^{11} Hz). The hysteresis curve (see Fig.) can be conditioned by the dependence of σ and ε on time (or ω) when there is no residual deformation; this is so-called *dynamic hysteresis*. In this case the internal friction does not depend on ε. This type of loss is usually called *relaxational internal friction*. A hysteresis loop may also appear (including the case of very small rates of deformation) due to microplastic deformation. This type of loss does not depend on ω; the losses grow with increasing ε, and this is called *hysteresis internal friction* (see *Amplitude-dependent internal friction*). The mechanisms of relaxation losses are various and combine a large group of the phenomena. For the most part internal friction involves a complex of different attenuation mechanisms. Each mechanism is most prominent over a specific frequency range, which causes the presence of a maximum in the curve $Q^{-1}(\omega)$. By varying the oscillation frequency from infrasound to hypersound under stable external conditions it is possible to separate contributions of different mechanisms in the internal friction. In perfect crystals the thermoelastic (see *Thermoelasticity*) losses (infrasound range of frequencies) and the interactions of elastic waves with the electron (low temperatures) and phonon waves (MHz–GHz) in the crystal are the main source of the internal friction. In real crystals the mechanisms of losses are conditioned by point, linear, surface and volume structural *defects*. Radiation in the infrasonic and sonic frequency ranges is often used in investigations of internal friction by studying the effects of regulation and reconstruction of the atomic structure, connected with the diffusion of vacancies, of impurity atoms, of pairs of point defects, etc. Internal friction involving the presence of *dislocations* in a crystal is studied the most. It appears over a wide range of frequencies, including infrasound and ultrasound. The concrete mechanism of adsorption is defined by the elastic and inertial properties of the dislocations, by operation of the *point defect–dislocation interaction* (*Snoek–Köster peak*, see *Snoek–Köster relaxation*, *Hasiguti peaks*), by potential barriers of the lattice itself [*Bordoni peaks* (P.G. Bordoni, 1949)], etc. *Stress relaxation* along *grain boundaries* and surfaces of *phase* separation is one of the important sources of internal friction. Other disturbances (*fractures*, *pores*, shrink holes) also contribute to the internal friction. It is necessary to take into account relaxation internal friction caused by micro-vortex currents, and by the magnetic field regulation in the ferromagnets.

INTERNAL STRESSES

Stresses existing in solids in the absence of external stresses, i.e. stresses caused by internal sources.

These internal stresses appear when the different parts of the body are interconnected in such a way that each part causes others to change their dimensions or shape. For instance, in crystals *defects* of the crystalline structure are sources of internal stresses; these defects can be of the *point*, *linear*, *planar* and *volume* types (and their ensembles). These stresses also exist in the presence of thermal gradients in solids. Stresses created by internal sources are calculated, as a rule, in the model of a *continuous*, *elastic*, *isotropic medium* (or an elastically anisotropic one). They are additive and can be described by the overall *internal stress tensor* $\sigma_{jk}(x_1, x_2, x_3)$ $(i, j, k = 1, 2, 3)$ (see

Stress tensor). At the free surface of a solid the internal stresses satisfy the boundary conditions $\sigma_{jk}n_k = 0$ $(i, k = 1, 2, 3)$, where n_k is the projection of the external normal to the surface on the coordinate axis x_k. If within the volume of the solid there are stresses caused by static external forces, they add to the internal stresses. They can appear in solids under the action of mechanical and thermal effects due to the nonuniformity of *plastic deformations*, and due to the possible occurrence of *phase transitions*. Internal stresses which appear after such processes are frequently called *residual stresses*. These stresses can have different degrees of localization, i.e. they can be counterbalanced within different regions of a solid. In *polycrystals* the following types of internal stresses are distinguished: (1) first-order residual stresses which are balanced within the overall volume of a solid, and produce a shift of an X-ray diffraction line, (2) second-order residual stresses (*microstresses*) which are balanced within a volume about the size of a crystallite or a mosaic block and broaden an X-ray diffraction line, and (3) third-order residual stresses which are balanced within a volume the size of a crystalline lattice *unit cell* and weaken X-ray diffraction lines.

X-ray diffraction methods (see *X-radiography*, *Nondestructive testing techniques*), and also (for first-order stresses) various mechanical methods based on the measurement of deformations, with subsequent recalculations that convert them into stresses with the aid of *Hooke's law*, are used for the determination of residual stresses. These (especially the first-order ones) can strongly influence the *strength* of solids.

INTERNATIONAL TABLES FOR CRYSTALLOGRAPHY

Reference guides for *crystallography* and structure analysis. "International Tables for Determination of Crystal Structures" were first published in 1935 (Berlin). The second generation comprises four volumes published between 1952 and 1974. Since 1983, the third generation "International Tables for Crystallography" has appeared as volumes A, B and C (Space Group Symmetry; Reciprocal Space; Mathematical, Physical and Chemical Tables, respectively) which provide comprehensive coverage of the status of modern crystallography. Particularly useful is Vol. A edited by T. Hahn which tabulates atom positions for the 230 space groups, and in addition has a great deal of background material on crystallography.

INTERPHASE IMPURITY DISTRIBUTION COEFFICIENT

The ratio of the *concentration* of a substance dissolved in a solid phase to that in a liquid phase. The equilibrium interphase impurity distribution coefficient is defined using a *phase diagram* as the ratio of abscissae of the solidus and liquidus points for a given impurity concentration. If this is smaller than 1, then the solid phase that forms as a result of *crystallization* contains fewer impurities, and vice versa. In actual crystallization schemes, the ratio of impurity content in the crystal to that in the melt is determined by the interphase impurity distribution coefficient that depends on the ratio of the rate of the crystallization front motion to the rate of equalization of impurity concentration in the melt. The latter process occurs by the diffuse transition of impurities in a thin stationary layer of melt that adjoins the crystallization front, and by forced or natural convective agitation of the remaining melt. Allowing for the above factors, the distribution coefficient can be calculated for actual crystallization schemes.

INTERSTITIAL ALLOYS

Alloys, one component of which consists of *interstitial atoms*. These alloys form for the ratio $r_{\mathrm{x}}/r_{\mathrm{m}} \leqslant 0.59$ of atomic radii of interstitial element r_{x} to metal r_{m}. Interstitial phases are divided to two classes: *interstitial solid solutions* of *metals*, and *interstitial compounds* or hybrids: carbides, nitrides, and some borides, oxides and silicides.

INTERSTITIAL ATOM

An *atom* that is localized in a *crystal* between *crystal lattice* sites. It is common for an interstitial atom to be situated at an *interstice*, i.e. in a cavity bounded by neighboring atoms at regular crystal lattice sites. However, in a number of cases more complicated configurations of an "extra" atom in a crystal can occur, such as when the atom intrudes into an atomic chain so that there are $n + 1$ atoms for n sites. These are *dumb-bell interstitial defect* ($n = 1$) and *crowdion* ($n \approx 5$–10) configurations. Intrinsic interstitial atoms appear in crystals as non-stoichiometric components (see *Stoichiometric composition*). This may happen, e.g., when a crystal is in thermodynamic equilibrium with the high-pressure vapor (see *High pressures*) of one of its components, or through the generation of equilibrium or nonequilibrium (during *nuclear radiations*, elastic *strain*) *Frenkel defects. Impurity atoms* of certain chemical elements can be localized in the interstices of various crystals. The following processes can take place: transitions of interstitial atoms between different types of interstices (e.g., between hexagonal and tetrahedral ones in a diamond-type lattice); trapping of an interstitial atom by a *vacancy*; dislodgment of an impurity atom from a site by an intrinsic atom; coagulation of intrinsic interstitial atoms with the formation of *dislocation loops*; and other reactions that involve intrinsic and impurity interstitial atoms, as well as *defects*. The *activation energy* of intrinsic interstitial atom diffusion is usually much lower than that of vacancies. The non-activation nature of the motion of interstitial Si atoms generated via nuclear radiation on a *silicon* crystal has been experimentally confirmed at low temperatures; the same is true of interstitial Ge atoms in *germanium* crystals. The presence of interstitial atoms can have a significant effect on the properties of crystals.

INTERVALLEY NOISE

Fluctuations of current flowing through a *many-valley semiconductor*, or of voltage across the latter. These result from population fluctuations of individual valleys. In the general case, the mobilities of *current carriers* in inequivalent valleys along the external electric field direction differ from each other. Therefore, random transitions of carriers from one valley to another cause corresponding changes of current (or of voltage), i.e. the appearance of intervalley noise. If the external field does not warm up the carriers, the spectral density of such noise is proportional to the square of the current flowing through the crystal. Its frequency characteristic is of the same form as *generation–recombination noise*, i.e. proportional to $(1 + \omega^2\tau^2)^{-2}$, where τ is the lifetime of an electron in a valley in *intervalley scattering*. Mobilities of carriers in isolated valleys are usually anisotropic. The partial currents of these groups may then be nonparallel, and intervalley noise in such crystals arises in the direction transverse to the direct current. The dependence of intervalley noise on the orientation of the electric field relative to the crystallographic axes allows varying the magnitude of intervalley noise, and facilitates its identification in experimental studies.

INTERVALLEY PHONONS

Phonons involved in interaction electron transitions from one valley to another (see *Many-valley semiconductors*). By virtue of quasi-momentum conservation (see *Conservation laws*) within an accuracy of a *reciprocal lattice* vector, a phonon quasi-wave vector which participates in an elementary transition between valleys is equal to $\boldsymbol{k}_{\text{phonon}} = \pm(\boldsymbol{k}_1 - \boldsymbol{k}_2)$, where, $\boldsymbol{k}_1$ and $\boldsymbol{k}_2$ are the *quasi-wave vectors* of the initial and final electron states, the "+" sign refers to emission of a phonon, and the "−" sign refers to absorption of a phonon. Since valleys are often situated near the *Brillouin zone* boundary (e.g., in *germanium*, *silicon*), the values of $|\boldsymbol{k}_1|$ and $|\boldsymbol{k}_2|$ and, accordingly, of $|\boldsymbol{k}_{\text{phonon}}|$ are close to the maximum value of a quasi-wave vector in the Brillouin zone. Such a value of $|\boldsymbol{k}_{\text{phonon}}|$ corresponds to the energy of intervalley phonons, which is close to the peak phonon energy in the crystal. See *Intervalley redistribution, Intervalley scattering.*

INTERVALLEY REDISTRIBUTION

Population change or readjustment of equivalent (or inequivalent) valleys of a *many-valley semiconductor* energy band. Typically, it occurs when electrons of different equivalent valleys obtain unequal energies from the electric field by virtue of the difference in their *effective masses* along the field direction due to *intervalley scattering*. Since the intervalley transition is related to the change of the electron *quasi-wave vector* (of the order of a *reciprocal lattice* constant), satisfying the *conservation law* of quasi-momentum requires participation of a *phonon* or an impurity in the intervalley scattering process. Intervalley redistribution has a significant effect on semiconductor properties: a high anisotropy of the conductivity arises from the misalignment of the directions of the current flow and the electric field strength

(*Sasaki effect*); when the intervalley distribution depends strongly on the electric field, then negative diffusive conduction can be observed. The Hall effect and magnetoresistance (see *Galvanomagnetic effects*) exhibit some characteristic features in their dependences on the electric and magnetic field intensities; absorption of light by free carriers decreases, and *birefringence* arises.

Intervalley redistribution may also occur as a result of a *spontaneous symmetry breaking*, which leads to a *multivalued electron distribution* (see *Sasaki effect*). This redistribution can also take place via the *Gunn effect* between valleys of different *energy bands.*

INTERVALLEY SCATTERING

Quantum transitions of electrons or holes between valleys of an energy band (see *Many-valley semiconductors*). The term is used mostly for describing scattering processes in *semiconductors* and *semimetals.* Intervalley scattering may take place in the interaction of carriers with *phonons*, impurities, or crystal surfaces. Since the distance between valleys in *momentum space* is comparable to the *Brillouin zone* size, this scattering involves transfer of a large *quasi-momentum* at each individual event. The probability of intervalley scattering is usually considerably lower than that of *intravalley scattering*, in which both initial and final states of the carrier are in the same valley. Yet, intervalley scattering plays an important role, since it controls relaxation processes of *level populations* of different valleys, and determines steady-state populations and the magnitude of the energy that *current carriers* absorb under the action of external fields (see *Intervalley redistribution*). Intervalley scattering is described by the probability of an intervalley transition, which (for an *electron–phonon interaction*) is

$$W^{\pm}_{\boldsymbol{k}_1\to\boldsymbol{k}_2} = C\left(N + \frac{1}{2} \pm \frac{1}{2}\right)\delta(\varepsilon_{\boldsymbol{k}_1} - \varepsilon_{\boldsymbol{k}_2} \mp \hbar\omega_{12})$$

where quasi-momenta $\boldsymbol{k}_1$ and $\boldsymbol{k}_2$ determine the initial and final states of electrons in valleys 1 and 2, respectively; $\varepsilon_{\boldsymbol{k}_1}$ and $\varepsilon_{\boldsymbol{k}_2}$ are the energies of these states; ω_{12} is the frequency of a phonon with a wave vector that is close to the distance between maxima 1 and 2 to within a *reciprocal lattice* vector; and N is the number of phonons with frequency ω_{12}. The upper sign refers to a transition with phonon emission, the lower refers to one with phonon absorption. The quantity C is defined with the help of the explicit form of the *Bloch wave functions* $|\boldsymbol{k}_{1,2}\rangle$ (see *Bloch theorem*).

The transition probabilities in scattering by impurities may be obtained from the above formula by setting $\omega_{12} = 0$, $N + 1/2 \pm 1/2 = 1$ in it. Then the quantity C is defined by the short-range part of the electron–impurity interaction potential. The change in electron momentum, required for an intervalley transition, may occur at its collision with the crystal surface. The possibility of such a transition is determined by geometric considerations, and depends on the orientation of the $\boldsymbol{k}_2 - \boldsymbol{k}_1$ vector relative to the surface. In this case the laws of the conservation of energy and of the projection of quasi-momentum on the surface (see *Conservation laws*) must be taken into account. See also *Current carrier scattering.*

INTRAMOLECULAR VIBRATIONS

Oscillations associated with relative displacements of atoms within a molecule or a molecular ion in a *molecular crystal.* The frequency range of the vibrations is 400–3000 cm^{-1}. For an impurity molecule in a *crystal lattice* the intramolecular vibrations are often *local vibrations*, significantly higher in frequency than the maximum frequency of the matrix crystal (e.g., NO_2^-, NO_3^-, SO_4^{2-}, etc. in an *alkali-halide crystal* lattice).

INTRINSIC CONDUCTIVITY

Electrical conductivity of a *semiconductor* due to the excitation of electrons from the *valence band* to the *conduction band*, and the generation of holes in the valence band, with the concentrations of electrons and holes equal to each other. Intrinsic conductivity predominates in intrinsic semiconductors. For more details, see *Electrical conductivity.*

INTRINSIC DEFECTS in crystals

Crystal lattice defects that are deviations from the regular arrangement of the atoms $A, B, \ldots$ of a formula unit $A_m B_n \ldots$ at preassigned sites. The simplest intrinsic defects are point *vacancies* V and *interstitial atoms* I of crystal components. One of the mechanisms for the production of both V and I is the thermal motion of the atoms.

If the presence of V or I is related to the excess or deficit of a certain component of a crystal compound, then the intrinsic defects are referred to as *nonstoichiometric defects*. *Frenkel defects*, *Schottky defects*, *antistructural defects* and *anti-Schottky defects* are called *stoichiometric intrinsic defects*, since the generation of these defects does not require particle exchange between a crystal and its environment, and does not upset the stoichiometric concentration ratios of the chemical constituents. More complex intrinsic defects are various aggregations of vacancies, intrinsic interstitial atoms, and *substitutional atoms*.

INTRINSIC LIGHT ABSORPTION EDGE

The increase of optical density in *semiconductors* and *insulators* due to photoexcitation that raises electrons from the *valence band* to unfilled states of the *conduction band*. The energy position of the intrinsic light absorption edge is directly related to the width of the *band gap*, E_g, while its spectral dependence is defined by the features of the *band structure* of the crystal. *Direct transitions* play a decisive role for those extrema in the conductivity. When the position ($\boldsymbol{k}$-value) of the minimum of the conduction band corresponds to the maximum of the valence band, $\boldsymbol{k}_c = \boldsymbol{k}_v$, then direct transitions take place and the initial and the final *quasi-momenta* of the electron practically coincide. The respective light *absorption coefficients* reach about 10^5 cm^{-1}. If $\boldsymbol{k}_c \neq \boldsymbol{k}_v$ then indirect transitions of electrons become possible, and the quasi-momentum *conservation law* is satisfied either through the absorption or emission of a *phonon*, or via an interaction with defects of the *crystal structure*. During indirect transitions the light absorption coefficient does not exceed 10^3 cm^{-1}. Optical interband transitions may be studied by the methods of *nonlinear optics*, using multiphoton excitation.

The position of an intrinsic light absorption edge depends on the temperature, pressure, density of free charge carriers, presence of impurities, and external electric and magnetic fields (see *Burstein–Moss effect*, *Franz–Keldysh effect*). The sensitivity to these factors may be utilized to analyze the structure of the crystal energy bands, and to develop regulating *optical filters* and modulators.

INVARIANCE in solid state physics

The invariability of the state of a solid under certain transformations that characterize its physical parameters. The basic properties of invariance arise from the invariability of the state as a result of changes of position and orientation in space and time. Invariance of other types is also possible (see *Gauge invariance*).

Invariance is described on the basis of *group theory*. The rules for a transition to other characteristics of an object related to its positions and orientations in a four dimensional space–time coordinate system contain only the parameters of these changes. The latter are considered as space–time transformations g, constituting the corresponding group G (in the nonrelativistic case the Galilean group or a subgroup of it). Physical quantities, among them state functions Ψ, realize unitary representations of the group G according to the law $T_g T_{g'} = T_{gg'}$, and satisfy the conservation of distance in representation space. Invariance requires the operators of physical quantities for the states Ψ and $T_g\Psi$ to have the same form. Average values of the operators must have a tensor form.

The invariance principle usually serves as a control tool, but can also be used in a constructive way. Let $G = H = \{h\}$ be the rotation-reflection group $O(3)$ in ordinary space (see *Continuous symmetry transformation groups*) with $x'_i = h_{ik}x_k$ (where x_i and x'_i are respectively the coordinates of a point of an object before and after a rotation h). The latter transforms $\Psi(\boldsymbol{x})$ to $T_h\Psi(\boldsymbol{x}) = \Psi(h^{-1}\boldsymbol{x})$, a vector field $A_i(\boldsymbol{x})$ to $A'_i(\boldsymbol{x}) = h_{ik}A_k(h^{-1}\boldsymbol{x})$, and the mean value $\overline{L}_i = (\Psi, L_i\Psi)$ of an axial vector operator to $\overline{L}_i = h_{ik}\overline{L}_k$ (under a proper rotation). The equality $(T_h\Psi, L_i T_h\Psi) = h_{ik}(\Psi, L_k\Psi)$ implies that $L_i T_h = h_{ik} T_h L_k$. It follows from the latter equality, combined with the condition of maximum simplicity of L_i, that L_i is proportional to a derivative with respect to the angle of rotation about the x_i axis. A more formal approach to this relation is associated with writing it in a matrix form in the basis of any irreducible representation $T(h)$ of the group H (it is important that all of them be known). Substituting $T(h)$ for T_h, we obtain the equation for determining the matrix forms $(L_i)_{\alpha\beta}$ ($\alpha, \beta = 1, 2, \ldots, j$, where j is the dimensionality of $T(h)$). In particular, if $j = 2$ and $T(h) = u(h)$

in the spinor representation, we obtain to within a similarity transformation that $L_i = \sigma_i$, the *Pauli matrices*. Invariance (or symmetry) with respect to time inversion means that together with a state Ψ of an object, which is not in a magnetic field, a state $\Theta\Psi$ is possible that is characterized by opposite directions of spins, and motions comparable to those in the state Ψ. Applying the Pauli principle, one can establish the antiunitarity property of Θ, its analytic forms, and the matrix forms of operators of the group $G + \Theta G$ (the so-called *co-representations*).

The *crystal field* is regarded as external with respect to the *quasi-particles* in a crystal, hence there arises an invariance of the potential energy $U(\boldsymbol{x}) = T_g U(\boldsymbol{x})$, where $\boldsymbol{x}$ is the set of coordinates, g is an element from $G_a = G + a_0 G$, G is the *space group* or *Shubnikov group* (magnetic space group), and a_0 is a certain antiunitary operator. Physical quantities, depending on their nature, implement representations or co-representations of the group G_a (see also *Space groups*). Most often, the term invariance is used with respect to the group H or the Euclidean group $G_e = \{g + (\boldsymbol{\alpha}/h)\}$, where $\boldsymbol{\alpha}$ is a *translation* vector, and $\Psi(\boldsymbol{x})$ is a scalar invariant if $T_g \Psi(\boldsymbol{x}) = \Psi(\boldsymbol{x})$; the space $\{\Psi_i(\boldsymbol{x})\}$ is invariant if $h_{ik} A_k(g^{-1}\boldsymbol{x}) = A_i(\boldsymbol{x})$; and $S(\boldsymbol{x})$ is a scalar invariant operator if $T_g S(\boldsymbol{x}) = S(\boldsymbol{x}) T_g$. Application of T_g to an expression containing x_i means substituting $h_{ik}x_k - h_{ki}\alpha_k$ for x_i.

INVARIANCE, SCALING THEORY

See *Scaling invariance hypothesis*.

INVARIANT, ELASTIC

See *Elasticity theory invariants*.

INVARIANTS in the theory of solids

Quantities conserved under operations of a symmetry group and composed of variables describing the state of a solid, e.g., *spin operators*, components of *magnetization* or *strain tensors*, etc., as well as the potentials of external fields. The special role of invariants arises from the fact that the Hamiltonian, the *thermodynamic potential*, the *dissipation function*, etc., can be represented as a sum of invariants. One can also obtain expressions to describe the dynamics of elementary excitations: the energy of a band state and its natural frequencies of vibrations, with invariants including quantities that determine the dynamics of a *quasi-particle* (*spin*, *quasi-momentum*).

With the help of *group theory* methods one can write down the invariants for a physical system. This permits the analysis of material characteristics: *electrical conductivity*, *dielectric constant*, and *magnetic susceptibility*, including those with *spatial dielectric dispersion* taken into account. For example, the conductivity tensor σ_{ij} enters the formula for Joule heat $Q = \sigma_{ij} E_i E_j$; therefore one can analyze the tensor σ_{ij} from the properties of the quantity Q represented as a sum of invariants. All variables and their combinations, including spin operators, can be classified in terms of their *irreducible representations of a group*, and depending on the required approximation one can consider not only the first but higher powers of parameters (e.g., E_i, E_j^2, etc.) as well. If the sets of functions f_i^α and φ_i^α simultaneously transform by the same (αth) representation, the invariant is constructed as $I_\alpha = \sum_i (f_i^\alpha)^* \varphi_i^\alpha$. This invariant enters the Hamiltonian, the thermodynamic potential, etc., with a certain phenomenological parameter determined for particular systems from comparison between theory and experiment, or using calculations based on basic principles. This scheme has been very successful for the construction of *spin Hamiltonians*, in providing a phenomenological description of magnetically ordered substances, and in *band theory* for the case of complicated forbidden gaps (e.g., *Luttinger Hamiltonian*).

INVERSE PHOTOELECTRON EMISSION

Emission of optical radiation by *solids* under bombardment of their surface by electrons. The principal mechanisms of inverse photoelectron emission are the radiative capture of the incident radiation at unoccupied states of the solid, and the radiative *relaxation* of collective (plasma) and single-particle excitations of the electron subsystem in the solid whose nonequilibrium concentration is maintained by the incoming electrons. The spectrum of inverse photoelectron emission is continuous, however, a single individual component is also present. The structure of the spectrum is determined by the nature of the material. In addition to the above-mentioned mechanisms, *radiative quantum transitions* stimulated by incident electrons can also occur with the particles adsorbed at the surface.

The principal physical characteristics of inverse photoelectron emission are the following: the intensity of the radiation (*photon yield* Y), its spectrum, polarization, and direction. The value of Y, which is defined as the ratio of the number of photons emitted in a certain wavelength range to the number of incident electrons, depends both on the nature and the structure of the material, the condition of its surface, and the kinetic energy E and angle of incidence of the electrons. In metals the photon yield is small ($Y \leqslant 10^{-4}$ photon/electron at $E = 10^3$ eV in the wavelength range 200–600 nm). The study of these physical characteristics provides information about properties of the surface (see *Electron–photon spectroscopy*). Inverse photoelectron emission of metals can serve as a secondary standard source of optical radiation. The luminescence produced by *luminophors* is widely used in practice (see *Cathodoluminescence*).

INVERSION ASYMMETRY

Lowering of symmetry of a *crystal structure* by removal of a *center of symmetry* with respect to a related centro-symmetric structure (e.g., transition from the *diamond* to the sphalerite structure). The term inversion asymmetry is used to characterize features arising in the electron energy spectra, e.g., term splitting due to the inversion asymmetry (see *Inversion splitting*) at a lattice site. Of particular importance is the lifting of the spin degeneracy of spin states at points of the *Brillouin zone*: in a crystal structure without an inversion center, only those states with oppositely directed wave vectors $\boldsymbol{k}$ and spins σ are found to be degenerate (owing to time inversion): $E(\boldsymbol{k}, \sigma) = E(-\boldsymbol{k}, -\sigma)$. There are features of electron energy spectra in *semiconductors* lacking inversion symmetry that produce physical effects that cannot be observed in crystals with an inversion center. Examples of these are: one type of *combined resonance*, the precession mechanism of electron spin *relaxation*, circular photogalvanic effects (see *Photoelectric phenomena*), and the linear *Zeeman effect* caused by the magnetic-field-induced shift of the valleys of a semiconductor close to a high symmetry point of the Brillouin zone (see *Many-valley semiconductors*). Phenomena caused by inversion asymmetry are observable in $A^{III}B^{V}$ type semiconductors, mercury selenide, and tellurium.

INVERSION AXIS

An element of symmetry, which includes rotation about a *rotation symmetry axis* (n, C_n) through a certain angle ($2\pi/n$) with a simultaneous reflection (m) in a perpendicular plane at the *center of symmetry* situated on this axis. In *crystals*, there can be inversion axes of the same orders as the ordinary rotation axes: of the first ($\bar{1}$, $C_i = S_2$), second ($\bar{2} = m$, $C_{1h} = C_s$), third ($\bar{3}$, $C_{3i} = S_6$), fourth ($\bar{4}$, S_4), and sixth ($\bar{6}$, C_{3h}) orders, where each inversion axis is listed in both the Hermann–Mauguin and the Schönflies notation. Only axes $\bar{3}$, $\bar{4}$ and $\bar{6}$ have an independent significance, since axis $\bar{1}$ is equivalent to a center of symmetry, and axis $\bar{2}$ corresponds to reflection (m) in a *symmetry plane*.

INVERSION-INEQUIVALENT POSITIONS

Two locations of an atom (an ion) that differ one from another by the inversion of the coordinates of all atoms of a *crystal* relative to a particular lattice site (see the figure). The inversion-inequivalent positions exist, for example, in tungsten, corundum, and Si. The coordinate systems for the inversion-inequivalent positions that are related to the crystal environment in the same

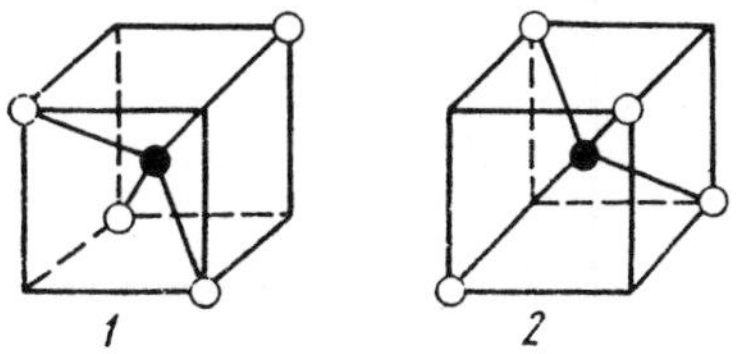

Two inversion-inequivalent positions (1 and 2) in cube centers of a diamond-type lattice: • inversion-inequivalent positions (sites or interstitial sites), o lattice atoms (only first coordination sphere is shown). The crystal field environment of position 1 will be the same as for its inequivalent counterpart 2, if the crystal is inverted in the coordinate system with origin centered at position 1.

way are mutually inverted. The crystal characteristics (in particular, the tensors corresponding to external fields) in these inversion-different coordinate systems are the same, whereas the external fields are inverted in the systems. The magnetic field $\boldsymbol{B}$ is, by virtue of its pseudovector nature, invariant under space inversion, while the electric field strength $\boldsymbol{E}$ changes its sign at the inversion operation. The inversion nonequivalence has no effect in many experiments. For example, the *electron paramagnetic resonance* spectra of inversion-inequivalent *paramagnetic centers* coincide. One can distinguish between these centers only upon application of the electric field along with the magnetic one (see *Electric field effects*). A similar situation exists in optics.

INVERSION LAYER

A charged subsurface region formed by minority *current carriers* in semiconductors. The inversion layer is a part of an *electric double layer* (see Fig.), which appears near the surface of the p-type (n-type) semiconductor where a positive (negative) charge is localized. This charge is due to the trapping of majority carriers at the Tamm states (see *Tamm levels*), i.e. levels of *surface defects* and adsorbed atoms. There is a widely used method for producing an inversion layer in a *metal–insulator–semiconductor structure* (or metal–oxide–semiconductor structure, MOS), where an electrical voltage of the same polarity as the free carrier bulk charge is applied to a metal electrode (gate or field electrode). The inversion layer can also arise near a semiconductor–metal interface. However, if the surface charge sign is opposite to the majority charge carrier sign, the *enriched layer* appears in the subsurface layer.

The potential distribution in an electrical double layer is determined by the *Poisson equation* (characteristic scale of its change is the *screening radius*). For a low density of surface charge the inversion layer does not appear. These charges are neutralized by a subsurface depletion layer in which the majority carrier concentration is reduced, and an uncompensated layer of acceptor (donor) impurities arises. As the surface charge (or the voltage across the MOS-structure gate) grows, this depletion layer shifts into the bulk semiconductor, and an inversion layer appears near the surface. Such charge redistribution leads to the change of the MOS-capacitor capacitance as a function of the electrical voltage V_g across the gate. This *capacitance–voltage characteristic* method allows one to study the charge distribution over the interface of a semiconductor and an insulator.

The inversion layer is separated from the semiconductor bulk by a nonconductive depletion layer, so the MOS-capacitor can transform into a transistor structure by establishing the source and sink contacts on the surface close to the field electrode. In this case, the conduction would not be

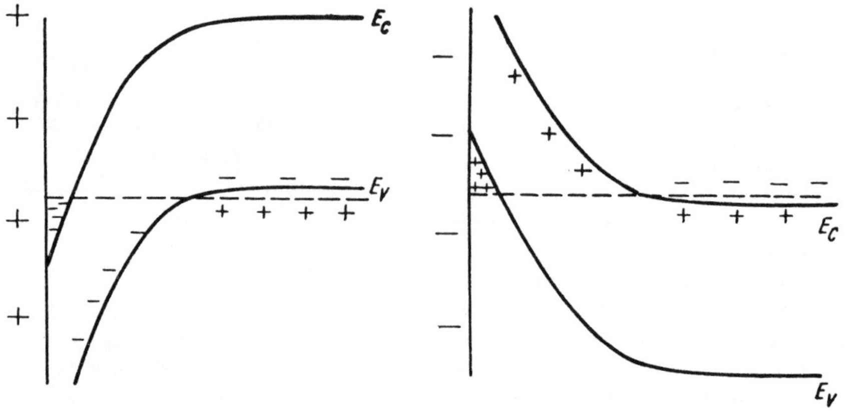

Energy diagrams of the subsurface region of p-type (left), and n-type (right) semiconductors.

shunted by the bulk. The resistance of the source-sink channel is modulated by the voltage V_g (*field effect*) due to the change of carrier concentration, and in the inversion layer the carrier mobility varies, e.g., with changing thickness (*size effect*). Studies of *galvanomagnetic effects* in the inversion layers not only yield information on the electronic properties of a semiconductor–insulator interface, but also enable optimization of the *field-effect transistor* parameters on the $Si–SiO_2$ structure that is a basic element of modern electronics.

Charge carriers in an inversion layer are localized in a potential well formed by a barrier on the semiconductor surface on one side, and by the bulk electrostatic charge on the other side. For strong inversion this potential well is rather narrow (comparable to the quasi-particle de Broglie wavelength), and the electron (hole) motion is quantized in the direction normal to the surface. However, in the direction along the inversion layer, the motion remains free, corresponding to a *quasi-two-dimensional electron gas*. For dimensionally quantized states to appear the scattering needs to be weak (attained at a high-quality interface at low temperatures when the thermal vibrations are ineffective). In the latter case the spacing of levels should exceed their impact broadening. It is also necessary to fill the lower subbands to provide low carrier concentrations in the inversion layer. Direct observation of the of two-dimensional nature of the energy spectrum is possible using the oscillations of the magnetic conduction of the inversion layer of $Si–SiO_2$ by the variation of V_g (due to the *Shubnikov–de Haas effect*), with the period of the oscillations depending on the magnetic field component normal to the surface. In strong magnetic fields the motion of two-dimensional carriers in the inversion layer is quantized (see *Quantizing magnetic field*), and their spectrum becomes discrete. In this case, the character of galvanomagnetic phenomena changes qualitatively and the *quantum Hall effect* appears. Using this effect, one can measure the value of the fine structure constant on a silicon field transistor, and design a resistance standard with an accuracy noticeably higher than that achieved with other methods. There are other specific electronic features of the inversion layer: the appearance of an *oriented superlattice* in the case of a surface parallel to a plane with a high *crystallographic index*; the method of controlling the doping of the interface $Si–SiO_2$ with Na^+ ions; and studying the two-dimensional plasma electron vibrations in the inversion layer.

INVERSION OF POPULATION

See *Level population.*

INVERSION SPLITTING, tunnel splitting

(I.B. Bersuker, 1960)

Splitting of energy levels in a multiatomic system in the presence of electron *degeneracy* or *pseudodegeneracy* where the *Jahn–Teller effect* produces two or more equivalent, quite deep minima of an *adiabatic potential* with tunneling between them (see *Tunnel effect*). This results in a lowering of the symmetry and a raising of some, or all, of the degeneracy. The value of the inversion splitting depends strongly on the ratio between the stabilization energy of the Jahn–Teller effect and the vibration quantum in a minimum energy state; it varies over a wide range from fractions to hundreds of cm^{-1}.

ION

An electrically charged atomic particle, or an atom that accepted (*negative ion*, called an *anion*) or gave up (*positive ion*, called a *cation*) one or several electrons (*singly charged ion* or *multiply charged ion*, respectively); also a negatively or positively charged molecule (*molecular ion*). The charge of an isolated ion such as Na^+ and F^- is a multiple of the electron charge e. In crystals with *ionic bonds* the electron density redistribution between the positively and negatively charged ions leads to fractional, i.e. non-integral values of the charge e, called *effective ionic charges*, which are related to their *ionic radii* (see *Atomic radii*). For instance, in an NaCl crystal, the number of electrons is approximately 10.2 (11 in the atom) in Na^+ and 17.8 (17 in the atom) in Cl^-, for an effective ionic charge of $0.8e$. In a LiF crystal the charge of Li^+ is $2.1e$ (3 in atom) and of F^- is $9.9e$ (9 in atom) so the effective ionic charge is $0.9e$.

ION BEAM CHANNELING

See *Channeling of ion beams.*

ION BOMBARDMENT of solids

Bombardment, or irradiation of solid targets by a beam of accelerated *ions*. This results in *backscattering* of ions, which is nearly total for the ion energy range 10^2–10^3 eV (for normal incidence). At higher energies the ions penetrate into the bulk, decelerate due to collisions with target *atoms*, and stop, thus accumulating in the material (*Ion implantation*, see also *Ion-mixing, Ionic synthesis*). *Sputtering* of target atoms, appearance of radiation-induced defects, and *phase transitions* take place in the irradiated matter at the ion bombardment. This is widely used in technology for treating *solid surfaces* of various types (see *Ion-stimulated surface modification, Ionic modification of films, Ion-plasma treatment, Ion lithography*), and for analysis of the substance composition (see *Secondary ion mass spectrometry*). The incidence of collimated ion beams at small angles to low-index directions in *monocrystals* results in *channeling* of the beams.

ION–ELECTRON EMISSION

Emission of electrons by a *solid surface* into a vacuum under ion bombardment. The ion–electron emission coefficient $\gamma = n_e/n_i$ is the ratio of the number of emitted electrons n_e to that of ions n_i incident on the surface. Ordinarily there is no energy threshold. For low-energy ions γ is almost independent of their energy and mass, but depends on their charge ($\gamma \sim 0.2$–0.3 for single-charged ions, and may exceed unity for multicharge ones).

Depending on the mechanism of energy transfer to an electron, *potential ion–electron emission* and *kinetic ion–electron emission* are known. The potential type takes place at the approach of a low-energy ion to the surface when an electron from the solid can transfer to the ion and neutralize it. The energy released at such a transition can be taken up by other electrons in the solid, which are then emitted into the vacuum. The kinetic type emission occurs under high-energy ion bombardment due to intensive electron exchange with the excitation of valence and inner level shells of the target atoms, and involves electrons with a wide energy range as well as Auger electrons.

The ion–electron emission depends on the ionization and excitation energies, as well as on the *work function* of the target material. The energy spectrum of the emitted electrons has a maximum at the energy 1–3 eV, and the Auger emission spectrum can exhibit fine structure. The principles of ion–electron emission underlie methods of analysis of the composition and electronic structure of matter, viz. *ion-neutralization spectroscopy* and *Auger ion spectroscopy*.

ION EMISSION

Emission of negative and positive *ions*, especially by *solids* or liquids to a vacuum or gaseous matter, but ionized gases (plasma) can also serve as ion emitters. Thermal-equilibrium ion emission arises at the thermal excitation of particles. This is the *surface ionization* of vaporizing atoms of the solid itself, of the atoms of impurity elements diffusing from the bulk, or of particles adsorbed from the gaseous medium, as well as the reaction products of the latter at the surface. Thermally nonequilibrium ion emission is stimulated by radiation of various kinds, these are ions (*ion–ion emission*), electrons (*ion–electron emission*), neutral particles (in ground or excited states), the photons of various energy ranges (*ion–photon emission*). Ion emission is used to study the composition and properties of solids.

ION EMISSION, SECONDARY

See *Secondary ion emission*.

ION ETCHING

A removal of material from a *solid surface* under action of *ion bombardment*. Ion etching results in cleaning the surface, exposing its structure, and forming the relief. The bombarding low-energy *ions* originate from various ion sources, glow discharge or radio-frequency plasma formed in a closed cavity containing the object with the surface to be etched (see *Ion-plasma treatment*). The main process involved in *etching* is cathode *sputtering* as well as ion-stimulated chemical reactions which produce volatile substances at the surface, as well as chemically active plasma components. The etching rate is determined by the ion flux density incident on the surface (up to a few mA/cm^2), and by the material properties. The bombardment of crystals reveals *grain boundaries, dislocation* outcrops, and so-called *etch figures* on the target surface. This allows one, in particular, to detect the regions with different states of stress (*selective etching*). Ion etching is used to cleanse oxidized or soiled details and blanks during the technological procedures of manufacturing *integrated circuits* in *microelectronics*, for revealing microstructure of metallography samples, for preparing objects for *electron microscopy* studies, and for obtaining *atomically clean surfaces* in ultrahigh vacuum.

ION EXCHANGER

See *Ionite*.

IONIC BOND, electrovalent bond

An interatomic *chemical bond* with an asymmetric charge distribution between the adjacent atoms. The ionic bond exhibits a molecular electric dipole moment; and its value can be regarded as a measure of the chemical bond ionicity. An ionic bond arises between atoms with a large difference in *electronegativity*.

The concept of ionic bond was introduced by W. Kossel (1916) who noted that it is energetically favorable to transfer an electron from a metal (M) to a halide (X) with the formation of a molecule M^+X^- consisting of ions with the electron shell configurations corresponding to those of noble gases. The resulting ionic bond is caused by the Coulomb attraction between oppositely charged ions. Typical compounds having ionic bonds are *alkali-halide crystals*. However, a perfect ionic bond is an idealization. The wave function is actually a superposition of those of covalent and ionic bonds, ψ_{co} and ψ_{ion}, respectively:

$$\psi = C_1\psi_{\text{co}} + C_2\psi_{\text{ion}}.$$

In this case the bond *ionicity* (β) is given by the expression

$$\beta = \frac{|C_2|^2}{|C_1|^2 + |C_2|^2}.$$

There are several methods for empirically estimating the value β. L. Pauling established that in the case of a predominantly *covalent bond* ($C_1 \gg C_2$) in a molecule A–B, one can calculate the bond energy as an arithmetic (or geometric) mean of the experimentally measured (covalent) bond energies of the molecules A–A and B–B. Pauling proposed to consider a deviation from this mean value as a measure of the contribution of the ionic component of the bond. According to Pauling the dependence of the parameter β on the difference between the electronegativities of atoms A and B, $X_{\text{A}} - X_{\text{B}}$, has the form

$$\beta = 1 - \text{e}^{-0.25}(X_{\text{A}} - X_{\text{B}})^2.$$

Even for typical alkali-halides β fails to reach unity; for example, for LiF $\beta \sim 0.92$, for KCl $\beta \sim 0.82$, and for NaCl $\beta \sim 0.75$.

IONIC CONDUCTIVITY

Electrical conductivity by ions as the charge carriers. Unlike electron conductivity, it involves the transfer of mass to electrodes. The substance deposition is described by the *Faraday law*: at the transfer of the electrical charge of 96,500 C, one gram-equivalent of ion components is carried to the electrodes. The substance decomposition under the action of electric current is called *electrolysis*, so ionic conductivity is frequently referred to as *electrolytic conductivity*. This is the prevailing type of conduction in many dielectrics, in particular, in *ionic crystals*, e.g., in *alkali-halide crystals*. Interstitial (intrinsic or impurity) ions can be current carriers such as Ag^+ in crystals AgCl, H^+ and OH^- in ice crystals, or charged vacancies (e.g., anion vacancies in alkali halide crystals). At low temperatures the ionic conductivity values are determined by the concentration of impurity or nonstoichiometric current carriers, and can differ from one sample to another made of the same material. The temperature dependence of the conductivity σ has an Arrhenius character, $\sigma = \sigma_0 \exp[-U_{\text{g}}/(k_{\text{B}}T)]$, where U_{g} is the *activation energy* of the current carrier jump process. At high temperatures, the ionic conductivity is determined by the concentration of stoichiometric current carriers that are the components of *Frenkel defects* or *Schottky defects*, and it is the same for every sample of a given substance. The activation energy of ionic conduction at high temperature is determined in this case by a sum $U = U_0 + U_{\text{g}}$ where U_0 is the energy of thermally-stimulated nucleation of a *stoichiometric defect*. The maximum conduction attained at a temperature close to that of melting is 10^{-3}–10^{-4} Ω·cm. *Solid electrolytes* exhibit abnormally high ionic conductivity.

IONIC CONDUCTIVITY COMPENSATION

Sublinear dependence of the *ionic conductivity* on the composition of multicomponent materials in the *condensed state of matter* for an equimolar interchange of any two of their components. Ionic conductivity compensation is called the *polyalkaline effect* when it involves relative replacement of alkali atoms in oxide-coated glasses. This compensation is an analogue of the sublinear dependence of the electric conductivity on the composition in some *alloys* for equimolar replacements of the components in them, and in semiconductors for the equimolar replacement of donor impurities by acceptor types. All these cases can be explained by a general model for the simultaneous transport of any two types of *current carriers* which makes use of a matrix defining their possible states with different migration abilities. There is a so-called bound case in which the mobility is very small or equal to zero, and the free case in which the mobility is large enough for electromigration observation.

IONIC-COVALENT BOND

A type of a *chemical bond* intermediate between the purely *ionic bond* and purely *covalent bond*. Crystals built of atoms with different degrees of *electronegativity* possess more or less differing average atomic charges when forming valence bonds. Therefore, in addition to a purely covalent bond, there appears also an admixture of ionic bond due to the attraction between charges of opposite sign. An example is the compounds $A^{III}B^{V}$ and $A^{II}B^{VI}$ where the elements of Groups V and VI acquire a small ($\sim$0.2–0.3) electron charge taken from the elements of Groups III and II, respectively. On the other hand, many crystals considered as *ionic crystals* have a partially covalent bond due to overlapping of the nearest neighbor wave functions. For such univalent crystals as *alkali-halide crystals* (and also silver- and thallium-halide ionic crystals) the charges fail to reach the limiting values ± 1, and amount to ± 0.96. This effect is even more noticeable for oxides, sulfides, and selenides (e.g., MgO, SiO_2, PbS and CdSe), since the ion O^{2-} (to say nothing of the ions S^{2-} and Se^{2-}) is unstable in the isolated state. Intermediate cases are possible when it is difficult to unambiguously establish average atomic charges that deviate greatly from those ascribed to their *valence*.

IONIC CRYSTALS

Crystals where the ionic type of *chemical bond* prevails. Ionic crystals consist of positively (*cations*) and negatively (*anions*) charged *ions* that can be monatomic or multiatomic. For example, ionic crystals of the first type are alkali halides, alkaline-earth metal halides and oxides where the cations are metal ions, and the anions are halogen or oxygen ions (NaCl, CaF_2, Al_2O_3, etc.). The ionic crystals of the second type are nitrates, sulfates, phosphates and other salts of metals where the anions are oxygen radical groups consisting of a few atoms (NO_3^-, SO_4^{2-}, PO_4^{3-}, etc.). A certain (small) *ionicity* is inherent in the chemical bonds of *covalent crystals*.

The lattice energy of ionic crystals consists mainly of an electrostatic part (*Madelung energy*) and a positive energy of interaction of the ion electron shells (*Pauli energy*), and it amounts to 6–10, 20–30, and 30–150 eV per pair of ions for alkali halides, alkaline-earth halides, and oxide crystals, respectively. As to the electrical properties, most ionic crystals at normal conditions are *insulators*, and some of them are superionic conductors (see *Solid electrolytes*). In view of a wide *band gap* (up to 13.6 eV in the case of LiF), ionic crystals are transparent to electromagnetic radiation in the visible and ultraviolet spectral ranges. The high-energy (up to 0.2 eV) optical *phonons* are well suited for determining the spectrum of atomic vibrations of ionic crystals. Physical properties of ionic crystals are strongly dependent on their *defect* content (anion and cation *vacancies* and *interstitial atoms* and ions, *dislocations*, *impurity atoms*). The latter result, e.g., in the appearance of selective absorption bands in the transparency range of pure crystals (see *Color centers*). Typical for many ionic crystals, e.g., halides, is the electron excitation decay (mostly *excitons*) with the formation of lattice defects.

IONICITY

Semi-quantitative characteristic of a *chemical bond*, which is intermediate between an *ionic bond* (with ionicity = 1) and a *covalent bond* (with zero ionicity). There is no unified and universally recognized definition of ionicity. The most commonly used definitions are those proposed by L. Pauling, C.A. Coulson, and J.C. Phillips. The first of these is based on the chemical bond energy, the second proceeds from a consideration of the coefficients of the expansion of single particle *orbitals* in terms of atomic states, and the third is based on the static *dielectric constant*. The parameters of the ground state of *crystals* correlate with ionicity, which allows its use for rough interpolations (see *Ionic bond*, *Ionic crystals*). More generally, the term ionicity can refer to the ionic properties of a solid.

IONIC MODIFICATION OF FILMS

An alteration of the structure and properties of *films* through *ion bombardment*. Irradiation is most often performed by ions of noble gases, hydrogen, and atoms involved in the film composition with the energy of about 10^2–10^3 eV. Subjected to the irradiation are either the substrate before the film deposition, or the entire system during the film synthesis. The ionic modification of carbon films has been investigated in detail, with irradiation by low-energy ions ($\leqslant$1000 eV), the most efficient for carbon film synthesis. The term "i-carbon" is used to denote the carbon film formed during the irradiation. It is distinct from the graphite-like carbon of films condensed without the irradiation. The "i-carbon" exhibits diamond-like properties: enhanced *hardness*, *density* and *transparency*, high electric resistivity, chemical resistance, the band gap width specific for diamond (see *Diamond-like carbon films*). Varying the ion energy one can obtain carbonic films with a wide spectrum of properties: density $d = 1.9$–4 g/cm^3; Mohs hardness 1–10; electric resistivity $\rho = 10^3$–10^{12} Ω·cm; *band gap* width 0.5–2 eV. The action of the ions causes the selective suppression of the graphite-like phase stable under normal conditions, and stimulates the formation of metastable phases, the foremost being diamond. In addition, formerly unknown crystal modifications of carbon were obtained: a face-centered cubic phase with lattice constant $a = 0.357$ nm (γ-phase) and a body-centered cubic phase with $a = 0.428$ nm (phase C_8). The ion-stimulation effect is caused by both elastic and inelastic processes. The important role of inelastic processes is apparent from the oscillating dependence of the ionic effect on the ion energy in the low energy range (<200 eV), which reflects the character of the energy dependence of the cross-section of ion resonance recharge at small graphite *clusters*.

In irradiation of already formed carbon films by high and medium energy ions, the *graphitization* process dominates, as is the case for heat treatment. Here, the film conductivity increases up to values more than 10^3 (Ω·cm)$^{-1}$.

IONIC RADIUS

A size characteristic of *ions* in ionic compounds. The ionic radius is defined so that the distance between the nearest neighboring ions in a *crystal lattice* of NaCl type would equal the sum of the radii of the corresponding ions. Historically there have been several internally consistent systems of ionic radii. For example, the radius of the ion O^{2-} is 0.132 nm in the Goldschmidt system; 0.136 in the Belov and Bokij system; 0.140 in the Pauling system; 0.146 in the Zachariasen system. Nowadays researchers use standard, accepted values tabulated in handbooks, and for O^{2-} this is 0.132 nm. When using ionic radii to determine the interionic distances in *crystals* with structures different from that of NaCl, one may have to introduce corrections close to 0.01 in order of magnitude (see also *Atomic radius*).

IONIC SYNTHESIS

A formation of new *phases* consisting of the atoms of a target and *atoms* introduced during *ion implantation* in the solid. As a rule, the energy of the implanted ions is between tens and hundreds of keV, the dose is more than 10^{17} ion/cm^2, the ion flux density is 10^{13}–10^{16} ion·cm^{-2}·s^{-1}; the target during the irradiation can be either at an elevated temperature (to ~1000 °C) or room temperature (or lower). The phases under synthesis are formed in a thin surface layer ($\leqslant$1 μm) in either an amorphous (see *Amorphization*) or a crystalline state directly during the implantation process. To stimulate the *crystallization* and to improve the synthesized compound structure, subsequent thermal, light pulse, electronic and other kinds of *annealing* are often used.

Phase transformations (see *Phase transitions*) that take place during the synthesis process under the ion implantation have some particular features:

(i) Because of essentially nonequilibrium conditions due to high-rate ($t \sim 10^{-10}$–10^{-11} s) nonthermal excitation of the lattice and the electron subsystem in local regions along the *ion* tracks, where very *high pressures* and temperatures arise, both equilibrium and metastable phases are synthesized (see *Metastable state*), as well as some compounds unobtainable by ordinary methods.
(ii) The phases under synthesis are formed by way of nucleation of elementary molecular complexes at the *crystal nuclei* (most often, various radiation-induced defects and defect clusters); such phases are usually finely dispersed.
(iii) The influence of the base material structure and orientation on the crystallization and type of synthesized compounds is predominant due to adaptation of the newly formed structural elements to details of the source phase.
(iv) The course of solid state chemical reactions during ionic synthesis is greatly affected by the high concentration of radiation-induced defects, structural disordering, and high local pressure.
(v) Important parameters of the ionic synthesis that influence the phase composition and the crystal structure of forming *films* are the ion flux density and the sample temperature during the irradiation. The ion energy determines only the embedding depth and the effective thickness of the synthesized layer.

The ionic synthesis method for *thin films* possesses some essential advantages compared to other traditional methods: purity of the process; opportunity to obtain a homogeneous gas mixture of strictly controlled composition; high reproducibility of results; opportunity to obtain deeply lying layers and multilayer systems, i.e. to govern the chemical and mechanical properties of the subsurface layers; opportunity to carry out the synthesis at low temperatures; opportunity to form metastable phases; high *adhesion* of the synthesized layers to the substrate; opportunity to synthesize the phases on local regions of the substrate.

The ionic synthesis method has prospects for solving problems of *microelectronics* (such as producing silicides, insulator and semiconductor layers), of *metal science* by modifiying surface properties (improvement of wear resistance, *hardness*, *corrosion resistance*, *thermal resistance*, etc. via thin films of carbides, borides, silicides, and other compounds on the surface), and of many other areas of science and engineering.

ION IMPLANTATION

Introduction of atoms into the surface layer of a *solid* by bombarding its surface with *ions* of energy from a few kiloelectronvolts to a few megaelectronvolts (see *Ion bombardment*). The penetration of the impurity atoms into the solid at implantation is due to the ion kinetic energy far exceeding the binding energy of the atoms in the crystal. The process does not need high temperatures as is the case with diffusion *doping* or *alloying*. Ion implantation is a nonequilibrium process so there are no thermodynamic restrictions on the concentration of implanted impurities, and this concentration can far exceed values limited by equilibrium *solubility*. This method allows one to introduce any element into any material at room temperature, and in a quantity strictly controlled by the ion current density and the exposure time. The penetration depth of ions of a given chemical type into the selected material is determined by their initial energy, and by the stopping ability of a target (see *Charged particle paths*). One can freely change the concentration profile by varying the ion doses and their energies in a controlled manner, and produce latent ("buried") doped layers by implanting ions of sufficiently high energy (hundreds of electronvolts). The use of electrical or mechanical scanning of the ion beam across the surface can assure high homogeneity doping. The ion implantation method allows one to carry out local doping by using a sharply focused ion beam or protective masks. Since the ions experience numerous collisions with the lattice atoms, a great number of *radiation-induced defects* appears in the implanted region. As a rule, the number of these defects is more than 100–1000 times that of the implanted ions. In *semiconductor materials*, the radiation-induced defects can mask the effect of chemical doping; so *annealing* is called for. *Pulsed annealing of semiconductors* is efficient after ion implantation (in particular, laser annealing). The dose range for *ion implantation doping* of semiconductors varies within the limits 10^{12} to 10^{15} ions/cm^2. Ion implantation into *metals* and *alloys* requires doses as large as 10^{15}–10^{18} ions/cm^2. This method allows one to control the solid surface properties over a broad range independently of the bulk properties, thus providing unique possibilities in the fields of solid-state electronics and *metallurgy*. Initially the ion implantation method was applied to semiconductors as a means for introducing electrically active impurities for producing *semiconductor junctions*.

Owing to the universality of the method, its complete compatibility with planar-epitaxial technology, and also its consistency with *diffusion*, ion implantation is used in the production processes of most *semiconductor devices* and integrated circuits. The guiding principles behind the development of solid-state electronics, viz., the increase of component packing density and speed and the improvement of quality, are well implemented by combining ion implantation with *ion lithography*. Ion implantation is also employed for introducing radiation-induced defects into the solid surface to modify its mechanical, electrical, electrophysical and optical properties. For example, the implantation of the ions H^+ and N^+ in *gallium arsenide* provides the disordered layers that are the basis of switching devices. The introduction of implantation disturbances into devices based on *magnetic bubble domains* allows one to control their static and dynamic characteristics. The ion implantation method is applied for governing the surface properties of metals and alloys. By appropriate selection of ions, it is possible to improve wear resistance, *corrosion resistance*, *radiation resistance*, fatigue *failure* resistance, and resistance to the surface *oxidation* of various materials, as well as to reduce their *friction* coefficients. The microscopic aspects of metallurgy (e.g., to find impurity diffusion rates in metals, *phase diagrams*, enthalpy and *entropy* of reactions, trapping of atoms of dissolved matter), as well as the mechanisms of *corrosion* and surface strengthening, can be understood better with the help of ion implantation data. In parallel with these applications, ion implantation is used to investigate metastable phases and obtain metallic amorphous alloys on the surface of metals and alloys. It also finds wider use for controlling the properties of *insulators*: *ceramics*, glasses (see *Vitreous state of matter*), and *polymers*. In particular, ion implantation into polymers allows transforming various polymers into conductors through increasing their electrical conductivity by many orders of magnitude.

ION IMPLANTATION DOPING of semiconductors

Method for the formation of controlled distributions of electrically active impurities in *semiconductors*. The method is based on *ion implantation* of atoms of chemical elements, which are capable of serving as *donors* or *acceptors* in the treated semiconductor, e.g., atoms of Groups V and III in silicon and germanium, respectively. Selection of the *ion* bombardment energy serves to control the doped layer depth (see *Charged particle paths*), while *lithography* helps to establish the layer distribution relief over the irradiated surface. However, many of the implanted atoms fail to occupy electrically active states during the implantation. For example, the atoms of Groups V and III in Si and Ge can be at interstitial positions or enter compounds with radiation-induced *defects* that appear at implantation, i.e. be in electrically inactive states, or in states with *deep levels*. This reduces the *mobility of current carriers*. At high doses the amorphization of semiconductors is also possible. Therefore, the technology of ion *doping* includes *annealing*. Thermal annealing can cause the diffusion spreading of the impurity distribution profile established at implantation. More efficient is *laser annealing* that does not lead to thermally active *diffusion*. Semiconductor ion doping imparts *diamond* layers with semiconductor properties and produces *semiconductor junctions* there, and doping *gallium arsenide* crystals and other semiconductor compounds provides them with donors and acceptors, etc. Due to the possibility of producing sharp-relief microprofiles of electric resistance in various materials, ion semiconductor doping is highly efficient in *integrated circuit* manufacture.

IONITE, ion exchanger

A substance enclosing ions which can be exchanged with *ions* in a surrounding electrolyte solution, in a melt, or in a liquid passing through it. Typically, ion exchangers or ionites are insoluble solids consisting of a polymeric space framework with fixed ions and counter-ions. The latter compensate the framework charge, and are capable of exchanging for equivalent quantities of ions of the same sign from the solution or the melt. According to their sign, the counter-ions are cationites (−), anionites (+) and ampholytes (±). According to the nature of the framework, there are inorganic, organic and mineral-organic ionites. A special group comprises *oxidation–reduction ionites* (*redoxites*) whose counter-ions are capable of reversible oxidation–reduction transformations. The ionites are substances of both natural (*zeolites*, clay, peat, cellulose, etc.) and synthetic (silica gel, synthetic zeolites, zirconylphosphates, ion-exchange resins) origin.

The important characteristics of ionites are the nature and the concentration of their functional groups. The types of functional groups containing fixed ions determine the selectivity of the ionites, i.e. their ability to interact selectively with various ions. The concentration of the functional groups influences the counter-ion concentration, i.e. the ionite exchange capability (calculated per unit of dry or swollen ionite).

Ionites are used in the form of grains, powders, fibers, tissues or *membranes*. Areas of application are: ion exchangers for the purification of substances from electrolyte admixtures, separation of mixtures, isolating and concentrating of substances (in water treatment, hydrometallurgy, recycling industry), catalysts (zeolites in petroleum chemistry), gas absorbers, and ion emitters in *mass spectrometry*.

IONIZATION, FIELD

See *Field ionization*.

IONIZATION, IMPACT

See *Impact ionization*.

IONIZATION SPECTROSCOPY

A variety of *characteristic energy loss spectroscopy*. Ionization spectroscopy deals with groups of electrons comprising the *secondary electron emission* energy spectrum, which expend some energy exciting electrons from inner K, L and M shells of atoms of a solid, or raising electrons from the *valence band* to upper (unoccupied) levels. These expended energies are directly related to ionization energies. There are transmission and reflection versions of ionization spectroscopy. In the former electrons that lose energy ΔE and are scattered at a small angle are detected. In the latter case the electrons are detected after a two-stage process: initial energy loss ΔE due to scattering at a small angle, and then elastic scattering at a wide angle. The scattering cross-sections decrease with growing primary electron energy E_{P}; to observe the ionization spectral lines, E_{P} is preferably selected close to the electron binding energy E_{B} for a given level ($E_{\mathrm{P}}/E_{\mathrm{B}} = 1$–$1.4$). Sometimes ionization lines and Auger lines (see *Auger effect*) fall into the same energy range. The ionizing lines can be differentiated by the dependence of the line position on E_{P} since the ionization line exhibits such a dependence, while the position of the Auger line is independent of E_{P}. Using ionization spectroscopy, one can obtain data similar to that from *X-ray photoelectron spectroscopy*. However, the binding energies detected in X-ray *photoelectron spectroscopy* are referenced to the *Fermi level*, whereas ionization spectroscopy determines the energy loss (see also *Electron spectroscopy*): $\Delta E = E_{\mathrm{B}} \pm \xi$, where ξ is associated with the maximum of the density of unoccupied states (see *Density of states*) relative to the Fermi level. The values ΔE pertain to the final relaxed and screened state of a *hole* at the ground level of the solid, and are also dependent on the chemical environment. One can estimate ξ by comparing E_{B} found from X-ray photoelectron spectroscopy to ΔE obtained from ionization spectroscopy.

The fine structure of the ionization spectrum contour should reflect the structure of the ground level and the distribution of the density of free electron states. The ground levels are broadened by the finite lifetime of the excited state with a hole (natural linewidth), Koster–Kronig transitions, *exchange interaction*, *Jahn–Teller effect*, *shake-up*, *shake-off*, and other many-particle effects. In a number of cases, e.g., for the L_{23}-level of silicon, phosphorus, the K-level of carbon, etc., the ground state level broadening does not exceed 1 eV, and the ionization line fine structure does reflect the distribution of the density of unoccupied electron states. The ionization spectroscopy of outer shells reflects the loss for transitions from the valence band to higher levels, and is sensitive to the condition of the *solid surface*. If independent photoelectron spectroscopy data on the valence band structure are available, then ionization spectroscopy provides information on unoccupied electron states.

Ionization spectroscopy is used for qualitative and quantitative analysis of the composition and the electronic structure of the crystal subsurface, and the adsorbed layers of solids. An advantage is its insensitivity with respect to the surface charging, its ease of implementation, the possibility of obtaining information on the density of free states, and convenient control of the probe depth by varying the primary electron energy E_P. This provides for the nondestructive elucidation of the composition and electronic structure of subsurface layers in the depth of a solid.

IONIZATION, SURFACE

See *Surface ionization.*

ION–LATTICE INTERACTION, ion–photon interaction

A dynamic characteristic of *noncentral ions* which arises from the dependence of the potential energy of these ions on lattice atom displacements from their equilibrium positions. The ion–lattice interaction is responsible for both the influence of the elastic stress in a crystal on the spectrum of tunneling states of noncentral ions, and the *ion–lattice relaxation*. Ion–lattice interaction data are obtained from experiments on *paraelectric resonance* in the presence of an applied axial pressure.

ION–LATTICE RELAXATION

A process of restoration of thermodynamic equilibrium in a system of *noncentral ions* after the action of some external disturbance (e.g., pulse saturation of the *paraelectric resonance* spectrum). By analogy with *paramagnetic relaxation*, two relaxation times describe the ion–lattice relaxation, namely: the *longitudinal relaxation time* (τ_1) and *transverse relaxation time* (τ_2). The time τ_1 is measured by electrocalorimetric means using pulse or continuous saturation, or it is estimated from the temperature broadening of the paraelectric resonance lines. The time τ_2 can be estimated from the uniform temperature-independent part of these lines. Both relaxation times depend on the concentration of the noncentral ions; and τ_1 also depends on the temperature (e.g., for Li^+ in KCl, $\tau_1^{-1} \sim T$ at $T < 4$ K, with the value $\tau_1 \approx 10^{-8}$–10^{-9} s at $T = 1.4$ K). See also *Relaxation time.*

ION LITHOGRAPHY

A method of formation of microstructures in the technology of large and very large *integrated circuits.* It includes ion beam irradiation (exposure) of a polymer *resist* (radiation sensitive coating) deposited on a *substrate*; obtaining a protective resist mask by a development process; *etching* the material through the mask, then removal of the mask (see *Lithography*). When using polymer resists, ion doses of 10^{11} to 10^{14} cm^{-2} are employed. The *negative resists* become polymerized, the molecular mass of the *polymer* increases and it becomes insoluble during development. The irradiation of *positive resists* results in the *polymer degradation*, which becomes soluble. The resist exposure is performed by a spatially modulated or computer-controlled sharply-focused ion beam.

Ion lithography is superior to other methods of *microlithography* (photolithography, electron and X-ray lithography) in submicron resolution for short exposure times, and is insensitive to *proximity effects*. This is due to the small lateral path of secondary electrons ($\sim$10 nm) during the exposure and to the efficient absorption of the radiation ($\sim$1 keV/nm) in the resist material. More descriptively, ion lithography is a method of direct microstructure formation in the technological and topological layers of integrated circuits. Layer formation results from the ion, chemical or ion-stimulated chemical etching of *insulator*, *metal*, or semiconductor layers by a controlled ion beam.

ION METALLURGY, implantation metallurgy

Production of *solid solutions* by the *ion implantation* method. A particular feature of the method is the possibility of creating materials that cannot be obtained under equilibrium conditions. Such metastable "alloys" can persist up to rather high temperatures. An example is the AgCu substitutional solid solution obtained by implantation of silver in copper (this decays at $T \geqslant 310\,^{\circ}\mathrm{C}$), the metastable substitutional solutions W–Cu or Ta–Cu appearing at the implantation of tungsten or tantalum, respectively, in copper, etc. See also *Ionic synthesis.*

ION-MIXING during irradiation

Redistribution of atoms in a *solid* during irradiation by an ion beam, caused by the appearance of *recoil atoms* and collision cascades (*recoil atom implantation*). This can involve distortion of the impurity concentration profiles at high-dose *ion implantation.* The most interesting processes for studies and practical applications are:

(1) Ion-mixing with cross-penetration of dissimilar atoms at the interface of the two-layer structure "thin film–substrate" aimed to produce thin ($\sim$10 nm) boundary layers with special properties (film thickness $d \sim 30$–50 nm, *ion* energy such that the maximum of energy loss is at the interface). At elevated irradiation temperatures (mostly $>$300 K), an increase of the atom penetration depth is observed due to *radiation-induced diffusion.*
(2) Ion-mixing via irradiation of multilayer structures formed of very thin ($d \sim 3$–5 nm) alternating layers aimed to produce *phases* (including metastable ones) or compounds with a certain structure and stoichiometry.

Ion-mixing is often studied via Rutherford *backscattering*, Auger microanalysis, and X-ray methods. Used for this purpose are universal ion sources of inert gases (less often P^+, N^+, etc.) of energy $E \sim 50$–600 keV with radiation doses $\Phi = 10^{15}$–10^{16} ions/cm^2. Using such sources is an advantage compared to direct ion implantation where a special ion source is needed for each implanted element, and high implantation doses are required (e.g., for metals $\Phi = 10^{17}$–10^{18} ions/cm^2). Dynamic ion-mixing occurs at the irradiation of the interface "film–substrate" during the film *condensation* (renewal), and this allows one to reduce sharply the ion energy ($E \sim 10$ keV). These methods of ion mixing are used to produce very thin *coatings* and surface layers with some special properties such as improved wear and *corrosion resistance*, *hardness*, high *adhesion*, surface *electrical conductivity*, etc.

ION-MOLECULAR CRYSTALS

Crystals formed by complex ions, and belonging to *heterodesmic structures*, i.e. those with more than one type of chemical bonding. One can group ion-molecular crystals into three types:

(i) those with a monatomic cation (positive ion) and a complex anion (negative ion; salts of oxyacids): perhalogenates, halogenates, nitrates, phosphates, sulfates, carbonates, niobates, tungstates and similar compounds of alkali and alkaline-earth metals;
(ii) those with a monatomic anion and a complex cation such as ammonium halides;
(iii) those with a complex cation and anion, e.g., NH_4NO_3, NH_4ClO_4, NO_2ClO_4, etc.

Ion-molecular crystals are generally low in symmetry and polymorphous. For example, five lattice modifications for NH_4NO_3 are known. The presence of low-symmetry complex ions can involve the orientational ordering or disordering of ion-molecular crystals, which is related to the possibility or impossibility of a free rotation of the complex ions at their lattice sites. It is for this reason that *order–disorder phase transitions* are common for ion-molecular crystals, as well as anisotropy of the crystal properties in most cases. In view of the purely *ionic bond* between the anion and the cation they have similarities with *ionic crystals* like *alkali-halide crystals*, while their status of being covalently bonded complex ions provides them with properties similar to those of *molecular crystals.* These features are graphically revealed in optical spectra where, along with the optical phonon bands of ion-molecular crystals, there are infrared bands typical of internal vibrations in complex ions (*intramolecular vibrations*). The ultraviolet spectra of ion-molecular crystals are complicated by intramolecular transitions of electrons from occupied orbitals to free ones. Therefore, a double-band model involving a *valence band* and a *conduction band* is not applicable to ion-molecular crystals, although there are intrinsic allowed electron energy levels (or quasi-bands) between these bands.

IONOLUMINESCENCE

Luminescence appearing during high-energy (0.001–10 MeV) ion bombardment of solids. Its salient features are a high density of excitation energy and the surface nature of the excitation owing to the small ion penetration depth in *solids* (from several to hundreds of angstroms depending on the beam energy). Hence, the ionoluminescence method yields information that pertains mostly to the subsurface layer; it underpins *ion–photon spectroscopy*. The ionoluminescence spectrum is determined by the composition of the target material and by the radiation-induced defects appearing under the *ion bombardment*. In addition, the intensity depends on the ion energy and density, and on the crystal orientation relative to the beam direction.

ION–PHOTON EMISSION

Emission of optical radiation by a *solid*, or by excited particles leaving its surface under bombardment by accelerated *ions* (or atomic particles). The radiation emitted by the surface itself is typical of insulators and semiconductors, and is widely investigated (see *Ionoluminescence*). There is another component of ion–photon emission which is specific for each substance, and related to photon emission by excited particles (sputtered and scattered *atoms*, molecules, *clusters* and their ions) noninteracting with the surface. The spectra of this component contain, in a general case, atomic and ionic lines as well as molecular bands, and in some cases continuous radiation is observable. The principal mechanisms of the excitation of sputtered atoms and ions are, first, the detachment mechanism when the electronic states of the particles are formed only during the process of their escape from the surface, and second, the kinetic mechanism when the particle excitation is caused by atomic collisions. Important in both cases are electron exchange processes in the solid, particle-emitting system. Quantitatively, the excited particle emission is characterized by the *photon emission factor* (*photon yield*) Y, which is the ratio between the number of electrons emitted in a given frequency range, and the number of ions incident at the surface. The value Y depends on the nature of the material, the condition of its surface, and the parameters of the bombarding ions (energy, angle of incidence, and ion type). In the case of metals, $Y \leqslant 10^{-4}$ photon/ion at the resonance transition frequency under bombardment of the target by ions of intermediate mass and energy. This parameter may increase noticeably ($\geqslant$10–100 times) when the surface is coated with atoms of chemically active gases. The principles of ion–photon emission underly the *ion–photon spectroscopy* method of elemental and *layer-by-layer analysis* of the surface.

ION–PHOTON SPECTROSCOPY

A physical method based on the *ion–photon emission* phenomenon for the study of a solid surface. Ion–photon spectroscopy employs measurements of the intensity and frequency range of optical emission from particles sputtered from the surface, the dependence of these characteristics on the incident ion position, the sputtered layer depth, the nature, and the surface condition of the target. This spectroscopy allows one to determine the elemental composition of the surface, and the distribution of elements over the depth and the surface. One can also study spectra of atoms and molecules, processes of oxidation, diffusion, surface particle adsorption, and mechanisms of the interaction of atomic particles with solids.

Quantitative elemental analysis is based on the correspondence between photon yield, Y, at the frequency of selected transitions of impurity atoms with concentration, c, and the atomization factor, s_{M} of the host atoms: $c = Y/(A\tau n s_{\mathrm{M}})$ (A is the probability of the transition; n and τ are respectively the probability of excitation and the atom lifetime in this state). The determination of c is facilitated if parameters n and s_{M} are known or can be calculated, and we note that the relation between c and Y is linear. In other cases one should use standards. The *layer-by-layer analysis* is performed by continuously measuring the dependence of the intensity of a spectral line of any impurity atom on the duration of the surface *etching*. The potentiality for quantitative analysis using ion–photon spectroscopy is almost the same as that of the related method of *secondary ion mass spectrometry*. Advantages of ion–photon spectroscopy are high sensitivity and accuracy of elemental identification, remote mode of data acquisition, and no need to apply an electric field to a target.

The ion–photon spectroscopy method uses ions (typically, Ar^+) with tens of keV in energy to probe the surface under conditions of high vacuum. The emitted particle radiation is analyzed by *optical spectroscopy* methods.

ION-PLASMA TREATMENT

Treatment of materials that consists in modifying the material surface and subsurface layers by gas plasma components, i.e. by high-speed electrons and *ions*. To this end, the material is placed in the immediate vicinity of or inside the region where a gas discharge occurs with its strong electric (constant or high frequency) and magnetic fields. A pressure of 10^{-2}–10^2 Pa is usually employed with a particle density of 10^{12}–10^{16} cm^{-3}, and a particle energy of 10–10^4 eV. Collisions of electrons and atoms with atoms of the irradiated material lead to the following effects: (i) surface *sputtering* (knocked out *atoms* and ions become plasma components and participate in the treatment); (ii) ion penetration into the material (*ion implantation*); (iii) deposition of particles from the plasma on to the *substrate*; (iv) activation of various heterogeneous chemical reactions at the treated surface. Using this ion-plasma treatment one can control material parameters over a wide range by changing the plasma chemical composition, the electric discharge parameters (energy of excited particles, ratios of excited and ionized plasma particles, multiplicity of ionization), and the magnitudes and directions of electric and magnetic fields. Sometimes the ion plasma component only is isolated for irradiation (see *Ion bombardment*, *Ion-stimulated surface modification*). Widely used ion-plasma treatments are:

(1) *plasma-chemical etching* and *ion-chemical etching*, including the chemical reactions of high-energy ("hot") plasma particles with the material;
(2) *ion-plasma deposition* and *ion-beam deposition* for producing films on various substrates. The atoms to be deposited are introduced into the gas phase by sputtering. Protective coatings obtained by this treatment have high *adhesion*; the process of their formation does not need high temperatures; application of several sputtering targets of various chemical composition allows one to govern the composition of the forming films;
(3) *ion-plasma cleaning*, polishing, producing the relief of *surface roughness*.

These treatments are efficient in combination with ion bombardment (so-called *reactive ion etching*, reactive ion-plasma deposition of materials). In these cases, an additional ion beam is used to accelerate the processes of forming volatile products of chemical reactions by incoming plasma particles on the irradiated surface, and the removal of the reaction products.

ION PROJECTOR, field-ion microscope

A lensless vacuum ion-optical apparatus for obtaining an image of a conducting material surface with a magnification of several million. A tip of a thin needle cooled by liquefied gas acts as a positive electrode, with its surface imaged on a luminescent screen. The noble gas atoms that fill the inner volume of the ion projector are ionized in the strong electric field ($\sim 5 \cdot 10^8$ V/cm) at a distance 0.5 to 1 nm from the needle tip that accepts their electrons. Under the action of the electric field, positive *ions* of the imaging gas are accelerated and bombard the screen which has a negative potential. The luminescence distribution of each screen element is proportional to the density of the incoming ionic flow, and thus reproduces the ion density distribution near the needle tip on an expanded scale. The magnification equals the ratio between the screen radius and the tip curvature radius. The contrast image of the needle tip appearing at the screen is determined by the presence of the local microscopic relief, and the imaging gas pressure that usually does not exceed 0.01 Pa. An ion projector is used in *field ion microscopy*.

ION-STIMULATED SURFACE MODIFICATION

Alteration of the structure and properties of the surface and subsurface layers of solids by *ion bombardment*. The bombardment of solids by ions with energy $E \sim 1$–10 keV leads mainly to *sputtering* and cleaning the *solid surface*, to modification of the *crystal morphology*, i.e. the formation of "cones", "craters", etc.; or, conversely, to ion polishing of the surface for special conditions. At high radiation doses Φ with inert gas ions ($\Phi \sim 10^{17}$–10^{20} cm^{-2}), gas blisters may form at the surface (see *Blistering*), and surface *swelling* and *scabbing* occurs, which imitates the processes connected with problems at the first wall of a thermonuclear reactor. The high-energy ions implant themselves into the irradiated material (see *Ion implantation*). Initial studies involved the *ion implantation doping of semiconductors* for the modification of their electrophysical characteristics, for the formation of p–n junctions, and for the design of devices. Implantation metallurgy (see *Ion metallurgy*) involves the modification of surface properties of metals and alloys to increase *microhardness*, wear resistance, ultimate fatigue strength, *corrosion resistance* and *cavitation resistance*, as well as modification of optical and other properties by high-dose implantation of various elements using beams of ions with energies of hundreds of keV, and $\Phi \sim 10^{16}$–10^{18} cm^{-2}. Mechanisms for the improvement of properties are diverse and not always clear. Depending on the ion–target combination and the regimes and conditions of the irradiation which take place simultaneously with the impurity implantation to the depth of a few thousand angstroms, there is observed the formation of equilibrium or metastable surface alloys, phase and/or structural transformations in thin subsurface layers (up to *amorphization*), the introduction of mechanical stresses, etc. When carried out in a technical vacuum, ion-induced *oxidation* and carbonization of the surface are also possible. A particular case is ion-nitriding, which improves some mechanical properties of the surface of these materials. See also *Ion-mixing*, *Ionic synthesis*, *Ionic modification of films*.

ION TREATMENT of surfaces

An action upon a *solid surface* by a flux of accelerated ions to change its profile, chemical composition, physicochemical and electrophysical properties, as well as to produce pits or to deposite foreign substances. The ion treatment of a surface is carried out with the aid of *ion etching*, ionic milling, *ion implantation*, ionic *surface passivation*, *ion lithography*, and ion-stimulated deposition. The main processes resulting from *ion bombardment* are cathode *sputtering* and ion implantation. This treatment is applied to metals, semiconductors, and insulators, and is one of the basic technological processes in the manufacture of *semiconductor devices* and *integrated circuits*. The ion treatment of a surface typically requires an ion source, ionic optics, a mass separator, and a source of accelerating voltage up to 200 kV.

IRIDIUM, Ir

Chemical element of Group VIII of the periodic system with atomic number 77 and atomic mass 192.22. Natural iridium is a mixture of two stable isotopes ^{191}Ir (38.5%) and ^{193}Ir (61.5%). Outer shell electron configuration is $4f^{14}5d^76s^2$. Successive ionization energies are 9.1, 17.0, $\approx$27, $\approx$39, $\approx$57 eV. Atomic radius is 0.136 nm; radius of Ir^{3+} ion is 0.073 nm, of Ir^{+4} ion is 0.068 nm. Electron affinity is 1.97 eV. It is chemically inert; oxidation state is +3, +4, +6, less often, +1, +2. Electronegativity is 1.55.

Iridium in a free form is a silvery-white *metal*. It has face-centered cubic lattice, space group $Fm\bar{3}m$ (O_h^5); $a = 0.3839$ nm at 299 K. Density is 22.65 g/cm^3 at 299 K (one of highest values for a simple substance); $T_{\text{melting}} = 2716$ K; $T_{\text{boiling}} \cong 4740$ K. Bonding energy is -6.93 eV at 0 K. Latent heat of melting is 26.38 kJ/mole, latent heat of evaporation is 629 kJ/mole, specific heat is 131 J·kg^{-1}·K^{-1} (at 298 K). Debye temperature is 420 K; the linear thermal expansion coefficient is $6.63\cdot10^{-6}$ K^{-1} (293 to 373 K); coefficient of thermal conductivity is 148 W·m^{-1}·K^{-1} (273 to 373 K); adiabatic coefficients of elastic rigidity of iridium are the highest of all cubic group crystals (except *diamond*): $c_{11} = 600$, $c_{12} = 260$, $c_{44} = 270$ (GPa) at room temperature. Isothermal bulk modulus of iridium is 355.3 GPa (at 298 K); Young's modulus is $\cong$514 GPa; Poisson ratio is 0.26; tensile strength is 0.23 GPa, relative elongation at the moment of rupture is 2%. Brinell hardness is 1.64 GPa, Vickers hardness is $\cong$220 in HV units for annealed iridium (at room temperature). Thermal neutron trapping cross-section is 440 barn; ion-plasma frequency is 16.02 THz. Linear low-temperature electronic specific heat term (Sommerfeld coefficient) is 3.2 mJ·mole^{-1}·K^{-2}. Electric resistivity is 51 nΩ·m (at 295 K); temperature coefficient of electrical resistivity is 0.003925 K^{-1} (273 to 373 K); Hall constant is $+3.18 \cdot 10^{-11}$ m^3/C; superconducting critical temperature T_c is 0.125 K; superconducting critical magnetic field is 1.9 mT (at 0 K). Iridium is a paramagnetic substance with molar magnetic susceptibility $+255.6 \cdot 10^{-6}$ CGS units (at 298 K); nuclear magnetic momentum of ^{193}Ir is 0.17 nuclear magnetons.

Iridium is used to manufacture electrodes and *thermocouples* (Ir–Rd alloy also used for thermocouples); special crucibles with high *corrosion resistance*; to apply protective *coatings*. Alloys of Ir with Pt and Pd are used to manufacture strain gages, resistors, and current brushes. Standards of length are made of Ir–Os alloy. The *Mössbauer effect* was discovered (1957) on the isotope ^{193}Ir.

IRON (Lat. *Ferrum*), Fe

A chemical element of Group VIII of the periodic table with atomic number 26 and atomic mass 58.847. Natural iron consists of four stable isotopes ^{54}Fe (5.84%), ^{56}Fe (91.68%), ^{57}Fe (2.17%), ^{58}Fe (0.31%). Five radioactive isotopes are known: ^{52}Fe, ^{53}Fe, ^{55}Fe, ^{59}Fe, ^{60}Fe. Outer electronic shell configuration is $3d^6 4s^2$. Successive ionization energies are 7.893, 16.18, 30.6 eV. Electronegativity is 1.64. Atomic radius is 0.126 nm, radius of Fe^{2+} ion is 0.076 nm, of Fe^{3+} ion is 0.067 nm. Oxidation state is +2, +3, and less often, +1, +4, +6, +8.

Pure iron is a shiny silvery-white *metal*. Iron crystallizes in four polymorphous modifications: α-Fe, β-Fe, γ-Fe, δ-Fe. α-Fe has a body-centered cubic lattice, space group $Im\bar{3}m$ (O_h^9); $a =$ 0.286653 nm at 298 K. β-Fe has the same structural type with $a = 0.2895$ nm (at 1173 K). γ-Fe has face-centered cubic lattice, space group $Fm\bar{3}m$ (O_h^5) with parameter $a = 0.36468$ at the temperature 1189 K. δ-Fe has body-centered cubic lattice, space group $Im\bar{3}m$ (O_h^9) with parameter $a = 0.29322$ nm at the temperature 1667 K. Below 1042 K, α-Fe is stable; above the *Curie point* 1042 K it preserves its crystalline structure but looses its ferromagnetic properties and transforms into paramagnetic β-Fe which is stable up to 1184 K. Between 1184 K and 1665 K, γ-Fe is stable. Above 1665 K, δ-Fe forms, which is stable up to the melting temperature $T_{\text{melting}} =$ 1809 K. Modifications of γ-Fe and δ-Fe are paramagnetic. There are indications that γ-Fe tends to be an *antiferromagnet*. The structure of low-temperature β-Fe is similar to that of the high-temperature modification of β-Fe, and is not usually regarded as an independent allotropic form. Physical properties of pure iron (technical purity, additions $\approx 10^{-3}$%; high purity, additions 10^{-4} to 10^{-7}%) are as follows: density of α-Fe is 7.874 g/cm^3 (at 293 K), of γ-Fe is 7.60 g/cm^3 (at 1203 K); bonding energy is 4.29 eV/atom at 0 K and 4.13 eV/atom at room temperature. Latent heat of $\beta \leftrightarrow \gamma$ transformation is 1.047 kJ/mole and of $\gamma \leftrightarrow \beta$ transformation is 1.40 kJ/mole; melting latent heat is 15.40 kJ/mole; sublimation latent heat 403.9 kJ/mole; boiling temperature $T_{\text{boiling}} = 3153$ K. Specific heat of iron depends on its structure and is an involved function of temperature; at 298 K and a constant pressure it is 450 $J \cdot kg^{-1} \cdot K^{-1}$ (298 K); Debye temperature is 467 K; linear thermal expansion coefficient of α-Fe is $11.7 \cdot 10^{-6}$ K^{-1} (at 293 K), for γ-Fe it is $21.5 \cdot 10^{-6}$ K^{-1}; coefficient of thermal conductivity at 298 K is 74.04 $W \cdot m^{-1} \cdot K^{-1}$; temperature of *martensitic transformation* is 843 K (technical purity iron), 1173 K (high purity iron). Rigidity is $\approx$85 GPa (at 298 K); Young's modulus is $\approx$219 GPa (at 298 K); Poisson ratio is $\approx$0.286 (at 298 K). Adiabatic coefficients of elastic rigidity are $c_{11} = 232.2$, $c_{12} = 135.6$, $c_{44} = 117.0$ for α-Fe crystal at 298 K; $c_{11} \approx 154$, $c_{12} \approx 122$, $c_{44} \approx 77$ GPa for γ-Fe at 1428 K. Temporary rupture strength is 300 MPa for high purity, 50 MPa for technical purity iron (at 298 K). Relative elongation is 50 to 22%, relative contraction is 93%. Brinell hardness is 530 MPa, microhardness is 600 MPa for iron of high purity (at 298 K), 1100 MPa for technical purity. Yield limit is 20.5 MPa for iron of high purity, 140 to 180 MPa for technical purity (at 293 K). Maximum impact strength is $\approx$345 MPa; temperature of transition to brittle state at impact tests is 188 K for iron of high purity and 285 K for iron of technical purity. Tensile strength of *thread-like crystal* (1.6 μm diameter) is 134 GPa. Self-diffusion coefficient is $1.04 \cdot 10^{-16}$ m^2/s for α-Fe (at 1043 K), $2.78 \cdot 10^{-15}$ m^2/s for β-Fe (at 1154 K) and $2.19 \cdot 10^{-15}$ m^2/s for γ-Fe (at 1425 K). Neutron fission relaxation length at energies >3 MeV is 6.5 cm; relaxation length for 1 MeV γ-quanta is 2.14 cm. Linear term in low-temperature molecular specific heat is 4.77 mJ·mole^{-1}· K^{-2}. Electrical resistivity is 98 nΩ·m (at 298 K); temperature coefficient of electrical resistance is 0.00651 K^{-1} (273 to 373 K). Hall constant is $+1.26 \cdot 10^{-9}$ m/C; thermal electromotive force coefficient is +17.0 μV/K; infrared reflectivity for a wave length of 5.0 μm is 90.8%. Work function of polycrystal is 4.31 V; Curie point is $T_C = 1042$ K. Initial magnetic susceptibility of α-Fe is about 1100; coercive force (for coarse crystalline structure) is (25–70)$\cdot 10^{-7}$ T (high purity) and $80 \cdot 10^{-7}$ T (technical purity). Maximum relative permeability (for coarse crystalline structure) 108,000–230,000 (high purity), 82,000 (technical purity). *Magnetostriction* is $4.0 \cdot 10^{-4}$, atomic magnetic moment is 2.22 of Bohr magnetons.

Iron is the most important metal of modern engineering. About 95% of all the metal production involves *iron alloys*.

IRON ALLOYS

Alloys based on *iron*. Carbon and nitrogen form with iron interstitial *solid solutions* of limited solubility, while all other alloying elements form substitutional solid solutions.

Iron alloys are the most common metal material of the modern period. They are conveniently distinguished as *steels*, *cast irons*, *precision alloys*, and *ferroalloys*. Iron alloys are classified according to the method of manufacture, level and quality of *alloying*, *crystal structure*, basic physical mechanical properties, and their purpose.

Depending on the concentration of the alloying elements, there are iron alloys of ferrite, *pearlite*, *martensite*, *austenite*, and intermediate classes. Iron alloys of the ferrite and austenite classes are single-phase. For the carbon content 2%, they are considered as cast irons. Steels and cast irons are classified: according to their structure as ferrite, ferrite-pearlite, *ledeburite*; according to their purpose as durable, nonmagnetic, forgeable; and according to the type of metal *fracture*, the latter determined by the extent of *graphitization* (e.g., white and gray cast irons). Iron alloys can be used as precision materials (magnetically soft, magnetically hard, welding) and so on. Iron alloys with a high content of alloying elements that are used to manufacture alloyed steels are classed as ferroalloys (e.g., ferrochrome, ferromanganese, ferronickel, ferrocerium, etc.). The properties of iron alloys are determined by their structure and phase composition which can vary depending on the treatment type: deformation, *heat treatment*, etc.

IRON, ARMCO

See *ARMCO iron*.

IRON, CAST

See *Cast iron*.

IRON GARNETS

Oxide monocrystals (*ferrites*) of composition $R_3Fe_5O_{12}$ (i.e. $3R_2O_3 \cdot 5Fe_2O_3$, where R is a *rare-earth element* or *yttrium*), isomorphous to the garnet mineral grossular or $Ca_3Al_2(SiO_4)_3$. They belong to the *cubic system*, space group $Ia\bar{3}d$ (O_h^{10}). The lattice constant of iron garnets varies within the range 1.228 ($Lu_3Fe_5O_{12}$) to 1.252 nm ($Sm_3Fe_5O_{12}$); density is 5.169 ($Y_3Fe_5O_{12}$) to 7.128 g/cm^3 ($Lu_3Fe_5O_{12}$). Fe^{3+} ions are located in the interstices of the oxygen matrix in tetrahedral (d) and octahedral $[a]$ coordinations, thus forming the corresponding tetrahedral and octahedral *magnetic sublattices*, and the R^{3+} ions have dodecahedral coordination, thus forming a dodecahedral $\{c\}$ magnetic sublattice (if R^{3+} is magnetic). The *magnetic moments* of the $[a]$- and (d)-sublattices are antiparallel, oriented towards each other, and they create resulting uncompensated moments, with the *magnetization* vector of the rare-earth sublattice oriented antiparallel to them. The magnetic moment per formula unit at $T = 0$ K varies from zero ($Yb_3Fe_5O_{12}$) to $18.2\mu_B$ ($Tb_3Fe_5O_{12}$), where μ_B is the Bohr *magneton*; the *Néel point* is from 539 K ($Lu_3Fe_5O_{12}$) to 578 K ($Sm_3Fe_5O_{12}$); the *magnetic compensation* point is in the range from 0 ($Yb_3Fe_5O_{12}$) to 286 K ($Gd_3Fe_5O_{12}$). The iron garnets of yttrium, samarium, europium, thulium and lutetium have no compensation point.

IRRADIATION ANNEALING

The *annealing* of defects in a *crystal* stimulated by *nuclear radiations*. Both impurity atoms and other *defects* formed prior to irradiation, as well as radiation-induced defects, take part in the process of irradiation annealing (see *Radiation physics of solids*). When radiation doses are large, annealing lowers the rate of accumulation of the defects, particularly if the intensities themselves are high enough. Irradiation annealing mechanisms are related to the processes of *radiation-induced diffusion* of defects, to atomic restructuring triggered by collisions of external particles with crystal atoms, and also to the *small-dose effect*.

IRRADIATION EMBRITTLEMENT

Depletion or total loss of *plasticity* of materials resulting from their irradiation with high-energy *neutrons*, *ions*, electrons, γ-quanta and other types of emissions. Processes of *low-temperature irradiation embrittlement* and *high-temperature irradiation embrittlement* can thereby develop.

Low-temperature irradiation embrittlement consists in raising the temperature of the brittle–viscous transition and lowering the relative elongation, transverse compression, and *impact strength* of irradiated *metals* and *alloys* within the temperature range of tests, that is below approximately $0.5T_{\text{melting}}$, where T_{melting} is the melting temperature. Low-temperature irradiation embrittlement is mainly caused by the same processes that are responsible for *irradiation hardening*. The *annealing* of irradiated metals at temperatures above $0.5T_{\text{melting}}$ lessens the effects of this embrittlement, and can lead to a complete recovery of plasticity back to its initial level.

High-temperature irradiation embrittlement constitutes a practically irreversible lowering of the plasticity parameters of irradiated metals and alloys at temperatures above $0.5T_{\text{melting}}$. The cause of this embrittlement involves a change of the ratio between the strength of the matrix to that of *grain boundaries*. The accumulation of helium generated during the course of nuclear reactions, and the *segregation* of adverse impurities at grain boundaries, both stimulated by irradiation, significantly lower the tear strength at grain boundaries.

IRRADIATION HARDENING

Technique of increasing strength and elasticity of crystalline bodies (*elastic modulus*, *proportionality limit*, *yield limit*, *ultimate strength*, *hardness*) by irradiating crystalline targets with high-energy particles and *photons* (*neutrons*, electrons, *ions*, γ-quanta, etc.) within the temperature range below $(0.4–0.5)T_{\text{melting}}$, where T_{melting} is the material melting temperature. Since the mechanical properties of *metals* and *alloys* are highly sensitive to structural and phase changes, irradiation hardening was found in them earlier than other radiation effects, and has been studied in more detail. The level of irradiation hardening depends on the nature of the irradiated material, the conditions of irradiation, and the dose. For pure crystals (*aluminum*, *copper*) the increase of the amount of stress required to induce plastic flow becomes noticeable even after irradiating with neutron doses up to 10^{12}–10^{14} n/cm^2. Moreover, when irradiating these materials with the much higher doses of 10^{21}–10^{22} n/cm^2 the resulting changes may be greater by a factor of 5 to 10. As a rule, irradiation hardening in metals and alloys is accompanied by *irradiation embrittlement*. Irradiation hardening in crystals results from the development of new centers of *dislocation* pinning, and of barriers for mobile dislocations. Beside the ordinary interaction forces with the lattice and other initial lattice defects, dislocations in an irradiated crystal, as the natural "carriers" of *plastic deformation*, have to penetrate through a whole spectrum of other barriers of radiation origin (irradiation defects).

IRRATIONAL TWINNING (a term introduced by N.A. Brilliantov, I.V. Obreimov, 1935)

Appearance of strip-like regions with a turned lattice structure in a monocrystal as a result of *plastic deformation*. This phenomenon is observable in *alkali-halide crystals* of the NaCl type. Unlike ordinary *twinning of crystals*, irrational twinning is characterized by an irregular angle of the lattice turn, blurred strip boundaries, and misalignment of the boundaries with crystallographic planes. The strip boundaries are close to planes of the $\{110\}$ type. Irrational twinning takes place under compression of planar prisms spalled off along planes of *crystal cleavage*. As a result of irrational twinning the lateral surface of prisms acquires an accordion-like profile with bevels at a normal to the loaded faces. X-ray, optical polarization, and selective *etching* methods have been used to establish that a *strain* emerges along various crystallographically equivalent *slip systems* in neighboring disoriented layers. The lattice disorientation angle in neighboring layers ordinarily does not exceed 2–3°, its maximum value of 6° being observed as a result of strain at elevated strain temperatures. Irrational twinning also occurs at stretching of NaCl type alkali halide crystals.

IRREDUCIBLE POLYNOMIAL BASIS

Finite set of polynomials $P(c_i)$ invariant under the symmetry group G of a physical system or problem. Any other invariant polynomials of the system can be expressed in terms of these polynomials. In the theory of *phase transitions* $G = \Phi$ is the crystallographic *space group* of the high-symmetry phase, and c_i are parameters of the transition, belonging to only one irreducible representation. In an expansion of the *thermodynamic potential* the coefficients of the invariants of different orders are considered to be comparable in their magnitudes. In this simplified model an irreducible polynomial basis $P(c_i)$ is used to connect coefficients of the invariants of various orders with possible low-symmetry phases. In this case the symmetry groups of the latter (subgroup of the Φ-group) can often be identified without computation by simple fitting, using a group representation (character) table or the Frobenius method. In other nonlinear theories of solid state physics the irreducible polynomial basis is constructed from the components of vectors, tensors (e.g., *strain tensor*), etc. Associated with a particular group G there can be a *point group*, a magnetic group, a space group, etc. The construction of an irreducible polynomial basis is founded on the *Noether theorem* (the degree of the polynomials from the basis does not exceed the order of the group).

IRREDUCIBLE TENSOR OPERATORS

Operators T_{lm} that satisfy certain commutation relationships, and experience the same transformations during rotation of a coordinate system as the lmth eigenfunction of the angular momentum operator. Spherical harmonics $Y_{lm}(\theta, \varphi)$ represent a particular case of such operators, where an irreducible tensor is formed from the components of a radius vector. Atomic and radio spectroscopy use irreducible tensor operators composed of the operators of the total angular momentum, $\boldsymbol{J}$, the *orbital angular momentum*, $\boldsymbol{L}$, the electron and nuclear *spins*, $\boldsymbol{S}$ and $\boldsymbol{I}$, as well as the magnetic and electric field vectors. Introduction of irreducible tensor operators simplifies calculations of matrix elements, which are then reduced to computing the simplest single reduced matrix element. Thus, the set of irreducible tensor operators $T_{lm}(S)$, where $l = 0, 1, \ldots, 2S$ and $m = -l, \ldots, l$, forms a complete orthonormal system of linearly independent matrices of rank $2S + 1$. In *magnetic resonance spectroscopy of solids*, *Stevens operators* (K.N. Stevens, 1953) also find wide use. They represent a set of linearly independent operators $G_l^m(J)$ composed of polynomials of mutually noncommuting components of the angular momentum (effective spin) vector $\boldsymbol{J}$. Such operators were initially derived as a generalization of spherical harmonics. Matrix elements of Stevens operators are easily calculated; and have been tabulated for $l = 1, \ldots, 7$, $m = -l, \ldots, l$. They are used as equivalent operators to calculate matrix elements of the operators of the interactions of electrons in the unfilled shells of transition ions in the presence of a crystalline electric field and an external magnetic field. The same matrix elements also serve as basis operators to construct the generalized spin Hamiltonian using the method of invariants; they are unnormalized linear combinations of irreducible tensor operators.

For radiospectroscopic spectra in solids, *equivalent operators* are used that display identical transformation properties during rotation of the coordinate system. According to the *Wigner–Eckart theorem* (E.P. Wigner, C. Eckart) for a set of states with a definite value of angular momentum, matrix elements of such operators possess a common factor, the so-called reduced matrix element, that is independent of the azimuthal quantum number. Thus, for electrons in an unfilled paramagnetic ion shell, one may skip the cumbersome calculation of the matrix elements of operators for the interaction of the electrons with the crystalline electric field or the Zeeman operator $(\boldsymbol{L} + 2\boldsymbol{S})\cdot\boldsymbol{B}$. Instead, one simply multiplies tabulated matrix elements of much simpler operators (polynomials of angular momentum components constructed from either Stevens operators or irreducible tensor operators) by the reduced matrix elements which also have been precalculated for all the transition ions.

The above operators underlie the *technique of irreducible tensor operators* used to describe both optical and radio-frequency spectra of paramagnetic ions by representing their Hamiltonians as expansions over the full system of irreducible tensor operators. Expressing a generalized *spin Hamiltonian* as a sum of these operators for the components of electron and nuclear spins, and electric and magnetic field vectors, is a powerful tool. It provides matrix elements of a generalized spin Hamiltonian in closed form without having to determine their wave functions (which is particularly important for multispin systems), thereby simplifying the calculation of the transformations of the generalized spin Hamiltonian during rotations of the coordinate system, and simplifying the procedure for excluding extra parameters from the Hamiltonian through an unitary gauge transformation. When applying this technique, one often resorts to the concept of *effective spin*. The latter is a fictitious angular momentum $\widetilde{S}$ that is introduced to calculate the energies E_M of a group (multiplet) of electron energy levels separated from other levels by a gap appreciably exceeding the level-to-level spliting within the group itself. The quantity $\widetilde{S}$ is chosen in such a way that the number of values $M = -\widetilde{S}, \ldots, \widetilde{S}$, which equals $2\widetilde{S}+1$, coincides with the multiplicity of the multiplet. However, it appears to be more convenient to use integer values of S to describe the doublet states of paramagnetic ions with even numbers of electrons, N. One then avoids misunderstandings that stem from the transformation properties of a half-integer effective spins. According to the Wigner–Eckart theorem, the elements of matrices describing interactions of electrons either with the electric fields of a crystal (or molecule) or with an external magnetic field may be expressed via the matrix elements of irreducible tensor operators or Stevens operators, constructed from components of the effective spin for different states of a multiplet. See also *Spin Hamiltonian.*

IRREVERSIBLE PROCESSES

Processes in a *thermodynamic system* accompanied by an increase of *entropy.* In the case of irreversible processes, the system cannot be brought back to its prior state, even if it is subjected to the action of the same factors such as applied forces in reverse order, or in reverse directions.

ISING MAGNET

A *magnetic material* with properties describable by the *Ising model* within an acceptable margin of error. Ising magnets include some rare-earth compounds, e.g., dysprosium–aluminum garnet (see *Iron garnets*) and *orthoferrites* at low temperatures. Typical of an Ising magnet is a pronounced anisotropy of its magnetic properties (see *Magnetic anisotropy*), which is related to the anisotropic character of the Ising model Hamiltonian. The discreteness of the energy spectrum in the Ising model results in an exponential temperature dependence of the additional *magnetization* along the selected axis, and of the magnetic part of the *specific heat* at low temperatures. Unlike the *Heisenberg magnet*, the *order parameter* in the Ising magnet differs from zero not only in a three-dimensional crystal, but also in a two-dimensional one.

ISING MODEL (E. Ising)

A spin model describing both magnetic (see *Ising magnet*) and nonmagnetic (*ferroelectrics*, *binary alloys* and others) macroscopic systems. The Ising model Hamiltonian has the form:

$$H_{\mathrm{I}} = -J \sum_{(\boldsymbol{m},\boldsymbol{n})} \sigma_{\boldsymbol{m}}\sigma_{\boldsymbol{n}} - B \sum_{\boldsymbol{m}} \sigma_{\boldsymbol{m}},$$

where J is the *exchange interaction* constant, each variable $\sigma_{\boldsymbol{n}} = 2S^z_{\boldsymbol{n}}$ assuming the values ± 1 (the spin $S = 1/2$), $\boldsymbol{m}$ and $\boldsymbol{n}$ are the vector indices of crystal lattice sites, and B is the external magnetic field. The first summation extends over nearest neighbor pairs. The Ising model is much simpler than other quantum spin models since the Hamiltonian has no noncommuting operators. All quantum states of this model are determined by sets of values $\sigma_{\boldsymbol{n}}$ at the lattice sites.

In the case of one-dimensional and two-dimensional lattices exact results have been obtained for some thermodynamic functions. For the one-dimensional Ising model there is no *phase transition*, and the *magnetization* vanishes as $B \to 0$. The Ising model free energy F for the two-dimensional square lattice at $B = 0$ and $J > 0$ is obtained using the *Onsager method*:

$$F = -k_B T\left[\frac{1}{2}\ln(2\sinh 2K) + \int_{-\pi}^{\pi}\frac{dq}{4\pi}\sinh^{-1}(\cosh 2K\coth 2K - \cos q)\right],$$

where $K = J/(k_B T)$. At the Curie temperature T_C a *second-order phase transition* takes place, and the *specific heat* c_p has a logarithmic singularity $c_p = A + B\ln|(T - T_C)/T_C|$. In the vicinity of the *Curie point*, the magnetization M tends to zero, $\langle M\rangle \propto (T - T_C)^{1/8}$; the *magnetic susceptibility* χ diverges as $(T_C - T)^{-7/4}$; and the *correlation length* r_c grows according to the law $r_c \propto |T - T_C|^{-1}$. Numerical calculations show that the three-dimensional Ising model also experiences a second-order transition, and has the *critical indices* $\alpha = 0.08$, $\beta = 0.33$, and $\nu = 0.64$.

ISOBAR (fr. Gr. ισος, equal, and βαρος, weight, heaviness)

A line on a compositional *phase diagram* showing an equilibrium *isobaric process* taking place at constant *pressure*. The isobar of an equilibrium process in a single component system is the plot of the function $V(T)$ at $P = \text{const}$ (where V, T and P are the volume, temperature, and pressure, respectively).

ISOBARIC PROCESS

A process occurring in a physical system at constant *pressure*, or, in a narrower sense, an equilibrium thermodynamic process involving the variation of system parameters with the *pressure* held constant. The dependence $V(T)$ at $P = \text{const}$, where V, T and P are the volume, temperature and pressure, respectively, is called an isobaric process equation, and its plot is called an *isobar*. The isobaric process equation can be of an involved form; but for a given *state of matter*, V is a continuous function of T and changes abruptly at the temperature of a *first-order phase transition*. The simplest form of the dependence $V(T)$ is that of an ideal gas, $V = RT/P$, where R is the gas constant (*Gay-Lussac law*, J. Gay-Lussac, 1802). For a solid at temperatures higher than the *Debye temperature*, the dependence $V(T)$ in the single-phase region is close to linear. An important parameter of an isobaric process is the *thermal volume expansion coefficient* $\alpha = (1/V)(\partial V/\partial T)_P$. For an *ideal gas* $\alpha = 1/T$. For a solid at temperatures higher than the Debye temperature, a good approximation is considered to be $\alpha = \text{const}$, unless the system undergoes a phase transformation. In an isobaric process, the system, owing to the thermal flow toward it, not only heats up but also produces work. Therefore, $C_P > C_V$ where C_P and C_V are the *specific heats* at constant pressure and constant volume, respectively.

ISOCHOR (fr. Gr. ισος, equal, and χωρα, region)

A line on a thermodynamic compositional *phase diagram* for a constant volume (isochoric) process; plots of the functions $P(T)$ and $V = \text{const}$ (P, T, and V are pressure, temperature and volume, respectively).

ISOCHORIC PROCESS

A process performed in a physical system at constant volume. Thermodynamic compositional *phase diagrams* represent an isochoric process by an *isochor*. In an equilibrium thermodynamic system without external fields, an isochoric process is described by the dependence $P(T)$ at $V = \text{const}$ (where P, T, and V are the system pressure, temperature and volume, respectively). For an *ideal gas*, $P = RT/V$ (*Charles' law*) where R is the gas constant. The function $P(T)$ is monotonic in the single phase region (solid, liquid, gaseous state of matter), and has discontinuities at the temperatures of *first-order phase transitions*.

ISOELECTRONIC

A situation when two different objects possess an equal number of electrons, e.g., the 48 electrons in the unit cell of *diamond* (8 carbon atoms) and boron azide BN (4 molecules), and the 20 electrons of the molecules NaF and MgO. For these cases the same or similar crystal structures are often observed: in the first case, the diamond and the sphalerite lattices; in the second case, the *rock salt* (NaCl, simple cubic) lattice. Sometimes the term isoelectronic is interpreted as an equal number of valence electrons, e.g., when substituting a Na atom for a K atom, or Br for Cl in an NaCl crystal, or C for Si in diamond, and so on.

ISOELECTRONIC SEQUENCE, isoelectronic series

A sequence consisting of a neutral atom and ions of neighboring elements in the periodic table having an equal number of electrons, e.g., H^-, He, Li^+, Be^{2+}, B^{3+} or Na, Mg^+, Al^{2+} and Si^{3+}, their electronic shells having the same configurations $1s^2$ and $1s^22s^22p^63s$, respectively. As the nuclear charge Z in each of these series grows, the absorption spectra corresponding to the outer electron transitions shift toward the violet (toward higher energies) while remaining very similar, and the ionization potentials increase.

ISOENERGETIC SURFACE, surface of constant energy

A surface in *momentum space* corresponding to a fixed value of the particle (*quasi-particle*) energy. This concept illustrates the geometric characteristics of the *dispersion law*. The normal to an isoenergetic surface determines the particle velocity direction. The shape of the isoenergetic surface for each branch of quasi-particles is energy-dependent: it is closed (sphere in isotropic case) near the edges of an admissible energy band. However, there always exists a range of values limited by *Van Hove singularities* where the isoenergetic surface is open and repeats periodically throughout reciprocal space. An example of an isofrequency surface is the Fermi surface of a metal or semiconductor.

ISOFREQUENCY SURFACE, surface of constant frequency

A surface in the space of wave or *quasi-wave vectors* corresponding to a fixed value of the frequency for the corresponding wave (*quasi-particle*). An isofrequency surface coincides with an *isoenergetic surface* after quantization of vibrations. The curvature of an isofrequency surface determines the nature of the decay of radiation, or of a wave scattered in the corresponding direction, at great distances from the radiator or scatterer.

ISOMER SHIFT

Shift of Mössbauer lines (see *Mössbauer effect*) that arises when the emitting and absorbing nuclei are in different chemical or crystallographic environments. The magnitude of the relative shift of emission and absorption lines is $\delta = A\{|\psi_a|^2 - |\psi_e|^2\}$ where A is a constant dependent on the parameters; ψ_a and ψ_e are the wave functions of s-electrons at the locations of the absorbing and emitting nuclei. The study of the isomer shift allows one to determine the values of $|\psi_a|^2$, needed for the analysis of the electronic structure and chemical bonding in solids.

ISOMORPHISM (fr. Gr. $\iota\sigma o\varsigma$, equal, and $\mu o\rho\varphi\eta$, form)

Perfect correspondence of the internal structure of crystals formed by materials with like chemical properties. Isomorphous substances possess the same *crystal structure*. *Isostructural* materials that form substitutional *solid solutions* are isomorphous. Still, this isomorphism is not present in all isostructural substances as some may differ in the nature and size of their particles. Isomorphic crystallization generally takes place in materials with atomic radii that differ by no more than 10–15%, and have the same valence and the same chemical bond type. Depending on the solubility of components, perfect (complete) and imperfect (limited) isomorphism is known. The *unit cell* dimensions of solid solutions of isomorphic substances exhibit linear dependences on the concentration (*Vegard rule*), although some violations of this law may occur.

ISOPERIODIC HETEROGENEOUS SYSTEMS, isoperiodic heterojunctions

Pairs or uninterrupted series of isomorphous (see *Isomorphism*) *semiconductor materials* that have the same lattice periods and, because of this, are suitable for forming defect-free *heterojunctions*. They are used in solid-state devices operating on the basis of *heterogeneous structures*, *heterojunction lasers*, heterophotodiodes, new types of *transistors*, and so on. The design of these isoperiodic systems involves evaluating the lattice period in *solid solutions* of related semiconducting compounds. Chemical diversity in these systems allows the variation of the heterojunction potential barriers, *refractive indices*, radiation wavelength, and the *intrinsic light absorption edge* of the materials in contact to optimize the device performance. The classical system $Ga_{1-x}Al_xAs/GaAs$ is based on the mutual replacements of Al and Ga that has almost no effect on the lattice dimensions. In a more general case, the *principle of isoperiodic substitution* applies more complicated substitutions of these types of multicomponent systems, such as a concurrent balanced substitution of two or more elements which compensate each other for their individual lattice distortions. For example, in the system $In_{1-x}Ga_xAs_yP_{1-y}$/InP the lattice constant retains the same value as that of InP since the partial substitution of Ga for In contracts the unit cell to the same extent that the substitution of As for P expands the unit cell if the condition $y \cong 2.2x/(1+0.06x)$ is satisfied. These two and other isoperiodic heterogeneous systems (GaAlSbAs/GaSb, InGaSbAs/GaSb, InGaAsP/GaAs, PbSnTeSe/PbTe) find their use in *optoelectronics*.

ISOSTRUCTURAL DECAY

A decomposition of an oversaturated *solid solution* that results in the formation of a *phase* whose structure is either identical (alloys Cu–Co, Cu–Ag, Cu–Fe, etc.) or close to an *alloy* structure with an ordered arrangement of atoms (alloys Ni–Ti, Ni–Si, Fe–Be, Al–Li etc.). During the *quenching* process of concentrated alloys, the precipitation of dispersion particles of the isomorphic phase tends to occur (see *Disperse structure*); and once hardened they prove to be two-phase alloys. Isostructural decay is present in most aging *high-temperature alloys* (see *Alloy aging*) based on nickel, iron, or cobalt, where the strengthening isomorphous phases are known as *gamma phases*.

ISOTHERM (fr. Gr. $\iota\sigma o\varsigma$, equal, and $\vartheta\varepsilon\rho\mu\eta$, heat)

A line on a compositional *phase diagram* showing an equilibrium *isothermal process* taking place at a constant temperature. For a homogeneous body in dynamic equilibrium the isotherms are the lines on plots in the coordinates (P, V), which correspond to the thermal equation of state at $T = \text{const}$. The isotherm for a solid is determined experimentally or within the framework of analytical models. The isotherm for an ideal gas is $PV = \text{const}$ (*Boyle's law*).

ISOTHERMAL PROCESS

A process taking place in a physical system at a constant temperature; in a narrower sense, an equilibrium thermodynamic process occurring at a constant temperature. Thermodynamic compositional *phase diagrams* show isothermal processes as *isotherms*. An isothermal process is accompanied by the supply/withdrawal of the required heat flux to/from the system. For instance, boiling a liquid or melting a solid at constant *pressure* constitute isothermal processes. *First-order phase transitions*, accompanied by a change in *entropy* $\Delta S = Q/T_0$ where Q and T_0 are the latent heat and the temperature of the transition, take place isothermally. An important characteristic of solids is the *isothermal compressibility* $\chi = -(1/V)(\partial V/\partial P)_T$ that is found during an isothermal process by measuring the matter volume change upon the application of external pressure. In an isothermal process the work done by the system is equal to the change in free energy, but with the opposite sign, that is $\mathrm{d}W = -\mathrm{d}F$.

ISOTOPE EFFECT in superconductors

A change of the *critical temperature of superconductors* T_c under substitution of one isotope of a constituent chemical element of the material (*metal*, *alloy*, compound) by another isotope of this same element. For most ordinary superconductors that are simple metals (Hg, Pb, Sn, In, etc.), the increase in the isotope mass, M, causes a lowering of T_c according to the power law $T_c \propto M^{-\alpha}$, where the isotope effect index α is close to 1/2 (or somewhat smaller). Since the dependence of the mean (Debye) *phonon* frequency ω_D and the *Debye temperature* Θ_D on M in the *harmonic approximation* has the form $\omega_D \sim \Theta_D \propto M^{-1/2}$, the dependence $T_c \propto M^{-1/2}$ is strong evidence for the *phonon mechanism of superconductivity* (Fröhlich, 1950). As T_c decreases, the value of α decreases (Zn, Al, Cd). Superconductors with low $T_c < 1$ K (Mo, Os, Ru, Zr, U) exhibit close to zero ($\alpha \approx 0$) or even negative ($\alpha < 0$) isotopic effects. This is caused, on the one hand, by a weakening of the *electron–phonon interaction* and, on the other hand, by a relative strengthening of the Coulomb repulsion, and follows directly from formulas for T_c and α in the weak-coupling approximation (see

Bardeen–Cooper–Schrieffer theory)

$$T_c = \Theta_D \exp\left\{-\frac{1}{\lambda_{ph} - \mu_c^*}\right\}, \quad (3)$$

$$\alpha \equiv \frac{\partial \ln T_c}{\partial \ln M} \approx \frac{1}{2}\left[1 - \left(\frac{\mu_c^*}{\lambda_{ph} - \mu_c^*}\right)^2\right], \quad (4)$$

$$\mu_c^* = \mu_c\left[1 + \mu_c \ln\left(\frac{\Theta_F}{\Theta_D}\right)\right]^{-1}, \quad (5)$$

where λ_{ph} is the dimensionless electron–phonon interaction constant; μ_c is the screened Coulomb interaction constant; and Θ_F is the Fermi temperature of the delocalized electron gas. As is easily seen, T_c decreases and α decreases (even down to a negative value at $\mu_c^* > \lambda_{ph} - \mu_c^* > 0$) with λ_{ph} decreasing or the Morel–Anderson Coulomb pseudopotential, μ_c^*, growing. In the vicinity of a structural instability of the *crystal lattice* phonon anharmonicity effects become prominent, and the dependence $\Theta_D \propto M^{-1/2}$ is violated, which can result in anomalies of the isotope effect.

In *high-temperature superconductors*, a reverse trend is observed: the isotope effect weakens as T_c increases. According to Eqs. (1) and (2), this could occur only when $\mu_c^* < 0$, being possible evidence of an additional (with respect to the electron–phonon interaction) attraction of a non-phonon nature between the electrons that form Cooper pairs. However, it cannot be excluded that the formulas like Eqs. (1) or (2) are invalid for these superconductors. This may be due either to an abnormally strong electron–phonon interaction or photon anharmonicity, or due to a fundamental difference between the *high-temperature superconductivity* mechanism and the ordinary mechanism of Cooper pairing of electrons (see *Cooper pairs*).

ISOTOPE EFFECT, KINETIC

See *Kinetic isotope effect*.

ISOTOPIC ORDERING

An ordered arrangement of atoms of several isotopes of one of the components of a compound over inequivalent sites in *substitutional alloys*, resulting in the separation of the isotopes into sublattices at low temperatures. Isotopic ordering is mainly due to a dissimilarity in the energy of *zero-point vibrations* of isotopes at nonequivalent positions in the lattice. Isotopic ordering is not accompanied by an *order–disorder phase transition*, but it can be detected using *inelastic neutron scattering*. Isotopic ordering was observed in the interstitial alloy CeD_2H, where the lowering of the temperature causes ordinary hydrogen atoms H to concentrate at the octahedral sites of the Ce face-centered cubic lattice, and deuterium atoms D at its tetrahedral sites.

ISOTROPY (fr. Gr. ισος, equal, and τροπος, turn)

Independence of the properties of a material on the direction. The *condition of isotropy* involves the absence of long-range order (see *Long-range and short-range order*) in a medium. Classical liquids and gases are isotropic. Polycrystals with randomly oriented equilibrium grains, and amorphous materials are macroscopically isotropic (see *Amorphous state*). Physical properties of an amorphous body cannot be characterized by tensors of odd rank.

ISOTROPY OF ELASTICITY

The independence of the elastic properties of a solid on the direction. For *elastically isotropic solids* the values of *Young's modulus*, the *shear modulus* and the *Poisson ratio* do not depend on the measurement direction. *Elastic moduli* c_{ik} for such solids are related by $c_{11} - c_{12} = 2c_{44}$ in a frame of the Voigt indices. Elastically isotropic solids include, e.g., amorphous solids (see *Amorphous state*). Crystalline solids are elastically anisotropic (see *Anisotropy of crystals*, *Anisotropy of elasticity*). Polycrystalline *metals* and *alloys* used in engineering consist of many fine randomly oriented crystal grains and behave as elastically isotropic, given the absence of *texture*. Their properties can be calculated from the elastic properties of *monocrystals* (see *Elasticity*, *Voigt averaging*, *Reuss–Voigt–Hill approximation*).

ISOVALENT IMPURITIES

Substitutional *impurity atoms* with the same effective charge and similar electronic structure as the atom that is replaced. Isovalent impurities add no long-range Coulomb potential. Their effect on the optical and electrical properties of a semiconductor material is determined by local levels induced by them in the *band gap*, and also by the elastic stress fields arising due to a mismatch between the covalent radii of the impurity and matrix atoms. Isovalent impurities can give arise to long-range strain fields.

The short-range potential of an isovalent impurity is due to a difference between the potentials of the matrix atoms and the impurities inside a unit cell, as well as to the electron polarization and local lattice distortion. If the solubility of impurity atoms is sufficiently high then their concentration could increase to the point where isovalent impurity atom pairs or even larger *clusters* form.

Jj

JACCARINO–PETER EFFECT (V. Jaccarino, M. Peter, 1962)

A reduction of the *paramagnetic limit* in superconductors with magnetic ions due to the mutual compensation of the external magnetic field and the internal exchange field of magnetic centers. The Jaccarino–Peter effect occurs in compounds with an antiferromagnetic electron–ion *exchange interaction*, results in an increase of the *upper critical field*, and can cause (e.g., in $Sm_{1-x}Eu_xMo_6S_8$) stimulation of superconductivity by magnetic fields.

JAHN–TELLER DISTORTION

Spontaneous distortion of the initial (highly symmetric) configuration of nuclei Q_0 of a polyatomic system in the presence of an electronic degeneracy to a lower symmetry arrangement due to the absence of a minimum of the *adiabatic potential* in this configuration (see *Jahn–Teller theorem*). The term Jahn–Teller distortion is used in a conventional sense since at points of electronic degeneracy, the adiabatic potential looses its meaning as the potential energy of nuclei in the field of electrons. The lack of a minimum of the adiabatic potential at the point Q_0 complicates the dynamics of the nuclear motion, and in the general case, may not lead to a static distortion of the configuration Q_0, but rather to a more complex set of properties. A static Jahn–Teller distortion can be observed either directly, in the presence of a low-symmetry perturbation provided by a vibronic interaction (so-called *vibronic amplification*), or in a cooperative Jahn–Teller effect (see *Jahn–Teller effect*).

JAHN–TELLER EFFECT (H. Jahn, E. Teller, 1937)

A set of phenomena, effects, and trends which appear in polyatomic systems which have a configuration differing from a linear configuration, in the state of electronic degeneracy due to *vibronic instability* (see *Jahn–Teller theorem*). Similar effects are observable in the state of *pseudodegeneracy* (*Jahn–Teller pseudoeffect*) and in linear systems (*Renner effect*). A direct consequence of the Jahn–Teller theorem is the formation of a complex *adiabatic potential*, and the impossibility of separating the electronic and nuclear motions in an *adiabatic approximation*. In contrast to the latter, for the determination of the energy spectrum and the wave functions (*vibronic states*) of a polyatomic system in an f-fold degenerate electronic state, it is necessary to solve a system of f coupled equations.

Doubly-degenerate E-term. For a doubly-degenerate orbital E-term, in a linear approximation with respect to the *vibronic interaction* (linear E–e problem), the adiabatic potential takes the form of a "*sombrero*" (see Fig. 1). Under free rotation along its groove, the system changes its configuration continuously in the space of symmetrized E-displacements of Q_θ and Q_ε which, for an octahedral configuration, are related to a tetragonal *Jahn–Teller distortion*. The groove depth (at the minimum), counted out from the degeneracy point, is called the stabilization energy of the Jahn–Teller effect; $E^E_{\text{JT}} = V^2/2k_E$ where V is the vibronic constant, and k_E is the force constant. By taking into account the quadratic terms of the vibronic interaction (quadratic E–e problem) along the groove of the adiabatic potential, there appear three minima M_i at points $\Phi = 0,\ 2\pi/3,\ 4\pi/3$ which alternate with three peaks C_i together with the Jahn–Teller deformations of octahedral and tetrahedral systems. In the case of low barriers between the minima, the free "rotation" becomes hindered, otherwise for a high barrier the nuclear motion acquires a tunneling character which leads

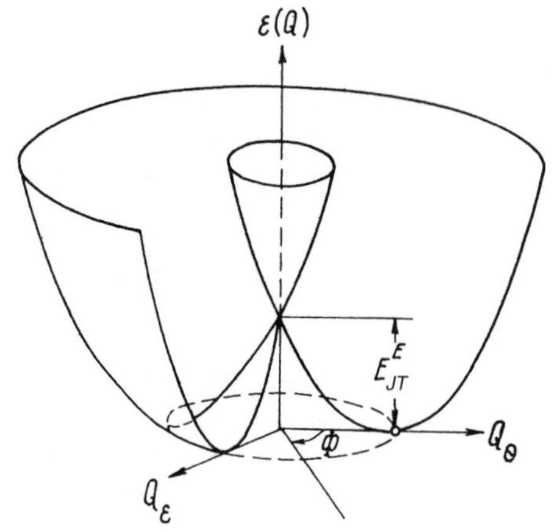

Fig. 1. “Sombrero” form of the Jahn–Teller adiabatic potential surface for the linear approximation of a doubly degenerate E-term.

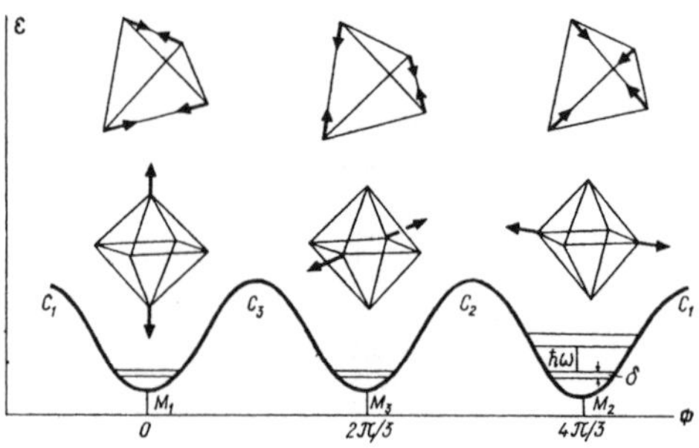

Fig. 2. Adiabatic potential (lower figure) for the quadratic E-problem showing the three Jahn–Teller diagonal (above figures) and tetragonal (middle figures) distortions of tetrahedral and octahedral configurations of atoms, respectively.

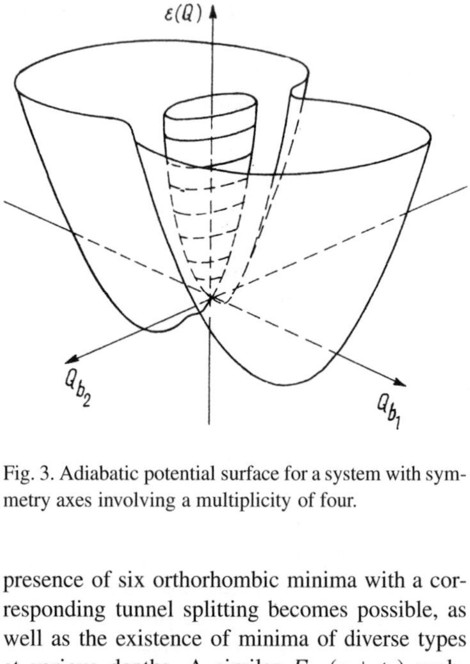

Fig. 3. Adiabatic potential surface for a system with symmetry axes involving a multiplicity of four.

to the *inversion splitting* of the energy levels δ (see Fig. 2). In all cases, the vibronic ground state remains doubly degenerate (*vibronic doublet*); this is followed by a nondegenerate *vibronic singlet*. From symmetry considerations and conservation of the multiplicity of the ground state, the physical values in the electron subsystem which depend only on the ground state, taking into account the vibronic interaction, are only reduced quantitatively (*vibronic reduction*). In the case of systems with symmetry axes of an exclusively even order multiple of four, the problem E–$(b_1 + b_2)$ appears with the adiabatic potential shown in Fig. 3.

Triply-degenerate T-term. For a triply-degenerate electronic T-term, in the general case, it is necessary to solve a T–$(e + t_2)$ problem. Neglecting the relation with T_2-vibrations (T–e problem), the adiabatic potential takes the form of three equivalent paraboloids intersecting at a single point; and the vibronic ground state is T. There are four minima of the adiabatic potential in a T–t_2 problem; the tunnel splitting leads to the ground *vibronic triplet* T and singlet A. In a full E–$(e + t_2)$ problem, the adiabatic potential surface has either three minima (of E-type) and four saddle points (of T-type) or vice versa, depending on the relation between the vibronic constants of proportionality, and the E and T_2 vibrations. By taking into account the quadratic relation, the presence of six orthorhombic minima with a corresponding tunnel splitting becomes possible, as well as the existence of minima of diverse types at various depths. A similar Γ_8–$(e + t_2)$ problem presents itself for the quadruplet electronic term Γ_8. The spectra of energies and wave functions as well as physically observable values for many types of Jahn–Teller problems have been calculated using various methods, mainly with the aid of numerical computations.

In those cases when not only a single, but rather several modes of a given symmetry are active in the Jahn–Teller effect (there is an infinite number

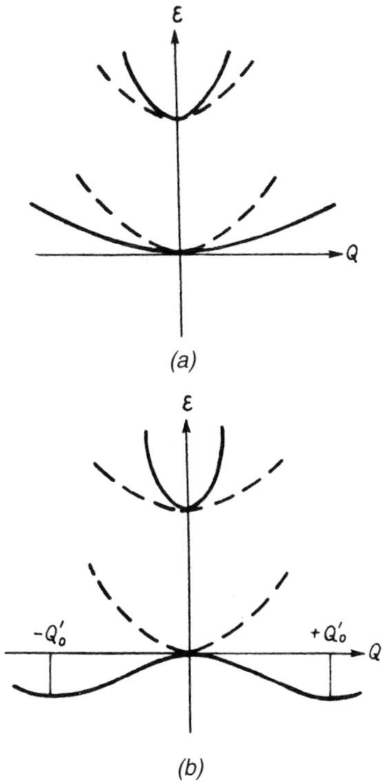

Fig. 4. One-dimensional adiabatic potential curves in the presence of vibronic mixing when there is (a) a reduction in curvature, or (b) a reversal of curvature with a resultant instability in the ground state.

of modes in the case of a crystal), we deal with the *Jahn–Teller multimode effect*. The interaction between single-center Jahn–Teller deformations in a crystal leads to an ordering that has the character of a *structural phase transition* (*Jahn–Teller cooperative effect*). In the case of a limited number of centers, a *Jahn–Teller cluster effect* takes place.

Jahn–Teller pseudoeffect. This effect emerges as a result of the vibronic interaction mixing of the ground electronic state with the fairly close (in energy Δ) excited state. Sometimes the less specific *second-order Jahn–Teller effect* term is used. A direct result of the vibronic mixing is either the reduction in curvature (Fig. 4(a)) or the appearance of an instability (negative curvature; Fig. 4(b)) of the ground state in a given direction Q (vibronic instability). The latter is observed for the condition $\Delta < V^2/k_0$, where V is the vibronic constant, and k_0 is the initial curvature without taking into account the vibronic interaction. Numerical calculations demonstrate that, depending on the values of V and k_0, the vibronic instability of the pseudoeffect can appear in the case of an appreciable energy separations Δ. The resulting maximum magnitudes, when the instability can still occur, are $\Delta_{\max} \sim 10$–15 eV for light atoms and strong *chemical bonds*; $\Delta_{\max} \sim 3$–5 eV for compounds of transition metals; $\Delta_{\max} \sim 0.1$–1 eV for rare-earth elements. The adiabatic potential in Fig. 4 is given by the formula $\varepsilon_{\pm}(Q) = (1/2)k_0Q^2 \pm [\Delta^2 + V^2Q^2]^{1/2}$. This demonstrates that, because of higher power exponents of Q (*vibronic anharmonicity*), the ground state at the points $\pm Q_0$ becomes unstable. It has been demonstrated that vibronic mixing with excited states is the main source of dynamic instability in the ground state of polyatomic systems. This conclusion is important for stereochemistry and crystal chemistry, the theory of phase transitions (see *Vibronic theory of ferroelectricity*), molecular transformations, transition states of chemical reactions, excimer *lasers*, and so on. Other observations of the Jahn–Teller pseudoeffect are analogous to those of the ordinary Jahn–Teller effect.

Jahn–Teller multimode effect. A complex of effects and trends in polyatomic systems possessing more than one set of Jahn–Teller-effect-active normal coordinates which transform with respect to the same irreducible representation. The Jahn–Teller multimode effect determines a series of properties of *point defects* in crystals, and this leads to the formation of a complex of electron–phonon states and bound states of a center with *phonons*. These appear in the form of localized and pseudolocalized *resonances* in spectral densities of IR absorption and *Raman scattering of light*, single-phonon satellites of optical bands, etc.

Jahn–Teller cooperative effect. This is the ordering or correlation of Jahn–Teller local deformations in crystals with a sublattice of ions (molecular groups) with degenerate (or pseudodegenerate) electronic ground states. This ordering, caused by the interaction of local distortions through the phonon field, can be either of the ferro-kind (same distortions in all *unit cells*), or of a more complex nature (antiferro-, incommensurate structures, etc.). The spectrum of elementary excitations of crystals with a cooperative Jahn–Teller effect is characterized by the presence of low-lying and temperature-dependent interacting electronic and phonon (acoustic) branches. With increasing temperature the correlations of local distortions weaken, and at the critical temperature T_c a phase transition to the distortion-disordered phase takes place. Since the phase transition is structural it induces anomalies of the elastic crystalline properties. At the same time, as the static lattice distortions below T_c induce substantial changes in the electronic structure which split the degenerate energy levels (bands), in the vicinity of T_c all the crystalline electronic parameters (magnetic, electric, etc.) also exhibit appreciable anomalies. The relationship between the electronic and phonon subsystems, which plays a determining role in crystals with a Jahn–Teller cooperative effect, leads to strong interrelationships between the elastic, electric, and magnetic properties of these crystals. This is the cause of some original effects in these crystals: appreciable influence of a magnetic field on a structural phase transition, *giant magnetostriction* and *electrostriction*, and so on. The Jahn–Teller cooperative effect has been found and studied in many compounds. The most widely studied are crystals containing rare-earth ions (like $TmVO_4$), spinel ($NiCr_2O_4$), *perovskites* ($KCuF_3$) with Jahn–Teller ions of transition metals, and *intermetallic compounds* (TmCd).

Examples of the Jahn–Teller cooperative and pseudoeffects are found in all major areas of the physics and chemistry of molecules and crystals, including spectroscopy over the entire electromagnetic spectrum, *crystal physics* and *crystal chemistry*, structural phase transitions, phenomena of photon transport, physics of impurity crystals, mechanisms of chemical reactions, chemical activation, and electron-conformational interactions in biology (see *Vibronic effects in optics*, *Vibronic effects in radio-frequency spectroscopy*, *Inversion splitting*).

JAHN–TELLER THEOREM

A theorem concerning the properties of a degenerate polyatomic system. It can be formulated as follows: if the *adiabatic potential* of a nonlinear polyatomic system has several branches f that coincide at the same point (f-fold degeneracy), then at least one of these branches lacks a minimum at this point. An exception is Kramers *spin degeneracy* for an odd number of electrons (total spin S is half-integer). A more limited theorem states that if a magnetic ion is at a crystal site of such a high symmetry that its ground state is not the Kramers minimum, then it will be favorable energetically for the crystal to distort and lower the symmetry to a sufficient extent to remove the degeneracy. There are some widespread simple, and in principle incorrect, formulations of the Jahn–Teller theorem which claim that in the presence of an electronic degeneracy a polyatomic system must spontaneously distort to remove the degeneracy. However, in the *Jahn–Teller effect* which follows from the theorem the degeneracy is not necessarily removed, but rather it can transform from an electronic degeneracy to a vibronic one (see *Vibronic states*), which lacks a *Jahn–Teller distortion* in the strict sense, since it is of a dynamic character.

JELLIUM MODEL

The simplest model of an interacting electron gas; it assumes that the positive charge compensating for the electronic charge is uniformly distributed in space. The only parameter that determines system properties is the electron density, or the average distance between electrons r_s. For a dense gas $r_s \ll 1$ (in Bohr radii) and in calculations of the total energy, the application of perturbation theory to the electron–electron interaction is justified. The zero-order term in the interaction yields the kinetic energy; the first-order term is the exchange energy, while higher-order terms yield the *correlation energy*. Calculations of the latter involve divergences, but their sum yields the correct final result: an expansion of the energy in powers of r_s, with a nonanalytic contribution proportional to $\ln r_s$. For a rarefied gas, $r_s \gg 1$, the uniform distribution of electrons fails to represent the ground state, and the so-called *Wigner crystallization* (see *Wigner crystal*) takes place. In this limiting case a strict calculation of energy is also possible; and $r_s{}^{-1}$ acts as the expansion parameter. In real *crystals* ($2 \leqslant r_s \leqslant 6$), calculations performed with the help of variational calculus or by application of various interpolation formulae often give very close results. The jellium model is often used for investigations of the properties and spectra of the ground state in *solids*, where the electron distribution is rather far from uniform. This application of the jellium model is based on the local density approximation of *density functional theory*. This approximation assumes the density of the exchange-correlation energy at every point of the inhomogeneous system equals that of the jellium model with the same average electron density.

JEWEL

See *Gems*.

JOSEPHSON COUPLING ENERGY

Change of the free energy (see *Thermodynamic potentials*) of two *superconductors*, caused by formation of a coherent state at their junction (see *Josephson effects*). The free energy of this *Josephson junction* is less than the sum of their energies based on the value of the Josephson coupling energy:

$$E_J = \frac{\hbar I_c}{2e}(1 - \cos\varphi),$$

where I_c is the Josephson critical current, and φ is the phase difference of the wave functions on the two sides of the contact. At $\varphi = 0$, when the *Josephson current* $I_s = I_c \sin\varphi$ goes to zero, the Josephson coupling energy is also equal to zero. At small φ ($I_s \to I_c\varphi$, and $\cos\varphi \to 1 - \varphi^2/2$) the energy grows as the square of the current I_s, which is similar to the way the energy of the superconducting condensate grows as the square of I_s in the case of ordinary *superconductivity*.

JOSEPHSON EFFECTS (B. Josephson, 1962)

The totality of quantum interference phenomena in superconducting *tunnel junctions* and also in other weakly coupled superconductors (*superconducting microbridges*, superconductor–normal metal–superconductor junctions, etc.). There are two types of Josephson effects, static (dc) and oscillatory (ac). The *dc Josephson effect* is nondissipative current flow through a junction between two superconductors, S_1 and S_2, separated by a very thin ($\sim 3{\cdot}10^{-7}$ cm) insulating layer (called a weak link) with zero voltage drop across it (see Fig. 1). The *Josephson current* is $I = I_c \sin\varphi$, where φ is the phase difference across the junction, and I_c is the critical current of the junction. The *ac Josephson effect* involves oscillations of the tunnel current when a voltage V is applied to the junction. The frequency of the oscillations is $\omega = d\varphi/dt = 2eV/\hbar$, where φ is the phase, and the frequency $f = \omega/(2\pi)$ to voltage V ratio is given by the expression $f/V = 2e/h = 4.8360{\cdot}10^{14}$ Hz/V. The dc Josephson effect was originally observed by J.M. Rowell (1963), and the effect of *Josephson radiation* from superconducting tunnel contacts was first recorded by I.K. Janson, V.M. Svistunov and I.M. Dmitrenko (1965). The Josephson effect is the result of *macroscopic quantum coherence* of the superconducting state. In the presence of electric (E) and magnetic (B) fields the phase across the junction φ changes according to the equations

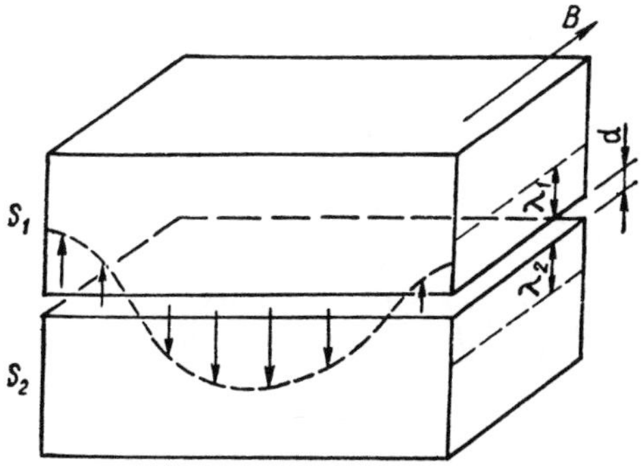

Fig. 1. Josephson current density oscillation in a junction located in an external magnetic field.

$$\frac{\partial\varphi}{\partial t} = \frac{2eV}{\hbar}, \qquad \frac{\partial\varphi}{\partial x} = \frac{2e\Lambda}{\hbar}B, \tag{1}$$

where Λ is the *penetration depth of magnetic field* into the superconducting junction. The first equation of Eq. (1) provides the ac Josephson effect. The second expression of Eq. (1) shows that in the presence of an external magnetic field the distribution of current in the junction varies according to a sinusoidal law with respect to the coordinate, and as a result, the maximum possible stationary current oscillates with the field, reaching zero at some of its values (*stationary quantum interference effect* analogous to Fraunhofer diffraction in optics, see Fig. 2). If an ac voltage $V(t) = V_0 \cos\omega t$ is applied, steps appear on the current–voltage characteristic of the tunnel contact (*Shapiro steps*, see

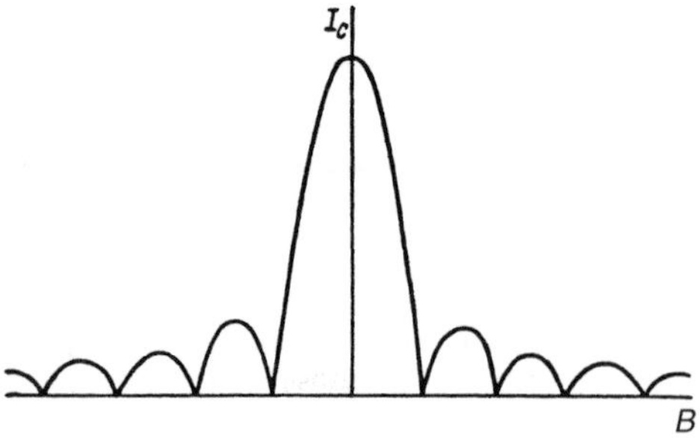

Fig. 2. A Fraunhofer diffraction picture of maximum current through a Josephson junction at zero-voltage.

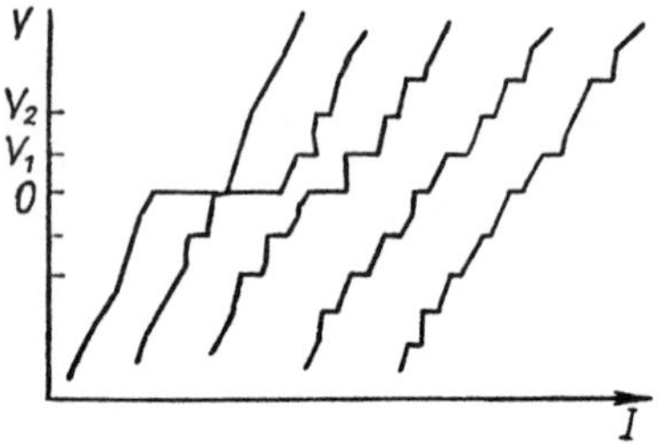

Fig. 3. Shapiro steps on the current–voltage characteristic of a tunnel junction.

Fig. 3) located at the voltages $V_n = n\hbar\omega/(2e)$. The height of the nth step oscillates in response to the alternating voltage amplitude as $J_n(2eV_0/(\hbar\omega))$, where $J_n(x)$ is an nth order Bessel function.

The inclusion of one or more weak links in a superconducting ring (see Fig. 4) induces a response (high frequency impedance, maximum super-current) which periodically changes with the period $\Delta B = \Phi_0/S$ in an applied field B, where $\Phi_0 = h/(2e)$ is the *flux quantum* and S is the ring area. Such a device is called a *Superconducting QUantum Interference Device* (SQUID). It allows one to measure extremely weak magnetic fields (down to 10^{-17} T), small voltages (10^{-15} V), and currents (10^{-10} A), and to detect minimal amounts of high frequency electromagnetic radiation.

The critical current of the tunnel junction is

$$I_c = \frac{\pi\Delta}{2eR}\tanh\left(\frac{\Delta}{2k_BT}\right),$$

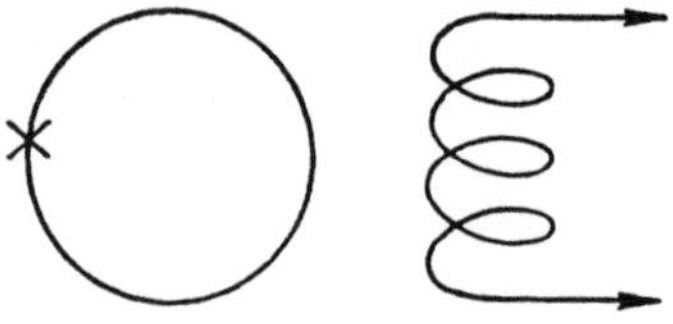

Fig. 4. Sketch of a SQUID with a single weak link (denoted by X), inductively coupled to an input coil.

where Δ is the superconductor *energy gap* ($\Delta \neq 0$ at $T < T_c$), and R is the resistance of the junction in the normal state (at $T > T_c$). For weak links with metallic conductivity (microcontacts) $I(\varphi)$ becomes non-sinusoidal:

$$I = \frac{\pi\Delta}{eR}\sin\left(\frac{\varphi}{2}\right)\tanh\left(\frac{\Delta\cos(\varphi/2)}{2k_BT}\right),$$

and the critical current at $T = 0$ appears to be twice as large (for the same contact resistance in the normal state). In high resistance contacts the *Josephson current* becomes negligibly small because thermodynamic *fluctuations* destroy the phase coherence. The maximum value of the resistance R for which the Josephson effect can still occur is determined by the condition $E_0 \geqslant k_BT$, where E_0 is *Josephson coupling energy*, $E_0 = \hbar I_c/(2e)$, which stabilizes the constant (coherent) value of the phase. At $T \sim 1$ K this criterion corresponds to a resistance not exceeding the order of 10^3 Ω. In addition to the thermodynamic fluctuations, quantum phase fluctuations (*macroscopic quantum tunneling*), which determine the probability of decay of the super-current at extremely low temperatures ($T < 0.1$ K), are observed in tunnel junctions. A *Josephson junction* under conditions of macroscopic quantum tunneling is a unique quantum system of macroscopic dimensions, whose behavior is similar to that of a single microparticle.

The space and time dependences of the phase φ of long superconducting tunnel junctions are provided by solutions of the sine-Gordon equation

$$\frac{\partial^2\varphi}{\partial x^2} - \frac{1}{c_0^2}\frac{\partial^2\varphi}{\partial t^2} = \frac{1}{\lambda_J^2}\sin\varphi, \qquad (2)$$

which has also been used to describe soliton motion. In this equation c_0 is the speed of the propagation of waves in the superconducting stripline

(*Swihart waves*) with the value $c_0 \sim 10^9$ cm/s; and λ_J is the *Josephson penetration depth* of a magnetic field into the junction ($\lambda_J \approx 10^{-3}$ to 10^{-1} cm). This equation has solutions in the form of the *Josephson vortices* in the tunnel junction, similar to *Abrikosov vortices*, but unlike the latter they do not possess a normal state core, and thus when moving only weakly dissipate energy. The reflection of vortices from the edges of junctions introduces stepwise singularities in the current-voltage characteristic (Shapiro or *Fiske steps*). These separate vortices may be regarded as the elementary carriers of information from the point of view of applications of *Josephson junctions* to digital electronics (computers). Eq. (2) also describes the weak oscillations of the phase with frequency $\omega_0 = c_0/\lambda_J \sim 10^{10}$ s^{-1} (the so-called *Josephson plasma frequency*). This mode exists due to the presence of the *Josephson inductance* $L_J = \hbar/(2eI_c \cos\varphi)$, conditioned by the dependence of the junction energy on the phase φ. The factor L_J is a nonlinear (parametric) inductance since it depends on the applied current.

JOSEPHSON JUNCTION

A *point contact* or *tunnel junction* between two superconductors which displays *Josephson effects*. For point contacts with direct conductivity, or for *superconducting microbridges*, the characteristic dimension of the constriction should not exceed the *coherence length* ξ. On the other hand, to prevent the destruction of the Josephson effect by *fluctuations* the contact resistance should not be too high (usually less than 100 Ω). According to their design, superconducting film microbridges, superconductor–insulator–superconductor tunnel junctions, superconductor–semiconductor–superconductor junctions, and others are singled out. Josephson junctions are used in *superconducting quantum interference devices* (SQUIDs), which measure magnetic flux and its space derivatives, and also any other electrical parameters (e.g., voltage, current), which can be transformed into a magnetic flux coupled to the superconducting circuit containing the Josephson junction. These junctions are used in low-temperature *thermometry*, as basic elements of superconductor digital electronics, and as nonlinear elements for detecting and mixing electromagnetic radiation over a wide range of frequencies.

JOSEPHSON MEDIUM

System of many superconducting regions connected by *Josephson junctions*. Josephson media can be regular or disordered. In regular two-dimensional lattices of niobium tunnel junctions, and of Josephson junctions based on the *proximity effect* of the type Pb–Bi/Cu and In/Au, a complex cooperative behavior is observed. In particular, the *Kosterlitz–Thouless transition* is observed with the dissociation of vortex–antivortex pairs at some temperature. Inhomogeneities purposefully introduced into Josephson media strongly influence the current–voltage characteristic. A *granular superconductor*, disordered to a greater or lesser extent, constitutes a special type of Josephson medium. For example, media based on the *ceramics* $BaPb_{1-x}Bi_xO_3$ display multiple Josephson switching with a voltage 0.1 to 1 V. A Josephson medium in an external magnetic field is described by the Hamiltonian

$$\widehat{H} = \sum_{\langle ij \rangle} I_{ij} \cos(\varphi_i - \varphi_j - A_{ij}),$$

$$A_{ij} = \frac{2e}{\hbar c} \int_i^j \boldsymbol{A}\, \mathrm{d}\boldsymbol{r},$$

where the amplitude of the *order parameter* is assumed to be constant within a grain or granule. Here φ_i is the phase of the order parameter in the ith granule; J_{ij} is the value of the weak bond between the granules i, j; $\boldsymbol{A}$ is the vector-potential of the magnetic field $\boldsymbol{B} = \nabla \times \boldsymbol{A}$.

JOSEPHSON RADIATION

Electromagnetic radiation from a *Josephson junction* due to the passage of an alternating superconducting current $j_s = j_c \sin\omega t$ upon applying a direct voltage to the electrodes. Josephson radiation had been predicted by B. Josephson (1962), and was detected by I.K. Janson, V.M. Svistunov, I.M. Dmitrenko (1965). The radiation frequency ω is proportional to the voltage V through the *Josephson relation* $\hbar\omega = 2eV$, which expresses the energy conservation law at the passage of a *Cooper pair* through the junction, with the simultaneous emission of a photon with energy $\hbar\omega$. For 1 μV the frequency $\omega/(2\pi)$ is 483.6 MHz. Josephson radiation may be to a large extent coherent and monochromatic (relative width of the frequency band $\leqslant 10^{-7}$). The power generated by a single Josephson junction is usually small ($\leqslant 10^{-10}$ W). In addition to the fundamental frequency, harmonics and subharmonics are often observed, as a result of the nonlinear properties of Josephson junctions.

The reciprocal effect, that is the excitation by the external radiation of steps on the current–voltage characteristic of the junction located at the voltages $V_n = n\hbar\omega/(2e)$, $n = 1, 2, \ldots$, was observed by S. Shapiro (1963) (see *Shapiro steps*). At a finite temperature noise due to fluctuations of the voltage on the junction [*Nyquist noise* and *shot noise* (H. Nyquist, 1927)], and also fluctuations of the *order parameter*, provide a finite width to the Josephson radiation. In the case of Nyquist noise the width $\Delta f = (2e/\hbar)^2 RT$, where R is the junction resistance in the normal state, corresponds to $\Delta f \sim 40$ kHz at $T = 4$ K and $R = 10^{-3}$ Ω.

JOSEPHSON VORTEX, fluxon

A localized distribution of current density in a superconducting tunnel junction with the characteristic dimension $\lambda_J = [\hbar c^2/(8\pi e \Lambda J_c)]^{1/2} \sim 10^{-3}$ to 10^{-1} cm, such that the total magnetic flux Φ created by this current is equal to the *flux quantum* $\Phi_0 = h/(2e)$, where λ_J is called the *Josephson penetration depth*, J_c is the Josephson critical current density, and Λ is the total depth of the magnetic field penetration. A Josephson vortex is a topological *soliton* solution of the sine-Gordon equation,

$$\frac{\partial^2\varphi}{\partial x^2} - \frac{1}{c_0^2}\frac{\partial^2\varphi}{\partial t^2} = \frac{1}{\lambda_J^2}\sin\varphi,$$

describing the distribution of phase φ in the Josephson junction. Josephson vortices, unlike *Abrikosov vortices* in *type II superconductors*, do not have a normal state core, so they only weakly dissipate their energy when moving along the junction. Therefore, in addition to stationary (immobile) Josephson vortices there are various related dynamic soliton structures. The transient formation of "soliton–antisoliton" (vortex–antivortex) pairs with a frequency ω in the energy gap for small oscillations of the phase, $\omega < \omega_0 = c_0/\lambda_J$ (ω_0 is Josephson plasma frequency), is one of them.

The reflections of Josephson vortices from the boundaries of the tunnel junction causes standing modes of oscillation. The motion of a one-dimensional chain of vortices may be regarded as a non-stationary *Josephson effect*, because the frequency of the field oscillations caused by this motion, and the respective average voltage across the barrier V, are related by the *Josephson frequency expression* $2eV = \hbar\omega$.

JOULE HEAT (J.P. Joule)

The heat emitted by an electrical conductor upon the passage of an electric current through it. If J is the current density and σ is the *electrical conductivity* then the Joule heat emitted per unit time per unit volume is equal to J^2/σ (*Joule–Lenz law*; established in 1841 by J.P. Joule and confirmed in 1842 by the experiments of E.H. Lenz).

JOULE–THOMSON EFFECT (J. Joule, W. Thomson, 1853)

Change of *temperature* of a gas at a change of *pressure* under adiabatic conditions. The Joule–Thomson effect is the basis of the most widely used methods of cryogenic cooling (see *Cryogenic temperatures*).

JUMP CONDUCTIVITY

See *Hopping conductivity*.

JUNCTION

See *Heterojunction*, *Homojunction*, *Josephson junction*, *Metal–semiconductor junction*, *Semiconductor junction*, *Tunnel junction*.

JUNCTION CAPACITANCE

See *Barrier capacitance*.

JUNCTION PHOTOEFFECT, gate photoeffect

An internal *photoeffect* which appears at the illumination of a junction or gate, i.e. a rectifying contact. It depends on the separation of the electron–hole pairs generated by the light (see *Semiconductor junction*) by means of the electric field at the junction. The cut-off layer (see *Metal–semiconductor junction*) at the semiconductor–metal boundary involves a p–n junction which possesses rectifying properties. As a result, the illumination of the rectifying contact causes a current to flow in the external circuit, and for the case of an open circuit a photoelectromotive force, *gate emf*, is generated at the junction (see *Surface photoelectromotive force*).

JUVENILE SURFACE, freshly prepared surface, virgin surface

A pure newly-prepared surface. In the case of *solids* such a surface is produced by making a *spalling*, by fracturing a bulk sample in a superhigh vacuum, or as a result of noble-gas-ion bombardment of the surface with subsequent *annealing* in a superhigh vacuum (see *Atomically clean surface*). In the case of liquids (e.g., mercury), a freshly prepared surface can be obtained by continuous formation and tearing off drops when pressing the liquid through a capillary.

Kk

KADOMTSEV–PETVIASHVILI EQUATION

See *Soliton*.

KANE MODEL

See *Semiconductors*.

KANE'S DISPERSION LAW (E.O. Kane, 1956)

Non-parabolic dependence of the electron energy E on the *quasi-momentum* $\boldsymbol{p}$ in the *conduction band* of a semiconductor. Kane's law is derived from a semiempirical theory that takes into account the interplay between the conduction and valence bands in $A^{\rm III}B^{\rm V}$ type semiconductors (see *Semiconductor materials*) with the bottom of the conduction band at the Γ point (center) of the *Brillouin zone* (InSb, GaAs, InAs, etc.). Such crystals are conventionally referred to as *InSb-type semiconductors*.

At the interaction of non-degenerate bands, under the condition $E(\boldsymbol{p}) - E(0) \ll E_{\rm g} + (2/3)\Delta$, we have

$$E(\boldsymbol{p}) - E(0) = \frac{p^2}{2m_0} - \frac{E_{\rm g}}{2}\left\{1 - \left[\frac{4p^2P^2}{3m_0E_{\rm g}^2} \times \left(2 + \frac{E_{\rm g}}{E_{\rm g} + \Delta}\right)\right]^{1/2}\right\},$$

where m_0 is the mass of the electron, $E_{\rm g}$ is the band gap, and P and Δ are parameters of momentum and energy. The *dispersion law* is non-parabolic, but isotropic with $E = E(p)$. The non-parabolic behaviour becomes negligible at large $E_{\rm g}$. The *band structure* model from which this equation was derived is called the *Kane model* (see *Semiconductors*).

KANZAKI FORCES (H. Kanzaki, 1957)

Fictitious forces that are to be applied to atoms of an *ideal crystal* in a solvent that is described in the *quasi-harmonic approximation* in order to create the same resulting displacements of these atoms in the unstressed crystal as those caused by the real forces of interaction between the forming crystal and the solvent.

KAPITZA LAW (P.L. Kapitza, 1928)

An empirical rule stating that the electrical resistance of a polycrystal of metal in a strong magnetic field grows in proportion to the magnetic field strength. It has been established for Cu, Au, Ag, and explained in the theory of *galvanomagnetic effects*.

KAPITZA TEMPERATURE JUMP (P.L. Kapitza, 1941)

A phenomenon in superfluid *helium* whereby the heating of a solid body immersed in liquid helium causes a temperature difference ΔT to arise at the interface between them. For small heat fluxes the temperature jump is proportional to the heat flux density Q:

$$\Delta T = R_k Q,$$

where the coefficient R_k is called the *Kapitza thermal boundary resistance*. The temperature dependence of R_k is described by the relation $R_k = A/T^n$, where $n = 1.4$–4.5, and at low temperatures $n \to 3$. The constant A depends on the acoustical properties of the bodies in contact, and on the quality of the surfaces. A weak dependence of R_k on the *Debye temperature* θ_D of a material is observed: $R_k \propto \theta_D^{0.6\text{–}0.8}$. The value R_k decreases insignificantly as the external pressure rises. For pure annealed copper, $R_k \sim 2.4\cdot 10^{-2}/T^3$ m$^2\cdot$K$^{-1}\cdot$W^{-1}. The Kapitza temperature jump is a universal physical phenomenon which arises at low temperatures at the interface between any heterogeneous media. The theory of this phenomenon is based on the assumption that the *heat transport* between the contacting bodies is carried by thermal *phonons*. For example, the main contribution to R_k is from the difference between the phonon spectra of He II and a solid body (I.M. Khalatnikov, 1952).

As the heat loading and the corresponding temperature gradient ΔT increase, processes in the liquid begin to play a more significant role. The value of R_k for large heat loading decreases as Q increases.

The Kapitza temperature jump hampers the cooling of bodies to extremely low temperatures, and must be allowed for in designing deep freezing systems for achieving *cryogenic temperatures*.

KAUZMAN PARADOX (W. Kauzman, 1948)

The property of a number of vitrifying systems (organic, oxide, polymerizing liquids) whereby the extrapolation of the physical characteristics of their liquid phase below the *vitrification* temperature T_V leads to nonphysical results. In particular, the extrapolated *entropy* of an overcooled liquid becomes smaller than that of the crystalline phase.

KELLERMAN MODEL (E.W. Kellerman, 1940)

This first serious model for computing *phonon optical spectra* takes into account inter-ionic interaction forces, and it served as the prototype for a number of subsequent models suited for simple *ionic crystals*. In addition to the Coulomb interaction between the ions regarded as point charges, the Kellerman model accounts for the effects of charge cloud overlapping between nearest neighbors. The contribution of the latter to the interaction energy is most often approximated as b/r^n or $Ae^{-\alpha r}$, where r is the distance between the atoms. The parameters of the overlap force, b and n, or A and α, are calculated by proceeding from known values of the *bulk modulus* (reciprocal of compressibility) and lattice constants. The potential energy of the crystal, U, is defined as the sum of *pair interaction* energies of ions in the lattice. The vibrational frequencies are solutions of the secular equation, elements of the determinant of the latter being expressed in terms of the second derivative d^2U/dr^2. Calculations following the Kellerman model for NaCl have indicated the existence of two types of phonon waves: acoustic and optical. The frequency *distribution function* is non-monotonic, in contrast to the Debye case. An important shortcoming of the Kellerman model is that it fails to take into account the polarization of ions displaced from their equilibrium positions. This shortcoming was overcome by the construction of the *model of a lattice of polarized ions* (K.B. Tolpygo), and the *shell model* that is equivalent to it (W. Cochran). See also *Phonon spectra*.

KELVIN EQUATION (W. Thompson, Lord Kelvin, 1871)

Thermodynamic equation describing the dependence of the vapor pressure P_r (or of the solubility C_r) on the radius of curvature r of the surface separating the *phases*:

$$\frac{P_r}{P_0} = \frac{C_r}{C_0} = \exp\left(\frac{2\sigma v}{rRT}\right),$$

where σ is the *surface tension* at the interface between the phases, v is the molar volume of the liquid or solid, R is the molar gas constant, and P_0 and C_0 are the values for a flat surface.

The Kelvin equation was obtained from the condition that the chemical potentials of the corresponding phases be equal to each other. It follows from this equation that the saturation vapor pressure (or solubility) on the boundary with drops of liquid or with *crystals* increases with the decrease of their dimensions, whereas the pressure inside a bubble or by the surface of a concave *meniscus* (i.e. when the radius of curvature is negative) is lower than that over a flat surface. Changes in the saturation vapor pressure (or solubility) only become significant at very small radii of curvature (usually, less than 10^{-5} cm). Thus, the Kelvin equation is usually applied when studying systems containing small objects (nuclei of a new phase, colloidal systems, etc.), as well as when dealing with *capillary phenomena*, and with processes associated with the formation of a new phase. For instance, according to the Kelvin equation, the *condensation* of a vapor in capillaries and microcracks occurs at vapor pressures lower than P_0. For the establishment of thermodynamic equilibrium, the Kelvin equation points out the direction of transport of the substance from smaller drops or crystals to larger ones. The nuclei of a new phase cannot start growing before the supersaturation defined by the Kelvin equation has been reached.

KERR EFFECT (J. Kerr, 1875)

A quadratic *electrooptical effect* involving the emergence of, or change in, *birefringence* resulting from the action of an electric field on a medium. It was first found in isotropic media (glass, polar liquids) by the enhanced transmittance in a transverse electric field of a polarization cell consisting of a crossed polarizer and analyzer with a glass plate between them. The Kerr effect occurs due to anisotropy of the *refractive index* exhibited by optically *uniaxial crystals*. The optical axis is aligned along the electric field, and the magnitude of the double refraction, Δn, is proportional to the square of the electric field intensity, $\Delta n = BE^2$, where B is the *Kerr constant* which depends on the temperature, light wavelength, molecular structure, and the phase *state of matter*.

The Kerr effect occurs both in static or low-frequency fields and in the high-frequency (optical) range; it may be treated in the terms of *nonlinear optics* as a manifestation of the nonlinear polarization of a medium in external electric fields. The Kerr effect occurs in crystals of all symmetry classes (see *Crystal symmetry*). In crystals that lack a *center of symmetry* and have a particular shape it is overshadowed by the much stronger linear electrooptic effect (the *Pockels effect*) when the applied electric field is oriented along the direction of the light propagation. The phenomenological description of this effect requires a polar fourth-rank tensor.

The Kerr effect in solids is usually associated with both lattice *strains* and a rearrangement of electronic states. The effect in a low-frequency electric field is usually accompanied by *electrostriction*, thus the Kerr effect for allowed and forbidden deformations should be distinguished. Because the Kerr effect "switches" almost instantaneously it is exploited for the high-frequency amplitude and phase modulation of light (see *Modulation spectroscopy*).

In addition to the above described electrooptic Kerr effect, there is also a *magnetooptical Kerr effect* (J. Kerr, 1876) which arises when linearly polarized light is normally incident upon the surface of a magnetized *ferromagnet*, and the reflected light is elliptically polarized, with the major axis of the polarization ellipse rotated through a certain angle relative to the plane of polarization of the incident light.

KHAIKIN OSCILLATIONS (M.S. Khaikin, 1960)

Sharp nonmonotonic dependence of the *surface impedance* of metals on a weak magnetic field $B \sim 0.01$ to 10 mT. Khaikin oscillations are observed at electromagnetic radiation frequencies $\omega \sim 10^{10}$–10^{11} s^{-1} in the course of experiments at liquid helium temperatures ($T < 10$ K), when the external magnetic field is applied parallel to the surface of a metal sample. Initially discovered by Khaikin in monocrystals of tin, cadmium and indium; later these oscillations were found and studied in a variety of *metals* and *semimetals* (bismuth, aluminum, copper, gallium, etc.). Khaikin oscillations are due to the resonance absorption of an electromagnetic wave at transitions of electrons passing from one *magnetic surface level* to another. Since these surface levels are not equidistant, Khaikin oscillations are not periodic with respect to reversals of the magnetic field, and are the result of superpositions of different resonance series. Resonance values of the magnetic field are proportional to $\omega^{3/2}$. The shape of a Khaikin oscillation resonance line is related to the dispersion law of *conduction electrons*, and to their scattering in the bulk and at the surface of the metal.

KIKOIN–NOSKOV EFFECT (I.K. Kikoin, M.M. Noskov, 1933)

A photomagnetoelectric effect involving the generation of an electric field under the action of light in a *semiconductor* placed in a magnetic field. The electric field is perpendicular to both the magnetic field and the direction of flow of the *current carriers*, the latter diffusing in the semiconductor from the illuminated side, where the absorbed photons produce *electron–hole pairs*, then drifting toward the non-illuminated side. The Kikoin–Noskov effect manifests itself through a pronounced non-uniform concentration of minority carriers brought about by the strong absorption of light.

KIKUCHI LINES (S. Kikuchi, 1928)

Lines found in the *electron diffraction* pattern (see *Electron diffraction analysis*) of thick crystals. The usual Laue–Bragg spots that appear on the photographic plate of an electron diffraction pattern from thin crystals vanish in the case of thick crystals, with dark and light lines (the Kikuchi lines) and bands (*Kikuchi bands*) observed in their stead. In the general case there is a light Kikuchi line parallel to each dark Kikuchi line. If the intensity between the pairs of Kikuchi lines differs from that of the background, then it is termed a Kikuchi band (it is a "surplus" band, if its intensity exceeds the background, and a "deficit" band in the opposite case). The Kikuchi lines are produced by electrons that underwent an inelastic scattering (on *phonons*, *plasmons*, or on valence or inner atomic electrons) by way of their diffraction at the reflecting planes of a *crystal lattice*. The structure of *Kikuchi patterns* depends upon the diffraction conditions and the crystal thickness. Almost all Kikuchi lines get split on crossing the Kikuchi bands. With increasing crystal thickness the dark Kikuchi lines turn into light ones. A rigorous quantitative description of the contrast of Kikuchi lines and bands is given by the theory of *dynamic radiation scattering*. One can utilize the Kikuchi lines to determine the crystal orientation and to obtain information on crystal lattice distortions. Kikuchi lines are an electron-optical analogue of *Kossel lines* from X-ray diffraction.

KILO... (fr. Gr. κιλιας, thousand)

Prefix to denote a decimal multiple equal to 1,000 times an original unit, abbreviated as k. An example: 1 km = 1,000 m.

KINETIC COEFFICIENTS

The quantities K_{ij} that define the linear dependence of thermodynamic flows or fluxes, I_i, on the thermodynamic forces, Y_j, that give rise to these fluxes in accordance with the expression

$$\frac{\mathrm{d}\gamma_i}{\mathrm{d}t} \equiv I_i = \sum_{j=1}^{n} K_{ij} Y_j \qquad (1)$$

for a physical system close to equilibrium. The parameters γ_i describing the nonequilibrium state, and the thermodynamic forces, Y_i, are related by the equations: $\partial S(\gamma)/\partial\gamma_i = -Y_i$, where $S(\gamma)$ is a nonequilibrium entropy. The kinetic coefficients obey the symmetry principle $K_{ij} = K_{ji}$, i.e. the *Onsager relations*. (Note that $K_{ij}(\boldsymbol{B}) = K_{ji}(-\boldsymbol{B})$ when the kinetic coefficients depend on a *magnetic field* $\boldsymbol{B}$.) Consider, for example, the case of thermodynamic forces arising from an electric field intensity $\boldsymbol{E}$, in the presence of gradients of the chemical potential μ, and the temperature T. To describe *transport phenomena* in a metal Eqs. (1) assume the forms

$$I_k = \sigma_{ki}\left(E_i - \frac{1}{e}\frac{\partial\mu}{\partial x_i}\right) + K_{ki}\frac{1}{T}\frac{\partial T}{\partial x_i},$$

$$Q_k - \frac{\mu}{e} I_k \qquad (2)$$

$$= -K'_{ki}\left(E_i - \frac{1}{e}\frac{\partial\mu}{\partial x_i}\right) + \kappa_{ki}\frac{1}{T}\frac{\partial T}{\partial x_i},$$

where I_k is an electric current, Q_k is an energy flux (thermal current), x_i are space coordinates, and e is the electric charge. The kinetic coefficients here are the *conductivity tensor* $\sigma_{ki} = \sigma_{ik}$, the *thermoelectric coefficient tensor*, the Peltier thermal e.m.f. coefficient (see *Thermoelectric phenomena*) $K_{ki} = K'_{ik}$, and the tensor $\kappa_{ki} = \kappa_{ik}$ which is related to the *thermal conductivity tensor*, $\hat{\chi}$, by the expression $\chi_{ik} = T^{-1}(\hat{\kappa} - \hat{K}\hat{\sigma}^{-1}\hat{K})_{ik}$. When $\boldsymbol{B} = 0$ the energy flow is expressed in terms of the thermal conductivity tensor, $\hat{\chi}$, under the conditions $I_k = 0$ and $Q_k = \chi_{ki}(\partial T/\partial x_i)$. The fact that the electric field $\boldsymbol{E}$ and the gradient of the chemical potential μ appear in the combination $\boldsymbol{E} - (1/e)(\partial\mu/\partial x)$ indicates the validity of the *Einstein relations* linking the conductivity with the *diffusion coefficient*.

When analyzing the equations of fluid dynamics applied to a non-ideal liquid, dissipative fluxes similar to the fluxes I_i in Eq. (1) appear. These dissipative fluxes are determined by such kinetic parameters as the first and second viscosity coefficients, the diffusion coefficient, the *thermodiffusion* coefficient, and the thermal conductivity.

KINETIC DECOUPLING OF COOPER PAIRS

The phenomenon of the decrease of the *energy gap* and of the *order parameter* in a superconductor brought about by the flow of supercurrent (*kinetic decoupling effect*). After carrying out a transformation to a moving reference frame, the energy of a *Cooper pair* traveling with a velocity $\boldsymbol{v}_s$ is found to be equal to $\Delta - \boldsymbol{p}\boldsymbol{v}_s$, where Δ is the value of the gap at $\boldsymbol{v}_s = 0$. Since the electron momentum assumes values along the direction of motion varying between $-p_F$ and p_F, where p_F is the Fermi momentum, the minimum energy gap is given by $\Delta_{min} = \Delta - p_F v_s$. The maximum v_s value for a gap different from zero is Δ/p_F corresponding to a maximum Cooper pair velocity $v_c \approx 10^4$ cm/s. This kinetic decoupling of Cooper pairs indicates the existence of a maximum dissipationless current density j_c in a *superconductor* (*depairing current density*) with a magnitude of the order of $j_c \approx n_s e v_c \approx 10^9$ A/cm^2 (see *Current states*).

KINETIC EQUATION

The basis for the statistical description of nonequilibrium systems comprising a great number of similar objects (particles, *quasi-particles*, etc.); it specifies the change with time, t, of a *distribution function*, $f(t)$, defined in the phase space of an individual object. The term was coined by L. Boltzmann in the latter part of the 19th century, and he also proposed the *Boltzmann equation*, a kinetic equation for the single-particle distribution function of a classical monatomic gas. A kinetic equation should satisfy the following condition: a change in the distribution function f during a short time interval, Δt, is completely determined by the function f itself, and by the external forces acting at that instant of time t, i.e. the history of the past evolution of a system in no way affects its current change. When this condition is satisfied with a sufficient precision then the system is referred to as being in the *kinetic stage of evolution*, which means being confined within a finite time period which depends both upon the system's individual characteristics (e.g., mean particle density, nature of their interactions), and its initial distribution. In the limit of short times the kinetic stage is restricted by the condition of a sufficiently small deviation from equilibrium when the magnitude of the particle *correlation length* is of the order of the radius of their interaction. On the other hand, the times must not be so long that weak, although long-lasting, time correlations can influence the system's evolution. Some systems consisting of a great number of particles, such as fluids, possess such internal conditions that the kinetic stage of evolution is completely missing.

Kinetic equations are widely used in statistical physics applied to nonequilibrium systems: in the theory of gases, theory of solids, theory of turbulence, plasma physics, theory of particle transit through matter, theory of radiation transfer, etc. A kinetic equation is ordinarily constructed, following L. Boltzmann, based on the phenomenological principle that relates a change in the function f during a short period of time to the balance of incoming and outgoing probability fluxes for an element in the single-particle phase space. Methods have already been developed for the self-consistent construction of a kinetic equation within the formalism of perturbation theory, with reference to some small parameter, proceeding from the macroscopic dynamic equations which describe the motion of particles in the system. For instance, one can derive the Boltzmann equation from the *Liouville equation* of classical mechanics written for the distribution of probabilities in the phase space of all the particles in a gas to the lowest approximation in the density. In quantum statistical physics an equation similar to the Boltzmann equation (see *Quantum kinetic equation*), which takes into account quantum-mechanical effective cross-sections and symmetry requirements, is obtained from the first non-vanishing terms in the approximation of the density, proceeding from the quantum Liouville equation for the system *density matrix*.

KINETIC EQUATION, QUANTUM

See *Quantum kinetic equation*.

KINETIC INSTABILITY OF SPIN WAVES

The exponential growth of the amplitudes of some groups of *spin waves* occurring in the case of highly nonequilibrium distributions in the spin wave spectrum. Such nonequilibrium distributions emerge, e.g., at the *parametric excitation of spin waves*, with their population growing by many orders of magnitude within a very narrow spin wave spectral region. When the parametrically excited waves reach a certain critical amplitude they give rise to kinetic instability at the lower end of their spectrum. The corresponding critical amplitude of parametrically excited spin waves is reached when the pumping amplitude exceeds its threshold value by 10–20 dB.

KINETIC ISOTOPE EFFECT

Influence of the isotopic composition of a reacting substance on the rate of a reaction, where the *reaction rate constants* of the molecules depend on the isotopes present. The kinetic isotope effect arises from the effects of atomic masses on the zero point vibrational levels of reacting molecules. The primary kinetic isotope effect depends upon the masses of the nuclei directly involved in the reaction, and the secondary effect involves masses of other nuclei. An abnormally large effect may be associated with tunneling reactions in solids, where the mass affects the probability of tunneling (see *Tunneling phenomena in solids*). A *magnetic kinetic isotope effect* has been observed, viz., the dependence of the rates of some chemical reactions on the nuclear magnetic moments associated with reacting *free radicals*. The kinetic isotope effect is of use in studies of the mechanisms of chemical transformations, and for the separation of isotopes.

KINETIC MOMENT

The same as *Angular momentum*.

KINETIC PHENOMENA

Physical phenomena occurring in non-equilibrium macroscopic systems under the action of external factors, or resulting from the collisions of particles or *quasi-particles*. These processes are described by the equations of macroscopic physics, including the *kinetic equations*, equations of fluid dynamics, Maxwell equations in a medium, etc. Kinetic phenomena can be classified into slow and fast varieties. The first class of *slow processes* involves those those for which the characteristic times of change of either the macroscopic parameters describing a given non-equilibrium system, or of the externally acting factors, are long compared to the characteristic microscopic times in this system. *Fast kinetic phenomena* are those for which the characteristic times of change of the external factors and of the macroscopic parameters are comparable to, or smaller than, the microscopic characteristic times. In the latter case kinetic phenomena are described by frequency-dependent *kinetic coefficients*, as distinct from the slow processes which involve constant or static kinetic coefficients. The distinction between fast and slow kinetic processes is largely a matter of convention, and depends on the choice of a characteristic microscopic time scale. If one takes for this scale the mean free time of particles or quasi-particles, using the kinetic equations with a *self-consistent field*, such as the *Vlasov equation*, it will be sufficient to deal with fast processes; with slow processes being described by equations of the fluid dynamics type in this case. Slow processes correspond to those associated with *transport phenomena*.

KINK

A particle-like excitation (*soliton*) in one-dimensional systems with a discrete degeneracy of its ground state energy. An example of a kink is a *domain wall* in a crystal with quasi-one-dimensional ordering (magnetic ordering, ferroelectric ordering, etc.). The simplest soliton solution of the sine-Gordon equation (see *Soliton*) describes a kink. Kinks play an important role in the thermodynamics of *quasi-one-dimensional crystals*; in particular, they determine the disturbance of long-range order (see *Long-range and short-range order*) known as the *central peak* of the correlation functions.

KINKS ON DISLOCATIONS

Elements of a dislocation line, which connect two of its segments, which lie in adjacent valleys of the potential *Peierls relief*. Depending on the kink width $\omega = a[U_0/(2U_P)]^{1/2}$, which is determined by the balance between the *Peierls energy* (U_P) and the *linear dislocation tension* energy (U_0), sharp and smooth kinks are distinguished. The kink width ranges from several lattice constants (a) in the case of *covalent crystals* to tens of lattice constants in the case of close-packed metals. The *kink energy* is given by the equation $U_k = (2a/\pi)(2U_0U_P)^{1/2}$. At zero temperature, the dislocation line may have only geometrical kinks if its end points are fixed in different valleys of the potential relief. At $T \neq 0$ the thermal fluctuations give rise to pairs of kinks; these pairs consist of kinks of opposite sign. The equilibrium concentration of paired kinks is defined as

$c_k^+ c_k^- = (1/a^2)\exp[-(2U_k)/(k_B T)]$. Kinks on a dislocation line are of crucial importance for describing dislocation mobility in the Peierls relief. Due to the translational symmetry along the dislocation line, the kink also exhibits a potential relief, which is called the *Peierls relief of the second kind*. The magnitude of this secondary relief is small in the cases of face-centered cubic and hexagonal close-packed metals, which exhibit smooth kinks, and is comparable to aU_P in the case of sharp kinks in *silicon* and *germanium*.

KIRKENDALL EFFECT (E.O. Kirkendall, 1947)

Transport of material in the *diffusion zone* during a *homogenization* process involving unequal partial *diffusion coefficients* D_A and D_B of two interdiffusing components A and B. The process is sustained by the self-maintaining action of a "source" of vacancies whose current has a magnitude g_v proportional to $[(D_A - D_B)/\omega]\nabla C_v$ (C_v is the vacancy concentration, and ω is an atomic volume), and by the *vacancy flow* itself. The flux which determines the effect involves *dislocations* whose *Burgers vectors* include an edge component. In phenomenological terms the Kirkendall effect is a *creep* under an effective stress in the diffusion zone maintained by supersaturation with vacancies, Δ/C_{v0} ($\Delta = C_v - C_{v0}$, where C_{v0} is an equilibrium vacancy concentration). The effect was experimentally discovered on brass–copper specimens (see Fig.) when inert markers that were placed on the plane of initial contact between the specimen components drifted due to the transport of matter in the diffusion zone. It follows from the simple Kirkendall effect theory for

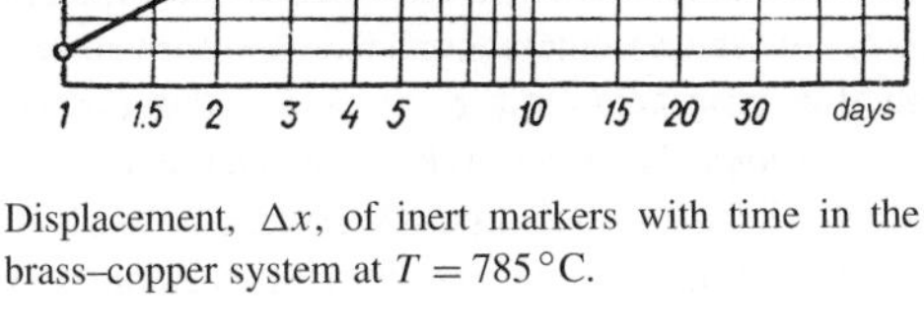

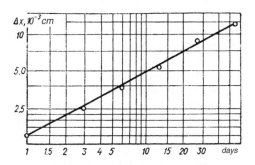

Displacement, Δx, of inert markers with time in the brass–copper system at $T = 785\,^\circ$C.

two-component substitutional *solid solutions* that the rate of marker displacement along the x axis, v, equals: $v = (D_A - D_B)\partial C_A/\partial x$. The markers move toward the component possessing the larger diffusion coefficient. The Kirkendall effect *per se* appears when the flow of vacancies involves only dislocations. Other fluxes (nucleating *pores*, *cracks*, etc.) can be removed, e.g., by applying a small *uniform compression* to the specimen (see *Barodiffusional effects*).

KIRKWOOD APPROXIMATION (J.G. Kirkwood, 1946)

One of the so-called *superposition approximations* that relate multiparticle *correlation functions* with each other. An example of the Kirkwood approximation is the following decomposition of a three-particle correlation function $f^{(3)}_{AAB}$ in terms of pairwise correlation functions $f^{(2)}_{AB}$ and $f^{(2)}_{AA}$:

$$f^{(3)}_{AAB}(\boldsymbol{r}_1, \boldsymbol{r}_2, \boldsymbol{r}'_1) \approx f^{(2)}_{AB}(|\boldsymbol{r}_1 - \boldsymbol{r}'_1|, t) f^{(2)}_{AB}(|\boldsymbol{r}_2 - \boldsymbol{r}'_1|, t) \times f^{(2)}_{AA}(|\boldsymbol{r}_1 - \boldsymbol{r}_2|, t),$$

with A and B designating the type of particles, $\boldsymbol{r}_1, \boldsymbol{r}_2$ and $\boldsymbol{r}'_1$, are the coordinates of the A, A and B particles, respectively; and t is time. The Kirkwood approximation can be employed to break up an infinite chain of coupled correlation functions that describe the spatial distribution of classical particles in statistical physics, and in the kinetics of dense gases and condensed media. Its use involves a transformation to a reduced description of the fluctuation spectrum of a multi-particle system on the level of two-particle correlation functions. This approximation facilitates the computation of a number of thermodynamic parameters. The validity and the regions of applicability of the Kirkwood approximation, which does not involve an expansion in powers of the density, depend upon the particular problem under consideration. They may be assessed by comparison with a *computer simulation*, or with an exact solution of the problem when one is available. For instance, the Kirkwood approximation applied to the diffusion-controlled transport of energy coincides with the exact solution for the case of immobile energy fluxes, it becomes an approximate solution when they become mobile, and exhibits a 40% error in the computation of the kinetics of the accumulation of *Frenkel defects* which are immobile at low temperatures.

KITTEL'S DOMAIN STRUCTURE (Ch. Kittel, 1946)

Plane-parallel (laminated) *magnetic domain structure* without domain closure, i.e. the *magnetic flux* does not close on itself within the sample. This domain structure occurs in crystals which have a single *easy magnetization axis* (*uniaxial magnetic crystals*). There is a square root dependence of the structure period, d, on the plate thickness, l, i.e. $d \propto (l l_0)^{1/2}$, where l_0 is a *characteristic length of magnetic material.* Kittel's domain structure proves advantageous when the *demagnetization field* energy is significantly smaller than the uniaxial anisotropy energy. This structure is only observed in sufficiently thin *plates.* With increasing thickness the structure becomes more complex, and *domain branching* becomes more pronounced near the surface, with $d \propto l^{2/3}$.

KKR METHOD

See *Korringa–Kohn–Rostoker method.*

KOHLER RULE (M. Kohler, 1938)

An empirically based assertion that the relative change of a *metal* magnetoresistance $\Delta\rho/\rho$ (see *Galvanomagnetic effects*) in a magnetic field $\boldsymbol{B}$ at various temperatures T can be represented by a universal relation:

$$\frac{\Delta\rho}{\rho} = \frac{\rho(B,T) - \rho(0,T)}{\rho(0,T)} = F\left[\frac{B}{\rho(0,T)}\right],$$

where $\rho(0,T)$ and $\rho(B,T)$ are the resistivities at $B = 0$ and $B \neq 0$, respectively. The function F is the same for samples of the same metal that differ from each other by their level of impurities and other defects. The Kohler rule is based on the assumption of similar *relaxation times*, τ, of the two types of *current carriers* traveling in a magnetic field. Observed deviations from this rule arise from a difference in the current carrier scattering mechanisms that affect τ.

KOHN ANOMALY

Peaks or breaks in the *phonon* dispersion law $\omega(\boldsymbol{q})$ (ω is the frequency) for wave vectors $\boldsymbol{q}$ coinciding with *Fermi surface* extremal diameters. The theoretical analysis of these singularities was carried out separately by A.B. Migdal (1958) and W. Kohn (1959) so they are often referred to as *Migdal–Kohn singularities*. They were observed experimentally in a series of metals and alloys with the help of *inelastic neutron scattering.*

The *phonon spectrum* is determined by the interatomic interaction and, therefore, the singularities in $\omega(\boldsymbol{q})$ are caused by anomalies in the electronic screening of this interaction, that is by anomalies in the *dielectric constant* $\varepsilon(q)$. For a spherical Fermi surface of radius k_F the first derivative of $\varepsilon(q)$ diverges logarithmically at $q = 2k_F$. In this case for the determination of the Kohn singularities the *Kohn construction* is applied, and a sphere of radius $2k_F$ is created around each site of the *reciprocal lattice.* Each point of the sphere that reaches the first *Brillouin zone* corresponds to a Kohn singularity, but experimentally the singularities are observed only for points belonging to several spheres simultaneously. To determine the locations of Kohn singularities in a real metal with a nonspherical Fermi surface a precise knowledge of its $\boldsymbol{k}$-space geometry is necessary. In *quasi-one-dimensional crystals*, *giant Kohn singularities* are observed which produce a network *Peierls instability*, and bring about a *second-order phase transition* to a distorted lattice state with decreasing temperature.

KOHN–LUTTINGER THEORY (W. Kohn, J. Luttinger, 1957)

The quantum theory of electric charge *transport phenomena.* A formalism that gives the complete *density matrix* of the charge carrier system in the stationary state has been developed within the Kohn–Luttinger theory. This theory uses a simplified but quite physical model of a real medium. A closed system with charge carriers (e.g., electrons) is examined. These electrons are considered to be free and noninteracting. They move in an applied electric field, and in the fields of randomly located *impurity atoms.* When the interaction between electrons and impurity centers is weak it is possible to show the validity of the *Boltzmann equation* for the diagonal elements of the density matrix in first order perturbation theory. The off-diagonal matrix elements are then expressed through the diagonal ones. Corrections have been calculated through terms of order λ^4, where λ is the dimensionless parameter characterizing the interaction force of impurities with electrons. For higher orders it was found that the distribution function equation does not reduce to the usual Boltzmann equation, but all correction terms are smaller than their "Boltzmann" counterparts.

KONDO EFFECT (J. Kondo, 1964)

Observation of a minimum electrical resistance R in experiments with nonmagnetic *metals* (Au, Ag, Cu, Al, Zn, etc.) as a function of temperature (T) caused by the conduction *electron scattering* off impurity atoms with unfilled inner electron shells and spin different from zero (*paramagnetic impurities*). For each material the effect is observed below a particular temperature T_K called the *Kondo temperature*. Such impurity atoms can be transition series (e.g., Cr, Mn, Fe, Co) and *rare-earth elements* (e.g., Ce, Tm, Yb).

The Kondo effect is a collective one involving an *indirect exchange interaction* of the conduction electrons with a paramagnetic impurity. As was shown by Kondo, corrections to the *Born approximation* for the amplitude of the *electron spin-flip scattering* from paramagnetic impurities depend on the Fermi–Dirac distribution function. In cases where the spin *exchange interaction* constant of the conduction electrons with the paramagnetic impurities is antiferromagnetic, the scattering amplitude increases logarithmically with the approach of the scattered electron energy to the *Fermi energy* of the metal. As a result, the spin component of resistance increases logarithmically with decreasing temperature, and a minimum in R appears through the competitive contribution of other scattering mechanisms (in particular, electron–phonon) resulting in the electrical resistance decreasing with the temperature. In the neighborhood of the Kondo temperature T_K (see *Abrikosov–Suhl resonance*) the temperature dependence of the electrical resistance changes sharply and tends to a finite limit at $T \to 0$. The growth of the electrical resistance with decreasing temperatures comes to a stop since the system of impurities transforms through the *Ruderman–Kittel–Kasuya–Yosida interaction* into the "*spin glass*" state in which the impurity spin orientations remain fixed, and the electron scattering channel through spin flips is suppressed. An applied magnetic field also fixes the spin orientation, and for $\mu_B B \gg k_B T$ (μ_B is the Bohr magneton) the minimum in R as a function of T disappears. Depending on the particular characteristics of the sample, switching on a rather strong magnetic field can either increase or decrease the electrical resistance.

KONDO LATTICE (J. Kondo)

A lattice of localized spins of paramagnetic ions (f-state ions) experiencing an antiferromagnetic *exchange interaction* with *conduction electrons*. In such a lattice the collective *Abrikosov–Suhl resonance* appears that causes a sharp increase of the *effective mass* of electrons transforming them into heavy electrons (heavy fermions).

KOOPMANS THEOREM (T.A. Koopmans, 1933)

This theorem provides a physical interpretation of energies of one-electron orbitals of a multielectron system in terms of the Hartree and Hartree–Fock approximations (see *Hartree–Fock method*). By Koopmans theorem the energy of every ith orbital taken with opposite sign is equal to the ionization energy of the system from the ith one-electron level. Koopmans theorem holds under the approximate assumption that removing an electron from the ith level does not change the orbitals of the remaining electrons. Some cases are known where Koopmans theorem is not valid, and the arrangement of one-electron levels by their energies is incompatible with the results of a Hartree–Fock calculation of the ionization energy as the difference between the ground state energies of $(N-1)$-electron and N-electron systems. In models different from those of Hartree or Hartree–Fock, Koopmans theorem may not hold.

KORRINGA–KOHN–ROSTOKER METHOD (J. Korringa, W. Kohn, N. Rostoker, 1947), KKR method

A non-linearized method, one of the *band structure computation methods* for crystals. The widely used KKR method has two equivalent formulations: one expressed in terms of multiple scattering theory is widely used in the theory of disordered systems, and the other led to the widespread linear muffin tin orbital approach (see *Linear methods of band structure computation*).

The solution of the Schrödinger equation in a crystal *muffin-tin potential* by the KKR method is carried out with the aid of a linear combination of trial basis functions. The solutions of the Schrödinger equation for a given energy inside the muffin tin sphere are used as the trial functions. They are matched continuously and smoothly to spherical waves outside the sphere which have the same angular momentum. The condition of canceling the "tails" of all "foreign" basis functions inside a given muffin tin sphere provides the secular equation of the KKR method; an equation which is non-linear in the energy, and has the form

$$\det\left\| B_{lmt,l'm't'}(E,\boldsymbol{k}) + \delta_{ll'}\delta_{mm'}\delta_{tt'}P_{lt}(E)\right\| = 0.$$

The so-called structure constants B depend only on the mutual arrangement of the atoms, and not on the lattice parameter, or on the muffin tin potential. The potential parameters P, on the contrary, do not depend on the structure. The order of the equation is determined by the number of spherical harmonics Y_{lm} in the wave function expansion (usually 9), and the number of atoms labeled by the index t in the cell.

An advantage of the KKR method, in comparison with the *augmented plane waves method*, is the small dimensionality of the matrix, and the possibility of calculating the structure constants for a given structure once and for all; a deficiency is the extremely inconvenient representation of the wave function in the regions between muffin tin spheres.

KORTEWEG–DE VRIES EQUATION

See *Boussinesq equation* and *Soliton.*

KOSSEL LINE (W. Kossel, 1935)

Diffraction pattern by X-rays generated within a single crystal. The interference lines (dark, light, mixed light-dark) are captured on a photoplate after irradiation by a beam of high-energy electrons. The arrangement of Kossel lines on a *kosselogram* is similar to the arrangement of interference lines on a diffractogram from a monocrystal exposed to a divergent monochromatic X-ray beam (see *Diffraction methods of investigation*). Rays which obey the *Bragg law* are diffracted on corresponding reflecting surfaces and give rise to Kossel lines on a photoplate. The Kossel line positions are determined by crossing the photoplate plane by *Kossel cones* with angular openings equal to $\pi - 2\theta$ (θ is the Bragg angle), and axes normal to the reflecting planes. The quantitative definition of Kossel line contrast is given by the theory of *dynamic radiation scattering.* Kossel lines are used to determine the lattice orientation relative to the crystal surface, to measure the lattice constant with a relative error $\sim 10^{-5}$, and to obtain information about crystal distortions, and degradations of crystal lattice symmetry (see *Crystal symmetry*).

KOSTERLITZ–THOULESS TRANSITION, called Berezinski–Kosterlitz–Thouless transition in the Russian literature

A *phase transition* occurring in planar (two-dimensional) degenerate systems with a two-component *order parameter* $\Psi e^{i\phi(x)}$ such as two-dimensional magnets, superconductors and superfluid helium. Upon tracing around a closed contour, the angular phase $\varphi(x)$ can change only by an integer multiple of 2π, i.e. $\Delta\varphi = 2\pi Q$. The quantity Q determines a resultant *topological charge* of the region confined by this contour, and equals the sum of topological charges of individual excitations (vortices) situated in this region (see *Topological inhomogeneity*). The contribution of each of the vortices to the system energy grows logarithmically with the increase of size R of the region: $\delta E = \pi J \ln(R/a)$ where a is the interatomic distance, and J is the interaction energy of nearest neighbor atoms. For several vortices with zero resultant topological charge, a similar contribution vanishes; therefore, the pairing of two vortices of opposite topological charge (i.e. opposite orientation) in a neutral vortex molecule produces a gain in energy (V.L. Berezinski, 1971). As the temperature increases, the number of "vortex molecules" and the distance between vortices in a newly formed molecule increase as well. At the temperatures $T \sim J/k_B$, the distances between molecules and between vortices in a molecule become equal, and the dissociation of "vortex molecules" occurs. This dissociation proceeds by the way of a *second-order phase transition*. The temperature of the transition T_c was calculated by J.M. Kosterlitz and D.J. Thouless (1973); hence its name.

The Kosterlitz–Thouless transition is associated with the appearance of a finite density of topological charge vortices, and exemplifies a *topological transition*. In contrast to phase transitions involving a change in the ground state symmetry, there is no long-range order (see *Long-range and short-range order*) in the two-dimensional system at any temperature $T > 0$, i.e. the average value of the order parameter is zero. At great distances, the correlation parameter changes from the exponential fall-off at $T > T_c$ to a power law at $T < T_c$. A transverse rigidity appears in the low-temperature phase; the susceptibility of such a phase (*Berezinski phase*) becomes infinite at all temperatures below T_c.

The temperature of the Kosterlitz–Thouless transition is found from the assumption that the change in the Gibbs free energy (see *Thermodynamic potentials*) is zero at the appearance of a vortex: $\delta G = \delta E - T\delta S = 0$ where $\delta S = \ln(R^2/a^2)$ is the vortex entropy, and R^2 is its area. Creating a single vortex increases the energy at $T > T_c = \pi J\rho_s/(2k_B)$. The value ρ_s determines the contribution of vortex-free thermodynamic fluctuations to the vortex energy; $\rho_s(T=0) = 1$. In films of superfluid ^{4}He, it is expressed via superfluid density $\overline{\rho}_s = J\rho_s m^2/(k_B\hbar^2)$ where m is the mass of a ^{4}He atom. The ratio ρ_s/T_c is independent of the film thickness, as confirmed by a number of experiments on films of ^{4}He.

KOSTER–SLATER MODEL (G.L. Koster, J.C. Slater, 1954)

The simplest variant of the Lifshits method of degenerate perturbation theory in which the crystal Hamiltonian and the perturbation induced by a defect are expressed in terms of *Wannier functions*. The Koster–Slater model is used for calculations of energy bands as an interpolation scheme.

KRAMERS DOUBLET

A pair of energy levels of half-integer spin having mutually conjugated eigenfunctions with respect to *time inversion*. A Kramers doublet is considered when taking into account the *spin–orbit interaction*. It follows from the *Kramers theorem* that the levels of a Kramers doublet coincide in the absence of an external magnetic field B. Magnetic dipole transitions between the levels of a Kramers doublet are always allowed, while electric dipole transitions are forbidden for small B.

KRAMERS ION

An ion with an odd number of electrons, and hence a half-integral spin. In accordance with the *Kramers theorem* the energy levels in a Kramers ion are doubly degenerate in the absence of an external magnetic field (*Kramers degeneracy*). *Non-Kramers ions* are also known: those with an even number of electrons and hence an integral spin which define their magnetic properties. In a crystalline electric field the energy levels of a free non-Kramers ion split either fully or in part. For example, when the value of the spin is $S = 1$, a non-degenerate term with *magnetic quantum number* $M = 0$, plus a doubly degenerate term with $M = \pm 1$ appear in a field of *axial symmetry*. Such *doublets* formed by non-Kramers ions in a crystal are traditionally called *non-Kramers doublets*. They may be described independently of other levels, particularly when the latter are sufficiently far removed from non-Kramers doublets. A *spin Hamiltonian* with effective spin $S = 1/2$ is then used.

KRAMERS–KRONIG RELATIONS (H. Kramers, R. Kronig, 1927)

Relations between the real, ε', and the imaginary, ε'', parts of the *dielectric constant* (permittivity) $\varepsilon(\omega)$:

$$\varepsilon' = \varepsilon(\infty) + \frac{1}{\pi}\int_{-\infty}^{+\infty} \frac{\varepsilon''(x)}{x - \omega}\,\mathrm{d}x, \tag{3}$$

$$\varepsilon'' = -\frac{1}{\pi}\int_{-\infty}^{+\infty} \frac{\varepsilon' - \varepsilon(\infty)}{x - \omega}\,\mathrm{d}x, \tag{4}$$

where the principal values of the integrals are taken, and $\varepsilon(\infty)$ is the high-frequency part of ε'.

The Kramers–Kronig relations were first obtained for the index of refraction $n(\omega)$ and the absorption coefficient $k(\omega)$ of an electromagnetic wave in the form:

$$n(\omega) = 1 + \frac{c}{\pi} \int_0^\infty \frac{k(\omega_1)\,d\omega_1}{\omega^2 - \omega_1^2}. \qquad (5)$$

The Kramers–Kronig relations are used to calculate one of the quantities n or k for any given ω using the available spectral dependence of the other. For example, one may calculate the frequency ω_m at which $n(\omega)$ undergoes its strongest variations due to the changes in the absorption $k(\omega)$ produced by various external influences on the medium. Such a possibility has important implications for *holography*.

These relations are valid for the real and imaginary parts of any analytical complex function. They follow from the principle of causality, and therefore, have an extremely general nature. In particular, they may be written for a material *susceptibility* of any type. The Kramers–Kronig relations can be generalized to account for *spatial dielectric dispersion*. Violations of these relations in the form (1) have been experimentally studied for a number of *molecular crystals*, such as CdS, etc., in the range of exciton absorption.

Historically the Kramers–Kronig relations provide the first example of dispersion relations. Generally speaking, these are the integral representations of response functions describing the reaction to external influences of a stationary physical system in equilibrium.

KRAMERS THEOREM (H. Kramers, 1929)

A theorem stating that the energy is at least twofold degenerate (more generally, has an even degeneracy) in systems with half-integer spin in the absence of an applied magnetic field. It is closely associated with *time inversion* (see *Kramers doublet*). In solids, the Kramers theorem is automatically satisfied for any values of the *quasi-momentum* $\boldsymbol{k}$, whenever $E(\boldsymbol{k}) = E(-\boldsymbol{k})$. When either $\boldsymbol{k}$ and $-\boldsymbol{k}$ differ by a *reciprocal lattice* vector or $\boldsymbol{k} = 0$, a double matching (superposition) of the energy bands takes place. In crystals possessing a *center of symmetry*, if spin interactions are taken into account, this occurs over the entire *k-space* (see *Group theory*).

KRIVOGLAZ–CLAPP–MOSS FORMULA
(M.A. Krivoglaz, 1957; P.S. Clapp, S.C. Moss, 1966).

A formula expressing the intensity I of the *diffuse scattering* of X-rays by non-distorted *solid solutions* in terms of the ordering energy of the solution, $w(\boldsymbol{\rho})$, for various atom-to-atom distances $\boldsymbol{\rho}$ among the pairs of interacting atoms. If $I(\boldsymbol{q})$ is expressed in electronic units ($\boldsymbol{q}$ is difference between wave vectors of the scattered and incident waves) then the Krivoglaz–Clapp–Moss formula for disordered solutions has the form

$$I(\boldsymbol{q}) = N(f_A - f_B)^2 \left[\sum_{\rho} \chi(\boldsymbol{\rho}) \cos(\boldsymbol{q} \cdot \boldsymbol{\rho}) \right]^{-1},$$

where N is the number of sites in the lattice, f_A and f_B are the atomic scattering factors for atoms A and B in the A–B solid solution, $\chi(\boldsymbol{\rho})$ are functions of $w(\boldsymbol{\rho})$, concentration c, and temperature T. Explicit expressions for $\chi(\boldsymbol{\rho})$ are obtained as expansions in powers of $w(\boldsymbol{\rho})/(k_B T)$, or of c. For example, $\chi(\boldsymbol{\rho}) = w(\boldsymbol{\rho})/(k_B T)$ for $\rho \neq 0$, and $\chi(0) = [c(1-c)]^{-1}$ in the high temperature limit.

Using experimental data on the diffuse scattering of X-rays by single crystals of alloys for various crystal orientations and various values of $\boldsymbol{q}$, and performing the Fourier transformation of $1/I(\boldsymbol{q})$ in a unit cell of the *reciprocal lattice*, one may use the Krivoglaz–Clapp–Moss formula to determine values of $w(\boldsymbol{\rho})$ for various values of $\boldsymbol{\rho}$. They permit one to calculate thermodynamic characteristics of alloys using statistical theories of *alloy ordering*.

KRONIG–PENNY POTENTIAL

See *Periodic potential*.

***k*-SPACE**, reciprocal space

Space used for representing wave propagation processes in a solid; its dimensionality coincides with that of a wave vector: $[k] = \mathrm{m}^{-1}$. In $\boldsymbol{k}$-space, propagating wave states (including quantum-mechanical ones), energy spectra of wave excitations, and scattering processes are featured. The coordinate space periodicity (*translational symmetry*) of an *ideal crystal* lattice is reflected in the reciprocal lattice of $\boldsymbol{k}$-space.

The wave vectors $\boldsymbol{k}$ in crystals of finite dimension take on discrete values. Using the *Born–von Kármán boundary conditions*, we obtain

$$\boldsymbol{k} = k_1\boldsymbol{b}_1 + k_2\boldsymbol{b}_2 + k_3\boldsymbol{b}_3 = 2\pi\left(\frac{m_1}{L_1}\boldsymbol{b}_1 + \frac{m_2}{L_2}\boldsymbol{b}_2 + \frac{m_3}{L_3}\boldsymbol{b}_3\right),$$

where $\boldsymbol{b}_1, \boldsymbol{b}_2, \boldsymbol{b}_3$ are the basis vectors of the reciprocal lattice; L_1, L_2, L_3 are the number of direct lattice unit cells in three coordinate directions; m_1, m_2, m_3 are integers taking on values, e.g., $0 \leqslant m_1 < L_1, 0 \leqslant m_2 < L_2, 0 \leqslant m_3 < L_3$. Thus, the total number of discrete values of $\boldsymbol{k}$ is equal to the number of crystal unit cells $N = L_1L_2L_3$. As the quantity N is very large, the distribution of wave vectors can be regarded as quasi-continuous, and summations over $\boldsymbol{k}$ can be replaced by integrations carried out inside a *Brillouin zone*

$$\sum_{\boldsymbol{k}} \cdots = \frac{V}{(2\pi)^3}\int \ldots \mathrm{d}\boldsymbol{k} = \frac{V}{(2\pi)^3}\int \mathrm{d}k_1\,\mathrm{d}k_2\,\mathrm{d}k_3 \ldots,$$

where V is the crystal volume. By correlating elementary or *collective excitations* of an ideal crystal with particular the wave functions $\psi_{\boldsymbol{k}}(\boldsymbol{r})$ it is possible to prove the *Bloch theorem*, which states that

$$\psi_{\boldsymbol{k}}(\boldsymbol{r}) = \mathrm{e}^{\mathrm{i}\boldsymbol{kr}}u_{\boldsymbol{k}}(\boldsymbol{r}),$$

where $u_{\boldsymbol{k}}(\boldsymbol{r}+\boldsymbol{l}) = u_{\boldsymbol{k}}(\boldsymbol{r})$. Therefore, both the wave function $\psi_{\boldsymbol{k}}(\boldsymbol{r})$ and any measured physical attribute of a crystal in $\boldsymbol{k}$-space are periodic functions of $\boldsymbol{k}$ with the period of the reciprocal lattice. For particular conditions (e.g., for small perturbations, for long wavelengths, and in certain other cases) the density of states, the energy spectrum, and other factors are very little altered for imperfect crystals, fluids and glassy-type materials.

KUBO THEORY

See *Linear response*.

KUBO–TOMITA METHOD (R. Kubo, K. Tomita, 1954)

A general formalism used to describe the *linear response* of physical variables to a weak external perturbation, $AF(t)$. The change in a physical variable $B(t)$ produced by an external force $F(t)$ is expressed as

$$\Delta B(t) = \int_{\infty}^{t} \varphi_{BA}(t-t')F(t')\,\mathrm{d}t'.$$

The function

$$\varphi_{BA}(t) = \frac{1}{\mathrm{i}\hbar}\mathrm{Tr}\,\rho\big[A, B(t)\big]$$

is called the *system response*, and ρ is the density matrix. The system susceptibility or admittance is a *kinetic coefficient* calculated as the limiting Fourier transform of the response function:

$$\chi_{BA}(\omega) = \lim_{\varepsilon\to+\infty}\int_0^{\infty} \varphi_{BA}(t)\exp[-\mathrm{i}\omega t - \varepsilon t]\,\mathrm{d}t.$$

As long as the reaction of the system to external perturbations remains linear, then calculating the susceptibility is reduced to calculating the time *correlation functions* for the *fluctuations* $A - \langle A\rangle$, $B - \langle B\rangle$ in the equilibrium state. In particular, this formalism provided the basis for developing the quantum theory of *magnetic resonance*, which does not make use of a kinetic equation for the *density matrix*.

KULIK FACTOR (I.O. Kulik)

A kinematic factor entering the microcontact function of the *electron–phonon interaction* that determines its dependence on the shape of the microcontact, on the orientation of crystallographic axes of the *metal* in the near-contact region with respect to the contact axis, and also on the average path length of elastic scattering of electrons. For a pure contact shaped like a circular hole in an infinitely thin screen that is opaque for electrons, the Kulik factor has the form

$$K(\boldsymbol{v}, \boldsymbol{v}') = \frac{4|v_z v_z'|}{|v v_z' - v' v_z|}\theta(-v_z v_z'),$$

where $\boldsymbol{v}$ and $\boldsymbol{v}'$ are the velocities of the electron before and after scattering, the z axis is parallel to the contact axis, and $\theta(x)$ is the theta function. For microcontacts made of an isotropic (polycrystalline) metal the Kulik factor has the form $(1 - \varphi/\tan\varphi)/2$ where φ is the scattering angle $(\boldsymbol{v}, \boldsymbol{v}')$. The Kulik factor accentuates the importance of large-angle scattering on the energy-dependent part of the microcontact resistance (see *Microcontact spectroscopy*).

KURCHATOVIUM, Ku

See *Transuranium elements.*

KURDYUMOV–SACHS ORIENTATION RELATIONS

See *Orientation relations.*

L

LABELED ATOMS

See *Tracers*.

LABYRINTHINE MAGNETIC DOMAIN STRUCTURE

A type of *magnetic domain structure* in a highly anisotropic magnetically uniaxial film, with a quality factor exceeding unity, and the *easy magnetization axis* normal to the growing surface. Such a structure is characterized by random orientation of the *magnetic domain walls*. It is observable in thin plates of *hexaferrites*, *iron garnets*, *ferrite spinels*, and in amorphous *magnetic films* of d- and f-metal alloys (see Fig. on the next page). The labyrinthine structure arises due to two circumstances: the lack of a preferred domain wall direction, and entropy production. Numerous experiments performed on various samples have demonstrated that a labyrinthine domain structure develops when a uniformly magnetized state loses its stability in an external magnetic field parallel to the *easy magnetization axis*, and the sample temperature is not too close to the *Curie point*. Moreover, if one artificially forms a regular striped domain structure in a film, the spatial ordering in the domains keeps gradually fragmenting under the effect of uniform external factors. These factors may include temperature effects, constant or alternating magnetic fields parallel to the easy magnetization axis, uniaxial extension or compression of the sample along that axis, etc., so that the magnetic domain structure eventually transforms to a labyrinthine type.

LAGRANGE COORDINATES (J.L. Lagrange, 1788)

The coordinates of a point in a continuous medium that characterize that point, are conserved, and remain specific for that point independent of any *strain* of the medium. They are often called *material coordinates* and are used in *continuum mechanics*. Lagrange coordinates were introduced by L. Euler. An example of the Lagrange coordinates of a point in a medium under deformation is the coordinates of that point in its initial state.

LAMBDA PHASE

See *Laves phases*.

LAMBDA POINT (λ-point)

Transition temperature $T_\lambda = 2.17$ K of liquid *helium* (^{4}He) from the normal (He I) to the superfluid (He II) state (see *Superfluidity*). This is a *second-order phase transition*. The term lambda point arises from the resemblence of the plot of the ^{4}He *specific heat* versus the temperature in the vicinity of T_λ to the Greek letter λ, as shown in the figure. Such anomalous behavior of the specific heat (logarithmic singularity at $T = T_\lambda$) is related to the large *critical fluctuations* over a broad region near the λ-point.

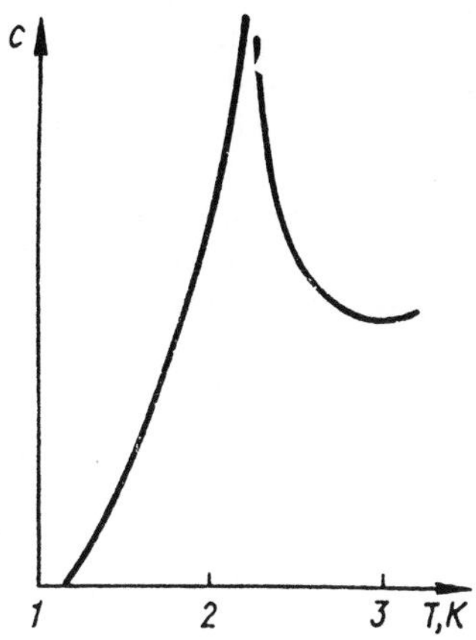

Lambda point of liquid ^{4}He.

Labyrinthine magnetic domain structure.

LAMBERT LAW (J. Lambert, 1760)

See *Reflection of light.*

LAMÉ COEFFICIENTS (G. Lamé, 1852)

Coefficients of the squared *dilatation*, u_{ii}^2, and of the sum of squares of all the components of the *strain tensor*, u_{ik}^2, in the expansion of the free energy of a deformed isotropic body. The free energy is represented in the form: $F = F_0 + (1/2)\lambda u_{ii}^2 + \mu u_{ik}^2$, where F_0 is a constant and λ and μ are the Lamé coefficients. The coefficient μ is the *shear modulus*, while the combination of two coefficients $K = \lambda + 2\mu/3$ is the *bulk modulus* (compressibility modulus). The quantities λ and μ are expressed via *Young's modulus*, E, and the *Poisson ratio*, ν, as follows:

$$\lambda = \frac{E\nu}{(1-2\nu)(1+\nu)}, \qquad \mu = \frac{E}{2(1+\nu)}.$$

LAMINA

See *Plates.*

LAMINATED CRYSTALS

See *Quasi-two-dimensional crystals.*

LANDAU DAMPING (L.D. Landau, 1946)

A collision-free damping of collective *plasma oscillations* (waves) in a classical (nondegenerate) electron–ion plasma, or in a quantum (degenerate) electron–hole *solid-state plasma.*

Classic Landau damping occurs when waves interact with resonance particles moving at velocities v that are close to the phase velocity of the wave, ω/k. Faster particles that catch up with the wave transfer some of their energy to it, while slower particles that lag behind extract energy from the wave. Since an equilibrium electron-ion plasma with a Boltzmann–Maxwell distribution $f(v)$ (see *Boltzmann distribution, Maxwell distribution*) has more slower particles (electrons, holes) than faster particles (see Fig. 1(a)), and the wave as a whole expends energy by accelerating slow particles, its amplitude falls exponentially with time, $e^{-\gamma_L t}$. The *decrement of Landau damping*, γ_L, is proportional to the derivative $\partial f/\partial v$ of the distribution function with respect to velocities

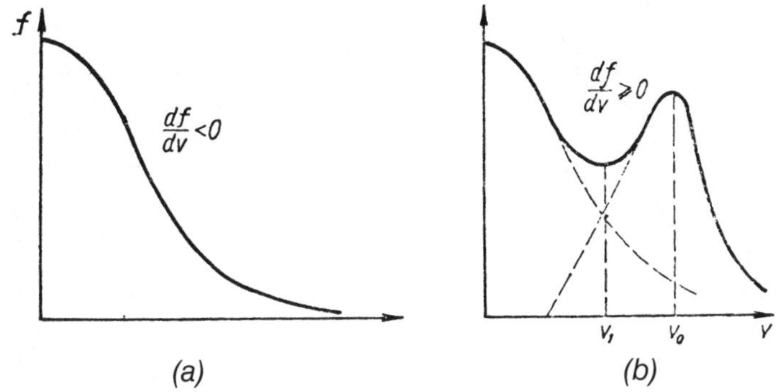

Fig. 1. Electron velocity distribution function (a) in an equilibrium Maxwell plasma, and (b) in a plasma pierced by an electron beam. The region $v_1 < v < v_0$ corresponds to the kinetic beam instability.

at $v = \omega/k$ and is positive for $\partial f/\partial v < 0$. The latter condition is satisfied for every v in an equilibrium plasma. However, we can have $\partial f/\partial v > 0$ in a certain velocity range in a highly nonequilibrium plasma (e.g., when an electric current or charged particle beam propagates through the electron-ion plasma (A. Akhiezer, Ya. Fainberg, D. Bohm, E. Gross, 1949; see Fig. 1(b)). In that case $\gamma_L < 0$, and an exponential build-up of plasma waves takes place, resulting in the so-called *kinetic beam* (or *current*) *instability*. The detection of this instability in a solid-state plasma is hampered by strong relaxation (scattering) of free charge carriers (*conduction electrons*, *holes*) or vibrations (*phonons*) and *point defects* (impurities, *vacancies*) of a crystal lattice. However, various *dissipation instabilities* become possible in a nonequilibrium solid-state plasma.

A quantum analog of the Landau damping takes place in the degenerate electron (hole) plasma of *metals* (*semimetals*). It results from the decay of collective (plasma) excitation quanta of the boson type, i.e. *plasmons*. These decay into one-particle electron excitations of the fermion type – the *electron–hole pairs* that follow the *conservation laws* of energy and momentum. For the cases of an isotropic three-dimensional (3D) degenerate electron plasma, and of a *two-dimensional electron gas* (2D) generated in *inversion layers* and stratified crystals (see *Quasi-two-dimensional crystals*), the range of Landau quantum damping with a nonvanishing imaginary part of the *dielectric constant* $\operatorname{Im}\varepsilon(q,\omega)$ is determined by the condition (J. Lindhard, 1954; F. Stern, 1967):

$$\max\left\{0, qv_F\left(\frac{q}{2k_F} - 1\right)\right\} < \omega < qv_F\left(\frac{q}{2k_F} + 1\right),$$

where v_F is the *Fermi velocity*, k_F is the electron Fermi momentum, and ω and q are the frequency and momentum, respectively (see Fig. 2(a) with the damping region hatched, where ω_p is the plasma frequency).

In a one-dimensional (1D) degenerate electron gas found in *thread-like crystals*, the range of Landau quantum damping is defined by the condition (P. Williams, A. Bloch, 1974):

$$q_\parallel v_F\left|\frac{q_\parallel}{2k_F} - 1\right| < \omega < q_\parallel v_F\left(\frac{q_\parallel}{2k_F} + 1\right),$$

where $q_\parallel$ is the longitudinal component of momentum in the direction of the chains. Due to shrinking of the phase volume available for the process of plasmon decay, there exists a "transparency window" (see Fig. 2(b)) with $\operatorname{Im}\varepsilon(q,\omega) = 0$ in the range $0 < \omega < q_\parallel v_F(1 - q_\parallel/2k_F)$ for $q_\parallel < 2k_F$.

Nonlinear Landau damping (R.K. Mazitov, 1965; O'Neil, 1965) takes place when the damping time, γ^{-1}, is longer than the oscillation period of the resonance particles in the traveling wave field, $-\varphi_0\cos(kx - \omega t)$. Phase trajectories of such particles are shown in Fig. 3. The region inside the separatrix corresponds to trapped particles, and that outside it to transient ones. The characteristic

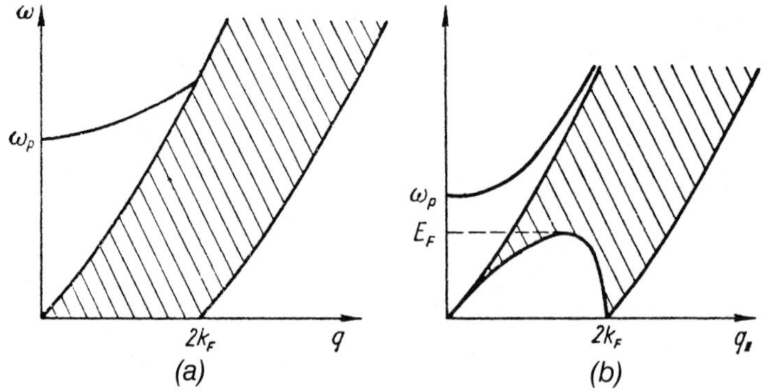

Fig. 2. Plasmon spectrum, $\omega(k)$, and Landau quantum damping region (hatched) of a degenerate electron plasma (a) in an isotropic 3D-metal, and (b) in a quasi-one-dimensional (chain-like) metal.

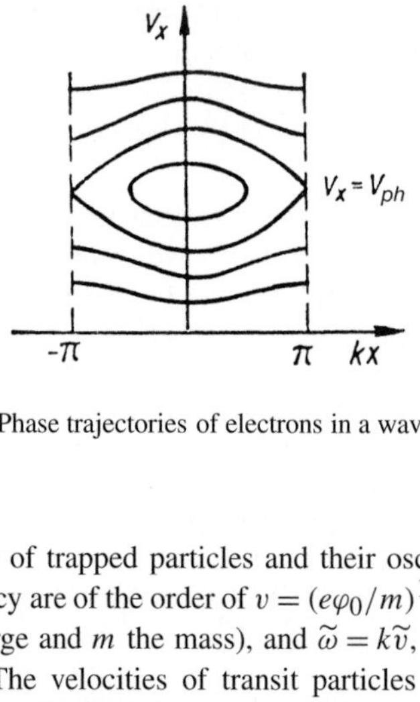

Fig. 3. Phase trajectories of electrons in a wave field.

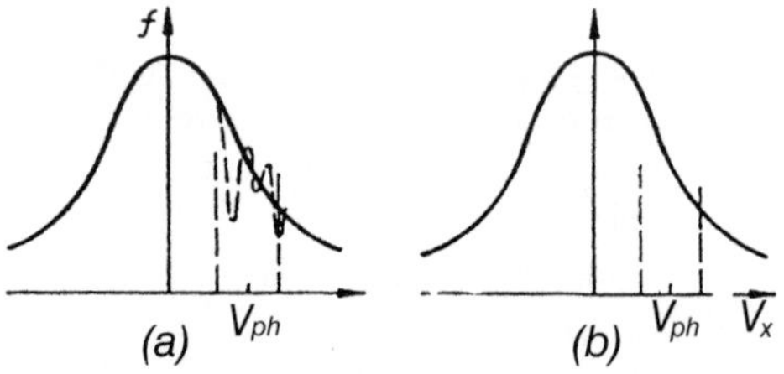

Fig. 4. Electron distribution function showing (a) formation of rapid oscillations, and (b) plateau region.

velocity of trapped particles and their oscillation frequency are of the order of $v = (e\varphi_0/m)^{1/2}$ (e is the charge and m the mass), and $\widetilde{\omega} = k\widetilde{v}$, respectively. The velocities of transit particles exceed $2\widetilde{v}$ in magnitude. The wave affects the electron trajectories, so collision-free damping weakens, compared to linear damping. The situation is that the particles with different energies have different oscillation frequencies in the wave field: while the frequency equals $\widetilde{\omega}$ at the bottom of the well, it tends to zero near the separatrix. Therefore, the points in the phase plane rotate so that a mixing of the initial equilibrium distribution occurs in the phase space. As a result, within several oscillation periods, a small-scale feature forms in the distribution function within a $4\widetilde{v}$ interval near the wave phase velocity $v_{\text{ph}} = \omega/k$ (see Fig. 4(a)). Due to electron collisions, fine oscillations of $f(v_x)$ vanish quickly and a plateau forms (see Fig. 4(b)).

The nonlinear damping decrement, $\gamma(t)$, oscillates with time. Let the field be switched on at $t = 0$. While the derivative $\partial f/\partial v_x$ is negative, the decrement is positive and is numerically equal to its linear value, γ_{L}. The resonance particles then extract the energy from the wave. During the next half-period, the resonance particles release, on the average, their energy and we have $\gamma < 0$. At $t \to \infty$, when a plateau has already formed in the distribution function, damping vanishes. The decrement never becomes zero in an electron plasma with collisions, but is proportional to $a\gamma_{\text{L}}$ for $t \gg 2\pi/\omega$, where $a = (\widetilde{\omega}\tau)^{-1} \ll 1$ is the ratio of the particle oscillation period to the time τ of electron scattering. The numerical value of $\gamma(t)$ may be found by equating the work done by the wave on the resonance particles to the depletion of wave energy.

The mechanism of nonlinear Landau damping explains the lowering of the *sound absorption* coefficient found in metals and semiconductors for growing wave amplitudes.

LANDAU DIAMAGNETISM (L.D. Landau, 1930)

Diamagnetism of free electrons in an external *magnetic field*, $\boldsymbol{B}$. The magnetic properties of an electron gas in the field result from the presence of the intrinsic spin magnetic moment of the electrons (see *Spin*), and from changes in the nature of the motion of free electrons in that field. The magnetic field bends electron trajectories in such a way that their projections onto the plane normal to $\boldsymbol{B}$ acquire the shape of closed trajectories (circular orbits in a uniform field $\boldsymbol{B}$). The evolving quasi-periodic motion of electrons in orbits is quantized and adds a diamagnetic component, χ_{LD} to the *magnetic susceptibility* of the electron gas:

$$\chi_{\mathrm{LD}} = -\frac{4m\mu_{\mathrm{B}}^2}{h^2}\left(\frac{\pi}{3}\right)^{2/3} n^{1/3},$$

where n is the electron gas density, m is the electron mass, and μ_{B} is the Bohr *magneton*. The spin magnetic moment of the electrons defines the paramagnetic part of the *susceptibility*, which exceeds the Landau diamagnetism by threefold. Measured susceptibilities of paramagnetic *metals* provide the algebraic sum of the diamagnetic and the paramagnetic contributions of both the electron gas and the ions of the crystal lattice. However, the technique of *electron paramagnetic resonance* allows determining the paramagnetic component alone, and hence the diamagnetic one as well. At low temperatures, the magnetic susceptibility of metals (both dia- and paramagnetic) follows an oscillatory dependence on $\boldsymbol{B}$ (see *De Haas–van Alphen effect*).

LANDAU LEVELS

See *Quantizing magnetic field*.

LANDAU–LIFSHITS DOMAIN STRUCTURE (L.D. Landau, E.M. Lifshits, 1935)

A plane-parallel (stratified) *magnetic domain structure* with *domains of closure* (see Fig.) that the authors suggested for ferromagnets with a single *easy magnetization axis*. Apparently it is never realized in uniaxial ferromagnets, but is found, as

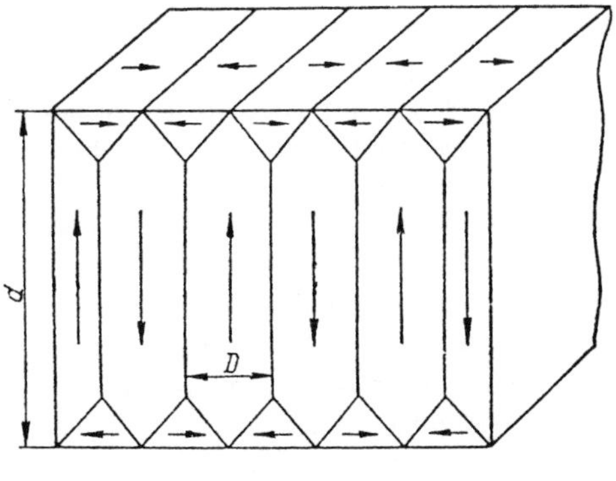

Domain structure of a plate of a cubic ferromagnet with its easy magnetization axis parallel to the (100) axis, and its surface normal to the easy magnetization direction. Arrows show directions of magnetization in the domains.

a rule, in cubic crystals with a positive *magnetic anisotropy* constant. The energy expenditure for the formation of domains of closure is related to the *magnetostriction*, and the equilibrium dimensions of the Landau–Lifshits domain structure are found from the condition of minimizing the magnetic *domain wall* plus magnetoelastic energy (see *Magnetoelastic interaction*). The domain size, D, is related to the size of the sample, d, and, in the case of thin enough *plates* it is given by the approximate relation $D \propto d^{1/2}$. With an increase in the plate thickness the domains fragment and branch (see *Magnetic domain structure*).

LANDAU–LIFSHITS EQUATION (L.D. Landau, E.M. Lifshits, 1935)

An equation derived within the phenomenological approach to describe the dynamics of the *magnetization*, $\boldsymbol{M}(\boldsymbol{r}, t)$, of a ferromagnet. It is formulated in terms of the normalized magnetization $\boldsymbol{m} = \boldsymbol{M}/|\boldsymbol{M}|$, and used to describe the *magnetic domain wall dynamics* at low velocities, the magnetic susceptibility of a ferromagnet with a domain structure, and ferromagnetic resonance. It is the basis of the phenomenological theory of magnetism, and is used to describe both ferromagnets and multi-sublattice magnetics (*antiferromagnets*, *ferrimagnets*). The Landau–Lifshits equations for the magnetization vectors of magnetic sublattices have the form:

$$\frac{\partial \boldsymbol{M}}{\partial t} = \gamma[\boldsymbol{M}\boldsymbol{H}_{\mathrm{eff}}] + \boldsymbol{R},$$

where $\gamma = g\mu_B/\hbar$, g is the gyromagnetic ratio ($g = 2$ for spin magnetism), μ_B is the Bohr magneton, $\boldsymbol{H}_{\text{eff}}$ is the effective field equal to $-\delta W/\delta \boldsymbol{M}$, and $W(\boldsymbol{M})$ is the magnetic energy expressed as a functional of the magnetization. For an n-sublattice magnetic material the equation for the magnetization of the ath sublattice, $\boldsymbol{M}_a$, is obtained by the substitution $\boldsymbol{M} \to \boldsymbol{M}_a$, $\boldsymbol{H}_{\text{eff}} = \boldsymbol{H}_{\text{eff}}^{(a)} = -\delta W(\boldsymbol{M}_1, \ldots, \boldsymbol{M}_n)/\delta \boldsymbol{M}_a$. The term $\boldsymbol{R}$ provides for relaxation of the magnetization field energy. Within the nondissipative approximation ($\boldsymbol{R} = 0$), the Landau–Lifshits equation has the following integrals of motion: energy W and momentum $\boldsymbol{P}$,

$$\boldsymbol{P} = \frac{\hbar}{2\mu_B} \int \frac{\mathrm{d}r(M_y \nabla M_x - M_x \nabla M_y)}{M_0 + M_z},$$

of the magnetization field. The length of the vector $\boldsymbol{M}$ is conserved, and has the Cartesian components

$$M_x = M_0 \sin\theta \cos\varphi,$$

$$M_y = M_0 \sin\theta \sin\varphi,$$

$$M_z = M_0 \cos\theta.$$

The Landau–Lifshits equations have the following forms in terms of the angular variables for $\boldsymbol{M}$:

$$\sin\theta \frac{\partial\theta}{\partial t} = -\frac{\gamma}{M_0}\frac{\partial W}{\partial\varphi},$$

$$\sin\theta \frac{\partial\varphi}{\partial t} = \frac{\gamma}{M_0}\frac{\partial W}{\partial\theta},$$

where $W = W(\theta, \varphi)$ is the functional of magnetic energy expressed in spherical coordinates θ and φ. In that form, the Landau–Lifshits Hamiltonian equations are: $\partial q/\partial t = \delta H/\delta p$, $\partial p/\partial t = -\delta H/\delta q$ for $q = \cos\theta$, $p = \varphi$ and $H = W$. In a one-dimensional case, when $\boldsymbol{M} = \boldsymbol{M}(x, t)$, the Landau–Lifshits equation for a ferromagnet, with its energy quadratic in powers of the magnetization, can be integrated explicitly if dissipation is ignored. It is equivalent to the method of the inverse problem of scattering theory, has an infinite series of *conservation laws*, and admits the existence of many-soliton solutions satisfying the asymptotic principle of superposition (see *Soliton*).

The dissipation term $\boldsymbol{R}$ is often written in the Landau–Lifshits or the Gilbert (T.I. Gilbert, 1955) form:

$$R_{\text{LL}} = \lambda_\gamma M_0\left[H_{\text{eff}} - m(mH_{\text{eff}})\right],$$

$$R_{\text{G}} = \frac{\alpha}{M_0}\left[M, \frac{\partial M}{\partial t}\right],$$

where λ and α are dimensionless relaxation constants. For $\boldsymbol{R} = \boldsymbol{R}_{\text{LL}}$ and $\boldsymbol{R}_{\text{G}}$, the Landau–Lifshits equations become equivalent under the substitution $\lambda \to \alpha$ and $g \to g' = g(1 + \alpha^2)$. The Landau–Lifshits equations containing $\boldsymbol{R}_{\text{LL}}$ or $\boldsymbol{R}_{\text{G}}$ exactly conserve the length of the magnetization vector $|\boldsymbol{M}| = \boldsymbol{M}_0 = \text{const}$, and provide a way to write these equations in angular coordinates. In that case, the right-hand sides of the equations for $\partial\theta/\partial t$ and $\partial\varphi/\partial t$ involve the terms $\alpha \sin^2\theta(\partial\varphi/\partial t)$ and $\partial\theta/\partial t$, respectively. However, the variable $\boldsymbol{R}$ in those forms describes the phenomenon of *relaxation* only qualitatively, coming into conflict with both microscopic calculations based on the method of *Green's functions*, and experimental data. Since this variable $\boldsymbol{R}$ fails to account for the dynamic symmetries of interactions of the *Heisenberg magnet*, V.G. Bar'yakhtar (1984, 1986) proposed a relaxation term in the form $\boldsymbol{R}_{\text{B}} = \boldsymbol{R}_{\text{e}} + \boldsymbol{R}_{\text{r}}$, where $\boldsymbol{R}_{\text{e}}$ and $\boldsymbol{R}_{\text{r}}$ originate from exchange and relativistic interactions, respectively.

The term $\boldsymbol{R}_{\text{e}} = -\lambda_{\text{e}} a^2 \boldsymbol{M}_0 \nabla^2 \boldsymbol{H}_{\text{eff}}$ describes relaxation when the resultant *magnetic moment* of a crystal, $\boldsymbol{\mu} = \int \boldsymbol{M}\, \mathrm{d}r$ is conserved, where λ_{e} is the exchange relaxation constant, and a is the lattice constant. The structure of $\boldsymbol{R}_{\text{r}}$ is essentially defined by the magnetic symmetry.

The Landau–Lifshits equation is widely used to describe both linear and nonlinear dynamic phenomena when a microscopic quantum analysis is either difficult or impossible, such as for *spin waves* in inhomogeneous or bounded magnetics (see *Walker modes*, *Magnetostatic oscillations*), and nonlinear magnetization waves (see *Magnetic domain wall dynamics*, *Magnetic soliton*).

LANDAU SUPERFLUIDITY CRITERION

The most general condition imposed upon the *dispersion law* for single-particle and collective elementary excitations in a superfluid quantum

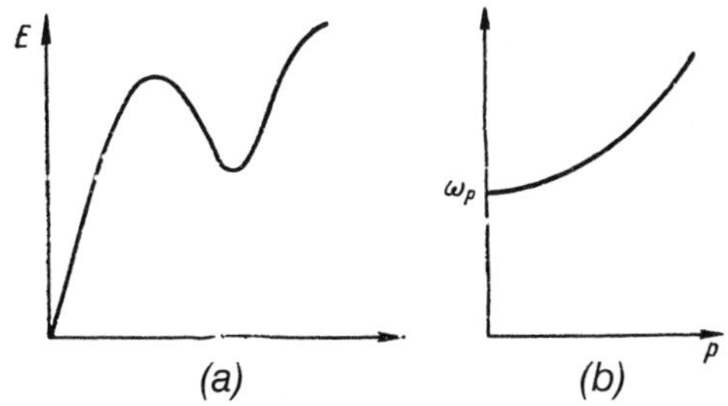

Fig. 1. Spectrum of quasi-particles in (a) neutral and (b) charged Bose-liquids.

particle system on the basis of the *conservation laws* of energy, $E(p)$, and momentum (*quasi-momentum*), p. The Landau superfluidity criterion has the form $v_c = \min\{E(p)/p\} > 0$ and constitutes a sufficient condition for the $E(p)$ spectrum to satisfy. Provided the flow rates of the quantum system as a whole fulfill the condition $v < v_c$, then all its intrinsic excitations resulting from interactions with other external systems (such as capillary walls, the *crystal lattice*) remain forbidden. In other words, no exchange of energy or momentum as well as no transitions of the system into excited states are possible. The Landau superfluidity criterion characterizes the degree of stability of the ground state of a superfluid component in a quantum system (the *Bose condensate*) with respect to external excitations.

In a neutral Bose-liquid (^{4}He), the superfluidity criterion is also met because, on the one hand, the acoustic dispersion law is valid for both single-particle oscillations, and many-particle excitations hybridized with the former at $p \to 0$, while on the other hand, a minimum in the $E(p)$ plot is present at $p \neq 0$ (see Fig. 1(a)). In a system of charged bosons (e.g., *bipolarons*), The Landau criterion is satisfied for a finite gap, equal to the plasma frequency ω_p, in the spectrum of elementary excitations at $p = 0$ (see Fig. 1(b)). Such a gap develops due to the Coulomb interaction. For a Fermi system (e.g., *conduction electrons*), the Landau superfluidity criterion is met when an energy gap of 2Δ appears in the spectrum of single-particle excitations (*electron–hole pairs*) at the *Fermi level* (see Fig. 2). The critical velocity then equals $v_c = \Delta/p_F$, where p_F is the Fermi momentum, defining the maximum *critical current* in *superconductors*.

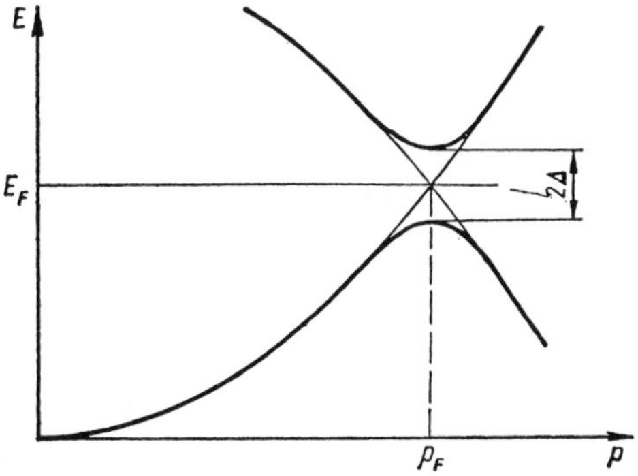

Fig. 2. Spectrum of quasi-particles in superconductors (2Δ is the energy gap).

LANDAU THEORY OF SECOND-ORDER PHASE TRANSITIONS (L.D. Landau, 1937)

A *phenomenological theory of phase transitions* based on the hypothesis that the nonequilibrium *thermodynamic potential* Φ may be expanded in a power series of a variable η called the *order parameter*:

$$\Phi = \Phi_0 + A\eta^2 + B\eta^4. \qquad (1)$$

The value $\eta = 0$ is attained in the phase of the highest symmetry, while for any $\eta \neq 0$ the symmetry of the state of matter is lower than that of the symmetric phase, and corresponds to an unsymmetric phase. The expansion coefficients in Eq. (1) are temperature dependent with $A = a(T - T_c)$, where T_c is the temperature of a *second-order phase transition.* The equilibrium value of η is obtained by minimizing Eq. (1), from which it follows that the symmetric phase ($\eta = 0$) is stable at $T > T_c$, while for the unsymmetrical one with $\eta \neq 0$ we have $\eta^2 = -A/(2B) \propto (T_c - T)$ at $T < T_c$.

In a general case, the order parameter is a multicomponent variable (see *Structural phase transition*). When formulating the Landau theory, the function of the change of crystal density $\delta\rho$ (or some other physical variable) in an unsymmetric phase is assumed to be expandable in the functions $\varphi_\alpha^{(m)}$ that transform according to the αth row of the mth irreducible representation of the symmetric phase spatial symmetry group: $\delta\rho = \sum_{\alpha,m} \eta_\alpha^{(m)} \varphi_\alpha^{(m)}$. All coefficients $\eta_\alpha^{(m)}$ are zero for the symmetric phase, while some of them may be nonzero for an unsymmetric phase. Therefore, the $\eta_\alpha^{(m)}$ coefficients play the roles of order parameters. Since the potential should be invariant with respect to the symmetric phase symmetry group, the expansion in the powers of the order parameter assumes the form

$$\Phi = \Phi_0 + \sum_m A_m \sum_\alpha |\eta_\alpha^{(m)}|^2 + \Phi_3 + \Phi_4 + \cdots. \tag{2}$$

Expansion (1) is a particular case of Eq. (2) for a second-order transition under a one-dimensional irreducible representation. Two limitations exist for a second-order transition under a given irreducible presentation. The continuity of the transition requires that Eq. (2) contain no invariants cubic in $\eta_\alpha^{(m)}$ that would correspond to one and the same irreducible representation. The stability of the emerging unsymmetric phase with respect to the presence of nonuniform *strains* requires that no invariants linear in the first spatial derivatives of the form $\eta_\alpha^{(m)}(\partial\eta_\beta^{(m)}/\partial x_i) - \eta_\beta^{(m)}(\partial\eta_\alpha^{(m)}/\partial x_i)$ be present. The Landau theory provides a qualitatively correct description of a second-order transition. However, in the close vicinity of T_c (the *critical region*), nonuniform fluctuations of the order parameter (*critical fluctuations*) start to play a significant role, and may qualitatively alter the results (see *Ginzburg number*).

LANDÉ g-FACTOR, magnetic splitting factor (A. Landé, 1921)

The dimensionless gyromagnetic ratio factor g_L which is a measure of the ratio of the magnetic moment to the angular momentum of an atom or subatomic particle in the presence of spin–orbit coupling. It is given by the Landé formula

$$g_L = \frac{3}{2} + \frac{S(S+1) - L(L+1)}{2J(J+1)}$$

which provides the Zeeman splitting of atomic energy levels in a magnetic field (see *Zeeman effect*). The values of the Landé g-factor depend on the electron configuration, and are valid for Russell–Saunders coupling. For a purely orbital moment ($S = 0$, $J = L$), the Landé factor equals unity, and for a purely spin moment ($L = 0$, $J = S$), it is 2. A free electron has the value $g = 2.0023$. Crystalline electric fields in solids act on the orbital motion of magnetic ions such as transition ions and cause the g-factor to deviate from the Landé value, sometimes to a considerable extent.

LANGEVIN–DEBYE FORMULA (P. Langevin, P. Debye, 1912)

Formula describing the nonlinear orientation polarization of noninteracting electric dipoles μ_0 under the effect of an applied electric field $\boldsymbol{E}$. For low fields a linear dependence on the field is expected for the average moment $\overline{\mu} = \alpha_D E$, where α_D is the dipole polarizability given by the *Debye formula*, $\alpha_D = \mu_0^2/(3k_BT)$. In relatively strong electric fields the polarization saturates, becoming nonlinear and described by the Langevin–Debye function:

$$\frac{\overline{\mu}}{\mu_0} = L\left(\frac{\mu_0 E}{k_B T}\right),$$

where $L(x)$ is the *Langevin function*. Meanwhile, the *dielectric constant* ε of a polar *insulator* in a strong field becomes weaker. For small x (in weak fields), we have $L(x) \approx x/3$, and the Debye formula follows. The Langevin–Debye formula is a generalization of the *Clausius–Mossotti equation* for polar insulators, and it accounts for the temperature dependence of ε.

LANGEVIN FUNCTION (P. Langevin, 1906)

A function defining the value of the magnetization of *paramagnets* in the classical limit, i.e. of paramagnets with large values of *magnetic moments* in their atoms (molecules)

$$L(x) = \coth x - \frac{1}{x}, \quad x = \frac{j\mu_B g H}{k_B T},$$

where μ_B is the Bohr magneton, g is the *gyromagnetic ratio*, and H is the magnetic field. The Langevin function may be obtained as the classical limit of the *Brillouin function* $B_j(x)$ for $j \to \infty$.

LANGMUIR–BLODGETT FILMS (I. Langmuir, K. Blodgett, 1937)

Films of organic materials of a specific type formed by successive deposition of monolayers on a substrate. Usually, the molecules of such materials are elongated, and their opposing ends have different chemical compositions, and differ in the nature of their interaction with the substrate.

For example, a number of fatty acids and their derivatives have molecules in the shape of long hydrocarbon chains built of $(CH_2)_n$ groups that exhibit hydrophobic properties (see *Lyophilic and lyophobic behavior*), one end of such a molecule having a hydrophilic group of atoms (amphiphilic molecules). When placed on a water surface, such a substance spreads into a monomolecular layer. This layer may be in various states of matter depending on the ambient conditions (see *Monolayer*). Molecules in a solid monolayer sit close to each other in a parallel arrangement so that their hydrophilic ends face the water surface and their hydrophobic ends look outside. A monolayer structured this way may be transferred to a solid substrate. By repeating the operation one may form Langmuir–Blodgett films of prescribed thickness that consist of monolayers alternating in a prescribed order. The structure, and the optical, magnetic, electric and other properties of Langmuir–Blodgett films have been widely studied, and their applications to *microelectronics* are well developed, especially those associated with the task of producing molecular level microelectronics.

LANTHANIDES (fr. lanthanum and Gr. $\varepsilon\iota\delta o\varsigma$, type), general symbol Ln

Family of 14 chemical elements with atomic numbers from 58 to 71, coming after La in the periodic system; belonging, as does La, to the *rare-earth elements*. Lanthanides include the elements: *cerium* Ce, *praseodymium* Pr, *neodymium* Nd, artificially produced radioactive *promethium* Pm, *samarium* Sm, *europium* Eu, *gadolinium* Gd, *terbium* Tb, *dysprosium* Dy, *holmium* Ho, *erbium* Er, *thulium* Tm, *ytterbium* Yb, *lutetium* Lu. These elements occur rarely and widely scattered in nature. Based on their properties, lanthanides are subdivided into the light cerium (from Ce to Gd) and the heavy (Tb to Lu) subgroups. Electronegativity ranges from 1.2 to 1.3.

Physical and chemical properties of all lanthanides are similar to each other as a result of the similarities of their electronic shell structures. The configurations of the outer shells of the majority of lanthanides are $5s^2p^66s^2$ (except Ce, Gd and Lu: each of which has one $5d$ electron). Electrons added to atoms of the lanthanides with increasing atomic number enter the $4f$-sublevel, hence the lanthanides are called f-elements). In chemical compounds, the atoms of lanthanides usually loose three electrons for the oxidation state +3, and some also form oxidation states +2 and +4. With increasing atomic number there is a gradual reduction of the radius of neutral atoms and ions from 0.102 nm (Ce^{3+}) to 0.080 nm (Lu^{3+}), the so-called *lanthanide contraction*.

In free form the lanthanides are silvery *metals*, T_{melting} decreases from 1073 (Ce) to 936 K (Lu). The metals possess comparably low electrical conductivity, and most of them are paramagnetic. Some lanthanides (Gd, Dy, Er) at low temperatures possess ferromagnetic properties. Upon fusion, lanthanides mix with each other yielding *solid solutions*, the so-called *mischmetal*. Many lanthanide compounds are used in the manufacture of special *optical glasses*, *luminophors*, crystallophosphors, laser materials; some lanthanides are components in *high-temperature superconductors*.

LANTHANUM, La

A chemical element of Group III of the periodic system with atomic number 57 and atomic mass 138.91, belongs to *rare-earth elements*. A natural mixture consists of a stable isotope ^{139}La (99.911%) and a radioactive isotope ^{138}La (0.089%), which decays by K-capture with half-life $T_{1/2} = 1.1 \cdot 10^{11}$ years. Isotope ^{139}La is formed by fission of uranium (6.3% of mass of all fission fragments) and is "the reactor poison". Configuration of outer electronic shells is $5d^16s^2$. Successive ionization energies are 5.577, 11.06, 19.176 eV. Atomic radius is 0.187 nm; radius of La^{3+} ion is 0.114 nm. Oxidation state is +3, and electronegativity is ≈ 1.1.

In a free form, lanthanum is a silvery-gray *metal*. It has three allotropic modifications: α-La, β-La, γ-La. The α-La form with a double hexagonal close-packed lattice and parameters $a = 0.3770$ nm, $c = 1.2159$ nm is stable at temperatures below 533 K. In the range 533 to 1150 K β-La forms a face-centered cubic lattice with parameter $a = 0.5304$ nm. At temperatures above 1150 K β-La transforms to body centered γ-La with $a = 0.426$ nm. Density of α-La is 6.162, of β-La 6.190, of γ-La 5.97 g/cm^3; $T_{\text{melting}} = 1193$ K; $T_{\text{boiling}} = 3743$ K. Heat of melting is 6.19 kJ/mole; heat of evaporation is 412.15 kJ/mole; and specific heat $c_p =$ 27.8 J·mole^{-1}·K^{-1}. Coefficient of thermal expansion is 4.9·10^{-6} K^{-1} (at 298 K). Thermal conductivity is 13.8 W·m^{-1}·K^{-1} (at 300 K). Debye temperature of β-La is 146 K. Electric resistivity of α-La is 0.568 μΩ·m (at 298 K), of β-La 0.98 μΩ·m (at 830 K), of γ-La 1.26 μΩ·m (at 1160 K). Temperature coefficient of electric resistance of α-La is 2.18·10^{-3} (at 273 K). Superconducting transition temperature of α-La is 4.90 K, of β-La 5.85 K. Work function is 3.3 eV.

Lanthanum is a paramagnetic substance. Thermal neutron cross-section of ^{139}La is 9 barn. At room temperature, normal elastic modulus is $\approx$6·10^4 MPa; Poisson ratio is 0.288. Yield limit of a cast sample at the temperature 293 K is 125 MPa, ultimate strength is 130 MPa, relative elongation is 8%. Vickers hardness (at 293 K): of a cast sample 490, of an annealed sample 363, of a forged sample 1180 to 1740 MPa. At room temperature, sufficiently pure lanthanum is forgeable (manufacturing sheets is possible) and pressable, but it does not possess sufficient toughness. Ion-plasma frequency is 8.70 THz. Linear low-temperature electronic specific heat (Sommerfeld coefficient) is 10 mJ·mole^{-1}·K^{-2}. Lanthanum is used in optical glasses, as a getter in alloys with Ni, as a component of mischmetal.

LARMOR PRECESSION (I. Larmor, 1895)

Precessional rotation of a stable system of identical (usually charged) particles with magnetic moments located in a uniform constant *magnetic field*, the field direction serving as the rotation axis. The *Larmor theorem* states that when a uniform magnetic field B is impressed upon a system of such particles (e.g., electrons or atomic nuclei), the equations of motion retain their form when transformed to a system of coordinates in uniform rotation about the direction of that field at the *Larmor frequency* $\omega_L = eB/(2m)$ (SI units). Here e and m are the charge and mass of an electron, respectively. The magnetic field thus produces a uniform precession of the orbit of each charged particle around the field direction. The Larmor precession is caused by the magnetic component of the *Lorentz force* upon charged particles, and is similar to the precession of a *gyroscope* axis in a uniform gravitational field. The Larmor theorem holds if ω_L remains small compared with the normal frequencies of orbiting particles in the absence of a magnetic field. We have $\omega_L \propto 3\cdot 10^{10}$ s^{-1} for electrons in a very strong magnetic field (1 T), while the electron orbiting frequency in an atom is of the order of 10^{15} s^{-1}. Thus, the Larmor theorem has an extremely wide application range. As the result of this precession of charged particles, a macroscopic system acquires a *magnetic moment*. Therefore, the Larmor theorem helps explain the phenomena of *diamagnetism*, the normal *Zeeman effect*, and the *Faraday effect* (magnetic *rotation of light polarization plane*).

LASER (acronym for Light Amplification by Stimulated Emission of Radiation)

A source of coherent radiation generated during stimulated emission or *light scattering* by an active medium located in a resonant cavity. Every laser contains three principal elements: a *laser active medium*, a system for pumping the active medium (pumping system), and an optical resonant cavity (see *Resonators*). The active medium can be either a gas, a liquid or a solid. To pump this medium light from nonlaser sources can be used, as well as an electric discharge, thermal heating, etc. The simplest *optical resonant cavity* consists of two flat mirrors placed parallel to each other. To extract the radiation, one of the mirrors is made semitransparent or has a special opening. Typical lasers generate radiation within a range from the UV to submillimeter frequencies, and operate either by continuously emitting power, or in a pulse mode. Some lasers operate at a fixed frequency, and others are tunable in their radiation frequency (*dye* lasers, etc.). Some continuous power lasers can generate in excess of 1 MW, and some pulse lasers can produce output powers exceeding 10 TW, providing energy flux densities at targets that exceed 10^{15} W/cm^2, with their pulse energy reaching 10 kJ.

Lasers are used in research (spectroscopy, *holography*, thermonuclear fusion, etc.), information technology (communications, data storage, etc.), and technological applications (laser chemistry, isotope separation, semiconductor production, cutting, *welding*, smelting, etc.). Three directions of laser applications in the physics of solids are: (a) studying interactions of laser radiation with solids; (b) using the effects of such interactions to study the composition and properties of solids (spectral analysis, *mass spectrometry*); and (c) using laser radiation to process various materials and alter or improve their properties (surface cleansing, *surface hardening*, *surface doping* and cladding, deposition of *thin films*, etc.). Laser types include *solid-state lasers*, gas lasers, excimer, plasma, and free-electron ones.

See also *Electron beam pumped laser*, *Heterojunction laser*, *Holographic laser*, *Injection laser*, *Lasers with distributed feedback*, *Semiconductor lasers*, *Solid-state laser*, *Streamer laser*, *Tunable lasers*, *X-ray laser*.

LASER ACTIVE MEDIUM

A medium capable of amplifying electromagnetic radiation by the use of a laser energy transition. Active laser media are known in every *state of matter*, including plasma. Among the most important such media are various solids including crystals and glasses (see *Active solid-state laser materials*) activated with ions of the transition and rare-earth elements (ruby (Cr^{+3}), yttrium–aluminum garnet, Nd^{+3}, neodymium glass, etc.); *alkali-halide crystals* with *color centers*, e.g., LiF: F_2^+; artificially layered structures (heterostructures) for charge recombination, inhomogeneously doped semiconductor crystals, and *heterogeneous structures*. They are used in *lasers* operating in every mode of generation from continuous emission to supershort pulses.

LASER ELEMENT, ACTIVE

See *Active laser element*.

LASER EPITAXY

A technique used to produce monocrystalline *semiconductor* layers on oriented *substrates* using *laser* pulse or scanning heating. A laser beam incident on a target evaporates atoms which fall onto a substrate to form a *thin film*. Laser epitaxy was first observed during *pulsed annealing of semiconductor* surface layers (Si, Ge, GaAs) transformed into the amorphous state by *ion implantation*. Without affecting the lower layers of a crystalline material, *pulse heating* can bring the temperature up to $T_{\text{melt.a}}$ at which the amorphous layer starts to melt. At $T < T_{\text{melt.a}}$, the process of laser epitaxy takes place in the solid phase, and when $T = T_{\text{melt.a}}$, it proceeds through the liquid phase (see *Liquid-phase epitaxy*). *Solid-phase laser epitaxy* is triggered when amorphous layers are pulse-heated within an interval of several microseconds, and the liquid phase type within nanoseconds. Heating the semiconductor by 6–8 ms long ruby laser pulses at power densities of 40–60 J/cm^2 brings the layer up to the temperature 900–1100 K, and restores the monocrystalline structure. Heating with 50 ns long pulses at power densities over 3 J/cm^2 melts the amorphous layer all the way down to the substrate.

The mechanism of laser epitaxy is described satisfactorily by the *thermal model of pulse heating* which assumes that the *mean free path* of the charge carriers produced in the semiconductor by the light pulse is shorter than the thermal diffusion length, and the carrier lifetime is shorter than the pulse duration. These conditions provide for the transformation into heat of energy of the electron–hole plasma produced by the radiation. Plasma phenomena appear during picosecond duration pulses. High *crystallization* rates attained during laser epitaxy with melting impart semiconductor layers with concentrations of impurities 10–100 times higher than those acquired by crystallization under equilibrium conditions.

LASER MATERIAL, ACTIVE

See *Active solid-state laser materials*.

LASER RESONATOR, DISPERSION

See *Dispersion laser resonator*.

LASERS WITH DISTRIBUTED FEEDBACK

A type of *laser* with enhanced spectral selectivity of the resonator cavity attained either by periodic modulation of the laser *active medium*, or from the *optical waveguide* along the radiation direction of propagation. Such a modulation by diffraction provides feedback resulting in a distributed radiation output from the cavity. The periodicity of the medium may be caused by changing its volume or by modulating the transverse surfaces of the waveguides, which are then corrugated by etching. First used in dye lasers (see *Solid-state quantum electronics*), the principle of distributed feedback has found its widest application in *injection lasers*, in those with optical pumping, and in those excited by an electron beam. Periodic modulation of the waveguide thickness is effected either inside the active zone with gain g (*distributed feedback lasers*) or outside it (*distributed Bragg reflection lasers*). Distributed reflection may be combined with a cavity having concentrated end-face mirrors. Distributed output makes it possible to obtain a collimated beam with low divergence determined by the length of the periodic output structure. The direction of the scattered wave depends on the order m of the dispersing grating: $\sin\theta_i = 2i/m - 1$, where θ_i is the angle of the scattered wave front relative to the direction of its propagation; and i is an integer. The separation of longitudinal normal frequencies (modes) of the resonant cavity of a distributed feedback laser depends on the length of the cavity L, i.e. the distance between the boundaries of the homogeneous and the periodic media.

LASER TECHNOLOGY

Processing and *welding* technology applications of *laser* radiation in various materials.

The high directionality of laser radiation allows focusing a ray on a spot with a diameter comparable to the wavelength of the radiation. The energy density attainable is sufficient to heat, melt and evaporate any natural or artificial material, so welding, boring, cutting and thermal processing are practable. *Solid-state lasers* and gas lasers operating in both the continuous wave (CW) and pulse modes are used.

The advantages of laser technology are manifold. The effects inflicted on materials are local in nature; the heating remains very localized, and there is no physical contact during processing. It is also possible to process materials in any transparent medium, or to work through transparent shields. These applications are widespread in the electronic and radio industries, and lasers with output powers of several dozen kilowatts are routinely used in the machine building industry (automobile manufacture and tool making).

At power densities of 10^5–10^6 W/cm^2 metals melt, which is the basis for *laser welding*, and provides high-quality joints between stainless steel, nickel, molybdenum, Kovar, etc. It is possible to weld metals with high thermal conductivities (aluminum, silver) as well as materials of sharply differing properties, such as nonmetals which resist welding by other techniques. At pulse energies between 0.1 and 30 J, typical for solid-state lasers, the depth of the material melted is 0.05–1 mm. The efficiency of point welding is several dozen operations per minute, and a continuous seam output is 1.5 m/min (for a depth of 0.5 mm). *Pulse laser welding* applications include the manufacture of vacuum electronic and semiconductor devices, sealing cases for integrated circuits and quartz resonators, producing high precision mechanical devices, etc. Using 400 W/cm^2 garnet laser output power, continuous seam welding of 3 m/min is achievable at melt depths of 0.3–0.4 mm. Carbon dioxide (CO_2) laser output beams with power levels of 10–20 kW can weld steel up to 50 mm thick.

At power densities of 10^7 W/cm^2 substances begin to evaporate, making possible *high-precision laser processing*. Laser drilling and cutting have undergone rapid development. One can produce holes from 0.005 mm in diameter in any material, including those extremely resistive to processing by other techniques (*ceramics*, *diamond*, corundum). Using *laser drilling* to produce diamond dies, ruby watch stones, and components for electronic devices increased productivity by more than tenfold. *Laser cutting* is widely used with ceramic components, in particular for inscribing substrates for semiconductor and hybrid integrated circuits. *Gas-laser cutters* are actively employed in the machine building industry; the substance melted by laser radiation being removed by a gas jet (usually air or oxygen). Garnet lasers with 100–500 W output power, and CO_2 lasers with 300–2000 W output, function as power tools in laser cutters. The material cut in this manner may be up to 10 mm thick, with cutting rates 0.5–10 m/min, depending on the thickness.

The manufacture of integrated circuitry and other electronic devices involves the formation of thin film patterns, and the control of various film parameters through selective ablation of film material.

Lasers are used for the *laser heat treatment of surfaces*, predominantly for thermal strengthening of steels and cast irons by *quenching*, as well as for *annealing* and thermal-chemical processing. Radiation output power densities for *heat treatments* reach 10^3–10^4 W/cm^2. During *laser hardening* it is possible to start cooling at extremely high temperatures, including the *melting temperature* at the surface. Note that a high temperature gradient in a thin surface layer provides extremely fast cooling rates (up to 10^6 deg/s) that result in a fine crystalline structure with a heavy concentration of *martensite* and a high *dislocation* density. Cutting tools thus processed increase their strength twofold, and stamping equipment increases in strength by a factor of three or more. A CO_2 laser with 1–10 kW output provides high rate continuous hardening of large-area surfaces to a depth of 0.7–1.0 mm. It also finds use in machine construction to attain higher wear resistance of rubbing surfaces of machine parts.

Applications of *laser heating* in technology include *doping*, producing amorphous metal layers, local activation of *etching* processes, thermochemical modification of surface properties, evaporating substances in vacuo to produce *thin films*, communication and manufacturing plates in printing, eye surgery, cosmetic surgery, other medical applications, etc.

LATENT PHOTOGRAPHIC IMAGE

The immediate result of light exposure on *photographic material*; modification of the photographic material with no visible change. For visualization the latent photographic image must be converted to a visible one with the help of some process for which the latent image plays the role of the initiator, catalytic agent, etc. This process cannot occur in the absence of this hidden image. In the AgX (X = F, C1, Br, I) emulsion layers of the silver halide process the latent image involves small atomic *clusters* generated by the light exposure (most sensitive emulsions have only 3 or 4 Ag atoms per microcrystal AgX). During the development the complete reduction of Ag in the microcrystals takes place. The physical mechanism of latent image formation in AgX microcrystals offered by R. Gurney and N. Mott (1938) is still accepted with some modifications and additions (*Gurney–Mott model*). According to this model, the high AgX light sensitivity (ability to form latent photographic image) is based on the unique combination of two physical features: high monopolar dark conductivity due to intersite cations Ag^+ (see *Frenkel defect*) and significantly smaller mobility of photo-generated holes relative to photoelectrons from the Ag^+ ions. In such a compound a sequence of elementary reactions is initiated, such as $X^- + h\nu \to X^0 + e^-$; $e^- + T \to Te^-$ (T is a deep trap); $T^- + Ag_i^+ \to TAg^0$; $X^- + h\nu \to X^0 + e^-$; $e^- + TAg^0 \to TAg^-$, $TAg^- + Ag_i^+ \to TAg_2^0$, etc. This sequence leads via quantum absorption to the formation of Ag_n^0 clusters of sizes sufficient for catalyzing the development through the capture of electrons from the developer without further participation of light. See also *Nonsilver photography*.

LATERAL INTERACTION of adsorbed particles

Interaction of adsorbed atoms and molecules on a solid surface, either immediate (direct for high adsorbate density) or indirect (proxy, via adsorbent or substrate) (see *Adsorption*). Several types of lateral interaction are identified: van der Waals (dispersion), deformation (distortions in crystal structure of solid surface via *chemisorption* forces), electrostatic (charged particles on *insulator* surfaces), direct exchange (overlap of particle wave functions) and indirect exchange through the electron gas of a metal or a semiconductor substrate (type of oscillating *Ruderman–Kittel–Kasuya–Yosida interaction*, RKKY interaction), *dipole–dipole interaction* (*image forces* on *metal* surface), and screened Coulomb interaction (see *Friedel oscillations*).

LATTICE

A totality of points (sites) arranged regularly in space with integer coordinates relative to a fixed coordinate system with its origin (*main reference point*) at one of the lattice sites. Each of the three vectors forming a primitive unit cell is called a *principal vector of the lattice*. A vector connecting any two sites of the lattice is called a *lattice vector*. A parallel displacement through any lattice vector converts the lattice into itself, i.e. it is a symmetry transformation of this lattice. Each reference point unambiguously defines the lattice, i.e. it may be regarded as the main reference point of the specific lattice. The main reference point in a given lattice may be selected in an infinite number of ways (see *Primitive parallelepiped*). The importance of the concept of a lattice is associated with the fact that the centers of particular atoms in any *crystal structure* form one or several lattices, metrically similar, and located in parallel, i.e. they form a *crystal lattice*. For example, the compound CaF_2 has a crystal structure, with the Ca and F atoms each on their own individual lattice sites, i.e. on sites of their own lattices.

LATTICE CONSTANT

See *Crystal lattice constants*.

LATTICE DESTABILIZATION

Approach to zero of the *shear* resistance for a special plane and direction (correspondingly the combination of elastic moduli $(C_{11} - C_{12})$ tends to zero) during the variation of thermodynamic parameters (temperature, pressure, magnetic or electrical fields, etc.). The lattice becomes mechanically unstable, and its rearrangement to another structure takes place. Almost complete lattice destabilization towards shear $\{110\}\langle 1\bar{1}0\rangle$ was discovered in the compounds V_3Si and Nb_3Sn in the neighborhood of the temperatures of *structural phase transitions*, and in the alloys InTl (28 to 39 at.% Tl) at the transition from a face-centered cubic to a tetragonal lattice. In other alloys (with the body-centered cubic lattice) during a temperature change only a partial reduction of the shear modulus and an increase of the degree of *anisotropy of crystals* take place.

LATTICE GAS

A mathematical model that deals with the behavior of a system of N particles distributed among N_0 (where $N < N_0$) sites of an undistorted lattice. Only one particle may be located at an individual site, every particle is at some site, but not all sites are occupied. The state of the system is determined by the distribution of the particles among the sites. It is impossible to determine the kinetic energy in such a system; yet, for a noninteracting lattice gas there is an ideal gas *equation of state* $PV = RNT$. If the energy of the particle interaction is known, one can simulate some structural properties of *alloys*, and of substitutional and interstitial *solid solutions*.

LATTICE STATICS METHOD

A method for calculating *static displacements of atoms* from their regular or ideal lattice sites in a crystal containing defects. The lattice statics method involves the introduction of fictitious forces (*Kanzaki forces*). When applied to the atoms of a defect-free crystal, these forces bring about atomic displacements identical to those caused by real defects. The Fourier components of the distribution of the fields of static displacements characterize the processes of *diffuse scattering of X-rays* and neutrons by crystals that contain *point defects*, *dislocations*, etc.

LATTICE, SURFACE

See *Surface lattices*.

LATTICE VIBRATIONS, SILENT

See *Silent lattice vibrations*.

LAUE CLASSES

Eleven classes of point *crystal symmetry* with a *center of symmetry*. Classified by crystal system they are: triclinic ($\bar{1}$), monoclinic ($2/m$), orthorhombic (mmm), tetragonal ($4/m$, $4/mmm$), trigonal ($\bar{3}$, $\bar{3}/m$), hexagonal ($6/m$, $6/mmm$), and cubic ($m\bar{3}$, $m\bar{3}m$). Laue classes are classes of diffraction symmetry (deduced from diffraction patterns) higher than its point symmetry, as follows from Friedel's law. Each of the 32 *point groups* falls into one of the Laue classes that results when a center of symmetry and certain derivative elements of symmetry are added. The Laue classes and diffraction symmetry are directly manifested in Laue patterns (see *Laue method*) and in electron diffraction patterns (see *Electron diffraction analysis*) of single crystals.

LAUE CONDITIONS, Laue equations

The relations between the directions along the interference maxima for the *X-ray diffraction* pattern of a monocrystal:

$$a(\cos\alpha - \cos\alpha_0) = H\lambda,$$

$$b(\cos\beta - \cos\beta_0) = K\lambda,$$

$$c(\cos\gamma - \cos\gamma_0) = L\lambda,$$

where a, b, c are the periods of the *crystal lattice* of a monocrystal along the X, Y, Z axes; α_0, β_0,

γ_0 are the angles of the X, Y, Z axes with the direction of the incident (primary) X-ray beam of wavelength λ; α, β, γ are the angles of X, Y, Z to the direction of the diffracted beam; H, K, L are integers.

LAUE DIFFRACTION, Laue geometry (M. Laue, 1912)

A scheme of X-ray diffraction of *crystals* (also electrons, etc.) with the diffracted beam exiting on the side opposite to its entry face (see *X-ray diffraction*). Two cases of Laue diffraction are identified; symmetric and asymmetric ones, when the reflecting planes (hkl) are and are not perpendicular to the entry face of the crystal, respectively. Several important effects have been observed during Laue diffraction. One is the periodic pumping of energy of the passing wave into the diffracted wave and back across the crystal depth with a period equal to the *extinction length* (this occurs for thicknesses, t, less than the absorption path length, t_{abs}, the so-called *Ewald pendulum solution*). Another effect is the *anomalous passage of X-rays* (the *Borrman effect*) for thicknesses much greater than t_{abs}. Laue diffraction is used in solid state physics to determine the diffraction symmetry of a material, and is often employed to orient single crystals. Using fast electrons in the Laue diffraction regime makes it possible to monitor the structural state of extremely thin layers or *films* ($t = 1$ μm), while X-rays probe less thin layers ($t = 0.1$–1 mm). Thick samples ($t \geqslant 1$–10 mm) may be studied using γ-quanta (see *Gamma diffractometry*) and neutrons.

LAUE METHOD (suggested by M. Laue and implemented by W. Friedrich and P. Knipping, 1912)

A technique of retrieving a diffraction pattern formed by spectrally continuous radiation diffracted by a *monocrystal*. Laue conditions are met in an experiment with polychromatic X-rays that vary continuously in wavelength from λ_{max} to λ_{min} for any given set of single *crystal lattice constants* a, b, c with orientations α, β, γ, α_0, β_0, γ_0 (see *Laue conditions*). The X-ray pattern is recorded on a planar X-ray sensitive film positioned normally to the incident beam behind a single crystal (*Laue pattern*) or in front of it. The Laue method is used to determine the *crystal symmetry* (i.e. its *Laue class*), the level of single crystal perfection, its orientation, etc. The method is used for the routine orientation of single crystals.

LAVES PHASES, lambda phases (λ-phases) (after Swiss scientist F. Laves)

A large group of *intermetallic compounds* with the formula AB_2 that form between elements (*metals*) with the ratio of atomic diameters d_{AA} and d_{BB} within the range 1.1–1.6. Arguing from geometric requirements, an ideal ratio is $d_{AA}/d_{BB} = 1.225$. Lambda phases are *crystals* with close-packed lattices. They fall into one of three structure types: λ_1 has structure type $MgZn_2$ with a hexagonal lattice, λ_2 has structure type $MgCu_2$ with a face-centered cubic lattice, and λ_3 has structure type $MgNi_2$ with a hexagonal lattice. Considering the positions of the large atoms, all three structures involve an alteration of doubled layers with vertical pairs of A atoms forming a hexagonal lattice. The difference between these structures consists in the angle of the shift between the alternating doubled layers relative each other. The *phases* exhibit metallic conduction and high *hardness*, but are brittle (see *Brittle failure*). Although the main factor responsible for the lambda phase formation is geometric, the stability of one or another modification is attributed in part to the interaction between the *Fermi surface* and *Brillouin zone* boundary. See also *Brass*.

LAW OF CONSTANCY OF ANGLES, Steno–Romé de L'Isle law

An empirical law of *crystallography* stating that, in one *crystal* or in different crystals of the same polymorphic modification of a material, the area and shape of facets, their mutual arrangement and number can change, but the angles between corresponding facets (i.e. those which are parallel in compared crystals) remain constant. The law holds with a great accuracy, up to fractions of minutes. The law was discovered in quartz (SiO_2) and haematite (Fe_2O_3) crystals. In 1772, J. Romé de L'Isle established the law's validity for crystals of all materials after carrying out a very large number of measurements.

LAW OF RATIONAL INDICES, Haüy law, law of integers (R. Haüy, 1783)

An empirical law of geometrical *crystallography* stating that, if any triad of non-coplanar *crystal* edges is assumed as coordinate axes a, b, c, the ratios of the intercepts on the axes by any two crystallographic (cleavage) planes are equal to the ratio of small (usually not exceeding five) prime integers: $(a_1/a_2):(b_1/b_2):(c_1/c_2) = h : k : l$. There are rules of "alignment" adopted for each symmetry type (*crystal system*) for the selection of coordinate axes and individual planes with parameters a_1, b_1, c_1. The validity of the law of rational indices follows from the crystal lattice structure and from the parallelism of edges and planes to rows and networks of the lattice. The smallness of the h, k, l integers is explainable by physical factors: planes with high reticular (atom) density, form a body with minimal surface *free energy* (see *Surface energy*). The law holds exactly for microscopic portions of a surface; macroscopic faces of a real crystal may fail to obey it due to an atomic step-shaped relief or to structural *defects* (see *Atomic steps on a surface*).

LAW OF ZONES, zone law, Weiss' zone law

An empirical law of *crystallography* stating that each plane, parallel to two actual or possible edges or lines of a *crystal*, is an actual or possible crystal face, while each direction parallel to the zone axis is an actual or possible crystal edge. The zone law is expressible via *crystallographic indices* of planes and directions as $hu + kv + lw = 0$, where h, k, l are Miller indices of a plane lying in a crystal zone, and u, v, w are zone indices of directions (edges) of the zone.

LAWRENCIUM, Lr

See *Transuranium elements*.

LAYER-BY-LAYER ANALYSIS

Method of investigating the structural state, phase, elemental composition, and also concentration distribution of atoms of different elements according to the depth of material. It is used in the physics of *metals*, *metal science*, and another areas of solid state physics. Layer-by-layer analysis is performed by the successive removal of layers of the required thickness from the investigated object, by the investigation of an oblique cut made at an angle from several minutes to several degrees relative to the surface (destructive methods of layer-by-layer analysis), or by "looking through" with the help of physical measuring tools at the material state at fixed depths without any destruction of the sample. Layer-by-layer analysis is used in combination with the methods of radioactive isotopes, *Mössbauer spectroscopy*, *microprobe X-ray spectrum analysis*, X-ray analysis, etc. The removal of layers is performed mechanically, chemically, electrolytically, by ion and electron etching, by *vacuum evaporation*, etc. The accuracy of determining the thickness of a removed layer in case of destructive layer-by-layer analysis does not exceed 0.1 μm, while *nondestructive testing techniques* allow one to investigate layers from 1 nm thickness and higher.

LAYERED STRUCTURE, ORDER–DISORDER

See *Order–disorder layered structure*.

LEAD (Lat. *plumbum*), Pb

A chemical element of Group IV of periodic system with atomic number 82 and atomic mass 207.19. Natural lead consists of four stable isotopes ^{204}Pb, ^{206}Pb, ^{207}Pb, ^{208}Pb. Electronic configuration of outer shell is $6s^2 6p^2$. Successive ionization energies are 7.37, 14.91, 31.97, 42.32, 68.8 eV. Atomic radius is 0.173 nm; radius of Pb^{4+} ion 0.084 nm. Oxidation states are +2, +4. Electronegativity is 1.55.

In a free form, lead is a bluish-gray *metal*. It has face-centered cubic lattice with parameter $a =$ 0.49495 nm, space group $Fm\bar{3}m$ (O_h^5). Density is 11.344 g/cm^3; $T_{melting} = 600.4$ K, $T_{boiling} =$ 2013 K. Heat of melting is 4.81 kJ/mole, heat of evaporation is 178 kJ/mole; specific heat is 129 kJ·kg^{-1}·K^{-1} (at 273 to 373 K). Thermal conductivity factor is 34.8 W·m^{-1}·K^{-1}; coefficient of linear thermal expansion is $27.56 \cdot 10^{-6}$ K^{-1} (at 293 K). Coefficient of self-diffusion is $7.32 \cdot 10^{-16}$ m^2/s (at 500 K). Debye temperature is 95 K. Adiabatic elastic moduli: $c_{11} \approx 48.1$, $c_{12} \approx 40.7$, $c_{44} \approx 14.7$ (GPa) at 298 K. Young's modulus is 28.2 GPa, Brinell hardness is ≈40 MPa. Tensile strength is ≈12.5 MPa, relative elongation is 55%. Coefficient of linear term in low temperature molar electronic specific heat is 2.98 mJ·mole^{-1}·K^{-1}.

Electrical resistivity is $20.68 \cdot 10^{-8}$ Ω·m. Superconducting transition temperature is 7.17 K. Nuclear magnetic moment of ^{207}Pb is 0.584 nuclear magnetons. Lead is used in the chemical industry, in electronics, in semiconductor engineering, for protection against radiation. Cases of wires and cables are manufactured from lead. See also *Lead alloys*.

LEAD ALLOYS

Alloys based on *lead*. They are characterized by high *density*, by low *strength*, *hardness*, *melting temperature*, by good antifriction properties, by relatively high acid strength towards some dilute acids (sulfuric, sulfurous, chromic, etc.). Dilute and concentrated lead alloys (see *Alloying*) are known. The first group comprises lead alloys with small additions of Fe, Cu, Sb, Cd or Ca, which do not reduce, and in some cases increase the *corrosion resistance* of lead, and therefore increase its *creep* limit and *long-term strength*. The second group comprises lead alloys which contain considerable quantities of elements which increase their strength, hardness and antifriction properties, and lower the melting temperature and shrinkage upon *casting*. Lead alloys are obtained by the melting of primary or secondary metals, and also of their *alloying elements*. Those with *sodium*, *potassium* and *calcium* are formed by an electrochemical method, i.e. by electrolysis of the melt salts together with a liquid lead cathode. There are antifriction alloys (see *Antifriction materials*), typographic alloys and solders. The majority of alloys are soft, easily subjected to mechanical processing: to rolling, stamping, *forging*, etc. Lead alloys are used as bearing materials, typographic and other low melting temperature alloys.

LEAD PLATING

Coating of various pieces and units with layers of *lead*. It is used to manufacture lead storage (automobile) batteries, to screen X-rays, and to protect ferrous and nonferrous metals from *corrosion* that would otherwise result from the effect of solutions of sulfuric acid or chromic acid and their salts in chemical installations. To produce lead plating the pieces are immersed into molten lead (hot technique), or liquid lead is sputtered onto their surfaces (metallization). To improve the wetting of the leaded surface the melted lead should contain antimony, tin, copper, or zinc. Electrochemical lead plating provides high purity coatings with better physical and mechanical properties. A reliable protection of ferrous metals in corrosion-active media is provided by pore-free lead plating at least 200–300 μm thick. Electrolytes that contain lead salts, and are usable for *electrochemical lead plating*, include fluorine borite, phenolsulfone, and other electrolytes containing lead salts.

LEBEDEV CRYSTALLITE HYPOTHESIS,

Lebedev glass structure hypothesis
(A.A. Lebedev, 1921)

A hypothesis on the structure of glasses. Initially, having measured the temperature dependencies that the coefficients of *refraction of light* and *thermal expansion* of dense silicate glasses feature within narrow regions of temperatures near structural transformations in crystals of quartz, tridymite and cristobalite, Lebedev supposed that glasses had a finely dispersed crystal structure. His later analysis (1933, 1948) included numerous experimental data on temperature dependences of their *internal friction* and data from *X-ray structure analysis*. Summing it up, he concluded that glasses contain regions of perfect short-range ordering, 0.7–1.5 nm in size, surrounded by molecular associations with less perfect ordering. That is, the glass is structurally a *random continuous network*

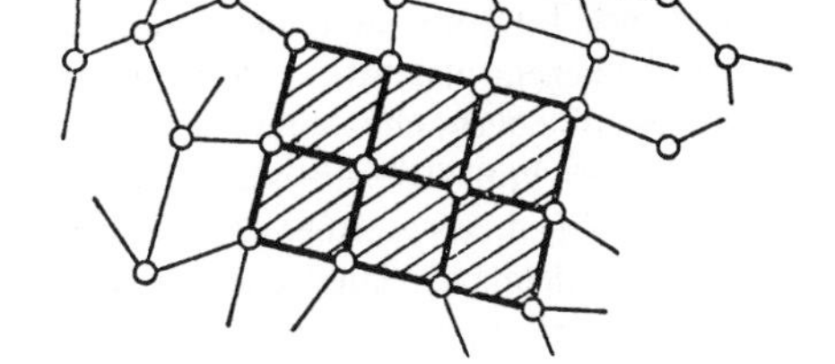

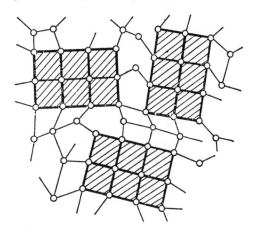

Crystallite model of disordered solid (ordered phase areas are hatched).

featuring nonuniform shifting *intermediate order* (see Fig.), in agreement with diverse experimental data on the structure of glass. See also *Vitreous state of matter, Polycluster amorphous solids.*

LED

Acronym for *Light emitting diode.*

LEDEBURITE (after German metallurgist A. Ledebur)

Structural compound of *iron*–carbon *alloys* (mostly, *cast irons*) that is a eutectic mixture (see *Eutectics*) of *austenite* and *cementite.* It forms at a temperature below 1145 °C (*carbon* content is 4.3%). Ledeburite at room temperature consists of cementite and *pearlite.* In steels subjected to *alloying* with chromium, tungsten, and other carbide-forming elements, ledeburite appears at low carbon content (0.7 to 1%) and is a eutectic mixture of alloyed austenite and *carbides* of alloying elements.

LENNARD-JONES POTENTIAL
(J.E. Lennard-Jones, 1924)

A *pairwise interaction* potential between atoms often used to describe the structure of the *condensed state of matter* and its dynamics:

$$U(r) = 4\varepsilon\left\{\left(\frac{\sigma}{r}\right)^{1/2} - \left(\frac{\sigma}{r}\right)^{6}\right\},$$

where r is the interatomic distance, the parameter σ characterizes the range of the interaction, and ε is its magnitude (see also *Van der Waals forces*). The attractive r^{-6} term arises from a fluctuating induced dipole–dipole interaction, and the r^{-12} term represents the ion core repulsion. The potential works well for inert gas (Ne, Ar, Kr, Xe) crystals.

LEVEL POPULATION

The number of particles (atoms, molecules, nuclei, paramagnetic impurities, etc.) occupying energy levels of a quantum system characterizing its various states. Provided there is no degeneracy, the number of particles satisfies the Boltzmann distribution for thermodynamic equilibrium at the ith level, given by

$$n_i = N\exp\left(-\frac{E_i}{k_B T}\right)\left[\sum_j \exp\left(-\frac{E_j}{k_B T}\right)\right]^{-1} \qquad (1)$$

where N is the total number of particles, and E_i is the energy of the ith level. Thus, according to Eq. (1), the level population decreases exponentially as its energy increases. An opposite situation is found in the case of *population inversion* where higher levels are more populated than lower ones. This inversion is a necessary precondition for the functioning of *lasers* and *masers.* The most typical way to produce inverted populations consists in equalizing the populations of a pair of levels i–k via external actions such as high power electromagnetic wave pumping (see Fig. in *Quantum amplifier*). It may then happen that for a different pair of levels, i–j, such that $E_i > E_j > E_k$, one obtains $n_i > n_j$. Inverted populations during *magnetic resonance* can be produced by causing the system to undergo an *adiabatic fast passage* through a resonance transition. The extent of inversion, viz., the number of particles at inverted higher levels, depends on various external factors: energy gaps, probabilities of relaxation transitions, etc.

When the particles possess a *magnetic moment* (i.e. *spin*), their distribution over magnetic sublevels is characterized by a *spin temperature*, T_s, which is determined via the Boltzmann distribution:

$$\frac{n_i}{n_j} = \exp\left(-\frac{E_i - E_j}{k_B T_s}\right). \qquad (2)$$

Provided such particles (the spin system) are in equilibrium with the lattice, one finds from Eqs. (1) and (2) that $T_s = T$. Applying an external action, one may either cool the spin system ($T_s < T$) or heat it ($T_s > T$). A spin system is cooled when its energy gap $E_i - E_j$ is adiabatically narrowed (see *Adiabatic demagnetization cooling*) through application of alternating magnetic (electric) fields to a system of magnetic (electric) dipoles (see *Magnetoelectric effect*), etc. A characteristic example of a heated spin system involves populations n_i and n_j equalized through *saturation effects.* Indeed, if $n_i = n_j$ (while $E_i \neq E_j$), then, according to Eq. (2) we have $T_s = \infty$. In the case of an inverted population, the distribution of particles (not necessarily having spin) over the energy levels is characterized by a *negative temperature.* If $n_i > n_j$ (while $E_i > E_j$), then we have $T_s < 0$ in accordance with Eq. (2). The concepts of spin temperature and negative temperature are used to describe effects resulting from deviations of level populations from equilibrium values.

LEVELS, SHOCKLEY

See *Shockley levels.*

LEVITATING ELECTRONS

Electrons localized at *surface levels* formed near a vacuum-dielectric interface with negative *electron affinity.* They are produced under the effect of electrostatic *image forces* with an attractive potential $V(z) = -\Lambda/z$, where

$$\Lambda = \frac{e^2(\varepsilon - 1)}{4(\varepsilon + 1)}, \qquad (1)$$

ε is the relative *dielectric constant*, and z is the distance measured from the *insulator* surface into vacuum. The surface state spectrum has the form:

$$\mathcal{E}_l(p_\parallel) = -\frac{\Delta}{l^2} + \frac{p_\parallel^2}{2m}, \qquad (2)$$

$$\Delta = \frac{m\Lambda^2}{2\hbar^2}, \quad l = 1, 2, 3, \ldots, \qquad (3)$$

where $p_\parallel = (p_x^2 + p_y^2)^{1/2}$ is the longitudinal two-dimensional momentum in the surface plane. Levitating electrons with their spectrum (2) may exist on the condition that the width of their localization region, $\lambda_l = \hbar^2 l^2/(\Lambda m)$, considerably exceeds that of the interface, the latter being of the order of an interatomic distance ($a \propto 10^{-8}$ cm). Such a condition is met, e.g., when electrons are localized above the surface of liquid helium, for which $\Lambda \sim 0.01$. In that case, the electrons would "soar" above the liquid surface in the gas phase (vapor) at the quasi-microscopic height $\lambda_l \geqslant 10^{-6}$ cm, justifying the term "levitating". The lifetime of levitating electrons increases when an additional electric field is switched on to press them to the liquid *helium* surface. Fig. 1 shows the resonance absorption of radiofrequency energy at a fixed frequency. This occurs by varying the clamping electric field strength, E, forcing the electrons to jump between discrete surface levels, the distance between the latter depending on E. Electrons scatter off the helium atoms (in the vapor) and off thermal excitations at the liquid helium surface

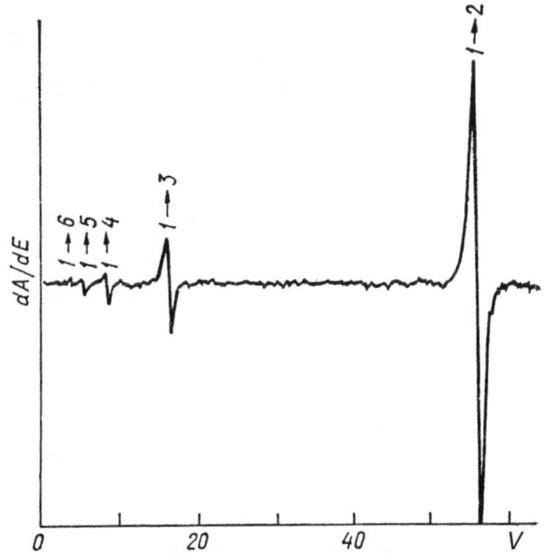

Fig. 1. First derivative dA/dE of absorbed radiofrequency power, A, with respect to electric field E versus the potential V of the external clamping field. Resonances correspond to electron transitions from the ground energy level ($l = 1$) to excited levels ($l \geqslant 2$) as indicated by arrows.

(*ripplons*). They dissipate energy in the process to produce a finite resonance width.

The two-dimensional electron system of finite surface density, $n_s \propto 10^4$–10^9 cm^{-2}, above the liquid helium is an ideal target for studying various collective effects related to the Coulomb interaction. It was in the system of levitating electrons that two-dimensional *plasma oscillations* were first observed, and instabilities of the charged surface of liquid helium were studied, including the formation of a two-dimensional *honeycomb crystal* (Fig. 2), and of charged bubbles (*bublons*) in strong electric fields, and where the two-dimensional electronic *Wigner crystal* (crystal of levitating electrons) was first discovered, with its existence region corresponding to particular values of the dimensionless parameter Γ satisfying the condition $\Gamma \equiv e^2(\pi n_s)^{1/2}/(k_B T) > 137$ (Fig. 3). At $\Gamma = 137$ during the liquid–crystal *phase transition* a two-dimensional triangular lattice of identically charged electrons forms, the first known example of *Coulomb crystallization.* The formation of the Wigner (electron) crystal is detected by the appearance of specific energy absorption resonances in the electromagnetic field,

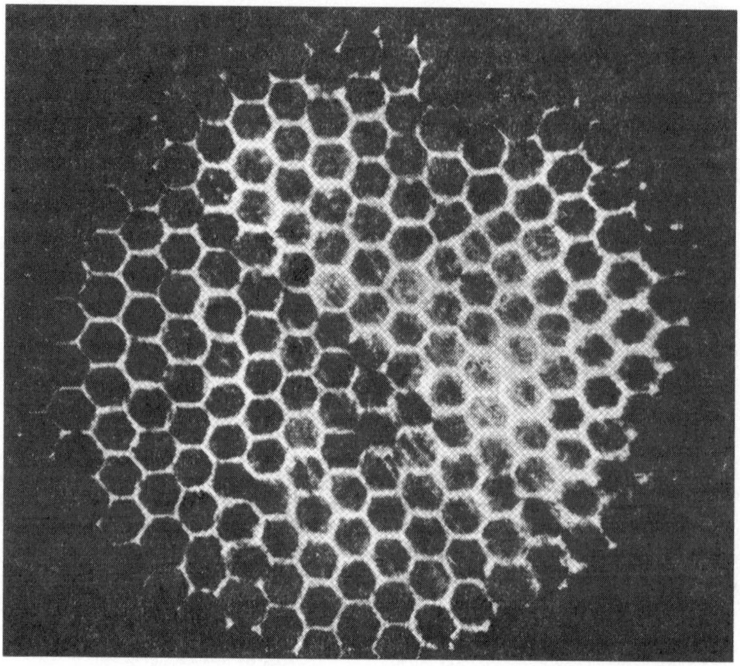

Fig. 2. A hexagonal honeycombed crystal on the surface of liquid helium in the clamping field E (under lateral illumination with a laser).

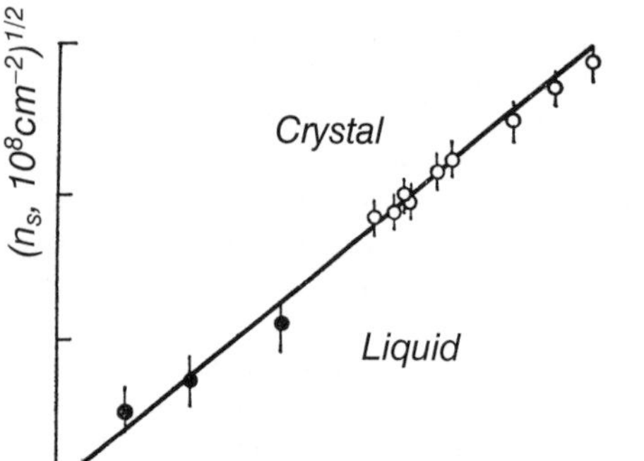

Fig. 3. Phase diagram of a two-dimensional system of levitating electrons above the surface of liquid helium in the $n_s^{1/2}$ versus T plane. The straight line indicates the boundary of the Wigner crystal–liquid phase transition from the liquid phase below to the crystalline state above.

and by other techniques. Specific *electron–ripplon* interactions define the structure of resonances in the crystalline state. There are grounds to believe that a system of levitating electrons at intermediate values of Γ can form a correlated state with short-range ordering similar to that found in a crystalline phase (see *Long-range and short-range order*).

LEVITATION CRYSTALLIZATION

Technology of growing crystals in the regime of microgravity ($10^{-3}g$) is achievable on board long-time orbital space stations. Levitation crystallization practically eliminates mixing of liquid layers by natural convection, and makes it easier to obtain perfect crystals. It features stronger laminar *mass transport* to the *crystallization* front, and prevents contamination of the growing crystals by container wall material. Levitation suppresses the effect of *segregation* when alloys with differing component densities crystallize, and broadens the possibilities of capillary formation of various complex prescribed profiles (see *Monocrystal growth*), even hollow shapes. See also *Levitation melting*.

LEVITATION MELTING

Melting of solid phase when gravity is either totally absent or reduced to the levels of microgravity achievable onboard long-term orbital space stations ($10^{-3}g$). To "suspend" the melting briquette of stock and make it levitate, different techniques are used. Ultrasound emission by appropriately oriented piezoelectric emitters can be focused on the target. Melted metal can be suspended in an electromagnetic field or placed in an electrostatic field generated by spherical electrodes around a spherical liquid droplet. Sufficient light pressure can be produced in the focus of a parabolic or elliptic light concentrator (preferably, solar radiation in free outer space should be used). Levitation melting allows smelting glass and growing *monocrystals* at higher temperatures, and provides for strict conformity to the prescribed stoichiometry, with preclusion of contamination of the melt with trace amounts of container (crucible, ampoule) material, or process gas phase.

LIBRATIONS, librons

Vibrations of a molecule (ion) or a group of atoms in a *crystal lattice* that involve hindered rotation, analogous to the oscillations of a pendulum; the orbit is closed in the (q, p) (position, momentum) plane. An example is the oscillation of the NO_2^- ion in an *alkali-halide crystal* lattice about its [001] axis, such that its C_2 axis does not leave the (110) plane. Another example is C_{60} molecules in K_3C_{60} crystals. Typical libron vibrations lie within the spectral range 100–300 cm^{-1}.

LIE GROUP

See *Continuous symmetry transformation groups*.

LIFETIME of a system in an excited state

The average time τ of the existence of a system in a given state, determined by the probabilities of different processes of transitions to other states: $\tau^{-1} = \sum_i \tau_i^{-1}$, where τ_i is lifetime of the system with respect to independent transition channels designated by the index i. The set of possible transitions and, correspondingly, the lifetimes, are specific for each type of excitation in solids. There is a quantum-mechanical probability for a spontaneous transition from each excited state to all states which lie lower in energy. For an electron in the excited level of a local electron center the lifetime for a spontaneous transition within the center accompanied by photon emission usually lies in the range $\tau_n \sim 10^{-6}$–10^{-9} s. Such *radiative quantum transitions* may occur with the participation of *phonons* (*indirect transitions*). *Nonradiative quantum transitions* are possible, where the energy of the excited electron is released in the form of phonons (*multiphonon transition*); their characteristic lifetimes can be much longer than τ_n. The thermal excitation of an electron to the *conduction band* is also possible, and accounts for the intrinsic conductivity of semiconductors.

An *exciton* lifetime is determined by the probability of its annihilation (electron–hole *recombination* of *Wannier–Mott exciton*), as well by the probability of its dissociation into an electron and a hole. The lifetime of electrons in the conduction band of semiconductors is characterized by the probabilities of their recombination with holes in the *valence band*, or their capture by defects (*recombination centers*); it depends on the concentration of holes and unoccupied defects sites, which in turn depend on the efficiency of electron and hole generation (see *Generation of current carriers*). In the stationary non-equilibrium state of a semiconductor the electron lifetime is given by $\tau_n = \Delta n/\lambda_n$, where Δn is the excess of electron concentration over the equilibrium value of n, and λ_n is the rate of nonthermal generation of electrons. A similar relation also applies to non-equilibrium holes, $\tau_p = \Delta p/\lambda_p$. The quantities τ_n and τ_p are called *relaxation lifetimes*. The lifetime of minority *current carriers*, e.g., of the holes in n-type semiconductors, under conditions when it is possible to neglect variations of the majority carrier concentration and the degree of electron filling of the recombination centers, then $\tau_p^{-1} = v \sum_i \sigma_i n_i$, where i designates the different types of recombination centers, σ_i is the cross-section of hole trapping by ith type center not occupied by holes, n_i is the concentration of such centers, and v is the thermal speed of the holes in the valence band. The time τ_p is an important material parameter for many electronic devices, and varies for different semiconductor materials from 10^{-3} to 10^{-8} s. The *relaxation* kinetics of the non-equilibrium concentration of carriers from its initial value Δp_0 is determined by the relation $\Delta p = \Delta p_0 \exp(-t/\tau_p)$. This expression becomes more complicated in the presence of traps. One should not confuse the lifetime of an electron (hole) in a band with their *relaxation times* within the band. The latter are determined by processes of energy and momentum relaxation. In metals the lifetime of an excited electron (time of thermalization) is determined by the electron–electron scattering time (generation of electron–hole pairs), and by the emission of phonons. The rate of both processes strongly depends on the temperature, and is also sensitive to the amount of static disorder (impurity scattering). In a similar way the lifetime for other non-equilibrium particles and quasi-particles in solids can be introduced, e.g., for *phonons, magnons, vacancies, interstitial atoms*, etc. For high concentrations of photons or phonons, induced transitions (see *Induced radiation*) play the main role in determining the value of the lifetime.

LIFSHITS MODEL (I.M. Lifshits, 1963)

A model of structural disorder (in amorphous materials or liquids) which describes the motion of electrons in a field produced by randomly arranged point scatterers (impurity centers). The Lifshits model is employed to study the *impurity band* in the energy spectra of electrons in disordered materials, a band arising due to the broadening of a *local electronic level.* In the leading approximation (in the concentration of impurities) the energy levels and quantum states of the impurity band acquire in this model a descriptive geometrical formulation, in which wave functions are localized at one or two impurity centers (although the energy levels also depend on the positions of other centers), while the *density of states* and the spatial correlations are represented on a certain scale by universal functions. In the immediate vicinity of the localized level, where within the framework of the geometrical formalism the density of states has a dip, other states become important, collectivizing a large number of centers, as a result of which the dip fills up to a significant extent. The structure of the spectrum and of the states in the vicinity of the edge of the initial spectrum was studied on the basis of this model, and an expansion in terms of the powers of the impurity concentration was carried out for the density of states, and therefore, for all thermodynamic quantities. All of the results mentioned were based on the idea of the existence over a definite region of the spectrum of representative *fluctuations*, which provide the main contribution to the density of states. This idea proved to be especially fruitful for the construction of *fluctuation levels*, where it developed into an effective computational method for the optimal fluctuation, which found wide application in the theory of *highly-doped semiconductors.*

LIFSHITS–ONSAGER HAMILTONIAN
(I.M. Lifshits, L. Onsager, 1952)

A Hamiltonian describing the dynamics of *conduction electrons* in an external magnetic field. The Lifshits–Onsager Hamiltonian $(\widehat{H})$ is constructed on the basis of the *dispersion law* for free electrons, $\varepsilon(\boldsymbol{p})$ ($\boldsymbol{p}$ is the *quasi-momentum*), according to the following correspondence principle: $\varepsilon \to (\widehat{H}) \equiv \varepsilon(\widehat{\boldsymbol{p}})$, $\widehat{\boldsymbol{p}} = \widehat{\boldsymbol{P}} - e\boldsymbol{A}(\widehat{\boldsymbol{r}})$. Here $\widehat{\boldsymbol{P}}$ is the generalized momentum operator, and $\boldsymbol{A} = \nabla \times \boldsymbol{B}$ is the vector potential. The order of operation of the mutually noncommutative components of the kinematic momentum operator, $\widehat{\boldsymbol{p}}$, is defined by the symmetrization rule: the periodic function $\varepsilon(\boldsymbol{p})$ is represented as a Fourier series, and the substitution $\boldsymbol{p} \to \widehat{\boldsymbol{P}}$ is made in the exponent of each term of that series. The Lifshits–Onsager Hamiltonian is approximate, with terms of the order of $\chi = eBa^2/\hbar$ (a is the *reciprocal lattice* parameter) dropped from it. In fields of 1–10 T we have $\chi \propto 10^{-5}$–10^{-4}. A limit to the applicability of this Hamiltonian is related to the emergence of band-to-band transitions in strong magnetic fields (see *Magnetic breakdown*).

LIFSHITS–ONSAGER QUANTIZATION RULES

Rules of quasi-classical quantization of energy levels of *conduction electrons* in a uniform magnetic field $\boldsymbol{B}$. Lifshits–Onsager quantization rules appear when the quasi-classical Bohr–Sommerfeld quantization rules are applied to the *Lifshits–Onsager Hamiltonian*, and provide the relation $S(E, p_z) = 2\pi eB(n + 1/2)$. Here $S(E, p_z)$ is the area bounded by a closed curve in *momentum space* that a classical electron traces in a magnetic field $\boldsymbol{B}$ at its energy E, and the projection p_z of its momentum on the magnetic field direction, where $n \gg 1$ is a positive integer (I.M. Lifshits, 1951; L. Onsager, 1952). For conduction electrons in a magnetic field at fixed n, this expression prescribes an approximately equally spaced set of energy levels (each level is infinitely degenerate). The existence of these levels is observed as macroscopic oscillations of various kinetic and thermodynamic characteristics of a *metal*, with the inverse magnetic field having the period $e\hbar/S_{ext}$, where S_{ext} is the extremal cross-section of the *Fermi surface* in the $p_z = \text{const}$ plane.

LIFSHITS POINT (I.M. Lifshits, 1930)

A point in the *phase diagram* of a crystal at which three *phases* are in equilibrium: disordered, ordered (uniform), and ordered incommensurate (modulated). At the Lifshits point it is found that thermodynamic potentials expressed as functions of the *order parameter* involve terms linear in the gradient of the order parameter. When moving away from the Lifshits point toward the ordered phase, a superstructure with varying wave vector $\boldsymbol{K}_S$ develops in the crystal. Changes in $\boldsymbol{K}_S$ sometimes occur stepwise, resulting in a sequence of transitions of the order–disorder type, forming the so-called *"devil's staircase"*. Various types of Lifshits points are possible, differing from each other in the type of anomaly in the physical quantities (*susceptibility*, *specific heat*, structure period, etc.) at these points. Lines of various types of *phase transitions* meet at Lifshits points. Usually, the transitions from the disordered to the uniform ordered or modulated phases are *second-order phase transitions*, while those of the order–disorder type between differing ordered phases are *first-order phase transitions*. Lifshits points are observable in ordered alloys, magnetically ordered crystals, systems with *charge density waves*, etc. Examples of substances that exhibit Lifshits points in their phase diagrams are $BaMnF_4$, CeSb, K_2SeO_4, MnP, $NaNO_2$, $NbSe_2$, and VO_2.

LIFSHITS TRANSITION (I.M. Lifshits, 1960)

A transition of the $2\frac{1}{2}$th order, an electronic topological transition. A change in the topology of the *Fermi surface* (breaking of a bridge, appearance of a new split-off cavity) under the effect of various external factors such as a *uniform compression*. For low enough temperatures in a metal that undergoes the Lifshits transition from the side of lower Fermi surface connectivity, its *thermodynamic potentials* involve singular terms of the form $|z|^{2.5}$, where z is a continuously changing parameter that characterizes the proximity of the system to its transition point. If we let $T \to 0$ in the absence of electron scattering, there should be square root singularities in the *density of states*, *specific heat*, *magnetic susceptibility*, conductivity, superconducting transition temperature, and certain other characteristics of a *metal* at the transition point ($z = 0$). Meanwhile, the linear expansion coefficient, thermal pressure coefficient, and thermal electromotive force should feature singularities of the form $|z|^{-1/2}$ due to the appearance of groups of electrons with abnormally low velocities near the Lifshits transition. At finite temperatures, and in the absence of electron scattering, the Lifshits transition ceases to be a *phase transition* in the strict sense of the word, since the Fermi surface of the metal is "eroded", and the anomalies in the above characteristics appear as broadened asymmetric steps and peaks. A typical example of the Lifshits transition is found during anisotropic *strain* of single crystals of $Bi_{0.9}Sb_{0.1}$ in which the electrons from one ellipsoid of the Fermi surface "pour over" into the other two ellipsoids. In the vicinity of the critical deformation point, when the ellipsoid is completely empty, smoothed breaks are observed in the conductivity, and in the coefficient of *sound absorption*, and a peak appears in the thermal electromotive force at low enough temperatures.

LIGANDS

Donor particles, anions (Cl^-, CN^-, $C_2O_4^{2-}$, etc.), or neutral molecules (H_2O, NH_3, etc.) located around the central atom in a particular order within an inner coordination sphere of a *coordination compound*. Ligands are usually attached to the central atom via donor atoms, and their number can be one (*monodentate ligands*, Cl^-, H_2O, NH_3), two (*bidentate ligands*, $C_2O_4^{2-}$, $NH_2(CH_2)_2NH_2$), or many (*polydentate ligands*). Bidentate and polydentate ligands form complexes that are more stable than the monodentate type (the so-called *chelate effect*). A ligand may act as a bridge if linked to two metal atoms, e.g., as is the case of two Cl^- ions in the molecule

$$\begin{array}{c} Cl \\ \diagdown \\ NH_3 \end{array} Pt \begin{array}{c} Cl \\ \diagup \diagdown \\ Cl \end{array} Pt \begin{array}{c} NH_3 \\ \diagup \\ Cl \end{array}$$

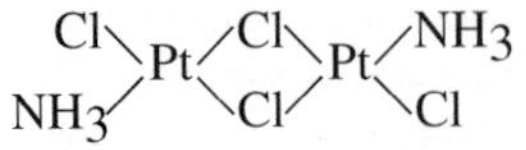

The nature of the chemical bond between ligands and the central atom and its effects upon the magnetic, spectral, and other properties of complex compounds are the subject of *ligand field theory* and *crystal field theory*. The reaction capability of complex compounds depends significantly on the mutual effects of ligands within their inner sphere; in particular those produced by ligands in a straight line, with the central atom on its opposite sides.

LIGHT ABSORPTION by solids

Decrease in intensity of radiation due to transfer of radiation energy to the medium, through which this radiation propagates. Quantum-mechanical processes of light absorption by solids involve the of annihilation of light quanta, with the energy transferred to *electrons*, *phonons*, *excitons* and other *quasi-particles*, and to *defects*. Therefore, the investigation of light absorption by solids can provide information on energy bands of crystals, the *density of electron states*, the energy spectrum of phonon and exciton excitations (see *Band theory of solids*), on electronic and vibrational states of impurity centers and other defects, and on the interaction between quasi-particles. At high radiation intensities, processes of many-photon absorption are possible (see *Optical spectroscopy*, *Nonlinear optics*, *Two-photon light absorption*, *Infrared spectroscopy*).

LIGHT DEFLECTORS

Devices for the smooth or discrete deviation of a light beam direction to a given angle. Many light deflectors use the natural or induced (electrical or mechanical) field variation of birefringence of electro-acousto-optical *crystals*, or of transparent *ceramics*. They are employed in projection devices for the optical processing of information, for the synthesis of electronic integrated circuit topology, in computer optical storage devices, etc.

There exist *electrooptical deflectors* of the analogue type (deviation changes smoothly with variation of control voltage) and *discrete deflectors* (deviation changes in steps determined by the number of the applied cascades of *polarization switching*, each consisting of a separating element, a crystal with high birefringence (e.g., calcite $CaCO_3$, calomel Hg_2Cl_2) and an electrooptical polarization switch, see *Electrooptical effect*). Acoustooptical deflectors produce the light deviation at a dynamic *diffraction grating*, formed within the acoustooptical element by a connected piezoelectric vibrator operating at a tunable ultrasonic radiation frequency (see *Acousto-optics*).

Traditional light deflectors meet most modern requirements except for fast response. The latter has been considerably improved (up to the order of nanoseconds) by the transition to integral-optical versions of light deflector design, which are characterized by the simultaneous reduction of energy consumption by three or more orders of magnitude.

LIGHT DISPERSION

See *Dispersion of light*.

LIGHT EMITTING DIODE (LED)

An optical (visible wavelength) transformer of electrical energy into incoherent light emission. It consists of a *semiconductor junction* to which a voltage ($V_g = 1.2$–2.5 V) is applied in the conducting direction. To operate the unit one has to *inject* secondary *current carriers* (e.g., electrons) into the contact via a homogeneous or heterogeneous p–n junction, so that the injected carriers *recombine* with the principal carriers at the base (e.g., holes). The energy thus released, eV_g, is partially dissipated by the diode body as heat (*emission-free recombination*, see *Nonradiative quantum transition*), and most of the remainder transforms into light (*emission recombination*). This emitted radiation $P_{h\nu}$ has an ampere–watt (or ampere–brightness) profile $P_{h\nu}(j)$, that depends on the injection current density. Typical values of j and $P_{h\nu}$ lie within the limits 5–500 A/cm^2 and 0.2–0.5 mW, respectively. Light emitting diodes are characterized by their internal, η_i, and external, η_e, efficiencies. The internal efficiency η_i gives the relative share of emission recombination, while $\eta_e = \eta_i F$. Here $F = 0.04 \div 0.06$ is the aperture coefficient accounting for radiation losses during emission from the LED. Depending on the design and the material efficiency, the ratio $P_{h\nu}/(I_g V_g)$ (here I_g is the LED current) may reach 0.5–2, the corresponding values of η_e being 10–12% for $50\% < \eta_i < 100\%$. The probability of emission recombination is higher in semiconductors with direct band gaps, compared to that for semiconductors with an indirect band gap (see *Indirect transitions*). Therefore, such LEDs are normally more efficient. The principal materials used for LEDs are of the $A^{III}A^{V}$ type. These are binary (GaAs, GaP, GaN), ternary ($Al_xGa_{1-x}As$, $GaAs_{1-x}P_x$, $Ga_xIn_{1-x}As$), and quaternary ($Ga_xIn_{1-x}As_yP_{1-y}$) compounds, as well as $A^{IV}B^{IV}$ compounds (e.g., SiC). They are doped with corresponding impurities to achieve n- and p-type conductivity and the desired radiation spectra from blue (470–483 nm, GaN, SiC) to red (690–780 nm, GaP, ZnO, AlGaAs) and IR (from 820–910 to 1300–1700 nm, AlGaAs, GaInAs, GaInAsP). To obtain the required emission color a double transformation may also be used, when one combines an IR LED and an anti-Stokes *luminophor*. To increase η_e and color contrast, and to form the desired phase function, both polymer and glass lenses are used. The LED current $I_g \propto \exp[eV_g/(k_B T)]$ is controlled by an external limiting resistance and remains within 10–100 mA for crystals 400 × 400 μm^2 in size. Light emitting diodes on the market are offered both with and without special casing. They serve as indicators, small-scale data presentation devices, are used in *optoisolators*, function in read-out devices that use magnetic tapes for data storage, in short-range fiber optic communication lines, sensor elements, etc.

LIGHT GENERATION

See *Parametric light generation*.

LIGHT PIPE

Dielectric waveguide (see *Integrated optics*) in the optical range. See also *Fiber optics*.

LIGHT POLARIZATION

See *Polarization of light*, *Self-induced light polarization change*, *Rotation of light polarization plane*.

LIGHT REFRACTION

See *Refraction of light*.

LIGHT SCATTERING

Phenomenon taking place during the propagation of light through various optically inhomogeneous media. The light wave experiences *diffraction of light* by inhomogeneities, thereby deviating from its initial direction of propagation. The principal quantitative characteristics of light scattering are the *differential scattering cross-section* $d s_s$ (ratio of the radiation flux dI_s, scattered into an elementary solid angle $d\Omega$, to the incident flux I_0: $d s_s = dI_s/I_0$); the *scattering phase function* (dependence of scattered intensity on scattering angle), the *frequency spectrum*, and *polarization of light*.

Physical causes producing optical inhomogeneities are quite diverse. The medium may become inhomogeneous, e.g., when one material, differing in its *refractive index*, is included in another. The scattering phase function depends on the ratio between the inhomogeneity size and the light wavelength λ_0. Light scattering may also be observed in macroscopically homogeneous media (i.e. containing no impurities) due, e.g., to random thermal motion of the material's particles. In that case, the reason for the optical inhomogeneity is fluctuations of the *dielectric constant* produced by different excitations of the medium. When the light wavelength λ_0 differs considerably from the wavelengths of the normal mode frequencies of the medium, then we have $\sigma_s \propto \lambda_0^{-4}$ (*Rayleigh law*). However, as λ_0 approaches the wavelengths of the normal modes of the medium, the Rayleigh law is no longer satisfied, and the intensity of the scattered light may be strongly enhanced (*resonance light scattering*).

From the quantum point of view light scattering is treated as the absorption of the incident photon, its energy $\hbar\omega_0$, momentum $\hbar k_0$ and polarization e_0, by the medium, a process accompanied by the simultaneous emission of another photon with energy $\hbar\omega_s$, momentum $\hbar k_s$ and polarization e_s. Provided the emitted photon frequency is equal to that of the absorbed one ($\omega_0 = \omega_s$) the light scattering is called elastic or *Rayleigh scattering of light*, and when $\omega_0 \neq \omega_s$ it is inelastic scattering. Meanwhile, the corresponding energy difference $|\hbar\omega_0 - \hbar\omega_s| = \hbar\omega$ is transferred to the medium, so the medium transforms to a different quantum state. Light scattering may take place

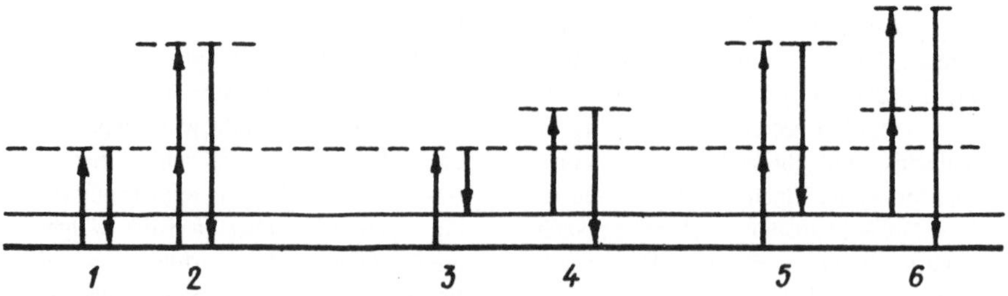

Scattering processes: (1) Rayleigh; (2) nonlinear Rayleigh (hyper-Rayleigh); (3) inelastic Stokes; (4) inelastic anti-Stokes; (5) nonlinear inelastic (hyper-Raman) Stokes; and (6) nonlinear inelastic (hyper-Raman) anti-Stokes.

when the scattered photon energy is either lower than the energy of the incident photons, $\hbar\omega_S < \hbar\omega_0$ (*Stokes scattering*), or greater than that energy, $\hbar\omega_S > \hbar\omega_0$ (*anti-Stokes scattering*). Inelastic light scattering in a solid may be accompanied by either the generation or annihilation of various types and numbers of *quasi-particles*. Typical examples of such processes are light scattering by acoustic phonons (*Brillouin scattering*), optical *phonons*, *magnons*, surface and bulk *polaritons*, etc. (see *Raman scattering of light*).

With the development of powerful sources of coherent radiation (*lasers*), it also became possible to study *nonlinear light scattering* processes during which a single quantum of scattered light is produced by two or more light quanta incident upon the medium (see Fig.). Lasers make it possible to use the techniques of *active spectroscopy* as well. In addition, as is also the case with light emission, *induced light scattering* becomes possible along with spontaneous light scattering during intense laser excitation involving already scattered photons.

See also *Induced light scattering*, *Photoinduced light scattering.*

LIGHT SCATTERING IN LIQUID CRYSTALS

This process strongly differs from light scattering in other media due to the presence of the large-scale *orientational order* of molecules, and the absence of translational ordering. A layer of a *nematic liquid crystal* only several millimeters thick is already opaque. Light scattering in it is approximately 10^6 times stronger than in a layer of ordinary liquid of identical thickness. The causes of such strong scattering are thermal fluctuations of the orientations of the *liquid crystal* molecules. The direction of the optical axis of the liquid crystal, as defined by the predominant orientation of its molecules, then changes. Scattering in *smectic liquid crystals* is usually weaker than in nematic types, since the stratified structure of the smectics hinders the development of fluctuations of the optic axis direction. At the same time inhomogeneities at the surface of liquid crystals and *defects* (e.g., *disclinations*, *confocal domains*) introduce perturbations into their structure, and result in light scattering that exceeds the scattering from thermal fluctuations. Besides, smectics of the C-type, in which the molecules are tilted with respect to the layer planes, are as opaque as nematics, because of fluctuations of the azimuth of the molecular inclination. Studies of the characteristics of light scattering provide information on the most important parameters of liquid crystals, their elastic moduli, viscosity factors, energy of interaction with the substrate, etc. These parameters are important for the design of liquid crystal displays (LCDs).

LIGHT WAVEGUIDE

See *Optical waveguide.*

LIGHT WAVE SELF-ACTION

See *Self-action of light waves.*

LIMITING SYMMETRY GROUPS, Curie groups

Point groups of symmetry which have infinite *symmetry axes* (∞). These limiting groups are subgroups of the (full) orthogonal group in three dimensions $O(3) = \infty\infty\bar{1} = \infty\infty m$ which describes the symmetry of uniform tensor fields of homogenious material media, and of geometrical figures in three-dimensional Euclidean

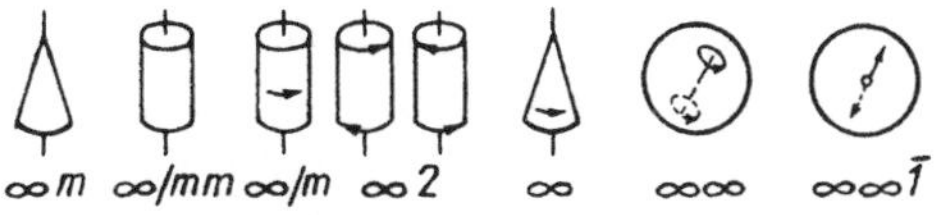

Reference figures with symmetries of Curie groups.

space. A uniform isotropic field and a sphere cut from it, whereby any diameter is a rotation axis for an arbitrary angle φ, including an infinitely small φ, $\infty = (2\pi/\delta\varphi)|\delta\varphi \to 0$, corresponds to the group $O(3)$. The inversion operation at the *center of symmetry* $\bar{1}$ is the third generator of the $O(3)$ group. The special subgroup $O^+(3) = SO(3) = \infty\infty$ of pure or proper rotations (no inversion) describes the symmetry of a continuous medium which allows *rotation of light polarization plane* in any direction.

Other subgroups of $O(3)$ (see Fig.) have a selected rotation direction ∞ or $\overline{\infty} = \infty \times \bar{1} = \bar{1} \times \infty$, which is combined with operations of mirror reflections in symmetry planes m, and with rotations around second-order axes 2 (180° rotation), such as: the group ∞ of a rotatable cone (of a collinear superposition of electric and magnetic fields), ∞m of a resting cone (of an electrical field), ∞/m of a rotatable cylinder (of a magnetic field), $\infty 2$ of a "twisted cylinder", which allows turning in opposite directions (the selected direction allows the rotation of the plane of polarization), ∞/mm of a resting cylinder (uniform uniaxial field of a mechanical stress).

LIMITING VELOCITY

See *Slonczewski limiting velocity.*

LINEAR COMBINATION OF ATOMIC ORBITALS, LCAO method

One of the techniques used to calculate the wave functions and energies of *electrons* in crystals. It is close to the *tight binding method* in the theory of molecules. In the LCAO method, the crystal potential is represented as a superposition of individual site potentials

$$V(r) = \sum_i V_i(r - R_i),$$

where R_i is the coordinate of the ith site in the lattice. The basis functions used for *band structure* calculations are Bloch sums composed of site-centered functions similar to *Wannier functions.* The LCAO method proper uses solutions of the Schrödinger equation for each respective Bloch sum and $V(r)$ term. As site functions become delocalized, the complexity of the solutions yielded by the LCAO method quickly builds up. Therefore, use is made of various modifications and approximations of the LCAO method when the site functions are artificially localized, as well as related techniques, such as linear combinations of Gaussian, Slater, and other *orbitals.*

LINEAR DICHROISM

See *Polaroid.*

LINEAR DISLOCATION TENSION

A force acting upon the elements of a *dislocation loop* that is defined by the dependence of the intrinsic energy of the loop on its length. To compute the energy of dislocation configurations, the *linear dislocation tension approximation* is used, adapting the model of a stretched string (see *Dislocation string*). The energy increment per unit increment of length of such a string is of the order of the *Burgers vector* squared multiplied by the *shear modulus.* More accurate expressions include a factor determined by the local value of the angle of the Burgers vector to the dislocation line, and also a logarithmic factor depending on the length of the curved section or curvature radius. The approximation of linear dislocation tension yields good results if these values are much larger than the magnitude of the Burgers vector. By minimizing expressions for the energy that are found within this approximation one may obtain equilibrium configurations of *dislocations* under their intersection or in the field of external stresses, the scale of the inhomogeneity of the latter along the line of dislocations being much larger than the Burgers vector length.

LINEAR HARDENING OF CRYSTALS

Strain hardening that manifests itself in larger values of the deforming stress σ needed to develop *plastic deformation* ε of a prescribed rate $\dot{\varepsilon} = \text{const}$ in a crystal. Three ranges (stages) of hardening describe the $\sigma(\varepsilon)$ dependence for *monocrystals* with face-centered cubic (FCC), body-centered cubic

Table 1. Coefficient of hardening

Metal	Lattice type	θ_{lhc} per unit modulus
Cu	FCC	$\approx 3\cdot 10^{-3}$
Fe	BCC	$\approx 2\cdot 10^{-3}$
Zn	HCP	$\approx 2\cdot 10^{-4}$

(BCC) and hexagonal close-packed (HCP) lattices: *easy slip*, linear hardening of crystals, and *dehardening*. As compared to others, the stage of linear hardening of crystals features a linear dependence of σ on ε, and the highest value of the coefficient of *hardening*, $\theta_{\text{lhc}} = \mathrm{d}\sigma/\mathrm{d}\varepsilon$. Crystals with FCC, BCC and HCP lattices have different values of θ_{lhc}, some of which are cited in Table 1 (at $T = 300$ K). The value of θ_{lhc} depends significantly on the type of *strain* (stretch, compression, *shear*), on the crystal orientation, and only weakly on T and $\dot{\varepsilon}$. The duration of the stage of linear hardening of crystals depends on the temperature, T. The transition from the stage of easy slip to that of linear hardening is related to activation of intersecting *slip systems*, to the appearance of *moving dislocations*, and to the fact that interaction with the latter needs much higher σ for *dislocations* to move. Long-range fields of σ control the resistance to the motion of dislocations at the stage of linear hardening.

LINEAR METHODS OF BAND STRUCTURE COMPUTATION

A group of methods differing from classic *band structure computation methods* (*Korringa–Kohn–Rostoker method*, *Augmented plane waves method*, etc.) in that their basis functions are energy independent. The energy dependence is either ignored, or an expansion is carried out about one or several energy reference points. Different variants of the method are often applied to different regions within the crystal (e.g., close to or far from nuclei). These methods provide an accuracy of *band structure* of several mRy, i.e. they stay within an error margin typical for constructing the crystalline potential. However, linear methods are computationally 10–100 times faster than classical ones. Most computations are done with the help of, first, the *linear muffin-tin orbital (LMTO) method* and, second, the *augmented spherical wave method*. The LMTO method is derived by linearizing the Korringa–Kohn–Rostoker method with some additional changes. Several modifications of the *linearized augmented plane waves (LAPW)* method are obtained by different forms of linearization. The LMTO method is faster, conceptually simpler, and more convenient for self-consistent operations. The advantage of the LAPW method is that it may be more easily generalized to cover potentials of an arbitrary shape (see *Muffin-tin potential*).

LINEAR PYROELECTRIC

A crystal with spontaneous polarization (see *Ferroelectricity*) that exists in the absence of an external electric field. The primitive cell has a nonvanishing electric dipole moment, and the direction of the spontaneous polarization cannot be reversed by applying an external electric field, even if it is as strong as the breakdown value. Examples of linear pyroelectrics are the mineral tourmaline and lithium sulfate Li_2SO_4. Their pyroelectric coefficients exhibit a very weak temperature dependence over a wide range of temperatures. Since pyroelectric coefficients in linear pyroelectrics are much smaller in magnitude than those in nonlinear ones, the former find practically no use as sensitive elements in *pyroelectric radiation detectors*. In some cases (e.g., tourmaline) the polarization is masked by neutralizing ions that collect on the surface from the atmosphere, and heating causes these ions to evaporate, with an observed increased in polarization, hence the origin of the name pyroelectricity.

LINEAR RESPONSE

A reaction of a physical system to a weak perturbation. External forces F_i (e.g., electric or magnetic fields) that affect the motion of individual particles may induce a linear response, so-called *mechanical perturbations*. An electric current may originate in a solid under their effect, its dielectric polarization (see *Polarization of insulator*) or *magnetization* may change. For small values of F_i, a linear relation holds between small deviations of the respective variables from their equilibrium values $A_i(t)$ and $F_i(t)$ (t is time):

$$A_i(t) = \sum_j \int_{-\infty}^{t} \varphi_{ij}(t,t')F_j(t')\,\mathrm{d}t,$$

where φ_{ij} is the linear response function, independent of F_i. The linear relation between $A_i(t)$ and $F_i(t)$ is a cause and effect one, i.e. $A_i(t)$ depends on values of $F_i(t')$ only if $t' < t$. In other cases the linear response may be caused by a weak spatial nonuniformity of the system, e.g., a nonuniform temperature or concentration of its current carriers. Linear response may then be expressed as fluxes of matter, momentum or energy that act to restore a uniform state of the system. Such perturbations are called *thermal perturbations*. Although the latter differ quite noticeably from mechanical perturbations on a microscopic level, the linear response itself has the same form. Spatial derivatives of the perturbed parameters of the system, e.g., temperature, particle density, etc., can act as external forces. The *Kubo linear response theory* relates the system response function to mechanical perturbations to the *correlation functions* of equilibrium *fluctuations* of microcurrents in a solid (see *Kubo–Tomita method*). For the case of a perturbation of the form $V(t) = \sum_i B_i F_i(t)$, where B_i are functions of the coordinates and momenta of the particles, or of the corresponding operators, Kubo theory introduces a response function expressed in terms of a two-time retarded *Green's function*

$$\varphi_{ij}(t,t') = \langle\langle A_i(t) B_i(t')\rangle\rangle$$
$$\equiv \theta(t-t')\langle\{A_i(t), B_j(t')\}\rangle,$$

where $\langle\ldots\rangle$ denotes averaging over the equilibrium distribution,

$$\theta(t) = \begin{cases} 1 & \text{for } t > 0, \\ 0 & \text{for } t < 0, \end{cases}$$

and $\{\ldots\}$ are Poisson brackets (commutator $1/(\mathrm{i}\hbar)[A(t), B(t')]$ in the quantum case). A remarkable feature of the linear response function is that the value of the linear response (a nonequilibrium characteristic) is expressed via the average over the state of statistical equilibrium. No such general theory exists for thermal perturbations, although relationships, similar to the *Kubo formulae*, may be obtained for specific cases through solving the *kinetic equations* or the Liouville equation.

LINE BROADENING

See *Inhomogeneous broadening*, *Homogeneous broadening*.

LINE INTENSITY

A spectroscopic parameter (ordinate of a curve) characterizing the absorption (transmission) intensity of a sample (see *Absorption index*). The line intensity versus frequency dependence determines the *line shape*, and this provides the *spectral line intensity* (i.e. line intensity at a given frequency), and the *integrated line intensity* (i.e. the area under the line contour).

LINE (LEVEL) WIDTH, level broadening

A measure of the energy uncertainty of a quantum-mechanical system interacting with an ensemble of particles (quasi-particles) that have a continuous or quasi-continuous energy spectrum, which makes the quantum states of the system not rigorously determined. The concept of level width permits the approximate description of the states of a closed system by exact wave functions when, strictly speaking, the system does not possess them. For the energy E_j of a closed system, the stable state wave function is $\psi_j(t) = \psi_j \exp[\mathrm{i}t(E_j/\hbar)]$. The interaction broadening of the level may be taken into account by introducing the modified wave function $\psi_j(t) = \psi_j \exp[\mathrm{i}t(E_j/\hbar + \mathrm{i}/\tau_j)]$, where the energy E has a random magnitude described by the *distribution function* $g_j(E)$ with its maximum near $E = E_j$, and τ_j is the average duration (*lifetime*) of the system in the given state j. The fluctuations of E_j for the quantum system ensemble are determined by the statistical distribution of external fields, which may either be stationary or fluctuate with time. If the interaction mentioned above is described as a random process then τ_j^{-1} is determined by the sum of probabilities of quantum transitions from the state with the energy E_j to all states with $E_k \neq E_j$. If the uncertainty of the energy $\Delta E_j \approx g_j(E_j)^{-1}$ due to the distribution $g_j(E)$ exceeds $\hbar\tau^{-1}$, then there is an *inhomogeneous level width* ΔE_j. If $\hbar\tau^{-1} > \Delta E_j$, the level width is determined by the uncertainty principle involving energy and time (*homogeneous broadening*). The mechanisms involving homogeneous and inhomogeneous (see *inhomogeneous *broadening*) contributions to the level width are determined, as a rule, by different interactions, but they could involve the same interaction (e.g., *electron–phonon interaction* for electron impurity states).

The intrinsic level widths are related to the experimentally measured line width ΔE_{jk} for transitions between energy levels E_j and E_k in accordance with the expression $\Delta E_{jk} \leqslant \Delta E_j + \Delta E_k$. The spectral line recording always involves an additional (applied perturbation) interaction $V(t)$ giving rise to the transitions. For example, for electric dipole transitions $V(t)$ is the dipole moment operator $e\boldsymbol{r}$ interacting with the oscillating electric field vector of an incoming electromagnetic wave. There are also magnetic dipole, electric quadrupole, and higher-order multipole transitions. The width depends on the method of recording the response to $V(t)$, such as the scanning frequency or magnetic field, and it can also depend on the strength or intensity of the exciting radiation. Some typical situations are:

(1) The spectral line width reflects the level width only partly when $V(t)$ excites transitions in such a manner that transition frequencies $\hbar\omega_{jk}$ do not depend on the parameters broadening the energy levels. Electron paramagnetic resonance (EPR) of a shallow donor, whose levels are broadened by the fluctuating electric fields of charged impurities, may be an example of this type. Such a broadening may mount up to several meV and be observed in optical spectra, while the EPR spectral lines of shallow donors are broadened by only several μeV, through much weaker magnetic interactions with nuclear and electronic spins.

(2) If a frequency of the quantum transition responsible for the spectral line is exposed to the influence of time varying (fluctuating) parameters (fields), then the spectral line width will reflect their distribution function $g_j(E)$ only when the transition frequency $\omega_{jk} = |E_j - E_k|/\hbar$ exceeds the reciprocal of the correlation time τ_c of the fluctuating fields. For the case $\omega_{jk} < \tau_c^{-1}$ an averaging of the fluctuating action takes place, the so-called *dynamic line narrowing effect*. This case, which is not unusual, occurs in various types of magnetic resonance.

(3) If $V(t)$ strongly disturbs the equilibrium of the quantum system, then the spectral line width may reflect changes in the spectral line due to the new equilibrium. For instance, optical transitions associated with the recharging of impurity centers may affect the quantity of the recharged ions, and hence result in an additional broadening of the spectral line. The inverse situation is possible (e.g., in high-resolution NMR), when the disturbance cools the quantum subsystem, which is responsible for the spectral line, leading to the suppression of fluctuations which cause broadening. The equilibrium can also be disturbed when the exciting signal is a high-power pulse.

Experimentally the spectral line width is determined by the energy spacing (or that of another parameter such as wavelength or magnetic field B which is scanned) between the points, where the *line intensity* has dropped to one-half of its maximum value ($\Delta_{1/2}$). When the spectrum is recorded as a first derivative, the spectral line width is characterized, as usual, by its peak-to-peak value (Δ_{PP}) which is the spacing between the extrema of the derivative recording. For most lineshapes the peak-to-peak width is less than the half amplitude width ($\Delta_{1/2}/\Delta_{\mathrm{PP}} = 3^{1/2} = 1.7321$ for Lorentzian shaped lines, and $\Delta_{1/2}/\Delta_{\mathrm{PP}} = (2\ln 2)^{1/2} = 1.177$ for Gaussian lines).

The proper theoretical calculation of the spectral line width requires a knowledge of its *line shape*. However, in the case of homogeneous broadening, its width is calculated from the knowledge of its corresponding *relaxation times*. In the case of inhomogeneous broadening the spectral lines are sometimes characterized by *moments of spectral lines*.

LINE SHAPE

Distribution of spectral *line intensity* as a function of the variable according to which the spectrum is recorded (wavelength λ, frequency ν, magnetic field B, etc.); it corresponds to the *form factor* of the line. Experimental line shapes are affected by instrumental settings as well as by physical factors, so line shape studies can provide information about the nature of the broadening mechanisms. The line shape may be characterized by a mathematical expression, or by a set of moments. The most common line shapes $F(x)$ are a *Gaussian shape*, a *Lorentzian shape*, and also their convolution (see *Line shape convolution*):

$$F_G(x) = \exp(-x^2 \ln 2) \quad \text{(Gaussian)},$$

$$F_L(x) = \frac{1}{1+x^2} \quad \text{(Lorentzian)},$$

where $x = (\omega_0 - \omega)/\Delta$, the center frequency of the resonance is ω_0, and Δ is the half-width at half-amplitude.

Specific line shapes can be due to *orientational line broadening*, and also due to the action of specific physical mechanisms (*Dyson shape*, *Holtsmark shape*). A line shape can arise from several sources of *inhomogeneous broadening* (*hyperfine interaction*, superhyperfine interaction, unresolved fine structure, etc.), or it can result from the superposition of overlapping lines. For the best resolution and analysis the first, second and third derivatives of the line are recorded. Also mathematical analogues of complex lines, calculated with the help of a computer, can be fit to experimental spectra with the proper choice of one or more variable parameters.

LINE SHAPE CONVOLUTION

A composite *line shape* that results from the simultaneous effect of two or more spectral distributions. The line shape function of a resonance absorption line in a paramagnetic system is defined by the spectral distribution of absorption probability for each separate *paramagnetic center*, $I_1(\omega' - \omega)$, and the spectral distribution of the number of particles with a given energy, $I_2(\omega' - \omega_0)$. Depending on the nature of the interaction of a paramagnetic particle with its environment, the function $I_1(\omega' - \omega)$ may be described by various distribution laws. The ones most often encountered in practice are the *Lorentzian shape*

$$L(\omega' - \omega) = \pi^{-1}\left[1 + \frac{(\omega' - \omega)^2}{\Gamma^2}\right]^{-1}$$

and the *Gaussian shape*

$$G(\omega' - \omega) = \pi^{-1/2} \exp\left[-\frac{(\omega' - \omega)^2}{\Gamma^2}\right],$$

where Γ is the linewidth. The function $I_2(\omega' - \omega_0)$ is usually Gaussian, $G(\omega' - \omega_0)$. The resulting line shape function is a convolution of I_1 and I_2, so that

$$I(\omega - \omega_0) = I_1 \times I_2 = \int_{-\infty}^{\infty} I_1(\omega' - \omega) I_2(\omega' - \omega_0)\, d\omega'.$$

In the most common and simplest case $I(\omega - \omega_0)$ is a convolution of a Gaussian and a Lorentzian shape

$$I(\omega - \omega_0) = G \times L = \int_{-\infty}^{\infty} G(\omega' - \omega) L(\omega' - \omega_0)\, d\omega'.$$

LINE SHAPE, HOLTSMARK

See *Holtsmark shape.*

LINE SHAPE, STATISTICAL THEORY

See *Statistical theory of line shape.*

LIQUATION

Separation (*segregation*) of a homogeneous (single-phase) multicomponent liquid (solution, melt) into two or more *phases* differing in composition. This can be done, e.g., by heating an alloy to melt the more fusible matter. New phases that form during liquation may be either liquid or solid (vitreous or crystalline). Liquation plays an important role during *vitrification* of multicomponent supercooled (metastable) highly viscous liquids. Sometimes, though rarely, the term "liquation" is used to denote the process of separating new *phases* from the initial one that differ from it only in structure.

LIQUID CRYSTAL, CHIRALITY

See *Chirality in liquid crystals.*

LIQUID-CRYSTAL DISPLAYS, LCD

Devices for visualizing information in the form of letters, numbers, symbols, lines, etc. based on a *liquid crystal* as an operating agent. The action of a liquid crystal display is based on the ability of a liquid crystal to change its optical properties under the action of an external electric field (or temperature). The field causes either a coordinated rotation of liquid crystal molecules (see *Electrooptical effects in liquid crystals*) or a strong liquid crystal flow (see *Electrohydrodynamic instabilities*). In the first case, the optical axis changes its direction; in the second case, a strong spatial nonuniformity of the effective *refractive index* develops. Both effects are reversible; they can be used for controllable modulation of the intensity of light passing through the liquid crystal layer. *Dyes* are introduced into a liquid crystal to bring about a color change.

Two types of liquid-crystal media suitable for displays are *ferroelectric liquid crystals* (providing a high switching rate) and liquid crystals dispersed into polymer arrays. These media have good prospects for the creation of TV screens, and of flexible wide-angle displays. There are also displays based on a thermal method of *information recording*. A laser beam that scans across the layer of a *smectic liquid crystal* introduces local changes due to melting, whereby its refractive index is changed. The stored information is erased by an electric field which restores the initial structure of the liquid crystal. Due to their simple manufacturing technology, high economic efficiency and functional variability, liquid crystal displays are widely used as alphanumeric indicators, TV screens, as well as in electronic watches, calculators, etc.

LIQUID CRYSTAL, LIGHT SCATTERING

See *Light scattering in liquid crystals*.

LIQUID CRYSTAL MODULATION

See *Wave-like modulation*.

LIQUID CRYSTAL ORDER

See *Degree of order, Order parameters in liquid crystals*.

LIQUID-CRYSTAL POLYMERS

Macromolecular analogs of common *liquid crystals*: typical of molecules or their fragments (e.g., lateral groups of flexible-chained comb-like molecules) in liquid-crystal polymers is the presence of *orientational order*. Liquid-crystal polymers occupy an intermediate position between amorphous and crystalline polymers. They are divided, like low molecular weight liquid-crystal polymers, into thermotropic (formed by change in temperature) and lyotropic (formed by change in concentration) types, and they form smectic, nematic and cholesteric *mesophases*. Liquid-crystal polymers might be used for film dichroic polaroids, and color indicators based on them.

LIQUID CRYSTALS

Thermodynamically stable states of matter with physical properties intermediate between those of a liquid and a solid *crystal*. Liquid crystals are composed of molecules or groups of molecules (e.g., *micelles*) often unsymmetrical in shape. Their most general features are the presence of long range *orientational order* in the arrangement of the asymmetric molecules, and the complete or partial absence of translational order of the locations of the centers of mass of the molecules.

Liquid crystals can be thermotropic and lyotropic. *Thermotropic liquid crystals* form during a temperature change: during melting a solid crystal may, before entering the isotropic liquid state, pass through one or more mesomorphic (intermediate) phases or *mesophases*. *Lyotropic liquid crystals* form when dissolving asymmetric objects in suitable solvents. Proceeding from rod-like to disc-like shaped component molecules, liquid crystals are subdivided into *calamitics* and *discotics*; and those formed from polymer macromolecules are called *liquid-crystal polymers*. According to the manner of ordering, mesophases are divided into nematic, cholesteric, smectic and columnar (these do not exhaust the classification, see also *Blue phases*). The most widely studied types are thermotropic calamitics of the *nematic liquid crystals*, *cholesteric liquid crystals* and *smectic liquid crystals* consisting of low-molecular weight organic materials with the common molecular formula

$$R-\langle\!\!\!\bigcirc\!\!\!\rangle-A = B-\langle\!\!\!\bigcirc\!\!\!\rangle-R',$$

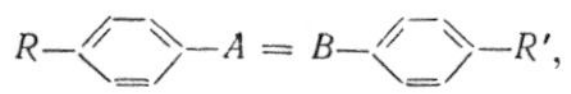

i.e. molecules with two benzene rings connected by a double or triple $A-B$ bond, terminated by short chains R and R' at the ends. Such molecules are about 2 nm long and about 0.5 nm thick. Nematics are characterized by the total absence of translational order in the arrangement of the molecular mass centers, and the presence of a preferable orientation of the longer axis of rod-like molecules (or of normals to the planes of disc-like molecules) along some direction denoted by the unit vector $\boldsymbol{n}$, which is called the *director.* A cholesteric can be pictured as a twisted nematic: when moving along an axis perpendicular to $\boldsymbol{n}$ the director $\boldsymbol{n}$ traces a spiral. Cholesterics consist of *chiral molecules*, i.e. those without planes of symmetry. Smectics have a unidimensional translationally ordered laminated structure. The best known are smectics A, B and C. The longer axes of the molecules are perpendicular to the plane of the layer in *smectics A* and inclined to it in *smectics C*. The mass centers of the molecules within the layer in both cases are positioned randomly. The layers can slip relative to one another, so smectics A and C resemble crystals in one direction, and liquids in the other two directions. The orientation of the molecules in *smectics B* is the same as that in the A-phase, but there is some translational and orientation ordering (see *Smectic liquid crystals*). *Columnar phases* are two-dimensionally ordered columns of disc-like molecules. There is no translational order in the arrangement of molecular mass centers within the columns; i.e. there is two-dimensional crystalline order in combination with one dimensional liquid order (see *Discotics*).

During an increase in temperature the best known mesophases replace each other in the following sequence: Crystal ↔ Smectic B ↔ Smectic C ↔ Smectic A ↔ Nematic (or Cholesteric) ↔ Isotropic liquid (some mesophases can be absent for a particular substance). The regions of the existence of mesophases (from fractions to some tens of degrees) may fall into the temperature range 60–400 °C. *Phase transitions* between the mesophases can be either of first or second order. Between cholesteric and isotropic liquids (in a narrow temperature range) sometimes blue phases occur with exotic and seldom-investigated properties. The types of ordering of lyotropic liquid crystals are mostly the same as in the case of thermotropic liquid crystals. However, the macromolecules or micelles can act as structural units of lyotropic liquid crystals, which renders some specifics to their ordering. For instance, an analog of smectic phases is *lamellar phases*, where the lipid structural units form, within the solvent (water), double layers similar to biological *membranes.* This similarity establishes the connection between liquid crystals and biological substances. Thus, the term "liquid crystal" encompasses a wide class of systems – from single-component, low molecular weight organic substances to complex biological structures. The term is attributed to the German physicist O. Lehmann (1890) who, at the request of the Austrian botanist F. Reinitzer (1888), the discoverer of liquid crystals, carried out the first systematic studies of liquid crystals.

The orientational order in liquid crystals underlies the anisotropy of their physical properties: optical, electrical, magnetic, elastic, etc. (see *Anisotropy of liquid crystals*). The theoretical description of liquid crystals employs the methods of statistical physics and hydrodynamics as well as the language of the *order parameter*, accepted in the theory of ordered systems (see *Order parameters in liquid crystals* and *Elasticity of liquid crystals*). Liquid crystals are systems with interesting topological properties giving rise to various *defects* and *textures* that exist in the field of the order parameter (director), and can be easily observed experimentally (see *Defects in liquid crystals*, *Domain wall in liquid crystals*). Thus, for example, the term "nematic" (fr. Gr. $\nu\eta\mu\alpha$, a thread) originated from the microscopic observation of *disclinations* looking like dark threads floating in the sample. Numerous electrooptic, hydrodynamic, and other effects are observed in liquid crystals (see *Wave-like modulation in liquid crystals*, *Twist effect*, *Frederiks transition*, *Electrohydrodynamic instabilities in liquid crystals*, *Guest–host effect*). These effects underlie different practical applications (see *Defectoscopy*, *Liquid-crystal displays*, *Thermography*, *Thermal indicators*). Liquid-crystal polymers serve as the source material for the manufacturing of super-durable fibers. Some lyotropic liquid crystals are suitable for the simulation of biological objects such as membranes.

See also *Cholesteric liquid crystal*, *Lyotropic liquid crystals*, *Smectic liquid crystal*, *Ferroelectric liquid crystal*, *Nematic liquid crystal*.

LIQUID CRYSTALS, MAGNETOOPTICAL EFFECTS

See *Magnetooptical effects in liquid crystals*.

LIQUID–GLASS TRANSITION

See *Vitrification*.

LIQUID METALS

Opaque liquids with electrical conductivity $\sigma \geqslant 5 \cdot 10^5\ (\Omega \cdot \mathrm{m})^{-1}$. Liquid metals are the melts of *metals* and their *alloys*, as well as melts of some *intermetallic compounds*, *semimetals*, and *semiconductors*.

LIQUID, QUANTUM

See *Quantum liquid*.

LIQUID-PHASE EPITAXY

An epitaxial deposition of a material from a liquid phase, i.e. from melt or solution (see *Epitaxy*). Liquid-phase epitaxy was originally used by H. Nelson (1963) in manufacturing germanium *films* for *tunnel diodes*. Now it is widely used to produce epitaxial films of III–V compounds and their *solid solutions* when creating *heterogeneous structures* to be the basis for *lasers*, *light emitting diodes* and photodetectors, and for the deposition of *magnetic films*. The essence of the technique is the material deposition on a *substrate* from a saturated melt or from a solution during its cooling. Compared to other methods of epitaxial growth, this method has several advantages, such as: simple equipment, high deposition rates, broad possibilities of doping films with different impurities, and the absence of the chemically active, toxic and explosion-prone gases used in *gas phase epitaxy*.

To deposit these epitaxial films, installations with an inclining furnace as well as those with a vertical one are used. In the former, the contact of the substrate with the solution is achieved by tilting; while in the latter, the substrate is immersed into the solution. To obtain multilayer epitaxial structures a multisection arrangement places the substrate into the depression of the movable graphite plate which serves as a bottom of cells filled with different melts. When depositing solid solution layers from a melt, the composition of the layer does not necessarily correspond to that of the dissolved substance. This is because the composition not only depends on the minimum system free energy arising from the chemical components, but it also depends on the mismatch of the of film and substrate crystal lattices, and this gives rise to internal stresses and an associated elastic energy. Therefore, as a result of the correlation between the elastic and chemical energies, the composition of the phase deposition changes when the liquid phase composition is varied (*effect of composition stabilization*). Besides, the composition of the layer being deposited depends on the presence of a region of restricted solubility of the solid solution components. These circumstances impose restrictions on the possibility of producing films of solid solutions of predetermined compositions.

LIQUID SEMICONDUCTORS

Materials possessing semiconductor properties in the liquid state. The process of *melting* oxides, sulfides and selenides (e.g., Bi_2O_3, Sb_2O_3, Sb_2S_3, Sb_2Se_3, etc.) does not cause them to lose their semiconductor properties, but their electrical conductivity does continue to increase with increasing temperature. *Semiconductors* with the diamond and zinc blend structures (e.g., Si, Ge, III–V compounds) do lose their semiconducting properties upon melting, since they experience a sharp increase of their *electrical conductivity* to values typical for *liquid metals*. With increasing temperature, some liquid semiconductors (e.g., Te–Se alloys, rich in Te) gradually lose semiconducting properties and acquire those of metals, while others, such as Se-rich Te–Se alloys, retain their semiconductor properties over a broad temperature range.

In liquid semiconductors, the energy region near the minimum of the *density of states* in the electron energy spectrum acts as a *band gap*. Given a sufficiently deep minimum, there appears in its vicinity a region of almost localized charge carrier states with low mobility (*pseudogap*). With an increase in temperature the pseudogap may disappear, with the liquid semiconductor transforming into a *metal*.

LIQUID STATE MODELS

Descriptions of liquids as systems of ordered microscopic regions. Either a solid or a gas can be taken as a basis for elaborating liquid state theory. Both approaches attempt to explain the ordered state of microscopic regions, or the short-range order (see *Long-range and short-range order*) of atomic arrangements, which have been proven by numerous *X-ray structure analyses*. Liquid state models generally follow the first approach. The second approach deals with statistical relationships for the model of a dense gas, and lacks much substantiation.

Problems of liquid state models that remain to be solved concern the structure of microscopic regions, their lifetime, nature of boundaries, and the formation mechanism for the regions (melting). There are two points of view regarding the structure of microscopic regions. One hypothesis asserts that the liquid consists of crystal "fragments" (*microcrystallites*) that retain a crystal structure. According to the second concept, a liquid is composed of *cybotaxes* (three-dimensional arrangements of molecules) that are intermediate between crystalline and gaseous states in their structural aspect.

There are no boundaries between microscopic regions in the cybotaxic model. In the microcrystallite model, crystal-like regions must have boundaries, but there is scarcely any evidence for their presence. It is clear that much of the heat of melting in this case goes into the generation of these boundaries. Frenkel thoroughly examined plausible reasons for the generation of boundaries and found them reducible to density *fluctuations*, generation of *vacancies*, and systems of microcavities and microcracks (see *Cracks*). Yet, the monotonic increase of the number of crystal *defects* with increasing temperature is inconsistent with the abrupt change of *viscosity* and *density* at *melting*. Microcrystallite boundaries are likely to arise as a result of the *anomalous thermal expansion* of a crystal surface (and at locations of defects). There, the lattice parameters increase more rapidly than they do inside the bulk of the crystal, and *shear* resistance vanishes at the melting temperature. The crystal turns into a liquid owing to the small value of the *shear modulus* along boundaries.

LITHIUM, Li

Element of Group I of the periodic system with atomic number 3 and atomic mass 6.941; it is an *alkali metal*. Natural lithium is a mixture of stable isotopes ^{6}Li (7.42%) and ^{7}Li (92.58%). Electronic configuration is $1s^2 2s^1$. Successive ionization energies are 5.39, 75.64, 122.42 eV. Electron affinity energy is 0.59 eV. Atomic radius is 0.157 nm; ionic radius of Li^+ is 0.068 nm. Oxidation state is +1, electronegativity is 0.95.

In a free form, lithium is a plastic, very soft silvery-white *metal*. It has a body-centered cubic lattice with parameter $a = 0.35023$ nm (at 20 °C), at 78 K it transforms to a hexagonal lattice. Density is 0.539 g/cm^3 (the lowest among all metals) at 273 K. $T_{\text{melting}} = 453.7$ K, $T_{\text{boiling}} = 1600$ K. Heat of melting is 3.0 kJ/mole; heat of evaporation is 133.7 kJ/mole; specific heat $c_p = 24.85$ J·mole^{-1}·K^{-1}, coefficient of thermal conductivity is 71 W·m^{-1}·K^{-1} (from 273 to 373 K). Debye temperature is 370 K. Resistivity is 0.0812 μΩ·m (at 273 K), the average coefficient of resistance is $4.5\cdot10^{-3}$ K^{-1}. Mohs hardness is 0.6, Brinell hardness is 5 MPa; tensile strength is 115 MPa; relative elongation is 50 to 70%. Adiabatic elastic moduli of lithium: $c_{11} = 13.42$, $c_{12} = 11.30$; $c_{44} = 8.89$ GPa at 298 K.

Metallic lithium is paramagnetic and its compounds are diamagnetic; magnetic susceptibility is $+2.04\cdot10^{-9}$ (at 293 K). It is used as an alloying addition to various *alloys*. The most important area of lithium application is nuclear power. Lithium compounds are used in manufacturing special glasses, thermostable porcelain and *ceramics*, etc.

LITHOGRAPHY (fr. Gr. $\lambda\iota\theta o\varsigma$, stone, and $\gamma\rho\alpha\varphi\omega$, am writing)

Historically, a printing technique that uses patterns etched in stone. Modern lithography techniques are used to etch systems of conducting and insulating regions on the surfaces of semiconductor plates to form *integrated circuits*. The surface of the plate is coated with a uniform layer (or several layers) of a *resist* sensitive to various types of radiation (accordingly, one speaks of *X-ray photolithography*, *electron lithography* and *ion lithography*). The intensity and spacial distribution of the irradiation controls the configuration of the protective mask remaining on the surface after processing the plate. It also controls the distribution of impurities in subsequent *doping* of the plate through the processed surface or the surface relief resulting from *etching*. The mask then has to be removed.

LITHOGRAPHY, ION

See *Ion lithography*.

LITTLE–PARKS EFFECT (W. Little, R. Parks, 1962)

A periodic variation ΔT_c in the *critical temperature*, T_c, of a thin walled superconducting cylinder or thin film wire loop that appears during variations of a magnetic field B applied parallel to the cylinder axis. The conditions on the wall thickness d for observing the Little–Parks effect are $d \ll R$, $\xi(T)$, $\lambda(T)$, where R is the cylinder radius; $\xi(T)$ is the *coherence length*, and $\lambda(T)$ is the *penetration depth of magnetic field* into the superconductor. The Little–Parks effect results from the *quantization of flux*. The value $\Delta T_c(B)/T_c$ is proportional to $(n - \Phi/\Phi_0)^2$, where Φ is the *magnetic flux* through the cylinder cross-section, $\Phi_0 = h/2e$ is the fluxoid or *flux quantum*, and n is an integer. For a given B, there is a value of n that minimizes the *superfluid velocity* v_s. According to estimates, $\max\{\Delta T_c(B)\} \approx 10^{-3}$ K.

LOADING, MICROSHOCK

See *Microshock loading*.

LOCAL ANALYSIS

Procedures for determining the chemical composition of microvolumes or thin layers of a *solid*. They are usually applied to retrieve the elemental composition of microphases, of individual inclusions, of micropatches or of thin layers of the *solid surface*. Both the basic components of the sample and impurities (10^{-2}–10^{-4}%) are identified. Various instrumental techniques are used that feature low detection thresholds (10^{-11}–10^{-14} g) and allow carrying out a multicomponent analysis. These techniques are characterized by a certain locality L (i.e. size of area or volume within which the analysis may be conducted with a given accuracy). One may distinguish longitudinal locality $L_{\parallel}$ (depth of analyzed region), and transverse locality $L_{\perp}$ (measured along analyzed surface).

Highly sensitive local analysis techniques ($L < 1$ μm^3) are the X-ray spectrum, laser, cathode luminescence, ionic probing microanalysis, Auger spectroscopy, *characteristic energy loss spectroscopy*, etc. *Layer-by-layer local analysis* is conducted at various depths with a layer step of about 0.1 μm. Beside the above techniques, various others are also applicable, such as atomic absorption, luminescence, radiation activation, etc. They are employed after thin layers of the sample are subjected to chemical or electrochemical dissolution, and the components to be identified are concentrated.

LOCAL BARRIERS for dislocations

Obstacles (*stoppers*) to interactions which result in a haphazard quality of displacements, and slow down the velocity of *dislocation* motion. Local barriers are typical of materials with low Peierls barriers (see *Peierls relief*), especially for *alkali-halide crystals*, and they also have an effect upon the mobility of *dislocations* in crystals with high Peierls barriers. The local barriers usually include vacancy–impurity complexes (see *Vacancy*, *Impurity atoms*), small-size impurity and vacancy *clusters*, peaks of internal uncompensated stress fields (see *Internal stresses*) of dislocations and sometimes of vacancies, *interstitial atoms*, impurity atoms, precipitates of foreign *phases*, and microscopic *dislocation loops*. The nature of local barriers has been studied most thoroughly in *silicon*. The so-called A and B *swirl defects* were found there. According to electron microscopy data, *B-defects* form during crystal growth, and are agglomerates of implanted silicon atoms with an admixture of other elements, e.g., *carbon*, which regulate the growth kinetics and stability. *A-defects* form by way of restructuring the B-defects, and are dislocation loops of an intrusion type. One may consider interactions of dislocations with both mobile and immobile local barriers. The immobility of local barriers stems from their large size and the low temperature of the experiment, when the diffusional motion of atoms is hindered (see *Diffusion*). Local barriers related to peaks of internal dislocation stresses are immobile in principle. Immobile local barriers usually have random positions throughout the crystal bulk, although some ordering can occur. Dislocations penetrate local barriers in either an activated or nonactivated manner during a finite time. The actual physical mechanisms of penetration depend on the barrier nature, and in many cases have not yet been studied. Spatially large local barriers (e.g., dislocation fields of internal stresses) are often avoided by being by-passed (*Orowan mechanism*). According to electron microscopy data, small local coagulation barriers may disintegrate under the forceful action of dislocations.

The mathematical formalism of local barrier penetration is based on geometrical statistics, i.e. on statistical characteristics of a dislocation line configuration. They include probability distributions of the lengths of dislocation segments, and of the angles of coverage of individual local barriers by a dislocation. Simultaneous interactions of a dislocation with a large number of randomly positioned local barriers introduce statistical features to its motion.

LOCAL CORROSION

The *corrosion* of spatially localized regions of a material. Corrosive *failure* may, depending on the properties of the metallic material and the level of hostility of the environment, localize at separate regions of the surface, with other areas remaining practically intact. The characteristic types of local corrosion are contact, crevice, pitting and intercrystalline. *Contact corrosion* appears in a device constructed from *alloys* differing in their electrochemical properties. It may also be observed in a single homogeneous material containing welds or areas of different structural states (stressed, tempered, etc.). *Crevice corrosion* involves an intense selective destruction in a crevice (slit) at the juncture of different parts of a construction, between gaskets and metal, etc. During *pitting (point) corrosion*, some metal spots dissolve at high rates, deeply enough to form penetrating sores. *Intercrystalline corrosion* (the most dangerous of all) develops selectively along the *grain boundaries* and results in the loss of *strength* and *plasticity* of the construction material.

LOCAL DENSITY APPROXIMATION

An approximate method for calculating the exchange-correlation energy E_{XC} of an inhomogeneous electron gas, assuming that it depends on the local charge *density* $\rho(r)$. More specifically, the value of E_{XC} is expressed through the exchange-correlation energy density ε_{XC} of a homogeneous gas at the given point: $E_{XC} = \int \mathrm{d}r \rho(r)\varepsilon_{XC}(\rho(r))$. The local density approximation is at present the main one employed in the practical application of *density functional theory*. Formally, the local density approximation is of zero order in the expansion of E_{XC} in the gradients of the electron density, but the overall area of its applicability is far broader. A drawback of this approximation is the irregular (nonzero) limit for the electron–electron interaction energy when the number of electrons converges to one, and also the exponential, rather than power law, fall off of the effective potential at large distances from the system (the so-called *problem of image forces*). This approximation often gives good results for ground state properties like cohesive energies and charge densities of valence electrons, but the results are poorer for excitation energies.

LOCAL ELECTRIC FIELD

The internal field produced by a sample polarization that affects an atom (or ion). Calculating the local field (E_{loc}) is an important problem in the theory of *insulators*, in particular of *ferroelectrics*, and also the theory of *magnetism*. This electric field arises from fixed charges outside the body (external electric field, E_0) and the sum of all fields of charges within the body itself, taking into account the sample polarization (see *Polarization of insulator*). The local field acting upon an atom at a point r_0 differs significantly from the macroscopic electric field, E, the latter being derived from the values of electric fields averaged over the volume of a *unit cell*. For instance, consider the center of symmetry point in a cubic environment assuming that the crystal itself is spherical in shape. We then have $E = E_0 - P/(3\varepsilon_0)$ (the negative term represents the depolarization arising from the crystal shape). Meanwhile, $E_{loc} = E_0$ owing to the mutual compensation of dipole fields in the sample, so that $E_{loc} = E + P/(3\varepsilon_0)$. Since $P = (\varepsilon - 1)\varepsilon_0 E$, we have $E_{loc} \gg E$ for certain crystals with large relative dielectric constants ε (e.g., ferroelectrics). For a lattice point with an environment other than cubic, the local field is $E_{loc} = E + \gamma P/(3\varepsilon_0)$, where γ is the *Lorentz factor* (see *Lorentz field*). Such a field may strongly differ both in value and sign from site to site in a crystal. The Lorentz model used to derive the above expressions for E_{loc} fails to account for the field produced by the molecule under consideration, i.e. for the self-action of the molecule on its environment (reactive field in the *Onsager theory*). Taking into account the reactive field causes the formula for E_{loc} to differ from the above. Both models yield consistent results only when the electric *dipole moments* are induced by an external electric field. Note that the field acting upon a free carrier always agrees with the macroscopic field, whereas $E_{loc} \neq E$ in the general case for localized electrons (e.g., electrons associated with impurities). That field depends on the radius of the localized electronic state.

LOCAL ELECTRONIC LEVELS

Localized states in the energy spectrum of a semiconductor, which correspond with certain discrete levels in the *band gap* and resonance (quasi-discrete) levels in allowed bands. As to their capability for either accepting an electron from the *valence band* or releasing it into the *conduction band*, the local electronic levels are divided into *acceptors* and *donors*. Certain impurities are *amphoteric* (see *Amphoteric centers*) and may behave either as donors and acceptors. The local electronic levels in crystalline semiconductors are generated by atoms at substitutional or interstitial positions in the lattice as well as by complexes comprising several *impurity atoms*, and by formations involving other lattice *defects* (*vacancies*, *stacking faults*) beside impurities. *Impurity levels* are divided into *shallow levels* and *deep levels* depending on their positions relative to the edges of the band gap, and of the properties of the impurity potential.

LOCALIZATION, ANDERSON

See *Anderson localization*.

LOCALIZATION OF PLASTIC DEFORMATION

Result of a nonuniform distribution of plastic shape changes within the bulk of a deformed material. Localization of plastic deformation accompanies *plastic deformation* of most crystalline bodies, and reflects the deformation behavior of the regions with highest *plasticity*. Different scales of localization of plastic deformation are found, characterized by the corresponding linear dimensions:

(1) submicroscopic (10^{-9}–$5\cdot10^{-9}$ m);
(2) microscopic ($5\cdot10^{-9}$–$5\cdot10^{-8}$ m);
(3) mesoscopic ($5\cdot10^{-8}$–$3\cdot10^{-7}$ m);
(4) quasi-macroscopic (optical) ($3\cdot10^{-7}$–10^{-5} m);
(5) macroscopic (above 10^{-5} m).

Electron microscopy techniques are commonly used to observe directly the localization of plastic strain at the first three levels. Using transmission *electron microscopy*, nonuniformities in the distribution of *defects* are commonly observed, while *replicas* and *scanning electron microscopes* make it possible to analyze microrelief that develops at the free surface of a target under plastic deformation. The localization of the fourth class, observable by an ordinary optical microscope, usually appears as bands that vanish during *polishing* and further *etching*, the so-called *shear bands* or *persistent slip bands*. The latter type is typical of nonstationary regimes of loaded targets. The fifth category of localization, called geometric localization of plastic deformation, appears as patches of more pronounced changes of shape observable by the naked eye. Such patches in the case of stretching are usually called *necks*. If these deformations develop, substances may deform over great distances without significant strain localization. *Superplasticity* is often observed in such substances. When plastic strain localizes as a result of changes in material structure or loading regime, or when a cellular dislocation structure appears (see *Fragmentation*), there is usually an accompanying loss of superplasticity. The most dangerous case is the fifth degree of localization since the appearance of a neck sharply lowers the level of deformation which leads to *failure*. The cause of the localization lies in the spatial heterogeneity of crystalline materials, which results not only from the presence of defects and impurities, but is also closely connected with the crystalline structure of a *perfect crystal*.

LOCALIZATION, SCALING THEORY

See *Scaling theory of localization*.

LOCALIZATION, SELF-

See *Self-localization*.

LOCALIZATION THRESHOLD

The same as *Mobility edge*.

LOCALIZATION, WEAK

See *Weak localization*.

LOCAL SYMMETRY

Relations of a lattice defect with its surrounding *ideal crystal* (molecule) which remain unchanged during particular rotations, reflections, inversions and rotations with inversions. The set of such transformations constitutes a *local symmetry group* (see *Defect symmetry group*).

LOCAL SYMMETRY GROUP

A symmetry *point group* with operations which leave the interactions of a lattice defect with a crystal (molecule) unchanged. The local symmetry group for a *point defect* coincides with the point group of the lattice site where the defect is located. For more complex defects the operators of the local symmetry group provide for the invariance of interactions within the complex, as well as those of the complex as a whole with the crystal. The local symmetry groups of interactions of a defect with nuclei (ions) of the environment retain the direction of the defect–nucleus axis, and are always subgroups of its symmetry *space group*, but not always those of its point group or symmetry class. All local symmetry groups for each of the 230 space groups are systematized in the *International Tables for Crystallography* (see also *Local symmetry*).

LOCAL VIBRATIONS

Impurity-induced modes of vibrational states in imperfect crystals. The frequencies of such modes lie outside the spectral vibrational bands of the unperturbed crystal. When the frequencies lie in the forbidden region between the optical and the acoustic vibrations, they are called *gap vibrations*. The amplitudes of local vibrations fall off exponentially as the distance from an *impurity atom*, while the radius of its state grows as the frequency of local vibration nears one of the band edges. In three-dimensional crystals local vibrations occur when the masses of the impurity atoms or their interatomic interactions differ significantly from the corresponding values for the surrounding lattice or matrix. When the spring constants describing the interaction are similar, the impurity atom must be lighter than the lattice atom).

Vibrational, rotational, and librational modes of impurity molecules in a crystal may also be treated as local vibrations. The basic mechanism to control the spectral broadening of local vibration frequencies is the decay of local vibrations into acoustic *phonons* of the unperturbed crystal, and also *phonon scattering* by local vibrations (*modulation broadening*).

If the vibrational frequencies of the impurity fall within the phonon spectrum of the unperturbed crystal, their broadening is small, and the respective modes are called *quasi-local vibrations* or *resonances*. Such vibrations result in the appearance of rather sharp peaks in the *density of states*. Provided the concentration of quasi-local vibrations is high enough, a complete restructuring of the phonon spectrum in the corresponding frequency range may take place.

Local and quasi-local vibrations can result in the appearance of new effects in imperfect crystals (e.g., maxima in the impurity *specific heat*), and are found in many diverse types of crystals during studies of the cross- section of *inelastic neutron scattering*, IR absorption, and *Raman scattering of light*.

LOMER–COTTRELL BARRIER (W.M. Lomer, A.H. Cottrell)

A sedentary *dislocation* consisting of three *partial dislocations*, b_1'', b, b_2'' linked by *stacking faults* (wavy lines in the figure) in intersecting planes P_1 and P_2. Such a barrier forms due to the reaction between two parallel dislocations with *Burgers vectors* b_1' and b_2' belonging to two different split dislocations $b_1 = b_1' + b_1''$ and $b_2 = b_2' + b_2''$. The product of that reaction, the dislocation $b = b_1' + b_2'$, is called a *vertex dislocation* or *stair-rod dislocation*. For example, in a face-centered cubic lattice, the Thompson tetrahedron notation for the reaction of dislocations $b_1 = A\delta + \delta B$ in the d plane and $b_2 = B\alpha + \alpha D$ in the a plane yields a

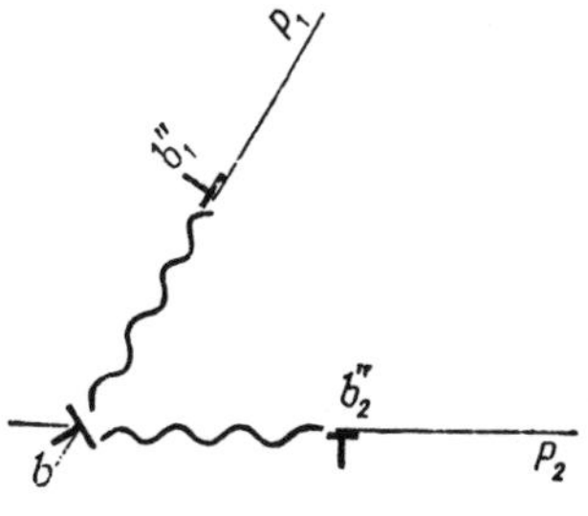

Lomer–Cottrell barrier.

Lomer–Cottrell barrier $A\delta + \delta\alpha + \alpha D$, the respective stair-rod dislocation $\delta\alpha$ having its Burgers vector $b = (a/6)[\bar{1}0\bar{1}]$. Since the Lomer–Cottrell barrier cannot slip in either of the planes P_1 or P_2, it blocks both planes. The formation of this barrier is an efficient means of *strain hardening* of crystals for the case of multiple slips.

LONDON PENETRATION DEPTH

See *Penetration depth of magnetic field.*

LONG-RANGE AND SHORT-RANGE ORDER

Order and symmetries (or correlations) in the spatial location of the elements of a solid structure (atoms, molecules, atomic groups of particular configurations, etc.) or the space correlations of values of physical parameters (e.g., magnetization $\boldsymbol{M}$, polarization $\boldsymbol{P}$). The pair *correlation function* in coordinate space $R_\Phi(\boldsymbol{r}) = \langle\Phi(\boldsymbol{x}), \Phi(\boldsymbol{x}+\boldsymbol{r})\rangle$, where $\Phi(\boldsymbol{x})$ is a physical quantity (atomic equilibrium position, local value of magnetic moment, etc.) and the angular brackets mean averaged over the coordinates x, is often used for describing the order. The degree of ordering is characterized by the *correlation radius* r_c, by the symmetry properties of $R_\Phi(\boldsymbol{r})$, and by the values of the *order parameters*. The correlation radius determines the average size of the area where $R_\Phi(\boldsymbol{r})$ is not negligibly small. If $\lim_{r\to\infty} R_\Phi(\boldsymbol{r}) \neq 0$ then $r_c \to \infty$, and the structure has *long-range order* with respect to the property Φ. If $r_c \sim a$ (a is the average interatomic distance) then only short-range (or local) ordering exists. If r_c has a finite value, considerably exceeding a, then the structure has *intermediate order*. If $\Phi(\boldsymbol{x})$ has some symmetry properties, then $R_\Phi(\boldsymbol{r})$ inherits these properties. The structure with lower symmetry is considered to be more ordered. The structure of $R_\Phi(\boldsymbol{r})$ with a single-axis of symmetry, e.g., where there is invariance with respect to rotations around some space axis, is more ordered, than an isotropic structure ($R_\Phi(\boldsymbol{r})$ which depends only on $|\boldsymbol{r}|$). The part of $R_\Phi(\boldsymbol{r})$ which does not disappear at $r \to \infty$ (let us designate it as $R_\Phi^a(\boldsymbol{r})$) may be the superposition of functions which differ by their symmetry types, or belong to the same symmetry types but with different parameters of the symmetry group $R_\Phi^a(\boldsymbol{r}) = a_1 R_{\Phi 1}^a(\boldsymbol{r}) + a_2 R_{\Phi 2}^a(\boldsymbol{r}) + \cdots$, where $R_{\Phi k}^a$ $(k = 1, 2, \ldots)$ are normalized functions. The coefficients a_k are called *long-range order parameters*, or simply *order parameters*. The structural order is usually determined by the function $R_{\Phi k}^a(\boldsymbol{r})$ of the lowest symmetry. The values of the order parameters a_k are measures of ordering. Among the structures with the same symmetry properties the one with the larger order parameter is more ordered. In the case of the same composition but different symmetries the structural states of the solid are called *phases*. In addition to long-range order parameters, sometimes *short-range order parameters* are introduced which reflect the measure of correlations over short distances (usually within the first *coordination sphere*). The order parameters are usually determined by atom-to-atom interactions, and depend on temperature (T), pressure (P), specific volume (V) and external forces. A jumpwise changing of the order parameters (or their derivatives) with variations of P, T, V or external forces is accompanied by jumpwise variations of different physical values, and this is called a *phase transition*. The values of P, T, V or those of applied external forces are called *critical parameters*. At a *first-order phase transition* a jumpwise variation of the order parameter occurs, and at a *second-order phase transition* some order parameter equals zero at one side of the *critical point*, differs from zero to the other side of it, and tends toward zero at the critical point itself.

Topological order is described by the correlations in the mutual locations of atoms. If $\Phi(\boldsymbol{x})$ is a function of the distribution of atoms, then $R_\Phi(\boldsymbol{r}) = \bar{\rho}^2 + f_\rho(\boldsymbol{r})$, where $\bar{\rho} = \langle\Phi(\boldsymbol{x})\rangle$ is the average density of the atoms, and the function $f_\rho(\boldsymbol{r}) = \langle\Phi(\boldsymbol{x}) - \bar{\rho}, \Phi(\boldsymbol{x}+\boldsymbol{r}) - \bar{\rho}\rangle$ describes the correlations of deviations from the average value. In liquid and amorphous solid phases the second term tends toward zero as $r \to \infty$, and its behavior at low r determines the short-range and intermediate order. In the ordered crystalline phase $f_\rho(\boldsymbol{r}) \neq 0$ at $r \to \infty$, and its symmetry is determined by the lattice symmetry. The variation of this symmetry is associated with a phase transition. For the analysis of order in solids the isotropic part of $R_\Phi(\boldsymbol{r})$, or the radial atomic *distribution function*, is often used.

Compositional order is determined by the correlations of the mutual ordering of atoms of different types in polycomponent solids. The variations in compositional order may occur without variations or with negligibly small variations of topological order, e.g., in ordered alloys (see *Alloy ordering*). Let us consider the alloy Cu–Zn with equal concentrations of copper and zinc atoms, with a simple cubic lattice and therefore topologically ordered at temperatures below the *melting temperature*. At a temperature close to zero the alloy is compositionally ordered: each atom of copper is surrounded by six atoms of zinc and each atom of zinc is surrounded by six atoms of copper, so there is *short-range order*. Besides this, the atoms of copper and zinc individually form cubic sublattices of a doubled (compared to the source lattice) period, i.e. there is also long-range compositional order. It is possible to introduce parameters to specify the long-range and short-range orderings. Let us designate by indices α and β the values related to the sublattices of copper and zinc (respectively). Let us assume that $w_{\alpha\mathrm{Cu}}, w_{\alpha\mathrm{Zn}}$ are the probabilities of locations of copper and zinc atoms, respectively, on the α-sublattice, with $w_{\beta\mathrm{Cu}}, w_{\beta\mathrm{Zn}}$ those values for the β-sublattice. The parameter for the long-range order η is determined in the following way:

$$\eta = 2\left(w_{\alpha\mathrm{Cu}} - \frac{1}{2}\right) = 2\left(w_{\beta\mathrm{Zn}} - \frac{1}{2}\right).$$

It can be seen that for $w_{\alpha\mathrm{Cu}} = 1$ and $w_{\beta\mathrm{Zn}} = 1$ we have $\eta = 1$ so all the points of the α-sublattice are occupied by the copper atoms, and the points of the β-sublattice are occupied by the zinc atoms. This might be the state of maximum long range order at absolute zero temperature. With increasing of temperature there will be interchanges of atoms of both types from one sublattice to the other, and at temperatures above some value T_c (the critical point) $w_{\alpha\mathrm{Cu}} = w_{\alpha\mathrm{Zn}} = w_{\beta\mathrm{Cu}} = w_{\beta\mathrm{Zn}} = 1/2$, so η becomes zero and the alloy becomes a compositionally disordered *solid solution* with the conservation of topological order.

Magnetic ordering. The magnetic moment of the atom $\boldsymbol{M}(\boldsymbol{x})$ located at point $\boldsymbol{x}$ interacts with the magnetic moments of the other atoms, due to which magnetic ordering can occur. When the correlation radius of the magnetic moment tends to infinity and $R_M(\boldsymbol{r}) = \langle \boldsymbol{M}(\boldsymbol{x}), \boldsymbol{M}(\boldsymbol{x}+\boldsymbol{r})\rangle \neq 0$ there exists long-range magnetic order. For uniform *magnetization* $R_M^a = \langle \boldsymbol{M}(\boldsymbol{x})\rangle^2 = S^2$ in periodic *modulated magnetic structures* $R_M^a(\boldsymbol{r}) = S^2 + a_k^2 f_k(\boldsymbol{kr})$, where f_k is a periodic function with period 2π. The quantities S and a_k are order parameters, and $\boldsymbol{k}$ is a parameter of the translation group. At $S = a_k = 0$ the body is magnetically disordered. The temperature at which S becomes different from zero is called the *Curie point*. Magnetic ordering can occur in topologically ordered crystalline phases, as well as in amorphous materials which lack topological order.

The ordering of the polarization vector $\boldsymbol{P}(\boldsymbol{x})$ in *ferroelectrics* is similar to the ordering of $\boldsymbol{M}(\boldsymbol{x})$ in *magnetic substances*. *Orientational order* is displayed in the correlations of vectors which determine the orientations of structural elements (e.g., molecules) which form a solid. Orientational order may be present together with topological disorder, as occurs in *quasi-crystals*. Besides this, topological ordering of molecules may coexist with their orientational disordering (e.g., the orientationally disordered *molecular crystal* N_2).

LONG-TERM HARDNESS, long-term high-temperature hardness

An auxiliary method for estimating the potential *high-temperature strength* and other high-temperature mechanical properties of a material from the time dependence of the imprint size during the measurement of high temperature *hardness* under permanent load at constant temperature. This depends on the surface state of the material. For measurements below 2300 K, the *method of static indention* is applied. An indenter in the form of the regular four-face pyramid with the angle 136° between opposite faces is made of sapphire (Al_2O_3) for use up to 2000 K, and of boron carbide (B_4C) for higher temperatures, where for the latter HV $= 1.854P/b^2$, where P is the load at the indenter, and b is the arithmetic mean of the diagonals of imprint after release. Above 2300 K the hot strength is determined by the method of one-sided flattening of a conical sample (cone angle 120°) on the plane surface of the punch. $H_c = 1.273P/d^2$, where d is the mean diameter of the imprint. The long-term hardness H was found to vary linearly with the *long-term strength* for a series of alloys. Three subsequent tests of high temperature hardness, performed with a factor of ten increase in exposure time at each sequential test (0.5, 5 and 50 min), provided indention curves or curves of hardness versus time. If $H = ar^n$, where a and n are the material constants dependent on its nature, n being the rate index, then one can compare the characteristics of long-term strength using the data on the long-term hardness.

LONG-TERM STRENGTH

The material *strength* sustained under application of permanent, long-term stress. As a characteristic of long-term working ability of, mainly, metallic materials, this is particularly important in thermal power engineering, gas-turbine construction, materials for aircraft and spacecraft engineering, and construction (concrete). A principal working characteristic of materials for long-term service life is their ultimate long-term strength σ_l^T (here T means the temperature, l is longevity) which provides a numeric value of stress that produces a fracture at working temperature within a given time. There is no theoretical limit of long-term strength. Its characteristics are determined from laboratory tests of standard samples under the conditions of single-axis tension by special test machines which maintain constant load and temperature during the test (10^4 h and more). The test results are approximated graphically by curvilinear or rectilinear (with turning points) diagrams in the coordinates "stress versus logarithm of time prior to *fracture*", and they are then extrapolated using temperature–time techniques from the theory of *plasticity* to the longevity corresponding to the working conditions (10^5 h and more). During the process of *creep* that precedes fracture one can observe a significant change in the structure, the phase composition, and the dislocation substructure of a *metal*. This makes it difficult to construct physically justifiable equations for the conditions of long-term loading and elevated temperatures. Therefore, the predictions of long-term strength made on the basis of shorter duration tests remain very approximate, and can be misleading.

LONSDALITE (after English crystal-chemist Dame Kathleen Lonsdale)

An allotropic form of *carbon* which was synthesized in 1965 and found in an iron meteorite in 1967. Density of lonsdalite is 3.51 g/cm^3. A framework of tetrahedrically coordinated carbon atoms bound to each other by sp^3-hybrid *orbitals* may be obtained in two ways. One of these yields a diamond-like cubic lattice where the packing order of atomic layers in the direction $\langle 111 \rangle$ follows the sequence ABCABC (Fig. 1(a)).

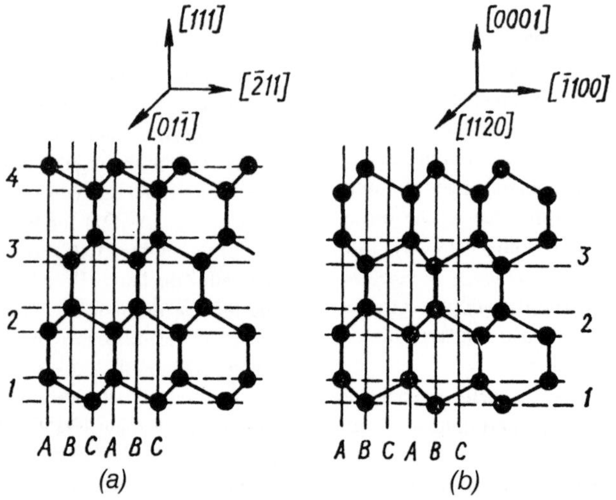

Fig. 1. Projection of atomic grids of the cubic diamond (a) and lonsdalite (b) lattices.

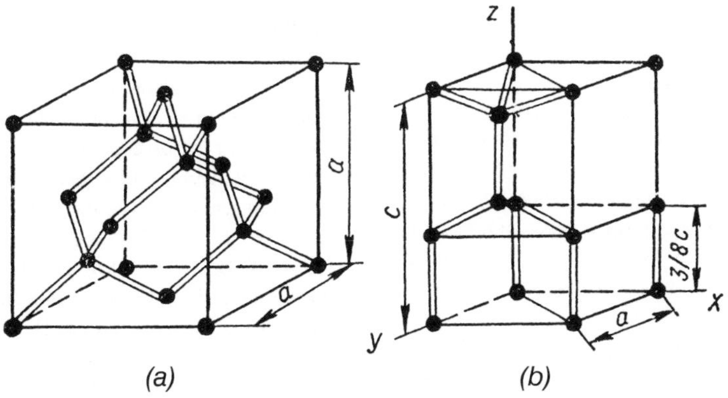

Fig. 2. Unit cells of the cubic diamond (a) and lonsdalite (b) lattices.

By changing the sequence of atomic layer packing to ABAB (Fig. 1(b)), the cubic structure is transformed to a "hexagonal" wurtzite-like modification. Concerning its interatomic bonds and their spatial arrangements lonsdalite is close to the cubic modification of *diamond* (Fig. 2(a)), but in the cubic modification all layers built of tetrahedra are oriented in the same direction, while in lonsdalite each successive tetrahedric layer is turned by 60° relative to the previous one. The lattice of lonsdalite may be thought of as a combination of two hexagonal close-packed sublattices displaced relative to one another by 3/8 of the lattice parameter c (Fig. 2(b)).

Lonsdalite, with symmetry *space group* $P6_3/mmc$ (D^4_{6h}), *crystal lattice constants* $a = 2.51$ Å, $c = 4.12$ Å, is unobtainable in pure form. Its content in natural impact diamonds is as high as 50%, in compression-synthesized ones it is up to 50%, and in those formed by static the compression of *graphite* it is up to 90%. Lonsdalite transforms to graphite at far lower temperatures than diamond; hence, to obtain lonsdalite through synthesis under dynamic conditions, effective heat withdrawal is necessary.

LORENTZ FIELD (H.A. Lorentz, 1934)

An electric field that acts upon an atom of an *insulator*, and is derived from polarization charges at the surface of a fictitious spherical cavity centered at the atom under consideration. Introducing such a cavity with a radius of about 10 lattice constants is an artificial mathematical tool which enables one to calculate the fields from the dipoles present inside the cavity, and the fields that result from the remainder of the sample by integrating over the effective surface charge density. The Lorentz field $E_{\rm L}$ turns out to equal to $P/(3\varepsilon_0)$ for the simple case when the neighbors around a given point in the lattice have cubic symmetry (see *Cubic system*). Here P is the polarization or dipole moment per unit volume of the crystal. Then the *local electric field* acting upon the atom is $E_{\rm loc} = E + P/(3\varepsilon_0)$ (*Lorentz formula*, where E is the average macroscopic field). In view of the relation $P = (\varepsilon - 1)\varepsilon_0 E$ between P and E, we find that $E_{\rm L} = (\varepsilon - 1)E/3$, and $E_{\rm loc} = (\varepsilon + 2)E/3$. For highly polarizable crystals ($\varepsilon \gg 1$), $E_{\rm L} \gg E$ and $E_{\rm loc} \approx E_{\rm L}$. For crystal field symmetries lower than cubic the Lorentz formula is more complicated: $E_{\rm loc} = E + \gamma P/(3\varepsilon_0)$, where γ is the *Lorentz factor*, with sign and magnitude depending on the lattice angle. The Lorentz field plays an important role in explaining the nature of *ferroelectric phase transitions*. Thus, the interactions of ions from different cells that seek to displace the atoms in a single direction reduce to the difference between the total field acting upon an ion and the macroscopic field averaged over the cell in the Lorentz field potential.

LORENTZ FORCE

The force $\boldsymbol{F}$ acting upon a particle of charge e that moves at the velocity $\boldsymbol{v}$ in an electromagnetic field:

$$F = e(E + v \times B),$$

where E is the electric field strength and B is the *magnetic flux density*. The effect of magnetic fields upon moving charged particles results in the redistribution of current across the conductor cross-section, giving rise to various *thermomagnetic phenomena* and *galvanomagnetic effects*.

LORENTZIAN SHAPE, Lorentzian

A spectral line profile described by the expression $I(x) = I_m\delta^2/(\delta^2 + x^2)$, where I_m is the maximum amplitude (for $x = 0$), and δ is the half-width at half-maximum. When presenting the line profile via first or higher order derivatives, the Lorentzian shape is described by the corresponding derivatives of the above expression. As a rule, various physical mechanisms result in a Lorentzian shape and produce *homogeneous broadening of lines* (e.g., *exchange interactions* in solids, and Brownian motions in liquids). Odd moments, $n = 1, 3, 5\ldots$, vanish from symmetry, and even moments $\langle(x - x_0)^n\rangle$ are infinite for $n \geqslant 2$. To avold this infinity one sometimes uses a so-called *truncated Lorentzian shape* which is finite within the interval $-a \leqslant x \leqslant a$, where $a \gg \delta$, and zero otherwise.

LOSCHMIDT NUMBER, Loschmidt constant (J. Loschmidt)

The number, $N_L = 2.68676{\cdot}10^{25}$ m^{-3}, of molecules in 1 m^3 of a substance which form an ideal gas under normal conditions (pressure $P = 101325$ Pa $= 760$ mm Hg $= 1$; temperature $T = 273.15$ K $= 0\,°$C). The Loschmidt number is proportional to the *Avogadro number*, $N_L = N_A/V_0$, where $V_0 = 22414.1$ cm^3 is the molar volume, or the volume of one mole of an ideal gas under normal conditions.

LOSSEV EFFECT (O.V. Lossev, 1923)

The effect of *electroluminescence* produced by radiative *recombination* during *injection* of nonequilibrium current carriers into a semiconductor via a *metal–semiconductor junction*, or via a *semiconductor junction*.

LOW-ENERGY ELECTRON DIFFRACTION (LEED)

The phenomenon of coherent scattering of electrons with energy of about 10 to 100 eV (the length of the *de Broglie wave* is approximately 0.1 nm), widely used for the structural analysis of *solid surfaces*. The physical cause of its high sensitivity is the fact that the *mean free path* of the electrons due to their low energy is 0.5 to 1 nm due to inelastic scattering (with the formation of *plasmons*). For observation of low-energy electron diffraction usually the reflection system of an electronograph (see *Electron diffraction analysis*) with grids and screen is used (see Fig.). Electrons which lose energy are filtered out and do not reach the screen. Thus the observed diffraction picture is formed by the electrons that are scattered elastically from the several atomic planes closest to the surface. Like the case of diffraction from a planar lattice, the LEED picture is observed for any value of the electron energy. However, the intensity of diffraction beams is a maximum when the scattering angles and energy satisfy the Laue conditions for a three-dimensional lattice. Analy-

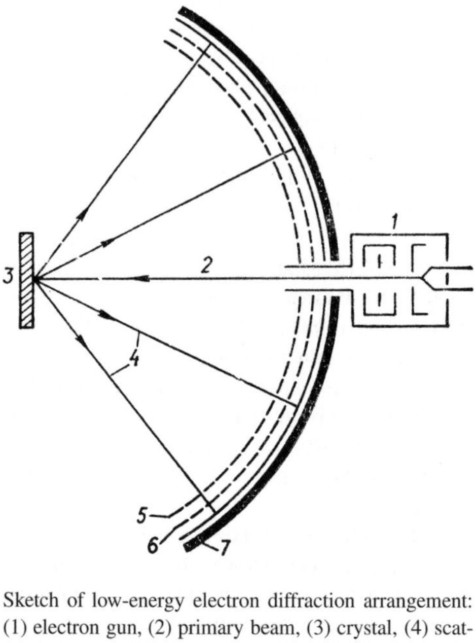

Sketch of low-energy electron diffraction arrangement: (1) electron gun, (2) primary beam, (3) crystal, (4) scattered beams, (5) screening grid, (6) filtering grid, (7) screen.

sis of the geometry of LEED pictures allows a simple determination of the symmetry and the periods of the surface lattice. To obtain the chemical composition of the *unit cell* and of the coordinates of its atoms in the general case requires the analysis not only of the positions but also of the intensities of the diffraction beams. This may be performed only on the basis of the dynamic theory of LEED because, due to the high probability of elastic scattering of low-energy electrons and the effects of multiple scattering, the usual kinematic theory of diffraction which neglects these effects is practically inapplicable. The measured dependences of the diffracted beams on the electron energy is compared with those calculated for various hypothetical positions of atoms in the surface lattice, while the generation of the theoretical curves requires time-consuming calculations that can be carried out only for the simplest cases.

In spite of these difficulties, the LEED method finds its wide application in studies of the physics and chemistry of surfaces. With its help the phenomenon of *surface reconstruction* has been developed, namely the formation at the surface of a lattice, that differs from the one in the bulk. Also surface *phase transitions* are successfully investigated (at clean surfaces, and in adsorbed layers), as well as the dynamics of thermal vibrations of surface atoms, the processes of formation of *thin films*, and other *surface phenomena*. The discovery of LEED by C.J. Davisson and L.H. Germer (1927) was the first experimental proof of the existence of the wave properties of microparticles.

LOW-ENERGY IONS

Ions with energies $\leqslant 10^4$ eV used in the *ion implantation doping* method for modifying the properties of thin subsurface layers to produce contacts, superfine p–n junctions (in radiation detectors, solar photoconverters, etc.), elements of chips, conducting and insulating coatings, and control for the height of *Schottky barriers*. The low-energy ions possess quite short mean free paths (100–10 nm or less) and, accordingly, small statistical dispersion of the paths. This is of importance when producing miniature *integrated circuits* with high packing density by ion doping. In view of the reduced probability of formation of a so-called disordered region, *point defects* are the prevailing type of radiation-induced damage below a certain value of the energy transferred at collisions of ions with target atoms. In the case of crystalline targets, the decrease of the bombarding *ion* energy lowers the probability of the ions to enter lattice channels, and increases the critical angles for *channeling*. There is interest in low-energy ion doping because of the need to produce submicron elements with gaps between them in modern chips, and also in connection with the development of molecular beam *epitaxy*. In some cases (e.g., manufacture of low-cost solar photoconverters, or doping of *metals* and *amorphous semiconductors*), low-energy ion implantation can be carried out with a gas discharge, or with the help of the simplest high-current ion guns without mass separation, which makes the technology easier and cheaper. Low-energy ion bombardment of heated targets causes an accelerated impurity *diffusion* which can be used as an independent method of ion doping (*ion injection*).

LOWER CRITICAL FIELD, first critical field

The temperature dependent maximum value of the external magnetic field strength, at which the full *Meissner effect* may still be observed in *type II superconductors*. The value of the lower critical field (B_{c1}) at absolute zero ($T = 0$ K) is given in the BCS theory by the expression $B_{c1} = [\Phi_0 \ln(\lambda/\xi)]/(4\pi\lambda^2)$, where Φ_0 is the *flux quantum*, λ is the *penetration depth of magnetic field* into the superconductor, and ξ is the *coherence length*. Since $\lambda > \xi$ in type II superconductors, B_{c1} is smaller than the *thermodynamic critical magnetic field* $B_c = \Phi_0/(2\sqrt{2}\pi\lambda\xi)$, which in turn is smaller than the *upper critical field* $B_{c2} = \Phi_0/(2\pi\xi^2)$.

LOW-FREQUENCY ENHANCED MAGNETIC RESONANCE ABSORPTION

A combined technique of resonance and *nonresonance absorption*. When a *paramagnetic resonance* is saturated with a high-frequency magnetic field B_1 (while energy is simultaneously absorbed from a low-frequency magnetic field parallel to the external static *magnetic field* B_0), there can develop, due to a change in the spin–spin reservoir temperature of the paramagnetic system (see *Dipole–dipole reservoir*), quite a strong amplification effect in the low-frequency *absorption* and dispersion that can reach two to three orders of magnitude. This technique can be employed to study low-frequency internal motion (*spin–spin interactions*, energy level crossings, *cross-relaxation*, etc.), and also to determine the temperatures of low-frequency *heat sinks* associated with local fields.

LOW-SPIN COMPLEXES

See *High-spin complexes.*

LOW-SYMMETRY EFFECTS in magnetic resonance

The totality of low-symmetry effects observed in the angular dependences (see *Angular dependence of spectra*) of the generalized *spin Hamiltonian* of *electron paramagnetic resonance*, *electron–nuclear double resonance*, *nuclear magnetic resonance*, and nuclear gamma ray resonance (see *Mössbauer effect*) spectroscopies. The signs of low symmetry in electron paramagnetic resonance are as follows: transition-to-transition disagreement between the extrema in the angular dependences of the resonance magnetic field, B_0; nonorthogonal orientations of the constant magnetic fields corresponding to extrema of one and the same transition; asymmetry in the angular dependences of B_0; frequency dependences of the asymmetry parameter and of the shifts of extrema; mismatch between the extrema found in the angular dependences of B_0 that correspond to one and the same transition for different centers occupying identical *crystal lattice* sites; mismatch between the extrema in the angular profiles of B_0 and those of line widths, intensities or *spin–lattice relaxation* times; mismatch between the extrema in the angular dependences of any spectral characteristic, and the specific crystallographic directions determined through either radiography or optical techniques. The resonance frequencies of electron–nuclear double resonance and nuclear magnetic resonance also display similar features. The study of low symmetry effects makes it possible to determine all the parameters of a generalized spin Hamiltonian. It also permits *magnetic resonance spectroscopy* techniques to evaluate which of 11 sets corresponds to the *symmetry group* of the spin Hamiltonian of a given *paramagnetic center.*

LOW-TEMPERATURE ANOMALIES in amorphous solids

Significant differences of thermal properties, and such characteristics as propagation velocity, *sound absorption*, and *internal friction* in amorphous solids compared to related parameters in crystalline *solids* at low temperatures (usually below 1 K). Low-temperature anomalies, observed in both insulating and metallic amorphous materials, point to certain structural features that are typical of all *disordered solids*, and that remain insensitive to the nature of their interatomic bonds.

The *specific heat* c_p of dielectric (insulating) glasses and *metallic glasses* in their superconducting state (when the contribution of electrons to the specific heat is negligible) is proportional to the first power of the temperature, T, rather than to T^3, as is the case for crystals in which phonons provide the main contribution to the specific heat. The *thermal conductivity* is proportional to T^2. The temperature dependence of the *sound* velocity (longitudinal and transverse) follows a $\ln T$ relation. In addition, resonance absorption of ultrasound is observed, and at low sound intensities the absorption factor is proportional to $\nu \tanh[h\nu/(2k_B T)]$ (ν is the sound frequency). Resonance absorption becomes saturated at high sound intensities (absorption factor decreases with growth of intensity). The characteristic value of the intensity at which sound is absorbed is noticeably higher for metals than it is for dielectric (insulating) glasses. *Nonresonance absorption* in metallic glasses is proportional to ν and independent of T, while that in dielectric glasses is proportional to T^3.

The above low-temperature anomalies can be explained by the presence of low-energy excitations of a specific type in amorphous materials, and such excitations are related to *two-level systems* (TLS). The latter are atomic configurations with closely lying energy levels (spaced by 10^{-4} eV) capable of communicating with each other via tunneling. Differences between the temperature dependences of the nonresonance absorption factors of sound in metallic and dielectric glasses are explained by the fact that in metals *conduction electrons* play a significant role in interacting with TLS. The features of the atomic structure responsible for various low temperature anomalies in amorphous materials remain mostly obscure. The *free volume model* relates the presence of TLS to the appearance of regions containing free volume, while *polycluster amorphous solids* may have cluster boundaries with unmatched cross links partially occupied by atoms.

LOW-TEMPERATURE DEFORMATION

Deformation at temperatures below the Debye temperature (usually at 77, 20.4, 4.2 K), carried out under conditions of suppression of dynamic recovery, involving an appreciable accumulation of *crystal lattice* defects, and a build-up of the latent energy of *strain*. Low-temperature deformation is a source of information on the mechanical properties of *solids* at low temperatures (see *Low-temperature physics*), and a powerful tool used to influence the defect structure and physical properties of samples (including *high-temperature strength*). With temperatures decreasing and approaching absolute zero, the mechanical properties of certain metals (Al, Cu, etc.) improve, while other metals (Fe, Mo, etc.) become brittle (see *Brittle failure*). Besides, such features as mechanical twinning, stepwise flow, interaction of dislocations with the electron system, all become more prominent. Finally, the speed of dislocation motion decreases, *plastic deformation* becomes more and more controlled by quantum mechanical tunneling (see *Tunneling phenomena in solids*) and *zero-point vibrations* of dislocations, rather than by thermally activated processes. Related to low-temperature deformation are the conditions for the development of a high-strength crystalline state with a *strength* 1/3 of its theoretical limit, attained with the ability to withstand significant elastic strains (up to 1%). Certain phenomena are typical for low-temperature deformations: *polymorphism*, *low-temperature recrystallization*, depletion of *dislocations* during *creep*, anomalies in the *yield limit*, *dehardening* during transition to the superconducting state (see *Superconductivity*), slip along the boundaries of twins (see *Twinning of crystals*), etc.

LOW-TEMPERATURE LUBRICANT

A layer of material spread over the *friction* surfaces of moving joints in machines and mechanisms operating at low temperatures (below 120 K). It has low *shear* resistance at low temperatures, decreases the friction force and rate of wear of joined surfaces within the full operating temperature range (from room to low), and withstands multiple cooling and heating. Widely used are: films of soft metals, mostly those having a face-centered cubic lattice (gold, silver, nickel, etc.); solid lubricating composite *coatings* with a hexagonal close-packed lattice (graphite, disulfides and diselenides of Mo, W, Nb. etc., and others), in combination with polymer binders; coatings produced by *chemical heat treatment* of the surface (sulfurizing, selenizing, etc.); metal–polymer coatings (porous metal layer saturated with fluoroplastic or other *polymers*); magnetic powder materials (mixtures of antifriction and magnetic powders), etc.

LOW-TEMPERATURE LUBRICATION

Techniques for transferring lubricants to low-temperature friction joints in order to prevent gripping and to ensure reliable operation of machines and mechanisms at low temperatures. Widely used are the preliminary application of a lubricant to a friction surface; offset-duplicator feeding of lubricant during the operation proper; manufacturing elements of friction joints from *composite materials* containing a lubricant; bringing the lubricant to and applying it at the friction surface by magnetic forces.

LOW-TEMPERATURE PHYSICS

A branch of condensed matter physics involving the investigation of phenomena and properties of materials at *low (cryogenic) temperatures*. (The 13th Congress of the International Low Temperature Institute (1971) recommended calling temperatures below 120 K cryogenic.) Investigations of liquid and solid *helium*, of *superconductivity*, of *cryocrystals* and cryogenic liquids, of electronic properties of *metals* and *semiconductors*, of low-temperature *magnetism* etc. are considered as low-temperature physics. The boundaries of low-temperature physics have not been universally established, but depend on the phenomenon under consideration. Very often low-temperature properties are observed under conditions when the investigated interaction energy appears to be higher than, or at least comparable to, thermal energy. Experiments carried out under this condition often provide the most dependable data for understanding many phenomena.

The reduction of the temperature leads to a lessening of the thermal motion of atoms and molecules, and at absolute zero, all materials appear to be in the ground state, characterized by zero entropy, and by a minimum value of their internal energy. The approach to this basic state creates conditions for the direct manifestation of the quantum nature of matter in the properties of various materials. The discoveries of many unusual phenomena, caused by the action of quantum effects at the microscopic level: *superfluidity*, *superconductivity*, *Josephson effects*, *quantum diffusion*, etc., result from this fact.

Increasing the temperature can be accompanied by the appearance of strong excitations. The main methods of low-temperature physics are based on the involvement of weakly excited states, the thermal behaviour of which can be close to the properties of elementary excitations (*quasi-particles*). For the description of particular low-temperature systems corresponding types of quasi-particles are often used: *phonons* and *rotons* in liquid helium; electrons and *holes* in *metals* and *semiconductors*; *excitons* and *librations* (see *Cryocrystal*) in *insulators*; *magnons* in *magnetic materials*, etc. Their energy spectrum is the main parameter of the quasi-particles, i.e. the dependence of the energy on the momentum (*quasi-momentum*). If the energy spectrum of all the quasi-particles of a system is known, its thermodynamic properties may be calculated directly by the methods of statistical physics of ideal or weakly non-ideal gases. Taking into account the interactions of quasi-particles clarifies our understanding of the kinetic properties of low-temperature systems.

Many methods developed in low-temperature physics often have a very general character, and they are widely used in other areas of physics.

LOW-TEMPERATURE PUMPING METHODS

Pumping methods for producing high vacuum that are based on the capability of cooled surfaces to retain incident gas molecules by way of *condensation* or *adsorption* (*cryopumping*). Working elements of such pumps are *cryopanels* cooled to low temperatures. In the simplest cases, acting as cryopanels are the container walls made of high *thermal conductivity* materials (*copper* is often used) and filled with the necessary *cooling agent*. As for condensation pumps, the temperature of their cooling agent should be such that saturation vapor *pressure* of the evacuated gases at that temperature remains lower than the final residual pressure prescribed for the evacuated volume. To reach a vacuum of about 10^{-5} Pa, cooling with liquid nitrogen ($\sim$77 K) suffices when water vapor, carbon dioxide or ammonia are evacuated, while N_2, O_2, CH_4, Ar require liquid hydrogen ($\sim$20.4 K), and H_2 and Ne need liquid helium ($\sim$4.2 K) cooling. Cryopanels used in adsorption pumps are coated with a layer of adsorbent, with the needed temperature and amount retrieved obtained from *adsorption isotherms*. The latter are the dependences yielding vapor pressure above adsorbent surface versus the amount of gas adsorbed for a given temperature. Highly active adsorbents (such as charcoal, silica gel, *zeolites*) quite actively adsorb the majority of gases even at liquid nitrogen temperature, and only when one needs to evacuate Ne, H_2 or He are lower temperatures required.

A combined *adsorption-condensation pumping method* is also available. It employs *cryogenic trapping* that involves easily condensable gases and high saturation vapor pressure gases condensing together at a given temperature, e.g., He or H_2. Layers of the forming condensate are porous and capable of holding volatile molecules, which are then confined inside these pores while precipitation proceeds at the surface. Thus, a high vacuum may be achieved even when noncondensable components are present in the gas mixture.

Maximum evacuation rates are achieved when a cryopanel is placed directly into the evacuated volume. The pumping rate V of a condensation pump depends on the area of its cryopanel, the gas temperature, T, the molecular mass, M, and the probability for the gas molecule to stick to the surface (*accommodation* factor), α. In most cases $\alpha \approx 1$, and the rate of pumping per unit area is approximated by the expression $V = 36(T/M)^{1/2}$ m/s, and pumping rates up to ~100 m/s appear quite feasible. Estimates show that at such rates cryogenic pumps fitted out with built-in liquefiers become economically superior to traditional pumps. An additional merit of low-temperature pumping techniques is the attainability of a "clean" vacuum, i.e. pumping may be done under conditions that fully preclude external impurities from entering the evacuated volume.

LOW-TEMPERATURE RECRYSTALLIZATION

The appearance of new grains in a *metal* deformed at low (4.2 K) temperatures (see *Low-temperature deformation*) after being heated to $T < 0.4T_{melt}$. Lowering the temperature of deformation to that of liquid helium (4.2 K) noticeably changes the characteristics of the *recrystallization* process: it lowers the initial recrystallization temperature (even down to room temperature) and the effective *activation energy* of the process, while accelerating grain growth. Low-temperature recrystallization stems from a higher latent energy of *strain* and more pronounced metastability of the *crystal lattice* produced by deformation at *cryogenic temperatures*. The mechanism of *nucleation* during low-temperature recrystallization, the process completeness, the nature and properties of the structure resulting from *annealing*, all depend on the type of deformation, the level of its uniformity across the volume, and the rate of the processes that restructure the dislocations. The energy of *stacking faults*, the level of internal strains, and the homologous temperature of annealing control them. The nucleation during low-temperature recrystallization is believed to result from either the *polygonization* that occurs when mobile large-angle boundaries are formed, or from fragmentation of the complex *dislocation walls*. A fine grain structure results from low-temperature recrystallization (1–3 μm grains) that features high mechanical characteristics and thermal stability. The phenomenon of low temperature recrystallization opens a way to study recrystallization diagrams of metals and their physical properties, taking into account the low temperature regime of activated deformation.

LOW-TEMPERATURE X-RADIOGRAPHY

Application of X-ray diffraction techniques to the study of *solids* at low temperatures (see *Low-temperature physics*). Currently, the principal objects of low-temperature radiography are the structure of *cryocrystals*, low-temperature *first-order phase transitions*, and *second-order phase transitions* (polymorphic, magnetic, ferroelectric, etc.), structural instabilities in *superconductors*, electric and magnetic effects in solids, certain aspects of *crystal lattice dynamics*, radiation damage of crystals, etc. The scope of tasks of low temperature radiography noticeably expands when crystals are simultaneously exposed to low temperatures and/or electric and magnetic fields (plotting E–T and B–T phase diagrams, studying phase transitions, the evolution of *magnetic domain structures*, etc.). Using powerful X-ray sources for low-temperature radiography purposes (such as synchrotron radiation, rotating anode tubes) introduces the possibilities of studying the *atomic magnetic structure* of solids with the aid of radiography.

LOW-TEMPERATURE YIELD LIMIT ANOMALIES

A significant deviation of the temperature dependence of the *yield limit* of crystals from the dependence that results from thermally activated elementary processes usually operative in *plastic deformation* (see Fig.). Low-temperature yield limit

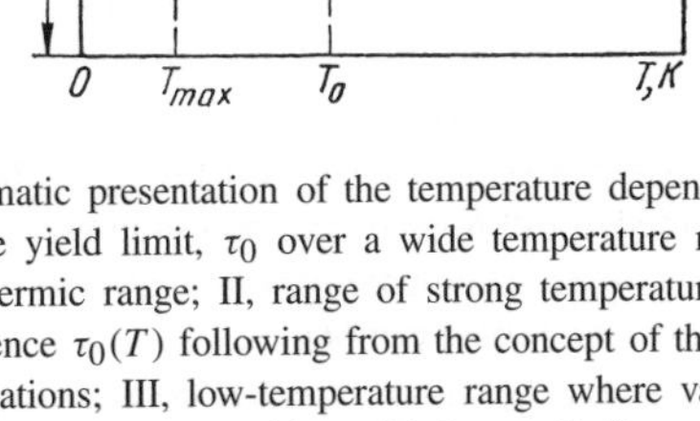

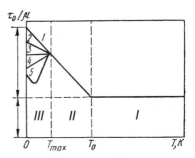

Schematic presentation of the temperature dependence of the yield limit, τ_0 over a wide temperature range: I, athermic range; II, range of strong temperature dependence $\tau_0(T)$ following from the concept of thermal fluctuations; III, low-temperature range where various anomalies are observed in $\tau_0(T)$ (curves 2–5).

anomalies were experimentally discovered in the late 1950s, and have been observed in numerous *metals*, *alloys*, and *ionic crystals* at temperatures below that of liquid nitrogen ($T < 77$ K). Detailed studies of such anomalies in face-centered cubic (FCC) and hexagonal close-packed (HCP) metals and alloys have demonstrated that their appearance and properties depend significantly on the concentration of impurities (see *Impurity atoms*) and on the value of the initial *strain*. Also of importance is the state of the sample surface. Theoretical explanations of those anomalies differ significantly for various specific mechanisms of plastic deformation. When the decisive factor controlling the motion of *dislocations* is local barriers (FCC and HCP crystals, ionic crystals with a CsCl lattice), the appearance of the anomaly is explainable by the role of thermal inertial properties of dislocations (see *Dislocation model of plastic deformation*). This means that the importance of thermal *fluctuations* in those crystals decreases with a drop in temperature, while the influence of dynamic effects increases, and the sensitivity of deformations to structural inhomogeneities grows. When the role of Peierls barriers is decisive (see *Peierls relief*) (body-centered cubic metals and alloys, and ionic crystals with the NaCl lattice), the anomaly is explainable by quantum-mechanical tunneling of dislocations and other quantum effects of dislocation motion. From a practical point of view, these anomalies are quite detrimental to *cryogenic technologies*.

LS-COUPLING

See *Russell–Saunders coupling*.

LUBRICANT, LOW-TEMPERATURE

See *Low-temperature lubricant* and *Low-temperature lubrication*.

LUMINESCENCE

Radiation that exceeds the thermal radiation of a medium and lasts significantly longer than the light frequency period. The phenomenon of luminescence is typical of optical processes taking place in various *states of matter*. Luminescence in *solids* is observed in *luminophors*, materials which contain *luminescent centers*.

The process of luminescence proceeds in three successive stages: excitation, energy migration or relaxation of the center from its excited state, and emission of a quantum of light. As to the excitation type, one may identify *photoluminescence* (excited by light); *radioluminescence* (excited by penetrating radiation, e.g., by an electron beam, see *cathodoluminescence*); *electroluminescence* (excited by electric field); *triboluminescence* (excited under mechanical action); *chemiluminescence* (generated during chemical reactions); *acoustoluminescence* (excited by an acoustic wave).

Of importance in solids is the second stage of luminescence, when the excitation energy absorbed by a luminophor is transferred (migrates to) a luminescent center. Mechanisms of energy migration in which free energy carriers (electrons or holes) take an active part are as follows: exciton mechanism, Auger mechanism involving bound current carriers (see *Auger effect*); intercenter tunneling of immobilized carriers. Energy migration results in a time lag between the processes of excitation and emission, which may be studied by measuring the kinetics using pulse excitation, and monitoring the thermal and spectral emissions. Included in the second stage of luminescence are *relaxation* processes during emission from the excited state. This involves the partial transformation of the excitation energy into thermal energy of the *crystal*, i.e. emission in the form of *phonons*. The relaxation process may be concurrent with the emission of an optical quantum of luminescent radiation, so that *hot luminescence* is observed. In many cases, the migration of energy to the luminescent center may only occur at high enough temperatures. That is linked with the trapping of free electrons (holes) by *attachment centers*, the probability of their ionization being proportional to $\exp[-E_a/(k_BT)]$, where E_a is the activation energy. When the ionization temperature of attachment centers increases, it results in *thermoluminescence*. The phenomenon of energy migration makes luminescence qualitatively distinct from various kinds of light scattering and reflection, or the parametric transformation of the radiation wavelength that is encountered in nonlinear optics (see *Parametric light generation*). The third stage of luminescence consists in the act itself of light emission. Depending on the mechanism of the elementary emission we can identify the following processes: band-to-band luminescence (*radiative recombination* of free electrons and holes), involvement of excitons (electrons coupled to holes via the Coulomb interaction), and impurity luminescence. In the latter case, the emission may occur via three types of mechanisms. The intracenter mechanism consists in a direct electron transition from the excited state of a luminescent center to its ground state. The intercenter mechanism involves a recombination of the nonequilibrium electron and hole localized at spatially separated *defects in crystals* (e.g., *donor–acceptor pair* in a semiconductor). Under a recombination mechanism, one of the carriers is free, i.e. it is in the *conduction band* (electron) or in the *valence band* (hole).

To identify the type of luminescence, one may study simultaneously the static and kinetic properties of luminescence and *photoconductivity* of a crystal, and also quantitatively analyze the basic characteristics of the radiation. The latter includes emission and excitation spectra (in the case of photoexcitation), the kinetics of afterglow (its temporal dependence), the temperature dependence, the dependence on the intensity of exciting radiation, and the polarization. For example, the kinetics of intracenter luminescence is described by an exponential dependence: $I = I_0 \exp(-t/\tau)$, where I_0 is the initial radiation intensity, τ is the inverse probability of an optical transition, t is the afterglow duration. The kinetics of *recombination luminescence* may be described by a hyperbolic excitation law $I = I_0/(1 + pt)^\alpha$, where p and α are constants. The specific type of kinetics may vary depending on the experimental conditions, e.g., the temperature, and the intensity of the excitation. However, only through a combination of techniques can one obtain reliable data on the elementary mechanisms that are operative.

Stokes' rule usually holds for photoluminescence of solids. It states that the energy of the excitation quantum exceeds that of the luminescence one, since a part of the excitation energy is transferred to the vibrational energy of crystal lattice atoms. Violation of that rule results in the so-called *anti-Stokes luminescence* in which the energy deficit is compensated by lattice phonons. A particular form of anti-Stokes luminescence is *cooperative luminescence* when several coherently excited centers accumulate energy to transfer it to a single emission center. To determine the symmetry and structure of luminescent centers in a solid one can analyze polarization properties of luminescent emission arising from an external anisotropic factor acting upon the crystal, and several such external factors are known. These are *piezospectroscopy* (uniaxial compression along certain crystallographic directions); the *Zeeman effect* or *Faraday effect* (a magnetic field) and the *Stark effect* (an electric field); polarized luminescence (excited by either linearly or circularly polarized light). Applying these techniques, one may identify the orientation of the absorbing and emitting dipoles relative to crystallographic axes, the multipolarity of transitions, and the point symmetry group of the luminescent center. A technique, especially informative from the point of view of the chemical nature and microstructure of the luminescent center, is that of *optically detected magnetic resonance* (ODMR). It consists in recording variations in the intensity of photoluminescence (absorption) under simultaneous action of optical excitation and microwave radiation upon a crystal placed in a magnetic field. The ODMR technique combines the principal advantages of luminescence spectroscopy and the methods of radiofrequency or microwave spectroscopy for studying crystalline defects, namely, their high selectivity and concentration sensitivity, as well as the possibility to obtain qualitatively new results. For example, one can study the electron paramagnetic resonance in both the ground and excited states of the luminescent center, analyze the spatial distribution of *paramagnetic centers*, and study the chemical nature of luminescent centers in thin film structures.

The phenomenon of luminescence in solids has found wide engineering applications. It underlies the functioning of visible and IR *semiconductor lasers* in the field of *optoelectronics*.

Electroluminescence is used in light emitting diodes fabricated on the basis of epitaxial films of GaAs, GaAlAs, GaP, SiC, ZnSe, etc. Radioluminescence is used in scintillation counters of α-, β-, and γ-rays to convert the energy of these rays into the optical spectral range. Anti-Stokes luminophors are used to convert IR range images into visible light in order to prepare visual images of variously heated targets (night viewing devices). The high sensitivity and the relative simplicity of implementation resulted in developing a technique of luminescence analysis designed for detecting minor concentrations of luminescent materials. This technique found various applications in *microelectronics* to measure impurities and defects in semiconductor crystals.

See also *Acoustoluminescence*, *Anti-Stokes luminescence*, *Cooperative luminescence*, *Electroluminescence*, *Exciton luminescence*, *Hot luminescence*, *Radioluminescence*, *Negative luminescence*, *Secondary luminescence*, *Thermally stimulated luminescence*, *Triboluminescence*, *X-ray luminescence*.

LUMINESCENCE ACTIVATOR

See *Luminescent centers*.

LUMINESCENCE DURING RADICAL RECOMBINATION

A particular type of *chemiluminescence* that takes place during the *recombination* of free radicals. In solids this effect occurs mainly in *semiconductors* at the material surface after the *adsorption* of free radicals. Their recombination process develops mainly as a monomolecular reaction, with an *activation energy* specific for each given material. To excite this luminescence there should be a source of *free radicals*, such as a chemical reaction, e.g., a reaction of burning (cathode luminescence), a photolytic electric discharge, thermal dissociation of molecules. Beside the principal bands which are typical for other types of *luminescence*, the radical recombination spectra also contain additional bands specifically related to surface emission centers of *luminophors*, to centers formed during adsorption of the radicals, and to chemiluminescence. The intensity of radical recombination luminescence grows almost linearly with the partial pressure of the radicals, and passes through a characteristic maximum with increasing temperature. The quantum yield (ratio of the number of *photons* to the number of molecules formed during the recombination of radicals per second) is low ($3\cdot10^{-3}$–10^{-6}). Radical recombination luminescence is used during studies of adsorption, catalysis, and electronic processes that take place at *solid surfaces* (e.g., with a crystal phosphor), and also for physico-chemical control of luminophors.

LUMINESCENCE QUENCHING, extinction of luminescence

The decrease or quenching of the *luminescence yield*, which occurs under the action of various factors: heating (*temperature quenching*), increased concentration of luminescence centers (*concentration quenching*), application of an electric field, exposure to light (*optical quenching*), friction, ultrasonic radiation, etc. The phenomenon of quenching of luminescence is due to the fact that exposure of the system to these above mentioned factors causes or enhances competitive processes involving nonradiative dissipation of excitation energy, or decreases the likelihood of *luminescent centers* becoming excited.

LUMINESCENCE QUENCHING BY METAL

Decrease of intensity, broadening, and shift of exciton *luminescence* band in an insulator or a semiconductor, which borders a metal film. In addition to the processes of absorption and *light scattering* in the metal, the metal quenching of luminescence is caused by the destruction of excitons near the interface, with the excitation of *plasma oscillations* and *electron–hole pairs* in the metallic film.

LUMINESCENCE YIELD

The ratio of some characteristic factor of the light emitted during *luminescence*, to the same characteristic of the exciting light. Ordinarily energy is chosen as the characteristic factor, and the luminiscence yield is called the *energy yield of luminescence*. In other cases the excitation can be characterized by the value of the current (current yield), by the number of ionizing particles (yield per particle), etc. See also *Quantum yield*.

LUMINESCENT ANALYSIS

A combination of research techniques, monitoring and testing methods applicable to various substances and structures that are based on recording *luminescence* (intensity, spectral features, dependencies on various factors, time, spatial distribution, etc.). Typical problems of luminescent analysis include those of detecting small amounts of specific impurities and structure defects, monitoring substance inhomogeneity and geometric flaws. Luminescent analysis combined with scanning *electron microscopy* offers wide capabilities and diverse approaches to solving such problems.

LUMINESCENT CENTERS

Local structural elements in solids (*impurity atoms*, *dislocations*, etc.) which cause *radiative quantum transitions* between their states to be observed as *luminescence*. When producing *luminophors*, luminescent centers are intentionally generated in them using various techniques, e.g., by introducing particular selected impurities (*luminescence activators*).

LUMINESCENT INDICATOR

A material with *luminescence* sensitivity to some physical or chemical factor, so that one may detect that factor and assess its characteristics (see *Luminescent analysis*). The term is also applied to a device for data presentation based on luminescence (usually *electroluminescence*).

LUMINOPHORS (fr. Lat. *lumen*, light, and Gr. $\varphi o \rho o \varsigma$, carrying)

Materials which produce luminescent radiation (see *Luminescence*), and are used in various lighting, visualization, and X-ray recording devices, in indicators, in systems for transmission and transformation of video information (oscilloscopes, television, computer displays, etc.). Many luminophors are synthetic solid-state materials which can be either inorganic or organic in composition. In widest use are inorganic crystalline luminophors (*phosphors*), that are found in fluorescent lamps, TV and oscilloscope cathode-ray tubes, electric luminescent panels, etc. Depending on their specific purpose, luminophors have to meet widely differing requirements in their radiation color, afterglow time, sensitivity to this or that excitation method, operating temperature range, etc. These requirements are met by appropriate selection of luminophor composition and methods of activation for particular *luminescent centers*.

LUTETIUM, Lu

A chemical element of Group III of the periodic system with atomic number 71 and atomic mass 174.97; it belongs to the *lanthanides*. Known isotopes have mass numbers 151, 153 to 156, 162, 164 to 180, the most abundant in nature (97.40%) being the stable isotope ^{175}Lu. Electronic configuration of filling outer shells is $4f^{14}5d^{1}6s^{2}$. Successive ionization energies are 5.426, 13.9, 20.96, 45.19 eV. Atomic radius is 0.1737 nm; radius of Lu^{3+} ion is 0.0848 nm. Oxidation state is +3, and electronegativity is $\approx$1.28.

In the free form lutetium is a silvery-white *metal*. It has a hexagonal close-packed crystal lattice with parameters $a = 0.35052$ nm, $c = 0.55494$ nm at room temperature, space group $P6_3/mmc$ (D_{6h}^4). There is a polymorphic transformation at $\approx$1820 K to a body-centered cubic modification with parameter $a = 0.390$ nm upon hardening, space group $Im\bar{3}m$ (O_h^9). Density is 9.84 g/cm^3, $T_{\text{melting}} = 1935$ K, T_{boiling} is about 3685 K. Bonding energy is -4.4 eV/atom at 298 K. Heat of melting is 18.85 kJ/mole, heat of boiling is 356.6 kJ/mole, heat of sublimation is 377 kJ/mole; specific heat $c_p = 26.56$ J·mole^{-1}·K^{-1}. Debye temperature is $\approx$185 K; linear thermal expansion coefficient of lutetium monocrystal is $19.0\cdot10^{-6}$ K^{-1} along the $\underline{6}$ principal axis and $5.2\cdot10^{-6}$ K^{-1} perpendicular to this axis at 273 to 301 K. Adiabatic elastic moduli of lutetium monocrystal: $c_{11} = 86.23$, $c_{12} = 32.0$, $c_{13} = 28.0$, $c_{33} = 80.86$, $c_{44} = 26.79$ (in GPa) at 300 K; isothermal bulk modulus is 41.1 GPa, Young's modulus is 84.3 GPa, shear modulus is 33.8 GPa, Poisson ratio is 0.233. Vickers hardness of annealed lutetium is 77 HV. Lutetium is easily amenable to mechanical treatment. Effective thermal neutron capture cross-section is 108 barn. Electrical resistivity of lutetium polycrystal is $\approx$73 nΩ·m at 298 K, thermal coefficient of the electrical resistance is 0.0014 K^{-1}. Electron work function of polycrystal is $\approx$3.0 eV. Lutetium is a paramagnet within the overall temperature range because its atomic $4f$-shell is completely full; the magnetic susceptibility is $0.102\cdot10^{-9}$; nuclear magnetic moment of ^{175}Lu is 2.9 nuclear magnetons. Lutetium is a prospective material as a getter (see *Gettering*).

LUTTINGER HAMILTONIAN (J.M. Luttinger, 1956)

A Hamiltonian involving the method of *effective mass* generalized to include the case of degenerate bands. It is often employed to represent the valence band dispersion in diamond and in zinc-blende (ZnS) semiconductors. The band energy extremum in cubic *semiconductors* is attained in their *valence band* at the center of the *Brillouin zone*. Bloch functions (see *Bloch theorem*) at that point transform as p-functions, their angular momentum being $j = 1$, or $j = 3/2$ (taking into account the *spin–orbit interaction* in the latter case). A matrix Hamiltonian describes the motion of *holes* in the vicinity of the zone center:

$$\begin{aligned}\mathcal{H}(\boldsymbol{k}) = &-\left(\gamma_1 + \frac{5}{2}\gamma_2\right)\boldsymbol{k}^2 I \\ &+ 2\gamma_2\left(J_x^2 k_x^2 + J_y^2 k_y^2 + J_z^2 k_z^2\right) \\ &+ 4\gamma_3\big([J_x J_y]k_x k_y + [J_y J_z]k_y k_z \\ &\quad + [J_z J_x]k_z k_x\big) \\ &+ \boldsymbol{k}(\boldsymbol{J}\cdot\boldsymbol{B}) + q\left(J_x^3 B_x + J_y^3 B_y + J_z^3 B_z\right),\end{aligned}$$

where the $\boldsymbol{J}$ are matrices that satisfy the *angular momentum* commutation relation $\boldsymbol{J} \times \boldsymbol{J} = \mathrm{i}\boldsymbol{J}$, $\boldsymbol{k}$ is the operator of *quasi-momentum* of a hole in the magnetic field $\boldsymbol{B}$, while the constants define the effective masses and the g-factor. The Luttinger Hamiltonian gives a good description of *cyclotron resonance*, acceptor states (see *Acceptor*) in p-Ge, p-Si, and intermetallides of the $A^{III}B^{V}$ type. This is the first example of a Hamiltonian applicable to the method of effective masses and written in the form of a sum of invariants. This approach has proved fruitful for constructing the equations of motion in other *semiconductors* (see *Group theory*).

LYDDANE–SACHS–TELLER RELATION

(R. Lyddane, R. Sachs, E. Teller)

The relationship between the frequencies of the long-wave ($k \to 0$) transverse, ω_{TO}, and longitudinal, ω_{LO}, optical *phonons* in cubic crystals:

$$\prod_i \frac{\omega_{\mathrm{LO}_i}^2}{\omega_{\mathrm{TO}_i}^2} = \frac{\varepsilon_0}{\varepsilon_\infty},$$

where ε_0 is the low-frequency (static) and ε_∞ is the high-frequency (optical) relative *dielectric constant* (permittivity) of a crystal. The product in the above relation is taken over all the optically active dipole oscillations.

LYOPHILIC AND LYOPHOBIC BEHAVIOR (fr. Gr. $\lambda\upsilon\omega$, am dissolving, $\varphi\iota\lambda o\varsigma$, love, $\varphi o\beta o\varsigma$, fear)

A characteristic of the interaction of molecules of a solid substance (or particles of a colloidal system) with a contacting liquid. Depending on the sign of the interaction one may identify substances which are *lyophilic* (strongly attracting), and *lyophobic* (repelling) suspensions in a liquid. Materials displaying lyophilic properties with water are called *hydrophilic*, and those displaying lyophobic behavior are called *hydrophobic*. Materials exhibiting such behavior in oil are called *oleophilic* (*lipophilic*) and *oleophobic* (*lipophobic*), respectively. Substances lyophilic with respect to one liquid may be lyophobic with respect to another. One may quantitatively characterize the degree of lyophilic versus lyophobic behavior by the *edge angle*, or by the heat of swelling (or *heat of solution*). Changes in lyophilic versus lyophobic behavior (*lyophilization* or *lyophobization* of substances, bodies, or surfaces) may occur during chemical transformations or physicochemical interactions. For example, *modification* by adsorption, a process involving *surface-active agents*, makes it possible to intentionally switch between the lyophilic and the lyophobic states during *flotation* (separation of fine particles of different substances) and during the manufacture of *composite materials*.

LYOTROPIC LIQUID CRYSTALS

Crystals formed during the course of dissolving certain organic compounds in water or in nonpolar liquids. Molecules in low molecular weight lyotropic liquid crystals are *amphiphilic*, i.e. they have a polar and a nonpolar part. Salts of long chain acids, such as potassium laurate, exemplify such molecules. A lyotropic liquid crystal structural unit is a *micelle*, a conglomerate of 10^2–10^3 molecules. Micelles may be normal, with polar parts of their molecules at the outside surface, or inverted with the polar parts inside. High molecular weight compounds have structural units that may be relatively short rigid fragments of a chain, e.g., a section of a DNA molecule about 300 nm long. The rigidity of such a fragment results from its interaction with the solvent.

In terms of structural organization, lyotropic (as well as thermotropic) *liquid crystals* are divided into nematics, cholesterics, and smectics. The latter are *lamellar phases*. Low molecular weight nematics are built of anisometric micelles (cylinders or disks). Cholesterics have twisted cylindrical or discotic micelles due to the presence of chiral molecules (see *Chirality*). Lamellar phases often feature double layers, with molecules of the solvent located between them. Textures and physical properties of lyotropic liquid crystals are similar to those found in thermotropic ones. However, long *relaxation times* (minutes to hours) are typical for lyotropic liquid crystals owing to their considerable viscosity, $1–10^3$ Pa·s. The lyotropic liquid crystal type structure is an attribute of many systems of biological origin.

Mm

MACHINING, ELECTRIC PULSE

See *Electric pulse machining.*

MACROCYCLIC COMPOUNDS

Compounds that contain cyclic arrangements of nine or more atoms in their molecules. Of most importance are *macroheterocyclic compounds* whose cycles includes heteroatoms (atoms of nitrogen, oxygen, sulfur, etc.) besides carbon atoms. Typical of macrocyclic compounds is the inclusion of neutral or charged particles into the macro-ring cavity, with the resulting formation of complexes, compounds of the "host–guest" type (see *Guest–host effect*), *clathrate compounds*. According to the type of *donor* atoms incorporated into the macro-ring, macrocyclic compounds may be classified under oxamacrocycles (crown ethers, O), azamacrocycles (N), thiamacrocycles (S), etc. Inclusion of donor atoms of more than one type is possible (e.g., azathiamacrocycles). *Crown ethers* form strong complexes with alkali and alkaline-earth ions, as well as compounds of the "host–guest" type. Aza- and thiamacrocyclic compounds bind ions of *transition metals*, i.e. Pb, Hg, Zn. Macrocyclic complexes of transition elements can form sandwich compounds exhibiting metallic conduction. Since many macrocyclic compounds favor the solid–liquid phase transition of ionic compounds, they serve as phase-transfer catalysts. Macroscopic compounds include some naturally occurring substances, e.g., porphyrins, certain antibiotics, and cyclodextrins.

MACROHARDNESS

Hardness measured when an *indenter* carries an appreciable load (10 N or more). Macrohardness H characterizes the resistance of a material to plastic elastic strain (see *Elasticity*, *Plastic deformation*) under local loading. The main ways to determine macrohardness are as follows: *Meyer hardness* (E. Meyer), or contact pressure $H = P/A$, where P is the load on the indenter, A is the projected area of the indentation; *Brinell hardness* (J.A. Brinell) is determined when penetration is carried out with a steel ball; *Vickers hardness* is measured using a regular tetrahedral diamond pyramid with an angle of 136° between opposite faces. In both latter cases, $H = P/A$. *Rockwell hardness* (S.P. Rockwell) is determined using a diamond cone by the depth of plastic impression at a given indenter load. Hardness of *minerals* is often determined with the help of *Mohs' scratch hardness test* (F. Mohs). *Mohs hardness* scale contains ten standard hardness values, with diamond the hardest (10) and talc the lowest value (1).

MACROLATTICE, imperfect superlattice

A regular arrangement of the particles of coherent precipitate *phases* in the bulk of an aged *alloy* crystal, which forms a *modulated structure* resulting from a diffusive *phase transition* (*alloy aging*, *alloy ordering*, an eutectoid *alloy decomposition*). Dimensions of the macrolattice cells amount to 0.5–5 nm, while the orientation of their axes depends upon the matrix of *anisotropy of elasticity*; for cubic crystals, it may be either $\langle 100 \rangle$ or $\langle 111 \rangle$. Unlike a *superlattice*, the macrolattice is imperfect: there is no strict periodicity in the positions of particles along fixed directions.

Depending on the difference in the *crystal structure* matrix and the phase separation, different types of macrolattices may develop. Under an isomorphic (see *Isomorphism*) or an isostructural phase transformation in a high-temperature superalloy (see *High-temperature alloys*) or a hard magnetic alloy of the *cunife*, *alnico*, etc., type, it is generally a primary primitive cubic lattice that later turns into a tetragonal one. If the transformation is not isomorphic (or isostructural) (in high-carbon and high-nitrogen *steels*, in beryllium bronze, in hard magnetic alloys Co–Pt, etc.), it is a tetragonal or rhombic lattice. In the latter case, a many-step twinning (see *Twinning structure*) of the macrolattice fragments of different orientations springs up, which leads to relaxation of elastic distortions.

MACROSCOPIC DEFECTS

Defects in crystals that have a characteristic linear size greater than the interatomic distance in all dimensions. Macroscopic defects have extra energy both in the form of *surface energy*, and from the field of the stresses that they induce. Macroscopic surface defects involve details of surface relief, including the shape of scratches, and elements of statistical *surface roughness*. These are smoothed by the molecular mechanism of *mass transport*. Three-dimensional macroscopic defects include *pores* and *cracks*. Pores may arise in a crystalline body during its growth from the melt as a result of negative jumps of volume at *crystallization*, due to the *interdiffusion* that occurs through differing partial *diffusion coefficients* of component atoms (see *Diffusion porosity*), during *sintering* of powders, and under the effect of nuclear radiation. Cracks are formed during *brittle failure* of a solid. An estimate of the critical size l^* of a crack capable of further growth is obtained by equating the elastic energy released during crack formation ($\sim\sigma^2 l^3/2E$, where E is the *elastic modulus*) to the crack surface energy ($\sim l^2\alpha$, where α is the surface energy density): $l^* \approx 2\alpha E/\sigma^2$. Macroscopic defects have a considerable effect on many physical features and application properties of solids.

MACROSCOPIC DESCRIPTION in solid state physics

Summarizing statements of solid state experimental results, often involving averaging procedures, and sometimes including a *phenomenological description*. Thermodynamic relationships, which establish a correspondence between various thermodynamic quantities, may serve as the conceptually simplest example of a macroscopic description. Such a description proceeds from the proper selection of physical quantities (scalar, vector, tensor) to describe the phenomena, and from expressions of mathematical relations between the said quantities. A macroscopic description is exemplified by *Ohm's law* (G. Ohm, 1826) which provides a linear relationship between the current density $\boldsymbol{j}$ in a conductor and the electric field intensity $\boldsymbol{E}$. Both quantities are vectors. This linear relationship is formulated through the application of the electric conductivity tensor $\widehat{\sigma}$ (see *Electrical conductivity*) or resistivity tensor ($\widehat{\rho}=\widehat{\sigma}^{-1}$):

$$j_i=\sigma_{ik}E_k,\quad E_i=\rho_{ik}j_k\quad (i,k=x,y,z).\quad (1)$$

These second-rank tensors σ_{ik} and ρ_{ik} provide the macroscopic description (phenomenological description) of conduction phenomena.

Invoking geometrical considerations and general laws of nature often allows one to reach important conclusions about the components of a macroscopic description. Thus, for a body in the state of equilibrium, Onsager–Casimir relations based on time reversal symmetry considerations require the conductivity and resistivity tensors to be symmetric ($\sigma_{ik}=\sigma_{ki}$, $\rho_{ik}=\rho_{ki}$), whereas *crystal symmetry* limits the number of independent components of the tensors σ_{ik} and ρ_{ik}. In a cubic crystal, e.g., the tensors σ_{ik} and ρ_{ik} reduce to scalars ($\sigma_{ik}=\sigma\delta_{ik}$, $\rho_{ik}=\rho\delta_{ik}$, $\rho=\sigma^{-1}$). In a hexagonal crystal, σ_{ik} and ρ_{ik} each have two independent principal values (one along the *symmetry axis* and the other in the plane, perpendicular to this axis). In accordance with the second law of thermodynamics the principal values of the conductivity and resistivity tensors (see *Principal axes of a tensor*) must be positive. The analytical results obtained from the application of general laws of nature to constituent quantities are inter-related through *Kramers–Kronig relations* which link together real and imaginary parts of generalized *susceptibilities* through relationships involving integrals over the frequency.

Included in a macroscopic description is the classification of solids by their characteristic features. For example, *metals* (normal metals and *superconductors*), *insulators* and *semiconductors* are classified according to their electrical conductivities; *paramagnets* (*ferromagnets* or *antiferromagnets* at low temperatures) and *diamagnets* are classified according to magnetic properties; brittle and plastic bodies (see *Brittle failure*) are classified according to their strength and elasticity; *crystals* and amorphous bodies (see *Amorphous state*) are classified according to structure, etc. An important component of a macroscopic description is based on equilibrium thermodynamics, as well as the thermodynamics of nonequilibrium processes. The limits within which the description holds true are often clear from the description itself, although the precise formulation of the limitations of a macroscopic description necessitates, as a rule, a microscopic treatment (see *Macroscopic defects*, *Macroscopic inclusions*).

MACROSCOPIC INCLUSIONS

Particles of a foreign *phase* in the bulk of a solid phase matrix that have a characteristic linear size exceeding the interatomic spacing in all three dimensions. They can either form because of the prior history of the solid phase, or be purposely introduced into this phase. Solid, liquid, gas, and combined *inclusions* can be distinguished. Macroscopic inclusions include hollow *pores*, gas-filled cavities, and solid phase particles insoluble in the matrix. The latter are generated through internal oxidation, through decomposition of oversaturated solutions (see *Alloy decomposition*), or via liquid inclusions captured from the liquid phase during crystallization. In many actual situations macroscopic inclusions give rise to fields of elastic tension, whose existence is responsible for some interactions of these intrusions. In anisotropic media such interactions may occur in both attractive and repulsive modes (see *Anisotropy of elasticity*). The presence of macroscopic inclusions may be responsible for consolidation of a crystalline matrix through mobile *dislocations* decelerated by inclusions. In various force fields, external or internal with respect to the matrix, macroscopic intrusions may move as an integral whole due to the existence of directed *mass transport* maintained by a corresponding force field. The presence of macroscopic inclusions in a solid affects the kinetics of both *recrystallization* and diffusive homogenization, decomposition of oversaturated solutions, *long-term strength*, and propagation of elastic vibrations. All the above features of macroscopic inclusions find wide use in technological practice.

MACROSCOPIC QUANTUM COHERENCE

in superconductors

A collective (cooperative) quantum phenomenon, whereby a superfluid or superconducting electron-pair (*Cooper pair*) *Bose condensate* can be described by a unified complex function $\psi(r,t)$ called the *order parameter* of the superconducting state. The existence of such a function for superfluid electrons in the bulk of a superconductor implies that their motion is highly correlated on a macroscopic scale. That is, the whole ensemble of electrons forming the superfluid condensate exhibits behaviors similar to a single quantum-mechanical particle. Related to this macroscopic quantum coherence are, e.g., the *quantization of flux* and *Josephson effects*.

MACROSCOPIC QUANTUM PHENOMENA

in superconductors

Phenomena related to the wave function phase $\chi(\boldsymbol{r},t)$ of *Cooper pairs* in superconductors: *Josephson effects*, *quantization of flux* (fluxoid), *macroscopic quantum tunneling*, and *macroscopic quantum coherence*. According to the BCS microscopic theory of superconductors, the *superconducting phase transition* is described by a complex *order parameter* $\psi = \Phi e^{i\chi}$, which acts as the wave function of a Cooper pair. By virtue of the *Bose–Einstein condensation* of pairs, the wave function of the whole system is $\psi_T \sim \psi^N$, where N is the number of Cooper pairs. Therefore, ψ can be considered as a macroscopic characteristic to describe not only a single particle, but the superconductor as a whole.

The phase $\chi(\boldsymbol{r},t)$ is coherent, i.e. it retains fixed values at different points $\boldsymbol{r}_1, \boldsymbol{r}_2$ of a sample, including those spaced at macroscopic distances, as well as at instants of time t_1 and t_2 separated by macroscopic time intervals. Observable physical quantities, such as charge density and current density, are expressed in terms of derivatives $\partial\chi/\partial\boldsymbol{r}$ and $\partial\chi/\partial t$ in gauge-invariant combinations with vector ($\boldsymbol{A}$) and scalar (φ) potentials:

$$\boldsymbol{j} = \left(\frac{n_S e}{2m}\right)(\hbar\nabla\chi - 2e\boldsymbol{A}),$$

$$\rho' = N_S e\left(\frac{\hbar\partial\chi}{\partial t} + 2e\varphi\right).$$

The uniqueness condition of a wave function in a closed superconducting ring leads to a phase change by a multiple of 2π when tracing round the ring. This is in accordance with the quantization of magnetic flux in *flux quantum* units $\Phi_0 = h/(2e)$ ($2e$ results from the electron pairing).

An outcome of this for a bulk hollow cylinder is the quantization of the flux, frozen in a hole, in units of Φ_0. In thin-walled shells (with wall thickness less than London *penetration depth of magnetic field*), the flux is not quantized. Yet, a circulating current arises with a magnitude and direction periodically changing in flux with increments of Φ_0 (*Aharonov–Bohm effect*). The flux, which penetrates into a superconductor in the form of *Abrikosov vortices*, is also quantized in fluxoid units Φ_0. In the case of superconductors separated by a thin insulating layer (*tunnel junction* or *weak link*), the phase is coherent on either side of the insulating barrier. The phase difference χ determines the superconducting current I that runs through the junction in accordance with the Josephson relation $I = I_c \sin\chi$. Given the voltage difference across the junction, the edges remain at equilibrium, i.e. $\rho' = 0$. This leads to the relationship $\partial\chi/\partial t = 2eV/\hbar$ for the phase difference, with consequent oscillations of the Josephson current in the course of time at a frequency (the Josephson frequency) determined by the relationship $\hbar\omega = 2eV$. A tunnel junction of small area may undergo transitions between states with different coherent phases by means of quantum tunneling through a Josephson barrier $\hbar I_c/(2e)$, which separates these states (*macroscopic quantum tunneling*). In the latter case, the phase χ ceases to be a classical variable, and the system is described by a "second quantization" wave function $\varphi(\chi)$.

MACROSCOPIC QUANTUM TUNNELING

Exhibition of quantum behavior by a macroscopic system, with a collective dynamic variable for a degree of freedom (see *Collective variables method*). The phenomenon of macroscopic quantum tunneling occurs with *Josephson junctions* and in *superconducting quantum interference devices* (SQUIDs), where the collective variable is the phase difference across the junction, or the total magnetic flux in the squid loop. These systems perform *zero-point vibrations* ω_0 with frequencies of 10^{11}–10^{12} Hz at metastable minima of the potential energy. This allows the inequality $\hbar\omega \gg k_B T$ to be realized, and the observation of a transition from the thermally activated penetration of the barrier to a temperature-independent macroscopic quantum tunneling mechanism at temperatures of about 1 K. Macroscopic quantum tunneling disappears not only with an increase in T, but also with increased energy dissipation in the system. Yet more sensitive to dissipation is the *macroscopic quantum interference* of more than one quantum state of a macroscopic system. A quantum-mechanical description of the phenomenon takes dissipation into account. Also classified as macroscopic quantum tunneling phenomena are quantum features of the low-temperature *plasticity* of solids (of *dislocation* motion).

MACROSTRESSES

First-order *residual stresses* in solids. Macrostresses appear as a result of nonuniform heating and cooling (*welding*, flame cutting, casting complicated shapes), pressure shaping and dressing of finished products, *heat treatment* (in particular, at the surface) and *sintering*, the application of different *coatings*, chemical or mechanical surface processing (turning, grinding, *polishing*, shotblasting), exposure of the surface to energy fluxes (laser or electron irradiation), *ion treatment*, diffusive surface saturation, etc. Macrostresses can be tensile or compressive. They generally concentrate in the subsurface layer, a few microns to 2 mm thick. These stresses may originate from such physical causes as: *pile-up of dislocations* of a similar sign; changes in *crystal lattice constants* due to a nonuniform distribution of *impurity atoms*; elastic strain of the lattice at its coherent bonding with a coating or at a coherent phase boundary interaction, etc. The magnitude and sign of macrostresses depend upon the material of a mechanical part, upon the manner of its treatment, upon the manner of fixing, way of cooling, etc. If a macrostress exceeds the *yield limit* in a plastic material, then relaxation occurs by way of *plastic deformation*. In low *plasticity* materials macrostress may exceed the *ultimate strength*, leading to the formation of *cracks*, spallings, and other surface defects. Macrostresses and their changes during the course of operation strongly affect the performance characteristics of mechanical parts.

MADELUNG CONSTANT (E. Madelung, 1918)

A dimensionless constant, α, dependent solely on the structure of a binary *ionic crystal*, which allows evaluating the potential φ arising from the Coulomb interaction at a lattice site produced by all the other ions from the equation: $\varphi = \alpha e/r_0$, where $\pm e$ are ionic charges, the sign of the potential φ is opposite to the sign of the charge at the site considered, and the characteristic length r_0 is the distance between nearest neighbor ions. The value of α is calculated by summing over all the atoms in the lattice, and it is a universal constant for a given lattice type. In crystals with several ions carrying different charges in their unit cell, the *Madelung potential* at the sth site is given by the equation:

$$\varphi(r_s) = \sum_{s'=1}^{p} \alpha_{ss'} \frac{e_{s'}}{r_0},$$

where the ionic charges are designated by e_s, $s = 1, 2, \ldots, p$ (p is the number of atoms in the *unit cell*), and the p^2 quantities $\alpha_{ss'}$ form a symmetric square matrix.

MAGGI–RIGHI–LEDUC EFFECT

See *Thermomagnetic phenomena.*

MAGNESIUM, Mg

A chemical element of Group II of the periodic system with atomic number 12 and atomic mass 24.305. Natural magnesium consists of the stable isotopes ^{24}Mg (78.60%), ^{25}Mg (10.11%) and ^{26}Mg (11.29%). Outer shell electronic configuration is $3s^2$. Successive ionization energies are 7.645, 15.035, 80.144 eV. Atomic radius is 0.1602 nm, radius of Mg^{2+} ion is 0.066 nm. Oxidation state is +2. Electronegativity is ≈1.29.

Magnesium in a free form is a silvery-white *metal.* It has a hexagonal close-packed lattice with the parameters $a = 0.32088$ nm, $c = 0.52095$ nm (at 298 K), space group $P6_3/mmc$ (D_{6h}^4). Density is 1.74 g/cm^3. $T_{\text{melting}} = 922$ K; $T_{\text{boiling}} = 1376$ K. Bonding energy of pure magnesium is 1.53 eV/atom at 0 K. Heat of melting is 8.5 kJ/mole, heat of evaporation is 130 kJ/mole, heat of sublimation is 152.6 kJ/mole. Specific heat is $c_p = 1.02$ kJ·kg^{-1}·K^{-1} (298 K). Debye temperature is 405 K; linear thermal expansion coefficient of magnesium monocrystal is 26.8·10^{-6} K^{-1} along the main sixfold axis $\underline{6}$, and 24.7·10^{-6} K^{-1} perpendicular to this axis at 280 K; coefficient of the thermal conductivity is 138 W·m^{-1}·K^{-1} at room temperature and normal pressure. Adiabatic elastic moduli of magnesium monocrystal: $c_{11} = 59.28$, $c_{12} = 25.90$, $c_{13} = 21.57$, $c_{33} = 61.35$, $c_{44} = 16.32$ GPa at 298 K; bulk modulus is ≈35 GPa, Young's modulus is ≈45.5, shear modulus is ≈18 GPa; Poisson ratio is ≈0.28 (at 298 K); at 293 K the properties of cast and deformed magnesium are characterized by the following parameters: ultimate strength is 0.113 and 0.196 GPa, relative elongation is 8.0 and 12.0%, yield limit is 0.025 and 0.088 GPa, respectively. Brinell hardness is 0.294 and 0.353 GPa; fatigue limit of annealed magnesium with basis 5·10^8 is 0.062 GPa; recrystallization temperature is 423 K. The strength of magnesium is increased by cold hardening. Self-diffusion coefficient of magnesium is 3.04·10^{-14} m^2/s along the axis $\underline{6}$ and 3.75·10^{-14} m^2/s perpendicular to it at 738 K. Ion-plasma frequency of magnesium is 17.78 THz; Sommerfeld coefficient of linear low-temperature electronic specific heat is 1.32 mJ·mole^{-1}·K^{-2}. Resistivity of monocrystal is 3.48·10^{-8} Ω·m along the principal axis $\underline{6}$ and 4.18·10^{-8} Ω·m perpendicular to it, temperature coefficient of electrical resistivity is 0.00408 K^{-1} along the principal axis $\underline{6}$ and 0.00390 K^{-1} perpendicularly to it, Hall constant is −8.42·10^{-11} m^3/C (room temperature, normal pressure); coefficient of the absolute thermal electromotive force is +3.3 μV/C along the principal axis $\underline{6}$ and +3.5 μV/C perpendicular to it. Work function of electron from polycrystal is 3.64 eV. Magnetic susceptibility of paramagnetic magnesium is +13.25·10^{-6} cm^3/mole in CGS units (300 K). Nuclear magnetic moment of ^{25}Mg isotope is +0.855 nuclear magnetons. Magnesium is mainly used in production of superlight alloys (see *Magnesium alloys*). Metallic magnesium is used for reduction of Zn, Th, U and other metals and their compounds.

MAGNESIUM ALLOYS

Alloys based on *magnesium*. They are the lightest known *construction materials*. They feature high impact *strength*, ability to absorb impact energy and vibration, as well as excellent tooling quality. Magnesium alloys are protected against *corrosion* by surface oxidation, or by *coating* with layers of varnish and paint. As to the initial processing conditions, these alloys can be cast or ductile types, and with respect to chemical composition they are alloyed with manganese (see *Alloying*); with *aluminum*, *zinc* and *manganese*; with *zirconium* and *zinc*; with less common and rare-earth metals; and with *lithium*. Mn enhances the *corrosion resistance* of an alloy and augments its strength; Al improves the strength and modifies the structure of cast Mg, Zn reduces the grain and enhances the alloy *plasticity* through refining. The *rare-earth elements* and less-common metals enhance the *creep* strength of an alloy at elevated temperatures (up to 250 °C), diminish its microporosity, and compensate for brittleness (see *Alloy brittleness*) arising from the presence of Zn. Lithium (more than 10%) substantially enhances the plasticity. The additional elements Ag, Be, Ca, Cd, Sn, Th, etc. are also added to some magnesium alloys. Mg forms *intermetallic compounds* called magnides with alloying elements, which significantly affect the alloy properties. The density of magnesium alloys varies, depending on the composition, within the range 1.36–2.0 g/cm^3, with magnesium–lithium alloys having the lowest density. Parts made of magnesium alloys display, due to their low density, large stiffness, and high values of specific heat. The coefficients of *thermal expansion* of magnesium alloys are 10–15% larger on the average than those of *aluminum alloys*. The *elastic modulus*, *yield limit* and *ultimate strength* grow at low temperatures, while the elongation and *impact strength* decrease, but there is no sharp drop in plasticity. Magnesium alloys are suitable for operation at cryogenic, normal and elevated temperatures. They find applications in the aerospace and automobile industries, for the manufacture of cinematic and photographic equipment, etc.

MAGNETIC AFTEREFFECT

The same as *Magnetic viscosity*.

MAGNETIC AGING

Deterioration of *magnetic material* properties with time. It occurs when a material changes from a *metastable state* to one closer to equilibrium. There are two types of magnetic aging: *reversible magnetic aging* due to *magnetic domain structure* reordering under the influence of an external magnetic field, of temperature variations, mechanical vibrations, irradiation, etc., and *irreversible magnetic aging* resulting from physico-chemical processes associated with the destruction of unstable crystallochemical configurations, with the redistribution of foreign inclusions, impurities, and structural defects (see *Defects in crystals*), and with the alleviation of inner stresses. Raising the temperature accelerates these processes markedly. Natural magnetic aging may last for months. To reduce aging the material is subjected to an *artificial aging*. The simplest way to stabilize the domain structure of a material operating near *remanence* is partial *demagnetization* with an alternating or reversed constant magnetic field. This stabilization is achieved by conditioning at elevated temperatures, and can be aided by annealing and work hardening. The maximum stability of the magnetization of a ferromagnet can be achieved by artificial aging, or by the application of a demagnetizing action similar to what is experienced during operation.

MAGNETIC ANISOTROMETER

An instrument for measuring the *magnetic anisotropy* constants, and determining preferred directions (*easy magnetization axes* or *hard magnetization axes*, crystallographic axes, etc.).

Underlying the operation of a magnetic anisotrometer are such phenomena as dependence of the components of the *magnetization* vector $\boldsymbol{M}$ or *magnetic susceptibility* χ on the orientation and strength of the applied *magnetic field* (determined by magneto-optic or radiophysical methods); the mechanical action of magnetic fields on the sample under study; ferromagnetic *resonance* (ferrimagnetic, antiferromagnetic, spin-wave, etc.), a *magnetic phase transition* from the uniformly magnetized state to that with *magnetic domain structure*, *galvanomagnetic effects*, etc.

There are no direct methods for measuring magnetic anisotropy constants. Indirect methods record the angle of rotation of the plane of polarization or the electric signal proportional to some component of $\boldsymbol{M}$ or χ, as well as the rotational mechanical moment acting on the sample in the magnetic field; the frequency of *ferromagnetic resonance* (at fixed magnetic field strength), or the resonant magnetic field (at fixed frequency); the strength and orientation of the magnetic field of a phase transition; the Hall voltage, etc.

The most common type of magnetic anisotrometer employs the *torsional moment method*. The sample under study (typically, having the shape of a disk) is fastened on an elastic suspension and placed in a uniform magnetic field $\boldsymbol{B}$. Since the anisotropic magnetic energy depends on the orientation of the magnetization vector $\boldsymbol{M}$ relative to preferred directions, a sample placed in the magnetic field starts experiencing the torque T that seeks to turn the sample to a position corresponding to the minimum total energy of the sample-suspension system. With the magnetic field vector $\boldsymbol{B}$ turning through an arbitrary angle φ (varying from 0 to 2π relative to the initial orientation), the torque (torsional moment) T also changes. To calculate magnetic anisotropy constants and determine the preferred directions, the coefficients of the Fourier expansion of $T(\varphi)$ are used. Typically, $T(\varphi)$ includes only even harmonics. The presence of odd harmonics indicates the presence of unidirectional anisotropy. The torsional moment in the first models of magnetic anisotrometers of the type described was measured directly by the torsional twist angle of the elastic suspension. Modern modifications use a compensation scheme of measurement, with the mechanical moment automatically balanced by means of, e.g., a compensation electromagnet. The field of the latter acts upon a small magnet fastened to the elastic suspension, holding it in a fixed position. The disadvantages of a magnetic anisotrometer using the torsional moment method are the high sensitivity of the device to mechanical vibrations, and the need to measure small torsional moments. In anisotrometers relying on other physical phenomena (ferromagnetic resonance, phase transitions, etc.), the sample under study has a rigid attachment system, so the first disadvantage is eliminated (see *Magnetometry*).

MAGNETIC ANISOTROPY

Nonequivalence of magnetic properties of materials in different directions.

Magnetic anisotropy plays an important role in magnetically ordered *crystals* (*ferromagnets*, *ferrimagnets*, *antiferromagnets*) where it determines, to a great extent, such characteristics as the frequency and width of the lines of *ferromagnetic resonance* or *antiferromagnetic resonance*, the *hysteresis loop* shape, *coercive force*, and the field of *magnetic saturation*. Magnetic anisotropy exists also in polycrystalline and amorphous materials, but is absent in liquids and gases.

Magnetic anisotropy is caused by the ordered arrangement of *magnetic moments* of particles of matter (atoms, ions), and the direction dependent character of the spin–orbit and dipolar interactions. The *spin–orbit interaction* determines the mutual orientations of the spin and orbital moments of a magnetic atom, and the latter is determined by the symmetry of the crystal field. The relative positioning of magnetic atoms in a *crystal lattice* causes the *dipole–dipole interaction* energy to depend on the magnetic symmetry of the crystal. Thus, both interactions depend on the orientation of atomic magnetic moments relative to crystallographic directions. A measure of the anisotropy is the energy density of magnetic anisotropy W_a, which for the n-sublattice magnetic crystal has the form of a series in even powers of components of magnetization vectors $\boldsymbol{M}_j$ of *magnetic sublattices*, where j is the index of a particular sublattice:

$$W_a = K_{ij}^{\alpha\beta} M_i^\alpha M_j^\beta + K_{ijkl}^{\alpha\beta\gamma\delta} M_i^\alpha M_j^\beta M_k^\gamma M_l^\delta + \cdots. \tag{1}$$

Due to the relative smallness of the spin orbit and dipolar energies compared to the exchange energy, the terms of the series (1) diminish, as a rule, rapidly as the power of $\boldsymbol{M}$ increases. The expansion coefficients $\widehat{K}$ are referred to as *anisotropy constants*. If tensors of anisotropy constants $\widehat{K}$ are determined only by crystallographic and magnetic anisotropy, such a magnetic anisotropy is referred to as *natural magnetic anisotropy*. In ferromagnets, the magnetic anisotropy manifests itself in the existence of easy and difficult directions of magnetization. The external magnetic field required to bring the magnetization to saturation has minima along the *easy magnetization axis*, and maxima along the *hard magnetization axis*. Some ferromagnets have one easy magnetization axis (e.g., hexagonal Co) and others have more than one (Fe, Ni, etc.). *Magnetic materials* having either one easy magnetization axis and several difficult ones, or one of the latter and several of the former, are referred to as *uniaxial magnetics*. A *biaxial magnetic* material has one easy and one difficult magnetization axis, and multiaxial materials have more than one of each type. In antiferromagnets, the direction along which the magnetic moments of sublattices are aligned is referred to as the *easy axis*. The effective field $\boldsymbol{B}_{\mathrm{A}j} = -\partial W_{\mathrm{A}}/\partial \boldsymbol{M}_j$ is referred to as the *anisotropy field* acting upon the jth sublattice. Sometimes the maximal value of $|\boldsymbol{B}_{\mathrm{A}j}| - B_{\mathrm{A}}$ is also referred to as the *anisotropy field*, and is used as a measure of the magnetic anisotropy energy.

Induced magnetic anisotropy arises under the action of external forces (e.g., pressure) and during the processing of a material (e.g., rolling, annealing in a magnetic field, etc.). It also arises during epitaxial growth of crystalline and amorphous *magnetic films* on a monocrystalline substrate (see *Growth anisotropy*). Several mechanisms exist that cause the induced magnetic anisotropy to appear. Under the magnetostriction mechanism, the magnetic anisotropy is due to the presence of mechanical stresses inside the sample. Another mechanism is connected with the formation and ordering of *defects* (impurities, *vacancies*, *dislocations*) during the course of the material processing. *Unidirectional* or *exchange magnetic anisotropy* arises at the interface between ferromagnetic and antiferromagnetic phases. Very often the antiferromagnetic phase is formed as a result of oxidation of a ferromagnet surface, for example, particles of Co coated with a CoO film. A particular feature of magnetic films with unidirectional magnetic anisotropy is the shift of hysteresis loops along the magnetic field direction.

MAGNETIC ANISOTROPY DISPERSION

See *Dispersion of magnetic anisotropy*.

MAGNETIC ANTIRESONANCE

A variety of phenomena arising from the vanishing of the real part μ' of the *magnetic permeability* $\mu(\omega) = \mu'(\omega) + \mathrm{i}\mu''(\omega)$, of a *magnetic substance* at a certain frequency ω_{A} called the frequency of antiresonance.

The most interesting manifestation of antiresonance is an essential (manifold) increase of the skin layer depth

$$\delta = \left[\frac{c^2}{2\pi\sigma\omega(\mu'' + (\mu'^2 + \mu''^2)^{1/2})}\right]^{1/2}$$

(cgs units) of a magnetic metal (see *Skin-effect*), i.e. the penetration depth of an electromagnetic wave (σ is the electrical conductivity). As a result, the *metal* has a *selective transparency* at the antiresonance frequency (predicted in 1959, discovered in 1969). The antiresonance frequency $\omega_{\mathrm{A}} = \gamma B$, where γ is the gyromagnetic ratio and B is the magnetic flux density ($\boldsymbol{B} = \boldsymbol{H} + 4\pi\boldsymbol{M}$; $\boldsymbol{H}$ is the magnetic field strength, and $\boldsymbol{M}$ is the magnetization). To observe antiresonance, it is necessary that the difference between ω_{A} and the frequency of *ferromagnetic resonance* (or para-, *antiferromagnetic resonance*) ω_{R} would considerably exceed the resonance line width $\Delta\omega_{\mathrm{R}}$. Since $\gamma H \leqslant \omega_{\mathrm{R}} \leqslant \gamma(HB)^{1/2}$, the condition $\Delta\omega_{\mathrm{R}} \ll 4\pi\gamma M$ is required. Due to this condition, the most convenient objects for investigating antiresonance are *ferromagnets* in applied fields of several tesla. Antiresonance is useful for the study of relaxation processes (dependence of $\mu''(\omega)$ of ferromagnetic metals). The transmission coefficient T of the electromagnetic wave of a frequency $\omega \approx \omega_{\mathrm{A}}$ through a ferromagnetic metal plate of thickness d is

$$T = \frac{c^2}{4\pi\sigma^2 d^2}\frac{2x^2}{\sin^2 x + \sinh^2 x},$$

$$x = \left(\frac{\sigma d^2}{\beta c^2}\frac{B}{M}|\omega - \omega_{\mathrm{A}}|\right)^{1/2},$$

where β is a factor on the order of unity, depending on the polarization of the electromagnetic wave incident on the plate.

MAGNETIC AXIS

The common principal axis of the g-tensor and of the crystal field tensor of the *quadrupole* and *hyperfine interactions* (see *spin Hamiltonian parameters*). The axis is parallel to one of the *crystal symmetry* directions, or perpendicular to one of the symmetry planes. The magnetic axis direction coincides with the applied magnetic field orientation corresponding to extrema in *angular dependences of spectra*. If *low-symmetry effects* appear in the spectra, viz., if the principal tensor axes are dissimilar, pseudoaxes averaged over all the resonance transitions will sometimes be conventionally selected. Such axes may deviate from the symmetry axes and from normals to the symmetry planes by as much as tens of degrees, and, with their orientation depending on the observation frequency (field), they have little physical significance.

A coordinate system that includes a magnetic axis is called a *magnetic coordinate system*. For high-symmetry spectra its axes coincide with specific crystallographic directions like $\langle 100\rangle$, $\langle 110\rangle$, $\langle 111\rangle$, $\langle 112\rangle$, etc. This approach is useful if the crystal has several equivalent *paramagnetic centers*, since describing the spectrum of each such center in its own magnetic coordinate system is much simpler than presenting it in a general *crystallographic system* of coordinates.

MAGNETIC BREAKDOWN

Quantum tunneling of *conduction electrons* in *metals* in a magnetic field, $\boldsymbol{B}$, between classical electron orbits of different *energy bands*. The term "magnetic breakdown" reflects the classically forbidden aspect of the *interband tunneling* in a magnetic field. Because of the small magnitude of the dimensionless quasi-classical approximation parameter, $\kappa = e\hbar B/p_{\mathrm{F}}^2 \ll 1$ (p_{F} is the electron Fermi momentum and $\omega_c = eB/m$ is the cyclotron frequency), the band-to-band transitions only occur within narrow classically forbidden regions of *momentum space* (magnetic breakdown regions). There the gap Δ between the bands is so small that the classical electron orbits from different bands approach one another to within distances comparable to those estimated from the quantum uncertainty of the transverse momentum: $(e\hbar B)^{1/2} \approx \kappa^{1/2} p_0$ (Fig. 1). The gap is usually

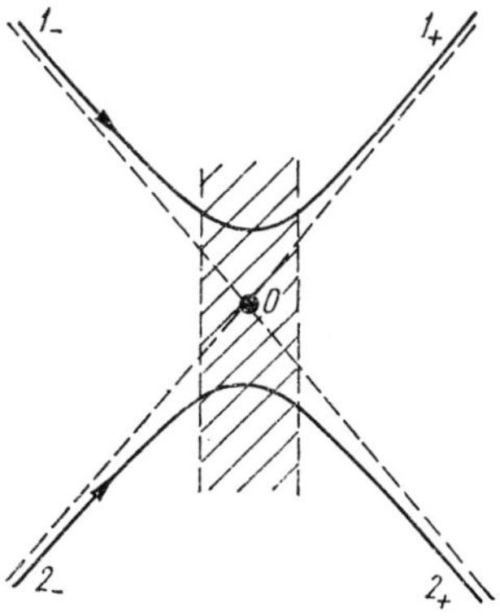

Fig. 1. Classical orbits (1 and 2) of different energy bands approaching one another, having the same momentum p_z of an electron of energy E in the form of two hyperbolae whose asymptotes intersect at the center point 0. The region of magnetic breakdown is indicated by hatching. 1_-, 2_- and 1_+, 2_+ are sections of classical orbits respectively entering and exiting the breakdown region. Arrows indicate directions of classical motion.

at its minimum in the Bragg reflection planes (see *Bragg diffraction*). Therefore, the band-to-band tunneling probability (magnetic breakdown probability) depends only on the orientation of the magnetic field, $\boldsymbol{B} = (0, 0, B)$, relative to the crystallographic axes, and on the value of the projection of the electron *quasi-momentum* (p_z) along the $\boldsymbol{B}$ direction. Since the magnetic breakdown regions are rather narrow they can be considered as point centers of two-channel quantum scattering: with or without a change of energy band number (see Fig. 1). Corresponding to the former is a magnetic breakdown scattering through the channels $1_- \to 2_+$ and $2_- \to 1_+$, and corresponding to the latter, through the channels $1_- \to 1_+$ and $2_- \to 2_+$. A closed configuration of quasi-classical trajectories corresponds to an arbitrarily aligned $\boldsymbol{B}$. In a field $\boldsymbol{B}$ parallel to the reciprocal lattice vector $\boldsymbol{b}$, an open periodic configuration with period b often results (Fig. 2). Generally the electron dynamics at magnetic breakdown is of a quantum nature, determined by the coherent interference of quasi-classical waves arising from multiple electron scattering at breakdown centers. The energy spectrum of electrons (magnetic breakdown spectrum) is defined in this quantum interference situation by the quasi-classical phase

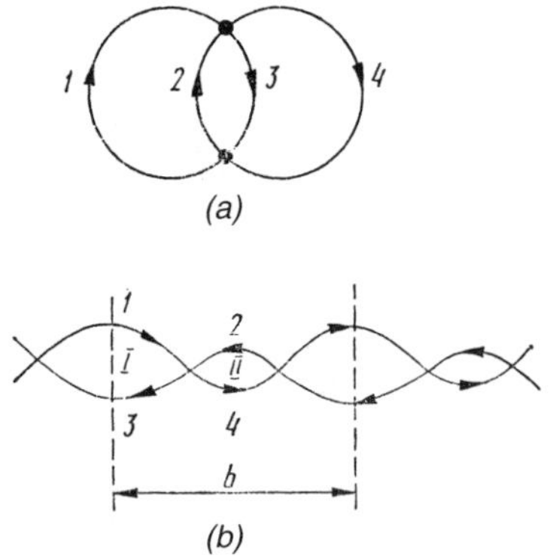

Fig. 2. Examples of magnetic breakdown configurations, where the curves are at the breakdown regions, and arrows indicate directions of classical motion. The upper curves show a closed configuration splitting up into orbits (1, 4) and (2, 3) at $W = 0$, and into orbits (1, 3) and (2, 4) at $W = 1$. The lower curves show a one-dimensional configuration with a period b splitting up into closed orbits (1, 3) and (2, 4) at $W = 0$, and into open orbits $(1, 4, \ldots)$ and $(2, 3, \ldots)$ at $W = 1$.

difference "accumulated" by electrons traveling across the configuration sections. The magnetic breakdown spectrum acquires a random (more precisely, quasi-stochastic) nature due to the incommensurability of these configurations with one another. In the case of closed configurations (Fig. 2(a)) it, like the Landau spectrum (see *Quantizing magnetic field*), emerges as a discrete set of terms $E_n(p_z)$ (n is the term number) separated by $\sim\hbar\omega_L$ (ω_L is a characteristic Larmor frequency), but not equally spaced; the E_n versus p_z dependence is irregular and rapidly oscillating, with an abnormally small (quantum) scale of the characteristic variation $\sim\kappa p_F$. Associated with the periodic configurations (Fig. 2(b)) is a spectrum of "magnetic bands" (broadened Landau levels).

The above features of the magnetic breakdown spectrum mean that the overall kinetics in a metal is drastically restructured at $B \geqslant B_0$ and $\omega_L\tau \gg 1$ (τ is the characteristic electron *relaxation* time at $B = 0$). Magnetic breakdown may reveal itself through two qualitatively different types of macroscopic phenomena, namely "coherent" and "random". This unusual dichotomy is of special importance for interpreting experimental data since κp_0 rather than p_0 (see above) plays the role of a characteristic momentum in the magnetic breakdown spectra. Consequently, even a minimal ($\sim\kappa p_F$) momentum transfer for small angle electron scattering from weak inhomogeneities of the *dislocation* strain field (or from other macroscopic lattice *defects*), and from *phonons*, can become effective. The relaxation processes are not characterized in this case by the "transport" time $\tau_{tr} \approx \tau_Q(p_0/q)$ customary for classical kinetics, but rather by the quantum lifetime of single-electron states, τ_Q, which is smaller than τ_{tr} by several orders of magnitude. Since τ_Q may have any relation to τ_B and to ω_L^{-1}, two basic regions of the parameter values are distinguished: (1) $\tau_B, \tau_Q \gg \omega_L^{-1}$, and (2) $\tau_B \gg \omega_L^{-1} \gg \tau_Q$. In the first case the collision-induced broadening of the magnetic breakdown terms is much greater than $\hbar\omega_L$. This corresponds to *coherent magnetic breakdown* "accumulated" by an electron traveling across the configuration sections.

The inequalities in the second case define the situation of a *random magnetic breakdown*. The small-angle scattering completely destroys the breakdown spectrum in the latter case, with the electrons traveling across the magnetic breakdown configurations as classical particles which make stochastic jumps with probabilities W and $1 - W$ between the sections when passing magnetic breakdown centers. Typical for a random magnetic breakdown are dissipation effects independent of the characteristic times τ_B and τ_Q, which persist at zero temperature, and are characterized by a *relaxation time* $\sim\omega_L^{-1}$. With the coherent magnetic breakdown situation, the kinetics in metals is defined by the quantum interference (Feynman's) sum over the magnetic breakdown paths, viz., over all the possible continuous trajectories traced out by an electron during an arbitrary time. The quantum oscillations with a magnetic field of any *kinetic coefficients* will be defined under these circumstances by the specific interference harmonics $\exp[i(S_1 - S_2)/\hbar]$, where the indices 1 and 2 mark two arbitrary breakdown paths, and $S_{1,2}$ are increments of the classical action along these paths. The relative amplitude of kinetic oscillations often may reach values ~ 1 (referred to as *giant magnetic breakdown oscillations*). As distinct from the kinetic oscillations, oscillations of the number density of states, and of the thermodynamic parameters coupled to them (*de Haas–van Alphen effect*), contain only harmonics with $S_{\text{cl}}/\hbar$ phases, where S_{cl} is the action increment for all the possible closed paths of the breakdown.

Also typical for a coherent magnetic breakdown is a variety of nonlinear effects in any external field capable of changing the electron momentum by a rather small value, of the order of $\sim\kappa p_F$, during a time $\sim\tau_B$ (or τ_Q). Coherent magnetic breakdown exhibits an abnormal sensitivity of kinetic coefficients to small inclinations of $\boldsymbol{B}$ to the symmetry axes (planes), or to the planes perpendicular to *reciprocal lattice* vectors. A very unusual situation shows up in the latter case: a pure metal behaves like an *incommensurate structure* with the quantum localization of conduction electrons. The above characteristics of the magnetic breakdown kinetics open a way to design fundamentally innovative electronic devices.

MAGNETIC BUBBLE DOMAIN, cylindrical magnetic domain

A small solitary *magnetic domain* shaped like an upright (perpendicular to the surface) circular cylinder. Magnetic bubble domains are observed in single-crystal films 0.1–100 μm thick with the easy magnetization axis normal to the sample surface, as well as in the presence of a *magnetic phase transition*. Examples of such films are *orthoferrites*, *iron garnets* of *rare-earth elements*, *amorphous magnetic substances*, alloys of rare-earth and transition metals, and other *magnetic substances*. The diameter of a magnetic bubble domain cross-section has a magnitude of the same order as the film thickness, and decreases as a constant *magnetic field* normal to the sample surface increases. An ordinary, solitary magnetic bubble domain is found only when such a field plays a stabilizing role. In this case, the film *magnetization* is oriented along the magnetic field, while the magnetic bubble domain magnetization is in the opposite direction. As the field increases to some critical value B_c (magnetic bubble domain collapse field) the domain size decreases, and then the domain disappears (*magnetic bubble domain collapse*). In the case when the field goes to the value B_2 (a *field of elliptic instability*), the magnetic bubble domain converts to a *band-type domain* or, more often, to the *labyrinthine magnetic domain structure*. In samples where the magnetic characteristics are substantially nonuniform in thickness over their surface area, *internal domains* in the magnetic bubble domain may exist, which are partly or entirely located inside of the film rather than on its surface, as in case of ordinary domains. Under the influence of an inhomogeneous magnetic field, temperature gradients, and other factors, a magnetic bubble domain may move over the film plane. When present with a fairly high density, due to magnetic dipole repulsive forces, they form a *magnetic bubble domain lattice*. The possibility for controlled magnetic bubble domain creation and annihilation at arbitrary sample points, as well as their motion by means of a magnetic field, permits their use in various devices (see *Magnetic bubble domain devices*).

MAGNETIC BUBBLE DOMAIN DEVICES

Solid-state computer engineering devices which utilize a magnetic film with *magnetic bubble domains*. The magnetic bubble domain devices are generally used for primary and auxiliary memory functions, but they are also capable of performing logic operations. These devices are constructed using a set of integrated micro-assemblies of different capacities (up to several megabits) controlled by an external electronic system. The micro-assembly contains: films of magnetic bubble domain materials with functional modules applied on the surface; a set of *permanent magnets* for establishing the bias field; a system for establishing the control magnetic field (control current). The diameters of the magnetic bubble domains used in these devices range from 5 μm to less than 1 μm.

The memory performance is based on the operation of the following elements: *magnetic bubble domain forwarding circuits*, which are often series–parallel connected for the purpose of increasing the data exchange rate; switches responsible for connections between the shift memory registers formed by elements of forwarding circuits; generator (source) of magnetic bubble domains; detector (sensing element) which takes the readings from magnetic bubble domains (generally by application of the *magnetoresistance* effect). In order to carry out the information read-out the magnetic bubble domain must be expanded into a film strip (see *Magnetic films*), so the data read-out module occupies a rather substantial part of the film surface. A magnetic bubble domain annihilator eliminates unnecessary domains. The domain micro-assembly, as well as the domain forwarding circuits, are characterized by a region of stable operation located near the functional units mentioned above. The magnetic bubble domain memory is nonvolatile, involves low energy consumption, is insensitive to external actions, and has high reliability. It has been demonstrated that memory devices may be created on the basis of *magnetic bubble domain lattices*, which provides a nearly tenfold increase of recording density compared to systems utilizing ordinary forwarding circuits.

MAGNETIC BUBBLE DOMAIN FORWARDING CIRCUITS

System of operating elements located on a film surface, which induces energy wells for *magnetic bubble domains* (MBD). Also it is intended for the bubble domain storage and transport. As a rule, an MBD comprises an associated system of memory shift registers, wherein the information is contained as code sequences: the logic unit "1" indicates the presence, and "0" denotes the absence of the MBD storage. The *forwarding circuits* are fabricated by photolithographic methods, and characterized by a variety of spatial shapes and types. Schemes of three types (and their combinations) are used: those in applications of *soft magnetic materials* (commonly, permalloy), current conductors, and ion-implanted structures. In the first and third cases the MBD shifting per single space period of the scheme is handled through the *magnetic reversal* of the scheme elements, or by means of a magnetic field (*operating field*). This field rotates at a fixed (tact or clock) frequency on the plane of the *magnetic film* as the field turns through an angle 360°. In the second case the *operating currents* perform the analogous function.

Magnetic bubble domain forwarding circuits occupy the major part of the film area defining the density of *information recording*. The magnetostatic repulsion among MBDs limits the scheme spacing period to 4–5 average diameters of a domain. Using permalloy, circuit densities $\sim 10^6$ MBD/cm^2 are achieved. Still higher densities are reached through ion-implanted structures (see *Ion implantation*). Clock frequencies are restricted by the *domain wall peak velocity*, and also by technical problems associated with the control field. For the first and third types of circuits the tact frequency is, commonly, ~200 kHz, for current conductors it is well above this.

The most important feature of the MBD forwarding circuits is the region of stable operation on the "shift field versus control field (current)" phase plane. This region should be within a narrow interval of the control field (current) value, and at the same time, it should have a maximum width with respect to the shift field value. The theoretical construction of the stable region requires an examination of the closed system of equations for the magnetic bubble domain and the scheme elements.

MAGNETIC BUBBLE DOMAIN LATTICE

(magnetic cylindrical domain lattice in Russian terminology)

A periodic two-dimensional *magnetic domain structure* of a thin *magnetic film*, its lattice sites having *magnetic bubble domains*. In a film with the *easy magnetization axis* perpendicular to its surface a *hexagonal lattice of magnetic bubble domains* can form. The equilibrium diameters of the domains of this lattice, and its period, depend on the value of the external *magnetic field* $\boldsymbol{B}$ perpendicular to the surface of the sample. Since the generation and annihilation of a separate magnetic bubble domain lattice are often impeded by the presence of an energy barrier, this lattice is often not in equilibrium, and in particular ranges of $\boldsymbol{B}$ values its period remains constant. The dynamical properties of a magnetic bubble domain lattice are determined by the set of natural oscillations of the *magnetization*, which are accompanied by small displacements of the bubble domains from their equilibrium positions at the lattice sites, and also by variations of their dimensions and shape. In the spectrum there are two activation-free branches of vibrations of the acoustic *phonon* type corresponding to displacements of the magnetic bubble domain lattice from the equilibrium sites, and there is also a series of branches of vibrations whose frequencies have finite values of the order of 10^8 Hz in the limit of infinite wavelengths. The propagation velocities of the lattice vibrations are usually of the order of 10^4 to 10^5 cm/s. These vibrations may be excited by an alternating magnetic field.

MAGNETIC BUBBLE DOMAIN, RIGID

See *Rigid magnetic bubble domain.*

MAGNETIC CHARACTERISTIC LENGTH

See *Characteristic length of magnetic material.*

MAGNETIC CHARGE, magnetic pole strength

An auxiliary concept introduced in *magnetostatics* to calculate the *demagnetization field* (analogue of concept of electric charge generating an electrostatic field). In contrast to electric charges, magnetic charges do not actually exist (see *Magnetic monopole*), since according to classical *magnetism* theory the magnetic field has no special sources besides electric currents. For media exhibiting a *magnetization* $\boldsymbol{M}$ one can introduce the concepts of a volume density, ρ_{m}, and of an area density, σ_{m}, of magnetic charge. The former is defined as the divergence of the magnetization, i.e. $\rho_{\mathrm{m}} = \operatorname{div} \boldsymbol{M}$, and the latter as a jump of the normal component of the magnetization at a discontinuity interface: $\sigma_{\mathrm{m}} = \Delta M_{\mathrm{n}}$ (e.g., $\sigma_{\mathrm{m}} = M_{\mathrm{n}}$ at a medium–vacuum boundary, where M_{n} is the normal component of the surface magnetization). See also *Magnetic pole.*

MAGNETIC CIRCUIT

A series of connected *magnetic substances* conducting *magnetic flux.* If the flux in a magnetic circuit is generated by a *permanent magnet* then such a circuit is called *magnetized circuit*. A magnetic circuit without permanent magnets is known as a *neutral circuit*, with the magnetic flux in it generated by an electric current flowing through coils enclosing part of the circuit. Depending on the source of flux there are magnetic circuits of the permanent, alternating, or pulsed types. Because of the formal analogy between electric and magnetic circuits, the same mathematical formalism is applicable to both. For instance, a counterpart of the Ohm's law emf $= IR$ is the equation mmf $= \Phi R_{\mathrm{m}}$, where Φ is the magnetic flux, R_{m} is the *reluctance* which limits the flux, and mmf is the *magnetomotive force* which generates the flux. In this circuit mmf is the analogue of the applied voltage or emf; Φ is the analogue of the current I, and R_{m} is the analogue of the resistance (impedance). Kirchhoff's laws are applicable to magnetic circuits. However, a fundamental difference between magnetic and electric circuits is the absence of dissipation of electromagnetic energy (*Joule heat*) in a magnetic circuit with magnetic flux Φ that is constant in time. The magnetic circuit approach is useful in designing permanent magnets, electromagnets, magnetic amplifiers, relays, electric measurement instruments, etc.

MAGNETIC COLLISION-FREE DAMPING of waves

The principal mechanism of damping of electromagnetic oscillations propagating in *metals.* Like *Landau damping*, magnetic collision-free damping is associated with the transfer of the electromagnetic field energy to particles (e.g., electrons) that move in phase with the wave. Yet, magnetic collision-free damping is not determined by the electrical force $e\boldsymbol{E}$, but rather by the magnetic force $\nabla(\boldsymbol{\mu}\boldsymbol{B})$ with which the alternating field of the wave $\boldsymbol{B}$ acts upon an electron rotating in an external field $\boldsymbol{B}_{\mathrm{ext}}$ ($\boldsymbol{\mu}$ is the mean *magnetic moment* of a rotating electron). Therefore, magnetic collision-free damping is often called *magnetic Landau damping*. Due to the Larmor rotation of electrons in the magnetic field (see *Larmor precession*), the scenarios of Landau damping and magnetic collision-free damping differ from each other. Landau damping involves particles whose velocity projections on the wave propagation direction coincide with the *phase velocity*, whereas in the magnetic collision free case projections of the particle velocity and of the wave phase velocity along the direction of $\boldsymbol{B}$ should coincide.

MAGNETIC COMPENSATION

Vanishing of the resultant overall magnetization of a *ferrimagnet* when at least one of the *magnetic sublattices* has a finite *magnetization* in the absence of an applied *magnetic field.* It occurs at a temperature, T_{c}, called the *compensation point*. The appearance of magnetic compensation is due to the nonequivalence of the *magnetic ions* (atoms) belonging to different sublattices, and to the resulting different temperature dependences of the sublattice *magnetic moments* in their antiparallel orientation. The occurrence of magnetic compensation is typical of most rare-earth *iron garnets*, of some *orthoferrites*, of *metallic glasses* based on d- and f-metal alloys, and also of a large number of mixed (or dilute) *ferrites* with diverse structures and compositions. Anomalies of thermodynamic and kinetic quantities such as *susceptibility*, *specific heat*, *coercive force*, etc., are observed at the compensation point. In addition, the temperature interval near T_C is particularly advantageous for studying noncollinear magnetic states due to a dramatic decrease in the critical magnetic field at *magnetic phase transitions* into such states.

MAGNETIC CONCENTRATION EFFECTS

Effects governed by matched changes of the electron and hole concentrations in a bipolar plasma in a semiconductor placed in crossed electric $\boldsymbol{E}$ and magnetic $\boldsymbol{B}$ fields. Such effects were discovered and studied in plasmas of Ge, Si and InSb. The electrons and holes drifting in the field $\boldsymbol{E}$ (see *Current carrier drift*) are deflected by the *Lorentz force* $\boldsymbol{F} = e(\boldsymbol{E} + \boldsymbol{v} \times \boldsymbol{B})$, which results in their being in excess on one surface of the material and depleted on the other surface (see Fig.). If the *recombination* rate of electrons and holes on the first side exceeds the rate of their *generation* on the second side, then the specimen is depleted of *current carriers*, whereas it is enriched with them otherwise. The specimen resistance, R, changes in both cases; this effect is termed the *galvanomagnetic recombination effect*, or *Welker effect* (H. Welker, 1951).

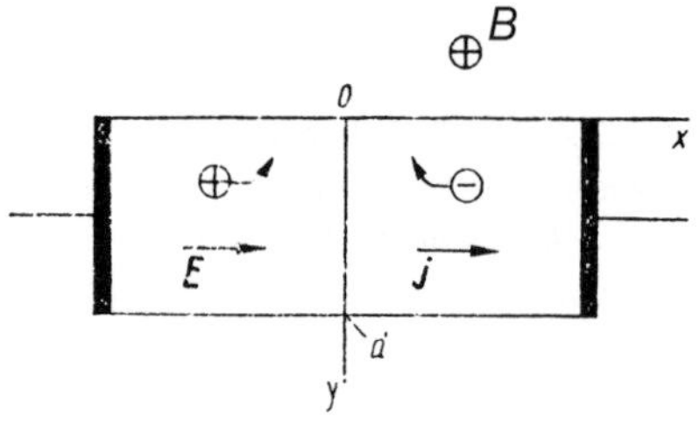

Magnetic concentration effects.

One can control the rates of recombination and generation by changing the *surface recombination* rates, I, at the surfaces involved. For $I_0 < I_d$, and a current density $j_x > 0$, the specimen is enriched with current carriers and its resistance R drops; for $j_x < 0$, the specimen is depleted of carriers and R increases. The emergence of such a rectification property is termed the *magnetic rectifying effect*, or *magnetic barrier effect*. The deflection of the injected current carriers (see *Injection*) by the magnetic field, and the change of their recombination rate, is called the *Suhl effect* (H. Suhl).

The intrinsic magnetic field of a high current flowing through a sample (e.g., wire) draws the electrons and holes into the central part of the specimen, which is known as the *pinch effect*. The magnitude of the current causing a marked current carrier redistribution is often evaluated based on the *Bennett condition* (W.H. Bennett, 1934) borrowed from gas plasma physics: it involves a parity between the kinetic pressure produced by electrons and holes, and the pressure of the magnetic field. As distinct from the gas plasma case, the surface generation of carriers has a significant bearing on the pinch effect in semiconductors. A longitudinal magnetic field $\boldsymbol{B}$ suppresses the pinch effect due to the developing Kadomtsev–Nedospasov instability which is employed to diagnose the latter. Flows of electrons and holes across the specimen may appear owing to spontaneous or deformation-induced anisotropy of the current carrier mobility. In this case, the deflection of electrons and holes toward one surface of the specimen is termed *electric pinch*. The above effects are used to make magnetic field and pressure gauges, and radiation modulators; they also occur in magnetic *diodes* and triodes.

MAGNETIC COOLING

See *Adiabatic demagnetization cooling*.

MAGNETIC DEFECTOSCOPY

See *Defectoscopy*.

MAGNETIC DIPOLE

A particle such as an unpaired electron which possesses a *magnetic moment* whose interactions with other magnetic moments in a medium takes place via the *magnetic dipole interaction*. Magnetic dipoles present in crystals are: nuclei (with non-zero *spins*), paramagnetic ions (see *Magnetic ion*), free radicals, and *triplet excitons*.

MAGNETIC DIPOLE INTERACTION, magnetic dipole–dipole interaction

Interaction between two magnetic moments $\boldsymbol{\mu}_1$ and $\boldsymbol{\mu}_2$ with each *magnetic moment* attracted by the magnetic field of the other. The resulting energy arising from this interaction is given by the equation:

$$E_{\mathrm{mdi}} = \frac{\mu_0}{4\pi} \times \frac{(\boldsymbol{\mu}_1 \cdot \boldsymbol{\mu}_2)|\boldsymbol{R}_{12}|^2 - 3(\boldsymbol{\mu}_1 \cdot \boldsymbol{R}_{12})(\boldsymbol{\mu}_2 \cdot \boldsymbol{R}_{12})}{|\boldsymbol{R}_{12}|^5},$$

where $\boldsymbol{R}_{12}$ is the radius vector connecting the two magnetic moments. If magnetic *dipole moments* are associated with particles with *spins* $\boldsymbol{S}_1$ and $\boldsymbol{S}_2$, the Hamiltonian terms for the magnetic dipole interaction is obtained by the direct generalization of the above equation through the replacement of $\boldsymbol{\mu}_1$ and $\boldsymbol{\mu}_2$ by their quantum analogs, $\gamma_1\hbar\boldsymbol{S}_1$ and $\gamma_2\hbar\boldsymbol{S}_2$, where γ_1 and γ_2 are the *gyromagnetic ratios*.

MAGNETIC DOMAIN, ferromagnetic domain

A region a *ferromagnet* (or other magnet) magnetized uniformly to saturation. A ferromagnetic sample in a sufficiently weak external field. is irregularly partitioned into magnetic domains present in thermodynamic equilibrium below the *Curie point*. The concept of magnetic domain was introduced by P. Weiss (1907) (see *Weiss theory*). In the absence of any present or prior external magnetic field, the *magnetization* vectors of each domain are oriented differently, so that the overall magnetization of a ferromagnetic crystal is close to zero. The shape and the dimensions of thermodynamically equilibrium magnetic domains are determined by the condition of minimum free energy of the ferromagnet. They depend on the shape and dimensions of the body, and are significantly larger than the *magnetic domain wall* thickness between them. In magnetic thin films *strip domains* and *magnetic bubble domains*, etc., can occur. In large-size samples of ferromagnets the domain shape can be more complicated.

MAGNETIC DOMAIN CHARACTERIZATION

Experimental studies of the size, shape, positioning, and restructuring features of *magnetic domains* in magnetic materials with a *magnetic domain structure*. Techniques of characterizing magnetic domains are based on various physical phenomena. Historically the first such technique (1919) was based on the *Barkhausen effect*, which consisted in recording Barkhausen jumps that result from remagnetization of separate domain groups. In the 1930s the technique of *magnetic powder patterns* was introduced (F. Bitter, 1931; N. Akulov, M.D. Detgyar, 1932; W.C. Elmore, 1936), making it possible for the first time to identify the size and shape of domains in various *ferromagnets*. It involved a special treatment of the specimen surface, including chemical and mechanical polishing. These techniques made it possible to judge the direction of the tangent to the projection plane of the domain magnetization: a scratch normal to that projection becomes richly decorated with the powder, while one parallel to it is not so affected (*scratch technique*). K.I. Sixtus and L. Tonks (1931) studied the *magnetic domain wall dynamics* in ferromagnetic wires by detecting a signal in an electromagnetic induction solenoid (induction technique). After the synthesis in the 1950s of magnetic materials transparent in the IR and visible regions, techniques based on various *magnetooptical effects* came into wide use.

Currently the leading methods for delineating magnetic domains are magneto-optic techniques that employ the *Faraday effect* and the Cotton–Mouton effect (see *Voigt effect*) for transparent materials, and the *Kerr effect* for opaque ones. The powder pattern method is often used in combination with magnetooptical techniques, making it possible to study both the bulk and the surface properties of magnetic domains. *Electron microscopy* based on differences between the *Lorentz force* deflections of electron beams in various domains with differing magnetizations is also used (*Lorentz microscopy*), as well as X-ray and neutron scattering. To study the dynamics of domain wall motions various induction and magnetooptical methods may also be used, including instantaneous and sequential superrapid photography.

MAGNETIC DOMAIN, LABYRINTHINE

See *Labyrinthine magnetic domain structure.*

MAGNETIC DOMAIN STRUCTURE

A spatial arrangement over a *magnetic material* sample of *magnetic domains* with various orientations of spontaneous *magnetization.* The main cause of the formation of a domain structure is the fact that in *ferromagnets*, in addition to the *exchange interaction* energy that produces the magnetization $\boldsymbol{M}$ and the *magnetic anisotropy* energy that gives the direction to $\boldsymbol{M}$, there exists a *magnetic static interaction* related to the existence of the *demagnetization field.* For uniform magnetization a strong demagnetization field appears around the sample with a large positive energy, and this energy can be reduced by dividing the ferromagnet into *domains* as shown in Fig. 1. It is possible to bring about this subdivision so the *magnetic flux* is closed inside the sample rather than outside it, and the demagnetization field outside diminishes almost to zero. This occurs with the appearance of *domains of closure* with triangular ends, as shown in Fig. 2. In crystals with a single *easy magnetization axis* (magnetic single-axis ferromagnets), the magnetization of the domains of closure is directed along a *hard magnetization axis*, and the energy expenditures for their formation are related to the magnetic anisotropy energy. In crystals (e.g., cubic) with several easy axes the magnetization of closure is directed along one of these axes. The energy expenditure for forming these closure domains is related to the magnetoelastic energy due to *magnetostriction.* All observable domain structures can be divided into two classes: those with completely closed magnetic flux (e.g., *Landau–Lifshits domain structure*), and those with open magnetic flux (e.g., *Kittel's domain structure*). The sample continues subdividing until the energy needed for forming new domain walls exceeds the decrease of the magnetostatic field energy associated with the subdivision. Eventually an equilibrium magnetic structure results from this competition between various kinds of energy (exchange, magnetostatic, magnetic-anisotropic energy), and this is brought about by minimizing the total *thermodynamic potential.* In crystals with more than one axes of easy magnetization one should take into account a contribution from the magnetoelastic energy (see *Magnetoelastic interaction*). The type of domain structure strongly depends on the dimensions and shape of the material. When the size decreases below a certain critical value the existence of a domain structure can become energetically unprofitable since the domain wall would occupy too large a part of the bulk. Sufficiently small grains become single-domain particles, and a collection of single-domain granules is called a superparamagnet. The most widely studied domain structures are those of thin *magnetic films.* There is a great diversity in the domain structures of thin films since their formation depends on the weak *magnetic dipole interaction.* This structure affects a significant number of crystal physical properties, such as kinetic, resonance, and optical

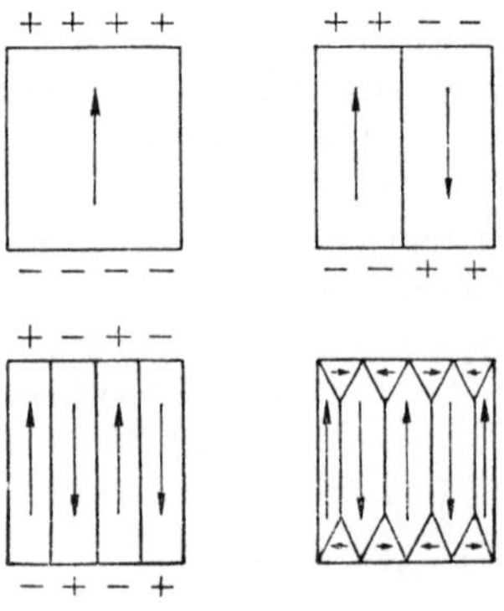

Fig. 1. Different forms of domains.

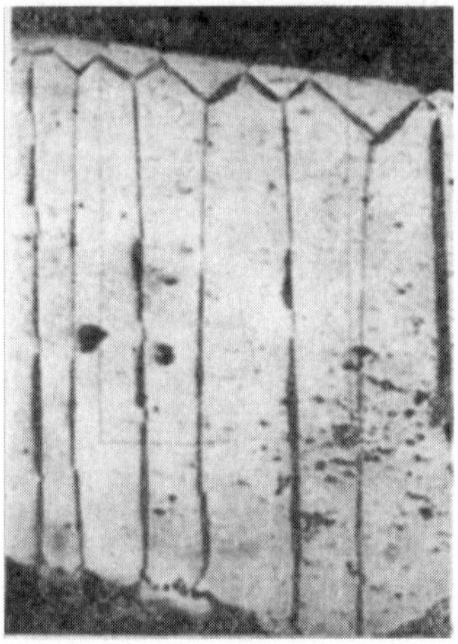

Fig. 2. Magnetic domain structure.

properties; this involves properties of ferromagnets which are of great importance in engineering, such as a high magnetic susceptibility (see *Soft magnetic materials*) and a large coercive force (see *Hard magnetic materials*). The domain structure finds a practical application, e.g., for storage elements in computers (see *Magnetic bubble domain devices*). The existence of a domain structure is confirmed by particular features of *magnetization curves*. In addition, there are several methods for the direct observation of domain structure (see *Magnetic domain characterization*).

MAGNETIC DOMAIN WALL, magnetic domain boundary

A layer between neighboring *magnetic domains* where there is an abrupt spatial change in the direction of the *magnetization* $\boldsymbol{M}$. The equilibrium orientation of $\boldsymbol{M}$ in different parts of a domain wall in a *ferromagnet*, subject to the condition $\boldsymbol{M}^2 = \text{const}$, is designated by the polar θ and azimuth φ angles, where θ is measured from the *easy magnetization axis* of the crystal. If both θ and φ are dependent on the same coordinate directed along a normal $\boldsymbol{n}$ to the surface of a domain wall, this corresponds to a domain wall with a one-dimensional $\boldsymbol{M}$-distribution. There also exist walls with multi-dimensional $\boldsymbol{M}$-distributions, in which the magnetization direction also changes along the surface of the wall, i.e. along z (see *Bloch line, Bloch point, Domain wall twisting*). The turning of $\boldsymbol{M}$ in a 180°-domain wall is sketched in Fig. 1. Such walls exist in ferromagnets with a single as well as with a few easy magnetization axes. In addition, there exist boundaries in magnetic-multi-axial crystals, at which the rotation of $\boldsymbol{M}$ is through an angle less than 180°. In cubic magnets with three easy magnetization axes of the type (100) (e.g., iron), 90°-domain walls can exist, while 70.5° and 109.5° walls exist with four axes of the type (111) (e.g., nickel). If $\operatorname{div}\boldsymbol{M} \neq 0$ in the wall then a demagnetization field $\boldsymbol{B}_m$ appears in it. If $\operatorname{div}\boldsymbol{M} = 0$ then $\boldsymbol{B}_m = 0$. A turning of the magnetization with $\boldsymbol{B}_m = 0$ occurs in a *Bloch wall*. Domain walls can form with different values of the magnetization projection on the wall plane from opposite sides; characterized by a non-zero *magnetic charge* density (*charged domain walls*). The wall structure depends on the exchange energy, the *magnetic anisotropy* energy, and the demagnetization fields. Exchange forces hinder the jump-like change of $\boldsymbol{M}$ when passing from one domain to another, and a gradual turn of $\boldsymbol{M}$ over the whole crystal is also impossible as it leads to an increase of the magnetic anisotropy energy. As a result, the main turn of $\boldsymbol{M}$ takes place over the thickness δ of the wall. In sufficiently large crystals the 180° wall is a *Bloch wall* with $\boldsymbol{B}_m = 0$, with the turn of $\boldsymbol{M}$ determined by the *exchange interaction* and the anisotropy energy. For a single-axis magnet this is described by Eq. (1) which is plotted in Fig. 2:

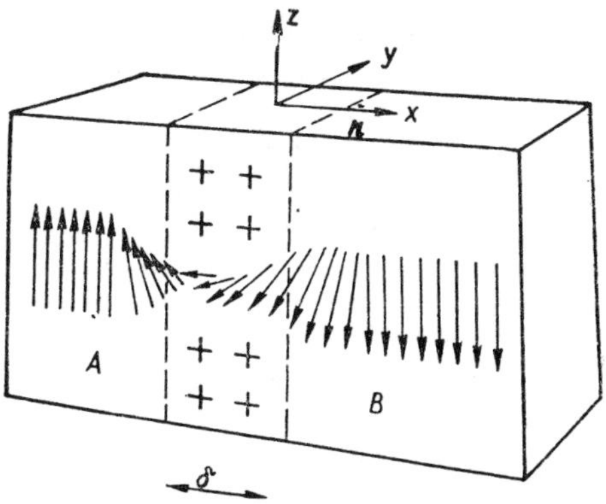

Fig. 1. Reversal of magnetization (designated by arrows) in the domain wall at the boundary between domains A and B. This wall has a positive sense of magnetic charge density.

$$\cos\theta = \mp\tanh\left(\frac{x}{\delta}\right), \qquad \sin\theta = \frac{1}{\cosh(x/\delta)}, \tag{1}$$

where the coordinate x is aligned along the normal to the Bloch wall; the wall thickness $\delta = (A/K)^{1/2}$, A is the exchange constant, and K is the constant of single-axis anisotropy. For Co with $A = 2\cdot10^{-11}$ J/m and $K = 9\cdot10^5$ J/m^3, we have $\delta = 15$ nm. Thus it is evident that although $\boldsymbol{M}$ varies continuously over the entire crystal, these pronounced changes mainly take place over relatively short distances, although these distances may involve hundreds of interatomic spacings. From Eq. (1) two Bloch walls can have a turn of θ both clockwise and counter-clockwise relative to x, with the same energy $(AK)^{1/2}$ (per unit

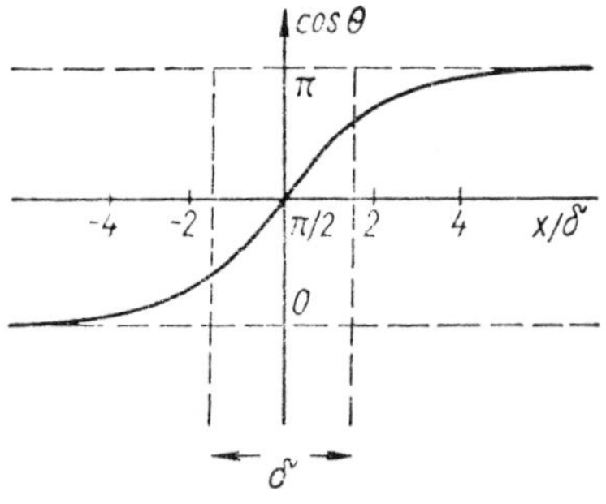

Fig. 2. Variation of the angle θ through a Bloch wall of thickness δ.

wall area) and the same thickness. In thin *magnetic films* with the easy magnetization axis parallel to the surface, static magnetic poles can emerge at the intersection of the Bloch wall with the film surface (zx). In the case of $\delta > h$ (h is film thickness), in spite of the turning of $\boldsymbol{M}$ in the plane perpendicular to the thin magnetic film, the turn in the film plane might be more favorable. In this case the domain wall contains a double layer of magnetic charges, and is called a *Néel wall.* In magnetic thin films of a certain thickness the formation of Bloch lines in Bloch walls is energetically profitable because positive magnetic charges from the walls on the film surfaces alternate with negative ones (*alternate polarity walls*). In thin permalloy films domain walls with short transverse branches are observable (*domain walls with transverse coupling*). In thin magnetic films with the surface normal to the easy magnetization axis and with $K > 2\pi M_s^2$ (M_s is the saturation magnetization), so-called *domain wall twisting* takes place. In these walls, in the vicinity of the central portion of the film, a turn of $\boldsymbol{M}$ is performed in the same fashion as at a Bloch wall; when approaching the wall surfaces, the plane of the turn of $\boldsymbol{M}$ is close to that at the Néel wall. If the $\boldsymbol{M}$ distribution is symmetric with regard to the central plane (that is parallel to the film surface) then such boundaries are called *symmetric twisted walls*. There are also asymmetric (in the sense indicated) walls. With the breakdown of the symmetry of a domain wall one can exclude the leakage field from the wall (*domain walls without leakage fields*).

Domain walls play an important role in the formation of the domain structure, and they also exert a significant influence on a range of dynamic phenomena exhibited by ferromagnets.

MAGNETIC DOMAIN WALL DYNAMICS

Variations of the positions or shapes of *magnetic domain walls* through the action of an external factor which is usually an applied *magnetic field* B. The variations involve changes in the volume of the different *magnetic domains* within the *magnetic domain structure.* The forward motion of a solitary domain (e.g., a *magnetic bubble domain*) can be described in terms of the dynamics of the domain wall that encloses it. There are experimental procedures for studying the details the motion of a separate domain wall, or of the average parameters of the wall in a domain structure, based on magnetooptical effects, inductance procedures, etc.

Studies of domain wall motion under the effect of B can provide the wall speed (V) through the friction force of the domain wall $F_{\text{fr}}(V)$ by using the expression $F_{\text{fr}}(V) = B\Delta M$, where ΔM is the difference between the *magnetizations* within the domains separated by the wall. For the condition $B > B_{\text{sf}}$, where $B_{\text{sf}} = F_{\text{sf}}/\Delta M$ and F_{sf} is the domain wall static friction force, the speed V grows at first according to the linear law $V = \mu(B - B_{\text{sf}})$, where μ is the *domain wall mobility*, then it grows more slowly, and finally stops growing when it reaches the *domain wall velocity limit* V_{sf} (see Fig.). Similar behavior is also observed for the domain wall radial velocity during the collapse of a magnetic bubble domain. The dynamics of the domain wall may be approximately described by an effective dynamic equation for the coordinate of the wall center u, and for a planar one-dimensional domain wall this equation has the form

$$m_* \frac{\partial^2 u}{\partial t^2} - \sigma\left(\frac{\partial^2 u}{\partial x^2} + \frac{\partial^2 u}{\partial y^2}\right) = F_{\text{fr}}\left(\frac{\partial u}{\partial t}\right) + B\Delta M, \quad (1)$$

where m_* and σ are the *effective mass* and energy of a unit area of the wall, respectively, and x, y are the coordinates of the plane of the undisturbed wall. Sometimes to describe the wall dynamics it is necessary to take into account the phenomenon

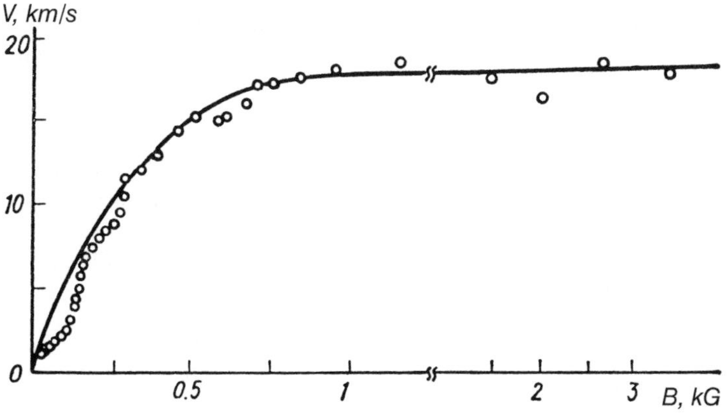

Dependence of the domain wall velocity in yttrium orthoferrite $YFeO_3$ on the external magnetic field.

of the magnetic after-effect (see also *Néel equation, Magnetic viscosity*).

At $B = 0$ Eq. (1) describes the free attenuating *flexural waves* of the domain wall with frequency $\omega = (\sigma/m_*)^{1/2}|\boldsymbol{k}_\perp|$, where $\boldsymbol{k}_\perp = (k_x, k_y, 0)$. If the wall is not one-dimensional (see *Domain wall twisting*), then its dynamics are more complex. For domain walls with vertical *Bloch lines* the equations for coordinates of the wall and the Bloch lines must take into account their mutual influence via the gyroscopic force F_G. The analysis may be performed on the basis of the *Slonczewski equations* for the wall coordinate u and the angle ψ that the magnetization makes with the domain wall plane. The wall dynamics may become multidimensional for high values of the field B, or within the region of negative differential wall mobilities.

The domain wall dynamics of the boundary of a bubble domain are associated with deformations, and with the forward movement of the domain. The small wall oscillations caused by non-uniform radial deviations $\Delta r(\varphi)$ of the bubble domain radius with respect to the angle φ are conditioned by deformations. These oscillations are characterized by a discrete set of the natural frequencies ω_n which depend on the component of B that is normal to the surface of the sample, and on the parameters of the *magnetic film*. In the field B_k of a magnetic bubble domain collapse, the frequency ω_0 corresponding to the uniform oscillations reduces to zero; in the field of an elliptical instability B_2 the frequency ω_2, of the elliptical deformations of the cylindrical cross-section, also reduces to zero. The characteristics of the forward motion of the domain due to the non-uniform field arise from the presence of the *gyroscopic force* F_G of the domain $F_G = A[M_0 \times V_c]$, where M_0 is the magnetization matrix, V_c is the domain velocity, the coefficient $A = (2\pi h/\gamma)S$, the *topological index* S is equal to the number of complete rotations of the magnetization vector in the center of domain wall with the turning of the side surface of the domain, h is the thickness of the film, and γ is the *gyromagnetic ratio*. The action of F_G deviates the bubble domain trajectory from the direction of the field gradient, making its dynamics similar to that of a charged particle in crossed magnetic and electrical fields. At high domain velocities the number and shape of the Bloch lines may change, leading to the *dynamic transformation of magnetic bubble domains*.

MAGNETIC DRIFT, unsymmetrical hysteresis loop drift

An effect occurring during the *magnetic reversal* of magnetic materials in an external *magnetic field* that periodically varies between certain values B_1 and B_2, with $B_1 \neq -B_2$, i.e. the magnetic reversal cycle is asymmetric. The magnetic drift occurs because at the end of the cycle, when the applied field has returned to its initial value B_1, the magnetization M has grown somewhat, instead of coming back to it initial value. Then the magnetic *hysteresis loop* is not closed on the $M(B)$ plane, but permanently drifts in the direction of increasing M. In some *magnetic substances* such a process may persist over a great number (up to 10^6) of magnetic reversal cycles. The drift results from minor random irreversible changes in the *atomic magnetic structure* of the specimen during the cyclic switching of the magnetic field, and it appears most prominently if the field magnitude varies near the value of the *coercive force*. Such changes in M may amount up to 0.01 mT for the magnetic drift in *ferromagnets*.

MAGNETIC FIELD

A field exerting a force on moving electric charges and bodies that possess a *magnetic moment* (whatever their state of motion). Its principal characteristics are: the vector of *magnetic flux density*, $\boldsymbol{B}$, the *magnetic field strength*, $\boldsymbol{H}$, where $\boldsymbol{B} = \mu\boldsymbol{H}$, and the permeability of the medium μ which, in general, is a tensor quantity. The Lorentz force $\boldsymbol{F}$ acting on a charge e moving at the velocity v is given by $\boldsymbol{F} = e\boldsymbol{v} \times \boldsymbol{B}$. The magnetic field action causes curving of charged particle trajectories, attraction or repulsion of current carrying conductors, and splitting of energy levels in atoms, molecules and crystals (*Zeeman effect*), *magnetic resonance* being a particular case of the Zeeman effect. Sources of $\boldsymbol{B}$ are magnetized bodies, current-carrying conductors, and electrically charged bodies in motion. The nature of all these sources is the same: the magnetic field originates from moving microscopic charged particles (electrons, protons, ions, etc.) and from intrinsic (spin) magnetic moments inherent to the particles.

It is clear from Maxwell's equations that an alternating magnetic field is generated when an electric field is changing with time, and a time-varying magnetic field in its turn generates an electric field.

To describe a magnetic field, field lines of force (flux density lines) are often invoked. The vector $\boldsymbol{B}$ is aligned tangent to each point of such a line. The flux density lines come closer together at locations with a higher magnitude of $\boldsymbol{B}$, and they spread apart wherever the field is weaker. The magnetic flux Φ through an area A perpendicular to the $\boldsymbol{B}$ direction is given by the expression $\Phi = BA$. If the field is not uniform then $\boldsymbol{B}$ must be integrated over the area to obtain the flux.

A magnetometer is used to measure the strength of a magnetic field (see *Magnetometry*).

MAGNETIC FIELD, CRITICAL

See *Critical magnetic fields.*

MAGNETIC FIELD PENETRATION DEPTH

See *Penetration depth of magnetic field.*

MAGNETIC FIELD, QUANTIZED

See *Quantizing magnetic field.*

MAGNETIC FIELD, SUPERSTRONG

See *Superstrong magnetic fields.*

MAGNETIC FIELD, THERMODYNAMIC

See *Thermodynamic critical magnetic field.*

MAGNETIC FILMS

Thin layers of a *magnetic substance* deposited on a planar or cylindrical non-magnetic *substrate*. Magnetic films are not only used as objects for studies or as working media alongside bulk magnetic specimens, but they often replace the latter for the sake of technological convenience and improved performance in *microelectronics*. Magnetic films may be monocrystalline, polycrystalline or amorphous, made of *soft magnetic materials* or of *hard magnetic materials*, may be strongly anisotropic or virtually isotropic, and they display a notable variety in their properties. There are magnetic films that exhibit a small ferromagnetic resonance *line width* (~0.02 mT), a very large Faraday rotation of $\sim 10^4$ deg/cm (see *Faraday effect*) that better reveals itself in the optical and IR parts of the spectrum, a very large *domain wall mobility* (up to 10^5 cm·s^{-1}·mT^{-1}), a small *magnetic reversal* time (~1 ns), a large (~1°) Kerr rotation of the plane of light polarization (see *Kerr effect*), etc.

Magnetic films generally differ considerably in their properties from bulk magnetic specimens. The dominant role in their anisotropy is usually played by the induced *magnetic anisotropy* (e.g., *epitaxial iron garnet films* exhibit a strong (growth) induced perpendicular uniaxial anisotropy, with the uniaxial anisotropy constant, K_u at least an order of magnitude greater than the cubic anisotropy

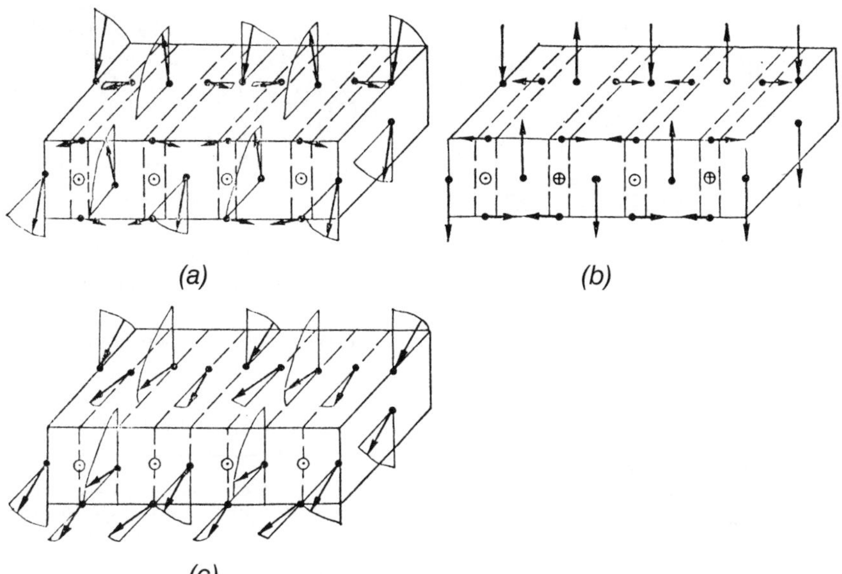

Fig. 1. Magnetic films. Perpendicular anisotropy case: (a) striped domain structure for unipolar twisted domain walls and $q \gg 1$, (b) corresponding case with $q \gg 1$ for heteropolar domain walls, and (c) unipolar domain walls with $q \ll 1$.

constant of bulk specimens). Magnetic films that are not too thick and are uniform in thickness possess a passing-through *magnetic domain structure*; *incomplete domains* may exist in non-uniform films. The good permeability of the magnetic films, in combination with their high specific Faraday rotation, permits employing magneto-optical techniques for the visual observation and study of their domain structure in transmitted light; other methods of *magnetic domain characterization* are also used. The domain structures of magnetic films are differentiated by the orientation of the *easy magnetization axis* (EMA), and by a parameter known as the quality factor $q = K_u/2\pi \boldsymbol{M}^2$, where $\boldsymbol{M}$ is the magnetization of the film:

(a) *The "perpendicular" anisotropy case* (EMA $\parallel \boldsymbol{n}$, where $\boldsymbol{n}$ is normal to the magnetic film surface). With $q \gg 1$, a *striped domain structure*, or *striped structure*, is formed with uni- or heteropolar twisted domain walls (see *Domain wall twisting*). In the structure with unipolar domain walls (when the vectors $\boldsymbol{M}$ in the centers of all domain walls in the magnetic film median plane are collinear), a small deviation of vectors $\boldsymbol{M}$ from the normal takes place in the domain centers on the film surfaces (Fig. 1(a)). There is no such deviation in the structure with heteropolar domain walls (Fig. 1(b)). The structure with heteropolar domain walls possesses a higher energy compared to that of the structure with unipolar boundaries, with the energy difference for these two structures tending to zero at $q \to \infty$. Since the films with EMA $\parallel \boldsymbol{n}$ exhibit rotational invariance, they also may contain a *labyrinthine magnetic domain structure*, isolated *magnetic bubble domains*, as well as mixed type structures composed, e.g., of randomly arranged fragments of the stripe and bubble domains. The bubble domains containing an even number of pairs of vertical *Bloch lines*, or containing no such lines at all (Fig. 2(a)), are composed of the striped domains with heteropolar domain walls; the striped domains with unipolar domain walls display a trend of turning into bubble domains having an odd number of pairs of Bloch lines (Fig. 2(b)). With the quality factor q decreasing, the energy of the *demagnetization field* starts playing the dominant role in the magnetic film energy, hence the angles between the vectors $\boldsymbol{M}$ and

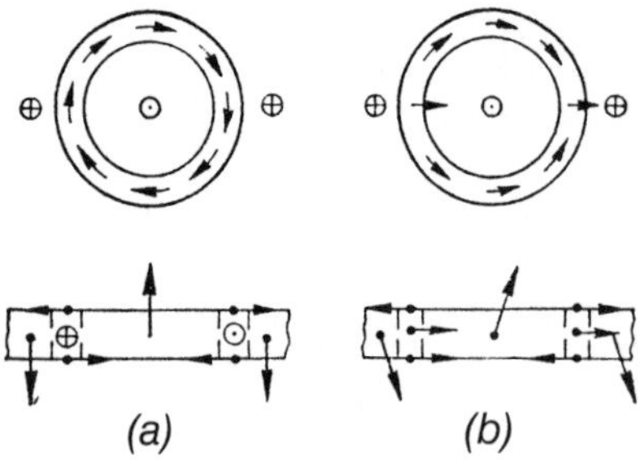

Fig. 2. Bubble domains.

the normal to the magnetic film surface display a tendency of decreasing. For $q \ll 1$, the widths of the domains and of the domain walls become comparable in their magnitude. The domain structure with unipolar domain walls (see Fig. 1(a)) acquires the form shown in Fig. 1(c); the structure with heteropolar domain walls (Fig. 1(b)) becomes disadvantageous in terms of the energy, and is not realized in practice for $q \ll 1$.

(b) *The "parallel" anisotropy case* (EMA$\perp \boldsymbol{n}$). In a magnetic film of infinite transverse dimensions, the ground state is a single-domain one with the $\boldsymbol{M}$ vectors everywhere parallel to the surface. The domain structure only appears when the transverse dimensions become limited due to magnetic poles emerging at the magnetic film edges. A domain structure with Bloch domain walls (Fig. 3(a)) appears in thick magnetic films, and the one with Néel walls in thin films (Fig. 3(c)). In the case of an intermediate thickness structures with mixed domain walls or with "transverse links" have been observed (Fig. 3(b)). The Bloch and Néel wall energies become comparable to each other at a film thickness $D \approx 0.14(A/K_u)^{1/2}$ where A is the *exchange interaction* constant, and, e.g., for permalloy we have $D \approx 35$ nm.

The domain structure pattern becomes appreciably more complex when the axis of easy magnetization deviates from the normal, as well as for types of anisotropy more complex than uniaxial. The remagnetization of magnetic films may occur either by way of *domain wall displacement processes*, or through coherent or non-coherent processes of magnetization vector rotations (see *Magnetization rotation processes*). This depends on the initial distribution of the *magnetization* vector $\boldsymbol{M}$, on the rate at which the remagnetizing field

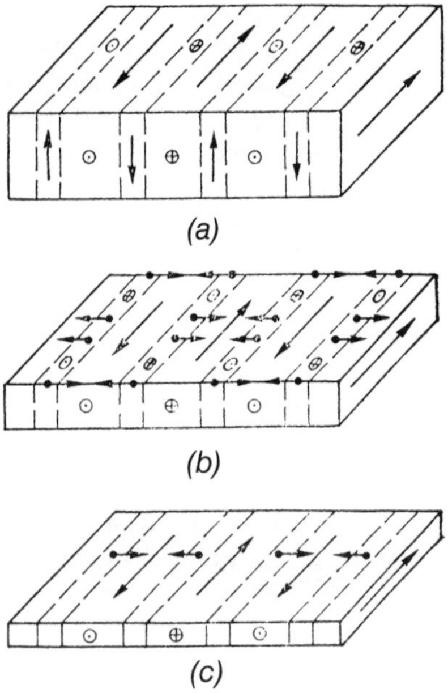

Fig. 3. Magnetic films. Parallel anisotropy case: (a) Bloch domain walls in thick magnetic films, (b) mixed domain walls in intermediate thickness films, and (c) Néel domain walls in thin films.

changes, and on its strength and degree of uniformity. The most important parameters defining the *magnetic domain wall dynamics* are the *domain wall mobility* and *domain wall velocity limit*; processes of vector $\boldsymbol{M}$ rotation are described in terms of the remagnetization time. The spectrum of elementary excitations of magnetic films includes magnetostatic waves (spin interaction without an exchange interaction) and *spin waves* (involving the exchange interaction); these may be direct or reversed, and surface or volume waves. Of great interest also are nonlinear spin waves and *magnetic solitons* in the films. Magnetic films have been manufactured by thermal evaporation in vacuo, by cathode *sputtering*, by chemical reaction techniques, by *epitaxial growth* from the liquid or gas (vapor) phase, etc.

The main fields of application:

(1) Computer engineering: *magnetic bubble domain forwarding circuits* in memory; working media in storage devices based on magnetic bubble domains or on vertical Bloch

lines (epitaxial films of mixed ferrite garnets); working media in storage devices with *thermomagnetic information recording* and magneto-optical reading (*amorphous films* of d-metal and f-metal alloys), etc.

(2) *Magnetooptics* and *integrated optics*: controlled transparencies, light modulators and group switches, transducers of optical modes (orthoferrite plates and epitaxial films of the bismuth- and praseodymium-containing ferrite garnets), etc.

(3) SHF, UHF and EHF engineering: guides, reciprocal and non-reciprocal devices using magnetostatic waves or *ferromagnetic resonance* (epitaxial iron–yttrium garnet films and *hexaferrites*).

(4) Household electronics: magnetic film heads, magnetooptical laser-player disks (permalloy and amorphous films of d-metal and f-metal alloys), etc.

MAGNETIC FLUX, flux of magnetic field

The flux Φ of the vector of *magnetic flux density* (magnetic field) $\boldsymbol{B}$ through a given surface area S. An element, $\mathrm{d}\Phi$ of magnetic flux through an infinitely small surface element $\mathrm{d}S$ is $\mathrm{d}\Phi = B_n\,\mathrm{d}S$, where B_n is the projection of $\boldsymbol{B}$ onto the direction of the unit vector $\boldsymbol{n}$, normal to $\mathrm{d}S$. The magnetic flux through an arbitrary surface S is defined by the integral $\Phi = \int B_n\,\mathrm{d}S$. For a closed surface the magnetic flux is zero, and this can be expressed mathematically with the aid of divergence, $\operatorname{div}\boldsymbol{B} = 0$, and the integral, $\oint B_n\,\mathrm{d}S = 0$, expressions. This situation results from the absence of *magnetic charges* in Nature (see *Magnetic monopole*). For a uniform magnetic field one can write $\Phi = BS$. The units of magnetic flux are: Wb (weber) $=$ Tm^2, in SI and Mx (maxwell) in CGS, where 1 Mx $= 10^{-8}$ Wb.

MAGNETIC FLUX DENSITY

The vector of magnetic flux density, $\boldsymbol{B}$, is a dynamic parameter of the *magnetic field* in a medium that averages the fields generated by individual electrons and other fundamental particles. The magnetic flux Φ passing through an area S normal to the direction of a uniform field $\boldsymbol{B}$ is given by the expression $\Phi = BS$. The magnetic flux density $\boldsymbol{B}$ can usually be expressed in terms of the magnetic field vector, $\boldsymbol{H}$, and the *magnetization* vector, $\boldsymbol{M}$. In the SI system of units we have

$$\boldsymbol{B} = \mu\boldsymbol{H} = \mu_0(\boldsymbol{H} + \boldsymbol{M}), \qquad (1)$$

where μ is the *magnetic permeability*, $\mu_0 = 4\pi\cdot 10^{-7}$ N/A^2 (or H/M) is the *magnetic permeability of vacuum*, and the magnetization $\boldsymbol{M}$ is the *magnetic moment* per unit volume. The latter is directly proportional to $\boldsymbol{H}$ in weak magnetic fields and isotropic media:

$$\boldsymbol{M} = \chi\boldsymbol{H}, \qquad (2)$$

where χ is the dimensionless *magnetic susceptibility*. Upon substitution of Eq. (2) into (1) we obtain $\boldsymbol{B} = \mu_0(1+\chi)\boldsymbol{H}$ and $\mu = \mu_0(1+\chi)$. (In the CGS system $\mu_0 = 1$ and $\boldsymbol{B} = \boldsymbol{H} + 4\pi\boldsymbol{M}$.) The units of magnetic flux density are tesla (T) in SI and gauss (G) in CGS; 1 T $= 10^4$ G.

MAGNETIC HARDNESS

The property of *magnetic materials* whereby the application of a very strong, reversed direction, magnetic field is required to bring the material back to its unmagnetized state (see *Demagnetization*). The extent of magnetic hardness is characterized by the *coercive force*, H_c, and remanence, B_r. Materials with large values of H_c and B_r, or those with the broadest *magnetic hysteresis* loops, are called *hard magnetic materials*; they are mainly used for *permanent magnets*.

MAGNETIC HYSTERESIS

A nonunique irreversible dependence of the *magnetization* $\boldsymbol{M}$ of a magnetically ordered material, e.g., a *ferromagnet*, an *antiferromagnet* or an *amorphous magnetic substance* during a variation (increasing or decreasing) of an external magnetic field on its prior history. A general cause of magnetic hysteresis is the presence within a certain range of applied field of some *metastable states* (side by side with stable ones), with irreversible transitions between them.

For a detailed explanation of this phenomenon in ferromagnets, the individual *magnetic domain structure* must be taken into account, both its formation through nucleation and incipient growth, as well as its variation during *magnetization* and *magnetic reversal* processes. These changes can proceed via *domain wall displacement processes*, as well as *magnetization rotation processes* during magnetization. Anything that hinders these processes and promotes the magnetically ordered material to be in a metastable state serves as a cause of the *hysteresis*. We can distinguish magnetic hysteresis associated with impeding the processes of rotation and *domain wall* displacement, as well as with hindering nucleation and the development of nuclei of remagnetization. The retardation of domain wall displacements can result from the interaction of the latter with *defects* (e.g., non-magnetic inclusions, intergrain boundaries, and others). The lag of the nucleation and the processes of rotation of $\boldsymbol{M}$ are caused by high potential barriers separating the stable states from the unstable ones.

The magnetic hysteresis can be characterized quantitatively by the parameters of a *hysteresis loop*, i.e. a closed curve plotting the dependence of magnetization $\boldsymbol{M}$ (or magnetic induction $\boldsymbol{B}$) on the applied field $\boldsymbol{H}$. A ferromagnetic hysteresis loop is shown in the figure, where curve (a) (*virgin curve*) plots the variation of $\boldsymbol{M}$ from the fully demagnetized state ($\boldsymbol{M} = 0$, $\boldsymbol{H} = 0$) to a state with practically constant $\boldsymbol{M} = \boldsymbol{M}_s$ (*magnetic saturation*) appearing at sufficiently large $\boldsymbol{H} = \boldsymbol{H}_m$. The subsequent course of the variation of $\boldsymbol{M}$ is indicated by arrows. When $\boldsymbol{H} = 0$, $\boldsymbol{M} = \boldsymbol{M}_r$ (*remanence*) and the value $\boldsymbol{M} = 0$ occurs at the applied field $\boldsymbol{H} = \pm\boldsymbol{H}_c$ (*coercive force*). Hysteresis loop 1 is called a *maximal loop*. Magnetic hysteresis loops called *low field loops* (curve 2) appear when $\boldsymbol{H}$ changes over a more limited range $-\boldsymbol{H}_1$ to $\boldsymbol{H}_1 \leqslant \boldsymbol{H}_m$. A differential loop over a very small range from $\boldsymbol{H}_1$ to $\boldsymbol{H}_2$ subject to the condition $|\boldsymbol{H}_2 - \boldsymbol{H}_1| \ll \boldsymbol{H}_1, \boldsymbol{H}_2$ is generally asymmetric. A loop exhibits a lag of $\boldsymbol{M}$ from $\boldsymbol{H}$ with the result that the energy acquired during magnetizing is not released completely during demagnetizing. The energy lost in one cycle (*hysteresis loss*) (see *Magnetic loss*) is described by the integral $\oint \boldsymbol{H}\,\mathrm{d}\boldsymbol{M}$ (the loop area). Eventually, this loss appears as heat. Loops of magnetic hysteresis exist for both quasi-static and dynamic variations of $\boldsymbol{H}$. In the latter case they are broader, due to the extra *dynamic loss* associated with the *magnetic viscosity*, eddy currents (in conductors), and other mechanisms. The loop shape, the energy dissipated per cycle, and other parameters (loss, $\boldsymbol{H}_c$, $\boldsymbol{M}_r$, etc.) depend on the type and value of the *magnetic anisotropy*, the chemical composition of the material, its structural condition, temperature, defect distribution, and method of preparation. For some treatments (*thermomechanical treatment*, *thermomagnetic treatment*, and so on), the hysteresis is observable not only at sign changes of $\boldsymbol{H}$, but also during its rotation (*rotational hysteresis*).

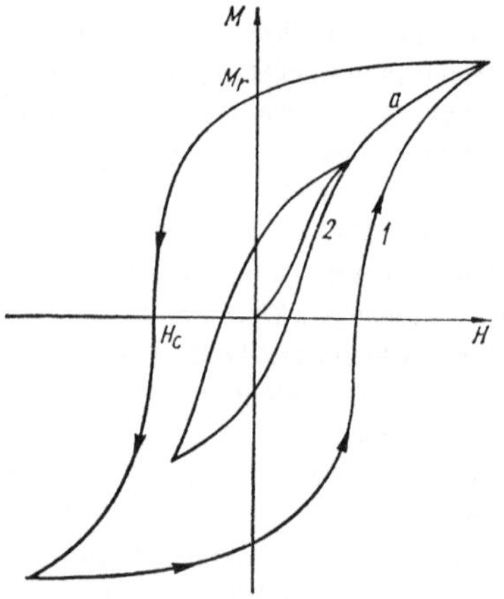

High field (1) and low field (2) hysteresis loops showing the nonunique dependence of the magnetization $\boldsymbol{M}$ on the strength of the applied magnetic field $\boldsymbol{H}$. The coercive field $\boldsymbol{H}_c$ and remanent magnetization $\boldsymbol{M}_r$ are indicated.

The source of hysteresis in other magnetically ordered materials (antiferromagnets, *ferrimagnets* in strong magnetic fields) is the same: the presence of metastable states and a domain structure (see *Intermediate state*). The magnetic hysteresis of amorphous magnets (*spin glasses*, *asperomagnets*) is caused by a nonergodicity of these materials.

Other physical properties of magnetically ordered materials can also exhibit a magnetic hysteresis behavior under a cyclic variation of $\boldsymbol{H}$, and examples are hysteresis of *magnetostriction*, of *galvanomagnetic effects*, of *magnetooptical effects*, etc. The dependence of $\boldsymbol{M}$ is also nonunique under a cyclic variation of temperature (*temperature hysteresis*) and stress (*magnetoelastic hysteresis*). Magnetic hysteresis phenomena have many features in common with the *hysteresis in phase transitions*.

MAGNETIC IMPURITIES in superconductors

Impurity atoms with incomplete inner shells possessing a non-zero *spin* and *magnetic moment.* Unlike non-magnetic impurities, magnetic impurities in a *superconductor* generally bring about the *destruction of Cooper pairs* in the singlet pairing case, and a change of the whole set of thermodynamic properties of the superconductor, including the values of the *critical temperature*, *energy gap*, and Josephson *critical current.* The destructive effect of magnetic impurities on the pairing arises from the *exchange interaction* (which violates time reversal symmetry) between the *conduction electron* spins and the magnetic impurity spin; this leads, over a particular range of impurity concentration, to the emergence of zero-gap superconductivity (see *Gapless superconductors*).

There are exceptions to the destructive effect of magnetic ions on the superconducting state. For example, magnetic order and superconductivity can coexist in Chevrel phases, $A_xMo_6X_8$, when A is a magnetic rare-earth ion which occupies a site remote from the the Mo ions of the MoX_8 groups. Some cuprate (i.e. high-temperature) superconductors become antiferromagnetic with a Néel temperature far below the superconducting transition temperature. An example is $ErBa_2Cu_3O_{7-\delta}$ with $T_N = 0.5$ K and $T_C = 88$ K.

MAGNETIC IMPURITY INTERACTION,
sometimes called carrier–impurity interaction

Interaction of *current carriers* with magnetic impurities in *metals* and *semiconductors.* The term carrier–impurity interaction was introduced in conjunction with studies of *exchange interactions* and spin correlations in magnetically doped semiconductors. This interaction develops in the presence of the electric fields of *defects*, with exchange forces included, and also as a result of the magnetic interaction of current carriers with impurities. The carrier–impurity interaction results in the dependence of internal degrees of freedom (electron, spin, vibron, etc.) of carriers and impurity centers on those electromagnetic forces that are switched on when carriers come close to impurities. The most extensively studied types of magnetic–impurity interactions are the exchange interaction that results in the *indirect electron spin–spin coupling* of the magnetic ions through the conduction electrons (*Ruderman–Kittel–Kasuya–Yosida interaction*, RKKY interaction), and in the *Kondo effect*, the giant spin splitting of *conduction bands*, during spin polarization and *relaxation* of the impurity centers through the *exchange scattering* of the current carriers. Exchange scattering is understood as the processes of the scattering of carriers by *paramagnetic centers*, which are due to their exchange interaction. Quantitatively, the efficiency of the exchange scattering process is characterized by its cross-section, σ. The value of σ in semiconductors depends on the properties of its impurities, and is at its maximum for the so-called *shallow centers* that have a larger radius. Thus, in the case of shallow *donors* of phosphorus P^0 in Si, the exchange scattering cross-section of its *conduction electrons* is $\sigma = 2\cdot10^{-12}$ cm^2 at $T = 1.25$ K, while for the deep centers of Fe^0 in Si we have $\sigma = 4\cdot10^{-15}$ cm^2. As for the charged impurities, spin-dependent processes involving the *trapping of current carriers* prove to be effective in the carrier–impurity interaction, which may lead to the spin orientation of the impurity centers (e.g., in $Si{:}Cr^+$).

MAGNETIC ION, magnetic atom

An ion (atom) possessing a *magnetic moment* which mainly arises from the *spin* and *orbital angular momenta* of electrons and, to a lesser degree (by three orders of magnitude) from the magnetic moments of atomic nuclei. The electronic magnetism of magnetic ions is influenced by the internal electric field in the crystal, which causes the magnetic properties of the ions to be anisotropic. Typical magnetic ions are transition series elements whose *magnetism* is due to the electrons in incompletely filled d-orbital shells, and rare-earth elements with incompletely filled $4f$-orbital shells. The *spin Hamiltonian* method is employed to characterize the properties of such magnetic ions. A magnetic ion giving rise to *paramagnetism* in a crystal, called a *paramagnetic ion*, exemplifies a broader concept of a *paramagnetic center.*

MAGNETIC LIQUID

See *Ferrofluid.*

MAGNETIC LOSS

Partial transformation to heat of electromagnetic energy transferred to magnetically ordered materials during the course of their cyclic *magnetic reversal* by an alternating *magnetic field.* The overall magnetic loss arises from *hysteresis loss* P_h, *vortex-current loss* P_v and loss P_r associated with inherent processes of spin system relaxation (see *Landau–Lifshits equation*). The loss P_h is present in almost all ordered materials; it is determined from the area of the *hysteresis loop*, and appears under dynamic, as well at quasi-static conditions of remagnetization of the samples. From a calculation per remagnetization cycle, the energy density associated with P_h may be determined as $Q_h = P_h/V = \oint \boldsymbol{H}\,d\boldsymbol{M}$, where the integration is performed over the area of the hysteresis loop. Often one also introduces the values $W_h = fQ_h$ of *power of loss* per unit volume of material and the *specific loss* $P_h = W_h/\rho$, where ρ is the *density* of the material, and f is the linear frequency variation of the magnetic field. Such factors are also introduced for the other magnetic loss components, and also for the total magnetic loss.

The vortex-current of the magnetic loss has a major significance only in ferromagnetic conductors; in particular it constitutes the major contribution to P in highly-textured electrical sheet *steels.* An explanation for the presence of this component involves the mechanism of the appearance of a vortex current arising from the influence of the *magnetic flux density* on the motion of *magnetic domain walls.* The dynamic behavior of the *domain structure*, reducing upon remagnetization to the vibration and bending of the walls, to fracturing of *domains* or to the radical reconstruction of the structure, crucially effects the formation of P_v. The calculation of the vortex current component is complex, but the results obtained in the simplest situations show that P_v is inversely proportional to electrical resistance, it depends nonlinearly on the frequency and width of domains, in a nonmonotonic manner on the angle between *easy magnetization axis* and direction of $\boldsymbol{H}$, and quadratically on the amplitude of the magnetic induction. This account of domain structure allows one to explain the experimentally observed mechanisms and the value of P, and thus to solve the problem of the so-called *additional loss* (the difference between $P_h + P_v^{(k)}$ and the experimentally measured value P, where $P_v^{(k)}$ is the vortex-current loss calculated without taking account of the domain structure). Usually the notation of P with fractional subscripts is used for characterizing of the magnetic loss. Their numerator designates the amplitude of the induction $B_m(T)$, and the denominator the remagnetization frequency f (Hz). Thus $P_{1.5/50}$ means that specific loss is measured at $B_m = 1.5$ T at the frequency $f = 50$ Hz. In basic *soft magnetic materials* the value of P_h is insignificant and $P = P_r + P_v$, whereas in polycrystalline sheets of electrical sheet steel P_r plays the main role, and in highly-textured sheets P_v plays this role. To reduce P_v, magnetically active *coatings* are applied to steel sheets which not only act as insulators, but by an appropriate selection of the heat expansion ratio they stretch the steel sheets, which leads to the reduction of the domain width, and to the lowering of P_v. In nonconducting ferromagnets $P_v = 0$, and the magnetic loss arises mainly from P_h and P_r.

MAGNETIC MATERIALS

Ferro- or ferrimagnetic materials whose magnetic properties determine their practical applications. Magnetic materials are an indispensable part of equipment related to the production, distribution, or consumption of electric power; they also play an important role in radio engineering and electronics, in telecommunications, automation, and computer engineering. The industry produces a wide range of these materials with a great diversity of combinations of physical properties, and they are commonly divided into the three main classes:

(1) *soft magnetic materials* used to establish *magnetic fluxes* of high density in definite space volumes. Typical devices using soft magnetics are electric generators, electric motors, and transformers. The magnetic fluxes in these devices change in magnitude and/or direction with time. Along with high *magnetization*, soft magnetics should possess small values of the *coercive force*, B_c, below 10^{-6}–10^{-3} T, and a high *magnetic permeability*;

(2) *hard magnetic materials*, mainly used to manufacture stable magnetic field sources: *permanent magnets*; these should posses a high *remanence* and $B_c > 10^{-2}$–1 T;

(3) *special purpose magnetic materials*, comprising those with a rectangular *hysteresis* loop, *magnetostrictive materials* and *thermomagnetic materials*, magnetic recording materials (see *Thermomagnetic information recording* and *Magnetic bubble domain devices*), materials for microwave equipment, and microelectronics, etc.

See also *Amorphous magnetic materials, Anisotropic magnetic materials, Hard magnetic materials, High-permeability magnetic materials, Soft magnetic materials.*

MAGNETIC MOMENT

A vector quantity characterizing the magnetic properties of closed *electric currents* in bodies or material particles. The magnetic moment, $\boldsymbol{\mu}$, of a closed circuit carrying a current I equals: $\boldsymbol{\mu} = I\boldsymbol{S}$, where $\boldsymbol{S}$ is a vector numerically equal to the area encompassed by the circuit, and aligned along the normal to the circuit plane in the right-handed sense. The SI unit for the magnetic moment is $\mathrm{A{\cdot}m^2}$. The intrinsic magnetic moments of fundamental particles arise from their spin. The magnetic moments of atoms are vector sums of their spin and orbital magnetic moments stemming from the spin and orbital motion of the electrons, and of the nucleons in their nuclei. The unit magnetic moments in atomic and nuclear physics are, respectively, the Bohr *magneton* $\mu_B = e\hbar/(2m_e) = 9.274{\cdot}10^{-24}$ J/T, and the *nuclear magneton*, $\mu_N = e\hbar/(2m_p) = 5.051{\cdot}10^{-27}$ J/T, where m_e and m_p are, respectively, the masses of the electron and the proton. The *magnetization*, which is the magnetic moment per unit volume of a material, serves as a measure of the state of magnetization for macroscopic bodies.

MAGNETIC MONOPOLE, Dirac monopole (P. Dirac, 1931)

Hypothetical particle possessing positive or negative *magnetic charge*, in analogy with electric charge.

The laws of classical electrodynamics leave room for particles possessing a single magnetic monopole to exist, and satisfy certain field equations and equations of motion. These laws incorporate no postulates that forbids the possibility of the existence of magnetic monopoles.

The situation in quantum mechanics is somewhat different. Consistent equations of motion for a charged particle moving in the field of a magnetic monopole, and those for a monopole moving in the field of a charged particle, can only be constructed if the electric charge, e, of the particle and the magnetic charge, g, of the magnetic monopole, are related through the expression:

$$eg = 2\pi n\hbar = nh,$$

where e is the charge of an electron, and n is a positive or negative integer. This condition arises when the particles are represented by waves in quantum mechanics, and interference effects emerge in the motion of particles of one type influenced by particles of another type. If the magnetic monopole were to exist, the above equation stipulates that all the charged particles located in its vicinity have a charge e equal to a multiple of the quantity h/g. Thus, the electrical charges must be quantized. Still, it is one of the fundamental laws of Nature that all existing electrical charges are multiples of the electron charge. Had the magnetic monopole existed, such a law would have a natural justification. No other explanation exists for the quantization of electrical charge. On assuming e as the charge of an electron whose magnitude is defined by the relation $e^2/(4\pi\varepsilon_0\hbar c) = 1/137$, one can derive from the above equation the smallest possible magnetic charge, g_0, of a magnetic monopole, called a Dirac monopole, defined by the relation $g_0^2/(4\pi\mu_0\hbar c) = 137/4$, so g_0 has a value much larger than e. It follows that the track of a fast moving magnetic monopole, e.g., in a Wilson chamber, must stand out sharply on the background of other particle tracks, but meticulous searches for them have been without success. The magnetic monopole is stable and cannot disappear until it annihilates with another monopole with a charge equal in magnitude and opposite in sign. Searches for magnetic monopoles generated by high-energy cosmic rays continuously incident on the Earth all turn out negative.

MAGNETIC NEUTRON DIFFRACTOMETRY

A collection of methods to investigate the *atomic magnetic structure* of materials (mostly, of *crystals*) based on the measurement of *thermal neutron diffraction* associated with the magnetic scattering of neutrons. The experiment is based on measuring the intensities of magnetically diffracted neutron beams using special diffractometers installed in the immediate vicinity of a nuclear reactor. One of the main experimental difficulties is distinguishing between the magnetic and the nuclear scattering, and this can be accomplished by using a neutron diffractometer and polarized *neutrons.* The most efficient separation is possible with a device known as a *spin-spectrometer* which gauges the extent of the depolarization of the neutrons interacting with the sample. This provides a set of magnetic diffraction reflection intensities that may be converted into magnetic *structure amplitudes* by ordinary structure computation techniques.

An important aspect of magnetic neutron diffractometry is the determination of the dependence on the scattering angle, θ, on the atomic magnetic amplitude (*form factor*) that is governed by the spatial distribution of the electrons responsible for the atomic magnetic moments (intrinsic, or induced by an external magnetic field). One seeks coincidence between the experimentally measured form factor values and those computed for an a priori assumed distribution of d- or f-electrons, taking into account *ligand* field energy level splittings. One can also take into account the polarized closed inner atomic shells, and the delocalized electrons.

In the neighborhood of *magnetic phase transition* points, *critical magnetic neutron scattering* appears near the Bragg magnetic reflections. This phenomenon is associated with *spin density* fluctuation enhancements near the *Curie point* (*Néel point*). Critical neutron scattering data can provide the spin moment *correlation lengths* in *magnetic substances* which are directly related to the magnetic parameters of the scattering. There are structure analysis techniques which permit one, proceeding from the neutron data, to construct the spin density distribution (total and differential), as well as the *magnetization* in a *unit cell.* Investigating one and the same chemical compound with X-ray and neutron diffractometry in parallel, one can elucidate separately the particularities of the charge and the spin density redistributions that occur during *chemical bond* formation.

MAGNETIC ORDER

See *Magnetism* and *Atomic magnetic structure.*

MAGNETIC PERMEABILITY, permeability

A physical quantity that expresses how the magnetic flux density $\boldsymbol{B}$ in a material varies with the *magnetic field* $\boldsymbol{H}$. The magnetic permeability $\widehat{\mu}$ is defined through the equation $\boldsymbol{B} = \widehat{\mu}\boldsymbol{H}$; and in general $\widehat{\mu}$ is a tensor dependent on the substance properties, the magnetizing force, and the temperature. The intrinsic magnetic permeability of a material, $\widehat{\mu}$, and magnetic permeability μ' of a particular body composed of that material can be differentiated, depending on whether the internal or external magnetic field $\boldsymbol{H}$ is involved in the $\boldsymbol{B} = \widehat{\mu}\boldsymbol{H}$ relation. The magnetic permeability μ' of a body depends upon the specimen shape, and is related to the intrinsic magnetic permeability $\widehat{\mu}$ by the demagnetization factor (see *Demagnetization fields*). In the SI system, the magnetic permeability is related to the *magnetic susceptibility*, χ, by the equation: $\mu = \mu_0(1+\chi)$, where μ_0 is the *magnetic permeability of vacuum* (in CGS $\mu = 1 + 4\pi\chi$). In *diamagnets*, $\mu < 1$, in *paramagnets* and magnetically ordered materials, $\mu > 1$, and in *ferromagnets*, $\widehat{\mu}$ may greatly exceed 1. As in the case of the magnetic susceptibility, the dynamic and static magnetic permeabilities are often introduced, with the *initial magnetic permeability*, *reversible magnetic permeability* and *differential magnetic permeability* distinguished from them.

MAGNETIC PERMEABILITY OF VACUUM

A proportionality factor in SI units relating the *magnetic flux density* (magnetic induction) $\boldsymbol{B}$ and the *magnetic field* $\boldsymbol{H}$ in vacuo through the expression $\boldsymbol{B} = \mu_0\boldsymbol{H}$, where $\mu_0 = 4\pi\cdot 10^{-7}$ H/m $=$ $1.25664\cdot 10^{-6}$ H/m. In the CGS system $\mu_0 = 1$, and is dimensionless.

MAGNETIC PHASE TRANSITIONS

Phase transitions involving changes of the *atomic magnetic structure* (or *magnetic domain structure*) of a magnetic material. Magnetic phase transitions are induced by changes in an external parameter (temperature, magnetic field, pressure, etc.) and may occur as both *first-order phase transitions*, and *second-order phase transitions.* We can distinguish several basic types of magnetic phase transitions.

An *order–disorder type phase transition* is associated with the transition of a material from the paramagnetic state into a magnetically ordered state (see *Magnetism*). This transition is usually second order at a particular temperature (*Curie point* for *ferromagnets*, *Néel point* for *antiferromagnets* and *ferrimagnets*, *freezing point* for *spin glasses* and other *amorphous magnetic substances*). The fluctuation region is usually rather broad for such phase transitions, hence the *Landau theory* is inapplicable; the similarity theory, the *renormalization group method*, the *epsilon expansion*, and the $1/n$-expansion approach can be employed to describe these transitions. In some materials, e.g., in antiferromagnetic Cr and ferromagnetic MnAs, the order–disorder transition occurs as a clearly defined first-order one. This is attributed to the *order parameter* being strongly linked with the lattice *strain*. The order–disorder type of phase transition also can be changed by the effects of *critical fluctuations*.

There is a great variety of *order–order type phase transitions*. Thus, phase transitions involving a change of the magnetic ordering variety are known, e.g., the antiferromagnet $\leftrightarrow$ ferromagnet first-order type occurring in MnP at $T = 50$ K and in FeRh at $T = 350$ K, the first-order ferromagnetic $\leftrightarrow$ *modulated magnetic structure* type in some rare-earth metals (see *Magnetism of rare-earth metals*, etc.), as well as the *spin-reorientation phase transitions* associated with changes in spin alignment. Among the latter are the *spontaneous magnetic phase transitions* occurring due to temperature or hydrostatic pressure variations, and the *induced magnetic phase transitons* due to changes of the external magnetic field and occurring at a definite magnitude of the applied B. A phenomenological description links spontaneous magnetic phase transitions with a change in sign of the *effective anisotropy constant* and, consequently, with the change of orientation of the *easy magnetization axis* (see *Magnetic anisotropy*). The direction of *magnetization*, $\boldsymbol{M}$, in ferromagnets (or $\boldsymbol{M} = \boldsymbol{M}_1 + \boldsymbol{M}_2$ of sublattices with *antiferromagnetism vector* $\boldsymbol{L} = \boldsymbol{M}_1 - \boldsymbol{M}_2$, in two-sublattice antiferromagnets and ferrimagnets) changes stepwise for the first-order case, while for the second-order case what are termed *angular phases* emerge, with $\boldsymbol{M}$ or $\boldsymbol{L}$ making an angle $\Theta \neq 0, \pi/2$ with selected axes that depends on the temperature and pressure. For instance, a magnetic phase transition associated with the magnetization turning from the hexagonal axis towards the basal plane in the ferromagnets Co and Gd, and the *intermetallic compounds* of the RCo_5 type (R is a *rare-earth element* or *yttrium*) takes place by way of two second-order phase transitions. Also, in some *orthoferrites* ($SmFeO_3$, $EuFeO_3$, etc.), the reorientation of the weak ferromagnetic moment from the a-axis to the c-axis proceeds by a second-order transition, and the angular phase stretches over a 10–20 K region. Realignment of $\boldsymbol{L}$ in antiferromagnets characterized by the *Dzialoshinskii interaction* may be accompanied by a weak ferromagnetic moment appearing or disappearing, as exemplified by the *Morin transition* in α-Fe_2O_3 (haematite), by the first-order transition in $DyFeO_3$ at 42 K, etc. Best known among induced magnetic phase transitions are *spin-flop phase transitions* in antiferromagnets (see *Antiferromagnetism*) occurring in a magnetic field parallel to the antiferromagnetism axis at a field magnitude $B \approx (2B_A B_E)^{1/2}$, where B_A is the anisotropy field, and B_E is the *exchange field*. *Spin-flip magnetic phase transitions* correspond to the transformation to parallel alignment of the sublattice magnetizations with the applied field increasing to $B \approx B_E$, as well as *metamagnetic phase transitions*. Induced magnetic phase transitions in finite specimens may be accompanied by the emergence of a particular thermodynamically stable domain structure: the *intermediate state* of a magnetic substance. The fluctuation region of these order–disorder transitions is usually not broad, so they can be described by the Landau theory.

Another group of magnetic phase transitions comprises transformations from a homogeneous state (a paramagnetic or a magnetically ordered one) to a macroscopically non-homogeneous state (having a domain structure or a long-period modulated magnetic structure).

MAGNETIC PHONONS

Acoustic-type excitations of an electron–ion system in a *metal* which exist only in a strong *quantizing magnetic field*. The magnetic phonons appear due to the interaction of *crystal lattice vibrations* with a quantum wave whose phase velocity differs least from the *sound velocity* in the metal. Since the spectrum of quantum waves is linear in this case, the divergence of dispersion curves at the interaction of the waves with acoustic *phonons* does not take place in the vicinity of a single point (intersection), but rather over the whole of their expanse. As a result, two types of tightly bound excitations exhibiting linear spectra are generated in a metal in the general case, termed magnetic phonons to discriminate them from other quantum waves that only weakly interact with lattice vibrations.

MAGNETIC PLASMA WAVES

A variety of weakly damped electromagnetic waves that propagate in *metals* in the presence of a constant external magnetic field. Magnetic plasma waves occur in a compensated metal when the spatial dispersion of the conductivity is absent, while frequency dispersion exists. Underlying the propagation of magnetic plasma waves is the fact that the *electrical conductivity* in a strong magnetic field dramatically decreases in directions transverse to the field, a metal behaving like an *insulator* along these directions. There are two types of magnetic plasma waves: those similar to the fast magnetoacoustic waves, and Alfvén waves in a plasma. The spectrum of both types of plasma waves, as distinct from the spectrum of *helicons*, is linear, with the velocity of the plasma wave proportional to the strength of the applied magnetic field. The polarization of magnetic plasma waves is linear, the vector of electric intensity of the wave being perpendicular to the external magnetic field. In experimentally attainable magnetic fields, the conditions required to observe magnetic plasma waves are present in *semimetals* with a small current carrier density (e.g., in *bismuth*).

MAGNETIC POLARON

An elementary excitation of electronic origin in a magnetically ordered crystal (*ferromagnet* or *antiferromagnet*) due to the *exchange interaction* between the spins of an electron (hole) and of an atom (molecule) in a crystal. A magnetic polaron can be considered as a compound *quasi-particle* constituted by an electron, a *hole* or an *exciton* enveloped by a region of nonuniform magnetization of a magnetic subsystem (a *magnon* "cloud"). Like an ordinary *polaron* (electron plus induced lattice polarization strain field) in an *ionic crystal*, a magnetic polaron is enveloped by a nonuniform polarization (strain) of the lattice (a *phonon* "cloud"). Such a configuration can travel over the crystal as an integral whole, and be a *current carrier* (charge carrier) or an energy carrier. The effective mass of a magnetic polaron may substantially exceed that of a free electron, and the *self-localization* of a magnetic polaron is possible in some cases. Depending on the sign of the exchange interaction, the nature of the ground state of a magnetic polaron will differ. If the exchange integral is negative then a magnetic polaron may exist at zero temperature. Some varieties of a magnetic polaron are: *ferron*, a ferromagnetic region enclosing a charge carrier in an antiferromagnetic semiconductor; *fluctuon*, a region of magnetically ordered phase enclosing an electron in a paramagnetic metal or alloy in the vicinity of a ferromagnetic phase transition; a *magnon bubble* which is a singlet state region enclosing an itinerant or localized charge in an antiferromagnet, etc.

MAGNETIC POLE

A section of the surface of a magnetized specimen (body) where the normal component of the *magnetization* vector M_n has a maximum value. The term North-seeking (N) or positive magnetic pole refers to the area where magnetic flux or magnetic induction ($\boldsymbol{B}$ field) lines exit the specimen (see *Magnetic flux density*), with flux lines entering the material at the South-seeking (S) or negative magnetic pole. In analogy with the electrostatic case one can ascribe to magnetic poles a non-zero surface density of *magnetic charge*, $\sigma_m = M_n$. The total absence of magnetic charges (see *Magnetic monopole*) in Nature means that magnetic induction $\boldsymbol{B}$ lines can never be discontinuous inside a specimen (body), so a magnetic pole of one polarity must always coexist with an equivalent magnetic pole of the opposite polarity on a magnetized specimen.

MAGNETIC POWDER PATTERN, Bitter pattern (F. Bitter, 1931), Akulov–Bitter pattern

A picture formed by the nonuniform distribution of small ferromagnetic particles on the surface of a *magnetic substance* with a *magnetic domain structure*. The nonuniformity is caused by the preferable aggregation of particles in regions with maximum magnetic field strength, close to where magnetic *domain walls* reach a sample surface, and also of *domains* which have a projection of their *magnetization* which is normal to the surface. F. Bitter (1931), and also N.S. Akulov and M.V. Degtyar (1932), suggested the use of powder patterns formed by applying ferromagnetic particles to the surface of a *ferromagnet* in the form of a liquid suspension, for *magnetic domain characterization.* A magnetic powder pattern can detect the distribution of the ends of vortices in superconductors.

MAGNETIC PRESSURE

An action exerted by a frozen-in *magnetic field* on a conducting liquid (e.g., on a plasma, including a *solid-state plasma*) and directed perpendicular to the field lines. The magnetic pressure, p_M, is proportional to the square of magnetic flux density, $\boldsymbol{B}$, viz., $p_M = B^2/2\mu$ (in SI units). It can be counterbalanced by the kinetic plasma pressure. If the magnetic pressure exceeds the kinetic one, a discharge compression (known as the *pinch-effect*) occurs; it was first described by W.H. Bennett (1934). The term was introduced by L. Tonks (1937). Magnetic pressure is a crucial factor in the operation of a tokamak nuclear fusion reactor.

MAGNETIC QUANTUM NUMBER (m, M)

A quantum number characterizing eigenvalues of the projection of the *angular momentum* operator $\hbar J$ along a magnetic field direction. For orbital motion the projection of the orbital angular momentum $\hbar L$, in the units of $\hbar$, can take on the $2L+1$ integer values $m_L = 0, \pm 1, \ldots, \pm L$, where L is the orbital quantum number. For the case of spins a spin magnetic quantum number, m_S, is introduced. In the case of an integer spin S, m_S may equal $0, \pm 1, \ldots, \pm S$, and for half-integer spin m_S takes on the values $\pm 1/2, \pm 3/2, \ldots, \pm S/2$.

The magnetic quantum number is used in optics, microwave, and radiowave spectroscopy to label energy levels, and to identify transitions between them. The hybridization of states with different m values may occur in crystals because of the atomic level splitting in the crystaline fields, and sometimes there is complete or partial *orbital angular momentum* decoupling or cancellation (see *Quenched (frozen) orbital angular momentum*). From a group theory perspective the states might be characterized by quantum numbers associated with the irreducible representations of *point groups.* The parameter (index) describing the states within the given representation is an analogue of the magnetic quantum number (m_k) for a crystal; it may take on $2l_k + 1$ values, where l_k is the dimension of the associated irreducible representation.

MAGNETIC RELAXATION

Process of attaining equilibrium in the spin subsystem of a body. Because of the weak coupling of the *spins* of atoms and subatomic particles with the particle motion, equilibrium within the spin system of a magnetically ordered medium (*ferromagnets* and *antiferromagnets*) is reached, as a rule, prior to the entire body coming to the state of equilibrium. A *spin temperature* can be assigned to the spin subsystem under such conditions (see *Level population*).

The magnetic relaxation becomes more involved due to the forces of different origin acting between the spins. The exchange forces (see *Exchange interaction*), largest in magnitude, cannot change the mean *magnetic moment* of the system, even if its value is a nonequilibrium one, but these forces can readjust individual spin directions to equilibrate the spin subsystem temperature. Spin interaction forces (see *spin–orbit interaction, magnetic dipole interaction*, etc.) are responsible for the *relaxation* of the mean magnetic moment, so its individual components can return to equilibrium with different rates.

Magnetic relaxation of the magnetic moment (magnetization) component perpendicular to the applied magnetic field in *paramagnets* is associated with the *spin–spin interaction*, and is known as the *spin–spin relaxation* time τ_2 (transverse relaxation time), while relaxation of the longitudinal component (component parallel to the field) of the magnetization to the lattice is known as the *spin–lattice relaxation* time τ_1 (longitudinal relaxation time). Typically, $\tau_1 \geqslant \tau_2$, with different mechanisms operating, so the magnitudes and the temperature dependences of the two *relaxation times* generally differ from each other. The magnetic relaxation of nuclear spins is conditioned by their relatively weak interactions both with other degrees of freedom in the solid, and with each other. Therefore, the time constant of nuclear magnetic relaxation is generally much longer than other magnetic relaxation times such as those due to electronic spins (see *Nuclear quadrupole relaxation*).

Magnetic relaxation affects magnetization and remagnetization processes (see *Magnetic viscosity*), and it determines the width of a *magnetic resonance* line, as well as the *magnetic susceptibility* dispersion. Magnetic relaxation puts limits to the employment of magnetic materials in engineering and in physical experiments. Since this relaxation (like other relaxation processes) depends on the structure of the solid and on the presence of *dislocations* and other *defects* (especially paramagnetic ones), the magnetic relaxation time can be controlled by technological treatments (*alloying*, *quenching*, etc.).

MAGNETIC RESONANCE

Absorption of electromagnetic waves by *magnetic moments* precessing in an external magnetic field. The absorption involves transitions ΔE between Zeeman energy splittings in an applied magnetic field $\boldsymbol{B}$, with the resonance condition expressed as $\Delta E = \hbar\omega = g\mu_{\mathrm{B}}B$ for electron spins, and as $\omega = \gamma B$ for nuclear spins, where g is the dimensionless g-factor, γ is the gyromagnetic ratio, and $g\mu_{\mathrm{B}} = \hbar\gamma$. The dynamics of the phenomenon is described by the Bloch equations for the time dependence of the magnetization vector $\boldsymbol{M}$:

$$\frac{\mathrm{d}M_x}{\mathrm{d}t} = \gamma(\boldsymbol{M}\times\boldsymbol{B})_x - \frac{M_x}{T_2},$$

$$\frac{\mathrm{d}M_y}{\mathrm{d}t} = \gamma(\boldsymbol{M}\times\boldsymbol{B})_y - \frac{M_y}{T_2},$$

$$\frac{\mathrm{d}M_z}{\mathrm{d}t} = \gamma(\boldsymbol{M}\times\boldsymbol{B})_z - \frac{M_z - M_0}{T_1},$$

where M_0 is the equilibrium magnetization in the absence of an applied radiofrequency signal, and T_1 and T_2 are, respectively, the spin–lattice and spin–spin relaxation times. This energy absorption by nuclei is termed *nuclear magnetic resonance* (NMR), and the corresponding absorption by magnetic moments of unpaired electrons in paramagnets is termed *electron paramagnetic resonance* (EPR). Ordinarily EPR resonant frequencies exceed NMR frequencies by three orders of magnitude. In magnetically ordered materials we have *ferromagnetic resonance*, *antiferromagnetic resonance* and *ferrimagnetic resonance*, respectively. Magnetic resonance imaging in medicine is an important application of NMR.

MAGNETIC RESONANCE IMAGING (MRI), MR tomography, NMR endoscopy (title in Russian)

A set of methods for investigating the internal structure of subjects by utilizing *nuclear magnetic resonance.* The resonance frequency ω_0 and permanent magnetic field B_0 in NMR are related by the expression $\omega_0 = \gamma B_0$ (here γ is the *gyromagnetic ratio* of the resonant nuclei). If B_0 is spatially nonuniform, i.e. $B_0 = B_0(x, y, z)$, then also $\omega = \omega_0(x, y, z)$. This means that each resonance signal with the frequency $\omega_0(x, y, z)$ arises from a particular volume element $\mathrm{d}V$ that produced this signal. However, according to Maxwell's equations there exist two-dimensional surfaces for any configuration of $B_0(x, y, z)$ where the value B_0 (and consequently, ω_0) does not vary. The reconstruction of the spatial structure of a subject under study is not simple. This problem was first resolved by P.C. Lauterbur (1973) who, using the principles of *tomography*, developed and implemented the method of obtaining NMR images, a procedure later given the name magnetic resonance imaging (MRI).

A principle of spatial focusing of *magnetic resonance*, which in various versions is often applied in *magnetic resonance spectroscopy* and for purposes of NMR, was first performed by R. Damadian (1976). The essence of the focusing was the creation of such a configuration of the field B_0 that it has sufficient uniformity and satisfies the resonance condition only over a limited region of space, with nuclei beyond this region either out of resonance or broadened beyond detection. By scanning over a sequence of these uniform resonant regions, one can study sequentially the spatial characteristics of the subject. Later more refined and sophisticated methods of preparing systematic magnetic field gradiants, and thereby obtaining improved NMR images, were developed.

The resolution of MRI is determined by the *line width*, and by the magnitude and regularity of the gradient of the field B_0. In the first experiments of Lauterbur, the resolution was 0.2 mm. MRI brought substantial changes to the methods of medical diagnostics. The application of MRI to solids is limited by the appreciable (compared to liquids) width of the NMR lines, and this has led to the study of model subjects. The concepts of the MRI have been adopted by other areas of magnetic resonance spectroscopy (see *EPR imaging*, *Tomography*).

MAGNETIC RESONANCE LINE NARROWING

See *Narrowing of magnetic resonance line.*

MAGNETIC RESONANCE, PHONON

See *Phonon magnetic resonance.*

MAGNETIC RESONANCE SPECTROMETERS

Instruments for recording the spectra of *magnetic resonance spectroscopy*. The principal spectrometer elements include a source of radiofrequency or microwaves, an absorbing cell (coil, cavity), and a detector. As a rule, magnetic resonance spectrometers also include sources of magnetic fields. They are most often used to observe and study the phenomena of *electron paramagnetic resonance* (EPR), *nuclear magnetic resonance* (NMR), EPR with *optical detection of magnetic resonance* (ODMR), and *electron–nuclear double resonance* (ENDOR).

EPR radiospectrometers are usually designed so that the microwave absorption in a specimen is observed while scanning a modulated magnetic field (or frequency). Various instrumental designs are available. In the simplest direct absorption EPR spectrometer the power from a microwave source (a klystron) is fed through a waveguide to a *resonator* (cavity) containing the substance under study and located in a strong magnetic field, and the wave passing through (or reflected from) the cavity goes to the detector. The *resonance* is identified from the change in the detector current. The spectrometers use magnetic field modulation at frequencies from tens of Hz to 1 MHz, and often employ synchronous detection to significantly improve the sensitivity. Another type of EPR spectrometer is a superheterodyne one in which a separate microwave source of a different frequency is introduced and the signal amplification is carried out at an intermediate difference frequency. Spectrometers operating in the 3 cm (X), 1.5 cm (K) and 8 mm (Q) bands find the widest use, although both longer and shorter wavelength (down to 2 mm) instruments are available. These spectrometers are also used to record *ferrimagnetic resonance*, *ferromagnetic resonance* and *antiferromagnetic resonance* in which the sample shape should be specified explicitly (usually a thin rectangular plate or a sphere). For these studies it is advantageous to vary the temperature through the phase transition point.

In the case of NMR radiospectrometers, the analogue of the absorbing cavity is a solenoid in the generator tuned circuit. where the specimen under study is placed. The solenoid is positioned in an external magnetic field $\boldsymbol{B}_0$ so that the radiofrequency field $\boldsymbol{B}_1$ is perpendicular to $\boldsymbol{B}_0$. During resonance, one records a small voltage drop in the tuned circuit. An increase in sensitivity is attained by modulating $\boldsymbol{B}_0$ and operating in a synchronous detection mode. To study NMR in solids, a number of high-resolution designs are used in which the sample is rotated rapidly at the "magic" angle. Also in use are multipulse techniques to accumulate signals, average over various NMR broadening mechanisms, and then retrieve the spectrum after its Fourier transformation.

When studying ODMR various optical devices are used to detect the resonance. They measure the angle of rotation of the polarization plane (*Faraday effect*) or the magnetic circular dichroism. Both effects are sensitive to the passage through the resonance magnetic field. ENDOR spectrometers are EPR spectrometers with an additional radiofrequency field introduced into their cavities to induce transitions between the closer spaced nuclear sublevels (NMR levels). The ENDOR spectrum is obtained by scanning the NMR frequency band while recording the EPR response. A nuclear double resonance radiospectrometer has an absorbing cell similar to those used in NMR. Its circuit operates at a tunable frequency, with pulses fed into the cavity to observe *spin echoes*. Scanning the frequency band and observing the echo with the help of superheterodyne signal detection provides nuclear double resonance spectra. Many radiospectrometers include Fourier analysis and signal accumulating units. EPR and NMR are also used to measure *relaxation times*. For continuous wave (CW) saturation the EPR spectrometer is fitted with a microwave generator covering a broad power range to obtain a series of EPR spectra at various levels of saturation. These data provide relaxation times. Alternatively, a pulse saturation technique uses high power pulses to saturate the specimen, and relaxation times are obtained from the time dependence of the EPR signal that follows each pulse. One may also use multipulse techniques to record spin echoes. Relaxation times in NMR spectroscopy are ordinarily measured using multipulse techniques.

The sensitivity of radiospectrometers depends on many factors. EPR spectrometers achieve a sensitivity of 10^{10}–10^{11} spin at a line width of 0.1 mT. NMR spectroscopy studies of solids often deal with rare isotopes reaching only 1% of the basic nuclei. Many spectrometers are fitted with additional devices to obtain spectra at high and low temperatures, at high pressure, during irradiation, etc.

Other magnetic resonance spectrometers include those used in *nuclear quadrupole resonance* (NQR), and *magnetic resonance imaging* (MRI). The latter technique, originally referred to as nuclear magnetic resonance imaging, is widely used in medicine for diagnosing cancer and other diseases. It employs a strong constant magnetic field provided by a superconducting magnet, with a superposed magnetic field gradient which makes the protons in the various parts of the body under examination come to resonance at slightly different frequencies. It is capable of providing three-dimensional images of the brain and other body organs.

MAGNETIC RESONANCE SPECTROSCOPY of solids

A part of the physics of solids dedicated to studying substances by their spectra in the radio and microwave frequency ranges, 10^6–10^{12} Hz. The majority of these techniques involve *electron paramagnetic resonance* (EPR), *nuclear magnetic resonance* (NMR), *ferromagnetic resonance, antiferromagnetic resonance, ferrimagnetic resonance, nuclear quadrupole resonance, cyclotron resonance, paraelectric resonance*, and various types of double and even triple *resonances*, with EPR and NMR by far the most widely utilized. Instruments used for magnetic resonance are called *magnetic resonance spectrometers.*

The high information yield of these measurements arises from the high sensitivity and resolution of the spectrometers, the capability of measuring spacings between the closely lying energy levels, and the ability to analyze the *line shape* with high precision. It is possible to identify the *local symmetry* of paramagnetic centers, and to deduce the crystallographic site and its neighborhood. Hence one may understand the nature and define the number of *paramagnetic centers* and defects surrounding them. Thus, magnetic resonance is a technique for crystal *defectoscopy.* It is also a method for studying *phase transitions* in materials that undergo electronic and molecular ordering, and to determine the characteristics of paramagnetic centers and their environment (the spin density distribution, effective magnetic and electric *dipole moments, electric field gradients*, nuclear *magnetic moments*, etc.). Magnetic resonance can also provide information on *band structure* parameters and the nature of *chemical bonds.* One may study *tunneling phenomena in solids* with its help. Magnetic resonance prompted the development of *masers*, and then *lasers*, which became solid-state instruments, and eventually MRI or *magnetic resonance imaging* in medicine.

MAGNETIC RESONANCE, SPIN PHONON

See *Spin phonon magnetic resonance.*

MAGNETIC RESONANCE, ZERO-FIELD

See *Zero-field magnetic resonance.*

MAGNETIC REVERSAL, remagnetization

The process of decreasing the magnitude and reversing the direction of the *magnetization* $\boldsymbol{M}$ of a magnetically ordered sample, which takes place at the reversal in direction of the external *magnetic field* $\boldsymbol{B}_{\text{app}}$ which had previously magnetized the material to saturation. In multi-domain *ferromagnets* and other materials exhibiting spontaneous magnetization, the reversal process takes place through the generation and growth of magneticly reversed regions, followed by the redistribution of magnetic phases through *domain wall* displacements. As this occurs, those phases with their magnetization pointing closest to the $\boldsymbol{B}_{\text{app}}$ direction grow in volume. The final stage of magnetic reversal involves the complete alignment of $\boldsymbol{M}$ in these domains in the $\boldsymbol{B}_{\text{app}}$ direction (see *Domain wall displacement processes*). Magnetic reversal of small magnetically ordered particles may proceed through irreversible jumps of $\boldsymbol{M}$ followed by reversible alignment of $\boldsymbol{M}$ with $\boldsymbol{B}_{\text{app}}$.

MAGNETIC SATURATION

The state of a material with the maximum possible *magnetization* value, $\boldsymbol{M}_\infty$, called the *saturation magnetization*, that undergoes no further changes with increases in the applied magnetic field. In the case of *ferromagnets*, $\boldsymbol{M}_\infty$ is attained at the completion of several processes: (a) the growth of the *domains* with their magnetic moments aligned along an *easy magnetization axis* as a result of the *domain wall displacement process*, (b) rotation of the magnetization vectors towards the direction of the applied field (*magnetization rotation process*), and what is called a *paraprocess*, viz., the growth of the number of *spins* aligned along the field under the action of a strong applied field, at the expense of those spins with antiparallel orientations (see *Magnetization curves*). In practice, this saturation is obtained in fields from a few millitesla up to 10 T at 20 °C. For *antiferromagnets* and *ferrimagnets* magnetic saturation is reached at a much higher magnitude of the applied magnetic field, close to that of the exchange field. For some *paramagnets*, a state approaching magnetic saturation is reached with a field $\sim$1 T at temperatures of $\sim$1 K.

MAGNETIC SEMICONDUCTORS

Semiconductor materials having as their main components ions with partially filled d- or f-orbital levels whose magnetic moments become ordered at $T \to 0$. Almost all known types of ordering can occur in magnetic semiconductors. For instance, there is ferromagnetic ordering in EuO and EuS, antiferromagnetic ordering in EuFe, and helical ordering in $HgCr_2S_4$. *Magnetic polymorphism* is found in EuSe: the two-sublattice antiferromagnetic structure is replaced by a three-, and then by a four-sublattice one with increasing temperature (the latter two are forbidden in the Heisenberg model, see *Heisenberg magnet*). One can easily control EuSe properties by external factors: a moderate pressure or a magnetic field transforms the antiferromagnetic ordering into a ferrimagnetic or ferromagnetic type (magnetoisotropic metamagnetic, see *Antiferromagnetism*). Unique EuSe properties are due to higher-order exchange interaction terms comparable to the bilinear Heisenberg exchange.

The magnetic properties of pure magnetic semiconductors are, due to a small intrinsic *current carrier* density, n, similar to those of *insulators*. A substantial difference only emerges via *doping* with electrically active impurities. In non-degenerate ferromagnetic semiconductors, the electrons of non-ionized donors enhance the local ferromagnetic ordering at $T \neq 0$ by way of indirect exchange between the localized d- or f-moments in the neighborhood of defects. As a result, the paramagnetic *Curie point*, Θ, may be considerably raised, even though the magnetic disordering temperature, T_C, does not substantially change. Donors added to antiferromagnets may even cause Θ to change sign, although the antiferromagnetic ordering remains.

In degenerate ferromagnetic semiconductors, the donor electrons get delocalized to effect indirect exchange over the whole crystal; thus T_C may be appreciably increased (doubled in EuO and EuS at $n \approx 5 \cdot 10^{26}$ m^{-3}). Unlike the case of *metals*, indirect exchange in magnetic semiconductors cannot be described by a *Heisenberg Hamiltonian* because of the relatively low *Fermi energy*.

Distinctive electrical and optical properties of magnetic semiconductors are also influenced by the strong *exchange interactions* of current carriers with localized d- or f-moments. Therefore, these properties are highly sensitive to magnetic ordering, even being qualitatively different in many cases from the properties of non-magnetic semiconductors. For instance, a huge long-wave (red) shift of the *intrinsic light absorption edge* occurs in ferromagnetic semiconductors at decreasing temperature or at increasing magnetic field strength B, amounting to 0.5 eV in $HgCr_2Se_4$. Such a shift results from a decrease of the current carrier energy at the onset of ferromagnetic ordering. The temperature dependences of the *photoconductivity* edge and light absorption edge are similar in ferromagnetic semiconductors. Typically, the absorption edge temperature shift is weak and occurs toward the short-wave side in antiferromagnetic semiconductors. An enormous spontaneous Faraday rotation (see *Faraday effect*) takes place in ferromagnetic semiconductors, exceeding in EuS, e.g., a value of 10^6 deg/cm. The Faraday quality factor also reaches gigantic values in the ferromagnetic semiconductors, amounting to 10^4 deg/dB in EuO. The n-type ferromagnetic semiconductors, unlike the non-magnetic ones, display, as a rule, a temperature dependent resistivity, ρ: with a high ρ peak in the vicinity of T_C. Ferromagnetic semiconductors exhibit a large *negative magnetoresistance* in the neighborhood of this peak. In non-degenerate ferromagnetic semiconductors, the ρ peak is mainly governed by the difference in the temperature dependences of the donor level depths and of the position of the *conduction band* bottom. Since the local magnetic ordering at a non-ionized donor is higher than that averaged over the crystal, the donor level rises slower than the band bottom with temperature, with the distance between them reaching its maximum not far from T_C. Accordingly, n is at its minimum there.

The ρ peak in degenerate ferromagnetic semiconductors is due to the temperature dependent scattering of conduction electrons (see *Current carrier scattering*) from ionized donors. The randomly positioned donors give rise to a non-uniformity in the conduction electron density distribution, and, consequently, to a non-uniformity of the *indirect exchange interaction* mediated through them. A component of static magnetization fluctuating in space appears, with the maximum scattering on this component occurring near T_C. A bend, instead of a peak, at the *Néel point* on the dependence of $\rho(T)$ on the temperature is found for pure antiferromagnetic semiconductors, while $d\rho/dT$ displays a singularity of the *specific heat* type.

Magnetic semiconductors are the only materials where the *heterophase self-localization* of current carriers has been observed; the phenomenon consists in the conduction electrons forming regions of normal unstable phase and stabilizing them by their localization inside such regions. That occurs when the electron energy is lower in the unstable phase than in the stable one. Hence, the conduction electrons in non-degenerate antiferromagnetic semiconductors create, independently of each other, ferromagnetic microregions and become self-localized in them. Such regions may consist of a few thousand atoms under the most favorable conditions. These *quasi-particles* (*ferrons*) are destroyed by the magnetic field that converts the whole crystal into the ferromagnetic state. Since the ferrons are immobile, their destruction is accompanied by an increase in the crystal conductivity. The existence of ferrons in non-degenerate EuSe and EuFe is proven by experimental data on the B dependences of their photoconductivity and *luminescence*.

Collective ferron states are possible in degenerate antiferromagnetic semiconductors, consisting in the spontaneous division of the crystal into ferromagnetic and antiferromagnetic regions, with the conduction electrons passing from the antiferromagnetic crystal portion into the ferromagnetic one. With n not too high, the ferromagnetic part consists of separate droplets, each containing a considerable number of electrons. With n rising, the number and dimensions of ferromagnetic droplets grow, and, starting from a certain density n_p, the ferromagnetic droplets start touching one another to form a simply connected ferromagnetic region. At $n < n_p$, the crystal behaves like an insulator, as all the electrons are trapped, each in its own droplet, and cannot carry current. Electron percolation occurs at n_p, i.e. the crystal behaves like a metal for $n > n_p$. When converted to the homophase state (by increasing the temperature to make the whole crystal paramagnetic, or by increasing the magnetic field to make the crystal ferromagnetic), a heterophase insulator ($n < n_p$) simultaneously undergoes a transition into the metallic state. At $n > n_p$, the transition into the homophase affects ρ relatively weakly. Such tendencies were observed in both degenerate EuSe and EuFe.

A related phenomenon was discovered in the degenerate ferromagnetic semiconductor EuO in a narrow n interval: it displays metallic type conductance at $T \to 0$; with increasing temperature, a transition to an insulating state occurs, the jump in ρ amounting up to 17 orders of magnitude. The high conductivity of ferromagnetic semiconductors at $T \to 0$ ensues from their inability to contain ferrons at $T = 0$. However, when the temperature rises, with the ferromagnetic ordering disrupted, the electrons reveal a trend to *self-localization* into regions with enhanced ferromagnetic ordering. Therefore, the electrons pass from delocalized levels to localized levels on the donors, with an enhanced ferromagnetic ordering taking place in their neighborhood (see *ferrons*).

MAGNETIC SHIELDING

Protection of a volume of space (or an object) against penetration by a *magnetic field* (as well as of its variations) arising from sources located outside this volume (or object). Achieved with no expenditure of electric energy, magnetic shielding is implemented by isolating the space to be shielded using special envelopes called *magnetic shields*. The basic principle behind magnetic shielding is the compensation (within the protected space) of the external magnetic field by an oppositely directed field produced by *magnetic poles* that develop along the shield edges at the places where the external field enters or exits. This compensation causes the magnetic field in the shielded space to be much smaller than it is without the magnetic screen. With increasing frequency of an applied electromagnetic field the magnetic shielding transforms into electromagnetic shielding because of the energy expended on generating eddy currents (see *Magnetic loss*) in the screen.

MAGNETIC SOLITON

A *soliton* of the field of magnetization, $\boldsymbol{M}$, in magnetically ordered crystals; a solitary solution of the nonlinear *Landau–Lifshits equations*. The simplest magnetic soliton in an easy-axis ferromagnet is a magnetic *domain wall*, and in an easy-plane magnetic material, it is a *rotation wave* or helical vortex. A magnetic soliton in the most restricted sense (sometimes called a *bion*) is associated with a uniform distribution of *magnetization* well away from the soliton (to the left and to the right of the soliton in the one-dimensional case). A so-called bion is characterized by two parameters: the velocity of the center of gravity, and the frequency of the internal precession of the vector $\boldsymbol{M}$. Domain walls, bions and *spin waves* constitute independent nonlinear modes of magnetic dynamics of a one-dimensional *ferromagnet* with an equilibrium magnetization. These objects determine the contribution of the magnetic degrees of freedom to the scattering of penetrating radiation, as well as the thermodynamics of quasi-one-dimensional *magnetic substances*.

In many cases a magnetic soliton behaves like a particle with an effective mass. A domain wall is an analogue of an unstructured particle, and a bion is an analogue of a particle having an internal degree of freedom (precession of the vector $\boldsymbol{M}$). A quantum analogue of a bion is the bound state of a large number of *magnons*, elementary excitations of a magnetic material (see *Magnon bubble*). If E and $\boldsymbol{P}$ are the energy and the momentum of a bion, respectively, and N is the number of magnons localized at it, then,

$$\mathrm{d}E = \boldsymbol{v}\,\mathrm{d}\boldsymbol{P} + \hbar\omega\,\mathrm{d}N,$$

where $\boldsymbol{v}$ and ω are the velocity and the frequency, respectively, of the bion internal precession. The bion frequency is always below the limit of linear spin wave frequencies, and is dependent on N. The requirement $\omega\,\mathrm{d}\omega/\mathrm{d}N < 0$ is usually the necessary condition for bion stability.

We can distinguish *topological magnetic solitons* whose stability is governed by the internal structure of the magnetization (e.g., a 180° domain wall in an easy-axis ferromagnet), and *dynamic magnetic solitons* whose stability depends on the existence of phase integrals of the vector $\boldsymbol{M}$ field equations (total energy, momentum, projection of momentum onto anisotropy axis, etc.). The bion is a dynamic magnetic soliton.

A two-dimensional magnetic soliton may exist in the form of a cylindrical magnon bubble (dynamic magnetic soliton), or of a *magnetic vortex* (topological magnetic soliton) with a topological charge determined by the circulation of the magnetization vector around the soliton axis. A magnetic vortex in an easy-plane ferromagnet is analogous to a vortex in superfluid ^{4}He, while that in an easy-axis one has no analogues in fluid physics, exhibiting a specific ω versus N dependence, and a sensitivity to the topological charge. A particular type of two-dimensional magnetic soliton is the *Bloch line* separating the segments of a planar domain wall having different directions of the turning vector $\boldsymbol{M}$.

A three-dimensional magnetic soliton in a single-axis ferromagnet may be thought of as a nucleus of a new *phase* differing by an inverse alignment of the vector $\boldsymbol{M}$. In an easy-axis ferromagnet such a soliton cannot be very small: it will bind a number N of magnons which exceeds a certain threshold value that depends on the energy of *magnetic anisotropy*, and decreases with increasing energy. This results from the fact that in the three-dimensional case a pair of particles, weakly attracted and obeying a quadratic *dispersion law*, is unable to form a *magnon bound state*. The Landau–Lifshits equations for an anisotropic magnetic substance allow the existence of three-dimensional topological magnetic solitons.

MAGNETIC SPECTROSCOPY

A seldom used term referring to the use of an applied magnetic field in spectroscopy, sometimes employed as a synonym for *magnetic resonance*. The magnetic field produces the *Zeeman effect* splitting of atomic and molecular energy levels into magnetic sublevels. The splitting can can be of the Paschen–Back type at very high applied fields.

MAGNETIC STRUCTURE

See *Atomic magnetic structure*.

MAGNETIC STRUCTURE ANALYSIS

Analysis of the structure of *magnetic materials* based on the dependence of their magnetic properties upon their structure. One can assess the changes in the structure of a solid at a *phase transition* by the changes that take place in its magnetic characteristics. Thus when *austenite* transforms into *martensite*, or into a mixture of *ferrite* and *carbides* in *steel*, then one can monitor the kinetics of such a transformation of a *paramagnet* into a *ferromagnet* by measuring the *magnetic saturation* of the materials. The composition of ferromagnetic *phases* can be evaluated by the *Curie point*. The transformation of a paramagnetic phase into an antiferromagnetic one in chromium-based alloys may be investigated via the specimen *susceptibility*. Measuring the *magnetic anisotropy* by the ballistic method, or by the so-called *magnetic dynamometer method*, permits one to investigate the *texture* of magnetic materials. *Dislocations* and dispersed inclusions in a ferromagnet produce changes in the *coercive force* that sometimes correlate with changes in the mechanical properties of the solid. Magnetic structure analysis is utilized on a large scale to check structural parts and components subjected to *plastic deformation* and *heat treatment*. Devices for this purpose are based on a differential inductance circuit design (see *Magnetometry*) or *coercimeters*. The scattering of neutrons by a magnetic material provides information on its magnetic structure, such as the relative alignments of the spin directions of atoms at various lattice sites in ferromagnets and antiferromagnets.

MAGNETIC STRUCTURE, MODULATED

See *Modulated magnetic structures*.

MAGNETIC SUBLATTICE

A system of identical magnetic atoms or ions positioned periodically in space, with *magnetic moments* of the same magnitude and direction. Magnetic sublattices are introduced to describe the *atomic magnetic structure* of *antiferromagnets* and *ferromagnets*. The translation periods of a magnetic sublattice may equal those of the *crystal structure*, or may be double or a multiple of them. In the latter case, the magnetic unit cell and the crystallographic one do not coincide, the former being a multiple of the latter. In a two-sublattice structure with individual magnetizations $\boldsymbol{M}_1$ and $\boldsymbol{M}_2$ one can write for the total magnetization $\boldsymbol{M} = \boldsymbol{M}_1 + \boldsymbol{M}_2$, and define the *antiferromagnetism vector* $\boldsymbol{L} = \boldsymbol{M}_1 - \boldsymbol{M}_2$. Structures of actual magnetic sublattices has been determined by the method of *magnetic neutron diffractometry*.

MAGNETIC SUBSTANCE, or sometimes simply magnetic

A term applied to materials which exhibit magnetic properties (see *Magnetism*). The classification of magnetic substances is based on the analysis of the magnitude and sign of their *magnetic susceptibility*, and of the nature of their *atomic magnetic structure*. Magnetic substances can be grouped into those without magnetic order (*diamagnets* and *paramagnets*), and *magnetically ordered substances*, viz., those revealing a magnetic ordering (*ferromagnets*, *antiferromagnets*, *ferrimagnets*, etc.).

MAGNETIC SUPERCONDUCTORS

Compounds where superconductivity and magnetic order occur together. The state of superconductivity in the presence of magnetic order is called the *magnetism/superconductivity coexistence phase*. The theory for the coexistence of these two types of ordering was first formulated by V.L. Ginzburg (1956). He pointed out the non-antagonistic nature of *ferromagnetism* and *superconductivity*, which makes the question of their possible coexistence nontrivial. Intense experimental investigations into the coexistence problem started in 1976, after the synthesis of rare-earth (R) containing ternary superconducting compounds of the Chevrel (RMo_6S_8) and RRh_4B_4 types. They possess a system of localized regularly positioned magnetic moments due to the f-electrons of the rare-earth ions. The *critical temperature of superconductors* for this class of compounds is about 10 K, and the *magnetic phase transitions* (of the ferro- or antiferromagnetic type) occur at temperatures of the order of a few degrees Kelvin. The magnetic ordering of rare earth ions in these ternary compounds involves the *indirect exchange interaction* of f-electrons via the conduction electrons (*Ruderman–Kittel–Kasuya–Yosida interaction* (RKKY)) and the *magnetic dipole interaction* of the moments. The mutual influence of the superconducting electrons and magnetic moments occurs due to the paramagnetic effect of the magnetic moment exchange field acting on the *Cooper pair* electron spins, and due to the magnetic field generated by localized moments and acting on the conduction electron orbital motion (exchange and electromagnetic mechanisms, respectively, of the interaction of superconductivity and magnetism).

Each of the above mechanisms results in an antagonism between superconductivity and magnetism, since the exchange and magnetic moment fields suppress the *Cooper effect*, while the superconductivity in its turn screens and reduces both the RKKY and the magnetic dipole interaction of the moments.

Theoretical and experimental studies have shown that *antiferromagnetism* and superconductivity weakly affect each other because of the rapid oscillations of the magnetic and exchange fields at the atomic distance, a, at which the magnetic moments change in an *antiferromagnet*. The superconductor feels an internal field averaged over a superconductivity *coherence length* $\xi \geqslant a$, the resulting field being weak. On the other hand, the competing types of ordering become modified in ferromagnets in the coexistence phase due to their strong mutual interaction (see *Reentrant superconductors*).

MAGNETIC SUPERSTRUCTURE

See *Modulated magnetic structures.*

MAGNETIC SURFACE LEVELS

Quantized energy values corresponding to the *magnetic surface states* of a charged *quasi-particle* when the particle motion in the direction perpendicular to crystal surface is limited by a potential barrier at the crystal boundary, and by the branch of the parabola of the effective potential energy in a magnetic field, U_{eff}, for this quasi-particle (see *Quantizing magnetic field*):

$$U_{\text{eff}} = \frac{1}{2} m\omega_H^2 (y - y_0)^2,$$

where $\omega_H = eB/m$ is the cyclotron frequency, $y_0 = -p_x/(eB)$ is the orbit center ($\boldsymbol{B} \parallel z$, $\boldsymbol{y}$ axis directed inside crystal). The nature of the spectrum of magnetic surface levels is effectively defined by the quantity y_0, i.e. by p_x. For the particular case of "skipping" or "sliding" electrons, and under the condition that that the Fermi energy, E_{F}, is much greater than $\hbar\omega_H$, the orbit center lies outside the crystal at a distance much larger than the dimensions of bulk level wave functions, $l_H = (\hbar/m\omega_H)^{1/2}$. Under such conditions, the particle is quantized in an almost triangular potential well as illustrated in Fig. 1:

$$E_N \approx \hbar\omega_H \left(\frac{E_{\text{F}}}{\hbar\omega_H}\right)^{1/3} N^{2/3} \gg N\hbar\omega_H.$$

Each bulk Landau level with a number N generates a continuous series of surface magnetic levels corresponding to different values of the longitudinal momentum p_x, and this cancels the bulk level degeneracy in p_x. The transitions between magnetic surface levels were detected by oscillations of the *surface impedance* in a weak magnetic field at liquid-helium temperature. The width of the magnetic surface levels is defined by the degree of perfectly specular reflection of quasi-particles

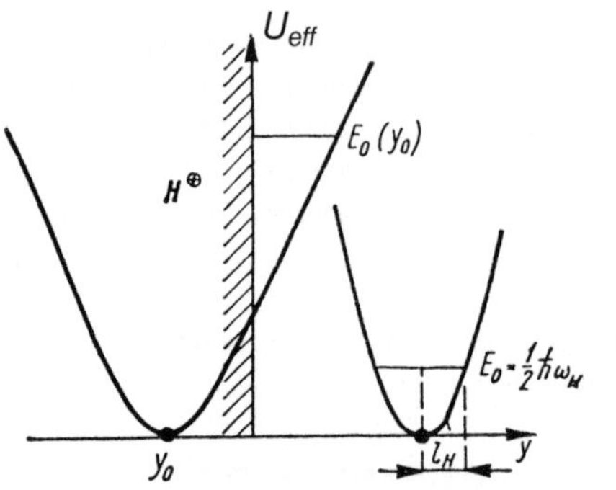

Fig. 1. Graphs of the effective potential energy of quasi-particles with different values of the longitudinal momentum p_x, where $l_H = [\hbar/(m\omega_H)]^{1/2}$ is the magnetic length, or the size of the wave function with $N = 0$, $E_0(y_0)$ is the surface magnetic level generated by the Landau level with $N = 0$ and $p_x = eB|y_0|$. The region outside the crystal is indicated by slopped lines.

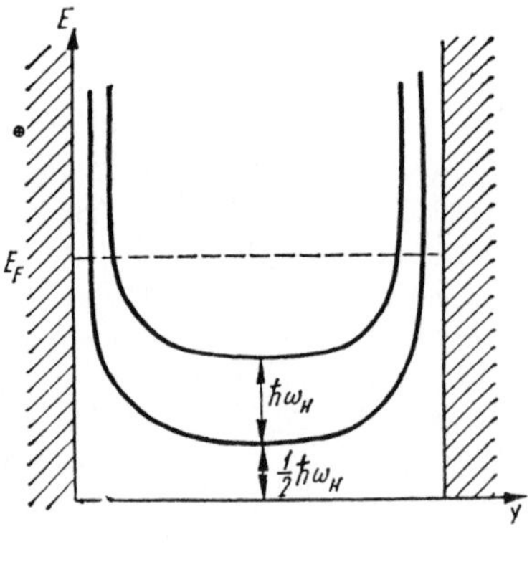

Fig. 2. Lower parts of the Landau bands with $N = 0$, 1 for two-dimensional electrons in a specimen of finite width.

from the surface, so that the perfection of the crystal surface is an important condition for one to be able to explore them.

The relative weakness of the necessary magnetic field permits the magnetic surface states to occur in *type I superconductors* where one can recognize them. A characteristic pattern of Landau quantum subbands emerges for two-dimensional electrons in a specimen confined over the y coordinate: the subband edges are raised at the approach to the specimen boundaries, as shown in Fig. 2. The motion along the level lines inside such a potential explains the *quantum Hall effect*, and the universal resistivity quantization of two-dimensional systems in a quantizing magnetic field.

MAGNETIC SURFACE STATES

Stationary quantum states of charged *quasiparticles* held at a crystal surface by a magnetic field parallel to it. In classical physics, such states correlate with the trajectories of particles "hopping" along the surface. Strictly speaking, the "hopping" trajectories occur only at specular reflection from the surface when each subsequent trajectory section duplicates the preceding one. While undergoing such trajectories, the motion of electrons along the normal to the surface is periodic, and hence is quantized. Since the trajectory of a hopping particle covers a smaller area than that of a particle moving in the bulk, and the energy distance between *magnetic surface levels* is larger than between the bulk Landau levels (see *Quantizing magnetic field*), therefore, magnetic surface states can only be observed in a weak magnetic field when the bulk Landau levels are usually "smeared" by the temperature. To see them, one should use the so-called sliding quasiparticles that approach the surface at small angles. Even a non-ideal surface is a mirror for them, as the only condition required for their specular scattering is that the transverse wavelength (i.e. the wavelength associated with electron motion along the normal to the surface) be much greater than the characteristic surface roughness size (see *Surface scattering of electrons*).

For the particles moving along the surface with momenta, p_x, comparable to the Fermi momentum p_F, the Larmor orbit center lies far beyond the crystal boundary. Their wave functions decrease into the bulk crystal to a distance of $l \approx l_H[\hbar/(l_H p_F)]^{1/3}$, where $l_H = [\hbar/(m\omega_H)]^{1/2}$ is the magnetic length, and $\omega_H = eB/m$ is the *cyclotron frequency*. Magnetic surface states of this kind were found experimentally via distinctive ($\propto B^{2/3}$) *surface impedance* oscillations in the conductors Sn, In, Cd, W and Bi at liquid-helium temperatures for $B < 10$ mT, with oscillation frequencies corresponding to the skin layer depth (see *Skin-effect*) comparable to the characteristic length l of the wave function decay into the bulk crystal.

MAGNETIC SUSCEPTIBILITY

A physical quantity characterizing the ability of a material to change its magnetic state (*magnetization*) with a changing applied *magnetic field*. Static and dynamic magnetic susceptibilities are known, the latter differing from the former for media exhibiting temporal dispersion. The *static susceptibility*, $\chi = M/H$, is the ratio of the magnetization, M, to the internal magnetic field, H. It is related to the *magnetic permeability*, μ, by $\mu = \mu_0(1+\chi)$ in SI units ($\mu = 1+4\pi\chi$ in CGS). The *dynamic susceptibility* $\chi(\omega)$ or $\chi(\boldsymbol{q},\omega)$ relates, in a similar manner, the Fourier components of the magnetization to the magnetic field (see *Susceptibility*). Static magnetic susceptibilities can differ in both magnitude and sign for different *magnetic substances*. For *diamagnets* ($\chi < 0$) and *paramagnets* ($\chi > 0$), the susceptibility is small, of the order 10^{-4}–10^{-6}, and it depends weakly on the external magnetic field. In *ferromagnets* and *ferrimagnets*, the susceptibility may be as high as 10^4 and strongly dependent on H. For that reason, the *differential susceptibility*, $\chi_d = dM/dH$, is often introduced to describe properties of magnetically ordered materials. This quantity χ_d is a property of the material that depends strongly on H; a dependence that is highly nonlinear for actual ferromagnets which exhibit a domain structure (see *Saturation magnetization curve*). At $H = 0$, χ_d has a certain (nonzero) magnitude called the *initial susceptibility*. The differential susceptibility grows with H, reaches a maximum, and then drops to a very small value with the onset of saturation at high fields (*saturation effect*). In such a high field, the differential susceptibility differs little from that of a paramagnet, and it is known as the *paraprocess susceptibility*. The so-called *reversible magnetic susceptibility*, χ_{rev}, is often singled out:

$$\chi_{rev} = \lim_{\Delta H \to 0} \frac{\Delta M}{\Delta H} \quad (\Delta H < 0).$$

Within a sufficiently small ΔH in the vicinity of any H the value χ_{rev} is determined mainly by reversible processes occurring in the magnetically ordered material with the variation of H. Hence, the difference $\chi_d - \chi_{rev}$ may serve as a gauge of the irreversibility of processes associated with *magnetic hysteresis* (see *Paramagnetism*, *Diamagnetism* and *Ferromagnetism*).

The magnetic susceptibility is a complicated, and sometimes hysteretic, function of the temperature. For instance, χ rises with T in ferromagnets, reaching a maximum (known as the *Hopkinson maximum*) near the *Curie point*, and then decreases over the paramagnetic region, following the *Curie–Weiss law*. An additional maximum of χ occurs in the neighborhood of various phase transitions, e.g., the reorientation-type *magnetic phase transition* in nickel, etc. Magnetic susceptibility is a structure-sensitive quantity in ferromagnets, so its magnitude may vary over a wide range for one and the same material, depending on its heat pretreatment. For anisotropic media, e.g., for magnetically ordered crystals where the *magnetic anisotropy* is taken into account, the susceptibility is a tensor quantity χ_{ik} depending on the direction in the material. The static susceptibility is a symmetric real-valued tensor, and the dynamic susceptibility $\chi_{ik}(\boldsymbol{q},\omega)$ will, in a nondissipative medium, be described by a Hermitian tensor, with the anti-Hermitian part determining the absorption of the alternating magnetic field energy. The $\chi_{ik}(\boldsymbol{q},\omega)$ tensor components obey general expressions following from *Kramers–Kronig relations* and the Onsager relation (see *Onsager method*). Components of the $\chi_{ik}(\boldsymbol{q},\omega)$ tensor have singularities (poles) at ω values close to natural oscillation frequencies, $\omega(\boldsymbol{q})$, of the magnetization (*spin waves*, *magnetostatic oscillations*, etc.).

MAGNETIC SWITCHING

Process of irreversibly changing the *magnetization* of a *magnetic substance*, with the change remaining after the remagnetizing field has been switched off. Magnetic switching is mainly due to jump-like displacements of the *magnetic domain walls* (see *Barkhausen effect*) and the reordering of the *magnetic domain structure*. Each Barkhausen jump is an elementary act of magnetic switching, while the magnetic *hysteresis loop* is the overall characteristic of magnetic switching. The basic parameters of magnetic switching (minimal force of switching, B_s, total polarization change, ΔM_s, time of magnetic switching, τ_s)

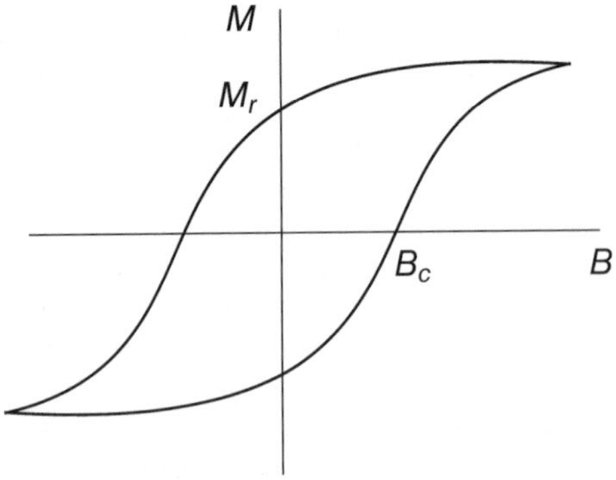

Hysteresis loop, or plot of the magnetization M versus the applied magnetic field B, showing the remanent magnetization M_r and the coercive field B_c.

are associated with the hysteresis loop characteristics, the coercive field B_c and the remanent magnetization M_r (see Fig.), and with the *domain wall mobility*, μ, in the following way: $B_s \geqslant B_c$, $\Delta M_s \leqslant M_r$, and $\tau_s \approx \mu d/B_s$ (d is an averaged domain wall displacement during the magnetic switching). Thus, the parameters of magnetic switching, which is a *magnetic reversal* process, effectively depend on the material composition, magnetic structure, manufacturing process, degree of imperfection, etc. Thermostable ferrite cores with highly rectangular hysteresis loops, which were formerly widely used in memory storage elements, have typical magnetic switching parameters: $B_s \approx 0.1$ mT, $\mu_0 \Delta M_s \approx 0.1$–$0.2$ T, $\tau_s < 10^{-6}$ s. Hard drives in personal computers often employ magnetic switching.

MAGNETIC SYMMETRY GROUP, magnetic symmetry Shubnikov group

A symmetry group for describing a magnetically ordered structure; a member of a *colored symmetry group.*

A crystallographic magnetic space symmetry group X comprises the elements (operators) g of space symmetry (rotation, mirror reflection, etc.) and the products gR, where R is the antisymmetry operation which reverses spin directions at each lattice site, and hence changes the sign of the *magnetic moment* density or magnetization $\boldsymbol{M}(\boldsymbol{r})$. If a space group is denoted by F or $F + gF$, then a Shubnikov group is $X = F + Rg\,F$, where $gR = Rg$, and $R^2 = e$, with e denoting the unit element. If the set RgF is empty ($X = F$), then X is called a *unicolor magnetic symmetry group*, equivalent to an ordinary *space group* (called Fedorov group in the Russian literature). If R is an element of the group X, i.e. $X = F \times \{e, R\}$, then it is a *neutral or gray magnetic symmetry group*, with $\boldsymbol{M}(\boldsymbol{r}) = 0$. These gray groups are possible for diamagnetic or paramagnetic crystals which have no time-averaged magnetic moments. All other Shubnikov groups are bicolor (or black-and-white) magnetic symmetry groups (see *Shubnikov groups*). Each of them is constructed as a group isomorphic to a certain (generator) space group by selecting a subgroup of index 2 in the latter, and substituting Rg for elements g in the coset.

The bicolor magnetic symmetry groups can be classified into 517 with an *antitranslation* operation (*AT-groups*) and 674 with an *antirotation* operation (*AR-groups*). Let $g = (a/h)$, where h is a rotation, and a is the vector of a subsequent *translation.* An AT-group includes, by definition, the antitranslation operation $(a/e)R = (a/R)$, viz., the elements (a_1/h) and $(a_2/h)R$ are both present (or both absent). The subgroup of translations in an AT-group is the translation group for the $\boldsymbol{M}(\boldsymbol{r})$ density, and there is also a subgroup comprising the usual $\rho(\boldsymbol{r})$ charge density translation group. An AR-group by definition cannot include the operations (a_1/h) and $(a_2/h)R = (a_2/hR)$ concurrently; the magnetic and electric periods coincide, and hR is an antirotation. There is a total of 1651 Shubnikov groups in all, 517 antitranslation groups, 674 antirotation groups, 230 gray or neutral groups, and 230 ordinary space groups.

The *crystallographic magnetic point symmetry groups* comprise all rotations and antirotations of the AR-group. There are in all 122 magnetic point groups, with 32 of them being the ordinary point groups, another 32 being the point group analogs of the gray space groups, and the remaining 58 being proper magnetic point groups.

When color is added to the 14 *Bravais lattices* of the ordinary crystallographic groups one obtains 22 additional such lattices, so there is a total of 36 Bravais lattices for the Shubnikov groups.

An *operator magnetic symmetry group* is constructed in conformity with a crystallographic group by substituting unitary operators for g operations and antiunitary operators for gR operations.

An operator magnetic symmetry group is isomorphic to the crystallographic one in coordinate function space, while in spinor space two operators, differing in sign, are associated with each crystallographic operation.

MAGNETIC TEXTURE

A preferential space orientation of the *easy magnetization axes* in a polycrystalline ferro- or ferrimagnetic specimen, which is indicative of the specimen *magnetic anisotropy.* Magnetic texture appears under: (i) application of directional mechanical stresses to the specimen, which brings about a preferential *crystallite* orientation (a crystallographic *texture*); (ii) *heat treatment* of the specimen below the *Curie point* in the presence of a magnetic field (*thermomagnetic treatment*), and (iii) thermomechanical treatment. The presence of magnetic texture can dramatically improve the magnetic properties in some *magnetic materials.* Classed into magnetically-textured *soft magnetic materials* are: crystallographically textured transformer steel, *perminvar,* permalloy, etc. Giving a magnetic texture to these materials decreases the magnitude of their *coercive force* and lowers the *magnetic loss.* In *hard magnetic materials* (magnico, *ticonal,* barium and cobalt *ferrites,* etc.), the magnetic texture results in an increase of their coercive force, remanence, etc.

MAGNETIC VISCOSITY

A measure of the time lag of *magnetization* changes in *magnetic substances* after variations in the applied *magnetic field.* Due to the magnetic viscosity, the steady-state magnetization is reestablished after a lag in time τ that can range from less than a millisecond to a few hours (see *Magnetic relaxation*). For example, it takes time for magnetic domains to grow and reorient. Magnetic viscosity is also observed in magnetic materials with internal degrees of freedom that result from the presence of mobile impurities and other *defects in crystals* interacting with the magnetic subsystem. There may be, e.g., a relaxation drift of the *magnetic susceptibility,* χ, in *ferromagnets* (or *ferrites*) towards its steady-state value through *domain walls* interacting with reorienting or diffusing defects. In the former case the *magnetic anisotropy* energy changes due to defects jumping from one lattice site to another, thus changing the symmetry of the local defect environment symmetry. In addition the mechanical defect strain field interacts with the magnetization (*magnetostriction*). In the latter case of defects diffusing in the lattice over distances of the order of the domain wall thickness, the *relaxation time* τ depends on the rate of *diffusion* and the type of diffusion (electronic, ionic, etc.), and it rises drastically with decreasing temperature; e.g., the carbonyl iron has $\tau \approx 10^3$ s at $-2\,°\mathrm{C}$, and $\tau \approx 10^{-2}$ s at $+100\,°\mathrm{C}$. The magnetic viscosity can be quantitatively described by introducing lagging terms (generally, nonlinear in domain wall displacements) into the *magnetic domain wall dynamics* equation (*Néel equation* and *Slonczewski equation*). The magnetic viscosity is particularly high in *hard magnetic materials* that are ensembles of *single-domain particles,* e.g., grains in polycrystalline *alloys* interacting with one another. This viscosity is governed by the irreversible magnetization rotation of such domains (see *Magnetization rotation processes*) due to thermal fluctuations. Some materials exhibit quite a range of lag times, τ, that can be determined from the number and positions of peaks appearing on the χ frequency dependence excited by a harmonically varying applied magnetic field. Magnetic viscosity appears in *spin glasses* and other nonergodic magnetic materials (see *Amorphous magnetic substances*).

MAGNETISM of solids (fr. Gr. $\mu\alpha\gamma\nu\eta\tau\eta\varsigma$, of Magnesia)

Branch of physics that studies *magnetic substances* or materials containing *magnetic moments,* generally arising from their constituent elements (atoms, molecules, ions, *conduction electrons*), including their interactions with each other or with external *magnetic fields* and external electric currents.

Basics. An electric charge e moving in a vacuum at the velocity $\boldsymbol{v}$ in the presence of a *magnetic flux density* (sometimes called magnetic induction) $\boldsymbol{B}$ experiences the *Lorentz force* $\boldsymbol{F}$ given by $\boldsymbol{F} = e\boldsymbol{v} \times \boldsymbol{B}$, so $\boldsymbol{B}$ is sometimes called the force per unit current. In free space we also define the associated magnetic field $\boldsymbol{H}$ through the expression $\boldsymbol{B} = \mu_0 \boldsymbol{H}$ where $\mu_0 = 4\pi \cdot 10^{-7}$ N/A^2 (or H/m) is the *magnetic permeability of vacuum.*

In a medium the difference between $\boldsymbol{B}$ and $\boldsymbol{H}$ involves the presence of magnetic field sources in the matter, viz. of moving electric charges (atomic electrons, conduction electrons, etc.) that give rise to microscopic currents (distinguishing between micro- and macroscopic currents is somewhat arbitrary), and of intrinsic (spin) moments of the substance particles: electrons and atomic nuclei. This relation between $\boldsymbol{B}$ and $\boldsymbol{H}$ is given by $\boldsymbol{B} = \mu_0(\boldsymbol{H} + \boldsymbol{M})$ in SI units ($\boldsymbol{B} = \boldsymbol{H} + 4\pi\boldsymbol{M}$ in CGS units) with the *magnetization* vector, $\boldsymbol{M}$, defined as the mean magnetic moment per unit volume. The theory of magnetism aims at establishing interrelations between $\boldsymbol{B}$, $\boldsymbol{H}$ and $\boldsymbol{M}$ in different magnetic materials under given external conditions (either steady-state, or time-dependent). The magnetization for a given $\boldsymbol{H}$ or $\boldsymbol{B}$ is defined by the material equations of a magnetic material, which result from (quantum-mechanical and statistical) averaging of the microscopic equations of motion for the sources of the magnetic field. More precisely, by virtue of the *Bohr–van Leeuwen theorem*, the magnetic properties of a substance can be consistently described within a quantum-mechanical framework. Nevertheless, ordinarily simple phenomenological equations are found to be adequate for describing the magnetization vector as a function of the coordinates and time: $\boldsymbol{M} = \boldsymbol{M}(\boldsymbol{r}, t)$ (see *Bloch equations* and *Landau–Lifshits equation*). In the static (or quasi-static) case, the relation between $\boldsymbol{M}$ and $\boldsymbol{B}$ (or $\boldsymbol{H}$) is determined by the *magnetization curve* $\boldsymbol{M}$ ($\boldsymbol{H}$) or $\boldsymbol{M}$ ($\boldsymbol{B}$), which differs in shape for different materials. In a number of cases, $\boldsymbol{M}$ is linear in $\boldsymbol{H}$:

$$\boldsymbol{M} = \chi\boldsymbol{H}, \tag{3}$$

where χ is the *magnetic susceptibility*, a tensor quantity for anisotropic bodies. But sometimes, e.g., for *ferromagnets*, the linearity of Eq. (1) may become invalid even in very weak fields.

Classification of magnets. The classification proceeds from the analysis of the nature of the atomic carriers of magnetism, and their interactions. *Diamagnets* and *paramagnets* are *weak magnetic materials* where the interaction between the magnetic moments is negligible in magnitude, or has no fundamental importance. Some specific magnetic properties are exhibited by *superconductors* (see *Magnetic superconductors*). The interactions of atomic magnetic moments (especially spin moments of unfilled d- and f-orbital shells in atoms of the Fe, Pd or Pt groups, rare earth elements, or actinides) are of major importance for magnetically ordered substances. For example, the *exchange interaction*, of electrostatic origin with energy $E_e = k_B\theta_W$ (where θ_W is the Weiss temperature, k_B is the Boltzmann constant) of the order of 10^{-20}–10^{-21} J, i.e. $\theta_W = 1000$–100 K, results in *magnetic ordering* when the spin orientations of neighboring atoms become correlated. This ordering occurs at a *magnetic phase transition* at a temperature T_c of the order of E_e/k_B, causing an *atomic magnetic structure* to appear in the material associated with a non-zero value of the mean atomic spin, even in the absence of an external magnetic field, $\boldsymbol{B}$. There is a great variety of magnetically ordered materials. Their main feature is the nature of the motion of the magnetic carriers, electrons; the magnetism of localized electrons, and that of delocalized electrons (see *Band magnetism*). In the simplest case of *ferromagnetism*, the *spins* of all the atoms (or of all the delocalized electrons contributing to ferromagnetism) acquire parallel orientations to form a *spontaneous magnetization* vector $\boldsymbol{M}_s$ (see also *Ferromagnetism of metals and alloys*). In *antiferromagnets* the spins of adjacent atoms acquire an antiparallel orientation, however, linear combinations of the spins (antiferromagnetic vectors $\boldsymbol{L}$) can still differ from zero. In *ferrimagnets*, both $\boldsymbol{M}$ and $\boldsymbol{L}$ are non-zero. For a two sublattice system with individual sublattice magnetizations $\boldsymbol{M}_1$ and $\boldsymbol{M}_2$ the total magnetization is $\boldsymbol{M} = \boldsymbol{M}_1 + \boldsymbol{M}_2$, and the *antiferromagnetism vector* is $\boldsymbol{L} = \boldsymbol{M}_1 - \boldsymbol{M}_2$.

Magnetic ordering of a more complex type is exemplified by spiral (helical) ordering (see *Modulated magnetic structures*). Amorphous magnetic materials (e.g., *spin glasses*) are also classified among the ordered magnetics. Typical of magnetically ordered materials is a complicated shape of their magnetization curve $M(B)$, which can exhibit hysteresis, i.e. depend on the sample history, and is sometimes indeterminate (see *Magnetic hysteresis* and *Nonergodicity*); and there is also a possibility of macroscopic nonuniformities being present (see *Magnetic domain structure*, and *Intermediate state* of a magnetic material).

There exist specific *resonances* that can arise from the magnetism of the particles of matter, such as: *nuclear magnetic resonance* (NMR) of nuclei, *electron paramagnetic resonance* (EPR) of paramagnets, and collective spin system excitations in magnetically ordered crystals which exhibit a special kind of weakly damped modes called *spin waves*, or *magnons* (see also *Ferromagnetic resonance* and *Antiferromagnetic resonance*).

Importance for science and technology. Magnetic methods (NMR, EPR, etc.) are employed to investigate the structure of materials in physics, chemistry, biology, and medicine. Engineering applications of magnetism are multifaceted: *magnetic materials* and *permanent magnets* are used in power and radio engineering; methods of magnetic *defectoscopy* are applied with success; devices based on the *magnetic resonance* phenomenon and on magneto-optical effects have been introduced into electronics, optoelectronics and computer engineering; magnetoresistance has been utilized; microwave devices using spin waves are employed, as well as *magnetic bubble domain devices*, and other magnetic computer memory elements.

MAGNETISM, AMORPHOUS

See *Amorphous magnetism*.

MAGNETISM AND GREEN'S FUNCTIONS

See *Green's functions and magnetism*.

MAGNETISM, BAND

See *Band magnetism*.

MAGNETISM OF RARE-EARTH METALS

A branch of the physics of magnetic phenomena which studies, both theoretically and experimentally, magnetic properties of the elements of the ^{57}La–^{71}Lu series (*rare-earth elements* or *lanthanides*). The strongly localized electrons in $4f$-orbitals act as carriers of magnetism in the rare earths, with the population of these orbitals increasing with the atomic number from 0 electrons for La up to 14 electrons for Yb and Lu. For the light rare-earth elements (Ce–Eu), $J = L - S$, and for the heavy ones (Gd–Tm), $J = L + S$, where J, L and S are the total, orbital, and spin *angular momenta*, respectively. The radii of the $4f$-orbitals are about 10 times smaller than the minimum interatomic distances in a solid which virtually precludes the possibility of any direct f–f *exchange interaction*. The *indirect exchange interaction* is effected via *conduction electrons*, and this is a long-range interaction of an oscillating nature (see *Ruderman–Kittel–Kasuya–Yosida interaction*). The electrostatic interaction of the *crystal field* with an anisotropic $4f$-orbital influences the *magnetic anisotropy* of rare-earth metals; the energy of magnetic anisotropy is comparable to the exchange interaction energy. These features of the magnetism of rare earths cause complex noncollinear magnetic structures to appear in magnetically ordered states, including helical ones (e.g., in Dy at $T = 85$–178 K), conical (Ho at $T < 20$ K), cycloidal (Er at $T = 20$–53 K), sinusoidal (Tm at $T = 25$–60 K) and some others (see *Modulated magnetic structures*). Theory can explain the emergence of noncollinear structures in rare earth metals through changes in the topology of the *Fermi surface* under magnetic ordering (see *Topological transition*). A variety of noncollinear ferro- and antiferromagnetic structures are found at low temperatures, and in strong magnetic fields (e.g., in Dy at $T < 85$ K, and in Sm at $T < 106$ K). Deformation and restructuring of magnetic structures is observed in relatively weak magnetic fields. Changes of magnetic ordering are associated with either first-order or second-order *magnetic phase transitions*; the transformation of a helical antiferromagnetic structure into a collinear ferromagnetic one is accompanied by a lowering of the symmetry from hexagonal to rhombic (see *Crystal symmetry*). Typical for the heavy rare-earth metals are high values of saturation magnetic moments (e.g., for Dy, $\mu = 10.65\mu_B$ along the a axis), as well as uniaxial anisotropy and *giant magnetostriction* (anisotropy and *magnetostriction* coefficients are greater by a factor of 10–100 for rare-earth metals than for the $3d$-iron group).

MAGNETISM OF TRANSITION METALS

Study of magnetic properties and phenomena of the *transition metals*. The atoms of transition metals possess partially filled inner electron shells (lacking some d, or d and f electrons); which makes these substances *paramagnets* at high temperatures.

Of the $3d$-metals, α-Fe, Co (hexagonal) and Ni are ferromagnets, with Co having the highest *Curie point* ($T_C = 1403$ K); Mn, γ-Fe and fcc Co display Néel-type antiferromagnetism. The antiferromagnetism of *chromium* (Cr) is due to a *spin density wave*, which results in incommensurate magnetic and crystallographic lattice constants. Chromium and some *chromium alloys* exhibit spin density waves and antiferromagnetism of delocalized electrons. The *ferromagnetism* of $3d$-metals and their alloys is intermediate between that of delocalized electrons (band or Stoner magnetism) and that of ions; this occurs because of the substantial overlap of $3d$-orbitals (see *Band magnetism*). Several intermetallics and $3d$-metal alloys (e.g., ScIn, Ni_3Ga, $Fe_{65}(Ni_{0.7}Mn_{0.3})_{35}$) are purely Stoner *ferromagnets*. Although the pure d-orbital metals of the following $4d$ and $5d$ periods exhibit no magnetic ordering, many of their alloys and intermetallic compounds are magnetically ordered materials, e.g., the Stoner ferromagnet $ZrZn_2$. Many oxides and sulphides of transition metals with metallic conductivity (e.g., Fe_3O_4) also exhibit magnetic ordering.

The $4f$-metals (except for Yb) are either ferromagnets (heavy $4f$-metals, from Gd to Tm) or antiferro-, or ferrimagnets (light $4f$-metals, from Ce to Eu) with antiferromagnetism possible at temperatures above the Curie point. The *atomic magnetic structure* of many $4f$-metals is a spiral type (or a superposition of spiral and collinear ferromagnetism). Magnetic electrons in rare earth metals belong to deep-lying $4f$-orbitals, and hence do not overlap. This makes them similar to magnetic insulators, describable in the terms of *magnetic ions*.

Besides ferro-, ferri- and antiferromagnetic ordering, a *spin glass* type ordering is possible in alloys of transition metals, e.g., in MnAu. Strong electron *paramagnets* may exist among the paramagnetic transition metals and alloys, e.g., $Cu_{93}Mn_7$. Their *magnetization curves* (in fields up to teslas) resemble those of the ferromagnets; yet, ferromagnetism has not been detected in such materials even at very low temperatures.

The *exchange interaction* in d-metals that gives rise to magnetism can be either direct (due to the overlap of d-orbitals) or indirect (due to s–d-exchange, or a *Ruderman–Kittel–Kasuya–Yosida interaction*, RKKY interaction); the magnetism of $4f$-metals is due to an indirect *exchange interaction* (via s–f-exchange).

MAGNETISM, SURFACE

See *Surface magnetism*.

MAGNETITE

Iron ferrite spinel $(FeO)Fe_2O_3$, or Fe_3O_4, collinear monocrystalline *ferrimagnet* with two sublattices. At $T > T_V = 125$ K (T_V is the *Verwey temperature*), it has cubic symmetry, and the inverse spinel structure (*space group* $Fd\bar{3}m$, O_h^7, symmetry *point group* $m\bar{3}m$) and, at $T < 125$ K, it has orthorhombic symmetry (space group $Imma$, D_{2h}^{28}). The lattice constant is 0.839 nm (cubic phase); the density is 5.24 g/cm^2; the *magnetic moment* per chemical formula unit is $4.1\mu_B$; and the *Néel point* is 858 K.

MAGNETIZATION

State of a *magnetic substance* being magnetized. The magnetization $\boldsymbol{M}$, also called magnetic polarization, is defined as the *magnetic moment* per unit volume. Weakly magnetic materials (*paramagnets, diamagnets*) are magnetized by an external *magnetic field* $\boldsymbol{B}/\mu_0 = \boldsymbol{H}$: $\boldsymbol{M} = \chi\boldsymbol{H}$, where χ is the *magnetic susceptibility* of the substance. Magnetically ordered materials may exhibit *spontaneous magnetization*, $\boldsymbol{M}_s$, which takes place below the *Curie point* even in the absence of an applied magnetic field (see *Ferromagnetism, Ferrimagnetism*). In the case of magnetic materials with complicated *atomic magnetic structures*, which includes several (n) *magnetic sublattices*, the term *sublattice magnetization* $\boldsymbol{M}_k$ ($k = 1, 2, \ldots, n$) is introduced. When evaluating this quantity, only magnetic moments of atoms, which belong to the kth sublattice, are taken into account. The dimensions of magnetization are A/m in the SI system of units, and oersted in CGS. The description of a magnetic state in terms of magnetization (or sublattice magnetization) as a function of coordinates and time $\boldsymbol{M} = \boldsymbol{M}(\boldsymbol{r}, t)$ is of basic importance in the phenomenological theory of *magnetism, micromagnetism*.

MAGNETIZATION CURVE

Dependence of the *magnetization* M on the strength of the applied *magnetic field*, B. Since B, H and M are uniquely related to each other, $B = \mu_0(H + M)$, it is sufficient to consider just two of them, e.g., $M = M(B)$. The magnetization curve for both a *diamagnet* and a *paramagnet* yields a unique relation $M = M(B)$. For magnetically ordered media which feature *magnetic hysteresis* such a relation becomes ambiguous, and the magnetization curve depends on the prehistory of the sample. For ferromagnetic media (*ferromagnets, ferrimagnets, amorphous magnetic substances*) one may identify the *curve of initial magnetization*. It gives the dependence $M = M(B)$ for the case when B increases from $B = 0$ (and $M = 0$), in contrast to those magnetic curves which describe a cycle of *magnetic reversal* (*hysteresis loop*). The profile of magnetization points to the existence of *magnetic phase transitions* induced by the applied field (see *Phase transitions in an external field*). These appear as jumps during first order phase transitions, or as discontinuities during second order transitions. Modern physical experiments to determine magnetization curves and their theoretical analysis, oftentimes computerized, are an important tool for studying the state of the magnetization of materials, and to define the fundamental properties of *magnetic materials.*

MAGNETIZATION PROCESS

Process of establishment of *magnetization* $\boldsymbol{M}$ in a material under the action of an external *magnetic field* $\boldsymbol{B} = \mu_0 \boldsymbol{H}$. This process in *diamagnets* consists in the generation of microscopic (atomic) currents and in *metals* and *semiconductors* in the realignment of the flow of *current carriers*; with the resulting magnetization antiparallel to the field. *Paramagnets* become magnetized along the field: *magnetic moments* of thermally disordered particles, which constitute a paramagnet, become oriented under the action of the field (see *Paramagnetism*). Processes of magnetization in magnetically ordered crystals are more complicated, and under different conditions are determined by different phenomena: not only by orientation of magnetic moments of particles by the magnetic field (see *Paraprocess*), but also by rotation of the vector of spontaneous magnetization as a whole (if the direction of magnetization does not coincide with an *easy magnetization axis*). Besides that, if a *magnetic domain structure* exists, the magnetization processes are governed by its rearrangement (shift of boundaries of domains, *magnetization rotation processes* in the magnetization direction, see *Saturation magnetization curve*). Of key importance for magnetic materials with several sublattices is the change of angles between the magnetizations of *magnetic sublattices*; this sometimes takes place through *magnetic phase transitions*, e.g., through a spin-flop transition in an *antiferromagnet*, transitions in metamagnetics, etc. Magnetization processes are significantly influenced by intrinsic magnetic fields of the sample (see *Demagnetization fields*). There is also a dependence of the nature of the magnetization on the sample shape, and this may lead to the generation of inhomogeneous states (*intermediate state*) of magnetic materials. The state of magnetization of magnetically ordered materials often depends on their prior history (see *Magnetic hysteresis*).

MAGNETIZATION ROTATION PROCESS

The stage of the magnetization process during which the angles between the vectors of the spontaneous *magnetization* $\boldsymbol{M}$ of different individual *domains* change in order to align the overall magnetization with the direction of the local *magnetic field* $\boldsymbol{B}$ acting within the body. For lower values of the internal magnetic field the main magnetization mechanism is domain wall displacement, and the rotation process takes place during the subsequent stage of magnetization buildup when practically all possible *domain wall displacement processes* have already taken place and only the orientation of domains in different regions of the sample volume may not yet coincide with the direction of $\boldsymbol{B}$. The rotation process occurs in magnetic fields sufficiently high to overcome the fields associated with the crystal magnetic anisotropy. At the end of the rotation the magnetization of a *ferromagnet* becomes saturated (see *Magnetization curve*). The contribution of the rotation process to the total magnetization depends on the *magnetic texture*, on the perfection of the *crystal lattice*, and on the shape and dimensions of the sample.

Rotation processes can be reversible or irreversible. *Irreversible* rotation processes can involve jumpwise changes of magnetization (see *Barkhausen effect*). If the displacements cannot occur in a crystal (e.g., *hard magnetic materials*), in a thin ferromagnetic film (see *Magnetic films*) or in a powder sample, then the magnetization increase is mainly due to rotations. A similar situation takes place when magnetizing along *hard magnetization axes.* Irreversible rotation processes in magnetically hard materials are the main cause of the *magnetic hysteresis* of ferromagnets.

MAGNETIZATION SATURATION

See *Saturation magnetization curve.*

MAGNETIZING FORCE

The same as *Magnetomotive force.*

MAGNETOACOUSTIC EFFECTS in metals

The oscillations with, and the resonance dependences on the magnetic field of *sound absorption,* and of the *sound velocity dispersion* in metals, due to the cyclotron pattern of the electron motion. Magnetoacoustic effects are most prominent in the region of a *quantizing magnetic field*: $\hbar\Omega \gg k_B T$ (Ω is the *cyclotron frequency*), where the electron motion is one-dimensional (see *Giant quantum oscillations*). In the case of $\hbar\Omega \ll k_B T$ magnetoacoustic effects are determined by the electrons that move over periodic trajectories in the magnetic field, $\boldsymbol{B}$, interacting with a periodic (in space and time) sound wave field, with the condition $kl, \tau\Omega \gg 1$ (l and τ are the mean free path of an electron and its corresponding transit time, and $\boldsymbol{k}$ is the wave propagation vector of the sound) guaranteeing the *ballistic regime* of electron motion. In the high-frequency region ($\omega\tau \gg 1$), the resonance absorption of sound occurs at $\omega = n\Omega$ ($n = 1, 2, \ldots$), assuming that the sound wave changes periodically with time (see *Acoustic cyclotron resonance*). The fact that the sound wave is periodic determines a number of geometrical *size effects* when the distance, $\boldsymbol{u}$, between the points of effective sound absorption (with $\boldsymbol{k} \cdot \boldsymbol{v} = 0$, where $\boldsymbol{v}$ is the electron velocity) is a multiple to the sound wavelength: $\boldsymbol{k} \cdot \boldsymbol{u} = 2\pi n$. The features of these effects depend upon the mutual orientation of the $\boldsymbol{k}$ and $\boldsymbol{B}$ vectors, as well as on the *Fermi surface* topology. For $\boldsymbol{k} \perp \boldsymbol{B}$ in *metals* with a closed Fermi surface, the magnitude of $\boldsymbol{u}$ represents the characteristic size of a closed electron trajectory (lying in a plane perpendicular to $\boldsymbol{B}$), with the condition $\boldsymbol{k} \cdot \boldsymbol{u} = 2\pi n$ defining oscillations of the sound absorption and of the speed of sound as functions of the reciprocal magnetic field (see *Geometrical resonance*). When open electron orbits exist, or when the vectors $\boldsymbol{k}$ and $\boldsymbol{B}$ are not orthogonal, there will be a drift motion of electrons along the directions of $\boldsymbol{k}$, and *magnetoacoustic resonance phenomena* occur. Such phenomena are pronounced when open electron orbits are present for which the $\boldsymbol{u}$ values do not depend on the electron momentum, the resonance condition being met for all electrons simultaneously (see *Magnetoacoustic resonance at open orbits*). When moving in closed orbits, the electrons can drift along the $\boldsymbol{k}$ vector only in the case $\boldsymbol{k} \cdot \boldsymbol{B} \neq 0$. The displacement $\boldsymbol{u}$ in one cyclotron period depends on the angle, θ, between vectors $\boldsymbol{k}$ and $\boldsymbol{B}$ and on the projection, $\boldsymbol{p}_B$, of the electron momentum on the magnetic field direction:

$$u = \frac{1}{eB}\frac{\partial S}{\partial \boldsymbol{p}_B}\cos\theta,$$

where S is the area of the section of Fermi surface cut by the plane $\boldsymbol{p}_B = \text{const}$. Those electrons for which the quantity $\partial S/\partial \boldsymbol{p}_B$ displays an extremum, $\partial S/\partial \boldsymbol{p}_B = (\partial S/\partial \boldsymbol{p}_B)_{\text{extr}}$, make a major contribution to the sound absorption This condition is realized at extremal points on the Fermi surface (where $\boldsymbol{p}_B$ attains a maximum) as well as at some other $\boldsymbol{p}_B$ points, if the Fermi surface is not convex. The input of the above electron groups determines a periodic (versus reciprocal magnetic field) system of asymmetric maxima in the sound absorption coefficient that emerge at the field magnitudes

$$B = \frac{k\cos\theta}{2\pi ne}\left(\frac{\partial S}{\partial \boldsymbol{p}_B}\right)_{\text{extr}}.$$

An additional condition determining the possibility of the resonance absorption of sound is the presence of points with $\boldsymbol{k} \cdot \boldsymbol{v} = 0$ on the $\boldsymbol{p}_B =$ const trajectory in *momentum space.* There are boundary points $\boldsymbol{p}_{\text{b}}$ which limit the regions of $\boldsymbol{p}_B$ values where intersection of the lines $\boldsymbol{p}_B = \text{const}$ and $\boldsymbol{k} \cdot \boldsymbol{v} = 0$ on the Fermi surface is impossible. Corresponding to the input of such points are jumps of the sound absorption coefficient at field values

$$B = \frac{k\cos\theta}{2\pi ne}\left(\frac{\partial S}{\partial \boldsymbol{p}_B}\right)_{\text{b}}.$$

Magnetoacoustic effects have been found in many metals, and they underlie the *ultrasound resonance spectroscopy method* employed to study energy spectra and kinetic parameters of electrons in metals.

MAGNETOACOUSTIC RESONANCE

A phenomenon arising from the interaction of spins and elastic waves that results in the generation of magnetoelastic waves. The coupling is characterized by a dimensionless parameter,

$$\zeta = \frac{\omega_{\text{me}}}{\omega_0} \approx \frac{B^2}{CM^2}, \qquad (1)$$

where M is the magnetization of the crystal (or of its *magnetic sublattice*), C and B are the elastic modulus (see *Elasticity theory*) and the *magnetostriction* coefficient, respectively, ω_0 is the *magnon* activation frequency, and ω_{me} is an input due to magnetostriction. Since $M \approx 10$–100 mT, $C \approx 10^8$–10^9 J/m^2, and B does not exceed 10^3–10^4 J/m^3 in most cases, Eq. (1) usually yields $\zeta \ll 1$. Therefore, magnons and *phonons* would be conventionally regarded as separate non-interacting *quasi-particles* in the first approximation, their interaction being considered in terms of perturbation theory as a cause of magnon–phonon scattering (see *Magnon–phonon interaction*). Under ordinary conditions such collisions obey the energy and momentum *conservation laws*, but only if three or more particles take part. That corresponds, in terms of the oscillation amplitudes of the *magnetization*, ΔM, and of the elastic strain, Δu_{ij}, to accounting for the third and higher power anharmonic terms of ΔM and Δu_{ij} in the energy. There is the singular case of magnetoacoustic resonance involving a magnon and a phonon with matching (or close to matching) magnitudes of energy and momentum. The interaction may be so strong that the individual nature of each quasi-particle vanishes, and a sort of hybrid quasi-particle comes forth instead. In terms of waves this means that coupled *magnetoelastic waves*, with ΔM and Δu making equivalent contributions to the amplitude, become normal waves to replace the elastic and *spin waves* with frequencies and wave vectors close to one another (with matching polarization).

Fig. 1 shows schematically the essence of magnetoacoustic resonance by plotting the branches of the frequency spectrum $\omega(k)$ of a *ferromagnet* as a function of wave vector $\boldsymbol{k}$ directed (for the sake of simplicity) along a *symmetry axis* of a high enough order so that the normal waves are circularly polarized. The dashed lines represent noninteracting elastic $\omega_{\text{p}}(k)$ and spin $\omega_{\text{s}}(k)$ waves in accordance with the *dispersion laws*:

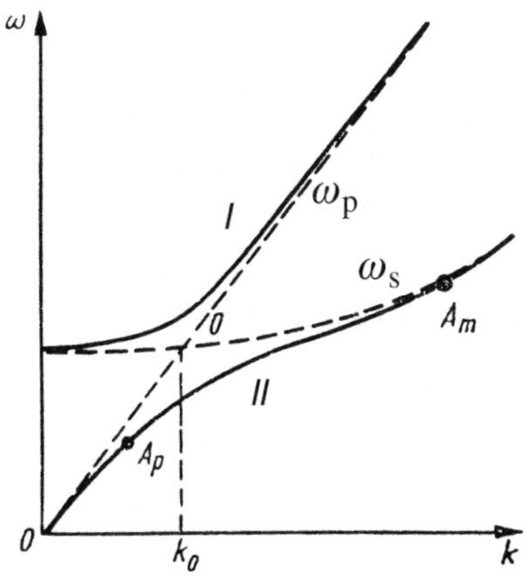

Fig. 1. Dispersion laws of uncoupled (dashed line) and coupled (solid line) elastic waves and spin waves, $\omega_{\text{p}}(k)$ and $\omega_{\text{s}}(k)$, respectively.

$$\omega_{\text{p}}(k) = \omega_{\text{D}}(ak), \qquad \omega_{\text{s}}(k) = \omega_0 + \omega_{\text{E}}(ak)^2, \qquad (2)$$

where $\omega_0 = \gamma B$ is the frequency of the uniform precession of magnetization ($\boldsymbol{k} = 0$) in the internal field B (comprising, in the general case, the effective fields of *magnetic anisotropy* and magnetostatic demagnetization of the specimen, that of magnetostriction, etc., aside from the external field); ω_{D} and ω_{E} are characteristic frequencies of the elastic (*Debye frequency*) and spin (exchange frequency) subsystems; and a is an interatomic spacing. The solid lines I and II on Fig. 1 correspond to coupled waves that take into account the *magnetoelastic interaction*.

The strongest magnetoacoustic coupling (*magnetoelastic resonance*) takes place at the intersection point O of the functions given by Eqs. (2), where the frequencies and wave vectors of the phonons and magnons become equal to each other. In the vicinity of this point the excitation of ΔM oscillations is necessarily accompanied by a Δu excitation, and vice versa, which permits the resonance generation of sound with an alternating microwave field, and the resonance generation of spin waves with *ultrasound.* The magnetoelastic coupling is practically absent at large distances from the point O, viz., if one travels, e.g., along the curve II, the wave is of a purely spin type over its upper portion (point A_m) and a purely elastic one over the lower portion (point A_p). For a given frequency ω, the point O may be positioned either below or above this frequency, depending on the magnitude of the field B. Thus, one can continuously transform the wave of a fixed frequency, ω, from a spin type into an elastic one, and back, by appropriately varying B. In practice such transformations may occur with a wave propagating in a highly nonuniform field B, which becomes either a spin or an elastic wave (with a corresponding change of velocity) to follow the B variations in space. This phenomenon is employed in microwave electronics to build delay lines controlled by the magnetic field (see *Acoustoelectronics*).

The detailed pattern of propagation of magnetoelastic waves effectively depends upon the dissipation processes both in the elastic and in the spin subsystems. The mutual influence of dissipation in the two subsystems is particularly prominent under the conditions of magnetoacoustic resonance.

The magnetoacoustic resonance phenomenon also occurs in *antiferromagnets.* A greater number of spectral branches, and, accordingly, a richer variety of magnetoacoustic phenomena, appear in this case. In antiferromagnets with a linear (or close to linear) dispersion law, magnetoacoustic resonance may occur over a broad frequency range. That occurs at an external field strength (or temperature) at which the *phase velocities* of magnons and acoustic phonons coincide with each other.

Investigations of magnetoacoustic phenomena have had a fundamental influence on the basic principles of *continuum mechanics.* Continuum dynamics equations are constructed based on the laws of conservation of momentum, of angular momentum, and of energy. The law of conservation of momentum leads to the appearance of a symmetrical *strain tensor* in the free energy (see *Thermodynamic potentials*) of the elastic medium, and in the equations of motion. Since neither orbital nor spin magnetic moments taken separately are conserved in magnetic materials, but only the

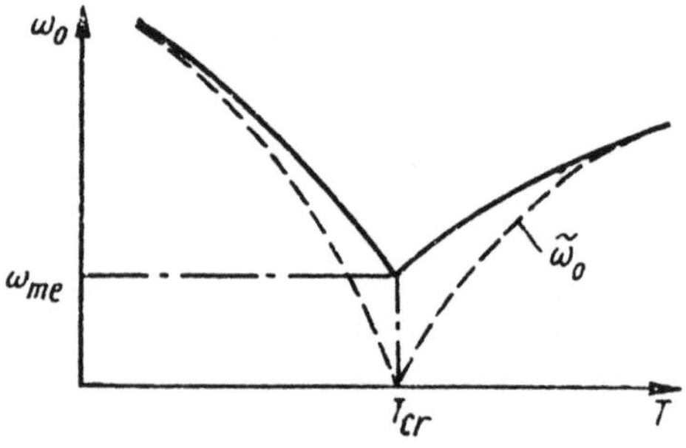

Fig. 2. Temperature dependence of the quasi-magnon activation frequency in the vicinity of a phase transition. The dashed line is the dependence of the magnon activation frequency, $\tilde{\omega}_0$, that does not take account of magnetoelastic coupling, the solid line is the same taking into account the coupling, and T_{cr} is the phase transition temperature.

total angular momentum, one should introduce an antisymmetric part of the strain tensor alongside the symmetric part to describe the magnetoelastic states of matter. Accordingly, the *stress tensor* turns out to be nonsymmetric also (contrary to classical elasticity theory). The antisymmetric part of the stress tensor defines an additional torque that appears due to the magnetic anisotropy, at a local elementary volume torsion of the body relative to the magnetization. Conditioned by the antisymmetric part of the stress tensor are reversal effects characteristic of magnetic media. A *nonreciprocity effect* was experimentally observed whereby the velocities of two transverse acoustic waves with mutually reciprocal wave vector alignments and polarizations differ from each other. The magnetoelastic interactions markedly increase in the vicinity of a magnetic *orientational phase transition*, if the magnetic mode of an activation frequency, ω_0, interacting with sound is a *soft mode* at this *phase transition*, so that without accounting for the magnetoelastic coupling, we find $\omega_0 = 0$ at the phase transition point. This could be a phase transition with respect to the temperature, magnetic field, or pressure. The gap for the quasi-magnon mode ω_0 does not go to zero at the transition due to the magnetoelastic coupling; it only reaches its minimum value $\omega_0 = \omega_{me}$ there, defined by the magnetostriction coefficients (Fig. 2). The magnetoelastic coupling

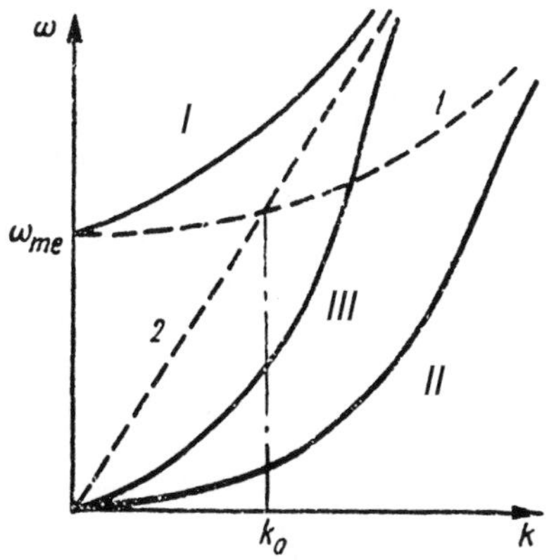

Fig. 3. Frequencies of coupled magnetoelastic waves, ω, plotted against the wave vector, $\boldsymbol{k}$, near the point of an orientational phase transition. The dashed lines 1 and 2 correspond to magnons and phonons, respectively, without taking into consideration the magnetoelastic coupling, and the solid lines correspond to: I , quasi-magnons, and II and III, quasi-phonons. The group velocity of quasi-phonons is zero at $k = 0$.

parameter, ξ, becomes unity at this point. The dynamic modulus of elasticity for the corresponding quasi-phonon mode goes to zero at the phase transition (Fig. 3) because of the magnetoelastic coupling, and so does the speed of sound defined by it. The first effect, the *magnetoelastic gap*, was experimentally discovered in the *antiferromagnetic resonance* of haematite, and a giant magnetoelastic gap (about 20 K) was found in terbium and dysprosium by *inelastic neutron scattering*. The second effect, reduction of the *sound velocity*, was observed in haematite (more than 50%) and in terbium. The impact of magnetoelastic effects at phase transitions on the general understanding of physics is associated with the fact that underlying these phenomena are fundamental properties of systems exhibiting *spontaneous symmetry breaking*.

MAGNETOACOUSTIC RESONANCE AT OPEN ORBITS

The resonance dependence on the magnetic field, $\boldsymbol{B}$, of the absorption of ultrasound in *metals* due to electrons drifting along the wave vector $\boldsymbol{k}$ while moving along open periodic orbits in *momentum space*. Unlike the *magnetoacoustic resonance* of closed electron orbits, this effect is observed even at $\boldsymbol{k} \perp \boldsymbol{B}$, occurring under the condition that the electron displacement, u, during one cyclotron period is an integer multiple of the sound wavelength: $ku = 2\pi n$. The quantity $u \sim b \sin\theta/(eB)$ does not depend on the electron momentum, and is determined by the period b of the *reciprocal lattice* along a direction where the Fermi surface is open, as well as by an angle, θ, between this direction and the wave vector $\boldsymbol{k}$. The sound absorption coefficient, Γ, is a function of the reciprocal magnetic field (see Fig. 4) with a period $\Delta(1/B) \sim 2\pi e/(kb \sin\theta)$ under the conditions of magnetoacoustic resonance with open orbits. Since the phenomenon is governed by the space (not time) periodicity of the acoustic wave, it is realized at low sound frequencies: $\omega\tau \ll 1$ (τ is the mean free path time of the electrons), if the condition $\Omega\tau \gg 1$ holds ($\Omega = eB/m^*$ is the *cyclotron frequency*). At $\omega\tau \gg 1$, the resonance condition takes the form $ku = 2\pi(n \pm \omega/\Omega)$ because of the *acoustic cyclotron resonance*, and each absorption maximum is split into two. The magnetoacoustic resonance at open orbits has been found in many metals, and is an effective means for the determination of Fermi surface parameters, and of the *effective masses* and *mean free paths* of electrons. Similar effects (e.g., $1/B$ periods) can occur as the *de Haas–van Alphen effect* and *Shubnikov–de Haas effect*.

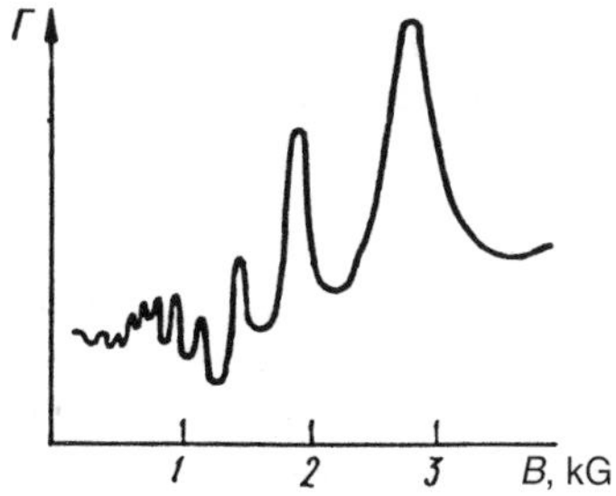

Fig. 4. Dependence of the ultrasound absorption coefficient Γ of tin on the magnetic field B.

MAGNETOACOUSTIC RESONANCE, NONLINEAR

See *Nonlinear magnetoacoustic resonance*.

MAGNETOACOUSTIC SIZE EFFECTS

The dependences of specific acoustic characteristics of a *metal* (e.g., mean coefficient of sound absorption Γ in the film) on the magnetic field strength $\boldsymbol{B}$ and on the material thickness, d. They develop due to the interaction of the ultrasound wave with *conduction electrons* when the mean free path l of the latter with respect to volume scattering is much larger than the specimen thickness ($l \gg d$, criterion of a "thin" specimen).

Magnetoacoustic size effects arise since in weak magnetic fields the electron orbits with their Larmor radii $r_{\rm L}$ do not fit into the specimen ($d < 2r_{\rm L}$) and are perturbed due to the surface specular scattering of the charge carriers (see Fig.). When the sound propagates in a direction normal to a thin metal layer placed in a magnetic field parallel to it, two *cyclotron resonances* develop along the grazing electron trajectories: a static and an oscillatory one. The first results in stronger sound absorption, Γ, and the second results in quasi-periodic oscillations of the function $\Gamma(\sqrt{d/B})$. At low enough temperatures one should observe quantum magnetoacoustic size effects in sound waves penetrating through thin conducting single crystal *films* ($qd \ll 1$, where q is the wave number). This effect involves both magnetic and spatial quantization of the carrier motion. Magnetoacoustic size effects in a strong *magnetic field* ($r_{\rm L} \ll d$) tilted with respect to the surface of the metal plate are similar to *Sondheimer oscillations*. Such oscillations of $\Gamma(B)$ are possible both during diffuse and specular reflection of these electrons off the boundary. In the range of strong magnetic fields the *helicon–phonon resonance* in a thin layer ($d \ll l$) also results in magnetoacoustic size effects.

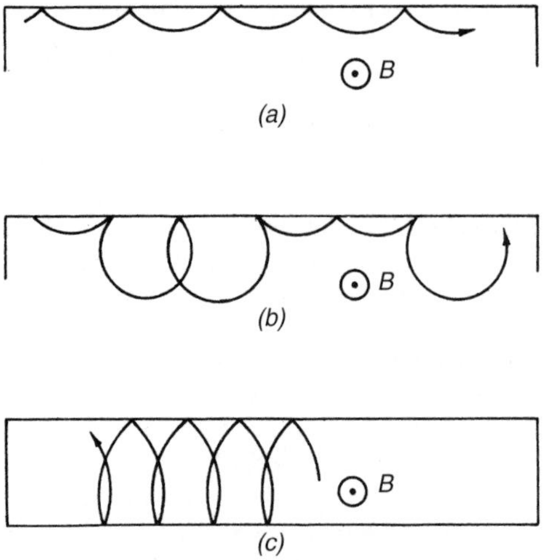

Types of surface electron trajectories in a magnetic field $\boldsymbol{B}$ applied parallel to the metal surface (no provision is made for surface scattering): (a) "sliding"; (b) "mixed" case of two-channel reflection from the boundary; and (c) orbit touching both specimen boundaries (in a weak magnetic field). Electrons effectively interact with the sound wave at sections of trajectories approximately parallel to the wave front.

Magnetoacoustic size effects carry important information on the energy spectrum of electrons (in particular on its local properties) which significantly complement descriptions of the *Fermi surface* obtained by other means. In addition, magnetoacoustic size effects are sensitive to both the integral and the differential parameters that characterize *surface scattering* of charge carriers, and may be used to study surface properties of crystalline conductors. The principal prerequisites for observing magnetoacoustic size effects are sufficient purity of the single crystal specimen to achieve the inequality $d < l$ at low temperatures, and a high quality of its surface. These conditions are practically the same as those necessary for observing *galvanomagnetic size effects* in metals, effects which have often been reported in the literature.

MAGNETOCALORIC EFFECT

Changes in the temperature of a *magnetic substance* with a variation in the applied *magnetic field* B under adiabatic conditions. The magnetocaloric effect is a consequence of the first law of thermodynamics. The change of the internal energy of the magnetic material equals the work done adiabatically by the magnetic field. Thus, the temperature of the material changes during adiabatic magnetization if the internal energy is temperature dependent. A small variation in the magnetizing field, ΔB, causes the temperature variation:

$$\Delta T = -\frac{T}{C_B}\left(\frac{\partial M}{\partial T}\right)_B \Delta B,$$

where M is the *magnetization*, and C_B is the *specific heat* at constant magnetic field. For a *paramagnet*, $M \propto B/T$ and thus $\Delta T \propto B\Delta B/T$; hence, the temperature of the magnetic material rises as the magnetic field increases, and falls as it decreases. The magnetocaloric effect is employed to achieve temperatures below 1 K by the adiabatic demagnetization of paramagnetic salts (see *Adiabatic demagnetization cooling*). For *ferromagnets* below the *Curie point*, $(\partial M/\partial T)_B < 0$, hence $\Delta T > 0$. The magnetocaloric effect attains its maximum value in the vicinity of the Curie point, and it is one of the most accurate indirect methods for determining the Curie temperature and the *magnetic saturation*. The magnetocaloric effect in *ferrimagnets* exhibits a change of sign in the neighborhood of the *magnetic compensation* point due to the change in sign of the temperature derivative of the magnetization.

MAGNETODIELECTRIC, magnetoinsulator

A powder *magnetic material* prepared by *pressing*, and consisting of ferromagnetic or ferrimagnetic particles of 1–100 μm size separated from each other by a non-magnetic *insulator* (dielectric). Employed as a powder in the magnetodielectrics are: the *soft magnetic materials* AlSiFe, carbonile iron, permalloy, etc., and the *hard magnetic materials*, various *ferrites* (e.g., the barium one), as well as alloys of the *alni*, *alnico*, etc., type. The insulator, acting simultaneously as a binder, may be: a "liquid glass", a glassy enamel, Bakelite, polystyrene, shellac, various resins, etc. *Rubbers* are employed as an insulator when using magnetodielectrics to fabricate elastic *permanent magnets*. The properties of a magnetodielectric can be varied over a wide range by selecting the composition and size of the magnetic powders, the type and content of the insulator, as well as by varying the pressing process techniques. Magnetodielectrics display high resistivity and small eddy-current loss (see *Magnetic loss*). They are used in radioelectronic and telecommunication equipment (coil cores, magnetic circuits of relays and chokes, etc.) at frequencies of up to 100 MHz, to manufacture permanent magnets, holding devices, etc. Magnetodielectrics successfully compete with ferrites in some fields of application.

MAGNETODYNAMIC NONLINEARITY

A principal cause of nonlinear electromagnetic effects that appear in pure *metals* at low temperatures. Due to the high *electrical conductivity* of a metal the electric field present in it is usually small, and the *magnetic field* generated by the current flowing through a sample proves to be the main source of nonlinearity. This magnetic field alters the electron dynamics, and thus influences the metallic conductivity. The mechanism of magnetodynamic nonlinearity works most effectively with a strong spatial dispersion when the conductivity is governed by a small group of *conduction electrons* confined to an effective potential well formed by the intrinsic field of the nonlinear wave. These electrons move along the electric current in trajectories that wind adjacent to the plane where there is a sign reversal of the intrinsic magnetic field. They exist only in a nonlinear regime. There is a pronounced spatial dispersion when the electron mean free path exceeds the specimen thickness. The magnetodynamic nonlinearity results in a sublinear *current–voltage characteristic* in a plate, and voltage stabilization in a wire. The external magnetic field controls the nonlinearity in such a manner that there is the possibility for a negative differential resistance to be present in a plate. For high frequencies the strong spatial dispersion is associated with the *anomalous skin-effect*. A magnetodynamic nonlinearity can lead to the dependence of the *surface impedance* on the amplitude of an incident electromagnetic wave (*nonlinear anomalous skin effect*), and the excitation of electric *current states in metals*.

MAGNETOELASTIC EFFECT

The same as the *Villari effect*.

MAGNETOELASTIC EXCHANGE INTERACTION

A *magnetoelastic interaction* caused by the dependence of the exchange integral J (see *Exchange interaction*) of atomic spins of a magnetically ordered crystal on the interatomic distance. The energy of magnetoelastic exchange is usually higher than that of ordinary quantum-mechanical spin interactions. However, magnetoelastic exchange possesses a higher symmetry (that does not vary for arbitrary uniform spin rotations of a magnet with unchanged atomic positions). By virtue of this, there are some magnetoelastic effects that do not arise from the magnetoelastic exchange interaction, e.g., to describe magnets it is necessary to allow for other types of interaction. The magnetoelastic exchange interaction controls, in particular, the volume change that results from magnetic ordering (this can alter the type of *magnetic phase transition*), but does not cause elastic *strains* to appear when the *magnetization* rotates with its length unchanged.

MAGNETOELASTIC GAP

An excitation in a *spin wave* spectrum of a magnetically ordered crystal whose magnetic symmetry would not suggest such an excitation spectrum (see *Goldstone theorem*); also a contribution to a spin wave excitation due to magnetoelastic interactions (see *Magnetoacoustic resonance*). The magnetoelastic gap appears most prominently in the vicinity of spin-realignment *magnetic phase transitions*, as the activation due to the *magnetic anisotropy* in the spin wave spectrum diminishes. Simultaneously vibrations of the elastic subsystem are transformed: the *dispersion law* for the sound (acoustic wave) interacting with the spin waves changes from linear to quadratic at the transformation point. Associated with the magnetoelastic gap is the invariance in the free energy of a *magnetic substance* which is linear in the deviations of the *magnetization* and *strain* from their equilibrium values, and determines the presence of an intrinsic ferroelastic phase transition in the material (see *Ferroelastic*).

MAGNETOELASTIC INTERACTION

Interaction between the *magnetization* and elastic displacements (of spin and elastic subsystems) in magnetically ordered solids. Underlying the magnetoelastic interaction is the dependence of the direct (dipole–dipole, spin–orbit), and the *exchange interactions* of magnetic atoms and *magnetic ions* upon the spacing between them. Of special importance are the *direct magnetoelastic interaction* and the *exchange magnetoelastic interaction*. These interactions can be described both microscopically, in the terms of a *spin–phonon Hamiltonian*, and through a phenomenological approach to the physics of *magnetism*, and they appear in both static and dynamic situations. The former (static) case results in *magnetostriction*, *mechanostriction*, in the magnetic anomalies in *thermal expansion*, in *elastic moduli*, in *piezomagnetism*, etc. The stronger exchange interaction associated with *magnetic phase transitions* may result in isomorphic changes of the structure of the *crystal lattice* and its symmetry. In the latter (dynamic) case the magnetoelastic interaction results in *magnetoacoustic resonance*, *spin–lattice relaxation*, etc. The magnetoelastic interaction is a relatively weak one, but its importance grows markedly in the vicinity of a magnetic phase transition where the *magnetic anisotropy* energy decreases. Near a second-order phase transition the frequency of the *soft mode* tends toward the value of the *magnetoelastic gap* rather than to zero. Here, as well, the *phonon* spectrum becomes highly distorted which, in turn, has a marked bearing on the static, dynamic, thermodynamic and other properties of the *magnetic material*.

MAGNETOELECTRIC

See *Magnetoferroelectrics*.

MAGNETOELECTRIC-ACOUSTIC PHENOMENA

See *Acoustomagnetoelectric phenomena*.

MAGNETOELECTRIC EFFECT (predicted by L.D. Landau and E.M. Lifshits, 1956)

Development (or change) of *magnetization* (electric polarization) in a solid under the action of an electric (magnetic) field. Several types of magnetoelectric effects are known, differing in their physical nature. The thermodynamic potential of some magnetically ordered crystals may include the terms $\Phi = -\alpha_{ik} E_i B_k$, where E_i and B_k are the electric and magnetic fields, respectively. Therefore, the electric field induces a magnetization in such materials: $M = -\partial\Phi/\partial B_k = \alpha_{ik} E_i$, and the magnetic field induces an electric polarization: $P = -\partial\Phi/\partial E_i = \alpha_{ik} B_k$. The classes of magnetic symmetry for which the tensor α_{ik} is nonzero were found by I.E. Dzialoshinskii. A magnetoelectric effect was first observed in the antiferromagnet Cr_2O_3. Besides the magnetoelectric effects induced by an external field, *spontaneous magnetoelastic effects* exist in *ferroelectrics*, which introduce some peculiar features into the picture of ferroelectric and *magnetic phase transitions*, and into the response of such a system to electric and magnetic fields. A *nonuniform magnetoelectric effect* may may appear in the presence of a macroscopic nonuniformity of the magnetization (e.g., one associated with *magnetic domain walls*) in magnetically ordered crystals, as well as in *modulated magnetic structures*; such phenomena may take place in crystals of any symmetry type.

Magnetoelectric effects based on the terms $\Phi = -\alpha_{ik} E_i B_k$ are impossible in crystalline *paramagnets* and *diamagnets* because the α_{ik} tensor vanishes in crystals without a magnetic structure, but whose symmetry operations include *time inversion*, T, by itself (T changes the sign of the magnetic field and magnetization). Still the thermodynamic potential of ferroelectric paramagnetic crystals involves the terms $\Phi = -\xi_{ijk} E_i B_j B_k$, where ξ_{ijk} is a third-rank tensor with the same symmetry as the *piezoelectric tensor.* Placing such crystals in a strong magnetic field, B_k, one can induce in them a magnetization along the j direction with a field E_i, or an electric polarization along the i direction with a field B_j. Such a magnetoelectric effect was first identified in a $NiSO_4 \cdot 6H_2O$ crystal.

The magnetoelectric effect is also possible in diamagnetic nonmetallic crystals with *paramagnetic centers* of impurity ions. This effect can be conveniently described in this case in terms of the *spin Hamiltonian* of a paramagnetic center. If the local symmetry of the paramagnetic center, as well as the symmetry of the crystal as a whole, is lacking a *center of symmetry* then the *spin Hamiltonian parameters* will be linear in the field E, and so will the magnetization of the crystal. Along with the purely static magnetoelectric effects caused by a change of spin Hamiltonian parameters, diamagnetic nonmetallic crystals with paramagnetic centers exhibit magnetoelectric effects associated with an unpaired electron traveling (as a rule, by way of tunneling) across different sites in the crystal, and with the effects of an electric field acting on such a motion. Under this type of magnetoelectric effect, an enormous enhancement of the magnetization of a specimen (by over an order of magnitude) is observable, arising from the action of an alternating electric field on the paramagnetic centers. This was first experimentally observed with $Al\text{–}O^-$ hole centers in quartz. An essential role belongs to the electric-field modulation of the processes of *relaxation* of electric and *magnetic dipoles.* One can interpret this magnetization enhancement as a quasi-stationary cooling of the system of paramagnetic centers down to a spin temperature (see *Level population*) below the lattice temperature.

The magnetoelectric effect was also predicted for non-magnetic conductors lacking an inversion center, where a magnetization must appear due to an electric current flowing: $M = \alpha E$, and an electric polarization must emerge in a magnetic field. The pseudotensor α includes the *relaxation time*; therefore, as in all *kinetic phenomena*, the time inversion *covariance* is not a limiting condition in this case.

Underlying the magnetoelectric effects is the fact that the electric and magnetic characteristics of a solid are determined by the same particles, namely, electrons that carry, in addition to an electric charge, a magnetic moment (spin and orbital). There are interactions (exchange and spin–orbit types) to interrelate the motion of charges in space, and the mutual alignment of their magnetic moments. External forces affect the energy of the electric and magnetic subsystems that interact both with one another and with their environment (e.g., with *phonons*). That results in the reciprocal influence of the electric field upon the magnetic characteristics, and of the magnetic field upon the electric characteristics, of a *solid.*

MAGNETOFERROELECTRICS, magnetoelectrics

Materials featuring simultaneous magnetic and ferroelectric (antiferroelectric) ordering. The natural magnetoelectrics are orthorhombic ericaite, $Fe_3B_7O_{13}Cl$ and chambersite, $Mn_3B_7O_{13}Cl$. However, the first studies of magnetoelectrics were conducted using synthetic crystals of the *perovskite* type. A large number of both ferroelectric and antiferroelectric crystals are known, which may be either *ferromagnetic, antiferromagnetic* or *ferrimagnetic.* Their ferroelectric and magnetic ordering occurs at different temperatures, θ_E and θ_M, respectively (see *Curie point*). Ordinarily, but not necessarily, $\theta_M < \theta_E$. A substance is a magnetoelectric at $T < \min(\theta_E, \theta_M)$. Magnetoelectrics are known with θ_E up to 1000 K and θ_M up to 400 K.

The specific properties of magnetoelectric crystals result from the interaction of the ferroelectric and magnetic subsystems, otherwise known as the *magnetoelectric interaction* (MEI). Microscopically the nature of this interaction is the same as that of the *magnetoelastic interaction*. One should distinguish between the exchange and relativistic MEI. The energy of MEI in magnetoelectrics may be comparable to the intrinsic energy of either the spin or the ferroelectric subsystem. MEI produces specific phenomena typical for magnetoelectrics only, such as the spontaneous *magnetoelectric effect*, the change of *atomic magnetic structure* by an external electric field, and the change of the electric state by a magnetic field. The latter may possibly induce corresponding *phase transitions*. There also exist *magnetoelectric oscillations* and waves (mutually related spin and ferroelectric waves, see *Magnetoacoustic resonance*). The unusual properties of magnetoelectrics make it possible to apply them to the construction of unique instruments in solid state *microelectronics*.

MAGNETOGYRIC RATIO

The same as *Gyromagnetic ratio*.

MAGNETOINSULATOR

See *Magnetodielectric*.

MAGNETOMECHANICAL PHENOMENA,
gyromagnetic phenomena

A group of phenomena determined by the interplay between mechanical and magnetic moments of microscopic particles serving as carriers of *magnetism*. Every such particle (electron, proton, neutron, atomic nucleus, atom) with *angular momentum* has an associated *magnetic moment*. Therefore, an increase of the total angular momentum of the particles comprising a physical body results in the appearance of an additional magnetic moment, and the body acquires additional angular momentum at its *magnetization*. Thus there appears a magnetic moment in an initially unmagnetized iron rod when it is rotated (see *Barnett effect*), and reciprocally, a freely suspended ferromagnetic specimen rotates on being magnetized (see *Einstein–de Haas effect*).

MAGNETOMECHANICAL RATIO

The same as *Gyromagnetic ratio*.

MAGNETOMETRY

Process of measuring magnetic fields, and sometimes more generally magnetic properties of materials. *Magnetometers* are based on a variety of principles, and are designed to determine the characteristics of a *magnetic field* (magnetic field $\boldsymbol{H}$, *magnetic flux density* or induction $\boldsymbol{B}$, time and space dependences of the field characteristics, etc.) or the magnetic properties of materials (*susceptibility*, *magnetic moment*, *magnetization*, *hysteresis* loops and losses, *coercive force*, *remanence*, parameters of *magnetic anisotropy* (see *Magnetic anisotrometer*), etc.).

Such devices are often labeled with more specific names originating either from the measurement method (e.g., *magnetic balance*, a device to gauge magnetic moments and susceptibilities), from the quantity being determined (e.g., *coercimeter*, to measure the coercive force, or *fluxmeter*, to measure magnetic flux), or else from the unit in which the device is calibrated (e.g., *teslameter* and *gaussmeter* measure B in the units T and G, respectively, or *webermeter*, a magnetic fluxmeter calibrated in $\mathrm{Wb} = \mathrm{T{\cdot}m^2}$).

The main methods that have been employed in magnetometry over the years are as follows. The *ballistic method* employs a ballistic galvanometer to measure the amount of electric charge generated in a search coil for a change of the magnetic flux threading through it. The *magnetometric method* is based on the magnetizing action of a specimen upon a magnetic needle, while the *electrodynamic method* is based on its action on a bracket with electric current. The *induction method* measures the e.m.f. induced in a secondary coil by an alternating magnetizing current passing through a primary coil. The *vibration method* measures the e.m.f. induced by vibrations of one magnetized specimen relative to another or to a coil. The *ponderomotive method* gauges the mechanical force acting on a specimen in a magnetic field (a pendulum, torsion or lever balance). *Bridge and potentiometric methods* determine the complex *magnetic permeability* in an alternating magnetic field. The *wattmeter method* is employed to measure remagnetization losses. The *calorimetric method* measures the changes of a specimen temperature at its remagnetization. Most of these methods are now antiquated, and seldom employed.

More widely used in recent years are *resonance and quantum methods* which depend on the phenomena of *nuclear magnetic resonance*, *electron paramagnetic resonance*, *Josephson effect* (see *Superconducting quantum interference devices*), etc. *Galvanomagnetic methods* make use of the Hall effect, and the changes of the resistivity in a magnetic field (see *Galvanomagnetic effects*). *Optical methods* are associated with the *rotation of light polarization plane* (*Faraday effect* and *Kerr effect*).

MAGNETOMOTIVE FORCE

A scalar quantity characterizing the magnetizing action of electric currents. The magnetomotive force F equals the circulation of the magnetizing vector $\boldsymbol{B}$ over a closed circuit L enclosing the electric currents I_i that generate the magnetic field

$$F \oint_L B\,\mathrm{d}l = \oint_L B_l\,\mathrm{d}l = \sum_{i=1}^{n} I_i,$$

where B_l is the projection of $\boldsymbol{B}$ on to the direction of the element dl of the integration contour, and n is the number of conductors (turns) carrying a current I enclosing the circuit. This concept is useful for *magnetic circuit* calculations.

MAGNETON

A unit of *magnetic moment* with the value $\mu = e\hbar/2m_{\mathrm{fp}}$, where e is the electron charge and m_{fp} is the mass of a fundamental particle. Of particular importance are the *Bohr magneton*, $\mu_{\mathrm{B}} = 9.274\cdot 10^{-24}$ J/T (defined by the mass of an electron) used for atomic magnetism, and the *nuclear magneton*, $\mu_{\mathrm{N}} = 5.051\cdot 10^{-27}$ J/T, with m_{fp} equal to the mass of a proton. Thus, electronic magnetism is much stronger by three orders of magnitude ($m_{\mathrm{p}}/m_{\mathrm{e}} = 1836$) than the magnetism of nuclei and nucleons. The magnetic moment of an elementary particle (as well as that of a system of particles) is quantized, and sometimes it is a multiple of the corresponding magneton. The magneton plays the part of an elementary magnetic moment; which is closely associated with the quantization of the particle *angular momentum*. For orbital motion the latter is a multiple of $\hbar$, for spins it can be a multiple of either $\hbar$ or $\hbar/2$.

MAGNETOOPTICAL EFFECTS

Phenomena consisting in changes of light polarization or intensity in a material medium under the action of a *magnetic field*, $\boldsymbol{B}$, or of a *magnetization*, $\boldsymbol{M}$, that take place as light passes through a medium, or is reflected from its surface. Falling into the first group are: the *Faraday effect*, or *magnetic circular birefringence*, the *Cotton–Mouton effect* (see *Voigt effect*), *magnetic circular dichroism* (differential absorption of waves with left and right polarization in the Faraday effect geometry), and *magnetic linear dichroism* (differential absorption of intrinsic waves with linear polarization in the Cotton–Mouton effect geometry). The changes in the intensity and polarization of light on its reflection from the surface of a magnetized medium constitute the *magnetooptical Kerr effect*. It can be polar ($\boldsymbol{M} \parallel \boldsymbol{n}$, where $\boldsymbol{n}$ is a normal to the surface), longitudinal, or meridional (vector $\boldsymbol{M}$ parallel to the surface and lying in the plane of incidence), and transverse or equatorial (vector $\boldsymbol{M}$ parallel to the surface and perpendicular to the plane of incidence). The correlation between the optical and magnetic properties occurs not only in static, but also in dynamic, phenomena, e.g., in the scattering of light from *spin waves*. Magnetooptical effects are found in materials of all types: in *insulators*, *semiconductors* and *metals*, in *diamagnets* and *paramagnets*; but they are strongest in magnetically ordered materials: *ferromagnets*, *antiferromagnets*, etc. The cause underlying the appearance of these effects is the dependence of the energy spectrum of a *magnetic substance* on its magnetic state.

MAGNETOOPTICAL EFFECTS IN LIQUID CRYSTALS

Changes in optical properties of *liquid crystals* resulting from the orienting action of an applied magnetic field upon the *director* (optical axis of a liquid crystal) due to the diamagnetic anisotropy with $\chi_{\mathrm{a}} = \chi_{\parallel} - \chi_{\perp} > 0$ (see *Electrooptical effects*).

MAGNETOOPTICS

Study of the interrelations between magnetic and optical properties of matter. These magnetooptical phenomena can be classified into: *magnetooptical effects* (*Faraday effect*, *Kerr effect*, etc.) caused by changes in the optical properties of materials (*dielectric constant* and *magnetic permeability* tensors) under the action of a *magnetic field* or *magnetization* (or sublattice magnetization vectors in multisublattice *magnetic substances*); and *photomagnetic effects* arising from the impact of optical radiation on such magnetic parameters as magnetization and *magnetic anisotropy* (reciprocal Faraday effect, photoinduced magnetic anisotropy, etc.). Magnetooptical phenomena appear in all magnetic materials, but are most pronounced in magnetically ordered media.

MAGNETOPLASTIC EFFECT

Effects of a magnetic field on mechanical properties of ferromagnets and other materials. Changes occur in the *yield limit*, *ultimate strength*, permanent set and *creep* of materials, wear hardness (see *Wear*), and the *durability* of parts and tools. In an alternating field the *strain* is discontinuous. The magnetoplastic effect arises from the interaction of *dislocations* with *magnetic domain walls*, or from magnetostrictive stresses (see *Magnetostriction*). The *magnetic domain structure* becomes much finer under strain, which makes the interaction of domain boundaries with dislocations in a weak field more intense. In a stronger field, domain boundaries disappear, but the drag on dislocations due to their disaligning action upon the spin moments remains. When studying magnetoplastic effects experimentally, it is necessary to take into account specimen heating by eddy currents in an alternating magnetic field, or at the switching (on/off) of the permanent field. The magnitude and sign of the magnetoplastic effect depend on many factors: the magnitude and direction of the magnetic field; the sequence in which stress and field have been applied; the temperature and the rate of deformation; the number of *impurity atoms*, the structural homogeneity of the material including the grain and subgrain sizes; and magnetic characteristics (domain structure, saturation *magnetization*, magnetostriction, etc.).

MAGNETOPLUMBITE

A *mineral* of hexagonal (in the crystallographic sense) symmetry. $Pb_2Fe_{15}Mn_7(Al, Ti)\ O_{38}$. Oxide monocrystalline *ferrimagnets* of composition $n_1(Pb, Ba, Sr)O{\cdot}2n_2Fe_2O_3$ ($n_1 \geqslant 1$, $n_2 \geqslant 3$), isomorphous to the above, are referred to as *hexaferrites* of the magnetoplumbite structure. The *unit cell* comprises four alternating blocks: S blocks of the cubic spinel structure (two-layered), and R blocks of a hexagonal structure (three-layered), positioned perpendicularly to the c axis according to the pattern SRS^*R^* (where the asterisk denotes blocks turned through 180°). The *space group* is $P6_3/mmc$ (D_{6h}^4), and the *point group* has symmetry $6/mmm$ (D_{6h}). The best known members of the magnetoplumbite family are: $PbO{\cdot}6Fe_2O_3$, barium ferrite $BaO{\cdot}6Fe_2O_3$ and strontium ferrite $SrO{\cdot}6Fe_2O_3$. Typical magnetoplumbite parameters are: *crystal lattice* constants $a = 0.586$–0.589 nm and $c = 2.303$–2.319 nm; density 5.12–5.67 g/cm^3, *magnetic moment* per unit chemical formula $\sim 20\mu_B$, and the *Néel point* 430–460 K.

MAGNETORESISTANCE

Changes in resistivity of solid conductors under the action of an external *magnetic field*, $\boldsymbol{B}$. There are transverse and longitudinal magnetoresistive effects with the electric current, $\boldsymbol{I}$, running perpendicular to the magnetic field $\boldsymbol{B}$, or with $\boldsymbol{I} \parallel \boldsymbol{B}$, respectively. The underlying cause of this effect is the bending of current carrier trajectories in the magnetic field (see *Galvanomagnetic effects*). The relative change of the transverse resistivity at room temperature amounts to $(\Delta\rho/\rho)_\perp \approx 10^{-4}$ at $B \approx 1$ T in *metals*, with the exception of Bi which has $(\Delta\rho/\rho)_\perp \approx 2$ at $B = 3$ T. The latter justifies the employment of Bi to gauge the magnetic field (see *Magnetometry*).

Decreasing the temperature and increasing B result in an increase in $(\Delta\rho/\rho)_\perp$ (see *Kapitza law*). In a weak field, $(\Delta\rho/\rho)_\perp$ is proportional to B^2. The proportionality factor is usually positive, i.e. resistivity increases with magnetic field; the exception being ferromagnets (see *Kondo effect*, *Mesoscopic fluctuations*). Since resistivity is sensitive to the presence of impurities and defects in the crystal lattice, and to the temperature, measurements may yield different ρ versus B dependences. Nevertheless, different data referring to one and the same metal are found to fit a single straight line (see *Kohler rule*).

The magnetoresistive effect is used to investigate electronic energy spectra and mechanisms of *current carrier scattering*, and to gauge the magnetic field strength.

Larger changes in magnetoresistance called *giant magnetoresistance* (GMR) have been observed in films made with alternating layers of a ferromagnetic and a non-ferromagnetic material. This effect was first detected in alternating layers of Fe and Cr, and is more pronounced in Co/Cr layers. Other materials, such as $La_{0.67}Ca_{0.33}MnO_x$, exhibit resistance changes of 10^3 in a magnetic field of 6 T, an effect referred to as *colossal magnetoresistance.*

MAGNETORESISTANCE, ANOMALOUS

See *Anomalous magnetoresistance.*

MAGNETOSTATIC OSCILLATIONS

Free oscillations of *magnetization* of a magnetically ordered material with characteristic inhomogeneity dimensions of the order of the specimen size. The frequency of magnetostatic oscillations, ω, must satisfy the inequality $\omega \ll c/l$ (l is the characteristic specimen size), implying that one is allowed to neglect the *effects of retardation* (lag) between different parts of the material. At the same time, the specimen dimensions should be large compared to the characteristic *exchange interaction* length, $D^{1/2}$ ($l \gg D^{1/2}$, where $D \approx 10^{-12}$–10^{-14} cm^2). The magnetic oscillations can be described, for these simplistic circumstances, by *magnetostatics* equations and by the equations of motion of the magnetization of *magnetic sublattices* (see *Landau–Lifshits equation*). The structure of the magnetic field and magnetization, and the spectrum of magnetic oscillations, depend on the specimen shape, and the boundary conditions at its surface. There is a variety of characteristic magnetic oscillations (see *Walker modes*), including those corresponding to uniform precession of the magnetization. The resonant frequency of *ferromagnetic resonance* for a ferromagnetic ellipsoid is given by the *Kittel equation* (Ch. Kittel, 1948):

$$\omega = \gamma\big[B + \mu_0(N_x - N_z)M\big]^{1/2} \times \big[B + \mu_0(N_y - N_z)M\big]^{1/2},$$

where μ_0 is the *magnetic permeability of vacuum*, γ is the *gyromagnetic ratio*, M is the saturation magnetization, B is the magnetizing component of the external field ($B \parallel z$), and N_x, N_y and N_z are the demagnetizing factors of the ellipsoid. The magnetic oscillations are excited with a nonuniform microwave field by placing the specimen in a wave guide or resonator.

MAGNETOSTATICS

A division of *magnetism* and electrodynamics that studies the properties of a steady-state (or relatively slowly changing) magnetic field (from a *permanent magnet* or direct electric currents). Calculations of the field make use of the equations of magnetostatics, which correspond to a special case of Maxwell's equations:

$$\nabla \cdot \boldsymbol{B} = 0, \qquad \nabla \times \boldsymbol{H} = \boldsymbol{j},$$

where $\boldsymbol{B}$ is the *magnetic flux density*, $\boldsymbol{H} = \boldsymbol{B}/\mu$ is the magnetic field, and $\boldsymbol{j}$ is the conduction current density. The basic field computation methods in magnetostatics ensue from the *Ampère theorem*, which asserts that the magnetic fields of closed direct current loops can be regarded as the fields of paired *magnetic charges* corresponding to *magnetic dipoles*. Introduction of the magnetic charge concept allows the formal application of field computation methods employed in electrostatics. The magnetostatic field in the absence of conduction currents is a potential field, i.e. the work of the forces of a magnetostatic field acting on a probe charge is independent of the path. The magnetostatic equations describe a varying magnetic field adequately enough, provided the characteristic dimension of the field inhomogeneity is much smaller than the wavelength of the oscillatory changes in B. Satisfying this condition is called the *magnetostatic limit*.

MAGNETOSTATIC WAVES

See *Spin waves.*

MAGNETOSTRICTION

Strain of a material induced by an applied magnetic field, discovered by J. Joule (1842). It is characterized by relative changes of the linear dimensions of the body $\lambda = \Delta l/l$ (*linear magnetostriction*) or of its volume (*bulk magnetostriction*). The magnetostriction phenomenon is present in all *magnetic materials*, with differing values of λ in different cases. In *ferromagnets* and *ferrimagnets*, $\lambda \sim 10^{-5}$–10^{-4}, in *antiferromagnets*, *paramagnets* and *diamagnets*, the magnetostriction is ordinarily smaller (10^{-6}–10^{-7}). *Giant magnetostriction* with $\lambda \sim 10^{-3}$–10^{-2} is exhibited by *rare-earth elements*, *actinides* and their compounds. The magnitude of λ depends on the orientation of the size change direction relative to the external magnetic field $\boldsymbol{B}$ and the crystallographic axes. One can determine λ along the field $\boldsymbol{B}$ (*longitudinal magnetostriction*) and perpendicular to it (*transverse magnetostriction*). Underlying this phenomenon is the fact that the interactions defining the magnetic state of a crystal depend upon the distances between magnetic atoms or ions. Hence, any change in the magnetic state under the action of an applied magnetic field, temperature, elastic stress, etc., results in the atoms being displaced from their equilibrium positions during the ensuing deformation of the body.

Magnetostriction is known as *exchange magnetostriction* or *direct magnetostriction* according to the type of magnetic interactions involved (*exchange interaction*, or direct interaction such as spin–orbit or dipole–dipole). The direct magnetostriction occurs in the region of *saturation magnetization* of ferromagnets at $B < B_S$ (B_S is the saturation magnetizing field) and displays a marked anisotropy. At $B > B_S$, i.e. in the so-called *paraprocess* region, the exchange magnetostriction is typical. In this region, the magnetostriction is isotropic in cubic crystals and anisotropic in uniaxial ones. The longitudinal and transverse magnetostrictions are usually opposite in sign for $B < B_S$ in most magnetic materials, and have similar signs for $B > B_S$. The value of λ at $B = B_S$ is called the *saturation magnetostriction*. Magnetostriction is an even function of the *magnetization $\boldsymbol{M}$*.

Antiferromagnets exhibit, in addition to the usual variety, a magnetostriction linear in the magnetization, and a piezomagnetic effect (see *Piezomagnetism*) reciprocal to it. A strain due to changes of the magnetic state with temperature (spontaneous magnetization) is termed *spontaneous magnetostriction*. An additional strain due to the magnetic state changing under the action of external elastic stresses is termed *forced magnetostriction*, or *mechanostriction*.

Associated with magnetostriction is the reciprocal process or *Villari effect*, which is the change in the magnetic induction B of a ferromagnetic material resulting from an applied mechanical stress.

MAGNETOSTRICTION, GIANT

See *Giant magnetostriction*.

MAGNETOSTRICTIVE MATERIALS (fr. Gr. $\mu\alpha\gamma\nu\eta\tau o\varsigma$, magnet, and Lat. *strictio*, tightening, narrowing)

Soft magnetic materials exhibiting a marked *magnetostriction*. Some basic parameters of magnetostrictive materials are: the magnetomechanical coupling parameter, k, whose square equals the ratio of the transformed energy (mechanical or magnetic) to the supplied energy (magnetic or mechanical, respectively), neglecting losses; the dynamic magnetostriction parameter, a, defining the transducer sensitivity in the projector mode; the relative *magnetic permeability*; the *sound velocity*; the saturation magnetostriction determining the ultimate sound intensity emitted by the magnetostriction transducer; and the *coercive force* and electric resistivity responsible for the energy loss through *hysteresis* and eddy currents, respectively. Employed as magnetostrictive materials are iron–nickel (*permalloy*), iron–cobalt (*permendur*), and iron–aluminum (*alfer*) alloyed with chromium, vanadium and other metals as well as with some soft ferrimagnetic materials. What should be taken into consideration when choosing magnetostrictive materials are: their *corrosion* resistance; the *plasticity* permitting fabrication into thin sheets; an operational temperature range conditioned by the temperature dependences of the material characteristics, and by its *Curie point*; the ease of processing; and the cost. Typical of iron–cobalt alloys is the combination of very high values of the magnetostriction, saturation magnetization, and Curie temperature. For this reason, they are used to design high-power transducers and high intensity projectors. Some permendur types are remarkable for their good plasticity. Iron–aluminum alloys possess large magnetostriction and electric resistivity, and in addition mechanical *strength*, but they display low ductility and a pronounced tendency to corrode. The cores of magnetostriction transducers are assembled from sheets of magnetostrictive materials electrically isolated from one another to reduce eddy current loss.

Ferrite magnetostrictive materials exhibit a very high *corrosion resistance*, and they are used, due to their high resistivity and the ensuing low eddy current loss, for the conversion of radio frequencies, particularly, in magnetostriction filters. Magnetostrictive materials find applications as projectors, and as receivers of ultrasonic mechanical vibrations.

MAGNETOTHERMAL PHENOMENA

Changes in the thermal states of materials that take place during changes in their magnetic states (at their *magnetization* or *demagnetization*). Singled out are: the magnetothermal phenomena during an adiabatic magnetization and demagnetization (*magnetocaloric effect* when the body temperature changes), and isothermal magnetothermal phenomena associated with the evolution or absorption of heat. In principle, magnetothermal phenomena are observable in any material, since underlying them are general laws of thermodynamics. A change of the internal energy of a body (see *Thermodynamic potentials*) is associated with a change in its magnetic state.

Most prominent are magnetothermal phenomena in *ferromagnets*, *antiferromagnets* and *ferrimagnets*. The features of the phenomena in these materials stem from the nature of their magnetization processes: shifting of domain boundaries (see *Domain wall displacement processes*); rotation of magnetic moments of domains (see *Magnetization rotation processes*); *paraprocess*; or processes involving of the decay or induced build-up of a noncollinear *atomic magnetic structure* (in antiferromagnets and ferromagnets). Particularly important are thermal effects accompanying the two latter processes. Closely linked by thermodynamics with magnetothermal phenomena occurring during magnetization are the anomalies of the specific heat found in the vicinity of the *Curie point*, *Néel point*, and other *magnetic phase transition* points (e.g., near the switch point of the noncollinear magnetic structure in ferrimagnets). The magnetothermal phenomena in some *paramagnets* are used to achieve temperatures below 1 K (see *Adiabatic demagnetization cooling*). When the atomic or nuclear magentic moments of a spin system absorb microwave or radiowave energy during a magnetic resonance experiment (see *electron paramagnetic resonance*, *nuclear magnetic resonance*) they transfer this energy to the lattice vibrations with a time constant called the spin–lattice *relaxation time*. If this relaxation time is very long the spin system can become heated, and temporarily acquire a spin temperature which is higher than the lattice temperature.

MAGNET, POWDER

See *Powder magnet*.

MAGNET, SAMARIUM–COBALT

See *Samarium–cobalt magnet*.

MAGNON

A *quasi-particle* corresponding to a spin oscillation wave (*spin wave*) in a magnetically ordered crystal. The number of different magnon types coincides with that of spin wave branches; hence *acoustic magnons*, *optical magnons*, *nuclear magnons*, etc., are distinguished. The density of magnons at thermal equilibrium depends on the temperature, T, and decreases as $T \to 0$. Magnons manifest themselves in the thermodynamic properties of *magnetic substances*, viz., they determine the low-temperature behaviour of the *magnetization*, $M(T)$ (see *Bloch law*), contribute to the *specific heat*, *thermal conductivity* and other thermal parameters, and also feature prominently in the microwave and kinetic properties of matter. Describing three-dimensional magnetic materials in terms of a gas of weakly interacting magnons is valid, as a rule, up to a temperature of the order of half the temperature of the transition to the paramagnetic state (see *Paramagnetism*). Taking into consideration, in addition to magnons, nonlinear excitations such as *magnetic solitons*, *magnon bound states* and *magnon bubbles*, may be necessary in low-dimensional magnetic materials, as well as under the conditions of strong excitation. Direct investigations of the magnon *dispersion law* have been carried out using the methods of *inelastic neutron scattering* and the scattering of light, as well as with *parametric excitation of spin waves*.

MAGNON BOUND STATE

A stationary state of the *Heisenberg magnet* with two spin directions (*magnons*), their wave function exponentially diminishing with distance between the "flipped over" spins. This localized magnon state is defined by the value of its total *quasi-momentum* of magnons, $\boldsymbol{k}$, and the discrete quantum number that prescribes the possible number of localized magnon states for a given $\boldsymbol{k}$, a number which remains finite. The localized magnon state energy levels lie below the minimum energy of the continuous spectrum of two-magnon states prescribed by the same value of $\boldsymbol{k}$. With $\boldsymbol{k}$ decreasing, the gap separating these discrete energy levels from the continuous spectrum narrows so that the localized magnon state eventually vanishes at a certain threshold value $\boldsymbol{k}_0$. Both one- and two-dimensional Heisenberg magnets have $\boldsymbol{k}_0 = 0$, while the value is non-zero for a three-dimensional magnet. The nature of the localized magnon state is identical to that of bound states of a particle in a potential well. The depth of that well is controlled by the $\boldsymbol{k}$ vector for the case of a localized magnon state. Localized states of more than two magnons are also possible (*spin complexes*, *magnon bubbles*). These states may be compared to *magnetic solitons*.

MAGNON BUBBLE

Magnetic soliton of a sizeable amplitude, inside which the *magnetization* vector differs from, and performs a precessional motion around, its equilibrium value in the remainder of the crystal. Magnon bubbles are *magnon bound states*.

MAGNON, HOT

See *Hot magnons*.

MAGNON MECHANISM OF SUPERCONDUCTIVITY

A hypothetical mechanism for the formation of bound electron pairs in a singlet or triplet spin state (see *Cooper pairs*, *Spin*) in metals with magnetic ordering of the ferromagnetic or antiferromagnetic type (see *Ferromagnet*, *Antiferromagnet*) occurring through the exchange of virtual quanta of collective spin excitations, i.e. *magnons*. Magnon exchange in ferromagnets always results in repulsion, whereas in antiferromagnets an effective attraction is possible (due to the opposite sign of the *exchange interaction* constant) between electrons in triplet states. Such an attraction may aid *triplet pairing* in anisotropic crystals (see *Anisotropy of crystals*). The possibility of a magnon mechanism of superconductivity has been discussed in connection with *reentrant superconductors* based on *heavy fermions* and the superconducting cuprates (see *High-temperature superconductivity*).

MAGNON–PHONON INTERACTION

Interaction of two types of *quasi-particles*: those associated with spin waves (*magnons*) and with elastic waves (*phonons*), which results from the *magnetoelastic interaction* in *magnetic substances*. Different elementary processes of the magnon–phonon interaction determine different physical effects (e.g., a magnon transforming into a phonon in *magnetoacoustic resonance*). The processes of a magnon emitting a phonon, of a phonon scattering by a magnon, and of a phonon being converted into two magnons (and vice versa), play a significant part in the establishment of thermal equilibrium in magnetic materials; e.g., they govern the heat exchange between the magnetic subsystem and the crystal lattice. One should take these processes into account to describe the *relaxation* of spin waves and of sound (see *Magnetic relaxation*).

MANGANESE, Mn

A chemical element of Group VII of the periodic table with atomic number 25, atomic weight 54.9380. The one stable isotope is ^{55}Mn. Outer shell electronic configuration $3d^5 4s^2$. Ionization energies (eV): 7.43, 15.64. Atomic radius 0.130 nm. Ionic radii (nm): Mn^{2+} 0.080, Mn^{3+} 0.066, Mn^{4+} 0.060, Mn^{7+} 0.046 for coordination number 6. Possible oxidation states are from +2 to +7 (the most stable being +2, +4 or +7). Electronegativity of Mn(II), Mn(IV) and Mn(VII) is ≈1.42, ≈1.85 and ≈2.07, respectively.

In the free state Mn is a pale gray brittle *metal*. It has four allotropic forms (α-Mn, β-Mn, γ-Mn, δ-Mn). Temperatures of polymorphic transformations: 990 K for $\alpha \leftrightarrow \beta$, 1360 K for $\beta \leftrightarrow \gamma$, 1412 K for $\gamma \leftrightarrow \delta$. α-Mn is stable at room temperature, and β-Mn is easily fixed in its metastable state through hardening (hardening that begins at a temperature within the β-Mn stability interval), whereas γ-Mn and δ-Mn cannot be fixed on cooling to room temperature. α-Mn has a complex cubic crystal structure with $a = 0.8904$ nm at 298 K, space group $I\bar{4}3m$ (T_d^3). The conventional unit cell contains 58 Mn atoms, whereas the primitive cell contains 29. Packing fraction is 54.8%. The structure is considered as a system of interpenetrating body-centered cubic lattices similar to α-W containing four types of Mn atoms with different coordinations and sizes, assumed to be in different oxidation states with magnetic moments in different directions. β-Mn also has a complex cubic structure, space group $P\bar{4}3m$, (T_d^1), $a = 0.6470$ nm at 1023 K. For non-equilibrium β-Mn, $a = 0.6302$ nm at 298 K. Conventional unit cell has 20 Mn atoms in two types of sites with different coordinations. Packing fraction is 67.6%. Modifications of α-Mn and β-Mn, unlike other metals, have no center of symmetry. Crystal structure of γ-Mn is treated as face-centered cubic with $a = 0.3859$ nm at $T = 1370$ K with 4 atoms in the unit cell and packing fraction is 74.05%, space group $Fm\bar{3}m$, (O_h^5). There is evidence that, even on rapid cooling to room temperature, γ-Mn attains a pseudo-cubic face-centered tetragonal structure, slightly deformed along one of the cube edges, space group $I4/mmm$ (D_{4h}^{17}), $a = 0.3780$ nm, $c = 0.3520$ nm, packing fraction 68.7%. Crystal structure of δ-Mn is body-centered cubic, space group $Im\bar{3}m$, (O_h^9), containing two Mn ions, $a = 0.3081$ nm at 1410 K. Packing fraction is 68.2%. Density of α-Mn is 7.471 g/cm^3, of β-Mn 7.245 g/cm^3 (at 298 K), of γ-Mn 6.335 g/cm^3 (at 1368 K), of δ-Mn 6.239 g/cm^3 (at 1407 K); $T_{\text{melting}} = 1517$ K, $T_{\text{boiling}} = 223$ K, heat of melting 14.9 kJ/mole, heat of evaporation 228.9 kJ/mole, bond energy of α-Mn is -2.98 eV/atom (at 0 K). Transition heats: α-Mn $\to$ β-Mn $\approx$2.54 kJ/mole, β-Mn $\to$ γ-Mn 2.28 kJ/mole, γ-Mn $\to$ δ-Mn 1.8 kJ/mole; heat of sublimation 224.8 kJ/mole. Specific heat of α-Mn is 0.477 kJ·kg^{-1}·K^{-1}, of β-Mn 0.649 kJ·kg^{-1}·K^{-1} and of γ-Mn 0.502 kJ·kg^{-1}·K^{-1} at 298 K. Debye temperature is 410 K; linear thermal expansion coefficient in temperature range 273 to 293 K is $22.3\cdot10^{-6}$ K^{-1} for α-Mn, $24.9\cdot10^{-6}$ K^{-1} for β-Mn, $14.75\cdot10^{-6}$ K^{-1} for γ-Mn; thermal conductivity is 66.57 W·m^{-1}·K^{-1} (at 298 K) and 5.86 W·m^{-1}·K^{-1} (at 95 K). Isothermal bulk modulus is 59.65 GPa (at room temperature); Young's modulus is $\approx$175 GPa at 294 K, shear modulus is 76.52 GPa (at 298 K); Poisson's ratio is 0.24 (at 298 K). Rockwell hardness is 70 HRC for α-Mn and 20 HRC for γ-Mn. At room temperature, α- and β-Mn are hard and brittle; therefore not subject to plastic deformation. γ-Mn is plastic and easily deformed until it transforms into α-Mn. Linear low-temperature electronic molar heat capacity: α-Mn $\approx$15.3 mJ·mol^{-1}·K^{-2}, γ-Mn 9.20 mJ·mol^{-1}·K^{-2}. Electric resistivity (free of residual resistance): α-Mn 1390 nΩ·m (at 295 K), β-Mn 910 nΩ·m, γ-Mn 350–400 nΩ·m (at room temperature); resistivity temperature coefficient: α-Mn (2–3·10^4 K^{-1}, β-Mn 0.00136 K^{-1}, γ-Mn 0.0055–0.0065 K^{-1}. Work function of polycrystal 3.83 eV. Manganese is paramagnetic but at low temperatures α-Mn and β-Mn are metastable, whereas γ-Mn and δ-Mn turn into antiferromagnetic state; Néel temperatures: α-Mn 98 K, γ-Mn $\approx$515 K, δ-Mn 625 K. Entropy changes of antiferromagnet–paramagnet transformation amount to 0.21 J·mol^{-1}·K for α-Mn and < 4.187 J·mol^{-1}·K for nonequilibrium γ-Mn. β-Mn remains paramagnetic until extremely low temperatures (1.5 K). Molar magnetic susceptibility of α-Mn is $+527\cdot10^{-6}$ (at 293 K), of β-Mn $+483\cdot10^{-6}$ (at 293 K), of γ-Mn $+467\cdot10^{-6}$ (at 1367–1407 K). Nuclear magnetic moment of ^{55}Mn is 3.461 nuclear magnetons. Manganese is one of the most industrially important metals. It is used in the manufacture of *steel* and *cast iron* as a deoxidizing and desulfurizing agent which provides wear resistance, strength, corrosion resistance, as well as in manufacturing of nonferrous metal alloys. Some industrially-used manganese compounds are toxic.

MANY-CONFIGURATION APPROXIMATION

A natural extension and refinement of the *Hartree–Fock method*. Whereas the latter treats the wave function of many particles as a product or determinant that is composed of one-particle wave functions (the *single-configuration approximation*), the many-configuration approximation represents the many-particle wave function as a linear combination of several determinants that are constructed from different sets of single-particle wave functions. Each determinant corresponds to a certain configuration of electrons distributed over the single-particle functions. Hence, the use of the term "many-configuration distribution". These determinants must differ from one another by at least two wave functions, since the energy in the Hartree–Fock approximation is stationary with respect to single-particle excitations. Arbitrariness in the choice of the single-particle functions may be limited by symmetry considerations on the basis of the requirement that the state must be an eigenstate of the total *spin*, of the *angular momentum*, of its projection on one of the axes, etc. The form of the functions is refined by the use of a variational principle.

MANY-VALLEY SEMICONDUCTORS

Semiconductors with several minima (maxima) of electron energy (treated as a function of the electron *quasi-momentum* $\boldsymbol{p}$) in the neighborhood of the bottom of the *conduction band* (top of *valence band*). If the electron energy has an extremum at point $\boldsymbol{p}_1 \neq 0$, then there will be similar extrema at all points $\boldsymbol{p}_\alpha$ that result from applying operations of *point group of symmetry*. Such extrema are called equivalent. Since the concentration of *current carriers* in semiconductors is not high, they only fill those states that are very near extremum points, and such groups of states are called *valleys*. If there is no degeneracy of the energy spectrum at the extreme point $\boldsymbol{p}_\alpha$, the electron energy is expressible as a series in powers of the deviation of the quasi-momentum $\boldsymbol{p}$ from valley center $\boldsymbol{p}_\alpha$. With an accuracy up to the first nonzero terms, we obtain

$$E(\boldsymbol{p}) = E(\boldsymbol{p}_\alpha) + \frac{1}{2}\sum_{ij} \frac{(p_i - p_{\alpha i})(p_j - p_{\alpha j})}{m_{ij}} \qquad (1)$$

The quantities $1/m_{ij}$ are components of the reciprocal *effective mass* tensor. This tensor is symmetric; and when written in the principal axis system Eq. (1) assumes the form

$$E(\boldsymbol{p}) = E(\boldsymbol{p}_\alpha) + \frac{(p_1 - p_{\alpha 1})^2}{2m_1} + \frac{(p_2 - p_{\alpha 2})^2}{2m_2} + \frac{(p_3 - p_{\alpha 3})^2}{2m_3}. \qquad (2)$$

The quantities $m_{ij}(m_1, m_2, m_3)$ are called effective mass tensor components, although, strictly speaking, these quantities fail to form a tensor. *Isoenergetic surfaces* in *momentum space* near the band edge are ellipsoids, ellipsoids of revolution when $m_2 = m_3$, and spheres when $m_1 = m_2 = m_3$. If $\boldsymbol{p}_\alpha \neq 0$, these surfaces are ellipsoids even in a cubic crystal. Such a situation is realized for *germanium* and *silicon* (see Fig.). The isoenergetic surfaces of these same semiconductors in the conduction band are ellipsoids of revolution with effective mass ratios 19.3 and 5.16, respectively. The response of current carriers to external actions depends on the relative positions of the valleys and their orientation relative to external forces. For instance, with the application of a uniaxial crystal strain, or strong applied magnetic or electric fields, the relative populations of various valleys can be significantly changed, and the electrical conductuctivity of the crystal is thereby affected. *Intervalley redistribution* is responsible for the *Gunn effect*, the *Sasaki effect*, and the generation of *intervalley noise*.

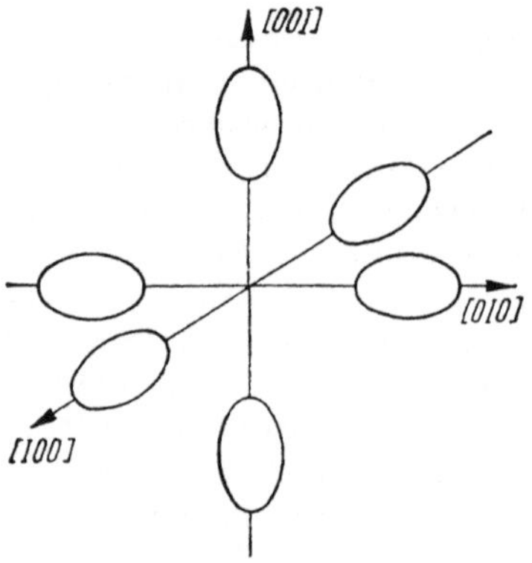

Isoenergetic ellipsoidal surfaces in the conduction band of Si.

MANY-WAVE APPROXIMATION

Method of *dynamic radiation scattering* theory (X-ray, neutron, or electron scattering), which allows for multiple scattering of an incident wave into several diffracted waves and vice versa, provided the *Bragg law* is completely or approximately satisfied for several systems of crystallographic planes. As a result, a unified wave field arises that is a coherent superposition of incident and diffracted waves with nodes and antinodes (standing wave); its interaction with atoms of the *crystal* is quite different from the ordinary case. In particular, the effects of an *anomalous passage of X-rays*, *indirect excitation* of forbidden reflections, etc. take place. For X-rays, thermal neutrons, and slow electrons, this occurs only if a crystal is aligned relative to the incident beam with a high degree of accuracy. In this case, wave amplitudes and refractive indices are found as eigenvectors and eigenvalues of a certain matrix, with elements that describe processes of individual scattering of waves one by another. The description of the passage of high-speed electrons or ions along a crystallographic *symmetry axis* (high-resolution *electron microscopy*, *channeling*) always requires the many-wave approximation, and the number of waves taken into consideration here is rather high. This approximation is also used in solving problems of X-ray optics. See also *Two-wave approximation*.

MANY-WELL POTENTIAL

Adiabatic potential of a polyatomic system, with two or more minima equivalent in energy and symmetry. In accordance with the theory of *vibronic interactions*, the latter are the only source of instability of highly symmetric configurations, and the formation of several minima of the many-well potential (see *Jahn–Teller effect*, *Renner effect*, *Vibronic instability*). The assumption of a many-well potential (in particular, of a *double-well potential*) is used in various descriptions of molecules, glassy systems, crystals, metastable states, and also in the theory of order–disorder *structural phase transitions*, the theory of noncentral impurities in crystals, in *magnetic resonance spectroscopy*, etc.

MARGIN OF PLASTICITY

Capability of a material to undergo *plastic deformation* without *failure*. The margin of plasticity is measured by the *plastic strain* before failure. For *metals* with a body-centered cubic lattice, the margin of plasticity correlates with the ratio of the values of *breaking stress* σ_F to yield stress (see *Yield limit*) σ_Y. The greater the ratio σ_F/σ_Y the larger the margin of plasticity. As σ_Y sharply increases with a decrease in temperature the margin of plasticity sharply drops upon approaching the temperature of the viscous–brittle transition (to the temperature of *cold brittleness* T_{CB}). That is, the margin of plasticity for body-centered cubic metals is larger, the greater the difference $T - T_{CB}$, where T is the testing or operating temperature. When a coarse-grained recrystal structure (see *Recrystallization*) with perfect *grain boundaries* is present, the margin of plasticity determined by the relative elongation at a temperature $T > T_{CB}$ is larger than that with a fine-grained or dislocational *cellular structure*, but the *plasticity* in the second case is retained down to lower temperatures.

MARGIN OF SAFETY

The relation between allowed and actually acting stresses. This is limited by the *safety factor* which is usually determined as the ratio of permissible stress (*yield limit*, *ultimate strength*, *endurance limit*, etc.) to the maximum stress that occurs during the course of construction. The safety factor takes into account the possibility of random overloads, the influence of crystal structure *defects* (*pores*, inclusions, *dislocations*, etc.), and the nonuniformity of their distribution. The selection of a value for the safety factor involves the necessity of conserving material and creating constructions of minimum weight. Minimum values of the safety factor are adapted for expendable products with short-term applications, and maximum values apply to constructions of long-term duration, especially under dynamic loads.

MARTENSITE (named after A. Martens)

1. In some crystalline materials: a low-temperature polymorphous structural modification. It results from a diffusion-free phase transition involving a lattice shift (see *Martensitic transformation*, *Polymorphism*), and has the appearance of individual crystallites that are smaller than the initial *crystal* grains. There is a strict crystallographic interrelation between the *crystal lattices* of the original and the martensitic *phases* (so-called *orientation relations*), which is responsible for the parallelism of some close-packed planes and directions of both lattices. As a consequence of the cooperative and natural (regular) shifting of atoms during the martensite crystal formation, there occurs a macroscopic shift of the specimen surface (surface relief arises in the form of ridges and valleys). Martensite crystals have the shapes of

needles, wafers and lenses. The internal structure is more complex, and sometimes includes a high density of dislocations (10^{11} cm^{-2}) such as twins, *stacking faults*, and sets of thin *plates*.

2. In *steel*: a structural component that is an oversaturated *solid solution* of carbon in a polymorphous modification of alpha *iron*. Since austenite dissolves up to 2.0% C at the hardening temperature, and the body-centered cubic α-iron lattice dissolves less than 0.002% C at room temperature, all the *carbon* goes into martensite as a result of a diffusion-free rearrangement of the face-centered cubic lattice into the body-centered cubic one. The carbon atoms are arranged along the c axis between Fe atoms in octahedral *pores*, and form a sublattice that is ordered with respect to C. In carbon steel martensite, however, filling all pores along the c axis never occurs (to accomplish this, the concentration of C would exceed its ultimate solubility in austenite). As a result of the inclusion of C atoms, the body-centered cubic Fe lattice becomes significantly distorted, i.e. a tetragonal or rhombic lattice arises, with *unit cell* parameters (a, b, c) that differ from each another. The degree of tetragonality (c/a ratio) increases linearly with the rise in C concentration, and when the carbon content reaches 1.8%, c/a equals 1.08. Martensite crystal lattice distortions promote high *strength* and *hardness* (high-carbon martensite hardness is about $9.8 \cdot 10^8$ Pa), and extremely low *plasticity* of hardened steel.

When steel is heated, C leaves the crystal lattice, forming disperse particles of iron *carbides*. As a result, the degree of tetragonality is reduced, crystal lattice distortions decrease, hardness and strength also decrease, whereas the plasticity increases. Martensite with such properties is called *tempered martensite*. The reverse martensitic transition in carbon steels is not observed because C, on heating, manages to leave the martensite lattice, and its reverse transition to austenite goes through diffusion. In carbon-free alloys (Fe–Ni), α- and γ-transformations proceed without diffusion. Besides α-martensite (see *Alpha phase*) with a body-centered cubic lattice, ε-martensite (see *Epsilon phase*) with hexagonal close-packed lattice and ε'-martensite with an 18-layered rhombohedral lattice form in manganese steels (>10% Mn). Like austenite, these phases possess close-packed structures, therefore they dissolve more C than the body-centered cubic lattice. Interstitial carbon atoms cause less distortions in ε'- and ε-martensite, so the hardness and strength of these steels are less than half those of hardened carbon steel. In the process of deformation, both ε'- and ε-martensite transform into α-martensite in the order: $\gamma \to \varepsilon \to \varepsilon' \to \alpha$. At 100 and 200 °C, respectively, ε'- and ε-martensite transform into austenite, i.e. the reverse $\varepsilon' \to \gamma$ and $\varepsilon \to \gamma$ martensite transformations take place. In other metals and alloys, which contain no *interstitial atoms* of C or N, the properties of martensite and the original phase differ very little. In nonferrous alloys (Cu–Al–Ni, Au–Cd) the possibility of *thermoelastic martensite* crystal formation exists. On cooling, the dimensions of such martensite crystals increase and, on heating, they decrease, which is due to forward and reverse martensite transformations taking place. The formation of martensite in steel is accompanied by volume increases of 4–5%. As a result, considerable *internal stresses* arise, leading to warping and *failure* of products. Besides, internal stresses result from directional cooperative *shear*, which contributes to crystal growth. These stresses are partially lowered due to the formation of internal twins (see *Twinning of martensite*) and *dislocations* inside martensite, as well due as to the formation of especially shaped martensite crystals and their distribution in the elastic matrix of the original phase.

MARTENSITE TWINNING

See *Twinning of martensite*.

MARTENSITIC POLYTYPES

Polytypes that arise at *martensitic transformations* (see *Polytypism*) in a diffusion-free manner, and are martensitic phases (see *Martensite*). Unlike equilibrium polytypes, martensitic polytypes are nonequilibrium (metastable) and turn under *plastic deformation* into other polytype structures. Martensitic polytypes are generated in metallic *alloys* with low energies of *stacking faults* ($\gamma < 30$ mJ cm^{-2}), and possess close-packed structures. In alloys with infinite solubility of the alloying element (see *Alloying*) (Fe–Mn, Co–Ni, Co–Cr, Co–Mn), high-temperature face-centered cubic and low-temperature hexagonal close-packed polymorphic modifications are polytypes (see *Polymorphism*). In alloys with limited solubility of the alloying element (Fe–Mn–C, Fe–Mn–N, Fe–Mn–Cu, Co–C, Co–Ti, Co–Al, Co–Nb, Co–Ta, Co–W), there arises a variety of martensitic polytypes with various multilayer *crystal lattices* that contain up to (and over) 100 close-packed layers in *unit cells*. A variety of martensitic polytypes arise on cooling nonferrous Cu-, Au- and Ag-based alloys as a result of a martensitic transformation, and from elastic deformation of Cu–Al–Ni *monocrystals* oriented along the [100] direction. As the external applied stresses are either increased or decreased, transitions between various martensitic polytypes take place, accompanied by an abnormally large elastic *strain* (see *Superelasticity*). Martensitic polytypes are generated also in a Ti–Ni alloy (*nitinol*).

MARTENSITIC TRANSFORMATION, martensitic phase transition

First-order phase transition in solids, which results in a martensite phase being formed from the original phase in a diffusion-free manner (see *Diffusion-free transformation*, *Martensite*). The martensite phase differs by *crystal lattice* type and by physical and mechanical properties. This transformation is one of the most common types of phase transition; it is a special case of a *polymorphic transformation*. According to the G.B. Kurdyumov theory a martensitic transformation involves a regular reorganization of a crystal lattice in which the atoms do not exchange places, but only displace over a distance which does not exceed an interatomic spacing. An example is a transformation between a hexagonal close-packed and a face-centered cubic structure by the sliding of close-packed planes. There is a close crystallographic interrelationship between the lattices of martensite and the original phase (involving so-called *orientation relations*); cooperative and directional atom displacement lead to the generation of relief on the specimen surface; in most cases, there is a high formation and growth rate of martensitic crystals ($\sim 10^{-7}$ s); a rapid decay of the transformation on reaching the specified temperature; a reverse martensitic transformation into the original phase on heating. Similar to other first-order phase transitions, a martensitic one proceeds through *nucleation* and nuclei growth; therefore, upon completion of the transformation, the solid is composed of a mixture of martensite crystals with different orientations.

The necessary condition for a martensitic transformation is an increase in the system free energy. However, the transformation, indicated by a downward pointing arrow on the figure, does not begin on cooling to the point of phase equilibrium T_0, where the free energies of original and martensite phases are equal, but rather at a lower temperature called the *martensitic point* M_s, and the transformation is completed at the temperature point M_f (see Fig.). On heating, the reverse martensitic transformation begins at a temperature A_s and ends at A_f, both higher than T_0: the gain in free energy must compensate for the energy outlay (in the form of elastic and surface energy) at the generation of martensite crystals. Martensitic transformations proceed over a wide range of temperatures: with the initial stage of the process at M_s, and the final stage at M_f. The temperatures of initial and final stages for reverse martensitic transformation are A_s and A_f, respectively. The kinetics of a martensitic transformation may be either athermic or isothermal. In the first case some martensite (up to 80%) is obtained at the temperature M_s during a short period of time, but further holding at this temperature does not lead to growth

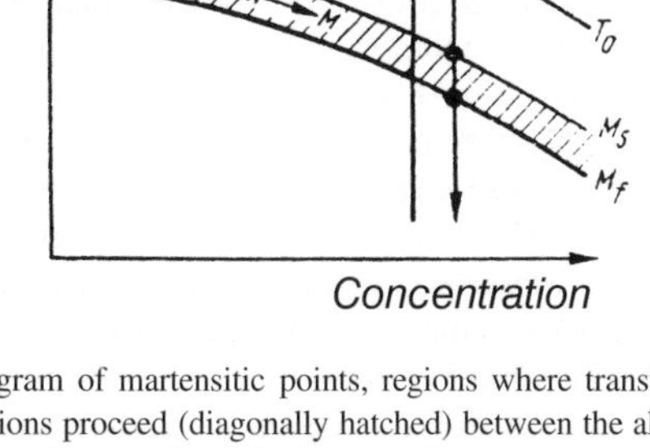

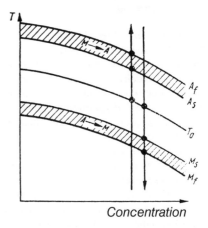

Diagram of martensitic points, regions where transformations proceed (diagonally hatched) between the alloy (A) and martensite (M) phases, and temperature versus alloying element concentration, as explained in the text.

of new martensite crystals. To obtain additional regions of martensite it is necessary to reduce the sample temperature. An isothermal transformation runs at a constant temperature with a varying rate, so the relationship between initial rate and temperature is in the form of a curve with a maximum.

Martensitic transformations were first discovered in *steels*, and then in pure *metals* (Co, Li, Ti, Zr), metal alloys (Cu–Al, Ag–Cd, Ni–Ti), *minerals*, organic crystals, and in other crystalline substances. The product may be thermally stable *phases* (e.g., in pure metals, in certain Ti-, Zr-, Co-based alloys) or metastable phases (martensitic steels, phases of Cu–Al, Ag–Cd and Ni–Ti alloys, etc.). This transformation finds its greatest practical application in steels (if hardened, steel gains high strength and hardness), and it underlies the *shape memory effect* and the *superelasticity* phenomenon.

MASER (acronym for Microwave Amplification by Stimulated Emission of Radiation)

Quantum generators and amplifiers of the radio frequency and microwave bands. The term *laser* (Light Amplification by Stimulated Emission of Radiation) came into use by analogy with maser. See *Solid-state quantum electronics*, *Quantum amplifier*.

MASS, EFFECTIVE

See *Effective mass*.

MASS OPERATOR OF QUANTUM MECHANICS, intrinsic energy part

Part of the *Dyson equation* for interacting fermions and bosons (see *Polarization operator*) that is irreducible (i.e. containing no bare connecting lines in *Feynman diagrams*) involving *Green's functions* (*propagators*) of particles and interactions. The quantum-mechanical mass operator (denoted by Σ or Π) determines the renormalization of the *quasi-particle* spectrum, in particular, of the effective electron mass in normal *metals*, due to the Coulomb interaction and the *electron–phonon interaction*, as well as the spectrum of *phonons*, *magnons*, and other quasi-particles. The anomalous quantum-mechanical mass operator Δ in the Gorkov–Dyson equations determines the *energy gap* in the elementary excitation spectrum of *superconductors*.

MASS SPECTROMETRY, mass spectroscopy

A technique for investigation of substances, based on determination of masses and quantities of ionized atoms and molecules from the material under study. A plot of mass abundance versus the mass values is called a *mass spectrum*. The essence of the technique is as follows: positively or negatively charged ions that have different masses and move in a vacuum are separated according to their mass-to-charge ratio under the effect of electric and magnetic fields of a mass-analyzer. To make use of the technique, the investigated sample is first ionized except for the cases when the substance under study exists in a plasma state (e.g., glow discharge, ionosphere). Mass spectrometry exhibits a very high absolute sensitivity (10^{-12}–10^{-13} Pa for gases, 10^{-11}–10^{-13} g for solids) and relative sensitivity (10^{-5}–10^{-8}% for solids). Mass spectrometry was initially used to determine the isotopic composition of atoms, and to make precise measurements of atomic masses. Later, it developed into an analytical method, used in nuclear power engineering, experimental physics, chemistry, geology, geological chemistry, biology, archeology, medicine, environmental monitoring, technology processes, space research, etc. Mass spectrometry allows one to find ionization, excitation, and dissociation energies of molecules, heats of evaporation and free formation energies of compounds, absolute age of geologic samples, and compositions of gas mixtures. Mass spectrometry underlies a wide variety of investigation and analysis methods which are used in solid state physics, and differ in the manner of the sample ionization. They are: mass spectrometry with surface ionization, *atomic probe method* (ionization in strong electric field), spark mass spectrometry (ionization in spark discharge), electronic-probe mass spectrometry (ionization of an electronic-beam evaporated sample by electron impact), *secondary ion mass spectrometry* (ionization by ion bombardment and *sputtering* of solids), mass spectrometry in rapid-atom bombardment, laser mass spectrometry (ionization during laser probing of a surface), thermodesorption mass spectrometry, and the method of electron- and irradiation-induced desorption of ions. Methods are also developed on the basis of mass spectrometry of neutral atoms, evaporated or sputtered along with ions from the surface of a solid and ionized by electron impact, in glow and high-frequency discharges, or by means of resonance and nonresonance multiphoton laser ionization. These methods find their use in the general analysis of elemental, isotopic, and molecular composition of the surface and bulk of a solid, in analysis of impurities (dopants) in gases and other substances, in local and layer-by-layer microanalysis of high spatial and layer resolution, in studies of processes of *diffusion*, *segregation*, *evaporation*, *adsorption*, *desorption*, *oxidation*, *corrosion*, catalysis, formation of *thin films*, sputtering, and other aspects of the interaction between energetic beams of atomic particles and solids. Of most widespread use in solid-state mass spectrometry are *mass-analyzers* with sector magnetic and electric fields, quadrupole magnets, and time-of-flight mass separation.

MASS SPECTOMETRY VIA SECONDARY IONS

See *Secondary ion mass spectometry.*

MASS TRANSFER, ANOMALOUS

See *Anomalous mass transfer.*

MASS TRANSPORT

Process of directional transport of material in a field that is external or internal with respect to a *crystal*. Volumetric and surface types of mass transport are recognized. Mass transport occurs during *creep*, *sintering*, changes in shape of *pores* and *cracks*, etc.

Volumetric mass transport may be of either a nonthreshold (diffusion) or threshold (dislocation) type. Diffusive mass flux in volumetric mass transfer is defined as

$$J_0 = -\frac{N_0 D_0}{k_B T} \nabla\mu,$$

where N_0 is the number of electrons per unit volume, D_0 is the coefficient of volumetric self-diffusion, T is the temperature, and $\nabla\mu$ stands for the atomic chemical potential gradient produced by the field to induce the flux. For a curved surface which bounds a hollow pore or a gas-filled cavity we can write $\nabla\mu_k = \alpha\omega\nabla k$ where α is the *surface tension*, k is the surface local curvature, and ω the atomic volume. The gradient $\nabla\mu_k$ directs the flux of crystal material into a pore cavity which causes pore healing (sintering). Foreign inclusions may move as a whole in external fields such as an electric field E ($\nabla\mu_E = ZeE$, where Ze is the effective atomic charge), a temperature field ($\nabla\mu_T = \theta\nabla T/T$, where θ is the *thermodiffusion coefficient*), or an inhomogeneous *vacancy* distribution field ($\nabla\mu_v = k_B T(\nabla C_v/C_v)$, where C_v is the vacancy concentration). For stresses that exceed a threshold the volumetric mass transport is determined by directional motion of *dislocation* ensembles, which may be responsible for crystal creep: $\dot{\varepsilon} = \rho b V(\sigma)$ (ρ is the mobile dislocation density, $V(\sigma)$ is the stress-dependent dislocation motion rate, b is the *Burgers vector*).

Surface mass transport may be carried out through the mechanism of surface self-diffusion and through material transport in a near-surface gas layer. In the first case, mass transport is described by the equation

$$J_S = -\frac{n_S D_S}{k_B T} \nabla_S\mu,$$

where $\nabla_S\mu$ is the surface gradient of atomic chemical potential, D_S is the coefficient of surface self-diffusion, and n_S is the surface atom density. For a gas mechanism surface mass transport may run in an evaporation-condensation mode involving *diffusion* in a near-surface gas layer. The corresponding flux equations have the following form:

$$J_{\text{e-c}} = \frac{P(k) - P_0}{(2\pi m k_B T)^{1/2}},$$

$$J_{\text{g-d}} = -\left(\frac{n_G D_G}{k_B T}\right)\nabla\mu_G,$$

where $J_{\text{e-c}}$ is the evaporating-condensing current, $J_{\text{g-d}}$ is the gas diffusion current, $P(k)$ is the pressure of the vapor in equilibrium with the portion of surface with local curvature k, P_0 is the equilibrium vapor buoyancy, D_G, N_G, $\nabla\mu_G$ are the *diffusion coefficient*, number of atoms per unit volume, and chemical potential gradient in the gas layer, respectively; m is the atomic mass.

MASS TRANSPORT, SURFACE

See *Surface mass transport.*

MASURIUM

See *Technetium.*

MATERIALS SCIENCE

A cross-disciplinary science dealing with properties of materials and factors responsible for these properties. Materials science endeavours to manipulate these properties during the process of material manufacture, and to predict changes that occur under operating conditions. The main components of materials science are the material as the subject of study, the production process, and technical applications. All technological procedures exert a direct influence on the material structure, whereas practical applications are determined by a whole complex of characteristics. Therefore, the physical basis of materials science (the subject of *physical materials science*) lies in the interrelation between the structure and properties of a material, while other matters are dealt with in applied or *technical materials science*.

Modern materials science is dependent on a number of disciplines such as *crystallography*, solid state physics, physical chemistry, electrochemistry, etc. Of fundamental importance for materials science are studies of *plastic deformation*, *recrystallization*, *phase transitions*, formation of *texture*, etc. Phase transition kinetics studies, which provided an opportunity to develop the foundations for the *thermomechanical treatment* of metals, have played an important role in the progress of *metal science*. A special importance is the increased understanding of the actual structures of materials, which differ from the ideal geometric abstraction of a *crystal lattice* by the presence of various kinds of *defects* (*point defects*, *dislocations*, interfaces, impurities, etc.). This explains how many processes occur under *strain*. The results obtained in this field, along with progress in the evolution of phase transition theory and that of electrochemical reactions during corrosion have, in their turn, promoted the development of manufacturing processes for materials with special operational characteristics: very strong (see *Strength*), wear-resistant (see *Wear*), corrosion-resistant (see *Corrosion*), etc. Whereas the growth of structural engineering industries stimulated the development of metallic materials with deformation characteristics, the progress of electrical engineering and electronics gave rise to materials with properties governed by their electronic structure (*semiconductors*, *insulators*, *superconductors*, *hard magnetic materials* and *soft magnetic materials*, etc.), which stimulated the development of the electronic theory of the solid state. Growing demands of *metallurgy* and electronics, as well as requirements of civil engineering and chemical branches of industry, are responsible for a new impetus toward developing new ceramics, a class of materials which has been in use for many centuries. Such inherent features of *ceramics*, as high *corrosion resistance* and high-temperature strength, compression strength, wear resistance, and outstanding dielectric properties have been under intense development.

The application of theories relating structure and physical properties, which were initially developed for *metals* and then extended to nonmetals including ceramic materials, has resulted in the creation of new engineering and *construction materials* based on extremely strong ceramics (*boron nitride*, *silicon carbide* and *boron carbide*, etc.) as well as in the application of glasses as semiconductors, laser working media (see *Lasers*), photosensitive materials, etc. Also of importance are metallic materials with a vitreous amorphous structure (*metallic glasses*, metglasses), possessing unique strength and magnetic characteristics.

MATRIX ISOLATION

Method of accumulation and study of atoms, molecules, ions, clusters, etc., in a matrix of solidified gases (see *Cryocrystals*). Typical matrices are noble gases, hydrogen, oxygen, nitrogen, carbon dioxide, clathrates, aerogels, etc. which solidify at low temperatures (*cryomatrices*); organic matrixes are used less often. An ideal matrix has no effect on the properties of particles that are frozen into it; it isolates them from collision and interaction with one another, and stabilizes complexes that are unstable under ordinary conditions. Test clusters for matrix isolation are obtained through joint *condensation* of particle and matrix gas fluxes on a cryogenic substrate. Use is also made of freezing of plasma components and photochemical synthesis directly into a matrix. It is through matrix isolation that frozen gases of numerous atoms and two-atom molecules were obtained. It permits the investigation of parameters of atoms, free radicals, and radical ions N, H, NH, C_2^-, CF_2, etc., those of *excimer* molecules ArO, ArN, KrF, XeBr, $XeCl_2$, etc., and complicated biomolecules. A number of compounds have been synthesized which had never been observed as a gas, e.g.: MO_2 (M = Li, Na, K, Ca, Sr), $M(CO)_n$ (M = Ni, Cu, Ag, Au, U), $M(N_2)_n$ (M = Ni, Pd, U). The main methods of investigations are spectroscopic: photoabsorption, *luminescence*, *Raman scattering of light*, NMR, EPR, etc. These methods allow determining the energy spectrum of particles, to diagnose processes of dissociation *recombination*, *diffusion*, formation of clusters (see *Cluster*), and accumulation of particles, and to study interactions of particles with one another and with elementary matrix excitations.

MATTHIAS RULE (B.P. Matthias)

An empirically found regular dependence of the *critical temperature of superconductors* in crystalline materials on the number of electrons per atom. A bell-shaped *Collver–Hammond curve* (M.M. Collver, R.H. Hammond) is an analogue of the Matthias rule for *amorphous superconductors*.

MAXIMUM OVERLAP

See *Pauling's maximum overlap principle*.

MAXWELL DISTRIBUTION (J. Maxwell, 1859)

The probability that a large number of identical classical particles in equilibrium at a temperature T, have particle velocity components in the range $v_i \to v_i + dv_i$ ($i = x, y, z$):

$$f(v)\,dv_x\,dv_y\,dv_z = \left(\frac{m}{2\pi k_B T}\right)^{3/2} \times \exp\left[-\frac{m(v_x^2 + v_y^2 + v_z^2)}{2k_B T}\right] dv_x\,dv_y\,dv_z,$$

where m is the particle mass, and k_B is the *Boltzmann constant*. The Maxwell distribution describes statistical ensembles of various *quasiparticles* subject to classical statistics, provided their *dispersion law* is quadratic (see *Boltzmann distribution*). This distribution was corroborated by O. Stern (1920) in his molecular beam experiments.

MAXWELL RELAXATION, dielectric relaxation

Relaxation of a space charge and its attendant electric field as a result of interactions of the charged particles that produce it. Some of the charge carriers must be mobile and satisfy the condition $\omega_p\tau \ll 1$, where τ is the carrier mean free path time, and ω_p is the *plasma oscillation* frequency. The characteristic field change time for Maxwell relaxation can be expressed in different ways: $\tau_M = \varepsilon/\sigma$, where ε is the *dielectric constant*, σ the electrical conductivity; $\tau_M = RC$, where R is the sample resistance and C is the capacitance; $\tau_M = l^2/D$, where l is the Debye *screening radius*, and D the charge carrier *diffusion coefficient*. During Maxwell relaxation the electric field is established by applying a voltage, and the charge results from fluctuating electron thermal motion. The *relaxation time* $\tau_M = \varepsilon/\sigma_d$, where σ_d is the diffusive conductivity, is introduced in the case of a nonlinear *current–voltage characteristic*. If $\sigma_d < 0$, $\tau_M < 0$, the charge grows gradually with time, corresponding to a *strong field* regime. If conditions opposite to those of Maxwell relaxation hold then the characteristic field change time is $1/\omega_p$.

MEAN FIELD APPROXIMATION

The common mean field approximation assumes that each magnetic ion in a material or crystal experiences a local magnetic field that is proportional to the magnetization that is present. See *Molecular field*, *Random phase approximation*, *Coherent potential method*, *Self-consistent field*, *Hartree–Fock method*, *Self-consistent field method* in superconductivity theory.

MEAN FREE PATH of particles

The average distance which a particle (quasiparticle in solids) traverses between two successive collisions with particles (or obstacles) of a medium. In the case of relatively rare collisions, its motion can be specified by giving values of the energy and momentum which change abruptly at the instant of collision. The path length between subsequent collisions depends on the particle energy; and to find its mean value it should be averaged over the particle energy distribution function. It can involve elastic, inelastic, or ionizing collisions. In the case of inelastic scattering, the interaction is accompanied by the generation or absorption of various elementary excitations (see *Quasi-particle*) in solids. The mean free path l is related to the cross-section σ for the corresponding process by the expression $l = 1/(\sigma N)$, where N is the concentration of scattering particles (*impurity atoms*, *defects*, *phonons*, *plasmons*, and so on). The parameters which determine the efficiency of transport (such as *electrical conductivity*, *thermal conductivity*, *viscosity* coefficient, etc.) depend on the corresponding mean free paths of mobile particles. In the case of an electron energy ≈ 75 eV, the mean free path of the electrons entering the solid before a collision event has a minimum value of about 0.5–1 nm. This allows the use of electron beams with this energy for the purpose of solid state surface analysis (see *Low-energy electron diffraction*, *Auger electron spectroscopy*).

MECHANICAL HEAT TREATMENT OF METALS

Strengthening treatment of *metals*, which includes a slight (10–15%) *plastic deformation* and polygonization stabilizing *annealing*. Mechanical heat treatment is an independent type of *thermoplastic technology*; it brings about the improvement of a complex of mechanical properties (*strength* and *plasticity*) through formation of specific dislocation substructures that limit the mean free path and mobility of linear *defects*.

The cause of the strengthening effect of a mechanical heat treatment is the formation of a higher density of intragrain boundaries in a *cellular structure*, or a polygonal fragmented structure (see *Fragmentation*), which interact with dissolved impurity atoms and mobile linear defects. For single-phase metals and *alloys* that undergo no polymorphic or phase transitions, mechanical heat treatment is the only effective method for varying technological and operating properties over a broad range by choosing corresponding dislocation distributions and substructures.

In order to increase the workability of brittle materials a cellular substructure is formed; while a polygonal (see *Polygonization*) substructure that contains an excess of *dislocations* of one sign is formed when resistance of high-temperature *creep* is to be increased. A polygonal substructure is thermally resistant under stress even up to premelting temperatures, which decreases the rate of creep by 1–2 orders of magnitude. Depending on the operating conditions, mechanical heat treatment requires optimization of preliminary *strain* in order to preclude any *recrystallization*.

For materials with complicated alloying conditions, mechanical heat treatment may be combined with other types of thermoplastic treatments, e.g., *thermomechanical treatment* that forms the hardened state (see *Hardening of materials*) as a result of an accumulation of dislocation-saturated structures at polymorphic and phase transitions, *ultrasonic treatment*, and subsequent aging (see *Alloy aging*). Under these, creep resistance, fatigue strength (see *Fatigue*), *impact strength* and resistance to *corrosion under stress* are increased. A mechanical heat treatment is usually carried out using standard pressing, rolling and drawing equipment, or using special equipment developed for new technologies. The application of a mechanical heat treatment for the hardening of products made of *nickel alloys*, *titanium alloys*, materials containing a metastable β-phase, *molybdenum alloys*, *tungsten alloys*, heat-resistant austenite steels (see *Austenite*), or rolled aging *aluminum alloys*, increases the economic efficiency of the use of these materials due to the improvement of material properties, the increase of reliability, and the saving in alloying elements.

MECHANICAL PROPERTIES OF SOLIDS

Ability of solids to resist *strain* and *failure*, combined with the capacity to undergo elastic deformation and *plastic deformation* under the action of external forces. Mechanical properties to characterize the elasticity of solids include *elastic moduli*, *proportionality limit* and *elastic limit*. *Strength* is described by *yield limit*, *ultimate strength*, *hardness*, and *breaking stress*. *Plasticity* is determined by relative elongation (shortening) and lateral contraction (spreading) in static testing for tension (compression), as well as by *impact strength* in dynamic testing (see also *Plastic limit*). In cyclic testing (see *Cyclic strength*), *fatigue limit* and *fatigue strength* (see *Fatigue*) are determined. The mechanical properties at elevated temperatures include *creep limit*, *long-term strength*, *long-term hardness* and *relaxation resistance*. At low temperatures, the mechanical properties of a solid with *metallic bonds* depend on the type of *crystal lattice*. Thus, *metals* and *alloys* with a face-centered cubic lattice (Cu, Ag, Au, Al, Pb, Ni, austenite *steels*, etc.) on lowering the temperature (down to 4.2 K) retain plasticity with a certain increase of strength. At low temperatures, these metals exhibit *creep*, its rate slightly depending on temperature. Relative elongation, lateral contraction, and impact strength of many metals and alloys with a body-centered cubic lattice (W, Mo, Cr, α-Fe, carbon and low-alloy steels) as well as of those with a hexagonal close-packed lattice (Be, Mg, Zn) tend to zero on cooling below the *viscous–brittle transition temperature* T_b (*cold brittleness*). On refining grains as well as on purifying the above metals of *impurity atoms*, in particular from interstitial impurities, T_b decreases. Typical of mechanical properties of solids with *covalent bonds* are high hardness and *brittleness* (see *Brittle failure*) even at the high purity pertaining to many semiconductors; their plasticity manifests itself at temperatures higher than 0.7 of the absolute melting temperature T_m. Mechanical properties of solids with *ionic bonds* (TiC, WC) are also typical for brittle materials, but plasticity manifests itself already at temperatures above $0.5T_m$. Mechanical properties of *polymeric materials* have certain specific features: an elastic strain that arises nearly instantly upon applying an external stress is rather small; a large reversible deformation (*high-elasticity state*), and irreversible viscous-fluid deformation take place during a period from a fraction of a second to many hours, depending on the temperature and polymer structure (see *Rubbers*). The *alloying* of metals leads to the generation of substitutional *solid solutions* (see *Substitutional alloys*), interstitial solid solutions (see *Interstitial alloys*) and/or a *heterogeneous system*. Such alloys are stronger than pure metals, but their plasticity is, as a rule, lower. Certain metal alloys with clearly defined cold brittleness provide an exception, e.g., alloys of *tungsten* and *molybdenum* with *rhenium*, their strength growing along with plasticity. *Heat treatment* and *thermomechanical treatment* enhance mechanical properties of solids that experience phase transformations (see *Polymorphism*, *Alloy ordering*, *Alloy aging*). Mechanical properties of solids change under irradiation. *Irradiation hardening* and *irradiation embrittlement* are observed; they are due to the generation of radiation-induced defects and to irradiation doping (at large irradiation doses). Mechanical properties of a solid in a *metastable state* change under irradiation because of the progress of radiation-induced processes, in particular, of *radiation-induced diffusion*. The environment has an effect on the mechanical properties. Thus, testing brittle materials, e.g., NaCl, in a saturated water solution causes a noticeable increase in strength (*Ioffe effect*), whereas *dehardening* takes place in the presence of *surface-active agents* (*Rebinder effect*, see *Strength reduction through adsorption*). *High pressures* increase the elastic moduli and strength of solids. Brittle solids exhibit a significant increase of the plasticity limit (see *Hydroextrusion*). The increase of plasticity is especially substantial when a phase with covalent or ionic bonding transforms under high pressure into a phase with metallic bonding owing to a *first-order phase transition*.

MECHANICAL STRESS

The force acting on an arbitrarily chosen unit region inside a body; a measure of internal forces arising during *strain*. Mechanical stresses are quantitatively characterized by the *stress tensor*. They stem from interatomic forces, whose range is of the order of interatomic distances. Forces that cause *internal stresses* are referred to in *elasticity theory* as *short-range forces* which act between neighboring points only. Forces that are exerted on a part of a body by surrounding regions act directly through the interface between them. This does not apply if the deformation of the body is accompanied by the generation of *space charges* and macroscopic electric or magnetic fields in it (semiconductors, *pyroelectric materials*, *piezoelectric materials*, *magnetostrictive materials*).

MECHANICAL TESTING OF MATERIALS

Tests consisting of determining, by mechanical means, those properties of materials that are responsible for the ability to resist deformation and *failure*, together with elastic and plastic behavior, under the action of external forces. The tests are carried out at room temperature, lowered and elevated temperatures. Mechanical testing of metals is divided into static types (sample loaded slowly and smoothly, or load remains constant during a long period), dynamic types (sample is loaded rapidly, i.e. with grippers moving faster than $1.6 \cdot 10^{-4}$ m/s), and cyclic types (sample undergoes repeated loads that vary in magnitude and possibly also in direction). The level of mechanical properties of a certain material depends on its nature, and on the nature of the applied stresses (type of the *state of stress*). Therefore, mechanical tests are also classified by types of loading (tension, compression, shearing, flexion, torsion, etc.) that permit testing in a linear, planar or spatial state of stress. Static testing evaluates such properties as *elasticity*, resistance to primary *plastic deformation*, resistance to major plastic deformation, *plasticity*. These properties are determined by tests for tension, compression, *flexure*, and torsion. The *hardness* of a specimen is determined by pressing into it a rigid tip (*indenter*) in the shape of a ball, cone or pyramid that experiences no appreciable deformation during the test. The state of stress thus attained makes it possible to use the hardness determination method for materials that are brittle in other kinds of testing. Micromechanical testing is used for assessing properties of a small amount of material (generally of a metal; e.g. welded joints, local *quenching* or *strain hardening*, degree of inhomogeneity, degree of *anisotropy of crystals*, etc.). Prolonged static loads are used to determine the resistance to plastic deformation (*creep limit*), and resistance to failure (*long-term strength*); these tests are carried out under special conditions (exposure to elevated temperatures, hostile media, etc.). *Fracture toughness* (short-term crack resistance) as an estimate of the resistance of a material to brittle failure is determined through static tests for tension or flexure on samples with purposefully induced fatigue *crack* (see *Fatigue*).

The principal objective of dynamic impact testing is to provide the simplest and least labor-consuming determination of the hazardous brittleness of metals for comparison purposes. Dynamic tests evaluate technical properties of metals at increased rates of deformation, when it is necessary for structures to experience high-rate loading under operating conditions.

Cyclic testing allows the determination of fatigue resistance, because most working parts of devices experience variable loads that cause the generation of fatigue cracks, their propagation and failure. The following schemes of loading are used to carry out tests for fatigue: pure bending at rotation; lateral flexure at rotation; the same operation in plane; tension–compression; variable torsion; and internal pressure. In addition, tests with combined loading are often used. Sometimes, along with tests for ordinary many-cycle fatigue, those for few-cycle, high-frequency, impact, thermal, thermomechanical and corrosion-mechanical fatigue are used.

MECHANICS

See *Analytical mechanics*, *Continuum mechanics*, *Physicochemical mechanics*, *Quantum mechanics*.

MECHANOCALORIC EFFECT (J.G. Daunt, K. Mendelssohn, 1938)

Heating of liquid *helium* ^{4}He in a vessel during a rapid transfer of He II from one vessel into another through a narrow capillary or a slot ($\sim$1 μm) at a temperature below that of the transition to the superfluid state (*lambda point*, 2.19 K at standard pressure), and the cooling of the escaping helium. The mechanocaloric effect has been observed below 0.4 K (P.L. Kapitza). For a slight temperature difference ΔT the flow-over process stops at the pressure difference $\Delta P = \rho S \Delta T$, where ρ is the helium density, and S is its *entropy*. The mechanocaloric effect is a macroscopic example of *quantum liquid* properties. The inverse phenomenon also occurs: spouting of He through a capillary at the addition of heat (*thermomechanical effect*). Both effects are explained by the quantum theory of *superfluidity*. The increase of temperature in a vessel raises the concentration of *quasi-particles* inside it. The behavior of He II is approximately described with the help of two sets of equations: one involving the normal component and the other the superfluid component of the interpenetrating liquids. The latter has zero entropy and does not interact with the normal component.

MECHANOCHEMISTRY

A branch of chemistry concerned with chemical and physicochemical transformations that take place when a material is exposed to external mechanical action: *polymer degradation* and synthesis of *polymers*, chemical reactions with *friction*, *strain* and *failure* of a solid (*tribochemistry*), reactions that take place under exposure to ultrasound and *high pressures*, etc. Methods of mechanochemistry are used in plasticization of *raw rubbers*, hydrolysis of cellulose, production of building materials, ultrasonic manufacturing of medicinal preparations, for prevention of chemical processes that cause aging and failure of assemblies and working pieces of machinery, etc.

MECHANOSTRICTION

Generalized term for a phenomenon related to *strain* of a body under the action of external electric or magnetic fields. According to the type of applied field, *electrostriction* and *magnetostriction* are distinguished. Electrostriction is observed in solid, liquid and gaseous *insulators* as a result of *polarization of insulator* in an electric field (displacement of ions of electric dipoles, or change of their orientation). Electrostriction applies to all solid dielectrics regardless of their structure and symmetry, as distinct from *piezoelectric materials* where the anisotropic structure (see *Anisotropic medium*) is of primary importance. Magnetostriction is strongly pronounced in *ferromagnets* and in certain *ferrites*, whereas it is very weak in *antiferromagnets*, and only rarely occurs in *diamagnets* and *paramagnets*. The deformation of bodies in a magnetic field is also called the *Joule effect* (J. Joule, 1842). An inverse phenomenon also occurs that involves the change of magnetic properties of a body under its forced deformation. In this case, an electromotive force arises in a coil that is wound, e.g., on a ferromagnetic core. The direct and inverse effects are used in sonic and ultrasonic emitters and receivers, respectively. The *Wiedemann effect* (G. Wiedemann, 1858) consists in the twisting of a current-carrying conductor in a longitudinal magnetic field.

MEDIUM, ACTIVE

See *Active medium*.

MEGA... (fr. Gr. $\mu\varepsilon\gamma\alpha\varsigma$, large)

Prefix for a physical unit to obtain a 10^6 multiple of the original unit, with the symbol: M. Example: 1 MW (megawatt) $= 10^6$ W.

MEISSNER EFFECT (W. Meissner, R. Ochsenfeld, 1933)

Phenomenon of *absolute diamagnetism* of superconductors (susceptibility $\chi = -1$ in SI units) involving complete expulsion of the magnetic field from the entire bulk superconductor except for a thin subsurface layer (see *Penetration depth of magnetic field*), in fields B not in excess of the *thermodynamic critical magnetic field* $B_c(T)$ for *type I superconductors*, and the *lower critical field* $B_{c1}(T)$ for *type II superconductors*. The Meissner effect is due to the shielding of the external field B by undamped surface super currents that exist via macroscopic quantum coherence and stability of the wave function of a superconducting condensate of *Cooper pairs*, with their lower energy state separated from the upper excited one-particle states by an *energy gap* of 2Δ (see *Landau superfluidity criterion*). Type II superconductors in the range of fields $B_{c1}(T) < B < B_{c2}(T)$ exhibit an incomplete Meissner effect in which the magnetic flux partially penetrates into the bulk superconductor in the form of *Abrikosov vortices* (see *Mixed state, Shubnikov phase*). In type I superconductors of an arbitrary shape and orientation relative to the external field $\boldsymbol{B}$, partial field penetration can occur through the coexistence of normal and superconducting regions (normal *domains*, see *Intermediate state*).

MELTING

The transition of a substance from the crystalline solid state to the liquid one (see *States of matter*). The process of melting involves absorption of heat (latent heat), and is a *first-order phase transition*.

MELTING HEAT

See *Heat of melting*.

MELTING, INCONGRUENT

See *Incongruent melting*.

MELTING TEMPERATURE, melting point

The *temperature* T_{melting} of the *phase transition* of a crystalline solid to the liquid state under a constant external pressure; a particular example of a *first-order phase transition*. As the melting arises from the breakage of some interatomic (intermolecular) bonds, T_{melting} depends

Table 1. Melting temperatures (in °C) of various high melting materials at atmospheric pressure

Metals and covalent crystals				Refractory compounds					
Material	$T_{melting}$	Material	$T_{melting}$	Material	$T_{melting}$	Material	$T_{melting}$	Material	$T_{melting}$
Ti	1670	Ir	2410	ThC	2625	TiN	2950	Al_2O_3	2050
Th	1750	Mo	2610	WC	2630	ZrN	2980	Y_2O_3	2410
Pt	1769	Os	2700	MoC	2692	TaN	3100	SrO	2450
Zr	1852	Ta	3000	SiC	2830	HfN	3310	BeO	2570
V	1900	Re	3180	VC	2830	BN	3700	CaO	2600
Cr	1915	W	3410	TiC	3250	TiB_2	2730	MgO	2800
Rh	1966			NbC	3480	NbB_2	3000	VO_2	2800
Hf	2222	Si	1410	ZrC	3530	ZrB_2	3040	ZrO_2	2800
Ru	2250	B	2070	TaC	3875	TaB_2	3120	CeO_2	2900
Nb	2415			HfC	3890	HfB_2	3250	ThO_2	3050

on the type of interatomic bond, the coordination number, and other characteristics of the *crystal lattice*. The melting temperature of a *metal* depends on the concentration of the electrons that determine the energy of the *metallic bond*. See Table 1.

MEMBRANE

Three types of membranes are as follows:

1. A flexible thin stretched *film* that serves as a sensor surface for a number of acoustic devices. The membrane elasticity arises from external forces that hold it stretched. In this regard it differs from a *plate*, since the elasticity of the latter is a function of its material and width. According to the shape of their border where the tension is applied, membranes are classified as rectangular, circular, etc. Natural vibrations of a membrane are systems of standing waves with particular patterns of nodal lines. Different systems of two dimensional standing waves correspond to different normal modes, the totality of the latter making up the discrete spectrum of membrane natural frequencies. Transverse vibrations of a homogeneous, uniformly stretched membrane are described by solutions of the wave equation

$$\frac{\partial^2\xi}{\partial t^2}=c^2\left(\frac{\partial^2\xi}{\partial x^2}+\frac{\partial^2\xi}{\partial y^2}\right),$$

where ξ is the lateral displacement of a point (x,y), t is time, $c=(T/\sigma)^{1/2}$ is the velocity of propagation of the vibrations, T is the tension, and σ is the membrane surface density. Solutions must satisfy the boundary condition $\xi=0$ at a fixed membrane edge, and initial conditions $\xi=f(x,y)$, $(\partial\xi/\partial t)=g(x,y)$ at $t=0$. The natural frequency spectrum of a rectangular membrane is given by the formula

$$\nu_{ik}=\frac{c}{2}\left[\left(\frac{m}{a}\right)^2+\left(\frac{n}{b}\right)^2\right]^{1/2},$$

($m,n=1,2,\ldots,a,b$ are membrane dimensions). Such an approach leaves out the scattering (emission) of energy, but this does not affect the calculation of natural frequencies. Allowing for emission loss is necessary only when calculating amplitudes of forced vibrations with frequencies that are close to resonance.

2. A membrane in engineering is a partition or diaphragm which exhibits selective penetrability for various compounds, and provides for the separation of those compounds, perhaps with the help of chemical catalytic processes. The membrane does not clog after long use, as is the case for filtration. Separation of a mixture flux into two components (one passing through the membrane, the other being retained) is due to the selective penetrability. The flux is caused by the application of gradients of pressure, concentration, electric potential, or temperature to the membrane.

3. A membrane in biology is a thin shell, often a bilayer which is two (perhaps elongated) molecules thick, confining a cell or subcellular formation. It provides sites for enzymes; ensures intercellular contacts; participates in processes of motion, secretion, absorption, protein synthesis,

and cell division; exhibits electrical activity (neurons) at the propagation of nerve impulses; detects and transmits information on minor environmental changes via special molecule-receptors and molecule-mediators, and facilitates storage of this information.

MEMBRANE MATERIALS

Polymeric film (see *Polymeric materials*), a thin metallic, ceramic or glassy *plate*, a layer of liquid, gas or vapor. Membrane materials are obtained through *pressing*, rolling, extrusion (stamping), leaching, *spraying*, electrophoresis, polymerization, casting, *etching*. Preferential and selective properties of membrane materials are governed by the structure of factors that arise during their formation: channels, slots, nets. The main function of membrane materials consists in separating the components of mixtures based on physical processes: *diffusion*, *evaporation*, ultra-filtration, reverse osmosis, etc. Membrane materials are widely used for purification and desalination of water, technological separation of gases, in chemical, food and microbiological industries, in medicine.

MEMORY, SOLID-STATE

See *Solid-state memory*.

MEMORY, STRUCTURAL

See *Structure memory*.

MENDELEVIUM, Md

See *Transuranium elements*.

MENISCUS (fr. Gr. $\mu\eta\nu\iota\sigma\kappa o\varsigma$, half-moon, crescent)

Shape of surface of liquid in a narrow vessel or a tube. Wetting liquids form a concave meniscus, and non-wetting liquids form a convex one (see *Wetting*).

MERCURY (Lat. *hydrargyrum*), Hg

A chemical element of Group II of the periodic system of elements; atomic number 80, atomic mass 200.59. Natural mercury has 7 stable isotopes: ^{196}Hg, ^{198}Hg, ^{199}Hg, ^{200}Hg, ^{201}Hg, ^{202}Hg and ^{204}Hg; 23 radioactive isotopes are known. Electronic configuration of filling outer shells is $4f^{14}5d^{10}6s^2$. Ionization energies are 10.43, 18.752, 34.3, 45.98 eV. Atomic radius is 0.150 nm; radius of Hg^{2+} ion is 0.110 nm. Oxidation state is +1, +2. Electronegativity is ≈1.6.

Mercury is silvery-white liquid *metal*, the only metal which remains liquid at low temperatures. It has a rhombohedral crystal lattice; $a = 0.2999$ nm, $\alpha = 70°45'$. Density is 13.5951 g/cm^3 (at 273 K) and 13.5459 g/cm^3 (at 293 K). $T_{\text{melting}} =$ 234.3 K, $T_{\text{boiling}} = 630$ K. Heat of melting is 2.352 kJ/mole; heat of evaporation is 58.38 kJ/mole; volume thermal expansion coefficient of solid mercury is $12.5\cdot10^{-5}$ to $17.1\cdot10^{-5}$ K^{-1} (in temperature range 183.2 to 233.7 K), temperature coefficient of liquid mercury is $1.823\cdot10^{-4}$ to $1.889\cdot10^{-4}$ K^{-1} (in temperature interval 234.3 to 623 K). Specific heat is 0.1419 kJ·kg^{-1}·K^{-1} (at 234.3 K), 0.1404 kJ·kg^{-1}·K^{-1} (at 273 K), 0.1396 kJ·kg^{-1}·K^{-1} (at 293 K), 0.1371 kJ·kg^{-1}·K^{-1} (at 373 K), 0.1357 kJ·kg^{-1}·K^{-1} (at 623 K). Electrical resistivity is 9.17 nΩ·m (at 273 K), 9.5833 nΩ·m (at 293 K). Mercury is diamagnetic with specific magnetic susceptibility $-0.168\cdot10^{-6}$ CGS units (at 293 K). Viscosity of liquid mercury is 0.001544 kg·m^{-1}·s^{-1} (at 293 K). Mercury vapor at low temperatures consists mainly of atoms, with increasing temperature the degree of association increases, and at the critical temperature it almost totally consists of diatomic molecules. Pressure of saturated mercury vapors is 0.00193 Pa (at 203 K), 27.273 mPa (at 273 K), 173.05 mPa (at 293 K), 1786.49 mPa (at 323 K). Adiabatic elastic moduli of mercury monocrystal: $c_{11} = 36.0$, $c_{12} = 28.9$, $c_{13} = 30.3$, $c_{14} = 4.7$, $c_{33} = 50.5$, $c_{44} = 12.9$ (in GPa) at 83 K. Mohs hardness of frozen mercury is 1.5. Compressibility of liquid mercury is 40.51 pPa^{-1} (at 303 K and pressure from 10^5 Pa to $5\cdot10^6$ Pa); compressibility coefficient of solid α-Hg is 35.2 pPa^{-1} (at 200 K and normal pressure). Mercury is widely used in the chemical industry, electronics, lighting engineering, etc.

MESIC ATOM, muonic atom

Atom of some chemical element where one of the electrons is replaced with a negatively charged *muon* (μ^-) which, like an electron, is also a lepton and behaves like a heavy electron. When the substance is irradiated with fast muons, mesic atoms form as the muons are decelerated: μ^- is trapped at a high-excitation state and replaces one of the atomic shell electrons. The transition to lower energy states is accompanied by radiation of characteristic *gamma quanta*, or by the *Auger effect.* Intensities of certain lines of characteristic radiation by a mesic atom depend on the nature of its bonding to nearest neighbors, which allows the determination of the type of chemical compound that is involved. As the transition of the muon to the K-shell takes place, the electron shell of a mesic atom with atomic number Z becomes similar to that of an ordinary atom with number $Z - 1$ because of the small Bohr radius of the bound muon. The small sizes of mesic atoms with hydrogen nuclei (radius of 10^{-11} cm) allow them to penetrate into electron shells, and even into nuclei of other atoms. This results in the generation of *mesic molecules*, and causes reactions that involve nuclei and muons (*muon catalysis*).

MESONIC CHEMISTRY

Chemistry of compounds containing *mesic atoms.* Replacement of one of the electrons with a negatively charged *meson* (*muon*, π- or K-meson) may take place as these particles are decelerated in the material. Since masses of mesons are hundreds of times that of an electron, the *Bohr radius* of a mesic atom appears to be significantly smaller than that of an ordinary *atom.* This results in the capability of the simplest mesic atom (mesonic hydrogen) to penetrate inside atomic shells of ordinary atoms and initiate nuclear transformations, formation of mesonic molecules, etc.

MESONS (fr. Gr. $\mu\varepsilon\sigma o\varsigma$, average, intermediate)

Unstable elementary particles that possess zero or integral *spin.* The first to be discovered and the best studied are π- and K-mesons that have masses between those of a proton and an electron; see also *muons* (at first erroneously called *μ-mesons*). Intense meson beams are obtained in proton accelerators, the so-called *meson factories*. This allows using mesons to study *solids.* The method of charge exchange of π-mesons on hydrogen nuclei is known; the probability of charge exchange depends on the atomic number of the nucleus that is bound to hydrogen, as well as on the nature of the chemical bond. This method allows identification of states of hydrogen in various media. Methods of *X-ray spectrum analysis* have been used for *mesic atoms. Channeling* behavior of π^+-mesons and μ^+-muons have been observed.

MESOPHASES

Thermodynamic *phases* of *liquid crystals* characterized by specific ranges of temperature, pressure, and concentration. In thermotropic liquid crystals, mesophases are found between the isotropic fluid (I) and solid crystal phases (S). The normal sequence of mesophases follows the decrease of degree of ordering with increase of temperature: $S \leftrightarrow$ smectic $B(H) \leftrightarrow$ smectic $C \leftrightarrow$ smectic $A \leftrightarrow$ nematic $N \leftrightarrow$ (cholesteric-Ch) $\leftrightarrow I$. In certain liquid crystals and their mixtures there can be deviations from this sequence, e.g., $C \leftrightarrow N \leftrightarrow A \leftrightarrow N \leftrightarrow I$. Melting points of mesophases range between -60 and $+400\,^\circ$C, and the width of mesophases varies from 0.01 to $\sim 100\,^\circ$C. Transitions $S \leftrightarrow N$ are *first-order phase transitions* with $\Delta H \sim (2\text{–}15)\cdot 10^3$ cal/mole. Phase transitions $N \leftrightarrow I$ and $N \leftrightarrow A$ are also first order with $\Delta H \sim 10^{-1}\text{–}10^{-2}$ and $1\text{–}10^{-2}$ kcal/mole, respectively. In *lyotropic liquid crystals*, thermal transitions $N \leftrightarrow I$ are first order phase transitions with $\Delta H \sim 10^{-3}$ kcal/mole. Transformations of mesophases are usually *second-order phase transitions.*

MESOSCOPIC FLUCTUATIONS

Scatter of observed parameters over an ensemble of specimens that exhibit identical macroscopic characteristics (dimensions, shape, impurity concentration, etc.) but differ in the specific configurations of static disorder.

It is reasonable to describe mesoscopic fluctuations in statistical terms. This approach seeks neither to depict atomic level motion of a particular specimen (microscopics, see *Microscopic description* of solids), or to limit itself to computing ensemble-averaged observables (macroscopics, see *Macroscopic description* of solids). Rather, its aim is to study the distributions of *mesoscopic* (intermediate length scale) fluctuations. The irregular dependence of these fluctuations on an external field allows one to observe them despite the fact that it is a particular specimen under experimental investigation, rather than the overall ensemble. For instance, the relationship between the total electrical conductivity of a specimen G and the magnetic field B has the form of a random process, but it is reproducible for a given specimen. It has been called the "*magnetic fingerprint*" of the specimen. Other "fingerprints" are relationships between G and electric field, or between G and the electron concentration.

The magnitude of mesoscopic fluctuations is unexpectedly high because of quantum effects. In the region of *weak localization* at zero temperature T, the root-mean-square mesoscopic conductivity fluctuation δG is of the order of the universal value $e^2/h \approx 4\cdot 10^{-5}\ \Omega^{-1}$ regardless of the size or shape of the sample, and decays rather slowly with increased T. The relative magnitudes of mesoscopic fluctuations of other physical quantities are often of the same order as $\delta G/G \ll 1$. However, if the mean value of X is small due to symmetry considerations (e.g., the thermal electromotive force is small because of electron–hole symmetry), then it is possible that $\delta X > X$ even in the region of weak localization. In an Anderson insulator, $\delta X \geqslant X$ at $T = 0$ (see *Anderson localization*).

An important feature of mesoscopic fluctuations consists in their sensitivity to minor changes in the disorder (configuration of impurities). For instance, a displacement of one impurity in a film of fixed width causes a finite change of G for any length of the film.

METAL, AMORPHOUS

See *Amorphous metals and metallic alloys.*

METAL AND ALLOY FERROMAGNETISM

See *Ferromagnetism of metals and alloys.*

METAL, ANTIFERROMAGNETIC

See *Antiferromagnetic metals.*

METAL CURRENT STATES

See *Current states in metals.*

METAL–INSULATOR INTERFACE

A transition layer between a metal and insulator in contact (see *Condensed matter interface*). The layer is heterogeneous in its chemical composition and structure. As a rule, the interface contains a large concentration of *surface electron states.* The quality of this interface to a great extent determines the electrophysical properties of *metal–insulator–semiconductor structures* and the characteristics of instrumental devices based on them.

METAL–INSULATOR–SEMICONDUCTOR STRUCTURE (MIS structure), metal–oxide–semiconductor structure (MOS structure)

A capacitor consisting of a *semiconductor* plate, an *insulator* layer (SiO_2) and a metal electrode. During the charging of the capacitor the electrical conductivity of the semiconductor changes in the neighborhood of the semiconductor–insulator interface due to a change in the concentration of charge carriers. This fact underlies the operation of a number of devices. The most common is a silicon based *metal–oxide–semiconductor field effect transistor* (MOSFET) (see Fig.). A thin insulating layer (width 100 nm) of silicon dioxide SiO_2 is applied by oxidation to a Si p-type *substrate*, then a metallic electrode (*gate*) is deposited. Two electron conduction regions are formed at a certain distance from each other under the surface of the oxide in the p-type Si; two metallic contacts (a source and a drain) are brought to these regions. If a positive voltage is

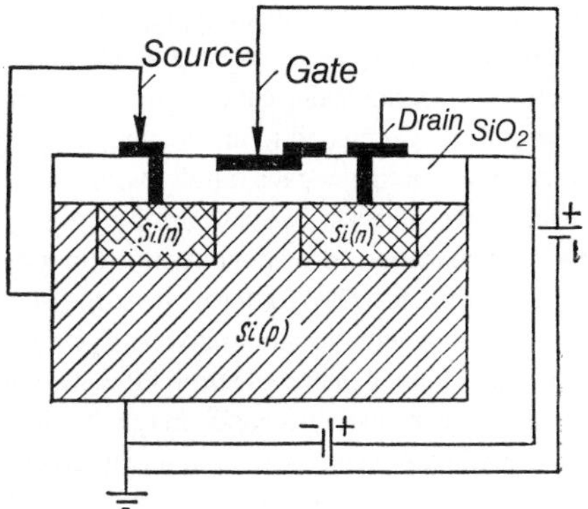

MOSFET transistor diagram.

applied to the gate, then all the electrons under it in p-Si will become attracted to the thin oxide layer, thereby creating a semiconducting *inversion layer* of n-type in it. This results in the generation of a current-carrying channel between the source and the drain. Such a system is equivalent to a vacuum triode (source is cathode, drain is anode, gate is grid). It may also serve as a memory unit. A two-layer insulator is used for this purpose: thin layers of SiO_2 and of silicon nitride. Electrical charge brought into Si may be transferred from Si to traps on the oxide–nitride interface. These traps remain charged for a relatively long time after removing the voltage between the gate and the substrate (memorization). This state may be read out by the change in properties of the subsurface area of the substrate. The MOS structure is a basic element of solid-state electronics; it is also used for studying surface properties of semiconductors (near the boundary with an insulator).

METAL–INSULATOR TRANSITION

See *Anderson localization*, *Mott metal–insulator transition*, and *Peierls transition.*

METALLIC BOND

Type of *chemical bond* in materials exhibiting metallic properties. Most atoms of the first three groups of the periodic table, as well as transition elements, have their electronic structure significantly modified when included in *crystal* lattices of *metals* and *alloys*. Their weakly bound outer s- and p-electrons lose their affiliation with their "own" atoms and move more or less freely through the whole crystal. The most typical example is Group I metals Li, Na, K as well as Cu and Ag, which serve as univalent cations immersed in an electron fluid. Compared to an isolated atom with a wave function that decays rapidly with distance, a nearly free electron has a wave function of a smoother character, corresponding to a kinetic energy far below the *Fermi energy*. It is this energy decrease, along with the Coulomb energy of an ion in a roughly homogeneous electron fluid, that is responsible for the metallic bond energy. Typical of metals are highly symmetric body-centered cubic, face-centered cubic and hexagonal close-packed lattices with high coordination numbers 8, 12 and 12, respectively. This explains their comparatively low melting points, high *forgeability* and *plasticity*. On the other hand, transition elements possessing several d-electrons (up to 8–9, as in Ni, Pd, Pt) exhibit a high metallic bond energy; and they are refractory and brittle, like, e.g., W. They have a large *covalent bond* contribution from highly localized d-electrons. Ions that differ very little in size easily replace one another, and this facilitates the formation of alloys over a wide range of composition, e.g., alloys of gold with copper and silver, bronzes, brasses, alloys of tin with lead, and many others. *Intermetallic compounds* InSb, InBi, HgTe have both metallic and covalent bond contributions. *Tin* can form different *phases*: a zero-gap *covalent crystal* α-Sn, and metallic β-Sn.

METALLIC CORROSION

See *Corrosion of metals.*

METALLIC GLASSES

Amorphous metallic alloys that are obtained through ultra-high-speed melt cooling. The rate of cooling must be about 10^6 K/s to prevent *crystallization* of the melt; this is attainable if at least one dimension of the specimen to be hardened is small. Therefore, metallic glass specimens have the form of foils, strips, wires with thickness of tens (more rarely, hundreds) of micrometers.

Types of metallic glasses. Melts capable of turning into glasses are called *glass forming*, and this capability depends on the melt composition. Several families of two-component metallic glasses are: *transition metal*–polyvalent metal (e.g., $Fe_{80}B_{20}$, $Pd_{80}Si_{20}$); polyvalent metal–polyvalent metal (e.g., $Mg_{70}Zn_{30}$); "heavy" transition metal–"light" transition metal (e.g., $Ni_{60}Nb_{40}$); rare-earth metal–polyvalent metal (e.g., $La_{70}Al_{30}$); rare-earth metal–monovalent noble metal (e.g., $La_{80}Au_{20}$); transition metal–"light" transition metal (e.g., $Cd_{70}Ni_{30}$); *uranium*–transition metal (e.g., $U_{70}Cr_{30}$). Multicomponent metallic glasses include atoms of more than one metal from a given group instead of only one of them; e.g., the alloy $Fe_{40}Ni_{40}P_{14}B_6$ belongs to the transition metal–polyvalent metal family.

Structure. Integral and partial radial distribution functions of atoms in metallic glasses, obtained through various diffractometry analysis methods, indicate the presence of short-range order (see *Long-range and short-range order*) in

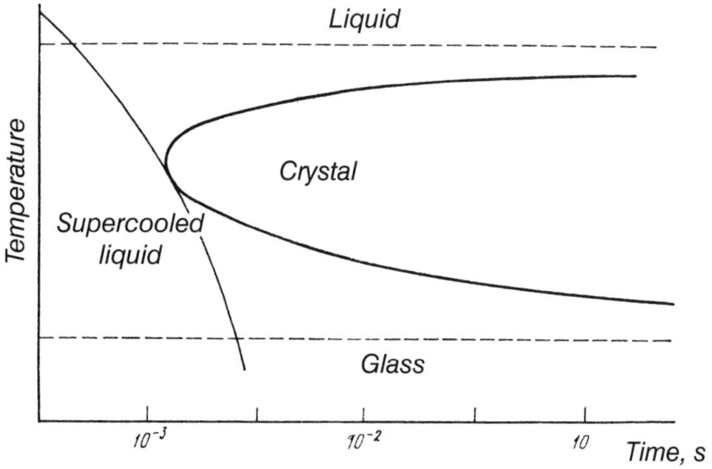

Diagram of alloy structure states.

the atomic arrangement extending over several interatomic distances. *Mössbauer spectroscopy* and other investigation methods indicate the presence of one or more types of local atomic ordering, and studies of surface structure of some metallic glasses using the scanning tunneling microscope (see *Scanning tunneling microscopy*) and the high-resolution *electron microscope* reveal ordered areas up to 2.5 nm in size. With the help of transmission electron microscopes, a domain-type structure was discovered in a number of glasses, the *domain* sizes being of the order of 10 nm. A number of models of metallic glass structure have been suggested. The *random close-packing model*, supplemented by concepts of free volume (see *Free volume model*), dates back to *Bernal's liquid model*, but fails to allow for intermetallic bonds and the existence of local topological and compositional order. In *polycrystal models* an amorphous body is considered as a dislocation- and disclination-disordered crystal, or a fine-grained *polycrystal*. The distances between extensive defects in *paracrystals* must extend over several interatomic distances, but as yet there is no experimental proof of their existence in metallic glasses. A polycluster model (see *Polycluster amorphous solids*) treats a metallic glass as an aggregation of locally matched, regular *clusters* with each atom in the cluster locally ordered. There may be several types of local order which may alternate randomly. Cluster boundaries are locally disordered and contain tunneling states responsible for *low-temperature anomalies* in amorphous solids.

Properties. The density of metallic glasses is, as a rule, 1–2% lower than that of the crystal alloy of the same composition. Metallic glasses owe their high strength to their lack of topological order (*ultimate strength* approaches the theoretical value). The *bulk modulus* is nearly the same and the *shear modulus* is 10–30% smaller than that of the crystal analogue. *Plasticity* is about 1% at temperatures below $0.7T_g$ (where T_g is the *vitrification* temperature) and up to several tens of percent at temperatures close to T_g. *Plastic deformation* is uniform and is described via Newton's equation at $T \approx T_g$. At low temperatures, plastic deformation is nonuniform; and *slip bands* about 20 nm in thickness are generated, separated by up to several μm. Uniform plastic deformation in polyclusters is a diffusely viscous flow controlled by *diffusion* and *slip* over cluster boundaries. At low diffusion mobility of atoms, this flow switches to a nonuniform deformation mode accompanied by the generation of slip bands. In the free-volume model, plastic deformation is associated with realignment of units containing free volume.

Due to the lack of thermodynamic equilibrium, metallic glasses undergo *structural relaxation* and *crystallization* with a typical crystallization enthalpy of 1–5 kJ/mole. The rate of structural relaxation and crystallization is determined by the speed of diffusion displacements of atoms, and the rate of formation of stable configurations and crystalline-phase nucleating centers by atoms. Instead of compositional *phase diagrams*, structural transformation diagrams in temperature–time coordinates are constructed for metallic glasses. The figure shows the *structural equilibrium diagram* of an alloy, with the thin line on the left corresponding to the curve of the melt cooling regime which ensures the formation of a metallic glass.

Diffusion coefficients of intrinsic atoms in metallic glasses are, as a rule, noticeably greater than those in crystals at the same temperature. Microscopic mechanisms of diffusion are not completely clear. In a free-volume model, the diffusion occurs along free cavities, the volumes of the latter being comparable to atomic volumes. In polyclusters two-dimensional diffusion occurs along cluster boundaries, while volume diffusion in clusters occurs mostly along vacancies.

The electrical resistivity of metallic glasses at temperatures near the *Debye temperature* is close to that of the melt. The thermal resistance coefficient may have a small positive, negative or zero value, depending on the composition of the glass. The *Mooij correlation* (J.H. Mooij, 1973) between the resistivity and the thermal resistance coefficient is observed. Some amorphous alloys are superconductors (see *Superconducting alloys*). The transformation temperatures of amorphous alloys into the superconducting state are, as a rule, lower than those of their crystal analogues, and usually do not exceed 9 K. Yet, amorphous superconductors find practical application because of their narrow range of transformation temperatures (about 0.05 K) and their high plasticity. Metallic glasses feature low-temperature anomalies in the heat capacity, heat conductivity, velocity of propagation and absorption of sound, which are typical of various types of glasses, and result from characteristics of their atomic structure, in particular, from the existence of *two-level systems*. At low temperatures, features of the resistance and magnetoresistance temperature dependences are found, which result from *weak localization* of electrons.

Many iron-, cobalt- and nickel-based metallic glasses of the "transition metal–polyvalent metal" type, which contain about 80% transition metals, are magnetically ordered media (see *Amorphous magnetic materials*, *Amorphous magnetic substances*). Their *Curie point* may be changed over a wide range (up to approximately 1000 °C) by varying the composition of the alloy. Corrosion properties of metallic glasses are, as a rule, noticeably better than those of their crystal analogues. Because of their extraordinary mechanical, electric and magnetic properties, metallic glasses find various practical applications. Their applicability is limited by comparatively low crystallization temperatures, insufficient thermal stability, and small thickness of the materials that are obtained in the form of strips and wires.

METALLIC HYDROGEN

A hypothetical high-pressure *phase* of solid hydrogen which exhibits metallic properties. According to general concepts of a metal–insulator (Peierls) transition, the change of electric properties is due to the broadening of *energy bands* and the disappearance of the *band gap* between metal and insulator, with strong compression of the material. Hydrogen becomes a conductor but retains the hexagonal close-packed structure of a *molecular crystal* by the gradual overlapping of the gap between the *valence band* and the *conduction band*. The further increase of pressure promotes the transition to the atomic metal phase with a body-centered cubic structure that is typical of alkali metals. Then, the appearance of one partially filled band to replace two overlapping ones provides high *electrical conductivity*. The experimental evidence indicates only a tendency for such a rearrangement into an atomic phase: as evidenced by *Raman scattering of light*, the vibration frequency of H_2 decreases under pressure above 0.3 Mbar, i.e. the strength of the bond decreases. The formation pressure of metallic hydrogen is theoretically estimated to be of the order of 2–7 Mbar. An important question is the stability of metallic hydrogen after removing the applied pressure. There is particular interest in the problem of metallic hydrogen because of its possible onset of *superconductivity* with a high critical temperature, perhaps $T_C \approx 100$ K.

METAL-LIKE COMPOUNDS

Compounds of *metals* with nonmetals (hydrogen, bromine, carbon, nitrogen, silicon, phosphorus, sulphur, germanium, arsenic, selenium, and tellurium). The nature of the *chemical bond* in metal-like compounds determines their properties. The presence of electrons delocalized over the crystal is responsible for their crystal chemistry characteristics: face-centered cubic, body-centered cubic, hexagonal close-packed or more complex structures, considerable *electrical conductivity* (with high negative temperature coefficient), *thermal conductivity*, and high *melting temperatures*. Unlike metals, metal-like compounds exhibit high *hardness*, small values of the *ultimate strength* and *plastic limit* at low temperatures, comparatively low values of the linear temperature expansion coefficient, and low resistance to thermal shock. Some metal-like compounds are actually *superconductors*. Most metal-like compounds are phases of variable composition, whose properties depend on the ratio of metal and nonmetal components. Transformation into semiconducting phases may also occur.

METAL, LIQUID

See *Liquid metals*.

METALLOGRAPHIC ANALYSIS

Analysis of *metal* microstructure by means of a metallographic microscope with the aid of reflected light. Before carrying out the metallographic analysis, a section of the metal specimen under study is prepared. The metal is cut, ground, polished and etched in such a way that there are no distortions of the structure as a result of a mechanical or thermal influence. Then, the metallographic specimen is examined under the microscope where the light beam is reflected from regions of the specimen according to their orientation. This is done so that those elements of the structure, which are perpendicular to the beam, appear as bright spots and those which are inclined with respect to the beam, appear as dark spots in the image.

These features reveal various aspects of the microstructure which were etched in different ways during the preparation of the specimen. Metallographic analysis uses a wide range of techniques and procedures. For bright-field illumination, the specimen is positioned perpendicular to the optical axis of microscope. Oblique illumination requires either displacement of the light source, or illumination of the specimen from under the objective lens; this allows increasing the contrast of the picture, but is difficult to implement when high magnification is needed. Dark-field illumination is achieved by placing the specimen in the ray that bypasses the objective lens and is directed by special reflectors. The contrast achieved by using dark-field illumination is opposite to that of bright-field illumination, i.e. those structure details, which appear dark under bright-field illumination, appear bright under dark-field illumination, and vice versa. The microscope resolution does not decrease with dark-field illumination.

Special methods of metallographic analysis include the phase contrast method, interference method, those of high-temperature and low-temperature analysis, those based on the application of polarized light, etc. Metallographic analysis is one of the main techniques for structure studies in the *physics of metals*, *metal science*, *materials science*, and *heat treatment*.

METALLOPORPHYRINS

Complexes formed by the porphyrin molecular structure with *metals*. According to chemical bond type, complexes are classified into ionic (labile) and covalent (stable), and according to coordination center N_4M geometry (N is a pyrrole ring nitrogen, and M is the metal), they are classified into centrally symmetric planar and non-centrally symmetric nonplanar types, into purely porphyrin complexes and porphyrin extracomplexes (depending on saturation state of central atom coordination number). There are also sandwich-type metalloporphyrins and various polymeric ones. The molecular structure of metalloporphyrins is highly specialized; it results from high covalence and strength of conjugated C=N and C=C bonds, and strong σ-electron overlap along the macrocycle which consists of four pyrrole rings connected by methine (–HC=) bridges. This results in the formation of very strong intramolecular compounds capable of extra-coordination. The diversity of metalloporphyrin structures is due to numerous chemical modifications of the molecules which do not effect the system of π-electrons. Metalloporphyrins of most metals (excepting alkalis, alkaline earths, lanthanides) are extra-stable complexes (stability constant in glacial acetic acid is 10^{35}–10^{45}). Materials produced on their basis exhibit a wide range of useful features. They are sensitive and selective catalysts for many chemical processes; when included into a thin film of *polymer* they act as photoelectrochemical energy converters. "Organic metals" exhibit properties of *semiconductors*; electrochromic materials have a long working life; photosensitive materials are used in phase three-dimensional *holography*, etc.

METALLURGY (fr. Gr. μεταλλουργια, metallurgy)

Study of *metals*, including processes of extraction from ores and other materials, and subsequent treatment in liquid and solid states, to provide them with specified shapes and properties. The most important metallurgic products are hot- and cold-rolled bar and sheet products, pipes, metal articles. By-products include chemical compounds, fertilizers, building materials, etc. Ferrous and nonferrous metallurgy are known. *Ferrous metallurgy* covers the manufacture of *steels* and *iron*-based *alloys* (*cast iron*, steel, ferrous alloys), while *nonferrous metallurgy* involves nonferrous metals and alloys based on them. *Powder metallurgy* involves the preparation of ferrous and nonferrous metal powders that are transformed into finished parts by *pressing* and subsequent *sintering*.

The production system of modern metallurgy includes mining complexes and concentrating mills, by-product coke plants, factories specializing in ferroalloys and refractory materials, metallurgical and metal-working combines and enterprises. There are four process stages:

(1) metallurgical production proper includes blast-furnace involvement,
(2) steel-making stage (open-hearth furnaces, converters, electric furnaces),
(3) rolling-mill procedures (roughing and billet mills, sheet, plate and section mills),
(4) there is an ever-increasing significance and wide application of operations of *heat treatment*, *thermomechanical treatment*, *chemical heat treatment*, and other kinds of rolled stock treatment, cold *plastic deformation*, application of protective, decorative and other kinds of *coatings*, and so on.

At present, the principal process is the two-stage (coke/blast-furnace) method of metal making that is based on the production of cast iron and reprocessing it into steel. A radical innovation was the development of continuous feed casting machines, which led to abandoning the reprocessing of ingots into feeds (blooming-slabbing mills) and reduced the consumption of rolling steel. However, there are disadvantages inherent in this two-stage method (high coke consumption with limited reserves of coking coal; minimal prospects of equipment productivity growth; unfavorable ecological aftereffects, etc.) that inspired a renewed interest in *direct processing metallurgy*. This involves the direct production of iron according to the scheme: shaft metallization furnace–electric furnace–continuous feed casting machine–rolling mill, without any blast-furnace or accompanying coke-chemical and agglomeration process stages.

It may also be possible, through improved methods of iron ore dressing which prepare practically pure iron oxides with economically acceptable performance, to employ deep reduction to obtain iron powders for the manufacture of work pieces, or to press compact feeds for rolling. This could provide finished products through bypassing the blast-furnace and steel-making process stages, without liquefying the metal (so-called *solid-state metallurgy*).

METALLURGY, ION

See *Ion metallurgy*.

METALLURGY, POWDER

See *Powder metallurgy*.

METAL MODIFICATION

See *Modification of metals*.

METAL OPTICS

Branch of physics that studies the interaction of *metals* with electromagnetic waves in the optical region (electrodynamic properties of metals). Metals typically have a high reflectivity R over a broad range of wavelengths λ due to the high concentration of *conduction electrons*. By interacting with an electromagnetic wave incident on the metal surface, conduction electrons produce alternating currents, and most of the energy acquired by electrons from the wave is emitted in the form of secondary waves that initiate an echo wave. A portion of the absorbed energy is transferred to lattice vibrations due to their interaction with the electrons. The conduction currents screen the external electromagnetic field, and attenuate the wave inside the metal (see *Skin-effect*). Conduction electrons can absorb arbitrarily small quanta $\hbar\omega$ of electromagnetic energy (ω is the radiation frequency). Therefore, they contribute to optical properties of metals, a contribution which is especially large in the infrared and radio-wave spectral regions. These properties are related to the complex *dielectric constant* $\varepsilon(\omega) = \varepsilon' - \mathrm{i}\sigma(\omega)/\omega$ (where ε' is the dielectric constant minus the contribution of conduction electrons, σ is the *electrical conductivity* of the metal), i.e. to its *refractive index* $n = n' - \mathrm{i}\kappa = \sqrt{\varepsilon}$ (where κ is *absorption index*). The imaginary part of n represents the exponential attenuation of the wave inside a metal. In the IR and optical frequency regions, the first approximation yields $\varepsilon(\omega) = \varepsilon(\infty) - (\omega_p/\omega)^2$ (here ω_p is the electron plasma frequency). At frequencies $\omega \geqslant \omega_p$, *plasma oscillations* of electrons are excited in the metal, and they bring about transparency at $\omega > \omega_p$. In the UV region, R falls off, and metals come close to *insulators* in optical properties. In the X-ray region, electrons of inner atomic shells determine the optical properties, and metals differ little from insulators. Like insulators, metals exhibit absorption bands that result from resonance excitation of transitions between different electron energy bands, and these resonances lead to variations in $\varepsilon'(\omega)$. By virtue of the strong interaction of electrons, absorption bands in a metal are considerably broader than in an insulator. Metals usually exhibit several bands situated mostly in the visible and near-UV, more rarely in the IR. There is a difference in phase between the waves reflected from a metal surface when polarized in the plane of incidence, and perpendicular to it, so plane polarized light becomes elliptically polarized upon reflection. As distinct from insulators, waves polarized in the plane of incidence always have $R \neq 0$.

METAL OVERHEATING

Heating of a *metal* above admissible operating temperatures (e.g., heating for *quenching* lies within the temperature range 850–950 °C for most types of *steel*). Heating to 1000 °C and above at hardening of steel is an example. Overheating of a metal results in an increase in grain size and a deterioration of mechanical properties. Hot plastic working of high-alloy steels is performed within the temperature range 1200–1000 °C (there is a specific narrow temperature range for each particular type of steel, but none of these individual ranges falls outside the above limits). Heating for *forging* to higher temperatures causes undesirable structural changes (e.g., *segregation* of carbon to *grain boundaries*), which may be responsible for the deterioration of *plasticity*. Operation (even short-term) of an article above a certain admissible temperature causes a decline of its mechanical properties, and a complete or partial drop in serviceability. The consequences of overheating are corrected through a repetitive *heat treatment* in accordance with the technological process adopted for a given article and a given steel type. If a metal is heated to temperatures which considerably exceed permissible ones, then the process of internal intergranular *oxidation* is observed (this kind of overheating is called *burn* in metallurgical practice); the mechanical properties of a work piece degrade drastically, and do not recover through repeated heat treatments.

METAL PHYSICS

See *Physics of metals*.

METALS

Materials exhibiting high *electrical conductivity*, characteristic luster, *plasticity* and *forgeability*. The high conductivity and high electromagnetic wave reflection factor in metals are due to the presence of free electrons with concentrations 10^{22}–10^{23} cm^{-3}. The free (delocalized) electrons influence various thermodynamic characteristics of metals, e.g., terms linear in temperature are added to the specific heat and thermal expansion coefficient. Direct proof of the electronic nature of metallic conduction is provided by electrical inertia experiments. Most crystals composed of identical atoms are metals, with only 19 elements forming nonmetallic crystals under equilibrium conditions. Under changes of temperature, pressure and magnetic field, crystals may undergo *structural phase transitions* (so-called *polymorphism*). *Crystals* that are not electrical conductors may, in this process, become metals. For instance, under *high pressure* the semiconductors *silicon* and *germanium* become metals, while metallic crystals may become nonmetals (e.g., white *tin* is a metal, and gray tin a *semiconductor*). Under elastic deformations that have no effect on the lattice symmetry, electronic topological transitions (see *Lifshits transition*) caused by abrupt change in dynamical properties of *conduction electrons* are observed in a number of metals. The concept of metallic structure as an ionic skeleton surrounded by an "electron fluid" that compensates for the forces of repulsion between ions and binds them in a *solid* is a fairly good representation of reality.

Electrical and thermal properties of metals. According to the electronic theory of metals, electrons are capable of moving through an ideal lattice without any resistance. The appearance of electrical resistance is due to *electron scattering* by faults in the lattice due to atomic thermal motion, the presence of impurity atoms, vacancies, dislocations, and other static *defects*. Conduction electrons in metals are charged low-energy excitations (*quasi-particles*) of an electronic *Fermi liquid*, and form a slightly imperfect Fermi gas with degeneracy temperature $T_0 = \varepsilon_F/k_B \sim 10^5$ K (ε_F is the Fermi energy). These values of T_0 correspond to characteristic electron velocities $\sim 10^8$ cm/s. With decreasing temperature, the amplitude of atomic oscillations decreases with the result that the electron *mean free path* l increases. This behavior of l causes an increase of conductivity with the decrease of temperature, which is typical of metals. At $T \to 0$ the electric resistance of actual metals saturates, that is it tends to a limit referred to as the *residual resistance* ρ_{residual} which is a measure of the lattice perfection. The criterion for this saturation is $k_F l > 1$ (*Ioffe–Regel rule*). The residual resistance of pure metals (e.g., W, Ga, Mg) is 10^5 times smaller than their resistance at room temperature. The electron mean free path l in these metals at $T \to 0$ is as long as several centimeters. At low temperatures samples of the same metal, differing in impurity content, follow the relationship $\rho = \rho_{\text{residual}} + \rho_{\text{ideal}}$ called the *Matthiessen rule* (L. Matthiessen, 1864) where ρ_{ideal} is the resistance due to scattering of electrons by thermal lattice vibrations. At $T \gg \Theta_D$ (Θ_D is the Debye temperature) $\rho_{\text{ideal}} \propto T$, while for $T \ll \Theta_D$ the temperature dependence has the typical form $\rho \propto aT + bT^5$. Here, a and b are constants that measure the scattering of electrons by electrons and thermal lattice vibrations (thermal *phonons*), respectively. Highly disordered metallic systems (*amorphous metals*) may experience *Anderson localization* that leads to a metal–insulator phase transition through *quantum interference phenomena*, the latter arising due to repeated scattering of electrons by static defects. *Superconductivity* occurs in 25 pure metals and numerous metallic alloys.

The superconducting state in metals is due to the formation of a condensate of *Cooper pairs* in the degenerate electron gas as a result of the *electron–phonon interaction*. The values of the superconducting transition temperature T_c for pure metals range from 0.026 K for Be to 9.25 K for Nb; the highest T_c for a classic metallic compound being 23 K for Nb_3Ge. Cuprate high temperature superconductors have transition temperatures as high as 133 K. The transition into the superconducting state in an external magnetic field is accompanied by expulsion of the field from the bulk metal (*Meissner effect*). The superconducting state is destroyed by an applied magnetic field that exceeds a certain critical value B_c. For pure metals, $B_c \approx$ 10–100 mT; for alloys, $B_c \approx$ 1–10 T. Electrons participate in the transfer of electric charge as well as heat.

There is a simple relationship

$$\frac{K}{\sigma T} = \frac{\pi^2}{3}\left(\frac{k_B}{e}\right)^2$$

between the thermal conductivity K and the electrical conductivity σ called the *Wiedemann–Franz law*. This law is valid in temperature regions where inelastic processes are negligible in electron scattering. Metals display a number of *thermoelectric phenomena*.

An external magnetic field bends electron trajectories, and thus affects transport processes in metals (*galvanomagnetic effects* and *thermomagnetic phenomena*). This effect is largest when the condition $r_L \ll l$ (here r_L is the ion Larmor radius) is satisfied. At $T = 4$ K, the resistance of some pure metals increases in a magnetic field by a factor of 10^4 to 10^5. Studying galvanomagnetic and thermomagnetic phenomena in monocrystalline metals at $r_L \ll l$ provides the topology of the *Fermi surface*, while the concentration of charge carriers in a metal is determined by the Hall effect. A magnetic field makes possible a nonuniform current distribution in a metal across the specimen cross section (*static skin-effect*). *Interband tunneling* of electrons in a magnetic field (*magnetic breakdown*) was detected in many metals. Under conditions of magnetic breakdown, the kinetics of electrons in pure metals is governed by quantum interference of electron waves that arise in band-to-band tunneling. Such interference may transform a pure metal into an incommensurate system with absolute quantum localization of a certain group of conduction electrons. Due to the high value of the metallic conductivity, a high-frequency electromagnetic wave cannot penetrate far into the metal depth, i.e. practically all the high-frequency current flows in a narrow skin-depth layer near the specimen surface (*skin-effect*). In typical metals, the depth of the skin-layer is $\delta \sim 10^{-6}$ cm at the frequency $\omega \sim 10^8$ Hz. Depending on the ratio between the values of l and δ, two types of skin-effect are distinguished: the *normal skin-effect* ($\delta \gg l$) and the *anomalous skin-effect* ($\delta \ll l$). Two types of phenomena are observed in a magnetic field under the conditions of the anomalous skin effect. The first is *cyclotron resonance* that arises when the Larmor frequency of some group of electrons coincides with the frequency of the applied high frequency field. The other includes various kinds of slightly decaying electromagnetic waves that cause an *anomalous penetration* of the high-frequency field into the metal to depths considerably exceeding δ.

At high temperatures, the evaporation or emission of a small number of electrons (*thermionic emission*) takes place. The number of departing electrons is proportional to $\exp[-U/(k_B T)]$, where U is the work function (*Richardson equation*). The escape of electrons from a metal may also take place without heating, through the *tunnel effect* that arises when an electric field is applied (*field electron emission*). Substantial field emission is observed in fields $E \approx 10^7$ V/cm. The magnitude of the field-emission current is an exponential function of E, and is practically independent of temperature. The electrical resistance of metals changes abruptly on melting (increases two- or threefold); three known exceptions being Bi, Sb and Ga: whose resistance decreases at melting.

Optical properties of metals; X-ray spectra. The characteristic luster of metals is caused by a high reflectivity (exceeding 99% in many metals) for electromagnetic waves with frequency ω lower than the plasma frequency ω_p (radio waves, IR, visible region). At $\omega > \omega_p$, metallic optical properties are close to those of insulators. In this case, of key importance are quantum effects – electron transitions between energy levels with the absorption of light quanta, and γ-quanta of X-radiation (*inner level photoemission effect*). For instance, the reflectivity of silver is 35% in the visible region and decreases to 4.2% in the UV (reflection by glass). In the X-ray region metals differ from insulators in optical properties by the structure of the absorption band edge. In a metal, the edge of the soft X-ray absorption band is sharply cut off, which reflects the degeneracy of the electron distribution in the *conduction band*. X-ray absorption bands contain information on the energy spectrum of a metal, i.e. they allow determining the overlap of energy bands by impurities

Mechanical properties of metals. Not only do free electrons define the electronic and thermophysical properties of metals, but they also noticeably affect the cohesive forces of atoms in a lattice, as well as the *elasticity*, *strength* and *plasticity*. A salient feature of a *metallic bond* consists in the decrease of the kinetic energy of a valence electron compared to its value in an isolated atom. The spatial directionality of p- and d-electron atomic orbitals leads to substantial anisotropy of elasticity, strength, and plasticity properties of *monocrystals*. Elastic properties of metals depend only slightly on the temperature. The role of conduction electrons makes itself evident in the change of mechanical properties of metals at a transition into superconducting states, e.g., the *elastic limit* of Pb becomes 5% lower, i.e. the process of *dehardening* takes place. According to the nature of the thermal change of plasticity metals can be divided into two groups: those that remain plastic down to liquid-helium temperatures, e.g., Al, Cu, Ni, Pb, and those that become brittle under cooling, the so-called cold brittle metals (see *Cold brittleness*), e.g., α-Fe, Mo, Cr. Plastic deformation of crystals occurs in the form of *slip*, *twinning of crystals*, *faulting*. The carriers of the elementary act of plastic deformation are *dislocations*. The theoretical strength of metals is observed on perfect (dislocation-free) crystals; the actual strength is smaller by factor of 100–1000 because of structural imperfections. Mechanical properties of metals depend on a number of external factors (temperature, force application rate, type of stressed state) and internal factors (chemical composition, microstructure). Application of *alloying* and various kinds of *heat treatment* and *thermomechanical treatment* (hardening, *programmed hardening* and other methods) substantially extend the range of practically important mechanical properties of metals.

Magnetic properties of metals. Essential to magnetic properties of metals is the division of metals into two groups. *Weak magnetic materials* form a group of metals exhibiting either *diamagnetism* or *paramagnetism*. *Strong magnetic materials* exhibit paramagnetism and, with decreasing temperature, attain magnetic ordering that results from the cooperative interaction of electrons from inner incomplete d- and f-orbitals of an atom (see *Ferromagnetism of metals and alloys* and *Antiferromagnetism*). In magnetic fields that exceed a certain, so-called critical value B_C, *antiferromagnetic metals* may turn into either ferromagnetic or paramagnetic states. The magnetic structure of an antiferromagnetic metal may be either commensurate or incommensurate with their crystallographic structure. Rare-earth metals (see *Magnetism of rare-earth metals*) exhibit complicated *atomic magnetic structures*: spiral (screw) structures with a pitch that is incommensurate with lattice parameters. At low temperatures in magnetic fields, metal single crystals exhibit complicated oscillating relationships between the magnetic moment and the field (see *De Haas–van Alphen effect*), which is due to the Landau quantization of the energy spectrum of conduction electrons in a magnetic field (see *Landau diamagnetism*). After investigating the de Haas–van Alphen effect along various crystallographic directions, one can reconstruct a Fermi surface, and find the *effective mass* of an electron in each of its extremal cross-sections.

METAL SCIENCE, physical metallurgy

Comprehensive study of the composition, structure and properties of *metals* (*alloys*), as well as their changes in response to thermal, chemical, mechanical, and other types of influences; a subfield of *materials science*. It provides the scientific basis for synthesizing metallic materials (e.g. alloys) with controlled properties. Metal science is closely related to the *physics of metals*.

METALS, CREEP

See *Radiation-induced creep*.

METAL–SEMICONDUCTOR JUNCTION

The area around the contact or junction between a *metal* and a *semiconductor* which includes a transition layer with a *space charge region*. Together with a *semiconductor junction* and a *metal–insulator–semiconductor structure* the metal–semiconductor junction is one of the main elements of semiconductor electronics, and is widely used as an *ohmic contact*.

The properties of this junction arise either from the difference between the work functions φ_M and φ_S of the metal and semiconductor, or are determined by the *surface states* on the *phase interfaces*. In the first case the surfaces in contact become oppositely charged due to the difference in the movement of electrons from the metal and semiconductor, and a space charge region is formed. This results in the energy *band bending* in the neighborhood of the contact (see figure). An n-type semiconductor with an upward turning band curvature has its space charge area depleted of majority carriers, has increased resistance, and is referred to as a *gate layer*. For downward band

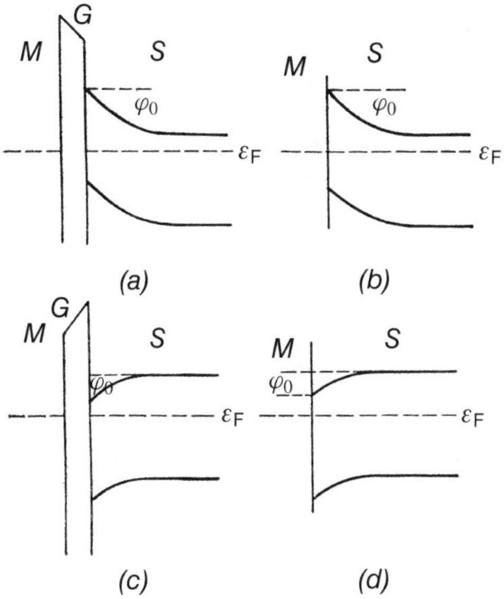

Energy plots for a metal (M)–semiconductor (S) junction with a gap (G) (a, c), and without a gap (b,d) for various band curvatures, where ε_F is the Fermi level.

curvature of an n-type semiconductor the space charge region is enriched with majority carriers, it has less resistance, and is referred to as an *antigate layer*. For a p-type semiconductor upward *band bending* corresponds to an antigate layer and downward bending to a gate layer.

The band curvature value is $\varphi_0 = f(\varphi_M - \varphi_S)$: for contact with a gap $\varphi_0 < \varphi_M - \varphi_S$, and for direct contact $\varphi_0 = \varphi_M - \varphi_S$. When the bending of the bands is caused by the surface conditions then only the bending direction and the type (gate or antigate layer) are determined by the charge, and $\varphi_0 \neq f(\varphi_M - \varphi_S)$. In an intermediate case the band bending can depend on both the work function difference and the surface charge conditions. The gate layer possesses rectifying properties, and the dependence of the current I on the voltage U usually follows the form

$$I = I_S\left[\exp\left(\frac{eV_1}{nk_BT}\right) - \exp\left(-\frac{eV_2}{n^*k_BT}\right)\right],$$

where $U = V_1 + V_2$, V_1 and V_2 are the voltage drops in the space charge region and at the gap, respectively; I_S, n and n^* are determined by the presence or absence of the gap, the involvement of one or two charge carrier types, and by their transition mechanisms through the contact (over-barrier current, tunneling current, generation–recombination current, current involving surface conditions, etc.). For direct contact we have

$$I_S = \frac{eV_n n_0}{4}\exp\left(-\frac{\varphi_0}{k_BT}\right),$$

where $n = 1$ and $n^* = \infty$. For low alternating signal injection the contact behaves as a resistance and a capacitance in parallel.

The capacitance value for direct contact in the case of a majority carrier current is determined by the space charge region recharging, and is equal to $C = \varepsilon\varepsilon_0/L$, where the width L of the space charge region is given by

$$L = \left[\frac{2\varepsilon\varepsilon_0(\varphi_0 - eU)}{e^2 n_0}\right]^{1/2},$$

and ε and ε_0 are dielectric constants of the semiconductor and vacuum, respectively. Such a capacitance is referred to as a *barrier capacitance*. The resistance value for direct contact is

$$R = \left(\frac{I_S e}{k_BT}\right)^{-1}\exp\left(-\frac{eU}{k_BT}\right).$$

If the current through the contact is determined by minority carriers, the metal–semiconductor contact capacitance includes also the *diffusion capacitance* C_D which for direct contact at low frequencies is equal to

$$C_D = \left(\frac{e\tau}{2k_BT}\right)I_S\exp\left(\frac{eU}{k_BT}\right),$$

where τ is the lifetime of minority carriers. The barrier capacitance does not depend on the frequency up to values $\omega \approx 1/\tau_M$, where τ_M is the Maxwell *relaxation time*. The diffusion capacitance decreases with increased frequency beginning with $\omega \approx 1/\tau$, and at $\omega \to \infty$, $C_D \to 0$.

A metal–semiconductor contact on the base of a gate layer is used as an active element in the following devices with a *Schottky barrier*: detector, mixer and parametric ultrahigh frequency *diodes*, *field-effect transistors*, *bipolar transistors* with bridging contact and collector on the base of metal–semiconductor contact, fast optical detectors, semiconductor *solar cells*, *charge coupled devices*, etc. (see *Semiconductor devices*). The advantages of these devices are operating speed, technological simplicity, small size of active region, etc.

METALS, HEAT TREATMENT

See *Mechanical heat treatment of metals.*

METAMAGNETISM

See *Antiferromagnetism.*

METASTABLE STATE

State of incomplete *thermodynamic equilibrium* of a macroscopic system, corresponding to a local minimum of its free energy, with a long lifetime before transformation to the equilibrium state. This transition usually occurs via the formation of nucleating centers of the new *phase* (see *Nucleation*). These centers may be, e.g., fragments of polar phases in *ferromagnets* and *ferroelectrics.* In quantum systems at low temperatures the metastable state may collapse by activation, or because of *macroscopic quantum tunneling.*

MICA

Minerals from the group of layered *aluminosilicates* with the common formula $AB_{2-3}[T_4O_{10}]$-$(OH,F)_2$ where A is K, Na, Ca; Ba, B is Al, Mg, Fe; T is Si, Al. The chemical composition is variable, and often one cation is exchanged for another. Micas can be divided in accordance with their chemical structure into subgroups: *black mica* $K(Mg,Fe)_3[Si_3AlO_{10}](OH,F)_2$ (magnesium–iron mica), *muscovite mica* KAl_2-$[AlSi_3O_{10}](OH)_2$ (aluminum mica), *lepidolite* $KLi_2Al[(Si,Al)_4O_{10}](F,OH)_2$ (lithium mica). A mica may be colored black, brown, reddish, colorless, white, or ruby. Hardness by Mohs scale is 2.5 and 4 parallel and perpendicular to the layers, respectively. Syngony (crystal class) is monoclinic. Structure consists of three layer sheets connected with each other by intercalated alkali cations. Each sheet consists of two silicon-alumino-oxygen tetrahedral layers and one inner octahedral layer with divalent or trivalent cations. Univalent cations (alkalis) are not in the sheets, but rather are situated between them. Physical properties are of the same type. Mica is an *insulator,* relative dielectric constant $\varepsilon = 9.7$–10.0 for black mica, 6.0–8.0 for muscovite, 6.7 for lepidolite. Magnetic susceptibility $\chi = (3\text{–}80)\cdot 10^{-6}$ cm^3/g for black mica, $(3\text{–}5)\cdot 10^{-6}$ cm^3/g for muscovite. Cohesion is rather perfect within layers. Micas are used in radio and electrical engineering as thermal and electrical *insulating materials.*

MICHELL PROBLEM (J.H. Michell, 1900)

The problem involves describing the *state of stress* of a prismatic *rod* with a uniformly loaded lateral surface. With the acceptance of the *Saint-Venant principle* (effects of a system of forces that add vectorially to zero applied to a region of a solid are only felt within that region), the problem reduces to the *Saint-Venant problem* involving forced *flexure* and *rod twisting,* and also to a certain *plane problem of elasticity theory.*

MICRO... (fr. Gr. $\mu\iota\kappa\rho o\varsigma$, small)

Prefix for a physical unit, which allows obtaining a multiple unit, equal to 10^{-6} of the initial unit. Symbol: μ. Example: 1 μs (microsecond) = 10^{-6} s.

MICROBRIDGE, SUPERCONDUCTING

See *Superconducting microbridges.*

MICROCRACK–DISLOCATION INTERACTION

This is the interaction of the elastic fields of *dislocations* with the vertex of a microfracture.

For a *crack* of size $l \ll d$ (d is a *crystallite* dimension) the elastic interaction of the crack with randomly distributed dislocations, where long-range fields introduce an additive term in the *elastic energy* (see *Elasticity*), is described by the effective tension $P_e = (P^2 + A\mu^2 b^2 n)^{1/2}$. Here P is the external tension, μ is the *shear modulus,* b is the *Burgers vector,* n is the dislocation density, A is a coefficient of order 1 for the case of a crack located within a grain, and of order 0.1 for a crack at a *grain boundary.* The effective increase in the loading and the consequent reduction of the critical size of the crack is caused by the *relaxation* of stresses from dislocations in the region with a size of order l in the vicinity of the crack cavity. For the continuum limit $d \ll l$ the above mentioned relaxation is described by introducing an effective specific surface energy $\gamma_e = \gamma_0 - B\mu b^2 nd/(24\pi)$

(γ_0 is the true specific *surface energy*, B is a coefficient of order 1). Relaxation of energies of randomly distributed dislocations located in the region of dislocation cell boundaries, can completely compensate for the increase of surface energy when the crack is being opened, which explains the *stratification-type failure*. The cracks can occur in equilibrium with different dislocation structures (e.g., with *pile-up of dislocations*). Changing the external conditions can shift this equilibrium either toward the healing (see *Healing of defects*) or toward the growth of fractures (see *Fatigue crack growth*). At high temperatures the elastic diffusion microcrack-dislocation interaction is significant (see *Diffusion*), and it can result in the healing of cracks with the formation of a limited *dislocation wall*.

MICROCRYSTALLITE PARAMETERS

The physical quantities: *surface energy*, molar surface, radius, content per mole, spacing between *microcrystallites* in amorphous bodies.

The surface energy of microcrystallites is not equal to the surface energy of a *crystal* in vacuo since microcrystallites are in contact with one another, and the external bonds of their surface atoms are partially saturated by their coupling with the surfaces of neighboring microcrystallites. The molar surface energy is given by the equation $\tilde{\sigma}_{ms} = \tilde{\sigma}_1 - \tilde{\sigma}_2$, where $\tilde{\sigma}_1 = [(k_V - k_S)/k_V]U$ is the total molar surface energy of microcrystallites (k_V and k_S are coordination numbers of atoms in the volume and at the lattice surface, U is the molar lattice energy). The quantity given by the equation $\tilde{\sigma}_2 = f\widetilde{N}_A^{1/3}\tilde{v}^{2/3}\sigma$ (coefficient f depends on the surface layer structure, $\widetilde{N}_A$ is the *Avogadro number*, $\tilde{v}$ is the molar volume, σ is the surface energy) is the part of the energy required by a certain region of a microcrystallite surface. The surface energy of microcrystallites in *metals* exhibiting body-centered cubic and face-centered cubic lattices is close to the value of the surface energy of the corresponding melts.

The *molar surface* (the total surface occupied by one mole of microcrystallites) is given by the equation $\widetilde{S} = \tilde{q}/\sigma_{ms}$, where $\tilde{q}$ is the heat of melting, σ_{ms} is the surface energy density of microcrystallites. The microcrystallite mean radius R and content per mole $\tilde{n}$ are determined from the following expressions: $R = 3\tilde{v}/\widetilde{S}$ and $\tilde{n} = \widetilde{S}/(4\pi R^2)$. In the case of metals with a body-centered cubic or face-centered cubic lattice, the area of a molar surface $\widetilde{S}$ amounts to thousands of square meters per mole, and the microcrystallite radius ranges from 10 nm (*alkali metals*) to 1.5 nm (metals with high *melting temperatures* $T_{melting}$); the product $RT_{melting}$ is thus roughly constant and amounts to $\sim 3\cdot 10^{-4}$ cm·K.

The mean intercrystalline distance is $a + \Delta a$, where a is the mean interatomic distance of the lattice, $\Delta a = \Delta\tilde{v}_{melting}/(1/2)\widetilde{S}$ ($\Delta\tilde{v}_{melting}$ is the volume increment due to melting). The ratio $\Delta a/a$ is nearly the same for different metals and amounts to ≈ 0.28. The distance between microcrystallites increases with increasing temperature: $\Delta a/a = B\exp(-T_{melting}/T)$, where B is a dimensionless numerical factor. The average value of B for monatomic materials (including alkali metals, noble gases in solid state, platinum, gold) and water is 0.78. Molecular solids may exhibit deviations of B from 0.78 toward both higher (1.3 for O_2 and PH_3) and lower (0.58 for CS_2) values.

MICROCRYSTALLITES

The smallest individual blocks or constituents of a polycrystal, with sizes of the order of several interatomic distances, which retain the structure of a *crystal*. In accordance with one of the *microcrystallite models* of the structure of the *amorphous state* of a solid, glasses (see *Vitreous state of matter*) are an aggregation of microcrystallites with random relative orientations. A modification of the microcrystallite model, known as the *paracrystallite model*, assumes that there are no distinct boundaries between individual microcrystallites; short-range distortions (see *Long-range and short-range order*) are not localized in the neighborhood of boundaries, but rather are smoothly spread over the entire bulk of microcrystallites (see Fig.). In the limiting case this model is equivalent to the *random continuous network* model of bonds, or *random close-packing model* of spheres with small *fluctuations* of short-range order parameters. The existence of microcrystallites in the majority of amorphous solids has not been confirmed by direct (diffraction) methods, since diffraction lines broaden as crystals become smaller in size. Broadening of lines may possibly result, not from

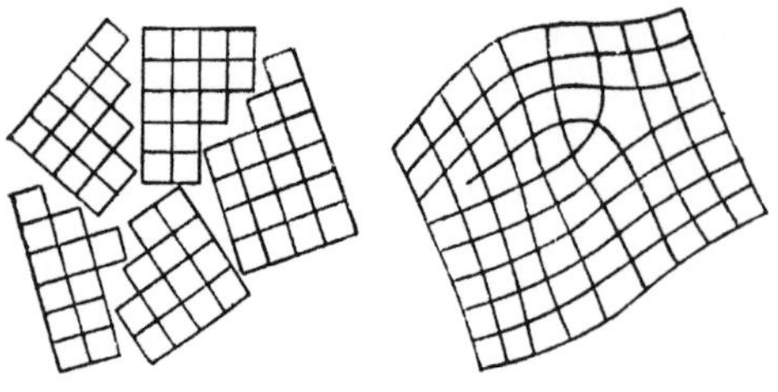

Microcrystallite (a) and paracrystallite (b) models of the structure of amorphous solids.

the lack of long-range order, but from the large growth of amplitudes of thermal *vibrations of surface atoms* on the microcrystallite boundaries. Evidence for a microcrystallite structure of *amorphous films* of Si, Ge, WO_3 has been recently obtained through direct observations with the help of a high-resolution electron microscope; the dimensions of individual microcrystallites are 1–3 nm. According to some *liquid state models* microcrystallites may also exist in liquids.

The mechanism for the decomposition of a *monocrystal* into microcrystallites at the *melting temperature* may be associated with the *anomalous thermal expansion* at the surface (and at the sites of *defects*) of a crystal. As a result of irregular growth of interatomic distances on cooling, these distances at sites of defects increase faster and reach critical values at the melting point when the *shear modulus* becomes closer to zero. Under these conditions, the growth of microcracks (see *Cracks*) takes place, the crystal decomposes into individual microcrystallites and becomes capable of flowing, as in the liquid state. Dimensions of microcrystallite cannot be smaller than certain limits in order to be compatible with the ordered structure of a *crystal lattice*, since the effect of anomalous thermal expansion of their boundaries would then die out. See also *Cybotaxes*.

MICROELECTRONICS, integrated circuit electronics

Branch of electronics which deals with the creation and application of extremely small-sized electronic devices with a high degree of integration. Such devices include *integrated circuits* (IC, or *chips*), functional microcircuits (see *Functional electronics*), as well as microcomponents of integrated devices. To produce an IC involves various actions taken on microscopic areas of a semiconductor: irradiation either with a narrow electron, X-ray, or laser beam, or through masks (protective *coatings* with specially prepared small windows and proper topology, see *Lithography*), selective *etching*, local *diffusion* of impurities and ionic *doping*, etc. As a result, various microareas of a crystal attain the properties of elements of an electronic circuit. Creation of a film IC involves the layer-by-layer deposition of *thin films* of various materials and with various configurations on a solid substrate, and the simultaneous generation of IC elements. Use is made of epitaxial build-up (see *Epitaxy*), film deposition, chemical or electrochemical film application, plasmochemical treatment, oxidation, metallization, methods of local surface treatment, etc. Combined application of film technology and micro-sized discrete devices produces a *hybrid IC*. An IC can include active elements to convert signals (*diodes*, *transistors*, *thyristors*, etc.) and passive elements to provide electric connections across the system (resistors, capacitors, inductors). Microelectronics involves a unified technological cycle of IC production, i.e. concurrent production of the entire set of circuit elements. Using a parallel method of production, a large number of devices are simultaneously formed on the surface of a plate that is several centimeters in diameter, and then the plate is sliced into individual crystals, each of them comprising a device (e.g., a transistor, or an entire IC). Development of microelectronics is accompanied by growth of the degree of integration and updating of the IC production technology (microtechnology). A chip can hold millions of individual elements with submicron sizes.

Integrated circuits may perform a wide variety of functions: applications to microcomputers, automatic control, robotics, data transmission and processing, etc. Functional operation principles are the same as those of the corresponding electronic circuits. Physical principles of the control of electron flows and electromagnetic waves are based on electronic processes operative in semiconductors used in transistors and semiconductor diodes.

Functional circuits have no direct analogs of electronic circuit elements. Functions of signal transformation in such a device are distributed over the bulk of a solid, and it is impossible to put individual functions in correspondence with certain parts of the crystal. The device matrix may include materials that exhibit semiconducting, ferroelectric or superconducting properties. Transfer and transmission functions in these circuits are realized not only by electric processes, but also by optical, acoustical, and magnetic processes that exert some influence on each other. *Optoelectronics* provides a connection between optic and electric signals. The optical part of conversions is based on processes of *real-time holography*, and on nonlinear optical effects (see *Nonlinear optics*). The relation between optical and electric processes is based on the phenomena of photoconductivity, electroluminescence, radiative recombination, etc. (see *Optoisolator*). *Acousto-electronics* uses the mutual influence between mechanical deformations of a crystal and its electric properties, e.g., in piezoelectric crystals. This allows transforming electrical signals into acoustic ones and vice versa. In opto- and acousto-electric transducers, certain circuit elements may be electrically insulated from one another. One of the rapidly developing trends of functional microelectronics is magneto-electronics, which uses magnetic materials (thin *magnetic films*, *magnetic semiconductors*, and possibly materials with *magnetic bubble domains*), as memory elements, transfer elements and other active circuit elements. Of possible future importance is the use in microelectronics of superconducting (see *Superconductivity*) and other low-temperature features of materials; this trend of functional microelectronics is sometimes called *cryoelectronics*. A switching element based on the property that an abrupt change takes place in the *electrical conductivity* of a superconductor under the action of a magnetic field exceeding the critical value B_{c2} is called a *cryotron*. Functional microelectronics is at a stage of rapid development.

MICROFIELD TOMOGRAPHY

See *Tomography of microfields*.

MICROHARDNESS

Hardness of microscopically small amounts of material. Typically, it is determined by the *static indentation method* or the *scratch technique*. As a rule, tests for macrohardness are carried out with specimens at least 5 μm in size. Modern *ultramicrohardness gauges* allow measuring the depths of *indenter* penetration as small as 20 nm. The method of microhardness is widely used for studies of mechanical properties of brittle solid materials, the anisotropy (see *Anisotropy of mechanical properties*) of these properties, for studies of *plasticity*, *Young's modulus*, and brittleness of *coatings*, epitaxial structures, *thread-like crystals*, various small-sized objects; it also provides estimates of the mobility of dislocation ensembles and starting stresses of *dislocations*. Measuring the length of a *crack* of the microhardness indentation can reveal the *fracture toughness* of the material. Recently, a method was developed for determining the brittleness of materials by measuring *acoustic emission* that arises in deformation by an indenter. Generation of an impression in microindentation on many materials is caused by dislocational plasticity (see *Dislocation model of plastic deformation*). However, of considerable importance for brittle solid crystals and covalent semiconductors can be the plasticity caused by the mobility of *interstitial atoms*. In semiconducting materials, high hydrostatic pressure (approximately equal to *hardness*) is generated under the indenter, which allows investigating *phase transitions* under pressure, in particular, transitions to a metallic *phase*.

MICROMAGNETISM

Branch of the phenomenological theory of *magnetism*, based on detailed calculations of the distribution of the *magnetization* of magnetically ordered bodies at the micrometer length scale with the help of the *Landau–Lifshits equations*, together with the Maxwell equations. In a narrow sense, the term is mainly limited to the investigation of the magnetization of ferromagnetic bodies of arbitrary shape and, with special reference, to cases involving *magnetic domain structure*. The term "micromagnetism" was introduced by W.F. Brown.

MICRON (μ)

Dated term, with symbol μ for a fractional length unit, equal to 10^{-6} m; the currently accepted name is micrometer (μm).

MICROPROBE (fr. Gr. $\mu\iota\kappa\rho o\varsigma$, small)

A narrowly-focused electron beam incident on a small area of a sample for the spectroscopic analysis of the emitted X-rays. The electron beam generated by an electronic optical device may be either stationary, or scanned in a raster pattern over the surface of the part of the solid under study. This principle is used in an *X-ray microanalyzer* (sometimes called *electron microprobe*), in a *scanning electron microscope* (raster microscope), an Auger electron microanalyzer and similar analytical electronic optical devices. The most important parameters of a microprobe are its minimal diameter d_{min} that controls the localization of the X-ray microanalysis, the resolution and depth of focus (varies from 0.1–1 μm for the former, and from 5–100 nm for the latter), and the maximal microprobe current i_{max} that controls analytic parameters of the X-ray microscopic analysis, and optic parameters of the scanning electron microscope.

MICROPROBE X-RAY DEFECT STRUCTURE ANALYSIS, X-ray diffraction microscopy

Use of *X-ray diffraction* methods for investigating *defects* in crystals. Microprobe X-ray structure analysis, which originated in works of W.S. Berg (1931) and C.S. Barret (1945), is based on determining the of extent of crystal perfection through the analysis of the intensity distribution of Bragg reflections.

There are many approaches to this defect structure analysis, which differ in observational geometry and use either monochromatic or polychromatic radiation to obtain the crystal image in the diffracted beam. The first group includes the Berg–Barret, Weissmann, Lambo methods, and others. Polychromatic radiation is used in the Schultz, Fujivara, Guinier–Tenevin methods, and others. With the help of these methods, sizes and angles of disorientation of grains (subgrains) are determined. Individual defects (*dislocations*, *stacking faults*, *segregations* of impurity atoms) are observed using topographic microprobe methods which are based on the phenomenon of extinction contrast (*Lang method*) and *anomalous passage of X-rays* (*Borrman method*). The Lang method is used for investigating dislocation structure of thin crystals ($\mu d < 1$, where μ is the linear absorption coefficient, d is the crystal thickness); the Borrman method is used for crystals with $\mu \sim 10$. Topographic methods of investigation of defects are applicable to comparatively perfect crystals (with density of dislocations $\leqslant 10^5$ cm^{-2}). See also *X-ray structure analysis*, *X-ray topography*.

MICROPROBE X-RAY SPECTRUM ANALYSIS

Nondestructive method for studying the elemental composition and distribution in microscopic volumes of solids. It is based on exciting and measuring the characteristic X-ray radiation of analyzed elements in the sample by an electron beam of energy 3–40 kV, which is focused to a thin probe (50 nm–1 μm in diameter). Resolution of the method for qualitative analysis is limited by scattering of probe electrons in the target material; hence when investigating bulk samples, the method provides quantitative data on the composition for microvolumes at least 1 μm in size. In studies of *thin films*, the resolution is determined by the diameter of the *electron probe*. Accuracy of analysis reaches 2% of actual element content in the volume under study, and sensitivity varies from 0.01 to 0.5% depending on the atomic number of the element.

Information on element distribution may be gained from recording the distribution of radiation intensity for linear motion of the object under the electron probe, or from raster-type images of the X-irradiated object via absorbed and back-scattered electrons. Raster-type images in absorbed, back-scattered and, especially, in secondary electrons also yield information on surface topography, and may be used for the investigation of surfaces with clearly defined relief. Here, the depth of focus is higher by 2–3 orders than that of optic microscopes. It should be noted that exact quantitative data on composition of a solid are not obtainable from investigation of solids with clearly defined relief. See also *X-ray spectrum analysis*.

MICRORADIOGRAPHY

A collection of methods for X-ray studies of the microstructure of *solids* (size and orientation of subgrains, impurity *segregations*, fine-dispersion precipitations, microporosity, distribution of *dislocations*, *stacking faults*, etc.). Microradiography includes *X-ray projection (dark field) microscopy* used when structural components of the material under study have different coefficients of X-ray absorption so that an X-ray beam, passing through the sample, forms a nonuniformly intense image of the object on the photographic film; and also *X-ray diffraction microscopy* based on obtaining an image of the *crystal* via diffracted X-rays. As a rule, all microradiography methods produce an image of natural size, or at a slightly enlarged scale. The necessity of subsequent optical magnification requires high-resolution photographic materials ($\sim$400 lines per mm).

The advantage of microradiography over other methods of observation of *defects* is its nondestructive nature; it requires no special *etching* of samples, and allows detection of the structure and concentration of inhomogeneities in the bulk of the sample (see also *X-radiography*, *Microprobe X-ray defect structure analysis*, *X-ray microscopy*).

MICROSCOPIC DESCRIPTION in solid state physics

Provides an insight into the atomic structure of solids, and of the motion of the atomic and subatomic particles in it. The complexity of this microscopic description stems from the necessity to coordinate the interaction of microscopic particles in the concentration range 10^{22}–10^{24} per cm^3. A major task of a microscopic description is to reveal the nature (and, if possible, to calculate the values) of quantities responsible for a *macroscopic description* in solid state physics (conductivity tensors, generalized *susceptibilities*, etc.). The points of departure for this description are the structure and energy spectrum of the solid, and the *quasi-particles* that implement this spectrum. It is important to find out how the particles that make up the body or the quasi-particles that describe elementary motions in the body contribute to the various properties and phenomena. As a rule, the microscopic description of a property (phenomenon) is achieved when macroscopic coefficients (e.g., conductivity) are expressed in microscopic terms. The determination of the temperature dependence of a physical quantity always calls for a microscopic approach since $k_B T$ is the measure of the thermal energy and motion of a single particle.

While formulating a microscopic description of properties of solids, it is highly important to state the "level of penetration" into the structure of the solid. Thus, when describing properties of *ionic crystals*, it suffices to treat ions with filled electron shells as structural units (e.g., the structural units for an NaCl crystal are Na^+ and Cl^- ions). Similarly, the structural units of *metals* are ions and conduction electrons, although every ion (in an ionic crystal or metal) consists, naturally, of a nucleus and many electrons, and a nucleus consists of nucleons. This restriction is substantiated by the fact that many properties of solids are not affected by the structure of individual ions or (nuclei). Moreover one can easily identify the properties that require a consideration of the structure of ions and nuclei for their description (e.g., X-ray absorption or *Mössbauer effect*).

A microscopic description often involves an intermediate (i.e. semiphenomenological) approach. An example is the *Drude–Lorentz–Sommerfeld* model of a metal, which postulates the existence of a degenerate (see *Degeneracy*) gas of *conduction electrons*. Macroscopic characteristics (*electrical conductivity*, *magnetic susceptibility*, etc.) are expressed via the electron density, charge, *mean free path*, *magnetic moment*. A self-consistent microscopic description should establish why there are free electrons in a metal, and evaluate their mean free path. As a rule, microscopic theory puts certain limits on various macroscopic descriptions. Thus, a linear relationship between $\boldsymbol{j}$ and $\boldsymbol{E}$ in electrical conductors (*Ohm's law*) is valid, if the energy that an electron acquires between collisions is small compared to its mean thermal energy. A microscopic description in solid state physics uses statistical thermodynamics and physical kinetics with their well developed theoretical physical tools (method of *Green's functions*, *kinetic equations*, etc.), and makes extensive use of numerical methods for comparison of various models with experiment (see *Pseudopotential method*, *Computer simulation*).

MICROSCOPIC THEORY OF FERMI LIQUID

Non-interacting particles obeying *Fermi–Dirac statistics* form a Fermi gas, and when these particles interact with each other in such a way that their properties remain close to those of a Fermi gas they constitute what is called a *Fermi liquid*.

The microscopic theory of a Fermi liquid (a liquid whose particles obey Fermi–Dirac statistics) provides an explanation for the appearance of zero-sound vibrations in a neutral Fermi liquid, and it also establishes a connection between the Landau correlation function and the forward scattering amplitude of quasi-particles at the angle zero. These considerations are based on the investigation of the properties of the vertex function

$$\Gamma_{\alpha\beta,\gamma\delta}(p_1,p_2;p_3,p_4) \equiv \Gamma_{\alpha\beta,\gamma\delta}(p_1,p_2;k), \tag{1}$$

where $k = p_1 - p_2 = p_4 - p_3$, $k \equiv (\boldsymbol{k},\omega)$ for neighboring values of the pairs of variables ($k \to 0$ for forward scattering). $\Gamma_{\alpha\beta,\gamma\delta}$ is obtainable by the summation of diagrams of the "ladder" approximation which have a singularity at $k \to 0$ that, in its turn, leads to the appearance of a singularity in $\Gamma_{\alpha\beta,\gamma\delta}$: namely that the result of taking the limit depends on the relation between ω and $|\boldsymbol{k}|$. By this the function

$$\Gamma^{\omega}_{\alpha\beta,\gamma\delta}(p_1,p_2) = \lim_{k\to 0,\ |\boldsymbol{k}|/\omega\to 0} \Gamma_{\alpha\beta,\gamma\delta}(p_1,p_2;k) \tag{2}$$

corresponds to the nonphysical limiting case of the scattering of quasi-particles on the Fermi surface with a small transfer of energy and a transfer of momentum strictly equal to zero, and the function

$$\Gamma^{k}_{\alpha\beta,\gamma\delta}(p_1,p_2) = \lim_{k\to 0,\ \omega/|\boldsymbol{k}|\to 0} \Gamma_{\alpha\beta,\gamma\delta}(p_1,p_2;k) \tag{3}$$

corresponds to real physical processes of the scattering of quasi-particles on the Fermi surface with a change of momentum, but no change of energy. The functions Γ^{ω} and Γ^{k} are connected by an integral relation. The poles of the function $\Gamma_{\alpha\beta,\gamma\delta}(p_1,p_2;k)$ characterize acoustic Bose excitations of a Fermi liquid. The equation describing these excitations coincides with the equation

$$(\boldsymbol{kv}-\omega)\,\delta\hat{n} - \boldsymbol{kv}\frac{\partial n_0}{\partial\varepsilon}\int \eta(\boldsymbol{p},\boldsymbol{p}')\,\delta n'\,\delta\tau' = 0 \tag{4}$$

for zero-sound vibrations, and the role of the kernel in the equation is played by the function $z_{\boldsymbol{p}} z_{\boldsymbol{p}'} \Gamma^{\infty}_{\alpha\beta,\gamma\delta}(\boldsymbol{p},\boldsymbol{p}')$ (where $z_{\boldsymbol{p}}$ is the renormalization constant of the quasi-particle *Green's function*), which allows one to identify it with the Landau correlation function

$$f_{\alpha\beta,\gamma\delta}(\boldsymbol{p},\boldsymbol{p}') = z_{\boldsymbol{p}} z_{\boldsymbol{p}'} \Gamma^{\infty}_{\alpha\beta,\gamma\delta}(\boldsymbol{p},\boldsymbol{p}'). \tag{5}$$

As in the case of Γ^{ω} and Γ^{k}, the function f^* is connected by an integral relation to the physical forward scattering amplitude:

$$g_{\alpha\beta,\gamma\delta}(\boldsymbol{p},\boldsymbol{p}') = z_{\boldsymbol{p}} z_{\boldsymbol{p}'} \Gamma^{k}_{\alpha\beta,\gamma\delta}(\boldsymbol{p},\boldsymbol{p}'). \tag{6}$$

The connection of the Landau correlation function with the vertex function allows one to obtain a sum rule for the Landau parameters A_l and B_l of a charged Fermi liquid, taking into account the long-range Coulomb interaction:

$$\sum_l (2l+1)\left(\frac{A_l}{1+A_l} + \frac{B_l}{1+B_l}\right) = -\frac{1}{1+A_0}. \tag{7}$$

In the case of a neutral Fermi liquid the sum rule is obtained by equating the left side of Eq. (7) to zero.

The isotropic model of a Fermi liquid is strictly applicable only for the alkali metals Na and K. The application of the Fermi liquid theory for describing the properties of other metals is complicated by the sharp anisotropy of the *Fermi surface*. Experimental values of the parameters $A_{2,3}$, $B_{0,1,2}$ for Na and K are small ($A_2 \approx (3\text{–}5)\cdot 10^{-2}$, $A_3 \approx 0$; $B_0 \approx -0.2$ to 0.3; $B_1 \approx 5\cdot 10^{-2}$, $B_2 \approx 0$). The only appreciably large Landau parameter is $A_0 \sim 1$. However, due to the complications involved in separating the electronic terms, it turns out that A_0, A_1 cannot be determined experimentally. Experimental Fermi liquid effects have been observed in ^{3}He, where they start to appear at $T < 0.1$ K. The values of the Landau parameters $A_{0,1}$, B_0 depend on the pressure P, thus: $A_1 = 2.1$, $A_0 = 10.8$, $B_0 = -0.67$ at $P = 3\cdot 10^4$ Pa; and $A_1 = 4.8$, $A_0 = 75.6$, $B_0 = -0.72$ at $P = 2.8\cdot 10^6$ Pa.

MICROSCOPY, ACOUSTIC

See *Acoustic microscopy*, *Thermoacoustic microscopy*.

MICROSCOPY, THERMAL

See *Thermal microscopy*, *Thermoacoustic microscopy*.

MICROSHOCK LOADING

Spatially nonuniform action of forces on a *solid surface*, which changes at a rate above $10^6\ \mathrm{s}^{-1}$, and is localized within several grains or structural components of the material. The stresses produced are comparable to or exceed the *yield limit* (or *ultimate strength*) of the material, which were obtained in static testing (see *Mechanical testing of materials*). Microshock loading occurs also under cavitation, at collisions between the solid surface and fine (1–100 μm) solid particles, drops of liquid (~1 μm diameter), with velocities ranging from tens to 1000 m/s (see *Cavitation resistance*). Microshock loading can produce *surface erosion* of material, which depends on the reaction of microstructures on the arising field of dynamic stresses.

MICROSPALLING

An ideally brittle spalling (break-off of fragments) of the crystal lattice of a body-centered cubic *metal*, which results from *nucleating crack* growth at the initial stage of its avalanche-like propagation. The specific crack propagation work at microspalling equals the ideal *surface energy* γ of a solid, since the lattice in submicrovolumes of the material surrounding the crack may be considered as ideal so plastic *relaxation* is prevented. Microspalling arises as an initial stage of macrofailure of body-centered-cubic *iron* and *steel* at the critical stress R_{MS} (*microspalling resistance*), when the *microspalling criterion* is met.

MICROSPALLING CRITERION

Strength loss through macrofailure of *iron* and *steels* with a body-centered cubic lattice, when such macrofailure begins at the submicrolevel via *microspalling*. Spalling itself is breakup into chips or bits. A necessary and sufficient condition for microspalling is the initial state of *failure* when the strain reaches its critical value R_{ms} (*microspalling resistance*) under the conditions of active *plastic deformation*: $s_{\mathrm{I}} = R_{\mathrm{ms}}$; $s_{\mathrm{i}} = s_{\mathrm{f}}$ (s_{I} is the maximum principal stretching strain, s_{i} is the intensity of strains at the microspalling site, s_{f} is the *yield limit* of the material). The first equality describes the Griffith opening of the *nucleating crack* (see *Griffith theory*), the second is the condition of fluidity in a complex strained state. The microspalling criterion may be represented in the form of dimensionless parameters: $j = K_{\mathrm{v}}$ for $s_{\mathrm{i}} = s_{\mathrm{f}}$ (here $j = s_{\mathrm{I}}/s_{\mathrm{f}}$ is the rigidity of the strained state, $K_{\mathrm{v}} = R_{\mathrm{ms}}/s_{\mathrm{f}}$ is the *viscosity* coefficient for steel). The dimensionless form of the microspalling criterion is convenient for describing the conditions of failure in a general case of complex strain, as given by the *strain tensor*. The microspalling criterion may be used for engineering computations of the extremal states of construction elements when evaluating the possibility of brittle failure of a unit, provided the phase diagram of its principal strains is known for those local volumes in which the *criterion of fluidity* of the substance $s_{\mathrm{i}} = s_{\mathrm{f}}$ is satisfied.

MICROSPALLING RESISTANCE

The critical (Griffith, see *Griffith theory*) stress, which results in the failure of a solid through an avalanche-type spread of *nucleating cracks*, as occurs in *slip bands* at *grain boundaries*, or at sections (*spallings*) of inclusions of microparticles of the second phase in body-centered cubic *iron* and *steel*. The magnitude of microspalling resistance R_{ms} is experimentally determined from low-temperature tension tests in the form of failure stress S_{f} during a drop of *plasticity*: $S_{\mathrm{f}} \approx \sigma_{\mathrm{u}}$, $\Psi \approx$ 2–3%, where σ_{u} is the ultimate resistance, Ψ is the necking (see *Neck*). The microspalling resistance may be theoretically estimated from general constants of the steel microstructure (in millimeters): the size of a *crystal grain* d_{gr} ($R_{\mathrm{ms}}^{\mathrm{gr}} \approx 18 d_{\mathrm{gr}}^{-1/2} \cdot 10^3$ Pa), the thickness of *cementite* laminas in *pearlite* d_{c} ($R_{\mathrm{ms}}^{\mathrm{c}} \approx 0.8 d_{\mathrm{c}}^{-1/2}$), the diameter of the cementite globule d_{cg} ($R_{\mathrm{ms}}^{\mathrm{cg}} = 2.5 d_{\mathrm{cg}}^{-1/2}$). The R_{ms} does not depend on the alloy intragranular submicrostructure, temperature, deformation rate, or type of *state of stress*, but rather it involves properties of a fundamental mechanical type (material constants), and may be used with calculations of the *strength* of structures undergoing composite stress. The magnitude of microspalling resistance changes substantially when a material undergoes *plastic deformation*.

MICROSTRAIN

Strain caused by microstresses in a solid, or a small strain (relative deformation $\leqslant 10^{-4}$) induced in a solid when examining its microplasticity (see *Plasticity*) by the application of small external loads which induce no macrofluidity of the sample. Microstrain is detected with the help of resistance transducers, inductive and capacitive pickups, and electromechanical and optic-mechanical systems.

MICROSTRESSES, microscopic stresses

Internal stresses that prevail in solids in the absence of external forces, and are balanced in small volumes (see *Stress scales of defect structures*). The causes of microstresses are crystal *defects*: *point defects*, linear, planar, volume defects (and ensembles of defects). In polycrystalline bodies (see *Polycrystal*), the term "microstresses" means stresses balanced in volumes of the order of crystal grains or mosaic blocks (see *Mosaic crystals*). Microstresses can have a pronounced effect on the physical properties of solids, and their behavior during *failure*.

MICROTOMOGRAPHY

See *Computational microtomography*.

MICROWAVE DEVICES

See *Microwave semiconductor devices*, *Photocontrolled microwave semiconductor devices*, *Superconducting microwave electronic devices*.

MICROWAVE FREQUENCY BANDS

The frequency intervals of the ultra-high frequency region of the electromagnetic spectrum are subdivided in accordance with radar standards established for waveguides. In solid-state physics the most widely used are the three-centimeter X-band, the one and a half-centimeter K-band and the eight-millimeter Q-band. In particular, the majority of electron paramagnetic resonance spectrometers are designed for operation in these bands. The frequencies within the range 8.2 to 12.4 GHz (wavelengths 2.4 to 3.7 cm) correspond to X-band propagating in rectangular waveguide with cross-section 22.9×10.16 mm^2; K-band from 18.0 to 26.5 GHz (wavelengths 1.1 to 1.7 cm) uses 10.63×4.32 mm^2 waveguide, and Q-band from 26.5 to 40.0 GHz (wavelengths 7.5 to 11.3 mm) employs waveguide with 7.11×3.56 mm^2 cross-section.

MICROWAVE SEMICONDUCTOR DEVICES

Instruments in which the interaction of *current carriers* with variable electromagnetic fields occurs at frequencies between 0.1 and 100 GHz (microwave range). Microwave semiconducting devices are used to generate, amplify, transmit, and transform signals. To generate and amplify microwave oscillations *generation type semiconductor diodes* and *transistors* are used. To transform and transmit signals use is made of generator diodes, transistors, *p–i–n-diodes*, *Schottky diodes*, varactors, etc. These various devices typically have small linear dimensions of active areas (spatial ranges in which electron fluxes interact with alternating electromagnetic fields). The carrier transit time through that range is comparable to the period of the generated oscillations. Further miniaturization for higher frequencies results in lower output power, and this becomes the principal limiting factor for their energy characteristics. Another factor limiting their energy capabilities is the thermal regime in which they operate. Moreover, durability of such devices strongly depends on the temperature of their active layers.

The most powerful source of continuous microwave oscillations at frequencies up to 10 GHz are *bipolar transistors* (output power reaching tens of watts). In the range from 10 to 33 GHz there are *impact ionization avalanche transit time diodes* (IMPATT diodes), built from *gallium arsenide* (output power up to several watts). As for the millimeter range, there are double IMPATT diodes built from indium phosphide (output power from tens to hundreds of milliwatts). An output power up to hundreds of watts may be obtained in the centimeter range from *trapped plasma avalanche triggered transit diodes* (TPATT diodes) operating in a pulse mode. IMPATT diodes are the most powerful emission sources in the millimeter band. For example, one diode operating in a pulse mode in the atmospheric transparency band, 94–96 GHz, puts out up to 30 W of power at an efficiency of about 6%. The drawback of IMPATT and TPATT diode generators is their high noise level. This is explained by processes related to the formation of the electron–hole plasma. Such a drawback is absent in generators built on *Gunn diodes* that typically have lower output power but high-quality spectral characteristics of their output signal. However, they have a relatively low frequency threshold of negative conductivity (up to 80–90 GHz in GaAs and 140–160 GHz in InP). To raise this upper frequency threshold, since the frequency is being limited by the inertia of electron transfer, one may operate Gunn diodes in a biharmonic mode, so the microwave power is derived from the second or even from the third harmonic of the fundamental frequency. In the mid-1980s GaAs Gunn diodes were putting out up to 80 mW at 94–96 GHz (power obtained from second harmonic), while the third harmonic yielded 3–5 mW at 140–150 GHz. Indium phosphide Gunn diodes yield up to 150 mW at 94–96 GHz (fundamental frequency) and up to 10 mW at 180–200 GHz (second harmonic).

Field-effect transistors made of GaAs are mainly used for low-noise amplifiers in the millimeter range. For example, a noise level of 4.5 dB is achieved at the frequency of about 10 GHz using these devices. The most powerful amplifiers are built on silicon bipolar transistors. Modern tendencies in building transistor amplifiers consist in merging several amplifying cascades into a monolith integrated circuit.

The principal materials used to produce microwave semiconducting devices are silicon, GaAs, and InP. Devices operating on the basis of GaAs, InP and solid solutions of these materials feature a high efficiency and low noise level. Electron inertia, however, affects the functioning of such devices in the millimeter range. The time required for electrons to reach their required distribution in microwave semiconducting devices is comparable to the microwave oscillation period, and this limits the potentially achievable frequencies and output power of the devices. Generally, microwave semiconducting devices are widely used to build various microwave instruments.

In parallel with perfecting semiconducting devices of the lumped type, solid state electronics is developing toward microwave devices functioning in the distributed operational mode. Dangling ultrasonic waves that interact with the electron flux resulted in the design and production of a new class of microwave semiconducting devices (see *Acousto-electronics*). See also *Photocontrolled microwave semiconductor devices.*

MIGRATION

Outward movement of objects away from locations in space; the physics of the *condensed state of matter* usually deals with migrations of small particles of microscopic (including atomic) sizes, and also of *quasi-particles*. Migration of *interstitial atoms*, *vacancies*, of autolocalized electron states (see *Self-localization*) of small radius, etc. represents a sequence of randomly oriented jumps with length of the order of interatomic distances; the *activation energy* of a jump is determined by the height of a potential barrier. Migration of conduction electrons and *holes*, *Wannier–Mott excitons*, etc. may be described as a motion of a particle with a *quasi-momentum* that changes when the particle interacts with *phonons*, *defects in crystals*, *current carriers*, etc. At low temperatures, the *mean free path* of a particle between two successive collisions generally considerably exceeds an interatomic distance. The mechanism of migration of particles with small masses may also involve tunneling of particles through a potential barrier (see *Tunneling phenomena in solids*). Mobile particles in a material experience random collisions with atoms and electrons of the medium as well as with one another, and, with no preferred direction present, execute *random walks* (*stochastic motion*). One should distinguish macro- and microscopic aspects of migration. If no reactions involving mobile identical particles take place, the initial macroscopic uniform spatial distribution of the latter is retained. But the microscopic distribution changes continually; and if the medium has some centers capable of undergoing reactions with particles (e.g., trapping at *traps for mobile particles*, *recombination*), the migration leads to the actualization of such reactions as the particles approach the centers. If the macroscopic distribution of centers is nonuniform, this will cause nonuniformity in the distribution of mobile particles. If the initial macroscopic spatial distribution of particles is nonuniform, then migration leads to equalization of the concentration. The change of concentration of particles $n(\boldsymbol{r}, t)$ per unit volume about point $\boldsymbol{r}$ at instant of time t that results from *diffusion* via a random walk is described by the equation $\partial n(\boldsymbol{r},t)/\partial t = \nabla^2 Dn$ (where ∇^2 is the Laplacian operator, D is the *diffusion coefficient*). Besides a diffusion component, migration includes a drift component that arises when forces act on the particle, e.g., those due to macroscopic electric, magnetic, elastic fields, or to the field of a defect of atomic scale. The force $-\nabla U(\boldsymbol{r})$ (where ∇ is the gradient operator, $U(\boldsymbol{r})$ is the particle potential energy) causes an increase of the hopping rate along the field, and its decrease in the opposite direction, when the migration occurs via a "hopping mechanism". If migration proceeds without activation, this force causes oppositely directed accelerations along and against the field. Drift results in a nonuniformity of particle distributions, or the establishment of a stationary current, depending on the boundary conditions. In the general case, $\partial n/\partial t$ is the sum of diffusion and drift terms, as well as terms that describe the generation and annihilation of particles.

Migration of a particle involves transport of mass; charge, *spin* and energy may also be simultaneously transferred in this process. Energy transport is also possible without mass transport, if the migrating particle is an excitation wave in a crystal, e.g., *exciton*, phonon, *plasmon*, etc. In particular, excitation of a certain region of a nonmetallic crystal under light absorption may cause *luminescence* from the dark region, induced by resonance transfer of an excited state between identical atoms. Migration of Wannier–Mott excitons, their subsequent trapping by *defects* of various types, and radiative recombination may cause luminescence at different frequencies. *Energy transfer* in condensed media may also occur directly by *photons*.

MILLER EFFECT (R.C. Miller, 1965)

Abrupt increase of the intensity of the second harmonic signal during the process of a nonlinear interaction between a laser beam and a ferroelectric crystal. It is observed in certain many-domain crystals with *domain* sizes comparable to the length of the coherent interaction. If domains are inclined at an angle of 90° to one another, as, e.g., in KDP crystals (potassium dihydrogen phosphate, KH_2PO_4), then the beam, in passing from one domain to another, changes its phase at the second harmonic frequency; growing in intensity by a factor of hundreds. The Miller effect finds application in the *nonlinear optics* of solids.

The term Miller effect also refers to the effective capacitance between terminals in a vacuum tube.

MILLER INDICES

See *Crystallographic indices.*

MILLER RULE (R.C. Miller, 1964)

A rule in *nonlinear optics*, which allows one to evaluate approximately the components of the nonlinear quadratic *susceptibility* tensor $d_{ijk}^{2\omega}$ that are responsible for the efficient *second harmonic generation* in anisotropic piezoelectric crystals and other media lacking a *center of symmetry*. The said components are determined from the components of the linear susceptibility χ_{ij} through a third-rank tensor $\delta_{ijk}^{2\omega}$, the *Miller tensor*,

$$d_{ijk}^{2\omega} = \chi_{ii}^{2\omega}\chi_{jj}^{\omega}\chi_{kk}^{\omega}\delta_{ijk}^{2\omega}.$$

This relationship may be extended to the case of sum-frequency generation $\omega_3 = \omega_1 + \omega_2$:

$$d_{ijk}^{\omega_3} = \chi_{il}^{\omega_3}\chi_{jm}^{\omega_1}\chi_{kn}^{\omega_2}\delta_{lmn}^{\omega_3}.$$

Since the components of the tensor δ_{ijk} differ from one material to another to a considerably lesser extent than the components of d_{ijk} (in practice one can set $\delta_{ijk} \approx$ const), an order-of-magnitude estimate of δ_{ijk} for any material may be found from the value of its *refractive index* n at a given frequency ($\chi = n^2 - 1$).

MILLI... (fr. Lat. *mille*, thousand)

Prefix for a physical unit, which gives the fractional part equal to 10^{-3} of the initial unit. Symbol: m. Example: 1 mA (milliampere) $= 10^{-3}$ A.

MINERALS

Naturally occurring chemical compounds (more rarely, native elements), which are products of various physicochemical processes taking place in the Earth's crust. A typical feature of minerals is their comparative homogeneity that results from a definite chemical composition and regular arrangement of atoms in a structure. The physical properties and characteristic geometrical shapes of minerals are related to the specifics of their composition and structure. Most geologists consider only solid compounds as minerals, but Vernadsky included some liquids and gases as minerals. There is disagreement on including certain compounds of organic origin (e.g., fossil resins), silicate glasses, and other compounds among minerals. Underlying the classification are differences in types of compounds, *crystal structures*, and features of the spatial arrangement of these structures (e.g., insular, chain or laminated structures). About 3000 types of minerals are known: silicates (25%), oxides and hydroxides (12%), sulfides and analogs (13%), phosphates, arsenates, vanadates (18%), halides, carbonates, sulfates and other oxygen-containing salts, native elements. Silicates include 6 structure types: Neso (individual SiO_4 tetrahedra), Soro (double tetrahedra Si_2O_7), Cyclo (rings $(SiO_3)_N$, $N = 2$, 3, 4, 6), Ino (single $(SiO_3)_N$ or double $(Si_4O_{11})_N$ chain), Phyllo (layered $(Si_2O_5)_N$), and Tekto (three-dimensional $(SiO_2)_N$ network).

Diagnostic attributes of minerals are: the shape of the crystal and precipitation, color, density, optical, mechanical, electrical, magnetic etc. properties. Products of laboratory or industrial processes with the composition and structure of natural minerals, are classed as artificial minerals. The latter may differ somewhat in their properties from natural minerals.

MIRROR GLIDE PLANE

See *Glide-reflection plane.*

MISHIMA ALLOYS (after T. Mishima)

High-coercivity alloys of the Fe–Ni–Al system (see *Alni*), which are used to manufacture *permanent magnets.*

MIXED CRYSTALS

Phases of variable or nonstoichiometric composition (*berthollides*) formed by two or more materials. Mixed crystals are widespread, and are found in many *mineral* and rock compositions. Mixed crystals may be separated from gas, liquid and solid phases. Mixing can vary over a wide range (from fractions to 100%). In the simplest cases it can be predicted from calculations based on *thermodynamic potentials* of mixed crystal formation. Two types of mixing, isovalent and heterovalent, are distinguished. At *isovalent mixing* one atom is replaced by another atom, or one ion by another, and true *solid solutions* are formed that are described by thermodynamic equations such as that of Gibbs–Duhen and others (e.g., K(Rb)Cl, KCl(Br), and so on). At *heterovalent mixing* it is possible to have two cases: without change of common atom numbers in formula units, but with different charge states of atoms (e.g., ZnS–GaAs, $BaSO_4$–$KMnO_4$, $2SnO_2$–$FeNbO_4$), or with a change in the number of atoms. This case leads to *vacancy* formation (in $CdCl_2$ added to NaCl there is one vacancy per Cd^{2+} ion); insertion of ions at interstices (see *Interstitial atom*) (in CdF_2–YF_3 one F^- ion is changed by one Y^{3+} ion), and filling of already existing vacancies (in $YO_{1.5}$–$PrO_{1.5+x}$ with x oxygen atoms ($x < 0.5$) filling vacancies in the oxygen sublattice of Y_2O_3). Such mixed crystals have certain features: *point defects*, irregular charge distribution, quasi-chemical complexes, ordering phenomena. Mixed crystals are increasingly used: ruby (Al_2O_3 with 0.05% Cr), $LiY(Nd)F_4$ and others are active elements in *solid-state lasers*; $YO_{1.5}$–$PrO_{1.5+x}$ is used for magnetohydrodynamic generators (due to electrical conductivity and *thermal resistance*), Na(Ti)I monocrystals are scintillators. *Ferrites*, and some *ferroelectrics*, are ordinarily mixed crystals (e.g., Fe_3O_4 is Fe_2O_3 plus FeO).

MIXED STATE of superconductors, vortex state

A condition involving a *type II superconductor* with a heterogeneous magnetic flux distribution and an average internal field $B = n\Phi_0$ when present in an applied field B_0 with a value between the *lower critical field* B_{c1} and the *upper critical field* B_{c2}. This mixed state is a result of the penetration of *Abrikosov vortices* into the superconductor. Each vortex has one *flux quantum* $\Phi_0 = h/2e$, and they are distributed with the density $n(B)$, with the value 0 below the field $B_0 = B_{c1}$ and the maximum value at the field $B_0 = B_{c2}$ ($n_{max} \propto 1/\xi^2$, where ξ is the *coherence length*). The dependence of the superconductor magnetization $M = (B - \mu_0 H)/\mu_0$ on the applied field B_0 has the form depicted in Fig. 1. The values of the critical fields B_{c1} and B_{c2} are given by the expressions: $B_{c1} = [B_c/(\kappa\sqrt{2})]\ln\kappa$, $B_{c2} = B_c\kappa\sqrt{2}$,

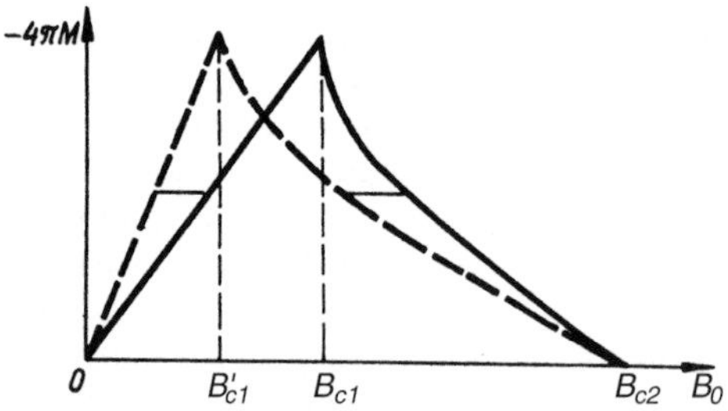

Fig. 1. Type II superconductor magnetization curve $M(B)$ dependence on the applied magnetic field B_0 in the absence of demagnetization effects (solid line), and for a demagnetization factor $n \neq 0$ (dashed line), where $B'_{c1} = (1-n)B_{c1}$.

Fig. 2. Lines of constant order parameter $|\psi|$ (dashed) and supercurrent flow (solid) in the case of a triangular vortex lattice.

where $B_c = \Phi_0/(2\sqrt{2}\pi\xi\lambda)$ is the *thermodynamic critical magnetic field*, $\kappa = \lambda/\xi$ is the *Ginzburg–Landau parameter*, and λ is the *penetration depth of magnetic field*.

At equilibrium, Abrikosov vortices form a regular triangular lattice on a plane perpendicular to the field (Fig. 2). Such a mixed condition structure is characteristic of an ideal type II superconductor. When there are defects or inhomogeneities, vortices form bundles located at defects. During current flow of density j through the superconductor the *Lorentz force* $\boldsymbol{f}_L = \boldsymbol{j} \times \boldsymbol{\Phi}_0$ ($\boldsymbol{f}_L$ is force on vortex per unit length) acts on Abrikosov vortices and makes them move in a direction perpendicular to $\boldsymbol{j}$ and $\boldsymbol{B}$. The Abrikosov vortex motion induces an electric field $\boldsymbol{E}$ in the superconductor, and as a result the superconductor enters a resistive state (see *Dynamic mixed state*) that is characterized by a nonlinear dynamic electrical resistivity $\rho(j) = \partial E/\partial j$ depending on the current density, temperature, magnetic field strength, and frequency. In real superconductors when transport current values are high enough, $j > j_p$ (j_p is the current density for the onset of vortex motion), then the Lorentz force overcomes the *vortex pinning* force that holds vortices at defects and inhomogeneities of the superconductor, and the Abrikosov vortex lattice is set into a coherent motion characterized by a linear dynamic resistance of viscous flow $\rho(j)|_{j>j_p} = \rho_f - \rho_n B/B_{c2}(T)$ (ρ_n is the metal resistivity in the normal state). For a current $j < j_p$ Abrikosov vortex motion may occur due to thermally activated processes (*magnetic flux creep*), and the dynamic resistance depends on the *phase state of the vortex ensemble* (position on the B versus T phase diagram of the mixed state), and also on the anisotropy and the type of superconductor defects. For example, the condition of a so-called *vortex liquid* state when current density values are small is characterized by the presence of a linear (ohmic) *thermally assisted flux flow* (TAFF) resistance $\rho(j) = \rho_{TAFF}(j) \propto \rho_f \exp[-U/(k_B T)]$, where $U(T,B)$ is the vortex activation energy, and the "*vortex glass*" or "*vortex crystal*" state exhibits a nonlinear dependence $\rho(j)$ of the type $\rho(j)_{j<j_p} \propto \exp[-(j_0/j)^{\mu} U/(k_B T)]$, where μ is a characteristic *critical index*. The role of thermally activated processes in the motion of Abrikosov vortices is very important for the electrodynamic characteristics of the mixed state in high-temperature superconductors in which the resistive state during moderate applied field values may be pulled into the temperature region where T is substantially lower than the *critical temperature* T_c.

MOBILITY EDGE

Term used for either the localization threshold (N.F. Mott, 1966) or the energy E_c that separates localized states from non-localized ones in *disordered solids*. Neither the *density of states* itself nor its derivatives undergo discontinuities at E_c (D.I. Thouless, 1970). The mobility edge concept is applicable to systems with spatial dimensionality $d \geqslant 2$, since all the states of a one-dimensional system remain localized, no matter how minimal the disorder (N.F. Mott, W.D. Twose, 1961). See also *Anderson localization*, *Amorphous semiconductors*.

MOBILITY OF CURRENT CARRIERS

Average rate of *current carrier drift* in an electric field of unit intensity; defined by the equation $\mu = |\boldsymbol{v}|/|\boldsymbol{E}|$, where $\boldsymbol{v}$ is drift rate, $\boldsymbol{E}$ is electric field intensity. The value of the mobility μ is determined by the balance between the acceleration of *current carriers* due to the electric field and the scattering of the carriers. The mobility μ of current carriers is related to the *diffusion coefficient* D by the *Einstein relation* $\mu = eD/(k_B T)$. In the absence of *degeneracy* of the carriers it follows from the solution of the *Boltzmann equation* that the electron (*hole*) mobility is given by $\mu = (e/m_n)\langle\tau_p(\varepsilon)\rangle$, where e is charge, m_n is *effective mass*, ε is energy, $\tau_p(\varepsilon)$ is the relaxation time of the quasi-momentum of scattered particles, the operator $\langle\ldots\rangle$ indicates averaging of current carriers over energy states with the help of the *distribution function*. If electrons (holes) are strongly degenerate, then $\mu = [e/m_n]\tau_p(\varepsilon_F)$, where ε_F is Fermi energy, because by virtue of the Pauli exclusion principle only electrons with $\varepsilon \approx \varepsilon_F$ are scattered. Anisotropy of energy spectra of electrons and *phonons* leads to anisotropy of the effective mass and relaxation time, which are described by the reciprocal effective mass tensor and the relaxation time tensor, respectively. If the crystal exhibits high microscopic *crystal symmetry* then, in spite of this anisotropy, the mobility of current carriers is a scalar quantity, which depends on components of the above tensors. The combination of reciprocal effective mass tensor components, which is included in the equation for the mobility instead of m_n, is sometimes called *conduction effective mass*, *Ohmic effective mass*, or *optical effective mass*. The last name owes its origin to the fact that the above combination of tensor components is included in the high-frequency mobility, which is introduced at the description of the motion of current carriers in an ac field of frequency $\omega \gg 1/\tau_p$. Often this is the region of visible radiation.

In *many-valley semiconductors* with cubic symmetry (*germanium*, *silicon*) the electrons of every valley have an anisotropic mobility. The inclusion of simultaneous electron drift in all valleys, which takes place due to the symmetrical location of electrons in reciprocal space (***k**-space*), leads to isotropy of the mobility. A change of the relative contributions to the current from different valleys, which takes place, e.g., at the *Sasaki effect* or at *strain*, results in anisotropy of mobility of current carriers. A low-symmetry crystal exhibits anisotropy of mobility of current carriers.

The temperature dependence of μ is determined by the mechanism of *current carrier scattering*. The mobility μ_H, which is obtained from measurements of electromotive force or Hall current (*Hall mobility*), differs from μ: $\mu_H = r\mu$, where the *Hall factor* r is a numerical parameter which depends on the mechanism of current carrier scattering, structure of energy bands and the magnitude of the magnetic field (see *Galvanomagnetic effects*). The difference between μ and μ_H is related to the fact that μ describes the response of current carriers to the force $-e\boldsymbol{E}$, which has an equal action on all current carriers, whereas μ_H describes the response to the *Lorentz force*, which depends on the particle velocity, and therefore exerts a different action on every current carrier. For Hall mobility determinations the quantity μ is sometimes called *conduction mobility* (mobility in terms of conduction) or *drift mobility*. The latter term is also used in reference to mobility, which is determined from the time of drift of current carriers between two separated points of the sample. Ambipolar drift (see *Ambipolar diffusion*) is described in terms of *ambipolar mobility*; *mobility in terms of magnetoresistance* is determined from magnetoresistance. *Surface mobility* of current carriers and *field-effect mobility* are introduced at consideration of the motion of current carriers at a semiconductor surface. Surface mobility of current carriers is the mean mobility of current carriers, which are situated in channels at different distances from the surface; it takes into account the surface scattering of current carriers. The field-effect mobility is the derivative of the current which flows through a sample of unit length and width, with respect to the charge of the field-effect electrode. The value of this quantity depends on the *trapping* of current carriers by electron *surface states*.

MOBILITY OF DOMAIN WALL

See *Domain wall mobility.*

MOBILITY ON A SURFACE

See *Surface mobility.*

MODEL OF CHARGES ON BONDS

Semiempirical model involving an approximate representation of the long-range part of the *adiabatic potential* for valence *crystal* atoms. The potential is the sum of a screened Coulomb interaction of nuclei, and the interaction of negative charges on bonds with one another and with nuclei. In semiconductors with a finite gap E_g, as distinct from *metals* where free electrons completely screen the potentials of ions at distances in excess of a Debye length, the screened potential of each nucleus (or lattice atom) decreases at infinity as $Ze/(\varepsilon_\infty r)$, where ε_∞ is the high-frequency *dielectric constant.* Yet, this is incompatible with the condition of crystal neutrality. J. Phillips proposed assigning compensation charges $-Ze/\varepsilon_\infty$ to regions of the overlap of electron wave functions of nearest covalent bonded atoms. In addition, the adiabatic potential must contain short-range terms to ensure stability of the system of charges.

MODIFICATION OF METALS

Introduction of *modifiers* (*magnesium*, ferrosilicon, *aluminum*, etc.) into metallic melts. Small amounts of modifiers (often a fraction of a per cent at most) promote *crystallization* of structural components of an *alloy* in a granular or spheroidal shape to improve its mechanical properties.

MODIFICATION OF POLYMERS

Purposeful modification of properties of *polymers* by regulating supramolecular structure (*nucleation* in crystallization, *heat treatment*), or by varying the molecular chemical composition (introducting reactive groups, etc.) in order, e.g., to increase the *impact strength* of plastics, or facilitate the dyeing of synthetic fibers.

MODULATED ELECTRIC FIELD

See *Electric field modulation method.*

MODULATED MAGNETIC STRUCTURES

Atomic magnetic structures with a periodic nonuniform distribution of spin density, the spatial constants λ of these structures being in a general case incommensurate with the *crystal lattice* spacing a, and dependent on the temperature. In various systems, λ varies by an order of magnitude from several a to hundreds (and even thousands) of a. If $\lambda \gg a$, modulated magnetic structures may be represented as long-period modulations of simple (spatially uniform) magnetic structures (ferromagnetic or antiferromagnetic). Sometimes the following terms are used to describe the same entities: *long-period magnetic structure*, *incommensurate magnetic structure*, *magnetic superstructure*. The main types of modulated magnetic structures are (see Fig.): SS-structure (*simple spiral*, *spiral magnetic structure*, *helical magnetic structure*); FS-structure (*ferromagnetic spiral*, *umbrella-shaped magnetic structure*); $\widetilde{\text{SS}}$-structure (its particular case is the *cycloidal magnetic structure*); FAN-structure (*fan shaped structure*, or *pseudo-ferromagnetic structure* existing only in an applied magnetic field); LSW- and TSW-structures (*longitudinal and transverse spin density waves*, respectively). They differ by the magnitude of the projections of the *magnetic moments* of the ions on the direction of the modulated magnetic structure vector $\boldsymbol{q}$ (direction of spatial gradients), and on the plane that is perpendicular to it.

The existence of modulated magnetic structures results from the presence of minima of a *thermodynamic potential* F of a magnetic material, the potential being considered a function of $\boldsymbol{q}$ at symmetry ("non-Lifshitz") points of the *Brillouin zone.* Modulated magnetic structures may appear in a phenomenological description of a magnetically ordered state, that includes exchange, when terms with *order parameter* derivatives are included in the magnetic part of the system potential F. Specific modulated magnetic structures can result from the competition of interactions described by the square of the first spatial derivative of the order parameter, or they can result from terms in F that are linear in the first spatial derivatives of the order parameter. The latter situation is characterized by a large value of the spatial constant λ ($\lambda \gg a$). An exchange-modulated magnetic structure of the SS-type has been identified

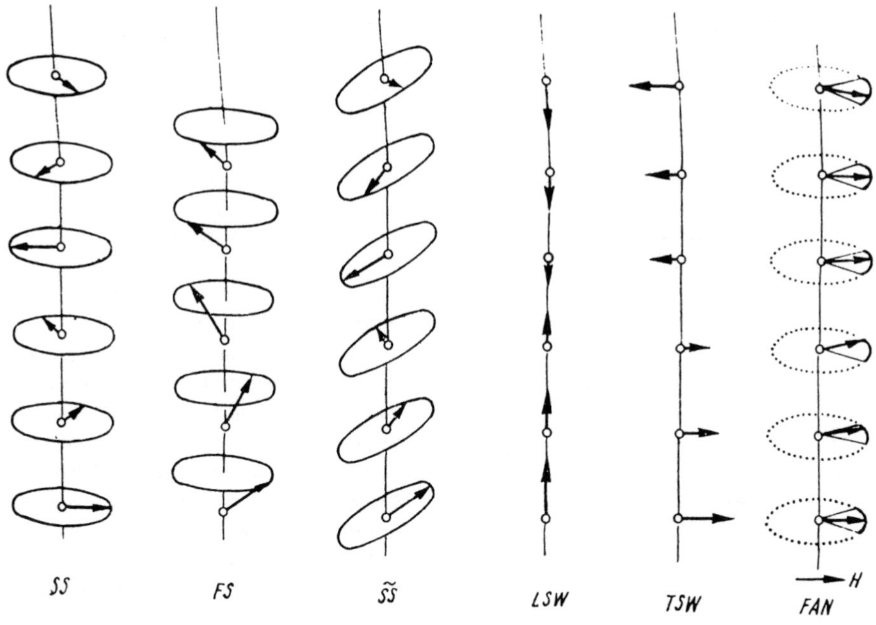

Modulated magnetic structures.

in the cubic crystals MnSi and FeGe which lack a center of inversion (*space group* $P2_13$, T^4). This structure owes its origin to the invariant $\boldsymbol{M}\nabla \times \boldsymbol{M}$ (V. Baryakhtar, E.S. Stefanovsky, 1968) that describes the nonuniform Dzialoshinskii–Moriya interaction (see *Dzialoshinskii interaction*).

MODULATED STRUCTURES

Regularly and quasi-periodically arranged regions that are, e.g., either *coherent precipitates* in aging (see *Alloy aging*), ordered, eutectically decomposing alloys (see *Eutectic alloys*, *Alloy decomposition*), individual (or group) *stacking faults* in one-dimensionally disordered alloys; or else antiphase *domains* in *antiferromagnets* and *ferroelectrics*. Complicated atomic or molecular complexes in *minerals*, in chemical organic or inorganic structures may also be considered as such regions. Modulated structures were first discovered by the X-radiographic investigation (see *X-radiography*) of an aged *cunife* alloy: the X-ray spectrum of this alloy has additional reflections (*satellite lines*) observed near the main Bragg reflections. The effect is explained by the model of a periodic change of the concentration of alloying elements (see *Alloying*) along one of the alloy's cubic directions. In diffuse *phase transitions* (aging, *alloy ordering*, eutectoid decomposition), a modulated structure is an ensemble of domains where the precipitates of the second phase form a *macrolattice*. One-, two-, and three-dimensional modulated structures are known. One-dimensional ones have the highest thermal stability in aging alloys as they correspond to a minimum of the elastic energy of the system. The formation of modulated systems is accompanied by a significant change in their physical and mechanical properties.

MODULATION SPECTROSCOPY

Investigation of optical spectra under conditions of modulation of material (sample) parameters (internal modulation), or of the light beam (external modulation). Used for producing modulation can be electric and magnetic fields, illumination intensity, temperature, uniform or uniaxial pressure. One can modulate such light beam parameters as wavelength λ, angle of incidence, polarization, etc. Since it is the change of light reflection ΔR (or light absorption) by the sample that is experimentally detected, a modulation spectroscopy method is called electro-, photo-, thermo-, piezo-, and wavelength-reflection (absorption). The main application field of this spectroscopy in solid state physics is the investigation of *band structure* and *surface phenomena*. Through the use of synchronous detection, modulation spectroscopy methods have a high sensitivity (in reflection up to $\Delta R/R \sim 10^{-6}$) and high resolution. In the region of weak fields, they are equivalent to the third derivative of the optical constants with respect to the *photon* energy. Such differentiation of an optical reflection (or absorption) spectrum cancels its constant component and emphasizes singularities of the spectrum, which are due to critical points in electron bands of a bulk crystal (*van Hove singularities*), or in the spectra of *surface electron states*. Electric modulation, the most convenient and widely used type, exhibits several types of modulation effects. Primary *electrooptical effects* due to the influence of the field on the wave function of electrons of the medium include (1) nonresonance effects: quadratic *Kerr effect* and linear *Pockels effect*, and (2) resonance effects: *Franz–Keldysh effect* for band-to-band and band-impurity transitions, field quenching, and energy shift of *excitons*, *phonons* and other *quasi-particles*. Related to electric modulation effects are electron concentration mechanisms for changes of *dielectric constant* (band filling effect, intraband transitions in *semiconductors* and *semimetals*, solid-state plasma; changes of concentration of adsorbed particles, changes of chemical composition of transition layer, etc.).

MODULATOR, ELECTROOPTICAL

See *Electrooptical modulators*.

MODULUS OF ELONGATION

See *Young's modulus*.

MODULUS OF RIGIDITY

See *Shear modulus*.

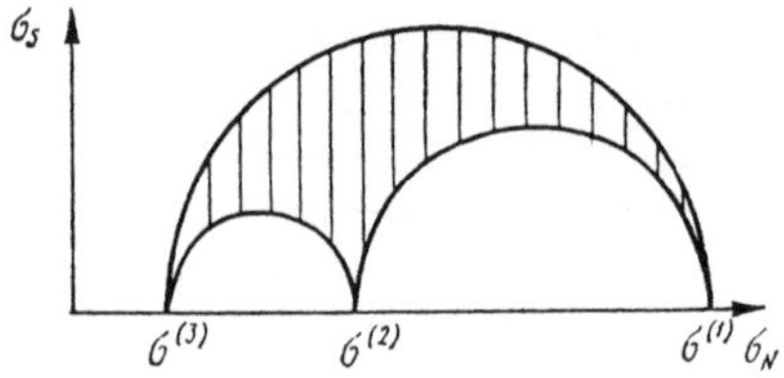

Mohr diagram.

MOHR DIAGRAM (O. Mohr, 1882)

Diagram (see Fig.), which provides a graphical determination of normal stress σ_{N} and shear stress σ_{S} for an arbitrarily selected direction, if the principal values σ^{i}, $i = 1, 2, 3$, and principal axes of the *stress tensor* (see *Principal axes of a tensor*) $\boldsymbol{e}^{(i)}$, $i = 1, 2, 3$, are known. If the unit vector in direction $\boldsymbol{e}$ has components v_1, v_2, v_3 ($v_1^2 + v_2^2 + v_3^2 = 1$) with respect to the tensor principal axes, then σ_{N} and σ_{S} in a Mohr diagram are coordinates of the point p that is located at the intersection of circumferences

$$\sigma_S^2 + \left[\sigma_{\mathrm{N}} - \frac{1}{2}\left(\sigma^{(i)} + \sigma^{(j)}\right)\right]^2 = \frac{1}{4}\left(\sigma^{(i)} - \sigma^{(j)}\right)^2 + v_k^2\left(\sigma^{(i)} - \sigma^{(k)}\right)\left(\sigma^{(j)} - \sigma^{(k)}\right),$$

where indices i, j, k form a cyclic sequence 1, 2, 3. The region of possible values of σ_{N} and σ_{S} on the Mohr diagram is shown by hatching. Another way to determine values of σ_{N} and σ_{S} is with the help of the *stress surface*. The Mohr construction is applicable to any symmetric second rank tensor S_{ij} by letting

$$\sigma_{\mathrm{N}} = S_{ij} v_i v_j,$$
$$\sigma_{\mathrm{S}}^2 = S_{ij} S_{jk} v_i v_k - (S_{ij} v_i v_j)^2.$$

Here, summation is meant to be taken over repeated indices.

MOLARITY

Measure of *concentration*, number of moles of a solute in 1 liter (1 dm^3) of solution. Alternate name: volumetric molar concentration. Measured in mole/liter (often denoted by M).

MOLECULAR BEAM EPITAXY

Epitaxial growth of *films* of chemical compounds from molecular or atomic beams of the constituent elements in a super-high vacuum (see *Epitaxy*). The advantage of this method is the following: (i) high purity is achieved through the use of highly purified initial materials and a super-high vacuum with negligible concentrations of H_2O, CO and O_2; (ii) low levels of surface defects result in a decreased density of *surface states*, plus an improvement of reproducibility and reliability; (iii) the possibility of obtaining very sharp ($\leqslant 1$ nm) variations of the composition and the level of *doping* through rigid control of the layer growth rate and the beam composition; (iv) the possibility of controlling the composition and the structure of layers by using analytical apparatus *in situ*. Molecular beam epitaxy is employed in the manufacture of high-technology *semiconductor devices* with multilayer crystals (*field-effect transistors*, *lasers*, *optical waveguides*, superhigh-frequency transducers and amplifiers, see *Microwave semiconductor devices*). A variation of the method is *ion beam epitaxy*.

MOLECULAR CRYSTALS

Class of solids whose intramolecular (i.e. internal) chemical bond energy significantly exceeds the intermolecular interaction energy. Owing to this, molecules in molecular crystals retain their individual identities and properties. In the case of polar molecules, the main contribution to the intermolecular interaction energy is from the *dipole–dipole interaction* (or the *quadrupole interaction*), while in the case of nonpolar molecules it arises from the *van der Waals interaction* (see *Van der Waals forces*). Molecular *crystals* also include a group whose molecules are linked together by *hydrogen bonds*. These bonds arise if molecules contain hydroxyl, OH, groups, or amino, NH_2, groups. Excited states give rise to new interactions; in particular, an important role is played by resonance interactions, their energy being inversely proportional to the cube of the distance.

In the solid state these molecular crystals are formed by noble gases (*atomic molecular crystals*, He, Ne, Ar); by molecules with strong chemical bonds H_2, O_2, HCl, CH_4, etc.; by halogens; by compounds with molecules consisting of nitrogen and phosphorus, of sulfur and phosphorus, of sulfur; by metal carbonyls. Crystal structures with molecular bonds are also formed by certain polymer chains (polyethylene), and by biological structures. Organic compounds form the largest group of molecular crystals.

In their electrical properties most molecular crystals are, as a rule, *insulators*, but some of them may under certain conditions exhibit noticeable electrical conductivity (see *Organic semiconductors*). Nearly all molecular crystals (except O_2 and NO_2) in the ground state are *diamagnets*. Absorption of electromagnetic radiation is caused by intramolecular electronic and electronic-vibrational transitions that are delocalized and generate an elementary excitation, an *exciton*. Lattice *phonons* also take part in the absorption of electromagnetic radiation.

Weak intramolecular bonds are responsible for the low mechanical strength of molecular crystals, low temperatures of melting (ranging from several tens to several hundreds °C) and often of evaporation. Their heat of sublimation is of the order of 10–20 kcal/mole, approximately 10–20 times smaller than isolated molecule dissociation energies.

Molecular crystals are widely used as scintillators and moderators of heavy particles in nuclear physics and industry, and as insulating materials in electrical engineering (molecular crystal resistivity $\approx 10^6\ \Omega\cdot$cm). Various organic compounds are found in medicines and *dyes*. A number of unique physical properties of molecular crystals should be useful in what might be called *molecular electronics*.

MOLECULAR ENERGY DIAGRAM

Scheme to illustrate the arrangement of energy levels of molecules according to their energies, in combination with the basic characteristics of the levels. For one-electron energy levels the scheme is usually called a *molecular orbital diagram*. The genetic relations of molecular orbitals with the single-electron levels of atoms which form a molecule are often indicated in the orbital diagram. These diagrams are constructed either from qualitative considerations, or using data calculated by, e.g., the *Hartree–Fock method*. For displaying multielectron energy levels, a scheme of molecular electron terms is utilized. Designations of the levels reflect their group theory classification, as a rule involving the irreducible representations of the symmetry *point group* of the molecule, and thus they provide the (orbital) degeneracy of the levels. For multielectron molecular terms, the multiplicity of the *spin degeneracy* of the state is also indicated. In analogy with electronic levels, similar diagrams are used for rotational, vibrational, and more complicated states, e.g., for electronic-vibrational states (*vibronic states*), etc.

MOLECULAR FIELD (B.L. Rozing, P. Weiss, 1907)

One of the most widely used approximations for explaining magnetic properties in a solid is to invoke a molecular field (*self-consistent field, average field*). The *molecular field approximation* assumes that the interaction of a certain spin S_i with N neighboring spins is approximated by introducting a molecular field B_{MF} that is proportional to the *magnetization* in a *ferromagnet*: $g\mu_{\mathrm{B}}B_{\mathrm{MF}} = \lambda M$. Here, λ is the molecular field constant, g is the *Landé g-factor*, μ_{B} is the Bohr *magneton*, and M is the magnetization. In the presence of an external magnetic field $B_{\mathrm{eff}} = \mu_0 H$, M satisfies the self-consistent equation $M = 2Ng\mu_{\mathrm{B}}SB_S[(\lambda M + g\mu_{\mathrm{B}}B)/(k_{\mathrm{B}}T)]$, where B_S is the Brillouin function, and N is the number of nearest neighbors. At $B = 0$, this equation describes the behavior of spontaneous magnetization that disappears in the paramagnetic state above the *Curie point*. In a microscopic approach, the local molecular field is expressed in terms of parameters J_{ij} of the exchange Hamiltonian and mean values of spins: $g\mu_{\mathrm{B}}B_{\mathrm{MF}}(i) = \sum_j J_{ij}\langle S_j\rangle$. This allows one to perform self-consistent calculations of spin behavior both in ferromagnets, and in more complicated *atomic magnetic structures*. This approximation is an asymptotically exact solution in models with an infinite *exchange interaction* radius R, and hence, it serves as a zero-order approximation in the perturbation theory of the thermodynamics of spin systems, which is based on an expansion in terms of $1/R$ or $1/N$. The molecular field is also used to describe other *phase transitions*: structural and ferroelectric types, those from the normal state to the superconducting state, and from the normal state to the superfluid state. When the role of fluctuations is significant (regions of critical behavior in the neighborhood of a phase transition temperature, one-dimensional systems, two-dimensional XY-models and Heisenberg model, dilute magnetic materials with short-range exchange in the neighborhood of flux thresholds, etc.), the molecular field approximation yields qualitatively incorrect results. In addition this approximation fails to describe collective spin excitations (*magnons*), and some low temperature features in *Heisenberg magnets*.

MOLECULAR INTEGRALS

The same as *Quantum chemistry integrals*.

MOLECULAR MASS DISTRIBUTION OF POLYMERS

Distribution of macromolecules of a given *polymer* sample expressed in molecular mass units. This molecular mass discussion pertains to polymers only, and results from the random nature of reactions of polymer generation and *polymer degradation*. The concept of molecular mass distribution loses its meaning for three-dimensionally linked polymers (see *Polymer cross-linking*), when the whole polymer body behaves as if it were a single "molecule". Two types of molecular mass distribution are known: proper distribution $\rho_n(M)$ of macromolecules in the number of chains with a given molecular mass M (numeric molecular mass distribution) and distribution $\rho_\omega(M)$ in masses of chains with a given molecular mass (mass molecular mass distribution). These are related through the expression $\rho_\omega(M) = \rho_n(M)M\overline{M}_n$, where $\overline{M}_n$ is the so-called number-average molecular mass given by $\overline{M}_n = \int_0^\infty \rho_n(M)M\,\mathrm{d}M$. There is also a mass-average molecular mass $\overline{M}_\omega = \int_0^\infty \rho_\omega(M)M\,\mathrm{d}M$ to characterize molecular mass distribution. The inequality $\overline{M}_\omega > \overline{M}_n$ is always true, and the deviation of the ratio $\overline{M}_\omega/\overline{M}_n$ from unity serves as a measure of the molecular mass distribution width (typically, $1 < \overline{M}_\omega/\overline{M}_n < 2$). The molecular mass distribution of a polymer is most often determined by sedimentation, fractionation, and chromatography experiments.

MOLECULAR ORBITAL

Single-electron wave function of a polyatomic (molecular) system. A molecular orbital is the generalization of *atomic orbital* to the case of several atoms. In the case of a *crystal* it is sometimes called a *crystal orbital*. The most common approximate method for determining a molecular orbital consists in its preliminary representation as a *linear combination of atomic orbitals* (LCAO), followed by finding the coefficients of the linear combination. The concept of molecular orbital is of fundamental importance in the theory of molecules and the theory of *solids*.

MOLECULAR SPECTRAL ANALYSIS

Determination of qualitative and quantitative composition of gaseous, liquid and solid compounds by absorption spectra of electronic-vibrational, vibrational, and rotational excited states, as well as by Raman spectra (see *Raman scattering of light*). Molecular spectral analysis of solids is the most efficient, using IR absorption spectra with their highly characteristic lattice vibration frequencies, *local vibrations* and *quasi-local vibrations*. The use of molecular spectral analysis by Raman spectra became as widespread as *optical absorption spectroscopy* after the invention of *lasers*, their radiation having no background. In certain cases Raman spectroscopy is used even more often, especially in the analysis of the composition of inorganic crystals, in particular, of *semiconductors*. To carry out the analysis, there is no need to record the entire spectrum or a major part of it. It suffices to use one or two *characteristic frequency band peaks* since they have only a weak dependence on the composition.

MOLECULE CHANNELING

Passage of fast molecules through thin *crystals* without dissociation. It is observed for hydrogen molecules that have energies 0.2–1 MeV/nucleon and pass through polycrystalline and monocrystalline foils about 1–10 μg/cm^2 (5–50 nm) thick. Despite the fact that the molecules lose their electrons upon entering the material (undergo complete "shelling"), the transit time through a *solid* is so short ($<10^{-16}$ s) that it is insufficient for *Coulomb explosion* of the molecule to separate protons by a large enough distance. The small spread in energy loss of protons ensures their escape from the solid with a small relative energy. It gives a molecule the possibility to be recreated through association and electron trapping during the exit. In the case of *monocrystals*, the protons move along the same or neighboring channels, and the molecules are capable, due to small energy loss and weak scattering, of passing through rather thick foils. Molecule channeling is sensitive to the defect structure of a solid, and can aid in its diagnostics.

MOLTER EFFECT (L. Molter, 1936)

Emission of electrons into a vacuum from a thin insulating layer on a conducting *substrate*, provided that there is a strong electric field (10^6 V/cm) in the layer. The emission current increases rapidly with an increase of the field. The Molter effect is due to the presence of a strong electric field in the layer, which causes *field emission of electrons* from the substrate into the layer, *electron heating* and *impact ionization* in the main bulk of the layer. Most of the voltage drop is in the neighborhood of the substrate. As a result, some of the high-speed electrons escape from the layer into the vacuum. In porous layers, the Molter effect is also caused by *Townsend impact ionization* that arises in *pores* (electrons escape mostly from pores).

MOLYBDENUM, Mo

Chemical element of Group VI of the periodic table with atomic number 42 and atomic weight 95.94. Natural Mo consists of 7 stable isotopes, and 10 radioactive ones are known. Outer shell electronic configuration $4d^5 5s^1$. Successive ionization energies (eV): 7.10, 6.15, 27.13, 40.53, 55.6, 71.7. Atomic radius 0.136 nm. Ionic radii (nm): Mo^{4+} 0.070, Mo^{6+} 0.062. Covalent radius is 0.130 nm. Oxidation states are +2, +3, +4, +5, +6. Electronegativity is 1.3.

Molybdenum is a light-gray *metal* having body-centered cubic structure, space group $Im\bar{3}m$ (O_h^9), $a = 0.314737$ nm at 298 K. Density is 10.22 g/cm^3; $T_{\text{melting}} = 2895$ K, T_{boiling} about 5100 K, heat of melting is 27.6 kJ/mole, heat of sublimation is 652.9 kJ/mole, heat of evaporation is 506 kJ/mole. Debye temperature is ≈464 K; coefficient of linear thermal expansion is ≈$6.0 \cdot 10^{-6}$ K^{-1} for unannealed state (at 298–2273 K). Thermal conductivity is 124–67 $W \cdot m^{-1} K^{-1}$ (at 1200–1900 K), 145 $W \cdot m^{-1} \cdot K^{-1}$ at 290 K and 183 $W \cdot m^{-1} \cdot K^{-1}$ at 90 K. Adiabatic elastic moduli: $c_{11} = 440.8$, $c_{22} = 172.4$, $c_{44} = 106.8$ GPa at 298 K. Young's modulus is ≈340 GPa, shear modulus is 119.7 GPa, Poisson ratio is ≈0.33. Breaking strength of wire 6.12 mm in diameter is 525–2500 MPa and depends on the degree of deformation; that of arc melted ingot is 571 MPa and that for monocrystalline molybdenum is 310 MPa. Hardness of molybdenum polycrystal is 1592 MPa at 293 K, 3730 MPa at 457.5 K and 126 MPa at 2130 K. Relative elongation 10–20%. Coefficient of self-diffusion is $3.1 \cdot 10^{-15}$ m^2/s at 2120 K. Electrical resistivity is 51.4 nΩ·m at 293 K. Temperature coefficient of electrical resistance is $479 \cdot 10^{-5}$ K^{-1} at 293–2893 K. Critical temperature of transition into superconducting state is 0.94 K. Critical magnetic field 9.65 mT. Work function is ≈4.3 eV. Reflection factor for optic waves with $\lambda = 5$ μm is 97.2%. Thermal neutron capture cross-section is 2.7 barn per atom. Molybdenum is a *paramagnet*, with molar magnetic susceptibility $+5.4 \cdot 10^{-6}$ CGS (at 291 K). Nuclear magnetic moment is 0.9 nuclear magnetons.

Molybdenum is used in the production of special *alloys* and alloyed *steels*. Molybdenum metal is a *construction material* for vacuum electronic industry and nuclear reactors. Liabilities are its high oxidizability and relatively high temperature of *cold brittleness*. Molybdenum is one of the most promising metals for creation of *high-temperature alloys* (see *Molybdenum alloys*).

MOLYBDENUM ALLOYS

Alloys based on *molybdenum*. Four main methods are singled out in molybdenum *alloying*.

(1) Introduction of highly active elements, exhibiting large affinity to interstitial impurities (rare earth metals, etc.).
(2) Alloying with a small quantity (about 1%) of transition metal to achieve competitive *segregation* with interstitial elements at crystal structure *defects* and to increase *high-temperature strength*.
(3) Joint alloying with an element (*titanium*, *zirconium*, hafnium, *vanadium*, *niobium*, *tantalum*) and *carbon* (nitrogen) to increase high-temperature strength through the mechanism of precipitation hardening (see *Precipitation-hardened materials*).
(4) Heavy alloying with *rhenium* (up to 50%) or with *tungsten*.

The production of most molybdenum alloys involves the simultaneous use of several alloying methods. Molybdenum alloys are melted in arc furnaces and electron beam furnaces, or produced through methods of *powder metallurgy*. Molybdenum alloys find use in electronics, rocket engineering, nuclear power industry, and in other areas.

MOMENT OF INERTIA

Sum of products of masses m_i of points in a solid that rotates about a fixed axis and the squares of their distances R_i from the axis: $I = \sum_i m_i R_i^2$. Knowledge of the moment of inertia allows expressing the *angular momentum* L of the body with respect to the fixed axis in terms of its *angular velocity*: $L = I\omega$; the rotational kinetic energy is expressible in terms of the moment of inertia and the square of the angular velocity: $K = (1/2)I\omega^2$. In the general case of an arbitrarily directed axis through the origin of coordinates, $I = \sum_{\alpha,\beta=1}^{3} I_{\alpha\beta} \cos\theta_\alpha \cos\theta_\beta$, where θ_α is the angle of this axis to the αth coordinate axis. The quantities $I_{\alpha,\beta} = \sum_i m_i (R_i^2 \delta_{\alpha\beta} - x_{i\alpha} x_{i\beta})$ form *moments of inertia* tensors and allow expressing L and K as $L_\alpha = \sum_{\beta=1}^{3} I_{\alpha\beta}\omega_\beta$; $K = (1/2)\sum_{\alpha,\beta=1}^{3} I_{\alpha\beta}\omega_\alpha\omega_\beta$. See also *Principal moments of inertia*.

MOMENTS OF SPECTRAL LINES

Characteristics of shapes of spectral (or some other) curves. For a resonance line (curve) that is described by a normalized (to unit area) *line shape* function $f(\omega)$, the nth moment M_n with respect to ω_0 is defined by

$$M_n = \int (\omega - \omega_0)^n f(\omega)\, d\omega.$$

If $f(\omega)$ is symmetrical with respect to ω_0 then all odd moments vanish. If all M_n are known, then $f(\omega)$ may be recovered. However, knowledge of only the first few moments provides important information on the shape of a resonance curve and, in particular, on the rate of its decay at shoulders away from ω_0. To illustrate, we consider the two most typical shapes of resonance curves, the symmetrical *Gaussian shape* and *Lorentzian shape*. The Gaussian shape is described by the normalized function

$$f(\omega) = \frac{1}{\Delta\sqrt{2\pi}} \exp\left[-\frac{(\omega-\omega_0)^2}{2\Delta^2}\right],$$

for which $M_2 = \Delta^2$, $M_{2n} = 1{\cdot}3{\cdot}5\cdots(2n-1)\Delta^{2n}$, and $M_4/M_2^2 = 3$. The half-width δ of the line at half-height is $\delta = \Delta(2\ln 2)^{1/2} \approx 1.18\Delta$. Thus, the value of the second moment for a Gaussian curve provides a satisfactory approximation for δ. The other common line shape is Lorentzian described by the normalized function

$$f(\omega) = \frac{\delta}{\pi} \frac{1}{\delta^2 + (\omega-\omega_0)^2}.$$

Since all even moments for a Lorentzian line, starting with the second one, are infinite, a more physically adequate approximation is that of a "cut-off" Lorentzian curve with $f(\omega) = 0$ at frequencies $|\omega - \omega_0| = \alpha \gg \delta$. Then, within the accuracy to terms of order δ/α, we may write

$$M_2 = \frac{2\alpha\delta}{\pi},$$

$$\frac{M_4}{M_2^2} = \frac{\pi\alpha}{6\delta} \gg 1,$$

$$\frac{\delta}{\sqrt{M_2}} = \frac{\pi}{2\sqrt{3}}\left(\frac{M_2^2}{M_4}\right)^{1/2} \ll 1,$$

i.e. δ of a Lorentz curve is much smaller than $\sqrt{M_2}$, which goes along with the slow decay of the shoulders. On the other hand, the assumption of a Gaussian curve may seem reasonable if the ratio M_4/M_2^2 is close to 3.

The advantage of the method of moments is that moments of lines may be calculated on the basis of general principles without finding the eigenvalues of a system Hamiltonian. In accordance with *linear response* theory, a function of shape $f(\omega)$ is expressible in terms of its autocorrelation function (see *Correlation function*) of magnetic or electric dipole moments of the system (in *magnetic resonance* or optical *light absorption*, respectively). This function depends on the system Hamiltonian term V that is responsible for the line broadening. Expanding the autocorrelation function in a series in powers of V, one can obtain closed form equations for moments of lines (e.g., the second moment equation is of the form $M_2 = -\mathrm{Tr}\{[V, S_x]^2\}/\mathrm{Tr}S_x^2)$. This procedure, referred to as *van Vleck's theory of moments*, was applied to magnetic resonance in calculating *line widths* caused by the magnetic *dipole–dipole interaction*. J.H. Van Vleck calculated M_2 and M_4 for a regular lattice of identical spins with dipole–dipole broadening and showed the resonance band shape to be close to Gaussian. In the case of diluted paramagnets, the ratio M_4/M_2^2 increases significantly, while the decrease of concentration of particles makes the shape of the line closer to Lorentzian. The *exchange interaction* has no effect on M_2, but increases M_4. The calculation of higher order moments is a cumbersome procedure; and describing a the shape of a line in terms of its moments is rarely done in practice.

MOMENTUM SPACE

Space of points that correspond to momentum values of masses or particles of a system. By momentum space in solid state physics one understands *quasi-momentum* space (i.e. ***k**-space*).

MONOCLINIC SYSTEM (fr. Gr. $\mu o\nu o\varsigma$, one, single, and $\kappa\lambda\iota\nu\omega$, bend)

Crystallographic system that is characterized by the presence of two right angles between coordinate axes ($\alpha = \gamma = 90° \neq \beta$) with unequal lengths ($a \neq b \neq c$). The angle β that is not a right angle is called the monoclinicity angle, and the axis b is aligned along the only twofold axis, or is perpendicular to the plane m. The so-called second setting is sometimes used: the c axis is along 2 or $\bar{2}$, $c < b < a$, $\gamma \neq 90°$. This system includes 2 *Bravais lattices* (primitive and body centered), 3 *point groups*, and 13 *space groups*.

MONOCRYSTAL, single crystal

Homogeneous *crystal* with a unified *crystal lattice* and anisotropic properties. The outward shape of a monocrystal results from its atomic structure and *crystallization* conditions: under equilibrium conditions, a monocrystal acquires a clearly pronounced natural *crystal faceting.* Examples of well-cut natural crystals are quartz, halite, Iceland spar, diamond, ruby. Some monocrystals may lack a regular cut (e.g., artificially grown rounded ruby crystals, *silicon* monocrystals).

Many monocrystals exhibit special physical properties: diamond and boron carbide are very hard, sapphire and fluorite are transparent over a wide wavelength range; quartz is piezoelectric (see *Piezoelectricity*); the strength of *thread-like crystals* of corundum is unmatched. Monocrystals are capable of changing their properties under exposure to external actions (light, mechanical stresses, electric and magnetic fields, radiation, temperature, pressure); hence, their use as transducers of various kinds in radio- and quantum electronics, acoustics, etc. As a result, there is a need for monocrystals of various sizes and shapes: from microcrystals, films and thread-like crystals several milligrams in mass to large specimens that weigh tens of kilograms. Initially, use was made of natural monocrystals, but reserves of these in nature are limited. The need for artificial ones grew; and many synthesized monocrystals (see *Monocrystal growth*) have no natural analog (see *Synthetic monocrystals*).

According to their degree of perfection, monocrystals are classified into non-mosaic (nearly perfect) and *mosaic crystals.* The most perfect contain only point defects (see *Defects in crystals*); the less perfect mosaic ones exhibit *dislocations* and *stacking faults.* Mosaic crystals with small block dimensions (10^{-5} cm and less) scatter X-rays in accordance with the theory of *elastic scattering of radiation.* Increase of disorientation angles between blocks up to a magnitude of tens of degrees results in the polycrystalline state (see *Polycrystal*). Many monocrystal properties (electrical, optical, strength, etc.) depend on the degree of perfection, and the impurity content. The most perfect ones are used in semiconductor engineering, quantum electronics, computer machinery, etc.

MONOCRYSTAL GROWTH, single crystal growing

Types of *crystallization* processes leading to the formation of *monocrystals.* If within the volume of the source phase there is only one nucleus capable of growing, then by the addition of particles to its surface a crystal of large dimensions might grow with all its atoms at regularly ordered lattice positions. Such a single crystal is called a monocrystal. Monocrystals are usually obtained with relatively small deviations from equilibrium in the source phase, since otherwise the addition of particles to the growing crystal surface occurs in a disordered manner, with the resulting formation of a *polycrystal.* To obtain a monocrystal, spatially uniform in its physical properties and structure, the crystallization conditions must be stable.

Crystallization conditions, nature of the material, and interface structure determine the different mechanisms of monocrystal growth. Depending on the concentration of surface defects (steps, etc.), the interface may be atomically smooth (see *Atomically clean surface*) or rough (see *Surface roughness*). The last case is realized when the distance between defects approaches unit cell dimensions. In the case of an atomically-smooth interface, the formation of a two-dimensional nucleus on its surface is necessary for monocrystal growth, and the probability of its appearance (as also for a three-dimensional nucleus) is associated with the energy outlay that depends on overcooling (oversaturation) of the source phase at the interface. Since this probability is low for small overcooling, the mechanism of bidimensional *nucleation* may be noticeably apparent at a comparably high deviation from equilibrium at the interface. The relation of the rate of monocrystal growth v to overcooling ΔT is determined by $v = A\exp(-B/\Delta T)$, where A and B are constants, weakly depending on the temperature T. The presence of a *screw dislocation* at the interface leads to the appearance of an echelon of steps, at which particles from the source phase may build. In this case, the dislocation mechanism of monocrystal growth is realized, which can proceed at small overcooling (oversaturation). The dependence of v on T under this mechanism is quadratic: $v \propto \Delta T^2$. If the crystal surface is rough, then practically every point of it is an active site for the addition of particles. The mechanism of crystal growth with a rough interface of phases is called normal. The relation of the growth speed to overcooling (oversaturation) for the normal mechanism is linear.

The presence of impurities and inclusions in the source phase affects the monocrystal growth.

Depending on the physicochemical conditions within the crystal, mechanical stresses and various defects (e.g., *vacancies, interstitial atoms, dislocations*, bloc boundaries, *pores*, inclusions) may appear which influence the crystal properties.

There are different methods for growing monocrystals: solutions, melts, and gas phase. They may be also obtained from the polycrystals and amorphous media by *recrystallization* and devitrification, respectively (see *Vitrification*). Selection of the method to use depends on the crystal physicochemical properties (melting temperature, volatility of components, their chemical activity, etc.). The most widespread are methods of growth from melts and solutions. Crystals are grown by drawing from the melt in a crucible with the seed in a region of the furnace where the temperature is lower than the crystallization temperature [*Czochralski method* (J. Czochralski, 1918)]; by placing the seed into the melt with slow cooling [*Kyropoulos method* (S. Kyropoulos, 1926)]; by moving nonmonocrystalline material through a zone with a temperature higher than the crystallization temperature (*zone crystallization, zone refinement*); by creating a layer of melt, held by the forces of surface tension, at the seed moved to the cold zone [*Verneuil method* (A. Verneuil, 1902)]; by moving the crucible plus melt to a zone below the crystallization temperature [*Stockbarger method* (D.C. Stockbarger, 1936)]; by removal of the latent heat of crystallization from the bottom of a stationary crucible with melt [*Stöber method* (F. Stöber, 1925)]; by drawing the crystal from the melt through a dye to form its profile [*Stepanov method* (A.V. Stepanov, 1969)], by molecular beam epitaxy growth on a substrate in vacuo, and by other techniques. Methods of growing crystals from solution are mainly associated with their growth at the seed upon cooling the solution as a result of reduction of the substance *solubility* in it, or by introduction into the solution, which is at a constant temperature, of an excessive (compared to equilibrium) concentration of the substance. See also *Crystallization from the gas (vapor) phase*.

MONOCRYSTAL, SYNTHETIC

See *Synthetic monocrystals*.

MONOLAYER, monatomic layer, monomolecular layer

A *film* of a foreign substance (1 atom or 1 molecule thick), adsorbed on the boundary of two phases. A monolayer may form on a *solid surface* owing to *adsorption* from the gaseous phase, or by precipitation (*segregation*) of impurities dissolved in a solid. The amount of material in a monolayer is designated by the absolute surface concentration of adsorbed particles, or by the ratio between the actual concentration and that of particles in a monolayer with close packing. Films with *coverage* less then 1 are called *submonolayer films*.

A monolayer is a quasi-bidimensional object that may exist in various *states of matter* (gas, liquid, liquid crystal or solid), depending on the conditions. Some monolayers form by a self-assembly process. A solid monolayer may possess its own lattice that is commensurate or incommensurate in structure with that of the *substrate*. The latter case can depend on the relationship between the depth of the potential relief of the substrate, and the energy of interaction of adsorbed particles (see also *Incommensurate structure*). At a change of coverage or temperature a *phase transition* can occur in a monolayer which differs in several ways from transitions in a three-dimensional system. Phase diagrams of monolayers generally display homogeneous regions as well as regions where various two-dimensional phases coexist.

Wide use is made of the ability of monolayers to change physicochemical properties of surfaces (*work function, surface energy*, electronic structure, etc.). Processes in monolayers determine the course of heterogeneous catalytic reactions, operation of emitters of many types, *crystal* growth, *wetting, corrosion*, etc. Specific properties are exhibited by monolayers of large organic molecules (see *Langmuir–Blodgett films*).

MONOPOLE EXCITATION

The same as *Shake-up*.

MONOPOLE IONIZATION

The same as *Shake-off*.

MONOPOLE, MAGNETIC

See *Magnetic monopole*.

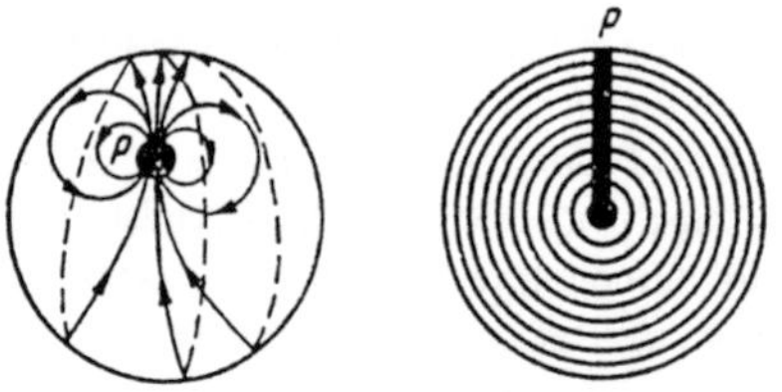

Left: monopole in laminated liquid crystal, showing distribution of vector field on a spherical layer which contains a singularity at point P. Right: semi-infinite disclination emergent from the center of a monopole, associated with a concentric spherical layer system.

MONOPOLE-TYPE DEFECT

A *defect in liquid crystals* which is a combination of a point defect and one or two linear defects. A monopole can form in a medium with a one-dimensional laminated structure where the orientation order of molecules within every layer is specified. It consists of a point *hedgehog* in a field normal to layers, with one or two *disclinations* attached to it, in the vector field that describes the ordering of molecules within layers (see Fig., left). If the hedgehog has a radial configuration, the monopole is a spherical concentric system of layers with a disclination emergent from its center (see Fig., right). Monopoles have been observed in cholesterics (see *Cholesteric liquid crystal*) and type C smectics (see *Smectic liquid crystal*). An analytical description of a monopole is similar to that of a Dirac monopole (*magnetic monopole*), which is a hypothetical isolated magnetic charge (see *Magnetic charge*). The stability of a monopole with respect to shortening the disclination length, and its transformation into a *boojum*, involves the presence of equal distances between layers.

MONOTECTIC ALLOYS (fr. Gr. $\mu o \nu o \varsigma$, one, and $\tau \eta \kappa \tau o \varsigma$, melted)

Alloys crystallizing from a liquid (melt) by simultaneous generation of a solid phase α, and another liquid that crystallizes through a eutectic (see *Eutectic alloys*) or some other reaction on subsequent cooling. Alloys are called *monotectoid alloys* if the initial *phase* is not a liquid, but rather a solid; with the latter transforming into two solid phases during the course of cooling. See also *Phase diagram*.

MONTE CARLO METHOD

Method of computer-aided solution of various problems of mathematics (statistical theory), which utilizes the generation of random numbers for simulation of probability distributions. Its efficiency is especially high when analytical models cannot be constructed for such distributions, or when approximations of an analytical approach are too rough. The Monte Carlo method is very close to the idea of a "computer experiment" (see *Computer simulation*).

Application of the Monte Carlo method in solid state physics gives significant results in investigations of *phase transitions* in *magnetic materials*, *ferroelectrics*, *binary alloys* (*Ising model*), for modeling processes of *diffusion* and self-diffusion of atoms, in the theory of *disordered solids*, *spin glasses*. The method makes use of one or another of the phenomenological models of matter, which must be close in properties to the system under study. Using the Monte Carlo method, "disorder" is introduced into the system, and averaging is performed on the results calculated for sample choices of random distributions. Their number must be large (from 10^3 to 10^9), with the upper bound determined by the computer capability, and the lower one by the desired accuracy. Another type of problem solved by Monte Carlo is related to simulations "from first principles". Fundamental physical laws and interactions of particles serve as starting points in this case. For instance, the solution of the Schrödinger equation is simulated for systems of particles, bosons or fermions (helium systems, neutron matter of pulsars, electron gas and liquid). This can provide, in particular, the most accurate results for the *correlation energy* of an electron gas in its dependence on the density, which are widely used in computational methods of solid state theory, especially within the framework of *density functional theory*.

MOREL–ANDERSON COULOMB PSEUDOPOTENTIAL (P. Morel, P.W. Anderson)

Dimensionless Coulomb repulsion constant in *superconductors*, renormalized through pairwise correlations of electrons scattering off each other far from the *Fermi surface*:

$$\mu^* = \frac{N(0)V_C}{1 + N(0)V_C \ln(E_m/\hbar\omega_D)},$$

where $N(0)$ is the electron *density of states* at the *Fermi level*, V_C is the screened Coulomb interaction matrix element averaged relative to the momentum transfers, ω_D is the *Debye frequency*, E_m is the maximum cutoff interaction energy which equals in order of magnitude the *Fermi energy* E_F or the plasma oscillation energy $\hbar\omega_p$ of electrons; thus, $\ln[E_m/(\hbar\omega_p)] \gg 1$ (see *Bogolyubov–Tolmachyov logarithm*).

MORIN TRANSITION (F.J. Morin, 1950)

Spontaneous (with no external *magnetic field*) *magnetic phase transition* from a weak ferromagnetic state to an antiferromagnetic state. It occurs in antiferromagnets that the exhibit the *Dzialoshinskii interaction*, and involves a change of spin orientation. The Morin transition results from the change of the equilibrium direction of the *antiferromagnetism vector* $\boldsymbol{L} = \boldsymbol{M}_1 - \boldsymbol{M}_2$ with changing temperature. The temperature where the Morin transition takes place is called the *Morin point*. This transition was first seen in haematite (α-Fe_2O_3) at $T_M = 258.5$ K, and has been detected in a number of *orthoferrites* and other *antiferromagnets*.

MOSAIC CRYSTALS, mosaic structure (fr. Ital. *mosaico*, picture, ornament)

Crystals that are aggregations of small monocrystalline blocks, slightly disoriented relative to one another. Practically every real crystal is, to a greater or lesser extent, a mosaic crystal. The sizes of mosaic blocks range from several hundreds to several thousands of crystallographic lattice constants, with disorientations of several angular minutes. For the larger size ranges the blocks are considered to be grains having ordinary low-angle grain boundaries. Thus, there is a continuous range of ordered states between mosaic structures and those characterized by ordinary small-angle *grain boundaries*. The *mosaic structure* forms during crystal growth (see *Crystallization*, *Block structure*) as a result of the appearance of various defects in the crystal (see *Defects in crystals*).

MÖSSBAUER EFFECT, nuclear gamma ray resonance (R. Mössbauer, 1958)

Emission and absorption of *gamma quanta* by atomic nuclei bound in a *solid*. These feature almost zero "recoil" energy of a nucleus (an effect not observed in liquids and gases). The underlying cause of the effect is that the nucleus recoil energy ratio $E_R = E/(2Mc^2)$ (here E is the energy of the γ-quantum, M is the mass of the nucleus) is small relative to *phonon* energies, so the *crystal lattice* as a quantum system is not excited, and the recoil is not absorbed by an individual nucleus but rather by the *crystal* as a whole. The probability of the Mössbauer effect is $f = \exp[-4\pi\langle\overline{x^2}\rangle/\lambda^2]$, where λ is the γ-quantum wavelength, and $\langle\overline{x^2}\rangle$ is the mean square of amplitude of nucleus oscillations in the γ-quantum direction. The absence of excited phonons causes the emission (absorption) *line width* to be determined solely by the width of the corresponding nuclear levels. The latter is much smaller than the energies of the *magnetic dipole interaction* and electric quadrupole interaction of a nucleus with electrons, and this permits studying the *hyperfine structure* of solids (see *Mössbauer spectroscopy*). Owing to the Mössbauer effect, the energy of γ-quanta is measured with an accuracy 10^{-13}–10^{-16} (ratio of line width to energy of γ-quantum) that is an all-time high for electromagnetic radiation. The Mössbauer effect finds its use in chemistry, solid state physics, metallurgy, geology, and biology.

MÖSSBAUER EFFECT, OPTICAL ANALOG

Generation of narrow *phononless lines* in optical spectra of imperfect crystals. Narrow absorption peaks in both γ-spectra and optical phononless lines result from quantum transitions between energy levels of *defects*, which involve neither emission nor absorption of *phonons*; from this point of view they are much alike. However, there are also important differences in the physical mechanisms for generating these spectral lines, in the conditions of their observation, as well as in applications of these phononless spectra.

MÖSSBAUER ELECTRON SPECTROSCOPY

Spectroscopy based on the detection of internal conversion electrons and Auger electrons generated as a result of resonance absorption of γ-quanta (*Mössbauer effect*). This technique permits the determination of the chemical state of Mössbauer atoms in surface layers of thickness ranging from several nm to hundreds of nm. Information in the form of the spectra is obtained with standard *Mössbauer spectrometers*. They consist of a radioactive Mössbauer source and an absorber that contains Mössbauer nuclei, a function generator, a vibrator with a feedback drive for the generation of a stable and linear relative Doppler shift between the source and the absorber, an electron detector with supply units for the amplifier of the differential amplitude discriminator, and a multichannel accumulator-analyzer operating in the mode of multichannel scaling. Parameters obtained from the spectra are similar to those of absorption *Mössbauer spectroscopy*.

By selecting the corresponding Mössbauer sources, one can investigate surface phenomena in samples that contain nuclei of ^{57}Fe, ^{119}Sn, ^{181}Ta, ^{151}Eu and other Mössbauer isotopes. Nuclear decay of the most common isotope ^{57}Fe after absorption of γ-quanta with energy of 14.4 eV from the radioactive ^{57}Co source is accompanied by the emission (in addition to γ-quanta and characteristic X-radiation) of K-shell conversion electrons (energy 7.3 keV, intensity 0.79, depth 10–400 nm); L-shell conversion electrons (13.6 keV; 0.08; 20 nm–1.3 μm, respectively); M-shell conversion electrons (14.3 keV; 0.01; 20 nm–1.5 μm); K–LL-Auger electrons (5.6 keV; 0.6; 7–200 nm); L–MM-Auger electrons (0.5 keV; 0.6; 1–2 nm). There are differential and integral methods for analyzing the distribution of electrons escaping from the surface of samples under study. The simplest and most frequently used is the integral analysis method which records in the scattering geometry all electrons of the sample that reach the cathode of a flow-type proportional counter (He + 5% CH_4). Differential Mössbauer electron spectroscopy of deep levels detects electrons of particular energy intervals. This requires the application of an ultrahigh vacuum, and the use of electronic spectrometers of high energy resolution (1–2%). Mössbauer electron spectroscopy is used for nondestructive analysis in solid state physics, chemistry and in *materials science* (*phase analysis*, *corrosion*, electrochemistry, implantation, metastable structures and amorphous alloys, structure of surface layers after physicochemical treatments, laser *alloying* and *hardening*, surface properties at different reactions, study of reactions *in situ*, etc.).

MÖSSBAUEROGRAPHY

Diffraction method of investigating the atomic, electric and magnetic structure of crystals, based on diffraction of Mössbauer γ-radiation (see *Mössbauer effect*). This is an analog of *X-radiography*, *electron diffraction analysis* and *neutron diffractometry*. The energy of low-temperature γ-transitions of Mössbauer nuclei corresponds to wavelengths $\lambda_\gamma \sim 0.01$–0.1 nm, therefore this radiation is diffracted by crystals with interplanar distance $d \sim \lambda_\gamma$. The diffraction of Mössbauer γ-radiation in resonance scattering by crystals that contain Mössbauer isotopes exhibits a number of features characteristic of the diffraction and scattering of X-rays, γ-quanta and neutrons. In diffraction by a crystal, coherent resonance scattering and *Mössbauer radiation Rayleigh diffraction* take place. Both channels of scattering in a crystal contribute to maxima in this process, so interference of both Rayleigh and nuclear (resonance) scattering appears.

Resonant γ-radiation is modulated by the *Doppler effect*, which allows one to easily vary the amplitude and phase of the scattered γ-quanta, facilitating the calculation of the phase of the Rayleigh amplitude and, by doing so, to solve the "phase" problem of structure analysis. The dependence of the amplitude of Mössbauer scattering on the direction of the magnetic field at the scattering nucleus, which is related to the orientation of atomic magnetic moments, makes it possible to investigate the *atomic magnetic structure* of materials. The relationship between the Mössbauer scattering amplitude and the gradient of the electric field at a nucleus is unique, in the sense that it provides the configurations of electric field gradients in crystals. This can show, e.g., that the axes of the electric field gradient tensor of the same atoms have different orientations at structurally equivalent points of a *crystal lattice*. The process of diffraction reveals these differences through the appearance of purely nuclear, magnetic, electric, or combined reflections that are "forbidden" by usual conditions of reflection extinction, i.e. reflections which fail to appear in *Rayleigh scattering of light*.

Gamma resonance diffraction by perfect crystals that contain Mössbauer nuclei is characterized by the collective interaction of nuclei with the γ-radiation. Provided the Bragg condition is met, the resonance value of the γ-quantum energy and the energy width at resonance differ from the energy and width of a Mössbauer level of an isolated nucleus. The collective nature of the interaction manifests itself also in the suppression of inelastic channels of nuclear reactions (*Kagan–Afanasyev effect*). Under conditions of *Bragg diffraction* by a perfect crystal, γ-radiation penetrates abnormally deeply compared to in a *mosaic crystal*, or to the absence of diffraction. This phenomenon is a nuclear analogue of the Borrman effect (see *Anomalous passage of X-rays*).

Diffraction of Mössbauer radiation by crystals that contain no Mössbauer nuclei also exhibits a specific feature. The use of extremely high energy resolution of Mössbauer detectors makes it possible to separate elastic and inelastic scattering in the region of a Bragg maximum. This permits the study of the dynamics of atomic motion with variations of the crystal temperature by observing some of the elastically scattered quanta. Kinematic and dynamic theories of the diffraction of Mössbauer γ-radiation by crystals have been developed, and the main physical processes of interaction of this radiation with matter have been studied. Mössbauerography experiments are under way in all the above directions. A significant obstacle to the widespead applicability of Mössbauerography is the unavailability of powerful sources of Mössbauer γ-quanta. There is hope for creating nondecaying sources of resonance scattering through synchrotron radiation, as well as through resonance γ-radiation, which arises when charged relativistic particles interact with crystals. The creation of a Mössbauer γ-laser would provide a powerful source for mössbauerography.

MÖSSBAUER RADIATION RAYLEIGH DIFFRACTION

Nuclear gamma resonance radiation (Mössbauer radiation) Rayleigh diffraction takes place at electronic shells of atoms and ions in condensed media. It is usually carried out in accordance with the X-ray diffraction scheme: gamma resonance radiation is scattered by the sample with subsequent energy analysis of the diffracted beam with the help of a Mössbauer absorber. As a rule, nuclei of Fe^{57} with a γ-quantum energy of 14.4 keV are used as the source. The main advantage of this procedure is the very high ($\sim 10^{-8}$ to 10^{-9} eV) relative resolution of the transferred energy, and the fact that it is not necessary to have Mössbauer nuclei in the sample. The majority of studies are with single crystals, and are devoted to determining the *Debye–Waller factor*, to the identification of thermal diffuse scattering, to the separation of static and dynamic diffuse scattering, and to the determination of the nature of the quasi-elastic, so-called *central peak*, in the diffraction study of *structural phase transitions*. Inelastic scattering from *crystal lattice* distortions around defects, relaxation motion in crystals with structural disorder, scattering in solid electrolytes, etc. is investigated. In diffraction studies of biological systems important results have been obtained involving the influence of water on the dynamical properties of proteins, etc. The low intensity of the diffracted beam, and therefore the low resolution of the momentum being transferred, is the main disadvantage of this technique. These difficulties are partly overcome by the application of coordinate-sensitive detectors, and also of ring detectors, and by synchronizing the motions of the source and absorber of the gamma resonance radiation. See also *Mössbauerography*.

MÖSSBAUER SPECTROSCOPY, nuclear gamma resonance spectroscopy

Method based on the *Mössbauer effect* of investigating *hyperfine interactions* between nuclei, and electric and magnetic fields, and hence various physical properties of *solids*. Mössbauer spectroscopy exhibits a remarkably high accuracy of measuring the energy of γ-quanta, which is defined by ratio of the resonance band width Γ_0 (10^{-10}–10^{-8} eV) to the energy E_0 of a nuclear transition $\Gamma_0/E_0 \approx 10^{-10}$–$10^{-16}$. This makes it possible to observe shifts of nuclear energy levels ($\leqslant 10^{-4}$ eV) caused by hyperfine interactions: electrostatic (Fig. 1(a)), electric quadrupole (Fig. 1(b)) and *magnetic dipole interactions* (Fig. 1(c)) between a nucleus and surrounding electrons by recording Mössbauer spectra (*nuclear gamma ray resonance spectra*). They represent the dependences of the number N of detected γ-quanta that pass through an absorber (transmission mode, Fig. 2(a)), or were scattered by the sample (scattering mode, Fig. 2(b)), or of X-rays and electrons that were emitted by the sample

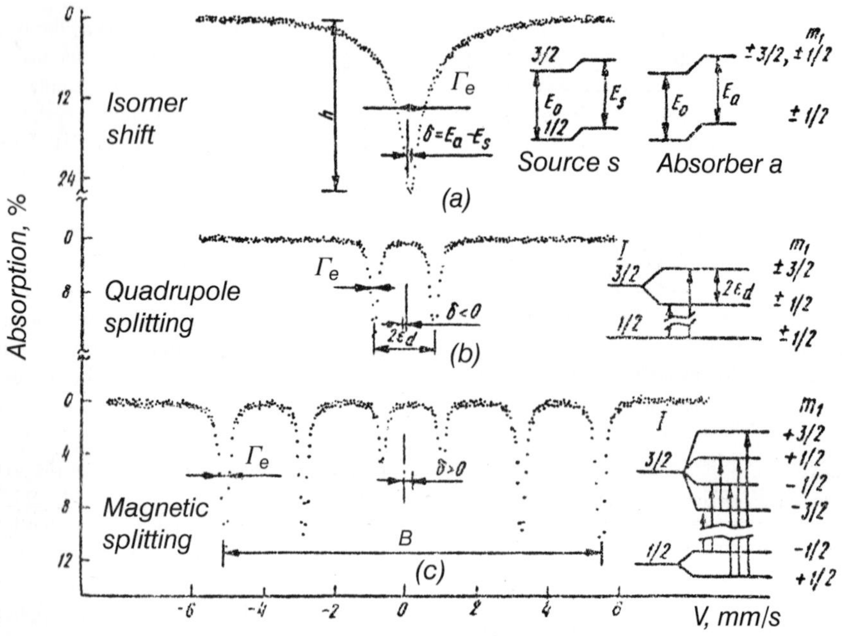

Fig. 1. Nuclear gamma resonance absorption spectra of ^{57}Fe, their energy levels, and their parameters.

due to internal conversion, on the extent of mismatch between the energies of the emitter and absorber (scatterer). The method of modulating γ-quanta energy, which results in the mismatch, is the most widespread version, and it is based on the *Doppler effect*. The γ-radiation source is set in motion past the absorber (scatterer) at a velocity relative to the absorber (scatterer) $v = \Gamma_0 c/E_0$, so that the γ-radiation energy changes by $\Delta E = \pm E_0 v/c$. Nuclear gamma resonance absorption spectra (similar to those of scattering) of conversion electrons exhibit *hyperfine structure*, which depends on the type of hyperfine interaction, and is described by the following parameters (Fig. 1):

(a) The magnitude of the Mössbauer effect h and area S depend (in general, in a nonlinear manner) on the number of nuclei n and on the Mössbauer effect probability f (probability of emission or absorption of γ-quanta without recoil; f is expressed via the mean square atomic oscillation amplitude, it characterizes the interatomic interaction in solids, and is a function of the temperature).

(b) *Line (level) width* Γ_e, where $\Gamma_e > 2\Gamma_0$ due to broadening caused by unresolved hyperfine structure, diffusive motion of atoms, *defects* of crystal structure, and self-absorption effects in the case of thick samples. Depending on the broadening mechanism, Γ_e yields information on the inhomogeneity of composition, the structural and magnetic states of solids, and the presence of defects.

(c) Shift $\delta = \delta_i + \delta_T$ of nuclear gamma resonance spectra relative to a standard. The *isomer shift* δ_i is caused by the interaction of the nucleus with the electrostatic field arising from the surrounding electrons, and depends on the difference in the charge density of s-electrons of the absorber and emitter nuclei. A temperature shift δ_T is associated with the second order relativistic Doppler effect, and characterizes vibratory motion of atoms in a solid. The quantity δ is an important characteristic of atom bonding in solids; it depends on the composition, structure, magnetic state, temperature, and pressure.

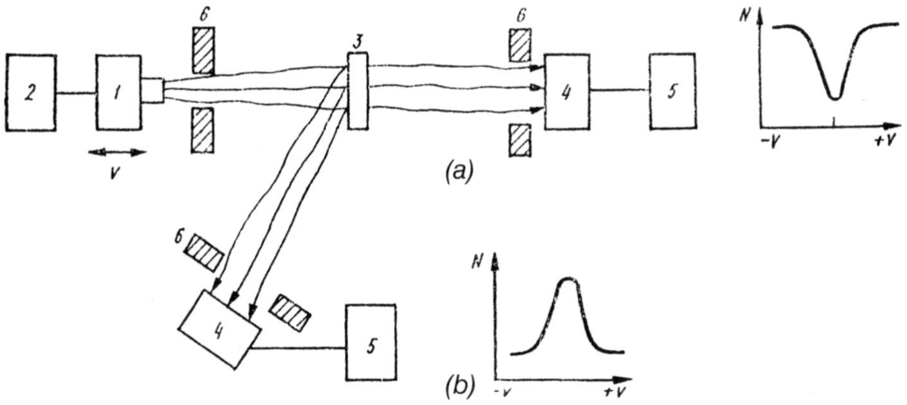

Fig. 2. Diagram of Mössbauer spectrometer: (a) transmission mode; (b) scattering mode. 1, modulator of γ-quanta energy with a radiation source; 2, modulation signal generator; 3, absorber or scatterer under study; 4, radiation detector; 5, pulse recording and analyzing unit; 6, collimator.

(d) Quadrupole splitting $2E_{\mathrm{d}}$ caused by the interaction of the nuclear quadrupole moment Q ($Q \neq 0$ for nuclei with spin $I > 1/2$) with the electric field gradient. It splits nuclear levels into two or more sublevels. For instance, nuclei of ^{57}Fe (or ^{119}Sn, ^{125}Te) have two sublevels, and accordingly, there are doublets in nuclear gamma resonance spectra (see Fig. 1(b)). The magnitude of the *quadrupole splitting* yields information on the asymmetry of the volume charge distribution, distortion of the *crystal lattice*, the presence of impurities.

(e) Magnetic splitting B typical of nuclear gamma resonance spectra of magnetically ordered materials. The interaction of the magnetic dipole moment of a nucleus $\boldsymbol{\mu}$ with the magnetic field $\boldsymbol{B}$ at a nucleus leads to the *Zeeman effect*. The nuclear gamma resonance spectrum in this case is a set of lines; their number equaling the transitions between sublevels that are allowed by the *selection rules*. For instance, the nuclear gamma resonance spectrum of ^{57}Fe exhibits six lines (see Fig. 1(c)) with the magnitude B measured from the distance between the extreme hyperfine components (for ^{57}Fe, there is a sextet of lines corresponding to transitions $\pm 3/2 \to \pm 1/2$). The magnetic hyperfine structure is sensitive to composition, temperature, pressure, and applied magnetic field, and is used to study magnetic and structural transformations, atomic ordering, relaxation phenomena, *texture*, *magnetic anisotropy*, *phase analysis*, etc.

MOTT–DAVIS MODEL of density of states (N.F. Mott, E.A. Davis, 1970)

Model of electronic *density of states* of amorphous semiconductors. According to the Mott–Davis model, tails of localized states of the *valence band* and *conduction band* are relatively narrow, and spread into the *band gap* to a depth of several tenths of an eV. There is a narrow band of compensated levels in the center of the gap which owes its existence to defects in a *random continuous network* of atoms; it "fixes" the *Fermi level* to the center of the gap. It follows from the Mott–Davis model that the *electrical conductivity* has three components, each of them dominating in its temperature region. At very low temperatures, the conductivity is due mainly to thermally activated jumps of carriers between states at the Fermi level, and is realized through a hopping mechanism (see *Hopping conductivity*). With a rise in temperature, the conductivity increases because of the excitation of carriers into localized states in the tails of the corresponding energy bands. These carriers also take part in charge transport through a hopping mechanism. On a further increase of temperature, the main mechanism responsible for the conductivity arises from carriers that are excited into delocalized states. See also *Amorphous semiconductors.*

MOTT LAW (N.F. Mott, 1969)

Temperature dependence of electrical conductivity due to jumps or hopping of *current carriers* over impurity- and defect-related localized states near the *Fermi level.* A model of localization centers randomly distributed in space explains why the temperature dependence of *hopping conductivity* is weaker than in an ordinary activation case. The Mott law may be written as $\sigma(T) = \sigma_0(T)\exp[-(T_0/T)^n]$, where $\sigma_0(T)$ is a weak function of the temperature, T_0 is a constant of the order of 10^8 K, the exponent n is always smaller than 1 and depends on particular physical properties of the object (relationship between *density of states* at the Fermi level; degree of *electron–phonon interaction*; presence of correlation effects; system dimensionality, etc.). In the three-dimensional case when the density of localized states is independent of energy and the electron-phonon interaction is negligible, we find $n = 1/4$ theoretically. Reported experimental studies of hopping conductivity of noncrystalline solids confirm the validity of the Mott law, but with values of n in the wider range: $0.2 \leqslant n \leqslant 0.55$.

MOTT–LITTLETON APPROXIMATION (N.F. Mott, M.J. Littleton, 1938)

A method of calculation of a static configuration of *defects* in a crystal lattice (often ionic) that takes into account long-range Coulomb forces and polarization forces. The physical significance of the approximation is that the electric dipole moments of ions, which are induced by a charged defect and located far from it, are determined by the macroscopic polarization.

MOTT METAL–INSULATOR TRANSITION (N.F. Mott, 1949)

Phase transition in an electron gas, which brings about an abrupt change of electrical conductivity. In the dielectric (insulator) state there is a gap between completely filled and empty energy bands. An electron transition results in the appearance of a free electron in the upper band and a hole in the lower band. If the concentration n of carriers is low, they form a bound state, an *exciton.* The electron gas screens the Coulomb interaction by decreasing the energy of an *electron–hole pair.* At $n > n_c$, the generation of excitons is no longer possible because the electrons delocalize and the system becomes a *metal.* Thus, strong electron–electron correlation leads to an abrupt change of free carrier concentration at the Mott metal–insulator transition. The value of n_c is determined by the distance d_0 between electron and hole. The Mott transition occurs in materials with carriers of the sp-type, e.g., transition-metal compounds (V_2O_3, NiS, etc.) and doped semiconductors. The last *impurity band* in the insulator state is split by the Coulomb interaction U. The *Density of states* at the *Fermi level* $N(E_F)$ becomes finite when the lower *Hubbard band* of width B (see *Hubbard model*) overlaps the upper one: $B \geqslant U$. This inequality roughly determines the condition for the Mott metal–insulator transition. Disordering effects may blur the upper band boundary, and form a density of states "tail". As only localized states contribute to this "tail", the transition takes place at considerable overlap of both bands and finite $N(E_F)$. Experimental data from many types of *semiconductors* fit the universal relation $n_c^{1/3} a_0 \approx 0.27$.

Mott (1978) showed that an energy gap decrease ΔE can result in an abrupt change of n from zero to n_c, the position of the energy minimum of the electron gas. Since this minimum equals the sum of the kinetic energies of electrons and holes ($\sim n^{2/3}$) plus the potential energy of their attraction $\sim(-n^{1/3})$, it is attained at $E_{min} < 0$. Thus, spontaneous generation of an electron-hole gas becomes energetically favorable before the two bands overlap at $\Delta E \leqslant -E_{min}$. In most cases, this gas has a lower energy than the alternative phase of a *Bose condensate* of excitons.

Since generation of a free electron and hole requires a finite energy, the Mott transition must be accompanied by an abrupt change of free energy F. Strong enough disorder cancels such jumps of n and F. This results from the fact that if an $N(E)$ "tail" exists, then the kinetic energy of the electron–hole gas depends less on n, which results in $E_{min} > 0$ (Mott and Kaveh, 1983). On further disordering, localization of electron states proceeds through an Anderson transition. This metal–insulator transition without an abrupt change of F was observed in such doped semiconductors, as Si:P (see *Anderson localization*).

MOTT PSEUDOGAP

See *Pseudogap*.

MOTT–WANNIER EXCITON

See *Wannier–Mott exciton*.

MOULDING

See *Pressing*.

MOUNTING OF CRYSTALS

Selection of *crystallographic axes* and *unit faces of crystals* for mounting purposes. Crystallographic axes are coordinate axes parallel to close-packed rows or planes of the space lattice. The angles of the intercepts cut by each crystal face on the axes serve as input data for calculating the indices of the crystal face. The *cubic system*, *tetragonal system*, and *orthorhombic system* are treated using Cartesian coordinates, whereas the crystallographic axes of other crystal systems are not all at right angles to one another, so they are more complicated to deal with. A mounting of the crystal in accordance with intrinsic crystallographic coordinates constitutes a structural mounting. In the case of the *monoclinic system*, the 2- (or $\bar{2}$)-fold axis of symmetry is often selected as the z-axis, the α and β angles for this nontypical mounting being chosen as 90°, with $\gamma \neq 90°$. A morphological mounting of a crystal, which utilizes convenient rather than intrinsic coordinates, is performed when there is a lack of information on the characteristics of the crystal structure, or for special purposes. It is carried out in accordance with certain rules, which are specified individually for each *crystal system*, and may be not be compatible with a structural mounting. See also *Goniometry*.

MRI

See *Magnetic resonance imaging*.

MUFFIN-TIN POTENTIAL, MT potential

A widely used approximation to a crystal potential. According to the MT-approximation, the entire crystal is divided into two regions: a region of MT-spheres circumscribed about nuclei and in contact with one another, and the region between the spheres. The MT-potential is chosen as constant (MT-zero) in the latter region, and is considered to be spherically symmetric inside the spheres. The MT-potential is used in *band structure computation methods* (*Korringa–Kohn–Rostoker method*, *augmented plane waves method*) and linear analogs of these methods (see *Linear methods of band structure computation*). A generalization of the MT-potential is the warped MT-potential which retains spherical symmetry inside the spheres, but allows for the potential to vary in the space between them. To represent an arbitrary potential as a generalized MT-potential, it can be expanded in terms of *spherical harmonics* inside the spheres, and in terms of plane waves outside them.

MULLIKEN EQUATIONS (R.S. Mulliken, J.K. Ruedenberg)

Approximate expressions for matrix elements of one- and two-particle interactions of localized wave functions $\varphi_\alpha(\boldsymbol{r})$, resulting from an approximation to the exchange density in the form

$$\varphi_\alpha(\boldsymbol{r})\varphi_\beta(\boldsymbol{r}') \approx \frac{1}{2}\left[\varphi_\alpha(\boldsymbol{r})\varphi_\alpha(\boldsymbol{r}') + \varphi_\beta(\boldsymbol{r})\varphi_\beta(\boldsymbol{r}')\right] \times \int \varphi_\alpha(\boldsymbol{r})\varphi_\beta(\boldsymbol{r})\,\mathrm{d}^3r.$$

The Mulliken equations lead to the *Mulliken–Ruedenberg method* which is one of the semi-empirical approaches of *quantum chemistry*. The Mulliken equations are used to calculate electronic bands and states of *defects in crystals*.

MULLIKEN–WOLFSBERG–HELMHOLZ METHOD (R.S. Mulliken, M. Wolfsberg, L. Helmholz)

A simplified modification of the *Hartree–Fock method* for application to calculations of the electronic structure of polyatomic complex *ions*. It is based on the determination of single-particle states of the complex as linear combinations of the valence *orbitals* of its constituent atoms. The coefficients in those linear combinations are found as the components of one-electron Hamiltonian H matrix eigenvectors from the eigenvalue equations of the form used in the *Hoffmann method*. The diagonal matrix elements are defined by the expression

$$H_{\mu\mu}^{A} = -I_{\mu}^{A} - \sum_{B(\neq A)} \frac{Z_B' + \Delta_\mu}{R_{AB}},$$

where I_μ^A is the electron energy of atom A in its one-particle state, taking into account the total charge and configuration of other electrons in the valence shell, Z_B' is the effective Mulliken charge of atom B, corrected by Δ_μ for *electric charge screening*, and R_{AB} is the interatomic distance. Off-diagonal elements are estimated from the formula

$$H_{\mu\nu} = -\frac{1}{2}k(H_{\mu\mu} + H_{\nu\nu})S_{\mu\nu},$$

where $k = 1.67$–2.0; other equations have also been proposed. An important feature of the Mulliken–Wolfsberg–Helmholz method is the procedure for self-consistency in charges and configurations that I_μ^A and Z_B' depend on. This procedure uses the results of the preceding iteration to determine the total charge on each atom, while I_μ^A values for the next iteration are calculated through interpolation from known ionization potentials of the atoms and corresponding ions. The Mulliken–Wolfsberg–Helmholz calculations allow determination of binding energies, ionization potentials, energies of electronic transitions, distributions of charge and *spin density*, as well as *conformations* of complexes.

MULTIAXIAL MAGNET

See *Magnetic anisotropy*.

MULTICRITICAL POINT, polycritical point

A point on the compositional *phase diagram* of a substance, where regions of several different *phases* come in contact, these regions (not necessarily all of them) being separated by *second-order phase transition* lines. *Bicritical points, triple critical points* and *tetracritical points* exemplify multicritical points. Schematic representation of corresponding regions of phase diagrams are given in the figure, where second-order phase transitions are indicated by dashed lines, first-order ones are indicated by solid lines; and the symbol I designates regions of phases which possess the highest symmetry.

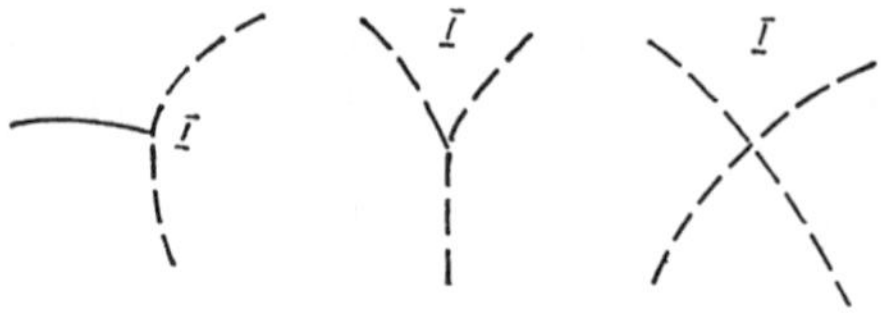

Types of multicritical points.

Tricritical points also include points where curves of first-order phase transitions turn into curves of continuous *phase transitions*. These points are also called *critical points of second-order transitions*, or *Landau critical points*.

A material can be brought to a multicritical point by changing more than one external parameter. These usually are temperature, pressure, electric or magnetic field, concentration of one of the components of a *solid solution*. Multicritical points have been found in compositional phase diagrams of many ferromagnets and antiferromagnets, ferroelectrics, and crystals with various types of structural ordering.

MULTIPHOTON PROCESSES

Interaction processes between electromagnetic radiation and atoms and molecules of matter when several *photons* simultaneously participate in each event (absorption, emission). The greater the number of photons that participate the less probable and the more difficult to observe are the processes. Multiphoton processes may either have no threshold, or exhibit a threshold value of radiation intensity where they appear.

Raman scattering of light is a two-photon process involving the simultaneous absorption of a photon of one frequency and the emission of a photon of another frequency in a *crystal*. The involvement of a phonon provides the energy balance. This process has no threshold; neither does *two-photon light absorption*. The coefficient of two-photon absorption is proportional to the wave intensity. Therefore, the larger the incident wave amplitude, the more rapid the attenuation of the wave in the crystal. In strong fields of laser radiation, nonlinear absorption of higher order is observed.

Three-photon processes include *second harmonic generation*, excitation of waves at sum and difference frequencies, *parametric light scattering*, and *parametric light generation*. The probability of multiphoton processes is proportional to product of the densities of photons that take part in the absorption and emission, i.e. the product of the intensities of waves of different frequencies. Besides that, this probability depends on the structure of the energy levels of the material plus the position of combination frequencies relative to these levels, and it increases sharply if one or several combination frequencies are close to those of intermediate transitions. The increase of the incident radiation intensity also causes the probability of multiphoton processes to increase. When passing from single-photon processes to multiphoton ones the *selection rules* change. Thus, in centrally symmetrical crystals a transition between states of like parity involves an even number of photons, and that between states with opposite parity involves an odd number (see *Mutual exclusion*). Therefore, transitions between different energy states that are forbidden for a certain set of photons are allowed for another number of photons. Investigations of multiphoton absorption spectra can involve studying transitions forbidden for single-photon excitation. This method finds wide use in the *nonlinear optics* of solids, and multiphoton processes accompanied by photon emission are used in *optical frequency converters*.

MULTIPLICATION OF DISLOCATIONS

An increase of the total length of dislocation lines in the bulk of a crystal developing during the course of *plastic deformation*. It results from the action of several sources, the main one being the *Frank–Read source*. To picture it one may consider a section of *dislocation* of length L fixed at its ends in the *slip plane*. It bends out under the applied stress σ, and upon reaching a certain critical configuration at $\sigma_s = Gb/L$ (G is the *shear modulus*, b is the *Burgers vector*), determined by the balance between the external force and the elastic stress force, the dislocation segment loses its stability. This loss of stability results in the formation of a closed *dislocation loop*, within which the initial position of the segment is recovered. Provided $\sigma > \sigma_s$ the Frank–Read source will continuously generate dislocation loops until its action is blocked by the long-range stresses produced by loops already formed. The action of other sources follows the same principle. However, in the case of a double transverse slip the pinning positions of a dislocation segment are the ends of sections of a *screw dislocation* that has executed its *cross slip* into a nearby plane, and then returned to a slip plane parallel to the initial one. In a pole source, only one end of the segment is fixed, and the other can eventually reach the crystal surface.

The *Bardeen–Herring sources* base their function on the processes of crawl of a fixed dislocation segment that proceed via the absorption of *point defects*, and this results in the formation of *helical dislocations* and *prismatic dislocations* in the crystal bulk.

MULTIPOLE

An aggregation of electric charges (e_i), which is characterized by a *multipole moment*. It may be introduced through the potential of a system of charges $\sum_i e_i/|\boldsymbol{R} \pm \boldsymbol{r}_i|$ at distances $\boldsymbol{R}$ from this system that far exceed the distance $\boldsymbol{r}_i$ to any ith charge (origin of coordinates is inside the multipole). For $R \gg r_i$, expanding $|\boldsymbol{R} - \boldsymbol{r}_i|^{-1}$ in a power series yields: $d_0 = \sum_i e_i$, zeroth-order moment or monopole moment; $\boldsymbol{d}_1 = \sum_i e_i \boldsymbol{r}_i$, *first-order moment* or *dipole moment*; $d_2 = \sum_i e_I(3r_{i\alpha}r_{i\beta} - \delta_{\alpha\beta})$, *second-order moment* or *quadrupole moment* (see *Quadrupole*), etc., where $r_{i\alpha}$ and $r_{i\beta}$ are projections of the vector $\boldsymbol{r}_i$ on the coordinate axes, and $\delta_{\alpha\beta}$ is the Kronecker delta. In the general case one speaks about the nth order moment.

Not only does a multipole moment determine fields at a distance R, but it also characterizes the behavior of the multipole in external fields and, in particular, there is the multipole interaction. Thus, the expression $\boldsymbol{d}_1 \cdot \boldsymbol{E}_0$ defines the interaction energy between a dipole and the uniform part of the electric field $\boldsymbol{E}_0$ at the dipole. Also, the expression $[(\boldsymbol{d}_1 \cdot \boldsymbol{d}'_1)R^2 - 3(\boldsymbol{d}_1 \cdot \boldsymbol{R})(\boldsymbol{d}'_1 \cdot \boldsymbol{R})]/R^5$ defines the *dipole–dipole interaction* between two dipoles $\boldsymbol{d}_1$ and $\boldsymbol{d}'_1$ spaced a distance R apart that far exceeds their sizes ($\boldsymbol{r}_i$). Also known are dipole–quadrupole, quadrupole–quadrupole and other interactions of higher order.

By analogy with electrical multipoles, magnetic, elastic and other kinds of multipoles can be introduced. The concept of multipole is used in solid-state physics for investigating processes of absorption (emission) of electromagnetic waves, phase transitions, broadening of resonance lines, etc. The concept of multipole has given rise to scientific disciplines and new concepts, such as quadrupole resonance, *dipole–dipole reservoir*, dipole instability (see *Vibronic instability*), etc.

MULTIWELL POTENTIAL

See *Many-well potential.*

MUONIUM

An atom (μ^+e^-), sometimes referred to as a light ("exotic") isotope of hydrogen, which is formed when a positively charged *muon* μ^+ forms a bound state with an electron e^-, an event that sometimes occurs in *semiconductors* and *insulators.* Muonium is unstable, and has a lifetime of 2.2 ms. All electrical and chemical interactions of the atom μ^+e^- in nonmetals are the same as those of a hydrogen atom; therefore, the investigation of the dynamics of the behavior of muonium in matter yields information on hydrogen behavior. Besides the normal state (μ^+e^-), there is also the state of *anomalous muonium* $(\mu^+e^-)^*$; in it, the *hyperfine interaction* is weaker than in μ^+e^-, it is anisotropic, and the precession frequency in a magnetic field is different. The $(\mu^+e^-)^*$ is a bound system located in the neighborhood of a lattice atom along a $\langle 111\rangle$ axis in silicon. Unlike $(\mu^+e^-)^*$, normal muonium (μ^+e^-) is an atom which rapidly diffuses in matter.

MUONS

Lepton elementary particles with spin 1/2, charge $+e$ or $-e$ (where e is the unit electric charge), rest energy 105.66 MeV, magnetic moment $M_\mu = 3.18\mu_p$ (μ_p is magnetic moment of proton). The former (incorrect) name was *μ-meson.* Intense beams of muons are generated by accelerators through the decay of *π-mesons* $\pi^\pm \to \mu^\pm + \nu$, where ν is a neutrino. This decay produces muons that are polarized along their momentum direction. The lifetime of a muon at rest is $\sim 2.2\cdot 10^{-6}$ s; a muon decays into an electron, neutrino and antineutrino: $\mu^\pm \to e^\pm + \nu + \overline{\nu}$. Since parity is not conserved in this decay, the angular distribution of *positrons* (or electrons) is asymmetrical relative to the direction of the muon spin. Positrons are emitted mostly in the same direction as the muon spin. This underlies the application of stopped muons μ^+ for the investigation of materials through the *method of muon spin rotation.* Muons are moderated in the material, and either become localized or diffuse. Observation of μ^+ (by detecting the decay of positrons) is feasible in the time interval from 10^{-7} to 10^{-5} s.

An external magnetic field B applied to the sample causes the muon spin to precess about the direction of the field. As this takes place, the number of positrons which escape in specific directions varies with time in a regular manner. The precession frequency is $\omega_\mu = eB/m_\mu$ with the value 13.55 MHz/T, where m_μ is the muon mass. The precession amplitude is proportional to the degree of polarization of the ensemble of muons. Depolarization caused by interaction of the muon spin and magnetic moment with atoms of the medium results in a decrease of the precession amplitude. This fact is used for studying properties of *solids*: *phase transitions* in magnetic materials and *superconductors*, and phenomena of *quantum diffusion* of muons in metals. When stopped in a material negatively charged muons (μ^-) form *mesic atoms.* Muons μ^+ form *muonium* atoms (μ^+, e^- bound state) in semiconductors and *insulators.* The phenomenon of *channeling* in monocrystals was observed for μ^+.

MUTUAL EXCLUSION, alternative exclusion

A *selection rule* for transitions involving the parity conservation law in quantum systems.

According to mutual exclusion, electric-dipole transitions are only possible between states of different parity. Magnetic-dipole and electric-quadrupole electron transitions only occur between states of the same parity. Mutual exclusion holds in the presence of a *center of symmetry*, relative to which the states are divided into odd and even types. A corollary of mutual exclusion is the *phonon optical spectrum* rule whereby transitions symmetric relative to the center of symmetry are forbidden in IR absorption spectra, while those that are antisymmetric are forbidden in spectra of the *Raman scattering of light.*

ISBN 0-12-561465-9

9 780125 614658

PERIODIC SYSTEM OF…

PERIOD	I	II	III	IV	V
	G…				
1	**H** 1; 1.00794±7; $1s^1$; 1; Hydrogen				
2	**Li** 3; 6.941±2; $2s^1$; 1 2; Lithium	**Be** 4; 9.01218±1; $2s^2$; 2 2; Beryllium	**B** 5; 10.811±5; $2s^2 2p^1$; 3 2; Boron	**C** 6; 12.011±1; $2s^2 2p^2$; 4 2; Carbon	**N**; 14.0067±1; $2s^2$…; Nitrogen
3	**Na** 11; 22.98977±1; $3s^1$; 1 8 2; Sodium	**Mg** 12; 24.305±1; $3s^2$; 2 8 2; Magnesium	**Al** 13; 26.98154±1; $3s^2 3p^1$; 3 8 2; Aluminum	**Si** 14; 28.0855±3; $3s^2 3p^2$; 4 8 2; Silicon	**P**; 30.97376±1; $3s^2$…; Phosphorus
4	**K** 19; 39.0983±1; $4s^1$; 1 8 8 2; Potassium	**Ca** 20; 40.078±4; $4s^2$; 2 8 8 2; Calcium	21 **Sc**; 2 9 8 2; $3d^1 4s^2$; 44.95591±1; Scandium	22 **Ti**; 2 10 8 2; $3d^2 4s^2$; 47.88±3; Titanium	23 …; 2 11 8 2; $3d^3 4s^2$; 50.…; Vana…
	29 **Cu**; 1 18 8 2; $3d^{10} 4s^1$; 63.546±3; Copper	30 **Zn**; 2 18 8 2; $3d^{10} 4s^2$; 65.39±2; Zinc	**Ga** 31; 69.723±4; $4s^2 4p^1$; 3 18 8 2; Gallium	**Ge** 32; 72.59±3; $4s^2 4p^2$; 4 18 8 2; Germanium	**As** …; 74.9216±1; $4s^2 4$…; Arsenic
5	**Rb** 37; 85.4678±3; $5s^1$; 1 8 18 8 2; Rubidium	**Sr** 38; 87.62±1; $5s^2$; 2 8 18 8 2; Strontium	39 **Y**; 2 9 18 8 2; $4d^1 5s^2$; 88.9059±1; Yttrium	40 **Zr**; 2 10 18 8 2; $4d^2 5s^2$; 91.224±2; Zirconium	41 **N**…; 1 12 18 8 2; $4d^4 5s^1$; 92.…; Nic…
	47 **Ag**; 1 18 18 8 2; $4d^{10} 5s^1$; 107.8682±3; Silver	48 **Cd**; 2 18 18 8 2; $4d^{10} 5s^2$; 112.41±1; Cadmium	**In** 49; 114.82±1; $5s^2 5p^1$; 3 18 18 8 2; Indium	**Sn** 50; 118.710±7; $5s^2 5p^2$; 4 18 18 8 2; Tin	**Sb** …; 121.75±3; $5s^2 5$…; Antimony
6	**Cs** 55; 132.9054±1; $6s^1$; 1 8 18 18 8 2; Cesium	**Ba** 56; 137.33±1; $6s^2$; 2 8 18 18 8 2; Barium	57 **La***; 2 9 18 18 8 2; $5d^1 6s^2$; 138.9055±3; Lanthanum	72 **Hf**; 2 10 32 18 8 2; $5d^2 6s^2$; 178.49±3; Hafnium	73 **T**…; 2 11 32 18 8 2; $5d^3 6s^2$; 180.…; Tan…
	79 **Au**; 1 18 32 18 8 2; $5d^{10} 6s^1$; 196.9665±1; Gold	80 **Hg**; 2 18 32 18 8 2; $5d^{10} 6s^2$; 200.59±3; Mercury	**Tl** 81; 204.383±1; $6s^2 6p^1$; 3 18 32 18 8 2; Thallium	**Pb** 82; 207.2±1; $6s^2 6p^2$; 4 18 32 18 8 2; Lead	**Bi** …; 208.9804±1; $6s^2 6$…; Bismuth
	Fr 87; …3.0197; $7s^1$; 1 8 18 32 18 8 2; …ncium	**Ra** 88; 226.0254; $7s^2$; 2 8 18 32 18 8 2; Radium	89 **Ac****; 2 9 18 32 18 8 2; $6d^1 7s^2$; 227.0278; Actinium	104 **Ku**; 2 10 32 32 18 8 2; $6d^2 7s^2$; (261); Kurchatovium	105 **N**…; 2 11 32 32 18 8 2; $6d^3 7s^2$; Nilsb…

* L…

…**r** 59; …7±1; $4f^3 6s^2$; 2 8 21 18 8 2; …mium	**Nd** 60; 144.24±3; $4f^4 6s^2$; 2 8 22 18 8 2; Neodymium	**Pm** 61; 144.9128; $4f^5 6s^2$; 2 8 23 18 8 2; Promethium	**Sm** 62; 150.36±3; $4f^6 6s^2$; 2 8 24 18 8 2; Samarium	**Eu** 63; 151.96±1; $4f^7 6s^2$; 2 8 25 18 8 2; Europium	**Gd**…; 157.25±3; $4f^7 5d$…; Gadolinium…

**…

… 2 9	**U** 92; 238.0289±1; $5f^3 6d^1 7s^2$; 2 9 21 32 18 8 2; …nium	**Np** 93; 237.0482; $5f^4 6d^1 7s^2$; 2 9 22 32 18 8 2; Neptunium	**Pu** 94; 244.0642; $5f^6 7s^2$; 2 8 24 32 18 8 2; Plutonium	**Am** 95; 243.0614; $5f^7 7s^2$; 2 8 25 32 18 8 2; Americium	**Cm**…; 247.0703; $5f^7 6d$…; Curium